Multicolour Edition

CELL BIOLOGY, GENETICS, MOLECULAR BIOLOGY, EVOLUTION AND ECOLOGY

[For Undergraduate and Postgraduate Students of Zoology, Botany and Biosciences]

Dr. P.S. VERMA
M.Sc., Ph.D., FESI, FAZ
Reader
Department of Zoology, Meerut College, Meerut

Dr. V.K. AGARWAL
M.Sc., Ph.D.
Reader
Department of Zoology, Meerut College, Meerut

S. CHAND PUBLISHING

S Chand And Company Limited

(ISO 9001 Certified Company)

S Chand And Company Limited

(ISO 9001 Certified Company)

Head Office: D-92, Sector–2, Noida – 201301, U.P. (India), Ph. 91-120-4682700

Registered Office: A-27, 2nd Floor, Mohan Co-operative Industrial Estate, New Delhi – 110 044, Phone: 011-49731800

S. CHAND PUBLISHING

www.**schandpublishing.com;** e-mail: **info@schandpublishing.com**

Marketing Offices:

Chennai	:	Ph: 23632120; chennai@schandpublishing.com
Guwahati	:	Ph: 2738811, 2735640; guwahati@schandpublishing.com
Hyderabad	:	Ph: 40186018; hyderabad@schandpublishing.com
Jalandhar	:	Ph: 4645630; jalandhar@schandpublishing.com
Kolkata	:	Ph: 23357458, 23353914; kolkata@schandpublishing.com
Lucknow	:	Ph: 4003633; lucknow@schandpublishing.com
Mumbai	:	Ph: 25000297; mumbai@schandpublishing.com
Patna	:	Ph: 4011400; patna@schandpublishing.com

First Edition 1974
Subsequent Editions and Reprints 1975, 76, 77, 78, 80, 81, 83 (Twice), 85, 86, 87, 89, 90, 91, 93, 94, 95, 97, 98, 99, 2001, 2002, 2003, 2004
First Multicolour Edition 2004; Reprints 2005, 2006, 2007, 2008 (Twice), 2009, 2010 (Twice), 2012 (Twice), 2013, 2014, 2016, 2018, 2019 (Twice), 2020 (Twice), 2021 (Twice), 2022 (Thrice), 2024, 2025

Reprint 2026

ISBN: 978-81-219-2442-9 **Product Code:** H6CGM68BIOL10ENZX0XO

PRINTED IN INDIA

By Shri Shyamlal Printing Press Private Limited, Plot 20/4, Site-IV, Industrial Area Sahibabad, Ghaziabad – 201 010 and Published by S Chand And Company Limited, A-27, 2nd Floor, Mohan Co-operative Industrial Estate, New Delhi – 110 044.

Preface

PREFACE

The multicoloured edition of the textbook of Cell Biology, Genetics, Molecular Biology, Evolution and Ecology is the outcome of sincere and combined efforts of the authors and editors (namely Shishir Bhatnagar, Shubha Pradhan, Malini Kothiyal) and, young and talented team of DTP of S. Chand & Company Ltd. Their main motive remained to provide relevant coloured diagrams explaining various intricate biological topics. Multicoloured figures and images of this edition would help our target readers to understand and fully appreciate the very gist of the subject matter. Authors and editors have remained quite choosy and vigilant regarding relevance and authenticity of each and every illustration/image finding its place in this textbook.

Authors earnestly hope that this multicoloured version of the fourteenth edition will enhance the curiosity of our target readers to know more and more about the subject. It will arm them with latest information for facing any type of exam quite adequately.

This book is meant for students of B.Sc., B.Sc. (Hons.) and M.Sc. of biological group. Students appearing in entrance exams of C.P.M.T., I.F.S., P.C.S. and I.A.S., etc, may be immensely benefited by this book.

Authors wish to express their thanks to Shri R.K. Gupta, the Managing Director, Mr. Navin Joshi, the General Manager of M/s S.Chand & Co. Ltd., New Delhi, for all their efforts to make this endeavour a pleasant surprise to the readers.

Authors

PREFACE TO THE FOURTEENTH EDITION

The revised edition of Cell Biology, Genetics, Molecular Biology, Evolution and Ecology comprises 84 chapters. The 21 new chapters, which have been added in this edition, are distributed in the five parts/sections of this textbook which are as follows :

1. Cell biology. Techniques in cell biology; Growth.

2. Genetics. Multiple genes (Quantitative genetics); Change in chromosome structure; Change in chromosome number; Human genetics; Transposable genetic elements (Jumping genes).

3. Molecular biology. Replication of DNA; Genetic engineering; Immunology; Genetic recombination and gene transfer (Bacterial conjugation, transformation and transduction).

4. Evolution. Direct evidences of evolution (Fossils); Examples of natural selection; Population genetics and evolution; Adaptive radiation; Barriers.

5. Ecology. Ecology in India; Ecological succession; Wild-life management; Biogeography; Adaptation.

Present edition of this book has been thoroughly revised, updated and enlarged. About 400 entirely new figures and data-packed tables have been added in this edition. All old chapters have been almost rewritten in the light of current researches. However, the old format of the book has been retained in order to familiarise the readers with the basic concepts. Revision questions (and problems) have been given at the end of each chapter to test the learning capacity of the readers. Answers to the problems have also been given at places where required.

In the revision of the book, the simplicity and clarity of the language has been maintained. Text of the book is accompanied with simple and self-explanatory diagrams. Every effort has been made to ensure that readers may get a balanced idea of the subject matter which may enlighten them regarding classical and modern concepts of the subject.

It is hoped that this textbook will serve the purpose of students of B.Sc., B.Sc. (Hons.), M.Sc. (Zoology, Botany and Biosciences) of various Indian Universities. This book can be used as a reference book by those students who are preparing for various competitive examinations/tests such as CPMT, CBSE (All India Medical Entrance Test), IFS, PCS, IAS and others.

Authors wish to express their thanks to Shri Ravindra Kumar Gupta and Shri T.N. Goel of M/s. S. Chand and Company Ltd., for their keen interest in the publication of this book.

Authors will feel highly obliged if suggestions for the improvement of the book are brought to their notice, so that future edition of the book may become more useful.

Authors

CONTENTS

CELL BIOLOGY

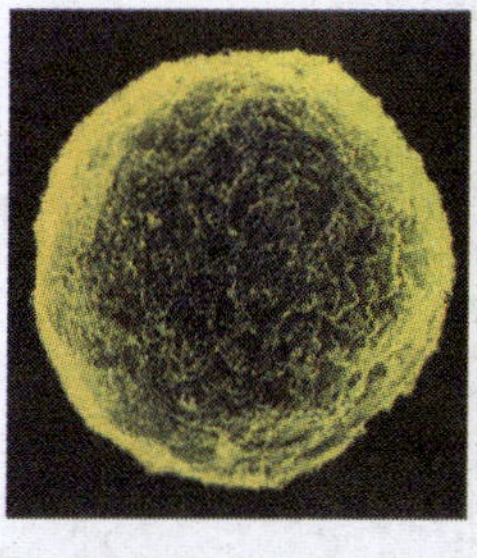

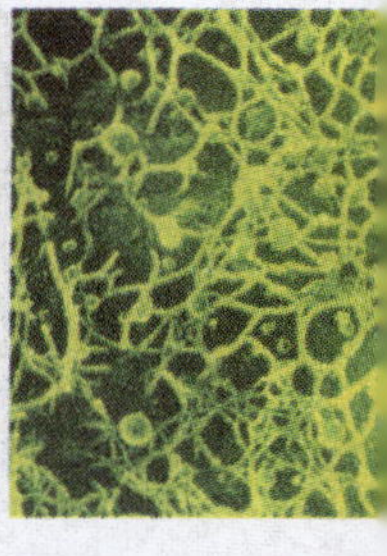

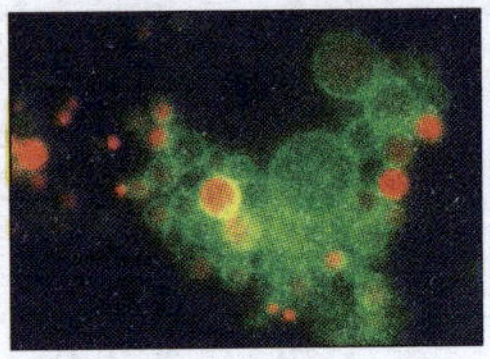

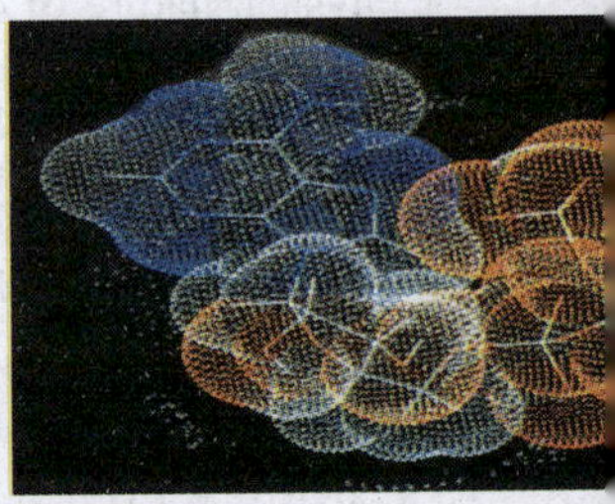

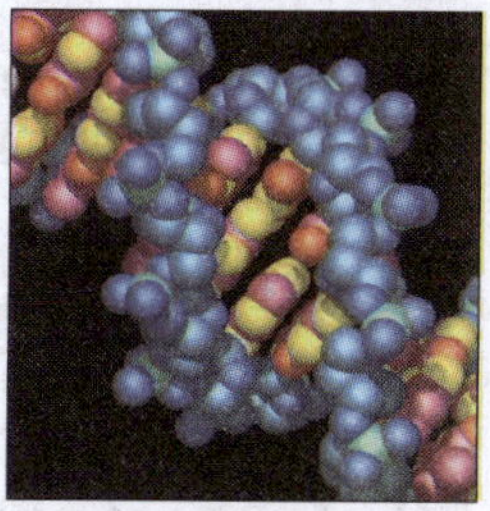

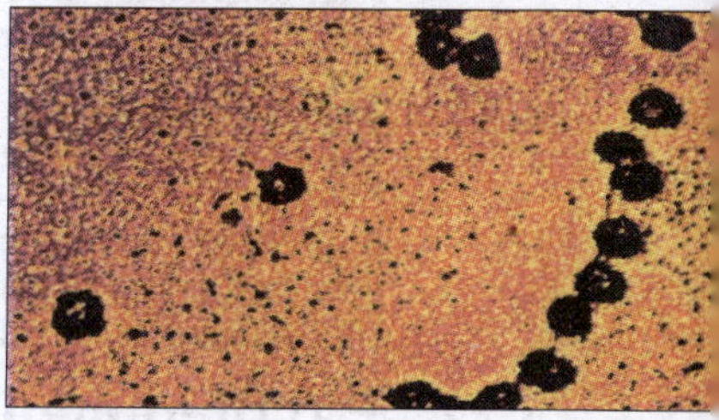

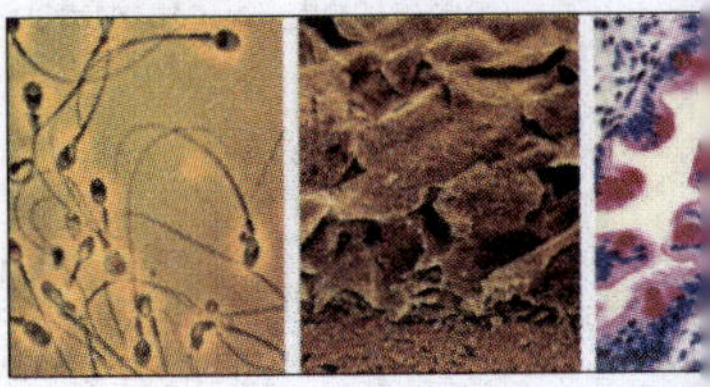

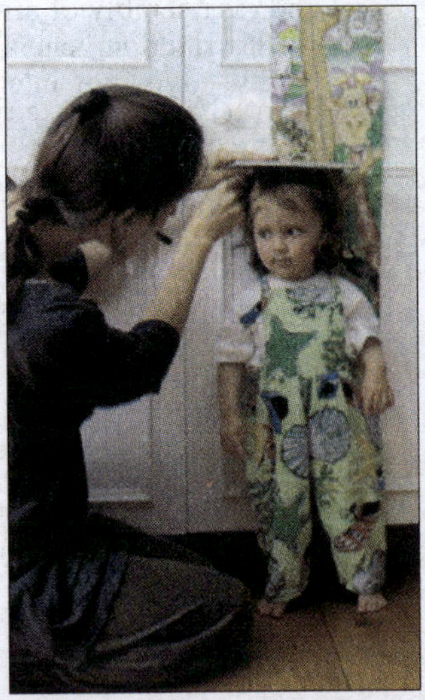

GENETICS

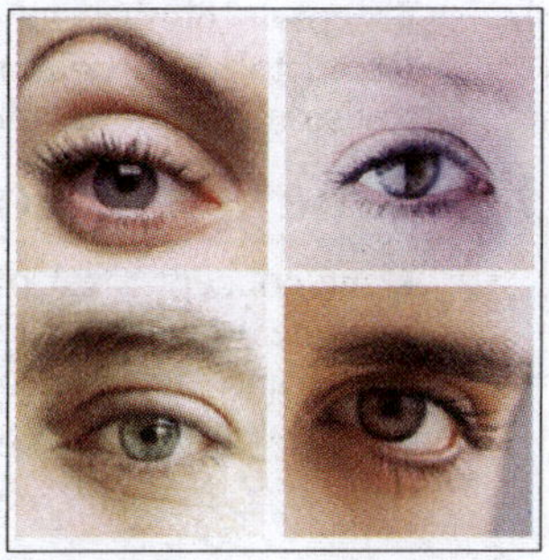

MOLECULAR BIOLOGY

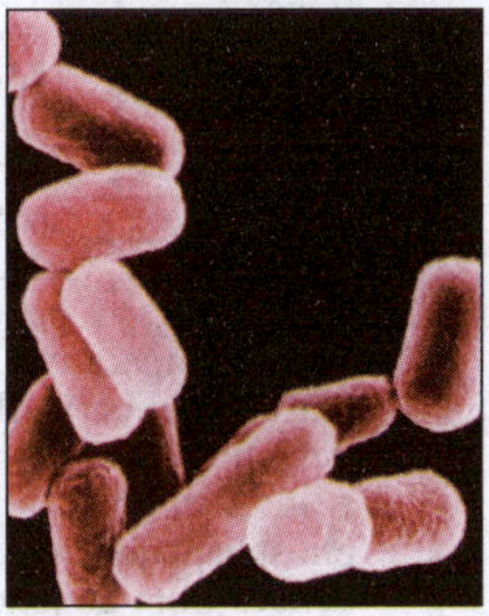

EVOLUTION

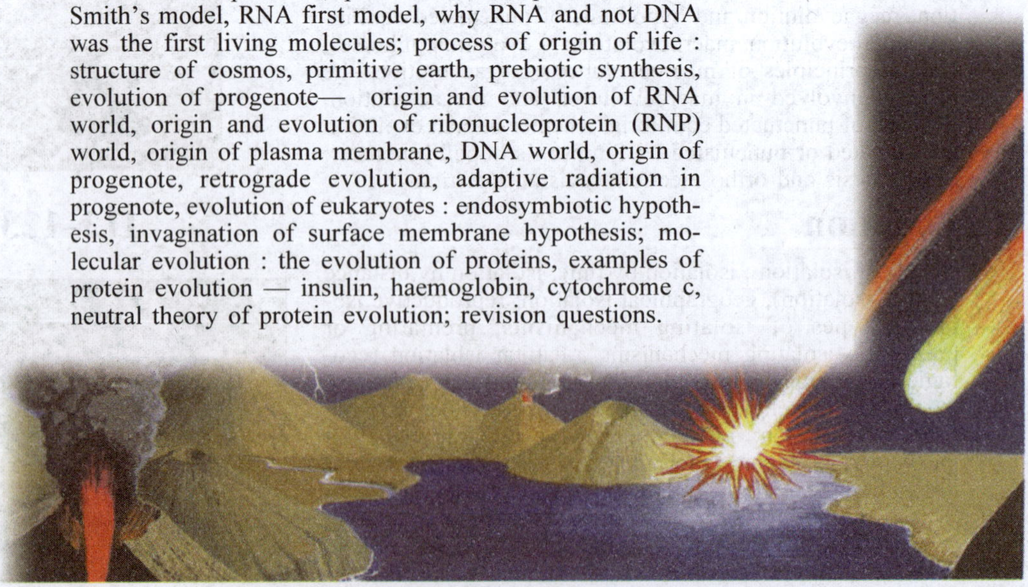

ECOLOGY

CELL BIOLOGY

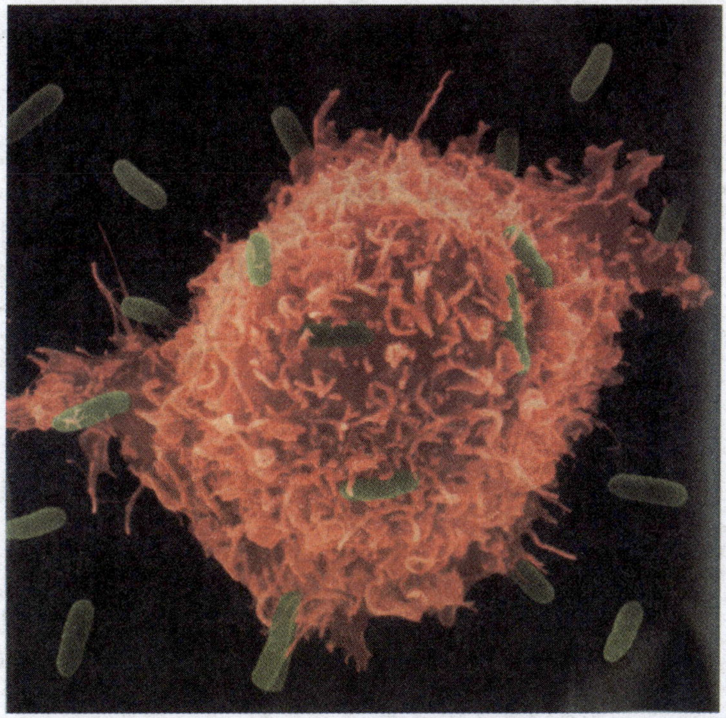

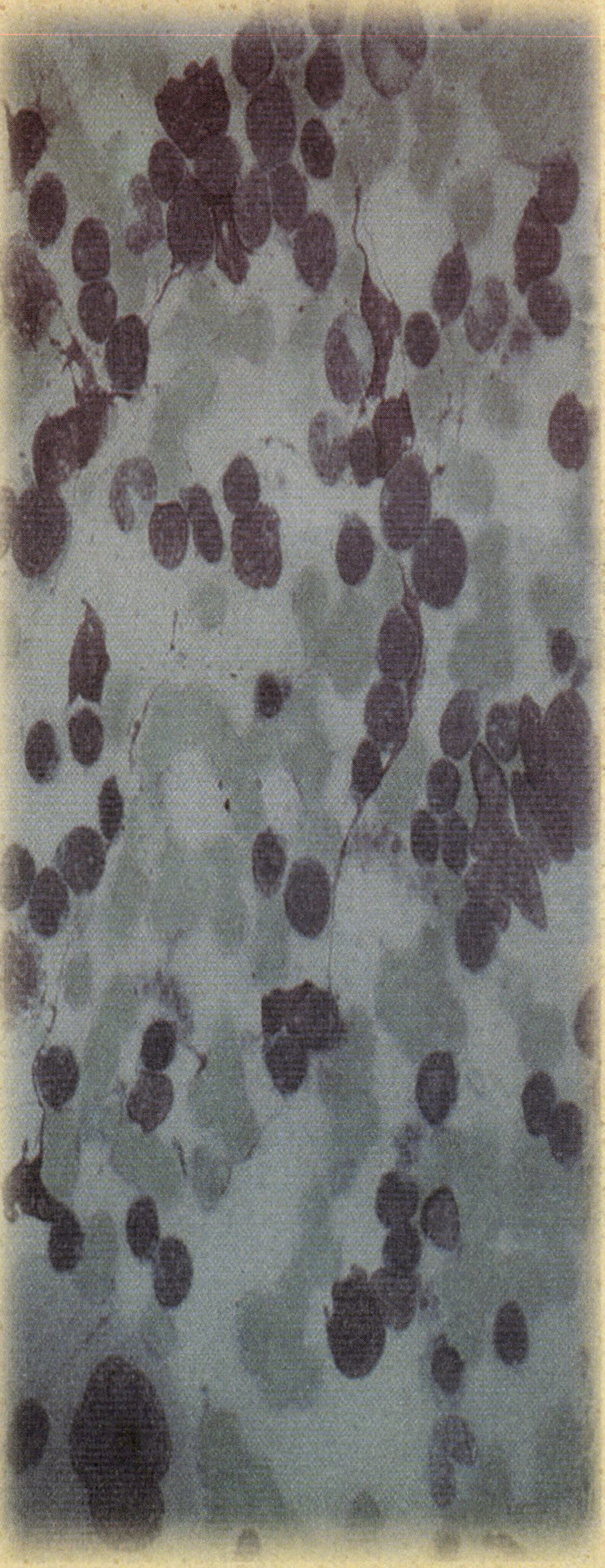

CHAPTERS

CHAPTER

1

Introduction

Nature's variety is boundless.

It seems to be an axiom of nature that where there is diversity, there is also similarity. Indeed, nature's variety is boundless. When walking through the woods, across a field, along a stream, through a zoo or wild life sanctuary, one is impressed with the diversity of life. Even looking through a microscope can be an elating experience. The universe of the cell too is complex and diverse. Like the world around us, the world of the cell is one of the forms specialized for a particular type of existence. And as is in the larger universe of the plant and animal kingdoms, where one can perceive basic life sustaining processes common to all organisms, in the cellular world many of the same processes and structures can be found in almost all cells. This generalization often leads to one of the most fundamental and obvious statement that the cell is the microscopic structural and functional unit of the living organisms. Thus, there are many cell types among fungi, protozoans and higher plants and animals. They differ in size, form and function, degree of specialization and average generation time. Yet at the ultrastructural level there is sameness about cells that is almost tedious. The same basic structures—nuclei, cytoplasmic matrix or cytsol, plastids, mitochondria, endoplasmic reticulum, Golgi apparatus, plasma membrane, etc.,—all appear with predictable regularity. Such a sameness can also be observed at the molecular level—all cell parts are made of highly organized groups of few types of molecules, *i.e.,* proteins, lipids, carbohydrates, nucleic acids, etc.

DEFINITION OF CELL BIOLOGY

The biological science which deals with the study of structure, function, molecular organization, growth, reproduction and genetics of the cells, is called **cytology** (Gr., *kytos* = hollow vessel or cell; *logous* = to discourse) or **cell biology**. Much of the cell biology is devoted to the study of structures and functions of specialized cells. The results of these studies are used to formulate the generalization applied to almost all cells as well as to provide the basic understanding of how a particular cell type carries out its specific functions. The cell biologist, without losing sight of the cell as a morphologic and functional unit within the organism, has to study biological phenomena at all levels of organization and to use all the methods, techniques and concepts of other sciences (Table 1-1).

Cytology versus Cell Biology

The cell biology has been studied by the following three avenues: **classical cytology** dealt with only light microscopically visible structure of the cell; **cell physiology** studied biochemistry, biophysics, and functions of the cell; and **cell biology** interpreted the cell in terms of molecules (macromolecules such as nucleic acids and proteins). In recent years distinction between classical cytology, cell physiology and cell biology has become blurred and outmoded and now two terms—cytology and cell biology are used as the synonyms (**Novikoff** and **Holtzmann**, 1970).

HISTORY OF CELL BIOLOGY

Ancient Greek philosophers such as **Aristotle** (384 —322 B.C.) and **Paracelsus** concluded that "all animals and plants, however, complicated, are constituted of a few elements which are repeated in each of them." They were referring to the macroscopic structures of an organism such as roots, leaves and flowers common to different plants, or segments and organs that are repeated in the animal kingdom. Many centuries later, owing to the invention of magnifying lenses, the world of microscopic dimensions was discovered. **Da Vinci** (1485) recommended the uses of lenses in viewing small objects. In 1558, Swiss biologist, **Conrad Gesner** (1516—1565) published results of his studies on the structure of a group of protists called **foraminifera.** His sketches of these protozoa included so many details that they could only have been made if he had used some form of magnifying lenses. Perhaps this is earliest recorded use of a magnifying instrument in a biological study.

Aristotle (384 —322 B.C.)

Table 1-1. Various levels of biological organization and instrumental resolving power (Source: De Robertis and De Robertis, Jr., 1987).

	Dimension	Biological field	Structures	Method of study
1.	0.1 mm or 100 µm or larger	Anatomy	Organs	Eyes and simple lenses
2.	100 µm to 10 µm	Histology	Tissues	Various types of light microscopes, X-ray microscopy
3.	10 µm to 0.2 µm (200 nm)	Cell biology	Cell, bacteria	
4.	200 nm to 1 nm	Submicroscopic morphology	Cell components, viruses	Polarization microscopy, electron microscopy
5.	Smaller than 1 nm (10A^0)	Ultrastructure, molecular and atomic structure.	Arrangement of atoms	X-ray diffraction

Further growth and development of cell biology are intimately associated with the development of optical lenses and to the combination of these lenses in the construction of the **compound microscopes** (Gr., *mikros* = samll; *skopein* = to see). Thus, the invention of the microscope and its gradual improvement went hand-in-hand with the development of cell biology.

1. Growth of Cell Biology during 16th and 18th Centuries

Fig. 1.1. Hooke's compound microscope.

The first useful compound microscope was invented in 1590 by **Francis Janssen** and **Zacharias Janssen**. Their microscope had two lenses and total magnifying power between 10X and 30X. Such types of microscopes were called **"flea glasses"**, since they were primarily used to examine small whole organisms such as fleas and other insects. In 1610, an Italian **Galileo Galilei** (1564 —1642) invented a simple microscope having only one magnifying lens. This microscope was used to study the arrangement of the **facets** in the compound eye of insects.

The Italian microanatomist **Marcello Malpighi** (1628—1694) was among the first to use a microscope to examine and describe thin slices of animal tissues from such organs as the brain, liver, kidney, spleen, lungs and tongue. He also studied plant tissues and suggested that they were composed of structural units that he called **"utricles"**. An English microscopist **Robert Hooke** (1635—1703) is credited with coining the term **cell** (L., *Cella* = hollow space) in 1665. He examined a thin slice cut from a piece of dried cork under the compound microscopes (Fig. 1.1) which were built by him. In 1665, **Hooke** published a collection of essays under the title *Micrographia*. One essay described cork as a honey comb of chambers or "cells". The chambers or cells are now recognized to be empty spaces left behind after the living portions of the cell had disintegrated. **Hooke** thought of the cells, he observed as something similar to veins and arteries of animals—they were filled with "juices" in living plants. But his crude microscopes did not permit the observation of any intracellular structure.

Anton van Leeuwenhoek (1632—1723)

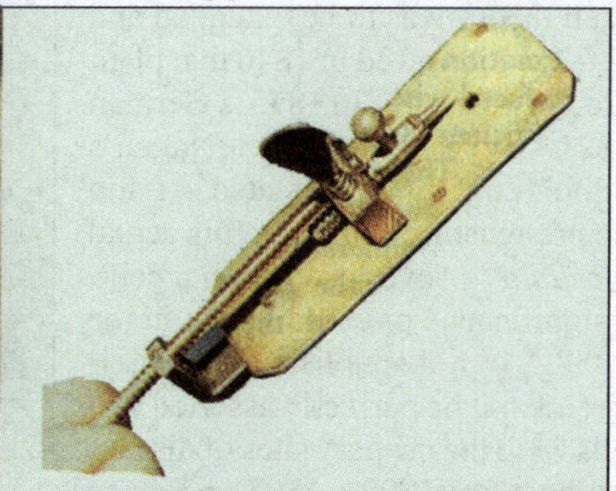

Fig. 1.2. Leeuwenhoek's microscope.

Dutch microscopist, **Anton van Leeuwenhoek** (1632—1723) had succeeded in greatly improving the art of polishing lenses of short focal length. He used his lenses in building numerous microscopes, some with magnifications approaching 300X (Fig. 1.2). **Leeuwenhoek** was the first to observe living free-living cells; he described in 1675, microscopic organisms in rainwater collected from tubes inserted into the soil during rainfall. His sketches included numerous bacteria (bacilli, cocci, spirilla and other Monera), protozoa, rotifers, and *Hydra*. **Leeuwenhoek** was also first to describe the sperm cells of humans, dogs, rabbits, frogs, fish and insects and to observe the movement of blood cells of mammals, birds, amphibians and fish, noting that those of fish and amphibians were oval in shape and contained a central body (the nucleus); while those of humans and other mammals were round. He also observed the striated muscles. **Leeuwenhoek's** observations were recorded in a series of reports that he sent during 1675—1683 to the Royal Society of London.

An English plant microanatomist **Nehemiah Grew** (1641—1721) published accounts of the microscopic examination of sections through the flowers, roots and stems of plants and clearly indicated that he recognized the cellular nature of plant tissues.

2. Growth of Cell Biology during 19th Century

Nineteenth century witnessed various cell biological inventions and formulations of various landmark theories such as cell theory and protoplasm theory. In 1807, **Mirbel** stated that all plant tissues were composed of cells. French biologist, **Rene Dutrochet** (1776—1827) correctly concluded in 1824, that all animal and plant tissues were "aggregates of globular cells." In 1831, an English botanist **Robert Brown** (1773—1858) discovered and named the **nucleus** in the cells (*e.g.*, epidermis, stigmas and pollen grains) of the plant *Tradescantia*. He established that the nucleus was the fundamental and constant component of the cells.

Cell Theory

In 1838, a German botanist **Mathias Jacob Schleiden** (1804—1881) put forth the idea that cells were the units of structure in the plants. In 1839, his coworker, a German zoologist, **Theodor Schwann** (1810—1882) applied Schleiden's thesis to the animals. Both of them, thus, postulated that the cell is the basic unit of structure and function in all life. This simple, basic and formal biological generalization is known as **cell theory** or **cell doctrine**. In fact, both **Schleiden** and **Schwann** are incorrectly credited for the formulation of the cell theory; they merely made the generalizations which were based on the works of their predecesors such as **Oken** (1805), **Mirbel** (1807), **Lamarck** (1809), **Dutrochet** (1824), **Turpin** (1826), etc., (see **Sheeler** and **Bianchi**, 1987). However, **Schleiden** was the first to describe the **nucleoli** and to appreciate the fact that each cell leads a double life—one independent, pertaining to its own development, and another as integral part of a multicellular plant. **Schwann** studied both plant and animal tissues and his work with the connective tissues such as bone and cartilage led him to modify the evolving cell theory to include the idea that living things are composed of both cells and the products or secretions of the cells. **Schwann** also introduced the term **metabolism** to describe the activities of the cells.

Louis Pasteur (1822—1895)

In the coming years, the cell theory was to be extended and refined further. **K. Nageli** (1817—1891) showed in 1846 that plant cells arise from the division of pre-existing cells. In 1855, a German pathologist **Rudolf Virchow** (1821—1902) confirmed the Nageli's principle of the cellular basis of life's continuity. He stated in Latin that the cells arise only from the pre-existing cells (*viz*., his actual aphorism was *"omnis cellula e cellula"* —every cell from a cell). **Virchow**, thus, established the significance of cell division in the reproduction of organisms. In 1858, **Virchow** published his classical textbook *Cellular Pathology* and in it he correctly asserted that as functional units of life, the cells were the primary sites of disease and cancer. Later, in 1865, **Louis Pasteur** (1822—1895) in France gave experimental evidence to support Virchow's extension of the cell theory.

The modern version of cell theory states that (1) All living organisms (animals, plants and microbes) are made up of one or more cells and cell products. (2) All metabolic reactions in unicellular and multicellular organisms take place in cells. (3) Cells originate only from other cells, *i.e.*, no cell can originate spontaneously or *de novo*, but comes into being only by division and duplication of already existing cells. (4) The smallest clearly defined unit of life is the cell.

The cell theory had its wide biological applications. With the progress of biochemistry, it was shown that there were fundamental similarities in the chemical composition and metabolic activities of all cells. **Kolliker** applied the cell theory to embryology—after it was demonstrated that the organisms developed from the fusion of two cells—the spermatozoon and the ovum. However, in the recent years, large number of sub-cellular structures such as ribosomes, lysosomes, mitochondria, chloroplasts, etc., have been discovered and studied in detail. Consequently, it may appear that cell is no longer a basic unit of life, because life may exist without cells also. Even then, the cell theory remains a useful concept.

Exception to cell theory. Cell theory does not have universal application, *i.e.*, there are certain living organisms which do not have true cells. All kinds of true cells share the following three basic characteristics: 1. A **set of genes** which constitute the blueprints for regulating cellular activities and making new cells. 2. A limiting **plasma membrane** that permits controlled exchange of matter and energy with the external world. 3. A **metabolic machinery** for sustaining life activities such as growth, reproduction and repair of parts. **Viruses** do not easily fit in these parameters of a true cell. Thus, they lack a plasma membrane and a metabolic machinery for energy production and for the synthesis of proteins. However, like any other cellular organism, viruses have (1) a definite genetically determined macromolecular organization; (2) a genetic or hereditary material in the form of either DNA or RNA; (3) a capacity of auto-reproduction; and (4) a capacity of mutation in their genetic substance. In consequence, viruses can only reproduce inside the host cells which may belong to animals, plants or bacteria. They use their own genetic programme for reproduction but rely on the raw materials (*i.e.*, amino acids, nucleotides) and biosynthetic machinery of the host cells (*i.e.*, ribosomes, tRNA, enzymes) for their multiplication. Thus, a virus may be defined as an infectious, subcellular and ultramicroscopic particle representing an obligate cellular parasite and a potential pathogen whose reproduction (replication) in the host cell and transmission by infection cause characteristic reaction in the host cells. Outside the host cells, viruses are just like non-living inert particles and like the salt or sugar, they can be purified, crystallized and placed into jars on a shelf for years. Due to this fact, viruses have been variously described such as "*naked genes that had somehow acquired the ability to move from one cell to another* (**Alberts** *et al.*, 1989), or as "*cellular forms that have degenerated through parasitism*", or as "*primitive organisms that have not reached a cellular state.*"

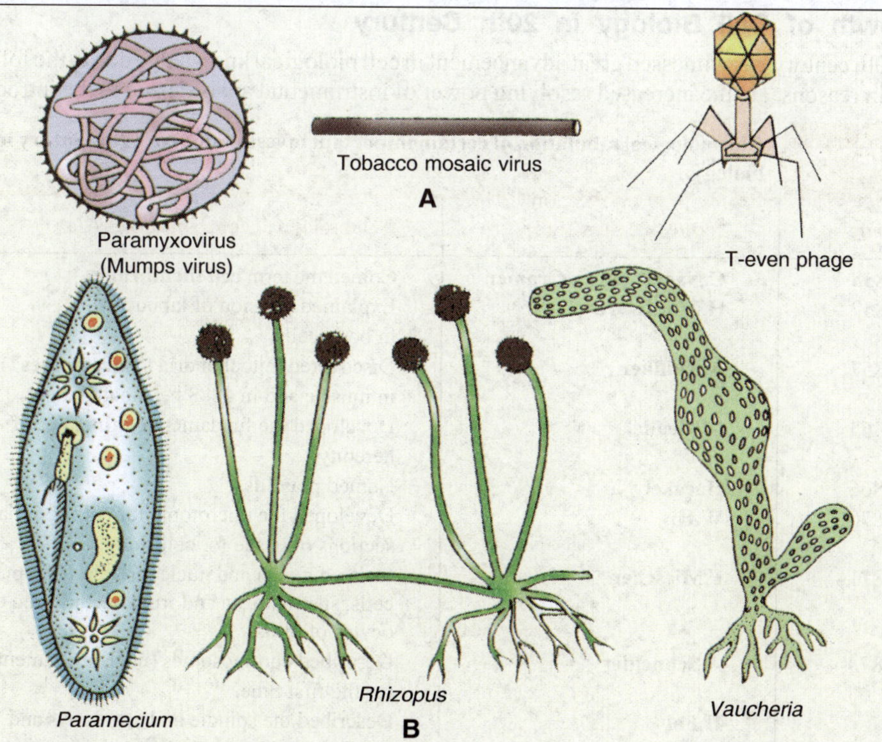

Fig. 1.3. Organisms forming exceptions to the cell theory : A—Three types of viruses; B—Three cases of cellular organization.

There are certain other organisms such as the protozoan *Paramecium*, the fungus *Rhizopus* and the alga *Vaucheria* (Fig. 1.3B) which do not fit into the purview of the cell theory. All of these organisms have bodies containing undivided mass of protoplasm which lacks cell-like organization and has more than one nucleus. They tend to raise the question that whether cell is a basic unit of structure in them.

Protoplasm Theory

Up to middle of the 19th century, greater emphasis was given to the cell wall and less to the cellular content. But soon cell biologists started to recognize the importance of "juicy" or "slimy" contents of the cells. In 1835, **Felix Dujardin** termed the jelly-like material within protozoans as **sarcode**. In 1835, **H.von Mohl** (1805—1875) described cell division. In 1839, the Czech biologist **J.E. Purkinje** (1787—1869) coined the term **protoplasm** to describe the contents of cells (animal embryos). **Von Mohl**, in 1846, applied the name protoplasm to the contents of embryonic cells of the plants. **Max Schultze**, in 1861, established similarity between sarcode and protoplasm of animal and plant cells and, thus, offering a theory which later on was improved and called **protoplasm theory** by **O.Hertwig** (1849—1922) in 1892.

Protoplasm theory holds that all living matter, out of which animals and plants are formed, is the protoplasm. The cell is an accumulation of living substance or protoplasm which is limited in space by an outer membrane and possesses a nucleus. The protoplasm which is filled in the nucleus is called **nucleoplasm** and that exists between the nucleus and the plasma membrane is called **cytoplasm**.

The last quarter of 19th century is usually considered as "**classical period of cell biology**". Since various significant cell biological discoveries have been made during this period. Certain landmark cell biological discoveries of second half of the 19th century have been tabulated in a chronological order in the Table 1-2.

3. Growth of Cell Biology in 20th Century

20th century has witnessed great advancement in cell biological knowledge due to the following two main reasons: (1) the increased resolving power of instrumental analysis due to the introduction

Table 1-2. Chronological tabulation of certain important investigations of 19th century in cell biology.

Year	Name of contributor	Cell biological contribution
1855	**C.Nageli** and **C. Cramer**	Coined the term **cell membrane.**
1857 – 1881	**H.M.Edwards**	Explained division of labour in body cells.
1857	**A.Kolliker**	Discovered mitochondria (**"sarcosomes"**) in muscle and in 1888 he isloated them.
1865	**G.Mendel**	Developed the fundamental principles of heredity.
1866	**Haeckel**	Named **plastids**.
1870	**W.His**	Developed the microtome for cutting serial sections of tissue for cell study.
1871	**F.Miescher**	Isolated nuclei and nucleoprotein from pus cells, spermatozoa and from haemolyzed erythr-ocytes of birds.
1873	**A.Schneider**	Described chromosomes (nuclear filaments) for the first time.
	H.Fol	Described the spindle and astral rays and showed in 1879 that only one sperm enters the egg in fertilization.

Year	Name of contributor	Cell biological contribution
1875	**E.Strasburger**	Described mitosis in plant cells and in 1882 introduced the terms **cytoplasm** and **nucleoplasm**. In 1884, he described fertilization in angiosperms.
	E. van Beneden	First observed the centriole.
1876	**O.Hertwig**	Studied reproduction in sea urchin and concluded that fertilization involves the union of sperm and egg pronuclei.
1878	**Schleicher**	Coined the term **karyokinesis.**
1879	**W.Flemming**	Introduced the term **chromatin** and described the longitudinal splitting of chromosomes during nuclear division of animal cells. In 1882, he coined the term **mitosis**.
1881	**Reinke** and **Rodewald**	Performed chemical analysis of protoplasm.
	E.G. Balbiani	Discovered the larval salivary gland chromosomes (*i.e.,* giant or polytene chromosomes) in *Chironomus.*
	Retzius	Described many animal tissues with a detail that has not been surpassed by any other light microscopist. In the next two decades, he, **Cajal** and other histologists developed staining methods and laid the foundations of microscopic anatomy.
1882	**W.Pfitzner**	Discovered chromomeres in the chromosomes.
1883	**E. van Benden**	Showed that in *Ascaris* the number of chromosomes in the gametes is half that of in the body cell.
	W. Roux	Proposed that chromosomes contain the units of heredity.
	Schimper	Introduced the term **chloroplast.**
	Meyer	Described details of chloroplast structure.
	E. Metchnikoff	Observed and named **phagocytosis**.
1886	**C.A. MacMunn**	Discovered cytochromes.
1888	**T.Boveri**	Described the structure of centrioles and coined the term **centrosome**. In 1892, he described spermatogenesis and oogenesis.
	W.Waldeyer	Introduced the term **chromosome.**
1890	**R.Altmann**	Stained mitochondria with a specific stain (1886), recognised their role in cellular respiration and considered them as autonomous organelles. In 1894, he coined the term **bioblasts** for mitochondria.
1892	**O.Hertwig**	Published his monograph—*Die Zelle und das Gewebe* (The cell and the tissue) in which he attempted to achieve a general synthesis of biological phenomena based on characteristics of the cell, its structure and function. He, thus, created cytology as a separate branch of biology.

1897	**C.Benda**	Coined the term **mitochondrion** and studied it in spermatozoa and other cells.
1898	**Camillo Golgi**	Described and coined the term **Golgi complex** for the reticular structure found in the cytoplasm of nerve cells of owls and cats. He used the silver staining method in studies.

of electron microscopy and X-ray diffraction techniques, and (2) the convergence of cytology with other fields of biological research, especially genetics (cytogenetics), physiology (cell physiology) and biochemistry (cytochemistry). Consequently new histochemical, cytochemical and immunocytochemical (using antibodies to localise antigens) techniques have been developed to detect various molecular components of the cell. Likewise, various cellular components have been separated by ultracentrifugation; different biochemical events of the cell could be known in detail by autoradiography; and methods of tissue culturing have made possible the study of living cells. Phase contrast microscopy and interference microscopy have been used to study the living cells. The ultrastructure of a cellular membrane could be observed by the techniques of freeze-fracturing and freeze-etching. Moreover, micromanipulators, micromanometric methods (*e.g.*, by Cartesian diver balance of **Zeuthen** weight of a single amoeba can be determined), chromatography, electrophoresis, spectrophotometry, etc., have provided new opportunities to cell biologists to investigate minute details of cell and its components. Due to the employment of various improved ultratechniques in the study of the cells, the validity of the cell theory and protoplasm theory has become vague. Therefore, presently, both of these theories have been replaced by another new theory called **organismal theory**.

Organismal Theory

The organismal theory holds that the body of all multicellular organisms is a continuous mass of protoplasm which remains divided incompletely into small centres, the cells, for the various biological activities. Thus, a multicellular organism is a highly differentiated protoplasmic individual, differing with a unicellular Protozoa only in size and degree of differentiation of the protoplasm. The differentiation involves separation of the protoplasm into subordinate semi-independent compartments, the so-called cells. Even the embryological development of a multicellular individual includes only growth and progressive internal differentiation of a small single protoplasmic individual (egg). Organismal theory too fails to ascertain the position of viruses.

Certain landmark cell biological discoveries and Nobel Prize winning investigations of 20th century have been tabulated in a chronological way in the Table 1-3.

Table 1-3. Chronological tabulation of certain important cell biological investigations of 20th century.

Year	Name of contributor	Cell biological contribution
1900	**C.Garnier**	Introduced the term **ergastoplasm.**
	J.Loeb	Discovered artificial parthenogenesis.
1901	**E.Strasburger**	Introduced the term **plasmodesmata.**
	T.H. Montgomery	Showed that homologous chromosomes undergo pairing or synapsis during the reduction division.
1902	**E.Fischer**	Got Nobel Prize for his pioneering studies of the proteins.
1903	**E.Buchner**	Discovered the enzymes and got Nobel Prize for it.
1904	**F. Meves**	Demonstrated the presence of mitochondria in plant cells.
1905	**J.B.Farmer** **J.E.Moore**	Coined the term **meiosis** for the reduction and cell division.

Year	Name of contributor	Cell biological contribution
1906	**M.Tswett**	Invented column chromatography.
	C.Golgi and **S.R. Cajal**	Got Nobel Prize for their contributions regarding the structure of nerve cells.
1907	**R.G. Harrison**	Developed the technique of tissue culture; cultivated amphibian spinal cord in a lymph clot.
1908	**E.Metchnikoff** and **P.Ehrlich**	Got Nobel Prize for their contributions on phagocytosis of bacteria during infection; staining procedures for bacteria, and studies on immunity.
1910	**A.Kossel**	Investigated the chemistry of the nucleus and got Nobel Prize for this contribution.
1915	**R.Wilstatter**	Got Nobel Prize for his studies on chlorophyll and other plant pigments.
1922	**A.V.Hill** and **O.Meyerhof**	Got Nobel Prize for their studies on the metabolism of muscle tissue and for relationship between muscle metabolism and lactic acid.
1924	**Lacassagne** and coworkers	Developed the first autoradiographic method to localize radioactive polonium in biological specimens.
1926	**T.Svedberg**	Got Nobel Prize for his studies on properties of colloids,especially proteins and for the development of analytical ultracentrifugation.
1930	**K.Landsteiner** 	Got Nobel Prize for the discovery of human blood groups and for studies of cellular agglutinins or antigens.
	Lebedeff	Designed and built the first interference microscope.
1931	**O.Warburg**	Got Nobel Prize for his studies on the nature and mode of action of respiratory enzymes and for the studies of oxidation and reduction in metabolism.
	E.Ruska and **M.Knoll**	Built the first transmission electron microscope.
	J.Q. Plowe	Coined the term **plasmalemma**.
	W.H.Lewis	Discovered pinocytosis.
1932	**F.Zernike**	Invented phase contrast microscope and got Nobel Prize for this invention in 1953.
1933	**T.H. Morgan** 	Got Nobel Prize for the discoveries concerning the role of chromosomes in the transmission of heredity.
	A.Tiselius	Introduced the technique of electrophoresis for separating proteins in solution.
1935	**J.Danielli** and **H.Davson**	Proposed protein-lipid-protein structure (sandwich model) of plasma membrane.
	M.Knoll	Demonstrated the feasibility of the electron microscope.
	M.W.Stanley	Isolated tobacco mosaic virus (TMV) in crystalline form.
1937	**A. von Szent-Gyorgyi**	Got Nobel Prize for his studies on biological oxidation and the involvement of vitamine C.

Year	Name of contributor	Cell biological contribution
	H.A.Krebs 	Discovered the citric acid cycle or tricarboxylic acid cycle and got Nobel Prize for this work in 1953.
1938	**Behrens**	Employed differential centrifugation to separate nuclei and cytoplasm from liver cells.
1939	**F.A.Lipman**	Proposed a central metabolic role for ATP.
1941	**Coons**	Used antibodies coupled to fluorescent dyes to detect cellular antigens.
1942	**Martin** and **Synge**	Developed partition chromatography, leading to paper chromatography two years later.
1943	**A.Claude**	Isolated cell components such as ribosomes, mitochondria and nucleus in relatively pure form by differential centrifugation.
1944	**Williams** and **Wyckoff**	Introduced the metal shadowing technique.
	C.F. Robinow	Demonstrated the nucleus (= nucleoid) in the bacteria.
1945	**K.R.Porter**	Discovered and named the **endoplasmic reticulum.**
	F.A. Lipman	Discovered coenzyme A (a key compound in cell metabolism) and got Nobel Prize in 1953 for his studies on this coenzyme.
1947	**C.F.Cori** and **G.T.Cori**	Got Nobel Prize for their studies of the metabolism of glycogen.
1948	**A.Tiselius**	Got Nobel Prize for his studies on the chemistry of proteins and for development of electrophoresis.
1948	**C.de Duve**	Isolated lysosomes and identified their enzymatic properties. In 1955, he coined the term **lysosome.**
	Grigg and **Hodge**	Studied fine structure of the flagellum of sperm.
1952	**G.E. Palade**	Described the ultrastructure of mitochondria.
	A. Morten and **R. Synge**	Got Nobel Prize for the development of chromatographic procedures for the separation of biological substances.
	Manton *et al.*	Studied fine structure of cilia of higher plants.
1953	**Robinson** and **Brown**	Reported ribosomes in the plant cells (*i.e.*, bean root).
1954	**J.Rhodin**	Described and named **microbody** in mouse kidney tissue.
	L.Pauling	Got Nobel Prize for his studies on the nature of chemical bonds, especially the peptide bond of proteins.
	Fawcett and **Porter**	Confirmed the 9 + 2 fibrillar arrangement of cilia and flagella.

Year	Name of contributor	Cell biological contribution
1955	**G.E. Palade**	Observed ribosomes in animal cells and in 1956 he detected RNA in the isolated ribosomes.
	F.Sanger	Completed the analysis of the amino acid sequence of bovine insulin; the first protein to be sequenced; got Nobel Prize in 1958 for this contribution.
1956	**J.H.Tjio** and **A.Levan**	Gave the first correct human chromosome count (46 chromosomes in diploid condition).
1959	**Brenner** and **Horne**	Used negative staining technique in visualizing viruses, bacteria and protein filaments.
	G.D. Robertson	Forwarded the concept of unit membrane.
1960	**Park** and **Pon**	Discovered **quantosomes** in the chloroplast.
	H.Ris and **M. Nass**	Independently verified the existence of DNA fibrils in mitochondria and chloroplasts.
1961	**M.Calvin**	Got Nobel Prize for his work on the assimilation of CO_2 by plants, photosynthesis, the "Calvin cycle."
1962	**M.F. Perutz** and **J.C.Kendrew**	Got Nobel Prize on their studies of the structure of globular proteins, especially myoglobin and haemoglobin.
1963	**J.Eccles, A. Hodkin** and **A. Huxley**	Got Nobel Prize for their work on the role of sodium and potassium ions in the conduction of nerve impulses along the nerve cell membrane.
1964	**K. Bloch** and **F.Lynem**	Got Nobel Prize for their studies on the metabolism of cholesterol and fatty acids.
	Kato and **Takeuchi**	Obtained a complete carrot plant from a single carrot root cell by tissue culture technique.
1965	**De Duve**	Coined the term **peroxisome** for catalase enzyme containing microbody.
	Harris and **Watkins**	Produced first **heterokaryons** of mammalian cells by the virus-induced fusion of human and mouse cells.
1967	**R.W.Breidenbach** and **H.Beevers**	Coined the term **glyoxisome** for the glyoxylate cycle containing microbody of plant cells.
1971	**E.A. Sutherland**	Got Nobel Prize for studies on the mechanism of action of hormones and role of cyclic AMP.
1972	**S.J. Singer** and **G.L. Nicolson**	Proposed the fluid mosaic model of cell membrane.
1974	**A.Claude, C.de Duve** and **G. Palade**	Got Nobel Prize for isolation and characterization of sub-cellular organelles and other particles.
1978	**P.Mitchell**	Got Nobel Prize for the studies of bioenergetics.
1981	**Allen** and **Inoue**	Perfected video-enhanced contrast light microscopy.
1982	**Aaron Klug**	Got Nobel Prize for his studies on the structure of complicated biological molecules (*e.g.*, proteins and nucleic acid in TMV and histone core of nucleosome) by using electron microscopy and X-ray crystallography.

Year	Name of contributor	Cell biological contribution
1984	**Schwartz** and **Cantor**	Developed pulsed field gel electrophoresis for the separation of very large DNA molecules.
	C.Milstein, J. F. Kohler and **J.K. Jerne**	Got Nobel Prize for his studies on molecular immunology.
1987	**S.Tonegawa**	Got Nobel Prize for primary discoveries in the field of antibodies (immunobiology).

UNIT OF MEASUREMENT OF CELL

The viruses and cells of most bacteria, blue green algae, animals and plants are minute in size and are measured by the fractions of **standard units**. The standard units are metres, grams, litres and seconds. The value of different units of measurements has been tabulated in Table 1- 4.

Table 1-4. Units of measurement used in cell biology (Avers, 1978).

A. Length

Metre (m)	Millimetre (mm)	Micrometre or Micron (µm)	Namometre or Millimicron (nm or mµ)	Angstrom (A^0)
1	1,000 (1×10^3)	1,000,000 (1×10^6)	1,000,000,000 (1×10^9)	1×10^{10}
0.001	1	1,000	1,000,000	1×10^7
0.000001	0.001	1	1,000	1×10^4
1×10^{-9}	1×10^{-6}	0.001	1	10
1×10^{-10}	1×10^{-7}	1×10^{-4}	0.1	1

B. Weight

Gram (g)	Milligram (mg)	Microgram (µg)	Nanogram (ng)	Picogram (pg)
1	1,000	1,000,000	1×10^9	1×10^{12}
0.001	1	1,000	1×10^6	1×10^9
1×10^{-6}	0.001	1	1×10^3	1×10^6
1×10^{-9}	1×10^{-6}	0.001	1	1×10^3
1×10^{-12}	1×10^{-9}	1×10^{-6}	0.001	1

CELL BIOLOGY AND OTHER BIOLOGICAL SCIENCES

The cell biology has helped the biologists to understand various complicated life activities such as metabolism, growth, differentiation, heredity and evolution at the cellular and molecular levels. Due to its wide application in various branches of biological science, many new hybrid biological sciences, have sprung up. Some of them are as follows:

1. Cytotaxonomy (Cytology and Taxonomy). Each plant and animal species has a definite number of chromosomes in its cells and the chromosomes of the individuals of a species resemble closely with one another in shape and size. These characteristics of the chromosomes help a taxonomist in determining the taxonomical position of a species. Further, cell biology furnishes strong support to the manner of origin of certain taxonomic units. Therefore, the cytotaxonomy can be defined as a cytological science which provides cytological support to the taxonomic position of any species.

2. Cytogenetics (Cytology and Genetics). Cytogenetics is that branch of cell biology which is concerned with the cytological and molecular bases of heredity, variation, mutation, phylogeny, morphogenesis and evolution of organisms. The Weismann's germ plasm theory, Mendel's laws of inheritance and the concept of gene could be well understood only after the application of cytological concept to the genetics.

3. Cell Physiology (Cytology and Physiology). The cell physiology is the study of life activities, *viz*., nutrition, metabolism, excitability, growth, reproduction or cell division and differentiation of the cell. The cell physiology has helped in understanding various complicated physiological activities at cellular level.

4. Cytochemistry (Cytology and Biochemistry). The cytochemistry is that branch of cytology which deals with the chemical and physico-chemical analysis of living matter. For example, the cytochemical analysis has revealed the presence of carbohydrates, lipids, proteins, nucleic acids and other organic and inorganic chemical compounds in the cells.

5. Ultrastructure and Molecular Biology. These are the most modern branches of biology in which the merging of cytology with biochemistry, physico-chemistry and especially macromolecular and colloidal chemistry become increasingly complex. Knowledge of the submicroscopic organization or ultrastructure of the cell is of fundamental importance because practically all the functional and physico-chemical transformations take place with the molecular architecture of the cell and at a molecular level. The recent discoveries in molecular biology such as the discovery of molecular model of DNA by **Waston** and **Crick** in 1953, molecular interpretation of protein synthetic mechanism, genetic code, etc., have an extraordinary impact on modern cell biology and biology.

Waston and Crick.

6. Cytopathology (Cytology and Pathology). The application of molecular biology to pathological science has helped in understanding various human diseases at molecular level. Because most diseases are caused due to disorder of genetic codes in DNA molecule which alter the synthetic process of enzymes and ultimately disturb metabolic activities of the cell.

7. Cytoecology (Cytology and Ecology). The cytoecology is the science in which one studies the effects of ecological changes on the chromosome number of the cell. The cytological studies on plants and animals have revealed that the ecological habitat and geographical distribution have the correlation with chromosome numbers.

REVISION QUESTIONS

1. Who had discovered the cell ? Explain, how is the growth of cell biology linked with the improvement in instrumental analysis ?
2. What is cell theory ? Describe the cell theory and explain the exceptions of cell theory.
3. What is meant by 'classical period of cell biology' ? Write about certain landmark discoveries of this period.
4. Write short notes on the following:
   (i) Protoplasm theory;
   (ii) Organismal theory;
   (iii) Branches of cell biology;
   (iv) Scope of cell biology.

CHAPTER

2

Techniques in Cell Biology

Work in a cytology laboratory where DNA, extracted from human cells, is analysec by techniques including electrophoresis and autoradiography.

Cells are tiny but complex bodies. It is difficult to see their structure; more difficult to understand their molecular composition and still difficult to find out the function of their various components. What one can learn about cells, depends on the tools at one's disposal and, in fact, major advances in cell biology have frequently taken place with the introduction of new tools and techniques to the study of cell. Thus, to gain divergent types of information regarding cell's structure, molecular organization and function, cell biologists have developed and employed various instruments and techniques. A basic knowledge of some of these methods is earnestly required.

MICROSCOPY

In the search for information about the structure and composition of cells, the cell biologists immediately face two limitations : the exceedingly small dimensions of cells and their component parts and the transparent nature of cells. The diameters of the majority of cells fall within a range of 0.2 and 50 μm. The human eyes have limited distinguishing or resolving power. The ability of an observational instrument such as a human eye or a microscope to reveal details of structure is expressed in terms of **limit of resolution** (*l*) which is defined as *the smallest distance that may separate two points on an object and still permit their observation as distinct separate points.* The unaided human eye under optimal conditions in green light (to which it is most sensitive) cannot distinguish between points less than about 0.1 mm or 100 μm apart. Structural details smaller than this, *e.g.*, cell, is unresolved unless some instrument capable of higher resolution is used. **Magnification**, *the increase in size of optical image over the size of the object being viewed*, is of no use unless the observational system can resolve the various parts of the structure being examined. Increased magnification without improved resolution results only in a large blurred image. The human eye has no power of magnification, so magnifying glasses may be used to magnify images up to about 10 times. A light compound microscope in which many lenses are combined together has a useful magnification of about 1,500 times.

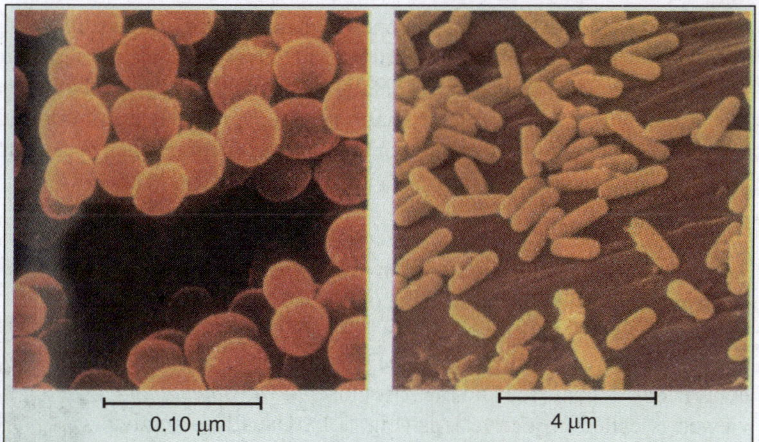

Microscopes magnify microorganisms manifold and help us to determine their shape and structure like spherical shaped cocci and rod-shaped, bacilli bacteria as shown here.

The limit of resolution (*l*) of any optical instrument (*i.e.,* eye or microscope) is given approximately by the Abbe's relationships :

$$\text{Resolution } (l) = \frac{\text{wavelength } (\lambda)}{\text{numerical aperture } (n \sin \alpha)}$$

where λ (lambda) is the wavelength ("colour") of the illumination or radiation used to form the image, *n* is refractive index (a function of density) of the material (*i.e.*, mostly air or water) between the specimen and the first lens (or objective lens), and $\sin \alpha$ is sine of the semi-angle of aperture of the first lens as viewed from the specimen. The quantity "$n \sin \alpha$" is often called the **numerical aperture** (NA).

Abbe's relationships make it clear that high resolution in a microscope can only be achieved by manipulating a small number of variables: the wavelength of the illuminating radiation, the refractive index and the aperture. The **aperture** is limited to something less than 90° since that would have the lens and specimen in contact with one another. In fact, 85° is about the limit in good optical microscopes. Such angles require an excellent lens. In most cases, the aperture is less because the edges of the lens introduce distortions and so cannot be used. **Refractive index** is easy to alter, but only within narrow limits. It can be increased by using oils to fill the space between the specimen and the objective lens. Transparent immersion oils used in today's microscopes (*i.e.,* **oil immersion lens**) have *n* up to about 1.6. Still 1.6 is big improvement over air or water (n=1). In a microscope, the smallest detectable detail is equal to about one-half the **wavelength** of light with which it is observed. The smaller the object, the shorter the wavelength of light required. Hence, the wavelength of light is the area which has great chances of improvement. One can, for example, use ultraviolet light instead of visible light, thus, improving resolution as much as twofold. In order to do that however, special lenses (*e.g.,* of quartz) must be used since ordinary glass blocks much ultraviolet light. In such a microscope, called **ultraviolet microscope,** the eyes cannot be used to view the image directly, for they are insensitive to ultraviolet light. Lastly, a specimen cannot absorb light of wavelength below 0.3 μm.

Thus, a good light microscope, with a numerical aperture of 1.4 and using light of short wavelength (0.4 µm) will resolve two points at about 0.17 µm separations. By such a microscope though, one can see considerable details in most cells, there is also a great deal that cannot be seen. For instance, ribosomes and chromatin threads of nucleus are about 0.02 µm in diameter and quite invisible to the light microscope. For them electron microscope is used. In cell biological studies, the following two types of microscopes are most extensively used :

Light Microscopy

The **compound light microscope** uses visible light for illuminating the object and contains glass lenses that magnify the image of the object and focus the light on the retina of the observer's eye. It consists of two lenses, one at each end of a hollow tube. The lens closer to eye is called **ocular lens** or **eyepiece** and the lens closer to the object being viewed is called **objective lens** (Fig. 2.1). Usually objective lenses of various magnifying powers are mounted on a revolving turret at the lower end of the tube. The object, supported by a glass slide under the objective lens, is illuminated by light beneath it. In ordinary microscopes light is reflected on the object by a mirror having concave and plane surfaces. In some microscopes, a third lens, called **condenser lens**, is located between the object and the light source and serve to focus the light on the object.

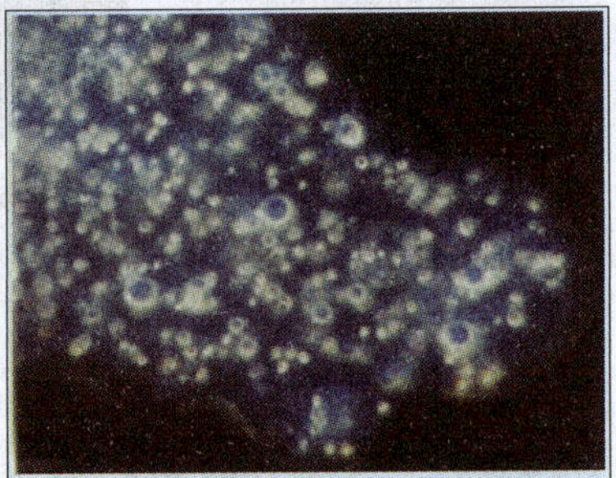

Dark - ground microscopy.

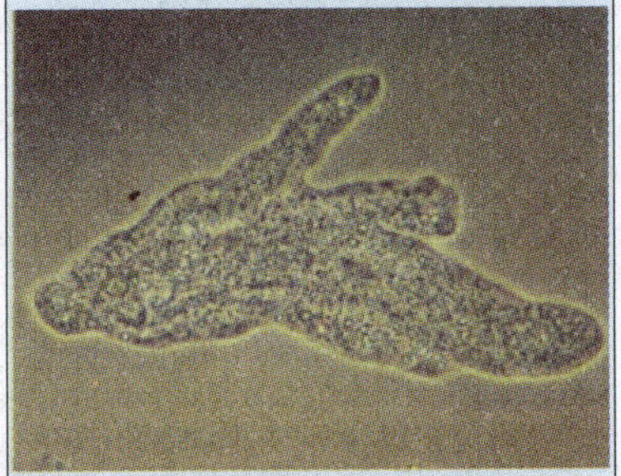

Phase-contrast microscopy.

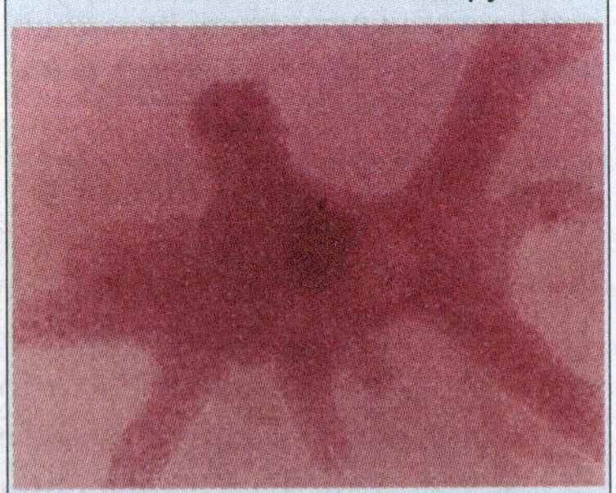

Oil-immersion microscopy.

In order to make full use of available resolving power of the compound light microscopes, special techniques have been designed to improve contrast. Certain improved types of light microscopes are of the following types :

1. Dark field microscope or Ultramicroscope. This type of light microscope is particularly useful for viewing suspensions of bacteria. In it, the object is viewed only with oblique rays and since one sees only those light rays that are scattered from objects, the images appear bright on a black background. The process is akin to seeing dust particles floating in a sunbeam.

2. Phase contrast microscope. This type of light microscope takes advantage of the fact that different parts of a cell have different densities and, hence, different refractive indices. Regions where the refractive index is changing they tend to bend light rays. In phase contrast microscope these bent rays are used to form patterns of destructive interference, yielding sharp contrasts. This technique is widely used to observe unstained and living cells (especially in mitotically dividing cultured cells).

The **interference microscope** is based on the principle of the phase contrast microscope and permits detection of small, continuous changes in refractive index. The variations of phase can be transformed into such vivid colour changes that a living cell looks like a stained preparation.

3. Polarization microscope. This type of light microscope is useful mainly for viewing highly ordered objects such as crystals or bundles of parallel filaments (*i.e.*, microtubules of mitotic spindle.) It is based on the behaviour of certain cellular materials when they are observed with polarized light. If the material is **isotropic**, polarized light is propagated through it with the same velocity, independent of the impinging direction. Such substances are characterized by having the same **index of refraction** in all directions. On the other hand, in an **anisotropic material** the velocity of propagation of polarized light varies. Such a material is also called **birefringent** because it presents two different indices of refraction corresponding to the respective different velocities of transmission. In a polarizing microscope, the specimen is placed between two closed polarizers and visible birefringent portions of the sample act like polarizing films and, hence, these portions of the sample are seen as bright objects on dark background.

Methods of Sample Preparation for Light Microscopy

Cells are transparent and optically homogeneous: so either they are viewed as such by instruments such as phase contrast microscope or to produce necessary contrast, the cells are passed through various steps of slide preparation such as killing, fixation, dehydration, embedding, sectioning, staining and mounting.

Thus, superior specimens for microscopic examination can be obtained by **killing** the cells and coagulating or **fixing** the protoplasm by preservatives, called **fixatives** such as alcohols, formaldehyde, mercuric chloride, picric acid, acetic acid and mixture of these. The process of fixation involves the following events — (1) The proteins and other macromolecules are precipitated. (2) The intracellular hydrolytic enzymes are denatured, preventing autolysis. (3) Cross links are formed between macromolecules, making the preparation more stable and minimizing shrinkage upon drying. (4) Substances are introduced which prevent attack by microorganisms. (5) The tissues become stiffer, making their sectioning easier. (6) The affinity of the tissue for dyes (stains) is increased.

Fixation is generally followed by **dehydration** (*i.e.,* gradual removal of water vapours from the tissue) by the organic solvents such as ethanol. The dehydrated specimens are **embedded** *i.e.*, they are infiltrated with molten paraffin which hardens upon cooling and provides enough support to allow thin sections to be cut with a **microtome**. By the microtome, serial sections, 5 to 10μm thick can be cut and placed on slides in the order of cutting and permitting a sequence of specimens for observation. These sections are stained with a non-vital-stain to increase the contrast.

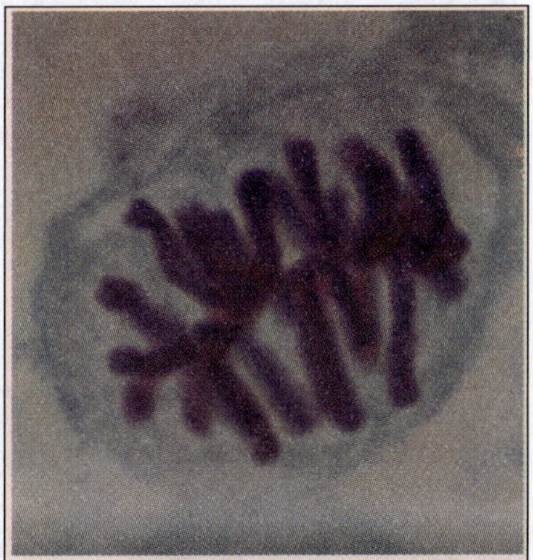

The Feuglen stain - This staining is specific for DNA.

Stains are the chemicals that can selectively attach to particular molecules of particular cellular structures and make them stand out from other parts of the cell. The non-vital stains fall into two main classses: **acid stains** such as eosin, orange G, aniline blue and fast green, all of which combine with basic molecules such as proteins of the fixed cells; and **basic stains** such as methylene blue, crystal violet, haematoxylin, basic fuchsin, etc., all of which combine with nucleic acids and other acidic molecules of the fixed cells. The cellular structures that stain with acid stains are called acidophilic and those that stain with basic dyes are called **basophilic**.

In addition, there are certain specific stains, called **cytochemical stains** that bind selectively to some specific groups of cellular macromolecules such as proteins, nucleic acids, polysaccharides and lipids. For example, Millon reaction, diazonium reaction and Naphthol Yellow 5 stain are used for the proteins; alkaline fast green is used for histone (basic protein); Feulgen reaction (using Schiff's reagent) is used for DNA; methyl green-pyronine stain (Unna-Pappenheim stain) is used in distinguishing between DNA and RNA and it stains DNA green and RNA red; acetocarmine and acetoorcein stains are used to stain chromosomes of dividing cells; periodic acid-Schiff (PAS) reaction is used for the demonstration of polysaccharide materials such as starch, cellulose, hemicellulose, and pectin in the plant cells and mucoproteins (glycoproteins), hyaluronic acid and chitin in animal cells; and fat soluble dyes such as Sudan Red and Sudan Black B are used for the lipids. The Sudan Black B is a specific stain for phospholipids and is used to stain Golgi apparatus.

Vital stains selectively stain the intracellular structures of living cells without serious alteration of cellular metabolism and function. For example, Janus green B selectively stains mitochondria; neutral red stains plant vacuoles and methylene blue stains Golgi apparatus and also nuclear chromatin of dividing cells.

All these steps often are time-consuming and cause artifacts in the cells. Hence, when speed is important and specimen is required for electron microscopy or for histochemical analysis, paraffin embedding may be replaced by fixation by freeze drying. **Freeze drying** is a method that avoids denaturation of enzymes and is particularly useful for histochemical staining. Tissue is frozen rapidly by plunging its small portions into liquid carbon dioxide or liquid nitrogen and, thus, required rigidity for sectioning by the **freezing microtome** is obtained. Frozen sections are stained and are dehydrated at low temperature (–30 to –40°C) in a high vacuum. At such low temperature ice crystals are of minimum size and few distortions or artifacts arise. Chemical composition and physical structure are maintained with little change. Another advantage is that fixation is rapid enough to arrest some cellular functions at their critical junctures which can then be observed and compared.

Electron Microscopy

The electron microscopy (Fig. 2.1) uses the much shorter wavelengths of electrons to achieve resolution as low as 3 A^0, with a usual working range between 5 to 12 A^0. In the electron microscope electromagnetic coils (*i.e.,* magnetic "lenses") are used to control and focus a beam of electrons accelerated from a heated metal wire by high voltages, in the range of 20,000 to 100,000 volts (new instruments are being developed that use 1,000,000 volts). The wavelength of an electron depends on the magnitude of the voltage and may be 0.01 A° or

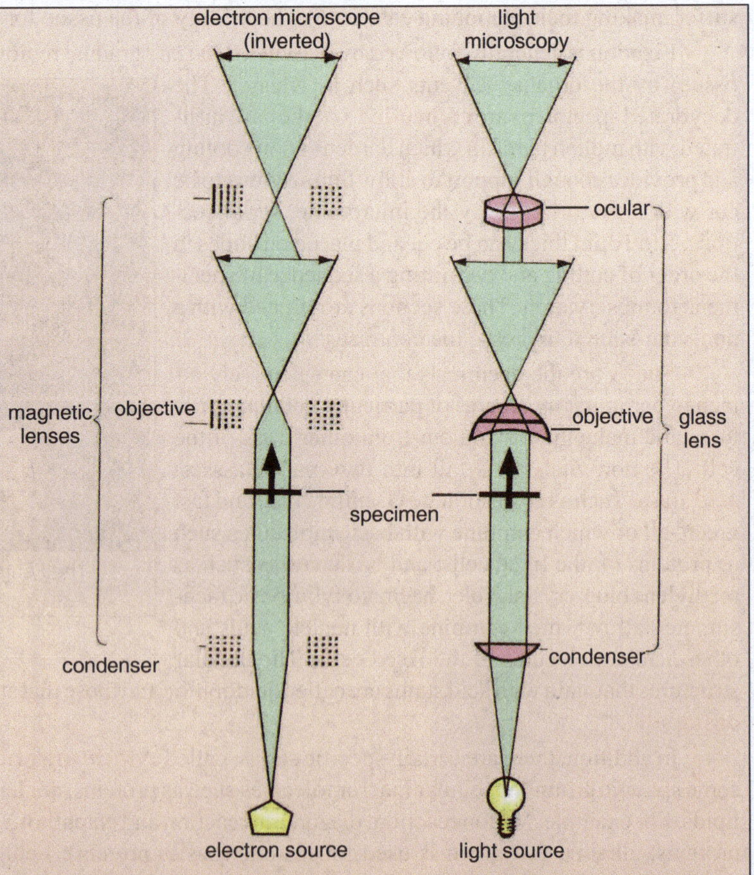

Fig. 2.1. Comparison of optical pathways in light and electron microscopes.

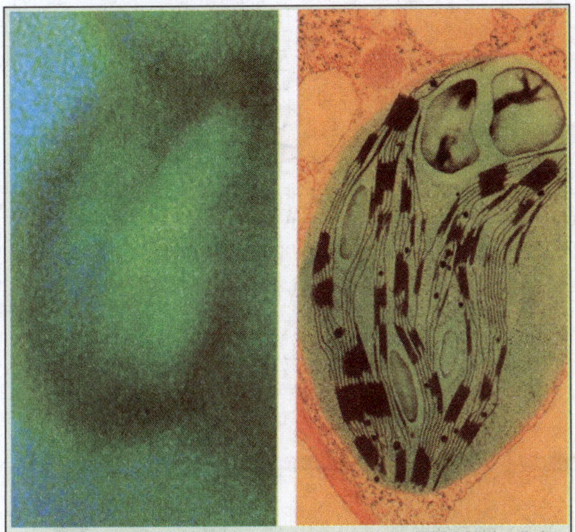

Chloroplasts at the same magnification (x25,000), by light microscopy (left), and by electron microscopy (right).

less. The electrons of the beam are scattered by a specimen placed in the path of the beam. Electrons that do manage to pass through the specimen are focused by an objective coil ('lens') and a final magnified image is produced by a projecter coil or 'lens'. The final image is viewed directly on the fluorescent screen or is recorded on photographic film to produce **electron micrograph**. This type of electron microscope is called **transmission electron microscope (TEM)**.

Unlike the compound light microscope, in which image formation depends primarily upon differences in light absorption, the electron mircroscope forms images as a result of differences in the way electrons are scattered by various regions of the object. Electrons have a very low penetrating power, that is, they are easily scattered by objects in their paths. The degree to which electrons are scattered is determined by the thickness and atomic density of the object: regions of high density(possessing atoms of high atomic number) scatter electrons more than regions of lesser density and consequently appear darker in the final image. Because electrons are scattered so easily, the specimen used in electron microscopy must be extremely thin (ultrathin, *i.e.,* 10 nm to 100 nm thick). If the sections were not extremely thin, most of the electrons would be scattered and a uniform dark image would result. Since, electrons are scattered even by gas molecules and so the electron beam must travel through the electron microscope in a very high vacuum and the samples must be completely dry and otherwise non-volatile. Thus, living cells which are wet cannot be viewed in electron microscope.

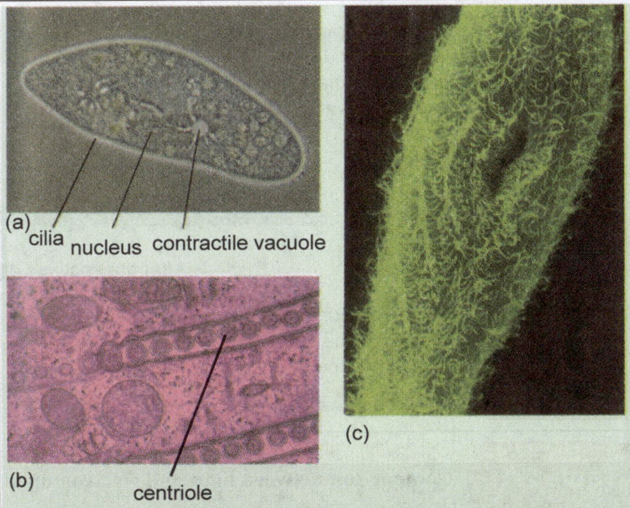

A comparison of microscope images
(a) A living *Paramecium* through light microscope
(b) A false - colour TEM photo showing centrioles
(c) A false colour SEM photo of a *Paramecium.*

Methods of Sample Preparation for Transmission Electron Microscopy

The standard procedure for the preparation of specimen for TEM entails fixation, dehydration, staining and sectioning similar to light microscopy (Table 2-1.). However, the most significant difference being the need for ultra-thin sections. Following techniques of sample preparation are generally used for different types of methods of studying ultrastructure of the cell:

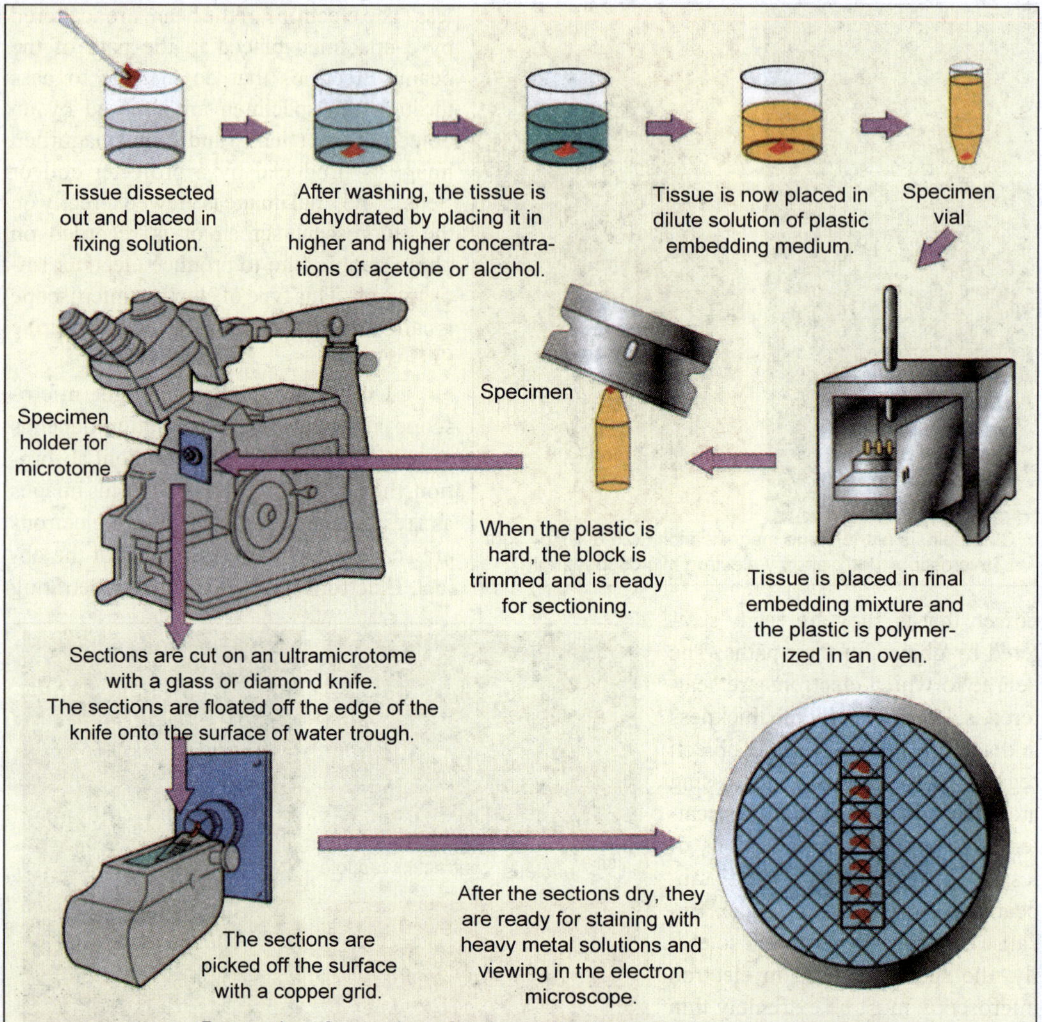

Preparation of a specimen for observation in the electron microscope.

Table 2-1. Comparison between light and electron microscopy.

Steps	Electron Microscopy	Light Microscopy
Fixation	Osmium tetroxide, potassium permanganate, formalin, glutaraldehyde.	Bouin's solution; formalin; Zenker's fluid.
Dehydration	Increasing concentration of ethanol (or acetone) followed by propylene oxide.	Increasing concentrations of ethanol followed by benzene.
Embedding	Araldite: Vestoplaw, Epan 812; Maraglas; Durcopan.	Paraffin.
Sectioning	Usually 10–100 nm thick sections cut with a glass or diamond knife on an ultramicrotome.	Usually 6μm thick sections cut with a razor blade on a microtome.

Steps	Electron Microscopy	Light Microscopy
Mounting	On a perforated metal disc (grid) usually covered with formvar or paralodian.	On a glass slide with an egg albumin adhesive. Deparaffinized in xylol for staining.
Staining	With salts of heavy metals such as lead acetate, lead citrate, lead hydroxide, uranyl acetate; phosphotungstic acid.	Selective chromatic stains (as haematoxylin and eosin), dehydrated in ethanol series, cleared in xylol and mounted for viewing in Canada balsam or Permount.
Viewing	Grid is placed between the condenser and objective lenses in a vacuum and the image is viewed on a phosphorescent screen.	Slide is placed between the condenser and objective lenses and the image viewed in the ocular lens.

1. Monolayer technique. Macromolecules such as DNA and RNA are studied by monolayer technique in which the macromolecules are extended on the air-water interface before being collected on a film.

2. Thin sectioning. This method uses a cutting device known as **ultramicrotome** to remove ultrathin (*i.e.,* 10 nm to 100 nm thick) sections from the specimen. To withstand the passage of ultrafine diamond or glass knife without tearing, the specimen is first embedded in a hard plastic such as, epoxy resin (Table 2-1). The resin is allowed to penetrate the sample before it is polymerized. Sections are floated from the knife of ultramicrotome onto the surface of water and picked up by touching them with a fine wire mesh or small circular copper grid (*i.e.,* small discs perforated with numerous openings). Prior to its use mesh or grid is coated with a thin monolayer film (7.5 to 15nm thick) of plastic (such as formvar or collodion) or carbon to provide a support to the sections (or sample) (Fig.2.2). The specimen is visualized through the holes of screen.

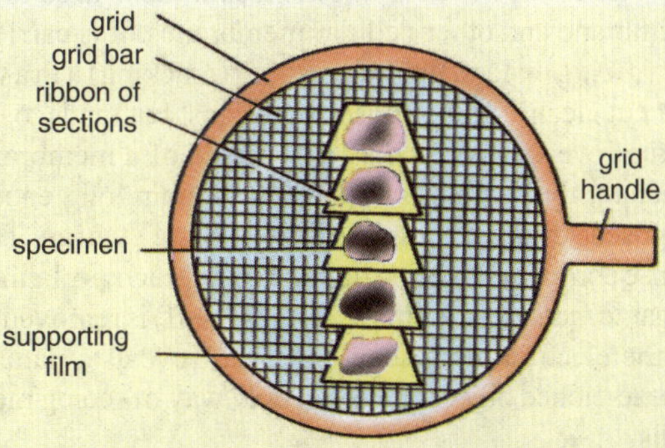

Fig. 2.2. Appearance of a ribbon of ultrathin sections on the grid (after Sheeler and Bianchi, 1987).

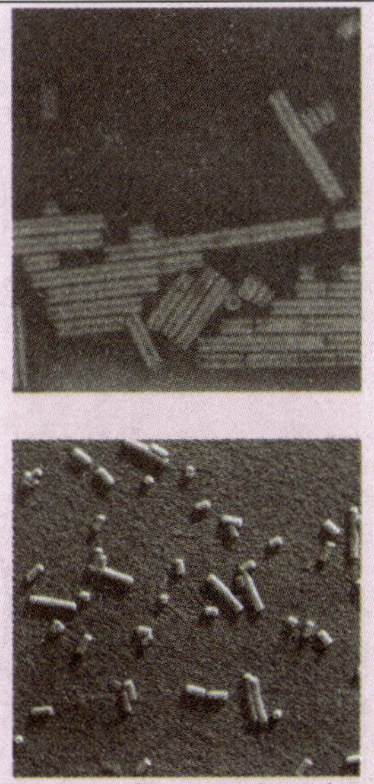
Examples of negatively stained and metal shadowed specimens.

Sections to be examined with the electron microscope are generally not stained (since no colours are seen with the electron microscope). However, contrast may be improved by "poststaining" with **electron stains** or electron-dense materials such as urynyl acetate, urynyl citrate, lead citrate, osmium tetroxide, etc. The method of thin sectioning is used to study morphology of cell.

3. Negative staining. This technique is used to study small particles such as viruses or macromolecules. Here, the specimen is embedded in a droplet of electron dense material, such as, phosphotungstic acid ($H_3PW_{12}O_{40}$). The electron stain penetrates into all the empty spaces (*i.e.,* openings and crevices) between the macromolecules. The spaces appear well defined in negative contrast. The portions of specimen that exclude stain transmit electrons readily, so their images can be seen.

4. Shadow casting or heavy metal shadowing. This technique is used to study three-dimensional appearance of viruses and certain macromolecules such as DNA molecules and collagen fibres. It involves placing of specimen in an evacuated chamber and evaporating at an angle, a heavy metal such as chromium, palladium, platinum or uranium from a filament of incandescent tungsten. The vapour of heavy metal is deposited on one side of the surface of the elevated particles; on the other side a shadow forms, the length of which permits determination of the height of the particle. In such a specimen, during electron microscopy, the electrons pass readily through the area of light metal content, less readily through the plane on which the particle sits, and are scattered more severely by the side of the particle on which metal has accumulated. Thus, by shadow casting, shape and profile of a particle can be discerned.

5. Tracers. Several biological processes such as pinocytosis, phagocytosis and transport of molecules across plasma membrane can be studied by the use of appropriate tracers (*e.g.,*gold, mercuric sulphide, iron oxide, etc.). These tracers are detected by their electron opacity. An ideal tracer should be non-toxic, physiologically inert, composed of small-sized particles of uniform and known size and preserved *in situ* during the processing of the tissue.

6. Freeze-fracture. This technique is used to study the molecular arrangement in the plasma membrane and other cellular membranes. It is carried out by rapidly cooling or freezing the sample (**cryofixation**) and then **fracturing** (cracking) it in a vacuum while it is still at —100ºC. The knife does not cut cleanly under those conditions, but tends to **fracture** (crack) the specimen along the lines of natural weakness, such as the middle of a membrane that runs parallel to the cut (Fig. 2.3). After fracture, the sample is left in the vacuum long enough to allow some water to evaporate from the exposed surfaces, a process called **freeze etching**. The exposed face is then shadowed with electron-dense combination of carbon and metal such as platinum to provide the necessary contrast, after which organic material (*i.e.,* the specimen itself) is removed by acids to leave a metal **replica** for examination in the electron microscope. Replica reveals a natural-looking representation of the surface of the freeze-etched object and is the only way of seeing membrane interior and certain other features of the cells.

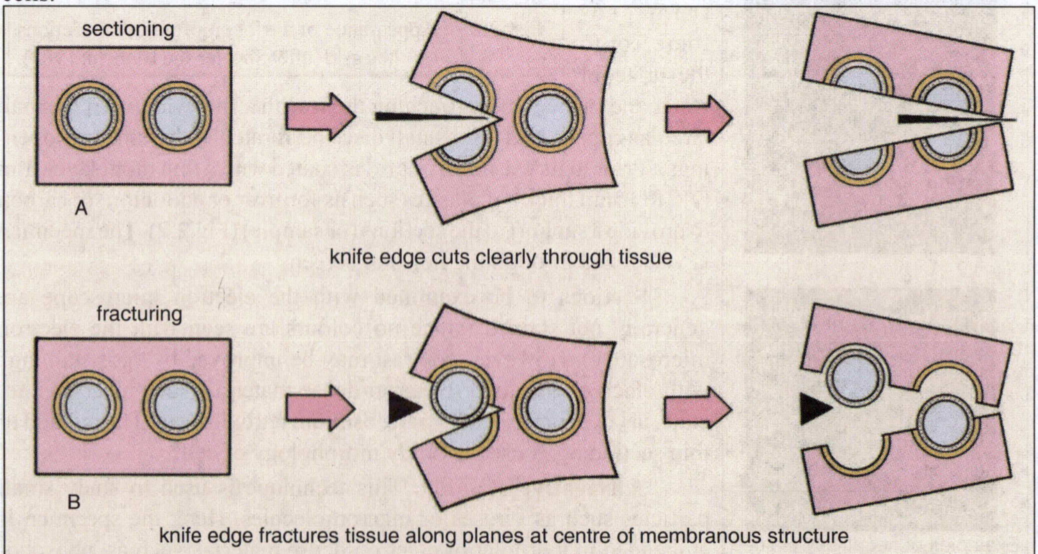

Fig. 2.3. A comparison of sectioning and fracturing tissues (after Sheeler and Bianchii, 1987).

7. Whole mounts. They are often used to examine chromosomes and other relatively thick objects that can be isolated free of debris. In these methods, the specimen is neither sectioned nor stained. Thick areas will scatter electrons more strongly than thin areas, providing enough contrast to form an image.

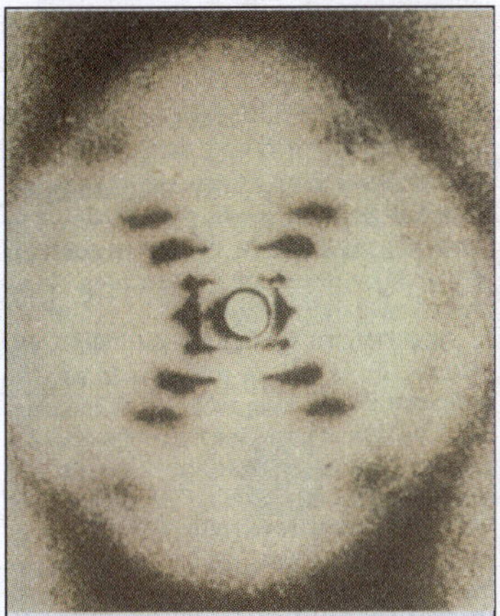

The X-ray diffraction pattern of DNA.

Scanning transmission electron microscopy (STEM). This electron microscope has less resolution power than the TEM (*i.e.*, about 200 A^0), yet is a very effective tool to study the surface topography of a specimen. In this instrument a narrow beam of electrons is scanned rapidly over a specimen and a three-dimensional image resulting from differential scattering of electrons by different parts of the surface of specimen is recorded on a cathode-ray oscilloscope and a photographic emulsion.

X- RAY DIFFRACTION ANALYSIS

This technique is used to analyze three-dimensional (tertiary) structure of DNA molecule and a variety of proteins such as myoglobin, haemoglobin, collagen, myelin sheath of nerve cells, myofibrils of striated muscles, etc. This method depends on the fact that X-rays are scattered or diffracted by the atoms of a substance. If the material has an ordered crystalline atomic structure, the resulting X-ray diffraction pattern is also ordered and reflects the three-dimensional arrangement of atoms in crystal.

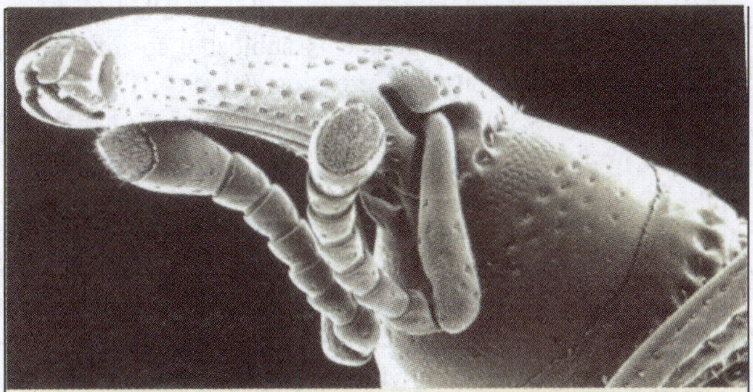

Scanning electron micrograph: the head and the mouthparts of a weevil.

CELL FRACTIONATION

Sometimes it becomes necessary to break up tissues and cells and to isolate various parts of the cell for structural or biochemical analysis. For this purpose, the technique of cell fractionation is employed. Cell fractionation method involves, essentially the homogenisation or destruction of cell boundaries by different mechanical or chemical procedures, followed by the separation of the subcellular fractions according to mass, surface and specific gravity by centrifuges.

Centrifuges in their various forms have become versatile tools of cell biology; they are used not only to characterize substances but to separate them. The **analytical ultracentrifuge** provides information concerning the mass and the shape of a molecule while **preparative ultracentri-**

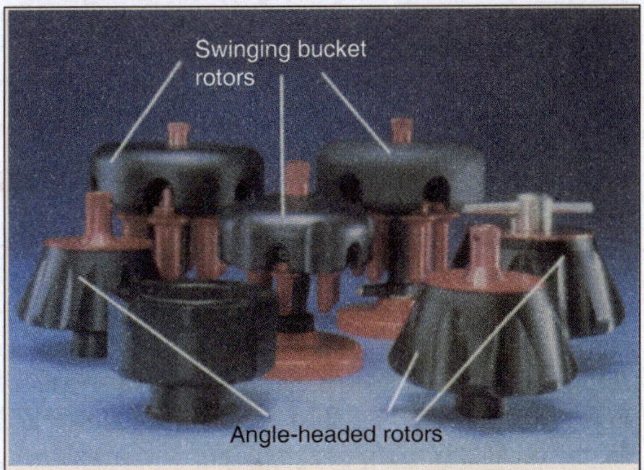

A variety of rotors used in a preparative ultracentrifuge.

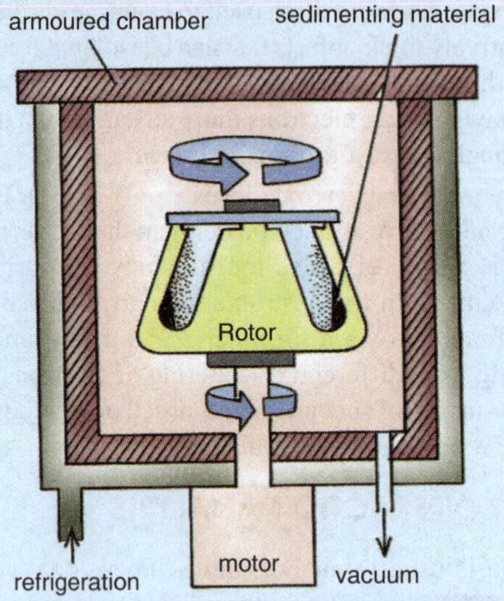

Fig. 2.4. The preparative ultracentrifuge. The sample (*i.e.*, homogenate) is contained in tubes that are inserted into a ring of cylindrical holes in a metal rotor. Rapid rotation of the rotor generates immense centrifugal forces, which cause particles in sample to sediment. The vacuum tends to reduce friction, preventing heating of the rotor and allowing the refrigeration system to maintain the sample at 4°C (after Alberts *et al.,* 1989).

fuge (Fig. 2.4) permits one to use these parameters to separate molecular types. An ultracentrifuge differs from other centrifuges only in attaining higher rotor velocities (*i.e.,* up to about 70,000 revolutions per minute or rpm). In addition, the analytical ultracentrifuge contains an optical system, allowing one to observe changes in the solute distribution as they occur in the sample. The rotors of all ultracentrifuge spin in a vacuum in order to prevent heating from air friction.

In the cell fractionation, the cells are gently broken by grinding a small piece of tisssue in a homogeniser having a moving close-fitting glass or plastic pestle within a tube that contains a medium such as sucrose solution to preserve the cellular organelles (Fig. 2.5). The solution containing homogenised or disrupted cells, is called **homogenate**. The homogenate is subjected to differential centrifugations of increasing velocity. The method depends on the principle that particles of different weight or sizes move at different rates through a solution in a centrifugal field. At each step larger particles form a gelatinous pellet at the bottom of the tube leaving smaller particles in the supernatant solution. By decanting the supernatant and spinning it harder, the next fraction can be brought down. Ultimately one is left with a supernatant solution having only soluble, molecular-sized components. The residual solution is called **cytosol**. The different molecules of cytosol are isolated by a variety of biochemical techniques such as chromatography, dialysis and electrophoresis.

The technique of cell fractionation has been improved through the use of **density gradient centrifugation.** In this method, centrifuge tube is loaded with layers of solution of varying densities of either sucrose, heavy water, cesium chloride or albumin, in a gradient from top to bottom. Once the gradient is formed, the homogenate is layered on the top and centrifuged until the particles reach equilibrium with the gradient. For this reason, this type of separation is called **equilibrium density** or **isopynic centrifugation** (**King**, 1986).

AUTORADIOGRAPHY

Autoradiography is a technique which is used to locate radioactive isotopes in cells, tissues, organs and whole organisms. A specimen is exposed to a solution containing molecules that have been made radioactive by the incorporation of radioactive isotopes, such as tritium (^{3}H), carbon 14 (^{14}C), phosphorus (^{32}P) and sulphur (^{35}S). The tagged molecules are often precursor molecules used by the cell in the synthesis of other needed molecules. At intervals, samples are removed from the solution; in case of smaller tissues, the samples are sectioned and mounted on glass slides or grids. The sections are then coated with a photographic emulsion and stored in the dark for periods ranging up to several months.

When a radioactive atom emits a beta particle (*i.e.*, electron) the photographic emulsion is affected in a manner similar to the exposure of a photographic emulsion to light. Over a period of time sufficient radioactive emissions occur to affect the silver grains of the emulsion. Black spots will appear at those sites when the emulsion is developed. Such spots will mark sites in the tissues where the radioactive atoms have accumulated. These sites can be identified by examining the stained tissue sections under the light microscope.

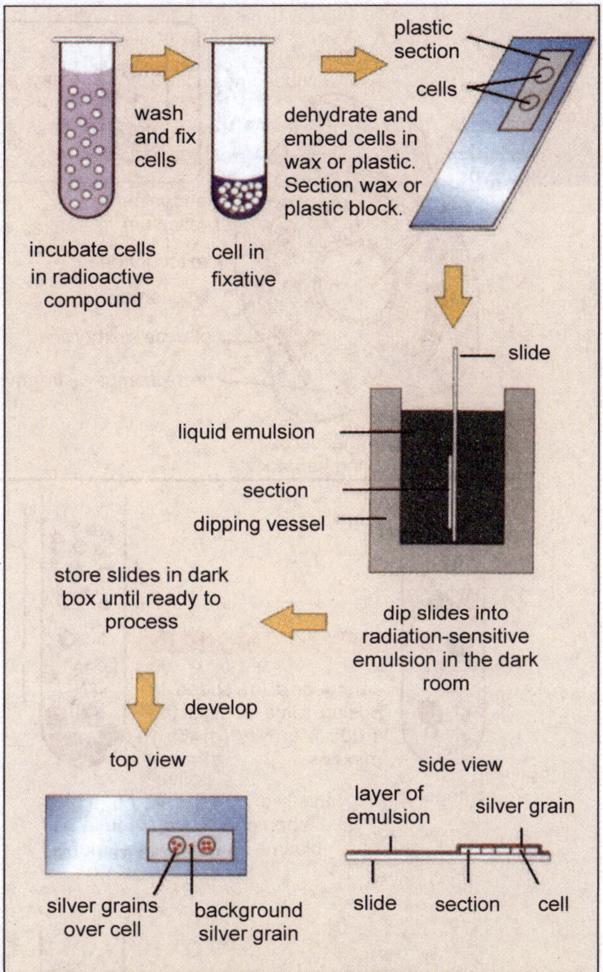

Step-by step procedure for the preparation of an autoradiograph.

In the technique of autoradiography, for the study of DNA metabolism of cell ^{3}H-thymidine is used; for RNA metabolism ^{3}H-uridine is used; for protein synthesis various tritiated (^{3}H-tagged) amino acids are used; and for polysaccharides and glycoproteins tritiated monosaccharides such as ^{3}H- mannose and ^{3}H-fucose are employed.

Pulse-labelling technique. This technique is used for those cases where biological molecules undergo considerable modifications after their synthesis (*e.g.*, ribosomal RNA). Here, actively growing cells are exposed to a radioactive precursor for a short period. The labelled precursor is then removed and replaced by 'cold' (unlabelled) precursor molecules. The unlabelled precursors are incorporated into the newly synthesized molecules and have the effect of 'chasing' the previously synthesized molecules containing the radioactive precursor through any maturation process. If the molecular species under investigation is sampled shortly after the start of the experiment, only the primary synthetic product will contain radioactivity. After longer time intervals, the original radioactive molecules will have been replaced with non-radioactive molecules. In this way the flow of radioactivity through a maturation process can be followed, together with any movement of molecular species within the cell.

CELL CULTURE

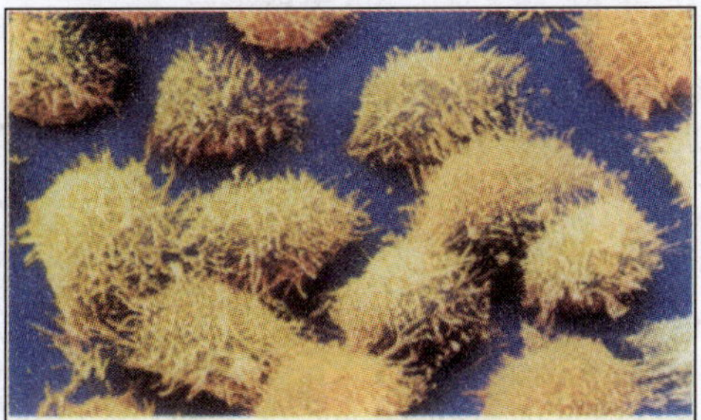
HeLa cells.

For cell biological observations sometimes it is needed to keep the animal and plant cell in living state outside the organism under favourable conditions. This process is called **cell culturing**. The cell cultures are of three main types: primary, secondary and those using established cell lines. **Primary cultures** are those obtained directly from animal tissue. The organ is aseptically removed, cut into small fragments and

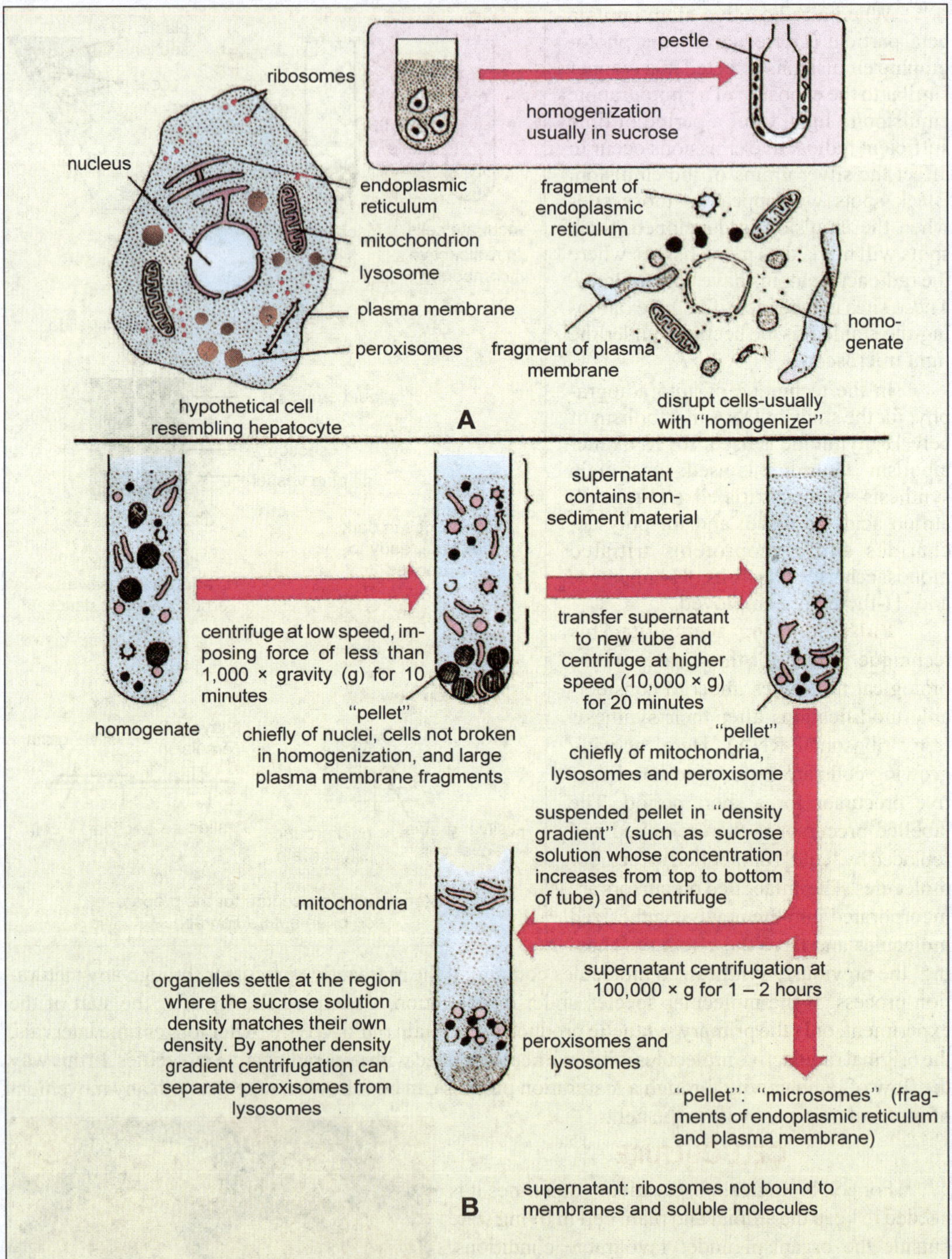

Fig. 2.5. Cell fractionation.

treated with trypsin enzyme to dissociate the cell aggregates into a suspension of viable single cells. These cells are plated in sterile petri-dishes and grown in the appropriate culture medium. When this culture is trypsinized and re-plated in a fresh medium then resultant culture is called **secondary culture**.

The other major type of cell culture uses **established cell lines**, which have been adapted to prolonged growth *in vitro*. Among the best known cell lines are **HeLa cells** (obtained from a human carcinoma), the **L** and **$3T_3$ cells** (from mouse embryo), the **BHK cells** (from baby hamster kidney) and the **CHO cells** (from Chinese hamster ovary).

CHROMATOGRAPHY

The chromatography is used to separate the molecules of different substances present together in a solution or cytosol. The solution is applied to an insoluble medium which has a different affinity for the individual molecules of the solution so that the molecules migrate through the medium at different rates. Following two types of chromatography are used in molecular biology :

(A) Paper chromatography. The paper chromatography (Fig. 2.6) is a smiple method for the separation of smaller molecules from one another. The molecules to be separated are applied to sheets of suitable paper, which are subsequently placed in a vessel which contains a suitable solvent. A distinction is made between ascending and descending paper chromatography according to whether the solvent migrates on the paper from below or above. Highly soluble components of the sample mixture will migrate at the solvent front; other substances move more slowly according to their solubility. From this it follows that the choice of solvent determines the speed of migration of the individual molecules and is crucial for a successful separation. The paper chromatography method is used for the separation of amino acids, nucleotides and other lower molecular-weight metabolic products.

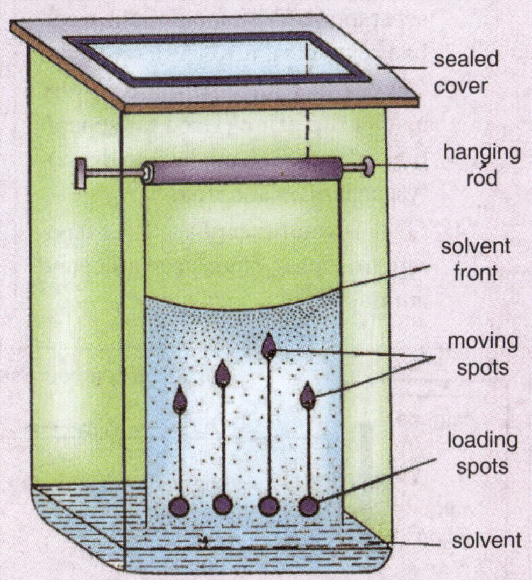

Fig. 2.6. Paper chromatography.

Thin layer chromatograph.

(B) Column chromatography. In column chromatography (Fig. 2.7), an insoluble medium is packed into a glass tube; the length and width of this tube influence the separation of the molecules. The molecules to be separated are applied to the top of the column and their migration is started by adding a solvent. The characteristic separation which results depends on the choice of solvent and carrier material. A positively charged carrier binds negatively charged molecules; other carriers contain pores which are penetrated by the smaller molecules, which are, therefore, slowed down. The solution which flows from the column is collected in small fractions, which contain the separated classes of molecules. Column chromatography is important for the separations of mixtures of proteins, that is, for the isolation of enzymes such as cytochrome C or RNA polymerase.

In a nutshell, presently the following types of chromatography are used for the isolation of different types of molecules (**King**, 1986; **Sheeler** and **Bianchi**, 1987):

A. Paper chromatography (used for separation of amino acids, nucleotides, and other low molecular weight solutes).

B. Thin layer chromatography or TLC (used for rapid separation of unsaturated and saturated fatty acids, triglycerides, phospholipids, steroids, peptides, nucleotides, etc.).

C. Column chromatography. It includes the following four types :

1. **Ion-exchange chromatography** (used for separation of proteins, RNA and DNA).
2. **Affinity chromatography** (used for separation of immunoglobulins, cellular enzymes, mRNAs).
3. **Gel permeation chromatography or Gel filtration** (Used for separation of proteins, nucleic acids, polysaccharides and lipids).
4. **Gas chromatography** (for the separation of lipids, oligosaccharides and amino acids).

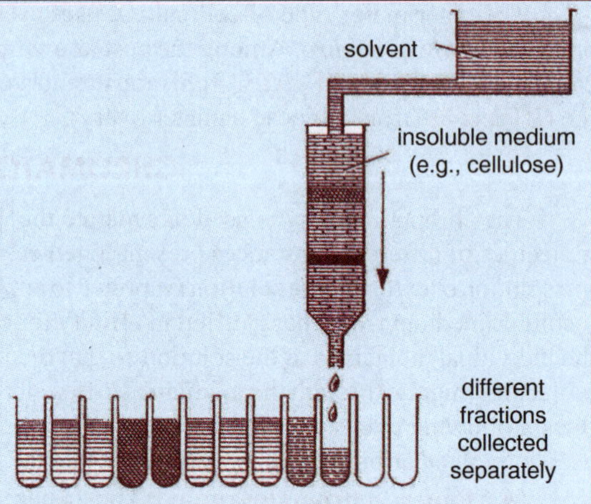

Fig. 2.7. Column chromatography.

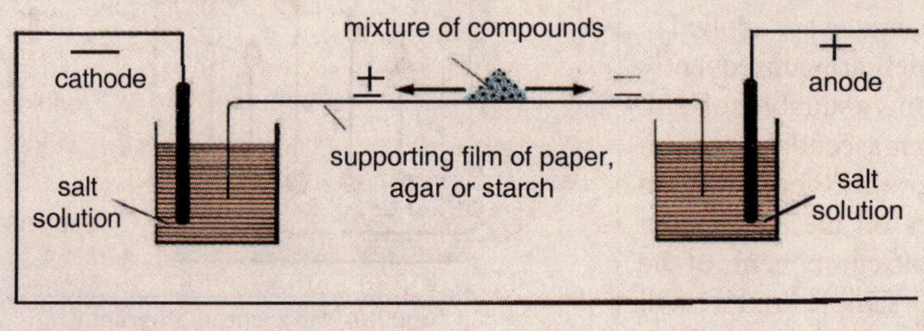

Fig. 2.8. Electrophoresis.

ELECTROPHORESIS

Molecules or macromolecules may be separated in an electric-field if they are charged to different extents. The mixture of compound is applied to supporting films which dip into two containers filled with a salt solution. One of the containers holds a cathode, the other an anode. On passing an electric current, the negatively charged molecules migrate to the anode and the positively charged molecules to the cathode. Paper, agar and starch are examples of substances which may be used as supporting films.

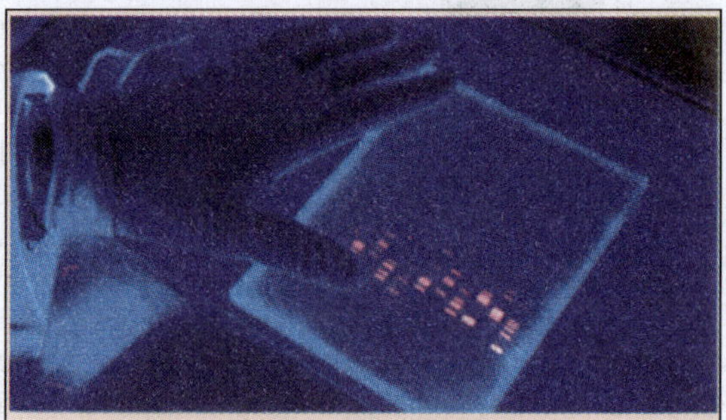
Separation of DNA restriction fragments by gel electrophoresis.

The rate of migration of a molecule in an electric field is determined by its size and the number of charged groups per molecule. The electrophoresis method is used in the separation of proteins, nucleic acids and their building blocks.

Some of the common types of techniques of electrophoresis, which are currently used in cell biology, are the following (**Sheeler** and **Bianchi**, 1987; **Alberts** *et al.*, 1989):

1. **Moving-boundary electrophoresis** (used for proteins).
2. **Gel or zone electrophoresis** (used for proteins).
3. **Discontinuous electrophoresis** (used for isolation of proteins of plasma membrane).
4. **SDS-PAGA** or **Sodium dodecyl sulphate-polyacrylamide gel electrophoresis** (used for separating and sizing macromolecules such as proteins, *e.g.,* membrane proteins, protein component of cytoskeleton, etc.).

5. Maxam - Gilbert technique (used for separation of polynuclotide fragments of RNA and DNA).

6. Immunoelectrophoresis (used for antigens and antibodies).

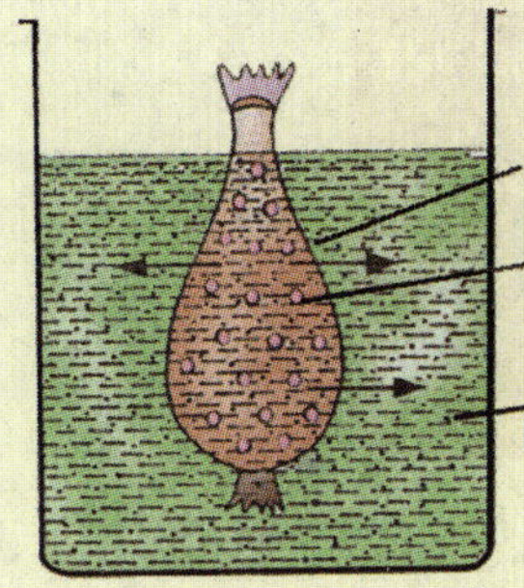

Fig. 2.9. Dialysis.

DIALYSIS

It is a sensitive method for separating lower molecular-weight component from macromolecules. A thin membrane in the form of a tube is filled with the solution containing the molecules to be separated. The pore size of the membrane allows the diffusion of small molecules such as salt or amino acids; larger molecules such as proteins or nucleic acids cannot pass through the pores and so remain inside the dialysis tube.

REVISION QUESTIONS

1. Enumerate various types of light microscopy. Describe the phase contrast microscopy.
2. Give a comparative account of light microscopy and electron microscopy.
3. Define the following terms : resolution, magnification, birefringent, vital stain, electron stain, cell fractionation, chromatography and electrophoresis.
4. What ways are available to improve contrast in light microscopy ?
5. Describe different techniques used in preparing sample for electron microscopy.
6. Write short notes on the following :

   (i) Cytochemical stains;
   (ii) Scanning electron microscope;
   (iii) X-ray diffraction analysis;
   (iv) Cell fractionation;
   (v) Autoradiography;
   (vi) Cell culture;
   (vii) Chromatography; and
   (viii) Electrophoresis.

CHAPTER

3

Cell

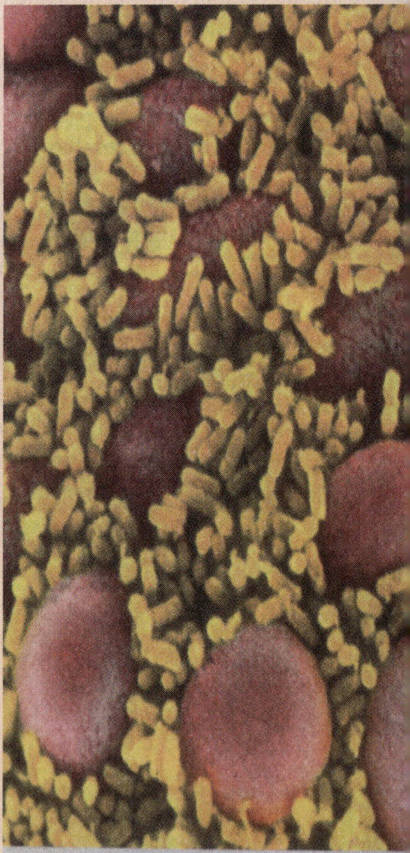

A close-up view of *E.coli* (yellow) in the human intestine (pink).

The cell is the basic unit of organization or structure of all living matter. Within a selective and retentive semipermeable membrane, it contains a complete set of different kinds of units necessary to permit its own growth and reproduction from simple nutrients. It has always been quite difficult to define a cell. Different cell biologists have defined the cell differently as follows : **A.G. Loewy** and **P. Siekevitz** (1963) have defined a cell as "*a unit of biological activity delimited by a semipermeable membrane and capable of self-reproduction in a medium free of other living systems*". **Wilson** and **Morrison** (1966) have defined the cell as "*an integrated and continuously changing system*." **John Paul** (1970) has defined the cell as "*the simplest integrated orgainization in living systems, capable of independent survival*."

All these definitions have excluded the viruses (see 'Exception of Cell Theory' in Chapter 1). A virus is neither an organism nor a cell, yet it consists of a core of nucleic acid (DNA or RNA) enclosed in an external mantle of protein. In the free state viruses are quite inert. They become activated only when they infect a living host cell and in the process only the nucleic acid core enter the host's cell. The nucleic acid which is the genetic substance, takes over the metabolic activity of the host cell and utilises the cell machinery for the formation of more viruses, ultimately killing the host cell. In a way, thus, viruses are cellular parasites that cannot reproduce by itself. But, because viruses are primitive and simpler units of life, therefore, they should be discussed prior to other cells.

VIRUSES

Viruses (L., venoum or poisonous fluid) are very small submicroscopic biological entities which though lack cellular organization (*viz.,* plasma membrane and metabolic machinery) possess their own genetic material, genetically determined macromolecular organization and characteristic mode of inheritance. For their multiplication, they essentially require the presence of some host cell, *i.e.,* they are obligate cellular parasites of either bacteria, plants or animals.

Structure

Viruses are quite a varied group (Fig. 3.1). They range in between 30 to 300 nm or 300 to 3000 A° in size, so they can be observed only by electron microscopy and X-ray crystallography. They have a regular geometrical and macromolecular organization. Basically an infectious virus particle (called **virion**) is composed of a **core** of only one type of nucleic acid (DNA or RNA) which is wrapped in a protective coat of protein, called **capsid**. The capsid consists of numerous **capsomeres**, each having a few **monomers** or **structural units**. Each structural unit is made up of one or more polypeptide chains. The capsomeres are of different shapes such as hollow prism, hexagonal, pentagonal, lobular or any other shape. The specific arrangement of capsomeres in the capsid determines the shape of a virion. Viruses have the following three different types of symmetry :

1. Icosahedral symmetry. Many viruses have spherical, cubical or polygonal shape which is basically **icosahedral** or 20-sided. Icosahedral symmetry depends on the fact that the assembly of the capsomeres causes the capsid of the virus to be at a state of minimum energy (**Caspar** and **Klug**, 1962). An icosahedral capsid comprises both **penta-meres** (*i.e.,* capsomeres containing 5 structural units) and **hexameres** (*i.e.,* capsomeres having 6 structural units). In an icosahedral virus the minimum number of capsomeres is 12 or its multiple such as 32, 42, 72, 92, 162, 252, 362, 492, 642 and 812. For example, the total number of capsomeres of different icosahedral viruses are : (1) Bacteriophage ϕ (phi) $\times$ 174 = 12 pentameres; (2) Turnip yellow mosaic virus or TYMV = 32 capsomeres; (3) Poliovirus = 32 capsomeres; (4) polyoma virus and papilloma virus = 72 capsomeres; (5) Reovirus

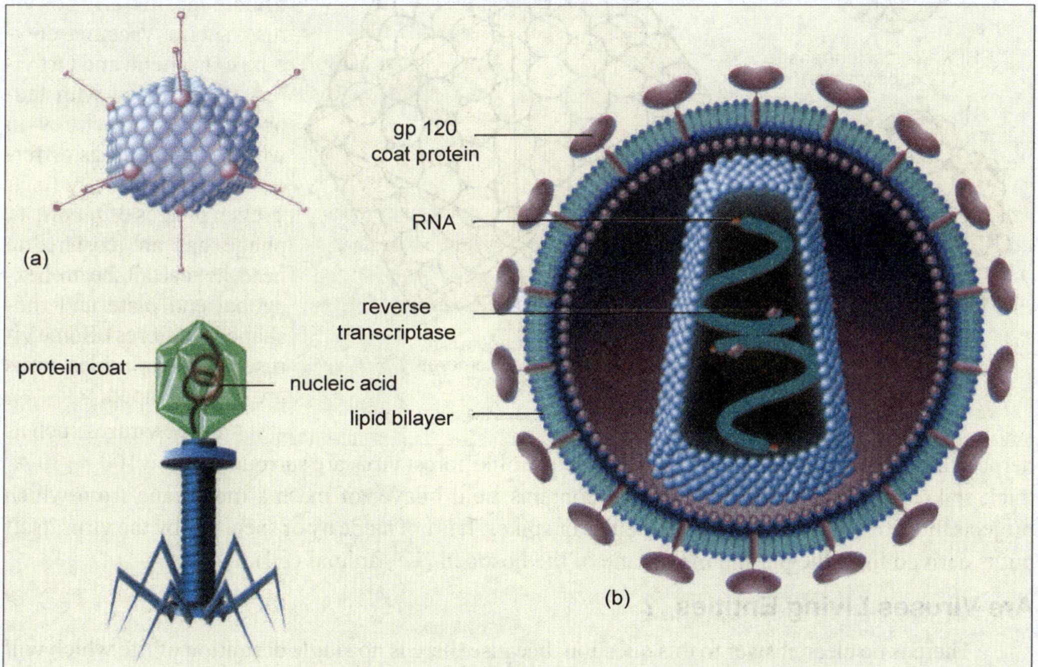

Virus diversity (a) Adenovirus, (b) HIV (c) T- even bacteriophage.

= 92 capsomeres; (6) Herpes virus = 162 capsomeres; (7) Adenovirus = 252 capsomeres; and (8) Tipula iridescent virus = 812 capsomeres. In all of these icosahedral viruses, only 12 capsomeres are pentameres, occupying 12 corners of five-fold symmetry, while the rest are hexameres (Fig. 3.2). Since a polyhedron of 20-sided icosahedron basically has triangular faces, it is also known as **deltahedron** (**Alberts** *et al.*, 1989).

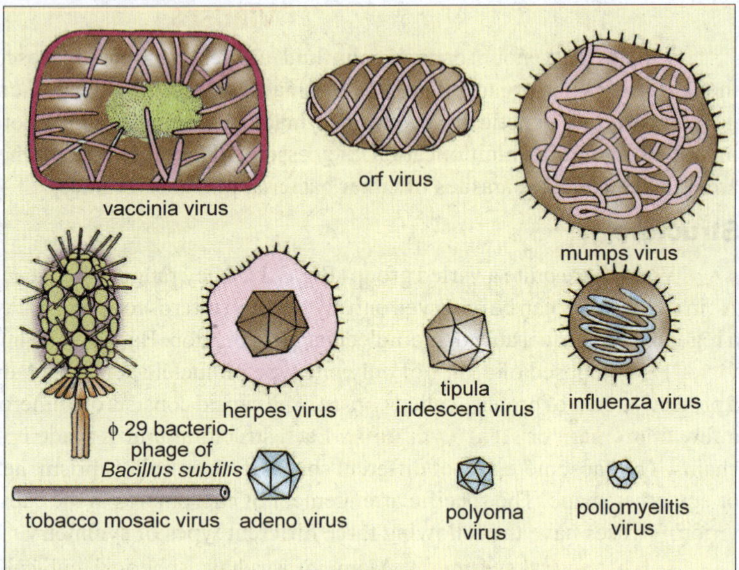

Fig. 3.1. Different kinds of viruses.

2. Helical or cylindrical symmetry. The rod-shaped helical capsid of viruses such as tobacco mosaic virus (TMV), bacteriophage M13 and influenza virus, consists of numerous identical capsomeres arranged into a helix because they are thicker at one end than the other.

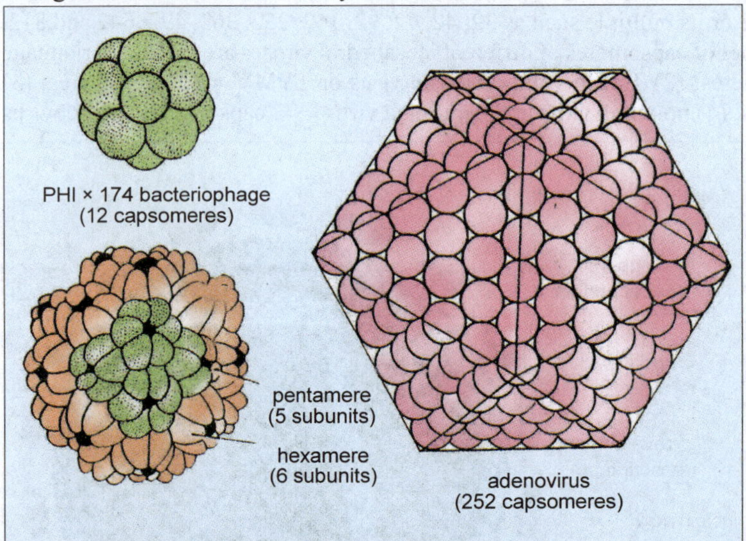

Fig. 3.2. Certain polyhedral (icosahedral) viruses.

3. Complex symmetry. Viruses with complex shaped capsids are of two shapes : those without identifiable capsids (*e.g.*, pox viruses such as vaccinia, cowpox, extromelia and Orf viruses) and those with tadpole-shaped structures in which each part has different sort of symmetry (*e.g.*, T-even phages of *E.coli*; T_2 phage has an icosahedral head, helical tail sheath, hexagonal end plate and rod-shaped tail fibres). Some viruses such as rabies virus are bullet-shaped.

Some viruses such as herpes virus, influenza virus, mumps virus and Semliki forest virus are surrounded by a 100 – 150 A^0 thick spiked membrane. This membrane contains lipid bilayer of plasma membrane from which projects the virus-specific protein molecules or spikes. It is not made by or specified by the virus itself but is derived from the plasma membrane of the host cell (*i.e.*, animal cell).

Are Viruses Living Entities ?

There is no clear answer to this question, because there is no single definition of life which will satisfy everyone. If life is defined as being cellular, then viruses are not alive. If life is defined as being capable of making new life directly through its own metabolic efforts, then viruses are not living. However, if life is defined as being able to specify each new generation according to its own genetic instructions, then viruses are living systems.

In fact, virus multiplication is very different from cell replication mechanisms. Cells produce their own chromosomes, proteins, membranes and other constituents and these materials are partitioned into progeny cells after a division process in the parent cell. As stated in the cell theory, cells arise only from other cells. Viruses do not give rise directly to new viruses. Instead, they must sabotage the biosynthetic machinery of their host cell so that virus-specific proteins and nucleic acids are made, according to viral genetic information. Eventually virus particles are assembled from newly-made molecules in the host cell and are released when the host cell bursts. They may then initiate new cycles of infection in other host cells. Thus, viruses borrow metabolism and a sheltering membrane from their host, but they provide the genetic instructions that ensure continuity of their species from generation to generation. Since viruses are entirely dependent on living cells for their replication, they cannot be a precellular form in evolutionary terms, but should be viewed as pieces of cellular genetic material which have gained some degree of individual autonomy (see **Bradbury** *et al.,* 1981).

Naming and Classification

Viruses are not named according to the method of binomial nomenclature like other organisms (Binomial nomenclature is the Linnean system of classification requiring the designation of a **binomen** (L., *bi* = twice + *nomen* = a name), the genus and species name, for every species of bacteria, blue green algae, plants and animals). Viruses tend to be named in a random fashion according to the disease caused (*e.g.,* poliomyelitis virus), the host organism (*e.g.,* bacterial viruses or bacteriophages, plant viruses and animal viruses), or some coded system (*e.g.,* T_1, T_2, P_1 phages). Recently, with increase in knowledge of viral biochemistry and molecular biology, various specific characteristics such as nature of nucleic acid (DNA or RNA), the symmetry of capsid, the number of capsomeres, etc., are now being used in viral classification. However, we will stick to the following conventional classification of viruses which is based on the type of the host cell :

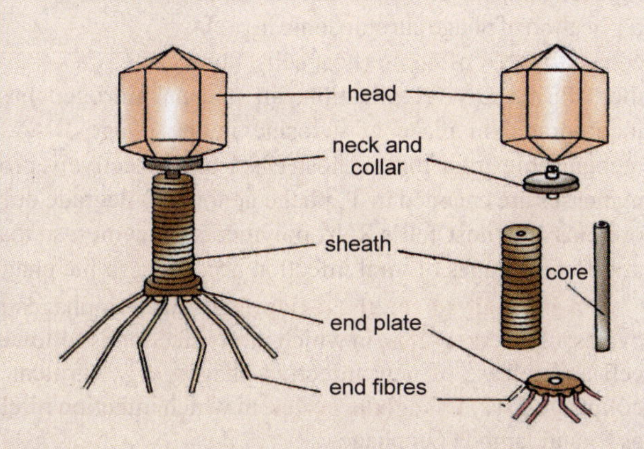

Fig. 3.3. A bacteriophage. A—External structure; B—Parts.

A. Bacterial viruses or bacteriophages. Viruses that parasitize the bacterial cells, are called **bacteriophages** or **phages** (phage means 'to eat'). The phages have specific hosts and they are of variable shapes, sizes and structures. The most widely studied phages are T-even bacteriophages such as T_2, T_4, T_6, etc., which infect the colon bacillus, *Escherichia coli* and are also known as **coliphages** (T for "type". The plural word phages refers to different species; the word phage is both singular and plural and in the plural sense refers to particles of same type. Thus, T_4 and T_7 are both phages, but a test tube might contain either 1T_4 phage or 100 T_4 phage; see **Freifelder,** 1985).

T_4 bacteriophage is a large-sized tadpole-shaped complex virus (Fig. 3.3). Its capsid comprises of an icosahedral **head** (1250 A^0 length and 850 A^0 width; 2000 capsomeres), a short neck with **collar** bearing 'whiskers' and a long helical **tail**. The tail is made up of a thick and hollow **mid-piece**, a hexagonal **base plate** or **end plate** to which are attached six **spikes** and six long **tail fibres**. The mid-piece consists of a central hollow **core** and a spring-like **contractile sheath** which comprises 24 rings of hexameres and remains helically arranged around the core. The T_4 genome or chromosome is a single DNA molecule which is 60 µm long, linear, double-stranded and tightly-packed within the head of the phage. Phage DNA contains more than 1,66,000 nucleotide pairs and encodes more than 200 different proteins (*i.e.,* proteins involved in DNA replication and in the assembly of head and tail). For example, T_4 phage DNA codes for at least 30 different enzymes (*e.g.,* helicases, topoisomerases, DNA polymerases, DNA ligases, etc.) all of which ensure rapid replication of phage chromosome in preference to DNA of *E.coli* (host cell). Further, during DNA replication, an unusual nitrogen base, called **5- hydroxymethylcytosine** is incorporated in place of cytosine in the phage DNA. This unusual base makes phage DNA recognisable from that of host DNA and selectively protects it from the nuclease enzymes. The nucleases are encoded in T_4 phage genome to degrade only the DNA of host cell. Some other phage proteins alter host cell's RNA polymerase enzymes, so that they transcribe different sets of T_4 genes at different stages of viral infection according to the phage's needs (see **Alberts** *et al.*, 1989).

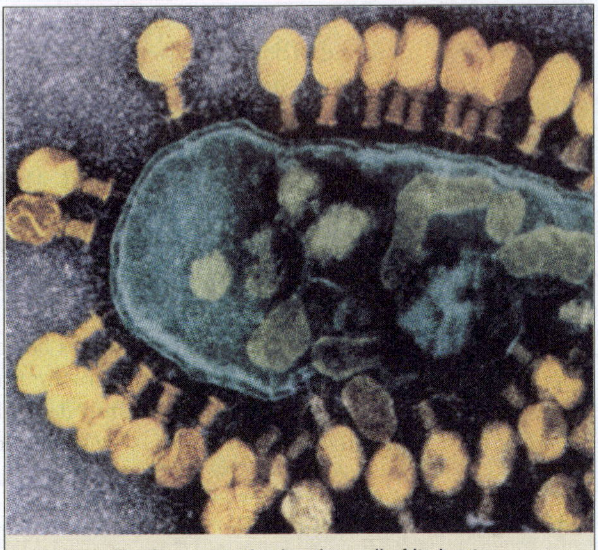

T_4 phage attached to the wall of its host.

Life cycle of the bacteriophage. Bacteriophages may have the following two types of life cycles: (1) **Lytic cycles**, in which viral infection is followed by **lysis** (bursting and death) of the host cell and release of new infective phages, *e.g.,* virulent phages such as T_4 and all other T-even coliphages. (2) **Lysogenic cycles**, in which infection rarely causes lysis, *e.g.,* temperate phages such as P_1 and lambda (λ) phages.

1. Lytic cycle of a virulent phage. Life cycle of a T_4 bacteriophage (Fig. 3.4) involves the following steps: 1. **Attachment** or **adsorption** of phage to bacterial (host) cell. 2. **Injection** or **penetration** of viral genetic material (DNA) into the host cell. 3. **Eclipse period**, during which synthesis of new phage DNA and protein coats takes place. 4. **Assembly** of phage DNA into protein coats.5. **Lysis** of host cell and **release** of the infective progeny phages. Such a phage is called **virulent** or **lytic phage** since it has infectiousness and it causes death of host cell by lysis.

The adsorption of the phage to its host is made possible by a reaction of chemical groups on the two during a random collision. Reactive groups (called **adsorption protein** or **pilot protein**; **Kornberg**, 1974) at the end of the tail of the phage can join with a complementary set of chemical groups (a **receptor site**) in the cell wall of the bacterium. During adsorption, long tail fibres of the phage are first to contact and attach to the cell. They help to position the phage's tail perpendicularly to the cell wall. Once the phage is attached to its prospective host, injection can take place involving a movement of phage DNA from its position inside the head of the phage through the hollow core of the tail into the bacterium. Entry is made possible by a hole punched in the bacterial cell wall, either by contraction of outer sheath of tail or by the action of enzymes carried by phage tail, or both. The protein coat or capsid of the phage remains outside the cell. Once inside the host cell, the phage DNA becomes a **vegetative phage**, *i.e.,* phage genes take over the metabolic machinery of the cell and direct it to produce replicas of the infecting virus. Although the cell continues to procure raw materials and energy from the environment, the phage genes allow only viral components to be built. Further, either the normal ability of the host DNA to control the cell is lost, or the host DNA is completely destroyed by early products of the viral genes. Thus, phage DNA is both replicated and transcribed; first the enzymes needed for synthesis of phage DNA are translated, then the capsid proteins are translated. Phage particles are assembled around condensed cores of the complete phage nucleic acid (by self-assembly method). At last lytic enzymes which have been coded by phage DNA, break open the bacterium and release the new phage particles which diffuse in the surrounding in search of new host (see **Mays**, 1981).

2. Lysogenic cycle. Certain bacteriophages such as P_1 and lambda (λ) phages, have entirely different pattern of life cycle than the virulent phages. This pattern is called **lysogeny** and is characterized by delayed lysis after phage infection. A virus with this capacity is called **temperate virus**. The infected host cell is said to be lysogenic because dormant virus may at any time become active and begin directing the synthesis of new virus particles.

In lysogeny, the process of adsorption and nucleic acid injection are quite similar to a lytic cycle of virulent phages, although different phages recognise different bacterial cell surface receptors. The next step, however, is unique to lysogeny. The nucleic acid is neither extensively replicated nor extensively transcribed. The virus generally expresses one or a few genes which code for a **repressor protein** that turns off (*i.e.*, represses) the expression of the other genes of the virus. In consequence, virus is not replicated, but phage DNA remains in the bacterium, being replicated in such a way that when the lysogenic bacterium divides, each daughter cell receives at least one phage genome in addition to the bacterial genome. There are two styles to this persistence of phage DNA : the phage chromosome may exist as a fragment of DNA outside the host's chromosome (*i.e.*, in

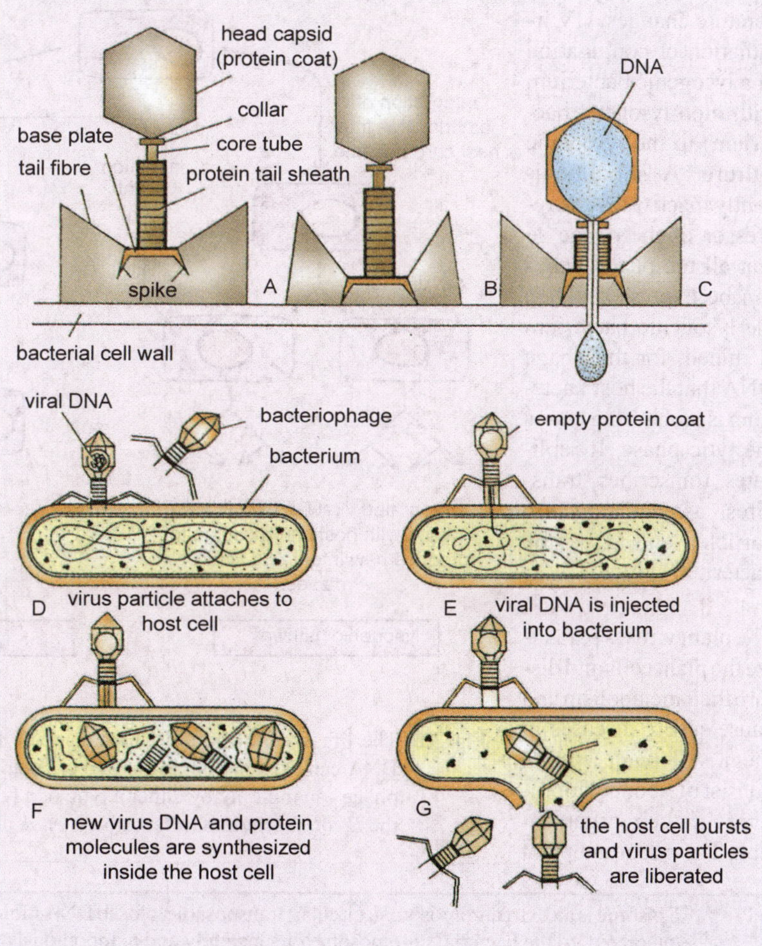

Fig. 3.4. A,B,C — Mode of attachment of a T-even phage (bacteriophage) on a bacterial cell wall and injection of DNA into the bacterium (host cell); D,E,F and G — Steps of viral reproduction inside the host cell.

the host's cytoplasm) essentially as a plasmid* (*e.g.,* P_1 bacteriophage) or it may attach itself to the host's chromosome as an episome* (*e.g.,* lambda phage). Thus, in case of lambda phage the DNA first becomes circular due to joining of its both cohesive ends and then is integrated into the circular DNA molecule of the bacterium. Such an integrated and dormant viral genome is often termed as **provirus** or **prophage**. The infection of *E. coli* cell with lambda phage and its consequent integration and adoption of lysogeny, renders that cell immune to further attack by phage of the same type.

The calm lysogenic period is ended by some type of shock (*e.g.,* temperature changes, UV irradiation or conjugation of a lysogenic bacterium with a non-lysogenic bacterium) to the lysogenic culture. A shock evidently inactivates the repressor of the phage so that all the phage genes can be expressed. Then the lysogenic bacterium is ruined, for the phage DNA that the host bacterium is harbouring enters the lytic phase. It replicates, transcribes, translates, assembles viron particles and lyses the bacterium (Fig. 3.5).

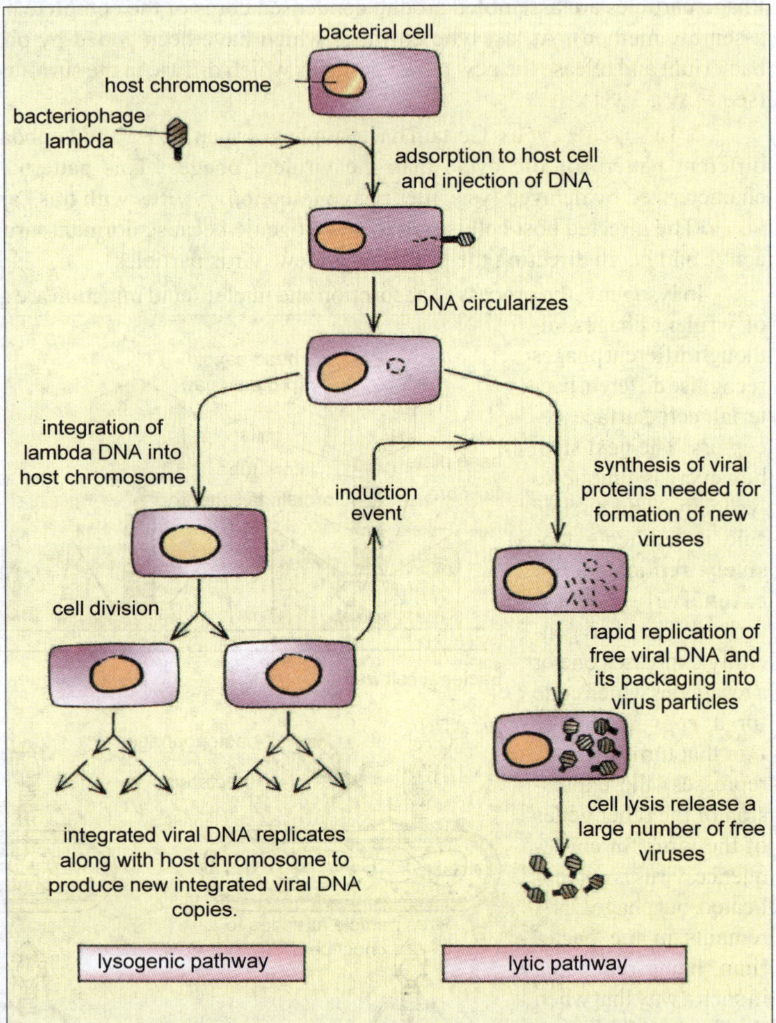

Fig. 3.5. The life cycle of bacteriophage lambda. Its double stranded DNA can exist in both linear and circular forms. The lambda phage can multiply by either a lytic or a lysogenic pathway in the *E.coli* bacterium (after Alberts *et al.,* 1989).

B. Plant viruses. The plant viruses parasitize the plant cells and disturb their metabolism and cause severe diseases in them. All plant viruses consist of ribonucleoproteins in their organization. The important plant

* **Episome** is an extrachromosomal, circular, transposable, closed DNA molecule which can exist either integrated ito the bacterial chromosome or separately and autonomously in the cytoplasm. **Plasmid** is that bit of autonomous genetic material of bacteria (*i.e.,* circular DNA) that exists only extrachromosomally and cannot be integrated into the bacterial chromosome (DNA) (see **Sheeler** and **Bianchi,** 1987). Now only the term plasmid is used for all kinds of extrachromosomal autonomous transposable circular DNA fragments (see **Alberts** *et al.,* 1989; **Burns** and **Bottino**, 1989).

viruses are tobacco rattle virus (TRV), tobacco mosaic virus (TMV), potato virus, beet yellow virus (BYV), southern bean mosaic virus (SBMV) and turnip yellow viruses (TYV). Among plants, few hundred viral diseases are caused, *e.g.,* mosaic diseases of tobacco, cabbage, cauliflower, groundnut and mustard; black-ring spot of cabbage; leaf roll of tomato; leaf curl of papaya, cotton, bean and soyabean; yellow disease of carrot, peach; little-leaf of brinjal. These diseases are spread mainly by insects such as aphids, leaf hoppers and beetles.

Tobacco mosaic virus (TMV). TMV is the most extensively studied plant virus. It was discovered by **Iwanowski** (1892) and obtained in a pure state (*i.e.,* in paracrystalline form) by **Stanley** (1935). **Bawden** and **Pirie** (1937) extensively purified TMV and showed it to be a nucleoprotein containing RNA. **H.Fraenkel-Conrat** experimentally demonstrated that RNA is the genetic substance of TMV.

TMV is a rod-shaped, helically symmetrical RNA virus (Fig. 3.6). Each virus particle is elongated, cigarette-like in shape having the length of 3000A^0 (300 nm) and diameter of 160 A^0 (16 nm; see **De Robertis** and **De Robertis, Jr.,** 1987). In each rod of TMV, there are about 2130 identical elliptical protein subunits or capsomeres. The capsomeres are closely packed and arranged in a helical manner around the RNA helix, forming a hollow cylinder. Thus, there is a hollow core (axial hole) of about 40A^0 (4nm) diameter which runs the entire length of the rod and contains the RNA molecule. The RNA molecule does not occupy the hole but is deeply embedded in the capsomeres. RNA of TMV is a single-stranded molecule consisting of 6500 nucleotides and is in the form of a long helix extending the whole length of viral particle. Lastly, there are about 16 capsomeres in each helical turn. Each capsomere contains about 158 amino acids and has a molecular weight of 18000 daltons. The whole TMV capsid has all amino acids found in other plant proteins.

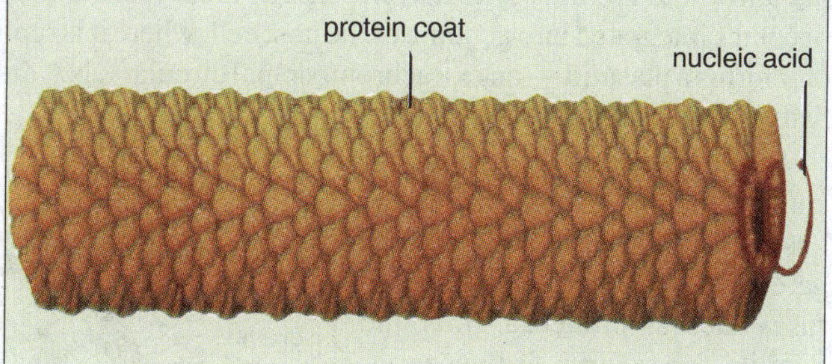

Fig. 3.6. Molecular organization of the tobacco mosaic virus (TMV) (after De Robertis and De Robertis, Jr., 1987).

TMV infects the leaves of tobacco plant. It is transmitted and introduced into the host cell by some vector or by mechanical means such as rubbing, transplanting and handling. Once inside the host cell, the viral RNA directs the metabolic systems of host to synthesize its own proteins and to replicate (multiply) its RNA molecule. All the raw materials for RNA replication and capsomere biosynthesis are derived from the host cell. Ultimately when numerous viral particles are formed by self-assembly method inside the host cell, they are released after lysis of the cell. Recently, it was found that plant viruses exploit the route of plasmodesmata to pass from cell to cell. For example, TMV is found to produce a 30,000 dalton protein called **P_{30}** which tends to enlarge the plasmodesmata in order to use this route to pass or spread its infection from cell to cell (see **Alberts** *et al.,* 1989).

C. Animal viruses. The animal viruses infect the animal cells and cause different fatal diseases in animals including man. Generally, they have a polyhedron or spherical shape and genetic material in the form of DNA or RNA. The protein coat or capsid of animal viruses is surrounded by an envelope.

Certain common viral infections of human beings are : common cold, influenza, mumps, measles, rubella (German measles), chickenpox, small pox, polio, viral hepatitis, herpes simplex, viral encephalitis, fever blisters, warts and some types of cancer. Among livestock and fowl, viruses cause encephalitis, foot and mouth disease, fowl plague, Newcastle disease, pseudorabis, hog cholera and a variety of warts and other tumors. A virus usually displays some specificity for a particular animal group.

Poliomyelitis is a most extensively studied RNA-containing animal virus. The polio virus has comparatively very simple organization. It consists of a protein shell built up of 60 structurally equivalent asymmetric protein subunits of approximately 60 A^0 diameter, packed together in such a way that they form a spherical shell of about 300 A^0 in diameter. The shell or capsid encloses a single stranded RNA molecule of 5,200 nucleotides.

Herpes virus is another most extensively studied animal virus. It possesses a DNA containing core embedded in a regular icosahedral **capsid** (162 capsomeres) and an outer **envelope** of lipids, proteins and carbohydrates. The DNA molecule of herpes virus is a single, linear, double-stranded having a molecular weight of 10^8 daltons and codes for about 100 average sized protein molecules.

Life cycle of animal viruses. Like the bacteriophages, animal viruses have two types of life cycles or growth : **1. Permissive growth** which permits an animal virus to multiply lytically and kill the host cell. **2. Non-permissive growth** which permits an animal virus to enter the host cell but does not allow it to multiply lytically. In some of these non-permissive cells, the viral chromosome either becomes integrated into genome of the host cell, where it is replicated along with the host chromosome or it forms a plasmid — an extrachromosomal circular DNA molecule — that replicates in a controlled fashion without killing the cell.

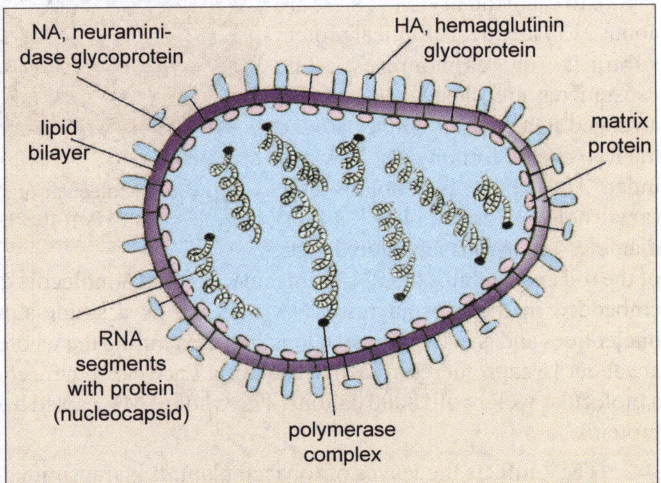

Fig. 3.7. The influenza viron (after Watson *et al.,* 1987).

The modes of infection or entry of animal viruses inside the host cells have been well investigated for the RNA-containing viruses such as influenza virus and Semliki forest virus. Influenza virus has segmented genome (Fig. 3.7). Each segment is a template for the synthesis of a different single mRNA. The RNA core is protected by an icosahedral capsid. The nucleocapsid (RNA + capsid) of influenza virus is ultimately surrounded by a phospholipid bilayer membrane in which are embedded the following two types of viral glycoproteins or **spikes** : **1. Large spikes** of trimers (*i.e.,* each unit of three monomers) of **hemagglutinin** or **Ha protein.** At the time of adsorption or attachment of the virus to the host cell, these HA spikes bind to virus-specific receptors on the surface or plasma membrane of the host cell. **2. Small spikes** of multimers of **neuraminidase** or **NA protein.** These NA spikes remove those virus-loaded receptors from the host's surface which have not entered the host cell through the process of receptor-mediated endocytosis. Thus, NA spikes cause the viruses to desorb from the host cell and make them free to infect other host cells.

Infection of influenza virus is initiated when the virus binds to specific receptor proteins on the plasma membrane of target host cell. These RNA viruses are taken in by the host cell by **receptor-mediated endocytosis** and are delivered to the **endosomes** (which are special vesicles having acidic medium or low pH between 5 to 5.5 and recycle the receptors; see Chapter 5). The low pH in endosome activates a **fusogenic protein** or **fusogen** in the viral envelope that catalyzes the fusion of membranes of the influenza virus and the endosome. This allows the escaping of intact viral particle into the cytosol of the host. Thus, both the viral protein and viral nucleic acid penetrate into the host cell. Ultimately viral RNA comes out from the viral capsid and gets attached to ribosomes to start the process of viral multiplication (Fig. 3.8).

Usually, the animal viruses are released from the host cell by rupturing (lysis) and subsequent death of host cells. However, sometimes viral particles are pinched off as buds from the cell surface and they retain the host's plasma membrane around them, *e.g.*, influenza virus, Semliki forest virus, retroviruses such as AIDS virus. AIDS or acquired immune deficiency syndrome is caused by **HIV-1** or Human immunodeficiency virus type -I.

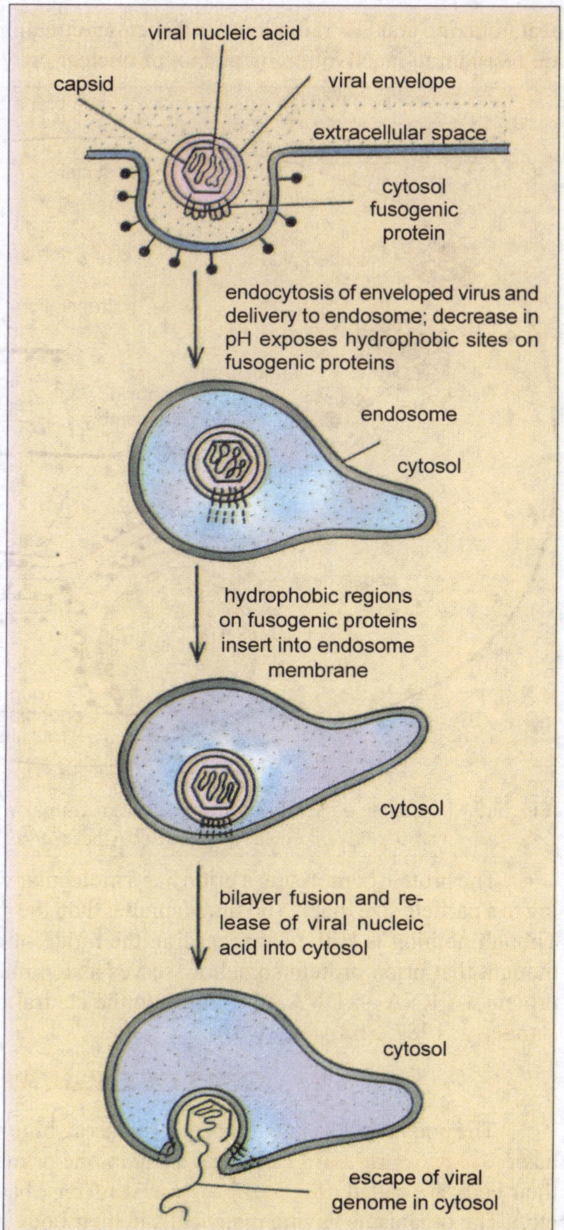

Fig. 3.8. Schematic diagram to show how fusogenic proteins on the surface of many enveloped viruses are thought to catalyze the fusion of viral and endosomal membranes (Alberts *et al.*, 1989).

Viroids. Viroids are small RNA circles, only 300 to 400 nucleotides long, lacking AUG codon (the signal for the start of protein synthesis). They are replicated autonomously despite the fact that they do not code for any protein. Having no protein coat or capsid, they exist as naked RNA molecules and pass from plant to plant only when the surfaces of both donor and recipient cells are damaged so that there is no membrane barrier for the viroid to pass (see **Alberts** *et al.,* 1989).

The term viroid was coined by **Diener** (1971) who discovered the first viroid, called **potato spindle tuber viroid** or **PSTV**. Viroids form a class of subviral pathogens which cause infections and diseases in many plants and also in animals (see **Sheeler** and **Bianchi**, 1987). Some of the plant diseases which are caused by the viroids are the following :

1. Potao spindle tuber (Its viroid contains 359 nucleotides in single and circular RNA molecule);
2. Citrus exocortis (Its viroid contains 371 nucleotides in RNA molecule);
3. Chrysanthemum stunt;
4. Chrysanthemum chlorotic mottle;
5. Cucumber pale fruit;
6. Hope stunt; and
7. Tomato plant macho (see **Sharma**, 1990).

Prions. Prions are described as 'rod-shaped' proteinaceous particles thought to cause a number of diseases in animals such as **Scrapie disease** of central nervous system of goats and sheep (in which animals scrape or scratch themselves against some gate post or similar object). Prions are also found to cause a Scrapie-like disease, called **Creutzfeldt-Jakob disease** of nervous system of humans and **Kuru** disease of brain of cannibalistic tribes of New Guinea. Prions were named by **S.B.Prusiner** (1984). They can survive heat, ionizing and UV radiations, and chemical treatment that normally inactivates viruses (*i.e.,* they are resistant to inactivation by phenol or nuclear enzymes).

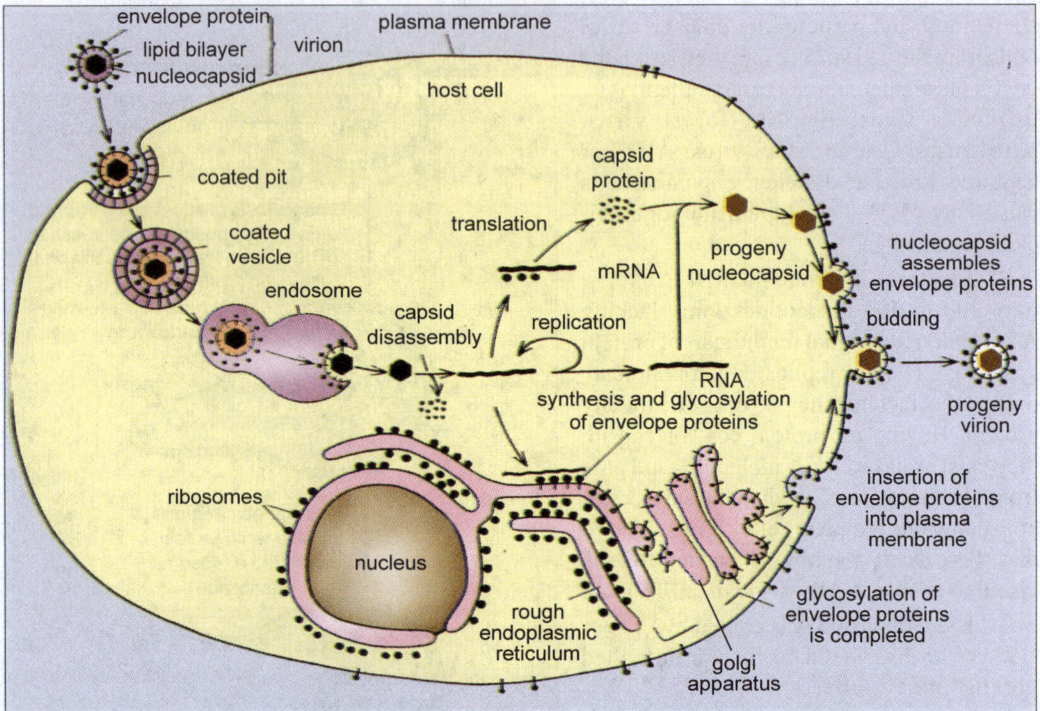

Fig. 3.9. Life cycle of Semliki forest virus : An animal virus containing a single-stranded RNA genome, capsid and spiked envelope (after Alberts *et al.,* 1989).

The protein comprising a prion has a molecular weight between 50,000 to 100,000, corresponding to a particle size that is 100 times smaller than the smallest virus (see **Sheeler** and **Bianchi**, 1987). Though nothing is still clear regarding the mode of replication, multiplication of prions, yet, it is thought that prion protein somehow serves as template to perform ' reverse translation', *i.e.,* from protein $\rightarrow$ RNA $\rightarrow$ DNA, thus, turning the central dogma of molecular biology on its head (see **Banerjee**, 1987; **Sharma**, 1990).

CELLS OF CELLULAR ORGANISMS

The body of all living organisms (bacteria, blue green algae, plants and animals) except viruses has **cellular organization** and may contain one or many cells. The organisms with only one cell in their body are called **unicellular organisms** (*e.g.,* bacteria, blue green algae, some algae, Protozoa, etc.). The organisms having many cells in their body are called **multicellular organisms** (*e.g.,* most plants and animals). Any cellular orgainsm may contain only one type of cell from the following types of cells :

A. Prokaryotic cells ; B. Eukaryotic cells.

The terms prokaryotic and eukaryotic were suggested by **Hans Ris** in the 1960's.

PROKARYOTIC CELLS

The prokaryotic (Gr., *pro* = primitive or before; *karyon* = nucleus) are small, simple and most primitive. They are probably the first to come into existence perhaps 3.5 billion years ago. For example, the **stromatolites** (*i.e.,* giant colonies of extinct cyanobacteria or blue green algae) of

Western Australia are known to be at least 3. 5 billion years old. The **eukaryotic** (Gr., *eu* =well; *karyon* = nucleus) cells have evolved from the prokaryotic cells and the first eukaryotic (nucleated) cells may have arisen 1.4 billion years ago (**Vidal,** 1983).

The prokaryotic cells are the most primitive cells from the morphological point of view. They occur in the bacteria (*i.e.,* mycoplasma, bacteria and cyanobacteria or blue-green algae). A prokaryotic cell is essentially a **one-envelope system** organized in depth. It consists of central nuclear components (*viz.,* DNA molecule, RNA molecules and nuclear proteins) surrounded by cytoplasmic ground substance, with the whole enveloped by a plasma membrane. Neither the nuclear apparatus nor the respiratory enzyme system are separately enclosed by membranes, although the inner surface of the plasma membrane itself may serve for enzyme attachment. The cytoplasm of a prokaryotic cell lacks in well defined cytoplasmic organelles such as endoplasmic reticulum, Golgi apparatus, mitochondria, centrioles, etc. In the nutshell, the prokaryotic cells are distinguished from the eukaryotic cells primarily on the basis of what they lack, *i.e.,* prokaryotes lack in the nuclear envelope, and any other cytoplasmic membrane. They also do not contain nucleoli, cytoskeleton (microfilaments and microtubules), centrioles and basal bodies.

Bacteria

The bacteria (singular bacterium) are amongst the smallest organisms. They are most primitive, simple, unicellular, prokaryotic and microscopic organisms. All bacteria are structurally relatively homogeneous, but their biochemical activities and the ecological niches for which their metabolic specialisms equip them, are extremely diverse.

Bacteria occur almost everywhere : in air, water, soil and inside other organisms. They are found in stagnant ponds and ditches, running streams and rivers, lakes, sea water, foods, petroleum oils from deeper regions, rubbish and manure heaps, sewage, decaying organic matter of all types, on the body surface, in body cavities and in the internal tracts of man and animals. Bacteria thrive well in warmth, but some can survive at very cold tops of high mountains such as Alps or even in almost boiling hot springs. They occur in vast numbers. A teaspoonful of soil may contain several hundred million bacteria. They lead either an autotrophic (photoautotrophic or chaemoautotrophic), or heterotrophic (saprotrophic or parasitic) mode of existence. The saprophytic or saprotrophic species of bacteria are of great economic significance for man. Some parasitic species of bacteria are pathogenic (disease producing) to plants, animals and man.

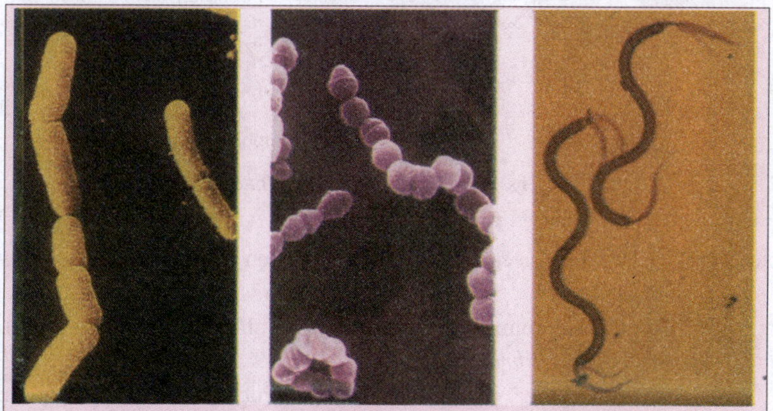

(a) Cells of *Pseudomonas* (cylindrical); (b) *Streptococcus* (spherical) (c) *Spirilla* (twisted shape)

Bacteria have a high ratio of surface area of volume because of their small size. They show high metabolic rate because they absorb their nutrients directly through cell membranes. They multiply at a rapid rate. In consequence, due to their high metabolic rate and fast rate of multiplication, bacteria produce marked changes in the environment in a short period.

1. Size of bacteria. Typically bacteria range between 1 μm (one micrometre) to 3 μm, so they are barely visible under the light microscope. The smallest bacterium is *Dialister pneumosintes* (0.15 to 0.3μm in length). The largest bacterium is *Spirillum volutans* (13 to 15μm in length).

2. Forms of bacteria. Bacteria vary in their shapes. Based on their shape, bacteria are classified into the following groups :

(1) Cocci (singular coccus). These bacteria are spherical or round in shape. These bacterial cells may occur singly (**micrococci**); in pairs (**diplococci**, *e.g.*, pneumonia causing bacterium, *Diplococcus pneumoniae*); in groups of four (**tetracocci**); in a cubical arrangement of eight or more (**sarcinae**); in irregular clumps resembling bunches of grapes (**staphylococci**, *e.g.,* boil causing bacterium, *Staphylococcus aureus* or in a bead-like chain (**streptococci**, *e.g.,* sore throat causing bacterium, *Streptococcus pyogenes*) (Fig. 3.10).

(2) Bacilli (singular, bacillus). These are rod-like bacteria. They generally occur singly, but may occasionally be found in pairs (**diplobacilli**) or chains (**streptobacilli**). Bacilli cause certain most notorious diseases of man such as tuberculosis (*Mycobacterium* or *Bacillus tuberculosis*), tetanus (*Clostridium tetani*), typhoid (*Salmonella* or *Bacillus typhosus*), diphtheria (*Corynebacterium diphtheriae*), leprosy (*Mycobacterium leprae*), dysentery and food poisoning (*Clostridium botylinum*). Certain well known diseases of the animals are also caused by bacilli, *e.g.,* anthrax (*Bacillus anthracis*) and black leg (*Clostridium chauvei*).

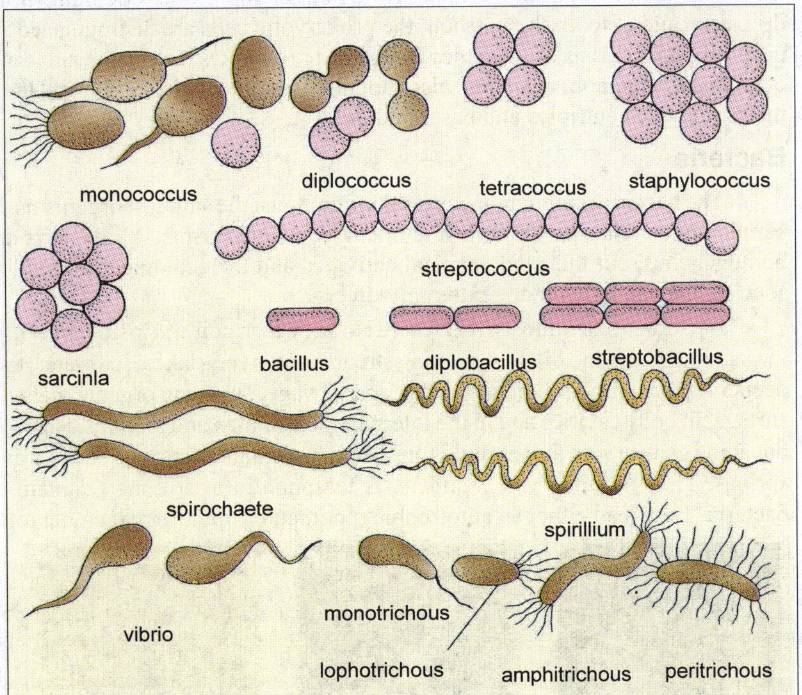

Fig. 3.10. Different forms of bacteria.

(3) Spirilla (singular, spirillum). These are also called **spirochetes**. These are spiral-shaped and motile bacteria (Fig. 3.10). Spirilla cause human disease such as syphilis (*Treponema pallidum*).

(4) Vibrios (singular vibrio). These are comma-shaped or bent-rod like bacteria (Fig. 3.10). Vibrios cause human disease such as cholera (*Vibrio cholerae*).

3. Gram negative and Gram positive bacteria. On the basis of structure of cell wall and its stainability with the Gram stain, the following two types of bacteria have been recognized : Gram positive and Gram negative bacteria. The Gram staining method is named after **Christian Gram** who developed it in Denmark in 1884. In this technique, when heat-fixed bacteria are treated with the basic dye, crystal violet, they become blue or purple. Such blue stained cells are treated with a mordant (*i.e.,* agent that fixes stains to tissues) such as iodine (*i.e.,* potassium iodide or KI solution) and ultimately washed with some organic solvent such as alcohol. Some bacteria retain the blue colour, while others lose it and stay colourless. The former are **Gram positive bacteria** (*e.g., Bacillus subtilis, Staphylococcus*, etc.) and the latter are **Gram negative bacteria** (*e.g., Escherichia coli, Simonsiella,* cyanobacteria, etc.). Colourless Gram negative bacteria may thereafter be stained pink with safranin stain for their better microscopic visibility.

The long search for the chemical basis of this differentiating staining reaction ended in 1950's when it was detected that cell wall of Gram negative bacteria has high lipid content which tends to be dissolved away by alcohol. The alcohol then can enter the cell and leach out the stain, whereas the cell walls of Gram positive bacteria form a barrier (*i.e.,* peptidoglycan layers) that prevent the penetration of the solvent inside the cell.

4. Structure of bacteria. Structural details of a bacterial cell can only be seen with an electron microscope in very thin sections. A typical bacterial cell has the following components:

A. Outer covering. The outer covering of bacterial cell comprises the following three layers:

I. Plasma membrane. The bacterial protoplast is bound by a living, ultrathin (6 to 8 nm thick) and dynamic plasma membrane. The plasma membrane chemically comprises molecules of lipids and proteins which are arranged in a **fluid mosaic pattern**. That is, it is composed of a bilayer sheet of **phospholipid** molecules with their polar heads on the surfaces and their fatty-acyl chains (tails) forming the interior. The **protein** molecules are embedded within this lipid bilayer, some spanning it, some exist on its inner side and some are located on its external or outer side. These membrane proteins serve many important functions of the cell. For example, the transmembrane proteins act as **carriers** or **permeases** to carry on selective transportation of nutrients (molecules and ions) from the environment to the cell or vice versa. Certain proteins of the membrane are involved in oxidative metabolism, *i.e.,* they act as enzymes and carriers for electron flow in respiration and photosynthesis leading to phosphorylation (*i.e.,* conversion of ADP to ATP). The bacterial plasma membrane also provides a specific site at which the single circular chromosome (DNA) remains attached. It is the point from where DNA replication starts. The first stage in nuclear division involves duplication of this attachment, followed by a progressive bidirectional replication of DNA by two replication forks.

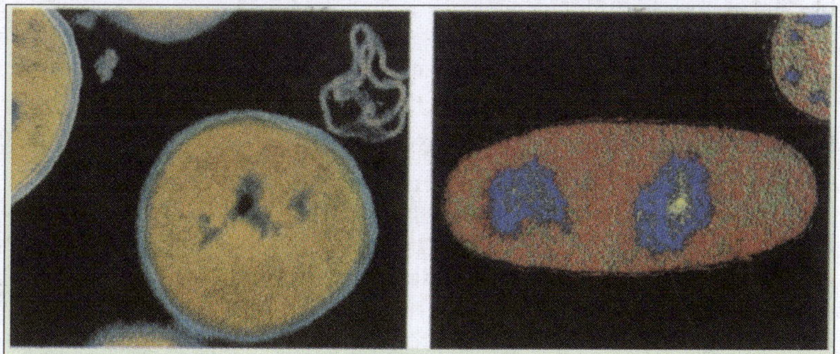

Pictures showing bacterial membrane -- blue in coci (left) and fine red outline in bacillus (right).

Plasma membrane intrusions. Infoldings of the plasma membrane of all Gram-positive bacteria and some Gram-negative bacteria give rise to the following two main types of structures:

(I) Mesosomes (or **chondrioids**). They are extensions of the plasma membrane within the bacterial cell (*i.e.,* cytoplasm) involving complex whorls of convoluted membranes (Fig. 3.11). Mesosomes tend to increase the plasma membrane's surface and in turn also increase their enzymatic contents. They are seen in chemoautotrophic bacteria with high rates of aerobic respiration such as *Nitrosomonas*, and in photosynthetic bacteria such as *Rhodopseudomonas* where they are the site of photosynthetic pigments. Mesosomes are involved in cross-wall (septum) formation during the division of cell.

(2) Chromatophores. These are photosynthetic pigment-bearing membranous structures of photosynthetic bacteria. Chromatophores vary in form as vesicles, tubes, bundled tubes, stacks, or thylakoids (as in cyanobacteria).

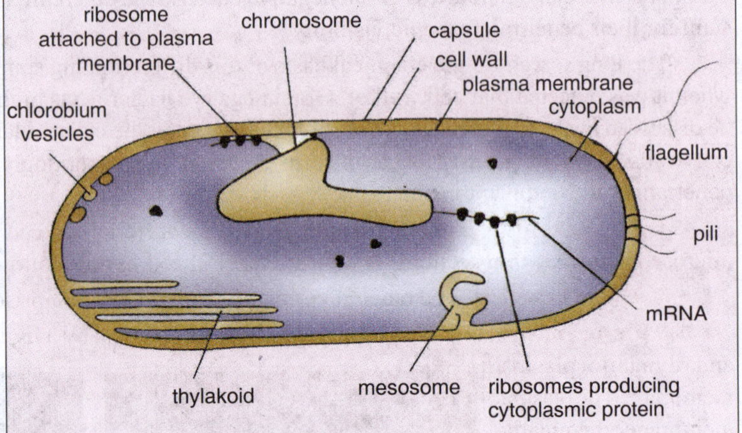

Fig. 3.11. A diagram of structures seen in the prokaryotic (bacterial) cells (after Sheeler and Bianchi, 1987; King, 1986).

II. Cell wall. The plasma membrane is covered with a strong and rigid cell wall that renders mechanical protection and provides the bacteria their characteristic shapes (the cell wall is absent in *Mycoplasma*). The cell wall of bacteria differs chemically from the cell wall of plants in that it contains proteins, lipids and polysaccharides. It may also contain chitin but rarely any cellulose.

Electron microscopy has revealed the fact that the cell wall of Gram-negative bacteria comprises the following two layers : 1. Gel, proteoglycan or peptidoglycan (*e.g.,* murein or muramic acid) containing **periplasmatic space** around the plasma membrane and 2. The **outer membrane** which consists of a lipid bilayer traversed by channels of **porin** polypeptide. These channels allow diffusion of solutes. The lipids of lipid bilayer are phospholipids and lipopolysaccharides (LPS). LPS have antigenic property and anchor the proteins and polysaccharides of the surrounding capsule (see **King**, 1986). The cell wall of Gram positive bacteria is thicker, amorphous, homogeneous and single layered. Chemically it contains many layers of peptidoglycans and proteins, neutral polysaccharides and polyphosphate polymers such as teichoic acids and teichuronic acids.

III. Capsule. In some bacteria, the cell wall is surrounded by an additional slime or gel layer called **capsule**. It is thick, gummy, mucilaginous and is secreted by the plasma membrane. The capsule serves mainly as a protective layer against attack by phagocytes and by viruses. It also helps in regulating the concentration, and uptake of essential ions and water.

B. Cytoplasm. The plasma membrane encloses a space consisting of **hyaloplasm, matrix** or **cytosol** which is the ground substance and the seat of all metabolic activities. The cytosol consists of water, proteins (including multifunctional enzymes), lipids, carbohydrates, different types of RNA molecules, and various smaller molecules. The cytosol of bacteria is often differentiated into two distinct areas : a less electron dense nuclear area and a very dense area (or dark region). In the dense cytoplasm occur thousands of particles, about 25 nm in diameter, called **ribosomes**. Ribosomes are composed of ribonucleic acid (RNA) and proteins and they are the sites of protein synthesis. Ribosomes of bacteria are 70S type and consist of two subunits (*i.e.,* a larger 50S ribosomal subunit and a small 30S ribosomal subunit). Non-functional ribosomes exist in the form of separated subunits which are suspended freely in the cytoplasm. During protein synthesis many ribosomes read the codes of single mRNA (messenger RNA) molecules and form **polyribosomes** or **polysomes**.

Reserve materials of bacteria are stored in the cytoplasm either as finely dispersed or distinct granules called **inclusion bodies** or **storage granules**. There are three types of reserve materials. First, there are organic polymers which either serve as reserves of carbon, as does **poly-β-hydroxybutyric acid**, or as stores of energy, as does a polymer of glucose, called **granulose** (*i.e.,* glycogen). Second, many bacteria contain large reserves of inorganic phosphate as highly refractile granules of metaphosphate polymers known as **volutin**. The third type of reserve material is elemental **sulphur**, formed by oxidation from hydrogen sulphide. It occurs as an energy reserve in the form of spherical droplets in certain sulphur bacteria.

C. Nucleoids. In bacteria the nuclear material includes a single, circular and double stranded DNA molecule which is often called **bacterial chromosome**. It is not separated from the cytosol by the nuclear membranes as it occurs in the eukaryotic cells. However, the nuclear material is usually concentrated in a specific clear region of the cytoplasm, called **nucleoid**. A nucleoid has no ribosome and nucleolus. The bacterial chromosome is permanently attached to the plasma membrane at one point, and when isolated often carries a number of membrane component with it. Bacterial chromosome does not contain histone proteins, however, chromosomes of some species are found to contain small quantities of a small heat-stable (HU) proteins that may be analogous to eukaryotic histones.

All three classes of RNA (*i.e.,* mRNA, tRNA, and rRNA) are formed (transcribed) by the activity of the single RNA polymerase (RNAP) species in prokaryotes. The messenger RNA formed at the chromosome is directly available for translation without processing, and so ribosomes may attach to the beginning of the mRNA strand and commence translation, while the other end of the mRNA is still being formed by transcription from DNA. Proteins for use within the cell are synthesized at cytoplasmic ribosomes; but ribosomes responsible for the synthesis of membrane proteins or proteins destined for export from the cell to form either the cell wall or secretory products, are attached to the plasma membrane. The resulting exportable polypeptides are ejected directly into or through the membrane as they are formed.

Plasmids. Many species of bacteria may also carry extrachromosomal genetic elements in the form of small, circular and closed DNA molecules, called **plasmids**. Some plasmids are merely **bacteriophage (viral) DNA** which may alternatively be incorporated within the chromosome. Other plasmids may be separated parts of the normal genome from the same or a foreign cell, and may recombine with the main chromosome. One function of some of these plasmids (called **colcinogenic factors**) is the production of antibiotically active proteins or **colicins** which inhibit the growth of other strains of bacteria in their vicinity. Some plasmids may act as **sex** or **fertility factors** (**F factor**) which stimulate bacterial conjugation. **R factors** are also plasmids which carry genes for the resistance to one or more drugs such as chloramphenicol, neomycin, penicillin, streptomycin, sulphonamides and tetracyclines.

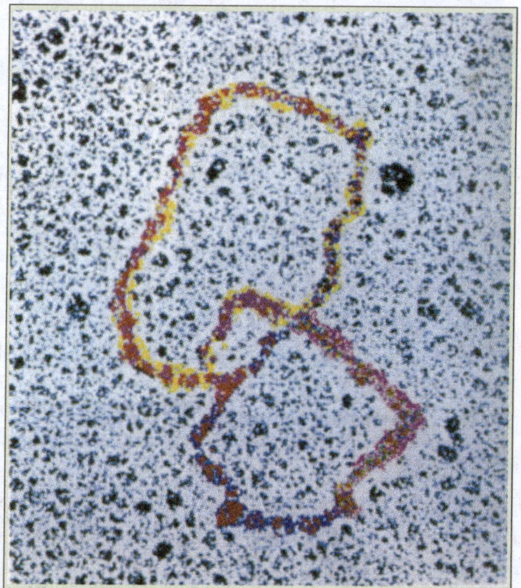

TEM of plasmids in bacterial cytoplasm.

D. Flagella and other structures. Many bacteria (*e.g., E. coli*). are motile and contain one or more **flagella** for the cellular locomotion (swimming). Bacterial flagella are smaller than the eukaryotic flagella (*i.e.,* they are 15 to 20 nm in diameter and up to 20 μm long) and are also simpler in organization. A bacterial flagellum consists of a helical tube containing a single type of protein subunit, called **flagellin**. The flagellum is attached at its base, by a short flexible **hook** that is rotated, like a propeller of ship, by the flagellar rotatory "**motor**" (*i.e.,* basal body; Fig. 3.12). The flagellar motor comprises four distinct parts : rotor (M ring), stator, bearing (S ring) and rod. The '**rotor**' is a protein disc integrated into the plasma membrane.

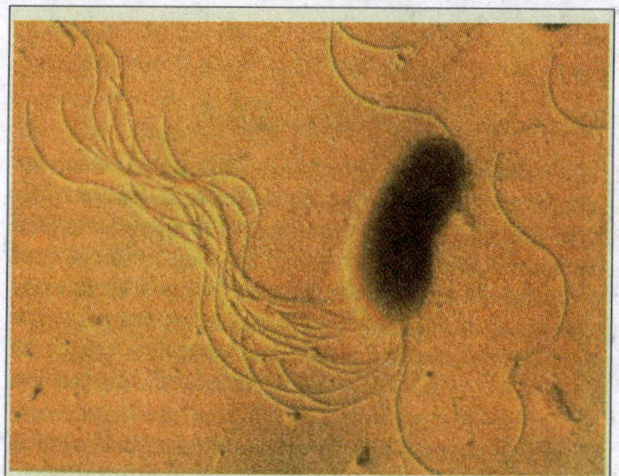

The bacterial flagellum.

It is driven by energy stored in the transmembrane proton H^+ gradient (not by ATP breakdown; see **Jones**, 1986) and rotates rapidly (~ 100 revolutions/second) in the lipid bilayer against another protein disc, called the '**stator**'. A **rod** links the 'rotor' to a hook and flagellum, thereby causing them to rotate. The protein "**bearing**" serves to seal the outer membrane of the cell wall as the rotating rod passes through it. The 'stator' and 'bearing' remains stationary (**Berg**, 1975; **Adler**, 1976).

According to the number and arrangement of the flagella in a bacterial cell, following four types of flagellation patterns have been recognized : **(1) Monotrichous**. There is a single flagellum at one pole of the cell. **(2) Lophotrichous**. There are several flagella at one pole. **(3) Amphitrichous**. The cell bears at least one flagellum at each pole. **(4) Peritrichous**. There are flagella all over the surface of cell (Fig. 3.10). Flagella-like **axial filaments** are the characteristics of some spirochetes which move like snakes through the environment. The axial filaments do not project away from the cell but are wrapped around the cell surface.

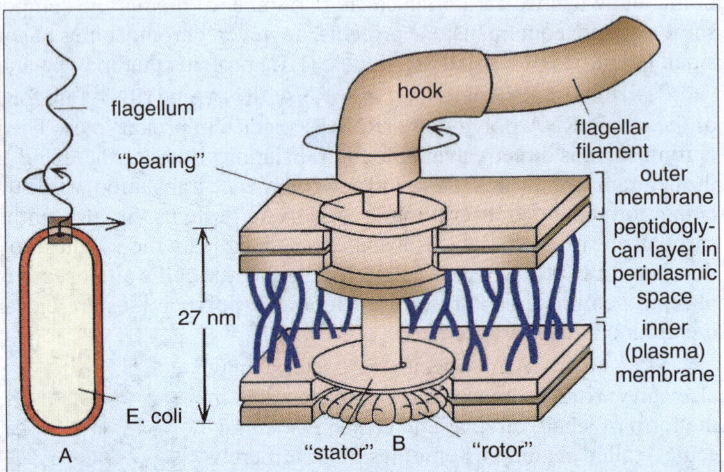

Fig. 3.12. Schematic drawing of the flagellar rotatory 'motor' of *E. coli* (after Alberts *et al.*, 1989).

Fimbriae or pili. Some bacteria (mostly Gram negative bacilli) contain non-flagellar, extremely fine, appendages called **fimbriae** (**Dugid** *et al.*, 1955) or **pili** (singular **pilus**; **Brinton**, 1959). Pili are non- motile but adhesive structures. They enable the bacteria to stick firmly to other bacteria, to a surface or to some eukaryote such as mold, plant and animal cells including red blood cells and epithelial cells of alimentary, respiratory and urinary tracts. Pili help in conjugation (*e.g.*, long F-pili or **sex pili** of male bacteria); in the attachment of pathogenic bacteria to their host cells (*e.g.*, attachment of gonorrhea- causing coccus, *Neisseria gonorrhoeae*, to the epithelial cells of the human urinary tract) and in acting as specific sites of attachment for the bacteriophages. Pili are known to be coded by the genes of the plasmid.

Spinae. Some Gram positive bacteria have tubular, pericellular and rigid appendages of single protein moiety, called **spinin**. They are called **spinae** and are known to help the bacterial cells to tolerate some environmental conditions such as salinity, pH, temperature, etc.

5. Nutrition in bacteria. Bacteria show wide diversity in their nutrition. Some are chemosynthetic, some are photosynthetic, but most of them are heterotrophic. Heterotrophic bacteria are mostly either saprophytic or parasitic. Parasitic bacteria live on the body of plants and animals and with few exceptions, most bacteria are pathogenic.

Bacteria inhabiting mouth – These bacteria possess a slime layer that allows them to cling to tooth enamel, where they can cause tooth decay unless they are removed by their chief antagonist, a toothbrush (seen here as green bristles).

Modes of respiration of bacteria are both aerobic and anaerobic. Some of the end products of bacterial anaerobic respiration are useful to man, so, they are used in the manufacture of various foods such as butter, cheese and vinegar. *Pseudomonas* is a gram negative heterotrophic aerobic form which can decompose (biodegrade) a wide variety of organic compounds such as hydrocarbons. So it is used in reducing water pollution due to petroleum spillage.

6. Reproduction in bacteria. Bacteria reproduce asexually by **binary fission** and **endospore** formation and sexually by **conjugation**. In the binary fission, the cell divides into two genetically identical daughter cells. During this process, the single circular chromosome first makes a copy of itself (*i.e.,* it duplicates) and daughter chromosomes become attached to the plasma membrane. They separate as the bacterial cell enlarges and ultimately the formation of a cross wall between the separating daughter chromosomes, divides the parent cell into two daughter cells.

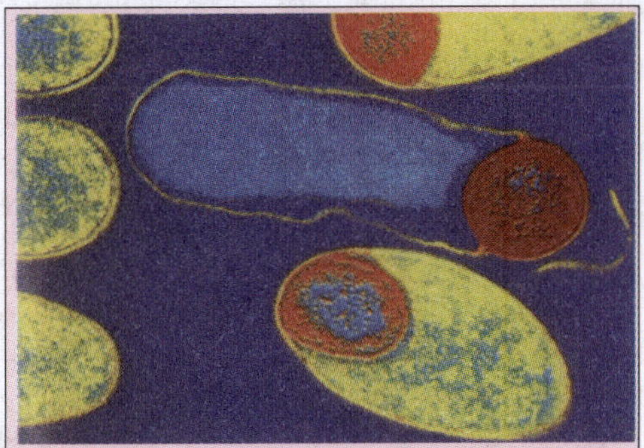

Resistant endospores, here coloured red, have formed inside bacteria of the germs (*clostridium*).

Under unfavourable ecological conditions, many bacteria (*e.g., Clostridium, Bacillus,* etc.) form spores which are not reproductive units but represent an inactive state. In endospore formation, a part of the protoplasmic material is used to form an impermeable coat or cyst wall around the chromosome along with some cytoplasm. The rest of the cell degenerates. The spore being metabolically inert can survive an unsuitable temperature, pH and drought. Under favourable conditions, spores imbibe water, become metabolically active again and germinate.

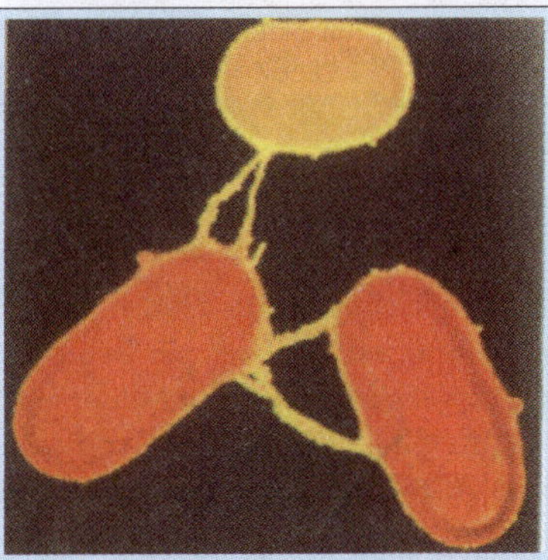

TEM of *E.coli* bacteria conjugating.

Bacterial conjugation is simplest form of sexual reproduction known. It was first of all observed in *E.coli* by **Laderberg** and **Tatum** in 1946. During the process of conjugation, a **F^+** or **donor** bacterium (equivalent to male) passes a piece of DNA or plasmid containing **fertility** or **F gene** to the **F^-** or **recipient** bacterium (equivalent to female). The donor's plasmid passes through the sex pilus of donor cell to the recipient. Following the conjugation, the progenies of the recipient express some of the characteristics of the donor. Thus, bacterial conjugation is a means of making new genetic combinations or **recombinations** which are expressed in the progeny.

Examples of Prokaryotic Cells

The following three types of prokaryotic cells are well studied ones :

1. Mycoplasma or PPLO. Among living organisms that have the smallest mass, are small bacteria called **mycoplasmas** which produce infectious diseases in animals including humans. Mycoplasmas are unicellular, prokaryotic, containing a plasma membrane, DNA, RNA and a metabolic machinery to grow and multiply in the absence of other cells (*i.e.,* they are capable of autonomous growth). They can be cultured *in vitro* like any bacteria, forming **pleomorphic** (Gr., *pleo* = many; *morphe* = forms) **colonies**, *i.e.*, depending on the type of culture medium, mycoplasmas tend to form different shaped colonies such as spheroid (fried - egg-shaped), thin, branching filaments, stellate, asteroid or irregular. They differ from the bacteria in the following respects :

1. Mycoplasmas are filterable through the bacterial filters (this fact was first demonstrated by **Iwanowsky** in 1892).

2. They do not contain cell wall and mesosomes.

3. Like the viruses and animal cells, they are resistant to antibiotics such as penicillin which kills bacteria by interfering with cell wall synthesis (see **Ambrose** and **Easty**, 1979).

4. Their growth is inhibited by tetracyclines and similar antibiotics that act on metabolic pathways.

Mycoplasmas were discovered by French scientists, **E. Nocard** and **E.R. Roux** in 1898 while studying pleural fluids of cattle suffering from the disease **pleuropneumonia** (*i.e.,* an infectious disease of warm blooded animals producing pleural and lung inflammation). Similar organisms were later isolated from other animals such as sheep, goats, dogs, rats, mice and human beings and were named as **pleuropneumonia - like organisms** (**PPLO**). PPLO were later on included under the genus *Mycoplasma* by **Nowak** (1929) and these organisms are now commonly called **mycoplasmas**. **W.V.Iterson** (1969) has placed PPLO in the group **Mycoplasmatceae** of bacteria. Currently mycoplasmas are considered as the simplest bacteria (see **Alberts** *et al.,* 1989). However, some cell biologists still prefer to place PPLO in between the viruses and bacteria (see **Sheeler** and **Bianchi**, 1987).

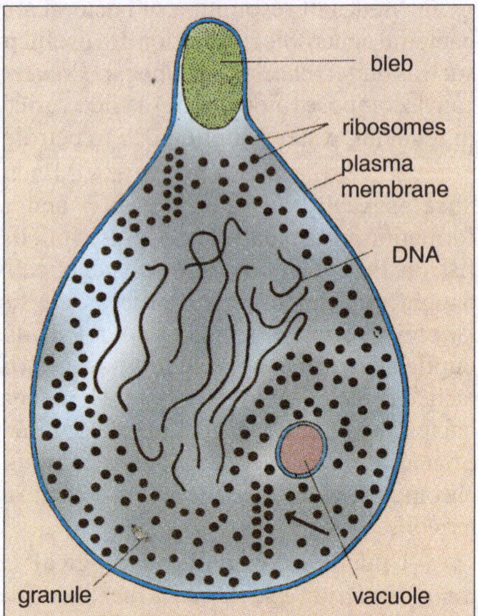

Fig 3.13. A schematic diagram of typical PPLO cell.

Mycoplasmas are mostly free-living, saprophytes or parasites. For example, *Mycoplasma laidlawii* (0.1 μm in diameter) is saprophytic and is found in sewage, compost, soil, etc. *Mycoplasma gallisepticum* (0.25 μm in diameter) is parasitic and pathogenic; it is the parasite of cells and cell exudates of respiratory organs of warm-blooded animals causing in them various chronic respiratory diseases.

Mycoplasmas range in size (diameter) from 0. 25 to 0.1 μm. Thus, they correspond in size to some of the large viruses. The spherical cell of a mycoplasma is bounded at its surface by a 75 A^0 thick plasma membrane which is composed of molecules of proteins and lipids, but there is no cell wall. Internally the cell's composition is more or less diffuse. The only microscopically discernible features within the cell are its genetic component, the DNA and the ribosomes. The DNA molecule is contained in a membraneless and clear nucleus-like region and it is a double helix which may exist either as the linear strands or a single circular molecule. The nuclear region is surrounded by numerous (50 to 100) 70S type ribosomes existing either freely or in the polysomes (Fig. 3.13). A variety of other cytoplasmic inclusions, such as vacuoles and granules, have also been detected, but their functions are not known. At one side of the cell occurs **bleb** (localised collection of fluid) of ill understood function.As in other prokaryotes, there is no intracellular membranous structure.

The PPLO cells contain many enzymes which may be required for DNA replication, the transcription of different kinds of RNA molecules and translation involved in protein synthesis, and also in the biosynthesis of adenosine triphosphate (ATP) by anaerobic breakdown of sugars. Unlike viruses, they are free living and do not require host cells for their duplication. PPLO reproduce by binary fission, budding, formation of small spore-like bodies and by growth of large branched filaments that ultimately fragment.

E.coli as seen in a electronmicrograph.

2. Escherichia coli. *E. coli* is a Gram negative, monotrichous, symbiotic bacillus of colon of human beings and other vertebrates. It is heterotrophic and non-pathogenic bacteria producing some vitamins (*e.g.*, vitamin K) for human use. Some strains of *E. coli* are known to recognise and bind specifically to sugar-containing target cells on the surface of gut lining of mammals (*e.g.,* D-mannose residues of epithelial cells of human gut or colon; **King**, 1986). *E. coli* is one of the best studied bacteria. It has served well in the field of molecular biology, since this bacterium is particularly easy to grow in an artificial medium where it divides every 20 minutes at 37^{0}C under optimal conditions. Thus, a single cell become 10^9 bacteria in about 20 hours.

The prokaryotic cell of *E. coli* (Fig. 3.14) is about 2µm long and 1µm wide. The cytoplasm of the bacterium is bounded by a typical fluid mosaic **plasma membrane.** External to the plasma membrane occurs the rigid and protective **cell wall** which has a complex organization; it comparises following two structures :

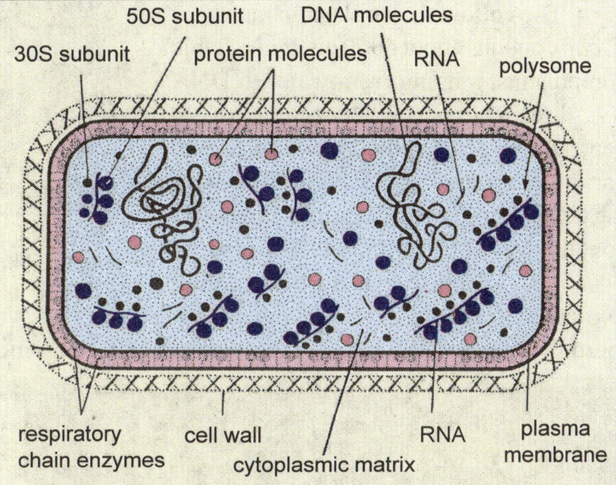

Fig. 3.14. A prokaryotic cell of *Escherichia coli.*

1. External membrane which is a lipid bilayer traversed by numerous **porin channels** that allow the diffusion of solutes. Each porin channel is formed by 6 to 8 subunits, each having three suspended hydrocarbon chains (Fig. 3.15). The porin is a polypeptide and it spans the full thickness of outer membrane. 2. Both membranes–the plasma membrane and external membrane of the cell wall – are separated by the **periplasmatic space.** This space contains a grid or reticulum of peptidoglycans. Some porin subunits remain attached to the peptidoglycan grid (Fig. 3.16).

The plasma membrane serves as a molecular barrier with the surrounding medium. It comprises a variety of transport proteins, called **permeases** which control the entrance and exit of small molecules and ions. It contributes to the establishment of bacterial protoplasm. *E. coli* has both oxygen-requiring (aerobic) and non-oxygen-requiring (anaerobic) respiratory machinery for the breakdown of sugar and contains a special group of proteins called the **electron transport chain** for the generation of stored energy in the form of ATP molecules. *E.coli* lacks mitochondria, and respiratory chain enzymes such as cytochromes, enzymes of Krebs cycle, NADH, acid phosphatase, etc.,are attached to inner face of the plasma membrane.

All genes of *E.coli* are contained on a single supercoiled, double-stranded, circular **DNA molecule**, which occurs in a clear zone of cytoplasm, called **nucleoid**, and is attached to the plasma membrane at one point. The total length of the DNA circle is about 1300 µm, comprising about 4.7 x 10^6 nucleotide pairs; this is enough DNA to code for about 4000 different proteins (see **Alberts** *et al.,* 1989). The DNA of *E.coli* is naked, lacking histones, but certain polyamines may be bound to some of its phosphates. Electron microscopy of isolated chromosome of *E. coli* has shown that DNA is folded into a series of **looped domains**, *i.e.,* about 45 loops radiate out from a dense proteinaceous **scaffold**. The DNA of loops is in the so-called supercoiled conformation in

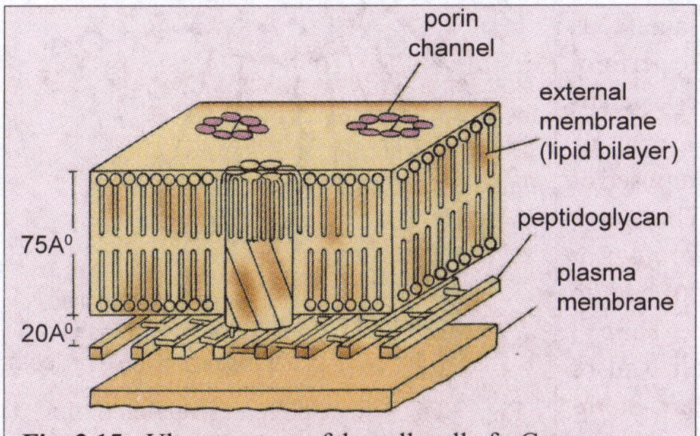

Fig. 3.15. Ultrastructure of the cell wall of a Gram-negative bacterium (after De Robertis and De Robertis, Jr., 1987).

which the double helix is itself twisted (**Schmid**, 1988). The enzyme **DNA gyrase** is responsible for the DNA supercoiling (it is inhibited by the drug called **coumermycin**; see **Freifelder**, 1985).

The colloidal cytoplasmic matrix of *E. coli* contains about 5000 distinguishable components, ranging from water to DNA (*i.e.*, three types of RNA, enzymes, glycogen, amino acids, monosaccharides and various other small molecules). Surrounding the DNA is dark dense region of matrix containing 20,000 to 30,000 70S type **ribosomes**, each existing in the form of their two subunits. During protein synthesis numerous complete ribosomes read the codes of mRNA molecules to form the polysomes (Fig. 3.14)

Cyanobacteria living inside the hairs of these polar bears are responsible for the unusual greenish colour of their coat.

Some bacteria thrive in extreme conditions like this hot spring.

3. Cyanobacteria or blue-green algae. The Gram-negative cyanobacteria or oxyphotobacteria (*i.e.*, oxygen yielding photosynthetic blue green algae) are one of the most successful and primitive (3.5 billion year old) groups of organisms on earth. They even inhabit the steaming hot springs and the undersides of icebergs. Cyanobacteria form another group of prokaryotes which include about 1500 species (85 genera and 750 species are found in India; see **Sharma**, 1992).

Cyanobacteria occur as individual cells, as small clusters or colonies of cells, or as long, filamentous chains. They lack flagella but are able to perform movement by rotatory motion or gliding over a gelatinous layer secreted through the cell surface.

A typical cell of a blue green alga is composed of outer cellular coverings and cytoplasm. The **outer cellular coverings** include an outermost **gelatinous** or **slimy layer**, **the capsule**, a middle **cell wall** and an innermost lipoproteinous **plasma membrane.** The cell wall of blue green algae resembles the cell wall of bacteria and contains an **outer bimolecular membrane** of phospholipids, lipoproteins and lipopolysaccharides, and a grid of peptidoglycans (muramic acid) in the **periplasmatic space** existing in between cell

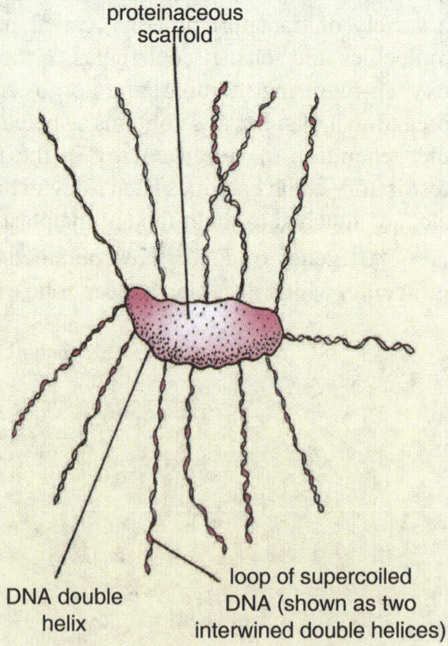

Fig. 3.16. Schematic representation of the chromosome of *E.coli*, showing only 12 of the 45 supercoiled loops (after King, 1986).

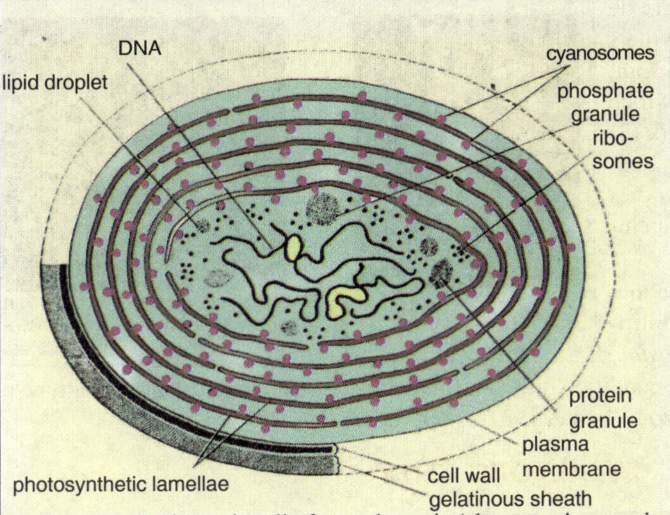

Fig. 3.17. A prokaryotic cell of cyanobacteria (electron microscopic view).

wall and plasma membrane. The **cytoplasm** of cyanobacteria appears more organized than that of other bacteria. The matrix extends throughout the cell. The cytoplasm (or protoplast) is differentiated in two regions : 1 Outer or peripheral pigmented region, the **chromoplasm** having photosynthetic lamellae or **thylakoids**. 2. Inner or central colourless region called **centroplasm** or **DNA plasm** having DNA and crystalline granules (Fig. 3.17).

Because the metabolism of the blue green algae is based on photosynthesis, therefore, the cells of them contain the photosynthetic pigment, *viz*., the **chlorophyll** and **carotenoid**. In addition to these pigments, these algae contain certain unique pigments collectively called **phycobilin**; one of the phycobilin is blue and called **phycocyanin**, while other type of phycobilin is red and called **phycoerythrin**. The photosynthetic pigments (chlorophylls and carotenoids) occur in flattened sacs called **lamellae** which remain arranged in parallel array. In between the lamellae occur certain granules of 400A^0 diameter. These granules contain phycobilin pigments and are called **cyanosomes** or **phycobilisomes**. They are attached to the outer lamellar membrane surface (**Berns**, 1983). Being earliest oxygenic photosynthesizers of earth, they made early earth's atmosphere aerobic providing the conditions favourable for the evolution of aerobic bacteria and eukaryotes.

The two subunits of 70S ribosomes of cyanobacteria are freely distributed in the cytoplasm and form polyribosomes during protein synthesis. As in all prokaryotes, the DNA molecule of blue green algae is circular, double- stranded helix and occurs in the centroplasm. This area (nucleoid) is not bound by the nuclear membrane and it does not contain a nucleolus.

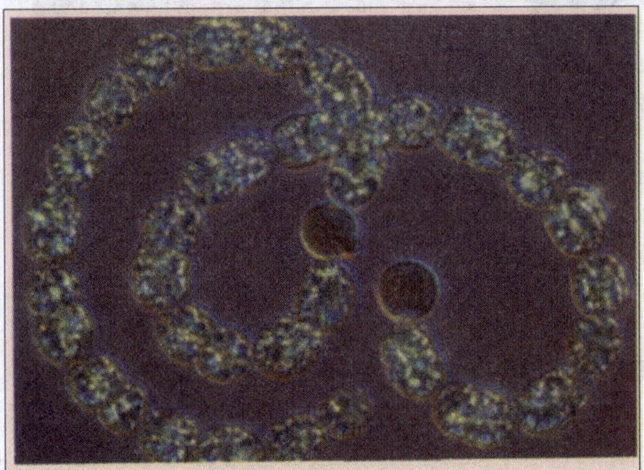

Nostoc - a cyanobacteria.

Cyanobacteria also contain a variety of **inclusions** in its cytoplasm. Membrane-bound inclusions are the gas vacuoles and the carboxysomes. **Gas vacuoles** are gas-filled cavities which are located in the inner part of chromatoplasm. They occur commonly in planktonic species such as *Nostoc, Anabaena, Phormidium, Calothrix, Galaeotrichia,* etc. Gas vacuoles serve the function of flotation or buoyancy. **Carboxysomes** contain enzymes involved in carbon dioxide fixation.

The cytoplasm of blue green algae also contains a variety of membrane-free inclusions such as **(1) cyanophycin granules** which are located in chromatoplasm and are protein storage products, containing large amount of arginine amino acid (**Fogg**, 1951) or copolymers of alanine and aspartic acid (**Simon**, 1971); **(2) myxophycean starch** which is the main food storage compound; (3) **polyglucon granules,** polyhedral bodies, lipid droplets, polyphosphate bodies etc., are some other cytoplasmic inclusions of cyanobacteria.

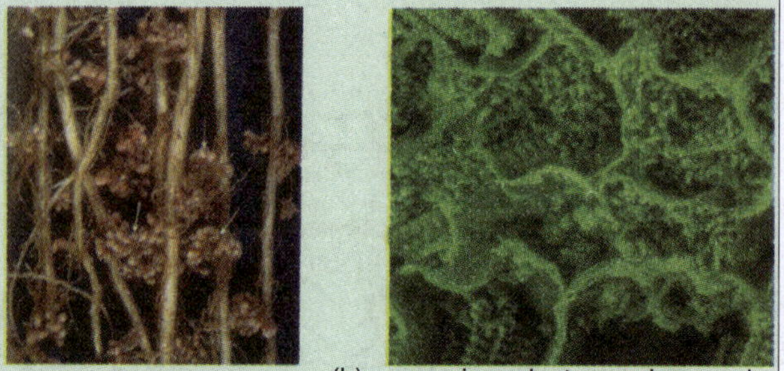

(a) Root-nodules of a leguminous plant.

(b) scanning electron micrograph shows the nitrogen-fixing bacteria inside cells within the nodules.

Lastly, many cyanobacteria (about 20 species) tend to fix atmospheric nitrogen as ammonia, *e.g., Anabaena, Nostoc, Mastigocladus,* etc. Under aerobic condition nitrogen fixation is done principally in special type of cells called **heterocysts,** as in *Nostoc* (**Donze**, 1971; **Carr**, 1976).

EUKARYOTIC CELLS

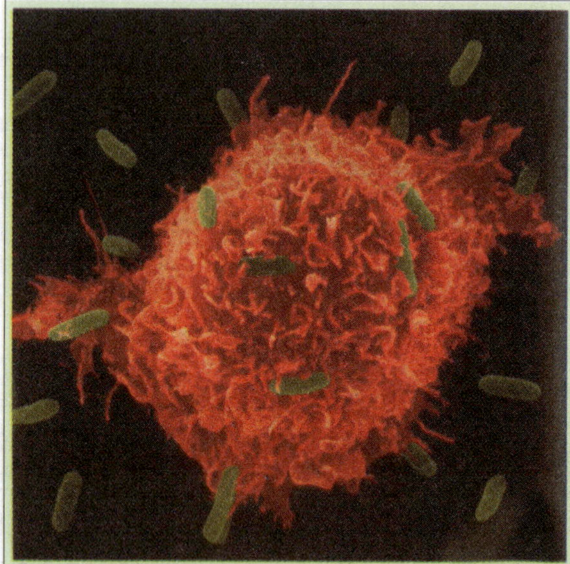

The white blood cell (in red) is eukaryotic and the bacterial cells (in green *E.coli)* are prokaryotic.

The eukaryotic cells (Gr., *eu*=good, *karyotic*=nucleated) are essentially **two envelope systems** and they are very much larger than prokaryotic cells. Secondary membranes envelop the nucleus and other internal organelles and to a great extent they pervade the cytoplasm as the endoplasmic reticulum. The eukaryotic cells are the true cells which occur in the plants (from algae to angiosperms) and the animals (from Protozoa to mammals). Though the eukaryotic cells have different shape, size, and physiology; all the cells are typically composed of plasma membrane, cytoplasm and its organelles, *viz.*, mitochondria, endoplasmic reticulum, ribosomes, Golgi apparatus, etc., and a true nucleus. Here the nuclear contents, such as DNA, RNA, nucleoproteins and nucleolus remain separated from the cytoplasm by the thin, perforated nuclear membranes. Before going into the details of cell and its various components, it will be advisable to consider the general features of different types of eukaryotic cells which are as follows:

Cell Shape

The basic shape of the eukaryotic cell is **spherical**, however, the shape is ultimately determined by the specific function of the cell. Thus, the shape of the cell may be **variable** (*i.e.,* frequently changing the shape) or **fixed**. Variable or irregular shape occurs in *Amoeba* and white blood cells or leucocytes (In fact, leucocytes are spherical in the circulating blood, but in other conditions they may produce pseudopodia and become irregular in shape). Fixed shape of the cell occurs in almost all protists (*e.g., Euglena, Paramecium*), plants and animals. In unicellular organisms the cell shape is maintained by tough plasma membrane and exoskeleton. In a multicellular organism, the shape of the cell depends mainly on its functional adaptations and partly on the surface tension, viscosity of the protoplasm, cytoskeleton of microtubules, microfilaments and intermediate filaments, the mechanical action exerted by adjoining cells and rigidity of the plasma membrane (*i.e.,* presence of rigid cell wall in plant cells). The shape of the cell may vary from animal to animal and from organ to organ. Even the cells of the same organ may display variations in the shape. Thus, cells may have diverse shapes such as **polyhedral** (with 8, 12 or 14 sides; *e.g.,* squamous epithelium); **flattened** (*e.g.,* squamous epithelium, endothelium and the upper layers of the epidermis); **cuboidal** (*e.g.,* in thyroid gland follicles); **columnar** (*e.g.,* the cells lining the intestine); **discoidal** (*e.g.,* red blood cells or erythrocytes);**spherical** (*e.g.,* eggs of many animals); **spindle shaped** (*e.g.,* smooth-muscle fibres); **elongated** (*e.g.,* nerve cells or neurons); or **branched** (*e.g.,* chromatophores or pigment cells of skin). Among plants, the cell shape also depends upon the function of the cell. For example, cells such as glandular hairs on a leaf, the guard cells of stomata and root hair cells have their special shape.

Cell Size

The eukaryotic cells are typically larger (mostly ranging between 10 to 100 μm) than the prokaryotic cells (mostly ranging between 1 to 10 μm). Size of the cells of the unicellular organisms is larger than a typical multicellular organism's cells. For example, *Amoeba proteus* is biggest among the unicellular organisms; its length being 1 mm (1000 μm). One species of *Euglena* is found up to 500 μm (0.5 mm) in length. *Euplotes* (a freshwater ciliate) is 120 μm in length. Another ciliate, *Paramecium caudatum* is from 150 to 300 μm (0.15 to 0.3 mm) in length. Diatoms have a length of 200 μm or more. The single-celled alga, *Acetabularia* which consists of a stalk and a cap is exceptionally large-sized and measures up to 10 cm in height.

The size of the cells of multicellular organisms ranges between 20 to 30 μm. Among animals, the smallest cells have a diameter of 4 μm (*e.g.,* polocytes); human erythrocytes being 7 to 8 μm in diameter. Largest animal cell is the egg of ostrich, having a diameter of 18 cm (its yolk or deutoplasm is about 5 cm in diameter); though, some nerve cells of human beings have a meter

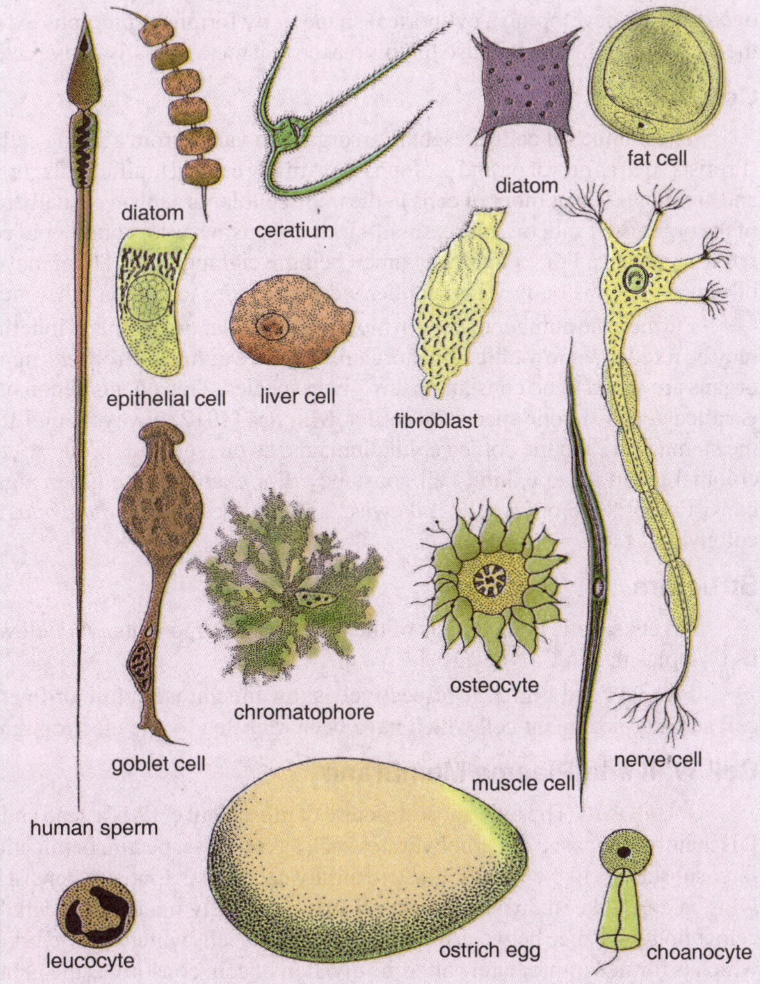

Fig. 3.18. Various types of eukaryotic cells showing different shapes.

long "tails" or axons. Among the multicellular plants, the largest cell is the ovule of *Cycas* (see **Dnyansagar**, 1988). The fibre cells (*i.e.,* sclerenchyma cells) of Manila hemp are over 100 cm in length.

Cell Volume

The volume of a cell is fairly constant for a particular cell type and is independent of the size of the organism. (This is called the **law of constant volume**.) For example, kidney or liver cells are about the same size in the bull, horse and mouse. The difference in the total mass of the organ or organism depends on the number, not on the volume of the cells. Thus, the cells of an elephant are not necessarily larger than those of other tiny animals or plants. The large size of the elephant is due to the larger number of cells present in its body.

If a cell is to be efficient, the ratio of volume to surface should be within a limited range. An increase in cell volume is accompanied by a much smaller expansion in the surface area of the cell (In fact, volume increases as cube of radius, while surface area increases as square of radius). In other words, a large cell has a proportionately smaller surface and a higher volume : surface ratio than a smaller cell. Further, a large cell volume has to accommodate many organelles simultaneously limiting the exchange of information and materials through the surface. This problem is partially overcome by developing a cylindrical shape or by forming numerous extensions (*e.g.,* microvilli) of the plasma membrane. It is also for this reason that metabolically active cells, tend to be smaller in size.

Cell Number

The number of cells present in an organism varies from a single cell in a **unicellular organism** (Protists such as protozoa and protophyta) to many cells in multicellular organisms (Most plants, fungi and animals). The number of cells in the multicellular organisms usually remains correlated with size of the organisms and, therefore, small-sized organism has less number of cells in comparison to large-sized organisms. For example, a human being weighing about 80 kg may contain about 60 thousand billion cells in his body. This number would be more in certain other multicellular organisms.

Further, the number of cells in most multicellular organisms is indefinite, but the number of cells may be fixed in some multicellular organisms. For example, in rotifers, number of nuclei in the various organs are found to be constant in any given species. This phenomenon of cells or nuclear constancy is called **eutely**. In one species of rotifer, **Martini** (1912) always found 183 nuclei in the brain, 39 in the stomach, 172 in the cornea epithelium, and so on (see **Hickman**, **Sr**., *et al.* 1979). Among plants, colonial green algae exhibit cell constancy. For example, the green alga, *Pandorina* has a colony consisting of 8, 16 or 32 cells. Likewise, another green alga, *Eudornia,* has 16, 32 or 64 cells in its colony.

Structure

An eukaryotic cell consists of the following components : A. Cell wall and plasma membrane; B. Cytoplasm; and C. Nucleus.

Fig. 3.19 and Fig. 3.21 respectively show the ultrastructure or finer details of a typical animal cell and a typical plant cell which have been revealed by the electron microscope.

Cell Wall and Plasma Membrane

1. Cell wall. The outermost structure of most plant cells is a dead and rigid layer called **cell wall**. It is mainly composed of carbohydrates such as cellulose, pectin, hemicellulose and lignin and certain fatty substances like waxes. Ultrastructurally cell wall is found to consist of a microfibrillar network lying in a gel-like matrix. The microfibrils are mostly made up of cellulose. There is a pectin-rich cementing substance between the walls of adjacent cells which is called **middle lamella**. The cell wall which is formed immediately after the division of cell, constitutes the **primary cell wall**. Many kinds of plant cells have only primary cell wall around them. Primary cell wall is composed of pectin, hemicellulose and loose network of cellulose microfibrils. In certain types of cells such as phloem and xylem, an additional layer is added to the inner surface of the primary cell wall at a later stage. This layer is called **secondary cell wall** and it consists mainly of cellulose, hemicellulose and lignin. In many plant cells, there are tunnels running through the cell wall called **plasmodesmata** which allow communication with the other cells in a tissue.

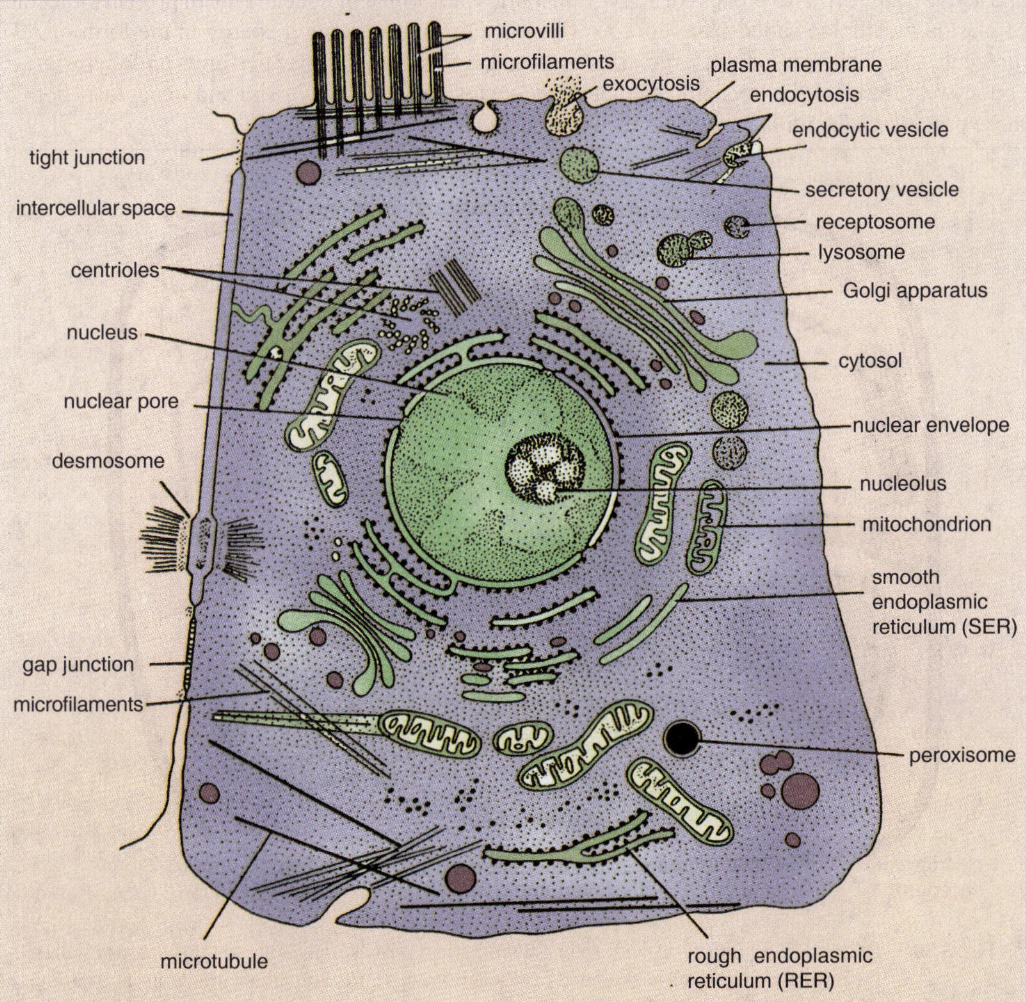

Fig. 3.19. Ultrastructure of a typical animal cell as seen in the electron microscope.

The cell wall constitutes a kind of exoskeleton that provides protection and mechanical support to the plant cell. It determines the shape of plant cell and prevents it from desiccation.

2. Plasma membrane. Every kind of animal cell is bounded by a living, extremely thin and delicate membrane called **plasmalemma, cell membrane** or **plasma membrane**. In plant cells, plasma membrane occurs just inner to cell wall, bounding the cytoplasm. The plasma membrane exhibits a tri-laminar (*i.e.,* three-layered) structure with a translucent layer sandwiched between two dark layers. At molecular level, it consists of a continuous bilayer of lipid molecule (*i.e.,* phospholipids and cholesterol) with protein molecules embedded in it or adherent to its both surfaces. Some carbohydrate molecules may also be attached to the external surface of the plasma membrane, they remain attached either to protein molecules to form **glycoproteins** or to lipids to form **glycolipids**. The plasma membrane is a **selectively permeable membrane**; its main function is to control selectively the entrance and exit of materials. This allows the cell to maintain a constant internal environment (**homeostasis**). Transport of small molecules such as water, oxygen, carbon dioxide, ethanol, ions, glucose, etc., across the plasma membrane takes place by various means such as osmosis, diffusion and active transport. The process of active transport is performed by special type of protein molecules of plasma membrane called **transport proteins** or **pumps**, consuming energy in the form of ATP molecules. For bulk transport of large-sized molecules, plasma membrane performs **endocytosis** (*i.e.,* endocytosis, pinocytosis, receptor-mediated endocytosis and phagocytosis) and **exocytosis** both of these processes also utilise energy in the form of ATP molecules.

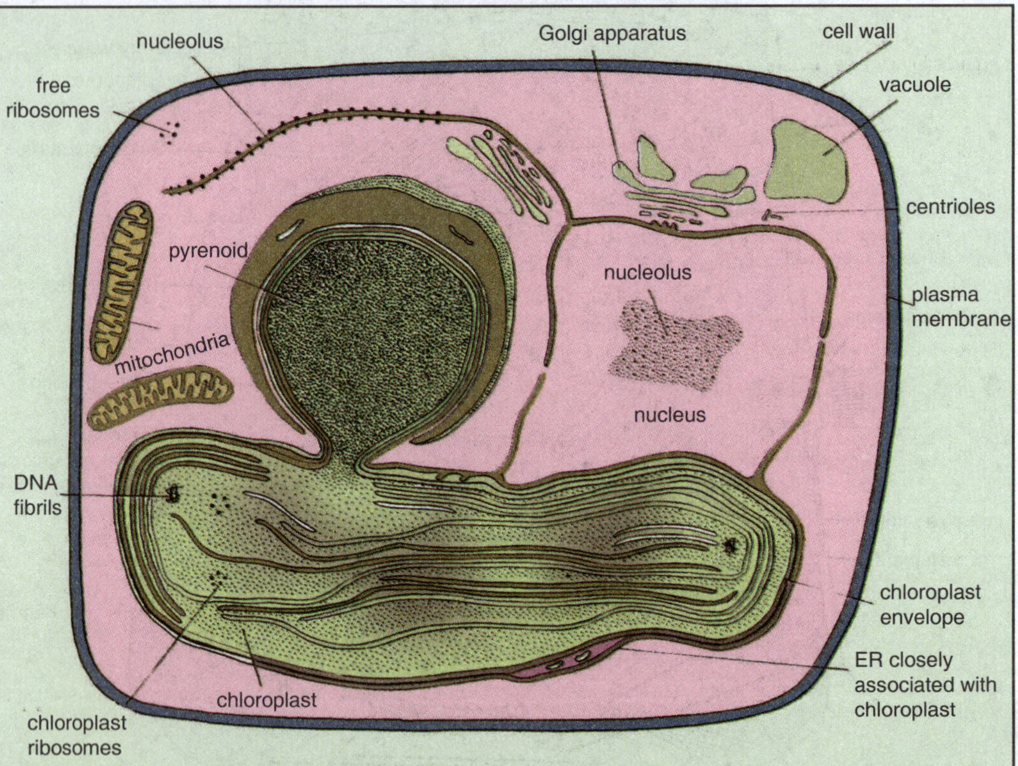

Fig. 3.20. Relationships among organelles in a hypothetical brown alga. Note that the flat sacs within the plastid are arranged in extended parallel arrays other than separated into grana and stroma systems as in higher plants.

Various cell organelles such as chloroplasts, mitochondria, endoplasmic reticulum and lysosomes are also bounded by membranes similar to the plasma membrane. All the cellular membranes have a basic trilaminar **unit membrane** construction. However, their structure and extent of activity are mainly depended on the relative proportion of their constituent protein and lipid molecules. Thus, membranes which are metabolically highly active, *e.g.,* those of mitochondria and chloroplasts have a greater proportion of proteins and more granular appearance than those membranes which are relatively less active, *e.g.,* myelin sheath of certain nerve fibres.

Cytoplasm

The plasma membrane is followed by the cytoplasm which is distinguished into following structures :

A. Cytosol. The plasma membrane is followed by the colloidal organic fluid called **matrix** or **cytosol**. The cytosol is the aqueous portion of the **cytoplasm** (the extranuclear protoplasm) and of the **nucleoplasm** (the nuclear protoplasm). It fills all the spaces of the cell and constitutes its true **internal milieu**. Cytosol is particularly rich in differentiating cells and many fundamental properties of cell are because of this part of the cytoplasm. The cytosol serves to dissolve or suspend the great variety of small molecules concerned with cellular metabolism, *e.g.*, glucose, amino acids, nucleotides, vitamins, minerals, oxygen and ions. In all type of cells, cytosol contains the soluble proteins and enzymes which form 20 to 25 per cent of the total protein content of the cell. Among the important soluble enzymes present in the matrix are those involved in glycolysis and in the activation of amino acids for the protein synthesis. In many types of cells, the cytosol is differentiated into following two parts : (i) **Ectoplasm** or **cell cortex** is the peripheral layer of cytosol which is relatively non-granular, viscous, clear and rigid. (ii) **Endoplasm** is the inner portion of cytosol which is granular and less viscous.

Cytoskeleton and microtrabecular lattice. The cytosol of cells also contains **fibres** that help to maintain cell shape and mobility and that probably provide anchoring points for the other cellular structures. Collectively, these fibres are termed as the **cytoskeleton**. At least three general classes of such fibres have been identified. 1. The thickest are the **microtubules** (20 nm in diameter) which consists primarily of the **tubulin** protein. The function of microtubules is the transportation of water, ions or small molecules, cytoplasmic streaming (cyclosis), and the formation of fibres or asters of the mitotic or meitotic spindle during cell division. Moreover, they form the structural units of the centrioles, basal granules, cilia and flagella. 2. The thinnest are the microfilaments (7 nm in diameter) which are solid and are principally formed of **actin** protein. They maintain the shape of cell and form contractile component of cells, mainly of the muscle cells. 3. The fibres of middle order are called the **intermediate filaments** (**IFs**) having a diameter of 10 nm. They having been classified according to their constituent protein such as **desmin filaments**, **keratin filaments**, **neurofilaments**, **vimentin** and **glial filaments**.

Recently, cytoplasm has been found to be filled with a three-dimensional network of interlinked filaments of cytoskeletal fibres, called **microtra-becular lattice** (**Porter** and **Tucker**, 1981). Various cellular organelles such as ribosomes, lysosomes, etc., are found anchored to this lattice. The microtrabecular lattice being flexible, changes its shape and results in the change of cell shape during cell movement.

B. Cytoplasmic structures. In the cytoplasmic matrix certain non-living and living structures remain suspended. The non-living structures are called **paraplasm** or **inclusions**, while the living structures are membrane bounded and are called **organoids** or **organelles**. Both kinds of cytoplasmic structures can be studied under the following headings :

(a) Cytoplasmic inclusions. The stored food and secretory substances of the cell remain suspended in the cytoplasmic matirx in the form of refractile granules forming the cytoplasmic inclusions. The cytoplasmic inclusions include oil drops, triacylglycerols (*e.g.*, fat cells of adipose tissue), yolk granules (or **deutoplasm**, *e.g.*, egg cells), secretory granules, glycogen granules (*e.g.*, muscle cells and hepatocytes of liver) and starch grains (in plant cells).

(b) Cytoplasmic organelles. Besides the separate fibrous systems, cytoplasm is coursed by a multitude of internal membranous structures, the organelles (literally the word organelle means a tiny organ). Membranes close off at specific regions of the eukaryotic cells performing specialized tasks : oxidative phosphorylation and generation of energy in the form of ATP molecules in mitochondria; formation and storage of carbohydrates in plastids; protein synthesis in rough endoplasmic reticulum; lipid (and hormone) synthesis in smooth endoplasmic reticulum; secretion by Golgi apparatus; degradation of macromolecules in the lysosomes; regulation of all cellular activities by nucleus; organization of spindle apparatus by centrosomes and so forth. Membrane-bound enzymes catalyze reactions that would have occurred with difficulty in an aqueous environment. The structure and function of some important organelles are as follows:

1. Endoplasmic reticulum (ER). Within the cytoplasm of most animal cells is an extensive network (reticulum) of membrane-limited channels, collectively called **endoplasmic reticulum** (or ER). Some portion of ER membranes remains continuous with the plasma membrane and the nuclear envelope. The outer surface of **rough ER** has attached ribosomes, whereas **smooth ER** do not have attached ribosomes. Functions of smooth ER include **lipid metabolism** (both catabolism and anabolism; they synthesize a variety of phospholipids, cholesterol and steroids); **glycogenolysis** (degradation of glycogen; glycogen being polymerized in the cytosol) and **drug detoxification** (by the help of the **cytochrome P-450**; **Darnell** *et al.*, 1986).

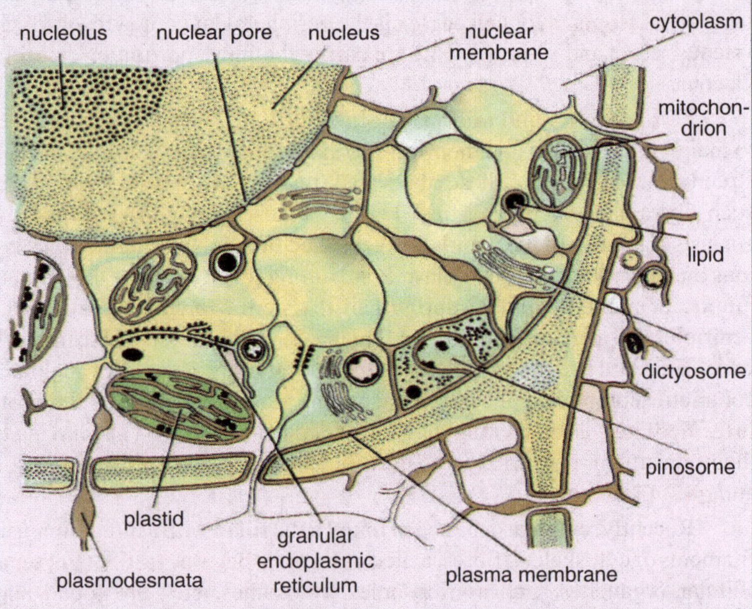

Fig. 3.21. Ultrastructure of plant cell (showing small portion of cell).

On their membranes, rough ER (RER) contain certain ribosome-specific, transmembrane glycoproteins, called **ribophorins I** and **II**, to which are attached the ribosomes while engaged in polypeptide synthesis. As a growing secretory polypeptide emerges from ribosome, it passes through the RER membrane and gets accumulated in the lumen of RER. Here, these polypeptide chains undergo tailoring, maturation, and molecular folding to form functional secondary or tertiary protein molecules. RER pinches off certain tiny protein-filled vesicles which ultimately get fused to cis Golgi. RER also synthesize membrane proteins and glycoproteins which are cotranslationally inserted into the rough ER membranes. Thus, endoplasmic reticulum is the site of biogenesis of cellular membranes.

2. Golgi apparatus. It is a cup-shaped organelle which is located near the nucleus in many types of cells. Golgi apparatus consists of a set of smooth **cisternae** (*i.e.,* closed fluid-filled flattened membranous sacs or vesicles) which often are stacked together in parallel rows. It is surrounded by spherical membrane bound **vesicles** which appear to transport proteins to and from it.

Golgi apparatus consists of at least three distinct classes of cisternae : **cis Golgi, median Golgi** and **trans Golgi**, each of which has distinct enzymatic activities. Synthesized proteins appear to move in the following direction : rough ER→ cis Golgi→ median Golgi →trans Golgi→secretory vesicles/ cortical granules of egg/ lysosomes or peroxisomes. Thus, the size and number of Golgi apparatus in a cell indicate the active metabolic, mainly synthetic, state of that cell. Plant cells contain many freely distributed sub-units of Golgi apparatus, called **dictyosomes**, secreting cellulose and pectin for cell wall formation during the cell division.

Generally, Golgi apparatus performs the following important functions : 1. The packaging of secretory materials (*e.g.*, enzymes, mucin, lactoprotein of milk, melanin pigment, etc.) that are to be discharged from the cell. 2. The **processing** of proteins, *i.e.*, glycosylation, phosphorylation,

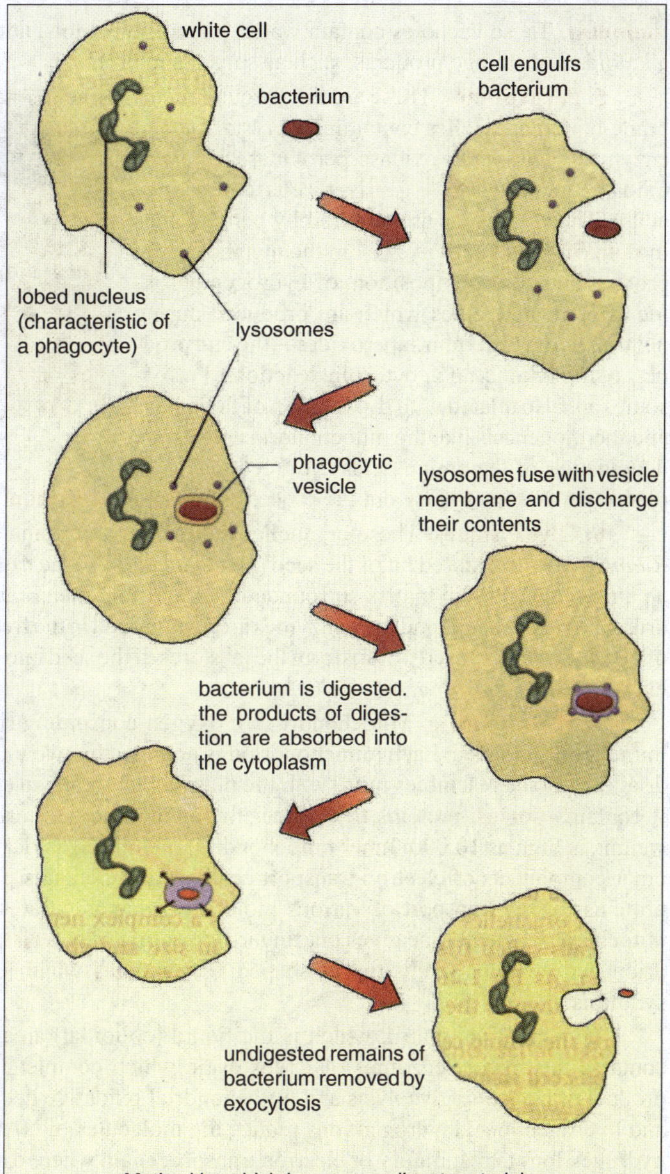

Method by which lysosomes digest material.

sulphation and selective proteolysis. 3. The synthesis of certain polysaccharides and glycolipids. 4. The sorting of proteins destined for various locations (*e.g.,* lysosomes, peroxisomes, etc.) in the cell. 5. The proliferation of membranous element for the plasma membrane. 6. Formation of the acrosome of the spermatozoa.

3. Lysosomes. The cytoplasm of animal cells contains many tiny, spheroid or irregular-shaped, membrane-bounded vesicles known as **lysosomes**. The lysosomes are originated from Golgi apparatus and contain numerous (about 50) hydrolytic enzymes (*e.g.,* **acid phosphatase** that is cytochemically identified) for intracellular and extracellular digestion. They digest the material taken in by endocytosis (such as phagocytosis, endocytosis and pinocytosis), parts of the cell (by autophagy) and extracellular substances. Lysosomes have a high acidic medium (pH 5) and this acidification depends on ATP- dependent **proton pumps** which are present in the membrane of lysosomes and which accumulate protons (H^+) inside the lysosomes. Lysosomes exhibit great **polymorphism**, *i.e.,* there are following four types of lysosomes : primary lysosomes (storage granules), secondary lysosomes (digestive vacuoles), residual bodies and autophagic vacuoles. The lysosomes of plant cells are membrane-bounded storage granules containing hydrolytic digestive enzymes, *e.g.,* large **vacuoles** of parenchymatous cells of corn seedlings, **protein** or **aleurone bodies** and **starch granules** of cereal and other seeds.

4. Cytoplasmic vacuoles. The cytoplasm of many plant and some animal cells (*i.e.,* ciliate protozoans) contains numerous small or large-sized, hollow, liquid-filled structures, the **vacuoles**. These vacuoles are supposed to be greatly expanded endoplasmic reticulum or Golgi apparatus. The **vacuoles** of animal cells are bounded by a lipoproteinous membrane and their function is the storage, transmission of the materials and the maintenance of internal pressure of the cell.

The vacuoles of the plant cells are bounded by a single, semipermeable membrane known as

tonoplast. These vacuoles contain water, phenol, flavonols, anthocyanins (blue and red pigment), alkaloids and storage products, such as sugars and proteins.

5. Peroxisomes. These are tiny circular membrane-bound organelles containing a crystal-core of enzymes (such as urate oxidase, peroxidase, D-amino oxidase and catalase, *e.g.*, liver cells and kidney cells). These enzymes are required by peroxisomes in **detoxification** activity, *i.e.*, in the metabolism or production and decomposition, of hydrogen peroxide or H_2O_2 molecules which are produced during neutralization of certain superoxides—the end products of mitochondrial or cytosolic reactions. Peroxisomes are also related with β-oxidation of fatty acids and thermogenesis like the mitochondria and also in degradation of the amino acids. In green leaves of plants, peroxisomes carry out the process of **photorespiration**.

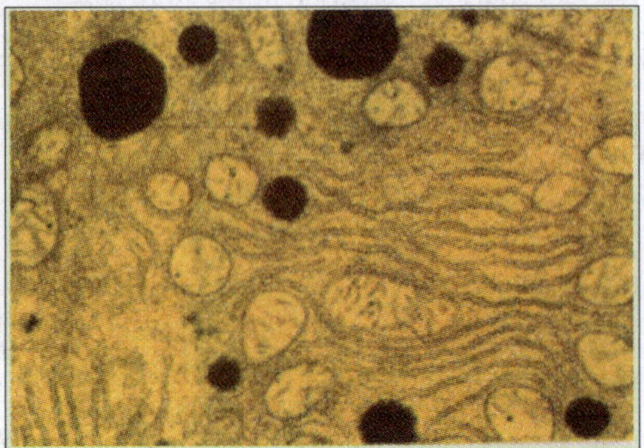

TEM of dark staining peroxisomes from the liver cell of a rat.

6. Glyoxysomes. These organelles develop in a germinating plant seed (*e.g.,* castor bean or *Ricinus*) to utilize stored fat of the seed (*i.e.*, to metabolise the triglycerides). Glyoxysomes consist of an amorphous protein matrix surrounded by a limiting membrane. The membrane of glyoxysomes originates from the ER and their enzymes are synthesized in the free ribosomes in the cytosol. Enzymes of glyoxysomes are used to transform the fat stores of the seed into carbohydrates by way of **glyoxylate cycle**.

7. Mitochondria. Mitochondria are oxygen-consuming ribbon-shaped cellular organelles of immense importance. Each mitochondrion is bounded by two unit membranes. The outer mitochondrial membrane resembles more with the plasma membrane in structure and chemical composition. It contains **porins**, proteins that render the membrane permeable to molecules having molecular weight as high as 10,000. Inner mitochondrial membrane is rich in many enzymes, coenzymes and other components of electron transport chain. It also contains **proton pumps** and many **permease** proteins for the transport of various molecules such as citrates, ADP, phosphate and ATP. Inner mitochondrial membrane gives out finger-like outgrowths (**cristae**) towards the lumen of mitochondrion and contains tennis-racket shaped **F_1 particles** which contain ATP-ase enzyme for ATP synthesis.

Mitochondrial matrix which is the liquid (colloidal) area encircled by the inner membrane, contains the soluble enzymes of Krebs cycle which completely oxidize the **acetyl-CoA** (an end product of cytosolic glycolysis and mitochondrial oxidative decarboxylation) to produce CO_2, H_2O and hydrogen ions. Hydrogen ions reduce the molecules of NAD and FAD, both of which pass on hydrogen ions to respiratory or electron transport chain where oxidative phosphorylation takes place to generate energy- rich ATP molecules. Since mitochondria act as the 'power-houses' of cells, they are abundantly found on those sites where energy is earnestly required such as sperm tail, muscle cell, liver cell (up to 1600 mitochondria), microvilli, oocyte (more than 300,000 mitochondria), etc. Mitochondria also contain in their matrix single or double circular and double stranded DNA molecules, called **mt DNA** and also the 55S ribosomes, called **mitoribosomes**. Since mitochondria can synthesize 10 per cent of their proteins in their own protein-synthetic machinery, they are considered as **semi-autonomous organelles**. Mitochondria may also produce heat (brown fat), and accumulate iron-containing pigments (Heme ferritin), ions of Ca^{2+} and HPO_4^{2-} (or phosphate; *e.g.,* osteoblasts of bones or yolk proteins).

8. Plastids. Plastids occur only in the plant cells. They contain pigments and may synthesize and accumulate various substances. Plastids are of the following types: **1. Leucoplasts** are colourless plastids of embryonic and germ cells lacking thylakoids and ribosomes. 2. **Amyloplasts** produce starch. 3. **Proteinoplasts** accumulate protein. 4. **Oleosomes** or **elaioplasts** store fats and essential oils. 5. **Chromoplasts** contain pigment molecules and are coloured organelles. Chromoplasts impart a variety of colours to plant cells, such as red colour in tomatoes, red chillies and carrots, various colours to petals of flowers and green colour to many plant cells. The green coloured chromoplasts are called **chloroplasts**. They have chlorophyll pigment and are involved in the photosynthesis of food and so act like the kitchens of the cell.

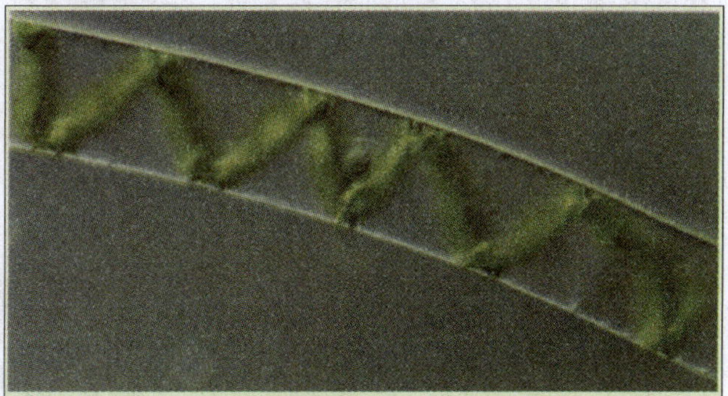

The ribbon - like chloroplast as seen in the *Spirogyra*.

Chloroplasts have diverse shapes in green algae but are round, oval or discoid in shape in higher plants. Like mitochondria, each chloroplast is bounded by two membranous envelopes, both of which have no chlorophyll pigment. However, unlike mitochondria there occurs third system of membranes within the boundary of inner membrane, called **grana**. The grana form the main functional units of chloroplast and are bathed in the homogeneous matrix, called the **stroma**. Stroma contains a variety of photosynthetic enzymes and starch grains. Grana are stacks of membrane-bounded, flattened discoid sacs, arranged like neat piles of coins. A chloroplast contains many such interconnected grana on which are located various photosynthetic enzymes and the molecules of green pigment chlorophyll and other photosynthetic pigments to trap the light energy. They contain DNA, ribosomes and complete protein synthetic machinery.

9. Ribosomes. Ribosomes are tiny spheroidal dense particles (of 150 to 200 A^0 diameter) that contain approximately equal amounts of RNA and proteins. They are primarily found in all cells and serve as a scaffold for the ordered interaction of the numerous molecules involved in protein synthesis. Ribosome granules may exist either in the **free state** in the cytosol (*e.g.*, basal epidermal cells) or **attached** to RER (*e.g.*, pancreatic acinar cells, plasma cells or antibodies-secreting lymphocytes, osteoblasts, etc.). Ribosomes have a sedimentation coefficient of about **80S** and are composed of two subunits namely **40S** and **60S**. The smaller 40S ribosomal subunit is prolate ellipsoid in shape and consists of one molecule of 18S ribosomal RNA (or rRNA) and 30 proteins (named as S_1, S_2, S_3, and so on). The larger 60S ribosomal subunit is round in shape and contains a **channel** through which growing polypeptide chain makes its exit. It consists of three types of rRNA molecules, *i.e.,* 28S rRNA, 5.8 rRNA and 5S rRNA, and 40 proteins (named as L_1, L_2, L_3 and so on).

10. Microtubules and microtubular organelles. With rare exceptions, such as human erythrocyte, microtubules are found in the cytoplasm of all types of eukaryotic cells. They are long fibres (of indefinite length) about 24 nm in diameter. In cross section each microtubule appears to have a dense wall of 6 nm thickness and a light or hollow centre. In cross section, the wall of a microtubule is made up of 13 globular subunits, called **protofilaments**, about 4 to 5 nm in diameter. Chemically, microtubules are composed of two kinds of protein subunits : **α-tubulin (tubulin A)** and **β-tubulin** (**tubulin B**), each of M.W. 55,000 daltons. The wall of a microtubule is made up of a helical array of repeating α and β tubulin subunits. Assembly studies have indicated that the structural unit is an **αβ dimer** of 8 nm length. Thus, in each microtubule, there are 13 protofilaments, each composed of αβ dimers that run parallel to the long axis of the tubule. The repeating unit is an αβ heterodimer which is arranged 'head to tail' within the microtubule, that is $\alpha\beta \rightarrow \alpha\beta \rightarrow \alpha\beta$. Thus, all microtubules have a defined **polarity** : their two ends are not structurally equivalent.

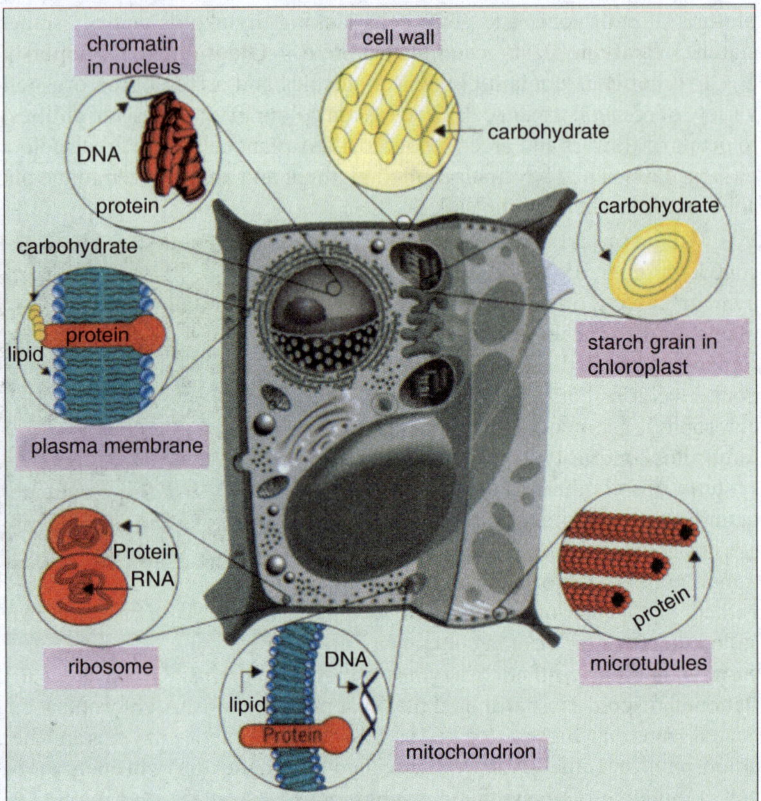

An overview of the types of biological molecules that make up various cellular structures.

Microtubules undergo reversible assembly-disassembly (*i.e.*, polymerization–depolymerization), depending on the need of the cell or organelles. Their polymerization is regulated by certain **MAPs** or **microtubule-associated proteins** (*e.g.*, Tau protein). The assembly of microtubules involves preferential addition of subunits (αβ dimers) to one end of tubule, called **A end** (or **net assembly end**); the other end of the tubule is called **D end** (or **net disassembly end**). Such an assembly involves the hydrolysis of GTP to GDP. Thus, assembly of tubulin in the formation of microtubules is a specifically oriented and programmed process. Centrioles, basal bodies and centromeres of chromosomes are the sites of orientation for this assembly. Calcium and **calmodulin** (an acidic protein having four Ca^{2+} binding sites) are some other regulating factors in the *in vivo* polymerization of tubulin. Certain drugs such as **colchicine** and **vinblastin**, are found to block the polymerization of tubulin.

The following cell organelles are derived from special assemblies of microtubules :

(1) Cilia and flagella. Ciliary and flagellar cell motility is adapted to liquid media and is executed by minute, specially differentiated appendices, called **cilia** and **flagella**. Both of these organelles have very similar structure; they differ mainly in size and number (*i.e.*, flagella are longer and fewer in number, while cilia are short and numerous). Cilia are used for locomotion in isolated cells, such as certain protozoans (*e.g.*, *Paramecium*). or to move particles in the medium, as in air passages and oviduct. Flagella are generally used for locomotion of cells, such as the spermatozoon and *Euglena* (protozoan). All cilia and flagella are built on a common fundamental plan : a bundle of microtubules called the **axoneme** (1 to 2 nm in length and 0.2 μm in diameter) is surrounded by a membrane that is part of the plasma membrane. The axoneme is connected with the basal body which is an intracellular granule lying in the cell cortex and which originates from the centrioles. Each axoneme is filled with **ciliary matrix**, in which are embedded two central **singlet** microtubules, each with the 13 protofilaments and nine outer pairs of microtubules, called **doublets**. This recurring motif is known as the 9 + 2 array. Each doublet contains one complete microtubule, called the **A subfibre,** containing all the 13 protofilaments. Attached to each A subfibre is a **B subfibre** with 10 protofilaments. Subfibre A has two **dynein arms** which are oriented in a clockwise direction. Doublets are linked together by **nexin links**. Each subfibre A is also connected to the central microtubules by **radial spokes** terminating in fork-like structures, called **spoke knobs** or **heads**.

Propulsion by both cilia and flagella is caused by bending at their base. Cilia move by a whip-like **power stroke** fueled by hydrolysis of ATP, followed by a **recovery stroke.** Flagellar movement is also powered by ATP hydrolysis. In contrast to cilia, they generally move by waves that emanate from the base and spread outward toward the tip.

2. Basal bodies and centrioles. Basal bodies and centrioles are similar in structure and function; both act as nucleating centres from which microtubules grow. **Centrioles** are cylinders that measure 0.2 µm × 0.5 µm. This cylinder is open on both ends, unless it carries a cilium or flagellum (then it is called **basal body** or **kinetosome**). The wall of a centriole has nine groups of microtubules arranged in a circle. Each group, called **blade** is a **triplet** formed of three tubules — A, B, and C that are skewed toward the centre. Tubule A has 13 protofilaments, while tubules B and C have only 10 protofilaments each. There are no central microtubules in the centrioles and no dynein arms like the cilia; however, triplets are linked by connectives. The **procentriole** (or daughter centriole) is formed at right angles to the centriole and is located near the proximal end of the centriole. Both centrioles are found in a specially differentiated region the **centrosome, cell centre** or **centrosphere**. The centrosome is juxtanuclear (L., juxta = near) and firmly attached to the nuclear envelope. At the time of cell division two pairs of centrioles are formed and form the spindle of microtubules which help in the separation and movement of chromosomes during concluding stages of cell divisions.

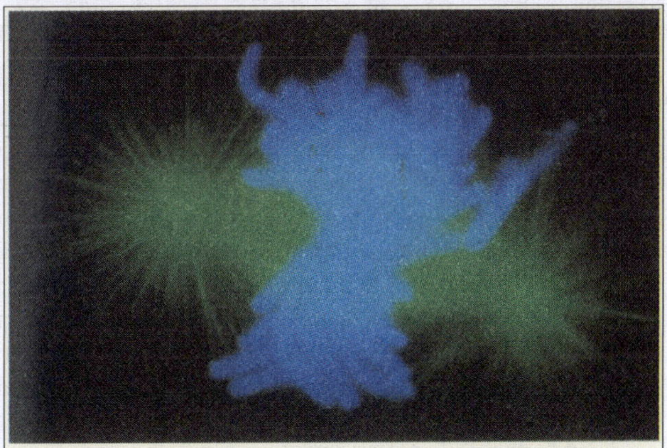

Mitotic spindle is constructed primarily of microtubules.

C. Nucleus

The nucleus is centrally located and spherical cellular component which controls all the vital activities of the cytoplasm and carries the hereditary material the DNA in it. The nucleus consists of the following three structures :

1. Chromatin. Nucleus being the heart of every type of eukaryotic cell, contains the **genes**, the hereditary units. Genes are located on the **chromosomes** which exist as **chromatin network** in the non- dividing cell, *i.e.,* during interphase. The chromatin has two forms : **1. Euchromatin** is the well-dispersed form of chromatin which takes lighter DNA-stain and is genetically active, *i.e.,* it is involved in gene duplication, gene transcription (DNA- dependent RNA synthesis) and **phenogenesis** or phenotypic expression of a gene through some type of protein synthesis. **2. Heterochromatin** is the highly condensed form of chromatin which takes dark DNA-stain and is genetically inert. Such type of chromatin exists both in the region of centromere (called **constitutive heterochromatin**) and in the sex chromatin (called **facultative heterochromatin**) and is late-replicating one.

Nucleus.

Chemically, the chromatin contains a single DNA molecule, equal amount of five basic types of histone proteins, some RNA molecules and variable amount of different types of acidic proteins. In fact, the chromatin has its unit structures in the form of **nucleosomes**. The chromatin binds strongly to the inner part of nuclear lamina, a 50 to 80 nm thick fibrous lamina lining the inner side of the nuclear envelope. Nuclear lamina is made up of three types of proteins, namely **lamin A**, **B** and **C**. Lamin proteins are homologous in structure to IF proteins and serve the following functions : 1. They anchor parts of interphase chromatin to the nuclear membrane. They tend to interfere with chromatin condensation during interphase of cell cycle. 2. Lamins may play a crucial role in the assembly of interphase nuclei after each mitosis.

2. Nuclear envelope and nucleoplasm. Nuclear envelope comprises two nuclear membranes— an **inner nuclear membrane** which is lined by nuclear lamina and an **outer nuclear membrane** which is continuous with rough ER. At certain points the nuclear envelope is interrupted by structures called **pores** or **nucleopores.** Nuclear pores contain octagonal **pore complexes** which regulate exchange between the nucleus and cytoplasm. The number of nucleopores is found to be correlated with the transcriptional activity of the cell. For example, in the frog *Xenopus laevis* oocytes (which are very active in transcription) have 60 pores/ μm^2 (and up to 30 million pore complexes per nucleus) , whereas frog's mature erythrocytes (inactive in transcription) have only about 3 pores/μm^2 (and a total of only 150 to 300 pores per nucleus) (**Scheer**, 1973).

The nuclear envelope binds the **nucleoplasm** which is rich in those molecules which are needed for DNA replication, transcription, regulation of gene actions and processing of various types of newly transcribed RNA molecules (*i.e.*, tRNA, mRNA and other types of RNA).

3. Nucleolus. Nucleus contains in its nucleoplasm a conspicuous, darkly stained, circular sub-organelle, called **nucleolus**. Nucleolus lacks any limiting membrane and is formed during interphase by the ribosomal DNA (rDNA) of **nucleolar organizer** (**NO**). Nucleolus is the site where ribosomes are manufactured. It is here where ribosomal DNA transcribes most of rRNA molecules and these molecules undergo processing before their step-wise addition to 70 types of ribosomal proteins to form the ribosomal sub-units.

Table 3-1. Differences between prokaryotic and eukaryotic cells (Source : Maclean and Hall, 1987).

Feature	Prokaryotic cell	Eukaryotic cell
1. Size	Mostly 1-10 µm	Mostly 10-100 µm
2. Multicellular forms	Rare	Common, with extensive tissue formation
3. Cell wall	Present in most but not in all cells	Present in plant and fungal cells only
4. Plasma membrane	Present	Present
5. Nucleus	Absent	Present
6. Nuclear membranes	Absent	Present
7. Chromatin with histone	Absent	Present
8. Genetic material	Circular or linear, double -stranded DNA : genes are not interrupted by intron *	Linear double-stranded DNA : genes frequently interrupted by intron sequences, especially in higher eukaryotes
9. Nucleoli and mitotic apparatus	Absent	Present
10. Plasmids	Commonly present	Rare
11. Cellular organelles :		
(i) Mitochondria	Absent	Present

* Intron is an intervening sequence of nucleotides in DNA, located within a gene that is not included in the mature mRNA.

Feature	Prokaryotic cell	Eukaryotic cell
(ii) Endoplasmic reticulum	Absent	Present
(iii) Vacuoles	Absent	Present
(iv) Lysosomes	Absent	Present
(v) Chloroplasts	Absent	Present (only in plants)
(vi) Centrioles	Absent	Present (absent in higher plants)
(vii) Ribosomes	Present (70S)	Present (80S)
(viii) Microtubules	Absent	Present
(ix) Flagellae	Simple structure composed of the protein flagellin	Complex 9 + 2 structure of tubulin and other protein
12. Respiration	Many strict anaerobes (oxygen fatal)	All aerobic, but some facultative anaerobes by secondary modifications
13. Metabolic patterns	Great variations	All share cytochrome electron transport chains, Krebs cycle oxidation, Embden-Meyerhof glucose metabolism or glysolysis
14. Photosynthetic enzymes	Bound to plasma membrane as composite chromatophores	Enzymes packaged in plastids bound by membrane
15. Sexual system	Rare : if present one way (and usually partial); transfer of DNA from donor to recipient cell occurs.	Both sexes involved in sexual participation and entire genomes transferred; alternation of haploid and diploid generations is also evident

Table 3-2. The cells of animals and plants have the following differences :

Animal cell	Plant cell
1. Animal cells are generally small in size.	1. Plant cells are larger than animal cells.
2. Cell wall is absent.	2. The plasma membrane of plant cells is surrounded by a rigid cell wall of cellulose.
3. Except the protozoan *Euglena* no animal cell possesses plastids.	3. Plastids are present.
4. Vacuoles in animal cells are many and small.	4. Most mature plant cells have a large central sap vacuole.
5. Animal cells have a single highly complex and prominent Golgi apparatus.	5. Plant cells have many simpler units of Golgi apparatus, called dictyosomes.
6. Animal cells have centrosome and centrioles.	6. Plant cells lack centrosome and centrioles.

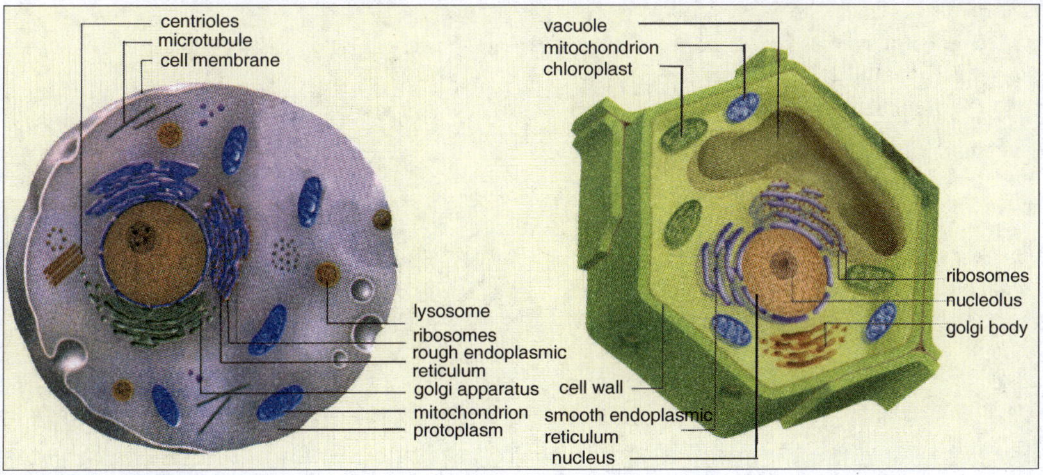

REVISION QUESTIONS

1. What are the viruses ? Write an essay on the viruses.
2. Give the life cycle of a virus.
3. What is a lysogenic phage ?
4. Describe the structural peculiarities of prokaryotic organization.
5. Write an essay on the bacteria.
6. Draw a well-labelled diagram of an animal cell as seen by the electron microscope. Comment upon the functions of nucleus, mitochondria, ribosome and microtubules.
7. Give an account of the structure of a typical animal cell.
8. Draw a labelled diagram of a typical plant cell as seen through an electron microscope. Describe the functions of specific structures of plant cells only.
9. Compare the characteristics of prokaryotic and eukaryotic cells.
10. Write short notes on the following :
    (i) Bacteriophage; (ii) Viroids; (iii) Prions; (iv) TMV; (v) PPLO; (vi) Bacteria; (vii) Blue green algae (Cyanobacteria).
11. Write differences between the following :
    (i) Viruses and bacteria;
    (ii) Prokaryotic cells and eukaryotic cells;
    (iii) Animal cells and plants cells.
12. "Structural complexity of eukaryotes is reflected in their subcellular structures". Discuss.

CHAPTER

4

Cytoplasmic Matrix

(Chemical Organization of the Cell)

All of life is conditioned by the chemistry of water.

Cells, tissues and organs are composed of chemicals, many of which are identical with those found in non-living matter, while others are unique to living organisms. The study of chemical compounds found in living systems and reactions in which they take part is known as **biochemistry**. Studies of the structure and behaviour of individual molecules constitute **molecular biology**. If the '*secret of life*' is to be found anywhere it is in these molecules (**Roberts**, 1986).

In fact, all living systems are subject to the same physical and chemical laws as are non-living systems. Within the cells of any organism, the living substance, or **protoplasm**, is itself comprised of a multitude of non-living constituents : proteins, nucleic acids, fats (lipids), carbohydrates, vitamins, minerals, waste metabolites, crystalline aggregates, pigments, and many others, all of which are composed of molecules and their constituent atoms. The *protoplasm is alive because of the highly complex organization of these non-living substances and the way they interact with one another*. This is just like a watch which is a timepiece only when all of its gears, springs,

Like small geometric units can be combined into higher-order patterns similarly the properties of a living thing emerge from the precise arrangement of component parts: atoms, molecules, cell parts, cells and so on.

and bearings are organized in a particular way and interact with one another. Neither the gears of a watch nor the molecules in protoplasm can interact in any way that is contrary to universal physical laws. Consequently, the more completely we can understand the functioning of protoplasm and its constituents on the basis of chemical principles, the more completely we can understand the phenomenon of life.

As already described in Chapter 3, the **cytoplasmic matrix** or **cytosol** is the fluid and soluble portion of the cytoplasm that exists outside the organelles (**Suzuki** *et al*., 1986). In this chapter the physical and chemical nature of the cytosol will be described.

PHYSICAL NATURE OF CYTOSOL (OR CYTOPLASMIC MATRIX)

The cytosol (cytoplasmic matrix) is a colourless or greyish, translucent, viscid, gelatinous or jelly-like colloidal substance. It is heavier than water and capable of flowing. In past, there has been a lot of controversy about the physical nature of the matrix. Different workers advanced different theories about the physical characteristics of the matrix. Their theories can be represented as follows :

1. Reticular theory suggests that the matrix is composed of reticulum of fibres or particles in the ground substances (Fig. 4.1 A).

2. Alveolar theory was proposed by **Butschili** in 1892 and according to it, the matrix consists of many suspended droplets or alveoli or minute bubbles resembling the foams of emulsion. (Fig. 4.1 B).

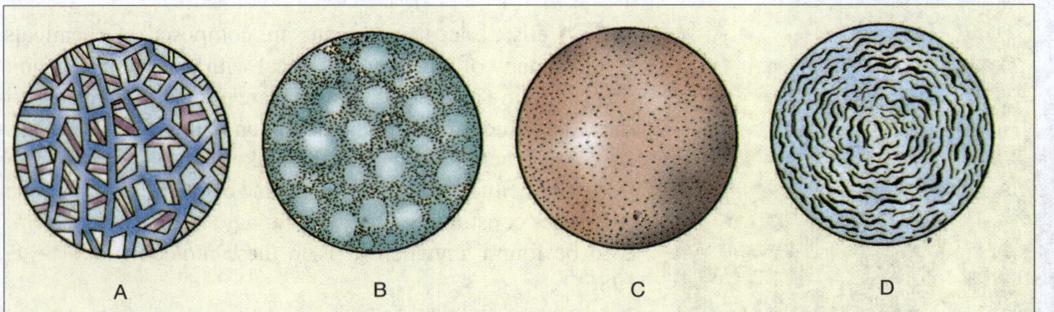

Fig. 4.1. Physical appearance of protoplasm. A—Reticular, B—Alveolar, C—Granular, and D—Fibrillar.

3. Granular theory was propounded by **Altmann** in 1893. This theory supports the view that the matrix contains many granules of smaller and larger size arranged differently. These granules were known as bioplasts (Fig. 4.1 C).

4. Fibrillar theory was proposed by **Fleming** and it holds that the matrix is fibrillar in nature (Fig. 4.1 D).

5. Colloidal theory has been forwarded very recently after the electron microscopical observations of the matrix. According to the recent concept, the matrix is partly a true **solution**, partly a **colloidal system**.

A **solution** is a mixture of liquid called **solvent** and any chemical substance in solid or liquid state, called **solute**. In a solution the particles of solute should be less than 1/10,000 millimetre in diameter. The solution part of the matrix consists of water as solvent in which various solutes of biological importance such as glucose, amino acids, fatty acids, electrolytes, minerals, vitamins, hormones and enzymes remain dissolved.

A **colloidal system** can be defined as a system which contains a liquid medium in which the particles ranging from about 1/1,000,000 to 1/10,000 millimetre in diameter, remain dispersed. Thus, the colloidal state is a condition in which one substance, such as protein or other macromolecule, is dispersed in another substance to form many small phases suspended in one continuous phase. In this way every colloidal system consists of two phases : a **discontinuous** or **dispersed phase** and a **continuous** or **dispersion phase**. Whole of the protoplasm (cytoplasm + nucleoplasm) is a colloidal solution, because the main molecular components of protoplasm— proteins—show all characteristics of the colloidal state. Proteins form stable colloids because, firstly, they are charged ions in solution that repel each other, and, secondly each protein molecule attracts water molecules around it in definite layers.

Phase Reversal

Cytosol (cytoplasmic matrix) like many colloidal systems, shows the property of **phase reversal**. For example, gelatin particles (discontinuous phase) are dispersed through water (continuous phase) in a thin consistency that is freely shakable (Fig. 4.2 A). Such a condition is called a **sol**. When the solution cools, gelatin now becomes the continuous phase and the water is in the discontinuous phase. Moreover, now the solution has stiffened and becomes semisolid and is called a **gel**.

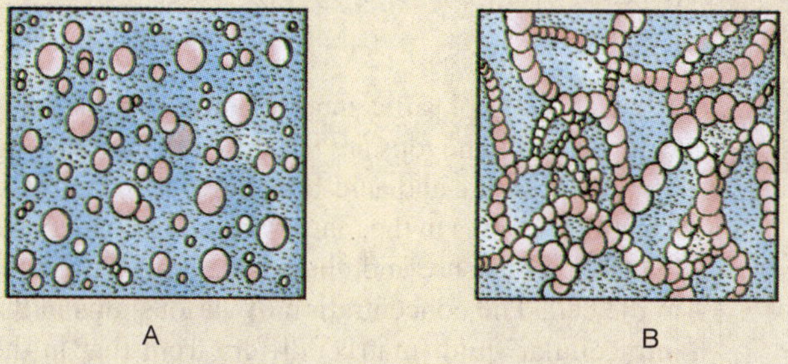

Fig. 4.2. Sol and gel state of the cytosol. A—Sol condition in which gelatin particles are the discontinuous phase, water the continuous phase; B—Gel condition in which gelatin particles form continuous phase (net work), enclosing water as discontinuous phase.

In gel state the molecules of colloidal substance remain held together by various types of chemical bonds or bond between H—H, C—H or C—N. The stability of gel depends on the nature and strength of chemical bonds. Heating the gel solution will cause it to become sol again, and the phases are reversed. Under the natural conditions, the phase reversal of the cytosol (cytoplasmic matrix) depends on various physiological, mechanical and biochemical activities of the cell.

CHEMICAL ORGANIZATION OF CYTOSOL (OR CYTOPLASMIC MATRIX)

Chemically, the cytoplasmic matrix is composed of many chemical elements in the form of atoms, ions and molecules.

Chemical Elements

Of the 92 naturally occurring elements, perhaps 46 are found in the cytosol (cytoplasmic matrix). Twenty four of these are considered essential for life (called **essential elements**), while others are present in cytosol only because they exist in the environment with which the organism interacts. Of the 24 essential elements, six play especially important roles in living systems. These **major elements** are carbon (C, 20 per cent), hydrogen (H, 10 per cent), nitrogen (N, 3 per cent), oxygen (O, 62 per cent), phosphorus (P, 1.14 per cent) and sulphur (S, 0.14 per cent). Most organic molecules are built with these six elements. Another five essential elements found in less abundance in living systems are calcium (Ca, 2.5 per cent), potassium (K, 0.11 per cent), sodium (Na, 0.10 per cent), chlorine (Cl, 0.16 per cent) and magnesium (Mg, 0.07 per cent). Several other elements, called **trace elements**, are also found in minute amounts in animals and plants, but are nevertheless essential for life. These are iron (Fe, 0.10 per cent), iodine (I, 0.014 per cent), molybdenum (Mo), manganese (Mn), Cobalt (Co), zinc (Zn), selenium (Se), copper (Cu), chromium (Cr), tin (Sn), vanadium (V), silicon (Si), nickel (Ni), fluorine (F) and boron (B).

All matter is composed of atoms. Photo of individual atoms on the surface of a silicon crystal developed by tunneling microscopy.

Ions

The cytoplasmic matrix consists of various kinds of ions. The ions are important in maintaining osmotic pressure and acid-base balance in the cells. Retention of ions in the matrix produces an increase in osmotic pressure and, thus, the entrance of water in the cell. The concentration of various ions in the intracellular fluid (matrix) differs from that in the interstitial fluid. For example, in the cell K^+ and Mg^{++} can be high, and Na^+ and Cl^- high outside the cell. In muscle and nerve cells a high order of difference exists between intracellular K^+ and extracellular Na^+. Free calcium ions (Ca^{++}) may occur in cells or circulating blood. Silicon ions occur in the epithelium cells of grasses. The free ions of phosphate (primary, $H_2PO_4^-$ and secondary, HPO_4^-) occur in the matrix and blood. These ions act as a buffering system and tend to stabilize pH of blood and cellular fluids. The ions of different cells also include sulphate (SO_4^-), carbonate (CO_3^-), bicarbonate (HCO_3^-), magnesium (Mg^{++}) and amino acids. Cellular functions of certain ions have been tabulated in Table 4-1.

(a) a person meditating; has a very distinctive brain wave pattern

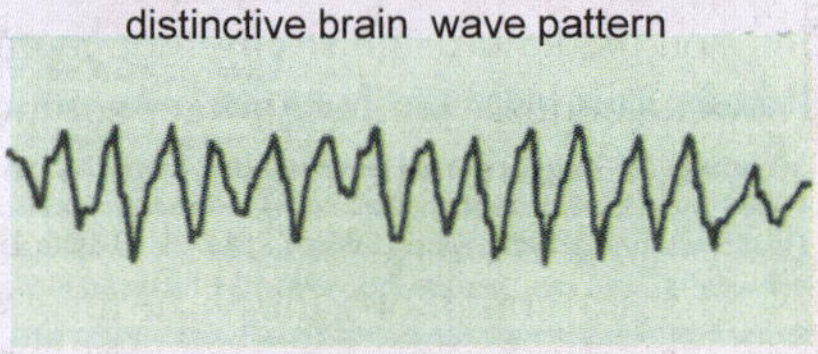

(b) The brain wave pattern of the same person while not meditating

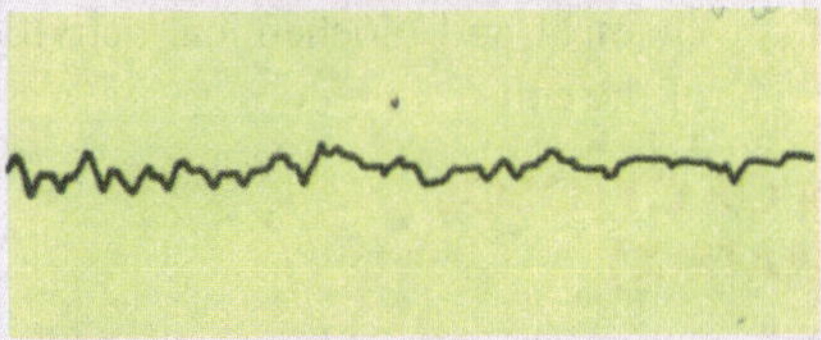

Ions, Nerve impulses and Meditation. Na^+ are crucial to electrical activity in the brain, which can be monitored and recorded.

Table 4-1. Cellular functions of certain ions. (Source : Sheeler and Bianchi, 1987).

	Element	Ionic form present	Functions
1.	Molybdenum	MoO_4^{2-}	Cofactor or activator of certain enzymes (*e.g.*, nitrogen fixation, nucleic acid metabolism, aldehyde oxidation).
2.	Cobalt	Co^{2+}	Constituent of vitamin B_{12}.
3.	Copper	Cu^+, Cu^{2+}	Constituent of plastocyanin and cofactor of respiratory enzymes.
4.	Iodine (Heaviest trace element)	I^-	Constituent of thyroxin, triiodothyronine and other thyroid hormones.
5.	Boron	BO_3^{3-}, $B_4O_7^{2-}$	Activates arabinose isomerase.
6.	Zinc	Zn^{2+}	Cofactor of certain enzymes (*e.g.*, carbonic anhydrase, carboxypeptidase).
7.	Manganese	Mn^{2+}	Cofactor of certain enzymes (*e.g.*, several kinases, isocitric decarboxylase).
8.	Iron	Fe^{2+}, Fe^{3+}	Constituent of haemoglobin, myoglobin and cytochromes.
9.	Magnesium	Mg^{2+}	Constituent of chlorophyll; activates ATPase enzyme.
10.	Sulphur	SO_4^{2-}	Constituent of coenzyme A, biotin, thiamine, proteins.
11.	Phosphorus	PO_4^{3-}, $H_2PO^-_4$	Constituent of lipids, proteins, nucleic acids, sugar phosphates, nucleoside phosphates.
12.	Calcium	Ca^{2+}	Constituent of plant cell walls; matrix component of bone tissue; cofactor of coagulation enzymes.
13.	Potassium	K^+	Cofactor for pyruvate kinase and K^+- stimulated ATPase.

Electrolytes and Non-electrolytes

The matrix consists of both electrolytes and non-electrolytes.

(i) Electrolytes. The electrolytes play a vital role in the maintenance of osmotic pressure and acid base equilibrium in the matrix. Mg^{2+} ions, phosphate, etc., are good examples of the electrolytes.

(ii) Non-electrolytes. Some of minerals occur in matrix in non-ionizing state. The non-electrolytes of the matrix are Na, K, Ca, Mg, Cu, I, Fe, Mn, Fl, Mo, Cl, Zn, Co, Ni, etc. The iron (Fe) occurs in the haemoglobin, ferritin, cytochromes and some enzymes as catalase and cytochrome

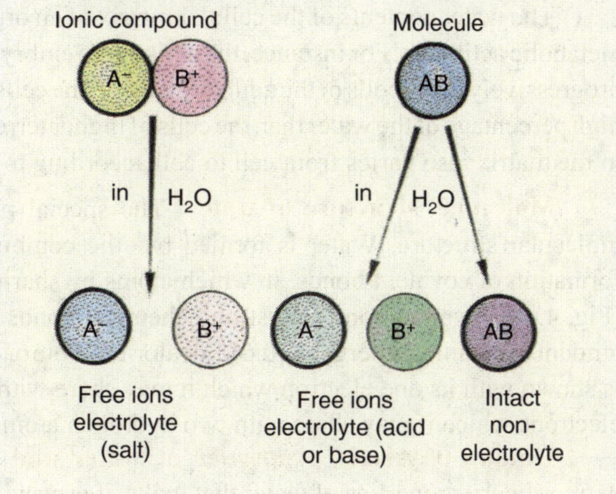

Fig. 4.3. Electrolytes and non-electrolytes.

oxidase. The calcium (Ca) occurs in the blood, matrix and the bones. The copper (Cu), manganese (Mn), molybdenum (Mo), zinc (Zn) are useful as cofactors for enzymatic actions. The iodine and fluorine are essential for the thyroid and the enamel metabolism, respectively.

TYPES OF COMPOUNDS OF CYTOSOL

Chemical compounds are conventionally divided into two groups : **organic** and **inorganic**. Organic compounds form 30 per cent of a typical cell, rest are the inorganic substances such as water and other substances.

Table 4-2. The approximate percentage composition of the human body (Source : Roberts, 1986).

Substance	Percentage
1. Water	65
2. Protein	18
3. Fat	10
4. Carbohydrate	5
5. Other organic	1
6. Inorganic	1

INORGANIC COMPOUNDS

The inorganic compounds are those compounds which normally found in the bulk of the physical, non-living universe, such as elements, metals, non-metals, and their compounds such as water, salts and variety of electrolytes and non-electrolytes. In the previous section, we have discussed a lot about the inorganic substances except the water which will be discussed in the following paragraph.

Water

The most abundant inorganic component of the cytosol is the water (the notable exceptions are seeds, bone and enamel). Water constitutes about 65 to 80 per cent of the matrix. In the matrix the water occurs in two forms, *viz*., **free water** and **bound water**. The 95 per cent of the total cellular water is used by the matrix as the solvent for various inorganic substances and organic compounds and is known as **free water**. The remaining 5 per cent of the total cellular water remains loosely linked with protein molecules by hydrogen bonds or other forces and is known as **bound water.**

The water contents of the cellular matrix of an organism depend directly on the age, habitat and metabolic activities. For instance, the cells of the embryo have 90 to 95 per cent water which decreases progressively in the cells of the adult organism. The cells of lower aquatic animals contain comparative high percentage of the water than the cells of higher terrestrial animals. Further the percentage of water in the matrix also varies from cell to cell according to the rate of the metabolism.

Molecular structure of water. The special physical properties of water are found in its molecular structure. Water is formed by the combination of hydrogen and oxygen through the formation of covalent bonds, in which atoms by sharing pairs of electrons, become linked together (Fig. 4.4). Covalent bonds are strong chemical bonds between atoms and contain a relatively large amount of chemical energy (110.6 kilocalories/Mole or 462 kilojoules/Mole). In Figure 4.4 hydrogen is shown with its one electron which it may share with an oxygen atom. Each oxygen atom has two electrons which it may share with two hydrogen atoms.

Unique physical properties of water and their biological utility. There are several extraordinary properties of water that make it especially fit for its essential role in the protoplasmic systems (*i.e.*, cytosol or matrix). Some of the unique properties of water are the following :

1. Water as a solvent. Water is most stable yet versatile of all solvents. Water's properties as a solvent for inorganic substances as mineral ions, solids, etc., and organic compounds such as carbohydrates and proteins, depend on water's dipole nature. Because of this polarity, water can bind electrostatically to both positively and negatively charged groups in the protein. Thus, each amino group in a protein molecule is capable of binding 2.6 molecules of water. The solvency is of great biological importance because all the chemical reactions that take place in the cells do so in aqueous solution. The water also forms the good dispersion medium for the colloidal system of the matrix.

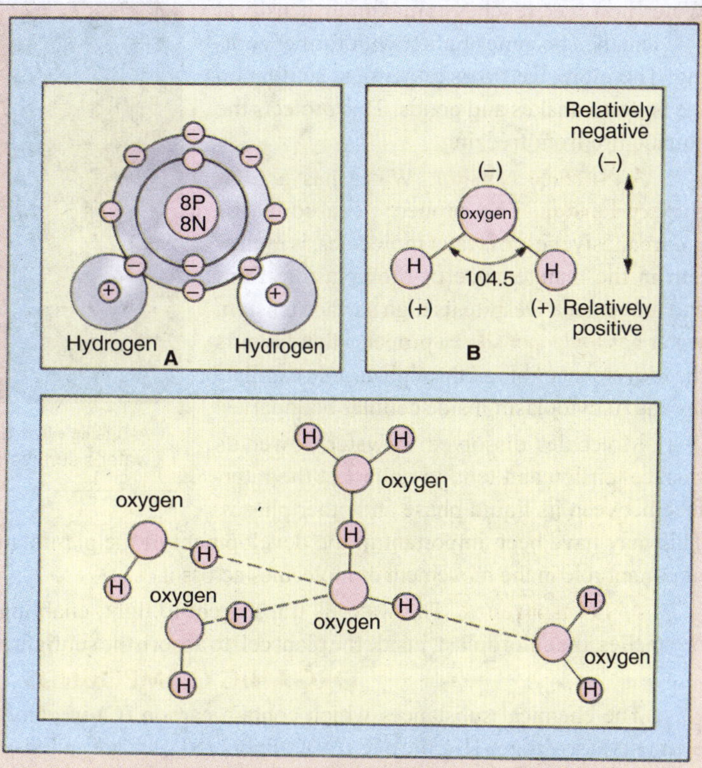

Fig. 4.4. Structure of a water molecule : A— How two hydrogen atoms share their single electrons with oxygen atom; B— The hydrogen atoms position themselves to one side of the oxygen, leaving a relatively negative cloud of electrons exposed on other side. The electrons of the hydrogen are maintained close to the oxygen, leaving the hydrogen relatively positive since its proton is exposed; C— A tetrahedron is formed due to formation of hydrogen bonds between four water molecules.

2. Water's thermal properties. Water is the only substance that occurs in nature in the three phases of solid, liquid and vapour within the ordinary range of earth's temperatures. Water has a **high specific heat** : it requires 1 calorie (4.185 joules) to elevate the temperature of 1 gram of water by 1°C (such as from 15 to 16° C). Such a high thermal capacity of water has a great moderating effect on environmental temperature changes and is a great protective agent for all life.

Water also has a **high heat of vaporization**. It requires more than 540 calories (2259 joules) to change 1 gram of liquid water into water vapour. Thus, water tends to have a remarkably high boiling point (100° C) for a substance of such low relative molecular mass. Were it not for this lucky accident, it is likely that liquid water would never have existed on earth and would have been lost to outer space. Further, for terrestrial plants and animals, cooling produced by the evaporation of water is an important means of getting rid of excess heat. Moreover, at the other temperature extreme, large amounts of energy (335 joules or 80 cal per gram) must be lost for water to be converted from the liquid to the solid state. This is called **heat of fusion.** Water's melting point being 0° C.

Another important property of water from a biological standpoint is its unique **density behaviour** during change of temperature. Most liquids become continually more dense with cooling.

Water, however, reaches its maximum density at 4° C and then becomes lighter with further cooling. Therefore, ice floats rather than settling on the bottom of lakes and ponds. This protects the aquatic life from freezing.

3. Surface tension. Water has a **high surface tension.** This property, caused by the great cohesiveness of water molecules, is important in the maintenance of protoplasmic form and movement. Despite its high surface tension, water has **low viscosity**, a property that favours the movement of blood through minute capillaries and of cytoplasm inside cellular boundaries.

A basilisk lizard runs across a pond, putting the water's surface tension to good use.

Molecules dissolved in water, lower its surface tension and tend to collect at the interface between its liquid phase and other phases. This may have been important in the development of the plasma membrane, and certainly plays an important role in the movement of molecules across it.

4. Transparency. The water is transparent to light, enabling the specialized photosynthetic organelles, the chloroplast, inside the plant cell to absorb the sunlight for the process of photosynthesis.

ORGANIC COMPOUNDS

The chemical substances which contain carbon (C) in combination with one or more other elements as hydrogen (H), nitrogen (N), sulphur (S), etc., are called **organic compounds**. The organic compounds usually contain large molecules which are formed by the similar or dissimilar unit structure known as the **monomers**. A monomer (Gr., *mono*=one, *meros*=part) is the simplest unit of the organic molecule which can exist freely. Some organic compounds such as carbohydrates occur in the matrix as the monomers. The monomers usually link with other monomers to form **oligomers** (Gr., *oligo*=few or little, *meros*=part) and **polymers** (Gr., *poly*=many, *meros*=part). The oligomers contain small number of monomers, while the polymers contain large number of monomers. The oligomers and polymers contain large-sized molecules or macromolecules. When a polymer contains similar kinds of monomers in its macromolecule it is known as **homopolymer** and when the polymer is composed of different kinds of monomers it is known as the **heteropolymer**.

The main organic compounds of the matrix are the carbohydrates, lipids, proteins, vitamins, hormones and nucleotides.

Simple sugars: Many animals consume sugar like this butterfly consuming nectar, a solution rich in glucose.

Carbohydrates

The carbohydrates (L., *carbo*=carbon or coal, Gr., *hydro*=water) are the compounds of the carbon, hydrogen and oxygen. They form the main source of the energy of all living beings. Only green part of plants and certain microbes have the power of synthesizing the carbohydrates from the water and CO_2 in the presence of sunlight and chlorophyll by the process of photosynthesis. All the animals, non-green parts of the plants (*viz.*, stem, root, etc.), non-green plants (*e.g.*, fungi), bacteria and viruses depend on green parts of plants for the supply of carbohydrates.

Chemically the carbohydrates are polyhydroxy aldehydes or ketones and they are classified as follows :

A. Monosaccharides (Monomers), B. Oligosaccharides (Oligomers), and C. Polysaccharides (Polymers).

A. Monosaccharides. The monosaccharides are the simple sugars with the empirical formula $Cn(H_2O)n$. They are classified and named according to the number of carbon atoms in their molecules as follows :

(i) Trioses contain three carbon atoms in their molecules, *e.g.*, glyceraldehyde and dihydroxy acetone.

(ii) Tetroses contain four carbon atoms in their molecules, *e.g.*,erythrulose and erythrose.

(iii) Pentoses contain five carbon atoms in their molecules, *e.g.*, ribose, ribulose, deoxyribose, arabinose and xylulose.

(iv) Hexoses contain six carbon atoms in their molecules, *e.g.*, glucose, mannose, fructose and galactose.

(v) Heptoses contain seven carbon atoms in their molecules, *e.g.*, sedoheptulose.

The monosaccharides usually exist as isomers. For example, three hexose sugars—glucose, fructose, and galactose, contain the same number of carbon, hydrogen and oxygen atoms (*i.e.*, $C_6H_{12}O_6$), but they are different sugars because of different arrangements of the atoms within the molecules. Glucose and galactose are **optical isomers** or **stereoisomers**. If a carbon atom is present in a molecule which has four different chemical groups bonded to it, the groups can be arranged in two distinct spatial arrangements about the carbon atom (such a carbon atom is often called **asymmetric carbon atom**). These two different arrangements are known as the **mirror-images** and a convenient example of such mirror-image structures are the two human hands which are identically structured but which cannot be superimposed on each other. The two isomers are designated as '**D**' or '**L**' by analogy to D– and L– glyceraldehyde, which are aldotrioses.

Simple carbohydrates: sugarcanes store large quantities of sucrose in special cells.

Most of the monosaccharides are optically active, meaning that their asymmetric carbon (s) cause the rotation of plane of polarised light. Molecules that rotate the plane of polarization to the right, as one faces the light source, are called **dextrorotatory** and are designated **d** or (+), while the opposite case is **levorotation**, designated **l** or (–). It is important to remember that the capitals D and L refer to structure, whereas the lower case **d** and **l** refer to optical activity established before the structure could be determined (see **Dyson**, 1978). Thus, one sees references to D (+) -glucose, also called dextrose. and D (–) -fructose, also called levulose.

Further, for the sake of simplicity, sugars can be represented in a linear straight chain form (Fig. 4.5). In fact, however, the more important configuration is the cyclic one; it is an isomer having an oxygen bridge between two of the carbons. Ring formation introduces a new asymetric carbon at position one. The stereochemistry of monosaccharides is such that the ring formed is either five- or six- membered; a seven-membered ring would involve too much strain. In pentose (five-carbon) sugars such as ribose, a five-membered **furanose ring** is formed. In hexoses such as fructose and glucose, a six-membered **pyranose ring** is formed (Fig. 4.6). A useful way of representing the ring-structures of sugars was proposed by **Haworth** (1927). The pyranose or furanose ring is considered to be in the plane perpendicular to the plane of the paper; thus, in gluco-pyranose, carbon atom 2 and 3 are in front of the paper, and carbon atom 5 and the ring oxygen lie behind the plane of the paper. The substitute groups are either above or below the plane of the ring (see **Ambrose** and **Easty**, 1977).

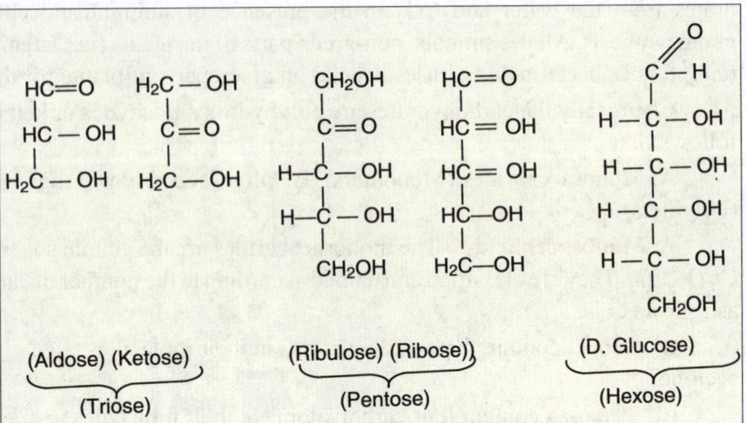

Fig. 4.5. Structure of some monosaccharides.

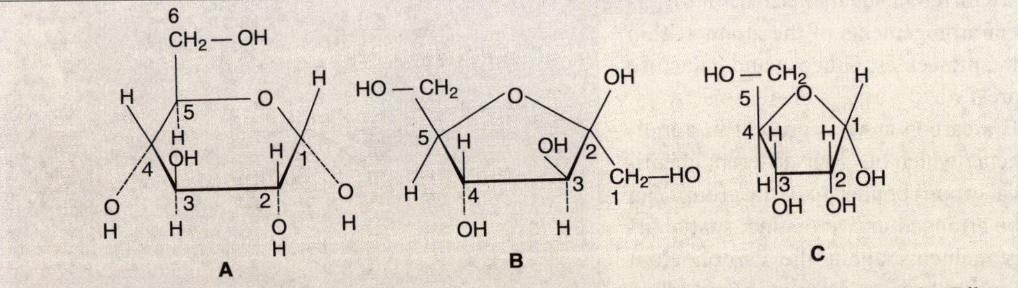

Fig. 4.6. Ring structures of monosaccharides proposed by Haworth. A—Glucose; B— Fructose; C—Ribose (after Ambrose and Easty, 1977).

The monosaccharides are the monomers and cannot split further or hydrolysed into the simpler compounds. The pentoses and hexoses are the most abundantly occurring monosaccharides of the matrix.

The pentose sugar, **ribose** is the important constituent molecule of the ribonucleic acid (RNA) and certain coenzymes as nicotinamide adenine dinucleotide (NAD), NAD phosphate (NADP), adenosine triphosphate (ATP) and coenzyme A (CoA). Another pentose sugar the **deoxyribose** is the important constituent of the deoxyribonucleic acid (DNA). The **ribulose** is a pentose sugar which is necessary for photosynthetic mechanism.

The **glucose**, a hexose sugar, is the primary source of the energy for the cell. The other important hexose sugars of the matrix are the fructose and galactose.

B. Oligosaccharides. The oligosaccharides consist of 2 to 10 monosaccharides (monomers) in their molecules. The monomers remain linked with each other by the **glycosidic bonds** or **linkages**. Certain important oligosaccharides are as follows :

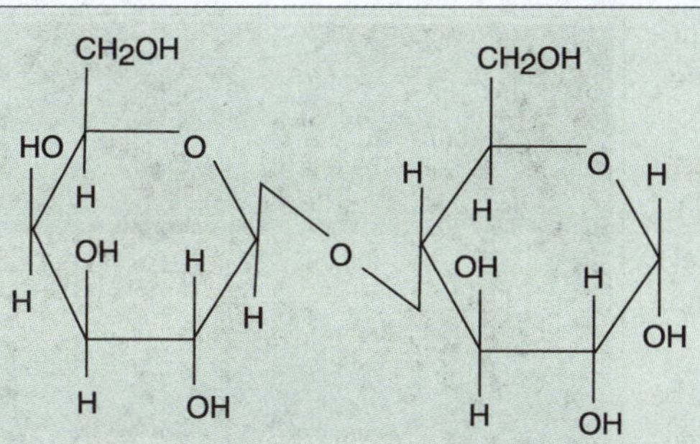

Fig. 4.7. Chemical formula of lactose.

(i) Disaccharides contain two monomers, *e.g.,* sucrose, maltose, lactose, etc.

(ii) Trisaccharides contain three monomers, *e.g.,*reffinose, mannotriose, rabinose, rhaminose, gentianose and melezitose.

(iii) Tetrasaccharides contain four monomers, *e.g.,*stachyose and scordose.

(iv) Pentasaccharides contain five monomers, *e.g.,* verbascose.

The most abundant oligosaccharides of the animal and plant cells are the disaccharides such as sucrose, maltose and lactose. The sucrose and maltose occur mainly in the matrix of plant cells, while the lactose occurs exclusively in the matrix of animal cells. The molecules of sucrose are composed of D-glucose and D-fructose. The molecules of maltose consist of two molecules of D-glucose. The molecules of lactose are composed of two monomers, *viz.*, D-glucose and D-galactose. Like monosaccharides all disaccharides are sweet, soluble in water and crystallizable.

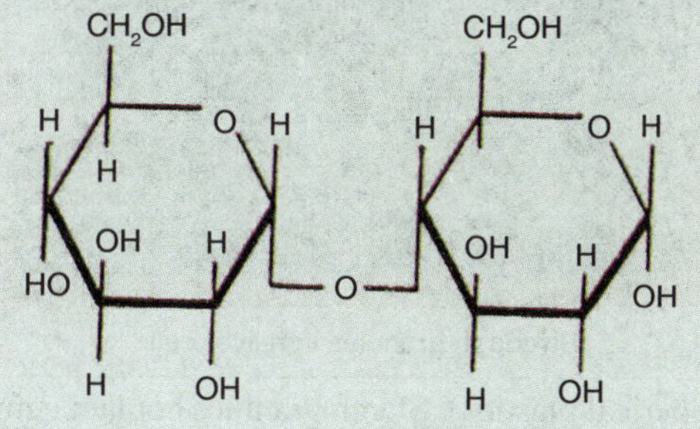

Fig. 4.8. Chemical formula of maltose.

C. Polysaccharides. The polysaccharides are composed of ten to many thousands monosaccharides as the monomers in their macromolecules. Their empirical formula is $(C_6H_{10}O_6)n$. The molecules of the polysaccharides are of colloidal size having high molecular weights. The polysaccharides can be hydrolysed into simple sugars.

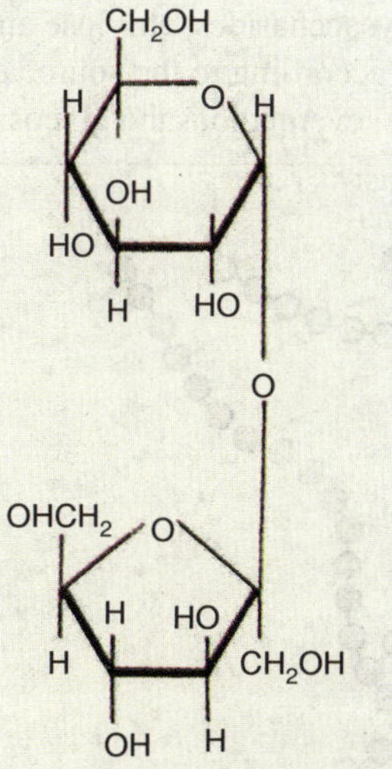

Fig. 4.9. Chemical formula of sucrose.

Polysaccharides can be divided into two main functional groups : the structural polysaccharides and the nutrient polysaccharides. The **structural polysaccharides** serve primarily as extracellular or intracellular supporting elements. Included in this group are **cellulose** (found in plant cell wall), **mannan** (a homopolymer of mannose found in yeast cell walls), **chitin** (in the exoskeleton of arthropods and the cell walls of most fungi and some green algae), **hyaluronic acid**, **keratin sulphate** and **chondroitin sulphate** (these three are found in cartilage and other connective tissues) and the **peptidoglycans** (in bacterial cell wall). The **nutrient polysaccharides** serve as reserves of monosaccharides and are in continuous metabolic turnover. Included in this group are **starch** (plant cells and bacteria), **glycogen** (animal cells), **inulin** (plants such as artichokes and dandelions) and

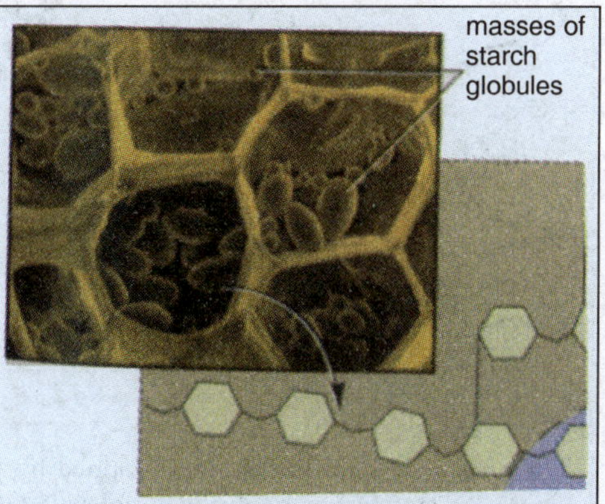

Starch is an energy-storage polysaccharide made of glucose subunits.

paramylum (an unbranched nutrient and storage homopolymer of glucose found in certain protozoa, *e.g.*, *Euglena*).

Molecules of some polysaccharides are **unbranched** (*i.e.*, linear) chains whose structure may be ribbon-like or helical (usually a left-handed spiral). Other polysaccharides are **branched** and, like many proteins, assume a globular form.

On the chemical basis, the polysaccharides can be divided into two broad classes : the homopolysaccharides and the heteropolysaccharides.

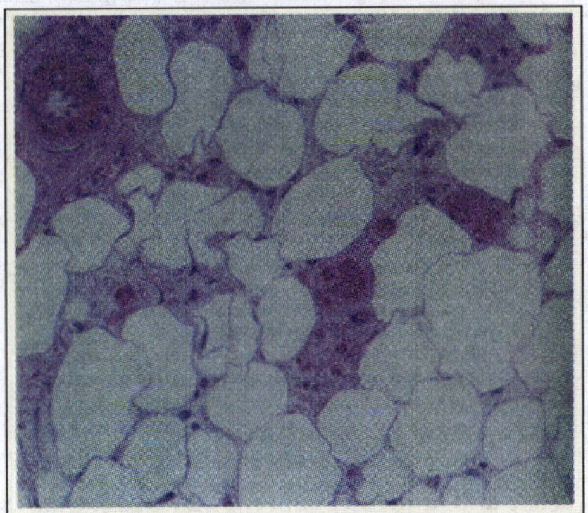
T.S of adipose cells of a mammal where fat is stored.

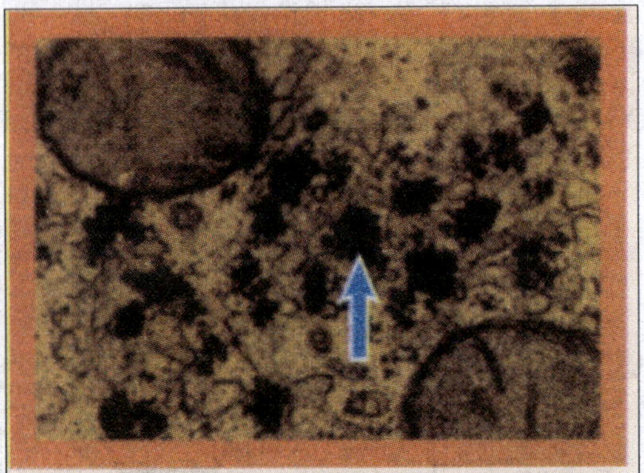
Glycogen granules in a liver cell.

Homopolysaccharides. The homopolysaccharides contain similar kinds of monosaccharides in their molecules. The most important homopolysaccharides of the matrix are the starch, glycogen, paramylum and cellulose.

(a) Starch. Starch is a nutrient, storage polysaccharide of plant cells (*e.g.*, potato tubers). It usually occurs in cells in the form of **grains** or **granules** (they are located inside the spherical plastids). Starch granules contain a mixture of two different polysaccharides, amylose and amylopectin, and the relative amounts of these two polysaccharides vary according to the source of the starch. **Amylose** is an unbranched 1→4 polymer of glucose and may be several thousand glycosyl

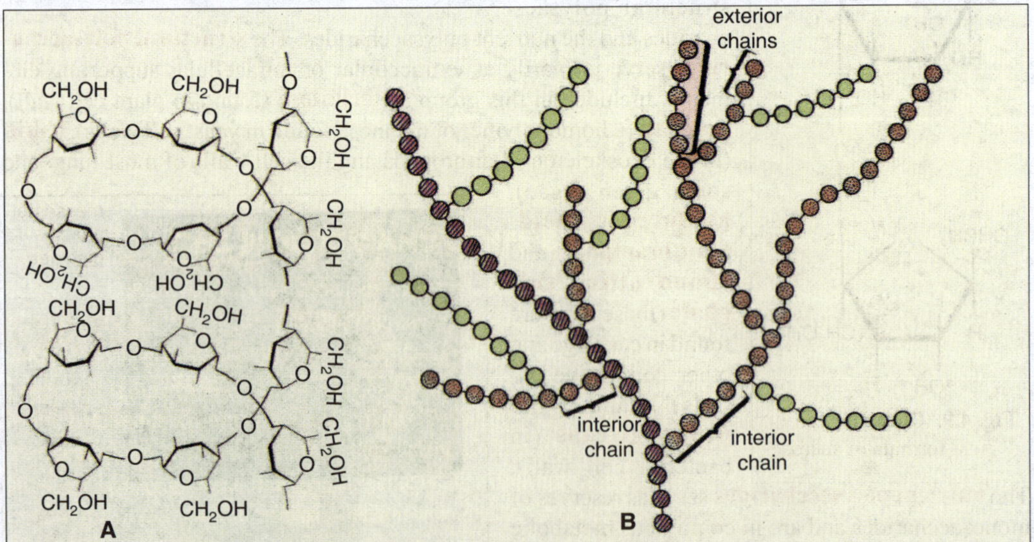

Fig. 4.10. A— Left-handed helix formed by the amylase polysaccharide; B—"Bush"- or "tree-like" structure of the glycogen molecule. Glucose units are represented by circles and the branch points (*i.e.*, 1→ 6 linkages) by heavier connections. The A chains are shown by open circles. The B chains are shown in the light shaded circles. The C chain is shown in dark shade. The reducing end (—OH group containing end) is denoted by the letter R.

units long. The polysaccharide chain exists in the form of a left-handed helix containing six glycosyl residues per turn (Fig. 4.10 A). The familiar blue colour that is produced when starch is treated with iodine is believed to result from the coordination of iodine ions in the interior of the helix. (In fact, such a colour reaction occurs when helix contains minimum six helical turns or 36 glycosyl units). **Amylopectin** is glycogen-like and is a branched polysaccharide containing many 1→4- and few 1→6-linked glucosyl units.

(b) Glycogen. Glycogen or animal starch is a branched, nutrient, storage homopolysaccharide of all animal cells, certain protozoa and algae. It is particularly abundant in liver cells and muscle cells of man and other vertebrates. Glycogen is more soluble than starch and exists in the cytoplasm as tiny granules. Glycogen molecules exist in a continuous spectrum of sizes, with the largest molecules containing many thousands (*e.g.*, 30, 000) of glucose or glycosyl units. Each glycogen molecule consists of long, profusely branched ('bush'-or 'tree-like' structure; Fig. 4.12B) chains of α-glucose molecules. The glycosidic bonds are established between carbon 1 and 4 of glucose (*i.e.*, α-1→4 linkages) except at the branching points, which involve linkages between carbon 1 and 6 (*i.e.*, α-1→6 linkages) (Fig. 4.11). A glycogen molecule contains three types of chains— A, B and C. There is only one **C chain** which bears many B and A chains and ends in the free reducing group (*i.e.*, carbon 1 of glucose at the end of C chain bears a hydroxyl or OH group). The **B chains** are attached directly to C chain and bear one or more **A chains**. The A chain may also be linked to the C chain.

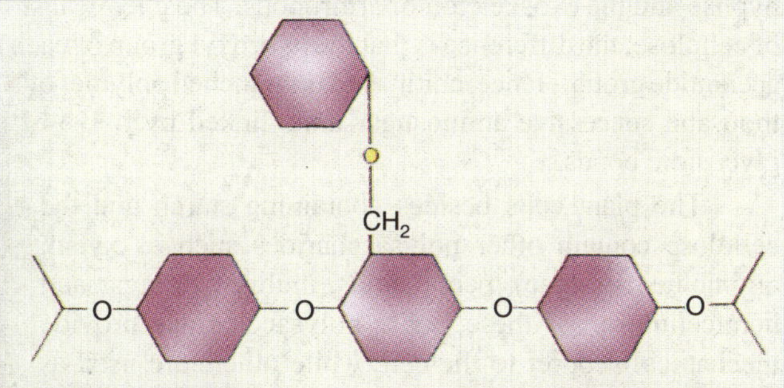

Fig. 4.11. Chemical formula of glycogen.

(C) Cellulose. Cellulose is most common and abundant biological product on earth. It is a major component of cell walls of plants and is also found in the cell walls of algae and fungi. Cellulose is an unbranched (straight) structural polysaccharide of glucose in which the neighboring monosaccharides are joined by β-1→4

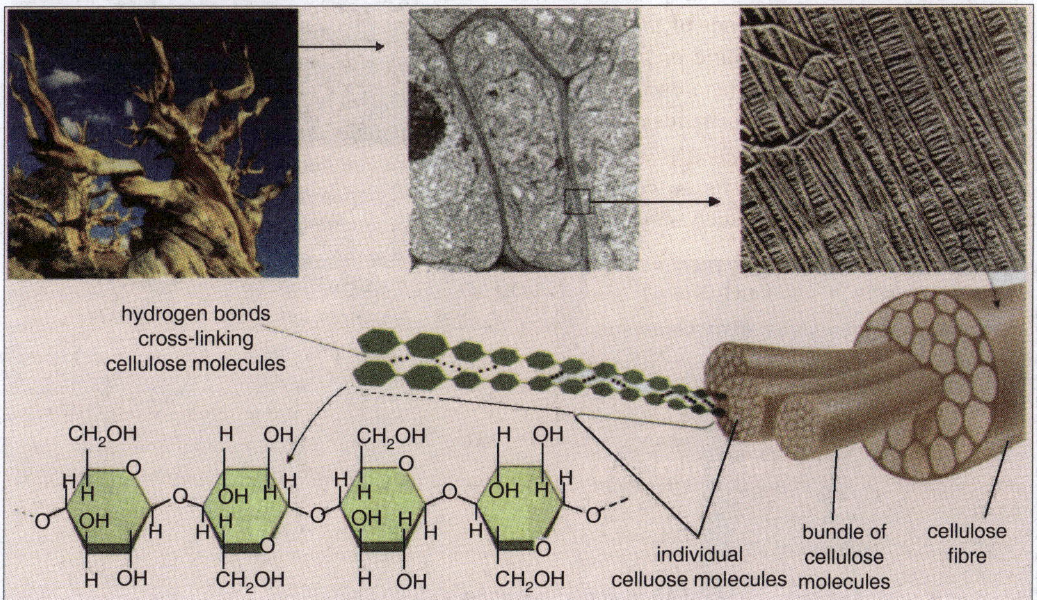

Cellulose structure and function.

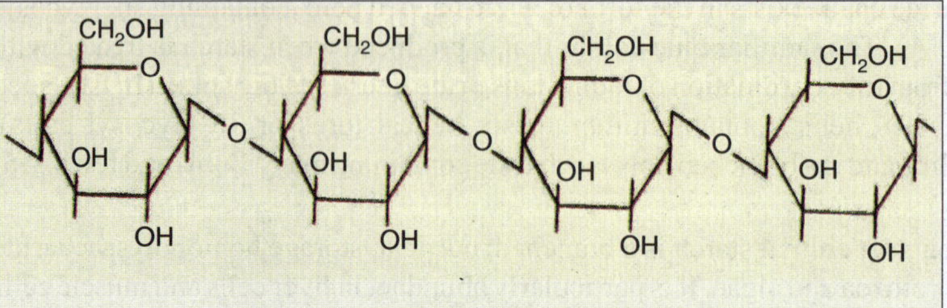

Fig. 4.12. Chemical formula of cellulose.

glycosidic bonds. Chain lengths vary from several hundred to several thousand glycosyl units (*e.g.*, in the algae *Valonia*, a single molecule of cellulose may contain more than 20,000 glycosyl units). In a cellulose molecule successive pyranose rings are rotated 180° relative to one another so that the chain of sugars takes on a "flip-flop" appearance (Fig. 4.12). Due to this, the OH groups of sugar molecules stick outwards from the chain in all directions which can form hydrogen bonds with OH groups of neighbouring cellulose chains, thereby establishing a kind of three-dimensional lattice. Thus, in plant cell walls 2000 cellulose molecules are organized into cross-linked, parallel **microfibrils** (having 25 nm diameter), whose long axis is that of the individual glucose chain.

(d) Chitin. Chitin is an extracellular structural polysaccharide found in the cell walls of fungal hyphae and the exoskeleton of arthropods. The chemical structure of chitin is closely related to that of cellulose; the difference is that the hydroxyl group of each number 2 carbon atom is replaced by an acetamide group. Hence, chitin is an unbranched polymer of **N-acetylglucosamine** containing several thousand successive aminosugar units linked by β-1→4 glycosidic bonds.

Tough, slightly flexible chitin supports the otherwise soft bodies of arthropods and fungi.

The plant cells besides containing starch and the cellulose contain other polysaccharides such as xylan, alginic acids (algae), pectic acids, inulin, agar-agar and hemicellulose. Of these, some polysaccharides provide mechanical support to the cell, while others are used as stored food material.

Heteropolysac-charides. The polysaccharides which are composed of different kinds of the monosaccharides and amino-nitrogen or sulphuric or phosphoric acids in their molecules are known as heteropolysac-charides. The most important heteropolysaccharides are as follows:

(a) Hyaluronic acid, keratin sulphate and chondroitin sulphate. Cartilage tissue contains the related acidic heteropolysaccharides such as hyaluronic acid, kera-

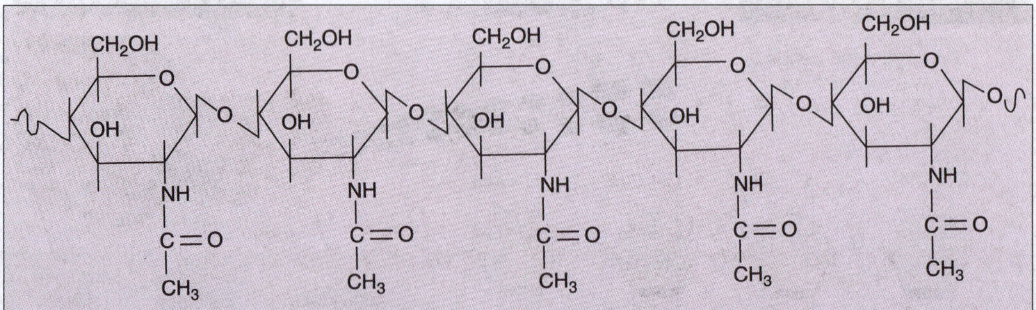

Fig. 4.13. A small segment of a chitin molecule.

tin sulphate and chondroitin sulphate. **Hyaluronic acid** is an unbranched heteropolysaccharide containing repeating disaccharides of **N-acetylglucosamine** (or D-glucosamine) and glucuronic acid. In addition to cartilage, hyaluronic acid is also found in other connective tissues, in the synovial fluid of joints, in the vitreous humor of the eyes, and also in the capsules that enclose bacteria.

Keratin sulphate, like hyaluronic acid, is a repeating disaccharide forming an unbranched chain. Each disaccharide unit of the polysaccharide consists of **D-galactose** and **sulphated N-acetylglucosamine**. It is found in cartilage and cornea.

Chondroitin sulphate is a repeating disaccharide consisting of alternating **glucuronic acid** and **sulphated N-acetyl galactosamine** residues. It is found in cartilage, bone, skin, notochord, aorta and umbilical cord.

Chitin is a primary component of the glistening outer skeleton of this grasshopper.

(b) Heparin. Heparin is a blood anticoagulant and found in the skin, liver, lung, thymus, spleen and blood. Its molecule contains the repeated disaccharide units, each having **D-glucuronic acid** and **D-glucosamine**.

(c) Proteoglycans, glycoproteins and glycolipids. Polysaccharides also occur in covalent combination with proteins and lipids, to form the following three types of molecules :

(i) Proteoglycans. The molecules of proteoglycans consist of much longer portion of polysaccharide and a small portion of protein. They are also called **mucoproteins** (**De Robertis** and **De Robertis**, **Jr**., 1987). The proteoglycans are amorphous and form gels which are able to hold large amounts of water.

The **cartilage proteoglycan** is found extracellularly in cartilage and bone. In its molecule, strands of protein, called **core protein**, extend radially from a long, central hyaluronic acid molecule. In each core protein strand, three carbohydrate bearing regions may be identified. The first region contains numerous oligosaccharides, the second region contains keratin sulphate chains and the third region contains chondroitin sulphate chains. This arrangement gives cartilage its resilience and tensile strength.

(ii) Glycoproteins (or **glycosaminoglycans** or **mucopolysaccharides**). In these molecules, the carbohydrate portion consists of much shorter chains which are often branched. Glycoproteins serve diverse roles in cells and tissues and include certain enzymes, hormones, blood groups, saliva, gastric mucin, ovomucoids, serum, albumins, antibodies or immunoglobins (see Table 4.3).

(iii) Glycolipids. These molecules are covalent combinations of carbohydrate and lipid. The carbohydrate portion may be a single monosaccharide or a linear of branched chain. Glycolipids form the component of most cell membranes, *e.g.*, cerebrosides and gangliosides.

Table 4-3. Carbohydrate content of glycoproteins (Source : Sheeler and Bianchi, 1987).

Glycoprotein	Percentage of carbohydrate	Function
1. Ovalbumin	1	Hens-egg food reserve
2. Follicle-stimulating hormone (FSH)	4	Hormone
3. Fibrinogen	5	Blood coagulation protein
4. Transferrin	6	Iron transport protein of blood plasma
5. Ceruloplasmin	7	Copper transport protein of blood plasma
6. Glucose-oxidase	15	Enzyme
7. Peroxidase	18	Enzyme
8. Luteinizing hormone	20	Hormone
9. Heptoglobin	23	Haemoglobin-binding protein of blood plasma
10. Erythropoietin	33	Hormone
11. Mucin	50–60	Mucus secretion
12. Blood-group glycoprotein	85	Unknown

Lipids (Fats)

The lipids (Gr., *lipos*=fats) are the organic compounds which are insoluble in the water but soluble in the non-polar organic solvents such as acetone, benzene, chloroform and ether. The cause of this general property of lipids is the predominance of long chains of aliphatic hydrocarbons or benzene ring in their molecules. The lipids are non-polar and hydrophobic. The common examples of lipids are cooking oil, butter, ghee, waxes, natural rubber and cholesterol. Like the carbohydrates, lipids serve two major roles in cells and tissues : 1. They occur as constituents of certain structural components of cells such as membranous organelles; plant pigments such as **carotene** found in carrots and **lycopene** that occurs in tomatoes; vitamins like A, E and K; menthol and eucalyptus oil; and (2) they may be stored within cells as reserve energy sources. Like the starch and glycogen, fat is compact and insoluble and provides a convenient form in which energy-yielding molecules (the fatty acids) can be stored for use when occasion arises.

(a) Fat is an efficient way to store energy.

(b) Wax is a highly saturated lipid.

Lipids are all made of carbon, hydrogen and sometimes oxygen. The number of oxygen atoms in a lipid molecule is always small compared to the number of carbon atoms. Sometimes small amounts of phosphorus, nitrogen and sulphur are also present. Natural fats and oils are compounds of **glycerol** (*i.e.,* glycerine or propane-1, 2, 3 triol) and **fatty acids**.

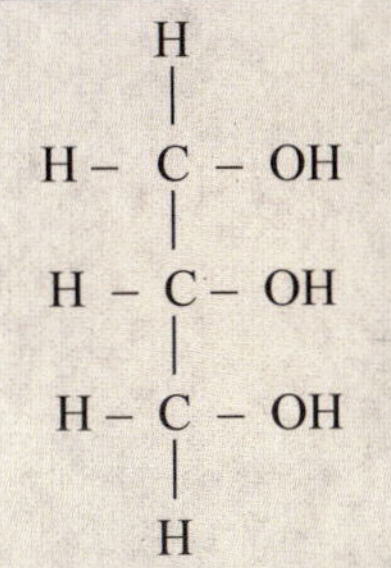

Glycerol

They are esters which are formed due to reaction of organic acids with alcohols. There is only one kind of glycerol : its molecular configuration shows no variation and it is exactly same in all lipids. The formula of glycerol is $C_3H_8O_3$ and following is its molecular structure :

Fatty acids. A fatty acid molecule is **amphipathic** and has two distinct regions or ends: a long **hydrocarbon chain**, which is **hydrophobic** (water insoluble) and not very reactive chemically, and a **carboxylic acid group** which is ionized in solution (COO^-), extremely **hydrophilic** (water soluble) and readily forms esters and amides. In neutral solutions, salts of the fatty acids form small spherical droplets or **micelles** in which the dissociated carboxyl groups occur at surface and the hydrophobic carbon chains project towards the centre. In cells, the fatty acids only sparingly occur freely; instead, they are esterified to other components and form the saponifiable lipids.

A fatty acid molecule may be either saturated or unsaturated. The **saturated fatty acids** consist of long hydrocarbon chains terminating in a carboxyl group and conform to the general formula :

$$CH_3 - (CH_2)_n - COOH$$

In nearly all naturally occurring fatty acids, *n* is an even number from 2 to 22. In the **saturated fatty acids**, most commonly found in animal tissues, *n* is either 12 (*i.e.*, **myristic acid**), 14 (*i.e.*, **palmitic acid**) or 16 (*i.e.*, **stearic acid**). In **unsaturated fatty acids**, at least two but usually no more than six of the carbon atoms of the hydrocarbon chain are linked together by double bonds ($-C=C-$), *e.g.*, **oleic acid**, **linoleic**, **linolenic, arachidonic** and **clupanadonic acids**. Double bonds are important because they increase the flexibility of the hydrocarbon chain, and thereby the fluidity of biological membranes. Unsaturated fatty acids predominate in lipids of higher plants and in animals that live at low temperatures. Lipids in the tissues of animals inhabiting warm climates contain larger quantities of saturated fatty acids.

Essential fatty acids. Some animals, especially mammals, are unable to synthesize certain fatty acids and, therefore, require them in their diet. They are called **essential fatty acids** and include linoleic acid, linolenic acid and arachidonic acid. Such essential fatty acids have to be obtained from plant material by the animal.

Types of lipids. The lipids are classified into three main types : 1. simple lipids, 2. compound lipids and 3. derived lipids.

1. Simple lipids. The simple lipids are alcohol esters of fatty acids :

$$\underset{\text{(Simple lipids)}}{\text{Triglyceride}} \xrightarrow[H_2O]{\text{Lipase}} \text{Glycerol} + 3 \text{ Fatty acids.}$$

Simple lipids are also of following two types :

(a) Neutral fats (Glycerides or triglycerides). They are triesters of fatty acids and glycerol. Neutral fats represent the major type of stored lipid and so accumulate in the cytoplasm.

(b) Waxes. Waxes have a higher melting point than neutral fats and are the esters of fatty acids of high molecular weight with the alcohol except the glycerol. The most important constituent alcohol of the molecules of waxes is the **cholesterol**, *e.g.*, bees wax.

2. Compound lipids. The compound lipids contain fatty acids, alcohols and other compounds as phosphorus, amino-nitrogen carbohydrates, etc., in their molecules. Some of the compound lipids are important structural components of the cell, in particular of cell membranes. The compound lipids of the cell are of the following types :

Waxes give plums their delicate whitish blush and help keep citrus fruit juicy.

(i) Phospholipids (or Glycerophos-phatides). Such type of lipids form the major constituent of cell membranes. In a molecule of phospholipid two of the —OH or hydroxyl groups in glycerol are linked to fatty acids, while the third —OH group is linked to phosphoric acid. The phosphate is further linked to a hydrophilic compound such as **etanolamine**, **choline**, **inositol** or **serine**. Each phospholipid molecule, therefore, has hydrophobic or water-insoluble tail which is composed of two fatty acid chains and a hydrophilic or water-soluble polar head group, where the phosphate is located. Thus, in effect the phospholipid molecules are detergents, *i.e.*, when a small amount of phospholipid is spread over the surface of water, there forms a monolayer film of phospholipid molecules; in this thin film, tail regions pack together very closely facing the air and their head groups are in contact with the water (Fig. 4.14). Two such films can combine tail to tail to make a phospholipid sandwich or self-sealing **lipid bilayer**, which is the structural basis of cell membranes.

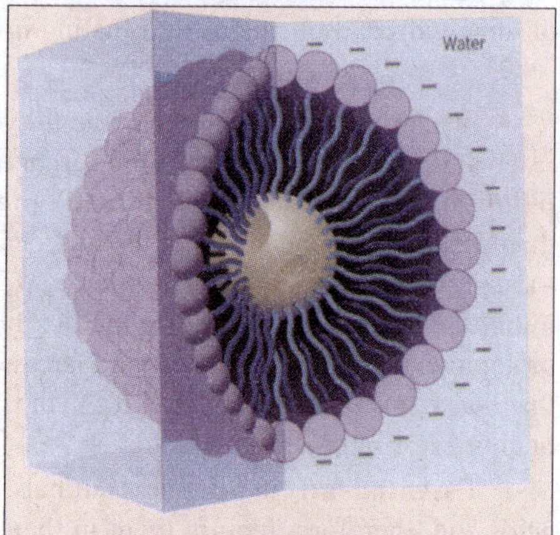

A soap micelle.

Various membranes of cell contain the following four types of phospholipids : **1. phosphatidyl choline** or **lecithin**; **2. phosphatidyl ethanolamine** or **cephalin**; **3. phosphatidyl serine**; and **4. phosphatidyl inositol**. The other important phospholipids of the matrix are the phosphoinositides (occur mostly in the cells of liver, brain, muscle and soyabean), plasmalogens and isositides. Plasmalogens are a special class of phospholipids which are especially abundant in the membranes of nerve and muscle cells and are also characteristic of cancer cells.

Liposomes. When aqueous suspensions of phospholipids are subjected to rapid agitation by using ultrasound (*i.e.*, insonation), the lipid disperses in the water and forms **liposomes** or **lipid vesicles**. Liposomes are small spherical bodies (25 nm to 1 μm in diameter) whose surface is formed by a bilayer of phospholipid molecules enclosing a small volume of the aqueous medium. They exhibit many of the permeability properties of natural membranes, *i.e.*, water soluble small molecules or ions can be enclosed by the liposomes and they can also traverse the lipid bilayer of latter. Recently, liposomes have been found to have great therapeutic promise, since, they can be used as vectors for the transfer of specific drugs, proteins, hormones, nucleic acids, ions or any other molecule into the specific types of animal cells. The contents of the liposomes can enter the target cells by two routes : 1. The liposomes can attach to the surface of target cells and may fuse with the plasma membrane, following which their contents are released into the cytosol or cytoplasmic matrix. 2. The entire liposomes may be endocytosed and degraded intracellularly (see **Sheeler** and **Bianchi**, 1987).

(ii) Sphingolipids. The sphingolipids occur mostly in the cells of the brain. Instead of the glycerol, they contain in their molecules amine alcohol (sphingol or sphingosine). For instance, the myelin sheaths of the nerve fibres contain a lipid known as **sphingomyelin** which contains sphingosine and phospholipids in its molecules.

(iii) Glycolipids. The glycolipids contain in their molecules the carbohydrates and the lipids. The matrix of the animal cells contains two kinds of glycolipids, *viz.*, cerebrosides and gangliosides.

(a) Cerebrosides. The cerebrosides contain in their molecules sphingosine, fatty acids and galactose or glucose. The cerebrosides are the important lipids of the white matter of the cells of brain and the myelin sheath of the nerve. The important cerebrosides are the kerasin, cerebron, nervon and oxynervons.

(b) Gangliosides. The gangliosides have complex molecules which are composed of sphingosine, fatty acids and one or more molecules of glucose, lactose, galactosamine and neuraminic acid. The gangliosides occur in the grey matter of the brain, membrane of erythrocytes and cells of the spleen. Gangliosides act as **antigens**.

One type of ganglioside, called **GM2**, may accumulate in the lysosomes of the brain cells because of a genetic deficiency that results in the failure of the cells to produce a lysosomal enzyme that degrades this ganglioside. This condition is called **Tay-Sachs disease** and leads to paralysis, blindness and retarded development of human beings.

Chlorophyll present in green plants are responsible for making them autotrophs.

3. Derived lipids (or Nonsaponifiable lipids). Some type of lipids do not contain fatty acids in their constituents and they are of following three types :

A. Terpenes. The terpenes include certain **fat-soluble vitamins** (*e.g.*, vitamins A, E and K), **carotenoids** (*e.g.*, photosynthetic pigments of plants), and certain **coenzymes** (such as coenzyme Q or ubiquinone). All the terpenes are synthesized from various numbers of a five-carbon building block, called **isoprene unit** (Fig. 4.17). The isoprene units are bonded together in a head-to-tail organization. Two isoprene units form a **monoterpene**, four form a **diterpene**, six a **triterpene**, and so on. The monoterpenes are responsible for the

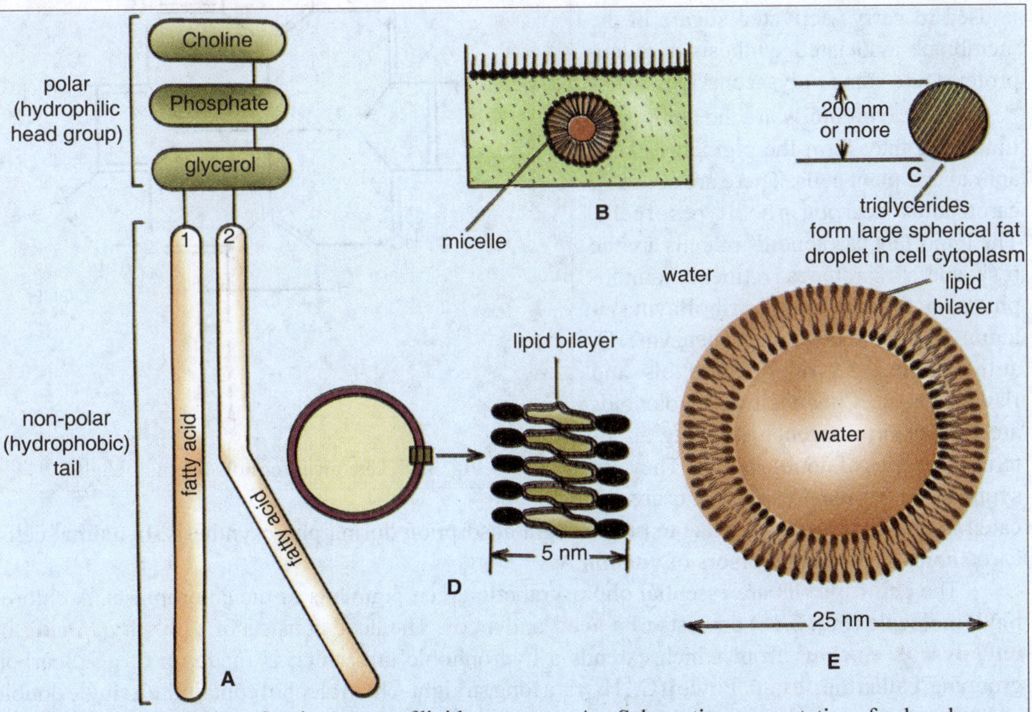

Fig. 4.14. Formation of various types of lipid aggregates. A—Schematic representation of a phospholipid molecule; B—Formation of micelle and monolayer film; C— Formation of a fat droplet by triglycerides; D—Formation of self - sealing lipid bilayer (*e.g.*, liposome); E—Cross section of a liposome (after Alberts *et al.*, 1989).

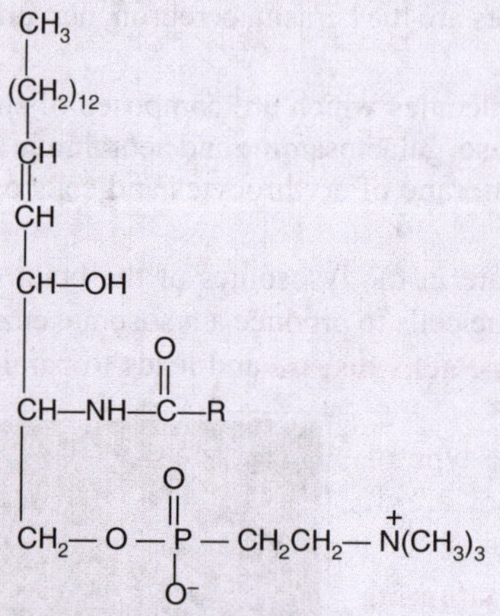

Fig. 4.15. General chemical formula of sphingomyelin.

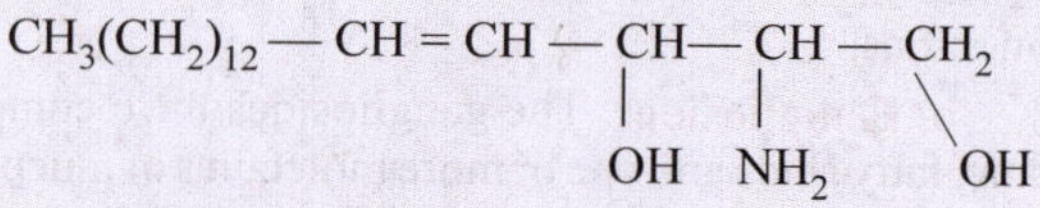

Fig. 4.16. Chemical formula of sphingosine.

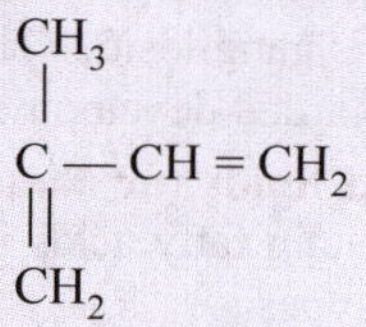

Fig. 4.17. Isoprene.

characteristic odours and flavours of plants (*e.g.*, **geraniol** from geraniums, **menthol** from mint and **limoneme** from lemons). **Dolicol phosphate** is a **polyisoprenoid** (*i.e.*, long chain polymer of isoprene) and is used to carry activated sugars in the membrane-associated synthesis of glycoproteins and some polysaccharides.

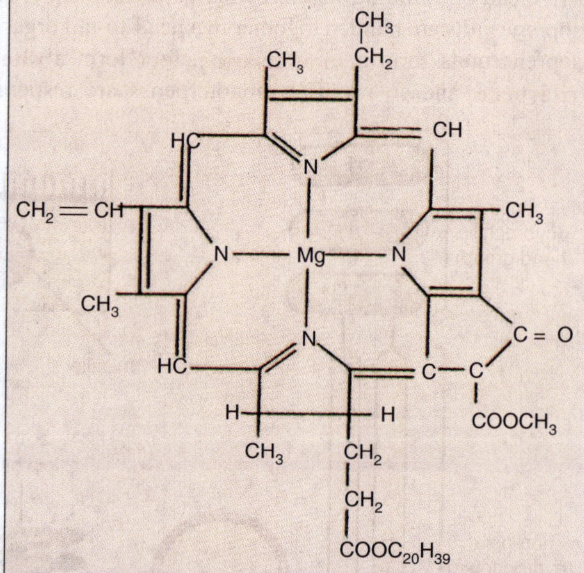

Fig. 4.18. Chemical formula of chlorophyll.

The **carotenoids** are the compound lipids and they form the pigments of the animal and plant cells. There are about 70 carotenoids occurring in both types of cells. The important carotenoids of cells are the α, β and γ carotenes, retinene, xanthophylls, lactoflavin in milk, riboflavin (vitamin B_2), xanthocyanins, coenzyme Q, anthocyanins, flavones, flavonols and flavonones, etc. Chemically all carotenoids are long-chain isoprenoids having an alternating series of double bonds. They are synthesized by plant tissues and are located in the chloroplast lamellae to help in light absorption during photosynthesis. In animal cells, carotenoids serve as precursors of vitamin A.

The **chlorophylls** are essential photosynthetic green pigments of the chloroplasts. A chlorophyll molecule (Fig. 6.18) consists of a **head** and a **tail**. The head consists of a **porphyrin ring** or **tetrapyrrole nucleus** from which extends a hydrophobic tail which is made up of a 20-carbon grouping, called the **phytol**. Phytol ($C_{20}H_{39}$) is a long straight- chain alcohol containing a single double bond. It may be regarded as a hydrogenated carotene (vitamin A). The porphyrins (Gr., *porphyra* = purple) are complex carbon-nitrogen molecules that usually surround a metal, *i.e.*, it is formed from four pyrrol rings linked together by methane bridges and metal atom (Mg or Fe) is linked to pyrrol rings. In chlorophyll molecule, the porphyrin surrounds a magnesium ion, while in haeme of **haemoglobin**, it surrounds an iron ion (Fig. 4.19). Many other pigments of animal cells such as **myoglobin** and **cytochromes** have porphyrin rings in their molecules.

B. Steroids. The steroids consist of a system of fused cyclohexane and cyclopentane rings. All are derivative of **perhydro - cyclopentano - phenanthrene**, which consists of three fused cyclohexane rings and a terminal cyclopentane ring (Fig. 4.20). Steroids have widely different physiological charac-

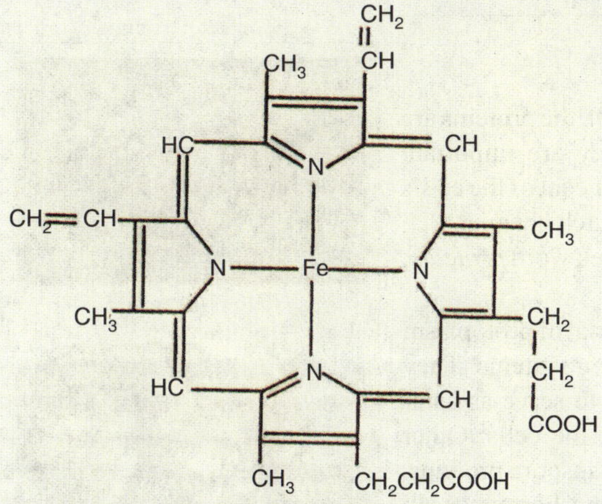

Fig. 4.19. Chemical formula of haeme portion of haemoglobin.

Some body builders endanger their health by taking 'steroids'.

teristics. For example, some steroids are **hormones** (*e.g.*, sex hormones such as estrogen, progesterone, testosterone and corticosterone) and affect cellular activities by influencing gene expression. Some steroids are **vitamins** (*e.g.*, vitamin D_2) and influence the activities of certain cellular enzymes. Some steroids (*e.g.*, cholic acid) are fat emulsifier found in the bile.

Alcohols of the steroids are called **sterols**. The common examples of the sterols are **cholesterol** found in animals and **ergosterol** and **stigmasterol** found in plants. Cholesterol (Fig. 4.21) is found in the plasma membrane of many animal cells and also in blood, bile, gallstone, brain, spinal cord, adrenal glands and other cells. It is the precursor of most steroid sex hormones and cortisones. **7- dehydro-cholesterol** is found in the skin where it is responsible for the synthesis of vitamin D in the presence of sunlight. **Ergosterol** is also a precursor of vitamin D.

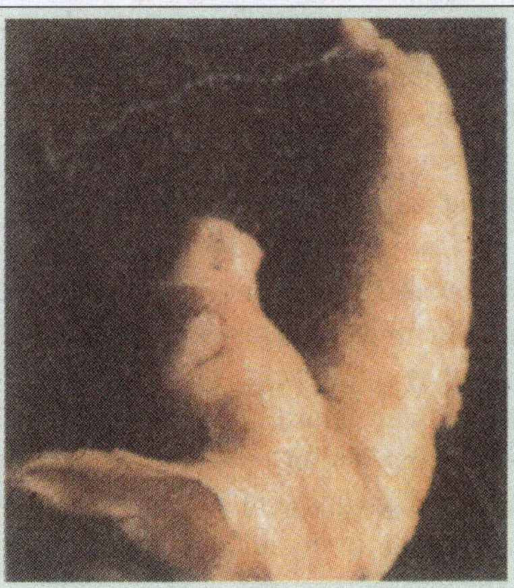
Cholesterol plug in artery

C. Prostaglandins. Hydroxy derivatives of 20-carbon polyunsaturated fatty acids are called **prostaglandins**. They are found in human seminal fluid, testis, kidney, placenta, uterus, stomach, lung, brain and heart. There are sixteen or more different prostaglandins, falling into nine classes (PGA, PGB, PGC .....PGI). Their main function is binding of hormones to membranes of the target

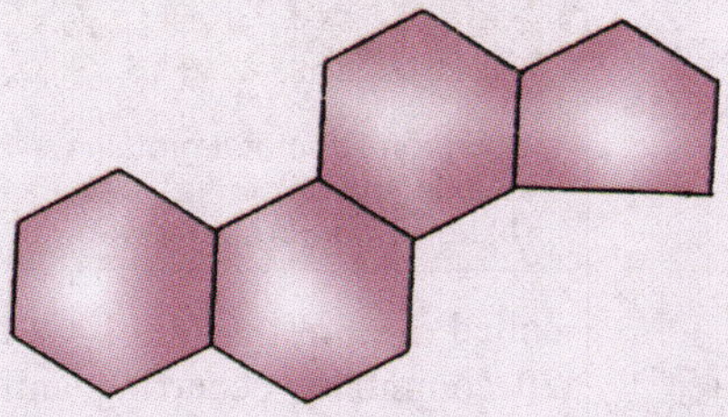
Fig. 4.20. Cyclopentano-perhydro-phenanthrene nucleus of the steroids.

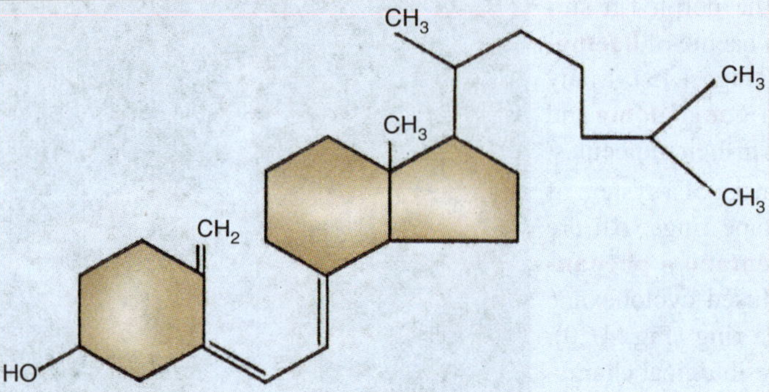

Fig. 4.21. Chemical formula of cholesterol.

cells. Being local chemical mediators, prostaglandins are continuously synthesized in membranes from precursors cleaved from membrane phospholipids by phospholipases. Their other important functions include initiation of contraction of smooth muscles (thus, helping in childbirth), aggregation of platelets and inflammation (*i.e.*, arthritis) (see **Alberts** *et al.*, 1989).

Proteins

Of all the macromolecules found in the cell, the proteins are chemically and physically more diverse. They are important constituents of the cell forming more than 50 per cent of the cell's dry weight. The term protein was coined by Dutch chemist **G.J. Mulder** (1802—1880) and is derived from Greek word *proteios*, which means "of the first rank".

Proteins serve as the chief structural material of protoplasm and play numerous other essential roles in living systems. They form enzymes—globular proteins specialized to serve as catalysts in virtually all biochemical activities of the cells. Other proteins are antibodies (immunoglobulins), transport proteins, storage proteins, contractile proteins, and some hormones. In every living organism, there are thousands of different proteins, each fitted to perform a specific functional or structural role. Indeed, a single human cell may contain more than 10,000 different protein molecules. Chemically, proteins are polymers of amino acids.

1. Amino acids. Nobel Laureate **Emil Fischer** (1902) discovered that all proteins consist of chains (linear sequence) of smaller units that he named **amino acids**. There are about 20 different amino acids (Table 4.4) which occur regularly as constituents of naturally occurring proteins. An organic compound containing one or more amino groups (–NH_2) and one or more carboxyl groups (—COOH) is known as amino acid. The amino acids occur freely in the cytoplasmic matrix and constitute the so called **amino acid pool**. Of the 20 commonly occurring amino acids, 19 may be represented by the following general formula (Fig. 4.22).

Common structural proteins include those of (a) hair, (b) horn and (c) spider web.

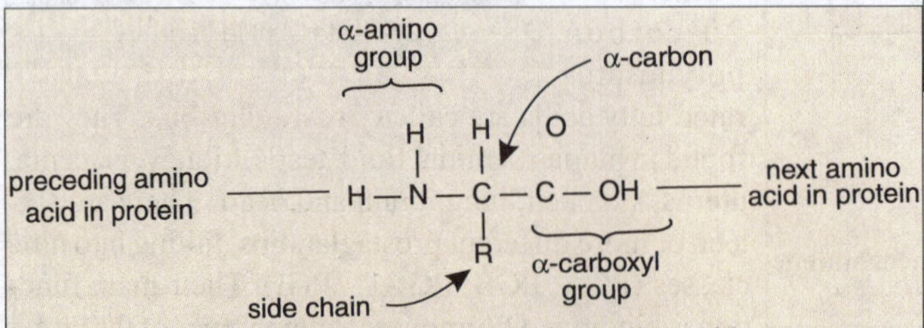

Fig. 4.22. Basic structure of an amino acid.

The sole exception is proline, where the amino group forms part of a ring structure. The central or **alpha** carbon atom of each amino acid is covalently bonded to four groups : (1) A hydrogen atom, (2) an amino group (—NH_2), (3) an acid (or carboxyl) group, and (4) a side chain called an **R-group**. It is the particular chemical structure of the R-group that distinguishes one amino acid from another. The name and structural formulae of the amino acids that regularly occur in proteins are given in Table 4-4.

Table 4-4. **The 20 - naturally occurring amino acids.**

Group of amino acid	Name of the amino acids, symbols and chemical formulae
A. Aliphatic amino acid	
I. Monoaminomonocarboxylic amino acids or simple amino acids.	1. Glycine (Gly, G) $H—CH(NH_2)—COOH$
	2. Alanine (Ala, A) $CH_3—CH(NH_2)—COOH$
	3. Valine (Val, V) $(CH_3)_2CH—CH(NH_2)—COOH$
	4. Leucine (Leu, L) $(CH_3)_2CH—CH_2—CH(NH_2)—COOH$
	5. Isoleucine (Ile, I) $CH_3—CH_2—CH(CH_3)—CH(NH_2)—COOH$
II. Monoamino-dicarboxylic or acidic amino acids.	6. Aspartic acid (Asp, D) $HOOC—CH_2—CH(NH_2)—COOH$
	7. Glutamic acid (Glu, E) $HOOC—CH_2—CH_2—CH(NH_2)—COOH$
III. Diamino-monocarboxylic or basic amino acids.	8. Lysine (Lys, K) $H_2N—CH_2—CH_2—CH_2—CH_2—CH(NH_2)—COOH$

Group of amino acid	Name of the amino acids, symbols and chemical formulae
	9. Arginine (Arg, R) $H_2N{-}C({=}NH){-}NH{-}CH_2{-}CH_2{-}CH_2{-}CH(NH_2){-}COOH$
	10. Histidine (His, H) 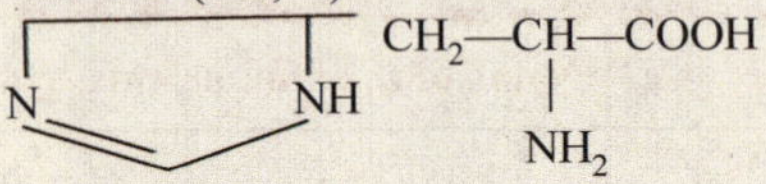
IV. Hydroxyl containing amino acids.	11. Serine (Ser, S) $HO{-}CH_2{-}CH(NH_2){-}COOH$
	12. Threonine (Thr, T) $CH_3{-}CH(OH){-}CH(NH_2){-}COOH$
V. Sulphur containing amino acids.	13. Cysteine (Cys, C) $HS{-}CH_2{-}CH(NH_2){-}COOH$
	14. Methionine (Met, M) $CH{-}S{-}CH_2{-}CH_2{-}CH(NH_2){-}COOH$
B. Aromatic amino acids.	15. Phenylalanine (Phe, F) 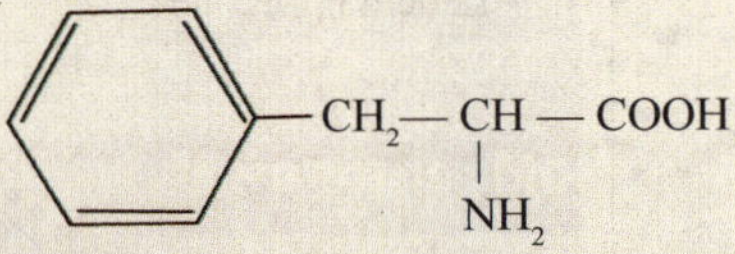
	16. Tyrosine (Tyr, Y) 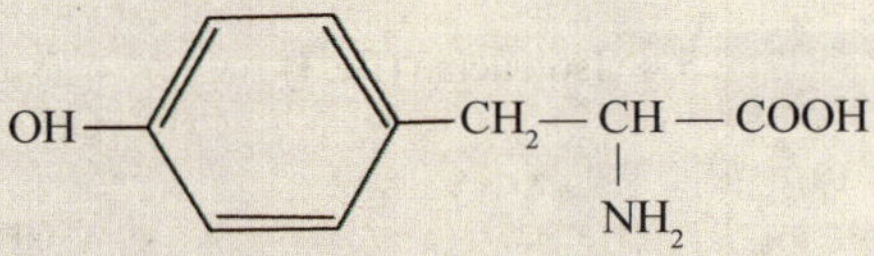
	17. Tryptophan (Try, W) 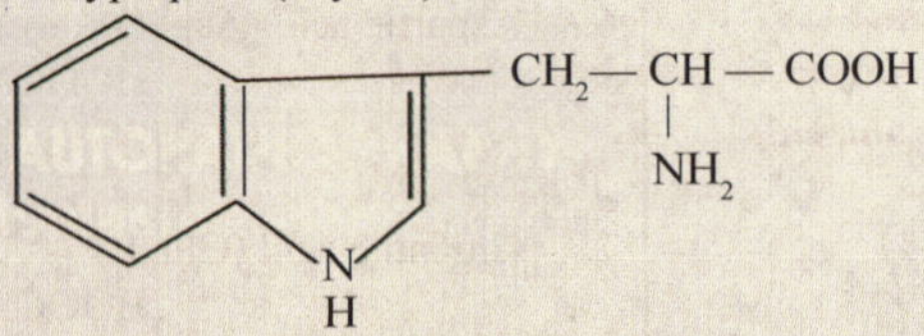
C. Secondary amino acids.	18. Proline (Pro, P) 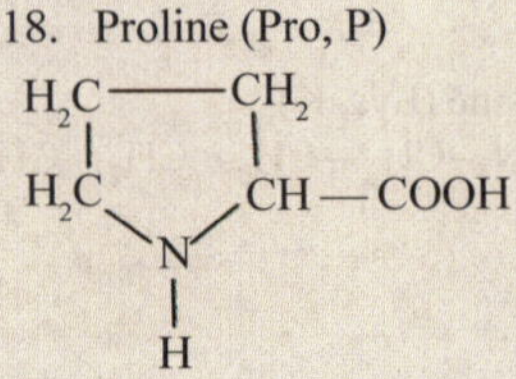

Group of amino acid	Name of the amino acids, symbols and chemical formulae
D. Amino acid amides.	19. Aspargine (Asn, N) 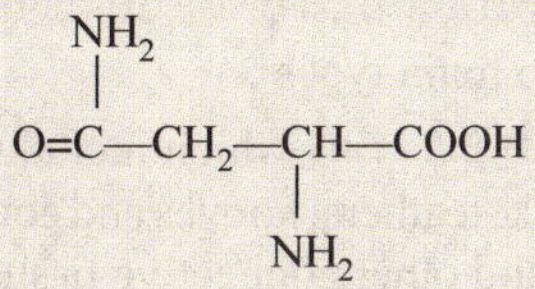 20. Glutamine (Glu, Q) 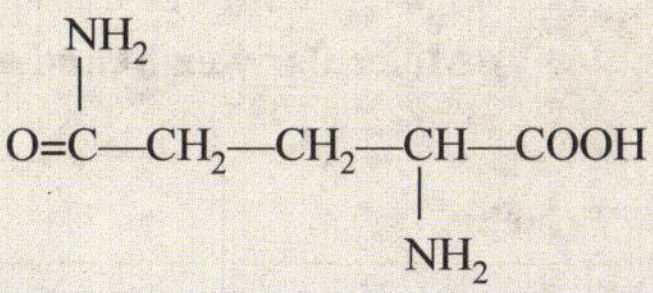

In certain amino acids R group is either a hydrogen atom (*e.g.*, glycine, the simplest amino acid) or a hydrophobic aliphatic (*e.g.*, leucine) or aromatic (*e.g.*, pheylalanine) hydrocarbon. In other cases, R group contains either an extra carboxyl group of an extra amino group or its equivalent. Glutamic acid and aspartic acid each have an extra carboxyl (– COOH) group. Lysine and arginine both contain an additional amino group or equivalent structure. Histidine also contains a N group. Other amino acids such as serine and tyrosine have hydroxyl groups in their side chains. Of a particular importance is the amino acid cysteine which possesses a thiol (SH) group.

2. Formation of proteins. Because a molecule of the amino acid contains both basic or amino (—NH_2) and acidic or carboxyl (—COOH) group, it can behave as an acid and base at a time. The molecules of such organic compounds which contain both acidic and basic properties are known as **amphoteric molecules**. Due to amphoteric molecules, the amino acids unite with one another to form complex and large protein molecules. When two molecules of amino acids are combined then the basic group (—NH_2) of one amino acid molecule combines with the carboxylic (—COOH) group of other amino acid and the loss of a water molecule takes place. This sort of condensation of two amino acid molecules by —NH—CO linkage or bond is known as **peptide linkage** or **peptide bond**. A combination of two amino acids by the peptide bond is known as **dipeptide**. When three amino acids are united by two peptide bonds, they form **tripeptide**. Likewise, by condensation of few or many amino acids by the peptide bonds the **oligopeptides** and **polypeptides** are formed respectively. The various molecules of polypeptides unite to form the **peptones**, **proteases** and **proteins**. Thus, protein macromolecules are the polymers of many amino acid monomers. The size (molecular weight), shape, and function of proteins are determined by the number, type and distribution of the amino acids present in the molecule. Proteins occur in a wide spectrum of molecular sizes from small molecules such as the hormone ACTH (or adrenocorticotrophic hormone) which consists of only 39 amino acids and has a molecular weight of 4500, to extremely large proteins such as haemocyanin (an invertebrate blood pigment) which consists of 8200 amino acids and has a molecular weight greater than 900,000 (see Table 4.5 for additional examples).

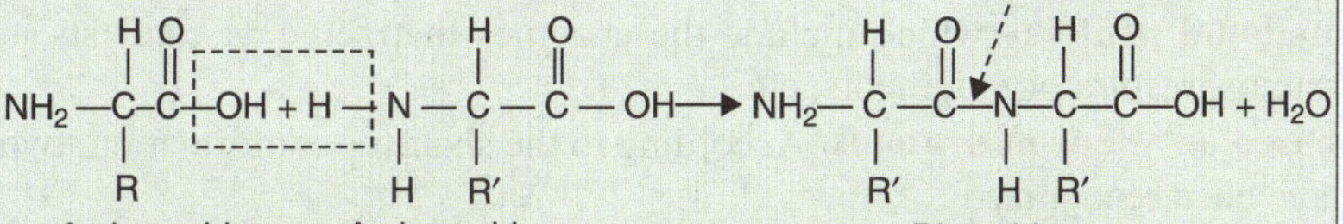

Fig. 4.23. A chemical reaction showing formation of a dipeptide.

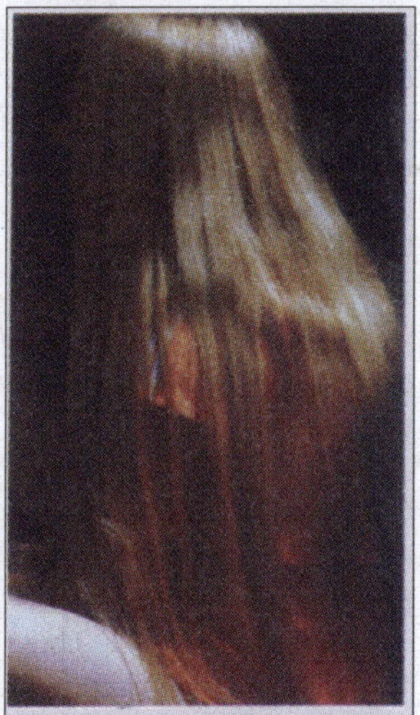
Keratin: major protein component of hair.

2. Types of proteins. Many different methods have been used to classify proteins, no method of their classification being entirely satisfactory :

(1) Classification based on biological functions. According to their biological functions, proteins are of two main types :

1. Structural proteins which include **keratin**, the major protein component of hair (cortex), wool, fur, nail, beak, feathers, hooves and cornified layer of skin; and **collagen**, abundant in skin, bone, tendon, cartilage and other connective tissues.

Table 4-5. **Molecular weight and amino acid content of some proteins (Source : Sheeler and Bianchi, 1987).**

Protein	Number of amino acids	Molecular weight
1. Adrenocorticotrophic hormone (ACTH)	39	4,500
2. Insulin	51	5,700
3. Ribonuclease	124	12,000
4. Cytochrome-c	140	15,600
5. Horse myoglobin	150	16,000
6. Trypsin	180	20,000
7. Haemoglobin	574	64,500
8. Urease	4,500	473, 000
9. Snail haemocyanin	8, 200	910,000

2. Dynamic or functional proteins which include the enzymes that serve as catalysts in metabolism, hormonal proteins, respiratory pigments, etc.

(2) Classification based on shape of proteins. According to the shape or conformation, two major types of proteins have been recognized :

(a) Fibrous proteins. Fibrous proteins are water-insoluble, thread-like proteins having greater length than their diameter. They contain secondary protein structure and occur in those cellular or extracellular structures, where strength, elasticity and rigidity are required, *e.g.*, collagen, elastin, keratin, fibrin (blood-clot proteins) and myosin (muscle contractile proteins).

(b) Globular proteins. Globular proteins are water-soluble, roughly spheroidal or ovoidal in shape. They readily go into colloidal suspension. They have tertiary protein structure and are usually functional proteins, *e.g.*, enzymes, hormones and immunoglobulins (antibodies). Actin of micro- filaments and **tubulins** of microtubules are also globular proteins (see **Alberts** *et al.*, 1989).

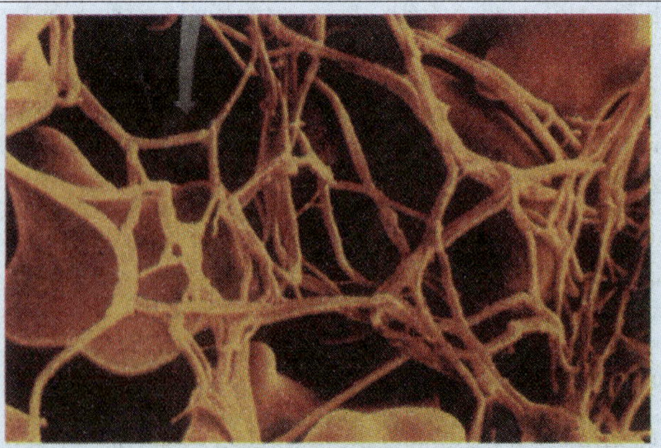
Fibrin threads and red blood cells are clearly visible in this blood clot.

(3) Classification based on solubility characteristics. According to this criterion proteins can be classified into two main types :

(A) Simple proteins. These proteins contain only amino acids in their molecules and they are of following types :

(i) Albumins. These are water soluble proteins found in all body cells and also in blood stream, *e.g.*, **lactalbu-**

min, found in milk and **serum albumin** found in blood.

(ii) Globulins. These are insoluble in water but are soluble in dilute salt solutions of strong acids and bases, *e.g.*, **lactoglobulin** found in milk and **ovoglobulin**.

(iii) Glutelins. These plant proteins are soluble in dilute acids and alkalis, *e.g.*, **glutenin** of wheat.

(iv) Prolamines. These plant proteins are soluble in 70 to 80 per cent alcohol, *e.g.*, **gliadin** of wheat and **zein** of corn.

(v) Scleroproteins. They are insoluble in all neutral solvents and in dilute alkalis and acids, *e.g.*, keratin and collagen.

(vi) Histones. These are water soluble proteins which are rich in basic amino acids such as arginine and lysine. In eukaryotes histones are associated with DNA of chromosomes to form nucleoproteins.

(vii) Protamines. These are water soluble, basic, light weight, **arginine** rich polypeptides. They are bound to DNA in spermatozoa of some fishes, *e.g.*, **salmine**, of salmon and **sturine** in sturgeons.

(B) Conjugated proteins. These proteins consist of simple proteins in combination with some non-protein components, called **prosthetic groups**. The prosthetic groups are permanently associated with the molecule, usually through covalent and/or non-covalent linkages with the side chains of certain amino acids. Conjugated proteins are of following types :

(i) Chromoproteins. Chromoproteins are a heterogeneous group of conjugated proteins which are in combination with a prosthetic group that is a pigment, *e.g.*, respiratory pigments such as **haemoglobin**, **myoglobin** and **haemocyanin**; **catalase**, **cytochromes**, **haemerythrins**; visual purple or **rhodopsin** of rods of retina of eye and yellow enzymes or **flavoproteins.**

(ii) Glycoproteins. Glycoproteins are proteins that contain various amounts (1 to 85 per cent) of carbohydrates. Of the known 100 monosaccharides, only nine are found to occur as regular constituents of glycoproteins (*e.g.*, glucose, galactose, mannose, fucose, acetylglucosamine, acetylgalactosamine, acetylneuraminic acid, arabinose and xylose). Glycoproteins are of two main types : 1. **Intracellular glycoproteins** which are present in cell membranes and have an important role in membrane interaction and recognition. They also serve as antigenic determinants and receptor sites. 2. **Secretory glycoproteins** are **plasma glycoproteins** secreted by the liver ; **thyroglobulin**, secreted by the thyroid gland ; **immunoglobulins** secreted by the plasma cells ; **ovoalbumins** secreted by the cells of oviduct of hen ; **ribonucleases** and **deoxyribonucleases. Mucus** and **synovial fluid** are also glycoproteins with lubricative properties.

(iii) Lipoproteins. Lipid containing proteins are called **lipoproteins.** Their lipid contents are 40 to 90 per cent of their molecular weight and this tends to affect the density of the molecule. There are four types of lipoproteins : 1. **High density lipoproteins (HDL)** or α-lipoproteins; 2. **Low density lipoproteins (LDL)** or β- lipoprotiens ; 3. **Very low density lipoproteins (VLDL)** or pre- β-lipoproteins; and 4. **Chylomicrons**. Lipoproteins include some of the blood plasma proteins, various types of membrane proteins, lipovitellin of egg yolk and proteins of brain and nerve tissue.

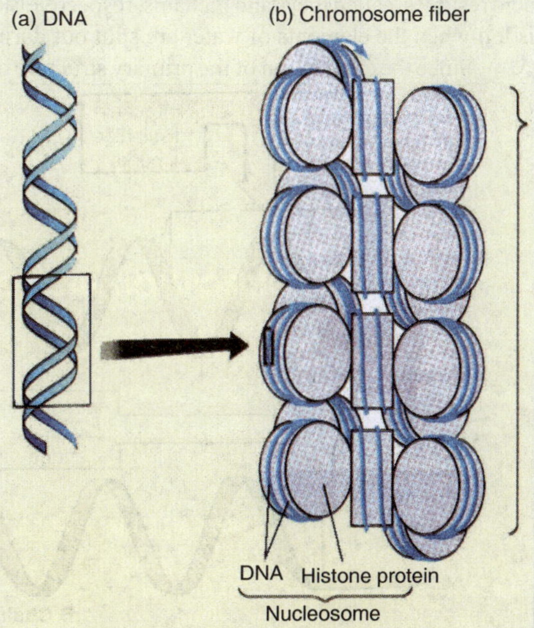

Position of histones in nucleic acids.

(iv) Nucleoproteins. Nucleoproteins are proteins in combination with nucleic acids (DNA and RNA). However, these proteins are not true conjugated proteins since the nucleic acid involved cannot be regarded as prosthetic groups. Nucleoproteins are of two types : 1. **Histones** which are quite similar in all plants and animals. Their highly basic nature accounts for the close associations histones form with the nucleic acids. Histones are involved in the tight packing of DNA molecules during the condensation of chromatin into chromosomes for the mitosis. 2. **Nonhistones** have great heterogeneous amino acid composition and are acidic in nature. They have selective combination with certain stretches of nuclear DNA and, thus, are involved in the regulation of gene expression.

(v) Metalloproteins. Metalloproteins are proteins conjugated to metal ions which are not part of the prosthetic group, *e.g.*, **carbonic anhydrase** enzyme contains zinc ions and amino acids in its molecule; **caeruloplasmin**, an oxidase enzyme containing copper; and **siderophilin** contains iron.

(vi) Phosphoproteins. Phosphoproteins are proteins in combination with a phosphate group, *e.g.*, casein of the milk and ovovitellin of eggs.

3. Structural levels of proteins. The protein as synthesized on the ribosome is a linear sequence of amino acids, polymerized by the elimination of water between successive amino acids to form the peptide bond, and existing as a randomly coiled chain without specific shape and possessing no biological (*i.e.*, catalytic) activity. Within seconds of synthesis being completed, the protein folds into a specific three-dimensional form, which is the same for all molecules of the same type of protein and which now is capable of doing catalysis. According to their mode of folding the following four levels of protein organization have been recognized :

(a) Primary protein structure. The primary protein structure is defined as the particular sequence of amino acids found in the protein. It is determined by the covalent peptide bondings between amino acids. Primary structure also includes other covalent linkages in proteins, for example the linkages that may exist between sulphur atoms of cysteine amino acids located in the chain of the protein insulin. The first protein to have its primary structure determined was of insulin, the pancreatic hormone that regulates glucose metabolism in mammals. Insulin has a molecular weight of 5,800 daltons and contains 51 amino acids. Insulin consists of two polypeptide chains of 21 and 30 amino acid residues, called the A and B chains, respectively (Fig. 4.24). (An **amino acid residue** is that which is left when the elements of water are split out during polymerization).

Since the elucidation of the primary structure of insulin in 1953 by **F. Sanger** (for which Sanger

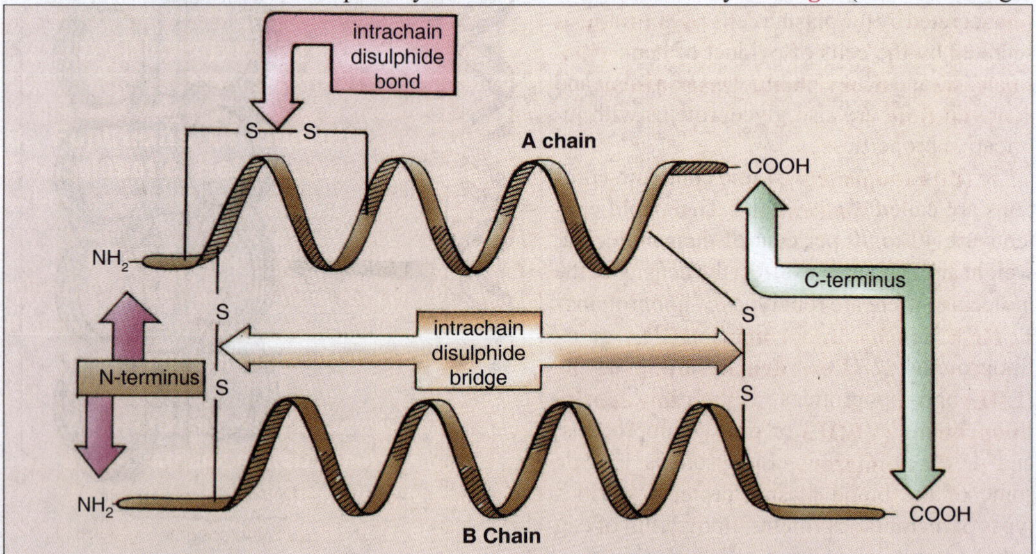

Fig. 4.24. Molecular structure of insulin (After Sheeler and Bianchi, 1987).

received a Nobel Prize), several hundred proteins have been fully sequenced. Among the fully sequenced proteins are ribonuclease and nearly 100 types of haemoglobin. For example, **Stein** and his coworkers established the amino acid sequence (*i.e.*, primary structure) of the enzyme **ribonuclease**. This enzyme is produced by the pancreas and secreted into the small intestine where it catalyzes the hydrolytic digestion of polyribonucleotide chains (RNA). The ribonuclease consists of a single 124 amino acid polypeptide having a molecular weight of about 12,000.

(b) Secondary protein structure. Secondary structure of the protein is any regular repeating organization of the polypeptide chain. There are three types of secondary protein structure : (1) **Helical structure** (*e.g.*, α-keratin and collagen); (2) **Pleated sheet structure** or **β- structure** (*e.g.*, fibroin of silk); and (3) **Extended configuration** (*e.g.,* stretched keratin). Most fibrous proteins have secondary structure. In globular protein, too, it is not uncommon for half of all the residues of each polypeptide to be organized into one or more specific secondary structures.

Collagen. The collagens (the source of leather, gelatin, glue, etc.) are a family of highly

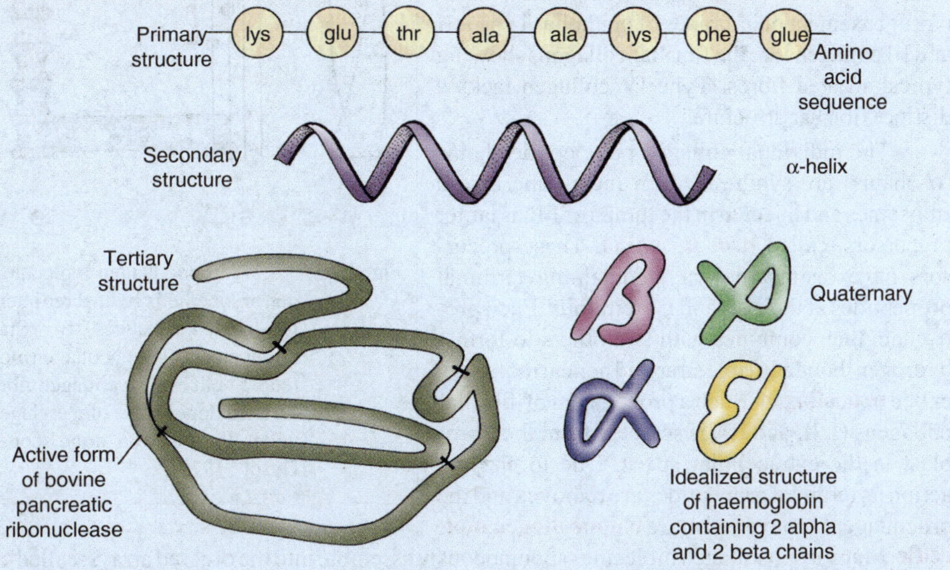

Fig. 4.25. The four levels of structural organization in protein molecules (after Stansfield, 1969).

characteristic fibrous proteins found in all multicellular animals (*e.g.*, in connective tissues). They are secreted by the fibroblasts constituting most abundant (up to 25 per cent of total body's proteins) proteins of mammals. The characteristic feature of collagen (or **tropocollagen**) molecules is their stiff, triple-stranded helical structure (which was discovered by **Rich**, **Crick** and **Rama-chandran**). Three collagen polypeptide chains are left-handed α-helices or alpha chains, each is about 1000 amino acid residues long. These chains are wound around one another in a regular superhelix to generate a rope-like collagen or tropocollagen molecule which is about 300 nm long and 1.5 nm is diameter (Fig. 4.26).

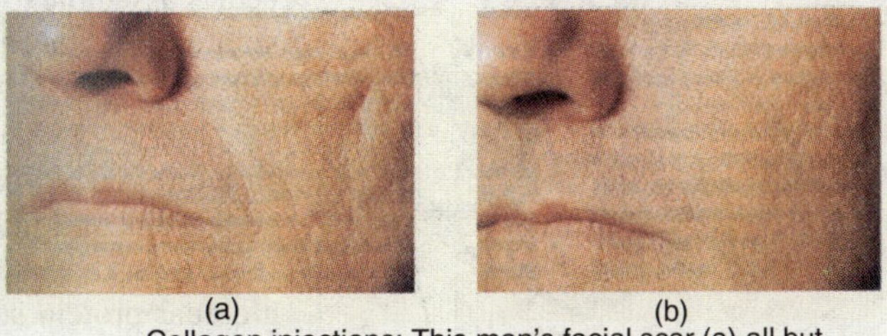

Collagen injections: This man's facial scar (a) all but disappears (b) after cosmetic collagen injections.

Collagens are exceptionally rich in **proline** (and **hydroxyproline**; both accounting for more than 20 per cent of collagen's amino acids) and **glycine**. Other dominant amino acids of collagens are **lysine** and **alanine**.

So far, about 20 distinct collagen-chains have been identified, each encoded by a separate gene. About 10 types of triplet-stranded collagen molecules have been found to assemble from various combinations of 20 types of α-chains. The best defined are types I, II, III and IV. **Type I collagen** is present in the dermis, tendons, ligaments, bone, cornea, dentine of teeth and internal organs and accounts for 90 per cent of body's collagen. **Type II collagen** is present mainly in cartilage, intervertebral disc, embryonic notochord and vitreous humour of eye. **Type III collagen** occurs in skin, cardiovascular system, gastro-intestinal tract and uterus. **Type IV collagen** is present in basal laminae or basement membranes of epithelia. Type I, II and III collagens are the fibrillar collagens showing typical striated fibres. Type IV collagen lacks a distinct fibrillar structure.

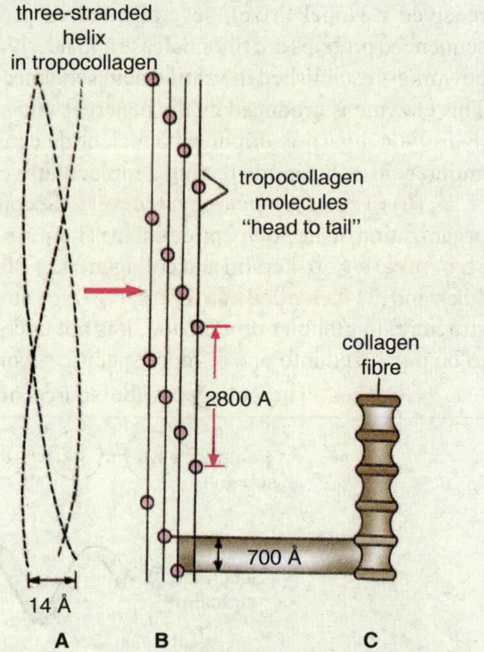

Fig. 4.26. A— A tropocollagen molecule (collagen or superhelix) with three intertwined left-handed a- helices ; B— Staggered arrangement of tropocollagen molecules (super helices) in a collagen fibril; C— Striated appearance of a collagen fibril under the electron microscope (after Dyson, 1978).

The individual collagen polypeptide chains (α-chains) are synthesized on membrane bound ribosomes and injected in the lumen of ER as larger precursors, called **pro- α-chains**. These precursors have distinct polarity, containing terminal propeptides at their N- and C- terminus. Each pro-α-chain then combines with two others to form a hydrogen-bonded, triple stranded helical molecule, called **procollagen**. Such a procollagen of fibrillar collagens (I, II, and III) is serceted from the fibroblast in the extracellular space. Due to enzyme action its terminal propeptides are removed and the procollagen is converted into a **tropocollagen molecule**. Many tropocollagen molecules spontaneously assemble into the ordered arrays, called **collagen fibrils**. The collagen fibrils are thin (10 to 300 nm in diameter), cable-like structures, many micrometres long, exhibiting cross-striations every 67 nm and are clearly visible in the electron microscope. The collagen fibrils often aggregate into larger bundles which can be seen in light microscope as **collagen fibres**. Type IV collagen molecules assemble to form a sheet-like **meshwork** that constitutes a major part of all basal laminae (**Martin** *et al*., 1985, **Burgeson**, 1988).

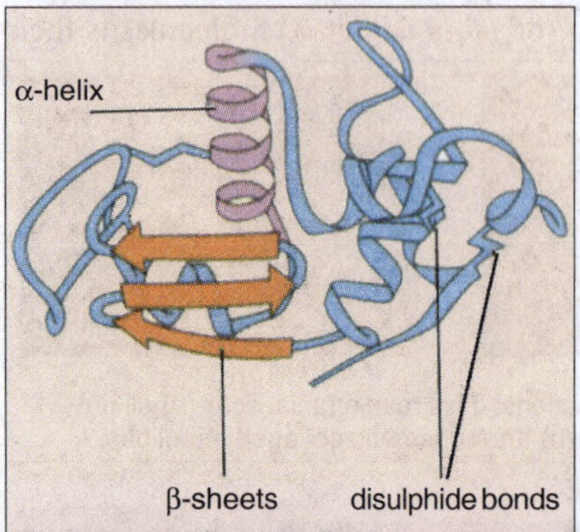

The tertiary structure of lysozyme, an antibacterial enzyme present in tears.

(C) Tertiary protein structure. Tertiary protein structure refers to a more compact structure in which the helical and non-helical regions of a polypeptide chain are folded back on themselves. This structure is typical of globular protein structure, in which it is the non-helical region that permits the folding. The folding of a polypeptide chain is not random but occurs in a specific fashion, thereby imparting certain steric (*i.e.*, three-dimensional) properties to the protein. For example, in enzymes folding brings together **active amino acids**, which are otherwise scattered along the chain, and may form a distinctive cavity or cleft in which the substrate is bound.

The complete tertiary structure of a protein can only be deducted by a laborious analysis of X-ray scattering patterns from crystals. The first protein to have its secondary and tertiary structure determined was **myoglobin**, a 153-amino acid, oxygen-binding protein found primarily in red muscle and largely responsible for the colour of that tissue. The work was done at Cambridge under the direction of **J.C. Kendrew** (1961). Although at some points the polypeptide chain does have secondary structure (alpha-helical structure), the chain is mainly characterized by seemingly random loops and folds.

In a tertiary protein the polypeptide chain is held in position by weak secondary bonds which are of different types such as **ionic bonds** (or electrostatic bonds or salt or salt bridges); **hydrogen bonds**; **hydrophobic bonds** and **disulphide bonds**.

(d) Quaternary protein structure. In proteins that are composed of two or more polypeptide chains, the quaternary structure refers to the specific orientation of these chains with respect to one another and the nature of the interactions that stabilize this orientation. The individual polypeptide chains of the protein are called **sub-units** and the active protein itself is called **multimer**. While multimeric proteins containing up to 32 subunits have been described, the most common multimers are **dimers**, **trimers**, **tetramers**, **pentamers** (*e.g.*, RNA polymerase) and **decamers** (*e.g.*, DNA polymerase III) (Table 4-6). If the protein consists of identical sub-units, it is called **homopolymers** and is said to have **homogeneous quaternary structure**, *e.g.*, the isozymes H_4 and M_4 of lactic dehydrogenase (LDH), enzyme phosphorylase and L-arabinose isomerase. The enzyme β-galactosidase consists of four identical polypeptide chains. Lastly, when the sub-units of the protein are different, the protein is called **heteropolymer** and is said to have a heterogeneous quaternary structure, *e.g.*, haemoglobin and immunoglobulins.Quaternary proteins are usually joined by hydrophobic forces. Hydrogen bonds, ionic bonds and possibly disulphide bonds may also participate in forming quaternary structures.

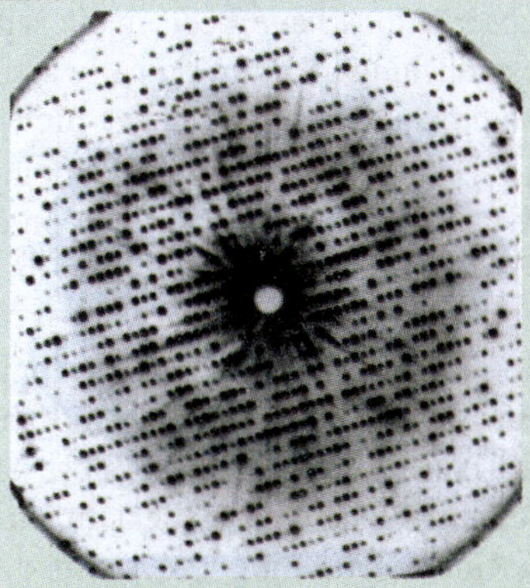

The X-ray diffraction pattern of myoglobin.

Table 4-6. Subunits and molecular weight of some multimer proteins (Source : Sheeler and Bianchi, 1987).

Protein	Molecular weight	Number	Subunits Designation	Molecular weight
1. Haemoglobin A (human)	64,500	4	Alpha chains (2)	15,700
			Beta chains (2)	16,500
2. Lactate dehydrogenase	135,000	4	A chain (0 to 4)	33,600
			B chain (4 to 0)	33,600
3. Immunoglobulin G	150,000	4	Light chains (2)	25,000
			Heavy chains (2)	50,000
4. Tryptophan synthetase (*E.coli*)	150,000	4	Alpha chains (2)	29,500
			Beta chains (2)	45,000
5. Aspartate transcarbamylase	306,000	12	C chains (6)	34,000
			R chains (6)	17,000
6. L-arabinose isomerase (*E.coli*)	360,000	6	(identical)	60,000
7. Apoferritin (iron storage protein)	456,000	24	(identical)	19,000
8. Thyroglobulin	670,000	2	(identical)	3,35,000

Some Examples of Tertiary and Quaternary Proteins

(i) Ribonuclease. C.B. Anfinsen initiated and confirmed the notion that, acting in concert, the specific primary structure of a polypeptide and the innate properties of the side chains of its amino acids cause the polypeptide to spontaneously assume its biologically active tertiary structure. In 1972, he got the Nobel Prize for this definitive work. **Anfinsen** identified four disulphide bridges in the ribonuclease protein, suggesting that the enzyme is highly folded (Fig. 4.25). As is the case with almost all enzymes, the catalytic activity of ribonuclease depends on the maintenance of a particular three-dimensional shape. In concentrated solutions of β-mercaptoethanol and urea, the disulphide bridges of the enzyme are broken and the resultant unfolding of the polypeptide chain is accompanied by a loss of enzyme activity. The enzyme is said to be **denatured**. If the β-mercaptoethanol and urea are removed by dialysis and the denatured ribonuclease reacted with oxygen, the four disulphide bridges re-form spontaneously, and essentially all the catalytic activity of the protein is restored. Similar observations have been made with other proteins, that is, they are capable of spontaneously re-establishing their biologically active tertiary (or even quaternary, *e.g.*, haemoglobin) structure after having undergone extensive molecular disorganization.

(ii) Haemoglobin. Haemoglobin is one of the fully sequenced protein. Our present understanding of the structure and function of haemoglobin is the outcome of 50 years of research of **M. F. Perutz**. He got the Nobel Prize in 1962, along with **J. C. Kendrew**, for their studies of haemoglobin and myoglobin.

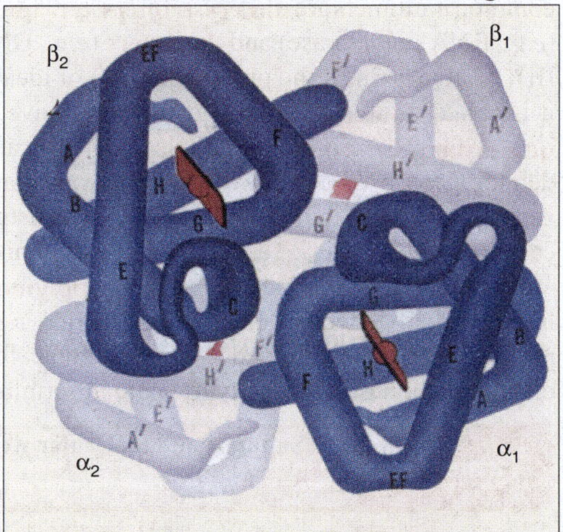

Haemoglobin molecule.

The haemoglobin is a conjugated globular protein, that is, it contains some non-protein part. In all but the lowest vertebrates, haemoglobin is a tetramer (a heteropolymer). In lampreys, however, haemoglobin is monomeric, that is, it contains a single globin chain like the myoglobin. In humans, most common type of haemoglobin is haemoglobin A (HbA), which consists of 574 amino acid residues and has a molecular weight of 64,500. Its secondary, tertiary and quaternary structure is typical of all higher vertebrate haemoglobins. The protein portion of the haemoglobin molecule, called **globin**, is composed of four polypeptide chains, each of which is also globular in shape. The four globin chains consist of two identical pairs : two **alpha chains** (141 amino acids each) and two **beta chains** (146 amino acids each). The non-protein portion of haemoglobin consists of four iron-containing haem groups, one associated with each of the four globin chains. Nineteen of the twenty biologically important amino acids are included in the globin of haemoglobin.

Haemoglobin molecule is highly symmetric; it can be divided into two identical halves, each consisting of an αβ- dimer. The complete tetramer is similar to a mildly flattened sphere having a maximum diameter of about 5.5 nm. The four polypeptide chains are arranged in such a way that unlike chains have numerous stabilizing interactions, whereas, like chains have few. A cavity about 2.5 nm long and varying in width from about 5 to 10 A° passes through the molecule along the axis. Each globin chain envelops its haem group in a deep cleft.

(iii) Immunoglobulins. The ability to resist infection by pathogens (viruses, bacteria and other unicellular parasites) and by multicellular endoparasites, is called **immunity**. Specific immune response function by recognizing particular chemical structures, known as **antigens**—on the surface of invading cells. An antigen can be a protein, lipid, carbohydrate or any other molecule. These antigens interact with protein molecules produced by the host, the **immunoglobulins**, which bind the antigen in much the same way as an enzyme binds its substrate. Specific **immune responses involve many different types of cells. One type, the B-lymphocyte** or **B-cell**, is capable of producing **free immunoglobulins**, called **antibodies**.

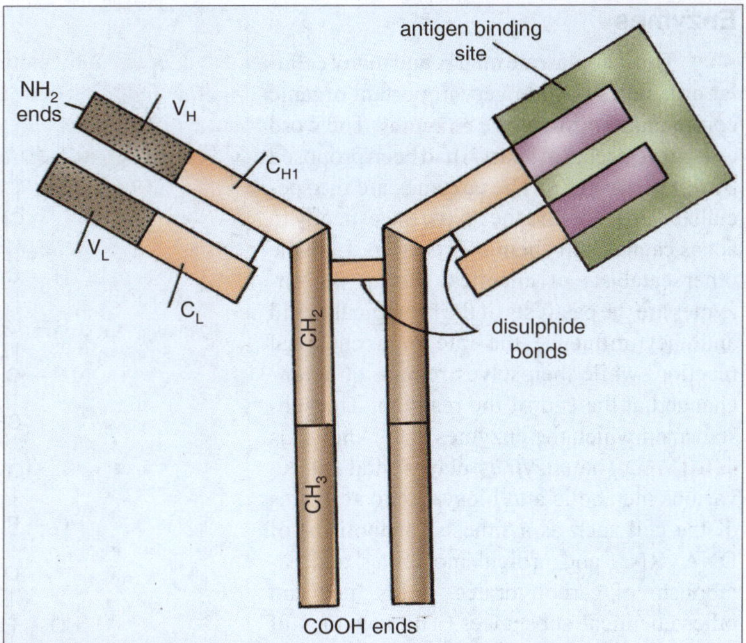

Fig. 4.27. Molecular structure of an immunoglobulin molecule (after Stansfield, 1986).

An immunoglobulin (Ig) molecule is a Y-shaped heteropolymer and is composed of two identical **H (heavy)** polypeptide chains and two smaller identical **L (Light)** polypeptide chains (Fig. 4.30). Heavy chains contain antigenic determinants in the "tail" (carboxyl) segments by which they can be classified as Ig G, Ig M, Ig A, Ig D or Ig E. Light chains can likewise be typed as kappa or lambda. Within a H chain class or L chain type, these segments exhibit very little variation in primary structure from one individual to another and are called **constant regions (C)**. The amino (—NH_2) ends, however, are extremely diverse in primary structure, even within a class and are called **variable (V) regions**. The V_H and V_L regions together form two **antibody-combining sites** (called **antigen - binding sites**) for specific interactions with homologous antigen molecules. The C_H region consists of three or four similar segments, presumably derived evolutionarily by duplication of an ancestral gene and subsequent modification by mutations; the similar segments are called **domains** and are labelled CH_1, CH_2, CH_3, etc. A mature lymphocyte (plasma cell) produces antibodies with a single class *H* chain and a single type of *L* chain, hence, also a single antigen-binding specificity. The first antibodies produced by a developing plasma cells are usually of class Ig M.

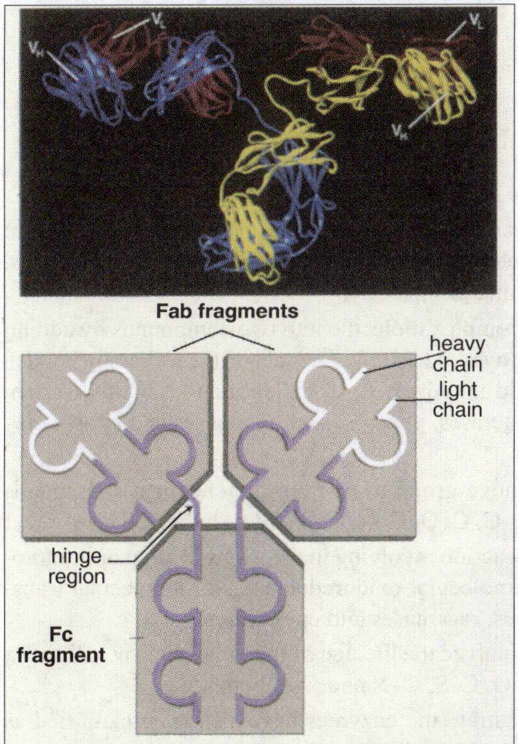

Antibody structure: (a) Ribbon model of an I_gG molecule (b) Schematic model showing the domain structure of an I_gG molecule.

Enzymes

The cytoplasmic matrix and many cellular organelles contain very important organic compounds known as the **enzymes**. The word enzyme (Greek. "in yeast") had been proposed by **Kuhne** in 1878. The enzymes are the specialized proteins and they have the capacity to act as catalysts in chemical reaction. Like the other catalysts of chemical world, the enzymes are the catalysts of the biological world and they influence the rate of a chemical reaction, while themselves remain quite unchanged at the end of the reaction. The substance on which the enzymes act is known as **substrate**. The enzymes play a vital role in various metabolic and biosynthetic activities of the cell such as synthesis (anabolism) of DNA, RNA and protein molecules and catabolism of carbohydrates, lipids, fats and other chemical substances. The enzymes of the matrix and cellular organelles are classified as follows :

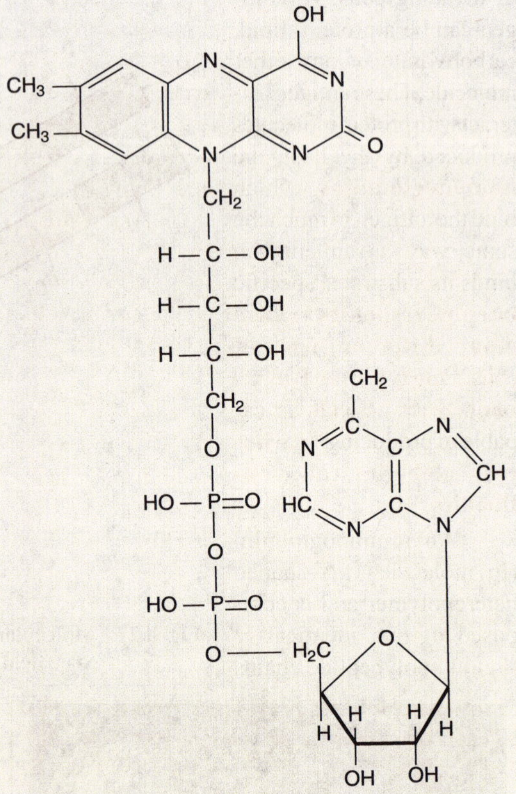

Fig. 4.28. Chemical formula of flavin adenine dinucleotide (FAD).

1. Oxireductases. The enzymes catalyzing the oxidation and reduction reaction of the cell are known as oxireductases. These enzymes transfer the electrons and hydrogen ions from the substrates, *e.g.*, hydrogenases or reductases, oxidases, oxygenases and peroxidases.

2. Transferases. The enzymes which transfer following groups from one molecule to other are known as transferases : one carbon, aldehydic or ketonic residues, acyl, glycosyl, alkyl, nitrogenous, phosphorus containing groups and sulphur containing groups.

3. Hydrolases. These enzymes hydrolyse a complex molecule into two compounds by adding the element of the water across the bond which is cleaved. These enzymes act on the following bonds– ester, glycosyl, ether, peptide, other C–N bonds, acid anhydride, C–C, halide and P–N bonds. Certian important hydrolase enzymes are the proteases, esterases, phosphatases, nucleases and phosphorylases.

4. Lysases. The lysase enzymes add or remove group to or from the chemical compounds containing the double bonds. The lysases act on C–C, C–O, C–N, C–S and C–halide bonds.

5. Isomerases. These enzymes catalyse the reaction involving in the isomerization or intramolecular rearrangements in the substrates, *e.g.*, intramolecular oxidoreductases, intramolecular transferases, intramolecular lysases, cis-trans-isomerases, racemases and epimerases.

6. Ligases or synthetases. These enzymes catalyze the linkage of the molecules by splitting a phosphate bond. The synthetase enzymes form C–O, C–S, C–N and C–C bonds.

According to the chemical nature of the substrate the enzymes have also been classified as follows:

1. Carbohydrases, 2. Proteases (endopeptidases and exopeptidases), 3. Amylases, 4. Esterases, 5. Dehydrogenases, 6. Oxidases, 7. Decarboxylases, 8. Hydrases, 9. Transferases, and 10. Isomerases.

The enzymes are **specific** in action and many factors such as pH, temperature and concentration of the substrate affect the rate of the activity of enzymes. Certain enzymes occur in the inactive form (called **proenzymes** or **zymogens**) and these are activated by other enzymes known as **kinases** to perform catalytic activities. Likewise, the enzyme **trypsinogen** of the pancreatic cells is activated in the intestine by the enzyme **enterokinase** and the enzyme **pepsinogen** of the Chief cells of the stomach is activated by the **hydrochloric acid** which is secreted by parietal cells.

Prosthetic Groups and Coenzymes

Certain enzymes such as **cytochromes** are the conjugated proteins and contain prosthetic group as **metalloporphyrins** complex in their molecules.

Certain enzymes cannot function singly but they can function only by the addition with the small molecules of other chemical substances which are known as **coenzymes**. The inactive enzyme (which cannot function singly) is known as the **apoenzyme**. The apoenzyme and coenzyme are collectively known as **holoenzyme**. For instance, the enzyme **hydrogenase** is an apoenzyme which can function either with the coenzyme NAD^+ or NADP.

Some important coenzymes or cofactors are as follows :

1. Nicotinamide adenine dinucleotide (NAD) or Diphosphopyridine nucleotide (DPN), 2. Nicotinamide adenine dinucleotide phosphate (NADP) or Triphosphopyridine nucleotide (TPN), 3. Flavin adenine mononucleotide (FAM), 4. Flavin adenine dinucleotide (FAD), 5. Ubiquinone (coenzyme Q or Q), 6. Lipoic acid (LIP or S_2), 7. Adenosine triphosphate (ATP), 8. Pyridoxyl phosphate (PALP), 9. Tetrahydrofolic acid (CoF), 10. Adenosyl methionine, 11. Biotin, 12. Coenzyme A (CoA), 13. Thiamine pyrophosphate (TPP), 14. Uridine diphosphate (UDP).

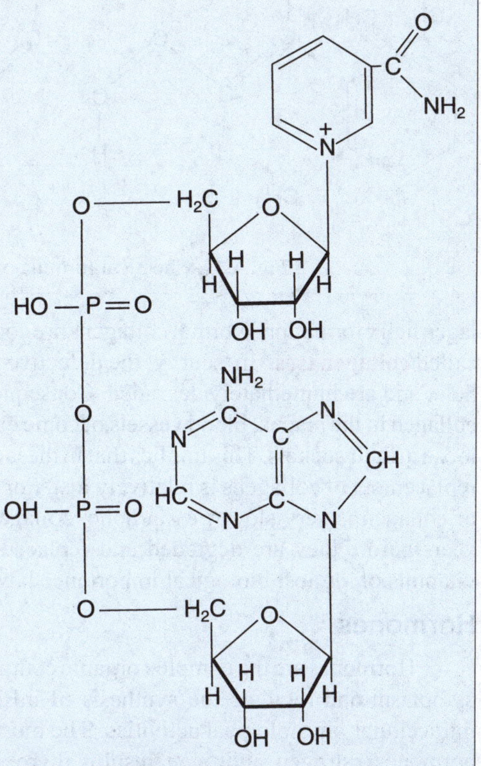

Fig. 4.29. Chemical formula of nicotinamide adenine dinucleotide (NAD^+).

Isoenzymes

Recently, it has been investigated that some enzymes have similar activities and almost similar molecular structures. These enzymes are known as **isoenzymes**. The isoenzymes have relation with the heredity (**Latner** and **Skillen**, 1969, and **Weyer** 1968). There are about 100 isoenzymes in the cell, *e.g.*, lactic dehydrogenase (LDH) occur in the form of five identical isoenzymes.

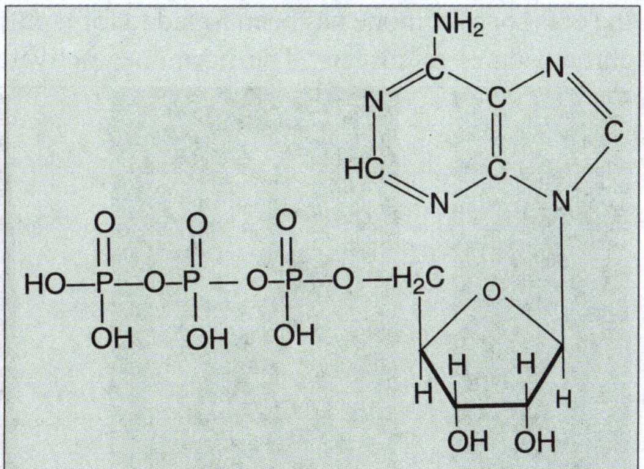

Fig. 4.30. Chemical formula of Adenosine triphosphate (ATP).

Vitamins

The vitamins are organic compounds of diverse chemical nature. They are required in minute amounts for normal growth, functioning and reproduction of cells. The vitamins play an important role in the cellular metabolism and act as the enzymes or other biological catalysts in the various chemical activities of the cell. Their importance for the animals has been reported by **Hopkins**, **Osborne**, **Mendal**, and **McCollum** (1912–1913). **Funk** (1912) demonstrated the presence of basic nitrogen in them and gave the name "vitamins" meaning vital amines to them.

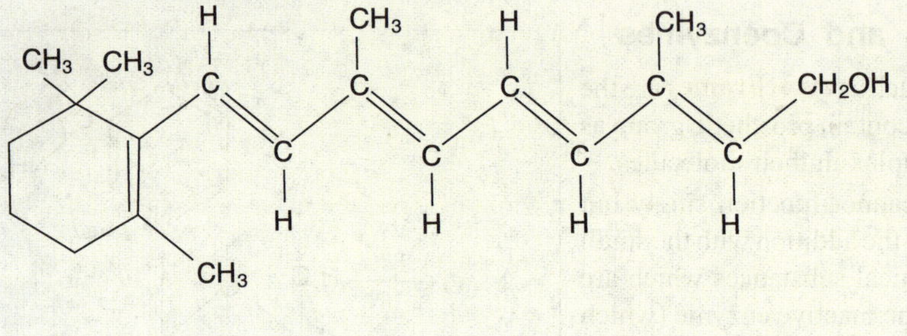

Fig. 4.31. Chemical formula of vitamin A.

The animal cell cannot synthesize the vitamins from the standard food and so they are taken along with the food. Their deficiency in the cell causes metabolic disorder and leads to various diseases. For example, the deficiency of ascorbic acid (Vitamin C) inhibits procollagen helix formation. Normal collagens are continuously degraded by specific extracellular enzymes, called **collagenases**. In scurvy, the defective pro-α-chains that are synthesized, fail to form a triple helix and are immediately degraded. Consequently, with the gradual loss of the pre-existing normal collagen in the matrix, blood vessels become extremely fragile and teeth become loose in their sockets. This implies that in these particular tissues degradation and replacement of collagens is relatively fast. For example, in bones, the 'turnover' of collagen is very slow, *i.e.*, in bone, collagen molecule persists for about 10 years before they are degraded and replaced (see **Alberts** *et al*., 1989). The vitamins of utmost biological importance have been tabulated in Table 4-7.

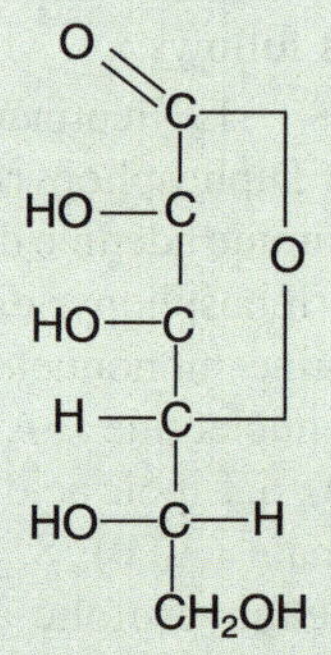

Fig. 4.32. Chemical formula of vitamin C or ascorbic acid.

Hormones

Hormones are the complex organic compounds which occur in traces in the cytoplasm and regulate the synthesis of mRNA, enzymes and various other intracellular physiological activities. The most important hormones are growth hormones, estrogen, androgen, insulin, thyroxine, cortisone, and adrenocortical hormones, etc. These hormones are synthesized by the ductless or endocrine glands and transported to various cells of multicellular organisms by blood vascular system. In cells they regulate various metabolic activities. For example, the ecdysone hormone has been found to form puffs (Balbiani rings) in the giant chromosomes of insects. The hormones activate or depress the gene at the

Hormones control metamorphosis in insects.

Table 4-7. Vitamins and their characteristics.

Vitamin	Daily requirement	Sources	Functions	Diseases and symptoms caused by lack of vitamin
Fat Soluble Vitamins 1. Vitamin A (Retinol)	750 μ gm	Animal fats (fish liver oil, egg-yolk, milk, butter, cheese); palm oils; red peppers; dark green leafy vegetables (spinach, methi, cabbage); yellow vegetables (carrot, pumpkin) and yellow fruits (mango, papaya).	Stored in liver; maintain general health and vigour of epithelial cells.	1. Skin becomes dry and scaly and so does cornea of eyes causing **xerophthalmia** or 'dry eye'; 2. **Night blindness or nyctalopia** (inability to see in dimlight).
2. Vitamin D (Calciferol)	200 IU (5 μ gm)	Fish liver oils, liver, egg-yolk, butter, fresh milk; also produced by our body when skin-cholesterol is exposed to ultra-violet rays of sunlight.	Involved in intestinal absorption of calcium and phosphorus and in calcium metabolism and bone formation.	1. **Rickets** in children; 2. **Osteomalacia** in adults.
3. Vitamin E (Tocopherol)	Trace amounts (15 IU)	Vegetable oils (especially polysaturated fatty acids); wheat germ oil; egg-yolk; green leafy vegetables; tomato; milk and butter.	Inhibit catabolism (*i.e.*, oxidation) of certain fatty acids of cellular membranes.	**Hemolytic anemia** due to oxidation of unsaturated fats resulting in abnormal structure and function of mitochondria, lysosomes and plasma membrane of cells.
4. Vitamin K (Naphtoquinone)	Trace amounts	Naturally produced by intestinal bacteria; liver; fresh green vegetables (spinach, cabbage, cauliflower).	Required for the formation of prothrombin (an essential component of blood-clotting).	**Haemorrhage or bleeding** in new-born infants; scurvy-like symptoms (blood takes longer to clot).
Water Soluble Vitamins 5. Vitamin B (Thiamine)	1.3 mg (boys) 1.2 mg (girls)	Yeast; whole cereals (*e.g.*, unpolished rice); pulses; oil-seeds, soyabean; nuts (especially groundnut); liver, pork, sea food, green leafy vegetables.	1. Essential for synthesis of acetylcholine; 2. Rapidly destroyed by heat. 3. Carbohydrate metabolism.	1. **Beriberi.** Partial paralysis of smooth muscle of gastrointestinal tract; paralysis of skeletal muscles, atrophy of limbs. 2. **Polyneuritis.**
6. Vitamin B_2 (Riboflavin)	1.6 mg (boys) 1.4 mg (girls)	Peas, beans, milk, egg-white, liver, kidney, germinated cereals and pulses, growing green leafy vegetables.	Forms the coenzyme FAD which is involved in metabolism of carbohydrates and proteins.	1. Blurred vision, cataract and corneal ulceration; 2. **Dermatitis** (inflammation of skin); 3. **Cheilosis** (cracking of skin at the corners of mouth and scaling of lips).

Vitamin	Daily requirement	Sources	Functions	Diseases and symptoms caused by lack of vitamin
7. Niacin (Nicotinic acid)	18 mg (boys) 15 mg (girls)	Meat, liver, fish, chicken, yeast, whole grain, peas, beans, pulses, nuts (groundnuts) potato, tomato, green vegetables, germinated seeds and milk (Maize is deficient in niacin).	1. Forms the coenzyme NAD which is involved in energy-releasing reactions; 2. In lipid metabolism inhibits production of cholesterol and help in fat breakdown.	**Pellagra**, a disease characterised by three D's-dermatitis, diarrhoea and dementia (psychological disturbance).
8. Vitamin B_6 (Pyridoxin)	1.5–2 mg	Liver, meat, fish (salmon), whole cereals, yellow corn, legumes, tomatoes, yoghurt.	1. Forms coenzymes involved in amino acid metabolism in brain; 2. Involved in fat metabolism.	1. Convulsions; 2. Dermatitis of eyes, nose and mouth; 3. Retarded growth.
9. Folic acid	50–100 mg	Green leafy vegetables, germinated pulses, eggs, liver.	1. Essential for synthesis of DNA. 2. Overcooking destroys it.	**Macrocytic anaemia** (production of abnormally large red blood cells).
10. Biotin (Vitamin H)	0.3 mg	Yeast, liver, egg-yolk, milk, kidneys.	Acts as coenzyme in metabolism of carbohydrates, fatty acids and nucleic acid.	1. Mental depression; 2. Muscular pain, fatigue; 3. Dermatitis; 4. Nausea.
11. Vitamin B_{12} (Cyanocobalamin)	0.2–1.0 μ gm	Liver, kidney, meat, fish, eggs, milk, cheese	Acts as coenzyme necessary for DNA synthesis, red blood cell formation, growth and nerve function.	1. **Pernicious anaemia;** 2. Malfunctioning of nervous system.
12. Vitamin C (Ascorbic acid)	40 mg	Amla, guava, citrus fruits (lime, lemon, orange), tomatoes, green leafy vegetables (cabbage, *chauli* and cauliflower).	1. Promotes protein synthesis (collagen), wound healing and iron absorption; 2. Protects body against infections; 3. Rapidly destroyed by heat.	**Scurvy.** A disease characterised by swelling of gums, multiple haemorrhages, anaemia and weakness.

particular locus on the chromosome. Thus, hormones serve to coordinate the various activities concerned with a particular function, *e.g.*, the hormone ecdyosone controls moulting and metamorphosis in insects (**Beermann**, 1965).

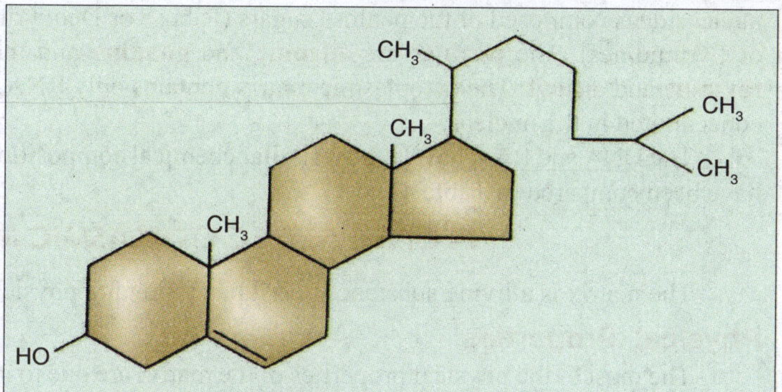

Fig. 4.33. Chemical formula of vitamin D.

In mammalian liver cells, the enzymes which convert glucose into glycogen are regulated by the hormone insulin which is synthesized by β-cells of the islets of Langerhans in the pancreas. Moreover, the hormone thyroxine, a secretion of thyroid gland, activates the enzyme phosphorylase to form glucose phosphate from the glycogen.

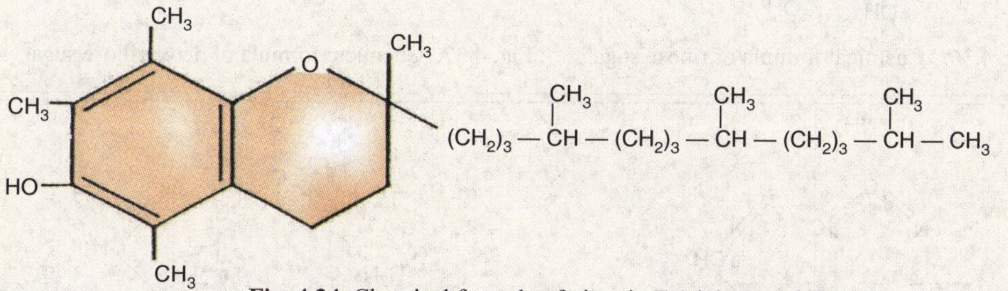

Fig. 4.34. Chemical formula of vitamin E (alphatocopherol).

Nucleic Acids

The nucleic acids are the complex macromolecular organic compounds of immense biological importance. They control the important biosynthetic activities of the cell and carry hereditary informations from generation to generation. There occur two types of nucleic acids in living organims, *viz.*, **Ribonucleic acid (RNA)** and **Deoxyribonucleic acid (DNA)**. Both types of nucleic acids are the polymers of the nucleotides. A nucleotide is composed of nucleoside and phosphoric acid. Even the

Photograph of DNA made by tunneling microscopy.

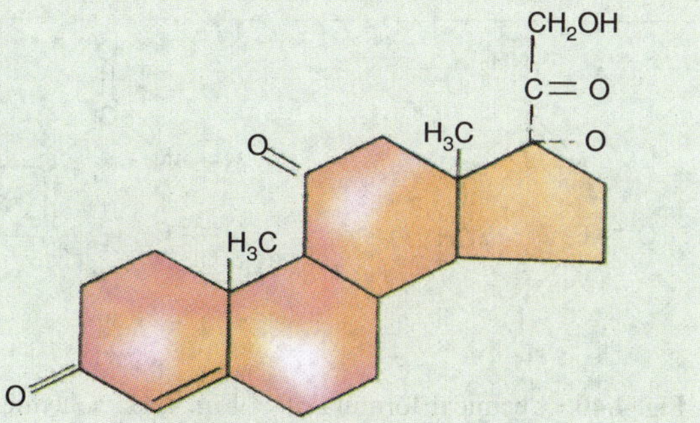

Fig. 4.35. Chemical formula of cortisone.

nucleoside is composed of the pentose sugars (Ribose or Deoxyribose) and nitrogen bases (Purines or Pyrimidines). The **purines** are **adenine** and **guanine** and the **pyrimidines** are the **cytosine**, **thymine** and **uracil**. The cytoplasmic matrix contains only RNA, while DNA exclusively remains concentrated in the nucleus.

The DNA and RNA have almost similar chemical compositions except a few differences. Both have been compared in Table 4-8.

PROPERTIES OF CYTOPLASMIC MATRIX

The matrix is a living substance and it has following physical and biological properties :

Physical Properties

The most of the physical properties of the matrix are due to its colloidal nature and these are as follows :

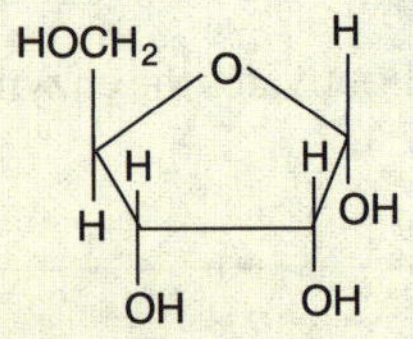

Fig. 4.36. Chemical formula of ribose sugar.

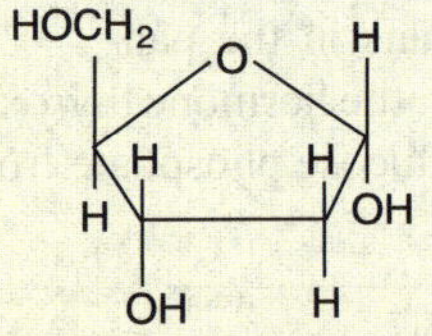

Fig. 4.37. Chemical formula of deoxyribose sugar.

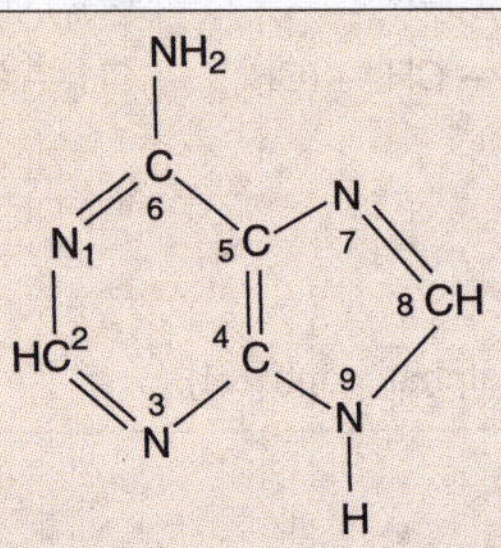

Fig. 4.38. Chemical formula of adenine.

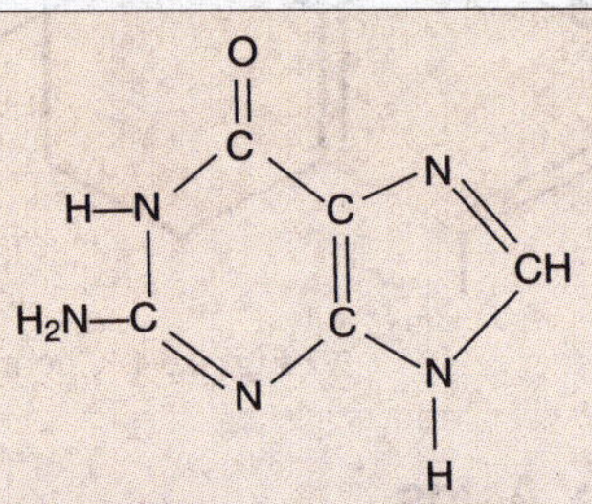

Fig. 4.39. Chemical formula of guanine.

1. Tyndall's effect. When a beam of strong light is passed through the colloidal system of the matrix at right angles in the dark room, the small colloidal particles which remain suspended in the colloidal system, reflect the light. The path of the light appears like a cone. This light cone is known as **Tyndall's cone** because this phenomenon has been first of all reported by **Tyndall** (1820—1893) in colloids.

2. Brownian movement. The suspended colloidal particles of the matrix always move in zig-

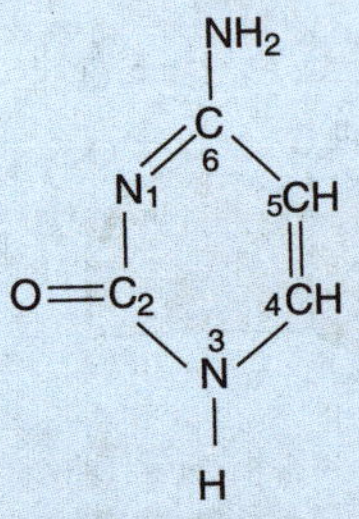

Fig. 4.40. Chemical formula of cytosine.

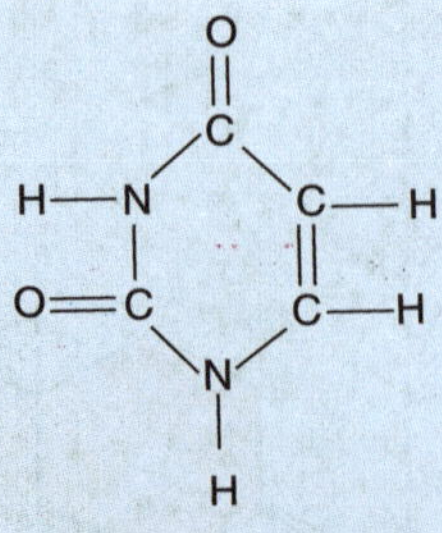

Fig. 4.41. Chemical formula of uracil.

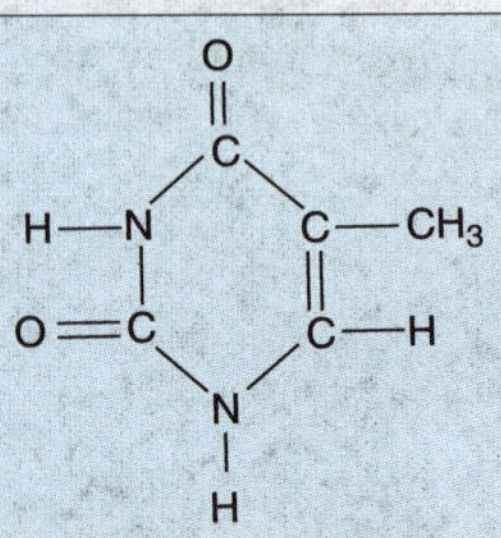

Fig. 4.42. Chemical formula of thymine.

zag fashion. This movement of molecules is caused by moving water molecules which strike with the colloidal molecules to provide motion to them. This type of movement was first of all observed by Scottish botanist **Robert Brown** in 1827 in the colloidal solution. Therefore, such movements are known as **Brownian movement**. The Brownian movement is the peculiarity of all colloidal solutions and depends on the size of the particles and temperature.

Table 4-8. Comparison of DNA and RNA.

DNA	RNA
1. It contains pentose sugar known as deoxyribose.	1. It contains pentose sugar called the ribose.
2. The molecule contains the phosphoric acid (phosphate) molecule which connects various sugars with one another.	2. The molecule contains the phosphoric acid (phosphate) molecule which connects various sugars with one another.
3. The nitrogen bases are : (i) **Purines**—adenine and guanine. (ii) **Pyrimidine**—cytosine and thymine.	3. The molecule contains following nitrogen bases in its molecule: (i) **Purines**—adenine and guanine. (ii) **Pyrimidines**—cytosine and uracil.
4. Molecules have four nucleotides as deoxyadenosine monophosphate, deoxyguanosine monophosphate, deoxycytidine monophosphate and thymidine monophosphate.	4. Molecules have four nucleotides as adenosine monophosphate, guanosine monophosphate, cytidine monophosphate and uridine monophosphate.
5. The molecule contains a double stranded helix structure in which many nucleotides remain arranged in pair.	5. The molecules consist of single chain of polynucleotides.
6. DNA is a genetic material and occurs in chromosomes, nucleoplasm and mitochondria, etc.	6. RNA is a carrier of genetic informations and it plays very significant role in the mechanism of protein synthesis. It mostly occurs in nucleolus, nucleoplasm and cytoplasm.

3. Cyclosis and amoeboid movement. Due to the phase reversal property of the cytoplasmic matrix, the intracellular streaming or movement of the matrix takes place. This property of intracullular movement of matrix is known as the **cyclosis**. The cyclosis usually occurs in the sol-phase of the matrix and is effected by the hydrostatic pressure, temperature, pH, viscosity, etc. The intracellular movements of the pinosomes, phagosomes and various cytoplasmic organelles such as the lysosomes, mitochondria, chromosomes, centrioles, etc., occur only due to cyclosis of the matrix. The cyclosis has been observed in most animal and plant cells.

The amoeboid movement depends directly on the cyclosis. The amoeboid movement occurs in the protozoans, leucocytes, epithelia, mesenchymal and other cells. In the amoeboid movement the cell changes its shape actively and gives out cytoplasmic projections known as **pseudopodia**. Due to cyclosis matrix moves these pseudopodia and this causes forward motion of the cell.

4. Surface tension. The molecules in the interior of a homogeneous liquid are free to move and are attracted by surrounding molecules equally in all directions. At the surface of the liquid where it touches air or some other liquid, however, they are attracted downward and sideways or inward, more than upward; consequently they are subjected to unequal stress and are held together to form a membrane. The force by which the molecules are bound is called the **surface tension** of the liquid. The cytoplasmic matrix being a liquid possesses the property of surface tension. The proteins and lipids of matrix have less surface tension, therefore, occur at the surface and form the membrane, while the chemical substances such as NaCl have high surface tension, therefore, occur in deeper part of the matrix.

5. Adsorption. The increase in the concentration of a substance at the surface of a solution is known as adsorption (L.,*ad*=to, *sorbex*=to draw in). The phenomenon of adsorption helps the matrix to form protein boundaries.

6. Other mechanical or physical properties of matrix. Besides surface tension and adsorption, the matrix possesses other mechanical properties, *e.g.*, elasticity, contractility, rigidity and viscosity which provide to the matrix many physiological utilities.

7. Polarity of the egg. The colloidal system due to its stable phase determines the polarity of the cell matrix which cannot be altered by centrifugation of other mechanical means.

8. Buffers and pH. The matrix has a definite pH value and it does not tolerate significant variations in its pH balance. Yet various metabolic activities produce small amount of excess acids or bases. Therefore, to protect itself from such pH variation the matrix contains certain chemical compounds as carbonate-bicarbonate system known as **buffers** which maintain a constant state of pH in the matrix.

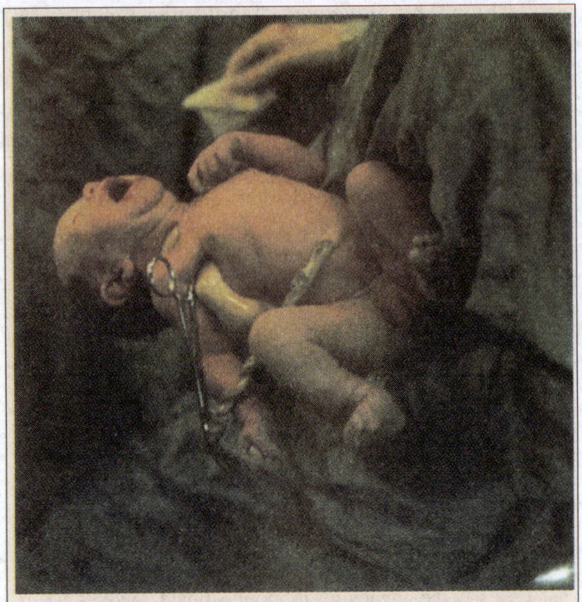
Surface tension: A baby's first breath is facilitated by a special coating called a surfactant, which the lungs secrete. This material acts much like a detergent to decrease the surface tension of the fluid layer lining the lungs. Without the surfactant hydrogen bonds in the water lining the small sacs of the lungs would pull water molecules together so tightly that the sacs would collapse.

Biological Properties

The matrix is a living substance and it has following biological properties :

1. Irritability. The irritability is the fundamental and inherent property of the matrix. It possesses a sensitivity to stimulation, an ability to transmission of excitation and ability to react according to stimuli. The heat, light, chemical substances and other factors stimulate the cytoplamic matrix to contract.

2. Conductivity. The conductivity is the process of conduction or transmission of excitation from the place of its origin to the region of its reaction. The matrix of nerve cells possesses the property of the conductivity.

3. Movement. The cytoplasmic matrix can perform movement due to cyclosis. The cyclosis depends on the age, water contents, heredity factors and composition of the cells.

4. Metabolism. The matrix is the seat of various chemical activities. These activities may be either constructive or destructive in nature. The constructive processes such as biosynthesis of proteins, lipids, carbohydrates and nucleic acids are known as **anabolic processes**, while the destructive processes such as oxidation of foodstuffs, etc., are known as **catabolic processes**. The anabolic and catabolic processes are collectively known as **metabolic process**.

5. Growth. Due to the secretory or anabolic activities (Gr., *anabolism*= a throwing up) of the cell, new protoplasm continuoulsy increases in its volume. The increase in the volume of the matrix causes into the growth of the cell which ultimately divides into daughter cells by the cell division.

6. Reproduction. The cytoplasm has the property of asexual and sexual reproduction.

REVISION QUESTIONS

1. What is cytoplasmic matrix ? Describe various theories regarding the physical nature of the matrix. Also discuss various properties of the matrix.
2. What is the structural basis for the unique properties of the water molecule ? How do the properties of water make it of importance to living systems ?
3. How would you define carbohydrates ? What are the major classes of carbohydrats ? How do the polysaccharides differ from the protein macromolecules ?
4. What are the major classes of lipids and what types of functions are served by them ?
5. Give the structure of triglyceride.
6. Name two porphyrins and give their functions.
7. What is the general formula of amino acids ?
8. What are primary, secondary, tertiary, and quaternary levels of protein structure ? What types of chemical bondings are responsible for each of these structural levels ? Which of the structural levels play a fundamental direct role in protein functioning ?
9. Describe the molecular structure of the following proteins : collagens, haemoglobin and immunoglobulin.
10. Write an essay on immunity and immunoglobulins.
11. How would you define an enzyme ? Describe some main types of enzymes.
12. Give structural formula of ribose and deoxyribose sugars.
13. What are three components of a nucleotide ?
14. List the nitrogenous bases which occur in DNA and RNA.
15. Enumerate the differences between DNA and RNA.
16. Write short notes on the following :
    (i) Amino acids ; (ii) ATP ; (iii) Vitamins ; (iv) Hormones ;
    (v) Nucleic acids ; (vi) Brownian movement ; and (vii) Cofactor.

CHAPTER

5

Plasma Membrane and Cell Wall

Diatoms. A glassy outer shell and a selectively permeable plasma membrane help cells maintain relatively constant internal conditions.

A plasma membrane encloses every type of cell, both prokaryotic and eukaryotic cells. It physically separates the cytoplasm from the surrounding cellular environment. Plasma membrane is a ultrathin, elastic, living, dynamic and selective transport-barrier. It is a fluid-mosaic assembly of molecules of lipids (phospholipids and cholesterol), proteins and carbohydrates. Plasma membrane controls the entry of nutrientes and exit of waste products, and generates differences in ion concentration between the interior and exterior of the cell. It also acts as a sensor of external signals (for example, hormonal, immunological, etc.) and allows the cell to react or change in response to environmental signals. The cells of bacteria and plants have the plasma membrane between the cell wall and the cytoplasm. For cells without cell walls (*e.g.*, mycoplasma and animal cells), plasma membrane forms the cell surface.

All biological membranes including the plasma membrane and internal membranes of eukaryotic cells (*i.e.*, membranes bounding endoplasmic reticulum or ER, nucleus, mitochondria, chloroplast, Golgi apparatus, lysosomes, peroxisomes, etc.) are similar in structure (*i.e.*, fluid-mosaic) and selective permeability but differing in other functions.

The plasma membrane is also called **cytoplasmic membrane**, **cell membrane**, or **plasmalemma**. The term cell membrane was coined by **C. Nageli** and **C. Cramer** in 1855 and the term plasmalemma has been given by **J. Q. Plowe** in 1931.

ISOLATION AND ANALYSIS

The plasma membrane is so thin that it cannot be observed by the light microscope. Structure of the plasma membrane of various cells has been studied by their isolation from the living systems and also by their artificial synthesis by using their constituent molecules (*e.g.*, liposome, see Chapter 4). The pure and isolated membranes are then studied by biochemical and biophysical methods. The purity of isolated membranes is controlled by **electron microscopy**, **enzyme analysis** and the **study of surface antigens**. A variety of cells such as mammalian red blood cell (erythrocytes), medullated nerve fibres, Ehrlich mouse ascites tumor cells, liver cells, striated muscle, *Amoeba proteus*, sea urchin eggs and bacteria, have been used in studying the ultra-structure of the plasma membrane. The mammalian erythrocytes and the myelin sheath of the nerve fibre, however, have provided the bulk of information regarding the structure and properties of the plasma membrane. For such experiments, human red blood cells or erythrocytes have been selected by **E. Gorter** and **F. Grendel** (1925) for following advantages : these cells are easy to obtain and are known to be extremely simple. Since these cells contain no intracellular organelles or membrane, so the only membrane structure to be considered is almost entirely that of the cell surface. Lastly, the plasma membrane of erythrocytes is relatively tough and does not readily fragment (See **Lucy**, 1975)

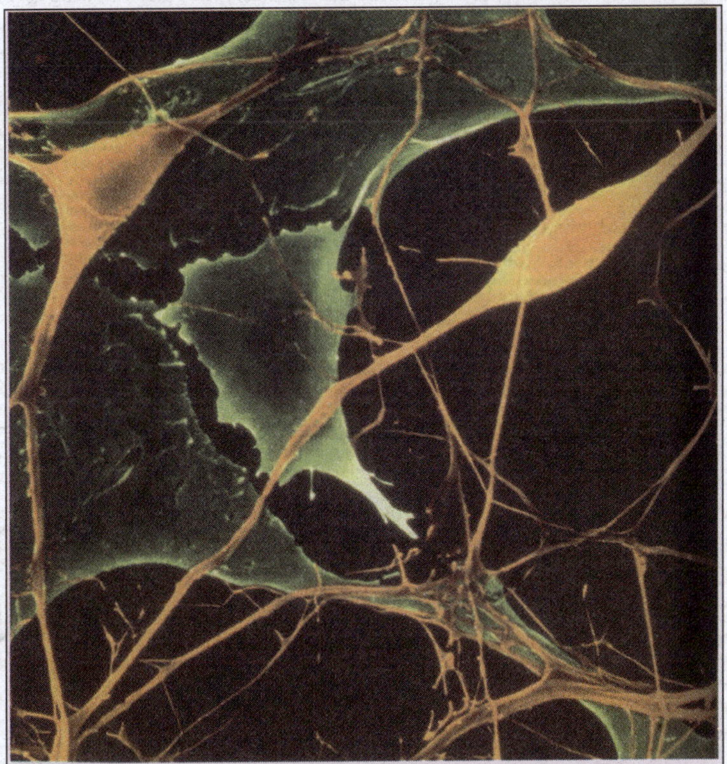

SEM of neurons. These have been used extensively in studying the ultra-structure of the plasma membrane.

Plasma membranes are more easily isolated from erythrocytes subjected to haemolysis. The cells are treated with hypotonic solutions (to be discussed elsewhere in the chapter) that due to endosmosis produce swelling and then loss of the heamoglobin content (*i.e.*, haemolysis). The resulting membrane is called a **red cell ghost**. If haemolysis is mild, permeability functions of the membrane can be restored by certain treatment, such a ghost is called **resealed ghost**. But if heamolysis is more drastic (*i.e.*, there is complete removal of the haemoglobin) and there is no chance of its resealing, the resulting membrane is called **white ghost**. While the resealed ghosts can be used for the study of physiological as well as biochemical properties, white ghosts can only be used for the study of biochemical properties.

The cell wall of yeast, *Saccharomyces cerevisiae*, can be enzymatically removed by the help of a snail gut enzyme, and the resultant protoplast serves as a source of plasma membrane in a manner similar to that of mammalian erythrocytes.

CHEMICAL COMPOSITION

Chemically, plasma membrane and other membranes of different organelles are found to contain proteins, lipids and carbohydrates, but in different ratios (Table 5-1). For example, in the plasma membrane of human red blood cells proteins represent 52 per cent, lipids 40 per cent and carbohydrates 8 per cent.

Table 5-1. Chemical composition of some purified membranes (in percentages) (Source : Darnell *et al.*, 1986).

Membrane	Protein	Lipid	Carbohydrate
1. Myelin (Nerve cell)	18	79	3
2. Plasma membrane :			
(i) Mouse liver	44	52	4
(ii) *Amoeba*	54	42	4
(iii) Human erythrocyte	52	40	8
3. Spinach chloroplast lamellae	70	30	0
4. Mitochondrial inner membrane	76	24	0

1. Lipids

Four major classes of lipids are commonly present in the plasma membrane and other membranes : **phospholipids** (most abundant), **sphingolipids**, **glycolipids** and **sterols** (*e.g.*, **cholesterol**) (For more details see Chapter 4). All of them are amphipathic molecules, possessing both hydrophilic and hydrophobic domains. The relative proportions of these lipids vary in different membranes. Phospholipids may be **acidic phospholipids** (20 per cent) such as **sphingomyelin** or **neutral phospholipids** (80 per cent) such as **phosphatidyl choline**, **phosphatidylserine**, etc. Many membranes contain cholesterol. Cholesterol is especially abundant in the plasma membrane of mammalian cells and absent from prokaryotic cells. **Cardiolipin** (diphosphatidyl glycerol) is restricted to the inner mitochondrial membrane (see **Darnell** *et al*., 1986).

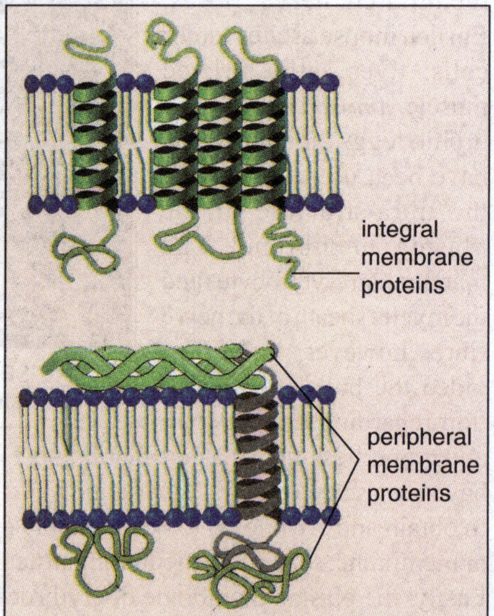

Classes of membrane proteins.

2. Proteins

The amount and types of proteins in the membranes are highly variable : in the myelin membranes which serve mainly to insulate nerve cell axons, less than 25 per cent of the membrane mass is protein, whereas, in the membranes involved in energy transduction (such as internal membranes of mitochondria and chloroplasts), approximately 75 per cent is protein. Plasma membrane contains about 50 per cent protein.

According to their position in the plasma membrane, the proteins fall into two main types : **integral** or **intrinsic proteins** and **peripheral** or **extrinsic proteins**, both of which may be either **ectoproteins**, lying or exposing to external or extracytoplasmic surface of the plasma membrane or **endoproteins**, lying or sticking out at the inner or cytoplasmic surface of the plasma membrane. The intrinsic proteins tend to associate firmly with the membrane, while the extrinsic proteins have a weaker association and are bound to lipids of membrane by electrostatic interaction. On the basis of their functions, proteins of plasma membrane can also be classified into three main types : structural

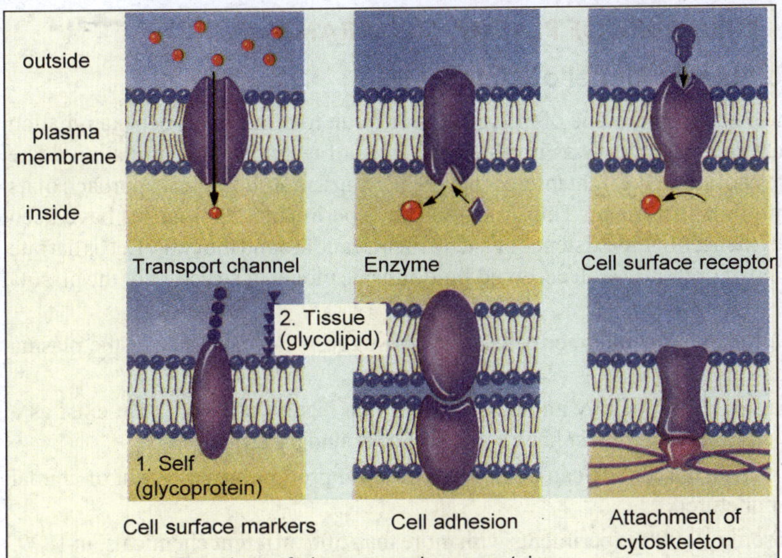

Functions of plasma membrane proteins.

proteins, enzymes and transport proteins (permeases or carriers). Some of them may act as **antigens**, **receptor molecules** (*e.g.*, insulin-binding sites of liver plasma membrane), **regulatory molecules** and so on. **Structural proteins** are extremely lipophilic and form the main bulk (*i.e.*, backbone) of the plasma membrane. **Enzymes** of plasma membrane are either **ectoenzymes** or **endoenzymes** and are of about 30 types (Table 5-2). **Transport proteins** transport specific substances across the plasma membrane and other cellular membranes.

3. Carbohydrates

Carbohydrates are present only in the plasma mambrane. They are present as short, unbranched or branched chains of sugars (**oligosaccharides**) attached either to exterior ectoproteins (forming **glycoproteins**) or to the polar ends of phospholipids at the external surface of the plasma membrane (forming **glycolipids**). No carbohydrate is located at the cytoplasmic or inner surface of the plasma membrane. All types of oligosaccharides of the plasma membrane are formed by various combinations of six principal sugars (all of which are glucose-derivatives) : **D-galactose**, **D-mannose**, **L-fucose**, **N-acetylneuraminic acid** (also called **sialic acid**), **N-acetyl-D-glucosamine** and **N-acetyl-D-galactosamine**.

Table 5-2. Some important enzymes present in the plasma membrane (Source : Sheeler and Bianchi, 1987).

1.	Acetyl phosphatase	11.	Cholesterol esterase
2.	Acetyl cholinesterase (Ectoenzyme of erythrocyte)	12.	Guanylase cyclase
3.	Acid phosphatase	13.	Monoglyceride lipase
4.	Adenosine triphosphatase	14.	NAD-ase (Ectoenzyme of erythrocyte)
5.	Mg^{2+} ATPase (Endoenzyme of erythrocyte)	15.	Protein kinase (Endoenzyme of erythrocyte)
6.	Na^+-K^+ ATPase (Ectoenzyme of erythrocyte)	16.	Phospholipase A
7.	Adenylate cyclase (Endoenzyme of erythrocyte)	17.	Lactase
8.	RNAase	18.	Maltase
9.	Alkaline phosphatase	19.	Sialidase
10.	Aminopeptidase	20.	UDP glycosidase

STRUCTURE OF PLASMA MEMBRANE

1. Evolution of Fluid Mosaic Model of Membrane

The existence of the plasma membrane of the cell was difficult to prove by direct examination before 1930's (when electron microscopy was invented) because of technological limitations. The membrane is beyond the resolution of the light microscope, rendering a morphological approach of its study quite unfeasible with this instrument. Thus, most of the experimental approaches have been provided by only indirect evidences of the existence of such a membrane around the cells. Let us narrate in brief the saga of evolution of presently well accepted fluid-mosaic model of structure of the plasma membrane :

1. The plasmolysis of plant cells in hypertonic solutions suggests the existence of the plasma membrane in the plants.

2. The very fact that a cell, especially an animal cell which has no cell wall, can exist as a physically defined entity suggests that it must have some sort of boundary around it.

3. The presence of plasma membrane can be inferred because protoplasm leaks out of animal cells when cell surface is punctured.

4. After performing some 10,000 experiments with more than 500 different chemicals, in 1899. **Overton** concluded that the peculiar osmotic properties of living protoplasts are due to a **selective**

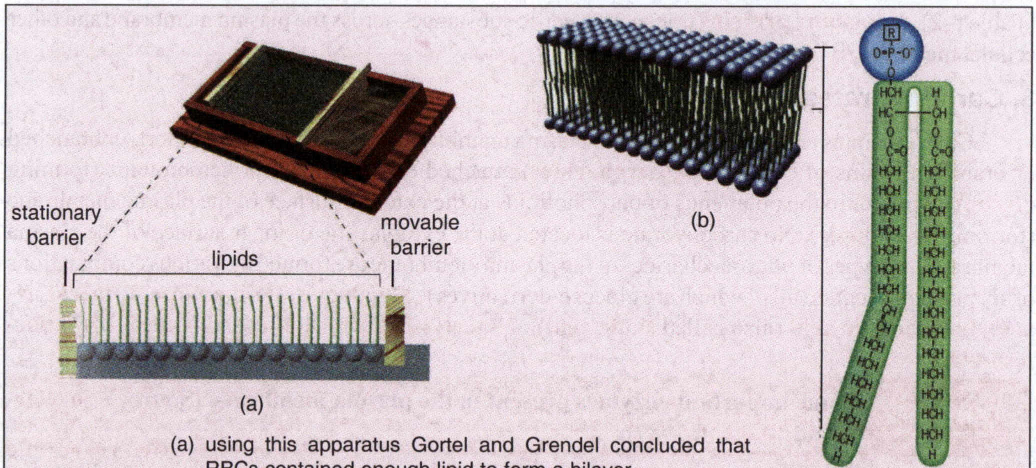

(a) using this apparatus Gortel and Grendel concluded that RBCs contained enough lipid to form a bilayer.
(b) bimolecular layer of phospholipids.

solubility mechanism. Hydrophobic compounds entered cells more rapidly than hydrophilic ones. **Overton** believed this was because of an outer lipoid layer in which hydrophobic compounds were more soluble. He correctly speculated that this layer might contain cholesterol, lecithin and fatty oils.

5. **Hober** (1910) and **Fricke** (1925) found that the intact cell had low electrical conductivity, indicating the presence of a lipid layer around it.

6. If a lipid containing **hydrophilic groups** (such as the carboxyl groups of fatty acids or the phosphate groups of phospholipids) is dissolved in a highly volatile solvent (*e.g.*, benzene) and several drops of it are then carefully applied to the surface of the water, the lipid spreads out to form a thin, one-molecule-thick or **monomolecular film**. In this film, it is found that the hydrophilic parts of each molecule project into the water surface and the hydrophobic parts are directed up, away from the water.

7. In 1917, **Langmuir** (Nobel Laureate of 1932 in chemistry) fabricated a **trough** or **film balance** (Fig. 5.1) for measuring the specific minimum surface area occupied by a monomolecular film

of lipid and the force necessary to compress all the lipid molecules into this area. Langmuir trough consists of a shallow trough filled with water on which lipid substance can be spread to make a monomolecular film. A barrier can be pushed across the trough to compress the film.

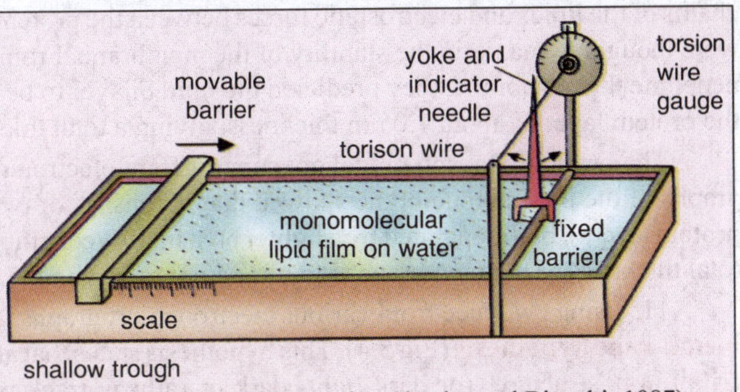

Fig. 5.1. Langmuir trough (after Sheeler and Bianchi, 1987).

8. In 1925, **Gorter** and **Grendel** extracted the lipids from erythrocyte ghosts of a variety of mammals (such as dogs, sheep, rabbits, guinea pigs, goats and humans) and spread them out on monolayers in the Langmuir trough. These investigators discovered that the area covered by the lipid monomolecular layer film was twice than what was needed to cover the surface of the cells from which the lipid was extracted. Consequently, they safely concluded that *erythrocytes were covered by a layer of lipids two molecules thick* (**lipid bilayer** or **bimolecular lipid layer**) oriented with polar groups toward the inside and outside of the cell.

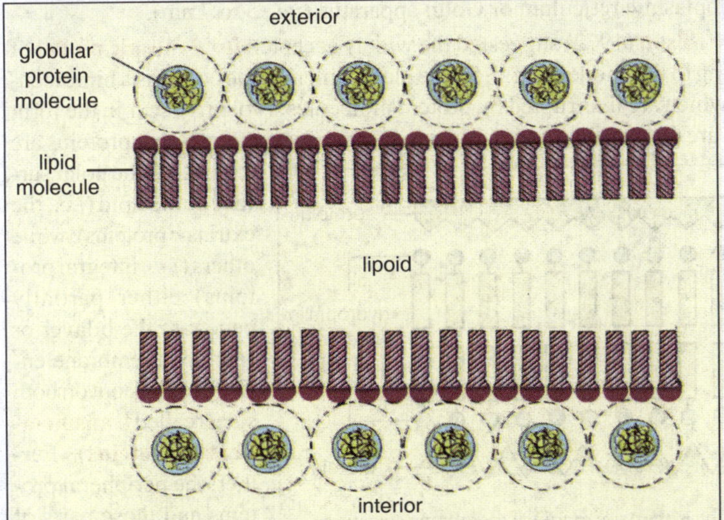

Fig. 5.2. The original Danielli-Davson model (1935) of membrane structure. The bimolecular layer of lipid molecules is of undefined thickness and is covered on each side by a continuous layer of globular proteins (after De Witt, 1977).

9. By studying the surface tension of cells (**Harvey** and **Cole**, 1931, **Danielle** and **Harvey**, 1935) suggested the presence of proteins in the plasma membrane, in addition to the lipids.

10. In 1935, **Danielli** and **Davson**, proposed a model, called **sandwich model**, for membrane structure in which a lipid bilayer was coated on its either side with hydrated proteins (globular proteins). Mutual attraction between the hydrocarbon

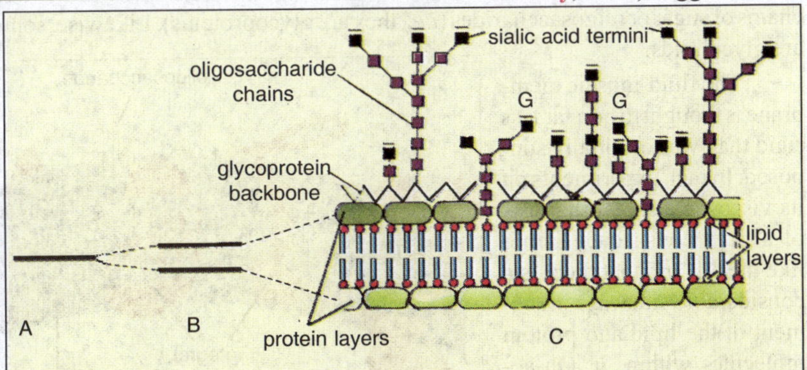

Fig. 5.3. The structure of plasma membrane as observed A — at low magnification of electron microscope; B — at high magnification of electron microscope. C — trilaminar model of plasma membrane showing possible arrangement of the lipid, protein and oligosaccharide molecules in the plasma membrane (after De Robertis *et al.*, 1970).

chains of the lipids and electrostatic forces between the protein and the "head" of the lipid molecules, were thought to maintain the stability of the membrane. From the speed at which various molecules penetrate the membrane, they predicted the lipid bilayer to be about 6.0 nm in thickness, and each of the protein layer of about 1.0 nm thickness, giving a total thickness of about 8.0 nm.

The Danielli-Davson model got support from electron microscopy (Fig. 5.3). Electron micrographs of the plasma membrane showed that it consists of two dark layers (electron dense granular protein layers), both separated by a lighter area in between (the central clear area of lipid bilayer). The total thickness of the membranes too turned out to be about 7.5 nm.

11. Using evidence from various electron micrographs, **Robertson** in 1960, proposed the **unit memb-rane hypothesis** (Fig 5.4). This hypothesis states that all cellular membranes have an identical **trilaminar** structure (or dark-light- dark or railway track pattern, see **Thorpe**, 1984). However, thickness of the unit membrane has been found to be greater in plasma membrane (10 nm) than in the intracellular membranes of endoplasmic reticulum or Golgi apparatus (*i.e.*, 5 to 7 nm).

12. **S.J.Singer** and **G.L.Nicolson** (1972) suggested the widely accepted **fluid mosaic model** of biological membranes. According to this model (Fig. 5.5), the plasma membrane contains a bimolecular lipid layer, both surfaces of which are interrupted by protein molecules. Proteins occur in the form of globular molecules and they are dotted about here and there in a mosaic pattern. Some proteins are attached at the polar surface of the lipid (*i.e.*, the extrinsic proteins); while others (*i.e.*, integral proteins) either partially penetrate the bilayer or span the membrane entirely to stick out on both sides (called **transmembrane proteins**). Further, the peripheral proteins and those parts of the integral proteins that stick on the outer surface (*i.e.*, ectoproteins) frequently contain chains of sugar or oligosaccharides (*i.e.*, they are glycoproteins). Likewise, some lipids of outer surface are glycolipids.

Fig. 5.4. Schematic diagram of the Robertson model of membrane structure. The lipid layer is defined as bimolecular, and the protein is extended but different on the two faces of the membrane (after Thorpe, 1984).

The fluid-mosaic membrane is thought to be a far less rigid than was originally supposed. In fact, experiments on its viscosity suggest that it is of a fluid consistency rather like the oil, and that there is a considerable sideways movement of the lipid and protein molecules within it. On account of its fluidity and the mosaic arrangement of protein molecules, this model of membrane structure is known

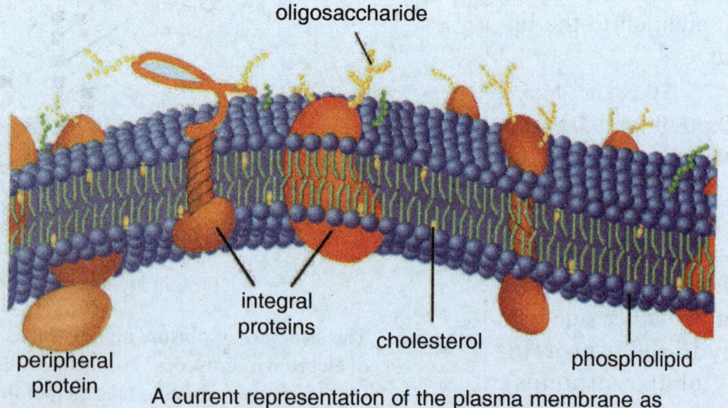

A current representation of the plasma membrane as proposed by Singer and Nicolson.

as the "fluid mosaic model" (*i.e.,* it describes both properties and organization of the membrane). The fluid mosaic model is found to be applied to all biological membranes in general, and it is seen as a dynamic, ever-changing structure. The proteins are present not to give it strength, but to serve as enzymes catalysing chemical reactions within the membrane and as pumps moving things across it.

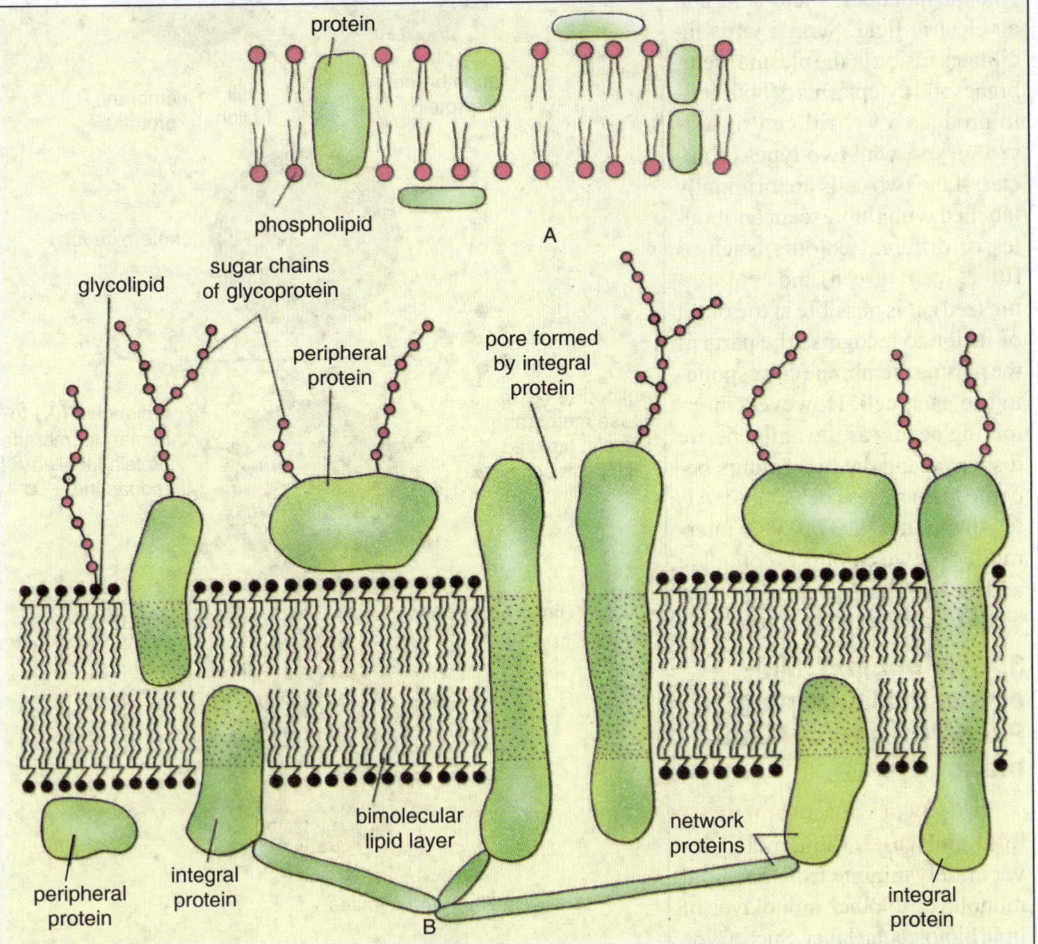

Fig. 5.5. Fluid mosaic model of the plasma membrane. A — Simplistic view; B— Complex view (after Berns, 1983; Sheeler and Bianchi, 1987).

2. Experimental Evidence in Support of Fluid Mosaic Model of Plasma Membrane

There is a good deal of evidence to support the fluid mosaic model of the plasma membrane:

A. Evidence in support of mosaic arrangement of proteins. Freeze-fracture electron microscopy of the plasma membrane by **Branton** (1968) revealed the presence of bumps and depressions (7 to 8 nm in diameter) which are randomly distributed. These were later shown to be transmembrane integral protein particles (Fig. 5.7). (For details of freeze-fracture technique, see Chapter 2).

B. Evidence in support of fluid property of lipid bilayer. Mobility of membrane proteins due to fluid property of lipid bilayer was demonstrated by a classical experiment of **D. Frye** and **M. Edidin** (1970). They fused two different types of cultured cells having different surface antigens (proteins). The **cell fusion** is achieved by the use of some fusogen such as an inactivated parainfluenza virus, called Sendai virus (named after a city of Japan). A **fusogen** is a membrane fusion promoting factor such as Sendai virus, lysophosphatides, oleic acid and an electric field. Sendai virus facilitates fusion of the plasma membranes and cytoplasms of both cells to produce a **hybrid cell** or **heterokaryon** with two types of nuclei. If the two cells are originally labelled with fluorescent antibodies of different colours, such as **fluorescein** (green) and **rhodamine** (red), it is possible at the onset of fusion to recognise the parts of the plasma membrane corresponding to each cell. However, intermixing occurs as the antigens are dispersed and the two colours become less and less detectable. After 40 minutes (at 37° C) the intermixing of two colours is complete and the two antigens can no longer be distinguished (Fig. 5.6).

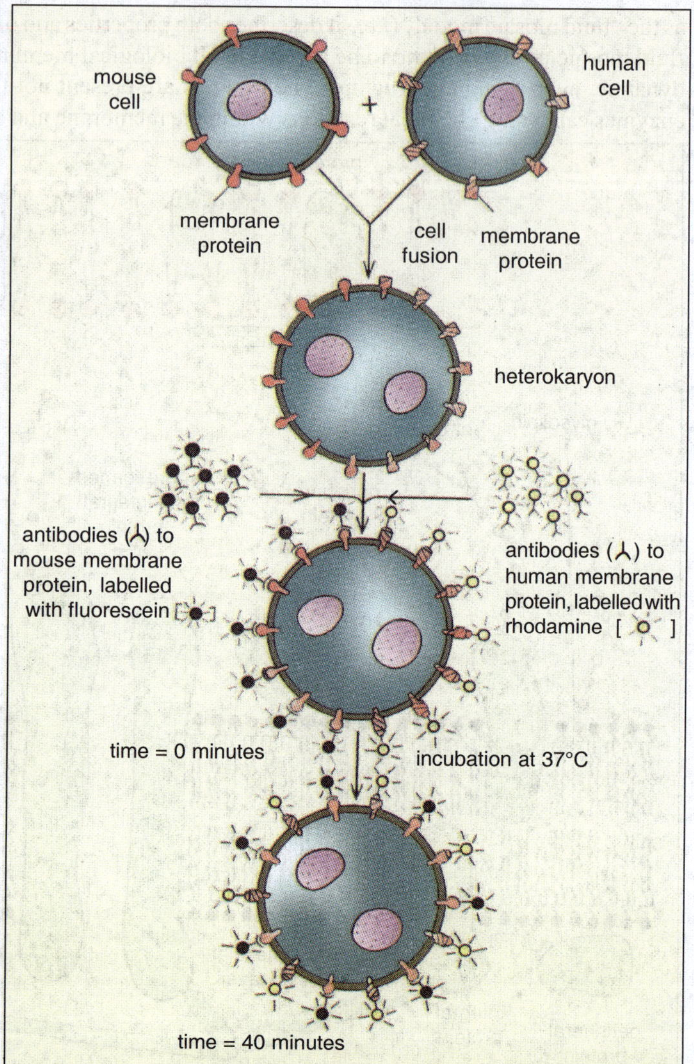

Fig. 5.6. Frye and Edidin's experiment demonstrating the mixing of plasma membrane proteins on mouse-human hybrid cells. In the plasma membrane of newly formed heterokaryon, the mouse and human proteins are intially confined to their own halves, but they intermixed with time (after Alberts *et al.*, 1989).

3. Role of Lipid Molecules in Maintaining Fluid Property of Membrane

(i) Types of movements of lipid molecules. Lipid molecules very rarely migrate from one lipid monolayer to other monolayer of lipid bimolecular layer. Such a type of movement is called **flip-flop** or **transbilayer movement** and occurs once a month for any individual lipid molecule. However, in membranes where lipids are actively synthesized, such as smooth ER, there is a rapid flip-flop of specific lipid molecules across the bilayer and there are present certain membrane-bound enzymes, called **phospholipid translocators** (*e.g.*, flippase) to catalyze this activity (**Bishop** and **Bell**, 1988). On the other hand, lipid molecules readily exchange places with their neighbours within a monolayer (~ 10^7 times a second). This results in their rapid **lateral diffusion**. Individual lipid molecules **rotate** very rapidly about their long axes and their hydrocarbon chains are flexible, the greatest degree of **flexion** occurring near the centre of the bilayer and the smallest adjacent to the polar head groups (Fig. 5.8).

(ii) Role of unsaturated fats in increasing membrane fluidity. A synthetic bilayer made from a single type of phospholipid changes from a liquid state to a rigid crystalline or gel (viscous) state at a characteristic freezing point. This change of state is called a **phase transition** and the temperature at which it occurs becomes lower if the hydrocarbon chains are short or have double bonds. Double bonds in unsaturated hydrocarbon chains tend to increase the fluidity of a phospholipid bilayer by making it more difficult to pack the chains together. Thus, to maintain fluidity of the membrane, cells of organisms living at low temperatures have high proportions of unsaturated fatty acids in their membranes, than do cells at higher temperatures.

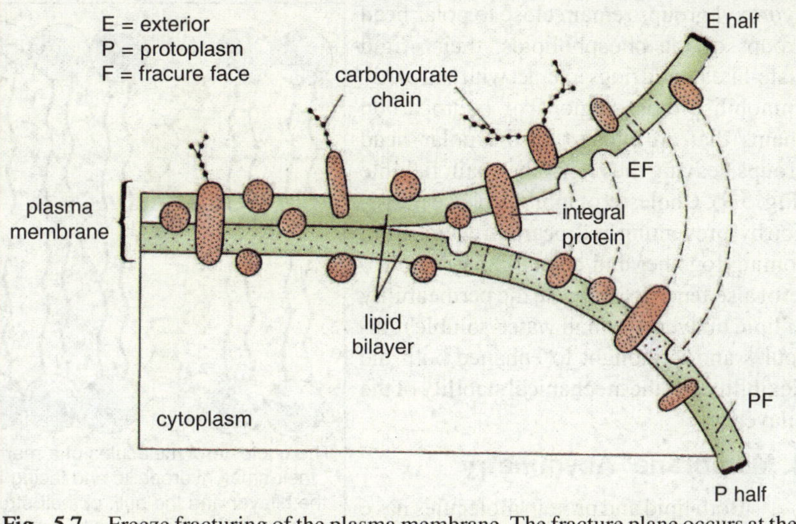

Fig. 5.7. Freeze fracturing of the plasma membrane. The fracture plane occurs at the centre of the lipid bilayer and passes over (or under) the integral membrane proteins. E=exterior, P=protoplasmic (cytosolic) side, F=fracture face (after Sheeler and Bianchi, 1987).

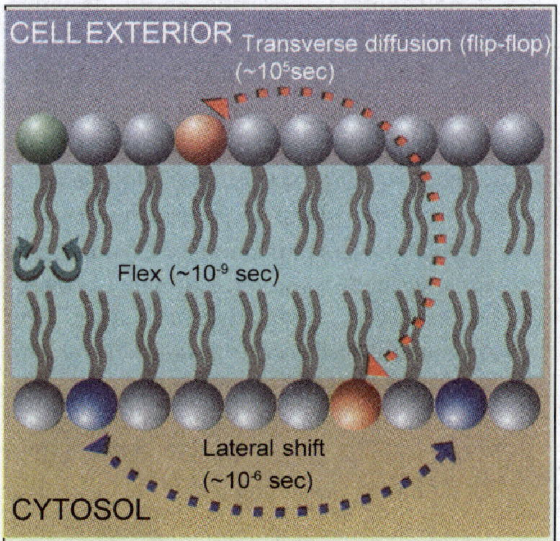

The possible movements of phospholipids in a membrane.

In fact, certain membrane **transport processes** and **enzyme activities** are found to cease when the lipid bilayer's viscosity increases beyond a threshold level (**Kimelberg**, 1977). In contrast, if lipid bilayer's fluidity is increased, the membrane's receptors for the hormone are withdrawn from the cell surface, thereby hampering hormone action (see **Sheeler** and **Bianchi**, 1987).

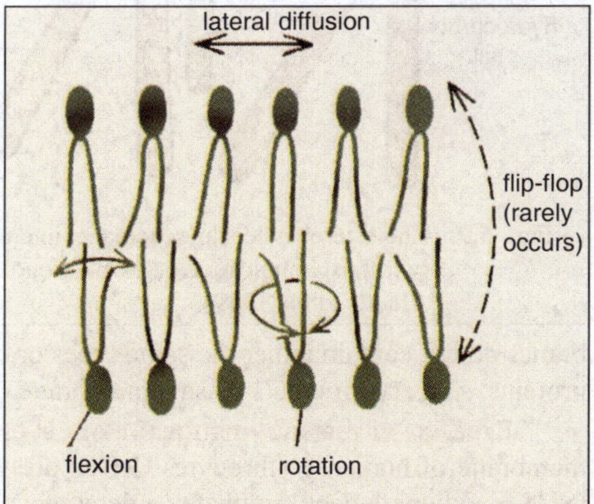

Fig. 5.8. The types of movement possible for phospholipid molecules in a lipid bilayer (after Alberts *et al*, 1989).

(iii) Role of cholesterol in maintaining fluidity of membrane. Eukaryotic plasma membranes are found to contain a large amount of cholesterol; up to one molecule for every phospholipid molecule. Cholesterol molecules orient themselves in the lipid bilayer in such a way that their hydroxyl groups remain close to polar head groups of the phospholipids, their rigid plate-like steroid rings interact with and partly immobilise those regions of hydrocarbon chains that are closest to the polar head groups, leaving the rest of the chain flexible (Fig. 5.9). Cholesterol inhibits phase transition by preventing hydrocarbon chains from coming together and crystallizing. Cholesterol also tends to decrease the permeability of lipid bilayers to small water-soluble molecules and is thought to enhance both the flexibility and the mechanical stability of the bilayer.

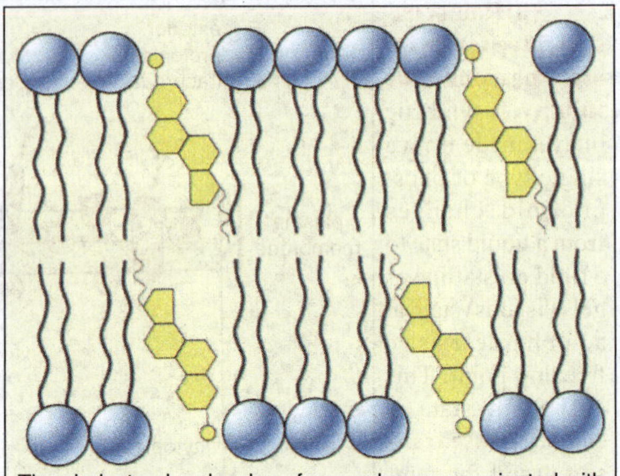

The cholesterol molecules of a membrane are oriented with their small hydrophilic end facing the external surface of the bilayer and the bulk of their structure packed in among the fatty acid tails of the phospholipids.

4. Membrane Asymmetry

Both lipid and protein molecules have irregular distribution in both monolayers of the lipid bilayer, this is called **membrane asymmetry.**

A. Phospholipid asymmetry in plasma membrane. The lipid composition and state of fluidity of two halves of the lipid bilayer are found to be strikingly different. For example, in human erythrocyte's plasma membrane, outer half contains those phospholipids which have more saturated fatty acid chains, and inner half contains those phospholipids which contain terminal amino groups and less saturated fatty acid chains. As a result, inner monolayer is more fluid than the outer lipid monolayer. Such a phospholipid asymmetry is generated in smooth ER. The asymmetry of glycolipids such as galactocerebroside, ganglioside, etc., in myelin sheath of nerves (*i.e.*, they are found only in the outer half of lipid bilayer) is found to be originated in lumen of Golgi apparatus. The specific role of lipid asymmetry of the membrane is still not clear.

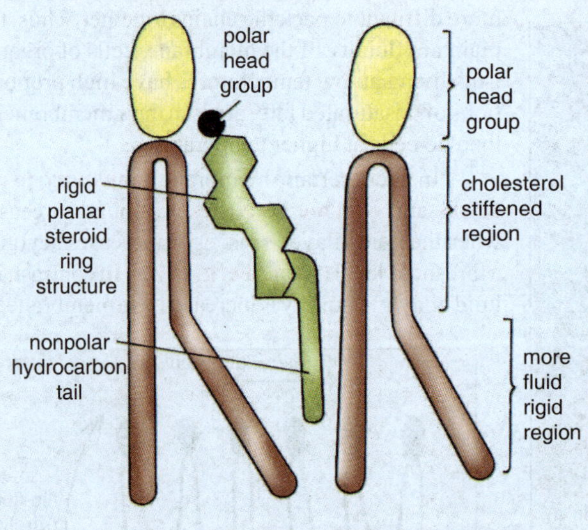

Fig. 5.9. Cholesterol molecule (schematic) interacting with two phospholipid molecules in a monolayer (after Alberts *et al.*, 1989).

B. Protein asymmetry in plasma membrane. The outer and inner sides of the plasma membrane and other membranes do not contain either the same types or equal amounts of the various peripheral and integral proteins, *e.g.*, erythrocyte's plasma membrane.

Proteins of plasma membrane of erythrocytes. When the extracted proteins of the plasma membrane of human erythrocytes (RBC) are studied by SDS polyacrylamide-gel electrophoresis (SDS = sodium dodecyl sulphate ; a detergent), approximately 15 major protein bands are detected, varying in molecular weight from 15,000 to 25,000. Most of these proteins are found to be peripheral proteins of cytosolic face of the plasma membrane. Important properties of some of these proteins are the following :

(i) Spectrin and other cytoskeleton proteins. Spectrin is the principal component of the protein meshwork (cytoskeleton) that underlies the erythrocyte's plasma membrane (Fig. 5.10). It, thus, maintains the structural integrity and biconcave shape of this membrane (**Branton** *et al.*, 1981). Spectrin is long, thin, flexible rod about 100 nm in length. It constitutes about 25 per cent of the membrane associated protein mass (about 2.5 x 10^5 copies per cell). Spectrin is a **heterodimer** and consists of two non-identical, antiparallel, loosely intertwind, flexible polypeptide chains, *i.e.*, **α–spectrin** (~ 240,000 daltons M.W.) and **β-spectrin** (~220,000 daltons M.W.), both being attached non-covalently to each other at multiple points including their ends (*i.e.*, phosphorylated 'head' and 'tail').

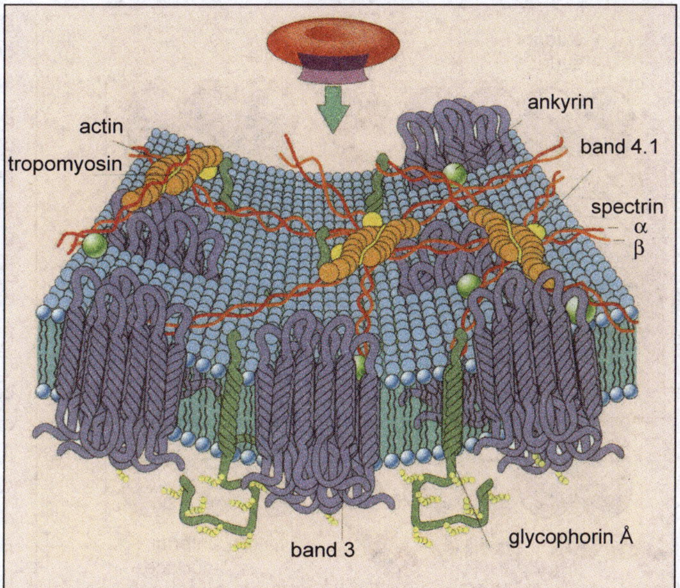

The plasma membrane of the human erythrocyte.

The spectrin heterodimers self-associate head-to-head to form 200 nm long **tetramers**. The tail ends of five or six spectrin tetramers are linked together by binding to short **actin filaments** (also called **band 5 proteins**; with 43,000 dalton M.W.) and each with 15 actin monomers and to another protein, called **band 4.1 protein** (82,000 dalton M.W.). These three proteins form the "**junctional complex**" of deformable, net-like meshwork of the cytoskeleton. Further, the binding of spectrin cytoskeleton to the cytosolic face of the erythrocyte's plasma membrane depends on a large intracellular attachment protein, called **ankyrin** (or **band 2.1 protein**; 210,000 dalton M.W.). Ankyrin tends to bind to both β-spectrin and to the cytoplasmic domain of a transmembrane protein, called **band 3 protein** (**Shen** *et al.*, 1986).

(ii) Glycophorin. It is a small transmembrane glycoprotein (single-pass membrane protein) having molecular weight of 55,000 daltons and 131 amino acid residues. This protein bears about 100 sugars on 16 separate oligosaccharide side chains (90 per cent of which is sialic acid). Despite there being more than 6 × 10^5

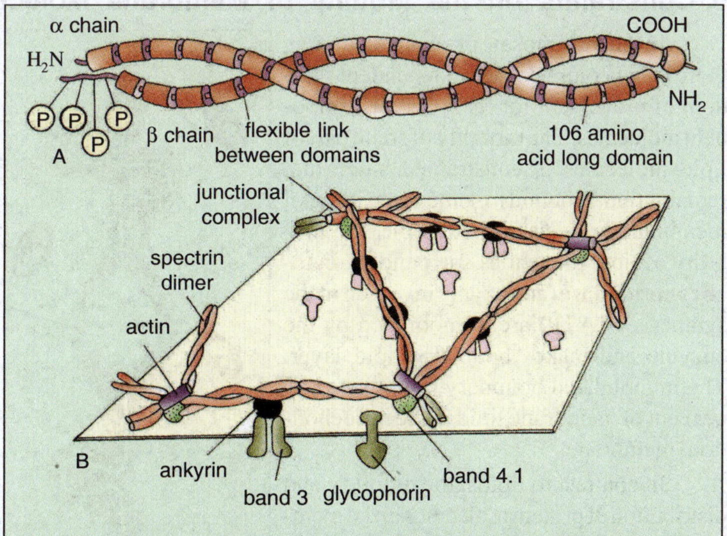

Fig. 5.10. A—One spectrin molecule from human red blood cell; B—Schematic drawing of the spectrin based cytoskeleton on the cytoplasmic side of the human red blood cell membrane.

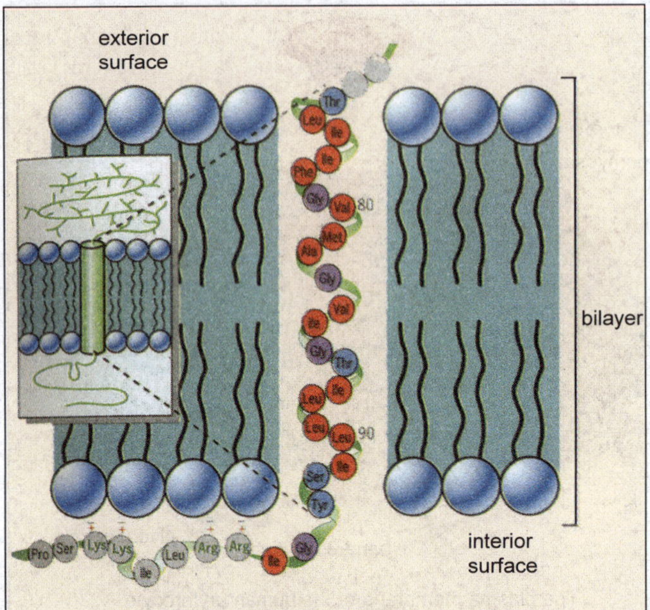

Glycophorin A, an integral protein with a single transmembrane domain.

glycophorin molecules per cell (*i.e.*, erythrocyte), their exact function is still not known. However, glycophorins are found to contain certain antigenic determinants (carbohydrates) for the **ABO blood groups** and **MN blood groups**. Further, sialic acid confers a high negative charge to the cell surface of erythrocyte. This sugar may be important in the life cycle of the erythrocytes as it has been shown that cells lose sialic acid as they age in the circulatory system. Correlated with this is the observation that *loss of sialic acid is a signal for removal and destruction of an erythrocyte by the spleen and liver.* In this way the life span of red blood cells may be regulated (see **King**, 1986).

(iii) Band 3 protein. Like the glycophorin, **band 3 protein** (93,000 daltons M.W.) is a transmembrane protein, but it is a **multipass membrane protein**, *i.e.*, its highly folded polypeptide chain (about 930 amino acid long) extends across the lipid bilayer at least 10 times. Each human erythrocyte contains about 10^6 and band 3 proteins, each of which forms either a dimer or tetramer in the membrane. Band 3 protein acts as the **anion exchange channels** in the membrane. As the erythrocytes pass through the lungs, they exchange bicarbonate (HCO^-) for chloride (Cl^-) through these hydrophilic channels during the process of CO_2 release (**chloride shift**).

5. Constraints on the Motility of Membrane Molecules

In the fluid mosaic plasma membrane, there is not complete and independent freedom of movement for its different component molecules. The mobility of some part of lipid molecules is constrained since that remains tightly bound to some of the integral membrane proteins. For example, the mobility of lipid molecules surrounding **cytochrome oxidase** (an enzyme involved in the synthesis of ATP) are immobilized by the enzyme and makes boundary lipid layer. The immobilized boundary lipid makes 30 per cent of membrane lipid in the mitochondrial membrane.

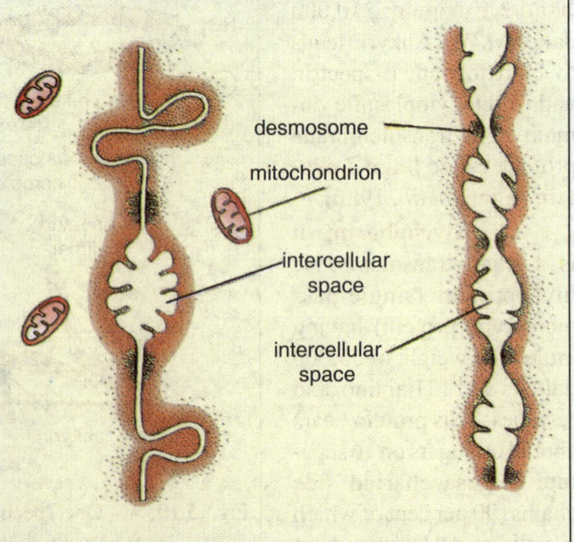

Fig. 5.11. Inter-cellular space.

In contrast to lipids, the mobility and distribution of protein molecules in the membrane is controlled by various ways : (1) Certain proteins of membrane are constrained by protein-protein interactions to form specialized ordered regions, representing 2 to 20 per cent of the membrane of a system, *e.g.*, gap junctions, synapsis of neurons and plaques of halobacteria. (2) Certain peripheral proteins (endoproteins) may form a bridge-like lattice work between integral proteins and restrict their lateral mobility, *e.g.*, spectrin-ankyrin-actin cytoskeletal meshwork provides a rigidity to the membrane of human erythrocytes and does not permit the clustering or **capping** of integral proteins when the appropriate antibodies or lectins are added. (3) In nucleated eukaryotic cells, the mobility of the peripheral endoproteins and integral proteins is restrained by their attachment to the ectoplasmic cytoskeleton. The cytoskeleton is extensive, including **myosin** filaments, **actin** filaments and **microtubules** (Fig. 5.10). Rearrangement of cytoskeletal components just below the cell surface manifests in the distribution of integral membrane proteins and also in the cellular motions, endocytosis and exocytosis.

The inter-cellular space. In the tissues of multicellular animals, the plasma membranes of two adjacent cells usually remain separated by a space of 10 to 150 A° wide. This inter-cellular space is uniform and contains a material of low electron density which can be considered as a cementing substance. This substance is found to be a mucopolysaccharide (Fig. 5.11).

ORIGIN OF PLASMA MEMBRANE

There is hardly any cell structure more important to the immediate health of the cell than the plasma membrane. If it is weakened or injured, the cell loses its ability to maintain gradients, to carry out the selective transport of nutrients, and to contain the pool of enzymes and organelles essential for the homeostasis. In consequence, new membranes may be added to existing membranes without altering the functions as a barrier and selective transporter. Also for maintaining the characteristic membrane asymmetry, the membrane must be assembled with precisely the correct moleular topography.

Thus, all cellular membranes grow from pre-existing membranes which act as **templates** for the addition of new precursors. All cells divide, daughter cells receive a full complement of membrane systems which undergo growth until the next division, to be passed on to subsequent progeny. Meanwhile the molecules within the membrane undergo continuous replacement.

The protein molecules of the plasma membrane are synthesized on both attached and free ribosomes. Proteins synthesized by free ribosomes may be inserted into the plasma membrane following their completion and release from the ribosomes. Proteins of plasma membrane synthesized on attached ribosomes of rough ER are **inserted** first into the membrane of RER and then **transferred** to the Golgi apparatus, **processed** there (*e.g.*, glycosylation) and ultimately are dispatched to the plasma membrane via the secretory vesicles. Likewise, the synthesis of phospholipid molecules of the plasma membrane takes place by the smooth ER (SER). Like the proteins, newly synthesized lipids are inserted into SER membranes, then they are passed to Golgi apparatus for the processing and ultimately are dispatched to the plasma membrane via small secretory vesicles. The cytosol also contains a number of **phospholipid transport proteins** that function to transfer phospholipid molecules from one cellular membrane to another (*e.g.*, from ER membranes to plasma membranes) (see **Sheeler** and **Bianchi**, 1987).

In fact, the process of glycosylation (or glycosidation, *i.e.*, addition of oligosaccharides containing the sugars such as galactose, fucose and/ or sialic acid, to the molecules of proteins and phospholipids of the plasma membrane) is completed at the level of Golgi apparatus. However, some sugars are added to the proteins in the lumen of RER.

FUNCTIONS OF PLASMA MEMBRANE

The plasma membrane acts as a thin barrier which separates the intra-cellular fluid or the cytoplasm from the extra-cellular fluid in which the cell lives. In case of unicellular organisms (Protophyta and Protozoa) the extra-cellular fluid may be fresh or marine water, while in multicellular organisms the extra-cellular fluid may be blood, lymph or interstitial fluid. Though the plasma membrane is a limiting barrier around the cell but it performs various important physiological functions which are as follows :

1. Permeability. The plasma membrane is a thin, elastic membrane around the cell which usually allows the movement of small ions and molecules of various substances through it. This nature of plasma membrane is termed as permeability. According to permeability following types of the plasma membranes have been recognised :

(i) Impermeable plasma membranes. The plasma membrane of the unfertilized eggs of certain fishes allows nothing to pass through it except the gases. Such plasma membranes can be termed as impermeable plasma membranes.

(ii) Semi-permeable plasma membranes. The membranes which allow only water but no solute particle to pass through them are known as semi-permeable membranes. Such membranes have not so far been recognised in animal cells.

(iii) Selective permeable plasma membranes. The plasma membrane and other intra-cellular membrane are very selective in nature. Such membranes allow only certain selected ions and small molecules to pass through them.

(iv) Dialysing plasma membranes. The plasma membranes of certain cells have certain extraneous coats around them. The basement membranes of endothelial cells are the best examples of extraneous coats. This type of plasma membrane having extraneous coats around it, acts as a dialyzer. In these membranes the water molecules and crystalloids are forced through them by the hydrostatic pressure forces.

Mode of Transport Across Plasma Membrane

The plasma membrane acts as a semipermeable barrier between the cell and the extracellular environment. This permeability must be highly **selective** if it is to ensure that essential molecules such as glucose, amino acids and lipids can readily enter the cell, that these molecules and metabolic intermediates remain in the cell, and that waste compounds leave the cell. In short, the **selective permeability** of the plasma membrane allows the cell to maintain a constant internal environment (**homeostasis**). In consequence, in all types of cells there exists a difference in ionic concentration with the extracellular medium (Table 5-3). Similarly, the organelles within the cell often have a different internal environment from that of the surrounding cytosol and organelle membranes maintain this difference. For example, in lysosomes the concentration of protons (H^+) is 100 to 1000 times that of

Table 5-3. Comparison of ion concentration inside and outside of a typical mammalian cell (Source : Maclean and Hall, 1987).

Component	Intracellular concentration (mM)	Extracellular concentration (mM)
Cations :		
Na^+	5-15	145
K^+	140	5
Mg^{2+}	30	1–2
Ca^{2+}	1-2	2.5–5
H^+	4×10^{-5} (pH 7.4)	4×10^{-5} (pH 7.4)
Anions* :		
Cl^-	4	110

* Since the cell must contain equal positive and negative charge (*i.e.*, be electrically neutral), the large deficit in intracellular anions reflects the fact that most cellular constituents are negatively charged, *e.g.*, HCO_3^-, PO_4^{3-}, proteins, nucleic acids, metabolites carrying phosphate and carboxyl groups, etc.

the cytosol. This gradient is maintained solely by the lysosomal membrane. Transport across the membrane may be passive or active. It may occur via the phospholipid bilayer or by the help of specific integral membrane proteins, called **permeases** or **transport proteins.**

A. Passive transport. It is a type of **diffusion** in which an ion or molecule crossing a membrane moves down its electrochemical or concentration gradient. *No metabolic energy is consumed in passive transport.* Passive transport is of following three types :

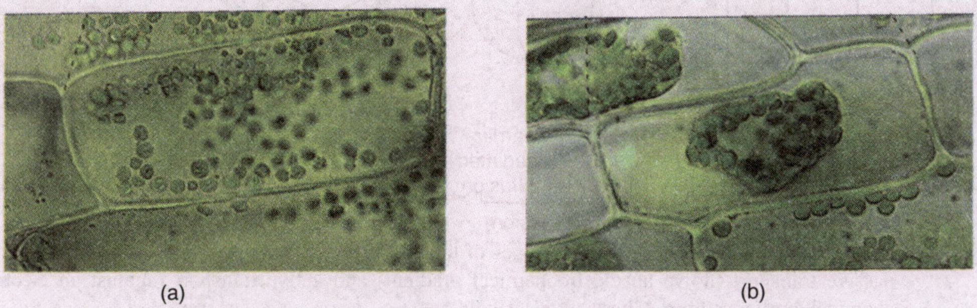

The effects of osmosis on a plant cell. (a) hypotonic : normal turgor pressure; (b) hypertonic : no turgor pressure.

1. Osmosis. The plasma membrane is permeable to water molecules. The to and fro movement of water molecules through the plasma membrane occurs due to the differences in the concentration of the solute on its either sides. The process by which the water molecules pass through a membrane from a region of higher water concentration to the region of lower water concentration is known as **osmosis** (Gr., *osmos*=pushing). The process in which the water molecules enter into the cell is known as **endosmosis,** while the reverse process which involves the exit of the water molecules from the cell is known as **exosmosis**. In plant cells due to excessive exosmosis the cytoplasm along with the plasma membrane shrinks away from the cell wall. This process is known as **plasmolysis** (Gr., *plasma*=molded, *lysis*=loosing) (Fig. 5.13).

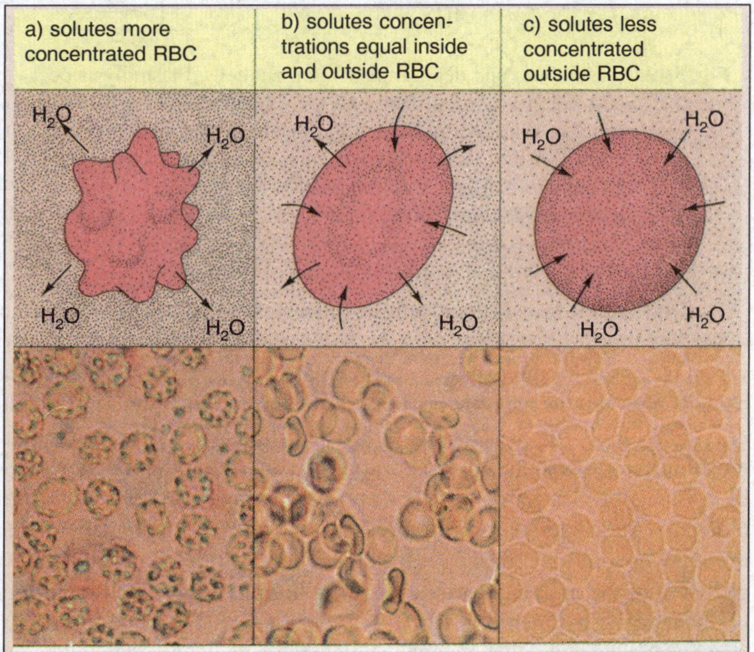

Osmosis and the RBC.

A cell contains variety of solutes in it, for instance, the mammalian erythrocytes contain the ions of potassium (K^+), calcium (Ca^+), phosphate (PO_4^-), dissolved haemoglobin and many other substances. If the erythrocyte is placed in a 0.9% solution of sodium chloride (NaCl), then it neither shrinks nor swells. In such case, because the intra-cellular and extra-celluar fluids contain same concentration and no osmosis takes place. This

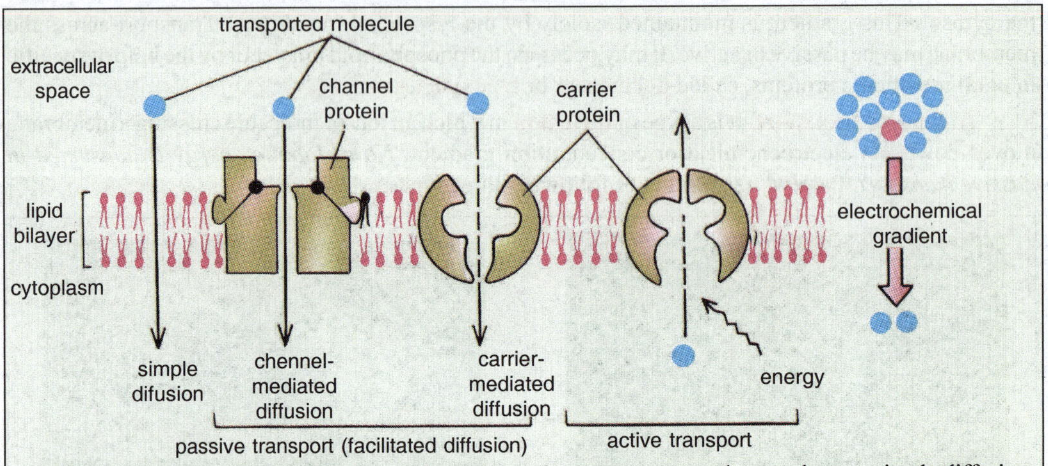

Fig. 5.12. Schematic diagram showing various types of transports across the membrane : simple diffusion, passive transport (down an electrochemical gradient) and active transport (against an electrochemical gradient) (after Alberts, *et al.*, 1989).

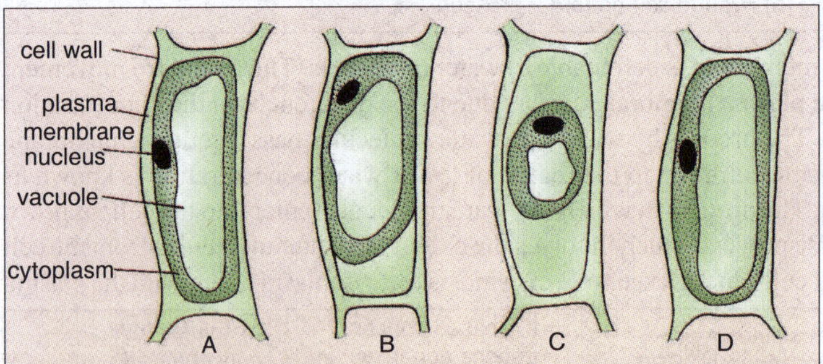

Fig. 5.13. Plasmolysis and deplasmolysis in plant cells. Plasmolysis occurs when a normal plant cell (A) is placed in a hypertonic solution. Water leaves the cell and the plasma membrane shrinks away from the cell wall (B,C). If solutes can penetrate the plasma membrane, the cell will eventually regain water—a process termed **deplasmolysis** (D) (after De Witt, 1977).

type of extra-cellular solution or fluid is known as **isotonic solution** or **fluid.** If the concentration of NaCl solution is increased above 0.9% then the erythrocytes are shrinked due to excessive exosmosis. The solutions which have higher concentrations of solutes than the intracellular fluids are known as **hypertonic solutions.** Further, if the concentration of NaCl solution decreases below 0.9% the erythrocytes will swell up due to endosmosis. The extra-cellular solutions having less concentration of the solutes than the cytoplasm are known as **hypotonic solutions.**

Due to endosmosis or exosmosis the water molecules come in or go out of the cell. The amount of the water inside the cell causes a pressure known as **hydrostatic pressure.** The hydrostatic pressure which is caused by the osmosis is known as **osmotic pressure.** The plasma membrane maintains a balance between the osmotic pressure of the intra-cellular and inter-cellular fluids.

2. Simple diffusion. In simple diffusion, transport across the membrane takes place unaided, *i.e.*, molecules of gases such as oxygen and carbon dioxide and small molecules (*e.g.*, ethanol) enter the cell by crossing the plasma membrane without the help of any permease. During simple diffusion, a small molecule in aqueous solution dissolves into the phospholipid bilayer, crosses it and then dissolves into the aqueous solution on the opposite side. There is little specificity to the process. The relative rate of diffusion of the molecule across the phospholipid bilayer will be proportional to the concentration gradient across the membrane.

3. Facilitated diffusion. This is a special type of passive transport, in which ions or molecules cross the membrane rapidly because specific permeases in the membrane facilitate their crossing. Like

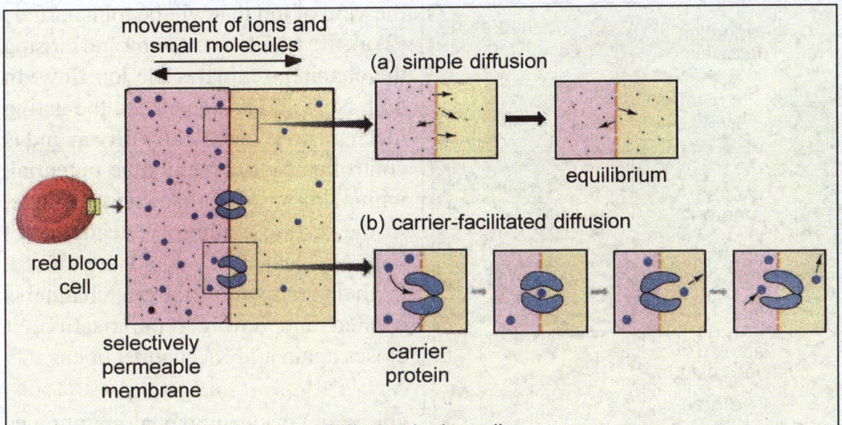

Diffusion in the cell.

the simple diffusion, facilitated diffusion does not require the metabolic energy and it occurs only in the direction of a concentration gradient. Facilitated diffusion is characterized by the following special features: (1) the rate of transport of the molecule across the membrane is far greater than would be expected from a simple diffusion. (2) This process is specific; each facilitated diffusion protein (called **protein channel**) transports only a single species of ion or molecule. (3) There is a maximum rate of transport, *i.e.*, when the concentration gradient of molecules across the membrane is low, an increase in concentration gradient results in a corresponding increase in the rate of transport. Currently, it is believed that transport proteins form the **channels** through the membrane that permit certain ions or molecules to pass across the latter (see **Darnell** *et al.*, 1986).

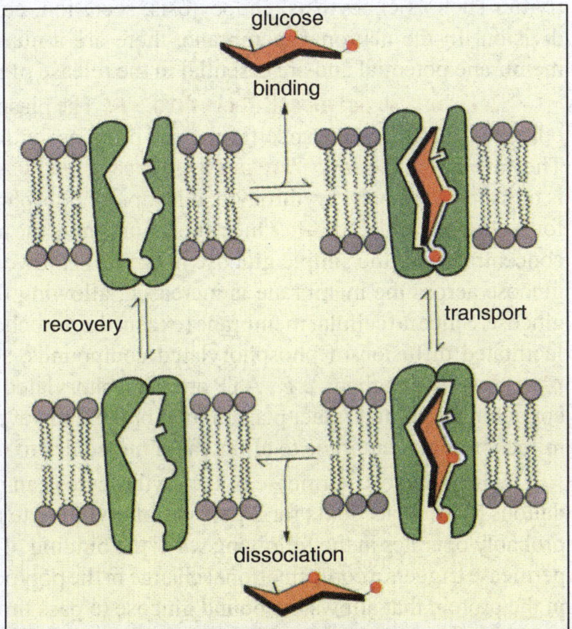

Facilitated diffusion.

Examples of Facilitated Diffusion

(i) Ionic transport through charged pores. Nerve conduction is propagated along the axonal membrane by **action potential** which regulates opening and closing of two main types of **ion channels** (*i.e.*, channel proteins with water filled pores) : **Na^+ channels** (or **voltage-gated Na^+ channels**) and **K^+ channels** (or **k^+ leak channels**). At the point of stimulation there is a sudden and several hundred fold increase in permeability to Na+, which reaches its peak in 0.1 millisecond (*i.e.*, the membrane potential may depolarise from –90 mV and overshoot to + 50 mV). At the end of the period, the membrane again becomes essentially impermeable to Na^+, but the K^+ permeability increases and this ion leaks out of the cell, repolarising the nerve fibre. In other words, during the rising phase of the spike, Na^+ enter through the Na^+ channels, and in descending phase K^+ is extruded through the K^+ channels.

Such ion channels also occur in other types of cells such as muscle, sperm and unfertilized ovum. They are not coupled to an energy source (ATP), so the transport they mediate is always passive ("down hill"), allowing specific ions mainly Na^+, K^+, Ca^{2+} and Cl^- to diffuse down their electrochemical gradient across the lipid bilayer (**Hille**, 1984). Further, an ion channel is made of integral proteins of neural membrane. This protein has two functional elements : (1) a **selective filter** which determines

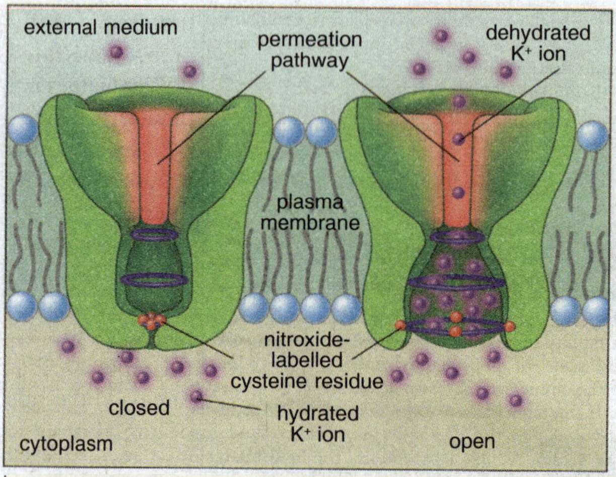

Use of EPR spectroscopy to monitor changes in conformation of a bacterial K^+ ion channel as it opens and closes.

the kind of ion that will be transported ; (2) a **gate** which by opening and closing the channel, regulates the ion flow. In both Na^+- and K^+- channels, the gating mechanism is electrically driven and is controlled by the membrane potential, without the need of other energy source. In the resting condition (steady state) both Na^+ and K^+ channels are closed. With depolarisation, the Na^+ channel is opened and during repolarisation, it closes again and K^+ channel opens.

Calcium ion channels (**Ca^{2+}-channels**) occur in axonal membranes and other membranes for the entrance of Ca^{2+} ions in the cell. Ca^{2+} ions have a fundamental role in many cellular activities such as exocytosis, endocytosis, secretion, cell motility, cell growth, fertilization and cell division. In the neuronal membrane, there are a number of Ca^{2+} channels that are driven by the membrane potential and are essential in the release of neurotransmitters (acetylcholine).

2. D-hexose permease of erythrocyte. The plasma membrane of mammalian erythrocytes and other body cells, contain specific channel proteins for the facilitated diffusion of glucose into the cells. They are called **glucose transporter**, **glucose permease** or **D-hexose permease**. After the glucose is transported into the erythrocyte, it is rapidly phosphorylated (by **hexokinase** enzyme and ATP) to form **glucose-6-phosphate**. Once phosphorylated, the glucose no longer leaves the cell; moreover, the concentration of the simple glucose in the cell is lowered. As a result, the concentration gradient of glucose across the membrane is increased, allowing the facilitated diffusion to continue to import glucose. Since no cellular membrane (except the mitochondrial membranes) contains any permease for facilitated diffusion of phosphorylated compounds, so a cell can retain any type of molecule by **phosphorylating** them, *e.g.*, ATP and phosphorylated nucleosides are never released from the cells containing a normal intact plasma membrane. However, permeases for ATP and ADP do exist in a mitochondrial membrane to allow these molecules to move across it.

The D-hexose permease of the erythrocyte is an integral and transmembrane protein of 45,000 daltons M.W. It accounts for 2 per cent of erythrocyte membrane protein. D-hexose permease, most probably operates in the following way : the binding of glucose to a site on the exterior surface of the permease triggers a conformational change in the polypeptide. This change somehow generates a pore in the protein that allows the bound glucose to pass through the membrane.

3. Anion exchange permease of erythrocyte. Band 3 polypeptide of plasma membrane of the erythrocytes and other cells is an **ion exchange permease protein** which catalyzes an one-for-one exchange of anions such as chloride (Cl^-) and bicarbonate (HCO_3^-) across the membrane (called **chloride shift**; erythrocyte has 100,000 times more permeability of Cl^- than other cells). The rapid flux of anions in the erythrocyte facilitates the transport in the blood of CO_2 from the tissues to the lungs. Waste CO_2 that is released from cell into the capillary blood, diffuses across the membrane of erythrocyte. In its gaseous from, CO_2 dissolves poorly in aqueous solutions such as blood plasma, but inside the erythrocyte the potent enzyme **carbonic anhydrase** converts it into a bicarbonate anion :

$$CO_2 + H_2O \xrightleftharpoons[\text{anhydrase}]{\text{carbonic}} \underset{\text{proton}}{H^+} + \underset{\text{bicarbonate}}{HCO_3^-}$$

This process occurs while the haemoglobin in the erythrocyte is releasing its oxygen into the blood plasma. The removal of oxygen from haemoglobin induces a change in its conformation that enables a globin histidine (amino acid) side chain to bind to the proton produced by carbonic anhydrase enzyme. The bicarbonate anion formed by carbonic anhydrase is transported out of the erythrocyte in exchange for a chloride (Cl^-) anions:

$$\underset{(in)}{HCO_3^-} + \underset{(out)}{Cl^-} \rightleftharpoons \underset{(out)}{HCO_3^-} + \underset{(in)}{Cl^-}$$

As the total volume of the blood plasma is about twice that of the total erythrocyte cytoplasm, this exchange triples the amount of bicarbonate that can be carried by blood as a whole. Without the presence of an anion exchange protein (*i.e.*, band 3 protein), bicarbonate anions generated by carbonic anhydrase would remain within the erythrocyte and blood would be unable to transport all of the CO_2 produced by tissue. The entire exchange process is completed within 50 millisecond (ms) during which time 5×10^9 HCO_3^- ions are exported from the cell. The process is reversed in the lungs : HCO_3^- diffuses into the erythrocyte in exchange for a Cl^-. Oxygen binding to haemoglobin causes release of the proton from haemoglobin. The CO_2 diffuses out of the erythrocyte and is eventually expelled in breathing. The exact mechanism of anion transport by the Band 3 protein is still unknown.

B. Active transport. Active transport uses specific transport proteins, called **pumps**, which use metabolic energy (ATP) to move ions or molecules against their concentration gradient. For example, in both vertebrates and invertebrates, the concentration of sodium ion is about 10 to 20 times higher in the blood than within the cell. The concentration of the potassium ion is the reverse, generally 20 to 40 times higher inside the cell. Such a low sodium concentration inside the cell is maintained by the sodium-potassium pump. There are different types of pumps for the different types of ions or molecules such as calcium pump, proton pump, etc.

Examples of Active Transport

1. Na^+- K^+- ATPase. It is an **ion pump** or **cation exchange pump** which is driven by energy

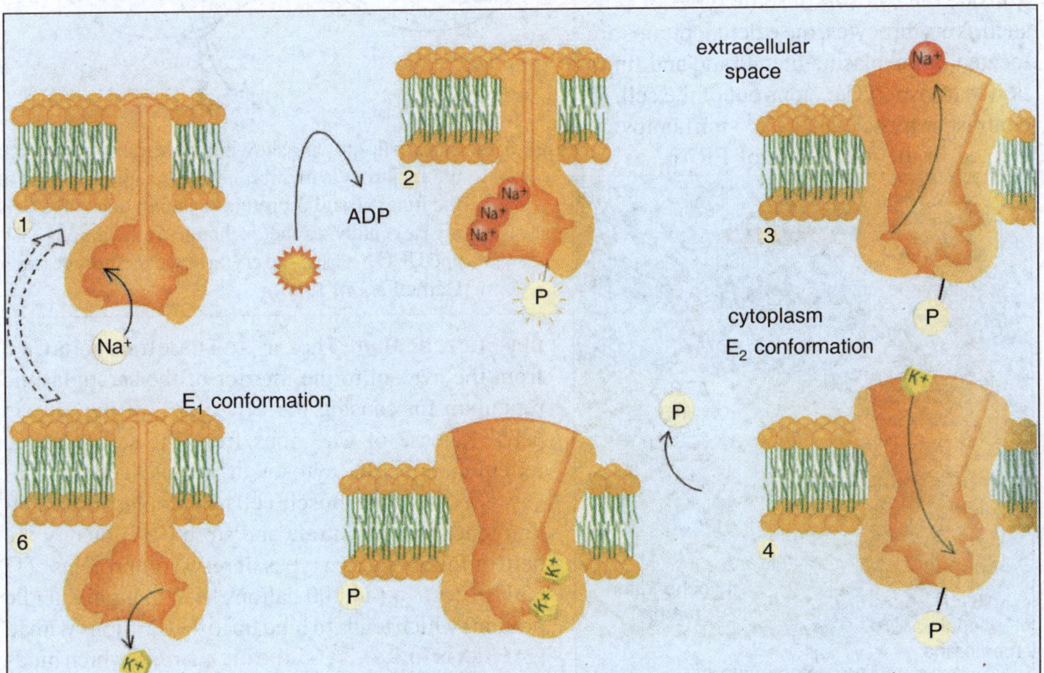

Schematic concept of the Na^+/K^+ –ATPase transport cycle.

of one ATP molecule to export three Na^+ ions outside the cell in exchange of the import of two K^+ ions inside the cell. Electrical organs of eels are found to be very rich in this enzyme or pump. N^+- K^+- ATPase is a transmembrane protein which is a dimer having two subunits : one smaller unit which is a glycoprotein of 50,000 daltons M.W., having a unknown function ; and another larger unit having 1,20,000 daltons M. W. The larger subunit of Na^+- K^+- ATPase performs the actual function of cation transport. It has three sites on its extracytoplasmic surface : two sites for K^+ ions and one site for the inhibitor **ouabain.** On its cytosolic side, the larger subunit contains three sites for three Na^+ ions and also has one catalytic site for a ATP molecule. It is believed that the hydrolysis of one ATP molecule somehow drives conformational changes in the Na^+- K^+- ATPase that allows the pump to transport three Na^+ ions out and two K^+ ions inside the cell (Fig. 5.15).

2. Calcium ATPase. Calcium pump or Ca^{2+}-ATPase pumps Ca^{2+}-ions out of the cytosol, maintaining a low concentration of it inside the cytosol. In some types of cells such as erythrocytes, the calcium pumps are located in the plasma membrane and function to transport Ca^{2+} ions out of the cell. In contrast, in muscle cells, Ca^{2+}-ion pumps are located in the membrane of ER or **sarcoplasmic reticulum**. The Ca^{2+}-ATPase transports Ca^{2+} from the cytosol to the interior of the sarcoplasmic reticulum for causing the **relaxation** of the muscle cells. Release of Ca^{2+} ions from the sarcoplasmic reticulum into the cytosol of muscle cells causes **contraction** of the muscle cells. Sarcoplasmic reticulum tends to concentrate and store Ca^{2+} ions by the help of following two types of reservoir proteins : (1) **Calsequestrin** (44,000 daltons M.W. ; highly acidic protein) which tends to bind up to 43 Ca^{2+} ions with it. (2) **High affinity Ca^{2+}- binding protein** which binds Ca^{2+} ions and also reduces the concentration of free

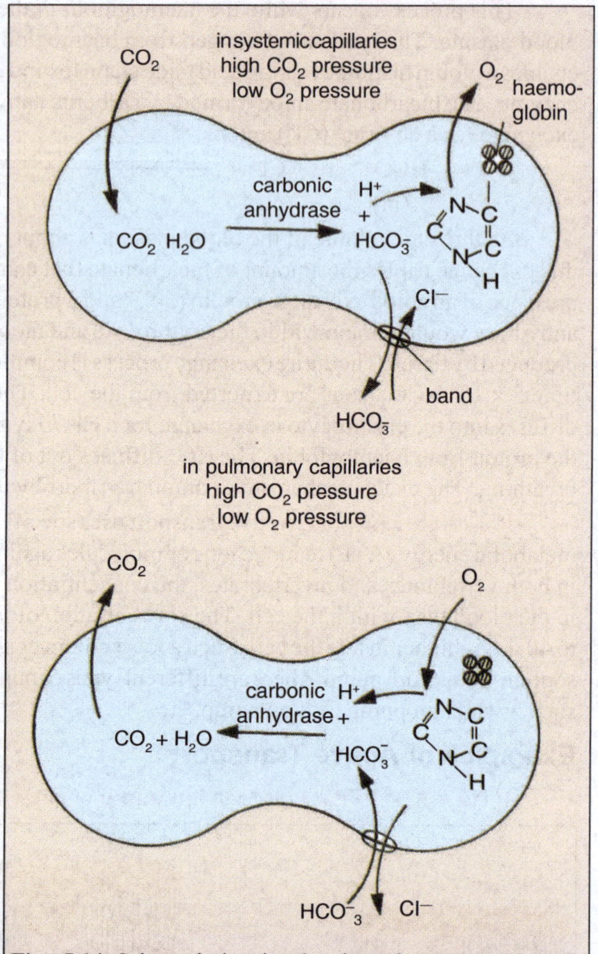

Fig. 5.14. Schematic drawing showing anion transport through the erythrocyte membrane in the capillaries and in the lungs. Band 3 protein (= anion exchange permease) catalyzes the exchange of the anions : Cl^- and HCO_3^- across the erythrocyte membrane (after Darnell *et al.*, 1986).

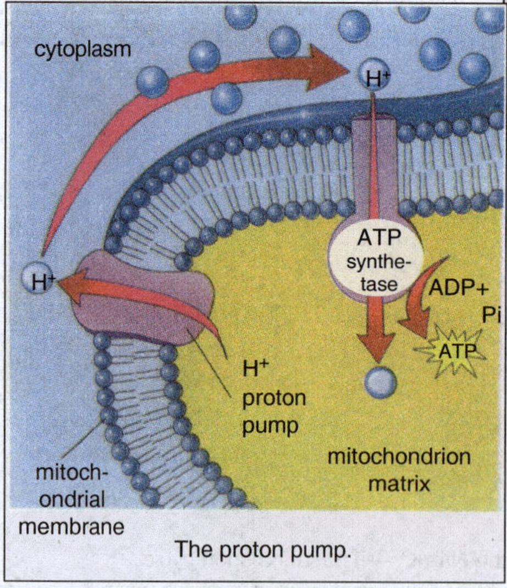

Ca^{2+} ions inside the sarcoplasmic reticulum vesicles and decreases the amount of energy needed to pump Ca^{2+} ions into it from the cytosol.

A calcium pump is a 100,000 M.W., polypeptide, forming 80 per cent of integral membrane protein of sarcoplasmic reticulum. In it, hydrolysis of one ATP molecule transports two Ca^{2+} ions in the counter-transport of one Mg^{2+} ion.

3. Proton pump or H^+-pump. The lysosomal membrane contains the ATP-dependent proton pump that transports protons from the cytosol into the lumen of the organelle, keeping the interior of lysosomes very acidic (pH 4.5 to 5.0). The pH of the cytosol is about 7.0 (Fig. 5.16).

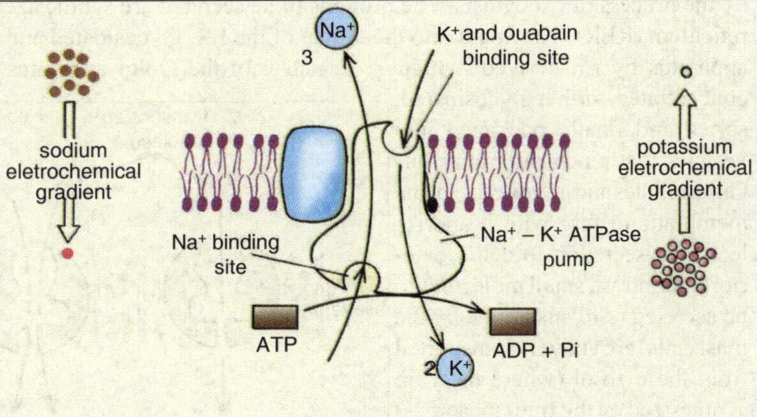

Fig. 5.15. The Na^+- K^+ ATPase in the plasma membrane actively pumps Na^+ out and K^+ into a cell against their electrochemical gradients. For every molecule of ATP hydrolyzed inside the cell, 3 Na^+ ions are pumped out and $2K^+$ ions are pumped in (after Alberts *et al.*, 1989).

Proton pumps also occur in mitochondria and chloroplasts where they participate in the generation of ATP from ADP. They also cause acidification of the mammalian stomach. In the apical membrane of a **parietal cell** or **oxyntic cell** (which sercete HCl or $H^+ Cl^-$) are located ATP-dependent proton pumps. Hydrolysis of ATP is coupled to the transport of H^+ ions out of the cell (into stomach lumen). HCl production, thus, involves three types of transport proteins : 1. anion-exchange protein; 2. chloride (Cl^-) permeases; and 3. ATP- dependant proton (H^+) pump.

Uniport, symport and antiport. Those carrier proteins which simply transport a single solute from one side of the membrane to the other ; are called **uniports.** Others function as **coupled transporters**, in which the transfer of one solute depends on the simultaneous transfer of a second solute, either in the same direction (**symport**) or in the opposite direction (**antiport**). Both symport and antiport collectively form the **cotransport**. Most animal cells, for example, must take up glucose from the extracellular fluid, where the concentration of the sugar is relatively high, by passive transport through the glucose carriers (such as **D-hexose permease**) that operate as the uniports. By contrast, intestinal and kidney cells must take up glucose from the lumen of the intestine and kidney tubules, respectively, where the concentration of the sugar is low. These cells actively transport glucose by symport with Na^+ ions whose extracellular concentration is very high. The anion exchange permease of human erythrocytes operates as an antiport to the exchange of Cl^- for HCO_3^- (**Alberts** *et al.*, 1989).

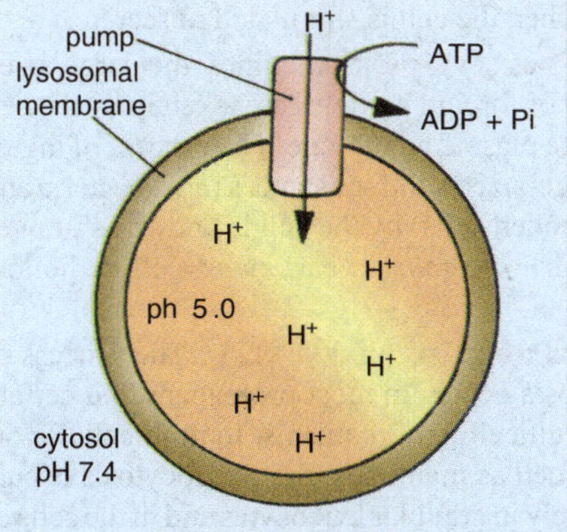

Fig. 5.16. The proton pump of the membrane of the lysosome (after Darnell *et al.*, 1986).

C. Bulk transport by the plasma membrane. Cells routinely import and export large molecules across the plasma membrane. Macromolecules are secreted out from the cell by **exocytosis** and are ingested into the cell from outside through **phagocytosis** and **endocytosis.**

1. Exocytosis. It is also called **emeiocytosis** and **cell vomiting**. In all eukaryotic cells, **secretory vesicles** are continually carrying new plasma membrane and cellular secretions such as proteins, lipids and carbohydrates (*e.g.*, cellulose) from the Golgi apparatus to the plasma membrane or to cell exterior by the process of exocytosis. The proteins to be secreted are synthesized on the rough endoplasmic reticulum (RER). They pass into the lumen of the ER, glycosidated and are transported to the Golgi apparatus by ER-derived **transport vesicles**. In the Golgi apparatus the proteins are modified, concentrated, further glycosidated, sorted and finally packaged into vesicles that pinch off from trans Golgi tubules and migrate to plasma membrane to fuse with it and release the secretion to cell's exterior. In contrast, small molecules to be secreted (*e.g.*, histamine by the mast cells) are actively transported from the cytosol (where they are synthesized on the free ribosomes) into preformed vesicles, where they are complexed to specific macromolecules (*e.g.*, a network of proteoglycans, in case of histamine; **Lawson** *et al.*, 1975), so that, they can be stored at high concentration without generating an excessive osmotic gradient.

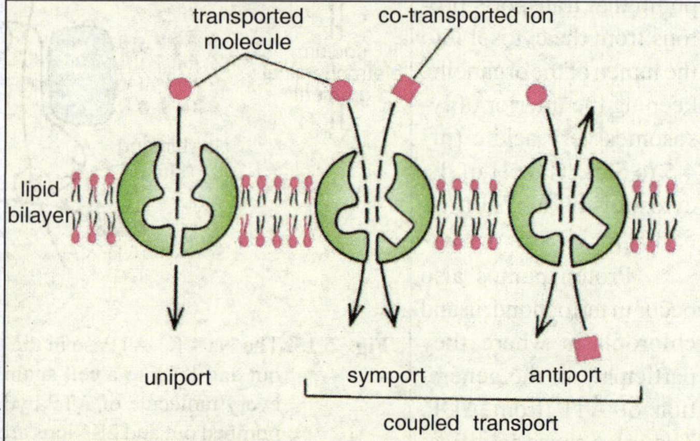

Fig. 5.17. Carrier proteins of membrane functioning as uniports, symports and antiports (After Alberts *et al.*, 1989).

During exocytosis the vesicle membrane is incorporated into the plasma membrane. The amount of secretory vesicle membrane that is temporarily added to the plasma membrane can be enormous : in a pancreatic acinar cell discharging digestive enzymes, about 900 μm^2 of vesicle membrane is inserted into the apical plasma membrane (whose area is only 30 μm^3) when the cell is stimulated to secrete.

The process of engulfment by a leucocyte ingesting a yeast particle.

2. Phagocytosis. Sometimes the large-sized solid food or foreign particles are taken in by the cell through the plasma membrane. The process of ingestion of large-sized solid substances (*e.g.*, bacteria and parts of broken cells) by the cell is known as **phagocytosis** (Gr., *phagein*=to eat, *kytos*=cell or hollow vessel).

Occurrence of phagocytosis. The process of phagocytosis occurs in most protozoans and certain cells of multicellular organisms. In multicellular organisms such as mammals, the phagocytosis occurs very actively in granular leucocytes and in the cells of mesoblastic origin. The cells of the mesoblastic origin are collectively known as the cells of **macrophagic** or **reticuloendothelial system**. The cells of macrophagic system are histiocytes of the connective tissue, the reticular cells of the hemopoietic organs (bone marrow, lymph nodes and spleen) and the endothelial cells which form the lining of capillary sinusoid of the liver, adrenal gland and hypophysis. The cells of macrophagic system can ingest bacteria, Protozoa, cell debris or even colloidal particles by the process of phagocytosis.

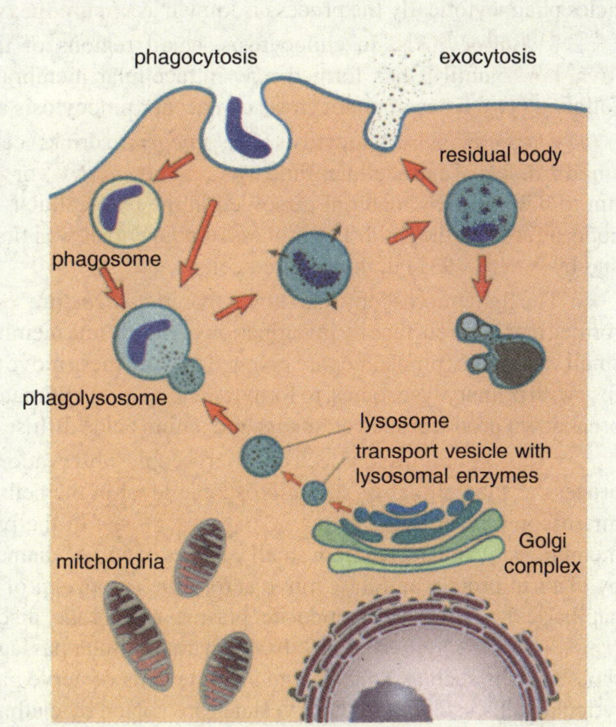

A summary of the phagocytic pathway.

Process of phagocytosis. In phagocytosis, first the target particle is bound, to the specific receptors on the cell's surface (process is called **adsorption**),then the plasma membrane expands along the surface of the particle and eventually engulfs it. Vesicle formed by phagocytosis is called **phagosome** and it is typically 1 to 2 μm or larger in diameter, much larger than those formed during pinocytosis and receptor-mediated endocytosis. The phagosomes migrate to the interior of the cell and fuse with the pre-existing lysosomes (to form phagolysosome). The food is digested by the hydrolytic enzymes (acid hydrolase) of the lysosomes and the digested food is ultimately diffused to the surrounding cytoplasm. In addition to the normal set of lysosomal hydrolases, macrophage's lysosomes contain enzymes that generate hydrogen peroxide (H_2O_2) and other toxic chemicals that aid in the killing of the bacteria. The undigested food is expelled from the plasma membrane by the process of **ephagy** or **egestion**. In macrophages, the undigested parts of ingested material such as the cell walls of micro-organisms, accumulate within lysosomes as residual bodies. Accumulation of residual bodies may be one reason why macrophages have a very short life time (*i.e.*, less than a few days).

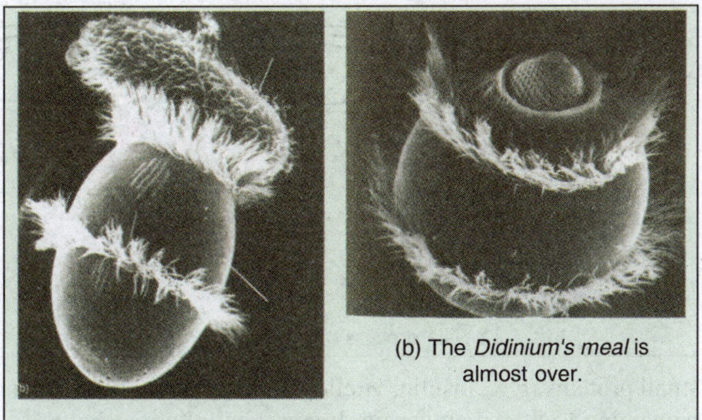
(b) The *Didinium's meal* is almost over.

(a) Here an egg-shaped protist *Didinum* illustrates phagocytosis by ingesting the smaller *paramecium*.

Kinds of phagocytosis. According to the physical and chemical nature of foreign substance following types of phagocytosis have been recognised :

(i) Ultraphagocytosis or colloidopexy. The process in which plasma membrane ingests smaller colloidal particles is known as **colloidopexy** or **ultraphagocytosis**, *e.g.*, leucocytes and the macrophagic cells of mammals.

(ii) Chromopexy. When the cell ingests colloidal chromogen particles phagocytotically the process is known as **chromopexy**, *e.g.*, some mesoblasitc cells.

3. Endocytosis. In endocytosis, small regions of the plasma membrane fold inwards or **invaginate**, until it has formed new intracellular membrane limited vesicles. In eukaryotes, the following two types of endocytosis can occur : pinocytosis and receptor-mediated endocytosis.

(i) Pinocytosis. Pinocytosis (Gr., *pinein* = to drink; 'cell drinkng') is the non-specific uptake of small droplets of extracellular fluid by **endocytic vesicles** or **pinosomes**, having diameter of about 0.1 μm to 0.2 μm. Any material dissolved in the extracellular fluid is internalized in proportion to its concentration in the fluid. The process of pinocytosis was first of all observed by **Edward** in *Amoeba* and by **Lewis** (1931) in the cultured cells.

The light microscopy has shown that in *Amoeba* tiny **pinocytic channels** are continually being formed at the cell surface by invagination of the plasma membrane. From the inner end of each channel small vacuoles or pinosomes are pinched off, and these move towards the centre of the cell, where they fuse with primary lysosomes, to form **food vacuoles**. Ultimately, ingested contents are digested, small breakdown products such as sugars and amino acids diffuse to cytosol.

Micropinocytosis. Electron microscopic observations have been made on the pinocytotic process at sub-cellular or sub-microscopic level in the cells. The pinocytosis which occurs at sub-microscopic level is known as **micropinocytosis**. In the process of micropinocytosis, the plasma membrane invaginates to from small vesicles of 650 A° diameter. These closed vesicles are not coated by clathrin protein and they move across the cytoplasm of endothelial cells (which line the blood capillaries) to fuse with opposite plasma membrane discharging their contents. This is called **transcytosis** (**Simionescue**, 1980). Such transcellular passage of fluids is also found to occur in other types of cells such as Schawn and satellite cells of nerve ganglion, macrophages, muscle cells and reticular cells, etc., but in them vesicles are coated by clathrin (see **Alberts** *et al.*, 1989).

(ii) Receptor-mediated endocytosis. In this type of endocytosis, a specific receptor on the surface of the plasma membrane "recognizes" an extracellular macromolecule and binds with it. The substance bound with the receptor is called the **ligand**. Examples of ligands may include viruses,

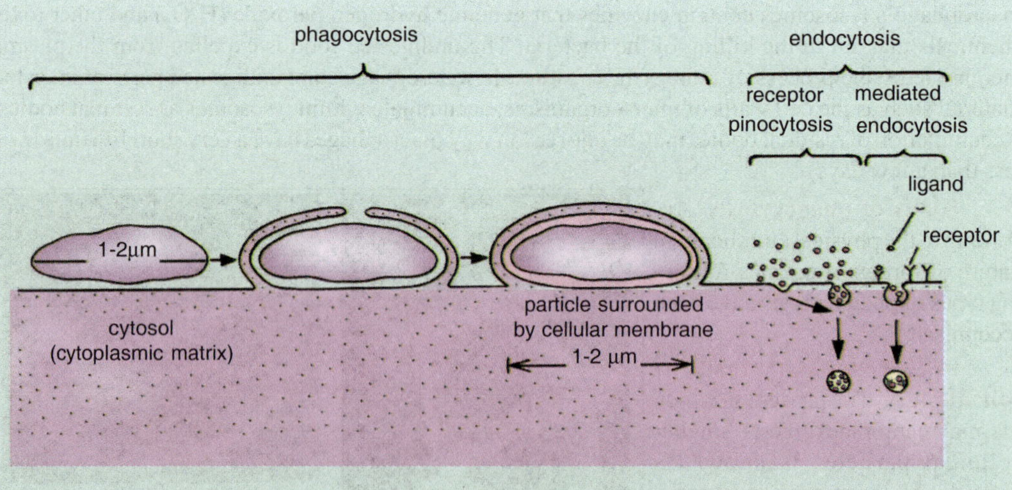

Fig. 5.18. The process of phagocytosis and endocytosis (after Darnell *et al.*, 1986).

small proteins (*e.g.*, insulin, vitellogenin, immunoglobin, transferrin, etc.), vitamin B_{12}, cholesterol containing LDL or low density lipoprotein, oligosaccharide, etc. The region of plasma membrane containing the receptor-ligand complex undergoes endocytosis. The whole process of receptor-mediated endocytosis, includes the following events :

1. Interaction of ligands and cell surface receptors. The macromolecules (ligands) bind to complementary cell-surface receptors. There are more than 25 different types of receptors which are involved in receptor-mediated endocytosis of different types of molecules. Such a receptor is a trans-membrane protein which contains two specific binding sites : (1) **ligand-binding site** at the external surface of plasma membrane ; and (2) **coated-pit binding site** at the inner or cytosolic face of the plasma membrane.

2. Formation of coated-pits and coated-vesicles. The endocytic cycle begins at specialized regions of the plasma membrane, called **coated-pits**. Coated-pits are depressions of plasma membrane having a coat of bristle-like structure towards their cytosolic side. The ligand-loaded receptors diffuse into these coated-pits. A coated-pit may accommodate about 1000 receptors of assorted variety. In fact, coated-pits serve as molecular filters and selective concentrating devices, since, they tend to collect certain receptors and leave others. They increase the efficiency of internalization of a particular ligand more than 1000-fold and also carry minor components of extracellular fluid. The life-time of each coated-pit is quite short—within a minute or so of being formed, it invaginates into the cell and pinches off to form the **coated-vesicles**. The coat of coated pits and coated vesicles is made up of protein, called **clathrin** and certain other proteins. A molecule of clathrin is composed of three large polypeptide chains and three smaller polypeptide chains, all of which together form a three-legged structure, called **triskelion**. A number of triskelions assemble into a basket-like network of hexagons and pentagons on the cytoplasmic surface of the membranes (**Pearse** and coworkers, 1981, 1987).

3. Fusion of endocytic vesicle and endosome. Once a coated vesicle is formed, the clathrin and associated proteins dissociate from the vesicle membrane and return to the plasma membrane to form a new coated-pit (**Schmid** and **Rothman**, 1985). The resultant **endocytic vesicle** gets fused with pre-existing endosomes and ultimately its contents are utilized by the cell.

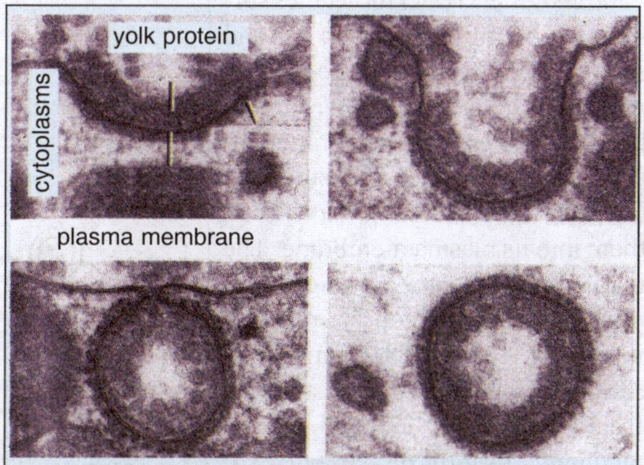

Receptor-mediated endocytosis.

Endosome or receptosome. Recently it has been found that in the cells exists a complex set of heterogeneous membrane-bound tubes and vesicles, called **endosome**, which extends from the periphery of the cell to the perinuclear region, where it lies quite close to Golgi apparatus. Thus, endosomes may be of two types : (i) **peripheral endosomes** just beneath the plasma membrane and (ii) **perinuclear** or **internal endosomes**. The interior of the endosome is acidic (pH 5-6) due to the presence of ATP-driven **proton (H^+) pumps** in its membrane that pumps H^+ ions into the lumen from the cytosol (**Sly** and **Doisy**, 1984). Endosomes lack in degradative enzymes.

Thus, via receptor-mediated coated-vesicles, the ligands are delivered to the peripheral endosomes which slowly move inward to become perinuclear endosomes. These perinuclear endosomes are converted into **endolysosomes** and then into **lysosomes** due to following three activities : 1. the fusion of transport vesicles from the Golgi apparatus, (**Note. Transport vesicles** capture a cargo of molecules, *e.g.*, proteins, from the lumen of one compartment as they pinch off from its membrane and then discharge that cargo into another compartment as they fuse with it. Thus, in such vesicular transport, the transported proteins do not cross any membrane and they are transferred from lumen to lumen). 2. continuous membrane retrieval, and 3. increased acidification (**Helenius** and coworkers, 1983, 1987). The endosomal compartment also acts as the main **sorting station** in the endocytic pathway. The acidic environment of the endosome causes dissociation of ligands from their receptors. Such ligands are destined for destruction in the lysosomes along with the other non-membrane-bound contents of the endosome. The receptor-proteins are either returned to the same plasma membrane domain from which they come or they go to lysosomes and are degraded.

Example of receptor-mediated endocytosis. Most animal cells are found to have a regulatory pathway for the uptake of cholesterol. Most cholesterol is transported in the blood in the form of particles of **low-density proteins** or **LDL**. Each of these large spherical particles (22 nm in diameter) contains a **core** of about 1500 cholesteryl ester molecules surrounded by a **lipid monolayer** and also contains a single large protein molecule (apoprotein). When the cell needs cholesterol for membrane synthesis, it synthesizes receptor proteins for LDL particles and inserts them into its plasma membrane. The human LDL receptor is a single-pass transmembrane glycoprotein which is composed of about 840 amino acid residues, only 50 of which stick out from cytoplasmic side of plasma membrane to form the **coated-protein-binding site**. The **LDL - binding site** of the receptor is exposed to cell surface. The LDL receptors move laterally within lipid bilayer, until they become associated to the newly formed coated-pits. Since coated-pits constantly pinch off to form coated-vesicles, the LDL particles are bound to receptors in the coated-pits and are rapidly internalized. After shedding their clathrin-coats the endocytic vesicles deliver their contents to endosomes. In endosomes, the LDL particles and their receptors are separated from each other ; the receptors are returned to the plasma membrane, while LDL ends up in the lysosomes. In the lysosomes, the cholesteryl esters in the LDL

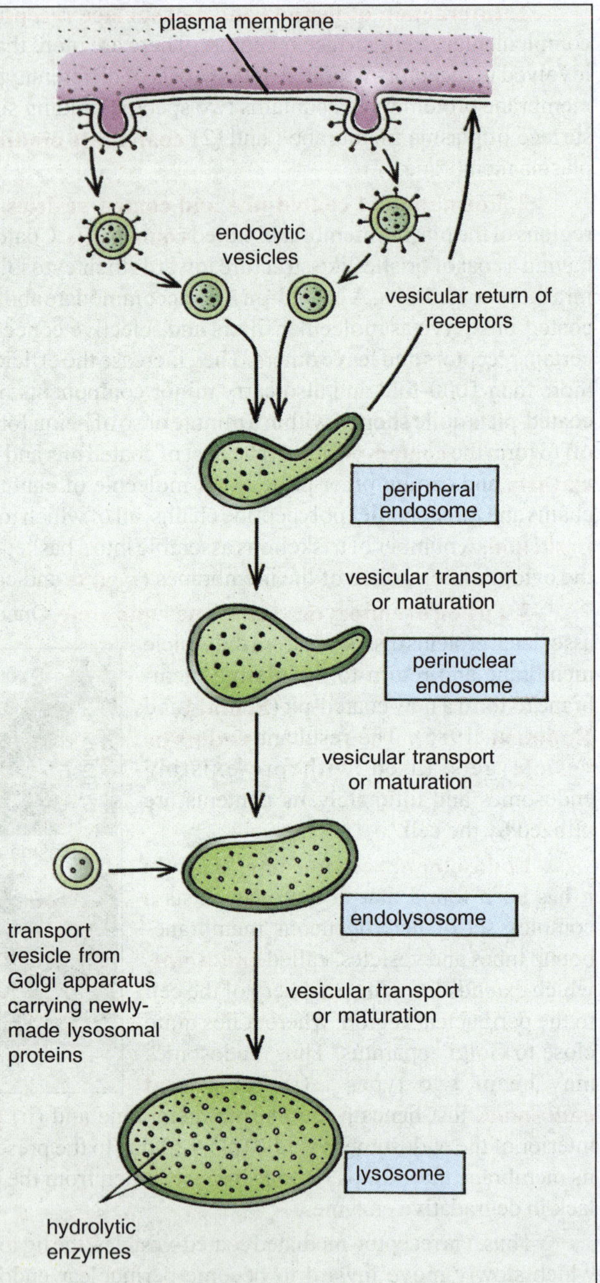

Fig. 5.19. An endocytic pathway : ligands which are endocytosed via coated pits, are delivered to the peripheral endosomes and then move sequentially to perinuclear endosomes, endolysosomes and lysosomes (Alberts *et al.*, 1989).

particles are hydrolyzed to free cholesterol molecules, which thereby become available to the cell for new membrane synthesis. If too much free cholesterol accumulates in a cell, this stops cell's own cholesterol synthesis and the synthesis of LDL-receptor proteins, so that less amount of cholesterol is made and less amount of it is taken up by the cell (**Brown** and **Goldstein**, 1984, 1986).

Energy utilisation by phagocytosis and endocytosis. Unlike pinocytosis, which is a constitutive process that occurs continuously, the phagocytosis is a triggered process in which activated receptors transmit signals to the cell interior to initiate the response (**Wright** and **Silverstein**, 1983). Both phagocytosis and pinocytosis are active mechanisms in the sense that the cell requires energy for

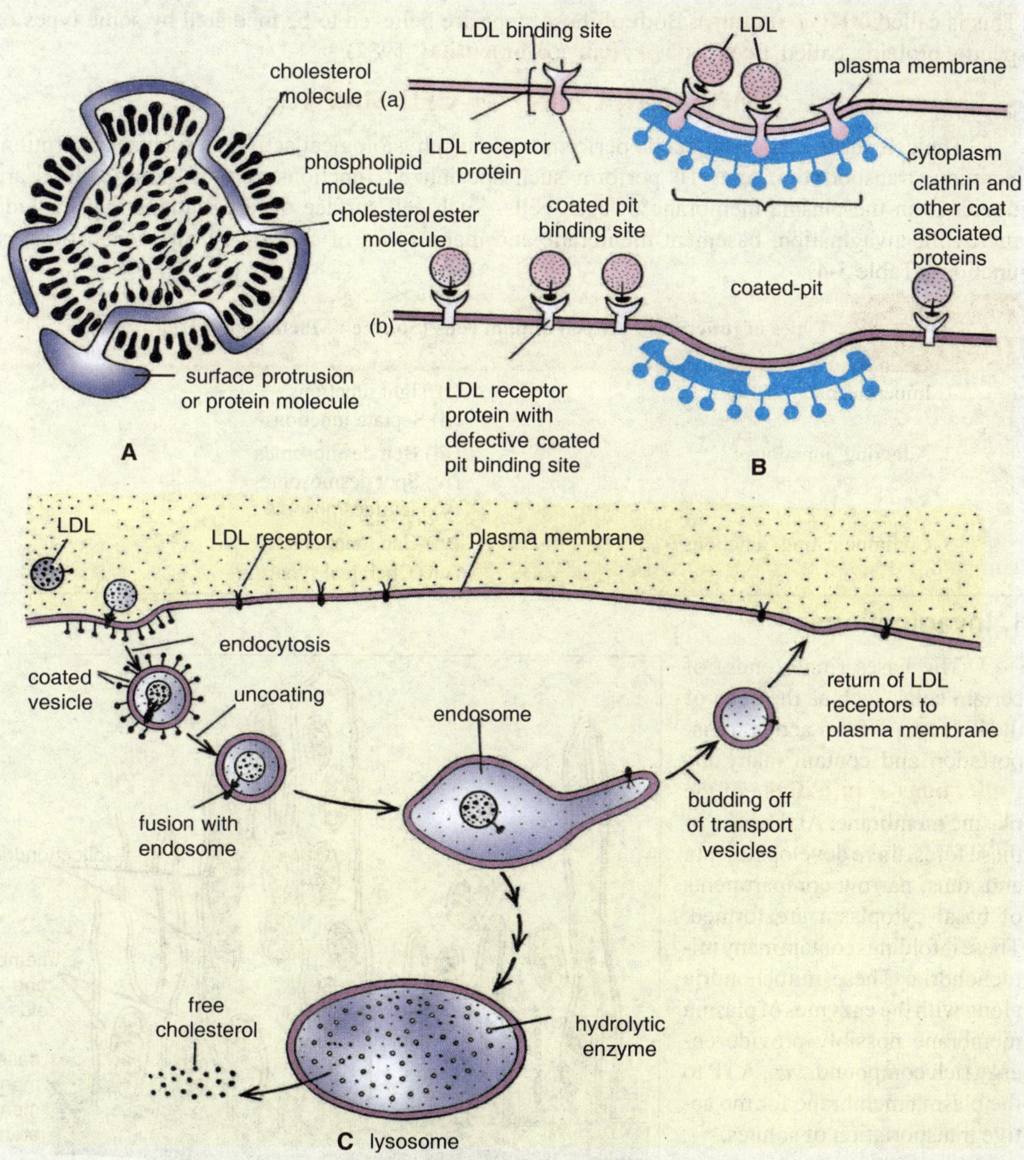

Fig. 5.20. A— A LDL particle in cross section ; B—LDL receptor proteins binding to a coated-pit in the plasma membrane ; C— Mechanism of receptor-mediated endocytosis of LDL particles (after Alberts *et al.*, 1989).

their operation. During phagocytosis by leukocytes oxygen consumption, glucose uptake and glycogen breakdown all increase significantly. The mechanism of endocytosis is found to involve the contraction of **microfilaments** of actin and myosin present in the peripheral cytoplasm (ectoplasm) which causes the plasma membrane to invaginate and to form the endocytic vacuole (pinosome/phagosome). Involvement of actin microfilaments is demonstrated by the action of the drug **cytochalasin B** which inhibits endocytosis and disorganizes actin microfilaments.

Membrane fusion during exocytosis and endocytosis. Both exocytosis and endocytosis involve the fusion of initially separate regions of lipid bilayer and occur in at least two steps : first the two bilayers come into close apposition, it is called **bilayer adherence**, and then they fuse together (This is called **bilayer joining**). Both of these steps are believed to be mediated by some types of specific proteins, called **fusogenic proteins** (**Blumenthal**, 1987).

DIFFERENTIATIONS OF CELL SURFACE

The cell surface of certain cells performs various physiological activities such as absorption, secretion, transportation, etc. To perform such specialized functions certain modifications are inevitable in the plasma membrane of such cells. Such cell surface differentiations may include microvilli, invagination, basement membrane and many types of cell-to-cell interconnections or junctions (Table 5-4).

Table 5-4. **Types of junctions between animal cells (Source : Maclean and Hall, 1987).**

1. Impermeable junctions :	(i) Tight junctions (ii) Septate junctions
2. Adhering junctions :	(iii) Belt desmosomes (iv) Spot desmosomes (v) Hemidesmosoms
3. Communicating junctions :	(vi) Gap junctions (vii) Chemical synapsis

1. Invaginations

The bases (inner ends) of certain cells, such as the cells of the kidney, perform active transportation and contain many **invaginations or infoldings** of the plasma membrane. At the base of these folds, there develops a septa and, thus, narrow compartments of basal cytoplasm are formed. These infoldings contain many mitochondria. These mitochondria along with the enzymes of plasma membrane possibly provide energy rich compound, *viz.*, ATP to the plasma membrane for the active transportation of solutes.

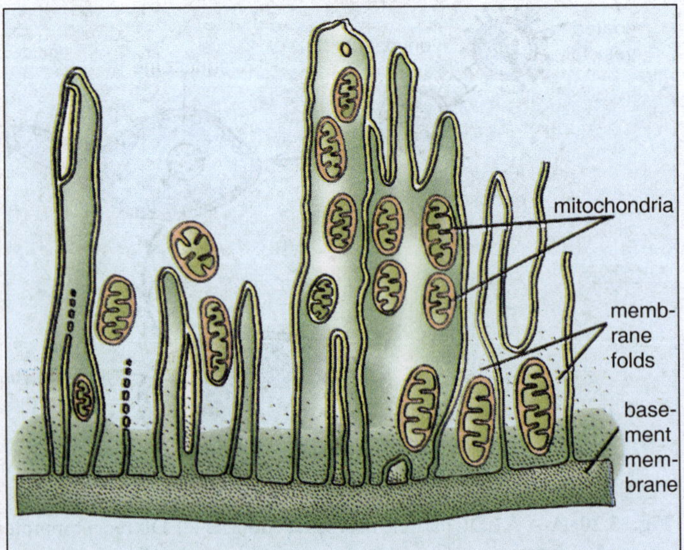

Fig. 5.21. Infoldings of plasma membrane.

2. Microvilli

Microvilli are finger-like,

slender projections of plasma membrane which are found in mesothelial cells, hepatic cells, epithelial cells of intestine (**striated border**), uriniferous tubules (**brush border**), gall bladder, uterus, growing oocyte and yolk sac. Microvilli increase the effective surface of absorption. For example, a single epithelial cell of intestine may have as many as 3000 microvilli and in a square millimetre of intestine there may be 200,000,000. These microvilli are 0.6 to 0.8 μm long and 0.1 μm in diameter. The narrow spaces between the microvilli form a kind of sieve through which substances may pass during the process of absorption. Within the cytoplasmic core of a microvillus fine microfilaments are observed which in the underlying cytoplasm form a **terminal web**. The microfilaments contain actin and are attached to the tips of the microvilli by α-actinin ; their function is to produce contraction of microvilli.

3. Basement Membrane

The interface between all epithelia and underlying connective tissue is marked by a non-cellular structure called **basement membrane**. This membrane comprises two basic layers : 1. **Basal lamia** which is in contact with the epithelial basal plasma membrane and is composed of fine feltwork of fibrils of collagen of Type IV that are embedded in an amorphous matrix. It is secreted by the epithelial cells. 2. **Reticular layer** exists just beneath the basal lamina and is composed of fine reticular fibres of reticulin protein. The reticular fibres are embedded in a ground substance. The reticular layer is synthesized by underlying connective tissue into which it is merged (**Wheater** *et al*., 1979). The basement membrane provides structural support for epithelia and may constitute an important barrier to the passage of materials between the epithelial and connective tissue compartments.

4. Tight Junctions (Zonula Occludens)

The cells of both vertebrate and invertebrate animals display junctions that are designed to prevent or reduce the flow of even small molecules between the lateral surfaces of adjacent cells. Such junctions are particularly characteristics of epithelial tissues. In higher animals these are termed **tight junctions** and in invertebrates these are called **septate junctions** (see **Maclean** and **Hall**, 1987).

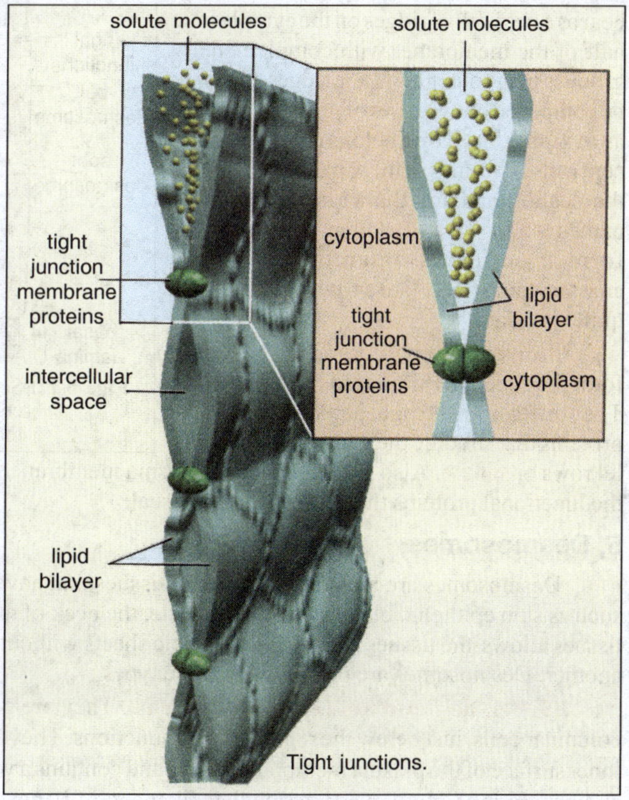

Tight junctions.

Tight junctions are situated below the apical border (often below the microvillar surface) of the epithelial cells and act as permeability barriers. Thus, all nutrients are absorbed from the intestine into one side of the epithelial cell and then released from the other side into the blood because tight intercellular junctions do not allow small molecules to diffuse directly from the intestine lumen into the blood. Also in pancreatic acinar tissue, they prevent the leakage of pancreatic secretory proteins, including digestve enzymes, into the blood.

Tight junctions are composed of thin bands that completely encircle a cell and are in contact with thin bands of

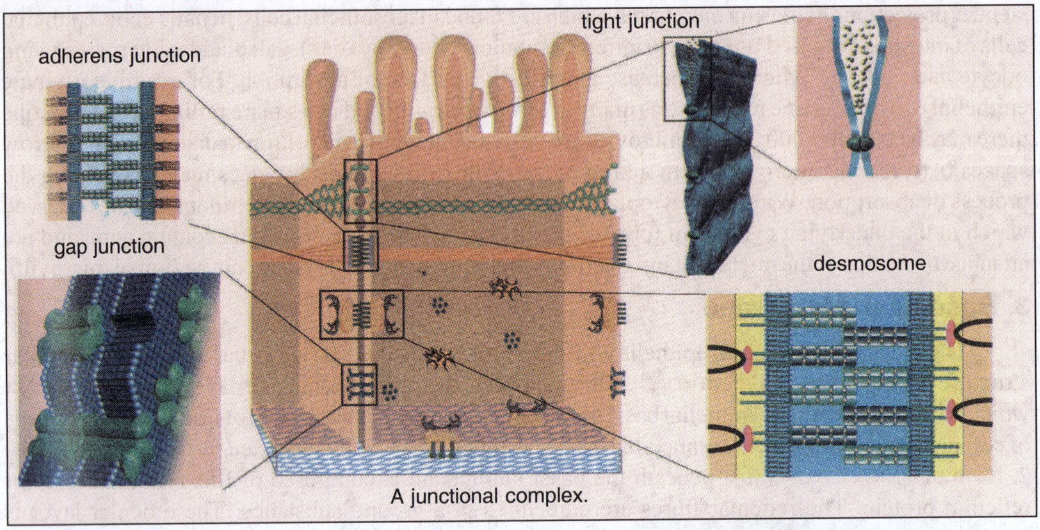

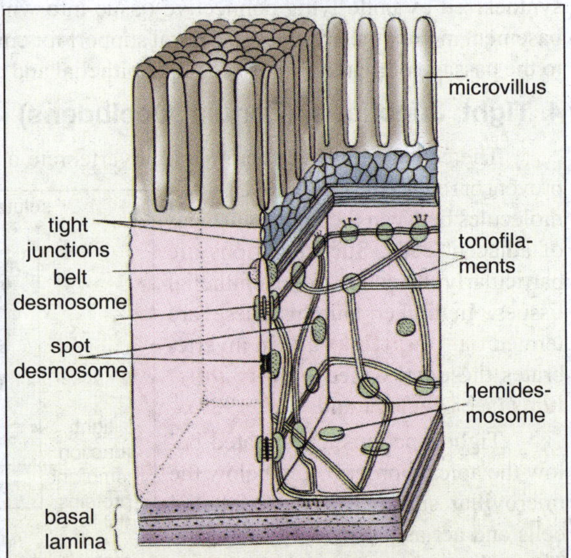

Fig. 5.22. Schematic diagram of the principal types of cell junctions, as found in the intestinal epithelial cell (after Darnell, *et al.*, 1986).

adjacent cells. In a thin section, in the tight junction two adjacent plasma membranes appear to be fused at a series of points. However, in three-dimensional structure, revealed by freeze-fracture technique (**Pinto da Silva** and **Kachar**, 1982), the tight junctions appear as a network of ridges on the cytoplasmic half of the membrane, with complementary grooves in the outer half. The ridges appear to be composed of two rows of protein particles, as in zipper, each one belonging to the adjacent cells. The lines of these particles produce the sealing and for this reason have been named **sealing strands**. Often, sealing strands form a series of interconnected and anastamosing lines, like a row of stitches in a quilted surface.

In invertebrates, **septate junctions** perform the functions similar to tight junctions. They differ from the tight junctions in that the proteins that straddle the gaps, occur in parallel rows or **septae**. Also in them adjacent plasma membrane surfaces are not in direct contact, so that the junctional proteins themselves form the seal.

5. Desmosomes

Desmosomes are abundantly found in tissues that have to withstand severe mechanical stress, such as skin epithelia, bladder, cardiac muscle, the neck of uterus and vagina. Their presence in such tissues allows the tissues to function as elastic sheets without the individual cells being torn one from another. Desmosomes are of following three types :

(i) Belt desmosomes (Zonula adherens). They are generally found at the interface between columnar cells, just below the region of tight junctions. They form a band that form a girdle around the inner surface of the plasma membrane. This band contains a web of 6 to 7 nm actin microfilaments and another group of interwoven intermediate filaments of 10 nm. Actin microfilaments are contractile and intermediate filaments play a structural role. At belt desmosome, the plasma membranes of adjacent cells are parallel, thicker than usual and 15 to 20 nm apart. The intercellular space between them is filled with an amorphous material.

(ii) Spot desmosomes (Macula adherens). The spot desmosomes act like rivets or "spot welds" to hold epithelial cells together at points of contact. They represent localized circular areas of contact about 0.5 µm in diameter, in which the plasma membranes of two adjacent cells are separated by a distance of 30 to 50 nm. The intercellular core or **central stratum** between the two membranes consists of specific desmosomal material rich in proteins and mucopolysaccharides. Under each facing plasma membrane of the spot desmosome, there is a discoidal intracellular **plaque**, 15 to 20 nm thick, having non-glycosylated proteins such as **desmoplakins I**, **II**, and **III** (**Muellar** and **Franke**, 1983). Numerous 10-nm thick intermediate filaments of keratin protein, called **tonofilaments**, converge towards the plaque. These filaments form a loop in a wide arc and course back into the cytoplasm. In addition, there are thinner filaments that arise from each dense plaque and traverse the plasma membrane to form "**trans-membrane linkers**" in the intercellular space. These linkers provide mechanical coupling and chemically are made of glycosylated proteins, called **desmogleins I** and **II**, with the carbohydrate moiety exposed toward the intercellular space (**Steinberg**, 1984). While the tonofil-aments provide the intracellular mechanical support, the cellular adhesion at the desmosome depends mainly on the extracellular coating material.

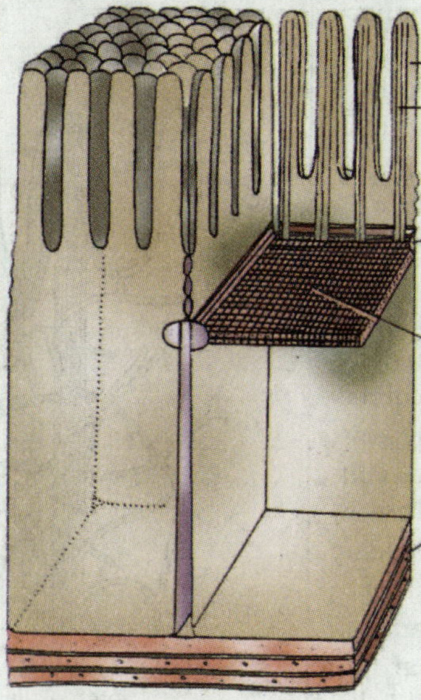

Fig. 5.23. Microvilli and terminal web (after Darnell, *et.al.*,1986).

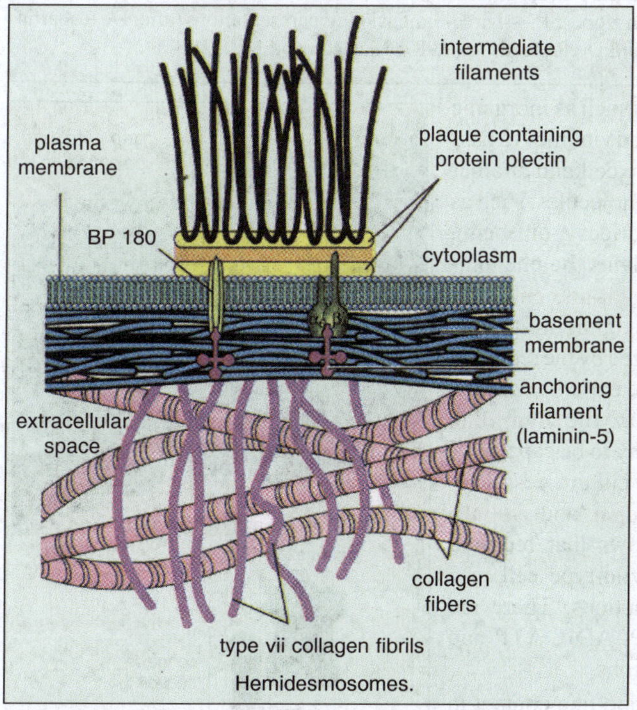

Hemidesmosomes.

(iii) Hemidesmosomes. They are half desmosomes which resemble spot desmosomes but join the basal surface of an epithelial cell to a basal lamina. They anchor extracellular proteins such as collagen and other proteins to the cell.

6. Gap Junctions (Nexus)

Many cells of the tissues of higher animals are coupled together by interconnecting **gap junctions**, **nexus** or **communicating junctions**. The presence of gap junctions explains the ionic or electronic connections between adjacent cells, *i.e.*, there are some cells which are **electrically coupled** and have regions of low resistance in the membrane through which there is a rather free flow of electrical current carried by ions. Such electrical coupling is found extensively in embryonic cells. In adult tissues it is usually found in epithelia, cardiac cells and liver cells. Skeletal muscles and most neurons do not show electrical coupling.

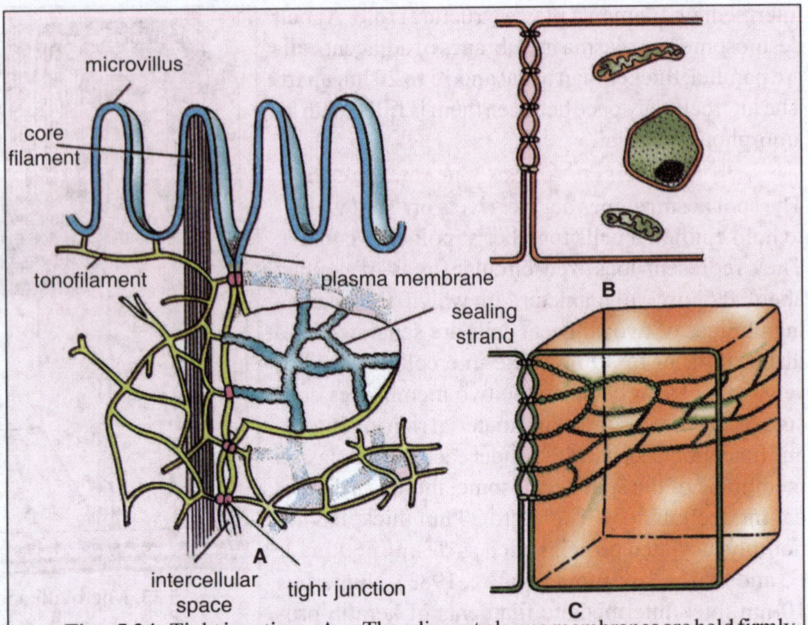

Fig. 5.24. Tight junctions: A—The adjacent plasma membranes are held firmly at the sealing strand which is composed of two rows of particles, as in zipper; B—Three-dimensional representations (after De Robertis and De Robertis Jr., 1987; Maclean and Hall, 1987).

In fact, gap junctions are found to permit molecules such as inorganic ions, sugars, amino acids, nucleotides and vitamins to pass with comparative freedom between one cell and another within a tissue, but they prevent larger molecules, such as proteins, nucleic acids and polysaccharides from being transferred. This observation also explains the phenomenon of **metabolic cooperation** or **metabolic coupling** between cells, *i.e.*, cells can transfer to neighbouring cells, the molecules which cannot be synthesized by the recipient cells. For example, in the tissue-culture experiments, the mutant cells which are deficient in the enzyme **thymidine kinase** can be shown by autoradiography to be capable of DNA synthesis only when grown in a culture vessel together with the wild type cells (**Hooper** and **Subak-Sharpe**, 1981). This observation shows that required thymidine has been passed from a wild-type cell to a mutant cell, presumably via gap junctions. There are certain other molecules such as AMP, ADP, ATP and cAMP that can pass through gap junctions.

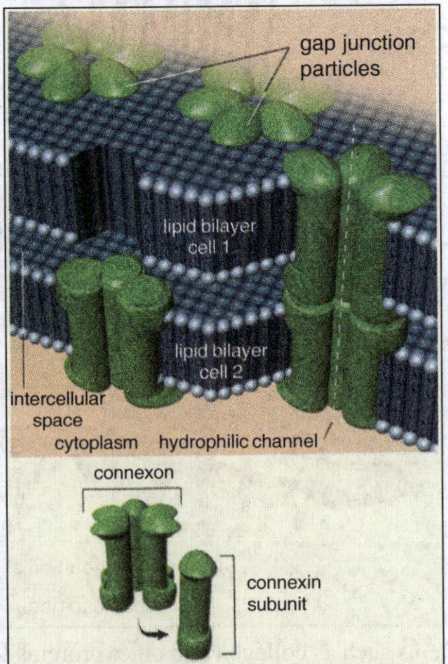

The structure of a gap junction.

A gap junction appears as a plaque-like contact in which the plasma membranes of adjacent cells are in close apposition, separated by a space of only 2 to 4 nm. Structurally, gap junctions consist of hollow channels round which a series of six protein subunits are located ; a channel has a diameter of about 1.5 to 2 nm. A single major protein (a macromolecular unit, called **connexon**) of 27000 daltons has been isolated from rat liver preparations consisting of almost pure gap junction material (**Hertzberg** *et al.*, 1981), A connexon appears as an annulus of six subunits surrounding the channel. It is believed that the sliding of the subunits caused the channel to open and close. The permeability of channel of gap junction is regulated by Ca^{2+} ions; if the intercellular Ca^{2+} ion level increases, the permeability is reduced or abolished. The gap junctions or connexons of adjacent cells are believed to line up to provide a continuous channel, made up of two connexons opposed end to end (Fig. 5.26).

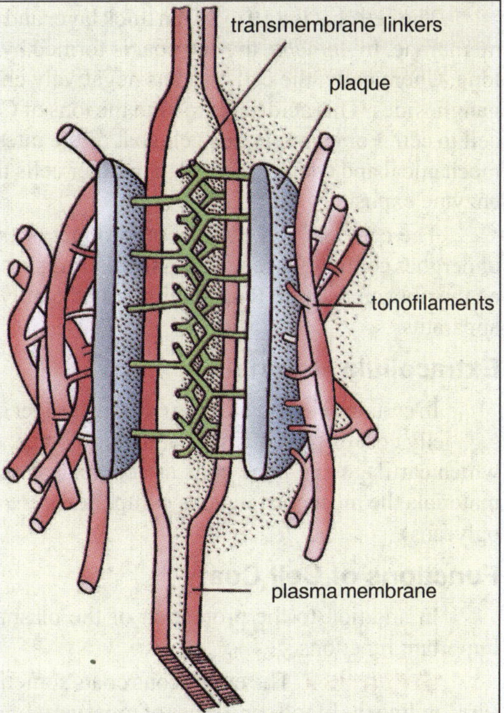

Fig. 5.25. Spot desmosome (after De Robertis and De Robertis Jr., 1987).

CELL COAT

The plasma membrane is surrounded and protected by the **cell coat**. Sometimes, the cell coat is also called **glycocalix**, because it contains sugar units in glycoproteins and polysaccharides. The cell coat is found to be equivalent to the oligosaccharide side chains of glycolipids and glycoproteins that stick out from the cell surface and are covalently attached to protein moieties. However, in many type of cells, there is a separate "fuzzy layer", beyond the cell coat, which is composed mainly of carbohydrates and is secreted by the cell.

The cell coat can be stained with PAS (see Chapter 2) or Alcian blue for the light microscopy and with **lanthanum** or **ruthenium red** for the electron microscopy. By the use of **lectins**, the carbohydrates can be specifically observed. Lectins are proteins which are normally derived from the plants and they tend to bind to the cell surface and cause agglutination. They are used as recognition molecules for the sugar components of glycoproteins. **Concanavalin A** is a lectin which has been isolated from jack beans and is specific for the glucose and mannose residues. Another lectin, called **germ agglutinin**, is specific for N-acetylglucosamine. These lectins may be labelled with fluorescent dyes or with electron-dense materials for electron microscopic observation.

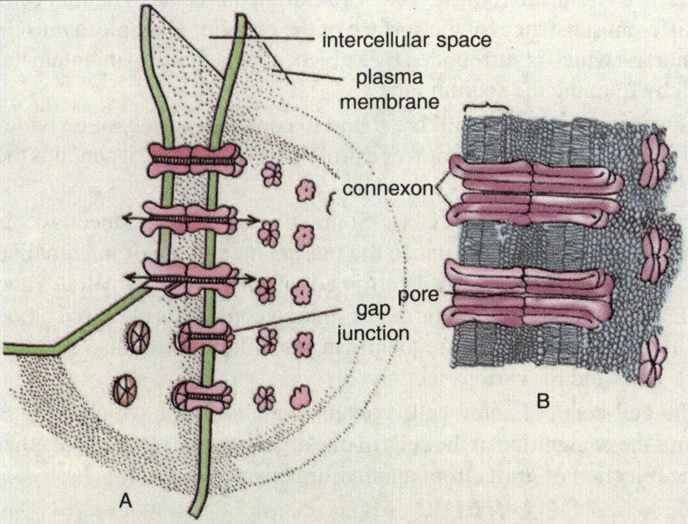

Fig. 5.26. Gap junctions. A— Location of gap junctions between two cells. The channels are made by particles in each membrane that traverse the intercellular space. The flow of fluid between the cells is indicated by arrows. B—Finer structure of unit structure or connexion of gap junction. The channel has a pore about 2 nm and is formed by two hexamers (six subunits) traversing the lipid bilayers of two plasma membranes (after De Robertis and De Robertis Jr., 1987).

The cell coat is a 10 to 20 nm thick layer and is in direct contact with the outer leaflet of the plasma membrane. In *Amoeba*, the cell coat is formed by fine filaments — 5 to 8 nm thick and 100 to 200 nm long. Chemically, the cell coat has negatively charged sialic acid termini, on both glycoproteins and gangliosides. This acid tends to bind the ions of Ca^{2+} and Na^{+}. The strength of the cell coat varies from cell to cell. For example, the cell coat of the intestinal epithelium is quite strong—it resists vigorous mechanical and chemical attacks; in other cells the coat is labile and may be depleted by washing or enzyme exposure (**Luft**, 1976).

The cell coat is the secretion product of the cell that is incorporated into the cell surface and undergoes continuous renewal. As already discussed, the glycoproteins of glycocalyx are synthesized on the ribosomes of RER and their final assembly with oligosaccharide moiety is attained in the Golgi apparatus.

Extracellula Materials

In certain cells, outside the cell coat proper and the fuzzy layer exist the **extracellular materials**, *e.g.*, jelly coat of eggs of fishes and amphibians, the basal laminae of epithelia, the matrix material in which cartilage and bone cells are embedded and the cell wall of plant cells. In these extracellular materials the most conspicuous components are **collagens** and **mucopolysaccharides** (glycosaminoglycans).

Functions of Cell Coat

In addition to the protection of the plasma membrane, the cell coat performs the following important functions :

(i) Filtration. The extraneous coats sometimes act as filters. For instance, the extraneous coats surrounding the blood capillaries of most vertebrates, especially the kidney glomerulus act as filter and regulate the passage of molecules through it. The extracellular coats of connective tissues contain the chemical compound **hyaluronate** which controls the diffusion.

(ii) Maintenance of the micro-environment of the cell. The extraneous coats of animal cells can affect the concentrations of different substances at the surface of the cell. For example, a muscle cell with its excitable plasma membrane which is surrounded by a glycocalyx is found to maintain the micro-environment of muscle cell by trapping the sodium ions.

(iii) Enzymes. The cell coat of intestinal microvilli are found to contain a variety of enzymes which are involved in the terminal digestion of carbohydrates and proteins. For example it contains the enzyme alkaline phosphatase.

(iv) Immunological properties of the extraneous coats. Some substances of extraneous coats provide immunological properties to the cell. As for instance the plasma membrane of mammalian erythrocytes is found to contain some specific, genetically determined substances (carbohydrates and proteins) corresponding to the A, B and O blood groups. The major sialoglycoproteins of the red blood cell membrane carry the M and N antigens that appear infrequently in man. The cell coat also contains the receptor sites for the influenza virus and for various lectins.

(v) Histocompatibility. The cell coats of some cells contain some antigens which provide **histocompatibility**, *i.e.*, they permit the recognition of the cells of one organism and rejection of other cells that are foreign to it (*e.g.*, the rejection of grafts from another organism).

CELL WALL

The plant cell is always surrounded by a **cell wall** and this feature distinguishes them from animal cells. The cell wall is a non-living structure which is formed by the living protoplast (A plant cell without its cell wall is called a **protoplast** ; **Alberts** *et al.*,1989). In most of the plant cells, the cell wall is made up of cellulose, hemicellulose, pectin and protein. In many fungi, the cell wall is formed of chitin and in bacteria, the cell wall contains protein-lipid-polysaccharide complexes. Thus, the cell wall is a rigid and protective layer around the plasma membrane which provides the mechanical support to the cell. The cell wall also determines the shape of plant cells. Due to the shape of cell walls many types of plant cell as the parenchymatous, collenchymatous, etc., have been recognised.

Chemical Composition

Chemically speaking, the plant cell wall is composed of a variety of polysaccharides (carbohydrates), lipids, proteins and mineral deposits, all exhibiting distinct staining reactions (Table 5-5).

Table 5-5. Chemical nature and staining reaction of various components of plant cell wall.

Substance	Chemical unit	Staining reaction
1. Cellulose	Glucose	Chlorzinc iodide (stains violet)
2. Hemicellulose	Arabinose, xylose, mannose, glucose and galactose	No specific stain
3. Pectin	Glucuronic and galacturonic acid	Ruthenium red
4. Lignin	Coniferyl alcohol (*e.g.*, hydroxy-phenyl propane)	Phloroglucinol hydrochloride (stains rose); chlorzinc iodide (stains yellow)
5. Cuticular substances	Fatty acids	Sudan III (stains orange)
6. Mineral deposits	Calcium and magnesium as carbonates and silicates	–

The polysaccharides of cell wall include cellulose, hemicelluloses, pectin compounds and lignins.

(1) **Cellulose** is a linear, unbranched polymer, consisting of straight polysaccharide chains made of glucose units linked by 1-4 β- bonds (called **glycosidic bonds** ; see Chapter 4) (**Note :** Complete hydrolysis of cellulose yields D-glucose and its partial hydrolysis yields disaccharide units, **cellobiose**.) These are the glucan chains which by intra-and intermolecular hydrogen bonding produce the structural units known as **microfibrils**, observable under electron microscopy and having toughness like the rubber. Each microfibril is ribbon-like flat fibre being 10 nm wide and 3 nm thick (or 25 to 30 nm in diameter) and is composed of about 2000 glucan chains in it. According to a classical estimate, each cellulose microfibril comprises three **micelles** or **elementary fibrils** : each elementary fibril contains about 100 cellulose molecules and each cellulose molecule is made up of 40 to 70 glucan chains (see **Thorpe**, 1984) (*i.e.*, One microfibril = 3 × 100 × 70 = 21000 glucan chains). Often numerous microfibrils get associated to form the **macrofibrils** having up to 0.5 μm diameter and observable under the light microscopy. Cellulose is synthesized by a wide variety of cells that include bacteria (*e.g.*, acetobacter,agrobacter and rhizobium), algae, fungi, cryptogams and seed plants. (2) **Hemicelluloses** are short but branched heteropolymers of various kinds of monosaccharides such as arabinose, xylose, mannose, galactose, glucose and uronic acid. Some of the common hemicelluloses go under the names xylans, arabinoxylans, glucomannans, galactomannans and xyloglucans.

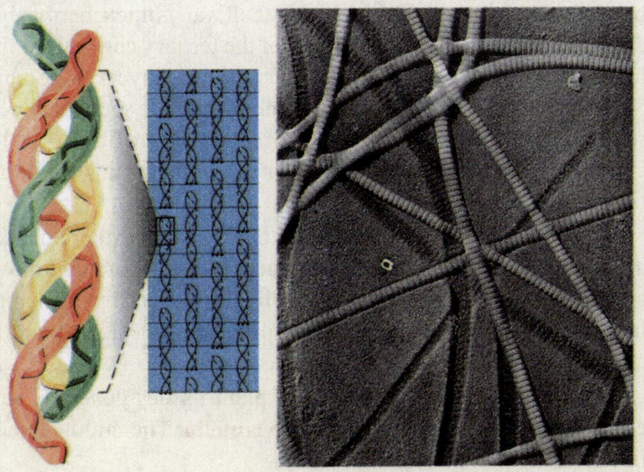
The structure of collagen.

(3) **Pectins** are water soluble, heterogeneous branched polysaccharides that contain many negatively charged D-galacturonic acid residues along with D-glucuronic acid residues. Because of their negative charge, pectins are highly hydrated and intensely bind cations. When Ca^{2+} is added to a solution of pectin molecules, it cross-links them to produce a semirigid gel. Such Ca^{2+} cross-links are thought to help hold cell-wall component together. (4) **Mannan** is a homopolysaccharide of mannose and is found in the cell wall of yeast, fungi and bacteria. (5) **Agar** is a polysaccharide, found in the cell wall of sea weeds and containing D-and L-galactose residues. (6) **Lignin** is a biological plastic and non-fibrous material. It occurs only in mature cell walls and is made of an insoluble hydrophobic aromatic polymer of phenolic alcohols (*e.g.*, hydroxyphenyl propane). (7) The **chitin** is a polymer of glucosamine.

Glycoproteins (present up to 10 per cent in primary cell wall) are hydroxyproline- rich proteins (like the collagen). In them, many short oligosaccharide side chains are attached to hydroxyproline and serine side chains. Thus, more than half the weight of glycoprotein is carbohydrate. These glycoproteins are known to act like the glue to increase the strength of the wall.

Cutin is also a biological plastic and is made of fatty acids (waxes). **Suberin** is a water-resistant substance, comprisig of fatty acids and found in the cork and cell wall of many plants. **Sporopollenin** is a lipoidal polymer forming tough wall (with species-specific patterns) of pollen grains.

Mineral deposits occur in cuticle in the form of calcium and magnesium carbonates and silicates. Deposits of calcium compounds are found in the cell wall of cruciferous and cucurbitaceous plants. Silicate deposits are common in the cell wall of Graminae family.

Structure

The cell wall is complex in nature and is differentiated in the following layers :
(i) Primary cell wall; (ii) Secondary cell wall; (iii) Tertiary cell wall.

(i) Primary cell wall. The first formed cell wall is known as primary cell wall. It is the outermost layer of the cell and in the immature meristematic and parenchymatous cells it forms the only cell wall. The primary cell is comparatively thin and permeable. Certain epidermal cells of the leaf and the stem also possess the cutin and cutin waxes which make the primary cell wall impermeable. The primary cell wall of the yeast and the fungi is composed of the chitin.

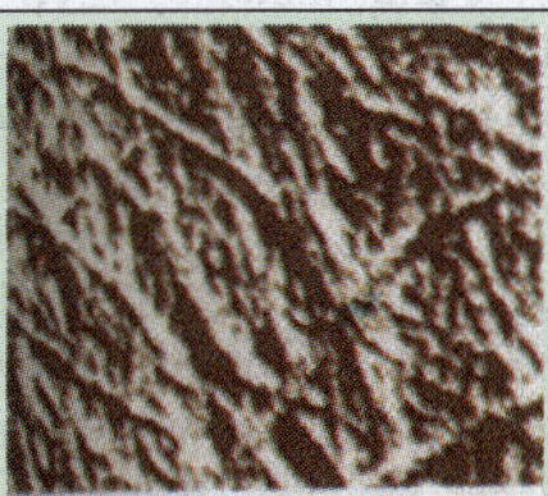

SEM of primary cell wall

SEM of a secondary cell wall

(ii) Secondary cell wall. The primary cell wall is followed by secondary cell wall. The secondary cell wall is thick, permeable and lies near the plasma membrane of the tertiary cell wall, if the latter occurs. It is composed of three concentric layers (S_1, S_2 and S_3) which occur one after another. Chemically the secondary cell wall is composed of compactly arranged macrofibrils of the cellulose, in between which sometimes occurs lignin as a interfibrillar material.

(iii) Tertiary cell wall. In certain plant cells, there occurs another cell wall beneath the secondary cell wall which is known as tertiary cell wall. The tertiary cell wall differs from the primary and secondary cell wall in its morphology, chemistry and staining properties. Besides the cellulose, the tertiary cell wall consists of another chemical substance known as the xylan.

Middle lamella. The cells of plant tissues generally remain cemented together by an intercellular matrix known as the **middle lamella**. The middle lamella is mainly composed of the pectin, lignin and some proteins.

Ultrastructure

Electron microscopy has shown that the cell wall is constructed on the same architectural principle which applied well in the construction of animal bones and such common building materials as fibre glass (plastic + glass fibres) or reinforced concrete (concrete + metal framework), *i.e.*, strong fibres (*e.g.*, cellulose microfibrils) resistant to tension embedded in an amorphous matrix (comprising hemicellulose, pectin and proteins) resistant to compression. In the primary cell wall, the fibres and matrix molecules are cross-linked by a combination of covalent bonds and non-covalent bonds to form a highly complex structure whose composition is generally cell-specific (Fig. 5.28). In fact, hemicellulose molecules (*e.g.*, **xyloglucans**) are linked by hydrogen bonds to the surface of the cellulose microfibrils. Some of these hemicellulose molecules are cross-linked in turn to acidic pectin molecules (*e.g.*, rhamnogalacturonans) through short neutral pectin molecules (*e.g.*, arabinogalactans). Cell wall glycoproteins are tightly woven into the texture of the wall to complete the structure of matrix.

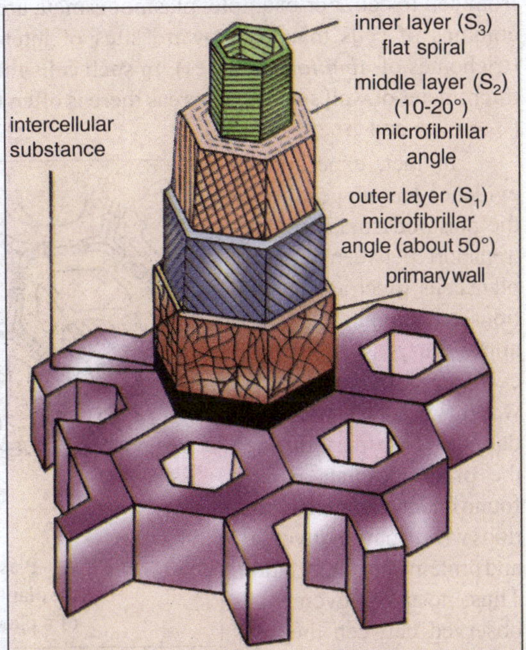

Fig. 5.27. Structure of a cell wall showing middle lamella, primary cell wall and three regions (S_1, S_2 and S_3) of the secondary cell wall (after Thorpe, 1984).

In the multilamellar secondary cell wall, cellulose microfibrils are laid down in layers, the microfibrils of each layer running roughly parallel with each other but at an angle to those in other layers (Fig. 5.27).

Plasmodesmata. Every living cell in a higher plant is connected to its living neighbours by fine cytoplasmic channels, each of which is called a **plasmodesma** (Gr., *desmos* = ribbon, ligament; plural, **plasmodesmata**) which pass through the intervening cell walls. The plasma membrane of one cell is continuous with that of its neighbour at each plasmodesma. A plasmodesma is a roughly cylindrical, membrane-lined channel with a diameter of 20 to 40 nm. Running from cell to cell through the centre of most plasmodesmata is a narrower cylindrical structure, the **desmotubule**, which remains, continuous with elements of the SER membranes of each of the connected cells. Between the outside of the desmotubule and the inner face of the cylindrical plasma membrane is an **annulus of cytosol**, which often appears to be constricted at each end of the plasmodemata. These constrictions may regulate the flux of molecules through the annulus that joins the two cytosols (**Gunning**, 1976, 1983).

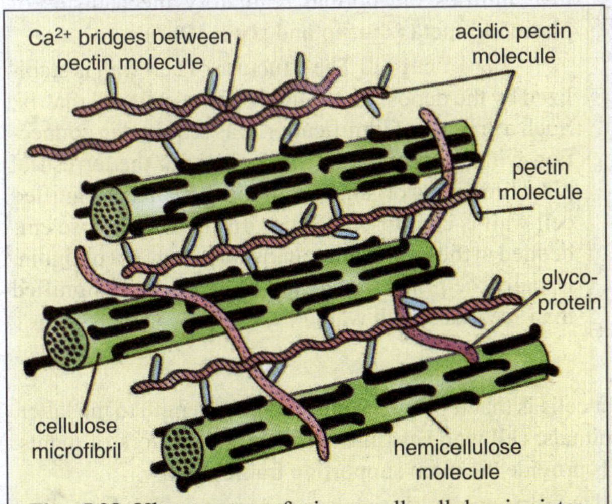

Fig. 5.28. Ultrastructure of primary cell wall showing interconnections between the two major components of the primary cell wall : the cellulose microfibrils and the matrix (after Alberts *et al.*, 1989).

Plasmodesmata are formed around the elements of smooth endoplasmic reticulum that become trapped during cytokinesis (of mitotic cell division) within the new cell wall that will bisect the parental cell. Plasmodesmata function in intercellular communication, *i.e.*, they allow molecules to pass directly from cell to cell. For example, plasmodesmata are especially common and abundant in the walls of columns of cells that lead toward sites of intense secretion, such as in nectar-secreting glands (trichomes of *Abutilon* nectaries). In such cells there may be 15 or more plasmodesmata per square micrometer of wall surface, whereas there is often less than 1 per square micrometer in other cell wall (**Gunning** and **Hughes**, 1976).

In fact, experimental evidence has suggested that the plasmodesmata mediate transport between adjacent plant cells, much as gap junctions of animal cells. They allow the passage of molecules with molecular weights of less then 800 daltons. Transport through the plasmodesmata is also found under complex regulations which may involve Ca^{2+} and protein phosphorylation. Thus, no dye movement is observed between the cells of the rooot cap and the root apex or between epidermal and cortical cells in either roots and shoots. However, certain plant viruses such as TMV can enlarge plasmodesmata in order to use this route to pass from cell to cell. Tobacco mosaic virus is known to synthesize a protein, called **P30** (30,000 dalton M.W.) that nullifies the normal regulatory mechanisms of plasmodesmata (**Zaitlin** and **Hull**, 1987).

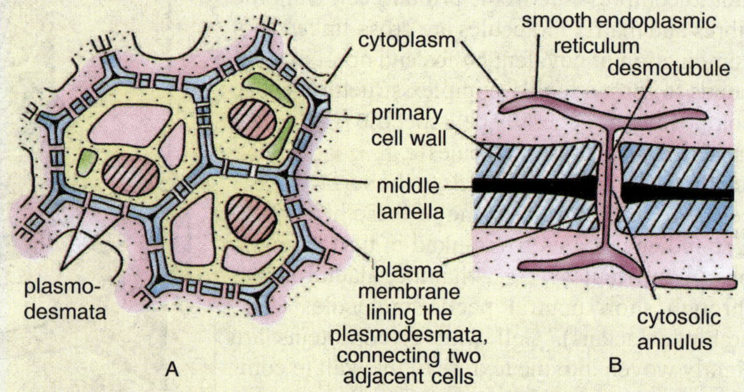

Fig. 5.29. A—Plasmodesmata pierce the cell wall and connect all cells in the plant together to form symplast ; B—Details of the structure of a plasmodesma (after Alberts *et al*., 1989).

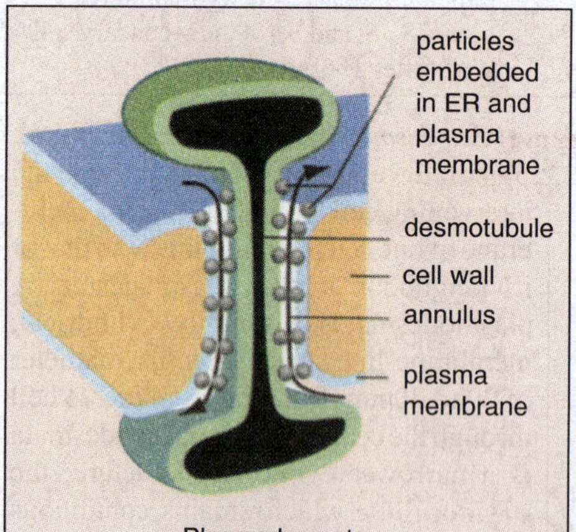

Plasmodesmata.

Lignification. The structure of cell wall is stabilized by the deposition of lignin in the cell wall matrix. Such a process of lignification was required in connection with the transition from aquatic to the terrestrial plant life during organic evolution of plants. A lignified cell wall is composed of microfibrils of cellulose embedded in the matrix containing large amount of lignin. Usually the primary cell wall becomes more lignified than secondary cell wall.

Functions of Cell Wall

The chief function of cell wall in plant cells is that it provides mechanical strength to the latter. Like the exoskeleton or endoskeleton of animals, cell wall acts like a skeletal framework of plants. Particularly in vascular plants, the cell walls provide the main supporting framework.

Despite its strength, the plant cell wall is fully **permeable** to water and solutes. This is because the matrix is riddled with minute water-filled channels through which free diffusion of water and water soluble substances such as gases, salts, sugars, hormones and like can take place. Moreover, the molecules of the matrix are strongly hydrophilic ("water-loving") with the result that in normal circumstances the cell wall is saturated with water like a sponge (*e.g.*, primary cell wall is 60 per cent water by weight). The cross-linked structure of the cell wall is, however, found to slightly impede the diffusion of small molecules such as water, sucrose and K^+. The average diameter of the spaces between the cross-linked macromolecules in most cell wall is about 5 nm, this is small enough to make the movement of any globular macromolecules with a M.W. much above 20,000 daltons extremely slow. Therefore, plants must subsist on molecules of low molecular weight, and any intercellular signaling molecules that have to pass through the cell wall must also be small and water soluble. In fact, most of the known plant signaling molecules, such as growth regulating substances—auxins, cytokinins and gibberellins—have molecular weights of less than 500 daltons.

During lignification, lignin is deposited in spaces between the cellulose molecules, making the cell wall much more rigid, and rendering it **impermeable**. Once lignification is complete the protoplasm can no longer absorb materials from outside the cell, which, therefore, dies. Hence, *lignified tissue is always dead*. Thus, a lignified tissue becomes well adapted for two types of functions: (1) It provides the mechanical strength due to its ligno-cellulose composition. (2) It transports water and salts, since, lignification involves loss of the protoplasm resulting in the formation of a hollow waterproof tube.

Origin and Growth of Cell Wall

The mechanism of cell wall formation includes the following steps :

1. Formation of matrix. Most cell wall matrix components are transported via vesicles derived from the Golgi apparatus and secreted by exocytosis at the plasma membrane. Golgi apparatus of the plant cells is involved mainly in producing and secreting a very wide range of extracellular polysaccharides, rather than the glycoproteins typical of animal cells. This is perhaps due to following facts : (1) Each of the polysaccharides of cell wall matrix is made of two or more sugars ; (2) At least 12 types of monosaccharides are used in their polymerization ; (3) Most of these polysaccharides are branched ; and (4) Many covalent modifications are introduced in the polysaccharides after they are synthesized. It is estimated that several hundred different enzymes are engaged in the assembly of the polysaccharide components needed to form a typical primary cell wall. Most of these enzymes are found in the endoplasmic reticulum and Golgi apparatus. Some enzymes, which are concerned with later covalent modifications of the polysaccharides, are found in the cell wall itself.

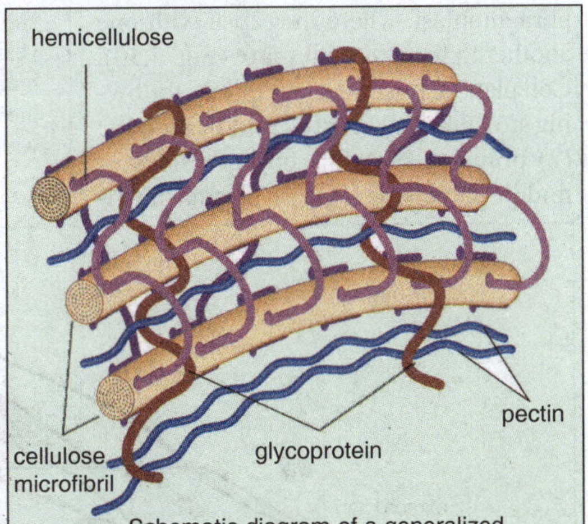

Schematic diagram of a generalized plant cell wall.

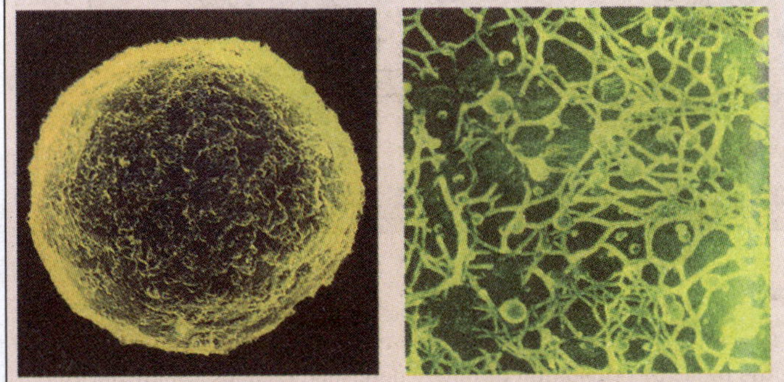
Cell wall growth studied in a moss leaf cell; progressive growth of cell wall microfibrils after 4 hours (left) and after 10 hours (right).

Since, the cell wall varies in composition and morphology at different locations around the cell, the Golgi-derived vesicles are directed at specific regions of the plasma membrane by the help of cytoskeleton (*i.e.*, microtubules and microfilaments). For example, we can consider the case of formation of new primary cell wall that separates two daughter cells after the karyokinesis of mitosis. At the end of telophase, a barrel-shaped or disc-like region, called **phragmoplast** (Gr., *phragma* = hedge, enclosure; *plasso* = to form) forms in the plane of former spindle equator. The phragmoplast comprises a double-ring of short microtubules on either side of, and terminating at, the division plane, and a set of microfilaments are coaligned with the microtubules. Golgi derived vesicles containing cell wall precursors, especially **pectin**, are guided inward along these oriented microtubules until they reach the phragmoplast, where they fuse with one another to form the **cell plate** (Fig. 5.30). Cell plate, at this stage, comprises following structures : (1) central middle lamella; (2) primary cell walls on both sides of middle lamella ; and (3) plasma mem-

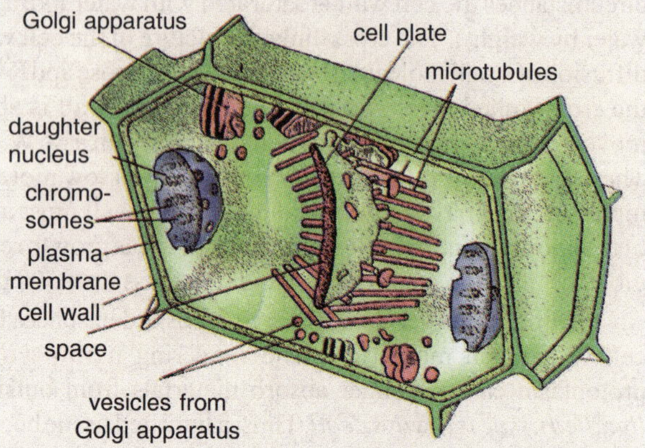

Fig. 5.30. Schematic representation of a cell of a higher plant as seen at telophase in mitosis. In the phragmoplast region, a region of membranes and microtubules, a cell plate forms and grows until it separates the cytoplasm into two daughter cells. The cell plate develops as a membrane-delimited structure enclosing a space in which new cell-wall will form. The Golgi apparatus contributes many vesicles to the phragmoplast membrane; the vesicle membranes apparently are incorportated into the membrane of the cell plate and the vesicle contents enter the forming cell wall (after Novikoff and Holtzman, 1970).

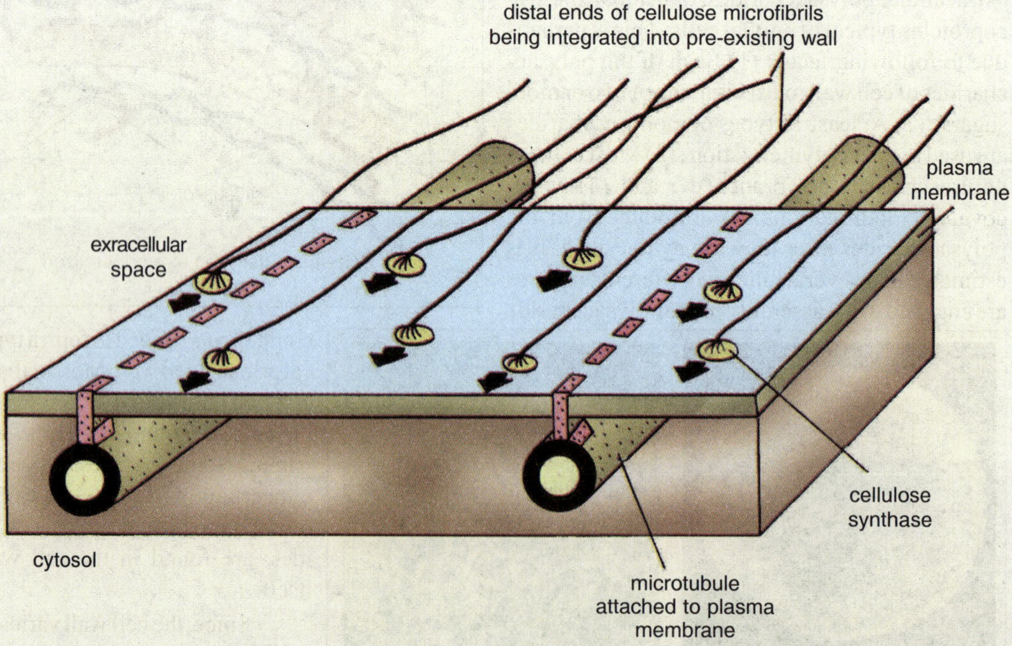

Fig. 5.31. A model explaining the mode of orientation of newly deposited cellulose microfibrils according to orientation of cortical microtubules (after Alberts *et al*, 1989).

brane lining cytoplasm of each daughter cell. Ultimately, the microtubular ring moves centrifugally outward as Golgi vesicles continue to add precursors to the growing cell plate. The cell plate fuses with the mother cell wall to create two separate daughter cells. It is not clear which component of the phragmoplast—the microtubules or the actin filaments (or both)—are responsible for the movement and guidance of the Golgi derived vesicles (see **Alberts** *et al*, 1989).

2. Synthesis and orientation of cellulose microfibrils. In most plants, cellulose is synthesized on the external surface of the cell by a plasma membrane bound enzyme complex, called **cellulose synthetase** which uses a sugar nucleotide precursor supplied from the cytosol, probably UDP- glucose (**Delmer**, 1987). As they are being synthesized, the nascent cellulose chains spontaneously assemble into **microfibrils** that form a layer on the surface of the plasma membrane (a lamella) in which all the microfibrils have more or less the same alignment (**Note.** Each cellulose molecule has a polarity, having a 1' and a 4' end). Cellulose synthetase complexes are thought to be associated with the ends of growing microfibrils and the sugars present in the extracellular matrix are polymerized into cellulose at these "terminal complexes". Extension of a cellulose microfibril is presumably achieved by lateral movement of the enzyme complex in the fluid phase of plasma membrane, with the microfibril "spun out" on the outer surface of the membrane behind the moving enzyme complex (Fig. 5.31). The direction in which the complex moves and the orientation of the microfibril depend on some interactions between the membrane complex and the underlying cytoplasmic microtubules (*i.e.*, microtubules of cell, cortex, **Herth**, 1985). Because the cellulose is synthesized at the plasma membrane, each new wall lamella forms internally to the last formed lamella. The cell wall, therefore, consists of concentrically arranged lamellae, with the oldest on the outside (**Brown**, 1985).

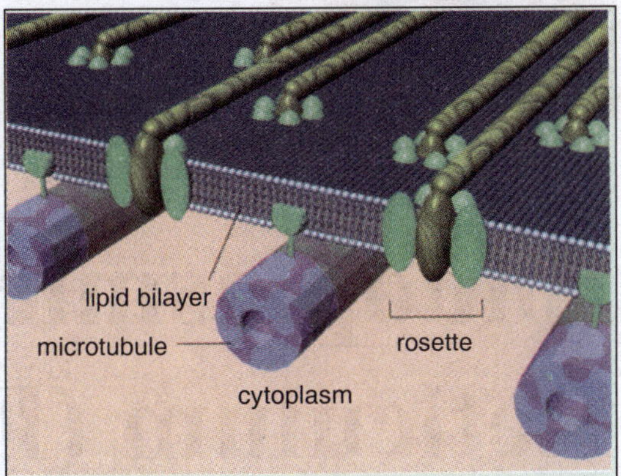

A model of cellulose fibril deposition

REVISION QUESTIONS

1. Describe various models of plasma membrane and explain which of these models is dynamic and why ?
2. Describe the 'Fluid mosaic model' of the plasma membrane. On the basis of this model explain different functions of the plasma membrane.
3. Describe the chemical composition of plasma membrane.
4. Discuss in detail various functions of the plasma membrane.
5. What is cell wall ? Describe the chemical composition, structure, origin and function of the plant cell wall.
6. Write short notes on the following : 1. Cell wall : its structure and function ; 2. Desmosome ; 3. Active transport ; 4. Endocytosis ; 5. Ion pumps ; 6. Nexus ; 7. Receptor-mediated endocytosis ; 8. Plasmodesmata.
7. Differentiate between the following :
   (i) Phagocytosis from pinocytosis ;
   (ii) Passive transport from active transport ;
   (iii) Macula adherens from macula occludens ;
   (iv) Primary cell wall from secondary cell wall.
8. Draw a well labelled diagram of "Fluid mosaic model" of the plasma membrane.
9. Describe the mode of origin and growth of cell wall.

CHAPTER

6

Endoplasmic Reticulum (ER)

The cytoplasmic matrix is traversed by a complex network of inter-connecting membrane bound vacuoles or cavities. These vacuoles or cavities often remain concentrated in the endoplasmic portion of the cytoplasm; therefore, known as **endoplasmic reticulum**, a name derived from the fact that in the light microscope it looks like a "net in the cytoplasm." (Eighteenth-century European ladies carried purses of netting called **reticules**).

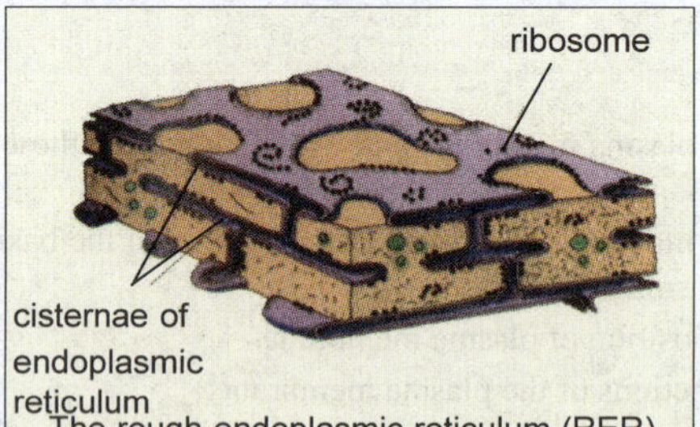

The rough endoplasmic reticulum (RER).

The name "endoplasmic reticulum" was coined in 1953 by **Porter**, who in 1945 had observed it in electron micrographs of liver cells. **Fawcett** and **Ito** (1958), **Thiery** (1958) and **Rose** and **Pomerat** (1960) have made various important contributions to the endoplasmic reticulum.

OCCURRENCE

The occurrence of the endoplasmic reticulum varies from cell to cell. The erythrocytes (RBC), egg and embryonic cells lack in endoplasmic reticulum. (**Note**. In the reticulocytes (immature red blood cells) which produce only proteins to be retained in the cytoplasmic matrix (cytosol) (*e.g.*, haemoglobin), the ER is poorly developed or non-existent, although the

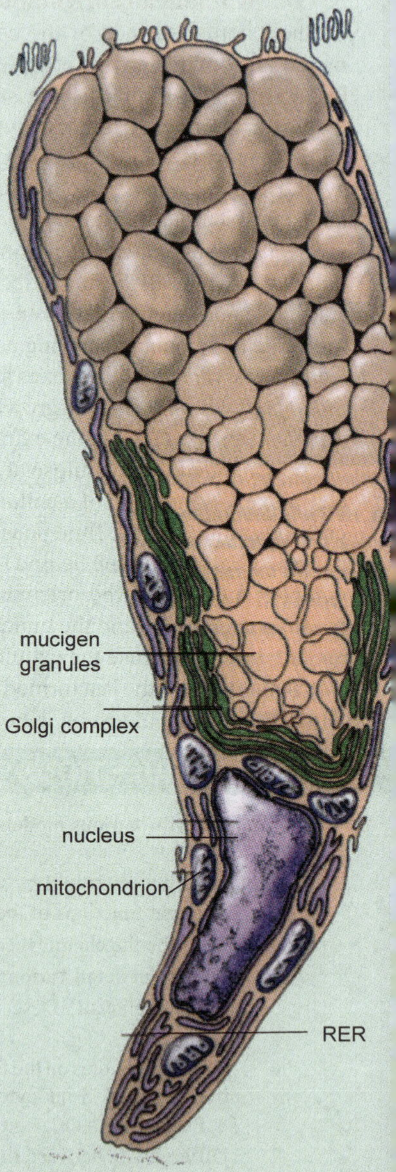

Drawing of a mucus - secreting goblet cell from the rat colon showing presence of RER.

cell may contain many ribosomes). The spermatocytes have poorly developed endoplasmic reticulum. The adipose tissues, brown fat cells and adrenocortical cells, interstitial cells of testes and cells of corpus luteum of ovaries, sebaceous cells and retinal pigment cells contain only **smooth endoplasmic reticulum (SER)**. The cells of those organs which are actively engaged in the synthesis of proteins such as acinar cells of pancreas, plasma cells, goblet cells and cells of some endocrine glands are found to contain **rough endoplasmic reticulum (RER)** which is highly developed. The presence of both SER and RER in the hepatocytes (liver cells) is reflective of the variety of the roles played by the liver in metabolism.

ER AND ENDOMEMBRANE SYSTEM

The endoplasmic reticulum is the main component of the **endomembrane system**, also called the **cytoplasmic vacuolar system** or **cytocavity network**. This system comprises following structures: (1) The **nuclear envelope**, consisting of two non-identical membranes, one opposed to the nuclear chromatin and other separated from the first membrane by a perinuclear space (both forming a cisternae), the two membranes being in contact at the nuclear pores; (2) The **endoplasmic reticulum**; and (3) the **Golgi apparatus,** which is mainly related to some of the terminal processes of cell secretion. **GERL** (or Golgi, ER and lysosome) refers to a special region of endomembrane system, which is more related to the Golgi apparatus and is involved in the formation of lysosomes.

The entire endomembrane system represents a barrier separating cytoplasmic compartments. The membrane of each component of this system has two faces : (i) the **cytoplasmic** or **protoplasmic face** and (ii) the **luminal face** (Fig. 6.1). The luminal face borders the perinuclear cisternae, the cavities of ER and SER, and the Golgi elements. It also corresponds to the interior of the secretory granules, the lysosomes and peroxisomes and also to faces of mitochondrial membranes confronting to outer mitochondrial chamber.

MORPHOLOGY

Morphologically, the endoplasmic reticulum may occur in the following three forms : 1. Lamellar form or cisternae (A closed, fluid-filled sac, vesicle or cavity is called **cisternae**) ; 2. vesicular form or vesicle and 3. tubular form or tubules.

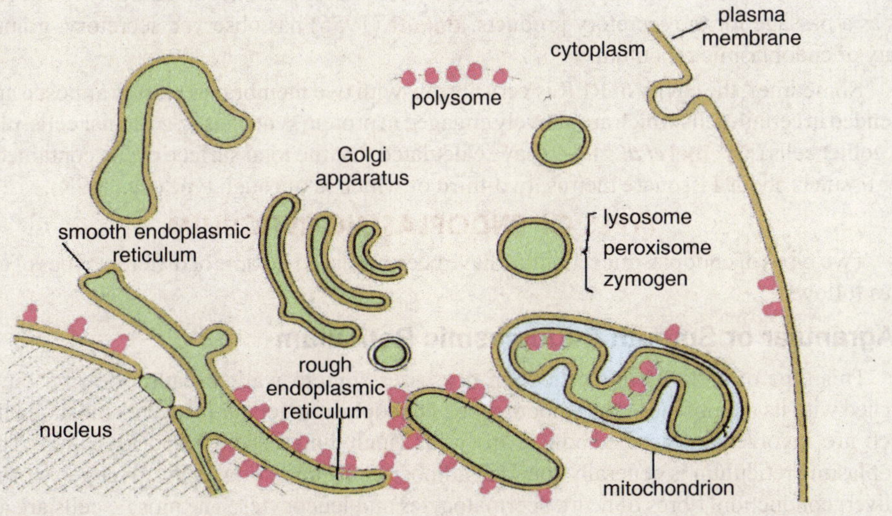

Fig. 6.1. Two faces of membranes of endomembrane system. In each membrane the luminal faces are shown in thick lines, while the cytoplasmic faces are depicted by thin lines. Ribosomes are always located on the cytoplasmic or matrix side (after De Robertis and De Robertis, Jr., 1987).

1. Cisternae. The cisternae are long, flattened, sac-like, unbranched tubules having the diameter of 40 to 50 µm. They remain arranged parallely in bundles or stakes. RER usually exists as cisternae which occur in those cells which have synthetic roles as the cells of pancreas, notochord and brain.

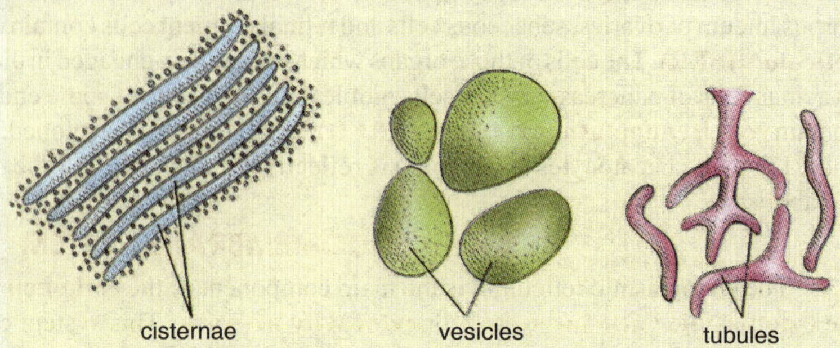

Fig. 6.2. Various components of the endoplasmic reticulum.

2. Vesicles. The vesicles are oval, membrane-bound vacuolar structures having the diameter of 25 to 500 µm. They often remain isolated in the cytoplasm and occur in most cells but especially abundant in the SER.

3. Tubules. The tubules are branched structures forming the reticular system along with the cisternae and vesicles. They usually have the diameter from 50 to 190 µm and occur almost in all the cells. Tubular form of ER is often found in SER and is dynamic in nature, *i.e.*, it is associated with membrane movements, fission and fusion between membranes of cytocavity network (see **Thorpe**, 1984).

ULTRASTRUCTURE

The cavities of cisternae, vesicles and tubules of the endoplasmic reticulum are bounded by a thin membrane of 50 to 60 A° thickness. The membrane of endoplasmic reticulum is fluid-mosaic like the unit membrane of the plasma membrane, nucleus, Golgi apparatus, etc. The membrane, thus, is composed of a bimolecular layer of phospholipids in which 'float' proteins of various sorts. The membrane of endoplasmic reticulum remains continuous with the membranes of plasma membrane, nuclear membrane and Golgi apparatus. The cavity of the endoplasmic reticulum is well developed and acts as a passage for the secretory products. **Palade** (1956) has observed secretory granules in the cavity of endoplasmic reticulum.

Sometimes, the cavity of RER is very narrow with two membranes closely apposed and is much distended in certain cells which are actively engaged in protein synthesis (*e.g.*, acinar cells, plasma cells and goblet cells). **Weibel** *et al.,* 1969, have calculated that the total surface of ER contained in 1ml of liver tissue is about 11 square metres, two-third of which is of rough type (*i.e.*, RER).

TYPES OF ENDOPLASMIC RETICULUM

Two types of endoplasmic reticulum have been observed in same or different types of cells which are as follows:

1. Agranular or Smooth Endoplasmic Reticulum

This type of endoplasmic reticulum possesses smooth walls because the ribosomes are not attached with its membranes. The smooth type of endoplasmic reticulum occurs mostly in those cells, which are involved in the metabolism of lipids (including steroids) and glycogen. The smooth endoplasmic reticulum is generally found in adipose cells, interstitial cells, glycogen storing cells of the liver, conduction fibres of heart, spermatocytes and leucocytes. The muscle cells are also rich in smooth type of endoplasmic reticulum and here it is known as **sarcoplasmic reticulum**. In the pigmented retinal cells it exists in the form of tightly packed vesicles and tubes known as **myeloid bodies**.

Glycosomes. Although the SER forms a continuous system with RER, it has different morphology. For example, in liver cells it consists of a tubular network that pervades major portion of the cytoplasmic matrix. These fine tubules are present in regions rich in glycogen and can be observed as dense particles, called **glycosomes**, in the matrix. Glycosomes measure 50 to 200 nm in diameter and contain glycogen along with enzymes involved in the synthesis of glycogen (**Rybicka**, 1981). Many glycosomes attached to the membranes of SER have been observed by electron microscopy in the liver and conduction fibre of heart.

2. Granular or Rough Endoplasmic Reticulum

The granular or rough type of endoplasmic reticulum possesses rough walls because the ribosomes remain attached with its membranes. Ribosomes play a vital role in the process of protein synthesis. The granular or rough type of endoplasmic reticulum is found abundantly in those cells which are active in protein synthesis such as pancreatic cells, plasma cells, goblet cells, and liver cells. The granular type of endoplasmic reticulum takes basiophilic stain due to its RNA contents of ribosomes. The region of the matrix containing granular type of endoplasmic reticulum takes basiophilic stain and is named as **ergastoplasm, basiophilic bodies, chromophilic substances** or **Nissl bodies** by early cytologists.

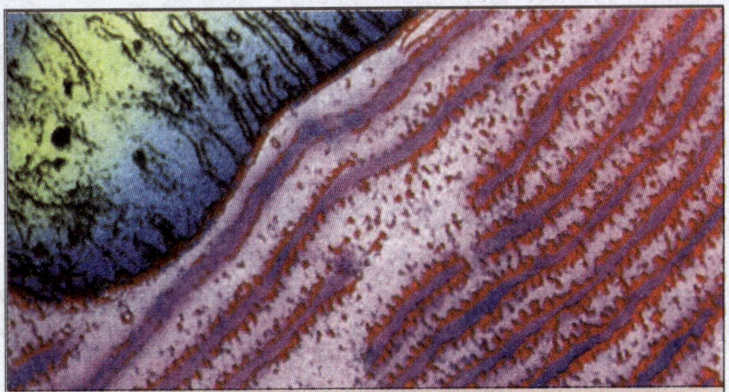

The ER extends throughout the cell. Note the clusters of ribosomes (0.02 μm wide) and the mitochondrion at the top of the photo.

In RER, ribosomes are often present as polysomes held together by mRNA and are arranged in typical "rosettes" or spirals. RER contains two transmembrane glycoproteins (called **ribophorins I** and **II** of 65,000 and 64,000 dalton MW, respectively), to which are attached the ribosomes by their 60S subunits.

Annulate Lamellae

Usually the endoplasmic reticulum has no pores or annuli in it but in certain cases the pores or annuli have been reported, *e.g.*, ER of invertebrates, ovocytes and spermatocytes of the vertebrates. These annuli resemble with the pores or annuli of the nuclear membranes. Like the annuli of nuclear membranes it contains a diaphragm across it (**Ward** and **Ward**, 1968) and possesses an octagonal symmetry (**Maul**, 1968). The annulate lamellae (pores) of the ER arise by the evagination from the nuclear envelope and have their association with the ribosomes (**Merriam** 1959; **Kessel**, 1963).

ISOLATION AND CHEMICAL COMPOSITION

The membranes of the endoplasmic reticulum can be isolated by subjecting homogenized tissues to differential centrifugation. Electron microscopy of such ER preparations reveals that the membranes disrupt to form closed vesicles (~100 nm diameter) of either a rough or a smooth form. These membranous entities were coined the term "**microsomes**" by **Claude** in 1940, and the relationship between microsomes and the elements of endoplasmic reticulum in the intact cell was established by **Palade** and **Siekevitz** in 1956.

Microsomes derived from rough ER are studded with ribosomes and are called **rough** or **granular microsomes**. The ribosomes are always found on the outside surface, the interior being biochemically equivalent to the luminal space of the ER. Homogenate also contains **smooth** or **agranular micro-somes** which lack attached ribosomes. They may be derived in part from smooth portion of the ER and in part from fragments of plasma membrane, Golgi apparatus, endosomes and mitochondria. Thus, while rough microsomes can be equated with rough portions of ER, the origin of smooth microsomes cannot be so easily assigned. However, since the hepatocytes of liver contain exceedingly large quantities of smooth ER, therefore, most of the smooth microsomes in liver homogenates are derived from smooth ER (see **Alberts** *et al.*, 1989).

As rough microsomes can be readily purified in functional form, they are especially useful for studying many biochemical processes carried out by the ER, *e.g.*, protein synthesis, glycosylation and lipid synthesis.

In rat liver, the membranes of microsomes are 60 to 70 per cent protein and 30 to 40 per cent phospholipid by weight. Thus, ER membranes contain more proteins, both in amount and kind (having about 33 types of polypeptides) than the plasma membrane. They are also richer in phosphotidyl- choline and poorer in sphingomyelin (**Thorpe**, 1984).

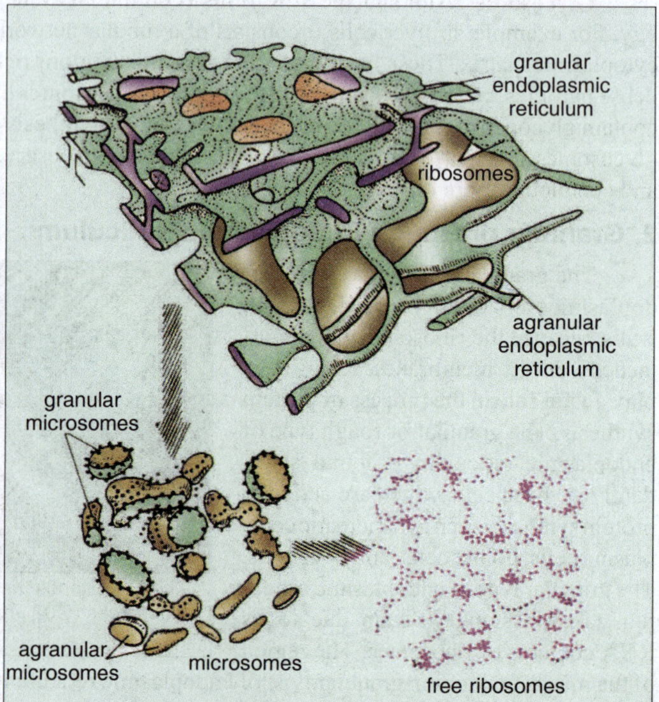

Fig. 6.3. Three-dimensional structure of endoplasmic reticulum showing microsomes and ribosomes.

ENZYMES OF THE ER MEMBRANES

The membranes of the endoplasmic reticulum are found to contain many kinds of enzymes which are needed for various important synthetic activities. Some of the most common enzymes are found to have different transverse distribution in the ER membranes (Table 6-1). The most important enzymes are the stearases, NADH-cytochrome C reductase, NADH diaphorase, glucose-6-phosphotase and Mg++ activated ATPase. Certain enzymes of the endoplasmic reticulum such as nucleotide diphosphate are involved in the biosynthesis of phospholipid, ascorbic acid, glucuronide, steroids and hexose metabolism. The enzymes of the endoplasmic reticulum perform the following important functions :

1. Synthesis of glycerides, *e.g.*, triglycerides, phospholipids, glycolipids and plasmalogens.
2. Metabolism of plasmalogens.
3. Synthesis of fatty acids.
4. Biosynthesis of the steroids, *e.g.*, cholesterol biosynthesis, steroid hydrogenation of unsaturated bonds.
5. $NADPH_2$+O_2—requiring steroid transformations: Aromatization and hydroxylation.
6. $NADPH_2$+O_2—requiring steroid transformations : Aromatic hydroxylations, side-chain oxidation, deamination, thio-ether oxidations, desulphuration.
7. L-ascorbic acid synthesis.
8. UDP-uronic acid metabolism.
9. UDP-glucose dephosphorylation.
10. Aryl-and steroid sulphatase.

Table 6.1. Transverse distribution of various enzymes in the membranes of endoplasmic reticulum (Source: Thorpe, 1984).

	Enzymes	Surface localization
1.	Cytochrome b_5 (involved in synthesis of unsaturated fatty acids)	Cytoplasmic face
2.	NADH- Cytochrome b_5 reductase	Cytoplasmic face
3.	NADP- Cytochrome c reductase	Cytoplasmic face
4.	Cytochrome P-450 (most abundant)	Both on cytoplasmic and luminal face
5.	ATPase	Cytoplasmic face
6.	5′-nucleotidase	Cytoplasmic face
7.	Nucleoside pyrophosphatase	Cytoplasmic face
8.	GDP-mannosyl transferase	Cytoplasmic face
9.	Nucleoside diphosphatase	Luminal face
10.	Glucose -6- phosphatase (histochemical marker enzyme)	Luminal face
11.	Acetanilide-hydrolysing esterase	Luminal face
12.	β- glucuronidase	Luminal face

ORIGIN OF ENDOPLASMIC RETICULUM

The exact process of the origin of endoplasmic reticulum is still unknown. But because membranes of ER resemble with the nuclear membrane and plasma membrane and also at the telophase stage the ER membranes are found to form the nuclear envelope. Therefore, it is normally assumed that the ER has originated by evagination of the nuclear membranes. **Seikevitz** and **Palade** (1960) have reported that the granular type of ER has originated first and later it synthesizes the agranular or smooth type of endoplasmic reticulum.

The synthesis of membranes of ER is found to proceed in the following direction : RER → SER. In fact, membrane biogenesis is a multi-step process involving, first, the synthesis of a basic membrane of lipid and intrinsic proteins and thereafter the addition of other constituents such as enzymes, specific sugars, or lipids. The process by which a membrane is modified chemically and structurally is called **membrane differentiation**. The ER (especially SER) is the organelle containing the main phospholipid synthesizing and translocating enzymes (*i.e.*, there occurs an intense flip-flop of lipid components). The insertion of proteins into ER membranes occurs at the level of RER. Most of these proteins are formed on membrane-bound ribosomes. However, some of these are synthesized by free ribosomes in the cytosol (cytoplasmic matrix) and then are inserted into the membrane. For example, the enzyme **NAD-cytochrome-b5-reductase** is synthesized in the cytosol (cytoplasmic matrix) and then becomes incorporated in various parts of the endomembrane system (*i.e.*, RER, SER and Golgi apparatus) and in the outer mitochondrial membrane (**Borghese** and **Gaetani**, 1980).

FUNCTIONS OF ENDOPLASMIC RETICULUM

The endoplasmic reticulum acts as secretory, storage, circulatory and nervous system for the cell. It performs following important functions:

A. Common Functions of Granular and Agranular Endoplasmic Reticulum

1. The endoplasmic reticulum provides an ultrastructural skeletal framework to the cell and gives mechanical support to the colloidal cytoplasmic martix.

2. The exchange of molecules by the process of osmosis, diffusion and active transport occurs through the membranes of endoplasmic reticulum. Like plasma membrane, the ER membrane has permeases and carriers.

3. The endoplasmic membranes contain many enzymes which perform various synthetic and metabolic activities. Further the endoplasmic reticulum provides increased surface for various enzymatic reactions.

Some Bedouin women have a smooth ER problem. Because this woman's clothing leaves little or no skin exposed to sunlight, her smooth ER may not be able to make enough of the vitamin D to maintain strong, healthy

4. The endoplasmic reticulum acts as an intracellular circulatory or transporting system. Various secretory products of granular endoplasmic reticulum are transported to various organelles as follows: Granular ER→ agranular ER → Golgi membrane→lysosomes, transport vesicles or secretory granules. Membrane flow may also be an important mechanism for carrying particles, molecules and ions into and out of the cells. Export of RNA and nucleoproteins from nucleus to cytoplasm may also occur by this type of flow (see **De Robertis** and **De Robertis**, **Jr**., 1987).

5. The ER membranes are found to conduct intra-cellular impulses. For example, the sarcoplasmic reticulum transmits impulses from the surface membrane into the deep region of the muscle fibres.

6. The ER membranes form the new nuclear envelope after each nuclear division.

7. The sarcoplasmic reticulum plays a role in releasing calcium when the muscle is stimulated and actively transporting calcium back into the sarcoplasmic reticulum when the stimulation stops and the muscle must be relaxed.

B. Functions of Smooth Endoplasmic Reticulum

Smooth ER performs the following functions of the cell :

1. Synthesis of lipids. SER performs synthesis of lipids (*e.g.*, phospholipids, cholesterol, etc.) and lipoproteins. Studies with radioactive precursors have indicated that the newly synthesized phospholipids are rapidly transferred to other cellular membranes by the help of specific cytosolic enzymes, called **phospholipid exchange proteins.**

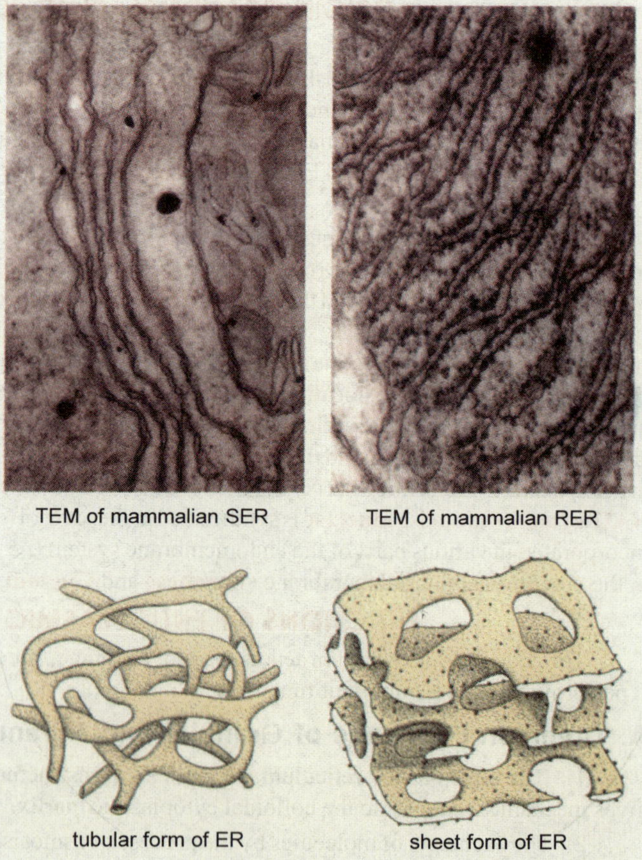

2. Glycogenolysis and blood glucose homeostasis. The process of glycogen synthesis (glycogenesis) occurs in the cytosol (in glycosomes). The enzyme **UDPG-glycogen transferase**, which is directly involved in the synthesis of glycogen by addition of **uridine diphosphate glucose (UDPG)** to primer glycogen is bound to the glycogen particles or glycosomes.

SER is found related to **glycogenolysis** or breakdown of glycogen. An enzyme, called **glucose-6- phosphatase** (a marker enzyme) exists as an integral protein of the membrane of SER (*e.g.*, liver cell). Generally, this enzyme acts as a glucogenic phosphohydrolase that catalyzes the release of free glucose molecule in the lumen of SER from its phosphorylated form in liver (Fig. 6.4). Thus, this process operates to maintain homeostatic levels of glucose in the blood for the maintenance of functions of red blood cells and nerve tissues.

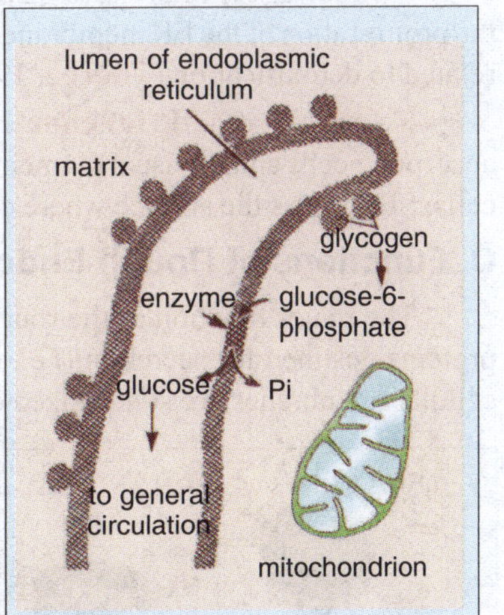

Fig. 6.4. Diagram of the intervention of the smooth endoplasmic reticulum in glycogenolysis with the consequent release of glucose. The enzyme (E), glucose-6- phosphatase, is present in the membrane and has a vectorial deposition by which it receives the glucose-6-phophate from the matrix surface. The product glucose penetrates the lumen of the endoplasmic reticulum (after De Robertis *et al.*, 1975).

3. Sterol metabolism. The SER contains several key enzymes that catalyze the synthesis of **cholesterol** which is also a precursor substance for the biosynthesis of two types of compounds— the steroid hormones and bile acids :

(i) Cholesterol biosynthesis. The cholesterol is synthesized from the acetate and its entire biosynthetic pathway involves about 20 steps, each step catalyzed by an enzyme. Out of these twenty enzymes, eleven enzymes are bounded to SER membranes, rest nine enzymes are the soluble enzymes located in the cytosol and mitochondria. Examples of SER-bound enzyme include **HMG-Co A reductase** and **squalene synthetase** (see **Thorpe**, 1984).

(ii) Bile acid synthesis. The biosynthesis of the bile acids represents a very complex pattern of enzymes and products. Enzymes involved in the biosynthetic pathway of bile acids are hydroxylases, mono-oxygenases, dehydrogenases, isomerases and reductases. For example, by the help of the enzyme **cholesterol 7α-hydroxylase**, the cholesterol is first converted into 7α-hydroxyl cholesterol, which is then converted into bile acids by the help of hydroxylase enzymes. The latter reaction requires NADPH and molecular oxygen and depends on the enzymes of Electron transport chains of SER such as **cytochrome P-450** and **NADPH-cytochrome-c-reductase**

(iii) Steroid hormone biosynthesis. Steroid hormones are synthesized in the cells of various organs such as the cortex of adrenal gland, the ovaries, the testes and the placenta. For example, cholesterol is the precursor for both types of sex hormones—estrogen and testosterone—made in the reproductive tissues, and the adrenocorticoids (*e.g.*, corticosterone, aldosterone and cortisol) formed in the adrenal glands. Many enzymes (*e.g.*, dehydrogenases, isomerases and hydroxylases) are involved in the biosynthetic pathway of steroid hormones, some of which are located in SER membranes and some occur in the mitochondria. This biosynthetic pathway has the following steps :

Acetate (Cytosol) → **Cholesterol** (SER) → **Pregnenolone** (Mitochondria) → **Progesterone** (SER) → **11-Deoxycortisol** (SER) → **Cortisol** (Mitochondria)

Progesterone (SER) → **11- Deoxycorticosterone** (SER) → **Aldosterone** (Mitochondria)

4. Detoxification. Protectively, the ER chemically modifies **xenobiotics** (toxic materials of both endogenous and exogenous origin), making them more hydrophilic, hence, more readily excreted. Among these materials are drugs, aspirin (acetyl-salicylic-acid), insecticides, anaesthetics, petroleum products, pollutants and carcinogens (*i.e.*, inducers of cancer; *e.g.*, **3-4- benzopyrene** and **3-methyl cholanthrene**).

The enzymes involved in the detoxification of aromatic hydrocarbons are **aryl hydroxylases**. It is now known that benzopyrene (found in charcoal-broiled meat) is not carcinogenic, but under the action of aryl hydroxylase enzyme in the liver, it is converted into **5, 6-epoxide**, which is a powerful carcinogen (see **De Robertis** and **De Robertis, Jr**., 1987).

A wide variety of drugs (*e.g.*, phenobarbital), when administrated to animals, they bring about the proliferation of the ER membranes (first RER and then SER) and/ or enhanced activity of enzymes related to detoxification (**Thorpe**, 1984).

5. Other synthetic functions. SER plays a role in the synthesis of triglycerides in intestinal absorptive cells and of visual pigments from vitamin A by pigmented epithelial cell of retina. In plant cells, SER forms the surface where cellulose cell walls are being formed.

C. Functions of Rough Endoplasmic Reticulum

The major function of the rough ER is the synthesis of protein. It has long been assumed that proteins destined for secretion (*i.e.*, export) from the cell or proteins to be used in the synthesis of cellular membranes are synthesized on rough ER-bound ribosomes, while cytoplasmic proteins are

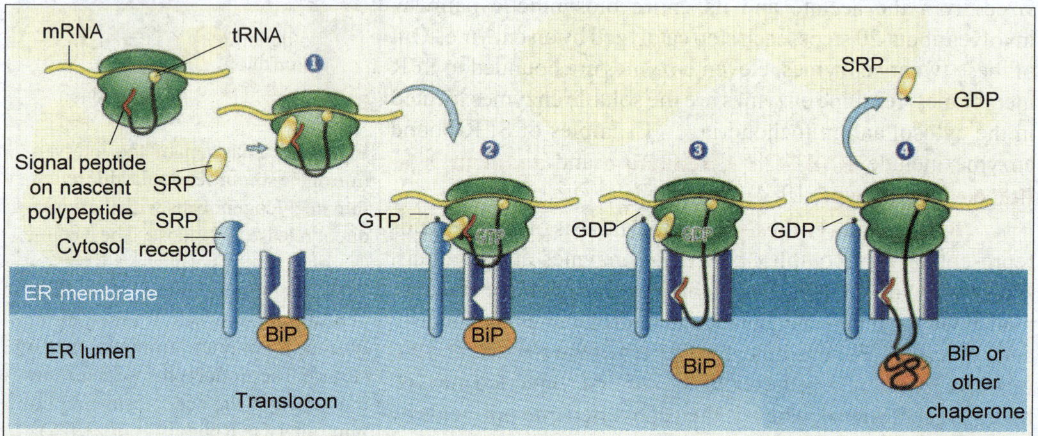

A schematic model for the synthesis of a secretory protein on a membrane - bound ribosome of the RER.

translated for the most part on free ribosomes. In fact, the array of the rough endoplasmic reticulum provides extensive surface area for the association of metabolically active enzymes, amino acids and ribosomes. There is more efficient functioning of these materials to synthesize proteins when oriented on a membrane surface than when they are simply in solution, mainly because chemical combinations between molecules can be accomplished in specific geometric patterns.

The membrane-bound ribosomes are attached with **specific binding sites** or **receptors** of rough ER membrane by their large 60S subunit, with small or 40S subunit sitting on top like a cap. These receptors are membrane proteins which extend well into and possibly through the lipid bilayer. The receptor proteins with bound ribosomes can float laterally like other membrane proteins and may facilitate formation of the polysome and probably translation which requires that mRNA and ribosome move with respect to each other.

Further, the secretory proteins, instead of passing into the cytoplasm, appear to pass instead into the cisternae of the rough ER and are, thus, protected from protease enzymes of cytoplasm. It is calculated that about 40 amino acid residues long segment at the— COOH end of the nascent protein remains protected inside the tunnel of 'free' or 'bound' ribosomes and rest of the chain, with—NH_2 end, is protected by the lumen of RER. The passage of nascent polypeptide chain into the ER cisterna takes place during translation leaving only a small segment exposed to the cytoplasm at any one time.

How the polypeptide chain gets through the lipid bilayer is not so clear, but it is quite reasonable to propose that the membrane proteins serving as ribosomal receptors also has a very fine channel through its core that opens into the cisterna of the rough ER. The chain may have great flexibility, permitting the amino acids to snake their way single file through the proposed pore. As soon as growing polypeptide chain reaches the cisterna, it folds into its secondary and tertiary structures and thus trapped in the cisterna of the rough ER.

Protein glycosylation. The covalent addition of sugars to the secretory proteins (*i.e.*, glycosylation) is one of the major biosynthetic functions of rough ER. Most of the proteins that are isolated in the lumen of RER before being transported to the Golgi apparatus, lysosomes, plasma membrane or extracellular space, are **glycoproteins** (a notable exception is albumin). In contrast, very few proteins in the cytosol (cytoplasmic matrix) are glycosylated and those that carry them have a different sugar modification.

The process of **protein glycosylation** in RER lumen is one of the most well understood cell biological phenomena. During this process, a single species of **oligosaccharide** (which comprises N-acetyl-glucosamine, mannose and glucose, containing a total of 14 sugar residues) is transferred to

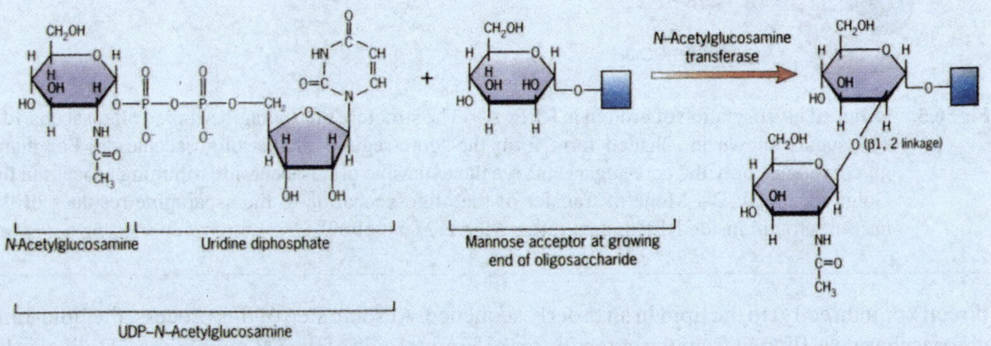

Example of a reaction catalyzed by a glycosyltransferase.

proteins in the ER. Because it is always transferred to the NH_2 group on the side chain of an asparagine residue of the protein, this oligosaccharide is said to be **N-linked** or **asparagine-linked** (Fig.6.5 A). The transfer is catalyzed by a membrane-bound enzyme (*i.e.*, **glycosyl transferase**) with its active site exposed on the luminal surface of the ER membrane. The preformed precursor oligosaccharide is transferred *en bloc* to the target asparagine residue in a single enzymatic step almost as soon as that residue emerges in the lumen of ER during protein translocation (Fig.6.5 B). Since most proteins are co-translationally imported into the ER, N-linked oligosaccharides are almost always added during protein synthesis, ensuring maximum access to the target asparagine residues, which are present in the sequences–*Asn-X-Ser* or *Asn-X-Thr* (where *X* is amino acid except proline). *These two sequences, thus, function as signals for N-linked glycosylation.*

The precursor oligosaccharide is held in the ER membrane by a special lipid molecule, **dolicol** (the carrier). The oligosaccharide is linked to the dolicol by a high-energy **pyrophosphate bond** which activates the oligosaccharide for its transfer from the lipid to an asparagine side chain (*i.e.*, it provides activation energy for the glycosylation reaction). The oligosaccharide is built up sugar by sugar on the membrane-bound dolicol (towards the cytosolic side) prior to its transfer to a protein. Sugars are first activated in the cytosol (cytoplasmic matrix) by the formation of **nucleotide-sugar intermediates** (*e.g.*, UDP-glucose, UDP-N-acetylglucosamine, and GDP-mannose), which then donates their sugar

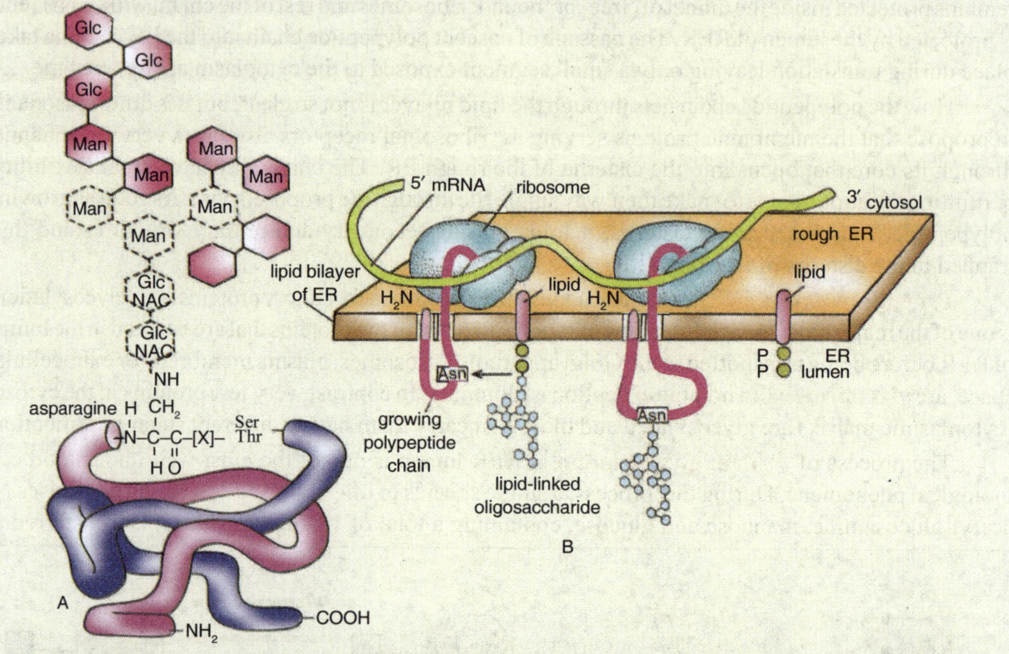

Fig. 6.5. N-linked glycosylation of protein in RER. A—The structure of asparagine-linked oligosaccharide. The sugars shown in shaded form form the 'core-region' of this oligosaccharide. For many glycoproteins, only the core sugars survive the extensive oligosaccharide trimming process in the Golgi apparatus; B—Mode of transfer of the oligosaccharide to the asparagine residues of the nascent protein inside RER lumen (after Alberts *et al.*, 1989).

(directly or indirectly) to the lipid in an orderly sequence. At some step of this process, the lipid-linked oligosaccharide is flipped from the cytosolic to the luminal side of the ER membranes. Dolicol is long and very hydrophobic : its 22 five-carbon units can span the thickness of lipid bilayer more than three times, so that the attached oligosaccharide is firmly anchored to the membrane.

While still in RER lumen, three glucose residues and one mannose residue are quickly removed from the oligosaccharides of most glycoproteins. Such oligosaccharide "trimming" or "processing" continues in the Golgi apparatus (Hirschberg and Snider, 1987; Kornfeld and Kornfeld, 1985). If a glycoprotein is to contain a

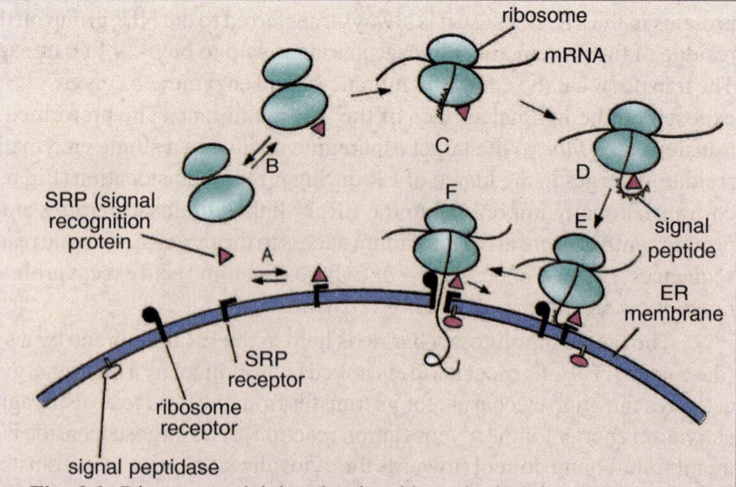

Fig. 6.6. Diagram explaining the signal hypothesis (after De Robertis and De Robertis, Jr., 1987).

terminal glucose, fucose or sialic acid, then those sugars are added in the Golgi apparatus where the appropriate sugar transferase enzymes are localized.

The signal hypothesis. The proteins for the secretion, the lysosomes and the membrane formation, are synthesized on the membrane bound ribosomes. The free and bound ribosomes were found to be continuously interchanging and show no differences between them. The **signal hypothesis** was proposed by **Blobel** and **Sabatini** (1971) to explain how the ribosomes which are meant for the biosynthesis of secretory type proteins get specifically attached to RER membranes. According to this hypothesis, the mRNA is able to recognize free or bound ribosomes. It is postulated that the mRNA for secretory proteins contain a set of **special signal codons** localized after the initial codon AUG. Once the ribosome "recognizes" the signal the ribosome becomes attached to the membrane of ER and the polypeptide penetrates. It is also postulated that at the luminal surface there is a **signal peptidase enzyme** that removes the signal peptide. Thus, the mRNA produces a **preprotein** of larger molecular weight than the final protein. This signal peptide has between 15 to 30 amino acids which are generally hydrophobic. Such a signal peptide probably establishes the initial association of the ribosome with the membrane, but some protein factors are involved. A **signal recognition protein** (SRP) complex binds to the nascent signal peptide and stops the translation until it reaches the ER membrane. It is suggested that a **SRP receptor** or **docking protein** which is a pore-containing integral membrane protein of ER, removes the SRP block, allowing for the translocation of the polypeptide into lumen of RER.

REVISION QUESTIONS

1. What is endoplasmic reticulum ? Describe the types, structure and functions of the endoplasmic reticulum ?
2. What functions seem relegated mostly to smooth endoplasmic reticulum ?
3. Proteins destined for secretion are translated primarily by the rough endoplasmic reticulum instead of by free ribosomes. What factors probably account for this selectivity ?
4. Write short notes on the following :
   (i) Endomembrane system; (ii) Signal theory; (iii) Microsomes ; (iv) Glycosomes; (v) Enzymes of ER; and (vi) Origin of endoplasmic reticulum.

CHAPTER

7

Golgi Apparatus

Schematic model of a portion of a Golgi complex from an epithelial cell

For the performance of certain important cellular functions such as biosynthesis of polysaccharides, packaging (compartmentalizing) of cellular synthetic products (proteins), production of exocytotic (secretory) vesicles and differentiation of cellular membranes, there occurs a complex organelle called **Golgi** complex or **Golgi** apparatus in the cytoplasm of animal and plant cells. The Golgi apparatus, like the endoplasmic reticulum, is a canalicular system with sacs, but unlike the endoplasmic reticulum it has parallely arranged, flattened, membrane-bounded vesicles which lack ribosomes and stainable by osmium tetraoxide and silver salts.

HISTORICAL

An Italian neurologist (*i.e.*, physician) **Camillo Golgi** in 1873 discovered and developed the **silver chromate method** (termed *la reazione nera*) for studying histological details of nerve cells. He, thus, opened a new field of scientific inquiry, called **neuromorphology**. In 1898, Golgi found that Purkinje cells (*i.e.*, nerve cells of cerebral cortex of brain) of barn owl contained an internal reticular network which stains black with the silver stain. He called this structure *apparato reticolare interno* (= internal reticular apparatus). By reporting the existence of such an organelle inside cell, he inadvertently raised a storm of controversy in the scientific world, which is commonly known as the Golgi controversy.

Camillo Golgi
(1844 - 1926)

The Golgi controversy. Since the refractive index of the Golgi apparatus is similar to that of cytosol (cytoplasmic matrix), the Golgi apparatus in the living cell was difficult to observe with the light microscopy, and this led to many controversies regarding its true nature. For years, it was thought to be an artifact of various fixation and staining procedures. In other words, many scientists believed that structure observed during numerous microscopy procedures and termed the Golgi, did not actually exist in the living cells (see **Berns**, 1983). For instance, **Holmgren** (1900) described a clear system of clear canals, which he called **trophospongium**. Though this structure was earlier described as being homologous with internal reticular apparatus, this comparison was later dropped. **Parat** and **Painleve** (1924) suggested **vacuome theory** : they believed that all plant and animal cells have only two fundamental but morphologically independent cytoplasmic components, *i.e.*, **vacuome** (watery vacuoles) or canals stainable with neutral red, and the **chodriome**, consisting of lipoidal mitochondria. They mixed up with Golgi apparatus and Holmgren's canalicular system and thought these were formed by the deposition of metallic silver or osmium on the vacuoles (see **Purohit**, 1980). **S.R. Cajal**, a contemporary of Golgi and Spanish histologist, was a solid supporter of Golgi during the years of controversy. Cajal referred to Golgi nets as the **Golgi-Holmgren canals**. He refined the Golgi's method of staining and became a pioneer student of the nervous system. Cajal verified Golgi's finding of a special internal cell complex and observed its morphology and behaviour under a variety of metabolic states. In the year 1906, **Camillo Golgi** and **S. R. Cajal** were jointly awarded the Nobel Prize. **Nath** (1930) using fresh eggs of the frog and later on, **Nath** and **Nangia** (1931) using telostean fish eggs, demonstrated that the vacuome and the Golgi apparatus were independent cytoplasmic organelles as were the mitochondria. Thus, not until electron microscopic studies were performed in the 1950's was the Golgi recognized and accepted as a legitimate cell organelle.

Due to their presumed high lipid contents, Golgi apparatuses were called **lipochondria (Baker**, 1951, 1953). Since originally these were known to be networks, they were also called "**dictyosomes**" (Gr., *dictyes*=net). Currently, the term **Golgi apparatus** is more prevalent one, than many other names such as **Golgi complex**, **Golgiosome**, **Golgi bodies**, **Golgi material**, **Golgi membrane**, etc. The Golgi apparatus of the cells of plants and lower invertebrates is usually referred to as **Golgi body** or **dictyosome**.

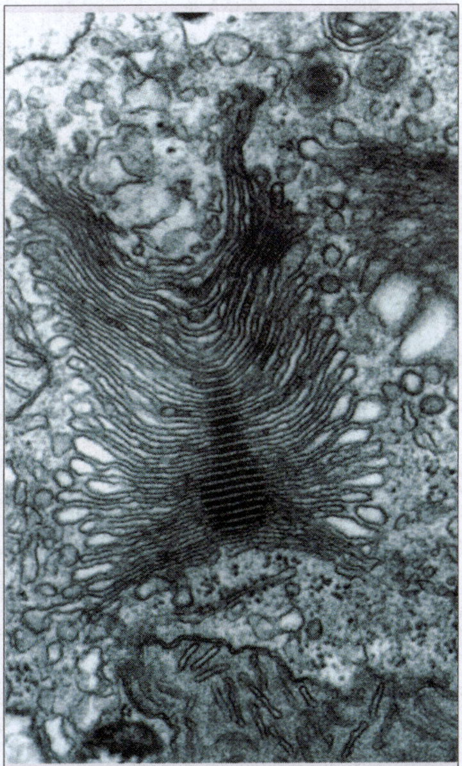
The various functions of the Golgi complex are summarised in this diagram.

OCCURRENCE

The Golgi apparatus occurs in all cells except the prokaryotic cells (*viz.*, mycoplasmas, bacteria and blue green algae) and eukaryotic cells of certain fungi, sperm cells of bryophytes and pteridiophytes, cells of mature sieve tubes of plants and mature sperm and red blood cells of animals. Their number per plant cell can vary from several hundred as in tissues of corn root and algal rhizoids (*i.e.*, more than 25,000 in algal rhizoids, **Sievers**, 1965), to a single organelle in some algae. Certain algal cells such as *Pinularia* and *Microsterias*, contain largest and most complicated Golgi apparatuses. In higher plants, Golgi apparatuses are particularly common in secretory cells and in young rapidly growing cells.

In animal cells, there usually occurs a single Golgi apparatus, however, its number may vary from animal to animal and from cell to cell. Thus, *Paramoeba* species has two Golgi apparatuses and nerve cells, liver cells and chordate oocytes have multiple Golgi apparatuses, there being about 50 of them in the liver cells.

DISTRIBUTION

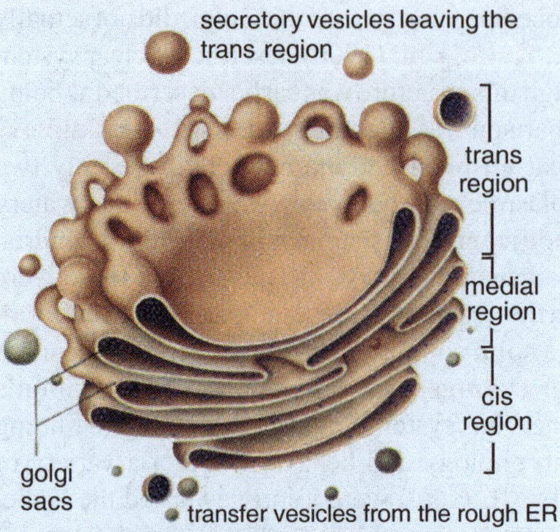

Cisternae are central, flattened closed compartments.

In the cells of higher plants, the Golgi bodies or dictyosomes are usually found scattered throughout the cytoplasm and their distribution does not seem to be ordered or localized in any particular manner (**Hall** *et al.*, 1974). However, in animal cells the Golgi apparatus is a localized organelle. For example, in the cells of ectodermal or endodermal origin, the Golgi apparatus remains polar and occurs in between the nucleus and the periphery (*e.g.*, thyroid cells, exocrine pancreatic cells and mucus-producing goblet cells of intestinal epithelium) and in the nerve cells it occupies a circum-nuclear position.

MORPHOLOGY

The Golgi apparatus is morphologically very similar in both plant and animal cells. However, it is extremely **pleomorphic** : in some cell types it appears compact and limited, in others spread out and reticular (net-like). Its shape and form may vary depending on cell type. Typically, however, Golgi apparatus appears as a complex array of interconnecting *tubules*, *vesicles* and *cisternae*. There has been much debate concerning the terminology of the Golgi's parts. The classification given by **D.J. Morre** (1977) is most widely used. In this scheme, the simplest unit of the Golgi apparatus is the **cisterna**. This is a membrane-bound space in which various materials and secretions may accumulate. Numerous cisternae are associated with each other and appear in a stack-like (lamellar) aggregation. A group of these cisternae is called the **dictyosome**, and a group of dictyosomes makes up the cell's Golgi apparatus. All dictyosomes of a cell have a common function (see **Berns**, 1983).

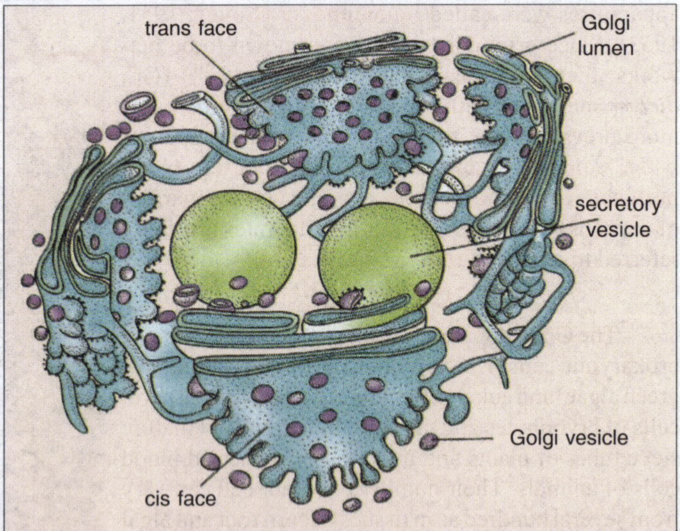

Fig. 7.1. Three-dimensional drawing of a Golgi apparatus ; it is drawn from electron micrographs of a secretory animal cell (after Alberts *et al.*, 1989).

The detailed structure of three basic components of the Golgi apparatus can be studied as follows (Fig.7.1) :

1. Flattened Sac or Cisternae

Cisternae (about 1 μm in diameter) are central, flattened, plate-like or saucer-like closed compartments which are held in parallel bundles or stacks one above the other. In each stack, cisternae are separated by a space of 20 to 30 nm which may contain rod-like elements or fibres. Each stack of cisternae forms a dictyosome which may contain 5 to 6 Golgi cisternae in animal cells or 20 or more cisternae in plant cells. Each cisterna is bounded by a smooth unit membrane (7.5 nm thick), having a lumen varying in width from about 500 to 1000 nm (see **Sheeler** and **Bianchi**, 1987).

Polarity. The margins of each cisterna are gently curved so that the entire dictyosome of Golgi apparatus takes on a bow-like appearance. The cisternae at the convex end of the dictyosome comprise **proximal, forming** or **cis-face** and the cisternae at the concave end of the dictyosome comprise the **distal, maturing** or **trans-face.** The forming or cis face of Golgi is located next to either the nucleus or a specialized portion of rough ER that lacks bound ribosomes and is called **"transitional" ER**. Trans face of Golgi is located near the plasma membrane. This polarization is called **cis-trans axis** of the Golgi apparatus.

2. Tubules

A complex array of associated **vesicles** and anastomosing **tubules** (30 to 50 nm diameter) surround the dictyosome and radiate from it. In fact, the peripheral area of dictyosome is fenestrated (lace-like) in structure.

3. Vesicles

The vesicles (60 nm in diameter) are of three types :

(i) Transitional vesicles are small membrane limited vesicles which are thought to form as blebs from the transitional ER to migrate and converge to cis face of Golgi, where they coalasce to form new cisternae.

(ii) Secretory vesicles are varied-sized membrane-limited vesicles which discharge from margins of cisternae of Golgi. They, often, occur between the maturing face of Golgi and the plasma membrane.

(iii) Clathrin-coated vesicles are spherical protuberances, about 50 μm in diameter and with a rough surface. They are found at the periphery of the organelle, usually at the ends of single tubules, and are morphologically quite distinct from the secretory vesicles. The clathrin-coated vesicles are known to play a role in intra-cellular traffic of membranes and of secretory products, *i.e.*, between ER and Golgi, as well as, between GELR region and the endosomal and lysosomal compartments.

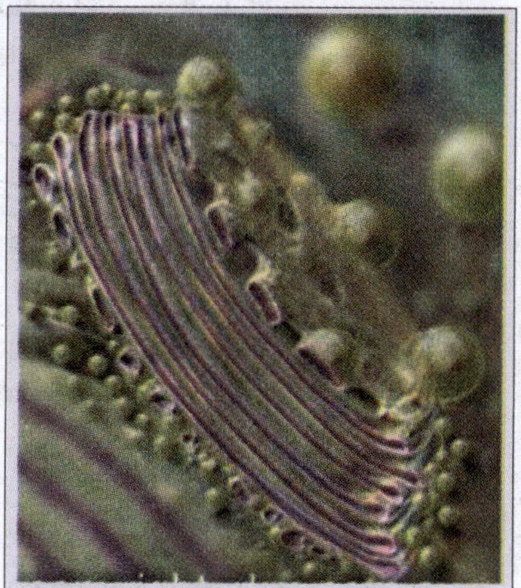

Secetory vesicles leaving the trans region.

The GERL Region

Golgi apparatus is a differentiated portion of the endomembrane system found in both animal and plant cells. This membranous component is spatially and temporally related to the endoplasmic reticulum (ER) on one side and by way of secretory vesicles, may fuse with specific portions of the plasma membrane. To the trans face of Golgi is associated the **trans-reticular Golgi**, **TGN** (=trans-Golgi-network ; **Alberts** *et al.*, 1989) or **GERL** (=Golgi + smooth ER + lysosomal), in which acid phosphatase enzyme (a characteristic lysosomal enzyme) makes its first appearance. GERL is found to be involved in the origin of **primary lysosomes** and of **melanin granules** ; in the processing, condensing and packaging of secretory material in endocrine and exocrine cells; and in lipid metabolism (**Novikoff**, 1976). GERL is also a region of sorting of cellular secretory proteins.

Zones of Exclusion

A Golgi body or Golgi apparatus is surrounded by a differentiated region of cytoplasm where ribosomes, glycogen, and organelles such as mitochondria and chloroplasts are scarce or absent. This is called **zone of exclusion** (**Morre** *et al.*, 1971) or **Golgi ground substance** (**Sjostrand** and **Hanzon**, 1954). Endoplasmic reticulum within the zone of exclusion has a smooth surface (lacking ribosomes), and coated vesicles of the Golgi apparatus are restricted to this region. Similar zones of exclusion are associated with microtubules (**Porter**, 1966), centrioles (**Bainton** and **Farquhar**, 1966), and regions of centriole formation (**Sorokin**, 1968).

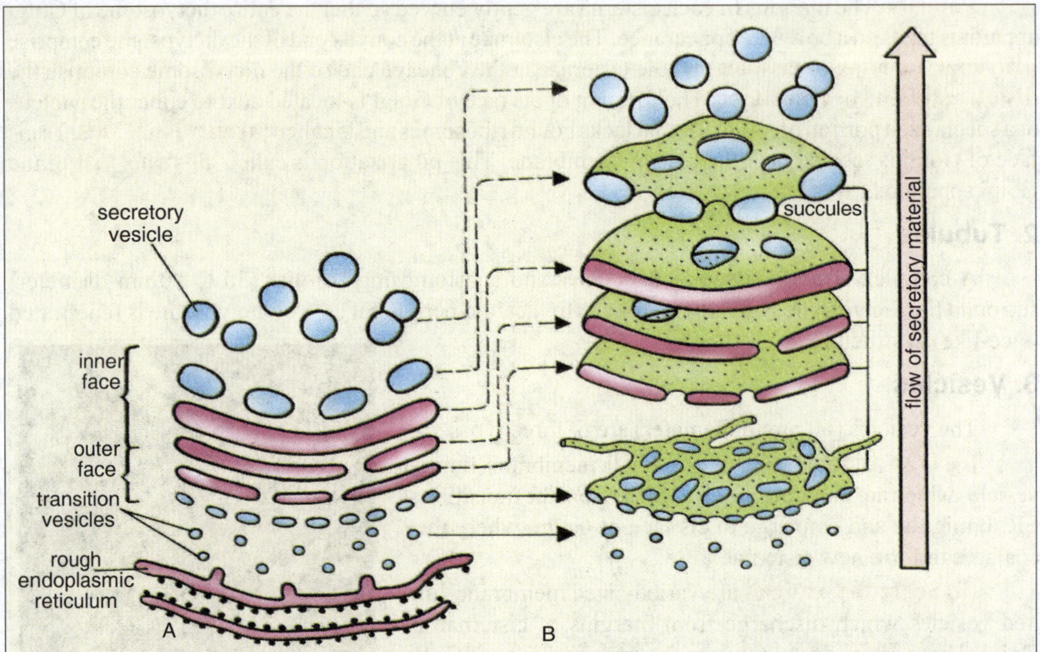

Fig. 7.2. The position and orientation of the Golgi apparatus in the secretory pathway. A—The components as they might be seen by electron microscopy of thin sections ; B—A three-dimensional reconstruction of Golgi apparatus (after Thorpe, 1984).

ISOLATION AND CHEMICAL COMPOSITION

Initially, Golgi apparatus was isolated only from cells of the epididymis, however, in recent years, it has been isolated from number of plant and animal cells. The isolation of Golgi apparatus is brought about mainly by gentle homogenization followed by differential and gradient homogenization. Gentle homogenization is preferred to preserve the stacks of cisternae. Due to its low density, Golgi apparatuses tend to form a distinct band in gradient centrifugation. The isolated Golgi apparatus is washed with distilled water for purifying it, though, its secretory components are lost (see **Thorpe**, 1984).

Chemically, Golgi apparatus of rat liver contains about 60 per cent lipid material. The Golgi apparatus of animal cells contains phospholipids in the form of **phosphatidyl choline**, whereas, that of plant cells contains **phosphatidic acid** and **phosphatidyl glycerol**. The Golgi apparatus also contains a variety of enzyme (Table 7-1), some of which have been used as cytochemical markers.

Cytochemical Properties of Golgi Apparatus

Different parts of Golgi apparatus have been histochemically identified by specific staining properties (**Thorpe**,1984 ; **Alberts** *et al*., 1989) :

1. Osmium tetroxide (O_sO_4) selectively impregnates the outer face (cis face) of the Golgi apparatus. This stain adheres well to lipids, especially phospholipids and unsaturated fats.

2. Phosphotungstic acid ($H_3PO_4 . 12 WO_3 . 24 H_2O$) selectively stain the maturing or trans face of Golgi stack. This stain is an anionic stain having special affinity for polysaccharides and proteins.

3. Glycosyl transferase and **thiamine pyrophosphatase** can be localized cytochemically in the trans cisternae of Golgi apparatus. Transferase enzymes are found to be located in the membranes of Golgi, not in the lumen of cisternae (**Thorpe**, 1984).

4. Acid phosphatase enzyme is cytochemically marked in the GERL region.

Table 7.1. Some important enzymes of the Golgi apparatus of animal cells (Source: Thorpe, 1984; Rastogi, 1988).

Enzymes : class and types	Function
A. Glycosyl transferases : Glycoprotein biosynthesis	
1. Sialyl transferases	Transfers sialic acid from CMP-sialic acid
2. Galactosyl transferases	Transfer galactose to lipids or proteins
B. Sulpho-and glycotransferases : Glycolipid biosynthesis	
3. Sulphotransferase	Transfer of sulphate from activated donor
4. Lysolecithin acetyltransferase	Transfer of acyl groups to phospholipid
5. Glycero-phosphate phosphatidyl transferase	Transfer of phosphatidyl group
C. Oxireductases : Oxidation and reduction	
6. NADH- cytochrome c-reductase	Removal or addition of hydrogen
7. NADPH- cytochrome c-reductase	Removal or addition of hydrogen
D. Phosphatases : Hydrolysis of phospholipids	
8. Glucose-6-phosphatase	Removal of phosphate
9. Thiamine pyrophosphatase (Nucleoside diphosphatase)	Hydrolysis of inorganic pyrophosphate
10. ATPase	Removal or addition of phosphate
11. Acid phosphatase	Removal of phosphate
E. Phospholipases : Hydrolysis of lipids	
12. Phospholipase A_1	Removal of non-specific fatty acid chains from phospholipids
13. Phospholipase A_2	Removal of fatty acid chains
F. Kinases : Phosphorylation	
14. Casein phosphokinases	Phosphorylation of casein
G. Mannosidases : Removal of mannose	
15. Mannosidase I and II	Removal of mannose residue from oligosaccharide

ORIGIN

Origin of Golgi apparatus involves the formation of new cisternae and there is great variation in shape, number and size of cisternae in each stack (dictyosome). The process of formation of new cisternae may be performed by any of the following methods: 1.Individual stacks of cisternae may arise from the pre-existing stacks by division or fragmentation. 2.The alternative method of origin of Golgi is based on *de novo* formation. In fact, various cytological and biochemical evidences have established that the membranes of the Golgi apparatus are originated from the membranes of the smooth ER which in turn have originated from the rough ER. The proximal Golgi saccules are formed by fusion of ER-derived vesicles, while distal saccules "give their all" to vesicle formation and disappear. Thus, Golgi saccules are constantly and rapidly renewed.

The cells of dormant seeds of higher plants generally lack Golgi apparatuses but they do display zone of exclusion having aggregation of small transition vesicles. Photomicrographs of cells in early stages of germination suggest progressive development of Golgi bodies in these zones of exclusion; and the development of Golgi apparatuses coincides with the disappearance of the aggregation of vesicles (see **Sheeler** and **Bianchi**, 1987).

FUNCTIONS

Golgi vesicles are often, referred to as the "**traffic police**" of the cell (**Darnell** *et al*., 1986). They play a key role in **sorting** many of cell's proteins and membrane constituents, and in **directing** them

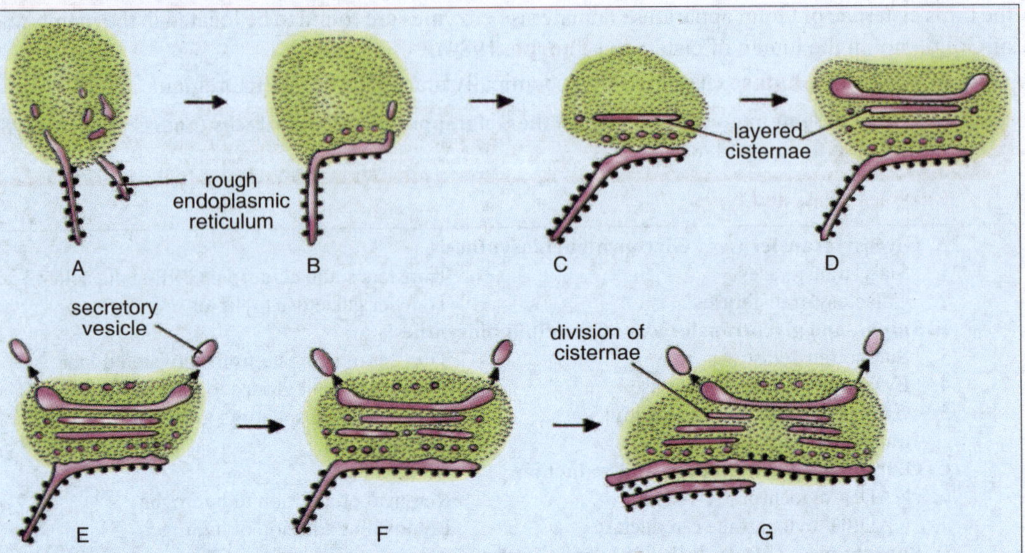

Fig. 7.3. A model of formation of Golgi apparatus from endoplasmic reticulum (A—C) and subsequent developmental stages : formation of stack of cisternae (C and D), formation of secretory vesicles (E) division (F,G) (after Sheeler and Bianchi, 1987).

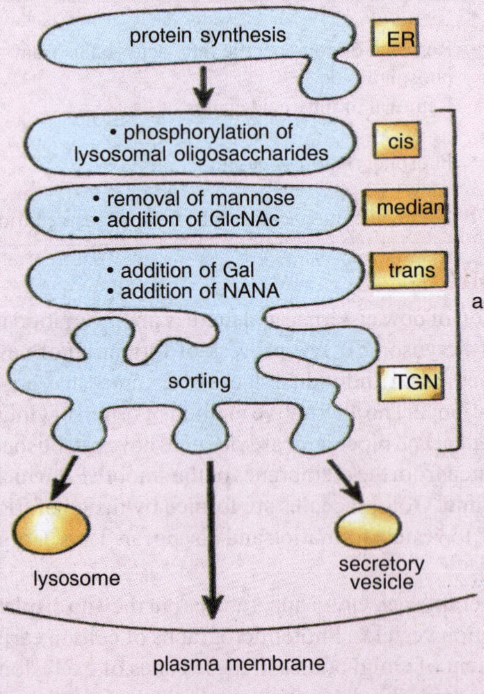

Fig. 7.4. The compartmentalization of the Golgi apparatus. GlcNAc = N-actyl-glucosamine galactose; NANA = N-acetyl neuraminic acid (sialic acid) (after Alberts *et al.*, 1989).

to their proper destinations. To perform this function, the Golgi vesicles contain different sets of enzymes in different types of vesicles—**cis, middle** and **trans cisternae**—that react with and modify secretory proteins passing through the Golgi lumen or membrane proteins and glycoproteins that are transiently in the Golgi membranes as they are *en route* to their final destinations (Fig. 7.4). For example, a Golgi enzyme may add a "signal" or "tag" such as a carbohydrate or phosphate residues to certain proteins to direct them to their proper sites in the cell. Or, a proteolytic Golgi enzyme may cut a secretory or membrane protein into two or more specific segments (*e.g.*, molecular processing involved in the formation of pancreatic hormone insulin : preproinsulin→ proinsulin→ insulin).

Recently, in the function of Golgi apparatus, subcompartmentalization with a division of labour has been proposed between the *cis* region (in which proteins of RER are sorted and some of them are returned back possibly by coated vesicles), and the trans region in which the most refined proteins are further separated for their delivery to the various cell compartments (*e.g.*, plasma membrane, secretory granules and lysosomes)

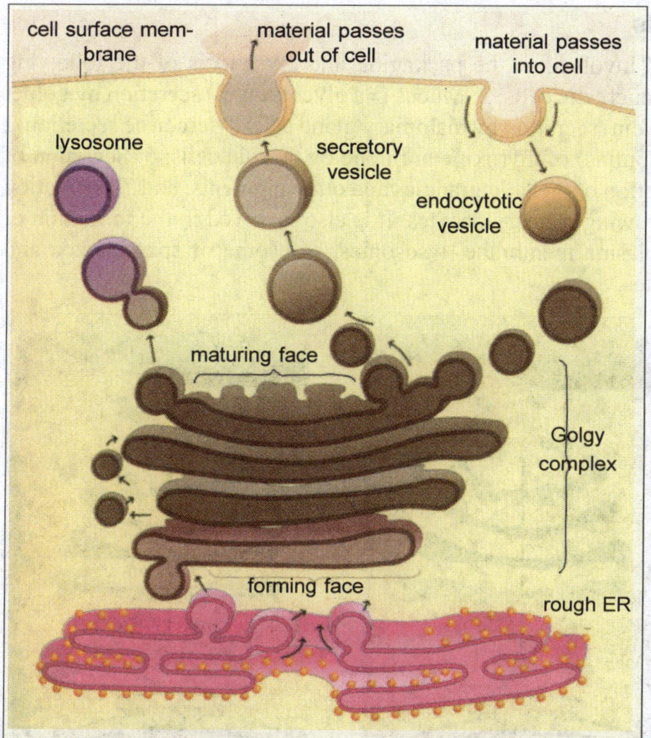

The EM of the Golgi complex with a very large number of vesicles.

(Fig.7.5; **Rothman** 1981; **Rothman** and **Leonard**, 1984).

Thus, Golgi apparatus is a centre of *reception*, *finishing*, *packaging*, and *dispatch* for a variety of materials in animal and plant cells:

1. Golgi Functions in Plants

In plants, Golgi apparatus is mainly involved in the secretion of materials of primary and secondary cell walls (*e.g.*, formation and export of glycoproteins, lipids, pectins and monomers for hemicellulose, cellulose, lignin, etc.). During cytokinesis of mitosis or meiosis, the vesicles originating from the periphery of Golgi apparatus, coalesce in the phragmoplast area to form a semisolid layer, called **cell plate**. The unit membrane of Golgi vesicles fuses during cell plate formation and becomes part of plasma membrane of daughter cells (For details see Chapter 5).

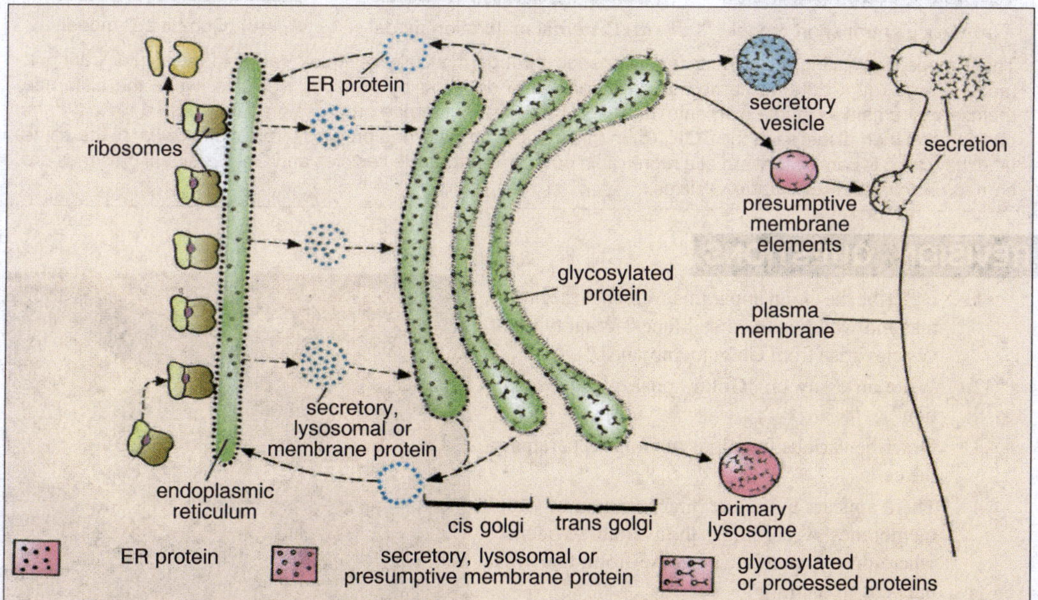

Fig. 7.5. Diagram illustrating the hypothetical dual function of cis and trans cisternae of the Golgi apparatus. The closed circles correspond to ER proteins that are removed from the rims of cis and middle Golgi cisternae ("refiners") and return to the ER (dashed arrow). The open circles represent secretory proteins which are destined for secretion or incorporation into organelles (lysosomes, plasma membrane) (after Sheelar and Bianchi, 1987).

2. Golgi Functions in Animals

In animals, Golgi apparatus is involved in the packaging and exocytosis of the following materials : 1. Zymogen of exocrine pancreatic cells; 2. Mucus (=a glycoprotein) secretion by goblet cells of intestine ; 3. Lactoprotein (casein) secretion by mammary gland cells (Merocrine secretion) ; 4. Secretion of compounds (thyroglobulins) of thyroxine hormone by thyroid cells; 5. Secretion of tropocollagen and collagen ; 6. Formation of melanin granules and other pigments; and 7. Formation of yolk and vitelline membrane of growing primary oocytes. It is also involved in the formation of certain cellular organelles such as plasma membrane, lysosomes, acrosome of spermatozoa and cortical granules of a variety of oöcytes.

(a) Vesicular transport model (b) Cisternal maturation model (c) "Combined" model

Three models depicting the dynamics of transport through the Golgi complex. (a) In the vesicular transport model, cargo (dark dots) is carried in an anterograde direction by transport vesicles, while the cisternae themselves remain as stable elements. (b) In this model, the cisternae progress gradually from a *cis* to a *trans* position and then disperse at the TGN, (c) In this model, the cisternae progress from *cis* to *trans*, at the same time that cargo is carried forward at a more rapid pace in anterograde vesicles and Golgi resident enzymes are carried backward in retrograde vesicles.

REVISION QUESTIONS

1. Describe the Golgi apparatus. Which is the proximal and which is the distal face ? What types of vesicles arise from Golgi membranes ?
2. Write an essay on "Golgi apparatus and secretion".
3. Describe various functions of Golgi apparatus in the cells.
4. There appears to be a regular turnover of Golgi membranes. According to the available evidence, where do the membranes come from, and what happens to them.
5. Write short notes on the following :

   (i) The Golgi controversy : (ii) Morre's classification of Golgi ; (iii) GERL region ; (iv) Isolation of Golgi apparatus ; (v) Enzymes of Golgi apparatus ; (vi) Compartmentalization of Golgi apparatus.

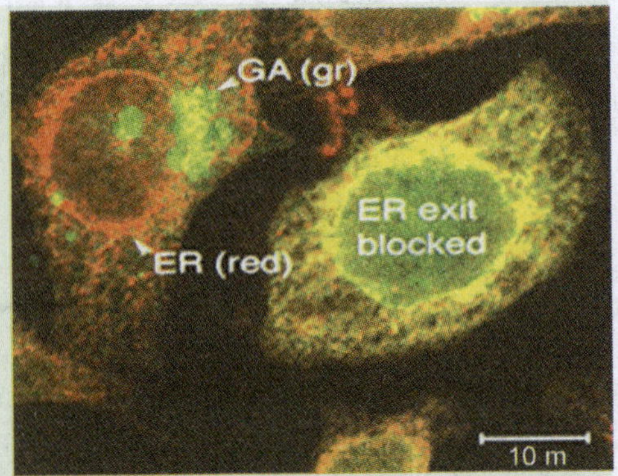

Photograph showing how proteins are delivered to the cell's Golgi apparatus for processing in vesicles that bud from the ER.

CHAPTER

8

Lysosomes

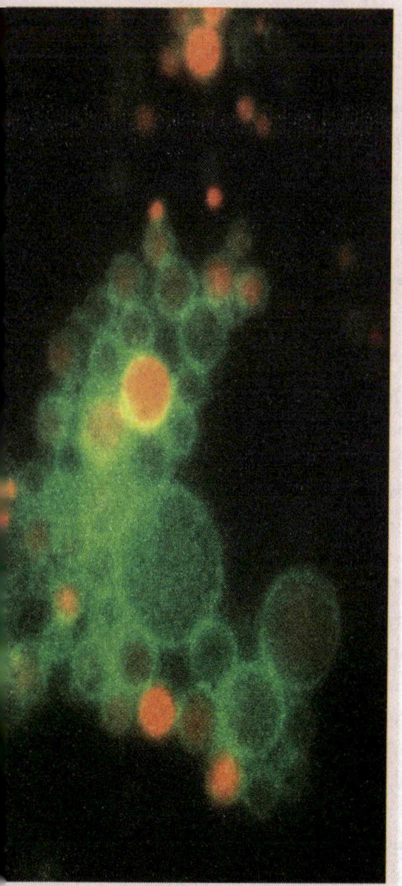

Lysosomes.

The lysosomes *(Gr., lyso*=digestive + *soma*=body) are tiny membrane-bound vesicles involved in intracellular digestion. They contain a variety of hydrolytic enzymes that remain active under acidic conditions. The lysosomal lumen is maintained at an acidic pH (around 5) by an ATP-driven proton pump in the membrane. Thus, these remarkable organelles are primarily meant for the digestion of a variety of biological materials and secondarily cause aging and death of animal cells and also a variety of human diseases such as cancer, gout, Pompe's disease, silicosis and I-cell disease.

HISTORICAL

During early electron microscopic studies, rounded dense bodies were observed in rat liver cells. These bodies were initially described as **"perinuclear dense bodies". C. de Duve,** in 1955, renamed these organelles as 'lysosomes' to indicate that the internal digestive enzymes only became apparent when the membrane of these organelles was lysed (See **Reid** and **Leech**, 1980). However, the term lysosome means lytic body having digestive enzymes capable of lysis (*viz.*, dissolution of a cell or tissue; (**De Robertis** and **De Robertis, Jr.**, 1987).

Christian de Duve (Born 1917, won Nobel Prize in 1974).

Lysosomes were investigated according to following two schools : (1) **C. de Duve** and his coworkers (1963, 1964, 1974) worked in Belgium and their approach was biochemical one. (2) **Alex Novikoff** and his research group (1962, 1964) worked in United States and their approach was morphological and cytochemical. For the discovery of lysosomes and a brilliant series of experiments on them, **de Duve** shared the 1974 Nobel Prize for physiology with **Palade** and **Claude**, both were pioneer cell biologists.

OCCURRENCE

The lysosomes occur in most animal and few plant cells (Table 8-1). They are absent in bacteria and mature mammalian erythrocytes. Few lysosomes occur in muscle cells or in acinar cells of the pancreas. Leucocytes, especially granulocytes are a particularly rich source of lysosomes. Their lysosomes are so large-sized that they can be observed under the light microscope. Lysosomes are also numerous in epithelial cells of absorptive, secretory and excretory organs (*e.g.*, intestine, liver, kidney, etc.). They occur in abundance in the epithelial cells of lungs and uterus. Lastly, phagocytic cells and cells of reticuloendothelial system (*e.g.*, bone marrow, spleen and liver) are also rich in lysosomes.

STRUCTURE

The lysosomes are round vacuolar structures which remain filled with dense material and are bounded by single unit membrane. Their shape and density vary greatly. Lysosomes are 0.2 to 0.5μm in size. Since, size and shape of lysosomes vary from cell to cell and time to time (*i.e.* they are polymorphic), their identification becomes difficult. However, on the basis of the following three criteria, a cellular entity can be identified as a lysosome: (1) It should be bound by a limiting membrane; (2) It should contain two or more acid hydrolases; and (3) It should demonstrate the property of enzyme latency when treated in a way that adversely affects organelle's membrane structure.

Table 8-1. Examples of plant and animal cells, tissues and organs containing lysosomes (Source : Sheeler and Bianchi, 1987).

A. Animal tissues	B. Protozoa
1. Liver	15. Leucocytes
2. Kidney	16. *Amoeba*
3. Nerve cells	17. *Tetrahymena*
4. Brain	18. *Paramecium*
5. Intestinal epithelium	19. *Euglena*
6. Lung epithelium	**C. Plants**
7. Macrophages	20. Onion seeds
(of spleen, bone marrow, liver and	21. Barley seeds
connective tissue)	22. Corn seedlings
8. Thyroid gland	23. Yeast
9. Adrenal gland	24. *Neurospora*
10. Bone	**D. Tissue culture cells**
11. Urinary bladder	25. HeLA cells
12. Prostate	26. Fibroblasts
13. Uterus	27. Chick cells
14. Ovaries	28. Lymphocytes

ISOLATION AND CHEMICAL COMPOSITION

Lysosomes are very delicate and fragile organelles. Lysosomal fractions have been isolated by **sucrose-density centrifugation** (or Isopycnic centrifugation) after mild methods of homogenization. Since the original de Duve's isolated lysosomal fractions were having contaminations of mitochondria, microsomes and microbodies, so, in 1960's it was investigated that rats injected with **dextran** or **Triton WR-1339**, incorporated these compounds into their lysosomes, thereby altering their density and making their cleaner separation possible by differential centrifugation and density gradients (see **Reid** and **Leech**,1980).

Lysosomes tend to accumulate certain dyes (vital stains such as Neutral red, Niagara, Evans blue) and drugs such as anti-malarial drug **chloroquine**. Such 'loaded' lysosomes can be demonstrated by fluorescence microscopy.

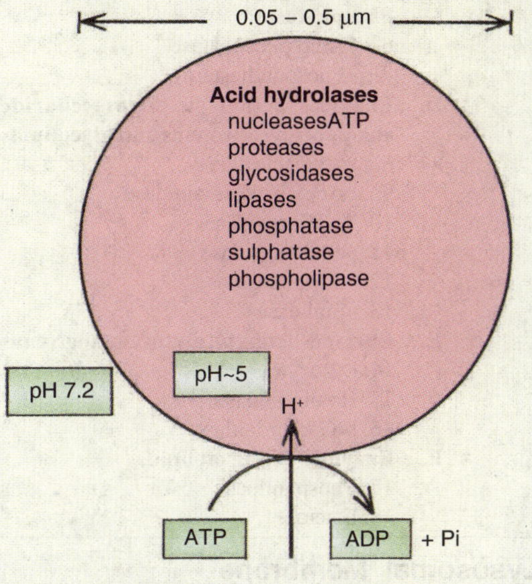

Fig. 8.1. A lysosome showing its various characteristics (after Albert *et al.*, 1989).

The location of the lysosomes in the cell can also be pinpointed by various histochemical or cytochemical methods. For example, lysosomes demonstrate the property of **metachromasia** with toluidine blue and give a positive acid Schiff reaction (see Chapter 2). Metachromasia is the property exhibited by certain pure dyestuffs, chiefly basic stains,of colouring certain tissue elements in a different colour. Certain lysosomal enzymes are good histochemical markers. For example, **acid phosphatase** is the principal enzyme which is used as a marker for the lysosomes by the the use of **Gomori'staining technique** (**Gomori**,1952). Specific stains are also used for other lysosomal enzymes such as **B- glucuronidase**, **aryl sulphatatase**, **N-acetyl-B-glucosaminidase** and **5-bromo-4-chloroindolacetate esterase**.

Lysosomal Enzymes

According to a recent estimate,a lysosome may contain up to 40 types of hydrolytic enzymes (see **Alberts** *et al*.,1989). They include **proteases** (*e.g.*, cathepsin for protein digestion), **nucleases**, **glycosidases** (for digestion of polysaccharides and glycosides), **lipases**, **phospholipases**, **phosphatases** and **sulphatases** (Table 8-2). All lysosomal enzymes are acid hydrolases, optimally active at the pH5 maintained within lysosomes.The membrane of the lysosome normally keeps the enzymes latent and out of the cytoplasmic matrix or cytosol (whose pH is about ~7.2), but *the acid dependency of lysosomal enzymes protects the contents of the cytosol (cytoplasmic matrix) against any damage even if leakage of lysosomal enzymes should occur*.

The so-called **latency** of the lysosomal enzymes is due to the presence of the membrane which is resistant to the enzymes that it encloses. Most probably this is due to the fact that most lysosomal hydrolases are membrane-bound, which may prevent the active centres of enzymes to gain access to susceptible groups in the membrane (see **Reid** and **Leech**,1980).

Table 8-2. Some lysosomal enzymes and their substrates (Source : Sheeler and Bianchi, 1987).

Enzyme	Substrate
A. Proteases and peptidases	
1. Cathepsin A,B,C,D and E	Various proteins and peptides
2. Collagenase	Collagen
3. Peptidases	Peptides
B. Nucleases	
4. Acid ribonuclease	RNA
5. Acid deoxyribonuclease	DNA

Enzyme	Substrate
C. Phosphatases	
6. Acid phosphatase	Phosphate monoesters
7. Phosphodiesterase	Oligonucleotides, phosphodiesters
D. Enzymes acting on oligosaccharide chains of glycoproteins and glycolipids	
8. b-galactosidase	b-Galactosides
9. Acetylhexosaminidase	Acetylhexosaminides, heparin sulphate
10. b-Glucosidase	b-Glucosides
11. a- Glucosidase	Glycogen
12. a-Mannosidase	a-Mannosidase
13. Sialidase	Sialic acid derivatives
E. Enzymes acting on glycosaminoglycans	
14. Lysozyme	Mucopolysaccharides, bacterial cell wall
15. Hyaluronidase	Hyaluronic acid, chondroitin sulphates
16. b-Glucuronidase	Polysaccharides, mucopolysaccharides
F. Enzymes acting on lipids	
17. Phospholipase	Lecithin, phosphatidyl ethanolamine
18. Esterase	Fatty acid esters

Lysosomal Membrane

The lysosomal membrane is slightly thicker than that of mitochondria. It contains substantial amounts of carbohydrate material, particularly **sialic acid**. In fact, most lysosomal membrane proteins are unusually highly glycosylated, which may help protect them from the lysosomal proteases in the lumen. The lysosomal membrane has another unique property of fusing with other membranes of the cell. This property of fusion has been attributed to the high proportion of membrane lipids present in the micellar configuration (**Lucy**,1969). Surface active agents such as liposoluble vitamins (A,K,D and E) and steroid sex hormones have a destabilizing influence, causing release of lysosomal enzymes due to rupture of lysosomal membranes. On the contrary, the cortisone, hydrocortisone and other drugs tend to stabilize the lysosomal membrane and have an anti-inflammatory effect on the tissue.

The entire process of digestion is carried out within the lysosome. Most lysosomal enzymes act in an acid medium. Acidification of lysosomal contents depends on an ATP-dependent proton pump which is present in the membrane of the lysosome and accumulates H^+ inside the organelle (**Reijngond**,1978). Lysosomal membrane also contains transport proteins that allow the final products of digestion of macromolecules to escape so that they can be either excreted or reutilized by the cell.

KINDS OF LYSOSOMES (POLYMORPHISM IN LYSOSOMES)

Lysosomes are extremely dynamic organelles, exhibiting polymorphism in their morphology. Following four types of lysosomes have been recognized in different types of cells or at different times in the same cell. Of these, only the first is the **primary lysosome**, the other three have been grouped together as **secondary lysosomes**.

1. Primary Lysosomes

These are also called **storage granules**, **protolysosomes** or **virgin lysosomes**. Primary lysosomes are newly formed organelles bounded by a single membrane and typically having a diameter of 100 nm. They contain the degradative enzymes which have not participated in any digestive process. Each primary lysosome contains one type of enzyme or another and it is only in the secondary lysosome that the full complement of acid hydrolases is present.

2. Heterophagosomes

They are also called **heterophagic vacuoles**, **heterolysosomes** or **phagolysosomes**. Heterophagosomes are formed by the fusion of primary lysosomes with cytoplasmic vacuoles containing **extracel-**

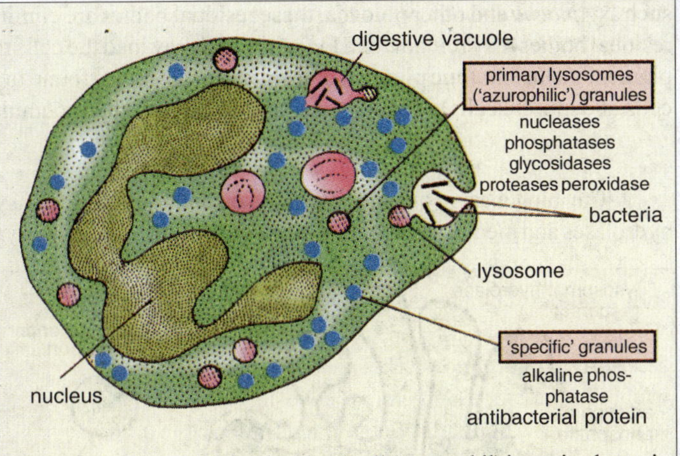

Fig. 8.2. Diagram of a white blood cell (neutrophil) ingesting bacteria. Two types of granules fuse with the phagocytotic vacuoles and contribute digestive enzymes and other components.

lular substances brought into the cell by any of a variety of endocytic processes (*e.g.*, pinocytosis, phagocytosis or receptor-mediated endocytosis, see Chapter 5). The digestion of engulfed substances takes place by the enzymatic activities of the hydrolytic enzymes of the secondary lysosomes. The digested material has low molecular weight and readily passess through the membrane of the lysosomes to become the part of the matrix (Fig. 8.2).

3. Autophagosomes

They are also called **autophagic vacuole**, **cytolysosomes** or **autolysosomes**. Primary lysosomes are able to digest **intracellular structures** including mitochondria, ribosomes, peroxisomes and glycogen granules. Such autodigestion (called **autophagy**) of cellular organelles is a normal event during cell growth and repair and is especially prevalent in differentiating and dedifferentiating tissues (*e.g.*, cells undergoing programmed death during metamorphosis or regeneration) and tissue under stress. Autophagy takes several forms. In some cases the lysosome appears to flow around the cell structure and fuse, enclosing it in a double membrane sac, the lysosomal enzymes being initially confined between the membranes. The inner membrane then breaks down and the enzymes are able to penetrate to the enclosed organelle.In other cases, the organelle to be digested is first encased by smooth ER, forming a vesicle that fuses with a primary lysosome (Fig. 8.4). Lysosomes also regularly engulf bits of cytosol (cytoplasmic matrix) which is degraded by a process, called **microautophagy**.

As digestion proceeds, it becomes increasingly difficult to identify the nature of the original secondary lysosome (*i.e.*, heterophagosome or autophagosome) and the more general term **digestive vacuole** is used to describe the organelle at this stage.

4. Residual Bodies

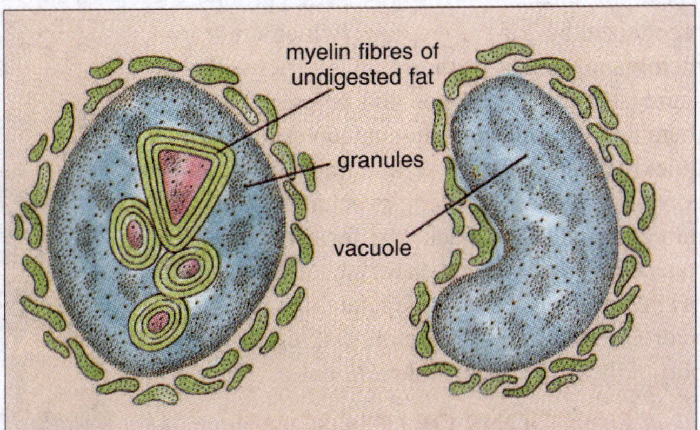

Fig . 8.3. Lysosomes of the kidney cells of rat, showing the presence of residues (A).

They are also called **telolysosomes** or **dense bodies**. Residual bodies are formed if the digestion inside the food vacuole is incomplete. Incomplete digestion may be due to absence of some lysosomal enzymes. The undigested food is present in the digestive vacuole as the residues and may take the form of whorls of membranes, grains, amorphous masses, ferritin-like or myelin figures (Fig.8.3).

Residual bodies are large, irregular in shape and are usually quite electron-dense. In some cells, such as *Amoeba* and other potozoa, these residual bodies are eliminated by **defecation**. In other cells, residual bodies may remain for a long time and may load the cells to result in their aging. For example, pigment inclusions (age pigment or **lipofuscin granules**) found in nerve cells (also in liver cells, heart cells and muscle cells) of old animals may be due to the accumulation of residual bodies.

ORIGIN

The biogenesis (origin) of the lysosomes requires the synthesis of specialized lysosomal hydrolases and membrane proteins. Both classes of proteins are synthesized in the ER and transported

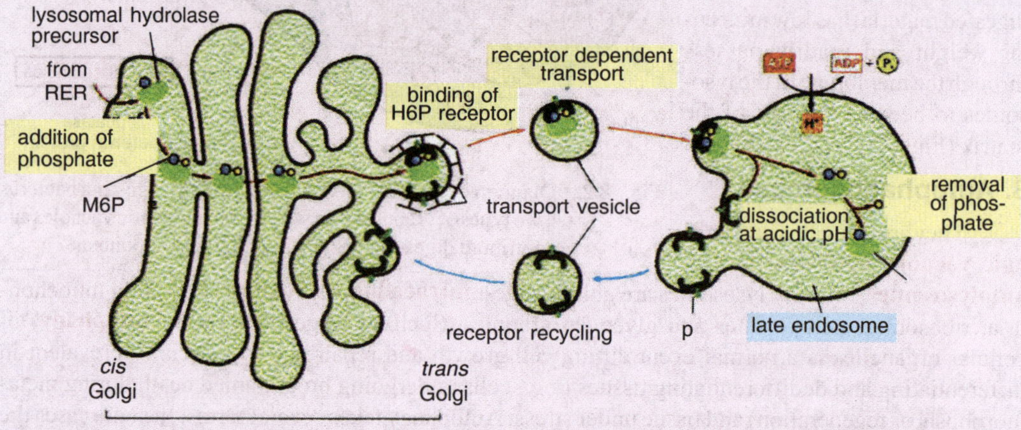

A summary of the autophagic pathway.

through the Golgi apparatus, then transported from the trans Golgi network to an intermediate compartment (an **endolysosome**) by means of **transport vesicles** (which are coated by clathrin protein; Fig. 8.4). The lysosomal enzymes are glycoproteins, containing N-linked oligosaccharides that are processed in a unique way in the cis Golgi so that their **mannose** residues are phosphorylated. These **mannose 6-phosphate (M6P)** groups are recognized by **M6P-receptors** (which are transmembrane proteins) in the trans Golgi network that segregates the hydrolases and helps to package them into budding clathrin-coated vesicles which quickly lose their coats. These transport vesicles containing the M6P-receptors act as shuttles that move the receptors back and forth between the trans Golgi network and endolysosomes. The low pH in the endolysosome dissociates the lysosomal hydrolases from this receptor, making the transport of the hydrolases unidirectional.

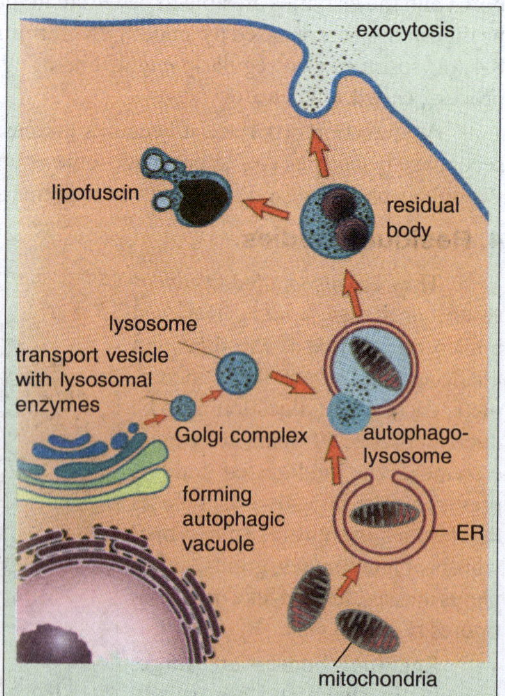

A summary of the autophagic pathway.

FUNCTIONS OF LYSOSOMES

The important functions of lysosomes are as follows:

1. Digestion of large extracellular particles. The lysosomes digest the food contents of the phagosomes or pinosomes. The lysosomes of leucocytes enable the latter to devour the foreign proteins, bacteria and viruses.

2. Digestion of intracellular substances. During the starvation, the lysosomes digest the stored food contents, *viz.*, proteins, lipids and carbohydrates (glycogen) of the cytoplasm and supply to the cell neccessary amount of energy.

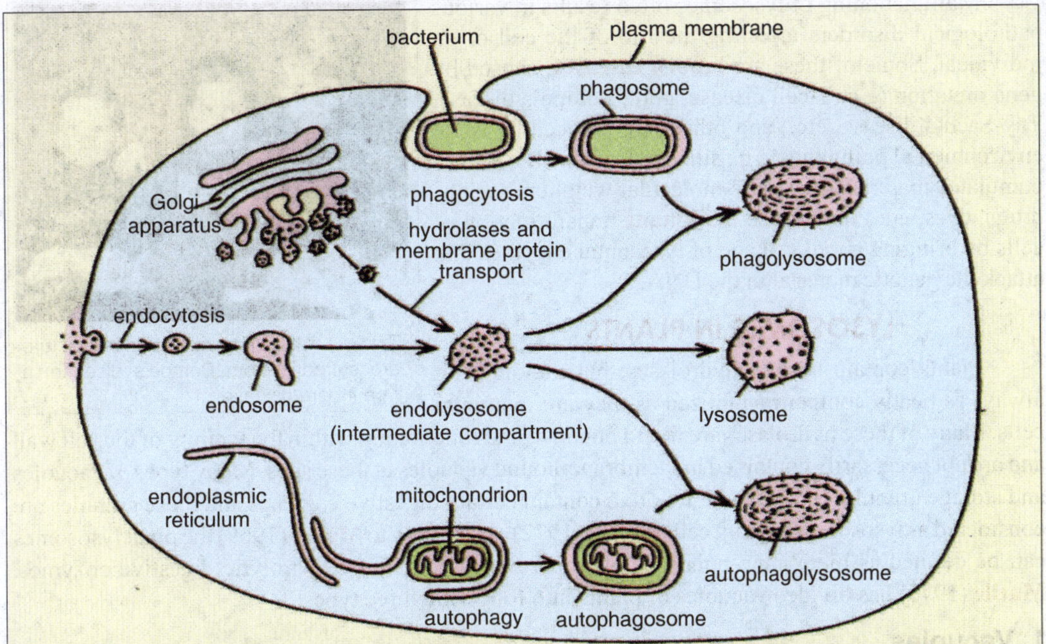

Fig. 8.4. Recently understood mechanism of the origin of three types of lysosomes : phagolysosome, lysosome (the classical secondary lysosome) and autophagolysosome. Transport vesicles (the classical primary lysosomes) originate from trans Golgi network to fuse with endolysosome which contains already endocytosed materials for digestion (after Alberts *et al*., 1989).

3. Autolysis. In certain pathological conditions the lysosomes start to digest the various organelles of the cells and this process is known as **autolysis** or **cellular autophagy**. When a cell dies, the lysosome membrane ruptures and enzymes are liberated. These enzymes digest the dead cells. In the process of metamorphosis of amphibians and \tunicates many embryonic tissues, *e.g.*, gills, fins, tail, etc., are digested by the lysosomes and utilized by the other cells.

4. Extracellular digestion. The lysosomes of certain cells such as sperms discharge their enzymes outside the cell during the process of fertilization. The lysosomal enzymes digest the limiting membranes of the ovum and form penetration

The lysosomes of sperms discharge their enzymes outside the cell during the process of fertilization. Here a human sperm is being seen fertilizing an egg.

path in ovum for the sperms. Acid hydrolases are released from **osteoclasts** and break down bone for the reabsorption; these cells also secrete lactic acid which makes the local pH enough for optimal enzyme activity. Likewise, preceding ossification (bone formation), **fibroblasts** release cathepsin D enzyme to break down the connective tissue (**Dingle**, 1973).

LYSOSOMES AND DISEASE

Malfunctioning of lysosomes often results in various pathological disorders affecting the life of the cell or an individual. Some of these are **inborn diseases**, caused by gene mutation (*e.g.*, I-cell disease, gout, Pompe's disease, Tay-Sach's disease, etc.) and others are induced by some environmental pollutants (*e.g.*, silicosis). Typically, the accumulated materials (*e.g.*, low-molecular weight materials, drugs, dyes, etc.) may cause **malignant** transformation of cells by bringing about leakage of lysosomal enzymes that attack the genetic material in the DNA.

Two out of the three siblings in this picture are suffering from Pompe's disease, a rare inborn disease.

LYSOSOMES IN PLANTS

Plants contain several hydrolases, but they are not always as neatly compartmentalized as they are in animal cells. Many of these hydrolases are found bound to and functioning within the vicinity of the cell wall and are not necessarily contained in membrane-bound vacuoles at these sites. Many types of vacuoles and storage granules of plants are found to contain certain digestive enzymes and these granules are considered as lysosomes of plant cell (**Gahan**, 1972). According to **Matile** (1969) the plant lysosomes can be defined as membrane-bound cell compartments containing hydrolytic digestive enzymes. **Matile** (1975) has divided vacuoles of plants into following three types:

1. Vacuoles

The vacuole of a mature plant cell is formed from the enlargement and fusion of smaller vacuoles present in meristematic cells; these **provacuoles**, which are believed to be derived from the ER and possibly the Golgi and contain acid hydrolases. These lysosomal enzymes are associated with the tonoplast of large vacuole of differentiating cells. Sometimes, mitochondria and plastids are observed inside the vacuole suggesting autophagy in plants (**Swanson** and **Webster**, 1989).

2. Spherosomes

The spherosomes are membrane-bounded, spherical particles of 0.5 to 2.5 μm diameter, occurring in most plant cells. They have a fine granular structure internally which is rich in lipids and proteins. They originate from the endoplasmic reticulum (ER). Oil accumulates at the end of a strand of ER and a small vesicle is then cut off by constriction to form particles, called **prospherosomes**. The prospherosomes grow in size to form spherosomes. Basically, the spherosomes are involved in lipid synthesis and storage. But, the spherosomes of maize root tips (**Matile**, 1968) and spherosomes of tobacco endosperm tissue (**Spichiger**, 1969) have been found rich in hydrolytic digestive enzymes and so have been considered as lysosomes. Like lysosomes they are not only responsible for the accumulation and mobilization of reserve lipids, but also for the digestion of other cytoplasmic components incorporated by phagocytosis.

3. Aleurone Grain

The aleurone grains or protein bodies are spherical membrane-bounded storage particles occurring in the cells of endosperm and cotyledons of seeds. They are formed during the later stages of seed ripening and disappear in the early stages of germination. They store protein (*e.g.*, globulins) and phosphate in the form of phytin. **Matile** (1968) has demonstrated that aleurone grains from pea seed contain a wide range of hydrolytic enzymes including protease and phosphatase which are required for the mobilization of stored protein and phosphate, although the presence of other enzymes such as β-amylase and RNAase suggest that other cell constituents may also be digested. Thus, like spherosomes, aleurone grains store reserve materials, mobilize them during germination and in addition form a compartment for the digestion of other cell components (**Hall** *et al.*, 1974). The aleurone grains are derived from the strands of the endoplasmic reticulum.

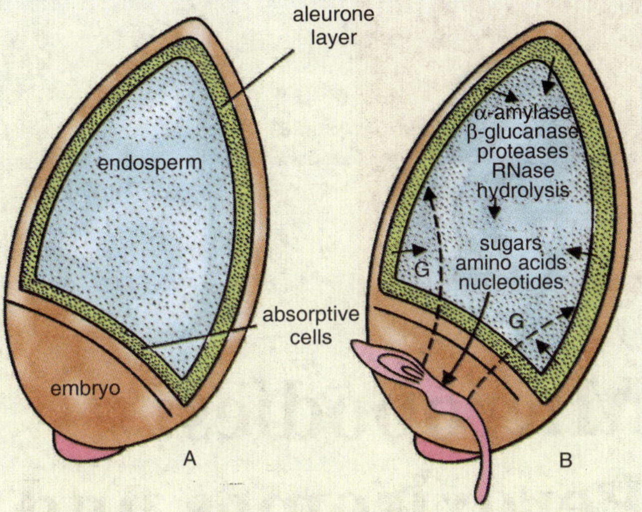

Fig. 8.5. Role of hormone and hydrolytic enzymes in seed germination. A—Ultrastructure of seed before germination; B—During germination, gibberellins (G) pass from the embryo to the aleurone layer, where the *de novo* synthesis of hydrolytic enzymes is induced. These enzymes break down the macromolecular stored reserves and the low molecular weight products are transported to the embryo, where they function as nutrients (after Thorpe, 1984).

During germination of barley seed, the activity of hydrolases is found to be controlled by hormones such as **gibberellic acid** (Fig. 8.5). Gibberellic acid, a plant growth hormone, is released by the embryo to the aleurone layer where, in turn, the hydrolases are released to the endosperm. This hormone operates by derepressing appropriate genes in the aleurone cells, which then begin to crank out new hydrolytic proteins (see **Thorpe**, 1984).

Extra-cellular Digestion by Plants

Plant cells are generally unable to engulf large particles, presumably because of the restrictions imposed on the cell by cell wall. The secretion of hydrolases to carry out extracellular digestion, therefore, becomes an important process. Hydrolases are commonly secreted by fungi, enabling the organism to degrade and grow on macromolecules it cannot transport into the cell. Higher plants also secrete hydrolases, a notable example being the insectivorous pitcher plants, which produce a proteinase-containing liquid in which victims are trapped and digested.

REVISION QUESTIONS

1. What are the lysosomes ? Describe their origin, structure and function.
2. Describe the method of isolation of lysosomes in the cells of plants and animals. Add a note about histochemical marking of lysosomal components : membrane and enzymes.
3. Describe the process of autophagy. What is the ultimate fate of the digestive vacuole ?
4. Write short notes on the following :
   (i) Polymorphism in lysosomes;
   (ii) Lysosomes and disease;
   (iii) Lysosomal enzymes;
   (iv) Lysosomes of plants.

CHAPTER

9

Microbodies: Peroxisomes and Glyoxysomes

The cells of Protozoa, fungi, plants, liver and kidney of vertebrates contain membrane-bound, spherical bodies of 0.2 to 1.5 μm diameter in close association of endoplasmic reticulum, and mitochondria or chloroplast, or both. These organelles have a central granular or crystalloid core containing some enzymes and are called **microbodies**. Microbodies use molecular oxygen like mitochondria, but instead of having cytochromes and capacity of ATP synthesis like them, they contain flavin-linked oxides and catalases for the hydrogen peroxide metabolism and also enzymes for fatty acid metabolism. Peroxisomes differ from mitochondria and chloroplasts in many ways. Most notably, these organelles are surrounded only by a single membrane, and they do not contain DNA (genome) or ribosomes. However, they resemble ER in being self-replicating membrane bound organelle.

HISTORICAL

Since the mid-1950s electron microscopists have observed small structures or bodies in cells that on morphological grounds have been aptly termed **microbodies**. **C.de Duve** and **P. Baudhuin** (1966) coined the term **peroxisome** for the microbodies of mammalian systems and studied their structure and function. **Leaf peroxisomes** were first isolated from spinach leaf homogenate (*i.e.*, from mesophyll cells) by **Tolbert's** group in Michigen in 1968. Glyoxylate cycle containing peroxisomes,

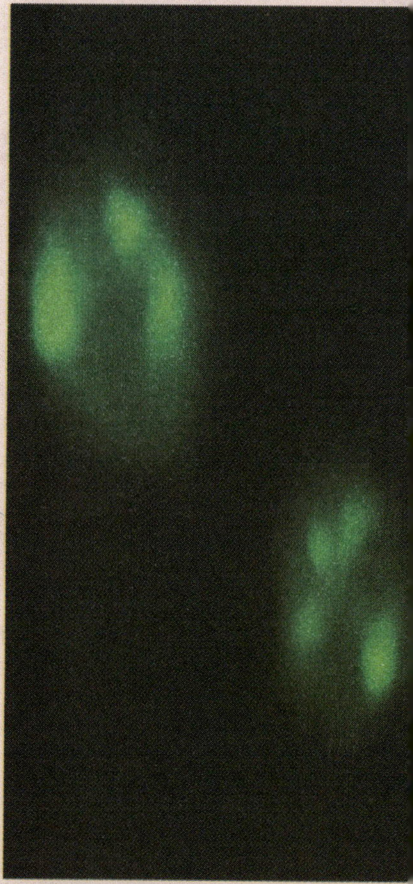

Peroxisomes illuminated by fluorscent pro

called **glyoxysomes**, were discovered in 1969 by **Beevers** in the endoplasm cells of germinating castor bean (*Ricinus*).

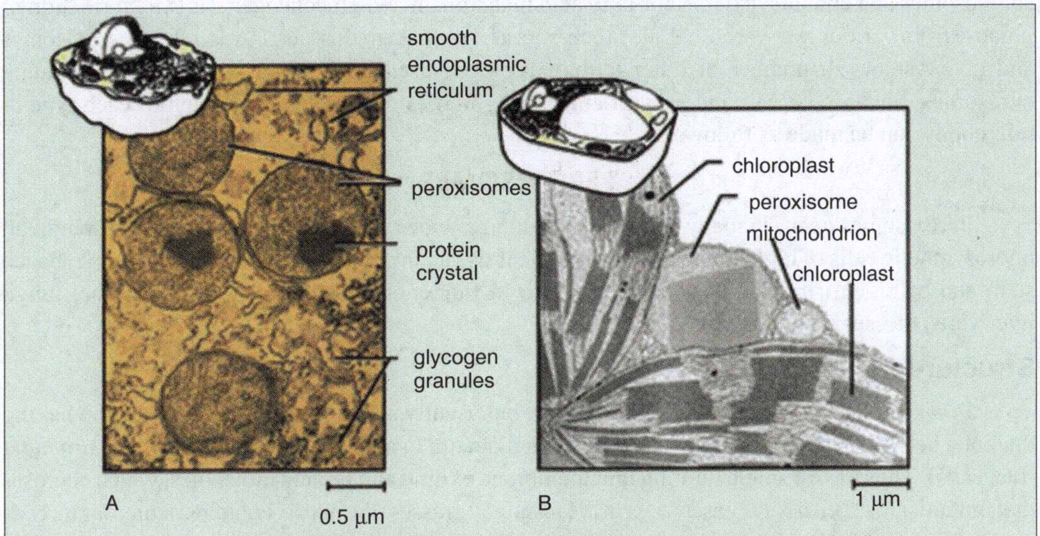

Peroxisome showing central crystalloid core in plant and animal cell.

MICROBODIES : STRUCTURE AND TYPES

Microbodies are spherical or oblate in form. They are bounded by a single membrane and have an interior or matrix which is amorphous or granular. Microbodies are most easily distinguished from other cell organelles by their content of **catalase** enzyme. Catalase can be visualized with the electron microscope when cells are treated with the stain **DAB** (*i.e.*, 3, 3′-diaminobenzidine). The product is electron opaque and appears as dark regions in the cell where catalase is present. By applying this technique microbodies have been observed by electron microscopy and subsequently isolated from various mammalian tissues such as liver, kidney, intestine and brain.

The technique of **isolation** of microbodies from animal and plant tissues includes the following steps : 1. Tissues are ground very carefully to save microbodies from disruption. 2. The homogenate is treated with differential centrifugation to obtain a fraction of the cell homogenate which is rich in microbodies. 3. The enriched fraction is subjected to isopycnic ultra-centrifugation on discontinuous or continuous sucrose density gradient.

Recent biochemical studies have distinguished two types of microbodies, namely **peroxisomes** and **glyoxysomes**. These two organelles differ both in their enzyme complement and in the type of tissue in which they are found. **Peroxisomes** are found in animal cells and in the leaves of higher plants. They contain catalases and oxidases (*e.g.*, D-amino oxidase and urate oxidase). In both they participate in the oxidation of substrates, producing hydrogen peroxide which is subsequently destroyed by catalase activity :

1. Reduced substrate → Oxidized substrate, coupled with FAD → $FADH_2$, coupled with $FADH_2$ + O_2 → FAD + H_2O_2

$$\text{Reduced substrate} \rightarrow \text{Oxidized substrate};\quad FAD \rightarrow FADH_2;\quad O_2 \rightarrow H_2O_2$$

2. $H_2O_2 \rightarrow H_2O + \quad O_2$

In plant cells, peroxisomes remain associated with ER, chloroplasts and mitochondria and are involved in photorespiration. **Glyoxysomes** occur only in plant cells and are particularly abundant in germinating seeds which store fats as a reserve food material. They contain enzymes of glyoxylate cycle besides the catalases and oxidases. Glyoxysomes remain intimately associated with lipid bodies, the spherosomes and contain enzymes for fatty acid metabolism and gluconeogenesis (*i.e.*, formation of glucose from various non-carbohydrate precursors as succinate in this case). Apart from peroxisomes and glyoxysomes, a number of other terms have been used to describe microbodies, including **cytosomes**, **phragmosomes** and **crystal-containing bodies**. A detailed discussion of each type of microbody can be made as follows:

PEROXISOMES

Peroxisomes occur in many animal cells and in a wide range of plants. They are present in all photosynthetic cells of higher plants in etiolated leaf tissue, in coleoptiles and hypocotyls, in tobacco stem and callus, in ripening pear fruits and also in Euglenophyta, Protozoa, brown algae, fungi, liverworts, mosses and ferns.

Structure

Peroxisomes are variable in size and shape, but usually appear circular in cross section having diameter between 0.2 and 1.5μm (0.15 to 0.25 μm diameter in most mammalian tissues; 0.5 μm in rat liver cells). They have a single limiting unit membrane of lipid and protein molecules, which encloses their granular matrix. In some cases (*e.g.*, in the festuciod grasses) the matrix contains numerous threads or fibrils, while in others they are observed to contain either an amorphous nucleoid or a dense inner core which in many species shows a regular crystalloid structure (*e.g.*, tobacco leaf cell, **Newcomb** and **Frederick**, 1971). Little is known about the function of the core, except that it is the site of the enzyme urate oxidase in rat liver peroxisomes and much of the catalase in some plants (see **Hall** *et al*., 1974).

Functions of Peroxisomes

Peroxisomes are found to perform following two types of biochemical activities :

A. Hydrogen peroxide metabolism. Peroxisomes are so-called, because they usually contain one or more enzymes (*i.e.*, D-amino acid oxidase and urate oxidase) that use molecular oxygen to remove hydrogen atoms from specific organic substrates (R) in an oxidative reaction that produces hydrogen peroxide (H_2O_2) :

$$RH_2 + O_2 \rightarrow R + H_2O_2$$

Catalase (which forms 40 per cent of total peroxisome protein) utilizes the H_2O_2 generated by other enzymes in the organelle to oxidize a variety of other substances—including alcohols, phenols, formic acid and formaldehyde—by the "peroxidative" reaction:

$$H_2O_2 + R'H_2 \rightarrow R' + 2H_2O$$

This type of oxidative reaction is particularly important in liver and kidney cells, whose peroxisomes **detoxify** various toxic molecules that enter the blood stream. Almost half of alcohol one drinks is oxidized to acetaldehyde in this way. However, when excess H_2O_2 accumulates in the cell, catalase converts H_2O_2 to H_2O :

$$2H_2O_2 \rightarrow 2H_2O + O_2$$

H_2O_2 and aging. Most cytosolic H_2O_2 is produced by mitochondria and membranes of ER, although there are also H_2O_2-producing enzymes localized in the cytoplasmic matrix. Catalase acts as a "**safety valve**" for dealing with the large amounts of H_2O_2 generated by peroxisomes; however, other enzymes such as **glutathione peroxidase**; are capable of metabolizing organic hydroperoxides and also H_2O_2, in the cytosol (cytoplasmic matrix) and mitochondria. The production of superoxide anion (O_2^-) in mitochondria and cytosol (cytoplasmic matrix) is regulated mainly by the enzyme **superoxide dismutase**. All of these protective enzymes are present in high levels in aerobic tissues.

Recently, a possible relationship has been stressed between peroxides and free radicals (such as superoxide anion—O_2^-)with the process of aging. These radicals may act on DNA molecule to produce **mutations** altering the transcription into mRNA and the translation into proteins. In addition, free radicals and peroxides can affect the membranes by causing peroxidation of lipids and proteins. For these reasons reducing compounds such as vitamin E or enzymes such as superoxide dismutase could play a role in keeping the healthy state of a cell.

B. Glycolate cycle. Peroxisomes of plant leaves contain catalaze together with the enzymes of **glycolate pathway**, as glycolate oxidase, glutamate glyoxylate, serine-glyoxylate and asparate-α-ketoglutarate aminotransferases, hydroxy pyruvate reductase and malic dehydrogenase. They also contain FAD, NAD and NADP coenzymes. The glycolate cycle is thought to bring about the formation of the amino acids–glycine and serine–from the non-phosphorylated intermediates of photosynthetic carbon reduction cycle, *i.e.*, glycerate to serine, or glycolate to glycine and serine in a sequence of reactions which involve chloroplasts, peroxisomes, mitochondria and cytosol (**Tolbert**, 1971). The glycolate pathway also generates C_1 compounds and serves as the generator of precursors for nucleic acid biosynthesis.

Photorespiration. In green leaves, there are peroxisomes that carry out a process called **photorespiration** which is a light-stimulated production of CO_2 that is different from the generation of CO_2 by mitochondria in the dark. In photorespiration, **glycolic acid** (**glycolate**), a two-carbon product of photosynthesis is released from chloroplasts and oxidized into **glyoxylate** and **H2O2** by a peroxisomal enzyme called **glycolic acid oxidase**. Later on, glyoxylate is oxidized into CO_2 and **formate**:

$$CH_2OH.\,COOH + O_2 \longrightarrow CHO - COOH + H_2O_2$$
$$CHO - COOH + H_2O_2 \longrightarrow HCOOH + CO_2 + H_2O$$

Photorespiration is so-called because light induces the synthesis of glycolic acid in chloroplasts. The entire process involves intervention of two basic organelles : chloroplasts and peroxisomes.

Lastly, photorespiration is driven by atmospheric conditions in which the O_2 tension is high and the CO_2 tension low. Apparently O_2 competes with CO_2 for the enzyme **ribulose diphosphate carboxylase** which normally is the key enzyme in CO_2 fixation during photosynthesis. When O_2 is used by the enzyme, an unstable intermediate is formed which breaks down into **3-phosphoglycerate** and **phosphoglycolate**. The latter tends to increase the glycolate concentration by removal of its phosphate group and, therefore, more glycolate is available for additional oxidation and CO_2 release.

Photorespiration is a wasteful process for the plant cell, since, it significantly reduces the efficiency of the process of photosynthesis (*i.e.*, it returns a portion of fixed CO_2 to the atmosphere). It is a particular problem in C_3 plants that are more readily affected by low CO_2 tensions ; C_4 plants are much more efficient in this regard (see Chapter 11).

C. β-oxidation. Peroxisomes of rat liver cells contain enzymes of β-oxidation for the metabolism of fatty acids. They are capable of oxidizing palmitoyl-CoA (or fatty acyl-CoA) to acetyl-CoA, using molecular oxygen and NAD as electron acceptors (**Lazarow** and **de Duve**, 1976). The acetyl-CoA formed by this process is, eventually, transported to the mitochondria where it enters into the citric acid cycle. If, alternatively, acetyl-CoA remains in the cytosol, it is reconverted into fatty acids and ultimately to neutral fats. β-oxidation pathway of the peroxisomes is very similar to the one that occurs in mitochondria with one very important exception. In mitochondria, the flavin dehydrogenase donates its electrons to the respiratory chain. It does not react with molecular oxygen. In peroxisomes, the dehydrogenase reacts directly with O_2 and in so doing generates H_2O_2. Mitochondria contain no catalase and, therefore, cannot deal with the formation of toxic hydrogen peroxide. For peroxisomes this is not a problem.

D. Other functions. Mammalian cells do not contain D-amino acids, but the peroxisomes of mammalian liver and kidney contain **D-amino acid oxidase**. It is suggested that this enzyme is meant for D-amino acids that are found in the cell wall of the bacteria. Thus, the presumed role of this enzyme is to initiate the degradation of D-amino acid that may arise from breakdown and absorption of peptidoglycan material of intestinal bacteria.

Uric acid oxidase (uricase) is important in the catabolic pathway that degrades purines. Thus, peroxisomes are unusually diverse organelles and even in different cells of a single organism may contain very different sets of enzymes. They can also adapt remarkably to changing conditions. For example, yeast cells grown on sugar have tiny peroxisomes. But when some yeasts are grown on methanol, they develop large-sized peroxisomes that oxidize methanol; when grown on fatty acids, they develop large peroxisomes that break down fatty acids to acetyl-CoA (**Veenbuis**, *et al.*, 1983).

Biogenesis of Peroxisomes

At one time it was thought that the membrane 'shell' of the peroxisomes is formed by budding of the endoplasmic reticulum (ER), while the 'content' or matrix is imported from the cytosol (cytoplasmic matrix). However, there is now evidence suggesting that new peroxisomes always arise from pre-existing ones, being formed by growth and fission of old organelles similar to mitochondria and chloroplasts.

Thus, peroxisomes are a collection of organelles with a constant membrane and a variable enzymatic content. All of their proteins (both structural and enzymatic) are encoded by nuclear genes and are synthesized in the cytosol(cytoplasmic matrix) (*i.e.*, on the free ribosomes). The proteins present in either lumen or membrane of the peroxisome are taken up post-translationally from the cytosol (cytoplasmic matrix). For example, **catalase** enzyme is a tetrameric haeme-containing protein that is made in the cytosol (cytoplasmic matrix) as the haeme-free monomers ; the monomers are imported into the lumen of peroxisomes, where they assemble into tetramers in the presence of haeme. Catalase and many peroxisomal proteins are found to have a **signal sequence** (comprising of three amino acids) which is located near their carboxyl ends and directs them to peroxisome (**Gould**, **Keller** and **Subramani**, 1988). Peroxisomes contain **receptors** exposed on their cytosolic surface to recognize the signal on the imported proteins. All of the **membrane proteins** of the peroxisomes, including signal receptor proteins, are imported directly from the cytosol (cytoplasmic matrix). The lipids required to make new peroxisomal membrarne are also imported from the cytosol (cytoplasmic matrix), possibly being carried by **phospholipid transfer proteins** from sites of their synthesis in the ER membranes (**Yaffe** and **Kennedy**, 1983).

Symbiotic origin. One hypothesis (**de Duve**, 1969) holds that the peroxisomes is a vestige of an ancient organelle that carried out all of the oxygen metabolism in the primitive ancestors of eukaryotic cells. When the oxygen, produced by photosynthetic bacteria, first began to accumulate in the primitive atmosphere of earth, it would have been highly toxic to most types of cells. Peroxisomes may have served to lower the concentration of oxygen in such cells while also exploiting its chemical reactivity to carry out useful oxidative reactions. This hypothesis also holds that the later development of mitochondria rendered the peroxisome largely obsolete because many of the same reactions—which had formerly been carried out in peroxisomes without producing energy—were now coupled to ATP formation by means of oxidative phosphorylation.

GLYOXYSOMES

Glyoxysomes are found to occur in the cells of yeast, *Neurospora*, and oil rich seeds of many higher plants. They resemble with peroxisomes in morphological details, except that, their crystalloid core consists of dense rods of 6.0 µm diameter. They have enzymes for fatty acid metabolism and gluconeogenesis, *i.e.*, conversion of stored lipid molecules of spherosomes of germinating seeds into the molecules of carbohydrates.

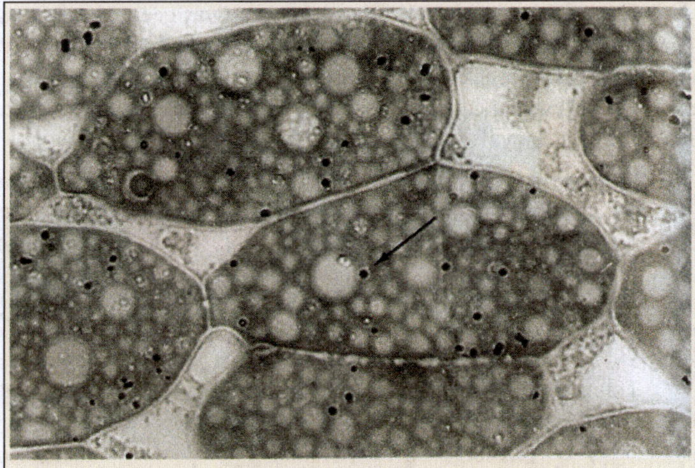

Glyoxysome localization within plant seedlings.

Functions

Glyoxysomes perform following biochemical activities of plants cells :

1. Fatty acid metabolism. During germination of oily seeds, the stored lipid molecules of spherosomes are hydrolysed by the enzyme lipase (glycerol ester hydrolase) to glycerol and fatty acids. The phospholipid molecules are hydrolysed by the enzyme phospholipase. The long chain fatty acids which are released by the hydrolysis are then broken down by the successive removal of two carbon or C_2 fragments in the process of **β-oxidation**.

β-Oxidation. During β-oxidation process, the **fatty acid** is first activated by enzyme fatty acid thiokinase to a **fatty acyl-CoA** which is oxidized by a FAD-linked enzyme fatty acyl-CoA dehydrogenase into **trans-2-enoyl-CoA**. Trans-2-enoyl-CoA is hydrated by an enzyme enoyl hydratase or crotonase to produce the **L-3-hydroxyacyl-CoA**, which is oxidized by a NAD linked L-3-hydroxyacyl-CoA dehydrogenase to produce **3-Keto acyl-CoA**. The 3-keto acyl-CoA looses a two carbon fragment under the action of the enzyme thiolase or β-keto thiolase to generate an **acetyl-CoA** and a new **fatty acyl-CoA** with two less carbon atoms than the original. This new fatty acyl-CoA is then recycled through the same series of reactions until the final two molecules of acetyl-CoA are produced. The complete β-oxidation chain can be represented as follows :

Fatty acid —(Activation; Fatty acid thiokinase+CoA)→ **Fatty acyl - CoA** —(Oxidation; Fatty acyl-CoA dehydrogenase + FAD)→ **trans-2-Enoyl-CoA**

trans-2-Enoyl-CoA —(Hydration (Enoyl-hydratase))→ **L-3-hydroxyl -CoA**

L-3-hydroxyl -CoA —(Oxidation; L-3-hydroxy acyl-CoA dehydrogenase +NAD)→ **3 keto acyl-CoA** —(Thiolytic cleavage; β-keto thiolase)→ **Fatty acyl-CoA** + **Acetyl-CoA**

In plant seeds β-oxidation occurs in glyoxysomes (**Cooper** and **Beevers**, **1969**). But in other plant cells β-oxidation occurs in glyoxysomes and mitochondria. The glyoxysomal β-oxidation requires oxygen for oxidation of reduced flavorprotein produced as a result of the fatty-acyl-CoA dehydrogenase activity. In animal cells β-oxidation occurs in mitochondria.

In plant cells, the acetyl-CoA, the product of β-oxidation chain is not oxidized by Krebs cycle, because it remains spatially separated from the enzymes of Krebs cycle, instead of it, acetyl-CoA undergoes the **glyoxylate cycle** to be converted into succinate.

2. Glyoxylate cycle. The glyoxylate pathway occurs in glyoxysomes and it involves some of the reactions of the Krebs cycle in which citrate is formed from **oxaloacetate** and **acetyl-CoA** under the action of citrate synthetase enzyme. The citrate is subsequently converted into **isocitrate** by aconitase enzyme. The cycle then involves the enzymatic conversion of isocitrate to **glyoxylate** and **succinate** by isocitratase enzyme:

$$\text{Isocitrate} \xrightarrow{\text{Isocitratase}} \text{Glyoxylate} + \text{Succinate}$$

The glyoxylate and another mole of acetyl-CoA form a mole of **malate** by malate synthetse:

$$\text{Acetyl CoA} + \text{Glyoxylate} \xrightarrow{\text{Malate synthetase}} \text{Malate}$$

This malate is converted to oxaloacetate by malate dehydrogenase for the cycle to be completed. Thus, overall, the glyoxylate pathway invloves :

$$\text{2 Acetyl-CoA} + NAD^{+} \longrightarrow \text{Succintiate} + NADH + H+$$

Succinate is the end product of the glyoxysomal metabolism of fatty acid and is not further metabolized within this organelle.

The synthesis of hexose or gluconeogenesis involves the conversion of succinate to oxaloacetate, which presumably takes place in the mitochondria, since the glyoxysomes do not contain the enzymes fumarase and succinic dehydrogenase. Two molecules of **oxaloacetate** are formed from four molecules of acetyl-CoA without carbon loss. This oxaloacetate is converted to **phosphoenol pyruvate** in the phosphoenol pyruvate carboxykinase reaction with the loss of two molecules of CO_2:

$$\text{2 Oxaloacetate} + \text{2ATP} \rightleftharpoons \text{2 Phosphoenol pyruvate} + 2CO_2 + \text{2ADP}$$

The phosphoenol pyruvate is converted into monosaccharides (*e.g.*, glucose, fructose), disaccharide (sucrose) and polysaccharide (starch) by following reation :

Phosphoenol pyruvate ⟶ **Triose-phosphate** ⟶ **Fructose-1, 6 - diphosphate**
↓
Fructose-6-phosphate ⟶ **Glucose-6-phosphate** ⟶ **Glucose-1-phosphate** ⟶ **UDP Glucose**

UDP Glucose **+Fructose-6 phospate** ⟶ **Sucrose**

Glucose-1-phosphate ⟶ **ADP Glucose** ⟶ **(Starch)**

Glucose-6-phosphate ⟶ **(Glucose)**

Fructose-6-phosphate ⟶ **(Fructose)**

REVISION QUESTIONS

1. What are microbodies ? Describe the symbiotic origin theory of the peroxisomes.
2. Define the terms peroxisome and glyoxysome.
3. What would be the most likely role of peroxisome in cells lacking mitochondria ?
4. What is the function of the glyoxylate cycle ?
5. What is photorespiration and how do peroxisomes contribute ?
6. Write an account of structure, synthesis, function and histochemical localization of the catalase enzyme in eukaryotic cells.
7. Describe the mode of biogenesis and various functions of peroxisomes in animal cells.

CHAPTER

10

Mitochondria

With wings beating 60 times per second, the ruby-throated hummingbird has a metabolic rate 50 times that of a human. The muscles of its wings are packed with mitochondria, which supply the ATP needed to meet the bird's energy demands.

The mitochondria (Gr., *mito*=thread, *chondrion* =granule) are filamentous or granular cytoplasmic organelles of all aerobic cells of higher animals and plants and also of certain micro-organisms including Algae, Protozoa and Fungi. These are absent in bacterial cells. The mitochondria have lipoprotein framework which contains many enzymes and coenzymes required for energy metabolism. They also contain a specific DNA for the cytoplasmic inheritance and ribosomes for the protein synthesis.

HISTORICAL

The mitochondria were first observed by **Kolliker** in 1850 as granular structures in the striated muscles. In 1888, he isolated them from insect muscles (which contain many slab-like mitochondria ; Fig. 10.1) and showed that they swelled in water and contain a membrane around them. In 1882, **Flemming** named them as **fila**. **Richard Altmann** (1890) developed a specific stain that had useful specificity for the mitochondria. He named this organelle, the **bioblast**. Altmann correctly speculated that bioblasts were autonomous elementary living particles that made a genetic and metabolic impact on the cells. The present name mitochondria was assigned by **Benda** (1897–98) to them. He stained mitochondria with alizarin and crystal violet. **Michaelis** (1900) used the supravital stain Janus green to demonstrate that mitochondria were oxidation-reduction sites in the cell. In 1912, **Kingsbury** suggested that the oxidation reactions mediated by mitochondria were normal cellular processes. **Otto Warburg** (1883–1970), who is considered as 'the father of respirometry', in 1910 isolated mitochondria ("large granules") by low-speed centrifugation of tissues disrupted by grinding.

He showed that these granules contained enzymes catalyzing oxidative cellular reactions.

Various steps of **glycolysis** were discovered by two German biochemists **Embden** and **Meyerhof** [**Gustav G. Embden** (1874–1933); **Otto F. Meyerhof** (1884–1951)]. Meyerhof got Nobel Prize in 1922 along with English biophysicist **A.V. Hill**, for the discovery of oxygen and metabolism of lactic acid in muscle (*i.e.*, production of heat in muscle). **Lohmann** (1931) discovered **ATP** in muscle. **Lipmann** (German biochemist in U.S. ; born 1899) discovered **coenzyme A** and showed its significance in intermediary metabolism. In 1941, he introduced the concept of "high energy phosphates" and "high energy phosphate bonds" (*i.e.*, ATP) in bioenergetics. **Warburg** linked the phenomenon of ATP formation to the oxidation of glyceraldehyde phosphate. **Meyerhof** showed the formation of ATP from phosphopyruvate and **Kalckar** related oxidative phosphoryalation to respiration. **Sir Hans Adolph Krebs** (German biochemist in England ; born 1900), in 1937, worked out various reactions of the **citric acid cycle** (or tricarboxylic acid or TCA cycle). His contribution was remarkable, because, up to that time radioactively labelled compounds were not available for biological studies and cellular sites of the reactions were not known with certainty. Krebs received the Nobel Prize in 1953 along with Lipmann for his discovery of the citric acid cycle.

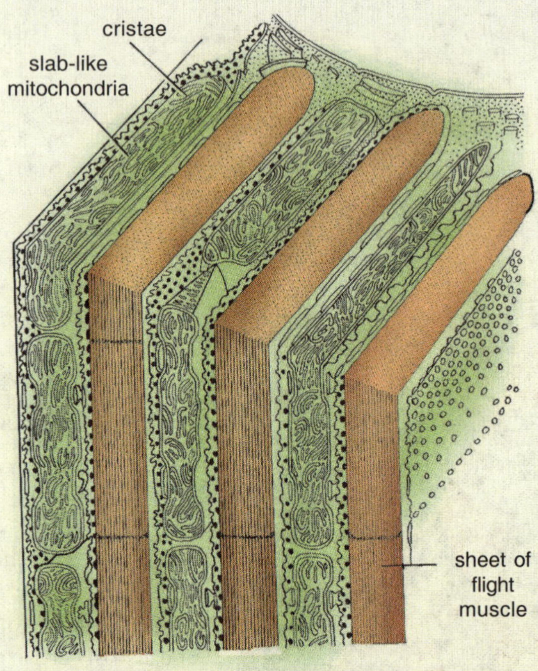

Fig. 10.1. Mitochondria of the flight muscle of a dragon fly, showing profuse cristae.

Kennedy and **Lehninger** (1948–1950) showed that the citric acid cycle (Krebs cycle), oxidative phosphorylation and fatty acid oxidation took place in the mitochondria. In 1951, **Lehninger** proved that oxidative phosphorylation requires electron transport. Among these early investigators of ETS the Nobel Prize recipients were **Warburg**, **Szent-Gyorgyi** and **Kuhn**. In 1961, **Mitchell** proposed the highly acclaimed "**chemiosmotic-coupling hypothesis**" for the ATP-production in mitochondria. He got the Nobel Prize in 1978 for the development of this model.

Palade (1954) described the ultra structure of cristale. In 1963, **Nass** and **Nass** demonstrated the presence of **DNA fibres** in the matrix of mitochondria of embryonic cells. **Attardi**, **Attardi** and **Aloni** (1971) reported the 70S type ribosomes inside the mitochondria.

Previously the mitochondria have been known by various names such as **fuchsinophilic granules**, **parabasal bodies**, **plasmosomes**, **plastosomes**, **fila**, **vermicules**, **bioblasts** and **chondriosomes**.

DISTRIBUTION OR LOCALIZATION

The mitochondria move autonomously in the cytoplasm, so they generally have uniform distribution in the cytoplasm, but in many cells their distribution is very restricted. The distribution and number of mitochondria (and also of mitochondrial cristae) are often correlated with type of function the cell performs. Typically mitochondria with many cristae are associated with mechanical and osmotic work situations, where there are sustained demands for ATP and where space is at a premium, *e.g.*, between muscle fibres, in the basal infolding of kidney tubule cells, and in a portion of inner segment of rod and cone cells of retina. Myocardial muscle cells have numerous large mitochondria called **sarcosomes**, that reflect the great amount of work done by these cells. Since the work of hepatic cells is mainly biosynthetic and degraditive, and work locations are spread throughout the cell, in these cells, it may be more efficient to have a large number of "low key" sources of ATP production distributed throughout the cell. Often mitochondria occur in greater concentrations at work sites, for example, in the oocyte of *Thyone briaeus*, rows of mitochondria are closely associated with RER membranes, where ATP is required for protein biosynthesis. Mitochondria are particularly numerous in regions where ATP-driven osmotic work occurs, *e.g.,* brush border of kidney proximal tubules, the infolding of the plasma membrane of dogfish salt glands and Malpighian tubules of insects, the contractile vacuoles of some protozoans (*Paramecium*). Non-myelinated axons contain many mitochondria that are poor ATP factories, since each has only a single crista. In this case, there is a great requirement for **monoamine oxidase**, an enzyme present in outer mitochondrial membrane that oxidatively deaminates monoamines including neurotransmitters (acetylcholine).

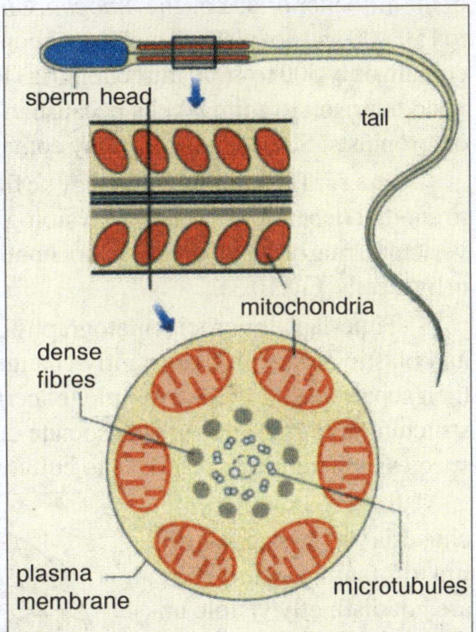

A cross-section of the tail of a human sperm showing the '9 + 2' arrangement of microtubule and mitochondria.

ORIENTATION

The mitochondria have definite orientation. For example, in cylindrical cells the mitochondria usually remain orientated in basal apical direction and lie parallel to the main axis. In leucocytes, the mitochondria remain arranged radially with respect to the centrioles. As they move about in the

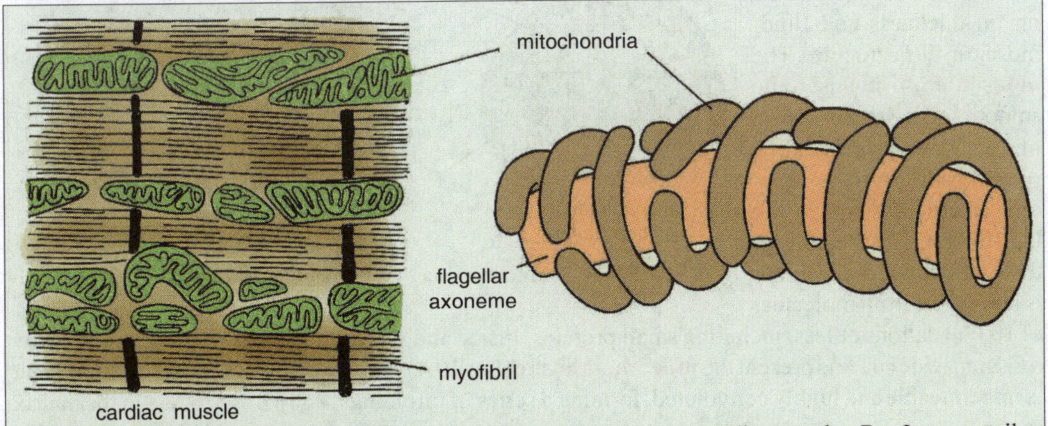

Fig. 10.2. Location of mitochondria near high ATP utilization : A— In cardiac muscles; B— In sperm tail (after Alberts *et al.*, 1989).

mitochondria form long moving filaments or chains, while in others they remain fixed in one position where they provide ATP directly to a site of high ATP utilization, *e.g.*, they are packed between adjacent myofibrils in a cardiac muscle cell or wrapped tightly around the flagellum of sperm (Fig.10.2).

MORPHOLOGY

Number. The number of mitochondria in a cell depends on the type and functional state of the cell. It varies from cell to cell and from species to species. Certain cells contain exceptionally large number of the mitochondria, *e.g.*, the *Amoeba*, *Chaos chaos* contain 50,000; eggs of sea urchin contain 140,000 to 150,000 and oocytes of amphibians contain 300,000 mitochondria. Certain cells, *viz.*, liver cells of rat contain only 500 to 1600 mitochondria. The cells of green plants contain less number of mitochondria in comparison to animal cells because in plant cells the function of mitochondria is taken over by the chloroplasts. Some algal cells may contain only one mitochondrion.

Shape. The mitochondria may be filamentous or granular in shape and may change from one form to another depending upon the physiological conditions of the cells. Thus, they may be of club, racket, vesicular, ring or round-shape. Mitochondria are granular in primary spermatocyte or rat, or club-shaped in liver cells (Fig.10.3).

Time-lapse microcinematography of living cells shows that mitochondria are remarkably mobile and plastic organelles, constantly changing their shape. They sometimes fuse with one another and then separate again. For example, in certain euglenoid cells, the mitochondria fuse into a reticulate structure during the day and dissociate during darkness. Similar changes have been reported in yeast species, apparently in response to culture conditions (see **Reid** and **Leech** 1980).

Size. Normally mitochondria vary in size from 0.5 μm to 2.0 μm and, therefore, are not distinctly visible under the light microscope. Sometimes their length may reach up to 7 μm.

Structure. Each mitochondrion is bound by two highly specialized membranes that play a crucial part in its activities. Each of the mitochondrial membrane is 6 nm in thickness and fluid-mosaic in ultrastructure. The **outer membrane** is quite smooth and has many copies of a transport protein called **porin** which forms large aqueous channels through the lipid bilayer. This membrane, thus, resembles a sieve that is permeable to all molecules of 10,000 daltons or less, including small proteins. Inside and separated from the outer membrane by a 6–8 nm wide space is present the **inner membrane** (Fig.10.4). The inner membrane is not smooth but is impermeable and highly convoluted, forming a series of infoldings, known as **cristae**, in the matrix space.

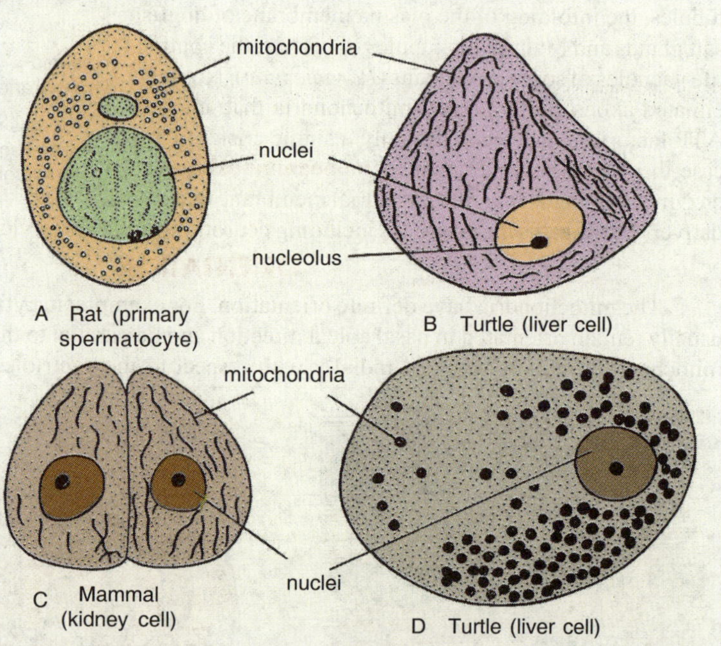

Fig.10.3. Mitochondria of different type of animal cells.

Thus, mitochondria are double membrane envelopes in which the inner membrane divides the mitochondrial space into two distinct chambers : 1. The **outer compartment, peri-mitochondrial space** or the **inter-membrane space** between outer membrane and inner membrane. This space is continuous into the core of the crests or cristae. 2. The **inner compartment**, **inner chamber** or **matrix space**, which is filled with a dense, homogeneous, gel-like proteinaceous material, called **mitochondrial matrix**. The mitochondrial matrix contains lipids, proteins, circular DNA molecules, 55S ribosomes and certain **granules** which are related to the ability of mitochondria to accumulate ions. Granules are prominent in the mitochondria of cells concerned with the transport of ions and water, including kidney tubule cells, epithelial cells of the small intestine, and the osteoblasts of bone-forming cells. Further, the inner membrane has an outer **cytosol** or **C face** toward the perimitochondrial space and an inner **matrix** or **M face** toward matrix.

In general, the cristae of plant mitochondria are tubular, while those of animal mitochondria are lamellar or plate-like (**Hall**, **Flowers** and **Roberts**, 1974), but, in many Protozoa and in steroid synthesizing tissues including the adrenal cortex and corpus luteum, they occur as regularly packed tubules (**Tyler**, 1973). The cristae greatly increase the area of inner membrane, so that in liver cell mitochondria, the cristae membrane is 3–4 times greater than the outer membranre area. Some mitochondria, particularly those from heart, kidney and skeletal muscles have more extensive cristae arrangements than liver mitochondria. In comparison to these, other mitochondria (*e.g.*, from fibroblasts, nerve axons and most plant tissues) have relatively few cristae. For example, mitochondria in epithelial cells of carotid bodies (or **glomus carotica** which are chemoreceptors, sensitive to changes in blood chemistry and lie near the bifurcations of carotid arteries) have only four to five cristae and mitochondria from non-myelinated axons of rabbit brain have only a single crista.

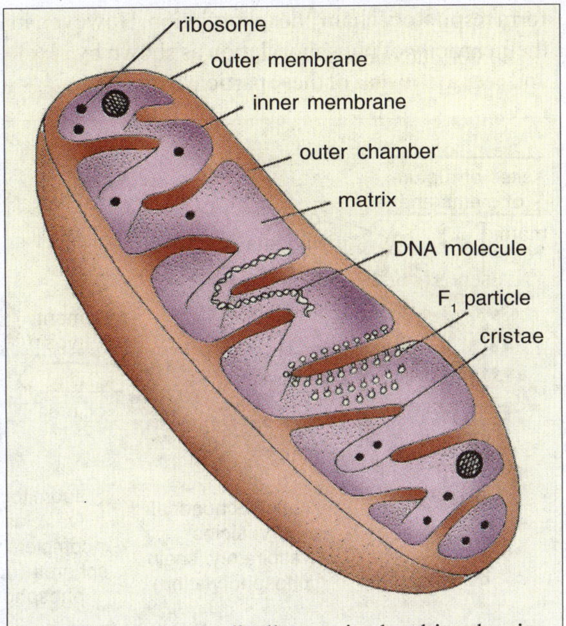

Fig. 10.4. A longitudinally cut mitochondrion showing its internal structure.

Attached to M face of inner mitochondrial membrane are repeated units of stalked particles, called **elementary particles**, **inner membrane subunits** or **oxysomes** (Fig.10.5). They are also identified as **F_1 particles** or **F_0-F_1 particles** and are meant for ATP synthesis (phosphorylation) and also for ATP oxidation (*i.e.*, acting as ATP synthetase and ATPase) (**Racker**, 1967). F_0-F_1 particles are regularly spaced at intervals of 10 nm on the inner surface of inner mitochondrial membrane. According to some estimates, there are 10^4 to 10^5 elementary particles per mitochondrion. When the mitochondrial cristae are disrupted by sonic vibrations or by detergent action, they produce **submitochondrial vesicles** of inverted orientation. In these vesicles, F_0-F_1 particles are seen attached on their outer surface (Fig.10.6). These submitochondrial vesicles are able to per-

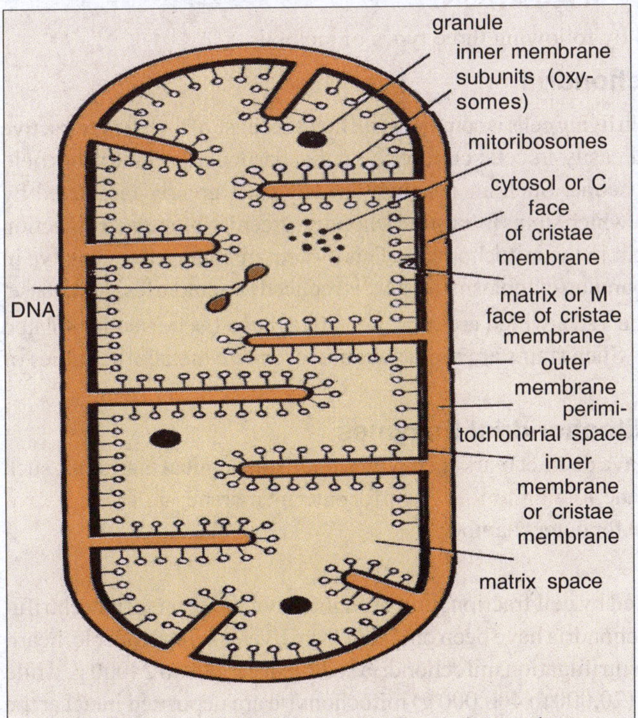

Fig. 10.5. A mitochondrion in sectional view to show its numerous oxysomes.

form respiratory chain phosphorylation. However, in the absence of F_0-F_1 particles, these vesicles lose their capacity of phosphorylation as shown by **resolution** (*i.e.*, removal by urea or trypsin treatment) and **reconstitution** of these particles (Fig.10.6).

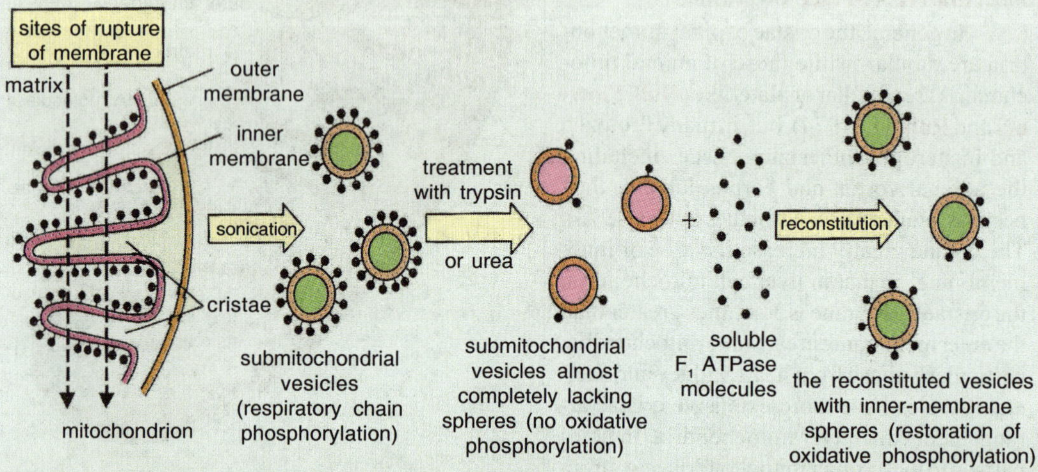

Fig. 10.6. Method of formation of small mitochondrial vesicles having oxysomes on their surface. Experimental results of resolution and reconstitution of oxysomes also have been shown (after Sheeler and Bianchi, 1987).

ISOLATION

Mitochondria have been studied by following three types of methods :

1. Direct Observation of Mitochondria

The examination of mitochondria in living cells is somewhat difficult because of their low refractive index. However, they can be observed easily in cells cultured *in vitro*, particularly under darkfield illumination and phase contrast microscope. Such an examination has been greatly facilitated by colouration with vital stain **Janus green** which stains living mitochondria greenish blue due to its action with cytochrome oxidase system present in the mitochondria. This system maintains the vital dye in its oxidized (coloured) state. In the surrounding cytoplasm the stain is reduced to a colourless leukobase.

Fluorescent dyes (*e.g.*, rhodamine 123), which are more sensitive, have been used in isolated mitochondria and intact cultured cells. Such stains are more suitable for *in situ* metabolic studies of mitochondria.

2. Cytochemical Marking of Mitochondrial Enzymes

Different parts of mitochondria have distinct marker enzymes for histochemical markings, such as **cytochrome oxidase** for inner membrane, **monoamine oxidase** for outer membrane, **malate dehydrogenase** for matrix and **adenylate kinase** for outer chamber.

3. Isolation

Mitochondria can be easily isolated by cell fractionation brought about by differential centrifugation. Homogeneous fractions of mitochondria have been obtained from liver, skeletal muscle, heart, and some other tissues. In differential centrifugation mitochondria sediment at 5000 to 24000 g, while in living cells at the ultracentrifugation (20,000 to 400,000 g) mitochondria are deposited intact at the centrifugal pole.

The two mitochondrial membranes have been separated by **density gradient centrifugation**. The outer membrane is separated by causing a swelling which can be brought about by breakage followed by contraction of inner membrane and matrix. Certain detergents such as digitonin and lubrol are often used for this purpose. Since outer membrane is lighter and much stronger, centrifugal force is needed to separate it. When outer membrane is removed with digitonin, the so-called **mitoplast** is formed. Mitoplast includes inner membrane with unfolded cristae and matrix. Mitoplast is found to carry out oxidative phosphorylation. The isolated outer membrane is revealed by negative staining and shows a "folded-bag" appearance (Fig.10.7). Such isolation of two membranes and compartments has enabled localization of various enzyme systems of mitochondria.

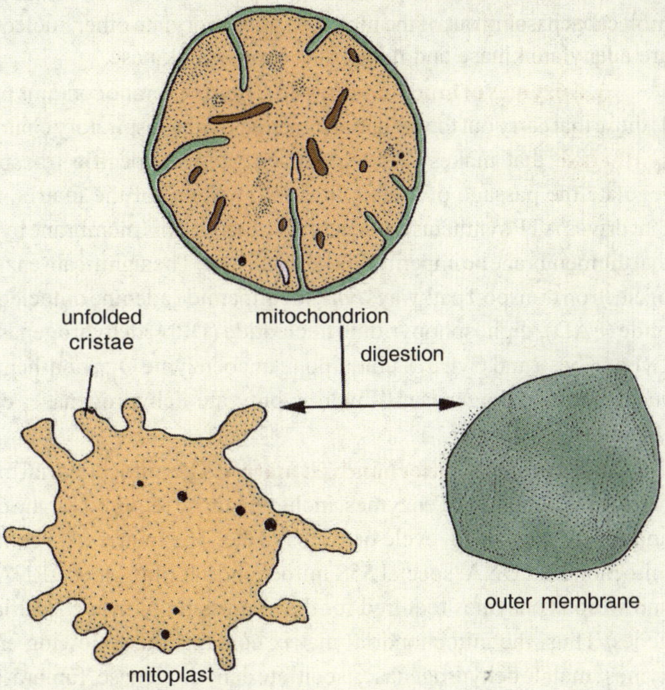

Fig. 10.7. Action of digitonin on liver mitochondrion to produce mitoplast and "folded-bag" of outer membrane (after De Robertis and De Robertis, Jr., 1987).

CHEMICAL COMPOSITION

The gross chemical composition of the mitochondria varies in different animal and plant cells. However the mitochondria are found to contain 65 to 70 per cent proteins, 25 to 30 per cent lipids, 0.5 per cent RNA and small amount of the DNA. The lipid contents of the mitochondria are composed of 90 per cent phospholipids (lecithin and cephalin), 5 per cent or less cholesterol and 5 per cent free fatty acids and triglycerides. The inner membrane is rich in one type of phospholipid, called **cardiolipin** which makes this membrane impermeable to a variety of ions and small molecules (*e.g.*, Na^+, K^+, Cl^-, NAD^+, AMP, GTP, CoA and so on).

The outer mitochondrial membrane has typical ratio of 50 per cent proteins and 50 per cent phospholipids of 'unit membrane'. However, it contains more unsaturated fatty acids and less cholesterol. It has been estimated that in the mitochondria of liver 67 per cent of the total mitochondrial protein is located in the matrix, 21 per cent is located in the inner membrane, 6 per cent is situated in the outer membrane and 6 per cent is found in the outer chamber. Each of these four mitochondrial regions contains a special set of proteins that mediate distinct functions :

1. Enzymes of outer membrane. Besides porin, other proteins of this membrane include enzymes involved in mitochondrial lipid synthesis and those enzymes that convert lipid substrates into forms that are subsequently metabolized in the matrix. Certain important enzymes of this membrane are monoamine oxidase, rotenone-insensitive NADH-cytochrome-C-reductase, kynurenine hydroxyalase, and fatty acid CoA ligase.

2. Enzymes of intermembrane space. This space contains several enzymes that use the ATP molecules passing out of the matrix to phosphorylate other nucleotides. The main enzymes of this part are adenylate kinase and nucleoside diphosphokinase.

3. Enzymes of inner membrane. This membrane contains proteins with three types of functions: 1. those that carry out the oxidation reactions of the respiratory chain; 2. an enzyme complex, called **ATP synthetase** that makes ATP in matrix ; and 3. specific transport proteins (see Table 10-1) that regulate the passage of metabolites into and out of the matrix. Since an electrochemical gradient, that drives ATP synthetase, is established across this membrane by the respiratory chain, it is important that the membrane be impermeable to small ions. The significant enzymes of inner membrane are enzymes of electron transport pathways, *viz.*, nicotinamide adenine dinucleotide (NAD), flavin adenine dinucleotide (FAD), diphosphopyridine nucleotide (DPN) dehydrogenase, four cytochromes (Cyt. b, Cyt. c, Cyt.c_1, Cyt. a and Cyt. a_3), ubiquinone or coenzyme Q_{10}, non-heme copper and iron, ATP synthetase, succinate dehydrogenase; β-hydroxybutyrate dehydrogenase; carnitive fatty acid acyl transferase (Fig. 10.8).

4. Enzymes of mitochondrial matrix. The mitochondrial matrix contains a highly concentrated mixture of hundreds of enzymes, including those required for the oxidation of pyruvate and fatty acids and for the citric acid cycle or Krebs cycle. The matrix also contains several identical copies of the mitochondrial DNA, special 55S mitochondrial ribosomes, tRNAs and various enzymes required for the expression of mitochondrial genes. Thus, the mitochondrial matrix contains the following enzymes : malate dehydrogenase, isocitrate dehydrogenase, fumarase, aconitase, citrate synthetase, α-keto acid dehydrogenase, β-oxidation enzymes. Moreover, the mitochondrial matrix contains different nucleotides, nucleotide coenzymes and inorganic electrolytes—K^+, HPO_4^-, Mg^{++}, Cl^- and SO_4^-.

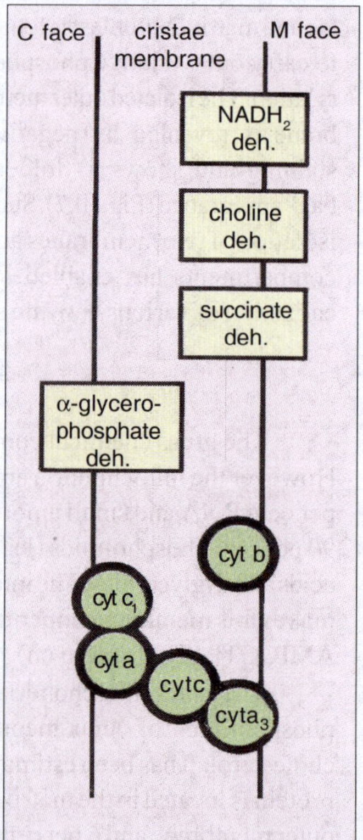

Fig. 10.8. A part of inner mitochondrial membrane (cristae) showing the distribution of different dehydrogenases and cytochromes on M face and C face.

MITOCHONDRIA AND CHLOROPLASTS AS TRANSDUCING SYSTEMS

In cells, energy transformation takes place through the agency of two main transducing systems (*i.e.*, systems that produce energy transformation) represented by mitochondria and chloroplasts. These two organelles of eukaryotic cells in some respects operate in opposite directions. For example, chloroplasts are present only in plant cells and especially adapted to capture light energy and to transduce it into chemical energy, which is stored in covalent bonds between atoms in the different nutrients or **fuel molecules**. In contrast, the mitochondria are the "power plants" or "power houses" that by oxidation, release the energy contained in the fuel molecules and make other forms of chemical energy (Fig.10.9). The main function of chloroplasts is **photosynthesis**, while that of mitochondria is **oxidative phosphorylation**. Finally, photosynthesis is an **endergonic** reaction, which means that it captures energy; oxidative phosphorylation is an **exergonic** reaction, meaning that it releases energy. Table 10-2 has enlisted some of the basic differences between these two transducing systems.

Table 8-1. Transport systems of inner mitochondrial membrane (Source: Sheeler and Bianchi, 1987).

System	Exchange
1. Decarboxylate carrier	Exchange for mole-for-mole basis of malate, succinate, fumarate and phosphate between matrix and cytosol.
2. Tricarboxylate carrier	Exchange on mole-for-mole basis citrate and isocitrate between matrix and cytosol.
3. Aspartate-glutamate carrier	Exchange aspartate for glutamate across membrane.
4. α-Ketoglutarate-malate carrier	Specifically exchange α-ketoglutarate for malate across membrane.
5. ADP-ATP carrier	Exchange of ADP for ATP.

FUNCTIONS

The mitochondria perform most important functions such as oxidation, dehydrogenation, oxidative phosphorylation and respiratory chain of the cell. Their structure and enzymatic system are fully adapted for their different functions. They are the actual respiratory organs of the cells where the foodstuffs, *i.e.*, carbohydrates and fats are completely oxidised into CO_2 and H_2O. During the biological oxidation of the carbohydrates and fats large amount of energy is released which is utilized by the mitochondria for synthesis of the energy rich compound known as **adenosine triphosphate** or **ATP**. Because mitochondria synthesize energy rich compound ATP, they are also known as "power houses" of the cell. In animal cells mitochondria produce 95 per cent of ATP molecules, remaining 5 per cent is being produced during anaerobic respiration outside the mitochondria. In plant cells, ATP is also produced by the chloroplasts.

Adenosine triphosphate or ATP

The ATP consists of a purine base **adenine**, a pentose sugar **ribose** and three molecules of the **phosphoric acids** (Fig. 4.30). The adenine and ribose sugar collectively constitute the nucleoside **adenosine** which by having one, two or three phosphate groups forms the adenosine monophosphate (AMP), adenosine diphosphate (ADP) and adenosine triphosphate (ATP) respectively. In ATP the last phosphate group is linked with ADP by a special bond known as "**energy rich bond**" because when the last phosphate group of the ATP is released the large amount of energy is released as shown by the following reaction :

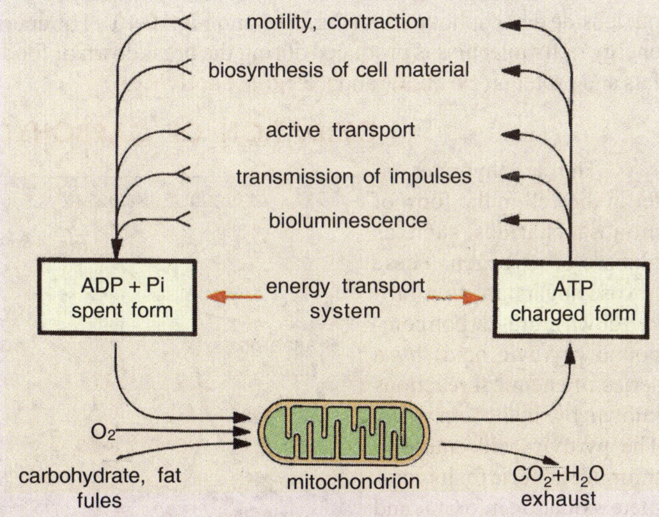

Fig. 10.9. Diagrammatic representation of the energy liberation in mitochondrion and its utilization in various cellular functions.

$$\text{A—P}\sim\text{P}\sim\text{P} \;=\; \text{A–P}\sim\text{P} + \text{Pi} + 7300 \text{ calories}$$

ATP　　ADP　　Phosphate group

In the above reaction, we have seen that by the breaking of the energy rich bond about 7300 calories of energy are released, while the common chemical bond releases only 300 calories of energy. The chemical reactions which synthesize the energy rich bond or~P bond require great amount of energy which is supplied by the oxidation of the foodstuffs in the mitochondria. The utility of energy rich phosphate bond (~P) of the ATP is that great amount of energy is kept stored in the ready state in a very limited space of the cell. The stored chemical energy is disposed of very quickly at the time of the need in various cellular functions such as respiratory cycle, protein and nucleic acid synthesis, nervous transmission, cell division, transportation and bioluminescence, etc.

Table 9-2. Differences between photosynthesis and oxidative phosphorylation (Source: De Robertis and De Robertis, Jr., 1987).

Photosynthesis	Oxidative phosphorylation
1. Occurs in the presence of light; thus, periodic	1. Independent of light; thus continuous
2. Uses H_2O and CO_2	2. Uses molecular O_2
3. Liberates O_2	3. Liberates CO_2
4. Hydrolyzes water	4. Forms water
5. Endergonic reaction	5. Exergonic reaction
6. $CO_2 + H_2O$ + energy $\rightarrow$ foodstuff	6. Foodstuff + $O_2 \rightarrow CO_2 + H_2O$ + energy
7. Takes place in chloroplast	7. Takes place in mitochondria

Because the terminal phosphate linkage in ATP is easily cleaved with release of free energy, ATP acts as an efficient phosphate donor in a large number of different phosphorylation reactions. In this way, ATP acts as a carrier molecule like the acetyl CoA and as coenzyme like the CoA or NAD.

Recently, besides ATP, certain other energy rich chemical compounds have been found to be active in the cellular metabolism. These are **cytosine triphosphate (CTP)**, **uridine triphosphate (UTP)** and **guanosine triphosphate (GTP)**. These compounds, however, derive the energy from the ATP by nucleoside diphosphokinases (Fig. 10.9 and Fig. 10.10). The energy for the production of ATP or other energy rich molecules is produced during the breakdown of food molecules including carbohydrates, fats and proteins (**catabolic** and **exergonic activities**).

OXIDATION OF CARBOHYDRATES

The carbohydrates enter in the cell in the form of monosaccharides such as glucose or glycogen. These hexose sugars are first broken down into 3-carbon compound (pyruvic acid) by a series of chemical reactions known by many enzymes. The pyruvic acid enters in the mitochondria for its complete oxidation into CO_2 and water. The reactions which involve in the oxidation of glucose into CO_2 and water are known to form the **metabolic pathways** and they can be grouped under the following heads :

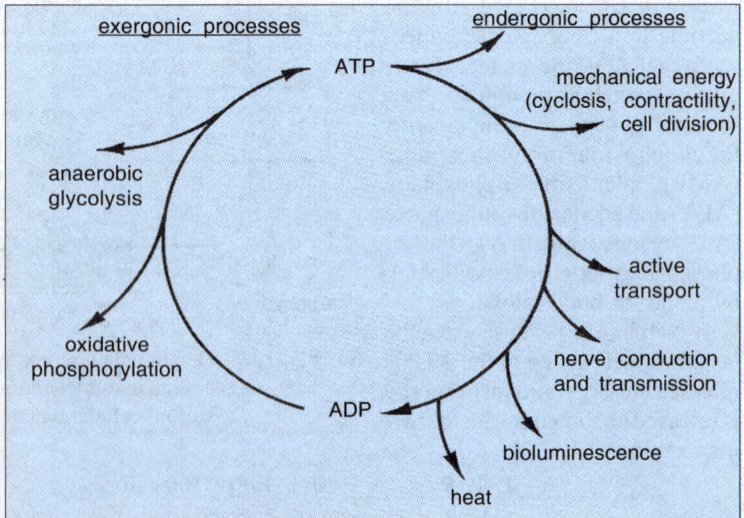

Fig. 10.10. Diagrammatic representation of uses and synthesis of ATP.

(i) Glycolysis or Embden-Meyerhof pathways (EMP) or Embden-Meyerhof-Parnas pathways (EMPP);
(ii) Oxidative decarboxylation;
(iii) Krebs cycle; citric acid cycle or tricarboxylic acid cycle;
(iv) Respiratory chain and oxidative phosphorylation.

1. Glycolysis

Under anaerobic conditions (*i.e.*, in the absence of oxygen) glucose is degraded into lactic acid or lactate by a process called **glycolysis** (*i.e.*, lysis or splitting of glucose), *e.g.*, it commonly occurs in vertebrate muscles when the energy demand in heavy exercise exceeds the available oxygen. [**Note**: According to Circular No. 200 of Committee of Editors of Biochemical Journals Recommendations (1975) the ending *ate* in lact*ate*, pyru*vate*, oxaloace*tate*, citr*ate*, etc., denotes any mixture of free acid and the ionized form(s) (according to pH) in which cations are not specified (see **Martin Jr.**, *et al.*, 1983). Most modern textbooks though have adopted this convention for all of the carboxylic acids, but we prefer to stick to the old pattern). If glycolysis is carried out under aerobic conditions the final products are pyruvic acid and coenzyme NADH. Glycolysis is achieved by a series of 10 enzymes all of which are located in the cytosol (cytoplasmic matrix). As shown in Figure 10.11, in this chain of reactions, the product of one enzyme serves as a substrate for the next reaction. To facilitate its analysis, the sequence of glycolysis can be subdivided into following three main steps : (i) Activation (stage I); (ii) Cleavage (stage II); and (iii) Oxidation (stage III).

(i) Activation. In reactions 1 to 3 the glucose molecule is converted into **fructose-1-6- diphosphate**. This step uses two molecules of ATP and involves the following enzymes: hexokinase, phosphoglucose isomerase (or phosphohexoisomerase) and phosphofructokinase.

(ii) Cleavage. In reactions 4 and 5, fructose -1-6-diphosphate splits into two (3-carbon) end products, **glyceraldehyde-3-phosphate** molecules. During the step of cleavage only two enzymes are used: aldolase (fructoaldolase) and triose isomerase (triosephosphate isomerase).

(iii) Oxidation. In reactions 6 to 10, two molecules of glyceraldehyde-3-phosphate are oxidized and ultimately converted into two molecules of **pyruvic acid**. This step produces four molecules of ATP by **substrate-level-phosphorylation** and involves the following enzymes: phosphoglyceric dehydrogenase (glyceraldehyde phosphate dehydrogenase), phosphoglyceric kinase, phosphoglyceromutase, enolase and pyruvic kinase.

The net energy yield of chain reactions of glycolysis is the production of two ATP molecules from one molecule of glucose. Under aerobic conditions, the end products of glycolysis are pyruvic acid and reduced coenzyme NAD (*i.e.*, NADH). NADH carries two electrons, taken from glyceraldehyde-3-phosphate and contains little energy. However, under anaerobic conditions, pyruvic acid remains in the cytosol (cytoplasmic matrix) and is used as a hydrogen acceptor and converted into lactic acid :

$$\text{Pyruvic acid} \xrightarrow[+\text{NADH}]{\text{Lactic dehydrogenase}} \text{Lactic acid} + \text{NAD} + \text{Energy}$$

In above case, the following equation represents the overall reaction of glycolysis :

$$\underset{\text{Glucose}}{C_6H_{12}O_6} + \text{Pi} + 2\text{ADP} \longrightarrow \underset{\text{Lactic acid}}{2C_3H_6O_3} + 2\,\text{ATP} + H_2O$$

Further, pyruvic acid is converted into ethyl alcohol via acetaldehyde by yeast, *Saccharomyces cerevisae* in the absence or deficiency of oxygen. This process is called **alcoholic fermentation :**

$$\underset{\text{Glucose}}{C_6H_{12}O_6} \xrightarrow{\text{Glycolysis}} \underset{\text{Pyruvic acid}}{CH_3CO.COOH} \xrightarrow{-CO_2} \underset{\text{Acetaldehyde}}{CH_3CHO} \xrightarrow{+2H} \underset{\text{Ethyl alcohol}}{CH_3CH_2.OH}$$

However, in the cells of higher plants under anaerobic conditions, pyruvic acid is converted into either ethyl alcohol or any organic acid such as malic acid, citric acid, oxalic acid and tartaric acid. The

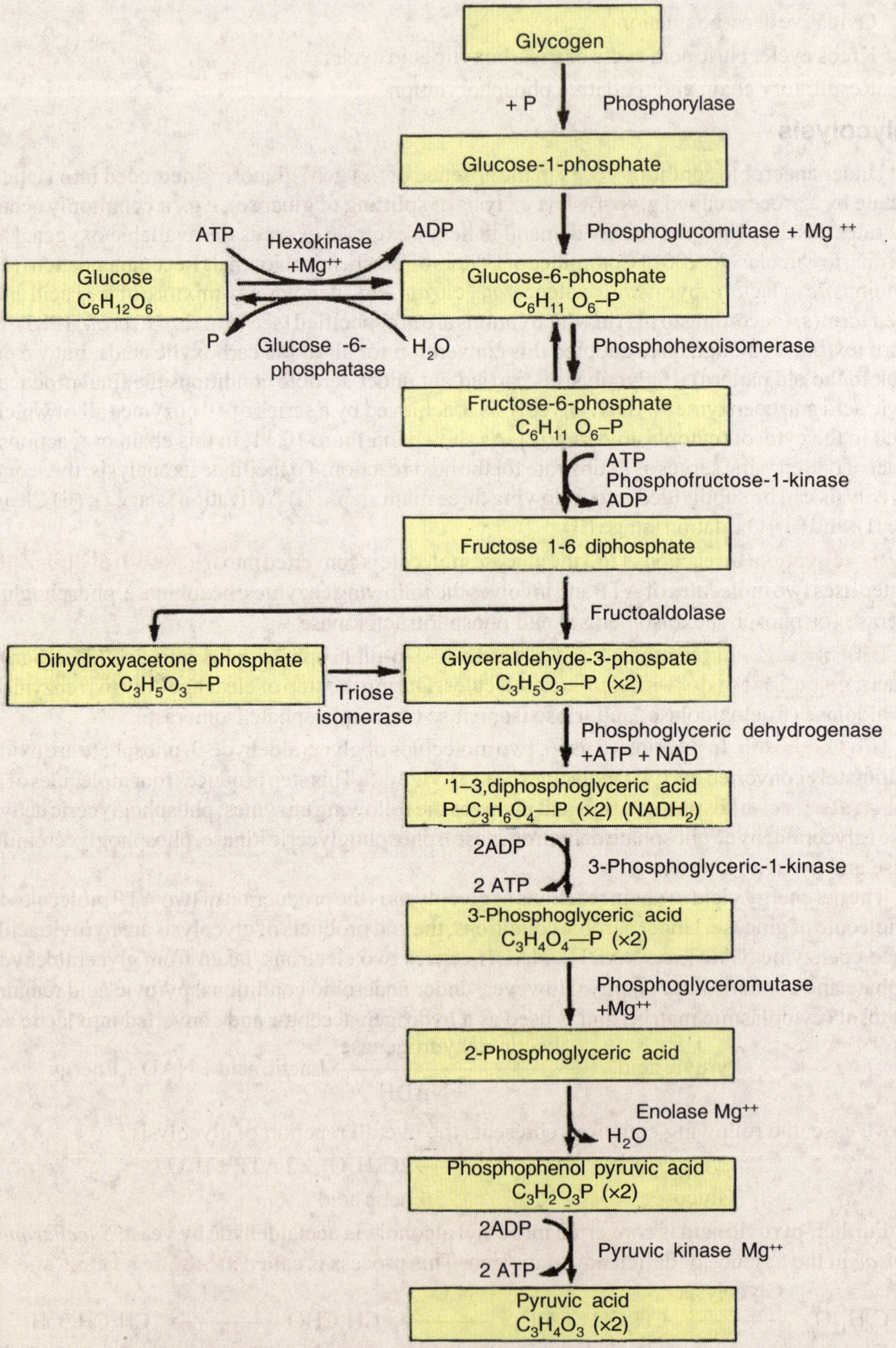

Fig. 10.11. Glycolysis or Embden-Meyerhof pathways showing the step-wise degradation of glucose to pyruvic acid.

anaerobic respiration of higher plants and fermentation of bacteria and yeast have the following differences :

(i) Fermentation is an extracellular process, *i.e.*, the respiratory substrate (*i.e.*, sugar, etc.) is present outside the cell and that too in the liquid medium, while the process of anaerobic respiration is intracellular, *i.e.*, the respiratory substrate is present inside the cell.

(ii) The enzyme (zymase) needed for fermentation is supplied from the micro-organism like yeast, while the enzymes (zymase-complex) required for the process of anaerobic respiration are present in the same cell of the higher plant in which the process is occurring.

However, both processes produce similar end-products as follows :

$$\underset{\text{Glucose}}{C_6H_{12}O_6} \xrightarrow[\text{or Fermentaion}]{\text{Anaerobic Respiration}} \underset{\text{Ethyl alcohol}}{2C_2H_5OH} + \underset{\text{Carbon dioxide}}{2CO_2} + \text{Energy}$$

2. Oxidative Decarboxylation

In aerobic organisms, since pyruvic acid still contains a large amount of energy, it must undergo further degradation, but this time inside the mitochondria. This is done in three consecutive steps : **oxidative decarboxylation** (removal of carboxyl or—COOH group), **Krebs cycle** and **oxidative phosphorylation**. Pyruvic acid directly enters the mitochondrial matrix and is converted into **acetyl-CoA** by the help of a huge enzyme, called **pyruvic acid dehydrogenase** (Fig. 10.12). The two NADH molecules (which are generated during glycolysis) cannot penetrate directly into the mitochondria, so their electrons are transferred to **dihydroxyacetone phosphate**, which shuttles them into the mitochondria. This process utilizes one ATP molecule for each NADH ; in all two ATP molecules are consumed for two NADH molecules. When both of these NADH pass through ETS, they tend to generate 6 ATP molecules.

Pyruvic acid dehydrogenase and its action. Sometimes two enzymes that catalyze sequential reactions form an **enzyme complex** and the product of the first enzyme does not have to diffuse through the cytoplasm to encounter the second enzyme. The second reaction begins as soon as the first is over. Some large enzyme (multienzyme) aggregates carry out whole series of reactions without losing contact with the substrate. For example, the conversion of pyruvic acid to acetyl CoA proceeds in *three* chemical steps, all of which take place on the same large multienzyme complex (*i.e.*, pyruvate dehydrogenase). Pyruvic acid dehydrogenase occurs in the mitochondrial matrix and is larger than a ribosome in size. It contains multiple copies of **three** enzymes namely **pyruvivc acid dehydrogenase**, **dihydrolipoyl transacetylase** and **dihydrolipoyl dehydrogenase**. It also contains **five** coenzymes (*e.g.*, NAD, coenzyme A, etc.) and **two regulatory proteins** (*e.g.*, protein kinase and protein phosphatase ; both regulating the activity of pyruvic acid dehydrogenase, turning it off whenever ATP levels are high).

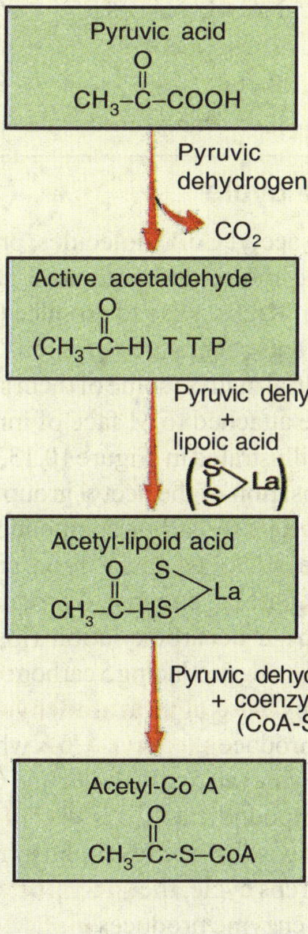

Fig. 10.12. The oxidative decarboxylation of pyruvic acid to acetyl coenzyme A.

Thus, during oxidative decarboxylation of one molecule of pyruvic acid, one mole of CO_2 is produced and one NAD is reduced to NADH. The end product of this reaction is a 2- carbon compound, the **acetyl group** which is attached to coenzyme A to produce the carrier molecule, called **acetyl CoA**.

Role of coenzymes in mitochondria. Some coenzymes have a central role in mitochondrial function. **Coenzyme A** (CoA) is part of a group (Table 10-3) that is derived from a nucleoside (adenine-D-ribose) and contains pantothenic acid (a vitamin of B complex) linked to the ribose by pyrophosphoric acid. CoA can be easily transformed into an ester at the thiol end(—SH) by acetyl group making acetyl-CoA. Acetyl-CoA is a carrier molecule in which acetyl group is linked by reactive bonds so that they can be transferred efficiently to other molecules. The same carrier molecule will often participate in many different biosynthetic reactions in which its group (*i.e.*, acetyl group) is needed, *e.g.*, growing fatty acid.

Other mitochondrial coenzymes are **nicotinamide adenine dinucleotide** (**NAD^+**) which contains the vitamin nicotinic acid of B complex and **flavin mononucleotide** (**FMN**) and **flavin adenine dinucleotide** (**FAD**), both of which contain riboflavin or vitamin B_2. NAD^+, FMN and FAD are important coenzymes not only in mitochondria but also in chloroplasts.

Table 10-3. **Some important coenzymes which act as carrier molecules in transfer of a group (Source: Alberts *et al.*, 1989).**

Coenzyme	Group transferred
1. ATP	phosphate
2. NADH, NADPH	hydrogen and electron (hydrogen ions)
3. Coenzyme A	acetyl
4. Biotin	carboxyl
5. S - Adenosyl-methionine	methyl

3. Krebs cycle

Two acetyl CoA molecules, produced above by oxidation of one molecule of glucose pass through a series of reactions of Krebs cycle to produce CO_2, H_2O and electrons. Enzymes and coenzymes of Krebs cycle are located in the mitochondrial matrix (some of them such as succinic dehydrogenase, are attached to M face of inner mitochondrial membrane). As illustrated in Figure 10.13, the Krebs cycle involves the condensation of the acetyl group with oxaloacetic acid to make **citric acid** (6-carbon compound). This step is directed by the enzyme **citrate synthase**. From citric acid, H_2O is released twice by **aconitase** enzyme to produce **isocitric acid**. This is followed by a decarboxylation (loss of CO_2) by **isocitric dehydrogenase**, producing 5 carbon **α- ketoglutaric acid**. CO_2 is released by **α- ketoglutarate dehydrogenase** in the presence of CoA to produce **succinyl - CoA** which changes by **succinyl kinase** enzyme (also called succinyl CoA synthetase) into a 4-carbon compound, the **succinic acid** (at this stage one GTP is generated by substrate level phosphorylation). The next enzyme of Krebs cycle, the **succinic dehydrogenase** converts succinic acid into **fumaric acid** and then **fumarase** enzyme produces **malic acid**. The mediation of **malate dehydrogenase** enzyme produces **oxaloacetic acid**, and thereby closes the Krebs cycle.

In this computer graphics image of ATP, adenosine is blue, pentose is white and the phosphate groups are red.

At each turn of the Krebs cycle, four pairs of hydrogen atoms are removed from the substrate intermediates by enzymatic dehydrogenation and two CO_2 molecules are released. These hydrogen atoms (or equivalent pairs of electrons) enter the respiratory chain, being accepted by either NAD^+ or FAD. Three pairs of hydrogen molecules are accepted by NAD^+, reducing it into NADH, and one pair by FAD, reducing it into $FADH_2$ (this pair of electrons comes directly from the succinic dehydrogenase reaction). Since it takes two turns of the cycle to metabolize the two acetyl groups that are produced by glycolysis from one molecule of glucose, a total of six molecules of NADH and two of $FADH_2$ are formed. During Krebs cycle are also produced two ATP molecules (*i.e.,* via GTP molecules).

Let us consider the specific function of the Krebs cycle. When a log is to be burned in a fire place, the log must first be chopped up into smaller chunks of fuel. Glucose, the fuel for metabolism, must also be broken into smaller pieces. In glycolysis, glucose is first split into two pyruvic acid molecules. Then, each pyruvic acid molecule is broken into three carbon dioxide fragments: One CO_2 molecule is given off during oxidative decarboxylation (*i.e.*, conversion of pyruvic acid into acetyl-CoA) and two CO_2 molecules are given off in the Krebs cycle. Thus, two turns of Krebs cycle will completely break up one glucose molecule. In fact, chopping off a log into kindling wood does not oxidize the wood. Nor does chopping glucose into smaller carbon fragment oxidize the glucose. The Krebs cycle itself releases no energy, but as the glucose breaks up, it also frees the hydrogen atoms attached to the carbons. Each hydrogen atom contains one proton (H^+) and one electron (e^-). The electrons which are released during complete oxidation of glucose (*i.e.*, glycolysis, oxidative decarboxylation and Krebs cycle) carry most of the energy of the glucose. In the final stage of cell respiration, the electron transport system, these electrons will at last release their energy to the cell.

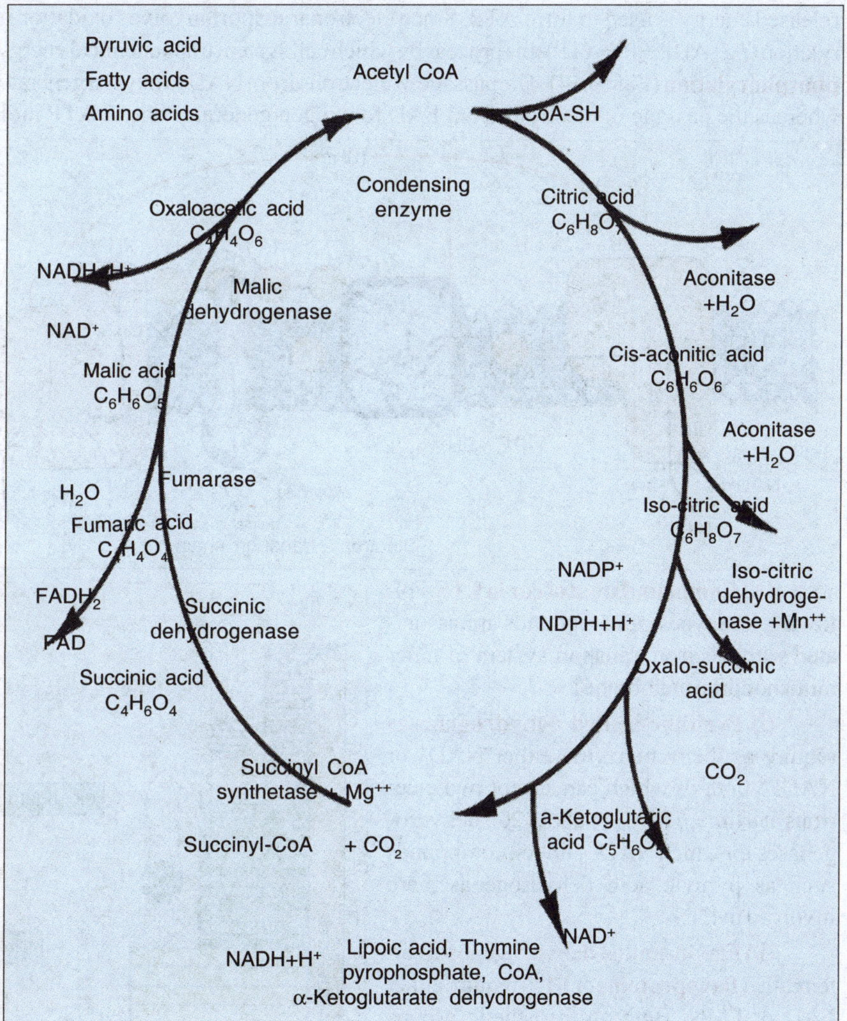

Fig. 10.13. Krebs cycle or tricarboxylic acid cycle.

4. Respiratory Chain and Oxidative Phosphorylation

Two molecules of $FADH_2$ and six molecules of NADH produced in Krebs cycle (from two molecules of acetyl-CoA) are oxidized by molecular O_2 in a **respiratory chain** or **electron transport system** or **ETS** involving a series of enzymes and coenzymes.

In the electron transport system, the successive electron acceptors are at lower and lower energy levels. With each transfer to a lower energy level, the electrons release some of their potential energy. That is why this series is called an **electron cascade,** like a cascade of falling water. At each stage, the released energy is used to form ATP. Since electron transport involves oxidation as well as phosphorylation (*i.e.*, ADP + P = ATP) this process by which cell system traps chemical energy is called **oxidative phosphorylation** (Fig 10.14). The passage of electrons from NAD to oxygen generates 3 ATP molecules, whereas the passage of electrons from FAD to oxygen generates only 2 ATP molecules.

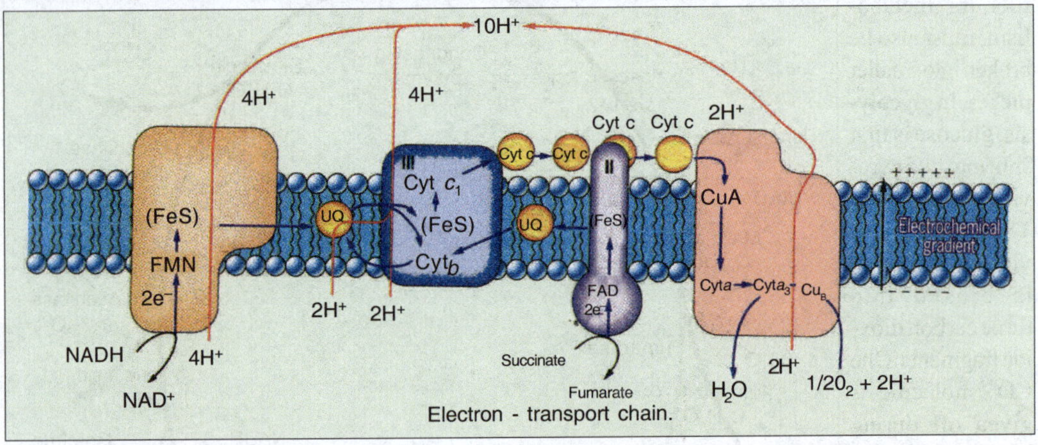

Electron - transport chain.

(A) Compounds that occur in ETS. Following five types of compounds are associated with electron transport system of inner mitochondrial membrane :

(i) Pyridine-linked dehydrogenases require as their coenzyme either NAD^+ or $NADP^+$ both of which can accept two electrons at a time. There are about 200 dehydrogenases for which NAD^+- linked compounds such as pyruvic acid dehydrogenase, are involved in ETS.

(ii) Flavin-linked dehydrogenases (often called **flavoproteins** or **FPs**) require either FAD or FMN. Both are prosthetic groups whose isoalloxazine ring can accept two hydrogen atoms. Flavin- linked enzymes are commonly involved in a number of enzyme systems such as fatty acid oxidation, amino acid oxidation and Krebs cycle activity (*e.g.*, succinic dehydrogenase or SDH).

(iii) Ubiquinones were so named because of their occurrence in so many different organisms and their chemical resemblance to

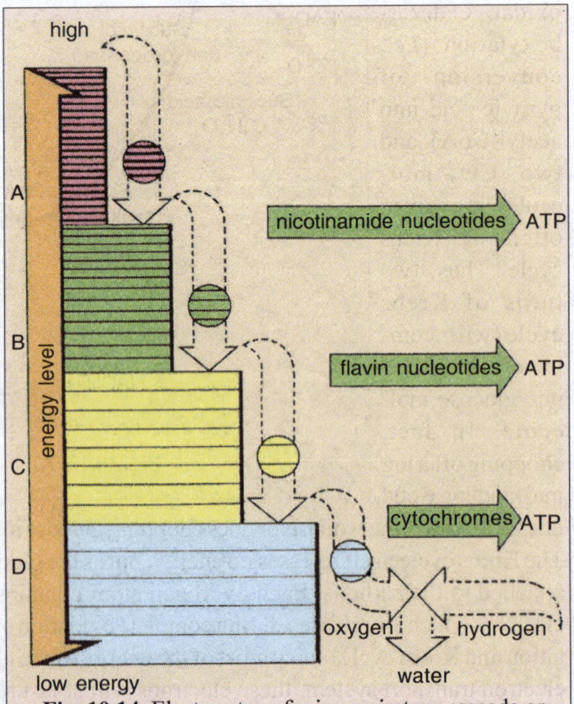

Fig. 10.14. Electron transfer in respiratory cascade or electron transport system.

quinone. They are found in several different forms including the **plastoquinones** of chloroplasts. The form of ubiquinones present in mitochondria is often called **coenzyme Q_{10}** (CoQ_{10} or Q). CoQ_{10} is a lipid soluble and accepts two hydrogen atoms (or two protons and two electrons) at a time.

(iv) Cytochromes are proteins containing iron-porphyrin (haem) groups. There are a large number of cytochromes in cells; most are found in mitochondria, although some also function in the ER and in chloroplasts. Mitochondria have five types of cytochromes which are arranged in the following order in an inner membrane: cyt. b, cyt. c_1., cyt. c, cyt. a and cyt. a_3. All of them transfer electrons by reversible valence changes of the iron atom (trivalent ferric or $Fe^{3+} \rightleftharpoons$ bivalent ferrous or Fe^{2+}).

(v) Iron-sulphur proteins (Fe_2S_2 and Fe_4S_4) are electron carriers of mitochondria containing iron and sulphur in equal amounts. The iron is reversibly oxidized during the electron transfer. Iron-sulphur proteins transfer one electron at a time.

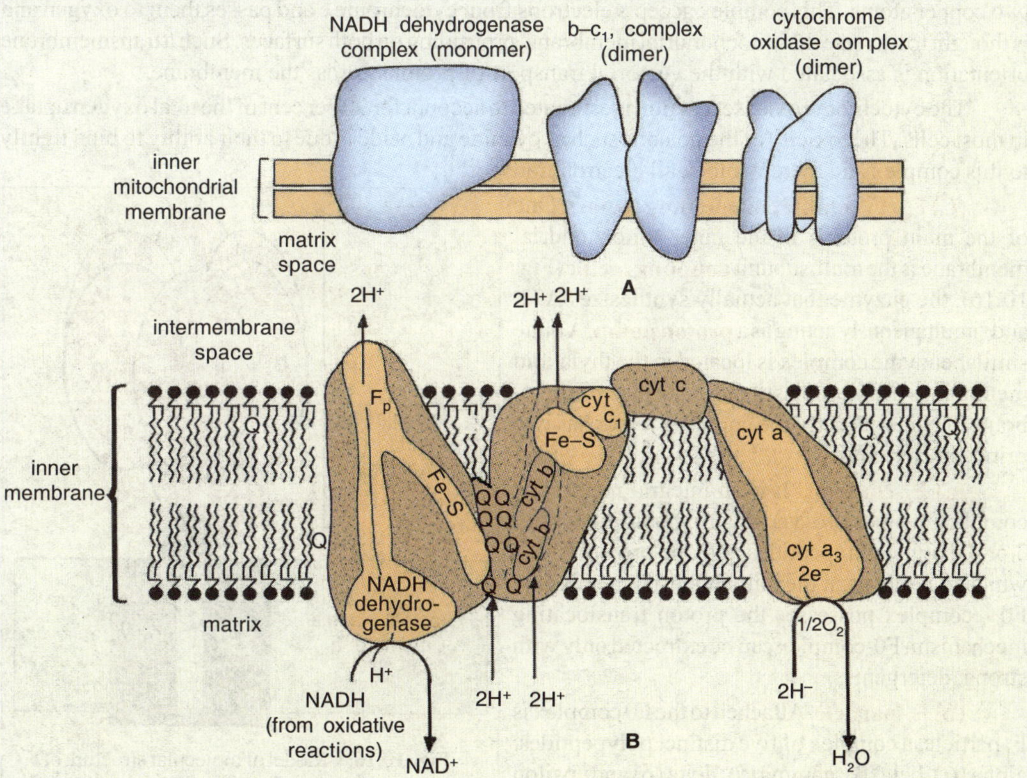

Fig. 10.15. Respiratory enzyme complexes. A—The relative sizes and shapes of the three respiratory enzyme complexes; B—Mode of transfer of H^+ through the three respiratory enzyme complexes of inner mitochondrial membrane (after Sheeler and Bianchi, 1987; Alberts *et al.*, 1989).

All these components of ETS are arranged in the inner mitochondrial membrane in the following sequence: NAD-linked succinic dehydrogenase (SDH), flavoprotein (FAD), non-haem iron protein or iron-sulphur protein, flavoprotein (FAD), cytochrome b, ubiquinone or coenzyme Q_{10}, cytochrome c_1, cytochrome c, cytochrome a, cytochrome a_3 and three coupling sites, where phosphorylation coupled with oxidation leads to production of ATP.

(B) Three complexes. Evidently above described components of ETS occur in the mitochondria in the form of following three complexes (**Green** *et al.*, 1967; **Capaldi** *et al.*, 1982; **Weiss** *et al.*, 1987):

(i) The NADH-dehydrogenase complex. It is the largest of the respiratory enzyme complexes, with a mass about 800,000 daltons and more than 22 polypeptide chains. It accepts electrons from NADH and passes them through a flavin and at least five iron-sulphur centres to ubiquinone (Q) that transfer its electrons to the next complex, the b-c_1, complex. This complex spans the inner mitochondrial membrane and is able to translocate protons across it from M side to C side (Fig.10.15).

(ii) The b-c1 complex. It contains at least 8 different polypeptide chains and is thought to function as a dimer of about 500,000 daltons. Each monomer contains three haemes bound to cytochromes and iron-sulphur protein. This complex accepts electrons from ubiquinone (Q) and passes them to cytochrome c, a small peripheral membrane protein that carries its electrons to the cytochrome oxidase complex. In the topology of this complex the Q-site may be in the middle of the membrane in the hydrophobic area and the cytochrome c-site on the C side.

(iii) The cytochrome oxidase complex. It comprises at least eight different polypeptide chains and is isolated as a dimer of about 300,000 daltons; each monomer contains two cytochromes (a, a_3) and two copper atoms. This complex accepts electrons from cytochrome c and passes them to oxygen and is thought to traverse the mitochondrial membrane, protruding on both surfaces. Such a transmembrane orientation is associated with the vectorial transport of protons across the membrane.

The cytochrome oxidase reaction is estimated to account for 90 per cent of the total oxygen uptake in most cells. The toxicity of the poisons such as cyanide and azide is due to their ability to bind tightly to this complex and thereby block all electron transport.

(C) F_0 - F_1 complex or coupling factors. One of the main proteins in the inner mitochondrial membrane is the multisubunit **coupling factor** (Fig. 10.16), the enzyme that actually synthesizes ATP and simultaneously acting as a **proton pump**. A quite similar enzyme complex is located in the thylakoid membranes of chloroplasts and in the plasma membrane of bacterial cell. The coupling factor has two principal components :

(a) F_0-complex. It is an integral membrane complex, composed of very hydrophobic proteins—3 or 4 distinct polypeptides and one proteolipid—which together span the mitochondrial membrane. F0 - complex possesses the proton translocating mechanism. F0-complex can be extracted only with strong detergents.

(b) F_1-particle. Attached to the F0 complex is F_1 particle, a complex of five distinct polypeptides: alpha (α), beta (β), gamma (γ), delta (δ) and epsilon (ε), with the probable composition of $\alpha_3\beta_3\gamma\delta\varepsilon$. F_1 forms the knob or 'tadpole' that protrudes on the matrix side of the inner mitochondrial membrane. F_1 particle can be detached from the membrane by mechanical agitation and is water soluble. When physically separated from the membrane, F_1 particle is capable only of catalysing the hydrolysis of ATP into ADP and phosphate. Hence, it is often called the **F_1-ATPase**. However, its natural function is the synthesis of ATP.

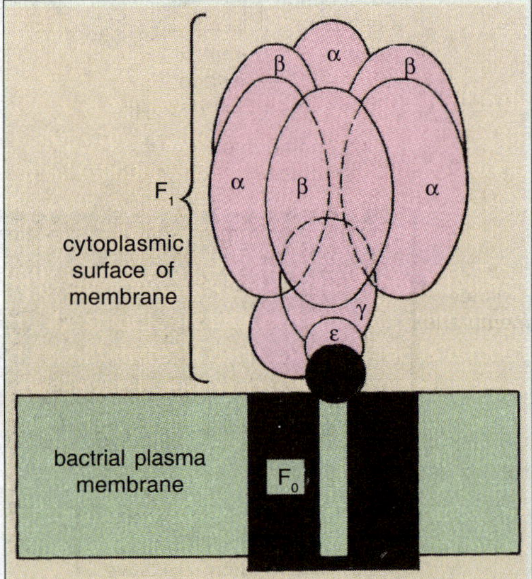

Fig. 10.16. Model of molecular structure of F_0 and F_1 partilces in the bacterial plasma membrane. The core of F_1 particle is an $\alpha_3\beta_3$ complex. The δ polypeptide of F_1 and probably the γ are involved in binding F_1 to the membrane-embedded F_0-particle. The role of ε subunit of F_1 complex is not known (after Darnell *et al.*, 1986).

(D) Redox reactions and redox couples. The movements of electrons between cellular reductants and oxidants represent a form of energy transfer in cells. A reductant (or **reducing agent**) is a substance that loses or donates electrons to another substance; the latter substance is the oxidant (or **oxidizing**

agent). Conversely, an oxidant is a substance that accepts electrons from another substance, the latter being the reductant. Reactions that involve the movement of electrons between reductants and oxidants are called **redox reactions**.

Different chemical substances have different potentials for donating or accepting electrons. The tendency of hydrogen to dissociate:

$$H_2 \rightleftharpoons 2H^+ + 2e^-$$

thereby releasing electrons, is used as a standard against which the tendencies of other substances to release or accept electrons is measured. The electron donor (*e.g.*, H_2 in the above reaction) and the electron acceptor (*e.g.*, $2H^+$ in the above reaction) are called a **redox couple** or **half cell**. The tendency of any chemical substance to lose or gain electrons is called the **redox potential** and is measured in volts(V). Measurements are made using an electrode that has been standardized against the $H_2 - 2H^+$ couple whose redox potential is set at 0.0 under standard conditions (pH 0.0, 1M(H^+), 25° C and 1 atmosphere pressure). This potential is noted by the symbol *Eo*. For biochemical reactions which normally occur at pH 7.0, the redox potential of the H_2–$2H^+$ couple is –0.421 V; standard redox potentials at pH7.0 are noted by the symbol *E′o*.

Any substance with a more positive *E′o* value than another has the potential for oxidizing that substance (*i.e.*, removing electrons from the substance with the more negative *E′o* value). The greater the difference in redox potentials, the greater the energy changes involved. The change in standard free energy changes, $\Delta G^{o\prime}$ is related to *E′o* as follows:

$$\Delta G^{o\prime} = nF\Delta E'o$$

where *n* is the number of electrons exchanged per molecule, *F* is the **Faraday** (96,406 J/V), and *E′o* is the difference in redox potential between the more positive and more negative members of the redox couple. For example, the oxidized form of cytochrome *c* (*E′o* value = $^+$0.254 V) can oxidize the reduced form of cytochrome *b* (*E′o* value = + 0.030V) by removal of two electrons. The difference between the redox potentials of the two is:

$$+0.254 - (+0.030) = +0.224\text{ V, therefore,}$$

$$\Delta G^{o\prime} = -2(96{,}406\text{ J/V})(0.224\text{ V})$$

$$= -43.19\text{ kJ (per mole of each cytochrome).}$$

(E) ATP synthesis. The potentials drop in three large steps, one across each major respiratory enzyme complex. The change in redox potential between any two electron carriers is directly proportional to the free energy released by an electron transfer between them. Each complex acts as an **energy-conversion-device** to harness this free-energy change, pumping H^+ across the inner membrane to create an electrochemical proton gradient as electrons pass through. The energy conversion mechanism underlying oxidative phosphorylation requires that each protein complex be inserted across the inner mitochondrial membrane in a fixed orientation, so that all protons are pumped in the same direction out of the matrix space. Such a **vectorial organization** of membrane proteins has been experimentally proved.

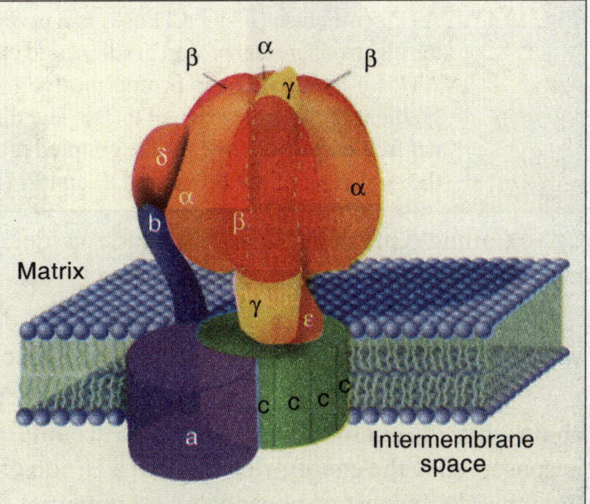

The structure of the ATP synthase.

Just as a flow of water from a higher to a lower level can be utilized to turn a water wheel or a hydroelectric turbine, the energy released by the flow of the protons down the gradient is utilized in the synthesis of ATP (see **Reid** and **Leech**, 1980). Similarly, resultant electrochemical proton gradient is harnessed to make ATP by F_0 - F_1 complex (acting as ATP synthetase and proton pump), through which the protons flow back into the matrix (Fig. 10.17). The ATP synthetase is a reversible coupling device that normally converts a back-flow of protons into ATP phosphate-bond energy, but it can also hydrolyze ATP (into ADP and phosphate) to pump protons in the opposite direction, if the electrochemical proton gradient is reduced.

The chemiosmotic theory. Several hypotheses have been proposed to explain the mechanism of electrochemical link between respiration and phosphorylation inside inner mitochondrial membrane. Most accepted one is that of **Mitchell's chemiosmotic coupling theory**, proposed in 1967. According to this theory, the inner membrane of the mitochondria acts as a transducer converting the energy which is provided by an electrochemical gradient, into the chemical energy of ATP. In this model (Fig. 10.18A), the membrane is **impermeable** to both H^+ and OH^- ions. For this reason, if pH differences are established across the membrane, they act as energy-rich gradients. The electron transport system is organized in "**redox loops**" within the membrane, and the electrons are passed from one carrier to another on the respiratory chain. At the same time, protons (H^+) are ejected toward the cytoplasmic side (C side), while OH^- remain on matrix side. This vectorial movement of protons creates a difference in pH (*i.e.*, lower pH on the C side and higher on the M side), which results in an electrical potential (Fig. 10.18B).

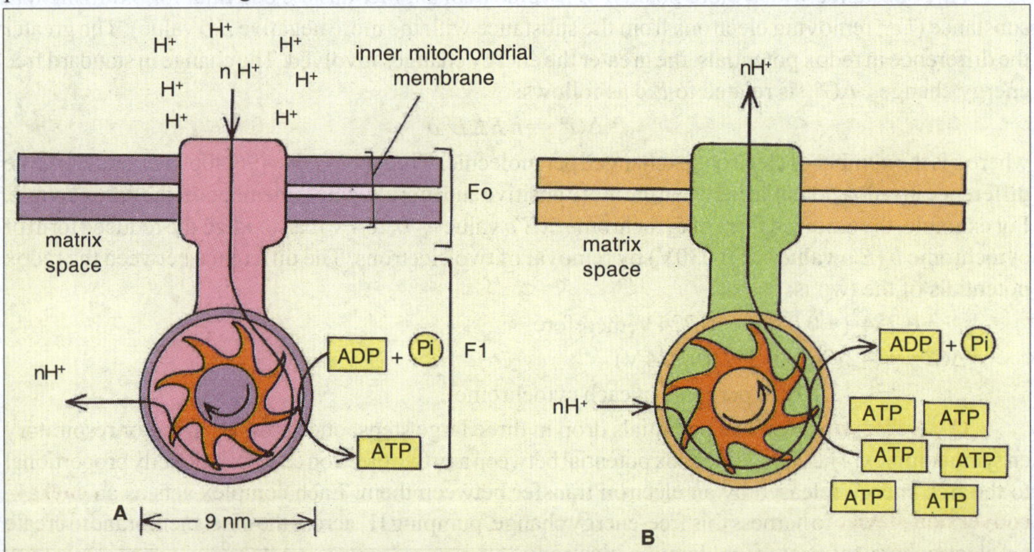

Fig. 10.17. ATP synthetase (F_0-F_1 ATPase) is a reversible coupling device that interconverts the energies of the electrochemical proton gradient and chemical bonds. The ATP synthetase can either synthesize ATP by harnessing the proton motive force (A) or pump protons against their electrochemical gradient by hydrolyzing ATP (B). The direction of operation at any given instant depends on the net free-energy change for the coupled processes of proton translocation across the membrane and the synthesis of ATP from ADP and Pi (after Albert *et al.*, 1989).

Calling Δ pH the **pH gradient** and $\Delta\psi$ (delta psi) the resulting **electrical gradient** in volts, the energy produced Δ P is the **proton motive force**:

$$\Delta P = \Delta\psi + 2.3\ RT/F\ \Delta pH$$

where R is the universal gas constant, T the absolute temperature, and F the Faraday constant.

The chemiosmotic theory postulates that the primary transformations occurring in the respiratory chain guide the **osmotic work** needed to accumulate ions. The energy generated by electron transport is conserved in the energy-rich form of a H^+ ion gradient. This gradient provides the driving force for the inward transport of phosphate and potential for generating ATP. ADP is brought into matrix in exchange for ATP, *i.e.*, cotransport. As indicated in Figure 10.18C, this gradient through the action of the proton pump of the F_0-F_1, drives the oxidative phosphorylation of ADP to form ATP by which mechanism free energy is conserved :

$$ADP + Pi \rightleftharpoons ATP + H_2O$$

In this reaction H_2O is also formed because of the dehydration, which leads to the removal of H^+ and OH^- ions.

In recent years, much experimental evidence has supported the validity of Mithchell's chemiosmotic theory. However, there is some dispute over the number of H^+ ions translocated out during two-electron transportation from NADH to oxygen. According to **Lehninger** and **Brand** (1979) these may be 9 to 12 in number instead of 6 as claimed by **Mitchell** (1967). However, **Mitchell** and **Moyle** (1979) have reasserted their original claim of 6 H^+ translocation.

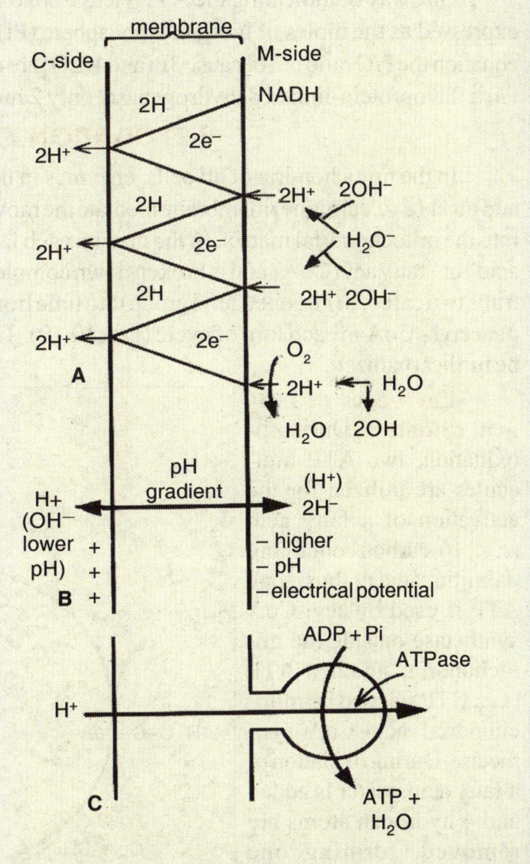

Fig. 10.18. Diagram explaining chemiosmotic coupling according to Mitchell. A—During electron transport H^+ ions are driven to the C side of inner mitochondrial membrane; B—This process produces a pH gradient and an electrical potential across the membrane; C—this gradient drives the proton pump of the ATPase and ATP is synthesized from ADP and Pi (after De Robertis and De Robrtis Jr., 1987).

(F) Energetics of glucose oxidation. Of the 686,000 calories contained in a mole of glucose, less than 10 per cent (*i.e.*, 58,000 calories) can be released by anaerobic glycolysis. The cell is able to store only 45 per cent of the chemical energy liberated by the combustion of glucose in the form of ATP (*i.e.*, only 36 ATP molecules). The rest of the energy is dissipated as heat or used for other cell functions.

At this stage, let us do stocktaking of ATP generation during aerobic respiration of one mole of glucose. We have seen that glycolysis and Krebs cycle can each generate 2 molecules of ATP per molecule of glucose by substrate level phosphorylation (total 4 ATP molecules). In addition 10NADH (*i.e.*, 2 NADH in glycolysis, 2NADH in oxidative phosphorylation and 6NADH in Krebs cycle) and $2FADH_2$ are produced which are equivalent to 34 ATP molecules. Thus, a total of 38 ATP molecules are produced per glucose molecule oxidized. However, in most eukaryotic cells 2 molecules of ATP are used in the transportation of 2 mole of NADH produced during glycolysis into the mitochondrion (*via* the malate shuttle) for their further oxidation (*via* ETS). Hence, the net gain of ATP is 38–2=36 molecules; since one high energy phosphate bond is equal to 36.8 kJ; so 36ATP = 1325 kJ or 36,000 calories:

$$C_6H_{12}O_6 + 36Pi + 36ADP + 6O_2 + 6H_2O \rightarrow 6CO_2 + 36\,ATP + 12H_2O$$

Glucose Inogranic phosphate (360,000 calories or 1325 kJ)

The P/O ratio. In the electron tranport chain, since the formation of ATP occurs in three steps, the equation can be written as follows:

$$NADH + H^+ + 3ADP + 3\,Pi + 1/2\,O_2 \rightarrow NAD^+ + 4H_2O + 3ATP$$

One way of indicating the ATP yield from oxidative phosphorylation is the **P/O ratio**, which is expressed as the moles of inorganic phosphate (Pi) used per oxygen atom consumed. Thus, in above equation the P/O ratio is 3 because 3Pi and 1/2 O_2 are used. On the other hand, when a substrate is oxidized *via* a flavoprotein-linked dehydrogenase, only 2 mol of ATP are formed , *i.e.*, P/O ratio is 2.

β-OXIDATION OF FATTY ACIDS

In the mitochondria of all cells, enzymes in the outer (*e.g.*, thiokinase or acyl - CoA synthetase) and inner (*e.g.*, carnitine) membrane mediate the movement of free fatty acids derived from fat molecules into the mitochondrial matrix. In the matrix, each fatty acid molecule exists in the form of "active fatty acid" or "fatty acyl CoA" and is broken down completely by a cycle of reactions, called **β-oxidation** that trims two carbons (*i.e.*, one acetyl group) at a time from its carboxyl end (β-end), generating one molecule of acetyl - CoA in each turn of cycle (Fig. 10.19). The acetyl- CoA produced is fed into Krebs cycle to be further oxidized.

Energetics of fatty acid oxidation. During β-oxidation, two ATP molecules are utilized for the activation of a fatty acid (*e.g.*, 16-carbon containing palmitic acid); thus, one ATP is used by acyl-CoA synthetase outside the mitochondria; another ATP (*i.e.*, GTP) is used by mitochondrial acyl-CoA synthetase. During oxidation of a fatty acid, water is added and 4 hydrogen atoms are removed, forming one $FADH_2$ molecule and one NADH molecule, from 2 carbon atoms nearest to CoA. When electrons of $FADH_2$ and NADH are passed through ETS, they release 5ATP molecules for each of the first 7 acetyl - CoA molecules formed by β-oxidation of palmitic acid, *i.e.*, 7 × 5 = 35 ATP molecules. β - oxidation of palmitic acid produces in total 8 mol of acetyl - CoA, each of which on oxidation by Krebs cycle produces 12ATP molecules; thus, making 8 × 12 = 96ATP molecules *via* this route. By deducting 2ATP used for initial activation of the fatty acid, a net gain of 129ATP (*i.e.*, 35 + 96 – 2 = 129) is achieved. In terms of energy 129 ATP molecules contain 4747 kJ (129 × 36.8 kJ). As the free energy of combustion of palmitic acid is 9791 kJ/mol, the process of β-oxidation captures as ATP molecules on the border of 48 per cent of the total energy of combustion of the fatty acid (see **Mayes** and **Granner**, 1985).

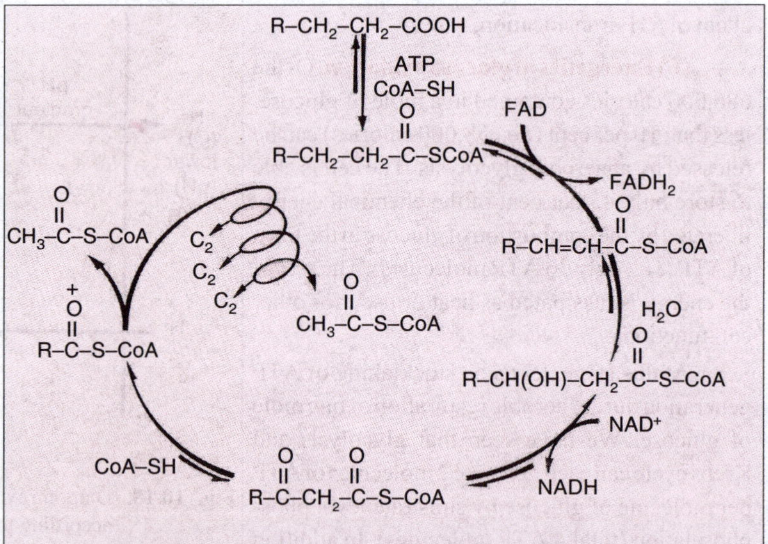

Fig. 10.19. The oxidation of fatty acids by the β-oxidation helical scheme.

OXIDATION OF PROTEINS

Before proteins can be introduced into the mainstream of metabolism (catabolism) they must be split into amino acids. The process is accomplished by protease enzymes similar to the process occurs in digestion. Each peptide bond is severed with the introduction of a water molecule, a hydrolytic reaction. Next, nitrogen is removed from amino acids by any one following two processes: oxidative deamination and transamination.

During **oxidative deamination,** the amino group of the amino acid is split off from the rest of the molecule, forming ammonia (NH_3). The remainder of the amino acid then enters the main metabolic stream as a keto acid. Water is required for this process and two hydrogens are removed by coenzyme NAD. The ammonia formed during deamination may be immediately excrerted or organized into another molecule before excretion; for example, in human being the ammonia is converted into urea molecules by the liver cells before being sent *via* the blood to the kidney. The energy derived from oxidative deamination depends upon the amino acid involved. For example, oxidative deamination of glutamic acid involves its conversion into α-ketoglutaric acid, which is oxidized by the Krebs cycle:

$$\underset{\text{(amino acid)}}{\text{Glutamic acid}} \xrightarrow[\swarrow H_2O \quad \nwarrow NH_3]{NAD \rightarrow NADH_2} \underset{\text{(keto acid)}}{\alpha\text{-Ketoglutaric acid}}$$

In this case, high-energy phosphates would be created by the transfer of hydrogen from $NADH_2$ (formed in deamination) through the cytochrome system.

Transamination reaction consists of an amino group being shifted from one molecule to another in exchange for an oxygen. For example, due to transamination amino acid glutamic acid being converted into α-ketoglutaric acid, as is shown by following reaction:

$$\underset{\text{(amino acid)}}{\text{Glutamic acid}} + \underset{\text{(keto acid)}}{\text{Oxaloacetic acid}} \longrightarrow \underset{\text{(keto acid)}}{\alpha\text{-Ketoglutaric acid}} + \underset{\text{(amino acid)}}{\text{Aspartic acid}}$$

In this case, the amino group is not lost completely but is transferred to one of the substrates of Krebs cycle. Oxaloacetic acid loses its oxygen and picks up the NH_2 group and becomes the amino acid, aspartic acid. Thus, one amino acid is converted into keto acid, while another keto acid is transformed into an amino acid. The usual purpose of such reactions is to maintain a particular balance among amino acids and substrates rather than providing grist for the metabolic mill.

Some amino acids such as alanine, cysteine, glycine, hydroxyproline, serine and threonine undergo enzymatic reaction to become pyruvic acid which enters mitochondria and is changed into acetyl-CoA and oxidized by Krebs cycle. Some other amino acids such as phenylalanine, tyrosine, tryptophan, lysine and leucine form acetyl-CoA directly without first forming the pyruvic acid.

OTHER FUNCTIONS OF MITOCHONDRIA

Besides the ATP production, mitochondria serve the following important functions in animals:

1. Heat production or thermiogenesis. As we have already discussed earlier that only 45 per cent of the energy released during the oxidation of glucose is captured in the form of ATP, the rest 55 per cent is either lost as heat or used to regulate body temperature of warm-blooded animals. In some mammals, especially young animals and hibernating species, there is a specialized tissue called **brown fat**. This tissue, typically located between the shoulder blades, is especially important in temperature regulation; it produces large quantities of body heat necessary for arousal from hibernation. The colour of brown fat comes from its high concentration of mitochondria, which are sparse in ordinary fat cells. The mitochondria appear to catalyze electron transport in the usual way but are much less efficient at producing ATP. Hence, a higher than usual fraction of the oxidatively released energy is converted directly to heat (called **non-shivering thermiogenesis**).

2. Biosynthetic or anabolic activities. Mitochondria also perform certain biosynthetic or anabolic functions. Mitochondria contain DNA and the machinery needed for protein synthesis. Therefore, they can make less than a dozen different proteins. The proteins so far identified are subunits of the ATPase, portions of the reductase responsible for transfer of electrons from CoQ to the iron of Cyt c, and three of the seven subunits in cytochrome oxidase. Altogether, no more than 5–10 per cent of mitochondrial components can be attributed to mitochondrial genes.

Some biosynthetic functions of mitochondria are of primary benefit to the rest of the cell. For example, the synthesis of **haeme** (needed for cytochromes, myoglobin and haemoglobin) begins with a mitochondrial reaction catalyzed by the enzyme, delta or δ-aminolevulinic acid synthetase. Likewise, some of the early steps in the conversion of cholesterol to steroid hormones in the adrenal cortex are also catalyzed by mitochondrial enzymes.

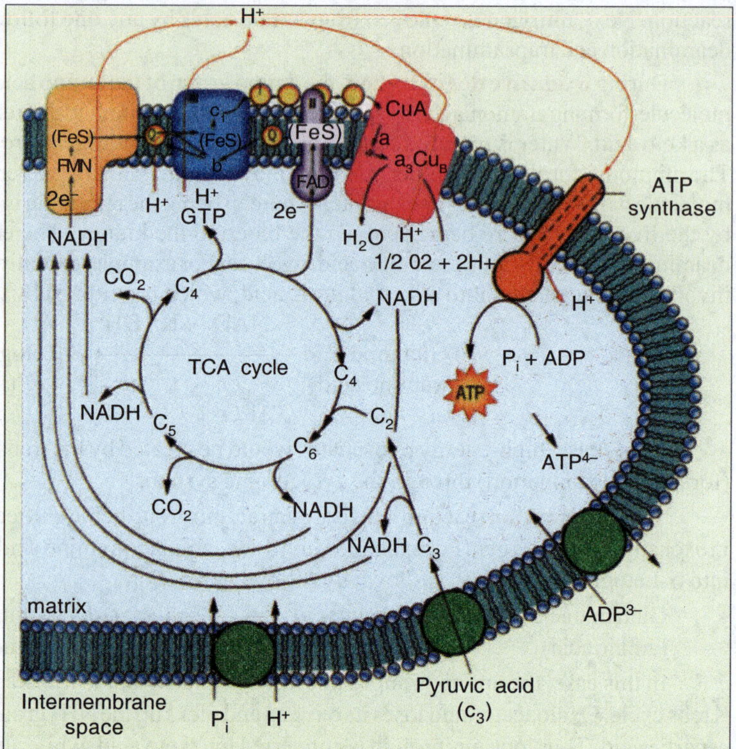

Summary of the major activities during aerobic respiration in a mitochondrion.

3. Accumulation of Ca^{2+} and phosphate. In the mitochondria of **osteoblasts** present in tissues undergoing calcification large amount of Ca^{2+} and phosphate (PO_4^-) tend to accumulate. In them microcrystalline, electrone-dense deposits may become visible. Sometimes, the mitochondria assume storage function, *e.g.*, the mitochondria of ovum store large amounts of yolk proteins and transform into yolk platelets.

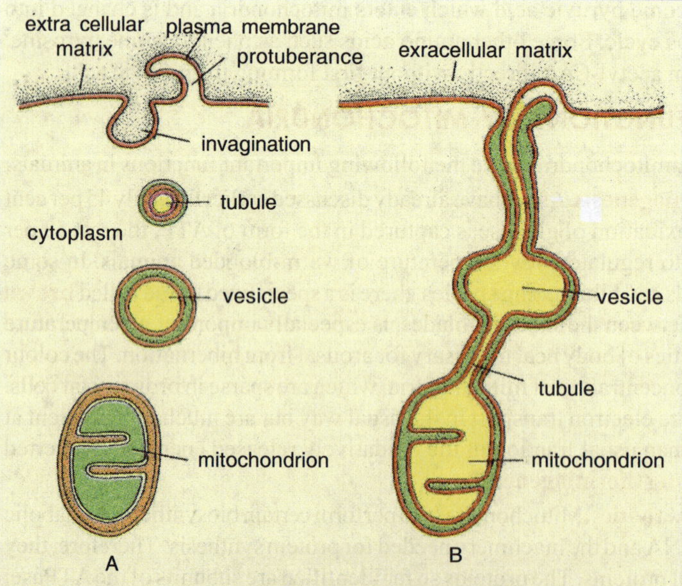

Fig. 10.20. Hypothetical diagrams showing the origin of mitochondria from plasma membrane.

BIOGENESIS OF MITOCHONDRIA

Regarding the origin of the mitochondria, several hypotheses have been postulated which are as follows:

1. "de novo" origin. According to this hypothesis, the mitochondria are originated "de novo" (L. anew) from the simple building blocks such as amino acids and lipids. But, there is no direct evidence in suppport of "de novo" hypothesis for the origin of the mitochondria therefore, it is discarded now.

2. Origin from the endoplasmic reticulum or plasma

membrane. According to **Morrison** (1966) the new mitochondria might have been originated from the endoplasmic reticulum or plasma membrane(Fig.10.20). This hypothesis also could not provide direct evidences, therefore, it is not well accepted at present time.

3. Origin by division of pre-existing mitochondria. The electron microscopic and radio-autographic observations of the culture cells have shown clearly that the new mitochondria are originated by the growth and division of pre-existing mitochondria. On average, each mitochondrion must double in mass and then divide in half once in each cell generation. Mitochondria are distributed between the daughter cells during mitosis and their number increase during interphase. Electron microscopic studies of *Neurospora crassa* (**Luck**, 1963) and HeLa cells (**Attardi** *et al.*, 1975) have suggested that organelle division begins by an inward furrowing of the inner membrane, as occurs in cell division in many bacteria (Fig. 10.21). After elongating, one or more centrally located cristae form a partition by growing across the matrix and fusing with the opposite inner membrane. This separates the matrix into two compartments. The outer membrane then invaginates at the partition plane, constricting until there is membrane fusion between the two inner membrane walls. Thus, two separable daughter mitochondria are formed.

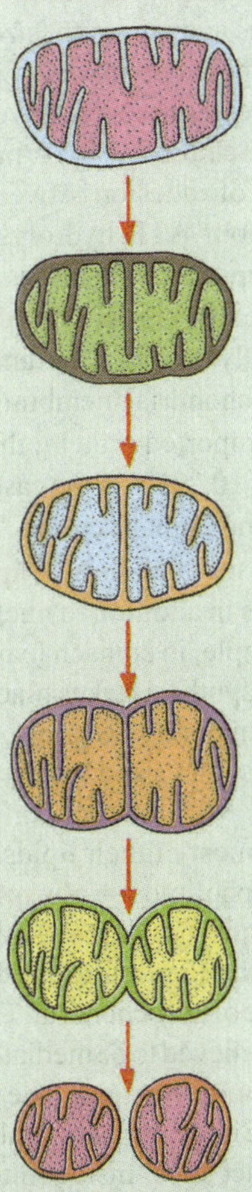

Fig. 10.21. Fission of a mitochondrion by partition formation (after Thorpe, 1984).

Mitochondria as semiautonomous organelles. Recently the study of mitochondrial and chloroplast biogenesis became of great interest because it was demonstrated that these organelles contain DNA as well as ribosomes and are able to synthesize proteins. The term **semiautonomous organelles** was applied to the two structures in the recognition of these findings. This term also indicated that the biogenesis was highly dependent on the nuclear genome and the biosynthetic activity of the ground cytoplasm. It is well established now that the mitochondrial mass grows by the integrated activity of both genetic systems, which cooperate in time and space to synthesize the main components. The mitochondrial DNA codes for the mitochondrial, ribosomal and transfer RNA and for a few proteins of the inner membrane. Most of the proteins of the mitochondrion, however, result from the activity of the nuclear genes and are synthesized on ribosomes of the cytosol (cytoplasmic matrix). The cooperation of two genomes has been greatly clarified by studies on the molecular assembly of cytochrome oxidase (**Saltzgaber** *et al.*, 1977). This cytochrome, as studied in *Saccharomyces cerevisiae* is made up of seven polypeptide subunits for a combined molecular weight of 139,000 daltons. Three of the polypeptides are coded by mt DNA and assembled on mitochondrial ribosomes. They are very hydrophobic and high in molecular weight (23,000–40,000 daltons). The remaining four subunits are coded by nuclear DNA and made on cytoplasmic ribosomes. These are hydrophilic polypeptides of lower molecular weight (4500–14,000 daltons).

Mitochondrial DNA. Mitochondrial DNA (mt DNA) molecule is relatively small, simple, double-stranded and except for the DNA of some algae and protozoans, it is circular. The size of mitochondrial genome is very much large in plants than in animals. Thus, mt DNA varies in length from about 5 μm in most animal species to 30 μm or so in higher plants. The mt DNA is localized in the matrix and is probably attached to the inner membrane at the point where DNA duplication starts. This duplication is under nuclear control and the enzymes used (*i.e.*, polymerases) are imported from the cytosol.

Mitochondrial ribosomes. Mitochondria contain ribosomes (called **mitoribosomes**) and polyribosomes. In yeast and *Neurospora*, ribosomes have been ascribed to a 70S class similar to that of bacteria ; in mammalian cells, however, mitoribosomes are smaller and have a total sedimentation coefficient of 55S, with subunits of 35S and 25S (**Attardi** *et al.*, 1971). In mitochondria, ribosomes appear to be tightly associated with the inner membrane.

Mitochondrial protein synthesis. As already described, mitochondira can synthesize about 12 different proteins, which are incorporated into the inner mitochondrial membrane. These proteins are very hydrophobic (*i.e.*, they are proteolipids). Thus, on the mitoribosomes are made the following proteins : three largest subunits of cytochrome oxidase (Fig. 10.6), one protein subunit of the cytochrome b-c_1 complex, four subunits of ATPase and a few hydrophobic proteins. One of the best known differences between the two mechanisms of protein synthesis (*i.e.,* in the cytosol and in the mitochondrial matrix) is in the effect of some inhibitors. The mitochondrial protein synthesis is inhibited by **chloramphenicol**, while synthesis in the cytosol (cytoplasmic matrix) is not affected by this drug. In contrast, **cycloheximide** has the reverse effect.

Import mechanism of mitochondrial proteins. Most mitochondrial proteins are coded by nuclear genes and are synthesized on free ribosomes in the cytosol (cytoplasmic matrix). The import of these polypeptides involves similar mechanism both in mitochondria, and chloroplasts. The transport processes involved have been most extensively studied in mitochondria, especially in yeasts (**Attardi** and **Schatz**, 1988). A protein is translocated into the mitochondrial matrix space by passing through sites of adhesion between the outer and inner membrane, called **contact sites**. Translocation is driven by both ATP hydrolysis and the electrochemical gradient across the inner membrane, and the transported protein is unfolded as it crosses the mitochondrial membranes. Only proteins that contain a specific **signal peptide** are translocated into mitochondria and chloroplasts. The signal peptide is usually located at the amino terminus and is cleaved off after import (Fig. 10.23A). Transport to the inner mitochondrial membrane can occur as a second step if a **hydrophobic signal peptide** is also present in the imported protein; this second signal peptide is unmarked when the first signal peptide is cleared (Fig. 10.23B). In the case of chloroplasts, import from the stroma into the thylakoid likewise requires a second signal peptide.

Mitochondrial lipid biosynthesis. The biogenesis of new mitochondria and chloroplasts requires lipids in addition to nucleic acids and proteins. Chloroplasts tend to make the lipids they require. For example, in spinach leaves, all cellular fatty acid synthesis takes place in the chloroplast. The major glycolipids of the chloroplast are also synthesized locally.

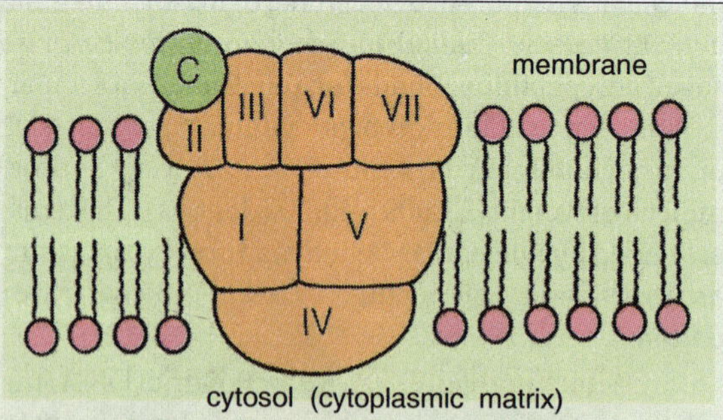

Fig. 10.22. Model of cytochrome oxidase of yeast showing topographical relationship of its seven subunits and its relation with cytochrome c (after Thorpe, 1984).

Mitochondria, on the other hand, import most of their lipids. In animal cells the phospholipids—**phosphatidyl-choline** and **phosphatidyl-serine**—are synthesized in the ER and then transferred to the outer membrane of mitochondria. The transfer reactions are believed to be mediated by **phospholipid exchange proteins**; the imported lipids then move into the inner membrane, presumably at contact sites. Inside mitochondria, some of the imported phospholipids are decarboxylated and converted into **cardiolipin** (diphosphatidyl glycerol). Cardiolipin is a " double" phospholipid that contains four fatty - acid tails; it is found mainly in the inner mitochondrial membrane, where it constitutes about 20 per cent of the total lipids.

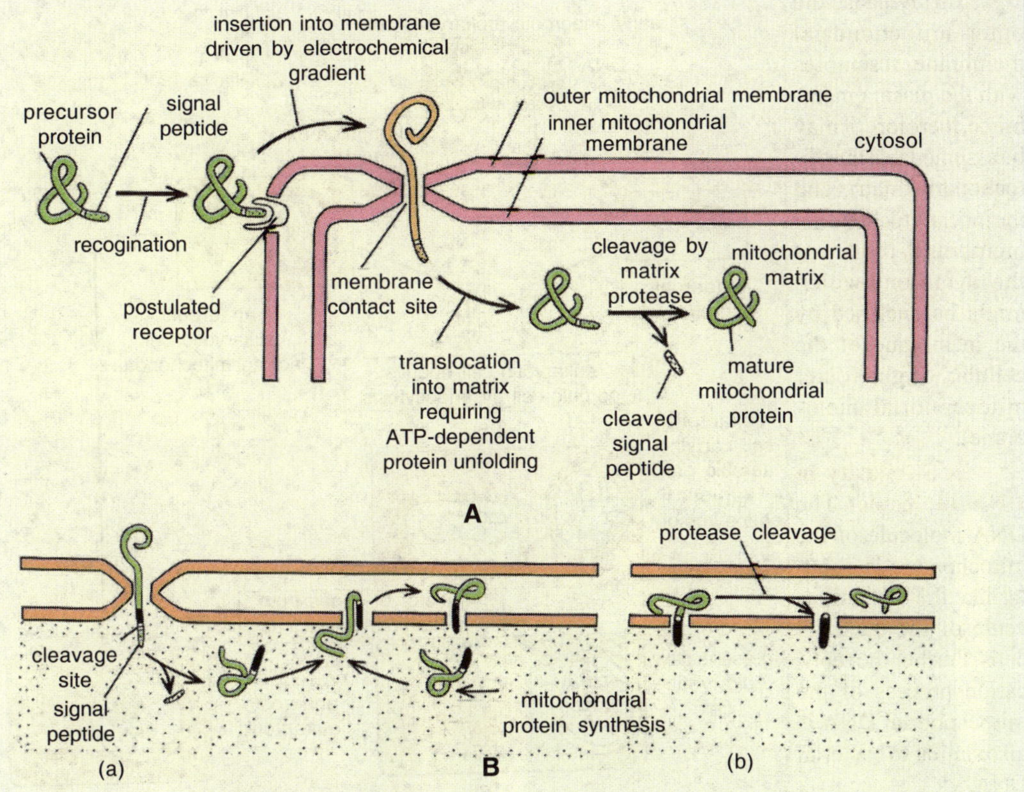

Fig. 10.23. A—Protein import by mitochondrial matrix through single signal; B—Import of proteins from the cytosol (cytoplasmic matrix) to the mitochondrial intermembrane space or inner membrane through two (or multiple) signals (after Albert *et al*; 1989).

Prokryotic Origin or Symbiont Hypothesis

Early cytologists such as **Altmann** and **Schimber** (1890) have suggested the possibility of origin of the mitochondria from the prokaryotic cells. According to their hypothesis, the mitochondria and chloroplasts may be considered as intra-cellular parasites of the cells which have entered in the cytoplasm of eukaryotic cells in early evolutionary days, and have maintained the symbiotic relations with the eukaryotic cells. The mitochondira are supposed to be derived from the bacterial cells (purple bacteria) while chloroplasts are supposed to be originated from the blue green algae (see **Margulis**, 1981). Due to these reasons **Altmann** suggested the name "**bioblasts**" to the mitochondria and he also hinted about their self-duplicating nature.

Recent cytological findings have also suggested many homologies between the mitochondria and the bacterial cells. The similarities between the two can be summarised as follows :

1. Similarity in inner mitochondrial membrane and bacterial plasma membrane. (i) In the mitochondria the enzymes of the respiratory chain are localized on the inner mitochondrial membrane like the bacteria in which they remain localized in the plasma membrane. The bacterial plasma membrane resembles with the inner mitochondrial membrane in certain respects.

(ii) The plasma membrane of certain bacterial cells gives out finger-like projections in the cytoplasm known as mesosomes. The mesosomes can be compared with mitochondrial crests. **Salton** (1962) has reported respiratory chain enzymes in the mesosomes.

(iii) Because the outer mitochondrial membrane resembles with the plasma membrane, therefore, it may be assumed that the mitochondrial matrix and the inner mitochondrial membrane represent the symbiont which might be enclosed by the membrane of the cellular origin (outer mitochondrial membrane).

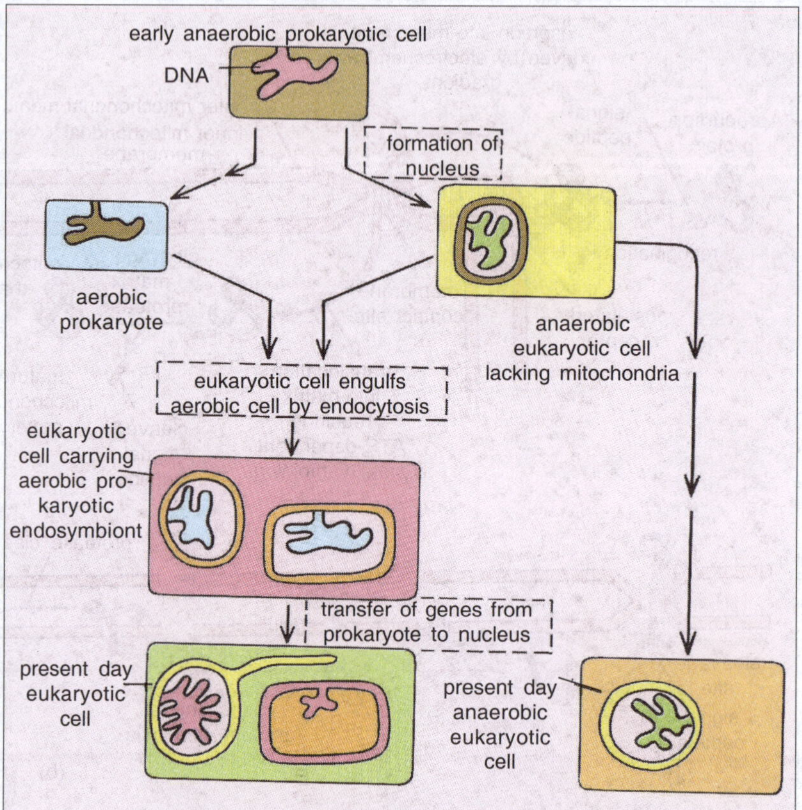

Fig. 10.24. Symbiotic origin of mitochondria and chloroplast (after Alberts *et al.*, 1989).

2. Similarity in DNA molecule. The DNA molecule of the mitochondria is circular like the DNA molecule of the bacterial cells. Further the replication process of the mitochondrial DNA is also similar to bacterial DNA.

3. Similarity in ribosomes. The mitochondrial ribosomes are small in size and resemble the ribosomes of the bacteria.

4. Similarity in the process of protein synthesis. The process of protein synthesis of both mitochondria and bacteria is fundamentally same because in both, the process of protein synthesis can be inhibited by same inhibitor known as chloramphenicol.

These degenerating muscle fibers are from a biospy of a patient and show accumulations of red staining "blotches" just beneath the cell's plasma membrane, which are due to abnormal proliferation of mitochondria.

Further, the mitochondria for the process of protein synthesis depend partially on the mitochondrial matrix and DNA and partially on the nucleus and cytoplasm of the eukaryotic cells. This shows the symbiotic nature of the mitochondria.

Due to the above-mentioned similarities between the bacteria and mitochondria, the symbiont hypothesis postulated that the host cell (eukaryotic cell) represented an anaerobic organism which derives the required energy from the oxidations of food by the process of glycolysis. While the mitochondria represent the symbionts which respire **aerobically** and contain the enzymes of **Krebs cycle** and **respiratory chain**. The symbionts seem to be capable to get the energy by **oxidative phosphorylation** from the partially oxidised food (pyruvic acid) of the host cell.

REVISION QUESTIONS

1. What are the characteristic structural features of mitochondira that aid in their identification ?
2. Why are mitochondria termed as the "power houses" of the eukaryotic cells ?
3. What is the major function of mitochondria ?
4. What differences exist in structure and function between the inner and outer membranes of the mitochondria ?
5. Would you expect plant cells to have the Embden-Meyerof glycolytic pathway ? Explain your answer.
6. What is the significance of acetyl-coenzyme A and where does it come from ?
7. What is an oxidative decarboxylation ? Give an example.
8. Of what value to the cell is a cyclic process such as the Krebs cycle ?
9. Define the terms redox reactions and redox potential. Explain Mitchell's chemiosmotic coupling theory.
10. Define the following : electron transport; respiratory chain; oxidative phosphorylation; cytochrome oxidase. Summarize various schemes for coupling electron transport to ADP phosphorylation.
11. What is a fermentation, and why is it needed by some cells ?
12. What is the special function of brown fat mitochondria and how is it carried out ?
13. Describe the β-oxidation of fats.
14. Describe the biogenesis of the mitochondria.
15. Explain why are the mitochondria considered as semiautonomous organelles.
16. Describe the energetics of glucose oxidation and compare it with that of fat oxidation.
17. Give a short account of symbiotic origin of mitochondria and chloroplasts.
18. Compare the functions of mitochondria and chloroplasts.

CHAPTER

11

Plastids

(Chloroplasts, Photosynthesis and Vacuoles)

Starch filling plastids in potato cells.

Plant cells are readily distinguished from animal cells by the presence of two types of membrane-bounded com partments– **vacuoles** and **plastids**. Both organelles are related to the immobile life-style of plant cells.

Plastids are present in all living plant cells and in *Euglena* (a protozoan). They are small bodies found free in the cytoplasm. Plastids are often more or less spherical or disc-shaped (1 µm to 1 mm in diameter), but may be elongated or lobed or show amoeboid characteristics. Their other identifying features are their double bounding membranes, the possession of **plastoglobuli** (spherical lipid droplets; store of lipids surplus to current requirements) and an internal membrane fretwork of many discrete internal vesicles. All plastids in a particular plant species contain multiple copies of same relatively small genome (DNA) and 70S-type ribosomes. They are self-replicating organelles containing a protein-synthesizing capacity comparable to that of mitochondria. They perform most important biological activities as the synthesis of food and storage of carbohydrates, lipids and proteins. Plastids are absent in the cells of fungi, bacteria, animals and male sperm cells of certain higher plants.

HISTORICAL

Chloroplasts were described as early as seventeenth century by **Nehemiah Grew** and **Antonie van Leeuwenhoek**. The term plastid was used by **Schimper** in 1885 ; he also classified the plastids of plants. **A. Meyer** in 1883, **F. Schmitz** in 1884 and **A.F.W. Schimper** in 1885 made detailed cytological studies of these cell organelles and showed that chloroplasts always arise from pre-existing chloroplasts. In 1918, **Wilstatter** and **Stoll** isolated and characterized the green pigments–chlorophylls *a* and *b*. **K. Porter** and **S. Granick** (1947) described the ultrastructure of grana of chloroplasts. The studies of **Julius Sachs** in the mid-nineteenth century show that chlorophyll was confined to the chloroplasts and was not distributed throughout the plant cell. He also showed that sunlight caused chloroplasts to absorb carbon dioxide and that chlorophyll is formed in chloroplasts only in the presence of light.

Dutrochet (1837) recognized that chlorophyll was essential to oxygen evolution by plants. **Liebig**, in 1845 indicated that carbon dioxide was the source of all organic compounds synthesized by green plants. In 1845, **von Mayer** recognized that green plants convert the solar energy into the chemical energy of organic matter :

$$CO_2 + H_2O \xrightarrow[\text{green plants}]{\text{sunlight}} \text{Organic matter} + O_2$$

In 1862, **Sachs** proved that starch was synthesized by plants in a light-dependent reaction (photosynthesis).

In 1931, an American biochemist, **Cornelius van Niel** observed that a certain type of photosynthetic bacteria fixed carbon dioxide in the presence of hydrogen sulphide. In this process no oxygen was evolved. Instead globules of sulphur were formed as a waste product. He concluded that during bacterial photosynthesis carbon dioxide was not split, rather hydrogen sulphide was broken down, the resultant hydrogen reduced carbon dioxide and sulphur was left behind :

$$6\,CO_2 + 12\,H_2S \longrightarrow C_6\,H_{12}O_6 + 6H_2O + 12S$$

This led van Niel to hypothesize that (1) oxygen produced during photosynthesis of higher plants comes from water and not from carbon dioxide. (2) Water is the hydrogen donor. (3) CO_2 molecules are incorporated intact into carbohydrates. He proposed the following reaction for all photosynthetic organisms :

$$\underset{\substack{\text{Hydrogen}\\\text{acceptor}}}{CO_2} + \underset{\substack{\text{Hydrogen}\\\text{donor}\\(H_2O\text{ or }H_2S)}}{2H_2A} \xrightarrow[\text{chlorophyll}]{\text{light}} \underset{\substack{\text{Reduced}\\\text{acceptor}}}{(CH_2O)} + \underset{\substack{\text{Dehydrogenated}\\\text{donor}\\(1/2\ O_2\text{ or }S)}}{2A} + H_2O$$

In 1932, **Emerson** and **Arnold** carried out the flashing light experiment and showed the existence of light and dark reactions. They introduced the concept of **photosynthetic unit** (or **PS I**) which is thought to be activated when light impinges on a photosynthetic unit.

An English biochemist, **Robert Hill** (1937) demonstrated **photolysis** of water by isolated chloroplasts in the presence of suitable electron acceptor (*e.g.*, ferricyanide)

$$2H_2O + \underset{\text{Ferricyanide}}{4Fe^{3+}} \xrightarrow{\substack{\text{illuminated}\\\text{chloroplast}}} O_2 + 4\,H^+ + \underset{\text{Ferrocyanide}}{4Fe^{2+}}$$

In 1941, **Ruben** and **Kamen** used O^{18} to show that in photosynthesis oxygen comes from water. **Calvin** and **Benson** (1948) showed that phosphoglycerate was an early product of CO_2 fixation. In 1954, **Arnon**, **Allen** and **Whatley** used $^{14}CO_2$ to show fixation of CO_2 by isolated chloroplasts. **Melvin Calvin** (1945–1954) made experiments with unicellular green alga *Chlorella* and used radioactive form of CO_2 ($^{14}CO_2$) to work out those anabolic reactions by which CO_2 is fixed into hexoses and other carbohydrates. These reactions are found to be independent of light and are called **dark reactions**,

biochemical reactions, **Calvin cycle** or **C_3 cycle** (since it involves the formation of a 3-carbon product). Calvin was awarded the Nobel prize in 1960.

Hill and **Bendall** (1960), proposed Z-sheme for electron transport from water to NADPH during photosynthesis. It linked the two photosystems—PS I and PS II.

In 1966, two Australian workers, **M.D. Hatch** and **G.R. Slack** suggested an alternative pathway for carbon fixation in corn and some other hot-weather plants. It is called **C_4 cycle**, since it involves a four carbon compound.

TYPES OF THE PLASTIDS

The term 'plastid' is derived from the Greek word "*plastikas*" (= formed or moulded) and was used by **A.F.W. Schimper** in 1885. **Schimper** classified the plastids into following types according to their structure, pigments and the functions :

1. Leucoplasts

The leucoplasts (Gr., *leuco* = white ; *plast* = living) are the colourless plastids which are found in embryonic and germ cells. They are also found in meristematic cells and in those regions of the plant which are not receiving light. Plastids located in the cotyledons and the primordium of the stem are colourless (leucoplastes) but eventually become filled with chlorophyll and transform into chloroplasts. **True leucoplasts** occur in fully differentiated cells such as epidermal and internal plant tissues. They never become green and photosynthetic. True leucoplasts do not contain thylakoids and even ribosomes (**Carde**, 1984). They store the food materials as carbohydrates, lipids and proteins and accordingly are of following types :

(i) Amyloplasts. The amyloplasts (L., *amyl*=starch ; Gr., *plast*=living) are those leucoplasts which synthesize and store the starch. The amyloplasts occur in those cells which store the starch. The outer membrane of the amyloplst encloses the stroma and contains one to eight starch granules. In some plant tissues such as potato tuber, amyloplasts can grow to be as large as an average animal cell. In them starch granules may become so large that they rupture the encasing membrane. Starch granules of amyloplasts are typically composed of concentric layers of starch.

(ii) Elaioplasts. The elaioplasts store the lipids (oils) and occur in seeds of monocotyledons and dicotyledons. They also include sterol-rich **sterinochloroplast**.

(iii) Proteinoplasts. The proteinoplasts are the protein storing plastids which mostly occur in seeds and contain few thylakoids (**Heinrich**, 1966).

2. Chromoplasts

The chromoplasts (Gr., *chroma*=colour; *plast*=living) are the coloured plastids containing **carotenoids** and other pigments. They impart colour (*e.g.*, yellow, orange and red) to certain portions of plants such as flower petals (*e.g.*, daffodils, rose), fruits (*e.g.*, tomatoes) and some roots (*e.g.*, carrots). Chromoplast structure is quite diverse ; they may be round, ellipsoidal, or even needle-shaped, and the carotenoids that they contain may be localized in droplets or in crystalline structures. The function of chromoplasts is not clear but in many cases (*e.g.*, flowers and fruits) the colour they produce probably plays a role in attracting insects and other animals for pollination or seed dispersal.

In general, chromoplasts have a reduced chlorophyll content and are, thus, less active photosynthetically. The red colour of ripe tomatoes is the result of chromoplasts that contain the red pigment **lycopene** which is a member of carotenoid family. Chromoplasts of blue-green algae or cyanobacteria contain various pigments such as **phycoerythrin**, **phycocyanin**, **chlorophyll a** and **carotenoids**. Chromoplasts are of following two types :

(i) Phaeoplast. The phaeoplast (Gr., *phaeo*=dark or brown; *plast*=living) contains the pigment **fucoxanthin** which absorbs the light. The phaeoplasts occur in the diatoms, dinoflagellates and brown algae.

(ii) Rhodoplast. The rhodoplast (Gr., *rhode*= red; *plast*=living) contains the pigment **phaeoerythrin** which absorbs the light. The rhodoplasts occur in the red algae.

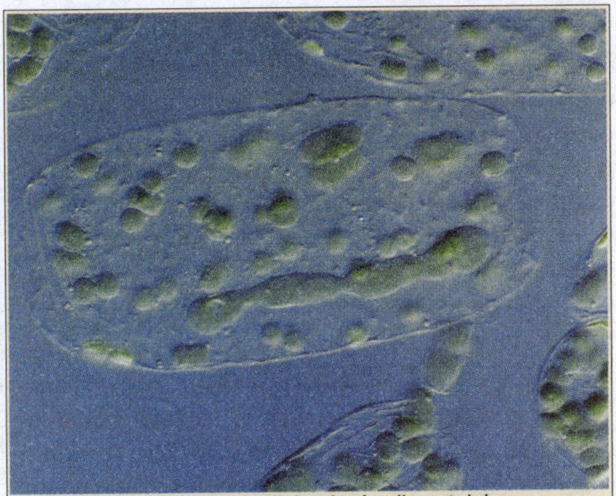

Light micrograph of a living leaf cell containing a number of chloroplasts.

3. Chloroplasts

The chloroplast (Gr., *chlor*=green; *plast*=living) is most widely occurring chromoplast of the plants. It occurs mostly in the green algae and higher plants. The chloroplast contains the pigment chlorophyll a and chlorophyll b and DNA and RNA.

According to **Schimper**, different kinds of plastids can transform into one another, as shown in following figure :

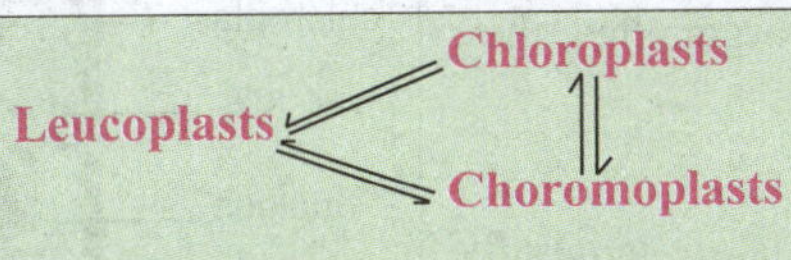

Fig. 11.1. Interconversion of three kinds of plastids.

Development of Plastids

All plastids, including chloroplasts, develop from **proplastids**, which are relatively small organelles present in meristematic cells. Proplastids develop according to the need of each differentiated cell. If a leaf is grown in darkness, its proplastids enlarge and develop into **etioplasts**. These have a semi-crystalline array of internal membranes that contain **protochlorophyll** (a yellow chlorophyll precursor) instead of chlorophyll. When exposed to light, the etioplasts rapidly develop into chloropalsts by converting protochlorophyll to chlorophyll and by synthesizing new membrane, pigments, photosynthetic enzymes and components of electron transport chain (**Thomson** and **Watley**, 1980 ; **Mullet**, 1988).

Light regulates many processes in plants. In ways that are still unclear, photoreceptors including phytochromes control the transcription of many light activated genes involved in chloroplast development, not only in the chloroplast itself but also in the nucleus. Approximately one-fifth of the 120 chloroplast genes are regulated in a light-dependent manner.

Phytochromes. Although there are several types of photoreceptor molecules in plants, the best characterized is the large (1,24,000 dalton) protein, called **phytochrome**. This protein contains a pigment that is responsive to light and can exist in two interconvertible forms : an **inactive form** produced by exposure to far red light (between visible red and infrared), and an **active form** produced by red light. Phytochrome is known to mediate many light-activated responses, including chloroplast rotation, plastid differentiation, seed germination, stem elongation, leaf initiation and flowering. For example, *Mougeotia* is an alga in which each cylindrical cell contains a single plate-like chloroplast. In response to light, the chloroplast rotates until it is perpendicular to the incident light. The photoreceptor molecules that initiate the response are phytochrome and a blue light receptor located on or very near the plasma membrane. Illumination of these receptors with a microbeam causes an influx of Ca^{2+} which binds to calmodulin. Calmodulin activates a network of actin filaments, attached both to the outer membrane of the chloroplast envelope and to the plasma membrane which mediates the rotation (**Wagner** and **Klein**, 1981).

CHLOROPLASTS

We will describe here only the chloroplasts in detail because of two reasons. Firstly they are most common plastids of many plant cells and secondly they perform the photosynthetic activity of greatest biological importance. By the process of photosynthesis the chloroplast synthesizes the carbohydrates which contain energy in the form of chemical energy. The chemical energy is utilized by all living beings to perform various life activities.

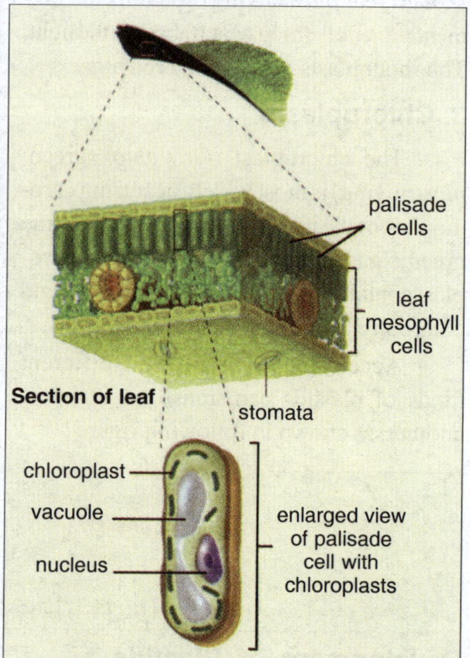

The functional organization of a leaf.

DISTRIBUTION

The chloroplasts remain distributed homogeneously in the cytoplasm of plant cells. But in certain cells, the chloroplasts become concentrated around the nucleus or just beneath the plasma membrane. The chloroplasts have a definite orientation in the cell cytoplasm. Since chloroplasts are motile organelles, they show passive and active movements.

MORPHOLOGY

Shape. Higher plant chloroplasts are generally biconvex or plano-convex. However, in different plant cells, chloroplasts may have various shapes, *viz*., filamentous, saucer-shaped, spheroid, ovoid, discoid or club-shaped. They are vesicular and have a colourless centre.

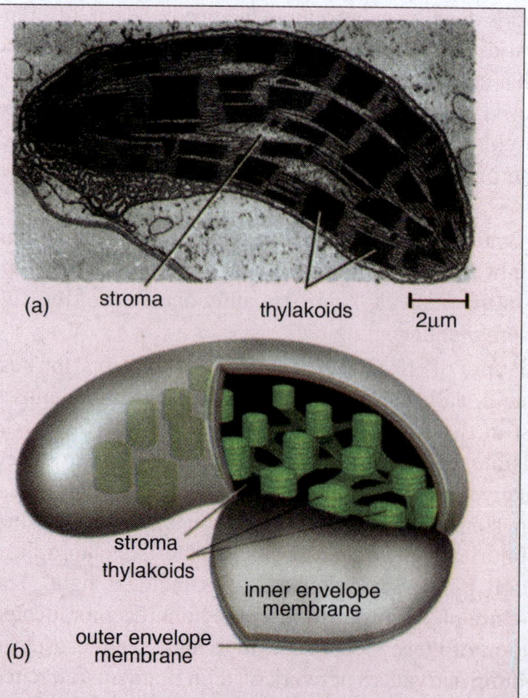

The internal structure of a chloroplast.

Size. The size of the chloroplasts varies from species to species. The chloroplasts generally measure 2–3μm in hickness and 5–10μm in diameter (*e.g.*, *Chlamydomonas*). The chloroplasts of polyploid plant cells are comparatively larger than the chloroplasts of the diploid plant cells. Generally, chloroplasts of plants grown in the shade are larger and contain more chlorophyll than those of plants grown in sunlight.

Number. The number of the chloroplasts varies from cell to cell and from species to species and is related with the physiological state of the cell, but it usually remains constant for a particular plant cell. The algae usually have a single huge chloroplast. The cells of the higher plants have 20 to 40 chloroplasts. According to a calculation, the leaf of *Ricinus communis* contains about 400,000 chloroplasts per square millimeter of surface area. When the number of chloroplasts is inadequate, it is increased by division; when excessive, it is reduced by degeneration.

Comparison of Chloroplasts and Mitochondria

Chloroplasts carry out their energy inter-conversions by chemiosmotic mechanisms in much the same way that mitochondria do and they are organized on the same principles. Thus, each chloroplast contains three distinct membranes which define three separate internal compartments—the intermembrane space, the stroma and the thylakoid space. The thylakoid membrane contains all the energy generating systems of chloroplasts.

Like the mitochondria, chloroplasts have a highly permeable outer membrane ; a much less permeable inner membrane, in which special carrier or transport proteins are embedded ; and a narrow intermembrane space in between. The inner membrane surrounds a large space called the **stroma**, which is analogous to the mitochondrial matrix and contains various enzymes, ribosomes, RNAs and DNA.

However, there is an important difference between the two : the inner membrane of the chloroplast is not folded into cristae and does not contain an electron-transport chain. Instead, the photosynthetic light-absorbing system, the electron-transport chain and an ATP synthetase are all contained in a third distinct membrane that forms a set of flattened disc-like sacs, the **thylakoids** (Fig. 11.2).

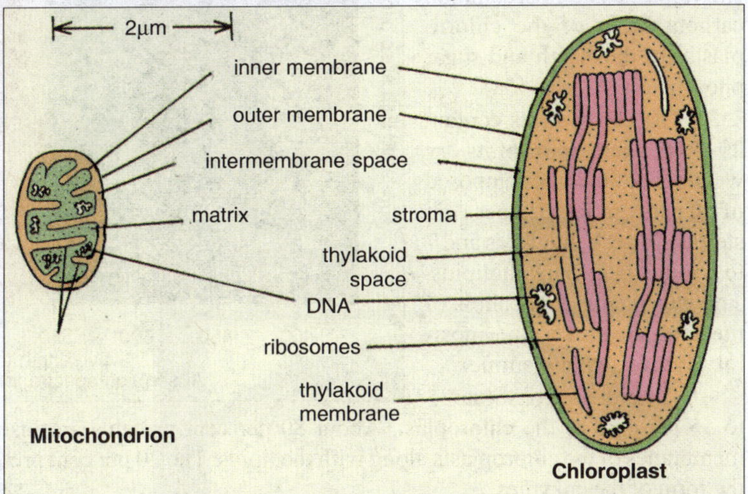

Fig. 11.2. Comparison of a mitochondrion and a chloroplast (after Alberts *et al.*, 1989).

In a general way, one might view the chloroplast as a greatly enlarged mitochondria in which the cristae are converted into a series of interconnected submitochondrial particles in the matrix space. The knobbed ends of the chloroplast ATP synthetases (F_0 - F_1 coupling factors), where ATP is made, protrude from the thylakoid membrane into the stroma, just as they protrude into the matrix from the membrane of each mitochondrial crista.

ISOLATION AND CHEMICAL COMPOSITION

Chloroplasts are routinely isolated from plant tissues by differential centrifugation following the disruption of the cells. Leaves are homogenized in an ice-cold buffered isotonic saline solution (*e.g.*, 0.35 M NaCl) at pH 8.0. The disruption is generally carried out with bursts of Waring blender. After filtration through a nylon gauze (20 μm pore size) to remove the larger particles of debris (*e.g.*, cell nuclei, tissue fragments and unbroken cells), the chloroplasts are separated by unbroken cells), the chloroplasts are separated by centrifugation of 200g for 1 minute. The chloroplast rich pellet is then resuspended and centrifuged again at 2000 g for 45 seconds to re-sediment the chloroplasts. Chloroplast preparations obtained by this method are generally mixtures of intact and broken organelles. Because the chemical composition, rate of photosynthetic activity and other properties of intact chloroplasts differ significantly from those of damaged organelles, it is often desirable to separate the two populations. This may be achieved by isopycnic density gradient centrifugation of the chloroplast preparation.

Details of ultrastructure of grana lamellae have been worked out by electron microscopy (fixation by gluteraldehyde and staining by osmium) and freeze-fracture technique.

The isolated chloroplasts of higher plants are found to contain the chemical composition shown in Table 11-1. The chloroplasts are composed of the carbohydrates, lipids, proteins, chlorophyll, carotenoids (carotene and xanthophylls), DNA, RNA and certain enzymes and coenzymes. The chloroplasts also contain some metallic atoms as Fe, Cu, Mn and Zn.

The carbohydrates occur in very low percentage in the chloroplasts. The most common carbohydrates of the chloroplasts are the starch and sugar phosphates.

The chloroplasts contain 20–30 per cent lipids of its dry weight. The lipids are composed of 50 per cent fats, 20 per cent sterols, 16 per cent waxes and 7 to 20 per cent phospholipids. The most common alcohols of the lipids are the choline, inositol, glycerol, ethanolamine.

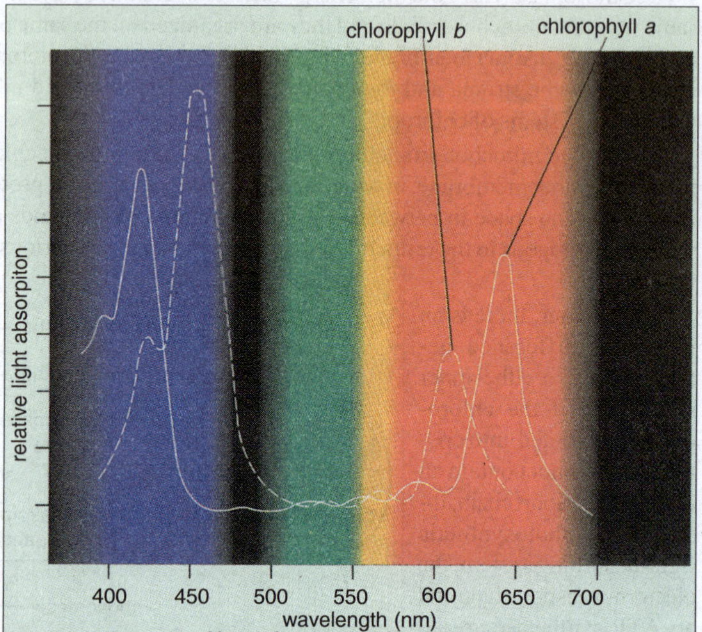

Absorption spectra for chlorophylls.

The proteins constitute 35 to 55 per cent of the chloroplast. About 80 per cent proteins are insoluble and forming the unit membranes of the chloroplasts along with the lipids. The 20 per cent proteins are soluble and occur in the form of the enzymes.

The chlorophyll is a green pigment of the chloroplasts. The chlorophyll contains an asymetrical molecule which has hydrophilic head of four rings of the pyrols and hydrophobic tail of the phytol chain. Chemically the chlorophyll is a porphyrin like the animal pigment haemoglobin and cytochromes except besides the iron (Fe), it contains Mg atom in between the rings of the pyrols which remain connected with each other by the methyl groups. The chlorophyll consists of 75 per cent **chlorophyll** *a* and 25 per cent **chlorophyll** *b*.

The carotenoids are carotenes and xanthophylls, both of which are related to vitamin A. The carotenes have hydrophobic chains of unsaturated hydrocarbons in their molecules. The xanthophylls contain many hydroxy groups in their molecules.

DNA of chloroplast of *Chlamydomonas* represents non-chromosomal genetic system and has been found to be related with cytoplasmic heredity. **Ruth Sager**, who is pioneer on nonchromosomal genes, was able to prepare genetic map of *Chlamydomonas* chloroplast. She has shown that the genome of the chloroplast of *Chlamydomonas* is circular like that of bacteria.

Table 11-1. Chemical composition of chloroplasts of higher plants.

Chemical constituents	Per cent dry weight	Components (per cent)
1. Proteins	35–55	Insoluble 80%
2. Lipids	20–30	Fats 50%
		Sterols 20%
		Wax 16%
		Phosphatides 2–7%

Chemical constituents	Per cent dry weight	Components (per cent)
3. Carbohydrates	Variable	Starch, sugar, phosphates 3–7%
4. Chorophyll	9.0	Chlorophyll *a* 75% Chlorophyll *b* 25%
5. Carotenoids	4.5	Xanthophyll 75% Carotene 25%
6. Nucleic acids		
RNA	3–4	
DNA	< 0.02–.01	

ULTRASTRUCTURE

A chloroplast comprises the following three main components (Fig.11.3) :

1. Envelope

The entire chloroplast is bounded by an **envelope** which is made of a double unit membranes. Across this double membrane envelope occurs exchange of molecules between chloroplast and cytosol (cytoplasmic matrix). Isolated membranes of envelope of chloroplast lack chlorophyll pigment and cytochromes but have a yellow colour due to the presence of small amounts of carotenoids. They contain only 1 to 2 per cent of the total protein of the chloroplast.

2. Stroma

The matrix or stroma fills most of the volume of the chloroplasts and is a kind of gel-fluid phase that surrounds the thylakoids (grana). It contains about 50 per cent of the proteins of the chloroplast, most of which are soluble type. The stroma also contains ribosomes and DNA molecules (*i.e.*, 80 DNA molecules per chloroplast per cell of *Chlamydomonas*; 20 to 40 DNA molecules per chloroplast per cell of leaf of maize), both of which are involved in the synthesis of some of the structural proteins of the chloroplast. The stroma is the place where CO_2 fixation occurs and where the synthesis of sugars, starch, fatty acids and some proteins takes place.

3. Thylakoids

The thylakoids (thylakoid = sac-like) consists of flattened and closed vesicles arranged as a membranous network. The outer surface of the thylakoid is in contact with the stroma, and its inner surface encloses an **intrathylakoid space** (the third compartment). Thylakoids may be stacked like a neat pile of coins, forming **grana** or they may be unstacked, **intergranal**, or **stromal thylakoids**, forming a system of anastomosing tubules that are joined to the **grana thylakoids**. There may be 40 to 80 grana in the matrix of a chloroplast. The number of thylakoids per granum may vary from 1 to 50 or more. For example, there may be single thylakoid (*e.g.*, red alga), paired thylakoids (*e.g.*, Chrysophyta), triple thylakoids and multiple thylakoids (*e.g.*, green algae and higher plants) (Fig. 11.4).

Molecular Organization of Thylakoids

Molecular organization of the membrane of thylakoids is based on the fluid-mosaic model of the membrane which represents following main characteristics: **fluidity**, **asymmetry** and **economy** (*i.e.*, lack of movement in the third dimension). Lipids represent about 50 per cent of the thylakoid membrane; these include those directly involved in photosynthesis (called **functional lipids**) such as **chlorophylls**, **carotenoids** and **plastoquinones**. **Structural lipids** of thylakoids include glycolipids, sulpholipids and a few phospholipids. Most of these structural lipids are highly unsaturated which confer to the membrane of thylakoids a high degree of fluidity.

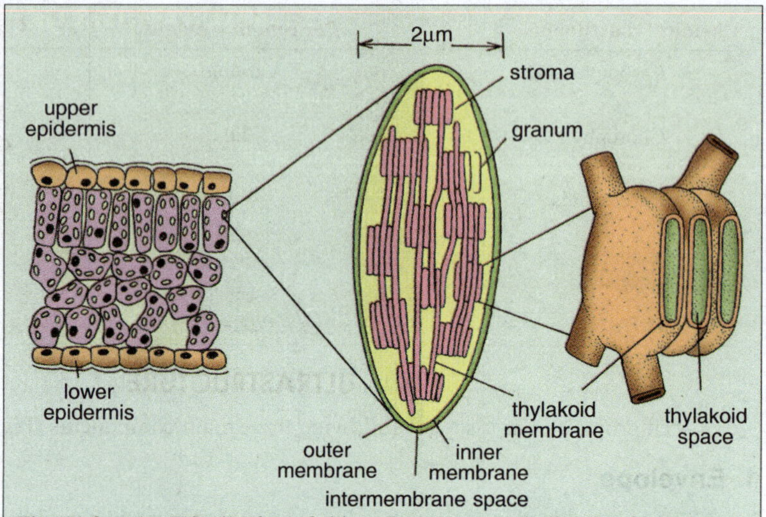

Fig. 11.3. A—Distribution of chloroplasts in mesophyll cells of a leaf; B—Ultrastructure of a chloroplast; C—Details of a granum (after Alberts *et al.*, 1989).

The protein components of thylakoid membrane are represented by 30 to 50 polypeptides which are disposed in the following five major supramolecular complexes (Fig. 11.5), which can be isolated with mild detergent :

1. Photosystem I (PS I). This complex contains a reactive centre composed of P700 (Type of pigment which is bleached at the wavelength of 700 nm), several polypeptides, a lower chlorophyll *a/b* ratio and β-carotene. It acts as a light trap and is present in unstacked thylakoid membranes. In it light induced reduction of $NADP^+$ takes place.

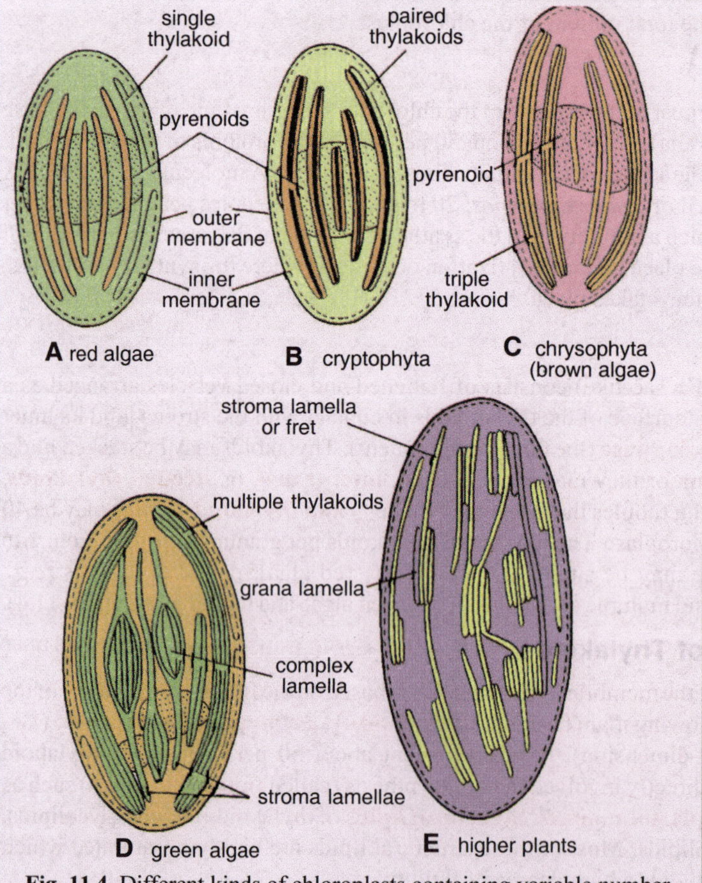

Fig. 11.4. Different kinds of chloroplasts containing variable number of thylakoids per granum.

2. Photosystem II (PS II). This complex comprises two intrinsic proteins that bind to the reaction centre of chlorophyll P680 (The pigment that bleaches when absorbing light at 680 nm). It contains a high ratio of chlorophyll *a/b* and β-carotene. Frequently, the PS IIs are associated with the light-harvesting complex and are involved in light induced release of O_2 from H_2O (*i.e.*, photolysis of water). PS II works as a light trap in photosynthesis and is mainly present in the stacked thylakoid membranes of grana.

3. Cytochrome b/f. This complex contains one cytochrome F, two cytochromes of b 563, one FeS centre and a polypeptide. It is uniformly distributed in the grana and acts as the electron carrier.

These three complexes are related to the electron transport and are linked by **mobile electron carriers** (*i.e.*, plastoquinone, plastocyanin and ferredoxin). Electron transport through PS II and PS I finally results in the reduction of the coenzyme $NADP^+$. Simultaneously, the transfer of protons from the outside to the inside of the thylakoid membrane occurs.

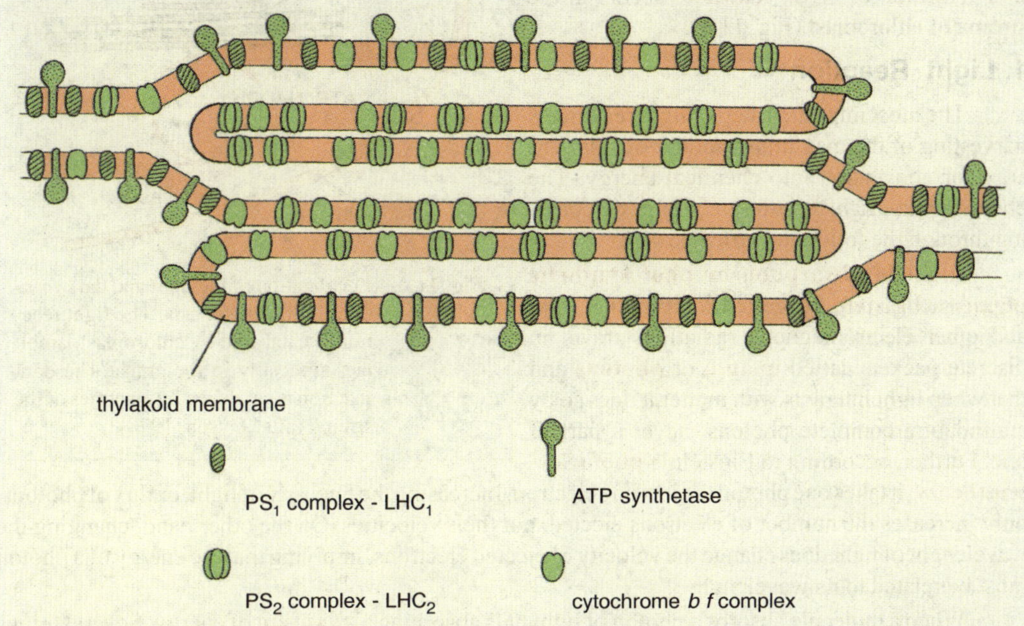

Fig. 11.5. Diagram showing the distribution of the main complexes within the thylakoid membranes both in the granal or stacked and stromal or unstacked regions (after De Robertis and De Robertis, Jr., 1987).

4. ATP synthetase. As in mitochondria, this complex consists of a CF_0 hydrophobic portion, a proteolipid that makes a proton channel, and a CF_1 (or coupling factor one) that synthesizes ATP from ADP and Pi, using the proton gradient provided by the electorn transport. ATP synthetase complexes are located in stacked membrane (grana).

5. Light harvesting complex (LHC). The main function of LH complex is to capture solar energy. It contains two main polypeptides and both cholorophyll *a* and *b*. LH complex is mainly associated with PS II, but may also be associated with PS I (**Anderson**, 1975). LHC is localized in stacked membranes and lacks photochemical activity.

Mutation and chloroplast structure. The organization of chloroplasts and other plastids is often modifed due to mutation. **D. Von Wettstein** (1956) reported that the plastids of normal barley plant have a well organized system of grana and stroma. But the plastids of an albino mutant of barley, fail to develop beyond a particular stage and there occurs no differentiation of grana and stroma. Further, the plastids of a yellow-green mutant of barley develop somewhat further than plastids of an albino plant.

FUNCTIONS OF THE CHLOROPLAST : PHOTOSYNTHESIS

It is well evident now that the process of photosynthesis consists of the following two steps:

1. **Light reaction**. It is also called **Hill reaction**, **photosynthetic electron transfer reaction** or **photochemical reactions.** In light reaction solar energy is trapped in the form of chemical energy of ATP and as reducing power in NADPH. During it, oxygen is evolved by photolysis or splitting of water molecule. Light reaction occurs in thylakoid membranes. 2. **Dark reaction**. It is also called **Calvin reaction**, **photosynthetic carbon reduction cycle** (**PCR cycle**), **carbon-fixation reaction** or

thermo-chemical reaction. In dark reaction, the reducing capacity of NADPH and the energy of ATP are utilized in the conversion of carbon dioxide to carbohydrate. Such a process of "**carbon fixation**" or "**CO_2-fixation**" occurs in the stroma of chloroplast (Fig. 11.6).

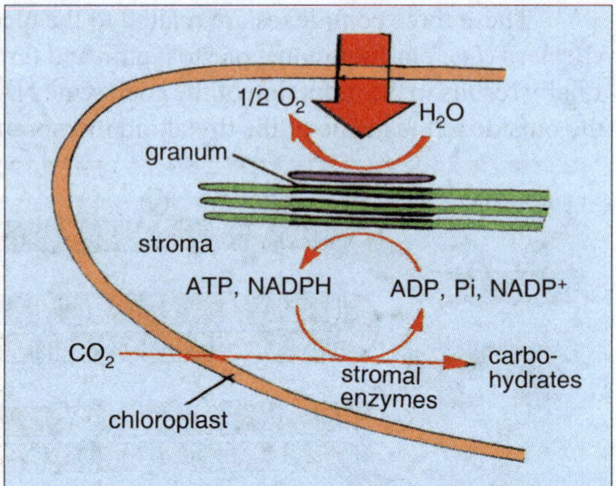

Fig. 11.6. Localization of the light and dark reaction of photosynthesis. The light reaction is catalyzed by chloroplast lamellae, especially in the grana. The dark reaction is catalyzed by enzymes of the stroma (after Dyson, 1978).

1. Light Reaction

The most important step of light reaction is harvesting of the maximum amount of solar energy for conversion into chemical energy. The photosynthetic light reaction is completed by passing through the following processes:

(i) Light absorption by photosynthetic pigment. Einstein suggested in 1905 that light and other electromagnetic radiations travel in discrete packets called **quanta** or **photons** and that when light interacts with matter it does so by annihilating complete photons, never a part of one. Further, according to Einstein's **photoelectric theory**, it takes one photon to eject one electron. Increasing the intensity of light, or flux of photons, only increases the number of electrons ejected, not their velocities. On the other hand, changing the wavelenght of light does change the velocity of ejected electrons, implying that the energy of a photon must be related to its wavelength.

When a molecule absorbs a photon of light, it is absorbing a **quantum** of energy. Several things can happen to this energy : (i) It can be dissipated in molecular motion, manifest as heat. (ii) It can be reemitted as a new photon of light at a longer wavelength, with the shift representing losses to other processes—if remission occurs very quickly, it is called **fluorescence** ; if there is a long lag (milliseconds to seconds) between absorption and reemission, the process is called **phosphorescence**. (iii) The energy of light can cause a chemical change in the compound that absorbs it. It is this latter possiblility that can happen when a molecule of chlorophyll absorbs a photon. In fact, chloroplast acts as an energy transducer, *i.e.*, it can convert light energy to chemical energy, much as a solar battery uses light to run a transistor radio.

Within a fraction of a second after light is absorbed by a photosynthetic pigment, the molecule is altered ; some electrons associated with the pigment are raised to new energetic heights, changing their spin or modifying their position. If enough energy is absorbed, an electron may even be ejected and, thus, oxidation occurs. When these events happen, we say that the molecule is in an **excited state**.

Photosynthetic units. The basic photosynthetic units seem to be groups of roughly 300 pigment molecules located in the chloroplast membranes (thylakoid disc). Although all the pigment molecules in the unit (carotenoids, chlorophyll *b*, etc.) are capable of capturing light energy, they must transfer it to a single key chlorophyll *a* molecule called the **reaction centre**. The latter then loses an electron to a series of electorn carrier molecules. Thus, the other 299 accessory pigment molecules are referred to as **antenna molecules** or **antenna pigments**, to designate their role in the capture of light energy.

Dual pigment systems. Higher plants are found to have two types of photosynthetic units, associated with two different pigment systems, which absorb light of different wavelenghts. **Pigment system I** or **photosystem I (PS I)** units occur in the thylakoid membrane in the form of small and densely packed particles. Each PS I unit consists of about 200 molecules of chlorophyll *a* and 50 carotene molecules. The reaction centre (chlorophyll *a* molecule) is called **P 700** because it has a maximum light absorption at 700 nanometers. Energy funneling into P 700 is responsible for the ejection of an electron from the chlorophyll. **Pigment system II** or **photosystem II (PS II)** units occur in the thylakoid membrane in the form of larger, more widely spaced particles or ES particles (or quantosomes). Each PS II unit consists of approximately 200 molecules of chlorophyll *a*, 200 molecules of carotenols, chlorophyll *b*, and chlorophyll *c* or *d*, depending upon the species. Its reaction-centre chlorophyll *a* is designated **P 690** or **shorter-wavelength trap**. In the thylakoid membrane PS I and PS II are probably arranged near one another forming the so-called **Z-scheme** (**Hill** and **Bendall**, 1960) because they are functionally related ; excitation energy originating in one can be shunted to the other system. However, two photosystems are coupled chemically rather than through direct energy transfer.

A sunburst through the leaves of a fern depicts the harvesting of light energy by photosynthesis, the ultimate source of energy for nearly all life on earth.

(ii) Electron transport systems and oxidation of water. When the P680 of photosystem II acquires a sufficient quantum of energy, it emits a pair of electrons. These electrons with high potential

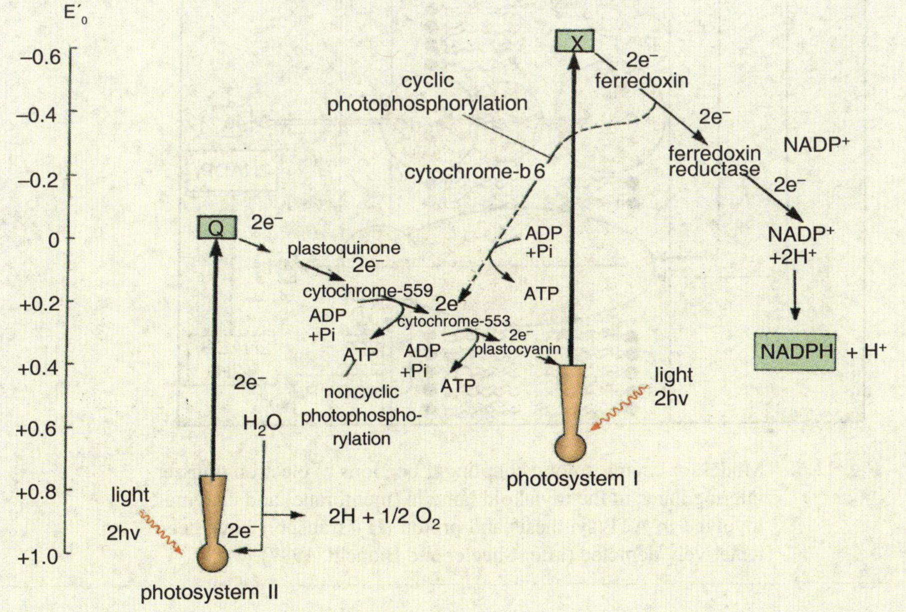

Fig. 11.7. Photosynthetic light reaction showing major events occurring in photosystem I and photosystem II of chloroplasts.

energy move down an electron transport chain (of thylakoid membrane) and during this process ATP molecules are formed (Fig. 11.7). Two electrons are passed through an electron acceptor **Q** (which is a quinone) to an electron transport chain involving four electron carriers (**plastoquinone**, **cytochrome-559**, **cytochrome-553** or **cytochrome f** and **plastocyanin**), before being passed on to PS I. The electrons are passed through four carriers successively at lower energy levels, so that at each step energy is released, which is harvested in the production of two ATP molecules (from ADP + Pi). The electron lost by P 680 is ultimately accepted by P700 of photosystem I. P700 (PS I) also captures light, and for absorbing each photon, it ejects an electorn. This electron is replaced by an electron from PS II and this flow of electron continues as long as the light is available. The electrons liberated from P700 are passed through acceptor x, to an electron transport chain (**ferredoxin**, **ferredoxin NADP reductase**) at successive lower energy levels. Finally these electorns reach NADP coenzyme, each molecule of which receives an electron, enabling it to pick up a H^+ ion (proton), thus, producing two molecules of NADPH from one molecule of H_2O used in PS II :

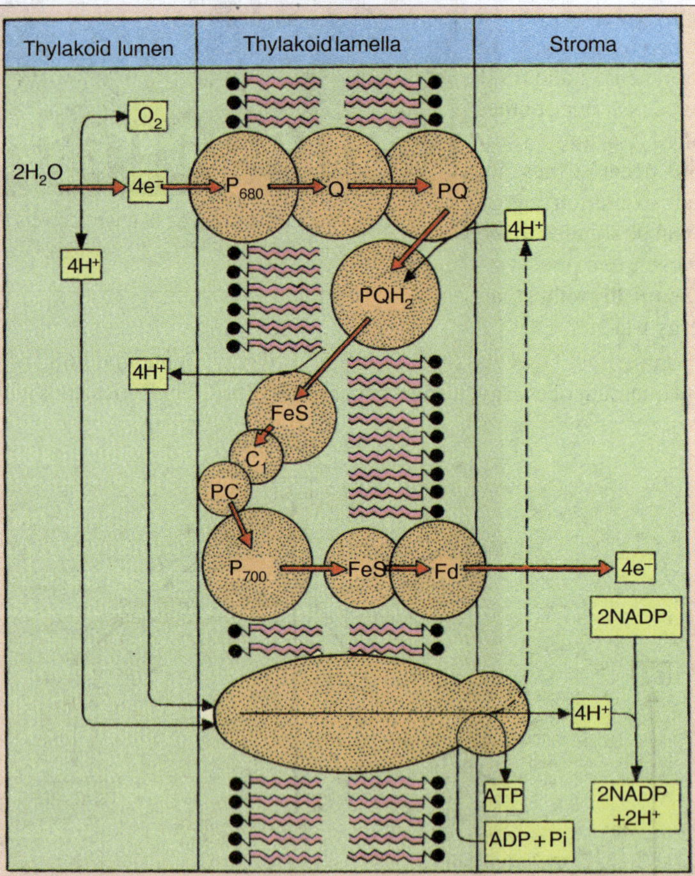

Fig. 11.8. Model for chemiosmotic coupling. Locations of electron transfer intermediates in the thylakoid lamella (membrane) and enzymes involved in ATP synthesis and proton (H^+) transport have been tentatively depicted (after Sheeler and Bianchi, 1987).

$$2e^- + 2H^+ + 2NaDP^+ \xrightarrow{2hv} 2NADPH$$

The oxidized P680 regains its electrons by the photolysis of water into $2H^+$, $2e^-$ and oxygen:

$$H_2O \xrightarrow{\text{2quanta or 2hv}} \tfrac{1}{2}O_2 + 2H^+ + 2e^-$$

Oxygen is given out by photosynthesizing plants. The protons (H^+) accumulate inside the thylakoid membrane resulting in a **proton gradient**. The energy released by the protons when they diffuse across the thylakoid membrane into the stroma (along H^+ concentration gradient) is used to produce ATP molecules, by **CF0 - CF1 ATP synthetase** in the membrane. As synthesis of ATP occurs in light and the process is not cyclic (*i.e.*, it needs a constant supply of water molecules to be oxidized and NADP to be reduced), the process is called **non-cyclic photophosphorylation**.

ATP production by **cyclic photophosphorylation** also occurs during the light reaction. This process involves another electorn transfer mechanism involving **cytochrome b6** and starting with P700; the ultimate acceptor of the de-energized electron is also P700 of photosystem I.

All the molecules of ATP and NADP generated in the light reaction of photosynthesis are used by soluble enzymes of stroma of chloroplast during the dark reaction. Total light reaction can be summarized as follows :

$$H_2O + 2NADP + 2\,ADP + 2Pi \xrightarrow{4hv} \tfrac{1}{2}O_2 + 2\,NADPH + 2ATP$$

2. Dark Reaction

The dark reaction is completed by passing through following three main phases:

(i) Phase 1 : Carboxylation. During this phase of dark reaction, three molecules of carbon dioxide (3C) are attached to three molecules of **ribulose 1,5, biphosphate** (**RuBP** ; this pentose sugar was previously termed RuDP or ribulose diphosphate ; 15C) to produce short- lived six-carbon intermediates. This process is called **carboxylation** and is catalyzed by the enzyme **RuBP carboxylase**, **carboxydismutase** or "**Rubisco**" (which is widely acclaimed as one of the most abundant proteins present on the planet Earth; see **Alberts** *et al.*, 1989). Rubisco is a large protein molecule (500,000 dalton MW) comprising 16 per cent of chloroplast protein. The six carbon intermediates are immediately broken down into six molecules of **PGA** or **3- phosphoglyceric acid** (*i.e.* a C-3 or three carbon compound, 6C×3C = 18C).

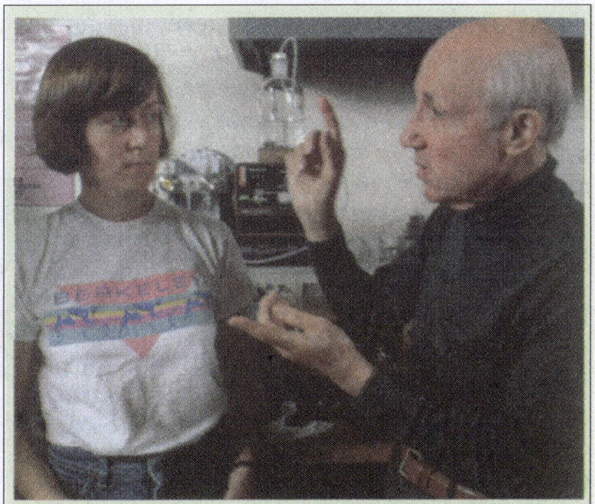

Calvin with a student. The sequence of steps in the light independent stage was investigated by a team led by Melvin Calvin.

(ii) Phase 2 : Glycolytic reversal. By utilizing six ATP molecules, these six molecules of PGA are transformed into six molecules of **1,3, diphosphoglyceric acid** (*i.e.*, 6C × 3C = 18C ; 3C from $3CO_2$ and 15C from 3 RuBP). These in turn get converted into six molecules of **glyceraldehyde-3-phosphate**, **3- phosphoglyceraldehyde** (**PGAL**) or **3-phosphoglyceric acid** (triose) by utilizing six NADPH molecules.

(iii) Phase 3 : Regeneration of RuBP. For the continuous running of Calvin cycle, there must be a regular supply of ATP and NADPH and also sufficient amount of RuBP. Three molecules of RuBP (3C × 5C = 15C) are regenerated by a complex series of reactions which utilize three ATP molecules and five molecules of 3-phosphoglyceric acid (5C × 3C = 15C). This leaves one molecule of PGAL as the net gain of one Calvin cycle. Two turns of Calvin cycle result in the production of one molecule of **glucose**. This glucose is used by the plant to form a large variety of organic compounds required for its structure and function (*e.g.,* starch, cellulose, lipids, amino acids and proteins, etc.). The dark reaction may be summed up as follows :

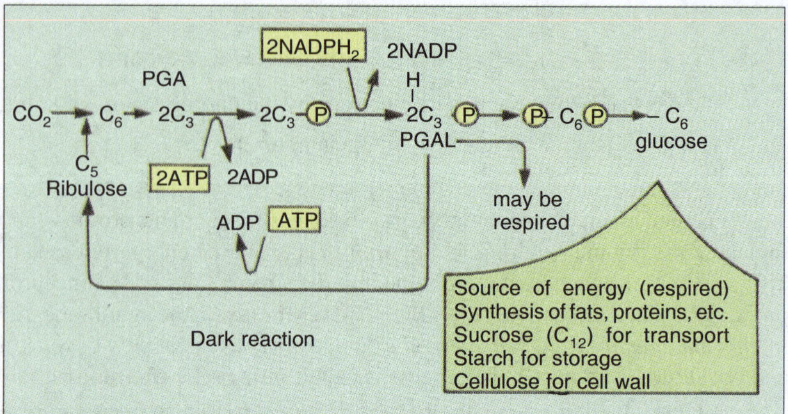

Fig. 11.9. Synthesis of carbohydrate during dark reaction (after Berns, 1983).

$$6\,RuBP + 6CO_2 + 18\,ATP + 12\,NADPH \longrightarrow 6RuBP + C_6H_{12}O_6 + 18ADP + 18Pi + 12\,NADP$$

Hatch and Slack Pathway or C_4 Pathway of CO_2 - fixation in Angiosperms

In many angiosperms (*e.g.,* maize, sugarcane, sorghum) having Krantz anatomy (*i.e.,*bundle sheath with chloroplasts), an alternative pathway of CO_2 fixation occurs called **C_4 pathway of carbon dioxide fixation**. In the mesophyll cells of leaf in the presence of **PEP carboxylase enzyme**, the CO_2 is assimilated by carboxylation of **PEP** (a 3-carbon acceptor, called **phosphoenol pyruvate**) to produce a 4-carbon compound, called **oxaloacetic acid** or **OAA** (OAA is an intermediate in the Krebs cycle of respiration). OAA is converted into another intermediate of Krebs cycles, the **malic acid** (4C) or **aspartic acid** in some cases. This acid is then transferred (probably by diffusion) to the cells surrounding the vascular bundle, the **bundle sheath cells** having enzymes of Calvin cycle. Here, malic acid undrgoes decarboxylation to produce pyruvic acid (3C) and CO_2. CO_2 enters the Calvin cycle (or dark reaction) to produce a molecule of 3-phosphoglyceric acid. Sugars formed in Calvin cycle are transported into the phloem.

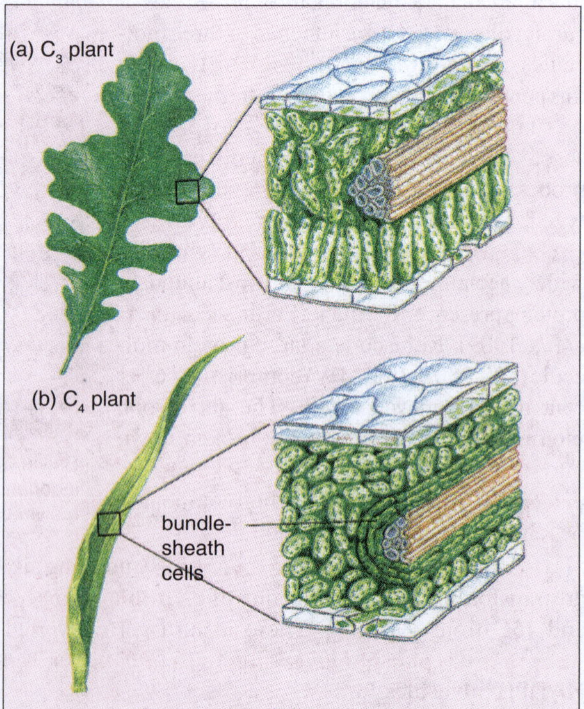

Comparision of C_3 and C_4 plants.

The pyruvic acid generated in the bundle sheath cells is transferred back to the mesophyll. It is converted to PEP by the

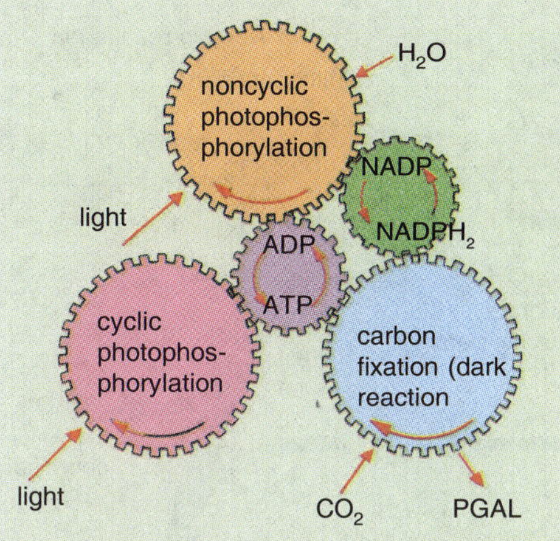

Fig. 11.10. Diagram showing relationship betwen photophosphorylation and carbon fixation. The entire photosynthetic process can be visualized as a series of interlocking gears, with the energy from light turning the two photophosphorylation gears. The turning of the cyclic-phosphorylation gear causes the gear of ATP synthesis to turn, and the turning of noncyclic-photophosphorylation gear causes both the ATP synthesis and NADPH- synthesis gears to turn. These two gears cause the carbon fixation gear to turn, with resultant production of carbohydrate (PGAL; triose) from CO_2 (after Berns, 1983).

expenditure of an ATP molecule. But because the conversion results in the formation of AMP (and not ADP), there remains a requirement of 2ATP for the regeneration of ATP from AMP.

The C-4 pathway is more energy-expensive than the C-3 pathway. While C-3 pathway requires 18 ATP for the synthesis of one molecule of glucose (*i.e.,* 9 ATP molecules per Calvin cycle: $9 \times 2 = 18$ ATP molecules), the C-4 pathway requires 30 ATP molecules (12 ATP more than the C-3 cycle). But realizing that many tropical plants would otherwise lose more than half of the photosynthetic carbon in photorespiration,the C-4 pathway is of adaptive advantage. Further, production of OAA in C-4 plants is a favourable step, since it permits closure of stomata and allows conservation of water.

CHLOROPLAST AS SEMIAUTONOMOUS ORGANELLE

Like the mitochondria, the chloroplasts have their own DNA, RNAs and protein synthetic machinery and are semiautonomous in nature.

1. DNA of chloroplast. Recently the chloroplasts of the algae and higher plants are found to contain DNA molecules. First of all **Ris** and **Plant** (1962) have reported DNA molecule in the chloroplast of the *Chlamydomonas*. Later on DNA molecule has been reported from the chloroplasts of other algae and higher plants. In general, chloroplasts have a double helical DNA circle with an average length of 45 µm (about 135,000 base pairs). The replication of chloroplast DNA has been followed with ^{3}H-thymidine. Maps of the location of genes (genetic maps) have been made in several chloroplast DNAs by the help of restriction enzymes. The gene for the large subunit of carboxydismutase enzyme has been fully sequenced and is found to contain 1425 nucleotides.

2. Ribosomes of chloroplasts. The chloroplasts contain the ribosomes which are smaller than the cytoplsmic ribosomes. The ribosomes of the chloroplast are of 70S type and resemble with the bacterial ribosomes. The ribosomes of the chloroplasts consist of two ribosomal RNAs, 23S rRNA and 16S rRNA (**Stutz** and **Noll**, 1967, **Bager** and **Hamilton**, 1967). **Lyttleton** (1962) has also separated polyribosomes or polysomes from the chloroplast. The chloroplasts also contain aminoacyl-tRNAs, aminoacyl-tRNA synthetases, methionyl-tRNA.

3. Protein synthesis. According to most recent studies (see **Hall**, *et al.*, 1974). the DNA of chloroplast codes for chloroplast mRNA, rRNA, tRNA, and ribsomal proteins. It also codes for certain structural proteins of thylakoid membranes. The synthesis of other chloroplast components as chlorophyll, carotenoids, lipids and photosynthetic and starch synthesizing enzymes, is controlled by nuclear genes. The 70S ribosomes of *Euglena* chloroplast are found to require Mg^{++} for their stability and also have a requirement for N-formy1 methionyl-tRNA in chain initiation protein synthesis like the bacteria. The protein synthetic mechanism of chloroplasts is inhibited by chloramphenicol like that of mitochondria and bacteria (**Ellis**, 1969).

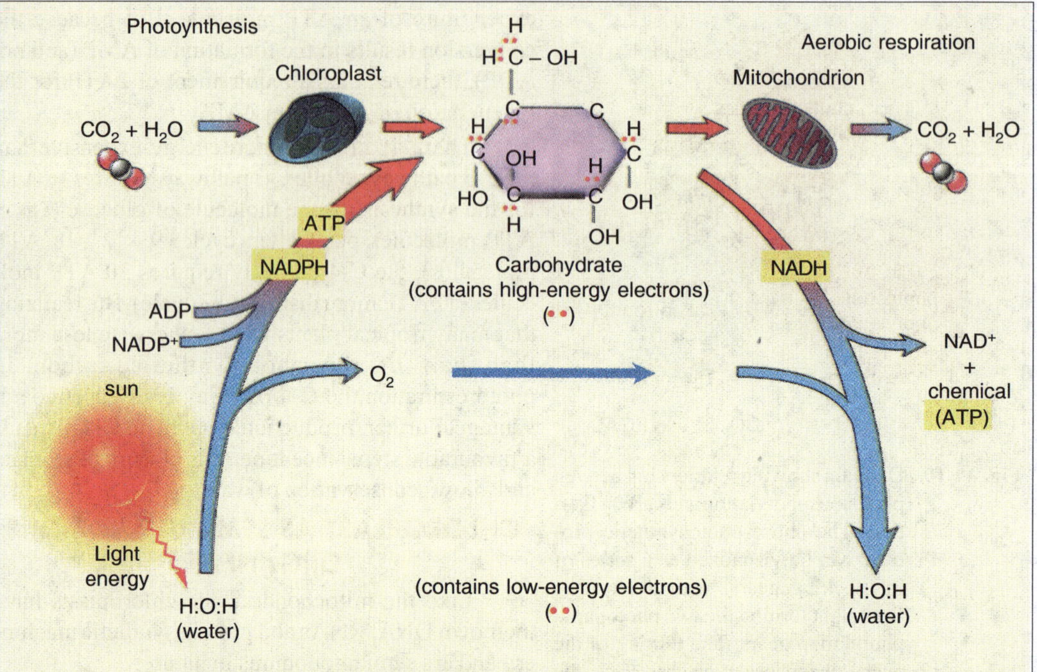

An overview of the energetics of photosynthesis and aerobic respiration.

The mode of synthesis of proteins of chloroplasts indicates towards their **semiautonomous** or **symbiotic nature**. For example, of the 30 known thylakoid polypeptides that function in photosynthesis, so far 9 have been demonstrated to be synthesized on chloroplastic ribosomes and 9 are coded by nuclear genes and synthesized on cytoplasmic ribosomes (**von Wettstein**, 1981). Synthesis of carboxydismutase (C Dase) presents a good case of cooperative action of two genetic systems (*i.e.,* chloroplastic and nuclear genetic systems). C Dase comprises 16 subunits : 8 subunits of high molecular weight (55,000 daltons) and 8 subunits of much smaller molecular weight (14,000 daltons). The large subunit is coded by genes present in chloroplastic DNA, while the small subunit is produced by nuclear genes. The small subunit (called P20) is synthesized as a precursor weighing 20,000 daltons on free ribosomes ; it then enters **post-translationally** into the stroma to be cleaved to attain its final size. It is postulated that the chloroplastic envelope has receptor sites that recognize the proteins that are to be incorporated into the organelle. The extra sequence (acting as the **signal**) that is present in P20 is composed of acidic amino acids, in contrast to the hydrophobic ones in the signal sequence of secretory proteins. After entering the chloroplast the signal sequences are removed by a protease enzyme, which is present in the envelope of chloroplast, and the small subunit of C Dase is released into the stroma (**Ellis**, 1981). Thus, chloroplast proteins may be synthesized by three avenues : (1) by an exclusive chloroplastic mechanism, (2) by a mechanism involving nuclear genes and chloroplastic ribosomes, and (3) by nuclear genes and cytoplasmic ribosomes.

Protein transport into chloroplasts resembles transport into mitochondria in many respects : both occur post-translationally, both require energy, and both utilize hydrophilic amino-terminal signal peptides that are removed after use. However, there is at least one important difference that while mitochondria exploit the electrochemical gradient across their inner membrane to help drive the transport, chloroplasts (which have an electrochemical gradient across their thylakoid but not their inner membrane) appear to employ only ATP hydrolysis to import across their double-membrane outer envelope.

Translocation of proteins into the thylakoid space of chloroplasts requires two signal peptides and two translocation events. The precursor polypeptide contains an amino-terminal chloroplast signal peptide followed immediately by a thylakoid signal peptide. The **chloroplast signal peptide** initiates translocation into the stroma through a membrane contact site by a mechanism similar to that used for translocation into mitochondrial matrix. The signal peptide is then cleaved off, unmasking the thylakoid signal peptide, which initiates translocation across the thylakoid membrane (Fig. 11.12).

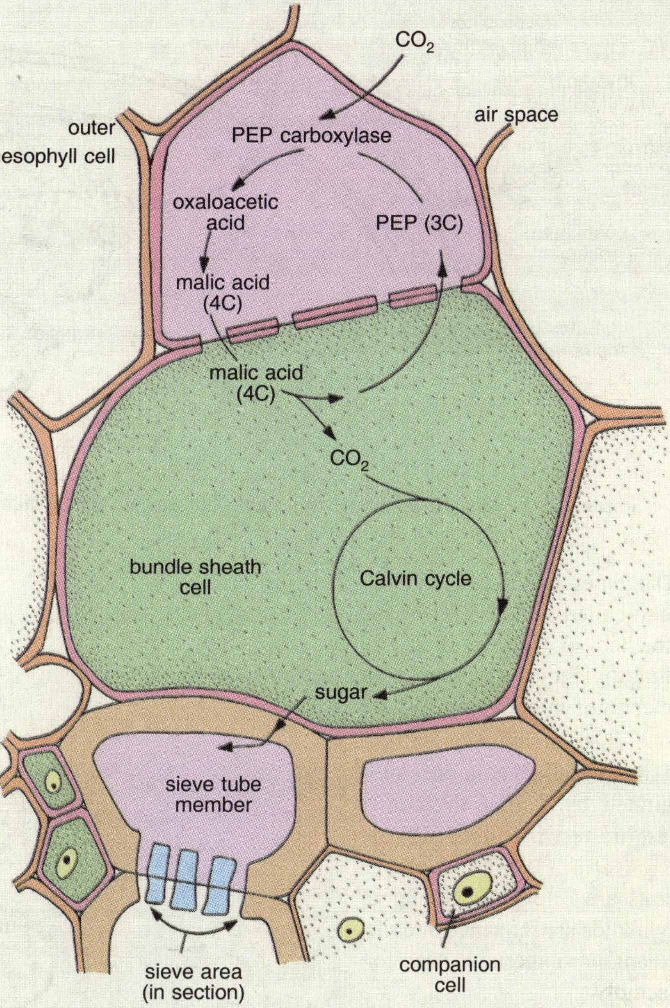

Fig. 11.11. The C_4 pathway.

BIOGENESIS OF CHLOROPLAST

The chloroplasts never originated *de novo*. Since the classic work of **Schimper** and **Meyer** (1883) it has been accepted that chloroplasts multiply by fission, a process that implies growth of the daughter organelles. This is easily observed in the alga *Nitella*, which contain a single huge chloroplast. In *Nitella* a division cycle of 18 hours has been cinematographically recorded for the chloroplast.

During the development of the chloroplast, the first structure to appear is the so-called **proplastid**, which has a double membrane. Development of proplastid into chloroplast takes place in the following steps :

1. In the presence of light, the inner membrane grows and gives off vesicles into the matrix that are transformed into discs (Fig. 11.13). These intrachloroplastic membranes are the thylakoids which, in certain regions, pile closely to form the grana. In the mature chloroplast the thylakoids are no longer connected to the inner membrane, but the grana remain united by intergranal thylakoids.

2. In the absence of light, a reverse sequence of changes takes place. This is the process of **etiolation**, in which the leaves lose their green pigment and the chloroplast membranes become disorganized. The chloroplasts are transformed into **etioplasts**, in which there is a paracrystalline

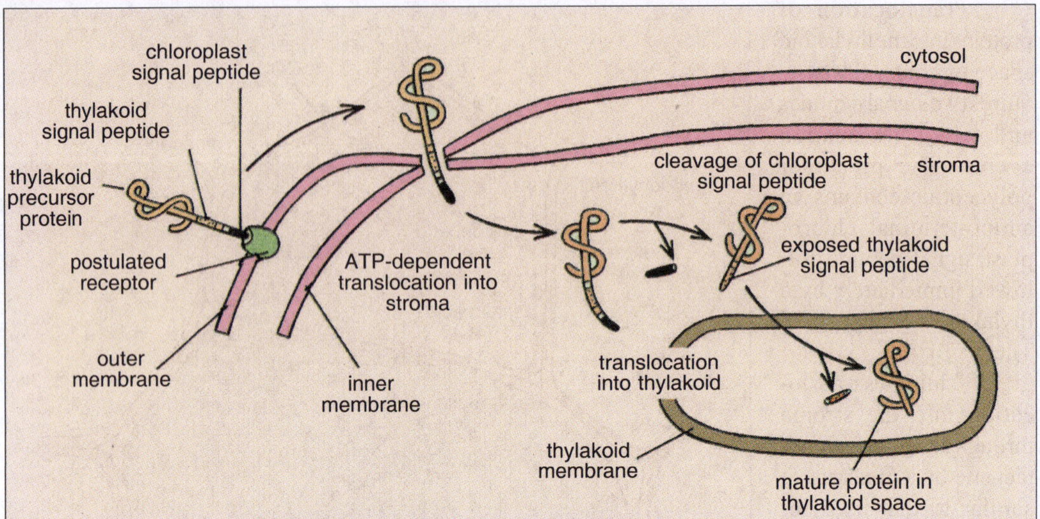

Fig. 11.12. Mode of translocation of proteins into stroma and thylakoid of chloroplast by the help of signal peptides (after Albert *et al.*, 1989).

arrangement of tubules forming the so-called **prolamellar body**. Attached to these bodies are young thylakoid membranes that lack photosynthetic activity.

The regular crystal lattice of two prolamellar bodies surrounded by young thylakoid membranes is observed by **Osumi** *et al.*, (1984). If etiolated plants are re-exposed to light, thylakoids are reformed and the prolamellar material is used for assembly.

The symbiotic origin of the chloroplast. In certain characteristics, the chloroplasts are comparable with that of a semi-autonomous or symbiotic organism living within the plant cells. They divide, grow and differentiate ; they contain circular DNA, ribosomal RNA, messenger RNA and are able to conduct protein synthesis. By visualizing these similarities between chloroplast and micro-organism, it has been suggested that chloroplast might have resulted from a symbiotic relationship between

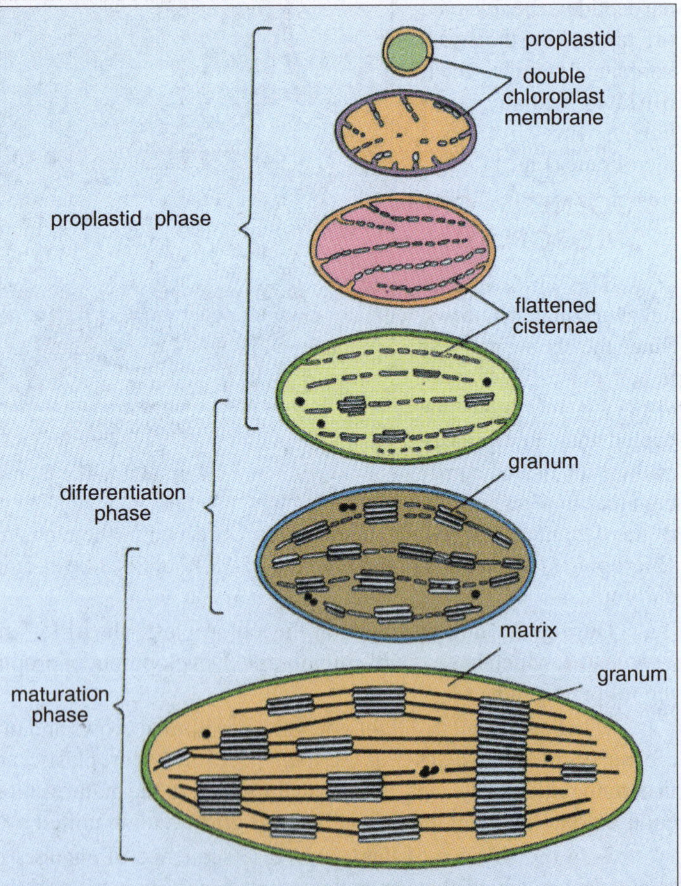

Fig. 11.13. Development of a chloroplast from a submicroscopic proplastid in the presence of light.

an autotrophic micro-organism, one which is able to transform radiant energy from sunlight and heterotrophic host cell. The symbiotic origin of the chloroplast appears very justified but **Kirk** (1966) has shown that certain important enzymes which are necessary for the development of the chlorophyll and for the photosynthetic mechanism are synthesized according to the codes of the nuclear DNA. There still exists certain doubt about the symbiotic origin of the chloroplast.

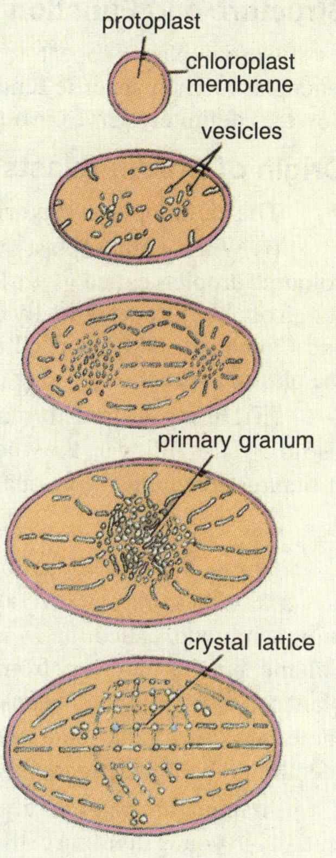

Fig. 11.14. Development of a chloroplast from submicroscopic proplastid in dark.

AMYLOPLASTS

Amyloplasts or starch granules are leucoplasts which lack any visible pigment and are involved in the synthesis and storage of various kinds of carbohydrates.

Structure and Function

The amyloplasts resemble proplastids and differ from them only in size. The contain less number of lamellae, but, can build up thylakoid structure found in chloroplast, in the presence of light. In the dark they synthesize and store starch. The conversion of glucose into starch is a chief characteristic of amyloplasts. However, chloroplasts can also synthesize starch and store it in stroma region. But, this starch in chloroplasts disappears quickly and is, therefore, known as **transitory starch**. The starch of amyloplasts on the other hand can be stored for longer periods and is, therefore, called **reserve starch**.

Origin of Amyloplasts

The starch development begins with the formation of a particle in the stroma. This particle consists of several tubuli and thylakoids and is surrounded by additional rings of starch. It grows till the amyloplast is filled with starch. Ultimately, the tubuli are pressed against the wall and gradually, starch granule exceeds the size of amyloplast. Finally, the membraneous wall of amyloplasts ruptures and withers away, so that only the starch granules can be seen.

CHROMOPLASTS

The plastids which contain pigments other than chlorophyll are the chromoplasts. They can be originated from chloroplasts and also from leucoplasts. The common example of their derivation from chloroplasts can be observed in the petals which are green initially but become coloured subsequently. Similarly in carrot roots, chromoplasts are derived from leucoplasts as is clear from the fact that the roots have no colour in the beginning but become coloured at a later stage.

Chromoplasts impart colour to roots (Carrot), fruit (Tomato) and petals (Rose).

Structure and Function

Chromoplasts appear to be products of degeneration or disintegration of chloroplasts or leucoplasts. Their specific function in plant cell is still little understood. However, their presence in floweres definitely serves in attracting insects for pollination and propagation.

Origin of Chromoplasts

The chromoplasts can originate either from chloroplasts or from leucoplasts :

(i) When a chloroplast transforms into a chromoplast, it has been observed that some yellow coloured droplets called **globulins** appear in the former. In the course of development chlorophyll and starch of chlorplasts gradually decrease, large globuli are formed, the lamellar structure breaks down and stroma is disorganized. Ultimately, globuli are arranged along plastid membrane and the centre of the plastid appears empty due to disintegration of stroma.

(ii) During the transformation of leucoplasts into chromoplasts, certain fibrils appear which give rise to crystals filling up the whole plastids. These crystals are normally found in the form of sheet-like structures containing large quantities of carotenoids.

VACUOLES

The most conspicuous compartment in most plant cells is a very large, fluid-filled vesicle called a **vacuole**. There may be several vacuoles in a single cell, each separated from the cytoplasm by a single unit membrane, called the **tonoplast**. Generally vacuoles occupy more than 30 per cent of the cell volume; but this may vary from 5 per cent to 90 per cent, depending on the cell type. Conventionally, plant cell biologists do not consider the vacuole to be part of the cytoplasm ; they tend to consider only three parts in a plant cell : nucleus, vacuole and cytoplasm—the latter containing all the other membrane-bounded organelles, including the plastids.

In immature and actively dividing plant cells the vacuoles are quite small. These vacuoles arise initially in young dividing cells, probably by the progressive fusion of vesicles derived from the Golgi apparatus. They are structurally and functionally related to lysosomes in animal cells and may contain a wide range of hydrolytic enzymes. In addition, they usually contain sugars, salts, acids and nitrogenous compounds such as alkaloids and anthocyanin pigments. The pH of plant vacuoles may be as high as 9 to 10 due to large quantities of alkaline substances or as low as 3 due to the accumulation of quantities of acids (*e.g.*, citric, oxalic and tartaric acids).

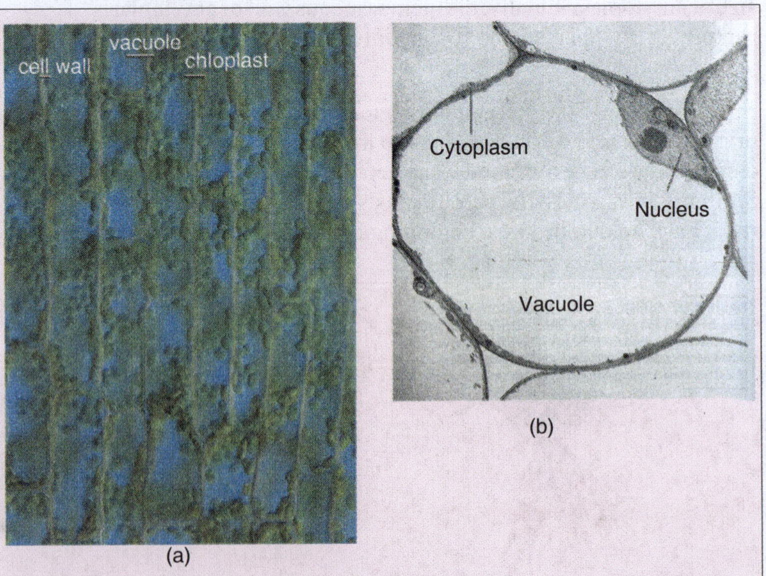

Plant cell vacuoles (a) Vacuoles seen in *Elodea* leaf (b) TEM of a cell showing vacuole.

Functions of Vacuoles

A plant vacuole has a variety of functions. It can act as a storage organelle for both nutrients and waste products ; as a lysosomal compartment (**Boller** and **Kende**, 1979), as a economical way of increasing the size of the cells, and as a controller of turgor pressure (To recall, turgor provides support for the individual plant cell and contributes to the rigidity of the leaves and younger parts of the plant). Different vacuoles with distinct functions (*e.g.*, lysosomal and storage) are often present in the same cell, we have already described the role of lysosomal vacuole, now, let us examine the storage function of a vacuole.

Storage functions of plant vacuoles. Plant vacuoles can store many type of molecules. In particular, they can sequester substances that are potentially harmful for the plant cell, if they are present in bulk in the cytoplasm. For example, the vacuoles of certain specialized cells contain such interesting products as rubber (in *Hevea brasiliensis*) or opium (in *Papaver somniferum*). Even Na+ ions are stored in these organelles, where their osmotic activity contributes to turgor pressure. Analysis of the giant cells of the alga *Nitella* indicates that **Na+ pumps** located in the tonoplast maintain low concentration of Na+ in the cytosol and four to five fold higher concentrations in the vacuole ; and since the vacuole occupies a much greater volume than the cytoplasm in *Nitella*, the greater bulk of cellular Na+ is in the vacuole.

The vacuole has an important **homeostatic function** in plant cells that are subjected to wide variations in their environment. For example, when the pH in the environment drops, the flux of H^+ into the cytoplasm is buffered by increased transport of H^+ into the vacuole. Similarly, many plant cells maintain turgor pressure at remarkable constant levels in the face of large changes in the tonicity of the fluids in their immediate environment. They do so by changing the somatic pressure of the cytoplasm and vacuole—in part by controlled breakdown and resynthesis of polymers such as **polyphosphate** in the vacuole, and in part by altering transport rate across the plasma membrane and the tonoplast. The permeability of these two membranes is partly regulated by turgor pressure and is determined by the distinct set of membrane transport proteins that transfer specific sugars, amino acids, and other metabolites across each lipid bilayer. The substances in the vacuole differ qualitatively and quantitatively from those in the cytoplasm. The tonoplast has little mechanical strength, however, the hydrostatic pressure must remain roughly equal in cytoplasm and vacuole, and two compartments must act together is osmotic balance to maintain turgor.

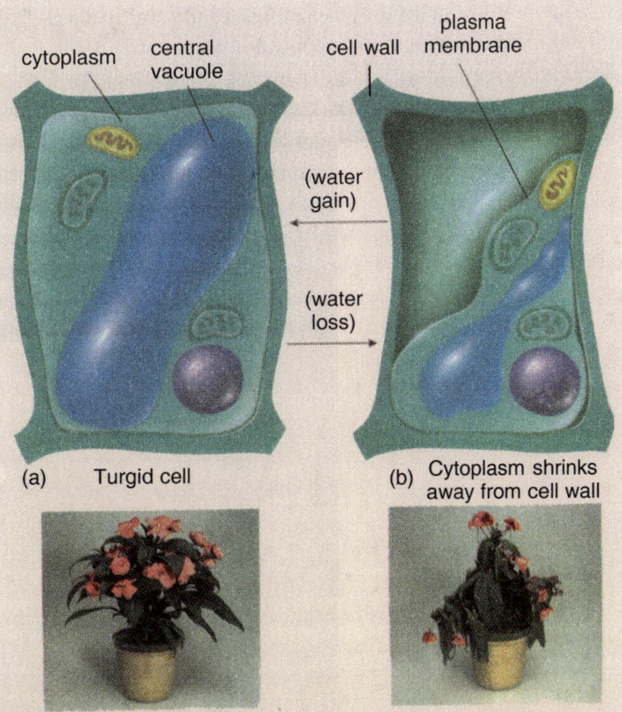

The central vacuole and turgor pressure in plant cells. (a) (top) when water is plentiful, it fills the central vacuole and help maintain the cell's shape (bottom). The pressure of water supports leaves (b)(top) when water is scarce, the central vacuole shrinks and cell becomes soft and shrunken (bottom) wilted plant.

Some of the products stored by vacuoles have a **metabolic function**. For example, succulent plants open their stomata and take up carbon dioxide at night (when transpiration losses are less than in the day) and convert it to **malic acid**. This acid is stored in vacuoles until the following day, when light energy can be used to convert it to sugar while the stomata are kept shut. As a second example, proteins can be stored for years as reserves for future growth in the vacuoles of the storage cells of many seeds, such as those of peas and beans. When the seeds germinate, the proteins are hydrolyzed and the amino acids are mobilized to form a food supply for the developing embryo.

Other molecules stored in vacuoles are involved in the interactions of the plant with animals or with other plants. The anthocyanin pigments, for example, colour the petals of some flowers so that they attract pollinating insects. Other molecules defend the plant against predators. Noxious metabolites released from vacuoles, when the cells are eaten or otherwise damaged, range from poisonous alkaloids to unpalatable inhibitors of digestion. The **trypsin inhibitors** commonly found in seeds and the **wound induced protease inhibitors** of leaf cells (to inhibit both insect and microbial proteases), both accumulate in the vacuole and are presumably designed to interfere with the digestive processes of herbivores.

REVISION QUESTIONS

1. What is the chloroplst? Describe the ultrastructure of the chloroplast. What are some of the more obvious similarities and differences between chloroplasts and mitochondria?
2. Define the following : granum, thylakoid, chromoplast, leucoplast, proplastid.
3. What part of chloroplast is associated with the light reaction ? Where dark reaction takes place?
4. Define the following : photon; a quantum of energy ; fluorescence; photoelectric effect. What are the possible fates of an absorbed photon ?
5. What is the Calvin cycle, and what is its purpose? Describe it.
6. Write an essay on "chloroplast and photosynthesis."
7. Write short notes on the following :

   (*i*) Chromatophores; (*ii*) Quantosome concept; (*iii*) Pyrenoid; (*iv*) Photosynthetic pigments of chloroplasts; (*v*) Dark reaction; and (*vi*) Origin of chloroplasts.
8. Describe the molecular organization of the thylakoids.
9. Write an account of structure and function of plant vacuoles.

CHAPTER

12

Nucleus

The nucleus (L., *nux* = nut) is the heart of the cell. It is here that almost all of the cell's DNA is confined, replicated and transcribed. The nucleus, thus, controls different metabolic as well as hereditary activities of the cell. A synonymous term for this organelle is the Greek word **karyon**. Nucleus serves as the main distinguishing feature of eukaryotic cells, *i.e.*, this is the true nucleus as opposed to the nuclear region, prokaryon or nucleoid of the prokaryotic cells (see Chapter 3). The following statement of **Vincent Allfrey** (1968) completely qualifies the central position of the nucleus in the affairs of an eukaryotic cell :

"*The cell nucleus, central and commanding, is essential for the biosynthetic events that characterize cell type and cell fraction; it is a vault of genetic information encoding the past history and future prospects of the cell, an organelle submerged and deceptively serene in its sea of turbulent cytoplasm, a firm and purposeful guide, a barometer exquisitely sensitive to the changing demands of the organism and its environment. This is our subject — to be examined in terms of its ultrastructure, composition and function.*"

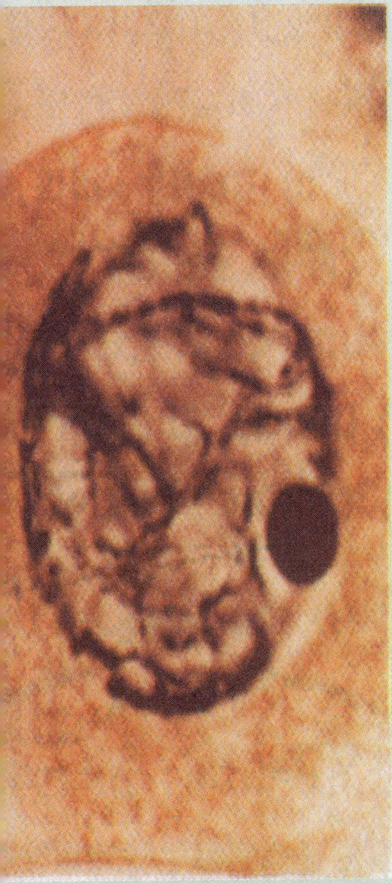

Interphase nucleus.

HISTORICAL

Nuclei were first discovered and named by **Robert Brown** in 1833 in the plant cells and were quickly recognized as a constant feature of all animal and plant cells. Nucleoli were described by **M.J. Schleiden** in 1838, although first noted by **Fontana** (1781). The term nucleolus was coined by **Bowman** in 1840. In 1879, **W. Flemming** coined the term chromatin for chromosomal meshwork. **Strasburger** (1882) introduced the terms cytoplasm and nucleoplasm. The existence of a membrane delimiting the nucleus was first demonstrated by **O.Hertwig** in 1893. In 1934, **Barbara McClintock** recognized and named nucleolar organizers in the chromosomes. In 1950, **Callan** and **Tomlin**, first observed the nucleopores in the nuclei of amphibian oocytes. The role of nucleus in heredity was firmly established by the grafting experiments of **Hammerling** (1953) with *Acetabularia*. Ultrastructure of nuclear envelope, pore complexes and nuclear lamina were worked out by **Kirschner** *et al*., (1977), **Schatten** and **Thoman** (1978), etc.

NUCLEO - CYTOPLASMIC INTERRELATIONSHIP

The evidences for nucleo-cytoplasmic communication as a factor in cell maintenance and development have been known before the rediscovery of Mendel's "genes". In the late nineteenth century, **Verworm**, **Balbiani** and others showed that following microsurgery, nucleated halves of various protozoans survived and grew, whereas the enucleated halves degenerated and died. Later, in the 1930s it was shown that insertion of nuclei into enucleated amoebae restored pseudopodial activity, feeding behaviour and growth. It was also shown that the nucleus was essential for the growth and regeneration of the morphologically complicated ciliate *Stentor*. In a classical series of experiments, spanning between 1934 and 1954, on the unicellular alga *Acetabularia*, **Hammerling** demonstrated by means of interspecific nuclear transplants, that morphological features, notably the shape of cap, were determined by the nucleus. He also showed that even after removal of the nucleus, the cell was able to continue morphogenesis for a time and proposed that the cytoplasm contained a store of morphogenetic material (later on rec-

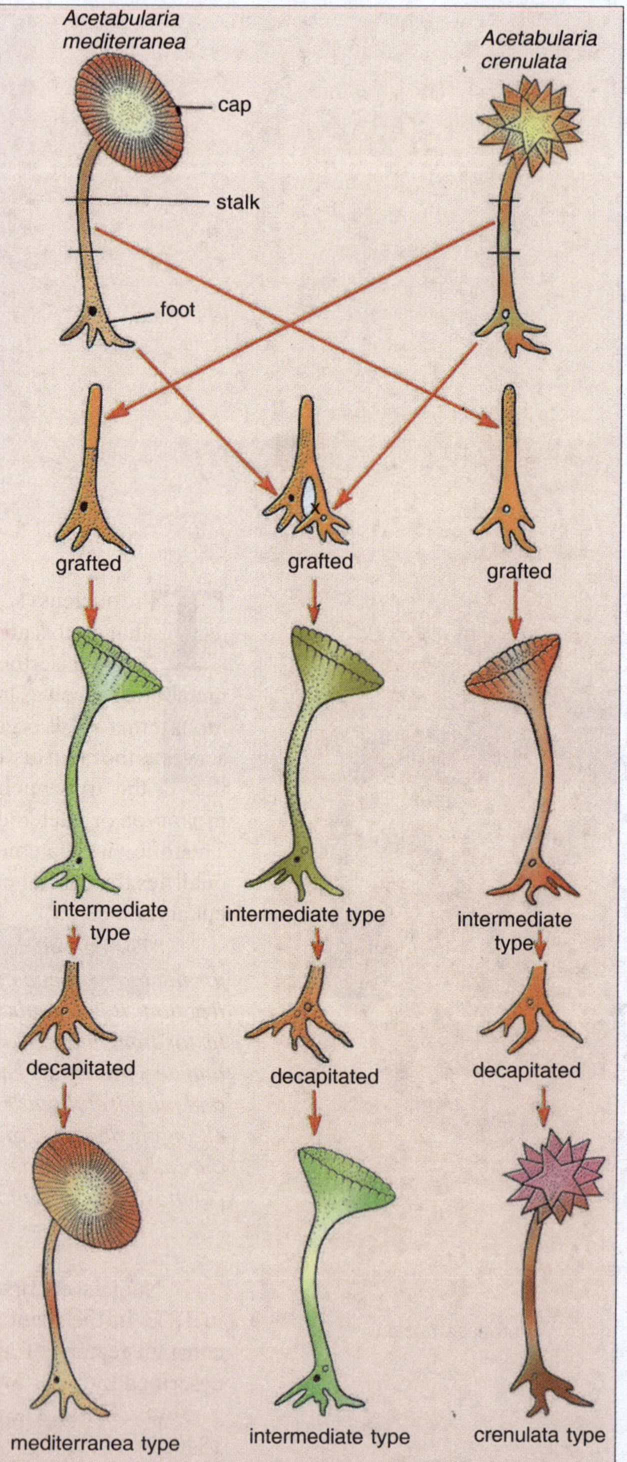

Fig. 12.1. Summary of the grafting experiments with *Acetabularia* to prove that the hereditary characters are determined by the nucleus and not by the cytoplasm.

ognized as mRNA molecules) that had been produced by the nucleus. Let us closely examine the Hammerling's classical nuclear transplantation experiments :

Hammerling's expe-riment. The body of an alga *Acetabularia* is about six centimeters long and is differentiated into a foot, a stalk and a cap. The cap has a characteristic shape for each species and is easily regenerated if removed. The single nucleus is situated in the rhizoid portion. *Acetabularia crenulata* has a cap, with about 31 rays, the tips of which are pointed, but *Acetabularia mediterranea* has about 81 rays with rounded tips. If the cap, stalk or even the nucleated portion of the rhizoid is removed, the remaining portion of the alga has the capacity to regenerate into a whole plant. The enucleated part loses the regeneration capacity after a few decapitations, but the nucleated portion always maintains this ability. When the stalk of one species is grafted on to the nucleated rhizoid of the other, an **intermediate type** of cap is formed. On decapitation, a second cap develops which resembles the cap of species which provides the nucleus (Fig. 12.1). When the nuclei of both the species are present in the same cytoplasm, an intermediate type of cap develops. Such experiments have clearly established that the nucleus is the storehouse for and the control tower of, all hereditary information.

OCCURRENCE AND POSITION

The nucleus is found in all the eukaryotic cells of the plants and animals. However, certain eukaryotic cells such as the mature sieve tubes of higher plants and mammaliam erythrocytes contain no nucleus. In such cells nuclei are present during the early stages of development. Since mature mammalian red blood cells are without any nuclei, they are called red blood "corpuscles" rather than cells (*L. corpus* = body, especially dead body or corpse).

The prokaryotic cells of the bacteria do not have true nucleus, *i.e.*, the single, circular and large DNA molecule remains in direct contact with the cytoplasm. The position or location of the nucleus in a cell is usually the characteristic of the cell type and it is often variable. Usually the nucleus remains located in the centre. But its position may change from time to time according to the metabolic states of the cell. For example, in the embryonic cells the nucleus generally occupies the geometric centre of the cell but as the cells start to differentiate and the rate of the metabolic activities increases, the displacement in the position of the nucleus takes place. In certain cells such as the glandular cells the nucleus remains located in the basal portion of the cell.

MORPHOLOGY

Number

Usually the cells contain single nucleus but the number of the nucleus may vary from cell to cell. According to the number of the nuclei following types of cells have been recognised :

1. Mononucleate cells. Most plant and animal cells contain single nucleus, such cells are known as **mononucleate cells.**

2. Binucleate cells. The cells which contain two nuclei are known as **binucleate cells**. Such cells occur in certain protozoans such as *Paramecium* and cells of cartilage and liver.

3. Polynucleate cells. The cells which contain many (from 3 to 100) nuclei are known as **polynucleate cells**. The polynucleate cells of the animals are termed as **syncytial cells**, while the polynucleate cells of the plants are known as **coeno-cytes**. The most common example of the syncytial

The nucleus surrounded by nuclear membrane. The pits are nuclear pores.

cells are the osteoblast (polykaryocytes of the bone morrow) which contain about 100 nuclei per cell and striated muscle fibres each of which contains many hundred nuclei. The siphonal algae *Vaucheria* contains hundreds of nuclei and certain fungi are the best example of the coenocytic plant cells.

Shape

The shape of the nucleus normally remains related with the shape of the cell, but certain nuclei are almost irregular in shape. The spheroid, cuboid or polyhedral cells (isodiametrical cells) contain the **spheroid** nuclei. The nuclei of the cylindrical, prismatic or fusiform cells are **ellipsoid** in shape. The cells of the squamous epithelium contain the **discoidal** nuclei. The leukocytes, certain infusoria, glandular cells of some insects and spermatozoa contain the irregular shaped nuclei. Nuclei of cells of silk glands of silk worm have finger-like extensions that greatly increase their surface area (Fig. 12.2).

Size

Generally nucleus occupies about 10 per cent of the total cell volume. Nuclei vary in size from about 3 μm to 25 μm in diameter, depending on cell type and contain diploid set of chromosomes. The size of the nucleus is directly proportional to that of the cytoplasm. **R. Hertwig** has given the following formula for the deduction of the size of the nucleus of a particular cell.

$$NP = \frac{V_n}{V_c - V_n}$$

Where NP is the nucleoprotein index, V_n is the volume of the nucleus and V_c is the volume of the cell.

Moreover, the size of the nucleus is related with the number of the chromosomes or ploidy. The haploid cells contain small-sized nuclei in comparison to the nuclei of the diploid cells. Likewise the polyploid cells contain larger nuclei than the diploid cells. Thus, the size of the nucleus of a cell depends on the volume of the cell, amount of the DNA and proteins and metabolic phase of the cell.

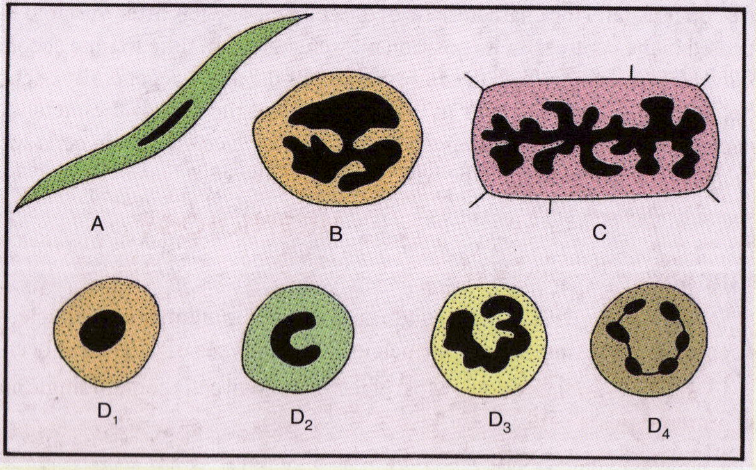

Fig. 12.2. Different shapes of the nucleus in animal cells. A—Elongated in muscle cell; B—Lobed in a human neutrophil cell; C—Branched in a silk spinning cell of an insect larva; D1 to D4—Variable shape in leucocytes.

ISOLATION TECHNIQUES

The number of nuclei can be measured easily in a chamber similar to that used for blood counts. For the isolation of nuclear envelope, nuclei are first of all separated from the rest of the cell. This is normally accomplished by disrupting tissue in homogenizers wherein the clearance is such that nuclei are not broken but the plasma membrane and endoplasmic reticulum are severely disrupted. Nuclei can then be harvested by differential centrifugation. They are then lysed by sonication and their envelopes separated on density gradient centrifugation.

Alternatively, DNAase digestion followed by extraction with salts releases envelopes which again can be banded on sucrose or cesium chloride gradients (see **Thorpe**, 1984).

The chemical organization of the nucleus has been investigated by two main approaches. The first, which is essentially biochemical, consists of isolating a large enough number of nuclei to permit analysis by biochemical methods. The second approach, which is essentially cytologic, uses the cytophotometric and radio-autographic methods. A nucleolar fraction may be obtained by treating the nuclei with highly ionic solutions and digesting the chromatin with DNAse (**Penman** *et al.,* 1966). The amount of DNA in an eukaryotic nucleus can be determined by microspectrophotometry, a technique that measures with precision the amount of Feulgen staining material each nucleus contains.

ULTRASTRUCTURE

The nucleus is composed of following structures : 1. The nuclear membrane or karyotheca or nuclear envelope; 2. The nuclear sap or nucleoplasm; 3. The chromatin fibres; and 4. The nucleolus.

1. Nuclear Envelope

The nuclear envelope (or perinuclear cisterna) encloses the DNA and defines the nuclear compartment of interphase and prophase nuclei. It is formed from two concentric unit membranes, each 5–10 nm thick. The spherical **inner nuclear membrane** contains specific proteins that act as binding sites for the supporting fibrous sheath of intermediate filaments (IF), called **nuclear lamina**. Nuclear lamina has contact with the chromatin (or chromosomes) and nuclear RNAs. The inner nuclear membrane is surrounded by the **outer nuclear membrane**, which closely resembles the membrane of the endoplasmic reticulum, that is continuous with it. It is also surrounded by less organized intermediate filaments (Fig. 12.3). Like the membrane of the rough ER, the outer surface of outer nuclear membrane is generally studded with ribosomes engaged in protein synthesis. The proteins made on these ribosomes are transported into space between the inner and outer nuclear membrane, called **perinuclear space**. The perinuclear space is a 10 to 50 nm wide fluid-filled compartment which is continuous with the ER lumen and may contain fibres, crystalline deposits, lipid droplets or electron-dense material (see **Thorpe**, 1984).

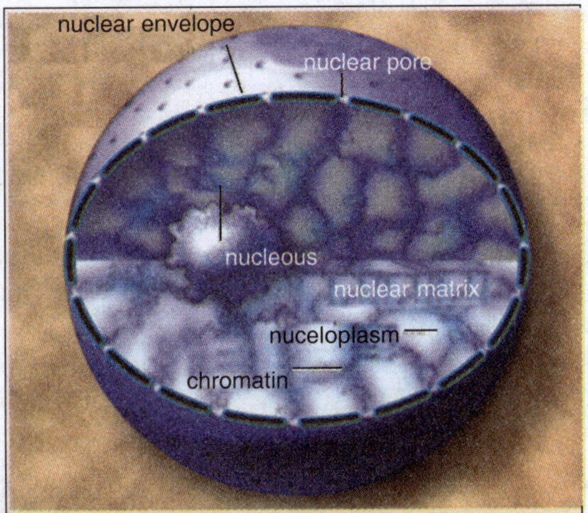

The cell nucleus.

Nuclear lamina. It is also called **fibrous lamina**, **zonula nucleum limitans**, **internal dense lamella**, **nuclear cortex** and **lamina densa**.

The nuclear lamina is a protein meshwork which is 50 to 80 nm thick (**DeRobertis** and **De Robertis, Jr**., 1987) or 10 to 20 nm thick (**Alberts** *et al.,* 1987). It lines the inside surface of the inner nuclear membrane, except the areas of nucleopores, and consists of a square lattice of intermediate filaments. In mammals, these intermediate filaments are of three types : **lamins A**, **B** and **C** having M.W. 74,000, 72,000 and 62,000 daltons, respectively. The lamins form dimers that have a rod-like domain and two globular heads at one end. Under appropriate conditions of pH and ionic strength, the dimers spontaneously associate into filaments that have a diameter and repeating structure similar to those of cytoplasmic filaments.

The nuclear lamina is a very dynamic structure. In mammalian cells undergoing mitosis, the transient phosphorylation of several serine residues on the lamins causes the lamina to reversibly disassemble into tetramers of hypophosphorylated lamin A and lamin C and membrane associated lamin B. As a result, lamin A and C become entirely soluble during mitosis, and at telophase they become dephosphorylated again and polymerize around chromatin. Lamin B seems to remain associated with membrane vesicles during mitosis, and these vesicles in turn remain as a distinct subset of membrane components from which nuclear envelope is reassembled at telophase. Inside an interphase nucleus, chromatin binds strongly to the inner part of the nuclear lamina which is believed to interfere with chromosome condensation. In fact, during meiotic chromosome condensation, the nuclear lamina completely disappears by the pachytene stage of prophase and reappears later during diplotene in oocytes, but does not reappear at all in spermatocytes.

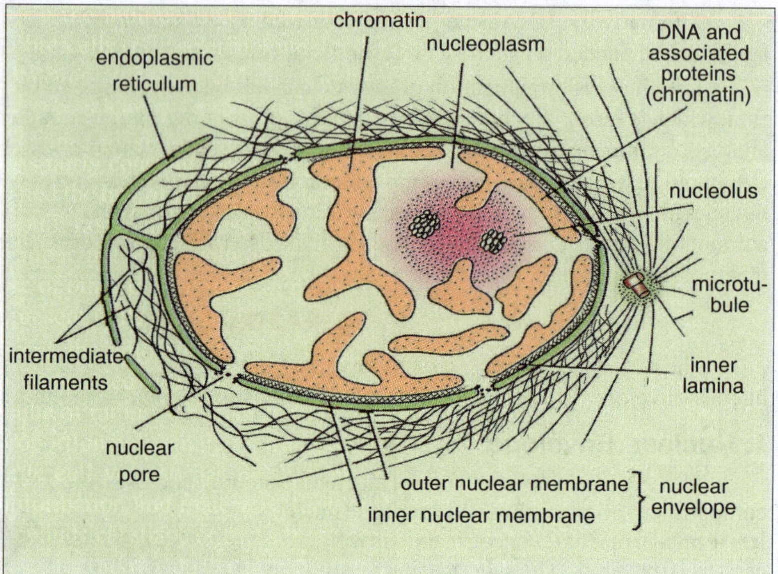

Fig. 12.3. Cross-section of a typical cell nucleus showing ultrastructure (after Alberts *et al.,* 1989).

The lamins may play a crucial role in the assembly of interphase nuclei. For example, when cells are left for a long time in colchicine (drug which arrests cells in metaphase), the lamins assemble around individual chromosomes, which then surrounded by nuclear envelopes give rise to micronuclei containing only one chromosome. A similar phenomenon occurs during normal amphibian development. In the first few cleavages of amphibian development, the nuclear envelope initially forms around individual chromosomes, forming several vesicles that then fuse together to form a single nucleus. This suggests that chromatin is the nucleating centre for the deposition of a nuclear lamina and envelope.

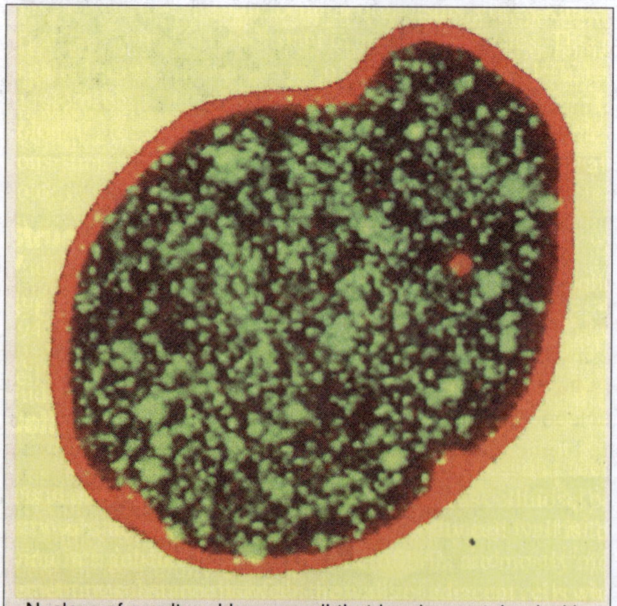

Nucleus of a cultured human cell that has been stained with fluorescently labeled antibodies to reveal the nuclear lamina (red) which lies on the inner surface of the nuclear envelope.

Nuclear pores and nucleocytoplasmic traffic. The nuclear envelope in all eukaryotic forms, from yeasts to humans, is perforated by **nuclear pores** which have the following structure and function :

1. Structure of nuclear pores.

Nuclear pores appear circular in surface view and have a diameter between 10nm to 100 nm. Previously it was believed that a diaphragm made of amorphous to fibrillar material extends across each pore limiting free transfer of material. Such a diaphragm called **annulus** has been observed in animal cells, but lacking in plant cells. Recent electron microscopic studies have found that a nuclear pore has far more complex structure, so it is called **nuclear pore complex**. Each pore complex has an estimated molecular weight of 50 to 100 million daltons. Negative staining techniques have demonstrated that pore complexes have an eight-fold or octagonal symmetry. More recent computerized image-processing techniques of **Unwin** and **Mulligan** (1982) have shown that the pore complex consists of two "**rings**" (**R** or annuli) at its periphery with an inside diameter of 80 nm, a large particle that forms a **central plug** (**C**) and radial '**spokes**' (**S**) that extends from the plug to the rings (Fig. 12.5 B and C). **Particles** (**P**) are anchored to cytoplasmic ring and are thought to be inactive ribosomes. The '**hole**' in the centre of the pore complex is an aqueous channel through which water-soluble molecules shuttle between the nucleus and the cytoplasm. This hole often appears to be plugged by a large central granule (**central plug**) which is believed to consist of newly made ribosomes or other particles caught in transit.

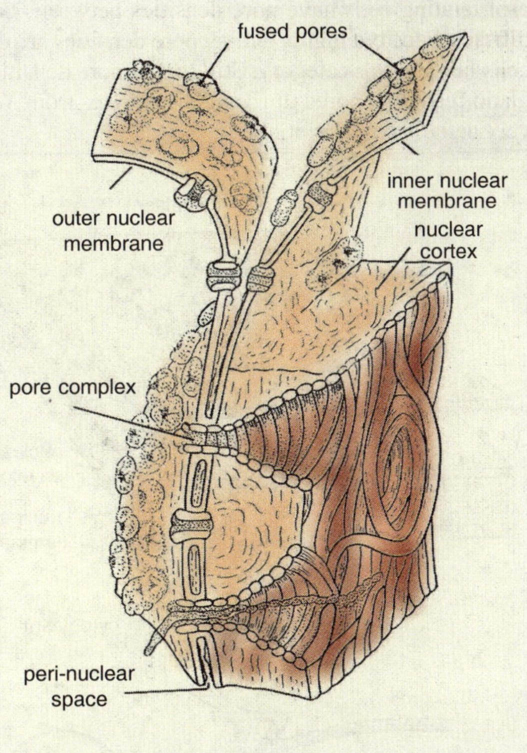

Fig. 12.4. A model of the nuclear surface complex as proposed by **Schatten** and **Thoman** (1978) (after Thorpe, 1984).

The pore complex perforates the nuclear envelope bringing the lipid bilayers of the inner and outer nuclear membrane together around the margins of each pore. Despite this continuity, which would seem to provide a pathway for the diffusion of membrane components between the inner and outer membranes, the two membranes remain chemically distinct.

Quite recently, following two proteins have been found to be associated to the nuclear pores : one is an integral membrane protein, a glycoprotein of 120,000 daltons that may anchor the annuli to the lipid bilayer (**Gerace** *et al.*, 1982). The second protein is a 63,000 dalton protein (that has covalently bound acetyl neuraminic acid) located on the cytoplasmic side of the electron-dense material that occludes the nuclear pores (**Davis** and **Blobel**, 1986). This protein may be involved in the transport of materials through the nuclear pores.

2. Number of nuclear pores (Pore density). In nuclei of mammals it has been calculated that nuclear pores account for 5 to 15 per cent of the surface area of the nuclear membrane. In amphibian oocytes, certain plant cells and protozoa, the surface occupied by the nuclear pores may be as high as 20 to 36 per cent. The number of pores in the nuclear envelope or **pore density** seems to correlate with the transcriptional activity of the cell (Table 12-1). Thus, pore densities as low as ~3 pores/μm^2 are seen in nucleated red blood cells and lymphocytes (which are inactive in transcription). These cells are highly differentiated but metabolically inactive and they are non- proliferating cells. The majority of proliferating cells have pore densities between 7 and 12 pores/μm². Among cells of a third type, differentiated but highly active, pore densities are often 15 to 20 pores/μm². Liver, kidney and brain cells fall into this category. Still higher pore densities are found in specialized cells, such as salivary gland cells (~40 pores/μm²) and the oocytes from *Xenopus laevis* (~50 pores/μm²), both of which are very active in transcription.

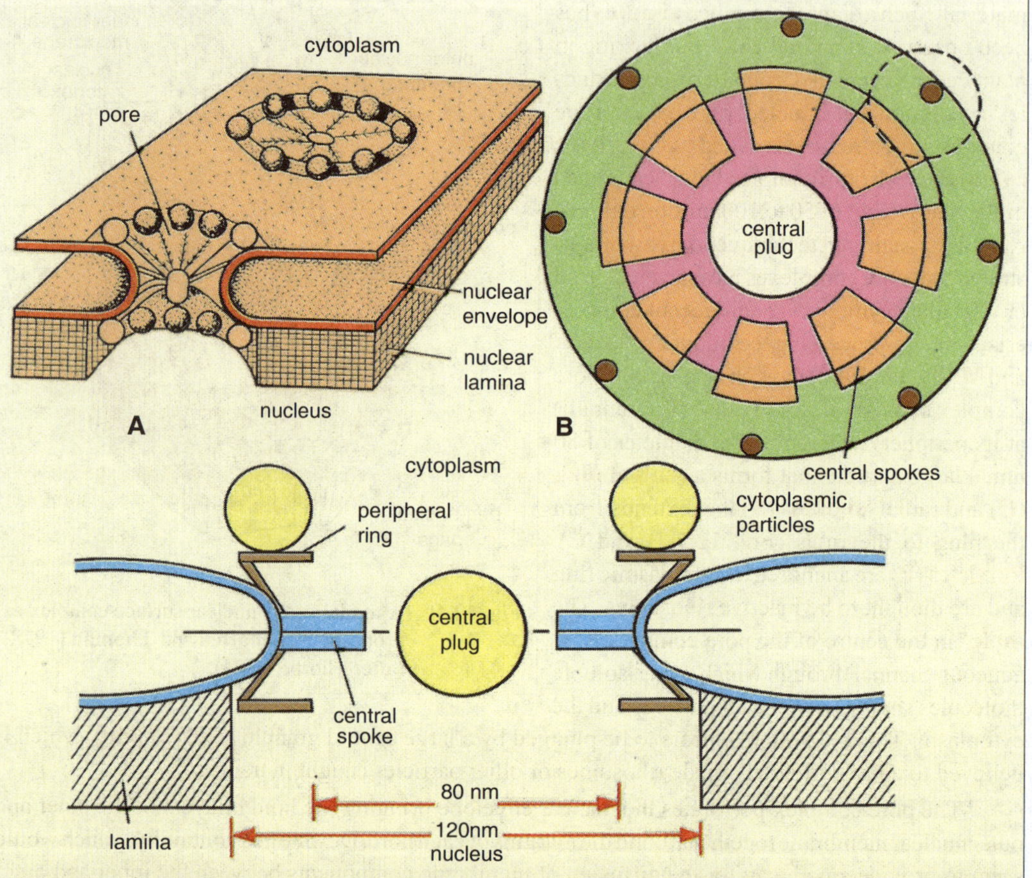

Fig. 12.5. Nuclear pore. A—A part of nuclear envelope with 80 nm pores occupied by pore complexes of octogonal radial symmetry; B—Nuclear pore in top view; C—A nuclear pore in cross section (after De Robertis and De Robertis, Jr., 1987).

Table 12-1. Pore densities and nuclear pores per nucleus in various cell types (source : Thorpe, 1984).

Cell type	Pore density (Pores/μm²)	Pores per nucleus
1. Human lymphocytes	3.2	405
2. Leopard frog embryo	5.6	1729
3. Human embryonic lung	8.5	2788
4. African green monkey kidney	8.6	4277
5. Newt heart	7.6	12,707
6. *Xenopus laevis* oocyte	51	37.7 x 10⁶
7. *Triturus alpestris* oocyte	50	57 x 10⁶

3. Arrangement of nuclear pores on nuclear envelope. In somatic cells, the nuclear pores are evenly or randomly distributed over the surface of nuclear envelope. However, pore arrangement in other cell types is not random but rather range from **rows** (*e.g.,* spores of *Eqisetum*) to **Clusters** (*e.g.,* oocytes of *Xenopus laevis*) to **hexagonal** (*e.g.,* Malpighian tubules of leaf hoppers) packing order (see **Thorpe**, 1984).

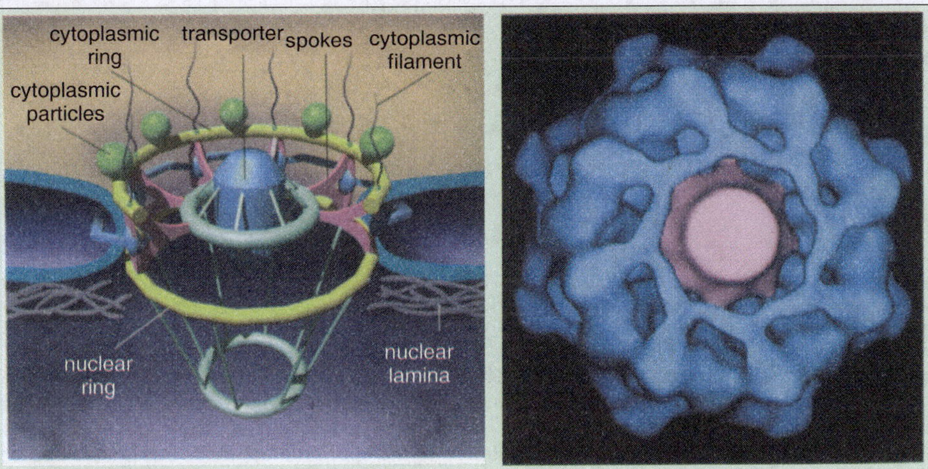

(a) A three dimensional representation of a vertebrate nuclear pore complex.

(b) A three dimensional model based on high resolution electron microscopy showing nuclear pore complex.

4. Nucleo-cytoplasmic traffic. Quite evidently there is considerable trafficking across the nuclear envelope during interphase. Ions, nucleotides and structural, catalytic and regulatory proteins are imported from the cytosol (cytoplasmic matrix); mRNA, tRNA and ribosome subunits are exported to the cytosol (cytoplasmic matrix) (see **Reid** and **Leech**, 1980). However, one of the main functions of the nuclear envelope is to prevent the entrance of active ribosomes into the nucleus.

The pore appears to function like a close fitting diaphragm that opens to just the right extent when activated by a signal on an appropriate large protein (having a diameter up to 20 nm). Recently, it has been investigated that the nuclear-specific proteins (called **karyophilic proteins**) have in their molecular structure some type of signals, called **karyophilic signals** or **nuclear import signals**, that enable them to accumulate selectively in the nucleus. For example, **nucleoplasmin** is an abundant, pentameric nuclear protein having distinct head and tail domains. Nucleoplasmins are actively transported through the nuclear pores, probably while still in their folded form. The karyophilic signal for such a nuclear import apparently resides in the tail domains and such an active nuclear transport requires energy which is derived from ATP hydrolysis. Similar signals are also noted in a short sequence (126-132 amino acids) of **simian virus 40 T** antigen molecule. These short sequences when attached to bigger molecules (even to metal particles such as gold) allow these bigger molecules to enter the nucleus via the nuclear pores.

5. Rate of transport through the nuclear pores. As we have already described, the nuclear envelope of a typical mammalian cell contains 3000 to 4000 pores (about 11 pores/ μm^2 of membrane area). If the cell is synthesizing DNA, it needs to import about 10^6 histone molecules from the cytoplasm every 3 minutes in order to package newly made DNA into chromatin, which means that on an average each pore needs to transport about 100 histone molecules per minute. Further, if the cell is growing rapidly, each nuclear pore needs to export about three newly assembled ribosomes per minute to the cytoplasm, since ribosomes are produced in nucleus but function in the cytoplasm. The export of new ribosomal subunits is particularly problematic since these particles are about 15 nm in diameter and are much too large to pass through the 9 nm channels of nuclear pores, it is believed that they are specifically exported through the nuclear pores by an active transport system. Similarly, mRNA molecules complexed with special proteins to form ribonucleoprotein particles, are thought to be actively exported from the nucleus.

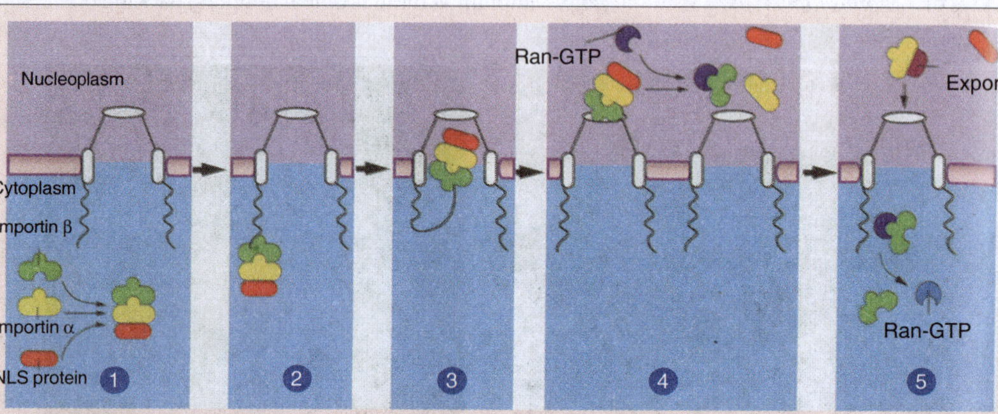

Importing proteins from the cytoplasm into the nucleus. The protein bearing a nuclear localization signal (NLS) binds to the heterodimeric receptor (importin α/β) (step 1) forming a complex that associates with a cytoplasmic filament (step 2). The protein complex is moved through the nuclear pore (step 3) and into the nucleoplasm where it interacts with Ran-GTP and dissociates (step 4). The importin β subunit is transported back to the cytoplasm in association with Ran-GTP (step 5), where the Ran-GTP is hydrolyzed. Ran-GDP is subsequently transported back to the nucleus and importin α back to the cytoplasm.

Lastly, nuclear pores are not the only avenues for nucleocytoplasmic exchanges (Fig. 12.6). For example, small molecules and ions readily permeate both nuclear membranes. Larger molecules and particles may pass through the membrane by formation of small pockets and vesicles that traverse the envelope and empty on the other side.

2. Nucleoplasm

The space between the nuclear envelope and the nucleolus is filled by a transparent, semi-solid, granular and slightly acidophilic ground substance or the matrix known as the **nuclear sap** or **nucleoplasm** or **karyolymph**. The nuclear components such as the chromatin threads and the nucleolus remain suspended in the nucleoplasm.

The nucleoplasm has a complex chemical composition. It is composed of mainly the nucleoproteins but it also contains other inorganic and organic substances, *viz*., nucleic acids, proteins, enzymes and minerals.

1. Nucleic acids. The most common nucleic acids of the nucleoplasm are the DNA and RNA. Both may occur in the macromolecular state or in the form of their monomer nucleotides.

2. Proteins. The nucleoplasm contains many types of complex proteins. The nucleoproteins can be categorized into following two types :

(i) Basic proteins. The proteins which take basic stain are known as the basic proteins. The most important basic proteins of the nucleus are **nucleoprotamines** and the **nucleohistones**.

The nucleoprotamines are simple and basic proteins having very low molecular weight (about 4000 daltons). The most abundant amino acid of these proteins is **arginine** (pH 10 to 11). The protamines usually remain bounded with the DNA molecules by the salt linkage. The protamines occur in the spermatozoa of the certain fishes. The nucleohistones have high molecular weight, *e.g.*, 10,000 to 18,000 daltons. The histones are composed of basic amino acids such as **arginine**, **lysine** and **histidine**. The histone proteins remain associated with the DNA by the ionic bonds and they occur in the nuclei of most organisms. According to the composition of the amino acids following types of histone proteins have been recognised, *e.g.*, histones rich in lysine, histones with arginine and histones with poor amount of the lysine.

(ii) Non-histone or Acidic proteins. The acidic proteins either occur in the nucleoplasm or in the chromatin. The most abundant acidic proteins of the euchromatin (a type of chromatin) are the **phosphoproteins**.

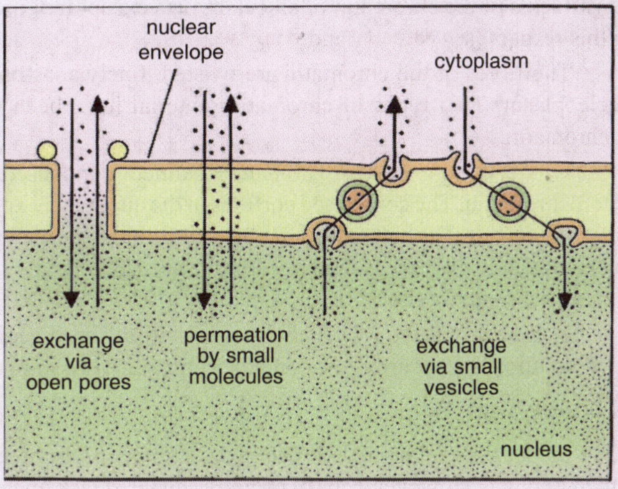

Fig. 12.6. Various avenues for transport of material from the nucleus to the cytosol (cytoplasmic matrix) (after Sheeler and Bianchi, 1987).

3. Enzymes. The nucleoplasm contains many enzymes which are necessary for the synthesis of the DNA and RNA. Most of the nuclear enzymes are composed of non-histone (acidic) proteins. The most important nuclear enzymes are the **DNA polymerase**, **RNA polymerase**, **NAD synthetase**, **nucleoside triphosphatase**, **adenosine diaminase**, **nucleoside phosphorylase**, **guanase**, **aldolase**, **enolase**, **3-phosphoglyceraldehyde dehydrogenase** and **pyruvate kinase**. The nucleoplasm also contains certain cofactors and coenzymes such as **ATP** and **acetyl CoA**.

4. Lipids. According to **Stoneburg** (1937) and **Dounce** (1955), the nucleoplasm contains small lipid content.

5. Minerals. The nucleoplasm also contains several inorganic compounds such as phosphorus, potassium, sodium, calcium and magnesium. The chromatin comparatively contains large amount of these minerals than the nucleoplasm.

3. Chromatin Fibres

The nucleoplasm contains many thread-like, coiled and much elongated structures which take readily the basic stains such as the basic fuchsin. These thread-like structures are known as the **chromatin** (*Gr., chrome*=colour) **substance** or **chromatin fibres**. Such chromatin fibres are observed only in the interphase nucleus. During the cell division (mitosis and meiosis) chromatin fibres become thick ribbon-like structures which are known as the **chromosomes**.

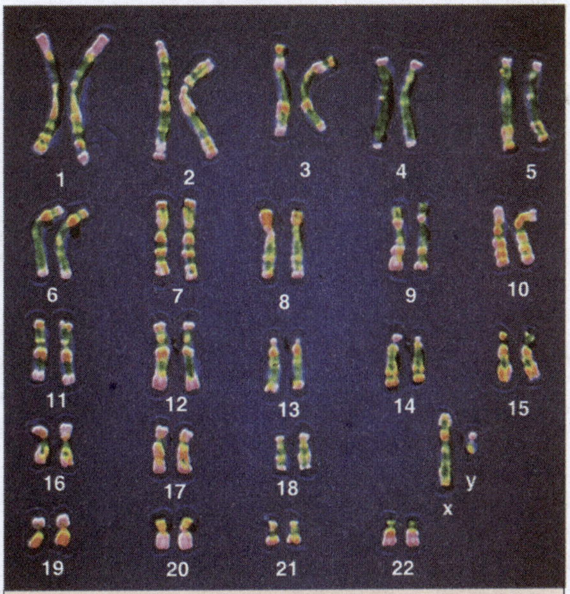

The 23 pairs of chromosomes in the human karyotype.

Chemically, chromatin consists of DNA and proteins. Small quantity of RNA may also be present but the RNA rarely accounts for more than about 5 per cent of the total chromatin present. Most of the protein of chromatin is histone, but "nonhistone" proteins are also present. The protein : DNA weight ratio averages about 1:1. Histones are constituents of the chromatin of all eukaryotes except fungi, which, therefore, resemble prokaryotes in this respect (see **Sheeler** and **Bianchi**, 1987).

The fibres of the chromatin are twisted, finely anastomosed and uniformly distributed in the nucleoplasm. Two types of chromatin material have been recognised, *e.g.*, heterochromatin and euchromatin.

A. Heterochromatin. The darkly stained, condensed region of the chromatin is known as heterochromatin. The condensed portions of the nucleus are known as **chromocenters** or **karyosomes** or **false nucleoli**. The heterochromatin occurs around the nucleolus and at the periphery. It is supposed to be metabolically and genetically inert because it contains comparatively small amout of the DNA and large amount of the RNA.

B. Euchromatin. The light stained and diffused region of the chromatin is known as the euchromatin. The euchromatin contains comparatively large amount of DNA.

4. Nucleolus

Most cells contain in their nuclei one or more prominent spherical colloidal acidophilic bodies, called **nucleoli**. However, cells of bacteria and yeast lack nucleolus. The size of the nucleolus is found to be related with the synthetic activity of the cell. Therefore, the cells with little or no synthetic activities, *e.g.*, sperm cells, blastomeres, muscle cell, etc., are found to contain smaller or no nucleoli, while the oocytes, neurons and secretory cells which synthesize the proteins or other substances contain comparatively large-sized nucleoli. The number of the nucleoli in the nucleus depends on the species and the number of the chromosomes. The number of the nucleoli in the cells may be one, two or four. The position of the nucleolus in the nucleus is eccentric.

A nucleolus is often associated with the **nucleolar organizer (NO)** which represents the secondary constriction of the nucleolar organizing chromosomes, and are 10 in number in human beings (Fig. 12.7). In corn, *Zea mays* chromosome 9 and 6 contain 'darkly staining knobs' or nucleolar organizers (**Heitz** and **McClintock**, 1930s). Nucleolar organizer consists of the genes for 18S, 5.8S and 28S rRNAs. The genes for fourth type of r RNA, *i.e.,* 5S rRNA occur outside the nucleolar organizer.

1. Chemical composition of nucleolus. Nucleolus is not bounded by any limiting membrane; calcium ions are supposed to maintain its intact organization. Chemically, nucleolus contains DNA of nucleolar organizer, four types of rRNAs, 70 types of ribosomal proteins, RNA binding proteins (*e.g.,* nucleolin) and RNA splicing nucleoproteins (U_1, U_2......U_{12}). It also contains phospholipids, orthophosphates and Ca^{2+} ions. Nucleolus also contains some enzymes such as acid phosphatase, nucleoside phosphorylase and NAD^+synthesizing enzymes for the synthesis of some coenzymes, nucleotides and ribosomal RNA. **RNA methylase** enzyme which transfers methyl groups to the nitrogen bases, occurs in the nucleolus of some cells.

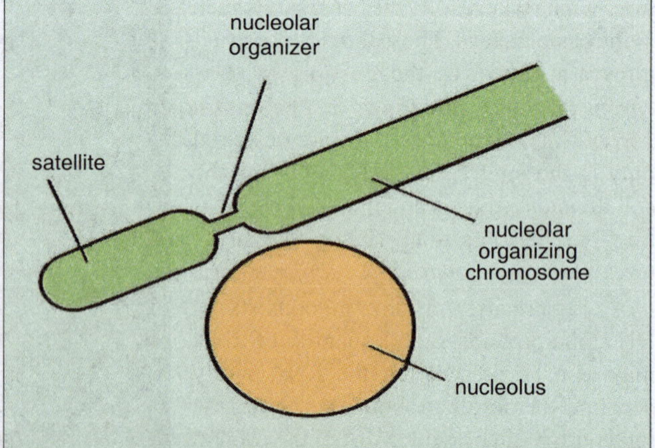

Fig 12.7. A satellited chromosome and an attached nucleolus.

2. Ultrastructure and function of nucleolus. Nucleolus are the sites where biogenesis of ribosomal subunits (*i.e.,* 40S and 60S) takes place. In it three types of rRNAs, namely 18S, 5.8S and 28S r RNAs, are transcribed as parts of a much longer precursor molecule (45S transcript) which undergoes processing (RNA splicing, for example) by the help of two types of proteins such as nucleolin and U3 sn RNP (U3 is a 250 nucleotide containing RNA, sn RNP represents small nuclear ribonucleoprotein). The 5S r RNA is transcribed on the chromosome existing outside the nucleolus and the 70 types of ribosomal proteins are synthesized in the cytoplasm. All of these components of the ribosomes migrate to the nucleolus, where they are assembled into two types of ribosomal subunits which are transported back to the cytoplasm. The smaller (40S) ribosomal subunits are formed and migrate to the cytoplasm much earlier than larger (60S) ribosomal subunits; therefore, nucleolus contains many more incomplete 60S ribosomal subunits than the 40S ribosomal subunits. Such a time lag in the migration of 60S and 40S ribosomal subunits, prevents functional ribosomes from gaining access to the incompletely processed heterogeneous RNA (hn RNA; the precursor of m RNA) molecule inside the nucleus (see **Alberts** *et al.,* 1989).

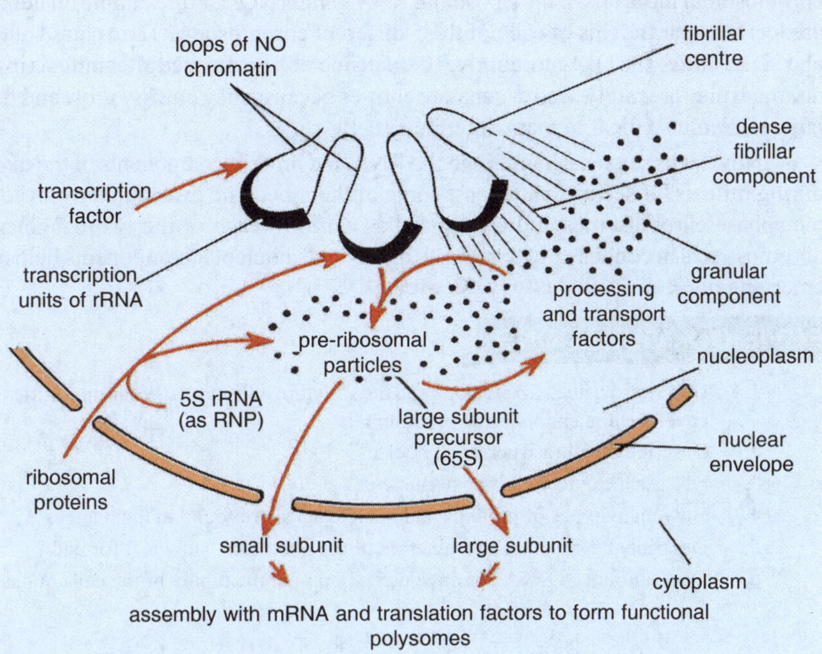

Fig. 12.8. Three different regions of the nucleolus and their involvement in ribosome assembly.

Different stages of formation of ribosomes are completed in three distinct regions of the nucleolus. Thus, their **initiation**, **production** and **maturation** seem to progress from centre to periphery. Following three regions have been identified in the nucleolus (Fig. 12.8) :

(i) Fibrillar centre. This pale-staining part represents the innermost region of nucleolus. The RNA genes of nucleolar organizer of chromosomes are located in this region. The transcription (*i.e.,* ribosomal RNA synthesis) of these genes is also initiated in this region.

(ii) Dense fibrillar component. This region surrounds the fibrillar centre and RNA synthesis progresses in this region. The 70 ribosomal proteins (rps) also bind to the transcripts in this region.

(iii) Cortical granular components. This is the outermost region of the nucleus where processing and maturation of pre-ribosomal particles occur.

3. Mitotic cycle of nucleolus. The appearance of nucleolus changes dramatically during the cell cycle. During meiosis as well as during mitosis the nucleolus disappears during prophase. As the cell. approaches mitosis, the nucleolus first decreases in size and then disappears as the chromosomes condense and all RNA synthesis stops, so that generally there is no nucleolus in a metaphase cell. When ribosomal RNA synthesis restarts at the end of mitosis (in telophase), tiny nucleoli reappear at the chromosomal locations of the ribosomal RNA genes (NOs). For example, in humans the r RNA genes are located near the tips of each of the 5 different chromosomes (*i.e.,* paired autosomes 13,14,15,21 and 22; **Franke**, 1981). Accordingly, 10 small nucleoli are formed after mitosis in a human diploid cell, although they are rarely seen as separate entities because they quickly grow and fuse to form the single large nucleolus typical of many interphase cells.

Now let us see what happens to the RNA and protein components of the disintegrated nucleolus during mitosis? It seems that at least some of them become distributed over the surface of all of the metaphase chromosomes and are carried as cargo to each of the two daughter cell nuclei. As the chromosomes decondense at telophase, these "old" nucleolar components help reestablish the newly emerging nucleoli (**Anastassova-Kristeva**, 1977).

REVISION QUESTIONS

1. Describe the nuclear envelope and the structure of its pores. What similarities occur between nuclear envelope and endoplasmic reticulum ?
2. Describe the ultrastructure of nucleus.
3. Discuss the cytochemistry of nucleus.
4. How many types of proteins and nucleic acids are found in the nucleus ?
5. Describe the structure and function of the nucleolus. How is it formed ?
6. Does amount of DNA has any correlation with the ploidy of the cell? What ?

CHAPTER

13

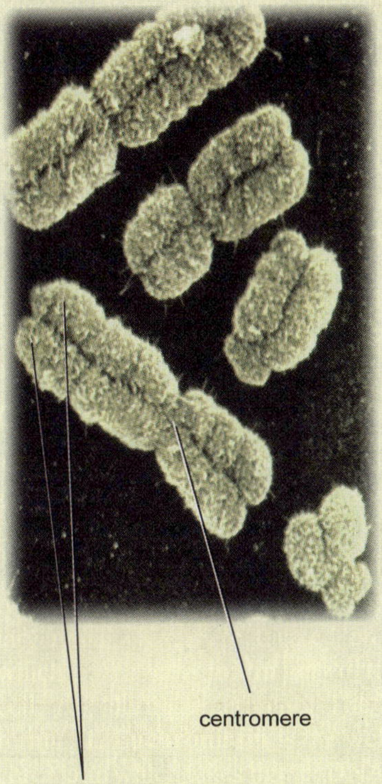

Human chromosomes during mitosis.

Chromosomes

The chromosomes are the nuclear components of special organisation, individuality and function. They are capable of self-reproduction and play a vital role in heredity, mutation, variation and evolutionary development of the species.

HISTORICAL

Karl Nagli (1842) observed rod-like chromosomes in the nuclei of plant cells. **E. Russow** (1872) made the first serious attempt to describe chromosomes. **A. Schneider** (1873) published a most significant paper dealing with the relation between chromosomes and stages of cell division. **E. Strasburger** (1875) discovered thread-like structures which appeared during cell division. **Walter Flemming** (1878) introduced the term chromatin to describe the thread-like material of the nucleus that became intensely coloured after staining.

W. Roux (1883) suspected the involvement of the chromosomes in the mechanism of inheritance. **Benden** and **Bovery** (1887) reported that number of chromosomes for each species was constant. The present name **chromosome** (Gr., *chrom*= colour, *soma*=body) was coined by **W. Waldeyer** (1888) to darkly stained bodies of nucleus. **W. S. Sutton** and **T. Boveri** in 1902 suggested that chromosomes were the physical structures which acted as messengers of

Theodor Boveri (1862-1915).

heredity. **Sutton** (1902) observed that the chromosome pair in synapsis is made up of one maternal and one paternal member. He believed that chromosomes, acting in this way, may be the physical basis for the Mendelian laws of heredity. He is credited as the originator of the theory of the chromosomal basis for heredity. **Thomas Morgan** and **Hermann Muller**, in the early 1900s, established the cytological basis for the laws of heredity. Working with *Drosophila* chromosomes, they located 2000 genetic factors on the four chromosomes of the fruit fly in 1922. In 1914, **Robert Feulgen** demonstrated a colour test known as **Feulgen reaction** for the DNA. In 1924, he showed that chromosomes contain DNA. In 1942, using cytochemical procedures, **Brachet** demonstrated the presence of another nucleic acid, RNA, and not long there after, **Mirsky** and **Pollister** (1946) showed that there were proteins associated with chromosomal material. **Heitz** (1935), **Kuwanda** (1939), **Geitter** (1940), and **Kaufmann** (1948) have described the morphology of chromosomes. **Dupraw** (1965) suggested 'folded fibre model' of the chromosome to suggest that it was made of a highly folded single molecule of DNA which is wrapped in chromosomal proteins. **R. D. Kornberg** (1974) proposed the 'nucleosome model' of the basic chromatin material. The term 'nucleosome' was coined by **P. Outdet** *et al.*, (1975).

CHROMOSOME NUMBER

The mumber of the chromosomes is constant for a particular species. Therefore, these are of great importance in the determination of the phylogeny and taxonomy of the species. The number or set of the chromosomes of the gametic cells such as sperms and ova is known as the gametic, reduced or **haploid** sets of chromosomes. The haploid set of the chromosomes is also known as the **genome**. The somatic or body cells of most organisms contain two haploid set or genomes and are knows as the **diploid cells**. The diploid cells achieve the diploid set of the chromosomes by the union of the haploid male and female gametes in the sexual reproduction. The suffix "—ploid" refers to chromosome "sets". The prefix indicates the degree of the ploidy.

The number of chromosomes in each somatic cell is the same for all members of a given species. The organism with the lowest number of the chromosomes is the nematode, *Ascaris megalocephalus univalens* which has only two chromosomes in the somatic cells (*i.e.*, 2n =2). In the radiolarian protozoan *Aulacantha* is found a diploid number of approximately 1600 chromosomes. Among plants, chromosome number varies from 2n = 4 in *Haplopappus gracilic* (Compositae) to 2n = >1200 in some pteridophytes. However,the diploid number of tobacco is 48, cattle 60, the garden pea 14, the fruit fly 8, etc. The chromosome number of some animals and plants is tabulated in Table 13-1. The diploid number of a species bears no direct relationship to the species position in the phylogenetic scheme of classification.

Table 13.1. **Chromosome number of some organisms.**

Group	Common name	Scientific name	Chromosome number
Animals :			
Protozoa	Paramecium	*Paramecium aurelia*	30–40
Cnidaria	Hydra	*Hydra vulgaris*	32
Nematoda	Round worm	*Ascaris lumbricoides*	24
Arthropoda	House fly	*Musca domestica*	12
	Mosquito	*Culex pipiens*	6
Chordata	Frog	*Rana esculenta*	26
	Pigeon	*Columba livia*	80
	Rabbit	*Oryctolagus cuniculus*	44
	Gorillia	*Gorilla gorilla*	48
	Man	*Homo sapiens*	46

Group	Common name	Scientific name	Chromosome number
Plants :			
Algae	Chlamydomonas	*Chlamydomonas reinhardii*	10?; 12? 16? (Haploid sets)
Fungi	Bread mold	*Mucor heimalis*	2
Gymnosperm	Yellow pine	*Pinus ponderosa*	24
Angiosperm	Cabbage	*Brassica oleracea*	18
	Coffee	*Coffea arabica*	44
	Potato	*Solanum tuberosum*	48
	Sugar cane	*Saccharum officinarum*	80
	Onion	*Allium cepa*	16

Lastly, while '*n*' normally signifies the gametic or haploid chromosome number, '*2n*' is the somatic or diploid chromosome number in an individual. In polyploid individuals, however, it becomes necessary to establish an ancestral primitive number, which is represented as '*x*' and is called the **base number**. For example, in wheat *Triticum aestivum* $2n = 42$; $n = 21$ and $x = 7$, showing that common wheat is a hexaploid ($2n = 6x$).

Autosomes and Sex chromosomes

In a diploid cell, there are two of each kind of chromosome (these are termed **homologous chromosomes**), except for the **sex chromosomes**. One sex has two of the same kind of sex chromosome and the other has one of each kind. For example, in human, there are 23 pairs of homologous chromosomes (*i.e.,* 2n = 46 ; a chromosome number which was established by **Tijo** and **Levan** in 1956). The human female has 44 non-sex chromosomes, termed **autosomes** and one pair of homomorphic (morphologically similar) sex chromosomes given the designation XX. The human male has 44 autosomes and one pair of heteromorphic or morphologically dissimilar sex chromosomes, *i.e.,* one X chromosome and one Y chromosome.

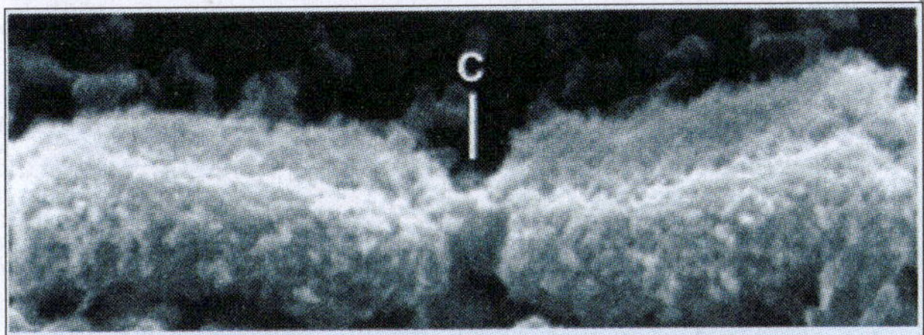

A mitotic chromosome; c = centromere.

MORPHOLOGY

Size

The size of chromosome is normally measured at mitotic metaphase and may be as short as 0.25 μm in fungi and birds, or as long as 30 μm in some plants such as *Trillium*. However, most metaphase chromosomes fall within a range of 3μm in fruitfly (*Drosophila*), to 5μm in man and 8μm to 12μm in maize. The organisms with less number of chromosome contain comparatively large-sized chromosomes than the chromosomes of the organisms having many chromosomes.

The monocotyledon plants contain large-sized chromosomes than the dicotyledon plants. The plants in general have large-sized chromosomes in comparison to the animals. Further, the chromosomes in a cell are never alike in size, some may be exceptionally large and others may be too small. The largest chromosomes are lampbrush chromosomes of certain vertebrate oocytes and polytene chromosomes of certain dipteran insects.

Shape

The shape of the chromosomes is changeable from phase to phase in the continuous process of the cell growth and cell division. In the resting phase or interphase stage of the cell, the chromosomes occur in the form of thin, coiled, elastic and contractile, thread-like stainable structures, the chromatin threads. In the metaphase and the anaphase, the chromosomes become thick and filamentous. Each chromosome contains a clear zone,known as **centromere** or **kinetocore**, along their length. The centromere divides the chromosomes into two parts, each part is called **chromosome arm**. The position of centromere varies form chromosome to chromosome and it provides different shapes to the latter which are following (Fig. 13.1) :

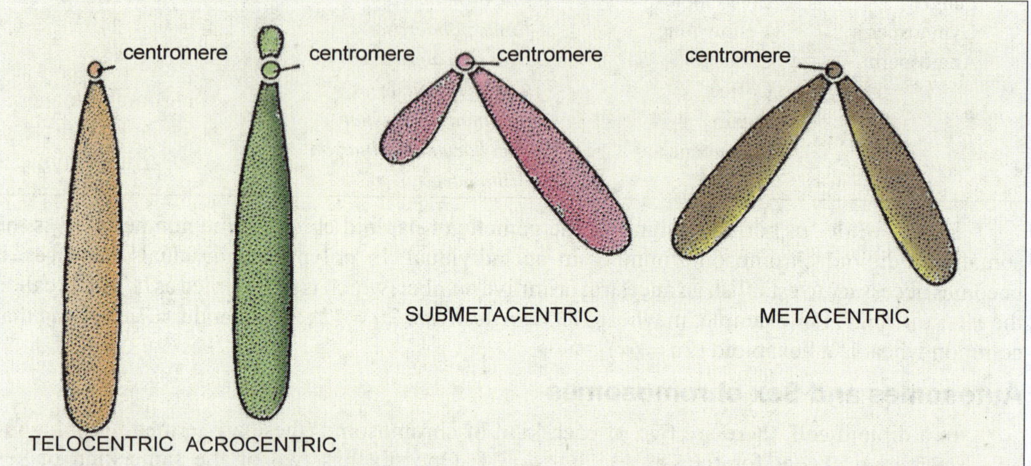

Fig. 13.1. The four morphological types of chromosomes according to the position of the centromere.

1. Telocentric. The rod-like chromosomes which have the centromere on the proximal end are known as the **telocentric chromosomes.**

2. Acrocentric. The acrocentric chromosomes are also rod-like in shape but these have the centromere at one end and thus giving a very short arm and an exceptionally long arm. The locusts (Acrididae) have the acrocentirc chromosomes.

3. Submetacentric. The submetacentric chromosomes are J- or L-shaped. In these, the centromere occurs near the centre or at medium portion of the chromosome and thus forming two unequal arms.

4. Metacentric. The **metacentric chromosomes are V-shaped** and in these chromosomes the centromere occurs in the **centre and forming two equal arms**. The amphibians have metacentric chromosomes.

Structure

While describing the structure of the chromosomes during various phases of cell cycle, cell biologists have introduced many terms for their various components. Let us become familiar with the following terms to understand more clearly the structure of the chromosomes (Fig. 13.2) :

1. Chromatid. At mitotic metaphase each chromosome consists of two symmetrical structures, called **chromatids**. Each chromatid contains a single DNA molecule. Both chromatids are attached to each other only by the centromere and become separated at the beginning of anaphase, when the sister chromatids of a chromosome migrate to the opposite poles.

2. Chromonema (ta). During mitotic prophase the chromosomal material becomes visible as very thin filaments, called **chromonemata** (a term coined by **Vejdovsky** in 1912). A chromonema represents a chromatid in the early stages of condensation. Therefore, 'chromatid' and 'chromonema' are two names for the same structure : a single linear DNA molecule with its associated proteins. The chromonemata form the gene-bearing portions of the chromosomes.

According to old view, a chromosome may have more than one chromonemata which are embedded in the achromatic and amorphous substance, called **matrix**. The matrix is enclosed in a sheath or **pellicle**. Both matrix and pellicle are non-genetic materials and appear only at metaphase

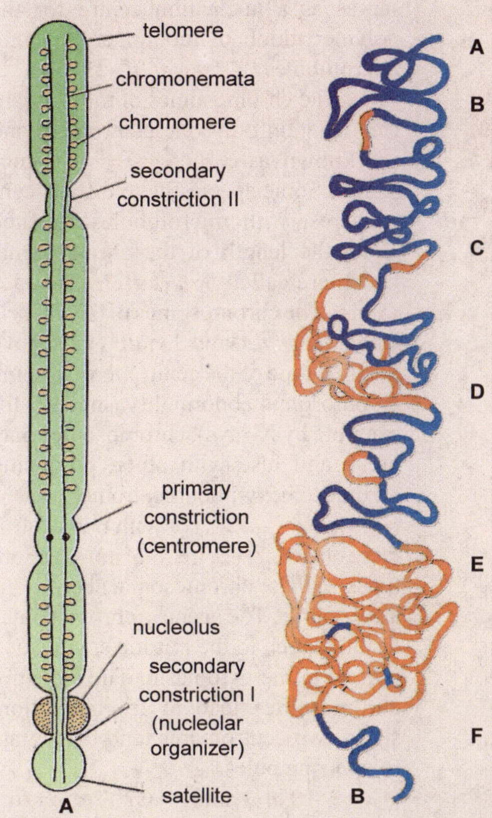

Fig. 13.2. A—Structure of a typical chromosome; B—Model of constitutive heterochromatin in a mammalian metaphase chromosome. A—Constitutive heterochromatin; B—Secondary constriction; I or nucleolar organizer; C—Primary constriction or centromere; D—Euchromatin; E—Secondary constriction. II—possible site of 5S rRNA cistrons; F—Telomere (after De Robertis *et al.,* 1975).

when the nucleolus disappears. It is believed that nucleolar material and matrix are interchangeable, *i.e.,* when chromosomal matrix disappears, the nucleolus appears and vice versa. Electron microscopic observations, however, have questioned the occurrence of pellicle and matrix in them.

3. Chromomeres. The chromomeres are bead-like accumulations of chromatin material that are sometimes visible along interphase chromosomes. The chromomere-bearing chromatin has an appearance of a necklace in which several beads occur on a string. Chromomeres become especially clear in the polytene chromosomes, where they become aligned side by side, constituting the chromosome beads (Fig. 13.2). At metaphase the chromosomes are tightly coiled and the chromomeres are no longer visible.

Chromomeres are regions of tightly folded DNA and have great interest for the cell biologists. They are believed to correspond to the units of genetic function in the chromosomes (see **De Robertis** and **De Robertis**, **Jr**., 1987). In fact, for long time most geneticists considered these chromomeres as genes, *i.e*., the units of heredity.

4. Centromere and kinetochore. Originally it was considered that the centromere consists of small granules or spherules. The centromere of the chromosome of the *Trillium* has the diameter of 3µm and the spherules have the diameter of 0.2 µm. The chromonema remains connected with the spherules of the centromere. Currently it is held that centromere is the region of the chromosome to which are attached the fibres of mitotic spindle. The centromere (a term much preferred by the geneticists) lies within a thinner segment of chromosome, the **primary constriction**. The regions of chromosome flanking the centromere contain highly repetitive DNA and may stain more intensely with the basic dyes. (*i.e.,* it is a constitutive heterochromatin, Fig. 13.2B). Centromeres are found to contain specific DNA sequences with special proteins bound to them, forming a disc-shaped structure, called **kinetochore** (a term that is much preferred by the cytologists). Under the EM, the kinetochore appears as a plate - or cup-like disc, 0.20 to 0.25 nm, in diameter situated upon the primary constriction or centromere. In thin electron microscopic sections, the kinetochore shows a trilaminar structure, *i.e.,* a 10 nm thick dense **outer proteinaceous** layer, a **middle** layer of low density and a dense **inner** layer tightly bound to the centromere (Fig. 13.3). The DNA of centromere does not exist in the form of nucleosome (**Ris** and **Witt**, 1981). Further, emanating from the convex surface of outer layer of kinetochore, in addition to the microtubules, a "corona" or "collar" of fine filaments has been observed. During mitosis, 4 to 40 microtubules of mitotic spindle become attached to the kinetochore and provide the force for chromosomal movement during anaphase. The main function of the kinetochore is to provide a centre

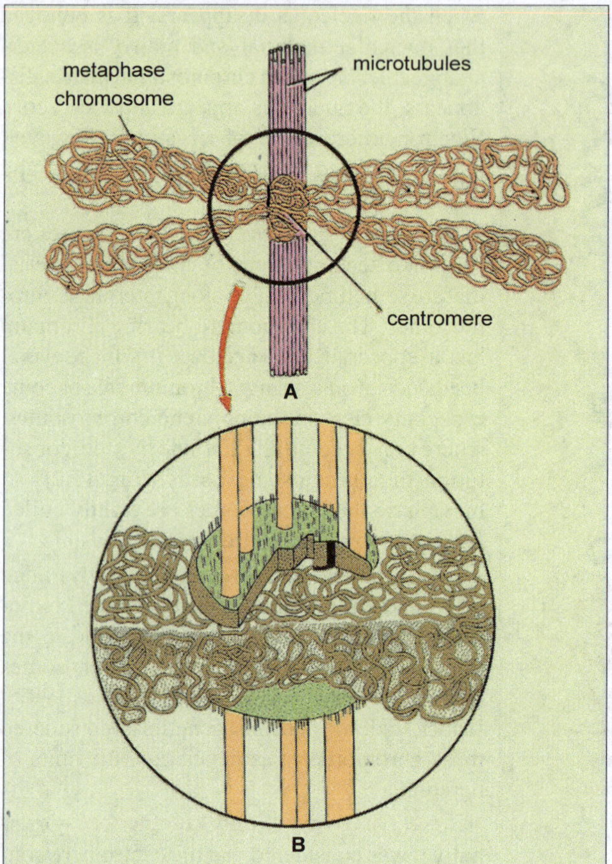

Fig. 13.3. A—A metaphase chromosome showing the folded fibre structure and the centromere with implanted microtubules; B—An inset a higher magnification, showing the convex electrone dense layer and the fibrillar material forming the "corona" of the centromere (after De Robertis *et al.,* 1975).

of assembly for microtubules, *i.e.,* it serves as a nucleation centre for the polymerization of tubulin protein into microtubules (**Telzer** *et al.,* 1975).

The chromosomes of most organisms contain only one centromere and are known as **monocentric** chromosomes. Some species have diffuse centromeres, with microtubules attached along the length of the chromosome, which are called **holocentric chromosomes**. The chromosomes of the *Ascaris megalocephala* and hemipterans have diffused type of the centromere. In some chromosomal abnormality (induced for example by X-rays), chromosomes may break and fuse with other, producing chromosomes without centromere (**acentric chromosomes**) or with two centromeres (**dicentric chromosomes**). Both types of these chromosomal aberrations are unstable. The acentric chromosomes cannot attach to the mitotic spindle and remain in the cytoplasm. The dicentric chromosomes lead to fragmentation, since, two centromeres tend to migrate to opposite poles.

5. Telomere. (Gr., *telo*=for; *meros*=part). Each extremity of the chromosome has a polarity and therefore, it prevents other chromosomal segments to be fused with it. The chromosomal ends are known as the **telomeres**. If a chromosome breaks, the broken ends can fuse with each other due to lack of telomeres.

6. Secondary constriction. The chromosomes besides having the primary constriction or the centromere possess secondary constriction at any point of the chromosome. Constant in their position and extent, these constrictions are useful in identifying particular chromosomes in a set. Secondary constrictions can be distinguished from primary constriction or centromere, because chromosome bends (or exhibits angular deviation) only at the position of centromere during anaphase.

7. Nucleolar organizers. These areas are certain secondary constrictions that contain the genes coding for 5.8S, 18S and 28S ribosomal RNA and that induce the formation of nucleoli. The secondary constriction may arise because the rRNA genes are transcribed very actively and, thus, interfering with chromosomal condensation. In human beings, the nucleolar organizers are located in the secondary constrictions of chromosomes 13, 14, 15, 21 and 22, all of which are acrocentric and have satellites.

8. Satellite. Sometimes the chromosomes bear round elongated or knob-like appendages known as **satellites**. The satellite remains connected with the rest of the chromosome by a thin chromatin filament. The chromosomes with the satellite are designated as the **sat chromosomes**. The shape and size of the satellite remain constant.

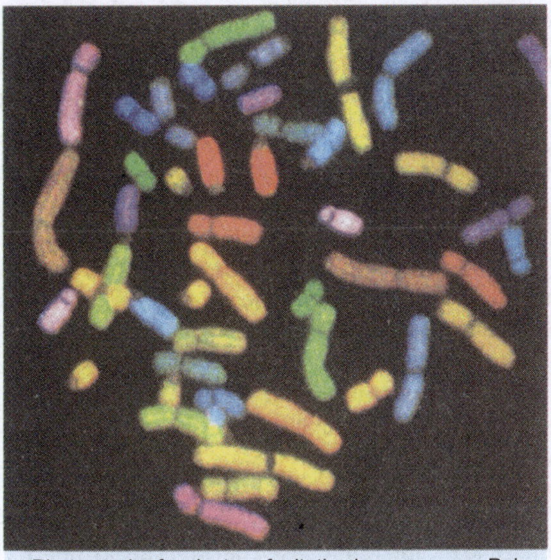

Photograph of a cluster of mitotic chromosomes. Pairs of homologous chromosomes can be identified and karyotypes can be prepared with their help.

Chromosome satellites are a morphological entity and should not be confused with satellite DNAs which are highly repeated DNA sequence.

Karyotype and Idiogram

All the members of a species of a plant or the animal are characterized by a set of chromosomes which have certain constant characteristics. These characteristics include the number of chromosomes, their relative size, position of the centromere, length of the arms, secondary constrictions and satellites. The term **karyotype** has been given to the group of characteristics that identifies a particular set of chromosomes. A diagrammatic representation of a karyotype (or morphological characteristics of the chromosomes) of a species is called **idiogram** (Gr., *idios* = distinctive; *gramma* = something written). Generally, in an idiogram, the chromosomes of a haploid set of an organism are ordered in a series of decreasing size. Sometimes an idiogram is prepared for the diploid set of chromosomes, in which the pairs of homologues are ordered in a series of decreasing size.

A karyotype of human metaphase chromosomes is obtained from their microphotographs. The individual chromosomes are cut out of the microphotographs and lined up by size with their respective partners. The technique can be improved by determining the so-called **centromeric index**, which is the ratio of the lengths of the long and short arms of the chromosome.

Some species may have special characteristics in their karyotypes; for example, the mouse has acrocentric chromosomes, many amphibians have only metacentric chromosomes and plants frequently have heterochromatic regions at the telomeres.

Uses of karyotypes. The karyotypes of different species are sometimes compared and similarities in karyotypes are presumed to represent evolutionary relationship. A karyotype also suggests primitive or advanced features of an organism. It may be symmetric or asymmetric. A karyotype exhibiting large differences in smallest and largest chromosomes of the set and containing fewer metacentric chromosomes, is called an **asymmetric karyotype** (Fig. 13.4). In comparison to a symmetric karyotype (*e.g., Pinus*; Fig. 13.4 A), an asymmetric karyotype (*e.g., Ginkgo biloba,* Fig. 13.4 B) is considered to be a relatively advanced feature. **Levitzky** (1931) suggested that in flowering plants there is a prominent trend towards asymmetric karyotypes. This trend has been well studied in the genus *Crepis* of the family Compositae. In many cases it was shown that increased karyotype asymmetry was associated with specialized zygomorphic flowers.

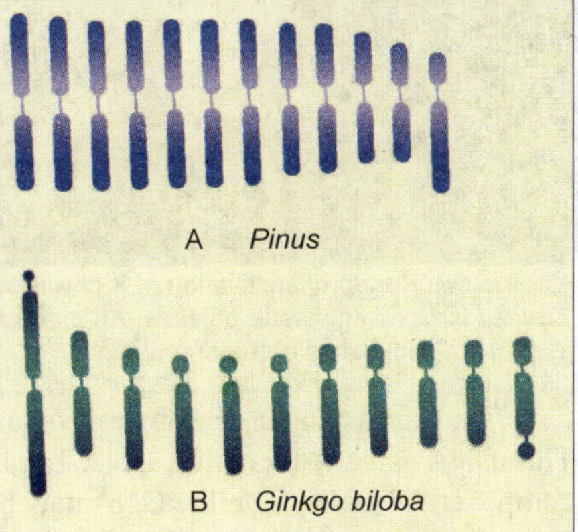

Fig. 13.4. Karyotypes : A–Symmetric; B– Asymmetric.

MATERIAL OF THE CHROMOSOMES

The material of the chromosomes is the chromatin, the structure of which has already been described in previous chapter (Chapter 12). Depending on their staining properties, the following two types of chromatin may be distinguished in the interphase nucleus :

1. Euchromatin. Portions of chromosomes that stain lightly are only partially condensed; this chromatin is termed **euchromatin**. It represents most of the chromatin that disperse after mitosis has completed. Euchromatin contains structural genes which replicate and transcribe during G_1 and S phase of interphase. The euchromatin is considered genetically active chromatin, since it has a role in the phenotype expression of the genes. In euchromatin, DNA is found packed in 3 to 8 nm fibre.

2. Heterochromatin. In the dark-staining regions, the chromatin remains in the condensed state and is called **heterochromatin**. In 1928, **Heitz** defined heterochromatin as those regions of the chromosome that remain condensed during interphase and early prophase and form the so-called **chromocentre**. Heterochromatin is characterized by its especially high content of repititive DNA sequences and contains very few, if any, structural genes (*i.e.,* genes that encode proteins). It is **late replicating** (*i.e.,* it is replicated when the bulk of DNA has already been replicated) and is not transcribed. It is thought that in heterochromatin the DNA is tightly packed in the 30 nm fibre.

Types of heterochromatin. In an interphase nucleaus, usually there is some condensed chromatin around the nucleolus, called **perinucleolar chromatin**, and some inside the nucleolus, called **intranucleolar chromatin**. Both types of this heterochromatin appear to be connected and together, they are referred to as **nucleolar chromatin**.

Dense clumps of deeply staining chromatin often occur in close contact with the inner membrane of the nuclear envelope (*i.e.,* with the nuclear lamina) and is called **condensed peripheral chromatin**. Between the peripheral heterochromatin and the nucleolar heterochromatin are regions of lightly staining chromatin, called **dispersed chromatin**. In the condensed chromosomes, the heterochromatic regions can be visualized as regions that stain more strongly or more weakly than the euchromatic regions, showing the so-called **positive** or **negative heteropyknosis** of the chromosomes (Gr., *hetero* = different + *pyknosis* = staining).

Heterochromatin has been further classified into the following types :

1. Constitutive heterochromatin. In such a heterochromatin the DNA is permanently inactive and remains in the condensed state throughout the cell cycle. This most common type of heterochromatin occurs around the centromere, in the telomeres and in the C-bands of the chromosomes. In *Drosophila virilis*, constitutive heterochromatin exists around the centromeres and such **pericentromeric heterochromatin** occupies 40 per cent of the chromosomes. In many species, entire chromosomes become heterochromatic and are called **B chromosome**, **satellite chromosomes** or **accessory chromosomes** and contain very minor biological roles. Such chromosomes comprising wholly constitutive heterochromatin occur in corn, many phytoparasitic insects and salamanders. In the fly *Sciara*, large metacentric heterochromatic chromosomes are found in the gonadal cells, but are absent in somatic cells. Entire Y chromosome of male *Drosophila* is heterochromatic, even though containing six gene loci which are necessary for male fertility (see **Suzuki**, *et al.*, 1986).

A calico cat. Random inactivation of X chromosome creates a mosaic of tissue patches. An example of facultative heterochromatin.

Constitutive heterochromatin contains short repeated sequences of DNA, called **satellite DNA**. This DNA is called satellite DNA because upon ultracentrifugation, it separates from the main component of DNA. Satellite DNA may have a higher or lower G + C content than the main fraction. For example, the mouse satellite DNA is a 240 base pair sequence that is repeated about 1000,000 (10^6) times in the mouse genome, constituting 10 per cent of the total mouse DNA. The exact significance of constitutive heterochromatin is still unexplained.

2. Facultative heterochromatin. Such type of heterochromatin is not permanently maintained in the condensed state ; instead it undergoes periodic dispersal and during these times is transcriptionally active. Frequently, in facultative heterochromatin one chromosome of the pair becomes either totally or partially heterochromatic. The best known case is that of the X-chromosomes in the mammalian female, one of which is active and remains euchromatic, whereas the other is inactive and forms at interphase, the **sex chromatin** or **Barr body** (Named after its discoverer, Canadian cytologist **Murray L. Barr**). Barr body contains DNA which is not transcribed and is not found in males. Indeed, the number of Barr bodies is always one less than the number of X chromosomes (*i.e.*, in humans, XXX female has two Barr bodies and XXXX female has three Barr bodies; **M.L. Barr**, 1959).

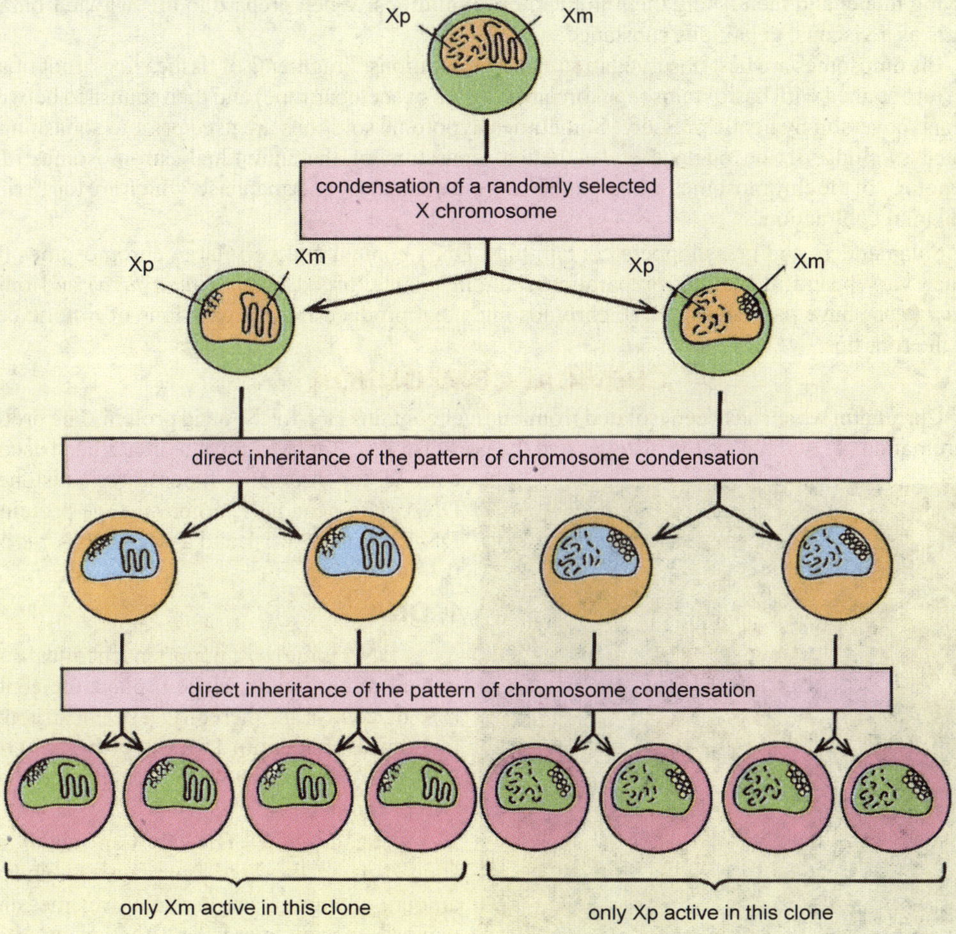

Fig. 13.5. The clonal inheritance of a condensed inactive X chromosome that occurs in female mammals (after Alberts *et al.*, 1989).

Dosage compensation and lyonization. In mammals all female cells contain two X chromosomes, while male cells contain one X and one Y chromosome. Presumably because a double dose of X chromosome products would be lethal, the female cells have evolved a mechanism for permanently inactivating one of the two X chromosomes in each cell (this process is called **dosage compensation**). Process of X chromosome inactivation is often termed **lyonization** after the name of British cytogeneticists **Mary Lyons**. In mice, this occurs between the third and the sixth day of development, when one or the other of the two X chromosomes in each cell is chosen at random and condensed into heterochromatin (or Barr body). Because the inactive chromosome is faithfully inherited, every female is a **mosaic** composed of clonal groups of cells in which only the paternally inherited X chromosome (Xp) is active and a roughly equal number of groups of cells in which only the maternally inherited X chromosome (Xm) is active (Fig. 13.5).

ISOLATION METHODS

Several methods have been developed over the years to prepare chromatin for microscopical examination. A relatively simple approach is to first prepare purified nuclei from thymus, liver, or any other desired source. Nuclei are then lysed with detergent and the chromatin pelleted at 20,000 rpm (about 50,000 g) in a preparative centrifuge. Chromatin can also be isolated biochemically by purifying nuclei and then lysing them in hypotonic solutions. When prepard in this way, chromatin appears as a viscous, gelatinous substance.

Chromosomes are best observed in **squash** preparations. Fragments of tissues (*e.g.*, root tips of onion) are stained with basic stains (*e.g.*, orcein, Giemsa or acetocarmine) and then squashed between slide and coverslip by gentle pressure. Sometimes hypotonic solutions are used prior to squashing to produce swelling of the nucleus and a better separation of the individual chromosomes. The morphology of the chromosomes is best studied during metaphase and anaphase, which are the periods of maximal contraction.

Salamanders and grasshoppers contain high DNA content and very large chromosomes that produce vary beautiful meiotic preparations; among plants, broad beans (*Vicia fava*) and onion (*Allium cepa*) have particularly large chromosomes that produce fine preparations of mitotic cells from the root tips.

CHEMICAL COMPOSITION

Chromatin which has been isolated from rat liver contains DNA, RNA and protein. The protein of chromatin is of two types : the histones and the non-histones. Rat liver chromatin has been used as a model for chromatin. It possesses a histone to DNA ratio near 1 : 1, a non-histone protein to DNA ratio of 0.6 : 1 and a RNA/DNA ratio of 0.1 : 1.

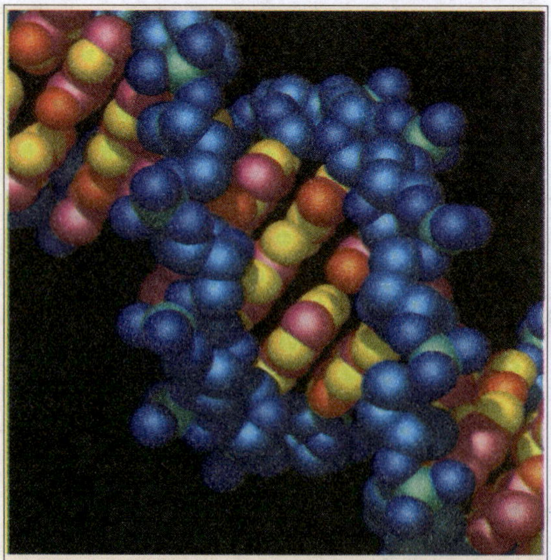

A computer - generated model of the structrue of DNA

1. DNA

DNA is the most important chemical component of chromatin, since it plays the central role of controlling heredity. The most convenient measurement of DNA is picogram (10^{-12} gm). DNA of chromatin represent the following two phenomena :

The C-value. The DNA in nuclei was stained using the **Feulgen reactions** and the amount of stain in single nuclei was measured using a special microscope (called **cytophotometer**). Both of these techniques demonstrated that nuclei contain a constant amount of DNA. Thus, all the cells in an organism contain the same DNA content (2C) provided that they are diploid. Gametes are haploid and, therefore, have half the DNA content (1C). Some tissues such as liver, contain occasional cells that are polyploid and their nuclei have a correspondingly higher DNA content (4C or 8C) (see Table 13-2).

Table 13.2. DNA content and chromosome component (after **De Robertis** and **De Robertis, Jr.**, 1987).

Cells	Mean DNA-Feulgen content	Presumed chromosome set
1. Spermatid	1.68 (1C)	Haploid (n)
2. Liver	3.16 (2C)	Diploid (2n)
3. Liver	6.30 (4C)	Tetraploid (4n)
4. Liver	12.80 (8C)	Octoploid (8n)

Thus, each species has a characteristic content of DNA which is **constant** in all the individuals of that species and has, thus, been called the **C-value**.

The C-value paradox. Eukaryotes vary greatly in DNA content but always contain much more DNA than prokaryotes. Lower eukaryotes in general have less DNA, such as nematode *Caenorhabditis elegans* which has only 20 times more DNA than *E.coli*, or the fruit fly *D. melanogaster* which has 40 times more DNA (*i.e.*., 0.18 pg or picogram per haploid genome). Vertebrates have greater DNA content (about 3 pg), in general about 700 times more then *E. coli*. One of the highest DNA content is that of the salamander *Amphiuma* which has 84 pg of DNA. Man has about 3 pg of DNA per haploid genome, or 3×10^9 base pairs, *i.e.*, the human genome could accommodate about 3 million average sized proteins if all the DNA were coding (or containing structural genes) and if this was true, salamanders would have 30 times more genes than human beings. This is called **C-value paradox** (**Gall**, 1981). It was detected quite early that there was little connection between the morphological complexity of eukaryotic organisms and their DNA content. For example, *E. coli* (containing 3,400,000 base pairs in its DNA) has about 3000 genes. Although it is difficult to estimate how many different genes exist in the human genome, there are probably not more than 20,000 to 30,000 genes (Note : According to a most recent estimate, there are 100,000 genes in human genome, see **Deviah**, 1994). There is no reason to believe that salamanders should have any more. From these facts, it can be easily concluded that most of the DNA in the eukaryotic genomes must be of a non-coding nature.

2. Histones

Histones are very basic proteins, basic because they are enriched in the amino acids arginine and lysine to a level of about 24 mole present. Arginine and lysine at physiological pH are cationic and can interact electrostatically with anionic nucleic acids. Thus, being basic, histones bind tightly to DNA which is an acid. There are five types of histones in the eukaryotic chromosomes, namely H1, H2A, H2B, H3 and H4.

One of the important discoveries that has come from chemical studies is that the primary structures of histones have been highly conserved during evolutionary history. For example, histone H4 of calf and of garden pea contains only two amino acid differences in a protein of 102 residues (**DeLange**, 1969). Likewise, the sequence of histone H3 from rat differs only in two amino acids from that of peas, out of 102 total amino acid residues. These organisms are estimated to have an evolutionary history of at least 600 million years, during which they diverged structurally. This conservation of structure suggests that over the eras, histones have had a very similar and crucial role in maintaining the structural and functional integrity of chromatin. Such an evolutionary conservation suggests that the functions of these two histones involve nearly all of their amino acids so that a change in any position is deleterious to the cell.

Histone H1 is the least rigidly conserved histone protein. It contains 210 to 220 amino acids and may be represented by a variety of forms even within a single tissue. H1 is present only once per 200 base pairs of DNA (in contrast to rest of the four types of histones each of which is present twice) and is rather loosely associated with DNA. H1 histone is absent in yeast, *Saccharomyces cerevisiae*.

Histones besides determining the structure of chromatin, play a regulatory role in the repression activity of genes.

3. Non-histones

In contrast to the modest population of histones in chromatin, non-histone proteins display more diversity. In various organisms, number of non-histones can vary from 12 to 20. Heterogeneity of these proteins is not conserved in evolution as the histones. These non-histones differ even between different tissues of the same organism suggesting that they regulate the activity of specific genes.

About 50 per cent non-histones of chromatin have been found to be structural proteins and include such proteins as **actin**, and α-and β-**tubulins** and **myosin**. Although for sometime these contractile proteins were thought to be contaminants, it is now believed that they are vital ingredients of the chromosome, functioning during chromosome condensation and in the movement of chromosomes during mitosis and meiosis (see **Thorpe**, 1984). Many of the remaining 50 per cent of non-histones include all the enzymes and factors that are involved in DNA replication, in transcription and in the regulation of transcription. These proteins are not as highly conserved among organisms, although they must carry out similar enzymatic activities. Apparently they are not as important as the histones in maintaining chromosome integrity.

ULTRASTRUCTURE

The field of ultrastructure of the chromatin is still the area where electron microscope had failed to provide us a clear picture of the organization of DNA in the chromatin. For the study of chromosomes with the help of electron microscope, whole chromosome mounts as well as sections of chromosomes were studied. Such studies had demonstrated that chromosomes have very fine fibrils having a thickness of 2nm—4nm. Since DNA is 2nm wide, there is a possibility that a single fibril corresponds to a single DNA molecule.

Single-stranded and Multi-stranded Hypotheses

When chromosomes are compared in related species which differ widely in DNA content, such differences may be attributed to one of two causes : (1) **lateral multiplication** of chromonemata leading to multiple or multi-strandedness, or (2) **tandem duplication** of DNA or chromonemata where lenghthwise duplication is responsible for chromatin differences. This latter condition will retain the single stranded feature of chromosomes.

Although multiple strandedness has been demonstrated in several cases of plants such as *Vicia faba* and animals such as dipteran salivary gland chromosomes, there are evidences against such hypothesis to become a generalization. In all these cases, however, tandem duplication of chromonemata (or DNA) evidently takes place. Indeed, there are many evidences to support the idea of single-stranded nature of chromatin. This was confirmed by the technique of pulsed gel electrophoresis that in yeast *Saccharomyces cerevisiae*, each chromosome is formed from a single linear DNA molecule (**Kavenoff** *et al.,* 1974).

Folded-fibre Model and Nucleosome Concept

If we presume that a single chromatid has a single long DNA molecule, we have no choice but to believe that DNA should be present in a coiled or folded manner. The manner of coiling and folding of DNA was a matter of debate and dozens of models were available for this purpose ; of them only two stand out and are important. A popular model was the **folded-fibre model**, proposed by **E.J. Dupraw** in 1965. According to it, the bulk of the chromosome is visualized to be composed of a tightly folded fibre which has a rather homogeneous diameter of 200 to 300 A°. This folded fibre is supposed to contain the DNA histone helix (of 30A° diameter) in a supercoiled condition (Fig. 13.6). Another model is most significant and universally accepted one and is called **nuclesome model** which was proposed by **R.D. Kornberg** (1974) (Fig. 13.7) and confirmed and christened by **P. Oudet** *et al.*, (1975). Thus, while in the folded-fibre model, it was proposed that the histones were bound on the outside of the DNA coils (*i.e.*, histone shell around DNA), the nucleosome model has proposed the converse (*i.e.*, histone particle with DNA round it). In other words, the earlier theory that basic chromatin fibre had DNA core surrounded by histones was incorrect (**Berns**, 1983). In fact, from a genetic perspective, a significant feature of packing mechanism through the nucleosomes lies in its topology : *at no point is the DNA buried* ; instead, it is freely exposed along the entire surface of the "spool", available for genetic expression. Nucleosomes seem to be universal device for compacting the long DNA molecules of eukaryotic cells.

Nucleosomes and Solenoid Model of Chromatin

In eukaryotes, DNA is tightly bound to an equal mass of histones, which serve to form a repeating array of DNA-protein particles, called **nucleosomes**. If it was stretched out, the DNA double-helix in each human chromosome would span the cell nucleus thousands of time. Histones play a crucial role in packing this very long DNA molecule in an orderly way (*i.e.*, nucleosome) into nucleus only a few micrometres in diameter. Thus, nucleosomes are the fundamental packing unit particles of the chromatin and give chromatin a "**beads-on-a-string**" appearance in electron micorgraphs taken after treatments that unfold higher- order packing (**Olins** and **Olins**, 1974).

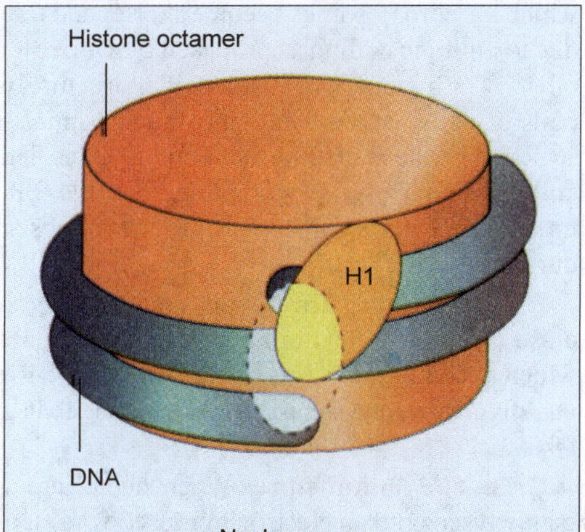

Nucleosome.

The nucleosome 'beads' can be removed from long DNA "string" by digestion with enzymes that degrade DNA, such as bacterial enzyme, **micrococcal nuclease**. After digestion for a short period with micrococcal nuclease, only the DNA between the nucleosome beads is degraded (Fig. 13.7). The rest is protected from digestion and remains as double-stranded DNA fragments 146 nucleotide pairs long bound to a specific complex of 8 nucleosome histones (the **histone octamer**). The nucleosome beads obtained in this way have been crystallized and analyzed by X-ray diffraction.

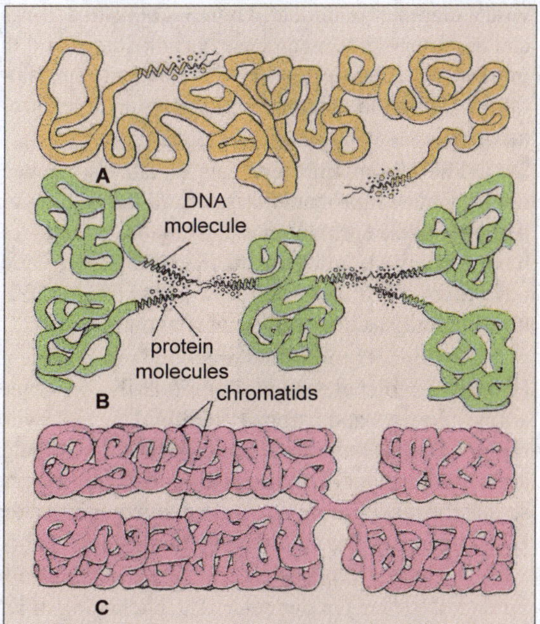

Fig. 13.6. Dupraw's folded fibre model of chromatin in interphase (A and B) and in metaphase (C).

Each nucleosome is a disc-shaped particle with a diameter of about 11 nm and 5.7 nm in height containing 2 copies of each 4 nucleosome histones–H2A, H2B, H3 and H4. This histone octamer forms a protein core [(*i.e.*, a core of histone tetramer $(H3, H4)_2$ and the apolar regions of 2(H2A and H2B)] around which the double-stranded DNA helix is wound 1¾ time containing 146 base pairs. In undigested chromatin the DNA extends as a continuous thread from nucleosome to nucleosome. Each nucleosome bead is separated from the next by a region of **linker DNA** which is generally 54 base pair long and contains single H1 histone protein molecule. Generally, DNA makes two complete turns around the histone octamers and these two turns (200 bp long) are sealed off by H1 molecules. (**Note** : In some organisms nucleosome DNA may vary from 162 base pairs (*e.g.*, rabbit cortical neurons) to 242 base pairs (*e.g.*, sea urchin sperm); **Reid** and **Leech**, 1980). Thus, on an average, nucleosomes repeat at intervals of about 200 nucleotides or base pairs. For example, an eukaryotic gene of 10,000 nucleotide pairs will be associated with 50 nucleosomes and each human cell with 6×10^9 DNA nucleotide pairs contains 3×10^7 nucleosomes.

Solenoid Models

H1 is reported to be phosphorylated just before mitotic and meiotic cell division to make possible the higher levels of coiling (see **Mays**, 1981). During mitosis or meiosis, the prophase is the stage during which the chromosomes become shorter and thicker due to multiple coiling as proposed by **Dupraw** and others. The hypothesis of a solenoidal structure, with coils of coils had renewal since nucleosomal sub-structure has been discovered. Thus, due to solenoid coiling of nucleosome containing fibre, the following types of chromosomal structures can be observed during the cell cycle (Fig. 13.12) :

1. The 10-nm fibre. When nucleosomes are in close apposition, they form the **10-nm filaments**, in which packing of DNA is about five-to seven-fold, *i.e.*, five to seven times more compact than free DNA.

2. The 30-nm fibre. When nuclei are very gently lysed onto an electron microscopy grid, most of the chromatin is seen to be in the form of a fibre, with a diameter of about 30 nm. Such **30-nm fibres** can be observed in metaphase chromosomes and in interphase nuclei and it probably represents the natural conformation of transcriptionally inactive chromatin.

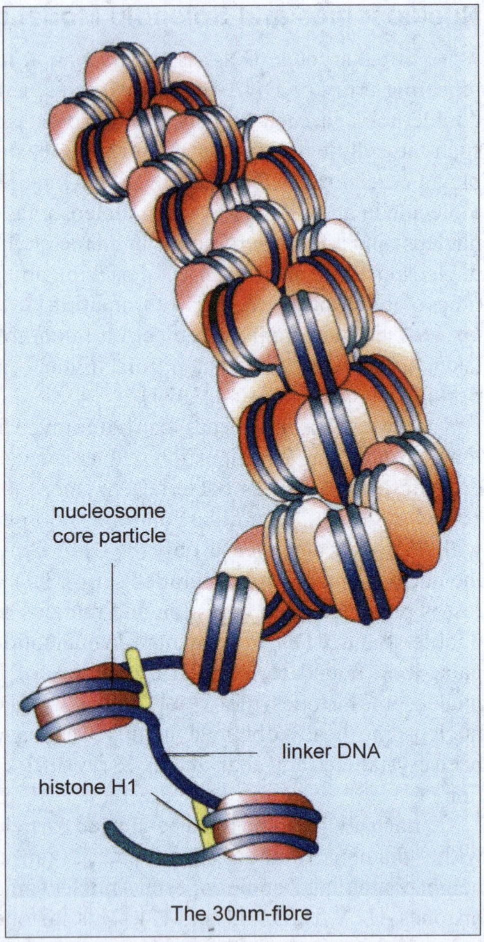

The 30-nm fibre consists of closely packed nucleosomes. It probably arises from the folding of the nucleosome chain into a **solenoid structure** having about six nucleosomes per turn (**Klug** and coworkers, 1976, 1979 and 1985). The DNA of 30-nm solenoid has a packing that is about 40-fold.

Histone H1 molecules are found responsible for packing nucleosomes into the 30-nm fibre. The H1 histone molecule has an evolutionarily conserved globular core or central region linked to extended amino-terminal and carboxyl-terminal "arms", whose amino acid sequence has evolved much more rapidly. Each H1 molecule binds through its globular portion to a unique site on a nucleosome and has arms that are thought to extend to contact with other sites on the histone cores of adjacent nucleosomes, so that the nucleosomes are pulled together into a regular repeating array (Fig. 13.10). The binding of H1 molecule to chromatin tends to create a local polarity that the chromatin otherwise lacks.

3. Radial loops of 30-nm fibre and metaphase chromosome. The nucleus is typically about 5µm (5×10^{-4} cm) in diameter. The packaging of DNA into a 30-nm chromatin fibre leaves a human chromosome about 0.1 cm long, so there must be several higher orders of folding. The probable nature of one further level of folding was originally suggested by the appearance of specialized chromosomes—the lampbrush chromosomes and polytene chromosomes. These two types of chromosomes

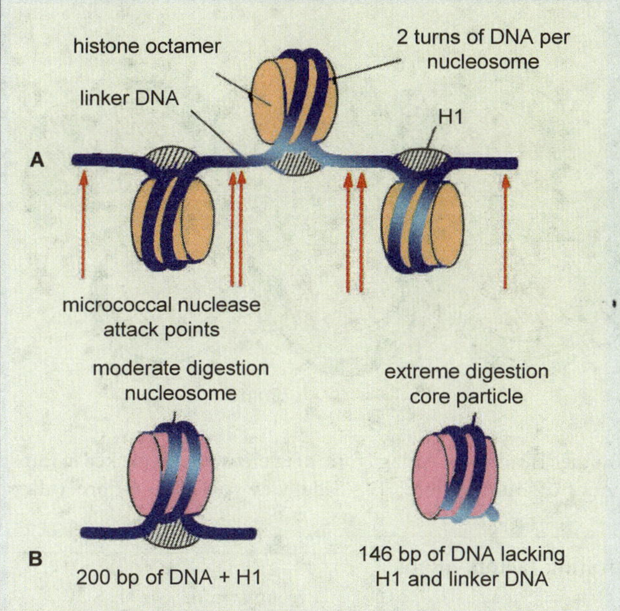

Fig. 13.7. Diagram showing the effect of nuclease enzymes on chromatin. A— Intact chromatin, note that the nucleo some is flat and that each histone octamer has two turns of DNA sealed off by histone H1. In living cells, the nucleosomes would be touching each other to form a 10 nm fibre and not stretched out (to form beads-on-a-string preparation) ; B— A nucleosome released by moderate digestion ; C— Core particle of nucleosome obtained by extensive digestion (after De Robertis and De Robertis, Jr. 1987).

seem to contain a series of **looped domains**— loops of chromatin that extend at an angle from the main chromosome axis. Since such loops do occur in *E. coli* chromosome, so the presence of loops may be a general feature of chromosomes.

In principle, looped domains in chromatin could be esta- blished and maintained by DNA binding proteins that clamp two regions of the 30- nm fibre together by recognizing specific DNA sequence that will form the neck of each loop. Alternatively, they could be formed by binding of DNA at the base of loop to a chromosome axis. Structural non-histone proteins could be involved in organizing the 30-nm fibres into loops. In an experiment, the histones are removed from the metaphase chromosome by adding the polyanion dextran sulphate. Histone-depleted chromosomes are found to have a central core of **scaffold** surrounded by a halo made of hoops of DNA (Fig. 13.11 A). The scaffold is made of non-histone proteins and retains the general shape of the metaphase chromosome. Each chromosome has two scaffolds, one for each chromatid, and they are connected together at the centromere region. When the histones are removed, the DNA which has packed about 40-fold in the 30-nm chromatin fibre, becomes extended and produces loops with an average length of 25μm (75,000 base pairs). In each loop the DNA exits from the scaffold and returns to an adjacent point. On the basis of these observations a model of chromosome structure has been proposed by **Laemmli** and coworkers (1979, 1984). In Laemmli's radial loop model DNA is arranged in loops anchored to the non-histone scaffold. Because the lateral loops have 25μm DNA, after contracting 40-fold in the 30-nm fibre, they would be only about 0.6μm long, a length consistent with the diameter of metaphase chromosome (1μm). Figure 13.11 B shows how the chromatin is arranged in loops which during metaphase, become arranged so that the base of the loops forms a scaffold in the centre of the chromatid. The base of the loop might be arranged on a helical coiled path (*e.g., Trillium* and *Tradescantia*).

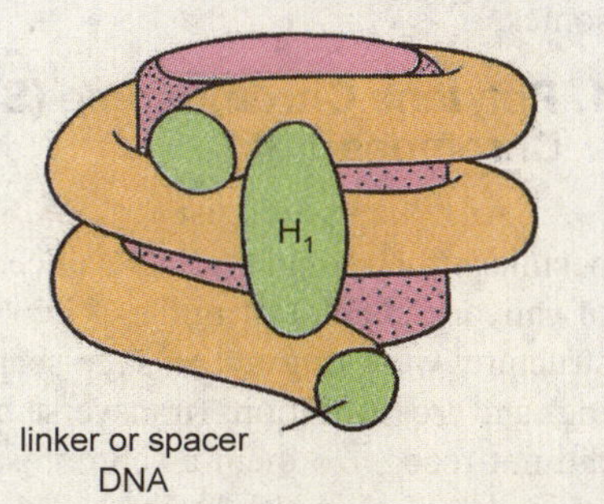

Fig. 13.8. A nucleosome (after Thorpe, 1984)

The chromatin in mitotic chromosomes is transcriptionally inert : all RNA synthesis ceases as the chromosomes condense. Presumably the condensation prevents RNA polymerase from gaining

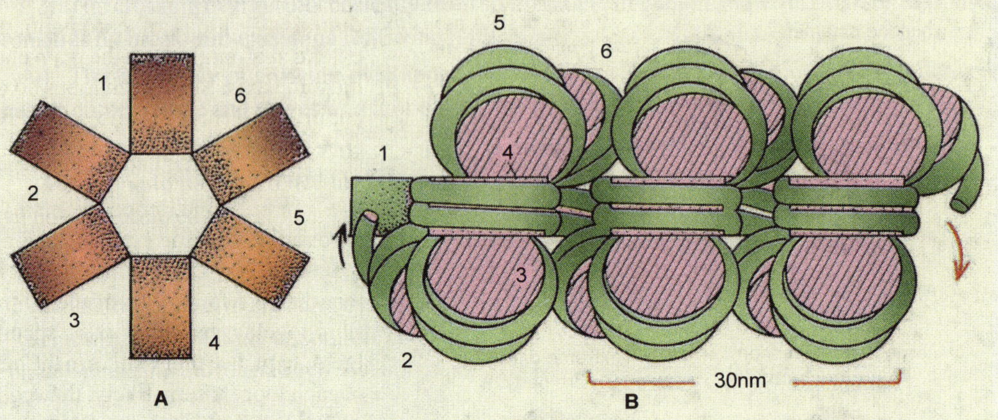

Fig. 13.9. A model suggested to explain how the "Beads-on-a-string" form of nucleosomes is packed to form the 30-nm fibre. A— Top view of 30-nm fibre; B—Side view of 30-nm fibre (after Alberts *et al,* 1989).

access to the DNA, although other controlling factors might also be involved.

FUNCTIONS

The function of chromosomes is to carry the genetic information from one cell generation to another. DNA being the only permanent component of chromosome structure, is the sole genetic material of eukaryotes.

GIANT CHROMOSOMES

Some cells at certain particular stages contain large nuclei with giant or large-sized chromosomes. The giant chromosomes are the **polytene** and **lampbrush** chromosomes.

1. Polytene Chromosome (Salivary Gland Chromosomes)

An Italian cytologist **E.G. Balbiani** (1881) had observed peculiar structures in the nuclei of certain secretory cells (*e.g.*, of salivary glands) of midge, *Chironomus* (Diptera). These structures were long and sausage-shaped and marked by swellings and cross striations (transverse bands). Unfortunately, he did not recognize them as chromosomes, and his report remained buried in the literature. It was not until 1933 that **Theophilus Painter**, **Ernst Heitz**, and **H. Bauer** rediscovered them in *Drosophila* and recognized them as the chromosomes. Since these chromosomes were discovered in the salivary gland cells, they were called **salivary gland chromosomes** (Fig. 13.13). The present name **polytene chromosomes** was suggested by **Kollar** due to the occurrence of many chromonemata (DNA) in them.

Thus, some cells of the larvae of the dipteran insects such as flies (*e.g., Drosophila*), mosquitoes and midges

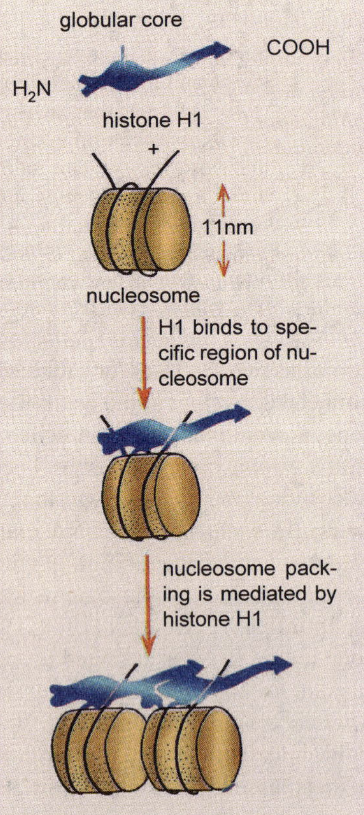

Fig. 13.10. The way histone H1 is thought to help to pack adjacent nucleosomes together (after Alberts *et al.*, 1989).

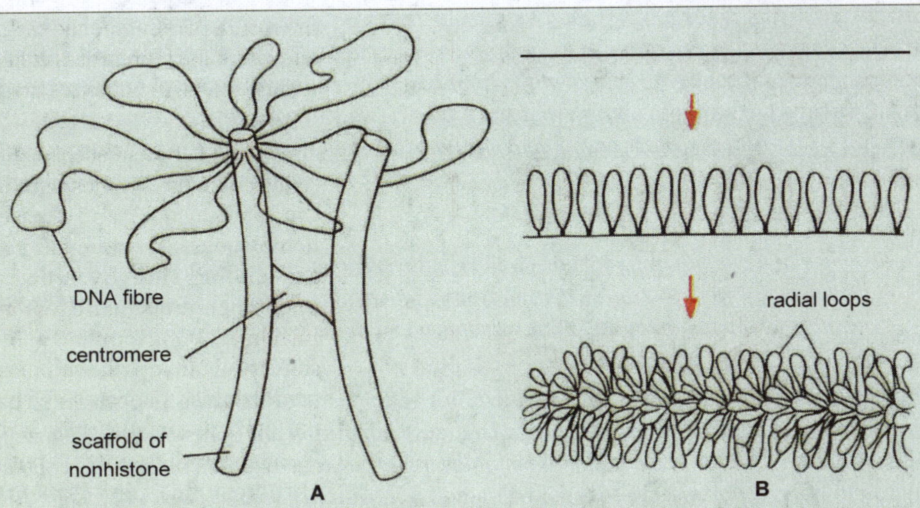

Fig. 13.11. Two methods by which DNA loops may form a metaphase chromosome. A—The non-histone proteins form two scaffolds, one per chromatid, while the naked DNA fibres form a halo around it; B—Laemmli's radial loop model of chromosome structure showing how 10-nm fibre form 30- nm fibre which further folds into radial loops (after De Robertis and De Robertis, Jr., 1987).

(*Chironomus*) become very large having high DNA content. These cells are unable to undergo mitosis and are destined to die during metamorphosis (Those cells of larva which are destined to produce the adult structures after metamorphosis, *i.e.,* imaginal discs remain diploid). Such polytenic cells are located most prominently in the salivary gland, but also occur in Malpighian tubules, rectum, gut, foot pads, fat bodies, ovarian nurse cells, etc. Polyteny of giant chromosomes is achieved by replication of the chromosomal DNA several times without nuclear division (endomitosis); and the resulting daughter chromatids do not separate but remain aligned side by side. In the process of endomitosis the nuclear envelope does not rupture and no spindle formation takes place. In fact, polyteny differs from polyploidy, in which there is also an excess DNA per nucleus, but in which the new chromosomes are separate from each other.

A giant chromosome of a midge (a small fly)

A polytene chromosome of *Drosophila* salivary gland has about 1000 DNA molecules which are arranged side by side and which arise from 10 rounds of DNA replication ($2^{10} = 1024$). Other dipteran species have more DNA, for example, *Chironomus* has 16000 DNA molecules in their each polytene chromosomes. Further, the polytene chromosomes are visible during interphase and prophase of mitosis. In them the chromomere (regions in which the chromatin is more tightly coiled) alternate with regions where the DNA fibres are folded more loosely (Fig. 13.15). The alignment of many chromomeres gives polytene chromosomes their characteristic morphology, in which a series of dark transverse **bands** alternates with clear zones called **interbands**. About 85 per cent of the DNA in polytene chromosomes is in bands and rest 15 per cent is in inter bands. The crossbanding pattern of each polytene chromosome is a constant characteristic within a species and helps in chromosome mapping during cytogenetic studies. For example, in *Drosophila melanogaster* there are about 5000 bands and 5000

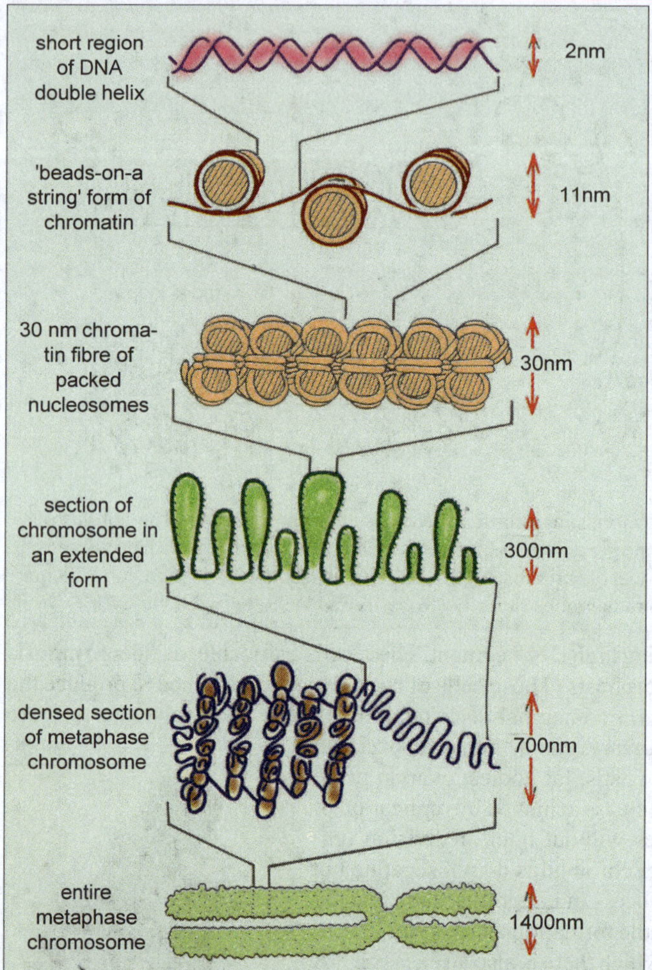

Fig. 13.12. Schematic diagram of some of many orders of chromatin packing which may give the highly condensed metaphase chromosome (after Alberts *et al.,* 1989).

interbands per genome and each band and interband represent a set of 1024 identical DNA sequences arranged in file.

Another peculiar characteristic of the polytene chromosomes is that the maternal and paternal homologous chromosomes remain associated side by side. This phenomenon is called **somatic pairing**. Consequently in the salivary gland cells the chromosome number always appear to be half of the normal somatic cells, *e.g., D. melogaster,* has only 4 polytene chromosomes. In *Drosophila*, pericentromeric heterochromatin of all polytene chromosomes also coalesces in a **chromocentre**.

The preparation of a slide of the polytene chromosomes of dipterans for light microscopy is rather easy. The larvae are taken at the third instar stage and the salivary glands are dissected out and squashed in aceto-carmine. In such preparations, these chromosomes in aggregate reach a length of as much as 2000 μm in *D. melanogaster*. In female *Drosophila,* the polytene chromosomes are found in the form of five long and one short strands radiating from a single more or less amorphous chromocentre (Fig. 13.15). One long strand corresponds

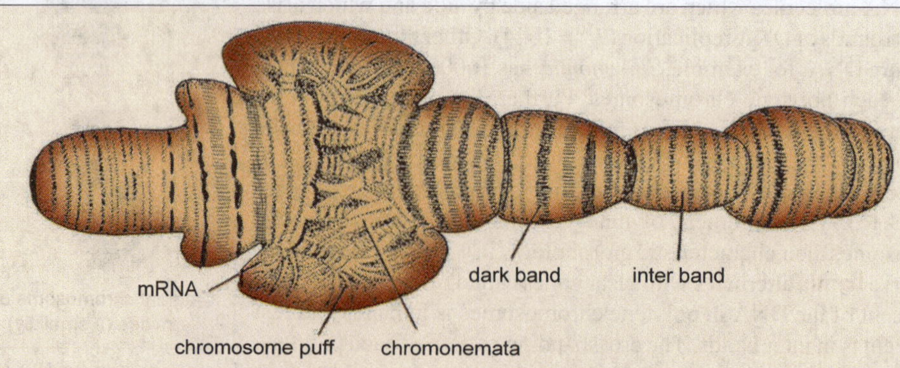

Fig. 13.13. Polytene chromosome of an insect, showing bands and interbands and a puff or Balbiani ring.

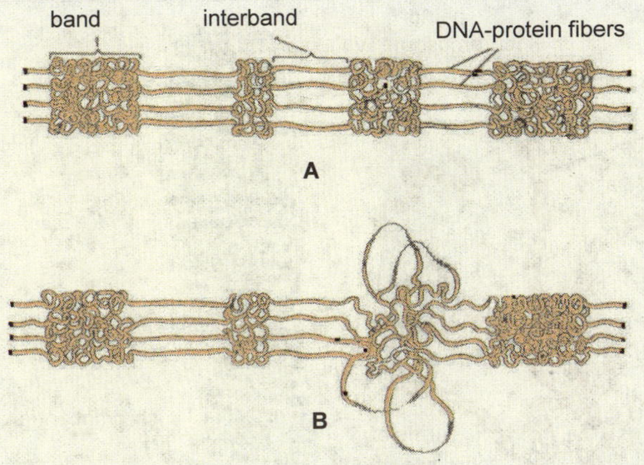

Fig. 13.14. A— Model of Dupraw and Rae showing band and interband structure in giant chromosome. B— The same model showing a possible method of "puffing" in one of the bands.

to the X chromosome and remaining four long strands are the left and right arms of II and III chromosomes. The shortest strand represents the small dot-like IV chromosome. Each of these chromosomes contains maternal and paternal homologues in somatic pairing which lacks in the sex chromosomes of male fruit flies. Thus, in male *Drosophila,* X chromosome remains single and thin and Y chromosome exists indistinctly fused with the chromocentre.

One-gene, one-band hypothesis. The fixed pattern of bands and interbands in a *Drosophila* polytene chromosome suggested the early cytologists such as **Painter** (1933) and **Bridges** (1936) that each band might correspond to a single gene. Accordingly, they concluded that *Drosophila* might contain only 5000 essential genes. It was also believed that bands were the sites of genes (DNA) and interbands were relatively inert linker regions. Recent data, however, have contradicted this simple "one-band, one-gene" hypothesis, now it is held that bands as well as inter-bands contain active genes and a band may even contain more than one genes.

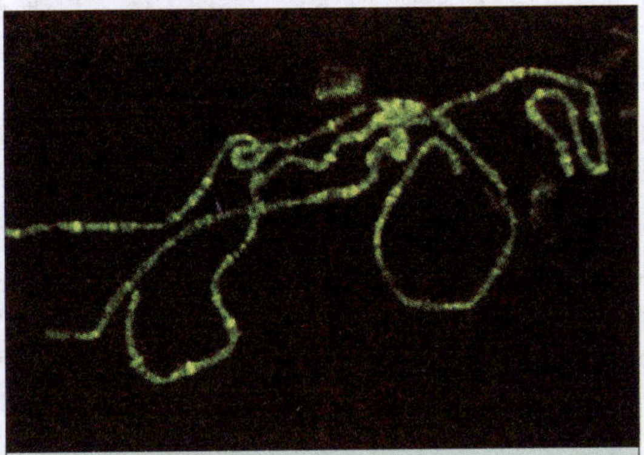

In the giant chromosomes of the *Drosophila* active genes appear as brighter bands.

At this juncture a question arises, why is the single long strand of chromatin in each chromosome subdivided into so many distinct regions ? Exact explanation of this question is still not known. However, **Alberts** *et al.,* (1989) believed that this type of organization (*i.e.,* banding and interbanding of chromosomes in general) may help to : (1) keep the DNA organized; (2) isolate genes from their neighbours and thereby prevent biological "crosstalk", or (3) regulate gene transcription for the cytodifferentiation, for example, constitutively expressed "**housekeeping**" **genes** could be located in interbands, whereas **cell-type-specific genes** could be confined to the bands.

Chromosome puffs or Balbiani rings. Chromosome puffs or **Balbiani rings** are the swellings of bands of the polytene chromosomes (Fig. 13.16) where DNA unfolds into open loops as a consequence of intense gene transcription (*e.g.*, mRNA formation). In 1954, **W. Beerman** compared the polytene chromosomes of different tissues of *Chironomus* larvae and showed that although the pattern of bands and interbands was similar in all tissues, the distribution of puffs differred from one tissue to another. **Beerman** and **Bahr** (1954) have studied the fine structure of these puffs. According to them these puffs represent regions where the tightly coiled chromosomal fibres open out to form many loops. In fact, puffing is a cyclic and reversible phenomenon : at definite times and in different tissues of the larvae, puffs may appear, grow and disappear. In salivary glands the appearance of some puffs has been correlated with the production of specific proteins which are secreted in large amounts in the larval saliva (**Grossbach**, 1977). The process of puffing involves several processes such as the accumulation of acidic proteins, despiralization of DNA, accumulation of **RNA polymerase II** (an

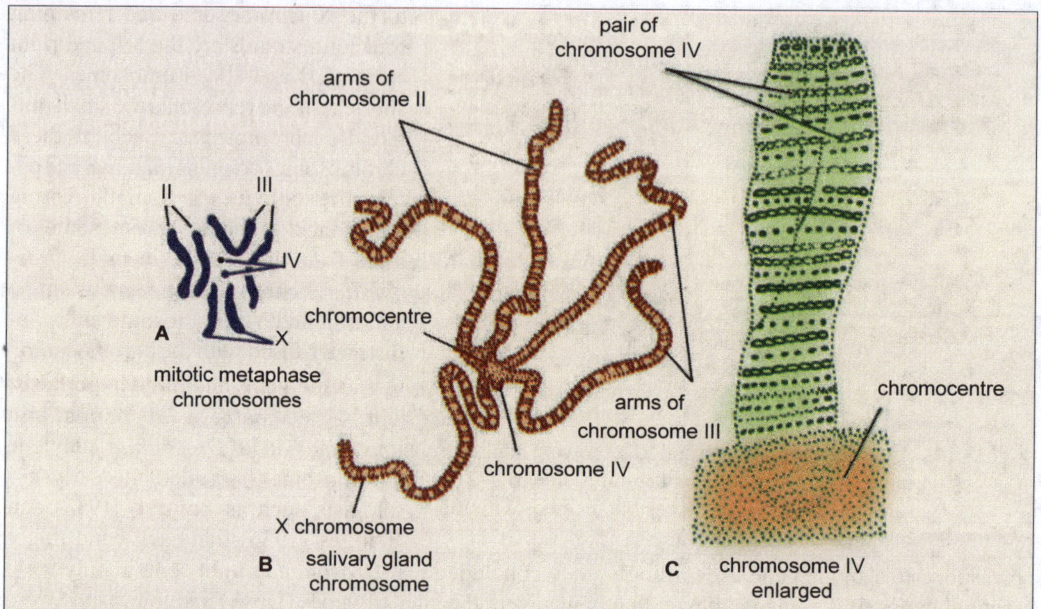

Fig. 13.15. Normal and polytene chromosomes of *Drosophila melanogaster*. A—Normal mitotic chromosomes; B—Polytene chromosomes of female; C—An enlarged IV chromosome.

enzyme involved in the transcription of m RNA molecules), synthesis of mRNA and release of newly synthesized mRNA in the cytoplasm.

2. Lampbrush Chromosomes

The lampbrush chromosomes were first observed in salamander (amphibian) oocytes in 1882. He coined the name because the chromosomes look like the brushes which were used for cleaning the glass chimneys of old-fashioned paraffin or kerosene lamps. They were described in detail in shark oocytes by **R. Ruckert** in 1892. **Thorpe** (1984) and **Burns** and **Bottino** (1989) preferred the term **test tube brush chromosomes** for them. However, due to recent investigations of **Gall** and coworkers (1962, 1983) the structure of these exceptionally large-sized chromosomes has been interpreted in

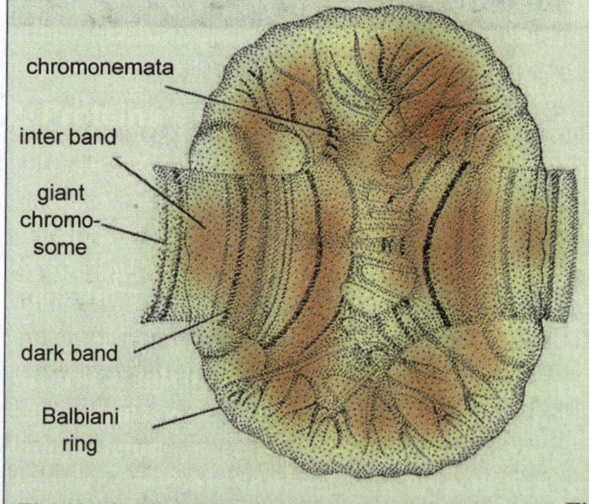

Fig. 13.16. A Balbiani ring of a polytene chromosome.

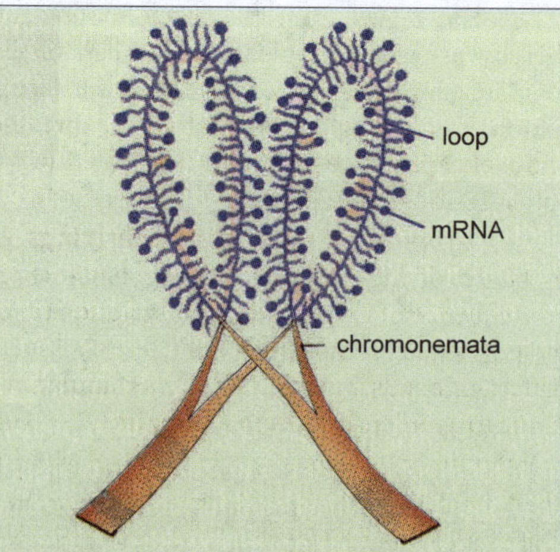

Fig. 13.17. Suggested form of the individual loops in chromosome puffs. Note the tiny fibrils of RNA and proteins attached to the loop.

functional terms, *i.e.*, now they are merely visualized as means of *"turning on and turning off"* of the genes.

The lampbrush chromosomes occur at the diplotene stage of meiotic prophase in the primary oocytes of all animal species, both vertebrates and invertebrates. Thus, they have been described in *Sagitta* (Chaetognatha), *Sepia* (Mollusca), *Echinaster* (Echinodermata) and in seveal species of insects, shark, amphibians, reptiles, birds and mammals (humans). Lampbrush chromosomes are also found in spermatocytes of several species, giant nucleus of *Acetabularia* and even in plants (**Grun**, 1958). Generally, they are smaller and "hairy" in invertebrates than in vertebrates. Lamphrush chromosomes are best visualized in salamander oocytes because they have a high DNA content. For example, the largest chromosome having a length up to 1 mm have been observed in urodele amphibian. Thus, lampbrush chromosomes are much larger (longer) than the polytene chromosomes of insects.

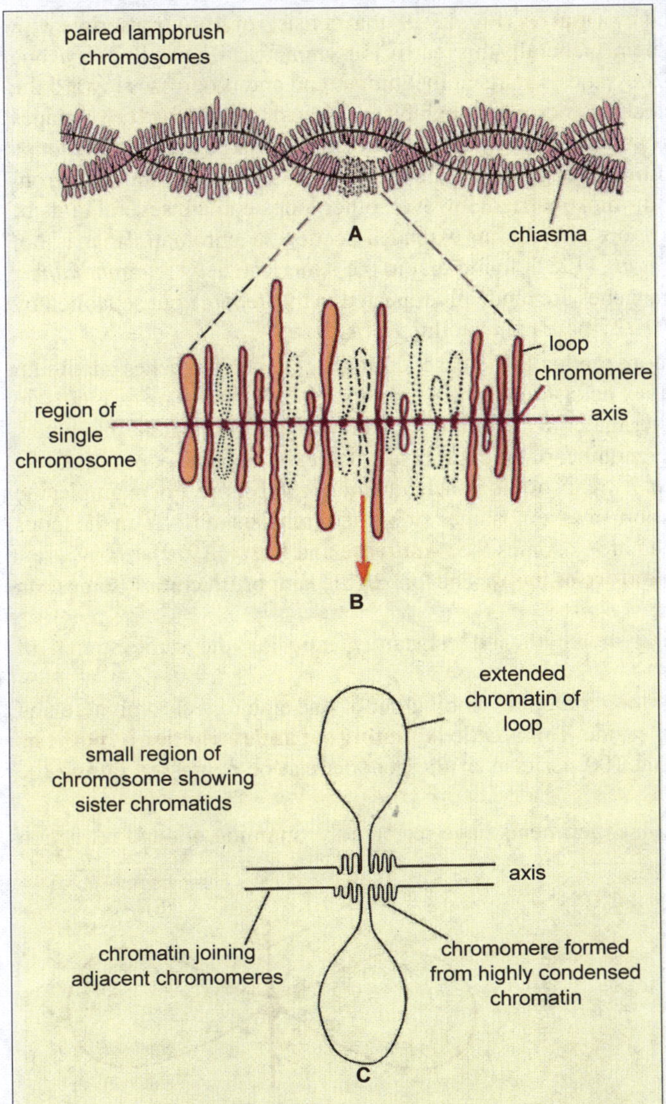

Fig. 13.18. Lampbrush chromosome structure. A—Bivalent or paired homologous chromosomes in pairing showing chiasmata, B—A part of one homologue showing paired loops given out by two chromatids; C—Single pair of loop (after Alberts *et al.*, 1989).

Since the lampbrush chromosomes are found in the prolonged diplotene stage of meiotic prophase I, they are present in the form of **bivalents** in which the maternal and paternal chromosomes are held together by **chiasmata**, at those sites where crossing over has previously occurred (Fig. 13.20). The paired homologues are not condensed as usual chromosomes would be; instead, they are very long and stretched out. Each bivalent has four chromatids, two in each homologue. The axis of each homologue consists of a row of granules or **chromomeres** from which **lateral loops** extend. The loops are always symmetrical, each chromosome having two of them, one for each chromatid. The loops can be categorized by size, thickness and other morphological characteristics. Each loop appears at a constant position in the chromosome ; this fact helps in the chromosome mapping. There are about 10,000 loops per chromosome set or haploid set (*e.g.*, oocytes of slamander *Triturus* ; see **Grant**, 1978). Each loop has an **axis** which is made of single DNA molecule that is unfolded from the chromosome for the intense RNA synthesis. Thus, about 5 to 10 per cent of the DNA exists in the lateral loops, the rest being tightly condensed in the chromomeres which are transcriptionally inactive. The centromeres of the chromosomes bear no loops.

Each loop of lampbrush chromosomes is found to perform intense transcription of **hn RNA** or **heterogeneous RNA** molecules. (*i.e.,* precursors of mRNA molecules for various ribosomal proteins and for five types of histone proteins). Electron microscopy of the loops has shown that **RNA polymerase** enzyme molecules are attached to the principal axis (DNA) of the loop from which RNA fibrils of increasing length extend. As transcription continues along the DNA strand of loop, the fibrils of RNA (*i.e.*, hnRNA) lengthen. Proteins get associated with these RNA fibrils as they are formed and ultimately ribonucleoprotein product is released.

Thus, each lateral loop is covered by a **matrix** (Fig. 13.19) that consists of RNA transcripts with hn RNA-binding proteins attached to them. Generally this matrix is asymmetrical, being thicker at one end of the loop than at the other. RNA synthesis starts at the thinner end and progresses toward the thicker end. Preparations spread for electron microscopy exhibit the typical '**christmas tree**" images with nascent ribonucleoprotein chains attached perpendicularly to the DNA axis. Many of the loops correspond to a single transcriptional unit (or single gene) and they are transcribed continuously from end to end; they form a continuous thin-thick matrix. However, other loops contain several units of transcription (or many genes); such loops include an extended section of chromatin that is not transcribed at all. For example, **Gall** *et al.*, (1981) found that in the American newt *Notophthalmus viridescens*, clusters of the five histone genes are tandemly repeated in the genome but separated by about 50,000 base pairs of repeats of a 225-base pair satellite DNA.

Further, the number of pairs of loops gradually increases during meiosis till it reaches maximum in diplotene. Such a lampbrush stage may persist for months or years as the primary oocyte builds up a supply of mRNA molecules and other materials required for its ultimate development into a new individual. As meiosis proceeds further, number of loops gradually decreases and the loops ultimately disappear either due to disintegration or by reabsorption back into the chromosome. For example, the addition of histone proteins to the lampbrush chromosomes stops the syntheisis of RNA on the loops and causes the loops to retract into the chromosomes (see **Ambrose** and **Easty**, 1979).

Certain hypotheses regarding nature of loops. The loop of the lampbrush chromosomes can be viewed in the following two ways :

1. It may be static, unchanging in length and constantly exposing the same stretch of chromosome fibre.

2. It may be dynamic, with new loop material spinning out of one side of a chromomere and returning to a condensed state on the other side. This is called **spinning out and retraction hypothesis** (**Gall** and **Callan**, 1962). It means that 100 per cent of the genome can be expressed during the lampbrush stage.

Recent, DNA-RNA hybridization experiments have rejected the spinning out and retraction hypothesis (see **Grant**, 1978).

Master and slave hypothesis. **Callan** and **Llyod** (1960) suggested that each loop pair and thus each chromomere is associated with the activity of one specific gene. Their master and slave hypothesis was postulated to explain the large size of the chromomere and of lampbrush loop; presently this hypothesis has become obsolete, but still holds interest. This hypothesis postulates that each loop consists not of one gene, but of a

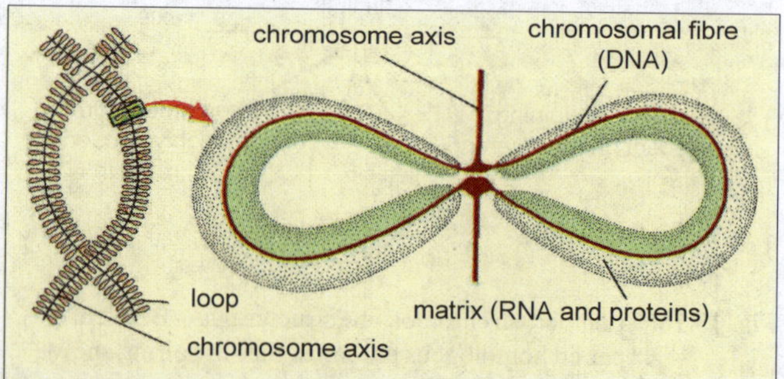

Fig. 13.19. Lampbrush chromosome. A—At low magnification; B—Loop magnified.

number of duplicate copies, linearly arranged, of one gene. There is a "**master**" **copy** at each chromomere and information is transferred from this to each of the "**slave**" **copies** which are matched against it to ensure that they are all identical to the master. The master copy of the gene does not take part in RNA synthesis, but the slave copies of the gene existing in the loop have a role in transcription. The advantage of having a number of duplicate copies of a gene is that a higher rate of RNA synthesis is possible.

Study of both polytene and lampbrush chromosomes provided the evidence that eukaryotic gene activity is regulated at the level of RNA synthesis (or transcription). Lampbrush chromosomes also show the possible way of gene amplification which is required during the growth phase of oocytes.

REVISION QUESTIONS

1. Give an account of the morphology, ultrastructure and chemistry of the chromosomes.
2. Distinguish between the members of each pair :
   (i) diploid-haploid; (ii) chromatid-chromosome; and (iii) euchromatin-heterochromatin.
3. Describe the basic structure of chromatin as we understand it. What is the role of histones in this structure ?
4. Describe the structue of the prokaryotic chromosome.
5. Write short notes on the following :chromosomal proteins ; nucleosome ; polytene chromosome; and lampbrush chromosomes.
6. Why the study of the chromosomes has become very significant in the field of biology ?
7. Write short account on the following : (i) Nucleosome concept and solenoid model ; (ii) Karyotype ; (iii) Kinetochore ; (iv) Heterochromatin ; (v) C-value and C-value paradox; (vi) Salivary gland

CHAPTER

14

Ribosomes

The ribosomes are small, dense, rounded and granular particles of the ribonucleoprotein. They occur either freely in the matrix of mitochondria, chloroplast and cytoplasm (*i.e.*, cytoplasmic matrix) or remain attached with the membranes of the endoplasmic reticulum and nucleus. They occur in most prokaryotic and eukaryotic cells and are known to provide a scaffold for the ordered interaction of all the molecules involved in protein synthesis.

HISTORICAL

Ribosomes are remarkable organelles of cell. They were studied before they were discovered. Thus, ribosomes were studied in the early 1930s, discovered and isolated in the early 1940s, scrutinized in 1950s and baptized in 1958. In 1960s they were dissociated and reconstituted; in 1970s sequenced and studied topographically; and in 1980s they continue to be object of considerable research.

Before 1930s, it was the prevailing view that DNA was found only in animal cells and RNA only in plant cells. In 1930s various direct studies, employing basic staining teachniques that discriminated between DNA and RNA, and spectrophotometric measurements of absorption in different cell regions, confirmed that RNA is present in the cytoplasm of both plant and animal cells and suggested that DNA is found exclusively

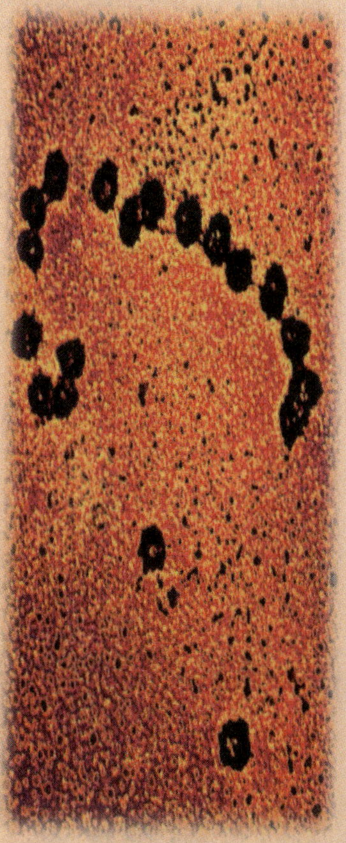

Ribosomes may be found free in the cytoplasm either singly or strung along messenger RNA molecules as they participate in protein synthesis. Ribosomes also stud the rough endoplasmic reticulum, giving it a rough appearance and allowing the synthesis of proteins within the ER.

in the nucleus (**Brachet** and **Caspersson**). Using quantitative techniques in 1940s, a very significant observation was made regarding the ribosome function. It was reported that cells were rich in RNA when they were active in protein synthesis. For example, secretory cells such as pancreas cells and silk gland cells, were noted to be RNA-rich, whereas cells of other types (non-secretory cells such as heart muscle cells which make little new proteins) are relatively RNA-poor. In 1940s **Albert Claude** homogenized chick and mammalian embryos and obtained a fraction containing what he called **microsomes**— particles of ribonucleoprotein and lipid visible with the dark field microscope. He, thus, showed that the cytoplasmic RNA was included in tiny particles of ribonucleoprotein later to be called "ribosomes'. In 1952, **G.E. Palade** described the ribosome. Their presence in both free and membrane attached form was confirmed by **Palade** and **Siekevitz** by the electron microscopy. In 1950s another technique, namely ultracentrifugal analysis was employed to study the ribosomes. This showed that ribosomes sedimented at discrete peaks in the 40S—70S range. When purified by centrifugation and electrophoresis, they were found to contain half RNA and half protein. As $MgCl_2$ was known to precipitate RNA, **Siekevitz** suggested that RNA might somehow be involved in protein synthesis. In 1952, **Siekevitz** and **Zamecnik** showed clearly that radioactive amino acids first were incorporated into proteins on ribosomes and then were released to the soluble portions of the cell.

In 1958, the papers presented at a meeting of the Biophysical Society at the Massachusetts Institute of Technology were published in a book form. **R.B. Roberts** edited this collection of papers and coind the name **ribosome** in his introductory comments. The term ribosome is due to rich RNA content of this organelle. **Tissieres** and **Watson** (1958) isolated 70S *E.coli* ribosomes and showed that they consist of two subunits, 50S and 30S. In 1960s ribosomes were subjected to exhaustive electrophoretic and chromatographic procedures, this time not to purify them but to examine their parts. It soon became clear that ribosomes contain three or four kinds of RNA and scores of proteins.

Recently, various workers such as **Lake**, **Nomura**, **Wittman**, **Traut** , **Stoffler**, **Kurland**, etc., have studied the relationship between rRNAs and ribosomal proteins to work out the topology of ribosomes (Topology includes study of detailed shape and positions of the individual proteins and rRNA molecules relative to each other; see **King**, 1986).

OCCURRENCE AND DISTRIBUTION

The ribosomes occur in cells, both prokaryotic and eukaryotic cells. In prokaryotic cells the ribosomes often occur freely in the cytoplasm. In eukaryotic cells the ribosomes either occur freely in the cytoplasm or remain attached to the outer surface of the membrane of endoplasmic reticulum. The yeast cells, reticulocytes or lymphocytes, meristamatic plant tissues, embryonic nerve cells and cancerous cells contain large number of ribosomes which often occur freely in the cytoplasmic matrix. The cells in which active protein synthesis takes place, the ribosomes remain attached with the membranes of the endoplasmic reticulum. Such cells are the pancreatic cells, plasma cells, hepatic parenchymal cells, Nissls bodies, osteoblasts, serous cells, or the submaxillary gland, chief cells of the glandular stomach, thyroid cells and mammary gland cells. The cells which synthesize specific proteins for the intracellular utilization and storage often contain large number of free ribosomes. Such cells are the erythroblasts, developing muscle cells, skin and hair.

METHOD OF ISOLATION

The ribosomes are usually isolated from the cell by the differential centrifugation method in which an analytical centrifuge is employed. The sedimentation coefficient of the ribosomes is determined by the various optical and electronic techniques. The sedimentation coefficient is expressd in the **Svedberg unit**, *e.g.,* **S** unit. The **S** is related with the size and molecular weight of the ribosomal particles.

TYPES OF RIBOSOMES

Recently accroding to the size and the sedimentation coefficient (S) two types of ribosomes have been recognised (Fig. 14.1).

1. 70S Ribosomes. The 70S ribosomes are comparatively smaller in size and have sedimentation coefficient 70S and the molecular weight 2.7× 106 daltons. (**Dalton** is the unit of molecular weight **(MW)**; one dalton equals the weight of hydrogen atom. For example, a water molecule weighs 18 daltons, see **De Robertis** *et al.*, 1970). According to the data of electron microscopy the dimension of the dry particles of 70S ribosomes are 170 × 170 × 200 A° (**Hall** and **Stayter**, 1959, **Huxley** and **Zubay**, 1960). They occur in the prokaryotic cells of the blue green algae and bacteria and also in mitochondria and chloroplasts of eukaryotic cells.

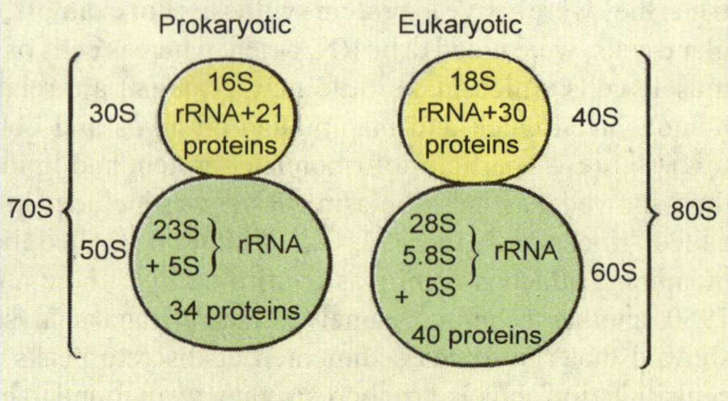

Fig. 14.1. Various components of prokaryotic (70S) and eukaryotic (80S) ribosomal subunits (after De Robertis and De Robertis, Jr. 1987).

2. 80S Ribosomes. The 80S ribosomes have the sedimentation coefficient of 80S and the molecular weight 40 × 10^6 daltons. The 80S ribosomes occur in eukaryotic cells of the plants and animals.

The ribosomes of mitochondria and chloroplasts are always smaller than 80S cytoplasmic ribosomes and are comparable to prokaryotic ribosomes in both size and sensitivity to antibiotics, although their sedimentation values vary in different phyla, *e.g.,* 77S in mitochondria of fungi, 60S in mitochondria of mammals and 60S in mitochondria of animals in general. The ribosomes of chloroplasts are 70S type.

NUMBER OF RIBOSOMES

An *E. coli* cell contains 10,000 ribosomes, forming 25 per cent of the total mass of the bacterial cell. In contrast, mammalian cultured cells contain 10 million ribosomes per cell, each of which is about twice as large as a prokaryotic ribosome.

STRUCTURE OF RIBOSOMES

The ribosomes are oblate spheroid structures of 150 to 250A° in diameter. Each ribosome is porous, hydrated and composed of two subunits. One ribosomal subunit is large in size and has a dome-like shape, while the other ribosomal subunit is smaller in size and occurring above the larger subunit and forming a cap-like structure.

The 70S ribosome consists of two subunits, *viz.*, 50S and 30S. The 50S ribosomal subunit is larger in size and has the size of 160 A° to 180 A°. The 30S ribosomal subunit is smaller in size and occurs above the 50S subunit like a cap (Fig. 14.1A).

The 80S ribosome also consists of two subunits, *viz.*, 60S and 40S. The 60S ribosomal subunit is dome-shaped and larger in size. In the ribosomes which remain attached with the membranes of endoplasmic reticulum and nucleus, etc., the 60S subunit remains attached with the membranes. The 40S ribosomal subunit is smaller in size and occurs above the 60s subunit forming a cap-like structure. Both the subunits remain separated by a narrow cleft (Fig. 14.1B).

The two ribosomal subunits remain united with each other due to high concentration of the Mg^{++}(.001M) ions. When the concentration of Mg^{++}ions reduces in the matrix, both ribosomal subunits get separated. Actually in bacterial cells the two subunits are found to occur freely in the cytoplasm and they unite only during the process of protein synthesis. At high concentration of Mg++ ions in the matrix, the two ribosomes (called **monosomes**) become associated with each other and known as the **dimer**. Further, during protein synthesis many ribosomes are aggregated due to common messenger RNA and form the **polyribosomes** or **polysomes**.

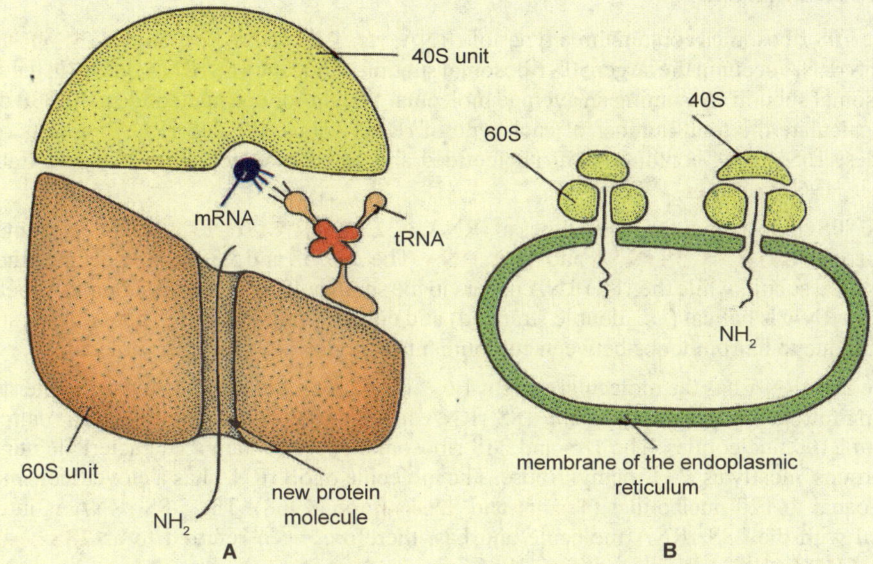

Fig. 14.2. Functional 80S ribosomes : A—Diagram of a ribosome showing the two subunits and the probable position of mRNA and tRNA. The nascent polypeptide chain passes through a kind of tunnel within the large subunit; B—Diagram showing the relationship between the ribosome and the membrane of endoplasmic reticulum and the entrance of the polypeptide chain into the centre of endoplasmic reticulum during the process of protein synthesis.

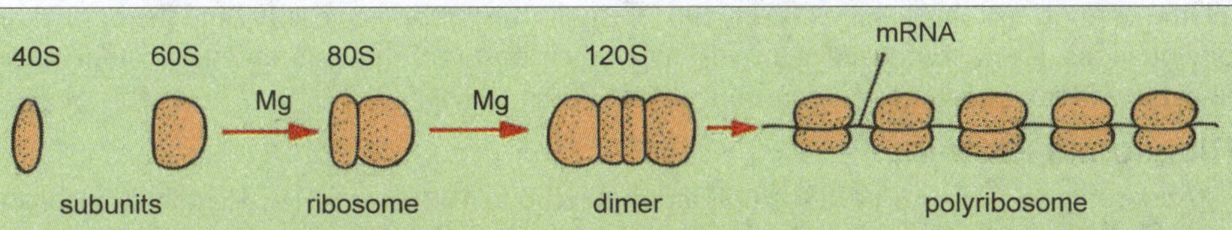

Fig. 14.3. Diagram of the subunit structure of the ribosome and the influence of Mg^{++} ions.

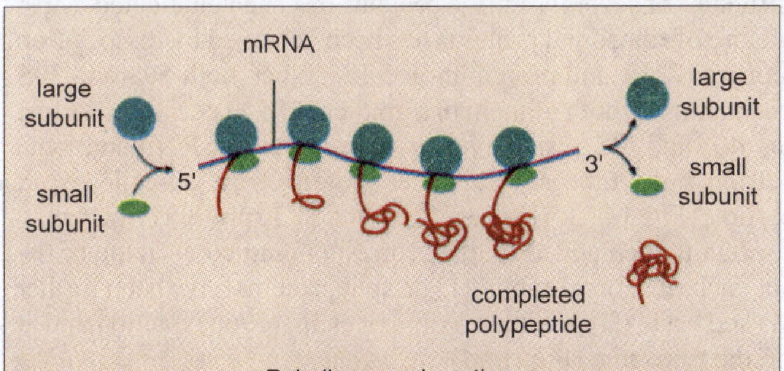

Polyribosome in action.

CHEMICAL COMPOSITON

The ribosomes are chemically composed of RNA and proteins as their major constituents ; both occurring approximately in equal proportions in smaller as well as larger subunit. However, the 70S ribosomes contain more RNA (60 to 40%) than the proteins (36 to 37%), *e.g.*, the ribosomes of *E. coli* contain 63% rRNA and 37% protein. While the 80S ribosomes contain less RNA (40 to 44%) than the proteins (60 to 56%), *e.g.*, yeast ribosomes have 40 to 44% RNA and 60 to 56% proteins ; ribosomes of pea seedling contain 40% RNA and 60% proteins. There is no lipid content in ribosomes.

1. Ribosomal RNAs

The 70S ribosomes contain three types of rRNA, *viz*., **23S rRNA**, **16S rRNA**, **5S rRNA**. The 23S and 5S rRNA occur in the larger 50S ribosomal subunit, while the 16S rRNA occurs in the smaller 30S ribosomal subunit. Assuming an average molecular weight for one nucleotide to be 330 daltons, one can calculate the total number of each type of rRNA. Thus, the 23S rRNA consists of 3300 nucleotides, 16S rRNA contains 1650 nucleotides and 5S rRNA includes 120 nucleotides in it (**Brownlee**, 1968 ; **Fellner**, 1972).

The 80S ribosomes contain four types of rRNA, *viz*., **28S rRNA** (or **25-26 rRNA** in plants, fungi and protozoa), **18S rRNA**, **5S rRNA** and **5.8S rRNA**. The 28S, 5S and 5.8S rRNAs occur in the larger 60S ribosomal subnit, while the 18S rRNA occurs in the smaller 40S ribosomal subunit. About 60 per cent of the rRNA is helical (*i.e*., double stranded) and contains paired bases. These double stranded regions are due to hairpin loops between complimentary regions of the linear molecule.

The 28S rRNA has the molecular weight 1.6×10^6 daltons and its molecule is double stranded and having nitrogen bases in pairs. The 18S rRNA has the molecular weight $0.6x10^6$ daltons and consists of 2100 nucleotides. The 18S and 28S ribosomal RNA contain a characteristic number of methyl groups, mostly as 2'-O-methyl ribose. The molecule of 5S rRNA has a clover leaf shape and a length equal to 120 nucleotides (**Forget** and **Weissmann**, 1968). The 5.8S rRNA is intimately associated with the 28S rRNA molecule and has, therefore, been referred to as **28S-associated ribosomal RNA** (**28S-A rRNA**) (**Avers**, 1976).

The 55S ribosomes of mammalian mitochondria lack 5S rRNA but contain **21S** and **12S rRNAs**. The 21S rRNA occurs in larger or 35S ribosomal subunits, while 12S rRNA occur in smaller or 25S ribosomal subunit. The sedimentation coefficient of ribosomes, ribosomal subunits, rRNAs and number of ribosomal proteins of certain representative organisms have been tabulated in Table 14.1.

It is thought that each ribosomal subunit contains a highly folded ribonucleic acid filament to which the various proteins adhere (**Hart**, 1965). But as the ribosomes easily bind the basic dyes so it is concluded that RNA is exposed at the surface of the ribosomal subunits, and the protein is assumed to be in the interior in relation to non-helical part of the RNA.

2. Ribosomal Proteins

According to **Nomura** (1968, 1973) and **Garett** and **Wittmann** (1973) each 70S ribosome of *E. coli* is composed of about 55 **ribosomal proteins**. Out of these 55 proteins, about 21 different molecules have been isolated from the 30S ribosomal subunit, and some 32 to 34 proteins from the 50S ribosomal subunit. The primary structure of several of these proteins has been elucidated. Most of the recent knowledge about the structure of ribosomal proteins has been achieved by dissociation of ribosomal subunits into their component rRNA and protein molecules. When both 50S and 30S ribosomal subunits are dissociated by centrifuging both of them in a gradient of 5 M cesium chloride, then there are two inactive core particles (40S and 23S, respectively) which contain the RNA and some proteins called **core proteins** (**CP**) at the same time several other proteins—the so-called **split proteins** (**SP**) are released from each particle (Fig. 14.3). There are SP50 and SP30 proteins which may reconstitute the functional ribosomal subunit when added to their corresponding core. Some of the split proteins are apparently specific for each ribosomal subunit. The split proteins have been further fractionated and divided into acidic (A) and basic (B) proteins. **Nomura** *et al*., (1968) fractionated at least six different groups of proteins in the ribosome (Fig. 14.4).

In all, 21 types of proteins have been isolated in smaller subunit (30S) of ribosome of *E. coli*. These are designated as S1 to S21. Similarly, in larger subunit (50S) 34 different proteins designated as L1 to L34, have been isolated. Thus, the 70S ribosome was thought to consist of 55 different proteins. However it was later shown that protein S20 is identical to L26, thus, the correct number of S proteins is 20. Likewise, L8 was shown to be an aggregate of proteins L7, L12 and L10; thus, the correct number of L proteins is 33. Thus, the prokaryotic 70S ribosome consists of 53 different proteins

(20S + 33L = 53 proteins). Similar organization of ribosomal proteins and RNA is found in 80S ribosomes (Fig. 14.5).

Table 14.1. Some characteristics of ribosomes of various organisms (Avers, 1976).

Source	Intact ribosomes	Ribosome subunits	rRNA in subunit	Number of proteins in subunit
Prokaryotes	70S	30S 50S	16S 23S, 5S	21 32-34
Eukaryotes	80S	40S 60S		~30 ~50
Animals		40S 60S	18S 28S,5S, 5.8S	
Plants		40S 60S	18S 25-26S, 5S, 5.8S	
Fungi		40S 60S	18S 25-26S, 5S 5.8S	
Protozoa (some other protists)		40S 60S	18S 25-26S, 5S 5.8S	

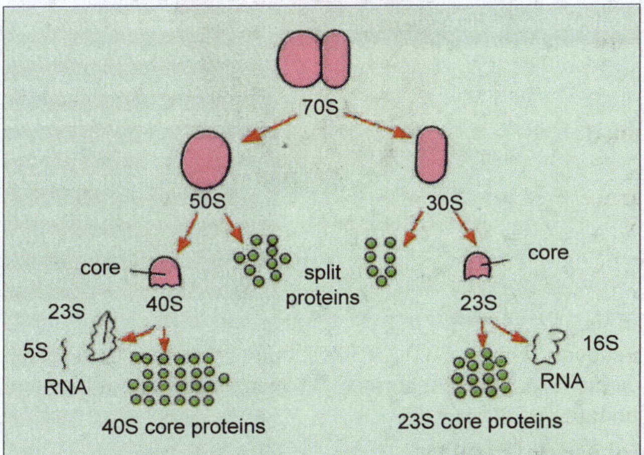

Fig. 14.4. Diagram showing the stepwise dismantling of the two subunits of 70S ribosome. Note that the proteins may be separated into split and core proteins. The 50S subunit contains 23S and 5S RNAs, and the 30S subunit has 16S RNA (after De Robertis and De Robertis, Jr., 1987).

Different rRNA molecules evidently play a central role in the catalytic activities of ribosomes in the process of protein synthesis. Various ribosomal proteins have been found to mainly enhance the catalytic function of the rRNA in the ribosomes (see **Alberts** *et al.*,1989).

3. Metallic Ions

The most important low molecular weight components of ribosomes are the divalent metallic ions such as Mg^{++}, Ca^{++} and Mn^{++}

ULTRASTRUCTURE

Molecular organization and function of ribosomes have been studied more intensively in prokaryotes than in eukaryotes. Fine or ultra-structure of 70S ribosome is very complex. In it, the RNA and proteins are intertwined and arranged in a complex manner in the two subunits. Since the positive protein charges are not sufficient to balance the many negative charges in the phosphates of the RNA, so ribosomes are strongly negative and bind cations and basic stains. Consequently, **negative staining** of ribosomes has led to better understanding

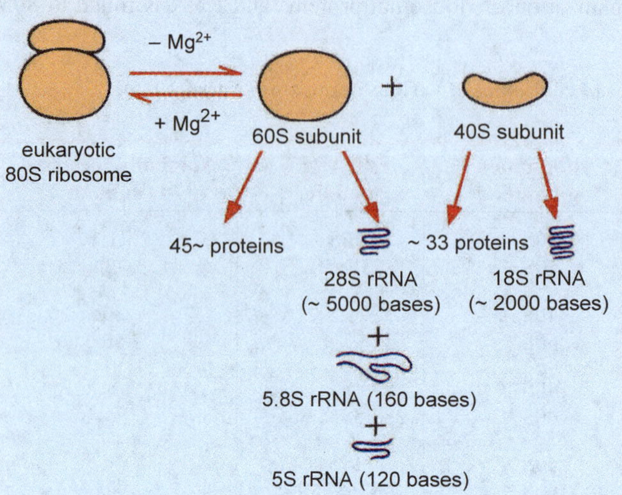

Fig. 14.5. Different RNA and protein components of eukaryotic ribosomes.

of the fine structure of these organelles. Recently following two models have been suggested to explain the three-dimensional structure of prokaryotic or 70S ribosomes :

1. Stoffler and Wittmann's Model (Quasi-symmetrical Model, 1977)

According to this model the 30S ribosomal subunit has an elongated, slightly bent prolate shape (Fig. 14.6). It is a bipartite structure. A transverse hollow or cleft divides the 30S subunit into two parts, a smaller **head** and larger **body**, giving it the appearance of a telephone receiver or embryo. In electron microscopy 50S ribosomal subunit showed various shapes depending on structure seen in different views such as frontal-maple leaf, lateral-kidney shaped or rear view-rounded. In a frontal view, the 50S subunit appears bilaterally symmetrical and shows three protuberances arising from a rounded base (maple leaf structure). The **central protuberance** being the most prominent. The 50S subunit is often compared with an armchair, with the rounded base forming a **vaulted seat**, the central protuberance forming the **back** and the lateral protuberances the **arms** of chair.

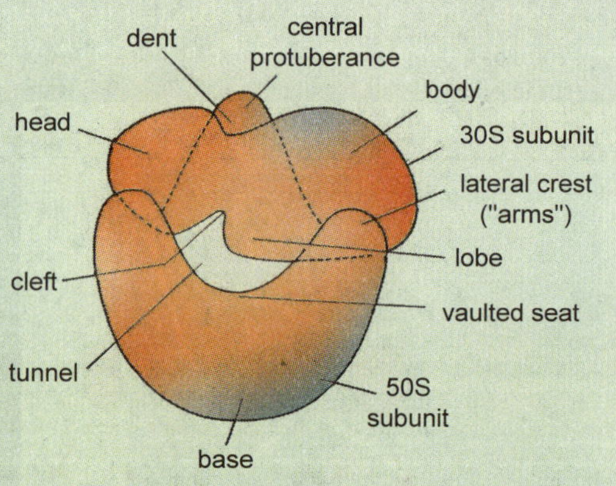

Fig. 14.6. Stoffler and Wittman's model of 70S ribosome.

When 30S and the 50S subunits become associated to form the 70S ribosome, the frontal face of the 30S subunit with its hollow faces the vaulted seat of the 50S subunit. The long axis of 30S subunit is oriented transversely to the central protuberance of the 50S subunit. A **tunnel** is formed between the hollow of the small subunit and vaulted seat of the large subunit.

2. Lake's Model (Asymmetrical Model, 1981)

This completely asymmetrical model of ribosome has been suggested by **James A. Lake** (1981). The smaller subunit has a **head**, a **base** and a **platform**. The platform separates the head from the base by the help of a **cleft**. This cleft is an important functional region ; it is suggested to be the site of codon-anticode interaction and as a part of binding site for initiation factors of protein synthesis.

The large subunit consists of a **ridge**, a **central protuberance** and a **stalk**. The ridge and the central protuberance are separated with the help of a valley (Fig. 14.7)

Three Dimensional Model of 80S Ribosome

In spite of the difference in overall sizes (as manifested in the greater molecular weights, sedimentation constants, sizes and numbers of rRNAs and proteins), the cytoplasmic ribosomes of eukaryotes (80S) are remarkably similar in morphology to those of prokaryotes. As in 30S subunits of prokaryote ribosomes, the 40S ribosomal subunit of eukaryotes is divided into **head** and **base** segments by a transverse groove (Fig. 14.8). The 60S ribosomal subunit is generally rounder in shape than the small subunit, although its one side is flattened ; this is the side that becomes confluent with the small subunit during the formation of the monomer or monosome (*i.e.*, functional 80S ribosome).

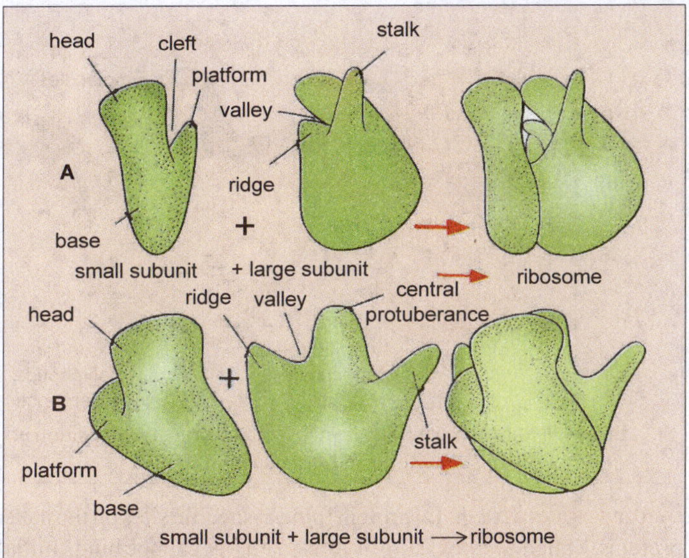

Fig. 14.7. Lake's model of the 70S ribosome. Note the three dimen sional structure of the ribosome in two different orientations.

Dissociation and reconstitution (self-assembly) of the ribosomes. To understand the three-dimensional organization of ribosomal proteins in the ribosomes and also for the investigation of interactions between the molecules of rRNA and proteins, following classical experiment of dissociation and reconstitution of **Nomura** and **Traub** (1968) can be considered. This experiment involves to take purified 30S ribosomal subunits, dissociate them by chemical means into their component RNAs and proteins and then allow them to reassociate under appropriate ionic conditions. Dissociation of 30S subunit may be achieved by treatment with four molar urea and two molar LiCl, which separate the proteins. If the 16S rRNAs previously extracted with phenol is placed in the presence of 20 different protein molecules of 30S ribosomal subunit, the reconstitution or self-assembly of 30S ribosomal subunit takes place in two steps :

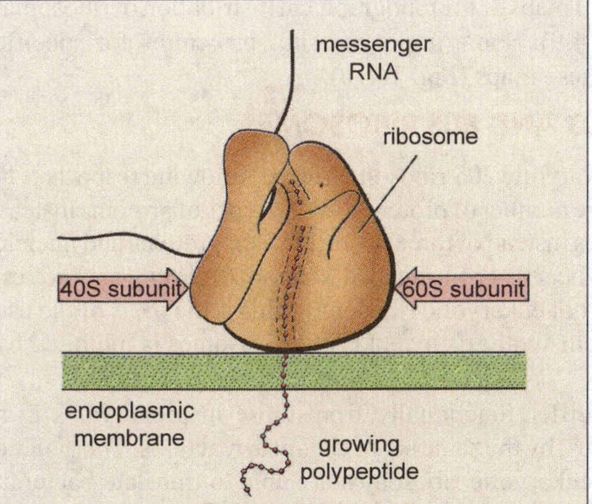

Fig. 14.8. A three dimensional model of eukaryotic cytoplasmic ribosome (after Sheeler and Bianchi, 1987).

$$\text{16S RNA} + \xrightarrow[\text{proteins}]{R_1} \text{RI particles} \xrightarrow{\text{heat}} \text{RI* particles} \xrightarrow{+\text{ S proteins}} \text{30S ribosomal subunit.}$$

In the first step, performed at a low temperature, the 16S RNA binds some of the 30S ribosomal proteins, forming an RI particles (*i.e.*, a reconstitution intermediate) that is inactive. In the second step, the RI particles are heated at 40ºC in the presence of the other proteins that have remained in the supernatant (*i.e.*, S proteins) thereby forming an excited intermediate, RI*, within 20 minutes fully active 30S ribosomal subunits are formed. The self-assembly of 30S subunits is highly specific. It can be achieved with 16S RNA of other bacteria, but not with 16S RNA from yeast or the 23S RNA from *E. coli.*

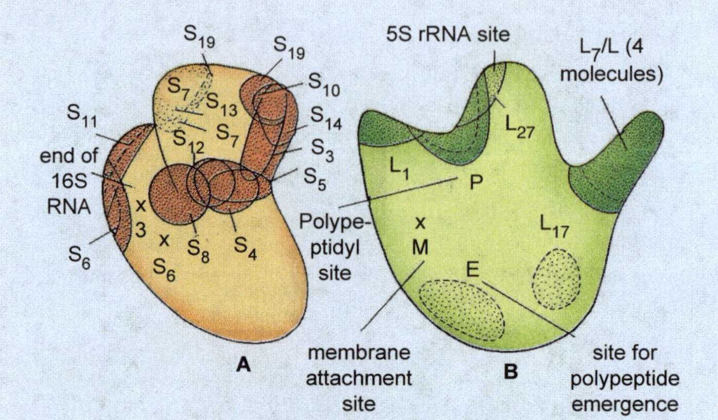

Fig. 14.9. Map of proteins in the ribosome showing their position in small subunit (A) and large subunit (B).

In similar manner, reconstitution of 50S ribosomal subunit is achieved. Finally, a complete functional ribosome is reconstituted spontaneously. In some of these experiments, when ribosomal protein is omitted (or modified) at a time, they show that certain ribosomal proteins require prior to the attachment of other proteins in order to become incorporated in a stepwise manner. For example, some ribosomal proteins, called **initial** or **primary binding proteins** (*e.g.*, L4 protein) bind at specific sites on the naked rRNA and without them the other proteins, called **secondary binding proteins** cannot bind. Initial binding proteins have also been found essential in the control of synthesis of ribosomal proteins.

Further, all ribosomal proteins of the 70S ribosomes also have been isolated and specific antibodies against them have been produced. Various immunological and chemical cross-linking procedures have made possible the construction of maps of the topographical distribution of ribosomal proteins within the ribosomal subunits (Fig. 14.9). Some important sites or centres for specific functions have also been indicated in some of these maps (Fig. 14.10).

COMPARISON OF 70S AND 80S RIBOSOMES

Eukaryotic 80S ribosomes differ from prokaryotic 70S ribosomes in the following respects : (1) they are considerbly larger; (2) they contain a large number of proteins (70–80 types of proteins instead of 53) ; (3) they have four types of RNA molecules instead of three types; (4) their proteins and nucleic acids are large-sized; (5) the RNA-protein ratio is near to 1 : 1 instead of 2 : 1 and (6) several antibiotics, such as **chloramphenicol**, inhibits bacterial but not eukaryotic ribosomes (this is the basis of the use of many antibiotics in medical treatment). Protein synthesis by eukryotic ribosomes is inhibited by **cycloheximide**.

However, eukaryotic ribosomes do not differ functionally from those in prokaryotes in a fundamental way; they perform the same functions, by the same set of chemical reactions. The genetic code is the same for all living organisms, and eukaryotic ribosomes are able to translate bacterial mRNAs efficiently, provided that a "cap" is added enzymatically (**Paterson** and **Rosenberg**, 1979).

Ribosomes from mitochondria and chloroplasts show resemblance to 70S ribosomes of bacteria. Their functions are also inhibited by chloramphenicol. Further, **hybrid ribosomes** containing one bacterial subunit and one subunit from the chloroplast ribosomes are found fully active in protein synthesis, but if hybrid ribosomes contain one subunit of bacteria and another subunit from any eukaryote, they are found to be inactive or non-functional in protein synthesis. However, ribosome constitution experiments have shown some homology between 70S and 80S ribosmes, *e.g.*, proteins L7 and L12 of *E. coli* can replace the homologous proteins in mammalian ribosomes.

BIOGENESIS OF RIBOSOMES

Ribosomes are not self-replicating particles. Synthesis of various component of ribosomes such as rRNAs and proteins, are under genetic control, *i.e.*, rRNAs and mRNAs (for various ribosomal proteins) are transcribed by genes (DNA). Since the mechanism of biogenesis of 70S and 80S ribosomes differ greatly, so can be studied separately as follows :

1. Biogenesis of 70S Ribosomes

Smith *et al.,* (1968) have suggested that in bacteria the RNA genes coding for the 5S, 23S, and 16S ribosomal RNAs are tightly clustered in a region of the chromosome and are present in only few copies. In other words, the ribosomal genes are in a single operon which is transcribed as a unit to synthesize a large molecule of RNA containing the 16S, 23S and 5S rRNA sequences. About 0.4 per cent of a *E.coli* chromosome is devoted to carrying rRNA sequences. In contrast, about 80 per cent of the RNA in an *E. coli* cell is rRNA. This discrepancy in percentages can be explained as follows—First, rRNA is greatly stabilized and protected by ribosomal proteins, and thus any rRNA molecule that is synthesized is expected to have a far longer lifetime than other RNA species (*i.e.,* tRNA, mRNA). Second, in contrast to other genes that may be transcribed only a few times or perhaps not at all during the life of a cell, the rRNA genes are in a state of perpetual transcription.

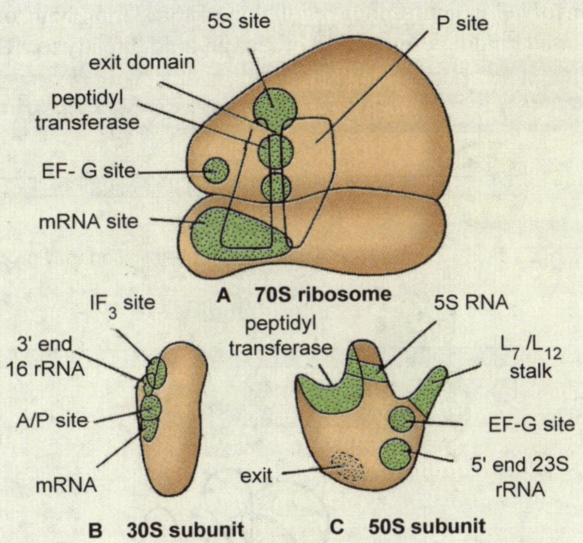

Fig. 14.10. Some important sites or centres for specific functions in the ribosome and its subunits.

Thus, in bacteria a single gene transcript containing the sequences of 16S, 23S and 5S rRNAs, is synthesized by a rRNA operon and this larger molecule is thought to undergo both tailoring and chemical modifications before each rRNA molecule assumes its mature form. During tailoring of larger rRNA molecule, 16S rRNA sequence is first of all cleaved off and is separated from the 23S and 5S sequences. The fragment containing 16S information is still larger than the mature 16S rRNA by at least 100 bases and is not methylated : both the methylation and the tailoring of this molecule takes place after it has associated with a number of proteins to form the precursor ribosomal subunits. The 5S rRNA is found not to undergo the processes of tailoring and methylation before it becomes mature (see **Good-enough** and **Levine**, 1974). The whole process of biogenesis of 70S ribosomes takes place in cytoplasm.

Regulation of synthesis of 70S ribosomal proteins. For *E. coli* the task of coordination of synthessis of 54 ribosomal proteins and 3 rRNAs may be quite difficult. Since synthesis of excess of proteins would be wasteful and if some of them were missing, ribosomes would not be assembled properly. Recent findings have indicated that control of synthesis of exact amount of proteins at the proper time is achieved mainly through blocking the translation of mRNAs for ribosomal proteins, when there is an excess of free ribosomal proteins.

For example, in *E. coli* mRNAs tend to be **polycistronic**, *i.e,* they contain the information for several proteins in a single mRNA molecule. Such a transcriptional unit for multiple proteins is called an **operon** and *E. coli* has six operons which contain genes for the ribosomal proteins. The longest of these operons contains genes for 11 ribosomal proteins. It has been shown that L4 ribosomal protein can inhibit the translation of several of these ribosomal proteins from the polycistronic mRNA (Fig., 14.11) (**Yates** and **Nomura**, 1980 ; **Dean** *et al.*, 1981). This is called **autogenous regulation of translation**, because, the protein blocks translation by binding to its own mRNA. This inhibition is overcome by addition of purified 23S rRNA, which binds to protein L4. *In vivo* experiments of **Lindahl** and **Zengel** (1982) have investigated that induction of L4 protein overproduction has greatly reduced the synthesis of all ribosomal proteins by this longest operon. Similar properties have been reported for the proteins of other operons (**Nomura**, 1986). All such ribosomal proteins which are involved in autogenous regulation of translation have one common property—they are primary or initial binding proteins, *i.e.*, they can bind directly to rRNA.

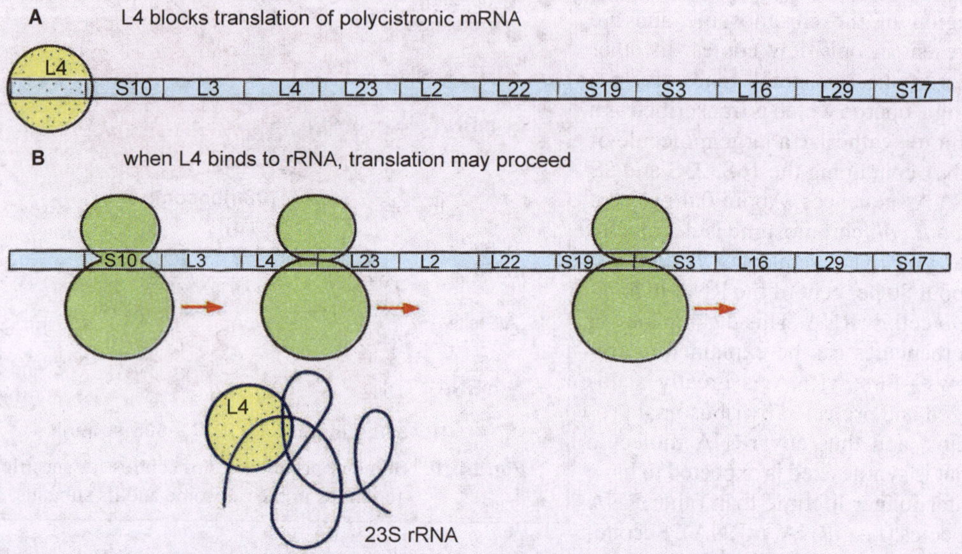

Fig. 14.11. Autogenous regulation of the translation of a polycistronic mRNA coding for 11 ribosomal proteins (after De Robertis and De Robertis, Jr., 1987).

2. Biogenesis of 80S Ribosomes

In eukaryotes, the biogenesis of ribosomes is much more complex and involves a long-lasting process in which several regions of cell are involved. The 5.8S, 18S and 28S rRNAs are transcribed as a much larger molecule in the **nucleolar organizer (NO)** which contains many copies 5.8S, 18S and 28S rRNA genes or **ribosomal DNA** (*i.e.*, there is gene redundancy or amplification). The DNA coding for the 5S rRNA is also highly repetitive, but the molecule is synthesized outside the nucleolus. It is in the nucleolus that newly synthesized rRNA accumulates and becomes associated, presumably by a self-assembly process, with 50 or more ribosomal proteins that have been synthesized in the cytoplasm by usual mechanism of protein synthesis and then migrate to the cytoplasm of cell, in the form of ribosomal subunits.

Biogenesis of 80S ribosomes involves the following three main events :

A. Ribosomal RNA synthesis by nucleolar organizer ;

B. Biosynthesis of ribosomal proteins ;

C. 5S RNA (or 5S rRNA) synthesis.

A. Ribosomal RNA synthesis inside nucleolus. Direct evidence that the nucleolus is responsible for the synthesis of rRNA was obtained in 1964, when it was discovered that an anucleolate mutant (O-nu) of the South African frog *Xenopus laevis*, was unable to synthesize rRNA **Brown** and **Gurdon**, 1964). Since, O-nu embryos were able to continue synthesizing 5S rRNA (**Miller**, 1973), it indicated that these genes were not located in the nucleolar organizer. In *Xenopus* chromosomes, the genes for 5S rRNA are found located at the telomeres.

(1) Ribosomal RNA genes. All organisms have multiple rRNA genes. In case of *Xenopus*, each nucleolar organizer contains 450 rRNA genes. These genes are **tandemly repeated** or **reiterated** along the DNA molecule (in a head to tail arrangement) and are separated from each other by stretches of **spacer DNA**, which is not transcribed. These rRNA genes are being actively transcribed and the

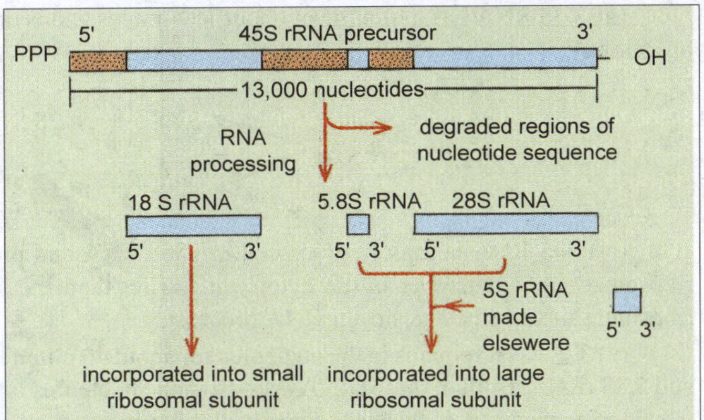

Fig. 14.12. The pattern of processing of 45S rRNA precursor molecule into three separate ribosomal RNAs (after Alberts *et al.*, 1989).

nascent RNA chains are spread perpendicularly to the DNA axis. Each gene is transcribed into a long RNA molecule (which varies in size from 40S to 45S according to species) which will eventually be processed to give rise to 18S, 28S and 5.8S rRNA. Because each rRNA gene has a fixed initiation site (promoter) and a fixed termination site, the transcripts adopt the characteristic "Christmas tree" or "fern leaf" configuration (Fig. 14.12). Nucleolar rRNA genes are transcribed by **RNA polymerase I** (about 100 enzymes per gene). RNA polymerase I molecules are found to remain bound to the nucleolar organisers during mitotic metaphase and anaphase (**Scheer** and **Rose**, 1984). During this period there is no RNA synthesis and so the enzyme molecules must remain in an inactive state.

The spacer DNA can be subdivided into a **nontranscribed spacer** and a **transcribed spacer**, the latter is copied into RNA but does not give rise to mature rRNA. Evidently spacer DNA provides multiple functional binding sites that attract factors needed to activate the promoter (**Busby** and **Reeder**, 1983). In electron microscopy, spacers have acted as **sinks** or storage areas for gene-specific binding proteins.

Gene amplification is the process by which a set of genes is selectively replicated. The rDNA in the amphibian oocyte undergoes this process to accumulate in the egg the huge number of ribosomes (10^{12}) that are used in the first stages of development. During pachytene there is an active replication of the nucleolar organizers and the rDNA is amplified 1000-fold. In *Xenopus* egg 25 pg of extra DNA with 2, 000,000 rRNA genes is accommodated by between 1000 to 1500 nucleoli. This amplified DNA is ultimately lost during development. The amplification of specific genes is not a common event. The rDNA is amplified in oocytes of amphibians, some beetles and spiders, as well as in the macromolecules of ciliate protozoons such as *Tetrahymena* and *Stylonichia*. Gene amplification also occurs in the DNA puffs of *Sciarid* dipterans and in chorion (egg shell) genes of *Drosophila*. In all these cases the amplified genetic material is not passed on the future cell generations.

(2) Processing of rRNAs inside nucleolus. As already described, rRNA genes are transcribed into a long precursor RNA (which is 40S in *Xenopus* and 45S in HeLa and other human cells) ; this precursor must be cleaved into 18S, 28S and 5.8 rRNA. In the cleavage process about 50 per cent of the precursor RNA is degraded. In HeLa cell, the processing of rRNA involves the following steps (**Weinberg** *et al*,. 1967 ; **Maden**, 1977) :

(i) The first ribosomal RNA in HeLa cells is a large 45S molecule of 14,000 nucleotides. Within this precursor molecule the rRNAs are separated by stretches of spacer RNA and the order of transcription is : 5' end — 18S—5.8S—28S—3' end. On a fully active gene about 100 RNA polymerase I enzymes (along with transcription factor I or TFI) are transcribing simultaneously on the rRNA gene.

(ii) In nucleolus, 45S RNA is rapidly **methylated**, even before transcription is completed. Methylations occur mostly on the ribose moiety (producing 2'-O- methyribose) and occur only in the 18S (46 methylations) and 28S (71 methylations) sequences that have to be conserved. Those segments of 45S which have to be degradde are not methylated.

(iii) 45S RNA has a lifetime of about 15 minutes and is then cleaved into smaller components as follows :

45 S → 41S → 32 S → 28 S + 5.8 S
41S → 20 S → 18 S

(iv) 20S RNA is rapidly processed in 18S rRNA and probably due to this reason the small ribosomal subunits appper in the cytoplasm earlier than the large ribosomal subunits. The large ribosomal subunits have a slower RNA processing.

(v) 32S RNA remains in the nucleolus for about 40 minutes and is then cleaved into 28S rRNA and 5.8S rRNA. Both of these rRNAs persist in the nucleolus for another 30 minutes before entering the cytoplasm as part of the large ribosomal subunit.

Thus, about half of 45S rRNA molecule is lost by the successive degradations. This degradation occurs in the regions that are non-methylated and have a higher content of GC. In this way processing of ribosomal RNA results into an increase in methyl groups and decrease in GC content.

Further, all these processing steps do not take place on naked RNA, but rather on RNA-protein complexes. Ribosomal proteins bind to rRNA at the nucleolus, and electron microscopic observations on "christmas tree" spreads stained with antibodies against specific ribosomal proteins have shown that the primary or initial ribosomal proteins bind before the synthesis of 45S rRNA is completed (**Chooi** and **Leiby**, 1981). In addition, a smaller nuclear ribonucleoprotein (U_3 sn RNP) becomes tightly bound to the 32S RNA precursor and is believed to be involved in its processing.

B. Biosynthesis of ribosomal proteins. There is a great possibility that *E. coli*-like translational regulation also exists in eukaryotes. For example, the early embryo of frog is found to contain the mRNAs for all the 70 ribosomal proteins, and except four of these mRNAs, all are not translated until the midblastula stage when synthesis of rRNAs is switched on in the nucleolus (**Pierandrei-Amaldi**, 1982). The remaining four mRNAs are translated at all times.

C. 5S RNA synthesis. The 5S rRNA is synthesized from 20,000 genes in the oocytes but only from 400 genes in the somatic cells, which differ slightly in sequence (six nucleotides out of 120). A gene for 5S rRNA contains an **internal control region** (**ICR**) in its middle region which is found essential for transcription. To this control region of gene remains attached a special protein, called **transcription factor IIIA** or **TF III A** which permits RNA polymerase III enzyme to recognize the promoter of a 5S rRNA gene (*i.e.*, TF III A initiates the synthesis of 5S rRNA).

FUNCTIONS

Ribosomes play a very significant role during biosynthesis of proteins and that will be discussed in a separate chapter.

REVISION QUESTIONS

1. What are the ribosomes ? What is meant by a 70S and 80S ribosome ? Describe the structure of both types of ribosomes.
2. How many types of RNA and proteins are found in the 70S and 80S ribosomes ?
3. Describe the process of biogenesis of ribosomes.
4. Describe the Nomura's experiment of ribosomal self-assembly.
5. Write short notes on the following :
   (i) Transcription factors ; (ii) Three-dimensional structure of ribosome.

CHAPTER

15

Cytoskeleton : Microtubules, Microfilaments and Intermediate Filaments

Eukaryotic cells are given shape and organization by the cytoskeleton, which consists of three types of proteins: microtubules, intermediate filaments, and microfilaments.

The ability of eukaryotic cells to adopt a variety of shapes and to carry out coordinated and directed movements depends on the **cytoskeleton**. The cytoskeleton extends throughout the cytoplasm and is a complex network of three types of protein filaments : **microtubules**, **microfilaments** (or actinfilaments) and **intermediate filaments** (**IFs**). The cytoskeleton is also can be referred to as **cytomusculature**, because, it is directly involved in movements such as crawling of cells on a substratum, muscle contraction and the many changes in the shape of a developing vertebrate embryo; it also provides the machinery for the cyclosis in cytoplasm. Cytoskeleton is apparently absent from the bacteria; it may have been a significant factor in the evolution of the eukaryotic cells.

The existence of an organized fibrous array or cytoskeleton in the structure of the protoplasm was postulated in 1928 by **Koltzoff**. He conceived of a cytoskeleton that determines both the shape of the cell and the changes in its form.

The main proteins that are present in the cytoskeleton are **tubulin** (in the microtubules),**actin**, **myosin**, **tropomyosin** and other (in the microfilaments) and **keratins**, **vimentin**,

desmin, **lamin** and others (in intermediate filaments). Tubulin and actin are globular proteins, while subunits of intermediate filaments are fibrous proteins. Great progress has been made in the isolation of these cytoskeletal proteins. In addition, by the production of specific antibodies against these proteins, it has been possible to examine under the light and the electron microscopes the disposition of the microtubules and microfilaments. The use of high-voltage electron microscopy on whole cells has also helped to demonstrate that there is a highly structured, three-dimensional lattice in the ground cytoplasm.

MICROTUBULES

Microtubules were fisrt of all observed in the axoplasm of the myelinated nerve fibres by **Robertis** and **Franchi** (1953). They called them **neurotubules**. The exact nature of microtubules was brought into light when **Sabatini**, **Bensch** and **Barnett** (1963) made use of the glutarldehyde fixative in the electron microscopy. Microtubeles of plant cells were first described in detail by **Ledbetter** and **Porter** (1963).

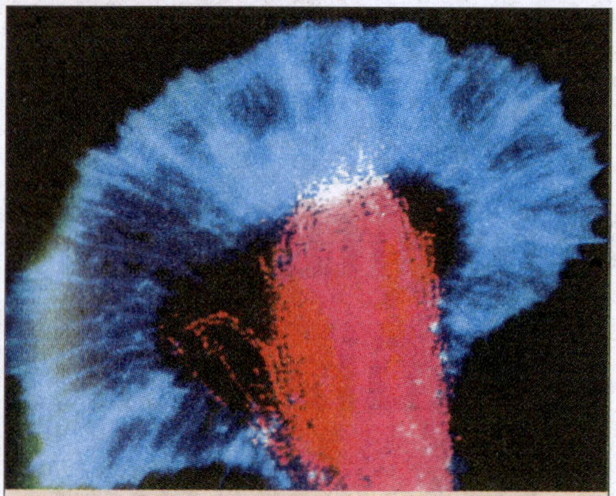

The terminus of a growing axon. Actin filaments are shown in blue, microtubules in red.

Occurrence

With rare exceptions such as the human erythrocytes, microtubules are found in all eukaryotic cells, either free in the cytoplasm or forming part of centrioles, cilia and flagella. The most abundant source of microtubules for the biochemical studies is vertebrate brain— high densities of microtubules exist in axons and dendrites of nerve cells. In the cytoplasm of animal and plant cells, microtubules occur at following seven sites :— 1. cilia and flagella, 2. centrioles and basal bodies, 3. nerve processes, 4. the mitotic apparatus, 5. the cortex of meristematic plant cells, 6. elongating cells such as during the formation of the lens or during spermatogenesis of certain insects. 7. selected structures in Protozoa such as the axostyle of parasitic flagellates, the axoneme of *Echinosphaerium*, the fibre systems of *Stentor*, and the cytopharyngeal basket of *Nassula*.

The stability of different microtubules varies. Cytoplasmic and spindle microtubules are rather labile structures, whereas, those of cilia and flagella are more resistant to various treatments.

Structure

Microtubules constitute a class of morphologically and chemically related filamentous rods which are common to both plant and animal cells. A microtubule consists of a long, unbranched, hollow tubules 24–25 nm in diameter, several micrometers long and with 6 nm thick wall having 13 subunits or **protofilaments**. Thus, the wall of the microtubule consists of 13 individual linear or spiralling filamentous structures about 5 nm in diameter, which in turn, are composed of tubulin. These protofilaments have a centre-to-centre spacing of 4.5 nm. Application of negative staining techniques has shown that microtubules have a lumen 14 nm wide and a protofilament or subunit structure in the wall (Fig. 15.1).

Chemical Composition

Biochemically, a protofilament of microtubule is made of a protein called **tubulin**. Tubulin is an acidic protein with a molecular weight of 55,000 and a sedimentation constant of 6S. It occurs in two different forms, called **α-tubulin** and **β-tubulin**, each containing about 450 amino acids. Both of these proteins have a distinct, though closely related, amino acid sequences and are thought to have evolved from a single ancestral protein. The two proteins show very little divergence from the lowest to the

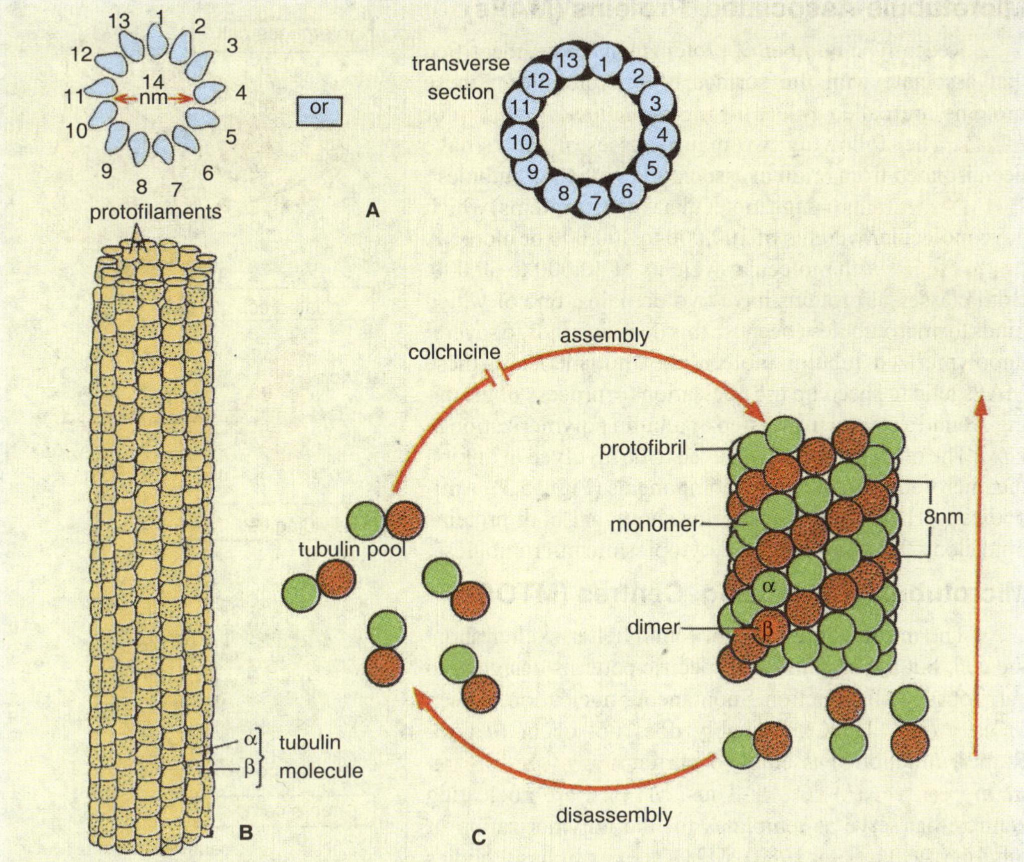

Fig. 15.1. Schematic diagrams of a microtubule, showing how the tubulin molecules pack together to formthe cylindrical wall. A—13 tubulin molecules (subunits of protofilaments) in cross section ; B—Side view of a short section of a microtubule, with the tubulin molecules aligned into rows, or protofilaments. Each of the 13 protofilaments is composed of a series of tubulin molecules, each with an α/β heterodimer ; C—Assembly and disassembly of the microtubule. The microtubule is being disassembled at the bottom while being simultaneoasly assembled at the top. Colchicine, by blocking the assembly process, produces depolymerization of the microtubules (after De Robertis and De Robertis, Jr., 1987 ; Alberts *et al.*, 1989).

highest eukaryotes ; for example, the β- tubulins of sea urchin flagella and chick brain cells differ only in one amino acid. Similarities such as this suggest that most mutations disrupt the functions of microtubules and are thus lethal and are eliminated by selection (see **King**, 1986).

Tubulin in the form of dimers (rather heterodimers of α- and β- tubulins ; each with 115,000 MW, see **Berns**, 1983) polymerizes into the microtubules. Thus, heterodimers of tubulins assemble to form linear "protofilaments" with the β- tubulin of one dimer in contact with the α- tubulin of the next. Since all the 13 protofilaments are aligned parallely with the same polarity, the microtubules are the **polar** structurs having a **plus** or **fast growing end** and **minus** or **slow-growing end**. The minus ends of cytoplasmic microtubules in cells are bound tightly to **microtubule organizing centres** (**MTOCs**) from which their assembly or polymerization starts. MTOCs also protect the minus ends of the microtubules from the disassembly. Generally, the plus ends of microtubules terminate near cell margins (Fig. 15.2) and are protected from disassembly by the **capping proteins** (see **Alberts** *et al.*, 1989).

Microtubule-Associated Proteins (MAPs)

Recently, a number of proteins have been identified that associate with the surface of microtubules ; these proteins are called **microtubule-associated proteins** or **MAPs**. The following two major classes of MAPs have been isolated from brain in association with microtubules : 1. **HMW proteins** (=high molecular weight proteins) which have molecular weights of 200,000 to 300,000 or more ; 2. **tau proteins**, with molecular weights of 40,000 to 60,000. Both classes of proteins have two domains, one of which binds to microtubules ; because this domain binds to several unpolymerized tubulin molecules simultaneously, these MAPs tend to speed up the nucleation (= process of grouping around a central mass) step of tubulin polymerization *in vitro*. The other domain is believed to be involved in linking the microtubule to other cell components (Fig. 15.3). Antibodies to HMW and tau proteins show that both proteins bind along the entire length of cytoplasmic microtubules.

Microtubule Organizing Centres (MTOCs)

The microtubules are not found helter-skelter about the cell, but are organized in specific patterns designed to carry out specific function. Spontaneous nucleation, as seen *in vitro* (Fig. 15.1C), probably does not occur *in vivo*. Rather, initiation of assembly occurs at **microtubule organizing centres (MTOCs)**. Thus, MTOCs are nucleating centres that serve as templates for the polymerization of tubulin (see **Thorpe**, 1983). MTOCs exist in basal bodies (*e.g. Chlamydomonas*); in centrioles (*e.g.*, most animal cells); at the poles of mitotic spindles in dividing cells that do not have centrioles (*e.g.*, most plant cells); on chromosomes (*i.e.*, **kinetochore**); in membranes and probably many other places as well. Recent studies have revealed that most cytoplasmic microtubules do not arise directly from the centrioles, but from a densely staining **pericentriolar material** that surrounds the centriole (see **King**, 1986).

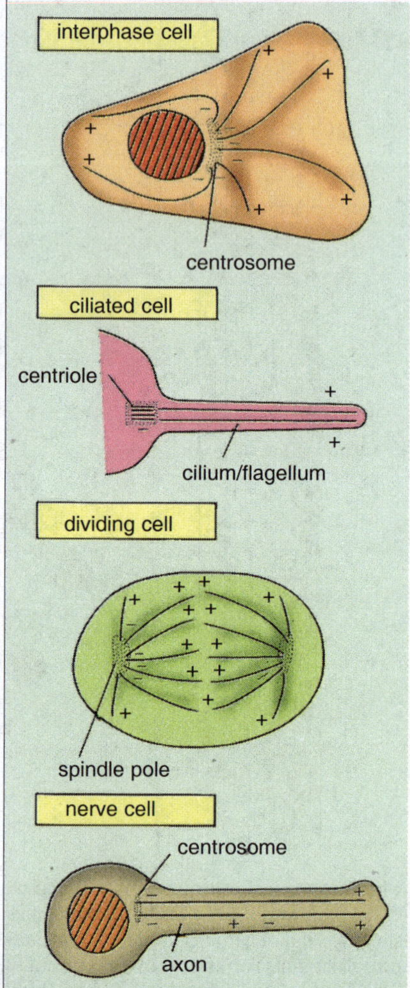

Fig. 15.2. The minus ends of microtubules in cells are generally embedded in a microtubule-organizing centre ; while the plus ends are often near the plasma membrane (after Alberts *et al.*, 1989).

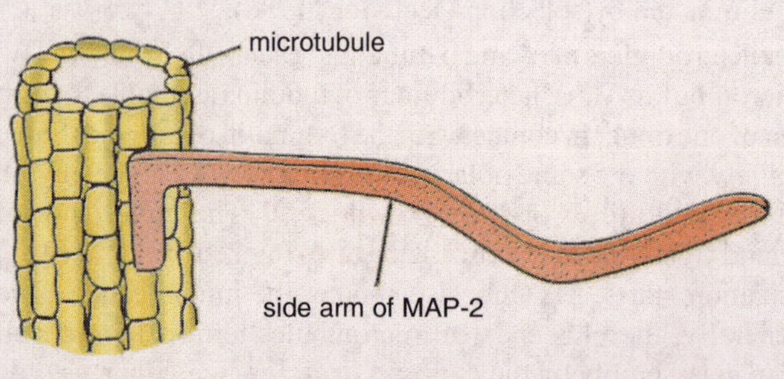

Fig. 15.3. A microtubule associated protein (known as MAP-2) showing its two domains (after Alberts *et al.*, 1989).

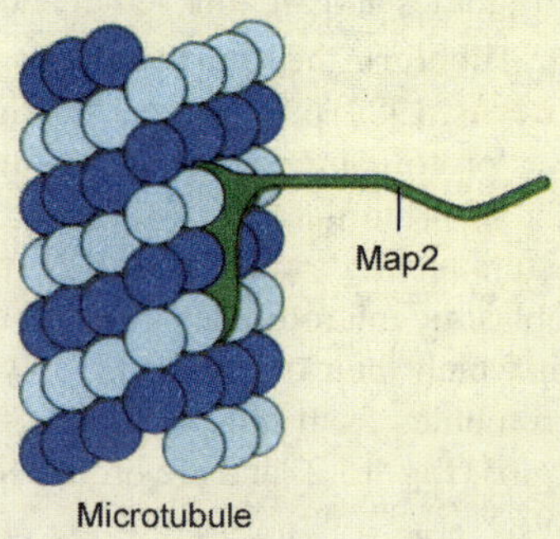

MAPs

Turning on and off of these organizing centres for microtubule assembly at different times in the cell's life are probably regulated by one or all the following factors : changes in nucleation centres, changes in Ca^{2+} concentration and involvement of MAPs.

Assembly and Disassembly of Microtubules

Cytoplasmic microtubules are highly dynamic structures, constantly forming and disappearing depending on cell activities. They, like the microfilaments, grow by the reversible addition of subunits, accompanied by nucleotide (GTP) hydrolysis and conformational change. The process of polymerization (assembly) and depolymerization (disassembly) of the microtubules appears to be a form of **self-assembly**. The assembly of microtubules from the tubulin dimers is a specifically oriented and programmed process. In the cell, the sites of orientation are MTOCs from which the polymerization is directed. The quantity of polymerized tubulin is high at interphase (cytoplasmic microtubules) and metaphase (spindle microtubules), but low at prophase and anaphase.

Microtubules as supporting rods. This protist has tentacles projecting from its cell body.

Within the cell, microtubules are in equilibrium with free tubulin. Phosphorylation of the tubulin monomers by a cyclic AMP-dependent kinase favours the polymerization. A definite relationship has been found between cell shape, the number and direction of microtubules and cAMP. The assembly and disassembly of tubulin constitute a polarized phemomenon. In a microtubule, the assembly of tubulin dimers takes place at one end, while disassembly is common at the other end (Fig. 15.1 C). If a cell is treated with certain drugs such as **colchicine**, **vincristine** or **vinblastine**, the assembly of the microtubules is inhibited, while the disassembly continues, leading to the disorganization of the microtubule. Further, the assembly is accompanied by the hydrolysis of guanosine triphosphate (GTP) to guanosine diphosphate (GDP) and lack of GTP stops the assembly. *In vivo* control of assembly and disassembly of tubulin involves Ca^{2+} and the calcium-binding protein **calmodulin**. The addition of Ca^{2+} inhibits polymerization of tubulin ; this effect is also enhanced by the addition of calmodulin.

The *in vivo* mechanism involved in self-assembly of the microtubules is still little understood, however, *in vitro* studies have revealed various interesting facts about it. Thus, in a classical study using isolated bovine brain tubulin, **Weingarten** *et al.*, (1975) demonstrated that tubulin alone was not sufficient to bring about *in vitro* assembly into microtubules. Under normal conditions, if brain microtubules are isolated and caused to depolymerize into tubulin subunits, the tubulin molecules will reassemble into microtubules if Mg^{2+} and GTP (an energy source) are added to the mixture. However, according to **King** (1986), *in vitro* assembly of microtubules can occur in the presence of low calcium concentration, MAPs, GTP, and a level of free tubulin monomers above a threshold concentration.

In vitro polymerization evidently involves two distinct phases, one of **initiation** and the other of **elongation**. The initiation event seems to involve the formation of some multimeric "**nucleating**" centre, following which the addition of more subunits proceeds rapidly during elongation. Thus, during *in vitro* polymerization of microtubules, α- and β- tubulins combine to form heterodimers (Fig. 15.4). The heterodimers associate to form multimeric **rings**, **spirals** and other intermediate structures which eventually open up to form strands or protofilaments. Side-by-side assembly of the protofilaments creates sheet-like structures that curl to form a tube. Elongation of this short cylinder occurs by direct addition of new heterodimers at one end of the tubule (*i.e.,* the plus end of tubule). It is believed that during anaphase, addition of dimers to one end of a microtubule is accompanied by the loss of dimers from the other end.

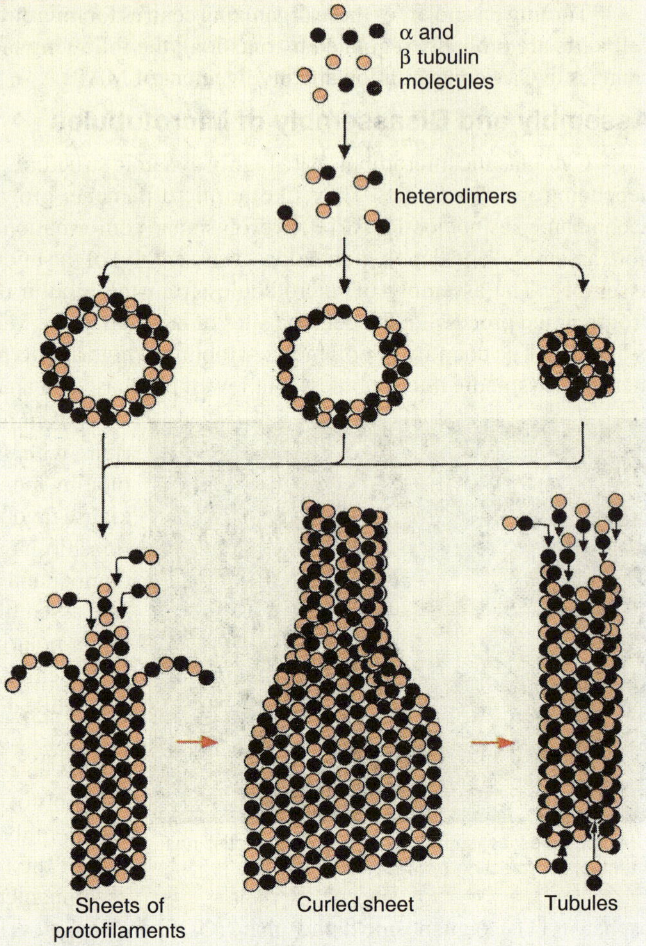

Fig. 15.4. Various steps in *in vitro* polymerization or assembly of the microtubules (After Thorpe, 1984).

Functions of Cytoplasmic Microtubules

Microtubules have several functions in the eukaryotic cells such as follows :

1. Mechanical function. The shape of the cell (*e.g.,* red blood cells of non-mammalian vertebrates) and some cell processes or protuberances such as axons and dendrites of neurons, microvilli, etc., have been correlated to the orientation and distribution of microtubules.

2. Morphogenesis. During cell differentiation, the mechanical function of microtubules is used to determine the shape of the developing cells. For example, the enormous elongation in the nucleus of the spermatid during spermiogenesis is accompanied by the production of an orderly array of microtubules that are wrapped around the nucleus in a double helical arrangement. Likewise, the elongation of the cells during induction of the lens placode in the eye is also accompanied by the appearance of numerous microtubules.

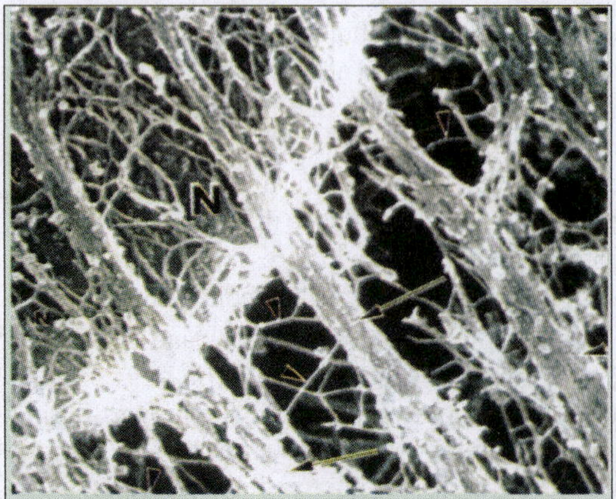

This photo shows the nucleus (N) of a cell being held in place by a network of cytoskeleton.

3. Cellular polarity and motility. The determination of the intrinsic polarity of certain cells is also related to the microtubules. Directional gliding of cultured cells is found to depend on the microtubules.

4. Contraction. Microtubules play a role in the contraction of the spindle and movement of chromosomes and centrioles as well as in ciliary and flagellar motion.

5. Circulation and transport. Microtubules are involved in the transport of macromol-

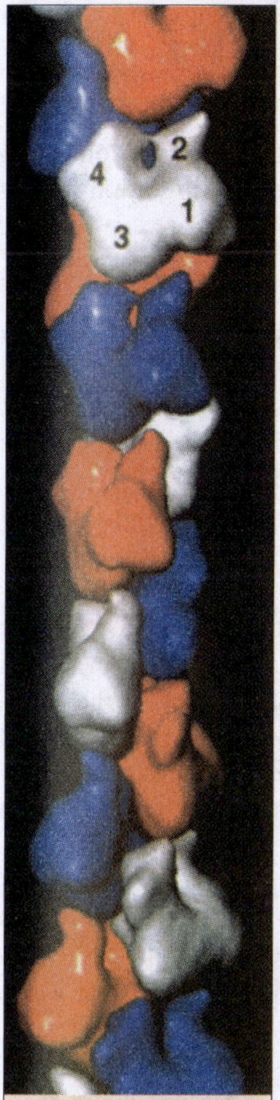

Actin filament structure.

ecules, granules and vesicles within the cell. **Examples** : 1. The protozoan *Actinosphaerium* (Heliozoa) sends out long, thin pseudopodia within which cytoplasmic particles migrate back and forth. These pseudopodia contain as many as 500 microtubules disposed in a helical configuration. 2. In the protozoan *Nassula*, microtubules drive the food in the gullet. 3. In melanocytes, melanin granules move centrifugally and centripetally with different stimuli. These granules have been observed moving between channels created by the microtubules in the cytoplasmic matrix. 4. In the erythrophores found in fish scales the pigment granules may move at a speed of 25 to 30 μm per second between the microtubules. 5. They have a role in axoplasmic transport of proteins, glycoproteins and enzymes.

MICROFILAMENTS

Thin,solid **microfilaments** of actin protein, ranging between 5 to 7 nm in diameter and indeterminate length, represent the active or motile part of the cytoskeleton. They appear to play major role in cyclosis and amoeboid motion. With high voltage electron microscopy a three-dimensional view of microfilaments has been obtained (*i.e.*, an image of **microtrabecular lattice**). These microfilaments are sensitive to **cytochalasin-B**, an alkaloid that also impairs many cell activities such as beat of heart cell, cell migration, cytokinesis, endocytosis and exocytosis. It is generally assumed that the cytochalasin-B-sensitive microfilaments are the contractile machinery of non-muscle cells.

Distribution

Microfilaments are generally distributed in the cortical regions of the cell just beneath the plasma membrane. In contrast, intermediate filaments and microtubules are found in subcortical and deeper regions of the cell. Microfilaments also extend into cell processes, especially where there is movement. Thus, they are found in the microvilli of the brush border of intestinal epitheliun and in cell types where amoeboid movement and cytoplasmic streaming are prominent.

Chemical Composition

Actin is the main structural protein of microfilaments. The concentration of actin in non-muscle cells is surprisingly high ; it may account up to 10 per cent of total cell protein. It can be extracted and *in vitro* settings will undergo polymerization reactions from G-actin monomer state to F-actin. In fact, the globular (=G actin) – fibrillar (=F-actin) transition is the basis of the classical sol-gel transition in the cytoplasm of moving cells. Further, there are present three types of actins— α , β and γ. The α- form of actin is found in fully mature muscle tissue. The other two forms are more characteristic of non-muscle cells.

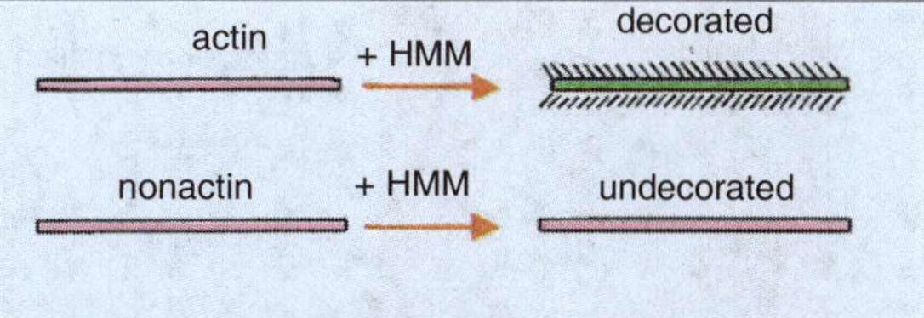

Fig. 15.5. Mode of "decoration" of actin filaments by HMM (after Thorpe, 1984).

In non-muscle cells, microfilaments, being of actin composition, can bind myosin (a contractile protein). *In vitro* and *in situ* mircrofilaments can be coated or "decorated" with heavy myosin (HMM) or S_1 heads. This binding results in an arrow-head pattern to the microfilaments in which the arrowheads all point in the same direction (Fig. 15.5). This pattern indicates that microfilaments possess a polarity, a property that is probably crucial to their role in mediating cell movements.

The HMM binding method has become a very useful method for identifying and localizing microfilaments in any type of cell. Intermediate filaments are not decorated by HMM.

Function

Microfilaments are found to be involved in movement associated with furrow formation in cell division, cytoplasmic streaming in plant cells (*e.g.*, *Nitella* and *Chara*) and cell migration during embryonic development.

INTERMEDIATE FILAMENTS

Intermediate filaments (IFs) are tough and durable protein fibres in the cytoplasm of most higher eukaryotic cells. Constructed like woven ropes, they are typically between 8 nm to 10 nm in diameter, which is "intermediate" between the thin and thick filaments in muscle cells, where they were first described ; their diameter is also between microfilaments (actin filaments) and microtubules. IFs are found resistant to colchicine and cytochalasin B and are sensitive to proteolysis.

In most animal cells IFs form a "basket" around the nucleus and extend out in gentle curving arrays to the cell periphery. IFs are particularly prominent where cells are subjected to mechanical stress, such as in epithelia, where they are linked from cell to cell at desmosomal junctions, along the length of axons, and throughout the cytoplasm of smooth muscle cells. Various names have been attached to the intermediate filaments that have a basis in the cell type in which they are observed. Thus, IFs in epidermal cells are called **tonofilaments**, in nerve cells they are referred to as **neurofilaments** and in neuroglial cells they are designated as **glial filaments**.

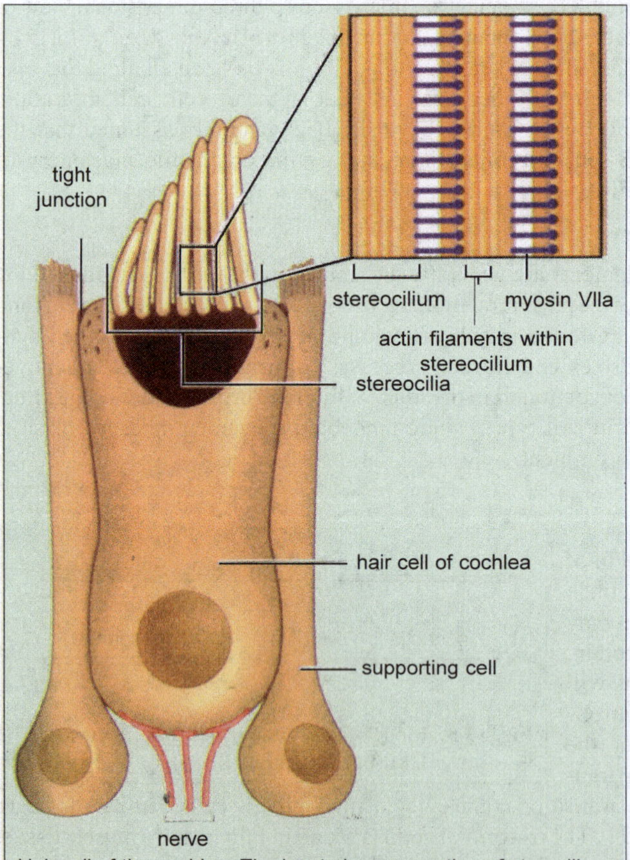

Hair cell of the cochlea. The inset shows a portion of sterocilia, which is composed of actin filaments.

In cross-section, intermediate filaments have a tubular appearance. Each tubule appears to be made up of 4 or 5 protofilaments arranged in parallel fashion (**Thorpe**, 1984). IFs are composed of polypeptides of a surprisingly wide range of sizes (from about 40,000 to 130,000 daltons).

Types of intermediate filaments. The intermediate filaments are very heterogeneous from the point of view of their biochemical properties, but by their morphology and localization can be grouped into following four main types (Table 15-1) :

1. Type I IF proteins. They are found primarily in epithelial cells and include two subfamilies of **keratin** (also celled **tono**, **perakeratin** or **cytokeratin**) : acidic keratin and neutral or basic keratin. Keratin filaments are always heteropolymers formed from an equal number of subunits from each of these two keratin subfamilies. The keratins are most complex class of IF proteins, with at least 19 distinct forms in human epithelia and 8 more in the keratins of hair and nails. Mammalian cytokeratin are α-fibrous proteins that are synthesized in cells of living layers of the epidermis and form the bulk of the dead layers of **stratum corneum**.

2. Type II IF proteins. They include the following four types of polypeptides: vimentin, desmin, synemin and glial fibrillary acidic protein (or glial filaments). **Vimentin** is widely distributed in cells of mesenchymal origin, including fibroblasts, blood vessel endothelial cells and white blood cells, **Desmin** is found in both striated (skeletal and cardiac) and smooth muscle cells. **Glial filaments** occur in some type of glial cells such as astrocytes and some Schwann cells, in the nervous system. **Synemin** is a protein of 230,000 daltons, which is also present in the intermediate filaments of muscle, together with desmin and vimentin. Vimentin and synemin containing IFs can be observed in the chicken erythrocytes.

Each of thse IF proteins tends to assemble spontaneously *in vitro* to form homopolymers and will also co-assemble with the other Types II IF proteins to form **co-polymers** and **heteropolymers**. In fact, co-polymers of vimentin and desmin, or of vimentin and glial fibrillary acidic protein, are found in some type of cells. For example, desmin remains concentrated in the Z-lines and T-tubule system of striated or skeletal system, together with vimentin, synemin and α- actinin. Since desmin links actin to plasma memberane, from this fact the name of desmin has been derived by **Lazarides** and coworkers in 1976 (in Greek desmin means link or bond).

Table 15.1. Characteristics of four types of intermediate filament proteins (Source : Alberts *et al.*, 1989).

Types of intermediate filaments	Component polypeptide (mass in daltons)	Cellular location
1. Type I	Acidic keratins (40,000—70,000) Neutral or basic keratins (40,000—70,000)	Epithelial cells and epidermal derivatives such as hair and nail
2. Type II	Vimentin (53,000) Desmin (52,000) Glial fibrillar acidic protein (glial filaments; 45,000) Synemin (230,000)	Many cells of mesenchymal origin Muscle cells Glial cells (astrocytes and some Schwann cells) Muscle cells
3. Type III	Neurofilament proteins (about 130,000, 100,000 and 60,000)	Neurons
4. Type IV	Nuclear lamins A, B and C (65,000—75,000)	Nuclear lamina of all cells

3. Type III IF proteins. These IF proteins assemble into **neurofilaments**, a major cytoskeletal element in nerve axons and dendrites, and consequently are called **neurofilament** proteins. In vertebrates, Type III IFs consist of three distinct polypeptides, the so-called neurofilament triplet.

4. Type IVIF proteins. They are the **nuclear lamins** which form highly organized two dimensional sheets of filaments. These filaments rapidly disassemble and reassemble at specific stage of mitosis.

General Structure of IFs

Despite the large differences in their size, all cytoplasmic IF proteins are encoded by members of the same multigene family. Their amino acid sequences indicate that each IF polypeptide chain contains a homologous central region of about 310 amino acid residues that forms an extended α -helix with three- short— α- helical interruptions (Fig. 15.6).

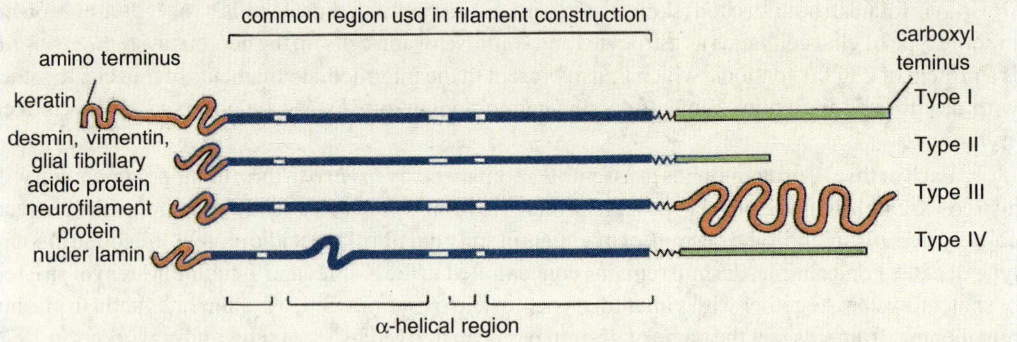

Fig. 15.6. All IF proteins share a similar central region (about 310 amino acid residues) that forms an extended helix with three short interruptions. The amino-terminal and carboxyl-terminal domains are non-helical and vary greatly in size and sequence in different IF proteins (after Alberts *et al.*, 1989).

Assembly of IFs

A current model of assembly of a intermediate filament includes the following steps : 1. Two identical **monomers** pair to form a **dimer** in which the conserved helical central regions are aligned in parallel and are wound together into a coiled coil. 2. Two dimers then line up side-by-side to form a 48 nm by 3 nm **protofilament** containing four polypeptide chains. 3.These protofilaments then associate in a staggered manner to form successively larger structures. 4. The final 10 nm diameter of the intermediate filament is thought to be composed of 8-protofilaments (*i.e.*, 32 polypeptide chains) joined end on end to neighbours by staggered overlap to form the long rope-like filaments (Fig.15.7). It is still not known whether IFs are polar structures (like actin and tubulin) or non-polar (like the DNA double helix).

IFs During Mitosis

Mitosis of cultured epithelial cells shows striking changes in intermediate filaments of cytokeratin and vimentin. During prophase the 10 nm filaments unwind into threads of 2 to 4 nm and into spheroidal aggregates containing both types of proteins. At metaphase and anaphase most vimentin and cytokeratin appear as spheroid bodies, while at telophase the filamentous cytoskeleton become gradually reestablished. From these experimental studies, **Franke** (1982) has concluded that the living cells contain factors that promote the reversible disintegration and restoration of intermediate filaments during mitosis.

Functions of IFs

The main function of most intermediate filaments is to provide mechanical support to the cell and its nucleus. IFs in epithelia form a transcellular network that seems designed to resist external forces. The neurofilaments in the nerve cell axons probably resist stresses caused by the motion of the animal, which would otherwise break these long, thin cylinders of cytoplasm. Desmin filaments provide mechanical support for the sarcomeres in muscle cells, and vimentin filaments surround and probably support the large fat droplets in the fat cells.

COMPARISON OF MICROTUBULES, INTERMEDIATE FILAMENTS AND MICROFILAMENTS

The three components of the cytoskeleton, namely microtubules, intermediate filaments and microfilaments have been compared in Table 15-2.

Table 15.2. Comparison of some properties of microtubules, intermediate filaments and microfilaments (Source : Thorpe, 1984).

Property	Microtubules	Intermediate filaments	Microfilaments
1. Structure	Hollow with walls made up of 13 protofilaments	Hollow with walls made up of 4 to 5 protofilaments	Solid made up of polymerized actin (F-actin)
2. Diameter (nm)	24 — 25	10	7 — 9
3. Monomer units	α- and β- tubulin	Five types of protein defining five major classes	G-actin
4. ATPase activity	Present in dynein arms	None	None
5. Functions	1. Motility of eukaryotes 2. Chromosome movement 3. Movements of intracellular materials 4. Contribute toward maintaining cell shape	1. Integrate contractile units in muscle 2. Cytoskeletal structural function in cytoplasm	1. Muscle contraction 2. Cell shape changes 3. Protoplasmic streaming 4. Cytokinesis

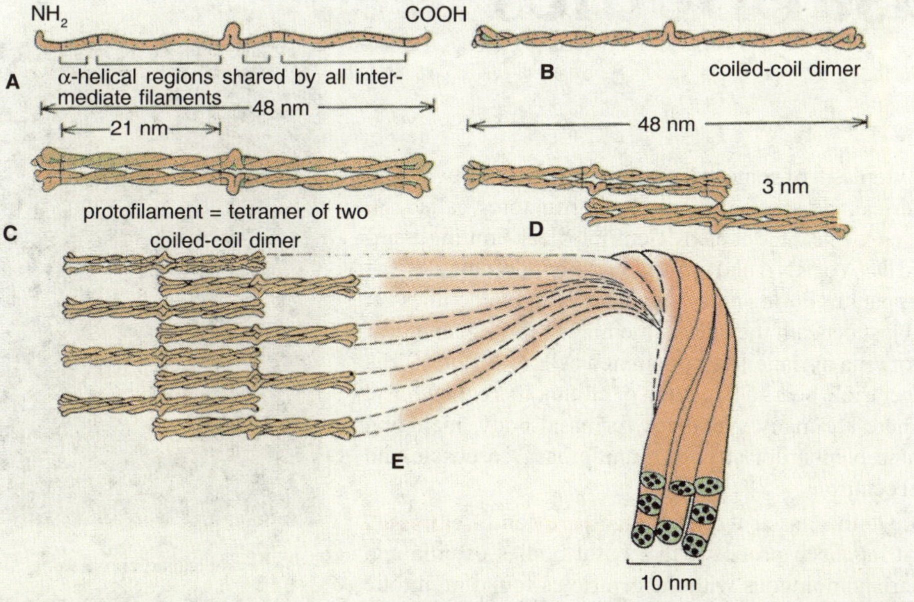

Fig. 15.7. A current model to explain the mode of assembly of a 10-nm thick intermediate filament (after Alberts *et al*., 1989).

REVISION QUESTIONS

1. What is cytoskeleton ? Write a short note on the cytoskeleton
2. What are microtubules ? Describe their structures, assembly, disassembly and functions.
3. What are the intermediate filaments ? Describe their types, structure and cellular functions.
4. Define the term microfilament. Describe the structure and function of microfilament in the cell.
5. Make a comparison of three main components of the cytoskeleton : microtubules, intermediate filaments and microfilaments.

CHAPTER

16

Centrioles and Basal Bodies

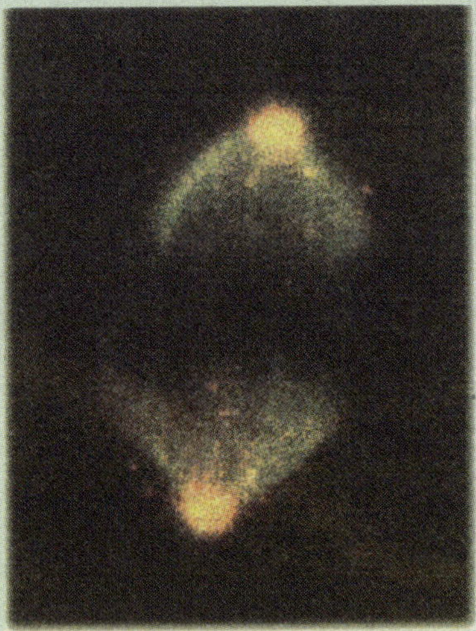

A dividing fibroblast.

Cytoplasm of some eukaryotic cells contains two cylindrical, rod-shaped, microtubular structures, called **centrioles**, near the nucleus. Centrioles lack limiting membrane and DNA or RNA and form a spindle of microtubules, the mitotic apparatus during mitosis or meiosis and sometimes get arranged just beneath the plasma membrane to form and bear flagella or cilia in flagellated or ciliated cells (**Fulton**, 1971). When a centriole bears a flagellum or cilium, it is called **basal body**. There are many synonyms for basal body, including kinetosome, blepharoplast, basal granule, basal corpuscle, and proximal centriole.

Due to the classic works of **Henneguy** and **Lenhossek** (1897), it has been proposed that basal bodies of cilia and flagella are homologous with the centrioles found in mitotic spindle.

OCCURRENCE

Centrioles occur in most algal cells (notable exception being red algae), moss cells, some fern cells and most animal cells. They are absent in prokaryotes, red algae, yeast, cone-bearing and flowering plants (conifers and angiosperms) and some non-flagellated or non-ciliated protozoans (such as amoebae). Some species of amoebae have a flagellated stage as well as an amoeboid stage ; a centriole develops during the flagellated stage but disappears during the amoeboid stage.

STRUCTURE

Centrioles and basal bodies are cylindrical structures which are 0.15–0.25μm in diameter usually 0.3–0.7μm in length, though, some are as short as 0.16μm and others are as long as 8μm (see **Fulton**, 1971). Both have following ultrastructural components :

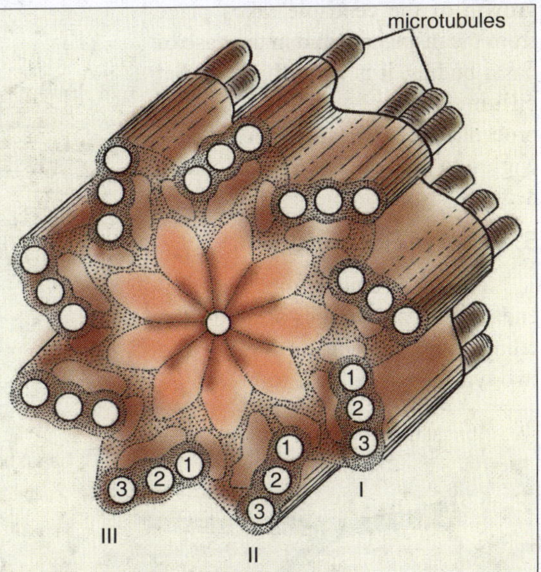

Fig. 16.1. Cross-sectional structure of a centriole. There are nine groups of three microtubules (after De Witt, 1977).

1. Cylinder Wall

The most striking and regular ultrastructural feature of centrioles and basal bodies is the array of nine triplet microtubules equally spaced arround the perimeter of an imaginary cylinder (Fig. 16.1). The space between and immediately around the triplet is filled with an amorphous, electron-dense material. In transverse section the triplets are arranged like vanes or blades of pinwheel or turbine. Each triplet or blade is tilted inward to the central axis at an angle of about 450 to the circumference; within each blade the tubules twist from one end to the other or describe a helical course. Since centrioles have no outer membrane, the triplets are considered to form the wall of the cylinder, and arbitrarily define the inside and outside of the centriole.

2. Triplets

The nine triplets that make up the wall are basically similar in centrioles and basal bodies. The three subunit microtubules have been designated A, B, and C, with the innermost tubule being A. Individual tubules are 200–260A° in diameter. Only the A tubule is round ; the others are incomplete, C-shaped and share their wall with the preceding tubule. At both ends the C tubule often terminates before the A and B tubules.

The substructure of A, B and C tubules, is similar to the structure of other microtubules (**Stephens**, 1970). The A tubule has 13, 40–45A° globular subunits around its perimeter. Three or four of these subunits are shared with the B tubules, which in turn share several of its subunits with the C tubules.

Often the triplets are thought to run parallel to one another and to the long axis of the cylinder, but this is not always the case. In the basal bodies of some organisms, the triplets get closer toward the proximal end, so the diameter of the cylinder gets smaller. In some centrioles the triplets are parallel to one another but turn in a long-pitched helix with respect to the cylinder axis (**Fulton**, 1971).

3. Linkers

The A tubule of each triplet is linked with C tubule of neighbouring triplet by protein **linkers** at intervals along their entire length. These linkers hold the cylindrical array of the microtubules and maintain the typical radial tilt of the triplets.

4. Cartwheel

There are no central microtubules in the centrioles and no special arms. However, often faint protein spokes are radiate out to each triplet from a central core, forming a pattern like a cartwheel. Such a cartwheel configuration determines the **proximal end** of a centriole and, thus, provides a structural and functional **polarity** to it. The growth of the centriole takes place from the distal end, and in the case of basal bodies, it is from this end that cilium is formed. Moreover, the procentrioles which are formed at right angles to the centriole, are located near the proximal end.

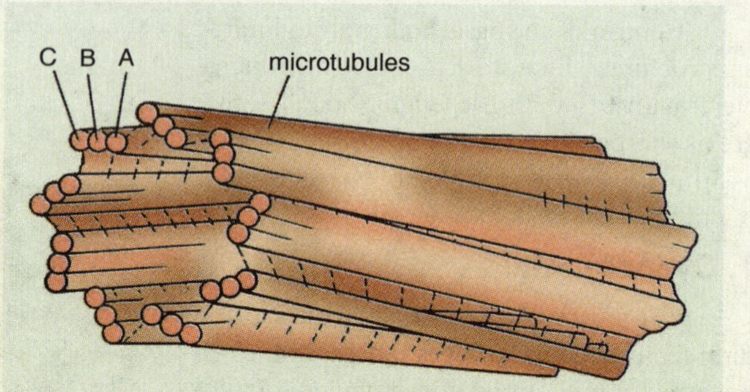

Fig. 16.2. Ultrastructure of a centriole (after Alberts *et al.*, 1989).

5. Ciliary Rootlets

In some cells, from the basal ends of the basal bodies originate the ciliary rootlets which are of following two types :

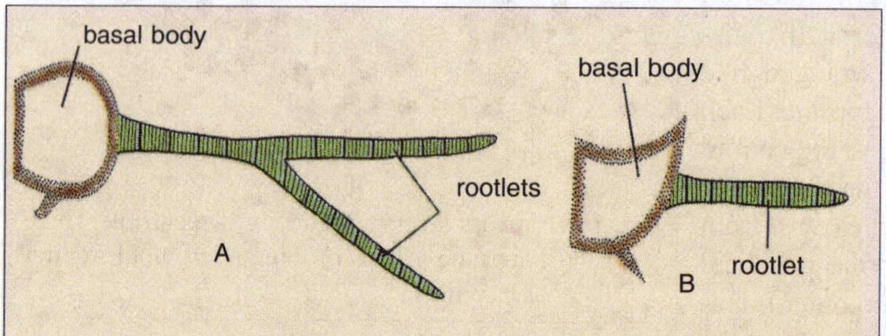

Fig. 16.3. Relation of rootlets with the basal body.

(i) Tubular root fibrils. The tubular root fibrils have the diameter of 200A°.

(ii) Striated rootlets. Most ciliary rootlets are striated, having a regular cross-banding with a repeating period of 55 to 70 nm. The striated fibres of rootlets are composed of parallel microfilaments; 3 to 7 nm in diameter, which in turn are formed of globular subunits. These fibres and filaments may have a structural role such as anchoring the basal body. Due to microfilaments, ciliary rootlets also have a contractile role. The rootlet may be double (*e.g.*, molluscs) or single (*e.g.*, the frog *Rana*).

6. Basal Feet and Satellites

The **basal feet** are dense processes that are arranged perpendicularly to the basal body. These processes impose a structural asymmetry on the basal body that is related with direction of the ciliary beat. A basal foot is composed of microfilaments that terminate in a dense bar. It may act as a focal point for the convergence of microtubules.

Satellites or **pericentriolar bodies** are electron-dense structures lying near the centriole that are probably nucleating sites for the microtubules.

CHEMICAL COMPOSITION

The microtubules of centrioles and basal bodies contain the structural protein, **tubulin**, along with lipid molecules (**Fulton**, 1971). The centrioles and basal bodies contain a high concentration of ATPase enzyme. There exists a controversy that whether centrioles and basal bodies have DNA and RNA. **Fulton** (1971) has doubted the presence of nucleic acids in these organelles.

ORIGIN OF CENTRIOLES AND BASAL BODIES

The idea prevalent years ago that new centrioles arise by the division of existing centrioles is no longer accepted. Rather it appears that new centrioles are either produced *de novo* or are synthesized using an existng centriole as some form of template (semi-autonomous replication).

1. Origin of centrioles by duplication of pre-existing centrioles. In cultured fibroblasts, centriole doubling begins at around the time that DNA synthesis begins (interphase). First the two members of a pair of centrioles separate ; then a daughter centriole, called **procentriole**, is formed perpendicular to each original centriole, the two organelles being separated from each other by a

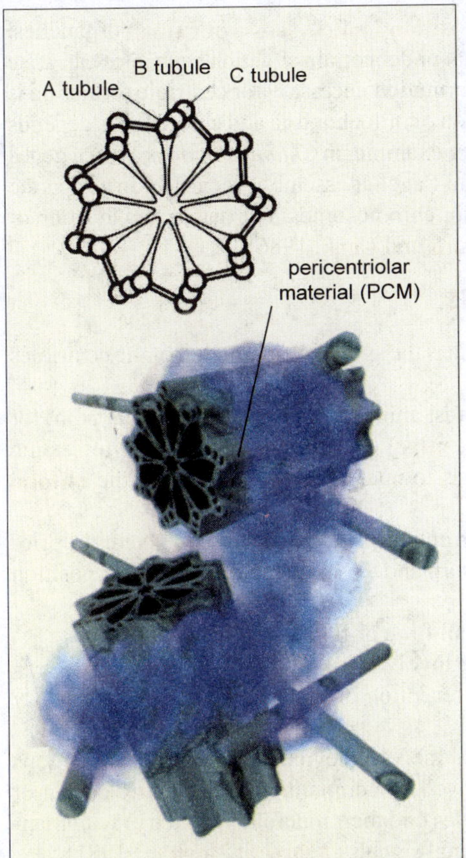

Structure of the centrosome.

distance of 50 to 100 nm. An immature centriole contains a ninefold symmetric array of single microtubules ; each microtubule then presumably acts as a template for the assembly of the triplet microtubules of mature centrioles. Each daughter centriole grows to mature size in late prophase maintaining their close proximity to and orientaton at right angles to the mother centriole. As a result, when the interphase nuclei reform at the end of nuclear division, a centrosome containing two centrioles exists beside each nucleus.

Development of centriole (or basal body) has been studied in the ciliates *Paramecium* and *Tetrahymena* and in tracheal epithelium of *Xenopus* and chicks. The stages of development are virtually the same in all of them. Development of a basal body begins with the formation of a single microtubule in an amorphous mass. Microtubules are added one at a time until there is an equally spaced ring of nine. As the microtubules appear, the amorphous mass is lost, as though it were being consumed in the production of the microtubules. There is some evidence that connectives exist between the microtubules, which could act to set the distance between them. Thus, a ring of nine complete microtubules (*i.e.,* A tubules) is formed. The C-shaped B microtubule develop next and finally the C mirotubules are added. The hub and cartwheel are added in the centre. The A-C links are not established until the end of development (see **Sheeler** and **Bianchi**, 1987).

2. Origin of basal bodies. In a ciliated vertabrate cell, which may contain hundreds of cilia, the centrioles of the precursor cell give rise to the many basal bodies required to nucleate the cilia in the mature cell. For example, during the differentiation of the ciliated epithelial cells that line the oviduct and the trachea, the centriole pair migrates from its normal location near the nucleus to the apical region of the cell where the cilia will form. There, instead forming a single daughter centriole in the typical manner, each centriole in the pair forms numerous electron-dense fibrogranular satellites. Many basal bodies then arise from these satellites and migrate to the membrane to initiate the formation of cilia.

3. The de novo origin of centrioles and basal bodies. There are certain cases where centrioles seem to arise *de novo*. For example, unfertilized eggs of many animals lack functional centrioles and use the sperm centriole for the first mitotic division (for cleavage), however, under certain experimental conditions— such as extreme ionic imbalance or electrical stimulation—the unfertilized egg can produce a variable number of centrioles. Each of these centrioles nucleates the formation of a small aster, one of which can be used by the egg for cleavage division, so that a haploid organism devlops by a process called **parthenogenesis**. In fact, centriole precursors are stored in the cytoplasm of unfertilized eggs and can be activated to form a new centriole under special situation.

Like the centrioles, the basal bodies are found to possess some capacity for self-assembly and they appear suddenly in *Naegleria* as it changes from its amoeboid form to a typical ciliate (see **Reid** and **Leech**, 1980).

The unusual mode of duplication of centrioles and their continuity over many generations led to the earlier suggestion that centrioles might be fully **autonomous**, **self-replicating** organelles. Although it is now known that this is not the case and that under certain situation centrioles can arise *de novo* in the cytoplasm. So, it is possible that some information necessary for centriole formation is usually carried in the centriole itself (just as the replication of mitochondria and chloroplasts depends on extrachromosomal genes carried in the organelles). For example, in *Chlamydomonas* a set of genes that encode proteins involved in basal body structure and flagellar assembly is carried on a separate genetic element that segregate independently of the major chromosomes. The nature and location of this genetic element have still to be investigated (**Ramanis** and **Luck**, 1986).

FUNCTIONS

1. Formation of basal bodies and ultimately the cilia is the specialized function of the centrioles in the cell.

2. The normal function of a pair of centrioles in most animal cells is to act as a focal point for the centrosome. The centrosome (also called the **cell centre**) organizes the array of cytoplasmic microtubules during interphase and duplicates at mitosis to nucleate the two poles of the **mitotic spindle**.

3. Sometimes centrioles can serve first one function and then another in turn : for example, prior to each division in *Chlamydomonas*, the two flagella resorb and the basal bodies leave their position to act as mitotic poles.

4. In spermatozoon one centriole give rise to the tail fibre or flagellum.

5. Centrioles and basal bodies are also found to be involved in ciliary and flagellar beat.

6. Centrioles and basal bodies have a role in the reception of optical, acoustic and olfactory signals.

7. Recently, it has been suggested that centrioles could serve as devices for locating the directions of signal sources. Such a role for them has been conceived by comparing the geometric design of centrioles (with their disposition in pairs at the right angles and their ninefold symmetry) with man-made devices such as radar scanners, that detect directional signals (**Albrecht-Buehler**, 1981).

REVISION QUESTIONS

1. Define the terms basal bodies and centrioles.
2. Describe the ultra-structure of the basal body and centrioles.
3. How are the centrioles and basal bodies (kinetosomes) are originated in the cell ?
4. What are the main functions of the basal bodies and the centrioles ?

CHAPTER

17

Cilia and Flagella

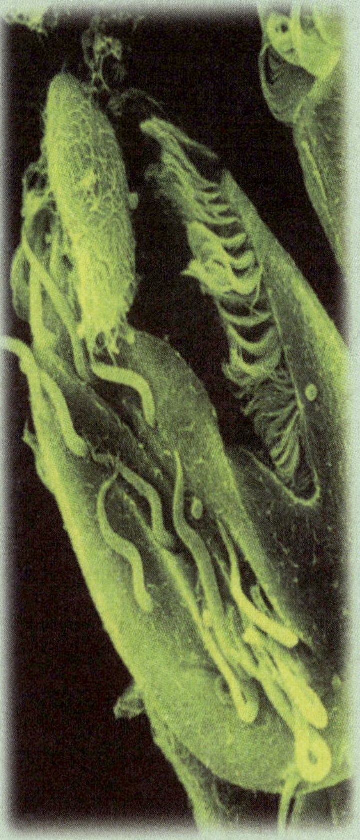

Euplotes with its fringe of waving cilia is attempting to consume *Paramecium* in the upper left corner.

The cilia (L.,*cili*=eye lash) and flagella (L., Little whip) are microscopic, contractile and filamentous processes of the cytoplasm which create food currents, act as sensory organs and perform many mechanical functions of the cell. Morphologically and physiologically, the cilia and flagella are identical structures but even then both can be distinguished from each other by their number, size and functions. Their distinguishing features are as follows :

1. The flagella are less (1 or 2) in number than the cilia which may be numerous (3000 to 14000 or more) in number.
2. The flagella occur at one end of the cell, while the cilia may occur throughout the surface of the cell.
3. The flagella are longer (up to 150 µm) processes, while the cilia are short (5 to 10 µm) appendages of the cytoplasm.
4. The flagella usually beat independently, while the cilia tend to beat in a coordinated rhythm.
5. The flagella exhibit undulatory motion, while the cilia move in a sweeping or pendular stroke.

STEROCILIA AND KINOCILIA

The cell sometimes gives out immobile cytoplasmic extensions known as **sterocilia**. The sterocilia differ from the true cilia which are known as **kinocilia**. The sterocilia occur in most epithelial cells of the epididymis and macula and crista of the internal ear. Sterocilia of the hair cells of the inner ear are responsible for the transduction of sound. These and other sterocilia do not contain microtubules. They contain, however, about 3000 microfilaments which are disposed longitudinally but have a definite polarity and a helical symmetry, with cross-bridges around the filaments (**De Rosier**, 1980).

DISTRIBUTION OF THE CILIA AND FLAGELLA

The flagella occur in the protozoans of the class Flagellata, choanocyte cells of the sponges, spermatozoa of the Metazoa and among plants in the algae and gamete cells. The cilia occur in the protozoans of the class Ciliata and members of other classes and ciliated epithelium of the Metazoa. The cilia may occur on external body surface and may help in the locomotion of such animals as the larvae of certain Platyhelminthes, Nemertines, Echinodermata, Mollusca and Annelida. The cilia may line the internal cavities or passages of the metazoan bodies as air passage of the respiratory system and reproductive tracts. The nematode worms and arthropods have no cilia.

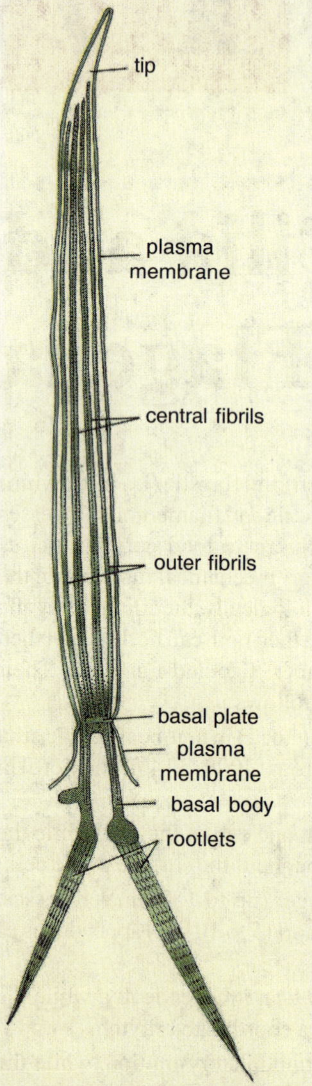

Fig. 17.1. A ciliary apparatus in L.S. showing fundamental structure of cilium or shaft, basal bodies and ciliary rootlets.

Except for sperm, the cilia in mammalian systems are not organelles of locomotion. But their effect is the same, that is, to move the environment with respect to the cell surface.

STRUCTURE OF THE CILIA AND FLAGELLA

The ciliary apparatus is composed of following basic components (1) the **shaft** or **cilium**, which is the slender cylindroid process that projects from the free surface of the cell; (2) the **basal body** or granule, a centriole like cellular organelle from which the cilium originates ; and (3) in some cells fine fibrils—called **ciliary rootlets**. Basal body remains separated from cilium by a **ciliary** or **basal plate** which has two functions : termination of the C Tubule of each triplet of basal body ; and beginning of two central microtubules. The cilia and flagella are extremely delicate, permanently formed, thread-like extension of cytoplasm and their thickness is often at the limit of the resolving power of light microscope.

ISOLATION AND CHEMICAL COMPOSITION OF CILIA AND FLAGELLA

The first detailed chemical analysis of the protein components of the cilia of *Tetrahymena pyriformis* was conducted by **I. R. Gibbons** (1963). Ciliary movements can be analyzed easily by scraping the pharyngeal epithelium of a frog or toad with a spatula and placing the scrapings in a drop of physiological salt solution between a slide and a coverglass. By certain recent techniques, a flagellum can be severed from a cell by a laser beam and ciliary membrane can be peeled off by detergent treatments.

Axoneme of cilia has a variety of proteins such as α and β **tubulins** in the micortubules, **dynein** (the microtubule ATPase), **nexin** and others (see Table 17-1).

ULTRASTRUCTURE OF THE CILIA AND FLAGELLA

An eukaryotic cilium or flagellum is composed of three major parts : a central axoneme or shaft, the surrounding plasma memberane and the interposed cytoplasmic matrix (Fig. 17.1).

1. Ciliary Membrane

Though the ciliary membrane (9.5 nm thick) is physically continuous with plasma membrane of the cell, but it contains far less amount of proteins than the latter (*i.e.*, it is atypically protein poor; **Satir**, 1977). Further, some of the proteins present in the ciliary membrane are specific to it and have a role as the barrier against the loss of ATP and certain essential ion that are required at appropriate concentrations to provide the energy for the ciliary movement.

Ciliary necklace. An unusual feature of the membranes of all somatic cilia is the presence of multiple strands (2 to 6 and up to 11) of particles, called **ciliary necklace**, at the base of the organelle. These particles can be seen in the electron micrographs of freeze-fractured cilium (see **Fumi Suzuki**). The ciliary necklace is found at a region in the cilium where microtubules and basal bodies make contact with the membrane. According to **Thorpe** (1984), ciliary necklace may have following two functions : 1. It may position the underlying basal body from which the cilium is originated. 2. It may help in the differentiation of ciliary membrane; *i.e.*, the rings of particles may retain proteins that would otherwise diffuse out and be incorporated into ciliary membrane.

2. Matrix

The bounded space of the cilium contains a watery substance known as **matrix**. In the ciliary matrix are embedded eleven microtubules of axoneme and other interconnecting proteins.

3. Axoneme

The axial basic microtubular structure of cilia and flagella is called **axoneme**. It is the essential motile element of these organelles. The axoneme is about 0.2 to 10 μm in diameter and may range from a few microns to 1 to 2 mm in length. The cilia may be thicker at the base and may become thinner gradually along the length.

Table 17.1. Major protein structures of the axoneme of the cilia and flagella (Source : **Alberts** *et al.*, 1989).

Axoneme component (periodicity along axoneme)	Function
1. Tubulin (8 nm)	Principal component of microtubules.
2. Dynein (24 nm)	Project from microtubule doublets and interact with adjacent doublets to produce bending.
3. Nexin link (86 nm)	Hold adjacent microtubule doublets together.
4. Radial spokes (29 nm)	Extend from each of the nine outer doublets inward to the central pair.
5. Sheath projections (14 nm)	Project as a series of side arms from the central pair of microtubules ; together with the radial spokes these regulate the form of the ciliary beat.

The axonemal elements of nearly all cilia and flagella (as well as the tails of sperm cells) contain the same 9 + 2 arrangement of microtubules. In the centre of the axoneme are two **singlet microtubules** or fibrils that run length of the cilium. Each of the central microtubules (25 nm in diameter) is composed of 13 protofilaments. The central fibrils, each has a wall of 6 nm thick and are located 35 nm away from each other. Both central fibrils are connected by a bridge and are enclosed in a common central **sheath**. A plane perpendicular to line joining the two central tubules divides the axoneme into a right and a left symmetrical half. It is generally accepted that the plane of the ciliary beat is perpendicular to this plane of symmetry.

Nine **doublet microtubules** (each 36 nm in diameter) surround the central sheath; they remain separated from each other by a distance of 20 nm and from the ciliary membrane by a distance of 25 nm. Each peripheral doublet consists of two microtubules or subfibres (18 to 25 nm in diameter), one is smaller (A) and complete, having 13 protofilaments of tubulin and lying closer to the axis; the other

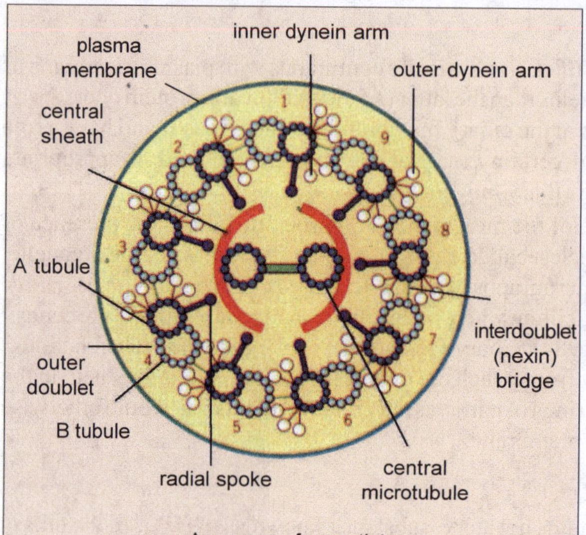

Axoneme of a protist.

subfibre (B) is larger and incomplete, having only 11 protofilaments. The B subfibre lacks the wall adjacent to A subfibre and is skewed at about 10° away from the axis. Other associated structures of the doublets are the following :

1. Dynein arms. Extending from each A subfibre are two **dynein arms**—an **outer arm** and an **inner arm**, that are oriented in the same direction in all microtubules (*i.e.*, peripheral doublets). This orientation is clockwise when the axoneme is viewed from base to tip. The arms contain **dynein**, which is large protein complex (nearly 2 million daltons) composed of 9 to 10 polypeptide chains, the largest of which are about 450,000 daltons. Dynein is a Mg^{2+} and Ca^{2+}-activated ATPase enzyme which after solubilization can recombine at the same position on the A microtubule. Dynein contains two or three elliptical or globular **heads** (depending on the source) linked to a common **root**, **foot** or **base** by the thin flexible **strands** or **stalks** (Fig. 17.3). Thus, the base of the dynein molecule attaches only to A subfibre, leaving the heads free to make contact with the adjacent B tubules of neighbouring doublet. It indicates that B tubule has different structure than the A tubule, so that, the base of dynein cannot attach to it. The resulting asymmetry is required to prevent a fruitless tug-of-war between the neighbouring microtubules, which presumably explains why each of nine outer microtubules is an A-B doublet (see **Alberts** *et al.*, 1989).

2. Nexin links. Adjacent doublets are joined or linked by **peripheral**, **interdoublet** or **nexin links** ; the nexin links have a periodicity of 86 nm. Nexin links extend from A tubule of one doublet to B tubule of adjacent doublet. Nexin protein has a molecular weight of about 150,000 to 160,000 daltons. Nexin links are highly elastic : their normal length is 30 nm, but they can be stretched to 250 nm without breaking (see **Darnell**, 1986). They are thought to function like the rubber bands

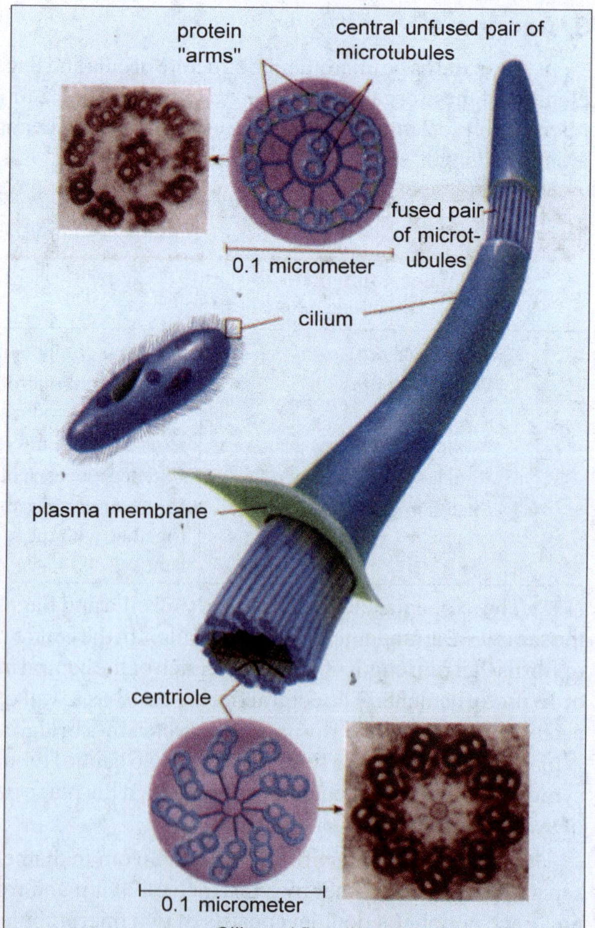

Cilia and flagella.

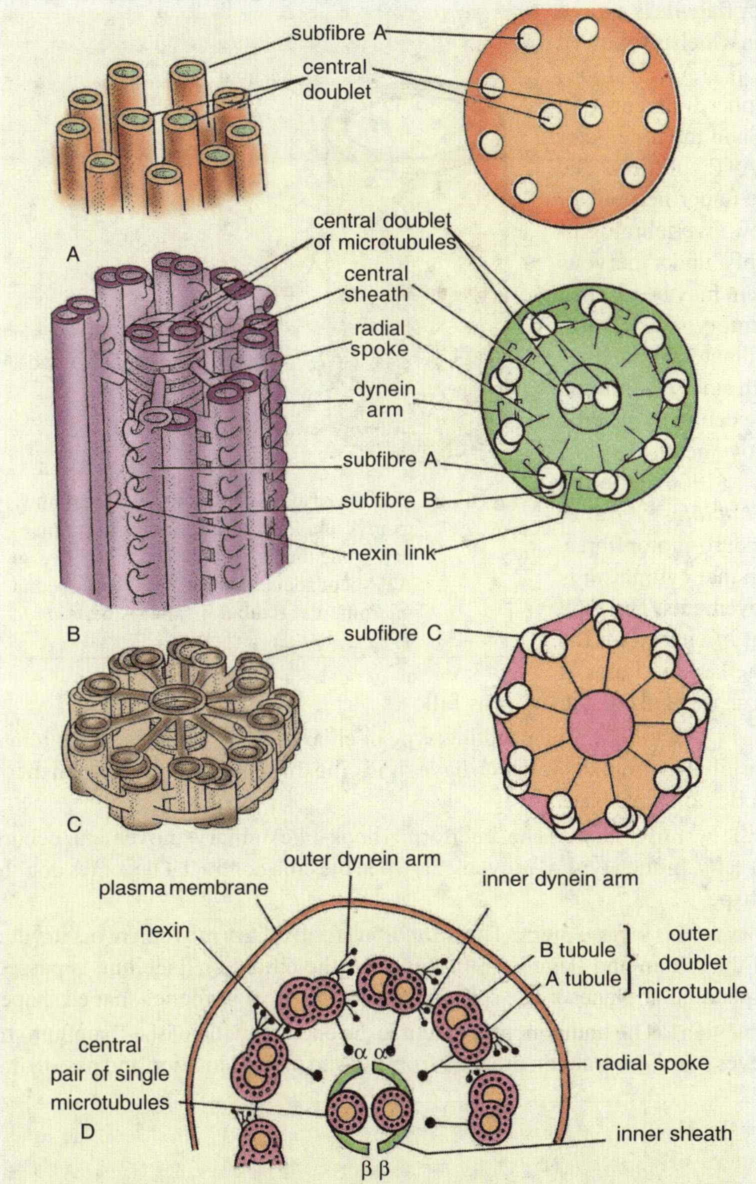

Fig. 17.2. Schematic diagram of a flagellar axoneme, showing relationships between its components at the flagellar tip (A), in the middle of the flagellum (B), and at the basal body (C). More detailed picture of flagellum in the cross section (after Berns, 1983 ; Alberts *et al.*, 1989).

to resist the sliding between adjacent double microtubules (*i.e.*, they maintain integrity of the axoneme during the sliding motion).

3. Radial spokes. There are 36 nm long radial bridges or links between the A subfibre and the sheath containing the central microtubules. These spokes terminate in a dense **knob** or **head**, which may have a fork-like structure. Earlier observation of **Warner** and **Satir** (1974) that the spokes are attached perpendicularly to the ciliary axis where it is straight and that they are relatively detached in bent or tilted regions of the axis has led to the hypothesis that *they may be active in the conversion of active sliding between outer doublets into local axial bending.*

Lastly, the structures of a cilium at the base and tip are slightly different from features described above (Fig. 17.2).

PHYSIOLOGY OF CILIARY MOVEMENT

The cilia and flagella serve many purposes and their movements propel the organism. The cilia are contractile structures and in them two types of rhythms known as **metachronic** and **isochronic** or **synchronous rhythms** produce the wave of contractions in the cilia. In the **metachronic** type of rhythm the cilia of a row beat one after the other, while in the **synchronous** or isochronic rhythm, all the cilia of a row beat simultaneously. In contrast to the cilia, the flagella exhibit undulant motion and beat independently.

The beating of cilia or flagella is caused by the intraciliary excitation which is followed by the interciliary conduction. **Hayashi** (1961) has reported that the two inner filaments of a cilium transmit excitation and the nine outer filaments are the seat of ATP splitting. The movement of cilia may be under nervous or cytoplasmic control. In a few invertebrate embryos the cilia are probably under nervous control since their movement may be stopped upon stimulation of the embryo. In ciliates they are thought to be coordinated by a neuromotor centre near the mouth since destruction of the fibres connecting the centre to the cilia results in uncoordinated movements (**Taylor**, 1920). However, **Okajima** (1966) reported coordinated movements in *Euplotes* even after complete dissection of the neuro-motor fibres. Recent studies have shown that cytoplasm is necessary for the ciliary movements. The ATP provides necessary amount of energy for the motion of the cilia and flagella. Four types of ciliary movements have been recognized which are as follows :

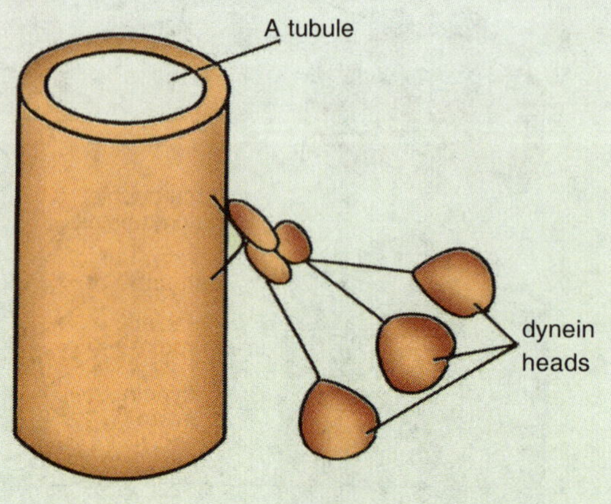

Fig. 17.3. The base of the dynein molecule binds tightly to an A tubule in an ATP-independent manner, while the large globular heads have an ATP-dependent binding site for a second microtubule (B tubule) (after Alberts *et al.*, 1989).

1. The pendulus ciliary movement. The pendulus type of ciliary movement is carried out in a single plane. It occurs in the ciliated protozoans which have rigid cilia. In such cases the movement of the cilia is carried out by a flexion at its base.

2. The unciform ciliary movement. The unciform (hook-like) ciliary movement occurs commonly in the metazoan cells. In such type of movement, when the cilia contract it becomes double and acquires a hook-like shape.

3. The infundibuliform ciliary movement. The infundibuliform ciliary movement occurs due to the rotary movement of the cilium and flagellum. In this case, the cilium or flagellum is passed through three mutually perpendicular planes in the space and makes conical or funnel-shaped shape.

4. The undulant movement. The undulant movement is the characteristic of the flagellum. In undulant movement the waves of the contraction proceed from the site of implantation and pass to the border.

Each beat of cilium or flagellum involves the same pattern of microtubule movement. Each cilium moves with a whip-like motion and its beat may be divided into two phases : 1. The fast **effective stroke** (or forward active stroke or power stroke) in which the cilium is fully extended and beating against the surrounding liquid (*i.e.*, it is like the action of an oar in a rowboat ; Fig. 17.4B). 2. The slow **recovery stroke**, in which the cilium returns

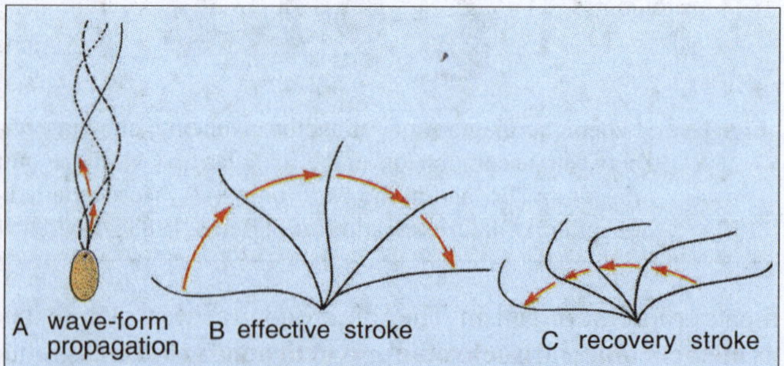

Fig. 17.4. Movement of cilia and flagella. A—Typical wave-form propagation of a flagellum, such as that found in sperm ; B—Effective stroke common with cilia of all types ; C—Recovery of cilium prior to the next effective stroke (after Dyson, 1978).

to its original position with an unrolling movement that minimizes viscous drag (Fig. 17.4B). The cycles of adjacent cilia are almost but not quite in synchrony, creating a wave-like pattern.

The flagellum instead of making whip-like movements, propagates quasi-sinusoidal waves (Fig. 17.4A), *i.e.*, successive waves move toward the tip of the flagellum, propelling the cell (*e.g.*, sperm) in the opposite direction.

The mechanism of force and movement (bending) by the flagellum has recently been studied extensively. It is well established now that the ciliary movement is generated by the microtubules and the associated structures of the flagellum. It was shown that the cell free flagella can be caused to move by adding an energy source such as ATP. Even broken pieces of cilia or isolated axoneme itself continue to beat, suggesting the role of microtubules in the movement. The contractile **axostyle** of some

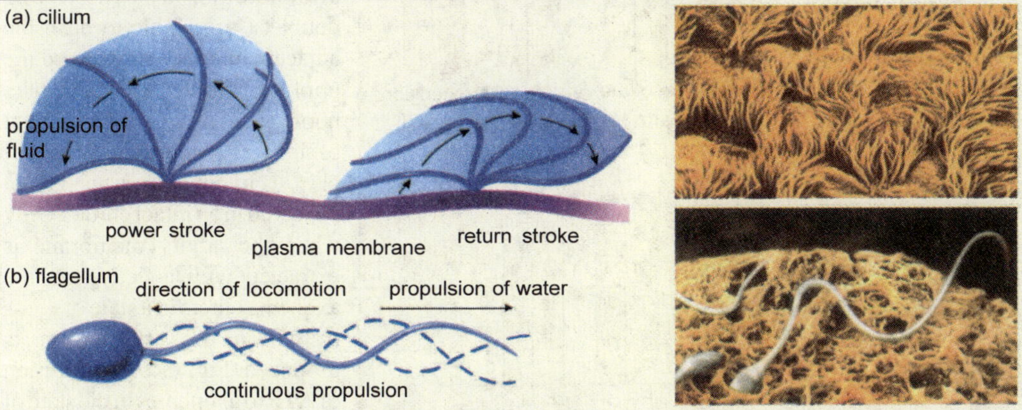

How cilia and flagella move.

microorganisms such as *Metamonadida* (a dinoflagellate that lives in the gut of termite; Fig. 17.5) is another example of microtubule mediated motile process (see **Berns**, 1983). In fact, bending force is produced by the sliding of microtubules. This has been shown by exposing isolated axonemes to proteolytic enzymes, which disrupts both the nexin links and the radial spokes but leaving the dynein arms and the microtubules themselves intact. If such a partially digested structure is exposed to as little as 10 µM ATP, the axoneme elongates until it is up to nine times its original length, the component microtubules in the axoneme telescoping out of the loosened structure. It seems that the adjacent peripheral doublets can actively slide against each other once they have been freed of their lateral cross-links (such as those made of nexin). Apparently, in intact structure this sliding movement is converted to bending. Further, since the adjacent outer doublets actively slide against each other, a force must be generated between them. This force is

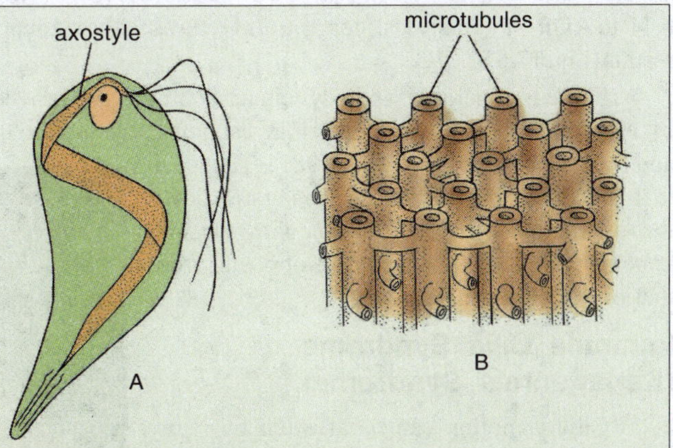

Fig. 17.5. Axostyle of *Metamonadida* flagellate. A—The relationship of the axostyle with the cell body; B—Schematic representation of the arrangement of microtubules into parallel sheets with protein cross bridges between them as seen in a T. S. of axostyle (after Berns, 1983).

apparently generated by dynein arms which 'walk' along the doublets, as has been suggested by **Peter Satir's** (1968) sliding filament hypothesis.

Sliding Filament Hypothesis

Recent experimental work on ciliary motion has shown notable similarities with the sliding mechanism involved in the interaction of actin and myosin in muscle. The dynein arms attached to subfibre A have been compared with the cross bridges of myosin and it has been postulated that they form intermittent attachments, by which one doublet (N_1) is able to push the adjacent one ($N_1 + 1$) toward the tip of the axoneme (Fig. 17.6). Under normal conditions, the attachment of subfibre A of N to subfibre B of N + 1 by dynein arms is not observed in an intact cilium. Only when the ciliary membrane is extracted with a detergent, the axoneme enters in a state of **rigor** in which the attachment is produced (Fig. 17.6 A). Addition of ATP to axonemes in the state of rigor restores motility and causes release of the dynein arm (Fig. 17.6 B). In this mechano-chemical cycle, the next step would be **reextension** of the dynein arm (Fig. 17.6C) and its **rebinding** at an angle, with a new, more proximal site on subfibre B (Fig. 17.6 D). This step involves the hydrolysis of ATP to ADP + Pi. In the last step, the arm returns to the rigor position and displacement of the doublets results (Fig. 17.6 E).

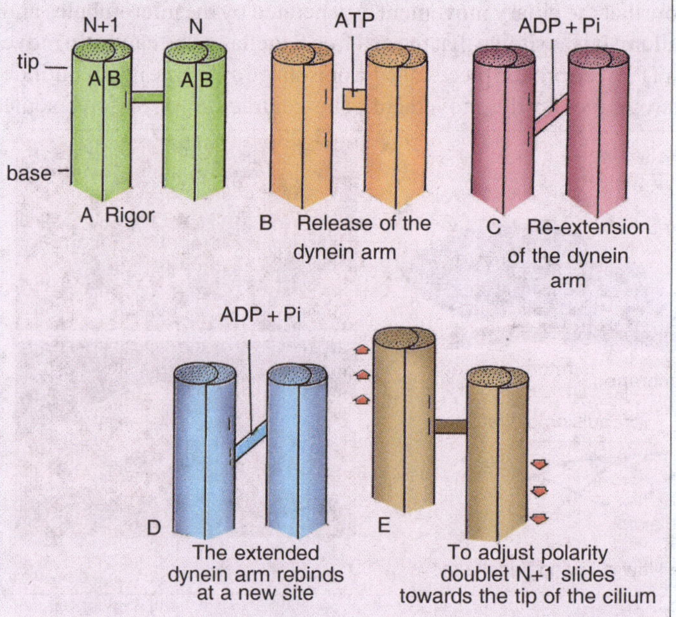

Fig. 17.6. Schematic representation of the mechanochemical cycle involved in sliding of filament in ciliary movement (after De Robertis and De Robertis, Jr., 1987).

Force is generated when dynein arms move. The movement of sliding is converted to bending by virture of radial spokes that bridge each other doublet to the inner pair of microtubules (**Warner** and **Satir**, 1974 ; **Huang** *et al.*, 1981). The wave that is generated by sliding is propagated down the organelle from base to tip, with the cell generally moving in a direction opposite from that of wave propagation.

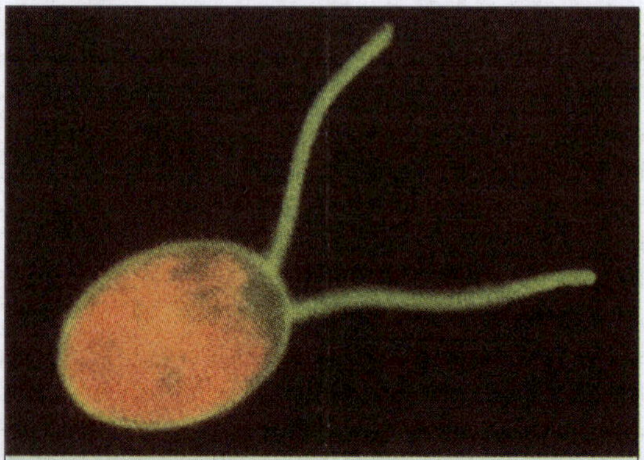

An eukaryotic flagellum.

Immotile Cilia Syndrome (Kartagenre's Syndrome)

Ciliary motion can be affected by many deficiencies in the protein composition of the organelle. For example, in immotile cilia syndrome, a condition characterized by severe respiratory difficulty (chronic bronchitis and sinusitis) and male sterility, the underlying genetic defect is the absence of inner and outer dynein arms on the peripheral doublets of both cilia and flagella. The symptoms of this syndrome result from the immobility of cilia in the respiratory tract and of the flagella in the sperm.

In *Chlamydomonas* several mutational defects have been studied in the axoneme of flagellum which may lead to paralysis of the flagellar function (**Luck**, 1984).

Other Functions of the Cilia and Flagella

1. The ciliary or flagellar movement provides the locomotion to the cell or organism.

2. The cilia create food currents in lower aquatic animals.

3. In the respiratory tract, the ciliary movements help in the elimination of the solid particles from it.

4. The eggs of amphibians and mammals are driven out from the oviduct by the aid of vibratile cilia of the latter.

Thus, the cilia and flagella serve many physiological processes of the cell, such as locomotion, alimentation, circulation, respiration, excretion and perception of sense.

ORIGIN OF CILIA

The newly formed basal bodies become aligned in rows beneath the apical plasma membrane and each basal body may then produce satellites from the side, a root from its base and a cilium from its apex.

The formation of the cilia from the basal bodies is started by the formation of a vesicle-like structure of the cytoplasm towards the distal end of the basal bodies. The walls of the vesicle are invaginated due to rapid growth of ciliary shaft. The walls of the vesicle are temporary and are replaced by the new and permanent ciliary sheaths.

DERIVATIVES OF CILIA

The cilia are modified into a variety of structures such as the rods and cones of the retina, crown cell of saccus vasculosus of third ventricle of fishes, primitive sensory cells of the pineal eye and cnidocil of the nematocysts of the coelenterates.

REVISION QUESTIONS

1. Differentiate between cilia and flagella. Describe the structure of the axoneme.
2. Explain sliding microtubule hypothesis of ciliary movement.
3. Flagella and cilia, though identical in structure, commonly exhibit a quite different pattern of movement. Describe the two patterns and conditions under which one or the other would be more appropriate.
4. Write short notes on the following :

   (a) axoneme ; (b) sterocilia ; (c) ciliary necklace ; (d) axostyle ; (e) sliding filament hypothesis ; (f) Kartagenre's syndrome

CHAPTER

18

Cell Growth and Cell Division

(Cell Cycle, Mitosis and Meiosis)

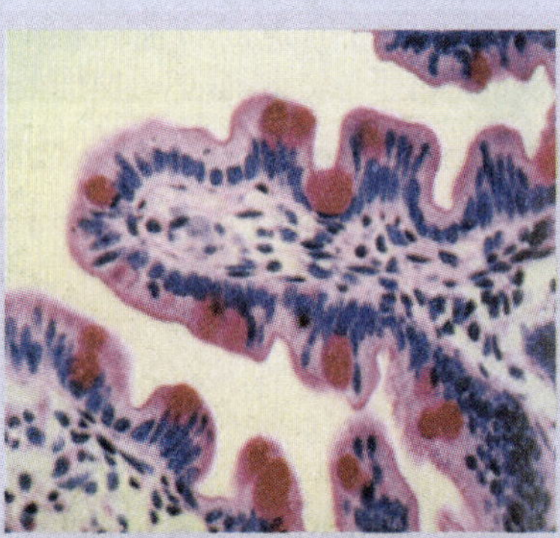

(a)

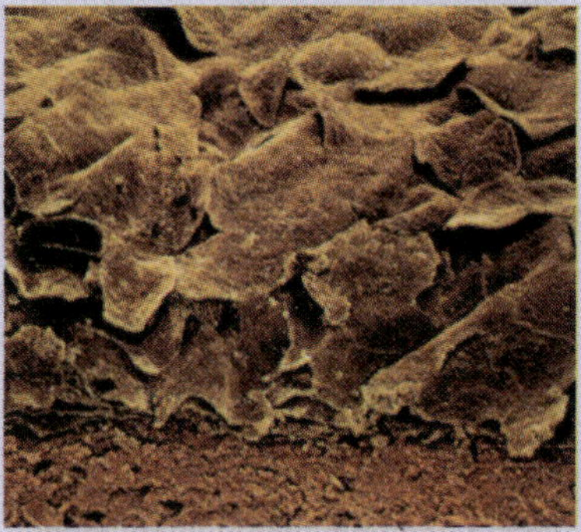

(b)

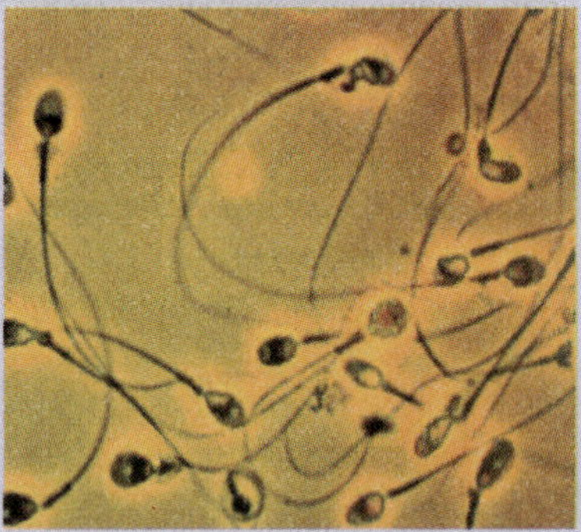

(c)

(a) Intestinal epithelial cells have a very rapid turnover; the entire gut lining is replaced every few days.

(b) SEM of human epidermis. Dead skin cells are constantly being lost and replaced by mitosis.

(c) Human sperms.

Growth— an increase in size or mass of a developing/ living system— is an irreversible process that occurs at all organizational levels. Often, it is difficult to define, because, it is. multifactorial, that is, growth embodies following three interacting growth patterns : (1) **auxetic growth**— an increase in cell mass or **auxesis** ; (2) **multiplicative growth**— an increase in cell number due to cell division; and (3) **accretionary growth**— growth due to accumulation of extracellular products (accretion means increase by addition on the surface of the material of same nature as that is already present, *e.g.*, the manner of growth of crystal). Generally, when rate of anabolism (*i.e.*, photosynthesis, protein synthesis, etc.) far exceeds the rate of catabolism (*i.e.*, respiration), the growth of protoplasm (*i.e.*, auxetic growth) takes place.

CELL CYCLE AND MITOSIS

All cells are produced by divisions of pre-existing cell. Continuity of life depends on cell division. A cell born after a division, proceeds to grow by macromolecular synthesis, reaches a species-determined division size and divides. This cycle acts as a unit of biological time and defines life history of a cell. **Cell cycle** can be defined as the entire sequence of events happening from the end of one nuclear division to the beginning of the next. The cell cycle involves the following three cycles (see **Albert** *et al.*, 1989).

1. Chromosome cycle. In it **DNA synthesis** alternates with **mitosis** (or karyokinesis or nuclear division). During DNA synthesis, each double-helical DNA molecule is replicated into two identical daughter DNA molecules and during mitosis the duplicated copies of the genome are ultimately separated.

2. Cytoplasmic cycle. In it **cell growth** alternates with **cytokinesis** (or cytoplasmic division). During cell growth many other components of the cell (RNA, proteins and membranes) become double in quantity and during cytokinesis cell as a whole divides into two. Usually the karyokinesis is followed by the cytokinesis but sometimes the cytokinesis does not follow the karyokinesis and results into the multinucleate cell, *e.g.*, cleavage of egg in *Drosophila*.

3. Centrosome cycle. Both of the above cycles require that the **centrosome** be inherited reliably and duplicated precisely in order to form the two poles of the mitotic spindle ; thus, centrosome cycle forms the third component of cell cycle.

Howard and **Pelc** (1953) have divided cell cycle into four phases or stages : G_1, S, G_2 and M phase. The G_1 phase, S phase and G_2 phase are combined to from the classical **interphase**.

1. G_1 Phase. After the M phase of previous cell cycle, the daughter cells begin G_1 of interphase of new cell cycle. G_1 is a resting phase. It is called **first gap phase**, since no DNA synthesis takes place during this stage; currently, G_1 is also called **first growth phase**, since it involves synthesis of RNA, proteins and membranes which leads to the growth of nucleus and cytoplasm of each daughter cell towards their mature size (see **Maclean** and **Hall**, 1987).

During G_1 phase, chromatin is fully extended and not distinguishable as discrete chromosomes with the light microscope. This is a time of resumption of normal cell metabolism which has slowed down during the previous cell division. Thus, G_1 involves transcription of three types of RNAs, namely rRNA, tRNA and mRNA ; rRNA synthesis is indicated by the appearance of nucleolus in the interphase (G_1 phase) nucleus. Proteins synthesized during G_1 phase (1) regulatory proteins which control various events of mitosis ; (2) enzymes (*e.g.*, DNA polymerase) necessary for DNA synthesis of the next stage ; and (3) tubulin and other mitotic apparatus proteins.

G_1 phase is most variable as to duration (Table 18-1) ; it either occupies 30 to 50 per cent of the total time of the cell cycle or lacks entirely in rapidly dividing cells (*e.g.*, blastomeres of early embryo of frog and mammals). Terminally differentiated somatic cells (*i.e.*, end cells such as neurons and striated muscle cells) that no longer divide, are arrested usually in the G_1 stage; such a type of G_1 phase is called **G_0 phase.**

Table 18.1. Different stages of a mitotic cell cycle and their duration in hours.

Parts of cell cycle	Phases	Description of phases	Duration in hours *Vicia faba*	Mouse L cells	Human HeLa cells
Interphase	G_1	Pre-DNA-synthesis phase	12	12	12
	S	DNA-synthesis phase	6	6–8	10
	G_2	Post-DNA synthesis phase	12	3–4	3
Mitosis	M	Mitotic phase	1	1	1

2. S phase. During the S phase or **synthetic phase** of interphase, replication of DNA and synthesis of histone proteins occur. New histones are required in massive amounts immediately at the beginning of the S period of DNA synthesis to provide the new DNA with nucleosomes. Thus, at the end of S phase, each chromosome has two DNA molecules and a duplicate set of genes. S phase

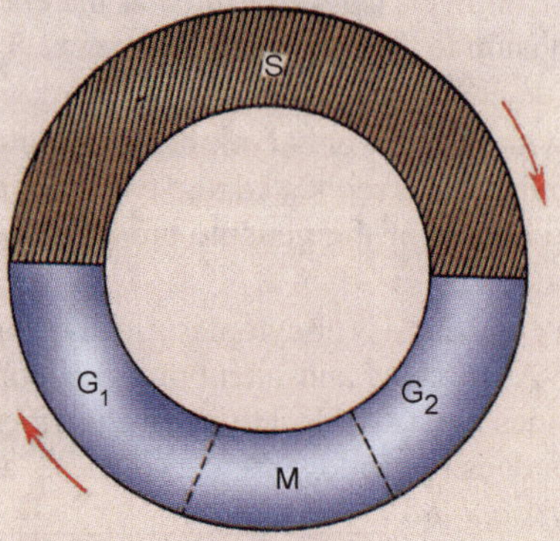

Fig. 18.1. The cell cycle or mitotic cycle, showing relative duration of phases (*e.g.,* interphase and mitotic phase) in a growing cell. S, synthesis of DNA ; G_1 the first gap or growth phase ; G_2, the second gap or growth phase ; and M, mitotic phase.

occupies roughly 35 to 45 per cent of cell cycle.

3. G_2 phase. This is a **second gap** or **growth phase** or resting phase of interphase. During G_2 phase, synthesis of RNA and proteins continues which is required for cell growth. It may occupy 10 to 20 per cent time of cell cycle. As the G_2 phase draws to a close, the cell enters the M phase.

General Events of Interphase

The interphase is characterized by the following features :

The nuclear envelope remains intact. The chromosomes occur in the form of diffused, long, coiled and indistinctly visible chromatin fibres. The DNA amount becomes double. Due to accumulation of ribosomal RNA (rRNA) and ribosomal proteins in the nucleolus, the size of the latter is greatly increased. In animal cells, a daughter pair of centrioles originates near the already existing centriole and, thus, an interphase cell has two pairs of centrioles.

In animal cells, net membrane biosynthesis increases just before cell division (mitosis). This extra membrane seems to be stored as **blebs** on the surface of the cells about to divide.

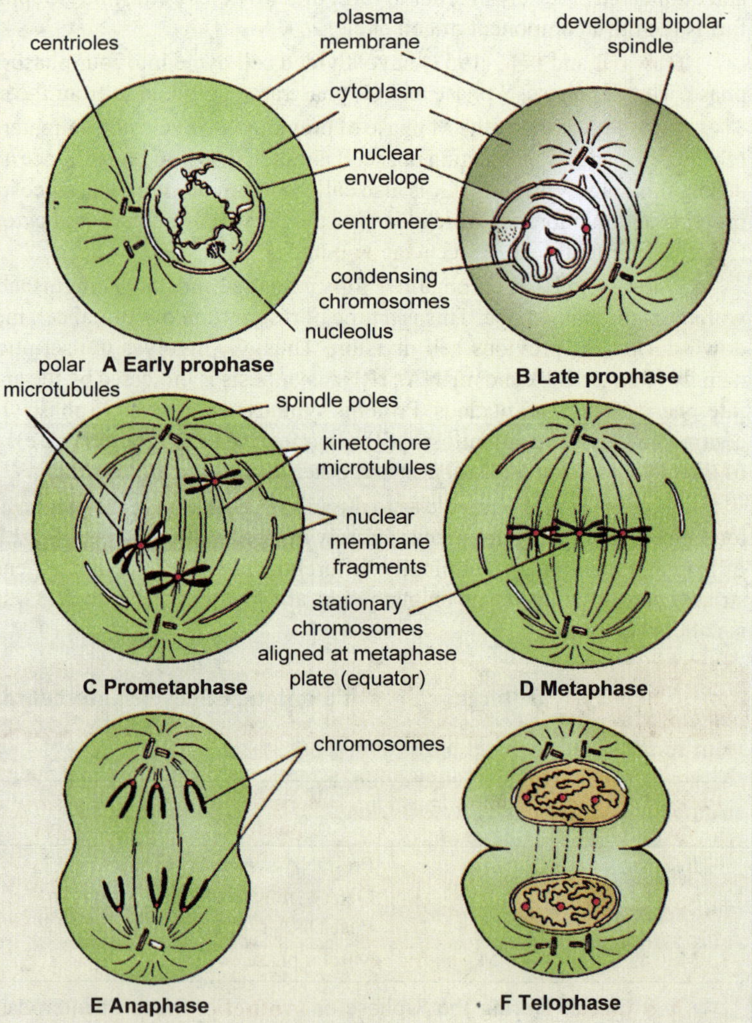

Fig. 18.2. Diagrammatic summary of mitosis in the animal cell (after Burns and Bottino, 1989).

4. M phase or Mitotic phase. The mitosis (Gr., *mitos*=thread) occurs in the somatic cells and it is meant for the multiplication of cell number during embryogenesis and blastogenesis of plants and animals. Fundamentally, it remains related with the growth of an individual from zygote to adult stage. Mitosis starts at the culmination point of interphase (*i.e.,* G_2 phase). It is a short period of chromosome condensation, segregation and cytoplasmic division. Mitosis is important for replacement of cells lost to natureal friction (**attrition**), wear and tear and for wound healing.

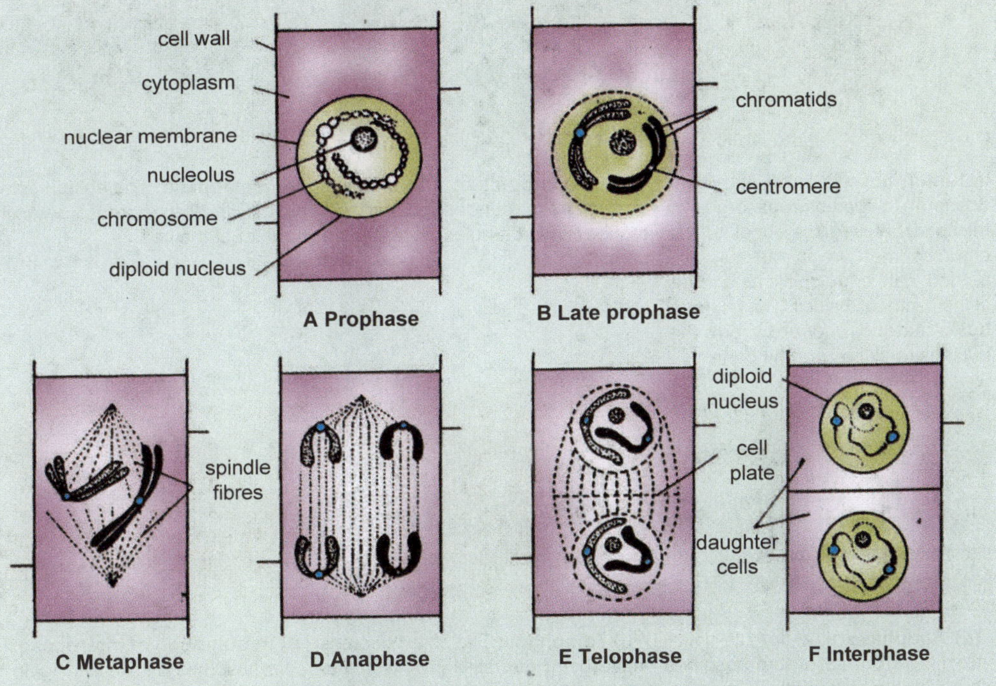

Fig. 18.3. Diagrammatic summary of mitosis in the higher plant cells.

As a process, mitosis is remarkably similar in all animals and plants. It is a smoothly continuous process and is divided arbitrarily into following stages or phases for convenient reference (Fig. 18.2 and Fig. 18.3) :

1. Prophase. The appearance of thin-thread like condensing chromosomes marks the first phase of mitosis, called **prophase** (Gr., *pro*=before ; *phasis*=appearance). The cell becomes spheroid, more refractile and viscous.

Each prophase chromosome is composed of two coiled filaments, the **chromatids**, which are the result of the replication of DNA during the S phase. As prophase progresses, the chromatids become shorter and thicker and two sister chromatids of each chromosome are held together by a special DNA-containing region, called the **centromere** or **primary constriction**. During prophase, proteins of the trilaminar **kinetochores** (one for each chromatid) start depositing or organizing on the centromere of each chromosome (see **Darnell** *et al.*, 1986). Further, during early prophase, the chromosomes are evenly distributed in the nuclear cavity ; as prophase progresses, the chromosomes approach the nuclear envelope, causing the central space of the nucleus to become empty.

In the cytoplasm, the most conspicuous change is the formation of the spindle or **mitotic apparatus**. In the early prophase, there are two pairs of centrioles, each one surrounded by the so-called **aster** which is composed of microtubules radiating in all directions. The two pairs of centrioles migrate to opposite poles of the cell along with the asters and become situated in antipodal positions. Between the separating centrioles forms a spindle. The microtubules of the spindle are arranged like two cones base to base, broad at the centre or equator of the cell and narrowing to a point at either end or pole. Mitotic spindle contains three main types of fibres (Fig. 18.7) : (1) **polar fibres**, which extend from the two poles of the spindle toward the equator ; (2) **kinetochore fibres**, which attach to the

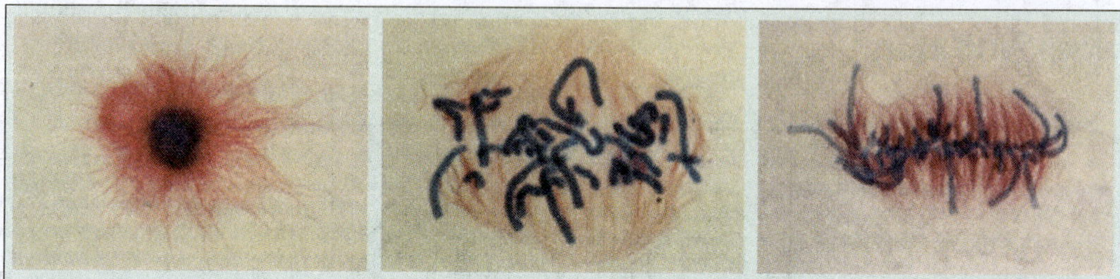

(a) Interphase in a cell of the endosperm (a food-storage organ in the seed) before mitosis begins: The chromosomes are in the thin, extended state and appear as a mass in the center of the cell. The microtubules extend outward from the nucleus to all parts of the cell.

(b) Late prophase: The chromosomes have condensed and attached to microtubules of the spindle fibers.

(c) Metaphase: The spindle fibers have pulled the chromosomes to the equator of the cell.

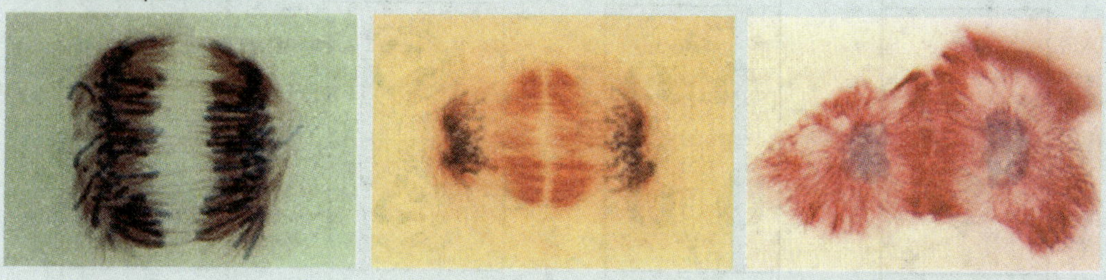

(d) Anaphase: The spindle fibers are moving one set of chromosomes to each pole of the cell.

(e) Telophase: The chromosomes have been gathered into two clusters, one at the site of each future nucleus.

(f) Resumption of interphase: The chromosomes are relaxing again into their extended state. The spindle fibers are disappearing, and the microtubules of the two daughter cells are rearranging into the interphase pattern.

kinetochores of centromeres of each mitotic chromosomes and extend toward the poles ; and (3) **astral fibres**, which radiate outward from the poles toward the periphery or cortex of cell. In cells of most higher plants, however, spindle forms without the aid of centrioles and lacks asters (Fig. 18.4).

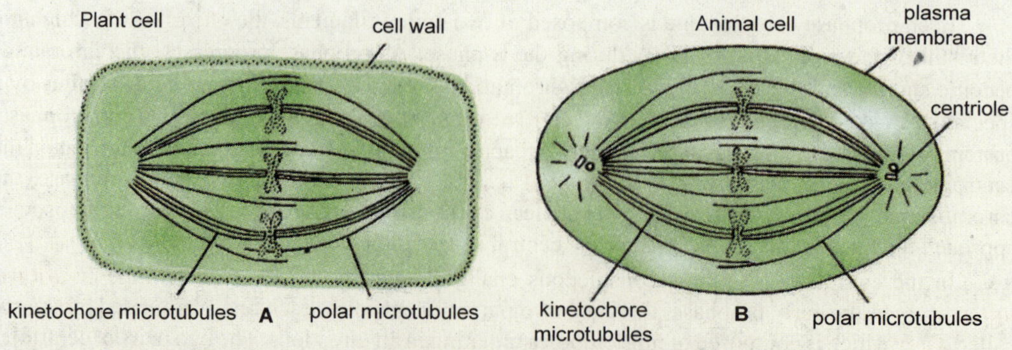

Fig. 18.4. Mitotic metaphase spindle structure in (A) a plant cell and (B) an animal cell. Higher plants lack centrioles and astral fibres (after Darnell *et al.*, 1986).

Lastly, during prophase, the nucleolus gradually disintegrates. Degeneration and disappearance of the nuclear envelope marks the end of prophase. This process is incompletely understood. However, following two factors may be involved in this process : 1. **Enzymatic action** either by some mitochondrial enzymes (see **Grant**, 1978), cytosolic MPF kinase (see **Alberts**, *et al.*, 1989) or nuclear RNA (or ribozyme ; **Burns** and **Bottino**, 1989). 2. **Physical action**, *i.e.*, physical stress exerted by microtubules which become attached to the nuclear envelope (see **Burns** and **Bottino**, 1989).

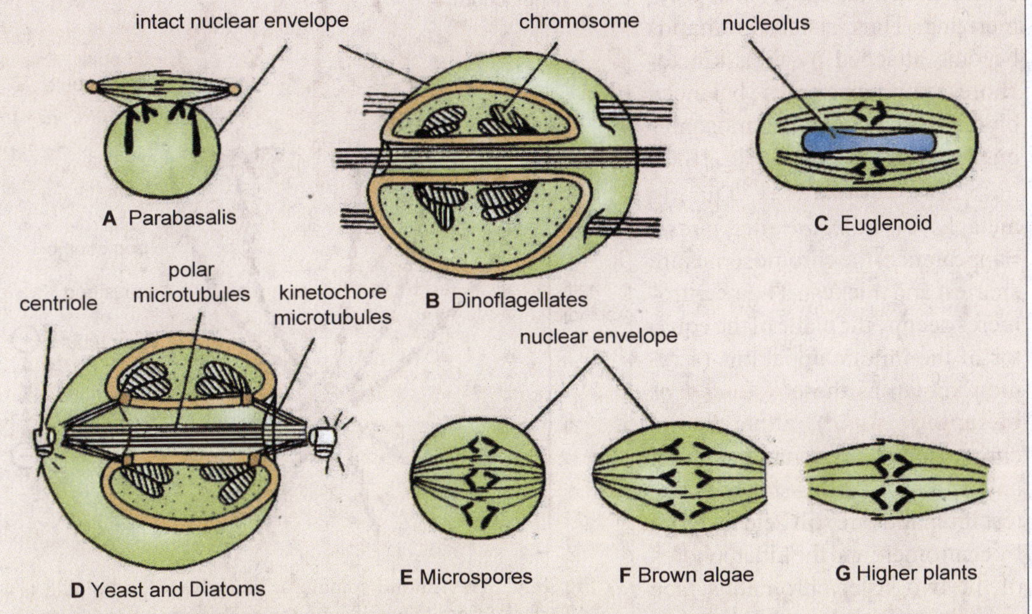

Fig. 18.5. Some variations of mitosis which have been observed during anaphase (after Albert *et al.*, 1989).

There are variations available with respect to the dissolution of nuclear envelope and the nucleolus. In a number of primitive classes of plants and animals the nuclear envelope does not dissolve during mitosis (Fig. 18.5).

Prometaphase. The breakdown of nuclear envelope signals the commencement of prometaphase and enables the mitotic spindle to interact with the chromosomes. This stage is characterized by a period of frantic activity during which the spindle appears to be trying to

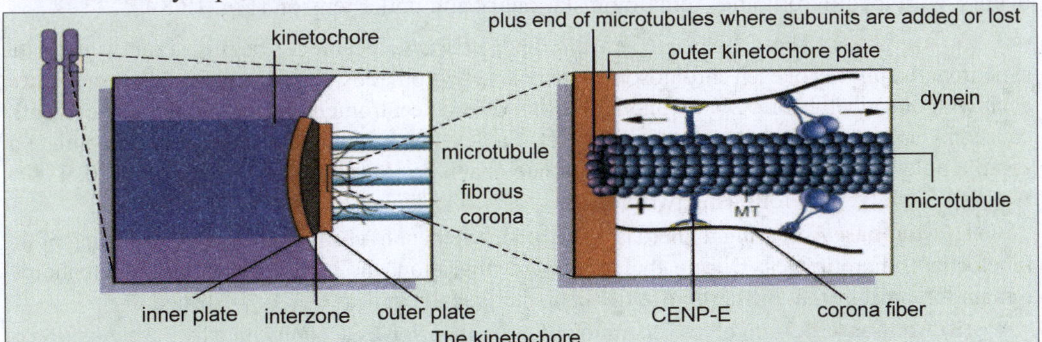

The kinetochore.

contain and align the chromosomes at the metaphase plate. In fact, at this stage the chromosomes are violently rotated and oscillated back and forth between the spindle poles because their kinetochores are capturing the plus ends of microtubules growing from one or the other spindle pole and are being pulled by the captured microtubules. The kinetochores thereby act as a "cap" that tends to protect the plus end from depolymerizing, just as the centrosome at the spindle pole tends to protect the minus end from depolymerizing. Thus, sister chromatids become attached by their kinetochores to opposite poles ; balanced bipolar forces hold chromosomes on the metaphase plate (Fig. 18.6).

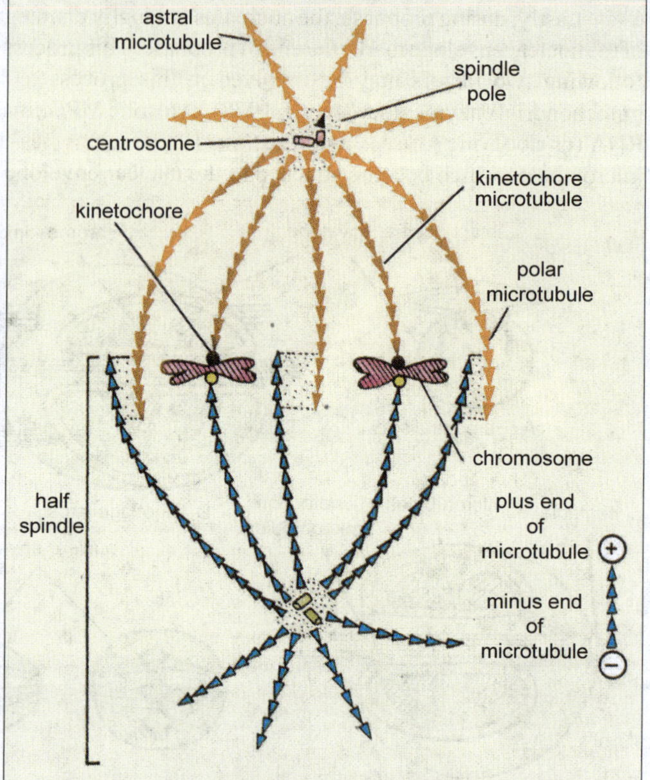

Fig. 18.6. The mitotic spindle at metaphae. The spindle is constructed from two half spindles, each composed of kinetochore, polar and astral microtubules. The polarity of the microtubules is indicated by the arrowheads (after Albert *et al*., 1989).

2. Metaphase. During metaphase (Gr., *meta*=after; *phasis* =appearance) the chromosomes are shortest and thickest. Their centromeres occupy the plane of the equator of the mitotic apparatus (a region known as the **equatorial** or **metaphase plate**), although the chromosomal arms may extend in any direction. At this stage the sister chromatids are still held together by centromere and the kinetochores of the two sister chromatids face opposite poles ; this would permit proper separation in the next phase (anaphase).

Metaphase occupies a substantial portion of the mitotic phase (see Table 18-2), as if the cell pause until all their chromosomes are lined up appropriately on the metaphase plate. At metaphase, subunits (tubulin dimers) are added to the plus end of a microtubule at the kinetochore and are removed from the minus end at the spindle pole. Thus, a poleward flux of tubulin subunits occurs, with the microtubules remaining stationary and under tension (Fig. 18.8A).

3. Anaphase. The anaphase (Gr., *ana*=up ; *phasis*=appearance) begins abruptly with the synchronous splitting of each chromosome into its sister chromatids, called **daughter chromosomes**, each with one kinetochore. Synchronous splitting of each centromere during prophase is evidently caused by an increase in cytosolic Ca^{2+}. In fact, Ca^{2+}-containing membrane vesicles accumulate at spindle poles and release calcium ions to initiate anaphase (**Hapler** and coworkers, 1980, 1987). Anaphase involves the following two steps :

(i) Anaphase A. During it, there is poleward movement of chromatids due to shortening of the kinetochore microtubules. During their poleward migration, the centromeres (and kinetochores) remain foremost so that the chromosomes characteristically appear U,V or J- shaped.

(ii) Anaphase B. It involves separation of poles themselves accompanied by the elongation of the polar microtubules. The astral microtubules also help in anaphase B by their attractive interaction with cell cortex.

Table 18.2. Duration of different phases of mitosis in certain plants and animals.

Organism	Duration in minutes			
	Prophase	Metaphase	Anaphase	Telophase
1. Mouse (spleen)	21	13	5	4
2. Grasshopper (neuroblasts)	102	13	9	57
3. Pea (root tip)	78	14.4	4.2	13.2
4. Onion (root tip)	71	6.5	2.4	3.8

4. Telophase. The end of the polar migration of the daughter chromosomes marks the beginning of the telophase ; which in turn is terminated by the reorganization of two new nuclei and their entry into the G_1 phase of interphase. In general terms, the events of prophase occur in reverse sequence during this phase. A nuclear envelope reassembles around each group of chromosomes to form two daughter nuclei. The mitotic apparatus except the centrioles disappears ; high viscosity of the cytoplasm decreases; the chromosomes resume their long, slender, extended form as their coils relax; and RNA- synthesis restarts causing the nucleolus to reappear.

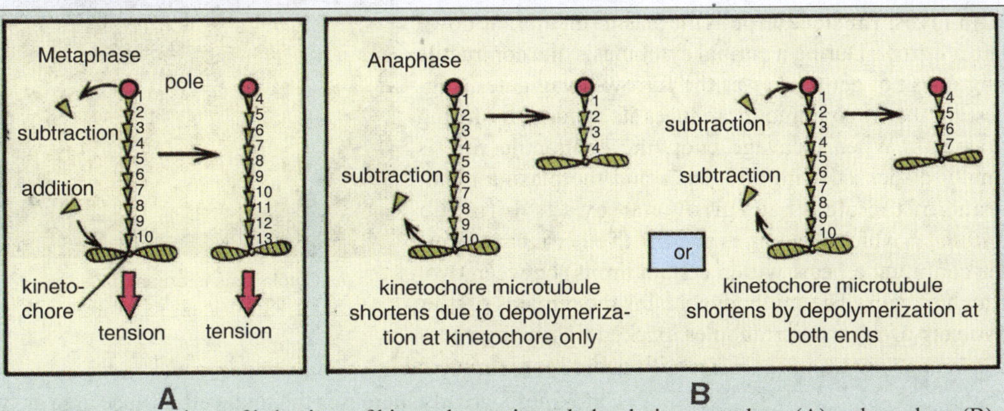

Fig. 18.7. Comparison of behaviour of kinetochore microtubules during metaphase (A) and anaphase (B). A—At metaphase, subunits are added to the plus end of a microtubule at the kinetochore and are removed from the minus end at the spindle pole. Thus, a constant poleward flux of tubulin subunits occurs, with the microtubules remaining stationary and under tension ; B—At the anaphase the tension is released, and the kinetochore moves rapidly up the microtubule, removing subunits from its plus end as it goes (BI). Its attached chromatid is thereby carried to a spindle pole. In some organisms, part of the chromatid movement is due to the simultaneous shortening of the microtubules (B II) (after Alberts *et al.*, 1989).

Cytokinesis

Both DNA synthesis and mitosis are coupled to cytoplasmic divison, or cytokinesis—the constriction of cytoplasm into two separate cells. During cytokinesis, the cytoplasm divides by a process, called **cleavage**. The mitotic spindle plays an important role in determining where and when cleavage occurs. Cytokinesis usually begins in anaphase and continues through telophase and into interphase. The first sign of cleavage in animal cells is **puckering** and **furrowing** of the plasma membrane during anaphase. The furrowing invariably occurs in the plane of the metaphase plate, at right angles to the long axis of the mitotic spindle. A cleavage furrow tends to form midway between asters originating from two centrosomes in fertilized sand dollar eggs.

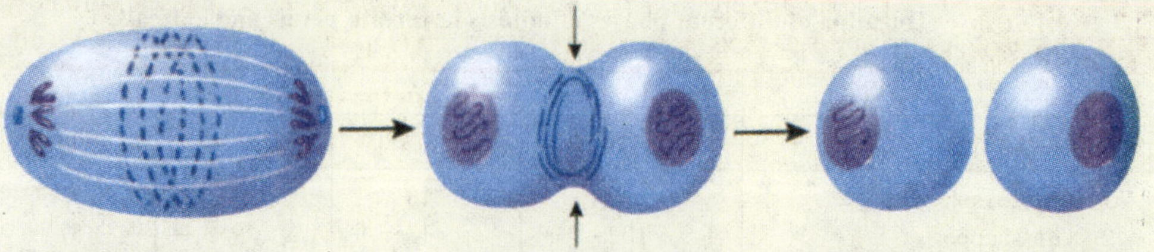

Microfilaments form a ring around the cell's equator

The microfilament ring contracts, pinching in the cell's "waist"

The waist completely pinches off, forming two daughter cells

Cytokinesis in an animal cell.

Cleavage is accomplished by the contraction of a ring composed mainly of actin filaments. This bundle of filaments, called **contractile ring**, is bound to the cytoplasmic face of the plasma membrane by unidentified attachment proteins. The contractile ring assembles in early anaphase, once assembled, it develops a force large enough to bend a fine glass needle inserted into the cell. Evidently this force is generated due to muscle-like sliding of actin and myosin filaments in the contractile ring. The actin-myosin interaction pulls the plasma membrane down into a furrow. During a normal cytokinesis, the contractile ring does not get thicker as the furrow invaginates, suggesting that it continuously reduces its volume by losing filaments. When cleavage ends, the contractile ring is finally dispensed with altogether and the plasma membrane of the cleavage furrow narrows to form the **midbody**, which remains as a tether (Tether means a rope for confining a beast within certain limits) between two daughter cells. The midbodycontains the remains of the two sets of polar microtubules, packed tightly together with dense matrix material.

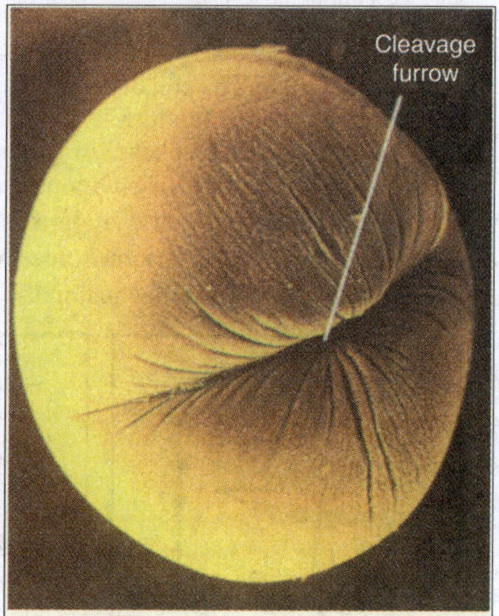

Cytokinesis in an animal egg.

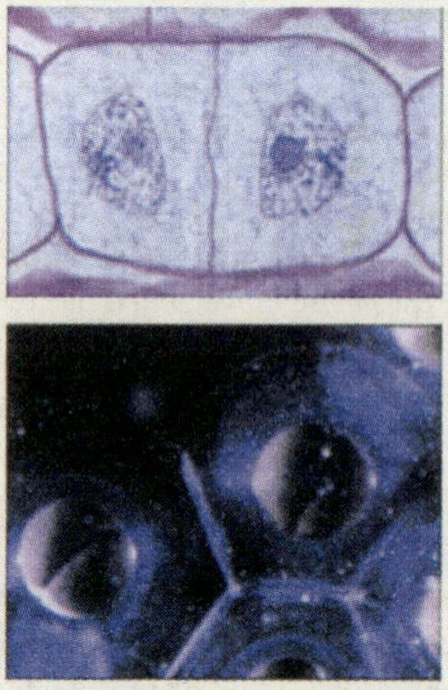

A furrow develops. It pinches the cell membrane in. As the furrow deepens the animal cell divides into two.

Cytokinesis greatly increases the total cell-surface area as two cells form from one. Therefore, the two daughter cells resulting from cytokinesis require more plasma membrane than in the plant cell. Lastly, prior to cytokinesis, in M phase large membrane-bounded organelles such as Golgi apparatus and the endoplasmic reticulum break up into smaller fragments and vesicles ; this may ensure their even distribution into daughter cells during cytokinesis. (Note : For cytokinesis in plants see Chapter 5).

Physiology of Cell Cycle and Mitosis

Follwing aspects of cell cycle and mitosis need somewhat more detailed explanation :

1. Regulation of mitotic chromosome cycle. Mitotic chromosome cycles is found to be regulated by the following three control factors (*i.e.*, diffusible proteins) : 1. The **S-phase activator** that normally appears in the cytoplasm only during S-phase and 'switches on' DNA synthesis (**Rao** and **Johanson**, 1970). 2. The **M-phase-**

promoting factor (**MPF**) that is present only in M-phase cytoplasm and causes chromosome condensation (**Johanson** and **Rao**, 1970). 3. The DNA-dependent **M-phase delaying factor** that is present in S-phase cytoplasm and inhibits the process leading to onset of MPF production.

The abrupt appearance and disappearance of these diffusible factors in the cytoplasm are landmark events in the cell cycle. The causal relationships among these factors guarantee that the events of the chromosome cycle will always occur in a fixed order, prohibiting such fatal accidents as chromosome condensation occurring in the middle of DNA synthesis. Each successive step depends on a preceding one (*i.e.* all processes of chromosome cycle are linked together as **dependent sequence**). Thus, (1) the cell cannot pass through mitosis until MPF has been produced ; (2) MPF cannot be produced until the M-phase-delaying-factor has disappeared ; (3) the M-phase-delaying factor and S-phase activator cannot disappear until DNA-synthesis has ended ; (4) DNA synthesis cannot end until all of the DNA has replicated ; (5) the DNA cannot begin to replicate until DNA re-replication block has been removed by passage through mitosis into G_1 ; and lastly (6) a cell cannot progress from mitosis into G_1, until the chromosomes have separated on the mitotic spindle.

MPF is a large-sized protein comprising two subunits– an inert subunit and a **kinase** subunit which can phosphorylate (and activate) the inert subunit (called **self-activation**) and other molecules (**Lohka** *et al.*, 1988). Thus MPF kinase directly phosphorylates several substances, including histone H1, thereby promoting chromosome condensation; and it may be through a cascade of phosphorylation that MPF triggers all the complex events of mitosis such as nuclear envelope breakdown and cytoskeletal change (*e.g.*, formation of mitotic spindle).

2. Dissolution and formation of nuclear envelope during mitosis. At least three parts of the nuclear envelope complex must be considered during its breakdown (at prophase) and reassembly (at telophase) : 1. outer and inner nuclear membranes; 2. the underlying nuclear lamina of lamin proteins and ; 3. the nuclear pores.

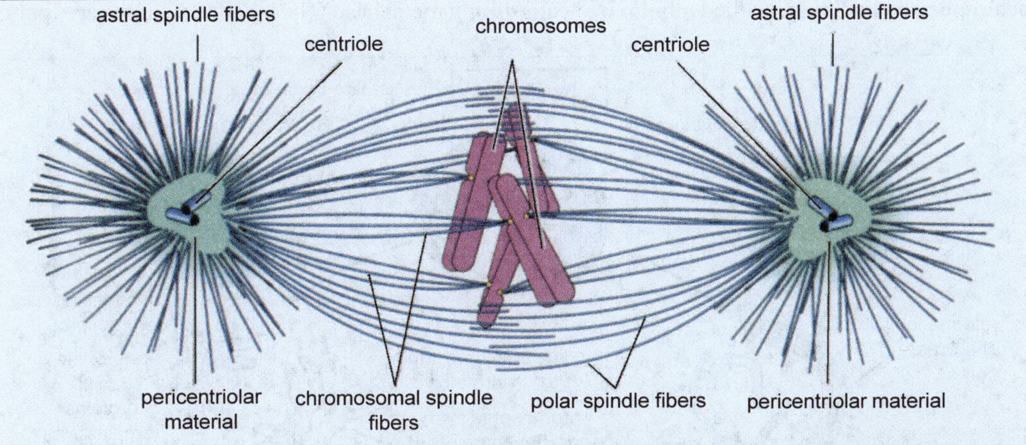

The mitotic spindle of an animal cell.

At prophase many proteins become phosphorylated by MPF and phosphorylation of nuclear lamins help regulate the disintegration and reconstruction of the nuclear envelope. The phosphorylation of the lamins occurs at many different sites in each polypeptide chain and causes them to disassemble, thereby disrupting the nuclear lamina. Subsequently, perhaps in response to a different signal (see Telophase), the nuclear envelope proper breaks up into small membrane vesicles. **Maul** (1977) has reported that in less than an hour (prophase to prometaphase) almost the entire 4000 pores disappear from the nuclear membranes of cultured mammalian cells. These pore complexes have been found on chromosomes during mitosis.

The sudden transition from metaphase to anaphase is thought to initiate dephosphorylation of many proteins, including histone H1 and the lamins, that were phosphorylated at prophase. Shortly thereafter, at telophase, nuclear membrane vesicles associated with the surface of individual chromosomes are fused to re-form the nuclear membranes which partially enclose clusters of chromosomes before coalescing to re-form the complete nuclear envelope (Fig. 18.8). During this process the presynthesized nuclear pores reassemble and the dephosphorylated lamins reassociated to form the nuclear lamina ; one of the lamina protein (lamin B) remains with the nuclear membrane fragments throughout mitosis and may help in nuclear assembly.

3. Role of cytoskeleton in mitosis. It has been said that the chromosomes in mitosis are like the corpse at a funeral : *they provide the reason for the proceedings but play no active part in them* (see **Alberts** *et al.,* 1989). The active role is played by their distinct cytoskeletal structures that appear short-termly in M phase. The first to form is a bipolar **mitotic spinde**, composed of microtubules and their associated proteins ; it is meant for poleward migration of daughter chromosomes during mitosis. The second cytoskeletal structure required in M phase in animal cells is a **contractile ring** of microfilaments and myosin that forms slightly later just beneath the plasma membrane ; it is meant for cytokinesis of the cell. The third cytoskeletal component is **meshwork** of intermediate filaments that surrounds the interphase nucleus ; it elongates during mitosis to enclose the two daughter nuclei and finally divides in half by the cleavage furrow.

Working of mitotic spindle during anaphase. During anaphase A, a surprisingly large force acts on a chromosome as it moves from the metaphase plate to the spindle pole. By hydrodynamic analysis it has been calculated that to move a chromosome, a force about 10^{-11} dynes is needed, and that the entire displacement— from equator to the pole of a chromosome— may require the use of about 30 ATP molecules. As each chromosome moves polewards, its kinetochore microtubules disassemble, so that they have nearly disappeared at telophase. The site of subunit loss can be determined by injecting labelled tubulin into cells during metaphase. The labelled subunits are found

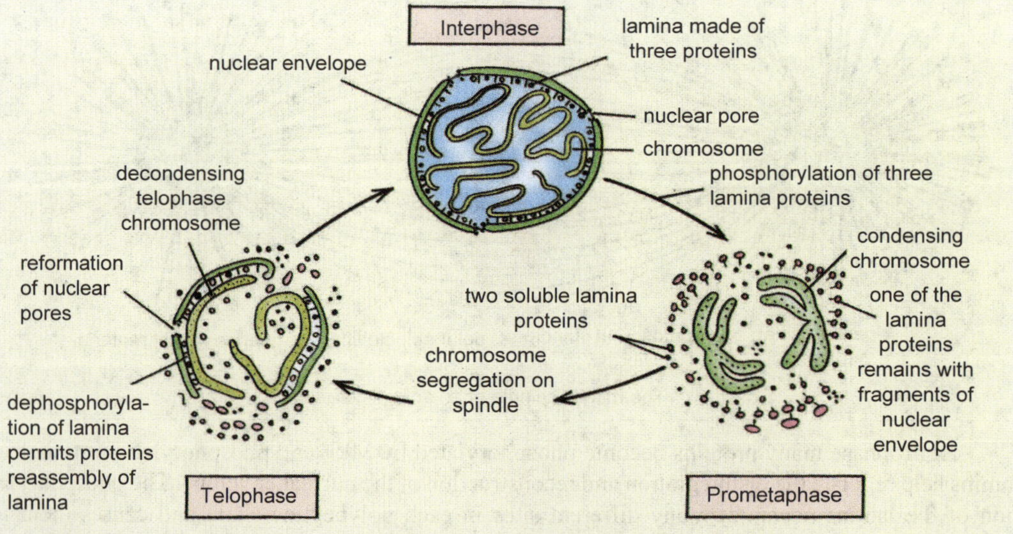

Fig. 18. 8. Schematic view of the nuclear-envelope cycle that occurs during mitosis (after Alberts *et al*., 1989).

to be added to the kinetochore end of kinetochore microtubules and then lost as anaphase A proceeds, indicating that kinetochore "eats" its way poleward along its microtubules at anaphase (Fig. 18.7) However, typically microtubule disassemble at kinetochores, poles or at both sites is probably necessery for equator- to- pole movement.

Further, the mechanism by which kinetochore of the chromosome moves up the spindle during anaphase A is still unknown. However, the following three models throw some light on it :

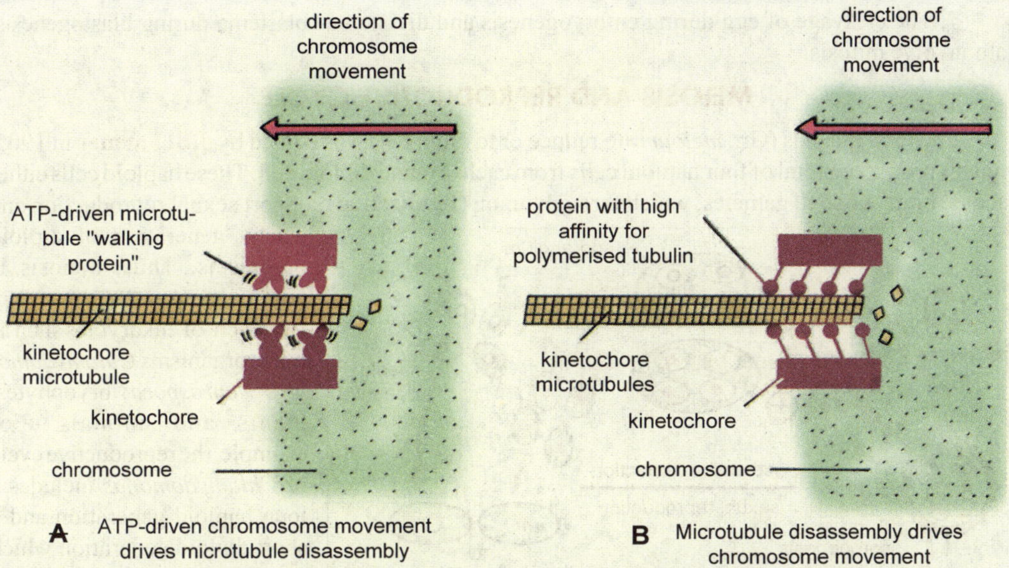

Fig. 18.9. Two alternative models to explain the generation of poleward force by the kinetochore for its chromosome during anaphase. A—Microtubule-walking proteins that resemble dynein or kinesin are part of the kinetochore, and they use the energy of ATP hydrolysis to pull the chromosome along its bound microtubules; B—Chromosome movement is driven by microtubule disassembly : as tubulin subunits dissociate, the kinetochore tends to slide poleward in order to restore its binding to the walls of the microtubule (After Alberts *et al.,* 1989).

1. The kinetochore hydrolyzes ATP to move along its attached microtubule, with the plus end of the microtubule depolymerizing as it becomes exposed (Fig. 18.9A). 2. The depolymerization of the microtubule itself causes the kinetochore to move passively to optimize its binding energy on the microtubule (Fig. 18.9B). 3. A system of elastic protein filaments might connect the kinetochore to the pole and pull the kinetochore steadily poleward. Thus, in this case, microtubules merely regulate movements of the chromosomes.

In mammalian cells, anaphase B begins shortly after the chromatids have begun their voyage to the poles and stops when the spindle is about 1.5 to 2.0 times its metaphase length (15 times increase in certain protozoa). Thus, anaphase B increases the distance between two spindle poles and in contrast to anaphase A, is accompanied by the polymerization of polar microtubules at their plus ends. Further, the polar microtubules from each half-spindle overlap in a central region near the spindle equator (*e.g.,* diatoms). During anaphase B these two sets of antiparallel polar microtubules appear to slide away from each other in the region of overlap. Dynein-like force generating protein may be involved in such a directed movement of chromosomes.

Significance of Mitosis

The mitosis has the following singificance for living organisms :

1. The mitosis helps the cell in maintaining proper size.

2. It helps in the maintenance of an equilibrium in the amount of DNA and RNA in the cell.

3. The mitosis provides the opportunity for the growth and development to organs and the body of the organisms.

4. The old decaying and dead cells of body are replaced by the help of mitosis.
5. In certain organisms, the mitosis is involved in asexual reproduction.
6. The gonads and the sex cells depend on the mitosis for the increase in their number.
7. The cleavage of egg during embryogenesis and division of blastema during blastogenesis, both involve mitosis.

MEIOSIS AND REPRODUCTIVE CYCLE

The term meiosis (Gr., *meioum*=to reduce or to diminish) was coined by **J.B. Farmer** in 1905. Meiosis produces a total of four haploid cells from each original diploid cell. These haploid cells either become or give rise to gametes, which through union (fertilization) support sexual reproduction and a new generation of diploid organisms. Thus, meiosis is required to run the **reproductive cycle** of eukaryotes such as microorganisms *Chlamydomonas, Neurospora*; bryophytes; plants and animals. For example, the reproductive cycle of *Chlamydomonas* includes a long haploid generation and a short diploid generation which involves the zygote formation. The zygote undergoes reduction division (*i.e.*, meiosis) resulting in the formation of haploid spores (Fig. 18.10). In higher plants, however, the reproductive cycle includes a long dominant diploid and multicellular generation (called **sporophyte**) and a short, multicellular haploid generation, called **gametophyte** generation. The tiny gametophyte is nurtured in specialized tissues of sporophyte. Male and female haploid cells called **spores**, are produced by meiosis in the diploid (sporophyte) organism. Spores grow into multicellular male and female haploid (gametophyte) structures, which through meiosis produce haploid cells corresponding to the actual gamete.

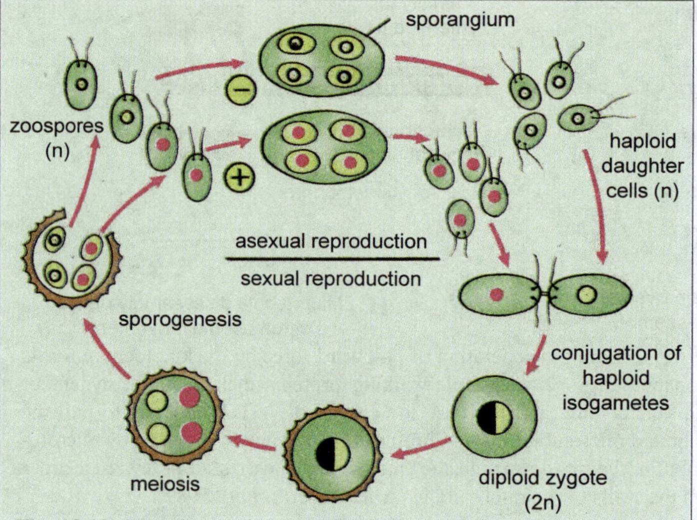

Fig. 18.10. Sexual reproduction cycle in *Chlamydomonas* sp. (after Burns, 1969).

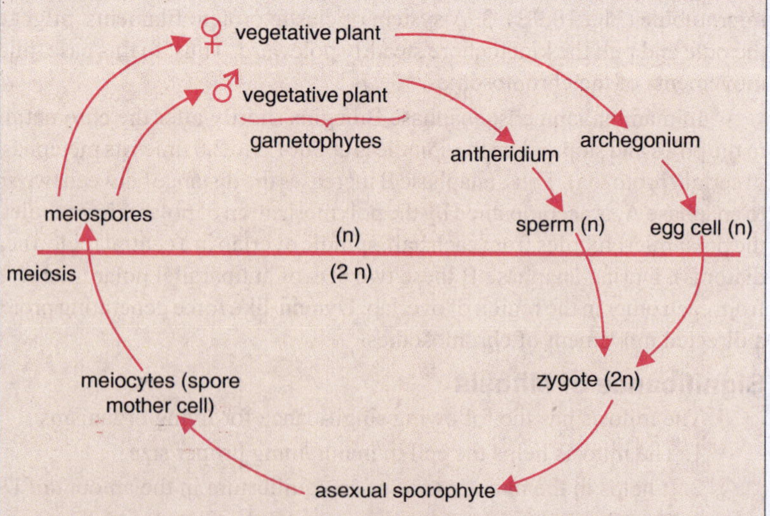

Fig. 18.11. Alternation of generation in Bryophyta (after Burns, 1969).

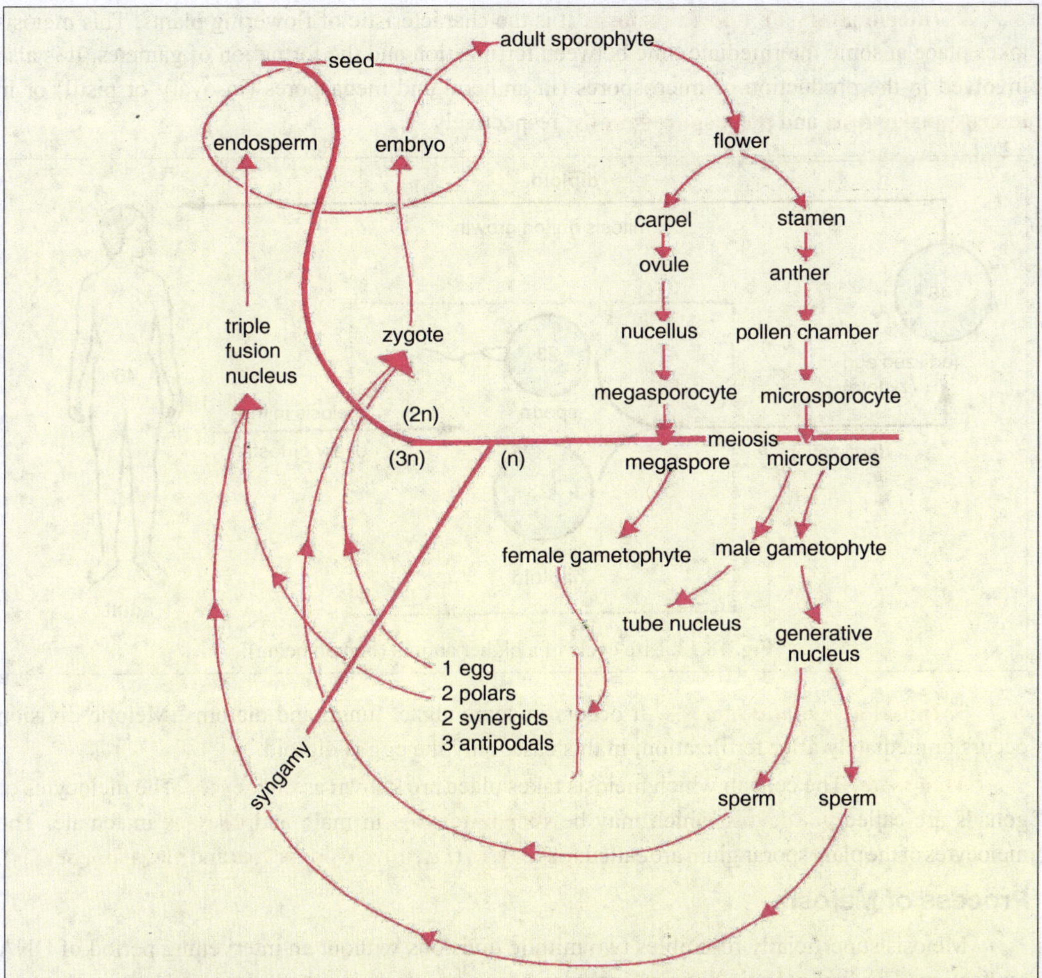

Fig. 18.12. Schematic diagram of the life cycle of an angiosperm (after Burns, 1969).

In both animals and plants, male and female gametes unite during fertilization to produce a **zygote** in which the diploid chromosome number is restored. In animals and simpler plants, the zygote matures to a new diploid organism. In the seed-producing plants, development is arrested at an early multicellular stage as a seed, which may remain stable for long time before germination permits a continuation of growth. Thus, reproductive cycle includes alternation of two generations : haploid and diploid and involves meiosis. (Fig. 18.11, Fig. 18.12 and Fig. 18.13).

Kinds of Meiosis

Meiosis occurs in the germ cells of sexually reproducing organisms. In both plants and animals, germ cells are localized in the gonads. The time at which meiosis takes place varies among different organisms, and on this basis the process can be classified into : **terminal, intermediate** or **initial**.

1. Terminal meiosis. It is also called **gametic meiosis** and is found in animals and a few lower plants. In terminal meiosis, the meiotic division occurs immediately before the formation of gametes or **gametogenesis** and will be discussed in detail in Chapter 20.

2. Intermediary or sporic meiosis. It is the characteristic of flowering plants. This meiosis takes place at some intermediate time between fertilization and the formation of gametes. It is also involved in the production of microspores (in anthers) and megaspores (in ovary or pistil) or in **microsporogenesis** and **megasporogenesis**, respectively.

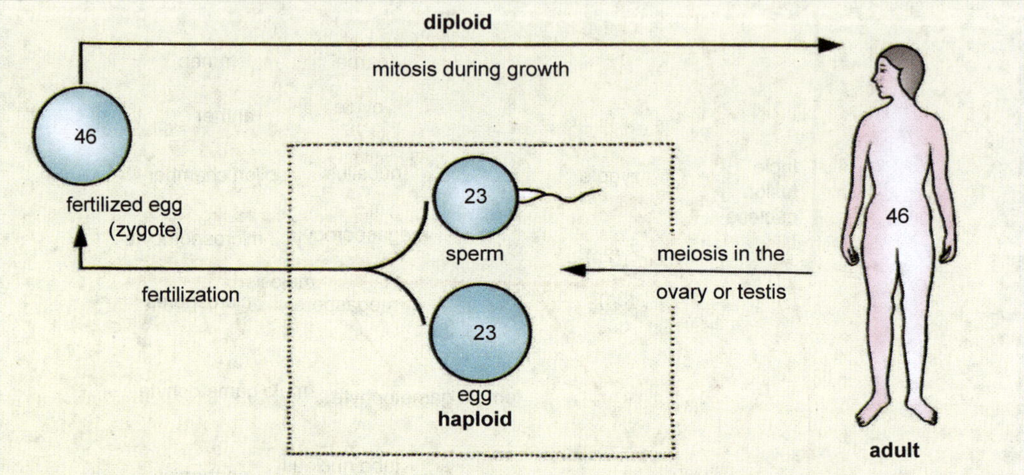

Fig. 18.13. Life cycle of a higher animal (human being).

3. Initial or zygotic meiosis. It occurs in some algae, fungi, and diatoms. Meiotic division occurs immediately after fertilization; in this case, only the egg is diploid.

Meiocytes. The cells in which meiosis takes place are known as **meiocytes**. The meiocytes of gonads are called **gonocytes** which may be **spermatocytes** in male and **oocytes** in female. The meiocytes of the plant sporangium are called **sporocytes** (*i.e.*, **microsporocytes** and **megasporocytes**).

Process of Meiosis

Meiosis superficially resembles two mitotic divisions without an intervening period of DNA replication. The first meiotic division includes a long prophase in which the homologous chromosomes become closely associated to each other and interchange of hereditary material takes place between them. Further, in the first meiotic division the reduction of chromosome number takes place and, thus, two haploid cells are resulted by this division. The first meiotic division is also known as the **heterotypic division**. In the second meiotic division the haploid cell divides mitotically and results into four haploid cells. The second meiotic division is also known as the **homotypic division**. In the homotypic division pairing of chromosomes, exchange of the genetic material and reduction of the chromosome number do not occur.

Both the meiotic divisions occur continuously and each includes the usual stages of the meiosis, *viz.*, prophase, metaphase, anaphase and telophase. The prophase of first meiotic division is very significant phase because the most cytogenetical events such as synapsis, crossing over, etc., occur during this phase. The prophase is the longest meiotic phase, therefore, for the sake of convenience it is divided into six substages, *viz.*, proleptonema (proleptotene), leptonema (leptotene), zygonema (zygotene), pachynema (pachytene), diplonema (diplotene) and diakinesis. The successive meiotic substages can be represented as follows :

Meiosis
- **Division I** or **Heterotypic division**
  - Prophase I–
    - Proleptotene
    - Leptotene
    - Zygotene
    - Pachytene
    - Diplotene
    - Diakinesis
  - Prometaphase I
  - Metaphase I
  - Anaphase I
  - Telophase I
- **Interphase**
- **Division II** or **Homotypic division**
  - Prophase II
  - Metaphase II
  - Anaphase II
  - Telophase II

Heterotypic Division or First Meiotic Division

Meiosis starts after an **interphase** which is not very different from that of an intermitotic interphase. During the premeiotic interphase DNA duplication has occurred at the S phase. In the G_2 phase of interphase apparently there is a decisive change that directs the cell toward meiosis, instead of toward mitosis (**Stern** and **Hotta**, 1969). Further, in the beginning of the first meiotic division the nucleus of the meiocyte starts to swell up by absorbing the water from the cytoplasm and the nuclear volume increases about three folds. After these changes the cell passes to the first stage of first meiotic division which is known as prophase.

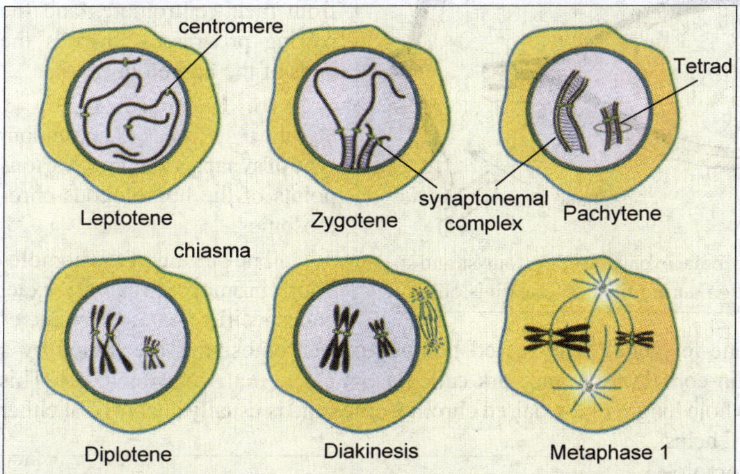

The stages of prophase I. The events at each stage are described in the tex.

Prophase I

The first prophase is the longest stage of the meiotic division. It includes following substages :

1. Proleptotene or Prolepto-nema. (Gr., *pro*=before; *leptas*= thin; *nema*= thread). The proleptotene stage closely resembles with the early mitotic prophase. In this stage the chromosomes are extremely thin, long, uncoiled, longitudinally single and slender thread-like structures.

2. Leptotene or Leptonema. In the leptotene stage the chromosomes become more uncoiled and assume a long thread-like shape. The chromosomes at this stage take up a specific orientation inside the nucleus; the ends of the chromosomes converge toward one side of the nucleus, that side where the centrosome lies (the **bouquet stage**). The centriole duplicates and each daughter centriole migrates towards the opposite poles of the cell. On reaching at the poles, each centriole duplicates and, thus, each pole of cell possesses two centrioles of a single diplosome.

3. Zygotene or Zygonema. (Gr., *zygon*=adjoining). In the zygotene stage, the pairing of homologous chromosomes takes place. The homologous chromosomes which come from the mother

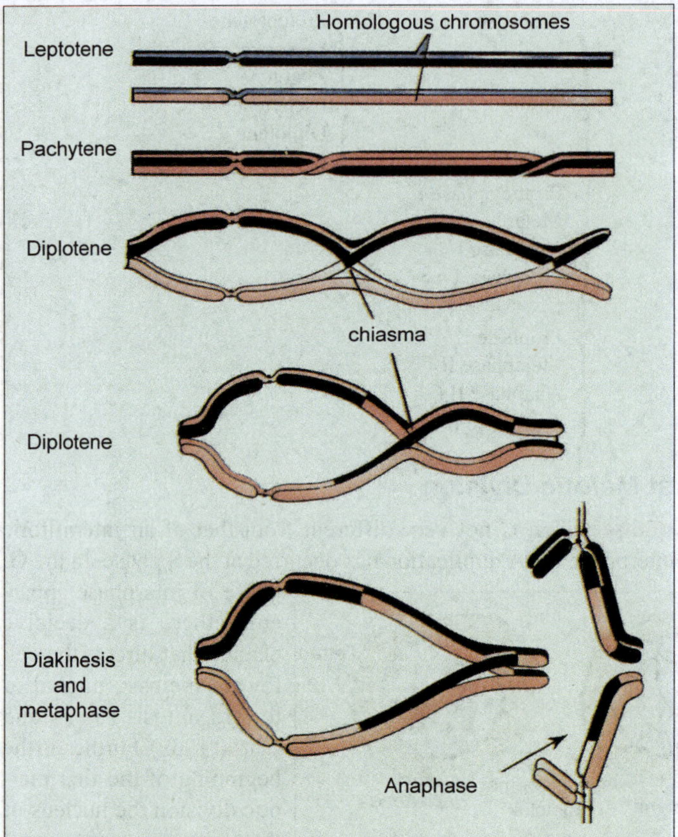

Fig. 18.14. Diagram of chromosomal exchange in the four-strand stage and of terminalization during first meiotic division.

(by ova) and father (by sperm) are attracted towards each other and their pairing takes place. The pairing of the homologous chromosomes is known as **synapsis** (Gr., *synapsis*=union). The synapsis begins at one or more points along the length of the homologous chromosomes. Three types of synapsis have been recognised.

(i) Proterminal synapsis. In proterminal type of synapsis the pairing in homologous chromosomes starts from the end and continues towards their centromeres.

(ii) Procentric synapsis. In procentric synapsis the homologous chromosomes start pairing from their centromeres and the pairing progresses towards the ends of the homologues.

(iii) Localized pairing or Random synapsis. The random type of synapsis occurs at various points of the homologous chromosomes.

The pairing of the homologous chromosome is very exact and specific (*i.e.*, alignment of chromosomes is exactly gene-for-gene). The paired homologous chromosomes are joined by a roughly 0.2-μm thick, protein-containing framework called a **synaptonemal complex** (**SC**). This complex extends along the whole length of the paired chromosomes and is usually anchored at either end to the nuclear envelope. SC helps to stabilize the pairing of homologous chromosomes and to facilitate the cytogenetical activity, called **recombination** or **crossing over** (occurring during pachynema). SC is not found in those organisms in which crossing over does not occur (*e.g.*, the male fruitfly, *Drosophila melanogaster* ; see **Burns** and **Bottino**, 1989).

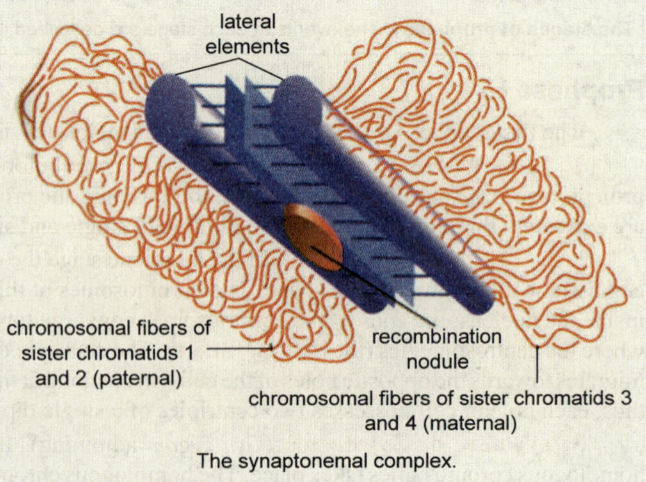

The synaptonemal complex.

4. Pachytene or Pachynema. (Gr., *pachus*=thick). In the pachynema stage the pair of chromosomes become twisted spirally around each other and cannot be distinguished separately. In the middle of the pachynema stage each homologous chromosome spilts lengthwise to form two chromatids. Actually, the doubling of the DNA molecule strands which is necessary for the subsequent duplication of chromosomes occurs earlier, before the beginning of meiotic prophase. Through the earlier part of the meiotic prophase, however, the DNA molecule in each chromosome behaves as a single body. In the pachynema stage, this is now changed, the two chromatids of each chromosome containing half of the DNA present in the chromosome at start, become partially independent of one another, although they still continue to be linked together by their common centromere. Each synaptonemal pair at this point is commonly referred to as **bivalent** or **dyads** because it consists of two visible chromosomes, or as a **quadrivalent** or **tetrad** because of the four visible chromatids.

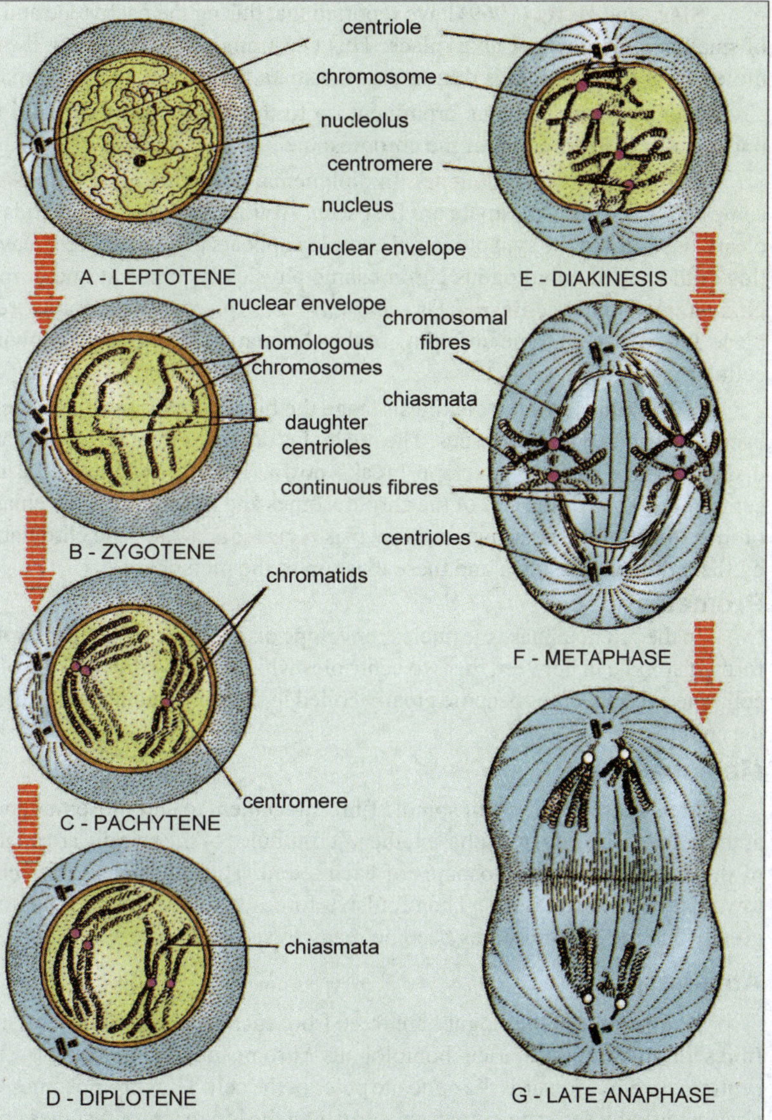

Fig. 18.15. Different stages of first meiotic division (after King 1965).

During pachynema stage an important genetic phenomenon called " crossing over" takes place. The crossing over involves reshuffling, redistribution and mutual exchange of hereditary material of two parents between two homologous chromosomes. According to recent views, one chromatid of each homologous chromosome of a bivalent may divide transversely by the help of an enzyme the **endonuclease** which is reported to increase in the nucleus during this stage by **Stern** and **Hotta** (1969). After the division of chromatids, the interchange of chromatid segments takes place between the non-sister chromatids of the homologous chromosomes. The broken chromatid segments are united with the chromatids due to the presence of an enzyme, **ligase** (**Stern** and **Hotta**, 1969). The process of interchange of chromatin material between one non-sister chromatid of each homologous chromosome is known as the **crossing over** which is accompanied by the **chiasmata formation**.

Stern and **Hotta** (1969) have reported that during the pachytene and zygotene stage, synthesis of small amount of DNA takes place. This DNA amount is utilized in the repairing of broken DNA molecule of the chromatids during the chiasmata formation and crossing over.

The nucleolus remains prominent up to this stage and it is found to be associated with the nucleolar organizer region of the chromosome.

5. Diplotene or Diplonema. In diplonema, unpairing or **desynapsis** of homologous chromosomes is started and chiasmata are first seen. At this phase the chromatids of each tetrad are usually clearly visible, but the synaptonemal complex appears to be dissolved, leaving participating chromatids of the paired homologous chromosome physically joined at one or more discrete points called **chiasmata** (singular, **chiasma**; Gr., *chiasma*= cross piece). These points are where crossing over took place. Often there is some unfolding of the chromatids at this stage, allowing for RNA synthesis and cellular growth.

6. Diakinesis. In the diakinesis stage the bivalent chromosomes become more condensed and evenly distributed in the nucleus. The nucleolus detaches from the nucleolar organizer and ultimately disappers. The nuclear envelope breaks down. During diakinesis the chiasma moves from the centromere towards the end of the chromosomes and the intermediate chiasmata diminish. This type of movement of the chiasmata is known as **terminalization**. The chromatids still remain connected by the terminal chiasmata and these exist up to the metaphase.

Prometaphase

In the prometaphase the nuclear envelope disintegrates and the microtubules get arranged in the form of spindle in between the two centrioles which occupy the position of two opposite poles of the cell. The chromosomes become greatly coiled in the spiral manner and get arranged on the equator of the spindle.

Metaphase I

Metaphase I consists of spindle fibre attachment to chromosomes and chromosomal alignment at the equator. During metaphase I, the microtubules of the spindle are attached with the centromeres of the homologous chromosomes of each tetrad. The centromere of each chromosome is directed towards the opposite poles. The repulsive forces between the homologous chromosomes increase greatly and the chromosomes become ready to separate.

Anaphase I

At anaphase I homologues are freed from each other and due to the shortening of chromosomal fibres or microtubules each homologous chromosome with its two chromatids and undivided centromere move towards the opposite poles of the cell. The chromosomes with single or few terminal chiasma usually separate more frequently than the longer chromosomes containing many chiasmata.

The actual reduction and **disjunction** occurs at this stage. Here it should be carefully noted that the homologous chromosomes which move towards the opposite poles are the chromosomes of either paternal or maternal origin. Moreover, because during the chiasma formation out of two chromatids of a chromosome, one has changed its counterpart, therefore, the two chromatids of a chromosome do not resemble with each other in the genetical terms.

Telophase I

The arrival of a haploid set of chromosomes at each pole defines the onset of telophase I, during which nuclei are reassembled. The endoplasmic reticulum forms the nuclear envelope around the chromosomes and the chromosomes become uncoil. The nucleolus reappears and, thus, two daughter chromosomes are formed. After the karyokinesis, cytokinesis occurs and two haploid cells are formed.

Both cells pass through a short resting phase of interphase. During interphase, no DNA replication occurs, so that chromosomes at the second prophase are the same double-stranded

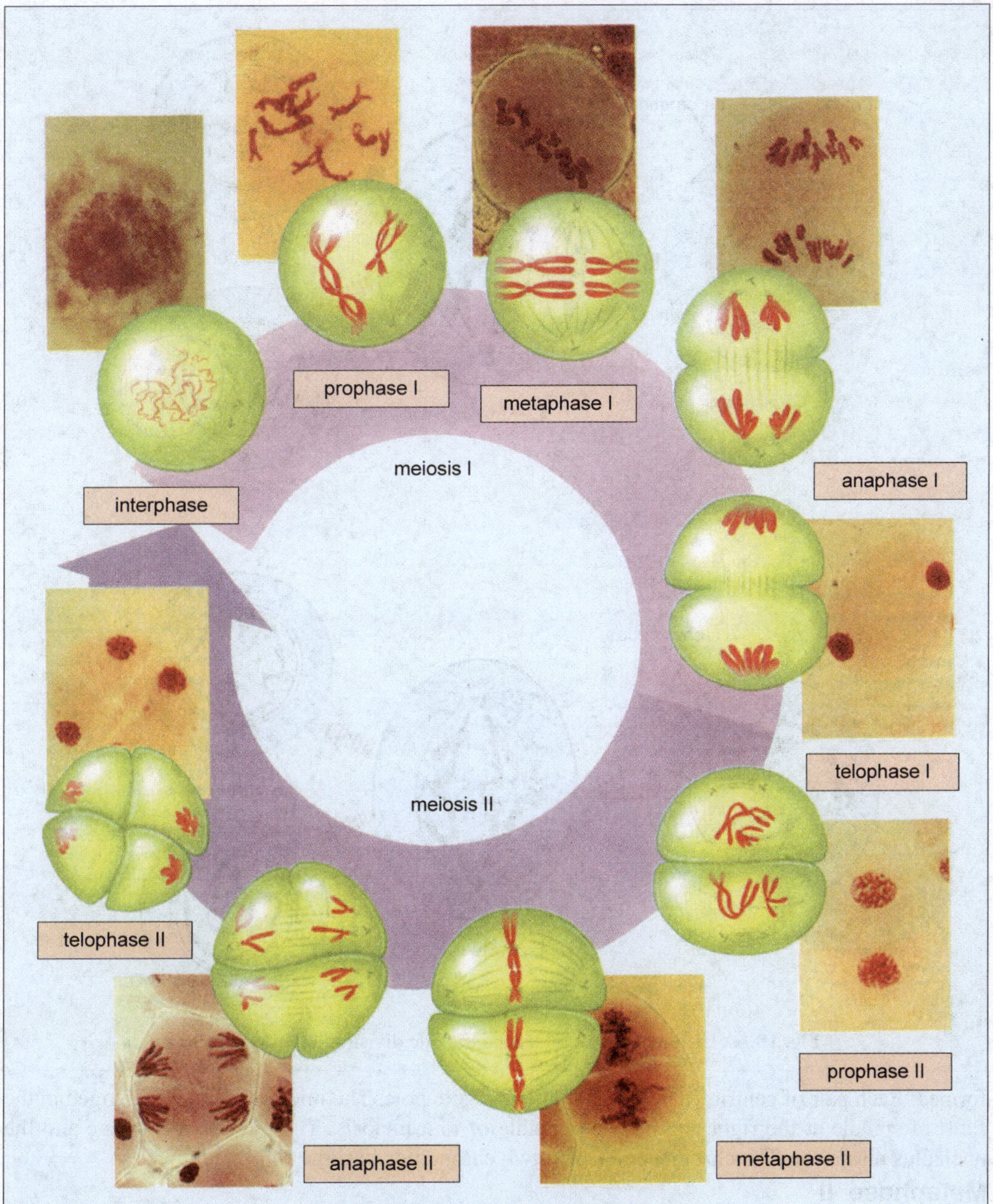

The stages of meiosis in animal cell.

structures that disappeared at the first telophase. In case of *Trillium* telophase I and interphase I do not occur and the anaphase I is followed by prophase II directly.

Homotypic or Second Meiotic Division

The homotypic or second meiotic division is actually the mitotic division which divides each haploid meiotic cell into two haploid cells. The second meiotic division includes following four stages.

Prophase II

In the prophase second, each centriole divides into two and, thus, two pairs of centrioles are

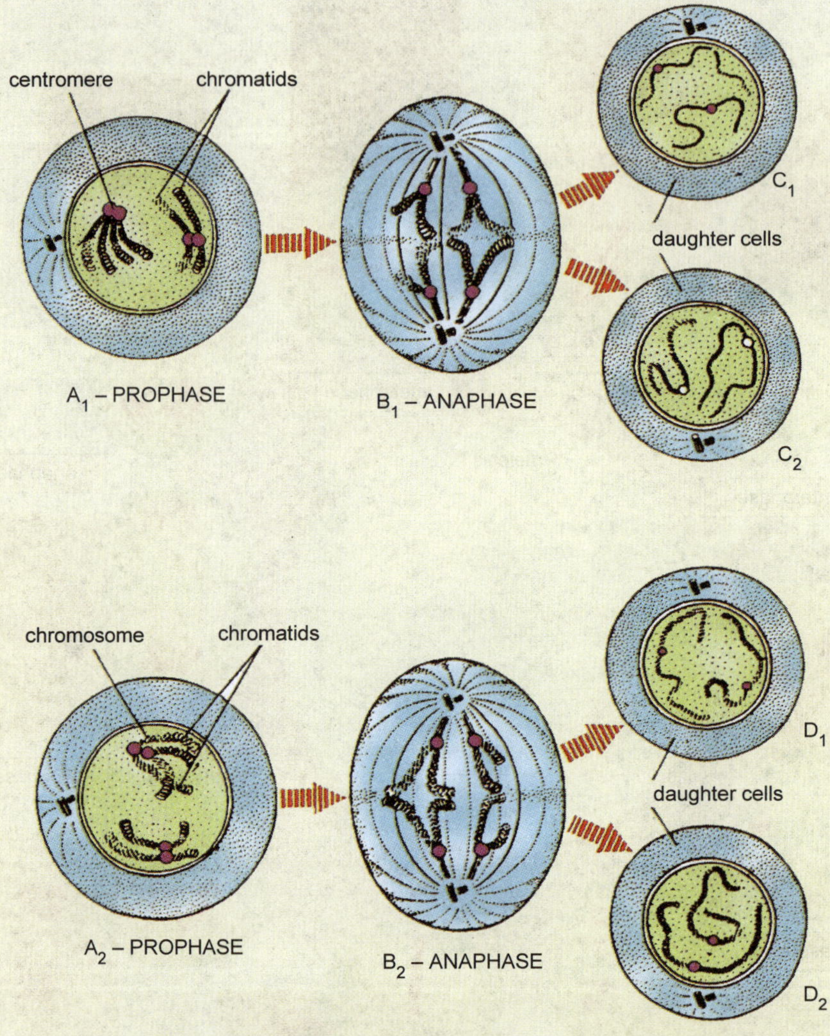

Fig. 18.16. Different stages of second meiotic division (after King, 1965).

formed. Each pair of centrioles migrates to the opposite pole. The microtubules get arranged in the form of spindle at the right angle of the spindle of first meiosis. The nuclear membrane and the nucleolus disappear. The chromosomes with two chromatids become short and thick.

Metaphase II

During metaphse II, the chromosomes get arranged on the equator of the spindle. The centromere divides into two and, thus, each chromosome produces two monads or daughter chromosomes. The microtubules of the spindle are attached with the centromere of the chromosomes.

Anaphase II

The daughter chromosomes move towards the opposite poles due to the shortening of chromosomal microtubules and stretching of interzonal microtubules of the spindle.

Telophase II

The chromatids migrate to the opposite poles and now known as chromosomes. The endoplas-

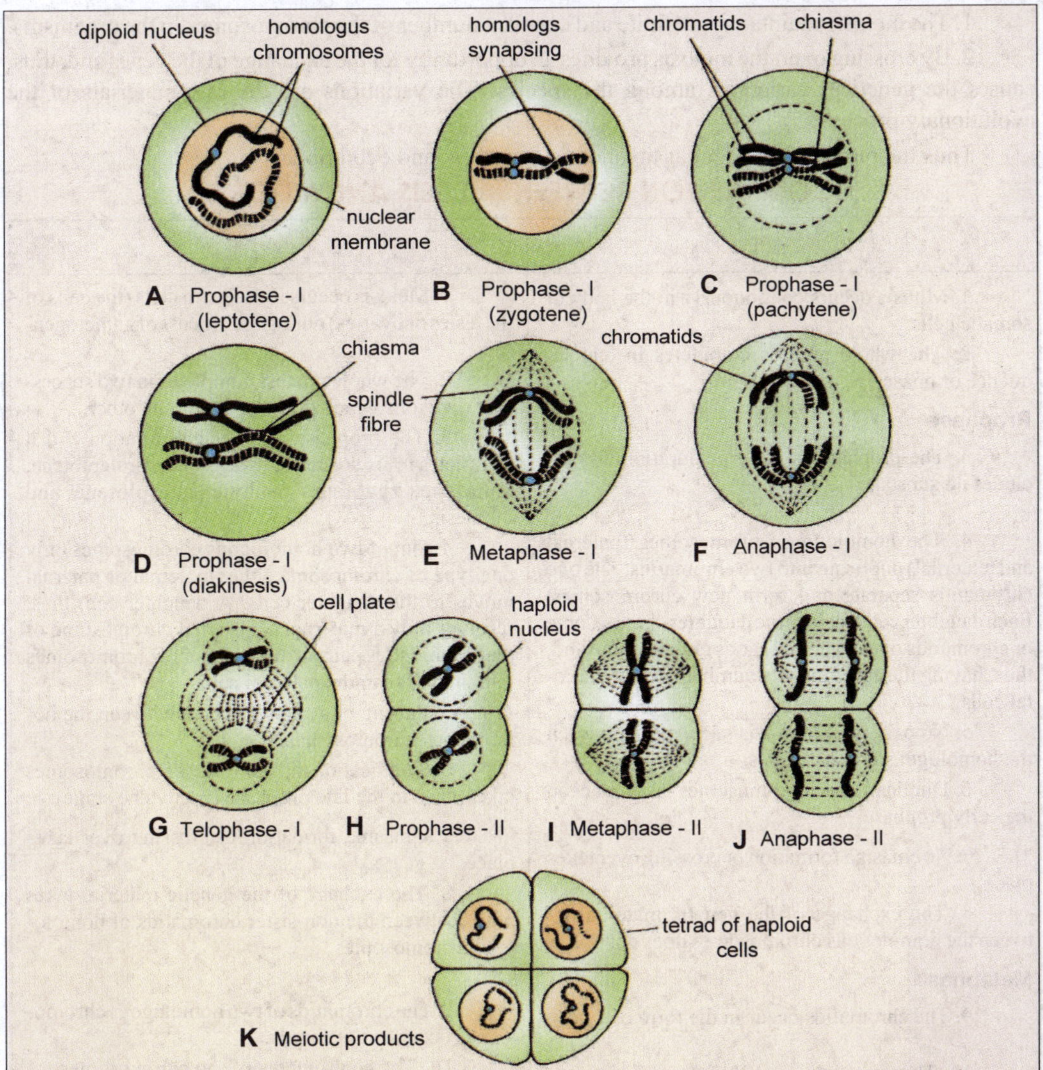

Fig. 18.17. Diagrammatic representation of meiosis in a plant cell showing one pair of homologous chromosomes (2n=2).

mic reticulum forms the nuclear envelope around the chromosomes and the nucleolus reappears due to synthesis of ribosomal RNA (rRNA) by rDNA and also due to accumulation of ribosomal proteins.

After the karyokinesis, in each haploid meiotic cell, the cytokinesis occurs and, thus, four haploid cells are resulted. These cells have different types of chromosomes due to the crossing over in the prophase I.

SIGNIFICANCE OF MEIOSIS

The meiosis has the greatest significance for the biological world because of its following uses :

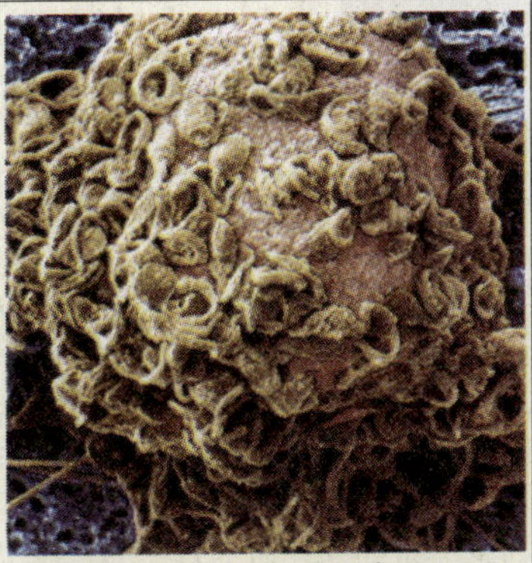

Meiosis produces egg and sperms.

1. The meiosis maintains a definite and constant number of the chromosomes in the organisms.

2. By crossing over, the meiosis provides an opportunity for the exchange of the genes and, thus, causes the genetical variations among the species. The variations are the raw materials of the evolutionary process.

Thus the meiosis is a peculiar taxonomic, genetical and evolutionary process.

COMPARISON BETWEEN MITOSIS AND MEIOSIS

Mitosis	Meiosis
1. Mitosis occurs continuously in the body or somatic cells.	1. Meiosis occurs in the germ cells (the cells of the testes or ovaries) during the process of gametogenesis.
2. The whole process completes in one sequence or phase.	2. The whole process completes in two successive divisions which occur one after the other.
Prophase	
3. The prophase is of short duration and includes no substage.	3. The prophase is of longer duration and it completes in six successive stages, *viz*., proleptotene, leptotene, zygotene, pachytene, diplotene and dikinesis.
4. The homologous chromosomes (paternal and maternal) duplicate into two chromatids. The two chromatids separate and form new chromosomes. Each daughter cell receives the daughter chromosome or chromatids of each homologous chromosome and, thus, having the chromosome number like the parental cells.	4. Out of two homologous chromosomes only one type of chromosome either maternal or paternal moves to the daughter cells. A daughter cell, thus, receives only a maternal or paternal chromosome of the homologous pair and the number of chromosomes remain half than the paternal cells.
5. No pairing or synapsis takes place between the homologous chromosomes.	5. Pairing or synapsis occurs between the homologous chromosomes.
6. Duplication of chromosomes takes place in the early prophase.	6. Duplication or splitting of chromosomes takes place in the late prophase (pachytene stage).
7. No chiasma formation or crossing over takes place.	7. Chiasma formation or crossing over takes place.
8. The exchage of the genetic material between the homologous chromosomes does not occur.	8. The exchage of the genetic material takes place between the non-sister chromatids of homologous chromosomes.
Metaphase	
9. The chromatids occur in the form of dyads.	9. The chromatids of two homologous chromosomes occur as the tetrads.
10. The centromeres of the chromosomes remain directed towards the equator and the arms of the chromosomes remain directed towards the poles.	10. The centromeres of the chromosomes remain directed towards the poles and the chromosomal arms remain directed towards the equator.
Anaphase	
11. The chromosomes are the monads, *i.e.*, having single chromatid.	11. The chromosomes are the diads, *i.e.*, having two chromatids and single centromere.
12. The chromosomes are long and thin.	12. The chromosomes are short and thick.
Telophase	
13. The telophase always occurs.	13. The first telophase is sometimes omitted.
Significance	
14. The chromosome number in each daughter cell remains the same like the parent cell.	14. In meiotic division the chormosome number is reduced to half in the daughter cells than the parental cells.
15. A diploid cell produces four diploid cells by a mitotic division.	15. A diploid cell produces four haploid cells by a meiotic division.

REVISION QUESTIONS

1. What is the cell division ? How many types of cell division occur in living organisms ? Discuss the use and biological significance of each type of cell division.
2. Define the terms : cell cycle and mitosis. Name the stages of cell cycle. Which is usually the longest stage? What are the major features of each mitotic phase ?
3. What basic activities occur during mitosis ? How does mitosis differ in animal and plant cells ?
4. What biochemical events take place in cells before visible cellular division occurs ? Compare the cytogenetic view of chromatin in interphase, in mitosis, and in meiosis.
5. Describe the behaviour and presumed role of centrioles during mitosis.
6. What is meiosis ? Describe the major features of each meiotic phase. Also discuss, why is meiosis needed for the production of gametes.
7. Summarize the event of the first meiotic prophase.
8. Which phases of meiosis are the same as the corresponding mitotic phase and which are different ? In what ways do they differ ?
9. Describe various roles of cytoskeleton during mitosis ; stress upon the function of mitotic spindle during anaphase A and anaphase B.
10. Describe the role of the microtubules in chromosome movement during mitosis and meiosis.
11. What is cytokiness ? Describe the process of cytokinesis in animal and plant cells.
12. What is a division furrow and contractile ring ? What influences appear to be instrumental in establishing contractile ring ?
13. If one creates two spheres out of one, additional surface would be needed. In cell division, where does this additional surface come from.
14. Describe the process of cell plate formation. What are plasmodesmata and how are they formed?
15. Write short notes on the following :
    (i) Auxetic growth ; (ii) Cell cycle ;
    (iii) G_0 phase ; (iv) Mitotic spindle ;
    (v) MPF ; (vi) Synaptonemal complex.

CHAPTER

19

Reproduction

Living things reproduce.

This is the inherent property of the living organisms to continue their race by the mechanism of reproduction. The reproduction is a process by which the living beings propagate or duplicate their own kinds. The reproduction may be of following two types :

1. Asexual reproduction ; 2. Sexual reproduction.

The word 'reproduction' implies replication, and it is true that biologic reproduction almost always yields a reasonable facsimile of the parent unit. However, sexual reproduction, performed by the majority of living organisms, produces the **diversity** which is required for survival in a world of constant change. The process, whether sexual or asexual, comprises a basic pattern : (1) the conversion of raw materials from the environment into the offspring, or sex cells that develop into offspring of a similar constitution,and (2) the transmission of a hereditary pattern or code (DNA of the genes) from the parent.

ASEXUAL REPRODUCTION

The development of new individuals without the fusion of the male and female gametes is known as **asexual reproduction**. The asexual reproduction usually includes amitotic or mitotic division of the body (somatic) cells, therefore, it is also known as **somatogenic** or **blastogenic reproduction**. The asexual reproduction is common only in lower plants and animals and it may be of following types:

1. Fission; 2. Budding; 3. Gemmule formation; 4. Regeneration.

1. Asexual reproduction by the fission. The fission is the most widely occurring type of asexual reproduction of the protozoans and various metazoans. In this method of reproduction the nuclear and cytoplasmic contents of the cell divide or split completely into smaller-sized daughter individuals. The fission itself may be of following types :

A. Binary fission. In the binary fission the animal body splits or divides in such a plane that two equal and identical halves are produced. It is most common in protozoans but it also occurs in certain lower metazoans. First of all the nucleus divides by amitotic or mitotic division and the division of the nucleus is followed by the division of the cytoplasm. According to the plane of fission following types of binary fission have been recognized in the organisms :

(i) Simple or orthodox type of binary fission. The simple or orthodox type of binary fission occurs in the irregular-shaped organims, *e.g., Amoeba* in which the plane of division is difficult to observe.

(ii) Transverse binary fission. The transverse binary fission occurs in some protozoans, *e.g., Paramecium* and some metazoans such as certain coelenterates, turbellarians and annelids. In transverse binary fission the plane of the division is always transverse to the longitudinal axis of the body of the organisms.

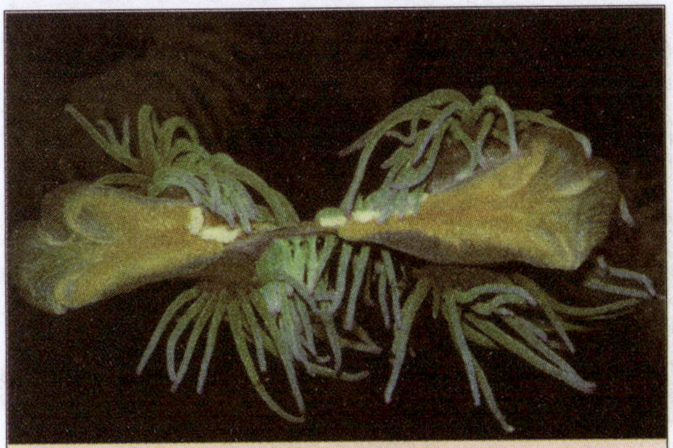

Binary fission in sea anemone.

(iii) Longitudinal binary fission. The longitudinal binary fission occurs in certain ciliates and flagellates, *e.g., Vorticella* and *Euglena* (Protozoa) and some corals (Anthozoa). In longitudinal binary fission the nucleus and the cytoplasm divide in the longitudinal plane.

(iv) Oblique binary fission. The oblique binary fission occurs in most dinoflagellates. In this type of fission the cell or body of the organism divides by the oblique division.

(v) Strobilation. In certain metazoan animals a special type of transverse fission known as the strobilation occurs. In the process of the strobilation several transverse fissions occur simultaneously and giving rise to a number of individuals which often do not separate from each other immediately. The strobilation occurs in the scyphozoan (*Aurelia*), certain polychaets and ascidians. In *Aurelia*, for instance, the strobilation occurs during the formation of ephyra larva.

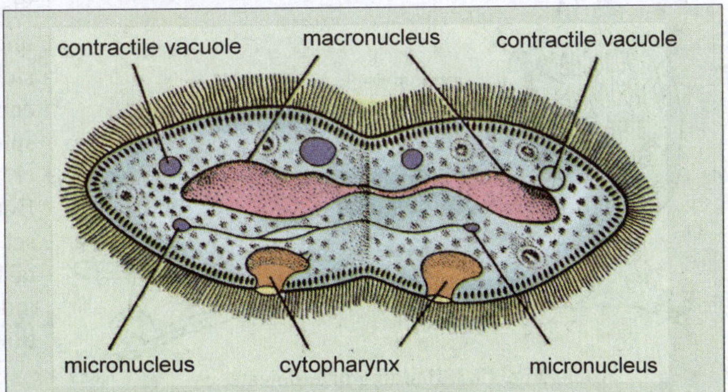

Fig. 19.1. Transverse binary fission in *Paramecium*.

B. Asexual reproduction by multiple fission. In the multiple fission, the nucleus of the cell divides very rapidly into many nuclei. Each daughter nucleus in later stage is surrounded by the little mass of the cytoplasm and forms the asexually reproducing body such as schizogont, gamont, spore, etc. The multiple fission occurs in most algae, fungi and some protozoans, *e.g., Amoeba, Plasmodium* and *Monocystis*, etc.

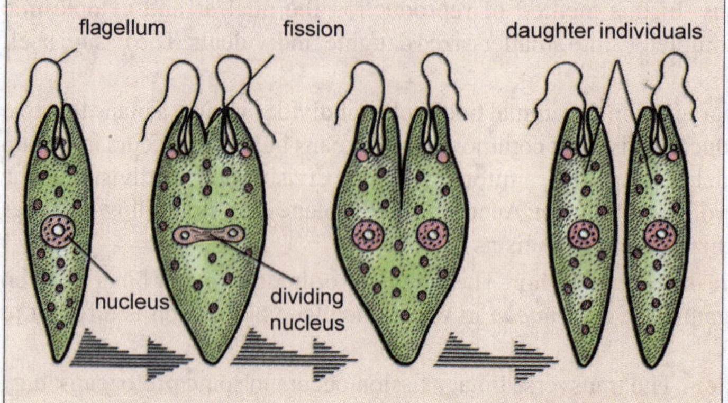

Fig. 19.2. Longitudinal binary fission in *Euglena*.

2. Asexual reproduction by budding or gemmation. In certain multicellular animals such as *Hydra* (coelenterates) and certain tunicates, the body gives out a small outgrowth known as the **bud**. The bud is supported by the parent body and it ultimately develops into a new individual. The process of development of a bud into an adult animal is called **blastogenesis**. The developing individual gets its food from the body of the parent and when it becomes fully mature it is detached from the body of the parent and leads an individual existence.

Budding in *Hydra*.

3. Asexual reproduction by gemmule formation. In certain metazoan animals the asexual reproduction is carried on by certain peculiar asexual bodies known as the **gemmules** and **statoblasts**. The gemmules occur in freshwater sponges (family Spongilidae) and the statoblasts occur in the bryozoans.

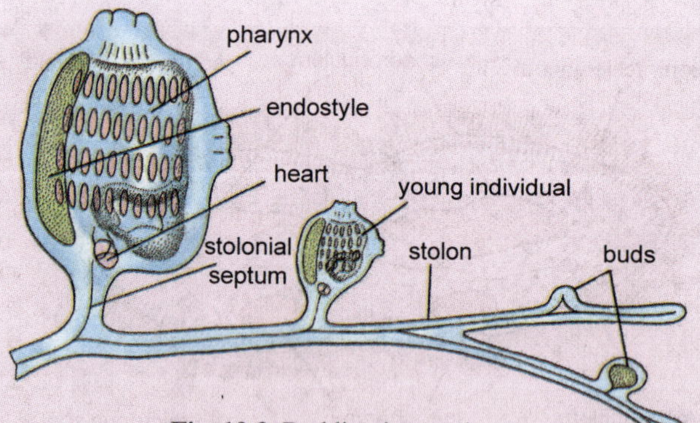

Fig. 19.3. Budding in a tunicate.

The gemmules and the statoblasts are composed of a group of undifferentiated cells which contain stored food material. These cells are enclosed and protected by the monaxon spicules in the gemmules and by the chitinous covering in the statoblasts. Both (gemmules and statoblasts) are set free by the destruction of the parental body and they develop into the new individuals in the favourable conditions.

4. Asexual reproduction by regeneration. The regeneration is a process by which the organisms develop or regenerate their lost or worn out parts. The regeneration is the best means of asexual reproduction in certain protozoans, sponges, coelenterates, planarians (Fig. 19.4) and echinoderms.

SEXUAL REPRODUCTION

In the sexual reproduction, the development of the new indivdual takes place by the fusion of the sex cells or male and female gametes (Fig. 19.5). The sexual reproduction is the most common type of reproduction among the plants and animals. It may be of following types :

A severed arm of starfish can regenerate into a whole new organism.

1. Syngamy. The syngamy is the most common type of sexual reproduction in the plants and animals. In syngamy (Gr., *syn*=together; *gam*=marriage) the fusion of two gametes takes palce completely and permanently. Following kinds of syngamy are prevalent among the living organisms :

(i) Autogamy. In autogamy (Gr., *auto*=self; *gam*=mariage) the male and female gametes are produced by the same cell or organisms and both gametes fuse together to form a zygote, *e.g.*, *Actinosphaerium* and *Paramecium.*

(ii) Exogamy. In exogamy (Gr. *exo*=external; *gam*=marriage) the male and female gametes are produced by different parents and both unite to form a zygote.

(iii) Hologamy. In the lower organisms, sometimes the entire mature organisms start to act as gametes and the fusion of such mature invividuals is known as the **hologamy**.

(iv) Paedogamy. Paedogamy is the sexual union of young individuals produced immediately after the division of the adult parent cell by mitosis.

(v) Merogamy. In the merogamy (Gr., *meros*=part; *gam*=marriage), the fusion of smaller-sized and morphologically different gametes (**merogametes**) takes place.

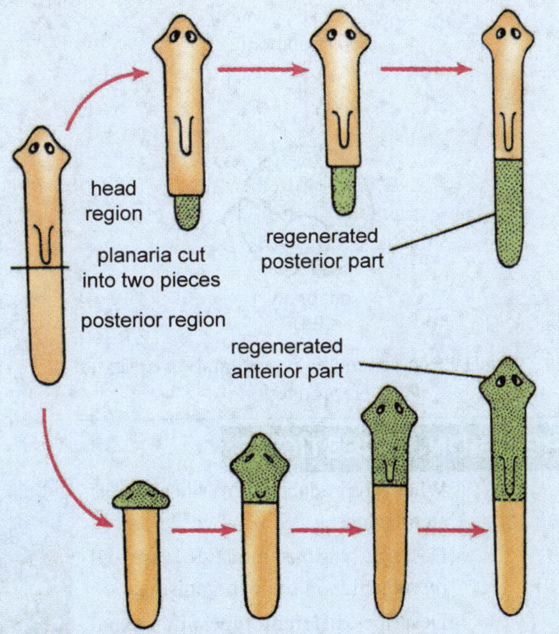

Fig. 19.4. Regeneration in *Planaria.*

(vi) Isogamy. In isogamy (Gr., *is*=equal; *gam*=marriage) the fusion of morphologically and physiologically identical gametes (**isogametes**) takes place.

(vii) Anisogamy. Some organisms produce two types of gametes. Both types of gametes differ from each other in their shape, size and behaviour and are collectively known as the **anisogametes** or **heterogametes**. The male gametes are motile and small in size and known as the **microgametes**. The female gametes are passive and have comparatively large size and known as the **macro**-or **megagametes**. The union of micro-and macrogametes is known as the **anisogamy** (Gr. *an*=without ; *is*=equal; *gam*= marriage). The anisogamy occurs in higher animals and plants but it is customary to use the term **fertilization** in them instead of the anisogamy or syngamy.

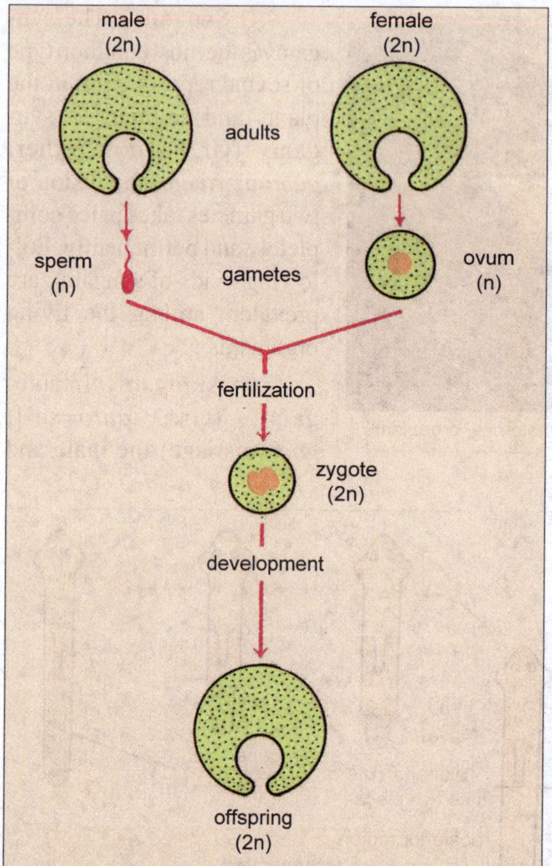

Fig. 19.5. A schematic representation of sexual reproduction.

(viii) Macrogamy. The syngamy or fusion of the macrogametes is known as **macrogamy** (Gr., *macro*=large; *gam*=marriage).

(ix) Microgamy. The microgamy (Gr., *micro*=small; *gam*=marriage) is common in certain protozoans, *e.g.,* foraminiferans and *Arcella.* In microgamy the fusion of microgametes takes place.

2. Conjugation. The conjugation is the temporary union of the two individuals of the same species. During the union both individuals known as **conjugants** exchange certain amount of nuclear (DNA) material and after which conjugants are separated. The conjugation is most common among the ciliates, *e.g., Paramecium* and bacteria.

3. Automixis. When the gamete nuclei of the same cell unite together to form new individuals this phenomenon is known as the **automixis**.

4. Parthenogenesis. The parthenogenesis (Gr., *parthenos*=virgin, *genesis*= birth) is the special type of sexual reproduction. In parthenogenesis, the eggs of an organism develop into the young individual without the fertilization of the eggs by the sperms. The parthenogenesis occurs in certain insects (wasps and bees, etc.) and rotifers.

REVISION QUESTIONS

1. What is reproduction ? Write about its significance.
2. Describe various modes of asexual reproduction in living organisms.
3. Describe different types of sexual reproduction.
4. Write short notes on the following:
   (i) Binary fission;
   (ii) Strobilation;
   (iii) Anisogamy.

Aphids give birth to live young produced by the process of parthenogenesis.

CHAPTER

20

Gametogenesis

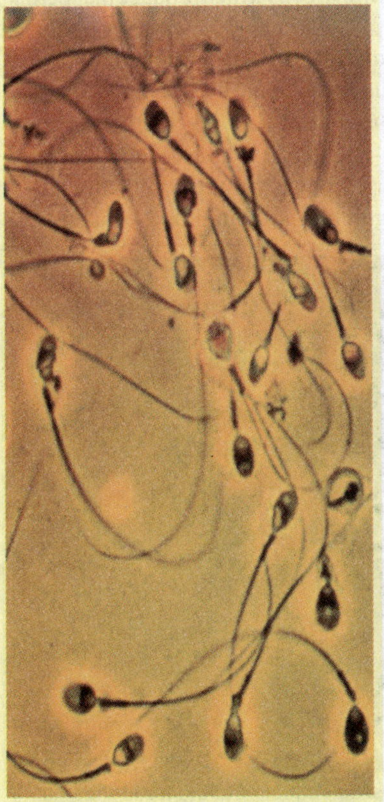

The average healthy human male produces about 1000 new sperms every second.

The gametogenesis (Gr., *gamos*=marriage; *genesis*= origin) is the process of gamete formation in the sexually reproducing animals. The sexually reproducing animals contain two types of cells in their body, *e.g.*, **somatic cells** and the **germinal cells**. Both types of cells have diploid number of chromosomes but each type has its different destiny. The somatic cells form various organs of the body and provide a phase for the maturation, development and formation of the germinal cells. The somatic cells always multiply by mitotic division.

The germinal cells form the gonads (testes and ovaries) in the animal body. These cells produce the gamete cells by successive mitotic and meiotic divisions. The male gamete is known as **spermatozoon** or sperm and the female gamete is known as **ovum** or **egg**. The process of sperm production is known as the **spermatogenesis** (Gr., *sperma*= sperm or seed; *genesis*= origin) and the process of production of ovum is known as **oogenesis** (Gr., *oon*= egg; *genesis*= origin). Both the processes can be studied in detail under separate headings.

SPERMATOGENESIS

The process of spermatogenesis occurs in the male gonads or testes. The testes of the vertebrates are composed of many seminiferous tubules which are lined by the cells of germinal epithelium. The cells of the germinal epithelium

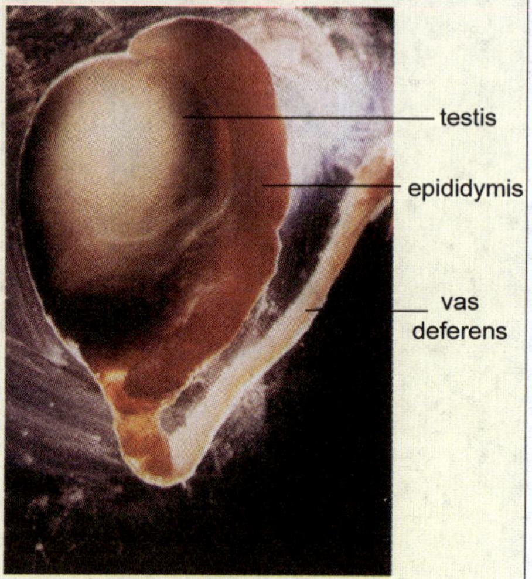

Human testis.

form sperms by the process of spermatogenesis. But in certain animals, *e.g.,* mammals and Mollusca, etc., there are somatic cells lying in between germinal cells, these somatic cells are known as **Sertoli cells.** The Sertoli cells anchor the differentiating cells and provide nourishment to the developing sperms. The insects do not possess Sertoli cells. The spermatogenesis is a continuous process and for the sake of convenience this process can be studied in two different stages.

1. Formation of spermatids; 2. Spermiogenesis.

1. Formation of Spermatids

The male germinal cells which produce the sperms are known as the **primary germinal cells** or **primordial cells**. The primordial cells pass through following three phases for the formation of spermatids :

(i) Multiplication phase. The undifferentiated germ cells or primordial cells contain large-sized and chromatin-rich nuclei. These cells multiply by repeated mitotic divisions and produce the cells which are known as the **spermatogonia** (Gr.,*sperma*=sperm or seed; *gone*=offspring). Each spermatogonium is diploid and contains 2X number of chromosomes.

(ii) The growth phase. In the growth phase, the spermatogonial cells accumulate large amount of nutrition and chromatin material. Now each spermatogonial cell is known as the **primary spermatocyte**.

(iii) The maturation phase. The primary spermatocytes are ready for first meiotic or maturation division. The homologous chromosomes start pairing (synapsis), each homologous chromosome splits longitudinally and by the chiasma formation the exchange of genetic material or crossing over takes place between the chromatids of the homologous chromosomes. The DNA amount is duplicated in the beginning of the division. By first meiotic division or homotypic division two **secondary spermatocytes** are formed. Each secondary spermatocyte is haploid and contains x number of chromosomes. Each secondary spermatocyte passes through the second maturation or second meiotic or heterotypic division and produces two **spermatids**. Thus, by a meiotic or maturation division a diploid spermatogonium produces four haploid spermatids. These spermatids cannot act directly as the gametes so they have to pass through the next phase, the **spermiogenesis**.

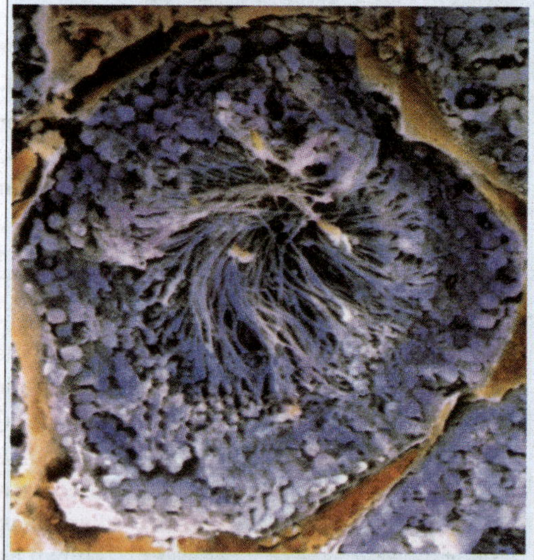
The microscopic structure of the testis showing section through a seminiferous tubule.

2. Spermiogenesis

The metamorphosis or differentiation of the spermatids into the sperms is known as **spermiogenesis**. Because the sperm or spermatozoon is a very active and mobile cell so to provide great amount of mobility to the sperm, the superfluous material of the developing sperms is discarded. For the reduction of the weight of the sperms following changes occur in the spermatids:

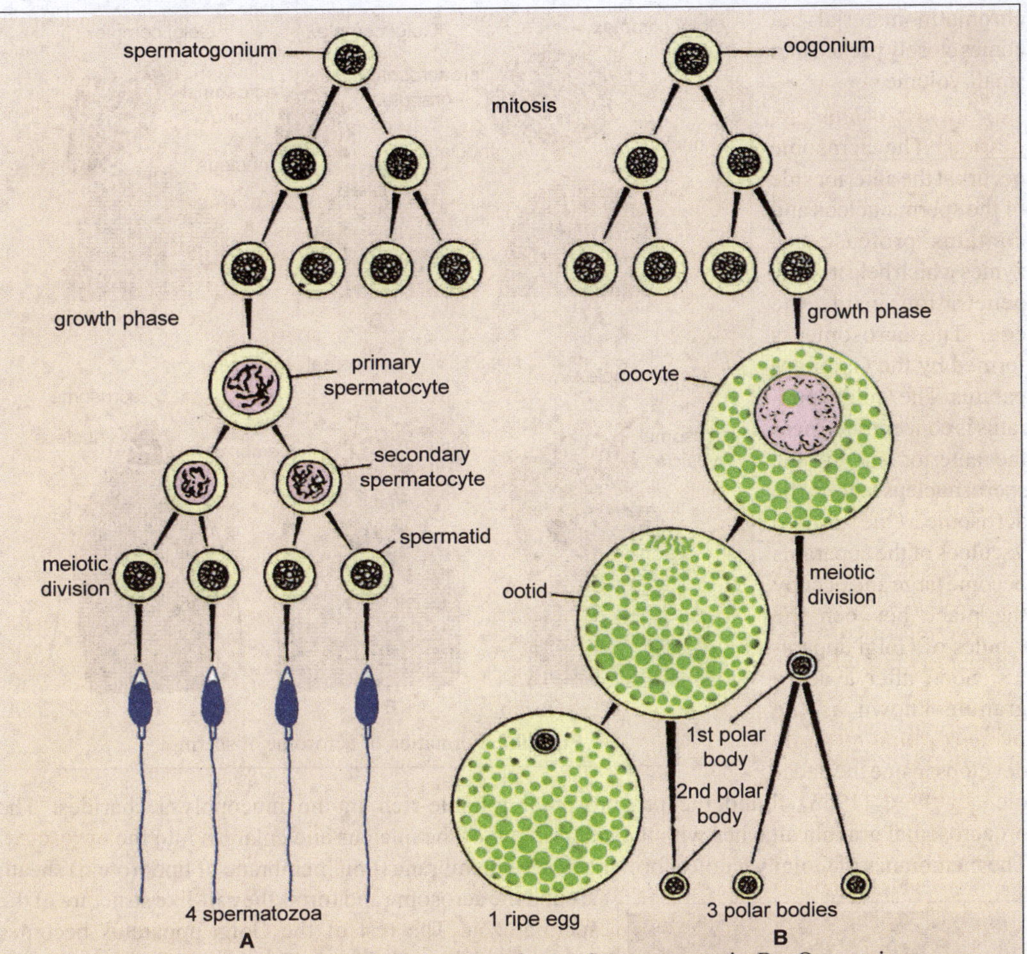

Fig. 20.1. Spermatogenesis and oogenesis. A—Spermatogenesis; B—Oogenesis.

(i) Changes in the nucleus. The nucleus loses water from the nuclear sap, shrinks and assumes different shapes in the different animals. The sperm nucleus in man and bull becomes ovoid and laterally flattened. In rodents and amphibians the sperm nucleus becomes scimitar-shaped with pointed tip. In birds and molluscs the nucleus becomes spirally twisted like a cork screw. The bivalve molluscs have the round sperm nucleus. The shape of the nucleus also determines the shape of the sperm head which becomes fully adopted for the active propulsion through the water. The RNA contents of the nucleus and the nucleolus are greatly reduced. The DNA becomes more concentrated and the

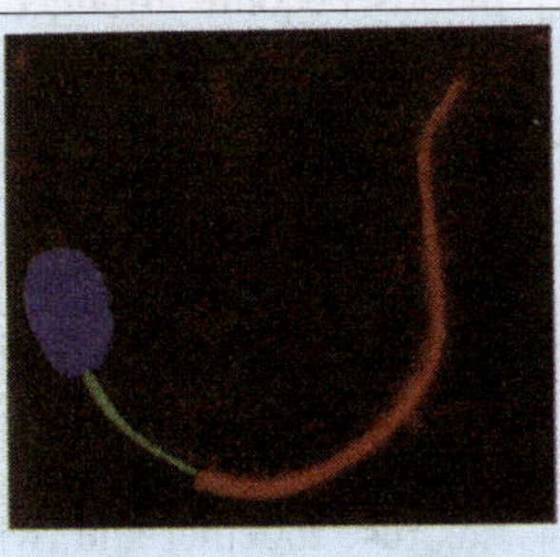

A–Mature bull sperm. The DNA is stained blue with DAPI; the mitochondria are stained green and tubulin of the flagellum is stained red.

B– Acrosome of mouse sperm, stained green by GFP.

chromatin material becomes closely packed into small volume.

(ii) Acrosome formation. The acrosome occurs at the anterior side of the sperm nucleus and contains protease enzymes which help its easy penetration inside the egg. The acrosome is formed by the Golgi apparatus. The Golgi apparatus is concentrated near the anterior end of the sperm nucleus to form the acrosome. One or two vaculoes of the apparatus become large and occupy the place between the tubules of Golgi apparatus. Soon after a dense granule known as the **proacrosomal granule** develops inside the vacuole. **Leblond** (1955) found the proacrosomal granule-rich in the mucopolysaccharides. The proacrosomal granule attaches with the anterior end of the nucleus and enlarges into the **acrosome**. The membranes of Golgi vacuoles form the double membrane (unit membrane of lipoprotein) sheath around the acrosome and forms the cap-like structure of the spermatozoa. The rest of the Golgi apparatus becomes reduced and discarded from the sperm as **Golgi rest**. In the sperms of certain animals an **acrosomal cone** or **axial body** also develops in between the acrosome and the nucleus.

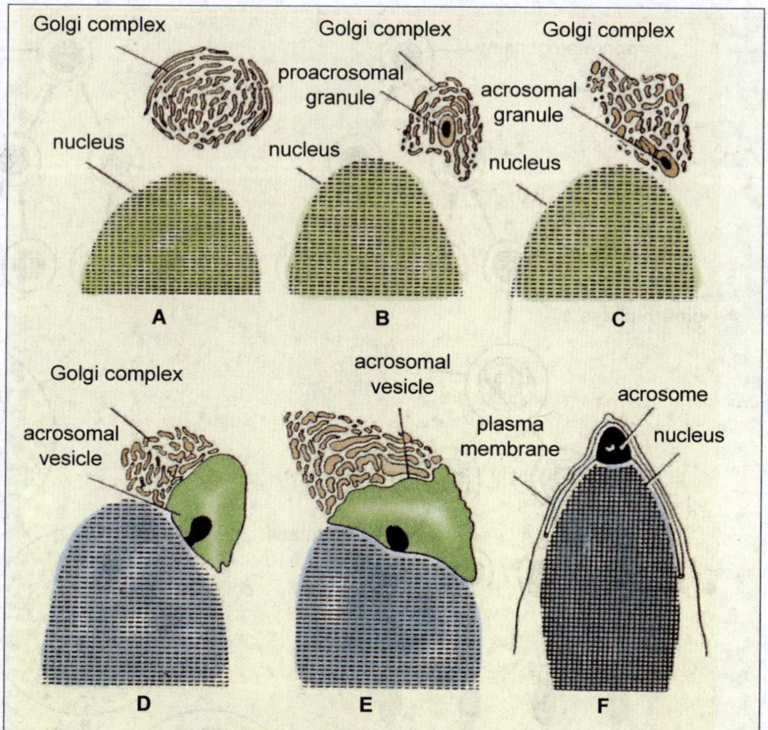

Fig. 20.2. Formation of acrosome of sperm.

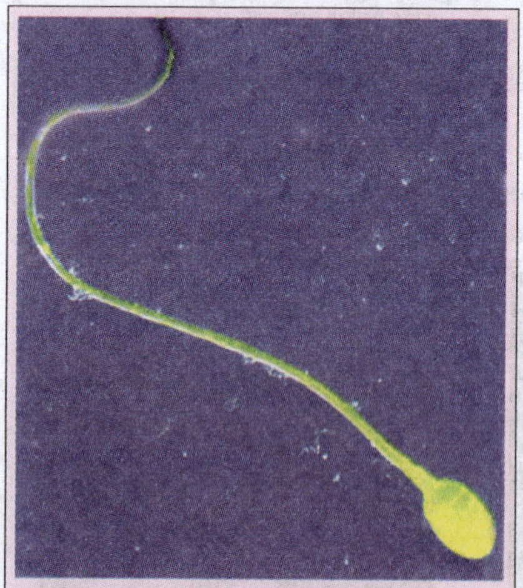
A human sperm.

(iii) The centrioles. The two centrioles of the spermatids become arranged one after the other behind the nucleus. The anterior one is known as the **proximal centriole** and the posterior one is known as the **distal centriole**. The distal centriole changes into the basal bodies and gives rise to the **axial filament** of the sperm. The axial filament or the flagellum is composed of a pair of central longitudinal fibres and nine peripheral fibres. The distal centriole and the basal part of the axial filament occur in the middle piece of the spermatozoa. The mitochondria of the spermatids fuse together and twist spirally around the axial filament.

Thus, most of the cytoplasmic portion of the spermatid except the nucleus, acrosome, centriole, mitochondria and axial filament is discarded during the spermiogenesis.

OOGENESIS

The process of oogenesis occurs in the cells of the germinal epithelium of the ovary, such cells are known as **primordial germinal cells**. The oogenesis is completed in the following three successive stages : 1. Multiplication phase; 2. Growth phase; 3. Maturation phase.

1. Multiplication Phase

The primordial germinal cells divide repeatedly to form the **oogonia** (Gr., *oon*=egg). The oogonia multiply by the mitotic divisions and form the **primary oocytes** which pass through the growth phase.

2. Growth Phase

The growth phase of the oogenesis is comparatively longer than the growth phase of the spermatogenesis. In the growth phase, the size of the primary oocyte increases enormously. For instance, the primary oocyte of the frog in the beginning has the diameter about 50 mm but after the growth phase the diameter of the mature egg reaches about 1000mm to 2000mm. In the primary oocyte, large amount of fats and proteins becomes accumulated in the form of **yolk** and due to its heavy weight (or gravity) it is usually concentrated towards the lower portion of the egg forming the **vegetal pole**. The portion of the cytoplasm containing the egg pronucleus remains often separated from the yolk and occurs towards the upper side of egg forming the **animal pole**.

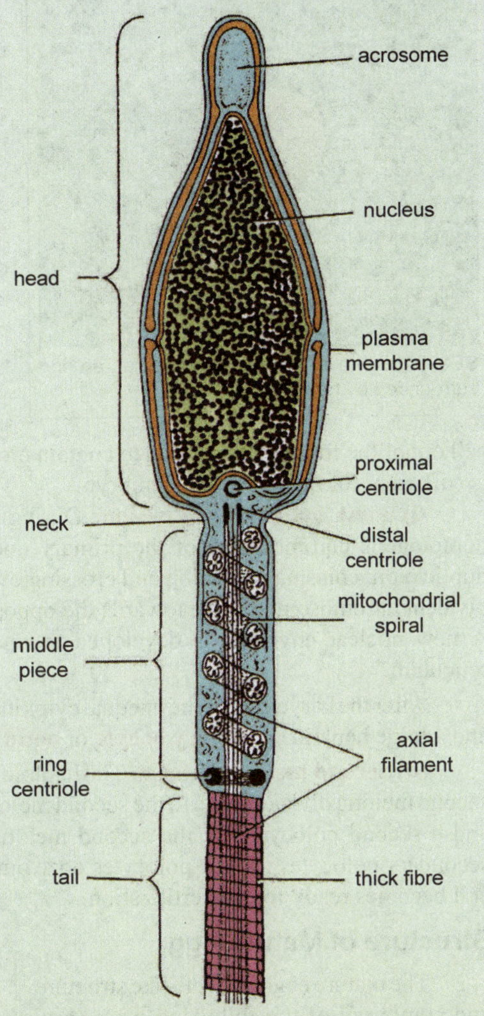

Fig. 20.3. Electron microscopic view of a flagellate spermatozoon.

The cytoplasm of the oocyte becomes rich in RNA, DNA, ATP and enzymes. Moreover, the mitochondria, Golgi apparatus, ribosomes, etc., become concentrated in the cytoplasm of the oocyte. In certain oocytes (Amphibia and birds) the mitochondria become accumulated at some place in the oocyte cytoplasm and forming the **mitochondrial clouds**.

During the growth phase, tremendous changes also occur in the nucleus of the primary oocyte. The nucleus becomes large due to the increased amount of the nucleoplasm and is called **germinal vesicle**. The nucleolus becomes large or its number is multiplied due to excessive synthesis of ribosomal RNA by rDNA of nucleolar organizer region of chromosomes. Thus, the nucleus or germinal vesicle of primary oocyte of *Triturus* has 600 nucleoli, of *Siredon* has 1000 nucleoli and *Xenopus* has 600 to 1200 nucleoli due to the synthesis of ribosomal RNA. The chromosomes change their shape and become giant **lampbrush chromosomes** which are directly related with increased transcription of mRNA molecules and active protein synthesis in the cytoplasm. When the growth of the cytoplasm and nucleus of the primary oocyte is completed it becomes ready for the maturation phase.

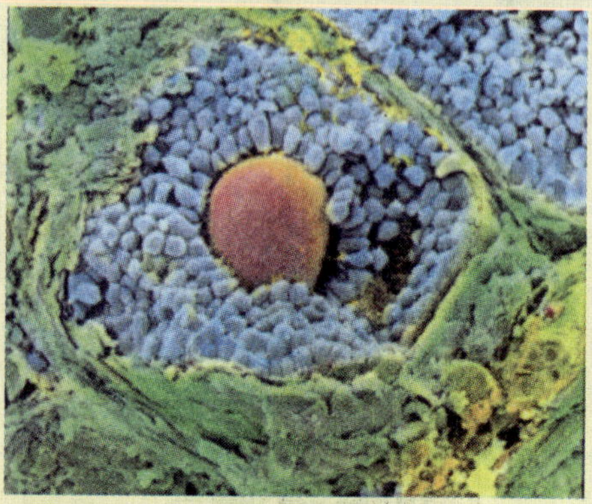

SEM through a primary oocyte showing an egg surrounded by follicle cell.

3. Maturation Phase

The maturation phase is accompanied by the maturation or meiotic division. The maturation division of the primary oocyte differs greatly from the maturation division of the spermatocyte. Here after the meiotic division of the nucleus, the cytoplasm of the oocyte divides unequally to form a single large-sized haploid egg and three small haploid **polar bodies** or **polocytes** at the end. This type of unequal division has the great significance for the egg. If the equal divisions of the primary oocyte might have been resulted, the stored food amount would have been distributed equally to the four daughter cells and which might prove insufficient for the developing embryo. Therefore, these unequal divisions allow one cell out of the four daughter cells to contain most of the cytoplasm and reserve food material which is sufficient for the developing embryo.

(i) First maturation division. During the first maturation division or first meiosis, the homologous chromosomes of the primary oocyte nucleus pass through the pairing or synapsis, duplication, chiasma formation and crossing over. Soon after the nuclear membrane breaks and the bivalent chromosomes move towards the opposite poles due to contraction of chromosomal fibres. A new nuclear envelope is developed around the daughter chromosomes by the endoplasmic reticulum.

After the karyokinesis the unequal cytokinesis occurs and a small haploid polar body or polocyte and a large haploid **secondary oocyte** or **ootid** are formed.

(ii) Second meiotic division. The haploid secondary oocyte and first polocyte pass through the second meiotic division. Due to the second meiotic division the secondary oocyte forms a mature egg and a second polocyte. By the second meiotic division the first polocyte also divides into two secondary polocytes : These polocytes ooze out from the egg and degenerate while the haploid egg cell becomes ready for the fertilization.

Structure of Mature Egg

The mature egg has a cell-like structure and composed of the following parts (Fig. 20.4) :

1. Plasma membrane. The mature egg is covered by a plasma membrane which is the unit membrane. It is composed of an outer and an inner layer of protein. Both the layers are 50A° in thickness. Between the proteinous layers there occurs a lipodous layer of 60A° thickness.

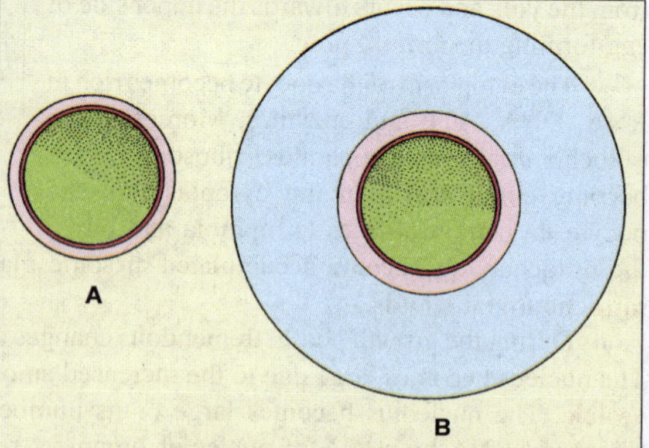

Fig. 20.4. Egg membranes of frog egg. A—Unfertilized egg as taken from oviduct; B—Fertilized egg with swollen jelly membrane (after Balinsky, 1981).

2. Primary egg membranes. In addition to the plasma membrane, the eggs of most animals except the sponges and certain coelenterates consist of certain other additional egg membranes. These membranes are known as the primary and secondary egg membranes. The primary egg membrane is secreted around the plasma membrane by the oocyte itself. In the insects, molluscs, amphibians and birds the primary egg membrane is known as the **vitelline**

membrane, while in tunicates and fishes this membrane is known as the **chorion**. The mammalian eggs contain similar membrane and in them this is known as the **zona pellucida**. The vitelline membrane is composed of mucoproteins and fibrous proteins. The vitelline membrane usually remains closely adhered to the plasma membrane but in later stages a space is developed between the plasma membrane and the vitelline membrane and this space is known as the **perivitelline space**.

3. Secondary egg membranes. The secondary egg membranes are secreted by the ovarian tissues around the primary egg membranes. They are composed of either jelly-coats in amphibians or chitinous shells in insects, ascidians and cyclostomes.

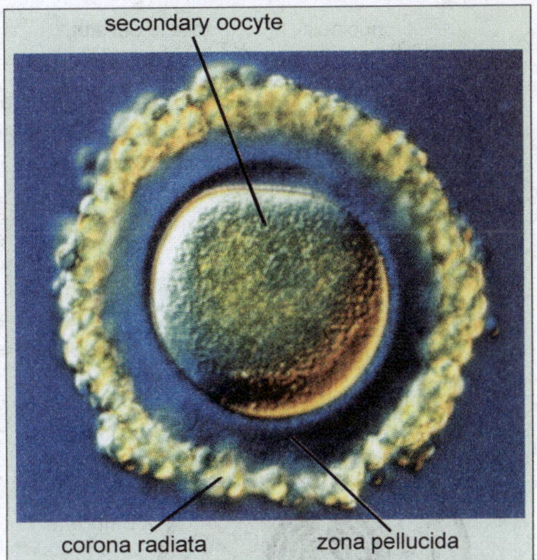

A human secondary oocyte shortly after ovulation.

4. Tertiary egg membranes. The tertiary egg membranes are formed by the oviduct or other accessory parts of female reproductive system. They may be composed of either jelly coats in amphibians, albumen and hard horny capsule in elasmobranch fishes, or albumen, shell membranes and calcareous shell in birds.

The egg, travels down the oviduct towards the uterus. It emits chemicals that attract sperm.

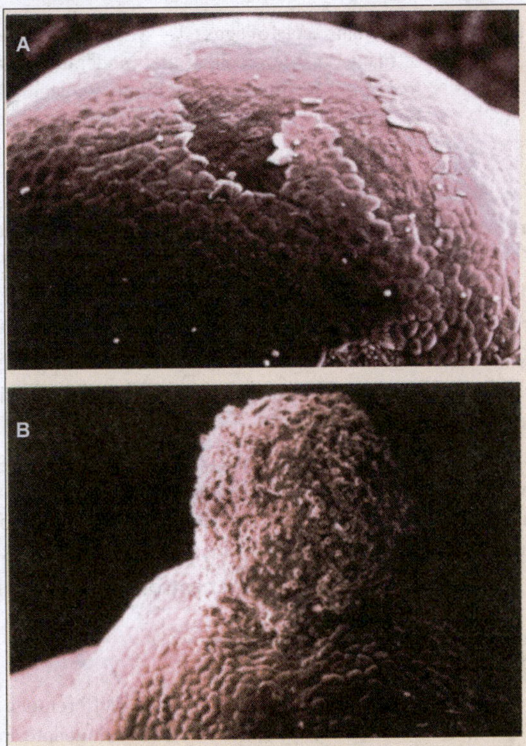

The mature follicle moves to (A) the surface of the ovary and (B) literally bursts through the ovary wall like a volcano.

5. The ooplasm. The cytoplasm of the egg cell is known as the ooplasm. The ooplasm consists of large amount of reserve food material in the form of **yolk**. It is also composed of a lipoprotein, pigment granules, water, RNA, ribosomes, mitochondria and various other cellular inclusions. The peripheral layer of the ooplasm is known as the **cortex** and it contains many microvilli and cortical granules. The microvilli are formed by the outpushings of the plasma membrane and they help in the transportation of the substances from the follicle cells to the egg during the development of the egg. The cortical granules are spherical bodies of various diameters. *e.g.,* 8.0μm in the sea urchin eggs and 2.0μm in the eggs of frog. The cortical granules are surrounded by the unit membranes and are originated from the Golgi apparatus. They contain homogeneous and granular acid mucopo-

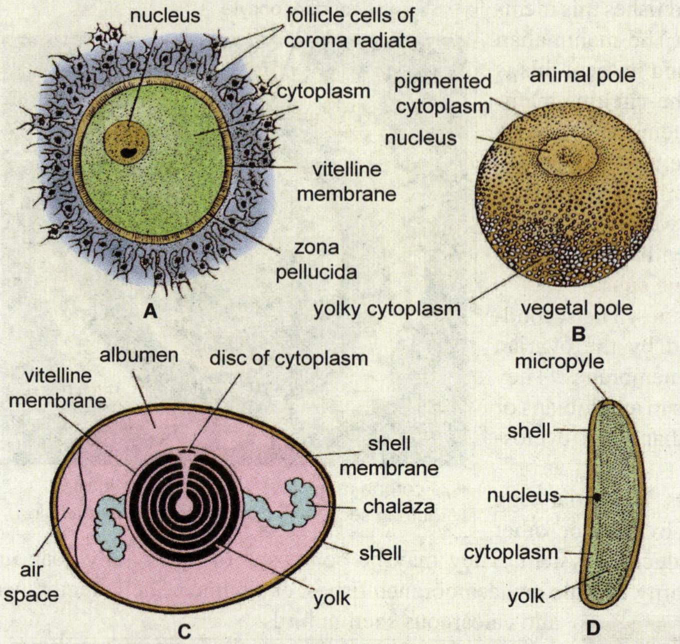

Fig. 20.5. Four main types of eggs : A—Microlecithal and isolecithal egg of man; B—Mesolecithal and moderately telolecithal egg of frog; C—Macrolecithal and highly telolecithal egg of the hen; D— Macrolecithal and centrolecithal egg of an insect, the fly.

lysaccharides and are present in the eggs of sea urchins, frogs, fishes, bivalve molluscs, some annelids and certain mammals. But they do not occur in the ova of man, rat, guinea pig, gastropod molluscs, urodele amphibians, insects and birds.

Yolk contents of the ooplasm. The amount of the yolk in the ooplasm varies from species to species. According to the amount of the yolk following types of egg cells have been recognised (Fig. 20.5).

1. Microlecithal. The eggs with very little amount of yolk are known as the microlecithal eggs, *e.g., Amphioxus*, eutherian mammals.

2. Mesolecithal. The ova or eggs containing moderate amount of yolk are called mesolecithal ova or eggs, *e.g., Peteromyzon,* Dipnoi and Amphibia.

3. Macrolecithal. The eggs with large amount of the yolk are known as the macrolecithal eggs, *e.g., Myxine,* cartilaginous and bony fishes, reptiles, birds and Monotremata. The eggs can also be grouped into two types on the basis of the distribution of the yolk in the ooplasm.

A. Homolecithal. The eggs with evenly distributed yolk contents in the ooplasm are known as the homolecithal eggs, *e.g.,* eggs of echinoderms.

B. Heterolecithal. The eggs in which the yolk is not evenly distributed in the ooplasm are known as the heterolecithal eggs. The heterolecithal eggs may be of following types :

(i) Telolecithal. When the amount of the yolk is concentrated in the one half of the egg to form the vegetative pole of the egg, then this condition is known as the telolecithal. Telolecithal eggs may be moderately telolecithal (*e.g.,* amphibians) or highly telolecithal (*e.g.,* hen and other birds). In macrolecithal and highly telolecithal eggs the amount of the yolk is very large and it occupies the largest portion of the egg except a small disc-shaped portion of the cytoplasm. The cytoplasm contains the zygote nucleus and is known as the germinal disc, *e.g.,* eggs of fishes, reptiles and birds.

(ii) Centrolecithal. In the centrolecithal eggs the yolk accumulates in the centre of the ooplasm, *e.g.,* eggs of insects.

REVISION QUESTIONS

1. Define the term gametogenesis. Differentiate between spermatogenesis and oogenesis.
2. Describe the ultrastructure of the flagellate spermatozoon.
3. Give an illustrated account of spermatogenesis.
4. What is oogenesis ? Describe the process of oogenesis.
5. What is an egg ? Give an account of various types of eggs of animals.
6. Write short notes on the following :
   (1) Spermiogenesis; (2) Growth phase of oogenesis (3) Egg membranes.

CHAPTER

21

Fertilization

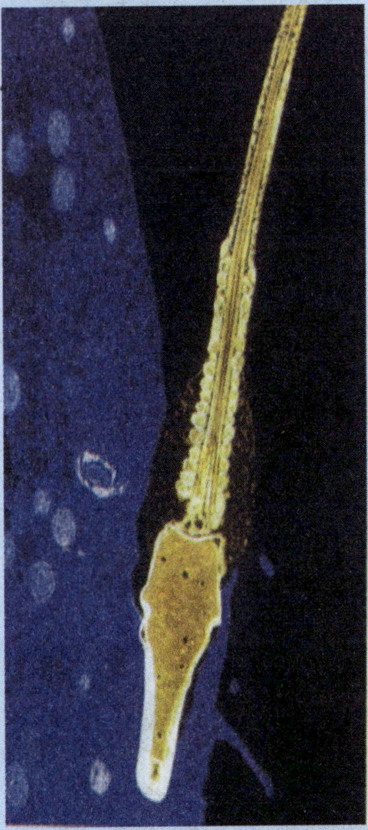

One sperm penetrates the membrane surrounding mammalian egg.

The union of the cytoplasm and pronuclei of the male and female gametes is known as the **fertilization** (L., *fertilis* = to bear). The fertilization is the most commonly used method for the production of the diploid zygotes in the sexually reproducing organisms of Metazoa and Metaphyta.

In the process of the fertilization the haploid male gamete (spermatazoa or pollen), which carries the paternal genetic information, unites with the haploid female gamete (ovum or egg), which carries genetic informations of female parent, to form a diploid zygote. The zygote ultimately, produces a diploid multicellular organism by the several repeated and organised mitotic divisions and cellular differentiation.

EXTERNAL AND INTERNAL FERTILIZATION

The fertilization always occurs in the aquatic media such as sea water, fresh water or intra-somatic (body) fluid of the maternal individual. If the fertilization occurs outside the body of the organism, it is known as **external fertilization** and if it occurs inside the body of the organism then it is termed as **internal fertilization**. The external fertilization is common in various invertebrates and chordates, while the internal fertilization occurs only in those animals which possess specialized sex organs for receiving and transmitting the sperms, *e.g.*, reptiles, birds, mammals and angiosperms, etc.

External fertilization is seen in fish.

FERTILIZIN AND ANTIFERTILIZIN

The process of fertilization is very specific. The sperms of one particular species fertilize the ova of the same species. This type of specificity of male and female gametes is of utmost biological importance and is achieved by the help of certain chemical compounds. It is found that the egg contains a chemical substance known as **fertilizin** (**Lillie**, 1919). The fertilizin is a glycoprotein which is composed of different types of amino acids and monosaccharides (glucose, fucose, fructose and galactose) according to the species. The molecular weight of the fertilizin is 300,000 and it contains large molecules.

The surface layer of sperm contains another proteinous substance known as **antifertilizin**. The antifertilizin is a protein which is composed of acidic amino acids. It has small molecules and the molecular weight is about 10,000.

The fertilizins of the eggs are supposed to attract the sperms which contain a particular type of antifertilizin. It has been found that egg fertilizin of any species reacts efficiently with the sperm antifertilizin of the same species. It has also been reported that the fertilizin in egg-water attracts the sperms of the same species and many sperms adhere together. This type of mutual adhesion of the sperms is known as the **agglutination**.

Internal fertilization is essential for reproduction on land.

PROCESS OF FERTILIZATION

The process of fertilization includes two successive steps which are as follows :

1. The activation of the egg; 2. The amphimixis.

1. Activation of the Egg

1. The process of activation of eggs is completed in following stages :

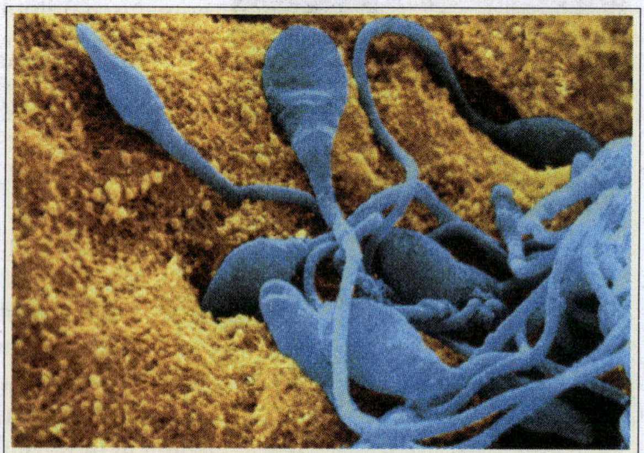

Sperms surround the oocyte, attacking its defensive barriers.

(i) Movement of the sperm towards the egg. The sperms which occur in the external or internal fluid media around the egg, swim towards the egg randomly. They collide with the egg by chance. The chance of colliding of the sperms with the egg occurs regularly in the nature and remains fruitful only due to the large number of the sperms and enoromously large size of the ovum. The fertilizins and antifertilizins become active after the chance collision of the sperms with the ova. The egg fertilizin usually occurs in the jelly surrounding the egg. It gradually dissolves in the surrounding water of the egg and forms the so called egg water.

(ii) Activation of the sperm. When a sperm with a specific antifertilizin comes in contact with the egg water of its own species then certain significant changes occur in the acrosome of the sperm.

The peripheral portion of the acrosome of sperm collapses and its enzymes the **lysins** are extruded and dissolve in the water. The central portion of the acrosome elongates and forms a 1 to 75μm long, thin tube known as the **acrosomal filament**. The acrosomal filament is the rigid tube which protrudes out from the sperm head. When the sperm possesses such an acrosomal filament it is said to be activated for the penetration in the unfertilized egg. When the activated sperms reach the egg the acrosomal filaments penetrate into the egg jelly and vitelline membrane by the help of dissolving action of the sperm lysins. As soon as the tip of the acrosomal filament touches egg membrane (plasma membrane), various important morphological and physiolgical changes are started in the egg.

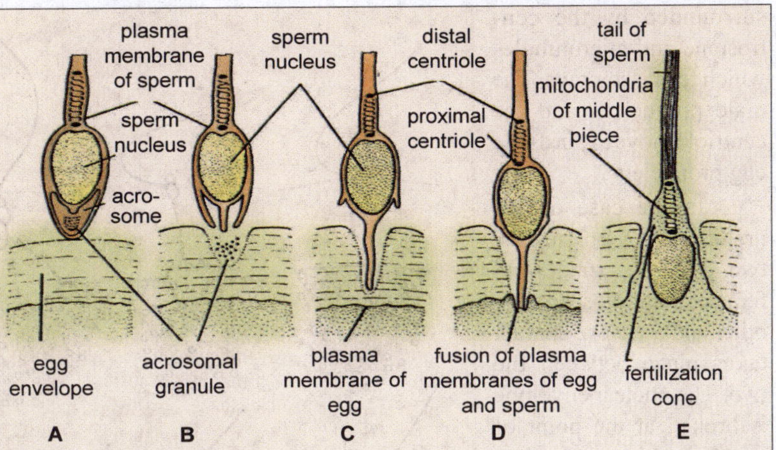

Fig. 21.1. Process of fertilization in animals.

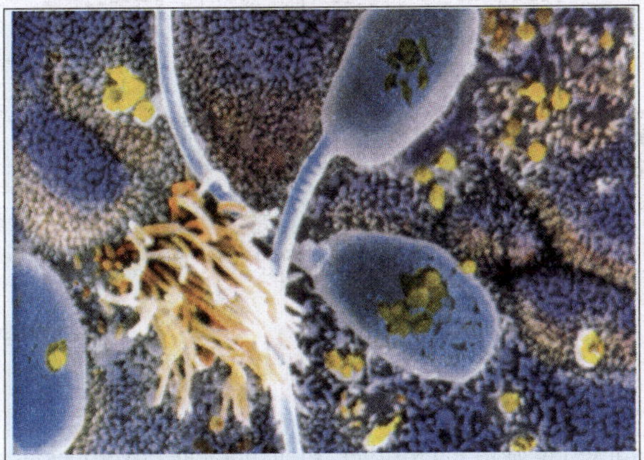

SEM of human sperm cells migrating inside the uterus travelling across a background of uterine mucosa cell.

(iii) The activation of egg insemination. As soon as the acrosomal filament touches the egg surface the ooplasm protrudes out at the point of contact into a cone-like process known as the **fertilization cone** (Fig. 21.2). The fertilization cone may be conical, cylindrical or irregular. When the fertilization cone is irregular in shape it contains many pseudopodia-like processes of the ooplasm. The fertilization cone is composed of the plasma membrane and hyaline cytoplasm. The fertilization cone engulfs the sperm and the sperm which is surrounded by the hyaline cytoplasm move inwards. The penetration of the sperm in the egg is known as the **insemination**. Immediately after the insemination a thin membrane known as the **fertilization membrane** is formed around the plasma membrane of the egg. The fertilization membrane prevents the entrance of further sperms in the egg.

2. Amphimixis

During the insemination the entire sperm may enter in the egg such as in the mammals or the tail of the sperm remains outside the egg such as in the echinoderms. In certain cases as in *Nereis,* the tail and middle piece of the sperm remain outside the egg and only the head and centrosome enter in the egg.

The nucleus of the sperm is known as the **male pronucleus.** The male pronucleus swells up by absorbing the water from the surrounding ooplasm and it becomes vesicular. The compactly arranged chromatin material of the male pronucleus becomes finely granular. The centriole of the sperm is surrounded by the centrosome and microtubules which form aster rays. The male pronucleus and the centriole move towards the egg pronucleus.

In the case of sea urchins and vertebrates, the two pronuclei (male and female) come close to each other and the close contact takes place between the two. The nuclear envelope is broken at the point of contact and the nuclear contents of both pronuclei are intermingled. The endoplasmic reticulum forms a new common nuclear envelope around the both pronuclei and, thus, forms a **zygote nucleus**. In case of *Ascaris,* annelids and molluscs this type of fusion of two pronuclei does not occur. In these animals the centrioles form the achromatic spindle from the microtubules of the ooplasm and both male and female pronuclei come close to each other, their nuclear envelopes are dissolved. The paternal and maternal homologous chromosomes get arranged on the equator of the achromatic figure and the first cleavage (mitotic) division of the egg occurs. After this division, the nuclear envelope is formed around the chromosomes of the daughter nuclei.

Fig. 21.2. Fertilization cone formation during fertilization of *Holothuria* (after Balinsky, 1976).

Post-Fertilization Changes in the Egg

After the fertilization following changes occur in the egg:

1. The zygote becomes ready for the cleavage and for the formation of the embryo.
2. The oxygen consumption of the zygote increases enormously.
3. The metabolic rate of the zygote increases greatly. For instance, the amount of amino acids and the permeability of the plasma membrane of the egg increases the volume of the egg, decreases the exchange of phosphate and sodium ions between the zygote and the surrounding media. Further, diffusion of the calcium ions from the egg starts and the hydrolysing activities of the proteolytic enzymes increase.
4. The protein synthesis is started.

KINDS OF FERTILIZATION

In the organisms following types of fertilization occur :

1. Monospermic fertilization. In most animals usually only one sperm enters in the egg, this type of fertilization is known as **monospermic fertilization**. The monospermic fertilization is common in the coelenterates, annelids, echinoderms, bony fishes, frogs and mammals.

2. Polyspermic fertilization. When many sperms enter in the egg, the fertilization is known as the **polyspermic fertilization**. It may be of two types :

(i) Pathological polyspermy. Under certain abnormal conditions when in a monospermic type of egg many sperms enter in the egg, the condition is known as the **pathological polyspermy**. This type of egg does not develop further and dies soon.

(ii) Physiological polyspermy. In the animals with large yolky eggs such as molluscs, selachians, urodels, reptiles and birds, the polyspermic fertilization usually occurs. Such polyspermic fertilization is known as **physiological polyspermy**. In these cases, many sperms enter in the egg but the pronucleus of only one sperm unites with the pronucleus of the egg and rest are degenerated soon. Such eggs are viable and develop further.

3. Polyandry. When two male pronuclei unite with a female pronucleus, the union is known as polyandry, *e.g.,* man and rat.

4. Polygamy. When two egg pronuclei unite with single male pronucleus, the phenomenon is known as polygamy, *e.g.,* sea urchins, polychaete worms, urodels and rabbits.

5. Gynogenesis. When only sperm activates the egg but its pronucleus does not unite with the egg pronucleus, *e.g.,* planarians and nematodes.

SIGNIFICANCE OF FERTILIZATION

1. The fertilization ensures the usual specific diploidy of the organisms by the fusion of the male and female pronuclei.
2. The fertilization establishes definite polarity in the eggs.
3. The fertilization provides new genetic constitution to the zygote.
4. The fertilization activates the egg for the cleavage.
5. The fertilization increases the metabolic activities and rate of the protein synthesis of the egg.
6. The fertilization initiates the egg to start cleavage and embryogenesis.

REVISION QUESTIONS

1. Define the term fertilization. What is the significance of fertilization ?
2. Describe the process and mechanism of fertilization in animals.
3. Write short notes on the following:
   (1) External and internal fertilization; (2) Post-fertilization changes in the egg; (3) Amphimixis; (4) Polyspermy; and (5) Fertilizin and antifertilizin reaction.

CHAPTER

22

Parthenogenesis

In spring and early summer, when food is abundant, aphid females reproduce parthenogenetically.

Usually an unfertilized ovum develops into a new individual only after the fertilization but in certain cases the development of the egg takes place without the fertilization. This peculiar mode of sexual reproduction in which egg development occurs without the fertilization is known as the **parthenogenesis** (Gr., *parthenos*=virgin; *genesis*=origin). An organism that has developed parthenogenetically is called a **parthenogenone** or **parthenote**. The phenomenon of the parthenogenesis occurs in different groups of the animals as in certain insects (Hymenoptera, Homoptera, Coleoptera), crustaceans, rotifers and also in some vertebrates such as several desert lizards, turkeys and some mammals.

There are certain conditions which are intermediate between parthenogenesis and fertilization, *e.g.,* partial fertilization, gynogenesis, androgenesis and merogony. 1. In **partial parthenogenesis**, the egg may be fertilized by only a part of sperm. For example, according to **Boveri** in sea urchin egg the fertilization of the egg (activation) takes place by the sperm aster. The sperm nucleus gets fused with the egg nucleus only in two cell stage. 2. In **gynogenesis**, the sperm penetrates the egg but takes no part in development. It degenerates in the egg without fusion with the egg nucleus, *e.g., Rhabditis aberrans.* 3. In **androgenesis**, the egg is activated by the sperm and development takes place without the participation of the egg nucleus. For example, if the **ova** of frogs and toads are treated with radium and then fertilized by normal sperms, the egg nucleus does not take part in development, but sperm (paternal) nucleus participates in normal development. 4. In **merogony**, egg fragments devoid of nucleus develop when fertilized by a normal sperm. If sea urchin eggs are shaken to produce small pieces, the fragments round up to form spheres. Some of these spheres are without nuclei. If such enucleated spheres are normally fertilized, they may develop into dwarf larvae.

The parthenogenesis may be of two types :

1. Natural parthenogenesis; 2. Artificial parthenogenesis.

NATURAL PARTHENOGENESIS

In certain animals the parthenogenesis occurs regularly, constantly and naturally in their life cycles and is known as the **natural parthenogensis**. The natural pathenogenesis may be of two types, *viz.,* complete or incomplete.

(i) Complete parthenogenesis. Certain insects have no sexual phase and no males. They depend exclusively on the parthenogenesis for the self-reproduction. This type of parthenogenesis is known as the **complete parthenogenesis** or **obligatory parthenogenesis**. It is found in some species of earthworms, badelloid rotifers, grasshoppers, roaches, phasmids, moths, gall flies, fishes, salamanders and lizards.

In the summer water fleas reproduce very rapidly by parthenogenesis. 10-12 unfertilized eggs are seen here inside the brood pouch of female.

(ii) Incomplete parthenogensis. The life cycle of certain insects includes two generations, the sexual generation and parthenogenetic generation, both of which alternate to each other. In such cases, the diploid eggs produce females and the unfertilized eggs produce males. This type of parthenogenesis is known as the **partial** or **incomplete** or **cyclic parthenogenesis**.

Cyclic parthenogenesis shows several variations in the alternation of sexual (S) and parthenogenetic (P) generations : (1) In gall flies (*e.g., Neuroterus*) there is an alternation of one sexual and one parthenogenetic generation per year (P, S,....P, S,.... P, S). (2) In aphids (plant lice), daphnids and rotifers, the sexual generation may come after many parthenogenetic generations during the summer of the year (P, P, P, P, P, P, S,.... P, P, P, P, P, P, S....). (3) In gall midge (*Miaster*) the larvae reproduce indefinitely by **paedogenetic parthenogenesis**. In this case, germ cells within the larvae develop parthenogenetically into parasitic larvae which feed on the mother larvae. These larvae tend to live under the bark of rotting logs and feed on fungus. Under favourable conditions winged males and females are produced. These stages reproduce sexually and help in dispersion. (4) In some groups there is no regularity between parthenogenetic and sexual generations.

The complete and incomplete type of natural parthenogenesis may be of following two types :

(a) Haploid or arrhenotokous parthenogenesis; (b) Diploid or thelytokous parthenogenesis.

(a) Haploid or arrhenotokous parthenogenesis. In the arrhenotokous parthenogenesis, the haploid eggs are not fertilized by the sperms and develop into the haploid individuals (Fig. 22.1). In these cases the haploid individuals are always males and the diploid individuals are the females *e.g.,* 1. Insects : (i) Hymenoptera (Bees and Wasps) (ii) Homoptera, (iii) Coleoptera (*Micromalthus debilis*), (iv) Thysanoptera (*Anthothrips verbasi*). 2. Arachnids, *e.g.,* ticks, mites and certain spiders (*Pediculoids ventricusm*). 3. Rotifers, *e.g., Asplanchne amphora.*

Thus, the queen bee is fertilized only once by one or many males (drones). She stores the sperm in her seminal receptacles and as she lays her eggs, she can either fertilize the eggs or allow them to pass unfertilized. The fertilized eggs become females (fertile queens or sterile workers depending up on the amount of royal jelly the developing young receives); the unfertilized eggs become fertile males or drones.

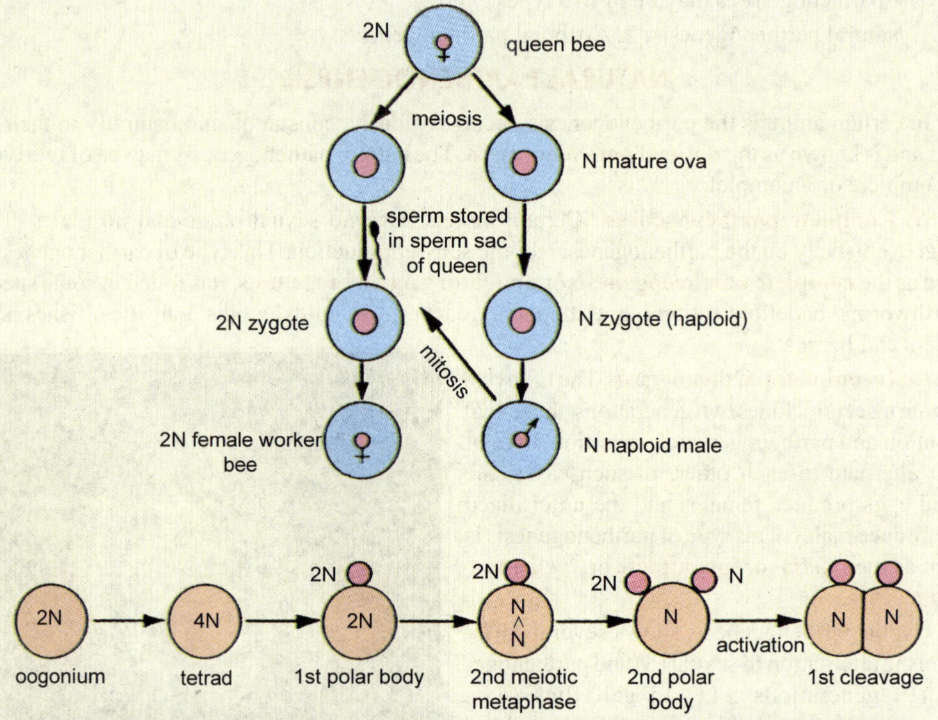

Fig. 22.1. Schematic representation of haploid parthenogenesis in bees (after Grant, 1978).

(b) Diploid or thelytokous parthenogenesis. In the diploid parthenogenesis, the young individuals develop from the unfertilized diploid eggs. The offspring of thelytoky could theoretically be either male or female; but normally it produces only diploid females (Fig. 22.2). For example, in aphids, females emerging in the spring produce several generations of females by diploid parthenogenesis resulting from suppression of first or second polar body. At summer's end some females produce sexual males and females by diploid parthenogenesis, males differing from females in lacking one sex chromosome. Males produce haploid gametes through normal meiosis which fuse to form diploid zygotes that emerge again in the spring as parthenogenetic females. Further, since thelytoky is also found in polyploid forms, it is also called **somatic parthenogenesis**.

Following types of the thelytoky have been recognised :

(i) Ameiotic parthenogenesis. Sometimes during the oogenesis, first meiotic or reduction divsion does not occur but second meiotic division occurs as usual. Such eggs contain diploid number of chromosomes and develop into new individuals without the fertilization. This type of parthenogenesis is known as **apomiotic** or **ameiotic parthenogenesis** and occurs in *Trichoniscus* (Isopoda), *Daphnia pulex* (Crustacea) *Compelona rufum* (Mollusca), weevils and long-horned grasshopper.

(ii) Meiotic parthenogenesis. Certain eggs develop by the usual process of oogenesis but at certain stages **diplosis** or doubling of chromosome number and production of diploid eggs occur. Such eggs develop into the diploid individuals and this phenomenon is known as the **meiotic parthenogenesis**.

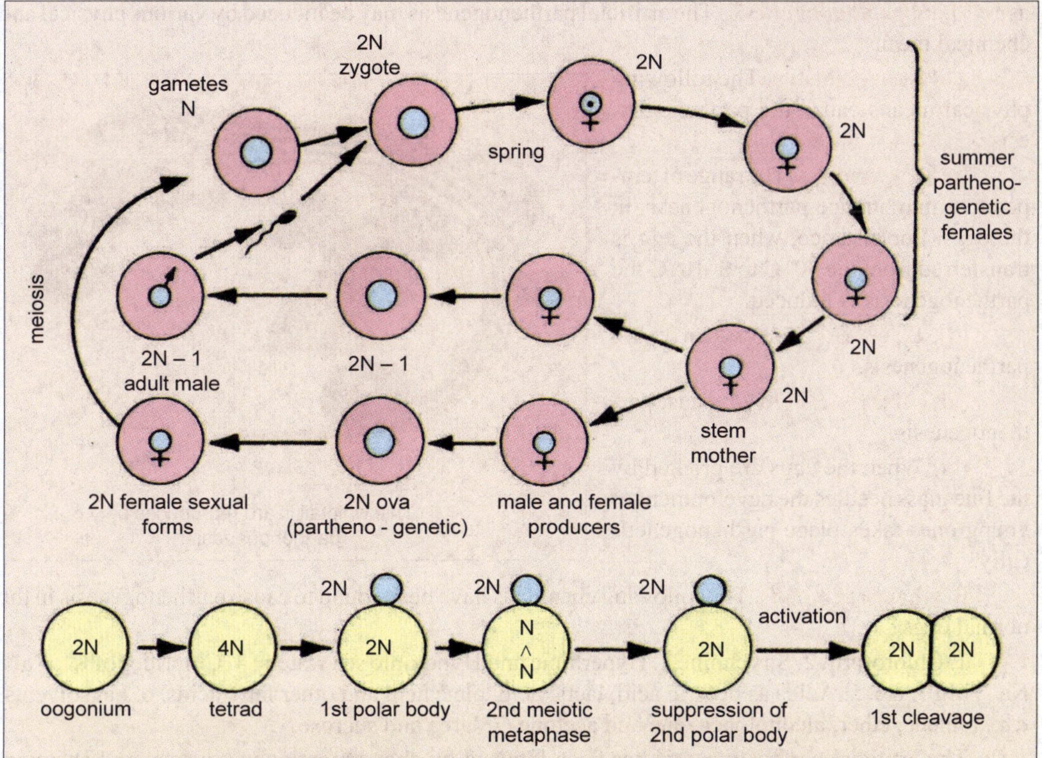

Fig. 22.2. Schematic representation of diploid parthenogenesis in aphids (after Grant, 1978).

The **diplosis** of the diploid thelytoky may occur by the following methods :

(a) By autofertilization. In certain cases the oocyte divides meiotically up to the formation of ootid and secondry polocyte. But the ootid and the secondary polocyte unite together to form a diploid egg which develops into a new individual, *e.g., Artemia salina* (Crustacea) and various other organisms.

(b) By restitution. Sometimes in primary oocyte karyokinesis forms a nucleus of the secondary oocyte and a nucleus of the first polocyte. But the karyokinesis is not followed by the cytokinesis. The chromosomes of both daughter nuclei are arranged on the equator and undergo second meiotic division to form a diploid ootid and a diploid polocyte. The dipioid ootid or ovum develops into a parthenogenetic diploid individual. This type of diplosis is known as the restitution, *e.g.,* insects of order Hymenoptera (*Nemertis conesceus*) and Lepidoptera.

Natural Parthenogenesis in Vertebrates

A few cases of natural parthenogenesis have also been reported in the vertebrates. The fish *Carassius auratus gibelio* is reported to consist of females only (**Lieder**, 1955). Likewise, males are found totally lacking in the lizard *Lacerta sexicola armeniaca* (**Lantz** and **Cyren**, 1936). In it females are reported to be originated by parthenogenesis. In turkeys 80 per cent of incubated eggs show early cleavage stages. Such parthenogentic forms have hatched and grown to reproducing adults which are found to be diploid male with ZZ sex chromosomes. In mammals too, up to 60 per cent of hamster eggs becomes spontaneously activated and develops up to two-cell stage (**Austin**, 1956).

ARTIFICIAL PARTHENOGENESIS

The eggs which always develop into the young individuals by the fertilization sometimes may develop parthenogenetically under certain artificial conditions. This type of parthenogenesis is known as **artificial parthenogenesis**. The artificial parthenogenesis may be induced by various physical and chemical means.

A. Physical means. The following physical means cause the parthenogenesis :

(i) Temperature. The range of temperature may induce parthenogenesis in the eggs. For instance, when the egg is transferred from the 30° C to 0–10° C the parthenogenesis is induced.

(ii) Electrical shocks can cause parthenogenesis.

(iii) Ultraviolet light can cause parthenogenesis.

(iv) When the eggs are pricked by the fine glass needles the development of young ones takes place parthenogenetically.

This whiptail lizard reproduces by parthenogenesis.

B. Chemical means. The following chemicals have been found to cause parthenogenesis in the normal eggs :

1. Chloroform 2. Strychnine 3. Hypertonic and Hypotonic sea waters. 4. Chlorides of K^+, Ca^{++}, Na^{++}, Mg^{++}, etc. 5. Acids as butyric acid, lactic acid, oleic acid and other fatty acids. 6. Fat solvents, *e.g.,* toulene, ether, alcohol, benzene and acetone. 7. Urea and sucrose.

The artificial parthenogenesis has been induced by above mentioned physical and chemical means by various workers in the eggs of most echinoderms, molluscs, annelids, amphibians, birds and mammals.

SIGNIFICANCE OF PARTHENOGENESIS

1. The parthenogenesis serves as the means for the determination of sex in the honey bees, wasps, etc., and it supports the chromosome theory of inheritance.

2. The parthenogenesis is the most simple, stable and easy process of the reproduction, *e.g.,* aphids (insects).

3. The parthenogenesis eliminates the variation from the population, but encourages development of the advantageous mutant characters.

4. The parthenogenesis causes the polyploidy in the organisms.

5. Due to the parthenogenesis, there is no need for the organisms to waste their energy in the process of mating but it allows them to utilize that amount of energy in the feeding and reproduction.

6. Honey bee and other social insects also control their sex ratio by parthenogenesis.

7. In aphids, parthenogenesis is a means of rapid breeding; the females reproduce by diploid parthenogenesis during summer.

REVISION QUESTIONS

1. What is parthenogenesis ? Describe different types of natural parthenogenesis in animals.
2. Give an account of artificial parthenogenesis.
3. What is the significance of parthenogenesis ?

CHAPTER

23

Growth

Living beings grow both in body mass and height.

Growth is an increase in size of an organism, reflecting an increase either in the number of its cells, or in its protoplasmic material or both. Cell number and protoplasmic content do not always increase together; cell division can occur without any increase in protoplasm, giving a larger number of smaller cells (e.g., cleavage). Alternatively, protoplasm can be synthesized with no cell division so that the cells become larger. Any increase in protoplasm requires the synthesis of cell components such as mitochondria, cell membranes, enzymes and other proteins.

Growth, thus, involves increase in size and weight of the organism due to the synthesis of new protoplasm. It also includes increase in amount of **apoplasmatic substances** such as the fibres and matrix of connective tissues of higher animals such as mammals. So the growth of body of a higher organism takes place by the addition of new substances, both protoplasmic and apoplasmatic, when the anabolic process dominates the metabolic activity. Conversely, when decomposition exceeds synthesis, first the internal food reserve (such as fat in the adipose tissues) is consumed to run the body machine, and then the energy is obtained at the expense of proteins of the protoplasm. This causes depletion of the living matter, resulting in **degrowth**.

LEVELS OF GROWTH

Among living organisms, growth can be recognised at the following two levels :

A. Cell growth. At the cellular level, the growth of all multicellular organisms is governed by two main activities. These are **reproduction** and **growth** of individual cells of the body. In Chapter 18 of Cell Division, we have already described that during the interphase stage, new materials such as nucleic acids (DNA, RNAs) and proteins are synthesized in the cells and, thus, the cells grow. The growth of individual cells comprising the body is the vital and the most essential factor of growth in all multicellular organisms. The rhythmicity of cell multiplication and growth can be studied well in tissue culture or in culture of unicellular organisms.

B. Growth of multicellular organisms. The growth of multicellular animals and plants in relation to growth and multiplication of their individual cells falls under the following three categories:

(1) Auxetic growth. (*Auxesis* = growth resulting from increase in cell size). In this type of growth, the volume of the body increases due to the growth of body cells without any increase in the number of cells. Examples of auxetic growth are quite rare and include nematodes, rotifers and tunicates.

(2) Multiplicative growth. This type of growth results due to the rise in the number of cells constituting the body. The increase in the number of cells is brought about by the mitotic divisions. In this case, however, the average size of the cells remains the same or increase insignificantly. This type of growth is called **multiplicative growth** and occurs in embryos during morphogenesis. It is also involved in prenatal growth of higher vertebrates.

(3) Accretionary growth. Generally post-embryonic growth of animals and plants occurs due to mitotic multiplication of some special types of cells occurring in specific locations of the body. The differentiated cells of organs and tissues of the body lose the capacity of division and growth (*e.g.*, muscles, nerve cells, osteocytes of bone, fat cells, xylem, phloem, parenchymal cells, etc.). These cells tend to perform physiological functions for the survival of the animal, whereas the special cells exist in an undifferentiated state as **reserve cells**, *e.g.*, **meristematic cells** in angiosperms, **stem cells** such as erythropoietic tissue of red bone marrow, periosteum cells of bone, ciliary body cells of vertebrate eye, and epidermal cells of stratum germinativum. In case of necessity, these reserve cells reinforce and replace the worn-out differentiated cells. In such an event they differentiate into the type of cells that they reinforce and replace. This type of growth is called **accretionary growth**.

According to **Green** and **Taylor** (1990), starting with an individual cell, growth of a multicellular organism can be divided into following three phases: (i) **cell division** or **hyperplasia**, *i.e.*, an increase in cell number as a result of mitotic division; (ii) **cell expansion** or **hypertrophy**, *i.e.*, an irreversible increase in cell size as a result of the uptake of water or the synthesis of living material: and (iii) **cell differentiation**, *i.e.*, the specialization of cells; in its broad sense, growth also includes this phase of cell development (*viz*., differentiation).

LIMITED AND UNLIMITED GROWTH

Studies of the duration of growth in plants and animals show that there are two basic patterns, called **limited** (definite or determinate) **growth** and **unlimited** (indefinite or indeterminate) **growth**. Growth in annual plants is limited and after a period of maximum growth, during which the plant matures and reproduces, there is a period of **negative growth** or **senescence** before the death of the plant.

Several plant organs show limited growth but do not undergo a period of negative growth, for example, fruits, organs of vegetative propagation, dicotyledonous leaves and stem internodes. Animals exhibiting limited growth include insects, birds and mammals.

Woody perennial plants on the other hand show unlimited growth. Unlimited growth also occurs in fungi, algae, monocotyledonous leaves, and many animals, particularly nonchordates, fishes and reptiles.

CELL GROWTH

The cell is a dynamic system that exhibits a unique phenomenon of growth. A cell grows at the expense of food materials that it draws from its environment and converts it into its cellular constituents. An increase in active cell mass is the result of synthetic and degenerative processes acting

An elephant's growth is limited while a crocodile shows unlimited growth.

simultaneously. When a cell has reached its maximum limiting size, it divided into two daughter cells. Such a cell is said to be in **growth-duplication cycle**. Single celled organisms such as *E.coli*, yeast, *Amoeba*, etc., and somatic cells in culture are examples of a growth-duplication cycle. A cell, born after a division, proceeds to grow by macromolecular synthesis, reaches species – specific division size and replicates to repeat the cycle. Mitosis (or nuclear division) marks the end of the growth-duplication cycle in eukaryotic cells and is usually preceded by DNA replication.

Kinetics of Cell Growth

There is a great problem to determine the kinetics of cell enlargement of the cells between divisions, since the criteria for measuring growth (for example, dry mass, volume and linear dimensions) do not behave consistently even within the same cell. Studies on fixed cells and on living cells have revealed the following two patterns of growth: linear and exponential. An **exponential growth pattern** means that growth rate is a function of total mass; as mass increases, growth rate increases accordingly. The curve is sigmoidal, behaving as an autocatalytic process with growth rate proportional to the amount of active protoplasm or replicating entities. **Linear growth pattern** means that growth rate is constant throughout the cell cycle and does not increase. Here, growth rate is independent of cell mass but is related to a constant number of elements (synthetic sites), the activities of which remain unchanged throughout the growth cycle.

Examples. 1. A growing cell not only increases in size but increases in weight too. This means that growth is a linear measure and can be studied as a function of time. *E.coli* has been largely used to study growth in laboratory conditions on a well-defined medium. The nutrient medium contains glucose as the carbon source and several inorganic ions dissolved in an aqueous medium. Growth is best studied at 37^0 C and it takes about 60 minutes to double the cell mass. However, growth of *E.coli* can be accelerated (20 minutes) by supplementing the medium with various amino acids, purine, and pyrimidine bases.

Observations on the growth of a single *E.coli* cell have shown that the cell grows and divides into daughter cells after a unit time, which is a constant factor for each generation. This unit time is called the **generation time**. The number of bacteria increases in an exponential phase, where growth and division are synchronized. Growth takes place as an exponent of 2, *i.e.*, $2^0, 2^1, 2^2, 2^3, 2^4$...cells, and the pattern is represented in the form of a growth curve (Fig. 23.1). The linear portion of the curve represents exponential growth, suggesting increase in number of cells as a function of time. Subsequently the growth curve flattens and this part of the curve represents a period of maximum stationary growth. During this period the cells divide slowly and growth no longer remains exponential. The cause of stoppage of exponential growth is attributed to several factors such as depletion of nutrients and lack of oxygen. Since the vitality of the genome is greatly reduced, so cells cannot divide infinitely.

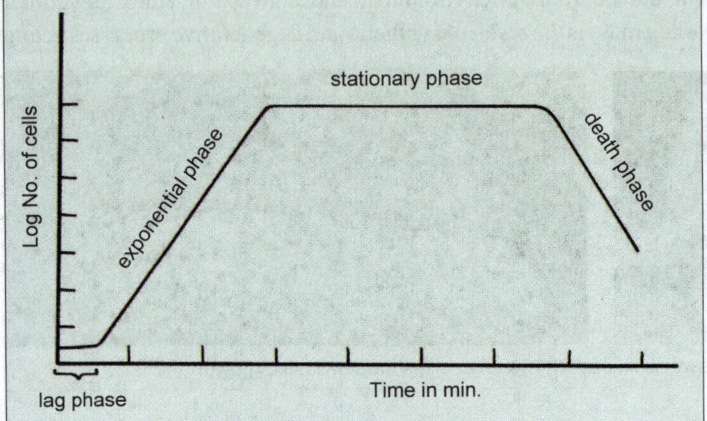

Fig. 23.1. Various patterns of growth in *E.coli*.

2. **Mitchison** (1963) measured the dry mass and volume of a budding yeast cell and demonstrated that mass increases **linearly** after division without a lag and continues at a constant rate until it doubles before the next division; that is, combined growth rate of daughter cells is double the original rate of the mother cell. On the other hand, cell volume follows a roughly **exponential curve** for the first three quarters of cell cycle, reaching a plateau prior to division.

Yeast growth kinetics emphasises various features common to the growth of cells that absorb nutrients through their surface : (1) growth rate is constant between divisions, and doubles immediately after; (2) cell mass doubles between divisions, (3) mass increase is correlated with a nuclear change rather than cytoplasmic; (4) there is no lag period – linear growth begins immediately after division. Exceptions to this pattern are found in *Amoeba*. Though mass in *Amoeba* doubles between divisions, its growth shows a diminishing rate of increase, reaching a plateau several hours before mitosis.

Evidently genetic factors determine growth patterns, but some differences are nutritional. Yeast cells absorb exogenous nutrients directly across the cell membrane while *Amoeba* engulfs solid food. Contractile vacuole changes in *Amoeba* introduce extraneous perturbations into cell mass measurements. Furthermore, a diminishing growth rate and a period of constancy at the end of *Amoeba* growth reflect a decline in feeding before division.

Mechanisms Involved in Cell Growth

In most cases, the kinetics of mass increase is usually matched by parallel synthesis of RNA, protein and membrane. The synthesis of DNA, being discontinuous, is not directly related to the kinetics of cell growth. It is however, involved in controlling the cell size.

1. RNA synthesis and cell growth. Generally, ribosomal RNA and tRNA are synthesized continuously throughout the eukaryotic cell cycle. The rate of synthesis may increase during cell cycle; in mammalian cells rate of rRNA synthesis becomes double after S phase. The pattern for mRNA is not known since different species of mRNA are synthesized at different periods of the cycle and at different rates. The correlation between overall cell growth patterns and rRNA synthesis suggests that the production of ribosomes might be an important site for growth regulation. Evidently, the total number of ribosomes in a bacterial cell controls the rate of synthesis of all proteins during growth, *i.e.*, the number of ribosomes per DNA genome is proportional to the rate of growth and protein synthesis (**O. Maaloe**). In eukaryotic cells, growth rate depends on total number of cytoplasmic ribosomes per cell and these are controlled by the nucleolus, the seat of ribosome synthesis.

2. Nucleolus and cell growth. Nucleolus shows cyclic changes during the cell cycle and is somehow related to cell growth. During interphase, when cells are actively growing, nucleoli are prominent and synthesize ribosomal RNA at a high rate. In prophase, when growth stops, nucleoli disappear, emptying their contents into the nucleoplasm. Nucleoli are absent in metaphase and anaphase but reappear early in telophase at twice their original number, as new nucleoli organize at nucleolar organizer sites in each daughter nucleus. The combined growth rate of the daughter cells increase to twice the rate of the original mother cell though total protoplasmic mass has not changed. The transition from a state of physiological "oneness" to that of physiological "twoness" is most closely coordinated with duplication of nucleolar organizers and formation of nucleoli. Thus, growth rate doubles after the nucleolar organizers are replicated during the S phase when twice the number of ribosomal RNA genes start to transcribe rRNA.

In fact, the nucleolus is a dynamic organelle, attuned to metabolic demands, responding rapidly to changing needs for new patterns and rates of growth. It becomes large and metabolically active in most growing and proliferating cells, such as tumor, and disappears in cells not active in protein synthesis. When there is a heavy demand for ribosomes, as in maturing oocytes, nucleolar function is greatly amplified by making many additional nucleoli (up to 1000 in some species), each equipped with a segment of DNA containing many copies of the rRNA genes.

Thus, growth regulation seems to hinge upon the nucleolus and its control over ribosome synthesis. Ribosome production and turnover may determine cell growth, with the nucleolus as the principal "flow-through" centre or "valve" regulating the entire process.

3. Protein synthesis and cell growth. Most eukaryotic cell growth results from total protein accumulation—the net balance between total protein synthesis and protein degradation. Both processes are subject to different modes of regulation, the former largely dependent upon the availability of the ribosomes. Further, though total cell protein may seem to increase continuously through the cell cycle, some individual proteins may be constant, others may be decreasing and still others may be increasing in a stepwise fashion. For example, during cell cycle the enzymes exhibit the following three patterns of their synthesis: 1. Synthesis may be periodic like the DNA synthesis and increase rapidly during one phase of the cycle, *e.g.*, enzymes involved in DNA synthesis, such as thymidine kinase do show a stepwise pattern of its synthesis. 2. Continuous synthesis of enzymes, either linear or exponential, is typical of many respiratory enzymes in mouse fibroblasts. 3. Some enzymes show a peak pattern. They increase rapidly during the cycle, and disappear, presumably as a result of degradation and turnover. Within the same cell, each protein may be regulated independently of others. However, some sets of proteins may be coordinately regulated, particularly those proteins that characterize a cell phenotype.

Two hypotheses or models have been proposed to explain various patterns of protein synthesis which are meant for cell growth:

1. Oscillatory repression model. According to this model, periodic enzyme production initiates end product repression by negative feedback. When the enzyme pool is high, enzyme synthesis is repressed; when low, enzyme synthesis is increased. This pattern will lead to stable oscillations which need not correspond in frequency to other events in the cell cycle, such as DNA synthesis.

2. Sequential gene expression model. According to this model, a chromosome is itself programmed for sequential gene expression of genes at different stages in the cell cycle. Here an ordered reading of genes is emphasized. The RNA polymerase moves along the genome transcribing genes in sequence. Hence, genes are available for transcription only at specific periods of the cell cycle. For example, linear reading of genes has been reported for synchronized growing yeast. The time of synthesis of 12 different enzymes in yeast corresponds to the position of their respective genes in the chromosomes. Each enzyme is synthesized in a stepwise manner with the position of gene seemingly dictating the order of its expression during the cell cycle.

REVISION QUESTIONS

1. Write an essay on the growh of living organisms.
2. What is growth? Describe the process of cell growth.

GENETICS, HUMAN GENETICS AND EUGENICS

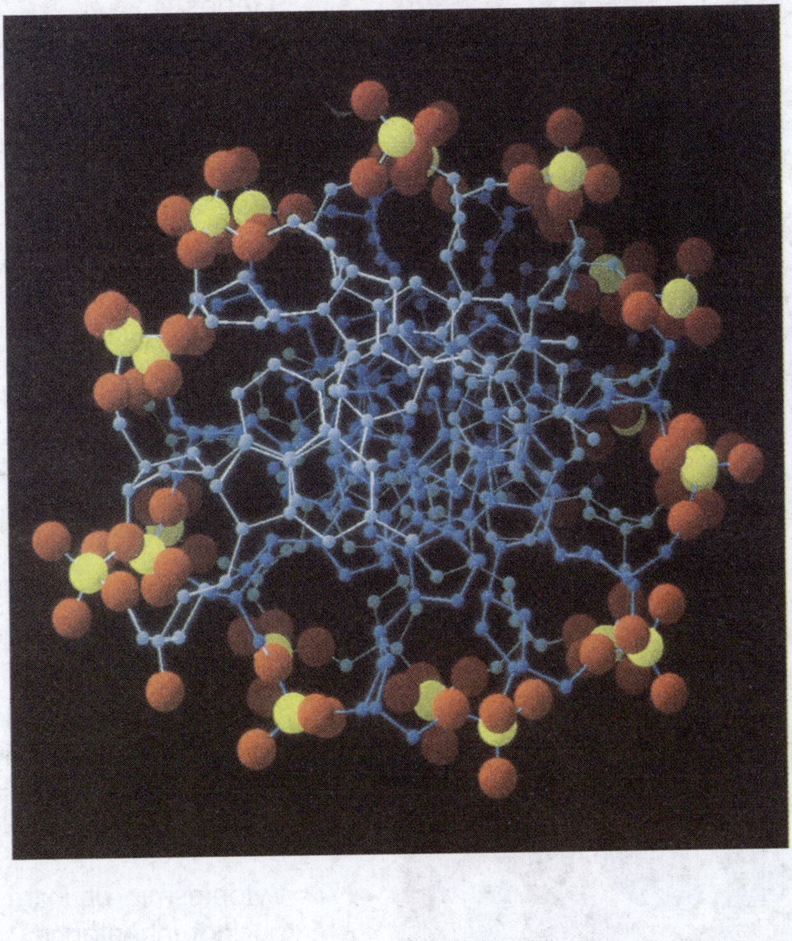

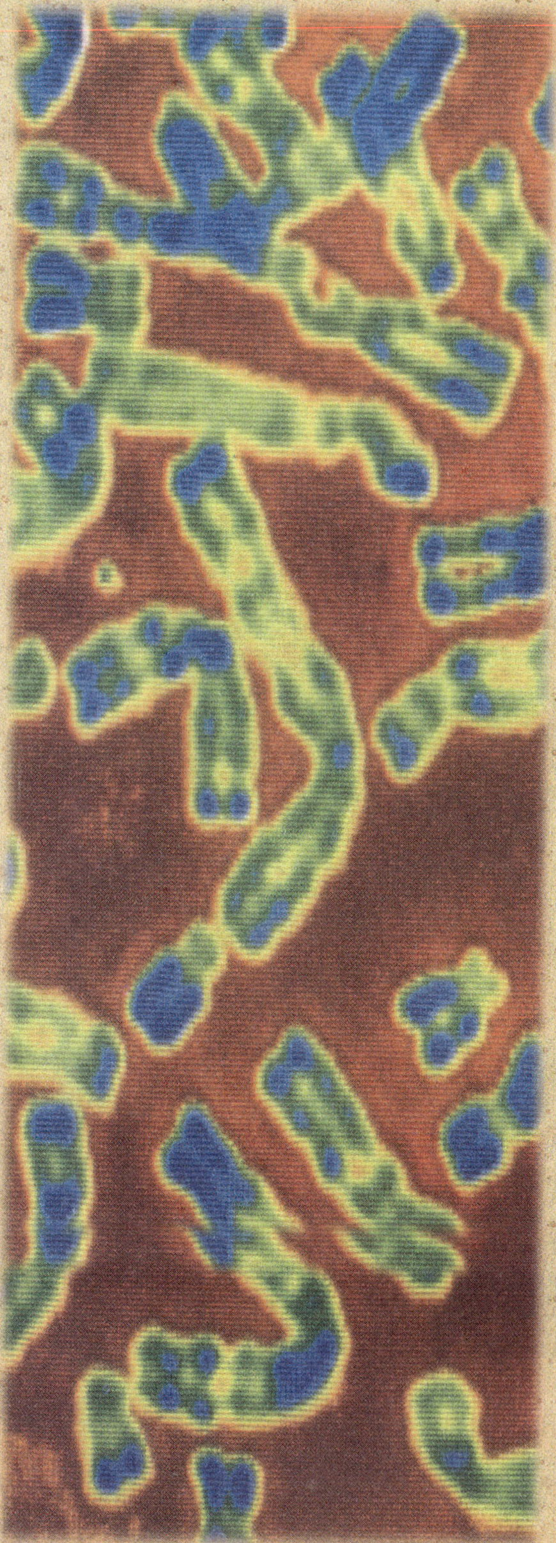

CHAPTERS

CHAPTER

1

Introduction

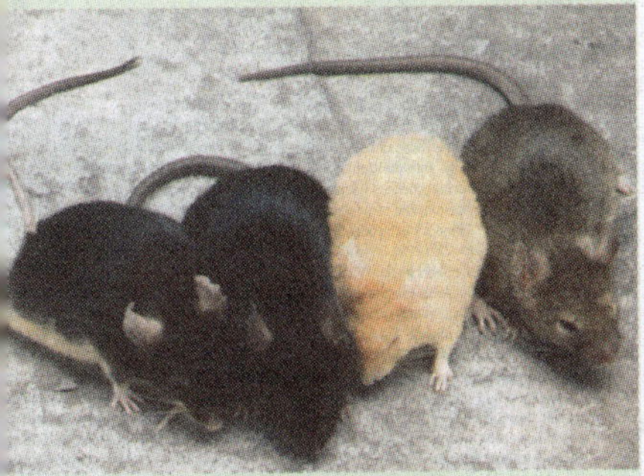

The science of genetics is the study of heredity which is the cause of similarities; and variation which is the cause of differences between individuals.

The science of heredity or genetics is the study of two contradictory aspects of nature : heredity and variation. The process of transmission of characters from one generation to next, either by gametes–sperms and ova–in sexual reproduction or by the asexual reproductive bodies in asexual reproduction, is called **inheritance** or **heredity**. Heredity is the cause of similarities between individuals. This is the reason that brothers and sisters with the same parents resemble each other and with their parents. **Variation** is the cause of differences between individuals. This is the reason that brothers and sisters who do resemble each other are still unique individuals. Thus, we have no trouble in recognizing the differences between sisters, for example, and even 'identical' twins are recognized as distinctive individuals by their parents and close friends. The science of genetics attempts to explain the mechanism and the basis for both similarities and differences between related individuals. It also tries to explain the phenomenon of evolution and cytodifferentiation.

The heredity and variations play an important role in the formation of new species (speciation). The biological science which deals with the mechanism of heredity and causes of variations in living beings (viruses, bacteria, plants and animals) is known as **genetics**. The word **genetics** was derived from the Greek root *gen* which means to become or to grow into and it was coined by **Bateson** in 1906 for the study of physiology of heredity and variations.

HISTORICAL

The history of most scientific disciplines including genetics are generally characterized by relatively long periods of stagnation punctuated by bursts of rapid progess. Most of these flurries of research are initiated by new technical developments. The science of genetics is a very young science in comparison to other biological sciences and its origin can be traced in the works of **Mendel** in the nineteenth century. But before Mendel's work men throughout the ages had some vague knowledge about genetics and more often have tried to explain the causes of heredity. About six thousand years ago men kept records of pedigrees of domestic animals such as horse and food plants as rice. The ideas or theories which have been forwarded from time to time to explain the phenomenon of inheritance can be categorized under the following headings : 1. Vapour and fluid theories; 2. Preformation theories; 3. Particulate theories.

1. Vapour and Fluid Theories

Early Greek philosophers such as **Pythagoras** (500 B.C.) proposed that every organ of animal body gives out some type of vapours. These vapours unite and form a new individual.

Hippocrates (400 B.C.) believed that the reproductive material is handed over from all parts of the body of an individual, so that the characters are directly handed over to the progeny.

Further, **Aristotle** (350.B.C.) thought that the semen of man has some "vitalizing" effect and he considered it as the highly purified blood. According to him the mother furnishes inert matter and the father gives the motion to the new life.

2. Preformation Theories

Leonardo da Vinci (1452–1519) proposed a theory that the male and the female parents contribute equally to the heredity of the offspring. **W. Harvey** (1578–1657) speculated that all animals arise from eggs and that semen only plays vitalizing role. **R. de Graaf** (1641–1673) observed that the progeny would have characteristics of father as well as of mother and, therefore, he proposed that both the parents should contribute to the heredity of progeny.

Malpighi (1673), the pioneer of preformationist school, concluded that development of any organism consisted simply of growth of preformed part. **A.V. Leeuwenhoek** in 1677 observed sperms of several animals (man, dog, rabbit and other mammals, frog, fish and insects) and also suggested their association with eggs. In 1679, **J. Swammerdam** studied development of insects and frog and suggested that development of an organism is a simple enlargement of a minute but preformed individual. The figure of **homunculus** (Fig. 1.1) or manikin, the miniature man in the sperm head, was published in 1695 by **Hartsoeker**. Such type of theories which advanced the concept of the presence of preformed embryo in the sex cells are known as **preformation theories**. Preformationists have, often, been divided into two schools: 1. **Ovists** who attached more importance to ova; they thought that "homunculus" was present in the ovum. 2. **Animalculists** or **spermatists** who attached more importance to sperm; they thought that a miniature but complete organism was present in the sperm.

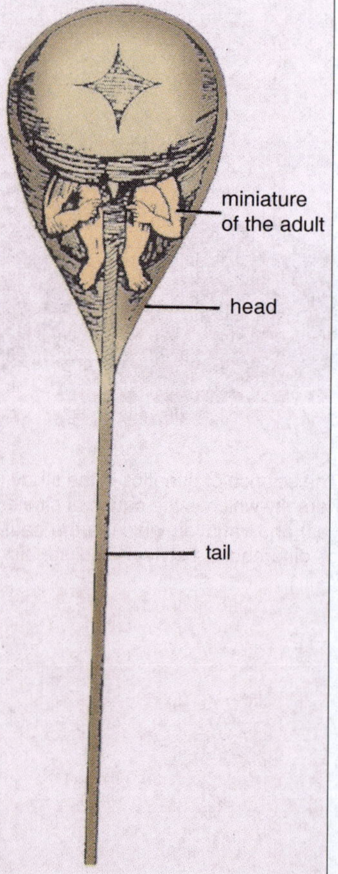

Fig. 1.1. Diagram of homunculus.

N. Grew in 1682 reported for the first time the reproductive parts of plants. **R. Camerarius** in 1694 described sexual reproduction in plants for the first time. He is also known to be first to produce a hybrid between two different plant species. In 1717, **Fairchild**

produced a hybrid having characteristics of both parents. This hybrid was called "**Fairchild's Sweet William**" or as "**Fairchild's mule**." This provides a means of artificial hybridization in plants. **J.G. Kolreuter** (1733–1806) obtained fertile hybrids from artificial crosses between two species of tobacco plants.

K.F. Wolff (1738–1794) finally refuted the preformation theory by proposing that neither egg nor sperm had a structure like homunculus but that the gametes contained undifferentiated living substance capable of forming the organized body after fertilization. Such an idea formed the very core of the theory of **epigenesis**. This theory suggested that many new organs and tissues which were originally absent, develop subsequently. However, Wolff believed that these tissues and organs developed *de novo* due to mysterious vital forces.

3. Particulate Theories

French biologist **Maupertuis** (1689–1759) has proposed that the body of each parent gives rise minute particles. In sexual reproduction, the particles of both individuals unite together to form a new individual. He thought that in certain cases the particles of the male parent might dominate on those of the female parent and produce the male individual. In the production of female individual the particles of female might dominate on particles of male. Thus, **maupertuis** proposed the concept of **biparental inheritance** by **elementary particles**. He studied the family pedigree of polydactyly and albinism in human beings.

The great biologist **Lamarck** (1744–1829) in 1809 proposed the phenomenon of "inheritance of acquired characters" among living organisms. But he failed to provide convincing evidences in support of his concepts.

In 1868 the well known naturalist **Charles Darwin** has given his famous **theory of pangenesis** which exclusively depends on the particulate theory. The central idea of pangenesis theory has been given first of all by **Hippocrates**. According to the pangenesis theory of Darwin each part of the animal body produces many minute particles known as **gemmules**. These gemmules are at first collected in the blood and later on are concentrated in the reproductive **organs**. When the animal reproduces into new individual, these gemmules pass on to it and it has blending of both parents. By this mechanism acquired characters would also be inherited because as the parts of the body changed so did the pangenes or gemmules they produced.

The theory of pangenesis was disapproved by **Galton** (1823–1911) and **Weismann** (1835–1934). **Weismann** in 1892 postulated the **theory of germplasm** to explain heredity. According to this theory the body of organisms contain two types of cells namely somatic cells and reproductive cells. The somatic cells form the body and its various organ systems, while the reproductive cells form sperm and ova. The somatic cells contain the **somatoplasm** and germinal or reproductive cells contain the **germplasm**. According to Weismann the germplasm can form somatoplasm but somatoplasm cannot form germplasm. Thus, the changes in the structure of somatic cells or somatoplasm which are caused by the environment (acquired characters) cannot influence the reproductive cells or germplasm. By cutting the tails of mice for many generations, Weismann always got tailed mice. So, by such experimental evidences he rejected the Lamarckism and pangenesis theory.

Though the particulate theory faced many problems in its beginning but its basic concept has formed the central core of the modern understanding of the genetics.

Augustinian Monk **Gregor Mendel** was the first investigator who laid the foundation of our modern concept of the particulate theory. He could understand the heredity problems more clearly than any one in the past, because his approach was simple, logical and scientific. By his famous experiments on pea plant he concluded that the inheritance is governed by certain factors which occur in the cells of each parent. He

Gregor Mendel (1822-1884).

thought that each parent has two such factors, while their sex cells (sperm or pollen and ovum or egg) have only one factor. However, he failed to explain the exact process by which these factors pass on the sex cells. **knight** (1799) and **Goss** (1824) conducted hybridization experiments on edible pea (*Pisum sativum*), but they failed to formulate any law of inheritance like the Mendel.

During 19th century and dawn of 20th century, the science of genetics have received solid support from landmark investigations in the field of cytology, embryology, biochemistry and genetics. **Von Baer** (1828) made discovery of the mammalian egg. **Pringsheim** (1855) first saw nuclear fusion in green algae (*Vaucheria*). Heredity transmission through the sperm and egg became known by 1860. **Ernst Haeckel**, noting that sperm consisted largely of nuclear material, postulated that the nucleus is responsible for heredity. **Oscar Hertwig** (1875) observed the entrance of the sperm into the sea urchin. He found nucleus to play an important role in hereditary mechanism. In 1884, **Hertwig** identified the hereditary substance with the chromatin of nucleus. **Strasburger** in 1875 discovered the chromosomes and he along with **Kolliker** and **Weismann** formulated the **nuclear theory of heredity**. **Flemming** (1882) investigated the process of mitosis.

Three plant breeders, namely **Hugo de vries** (Holland), **Karl Correns** (Germany) and **Erich Tschermak** (Austria), rediscovered the Mendel's laws in 1900. Each of them reached similar conclusions before they knew of Mendel's work. **Bateson** (1902) published a book "*The Principles of Heredity*." From 1902 to 1909 he introduced the terms **allelomorphs**, **homozygote**, **heterozygote**, **F1**, **F2** and **epistatic gene**. **Bateson** was the first to have Mendel's paper translated into English and the first to show that Mendel's theory also applied to animals. He coined the term **genetics** in 1905. In 1906, **Bateson** and **Punnett** reported first case of linkage in sweet pea, however, they failed to explain the phenomenon of linkage correctly. **R.C. Punnett** devised the **Punnett's square** for making gametic combinations theoretically. American cytologist, **Walter S. Sutton** in 1902 proposed the **chromosome theory of heredity** in his classic paper "*The Chromosomes in Heredity*," in which he postulated that the newly rediscovered Mendel's hereditary factors were physically located on chromosomes. This theory provided a mechanism of transmission to explain the behaviour of Mendel's factors and brought together two independent desciplines– the genetics and the cytology. Thus, the year 1903 is the year of birth of **cytogenetics**.

Archibald E. Garrod (1902) deciphered the inheritance pattern and metabolic nature of the human disease **alkaptonuria** (in which urine of patient turns dark to black upon exposure to air). In 1908, **Garrod** presented in a lecture nearly all the facts that we know today concerning this disease. He also postulated various enzymes involved in this metabolic error, but he could not identify them. In 1909 **Johannsen** formulated the **genotype-phenotype concept** to distinguish hereditary variations from environmental variations. According to him, the genotype of an individual represents the sum total of heredity, while phenotype of an individual represents the observable structural and functional properties which are produced by the interaction between genotype and environment. (In 1877, **Johannsen** coined the term **gene**). The hypothesis that "genes can change (mutate) to give rise to new genes (mutant genes)" was seriously tested, beginning in 1908 by American biologist **Thomas H. Morgan** and his young collaborators (Ph.D. students), such as, **Calvin B. Bridges**, **Hermann J. Muller** and **Alfred H. Sturtevant**. They worked on the fruit fly, *Drosophila melanogaster*. **(W.E. Castle** suggested the fruit fly to Morgan). In 1910, first white eye mutant was detected in *Drosophila* by this team of workers and it is first reported case of sex linkage.

Sir A.E. Garrod, father of biochemical genetics.

T.H. Morgan (1866–1945) proposed in 1911 the **theory of linkage**. He turned the chromosome theory of inheritance into the concept of genes being located in a linear array on each chromosome. In

Thomas Hunt Morgan 1866-1945.

1926, his book '*The Theory of the Gene*' was published and he got Nobel prize in 1934. Cytological basis of **crossing over** was first described by the Belgian cytologist **F.A. Janssens** in 1911. **H.J. Muller** and **L.J. Stadler** independently discovered that X-rays induce mutations. **H.J. Muller** got Nobel Prize in 1946 for the discovery of the induction of mutation in *Drosophila* by X-rays. In 1916, **Bridges** made discovery of the phenomenon of **non-disjunction** in *Drosophila*. In 1921, he proposed the **genic balance** mechanism of sex determination in *Drosophila*.

B.O. Dodge in the late 1920's and early 1930's first determined the genetics of *Neurospora*. **T.M. Jenkins** (1924) reported a case of **cytoplasmic inheritance**, called **Iojap striping** in maize. **Barbara McClintock** and **Harriet Creighton** working at Cornell University, USA, with the corn plant, *Zea mays,* devised an elegant demonstration of chromosome breakage and rejoining during crossing over. In 1937, **Richard Goldschmidt** stimulated exploratory questions on the chemical nature of gene (of *Drosophila*). In the 1940's, two significant discoveries were made concerning the chemical nature of the gene: 1. **Oswald Avery**, **C.M. MacLeod** and **M. McCarthy** (1944) were able to establish by experiments with pneumonia-causing (virulent) bacteria that *genes were composed of a specific type of nucleic acid, called* **deoxyribonucleic acid (DNA)**, *and not proteins.* 2. While studying the biochemical basis for the eye colour in *Drosophila,* **George Beadle** and **E.L. Tatum** were able to show that the lack of brown colour in various mutants was due to a defect in one step in the biosynthesis of the brown pigment. They proposed the **one-gene one-enzyme hypothesis** which suggested that the action of each gene is through the synthesis of a protein (enzyme) which in turn catalyzes a single chemical reaction. They proved this hypothesis through the use of multitude of mutants in the fungus, *Neurospora* (in 1941). In most cases, each mutation was due to a change in a single gene. Thus, they initiated the branch of **biochemical genetics**. The term **molecular biology** was first used in 1945 by **William Astbury**, who was referring to the study of the chemical and physical structure of biological macromolecules.. **Joshua Laderberg** (1946) first demonstrated the phenomenon of recombination in the bacteria *E.coli.* **Beadle**, **Tatum** and **Laderberg** got Noble prize in 1958. Recombination in phage was first demonstrated in 1948 by **Max Delbruck** and **Mary Delbruck**. The chemistry of DNA and RNA has been worked out by **A. Kornberg** and **S. Ochoa** ; both got Noble Prize in 1959.

Oswald Avery 1877-1955.

Max Delbruck 1906-1981.

Prior to discovery of the chemical structure of the genetic material, the 'gene' was an abstract, indivisible unit of heredity (comparable to old concept of the indivisible atom). This period in history is referred to as **classical** or **formal genetics**. The word "formal" pertains to the extrinsic aspect of something as distinguished from its substance or material. The era of **molecular genetics** followed the discovery of DNA structure (*i.e.,* 1953) when the fundamental unit of heredity was determined to be DNA nucleotide and the 'gene' was found to consist of an aggregate of nucleotides. In 1953, one of the most significant twentieth-century discoveries in biology was made by **James watson** and **Francis Crick**. Their paper published in the British Journal *Nature* in which they proposed the molecular structure of DNA, *i.e.,* the molecular composition of the gene. **Watson**, **Crick** and **Wilkins** got Nobel Prize in 1962 for the discovery of double helix model of DNA which opened the new vistas in the genetical world.

Francis Crick and James Watson.

Seymour Benzer performed extensive investigations on the genetics of T_4 bacteriophage of *E. coli* and in 1955, was able to define the gene in terms of function (**cistron**), recombination (**recon**) and mutation (**muton**) and to place an accurate molecular size estimate on the conceptual gene components. In simple terms, Benzer demonstrated that the linear array of genes on chromosomes as shown by Morgan, was extended down to the molecule of DNA making up the chromosome. In 1961, **Francois Jacob** and **Jacques Monod** provided genetic evidence for a method of gene regulation in bacteria, now called the **operon**. In 1965, **Jacob**, **Monod** and **Lwoff** were awarded Nobel Prize for their contribution to microbial genetics. **Gaulian** and **Kornberg** isolated, purified and utilized **DNA polymerase** of *E. coli*. **R.W. Holley** got Nobel Prize (1968) for the discovery of base sequence of tRNA. **Holley** died in 1993. During 1961–1968, the genetic code of DNA was solved by **M.W. Nirenberg**, **J.H. Matthaei**, **p.Leder**, and **H.G. Khorana**. They synthesized small RNA molecules (mRNAs) of known composition and observed which amino acid was incorporated into protein in a cell-free protein synthesizing system. In 1968 **Nirenberg** and **Khorana** discovered the complicated DNA code known as **genetic code**. Both scientists along with **Holly** received the Nobel Prize in 1968.

N.L. Dhawan and **R.L. Paliwal** (1964) studied the cytoplasmic inheritance in maize. The term **transposons** (*i.e.,* jumping genes) is used in 1974 by **R.W. Hedges** and **A.E. Jacob** of Hammersmith Hospital in London, for a DNA segment or genetic element which could move from one molecule to another and carried resistance for antibiotic ampicillin in the bacterial cells. These transposons, however, were originally discovered in maize plant by **Barbara McClintock** by the name **controlling elements** in 1956. During the late 1970's, the science of genetics entered a new era dominated by the use of **recombinant DNA technology** or **genetic engineering** to produce novel life forms not found in nature. Through this technology, it has been possible to transfer genes from mammals into bacteria, causing the microbes to become tiny factories for making (in relatively large quantities) proteins of great economic significance such as **hormones** (insulin, growth hormones) and **interferon** (lymphocyte proteins that prevent replication of a wide variety of viruses). These proteins are produced in such small quantities in humans that the cost of their extraction and purification from tissues has been very expensive, thus, restricting their medical use in **prophylaxis** (prevention) and **therapeutics** (treatment) of disease. By genetic engineering, it has become possible to produce various **blood clotting factors**, **complement proteins** (part of the immune system) and other substance for the improvement of genetic deficiency diseases (**euphenics**) other current fields of genetic research are oncogenes (cancer), antibody diversity (immunogenetics), homeotic mutation and behaviour.

Joshua Lederberg

SCOPE OF GENETICS

Geneticists study all aspects of genes. The study of the mode of gene transmission from generation to generation is broadly called **transmission genetics**; the study of structure and function of the gene forms the **molecular biology**, and the study of behaviour of genes in populations is called **population genetics**. These three major subdivisions of genetics are arbitrary and there is considerable overlapping. It is the knowledge of how genes act and how they are transmitted down through the generations that has unified biology; previously, specific set of biological phenomena had each been assigned to separate disciplines. An understanding of how genes act is now essential prerequisite for such biological fields of study as development, cytology, physiology and morphology. An understanding of gene transmission is a fundamental aspect of areas such as ecology, evolution and taxonomy. Further unification has resulted from the discovery that the basic chemistry of gene structure and function is very similar across the entire spectrum of life on the earth. Thus, not so long ago, biology was divided into many camps that rarely communicated with each other; today, however, every biologist must be a bit of a geneticist, because the findings and techniques of genetics are being applied and used in all fields. In fact, genetics contributed the modern prototype for all of the biology. It provides a unifying thread for the previously diverse fields of biology.

IMPORTANCE OF GENETICS

The cultural evolution of human beings is strongly influenced by knowledge of hereditary phenomena of early man. Civilizations itself become possible when nomadic tribes learned to domesticate plants and animals. Long before biology existed as a scientific discipline, people selected grains with higher yields and greater vigour and animals with better fur, meat or milk. They also were mystified about the inheritance of desirable and undesirable traits in the human population. Despite this long-standing concern with heredity and the practice of selective breeding, it was not until the discovery of Mendel's laws that we were able to explain the actual basis for inheritance.

Jacques Monod

Francois Jacob

Like other disciplines of science, the genetical insight has produced new challenges as well as solutions to some human problems. For example, early in this century a new wheat strain called **Marquis** was developed in Canada. This high-quality strain is resistant to disease; furthermore, it matures two weeks earlier than other commercially used strains—a very important factor where the growing season is short. The introduction of Marquis strain of wheat had opened up millions of square kilometres of fertile soil to cultivation in such northern countries as Canada, Sweden and the USSR. Likewise, **IR26** strain of rice was developed by geneticists in 1973; it has a wide range of desirable characteristics such as resistance to several viral and fungal diseases and protection from insects such as green leaf hopper, brown hopper and stem borer. In addition to improving crop varieties, geneticists have learned to change the genetic systems of insects to reduce their fertility. This technique has provided an important new tool in the age-old struggle to keep insects away of human crops and habitations.

In recent years, such successes led to the concept of "**Green Revolution**". Using sophisticated breeding techniques based on new knowledge about genes, geneticists created high-yielding varieties of dwarf wheat and rice. Large-scale planting of these crops around the world did provide new food supplies, but new problems quickly became apparent. These specialized crops require extensive cultivation and costly fertilizers. The use of the new high-yield varieties produced diversed social and economical problems for the poor/developing countries where they are most needed. Furthermore, the spread of **monoculture** (the extensive dependence on a single plant variety) left vast areas at the mercy of some newly introduced or newly evolved form of pathogen (*e.g.,* plant disease) or insect pest. With the huge population of humans on earth, our dependence on high yield varieties of crop plants and domestic animals has become increasingly clear. In fact, the stability of human society depends on the ability of geneticists to juggle the inherited traits that shape life forms, keeping a jump ahead of the destructive parasites and predators (Fig. 1.2).

In recent years, advances in **biotechnology** have led to the creation of special genetically engineered strains of bacteria and fungi that carry specific genes from unrelated organisms such as humans. (**Note**: Biotechnology means use of living organisms or processes to modify or make products and to improve plants or animals, see **Peter Funk**, 1995). As already has been mentioned under the head of historical, these microbes produce such useful compounds as insulin, human growth hormone and the antiviral (or anticancer) agent–the interferon.

Further, the most exciting and alarming application of genetic knowledge is to the human species itself. Genetic discoveries have had major effects on medicine. One can now diagnose hereditary or genetic disease before or soon after birth, and in some cases we can provide secondary treatments. Using **family pedigree analysis**, a genetic counsellor can give prospective parents the information they need to make intelligent decisions about the risks of genetic disease in their offspring. Some refined techniques as **amniocentesis** and **fetoscopy** provide information about possible genetic disease at early stages of pregnancy. A battery of post natal chemical tests can detect problems in the newborn infants, so that some corrective methods can be applied immediately to lessen the impact of many genetic diseases.

However, our new ability to recognize genetic disease poses an important moral dilemma. An estimated 5 per cent of our population survives with severe physical or mental genetic defects. This percentage probably will increase with extended exposure to various environmental factors and paradoxically, with improved medical technology. For example, of those patients admitted to paediatric hospitals in North America, 30 per cent estimated to have genetic diseases. This will certainly increase great financial burden on human society.

Lastly, knowledge of genetic mechanisms has made us aware of some new dangers as well. Primarily, some geneticists fear that there may be an accidental release from some laboratory of an artificial pathogen that has never existed on this planet before and that pathogen may cause havoc. Some geneticists even fear that increased exposure to chemical food additives and to vast array of chemicals in other commercial products may be changing the human genetic makeup in a very undesirable and haphazard manner. This type of random genetic changes can also be caused by such environmental agents as fallout from nuclear weapons, radioactive contamination from nuclear reactors, and radiation from various X-ray machines. These agents may be contributing to inherited disease, but they almost certainly are contributing to the incidence of cancer (which is a genetic disease of the somatic or body cells). Anyhow in modern era the genetics has revolutionised the agriculture, horticulture, animal husbandry and many other branches of science. The science of genetics has proved worthy in removing many faulty concepts of man about the inheritance.

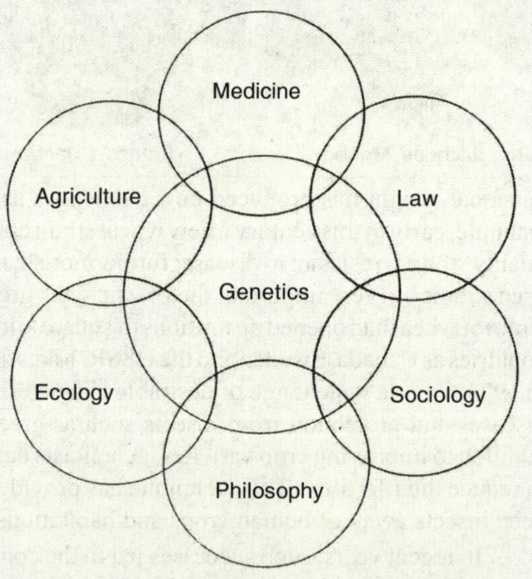

Fig. 1.2. Impact of genetics on different interacting areas of human ventures (after Suzuki *et al.,* 1986).

BRANCHES OF GENETICS

In recent years, the science of genetics has proliferated into numerous distinctive subdisciplines. Some of the significant branches of genetics are the following :

1. **Plant genetics.** The genetics of plants.
2. **Animal genetics.** The genetics of animals.
3. **Human genetics.** It involves the study of heredity of human traits, human disorders, betterment and correction of human disorders.

4. Microbial genetics. It deals with the genetics of microorganisms (*viz.,* viruses, bacteria, unicellular plants and animals).

5. Fungal genetics or mycogenetics. The genetics of fungi.

6. Viral genetics. Genetics of virus.

7. Drosophila genetics. Genetics of fruit fly, *Drosophila* sp.

8. Mendelian genetics. It involves study of heredity of both qualitative (monogenic) and quantitative (polygenic) traits and the influence of environment on their expressions.

9. Quantitative genetics. It involves the study of heredity of quantitative traits such as height, weight and IQ in human beings and milk production in cattle.

10. Morganian genetics. It includes study of recombination (crossing over) in all kinds of organisms such as higher plants, animals, fungi, bacteria and viruses. It also involves the preparation of linkage maps of chromosomes.

11. Non-Mendelian genetics. It involves a study of the role of cytoplasm and its organelles (particularly chloroplasts and mitochondria) in heredity.

12. Mutations. They involve study of heredity of both chromosomal changes (structural and numerical) and also gene mutation.

13. Cytogenetics. It provides the cytological explanations of different genetical principles.

14. Molecular genetics. It includes the study of structure and function of gene and regulation of its activity.

15. Transmission genetics. It includes the study of mode of gene transmission from generation to generation. The kind of studies that Mendel performed are now included in the discipline of transmission genetics.

16. Clinical genetics. Genetics involved in the detection of causes of diseases such as haemophilia, colour blindness, diabetes, phenylketonuria.

17. Immunogenetics. It deals with genetics of production of different types of antibodies; the diversity of antibodies has been found to be under control of genetic regulation.

18. Behavioural genetics. It involves the interaction of genes with the environment to produce a particular pattern of behaviour. In *Drosophila* many behaviour genes have been identified, *e.g.,* mutants described as sluggish, non-climbing, flightless, easily shocked, etc., and genes regulating sexual behaviour. In primates including humans, it has been found that IQ (intelligence quotient) is governed by genetics (parentage), environment (adopted parents) and developmental stage (age) of an individual.

19. Forward genetics and reverse genetics. During the last decade, the term **reverse genetics** has been used for physical mapping and isolation of genes whose protein products are unknown. The term **forward genetics** has been used for genes which are mapped on the basis of phenotype (or gene product or protein), using the technique of classical genetics. However, recently in 1991, the term reverse genetics has been redefined by **Paul Berg** (Nobel Laureate). According to him, the term reverse genetics should be restricted to those studies, where we start the study with a DNA segment with unknown phenotypic effect, introduce this DNA (without any alteration or other modification) into a plant or an animal and then follow the phenotypic effect.

REVISION QUESTIONS

1. Explain the following :
   Germplasm theory; pangenesis theory; preformation theory; particulate theory; and variation.
2. Why the study of genetics is important for human society?
3. Give a brief account of the scope and importance of genetics, outlining the newer areas of study in this subject.
4. Describe the historic growth of genetics in last two decades.
5. Enumerate and define various branches of genetics.

CHAPTER

2

Genetical Terminology

Like other sciences, the science of genetics has its specific terminology which minimizes the chances of confusion, inconvenience and unnecessary repetition of full sentences. We are giving here certain most common terms which are used more frequently in genetics.

Acquired character. The alteration in the morphology or physiology of an organism in response to its ecological factors (environment) is known as acquired character. Acquired characters are usually not heritable.

Albinism. Absence of colour in skin, hair and eyes or absence of chloroplast in a plant; an inherited trait.

Albino. The animal without pigmentation in skin, hairs and eyes is called albino.

Allele (Allelomorph). One of two or more forms that can exist at a single gene locus, distinguished by their differing effects on the phenotype. Alleles are genes controlling the same characteristic (*e.g.* hair colour) but producing different effects (*e.g.* black or red), and occupying corresponding positions on homologous chromosomes.

Amniocentesis. Puncture of the uterine wall with a needle for the purpose of obtaining amniotic fluid, which can be analyzed to determine whether the foetus has a genetic abnormality. Amniotic fluid contains sloughed foetal cells.

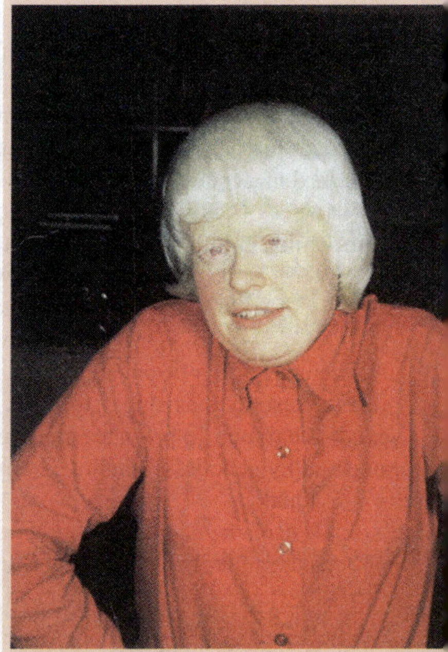

Albinos illustrate pleiotropy.

Aneuploidy. Karyotypic abnormality in which a specific chromosome(s) is present in too many or too few copies.

Animal breeding. The practical application of genetic analysis for development of pure-breeding lines of domestic animals suited to human purposes.

Autosome. The chromosomes which are not associated with sex are known as autosomes. Except the sex chromosomes (X) and (Y) other chromosomes are the autosomes.

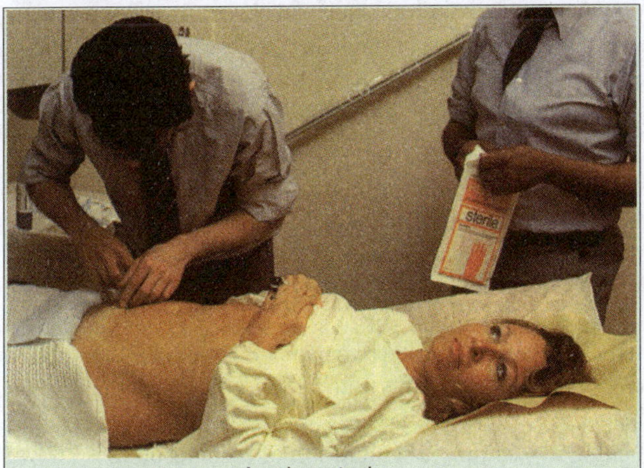

Amniocentesis.

Back cross. The cross of a progeny individual with its parents is known as back cross.

Barr body. A densely staining mass that represents an X-chromosome inactivated by dosage compensation.

Bead theory. The disproved hypothesis that genes are arranged on the chromosome like beads on a necklace, indivisible into smaller units of mutation and recombination.

Bivalent. A pair of synapsed homologous chromosomes is known as a bivalent.

Blending inheritance. A discredited model of inheritance suggesting that the characteristics of individual result from the smooth blending of fluid-like influences from its parents.

Carrier. A heterozygous individual. An individual who possesses a mutant allele but does not express it in the phenotype because of a dominant allelic partner; thus, an individual of genotype *Aa* is a carrier of *a* if there is complete dominance of *A* on *a*.

Chiasma (plural **chiasmata**). A cross-shaped structure commonly observed between nonsister chromatids during meiosis; the site of crossing-over.

Chromatin. A DNA, RNA, histone and non-histone protein containing thread-like coiled structure of interphase nucleus is called chromatin.

Chromosome. The nucleoprotein structure which are generally more or less rod-like during nuclear division. The genes are arranged on the chromosomes in a linear fashion. Each species has a characteristic number of chromosomes. Chromosomes play most important role in inheritance.

Cis arrangement. Linkage of dominants of two or more pairs of alleles on one chromosome and the recessive on the homologous chromosome.

Cistron. A term equated with the term gene. It is a region of DNA that encodes a single polypeptide (or functional RNA molecule such as tRNA or rRNA).

Clone. A group of genetically identical cells or individuals derived by asexual division from a common ancestor.

Cloning. Asexual production of a line of cells or organisms or segments of DNA genetically identical to the original.

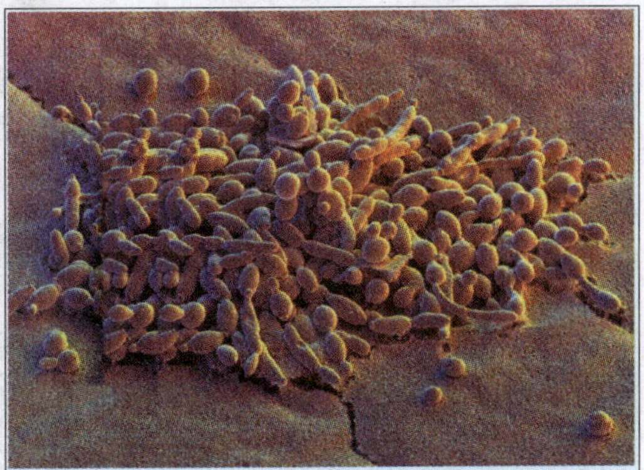
Cells of yeast multiply rapidly by budding, producing vast clones of genetically identical cells.

Codominance. When both the alleles (dominant and recessive) are equally expressed in the hybrid, the phenomenon is known as codominance, *e.g.*, when the red and white cattles are crossed, they produce a roan offspring which possess both red and white hairs on the skin.

Crisscross inheritance. Transmission of a gene from male parent to female child to male grandchild—for example, X-linked inheritance.

Cross. The deliberate mating of two parental types of organisms in genetic analysis.

Crossing over. Crossing over is a phenomenon in which exchange of chromosomal segment or genetic material occurs between the non-sister homologous chromosomes by breakage and union.

Culture. Tissues or cells multiplying by asexual division, grown for experimentation.

Cytoplasmic inheritance. Inheritance via genes found in cytoplasmic organelles.

Daltonism. Red and green colour blindness, a recessive trait known to be X-linked.

Deoxyribonucleic acid (DNA). It is a genetic material of many viruses, bacteria, plants and animals. It is a double stranded, helically coiled, macromolecule which is composed of phosphoric acid, deoxyribose sugar, two pyrimidenes (cytosine and thymine) and two purines (adenine and guanine), It is found to be most stable biological molecule which contains encoded genetic informations.

Dihybrid. An individual which is hybrid or hetrozygous in two pairs of alleles or allelomorphs is known as dihybrid. For instance, the cross between pea plants with yellow round seeds and green wrinkled seeds produces a dihybrid having yellow round seeds.

Diploid. An individual or cell containing two complete haploid sets of chromosomes is known as diploid.

Dominance. A phenomenon in which one member of a pair of allelic genes expresses itself as a whole (complete dominance) or in part (incomplete dominance).

Dominant allele. An allele that expresses its phenotypic effect even when heterozygous with a recessive allele; thus if *A* is dominant over *a*; then *AA* and *Aa* have the same phenotype.

Dominant phenotype. The phenotype of genotype containing the dominant allele; the parental phenotype that is expressed in a heterozygote.

Dominant trait. When out of two contrasting characters or traits only one expresses or appears in a generation. That trait is known as dominant trait, *e.g.*, in pea, round character of seed is dominant over wrinkled character of seed.

Dosage compensation. (1) Inactivation of X-chromosome in mammals so that no cell has more than one functioning X chromosome. (2) Regulation at some autosomal loci so that homozygous dominants do not produce twice as much product as the heterozygote.

Down's syndrome. An abnormal human phenotype including mental retardation, due to a trisomy of chromosome 21; more common in babies born to older mothers.

Environment. The combination of all the conditions external to the genome that potentially affect its expression and its structure.

Down's syndrome.

Episome. A genetic element

(closed, circular DNA molecule) in bacteria that can replicate free in the cytoplasm or can be inserted into the main bacterial chromosome and replicate with the chromosome, *e.g.*, F factor of *E. coli.*

Epistasis. A situation in which an allele of one gene obliterates the phenotypic expression of all allelic alternatives of another gene. The masked gene is said to be **hypostatic**.

Euchromatin. A type of chromatin that is non-condensed during interphase and condensed during nuclear division, reaching a maximum in metaphase. The banded segments of the polytene chromosomes of *Drosophila* larval salivary glands contain euchromatin.

Euploidy. Variation in chromosome number by whole sets or exact multiples of the monoploid (haploid) number, *e.g.* diploid, triploid. Euploids above the diploid level may be referred to collectively as **polyploids**.

Eugenics. Controlled human breeding based on notions of desirable and undesirable genotypes.

Exon. DNA sequence of a cistron that are transcribed into mRNA and are translated into protein.

Expressivity. The degree to which a particular genotype is expressed in the phenotype under a variety of environmental conditions.

F factor. The fertility factor in the bacterium, *Escherichia coli.* It is composed of DNA and must be present in a cell to function as a donor in conjugation.

Filial generations. Successive generations of progeny in a controlled series of crosses, starting with two specific parents (the P generation) and selfing or intercrossing the progeny of each new (F_1, F_2 ....) generation.

F_1 or First filial generation. The word filial is derived from the Latin word *filin*, meaning the son. The first generation of a given cross is known as F_1 generation.

F_2 or Second filial generation. The second generation which is resulted by interbreeding or selfing of F_1 offsprings is known as second filial generation.

Forward mutation. A mutation that converts a wild-type allele to a mutant allele.

Frame-shift mutation. The insertion or deletion of a nucleotide pair or pairs, causing a disruption of the translational reading frame.

Gamete. A sex cell having haploid set of chromosomes and arising due to meiotic cell division of diploid germ cell is known as gamete. The male gamete is known as pollen or sperm and female gamete is known as ovum or egg.

Gene. The fundamental physical and functional unit of heredity, which carries information from one generation to the next; a segment of DNA, composed of a transcribed region and a regulatory sequence, that makes possible transcription.

Genetically engineered animals like these sheep can be used to produce proteins.

Gene interaction. The co-ordinated effect of two or more genes in producing a given phenotypic trait.

Gene mutation. Change in the structure of a gene.

Genetic engineering. Array of techniques that facilitate the manipulation and duplication of pieces of DNA for industrial, medical and research purposes.

Genome. A complete set of chromosomes, or of chromosomal genes, inherited as a unit from one parent, or the entire genotype of a cell or individual.

Genotype. The genetic makeup or constitution of an individual, with reference to the traits under consideration, usually expressed by a symbol, *e.g.*, +, DD (tall), *dd* (short), etc.

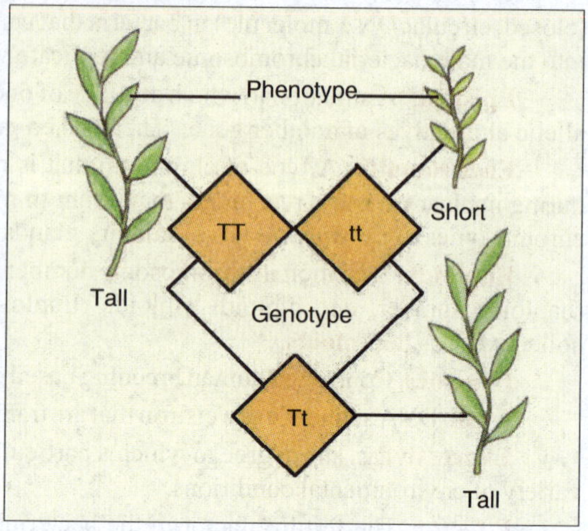

Gynandromorph or gynander. When the body of an individual exhibits both the male and female characters, this type of individual in known as gynandromorph. A sexual mosaic.

Haploid (Monoploid) . An individual or cell containing a single complete set of chromosomes is known as haploid.

Haemophilia. A metabolic disorder characterized by free bleeding even from slight wounds because of the lack of clot forming substances. It is associated with a sex- or X-linked recessive gene.

Heterochromatin. Condensed chromatin. **Constitutive heterochromatin** occurs in centromeric region of chromosomes of a given species; it includes certain important, genetically active regions. **Facultative heterochromatin** is that chromatin that makes up the genetically inactive whole chromosomes (*e.g.,* the X chromosome in human females).

Heterogametic sex. A sex containing either only one or two different sex chromosomes such as XO, XY or ZW produces two types of gametes half with X or Z chromosome and other half with Y or W or no chromosomes. This type of sex is known as heterogametic sex.

Heterozygote (Heterozygous). An individual containing both dominant and recessive genes or traits or characters of a allelic pair is known as heterozygous or hybrid.

Holandric gene. A gene which occurs only in Y chromosome is known as holandric gene.

Homogametic sex. The sex which possesses two identical sex chromosomes (XX or ZZ) and produces single type of gametes each with X or Z chromosome is known as homogametic sex .

Homologous chromosomes (Homolog). The identical male and female parent chromosomes occur in pairs and are known as homologous chromosomes. Each chromosome of a homologous pair is known as **homolog**.

Homozygote (Homozygous). The organism having two similar genes for a particular character in a homologous pair of chromosomes is known as homozygous or genetically 'pure' for that particular character.

Hybrid. (i) A heterozygote. (2) A progeny individual from any cross involving parents of differing genotypes.

Hybridoma. A cell hybrid between an antibody-producing B cells and a tumour cell, which divides indefinitely and produces a single antibody (monoclonal antibody).

Inbreeding. Mating between relatives more frequently than would be expected by chance.

Incomplete dominance. The condition in heterozygotes where the phenotype is intermediate between the two homozygotes. In some plants the cross of red and white produces pink-flowered progeny.

Independent segregation. The independent behaviour of genes occurs on different pairs of chromosomes. This phenomenon is known as independent segregation.

Intron. A sequence of nucleotides in DNA that does not appear in mRNA; probably excised from hnRNA in its processing.

Karyotype. The entire chromosome complement of an individual or cell, as seen during mitotic metaphase.

Klinefelter's syndrome. A genetic disease due to the XXY karyotype. It produces sterile males with some mental retardation.

Lawn. A continuous layer of bacteria on the surface of an agar medium.

Lethal gene. A gene whose phenotypic effect is sufficiently drastic to kill the bearer. Death from different lethal genes may occur at any time from fertilization of the egg to advanced age. Lethal genes may be dominant, incompletely dominant, or recessive.

Linkage. The occurrence of different genes on the same chromosome. They show nonrandom assortment at meiosis.

Linkage group. All of the genes located physically on a given chromosome.

Linkage map. A chromosome map; an abstract map of chromosomal loci, based on recombinant frequencies.

Locus. The position or place on a chromosome occupied by a particular gene or one of its alleles.

Antirrhinums showing incomplete dominance.

Maternal inheritance. A type of uniparental inheritance in which phenotypic differences in progeny occur due to factors such as chloroplasts and mitochondria transmitted by the female gamete.

Meiosis. Meiosis is the reduction division in which the diploid or somatic chromosome numbers are reduced to half. The meiosis often produces haploid gametes or individuals.

Mendelian ratio. A ratio of progeny phenotypes reflecting the operation of Mendel's laws.

Mendel's first law. The two members of a gene pair segregate from each other during meiosis; each gamete has an equal probability of obtaining either member of the gene.

Mendel's second law. The law of independent assortment; unlinked or distantly linked segregating genes pairs behave independently.

Mitosis. Process of cell division, whereby the genetic material is precisely divided and two new chromosome sets identical to the original are generated.

Monohybrid. When the cross takes place between the parents differing in a single pair of contrasting characters resulted into a monohybrid individual.

Monohybrid cross. The cross between the two parents differing in a single pair of contrasting characters is known as monohybrid cross.

Monosomic. An individual lacking one chromosome of a set ($2n - 1$).

Mutation. A spontaneous permanent change in a gene or chromosome which usually produces a detectable effect in the organism concerned and is transmitted to the offsprings.

Nondisjunction. The failure of homologs (at meiosis) or sister chromatids (at mitosis) to separate properly to opposite poles.

Norm of reaction. The pattern of phenotypes produced by a given genotype under different environmental conditions.

Oncogene. A gene that induces uncontrolled cellular proliferation (*i.e.,* cancer). Oncogenes may be either cellular or viral in origin.

Pedigree. A "family tree", drawn with standard genetic symbols, showing inheritance patterns for specific phenotypic characters.

Penetrance. The proportion of individuals of a particular genotype that show the expected phenotype under a certain set of environmental conditions.

Petite. A yeast mutation producing small colonies and altered mitochondrial functions. In **cytoplasmic petites** (neutral and suppressive petites), the mutation is a deletion in mitochondrial DNA, in **segregational petites**, the mutation occurs in nuclear DNA.

A fruitfly with a mutant gene, which has resulted in legs developing on its head in place of antennae.

Phenocopies. When two genotypes produce the same phenotype due to different environments, then each one is called the **phenocopy** of the other, because they differ genotypically. For example, in *Drosophila melanogaster,* the normal (natural or wild) body colour is brown and a hereditary variant has yellow colour, when the larvae of wild type *Drosophila* with brown body colour are raised on food containing silver salts, they develop into yellow bodied flies. Thus, these flies are the phenocopies of yellow mutant, but would give rise to wild type brown flies in normal environment.

Phenotype. The appearance or discernible character of an individual, which is dependent on its genetic makeup usually expressed in words, *e.g.,* "tall", "dwarf", "wild type", "albino", "prolineless".

Philadelphia chromosome. A translocation between the long arms of chromosome 9 and 22, often found in the white blood cells of patients with chronic myeloid leukemia.

Plant breeding. The application of genetic analysis to development of plant lines better suited for human purposes.

Plasmagene. A self-replicating, cytoplasmically located gene.

Plasmid. A closed, circular DNA molecule of restricted size (*i.e.,* a few tens of thousands of nucleotide pairs), existing only in the cytoplasm (of bacteria); incapable of integration into the bacterial "chromosome". Now plasmid is used to include both episomes and plasmids.

Pleiotropic mutation. A mutation that has effects on several different characters.

Pleiotropy. The influencing of more than one trait by a single gene.

Point mutation. Change in a single base of the DNA molecule.

Polygenes. Two or more different pairs of alleles, with a presumed cumulative effect that governs such quantitative traits as size, pigmentation, intelligence, among others. Those contributing to the trait are termed **contributing** (effective) alleles; those appearing not to do so are referred to as **non-contributing** or **non-effective alleles**.

Position effect. A phenotypic effect dependent on a change in position on the chromosome of a gene or group of genes.

Progeny. Offspring individuals.

Pseudoalleles. Non-alleles so closely linked that they are often inherited as one gene, but shown to be separable by crossover studies.

Pulse-chase experiment. An experiment in which cells are grown in radioactive medium for a brief period (the **pulse**) and then transferred to non-radioactive medium for a longer period (the **chase**).

Punnett square. A "checkerboard" grid designed to determine all possible genotypes produced by a given cross. Genotypes of the gametes of one sex are entered across the top, those of the other down one side. Zygote genotypes produced by each possible mating are then entered in the appropriate squares of the grid.

Pure line (pure breeding line). A strain of individuals homozygous for all genes being considered.

Recessive allele. An allele whose phenotypic effect is not expressed in a heterozygote.

Recessive phenotype. The phenotype of a homozygote for the recessive alleles; the parental phenotype that is not expressed in a heterozygote.

Reciprocal cross. A second cross of the same genotypes in which the sexes of the parental generation are reversed. The cross *AA* X *aa* is the reciprocal of the cross *aa* X *AA*.

Recombinant DNA. (1) A novel DNA sequence formed by the combination of two non-homologous DNA molecules. (2) A DNA molecule (in practice generally a bacterial plasmid) which has been enzymatically cut and DNA of another individual of the same or a different species inserted in the space so produced, then reannealed to the closed, circular form.

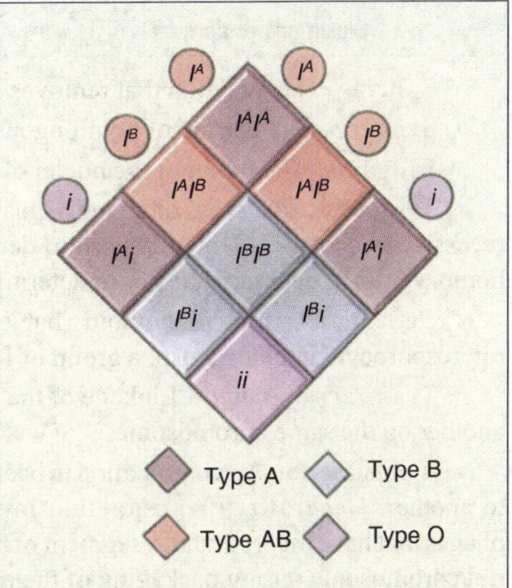

A punnett square for inheritance of ABO blood group in humans.

Recombination. The new association of genes in a recombinant individual; this association arises from independent assortment of unlinked genes, from crossing over between linked genes, or from intracistronic crossing-over.

Retrovirus. RNA viruses that replicate with reverse transcriptase. Since the enzyme produces a DNA copy of the viral RNA that is the reverse of transcription, the name *retro,* suggesting backward transcription, is used.

Reverse transcriptase. An enzyme, carried within the coat of the retroviruses, that makes double-stranded DNA from a single-stranded RNA template.

Segregation. The separation of allelomorphic (allelic) genes into different gametes at meiosis; or, separation in its offspring of traits which are combined in a hybrid.

Sex chromosomes. Heteromorphic chromosomes that do not occur in identical pairs in both sexes in diploid organisms; in humans and fruit flies these are designated as X and Y chromosomes, respectively; in fowl, as the Z and W chromosomes.

Sex linkage. The location of a gene on a sex chromosome.

Sexduction. Incorporation of bacterial chromosomal genes in the fertility plasmid, with subsequent transfer to a recipient cell in conjugation.

Sex-influenced trait. One in which dominance of an allele depends on sex of the bearer, *e.g.,* pattern baldness in humans in dominant in males, recessive in females.

Sex-limited trait. One expressed in only one of the sexes; *e.g.,* cock feathering in fowl is limited to normal males.

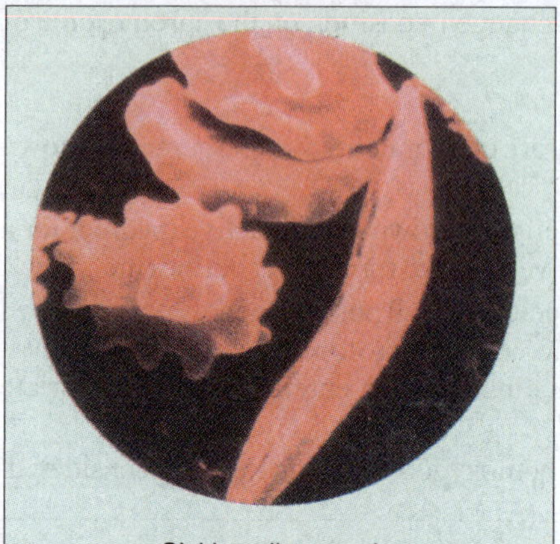
Sickle-cell anaemia.

Siblings or Sibs. The individuals having same maternal and paternal parents, *e.g.*, brother sister relationship.

Sickle-cell anaemia. Anaemia in humans inherited as an autosomal recessive and due to a single amino acid substitution in the beta-haemoglobin chain.

Soma (adj., **somatic**). The body, cells of which in mammals and flowering plants normally have two sets of chromosomes, one derived from each parent.

Somatic cell. A cell that is not destined to become a gamete; a "body cell", "whose genes will not be passed on to future generations.

Somatic-cell genetics. Asexual genetics, involving study of somatic mutation, assortment, and crossing-over, and of cell fusion.

Splicing. The reaction that removes introns and joins together exons in RNA.

Synapsis. The pairing of homologous chromosomes that occur in prophase-I of meiosis.

Syngamy. The union of the nuclei of sex cells (gametes) in reproduction.

Test cross. The cross of an individual (generally of dominant phenotype) with one having the recessive phenotype. Generally used to determine whether an individual of dominant phenotype is homozygous or heterozygous, or to determine the degree of linkage.

Tetrad. The four monoploid (haploid) cells arising from meiosis of a megasporocyte or microsporocyte in plants; also, a group of four associated chromatids during synapsis.

Trans arrangement. Linkage of the dominant alleles of one pair of gene and the recessive of another on the same chromosome.

Transduction. Recombination in bacteria whereby DNA is transferred by a phage from one cell to another. **Generalized transduction** involves phages that have incorporated a segment of bacterial chromosome during packaging of the phage; **specialized transduction** involves temperate phages that are always inserted into the bacterial chromosome at a site specific for that phage.

Trihybrid. The individual which is heterozygous for three pair of alleles is known as trihybrid.

Trisomic An individual with one extra chromosome of a set ($2n + 1$).

Turner's syndrome. A series of abnormalities in humans due to monosomy for the X chromosome (XO). Individuals are phenotypically female, but are sterile.

Uniparental inheritance. The transmission of certain phenotypes from one parental type to all the progeny; such inheritance is generally produced by organelle genes.

Variation. The differences among parents and their offspring or among individuals in a population.

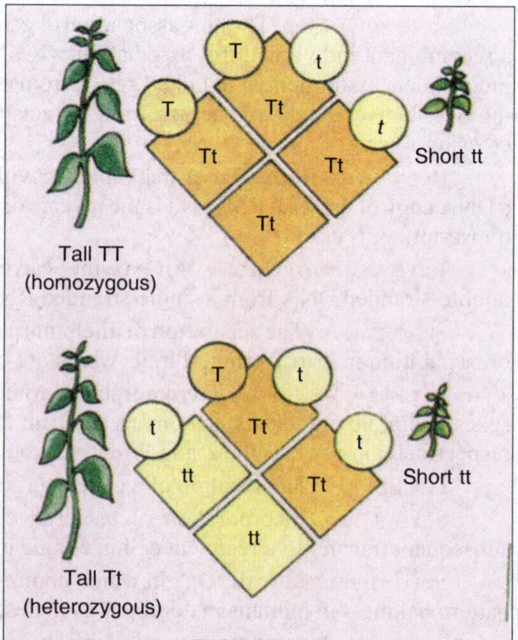

TEST-CROSS.

Vector. In cloning, the plasmid or phage chromosome used to carry the cloned DNA segment.

Wild type. The genotype or phenotype that is found in nature or in the standard laboratory stock for a given organism.

X linkage. The presence of a gene on the X chromosome but not on the Y.

X- and -Y linkage. The presence of a gene on both the X and Y chromosomes.

Y linkage. The presence of a gene on the Y chromosome but not on the X.

Zygote. The cell formed by the fusion of an egg and a sperm; the unique diploid cell that will divide mitotically to create a differentiated diploid organism.

SYMBOLS IN GENETICS

An organism with diploid (2n) cells has paired chromosomes and out of two contrasting characters each chromosome contains only one gene or trait for a single character. According to the **classical method of symbolization**, the dominant character is expressed in capital letter as the tall character is represented by 'T' and the recessive character is represented by 't'. Now homozygous tall plant will contain the genotype 'TT' and likewise a homozygous dwarf plant will have the genotype "tt". Because the sperm or ova contains only one chromosome of a homologous pair, therefore, it contains only single gene, *e.g.,* T or t. A heterozygous has both dominant and recessive characters, therefore, its genotype can be expressed by 'Tt' letters.

Recently, the classical method of symbolization has been modified a little. In the **modified classical method of symbolization**, like classical method, capital and small letters are commonly used to designate dominant and recessive alleles, but in contrast to that, the genetic symbol corresponds to the first letter in the name of the abnormal, recessive or mutant trait. For example, in man the recessive trait of albino is represented by letter 'a' while the normal trait is represented by capital letter 'A'.

In modern genetical literature one more method of symbolization called "**wild type symbolism**" is widely used. When out of two phenotypes, one phenotype is of much more common occurrence in the population than its alternative phenotype, the former is usually referred to as **wild type**. The phenotype which is rarely observed in the population is called the **mutant type**. In this system, the symbol plus (+) is used to indicate the normal allele for wild type. The base letter for the gene usually is borrowed from the name of the mutant or abnormal trait. If the mutant gene is recessive the symbol would be a lower case letter corresponding to the initial letter in the name of the trait. Its wild type or normal dominant would have the same lower case letter but with a^+ as a superscript. For example, black body colour in *Drosophila* is governed by a recessive gene b and wild type (gray body) by its dominant allele b^+.

REVISION QUESTIONS

1. Define the following terms : genotype; phenotype; hybrid; gene; allele; mutation; linkage and eugenics.
2. Describe the method of wild type symbolism.

CHAPTER

3

Mendel and His Work

Mendel's garden, as seen in the 1980's

Johann Mendel was the pioneer of classical geneticists. He was born in July 22,1822 in Heinzendorf in Austrian Silesia, where his father, Anton Mendel was the owner of a small farm. He graduated from the Gymnasium in 1840. In his youth, he led a disastrous, poor, difficult and sad life. In October 1843, Mendel was admitted to the Augustinian monastery at Brunn in Moravia (a Czechoslovakian town) where he took the name Gregor as novice and besides performing his other duties, he took keen interest in natural sciences. In the year 1846, Johann Gregor Mendel attended courses of agriculture, pomiculture and viniculture at the Philosophical Academy in Brunn (now called Brno in Czechoslovakia). After finishing his theological studies in 1848, he was appointed as a substitute teacher in the Imperial Royan Gymnasium in Znaim in the year 1849. From 1851 to 1853, he studied mathematics and natural science in the university of Vienna. In 1853, he took the membership of Zoological-Botanical Society of Vienna. On April 5, 1854, he wrote a letter to Vienna Zoological-Botanical Society about damage by pea-weevil, *Bruchus pisi* and, thus, showed his interest in peas. In May 1854, he was appointed to the post of supply teacher of physics and natural science in a higher secondary school of Brunn. He continued to hold this post until 1868 when he was elected abbot. In the spring of 1856, he began experimental crossing of pea varieties.

In 1862, Mendel became a founding member of the Brunn Natural Science Society. On February 8, 1865, he delivered his first lecture on pea experiments to Brunn Natural Science

Gregor Johann Mendel (1822-1884).

Society. In 1866 his paper "*Experiments on plant hybridization*" published in volume 4 of the proceedings of the Natural Science Society. In the same year, he began experiments with other plant species. In this paper, Mendel proposed some basic genetic principles. But unfortunately his remarkable piece of work remained unattended and unappreciated up to 1900. There were several reasons for the sad neglect of Mendel's work. These include (i) biologist's preoccupation with speculation concerning Darwin's theory of evolution "origin of species" which appeared in 1859; (ii) obscurity of the journal in which Mendel published his results; and (iii) unaccoustomedness of professional biologists of ninteenth century to think in the statistical manner which Mendel introduced in the study of hybridization. Further, Mendel himself made his work known only to some of the most famous hybridizers of his time such as **Carl von Naegeli** and **Anton Kerner von Marilaum**, but not to the younger generation of scientists who were perhaps less prejudiced against new ideas. **Naegeli** and **Kerner** knew of the Mendel's paper but they did not review it or discuss it, perhaps because they considered him an outsider and amateur. Moreover, Naegeli's negative approach in discouraging Mendel to pursue the right path becomes apparent in his insistent suggestions to Mendel to test his genetical principles on 26 species of hawk-weed (*Hieracium*), for which he supplied seed and plants. When Mendel crossed many varieties of hawk-weeds, the progeny did not show evidence of segregation of genes from parents, but rather were all like their mothers. Later on, these plants were known to be apomictic (*i.e.,* parthenogenetic) or capable of reproduction without fertilization. Mendel wasted his valuable six years on the hybridization experiments on this plant species, ruined his eye sight, but even then failed to confirm or even to test his theory. The results of these ill-fated experiments were published in 1869, in the Proceedings of the Natural Science Society, Brunn. His defeat with *Hieracium* led Mendel to withhold further publication and eventually to cease scientific work. Further, this neglect of his work and other economical and physical hazards made him greatly disappointed and bitter. His health progressively degenerated and he became far too obese and began to suffer from dropsy due to heart and kidney failure. The pioneer of classical genetics, thus, died unknowingly, amidst the feelings of despair on 6th January 1884 and buried in Brunn Central Cemetery (See **Dunn**, 1965; **Serra** 1965).

REDISCOVERY OF MENDEL'S WORK

Mendel's research paper remained dormant and unnoticed by the scientific world until 1900. During these intervening thirty four years many developments occurred in biology which prepared the way for the rediscovery of Mendel's work. For instance, during this period **Haeckel** (1866) recognised the active role of nucleus in heredity; **Weismann**, **Hertwig**, **Strasburger** and **Kolliker** suspected active participation of chromosomes in heredity transmission. **Roux** (1883) suggested that the chromosomes must contain qualitatively different hereditary determiners arranged in linear orders. **Weismann** supported the idea of **Roux** by propounding his germplasm theory. Further, some workers such as **Darwin** (1868) in England, **Vilmorin** (1879) in France, **Rumpau** (1891) in Germany and **Bohlin** (1897) in Sweden carried out hybridization experiments, very much like Mendel, on different plants and observed the phenomenon of dominance, but

Charles Darwin.

they failed to provide any conclusive explanation to their findings.

It was in the beginning of 20th century that three botanists, namely **Hugo de Vries,** working on *Oenothera*; **Carl Correns** working on *Xenia,* peas and maize and **Erich von Tschermak** working on various flowering plants, independently drawn the conclusions like Mendel. Later these botanists came across the research paper of Mendel and rediscovered it in 1900. Mendel's original paper was republished in Flora, 89, 364 (1901). **Bateson** confirmed Mendel's work by a series of hybridization experiments.

MENDEL'S SELECTION OF THE EXPERIMENTAL PLANT

For his hybridization experiments Mendel had certain consideration in his mind about the choice of a suitable material. Mendel's considerations about the material were as follows :

A pod of the garden pea which was selected by Mendel for his experiment.

1. Variation. The organisms which are to be chosen for the genetical experiments, should have a number of detectable differences and at a time only single detectable character should be considered.

2. Reproduction. The chosen organisms should be sexually reproducing (*i.e.,* by fusion of male and female sex cells) because only then the offsprings will be able to receive different characters from both the male and female parents.

3. Controlled mating. The chosen organisms should be able to mate in controlled or well-planned conditions. Because in genetical experiments sometimes we have to rear genetically pure parents by methods of controlled mating. One should maintain careful records of the offsprings of many generation.

4. Short life cycle. The chosen organisms should have very short life cycles.

5. Large number of offsprings. The organisms which have been chosen for the genetical experiments should produce large number of offsprings after each successive mating because it will help in deducing the correct conclusions.

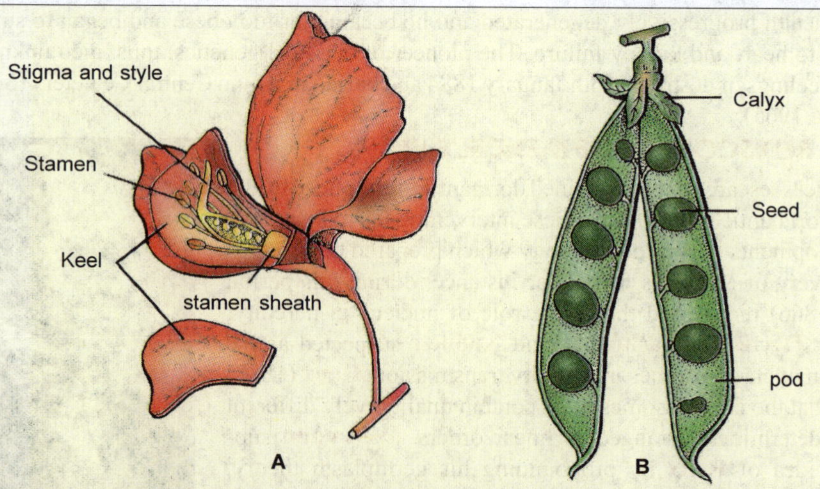

Fig. 3.1. A–A self-fertilizing flower of *Pisum sativum.* A–Portion of the keel which encloses the reproductive organs has been cut open to show the stamen and stigma; B–The mature fruit (pod) containing seeds that develop from the fertilized pea flower.

6. Convenience in handling. The experimental species should be of a type that can be raised and maintained conveniently and inexpensively in the laboratory. For instance, the elephant will prove entirely useless material for genetical experiments than the *Drosophila,* pea plants, tomato, rats, guinea pigs, etc., which have been generally used and are still used in hybridization experiments. *Arabidopsis thaliana* is a small, economically unimportant member of mustard family. In last four decades, it has became a most favorite research material for the plant geneticists and molecular biologists. *A. thaliana* is often nicknamed as *Drosophila melanogaster* (an insect) and *Caenorhabditis elegans* (a nematode) of the plant kingdom (**Gardner** *et al.,* 1991). This plant has the following advantages : 1. small size; 2. short generation time (about 5 weeks); 3. high seed production (up to 40,000 seeds per plant); 4. very small genome ($7X10^7$ nucleotide pairs); 5. very little interspersed repetitive DNA; and 6. natural self pollination.

MENDEL'S MATERIAL AND CROSSING TECHNIQUE

Mendel found edible pea (*Pisum sativum*) a best material for his hybridization experiments. The pea plant has various contrasting characters among its different varieties such as stem may be *tall* or *dwarf,* cotyledons may be *green* or *yellow*; seeds may be *round* or *wrinkled,* seed coat may be *coloured* or *colourless*; the unripe pods may be *green* or *yellow*; the ripe pods may be *inflated* or *constricted* between the seeds, flowers may have *axial* or *terminal* positions and the colours of flowers may be *red* or *white.* Besides these contrasting characters, the pea plant is a very satisfactory material for the hybridization experiments due to its flower structure. The flowers of pea plants are so constructed that the pollens of a flower normally fall on the stigma of the same flower and, thus, affects self-pollination or self-fertilization. For the required cross-pollination, the anthers have to be removed from the flower in bud stage (*i.e.,* before their maturity). This operation of removal of anthers is called **emasculation**. The stigma is protected against any foreign pollen with the help of its covering by a bag. The pollen, then at the dehiscence stage is brought from the plant to be used as a male parent and by the help of a brush is dusted on the feathery stigma of the **emasculated** flower. At the time of such **cross pollination**, the pollen should be mature and stigma should be receptive.

For each of seven pairs of characters listed in Table 3-1, plants with one alternative trait were used as female, and those with the other alternative as male. Reciprocal crosses were also made, *i.e.,* each of the crosses was made in two ways, depending on which phenotype is used as male or female. For example the following two crosses are reciprocal crosses :

phenotype A (*e.g.,* Tall) ♀ × phenotype B (*e.g.,* Dwarf) ♂

phenotype B (*e.g.,* Dwarf) ♀ × phenotype A (*e.g.,* Tall) ♂

The population obtained as a result of crossing plants showing contrasting characters is called **F_1 generation.** The progeny of F_1 plants was then obtained by self-fertilization and it forms the **F_2 generation**. Similarly, F_3, F_4, etc., generations can also be obtained.

Further, for getting the exact results in the breeding experiments, it was necessary for Mendel to rear genetically pure variety of pea plants for a single character. Mendel adopted self-fertilization technique for it. For instance, to get pure character for tallness, he self fertilized a tall pea plant for many generations till the resulted offsprings always produced only tall plants. Likewise, he got genetically pure variety for dwarf pea plants. Mendel cross pollinated these two varieties of pea plants which were differing in a pair of contrasting characters, *viz.,* tallness and dwarfness of the stem. When he made observations on the offsprings of first generation he found only tall plants. He allowed the self pollination in the offsprings of first generation and made further observations on the offsprings of second generation. He was astonished to note both tall and dwarf offsprings in the second generation. This showed him that the character of dwarfness disappeared in first generation but again reappeared in second generation. Further, the tall and dwarf plants of second generation were always in the ratio of 3:1 (3 tall : 1 dwarf). He self-pollinated the dwarf offsprings of second generation and found only dwarf plants in third generation. But when he self-pollinated the tall plants of second generation then he found that one-third (1/3) tall plants yield only tall plants in third generation, while rest two-third (2/3) tall plants yield tall and dwarf plants in the ratio of 3:1.

On the basis of the results of his experiments Mendel recognized the **phenomenon of dominance** and formulated following two laws :

1. Law of segregation ;
2. Law of independent assortment.

Actually Mendel himself did not postulated any genetical principle or laws, he simply gave conclusive theoretical and statistical explanations for his hybridization experiments in his research paper. However, it was **Correns**, the discoverer of Mendel's work, who thought that Mendel's discovery could be represented by these fundamental laws of heredity.

A mendelian trait wrinkled and round seeds.

PHENOMENON OF DOMINANCE

The cross between the pea plants differing in single pair of contrasting characters is known as **monohybrid cross**. As we have already noticed that when Mendel made a monohybrid cross (Fig. 3.2) between tall and dwarf pea plants then only tall pea plants appeared in the first filial generation (F_1). But when the F_1 progeny were allowed to be self-fertilized, both tall and dwarf characters appeared in the second filial generation or F_2. This shows that in F_1 hybrid the character of tallness dominates or conceals the character of dwarfness and so the character of dwarfness could not express itself in F_1 generation. The character which expresses itself (*i.e.,* tall) in F_1 generation is called by Mendel as **dominant character**, while the character which remained unexpressed or latent had been called **recessive**. According to these results Mendel described the phenomenon of dominance in following way : *in crossing between pure (homozygous) organisms for contrasting characters of a pair, only one character of the pair appears in the first filial generation.*

In pea plant Mendel found following characters to be dominant or recessive in various pairs of contrasting characters (see Table 3-1).

Certain Examples of Phenomenon of Dominance

After Mendel several geneticists tested the validity of the phenomenon of dominance on several plants and animals. They found its wide application in various plants and animals. A few important examples are described as follows :

1. Phenomenon of Dominance in Plants

Besides pea plant the phenomenon of dominance has also been observed in the following plants (see Table 3-2).

2. Application of Phenomenon of Dominance in Animals

A. The phenomenon of dominance is also applicable well to the animals. For instance, when homozygous black guinea pig is crossed with a homozygous white guinea pig (Fig. 3.3) then all hybrids of first filial generation (F_1) are found to be black. The black hybrids of F_1 when mated among themselves they produced black and white offsprings in 3:1 ratio. This shows that black coat colour dominates over white coat colour.

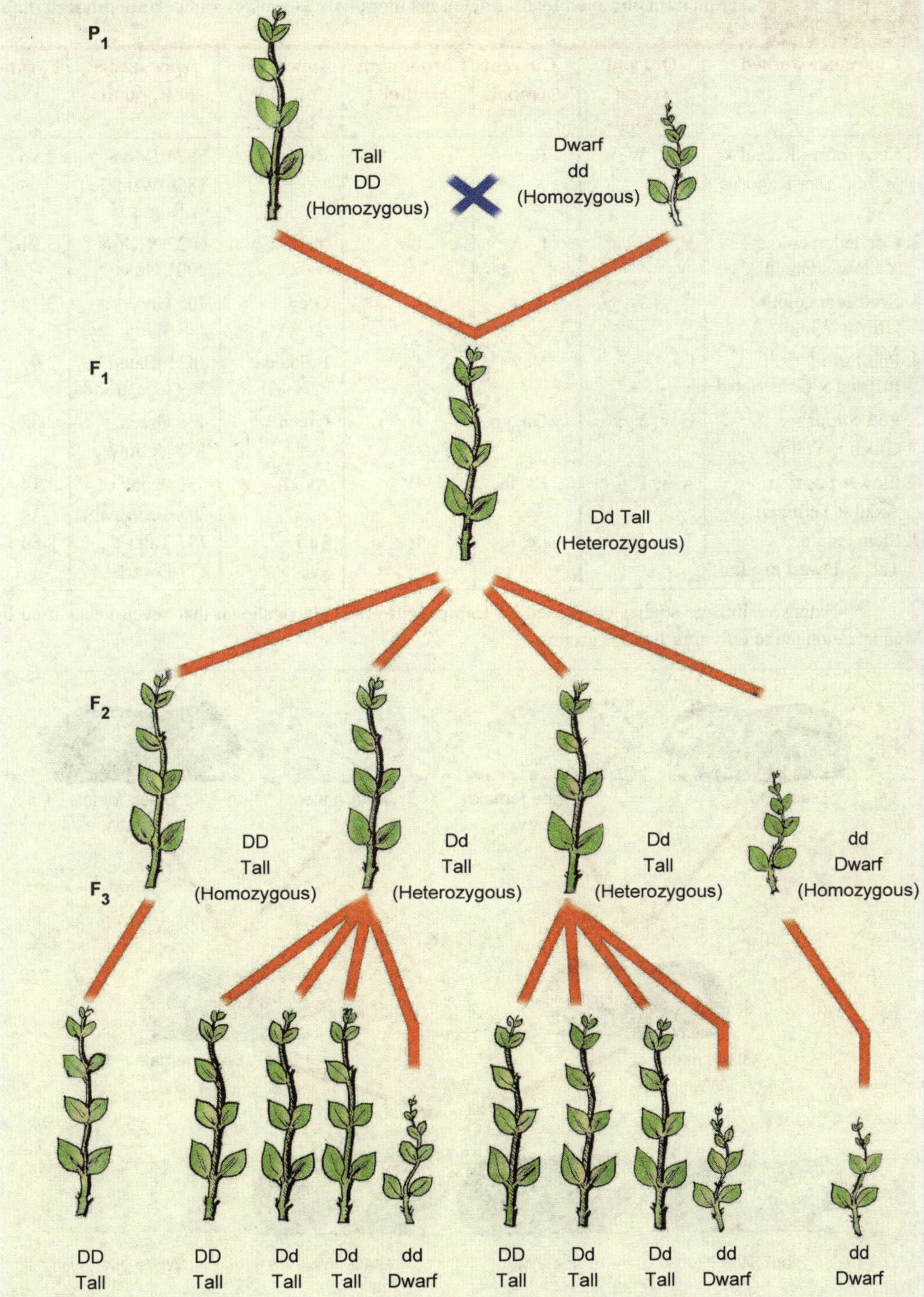

Fig. 3.2. A cross between a tall (TT) and a dwarf (tt) pea plant and their offsprings of F_1 and F_2 generations.

Table 3.1. Actual data obtained by Mendel in his monohybrid crosses and certain modern data.

Character studied	Old and recent symbols	Current symbols	Chromosome location*	Appearance of all F_1 hybrids	Appearance of F_2 plants	F_2 ratio
1. Seed form : Round × Wrinkled or Rugosus	R, r ; W, w	R, r	7	Round	5474 Round 1850 Wrinkled or Rugosus	2.96 : 1
2. Cotyledon colour : Yellow × Green	Y, y ; G, g	I, i	1	Yellow	6022 Yellow 2001 Green	3.01 : 1
3. Seed coat colour : Grey × White	G, g; W, w	A, a	1	Grey	705 Grey 224 White	3.15 : 1
4. Pod form : Inflated × Constricted	I, i; C, c	V, v	4	Inflated	882 Inflated 299 Constricted	2.95 : 1
5. Pod colour : Green × Yellow	G, g; Y, y	Gp, gp	5	Green	428 Green 152 Yellow	2.82 : 1
6. Flower position : Axial × Terminal	A, a; T, t	Fa, fa	4	Axial	651 Axial 207 Terminal	3.14 : 1
7. Stem length : Tall × Dwarf or Short	T, t; D, d	Le, le	4	Tall	787 Tall 277 Dwarf	2.84 : 1

* Extensive linkage studies conducted by **Lamprecht** (1961) have shown that seven genes used by Mendel belonged to only four linkage groups.

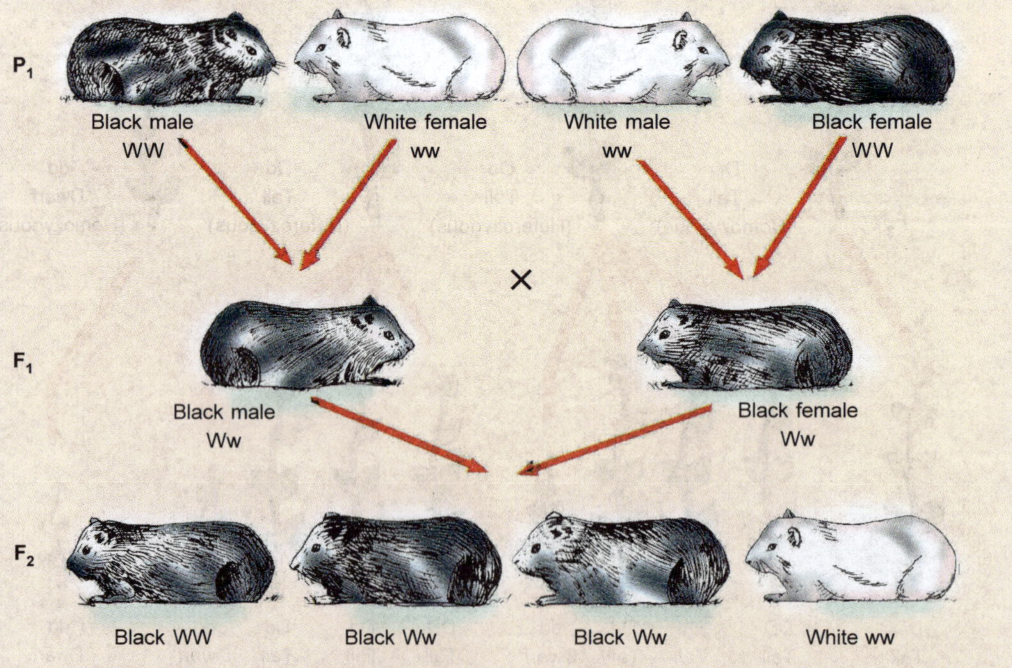

Fig. 3.3. A cross between black and white guinea pigs.

Table 3.2. The dominant and recessive characters in plants.

Name of the plant	Dominant	Recessive
1. Nettle	Serrated leaves	Smooth margined leaves
2. Sunflower	Branched habit	Unbranched habit
3. Cotton	Coloured lint	White lint
4. Maize	Round starchy kernel	Wrinkled sugary kernel
5. Snapdragon	Red flower	Non-red flower
6. Barley	Beardlessness	Beardness
7. Wheat	Susceptibility to rust	Immunity to rust
8. Tomato	Two celled fruit	Many celled fruit

B. Certain other examples of phenomenon of dominance in animals. The dominant and recessive traits or characters of some animals can be tabulated in following table :

Table 3.3. The dominant and recessive characters in animals.

Name of animal	Body character	Dominant	Recessive
1. Cat	Skin colour	Tabby	Black or blue
	Length of hair	Short hairs	Long hairs (Angora)
2. Dog	Skin colour	Grey	Black
	Tail	Stumpy	Normal tail
3. Cattle	Colour of face	White	Coloured
	Horn	Polled or Hornless	Horned
4. Horse	Skin colour	Black	Red
	Movement	Trotting	Pacing
5. Sheep	Hair or wool or fleece	White	Black
6. Swine or Pig	Skin colour	Black	Red
	Hoof	Uncleft	Normal
7. Salamander	Body colour	Dark	Light
8. Fruit fly (*Drosophila*)	Eye colour	Red	White
	Wings	Flat and yellow	Curled and white
	Body colour	Grey	Black
9. Land snail	Shape of shell	Unbanded shell	Banded shell

C. Dominant and recessive characters in man. In man various types of dominant and recessive characters have been reported. According to their nature these can be divided into normal, abnormal, and sex-linked characters.

(i) Normal characters. These characters always behave as dominant and recessive. These are tabulated as follows :

Table 3.4. Normal human traits.

Body part	Character	Dominant	Recessive
1. Hair	Form	Curly	Straight
	Colour	Dark	Light
2. Skin	Colour	Dark	Light
	Pigment	Normal (with melanin)	None (albinism, without melanin pigment)
3. Eyes or Iris	Colour	Brown	Blue

(ii) Abnormal characters. The abnormal characters occur unusually and can be tabulated as follows :

Table 3.5. Abnormal human traits.

Body part	Dominant	Recessive
1. Hair	Absent	Present
2. Skin epidermis	Thickened	Normal
3. Fingers	Short	Normal
	Webbed	Normal
	Extra digits	Normal
4. Ear	Normal hearing	Deaf mutism
5. Ear lobe	Free ear lobe	Attached ear lobe
6. Eyes	Opaque lens	Normal
	Glaucoma	Normal
7. Teeth	Absent	Present
8. Tongue	Ability to role	Inability to role
9. Taste of phenylthiocarbamide (PTC)	Taste bitter to PTC	Tasteless to PTC

(iii) Sex-linked characters. Certain dominant and recessive characters depend on the sex of individual and are sex-linked. These are as follows :

1. Colour vision : Normal (Dominant) and Colour blind (Recessive).
2. Blood clotting : Normal (Dominant) and Haemophilia (Recessive).

Mechanism of Dominance

Mendel carefully studied the results of his experiments and it became evident to him that there is a clear cut difference in between the actual visible character and that something which caused its production. Because the character cannot be present as such in sex cells (sperm and ova) which form the only link between a new individual and the parents. Therefore, there must be something which represents the characters and is responsible for their production. This something of Mendel is now called the **factor** or **gene**. The gene, thus, can be considered as the unit of inheritance that is transmitted in a gamete and determines or controls the development of a character by interaction with the other genes, the cytoplasm and the environment.

Blue eyes are considered a recessive trait in humans.

The cytological investigations have now established that the genes are the units of deoxyribonucleic acid (DNA) which along with ribonucleic acid (RNA) and nucleoproteins constitute the thread-like stainable structures the **chromosomes**. The chromosomes are specific in number, shape and size to a particular species.

A diploid cell has two sets of chromosomes which come from two different parents (male and female) via gametes (sperm and ova). The chromosomes of similar size and nature often form pairs during meiotic cell division and such identical chromosomes are known as **homologous chromosomes**. Each character of a pair of contrasting characters is represented by an **allele**. (When a gene for a unit character contains two or more alternative forms, they are called **allelomorphs** or **alleles**. All alleles of a gene are produced due to mutation of a wild gene (or normal gene). Thus, homozygous tall pea plant has two identical alleles TT on both gene loci of the homologous chromosomes; likewise, homozygous dwarf pea plant contains tt alleles.

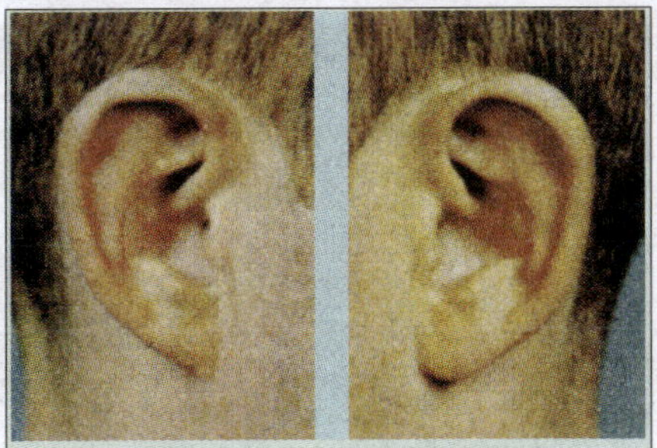

Attached and free earlobes.

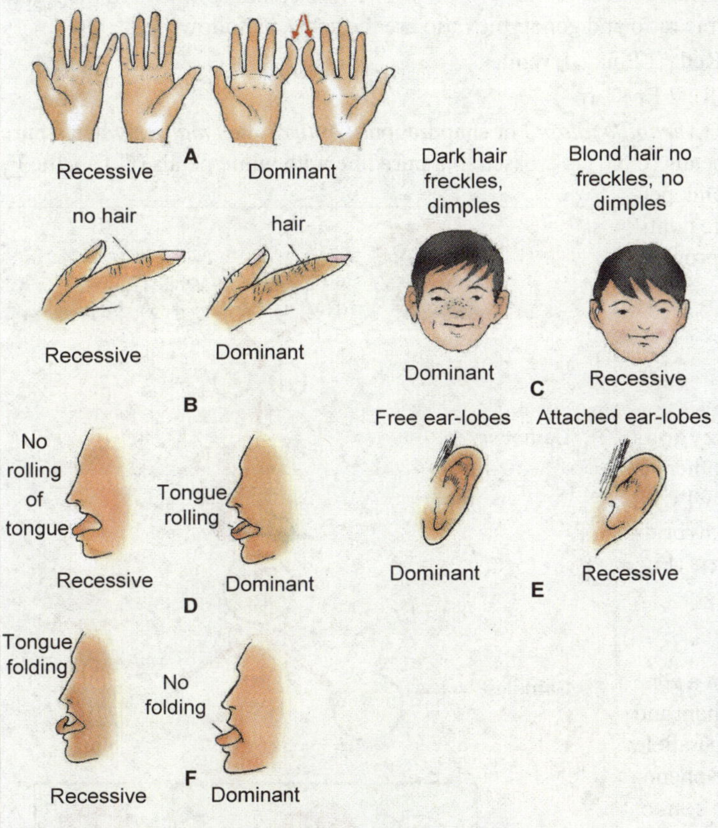

Fig. 3.4. Certain abnormal recessive and dominant characters of man.

During the gametogenesis, the homologous chromosomes with TT or tt genes are separated and each chromosome with T or t gene is passed to the gamete. The gametes of both parents unite during the process of fertilization and produce a new individual containing both tall (T) and dwarf (t) characters. This new individual of first generation (F_1) contains two different genes (*i.e.*, alleles) of a contrasting pair of characters, therefore, it is known as **heterozygote** or **hybrid**. Because the hybrids of F_1 have tall stems so the character of tallness (T) is considered as **dominant** and because the character for dwarfness could not express itself in F_1 generation, therefore, it is considered as **recessive**.

Variation in Dominance Relation

Mendel reported full dominance and recessiveness for all the seven gene pairs (or allele pairs) he studied (see Table 3-1). He may have been selective in his choice of pea characters to study, because variations on the basic theme appear quite often in analysis. For instance, in some cases the phenotypes of heterozygotes are found to be different from either of the homozygotes :

1. Incomplete Dominance

Sometimes in a heterozygote dominant allele does not completely mask the phenotypic

Incomplete dominance shown in the flower colour of snapdragons.

expression of the recessive allele and there occurs an intermediate phenotype in the heterozygote. This is called **incomplete dominance**.

Examples. 1. When a red flowered pea plant (RR) is crossed with white flowered pea plant (rr) then the F_1 hybrid pea plants are found to have pink flowers. It shows that gene for red colour could not completely dominate the gene for **white** colour as shown in the figure 3.5. In such a case, F_2 phenotypic ratio and genotypic ratio are the same, as follows :

F_2 phenotypic ratio = 1 Red : 2 Pink : 1 White

F_2 genotypic ratio = 1 RR : 2 Rr : 1 rr

2. In four-o'clock plants (*Mirabilis jalapa)* or snapdragons (*Antirrhinum majus*), when a pure line or homozygous with red petals (C_1C_1) is crossed to a pure line with white petals (C_2C_2), the F_1 progeny has no red petals but pink petals (C_1C_2). It shows incomplete dominance. If an F_2 is produced, its progeny exhibits the following results :

¼ red petals – $1C_1C_1$

½ pink petals – $2C_1C_2$

¼ white petals – $1C_2C_2$

3. When a homozygous Andalusian fowl with black feathers is crossed with a homozygous fowl with splashed white feathers the F_1 hybrids are found to contain blue feathers (Fig. 3.6).

2. Codominance

Sometimes both alleles of a gene in a heterozygote lack the dominant and recessive relationship, *i.e.,* each allele is capable of some degree of phenotypic expression. In a sense, codominance is no dominance at all, the heterozygote showing the phenotypes of both homozygotes. Hence, heterozygote genotype gives rise to a phenotype distinctly different from either of the homozygous genotypes.

Symbolism for codominant alleles. For codominant alleles, all upper case base symbols with different superscripts are used (see **Stansfield**, 1986; **Suzuki** *et al.,* 1986). The upper case

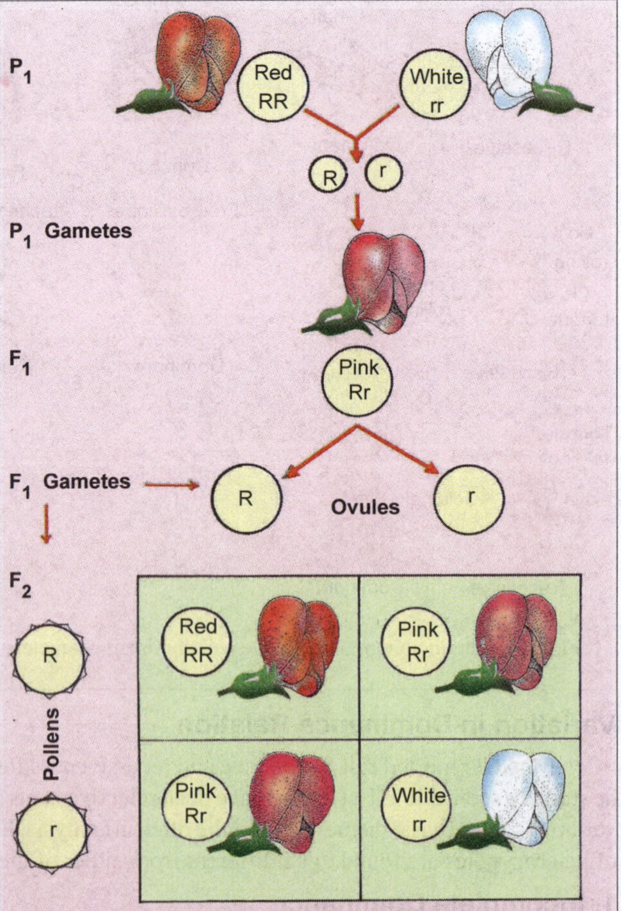

Fig. 3.5. A cross between a red flowered and a white flowered pea plant showing incomplete dominance.

letters indicate that each allele can express itself to some degree even when in the presence of its alternative allele (heterozygous).

Examples. 1. The coat colour of the Shorthorn breed of cattle represent a classical example of codominance. when a cattle of red coat (C^RC^R) is crossed with the cattle of white coat (C^wC^w), the F_1 heterozygote or hybrid is found to possess roan coat (C^RC^W) (Fig. 3.7). In roan coat the red and white hairs occur in definite patches but no hair has intermediate colour of red and white.

2. The alleles governing the **M-N blood group** system in humans are codominants and may be represented by the symbols L^M and L^N, base letter L being assigned in honour of its discoverers (**Landsteiner** and **Levine**). Here, three blood groups are possible–M, N and MN–and these are determined by the genotypes L^ML^M, L^NL^N, and $L^M L^N$, respectively. Blood groups actually represent the presence of an immunological antigen on the surface of red blood cells. People of L^ML^N genotype have both antigens. In the following summary chart, agglutination is represented by + and non-agglutination by – sign :

Genotype	Reaction with antisera:		Blood group
	Anti–M	Anti–N	(Phenotype)
1. L^ML^M	+	–	M
2. L^ML^N	+	+	MN
3. L^NL^N	–	+	N

3. The inheritance pattern of human disease sickle-cell anaemia shows, besides many other genetic phenomena, the incomplete dominance (at cellular or cell shape level) and codominance (at molecular, *i.e.*, haemoglobin level). The gene pair concerned Hb^A (for haemoglobin A) and Hb^S (for haemoglobin S) affects the oxygen transport molecule haemoglobin—the major constituent of red blood cells (erythrocytes). The three genotypes have different phenotypes, as follows :

$Hb^A Hb^A$: Normal. Red blood cells never sickled; they contain one type of haemoglobin, *i.e.*, haemoglobin A.

Hb^SHb^S : Severe, often fatal anaemia. Red blood cells sickled-shaped; contain one type of haemoglobin, *i.e.*, haemoglobin S.

Hb^AHb^S : No anaemia. Red blood cells sickle-shaped only under abnormally low oxygen concentration;

Fig. 3.6. A monohybrid cross between black and splashed white Andalusian fowl showing the incomplete dominance.

contain both types of haemoglobins, *i.e.,* haemoglobin A and haemoglobin S.

Thus, in regard to anaemia the Hb^A allele is dominant. In regard to blood cell shape there is incomplete dominance. And lastly, in regard to haemoglobin, there is codominance. The Hb^S allele in

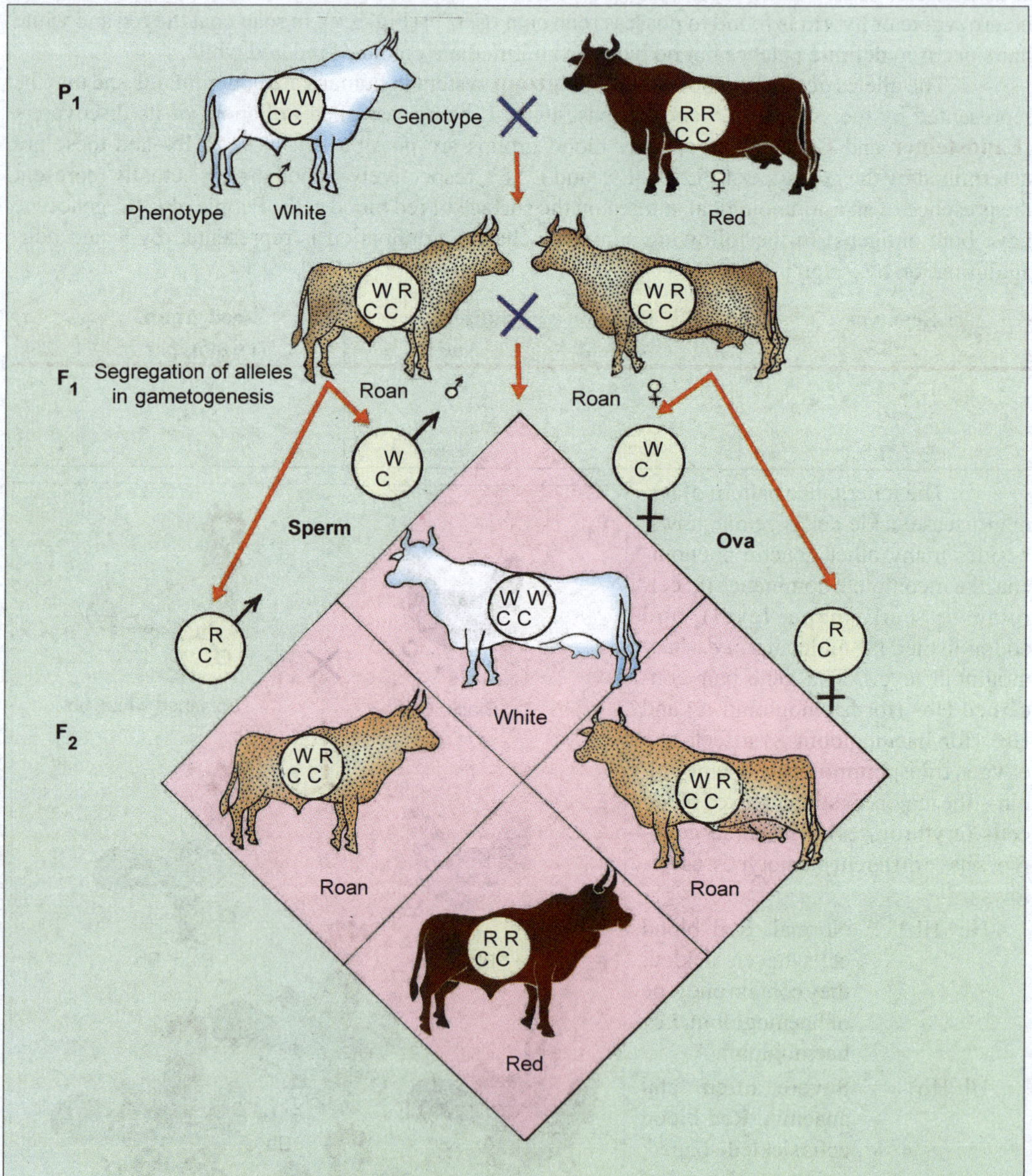

Fig. 3.7. A monohybrid cross in a white-coated and a red-coated cattle showing codominance.

homozygous condition (Hb^SHb^S) acts as a **lethal gene**, *i.e.,* it causes the death of its bearer; the homozygotes dies of fatal anaemia before they attain sexual maturity. A marriage between two carriers (*i.e.,* heterozygotes possessing a deleterious recessive allele hidden from phenotypic expression by the dominant normal allele) results in carriers and disease free children in the ratio of 3 : 1, that changes, later on, into the ratio of 2 : 1 due to the death of homozygotes :

Sickle cell carrier (Hb^A/Hb^S) × Sickle cell carrier (Hb^A/Hb^S)

Hb^A/Hb^A	Hb^A/Hb^S	Hb^A/Hb^S	Hb^S/Hb^S
Disease free (Normal) (Homozygous dominant)	Carrier	Carrier	Sickle-shaped erythrocytes; usually dies due to anaemia (Homozygous recessive)
	Normal (Heterozygotes with sickle cell trait)		

LAW OF SEGREGATION

Mendel's first law–the law of segregation is also known as law of purity of gametes. The law states that *the hybrids or heterozygotes of F_1 generation have two contrasting characters or allelomorphs of dominant and recessive nature. These alleles though remain together for long time but do not contaminate or mix with each other and separate or segregate at the time of gametogenesis, so that each gamete receives only one allele of a character either dominant or recessive.*

For the proper understanding of the Mendel's law of segregation it will be helpful to study one of the Mendel's monohybrid cross (Fig. 3.5). Mendel crossed a homozygous red flowered pea plant with a homozygous white flowered pea plant. The F_1 heterozygotes or hybrids were found to be pink or purple flowered, thus, showing the incomplete dominance of red colour over white colour. When the F_1 hybrids were allowed to be self-fertilized, they produce both coloured (red or purple) and white flowered pea plants in F_2 generation in the ratio of three and one (3 : 1). The reappearance of white colour in F_2 generation indicates towards the process of segregation.

Mechanism of Segregation

The mechanism of segregation in above mentioned monohybrid cross between red and white flowered pea plants can be understood by assuming that the homozygous red flowered pea plant has the allele RR for redness and white flowered pea plant has the alleles rr for whiteness. The pea plant with RR alleles produces the gametes with single allele R and pea plant with rr alleles produces the gametes with the allele r. The gametes of both united to form a hybrid or heterozygote having the alleles Rr both for redness and whiteness. Due to the phenomenon of incomplete dominance the allele R for red colour partially expresses itself in hybrids of F_1, while the allele r for white colour remains latent or recessive. Both the allele R and r remain together for long time but they do not effect each other. Neither they mix nor they contaminate each other. Because a gamete can contain only one chromosome of a homologous pair, therefore, each gamete can carry single allele R or r. At the time of gametogenesis two types of gametes are produced by F_1 hybrids in equal numbers. Half of the gametes carry the allele R and other half carry the allele r. These gametes during the process of fertilization can unite in three possible combinations, *viz.*, RR, Rr and rr, to produce three types of individuals in F_2 generation. Thus, in F_2 75% individuals have coloured flowers and 25% white flowers. The appearance of white colour in F_2 generation indicates that in the hybrid the allele (r) for white colour remains along with allele (R) for red colour but does not mix with it or contaminated by it and it separates or segregates during gametogenesis.

Certain Other Examples of Law of Segregation

The law of segregation is universal in its application and it has been found to occur in both plants and animals. The examples which are cited in phenomenon of dominance can also be considered for the law of segregation. However, to understand the mechanism of segregation more clearly in animals it will be helpful to consider an original experiment of well known genetist **T.H. Morgan** on *Drosophila.*

Morgan crossed a homozygous long-winged (wild type) *Drosophila* with a homozygous vestigial-winged (mutant) *Drosophila.* The F_1 heterozygotes or hybrids were found to be long winged. When the F_1 hybrids were allowed to mate among themselves, they produced long-winged and vestigial-winged hybrids in three and one ratio (3 : 1) in the F_2 generation (Fig. 3.8).

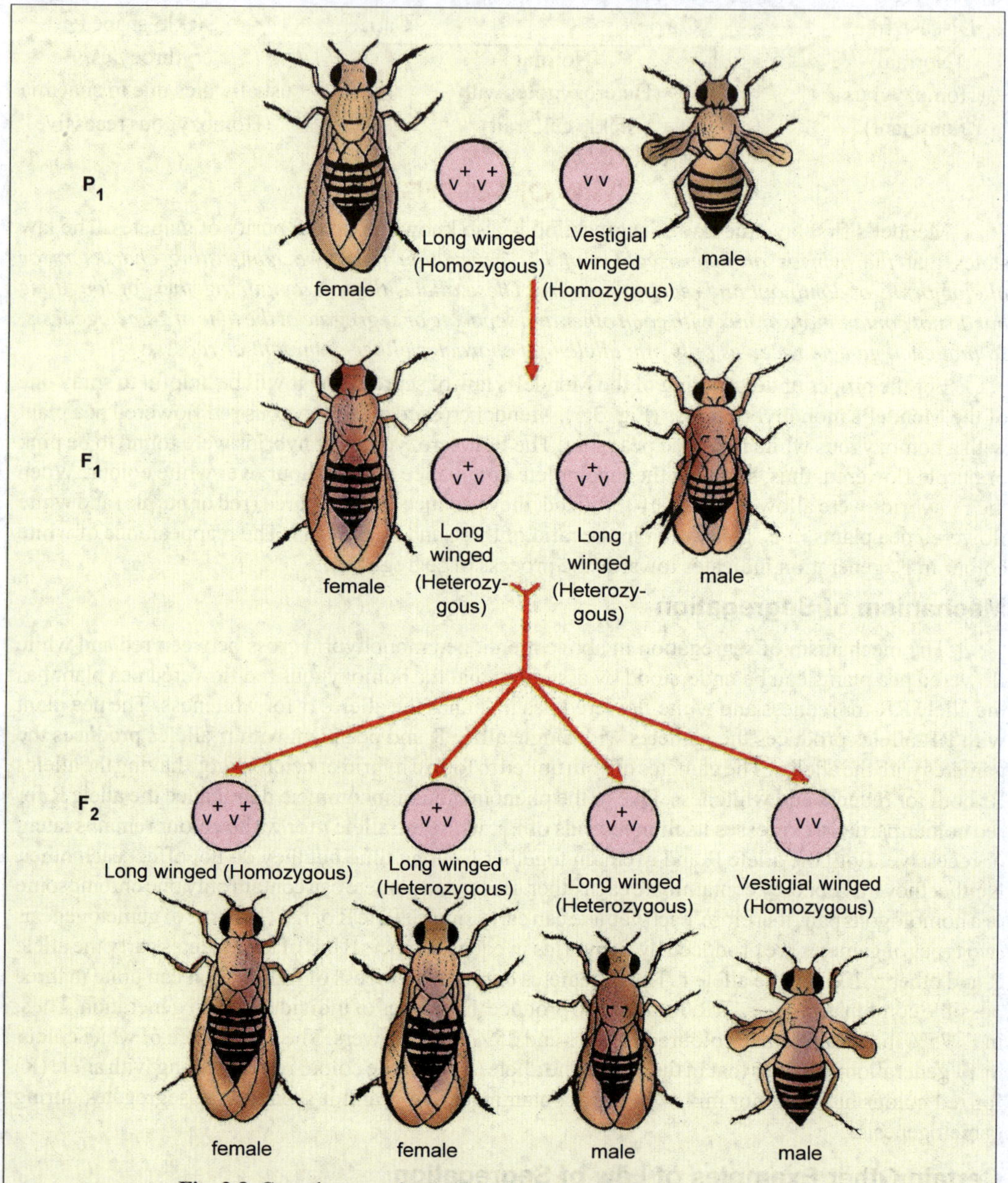

Fig. 3.8. Cross between long-winged and vestigial-winged *Drosophila.*

In this cross the mechanism of segregation can be well understood by assuming that the homozygous long-winged *Drosophila* has a pair of alleles v^+v^+ for longness of wing and similarly homozygous vestigial-winged *Drosophila* has the alleles vv for vestigial nature of wings. The long-winged *Drosophila* fly thus produces the gametes (sperm or ova) with the single allele v^+ and the vestigial-winged fly produces the gametes with the single allele v. The gametes of both parents unite in the process of fertilization and produce a hybrid with long wings in F_1 generation. The genotype of F_1 hybrid is v^+v and in it the allele v^+ is dominant over the alleles v which is recessive. At the time of gametogenesis these alleles (v^+ and v) are separated along with chromosomes to form two types of gametes, half of the gametes have the allele v^+ and other half have the allele v. These gametes unite in three possible combinations, *viz.*, v^+v^+, v^+v and vv, to produce genotypically three types of individuals v^+v^+, v^+v and vv in 1:2:1 ratio in F_2 generation. Phenotypically here occur only three long-winged and one vestigial-winged *Drosophila*. Thus, the dominant and recessive alleles remain together for long time without contaminating or mixing with one another and segregate during gametogenesis.

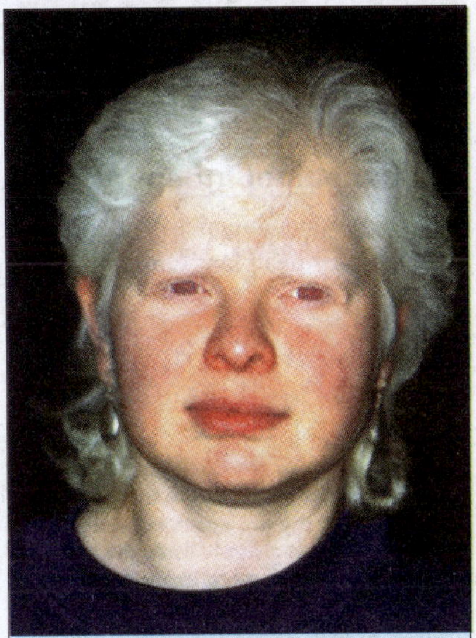

This person is an albino.

2. Albinism in humans. Lack of pigment production in the human body is an abnormal recessive trait called **albinism**. Using A and a to represent the dominant (normal = melanin production) allele and the recessive (albino = no melanin production) allele respectively, the following three genotypes and two phenotypes are possible:

Genotypes	Phenotypes
AA (Homozygous dominant)	Normal (Pigment)
Aa (Heterozygote)	Normal (Pigment)
aa (Homozygous recessive)	Albino (No pigment)

The person with recessive *aa* genes do not produce the tyrosinase enzyme which is needed by melanocytes for the synthesis of (black) pigment. As a result in an albino patient melanocytes are present in normal number in their skin, hairs, iris of eyes, etc., but lack in melanin pigment.

LAW OF INDEPENDENT ASSORTMENT

To formulate the law of dominance and law of segregation **Mendel** considered monohybrid crosses in which single pairs of contrasting characters were considered at a time. But he tried to find out how different characters would behave in relation to each other in their inheritance from generation to generation. For this purpose Mendel crossed two varieties of pea plants which were differing in two pairs of contrasting characters. Because such crosses yielded dihybrids and at a time two pairs of contrasting characters had been considered in them, therefore, these crosses were known as **dihybrid crosses**.

Mendel's Dihybrid Cross

In one of his hybridization experiment Mendel crossed a homozygous pea plant having **yellow round seeds** with the homozygous pea plant having **green wrinkled seeds** (Fig. 3.9). The F_1 hybrids were found to have yellow round seeds. When the F_1 hybrids were allowed to cross among themselves they produced four types of seeds in the ratio of 9 : 3 : 3 : 1 given as follows :

1. Yellow Round – 9
2. Yellow Wrinkled – 3
3. Green Round – 3
4. Green Wrinkled – 1.

Beside getting the ratio of 3 : 1 of the monohybrid crosses Mendel got the ratio of 9 : 3 : 3 : 1.

This irregularity in the ratio of F_2 offspring was explained by Mendel stating that "*when the parents differ from each other in two or more pairs of contrasting characters or factors then the inheritance of one pair of factors is independent to that of the other pair of factors*". This is the Mendel's law of independent assortment.

Mechanism of Independent Assortment

The mechanism of independent assortment can be understood easily by assuming that the homozygous pea plant with yellow round seeds has the alleles YY and RR for the yellow colour and

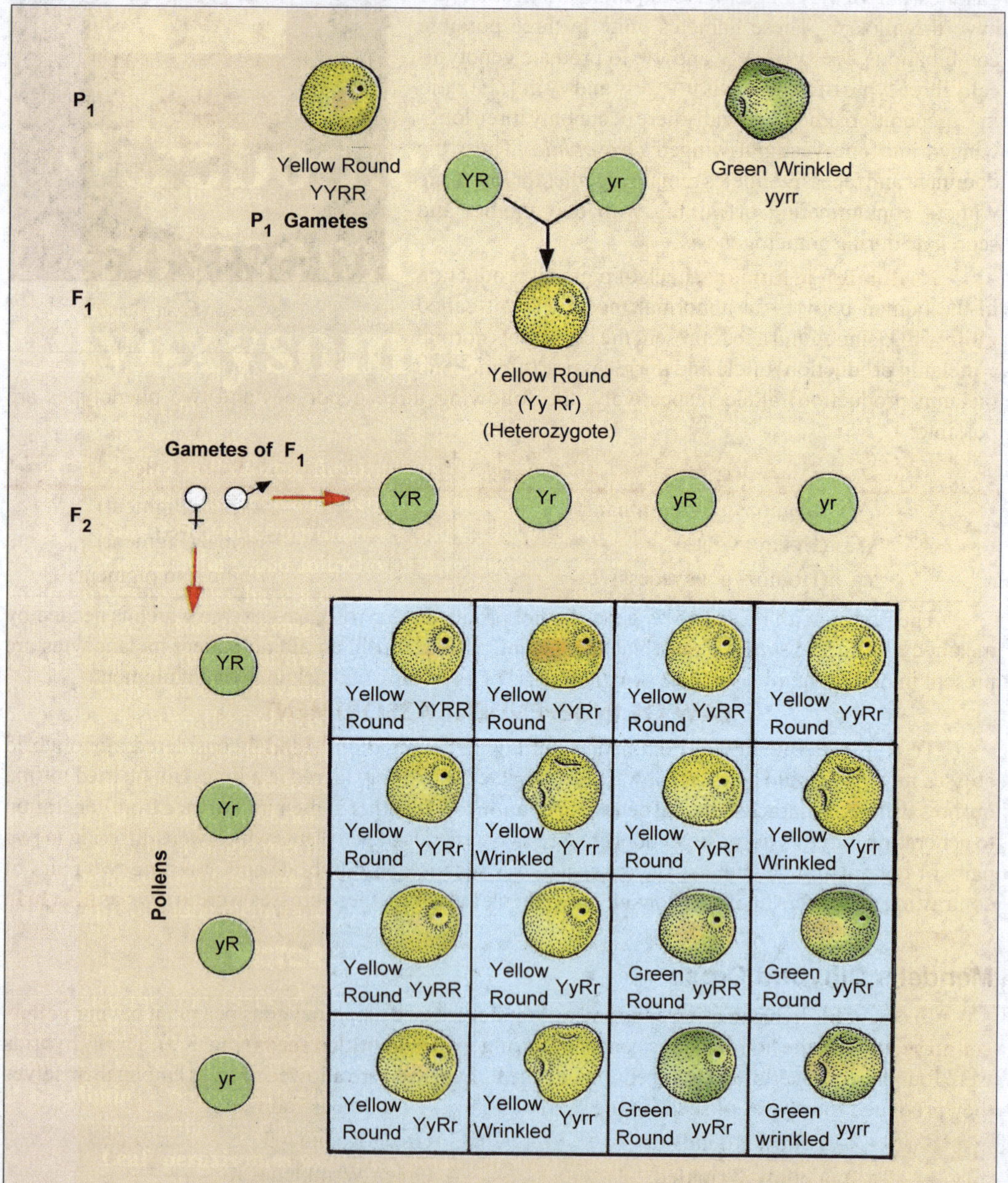

Fig. 3.9. A dihybrid cross between a yellow round and a green wrinkled pea plant.

roundness of the seed, respectively. Similarly the homozygous pea plant with green wrinkled seeds contains the alleles yy and rr for the green colour and wrinkledness of seeds. The gametes which are produced by YYRR and yyrr plants are YR and yr types respectively. When both parents are crossed the union of both types of gametes takes place to give the F_1 hybrid (Yy Rr). The F_1 hybrid have been found to contain yellow round seeds showing the dominance of allele Y for yellow colour over the recessive allele y for green colour and the dominance of allele R for roundness over the recessive allele r for wrinkledness of seed. Now the F_1 hybrids have four types of alleles, *viz.*, Y for yellow colour, y for green colour, R for round shape and r for winkledness of seed. During gametogenesis these four alleles, may combine in following four combination :

1. The allele Y may be associated with the allele R to give rise YR combination.
2. The allele Y may be associated with the allele r to give rise Yr combination.
3. The allele y may be associated with the allele R to give rise yR combination.
4. The allele y may be associated with the allele r to give rise yr combination.

Thus, four types of alleles are assorted independently to produce four types of gametes, *viz.*, YR, Yr, yR and yr. These four types of gametes (pollens or ovules) of F_1 hybrid unite at random in the process of fertilization and produce sixteen types of individuals in F_2 generation which are given in the table 3-6.

Table 3.6. **The phenotype and genotype of F_2 generation of a dihybrid cross between pea plants with yellow round and green wrinkled seeds.**

Number of individuals	Genotype class	Phenotype class
1	YY RR	Homozygous Yellow Round
2	YY Rr, YY Rr	Heterozygous Yellow Round
2	Yy RR, Yy RR	
4	Yy Rr, Yy Rr, Yy Rr, Yy Rr	
		Yellow Round=9
1	YY rr	Homozygous Yellow Wrinkled
2	Yy rr, Yy rr	Heterozygous Yellow Wrinkled
		Yellow Wrinkled=3
1	yy RR	Homozygous Green Round
2	yy Rr, yy rR	Heterozygous Green Round
		Green Round=3
1	yy rr	Homozygous Green Wrinkled
		Green Wrinkled=1
16	– –	9 : 3 : 3 : 1

Thus, the sixteen F_2 individuals have the ratio of 9 yellow round : 3 yellow wrinkled : 3 green round : 1 green wrinkled. These results have proved the law of independent assortment and showed that each pair of contrasting characters behaves independently and bears no permanent association or relation with a particular character. The allele Y was associated with allele R in parent but it does not always remain associated with it and it is also associated with the allele r.

A Case of Reverse Genetics in Mendel's Wrinkled Character

In the year 1990, a group of scientists, working at the John Innes Institute of Norwich (U.K), cloned the pea gene r (rugosus; old name wrinkled, w) which determines whether the seed is round or wrinkled. It was shown that an isoform of **starch branching enzyme (SBEI)** is present in round (RR or WW and Rr or Ww) seeds, but absent in wrinkled (rr or ww) seeds. The gene for SBEI is found to be located on r locus, but the gene is interrupted by a small DNA sequence (0.8 kilobases) called **transposon-like insertion**, due to which aberrant SBEI enzyme is produced. This defective enzyme leads to metabolic disturbances in the biosynthesis of starch, lipid and protein. This results in an increase of free sugars due to failure of starch formation. This probably leads to higher osmotic pressure and, hence, higher water content and larger cell volume in rr (ww) seeds. Seeds lose large proportion of their water on maturation leading to shrinkage in volume. Since testa does not shrink with the cotyledons, seeds become wrinkled.

Such a type of following of the molecular basis of a Mendelian trait comes under the purview of modern branch of genetics, called **reverse genetics** (from DNA to phenotype) as opposed to classical **forward genetics** (from phenotype to DNA).

Dihybrid Cross in *Drosophila*

In a dihybrid cross homozygous long-winged and black-bodied *Drosophila* fly is crossed with a vestigial-winged, grey-bodied fly, they produced in F_1 long-winged grey-bodied *Drosophila*. The

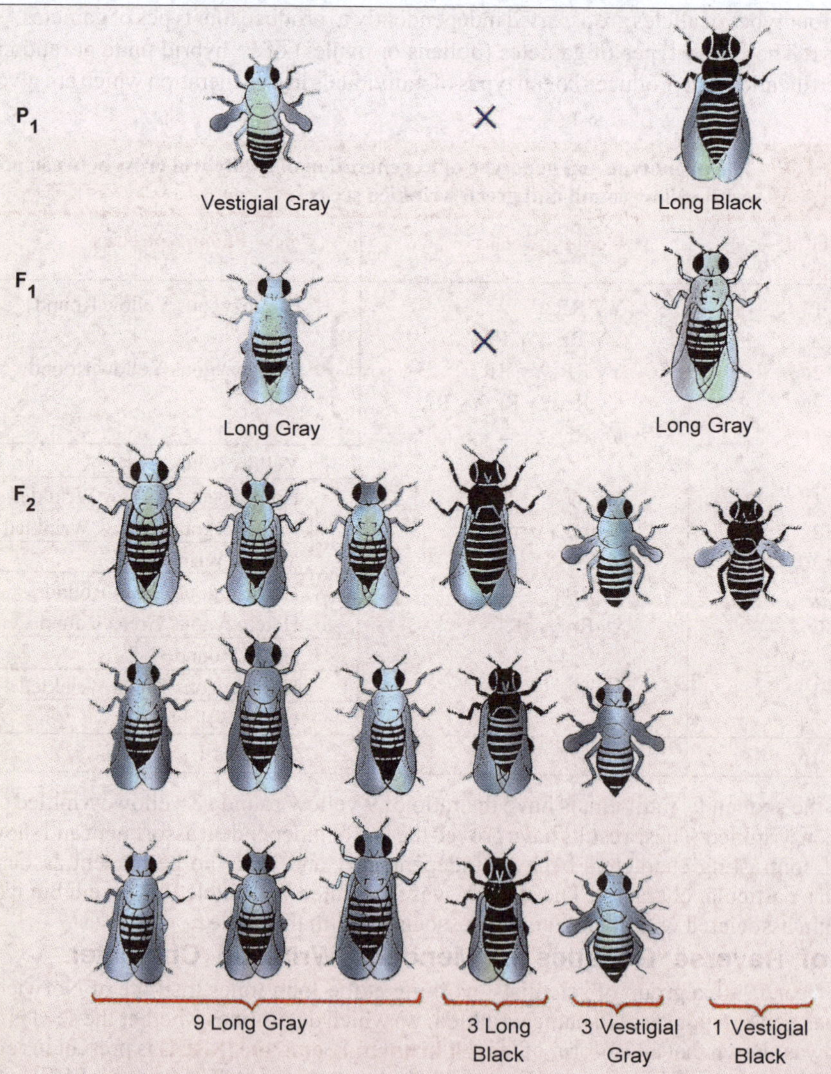

Fig. 3.10. A dihybrid cross between the vestigial-grey and long-black *Drosophila*.

hybrids of F_1 generation after mating with each other produced in F_2 generation 9 long-winged grey-bodied, 3 long-winged black-bodied, 3 vestigial-winged grey-bodied and 1 vestigial-winged black-bodied *Drosophila* (Total=16). Therefore, the 16 individuals of F_2 have the ratio of 9 : 3 : 3 : 1 (Fig. 3.10).

Back Cross and Test Cross

When F_1 individuals are crossed with one of the two parents from which they were derived, then such a cross is called **back cross**. In such back crosses, when F_1 is back crossed to the parent with dominant characters (phenotype), no recessive individuals are obtained in the progeny. On the other hand, when it is crossed with recessive parent, both phenotypes appear in the progeny. While both of these crosses are back crosses, only the cross with the recessive parent is known as **test cross**. It is called a test cross, because it is used to test whether an individual is homozygous (pure) or heterozygous (hybrid). For a monohybrid the test cross ratio remains 1 : 1, but, for a dihybrid test cross ratio becomes 1 : 1 : 1 : 1.

Example of Monohybrid Back Cross and Test Cross

In a monohybrid cross of homozygous tall (DD) and homozygous dwarf (dd) pea plants, when a F_1 haterozygous tall (Dd) plant is crossed either with its dominant parent to perform a back cross or with its recessive parent to perform a test cross following results are obtained :

P_1 :	Homozygous Tall DD	× ↓ Dd	Homozygous Dwarf dd
F_1 :		Heterozygous Tall	
A. Back cross :	F_1 Tall Dd	× ↓	P_1 Tall DD
Back cross progeny :	½DD Homozygous tall	:	½Dd or all tall. Heterozygous tall
B. Test Cross :	F_1 Tall Dd	× ↓	P_1 Dwarf dd
Test cross progeny :	½Dd Heterozygous tall	:	½dd or Test cross ratio = 1:1 Homozygous dwarf

Example of Dihybrid Test Cross

The test cross of a heterozygous yellow round seeded pea plant with a double recessive parent (green wrinkled, yyrr), yields the test cross genotypic and phenotypic ratio of 1:1:1:1 as follows :

Parents :	F_1 Heterozygous yellow round Yy Rr ↓	×	F_1 Homozygous green wrinkled yyrr ↓
Gametes :	YR, Yr, yR, yr		yr

Test cross progeny : ¼ Yy Rr : ¼ Yy rr : ¼ yy Rr : ¼ yy rr
or ¼ yellow round : ¼ yellow wrinkled : ¼ green round : ¼ green wrinkled
or 1:1:1:1.

Multihybrid Cross

The parents which differ in more than two pairs of contrasting characters then the cross between them is known as **polyhybrid** or **multihybrid cross**. The F_1 hybrids in these cases are known as **polyhybrids** or **multihybrids**. The law of independent assortment is also applicable to these crosses.

Example of A Multihybrid Cross

For example, we can consider a trihybrid cross of pea plant. The mating of a homozygous yellow, round, and tall pea plant (YY RR TT) with a homozygous green, wrinkled, and dwarf pea plant (yy rr tt) produces in F_1 yellow, round and tall (Yy Rr Tt) trihybrids. These F_1 trihybrids when self-crossed among themselves, they produce a F_2 progeny including 64 individuals in the phenotypic ratio of 27 yellow, round, tall : 9 yellow, round, dwarf : 9 yellow, wrinkled, tall : 9 green, round, tall : 3 yellow, wrinkled, dwarf : 3 green, wrinkled, tall : 3 green, round, dwarf : 1 green, wrinkled, dwarf or 27 : 9 : 9 : 9 : 3 : 3 : 3 : 1. The possible genotypes and phenotypes of this trihybrid cross have been summarized in Table 3-7.

DEVIATION FROM MENDEL'S DIHYBRID PHENOTYPIC RATIO

The Mendelian dihybrid phenotypic ratio of 9 : 3 : 3 : 1 is obtained only when the alleles at both gene loci display dominant and recessive relationship. If one or both gene loci have incompletely dominant alleles, or codominant alleles or lethal alleles, the dihybrid ratio becomes modified variously, such as follows :

Table 3.7. Genotypic and phenotypic ratios of a trihybrid cross in between a yellow, round, tall (YY RR TT) and a green, wrinkled, dwarf (yy,rr,tt) pea plant :

Genotype	Genotypic ratio	Phenotype	Phenotypic ratio
YYRRTT	1	Yellow, Round, Tall	27
YYRrTT	2		
YyRRTT	2		
YyRrTT	4		
YYRRTt	2		
YYRrTt	4		
YyRRTt	4		
YyRrTt	8		
YYrrTT	1	Yellow, Wrinkled, Tall	9
YyrrTT	2		
YYrrTt	2		
YyrrTt	4		
yyRRTT	1	Green, Round, Tall	9
yyRrTT	2		
yyRRTt	2		
yyRrTt	4		
YYRRtt	1	Yellow, Round, Dwarf	9
YYRrtt	2		
YyRRtt	2		
YyRrtt	4		
yyrrTT	1	Green, Wrinkled, Tall	3
yyrrTt	2		
YYrrtt	1	Yellow, Wrinkled, Dwarf	3
Yyrrtt	2		
yyRRtt	1	Green, Round, Dwarf	3
yyRrtt	2		
yyrrtt	1	Green , Wrinkled, Dwarf	1

1. 3 : 6 : 3 : 1 : 2 : 1 Ratio

When the dihybrid parent have dominant and recessive alleles at one gene locus and codominant alleles at second gene locus, the F_2 9 : 3 : 3 : 1 phenotyic ratio becomes 3 : 6 : 3 : 1 : 2 : 1.

Example. In cattles, hornless or polled (P) condition is dominant to horned (p) condition, and trait of white (W) coat is codominant to the trait for red (w) coat colour. The mating of homozygous white polled (PP WW) cattle with a homozygous red horned (pp ww) cattle produces F_1 heterozygotes (dihybrids) with the phenotype of hornless, roan and genotype of Pp Ww. These hornless roan F_1 dihybrids produce a F_2 progeny in the ratio of 3 : 6 : 3 : 1 : 2 : 1 as illustrated in following diagram:

P_1 :	Phenotype	Hornless White	×	Horned Red
	Genotype	PP WW		pp ww
		↓		↓
P_1	Gametes :	PW		pw
F_1 :			Pp Ww Hornless Roan	

F_2 : 3 Hornless White : 6 Hornless Roan : 3 Hornless Red : 1 Horned White : 2 Horned Roan : 1 Horned Red.

2. 1 : 2 : 1 : 2 : 4 : 2 : 1 : 2 : 1 Ratio

When each parent of a dihybrid cross has incompletely dominant alleles at both gene loci, then in F_2 large number of phenotypic classes are produced.

Example. In snapdragons, red flower colour (R) is incompletely dominant to white flower colour (r) and trait for broadness of leaf (B) is incompletely dominant to the trait for narrowness of leaf (b). The dihybrid cross between red, broad plant (RR BB) and white, narrow plant (rr bb) produces heterozygotes (dihybrids) having pink flowers and leaves of intermediate width and genotype of Rr Bb. These F_1 dihybrids produce the F_2 progeny in 9 phenotypic classes as follows :

P_1 :	Phenotype	Red Broad	×	White Narrow
	Genotype	RR BB		rr bb
		↓		↓
P_1	Gametes :	RB		rb
F_1 :	Phenotype		Pink Intermediate	
	Genotype		Rr Bb	

F_2 : 1 Red Broad : 2 Red Intermediate : 1 Red Narrow : 2 Pink Broad : 4 Pink Intermediate: 2 Pink Narrow : 1 White Broad : 2 White Intermediate : 1 White Narrow.

3. 3 : 1 : 6 : 2 Ratio

When the F_1 dihybrids have dominant-recessive alleles at one gene locus and recessive lethal alleles at second gene locus, the F_2 offsprings manifest the phenotypic ratio of 3 : 1 : 6 : 2.

4. 1 : 2 : 1 : 3 : 4 : 2 Ratio

When the F_1 dihybrids contain codominant alleles at first gene locus and recessive-lethal alleles at second gene locus, then their F_2 progeny display the phenotypic ratio 1 : 2 : 1 : 3 : 4 : 2.

5. 4 : 2 : 2 : 1 Ratio

This phenotypic ratio is obtained when at both gene loci of F_1 dihybrids occur the recessive lethal alleles.

REVISION QUESTIONS AND PROBLEMS

1. Give a brief life sketch of Mendel and state why his name is so significant for geneticists ?
2. Why did Mendel use pea as the experimental material in his hybridization experiments ? Give an account of his procedure and method of drawing the conclusions from the results of experiments.
3. "Individuals of identical phenotype may have different genotypes and vice versa." State whether this statement is true or false and why ?
   (a) Define and explain Mendel's law of segregation ?
   (b) State, giving evidence, whether or not segregation can occur at either of the meiotic division ?
   (c) Does segregation occur in asexual reproduction ? In sexually reproducing homozygotes ? Explain.
4. Explain why phenotypically identical, or at least very similar, parents may produce very different kinds of offspring.
   In rabbits certain **short-haired** individuals when crossed with long-haired ones produce only **short-haired** progeny. Other **short-haired** individuals when crossed with **long-haired** ones produce approximately equal numbers of **short-haired** and **long-haired** offspring. When **long-haired** individuals are inter-crossed, they always produce progeny like themselves.
   (a) Outline a hypothesis to explain these results and show the genotypes of all individuals ?
   (b) How would you proceed to test this hypothesis ? Show the results you would expect in the crosses you describe.
5. Why sexually reproducing organisms evolve more rapidly than the asexually reproducing ones ?
6. Human egg is much larger than the sperm ? Does it make the child to inherit more from the mother than from the father ? Explain fully.
7. What is the difference between a chromosome and a gene ?
8. A black mouse mates with a brown mouse, and all the offspring are black. Why are no brown offspring produced ? Explain your answer fully.

CHAPTER

4

Genetic Interaction and Lethal Genes

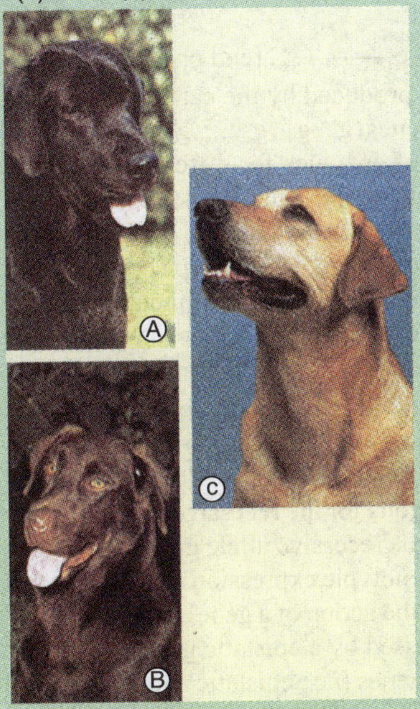

Masking Genes : Coat Colour in Labrador Retrievers. Labrador dogs can be (a) black, (b) chocolate, or (c) yellow, and their colouring is controlled by the black-coat gene (*B* or *b*) and the extension gene (*E* or *e*). When a dog receives two doses of the *e* allele, it will be yellow; *ee* masks the effects of the black-coat gene.

During the discussion of Mendel's monohybrid and dihybrid crosses, we encountered with the fact that for the determination of single phenotypic trait of an organism, two alleles or allelomorphs of a single gene interacted in various ways. Such as out of two allelomorphs of a single gene, one allelomorph might show simple (complete) dominance over the action of other which was recessive; or both allelomorphs might have partial or incomplete dominant relationship or both allelomorphs might have equal expression or codominant relationship. These kinds of genic or genetic interactions occur in between the two allelomorphs of a single type of gene and are usually referred to as **intra-allelic** or **allelic genetic interactions**. These kinds of genetic interactions give the classical ratios of 3 : 1 and 9 : 3 : 3 : 1. But, in addition to intra-allelic genetic interactions, **non-allelic** or **inter-allelic genetic (genic) interactions** also occur. In inter-allelic genetic interactions, the independent (non-homologous) genes located on the same or on different chromosomes interact with one another for the expression of single phenotypic trait of an organism. The discovery of the inter-allelic genetic interactions has been made after Mendel and they can be best understood by considering the way by which a phenotypic trait is goverened by a gene.

TIME OF GENETIC INTERACTION

The gene is a chemical determiner. Whereas a phenotypic trait results from the combined action of many genes and their products constantly interacting with the environment. The environment includes not only ecological factors such as temperature and the amount or quality of light, but also internal factors such as hormones and enzymes. The enzymes are proteins and the specific molecular organization of protein is determined by genes. The enzymatic proteins perform catalytic function in various cellular chemical (metabolic) reactions and causing the splitting or union of various molecules. Each cellular chemical reaction involves stepwise conversion of one substance called **precursor** into another, called **end product**. Each step being mediated by a specific enzyme. All the subsequent steps of a chemical reaction constitute the **biosynthetic pathway**.

Thus, a simplest biosynthetic pathway includes various steps, each step is catalyzed by a specific enzymatic protein and each enzymatic protein in its turn depends on a specific gene for its production. For example, we may consider a simple biosynthetic pathway which transforms a precursor substance 'P' into the end product 'C' in following three subsequent steps :

$$\text{P (precursor)} \longrightarrow \underset{\uparrow\, g_1^+}{e_1} \longrightarrow \text{A} - \underset{\uparrow\, g_2^+}{e_2} \longrightarrow \text{B} - \underset{\uparrow\, g_3^+}{e_3} \longrightarrow \text{C (end product)}$$

In this biosynthetic pathway each metabolite (A, B, C) is produced by the catalytic action of different enzymes (e_1, e_2, e_3.....e_x, specified by different wild type genes (g_1^+, g_2^+, g_3^+......g_x^+). When more than two or more genes become involved in the specification of enzymes for different steps of a common biosynthetic pathway, the phenomenon of genetic interaction occurs. If substance C is essential for the production of a normal phenotype and the recessive mutant alleles g_1, g_2, g_3 produce defective enzymes, then a mutant or abnormal phenotype would result from a genotype homozygous recessive at any of the three loci. If wild gene g_3^+ becomes mutant, the conversion of metabolite B to C does not occur and substance B tends to accumulate in excessive quantity; if g_2^+ becomes mutant, substance A will accumulate. Thus, the mutant genes caused "**metabolic blocks**" in synthetic pathway.

An organism with a mutation only in gene g_2^+ could produce a normal phenotype, if it was given either substance B or C, but an organism with a mutation in gene g_3^+ has a specific requirement for substance C for the production of normal phenotype. Thus, gene g_3^+ becomes dependent upon g_2^+ for its expression as a normal phenotype. If the genotype is homozygous for the recessive g_2 allele, then the biosynthetic pathway ends with substance A. Neither g_3^+ nor its recessive allele g_3 has any effect on the phenotype. Thus, genotype (g_2g_2) can hide or mask the phenotypic expression of alleles at the g_3^+ locus. Originally a gene or locus which suppressed or masked the action of a gene at another locus was termed **epistatic gene**. The gene or locus which was suppressed by a epistatic gene was called **hypostatic gene**. Later studies revealed the fact that both loci or genes (*i.e.,* epistatic and hypostatic) could be epistatic to one another. Presently, the term epistasis (Greek, standing upon) is used for almost any type of allelic genetic interaction.

Difference Between Dominance and Epistasis

The phenomenon of dominance involves intra-allelic gene suppression, or the masking effect which one allele has upon the expression of another allele at the same locus, while the phenomenon of epistasis involves inter-allelic gene suppression or the masking effect which one gene locus has upon the expression of another. The classical phenotypic ratio of 9 : 3 : 3 : 1 observed in the progeny of dihybrid parents becomes modified by epistasis into ratios which are various combinations of the 9 : 3 : 3 : 1 groupings.

NON-EPISTATIC INTER-ALLELIC GENETIC INTERACTIONS

In certain cases, two pairs of genes determine a same phenotype but assorted independently, produce new phenotypes by mutual non-epistatic interactions and the F_2 phenotypic ratio 9 : 3 : 3 : 1

remains unaltered. Two pairs of genes which interact to affect size and shape of comb but are independently transmitted exist in chicken.

Example

Combs in fowl (9 : 3 : 3 : 1). The classical case of genetic interaction of two genes is discovered by **Bateson** and **Punnett** (1905–1908) in fowls. There are many different breeds of domestic chicken. Each breed possesses a characteristic type of comb. The Wyandotte breed has a comb called "rose", the Brahmas breed has a comb called "pea", and the Leghorns have a comb called 'single'. Each of these types can be bred true.

A cross of chicken with a rose comb to one with a single comb produces ¾ rose and ¼ single, showing dominance of rose over single. Another cross between pea combed and single combed chickens produces pea and single combed chickens in the ratio of 3 : 1, showing dominance of pea over

Phenotypes	Walnut	Rose	Pea	Single
Genotypes	RRPP RRPp RrPP RrPp	RRpp Rrpp	rrPP rrPp	rrpp
Ratio	9	3	3	1
Explanation	Supplemental interaction of R and P	Dominance of R over r	Dominance of P over p	Recessiveness of rr and pp

Fig. 4.1. Supplementary inheritance. Inheritance of the shape of the comb in chickens.

single. But, when a rose combed chicken crossed with that of pea combed, the F_1 progeny was found with a different type of comb known as 'walnut' (Malay breed). When the F_1 walnut combed chickens were bred together, in F_2 all four types of combs, *i.e.,* 9/16 walnut, 3/16 rose, 3/16 pea and 1/16 single appeared (Fig. 4.1).

These peculiar results were interpreted by **Bateson** and **Punnett** as follows : The rose comb is caused by the combination of homozygous recessive genes "pp" and homozygous or heterozygous dominant genes 'RR' or 'Rr'. The pea comb is supposed to be produced by combination of a homozygous recessive condition (rr) and homozygous or heterozygous dominant condition (PP or Pp). While, the single type comb is produced by the double recessive, rrpp, genes. Thus, R gene determines the shape of rose comb and P gene determines the shape of pea comb, but when both genes happen to come together in single individual due to cross between rose and pea combed chickens, they interact to produce a walnut comb in F_1. In the cross of two walnut chickens, two genes interact variously and produces four types of offsprings in F_2.

Thus, here two pairs of genes interact to produce comb size and shape in fowl. During the inheritance of combs in fowls, the genes themselves do not determine the development of a character (presence or absense of comb) and simply modify a character determined by a basic gene and, therefore, known as **supplementary** or **modifying genes**.

KINDS OF EPISTATIC INTERACTION

When in dihybrid crosses, the epistatic interactions occur between two genes, less than four phenotypes appear in F_2. Such bigenic (two gene) epistatic interactions may be of following six types:

1. Dominant Epistasis (12 : 3 : 1)

When out of two genes, the dominant allele (*e.g.,* A) of one gene masked the activity of alleles of another gene (*e.g.,* B) and expressed itself phenotypically, then A gene locus is said to be epistatic to the B gene locus (Table 4-1). Because, the dominant allele A can express itself only in the presence of either B or b allele, therefore, such type of epistasis is termed as **dominant epistasis**. The alleles of hypostatic locus or gene B will be able to express themselves phenotypically only when gene locus A may contain two recessive alleles (aa). Thus, the genotype AA BB or Aa Bb and AA bb or Aa bb produce the same phenotype whereas the genotype aa BB or aa Bb and aa bb produce two additional phenotypes. The dominant epistasis modify the classical ratio of 9 : 3 : 3 : 1 into 12 : 3 : 1 ratio.

Table 4-1. **Mode of dominant epistasis.**

Epistatic alleles	Hypostatic alleles	Phenotypic expression of allele
1. AA, Aa	BB, Bb, bb	A
2. aa	BB, Bb	B
3. aa	bb	b

Example

Dominant epistatis in dogs. Among dogs, the colours of coats depend upon the action of two genes. One gene locus has a dominant epistatic inhibitor allele (I) of coat colour pigment (see Table 4-2). The allele I prevents the expression of colour allele at another independently assorting, hypostatic gene locus (B or b) and produces white coat colour. The alleles of hypostatic gene locus (BB, Bb, or bb) express only when two recessive alleles (ii) occur on the epistatic locus, *i.e.,* ii BB or ii Bb produces black and ii bb produces brown individuals. When two such white coat colour dogs are crossed, in F_1 the white, black and brown coat colours appear in 12 : 3 : 1 ratio as shown in (Fig. 4.2).

Table 4-2. **Mechanism of dominant epistasis in dogs.**

Epistatic alleles	Hypostatic alleles	Phenotypic expression of allele	F_2 Phenotypic ratio
1. II, Ii	BB, Bb, bb	I (no pigment)	White=12
2. ii	BB, Bb	B (Black)	Black=3
3. ii	bb	b (Brown)	Brown=1

P: White (Male) Ii Bb X White (Female) Ii Bb

P Male gametes →
P Female gametes ↓

F_1 :

	IB	Ib	iB	ib
IB	II BB White	II Bb White	Ii BB White	Ii Bb White
Ib	II Bb White	II bb White	Ii Bb White	Ii bb White
iB	Ii BB White	Ii Bb White	ii BB Black	ii Bb Black
ib	Ii Bb White	Ii bb White	ii Bb Black	ii bb Brown

F_1 Phenotypic ratio : 12/16 White : 3/16 Black : 1/16 Brown or 12 : 3 : 1.

Fig. 4.2. Checkerboard derived from a cross between two heterozygous white coated dogs showing 12 : 3 : 1 ratio due to dominant epistatic inhibitor genes.

2. Recessive Epistasis (9 : 3 : 4)

Sometimes the recessive alleles of one gene locus (aa) mask the action (phenotypic expression) of alleles of another gene locus (BB, Bb or bb alleles) (see Table 4-3). This type of epistasis is called **recessive epistasis**. The alleles of B locus express themselves only when epistatic locus has dominant alleles (*e.g.,* AA or Aa). Due to recessive epistasis the phenotypic ratio 9 : 3 : 3 : 1 becomes modified into 9 : 3 : 4 ratio.

Table 4-3. **Mode of action of recessive epistasis.**

Epistatic alleles	Hypostatic alleles	Phenotypic expression of allele
1. a a	BB, Bb, bb	a
2. AA, Aa	BB, Bb	B
3. AA, Aa	bb	b

Example

Recessive epistasis in mice. In mice various types of epistatic genetic interactions have been reported. The most interesting case is of recessive epistasis in coat colours. The common house mouse occurs in a number of coat colours, *i.e.*, agouti, black and albino. The agouti colour pattern is commonly occurred one (wild type) and is characterized by colour banded hairs in which the part nearest the skin is gray, then a yellow band and finally the distal part is either black or brown. The albino mouse lacks totally in pigments and has white hairs and pink eyes.

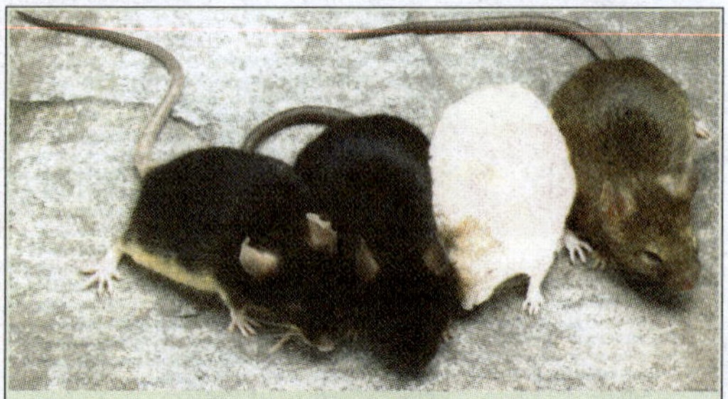

Different coat colours in mice.

Table 4-4. Mode of action of recessive epistasis in alleles for coat colour in mice.

Epistatic alleles	Hypostastic alleles	Phenotypic expression of allele	F_2 Phenotypic ratio
1. cc	AA, Aa, aa	c	Albino=4
2. CC, Cc	AA, Aa	A	Agouti=9 (due to supplementary genes)
3. CC, Cc	aa	a	Black=3

When a homozygous black (CC aa) is crossed with a homozygous albino (cc AA) in F_1 all agouti (Cc Aa) offsprings appear. When, the F_1 agouti are crossed among themselves in F_2 agouti, black and albino offsprings appear in the ratio of 9 : 3 : 4 as shown in the Fig. 4.3.

P_1 : Black CCaa X Albino ccAA

P_1 gametes : (Ca) ↓ (cA)

F_1 : Agouti CcAa

F_2 :

F_1 Male gametes → / F_1 Female gametes ↓	CA	Ca	cA	ca
CA	CC AA Agouti	CC Aa Agouti	Cc AA Agouti	Cc Aa Agouti
Ca	CC Aa Agouti	CC aa Black	Cc Aa Agouti	Cc aa Black
cA	Cc AA Agouti	Cc Aa Agouti	cc AA Albino	cc Aa Albino
ca	Cc Aa Agouti	Cc aa Black	cc Aa Albino	cc aa Albino

F_2 Phenotypic ratio : 9/16 Agouti : 3/16 Black : 4/16 Albino.

Fig. 4.3. Checkerboard derived from a cross between coloured (black) and albino mice showing 9 : 3 : 4 ratio due to recessive epistatic genes.

Supplementary genes. In the cross between black (CCaa) and albino (ccAA) mice, one thing becomes apparent that two independent pairs of genes (*i.e.*, C–c and A–a) have interacted in the production of the phenotypic trait (*i.e.*, coat colour) in such a way that one dominant (C) produces its effect whether or not the second (A) is present, but the second (A) gene can produce its effect only in the presence of the first. These genes (*i.e.*, C and A) have been termed as **supplementary genes** (see Villee *et al.*, 1973).

3. Duplicate Genes with Cumulative Effect (9 : 6 : 1)

Certain phenotypic traits (*e.g.*, coat colouration) depend on the dominant alleles of two gene loci. When the dominant condition (homozygous or heterozygous) at either locus (but not both) produces the same phenotype, the F_2 ratio becomes 9 : 6 : 1.

Table 4-5. **Mode of action of mutually supplementary genes.**

Epistatic alleles	Hypostatic alleles	Phenotypic expression of alleles
1. a a	bb	Neither a nor b
2. a a	BB, Bb	B only } single gene effect
3. AA, Aa	bb	A only } single gene effect
4. AA, Aa	Bb, Bb	A+B A and B alleles mutually supplement each other

Example

Cumulative effect in coat colour of pigs. In the Duroc-jersey breed of pigs, coat colour is influenced by two pairs of genes that interact in peculiar manner (Table 4-6). Sandy coat colour results from a dominant gene S, and the homozygous recessive (ss) is white in colour. Sandy coat colour may also result from a non-allelic dominant gene R; its homozygous recessive (rr) is also white. When a sandy pig (SS rr) is crossed with a second sandy pig (ss RR), the F_1 offsprings were found with red coloured coats. Such interactions are said to be the result of **mutually supplementary genes**. When F_1 red coated pigs cross bred among themselves they produce red, sandy and white coats in the ratio of 9 : 6 : 1 as shown in the Figure 4.4.

Duroc-jersey pig.

P : Sandy SS rr X Sandy ss RR

P gametes : (Sr) (sR)

F_1: Red Ss Rr

F_1 cross : Male Ss Rr X Female Ss Rr

↓

F_1 Male gametes → / F_2 Female gametes ↓	SR	Sr	s R	sr
SR	SS RR Red	SS Rr Red	Ss RR Red	Ss Rr Red
Sr	SS Rr Red	SS rr Sandy	Ss Rr Red	Ss rr Sandy
sR	Ss RR Red	SsRr Red	ss RR Sandy	ss Rr Sandy
sr	Ss Rr Red	Ss rr Sandy	ss Rr Sandy	ss rr White

F_2 : (above)

F_2 Phenotypic ratio : 9/16 Red : 6/16 Sandy : 1/16 White or 9 : 6 : 1.

Fig. 4.4. A cross between two strains of pig having sandy coats producing 9 : 6 : 1 ratio due to mutually supplementary genes.

Table 4-6. Mode of action of mutually supplementary genes for coat colour in pig.

Epistatic alleles	Hypostatic alleles	Phenotypic Expression of alleles	F_2 Phenotypic ratio
1. rr	ss	Neither r nor s (No pigment)	White=1
2. rr	SS,Ss	S }	
3. RR, Rr	ss	R }	
4. RR, Rr	SS, Ss	R+S (Mutually supplementary genes)*	Red=9

* Presence of dominant alleles on both epistatic and hypostatic loci produce cumulative phenotypic expression.

4. Duplicate Recessive Genes (or Complimentary genes) (9 : 7)

If both gene loci have homozygous recessive alleles and both of them produce identical phenotypes, the F_2 ratio 9 : 3 : 3 : 1 would become 9 : 7. In such case, the genotypes aa BB, aa Bb, AA bb, Aa bb, and aa bb produce one phenotype (Table 4-7). Both dominant alleles when present together, complement each other and are called **complementary genes** and produce a different phenotype. A case of such complemental inheritance, resulting from the combined action of complemental genes is known in sweet peas.

Table 4-7. Mode of action of complementary genes.

Epistatic allele	Hypostatic allele	Phenotypic expression of alleles
1. a a	BB, Bb, bb	} No Phenotype production
2. AA, Aa, aa	bb	}
3. AA, Aa	BB, Bb	Phenotype expression due to dominant alleles on both loci (complementation of gene action)

Example

When a pure line variety of white flowered sweet pea (*Lathyrus odoratus*) was crossed with another pure line variety of white flowered sweet pea, in F_1 purple or red flowered plants were produced (Table 4-8). The F_1 plants when self-pollinated or crossed among themselves, produced the F_2 generation with the phenotypic ratio of 9 coloured and 7 white flowered plants.

Purple, red and white flowers of pea.

Table 4-8. **Mode of action of complementary alleles in the production of coloured flowers in sweet pea.**

Epistatic alleles	Hypostatic alleles	Phenotypic expression of allele	F_2 Phenotypic ratio
1. c c	EE, Ee, ee	Neither c nor E or e	} write = 7
2. CC, Cc	e e	Neither C or c nor e	
3. CC, Cc	EE, Ee	Both C+E (complementation)*	Coloured=9

* Chromogen production due to complementation of C and E (*i.e.,* dominant epistatic and hypostatic alleles).

These surprising results could be understood by analysing the mechanism of colour production in flowers. A given enzyme (genetically controlled as to absence or presence in a given individual) acts upon chromogen (a colourless colour base whose absence or presence is also genetically controlled) to produce the purple or red colour of flowers. The dominant allele or alleles (CC or Cc) of gene C are responsible for the presence of chromogen, while the homozygous recessive alleles (cc) of this gene are responsible for the absence of chromogen. Likewise, the dominant alleles of gene E in homozygous (EE) or heterozygous (Ee) conditions caused the production of an enzyme which is necessary for colour production from chromogen, while homozygous recessive (ee) condition does not produce any such enzyme.

The appearance of 9 : 7 ratio instead of 9 : 3 : 3 : 1 ratio from the cross of two white flowered sweet pea plants (CC ee X cc EE) can be illustrated as follows :

P :	White flowers CC ee	X	White flower cc EE
	↓	↓	↓
P gametes :	(Ce)		(cE)
F_1 :	Purple flower Cc Ee	X	Purple flower CcEe

F_1 Male gametes → / **F_1 Female gametes ↓**	CE	Ce	cE	ce
CE	CC EE Purple	CC Ee Purple	Cc EE Purple	Cc Ee Purple
Ce	CC Ee Purple	CCee White	Cc Ee Purple	Cc ee White
cE	Cc EE Purple	Cc Ee Purple	cc EE White	cc Ee White
ce	Cc Ee Purple	Cc ee White	cc Ee White	cc ee White

F_2 :

F_2 phenotypic ratio : 9/16 purple : 7/16 white or 9 : 7.

Fig. 4.5. Complementary genes : production of 9 : 7 phenotypic ratio in sweet peas due to complementary genes.

5. Duplicate Dominant Genes (15 : 1)

If the dominant alleles of both gene loci produce the same phenotype without cumulative effect, the 9 : 3 : 3 : 1 ratio is modified into a 15 : 1 ratio (Table 4-9).

Table 4-9. **Mode of action of duplicate dominant genes.**

Epistatic alleles	Hypostatic alleles	Phenotypic expression of alleles
1. AA, Aa	bb	Same phenotype (A,B, A+B) (Presence of one dominant allele on both or either loci result in same phenotype)
2. AA, Aa	BB, Bb	
3. a a	BB, Bb	
4. a a	bb	Another phenotype (a or b) (Absence of dominant allele on both or either loci results in different phenotype).

Example

The seed capsules of shepherd's purse (genus *Capsella)* occur in two different shapes, *i.e.,* triangular and top-shaped. When a plant with triangular seed capsule is crossed with one having top-shaped capsule, in F_1 only triangular, character appears. The F_1 offsprings by self crossing produce the F_2 generation with the triangular and top-shaped seed capsules in the ratio of 15 : 1. Two independently segregating dominant genes (A and B) have been found to influence the shape of the capsule in the same way (Table 4-10). All genotypes having dominant alleles of both of these genes (A and B) would produce plants with triangular-shaped capsules. Only those with the genotype aa bb would produce plants with top-shaped capsules. The results of this example has been shown in Figure 4.6.

Capsella, flattened, triangular seedpods.

Table 4-10. **Mode of action of duplicate genes in producing triangular seed capsules in *Capsella*.**

Epistatic alleles	Hypostatic alleles	Phenotypic expression of alleles	Phenotypic ratio
1. AA, Aa	bb	Phenotype of either or both dominant alleles, *i.e.,* A or B or both	Triangular=15
2. AA, Aa	BB, Bb		
3. a a	BB, Bb		
4. a a	bb	Recessive phenotype of a and b alleles	Top-shaped=1

P : Triangular X Top-shaped
AA BB ↓ aa bb
P gamete : (AB) (ab)
F_1 : Triangular X Triangular
Aa Bb Aa Bb

F_1 Male gametes → F_2 Female gametes ↓		AB	Ab	aB	ab
	AB	AA BB Triangular	AA Bb Triangular	Aa BB Triangular	Aa Bb Triangular
	Ab	AA Bb Triangular	AA bb Triangular	Aa Bb Triangular	Aa bb Triangular
F_2 :	aB	Aa BB Triangular	Aa Bb Triangular	aa BB Triangular	aa Bb Triangular
	ab	Aa Bb Triangular	Aa bb Triangular	aa Bb Triangular	aa bb Top-shaped

F_2 Phenotypic ratio : 15/6 Triangular : 1/16 Top-shaped or 15:1.

Fig. 4.6. Duplicate dominant genes : a cross between two strains of *Capsella* having triangular and top-shaped seed capsules to get 15 : 1 F_2 dihybrid ratio.

6. Dominant and Recessive Interactions (13 : 3)

Sometimes, the dominant alleles of one gene locus (A) in homozygous (AA) and heterozygous (Aa) condition and the homozygous recessive alleles (bb) of another gene locus (B) produce the same phenotype, the F_2 phenotypic ratio becomes 13 : 3 instead of 9 : 3 : 3 : 1. In such case, the genotype AA BB, AABb, Aa BB, Aa Bb, AA bb and Aa bb produce same phenotype and the genotype aa BB, aa Bb and aa bb produce another but same phenotype (Table 4-11).

Table 4-11. **Mode of action of interaction between dominant and recessive alleles.**

Epistatic alleles	Hypostatic alleles	Phenotypic expression of allele
1. AA, Aa	BB, Bb, bb	A inhibits B or b (No phenotype of B or b)
2. a a	BB, Bb, bb	a does not inhibit B or b (Phenotype of B or b)

Example

White Leghorn and white Plymouth Rock.

In Leghorn type of fowl the white colour of feather is caused by the dominant genotype CC II, similarly the white colour of feathers of Plymouth Rock breed is caused by the recessive genotype cc ii (Table 4-12). When both white varieties of fowl are crossed, in F_1 white coloured hybrids appear. The F_1 hybrids in F_2 produce the white and coloured offsprings in the ratio of 13 : 3, as have been illustrated in Fig. 4.7.

Table 4-12. Method of interaction between dominant and recessive genes.

Epistatic alleles	Hypostatic allele	Phenotypic expression of allele	F_2 Phenotypic ratio
1. II, Ii	CC, Cc, cc	I (dominant inhibitor)	White = 12
2. ii	CC, Cc	C (due to recessive inhibitor i)	Coloured = 3
3. ii	cc	c (due to c and i)	White = 1

P : White Leghorn CC II X White Plymouth Rock cc ii

P gametes : (C I) ↓ (c i)

F_1 : White Cc Ii X White Cc Ii

F_1 male gametes → CI, C i, cI, ci

F_1 female gametes ↓

F_2 :

	CI	C i	cI	ci
CI	CC II White	CCIi White	Cc II White	Cc Ii White
C i	CC Ii White	CCii Coloured	Cc Ii White	Cc ii Coloured
cI	Cc II White	Cc Ii White	cc II White	cc Ii White
ci	Cc Ii White	Cc ii Coloured	cc Ii White	cc ii White

F_2 phenotypic ratio : 13/16 white : 3/16 coloured breeds or 13 : 3.

Fig. 4.7. A cross between two white coloured breeds of fowls to get 13 : 3 F_2 dihybrid ratio.

The various epistatic ratios can be summarized in the following table :

Table 4-13. Summary of various epistatic ratios.

Genotype	A–B– (AA BB, Aa BB, AA Bb, Aa Bb)	A–bb (AA bb, Aa bb)	aa B– (aa BB, aa Bb)	aa bb
Classical ratio	9	3	3	1
Dominant epistasis	12		3	1
Recessive epistasis	9	3	4	
Duplicate gene with cumulative effect	9	6		1
Duplicate dominant genes	15			1
Duplicate recessive genes	9	7		
Dominant and recessive interaction	13		3	

ATAVISM OR REVERSION

While we were discussing the cross between two white flowered sweet pea plants we observed that F_1 hybrid had purple colour unlike their immediate parents but like their remote ancestors. The appearance of such offsprings which resemble with their remote ancestors are called **throwbacks**, **atavisms** or **reversions**.

LETHAL GENES

Lethal genes are mutant genes and result in the death of the individual which carries them. Death of the individual occurs either in the prenatal or postnatal period prior to sexual maturity. A **fully** (completely) **dominant lethal allele** kills both in homozygous and heterozygous states. Individuals with a dominant lethal allele die before they can leave progeny. Therefore, the mutant dominant lethal is removed from the population in the same generation in which it arose. **Recessive lethal genes** kill only when they are in a homozygous state and they may be of two kinds : 1. one which has no obvious phenotypic effect in heterozygotes and 2. one which exhibits a distinctive phenotype when heterozygous.

The **completely lethal genes** usually cause death of the zygote, later in the embryonic development or even after birth or hatching. Complete lethality, thus, is the case where no individuals of a certain genotype attain the age of reproduction. However in many cases lethal genes become operative at the time the individuals become sexually mature. Such lethal genes which handicap but do not destroy their possessor are called **subvital**, **sublethal** or **semilethal** genes. The lethal alleles modify the 3:1 phenotypic ratio into 2 : 1.

Snapdragons.

Examples of Lethal Alleles

A. Lethal alleles in plants. In plants, recessive lethal alleles are known which produce **albinism**, where absence of chlorophyll is lethal (fatal) to them. Following two examples illustrate this fact :

1. In snapdragons (*Antirrhinum majus*) three types of plants occur : 1. green plants with chlorophyll ; 2. yellowish green plants with carotenoids, usually are referred as pale green, **golden** or **auria** plants and 3. white plants without any chlorophyll. The homozygous green plants have the genotype CC and the homozygous white plant has the genotype cc. The auria plants have the genotype Cc because they are heterozygotes of green and white plants. When two such auria plants are crossed, the F_1 progeny has identical phenotypic and genotypic ratio of 1 : 2 : 1 (*viz.*, 1 green (CC) : 2 auria (Cc) : 1white (cc). But the white plants because lack chlorophyll pigment, therefore, die to modify the ratio of 1 : 2 : 1 into 1 : 2 or 2 : 1. In this case the homozygous recessive genotype (cc) is lethal.

F_1 heterozygote : Auria X Auria
Cc Cc

F_2 : 1 Cc : 2 Cc : 1cc
Green Auria White (lethal)
or lCC : 2 Cc or 1 : 2.

Thus, c allele exhibits a lethal effect when homozygous and a distinctive phenotypic effect (*e.g.*, auria) when heterozygous.

2. In maize (*Zea mays*) the amount of chlorophyll is controlled by a recessive allele (g) which exhibits a lethal effect in homozygous (gg) and in heterozygous condition (Gg) has phenotype similar to homozygous condition for dominant gene GG. It modifies 3 : 1 phenotypic ratio into 2 : 1.

F_1 heterozygote : Green X Green
Gg Gg

F_2 : 1 GG : 2 Gg : 1 gg
Green Green White (lethal)
or 1 GG : 2 Gg or 1 : 2 or 2 : 1.

B. Lethal alleles in animals. Among animals, the following three examples exhibit the role of recessive lethal alleles :

1. The inheritance of mouse body colour was studied by the French geneticist, **L. Cuenot** in 1905. The coat colour of mice is governed by a multiple allelic series (see Chapter 10) in which A allele determines **agouti** or mousy-coloured coat, A^Y allele determines **yellow** coat and a allele forms **black** coat. The dominance hierarchy is as follows : $A^Y > A > a$. The A^Y allele also acts as a **recessive lethal**, since in the homozygous state (A^YA^Y), it kills the individual in early embryonic stage (*i.e.,* during gastrulation). Thus, when two yellow coated heterozygotes (A^YA) are crossed, they produce a progeny showing a ratio of 2:1 since homozygous yellow (A^YA^Y) individuals are never borned due to lethal effect of A^Y gene :

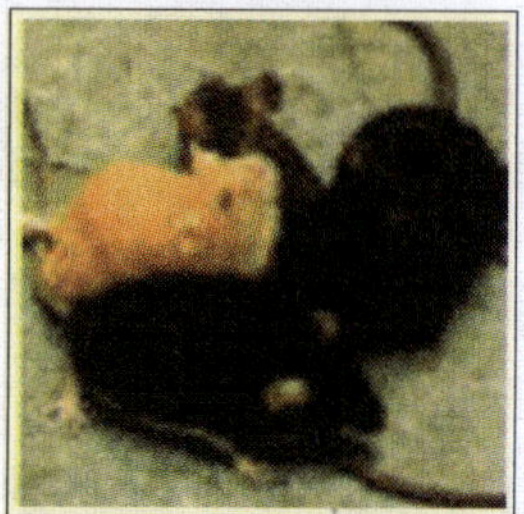

Coat colour in mice is governed by multiple alleles.

Parents : Yellow X Yellow
A^YA A^Ya
(Hybrid of yellow and agouti) ↓ (Hybrid of yellow and black)

Progeny : 1A^YA^Y : 2 A^YA : 1 Aa
Homozygous Yellow (die in uterus) Heterozygous Yellow Agouti
or 2 Yellow : 1 Agouti or 2:1.

2. In the chicken an incompletely dominant gene (cp) in heterozygous condition (cp/+) cause "creeper" condition. The creeper birds have much shortened and deformed legs and wings, giving them a squatty appearance and creeping gait. A cross of two creeper birds yields viable offsprings in the ratio of 2 creepers : 1 normal. The homozygous creepers having such a gross deformities that they die during incubation.

F_1 : cp/+ X cp/+
Creeper (Heterozygous) ↓ Creeper

F_2 : 1 cp/cp : 2cp/+ : 1+/+
Creeper Creeper Normal
(Homozygous; dies) (Heterozygous) (Homozygous)

3. In cattle, a recessive lethal gene in homozygous condition (aa) causes calves to born "amputated" which die soon after birth. The cross between two carriers (a+) produces the following result :

Parents : Normal X Normal
a+ a+

P gametes : Ⓐ⊕ Ⓐ⊕

Progeny : 1aa : 2a+ : 1 ++
Amputated (die) Normal

4. In *Drosophila* various sublethal and sex-linked lethal genes have been reported. The genes for vestigial wings and genes for white eyes are best examples of sublethal genes of *Drosophila.* Both of these genes reduce the viability of flies up to greatest extent. In *Drosophila,* certain recessive lethal genes like curly wings (Cy), plum eyes (Pm) and stubbles (Sb) influence the viability of the flies when present in homozygous condition.

C. Lethal alleles in human beings. In humans several hereditary diseases have lethal effects. Few important lethal genes of man are following :

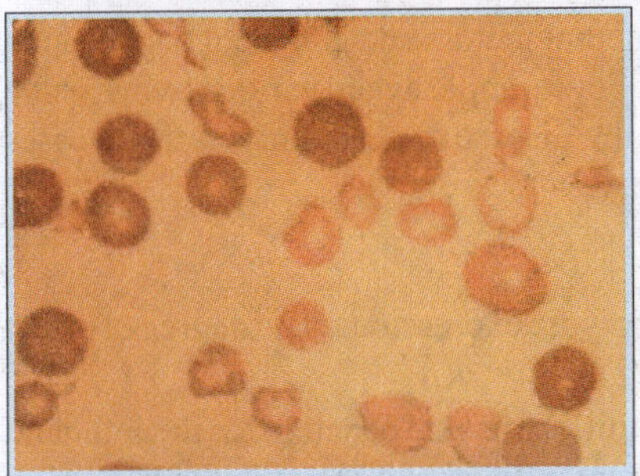

Small pale abnormally - shaped RBC are associated with thalassemia major. The darker cells likely represent normal RBCs, from a blood transfusion

1. Congenital ichthyosis. One of the most typical cases of a recessive lethal gene in man is expressed in congenital ichtyosis. At birth children afflicted with this disease have a crusted leathery skin with deep fissures down to the subcutaneous tissue; the fissures lead to bleeding, infection and death. Congenital ichthyosis occurs only when there occurs homozygous condition for its recessive lethal genes.

2. Infantile amaurotic idiocy. A recessive allele in homozygous condition causes a fatal disease called **infantile amaurotic idiocy** in juvenile stage. Bearers of this genotype begin to lose their eye sight between the age of four to seven years. The complete blindness is followed by mental degeneration and finally death before adolescence.

3. Thalassemia. Thalassemia or **Cooley's anaemia** is a haemoglobin disease somewhat similar to sickle cell anaemia. It occurs mostly in children (in India and other countries such as Italy, Greece and Syria) and is nearly 100 per cent fatal (lethal). Thalassemia is controlled by a single gene c which in homozygous condition (cc), produces the severe Cooley's anaemia or **thalassemia major** and causes death of the patient. The heterozygous condition of this lethal gene (Cc) results in a mild form of the disease called **thalasemia minor** or **microcythemia** (see **Gurdon**, 1968).

PENETRANCE

The ability of a given gene or gene combination to be expressed phenotypically to any degree is called **penetrance**. It is of following two kinds :

1. Complete Penetrance

Most dominant and recessive genes in homozygous conditions and many completely dominant genes even in heterozygous conditions give their complete phenotypic expressions. Such genes are called to have **complete penetrance.**

Examples of Complete Penetrance

1. In pea the alleles (RR) for red flowers and the alleles (rr) for white flowers have complete penetrance in homozygous conditions.

2. In *Drosophila* the recessive alleles for vestigial wings in homozygous conditions have complete penetrance.

3. In guinea pigs the dominant allele 'B' for black coat has complete penetrance both in homozygous and heterozygous conditions.

2. Incomplete Penetrance

Some genes in homozygous as well as in heterozygous conditions fail to provide complete (cent per cent) phenotypic expression of them. Such genes are called to have **incomplete penetrance**.

Examples of Incomplete Penetrance

(i) Polydactyly in man is thought to be produced by a dominant gene P. The normal condition with five digits on each limb is produced by the recessive genotype (pp). Some heterozygous individuals (Pp) are not polydactylus and, therefore, has a penetrance of less than 70%.

(ii) In man, the tendancy to develop diabetes mellitus (a condition in which there is an excess of sugar in the blood) is controlled by certain genes. However, not everyone carrying the genes for diabetes actually develops the condition, for these genes have incomplete penetrance.

Effects of environment on penetrance. The environmental factors and genetical background have some definite effect on the degrêe of penetrance of a gene. For example, when various twins which carry genes for diabetes mellitus are studied, it is found that the disease appears only in those cases which ate more carbohydrate foods (starch and sugars).

EXPRESSIVITY

A trait though penetrant, may be quite variable in its phenotypic expressions. The degree of effect produced by a penetrant genotype is called **expressivity**.

Example of expressivity. In man the polydactylous condition may be penetrant in the left hand (6 fingers) and not in the right (5 fingers); or it may be penetrant in the feet and not in the hands.

Effects of environment on expressivity. The expressivity of a given gene is often influenced by environmental conditions. For instance, the expressivity of completely penetrant gene for vestigial wings in *Drosophila* is influenced by the temperature at which the fly develops, with the effect being most obvious at lower temperature.

Other examples of environmental effects on the expressivity of a gene includes such cases as the differences in the severity of symptoms of an inheritable allergy, or the differences in height of identical twins who have been raised in different home (with different diets), or who have had different medical histories (one with a serious childhood disease, the other escaping this disease).

PLEIOTROPISM

Uptill now we have observed that a specific gene has a specific effect upon a specific phenotypic trait or in other words, each gene (allele) has its relation with a single phenotypic trait, but, this is not the case. A single gene often influences more than one phenotypic trait. However, it may be that one gene may cause evidently well marked expression of some phenotypic trait (**major effect**) then the others with less evident phenotype (**secondary effect**). Most genes have their multiple effects and are called **pleiotropic genes**. The phenomenon of multiple effect (multiple phenotypic expressions) of a single gene is called **pleiotropism**.

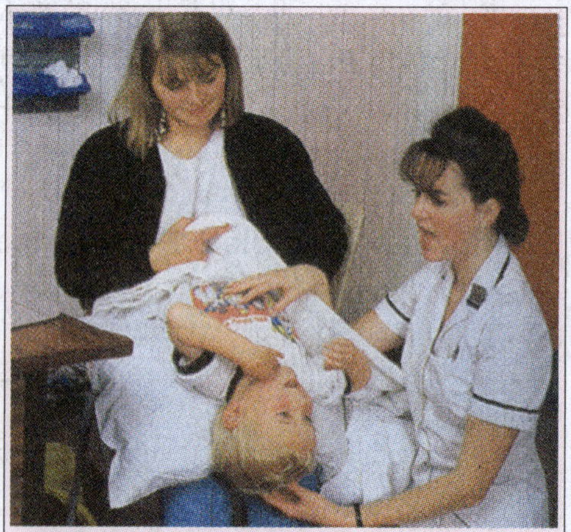

A young cystic fibrosis patient. This disease is a case of pleiotropy. These patients have problems with breathing & digestion. They produce rather viscous mucus, particularly in the gut, pancreas and lungs. Male patients are infertile.

Examples of Pleiotropism

1. In *Drosophila* the recessive gene for vestigial wings cause vestigial wings in homozygous condition. However, careful observations show that other traits as well are affected—(i) the tiny wing- like balancer behind the wings; (ii) certain bristles; (iii) the structure of the reproductive organs; (iv) egg production is lowered, and, (v) longevity is reduced.

2. In human, the gene for disease **phenylketonuria** has pleiotropic effect and produces various abnormal phenotypic traits, collectively called **syn-**

drome. For example, the affected individuals secrete excessive quantity of amino acid phenylalanine in their urine, cerebrospinal fluid and blood. They become short stature, mentally deficient, with widely spaced incisors, with pigmented patches on skin, with excessive sweating, and with non-pigmented hairs and eyes.

3. The A^Y allele for yellow coat in mice is also a good example of pleiotropic gene. It affects two characters : coat colour and survival. It is most probable that both effects of the A^Y allele are the result of same basic cause which promotes the yellowness of coat in a single dose and death in double dose. Genetic analysis has revealed that lethal pleiotropic A^Y allele basically affects the cartilage of mice and cause death.

REVISION QUESTIONS AND PROBLEMS

1. Distinguish between epistasis and dominance. What does gene interaction mean ?
2. Write short notes on the following :
   (i) Epistasis and hypostasis; (ii) Complementary genes; (iii) Supplementary gens; (iv) Inhibitors; (v) Mutually supplementary genes; (vi) Atavism; (vii) Pleiotropic genes; (viii) Expressivity; (ix) Penetrance.
3. What are the lethal genes ? Describe lethal genes by considering one example from each of the following—plants, animals and humans.
4. Four comb shapes namely rose, pea, walnut and single in poultry are known to be governed by two gene loci. The genotype R–P–produces walnut comb, characteristic of the Malay breed : R–pp produces rose comb, characteristic of the Wayandotte breed; rr P–produces pea comb, characteristic of the Brahma breed ; rr pp produces single comb, characteristic of the Leghorn breed. (a) If pure Wyandottes are crossed with pure Brahmas what phenotypic ratios are expected in the F_1 and F_2 ? (b) A Malay hen was crossed with a Leghorn cock and produced a dozen eggs, 3 of which grew into rose combed birds and 9 with walnut combs. What is the probable genotype of hen ? (c) Determine the proportion of comb types that would be expected in offspring from each of the following crosses: (1) Rrpp×RrPP, (2) rrPp×RrPp, (3) rrPP× RRPp, (4) RrPp×rrpp, (5) RrPp×RRpp, (6) RRpp × rrpp; (7) RRPP × rrpp ; (8) Rrpp × Rrpp; (9) rrPp × Rrpp; (10) rrPp × rrpp.
5. Listed below are 7 two-factor interaction ratios observed in progeny from various dihybrid parents. Suppose that in each of these cases one of the dihybrid parents is testcrossed (instead of being mated to another dihybrid individual). What phenotypic ratio is expected in the progeny of each testcrosses? (1) 9 : 6 : 1 ; (2) 9 : 3 : 4 ; (3) 9 : 7 ; (4) 12 : 3 : 1; (5) 9 : 3 : 3 : 1; (6) 13 : 3.
6. **Red** colour in wheat kernel is produced by the genotype R–B–, **white** by the double recessive genotype (rr bb). The genotypes R–bb and rr B– produce **brown** kernels. A homozygous red variety is crossed to a white variety what phenotypic results are expected in the F_1 and F_2.
7. In mice, a dominant allele C must be present in order for any pigment to be developed in the coat. The kind of pigment produced depends upon another locus, thus, B– produces black and bb produces brown. Individuals of epistatic genotype, cc, are incapable of pigment production and are called albinos. A homozygous black female is test crossed to an albino male. (a) What phenotypic ratio is expected in the F_1 and F_2 ? (b) If all the F_2 albino mice are allowed to mate at random, what genotypic ratio is expected in the progeny ?
8. An inhibitor of pigment production in onion bulbs (I–) shows dominant epistasis over another locus, the genotype ii R– producing red bulbs and ii rr producing yellow bulbs. (a) when a pure white strain is crossed to a pure red strain and produces an all white F_1 and an F_2 with 12/16 white, 3/16 red and 1/16 yellow. What were the genotypes of the parents? (b) If yellow onions are crossed to a pure white strain of a genotype different from the parental type in part (a), what phenotypic ratio is expected in the F_1 and F_2 generations ?
9. In *Drosophila*, a dominant gene (D) for a phenotype called **dichaete** alters the bristles and also makes the wings to remain extended from the body while the fly is at rest. It is homozygous lethal (*i.e.*, DD). (a) Diagram a cross between two dichaete (Dd) flies and summarize the expected results. (b) Diagram a cross between dichaete and wild type and summarize the expected results.

ANSWERS TO PROBLEMS

4. (a) F_1 : all walnut comb; F_2 : 9/16 walnut : 3/16 rose : 3/16 pea : 1/16 single ; (b) RR Pp

(c)

	R-P- Walnut	R-pp Rose	rrP- Pea	rrpp Single
(1)	3/4	–	1/4	–
(2)	3/8	1/8	3/8	1/8
(3)	all	–	–	–
(4)	1/4	1/4	1/4	1/4
(5)	1/2	1/2	–	–
(6)	–	all	–	–
(7)	all	–	–	–
(8)	–	3/4	–	–
(9)	1/4	1/4	1/4	1/4
(10)	–	–	1/2	1/2

5. (1) 1 : 2 : 1 ; (2) 1 : 1 : 2 ; (3) 1 : 3 ; (4) 3 : 1; (5) 2 : 1 : 1 ; (6) 1 : 1 : 1: 1; (7) 3 : 1.
6. F_1 : red; F_2 : 9/16 red : 6/16 brown : 1/16 white.
7. (a) F_1 : all black; F_2 : 9/16 black : 3/16 brown : 4/16 albino ;
   (b) 1/4 BB cc : 1/2 Bb cc : 1/4 bb cc.
8. (a) II rr × ii RR ; (b) F_1 : all white; F_2 : 12/16 white : 3/16 red : 1/16 yellow.
9. (a) **P :** Dd × Dd
   Gametes : (D) (d)   (D) (d)
   Progeny : 1 DD : 2 Dd : 1 dd
   Dies   Dichaete   Wild type
   Summary : 2 Dichaete : 1 Wild
   (b) **P :** D d × dd
   Gametes : (D) (d)   (d)
   Progeny : 1 Dd : 1 dd
   Summary : 1 Dichaete : 1 Wild.

CHAPTER

5

Quantitative Genetics

(Inheritance of Multiple Genes)

A field of pumpkins, where size is under polygenic control.

The phenotypic traits of the different organisms may be of two kinds, *viz.,* qualitative and quantitative. The **quali tative traits** are the classical Mendelian traits of **kinds** such as form (*e.g.,* round or wrinkled seeds of pea); structure (*e.g.,* horned or hornless condition in cattles); pigments (*e.g.,* black or white coat of guinea pigs); and antigens and antibodies (*e.g.,* blood group types of man) and so on. We have already discussed in previous chapters that each qualitative trait may be under genetic control of two or many alleles of a single gene with little or no environmental modifications to obscure the gene effects. The organisms possessing qualitative traits have distinct (separate) phenotypic classes and are said to exhibit **discontinuous variations**. The **quantitative traits**, however, are economically important measurable phenotypic traits of **degree** such as height, weight, shape, skin pigmentation, metabolic activity, reproductive rate, behaviour, eye-facet or bristle number in *Drosophila*, susceptibility to pathological diseases or intelligence in man; amount of flowers, fruits, seeds, milk, meat or egg produced by plants or animals, etc. Economically important traits such as body weight gains, mature plant heights, egg or milk production records, yield of grain per acre, etc., are also quantitative traits. The quantitative traits are also called **metric traits**. They do not show clear cut differences between individuals and forms a spectrum of phenotypes which blend imperceptively from one type to another to cause **continuous variations**.

Height in humans is under polygenic control.

In contrast to qualitative traits, the quantitative traits may be modified variously by the environmental conditions and are usually governed by many factors or genes (perhaps 10 or 100 or more), each contributing such a small amount of phenotype that their individual effects cannot be detected by Mendelian methods but by only statistical methods. Such genes which are non-allelic and affect the phenotype of a single quantitative trait, are called **polygenes** or **cumulative genes**. The inheritance of polygenes or quantitative traits is called **quantitative inheritance**, **multiple factor inheritance**, **multiple gene inheritance** or **polygenic inheritance**.

Multiple Factor Hypothesis

Two or more different pairs of alleles, with presumed cumulative effects, govern the quantitative traits. Those alleles which contribute to the trait involved are called **contributing**, **effective** or **active alleles**; those alleles which do not appear to do so are referred to as **non-contributing**, **non-effective** or **null alleles**. A gene, individually exerting a slight effect on the phenotype but along with a few or many other genes, controls a quantitative trait is called a **polygene** (a term coined by **K. Mather**). Since there are usually many genes of this kind for one quantitative trait, they are named **multiple factors**, each factor having too small an effect to be traced. This **multiple factor hypothesis** (*i.e.,* large number of genes, each with a small effect, are segregating to produce quantitative variation) has long been the basic model of quantitative genetics.

Differences Between Quantitative and Qualitative Genetics

The genetical studies of qualitative and quantitative traits are called **qualitative genetics** and **quantitative genetics**, respectively. The major differences between the two are following :

Table 5.1. Differences between qualitative and quantitative traits or genetics.

Qualitative genetics	Quantitative genetics
1. Characters of kind.	1. Characters of degree.
2. Discontinuous variation; distinct phenotypic classes.	2. Continuous variations; phenotypic measurements form a spectrum.
3. Single gene effects.	3. Polygenic control; effects of single genes too slight to be detected.
4. Concerned with individual matings and their progeny.	4. Concerned with population of organisms consisting of all possible kinds of matings.
5. Analyzed by making counts and ratios	5. Statistical analyses give estimates of population parameters such as the mean and standard deviation.

HISTORICAL

In 1760, **Joseph Kolreuter** inadvertently reported first case of continuous variation due to quantitative trait. He crossed the tall and dwarf varieties of tobacco, *Nicotiana*. The F_1 plants were intermediate in size between the two parent varieties. The F_2 progeny showed a continuous gradation from the size of the dwarf to that of the tall parent. Since the basic principles of genetics were yet not established then, these results could not be explained by Kolreuter. In post-mendelian era, there were two main groups of geneticists : 1. **Mendelians**, who believed that all evolutionary important heritable differences were qualitative and discontinuous (*e.g.*, **Bateson** and **de Vries**). 2. **Biometricians**, who proposed that heritable variation was basically quantitative and continuous and that genes did not exist as separate units (*e.g.*, **Galton**). Later on, it was resolved that both of these views were only partly correct. To prove that mendelians and biometricians were only partly correct, **Johannsen** (1903) used the character of seed weight in beans and analysed its breeding behaviour. He proved that continuous variation are heritable, but he could not explain genetics of the quantitative traits. It was **Yule** (1906) who suggested that quantitative variation may be controlled by large number of individual genes, each having a small effect. **H. Nilsson-Ehle** (1908), **East** (1910) and **C.B. Davenport** (1913) demonstrated segregation and independent assortment of genes controlling quantitative traits such as kernel colour in wheat, corolla length in tobacco and skin colour in negro and white population of USA, respectively.

CHARACTERISTICS OF MULTIPLE GENES

Multiple genes for quantitative traits have following characteristics :

1. Each contributing allele in the series of multiple genes produces an equal effect.
2. Effects of each contributing allele are cumulative or additive.
3. There is no dominance, rather, there exist pairs of contributing and non-contributing alleles.
4. There is no epistasis (masking of the phenotypes) among genes at different loci.
5. There is no linkage involved.
6. The environmental conditions have considerable effect on the phenotypic expression of poly- genes for the quantitative traits. The genotype determines the **range** an individual will occupy with regard to a given quantitative character; environment determines the **point** within the genetically determined range at which an individual's measurements will fall.

EXAMPLES OF QUANTITATIVE INHERITANCE

1. Kernel Colour in Wheat

A whole grain or seed of a cereal plant such as

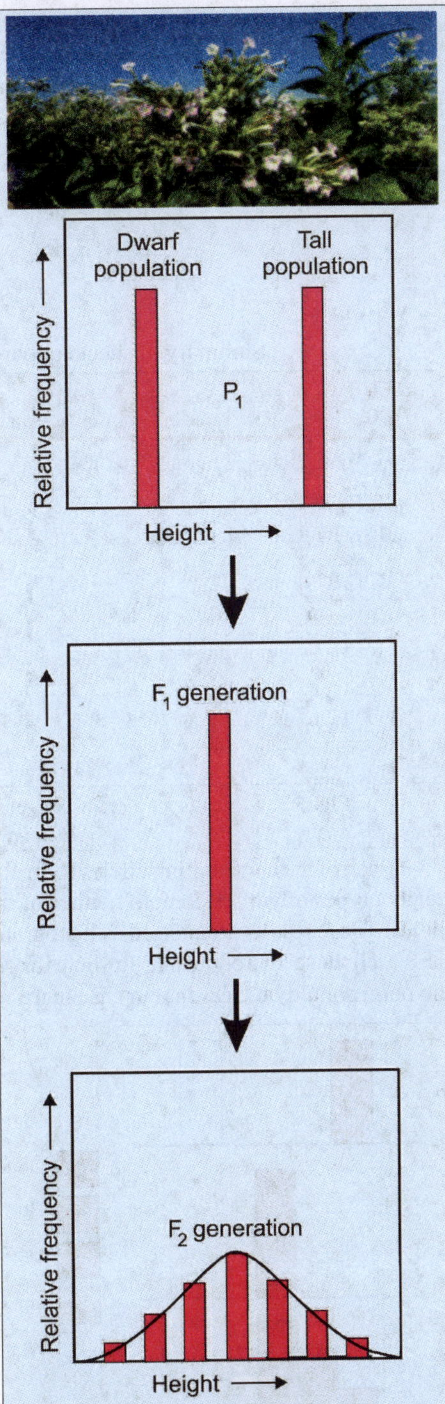

Histograms showing the relative frequency of individuals expressing height phenotypes derived from Kolreuter's cross between dwarf and tall tobacco plants carried to the F_2 generation. The photograph shows a tobacco plant.

corn, wheat, barley, etc., is called **kernel**. Kernel colour in wheat is a quantitative trait and its inheritance was studied by Swedish geneticist **H. Nilsson– Ehle** for the first time in 1908. When he crossed a certain red strain to a white strain, he observed that the F_1 was all light red and that approximately 1/16 of the F_2 was as extreme as the parents, *i.e.*, 1/16 was white and 1/16 was red. He interpreted these results in terms of two genes, each with a pair of alleles exhibiting cumulative effect (Fig. 5.1).

P : Red kernel × White kernel

$R_1 R_1 R_2 R_2$ $r_1 r_1 r_2 r_2$

F_1 : Light red

$R_1 r_1 R_2 r_2$

F_2 : Summary of checker board derived results, *i.e.*, F_2 genotypic and phenotypic ratios:

Genotype	Genotypic ratio	Number of contributing alleles	Phenotype	Phenotypic ratio
$R_1 R_1 R_2 R_2$	1	4	Red	1 } Colou-red (15/16)
$R_1 R_1 R_2 r_2$ $R_1 r_1 R_2 R_2$	2 2 }	3	Medium red	4
$R_1 r_1 R_2 r_2$ $R_1 R_1 r_2 r_2$ $r_1 r_1 R_2 R_2$	4 1 1 }	2	Light red	6
$R_1 r_1 r_2 r_2$ $r_1 r_1 R_2 r_2$	2 2 }	1	Very light red	4
$r_1 r_1 r_2 r_2$	1	0	White	1 } Colourless (1/16)

Fig. 5.1. Results of a cross between two varieties of wheat having red kernel and white kernel showing cumulative effect of alleles.

Each of the contributing alleles R_1 or R_2 adds some red to the phenotype of kernel colour, so that the genotypes of whites contain neither of these alleles and a red genotype contains only R_1 and R_2 alleles. These results are plotted as histograms in Fig. 5.2. Here five phenotypic classes are obtained in F_2; each 'dose' of a contributing allele for pigment production increases depth of colour. At this stage one point should be clear that in case there were two genes involved, there would be obtained 15 : 1 ratio (15 coloured : 1) (see Fig. 5.1).

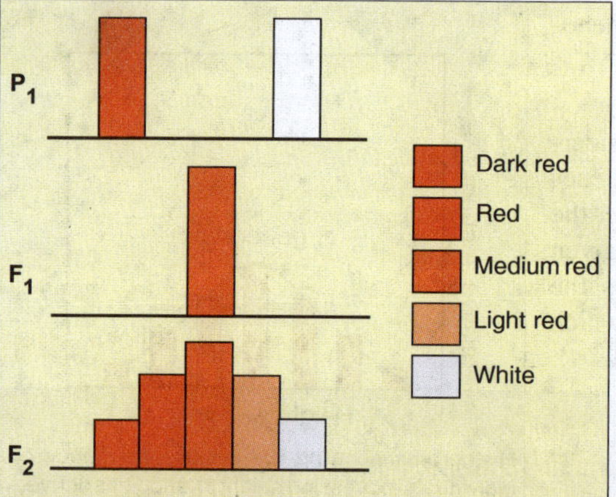

Fig. 5.2. Wheat colour as an example of a quantitative trait (after Stansfield, 1969).

Later on, when certain other strains of wheat with dark red kernels were crossed with whites exhibited an F_1 phenotype intermediate between the two parental types, but only 1/64 of the F_2 are white. In this case the F_1 is probably segregating for three pairs of genes and only the genotype r_1r_1 r_2r_2 r_3r_3 produce white. There are seven classes of phenotypes in a ratio of 1 : 6 : 15 : 20 : 15 : 6 : 1.

Above described ratios, *i.e.*, 1 : 4 : 6 : 4 : 1 and 1 : 6 : 15 : 20 : 15 : 6 : 1, can be easily obtained by the expansion of binomial equation $(1/2+1/2)^n$, where n is the number of alleles. In case of 2 genes, n=4, while in case of 3 genes n=6. This expansion can be obtained by the use of Pascal's triangle (see Table 5-2).

E.M. East (1913) extended the polygenic hypothesis to several cases in plants. For instance, in case of maize it was demonstrated that the ear size is controlled by multiple factors. Similarly, flower size in tobacco had the same pattern of inheritance. In these cases, the number of genes controlling the character were many and usually more than two or three.

Table 5.2. Expansion of binomial $(1/2+1/2)^n$ using different values of n (number of alleles) with the help of Pascal's triangle.

Number of alleles	Expansion of binomial $(1/2+1/2)^n$
1	1 : 1
2	1 : 2 : 1
3	1 : 3 : 3 : 1
4	1 : 4 : 6 : 4 : 1
5	1 : 5 : 10 : 10 : 5 : 1
6	1 : 6 : 15 : 20 : 15 : 6 : 1
7	1 : 7 : 21 : 35 : 35 : 21 : 7 : 1
8	1 : 8 : 28 : 56 : 70 : 56 : 28 : 8 : 1

2. Skin Colour in Man

Another classical example of polygenic inheritance was given by **Davenport** (1913) in Jamaica. He found that two pairs of genes, A-a and B-b cause the difference in skin pigmentation between negro and caucasian people. These genes were found to affect the character in additive fashion. Thus, a true negro has four dominant genes, AABB, and a white has four recessive genes aabb. The F_1 offspring of mating of aabb with AABB, are all AaBb and have an intermediate skin colour termed **mulatto**. A mating of two such mulattoes produces a wide variety of skin colour in the offspring, ranging from skins as dark as the original negro parent to as white as the original white parent. The result of this cross have been shown in Figure 5.3. The F_2 phenotypic and genotypic ratios have been tabulated in Table 5-3.

Human skin colour is a result of polygenic inheritance.

Flowers of a tobacco plant.

P_1 : AABB Negro × aabb White

↓ ↓

P_1 gametes : (AB) ↓ (ab)

F_1 : AaBb Mulattoes Intermediate skin colour

Intercross : Aa Bb Mulattoes × AaBb Mulattoes

♀ \ ♂	AB	Ab	aB	ab
AB	AABB Like negro	AABb Darker than mulattoes	AaBB Darker than mulattoes	AaBb Like mulattoes
Ab	AABb Darker than mulattoes	AAbb Like mulattoes	AaBb Like mulattoes	Aabb Lighter than mulattoes
aB	AaBb Darker than mulattoes	AaBb Like mulattoes	aaBB Like mulattoes	aaBb Lighter than mulattoes
ab	AaBb Like mulattoes	Aabb Lighter than mulattoes	aaBb Lighter than mulattoes	aabb Like white

F_2 :	Negro colour (1/16)	Colour between mulattoes and negro (4/16)	Colour of mulattoes (6/16)	Colour between mulattoes and white (4/16)	White skin colour (1/16)

Fig. 5.3. Checkerbord exhibiting segregation of skin colour in F_2 generation from a marriage between a negro and a white person.

Table 5.3. **Phenotypic and genotypic ratios of F_2 generation of cross shown in Figure 5.3.**

Phenotypes	Genotypes	Genotypic Frequency	Ratio
Black (Negro)	AABB	1	1
Dark	Aa BB,	2	4
	AA Bb	2	
Intermediate (Mulatto)	Aa Bb	4	6
	aa BB	1	
	AA bb	1	
Light	Aa bb	2	4
	aa Bb	2	
White	aa bb	1.	1

These results are clearly showing that A and B genes produce about the same amount of darkening of the skin and, therefore, the increase or decrease of A and B genes cause variable phenotypes in F_2 in the ratio of 1 Negro : 4 dark : 6 intermediate : 4 light : 1 white.

Other examples of quantitative traits of human beings include height, intelligence (I.Q.), hair colour (except for red versus non-red) and eye colour.

3. Eye Colour in Man

In human beings, the colour of eye is found to be determined by polygenes. These genes have been suggested by polygenes. These genes have been suggested to be X-linked (see **Burns** and **Battino**, 1989). At least 9 classes of eye colour can be recognized in humans. In order of increasing amount of melanin pigmentation, these eye colours can be designated as light blue, medium blue, dark blue, grey, green, hazel, light brown, medium brown and dark brown. The number of contributing alleles for these colours have been tabulated in Table 5-4.

Table 5.4. Number of contributing alleles for each type of eye colour of human beings (Source : Burns and Bottino, 1989).

Number of contributing alleles	Eye colour
0	Light blue
1	Medium blue
2	Dark blue
3	Grey
4	Green
5	Hazel
6	Light brown
7	Medium brown
8	Dark brown

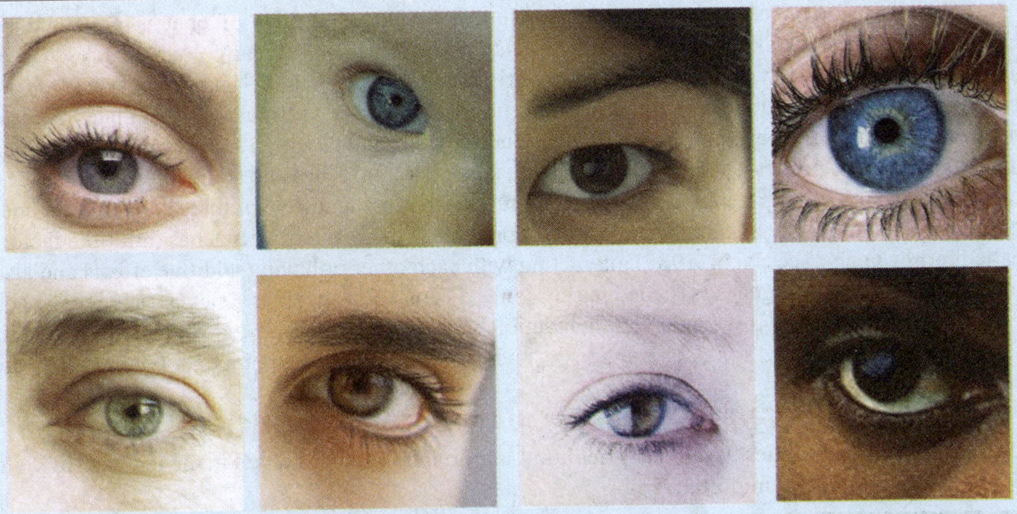
Different eye colours found in human beings.

TRANSGRESSIVE VARIATION

In some types of quantitative inheritances, sometimes, F_2 offsprings do not display continuous variation but some individuals exhibit great degree of variability and do not resemble with their either parent or even remote ancestors in that quantitative trait. Such a case in which the extremes of F_2 exceed those of the parent is called **transgressive variation**.

Example. Punnett and **Bailey** (1914, 1923) have reported first case of transgressive variation from a cross in between a large Golden Hamburg chicken with the smaller Sebright Bantam variety of chicken. The F_1 was intermediate in size between the parents and fairly uniform, but mean size of the F_2 was about the same as that of the F_1, but the variability of the F_2 was so great that a few individuals were found to exceed the size of either parental type (showing transgressive variation). In this case, four gene loci are thought to cause the transgressive variation as follows :

P :	aa BB CC DD Large Golden Hamburg chicken (6 contributing alleles)	X ↓	AA bb cc dd Small Sebright Bantam chicken (2 contributing allele)

F_1 : Aa Bb Cc Dd
Intermediate sized hybrid
(4 contributing allele)

F_2 : Some genotypes could segregate out in the F_2 with phenotypic values which exceed that of the parents. For example :

AA BB CC DD	8	Contributing alleles	Larger than Golden Hamburg
Aa BB CC DD	7	,, ,,	
Aa bb cc dd	1	,, ,,	Smaller than Sebright Bantams
aa bb cc dd	0	No contributing allele (physiological minimum)	

MODIFIERS OR MODIFYING GENES

The inheritance of certain traits is governed by a single pair of genes which determines the presence or absence of the trait plus a number of multiple genes which determine the extent of the trait. Thus, modifier is a gene that affects the expression of another nonallelic gene.

Example. In mice pie bald spotting pattern (or white spotting) of coat is governed by homozygous recessive alleles ss. Dominant allele S determines solid coat colour (or presence or absence of spots). Mice with ss genotype show variation in spotting pattern—white spots in some are present only on the belly, ranging to those with an entirely white coat with many intermediates. A cross between two animals differing in spot pattern produces F_1 individuals which are intermediate between the two parents. F_2 progeny shows variations with some parental types and some intermediates. The back cross of an F_1 with parental types with a frequency that is suggestive of three or four polygenes influencing the trait. The appearance of different patterns is the result of polygenes (multiple factors called **minor genes**, see **Sarin**, 1985) interacting with each other or having additive effects and also interacting with pie bald genes ss (called **major genes**). Thus, the polygenes s_1s_1, $s_2 s_2$, $s_3 s_3$, $s_n s_n$, each with small expression interact and exert their action by changing the magnitude of a major gene pair ss. The minor genes (*e.g.*, s_1 s_1, s_2 s_2, s_3 s_3, s_n s_n) which modify the effect of a major gene pair ss are called **modifiers**.

SIGNIFICANCE OF QUANTITATIVE GENETICS

Quantitative genetics has great agricultural importance and has helped in the increase of yield of various economically important crops.

REVISION QUESTIONS AND PROBLEMS

1. (a) Are quantitative characters restricted to sexually reproducing organism ?
   (b) Define the multiple-gene hypothesis. Evaluate its practical and theoretical significance and describe some of its limitations.
   (c) Describe briefly the evidence supporting the multiple-gene hypothesis.
   (d) Discuss the statement : No new principles of genetics have originated from the study of quantitative characters.
   (e) Why is it more different to study the inheritance of quantitative characters such as size, weight, and intelligence than qualitative ones such as ABO and Rh blood antigens ?
2. Distinguish between the term *polygene* and *modifying gene.* What have the two kinds of genes in common.
3. Describe the kinds of observations that would lead you to suspect that a certain human character was controlled by polygenes.
4. Which of the following human phenotypes would appear to be based on polygene inheritance : intelligence, absence of incisors, height, phenylketonuria, ability to taste phenylthiocarbamide, skin colour, cryptophthalmos (*i.e.,* failure of eyelids to separate in embryonic development), eye colour ?

5. Show, by means of appropriate genotypes, how parents may have children taller than themselves.
6. Skin colour in man is controlled by additive genes : If both mother and father have intermediate skin colour, can you expect children (i) with lighter skin, (ii) with darker skin ? Can you expect children with darker skin, if both parents have light skin ?
7. Two homozygous varieties of *Nicotiana longiflora* have mean corolla lengths of 40.5 mm and 93.3 mm. The average of the F_1 hybrids form these two varieties was of intermediate length. Among 444 F_2 plants, none was found to have flowers either as long or as short as the average of the parental varieties. Estimate the minimal number of pairs of alleles segregating from the F_1.

ANSWERS TO PROBLEMS

4. All traits showing practically continuous variation are likely to be due to polygenes. Here, the traits of intelligence, height, skin colour, and eye colour probably involve polygenes.
5. Any parental genotypes that can produce at least some progeny with a greater number of contributing alleles than they themselves have are possible, for example, Aa Bb Cc Dd × Aa Bb Cc Dd, Aa Bb Cc Dd × aa bb Cc Dd, etc.
7. If four pairs of alleles were segregating from the F_1, we expect $(1/4)^4 = 1/256$ of the F_2 to be as extreme as one of the other parental average. Likewise, if five pairs of alleles were segregating, we expect $(1/4)^5 = 1/1024$ of the F_2 to be as extreme as one parent or the other. Since none of the 444 F_2 plants had flowers this extreme, more than four loci (minimum of five loci) are probably segregating from the F_1.

CHAPTER

6

Inbreeding, Outbreeding and Hybrid Vigour

The word reproduction implies replication, and it is true that biologic reproduction almost always yields a rea sonable carbon copy of parent unit. However, sexual reproduction, practiced by the majority of animals, plants and microorganisms, produces diversity needed for survival in a world of constant change (*i.e.,* evolution). A sexually reproducing species may have individuals which are unisexual or bisexual. Bisexuality is common in plants and lower animals. Higher animals are mostly unisexual, *i.e.,* separate male and female sexes exist in these species. The essential feature of sexual reproduction, whether the individuals of a species are unisexual or bisexual, homogametic or heterogametic (*i.e.,* producing one type of gametes or more than one type of gametes) is a union of two gametes to form a zygote. Sexual reproduction performs the basic function of providing a great variety of genotypes than could arise under asexual reproduction (including apomixis and parthenogenesis); asexual reproduction produces only one type of genotype in the offspring. The production of an infinite variety of genotypes has much higher evolutionary significance in the sense that the better genotype can be selected and perpetuated by natural selection.

Among living organisms fundamentally following two systems of matings occur : **inbreeding** refers to the production of offspring through matings between related parents, where as **outbreeding** is the production of offspring through matings

Inbreeding in self fertilizing pea plants was a real advantage to Mendel.

between unrelated parents. Inbreeding results in an increase in homozygosity where as outbreeding results in an increase in heterozygosity.

INBREEDING

The process of mating of individuals which are more closely related than the average of the population to which they belong, is called **inbreeding**. For example, parthenogenesis in animals and apomixis and self-fertilization in plants are the most extreme type of inbreeding. Inbreeding in self-fertilizing pea plants was a real advantage to Mendel in his studies which provided pure lines of pea plants for his hybridization experiments. The term 'pure line' was coined by **W. Johannsen** in 1903 for the true breeding self-fertilized plants. (For obtaining pure lines Johannsen performed classical crosses on common garden bean plant or *Phaseolus vulgaris*). **Pure line** population is the one which breeds true when selfed without producing any genetic variability in the progeny.

Methods of Inbreeding

In plants, ova fertilized by the pollen of either the same plants (in case of bisexual plants) or of the other plant of the same genotype (in case of unisexual as well as bisexual plants), is called **self-fertilization**. However, in bisexual plants numerous structural and functional adaptations have been recorded (such as self sterility, see Chapter 10) which help plants with bisexual or hermaphrodite flowers avoid self-fertilization.

Normally, inbreeding is affected by restrictions in population size or area which brings about the mating between relatives. Since close relatives have similar genes because of common heritage, inbreeding increases the frequency of homozygotes, but does not bring about a change in overall gene frequencies. Thus, a mating between two heterozygotes as regards two alleles A and a will result in half of the population homozygous for either gene A or a and half of the population heterozygous like the parent but the overall frequencies of A and a remain unchanged :

Aa	X	Aa
1 AA :	2 Aa :	1 aa

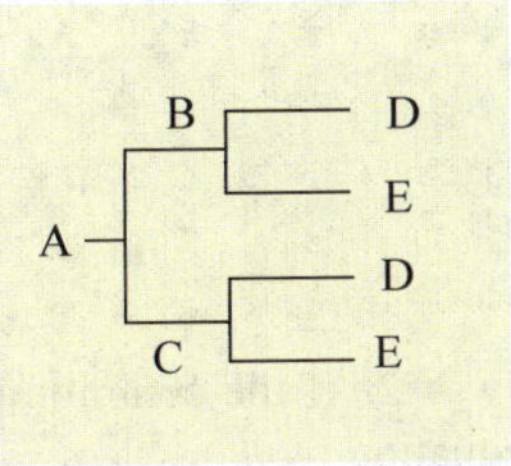

Thus, inbreeding brings about the recessive gene to appear in a homozygous state (aa). Once a recessive allele is in a homozygous state, natural selection can operate upon the rare recessives. Artificial selection is also possible as the homozygous recessives are phenotypically differentiated from the dominant population. The inbred pedigrees can be depicted as follows :

Here, B and C are full sibs, *i.e.*, have common parents. This pedigree can also be represented by the following arrow diagram :

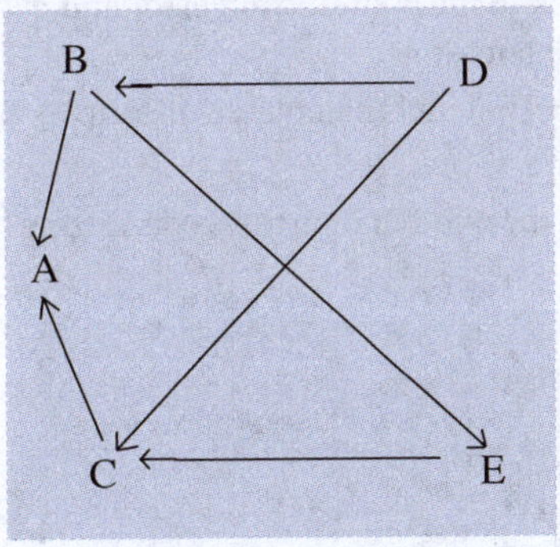

1. Coefficient of relationship (R). Coefficient is expression of the amount or degree of any quality possessed by a substance. It is also the degree of physical or chemical change normally occurring in that substance under stated conditions. The **coefficient of relationship (R)** characterises the percentage of genes held in common by two individuals due to their common ancestory. Each individual gets only a sample half of his genotype from one of his parent, each arrow in the above arrow diagram represents a probability of half. The sum (Σ) of all pathways between two individuals through common ancestors is the coefficient of relationship and is represented by R :

(i) R_{BC} = The coefficient relationship between the full sibs B and C and is calculated as follows :

i.e., individuals B and C contain ½×½ = ¼ of their genes in common through ancestor D.

(ii) *i.e.*, individuals B and C contain ½×½ = ¼ of their genes in common through ancestor E.

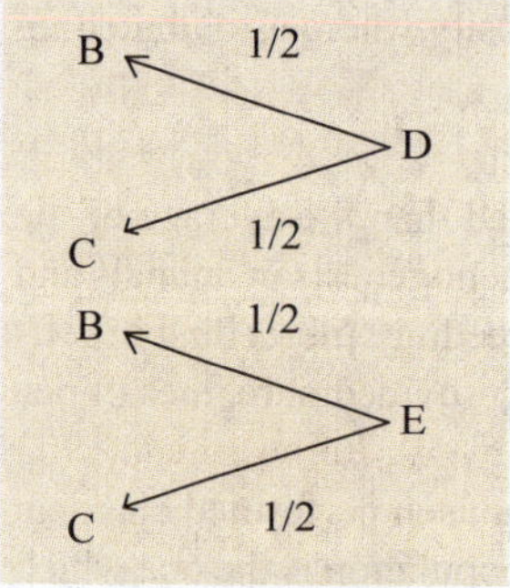

(iii) The sum of these two pathways, the coefficient of relationship, between the full sibs B and C = ¼+¼=½ = 50 per cent.

2. Inbreeding coefficient. In a diploid organism each gene has two alleles occupy the same locus. They are called **identical genes** if they have descended from the same gene; such genes are homozygous at the locus. Such a homozygosity is also caused when two alleles in a diploid organism are not descended from the common gene but the alleles of identical origin are brought together through mating between first cousins. Such alleles are called **similar alleles**. The fine difference between these two types of alleles becomes clear by the following chart : (see page 64).

The probability that the two alleles in a zygote are identical by descent, *i.e.*, are the replication product of the same gene of an ancestor is measured by the **inbreeding coefficient** (**F**) and is calculated as follows :

1. If the parents B and C are full sibs, *i.e.*, B and C parents are 50 per cent related, the inbreeding coefficient of individual (A) can be calculated by the equation $F_A = \frac{1}{2} R_{BC}$, where R_{BC} is the coefficient of relationship between the full sib parents (B and C) of A.

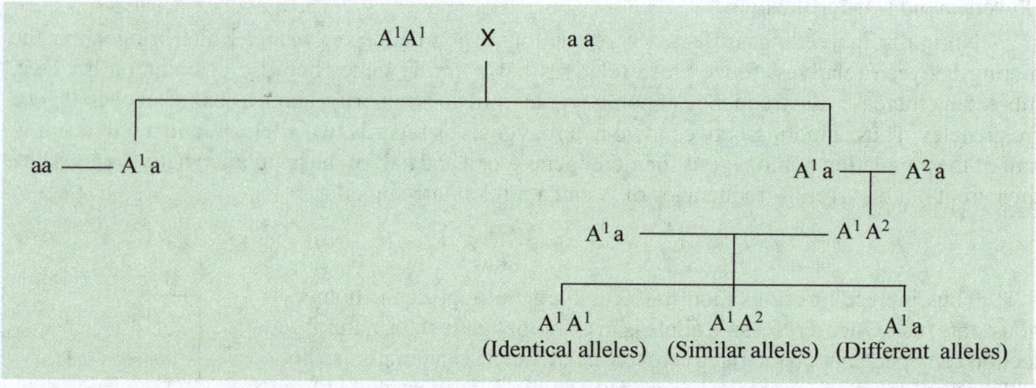

2. If the common ancestors are not inbred, the inbreeding coefficient is calculated by the equation :

$$F = \Sigma (\tfrac{1}{2})^{n_1+n_2+1}$$

where n_1, is the number of generations (arrows) from one parent back to the common ancestor and n_2 is the number of generations from the other parent back to the same ancestor.

3. In case the common ancestors are inbred, the inbreeding coefficient is calcutated as follows:

$$F = \Sigma (\tfrac{1}{2})^{n_1+n_2+1} \text{ (1+F ancestor)}$$

4. The coefficient of inbreeding is also calculated by counting the number of arrows connecting the individual through one parent back to the common ancestor and back again to his other parent by the following equation :

$$F = \Sigma (\tfrac{1}{2})^n (1+F_A)$$

n = number of arrows which connect the individual through one parent back to the common ancestor and back again to his other parent. F_A is the inbreeding coefficient of the common ancestor. For example, the inbreeding coeffi-

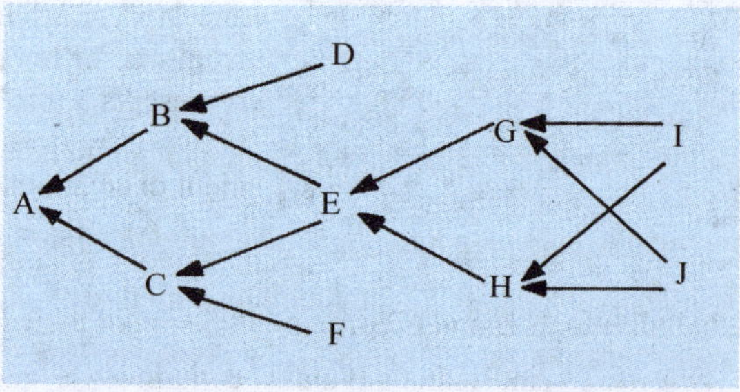

Range cattle usually breed panmicticly.

cient for A in the following arrow diagram can be calculated by following method:

B and C are the parents of A. There is only one pathway from B and C and that goes through ancestor E. Ancestor E is inbred, because its parents (G and H) are full sibs and are 50 per cent related. The inbreeding coefficient can be calculated as

$F_E = ½ R_{GH}$ (R = the coefficient of relationship between the full sibs G and H)

or $F_E = ½ (0.5) = 0.25$

$F_A = \Sigma (½)^n (1+F_{E\ (ancestor)})$

or $F_A = (½)^3 (1+0.25) = 0.156$

3. Panmixis (Random mating). If the breeder assigns no mating restraints upon the selected individuals, their gametes are likely to randomly unite by chance alone. This is commonly the case with outcrossing (non-self-fertilizing) plants. Wind or insect carry pollen from one plant to another in essentially a random manner. Even livestock such as sheep and range cattle are usually bred panmicticly. The males locate females as they came into heat, copulate with ("cover") and inseminate them without any artificial restrictions as they forage for food over large tracts of grazing land. This mating method is most likely to generate the greatest genetic diversity among the progeny.

4. Assortative and disassortative matings. In sexually reproducing organisms, the most rapid inbreeding system is that between brothers and sisters who share both parents in common. This type of mating is called **full-sib mating** and produces inbreeding coefficient of 25 per cent in the first generation of inbreeding (F_2 of Mendel). This rate is reduced in succeeding generations since some of the alleles are now already identical. Within 10 generations, full-sib matings can produce an inbreeding coefficient of 90 per cent. The other inbreeding systems are **half sib mating**, **parent-offspring mating**, **third-cousin mating** and so on. All these inbreeding systems are called **genetic assortative matings** since the parents of each mating type are sorted and mated together on the basis of their genetic relationship. Such a breeding method tends to increase the inbreeding coefficient. The assortative mating is also of the phenotypic type, *i.e.,* the mating between two like phenotypes, two like dominant phenotypes or between two like recessive phenotypes. If assortative selective mating is continued for many generations, the heterozygotes are eliminated and the resulting population consisted of homozygous dominants and homozygous recessives. If more than one locus is considered at a time, the rate of homozygosity achievement will be slower than for one locus. This is so because now the kind of heterozygotes produced will be more combinations of different loci, *e.g.,* Aa BB, AA Bb,........) and eliminating these will need more number of generations.

Disassortative mating refers to the mating of unlike phenotypes and genotypes and tends to maintain heterozygosity, as in the case of mating between unlike sexes. This preserves the dissimilarities both genetic as well as phenotypic. In primitive organism, sexual differences arose at a single gene locus, *i.e.,* one sex was homozygous and the other heterozygous for that locus, and the disassortative

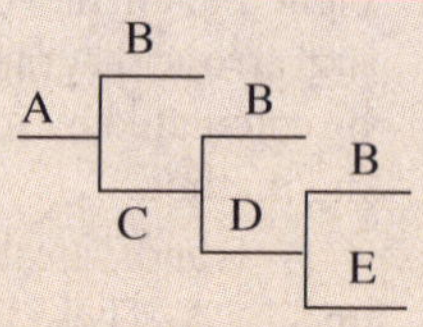

Fig. 6.1. Pedigree exemplifying close line breeding.

matings were the matings between an homozygous and an heterozygous individual for sex locus. Disassortative mating also results from dichogamy, (**Dichogamy** = producing mature male and female reproductive structures at different times); self-sterility in plants in which the mating of like phenotypes (inbreeding) is not possible and fertilization between plants with different genotype is favoured. This maintains heterozygosity within a diploid breeding population.

5. Line breeding. It is a special form of inbreeding utilized for the purpose of maintaining a high genetic relationship to a desirable ancestor. Figure 6.1 shows a pedigree in which close line breeding to B has been practiced so that A possesses more than 50 per cent of B's genes. D possesses 50 per cent of B's genes and transmits 25 per cent to C. B also contributes 50 per cent of his genes to C. Hence, C contains 50 per cent + 25 per cent = 75 per cent B genes and transmits half of them (37.5 per cent) to A. B also contributes 50 per cent of his genes to A. Therefore A has 50 per cent + 37.5 per cent = 87.5 per cent of B's genes.

Genetic Effects of Inbreeding

The continuous inbreeding results, genetically, in homozygosity. It produces homozygous stocks of dominant or recessive genes and eliminate heterozygosity from the inbred population. For example, if we start with a population containing 100 heterozygous individuals (Aa) as shown in figure (Fig. 6.2.), the expected number of homozygous genotype is increased by 50% due to selfing or inbreeding in each generation.

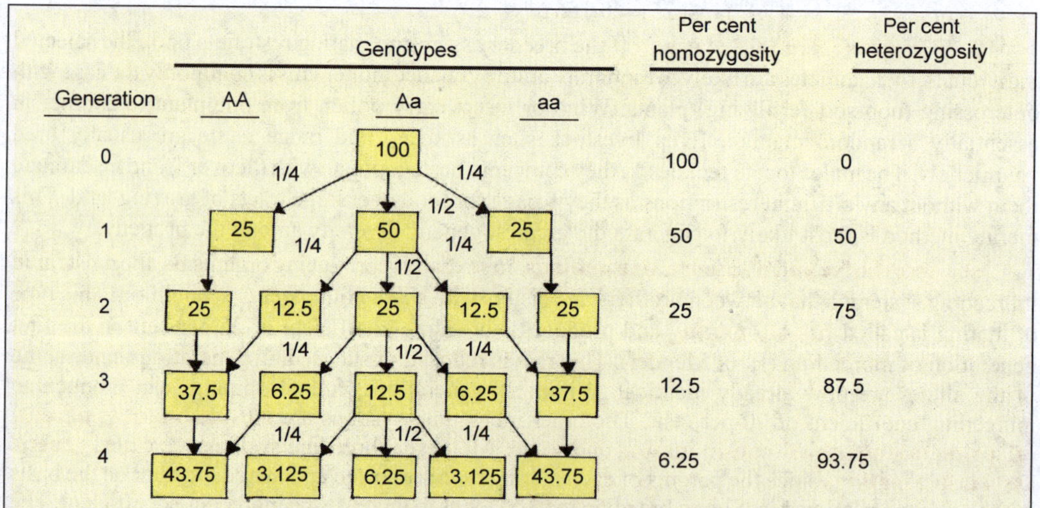

Fig. 6.2. Expected increase in homozygosity due to inbreeding.

Thus, due to inbreeding in each generation the heterozygosity is reduced by 50% and after 10 generation we can expect the total elimination of heterozygosity from the inbred line and production of two homozygous or pure lines. But, because a heterozygous individual possesses several heterozygous allelic pairs, therefore, we can conclude that inbreeding will operate on all genes loci to produce totally pure or homozygous offsprings. In man if inbreeding continued over a number of generations it results in increasing homozygosity, but some what slowly. The different types of inbreedings and their corresponding increase in homozygosity have been graphically illustrated in Fig. 6.3.

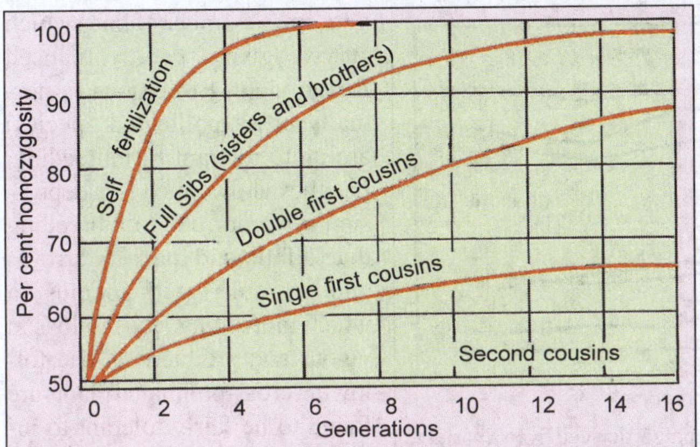

Fig. 6.3. Increase in percentage of homozygosity under various systems of inbreeding.

Decrease in vigour upon inbreeding.

Inbreeding Depression

In a heterozygote the inbreeding increases the probability of homozygosity of deleterious recessive alleles in a inbred population. In other words, one of the consequence of inbreeding is a loss in vigour (*i.e.*, less productive vegetatively and reproductively) which commonly accompanies an increase in homozygosity. This is called **inbreeding depression**.

Inbreeding depression is found to occur due to following four features of inbreeding : (1) increase in frequency of homozygotes, (2) increase in variability between different inbred families, (3) reduction in value of quantitative character in the direction of recessive values, and (4) the dependence of this reduction in value upon dominance. If this inbreeding effect is multiplied for many genes at many loci, there may be a large reduction in value for many traits, including those that affect fitness and survival. In corn (maize) for example, **E.M. East** (1908) and **G. H. Shull** (1909) studied the effects of inbreeding for 30 generations of inbreeding and found independently, that the yielding ability in these lines finally reduced to about one third of the open pollinated variety from which these samples were derived. Both of these authors draw the following important conclusions : (1) A number of lethal and sub-vital types appear in early generations of selfing. (2) The material rapidly separates into distinct lines, which become increasingly uniform for differences in various morphological and functional characteristics. (3) Many of the lines decrease in vigour and fecundity until they cannot be maintained even under the most favourable cultural conditions. (4) The lines that survive show a general decline in size and vigour.

Figure 6.4 shows the decline in size and vigour due to inbreeding in maize; here, the inheritance of two quantitative traits namely plant height and grain yield of three lines are shown for 30 generations of inbreeding. It can be noticed that fixation for plant height occurred after five generations of inbreeding. However, yield continued to decline for at least 20 generations until it reached one-third that of open-pollinated variety from which they were derived. Despite this conspicuous decline, maize was found more tolerant to inbreeding than some organisms where few strains survive two or three generations of inbreeding, *e.g.*, alfalfa and onions. Figure 6.5. illustrates the deterioration in yield of self-fertilized lines of alfalfa and of cross pollinated onions. In alfalfa, upon selfing many sub-vital and lethal types appear and the rate of deterioration of general vigour and productivity is alarming. The

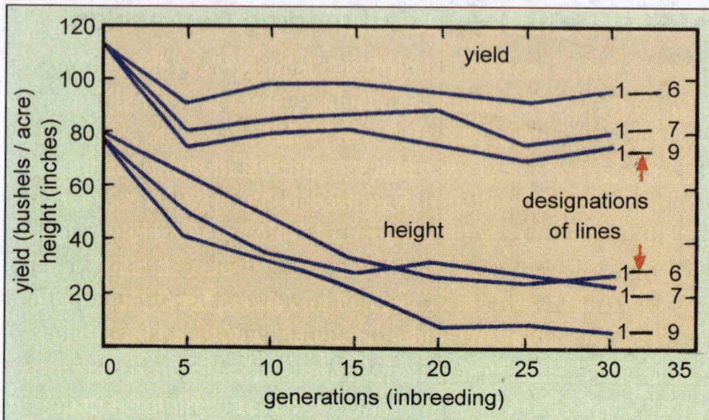

Fig. 6.4. A comparison of three lines of maize, derived from a variety, self-fertilized for 30 generations. Initially, there were four lines, but it became impossible to maintain one of them beyond 20 generations of inbreeding.

very small number of lines which survive give a greatly reduced forage yield. But onions (a normally cross pollinated species) are quite tolerant to inbreeding, *i.e.,* they show much less depression in vigour due to inbreeding than alfalfa and maize. Carrot is another cultivated species in which inbreeding leads to loss in vigour and production. The following cross-pollinated plants are found to be fairly tolerant to inbreeding : sunflowers, rye, timothy, smooth broomgrass and orchard grass. In certain self-pollinated species and normally cross fertilizing species such as cucurbits, inbreeding is found to be continued indefinitely with impunity.

In most animals inbreeding is found to have less remarkable effects on vigour. For example, in rats continuous brother-sister matings were performed for 25 generations, but no drastic deterioration was detected. In *Drosophila* inbreeding usually results in a rapid loss of vigour, but some strains compare favourably with outbred populations after long continued inbreeding. However, in certain breeds of cattle, intensive inbreeding has led to an unfortunate condition; for example, exhaustive inbreeding and selection of beef cattle breed (Hereford) produced dwarf calves of low economic value. These calves show characteristic head and body features of the **brachycephalic dwarfism** (*i.e.,* the characteristic short broad head, extra long lower jaw, bulging forehead, out of proportion abdomen and short legs). Breeding data indicate that a basic recessive gene is necessary for dwarfing, but additional modifier genes have been postulated to account for the different types of dwarfs (see **Gardner**, 1968).

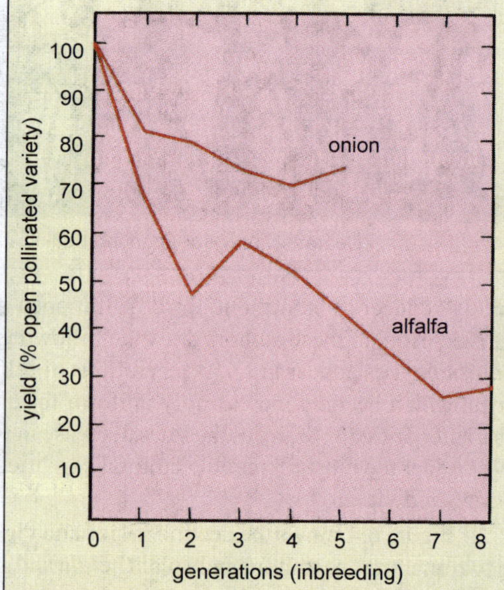

Fig. 6.5. Yields of self-fertilized lines of alfalfa and onions as per cent of open pollinated parental varieties.

Practical Applications of Inbreeding

The correlation of inbreeding and homozygosity exhibits that how inbreeding may cause deleterious effects. As we already know that in a heterozygous individual, the harmful recessive alleles remain masked by their normal dominant alleles. If a heterozygous individual undergoes inbreeding for various generations, there will be equal chances of homozygosity for dominant as well as recessive alleles. In homozygous condition recessive alleles will be able to express their deleterious phenotypic effects on the individual. On the other hand, the homozygosity for dominant alleles has equal opportunity to express their beneficial phenotypic effects on inbred races. The practical applications of inbreeding are following :

The animal breeder have employed the inbreeding to produce best races of horses.

1. Because inbreeding cause homozygosity of deleterious recessive genes which may result in defective phenotype, therefore, in human society, the religious ethics unknowingly and modern social norms consciously have condemned and banned the marriages of brothers and sisters. Further, the plant breeders and animal breeders too avoid inbreedings in the individuals due to this reason.

2. The inbreeding because, results in the homozygosity of dominant alleles, therefore, it is a best mean of mating among hermaphrodites and self-pollinating plant species of several families. The animal breeder have employed the inbreeding to produce best races of horses, dogs, bulls, cattles, etc. The modern race horses, for example, are all descendents of three Arabian stallions imported into England between 1689 and 1730 and mated with several local mares of the slow, heavy type that had carried the medieval knights in heavy armour. The fast runners of F_1 were selected and inbred and stallions of the F_2 appear as beginning points in the pedigrees of almost all modern race horses. This sort of inbreeding in also called **line breeding** which has been defined as the mating of animals in such a way that their descendents will be kept closely related to an unusually desirable individual.

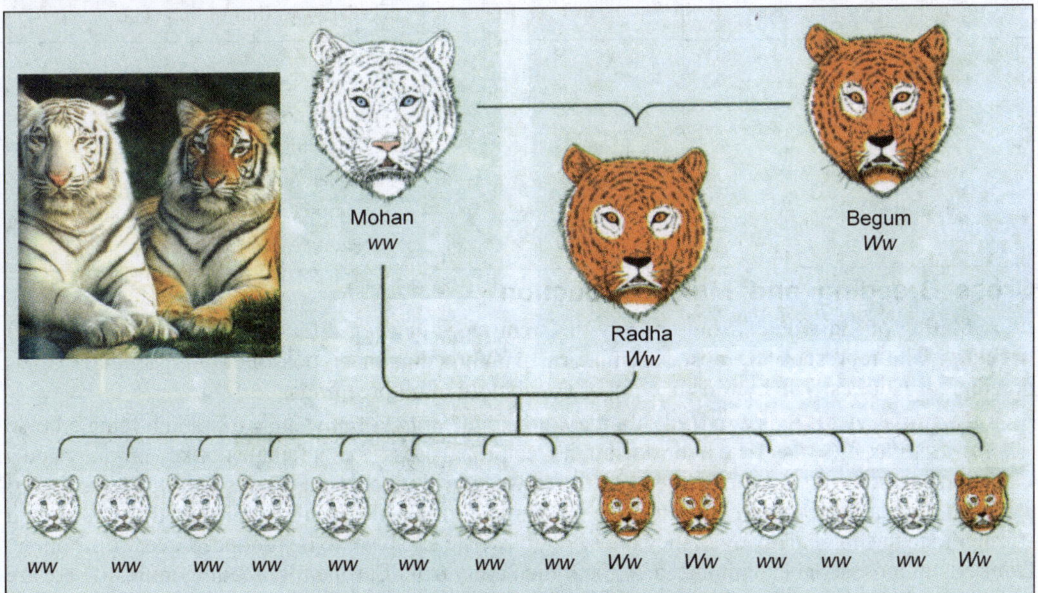

Inbreeding results in an increase in homozygosity.

Similarly, **merino sheep** are widely known as fine wool producers. They are the result of about 200 years of inbreeding. This strain was being developed in Spain in the 17th century by stock raisers. They observed that the ancestors of the present day merino sheep had two coats of wool, one composed of long, coarse fibres arising from primary follicles, and a second coat composed of short fine wool arising from clusters of secondary follicles. Intensive artificial selection was maintained for animals with more uniform production of fine wool and a lesser amount of coarse wool. For a time, Spain had a monopoly on the valuable merino sheep. When France invaded Spain, merino sheep were removed to France where they were maintained and eventually distributed to other parts of the world. Merino sheep were taken to South Africa and in 1796 they were introduced into Australia which has since become the world's largest producer of fine wool.

OUTBREEDING AND HYBRID VIGOUR

When a mating involves individuals that are more distantly related than the average of the selected group it is classified as **outcrossing** or **outbreeding** which is a **negative genetic assortative mating**. Outbreeding involves crossing individuals belonging to different families or crossing different inbred varieties of plants or crossing different breeds of livestock. Outbreeding increases heterozygosity and enhances the vigour of the progeny, *i.e.,* hybrid has superior phenotypic quality but often has poor breeding value than the parental populations. Thus, the two inbred parents homozygous for different genes, if crossed produce F_1 progeny heterozygous for all the genes. Such a F_1 progeny or hybrid may have improved general fitness, resistance to diseases and it may show remarkable growth and vigour. The superiority of the hybrid over the best parent is called **heterosis**, a term coined by **Shull** (1914) for describing **hybrid vigour**. In ordinary usage, the terms heterosis and hybrid vigour are used as synonyms, however, sometimes a distinction is made between these two terms. According to **Shull**, the developed superiority of the hybrids is the 'hybrid vigour' and heterosis is the mechanism by which this superiority is developed. Also, according to **Whaley** (1944) '*hybrid vigour denotes the manifest effects of heterosis*'. **Powers** regarded the heterosis as a phenomenon encountered in quantitative inheritance and he has shown following relationship between F_1 hybrids and their parents (see Table 6-1.).

Table 6.1. **Manifestation of heterosis (in terms of quantitative inheritance) in F_1 hybrids as suggested by Powers.**

Parents and their score	Score in F_1 hybrids	Manifestation of heterosis
	6	Heterosis
Parent A (5)	5	Complete dominance of large size
	4	Partial dominance of large size
	3	No dominance
	2	Partial dominance of smaller size
Parent B (1)	1	Complete dominance of small size
	0	Negative heterosis

Cross Breeding and Mule Production

Mating of individuals from entirely different races or even different species is called **cross breeding**. This represents the most extreme form of outbreeding that is possible among animals. Cross breeding produces sterile hybrids in comparison to normal outbreedings.

Example. A **mule** is a hybrid of a male donkey (*Equus asimus,* 2n = 62) and a female horse (*Equus caballus,* 2n = 64). The hybrid from the reciprocal cross (*i.e.,* a female donkey or jenny and a male horse or stallion) is called **henny**. Mule shows hybrid vigour and because of this it has served mankind as a patient beast of burden since time immemorial. Mules are larger than the donkey and sturdier than the horse. However, they are sexually sterile and have to be produced every time anew. Donkey stallions have been imported by the Indian army from Europe for breeding mules. There are two kinds of mules which are used by the Indian army : (1) general service type and (2) mountain artillary type. The latter are very important as they are firm-footed animals that can carry heavy loads on steep Himalayan mountain terrain.

Manifestation of Heterosis

In addition to increase in size and productiveness, heterosis is manifested in many ways. It may be either morphological or physiological in nature. For example, in some crosses of beans certain F_1

A Mule (a) and a Hinny (b).

hybrids contain greater number of nodes, leaves and pods than their parents; however, the gross size of plant remain unaffected. In some hybrids, the growth rate is increased but there occur no increase in size of mature plant. Further, earlier maturity of F_1 hybrids than in either parent is another manifestation of heterosis and is sometimes accompanied with actual decrease in total plant weight. In hybrids, greater resistance to diseases and to insect infestation and increased tolerance to erratic climatic condition are some other examples of heterotic effects in plants.

Sprague forwarded the following explanation for the hybrid vigour: "*it appears that hybrid superiority may result from a more efficient utilization of nutrient, increased rate of cell division, greater ability to synthesize required growth substances, and possibily from other as yet unrecognised causes.*"

Some Examples of Heterosis in Plants

G.H. Shull (1909) has shown that in corn or maize, hybrids between inbreds of diverse parentage generally give greater hybrid vigour than that shown by hybrids between inbreds derived from same or similar open pollinated varieties. However, such an exploitation of heterosis in maize posed one problem; most inbred lines were so infertile that they did not produce enough seeds for commercial plantings. Shull's method was modified by a method called **double cross method** (by **Jones**, 1917).

Double cross method for producing hybrid corn. Starting with four inbred lines (A, B, C, D), a single cross is made between A and B by growing the two lines together and removing the tassels from line A so that A cannot self-fertilize, and, thus, received only B pollen. In another locality the same method is followed for lines C and D. The yield of single cross hybrid seed is usually low because the inbred parent lacks vigour and produces small cobs. Plant that germinate from single cross seeds are usually vigorous hybrids with large cobs and many kernels. It is undesirable for the single-cross hybrid to self-fertilize, as this inbreeding process commonly produces less vigorous progeny. Therefore, a double cross is mode by using pollen from the CD hybrid on the AB hybrid (see Fig. 6.6).

Inbred A × *Inbred B*　　　*Inbred C* × *Inbred D*

↓ Single cross　　　↓ Single cross

AB hybrid × *CD hybrid*

↓ Double cross

ABCD hybrid

Fig. 6.6. The procedure of double crossing in maize.

Heterosis in other plants. In a number of cultivated plants such as onion, alfalfa and cabbage, it was shown that the intervarietal hybrids gave higher yields than the better of the parental varieties.

Genetical Basis of Heterosis

The genetical basis of heterosis is still a subject of controversy and following two hypotheses have been propounded to explain it :

1. Dominance hypothesis of heterosis. The dominance hypothesis of heterosis holds that increased vigour and size in a hybrid is due to combination of favourable growth genes by crossing two inbred races. In other words, the hybrid vigour is a result of action and interaction of dominant or fitness factors or cumulative (polygenic) effect of dominant genes.

Example. If we suppose, that a quantitative trait is governed by four genes. Each recessive genotype contributes one unit to the phenotype and each dominant genotype contributes two units to the phenotype. A out cross (outbreeding) between two inbred lines can produce more heterotic F_1 individuals than the parents, in following manner :

Parent :	AA bb CC dd	×	aa BB cc DD
Phenotypic value :	2+1+2+1= 6	↓	1+2+1+2 = 6
F_1 :		Aa Bb Cc Dd	
		2+2+2+2 = 8	

The dominance hypothesis of heterosis has been supported by certain experiments. For example, **Quinby** and **Karper** (1946) studied heterosis in *Sorghum* and observed that the heterozygote *Mama* is significantly late in maturity and produces a greater weight of grain than either of the homozygote parents, *Mama* or *mama.* **Keeble** and **Pellow** studied two varieties of pea both semi-dwarf, one with thin stem and long internodes and the other with thick stems and short internodes. The F_1 hybrid was much taller than either of the parents, combining the long internodes of one parent and many nodes of the other.

2. Over dominance hypothesis of heterosis. Overdominance hypothesis was proposed by **Shull** and **East**, independently in 1908. They considered that there is a physiological stimulus to development that increases with the diversity of the uniting gametes. In Mendelian terms, it means that there are loci at which the heterozygote is superior to either homozygote and that vigour increases in proportion to the amount of heterozygosity. The over dominance hypothesis is variously known as **single gene heterosis**, **cumulative action of divergent alleles**, or **stimulation of divergent alleles**. **Fisher** (1930) called it **superdominance**.

G.H. Shull.

Example. If we suppose that four gene loci are contributing to a quantitative trait, homozygous recessive genotype contribute 1 unit to the phenotype, heterozygous genotypes contribute 2 units to the phenotype and homozygous dominant genotypes contribute 1½ units. Then the results can be represented as follows :

Parents :	AA BB cc dd	×	aa bb CC DD
Phenotypic value :	1½+1½+1+1=5	↓	1+1+1½+1½ =5
F_1 :		Aa Bb Cc Dd	
		2+2+2+2=8	

Application of Heterosis

Heterosis has been exploited at commercial scale both in plants and animals. Among plants, it is applied to crop plants, ornamentals and fruit crops. It is found more important in vegetatively propagated perennial plants. Thus, in fruit plants and ornamental plants, if heterosis is once achieved, it may be maintained for long. Among cross pollinated crops, heterosis is exploited in the form of hybrids, composites and synthetic seeds.

Sometimes, intermediate phenotypes are preferred, in such a case heterosis involves mating of parents having opposite phenotypes. For example, general purpose cattle can be produced by crossing beef type with a dairy type. The offspring commonly produce an intermediate yield of milk and hang up a fair carcass when slaughtered. The same is true of the offspring from crossing an egg type (such as Leghorn breed of chicken) with a meat type (such as the Cornish). Crossing phenotypic opposite may also made to correct specific defects. For example, 'weedy' relatives of agriculturally important crops may carry genes for resistance to specific diseases. Hybrids from such crosses may acquire disease resistance, and successive rounds of selection combined with back crossing to the crop variety can eventually fix the gene or genes for disease resistance.

EVOLUTIONARY SIGNIFICANCE OF INBREEDING AND OUTBREEDING

The inbreeding and outbreeding, both, provide raw material to natural selection. Inbreeding allows natural selection to operate on recessive genes, but does not permit the introduction of good mutations from outside. While, outbreeding provides an opportunity for the accumulation of good traits of different races in one individual or line. It expresses good qualities of the races and masked the deleterious recessive alleles. Thus, it can be concluded at last that inbreeding and outbreeding, both, provide new allelic combinations which may be good or bad for the natural selection.

REVISION QUESTIONS AND PROBLEMS

1. (a) Describe what is meant by degree of inbreeding.
   (b) Describe, with the help of an illustration, the genetic effects of inbreeding and how they are related to the degree of inbreeding and to the viability of offspring. Are the genetic effects the same in both cross and self-fertilizing organisms?
   (c) State which kind of organism (cross-fertilizing or self fertilizing) is expected to suffer the harmful effect of inbreeding more severely and why ?
2. (a) Describe what is meant by outbreeding.
   (b) Describe the genetic effects of outbreeding and their relation to the viability of offspring.
   (c) State what the ultimate effect of continued outbreeding in a population will be and show what effect an occasional round of inbreeding would have on such a population.
3. (a) What effect does inbreeding have on (1) allele frequency and (2) heterozygosity ?
   (b) Self-fertilization results in a reduction in vigour in one species of plants but not in another. What conclusion could you draw regarding the form of breeding natural to the two species ?
   (c) Under what circumstances would inbreeding not have deleterious consequences ?
   (d) Why is selection within a pure line futile ?
   (e) How is heterosis related to effects of inbreeding ?
4. Define the terms inbreeding depression and inbreeding coefficient and describe them.
5. What is heterosis ? What are its major manifestations ? How has the phenomenon of heterosis been utilized for plant breeding ?
6. Differentiate between the following :
   (i) Heterosis and hybrid vigour ;
   (ii) Dominance hypothesis and overdominance hypothesis.
7. Demonstrate the following rates of reduction in heterozygosity per generation are correct :
   1. One-half for self-fertilization.
   2. One-fourth for brother-sister matings.
   3. One-eight for half–brother-half-sister matings.
   4. One-sixteenth for cousin matings.
8. Skin colour in man is controlled by additive genes. If both mother and father have intermediate skin colour, can you expect children (i) with lighter skin, (ii) with darker skin ? Can you expect children with darker skin, if both parents have light skin ?
9. The average plant heights of two inbred tobacco varieties and their hybrids have been measured following results : Inbred parent (P_1)=47.8 inches, inbred parent (P_2)=28.7 inches, F_1 hybrid ($P_2 \times P_2$)=43.2 inches. (a) Calculate the amount of heterosis exhibited by F_1. (b) Predict average height of the F_2.

ANSWERS TO PROBLEMS

9. (a) The amount of heterosis is expressed by the excess of the F_1 average over the midpoint between the two parental means :

   Heterosis of F_1 = $\bar{X}_{F1} - \frac{1}{2}(\bar{X}_{P1} + \bar{X}_{P2})$

   = 43.2–½ (47.8+28.7)

   = 43.2–38.25 = 4.95 inches

   (b) As a general rule the F_2 shows only about half the heterosis of the F_1 : ½ (4.95) = 2.48. Hence, the expected height of F_2 plants = 38.25 + 2.48 = 40.73 inches.

CHAPTER

7

Linkage

The hereditary units or genes which determine the characters of an individual are carried in the chromosomes and an individual usually has many genes for the determination of various different characters. As there are more genes than the chromosomes, it can be expected that each chromosome contains more than one gene. The genes for different characters may be either situated in the same chromosome or in different chromosomes. When the genes are situated in different chromosomes, the characters they control appear in the next generation either together or apart, depending on the chance alone. They assort independently according to Mendel's law of independent assortment. But if the genes are situated in the same chromosome and are fairly close to each other, they tend to be inherited together. This type of coexistence of two or more genes in the same chromosome is known as **linkage**. The difference between independent assortment and linkage can be understood by the following two examples :

Morgan and his co-workers by their investigation on the *Drosophila* found two types of linkage - complete linkage and incomplete linkage.

Example 1. Genes on different chromosomes assort independently giving a 1 : 1 : 1 : 1 test cross ratio which is as follows :

P_1 :	AA BB	×	aa bb
P_1 gametes :	(AB)	↓	(ab)
F_1 :		Aa Bb	
Test cross :	Aa Bb	×	aa bb
Gametes :	(AB) (Ab) (aB) (ab)		(ab)
F_2 : ¼ Aa Bb : ¼ Aa bb : ¼ aa Bb : ¼ aa bb or 1 : 1 : 1 : 1.			

Example 2. Linked genes do not assort independently but tend to stay together in the same combination as they were in the parents. In the following figure, the genes to the left of the

slash line (/) are on one chromosome and those on the right are on the homologous chromosome.

P_1 :	AB/AB	×	ab/ab
P_1 gametes :	(AB)	↓	(ab)
F_1 :		AB/ab	
Test cross :	AB/ab	×	ab/ab
Gametes :	(AB) (ab)		(ab)
F_2 :	½ AB/ab	:	½ ab/ab or 1 : 1.

This type of large deviation of F_2 results of example 2 from a 1 : 1 : 1 : 1 ratio in the test cross progeny of dihybrid is an evidence for linkage.

HISTORICAL

T.H. Morgan

The hypothesis that linked genes tend to remain in their original combinations because of their location in the same chromosome was advanced by **T.H. Morgan** in 1911. But prior to Morgan, **W. Sutton** and **T. Boveri** (1902), **Sutton** (1903) and **Bateson** and **Punnet** (1906) had given some hints about the phenomenon of linkage. It is strange that **Mendel** could not detect the phenomenon of linkage in pea plant, since extensive linkage studies of **Lamprecht** (1961) have demonstrated that seven genes used by Mendel belonged to only four linkage groups (chromosomes 1,4,5 and 7). The views of various geneticists about the phenomenon of linkage can be represented as follows :

1. Sutton-Boveri Chromosome Theory of Heredity

The formal statement of the chromosome theory of heredity is usually credited to both **walter Sutton** (an American graduate student who studied meiosis in grasshopper and confirmed **Montegomery's** conclusion (1901) that associations were brought about only between paternal and maternal chromosomes) and **Theodor Boveri** (a great German biologist, who in 1902 demonstrated in sea urchin that different chromosomes of a set possess different qualities. He also showed that presence of a complete set of chromosomes was important for survival). In 1902, these investigators recognised independently that the behaviour of Mendel's genes during the production of gametes in peas, precisely paralleled the behaviour of chromosomes at meiosis : genes are in pairs (so are chromosomes); the members of a gene pair segregate equally into gametes (so do

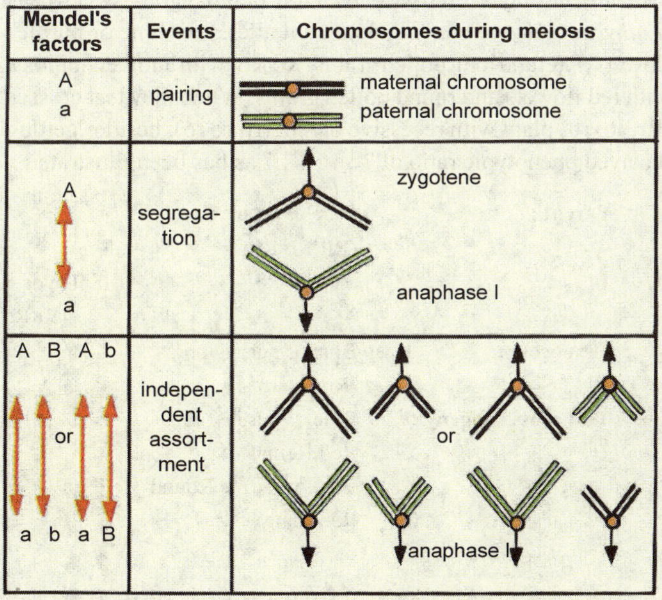

Fig. 7.1. Parallelism between Mendel's hypothetical particles (factors or genes) and chromosomes during meiosis (after Suzuki *et. al.,* 1986).

Walter Sutton (1877-1916).

the members of a pair of homologous chromosomes); different gene pairs act independently (so do different chromosome pairs) (Fig. 7.1.). It was, therefore, concluded that genes or Mendelian factors were located on chromosomes. This generalization is now known as **chromosome theory of inheritance (or of heredity)**. This theory provides a physical basis of heredity.

2. Sutton's View on Linkage

Sutton (1903) predicted that the chromosomes are the bearers of the units of heredity or genes and since the number of hereditary units is much larger than the number of chromosomes, therefore, each chromosome must contain a number of genes. He further stated that since the chromosomes move as units during meiosis to the gametes, all the genes which are situated in the same chromosome will be linked together. As a result each species of animals and plants would have a specific number of groups of linked genes which would correspond with the number of chromosomes found in that species. Unfortunately **Sutton** could not prove his predictions experimentally.

Theodor Boveri (1862 – 1915).

3. Bateson and Punnet's Coupling and Repulsion Hypothesis

Bateson and **Punnett** (1905–1908) formulated the 'hypothesis of coupling and repulsion' to explain the unexpected F_2 results of a dihybrid cross between a homozygous sweet pea (*Lathyrus odoratus)* having dominant alleles for blue or purple flowers (RR) and long pollen grains (Ro Ro) with another homozygous double recessive plant (rr, roro) with red flowers and round pollen grains. When they test crossed a heterozygous blue or purple long (Rr, Ro ro) plant with recessive parent (rr, ro ro), besides getting the 1 : 1 : 1 : 1 test cross ratio, they received phenotypic ratio of 7 : 1 : 1 : 7 as has been illustrated :

Parent :		Blue or purple Long (RR Ro Ro)	×	Red Round (rr ro ro)
F_1 :			All blue or purple long (Rr Ro ro)	
Test cross :	**F_1**	Blue or purple long (Rr Ro ro)	×	Red Round (rr ro ro)
Test cross progeny :		Blue or purple Long	=	192
		Red Round	=	182
		Blue or purple Round	=	23
		Red Long	=	30
				427

Test cross ratio : 7 Blue or purple Long : 1 Blue or purple Round : 1 Red Long : 7 Red Round or 7 : 1 : 1 : 7.

The 7 : 1 : 1 : 7 test cross ratio clearly indicated that there was a tendency in the dominant alleles (R Ro) to pass together to the same gamete. Similar was the case with recessive alleles (r ro). This tendency of dominant or recessive alleles to inherit together was explained as "**gametic coupling**" by **Bateson** and **Punnett**.

Further, when they crossed blue or purple round (RR ro ro) with red long (rr Ro Ro), the F_1 hybrids were found to be heterozygous blue or purple long (Rr Ro ro). The F_1 hybrid when test crossed with recessive (rr ro ro) parent, the test cross ratio was 1 blue or purple long : 7 red long : 7 blue or purple round : 1 red round, as has been illustrated in following figure :

Parent :	Blue or purple Round (RR ro ro)	×	Red Long (rr Ro Ro)
F_1 :	All Blue or purple Long (Rr Ro ro)		
Test cross :	F_1 Blue or purple Long (Rr Ro ro)	×	P_1 Red Round (rr ro ro)

Test cross progeny : 1/16 Blue or purple Long : 7/16 Blue or purple Round : 7/16 Red Long : 1/16 Red Round or 1 : 7 : 7 : 1.

R.C. Punnett (1875 – 1967)

William Bateson (1861 – 1926)

Hence, the two dominant pairs of alleles repelled each other. The tendency of both dominant or both recessive alleles to repel each other, so that the gametes of genotypes of R ro and r Ro are formed more frequently, was termed **repulsion**.

Bateson and Punnett could not explain the exact reasons of coupling and repulsion, and it was **T.H. Morgan** who while performing experiments with *Drosophila,* in 1910, found that coupling or repulsion was not complete. He further suggested that the two genes are found in coupling phase or in repulsion phase, because they are present on the same chromosome (coupling) or on two different homologous chromosomes (repulsion). Such genes are then called **linked genes** and the phenomenon of inheritance of linked genes is called **linkage** by Morgan. The terms **cis** and **trans** were employed later (**Haldane**, 1942), the former replacing coupling and the later, repulsion (Fig. 7.2).

Fig. 7.2. Cis and trans arrangement of two pairs of linked genes of *Lathyrus odoratus* (after Burns 1969).

4. Morgan's Views on Linkage

Morgan stated that the pairs of genes of homozygous parent tend to enter in the same gametes and to remain together, whereas same genes from heterozygous parent tend to enter in the different gametes and remain apart from each other. He further stated that the tendency of linked genes remaining together in original combination is due to their location in the same chromosome. According to him the degree or strength of linkage depends upon the distance between the linked genes in the chromosome. Morgan's concept about the linkage developed the theory of linear arrangement of genes in the chromosomes which helped the cytogeneticists in the construction of genetic or linkage maps of chromosomes.

5. Chromosome Theory of Linkage

Morgan along with **Castle** formulated the chromosome theory of linkage which is as follows:

1. The genes which show the phenomenon of linkage are situated in the same chromosomes and these linked genes usually remain bounded by the chromosomal material so that they cannot be separated during the process of inheritance.

2. The distance between the linked genes determines the strength of linkage. The closely located genes show strong linkage than the widely located genes which show the weak linkage.

3. The genes are arranged in linear fashion in the chromosomes.

W.E. Castle

KINDS OF LINKAGE

T.H. Morgan and his co-workers by their investigation on the *Drosophila* and other organisms have found two types of linkage, *viz.,* complete linkage and incomplete linkage.

1. Complete Linkage

The complete linkage is the phenomenon in which parental combinations of characters appear together for two or more generations in a continuous and regular fashion. In this type of linkage genes are closely associated and tend to transmit together.

Example. The genes for bent wings (bt) and shaven bristles (svn) of the fourth chromosome mutant of *Drosophila melanogaster* exhibit complete linkage.

Complete linkage in male *Drosophila*. In most of the organisms crossing-over takes place both in males and females. But in male *Drosophila* and female silkworm, *Bombyx mori* (see **Swanson**, 1957) crossing-over takes place either very rarely or not at all. This becomes clear from Morgan's experimental results from *Drosophila.* In 1919, **T.H. Morgan** mated gray bodied and vestigial winged (b^+vg/b^+vg) fruit flies with flies having black bodies and normal wings (bvg^+/bvg^+). F_1 progeny had gray bodies and normal long wings (b^+vg/bvg^+), indicating thereby that these characters are dominant. When F_1 males (b^+vg/bvg^+), were backcrossed (*i.e.,* test crossed) to double recessive females (bvg/bvg or black vestigial), only two types of progeny (one with gray bodies and vestigial wings, b+vg/bvg and the other with black bodies and normal wings, to bvg^+/bvg instead of four types of phenotypes were obtained (see **Sinha** and **Sinha**, 1990) :

Parents :	Gray, Vestigial	×	Black, Long
	b^+vg/b^+vg		bvg^+/bvg^+
Gametes :	(b^+vg)		(bvg^+)
F_1 :		All Gray, Long	
		b^+vg/bvg^+	
Test cross :	F_1 male Gray, Long	×	Female Black, Vestigial
	b^+vg/bvg^+		bvg/bvg
Gametes :	(b^+vg) (bvg^+)		(bvg)
	(only two types of gametes		
	due to complete linkage		

and lack of crossing over
in male *Drosophila*)

Test cross ratio : ½ Gray, Vestigial : ½ Black, Long or 1 : 1.
b^+vg/bvg bvg^+/bvg

Here, the use of the testcross is very important. Because one parent (the tester) contributes gametes carrying only recessive alleles, the phenotypes of the offspring represent the gametic contribution of the other double heterozygote parent. So the genetical analyst can concentrate on one meiosis and forget the other. This is in contrast to the situation in an F_1 selfing where there are two sets of meiotic divisions to consider one for the F_1 male parental gametes and one for the F_1 female.

2. Incomplete Linkage

The linked genes do not always stay together because homologous non-sister chromatids may exchange segments of varying length with one another during meiotic prophase. This sort of exchange of chromosomal segments in between homologous chromosomes is known as **crossing over** (Fig. 7.3). The linked genes which are widely located in chromosomes and have chances of separation by crossing over are called **incompletely linked genes** and the phenomenon of their inheritance is called **incomplete linkage**.

Example. The incomplete linkage has been reported in female *Drosophila* and various other organisms such as tomato, maize, pea, mice, poultry and man, etc. Here, the examples of incomplete linkage have been considered only for *Drosophila* and maize.

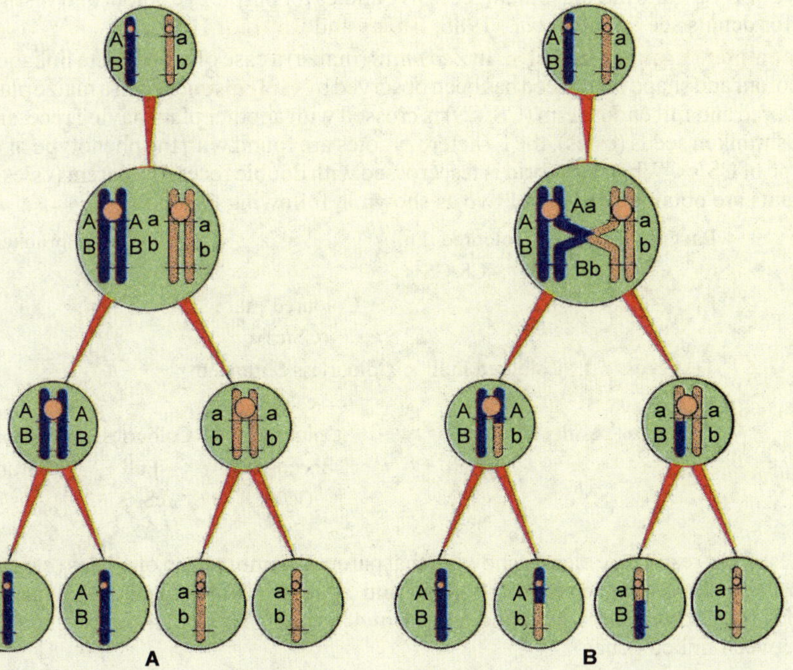

Fig. 7.3. A–Diagram of the segregation of two pairs of allelomorphic genes localized on the same pair of chromosomes without crossing over. The result is two types of gametes, AB and ab. A case of complete linkage; B–Diagram of the segregation of two pairs of allelomorphic genes on the same chromosome between which crossing over takes place during meiosis. Four types of gametes result : AB, Ab, aB and ab, A case of incomplete linkage (after De Robertis *et al.,* 1970).

Incomplete linkage in female Drosophila. When F_1 females of the Morgan's classical cross in *Drosophila* between gray, vestigial (b^+vg/b^+vg) and black, normal or long (bvg^+/bvg^+) were test-crossed to double-recessive (bvg/bvg) males, all four types of progeny were obtained in following ratio, showing occurrence of crossing-over :

Parents :	Gray, Vestigial b^+vg/b^+vg	×	Black, Long bvg^+/bvg^+
Gametes :	(b^+vg)		(bvg^+)
F_1 :		Gray, Long b^+vg/bvg^+	
Test cross :	F_1 Female Gray, Long b^+vg/bvg^+ ↓	×	Male Black, Vestigial bvg/bvg ↓
Gametes :	(b^+vg) (bvg^+) = Non-crossovers (b^+vg^+) (bvg) = Recombinants		(bvg)
Test cross ratio :	1. Gray, Vestigial; b^+vg/bvg = 41.5% 2. Black, Long; bvg^+/bvg = 41.5%	}	83% parental combination showing linkage
	3. Gray, Long ; b^+vg^+/bvg = 8.5% 4. Black, Vestigial; bvg/bvg = 8.5%	}	17% recombinants due to crossing over

Similar different test cross ratios (showing complete linkage in males and incomplete linkage in females) were obtained for F_1 males and females of *Drosophila* by **Bridges**, one of the student of Morgan. He made a cross between fruit flies having wild dominant alleles for red eye colour and normal wings (pr^+vg^+/pr^+vg^+) and having mutant recessive alleles for purple eye colour and vestigial wings (pr vg/pr vg) (for details see **Suzuki** *et al.*, 1986; **Burns** and **Bottino**, 1989).

2. Incomplete Linkage in maize. In *Zea mays* (maize) a case of incomplete linkage between the alleles for colour and shape of the seed has been observed by **Hutchison**. When a maize plant with seeds having coloured and full endosperm (CS/CS) is crossed with another plant having recessive alleles for colourless, shrunken seeds (cs/cs), the F_1 heterozygotes are found with the phenotype of coloured full and genotype of CS/cs. When F_1 hybrid is test crossed with double recessive parent (cs/cs) four classes of descendants are obtained instead of two as shown in following figure :

Parents :	Coloured Full CS/CS	×	Colourless Shrunken cs/cs	
F_1 :		Coloured Full (CS/cs)		
Test cross :	F_1 Coloured Full × Colourless Shrunken CS/cs cs/cs			
Test cross results :	Coloured : Full CS/cs 48%	Coloured : Shrunken Cs/cs 2%	Colourless : Full cS/cs 2%	Colourless Shrunken cs/cs 48%

The test cross results are clearly showing that parental combination of alleles (*e.g.,* CS/cs and cs/cs) are those expected from complete linkage and appear in 96% cases, the other two are new combinations (*e.g.,* Cs/cs and cS/cs) and appear in 4% cases. Thus, in 4% cases crossing over has occurred between linked genes.

LINKAGE GROUPS

All the linked genes of a chromosome form a **linkage group**. Because, all the genes of a chromosome have their identical genes (alleomorphs) on the homologous chromosome, therefore linkage groups of a homologous pair of chromosome is considered as one. The number of linkage group of a species, thus, corresponds with haploid chromosome number of that species.

Example. 1. *Drosophila* has 4 pairs of chromosomes and 4 linkage groups.
2. Man has 23 pairs of chromosomes and 23 linkage groups.
3. Corn (*Zea mays*) has 10 pairs of chromosomes and 10 linkage groups.

However, in organisms the female or male sex having dissimilar sex chromosomes (*e.g.*, human beings, *Drosophila,* fowl, etc.), one more linkage group occur than the haploid number (see **Burns** and **Bottino**, 1989).

Example. 1. Female human beings : = 22 pairs of autosomes or non-sex chromosomes + 1 pair of homomorphic X chromosomes
= 22 autosomal linkage groups + 1 X chromosomal linkage group
= 23 linkage groups.

2. Male human beings : = 22 pairs of autosomes + 2 heteromorphic sex chromosomes, *i.e.,* 1 X chromosome + 1 Y chromosome
= 22 autosomal linkage group + 1 X chromosomal linkage group + 1 Y chromosomal linkage group
= 24 linkage groups.

SIGNIFICANCE OF LINKAGE

The phenomenon of linkage has one of the great significance for the living organisms that it reduces the possibility of variability in gametes unless crossing over occurs.

REVISION QUESTIONS AND PROBLEMS

1. Describe the phenomenon of linkage by giving suitable examples. Why is linkage an exception to Mendel's second law ?
2. Linkage studies provided the first proof that a gene occupies a fixed locus on a specific chromosome. Discuss.
3. Mendel studied seven pairs of contrasting characters in the garden pea. Why do you suppose, did he not discover the principle of linkage ?
4. How many linkage groups are there in the (a) human male, (b) human female, (c) female grasshopper and (d) male grasshopper.
5. In a given plant long leaves (*S*) and green veins (*Y*) are dominant, respectively, over short leaves (*s*) and yellow vein syy. The cross *SSYY* × *ssyy* produced an F_1 *Ss Yy*. When F_1 plants were inbred, the F_2 consisted of 570 long, green individuals and 190 short, yellow. Are the genes *S* and *Y* linked.
6. The trihybrid *Aa Bb Cc* is test crossed to the triple recessive *aa bb cc,* and the following phenotypes are obtained in the progeny : *64 abc, 2 abC, 11 aBc, 18 aBC, 14 AbC, 17 Abc, 3 ABc, 71 ABC.*
   (a) Which of these loci are linked ?
   (b) What is the correct genotypes of each parent ?
7. Following two pairs of alleles are known in tomato :

   Cu = "curl" (leaves curled)

   cu = normal leaves

   Bk = "beakless" fruits

   bk = "beaked" fruits, having sharp pointed protuberance on blossom end of mature fruit.

   The cross of two doubly heterozygous "curl beakless" (*Cu Bk/cu bk)* plants yields four phenotypic classes in the offspring, of which 23.04 per cent are "normal beaked". Are these two pairs of genes linked ? How do you know ?

ANSWER TO PROBLEMS

3. Because genes for some of the traits, Mendel followed are on different chromosome pairs; linked traits with which he worked (seed colour/flower colour; pod colour/plant height, among others) are so far apart on their chromosomes that genes for these traits appear to segregate randomly. In one other case Mendel did not make the cross that would have revealed linkage. (Also consult the paper of **Blixt,** 1975).
4. (a) 24; (b) 23 (c) 12; (d) 12.
5. The genes *S* and *Y* are linked; the ratio is 3 : 1 indicating complete linkage, that is, no crossing over. If not linked, 9 : 3 : 3 : 1 ratio would have been obtained.
6. (a) All loci are linked.

   (b) Parental genotypes : ACB/abc and abc/abc.
7. Yes, they are linked. Normal beaked plants are double heterozygous in this case, and should produce about 6 per cent (1/16) of progeny with this phenotype. The number actually observed is about 3.68 time greater than is to be expected with unlinked genes.

CHAPTER

8

Crossing Over

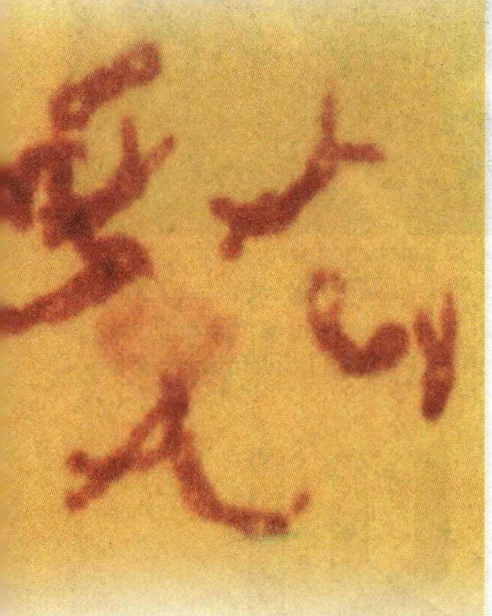

Prophase – I stage showing crossing over.

In the chapter of linkage, we have stated that the genes located in the same chromosome show linkage. These linked genes may either remain together during the process of inheritance and, thus, showing complete linkage or they may be segregated or separated during gametogenesis and, thus, displaying the incomplete linkage. The incomplete linkage takes place due to the occurrence of new combinations or recombinations of linked genes. The recombination in its turn is accomplished through a process known as **crossing over** in which the non-sister chromatids of homologous chromosomes exchange the chromosomal parts or segments. In another words "*the crossing over is a process that produces new combinations (recombinations) of genes by interchanging of corresponding segments between non-sister chromatids of homologous chromosomes*". The chromatins resulting from such interchanges of chromosomal parts are known as **cross overs**. The term **crossing over** was coined by **Morgan**. No crossing over occurs in male *Drosophila* and female silk worm, *Bombyx mori*. Certain external agents such as heat shock, chemicals, radiations, etc., have profound effect on crossing over.

Characteristics of Crossing Over

1. Crossing over or recombination occurs at two levels (i) at gross chromosomal level, called **chromosomal crossing over** and (ii) at DNA level, called **genetic recombination**.

2. A reciprocal exchange of material between homologous chromosomes in heterozygotes is reflected in crossing over.

3. The crossing over results basically from an exchange of genetic material between non-sister chromatids by break-and-exchange following replication.

4. The frequency of crossing over appears to be closely related to physical distance between genes on chromosome and serves as a tool in constructing genetic maps of chromosomes.

TYPES OF CROSSING OVER

According to its occurrence in the somatic or germ cells following two types of crossing over have been recognized :

1. Somatic or Mitotic Crossing Over

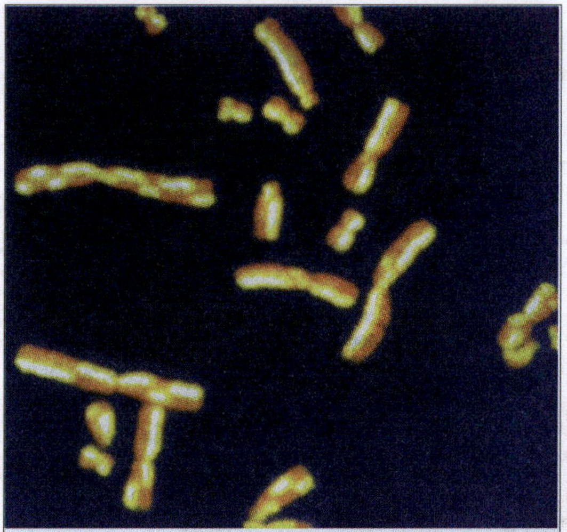

Demonstration of sister chromatid exchanges in mitotic chromosomes.

When the process of crossing over occurs in the chromosomes of body or somatic cells of an organism during the mitotic cell division it is known as **somatic** or **mitotic crossing over**. The somatic crossing over is rare in its occurrence and it has no genetical significance. The somatic or mitotic crossing over has been reported in the body or somatic cells of *Drosophila* by **Curt Stern** and in the fungus *Aspergillus nidulans by* **G. Pontecorvo**.

Example. Stern (1936) observed that *Drosophila* females which are heterozygous for sex-linked recessive genes, yellow body colour (Y) and singed (=burned or scorched) hairs and bristles (sn), are phenotypically like wild types but occasionally show yellow spots on their body. Homozygous recessive individuals for these genes are yellow bodied and with bent blunted bristles. Normally, all the cells of a heterozygous should be normal and should not express either of

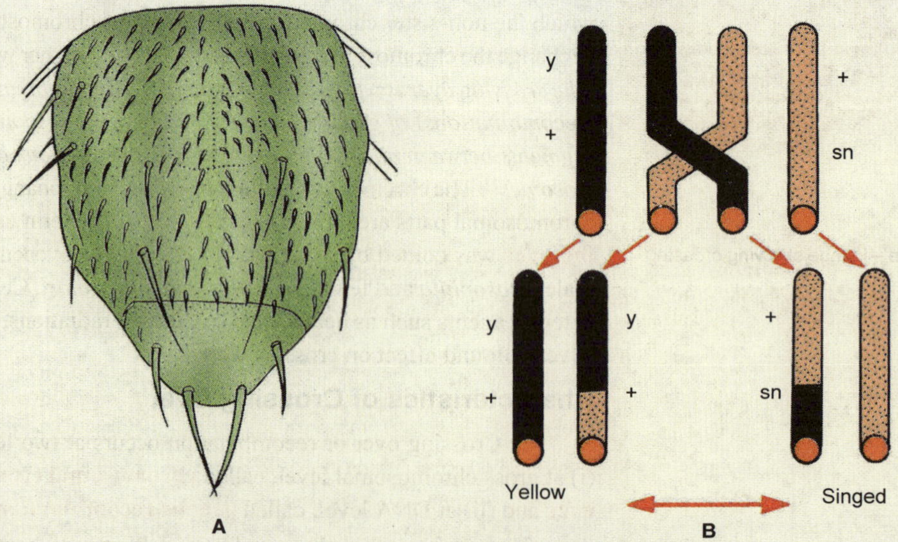

Fig. 8.1. Somatic or mitotic crossing over in *Drosophila.* A—Thorax of a fruit fly heterozygous for y (yellow body) and sn (singed hairs and bristles) with a twin spot; B—Mechanism of crossing over leading to homozygosity for distal genes.

these genes. Yellow skin spots can appear only if cells are homozygous for y, indicating thereby that whenever yellow spots appear in heterozygous flies, there is prior crossing over between Y and sn alleles. Stern found that equal sections of singed and yellow tissues lie adjacent to each other or form **twin spots**. This indicate that such tissues arise as a result of mitotic crossing over, rather than due to mutation. Somatic or mitotic crossing over occurs at a four strand stage and during this process there is pairing of homologous chromosomes (Fig. 8.1).

2. Germinal or Meiotic Crossing Over

Usually the crossing over occurs in germinal cells during the gametogenesis in which the meiotic cell division takes place. This type of crossing over is known as **germinal** or **meiotic crossing over**. The meiotic crossing over is universal in its occurrence and is of great genetic significance.

MECHANISM OF MEIOTIC CROSSING OVER

The process of crossing over includes following stages in it, *viz.,* synapsis, duplication of chromosomes, crossing over and terminalization. The chromosomes which tend to undergo recombination due to meiotic crossing over necessarily complete two functions : 1. 99.7 per cent replication of DNA and 75 per cent synthesis of histones, both of which take place prior to onset of prophase 1, and 2. attachment of each chromosome by its both ends (**telomeres**) to the nuclear envelope (*i.e.,* to nuclear lamina) via the specialized structure, called **attachment plaques** (see **Suzuki** *et al.,* 1986). This event occurs during the **leptotene** stage of prophase I and though each chromosome at this stage is visually long and thin thread, but contains material of two sister chromatids (*i.e.,* two DNA molecules plus almost duplicated amount of histones).

1. Synapsis

Synapsis or intimate pairing between the two homologous chromosomes (one maternal and another paternal) is initiated during **zygotene** stage of prophase I of meiosis I. Synapsis often starts when the homologous ends of the two chromosomes are brought together on the nuclear envelope and it continues inward in a zipper-like manner from both ends, aligning the two homologous chromosomes side by side (*e.g.,* mammals). In other cases, synapsis may begin in internal regions of the chromosomes and proceed toward the ends, producing the same type of alignment. By synapsis each gene is, thus, brought into juxtaposition (=being side by side) with its homologous gene on the opposite chromosome. Thus, synapsis is the phase of prolonged and close contact of homologous chromosomes due to attraction between two exactly identical or homologous regions or chromomeres. The resultant pairs of homologous chromosomes are called **bivalents**. The phenomenon of synapsis has always intrigued cytogeneticists. While **Darlington** tried to explain the cause of synapsis by proposing his precocity theory, **Moses** identified a factor in the formation of synaptonemal complex which aids in synapsis.

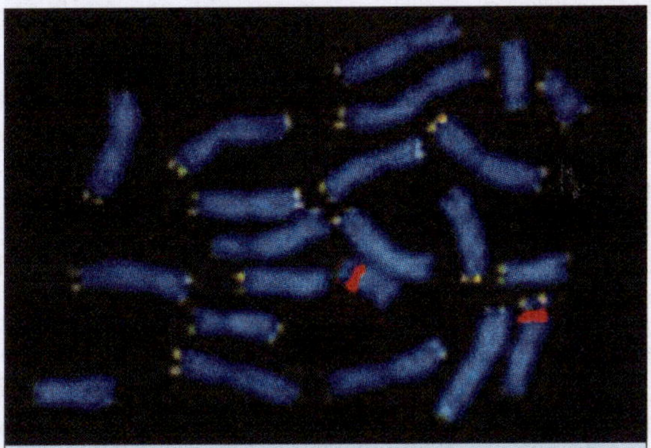

Telomeres (shown here in yellow colour).

(a) Precocity theory. To explain the question that why do homologous chromosomes, during synapsis, approach each other from a considerable distance and become closely associated, a British cytologist **C.D. Darlington** in 1937, proposed the **precocity theory of meiosis** which embraces well the cause of synapsis in it. According to the precocity theory of meiosis, chromosomes enter into meiotic prophase I, in contrast to mitotic prophase, as unreplicated structures, each of which consisting of a single chromatid which he considered as unbalanced or unsaturated state in electrostatical relations. In order to become saturated or balanced the chromosomes must pair. Thus, the sequences of meiotic pairing or synapsis are determined precociously.

Precocity theory is now untenable since it erroneously stressed upon the idea that DNA synthesis or chromosome duplication takes place later in pachytene or diplotene and then results in separation of homologous chromosomes.

(b) Synaptonemal complex. Montrose J. Moses (1956) has revealed a highly organized structure of filaments called **synaptonemal complex** in between the paired chromosomes of zygotene and pachytene stages in crayfish by electron microscopy. Synaptonemal complex has also been observed in a wide variety of species of plants and animals.

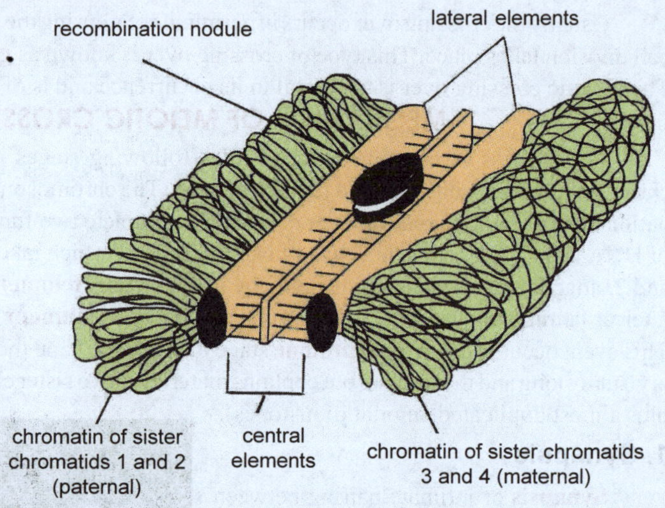

Fig. 8.2. A typical synaptonemal complex (after Alberts *et al.,* 1989).

In electron-micrographs the synaptonemal complex appears as three parallel dense lines that lie equally spaced in a plane and are flanked by chromatin. The elements of two lateral lines usually appear densest, while the element of central line is of variable prominence. Some fine transverse strands also cross between lateral elements, connecting them with the central element. Though, the morphology of lateral and central elements may vary from species to species, but the basic structure and the spacing of the synaptonemal complex is constant within the species.

Cytochemical studies have shown that the lateral elements of synaptonemal complex are rich in DNA, RNA and proteins, but that the central element of it contains mainly RNA, protein and little DNA.

Functions of synaptonemal complex. The synaptonemal complex is found to be concomitant of both chiasma formation and crossing over. For instance, the synaptonemal complex may serve crossing over by facilitating effective synapsis in one or more of the following ways—(a) to maintain pairing in fixed state for an extended period, (b) to provide a structural framework within which molecular recombination may occur, and (c) to segregate recombination DNA from the bulk of the chromosomal DNA.

Robert King (1970) suggested that the synaptonemal complex may orient the non-sister chromatids of homologs in a manner to facilitate enzymatically induced exchanges between their DNA molecules. More recently, **D.A. Comings** and **T.A. Okada** (1971), have shown electron microscopically that synapsis occurs at two levels one at chromosomal level and the other at the molecular level. According to them, the synaptonemal complex pulls homologous chromosomes into approximate association with each other but plays no role in molecular pairing of DNA strand.

Recently, each event of recombination due to crossing over is found to be performed by a large protein assembly of 90 nm diameter, called **recombination nodule**, which is placed on the synaptonemal complex (**Carpenter**, 1977, 1987). Each nodule marks the site of a large multienzyme 'recombination machine' which brings local regions of DNA on maternal and paternal chromatids together across the 100-nm wide synaptonemal complex.

2. Duplication of Chromosomes

The synapsis is followed by duplication of chromosomes (in **pachytene**). During this stage, each homologous chromosome of bivalents splits longitudinally and form two identical sister chromatids which remain held together by an unsplitted centromere. The longitudinal splitting of chromosomes is achieved by the separation of already duplicated DNA molecules along with certain chromosomal proteins. At this stage each bivalent contains four chromatids, so it is known as **tetrad**.

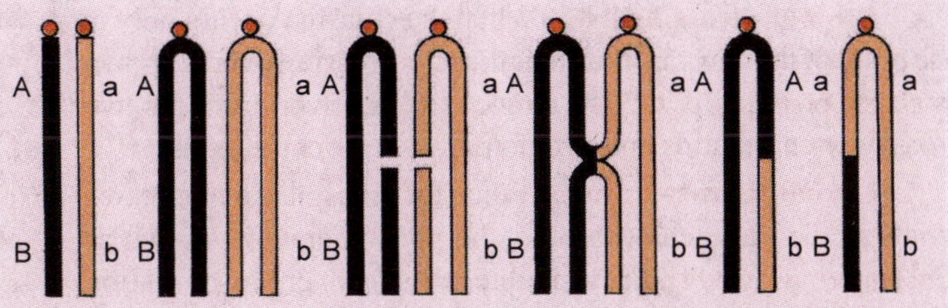

Fig. 8.3. Diagram showing the mechanism of crossing over.

3. Crossing Over by Breakage and Union

It is well evident that crossing over occurs in the homologous chromosomes only during the four stranded or tetrad stage. Homologues continue to stay in synapsis for days during pachytene stage and chromosomal crossing over occurs due to exchange of chromosomal material between non-sister chromatids of each tetrad. In pachytene, the recombination nodules become visible between synapsed chromosomes.

During the process of crossing over, two non-sister chromatids first break at the corresponding points due to the activity of a nuclear enzyme known as **endonuclease (Stern** and **Hotta**, 1969). Then a segment on one side of each break connects with a segment on the opposite side of the break, so that the two non-sister chromatids cross each other. At this stage .3 per cent synthesis of DNA (*i.e.,* P–DNA replication) occurs to fill the gap. The fusion of chromosomal segments with that of opposite one takes place due to the action of an enzyme known as **ligase (Stern** and **Hotta**, 1969). The crossing of two chromatids is known as **chiasma** (Gr., *chiasma*=cross) formation. The crossing over, thus, includes the breaking of chromatid segments, their transposition and fusion.

Chiasma frequency or percentage of crossing over. The crossing over may take place at several points in one tetrad and may result in the formation of several chiasmata. The number of chiasmata depends on the length of the chromosomes because the longer the chromosome the greater the number of chiasmata. In a species each chromosome has a characteristic number of chiasmata. The frequency by which a chiasmata occurs between any two genetic loci has also a characteristic probability. The more apart two genes are located on a chromosome, the greater the opportunity for a chiasma to occur between them. The closer two genes are linked lesser the chances for a chiasma occurring between them.

4. Terminalisation

After the occurrence of process of crossing over, the non-sister chromatids start to repel each other because the force of synapsis attraction between them decreases. During **diplotene**, desynapsis begins, synaptonemal complex dissolves and two homologous chromosomes in a bivalent are pulled away from each other. During **diakinesis**, chromosomes detaches from the nuclear envelope and each bivalent is clearly seen to contain four separate chromatids with each pair of sister chromatids linked at their centromeres, while non-sister chromatids that have crossed over are linked by chiasmata. The chromatids separate progressively from the centromere towards the chiasma and the chiasma itself moves in a zipper fashion towards the end of the tetrad (Fig. 8.4). The movement of chiasma is known as **terminalisation**. Due to the terminalisation the homologous chromosomes are separated completely.

KINDS OF CROSSING OVER

According to the number of chiasma following types of crossing over have been described.

1. Single crossing over. When the chiasma occurs only at one point of the chromosome pair then the crossing over is known as **single crossing over**. The single crossing over produces two cross over chromatids and two non-cross over chromatids.

2. Double crossing over. When the chiasmata occur at two points in the same chromosome, the phenomenon is known as **double crossing over**. In the double crossing over, the formation of each chiasma is independent of the other and in it four possible classes of recombination occur. In the double crossing over following two types of chiasma may be formed :

(i) Reciprocal chiasma. In the reciprocal chiasma the same two chromatids are involved in the second chiasma as in the first. Thus, the second chiasma restores the order which was changed by the first chiasma and it produces two non-cross over chromatids. The reciprocal chiasma occurs in two strand double crossing over in which out of four chromatids only two are involved in the double crossing over.

(ii) Complimentary chiasma. When both the chromatids taking part in the second chiasma are different from those chromatids involved in the first chiasma, the chiasma is known as **complimentary chiasma**. The complimentary chiasma produces four single cross overs but no non-cross over. The complimentary chiasma occurs when three or four chromatids of the tetrad undergo the crossing over.

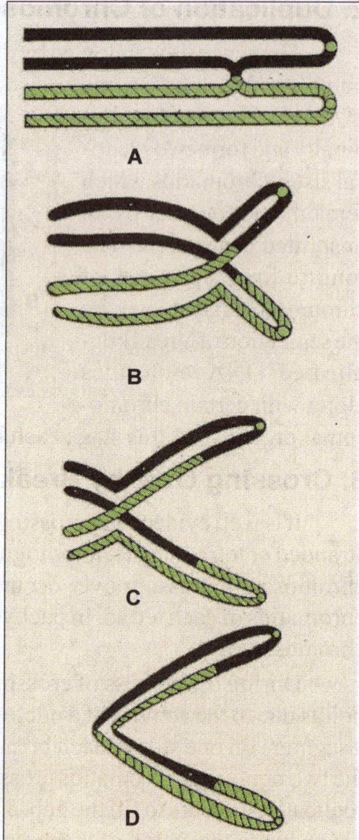

Fig. 8.4. Diagram showing the terminalisation.

3. Multiple crossing over. When crossing over takes place at more than two places in the same chromosome pair then such crossing over is known as **multiple crossing over**. The multiple crossing over occurs rarely.

THEORIES ABOUT THE MECHANISM OF CROSSING OVER

Following theories have been propounded to explain the mechanism of crossing over :

1. Duplication theory. This theory was proposed by **John Belling** (1928) while studying meiosis in some plant species. Belling believed that crossing over might occur during duplication of homologous chromosomes and might brought about due to novel attachments formed between newly synthesized genes. He visualized genes as beads (described as **chromomeres)**, connected by non-genic linking elements—the **interchromomeric regions**. During duplication of chromosomes, initially the chromomeres are duplicated and the newly formed chromomeres remain tightly juxtaposed to the old ones. When interchromomeric regions are synthesized to join these new genes or chromomeres, they may switch from a newly synthesized chromomere on one

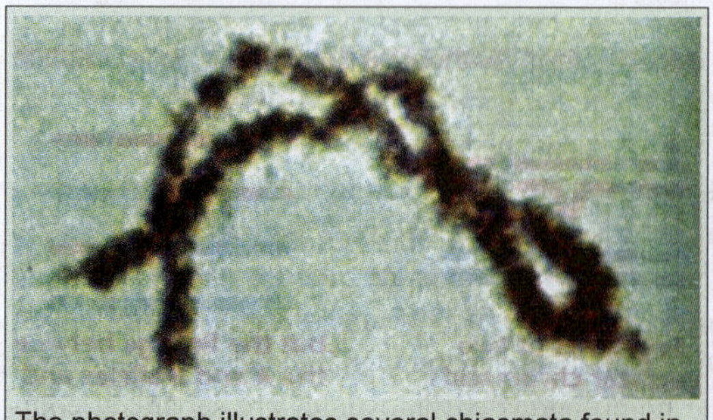

The photograph illustrates several chiasmata found in a tetrad isolated during first meiotic prophase stage.

homologous chromosome to an adjacent chromomere of other homologue. This results in the formation of recombinants or crossovers in new sets of chromatids (Fig. 8.6).

2. Copy choice hypothesis or Switch model. To explain recombination in micro-organisms, **J.**

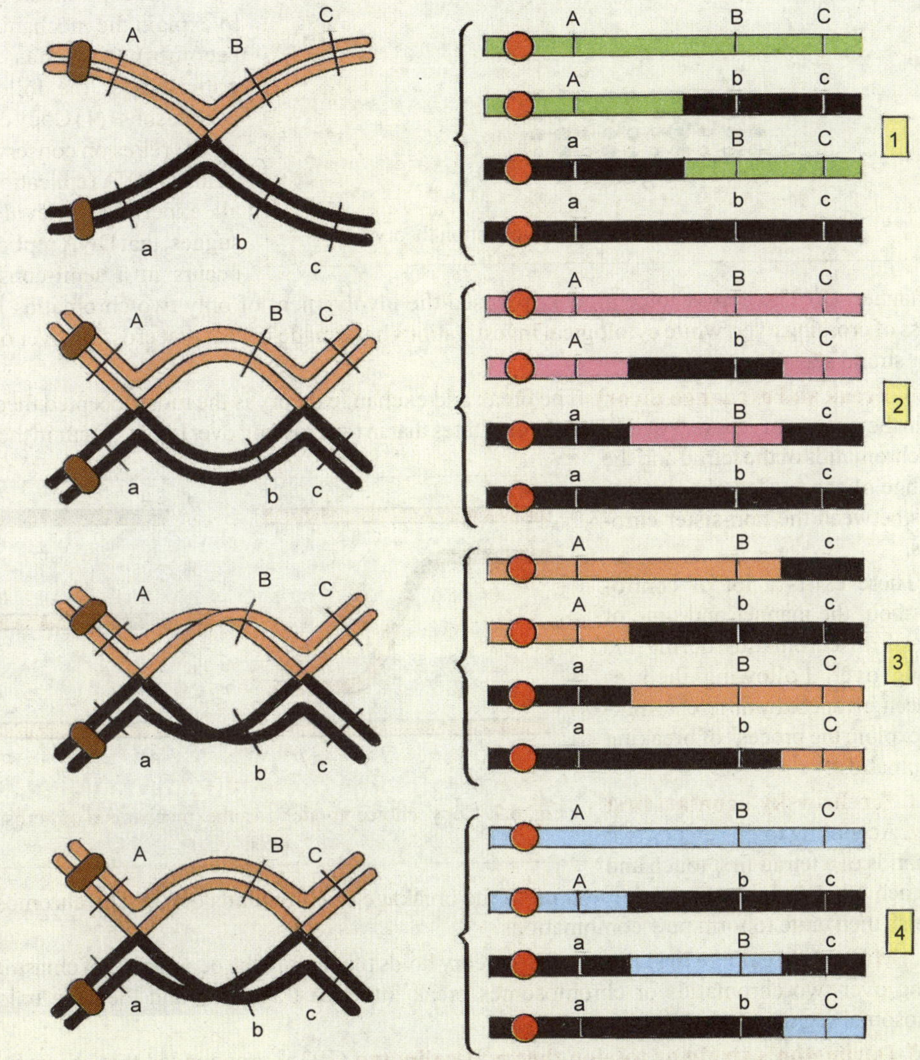

Fig. 8.5. Diagram showing the results of single and double crossing over. 1. Single chiasma. 2. Two strand double cross over. 3. Four strand double cross over. 4. Three strand double cross over.

Laderberg (1955) proposed a modified version of Belling's hypothesis. This is called copy choice mechanism of recombination or crossing over. According to this model a daughter chromosome is formed by alternate use of recipient and donor chromosome material as a kind of model or template. The daughter chromosome is then like the recipient chromosome except for portions "copied" from the donor chromosome segment. At the end of process a recipient cell would contain (1) original donor segment, (2) the originally recipient whole chromosome, and (3) a "hybrid" daughter chromosome. In

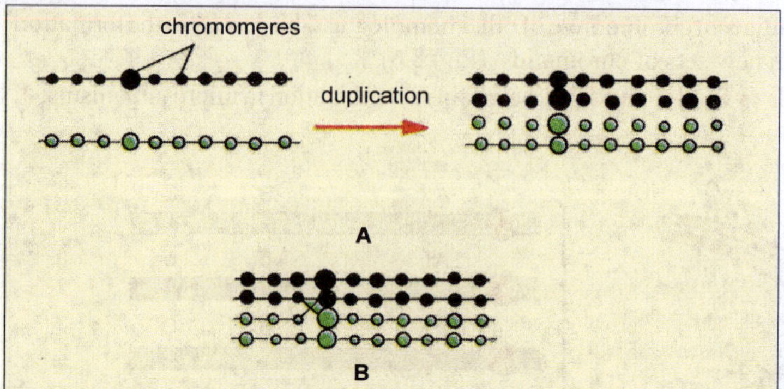

Fig. 8.6. Three steps of Belling's duplication theory.

succeeding divisions of the recipient cell, the donor segment is lost (see **Burns**, 1969).

The copy choice model to explain the mechanism of recombination has been criticised for the following two reasons : (1) Copy choice model relies on conservative mode of DNA replication, but all experimental evidences suggest that DNA replication occurs in a semi-conservative manner. (2) The copy choice model proposed the involvement of only two chromatids in the process of crossing over, while cytological investigations have made it clear that crossing over occurs at four strand stage.

3. Break and exchange theory. The break and exchange theory is the most accepted theory to explain the process of crossing over. This theory states that in the crossing over breaks occur in the non-sister chromatids of the tetrad and the exchange of chromosomal segments occurs between the non-sister chromatids.

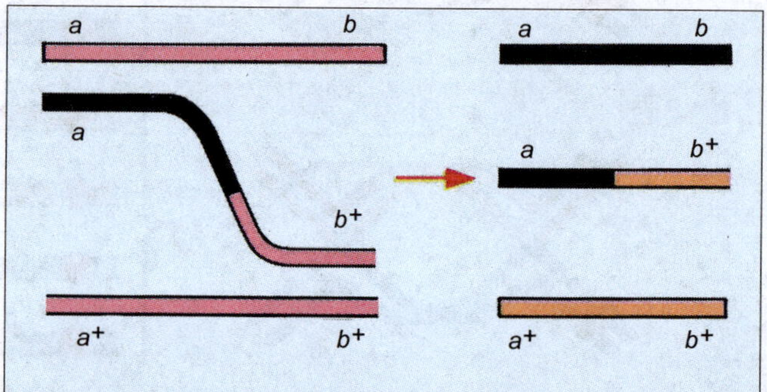

Fig. 8.7. Copy choice model for the mechanism of crossing over.

There exists a lot of controversy about the manner and time of the break of chromatids during the crossing over. Following theories have been advanced by different workers to explain the process of breaking of chromatids :

1. Serebrovsky's contact first theory. According to this theory two chromatids of a tetrad first touch and cross each other and then at the point of cross the breakage of chromatids occur. The chromosome segments then unite to form new combinations.

2. Muller's breakage first theory. This theory holds that before the occurrence of chiasma and crossing over two chromatids or chromosomes break into two fragments and then the union of chromosomal segments takes place.

3. Darlington's strain or torsion theory. Darlington (1935) propounded that chromosomal breakage occurs as a result of strain during pairing. According to this theory during the late pachytene of meiosis, the homologous chromosomes pair and become spirally twisted around each other. Due to twists the coils are formed and these are known as **relational coil**. Then there develops a sort of strain on one chromatid. As a result of the strain due to torsion of chromosomes, the chromatids of the chromosomes break at the point of contact and, thus, recombination results.

4. Hybrid DNA models. In the last three decades several hybrid DNA models have been proposed to explain the mechanism of recombination. In these models, only one strand in each of the two DNA duplexes belonging to non-sister homologous chromatids breaks. The single strands released from these breaks then pair crosswise with unbroken strands by complementary base pairing. This results in the formation of hybrid DNA segments. Some important hybrid DNA models have been proposed by **Whitehouse** (1963) and **Holliday** (1964) and these have been supported by (**Hotta** and **Stern**, 1977, 80, 81).

TETRAD ANALYSIS

Genetical proof that crossing over occurs during the tetrad stage has been provided by red bread mold (*Neurospora crassa*) and other fungi and algae. In a haploid such as *Neurospora,* study of linkage, crossing over and chromosome mapping is facilitated due to following reasons : 1. Since a single allele per locus is present in a haploid, products of recombination can be studied without any difficulty due to dominance. 2. The life cycle of certain fungi (*e.g., Neurospora*) enables us to analyse the products of a single meiosis. 3. In *Neurospora* meiotic products are linearly arranged in the form of **ordered tetrad**, because of which a distinction can be made between 'first division segregation' and 'second division segregation' which is used in detecting the crossing over between centromere and a gene.

Neurospora asci.

In *Neurospora,* for example, the products of a single meiosis can be recovered and examined. The products of a single meiosis may consist of four or eight spores (the meiospores, called **ascospores**), retained in a sac-like structure, the **ascus**, and are described as **tetrad**. Whenever, the products of meiosis consists of eight spores, they result from mitosis of four meiotic products. The group of eight spores is called **octad**, but for convenience it is also referred to as tetrad, since octad represents paired tetrad (Fig. 8.8). Further, the meiospores may be enclosed in two different ways: (1) **Unordered tetrads**, *e.g., Saccharomyces cerevisiae* (baker's yeast), *Chlamydomonas rheinhardi* (green alga), *Aspergillus nidulans* (green bread mold). (2) **Linear** or **ordered tetrads**, *e.g., Neurospora crassa* and *Ustilago hordei* (barley smut).

Tetrad analysis is the technique of using each of the individual spores of a tetrad for analysis to acquire information on linkage and recombination. Thus, for tetrad analysis in *Neurospora,* each ascospore is removed in order from ascus and germinated to determine the physiological phenotype or examined visually for such morphological traits as colour, pattern of growth, etc. Due to ordered tetrads in *Neurospora,* a cross between normal (a^+) and mutant (a) strain, will give rise to a linear arrangement of four normal spores (a^+) at one end followed by four mutant spores (a) at the other end (*i.e.,* 4 : 4 ratio). Such an arrangement would be disturbed if crossing over occurs between cen-

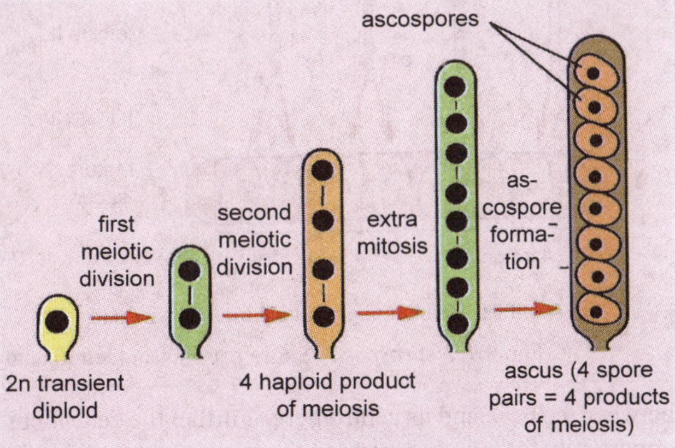

Fig. 8.8. Production of paired tetrad in an ascus, due to linear meiosis and subsequent mitosis to produce eight spores.

tromere and the gene, because crossing over occurs at four strand stage and not at two strand stage; it will produce (2 : 2 : 2 : 2 ratio). Thus, when crossing over is absent, it leads to 4 : 4 arrangement of spores and this is described as **first division segregation** and when crossing over takes place leading to paired arrangement (*e.g.,* 2 : 2 : 2 : 2), it is described as **second division segregation**.

1. First division segregation. A cross between a culture with a wild type (c^+) spreading form mycelial growth and one with a restricted form of growth, called **colonial** (c) has been shown in Fig. 8.9 A. If the ascospores are removed one by one from the ascus in linear order and each is grown as a separate culture, a linear ratio of 4 colonial : 4 wild type indicates that a first division segregation occurred, *i.e.,* during first meiotic anaphase both of the c^+ chromatids moved to one pole and both of the c chromatids moved to the other pole. The 4 : 4 ratio indicates *no crossing over* has occurred

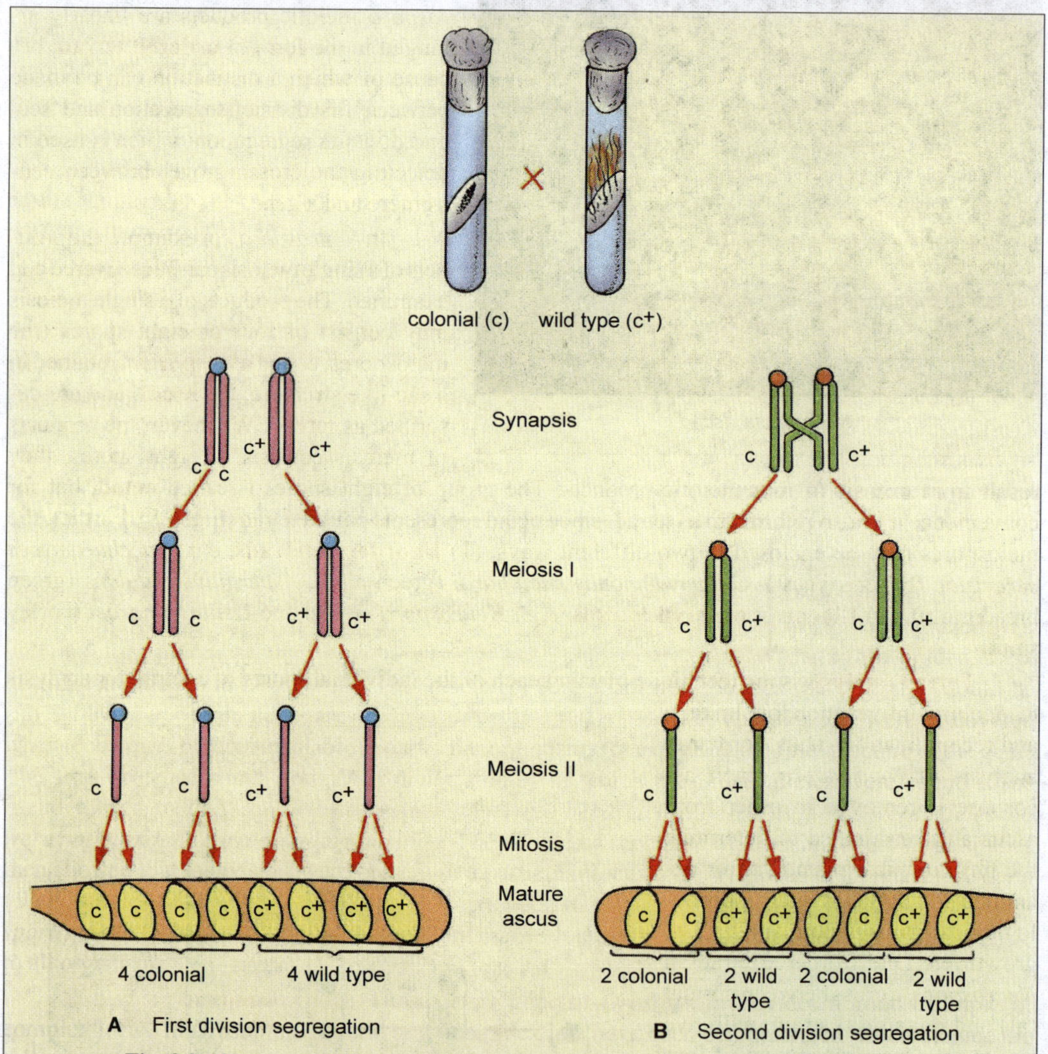

Fig. 8.9. *Neurospora* : Spore patterns and tetrad analysis (after Stansfield, 1986).

between the gene and its centromere. Further the gene locus is from the centromere, the greater is the opportunity for crossing over to occur in this region. So if the meiotic products of a number of asci are analyzed and most of them are found to exhibit a 4:4 pattern, then the locus of c must be close to centromere (this information is used in chromosome mapping).

2. Second division segregation. If crossing over occurs between the centromere and c locus (Fig. 8.9 B), it results in a c^+ chromatid and a c chromatid being attached to the same centromere. Hence, during first anaphase, c^+ and c fail to separate from each other. During second anaphase, sister chromatids move to opposite poles, affecting segregation of c^+ and c. The 2 : 2 : 2 : 2 linear pattern is indicative of a second division segregation ascus produced by crossing over between the gene and its centromere.

Lindegren (1933) studied the inheritance of two pairs of constrasting characters in *Neurospora crassa* and presented the first direct evidence to prove that crossing over occurs at a four-strand stage and involves only two of the four strands at any one place (Fig. 8.10).

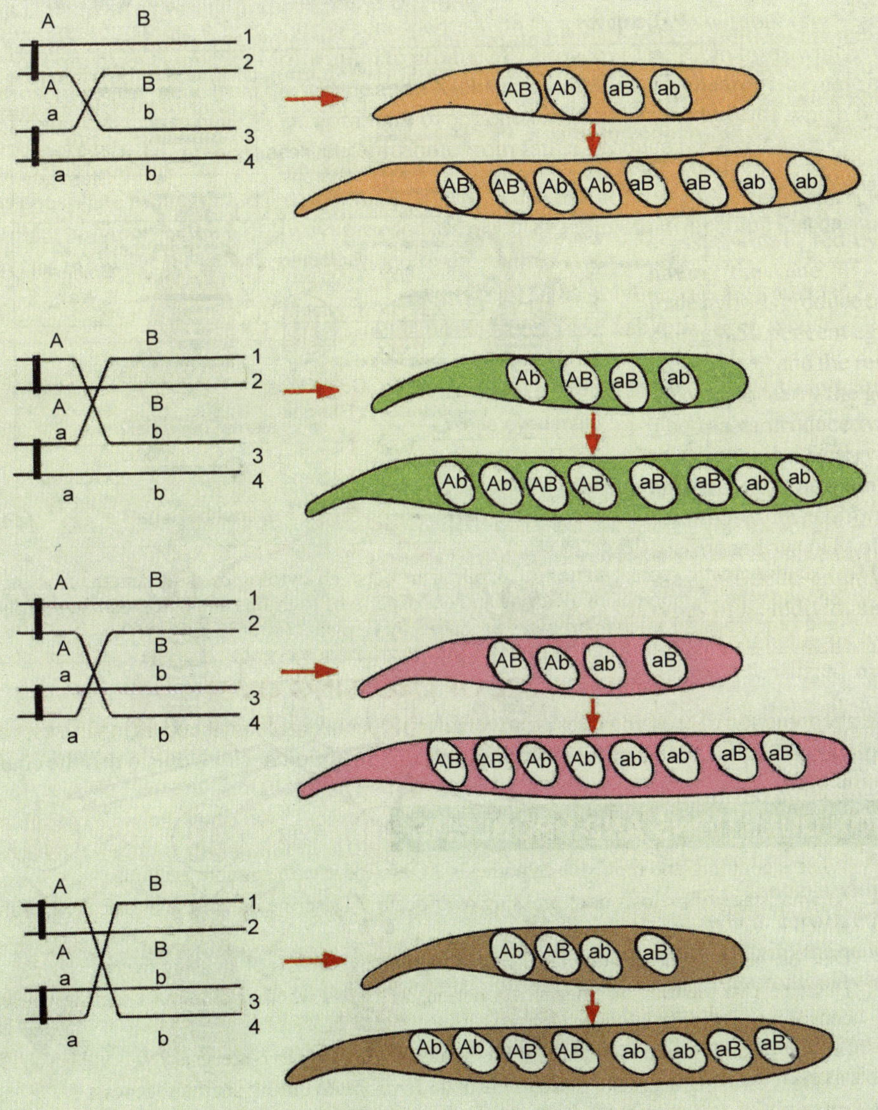

Fig. 8.10. *Neurospora.* Segregation patterns of ascospores in asci as a result of crossing over involving different strands (numbered 1 to 4) at a four-strand stage.

CYTOLOGICAL DETECTION OF CROSSING OVER

Creighton and McClintock's experiment. Creighton and McClintock like **Stera** made convincing correlation between cytological evidence and genetical results of crossing over in maize. They made the use of knob of 9th chromosome of maize which in two different strains might had one allele for either coloured aleurone (C) or colourless aleurone (c) and one allele for either starchy endosperm (Wx) or waxy endosperm (wx). The results of their experiment have been illustrated in Figure 8.11.

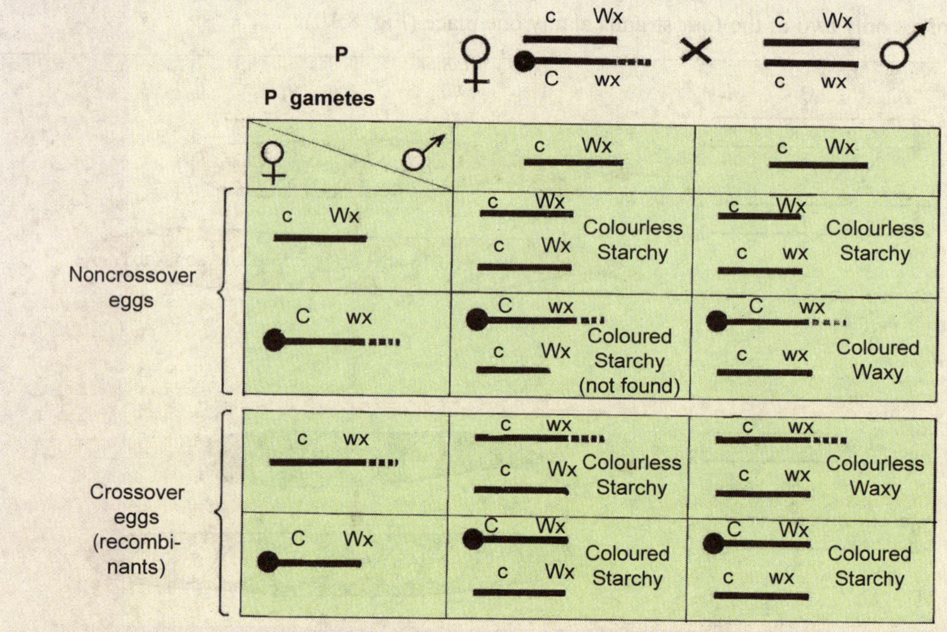

Fig. 8.11. A diagrammatic representation of parallelism between cytological and genetic crossing over showing the phenotypes of aleurone and endosperm of F_1 grains and chromosome morphology and genotypes for microsporocytes produced by F_1 plants (after Burns, 1969).

SIGNIFICANCE OF CROSSING OVER

The phenomenon of crossing over is universal in its occurrence and it occurs in viruses, bacteria, moulds, plants and animals. It is necessary for the natural selection because due to this the chances of variation increase.

REVISION QUESTIONS AND PROBLEMS

1. What is coupling and repulsion hypothesis ? Describe it with suitable example.
2. At what stage of meiosis does crossing over occur ? Describe one of the classical experiments in which it was detected cytologically.
3. Discuss the copy choice and breakage and reunion hypothesis of crossing over.
4. Describe two theories of chiasma formation ? Discuss whether chiasmata are the cause or the consequence of crossing over.
5. (a) Does crossing over take place in the female *Drosophila* in oogenesis ?
   (b) Does crossing over take place in the male *Drosophila* during spermatogenesis ?
6. What is tetrad analysis ? How was *Neurospora* found suitable for the study of crossing over and recombination. Using *Neurospora,* how can you show that crossing over takes place at four strand stage.

7. There is 21% crossing over between the gene locus of p and that of c in the rat. Suppose that 150 primary oocytes could be scored for chiasmata within this region of the chromosome. How many of these oocytes would be expected to have a chiasma between these two genes ?
8. What do you understand by first division segregation and second division segregation in *Neurospora.*
9. What are somatic crossing over and mitotic recombination ?
10. Write short notes on the following :
    (i) Chiasmata theory; (ii) Interference and coincidence;
    (iii) Copy choice theory; (iv) Tetrad analysis.
11. Assume that a crossover and resulting chiasma occured between two gene loci 100 per cent of the time. (a) What would be the percentage of recombinant chromosomes among the progeny. (b) If a chiasma occured 50 per cent of the time, what would this percentage be ?
12. In *Drosophila*, the recessive sex-linked genes abnormal eye facet (*fa*) and singed bristles (*sn*) show 18 per cent recombination.
    (a) If a singed male is crossed to fa^+/fa^+ female, what phenotypes are expected in the F_1 ? (singed = burned, charred or couterised).
    (b) If F_1 males and females are inbred, what phenotypic proportions would be expected to occur in F_2 males and females ?

ANSWERS TO PROBLEMS

5. (a) Yes, (b) No.
7. 63.
11. (a) 50 per cent, with results indistinguishable from independent assortment.
    (b) 25 per cent.
12. (a) F_1 females : *fa*+/+ *sn* (wild type); F_1 males : *fa* + / Y (facet).
    (b) From the data the crossover frequency between *fa* and *sn* is 18 per cent. Therefore, crossover eggs will be ++ (9 per cent) and *fa sn* (9 per cent) and non-crossover eggs will be *fa*+ (41 per cent) and +*sn* (41 per cent). Union of eggs with *fa*+, X- bearing sperm will produce females in the phenotypic proportions of 50 per cent wild type and 50 per cent facet. Union of eggs with Y- bearing sperm will produce males in the phenotypic proportion of 41 per cent facet, 41 per cent singed, 9 per cent facet, singed, and 9 wild type.

CHAPTER

9

Genetic and Cytological Mapping of Chromosomes

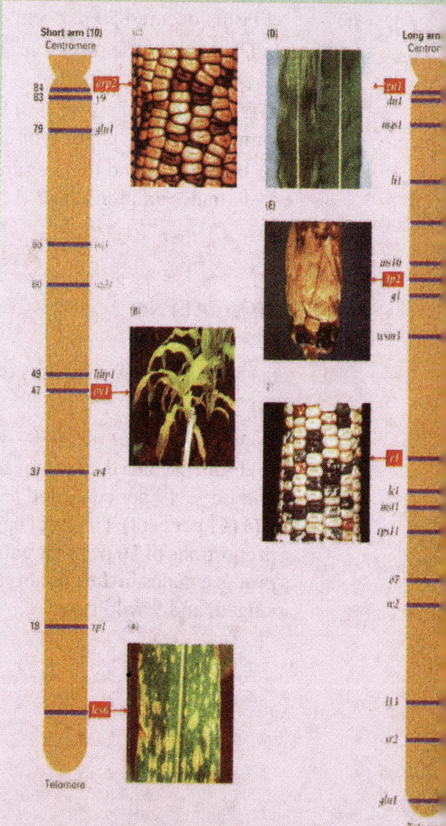

Genetic map of chromosome 10 of corn.

Our discussion of linkage and crossing over has made clear so far, that 1. Because the number of genes usually exceeds the number of chromosomes in different species, therefore, many genes have to be located on same chromosomes. 2. The genes remain arranged linearly on the chromosomes and they have no option before them, except to behave according to the chromosomes during gameto-genesis and inheritance, therefore, all the genes on a chromosome inherit together and are said to be linked with each other to form linkage groups. 3. The number of linkage groups corresponds to the number of homologous pairs of chromosomes or bivalents of the species. 4. The linked genes do not always remain linked, but, occasionally are departed from other members of their linkage groups by crossing over. 5. The closely linked genes have less chances of departure or frequency of crossing over than the widely located genes. 6. And each gene has definite order and location in a linkage group or chromosomes, as the crossing over frequency has been found constant for two given linked genes of a species. For example various genetic crosses have shown that in *Drosophila,* among three mutant sex-linked genes (*viz.,* genes for white eyes, yellow body and cut wings), the genes for white and yellow are always found to have 1% crossing over and similarly, white and cut genes are found to have 20% crossing over . The percentage or frequency of crossing over, thus, appears to be closely related to physical distance between genes, When, one knows all the genes, linkage groups and number of linkage groups of a species, it becomes possible for him that by adopting the crossing over as a tool, he may determine the relative distances between the genes in a linkage group and also their order and may give diagrammatic representation of chromosomes showing the genes as points separated by distances proportional to the amount of crossing over. Such a diagrammatic, graphical representation of relative distances between linked genes of a chromosome is called **linkage** or **genetic map**. Because, such a linkage or genetic map is the outcome of crossing over studies, is also called **cross over map**.

A. CONSTRUCTION OF A LINKAGE MAP OR GENETIC MAPPING

The method of construction maps of different chromosomes is called **genetic mapping**. The genetic mapping includes following processes :

1. Determination of Linkage Groups

Before starting the genetic mapping of the chromosomes of a species, one has to know the exact number of chromosomes of that species and then, he has to determine the total number of genes of that species by undergoing hybridization experiments in between wild and mutant strains. By the same hybridization techniques, it can also be easily determined that how many phenotypic traits remain always together or linked and consequently their determiners or genes during the course of inheritance. And thus, the different linkage groups of a species can be worked out.

2. Determination of Map Distance

The intergene distance on the chromosomes cannot be measured in the customary units employed in light microscopy, geneticists use an arbitary unit to measure the **map unit**, to describe distances between linked genes. A map unit is equal to 1 per cent of crossovers (recombinants); that is, it represents the linear distance along the chromosome for which a recombination frequency of 1 per cent is observed. These distances can also be expressed in **morgan units**; one morgan unit represents 100 per cent crossing over. Thus, 1 per cent crossing over can also be expressed as **1 centimorgan (1cM)**, 10 per cent crossing-over as **1 decimorgan** and so on. The Morgan unit is named in honour of **T.H. Morgan**; however, most geneticists prefer map units.

Quite interestingly, it is now possible to calculate the size of many genes, as well as distances separating them, and to photograph genes in the electron microscope (see **Burns** and **Bottino**, 1989).

Examples

1. If a F_1 hybrid having the genotype Ab/aB produces 8% of cross over gametes AB and ab, then the distance between A and B is estimated to be 16 map units or centimorgan.

2. If the map distance between the gene loci B and C is 12 centimorgan, then 12% of gametes of genotype BC/bc should be crossover types, *i.e.,* 6% bC.

Because, each chiasma produces 50% crossover products, 50 percent crossing over is equivalent to 50 map units or centimorgans. If the mean number of chiasmata is known for a chromosome pair, the total length of the map for that linkage group may be predicted :

Total length = mean number of chiasmata × 50

Two Point Test Cross

The percentage of crossing over between two linked genes is calculated by test crosses in which a F_1 dihybrid is crossed with a double recessive parent. Such crosses because involved crossing over at two points, so called **two point test crosses**. For example, a dihybrid having the genotype Ac/ac is test crossed with a double recessive parent (ac/ac), then among F_2 test cross hybrids we may get 37% dominant genes at both gene loci (AC/ac), 37% recessive genes at both gene loci (ac/ac), 13% dominant gene at first gene locus and recessive gene at the second gene locus (Ac/ac), and 13% recessive gene at first gene locus and dominant gene at second gene locus (aC/ac). The last two groups (*i.e.*, 13% Ac/ac and 13% aC/ac) were produced by crossover gametes (13+13) from the dihybrid parent. Thus, 26% of all gametes (13+13) were of cross over types and the distance between the loci A and C is estimated to be 26 centimorgans. Because, double crossovers usually do not occur between genes less than 5 centimorgans apart, so for genes further apart, the **three point test crosses** are used.

Three Point Test Cross

A three point test cross or trihybrid test cross (involving three genes) gives us information regarding relative distances between these genes, and also shows us the linear order in which these genes should be present on chromosome. Such a three point test cross may be carried out if three points or gene loci on a chromosome pair can be identified by marker genes. If, in addition to genes A and C indicated above, a third marker gene B is located in fairly close proximity in the same linkage group, all three markers may be used together in conducting a more precise analysis of the map distance and the relative position of the three points.

Suppose that we testcross trihybrid individuals of genotype ABC/abc and find in the progeny the following:

36% ABC/abc 36% abc/abc		9% Abc/abc 9% aBC/abc		4% ABc/abc 4% abC/abc		1% AbC/abc 1% aBc/abc
72% Parental type	:	18% Single crossovers between A and B. (region I)	:	8% Single crossovers between B and C. (region II)	:	2% Double crossover

To find the distance A–B we must count all crossovers (both singles and doubles) that occurred in region I = 18% + 2% or = 20% or 20 map units between the loci A and B. To find the distance B–C we must again count all crossovers (both singles and doubles) that occurred in region II = 8%+2% = 10% or 10 map units between the loci B and C. The A–C distance is, therefore, 30 map units when double crossovers are detected in a three point linkage experiment and 26 map units when double crossovers are undetected in the two-point linkage experiment above.

Without the middle marker (B), double crossovers would appear as parental types and hence we underestimate the true map distance (crossover percentage). In this case the 2% double crossovers would appear with the 72% parental types, making a total of 74% parental types and 26% recombinant types. Therefore, for any three linked genes whose distances are known, the amount of detectable crossovers between the two outer markers A and C when the middle marker B is missing is; (A–B crossover percentage) plus (B–C crossovers percentage) minus (2 X double crossover percentage).

3. Determination of Gene Order

After determining the relative distances between the genes of a linkage group, it becomes easy to place genes in there proper linear order. For example, if the linear order of three genes ABC is to be determined, then these three genes may be in any one of three different orders depending upon that which gene is in the middle. For the time being we may ignore left and right end alternatives. If double crossovers do not occur, map distances may be treated as completely additive units.

A scientist checking a genetic map of mice.

Now, if we suppose that the distance between the genes A–B = 12, B–C = 7 A–C = 5, we can determine the order of genes correctly in the following manner :

Case I. Let us assume that gene A is in the middle (*e.g.*, B–A–C) :

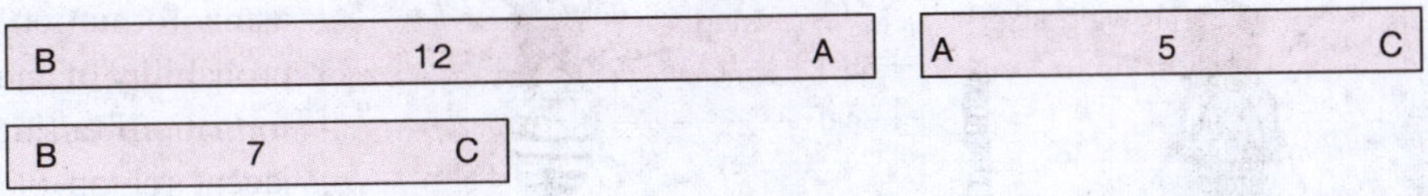

In this case because, the distances between B–C are not equitable, genes A cannot be in the middle.

Case II. Let us assume that gene B is in the middle (*e.g.*, A–B–C) :

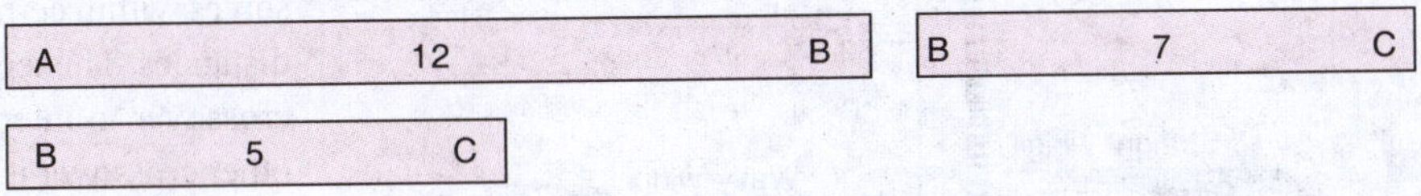

In this case, because the distance between A–C are not equitable, therefore, gene B cannot be in the middle.

Case III. Let us assume that gene C is in the middle (*e.g.*, A–C–B).

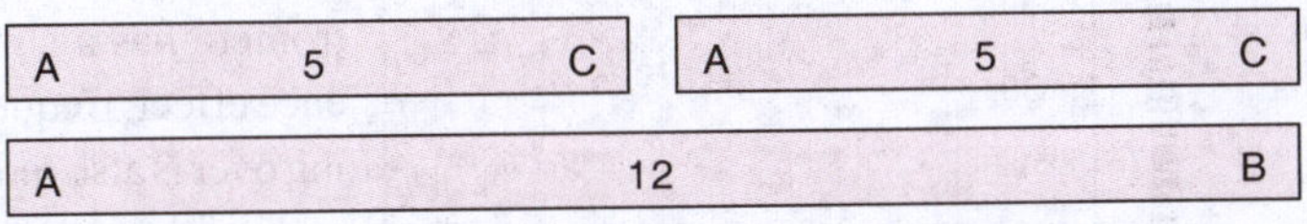

In this case, because the distances between A–B are equitable, therefore, gene C must be in the middle.

Thus, the relative distances and ordering of genes in a linkage group are determined in separate segments by two point test crosses or three point crosses, as the case may be.

4. Combining Map Segments

Finally, the different segments of maps of a complete chromosome are combined to form a complete genetic map of 100 centimorgans long for a chromosome.

Example. For example, suppose we have to combine following three map segments.

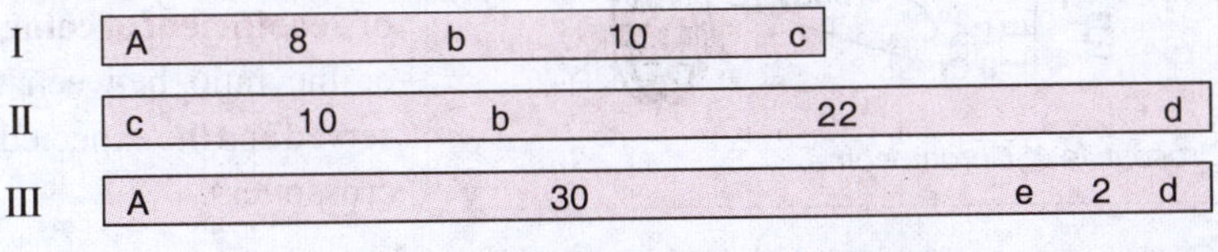

We can superimpose each of these segments by aligning the genes shared in common.

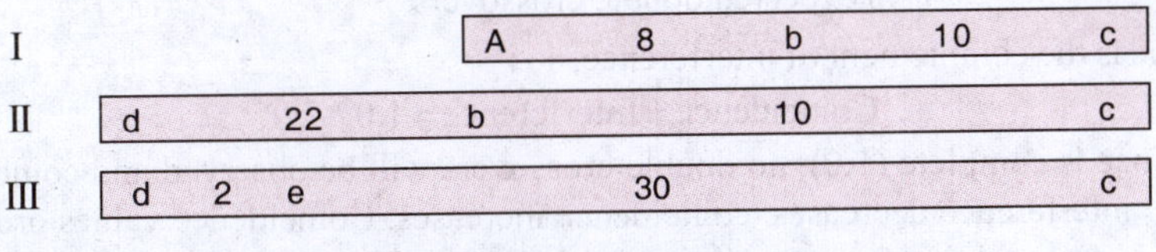

Then finally we may combine the three segments into one map :

The a to d distance = (d to b) – (a to b) = 22–8 = 14

The a to e distance = (a to d) – (d to e) = 14–2 = 12

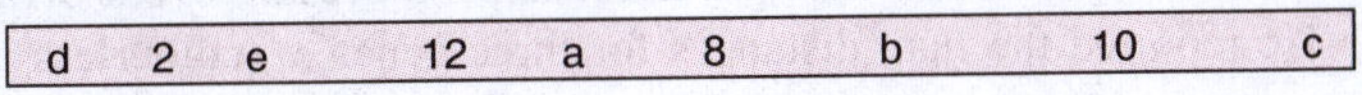

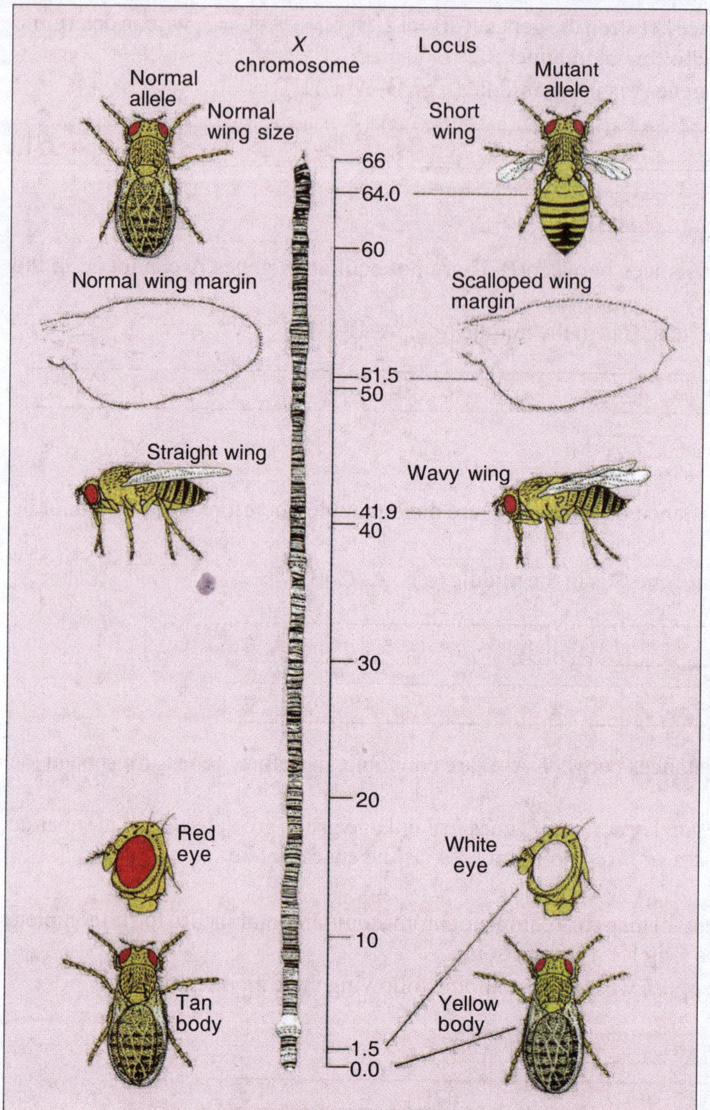

Map of the *Drosophila X Chromosome.*

Interference and Coincidence

In most higher organisms it has been found that one chiasma formation reduces the probability of another chiasma formation in an immediately adjacent region of the chromosome, probably because of physical inability of the chromatids to bend back upon themselves within certain minimum distances. The tendency of one crossover to interfere with the other crossover is called **interference**. Thus, the proximity of one crossover to another decreases the probability of another very close by. The centromere has a similar interference effect; frequency of crossing over is also reduced near the ends of the chromosome arms. The net result of this interference in the observation of fewer double crossover types than would be expected according to map distances. The strength of interference varies in different segments of the chromosome and is usually expressed in terms of a **coefficient of coincidence**, or the ratio between the observed and the expected double crossovers.

$$\text{Coefficient of coincidence} = \frac{\%\ \text{observed double crossovers}}{\%\ \text{expected double crossovers}}$$

The coincidence is the complement of interference, so :

$$\text{Coincidence} + \text{Interference} = 1.0$$

When interference is complete (1.0), no double crossovers will be observed and coincidence becomes zero. When, interference decreases, coincidence increases. Coincidence values ordinarily vary between 0 and 1. Coincidence is generally quite small for short map distance. There is no interference across centromere.

Example

For explaining interference and coincidence, we can consider the results of one of the experiment of **Hutchison** (1922). He reported the map distances for three genes, c (colourless aleurone), Sh (shrunken grains), and wx (waxy endosperm) of corn and observed following crossing over frequencies between these genes :

Table 9-1 Crossing over frequencies between genes c, Sh and wx of corn.

Regions	Genes	Percentage crossovers	Map-distances (in map units)
I	c–Sh	3.4	3.4 + 0.1= 3.5
II	Sh–wx	18.3	18.3 + 0.1= 18.4
Double crossover	c–Sh–wx	0.1	

Table 9-2. Some linkage groups and chromosome assignment in human beings (Source : *McKusick*, 1987).

Number of the chromosome	Gene	Number of the chromosome	Gene
1	Amylase (pancreatic)	13	Ribosomal RNA
1	Actin, skeletal muscle alpha chain	13	Collagen IV, alpha–1 chain
1	Xeroderma pigmentosum A	13	Collagen IV, alpha–2 chain
1	Rhesus blood antigen (Rh)	14	Ribosomal RNA
1	Histone cluster B : H3, H4	14	T-cell leukemia–1
1	Cystic fibrosis antigen	14	Immunoglobulin heavy chain gene cluster
1	Acid phosphatase	15	Ribosomal RNA
2	Elastin	16	Haemoglobin alpha
2	Interferon-1	16	Haemoglobin zeta
2	Collagen III	16	Nonhistone chromosomal protein-1
2	Collagen IV	17	Growth hormone
2	Glucagon	18	Gastric releasing peptide
2	Tubulin, alpha, testis specific	19	Bombay phenotype
3	Rhodopsin	19	Lewis blood group
3	Somatostatin	19	Green/blue eye colour
4	Huntington's disease	20	Growth hormone releasing factor, somatocrinin
4	MN blood group	21	Ribosomal RNA
4	Diabetes insipidus	22	Myoglobin
5	Fibroblast growth factor, acidic	22	Ribosomal RNA
6	Tubulin, beta M40	X	Polymerase DNA, alpha
6	Insulin - dependent diabetes mellitus	X	Gonadal dysgenesis, XY female type
7	Actin, cytoskeletal beta	X	Haemophilia A, factor VIII
7	Collagen I alpha 2	X	Haemophilia B, factor IX
7	Histone cluster A : H1, H2A, H2B	X	Deutan (green) colour blindness; green cone pigment
7	Non-histone chromosomal protein-2	X	Protan (red) colour blindness; red cone pigment
8	Polymerase, DNA, beta	X	Ocular albinism
8	Carbonic Anhydrase cluster	Y	Azoospermia–third factor
9	ABO blood group	Y	Stature
10	Hexokinase–1	Y	H–Y antigen
11	Catalase	Y	Testicular determining factor (TDF)
11	Insulin	Y	Pseudoautosomal segment (PAS)
12	Salivary protein complex		
12	Collagen II, alpha–1		

If crossing over in region I and II were independent, we should predict 0.035 X 0.184 = 0.6 per cent double crossovers, where as only 0.1 per cent was observed

$$\text{So, coincidence} = \frac{0.1}{0.6} = 0.167$$

Linkage Maps of Different Organisms

By adopting the above mentioned techniques, geneticists have constructed the linkage or genetic maps of various organisms, such as, viruses, bacteria, fungi, tomato, barley, wheat, rice, sorghum, morning glory, garden pea, maize, *Drosophila*, chickens, mice, man (Table 9-2), etc. The first linkage map has been constructed for two chromosomes of *Drosophila* by **Strutevant** in 1911. The linkage maps of other chromosomes of *Drosophila* have been constructed by **C.B. Bridges**. The linkage or genetic mapping in maize has been done by **McClintock** under the leadership of **R.A. Emerson**.

C.B. **Bridges** (1889-1938).

Syntenic Genes

If two or more specific human gene products and a given human chromosome are both present in the same hybrid cells, then those genes are located in the same chromosome; that is, they are **syntenic**. The term **synteny** refers to genes that are located on the same chromosome, whether or not they show recombination; **linkage** refers only to genetic loci that have been shown by recombination studies to be in the same chromosome. Syntenic genes may be so far apart in their chromosome that they seem to segregate independently; that is, they may show as much as 50 per cent recombination as would be exhibited by non-syntenic genes.

B. CHROMOSOME, PHYSICAL OR CYTOLOGICAL MAPPING

It has been found that linkage map distances between genes are not necessarily proportional to physical linear measurements. Special cytological techniques have been used to determine the physical locations of a gene in a chromosome. Localization has been accomplished by identifying a gene locus with relation to some visible landmark such as a chromomere or cross band. In this process of **chromosome**, **physical** or **cytological mapping**, the polytene chromosomes of salivary glands of *Drosophila* have been found very useful. **T.H. Painter** was the first geneticist who used polytene chromosomes of *Drosophila* in the cytological verification of genetic data. He related the bands on the giant chromosomes to genes, but he was more interested in the morphology of the chromosomes and applications concerning speciation than in the association of chromosome sections with particular genes. **Bridges**, beginning in 1934, made extensive and detailed investigation of the salivary gland chromosomes and, in the course of his investigations, developed a tool of practical usefulness in relating genes to chromosomes. In applying this method to *Drosophila melanogaster*, he prepared a series of cytological or chromosome maps to correspond with the linkage maps already available.

Cytological Mapping of Chromosomes of Drosophila

About 5000 single cross bands have been noted on the four pairs of salivary gland chromosomes in *D. melanogaster*. This number is considered a minimum approximation of the number of genes in that animal. Some genes have been associated with individual bands. Bridges system of designating parts of chromosomes with numbers, subdivision with letters, and numbering bands within subdivision had made it possible for investigators to discuss precise locations. In this system, fairly uniform divisions are numbered in order throughout each entire chromosome set from O at the beginning of the X chromosome to 102 at the end of chromosome 4. Subdivisions within the areas are identified with letters from A to F, and bands within subdivisions are numbered from left to right. For example, the gene (w) for white eyes is in bands 3C2. In linkage units this gene is located at 1.5 in the X chromosome. Linkage data do not correspond exactly with cytological locations, but the linear sequence of genes can be verified from salivary preparations.

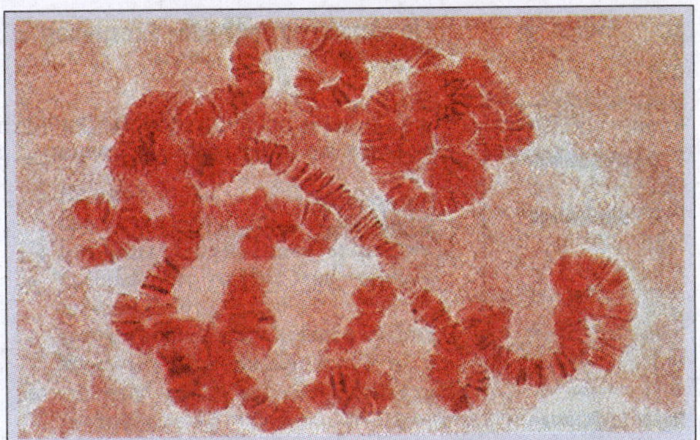

'Giant' chromosome from a cell in the salivary gland of *Drosophila*, showing banding.

Differences Between Genetic and Chromosome Maps

The frequency of crossing over usually varies in different segments of the chromosome but is a highly predictable even between any two genes loci. Therefore, the actual physical distances between linked genes bears no direct relationship to the map distance calculated on the basis of crossover percentages. However, the linear order of linked genes is identical in both maps.

Uses of Genetic Maps

The genetic maps have following uses :

(i) The chromosome maps display the exact location, arrangement and combination of genes in a linkage group or chromosomes.

(ii) They are useful in predicting results of dihybrid and trihybrid crosses.

REVISION QUESTIONS AND PROBLEMS

1. What is the difference between a linkage (genetic) map and a chromosome (cytological map) ?
2. How are genetic maps constructed ? What are the fundamental assumptions on which genetic (linkage)mapping is based ? Why are extremely short regions used in establishing genetic map ?
3. Barring chromosomal aberrations, the order of genes on a genetic map of a chromosome corresponds completely with the order on a cytological map. Why do the distances on the two maps not always correspond ?
4. In the mouse, the genes studied to date (approximately 300) fall into 20 linkage groups. How many chromosomes would you normally expect to find in a somatic cell of a mouse ? Explain.
5. Which of the following features must an organism possess to make the construction of genetic maps possible ?
   a. Morphologically distinct sexes;
   b. Many monogenically inherited pairs of traits;
   c. Individuals heterozygous for at least three pairs of alleles;
   d. Standard environmental conditions;
   e. Homozygous recessive stocks;
   f. More than two alleles per locus;
   g. X and Y chromosomes.
6. Loci for the human Duffy blood group, that is, production of either, both, or neither of the antigens Fy^a and Fy^b, and the rare Charcot-Marie Tooth disease (a severe sensory and motor neuropathy) are both located on autosome 1. Studies suggest a recombination frequency of about 0.15. Give the distance between the two loci in map units.
7. Two dominant mutants in the first linkage group of guinea pigs govern the traits pollex (*Px*), which is the ativistic return of thumb and little toe, and rough fur (*R*). When dihybrid pollex, rough pigs (with identical linkage relationships, were crossed to normal (wild type) pigs, their progeny fall into four phenotypes : 79 rough, 103 normal, 95 rough pollex and 75 pollex. (a) Determine the genotypes of the parents. (b) calculate the amount of recombination between *Px* and *R*.

8. The genes for two nervous disorders, waltzer (*v*) and jittery (*ji*) are 18 map units apart on chromosome 10 in mice. A phenotypically normal F_1 group of mice carrying these two genes in coupling phase is being maintained by a commercial firm. An order arrives for two dozen young mice each of waltzer, jittery and waltzer plus jittery. Assuming that the average litter size is seven offspring, and including a 10% safety factor to ensure the recovery of the needed number of females that need to be bred. Calculate?

9. Assume the presence in corn of the recessive *ds* and *mp*. The genes are linked and 20 map units apart. From the cross *ds mp*/++ × *ds*+/+*mp*, what percentage of the progeny would be expected to be both *ds* and *mp* in phenotype ?

ANSWERS TO PROBLEMS

6. 15 map units.

7. (a) **P** : *PxR/PxR* × *pxr/pxr*
   Pollex, rough   Normal

   (b) The 79 rough and 75 pollex types are recombinants, constituting 154 out of 352 individuals = 0.4375 or approximately 43.8% recombination.

8. F_1 : *vji* / + +

   F_2 : If 18% are crossover types, then 82% should be parental types.

		.41 *vji*	.41 + +	.09*v* +	.09+*ji*
82% Parerlal types	.41*vji*	.1681 *vji/vji* Waltzer, Jittery	.1681 *vji* / + + Wild type	.0369 *vji* / *v* + Waltzer	.0369 *vji* / + *ji* Jittery
	.41 + +	.1681 + + / *vji* Wild type	.1681 + + / + + Wild type	.0369 + + / *v* + Wild type	.0369 + + / + *ji* Wild type
18% Cross over types	.09 *v* +	.0369 *v* + / *vji* Waltzer	.0369 *v* + / + + Wild type	.0081 *v* + / *v* + Waltzer	.0081 *v* + / + *ji* Wild type
	.09 + *ji*	.0369 + *ji* / *vji* Jittery	.0369 + *ji* / *vji* Wild type	.0081 + *ji* / *v* + Wild type	.0081 + *ji* / + *ji* Jittery

Summary of the F_2 results :

1. Wild type = .6681 or 66.81%
2. Waltzer, Jittery = .1681 or 16.81%
3. Waltzer = .0819 or 8.19%
4. Jittery = .0819 or 8.19%

Waltzer or jittery phenotypes are least frequent and, hence, are the limiting factors. If 8.19% of all progeny are expected to waltzer, how many offspring need to be raised to produce 24 walterzs. .0819x = 24, x = 24/.0819 = 293.04 or approximately 293 progeny. Adding the 10% safety factor, 293+29.3 = 322.3 or approximately 322 progeny. If each female has 7 per litter, how many females need to be bred ? 322 / 7 = 46 females.

9. 40% of the gametes of ds mp/++ parent will carry both *ds* and *mp*; 10% of the gametes produced by the *ds*+/+*mp* parent will carry both recessives. Therefore, 0.4×0.1 = 0.04 or 4% of the progeny will be homozygous for these genes.

CHAPTER

10

Multiple Alleles

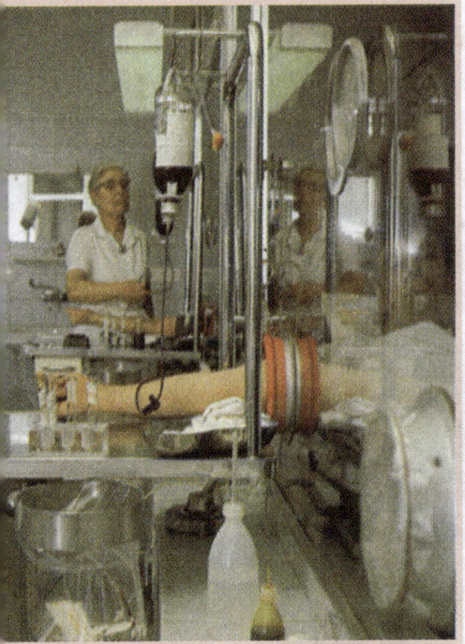

When transfusing blood, the blood groups of donor and recipient must be matched. Inheritance of A, B, AB and O blood types of man is determined by a series of three allelomorphic genes.

So far, it has been observed that a given phenotypic trait (character) of an individual depends on a single pair of genes, each of which occupies a specific position called the gene **locus**, on a homologous chromosome. Moreover, a particular gene has been found to occur in two alternative forms or **allelomorphs**, one being dominant and other recessive; one being wild form and the other mutant form. For example, a gene (L) for length of *Drosophila* wings may occur in two alternative forms : a allele or gene (L^+) for normal development of wings and another allele or gene (L^{vg}) for vestigial wings. Because, most flies have normally developed wings, so, it can be easily concluded that gene L^+ is the original form of gene or allele from which the other form of gene or allele (L^{vg}) might have originated by certain mutational event at sometime in past. The gene L^+ for normal development of wing is called the normal or **wild type allele** of the gene L and usually symbolized as L^+, while the mutated gene L^{vg} for vestigial wing is called reduced type or **mutant allele** of the gene L. A fly with normal wings, thus, has two wild type alleles (L^+L^+) and the vestigial winged fly has two mutant alleles ($L^{vg}L^{vg}$). Both of these allelic forms (L^+ and L^{vg}) of gene L occur at corresponding positions on genetically identical (homologous) chromosomes of same or different individual.

If the mutant allele has developed from the wild form of allele due to mutation, one may expect that the wild form of allele can mutate in more than one way. The mutant form of allele too can mutate once again to give rise to another mutant form of allele. Therefore, it is possible to have more than two allelic forms, *i.e.,* **multiple alleles**, of one kind of gene. Although only two actual alleles of a gene can exist in a diploid cell (and only one in a haploid cell), the total number of possible different allelic forms that might exist in a population of individuals is often quite large. This situation is termed as **multiple allelism**, and the set of alleles itself is called a **multiple allelic series**.

CHARACTERS OF MULTIPLE ALLELES

The most important and distinguishing features of multiple alleles are summarized below :

1. Multiple alleles of a series always occupy the same locus in the chromosome.

2. Because, all the alleles of multiple series occupy same locus in chromosome, therefore, no crossing-over occurs within the alleles of a same multiple allele series.

3. Multiple alleles always influence the same character.

4. The wild type allele is nearly always dominant, while the other mutant alleles in the series may show dominance or there may be an intermediate phenotypic effect.

5. When any two of the mutant multiple alleles are crossed, the phenotype is mutant type and not the wild type.

SYMBOLISM FOR MULTIPLE ALLELES

The dominance hierarchy is defined at the beginning of each problem involving multiple alleles. A capital letter is commonly used to designate the allele which is dominant to all other alleles in the series. The corresponding lower case letter designates the allele which is recessive to all others in the series. Other alleles which are intermediate in their degree of dominance between these two extremes, are usually assigned the lower case letter with some suitable superscript.

Fig. 10.1. Different coat colours in rabbits (cafter Burns 1969).

EXAMPLES

The best examples of multiple allelic system have been observed in coat colour of rabbits, wings of *Drosophila* and blood groups in man.

1. The C gene in Rabbit (Coat Colour)

The coat of rabbit may have following different colours :

(i) Full colour. The coat of the ordinary (wild type) rabbit is referred to as "**agouti**" or **full colour**, in which individuals have banded hairs, the portion nearest the skin being gray, succeeded by a yellow band, and finally a black or brown tip. The genes for full colour may be represented by capital letter C or c^+.

(ii) Chinchilla. In some individuals, the coat lacking the yellow pigment and due to the optical effect of black and gray hairs, have the appearance of silvery-gray. The gene for chinchilla is represented as c^{ch}.

(iii) Himalayan. The Himalayan type coat is white except for black extremities (nose, ears, feet and tail). The eyes are pigmented. The gene for Himalayan coat is represented by c^h.

(iv) Albino. The albino coat totally lacks in pigmentation and the eyes of an albino also remain pink due to lack of pigment in iris of eye. The gene for albino is represented by c.

Crosses of homozygous agouti (c^+c^+) and albino (cc) individuals produce a uniform agouti F_1; interbreeding of the F_1 produces an F_2 ratio of 3 agouti : 1 albino (Fig. 10.2). Two third of F_2 agouti are found to be heterozygous by testcrosses. Thus, it is a case of monohybrid inheritance, with agouti completely dominant to albino. Likewise, crosses between chinchilla and agouti (Fig. 10.3) produce all agouti individuals in the F_1 and a 3 agouti : 1 chinchilla ratio in the F_2. Such complete dominance of agouti also occurs on Himalayan (Fig. 10.4). Further crosses, reveal that c^{ch} allele for chinchilla though is recessive to c^+ allele for agouti coat or skin, is incompletely dominant over Himalayan (c^h) and albino (c) alleles (Fig. 10.5 and Fig. 10.6). Likewise, c^h allele for Himalayan coat is recessive to c^+ (agouti) and c^{ch} (chinchilla) but dominates over albino (Fig. 10.7). The results of all these crosses exhibit that c+ (agouti), c^{ch} (chinchilla), c^h (Himalayan) and c (albino) are allelic to each other and the alleles of this multiple allelic series have following dominance hierarchy :

$$c^+ > c^{ch} > c^h > c$$

The possible phenotypes and their associated genotypes of this multiple allelic series can be summarized in Table 10-1.

P₁ : Agouti (c^+c^+) X Albino (cc)
↓
F₁ : Agouti (c^+c)
F₂ : $1c^+c^+ : 2c^+c : 1cc$
3 Agouti : 1 Albino

Fig 10.2. A monohybrid cross between agouti and albino rabbits.

P_1 : Agouti ($c^+ c^+$) X Chinchilla ($c^{ch}c^{ch}$)
↓
F_1 : Agouti (c^+c^{ch})
F_2 : $1c^+c^+ : 2c^+ c^{ch} : 1 c^{ch}c^{ch}$
3Agouti : 1 Chinchilla

Fig. 10.3. A monohybrid cross between agouti and chinchilla rabbits.

P_1 : Agouti (c^+c^+) X Himalayan (c^hc^h)
↓
F_1 : Agouti (c^+c^h)
F_2 : $1c^+c^+ : 2 c^+c^h : 1 c^hc^h$
3 Agouti : 1 Himalayan

Fig. 10.4. A monohybrid cross between agouti and Himalayan (or Russian) rabbits.

P_1 : Chinchilla ($c^{ch} c^{ch}$) X Himalayan (c^hc^h)
↓
F_1 : Light gray ($c^{ch} c^h$)
F_2 : $1 c^{ch} c^{ch} : 2 c^{ch} c^h : 1 c^h c^h$
1 Chinchilla : 2 Light gray : 1 Himalayan

Fig. 10.5. A monohybrid cross between chinchilla and Himalayan rabbits. Showing incomplete dominance of chinchilla on Himalayan.

P_1 : Chinchilla ($c^{ch} c^{ch}$) X Albino (c c)
↓
F_1 : Light gray ($c^{ch} c$)
F_2 : $1c^{ch} c^{ch} : 2 c^{ch} c : 1cc$
1Chinchilla : 2 Light gray : 1 Albino

Fig. 10.6. A monohybrid cross between chinchilla and albino rabbits showing incomplete dominance of chinchilla over albino.

P_1 : Himalayan ($c^h c^h$) X Albino (c c)
↓
F_1 : Himalayan ($c^h c$)
F_2 : $1 c^h c^h : 2c^h c : 1 c c$
3 Himalayan : 1 Albino

Fig. 10.7. A monohybrid cross between Himalayan and albino rabbits.

Table 10-1. **The phenotypes and genotypes of multiple allelic series for coat colour in rabbit.**

Phenotypes	Genotypes
Full colour (Agouti)	c^+c^+, c^+c^{ch}, c^+c^h, c^+c
Chinchilla	$c^{ch}c^{ch}$
Light gray	$c^{ch}c^h$, $c^{ch}c$
Himalayan	c^hc^h, c^hc
Albino	cc

Karl Landsteiner (1868-1943).

It should be noted here that in this case four allelic forms of genes may produce at least ten genotypes, whereas in case of single pair of alleles at a given locus only three genotypes were produced where dominance is complete. Thus, as the number of genes in a series of multiple alleles increases, the variety of genotypes rises still more rapidly, such as exemplified on next p\age.

Alleles in series	Genotypes
2	3
3	6
4	10
5	15
n	n/2 (n+1)

2. A, B, AB and O Blood Groups in Humans **Landsteiner** in 1900 and 1902 discovered two kinds of **agglutinogens** or **antigens**, called **A** and **B antigens** from the surface of red blood cells of human blood. He found that out of A and B antigens, a person may contain either one (*i.e.,* A or B antigen) or neither of them. Accordingly, he recognised three kinds of **blood types** or **blood groups** : **type A**, **type B**, and **type O**. The fourth and most rare, the **A B blood group** or type, was discovered in 1902 by two of **Landsteiner's** students, **Von Decastello** and **Sturli**. For A and B antigens, there occur two **agglutinins** or **antibodies** : anti-A (or α) and anti-B (or β). Recent, chemical investigations have shown that A and B antigens are not proteins but, are mucopoly-

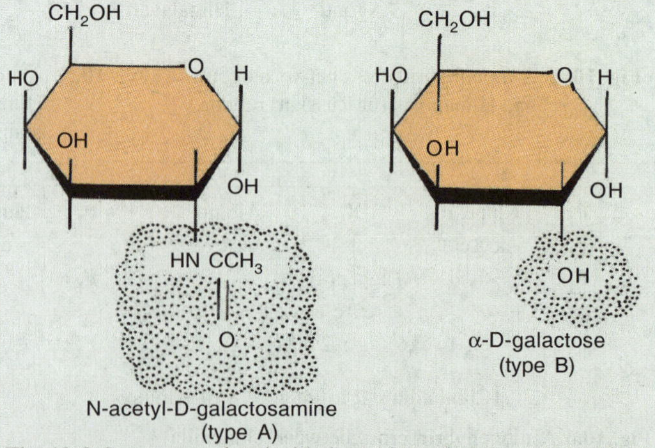

Fig. 10-8. Sugars associated with the surface of erythrocytes in blood type A and type B persons (after Goodenough and Levine, 1974).

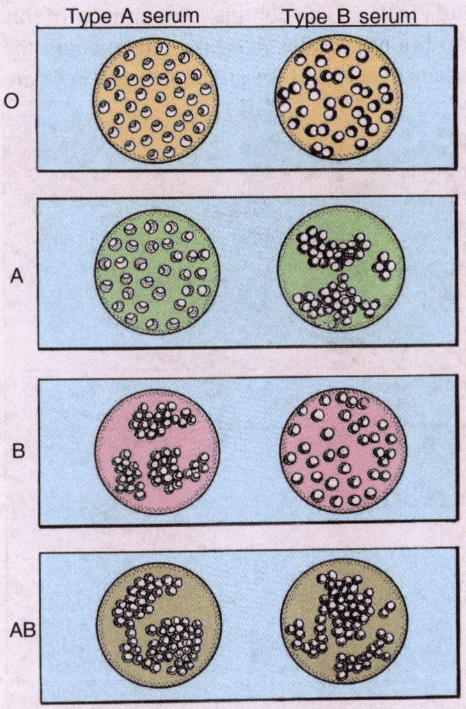

Fig. 10.9. Agglutination tests for A, B, AB and O human blood groups (after Gardner, 1968).

saccharides (sugars + aminoacids) of 300,000 molecular weights.

The four blood types have different agglutinizing properties. To determine the blood group types of different persons, an **agglutination test** is performed. On a glass slide is placed a drop of type A serum (containing anti-B antibodies) and a separate drop of type B serum (containing anti-A antibodies). When a drop of type O blood is added to each drop there is no agglutination or clumping of red blood cells in either drop takes place. This shows that O blood group has neither A nor B antigen. If a drop of type B blood is added agglutination occurs with type A serum; type A red blood cells are agglutinated by type B serum and type AB red blood cells are agglutinated by both sera. The agglutination tests for four types of human blood has been illustrated in Figure 10.9.

Each person, therefore, can use the blood of its own blood group in emergency, otherwise, clumping of red blood cells may take place, if blood group of different type is transfused in him. The clumped red blood cells occlude capillaries and, thus, deprive vital organs of normal blood supply and may lead to death. The characteristics of blood groups and the types of transfusions can be summarized in following table 10-2.

Table 10-2 **Human blood groups.**

Blood groups (phenotype)	Antigen in red blood cells	Antibodies in plasma	Can give blood to groups	Can receive blood from group	Genotype
O	None	Anti-A, Anti-B,	O, A, B, AB	O	ii
A	A	Anti-B	A, AB	O, A	I^AI^A or I^Ai
B	B	Anti-A	B, AB	O, B	$I^B I^B$ or I^B i
AB	A and B	None	AB	O, A, B, AB	$I^A I^B$

Multiple Allelic Inheritance of A, B, AB and O Blood Types

Bernstein (1925) proposed that inheritance of A, B, AB and O blood types of man is determined by a series of three allelomorphic genes. The gene controlling blood types has been labelled as **L** (after the name of discoverer, **Landsteiner**) or **I** (from **isoagglutination**, the technical term for the **agglutinogen** (antigen) or clumping of the red blood cells by an agglutinin or antibody. The prefix **iso** is derived from the greek *isos,* meaning equal and indicates that the agglutinations caused by a serum from the same species, man). The I gene exists in three different allelic forms : $\mathbf{I^A}$, $\mathbf{I^B}$ and **i**. The first two alleles produce characteristic antigens on the surface of erythrocytes. Thus, I^A allele specifies A antigen, I^B allele determines B antigen and allele specifies no antigen.

The pedigree analysis has shown that alleles I^A and I^B have dominance over i allele. Likewise the pedigree analysis of A and B parents revealed that their children have both A and B antigens on the erythrocytes, showing codominance between I^A allele and I^B allele. The dominance hierarchy of this allelic series can be depicted as follows : $I^A = I^B > i$. Additional studies that take into account the subgroups of the A antigen indicate that I^A allele may occur in at least four allelic forms. These are

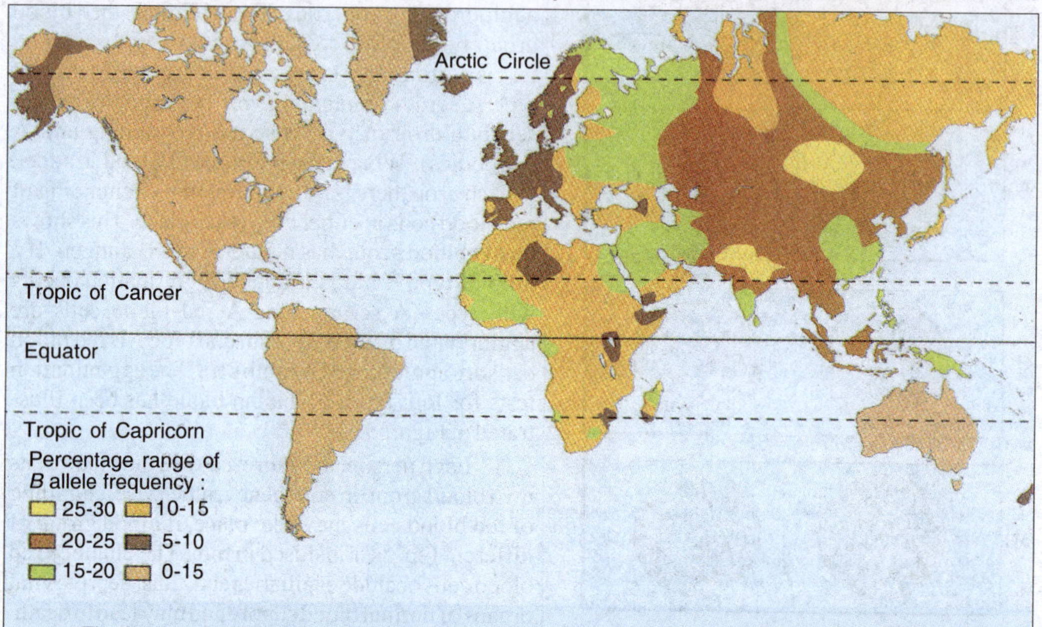

The frequency with which the I^B allele of the ABO blood group occurs throughout the world.

symbolized I^{A1}, I^{A2}, I^{A3}, and I^{A4}. I^{A1} is dominant to all other I^A alleles, I^{A2} is recessive to I^{A1} but dominant to the other two and so on. Considering all the six alleles of gene I, *i.e.,* four forms of I^A, one of I^B and one of i, dominance within the multiple allelic series can be shown in the following way :

$$[(I^{A1} > I^{A2} > I^{A3} > I^{A4}) = I^B] > i$$

Neglecting the very rare I^{A4} allele, this series of multiple alleles produces 15 genotypes and 8 phenotypes (see Table 10-3).

Table 10-3 **Summary of genotype and phenotypes of A, B, AB and O blood group (after Burns and Bottino, 1989).**

Genotype	Phenotype	Genotype	Phenotype
$I^{A1} I^{A1}$	A_1	$I^{A1} I^B$	A_1B
$I^{A1} I^{A2}$			
$I^{A1} I^{A3}$		$I^{A2} I^B$	A_2B
$I^A i$		$I^{A3} I^B$	A_3B
$I^{A2} I^{A2}$	A_2	$I^B I^B$	B
$I^{A2} I^{A3}$		$I^B i$	
$I^{A2} i$		ii	O
$I^{A3} I^{A3}$	A_3		
$I^{A3} i$			

The H Antigen and Bombay Phenotype

Antigens A and B of A, B and O blood phenotype are synthesized from a precursor mucopolysaccharide in the presence of the dominant allele of another pair designated as H and h. With genotypes HH or Hh the precursor is converted to an H antigen which, in turn, in the presence of IA and/or IB allele is partly converted to antigen A and/or antigen B. Gene h is termed amorph because it is producing no demonstrable product. So long as persons are of genotype H–(i.e., HH or Hh), A persons of group A produce antigens A and H, group B persons produce antigens B and H, and group AB persons produce antigens A, B and H. However, group O persons produce only antigen H if they are of the genotype ii H– (Fig. 10.10). On the other hand, blood of person of genotype- -hh does not react with anti-A, anti-B, or anti-H. This is very rare Bombay phenotype (i.e., one case in 13,000 persons; it is so named because it was first described in a family from the Bombay metropolitan. The allele h is found to be epistatic to the multiple alleles at the A-B-O locus. Erythrocytes of person having- -hh genotype give no reaction with anti-A or anti-B sera (even though they possess IA or IB genes); in fact, they contain no antigen of this multiple allelic series.

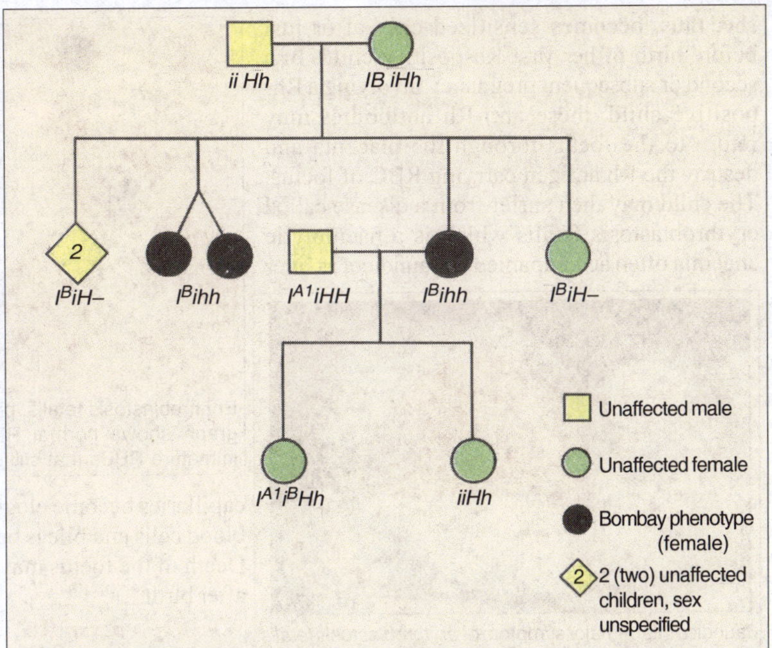

Fig. 10.10. Pathways leading to production of antigens on the red blood cells or erythrocytes showing how the Bombay phenotype is caused (after Burns and Bottino, 1989).

3. Rh Factor

The surface of erythrocytes (RBC) of some individuals contain one more type of antigen called Rh factor besides the A and B antigens. It is named after the Macaca rhesus monkey in which Rh factor was first discovered by Landsteiner and Wiener in 1940. Human beings are found to contain eight different types of Rh antigens.

The production of Rh antigen (Rh blood phenotype) depends on three closely set autosomal genes (pseudoallels). If any one of them is dominant, a Rh antigen is produced, but if all of them are recessive, no Rh antigen is formed. The individuals possessing the Rh antigen are called Rh-positive (Rh+) and those lacking it are Rh-negative (Rh–). Both of these types of persons are normal and none has natural anti-Rh antibodies in their blood plasma. However, a Rh-negative person can develop these antibodies on receiving Rh antigens through transfusion of Rh-positive blood. Such a blood transfusion will be safe only when the recipient had never been exposed to Rh-positive blood earlier. If already exposed, the previously developed anti-Rh antibodies will agglutinate the donor's RBC. In fact, the degree of RBC agglutination depends upon the amount of anti-Rh antibodies present. The high concentration of anti-Rh causes severe agglutination of RBC which sometimes proves fatal.

Erythroblastosis fetalis. The incompatibility of Rh-positive and Rh-negative bloods may also be noticed in death of the child before or soon after birth. If a Rh-negative women marries a Rh-positive man and bears a Rh-positive foetus, sometime due to some placental defect, some of the foetal RBC carrying Rh antigens may pass into her own blood stream and cause the production of anti-Rh antibodies. The concentration of anti-Rh antibodies is gradually built up in the mother and she, thus, becomes sensitized only at or just before birth of her first Rh-positive child. In a second or subsequent pregnancy involving a Rh-positive child, these anti-Rh antibodies may return to the foetus through the placenta and destroy the Rh antigen carrying RBC of foetus. The child may then suffer from a disease called erythroblastosis fetalis which is a haemolytic anaemia often accompanied by jaundice, as liver capillaries become clogged with the remains of red blood cells and bile is being absorbed by the blood. Death of the foetus may occur before birth or soon after birth.

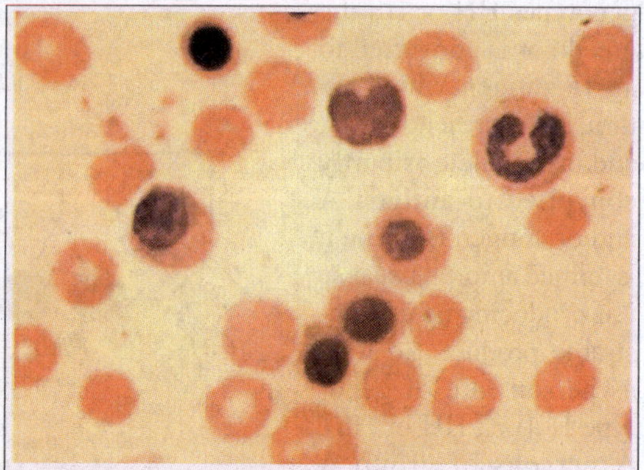

Erythroblastosis fetalis, photomicrograph : This photograph shows normal RBC's, damaged RBC's and immature RBCs that still contain nuclei.

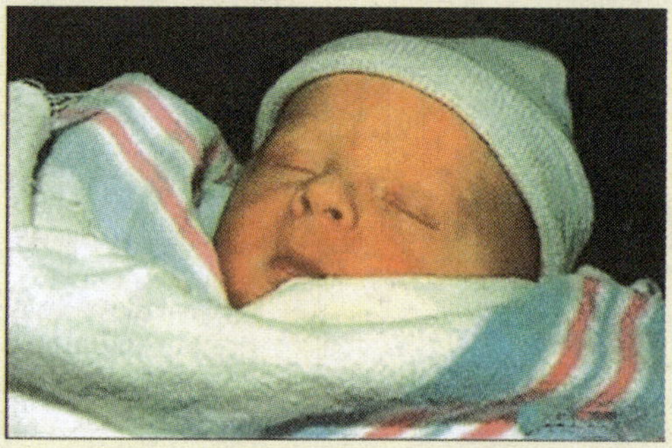

Jaundice infant-major symptom of *erythroblastosis fetalis.*

Genetics of Rh^+ blood type. There are ample evidences which show the genetic basis of Rh^+ and Rh– phenotypes. A single pair of genes, R and r was postulated for Rh+ and Rh– blood types respectively. The Rh+ blood type, later on, is found to be composed of several antigens such as C, c, D, d, E and e, all of which indicate towards the possibility of multiple allelism of gene R. Following two hypotheses have been forwarded to explain nature of inheritance of R gene:

1. Wiener's hypothesis. Wiener postulated a number of (at least eight) multiple alleles at a single locus. According to him, gene R contains eight alleles such as r, Ro, R′ R′′, R_1 , R_2 R_x or R_z and R_y.

2. Fisher's hypothesis. Fisher rejected the Wiener's concept of multiple allelism for R gene, instead of it, he proposed that a series of at least three pairs of pseudoalleles remain so closely linked with each other that they are usually inherited as a block. According to him, gene R is composed of three pairs of pseudoalleles or separate gene such as Cc, Dd and Ee. Recent genetical investigations have confirmed the Fisher's concept of pseudoallelism.

The concepts of Wiener and Fisher has been compared in Table 10-4.

Table 10-4 Comparison of Wiener's and Fisher's hypotheses about the genetics of Rh+ blood phenotype.

Gene symbol		Antigen produced	Phenotypes
Wiener	Fisher		
r	cde	none	Rh^-
Ro	cDe	Ro	Rh^+
R′	Cde	R′	Rh^+
R′′	cdE	R′′	Rh^+
R_1	CDe	Ro and R′′	Rh^+
R_2	cDE	Ro and R′′	Rh^+
Rx or Rz	CDE	R′ and R′′	Rh^+
Ry	CdE	R′and R′′	Rh^+

4. Eye Colour in *Drosophila*

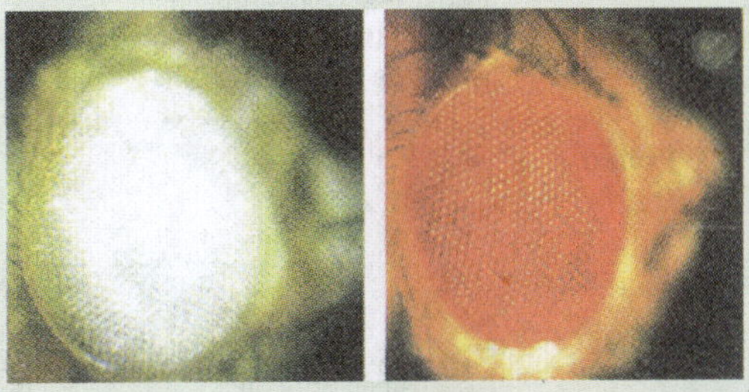

The photographs show white eye and the brick-red wild-type eye colour in *Drosophila*.

In *Drosophila*, normal red eye colour is determined by a X-linked wild type gene. White eyed Drosophila was one of the first mutants known in the fruit flies. The traits of red eye and white eye exhibited simple dominant recessive relationship. Subsequently, different shades between red and white were recovered. About a dozen different alleles are now known to occur at this locus; they are red or wild type (w^+) through coral (w^{co}), blood (b^{bl}), eosin (w^e), cherry (w^{ch}), appricot (w^a), honey (w^h), buff (w^{bf}), tinged (w^t), pearl (w^p) and ivory (w^i), to white (w). All of these were considered, on the basis of F_2 ratios, to form a multiple series, wild being dominant to all others and white recessive to all– $w^+ > w^{co} > w^{bl} > w^e > w^{ch} > w^a > w^{bf} > w^t > w^p > w^i > w$. When any two recessive alleles were brought together, intermediate types, called compound, is obtained. However, some of the members of this multiple allelic series have been found to be pseudoalleles by Lewis (1951) (see Chapter 11: Fine structure of Gene).

5. Self-sterility Alleles

A series of self-sterility alleles insures cross pollination in many plants. In fact, it is well known for long that some plants just will not self-pollinate in contrast to selfing in pea plant by Mendel.

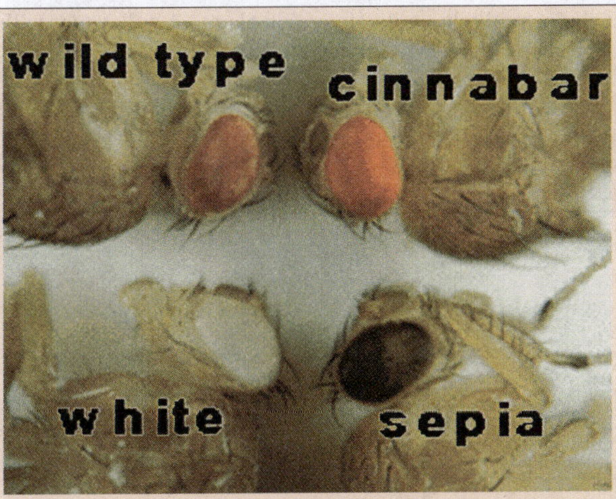

Eye colour in *Drosophila*.

Thus, a single plant may produce both male and female gametes but pollen grains of this plant fail to fertilize the ovules of the same plant; so as a result no seed will ever be produced. The same plants, however, will cross with certain other plants, so evidently they are not sterile. This phenomenon is called self-incompatibility or self-sterility. Kolreuter (1764) described self-sterility in tobacco plant (Nicotiana). Later on, the phenomenon of self-sterility is found to be common in many other dicot and monocot plants such as sweet cherries, petunias and evening primroses.

East and Mangelsdrof (1925) proposed a series of self-sterility alleles, labelled S_1, S_2, S_3, S_4 ..... S_n. A Nicotiana plant could have any two of these, but no more, since they are alleles located opposite each other in a pair of chromosomes. Fertilization could be accomplished only by a pollen grain with one of the alleles not present. For example, S_1/S_2 plants if pollinated by S_1/S_2 pollen would set no seed because neither S_1 nor S_2 could affect fertilization. Apparently pollen tubes will not grow down a style of the same genotype (Fig.10.11). Various combinations of crosses with their progenies have been tabulated in Table 10-5.

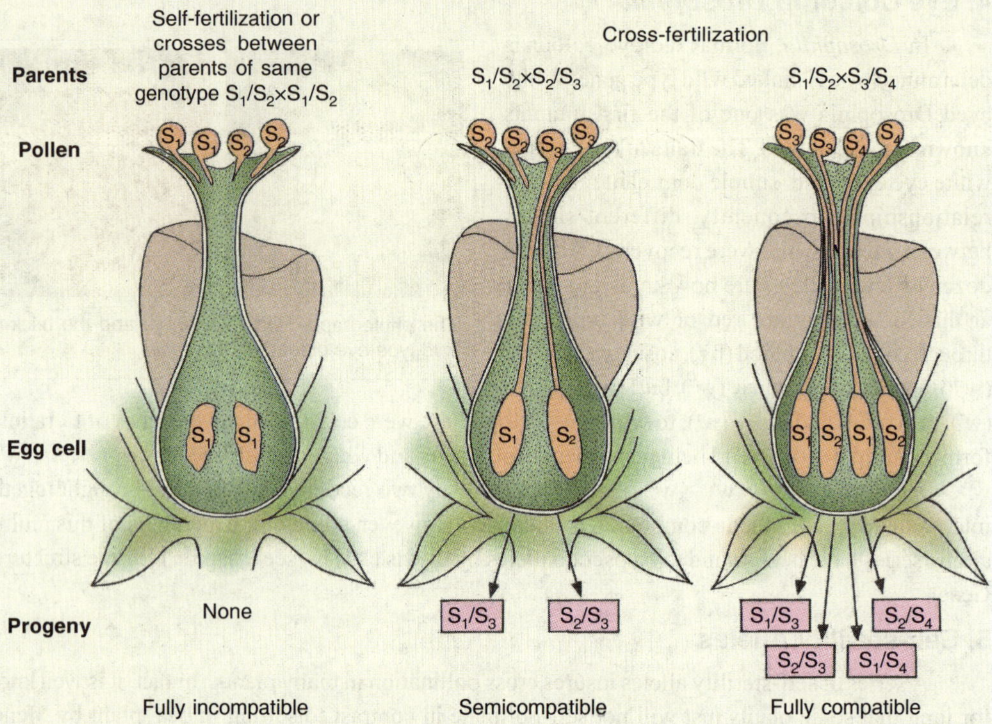

Fig.10.11. Diagram showing how multiple alleles control self-sterility in certain plants. A pollen tube will not grow if the S allele that it contains is present in the female parent (after Suzuki *et al.*, 1986).

Table 10-5 Functional pollen produced and progeny resulting from crosses of different genotypes of self-sterility alleles (Source: Singleton, 1967).

Genotype of parent	Functional pollen	Progeny
1. $S_1/S_2 \times$ self	None	0 seed
2. $S_1/S_2 \times S_1/S_3$	S_3	S_1/S_3 S_2/S_3
3. $S_1/S_2 \times S_2/S_3$	S_3	S_1/S_3 S_2/S_3
4. $S_1/S_3 \times S_1/S_2$	S_2	S_1/S_2 S_2/S_3
5. $S_1/S_3 \times S_2/S_3$	S_2	S_1/S_2 S_2/S_3
6. $S_2/S_3 \times S_1/S_2$	S_1	S_1/S_3 S_1/S_2
7. $S_2/S_3 \times S_1/S_3$	S_1	S_1/S_2 S_1/S_3

We are not limited to three alleles for self-sterility. Several more have been found in Nicotiana. In evening primrose (Oenothera), 37 different self-sterility alleles have been observed. Red clover contains more than 200 alleles for the self-sterility (Bateman,1949).

Molecular biology of self-sterility. Recently some biologists have tried to understand the mechanism of self-sterility at the molecular level. For example, Nasrallah et al., (1985) cloned DNA from S6 allele of Brassica oleracea and showed that the S6 allele causes the production of a S-specific glycoprotein which can be detected by an antibody. It is also shown that different S alleles may have different sequences.

REVISION QUESTIONS AND PROBLEMS

1. What are multiple alleles ? Give a brief account of multiple allelism ?
2. A hypothetical series of 20 multiple alleles is known for a certain locus. How many phenotypic and genotypic classes are possible ?
3. Is it possible to cross two agouti rabbits and produce both Chinchilla and Himalayan progeny ?
4. A multiple allelic series is known in the Chinese primrose where A (Alexandria type = white eye)> an (normal type = yellow eye)> a (primrose queen type = large yellow eye). List all of the genotypes possible for each of the phenotypes in this series.
5. Plumage colour in mallard ducks is dependent upon a set of three alleles : M^R for restricted mallard pattern, M for mallard, and m dusky mallard. The dominance hierarchy is M^R >M> m. Determine the genotypic and phenotypic ratios expected in the F_1 from the following crosses :
(a) $M^RM^R \times M^RM$, (b) $M^RM^R \times M^Rm$, (c) $M^RM \times M^Rm$, (d) $M^Rm \times Mm$, (e) $Mm \times mm$.
6. In guinea pigs the intensity of coat colour varies from black (variety A) to dark sepia (Variety B), to medium sepia (variety C) to white (variety D). P1 and F1 crosses were made between varieties with the following results :

P_1	F_1	F_2
1. A × B	Black	3 Black : 1 Dark sepia
2. A × C	Black	3 Black : 1 Medium sepia
3. A × D	Black	3 Black : 1 White
4. B × C	Dark sepia	3 Dark sepia : 1 Medium sepia
5. B × D	Dark sepia	3 Dark sepia : 1 White
6. C × D	Light sepia	1 Medium sepia : 2 Light sepia : 1 White.

Explain the inheritance of coat colour intensity.

7. A woman of blood group A marries a man of blood group O. There are three children in the family. The children's blood types are O, A and AB. Which child was definitely adopted ?
8 What genotypes and their proportions would be produced by the following crosses ?
(a) $I^A i \times ii$; (b) $I^AI^B \times I^Ai$;
(c) $I^B I^B \times I^Ai$; (d) $I^AI^B \times I^A I^B$.
9. Considering human blood group A to include three subtypes, and groups B and O to include one each how many phenotypes are included in the A-B-O series. ?
10. A woman whose father was of blood group AB and whose mother was A marries a man of blood group B. Of their two children one belongs to blood group O and the other to group A.
(a) What is the genotype of the woman ?
(b) What is the genotype of her husband ?
(c) What is the genotype of her mother ?
11. A paternity case involves these facts: the woman is A1, her child is O, and the alleged father is B. Could he be the father ? Explain.
12. What do you understand by self-incompatibility (= self-sterility) ? Explain this phenomenon by giving suitable example.
13. A number of self-incompatibility alleles is known in clover such that the growth of a pollen tube down the style of a diploid plant is inhibited when the latter contains the same self-incompatibility alleles as that in the pollen tube. Given a series of self-incompatibility alleles S^1, S^2, S^3, S^4, what genotypic ratios would be expected in embryos and in endosperms of seeds from the following crosses ?

	Seed parent	Pollen parent
a	$S^1 S^4$	$S^3 S^4$
b	$S^1 S^2$	$S^1 S^2$
c	$S^1 S^3$	$S^2 S^4$
d	$S^2 S^3$	$S^3 S^4$

14. Write short notes on the following:(i) Multiple alleles; (ii) ABO blood group; (iii) Universal donor, and universal recipient; (iv) Inheritance of Rh phenotype ; (vi) Bombay phenotype.

ANSWERS TO PROBLEMS

2. Phenotypic classes = 20; Genotypic classes = 210.
3. No, because of the dominance relationship of the four multiple allele: $c^+ > c^{ch} > c^h > c$.
4. Alexandria type (white eye) = A A, A a^n, A a ;
   Normal type (Yellow eye) = $a^n a^n$;
   Primrose queen (large yellow eye) = a a.
5. (a) 1/2 M^RM^R : 1/2 M^RM ; all restricted;
   (b) 1/2 M^RM^R : 1/2 M^RM ; all restricted;
   (c) 1/4 M^RM^R : 1/4 M^RM : 1/4 M^RM : 1/4 Mm; 3/4 restricted : 1/4 Mallard;
   (d) 1/4 M^RM : 1/4 M^Rm: 1/4 Mm : 1/4mm ;
   1/2 restricted : 1/4 Mallard : 1/4 Dusky.
   (e) 1/2 Mm : 1/2 mm ; 1/2 Mallard : 1/2 Dusky.
6. Multiple allelic series with A dominant over B, C and D; B dominant over C and D; C and D with intermediate dominance. Dominance hierarchy of these alleles is A>B>(C=D).
7. Child AB adopted.
8. (a) 1/2 I^{Ai} : 1/2 ii ; (b) 1/4 I^AI^A : 1/4 I^A i : 1/4 I^B I^A : 1/4I^B i;
   (c) 1/2 I^BI^A : 1/2 I^B i; (d) 1/4 I^AI^A : 1/2 I^AI^B : 1/4 I^BI^B.
9. 8.
10. (a) I^A i ; (b) I^B i ; (c) I^A i.
11. Yes,if the woman is IAi and the man IBi.
13. (a) Embryos = 1/2 S^1 S^3 : 1/2 S^3 S^4 ;
    (b) None ;
    (c) Embryos = 1/4 S^1 S^2 :1/4 S^1 S^4 ; 1/4 S^3 S^4;
    Endosperm : 1/4 $S^1S^1S^2$: 1/4 S^1 S^1 S^4 : 1/4 S^3 S^3 S^2 : 1/4 $S^3S^3S^4$;
    (d) Embryos = 1/2 S^2S^4 : 1/2 S^3S^4;
    Endosperms = 1/2 $S^2S^2S^4$: 1/2 $S^3S^3S^4$.

CHAPTER

11

Fine Structure of Gene

Some of what this boy will be as an adult will be influenced by what he learns from his grandfather, but much will reflect the genes he has inherited from his grandfather.

The hereditary units which are transmitted from one generation to the next generation are called **genes.** A gene is the fundamental biologic unit, like the atom which is the fundamental physical unit. **Mendel** while explaining the results of his monohybrid and dihybrid crosses, first of all conceived of the genes as particulate units and referred them by various names such as **hereditary factors** or **hereditary elements** (or **Merkmal).** In the beginning of present century Mendel's factor came to be known as **gene** (**Johannsen,** 1909). The presence of the gene is detected only when a mutation occurs in it. Initially genes were considered as beads and chromosomes as strings of beads (**Morgan,** 1911). Mutation was supposed to alter the bead structure and recombination (or crossing over) was regarded to involve a breakage between two beads followed by their exchange between paired chromosomes. Each bead was thought of to control one character by controlling some biochemical step. Thus, a gene was considered to be a unit of mutation, recombination and function.

Muller (1932) have compared the mutant genes with wild type genes (or standard genes) and classified them into the following types : **(1) Hypomorphs.** These are the genes which have the same effect as the standard gene but their phenotypic expression is less effective **(2) Amorphs.** These are genes which have very less effect than the standard gene, *i.e.*, they do not cause any significant change in phenotype. e.g., h antigen of Bombay phenotype. (3) **Hypermorphs.** These are genes which have greater effect than the standard gene, *i.e.*, they produce much phenotypic change. **(4) Antimorphs.** These are genes which result in opposite phenotypic effect than the standard gene, *i.e.*, they produce abnormal phenotype **(5) Neomorphs.** These are genes which produce a phenotypic effect that is qualitatively different from that of standard gene, *e.g.*, Bar locus of *Drosophila*.

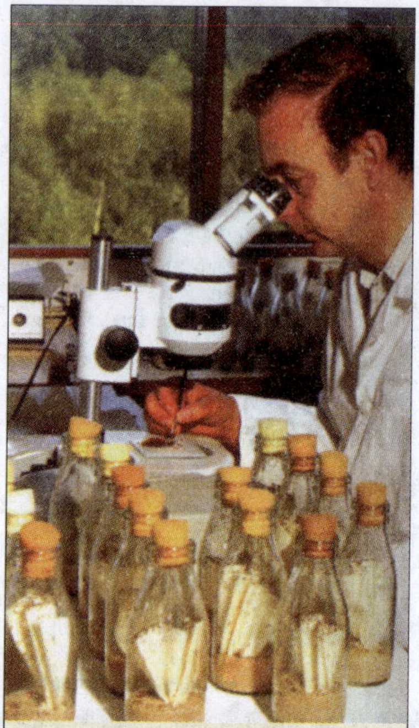

A breeding experiment using *Drosophila* : the photograph shows sterilised milk bottles containing nutritious growth medium for the culturing of *Drosophila*. Larvae feed in the medium but pupate on the filter paper; adult flies are seen above the medium. They are anaesthetised and examined.

Development in molecular genetics led to a fine analysis of the structure and function of the gene and has given a new concept (*i.e.*, chemical basis of gene) of the hereditary unit. It was soon realized that unit of mutation can range from a single base to a stretch of DNA molecule and that recombination can occur between two DNA bases. Thus, gene is neither a unit of mutation nor of recombination. One-*gene-one-character hypothesis* (**De Vries**) or *one-gene-one-function hypothesis* (**Horowitz** and **Leupold,** 1951) too became obsolete concepts. It was **Sturtevant** and **Morgan** (1923) who for the first time demonstrated that the **bar** locus of *Drosophila* contains more than one functional and recombinational units. In 1950, **Roper** found that crosses between different biotin requiring mutants of *Aspergillus* can yield wild type recombinants. **Benzer** (1950) reported fine structure of a genetic region of rII locus of T4 bacteriophage and recognized three types of genes—cistron, recon and muton, in it. This was followed by **E.B. Lewis**' (1952) demonstration that the white eye locus of **Drosophila** can be divided into many functional subunits. These findings led **Pontecorvo** (1952) to suggest that each gene has many linearly arranged mutable sites between which recombination can occur.

GENE CONCEPT

Gene is a conceptual unit. Its place on a chromosome is called a **locus (Demerec,** 1955). According to **Demerec,** a locus is a section of a chromosome that has a unitary function, and any mutation occurring within its boundaries produces an impairment of this function.

1. Test of Allelism (Allelism and Pseudoalleles)

If one wishes to find out whether two mutant alleles in question are allelic to each other or not, then from their crosses it is expected that F_1 individuals (which have mutant phenotypes) should not give rise to the wild type in the F_2 generation. This is called **test of allelism.** For example, if a and b are two alleles, the allelic test can be shown as in Fig. 11.1. This classical concept of allelism has to be modified due to experimental results of several workers working on *Drosophila*.

aa × bb
↓
F_1 : a/b

F_2 : { 1. If wild type appears a and b are non-allelic;
2. If no wild type appears, a and b are allelic.

Fig. 11.1. A test of allelism.

1 Bar locus in *Drosophila*. Bar locus in *Drosophila* provides experimental proof of two genetical phenomena—(1) intragenic or intralocus crossing over and (2) position effect. The bar gene of *Drosophila* is dominant and located on the X chromosome. The bar phenotype is characterized by

Alfred Henry Sturtevant
(1891-1970)

a narrower, oblong, bar-shaped eye with fewer facets. The Y chromosome of males has no allele of the bar gene. Wild type (B^+B^+) flies have 779 facets in their oval-shaped eyes. In the heterozygous bar (B^+B^+) flies the number of facets is reduced to 358, whereas in the homozygous recessive ones (BB) this number goes down to 68. Since the mutant allele produces a phenotypic effect that is qualitatively different from that of its wild type counter part, it is called **neomorph.** In a population of homozygous bar flies sometimes (1 in 1600 cases) normal eyed and sometimes flies with very narrow eyes arise. The latter are called **ultrabar** (B^u) and in homozygous (B^uB^u) condition they possess only 25 facets. Bar-eyed female flies sometimes produce normal eyed or ultrabar progeny, although with a very low frequency (about 1 in 1600). Similarly, ultrabar stocks produce some normal and some bar-eyed flies. The exceptional flies (*i.e.*, ultrabar) do not arise due to mutation because the mutation frequency of this locus is much lower than in 1600. In order to explain these results, **Sturtevant** and **Morgan** (1923) postulated that new types of flies arose as a result of intragenic rather than intergenic crossing over at the bar locus. They used strains in which ultrabar or bar locus is flanked by the marker genes f (forked bristles) and fu (fused wing veins), *i.e.*, the gene B is 57.0 crossover units from the centromere; the marker gene f is proximally situated at 56.7 and other marker gene fu is distally situated at 59.5. These three genes are so closed together that only one crossing over is normally possible in this region at a time. Whenever, there occurs an intragenic crossing over at the bar locus, it is accompanied by a separating of the flanking markers. Thus, $f^+B\ fu^+/\ f B\ fu$ females crossed to males with forked bristles and fused wing veins ($fB\ fu$ or fB^+fu) produced about 20,000 progeny, all of which were bar eyed except 7 which were normal eyed (B^+) and 2 which were ultrabar (Bu) (Fig. 11.2). It should be remembered that the *Drosophila* males are hemizygous for the bar locus and that there is no crossing over. Three of the normal-eyed and both the ultrabar flies had forked bristles but normal bristles, however, four of the normal eyed had normal bristles but fused wing veins, *i.e.*, in all the nine cases (Fig. 11.2) gene f and fu had separated as a result of recombination.

f	B	fu		f	B^+	fu^+	(3)
			→	f^+	B^+	fu	(4)
f^+	B	fu^+		f	B^u	fu^+	(2)

Fig. 11.2. Genotypes of the females flies (A) and of the seven and two abnormal progeny (B) produced by them as a result of crossing between marker genes f and fu.

On the basis of these and similar experiments **Sturtevant** proposed that exceptional flies arise as a result of **unequal crossing over.** An unequal crossing over in homozygous bar-eyed females will produce two types of gametes one with two bar genes and other with no bar genes (Fig. 11.3). In a similar way an unequal crossing over between bar an ultrabar produces nonviable triple bar and viable normal progeny.

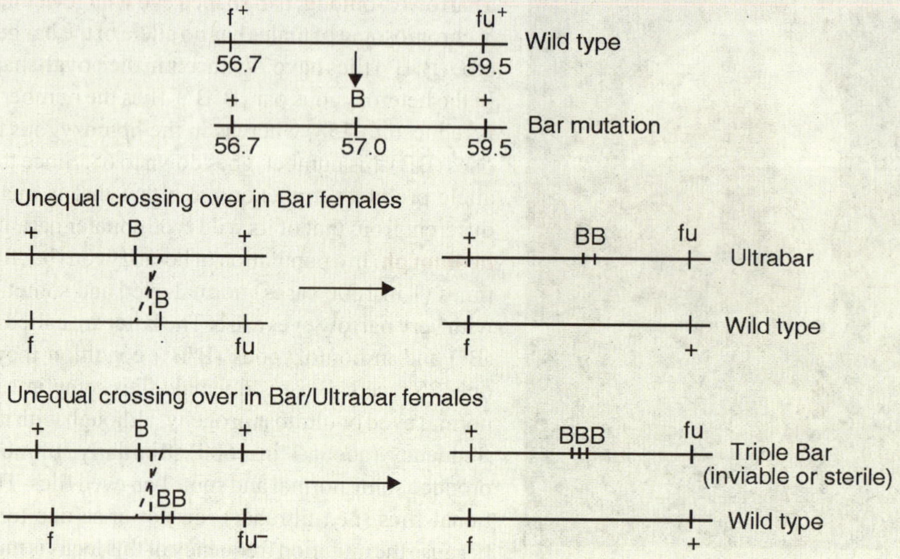

Fig. 11.3. Unequal crossing over and production of ultrabar and triple bar *Drosophila*.

2. Lozenge locus. In 1940 and later, **Oliver** and his student **Green** conducted experiments using different mutants and heterozygotes at the locus **lozenge** (l_z). The l_z locus is responsible for smaller darker and more elliptical (narrow ovoid) eyes (fig. 11.4) **Oliver** obtained unexpected results from a cross between *Drosophila* mutants carrying the genes l_zy and l_zs which were presumed to be alleles and which had been represented in exactly corresponding positions on a chromosome pair l_zg/l_zs. All progeny from this cross were expected to have the lozenge phenotype, but few were wild type. This indicated that these two alleles were not true alleles, i.e., two genes were not in identical positions on homologous chromosomes, but were located in slightly different areas—l_zg+/+ l_zs. Cross-overs had presumably occured, placing l_zg and l_zs on the same homologue l_zgl_zs/++ and each in heterozygous condition. This laid the foundation for work on the intragenetic recombination leading to the resolution of fine structure of gene.

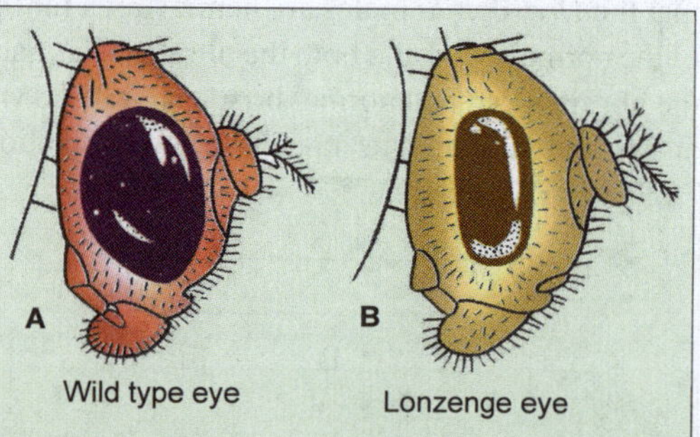

Fig. 11.4. Lozenge eye in *Drosophila*. A-Wild type eye; B-Lozenge eye (after Gardner, 1968).

3. Apricot eye colour in *Drosophila*. In 1951, **E.W. Lewis** reported results of one of his experiments where from a cross of apricot (w^{apr}) and white (w) eye he obtained F_1 having intermediate eye colour. In F_1 generation he expected segregation for apricot and white, but to his surprise, he recovered from F_2 wild type (red eyes) also, showing that the two presumed alleles were non-allelic, *i.e.*, both occupying different sites, **Lewis,** therefore, preferred to called them **pseudoalleles** and the phenomenon as **pseudoallelism.** Thus, pseudoalleles are those genes which are located almost at the same place on the linkage map but show recombination, indicating thereby that although they are functionally allelic, structurally they are non-allelic. **Pontecorvo** later preferred to use the term **Lewis effect** for the pseudoalleles. With accumulated crossover data, the distance between the two sites was placed at about 0.01.

Pseudoalleles show complementation, thereby, indicating their functional individuality. Thus, if a and b are non-allelic the parents should be represented as aa ++ and ++ bb and in F_2, wild type (++)

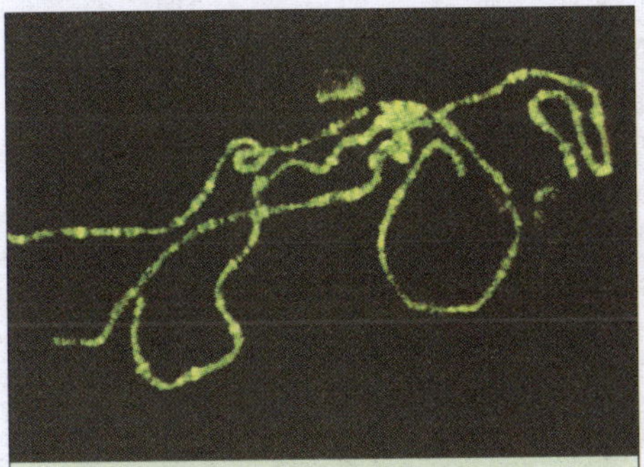

Genes on a chromosome.

would be recovered as expected. Thus, the appearance of wild type in F_2 generation could be explained. However, it was difficult to explain why F_1 progeny a+/+b was not wild type. For explaining this, another criterion, called **complementation,** was employed. Complementation means that in allelic forms there will be lack of complementation due to which F_1 individuals will not be wild type. But since intragenic interalleic crossing over (recombination) is possible, in the F_2 generation wild type individual appear. In F_1 individual a+/+b, the **position effect** (see also Chapter 16) does not allow the wild type to be expressed. As a result of crossing over, ++ and ab gametes are produced from the F_1 individuals and in the F_2 generation ++/++ or ++/ab will express the wild phenotype. Thus, it has been demonstrated that within the same gene ++/ab and a+/+b exhibit different phenotypes. This is known as the **cis-trans effect** (Fig. 11.5).

++ / ab	or	apr/w / +/+	a+ / +b	or	apr/+ / +/w
	Cis (wild)red			trans(mutant) apricot	

Fig. 11.5. Cis and trans arrangements of alleles a and b or apr and w alleles of *Drosophila* in heterozygous condition.

Why do the w and apr genes behave differently in different positions? There is no direct answer to this question at present, but one hypothesis is that the production of red pigment is a two-step chemical process. One gene may control one step, while the other gene is close proximity may control another step in the same reaction. When the two genes w^+ and apr^+ (+ +) occur next to each other on the same chromosome (cis position) they produce the normal red eye colour. When they are opposite to each other; however, chemical blocks occur that stop the reaction at one or more points. So cis-trans effect is a type of position effect of the genes with respect to each other rather than their presence or absence determines the end result (**Gardner,** 1968).

2. Cistron, Recon and Muton

When two genetical units exhibit cis-trans phenomenon, they are considered to belong to the same functional unit, called **cistron.** In case two units do not exhibit cis-trans effect, or in other words if the two units show complementation, they belong to two functional units or cistrons, through they may control the same character. Keeping in view these considerations, geneticists have to modify the classical concept of gene; they hold that a classical gene can consist of more than one functional units or cistrons and that crossing over or recombination within a gene or even within a cistron is possible.

With the help of recombination studies at intragenic level (*e.g.*, rII locus of T_4 bacteriophage, lozenge and white eye loci in *Drosophila*) **Micromaps** have been prepared for different genes in the same fashion as genetical maps of different chromosomes were initially prepared. Thus, a gene was further divided into smaller units called **mutational sites.** For instance, the white eye locus of *Drosophila* has been resolved into several sites. Likewise, a micromap of lozenge locus in *Drosophila melanogaster* show four groups of alleles located at four mutational sites (Fig. 11.6).

Seymour Benzer.

Moreover, ultimate in studies of **fine structure mapping** or **micromapping** was carried out by **Seymour Benzer** (1950, 1962) with phage T_4. For his experiments **Benzer** utilized mutations of the rapid lysis (r) character. The wild phage (r^+) is capable of forming plaques on *E. coli* strain B as well as another *E. coli* strain which is lysogenic for phage λ and is called K-12 λ. Mutations of r^+, called *rII* mutants, can lyse only strain B, and not strain K-12λ. The ability to form plaques on strain K-12λ was used as a selective procedure in the recovery of recombinant wild-type phage particles. By the use of double infections, **Benzer** performed complementation tests between pairs of *rII* mutants and found that all could be assigned to one or the other two functional groups, indicating the presence of two separate genes called *A* and *B* (Fig. 11.7).

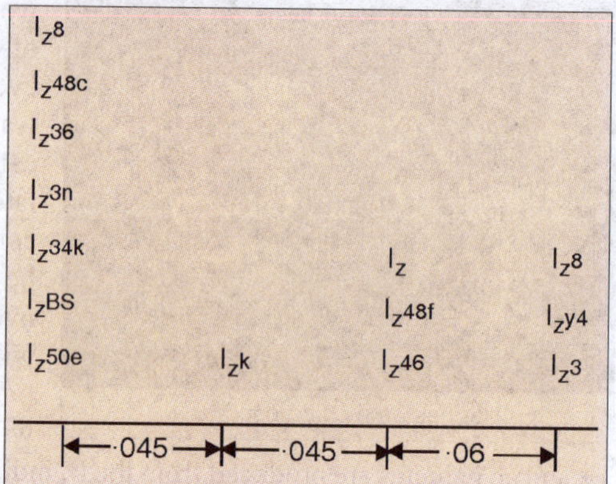

Fig. 11.6. A number of alleles of lozenge occupying four different mutational sites. Recombination frequencies between these four sites have also been indicated.

Benzer coined the term **cistron** to denote a genetic unit that has a single function as demonstrated by failure to complement in the trans configuration of the cis-trans test. Thus defined, a cistron is the equivalent of a functional gene. **Benzer** also applied the term **muton** to the smallest unit of a gene whose mutation can produce a mutant phenotype, and the term **recon** to the smallest distance within which recombination can occur. **C. Yanofsky** (1967) have shown that Benzer's muton is equivalent to a single nucleotide and his recon to the distance separating two adjacent nucleotides.

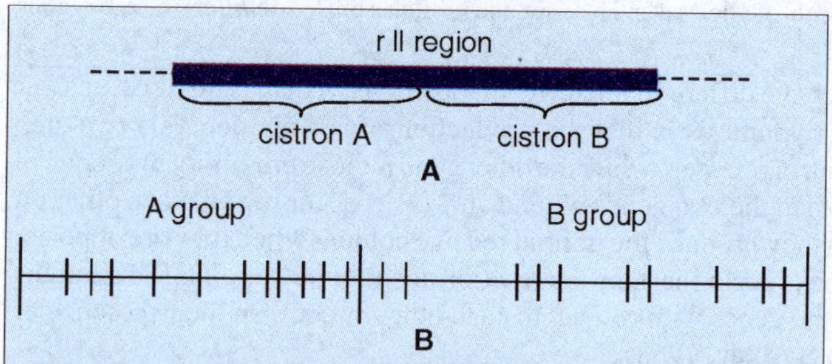

Fig. 11.7. A—rII region of T_4 phage showing two cistrons; B–Gene map of rII locus.

Benzer' work was important because it strengthened the concept of the gene as a unit of function and in addition permitted visualization of the gene in physical terms of DNA.

Complex Gene Loci

The Russian geneticist **N.P. Dubinin** in 1939 recognized **step allelism** (that is, the graded effect of different alleles) in *Drosophila*. This discovery fore shadowed the idea that "genes" could be subdivided, *i.e.*, certain genes are **compound** or **complex genes.** The complex loci were discovered when techniques became available in particular organisms for exploring the fine structure of the gene. The first evidence came from *Drosophila* studies in which presumed alleles were found to be pseudoalleles, that is, *closely linked genetic units that behaved ordinarily as if they were alleles but were not in exactly corresponding position.* They could be separated by crossing over. Virus studies then showed that many sites were present in a particular locus (*e.g.*, rII locus in T_4 phase). Around 1970s, studies of gene's fine structure were extended to bacteria Salmonella), yeast, mold (*Neurospora*), wasp (*Mormoniella*), flies, cotton and maize. It appears that complex gene structure is basic to all organisms.

Examples of complex loci. The lozenge gene of *Drosophila* provides a good example of complex locus. In *Drosophila* several other complex loci have been demonstrated which include the following loci : bx (bithorax), sb (stubble bristles), N (notch wing), g (garnet eye colour), sn (singed bristles), dp (dumpy wing) and s (star).

Complementary interaction between alleles indicates complex loci even when recombination between them has not been observed. By this technique, the Td region in *Neurospora crassa* which controls the final step in tryptophan synthesis, has been shown to be a complex of several sites in linear order.

In maize, complex loci have been reported at the A locus, which controls pigment production; the R locus, which control pigmentation; and the wx locus, which influence the nutrient content of the seed.

REVISION QUESTIONS AND PROBLEMS

1. What are the chemical units of DNA that correspond to the genetic units of replication, recombination, mutation, and function? Outline one form of evidence for each statement. Are the chemical units the same in both eukaryotes and prokaryotes? discuss data that support your answer.
2. Which of the following you not equate with the term (1) gene and (2) allele, and why? (a) Recon (b) Cistron (c) Muton.
3. Is there generally a correlation between locations of genes in a chromosome and their phenotypic effect? Give examples.
4. Discuss the current concept of the gene and the evolution of this concept beginning with Mendel's factors of inheritance.
5. How do complementation and recombination differe? Describe complementation and recombination tests for allelism.
6. What are pseudoalleles? Describe the concept of complex loci.
7. Giving suitable examples, discuss how did our concept of allelomorphism change during 1940-60.
8. Discuss the fine structure of gene as reveraled through the work on lozenge locus in *Drosophila* and rII locus in T_4 bacteriophage.
9. Write short notes on the following : (i) Position effect : (ii) Cis-trans effect : (iii) rII locus in T_4 bacteriophage; (iv) Lozenge locus.
10. Suppose you were given two recessive mutants of *Drosophila*, short wing (s) and reducing wing (r) both autosomal and both belonging to the same linkage group.
    (a) What one cross could you use to determine if s and r are alleles ?
    (b) What F_1 phenotype would indicate allelism?
    (c) Assuming allelism, what cross could you use to determine s and r occupied the same or different sites within the cistron?
    (*d*) What data would indicate that s and r occupied different sites ?
11. The recessive genes a and b are allelic, each producing a narrow eye phenotype in *Drosophila* when homozygous. Females heterozygous for these genes were crossed to b/b males. Of a total of 50,000 progeny, 10 wild-type individuals were recovered, the rest of the progeny showing a narrow eye phenotype.
    (a) Assuming reciprocal classes are equal, how many map units apart are these mutants?
    (b) If one map unit corresponds to 10,000 nucleotides of DNA, how many nucleotides separate these mutant sites.
12. The DNA of bacteriophage T_4 contains approximately 200, 000 nucleotide pairs. The rII region of the T_4 genome occupies about 1 per cent of its total genetic length. Benzer has found that about 300 sites are separable by recombination within the rII region. Determine the average number of nucleotides in each recon.
13. Suppose a certain cistron is found to consists of 1500 deoxyribonucleotides in sequence. (a) What is the maximum number of mutons of which this cistron could consist? (b) Is this number likely to be too high, too low, or about right for an actual organism.
14. Distinguish among these three concepts : cistron, muton and recon.

ANSWERS TO PROBLEMS

10. (a) s/s × r/r (b) A mutant wing phenotype would indicate allelism.
    (c) s/r × s/s or s/r × r/r (d) The appearance of a rare wild-type recombinant indicates that s and r occupied different sites within the cistron.
11. (a) 0.04 map units : (b) 4,000 nucleotides.
12. Approximately 7 nucleotides per recon.
13. (a) 1,500; (b) Too high.

CHAPTER

12

Sex-Linked Inheritance

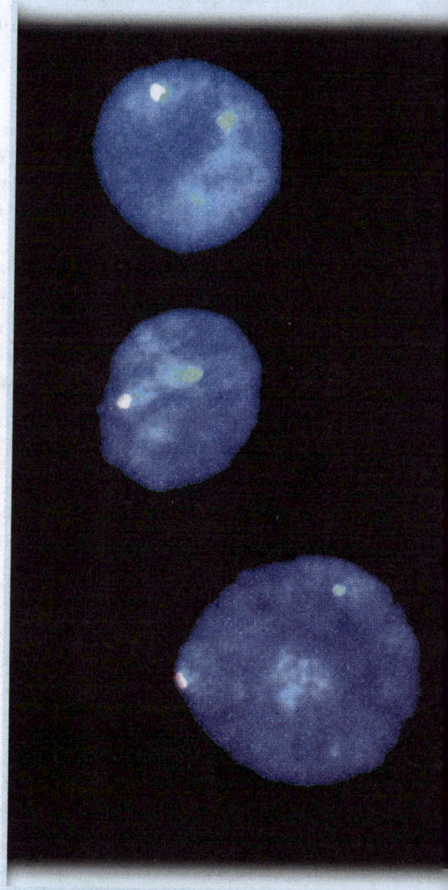

Demonstration of the X and Y chromosomes (the blue and the pink dots respectively) in mammalian fetal cells using fluorescent *in situ* hybridization (FISH).

In Mendel's crosses we have already observed that the progeny of a cross between two individuals of pure lines is the same regardless of which individual is female or which individual is male or we can say that sex makes no difference in Mendel's crosses. But the Mendel's laws are not applicable on those genes which are exclusively located either in X or Y chromosome. It has been observed that the genes occurring only in the X chromosomes are represented twice in female (because female contains 2X chromosomes) and once in male (because male has only one X chromosome). Moreover, if the recessive type of genes occur in X chromosomes of males, they express themselves phenotypically. Because in such case Y chromosome contains no dominant allelomorph or gene to overcome the recessive gene of X chromosome. The genes which occur exclusively on the X chromosome (mammals, *Drosophila*, *Melandrium*, etc.) or on the analogous Z chromosome (in birds and other species with ZO or ZW mechanism of sex determination) are called **X- or Z -linked genes**. The genes which exclusively occur in Y chromosome are called **holandric genes**. The inheritance of X- or Z-linked and holandric genes is called **sex-linked inheritance**. In XX–XY type organisms, sex-linked genes (see Fig.12.1) can be classified into following three types:

A. X-linked. The X-linked type sex-linked inheritance is performed by those genes which are localized in the non-homologous sections of X-chromosome, and that have no corresponding allele in Y chromosome. The X-linked genes are commonly known as **sex-linked genes**.

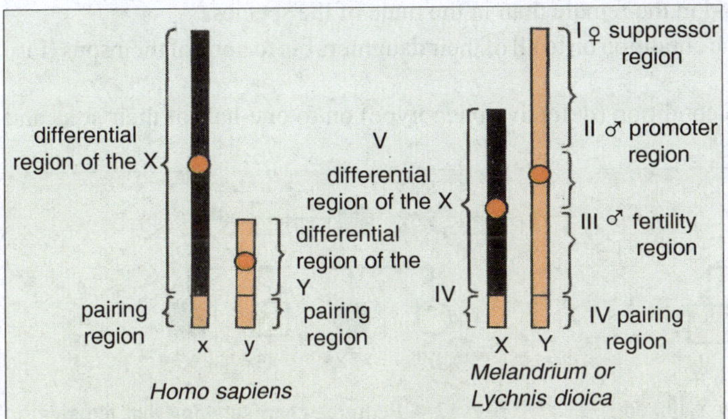

Fig. 12.1. Differential and pairing regions of sex chromosomes of humans and of the plant *Melandrium album* (after Suzuki *et al.*, 1986).

B. Y-linked. The Y-linked type sex-linked inheritance is performed by those genes which are localized in the non-homologous section of Y chromosome, and that have no alleles in X-chromosome. The Y-linked genes are commonly known as **holandric genes** (Greek, *holos* = whole, and *andros* = man).

C. XY-linked. The XY-linked type sex-linked inheritance is performed by those genes which are localized in homologous sections of X and Y chromosomes.

A. INHERITANCE OF X-LINKED (SEX-LINKED) GENES

Characteristics of Sex-linked Inheritance

The X-linked genes exhibit following characteristic patterns of inheritance :

1. The differential region of each chromosome (*i.e.*, X) contain genes that have no counterparts on the other kind of sex chromosome. These genes, whether dominant or recessive, show their effects in the male phenotype. Genes in the differential regions are called **hemizygous** ("half-zygous") in the males.

2. The X-linked recessive genes show the following two more peculiar features: **criss-cross** pattern of inheritance (*i.e.*, in criss-cross inheritance, a X-linked recessive gene is transmitted from P_1 male parent (father) to F_2 male progeny (grandsons) through its F_1 heterozygous females (daughters), which are called **carriers**) and different F_1 and F_2 results (ratios) in the reciprocal crosses.

3. The X-linked recessives can be detected in human pedigrees (also in *Drosophila*) through the following clues :

(i) The X-linked recessive phenotype is usually found more frequently in the male than in the female. This is because an affected female can result only when both mother and father bear the X-linked recessive allele (*e.g.*, $X^A X^a \times X^a Y$), whereas an affected male can result when only the mother carries the gene. Further, if the recessive X-linked gene is very rare, almost all observed cases will occur in males.

(ii) Usually none of the offspring of an affected male will be affected, but all his daughters will carry the gene in masked heterozygous condition, so one half of their sons (*i.e.*, grandsons of F_1 father) will be affected (Fig. 12.2).

(iii) None of the sons of an affected male will inherit the X-linked recessive gene, so not only will they be free of the defective phenotype; but they will not pass the gene along to their offspring.

4. Dominant X-linked genes can be detected in human pedigrees (also in *Drosophila*) through the following clues :

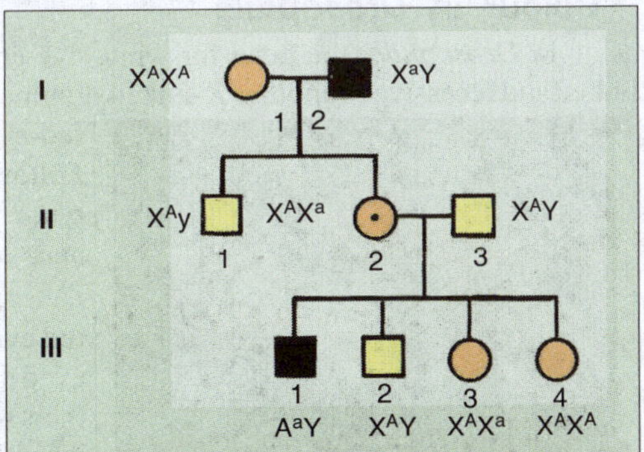

Fig. 12.2. Pedigree showing how X-linked recessive genes are expressed in males, then carried unexpressed by females in the next generation, to be expressed in their sons. II.3 and III.4 heterozygous or carrier females are not distinguished phenotypically (after Suzuki *et al.*, 1986).

(a) It is more frequently found in the female than in the male of the species.

(b) The affected males pass the condition on to all of their daughters but to none of their sons (Fig. 12.3).

(c) Females usually pass the condition (defective phenotype) on to one-half of their sons and daughters (Fig.12.4).

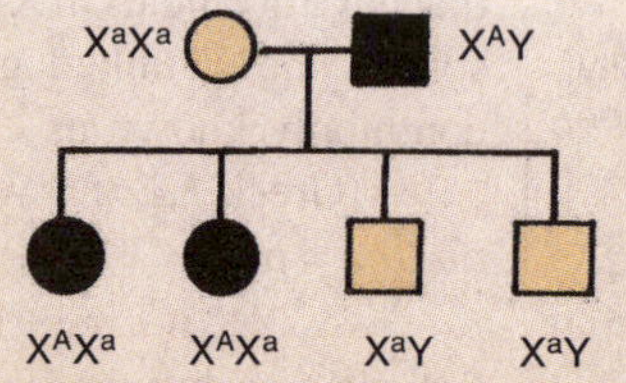

Fig. 12.3. Pedigree chart showing how X-linked dominants are expressed in all the daughters of affected males (after Suzuki *et al.*, 1986).

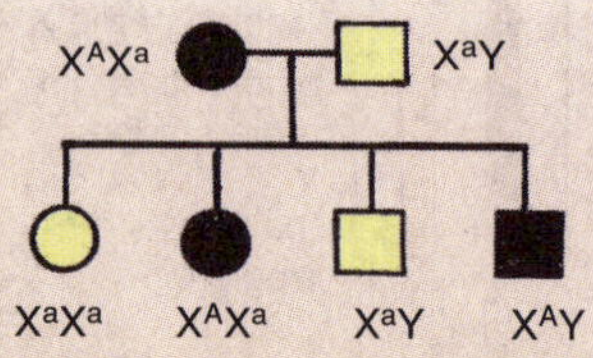

Fig. 12.4. Pedigree chart showing that females affected by an X-linked dominant condition usually are heterozygous and pass the condition to one-half of their progeny (after Suzuki *et al.*, 1986).

(d) A X-linked dominant gene fails to be transmitted to any son from a mother which did not exhibit the trait itself.

In humans, X-linked dominant conditions are relatively rare. One example is **hypophosphatemia** (vitamin D-resistant rickets). Another example includes hereditary enamel hypoplasia (**hypoplastic amelogenesis imperfecta**), in which tooth enamel is abnormally thin so that teeth appear small and wear rapidly down to the gums.

Red eyed male *Drosophila*.

Example of Inheritance of X-Linked Recessive Genes

The crisscross inheritance of recessive X- linked genes can be well understood by following classical examples in *Drosophila*, man, moth and chikens etc.:

1. Inheritance of X-Linked Gene for Eye Colour in *Drosophila*

In *Drosophila*, the gene for white eye colour is X-linked and recessive to another X-linked,dominant gene for red-eye colour. It is discovered by **Morgan** in 1910. Following crosses between white eyed and red eyed *Drosophila* will make clear the characteristic criss-cross inheritance of gene for white eyed colour in it :

Red eyed female *Drosophila*.

(a) Red eyed female × White eyed male. If a wild red eyed female *Drosophila* is crossed with a mutant white eyed male *Drosophila*, all the F_1 individuals irrespective of their sex have red eyes (Fig. 12.5). When the red eyed male and red eyed female individuals of F_1 are intercrossed, the F_2 progeny is found to include an exclusively red eyed female population and a male population with 50 per cent red eyed individuals and 50 per cent white eyed individuals. Thus, F_2 generation includes red eyed and white eyed individuals in the ratio of 3: 1.

(b) White eyed female × Red eyed male. When a white eyed female *Drosophila* is crossed with a red eyed male *Drosophila*, all the female individuals in the F_1 generation are red eyed (Fig. 12.6). When these red eyed female individuals and white eyed male individuals of F_1 are intercrossed, the female population of F_2 generation is found to include 50 per cent red eyed and 50 per cent white eyed flies. Similarly, the male population of F_2 includes 50 per cent, red eyed and 50 per cent white eyed flies.

The inheritance of X-linked recessive gene for white eyes can be understood more properly by considering the behaviour of X and Y chromosome, such as follows :

In the cross between red-eyed female and white-eyed male *Drosophila,* the red-eyed female contains the gene ++ for red colour of eye. The white-eyed male contains a single gene w for white colour of eye which remains located in X chromosome. This one allelic condition of male is termed **hemizygous** condition in contrast to the homozygous or heterozygous possibilities in the female. The female being homogametic produces only one type of gametes or eggs each with the gene + for red-coloured eyes. The male being heterogametic produces two types of gametes, 50 per cent sperms with X chromosome containing w gene, 50 per cent sperms with 'Y' chromosome without w or + gene. The gametes of both parents unite in fertilization to produce F_1 progeny. The F_1 hybrids which receive a X-chromosome with + gene from the female and a X-chromosome with w gene from the male becomes red-eyed female because gene '+' is dominant over gene 'w'. Further the hybrids which happen to receive a gene '+' from mother and Y chromosome from father produce red-eyed males.

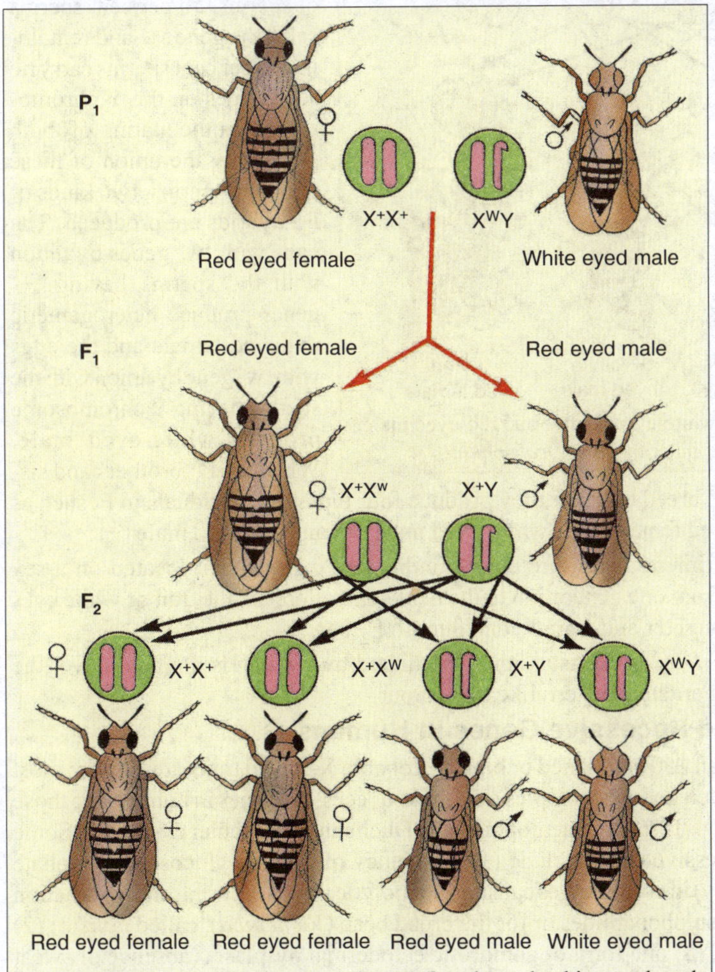

Fig. 12.5. A cross between red-eyed female and white-eyed male showing sex-linked inheritance in *Drosophila.*

The F_1 red-eyed female with the gene '+ w' when crossed with F_1 red-eyed male having the gene '+'. The female hybrids produce two types of eggs, 50 per cent eggs carry the gene '+' and the remaining 50 per cent carry the gene 'w'. The males produce two types of sperms, half carry the '+' and half carry no such gene on Y chromosomes. The union of sperms and ova of F_1 offsprings may produce four possible types of F_2 individuals :

1. The eggs with '+' genes if fertilizied by sperms with '+' genes produce homozygous red-eyed females.
2. The eggs with '+' gene if fertilized by the sperms with 'Y' chromosomes produce the red-eyed males.
3. The eggs with the gene 'w' when fertilized by the sperms having the gene '+' produce heterozygous red-eyed females.
4. The eggs with the gene 'w' when fertilized by the sperms having the Y chromosome white-eyed males are produced.

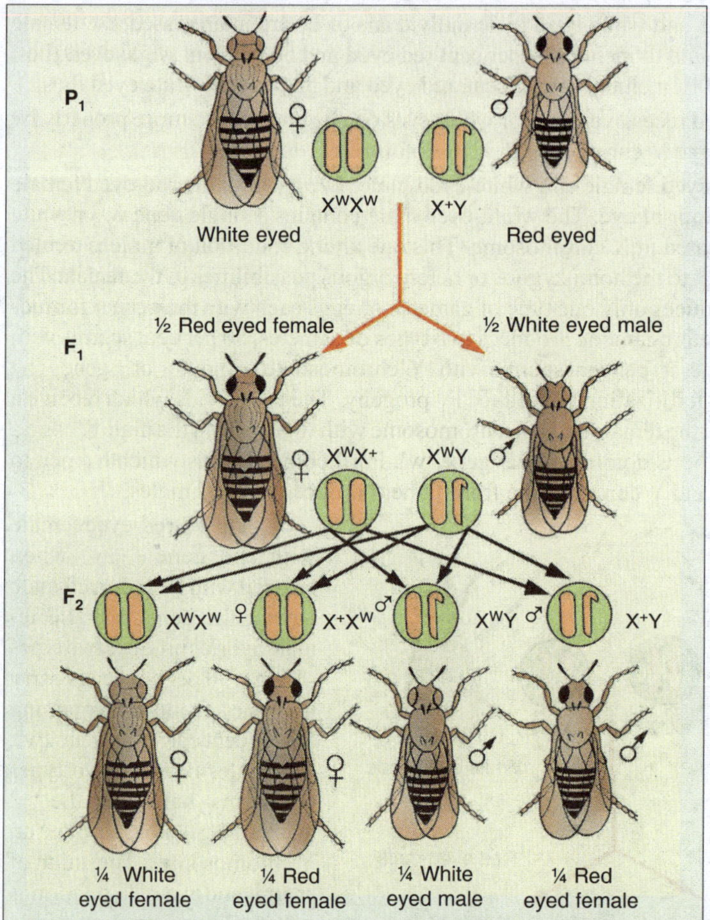

Fig. 12.6. A cross between white-eyed female and red-eyed male showing X-linked inheritance in *Drosophila.*

Likewise in other experiment in which a white-eyed female is crossed with red-eyed male, similar X-linked inheritance of recessive gene for white eye colour is revealed. The white-eyed female contains the gene 'ww' located on both X chromosomes. The red-eyed male contains the gene '+' located on single X chromosome. The female being homogametic produces single type of eggs with single 'w' gene for whiteness, while the male being heterogametic produces two types of sperms, 50 per cent sperms carry the gene '+' and remaining 50 per cent sperms carry no such genes on the Y chromosomes. In the mating of both parents by the union of these eggs and sperms, two kinds of F_1 hybrids are produced. The eggs with 'w' genes by union with the sperms having '+' genes produce heterogametic red-eyed female and the eggs with 'w' gene by union with the sperms having Y chromosome produce white-eyed male. When the F_1 brothers and sisters with the genes 'w' and '+w' breed together they produce four types of individuals in F_2 such as white-eyed female (ww), red-eyed female (+w),white-eyed male (w) and red-eyed male (+).

The results of these experiments, thus, are clearly indicating that the trait located on a sex chromosome alternates the sex from one generation to the next generation, *i.e*, the trait of white eyes transfers from P_1 father to F_1 daughter and from F_1 daughter to F_2 son.

The trait of barred eye in *Drosophila* is another sex-linked trait which is dominant over the normal eye shape but display inheritance pattern like eye-colour.

2. Inheritance of X-Linked Recessive Genes in Humans

In human beings more than 150 confirmed or highly probable X-linked traits are known; most of these are recessives. Certain well known examples of X-linked recessive genes in humans are those for red- green colour blindness or **daltonism**, haemophilia and Duchenne's muscular dystrophy. Some other examples of X-linked recessive traits include (1) deficiency of enzyme glucose-6 phosphate dehydrogenase (G6PD deficiency) in erythrocytes causing haemolytic anaemia during allergy reaction of persons for the drugs such as sulphonamides or for the broad bean (*Vicia faba*), called **favism**; (2) two forms of diabetes insipidus; (3) one form of anhidrotic ectodermal dysplasia (absence of sweat glands and teeth); (4) absence of central incisors; (5) certain forms of deafness ; (6) spastic parapelagia (*i.e.* tetanoid or partial paralysis of lower extremities with increased irritability and spasmodic contraction of the muscles); (7) uncontrollable rolling of the eye balls (nystagmus); (8) a form of cataract; (9) night blindness ;(10) optic atrophy; (11) juvenile glaucoma (hardening of eye ball); (12) juvenile muscular dystrophy and (13) white frontal patch of hair.

(1) Colour blindness. In human beings, a dominant X- linked gene is necessary for the formation of the colour sensitive cells, the **cones**, in the retina of eye. According to **trichromatic theory** of colour vision, there are three different types of cones, each with its characteristic pigment that react most strongly to red, green and violet light. The recessive form of this gene (*i.e.*, presence of recessive X-linked allele for colour blindness) is incapable of producing the colour sensitive cones and the homozy- gous recessive females ($X^c X^c$)and hemizygous recessive males ($X^c Y$) are unable to distinguish between these two colours (**Wilson,** 1911). The frequency of colour blind women is much less than colour blind man.

Detailed studies indicate that there are two closely linked gene loci each with several multiple alleles controlling the colour blind trait. Lack of the **chloroble pigment** in the retinal cones results in an inability to discriminate green colours. This defect is known as **deuteranopia** or **deutan colour blindness**. It affects about 8 per cent of human males but only about 0.7 per cent of females. The deutan or green colour blindness is first to be described in literature and is most commonly encountered sex-linked trait in human beings. Likewise, lack of **erythrolable pigment** which is necessary for discrimination in red end of the spectrum results in **protanopia** or **protan colour-blindness**. This abnormality is much less common than the deutan type, occurring in only about 2 per cent of males and in only 4 women out of 10,000. It is also caused by an X-linked recessive gene present quite close to the deuteranopia locus. Still other forms of colour blindness, some X-linked and some autosomal, are also known in humans.

In red-green colour blindness, a green table and red balls appear much the same colour.

The inheritance of colour blindness can be studied in the following two types of marriages :

(i) Marriage between colour-blind man and normal visioned woman. When colour-blind man marries with a normal visioned woman, then they will produce normal visioned male and female individuals in F_1. The marriage between a F_1 normal visioned woman and normal visioned male will produce in F_2 two normal visioned female, one normal visioned male and one colour-blind male (Fig. 12.7).

Parent : $X^+ X^+$ (Normal female) X $X^c Y$ (Colour-blind male)

Gametes : (X^+) (X^+) — (X^c) (Y)

F_1: ½ X^+X^c (Normal but carrier female) : ½ X^+Y (Normal but hemizygous male)

(Marriage between a carrier female and a normal male produces the progeny as follows —)

F_2 :
¼ X^+X^+ = Normal female
¼X^+X^c = Normal but carrier female
¼X^+Y = Normal male
¼X^cY = Colour-blind male
} 3 Normal : 1 Colour-blind or 3:1

Fig. 12.7. A marriage between a normal visioned female and a colour-blind man and result of the marriage of F_1 carrier female with a normal male.

(ii) Marriage between normal visioned male and colour-blind female. If a woman (XX) is colour blind and she happens to marry a normal visioned male (XY), then all F_1 sons will be colour-blind and daughters will be normal visioned. Because male receives one X-linked recessive gene for colour-blindness from colour-blind mother. The daughter receives one X-linked dominant gene for normal vision from father and one X-linked recessive gene for colour-blindness from the mother. The F_1 brother and sisters, if inbred or married they will produce in F_2 a colour-blind homozygous daughter, a normal visioned heterozygous daughter, a normal visioned hemizygous son and a hemizygous colour-blind son (Fig. 12.8).

(2) Haemophilia. Haemophilia is the most serious and notorious disease which is more common in men than women. This is also known as bleeder's disease. The person which contains the recessive gene for haemophillia lacks in normal clotting substance (thromboplastin) in blood so minor injuries cause continuous bleeding and ultimate death of the person due to haemorrhages. This hereditary disease was reported by **John Cotto** of Philadelphia in 1803 in man. Recently, two types of X-linked haemophilia have been recognized :

Parents :	X^cX^c	×	X^+Y	
	Colour-blind female		Normal male	
Gametes :	(X^c) $X^c)$		(X^+) (Y)	
F_1:	½ X^+X^c		½X^cY	
	Normal but carrie female		Colour-blind but hemizygous male	

(Marriage of F_1 carrier female with a colour blind male produces the following progeny—)

F_2 :	¼X^+X^c	= Normal but carrier female	
	¼ X^cX^c	= Colour-blind female	2 Normal : 2 Colour-blind or 1 : 1
	¼X^+Y	= Normal male	
	¼X^cY	= Colour-blind male	

Fig. 12.8. A marriage between a colour-blind female and a normal male and result of the marriage of F_1 carrier female with a colour-blind male.

The last Russian czar and his family. Tsarina Alexandra (granddaughter of Queen Victoria of England) was a carrier of haemophilia. She passed this disease on to her son Alexis.

(a) Haemophilia A. It is characterized by lack of antihaemophilic globulin (Factor VIII). About four fifths of the cases of haemophilic are of this type.

(b) Haemophilia B. It is also called **"christmas disease"** after the family in which it was first described in detail. Haemophilia B results from a defect in plasma thromboplastic component (factor IX). This is milder form of haemophilia.

Woodliff and Jackson in 1966 have found the occurrence of both types of haemophilia in an unusual family, both segregating independently; therefore, they concluded that the two loci (*i.e.* haemophilia A and haemophilia B) were far apart on the X chromosome (*i.e.*, having a distance of more than 40 map units between them.

Haemophilia is well known in the royal families of Europe, where it is traceable to

Queen Victoria of England, who must have been heterozygous (carrier). No haemophiliac is known in her ancestry; hence, it is concluded that her haemophilia allele arose from a mutant gamete. Since it is caused by recessive X-linked gene, a lady may carry the disease (*i.e.*, she is carrier but nonsufferer and would transmit it to 50 per cent of her sons, even if the father is normal (Fig. 12.9).

Parents :	X^+X^h Normal (Carrier) mother	×	X^+Y Normal father	
Gametes:	$(X^+)(X^h)$		$(X^+)(Y)$	
Progeny :	1/4 X^+X^+	=	Normal daughter	3 Normal :1 Haemophilic or 3 : 1
	1/4 X^+X^h	=	Normal but carrier daughter	
	1/4 X^+Y	=	Normal son	
	1/4 X^hY	=	Haemophilic son	

Fig. 12.9. Inheritance of haemophilia disease in human beings.

3. Inheritance of Z-linked Recessive Genes in Moths

In case of birds, moths, butterflies, etc., the females are heterogametic and males are homogametic quite unlike that of *Drosophila* and man. Here also sex-linked genes follow the "crisscross" pattern but from mother through heterozygous F_1 sons to grand-daughters of F_2. The common example of ZW-ZZ sex-linkage is plymouth rock chicken.

In moths, sex-linkage was discovered by the pioneer studies of **Doncaster** and **Raynor** in 1906. They were studying inheritance of wing colour in the magpie moth (*Abraxas*), using two different pure lines—one with light wings, the other with dark wings. If light winged females are crossed with dark winged males, all the progeny have dark wings, thus, showing that the allele for the light wings is recessive. However, in the reciprocal cross (dark female × light male), all the female progeny have light wings and all the male progeny have dark wings. Thus, this pair of receiprocal crosses does not give similar results, and in the second cross the wing phenotypes are associated with the sex of the moths.

Later on, these peculiar results were explained on the basis of characteristic pattern of inheritance of Z-linked genes (Fig. 12.10)

First Cross (reciprocal of second)	**Parents:**	Z^LZ^L Dark male	×	Z^lw Light female
	Gametes:	$(Z^L)(Z^L)$ sperms		(Z^l) (w) ova
	F_1:	1/2 Z^LZ^l Dark male or all dark	:	1/2 Z^LW Dark female
Second Cross: (reciprocal of first)	**Parents:**	Z^lZ^l Light male	×	Z^LW Dark female
	Gametes:	$(Z^l)(Z^l)$ sperms		(Z^L) (W) ova
	F_1:	Z^LZ^l Dark female or 1:1 ratio		1/2 Z^lW Light male

Fig. 12.10. Sex-linked inheritance in moth showing different result in the reciprocal crosses.

4. Sex Linkage in Poultry

In plymouth rock chicken the gene for barred feathers is dominant and the gene for black or red unbarred feathers is recessive. Both the genes are Z-linked. A barred male chicken contains two genes for barring because it has two sex chromosomes (ZZ). When this barred male with the gene BB is crossed with unbarred female containing single recessive gene 'b' in its Z chromosome (W chromosome contains no genes) produce in F_1 only barred males and females (Fig. 12.11). These F_1 barred males and females when inbred proeuce in F_2, a hemizygous non-barred female, a hemizygous barred female a heterozygous barred male and a homozygous barred male (*i.e.*, 3 Barred : 1 Non-barred).

In another cross when barred hens and non-barred cocks are crossed, the F_1 have half barred male and half non-barred females (Fig. 12.12). These by inbreeding produce in F_2 half barred males and females and half non-barred males and females (*i.e.*, 2 Barred : 2 Non-barred or 1:1 ratio).

B. INHERITANCE OF Y-LINKED GENES

Genes in the non-homologous region of the Y chromosome pass directly from male to male (Fig. 12.13). In man, the Y-linked or holandric genes such as **ichthyosis hystrix gravis hypertrichosis** (excessive development of hairs on pinna of ear) are transmitted directly from father to son. Recently, certain other holandric genes have been reported in humans, *e.g.*, genes for H-Y antigen, histocompatibility antigen, spermatogenesis, height (stature) and slower maturation of individual. The Y-linked gene of H-Y antigen is located on short arm of Y chromosome, while genes controlling spermatogenesis occur on the long arm of Y-chromosome. Y chromosome of *Drosophila* too contain Y-linked gene for male fertility.

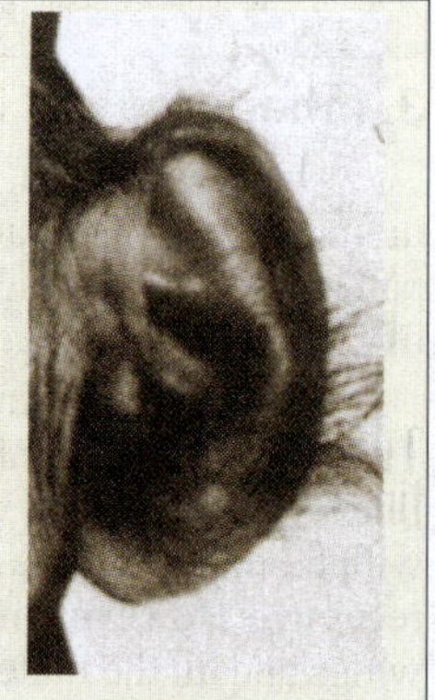

Excessive development of hair on pinna of ear.

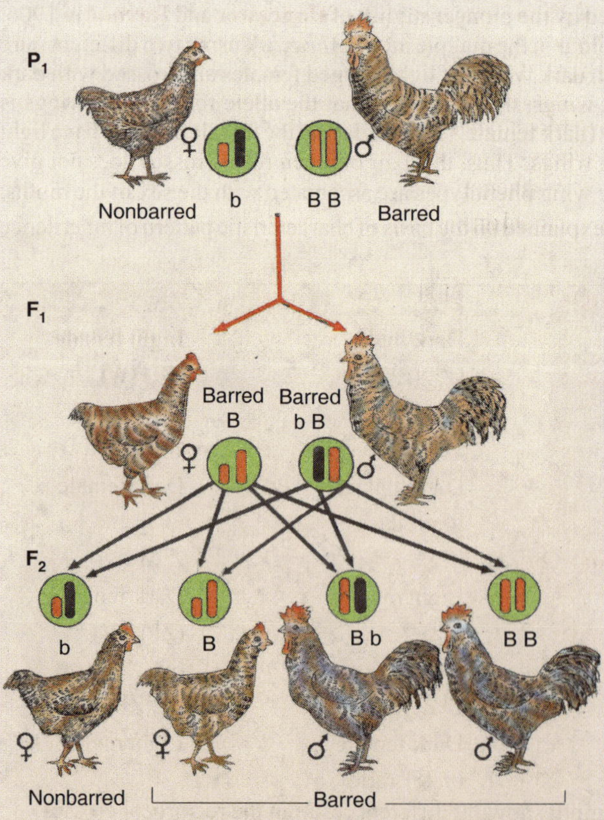

Fig. 12.11. A cross between a non-barred hen and a barred cock showing sex-linked (Z-linked) inheritance.

In the fish *Lebistes*, the Y chromosome contains a Y-linked gene **maculatus** that determines a pigmented spot at the base of dorsal fin of male individuals. This phenotype is passed only from father to son, and females never carry or express the gene.

Procupine man. An Englishman by the name of **Edward Lambert** was born in 1717. His skin was like thick bark which had to be shed periodically. The hairs on his body were quill-like and he subsequently has been referred to as the "**porcupine man**". He had six sons, all of which exhibited the same trait. The trait appeared to be transmitted from father to son through four generations. None of the daughters ever exhibited the trait. In fact, it has never been known to appear in females. So this trait was believed to be caused by a holandric or Y-linked gene (See Stansfield, 1986).

C. INHERITANCE OF X-Y-LINKED GENES

The genes which occur in homologous section of X and Y chromosome have inheritance like the autosomal genes. The X-Y linked genes are **partially** or **incompletely sex -linked**, because, sometime, the crossing over may occur in the homologous sections of X and Y chromosomes. In humans several diseases are XY-linked. Certain XY-linked genes of man are of total colour blindness, two skin diseases (*Xeroderma pigmentosum* and *Epipermolysis bullosa*), *Retinitis pigmentosa*, etc.

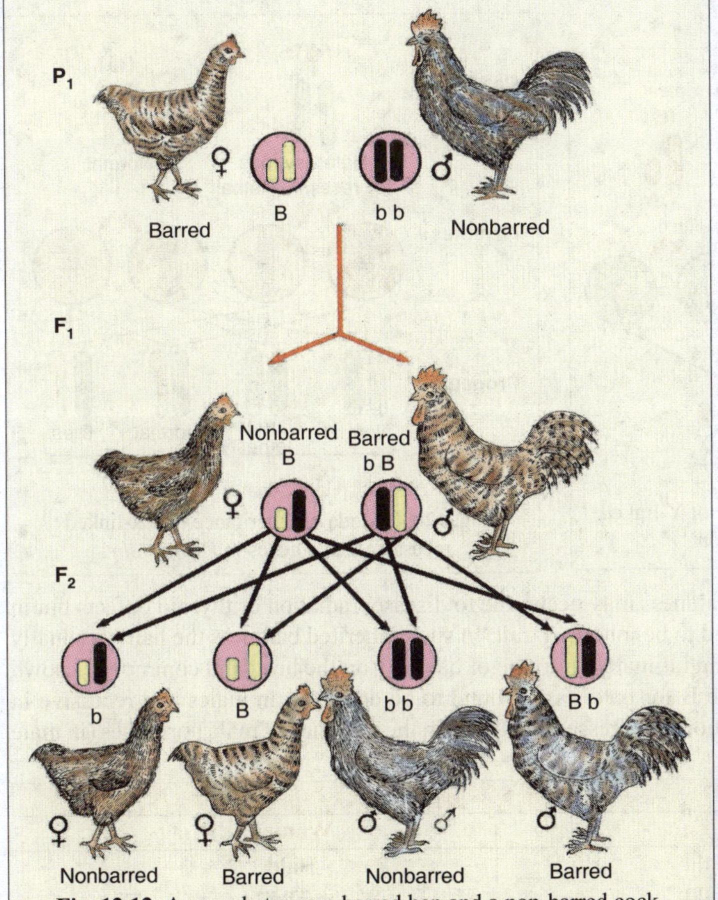

Fig. 12.12. A cross between a barred hen and a non-barred cock showing sex linked (Z-linked) inheritance.

SEX-LINKED LETHALS

Certain X-linked genes are **lethals**, *i.e.*, they cause death of an individual from egg up to sexually mature adult stage. X-linked lethals have been reported both in *Drosophila* and human beings.

1. Sex-linked lethals in *Drosophila*. In *Drosophila*, rare females regularly produce female and male progeny in the ratio of 2:1, which indicates that half of the male progeny is missing. This peculiarity can be explained due to the presence of a recessive sex-linked lethal gene in the heterozygous condition in the female parent (Fig. 12.14).

2. Sex-linked lethals in humans. In fact, the gene for haemophilia is a recessvie **sex-linked lethal,** since it may cuse death. Slight scratches, accidental injuries, or even bruises, which would not be serious in normal persons, may result in fatal bleeding for the haemophiliac. Often, internal bleeding (from bruises, internal lesions and so forth) is more important in producing lethality. By bringing about death, sex-linked lethals will alter the sex ratio in a progeny.

Children with Duchenne muscular dystrophy.

Duchenne (or progressive pseudohypertrophic) **muscular dystrophy** is another fatal disorder of humans. In it affected individual, though apparently normal in early childhood, exhibits progressive wasting away of the muscules, resulting in confinement to a wheel-chair by about the age of 12 years and death in teen years (in adolescence). The Y-linked recessive allele responsible for this disorder is a lethal and will change the sex ratio in a given group of offspring over time.

SEX-INFLUENCED GENES

Sex influenced genes are those whose dominance is influenced by the sex of the bearer. Thus, male and female individuals may be similar for a particular trait but give different phenotypic expressions of the same trait.

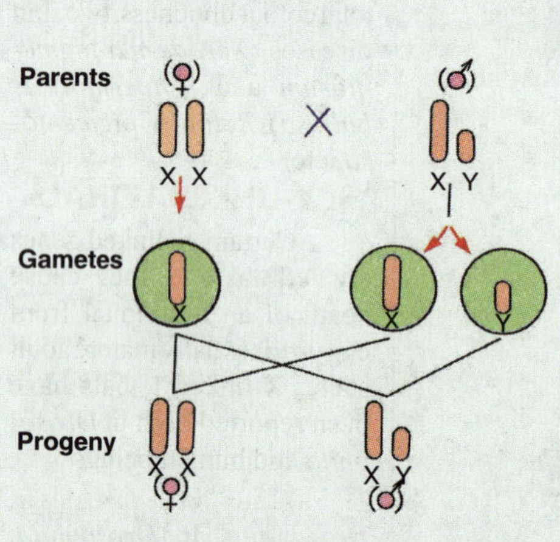

Fig. 12.13. Mode of inheritance of Y-linked genes from father to son.

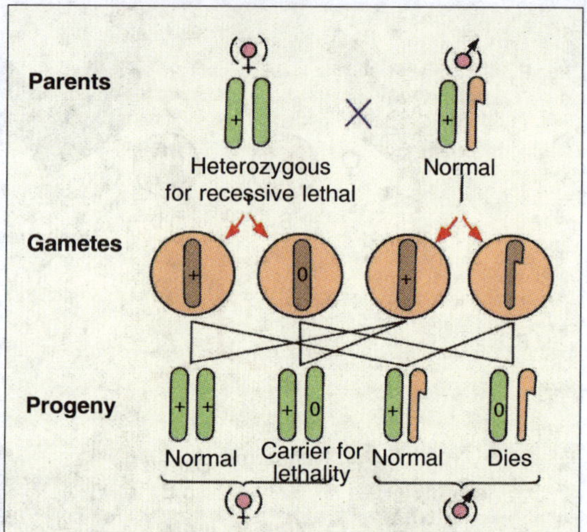

Fig. 12.14. Mode of inheritance of sex-linked recessive lethal genes in *Drosophila*.

Example 1. In man the baldness may occur due to disease, radiation or thyroid defects but in some families balldness is found to be inherited trait. In such inherited baldness the hairs gradually become thin on head top, leaving ultimately a fringe of hair low on the head and commonly known as **pattern baldness**. The gene B for baldness is found to be dominant in males and recessive in females. In heterozygous condition it expresses itself only in the presence of male hormones (in male sex):

Genotype	Phenotypes	
	Men	Women
BB	Bald	Bald
Bb	Bald	Non-bald
bb	Non-bald	Non-bald

2. In sheep, the genes for the development of horns is dominant in males and recessive in female (Table 12.1).

Table 12.1. Phenotypic expression of dominant sex-influenced gene for horned trait occurs only in male sheep.

Genotype	Phenotype	
	Males	Females
h^+h^+	Horned	Horned
h^+h	Horned	Hornless
hh	Hornless	Hornless

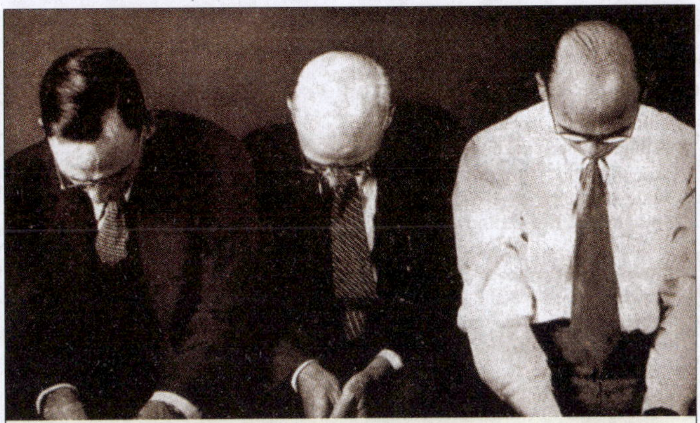

Pattern baldness in a father and one but not the other of his two sons. The father is in the center, with one son on each side.

Thus, here same allele (h^+) has different expressions in male and female individuals, but both alternative forms of the trait (*e.g.*, horned and hornless) are known in each sex, *i.e.*, both horned and hornless sheep are known as female and male.

3. Further in cattles also sex-influenced genes occur as spotting in cattle. The genes for mahogany and white spotting is dominant in males and the genes for red and white spotting is dominant in females.

SEX-LIMITED GENES

Sex-limited genes are autosomal genes whose phenotypic expression is determined by the presence or absence of one of the sex hormones. Their phenotypic effect is limited to one sex or other. In other words, the penetrance of a sex-limited gene in one sex remain zero. The sex-limited genes are mainly responsible for secondary sex characters in cattle, humans and fowl.

Example 1. The bulls have genes for milk production which they transmit to their daughters, but they or their sons are unable to express this trait. The production of milk is, therefore, limited to variable expression only in the female sex.

2. Beard development in human beings is a sex limited trait as men normally have beards, whereas women normally do not. Likewise, the genes for male voice, body hair and physique are autosomal in human beings, but they are expressed only in the presence of androgens which are absent in females.

3. In chicken the recessive gene (h) for cock feathering (Fig. 12.15) is male sex-limited (*i.e.*, it is penetrant only in male environment).

Genotypes	Phenotypes	
	Males	Females
HH	Hen-feathering	Hen-feathering
Hh	Hen-feathering	Hen-feathering
hh	Cock-feathering	Hen-feathering

In chicken, the recessive gene for cock feathering is male sex-limited.

NON-DISJUNCTION

The failure of homologs (at meiosis) or sister chromatids (at mitosis) to separate properly to opposite poles is called **non-disjunction**. The sex chromosomes pass through the phenomenon of non-disjunction and present various interesting situations. The non-disjunction may be of following types:

1. Primary Non-disjunction

The phenomenon of primry non-disjunction was discovered by **C.B. Bridges** (1916) during his classical matings of *Drosophila melanogaster*. Let us consider one of hisclassical crosses as follows:

The gene for wild type red eyes (+) is carried by the X chromosome; a recessive allele (v) produces vermilion eyes in homozygous females (X^vX^v) and in all males (X^vY). Normally vermillion-eyed females mated to red-eyed males produce only red-eyed daughters and vermilion eyed sons (Fig. 12.16).

Fig. 12.15. The hen and cock feathering in domestic fowl. Cock-feathering is characterized by long, pointed, curving neck and tail feathers.

Parents:	X^vX^v Vermilion female (or mother)		×	X^+Y Red male (or father)
Gametes :	**Eggs :**	(X^v)	(X^v)	
	Sperms :	(X^+)	(Y)	
F_1 :	1/2 X^+X^v Red female (or daughter)	:	1/2 X^vY Vermilion male (or son)	

Fig. 12.16. Results of a normal cross between vermilion-eyed female and red-eyed male.

Crosses of this type, however, in rare cases produce unexpected vermilion-eyed daughters and red-eyed sons with a frequency of one per 2,000 to 3,000 offspring. **Bridges** speculated that these unusual progeny are due to a failure of the X chromosomes in an XX female to disjoin during meiosis I of oogenesis. This phenomenon was termed **primary disjunction** by Bridges and this event tends to produce three types of eggs: the majority of eggs with normal single X chromosome and a small number of eggs with their two X chromosomes or no X at all. These eggs on normal fertilization produce the following results :

The metafemales (AA XXX) are weak and seldom live beyond the pupal stage; the AAOY individual die in the egg stage. Note that the AA XXY female indicate that the presence of a Y chromosome does not determine maleness itself, though males without it (AAXO) are sterile. From these results three facts became established : 1. In *Drosophila* sex determination occurs according to genetic balance theory of Bridges. 2. In *Drosophila*, Y chromosome contains genes for the spermatogenesis. 3. It provides strong support to chromosome theroy of inheritance (*i.e.*, Mendelian factors or genes are carried on the chromosomes).

2. Secondary Non-disjunction

In next cross, **Bridges** mated the exceptional vermilion-eyed females (AAX^vX^vY) that arose as a result of primary non-disjunction to normal red-eyed males (AAX^+Y). The progeny of this cross included 96 per cent female with red eyes and only 4 percent females had vermilion eyes indicating that there was 4 per cent secondary non-disjunction (Fig. 12.18).

Thus, primary non-disjunction may occur in either XX females or XY males. In the former it leads to the production of XX and O eggs. Occurrence in the first meiotic division of males produces XY

Parents: AAX^vX^v × $AA\ X^+Y$

Vermilion female — Red male

Gametes : Eggs: $(A\ X^v)$ + (AX^vX^v) + (AO)

(numerous) — (rare due to non-disjunction) — (rare called **nullo**)

Sperms : (AX^+) (AY)

F_1:
1. $AA\ X^+\ X^v$ = Red female (numerous; normal)
2. $AA\ X^+\ X^v\ X^v$ = Red "metafemales" (rare; die)
3. $AA\ X^+O$ = Red male (rare; sterile)
4. $AA\ X^vY$ = Vermilion male (numerous; normal)
5. $AA\ X^vX^vY$ = Vermilion female (rare; fertile)
6. AAOY = Die (rare)

Fig. 12.17. Inheritance of eye colour in *Drosophila* due to primary non-disjunction of sex chromosomes.

Parents : $AA\ X^vX^vY$ × $AA\ X^+Y$

Gametes : Vermilion females — Red males

Eggs : $(AX^v) + (AX^vY) + (AX^vX^v)$ + $(AY) + (AX^v) + (AX^vY)$

Due to normal disjunction — Due to secondary non-disjunction

Sperms : (AX^+) + (AY)

F_2 :

	Eggs	Sperms AX^+(50%)	Sperms AY(50%)	
	AX^vX^v (4%)	$AAX^+X^vX^v$ Metafemale (die) (2%)	AAX^vX^vY Vermilion female (2%)	4% secondary exceptional progeny phenotype
X-Y	AY (4%)	AAX^+Y Red male (2%)	AAYY (die) (2%)	96% regular (i.e., expected) progeny phenotypes
Pairing (16%)	AX^v (4%)	AAX^+X^v Red female (2%)	$AA\ X^vY$ Vermilion male (2%)	
Eggs :	AX^vY (4%)	AAX^+X^vY Red female (2%)	AAX^vYY Vermilion male (2%)	
	AX^v (42%)	AAX^+X^v Red female (21%)	AAX^vY Vermilion male (21%)	
X-X Pairing (84%)	AX^vY (42%)	AAX^+X^vY Red female (21%)	AAX^vYY Vermilion male (21%)	

Fig. 12.18. Results obtained from secondary non-disjunction.

and O sperms. When such non-disjunction takes place during the second meiotic division then XX, andXY and O sperms result. Secondary non-disjunction, on the other hand, occurs in XXY females,where it gives rise to XX, XY, X and Y eggs (Fig. 12.19). As the term non-disjunction implies, these aberrant gametes are produced only as a result of failure of the sex chromosomes to disjoin after synapsis; they are not physically attached.

Similar results were obtained by **Bridges** (1916) from the cross of red eyed male and white eyed female ***Drosophila.***

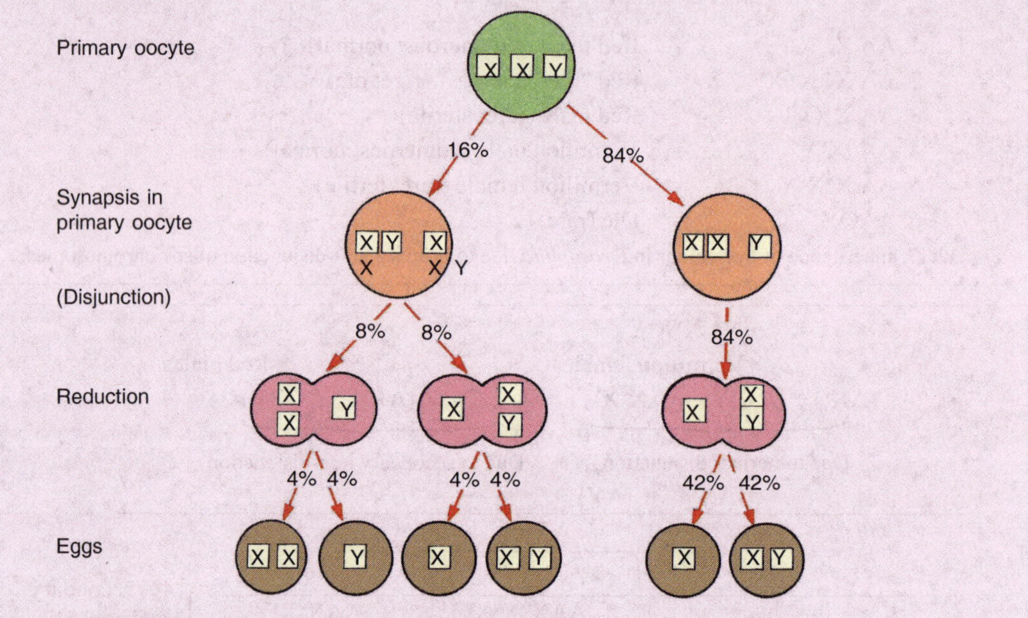

Fig. 12.19. Secondary non-disjunction in XXY female *Drosophila*, resulting in 46 per cent X, 46 per cent XY, 4 per cent XX, and 4 per cent Y eggs.

REVISION QUESTIONS AND PROBLEMS

1. In the domestic fowl (*Gallus domesticus*) the gene for plumage colour is sex-linked (Sturtevant, 1912). The dominant allele G determines *gold*-coloured plumage, and its recessive allele *g* determines *silver*-coloured plumage. A cross is made between a homozygous *gold*-coloured male and a *silver*-coloured female. The F_1 males are mated with F_1 females. Give the genotypes and phenotypes for each sex in the F_1 and F_2.
2. The gene for yellow body colour y in *Drosophila* is recessive and sex linked. Its dominant allele y^+ produces will type body colour. What phenotypic ratios are expected from the crosses (a) yellow male × yellow female, (b) yellow female × wild type male, (c) wild female (homozygous) × yellow male, (d) wild type (carrier) female × wild type male, (e) wild type (carrier) female × yellow male ?
3. Could a recessive mutant gene in humans be located on the X chromosomes if a woman exhibiting the recessive trait and a normal man had a normal son ? Explain.
4. A certain type of white forelock in man appears to follow the sex influenced mode of inheritance, being dominant in men and recessive in women. Using the allelic symbols w and w′, indicate all possible genotypes and the phenotypes thereby produced in men and women.
5. The male buffalo differs from the female in having a well-developed mane. What genetic explanations might be advanced to explain this dimorphism ?
6. Distinguish between the following, giving the location of the genes, the mode of transmission, and relationship to sex:

1. Sex-influenced and holandric characters;
2. Sex-limited and sex-influenced characters;
3. Sex-linked and sex-influenced characters.

7. Write shrot notes on the following: (i) Haemophilia; (ii) Colour blindness; (iii) Non-disjunction; (iv) Heterogametic sex; (v) Sex chromosomes and autosomes.
8. In *Drosophila* carnation (c) is a recessive eye colour mutant and short wings (s) is a recessive wing mutant. A carnation-eyed female, otherwise wild-type, was crossed to a normal-eyed, short-winged male. The male progeny all had carnation eyes, but half had short wings and half had normal wings. All female progeny had normal red-eye colour, but half had short wings and half had normal wings.
   a. What was the genotype of the parental female ?
   b. What was the genotype of the parental male ?
   c. What are the genotypes of the male and female progeny ?
9. Amale *Drosophila* with reduced eyes was crossed to a female with normal eyes. The F_1 consisted of a total of 67 males, all with normal eyes, and 65 females, all with reduced eyes. How would you explain the inheritance of this character ?
10. White (w) a sex-linked recessive gene of *Drosophila* which blocks eye pigment formation, while scarlet (st) is an autosomal recessive gene which results in bright red eyes when homozygous. Wild-type eye colour (+) is dull red and sex-linked. A white-eyed male heterozygous for scarlet was crossed to a scarlet-eyed female heterozygous for white. What phenotypes are expected in the F_1 males and in what proportions ?
11. In fowl, barring (B) is sex-linked and dominant, the recessive allele (b) producing solid black colour when homozygous. Silky feathers (s) is a recessive autosomal allele, as opposed to nonsilky (S). If black cocks, heterozygous for silky, are crossed to barred, silky hens, what genotypes and phenotypes will be produced and in what proportions?
12. 'Bent', a dominant sex-linked allele B, in the mouse, results in a short, crooked tail; its recessive allele, b, produces normal tails. If a normal-tailed female is mated to a bent-tailed male, what phenotypic ratio should occur in the F_1 ?
13. In the guppy a dominant allele, M, results in the presence of a black spot on the dorsal fin. Spotted males transmit the trait only to sons and not to daughters, and male and female progeny of such daughters do not show the trait. How is this trait inherited ?
14. Red-green colour blindness in humans is recessive and sex-linked. If a woman heterozygous for colour blindness marries a colour blind man, what is the probability that their first child will be a colourblind daughter ?
15. A married couple, both of whom had normal vision, produced a colour blind son. Examination of cell samples from the son showed the presence of a Barr body. What is the probable genotype of the son with respect to sex chromosomes and colour blindness ? What is the simplest explanation that will account for this genotype ?
16. A sex-linked recessive gene in humans produces colourblind men when hemizygous and colour blind women when homozygous. A sex-influenced gene for pattern baldness is dominant in men and recessive in women. A heterozygous bald, colour blind man marries a non-bald woman with normal vision whose father was non-bald and colourblind and whose mother was bald with normal vision. List the phenotypic expectations for their children.

ANSWERS TO PROBLEMS

2. (a) All offspring yellow; (b) All females wild type, all males yellow; (c) All offspring wild type; (d) all female wild type : 1/2 wild type males : 1/2 yellow males; (e) Females and males : 1/2 wild type : 1/ 2 yellow.
3. Yes, if it was incompletely sex-linked and the father carried the normal gene on the homologous portion of his Y chromosome.

4.

Genotype	Men	Women	Genotypes
w dominant in men:			w′ dominant in men:
ww	Forelock	Forelock	w′w′
ww′	Forelock	Normal	w′w
		(or)	
w′w′	Normal	Normal	ww

8. Carnation is a sex-linked; short wing is autosomal.
   (a) Female : c/c +/s ; (b) Male : +/Y s/s;
   (c) Progeny males : c/Y +/s and c/Y s/s;
   Progeny females : c/+ s/+ and c/+ s/s.
9. The trait of reduced eyes is sex-linked and dominant.
10. 1/2 White (w/Y st/+ and w/Y st/st) : 1/4 wild type (+/Y st/+) : 1/4 scarlet (+/Y st/st)
11. In birds females are XY and males are XX.
    Males : 1/2 B/b s/S, Barred, Non-silky;
    1/2B/b s/s, Barred, Silky;
    Females : 1/2 b/Y s/S, Black, Non-silky,
    1/2 b/Y s/s, Black, Silky.
12. All females 'bent'; all males normal.
13. Gene M is located on the Y chromosome, *i.e.*, it is a holandric or Y-linked gene.
14. The probability for a daughter is 1/2 and the probability for colour blindness is 1/2. Thus, the probability for a colourblind daughter is 1/2 × 1/2 = 1/4.
15. The mother was heterozygous for colourblindness; nondisjunction of the X chromatids, each bearing gene c, occurred at the second meiotic division to produce an egg of sex genotype X^cX^c. Fertilization by a Y-bearing sperm resulted in a son.

16.

	Phenotype	Daughters	Sons
1.	Bald, Normal vision	1/8	3/8
2.	Bald, Colour blind	1/8	3/8
3.	Non-bald, Normal vision	3/8	1/8
4.	Non-bald, colour blind	3/8	1/8

CHAPTER

13

Determination of Sex and Sex Differentiation

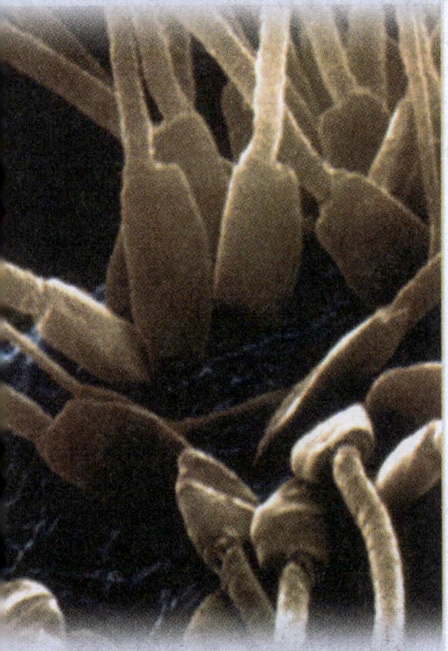

The average human ejaculate contains 175 million X – bearing spermatozoa and 175 million Y – bearing spermatozoa which determines the sex of the child.

Members of almost all species are often divided into two sections according to the kind of gamete or sex cell produced by them, *i.e.,* male sex and female sex. The word **sex** has been derived from Latin word *sexus* meaning section or separation. However, some of the lowest forms of plant and animal life are found to have several sexes. For example, in one variety of the ciliated protozoan *Paramecium bursaria* there are eight sexes or "mating types" all morphologically identical. Each mating type is physiologically incapable of conjugating with its own type, but may exchange genetic material with any of the seven other types within the same variety. Further, in organisms in which the number of sex reduced to just two, sexes may reside in different individuals or within the same individual. An animal possessing both male and female reproductive organs is usually referred to as a **hermaphrodite**. In plants where **staminate** (male) and **pistillate** (female) flowers occur on the same plant, the term of preference is **monoecious**. Most of our flowering plants have both male and female parts within the same flower (called **perfect flower**). The organisms in which both male and female gametes are produced by different individuals are called **dioecious**. The sex cells and reproductive organs form the **primary sexual characters** of male and female sexes. Besides these primary sexual characters, the male and female sexes differ from each other in many somatic characters known as **second-**

ary sexual characters. The phenomenon of molecular, morphological, physiological or behavioral differentiation between male and female sexes is called **sexual dimorphism**.

In comparison to most animals, relatively few angiosperms are dioecious, *e.g.,* asparagus, date palm, hemp, hops and spinach. In fact, among the higher plants, there are eight types of sex phenotype expression (Table 13-1).

Table 13.1. Eight main types of sex expressions in higher plants.

Types of sex expression	Flowers of different types
1. Hermaphrodites	All perfect (O) flowers.
2. Monoecious	Separate female (O) and male (O) flowers, but on the same plant.
3. Dioecious	Separate female (O) and male (O) flower on different plants.
4. Andromonoecious	Perfect (O) and male (O) flowers on the same plant.
5. Gynomonoecious	Perfect (O) and female (O) flowers on the same plant.
6. Trimonoecious	Perfect (O), female (O) and male (O) flowers on the same plant.
7. Androdioecious	Perfect (O) and male (O) flowers on separate plants.
8. Gynodioecious	Perfect (O) and female (O) flowers on separate plants.

The phenomenon of sexual dimorphism has been a biological riddle for the thinkers and biologists of all time. People always tried to know those factors which determine the male and female sexes of a species. Literally hundreds of mistaken hypotheses and wild guesses were proposed before 1900 in vain attempts to find out a solution to the problem of determination of sex. Modern geneticists have reported many different mechanisms of determination of sex in living organisms. Some important and common mechanisms of sex determination are following :

Clarance McClung (1870-1946)

A. GENETICALLY CONTROLLED SEX DETERMINING MECHANISMS

Most of the mechanisms of the determination of the sex are under genetic control and they may be classified into following categories :

1. Sex chromosome mechanism or Heterogamesis;
2. Genic balance mechanism;
3. Male haploidy or haplodiploidy mechanism;
4. Single gene effects.

A. Sex Chromosomal Mechanisms (Heterogamesis)

1. Discovery of sex chromosomes. In sexually dimorphic dioecious organisms besides morphological and behavioural differences between both sexes, the sexual diversity also occurs at the level of chromosomes. The chromosomal differences between the sexes of several dioecious species were found earlier in the course of cytological investigations. A German biologist, **Henking** in 1891 while studying spermatogenesis of the squash bug, *Pyrrhocoris,* noted that meiotic nuclei contained 11 pairs of chromosomes and an unpaired element is moved to one of the poles during the first meiotic division. **Henking** called this unpaired element a "**x body**" and interpreted it as a nucleolus.

The significance of X body was not immediately understood, but in 1902 an American geneticist **Clarance McClung** who had made extensive observations of gametogenesis in grasshoppers, suggested that the X body was involved in some way with the determination of sex. He reported that the somatic cells of the female grasshopper (*Xiphidium fasciatum*) contained 24 chromosomes

Edmund Wilson.

whereas those of the male had only 23. In 1905, **Edmund Wilson** noted that females of *Protenor* (another hemipteran bug) have 7 pairs of chromosomes, whereas males have 6 pairs and an unpaired chromosome, which **Wilson** called the **X chromosome**. Also in 1905, Miss **Nettie Stevens** found that males and females of the beetle *Tenebrio* have the same number of chromosomes, but one of the pairs in males is heteromorphic (of different size). One member of the heteromorphic pair appears identical to the member of a pair in the female; she called this the X chromosome. The other member of the heteromorphic pair is never found in females; she called this the **Y chromosome**. Stevens found a similar situation in *Drosophila melanogaster* which has four pairs of chromosomes, with one of the pairs being heteromorphic in males (Fig. 13.2). Likewise, **Stevens** and **Wilson** while working with the milkweed bug, *Lygaeus turicicus* found same number of chromosome in both the sexes, (*i.e.*, 14). In female all chromosomes were paired and the homologues were equal in size. In the male, all the chromosomes were paired, but the chromosome identified as the homologue to the X chromosome was distinctly smaller Y chromosome.

Nettie Stevens.

2. Types of sex chromosomes. In dioecious organisms, thus, two types of chromosomes were recognised which are as follows :

(i) Autosomes. The chromosomes which have no relation with the sex and contain the genes which determine the somatic characters of the individuals are known as **autosomes (A)**.

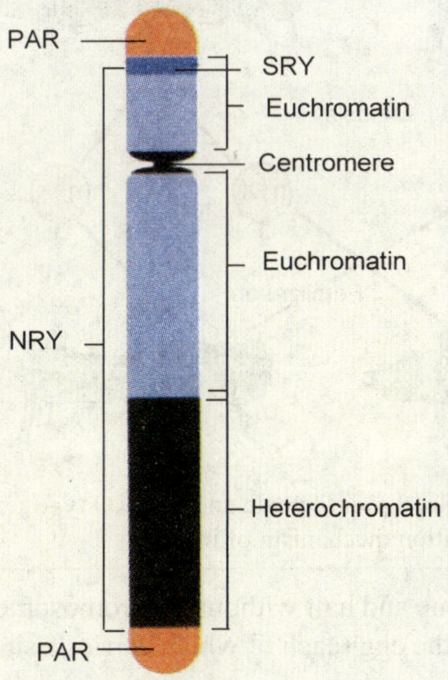

Human Y Chromosome
The various regions of the human Y chromosome

(ii) Sex chromosomes. The chromosomes which are responsible for the determination of sex are known as **sex chromosomes**, *e.g.,* X and Y chromosomes.

3. Structure of sex chromosomes. The X and Y sex chromosomes exhibit structural differences. The cytological studies have shown that the X chromosomes of most organisms are straight, rod-like and comparatively larger than Y chromosomes. The Y chromosome is smaller in size with one end slightly curved or bent to one side in *Drosophila*; in man and *Melandrium* no such curvature of Y chromosome occurs. The X chromosomes have large amount of euchromatin and small amount of heterochromatin. The euchromatin has large amount of DNA material, hence, much genetic information. The Y chromosome contains small amount of euchromatin and large amount of heterochromatin. The Y chromosome has little genetic information, therefore, sometimes it is referred to as genetically **inert** or **inactive**.

TYPES OF SEX CHROMOSOMAL MECHANISM OF SEX DETERMINATION

In dioecious diploidic organisms following two systems of sex chromosomal determination of sex have been recognized:

(a) Heterogametic males;
(b) Heterogametic females.

Heterogametic Males

In this type of sex chromosomal determination of sex, the female sex has two X chromosomes, while the male sex has only one X chromosome. Because, male lacks a X chromosome, therefore, during gametogenesis produces two types of gametes, 50 per cent gametes carry the X chromosomes, while the rest 50 per cent gametes lack in X chromosomes. Such a sex which produces two different type of gametes in terms of sex chromosomes, is called **heterogametic sex**. The female sex, because, produces similar type of gametes, is called, **homogametic sex**. The heterogametic males may be of following two types :

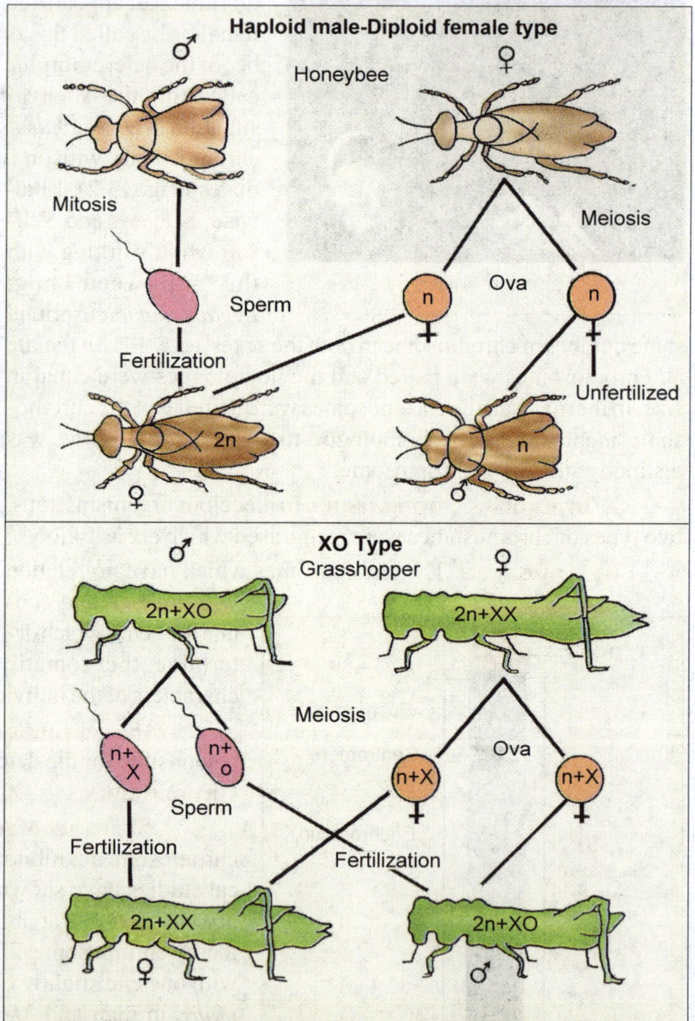

Fig. 13.1. Haploid male, diploid female and XX-XO type sex determination mechanism of insects.

(i) XX-XO type. In certain plants (*e.g., Vallisneria spiralis, Dioscorea sinuata*, etc.,) and insects specially those of the orders Hemiptera (true bugs) and Orthoptera (grasshoppers and roaches), the female having two X chromosomes (hence, referred to as XX) and are, thus, homogametic, while the male having only one X chromosome (hence, referred to as XO). The presence of an unpaired X chromosome determines the masculine sex. The male lacking in one X chromosome produces two types of sperms: half with X chromosome and half without X chromosome. The sex of the offspring depends upon the sperm that fertilizes the egg (each of which carries a single X chromosome) as shown in Fig. 13.1.

(ii) XX-XY type. In man, other mammals, certain insects including *Drosophila* and *Lygaeus turicicus* and in certain angiospermic plants such as *Melandrium album* (*Lychis*), *Humulus lupulus,*

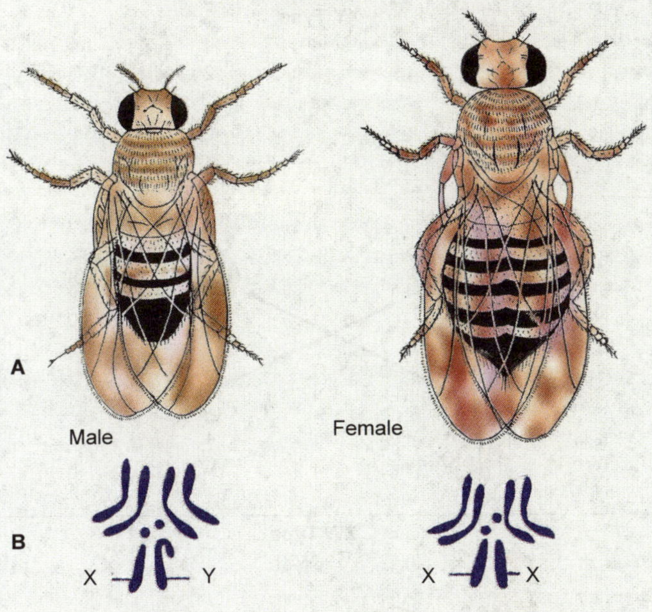

Fig 13.2. Male and female *Drosophila melanogaster* and their chromosomes.

Coccinia indica, the female possesses two homomorphic X chromosomes in their body cells (hence, referred to as XX) and they being homogametic, produce one kind of eggs, each with one X chromosome.The males of these organisms possess one X chromosome and one Y chromosome (hence, referred to as XY). The males having two heteromorphic sex chromosomes producce two kinds of sperms : half with X chromosome and half with Y chromosome. The sex of embryo depends on the kind of sperm. An egg fertilized by a X-bearing sperm, produces a female, but, if fertilized by a Y-bearing sperm, a male is produced (See Fig.13.3).

Heterogametic Females

In this type of sex chromosomal determination of sex, the male sex possesses two homomorphic X chromosomes, therefore, is homogametic and produces single type of gametes, each carries a single X chromosome. The female sex either consists of single X chromosome or one X chromosome and one Y chromosome. The female sex is, thus, heterogametic and produces two types of eggs, half with a X chromosome and half without a X chromosome (with or without a Y chromosome). To avoid confusion with that of XX-XO and XX-XY types of sex determining mechanisms, instead of the X and Y alphabets, Z and W alphabets are generally used respectively.

The heterogametic females may be of following two types :

(i) ZO-ZZ system. This system of sex determination is found in certain moths and butterflies. In this case, the female possesses single Z chromosome in its body cells (hence, is referred to as ZO) and is heterogametic, producing two kinds of eggs, half with a Z chromosome and half without any Z chromosome. The male possesses two Z chromosomes (hence, referred to as ZZ) and is homogametic, producing single type of sperms, each of which carries a single Z chromosome. The sex of the offspring depends on the kind of egg as shown below :

Parent :	Female	X	Male
	2A+ZO		2A+ZZ
Gametes :	(A+Z) (A+O)		(A+Z) (A+Z)
	Ova		Sperms
F_1 :	2A+ZZ		2A+ZO
	Male		Female

(ii) ZW-ZZ system. This system of sex determination occurs in certain insects (gypsy moth) and vertebrates such as fishes, reptiles and birds and plants such as *Fragaris elatior.* Here the female sex has one Z chromosome and one W chromosome. It is heterogametic and produces two types of ova, 50 per cent ova carry the Z chromosomes, while rest 50 per cent ova carry W chromosomes. The male sex has two homomorphic Z chromosomes and is homogametic producing single type of sperms, each carries a Z chromosome. The sex of the offspring depends on the kind of egg, the Z bearing eggs produces males but the W bearing eggs produce females as shown in Fig. 13.3.

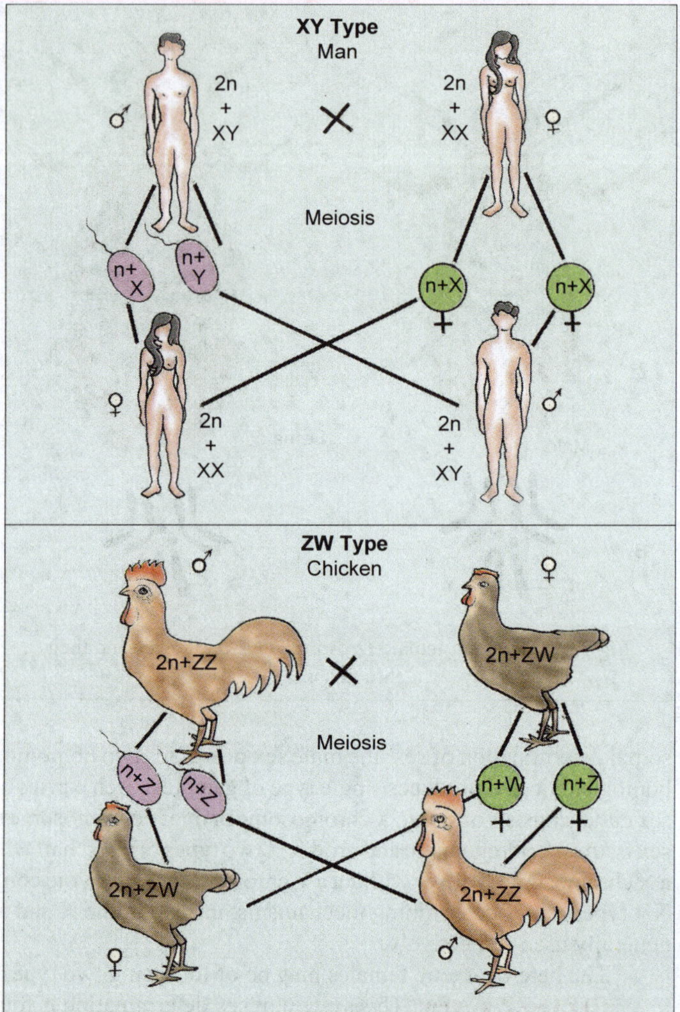

Fig 13.3. The XX-XY and ZW-ZZ type determination of sex in man and chicken.

B. Genic Balance Mechanism

By studying sex chromosomal mechanism of sex determination, it may appear at first glance that some genes carried by the sex chromosomes (X and Y) were entirely responsible for sex. But this is not the case. Extensive experimentation of different workers (**Wilson**, 1909 ; **Bridges**, 1921 and **Goldschmidt**, 1934) on different organisms have revealed the fact that most organisms generally have inherent potentialities for both sexes and each individual is found to be more or less intermediate between male and female sexes (Hence may be referred to as **intersex**). There seems to exist a delicate **balance** of masculine and feminine tendencies in the hereditary compliment of an individual and mechanisms like the XY, ordinarily serve to trip the

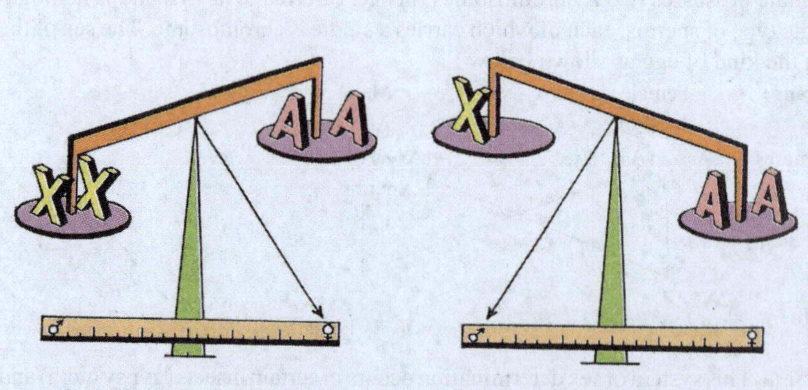

Fig. 13.4. A diagrammatic representation of genic balance theory in *Drosophila*.

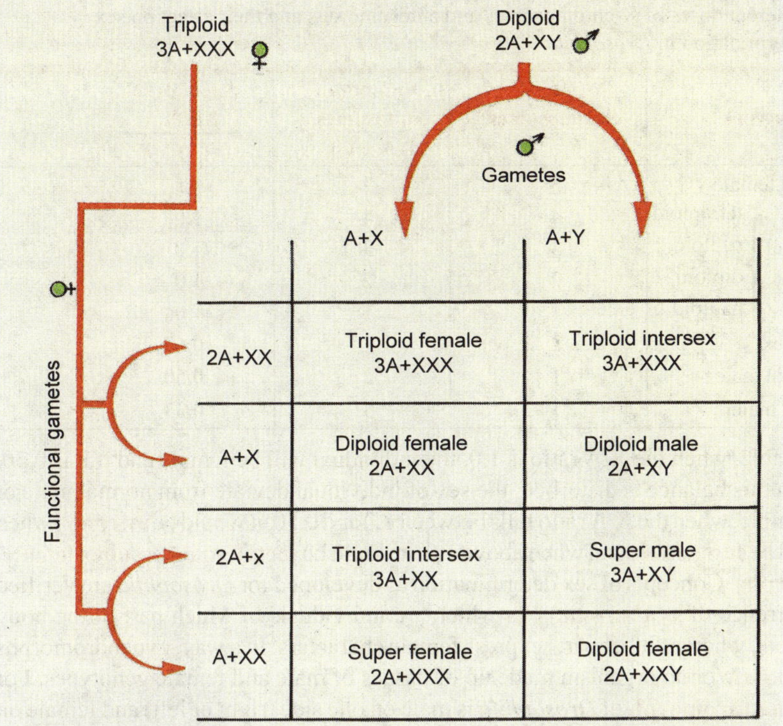

Fig. 13.5. Results obtained from a Bridge's classical cross of a triploid (3A+XXX) female fly and a diploid (2A+XY) male fly (*Drosophila.*)

balance in one direction or another. Such **genic balance** mechanism of determination of sex was first of all studied in *Drosophila* by **C.B. Bridges** in 1921.

Sex determination in Drosophila. In *Drosophila,* the presence of Y chromosome has been found essential for the fertility of male sex but that has nothing to do with the determination of male sex. In this fly, the sex is determined polygenically. The sex determining genes were so distributed that the net effect results in the autosomes determining maleness and the X chromosomes femaleness. The sex of an individual then depends upon the ratio of X chromosomes to autosomes. If each haploid set of autosomes carries factors with a male-determining value equal to one (1), then each X chromosome carries factors with a female determining value of one and half (1½). Let A represent a haploid set of autosomes. In a normal male (AAXY) the male and female determinants are in the ratio of 2: 1½ and, therefore, the genic balance is in the favour of maleness. A normal female (AAXX) has a male and female ratio of 2:3 and, therefore, the balance is in the favour of femaleness (See Fig. 13.4.)

Recent studies indicate that the determination in chickens and probably in birds in general, is similar to that of *Drosophila, i.e.,* it is dependent upon the ratio between the Z chromosomes and the number of autosomal sets of chromosomes. The W chromosome is not a strong female sex determining element (see **Stansfield**, 1986). Same system of genic balance in which the X/A ratio is critical, has been reported for the flowering plant (angiosperm) *Rumex acetosa.*

The genic balance theory of Bridges has been supported by the following two findings :

1. Polyploid flies. Bridges crossed the experimentally produced triploid (3n individual having three whole sets of chromosomes) female *Drosophila* (3A:3X) to diploid males (2A:XY). The results obtained from such a cross are shown in Fig. 13.5. From this cross he obtained normal diploid males, triploid females, intersexes, supermales and superfemales. The occurrence of triploid intersexes from such a cross, clearly established that the autosomes also carry genes for sex determination. These intersexes were sterile individuals and had phenotype in between male and female sexes. The occurrence of such intersexes, super males and super females were explained by him by genic balance mechanism. Different combination of X chromosomes and autosomes and corresponding sex expression in *Drosophila* can be summarized in Table 13-2.

Table 13.2. Different doses of X-chromosomes and autosome sets and their effect on sex determination in *Drosophila.*

Phenotypes		Number of chromosomes	Number of autosomes (A sets)	Sex index= No. X's / No. A sets
Super female		3	2	1.5
Normal female	tetraploid	4	4	1.0
	triploid	3	3	1.0
	diploid	2	2	1.0
	haploid	1	1	1.0
Intersex		2	3	0.67
Normal male		1	2	0.50
Super male		1	3	0.33

As shown in the table, when the X/A ratio is 1.0, the individual will be female and if it is 0.50, it would be male. When this balance is disturbed, the sex of individual deviate from normal male or normal female. For example, when the X/A ratio falls between 1.0 and 0.50, it would be **intersex**; when it is below 0.50, it would be **supermale** and when above 1.0, it would be **metafemale** or **super female**.

2. Gynandromorphs. Concepts of sex determination as developed for *Drosophila* are verified by the occasional occurrence of **gyandromorphs** which are individuals in which part of the body expressed male characters, whereas other parts express female characters. In a way, gynandromorphs represent one kind of **mosaic** or an organism made up of tissues of male and female genotypes. For example, a **bilateral gynandromorph** of *Drosophila* is male on one side (right or left) and female on the other (Fig. 13.6). It results due to the loss of an X-chromosome in a particular cell during development, *i.e.,* when the laggered X chromosome fails to be incorporated in a daughter nucleus and is lost forever (Fig. 13.7). If this event happens during first cleavage (or mitotic division) of the zygote, then one of the two blastomeres will have AAXX chromosomal complement and the other will have

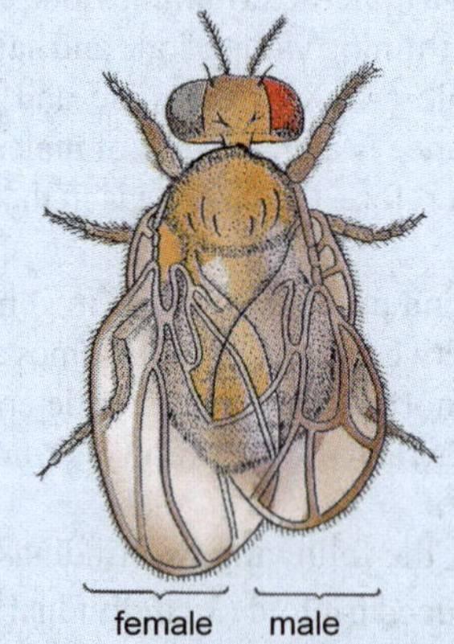

Fig. 13.6. A bilateral gynandromorph *Drosophila.*

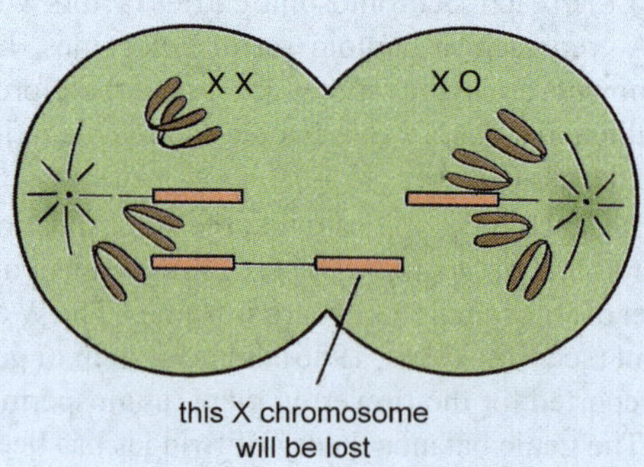

Fig. 13.7. The loss of an X chromosome during mitosis in a 2A+XX cell and formation of two types of cells—XX and XO.

AAXO. The portion of the body developing from AAXX blastomere will be normal female and the portion developing from the AAXO blastomere will be male. The cytological examination of gynandromorphs suggested that Y chromosome does not play any role in the determination of sex in *Drosophila.*

Sex Determination in Man

In man like *Drosophila* XX-XY type sex determining mechanism occurs but here the Y chromosome contains potent male sex-determining genes which can almost completely overcome the feminizing action of the rest of the genotype. The conclusive evidences that Y chromosome is a determiner of fertility and sex of male individual came from certain abnormal conditions (called syndromes) which contained aneuploidic sex-chromosomal abnormalities (for details see Chapter 18. Human Genetics). For instance, **Turner's syndromes** (XO) are sterile female individuals having certain abnormalities such as short stature, congenital malformations, shield chest, pronounced webbing of the neck, short fourth metacarpel, colour blindness, etc. Similarly, **Klinefelter's syndromes** (XX Y) are males, despite the presence of two X chromosomes. A person with extra one X and Y chromosome display true **hermaphroditism** having both ovarian and testicular tissues and variable degrees of intersexual development of the genitalia.

C. Male Haploidy or Haplodiploidy Mechanism

Male haploidy or haplodiploidy or arrhenotokous parthenogenesis is particularly common in the hymenopterous insects such as ants, bees, sawflies and wasps (*e.g., Bracon hebetor*). In these insects, since, fertilized eggs develop into diploid females and unfertilized ones into haploid males; so arrhenotoky is both a form of reproduction and a means of sex determination. Meiosis is normal in females, but crossing over and reduction in chromosome number fail to occur during spermatogenesis in males due to their haploidy (See Fig. 13.8).

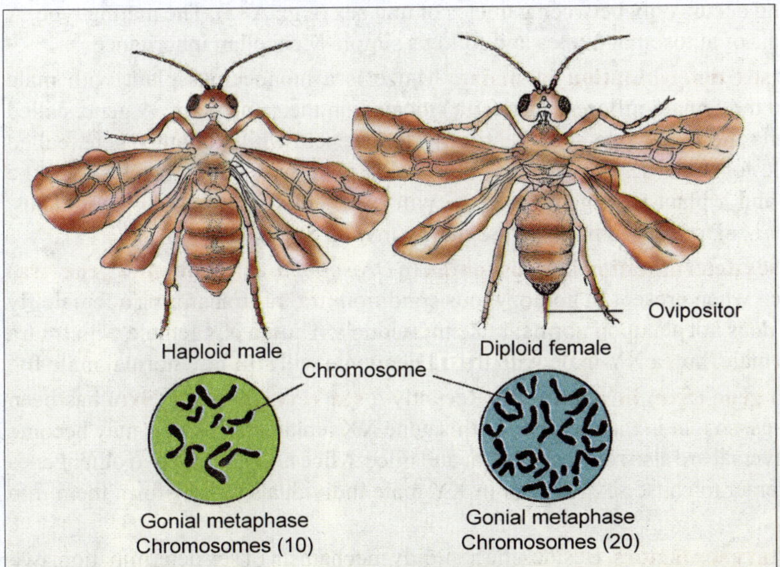

Fig. 13.8. Male and female *Bracon hebetor* and their gonial metaphase chromosomes.

For example, a honeybee queen (whose diploid number is 32) can lay two types of eggs. By controlling the sphincter of her sperm receptacle (which holds sperms previously obtained in matings with males during nupital flight), she produces a fertilized egg (a diploid zygote having 32 chromosomes and developing into a female) or an unfertilized egg (a haploid zygote having 16 chromosomes and developing into a male). The diploid female zygotes can differentiate into either workers (sterile) or queens (fertile) depending on the diet they consume during their development.

D. Single Gene Control of Sex

In certain organisms, for example *Chlamydomonas, Neurospora,* yeast, *Asparagus,* maize, *Drosophila,* etc., individual single genes are found to be responsible for the determination or expression of sex, following cases exemplified the single gene control of sex :

(a) Sex-determination in Asparagus. *Asparagus* is a dioecious plant, however, sometimes the female flowers bear rudimentary anthers and the male flowers bear rudimentary pistils. Thus, sometime it may happen that a rare male flower with poorly develop pistil may set seeds. In one of the experiment when the seeds of such a rare male flower were raised into plants, then, the male and female plants were found to be present in 3 : 1 ratio. When the male plants raised thus were used to pollinate the female flowers on female plants, only two third of them showed segregation indicating that the sex is controlled by a single gene.

Rare male plant
Pp
|
Selfed
↓

PP	Pp	pp
(25%)	(50%)	(25%)
Male (75%)		Female (25%)

Fig. 13.9. Segregation for sex in seed obtained from a rare bisexual flower in *Asparagus* showing monogenic control.

(b) Sex determination in Neurospora. *Neurospora* has two sexes exactly equivalent and designated A and a. Mating occurs only between gametes of unit sex (*e.g.*, A×a). The mating type, A or a is determined by a pair of autosomal alleles and follows simple Mendelian inheritance.

(c) Monogenic sex determination in maize. Maize is a monoecious plant with male inflorescence (**tassel**) and the female inflorescence (**silk**) located on the same plant. A gene, called **tassel seeds (ts)**, converts the tassel into seed bearing inflorescence, while, another gene called **silkless, sk**, is responsible for the absence of silks. Therefore, a plant with genotype sk/sk will be effectively a male plant and a plant with genotype ts/ts will be effectively a female plant. Thus, individual single genes (*viz.,* sk or ts) can impose bisexuality in maize.

(d) Monogenic sex determination in Drosophila. In *Drosophila,* a **transformer gene (tra)** has been recognized which when present in homozygous condition (tra/tra) transforms a female fly into a sterile male, but, it does not act upon normal male individuals. Thus, a XX female with tra/tra genotype will be a sterile male, but, a XY male with tra/tra genotype will still be a normal male fly.

(e) Sex reversal gene (Sxr) in mammals. Recently a **sex reversal gene (Sxr)** has been discovered in human beings, so that in the presence of this gene XX female individuals may become male. Such cases of sex reversal are also reported in goat and mice. Mice also contain two other genes Tdy and Tda-1 which interact to cause sex reversal in XY male individuals to transform them into females.

(f) Complementary sex factors. Besides the haploidy mechanism of sex determination, two hymenopteran insects – *Bracon hebetor* (a tiny parasitic wasp which is also called *Habrobracon*

Parents :	$s^a s^b$ Diploid	X	s^a Haploid		
Gametes :	(s^a) (s^b) Eggs		(s^a) Sperm		
F_1 :	1. When eggs are unfertilized	=	¼ (s^a) Haploid male	+	¼ (s^b) Haploid male
	2. When eggs are fertilized	=	¼ $s^a s^a$ Diploid male	+	¼ $s^a s^b$ Diploid female

Fig. 13.10. Determination of male sex in *Bracon* by heterozygosity.

juglandis) and honey bees–are known to produce males by **homozygosity** at a single gene locus. At least nine sex alleles are known at this locus in *Bracon* and may be represented by s^a, s^b, s^c, s^d.....s^i. All females must be heterozygotes such as $s^a s^b$, $s^a s^c$, $s^d s^f$, etc. If an individual is homozygous for any of these alleles such as $s^a s^a$, $s^c s^c$, etc., it develops into a diploid male (which is usually sterile). Haploid males contain only one of the alleles at this locus, *e.g.,* s^a, s^b, s^c, s^g, etc. (Fig. 13.10).

B. METABOLICALLY CONTROLLED SEX DETERMINING MECHANISM

Certain workers have seen the possibility of sex determination in the phenomenon of metabolism. **Crew** suggested that sex is a physiological equitable division between anabolic and catabolic individuals. **A.F. Shull** and **D.D. Whitney** have shown that by increasing metabolic rate in rotifers the occurrence of male individuals increases than females. Likewise, **Riddle** found that metabolism had some definite role in the determination of sex in pigeons and doves, because, increased rate of metabolism developed the male potency, while decreased rate of metabolism caused femaleness.

C. HORMONALLY CONTROLLED SEX DETERMINING MECHANISM

In many cases it has been observed that sexual differentiation is controlled by hormones. Following classical examples would make clear that how hormones control the sex of an individual :

(a) Sex in Bonellia. An excellent example of environmental determination of sexual phenotype is afforded by *Bonellia viridis,* a marine echiuroid worm studied extensively by **F. Baltzer** (1935). The adult female is about an inch (2.53 cm) long and has fairly complex anatomical organization. Male is of the size of large Protozoa and has rudimentary organs. The males live as parasites in the uterus of the females. All larvae of *Bonellia* are genetically and cytologically similar. If a particular larva settles on the proboscis of an adult female, it becomes a male individual. On the other hand, if a larva develops in isolation (*i.e.,* in water) it develops into a female. Further, if a incompletely developed male is detached from the proboscis of female, it becomes an intersex. Thus, it becomes evident that proboscis of adult female secretes some hormone-like substance which suppresses femaleness and initiating maleness in the larvae which are attached with it.

(b) The Crew's hen. In birds only one gonad of a normal female develops into a functional ovary. The other gonad remains rudimentary. If the functional ovary of a hen is destroyed, the rudimentary gonad develops into a testis (**Crew**, 1923). Thus, the female sex is reversed into male sex due to the phenomenon called **sex reversal**. Such a sex reversed hen can even father chicks which will be expected to show a sex ratio of two females to one male, since in birds the female is the heterogametic sex. The case of sex reversal of hen can be interpreted as follows :

During embryonic development, the XY (ZW) genotype stimulates

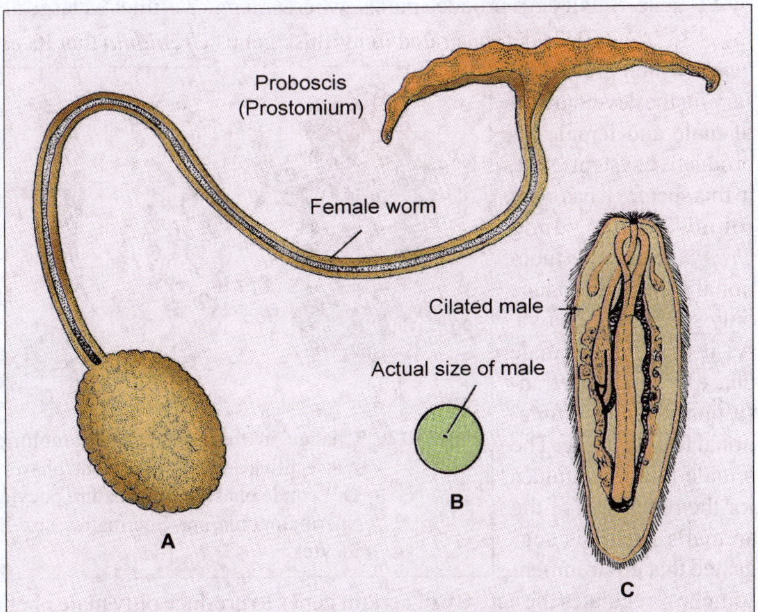

Fig. 13.11. The female and male of the echiuroid *Bonellia viridis.*

the pituitary gland to produce female hormones that cause the gonad of the hen to develop into an ovary. After the development of ovary, the pituitary ceases to produce female hormones, due to inhibition of the pituitary by hormones produced by the ovary, thus, acting as **developmental feed-back system**. The high level of female hormones secreted sequentially by the pituitary and the ovary is sufficient to suppress the action of male hormone producing cells of the body such as the steroid producing cells of the adrenal glands. When the ovary of a hen is removed, the steroid cells of adrenal become active and provoked the development of rudimentary gonad into testis which itself is an endocrine gland. Both endocrine glands (*e.g,* adrenal and testis) produce large amount of male hormones which sufficiently suppress the action of the female hormone cells of pituitary. Such cases of sex reversals have also been observed in amphibians (*e.g.,* in salamander by **E. Witschi**).

(ii) Hormones and freemartin. Another classical example of hormonal control of sex determination has been found in cattles. In cattle, when twin calves of different sexes occur, the female member is usually a sterile intersex called a **freemartin**. The freemartin has external female genitalia but internal sex organs are more or less like those of male. The male twin is usually normal. **F.R. Lillie** (1917) has suggested that the formation of a freemartin was due to a fusion of the foetal membranes of the twin calves, while they were in uterus of the mother. The fusion of the foetal membranes permitted the blood of each twin to circulate in the blood vessel of the other. The male hormones produced by the male twin are presumed to suppress the differentiation of the female internal sex organs of the co-twin.

D. ENVIRONMENTALLY CONTROLLED SEX DETERMINING MECHANISM

In some organisms, the environment determines the sexual phenotype of the individuals. Thus, some environmental factors as size of parent body or of egg, age of parent and temperature are found to determine the sex in following cases :

1. The marine annelid *Ophryotrocha* differentiates into a sperm producing male as a young animal and then changes into an egg-laying female when it gets older. If parts of an older female is amputated, the worm reverts to the male form, indicating that *size rather than age* is the important factor controlling the sex of the individual.

2. In the marine archiannelid *Dinophilus* sex appears to depend solely on size of the egg produced by a female. Small eggs produce males, where as eggs 27 times as large always develop into females.

3. **Coe** (1943) demonstrated in mollusc genus *Crepidula* that its every zygote contains all the genes which are necessary for the development of male and female reproductive systems. But, in this species it has been found that young *Crepidula* remains functional male and produce only sperms but no ova. As it matures the male phase is changed by transitional phase to a functional female phase. The female phase continues for the remainder of the animal's life. It is suggested that environment somehow regulates the activity of certain genes to produce only male phenotype during young stages, while inhibits them during adult stages and simultaneously initiating other genes to produce female phenotype.

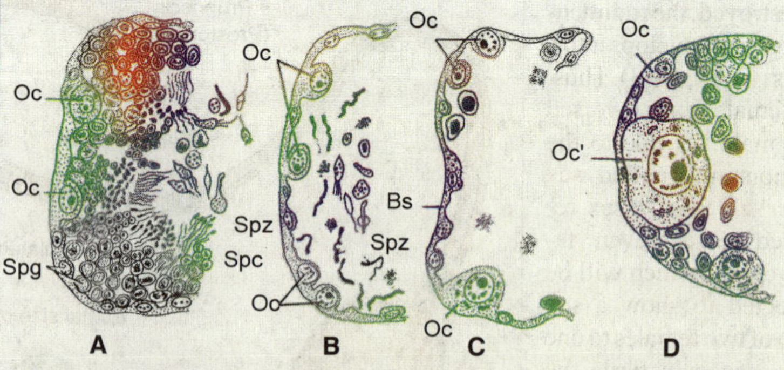

Fig. 13.12. Change in the gonad of the mollusc *Crepidula* experiencing consecutive sexuality. A–Male phase; B, C–Transitional phases; D–Female phase. oc–Immature oocyte, oc′–Mature oocyte, spg–Spermatogonia, spt–Spermatids, spz–Spermatozoa, spc–Spermatocytes.

4. The sex of some reptiles may depend upon the temperature at which the individual develops. For example, in most turtles, only females are produced at high temperature (30–35°C) and only males are produced at low temperatures (23–28°C). The reverse is true in crocodiles, alligators and some lizards, where males are produced at high temperature and females are produced at low temperature.

SEX DETERMINATION IN PLANTS

Most flowering plants are monoecious and, therefore, do no not have sex chromosomes. In fact, the ability of mitotically produced cells with exactly the same genetic dowry to produce tissues with different sexual functions in a perfect flower clearly indicate the bipotentiality of such plant cells. Well known examples of dioecious plants are found to be under the genetic control of a single gene. However, at least one extensively-investigated case of chromosomal sexuality is known in plants, *i.e.,* in the genus *Melandrium* (of pink family Caryophyllaceae). Here the Y chromosome determines a tendency to maleness just as it does in humans.

In *Melandrium* (*Lychnis*), staminate or male plants are XY and pistillate or female plants are XX. The X/A ratio bears no relation to "sex" but, through studies of plants with multiple sets of chromosomes, the X/Y ratio is found to be critical. Thus, X/Y ratios of 0.5, 1.0 and 1.5 are found in plants having only staminate flowers; in plants whose X/Y ratio is 2.0 to 3.0, occasional perfect flowers occur among otherwise all staminate flowers.

SEX DIFFERENTIATION

Although the primary determination of sex is made at fertilization, but the embryo acquires its definite sex characteristics by a more complex mechanism called **sex differentiation**. Thus, sex determination is not a single, once-and-for all decision. But rather it is a gradual *awakening* that spreads from the ovary to testis to the reproductive tract, to the general body tissues and even to the brain itself; finally almost the whole body is overtaken by the process of sexual differentiation. In human beings and other animals sex differentiation occurs in the following steps :

1. Genetic Sex

Normal females ordinarily have two X chromosomes; normal males have one X and one Y. The genes on these sex chromosomes determine femaleness or maleness. Thus, one can speak of females having the genetic sex designation XX and males having the genetic sex designation XY.

Further, since the X-chromosome carries much more genetic information in striking contrast to Y chromosome, one might wonder how it is that the female can carry a double dose of many vital X-linked genes, whereas the male has only a single dose of these X-linked genes. Such inequality in fact cannot be tolerated and so female mammals and male *Drosophila* seem to have developed their own types of dosage compensation mechanisms.

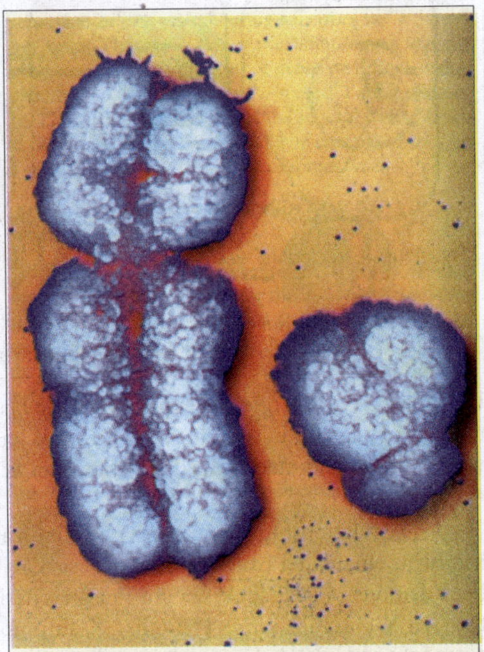
Photomicrograph of human sex chromosomes.

Dosage Compensation of Genes

The term dosage compensation was coined by **Muller** (1932) and this phenomenon was originally discovered in *Drosophila*. Dosage compensation of the genes is done either by **hypoproduction** due to inactivation of one X chromosome in homogametic female sex, as observed in mammals, or by the

hyperproduction due to hyperactivity of X-chromosome in the heterogametic male sex as observed in *Drosophila*.

(i) X-chromosome inactivation in mammals. It has been demonstrated that in homogametic XX female individuals, one X-chromosome gets characteristically condensed and inactivated. Such chromatin material is called **facultative heterochromatin**, since it becomes inactive in certain part of the life cycle and resumes activity before entering the germ line. The phenomenon of inactivation of X chromosome was confirmed by the observation of the Barr body.

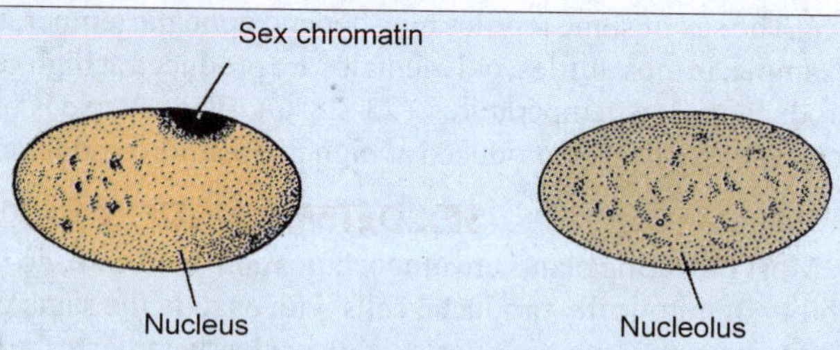

Fig. 13.13. Sex chromatin in the nucleus of female cat (A). Male cat (B) has no sex chromatin in its nucleus.

Barr body. Barr and **Bertram** (1949) reported a deeply stained chromatin body (*i.e.,* a chromocentre) in the nerve cells of female cat which was absent in the male. This chromatin body is called **sex chromatin** or **Barr body** after the name of its discoverer. Such Barr body has also been observed in most of the body cells (*e.g.,* skin, oral epithelium and blood cells) of man and other mammals. Human females have the Barr body in the nuclei of their body cells in higher proportion than males and are, therefore, referred to as **sex chromatin positive**. The human males are called **sex chromatin negative**.

The sex chromatin appears in the interphase nucleus as a small chromocentre, heavily stained with basic dyes. It gives a positive Feulgen reaction and has a relatively constant position in each tissue and species. It can be found in four position: (i) attached to the nucleus as in nerve cells of certain species; (ii) attached to the nuclear membrane as in cells of epidermis or of the oral mucosa; (iii) free in the nucleoplasm as in neurons after electric stimulation, and (iv) as a nuclear expansion. The best known example of nuclear expansion is that of the neutrophil leukocyte of female in which the sex chromatin (Barr body) appears as a small rod called the **drumstick**.

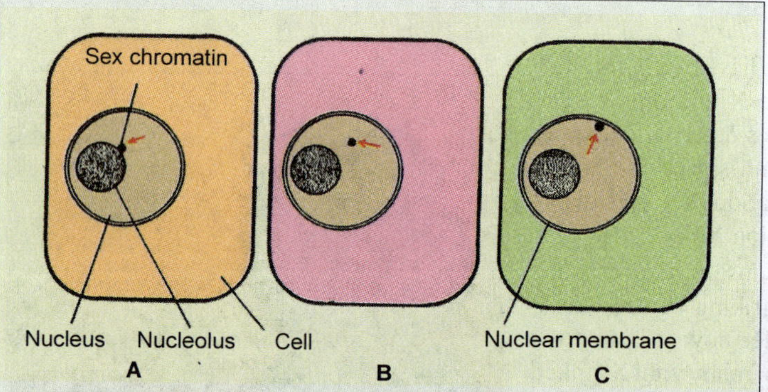

Fig. 13.14. A–Position of sex chromatin near nucleus; B–In nucleoplasm and C–Attached to the nuclear membrane.

The relationship between sex chromatin and sex chromosomes has been elucidated by **Ohno**, **Kaplan** and **Kinosita**, 1959. It has been demonstrated that the sex chromatin is derived only from one of the two X-chromosomes. The other X chromosome behaves like a autosome and is not heteropycnotic at interphase. Later, **Lyon** (1972)

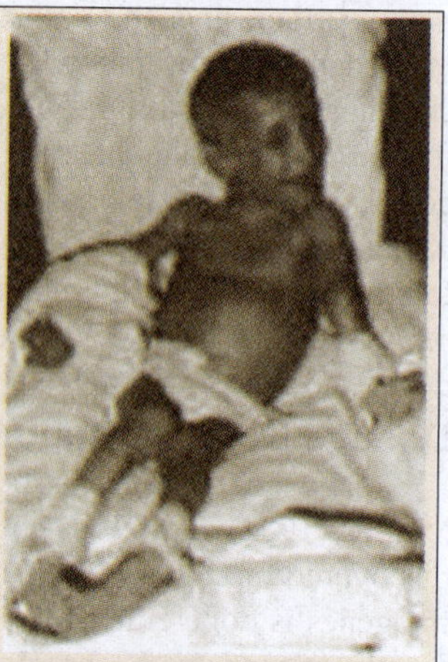

This child is suffering from Lesh – Nyhan syndrome

confirmed the existence of Barr body in normal females, metafemales or super females (XXX) and in Klinefelter males (XXY).

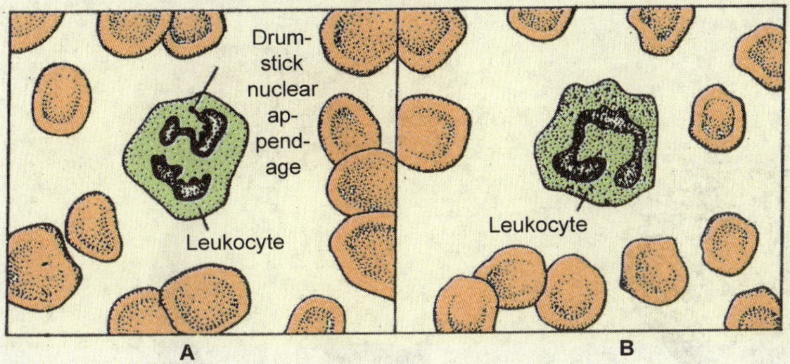

Fig. 13.15. A–Leukocyte of human female with drumstick-like sex chromatin; B–Leukocyte of human male without any drumstick.

Lyon's hypothesis. It was demonstrated by **M.F. Lyon** of U.K. that whenever the number of X chromosomes was two or more than two, the number of Barr bodies was one less than the number of X chromosomes (nX-1; *i.e.,* one Barr body in XX females and XXY males; two Barr bodies in XXXY males and XXX metafemales). Thus, in normal female only one active X chromosome is present.

Mary F. Lyon (1925—)

Which of the two X chromosomes remains active in female individuals, is determined at the early stages of development. It was observed by **Lyon** in 1961 that each of the paternal (P) and maternal (M) X chromosomes has a chance to become inactive (*i.e.,* Barr body). In other words, the inactivation of X chromosome is a random phenomenon. This fact has been demonstrated in human diseases linked to X chromosome. The **Lesch-Nyhan syndrome**, in which a deficiency of one enzyme of the purine metabolism (*i.e.,* hypoxanthineguanine phosphoribosyl transferase) produces mental retardation and increased uric acid levels results, from a recessive mutation in the X chromosome. This is shown as follows : if fibroblasts of these patients are cultured *in vitro,* two types of cell clones are obtained. Half the clones contain the enzyme, whereas the other half (in which the X carrying the normal gene is condensed) lack the enzyme.

In the human embryo X chromosome inactivation starts in the late blastocyst about the 16th day of life. Once the inactivation is established, it is irreversibly maintained in somâtic cells, however, in germ cell line reactivation occurs at a specific stage of germ cell development (*i.e.,* entry into meiotic prophase; **Martin**, 1982).

The good illustration of X chromosome inactivation is seen with tortoise-shell cats, where the coat is a mosaic patchwork of black and yellow hair. Black hair is produced by the dominant allele B, and yellow by its recessive allele b. This gene is

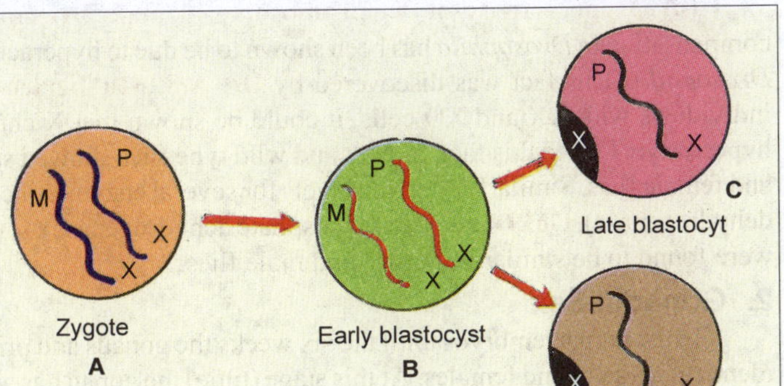

Fig. 13.16. Diagram of the evolution of XX chromosome. A–In the zygote both paternal (P) and maternal (M) X-chromosomes are euchromatic; B–In the early blastocyst the same is true as in A; C–In late blastocyst 50% of the cells have a maternal heterochromatic X chromosome and the other 50% have the paternal heterochromatic X chromosome. In diagram C formation of sex chromatin by one of the X chromosome has been shown.

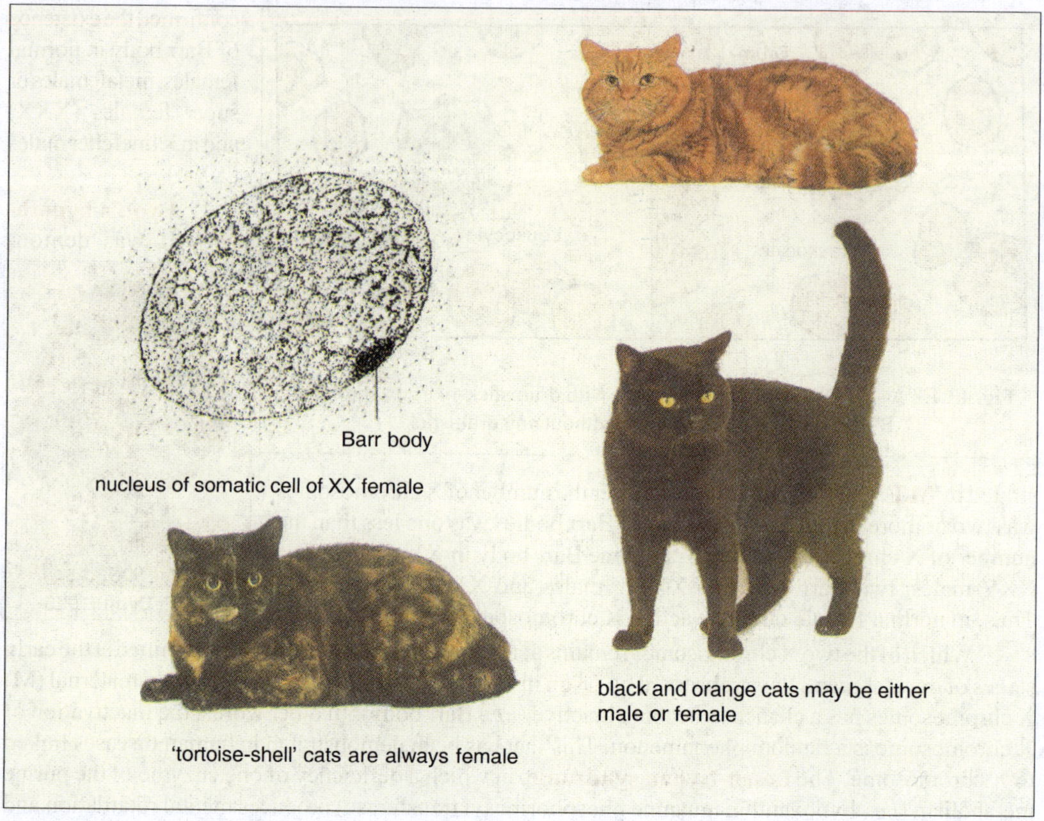

X-linked, so if one X chromosome contains the dominant allele B and other X chromosome the recessive allele b, random inactivation will allow both coat colours to be expressed. Male tortoise-shell cats are understandably rare, since it has only one X chromosome.

(ii) Hyperactivity of X chromosome in male Drosophila. The phenomenon of dosage compensation in *Drosophila* has been shown to be due to hyperactivity of one X chromosome in male *Drosophila.* This fact was discovered by **Dr. A.S. Mukherjee** of Calcutta University. In mosaic individuals with XX and XO cells, it could be shown that X chromosome in XO cells was always hyperactive. Due to this fact, mutant and wild type flies showed same intensity of eye colour in male and female flies. Similarly, enzyme levels for several enzymes including **6PGD** (6 phosphogluconate dehydrogenase), **G6PD** (glucose-6 phosphate dehydrogenase), **tryptophan pyrrolase** and **fumarase** were found to be similar in female and male flies.

2. Gonadal Sex

In the human embryos until the six weeks the gonads and primordia of the urinogenital tract are identical in males and females. At this stage (time) the gonad has already been invaded by the primary XX or XY cells. At this point, a gene or set of genes, called **testis-determining factor** or **TDF**, present in the Y chromosome causes the undifferentiated gonad to differentiate into a testis and the absence of this gene allows the gonad to become an ovary. The development into a testis starts as soon as the gonocytes (*i.e.,* primordial germ cells) from the yolk sac have finished their migration into the gonadal ridge. Gonocytes of the male (XY) migrate deeper into the gonadal blastema forming the medulla and female gonocytes (XX) remain at the periphery, forming a thick cortical layer. Hence, the XX genetic sex is ordinarily associated with **ovarian gonadal sex**, and XY is associated with **testicular gonadal sex**.

Further, identical TDF region of Y chromosome has been identified in gorillas, monkeys, dogs, cattle, horses and goats, further suggesting that this is a common sex-determining mechanism in mammals. The current theory is that the TDF gene is the **master switch** that when turned on, activate on entire series of genes whose function is sex differentiatiation (see **Burns** and **Bottino**, 1989).

H-Y antigen. It is known that the Y chromosome in mammals is a strong male inducer and directs the organogenesis of the testis. The short arm of the Y chromosome of human beings, near the centromere contains a gene responsible for the production of a protein, called **H-Y antigen**. This antigen occurs in the plasma membrane of all male tissues and possibly tends to regulate H-Y receptor cells. If these are defective, the undifferentiated gonadal cells fail to respond to H-Y antigen and testes do not form, leading the individual to mature as a female.

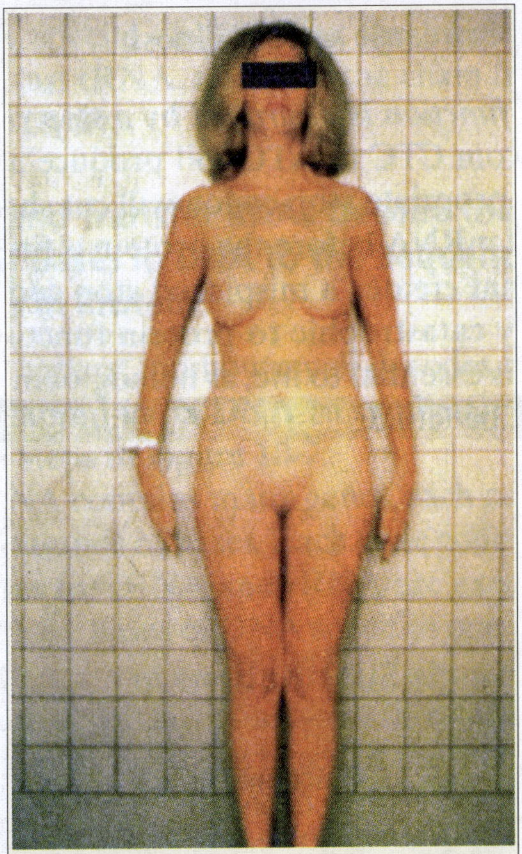

This individual has the male sex chromosome. As a result, she has testes in her abdominal cavity. Female appearance is the result of a mutation in a single gene for the receptor protein for androgens, so body cells don't respond to male hormones.

Cortico-medullary antagonism. Experiments in which male and female amphibians larvae (*e.g.,* toad, frog, salamander) have been surgically united to one another so that their circulation fused, gave rise to the concept of "**cortico-medullary antagonism**". If the medulla was stimulated to develop, it was thought to produce a substance that could act locally or via the circulation to inhibit the ingrowth of ovarian cortex and vice versa. For example, in case of **freemartin** which occurs in cows, sheep, pigs and goats, a female foetus that is twin to a male and shares a conjoined placental circulation undergoes a partial sex reversal of her ovaries, which interferes with the subsequent sexual differentiation of the reproductive tract.

3. Hormonal or genital sex. The embryonic gonads produce hormones that, in turn, determine the morphology of the external genitalia and the genital ducts. XX embryos normally develop ovaries, female external genitalia and Mullerian ducts. XY embryos, on the other hand, ordinarily develop testes, male external genitalia, and Wolffian ducts are suppressed; in XY embryos the Mullerian duct remain under developed. Thus, there is a distinction between male and female genital sex.

For example, the human gonad differentiates into a definite testicle at the seventh week, whereas the female gonad begins to develop into an ovary between the eight and nine weeks of development. At the time of differentiation an important epigenetic factor is the production of androgens (testosterone) by somatic cells in the embryonic male gonad. In mammals, administration of testosterone hormone to the mother produces in the foetus a shift in the differentiation of XX genitalia to a male type, producing **masculine pseudohermaphroditism**. This hormone locally accelerates the development of the testis, whereas in the female the absence of the hormone permits the slower development characteristic of the ovary (**Jost**, 1970; **Ohno**, 1976).

4. Somatic sex. Production of gonadal hormone continues to increase until **secondary sex characters** appear at puberty. These include amount and distribution of hair (*e.g.,* facial, body, axillary, pubic), pelvis dimensions, general body proportions, subcutaneous fats over hips and thighs and breast development in females, as well as, increased larynx size and deepening of the voice in the male.

5. Socio-psychological sex. In most individuals, genetic sex, gonadal sex, genital sex and somatic sex are consistent; XX persons, for example, develop ovaries, female genitalia and female secondary sexual characteristics. Ordinarily these persons are raised as females and adopt the faminine gender role under whatever culture pattern has been established in the society of which they are members. A similar consistency from genetic sex to socio-psychological sex is seen for XY individuals. Intersexuality occurs when the chromosomal contribution is inconsistent with the gonadal or other secondary sexual characters.

REVISION QUESTIONS AND PROBLEMS

1. What is sex determination ? Describe various examples of sex chromosomal mechanisms of sex determination.
2. Enlist various methods of sex determination in animals. Describe the genic balance theory of sex determination.
3. What is dosage compensation ? How is this achieved ?
4. What is sex differentiation ? Describe this phenomenon by considering the example of human beings.
5. Distinguish between :

   (a) Primary and secondary sex characters; (b) Sex determination and sex differeniation; (c) Hermaphrodites and monoecious individuals; (d) Heterogametic and homogametic sex; (e) Autosome and sex chromosome; (f) Freemartin and pseudohermaphrodite.
6. a. Cite evidence for the statement that the gonads of vertebrates are potentially dual in function.

   b. Discuss the evolutionary advantages of the haplodiploid scheme of sex determination.
7. Write short notes on the following :

   (i) Sex reversal; (ii) Barr body ; (iii) Lyon's hypothesis; (iv) Gyandromorphs; (v) Haplodiploidy ; (vi) Crew's hen ; (vii) Sex determination in plants ; (viii) *Bonellia*; (ix) Cortico-medullary antagonism.
8. Sex determination in the grasshopper is by the XO method. The somatic cells of a grasshopper are analyzed and found to contain 23 chromosomes. (a) What sex is this individual (b) Determine the frequency with which different types of gametes (number of autosomes and sex chromosome) can be formed in this individual ? (c) What is the diploid number of the opposite sex ?
9. If the diploid number of the honey bee is 16, (a) how many chromosomes will be found in the somatic cells of the drone (male) ? (b) How many bivalents will be seen during the process of gametogenesis in the male ? (c) How many bivalents will be seen during the process of gametogenesis in the female?
10. In *Drosophila,* the ratio between the number of X chromosomes and the number of sets of autosomes (A) is called the "sex index". Diploid females have a sex index (ratio X/A) = 2/2 = 1.0. Diploid males have a sex ratio of ½ = 0.5. Sex index values between 0.5 and 1.0 give rise to intersexes. Values over 1.0 or under 0.5 produce weak and inviable flies called super females (metafemales) and supermales (metamales), respectively. Calculate the sex index and the sex phenotype in the following individuals : (a) AAX ; (b) AA XX Y; (c) AAAXX; (d) AAXX; (e) AAXXX; (f) AAAXXX; (g) AAY.
11. Suppose that a female undergoes sex reversal to become a functional male and is then mated to a normal female. Determine the expected F_1 sex ratios from such matings in species with (a) ZW method of sex determination; (b) XY method of sex determination.
12. A completely pistillate inflorescence (female flower) is produced in the castor bean by the recessive genotype nn. Plants of genotype NN and Nn have mixed pistillate and staminate flowers in the

inflorescence. Determine the types of flowers produced in the progeny from the following crosses : (a) NN female x Nn male ; (b) Nn female x Nn male ; (c) nn female x Nn male.

13. In *Drosophila,* what function of the progeny of the cross Tra Tra XX × tra tra XY is "transformed"?
14. For humans, give the genetic sex of (a) most true hermaphrodites, (b) masculizing malepsuedohermaphrodites, (c) feminizing male pseudohermap hrodites; (d) female pseudohermaphrodites.

ANSWERS TO PROBLEMS

8. (a) Male; (b) ½ (11A + 1X) : ½ 11A ; (c) 24.
9. (a) 8; (b) None; meiosis cannot occur in haploid males ; (c) 8.
10. (a) 0.5 male; (b) 1.0 female; (c) 0.67 intersex ; (d) 1.0 female ; (e) 1.5 super female; (f) 1.0 female (triploid); (e) Lethal.
11. (a) 2 females : 1 male; (b) All females.
12. (a) All mixed ; (b) ¾ mixed : ¼ pistillate; (c) ½ mixed : ½ pistillate.
13. ¼
14. (a) XX; (b) XY; (c) XY; (d) XX.

CHAPTER

14

Chromosomal Mutation-I

(Cytogenetics : Changes in Structure of Chromosomes)

10 month old, Onya – Birri, the only albino koala in captivity with his mother. His condition resulted because of a mutation.

Genetics makes extensive use of deviations from the norms, and the study of chromosomes is no exception. The chromosomes of each species has a characteristic morphology (structure) and number. But, sometimes due to certain accidents or irregularities at the time of cell division, crossing over or fertilization some alterations in the morphology and number of chromosomes take place. The slightest variation in the organisation of chromosomes is manifested phenotypically and is of great genetical interest. The changes in the genome involving chromosome parts, whole chromosomes, or whole chromosome sets are called **chromosome aberrations** or **chromosome mutations**. Chromosome mutations have proved to be of great significance in applied biology—agriculture (including horticulture), animal husbandry and medicine.

Chromosome mutations are inherited once they occur and are of the following types :

A. Structural changes in chromosomes :

1. Changes in number of genes
   (a) Loss : deletion
   (b) Addition : Duplication
2. Changes in gene arrangement :
   (a) Rotation of a group of genes 180^0 within one chromosome : inversion

(b) Exchange of parts between chromosomes of different pairs : translocation.

B. Changes in number of chromosomes :

1. Loss, or gain, of a part of the chromosome set (aneuploidy)
2. Loss, or gain, of whole chromosome set (euploidy)
   (a) Loss of an entire set of chromosomes (haploidy)
   (b) Addition of one or more sets of chromosomes (polyploidy).

Both types of changes (structural and numerical) in chromosomes can be detected not only with a microscope (*i.e.*, cytologically) but also by standard genetic analysis. This gave birth to a hybrid science, called **cytogenetics** which attempts to correlate cellular events, especially those of chromosomes, with genetic phenomena.

STRUCTURAL CHANGES IN CHROMOSOMES

For better understanding of the abnormalities of chromosome structure, let us consider two important features of chromosome behaviour : (1) During prophase I of meiosis, homologous regions of chromosomes show a great affinity for pairing and they often go through considerable contortions in order to pair. This property results in many curious structures observed in cells containing one normal

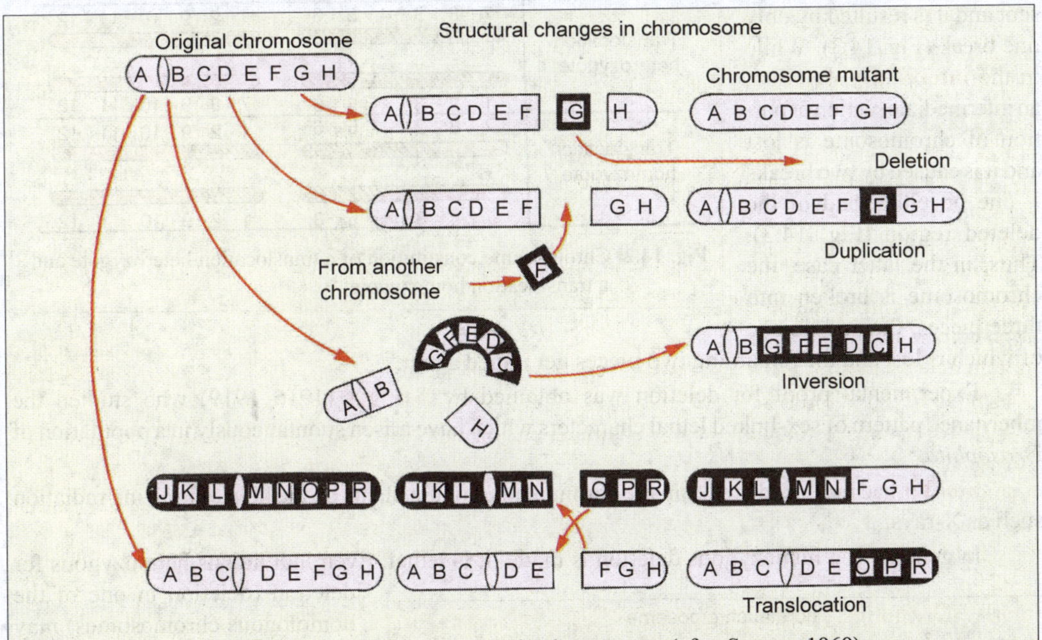

Fig. 14.1. Structural changes in chromosome (after Savage, 1969).

chromosome set plus an aberrant set. (2) structural changes usually involve chromosome breakage; the broken chromosome ends are highly "reactive" or "sticky", showing strong tendency to join with broken ends.

Types of Structural Changes in Chromosome

Structural changes in chromosome may be of the following types (Fig. 14.1) : 1. **deficiency** or **deletion** which involves loss of a broken part of a chromosome; 2. **duplication** involves addition of a part of chromosome (*i.e.*, broken segment becomes attached to a homolog which, thus, bears one block of genes in duplicate); 3. **inversion** in which broken segment reattached to original chromosome in reverse order, and 4. **translocation** in which the broken segment becomes attached to a non-homologous chromosome resulting in new linkage relations.

Further, structural abnormalities can occur in both homologous chromosomes of a pair or in only one of them. When both homologous chromosomes are involved, these are called **structural homozygotes**, *e.g.*, deletion homozygote, duplication homozygote, etc. When only one homologous chromosome is involved, it is called **structural heterozygote**. In Fig. 14.2 have been shown translocation heterozygote and translocation homozygote.

1. Deletion (or Deficiency)

The simplest result of breakage is the loss of a part of a chromosome. Portions of chromosomes without a centromere (called **acentric fragments**) lag in anaphase movement and are lost from reorganizing nuclei or digested by nucleases. Such loss of a portion of a chromosome (and of some genes) is called **deletion**. The chromosomes with deletions can never revert to a normal condition. If gametes arise from the cells having a deleted chromosome, this deletion is transmitted to the next generation. Further, a deletion can be terminal or intercalary (interstitial). In **terminal deletion** a terminal section of a chromosome is absent and it is resulted by only one break (Fig. 14.3). While in the **intercalary deletion**, an intermediate section or portion of chromosome is lost and it is caused by two breaks —one on either end of the deleted region (Fig. 14.3). Thus, in the latter case, the chromosome is broken into three pieces, the middle one of which is lost and the remaining two pieces get joined again.

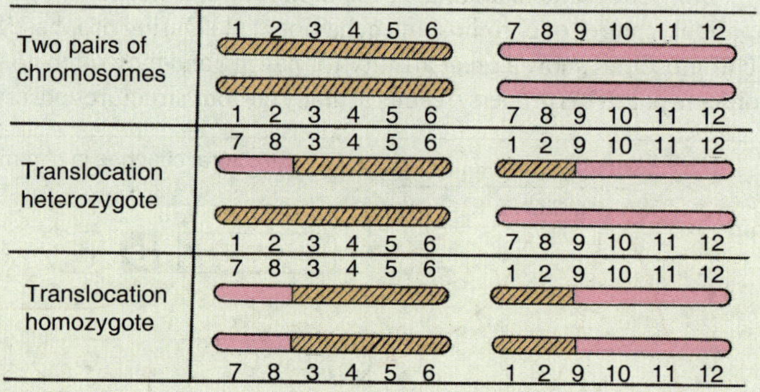

Fig. 14.2. Chromosome constitution of a translocation heterozygote and a translocation homozygote.

Experimental proof for deletion was obtained by **Bridges** (1916–1919) who studied the inheritance pattern of sex-linked lethal characters which have arisen spontaneously in a population of *Drosophila*.

Stadler had produced breaks in the chromosomes of the plants by the help of ionizing radiation such as X-rays.

In general, if a homozygous deletion is made, it is lethal. Even individuals heterozygous for deletion (deletion in one of the homologous chromosomes) may not survive. However, smaller deletion in heterozygous condition can be tolerated by the organisms. If meiotic chromosomes in such heterozygotes are examined, the region of deletion can be detected by the failure of the corresponding segment on the normal chromosome to pair properly; so a "**deletion-loop**" results. Deletion loops are also detected in polytene chromosomes of *Drosophila*,

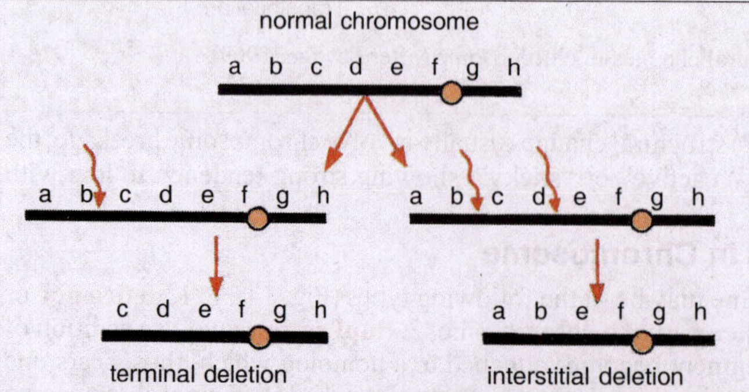

Fig. 14.3. Production of terminal and interstitial deletion. Chromosomes can be broken when struck by ionizing radiation (wavy arrows).

where the homologs exist in permanent state of pairing (Fig. 14.4). The cytological studies of pairing between normal and deleted chromosomes have helped a lot in finding out the relative positions of genes in chromosomes.

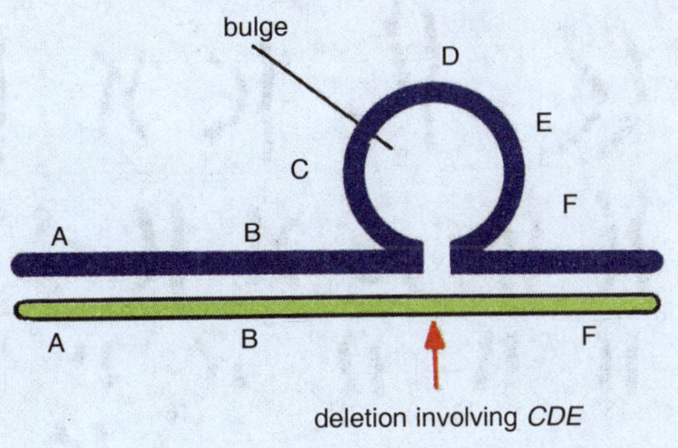

Fig. 14.4. Formation of a deletion loop during synapsis in a deletion heterozygote (after Stansfield, 1986).

Genetical effects of deletion. Deletion of some chromosome regions produce their own unique phenotypes. A good example of this is a dominant **notch-wing** mutation in *Drosophila.* In fact, this is a small deletion and acts as a recessive lethal in this regard. Further, in the presence of a deletion, a recessive allele of the normal homologous chromosome will behave like a dominant allele, *i.e.,* it will be phenotypically expressed, this phenomenon is called **pseudodominance** (Fig. 14.5).

The phenomenon of pseudodominance exhibited by deficiency heterozygotes has been utilized for the location of genes on specific chromosomes and in preparing **cytological maps** in *Drosophila,* maize, bacteriophage and other organisms. Such cytological maps are often used to verify the **genetic maps** (based on linkage analysis) of these organisms.

Examples of pseudodominance (deletion). 1. **Gates** (1921) demonstrated the effects of pseudodominance. Certain mice carrying the recessive allele (v) in homozygous condition move about erratically until exhausted. Such mice are called '**waltzing mice**'. When homozygous waltzing mice (vv) are mated with normal mice carrying the dominant allele (VV), all the affecting (Vv) are normal. However, in a cross between homozygous normal female and a waltzer male, one of the seven offspring was a waltzer. A deletion had resulted in the elimination of the dominant allele (V); so that the recessive allele (v) for waltzing had expressed itself.

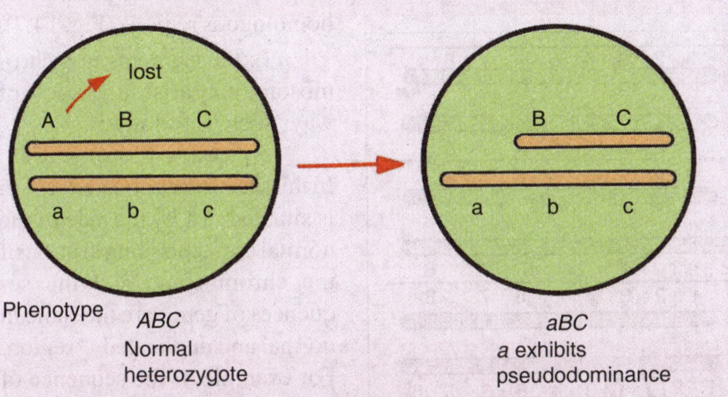

Fig. 14.5. Pseudodominance. A deficiency in the segment of chromosome bearing the dominant gene A allows the recessive allele a to become phenotypically expressed (after Stansfield, 1986).

2. Human babies missing a portion of the short arm of chromosome 5 (autosome) have a distinctive cat-like cry; hence, the French name "**cri du chat**" (cry of the cat) **syndrome** (first described by **Lejeune** *et al.,* 1963). They are also mentally retarded (IQ below 20), have malformation in the larynx, moon faces, saddle noses, small mandibles (micrognathia), malformed low-set ears and microcephally (small head).

The standard designation for the short arm of a non-metacentric chromosome is *p,* that for the longer arm is *q.* A deletion is indicated by a superscript minus sign, and added segments are indicated by supercript plus sign. Hence, the karyotype of a cri du chat patient is $5p^{-}$.

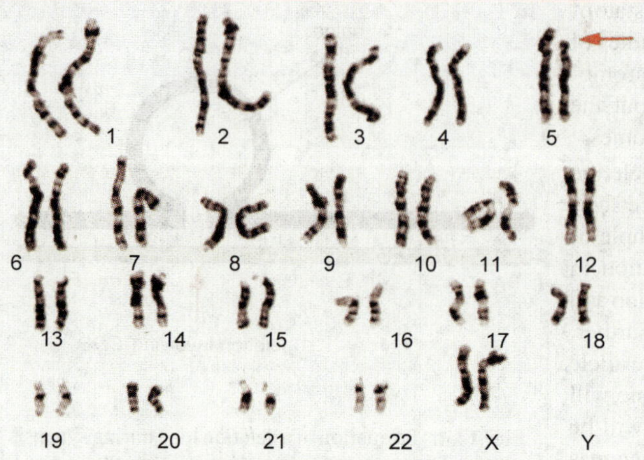

Karyotype of a child exhibiting cri-du-chat syndrome.

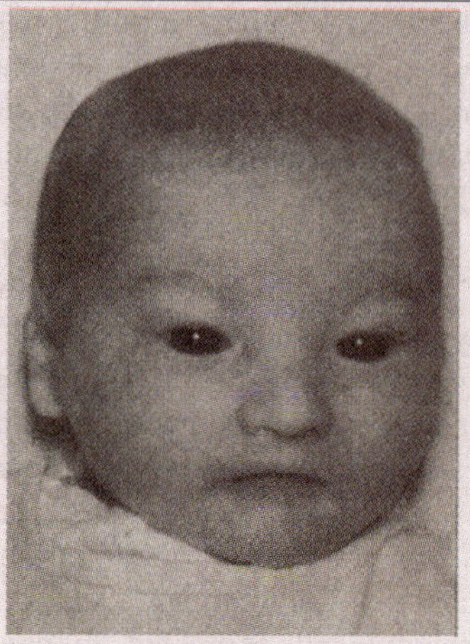
A child suffering from cri-du-chat syndrome.

3. Deletion of part of the short arm of one X chromosome produces a typical Turner syndrome (see chapter 18 of Human Genetics).

2. Duplication

The presence of a part of a chromosome in excess of the normal complement is known as **duplication**. Thus, due to duplication some genes are present in a cell in more than two doses. If duplication is present only on one of two homologous chromosomes, at meiosis the chromosome bearing the duplicated segment forms a **loop** to maximize the juxtaposition (during pairing) of homologous regions (Fig. 14.7).

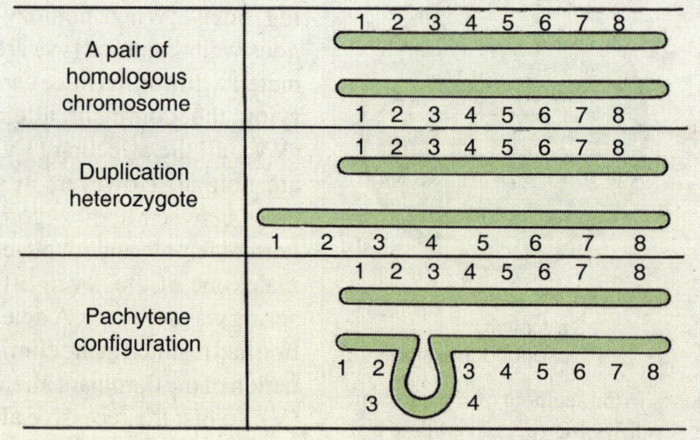

Fig. 14.6. Loop formation during chromosome pairing in a duplication heterozygote.

Extra segments in a chromosome may arise in a variety of ways such as follows :

1. Tandem duplication. In this case the duplicated region is situated just by the side of the normal corresponding section of the chromosome and the sequences of genes are the same in normal and duplicated region. For example, if the sequence of genes in a chromosome is ABC. DEFGH (The full stop depicts the centromere) and if the chromosomal segment containing the genes DEF is duplicated, the sequence of genes in tandem duplication will be ABC. DEF DEFGH.

2. Reverse tandem duplication. Here, the sequence of genes in the duplicated region of a chromosome is just the reverse of a normal sequence. In the above mentioned example, therefore, the sequence of genes due to reverse tandem duplication will be ABC. DEF...FED.GH.

3. Displaced duplication. In this case the duplicated region is not situated adjacent to the normal section. Depending on whether the duplicated portion is on the same side of the centromere as the original section or on the other side, the displaced duplication can be termed either **homobranchial** or **heterobranchial**.

Example. Homobranchial duplication = ABC. DEFG ..DEFH

Heterobranchial duplication = A ..DEFB.. C. DEFGH

4. Transposed duplication. Here, the duplicated portion of chromosome becomes attached to a non-homologous chromosome. For example, if ABC.DEFGH and LMNOPQ. RST represent the gene sequences of two nonhomologous chromosomes, a transposed duplication will result into chromosomes with gene sequence ABC.GH and LMNDEF. OPQ. RST. Such a transposed duplication may be either **interstitial** (*e.g.,* LMN ..DEF OPQ. RST) or **terminal** (*i.e.,* LMN OPQ. RST DEF).

5. Extra-chromosomal duplication. In the presence of centromere the duplicated part of a chromosome act as independent chromosome.

Genetical effects of duplication. Due to duplication, there occur unequal crossing over which results in deletion and reduplication which produce distinct phenotypes as shown by the following examples :

1. Bar eye in Drosophila. The Bar phenotype of *Drosophila* is characterized by narrower, oblong, bar-shaped eye with few facets. It is determined by a X-linked recessive allele B. The classical studies of **Bridges** (1936) showed that the bar trait of *Drosophila* is associated with the duplication of a segment of the X-chromosome, called **section 16A**, as observed in salivary gland chromosomes (Fig. 14.7). Each added section 16A intensifies the bar phenotype (*i.e.,* duplication behaves genetically as a dominant factor). However, the narrowing effect is greater if the duplicated segments are on the same chromosome (called **position effect**) (Table 14-1). Thus, cis and trans arrangements of the same number of 16A segments give different phenotypes (compare heterozygous ultrabar and homozygous bar eyes).

Some of the other well known duplications of *Drosophila* lead to following phenotypic effects : (1) a reverse repeat in chromosome 4 causes *eyeless* dominant (Ey); (2) a tandem duplication in chromosome 3 causes *confluens* (Co) resulting in thickened veins, and (3) another duplication causes *hairy wing* (Hw).

2. In humans, unequal crossing over between homologous chromosomes bearing σ (sigma) and β (beta) genes for σ and β subunits of adult haemoglobin (HbA), results in deletions and duplications of these genes. Deletions result in **Lepore** and **Kenya** variants of adult haemoglobin (HbA), both causing anaemia (*i.e.,* one type of thalassemia), while duplication result in Anti-Lepore and Anti-Kenya variants of haemoglobin A (see **Suzuki** *et al.*, 1986).

Genetic redundancy, of which duplication is one type, may protect the organism from the effects of a deleterious recessive gene or from an otherwise lethal deletion.

3. Inversion

Inversion involves a rotation of a part of a chromosome or a set of genes by 180^0 on its own axis. It essentially involves occurrence of *breakage* and *reunion.* The net result of inversion is neither a gain nor a loss in the genetic material but simply a rearrangement of the gene sequence. An inversion can occur in the following way : suppose that the normal order of segments within a chromosome is 1-2-3-4-5-6 ; breaks occur in regions 2-3 and 5-6 and broken piece is reinserted in reverse order. This results in an inverted chromosome having segments 1-2-5-4-3-6 (Fig. 14.8).

An **inversion heterozygote** has one chromosome in the inverted order and its homologue in the normal order. The location of the inverted segment can be detected cytologically in the meiotic nuclei of such heterozygotes by the presence of an **inversion loop** in the paired homologs. The location of the centromere relative to inverted segment determines the genetic behaviour of the chromosomes. If the centromere is not included in the inversion it is called **paracentric inversion** and when inversion includes the centromere it is called **pericentric inversion** (Fig. 14.9). Homologous chromosomes, with identical inversions in each member, pair and undergo normal distribution in meiosis. However, crossing over in inversion heterozygotes produce deletions, duplications and other curious configurations.

A. Crossing over in pericentric inversion. Crossing over in a heterozygous pericentric inversion result in deletions and duplications and also produces rod-shaped (acrocentric) chromosomes. The first meiotic anaphase figures appear normal, but the two chromatids of each chromosome

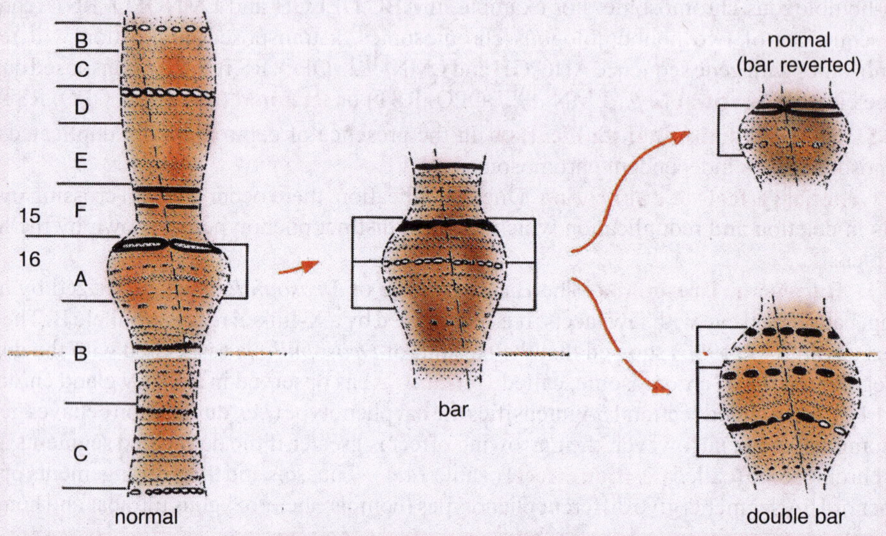

Fig. 14.7. Production of double-bar (or ultrabar) and bar-revertant (normal) chromosomes by asymmetric pairing and recombination in duplication homozygote. A–Diagrammatic representation of formation of unequal cross-over; B–Cytological representation (after Suzuki, *et al.,* 1986).

Table 14.1. Comparison of genotypes and phenotypes for bar eye in *Drosophila* females showing position effects of 16A segment (Source : **Burns** and **Bottino,** 1989).

X chromosomes	Phenotype		Mean number of facets
1. 16A/16A	Normal	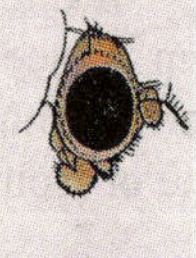	779
2. 16A, 16A/16A	Heterozygous bar eye	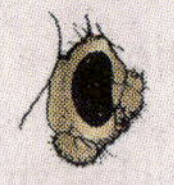	358
3. 16A, 16A/16A,16A	Homozygous bar eye		68
4. 16A, 16A, 16A/16A	Heterozygous ultrabar (=double bar)	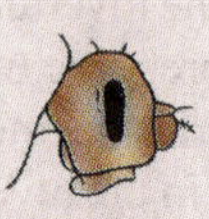	45
5. 16A, 16A, 16A/16A, 16A, 16A	Homozygous ultrabar	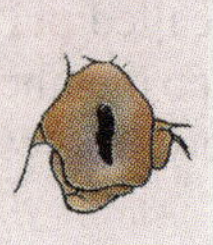	25

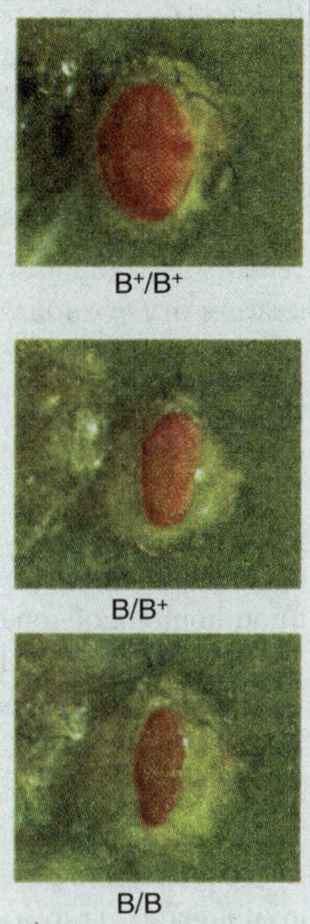

Photographs show two bar eye phenotypes and the wild type (B+/B+) in a *Drosophila.*

usually have arms of unequal length depending upon where the crossing over occurred (Fig. 14.10). Half of the meiotic products (gametes/pollen grains) are non-functional and inviable due to the presence of duplications and deletions in them. The other half of the gametes are functional : one-quarter have the normal chromosome order, one-quarter have the inverted arrangement.

B. Crossing over in paracentric inversion. A crossing over in the inverted region of a heterozygous paracentric inversion produces a **dicentric chromosome** (possessing two centromeres) which forms a bridge from one pole to the other during first anaphase. The bridge will rupture somewhere along its length and resulting fragments will contain duplication and/or deletion. In this case, an accentric fragment (without a centromere) is also formed and since it usually fails to move to either pole, it is not included in any meiotic products (gametes). Here also half of the meiotic products are non-functional, one-quarter are functional with a normal chromosome, and one-quarter are functional with an inverted chromosome (Fig. 14.11). Thus, heterozygotes for paracentric inversions are highly sterile and produce only parent-like progeny.

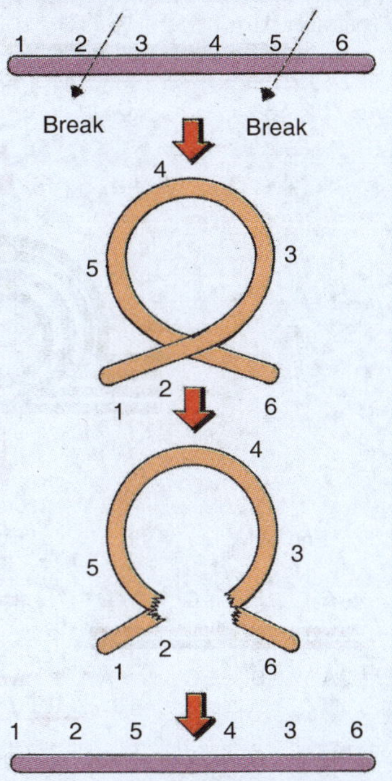

Fig. 14.8. Origin of an inversion in a chromosome.

Comparison of banded karyotype of humans and apes reveals numerous paracentric as well as pericentric inversions in humans as compared to apes. In female maize plants dicentric bridges are found to form and they undergo the bridge-breakage-fusion-cycle.

Further, inversion heterozygotes often have mechanical pairing problems in the area of the inversion; this also reduces crossing over and recombinant frequency in the vicinity (see **Suzuki** *et al.,* 1986). Due to this observation, inversions are called **cross-over-suppressors**. However, this reduction in crossing over is not actual reduction in cytological crossing over, but it is the result of lack of recovery of the products of single crossing over (see **Burns** and **Bottino**, 1989).

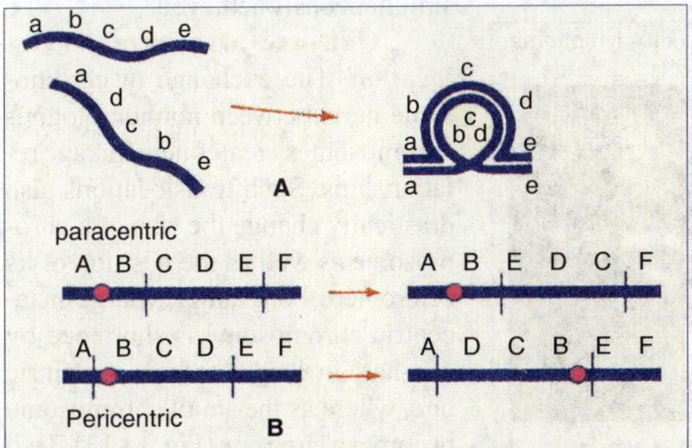

Fig. 14.9. A–An inversion loop in paired homologs of an inversion heterozygote; B–The location of the centromere relative to the inverted segment.

Advantages of inversions. Fertility of inversion homozygotes and sterility of inversion heterozygotes lead to establishment of two group (or varieties) which are mutually fertile but do not breed well with the rest of the species. Both varities evolve

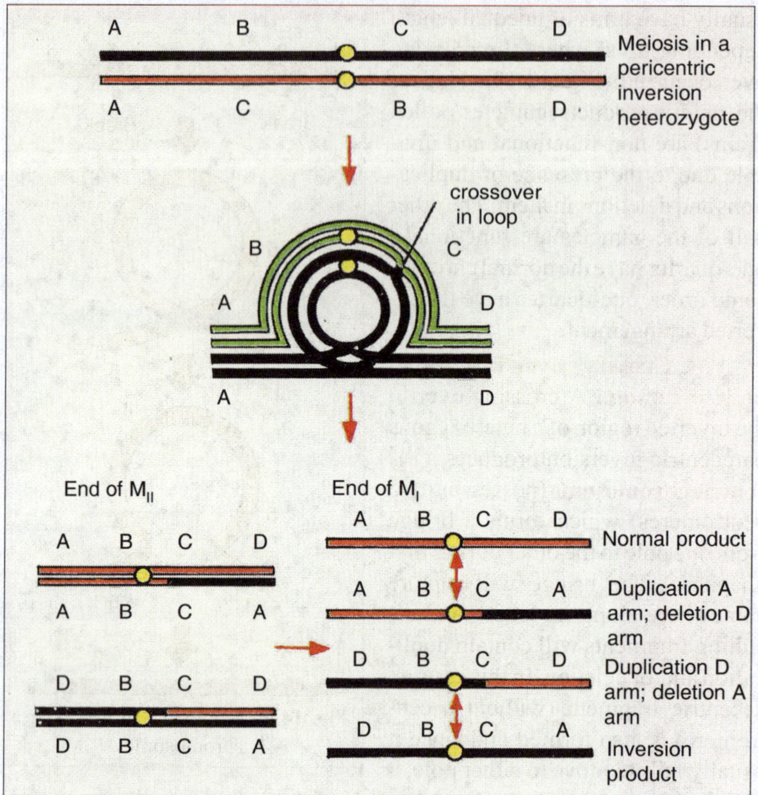

Fig. 14.10. Meiotic products resulting from a single crossover within a heterozygous pericentric inversion (after Suzuki *et al.,* 1986).

in different directions and later become reproductively isolated species. There is plenty of cytological evidence to prove that such evolutionary mechanisms have and are operating in *Drosophila* and a number of other organisms.

4. Translocation

The shifting or transfer of a part of a chromosome or a set of genes to a non-homologous one, is called **translocation**. There is no addition or loss of genes during translocations, only a rearrangement (*i.e.,* change in the sequence and position of a gene). Translocations may be of following three types (Fig. 14.12) :

1. Simple translocations. They involve a single break in a chromosome. The broken piece gets attached to one end of a nonhomologous chromosome.

2. Shift translocation. In this type of translocation, the broken segment of one chromosome gets inserted interstitially in a nonhomologous chromosome.

3. Reciprocal translocations. In this case, a segment from one chromosome is exchanged with a segment from another nonhomologous one, so that in reality two translocation chromosomes are simultaneously achieved.

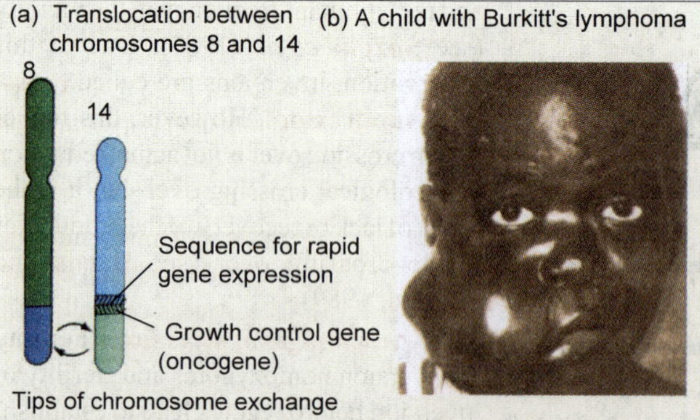

(a) In a reciprocal chromosome translocation, the tips of chromosome 8 and 14 are exchanged causing Burkitt's lymphoma. (b) A jaw tumour like this child's is a common symptom of Burkitt's.

Outcomes of reciprocal translocation. The exchange of chromosome parts between nonhomologous chromosomes creates new linkage relationships. Such translocations also drastically change the size of a chromosome as well as the position of its centromere. For example, a large metacentric chromosome is shortened by one-half in length to an acrocentric one, where as the small chromosome becomes a large one (Fig. 14.13). Two types of translocations have been recognized : homozygous and heterozygous (Fig. 14.14). The **translocation**

homozygotes may have normal meiosis and in fact, are difficult to detect cytologically unless morphologically dissimilar chromosomes are involved, or banding patterns differ markedly. The **translocation heterozygotes** produce both translocated and normal chromosomes and exhibit characteristic cytological and genetical effects.

Thus, translocation heterozygotes are marked by considerable degree of meiotic irregularity. In order to affect pairing of all homologous segments, peculiar and characteristic formations occur during synapsis. Typically, a cross-shaped configuration is seen in prophase-I. This structure opens into either a ring or a figure of 8, both comprising four chromosomes (Fig. 14.15). In case of 8 figure formation, both normal chromosomes move to one pole and both translocation chromosomes move to the other pole at anaphase-I. This is called **alternate segregation** (or **disjunction**) and in it functional meiospores (in higher plants) or gametes (animals) will be produced because the meiotic products will have a full gene complement. In case of ring formation, **adjacent type segregation** may occur and as a result each of translocation chromosomes and normal chromosomes move to opposite poles. In this case all of the gametes will contain some extra segments (duplications) and some pieces will be missing (deletion). Semisterility resulting from adjacent type segregation during reciprocal translocation is easily observed in such plants as corn (maize), wheat, pea and *Datura*. Ears of corn lack about half the kernels, and these are arranged irregularly. Abortive pollen is reduced in size (Fig. 14.16).

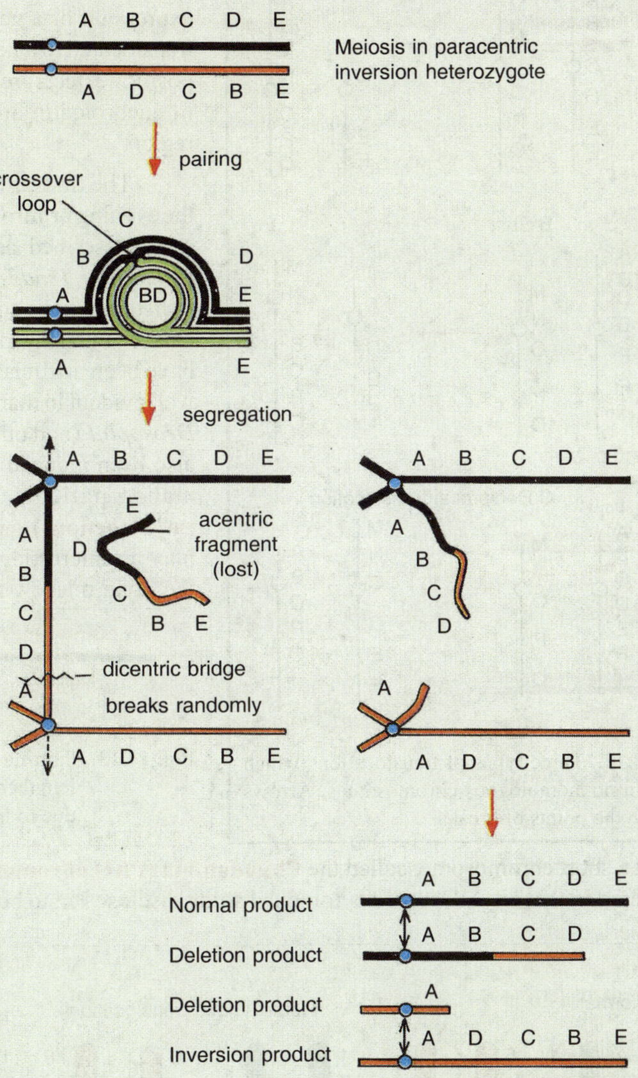

Fig. 14.11. Meiotic products resulting from a single crossing over within a heterozygous paracentric inversion loop (after Suzuki *et al.*, 1986).

Lastly, some genes which formerly were on nonhomologous chromosomes will no longer appear

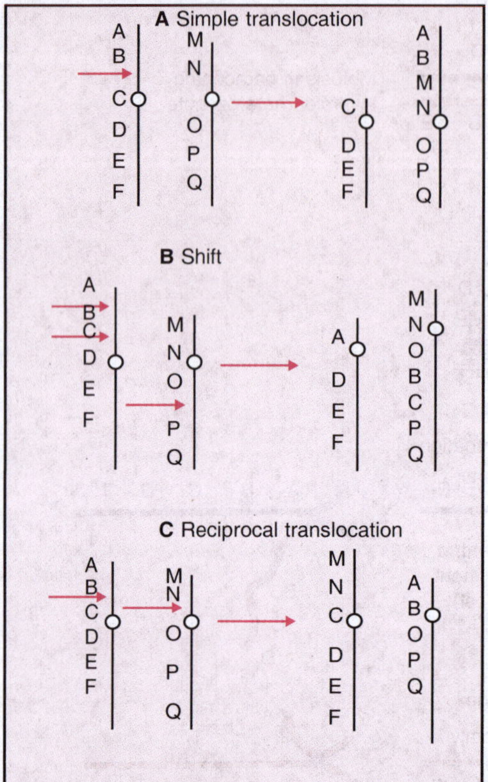

Fig. 14.12. Three types of translocations which occur in nonhomologous chromosomes. Arrows indicate the points of breaks.

to be assorted independently. And the phenotypic expression of a gene may be modified when it is translocated to a new position in the genome. Such position effects are particularly evident when genes in euchromatin are shifted near heterochromatin region.

The first case of translocation was studied in the evening primrose (*Oenothera*) which was originally described as a mutation by **de Vries**. In *Oenothera, Tradescantia* and *Rhoeo* translocations in heterozygous condition are frequently found in nature. In many other crop plants translocations have been artificially induced by X-rays. They are well evident in many animals including humans. In *Drosophila* and other animals, translocations have also been induced by X-rays. For example, in humans, patients with **chronic myelocytic (myologenous) leukemia** (a kind of cancer) display an interesting chromosomal abnormality. In the bone marrow and in cells derived from it, is

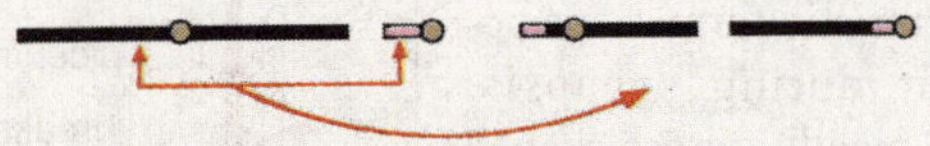

Fig. 14.13. Shorting of metacentric chromosome and lengthening of acrocentric chromosomes due to translocation.

present a short chromosome, called the **Philadelphia (Ph1) chromosome** (so named because it was discovered in that city). Detailed cytological study disclose Ph[1] to be a number 22 chromosome that

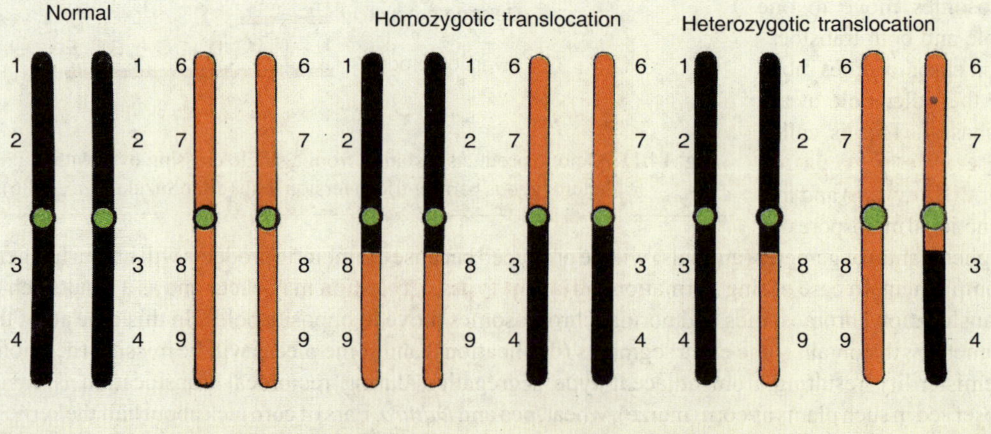

Fig. 14.14. Schematic represenation of homozygotic and heterozygotic reciprocal translocations compared with the normal arrangement.

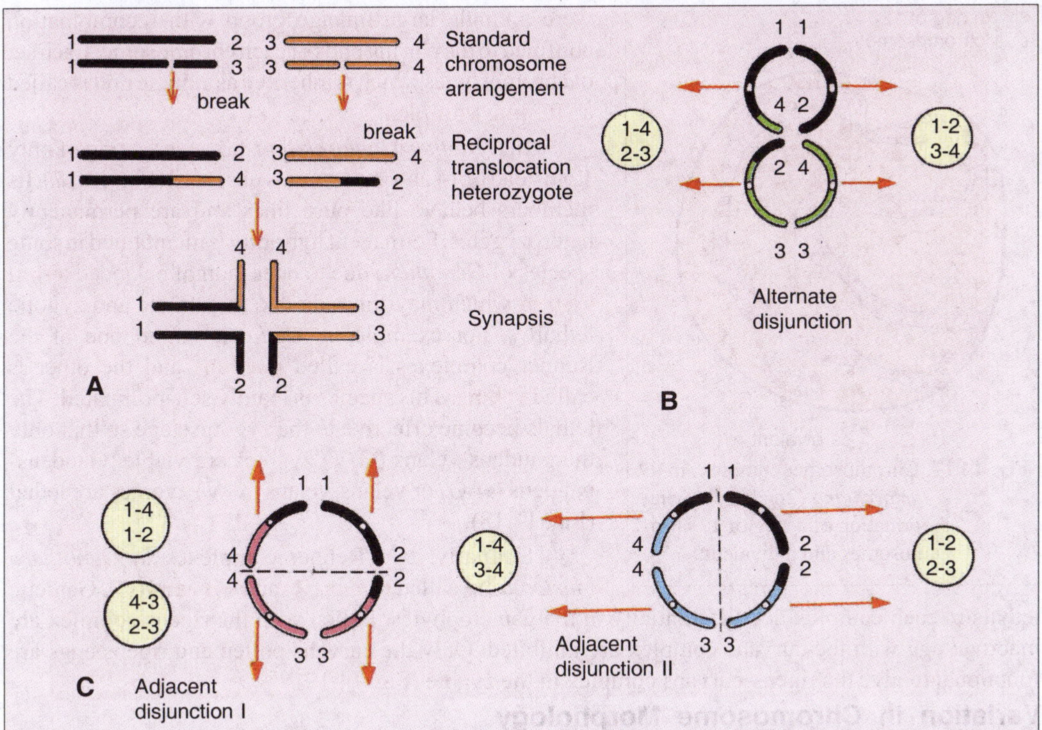

Fig. 14.15. A–Cross shaped figure formation during heterozygous reciprocal translocation; B–Alternate disjunction; C–Adjacent I and II disjunctions (after Stansfield, 1986).

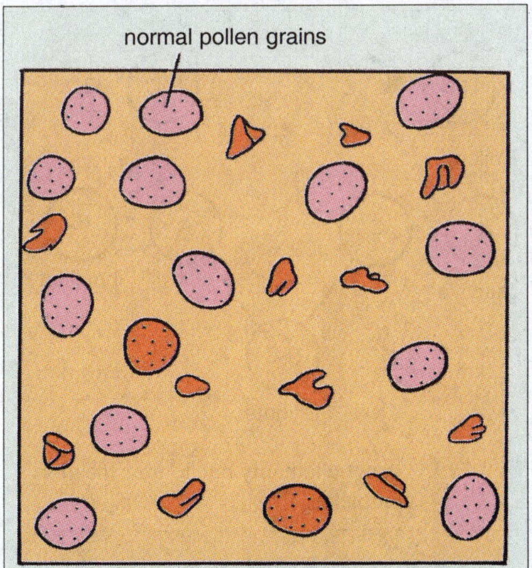

Fig. 14.16. Normal and aborted pollen in a semisterile corn plant. The small, shriveled pollen grains contain an euploid meiotic products of reciprocal translocation heterozygote (after Suzuki *et al.*, 1986).

has lost most of the distal part of its longer arm (22 q^-). The deleted part of autosome 22 is translocated to one of the larger autosomes (most frequently to the distal end of chromosome 9). This translocation exhibits position effect and it is not transmitted to offspring of persons having Philadelphia chromosome (Ph[1] does not appear in gametes of the patients).

Translocation complexes and lethality. In *Oenothera,* a rare series of reciprocal translocations have occurred which involve all 7 of its chromosome pairs. If each chromosome end is labelled with a different number, the normal set of 7 chromosomes would be represented as 1-2, 3-4, 5-6, 7-8, 9-10, 11-12, and 13-14; likewise a translocation set would be represented as 2-3, 4-5, 6-7, 8-9, 10-11, 12-13 and 14-1. Such a **multiple translocation heterozygotes** would form a ring of 14 chromosomes during meiosis. Different lethals in each of two haploid sets of 7 chromosomes administer structural heterozygosity. Since only alternate disjunction from the ring can form viable gametes, each group of 7 chromosomes behaves as though it

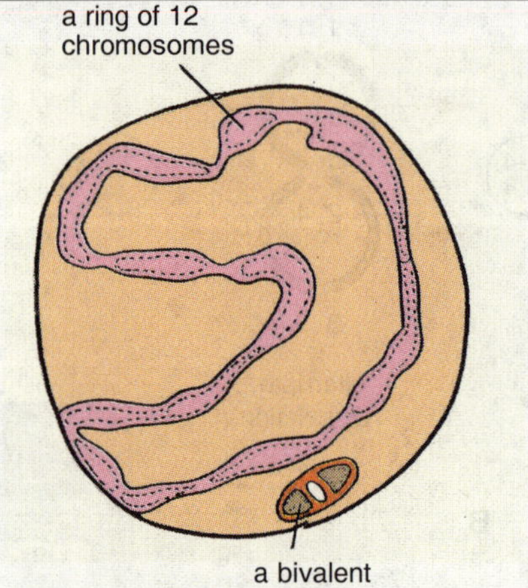

Fig. 14.17. Chromosome complex in *O. lamarckiana* (2n=14) showing formation of a ring of 12 chromosomes and a bivalent.

were a single large linkage group with recombination confined to the pairing ends of each chromosome. Each set of chromosomes which is inherited as a single unit is called a "**Renner complex**."

In *Oenothera lamarckiana,* however, a ring of only 12 instead of 14 chromosomes is observed (Fig. 14.17). Its members behave like pure lines and are permanently heterozygotes. Permanent hybridity is maintained in some species of *Oenothera* due to operation of balanced **lethal system** which may function due to gametic and zygotic lethality. For example, in *O. lamarckiana* one of the Renner complexes is called **gaudens** and the other is called **velans**. This species is mainly self-pollinated. The lethals become effective in the zygotic stage so that only the gaudens-velans (G-V) zygotes are viable. Gaudens-gaudens (G-G) or velans-velans (V-V) zygotes are lethal (Fig. 14.18).

Similarly, two Renner complexes in *Oenothera muricata* are called **rigens (R)** and **curvans** (**C**). Gametic lethals in each complex act differentially in the gametophytes. Pollen with the rigens complex are inactive; egg with the curvans complex are inhibited. Only the curvans pollen and rigens eggs are functional to give the rigens-curvans complex in the zygote (Fig. 14.19).

Variation in Chromosome Morphology

Various changes in chromosome structure often produce variation in chromosome morphology such as isochromosomes, ring chromosomes and Robertsonian translocation.

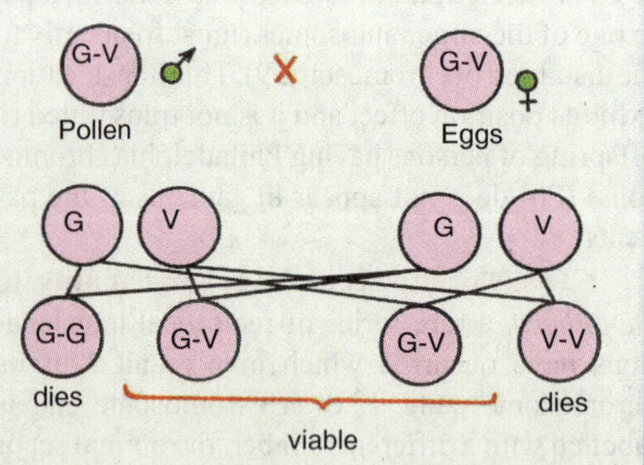

Fig. 14.18. Zygotic lethality due to translocation complexes.

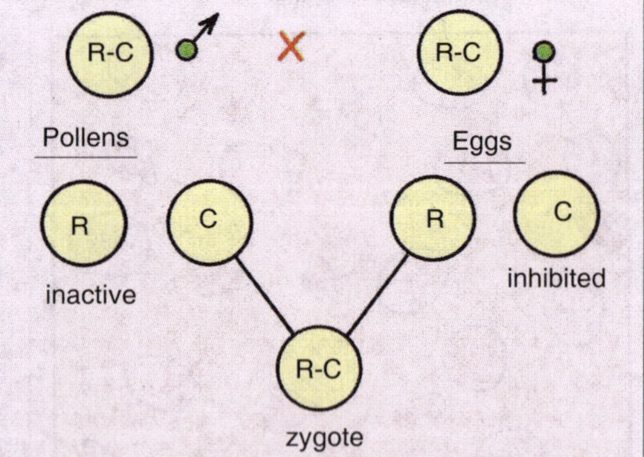

Fig. 14.19. Gametic lethality due to translocation complexes.

1. Isochromosomes. An isochromosome is a chromosome in which both arms are identical. It is thought to arise when a centromere divides in the wrong plane, yielding two daughter chromosomes, each of which carries the information of one arm only but present twice. For example, telocentric X chromosome of *Drosophila* may be changed into an "attached-X" which is formed due to misdivision of the centromere (Fig. 14.20).

2. Ring chromosomes. Chromosomes are not always rod-shaped. Occasionally ring chromosomes are encountered in higher organisms. Sometimes breaks occur at each end of the chromosome and broken ends are joined to form a ring chromosome. Crossing over between ring chromosomes can lead to bizarre anaphase figures (Fig. 14.21).

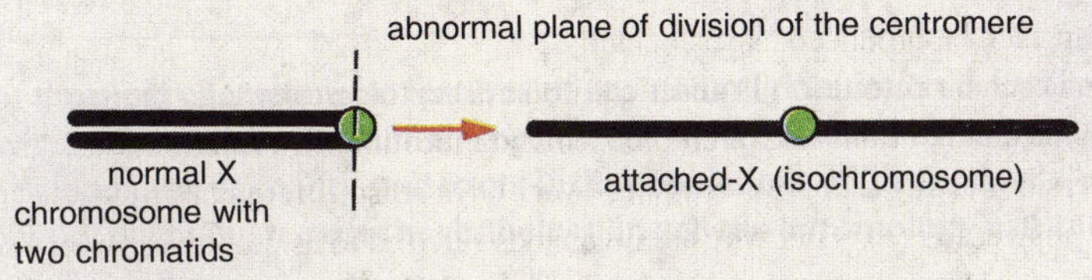

Fig. 14.20. Origin of an attached-X chromosome.

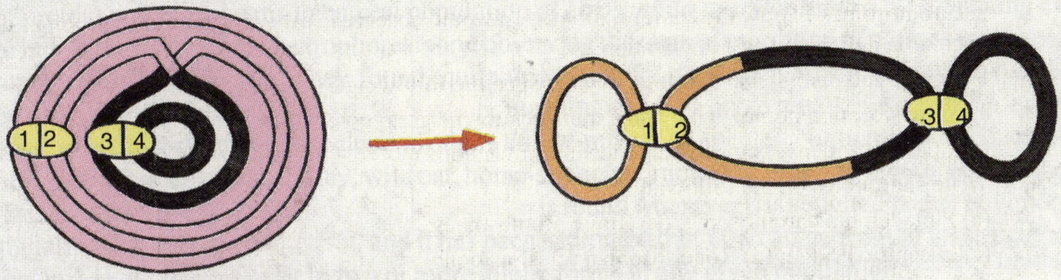

Fig. 14.21. Formation of double bridge during single crossing over in ring chromosomes (after Stansfield, 1986).

3. Robertsonian translocation. Sometimes whole arm fusions occur in the non-homologous chromosomes. It is called **Robertsonian translocation**. Thus, Robertsonian translocation is an eucentric reciprocal translocation where the break in one chromosome is near the front of the centromere and the break in the other chromosomes is immediately behind its centromere. The resultant smaller chromosome consists of largely inert heterochromatic material near the centromere; it normally contains no essential genes and tends to become lost. Thus, Robertsonian translocation results in a reduction of the chromosome number (Fig. 14.22).

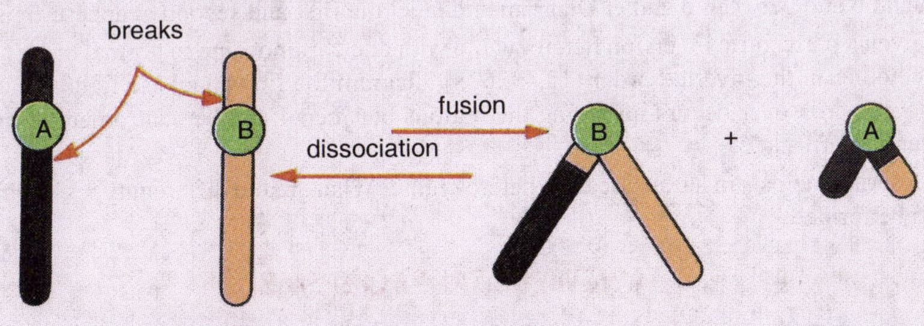

Fig. 14.22. Robertsonian translocation results in the formation of metacentric chromosome by fusion of two acrocentric chromosomes (after Stansfield, 1986).

The Robertsonian translocation is found to have a role in the evolution of human beings. Humans have 46 chromosomes whereas the great apes (Chimpanzees, Gorillas and Orangutans) have 48 chromosomes. Cytogeneticists believe that humans evolved from a common human/ape ancestor due to centric fusion of two acrocentric chromosomes to produce a single large chromosome containing the combined genetic content of two acrocentric chromosomes. It is suspected that structural rearrangements of chromosomes may lead to reproductive isolation and the formation of new species.

REVISION QUESTIONS AND PROBLEMS

1. Why are chromosomal aberrations considered to have less significance than gene mutations for subsequent generations.
2. a. Describe and illustrate how : (1) deletions, (2) inversions, and (3) reciprocal translocations arise in nature?

   b. How can each be produced experimentally ?

   c. How can each be detected : (1) genetically and (2) cytologically ?
3. What may be an important role of chromosomal duplications in evolution?
4. a. Do inversions always suppress crossing-over? Give reasons for your answer.

   b. If your answer to : (a) is no, why are inversions referred to as crossover-suppressors ?

   c. Show how : (1) paracentric and (2) pericentric inversions can act as crossover suppressors ?
5. Differentiate between paracentric and pericentric inversions.
6. Why are salivary gland chromosomes of *Drosophila* commonly used for study of structural changes in chromosomes ? Draw the configurations, you would observe in salivary gland chromosomes due to either a deficiency or a duplication.
7. How will you distinguish cytologically between the following ?

   a. A translocation homozygote and a translocation heterozygote ;

   b. A paracentric inversion and a pericentric inversion.
8. Write short notes on the following :

   (a) Philadelphia chromosome; (b) Ring chromosomes;

   (c) Dicentric bridge; (d) Robertosonian translocation;

   (e) Pseudodominance; (f) Deficiency; (g) Renner complex ; (h) *Oenothera.*
9. Assume a chromosome with the following gene sequence (the full stop (.) represents the centromere) :

   ABCD. EFGH

   You find the following aberrations in this chromosomes; for each identify the specific kind of aberration:

   (a) ABCD. EFH ;(B) ADCB. EFGH;

   (c) ABCDCD. EFGH.
10. A chromosome with segments in the normal order is (abcdefgh). An inversion heterozygote has the abnormal order (abfedcgh). A three-strand double crossover occurs involving the regions between a and b and between d and e. Diagram and label the first and second anaphase figures.
11. Given a pericentric inversion heterozygote with one chromosome in normal order (1234.5678) and the other in the inverted order (15.432678), diagram the first anaphase figure when a four-strand double-crossover occurs involving the regions between 4 and he centromere (.) and between the centromere and 5.
12. Describe meiosis in a translocation heterozygote. What are the different types of gametes expected to be formed?

ANSWERS TO PROBLEMS

9. (a) Deletion; (b) Inversion; (c) Duplication.
10. First anaphase: a diad, a loop chromatid and an acentric fragment; second anaphase: the diad splits into two monads, the loop forms a bridge and the acentric fragments becomes lost.

CHAPTER

15

Chromosomal Mutation-II

(Cytogenetics : Changes in Chromosomes Number)

Orangutans have 24 pairs of chromosomes. In the human lineage, two chromosomes underwent end-to-end fusion and produced human chromosome number 2, reducing the number of pairs of chromosomes to 23.

Each species has a characteristic number of chromo somes in the nuclei of its gametes and somatic cells. The gametic chromosome number constitutes a basic set of chromosomes called **genome**. A gamete cell contains single genome and is called **haploid**. When haploid gametes of both sexes (male and female) unite in the process of fertilization a **diploid** zygote with two genomes is formed. The diploid zygote undergoes embryological development and forms an adult animal which upon attaining sexual maturity produces haploid gametes. And this alternation of generation continues between haploidic and diploidic generation in most species. However, sometimes irregularities occur in nuclear division or "accidents" (as from radiations) may befall interphase chromosomes so that cells or entire organisms with aberrant genomes may be formed. Such chromosomal aberrations may include whole genomes and entire single chromosomes. Changes in number of whole chromosomes is called **heteroploidy** (see **Burns** and **Bottino**, 1989). Heteroploidy may involve entire sets of chromosomes (**euploidy**), or loss or addition of single whole chromosomes (**aneuploidy**). Each may produce phenotypic changes, modifications of phenotypic ratio, or alteration of linkage groups. Many are of some evolutionary significance.

A. EUPLOIDY

The term **euploidy** (Gr., *eu* = even or true; *ploid* = unit) designates genomes containing chromosomes that are multiples of some basic number (x). The euploids are those organisms which contain balanced set or sets of chromosomes in any number. The number of chromosomes in a basic set is called the **monoploid number**, x. Those euploid types whose number of sets is greater than two are called **polyploid**. Thus, $1x$ is **monoploid**, $2x$ is **diploid**; and the polyploid types are $3x$ (**triploid**), $4x$ (**tetraploid**), $5x$ (**pentaploid**), $6x$ (**hexaploid**) and so on. The **haploid** (**n**) refers strictly to the number of chromosomes in gametes. In most animals and many plants the haploid number and monoploid number are the same. Hence n or x (or $2n$ or $2x$) can be used interchangeably. However, in case of polyploids the usage of n may create confusion. For example, in modern wheat *x and n* are different. Wheat has 42 chromosomes, but careful study reveals that in hexaploid there are six rather similar but not identical sets of seven chromosomes. So, $6x = 42$, and $x = 7$. However, gametes of wheat contain 21 chromosomes, hence, 2n = 42 and n = 21 (see **Suzuki** *et al.*, 1986). A triploid hybrid of wheat, from which a hexaploid *Triticum spelta* has been obtained due to colchicine treatment, contains $2n = 3x = 21$. Such haploids, since are obtained from the polyploids (*i.e.*, cross of tetraploid emmer wheat and diploid goat grass), they are called **polyhaploids**, just to differentiate them from the normal monoploids.

The lower organisms such as bacteria and viruses are called **haploids** because they have a single set of genetic elements. However, since they do not form gametes comparable to those of higher organisms, the term monoploid would seem to be more appropriate (see **Stansfield**, 1986).

(1) Monoploidy

Monoploids have a single basic set of chromosomes, *e.g.*, 7 in barley and 10 in corn. Monoploidy is common in plants and rare in animals.

(i) Origin and production of monoploids. Monoploids in some cases are found normally and are produced due to parthenogenesis, as in male (drone) hymenopteran insects such as bees, wasps and ants. In these insects, queen and workers are diploid females. In angiosperms (flowering plants) monoploids may also originate spontaneously due to parthenogenetic development of egg. Such rare monoploids have been obtained in tomatoes and cotton under cultivation. Rarely monoploid plants may originate from the pollen tube, synergids and antipodals of the embryo sac and are called **androgenic monoploids** (or androgenic haploids).

Certain insect species, such as ants and bees produce males that are derived from unfertilized eggs. These organisms are known as monoploids and are quite rare.

Monoploids can be produced by artificial means by the following methods : (1) X-ray treatments, (2) delayed pollination, (3) temperature shock (cold treatment), (4) colchicine treatment, (5) distant (interspecific or intergeneric) hybridization), (6) anther or pollen culture. Among these techniques, the most important ones are distant hybridization and anther culture.

(a) Distant hybridization. Interspecific crosses in genera of *Solanaceae* (*e.g.*, *Solanum* and *Nicotiana*) have been employed for the production of both parthenogenetic and androgenic monoploids. By this technique monoploids have been obtained in large number in potato. **Kasha** and **Kao** (1970) have used this technique for producing monoploids in large number in barley. They discovered that when diploid barley, *Hordeum vulgare*, is pollinated using a diploid wild relative called *Hordeum bulbosum*, fertilization occurs, but during the ensuing somatic cell divisions, the chromosomes of *H. bulbosum* are preferentially eliminated from the zygote, resulting in a haploid embryo (such a **haplodization process** appears to be caused by a genetic incompatiability between the chromosomes of the different species). The resulting haploids can be doubled with colchicine treatment. This technique has resulted in the rapid production of new varieties of barley and applied to other plant species also.

(b) Anther or pollen culture. The production of monoploids in tobacco plant by anther and pollen culture was demonstrated for the first time in the laboratory of **Prof. S.C. Maheshwari** of Dept. of Botany of Delhi University (**Guha** and **Maheshwari**, 1967). In this technique, a cell which is destined to become a pollen grain, may be induced by cold treatment to grow instead into an **embryoid**, a small dividing mass of cells. The embryoid may be grown on agar to form a monoploid plantlet, which can then be potted in soil to mature. Subsequently, such monoploids were produced for various crop plants such as soyabean, rice, wheat, mustard, and tobbaco. Presently, this technique is regarded as a very potential source of monoploid production.

(ii) Morphology of monoploids. Monoploid plants have reduced size of all vegetative and floral parts. In monoploid *Nicotiana* **kostoff** reported that the leaves, flowers and over all plant size were smaller. The size of seed and stomata as well as diameter of pollen were found smaller in monoploids than in the diploids. Even the size of nucleus (or the nuclear volume) of a monoploid often was found to be just half than the nucleus of the diploid cell.

(iii) Cytology of monoploids. In monoploids each chromosome is represented only once due to which there is no zygotene pairing and all the chromosomes appear as **univalents** on the metaphase plate at the time of meiosis. During anaphase each chromosome move independently of the other and goes to either of the two poles. According to law of probability the chance that a particular chromosome will go to a particular pole as half and the chance that all the chromosomes of a monoploid set will go to the same pole is ½ X ½ X ½....n times, where n = number of chromosomes in the monoploid set. So, the frequency of gametes with the haploid set or n number of chromosomes will be $(½)^n$. This indicates that higher the number of chromosomes in a haploid set, lesser will be the frequency of all of them being included in the same gamete. Gametes containing less than the haploid number of chromosomes are normally not viable. Therefore, monoploid organisms are highly sterile. For instance, a monoploid in maize (2n = 20) will have 10 chromosomes and the number of chromosomes in a gamete can range from 0–10. Consequently, considerable sterility is found in such monoploid maize plants.

In contrast, in monoploid male honey bees during spermatogenesis the meiosis is bypassed by mitosis. As a result, their sperms are haploid and viable.

(iv) Uses of monoploids. In a monoploid, since there is only one copy of each chromosome and only one allele of each gene, so, in it each gene is expressed whether it is dominant or recessive. This facilitates genetic experiments and this is the reason why microorganisms have been able helpful in genetic studies. For the same reason, scientists are trying hard to develop haploid strains of the flowering plants. Success has been achieved in developing monoploid strains of *Nicotiana*, *Datura* and *Triticum*. From these monoploid strains have also been developed pure breeding strains which are resistant for the insecticides and also for toxic compounds normally produced by the parasites of these plants.

(2) Polyploidy

Any organism with more than two genomes (2*x*) is called a **polyploid**. Many plant genera include species whose chromosome numbers constitute a euploid series. For example, the rose genus *Rosa* includes species with the somatic numbers 14, 21, 28, 35, 42 and 56. These numbers are the multiples of 7. Therefore, this is a euploid series of the basic monoploid number 7, which gives diploid, triploid, tetraploid, pentaploid, hexaploid and octaploid species. Except diploids, rest of these belong to polyploid category. Ploidy levels higher than tetraploid are not commonly encountered in natural populations, but our most important crops and ornamental flowers are polyploids, *e.g.,* wheat (Hexaploid 6*x*), strawberries (octaploid, 8*x*), many commercial fruits and ornamental plants. Generally, polyploidy is common in plants (more common in monocots) but rare in animals.

Types of polyploidy. There are following three different kinds of polyploids : (i) autopolyploids, (ii) allopolyploids and (iii) autoallopolyploids. Suppose that there are four different haploid sets of chromosomes A, B_1, B_2 and C, in which B_1 and B_2 genomes are related. By using these genomes, all three types of polyploids can be derived as have been shown in Figure 15.1.

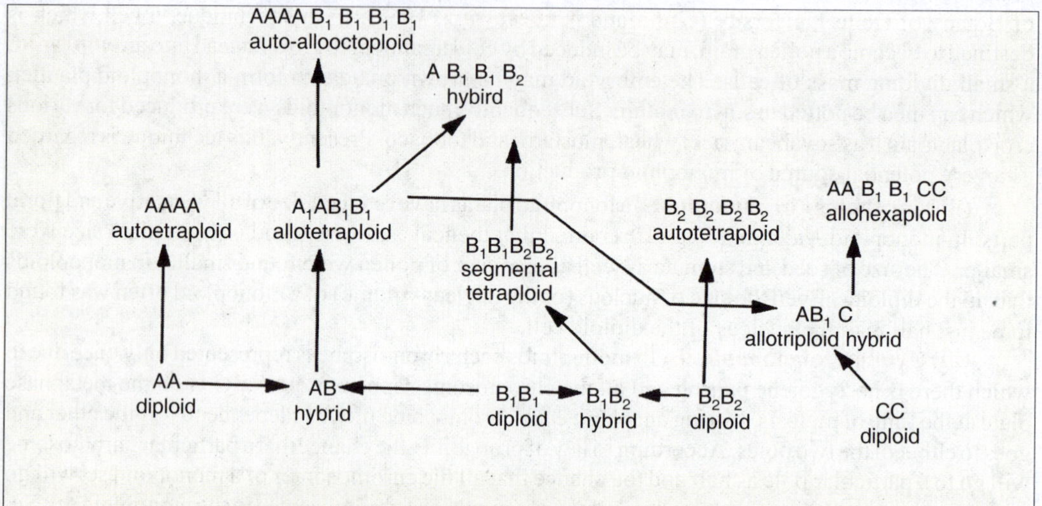

Fig. 15.1. Mode of formation of different kinds of polyploid.

Flower of one of the autoploid species of Chrysanthemum.

(a) Autopolyploids. The autopolyploids are those polyploids, which consist of same basic set of chromosomes multiplied. For example, if a diploid species has two similar sets of chromosomes or genomes (AA), an autotriploid will have three similar genomes (AAA), and an autotetraploid, will have four such genomes (AAAA).

(i) Origin and production of autopolyploids. The autopolyploids may occur in nature or may be produced artificially. When they are found in nature, their autopolyploidy nature is deduced by their meiotic behaviour. One of the common example of natural autopolyploidy is 'doob' grass (*Cynodon dactylon*) which is quite commonly cultivated in U.P. and Bihar. Its **autotriploid** status was established from its meiotic behaviour by **Prof. P. K. Gupta**, an eminent cytogeneticists of Northern India, working in Dept. of Agriculture Botany of Meerut University (**Gupta** and **Srivastava**, 1970). *Cynodon* is quite successful in the cultivation mainly due to its efficient way of vegetative propagation (since being triploid, it is sterile and setting no seed). Polyploids may arise naturally by following means : (i) in natural populations polyploidy may arise as a result of interference with cytokinesis, once chromosome replication has occurred; (ii) it may occur either in somatic tissues which give rise to tetraploid branches or during meiosis which produces unreduced gametes. All these natural inductions of polyploidy may occur due to chilling.

Some of common examples of autotriploid crop plants, which are mainly produced by artificial methods, are seedless varieties of watermelons, sugar beet, tomato, grapes and banana. Similarly, many important crop plants include **autotetraploids** such as rye (*Secale cereale*), corn (*Zea mays*), red clover (*Trifolium pratense*), berseem (*Trifolium alexandrium*), marigolds (*Tagetes*), snapdragons (*Antirrhinum*), *Phlox*, grapes, apples, *Oenothera lamarkiana* (which was recognized as mutation by **Hugo de Vries**).

Induced autopolyploidy. The autopolyploidy have been induced in many plant and animal cells by artificial means such as chemical (*e.g.,* chloral hydrate, colchicine, sulphanil amide, mercury chloride, hexachlorcyclohexane, etc.), radioactive substances, *e.g.,* radium and X-ray) and temperature shocks. These inducers usually disturb the mitotic or meiotic spindle and cause non-segregation of already duplicated chromosomes, during cell divisions.

Colchicine is a drug (*i.e.,* an alkaloid obtained from the corms of plants–*Colchicum autmunale* and *C. luteum*) and its aqueous solution is found to prevent the formation and organization of spindle fibres, so the metaphase chromosomes of the affected cells (called **C-metaphase** or **colchicine metaphase**) do not move to a metaphase plate and remain scattered in the cytoplasm. Even the process of cytokinesis is prevented by colchicine and with duplications of chromosomes the number goes on increasing. As colchicine interferes with spindle formation, its effects are limited to dividing and meristematic cells.

(ii) Effects of autopolyploidy. Autopolyploidy results in **gigantism** of plant cells, *i.e.,* leaves, flowers and fruits of an autopolyploid are larger in size than a diploid plant. For example, the size of lower epidermis of leaf of a tetraploid *Saxifraga pensylvanica* was found greater than the diploids (Fig. 15.2). Some of significant effects of autopolyploidy are as follows : (1) With the increase in cell size, the water content increases which leads to a decrease in osmotic pressure. This results into loss of resistance against frost, etc. (2) Due to slower rate of cell division, the plant's growth rate decreases. This leads to a decrease in **auxin** supply and a decrease in respiration. 3. Due to slow growth rate, the time of blooming of an autopolyploid is delayed. 4. At higher ploidy level, such as autooctoploids, the adverse effects become highly pronounced and lead to the death of the plants.

Polyploid varieties with an even number of genomes (*e.g.,* tetraploids) are often fully fertile (Fig. 15.3), whereas those with an odd number (*e.g.,* triploids) are highly sterile (Fig. 15.4).

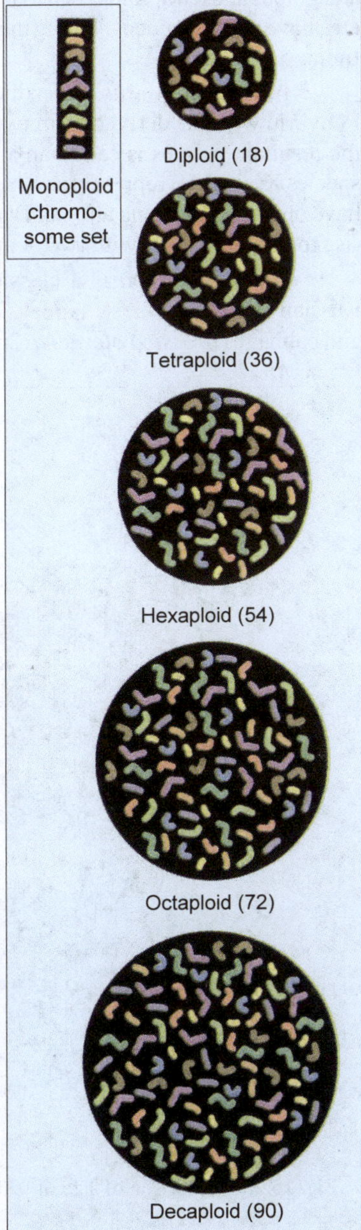

Chromosome numbers in diploid and polyploid species of *Chrysanthemum*. Each set of homologous chromosomes is depicted in a different colour.

Uses of induced polyploidy. Since in the induced polyploids, the fertility level and seed set are low, so seedless fruits can be produced by using triploids as in case of seedless watermelons which were produced by a Japanese scientist, **Dr. Hitoshi Kihara**. These triploids are obtained from seeds raised by a cross of tetraploid and diploid plants. The tetraploids have been produced from the diploids by colchicine treatment. By adopting these methods a variety of triploids such as sugar beet, tomato and grapes and tetraploids such as rye, barley, corn, apple, grapes, marigolds, snapdragons, lily, phlox, etc., have been obtained. Among the forage crops, tetraploid barseem is a very popular crop in Northern India.

(b) Allopolyploids. When the polyploidy results due the doubling of chromosome number in a F_1 hybrid which is derived from two distinctly different species, then, it is called **allopolyploidy** and the resultant species is called an allopolyploid. Let A represent a set of chromosomes (genome) in species X, and let B represent another genome in a species Y. The F_1 hybrids of these species then would have one A genome and another B genome. The doubling of chromosomes in the F_1 hybrids will give rise to allotetraploids with two A and two B genomes (see Fig. 15.5).

Raphanobrassica is a classical example of allopolyploidy or amphipolyploidy. In 1927, a Russian geneticist, **G.D. Karpechenko** performed a cross between radish (*Raphanus sativum,* 2n = 18) and cabbage (*Brassica oleracea,* 2n = 18) and in F_1 got sterile (diploid) hybrids. Among these sterile

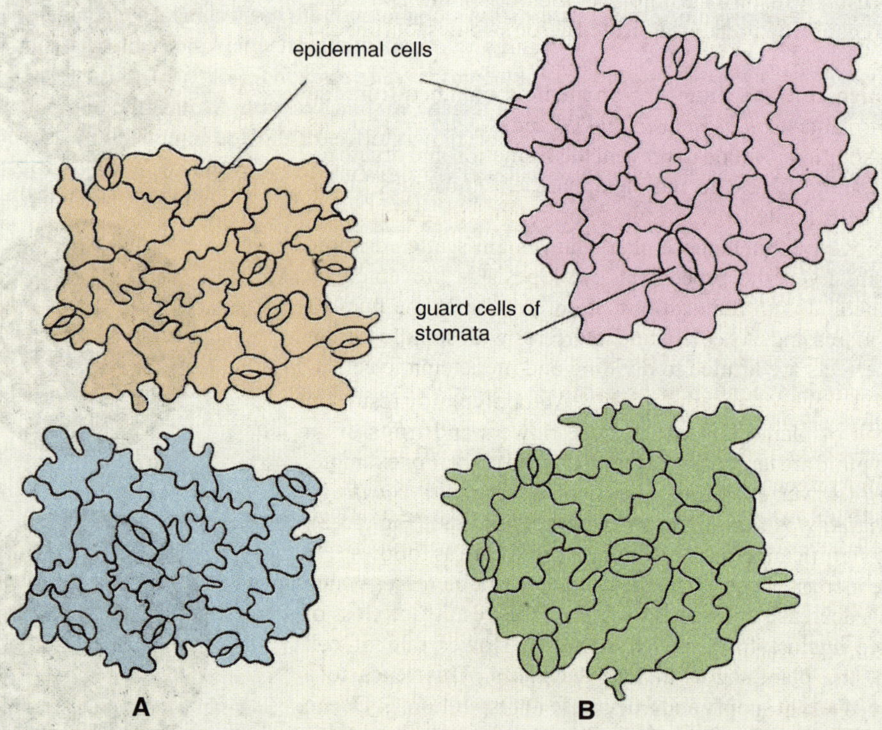

Fig. 15.2. Comparison of size of leaf epidermal cells of a diploid (A) and tetraploid (B) saxifrage (after Burns and Bottino, 1989).

F_1 hybrids, he found certain fertile plants which were found to contain 36 chromosomes. These fertile tetraploids were called *Raphanobrassica.*

Synthesized Allopolyploids

To find out the origin of naturally occurring allopolyploids some cytogeneticists produced certain allopolyploids in laboratory by employing artificial means. Common hexaploid wheat and tetraploid cotton furnish two such examples.

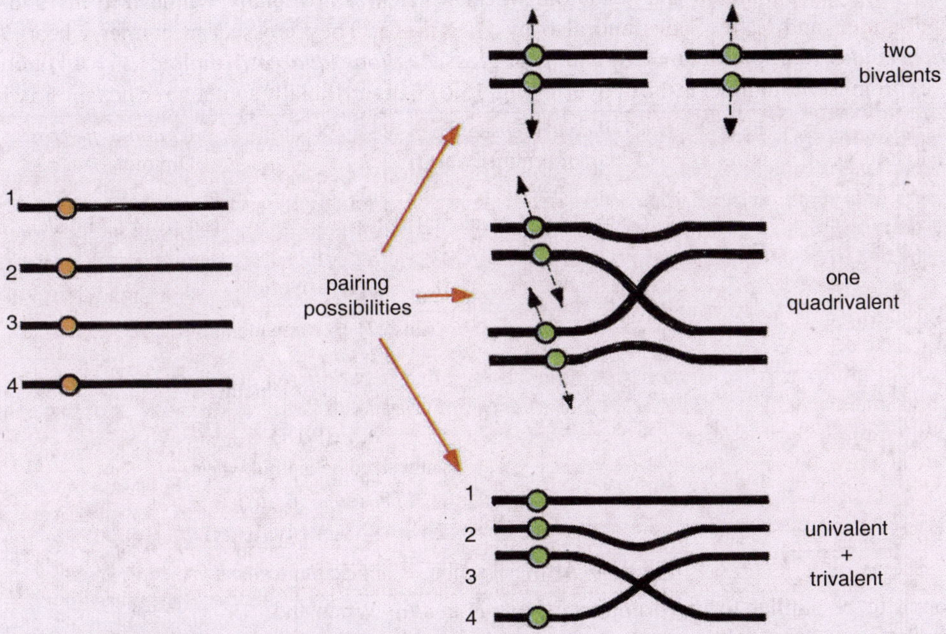

Fig. 15.3. Meiotic pairing possibilities in tetraploids (each chromosome is really two chromatids).

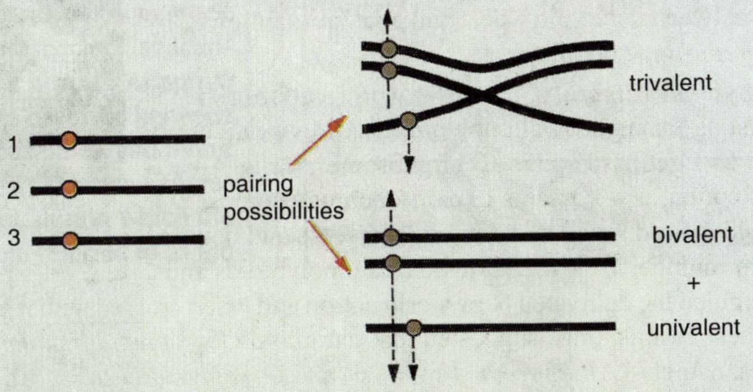

Fig. 15.4. Meiotic pairing possibilities in triploids (each chromosome is really two chromatids).

P_1 : Species X (AA) (Diploid) X Species Y (BB) (Diploid)

↓

F_1 : AB

Diploid sterile hybrid

↓ Colchicine

AA BB

Amphidiploid tetraploid

(Fertile)

Fig. 15.5. Formation of an amphidiploid tetraploid.

(i) *Triticum spelta* is a hexaploid wheat which was artificially synthesized in 1946 by **E.S. McFadden** and **E. R. Sears** and also by **H. Kihara**. They crossed an emmer wheat, *Triticum dicoccoides,* (tetraploid : 2n = 28) with goat grass, *Aegilops squarrosa* (diploid ; 2n = 14) and doubled the chromosome number in the F_1 hybrid (Fig. 15.6). This artificially synthesized hexaploid wheat was

P_1 : *Triticum dicoccoides* (Tetraploid emmer wheat) AA BB (2n = 28 ; 14 bivalents) X *Aegilops squarrosa* (Diploid goat grass) DD (2n = 14 ; 7 bivalents)

↓

F_1 : ABD
Triploid hybrid
(2n = 21 ; 21 univalents)

↓ Colchicine

AA BB DD
Synthesized hexaploid wheat
(*Triticum spelta*)
(2n = 42 ; 21 bivalents).

Fig. 15.6. Artificial synthesis of hexaploid wheat.

found to be similar to the primitive wheat *T. spelta.* When the synthesized hexaploid wheat was crossed with naturally occurring *T. spelta,* the F_1 hybrid was completely fertile. This suggested that hexaploid wheat must have originated in the past due to natural hybridization between tetraploid wheat and goat grass followed by subsequent chromosome doubling.

The pods of the amphidiploid form of *Gossypium,* the cultivated cotton plant.

(ii) *Gossypium hirsutum,* the New world cotton plant, is another interesting example of allopolyploidy. Old world cotton, *Gossypium herbaceum,* has 13 pairs of chromosomes, while, American or "upland cotton" also contains 13 pairs of chromosomes. **J.O. Beasley** crossed the old world and American cottons and doubled the chromosome number in the F_1 hybrids. The allopolyploids, thus, produced resembled the cultivated New world cotton and when crossed with it gave fertile F_1 hybrids (Fig. 15.7). These results, thus, suggested that tetraploid *Gossypium hirsutum* originated from two diploid species, namely *G. herbaceum* (2n = 26) and *G. raimondii* (2n = 26).

Gossypium herbaceum (Old world cotton) (2n = 26 ; 13 bivalents) X *Gossypium raimondii* (American or upland cotton) (2n = 26 ; 13 bivalents)

↓

F_1 hybrid
(2n = 26; 26 univalents)

↓ Colchicine

New world cotton
(*Gossypium hirsutum*)
(2n = 52 ; 26 bivalents)

Fig. 15.7. Artificial synthesis of new world cotton.

(iii) Triticale *(Triticosecale Wittmack)* is the first man made cereal which has been developed in recent years and is cultivated on about one million hectares of land throughout the Globe for the commercial use. Triticale is an artificial allopolyploid which has been derived by crossing wheat (*Triticum*) and rye (*Secale*). Depending upon whether *Triticum* is a tetraploid (2n = 4*x*= 28) or hexaploid (2n = 6*x* = 42), one would get hexaploid triticale (2n = 6*x* = 42; Fig. 15.8) or octaploid triticale (2n = 8*x* = 56; Fig. 15.9), respectively. In each case, only diploid rye (2n = 4*x* = 14) was used.

Triticum durum X *Secale cereale*
(Tetraploid wheat ; 2n = 28) (Diploid rye ; 2n = 14)
↓
F_1 hybrid (sterile)
(Triploid ; 2n = 21)
↓ (Chromosome doubling)
Hexaploid triticale
(2n = 42)

Fig. 15.8. Artificial synthesis of a hexaploid triticale.

(c) Segmental allopolyploids. Different genomes of some allopolyploids are not quite different from each other. Consequently in these polyploids chromosomes belonging to different genomes do pair together to some extent. This indicates that segments of chromosomes and not the whole chromosomes are homologous. Therefore, such allopolyploids are called **segmental allopolyploids (Stebbins**, 1943, 1950). The segmental allopolyploids are intermediate between autopolyploids and allopolyploids and can be identified by their peculiar meiotic behaviour.

Triticum aestivum X *Secale cereale*
(Hexploid wheat; 2n = 26) (Diploid rye; 2n = 14)
↓
F_1 hybrid (sterile)
(Tetraploid ; 2n = 28)
↓ (Chromosome doubling)
Octoploid triticale
(2n = 56)

Fig. 15.9. Artificial synthesis of a octoploid triticale.

It is generally believed that most naturally occurring polyploids are segmental allopolyploids. Our common hexaploid bread wheat too is found to be a segmental hexaploid.

Polyploidy in animals. Polyploidy is rare in animals but occur in flatworms, leeches and brine shrimp. In mice, also, 40 per cent liver cells are tetraploids, and about 5 per cent are octoploids. Polyploidy in humans have been found in liver cells and cancer cells. In them polyploidy is whether complete or as a mosaic, it leads to gross abnormalities and death.

Triticum aestivum.

Phenotypic Effects of Polyploidy

The increase in the genome's size beyond the diploid level is often caused following detectable phenotypic characteristics in a polyploid organism :

(i) Morphological effect of polyploidy. The polyploidy is invariably related with **gigantism**. The polyploid plants have been found to contain large-sized pollen grains, cells, leaves, stomata, xylem, etc. The polyploid plants are more vigorous than diploids.

(ii) Physiological effect of polyploidy. The ascorbic acid content has been reported to be higher in tetraploid cabbages and tomatoes than in corresponding diploids. Likewise corn meal of a tetraploid maize seed contain 40 per cent more vitamin A than cornmeal from a diploid plant.

(iii) Effect on fertility of polyploidy. The most important effect of polyploidy is that it reduces the fertility of polyploid plants in variable degrees.

(iv) Evolution through polyploidy. Interspecific hybridization combined with polyploidy offers a mechanism whereby new species may arise suddenly in natural populations.

B. ANEUPLOIDY

Changes that involve parts of a chromosome set results in individuals, called **aneuploids** (Gr. *aneu* = uneven ; *ploid* = unit). Aneuploidy can be either due to the loss of one or more chromosomes (**hypoploidy**) or due to addition of one or more chromosomes to the complete chromosome set (**hyperploidy**). Hypoploidy is mainly due to the substraction (or loss) of a single chromosome, called **monosomy** (2n–1) or due to the loss of one pair of chromosome called **nullisomy** (2n–2 ; two lost chromosomes are homologs). Likewise, hyperploidy may involve addition of either a single chromosome, called **trisomy** (2n + 1) or a pair of chromosomes, called **tetrasomy** (2n + 2). In the monoploid organisms, addition of single chromosome produces **disomy** (n + 1). All of these aneuploids are probably produced by nondisjunction during mitosis or meiosis (Fig. 15.10).

1. Monosomy

Diploid organisms which are missing one chromosome of a single pair are monosomic with the genomic formula 2n –1. A monosomic individual forms gametes of two types, (n) and (n – 1). The n –1 gametes do not survive in plants, but, in animals that may cause genetic imbalance, which is manifested by high mortality or reduced fertility of resulted organism (Fig. 15.11).

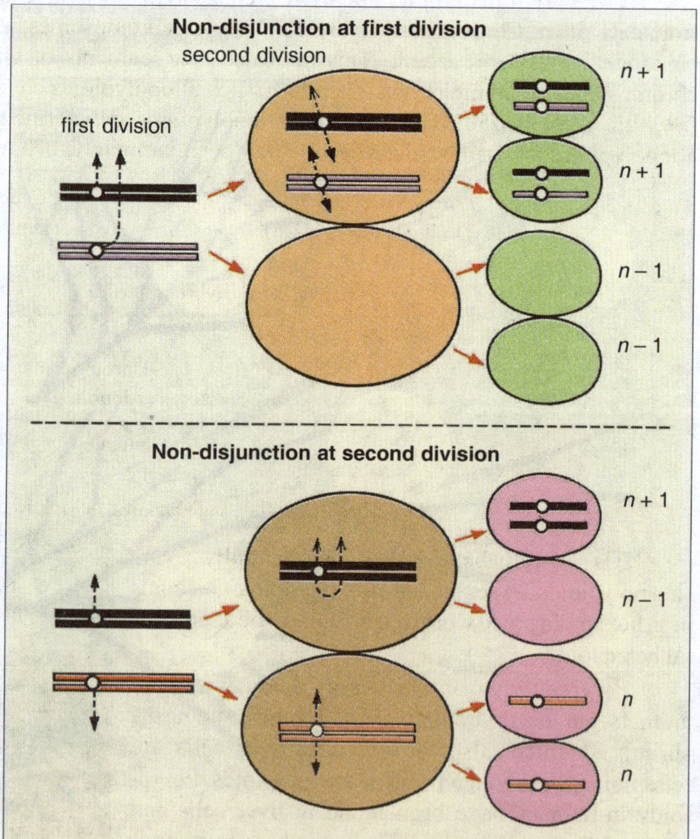

Fig. 15.10. The mode of origin of aneuploid gametes by non- disjunction at either the first or second meiotic division (after Suzuki *et al.* 1986).

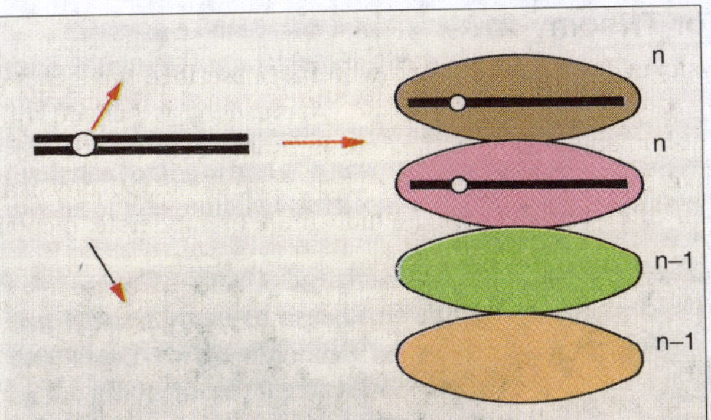

Fig. 15.11. Behaviour of a chromosome at meiosis (after Suzuki, *et al.*, 1986).

Monosomy in diploids is not tolerated, since it creates imbalance due to loss of one complete chromosome. However, in polyploids, monosomy has no apparent effect, since they have several chromosomes of same type and loss of one chromosome can be easily tolerated. The number of possible monosomics in an organism will be equal to the haploid chromosome number. For example, in common wheat, since 21 pairs of chromosomes are present, 21 possible monosomics are known. These 21 monosomics in wheat were produced by **E.R. Sears** in variety called "**chinese spring**" and being used for genetic studies all over the world. Monosomics were also reported in cotton (2n= 52) by **J. Endrizzi** and his coworkers and in tobacco (2n = 48) by **E.R. Clausen** and **D.R. Cameron**. Monsomics have also been produced in maize and tomato (2n = 24) despite their being diploids. **Double monosomics** (2n–1–1) or **triple monosomics** (2n–1–1–1) could also be produced in polyploids such as wheat. Double monosomic means that the chromosome number is 2n–2 like that in a nullisomic, but the missing chromosomes are non-homologous. The same explanation is applied to the triple monosomics.

2. Nullisomy

An organism which has lost a chromosome pair is a nullosomic. The nullosomic organism has the genomic formula (2n–2). A nullosomic diploid often does not survive, however, a nullosomic polyploid (*e.g.,* hexaploid wheat, 6*x*–2) may survive but exhibit reduced vigour and fertility.

3. Trisomy

Trisomics are those diploid organisms which have an extra chromosome (2n + 1). Since the extra chromosome may belong to any one of different chromosomes of a haploid complement, the number of possible trisomics will be equal to the haploid chromosome number. For example, haploid chromosome number of barley is 7, consequently in it seven trisomics are possible. Further, when the extra chromosome is identical to its homologs, such a trisomic is called **primary trisomic**. There are also secondary and tertiary trisomics. While the **secondary trisomic** means that the extra chromosome should be an isochromosome (*i.e.,* both chromosome arms genetically similar), a **tertiary trisomic** would mean that the extra chromosome should be the product of translocation (Fig. 15.12). Trisomics were obtained for the first time in jimson weed (*Datura stramonium*) by **A.F. Blakeslee** and **J. Belling** (1924). Since the haploid chromosome number of this species is n=12, so here, 12 primary trisomics, 24 secondary trisomics and a large number of tertiary trisomics are possible. Most of the trisomics were identified by the size, shape and other morphological features of the fruit of jimson weed (Fig. 15.13). In barley, such a trisomic series is produced and extensively studied by **T. Tsuchiya**.

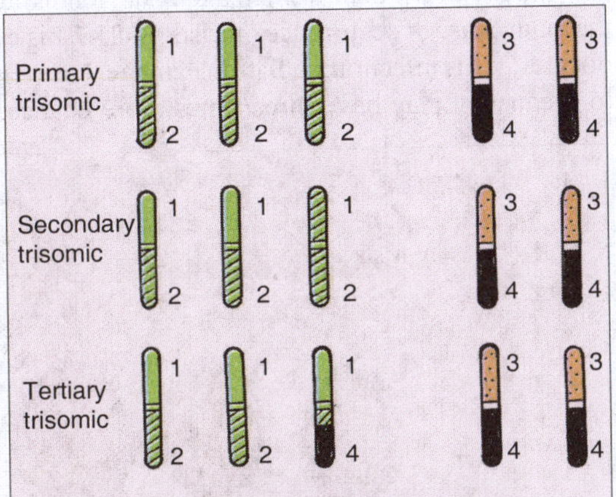

Fig. 15.12. Three kinds of trisomics.

Trisomy in humans. In human beings, the following three syndromes have been studied :

A. Down's syndrome (DS) or Trisomy-21.

Down's syndrome is named after the physician **J.Langdon Down** who first described this genetic defect in 1866 and it was formally called **mongolism** or **mongolian idiocy**. It is usually associated with a trisomic condition for one of the smallest human autosomes (*i.e.,* chromosome 21). It is the most common chromosomal abnormality in live births (1/650 births). There are about 50 physical characteristics shown by DS infants soon after birth. These include mild or moderate mental

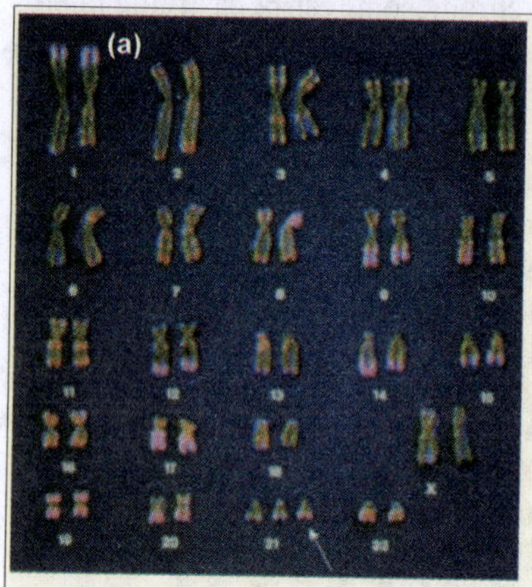

(a) Karyotype of a Down's syndrome child.

(b) Girls suffering from Down's syndrome.

retardation; eyes that slant up and out with internal epicanthal folds; a tongue that is large, swollen and protruding ; small and under developed ears; a single palmar crease; short stature; stubby fingers; an enlarged liver and spleen. Women over 45 years of age are about twenty times more likely to give birth to a child with DS than women aged 20. Nondisjunction of chromosome pair 21 during oogenesis is the main cause of occurrence of trisomy-21. This event is found to be affected either by senescence of oocytes, virus infection, radiation damage, etc. (*e.g.,* mothers who have had infectious hepatitis prior to pregnancy may have three times more chances to give birth to DS infants). Nondisjunction of

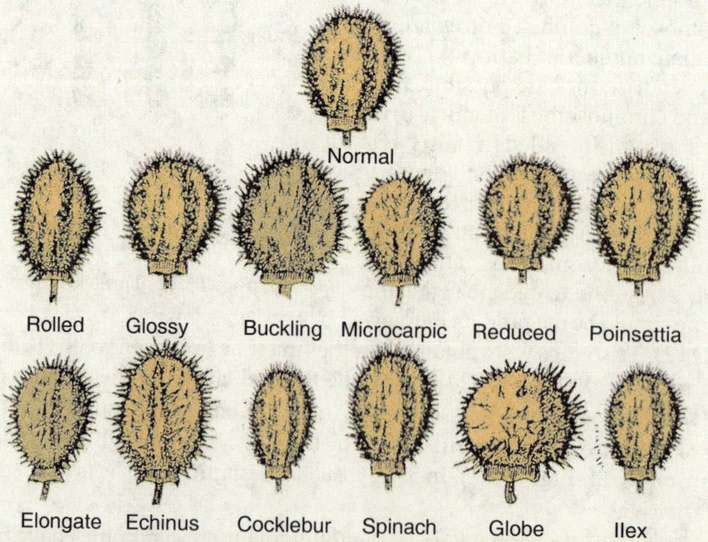

Fig. 15.13. Fruit capsules of the 12 primary trisomics of *Datura stramonium,* each with its particular phenotype (after Sybenga, 1972).

Plant of *Datura stramonium*.

chromosome pair 21 during spermatogenesis can also produce child with DS, but paternal age does not seem to be associated with its incidence.

Lastly, in about 2 to 5 per cent cases, the normal chromosome number is present (2n = 46), but the extra chromosome 21 is attached (translocated) to one of the larger autosomes (usually chromosome 14).

B. Edward's syndrome or Trisomy-18. First described in 1960 by **John H. Edwards** and his colleagues, **trisomy-18** is found to contain an incidence of about 0.3 per 1000 births. It is characterized by multiple malformations, primarily low-set ears; small receding lower jaw; flexed and clenched fingers; cardiac malformations; and various deformaties of skull, face and feet. Harelip and cleft palate often occurs. Death takes place around 3 to 4 months of age. Trisomy-18 children show evidence of severe mental retardation, which is more pronounced in females (the reason is still not clear). Like the Down's syndrome, occurrence of Edward's syndrome is too related with maternal age (*i.e.*, 35 to 45 year old mothers have more chance of giving birth to trisomy-18 infant).

C. Patau syndrome or Trisomy-13. This syndrome was described in 1960 by **Klaus Patau** and coworkers. Its incidence is about 0.2 per 1000 births. Individuals with Patau syndrome appear to be markedly mentally retarded; have sloping forehead, harelip and cleft palate. Polydactyly (both hands and feet) is almost always present; the hands and feet are deformed. Cardiac and various internal defects (of kidney, colon, small intestine) are common. Death usually occurs within hours or days, but the foetus may abort spontaneously. (**Note** : Sex chromosomal variations will be described in Chapter 18 of Human Genetics).

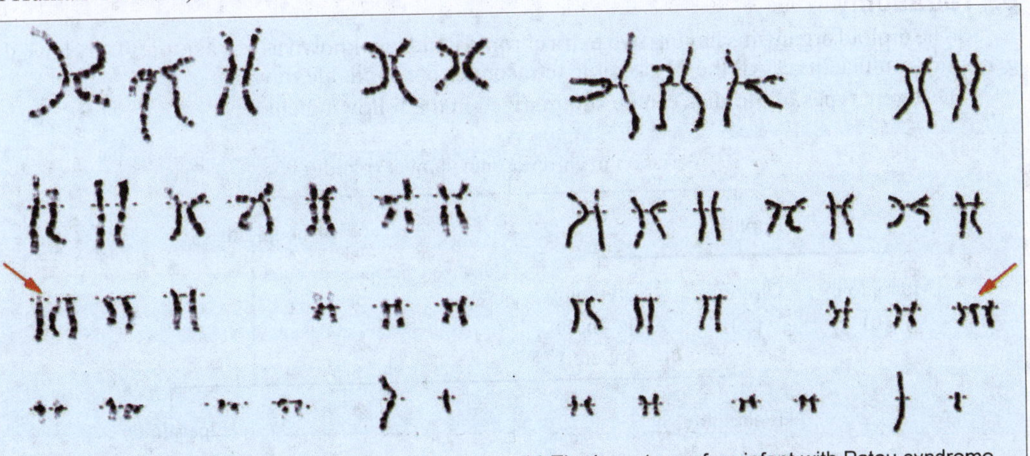

(a) The karyotype of Edwards syndrome. (b) The karyotype of an infant with Patau syndrome.

Trisomy in non-humans. Trisomy-22 has been reported in chimpanzees (**McClure** *et al.*, 1969); this shows Down syndrome-like phenotypic features. **Trisomy-21** has been reported in the gorilla.

Cytology of trisomics. The trisomics have an extra chromosome which is homologous to one chromosome of the diploid complement. Therefore, it forms **trivalent** which may take a variety of shapes in primary and secondary trisomics (Fig 15.14). In a tertiary trisomic a characteristic pentavalent is observed.

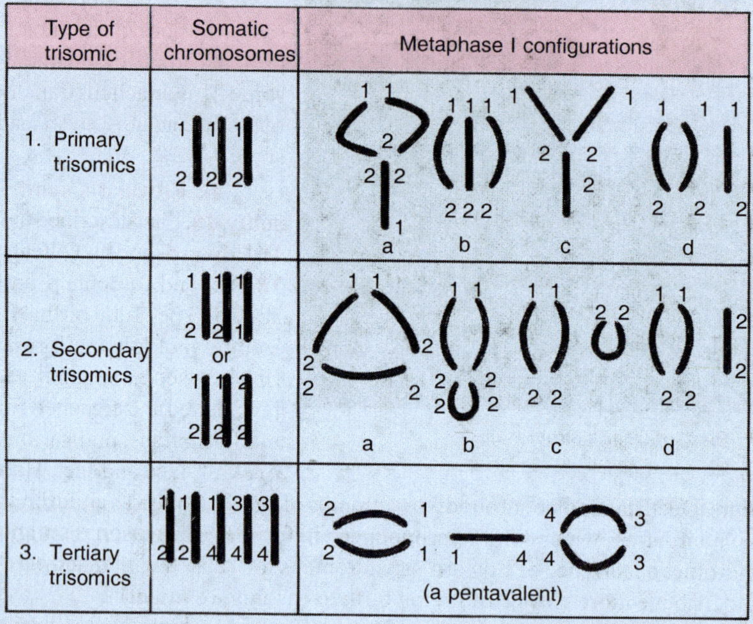

Fig. 15.14. Meiotic configurations formed at metaphase I in different types of trisomics.

4. Double Trisomy

In a diploid organism when two different chromosomes are represented in triplicate, the double trisomic is resulted. The double trisomic causes great genetic imbalance and has the genomic formula 2n+1+1.

5. Tetrasomy

The diploid organisms having two extra chromosomes are known as **tetrasomic**. They have the genomic formula 2n+2. All the 21 possible tetrasomics are available in wheat.

Different types of ploidies can be summarized in the following chart :

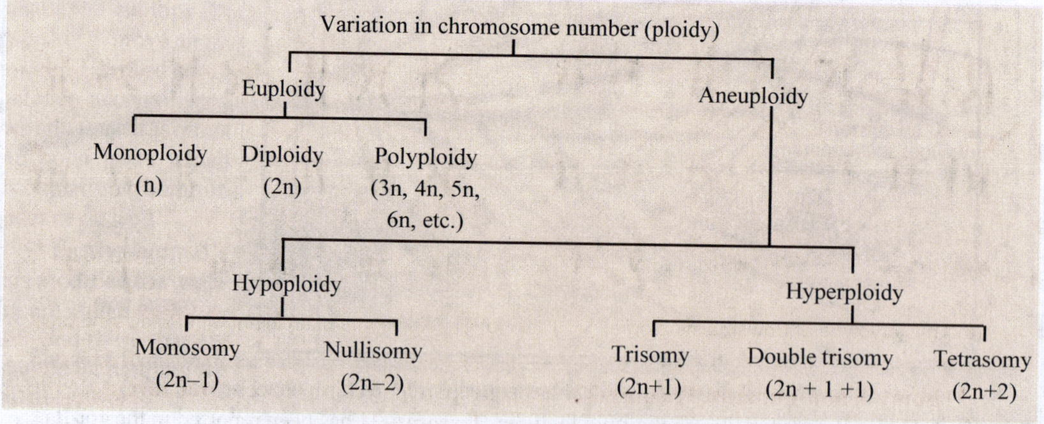

REVISION QUESTIONS AND PROBLEMS

1. How can you differentiate between the terms haploidy and monoploidy ? How can monoploids be produced and utilized in plant breeding?
2. What is polyploidy ? What are different kinds of polyploids ? How will you distinguish between autopolyploids and allopolyploids.
3. How can triploidy leads to seedlessness ? Discuss this by using examples of Kihara's seedless water-melons.
4. Some of the most desirable apples are triploids. If a desirable mutation occurred in a branch of triploid tree, how would you establish an orchard of trees with this mutation.
5. Briefly discuss the significance and role of polyploidy in evolution. Be sure to state whether autopolyploidy or allopolyploidy has been more important in speciation and why?
6. What is triticale ? What is the difference between primary and secondary triticales ?
7. How will you distinguish cytologically : (i) between a double monosomic and a nullisomic; (ii) between a primary trisomic and a secondary trisomic.
8. Write short notes on the following :
   (a) Aneuploidy; (b) Euploidy ; (c) Nullisomic ; (d) Evolution of wheat ; (e) *Raphanobrassica* ; (f) Colchicine treatment, (g) Down's syndrome ; (h) Trisomy-18; and (i) Patau syndrome.
9. Abyssinian oat (*Avena abyssinica*) appears to be a tetraploid with 28 chromosomes. The common cultivated oat (*Avena sativa*) appears to be a hexaploid in the same series. How many chromosomes does the common oat possess?
10. The diploid number of an organism is 12. How many chromosomes would be expected in (a) a monosomic, (b) a trisomic, (c) a tetrasomic, (d) a double trisomic, (e) a nullisomic, (f) a monoploid (g) a triploid, (h) an autotetraploid ?
11. Application of colchicine to a vegetative bud of a homozygous tall diploid tomato plant (DD) caused development of a tetraploid branch. What is the genotype of the somatic cells of this branch?
12. Different species of rhododendron have somatic chromosome numbers of 26, 39, 52, 78, 104 and 156. By what means does evolution appear to be taking place in this genus ?
13. How many sets are represented in the species with 156 chromosomes ?
14. Both autoploidy and allopolyploidy can result in a species with double the original chromosome number. If you had a plant with 4n complement. How would you determine cytologically if it were an autopolyploid or an allopolyploid ?
15. The loci of genes A and B are on different chromosomes. A dihybrid autotetraploid plant of genotype AA aa BB bb is self-pollinated. Assume that only diploid gametes are formed and that the loci A and B are very close to their respective centromere (chromosome segregation). Find the phenotypic expectations of the progeny.
16. The European raspberry (*Rubus idaeus*) has 14 chromosomes. The dewberry (*Rubus caesius*) is a tetraploid with 28 chromosomes. Hybrids between these two species are sterile F_1 individuals. Some unreduced gametes of the F_1 are functional in backcrosses. Determine the chromosome number and level of ploidy for each of the following : (a) F_1; (b) F_1 backcrossed to *R. idaeus ;* (c) F_1 backcrossed to *R. caesius;* to (d) chromosome doubling of F_1 (*R. maximus*).
17. If a plant were trisomic for one of its chromosomes and these chromosomes carried the alleles A, A^1, and A^2, respectively, what would be the genotypes of the gametes produced with respect of these genes ?
18. What types and proportions of eggs will be produced by a *Drosophila* female trisomic for chromosome 4 and of genotype +/+ +/ey ? (ey= eyeless is a recessive fourth chromosome gene).
    (a) If the female is crossed to an ey/ey male, what phenotypic ratio is expected in the offspring ?
    (b) What proportion of the diploid offspring will be eyeless ?

ANSWERS TO PROBLEMS

9. 42.
10. (a) 11; (b) 13 ; (c) 14 ; (d) 14 ; (e) 10 ; (f) 6 ; (g) 18.
11. DDDD.
12. Euploidy ; the different chromosome number are respectively 2n, 3n, 4n, 6n, 8n, and 12 n.
13. 12 sets (*i.e.,* 12 × 13).
14. Examine meiotic cells cytologically. If an autopolyploid, complex synaptic associations involving chromosomes (eight chromatids) will be evident. If an allopolyploid, tetrads composed of two chromosomes (four chromatids) each should occur.
15. 1225 AB : 35 Ab : 35 aB : 1 ab.
16. (a) 21, triploid ; (b) 28, tetraploid ;
    (c) 35, pentaploid ; (d) 42, hexaploid.
17. Six kinds of gametes could be produced : AA^1, A^2, AA^2, A^1, A^1A^2 and A.
18. (a) Eggs : 1 +/+ : 2 +/ey : 2 + : 1 ey
    Phenotypic ratio : 5 + : 1 ey (*i.e.,* 5 wild type : 1 eyeless) ;
    (b) 1/3 of the diploid progeny will be eyeless.

CHAPTER

16

Gene Mutation

A gene mutation is abrupt inheritable qualitative or quantitative change in the genetic material of an organism. Since in most organisms genes are segments of DNA molecule, so a mutation can be regarded as a change in the DNA sequence which is reflected in the change of sequence of corresponding RNA or protein molecules. Such a change may involve only one base/base pair or more than one base pair of DNA. Mutations occur in a **random** manner, *i.e.,* they are not directed according to the requirements of the organism. Most mutations occur **spontaneously** by the environmental effect, however, they can be induced in the laboratory either by radiations, physical factors or chemicals (called **mutagens**). A unicellular organism is more subjected to environmental onslaughts since it is at the same time a somatic or germ cell. In multicellular organisms the germ cells are distinct cells, and are relatively protected from the environment. Mutation has a significant role to play in the origin of species or evolution.

A chance mutation in a single gene of the speckled moth *Biston betularia* produced a fortunate change from the normal speckled colouration, seen in the moth on the left, to the melanic form seen on the right. In sooty areas, this new form was better camouflaged and so had a better chance of avoiding predators.

HISTORICAL BACKGROUND

The earliest record of point mutations dates back to 1791, when **Seth Wright** noticed a lamb with exceptionally short legs in his flock of sheep. Visualising the economic significance of this short-leged sheep. *i.e.,* short legged sheep could not cross the low stone fence and damage the crop fields in the vicinity, he produced a flock of sheeps, each of which having short legs by employing artificial breeding techniques. The short legged breed of sheep was known as **Ancon** breed. Later on, the trait of short legs was found to be resulted from a recessive mutation and the short legged individuals were found to be homozygous recessive.

Hugo de Vries was the first hybridist who used the term "mutation" to describe the heritable phenotypic changes of the evening primrose, *Oenothera lamarckiana.* Many mutations described by de Vries in *O. lamarckiana,* are now known to be due to variation in chromosome number or ploidy and chromosomal aberrations (*viz.* **gross mutations**). The first scientific study of mutation was started in 1910, when **Morgan** started his work on fruitfly, *Drosophila melanogaster* and reported white eyed male individuals among red eyed male individuals. The discovery of white eyed mutants in *Drosophila* was followed by an extensive search of other mutants of *Drosophila* by **Morgan** and his co-workers and other geneticists. Consequently about 500 mutants of *Drosophila* have been reported by geneticists all over the world. Later on, several cases of mutations have been reported in a variety of microorganisms (*e.g.,* bacteriophages, bacteria (*Escherichia coli*), *Neurospora,*etc., plants (*i.e.,* pea, snapdragon, maize, etc.) and animals, *i.e.,* rodents, fowls, man, etc.).

OCCURRENCE

Mutations occur frequently in the nature and have been reported in many organisms, *e.g., Drosophila,* mice and other rodents, rats, rabbits, guinea pigs and man. In the *Drosophila,* mutation causes white and pink eyes, black and yellow body colours, and vestigial wings. In rodents the mutuations are responsible for black, white and brown coats. In man, the mutations cause variation in hair colour, eye colour, skin pigmentation and several somatic malformations. Various genetical diseases of human beings such as haemophila, colourblindness, phenylketonuria, etc., form other examples of mutation in human beings.

How does a mutation act? Any change in sequence of nucleotides in the DNA will result in the corresponding change in the nucleotide sequence of mRNA. This may result in alignment of different tRNA molecules on mRNA (during protein synthesis). Thus, the amino acid sequence, and, hence, the structure and properties of the enzyme formed will be changed. This defective enzyme or structural protein may adversely affect the trait controlled by the protein. In consequence, a mutant phenotype makes its expression.

KINDS OF MUTATIONS

There exists a lot of controversy about the possible kinds of mutations among geneticists. They have been classified variously according to different criteria as follows :

1. Classificatjon of Mutation According to Type of Cells

According to their occurrence in somatic and germinal cells following types of mutations have been classified :

A. Somatic mutations. The mutations occurring in non-reproductive body cells are known as **somatic mutations**. The genetical and evolutionary consequences of somatic mutations are insignificant, since only single cells and their daughter cells are involved. If, however , a somatic mutation occurs early during embryonic life, the mutant cells may constitute a large proportion of body cells and the animal body may be a mosaic for different types of cells. Somatic mutations have been often related with malignant (cancerous) growth. Examples of somatic mutation have been reported in *Oenothera lamarckiana* (Hugo de Vries) and several other cases in-

In two kittens in this litter, the colour of the left eye doesn't match that of the right eye. These are examples of 'mosaic' phenotypes, often due to somatic mutation.

cluding man. In man somatic mutation causes several fatal diseases such as *paraoxysmal nocturnal hemoglobinura, circumscribed neurofibroma, unilateral retinoblastoma* and *heterochromia of the iris.*

B. Gametic mutations. The mutations occurring in gamete cells (*e.g.,* sperms and ova) are called **gametic mutations**. Such mutations are heritable and of immense genetical significance. The gametic mutations only form the raw material for the natural selection.

2. Classification of Mutations According to the Size and Quality

According to size following two types of mutations have been recognised :

A. Point mutation. When heritable alterations occur in a very small segment of DNA molecule, *i.e.,* a single nucleotide or nucleotide pair, then this type of **mutations** are called "**point mutations**". The point mutations may occur due to following types of subnucleotide change in the DNA and RNA.

1. Deletion mutations. The point mutation which is caused due to loss or deletion of some portion (single nucleotide pair) in a triplet codon of a cistron or gene is called **deletion mutation**. Deletion mutations have been frequently reported in some bacteriophages (Phage T_4).

2. Insertion or addition mutation. The point mutations which occur due to addition of one or more extra nucleotides to a gene or cistron are called **insertion mutations**. The insertion mutations can be artificially induced by certain chemical substances called **mutagens** such as acridine dye and proflavin. A proflavin molecule, it is believed, insert between two successive bases of a DNA strand, thereby stretching the strand lengthwise. At replication, this situation would allow the insertion of an extra nucleotide in the complementary chain at the position occupied by the proflavin molecule.

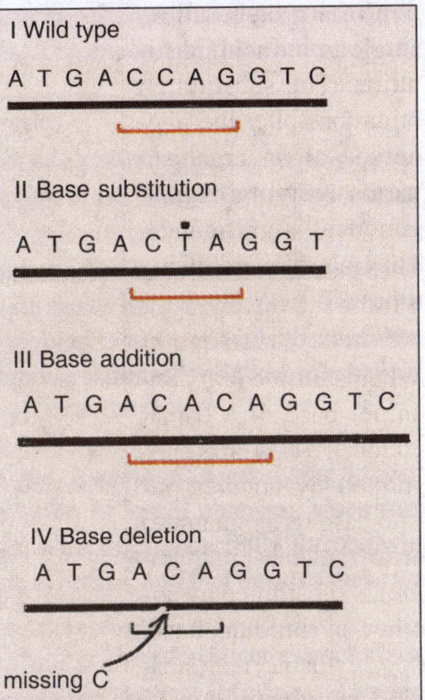

Fig. 16.1. Three types of point mutations. Only the base sequence of one DNA strand is shown. Changes are shown in square; the horizontal brackets indicate the affected segment (after Freifelder, 1985).

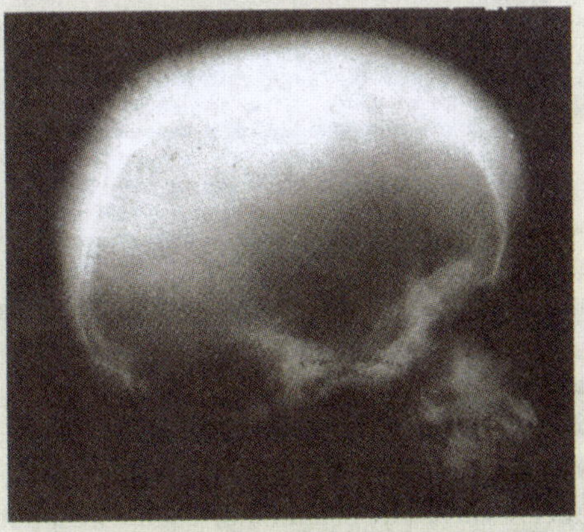

A child with thalassemia has a single base substitution of a gene for haemoglobin. The spleen removes altered RBC, eventually causing severe anaemia. RBC are replaced by excessive bone marrow. The rapid growth of bone marrow in the skull causes an enlarged head and a peculiar "hair on end" appearance of the bone when X-rayed.

The mutations which arise from the insertion or deletion of individual nucleotides and cause the rest of the message downstream of the mutation to be read out of phase, are called **frameshift mutations**. They result in the production of an incorrect, hence, inactive protein, due to which the death of the cell may occur.

3. Substitution mutation. A point mutation in which a nucleotide of a triplet is replaced by another nucleotide, is called **substitution mutation**. The substitution mutation affect only a particular triplet codon. Such an altered code word (triplet codon) may designate a different amino acid and may result in the production of a protein with a single amino acid substitution. The substitution mutations alter the phenotype of an organism variously and are of great genetical significance. They may be of following types :

(i) Transition. When a purine (*e.g.,* adenine) base of a triplet codon of a cistron is substituted by another purine base (*e.g.,* guanine) or a pyrimidine (*e.g.,* thymine) is substituted by another pyrimidine base, (*e.g.,* cytosine) then such kind of substitution is called **transition**. The transitional substitution mutations occur due to tautomerization.

Tautomerization

In a DNA molecule, normally, the purine, adenine (A) is linked to the pyrimidine, thymine (T), by two hydrogen bonds, while the purine guanine (G) is linked to the pyrimidine, cytosine (C) by three hydrogen bonds. Besides the common molecular configurations, each DNA base may have some altered uncommon molecular configuration, as has been shown in the figure 16.2.

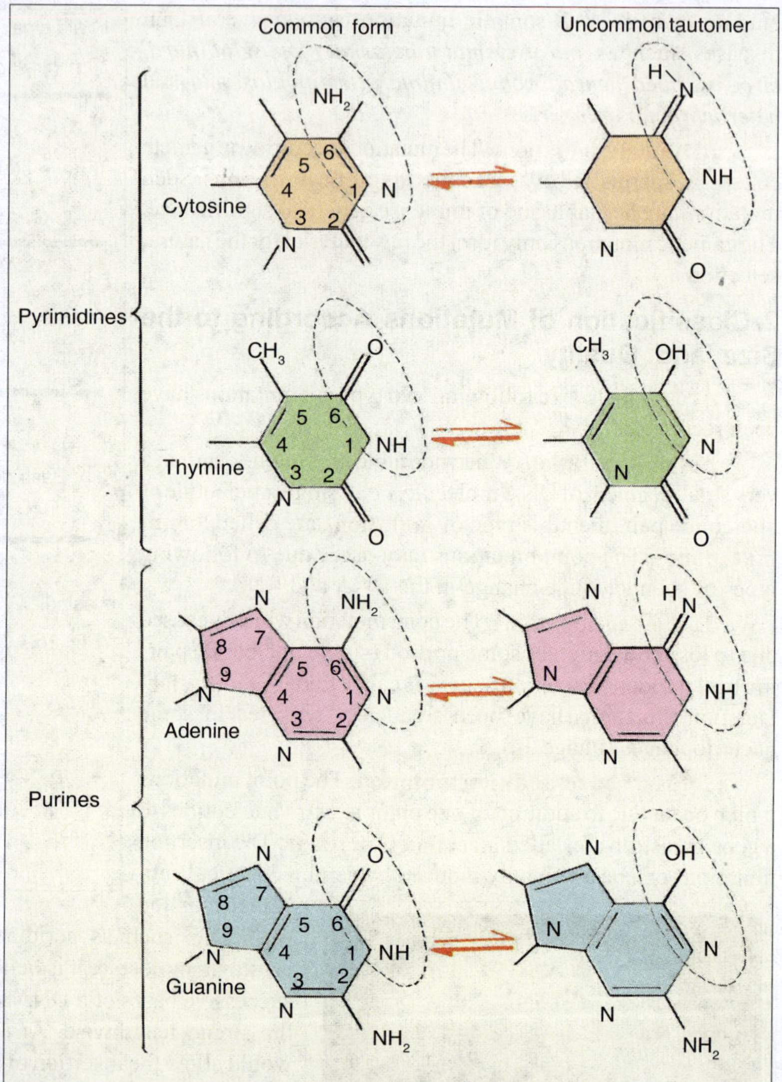

Fig. 16.2. The uncommon forms of DNA bases.

Such uncommon forms of DNA bases are generated by single proton shifts and are called **rare states** or **tautomers**. A tautomeric shift is believed to occur when the amino (NH_2) form of adenine is changed to an imino (NH) form. Similarly, a tautomeric shift may occur in thymine changing it form the keto (C=O) form to the rare enol (COH) form. When a base occurs in its rare or tatuomeric state, it cannot be linked to its normal partner. However, a purine, such as adenine can in its rare state forms a bond with cytosine (besides thymine), provided the cytosine is in its normal state.

Watson and **Crick** (1953) hypothesized that the occurrence of the bases in their rare states provides a mechanism for mutation during DNA replication. If, for example, adenine in an old chain is in its rare state at the moment that the complementary new chain reaches it, cytosine can pair with it (adenine) and be added to the growing end of the new chain. The result of this type of pairing is the formation of a DNA molecule that contains an exceptional base pair. This situation is not stable and at the next replication adenine is expected to return to its common state and to pair with thymine. Cytosine introduced into the complementary strand due to tautomeric shift in adenine, would then pair with guanine. Thus, there would be formed two kinds of DNA molecules, one that is identical to the original DNA and another that has undergone a base pair substitution of G-C for A-T. This transitionally substituted DNA molecule has altered coding at a point and results in recognizable mutation. Such mutations which formed during DNA replication are called **copy error mutations**. Such copy error mutations have been shown by Figure 16.4.

The abnormal pairing due to transitional substitution may also occur due to **ionization** of a base at the time of DNA replication. Ionization involves the loss of the hydrogen from the number one nitrogen of a base. For example, in its ionized state, thymine can pair with guanine, if the guanine is in its common form. In a similar fashion, guanine in its **ionized** state can pair with thymine in its common form. From any such unstable base pair, a transition will result following the steps outlined in Fig. 16.4 for A-T to G-C and G-C to AT.

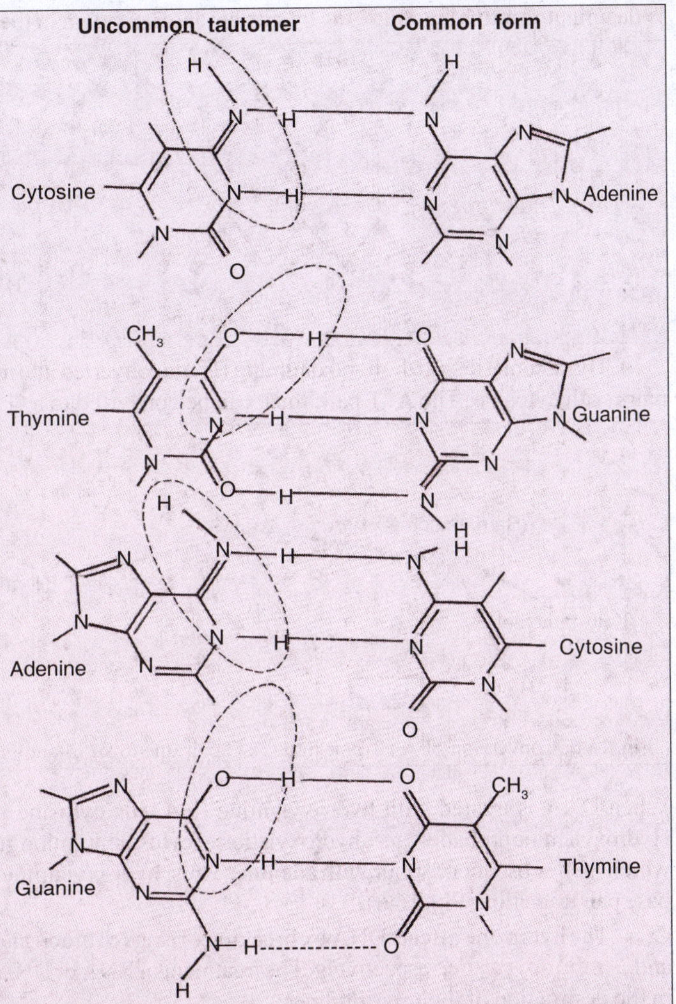

Fig. 16.3. Pairing qualities of rare tautomers of four DNA bases.

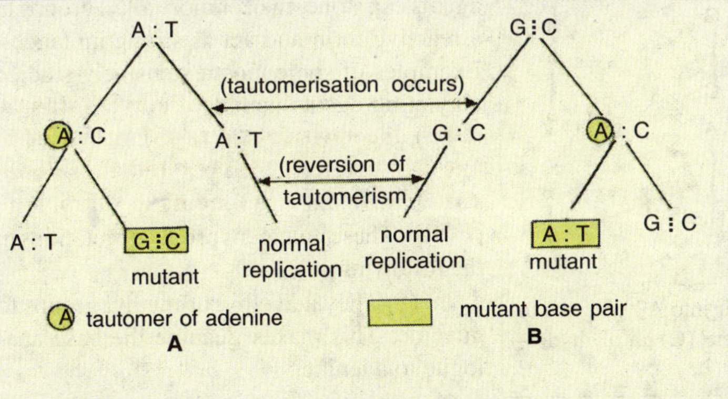

Fig. 16.4. Copy error mutation due to tautomerization of adenine.

Effect of Chemical Mutagens on Nucleotide Sequence

(a) Alteration in Resting Nucleic Acid

1. Deamination. Some chemical substances such as nitrous acid causes transitional mutation due to oxidative deamination of DNA bases. In the process of oxidative deamination, the amino group (NH_2) of a DNA base is replaced by hydroxyl (OH) group by the chemical mutagen. Thus, adenine is deaminated into **hypoxanthine** by nitrous acid as shown in the following figure :

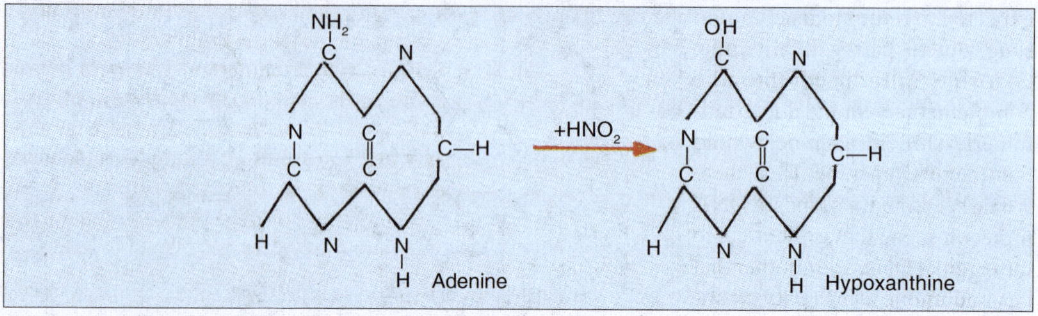

By tautomeric shift the hypoxanthine (HX) is converted into more common or keto-tautomer which pairs with cytosine. The A: T pair, thus, can be converted to a G : C pair.

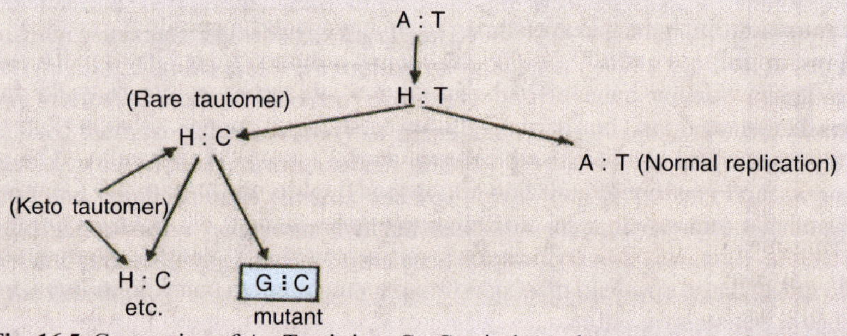

Fig. 16.5. Conversion of A : T pair into G : C pair due to keto-tautomer of adenine.

Similarly, deamination converts cysosine to uracil, which has pairing properties similar to thymine and in such a case G: C pair would be changed into A : T pair.

2. Hydroxylamine (HA = NH_2. OH) and hydrazine (HZ = NH_2 NH_2). When DNA is treated with hydroxylamine (HA), its cytosine base is the strongest reacting base. Hydroxylamine probably cause hydroxylation of cytosine at amino group giving rise to **hydroxylcytosine**, which then subsequently pair with adenine. Thus, hydroxylamine (HA) induces in DNA a GC → AT base pair transition (Fig. 16.6).

The hydrazine affects DNA by breaking of rings of uracil and cytosine giving rise to **pyrazolone** and **3-aminopyrasole**, respectively. The treatment of RNA or DNA with **anhydrous hydrazine** results in the destruction of their pyrimidines.

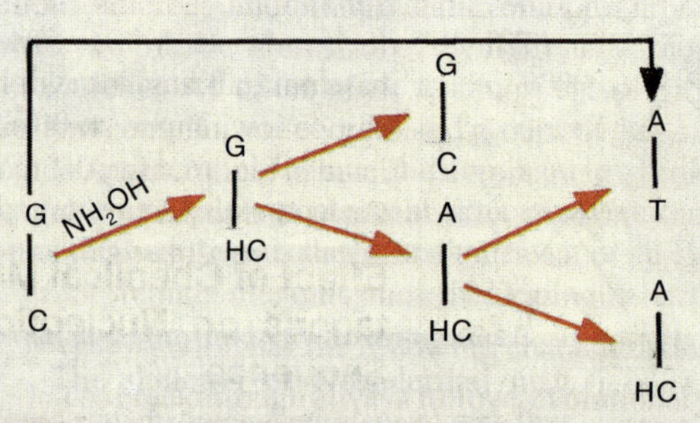

Fig. 16.6. Conversion of GC pair into AT pair due to conversion of cytosine (C) into hydroxylcytosine (HC).

3. Alkylating agents. Some alkylating agents carry one, two, or more alkyl groups in a reactive form and act as strong mutagens. Examples of some most extensively studied alkylating agents include **diethyl sulphate (DES)**, **dimethylsulphate (DMS)**, **methyl methane sulphonate (MMS)**, **ethyl ethane sulphonate (EES)** and **ethyl methane sulphonate (EMS)**. These mutagens produce mutations in the following ways :

(1) They add ethyl or methyl groups to guanine. This makes guanine the base analogue to adenine.

(2) They remove the alkylated guanine. This is known as **depurination**. The loss of the base produces gaps in the DNA chain which may be filled with a wrong base, thus, producing mutation.

(3) The gap may also produce a deletion, causing mutation.

(b) Alteration during Replication of Nucleic Acid

1. Base analogues. Certain chemical substances have molecular structure similar to the usual DNA bases that, if they are available, such **analogues** may be incorporated into a replicating DNA strand. For example, **5-bromouracil** (5BU) or its nucleoside **5-bromodeoxyuridine** (5-BUdR) in its usual (keto) form is a structural analogue of thymine (5-methyl uracil) and it will substitute for thymine. Thus, an A-T pair becomes and remains A-BU. There is some in vitro evidence to indicate the BU immediately adjacent to an adenine in one of DNA strands causes the latter to pair with guanine. But, in its rare (enol) state, 5BU behaves similar to the tautomer of thymine and pairs with guanine. This converts A : T to G : C as shown in Figure 16.7.

2–Aminopurine (2-AP) is another base analogue which is a relatively undifferentiated purine that apparently can pair with cytosine and thymine. It is thought that 2-AP acts by "switching" pyrimidines : for example, it may be incorporated opposite thymine during one round of replication and then pair with a cytosine at the next round to produce an AT → GC transition (see **Goodenough** and **Levine**, 1974).

2. Inhibition of precursors of nucleic acids. There are some mutagens which interfere with the synthesis of nitrogen bases of nucleic acids such as purines or pyrimidines. Often lack of one base either causes breaks or causes pairing mistakes. For example, **azaserine** (an potent alkylating agent) inhibits purine synthesis and **urethane** (a mild alkylating agent) is an inhibitor of pyrimidine synthesis. However, urethane induced chromosome breaks are inhibited by thymine.

3. Transversion. The substitution mutation when involves the substitution or replacement of a purine with a pyrimidine or vice versa then that type of substitution mutation is called **transversion mutation**. The existence of transversion mutation was first of all postulated by **E. Freese** in 1959. We have still poor information about the mechanism of induction, identification and characterization of transversion mutations. Moreover, it is extremely difficult to recognize transversion mutations genetically. However, they can be recognized only by analysis of amino acid substitutions in proteins.

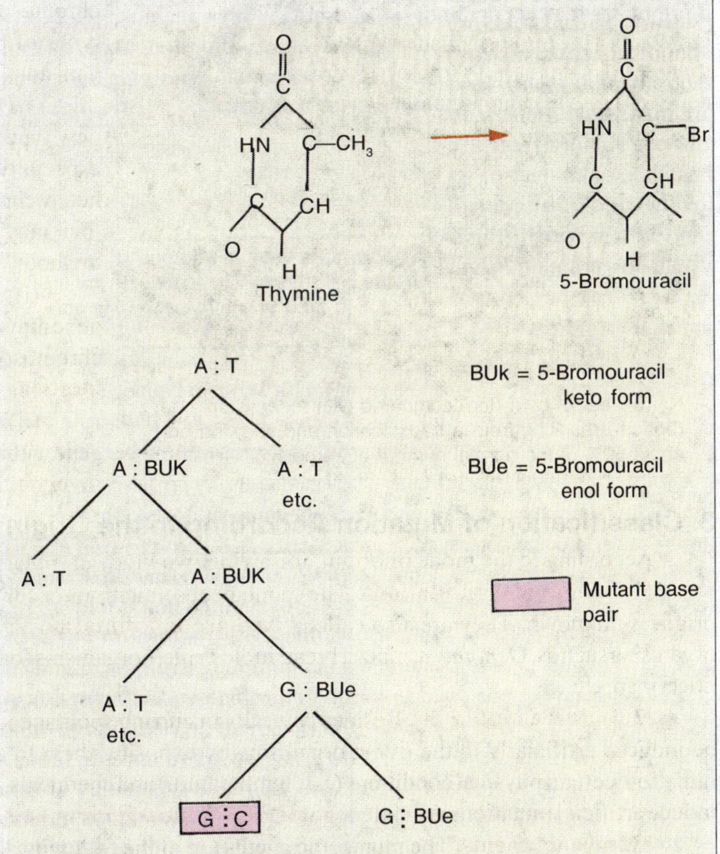

Fig. 16.7. Conversion of A : T base pair into G : C base pair due to keto and enol forms of bromouracil.

Effects of Physical Conditions on Nucleotide Sequence

High temperature and low pH value are known to affect depurination or loss of purine bases. The removal of a purine from a strand of DNA leaves a gap at that point. At the time of replication, it would be possible for any of the four bases to insert in the complementary newly formed strand. If the inserted nucleotide contained a purine, the complementary strand would contain a transversion.

B. Multiple mutations or gross mutations. When changes involving more than one nucleotide pair, or entire **gene**, then such mutations are called **gross mutations**. The gross mutations occur due to rearrangements of genes within the ganome and may be of the following types :

1. The rearrangement of genes may occur within a gene. Two mutations within the same functional gene can produce different effects depending on gene whether they occur in the cis or trans position.

2. The rearrangement of gene may occur in number of genes per chromosome. If the numbers of gene replicas are non-equivalent on the homologus chromosomes, they may cause different types of phenotypic effects over the organisms.

Heterozygous inversion and translocation in *Drosophila*.
(A) Pairing between two X chromosomes one of which contains an inverted section.
(B) Pairing between two normal chromosomes and two that have exchanged sections.

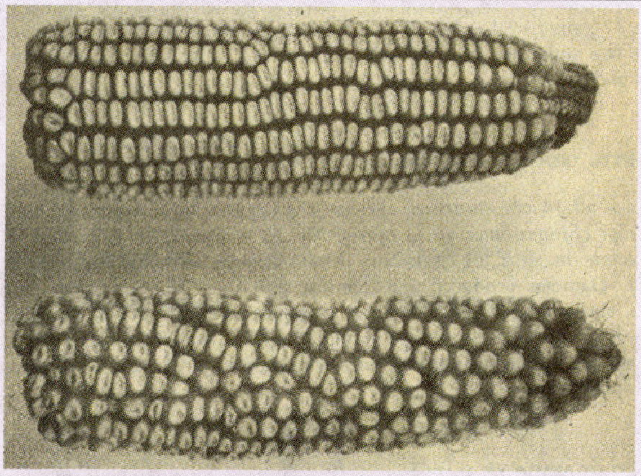

Normal ear of corn (top) compared with a semisterile ear (bottom) resulting from a translocation and showing only about 50% of the normal number of kernels.

3. Due to movement of a gene locus new type of phenotypes may be created, especially when the gene is relocated near heterochromatin. The movement of gene loci may take place due to following method:

(i) **Translocation.** Movement of a gene may take place to a non-homologous chromosome and this is known as **translocation**.

(ii) **Inversion.** The movement of a gene within the same chromosome is called **inversion**.

3. Classification of Mutation According to the Origin

According to the mode of origin, following two kinds of mutations have been recognised :

(1) Spontaneous mutations. The spontaneous mutations occur suddenly in the nature and their origin is unknown. They are also called "**background mutation**" and have been reported in many organisms such as, *Oenothera,* maize, bread molds, microorganisms (bacteria and viruses), *Drosophila,* mice, man, etc.

(2) Induced mutations. Besides naturally occurring spontaneous mutations, the mutations can be induced artificially in the living organisms by exposing them to abnormal environment such as radiation, certain physical conditions (*i.e.,* temperature) and chemicals. The substances or agents which induce artificial mutations are called **mutagenes** or **mutagenic agents**.

Mutagenic agents. The mutagenic agents are of the following kinds :

A. Radiations. The radiations which are important in mutagenesis are of two categories : one type is **ionizing radiations** such as X-rays and gamma rays; alpha and beta rays; electrons, neutrons, protons and other fast moving particles. The second type is **non-ionizing radiations** such as ultraviolet and visible light. Both types of radiations induce mutations by following methods :

Fig. 16.8. Radiation-induced chromosomal bridges and fragments in cells of X-rayed anthers of *Trillium*.

(i) Ionizing radiations as mutagens. Relatively little is known about the mechanism by which ionizing radiations cause mutation. As, we are already familiar that matter composed of atoms and atoms, in their turn, are made up of a positively charged atomic nucleus (with neutrons, protons) and a surrounding constellation of negatively charged electrons. The charges of atomic particles remain so balanced that normal atoms are electrically neutral. When ionizing radiations pass through matter, they dissipate their energy in part through the ejection of electrons from the outer shell of atoms and the loss of these balancing, negatively-charged particles (electrons) leaves atoms which are no longer neutral but are positively charged. The positively-charged atom is called **ion**. The ejected electrons move at high speed, knock other electrons free from their respective atoms and when their energy is dissipated, become attach to other atoms and convert the ion into negatively charged ions. To achieve their stable configuration (*i.e.,* neutral charge) ions undergo many chemical reactions and during these chemical reactions ionizing radiation is thought to cause mutation.

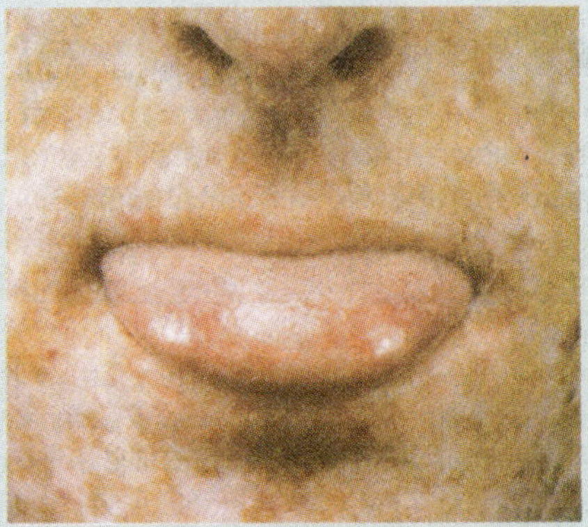

The inherited genetic disorder xeroderma pigmentosum. Those who have this disease are homozygous for a mutation that destroys body cells' ability to repair UV damage. These individuals develop extensive skin tumours on exposure to sunlight.

Further, ionizing radiations cause breaks in polysugar phosphate backbone of DNA and, thus, causing chromosomal mutations such as break, deletion, addition, inversion and translocation. During breakage of DNA molecule due to ionizing radiation the active role of oxygen is predicted. Because, oxygen is important in the formation of H_2O_2 and HO_2 in irradiated water and these products may induce breaks in DNA molecule.

(ii) Non-ionizing radiations as mutagens. The ultraviolet (UV) light is a non-ionizing radiation which may cause mutation. The most effective wave length of ultraviolet for inducing mutations is about 2,600 A°. This is a wave length that is best absorbed by DNA and a wave length at which proteins absorb little energy. When a substance absorbs sufficient energy from the ultraviolet light, some of their electrons are raised to higher energy levels, a state called **excitation**. The excited molecule becomes reactive and mutated and is called **photoproducts**.

Dimerization. The ultraviolet radiation produces several effects on DNA, one being the formation of chemical bonds between two adjacent pyrimidine molecules in a polynucleotide and particularly, between adjacent thymine residues as shown in Fig. 16.9. As the two thymine residues associate, or **dimerize** to form a **dimer**, their position in the DNA helix becomes so displaced that they can no longer form hydrogen bonds with the opposing purines and thus regularity of the helix becomes distorted. Thus, dimerization interferes with the proper base pairing of thymine with adenine, and may result in thymine's pairing with guanine. This will produce a T-A To C-G transition.

B. Temperature as mutagen. The rate of all chemical reactions are influenced by temperature. It is not surprising that temperature can be mutagenic. It is reported that the rate of mutation is increased due to increase in temperature. For example, an increase of 10°C temperature increases the mutation rate two or three fold. Temperature probably affects both thermal stability of DNA and the rate of reaction of other substances with DNA.

A study of swedish nudist indicated that the scrotal temperature of human males in ordinary clothing is about 3°C higher than that of nude male. The higher temperature could well increase the mutation rate nearly two fold, leading the investigators to suggest that the wearing of pants has possibly been much more unhygienic than fall out from testing of nuclear devices threatens to be. They suggested the wearing of kilts as one solution.

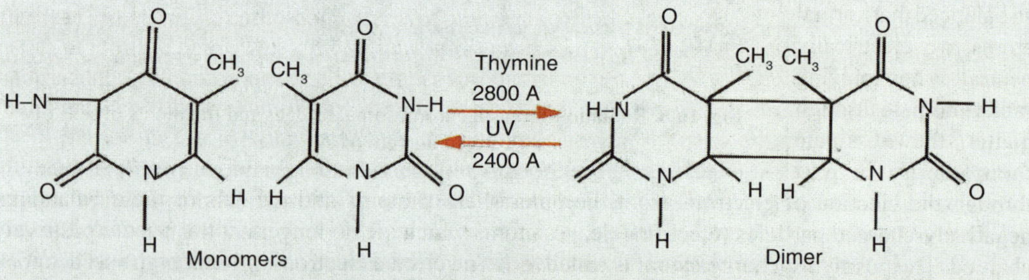

Fig. 16.9. Formation of a dimer of thymine.

C. Chemical mutagens. Many chemical substances have been responsible to increase the mutability of genes. The ability of chemicals to induce mutation was first of all demonstrated by **Auerbach** and **Robson** in 1947 using mustard gas and related compounds as the nitrogen and sulphur mustards, mustard oil and chloracetone in experiments with male *Drosophila melanogaster*. Since then many chemical compounds which are ordinarily considered to be non-toxic have been found to be mutagenic in certain specific situations. Any chemical substance that affects the chemical environment of chromosomes is likely to influence, at least indirectly, the stability of DNA and its ability to replicate without error. A chemical mutation can cause mutation only when it enters in the nucleus of the cell. It can affect the chromosomal DNA by following two ways :

(1) Direct gene change. Certain chemical mutagens affect DNA directly. They affect the constituents of DNA only when DNA is not replicating. For example, nitrous acid converts adenine into hypoxanthine and cytosine to uracil by deamination. Like the nitrous acid, nitrogen mustard, formaldehyde, epoxides, dimethyl and diethyl sulphonate, methyl and ethyl methanesulphonate (MMS and EMS) and nitrosoguanadine (NG) also have direct mutagenic effect on the DNA molecule.

(2) Copy error. Certain chemical compounds, called **base analogues** (*e.g.,* 5-bromouracil, 2-aminopurine, etc.) closely resemble with certain DNA bases and are, therefore, act as mutagens. During DNA replication, they are incorporated by DNA in place of the normal DNA bases. Certain other base analogues such as urethane triazine, caffiene (in coffee, tea and soft drinks), phenol and carcinogens, acridines (proflavin, etc.), have mutagenic effects. Certain inorganic substances such as manganese chloride is mutagenic for many organisms, as, they are the compounds which bind calcium and, thus, interfere with the integrity of the chromosome structure.

4. Classification of Mutation According to the Direction

According to their mode of direction following types of mutations have been recognised :

(A) Forward mutations. In an organism when mutations create a change from wild type to abnormal phenotype, then that type of mutations are known as **forward mutations**. Most mutations are forward type.

(B) Reverse or back mutations. The forward mutations are often corrected by error correcting mechanism, so that an abnormal phenotype changes into wild type phenotype. They may be of the following types :

(i) Single site mutation. Some reverse mutations change only one nucleotide in the gene and are called **single site mutations**. For example, due to forward mutation the adenine is changed into guanine and backward mutation change guanine into adenine :

$$\text{Adenine} \xrightarrow{\text{forward}} \text{Guanine} \xrightarrow{\text{reverse}} \text{Adenine}$$

(ii) Mutation suppressor. When a mutation occurs at a diffetent site from the site where already primary mutation occurred and that mutated gene reverse the effects of primarily mutated gene, then such (secondary) mutations are called **mutation suppressors**. They may be of following types :

(a) Extragenic suppressor. The extragenic suppressor mutation occurs in a different gene from that of the mutant gene. In *E. coli,* a gene mutation suppressor gene called **rec A** (*rec* for recombination) is known which is necessary for recombination and is found to repair ultraviolet-induced thymine dimers of a gene by a process called **postreplication recombinational repair** (see **Goodenough** and **Levine**, 1974).

(b) Intragenic suppressor. The intragenic suppressor mutation occurs in a different nucleotide within the same gene and shift the reading frame back into register.

This *Drosophila* is a mutant, bithorax. Due to a mutation in a gene regulating a critical stage during development it possesses two thorax segments and thus two sets of wings.

(c) Photoreactivation. In photoreactivation type reverse mutation reversal of ultraviolet induced thymine dimers takes place by specific enzymes in the presence of visible light waves. During ultraviolet radiation a particular enzyme is selectively bound to the bacterial DNA. During photoreactivation the enzyme is activated by visible light and that cleaves the pyrimidine or purine dimers into monomers and restores their original forms.

(d) Excision repair or Dark reactivation. In an ultraviolet (UV) induced mutation, the reverse mutation may also occur in the absence of light. According to **Howard Flanders** and **Boyce** (1964) dark reactivation includes following stages : (i) An enzyme possibly endonuclease makes a cut in the polynucleotide strand on either side of the dimer which may be formed due to ultraviolet radiation and excises a short, single strand segment of the DNA. (ii) Another enzyme, possibly exonuclease widens the gap produced by the action of the endonuclease. (iii) DNA polymerase resynthesizes the missing segment, using the remaining opposite strand as a template; and (iv) the final gap is closed by some enzymatic rejoining process, (*i.e.*, DNA ligase).

5. Classification of Mutation According to Magnitude of Phenotypic Effect

According to their phenotypic effects following kinds of mutations may occur :

1. Dominant mutations. The mutations which have dominant phenotypic expression are called dominant mutations. For example, in man the mutation disease **aniridia** (absence of iris of eyes) occurs due to a dominant mutant gene.

2. Recessive mutations. Most types of mutations are recessive in nature and so they are not expressed phenotypically immediately. The phenotypic effects of mutations of a recessive gene is seen only after one or more generations, when the mutant gene is able to recombine with another similar recessive gene.

3. Isoalleles. Some mutations alter the phenotype of an organism so slightly that they can be detected only by special techniques. Mutant genes that give slightly modified phenotypes are called **isoalleles**. They produce identical phenotypes in homozygous or heterozygous combinations.

4. Lethal mutations. According to their effects on the phenotype mutations may be classified as lethals, subvitals and supervitals. **Lethal mutations** result in the death of the cells or organisms in which they occur. **Subvital mutations** reduces the chances of survival of the organism in which they occur. **Supervital mutations**, in contrast, cause the improvement of biological fitness under certain conditions.

6. Classification of Mutation According to Consequent Change in Amino Acid Sequence

1. Missence mutations. They change the meaning of a codon, changing one amino acid into another.

2. Temperature sensitive mutations or Ts mutations. If the substitution produces a protein that is active at one temperature (typically 30°C) and inactive at a higher temperature (usually 40– 42°C).

3. Nonsense or chain termination mutations. They arise when a codon for an amino acid is mutated into a termination codon (UAG, UAA or UGA), resulting in the production of a shorter protein.

Since, temperature-sensitive and chain termination mutations exhibit the mutant phenotype only under certain conditions, they are called **conditional mutations**; they are the most versatile and useful mutations.

4. Silent mutations. They change a nucleotide but not the amino acid sequence because they affect the *third position* of the codon, which is usually less important in coding. This is a silent mutation because it leaves the protein sequence unchanged.

7. Classification of Mutation According to the Types of Chromosomes

According to the types of chromosomes, the mutations may be of following two kinds :

1. Autosomal mutations. This type of mutation occurs in autosomal chromosomes.

2. Sex chromosomal mutations. This type of mutation occurs in sex chromosomes.

MUTATION RATE

The frequency with which genes mutate spontaneously is called **mutation rate**. Most genes are relatively stable and mutation is a rare event. The great majority of genes have mutation rate of 1×10^{-5} to 1×10^{-5}, *viz.,* one gamete in 100,000 to one gamete in million would contain a mutation at a given locus. Mutations occur much more frequently in certain regions of the gene than in others. The favoured regions are called **hot spots**. The mutation rate is influenced by various factors which are as follows :

1. Genetic control of mutation rate. There are ample evidences which show that mutation rate is under genetic control, *viz.,* certain genes called **mutator genes** may increase the mutation rate in

Drosophila (**Demere**, 1937), maize (**Rhoades**, 1938) and *E. coli* (**Goldstein**, 1955). However, certain suppressor genes may decrease the rate of mutation.

In bacteria, as well as in eukaryotes, spontaneous mutations most frequently also caused by **transoposons** which are segments of DNA that have a tendency to jump around the genome.

2. Viral control of mutation rate. Virus reportedly affect the mutability of host's genes. **Sprague** (1963) experimented with maize suggested that virus may cause mutation. **Baumiliar** (1967) reported that viruses increase the mutation rate in *Drosophila melanogaster.* But, still we do not know how viruses increase the mutability of host genes.

3. Environmental control of mutation rate. There are three major environmental factors that affect mutation rates, *viz,* temperature, certain radiations and chemicals.

METHOD OF DETECTION OF SEX-LINKED LETHAL MUTATION

H.J. Muller devised an easy method for detecting lethal mutations in the sex chromosomes of *Drosophila.* This is called a ClB method in which a special type of female fly is employed which carries a normal X chromosome and an abnormal X chromosome. The abnormal X chromosome contains an inversion mutation C (which prevents the chromosome to do crossing over with the normal X chromosome, therefore, called **crossover suppressor**), a recessive lethal mutational gene, 1 and a dominant gene B for bar-eye. In Muller's ClB technique these ClB female flies are mated with males which are previously treated with some mutagenic agent (such as X-rays) to cause mutation in some of their sperms. The resulting zygotes are of four types and one of these, the ClB male, fail to survive because such embryos contain a recessive lethal which expresses itself when hemizygous. Thus, only one class of male (with ClB X chromosome) male remains to fertilize the F_1 females. Each heterozygous ClB female results from the fertilization of a ClB egg and an irradiated X-bearing sperm and some of these sperms will contain mutated X-chromosomes. Mated F_1 heterozygous ClB females are distributed individually into culture tubes in which each lays fertile eggs and so produces a single F_2 culture. The culture produced by females bearing an induced lethal mutation contain only females; whereas female bearing irradiated X-chromosomes in which no recessive lethal has been induced yield cultures containing some wild type males. Thus, if in a population of 1000 cultures 990 contained some males and 10 contained only females, the induced rate of sex-linked, recessive lethal mutations would be 1 per cent.

H.J. Muller. (1890–1967)

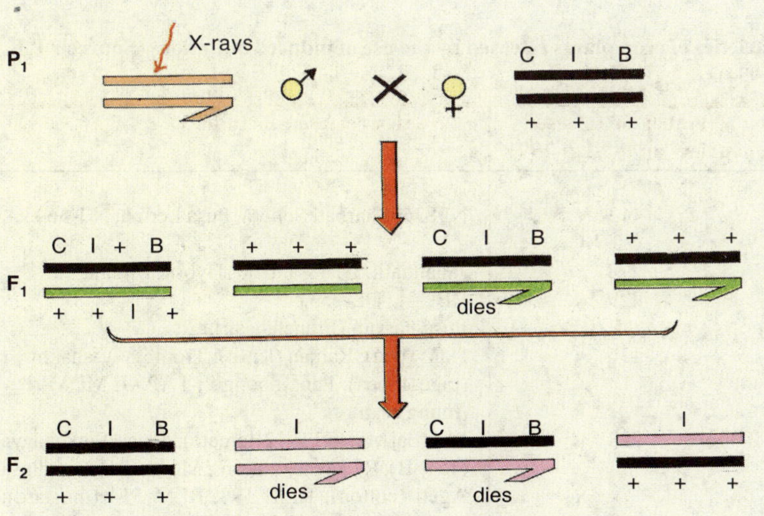

Fig. 16.10. Diagram of Muller's ClB technique for detecting sex-linked (X-linked), recessive lethal mutations.

Besides the ClB method there are many more methods such as **Muller-5 method** (for the detection of sex linked lethal mutations); **attached X- method** (for the detection of sex-linked visible mutations); **balanced lethal systems** (for the detection of autosomal mutations); and **Stadler's method** and **Singleton's method** (for the detection of specific loci in plants).

PRACTICAL APPLICATIONS OF MUTATIONS

Mutations are generally deleterious and recessive for the organisms, therefore, majority of them are of no practical value. **A. Gustafsson** has estimated that less than one in 1000 mutants produced may be useful in plant breeding. In India, several useful mutations of various cereals and other crop plants have been developed (see Table 16.1).

Dr. M.S. Swaminathan

1. Wheat. In bread wheat, many useful mutations have been obtained and utilized in plant breeding, *e.g.,* branched ears, lodging resistance, high protein and lysine content, amber seed colour and awned spikelets. **Dr. M.S. Swaminathan**, one of the most distinguished and legendary figure in the field of cytogenetics and plant breeding in Indian subcontinent, have utilized amber mutation of Mexican wheat variety to develop a new variety of wheat, called **Sharbati Sonora** while working at Indian Agriculture Research Institute (IARI), New Delhi. According to **Dr. N.E. Borlaug** (Nobel laureate), this variety of wheat pave the way for the Green Revolution in India (see **Gupta**, 1994).

2. Rice. In rice, one of the high yielding varieties **Reimei** was developed through mutations isolated after gamma irradiation. Certain developed mutants of rice are found to contain increased contents of proteins and lysine. In certain other mutant rice the duration of crop was reduced by as many as 60 days.

3. Barley. In barley, mutations called **erectoides** and **eceriferum** have been induced. These mutants had high yields including several useful characters.

SIGNIFICANCE OF MUTATION

The vast majority of mutations are deleterious to the organism and are kept at low frequency in the population by the action of natural selection. Mutant types are generally unable to complete equally with wild type individuals. Even under optimal environmental conditions many mutants appear less frequently than expected.

Table 16-1. **List of varieties of crop plants released by the use of induced mutations (Source : P.K. Gupta, 1994).**

Type of crop	Number of released variety up to Year 1991	Varieties released in India
I. Cereals		
1. Bread wheat	113	NP836; Sharbati Sonora, Pusa Lerma; N/I-5643.
2. Durum wheat	25	–
3. Rice	278	Jagannath, I/T48, I/T 60, Hybrid Mutant 95.
4. Barley	229	RBD–1, DL-253.
II. Legumes	15	Pusa Parvati (French bean).
	136	Hans (pea); Ranjan (lentil); Trombay Vishakhi (pigeon pea); Pant moong 2; TAP -7; MUM–2 (mung bean)
III. Other Crops	223	Pusa Lal Meeruti, S-12 (tomato); Aruna, Sowbhagya (157–B), RC8 (castor bean); MCU–7, Rasmi, Pusa Ageti (cotton); RLM 198, RLM 514 (mustard); New Hybrid Bajra 5 (NHB5); Pusa 46 (pearl millet); Co 997, Co 6608 (sugar cane); MDU1 (chili); JRC–7447 (jute).

REVISION QUESTIONS AND PROBLEMS

1. What are missense, nonsense, and frameshift mutations and what are the consequences of each.
2. What do you understand by spontaneous mutations and induced mutations? Discuss variation in mutation rates and frequencies at different loci within an organism.
3. (a) What critical evidence is needed to distinguish a gene mutation from a minute deletion, *i.e.*, one too small to be cytologically detectable?
   (b) Show how you would determine the rate of mutation from a recessive to a dominant allele in man.
   (c) Describe what is meant by a tautomeric shift?
   (d) Are the mutational consequences of tautomeric shifts base-pair transitions or base pair transversions? Explain, using a specific example.
4. Explain : (i) Why geneticists find most mutations to be deleterious?
   (ii) Why, nevertheless, the mutation process is considered to be the basis of evolutionary progress?
5. Answer each of the following questions as briefly as possible:
   (a) Which type of mutation, one induced by a base analog or one induced by proflavine, would you expect to be more deleterious to an organism and why?
   (b) What evidence is there that ionization caused by X-rays need not occur in the gene itself to cause mutation?
   (c) What are the possible mechanisms by which a gene may change to many different allelic forms?
   (d) Why are sex-linked lethal mutations easiar to detect than autosomal lethals?
6. Three repair mechanisms are known in *E. coli* for the repair of DNA damage (pyrimidine dimer formation) after exposure to ultraviolet light: (1) photoreactivation; (2) excision (dark) repair; and (3) postreplication repair. Compare and contrast these mechanisms, indicating how each achives repair and how the events occurring in each may lead to gene mutations.
7. Why are most mutations in structural genes recessive to their wild type alleles?
8. Describe different kinds of radiations and chemical mutagens utilized for induction of mutations.
9. Which of the following would be likely to suffer the greatest, and which the least genetic damage from radiation exposure : (a) a haploid, (b) a diploid, (c) a polyploid ?
10. If the mutation rate of a certain gene is directly proportional to the radiation dosage and the mutation rate of *Drosophila* is observed to increase from 3% at 1000 R to 6% at 2000R. What percentage of mutations would be expected at 3500 R ?
11. The X-linked recessive mutations are more easily studied in appropriate organisms than are autosomal ones. Why ?
12. Discuss the procedure used in the detection of sex-linked lethal mutations by ClB method.
13. Some individuals have a patch of blond hair in a head of brown hair. What types of mutation would this be ?
14. If a drastic alteration occurred in the structure of one of the genes for 28S rRNA, do you think that the translation of mRNA into protein would cease ? If not, why not ?
15. Explain the difference between a transition and a transversion and give an example of each.
16. What possible explanations can you offer for the reversion of a mutant to the wild-type phenotype?
17. What is the difference between intragenic and intergenic suppression ? Give an illustration of each.
18. How many base pairs would have to be deleted in a mutational event to eliminate a single amino acid from a protein and not change the rest of the protein ?
19. Compare the effects of nitrous acid, hydroxylamine and 5-bromouracil on DNA.
20. The "dotted" gene in maize (Dt) is a "mutator" gene influencing the rate at which the gene for colourless aleurone (a) mutates to its dominant allele (A) for coloured aleurone. An average of 7.2 coloured dots (mutations) per kernel was observed when the seed parent was dt/dt, a/a and the pollen parent was Dt/Dt, a/a. An average of 22.2 dots per kernel was observed in the reciprocal cross, How can these results be explained ?
21. Write short notes on the following :
   (a) Base substitution; (b) Ionizing radiation; (c) Balanced lethal; (d) Somatic mutation; (e) Acridine dye; (f) Lethal mutation.

ANSWERS TO PROBLEMS

7. Wild type alleles usually code for complete, functional enzymes or other proteins. One active wild type allele can often cause enough enzyme to be produced so that normal or nearly normal phenotypes result (dominance). Mutations of normally functioning genes are more likely to destroy the biological activities of proteins. Only in the complete absence of the wild type gene product would the mutant phenotype be expressed recessiveness.

9. Haploid greatest, polyploid least. Most mutations are recessive, and recessives have only a very low probability of being expressed in polyploids with their multiple sets of chromosomes bearing normal (dominant) alleles. Both dominant and recessive mutations are expressed at once in haploids.

10. 10.5%.

11. Recessive mutations are more easily detected in hemizygous males.

13. Somatic mutation.

14. Translation would not cease since numerous genes for rRNA are present in the genome. The mutation of one of these would probably not interfere with protein synthesis.

16. *Intragenic mutation within the same codon,* either restoring the original amino acid or resulting in the presence of a compatible amino acid; *intragenic mutation within the same cistron,* such as one that restores the normal reading frame; *intergenic direct suppression,* such as alteration in some component directly involved in protein synthesis, for example, tRNA; *intergenic indirect suppression* by an alteration in the cellular milieu.

18. Three, as any other number of deletion (or addition) would cause a frameshift and other amino acid changes.

20. Seed parent contributes two sets of chromosomes to triploid endosperm; one Dt gene gives 7.2 mutations/kernel, two Dt genes increase mutations to 22.2/ kernel.

CHAPTER

17

Cytoplasmic or Extra-Nuclear Inheritance

The variegated leaves of the shrub, *Acanthopanax*.

In preceding chapters, we have discussed different roles of genes of the nuclear chromosomes in inheritance, cellular metabolism, development and mutation of the organisms in which they occur. Though, the genes of nuclear chromosomes have a significant and key role in the inheritance of almost all traits from generations to generations, but they altogether cannot be considered as the sole vehicles of inheritance, because certain experimental evidences suggest the occurrence of certain extranuclear genes or DNA molecules in the cytoplasm of many prokaryotic and eukaryotic cells. For example, bacterial cells such as *E. coli* possesses a single main chromosome in the nucloid and often extra DNA elements called **plasmids** in the cytoplasm; the eukaryotic cells possess a main complement of chromosomes in the nucleus and extra-DNA molecules or chromosomes in their mitochondria and chloroplasts. Qualifying also as extra hereditary elements are certain viruses, bacteria, and algae (which live as endosymbionts inside the eukaryotic cells). These cytoplasmic extranuclear genes or DNA molecules of plasmids, mitochondria, chloroplasts, endosymbionts and cellular surfaces have a characteristic pattern of inheritance which does not resemble with that of genes of nuclear chromosomes and is known by different terms such as **non-Mendelian**, **non-chromosomal**, **uniparental**, **maternal**, **extra-chromosomal**, **cytoplasmic** and **extra-nuclear** inheritance.

EVIDENCES FOR CYTOPLASMIC FACTORS

Traits with extranuclear basis are identified by the accumulated evidence from a number of diagnostic criteria such as follows :

1. Because the female gamete contributes almost all of the cytoplasm to the zygote and male gamete (sperm or pollen) contributes only a nucleus, an inheritance pattern that differs between reciprocal crosses suggests a cytoplasmic involvement. This is clearly the basis for **uniparental** or **maternal inheritance** where the progeny always resemble one parent, most commonly the female parent (*e.g.,* shell coiling in *Limnaea peregra*).

2. Differences in reciprocal crosses which cannot be attributed to six-linkage or some other chromosomal basis tend to implicate extranuclear factors (*e.g.,* chloroplast inheritance in *Mirabilis jalapa*).

3. The uniparental inheritance of a trait which cannot be atributed to unequal cytoplasmic contributions from parental gametes may, however, involve cytoplasmic factors (*e.g.,* streptomycin resistance in *Chlamydomonas*).

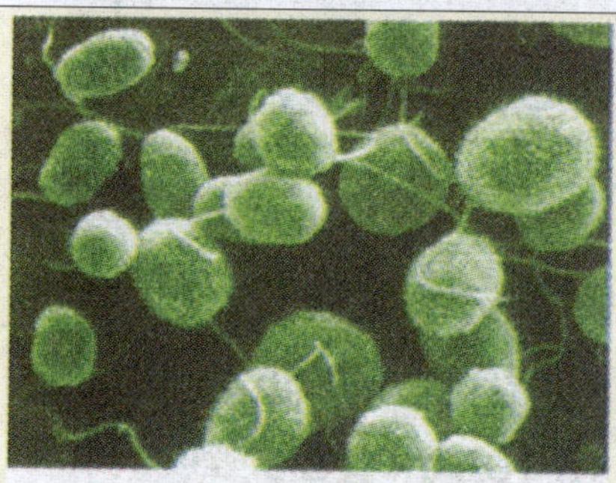

The green alga *Chlamydomonas.*

4. Whenever trait fails to demonstrate the classical segregation patterns and deviates from standard ratios, the conclusion is again a cytoplasm based type of inheritance (*e.g.,* mitochondrial inheritance in yeast).

5. When the trait fails to show linkage to any known nuclear linkage groups and assort independently from nuclear genes, a cytoplasmic mode of inheritance is suggested.

6. Many types of mutants that fit the above criteria will show segregation during mitotic division. This is very common in variegated plants that carry more than one type of plastid (chloroplast) per cell.

The cytoplasmic inheritance, therefore, will be understood to be based on cytoplasmically located, independent, self-replicating nucleic acids, which differ from chromosomal genes by their location within the cell, and have their own nucleotide sequences. The smallest heritable extra chromosomal unit is called a **plasmagene**. All of the plasmagenes of a cell constitute the **plasmon**.

A. EXTRA-NUCLEAR INHERITANCE IN EUKARYOTES

Many geneticists have studied various cases of extra-nuclear inheritance in different eukaryotes. Certain most important examples of extra-nuclear inheritance in eukaryotes are the following :

1. Maternal Inheritance

In certain cases, it has been observed that certain characteristic phenotypic traits of F_1, F_2 or F_3 progeny are not the expression of their own genes, but rather those of the maternal parents. Such phenotypic expressions of maternal genes (genotype) may be short-lived or may persist throughout the life-span of the individual. The substances which produce the maternal effects in the progeny are found to be transcriptional products (*i.e.,* mRNA, rRNA and tRNA) of maternal genes which have been manufactured during oogenesis and which exist in the ooplasm of unfertilized eggs in the form of inactive protein coated and late translating mRNA molecules (**informosomes**) or inactivated rRNA and tRNA. These transcriptional products of maternal genes produce their phenotypic effects during early cleavage and blastulation when there occur little or no transcription since, maternal and paternal genes of zygote remain engaged in mitotic replication or duplication of DNA. There may be other reasons of maternal effect which are still little understood. The maternal inheritance has been studied in some of the following cases :

The photograph illustrates mixture of right vs. left handed coiled snails.

(a) Shell coiling in Limnaea. In the snails (gastropods), the shell is spirally coiled. In most cases the direction of coiling of the shell is clockwise, if viewed from apex of the shell. This type of coiling is called **dextral**. However, in some snails the coiling of shell may be counter clockwise or **sinistral**. Both types of coilings are produced by two different types of genetically controlled cleavages, one being **dextral cleavage**, another being **sinistral cleavage** (Fig. 17.1).

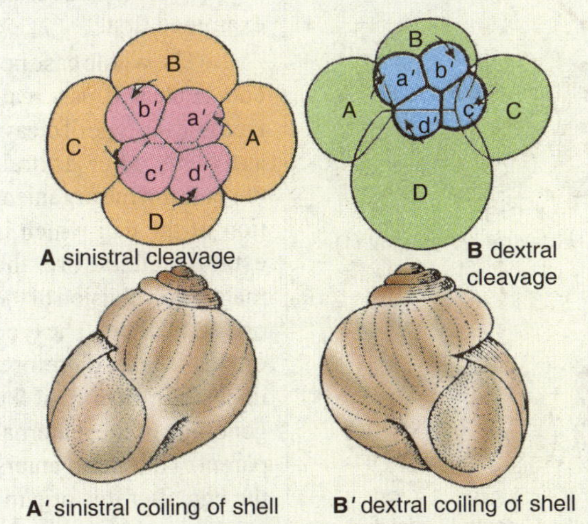

Fig. 17.1. Correlation of sinistral (A) and dextral (B) cleavage with sinistral (A′) and dextral (B′) coiling of the shell in gastropods.

There are some species of gastropods in which all the individuals are sinistral but the main interest attaches to a species in which sinistral individuals occur as a mutation among a population of normal dextral animals. Such a mutant was discovered in the freshwater snail *Limnaea peregra* (**A. Sturtevant**, 1923). Breeding and cross breeding of dextral and sinistral snails showed that the difference between the two forms is dependent on a pair of allelomorphic genes, the gene for sinistrality being recessive (S), and the gene for the normal dextral coiling being dominant (S^+). The two genes are inherited according to Mendelian laws, but the action of any genic combination is visible only in the next generation after the one in which a given genotype is found.

The eggs of a homozygous sinistral individual (SS) are fertilized by the sperm of a dextral individual (S^+S^+), the eggs cleave sinistrally and all the snails of this F_1 generation show a sinistral coiling of the shell. Thus, the gene of sperm (S^+) do not manifest themselve, although the genotype of the F_1 generation is S^+S. If a second generation (F_2) is bred from such F_1 sinistral individuals, it is all dextral, instead of showing segregation as would be expected in normal Mendelian inheritance. In fact, segregation does take place in the F_2 generation so far as the genes are concerned, but the new genic combinations fails to manifest themselves, since the coiling is determined by the genotype of the mother. The genotype of F_1 mother being S^+S, the gene for dextrality dominates and is responsible for the exclusively dextral coiling of the second generation. Only in the F_3 generation does segregation in the ratio of 3 : 1 becomes apparent, since the individuals of the F_2 generation had the genotypes —1S^+S^+; 2 S^+S, 1SS, 1/4 of them, on the average, produce eggs developing into sinistral individuals (Fig. 17.2).

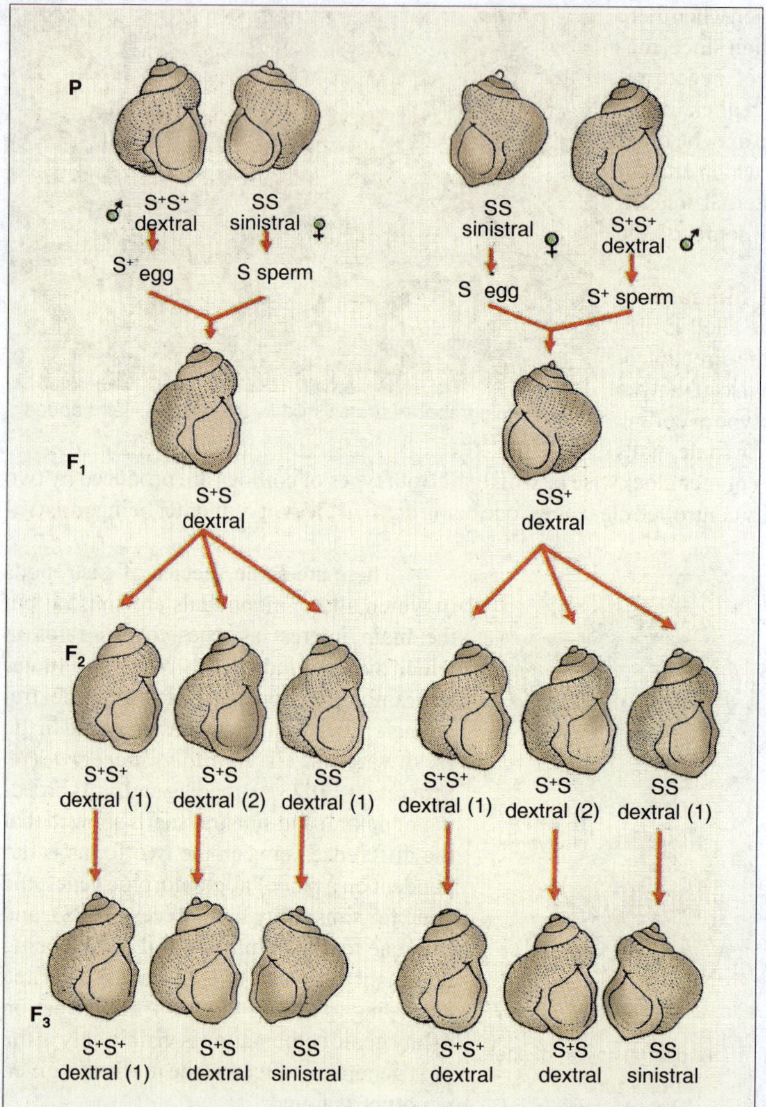

Fig. 17.2. Maternal effect in direction of coiling of the shell in *Limnaea.*

It is easy to understand that the results of a reciprocal cross that is, of the fertilization of the eggs of a homozygous dextral individual (S^+S^+) by the sperm of a sinistral individual (SS)—will lead to a somewhat different type of pedigree : the F_1 generation will be dextral (with genotype S^+S) and the F_2 generation again all dextral (with genotypic ratio of $1S^+S^+:2S^+S$: 1SS). The F_3 generation will show segregation among broods, just as in the cross examined first.

The whole case becomes clear if it is realized that the type of cleavage (sinistral or dextral) depends on the organization of the egg which is established before the maturation division of the oocyte nucleus. The type of cleavage is, therefore, under the influence of the genotype of the maternal parent. The sperm enters the egg after this organization is already established. Lastly, the direction of coiling of shell depends upon the orientation of the mitotic spindle of first cleavage of the zygote. If the spindle is tipped toward the left of the median line of the egg cell, the sinistral pattern will develop; conversely if the mitotic spindle is tipped toward the right of the median line of the cell, the dextral pattern will develop. The spindle orientation is, thus, controlled by the organization of ooplasm which becomes established during oogenesis and before fertilization.

(b) Eye pigmentation in water fleas and flour moths. Like the *Limnaea* and *Ambystoma mexicanum,* the maternal effect has also been observed in at least two very different invertebrates, the water flea (*Gammarus*) and the flour moth (*Ephestia kuhniella*). The normal colour of the eye in both invertebrates is dark due to the dominant gene (AA or KK) in which the dominant gene K directs the production of a hormone-like substance called **kynurenine** which is involved in the pigment synthesis. The recessive mutants do not possess pigment in the eye (*viz.,* kynurenineless) and have the genotype aa or kk. When aa or kk female is crossed with heterozygous male with Aa or Kk genes, only half of the larvae show dark pigment in the eye. But, a cross between Aa or Kk female and aa or kk male produces all larvae with dark eyes. On reaching the adult stage, half of the offsprings (those of the genotype aa) become light eyed. This indicates that some kynurenine diffuses from the Aa mother into all young (larvae), enabling them to manufacture pigment regardless of their genotype. The aa progeny, however, has no means of continuing the supply of kynurenine, with the result that their eyes eventually become light. This example suggests an ephemeral type of maternal effect.

2. Extra-nuclear Inheritance by Cellular Organelles

Chloroplasts and mitochondria are organelles that contain their own DNA and protein-synthesizing apparatus. A widely held theory concerning their origin proposes that they were once infectious endosymbiotic prokaryotes that evolved such a dependence on the gene products of the host that they are no longer able to function autonomously.

This theory has been supported by the fact that the genetic components of these organelle are often similar to those found in prokaryotes. For example, the chloroplasts of certain algae and *Euglena* contain 70S type small ribosomes and "naked" chromosomes or DNA which is circular. Their protein synthesis begins with the amino acid N-formyl methionine, as does prokaryotic protein synthesis, and their DNA-dependent RNA polymerase is sensitive to the inhibitor **rifampicin**. The genetic materials of chloroplasts and mitochondria will be transmitted to offspring almost exclusively via the egg. Maternal inheritance due to chloroplast and mitochondria is well illustrated by the following examples:

The four o'clock plant with a green branch.

Fig. 17.3. Leaf variegation in *Mirabilis jalapa,* the four-o'clock plant. Flowers may form on any branch (variegated, green, or white), and these flowers may be used in crosses.

(a) Chloroplast inheritance in variegated four o' clock plant. The cytoplasmic or extra-nuclear inheritance of colour in plant by plastids was first of all discribed by **C. Correns** in 1908 in the four o'clock plant, *Mirabilis jalapa.* In contrast to other higher plants, *Mirabilis* contains three types of leaves and parts : (1) Full green leaves or branches having chloroplast, (2) White (pale) leaves and branches having no chloroplast, (3) Variegated branches having leukoplast in white (pale) areas and chloroplast in green patches (Fig. 17.3). Because, the chlorophyll pigment of chloroplast is related with photosynthesis of food and leukoplasts are incapable to perform photosynthesis, so the white or pale parts of plant survive by receiving nourishment from green parts. **Correns** reported that flowers on green branches produced only green offsprings, regardless of the genotype and phenotype of pollen parent and likewise, flowers from the white or pale branches produced only white or pale seedings regardless of genotype and phenotype of pollen parent. The plants developing from the white or pale seedings die because they lack chlorophyll and cannot carry on photosynthesis. **Correns** further reported that flowers from the variegated branches yielded mixed progeny of green, white (pale) and variegated plants in widely varying ratios (Fig. 17.4). These results are summarized in Table 17-1.

The irregularity of transmission from variegated branches could be understood by considering cytoplasmic genes (plasmagenes) of plastids. A study of the egg during oogenesis in *Mirabilis* reveals that the ooplasm contains plastids like cytoplasm of other plant cells. If the egg cell is derived from green plant tissues, its ooplasm will contain coloured plastids; if derived from white plant tissues, its ooplasm will contain white plastids; if derived from variegated tissues, its cytoplasm may contain coloured plastids only, white plastids only or a mixture of coloured and white plastids. A study of the pollenogenesis, however, reveals that pollen contains very little cytoplasm which in most cases is devoid of plastids. Without the plastids, the pollen cannot affect this aspect of the offspring's phenotype.

Mitotic segregation. Variegated branches of *Mirabilis jalapa* produce three kinds of eggs : some contain only white chloroplasts, some contain only green chloroplasts and some contain both types of chloroplasts. In the subsequent mitotic divisions, some form of cytoplasmic segregation occurs that segregate the chloroplast types into pure cell lines, thus, producing the variegated phenotype in the progeny individual. This process of sorting might be described as "**mitotic segregation**"of this is a pure extra- nuclear phenomenon. In mitotic segregation since both segregation and recombination of organelle genotype takes place, so it is called **cytoplasmic segregation and recombination** (its acronym is **CSAR**).

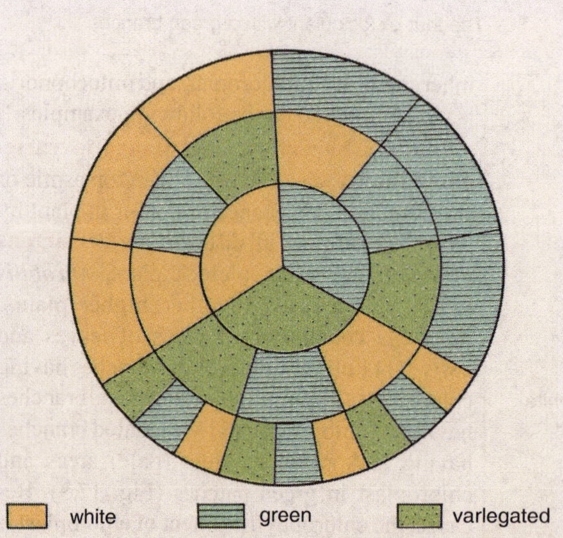

Fig. 17.4. Plastid inheritance of *Mirabilis jalapa.* The central circle represents the type of branch that produces flowers which are pollinated. Intermediate circle represents branch from which pollen is used and outer circle shows the progeny.

(b) Maternal inheritance by iojap gene of corn. Another example from higher plants also suggests the existence of plastid genes controlling plastid integrity. A gene in corn plant called **iojap** (**ij**) has been mapped by **M. Rhoades** (1946) to nuclear chromosomes 7. Plants homozygous for ij are either inviable white seedings or variegated with a characteristic white striping, the phenotype being known as **striped**. When the variegated plants serve as females in a cross, they give rise to green, white, and striped progeny, regardless of the nuclear genotype of the paternal parent. Thus, if the pollen derives from a normal green Ij/Ij plant as in Figure 17.5 b, the resulting progeny will be Ij/ij heterozygotes, but many will exhibit abnormal plastid pigmentation : the presence of the "normal" Ij gene has no curative effect. In the reciprocal Ij/Ij female X ij/ij male cross (Fig. 17.5). On the other hand, the Ij/ij progeny are all normally pigmented.

Table 17-1. **Chloroplast inheritance in variegated four o'clock plants.**

Branch of origin of the male parent	Branch of origin of the female parent	Progeny
Green	Green	Green
	Pale or white	Pale or white
	Variegated	Green, pale or white, variegated
Pale or white	Green	Green
	Pale or white	Pale or white
	Variegated	Green, pale or white, variegated
Variegated	Green	Green
	Pale or white	Pale or white
	Variegated	Green, pale or white, variegated

The iojap trait, thus, exhibits classical maternal inheritance once it has become established in an ij/ij plant. Moreover, once established, it becomes independent of the ij gene, as can be demonstrated by crossing F_1 Ij/ij variegated females to Ij/Ij normal males. As shown in Figure (17.5c), a mixture of green, striped and white progeny again results, even though some of the striped and white plants now have an Ij/Ij genotype. Thus, the iojap trait, once established, is permanent.

The iojap phenomenon has been explained by two hypotheses. One hypothesis holds that the ij/ij genetic constitution could bring about or permit, frequent mutations in the chloroplast genome that result in the production of lines of abnormal plastids. Another hypothesis suggests that certain cytoplasmic elements other then chloroplast mutations come into being or residence in ij/ij cells, are later inherited in the absence of this "susceptible" or "permissive" genotype, and bring about the bleaching of chloroplasts.

This type of maternal inheritance by plasmagenes of chloroplasts has been also studied in many other higher plants such as barley, *Oenothera* sp., rice, etc.

(c) Extra-nuclear inheritance by mitochondria. The most important work on the genetics of mitochondria done in yeast which was initiated by the discovery of petite mutants by **B. Ephrussi** (1953). Subsequently mt DNA was studied in several organisms including plants and animals.

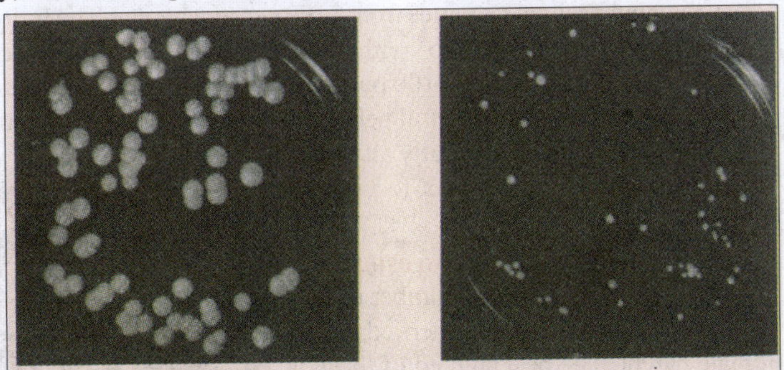

A comparison of normal vs. petite colonies in the yeast *Saccharomyces cerevisaiae*.

(i) Petite in yeast. Yeast, *Saccharomyces cerevisiae,* are single-celled ascomycetes fungi. In the life cycle, diploid and haploid adult alternates, the former reproducing by asexual meiospores called **ascospores**, the latter by **isogametes**. The **petite** mutants in yeast fail to grow on carbon source such as glucose and produce smaller colonies (the "littles") when grown on sugars such as glucose. Since this difference can be observed only when such yeast cultures are kept in a oxygen-containing environment; so it is concluded that petite mutants have a defective aerobic respiratory mechanism. In other words, slow growth of petite can be attributed to yeast cells utilization of less efficient fermentation process. These petites differ from wild type, called **grande** and are characterized by (i) their insensitivity to inhibitors of aerobic pathways (such as cyanide), (ii) absence of cytochromes a, a_3, b and a number of other changes in mitochondrial respiratory enzymes; (iii) incomplete development of mitochondria; and (iv) lack of stainability of petite mitochondria.

(a) Ij/Ij Green (Female) X ij/ij Striped (Male)
↓
Ij/ij All green

(b) ij/ij Striped (Female) X Ij/Ij Green (Male)
↓
Ij/ij Green | Ij/ij White | Ij/ij Striped

(c) Ij/ij Striped of F_1 of b (Female) X Ij/Ij Green (Male)
↓
Ij/Ij Green | Ij/Ij Striped | Ij/Ij White | Ij/ij Green | Ij/ij Striped | Ij/ij White

Fig. 17.5. (a) Cross between green (normal) and striped (iojap) plants. (b) Reciprocal of cross (a) (c). Cross of F_1 striped females (of cross b) to normal (green) males.

The petite mutants can be **segregational**, *i.e.,* they follow mendelian segregation and, therefore, presumably controlled by chromosomal genes. They may also be **vegetative**, *i.e.,* non-segregational or extra-chromosomal. The genetic basis of petite character is a cytoplasmic factor ρ+ (rho) which may be absent or defective in petites. Thus, a vegetative petite can be **neutral** (ρ^0) which completely lack ρ^+ or it may be **suppressive** (ρ^-) having a defective ρ^+. The neutral petites are not transmitted while suppressive petites are transmitted to a fraction of vegetative diploid progeny. In various strains of yeast, the suppressiveness varies from 1–99 per cent petites. The following two lines of evidences have suggested the association of ρ^+ with mitochondrial DNA (mt DNA); (1) **Ethidium bromide**, which induces petite mutations with 100 per cent efficiency, causes degradation of mt DNA after prolonged exposure of cells. In fact, neutral petites have been found lacking in mt DNA. (2) Supressive petites contain mt DNA which is greatly altered in base composition with respect to wild mt DNA.

(ii) Poky strain of Neurospora. In fungi, *Neurospora crassa* a number of mutations of mitochondria are inherited via the female parent. The best studied of these is the **poky strain** of *N. crassa,* first isolated by **Mitchell** and **Mitchell** (1952). A poky mutant differs from wild type strain of *Neurospora* in the following aspects : (1) it is slow growing; (2) it shows maternal inheritance, and (3) it has abnormal cytochromes. Of the three cytochromes—cyt a, b and c

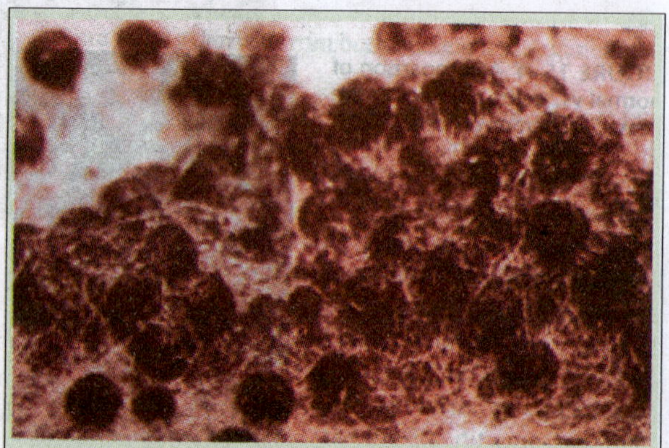

Micrograph illustrating the growth of the pink bread mold *Neurospora crassa.*

found in wild type, cyt a and cyt b are absent, and cyt c is in excess in poky mutant. In reciprocal crosses, poky character shows maternal inheritance:

poky (female) × wild type (male) → all poky

wild type (female) × poky (male) → all wild type

However, there are other marker nuclear genes (ad^+/ad^-) which show 1 : 1 mendelian segregation. The following evidences suggested that poky trait may be located in mitochondrial DNA: (i) slow growth may be due to lack of ATP energy and source of this energy is mitochondria; (ii) cytochromes in poky strain differ from those in wild type in quality and quantity and these cytochromes are found in mitochondria.

(iii) Male sterility in plants. In plants, the phenotype of male sterility is found to be controlled either by nuclear genes or plasmagenes (cytoplasm) or by both. Therefore, the trait of male sterility of plants is controlled by the following three methods :

(a) Genetic male sterility. In this type of male sterility, the sterility is controlled by a single nuclear gene which is recessive to fertility, so that the F_1 progeny would be fertile and in F_2 generation, the fertile and sterile individuals will be segregated in the typical 3 : 1 ratio (Fig. 17.6).

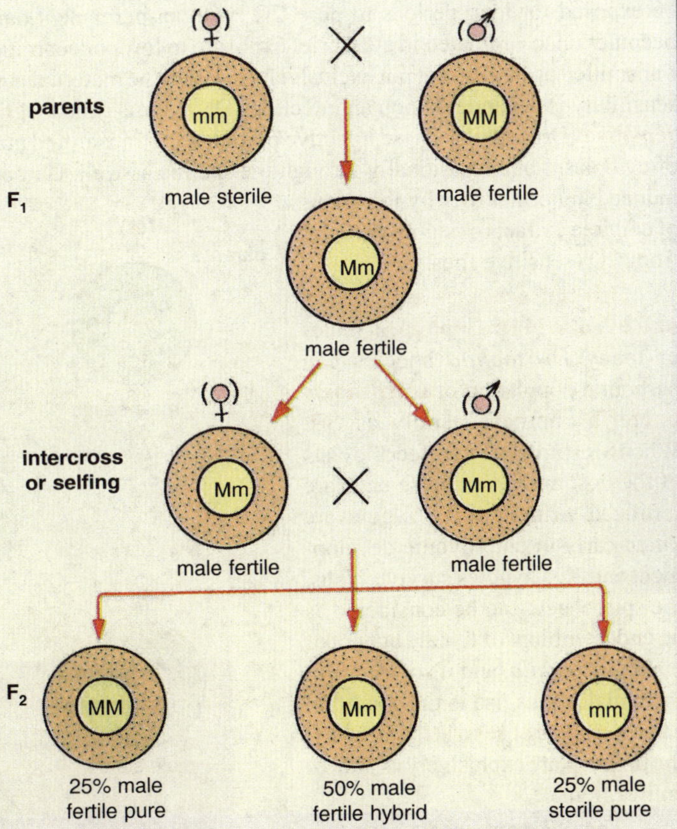

Fig. 17.6. Inheritance pattern of genetic male sterility.

(b) Cytoplasmic male sterility (CMS). In maize and many other plants, cytoplasmic control of male sterility is known. In such cases, if the female parent is male sterile (having plasmagene for male sterility), the F_1 progeny would always be male sterile, because the cytoplasm is mainly derived from the egg which is obtained from the male sterile female parent (Fig. 17.7).

(c) Cytoplasmic genetic male sterility. In certain plants, though the male sterility is fully controlled by the cytoplasm, but a **restorer gene** if present in the nucleus, will restore fertility. For example, if the female parent is male sterile (due to plasmagene of male sterility) then the nuclear genotype of the male parent will determine the phenotype of F_1 progeny. Thus, if male sterile female parent contains recessive nuclear genotype rr of restorer gene and male parent is RR, having homozygous dominant restorer genes. Their F_1 progeny would be male fertile Rr. However, if the male parent is male fertile rr, the F_1 progeny would be male sterile rr. If the F_1 male fertile heterozygote (Rr) is test crossed with male fertile rr male, a progeny with 50 per cent male fertile and 50 per cent male sterile will be obtained (Fig. 17.8).

Since, in maize expression of male sterility depends on an interaction between nuclear and extra-chromosomal genes. Male sterile lines can bear seeds only after cross-pollination. For this reason they are useful in raising hybrid seeds, especially on large scale.

Later on, in maize the following four types of cytoplasms have been recognized : the normal (N) cytoplasm and three types of male sterile cytoplasms (T, C and S). The recent studies of mitochondria in these cytoplasm revealed that the factors responsible for cytoplasmic male sterility are located in mitochondrial DNA (mt DNA) and mt DNA of N, T, C and S cytoplasms are found to be different. The cytoplasmic male sterility (CMS) of C and S type can be reversed by nuclear storer genes, however, the CMS-T cannot.

3. Extra-Nuclear Inheritance by Endosymbionts

Certain intra-cellular parasites such as bacteria and virus particles maintain symbiotic relationship with host cells. They are self-reproducing and look like the cytoplasmic inclusions. Sometimes they exhibit an infection like transmission with a hereditary continuity of their own. Generally such symbionts are coined by letters of the Greek alphabets (sigma, kappa, mμ, etc.). The various types of infective symbionts are as follows :

(i) Sigma virus in Drosophila. L. Heritier and **Teissier** found that a certain strain of *Drosophila melanogaster* shows a high degree of sensitivity to carbon dioxide, where as the wild type strain can be exposed for long periods to pure CO_2 without permanent damage, the sensitive strain quickly becomes unco-ordinated in even brief exposure to low concentrations. This trait (extra-sensitivity) is transmitted primarily, but not exclusively, through the maternal parent. Tests have disclosed that CO_2 sensitivity is dependent upon an infectious DNA virus called **sigma**, found in the cytoplasm of **CO_2 sensitives** Drosophila. These infective particles are transmitted normally via the egg's larger amount of cytoplasm but occasionally through the sperms as well. Carbon dioxide sensitivity may even be induced in normal flies by injections of cell free extracts of sigma particles from CO_2 sensitive flies.

(ii) Spirochaetes and maternal sex ratio in Drosophila. Females of many *Drosophila* species can harbour a population of **spirochaete bacteria** known generally as **SR**. When SR spirochaetes infect the eggs of the host and when these eggs are fertilized, virtually all XY zygotes are killed early in embroyonic development and XX zygotes survive. Thus, the spirochaete can be considered as an endosymbiont of female but not of male *Drosophila,* and its presence in the female gives rise to the condition called **maternal sex ratio**, in which the progeny are exclusively or almost entirely female.

The SR spirochaete is infectious, for when isolated from the haemolymph of female carriers and introduced into normal females the latter become carriers. Why the fe-

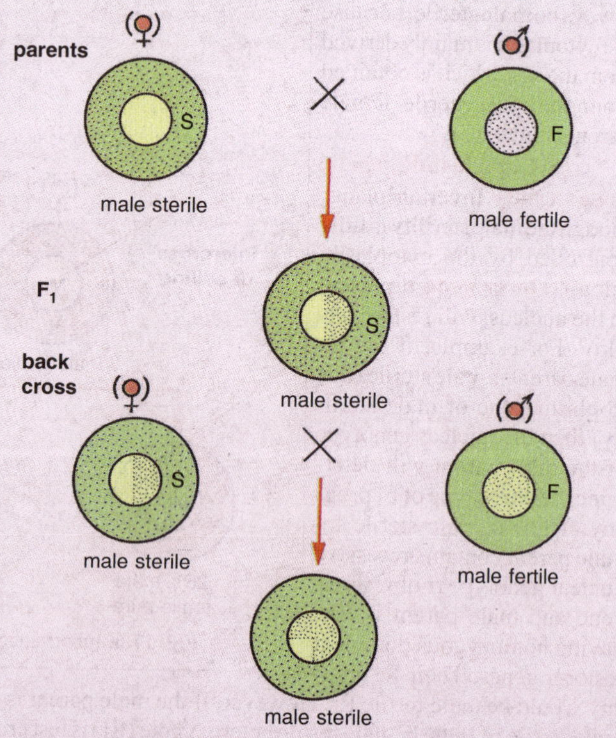

Fig. 17.7. Maternal inheritance of cytoplasmic male sterility.

male genotype permits their retention, and conversely, why XY cells are sensitive to their presence is not yet known. **K. Oishi** and **D. Poulson** (1970) have reported DNA-containing viruses in these endosymbiont spirochaetes of female *Drosophila.*

T.M. Sonneborn (1905-1981).

(iii) Kappa particles. In 1938, **T.M. Sonneborn** reported that some races (known as "**killers**" or **killer strain**) of the common ciliate protozoan, *Paramecium aurelia* produce a poisonous substance, called **paramecin** which is lethal to other individuals called "**sensitives**". The paramecin is water soluble, diffusible and depends for its production upon cytoplasmically located particles called **kappa**. Electron microscopic observations have shown that kappa particles are about 0.4μ long symbiotic bacteria, *Caedobacter taeniospiralis*; 20 per cent of kappa bacteria of the killer strain contain a refractile protein containing "**R body**" and are called "**brights**". They are infected with a virus that controls the synthesis of toxic viral protein, the paramecin (see **Gardner** *et al.,* 1991). A killer *Paramecium* may contain hundreds (*e.g.,* 400) of kappa particles. The presence of kappa particles in the killer *Paramecium* is dependent for their maintenance and replication on the chromosomal dominant gene K. Paramecia with nuclear genotype kk are unable to harbour kappa particles.

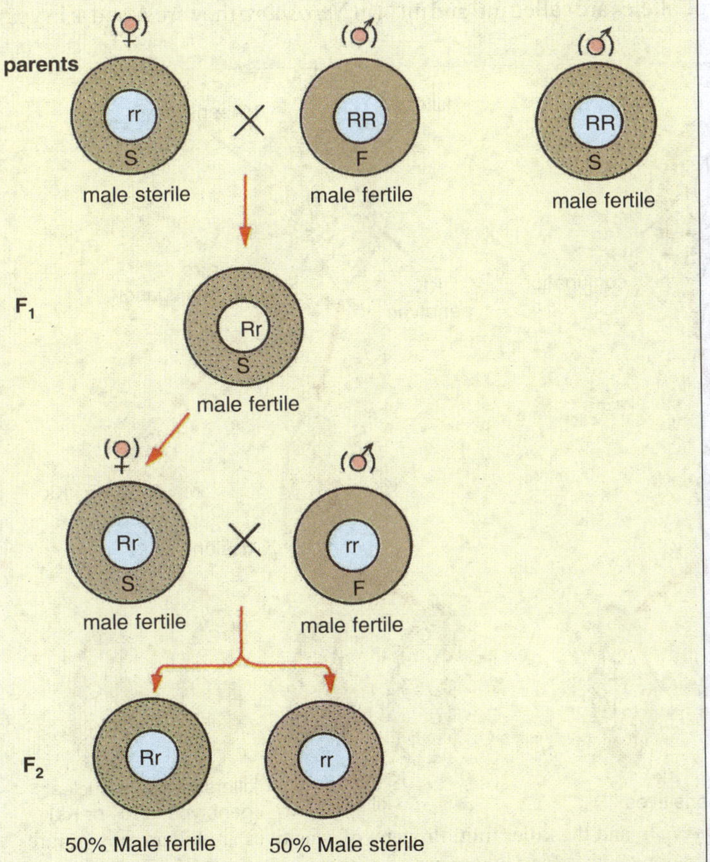

Fig. 17.8. Inheritance of cytoplasmic male sterility and effect of restorer gene in making male steriles into male fertiles.

When a *Paramecium* of killer strain having the genotype KK or (K$^+$) conjugates with the *Paramecium* of non-killer strain having the genotype kk, the exconjugants are all heterozygous for Kk genes (Fig. 17.9). The Kk genotype suggests that both exconjugant should be killers. But this is not the case. If conjugation is normal, *i.e.,* lasts only for a short time, and no exchange of cytoplasm takes place between the two, both killers and non-killers (sensitive) are produced. However, rare or prolonged conjugation (*i.e.,* lasting for long time) permits mixing of cytoplasm of both conjugants and results killers only. The killer trait is stable only in killer strain with KK genotype and is suitable in sensitive strain with kk genotype.

(iv) mμ particle. Another type of killer trait known as **mate killer** has been reported in *Paramecium* by **R.W. Siegel** in 1952. The mate killer trait is imparted by a cytoplasmic mμ particle and a *Paramecium* with a mμ particle is called **mate killer** because when it conjugates with a *Paramecium* without any mμ particle is called **mate sensitive**, then it kills the latter. The mμ particles exist only in those cells whose micronucleus contains at least one dominant gene of either of two pairs of unlinked chromosomal genes (M_1 and M_2). The mμ particles are composed of DNA, RNA and other substances and are symbionts.

(v) Milk factor in mice. Bittner found that females of certain lines of mice are highly susceptible to mammary cancer and this trait was found to be maternally transmitted trait. Results of reciprocal crosses between these and animals of low-cancer-incidence strain depend on the characteristic of the female parent. When the young mice of a low-incidence strain are allowed to be nursed by susceptible foster mothers produces a high rate of cancer in them. Apparently this is a case of infective agent transmitted in the milk. This so called milk factor resembles in many respects with a virus and has been discovered to be transmissible also by saliva and semen. The presence of milk factor also depends on nuclear genes.

4. Uniparental Inheritance in *Chlamydomonas reinhardi*

Like fungi, algae rarely have different sexes, but they do have mating types. In many algal and fungal species, there are two mating types that are determined by alleles at one locus. A cross can occur only if the parents are of different mating types. The mating types are physically identical but physiologically different. Such species are called **heterothallic** (literally "different bodied"). In *Chlamydomonas*, the mating type alleles are called mt^+ and mt^- (in *Nerospora* they are A and a; in yeast a and α).

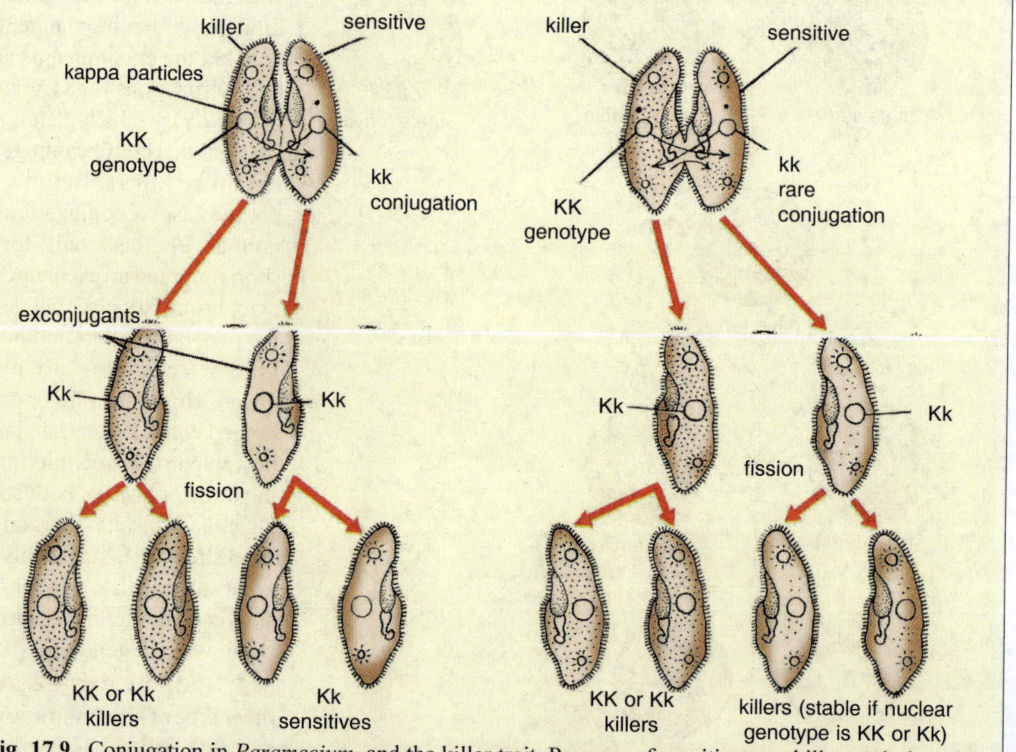

Fig. 17.9. Conjugation in *Paramecium* and the killer trait. Progeny of sensitives are killers only in rare situations where conjugation persists for a longer period so that kappa-containing cytoplasm is introduced into the conjugating sensitive. Kappa particles, however, are maintained only in the presence of K-nuclear genotype (after Burns, 1969 and Gardner, 1968).

In 1954, **Ms. Ruth Sager** isolated a **streptomycin - sensitive (sm-s)** mutant of *Chlamydomonas* with a peculiar inheritance pattern. In the following crosses, sm-r and sm-s indicate streptomycin resistance and streptomycin sensitivity, respectively, and mt is the mating-type gene :

$$mt^+ \text{ sm-r} \times mt^- \text{ sm-s} \rightarrow \text{progeny all sm-r}$$

$$mt^+ \text{ sm-s} \times mt^- \text{ sm-r} \rightarrow \text{progeny all sm-s}$$

Here, occurs a difference in reciprocal crosses; all progeny cell show the streptomycin phenotype of the mt^+ parent. Like the maternal inheritance this is a case of **uniparental inheritance**. In fact, **Sager** now refers to the mt^+ mating type as the female, using this analogy.

Ruth Sager.

REVISION QUESTIONS AND PROBLEMS

1. (a) What is the basic test by which cytoplasmic inheritance is distinguished from nuclear inheritance in almost all organisms ?
   (b) What specific properties do chromosomal genes possess?
2. Answer each of the following as briefly and completely as possible :
   (a) Which do you think would be easier to identify, the effects of plasmagenes or the effects of chromosomal genes? Explain
   (b) Why is it often difficult to distinguish between cytoplasmic gene (plasma genes) and viruses?
   (c) What conditions must be satisfied to prove that cytoplasmic genes are present in the chloroplast? Discuss.
3. Discuss the role of chloroplasts and mitochondria in the cytoplasmic inheritance.
4. Write short notes on the following :
   (i) Plasmagenes; (ii) Maternal inheritance; (iii) Male sterility; (iv) Kappa particle; (v) 'Petite' in yeast.
5. Suppose that a snail had a dextral coiling. Upon self-fertilization, it produces progeny all of which showed sinistral coiling. How do you explain results.
6. A male and female *Ephestia* moth, both coloured as larvae, were crossed. About half of the adult progeny were coloured, half were white. What colour did the male and female parents possess as adults?
7. Most strains of *Chlamydomonas* are sensitive to streptomycin(s). A strain is found which requires streptomycin in the culture medium for its survival (sd). How could it be determined whether streptomycin-dependence is due to a chromosomal gene or to a cytoplasmic element ?
8. Exposing a culture of white yeast to the mutagenic action of mustard gas produced some red individuals. When the red mutants were propagated vegetatively, some white cells frequently reappeared. How can these results be explained ?
9. Given seed from a male sterile line of corn, how would you determine if the sterility was genic or cytoplasmic ?
10. A four-o'clock plant with three kinds of branches (green, variegated and "white") is used in a breeding experiment. What kinds of progeny are to be expected from each of these crosses :
    (a) Green female X White male, (b) White female X Green male ;
    (c) Variegated female X Green male ?
11. Determine which of the three paramecial phenotypes (killer, unstable, or sensitive) is produced by the following combinations of genotype and cytoplasmic state.

	Genotype	**Cytoplasm**
(a)	KK	Kappa
(b)	Kk	No Kappa
(c)	kk	Kappa
(d)	KK	No Kappa
(e)	Kk	Kappa
(f)	kk	No Kappa

12. Some *Drosophila* flies are known to be very sensitive to carbon dioxide gas, rapidly becoming anesthetized under its influence. Sensitive individuals possess a cytoplasmic particle called sigma which has many of the attributes of a virus. Resistant individuals do not have sigma. This trait show strictly maternal inheritance. Predict the results of a cross between (a) sensitive female X resistant male; (b) sensitive male and resistant female.

ANSWERS TO PROBLEMS

6. Coloured female (Kk); white male (kk).
7. Cross mt^- ss (male) × mt^+ sd (female); if chromosomal, 25% of the sexual progeny should be mt^- ss, 25% mt^+ ss, 25% mt^- sd, 25% mt^+ sd; if cytoplasmic, almost all of the progeny should follow the maternal line (streptomycin dependent) while mating type segregates 1 mt^- : 1 mt^+.
8. Asexual reproduction cannot produce segregation of nuclear or chromosomal genes, but cytoplasmic constituents can be differentially distributed to fission products.
9. Plant the seed and pollinate the resulting plants with normal pollen from a strain devoid of male sterility. If the F_1 is sterile, then it is cytoplasmic; if the F_1 is fertile, it is genic.
10. (a) All green progeny because of chloroplast containing eggs from the pistillate parent.

    (b) All "white" because of all colourless (defective) plastids in the eggs of the pistillate parent.

    (c) Green, variegated, and "white" in irregular ratio, reflecting the three kinds of eggs produced by the pistillate parent.
11. Killer = (a) (c); Unstable = (c); Sensitive = (b) (d) (f).
12. (a) All progeny sensitive; (b) All progeny resistant.

CHAPTER

18

Human Genetics

Human Genetics.

For the study of inheritance in any species geneticists prefer (1) to have **isogenic strains**, standard stocks whose members are genetically identical, (2) to mate members of different isogenic strains or inbred lines and (3) to raise the offspring under carefully controlled conditions. Judged by these criteria, man is not a very favourable subject for studies of inheritance, because (i) members of the human race are genetically diverse, *viz.,* they are heterozygous for many genes and there are wide variations in their physical, biological and social environments; (ii) in him controlled matings under standardized environment are impossible and non-ethical : (iii) he has small individual progenies, so unfavourable to the use of certain standard research techniques of the genetics, and (iv) he has long time between successive generations. Despite these difficulties, a great deal has been learned about human inheritance and the field is progressing rapidly. To study the genetics of various human traits several techniques such as pedigree analysis, amniocentesis (and cell culture), twin's study, etc., have been utilized.

Further medically oriented investigations on man have provided detailed informations about biochemistry, morphology, anatomy and physiology to geneticists and they have made use of these in human genetics. Most of the genetical phenomena such as linkage, sex-linkage, sex-determination, etc., of man have already been discussed in previous chapters, therefore, in this chapter following topics of human genetics only, have been discussed.

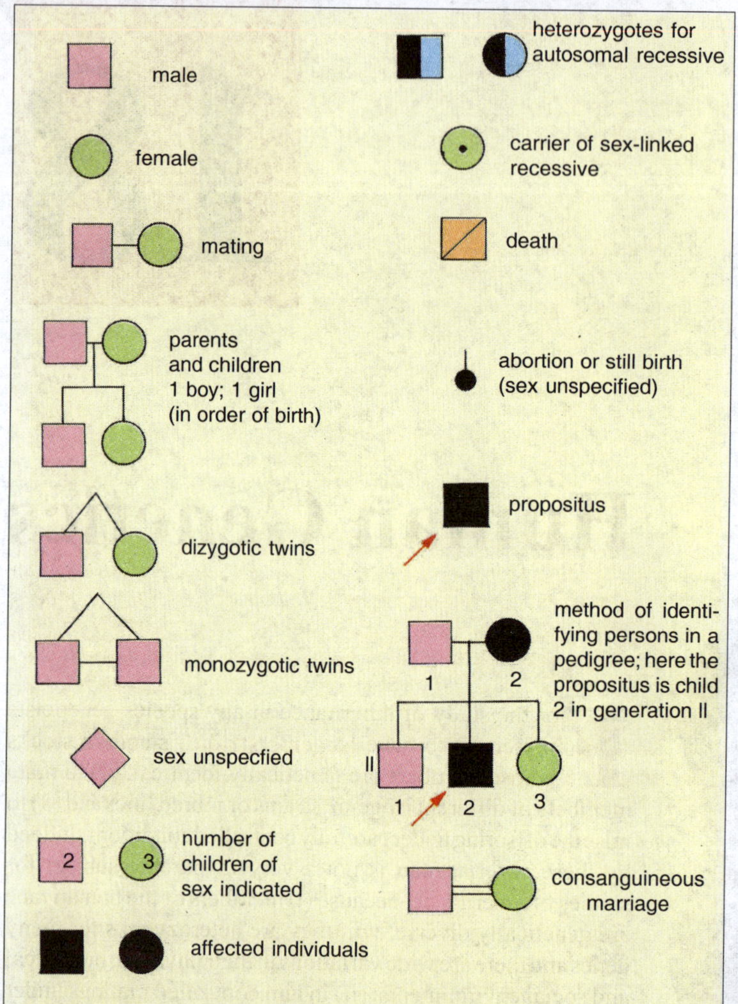

Fig. 18.1. Symbols commonly used in pedigree charts (after Bodmer and Cavalli-Sforza, 1976).

PEDIGREE ANALYSIS

Since, in human beings controlled crosses cannot be made, so human geneticists have to resort to a scrutiny of established matings in the hope that informative matings have been made by chance. The scrutiny of established matings is called **pedigree analysis**. A member of a family who first comes to the attention of a geneticist is called the **propositus**. Usually the phenotype of the propositus is exceptional in some way—for example, a dwarf. The investigator than traces the history of the character shown to be interesting in the propositus back through the history of family, and a family tree or **pedigree** or **pedigree chart** is drawn up using certain standard symbols (see Fig. 18.1).

A family pedigree chart conventionally has circles for female individuals and squares for male individuals. A marriage is indicated by a horizontal bar connecting a circle and square and the symbols for offspring are shown suspended from a line drawn perpendicular to the marriage bar. Individuals on the same line are from the same generation, and individuals in each generation are numbered in sequence (such as II·1, II·2, II·3, and so on). The unshaded circles or squares designate people who are normal for the character being studied, while solid black squares or circles depict "affected" individual. Heterozygotes customarilly are designated by colouring half of the symbol blacks (Fig. 18.1). Carrier of a sex-linked recessive gene is designated by a black dot in the middle of the symbol. Occasionally, an arrow pointing at a particular affected individual is added to indicate that this is the person who brought the trait to the geneticists attention.

Example. A case of polydactyly, the occurrence of extra fingers, in man can be considered as an example for pedigree analysis. A family pedigree for polydactyly has been illustrated in Fig. 18.2.

Here, we have a marriage between a polydactylous man and a normal woman (generation I). They reproduce three children, a polydactylous daughter, a polydactylous son, and a normal son (generation II). The first and third individuals of II generation, each marry normal (pentadactylous) persons, their children are shown in generation III. From the analysis of pedigree of polydactyly, one fact becomes clear that polydactylous trait is dominant on the normal pentadactylous trait which is recessive.

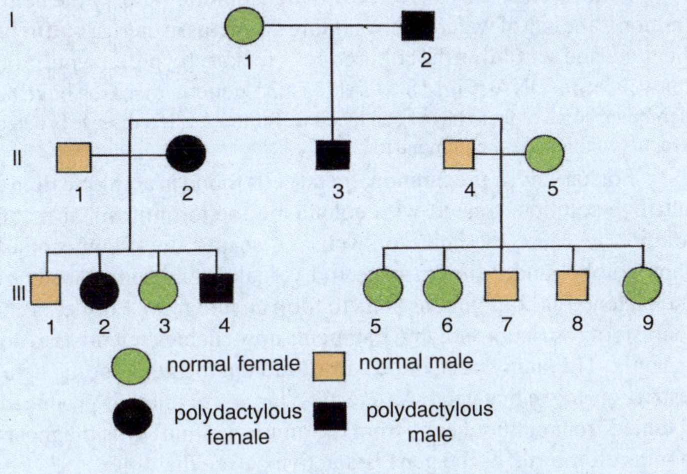

Fig. 18.2. A family pedigree for polydactyly in man (after Burns 1969).

How pedigree charts are read ? Many human diseases and other exceptional conditions are determined by simple mendelian recessive alleles. There are certain clues in the pedigree that must be sought. Characteristically the diseased condition appears in progeny of unaffected parents. Further more, two affected individuals cannot have an unaffected child. Quite often such recessive alleles are revealed by consanguineous matings–for example, cousin marriages. This is particularly true of rare conditions where chance matings of heterozygotes are expected to be extremely rare, *e.g.*, albinism, Tay-Sachs disease, cystic fibrosis and phenylketonuria (PKU).

There are also examples of exceptional conditions caused by dominant alleles. Once again, there are some simple rules to follow to ascertain from pedigrees a condition caused by a dominant allele : the condition typically occurs in every generation; unaffected individuals never transmit the condition to their offspring; two affected parents may have unaffected children; and the condition is passed, on average, to one-half of the children of an affected individual. Some examples of abnormal conditions caused by dominant alleles in humans are achondroplasia (a kind of dwarfism), Huntington's corea and brachydactyly (very short fingers).

AMNIOCENTESIS

Prenatal (=occurring before birth) screening of babies for gross chromosomal aberrations (such as polyploidy, aneuploidy, deletions, translocations, etc.), as well as, sex prediction, is now possible by the technique of amniocentesis.

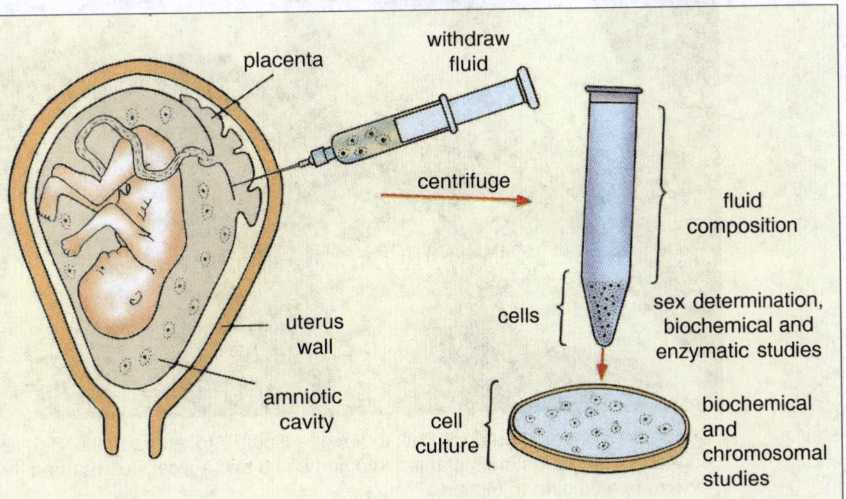

Fig. 18.3. Various steps of amniocentesis technique (after Suzuki *et al.*, 1986).

Amniocentesis involves **removing** amniotic fluid by the help of a hypodermic needle from the amnion ("the bag of water") surrounding the foetus in uterus; **culturing** the foetal cells contained within the fluid; and **screening** the cultured cells for karyotype, enzyme production and restriction site pattern analysis of its DNA (Fig. 18.3). At least 35 genetic diseases have been detected in human beings by this technique; some of them have been listed in Table 18.1. If a disease is detected by this method, abortion of such a foetus can then be recommended.

For karyotype preparation, foetal cells found in amniotic fluid can be cultured *in vitro* in a highly nutritive solution, treated with colchicine to stop mitosis at metaphase, subjected to a hypotonic solution to cause the cells to swell and scatter the chromosome, placed on a slide, stained and photographed under a microscope. Individual chromosomes are then cut from the resulting photograph and matched as homologous pairs to form an idiogram. Further, a combination of amniocentesis with analysis of restriction enzyme fragments now enables scientists to do prenatal screening for sickle cell anaemia. The single nucleotide substitution in the gene for the beta haemoglobin chain eliminates a restriction site in that gene. As a result a larger fragment is produced upon enzyme digestion of DNA obtained from cultured cells from the amniotic fluid. This diagnosis could not be made until several months after birth, as the gene is inactive before that time.

TWINS

Normally, a woman gives birth to only one young at a time, but sometimes more than one child are born to a woman at the same time. The delivery of more than one baby by a mother is called **multiple births**. Most commonly in the case of multiple births, the number of births is two and the individuals which are born in such a way are called **twins**. In case of multiple births, the number of infants may be three (**triplets**), four (**quadruplets**), five (**quintuplets**) or even more (*e.g.,* eleven). Twins are of two basic types :

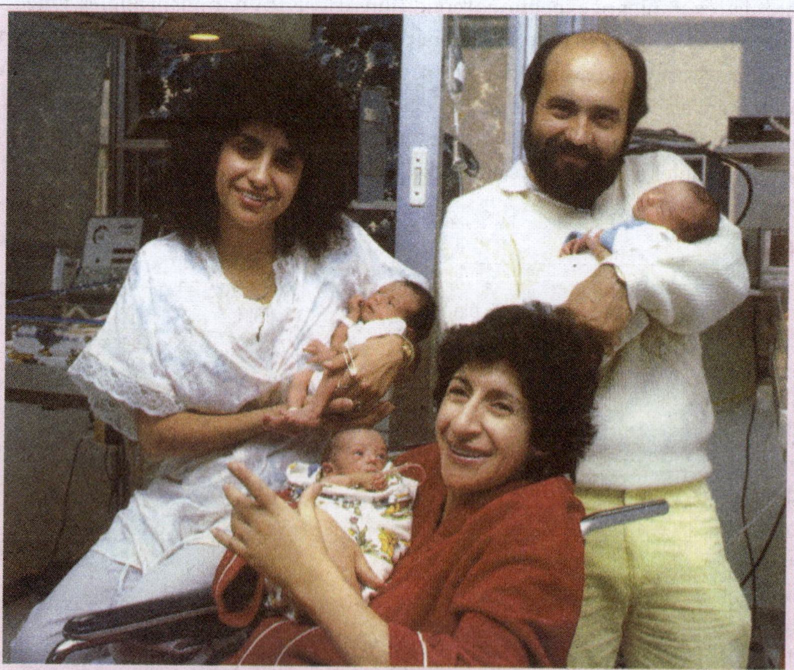

Surrogate grandmotherhood. Pat Anthony, a South African had taken fertilized eggs from her daughter, implanted into her womb and successfully carried them to term to give birth to triplets.

Table 18-1. Some common genetic diseases of humans which can be detected by amniocentesis (Source : Suzki et al., 1986).

Disease	Possible cause
I. Inborn errors of metabolism	
1. Cystic fibrosis	Mutated gene unknown
2. Duchenne muscular dystrophy	Mutated gene unknown
3. Gaucher's disease	Defective glucocerebrosidase enzyme
4. Tay–Sachs disease	Defective hexosaminidase A
5. Essential pentosauria	A benign condition
6. Classic haemophilia	Defective clotting factor VIII
7. Phenylketonuria	Defective phenylalanine hydroxylase
8. Cystinuria	Mutated gene unknown
9. Metachromatic leuko dystrophy	Defective arylsulphatase A
10. Galactosemia	Defective galactose 1-phosphate uridyl transferase
II. Haemoglobinopathies	
11. Sickle cell anaemia	Defective β-globin chain
12. β-thalassemia	Defective β-globin chain

1. Identical or Monozygotic Twins

Twins having no variability in their traits are called **identical twins**. They are produced from a single fertilized ovum. Sometimes, the two blastomeres resulting from the first cleavage (mitosis) of the fertilized egg or zygote separate from each other and develop into independent embryos (hence they are also called **monozygotic twins**). Since, the two embryos have arisen by mitosis, they have the same chromosome sets in their body cells. Thus, identical twins possess identical genes and they are like one person walking around in two bodies. These twins belong to the same sex. The situation is the same for identical triplets and other multiplets, but they are quite rare in human beings.

Identical twins.

Sometimes, identical twins fail to separate completely from each other, they are called **siamese twins**.

2. Fraternal or Dizygotic Twins

Non-identical or fraternal twins are produced by simultaneous fertilization of two separate ova by two separate sperms. Two foetuses so obtained from two zygotes are called **dizygotic twins**. They contain different sets of genes and are like any other brothers and sisters, but just happen to grow simultaneously in the same uterus. So fraternal twins differ from each other in their traits, development and often even sex.

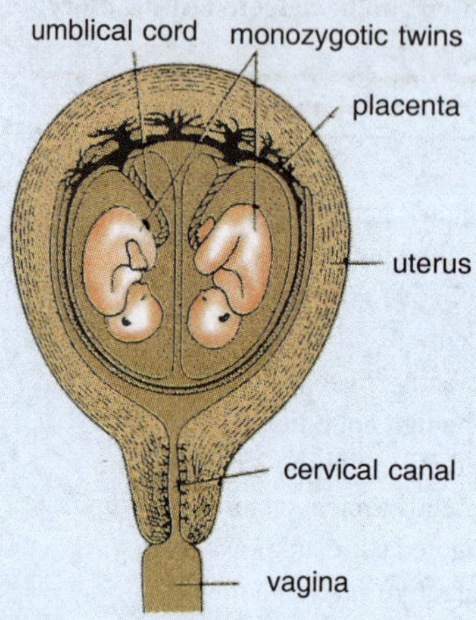

Fig. 18.4. Developing foetuses of monozygotic twins.

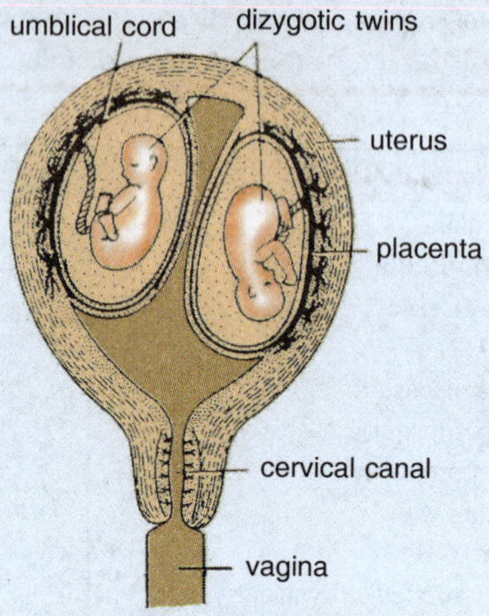

Fig. 18.5. Developing foetuses of dizygotic twins.

Twins are useful in detecting the relative effectiveness of heredity and environment upon the expression of a disease or trait. The hereditary trait is most likely shown by both members of the identical twins. If it is shown by one member of the identical twins, then the environmental factor is playing the major role since both possess the same genes. Very recently, studies of identical twins separated since birth revealed that genes (heredity) play a much greater role in determining behaviour than previously thought. Although these twins were not raised together, they exhibited many similar behavioural traits. In any behavioural characteristic studied genes are found to exert at least 50 per cent influence on the trait. The conclusion is that genes play a major role in shaping almost any type of behaviour including **alcoholism**, **criminality**, **intelligence (I.Q.)**, **political attitudes**, **scizophrenia** and **sociability**.

HUMAN TRAITS

Heredity affects a variety of human traits. They may involve physical appearance such as height, facial features, eye colour, skin colour, baldness, hair colour and form. Heredity also determines biochemical conditions such as genetic diseases, *e.g.,* brachydactyly, Huntington's chorea, porphyra, phenylketonuria, sickele-cell anaemia, and dibetes. According to **McKusick** (1987), in man there are

Table 18-2. Comparison of monozygotic and dizygotic twins (Source : Kapur and Suri, 1984).

Feature	Monozygotic twins	Dizygotic twins
1. Development	Result from a single fertilized ovum	Result from two fertilized ova
2. Heredity	They have the same heredity in similar environments	They have different heredity in similar environments
3. Reaction to grafts	Skin grafts between the monozygotic twins are not rejected	Skin grafts between the dizygotic twins are usually rejected

approximately 100,000 genes, however, definite positions of few genes on the chromosomes have been worked out. Thus, 1300 genes are found to be located on the autosomes, 107 on X-chromosomes and only one on the Y-chromosomes. More than 500 human genetic diseases are known to be caused by **single-gene defects** (**Albert** *et al.,* 1989). Certain well known dominant and recessive human genetic traits have been enlisted in Table 18-3.

Table 18-2. Some common phenotypic traits of humans.

	Dominant	Recessive
1.	Curly hair	Straight hair
2.	Dark brown hair	All other colours
3.	Coarse body hair	Fine body hair
4.	Pattern baldness (Dominant in males)	Baldness (recessive in females)
5.	Normal skin pigmentation	Albinism
6.	Brown eyes	Blue or grey eyes
7.	Near or far sightedness	Normal vision
8.	Normal hearing	Deafness
9.	Normal colour vision	Colour blindness
10.	Broad lips	Thin lips
11.	Large eyes	Small eyes
12.	Polydactylism (extra digits)	Normal digits
13.	Brachydactylism (short digits)	Normal digits
14.	Syndactylism (webbed digits)	Normal digits
15.	Hypertension	Normal blood pressure
16.	Diabetes insipidus	Normal excretion
17.	Huntigton's chorea	Normal nervous system
18.	Normal mentality	Schizophrenia
19.	Migraine headache	Normal
20.	Normal resistance to disease	Susceptibility to disease
21.	Enlarged spleen	Normal spleen
22.	Enlarged colon	Normal colon
23.	A or B blood group (antigen)	O blood group
24.	Rh blood factor	No Rh blood factor

DISORDERS DUE TO MUTANT GENES

1. PTC tasters. In human beings, there exists a autosomal dominant allele for the ability to taste the bitterness of the chemical, called **phenylthiocarbamide (PTC)**. This gene is inherited in simple Mendelian fashion to give a 3 : 1 monohybrid ratio (Fig. 18.6). Thus, tasters of PTC may have two genotypes (TT and Tt), whereas non-tasters will have only one genotype (tt).

P : TT (Taster) X tt (Non-taster)

↓

F_1 : Tt (Taster)

[Marriage of F_1 heterozygote (Tt) and a heterozygote for taster (Tt)]

F_2 : 1TT : 2 Tt : 1tt

3 Taster : 1 Non-taster

Fig. 18.6. Inheritance pattern of PTC taster trait.

2. Brachydactyly. In this disorder, the middle joints of the digits (phalanges) in hands or feet remain rudimentary and fuse with an adjacent joint (Fig. 18.7). Consequently, the fingers and toes become abnormally short. The defect of brachydactyly is caused by a dominant gene existing on an autosome. When an abnormal gene dominates the normal gene, this is called **dominant inheritance**.

3. Huntington's chorea. It is a fatal disease of man. It is characterized by uncontrolled jerking of body (due to involuntary twitching of voluntary muscles) and a progressive degeneration of the central nervous system (*i.e.,* brain and spinal cord), accompanied by gradual mental and physical deterioration. The mean age of onset of these symptoms is between 35 and 40. This disease is caused by an autosomal dominant gene.

4. Tongue rolling. The trait of **ability to roll one's tongue** is also controlled by a dominant allele.

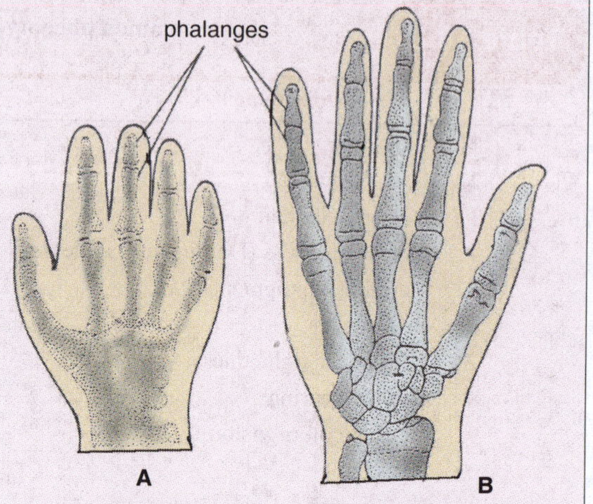

Fig.18.7. Comparison of a brachydactyl hand and a normal hand. A—A brachydactyl hand showing two phalanges in each finger; B—A normal hand with three phalanges in each finger.

5. Inborn errors of metabolism. In 1908, an English physician, **Archibald E. Garrod** published a book called *Inborn Errors of Metabolism.* He proposed that the genes produce enzymes whereas the mutant forms of these genes do not produce them, hence, resulting in physiological abnormalities. He studied the metabolism of phenylalanine (an essential amino acid) and showed that the metabolism of this amino acid proceeds in chains of enzyme-mediated reactions and a change or absence of an enzyme result an abnormality. There are three important diseases associated with metabolic breakdown of the phenylalanine :

The women in this photograph are sisters. Both are homozygous for the same mutant PAH gene. The bride is the younger of the two. She was diagnosed early and put on the PKU diet. Her elder sister was diagnosed too late to begin the diet. She is showing the symptoms of phenylketonuria.

(i) Phenylketonuria (PKU). Persons having autosomal homozygous recessive alleles pp fail to produce an enzyme called **phenylalanine hydroxylase**. This enzyme is required by the body to convert phenylalanine into tyrosine. As a result, the concentration of phenylalanine increases in blood. This amino acid is then partly converted into phenylpyruvate, phenylacetate, phenyllactate and other derivatives, all of which (including phenylalanine) are excreted in the urine. The patients of PKU have light body pigmentation and are physically and mentally retarded. They are called **phenyl pyruvate idiots** having mean IQ of 65 (**Burns** and **Bottino**, 1989). The idiocy of PKU is thought to be due to damage of the brain tissues by the phenylpyruvate in the cerebrospinal fluid. Here, the abnormal gene is suppressed in heterozygous individual, it is called **recessive inheritance**.

(ii) Alkaptonuria. The persons with autosomal homozygous recessive genes *hh* fail to produce the enzyme **homogentisic acid oxidase** which catalyzes the oxidation of alkapton (*i.e.* 2-5-dihydrophenylacetic acid or homogentisic acid). Therefore, in them, normal oxidation of alkapton into acetoacetic acid and ultimately into water and carbon dioxide does not take place and large amount of alkapton is accumulated in the blood and is excreted in the urine, which turns black upon exposure to

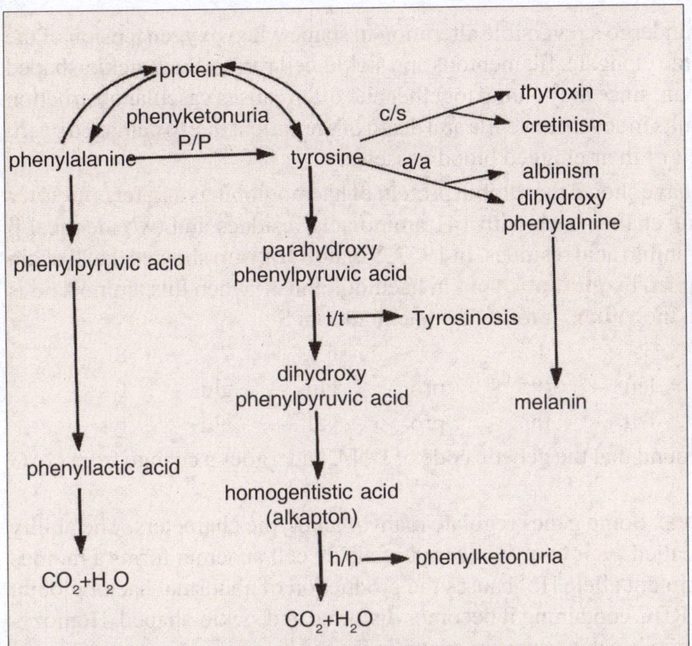

Fig. 18.8. Metabolism of phenylalanine in human body and various inborn errors of metabolism.

the air. In such alkaptonuric persons, the disease is manifested by the darkenings of cartilaginous regions (such as ear pinna) and a proneness to arthritis (a condition called **ochronosis**).

Urine test for alkaptonuria.

(iii) Albinism. The persons with recessive aa genes do not produce the **tyrosinase** enzyme which is needed by melanocytes for converting DOPA (3, 4- dihydroxy- phenyl- alanine) into melanin (a dark brown pigment). As a result in an albino patient melanocytes are present in normal number in their skin, hair, iris of eyes, etc., but lack in melanin pigment. This trait is inherited like a typical Mendelian monohybrid cross.

A person suffering from albinism.

6. Sickle-cell anaemia. The inheritance pattern of the trait of sickle-cell anaemia in humans presents an example of various genetical phenomena such as codominance, incomplete dominance, lethal effect, pleiotropism and polymorphism. When an allele results in a number of related changes, then such changes are collectively called **syndrome**. The syndrome, called sickle cell anaemia, is caused by a autosomal mutant allele, Hb^S, that in homozygous condition causes the production of an abnormal haemoglobin, called **haemoglobin S**; this indicates **primary effect** of mutant gene. In 1949, **Pauling** and his coworkers reported that the formation of an abnormal haemoglobin (*i.e.,* Hb^S) which differed in its electrophoretic mobility from normal haemoglobin (Hb^A or adult haemoglobin) was the cause of this hereditary disease.

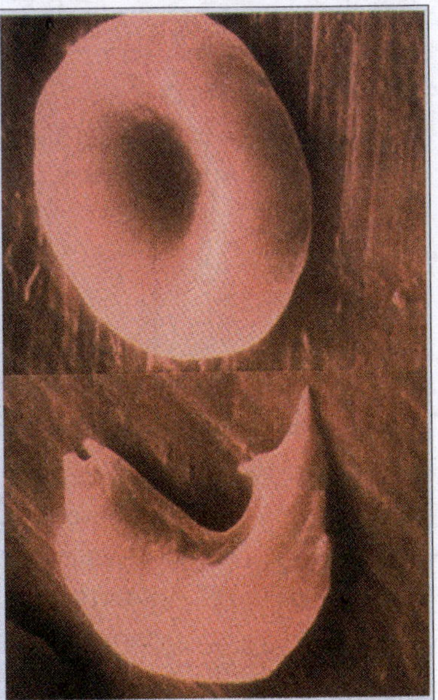

Top red blood cell is normal, bottom cell is sickled.

Further, due to **subsidiary effect** of abnormal haemoglobin (*i.e.,* Hb^S), red blood cell of some persons suffered from this fatal form of haemolytic anaemia undergo a reversible alteration in shape when oxygen tension of the plasma falls slightly and they assume elongate, filamentous and sickle-cell forms. Such sickle-shaped red blood cells show a lower life span, since they clump together and often causes vascular obstruction and are rapidly destroyed. This results in severe anaemia and death of the patient due to damaged heart, kidney, spleen and brain as a result of their clogged blood vesseles.

Biochemical investigations have shown that globin protein of haemoglobin is a **heteropolymer** having two identical α polypeptide chains, each with 141 amino acid residues and two identical β polypeptide chains, each with 146 amino acid residues. In 1957, **Vernon Ingram** showed that β chain amino acid of 6th position is occupied by **glutamic acid** in haemoglobin A, when this amino acid is substituted by another amino acid, the **valine**, it forms the haemoglobin S :

	1	2	3	4	5	6	7	8
Hb^A =	val -	his -	leu -	thr -	pro -	glu -	glu -	lys -
Hb^S =	val -	his -	leu -	thr -	pro -	val -	glu -	lys -

At the molecular level, it is found that the genetic code of DNA undergoes a change from GAG to GTC (see **Alberts** *et al.*, 1989).

Pleiotropy of sickle cell allele. Some genes regulate many phenotypic characters. The ability of a gene to have many effects is called **pleiotropy**. The trait of sickle cell anaemia forms a famous example of pleiotropy. Thus, the mutant allele Hb^S causes the production of abnormal haemoglobin. As a consequence the shape of the RBC containing it becomes distorted and sickle-shaped. Homozygous individuals normally die early in life due to severe anaemia caused by premature destruction of the sickled red blood cells. But heterozygotes who have both normal (haemoglobin A) and abnormal (haemoglobin S) haemoglobins and who also have mild anaemia, are naturally protected against contracting malaria as the parasite cannot live in these distorted cells. Heterozygotes may, therefore, survive better in region where malaria is endemic. Such population have both normal individuals and individuals heterozygous for the gene. The latter act as carriers of the gene from generation to generation as shown below :

$Hb^A\ Hb^S$ Sickle cell carrier × $Hb^A\ Hb^S$ Sickle cell carrier

↓

¼ $Hb^A\ Hb^A$ Normal; haemoglobin A; no sickling of RBC	½ $Hb^A\ Hb^S$ Sickle cell trait; carriers; produce both type of haemoglobin due to codominance; sickling under reduced O_2 tension	¼ $Hb^S\ Hb^S$ Sickle cell anaemia; normally die early in life; produce haemoglobin S only
↓	↓	↓
Heterosis : More susceptible to malarial infection in prone areas	Better adapted for malaria in prone area	suffer from lethal anaemia

Polymorphism of sickle cell trait. The existence of two or more genetically different classes in the same interbreeding population at the same time, in the same habitat, is known as **polymorphism**. These classes are controlled by two competing alleles in the same locus. In **balanced polymorphism**, there establishes an intermediate equilibrium between the frequencies of these alleles and two classes are maintained together in the same population. For example, in case of sickle-cell anaemia superiority retains the two types at equilibrium. Here, superiority of heterozygotes (Hb^A Hb^S; which is resistant to malaria) over the normal homozygotes (Hb^A Hb^A) maintains the two alleles of haemoglobin (*i.e.*, normal and sickling) in an equilibrium.

HUMAN CYTOGENETICS

A successful attempt to count the number of human chromosomes was made in 1912 by **Winiwarter** who proposed that human chromosomes are 48 in women and 47 in man; men having one X chromosome and women having two X chromosomes. **Painter**, in 1923, while examining the testicular material of man, observed a heteromorphic pair of sex chromosomes and proposed the XY mechanism of sex determination in man. **Tjio** and **Levan** (1956) cultured somatic cells from fibroblasts of human embryos and counted the human chromosome number as 46. This chromosome number was confirmed by **Ford** and **Hamerton** while working with testicular material in the same year. **Tjio** and **Levan** provided greatly improved techniques for chromosome preparations. **Moorhead** *et al.*, (1960) described a simple method of culturing of lymphocytes from human blood.

Karyotyping human chromosomes. For karyotyping of human chromosomes venous blood is taken and blood leucocytes are stimulated to divide (by mitosis) *in vitro* by the addition of **phytohaemagglutinin**. Colchicine is added to arrest cell division at metaphase stage. It is further treated with hypotonic saline solution which results in swelling of cells and dispersal and better clarity of chromosomes for counting and morphological study. There after, the material is stained (*e.g.*, with Giemsa technique) to demonstrate the banding patterns of chromosomes. Finally, a suitable metaphase spread is photographed through a high power microscope. The individual chromosomes are cut out from the photograph. The chromosomes are then arranged in an orderly fashion in homologous pairs, to produce a standard arrangement, the karyotype.

To characterize a chromosome in the karyotype, the following parameters are used :

1. Shape of chromosome ;
2. Length of chromosome ;
3. Centromeric index, *i.e.*, this index is expressed in the form of ratio of the short arm length to the total chromosome length :

$$\text{Centromeric index} = \frac{\text{Short arm length}}{\text{Total chromosome length}}$$

For example, centromeric index in a metacentric chromosome is 0.5.

4. Proportion of the arms, *i.e.*, it is ratio between the long arm and short arm of the chromosome. This ratio is 1 : 1 in a typical metacentric chromosome.

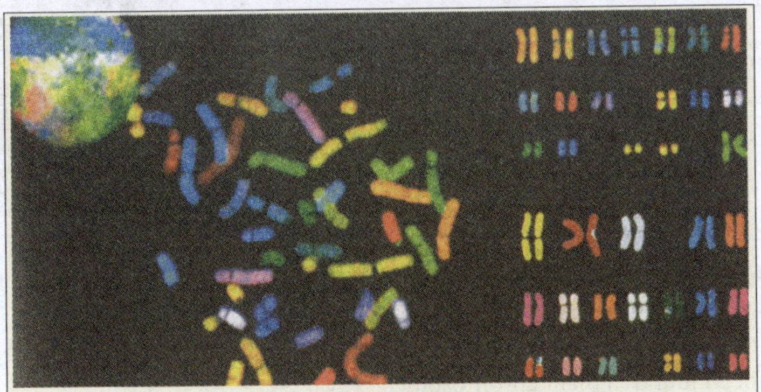

Spectral karyotyping of human chromosomes utilizing differentially labelled "painting" probes.

Classification. The human metaphase chromosomes were first of all classified by a conference of cytogeneticists at Denver, Colorado in 1960 and is known as the **Denver classification**. To follow this classification, each of the 22 pairs of autosomes has been numbered from 1 to 22 according to their decreasing size. **Patau** (1960) divided the human chromosomes into the following seven groups designated A to G :

1. A group : 1 to 3 pairs — Metacentric
2. B group : 4 to 5 pairs — Submetacentric
3. C group : 6 to 12 pairs — Submetacentric
4. D group : 13 to 15 pairs — Acrocentric
5. E group : 16 to 18 pairs — Submetacentric (16 is metacentric)
6. F group : 19 to 20 pairs — Metacentric
7. G group : 21 to 22 pairs — Acrocentric

Group A consists of longest metacentric chromosomes.

Group G consists of the shortest acrocentric chromosomes. These chromosomes have satellites that correspond to nucleolar organizers. Chromosomes of group D also contains satellites. In males, group G includes a variable Y chromosome which lacks the satellites.

The X chromosome is the member of group C and can be identified by special banding or staining methods.

Banding Techniques

Recently banding techniques reveals structural details of chromosomes. The main banding techniques are identified by letters such as Q, G, C, R, T, F and N bands:

1. Q banding. It uses fluorescent dyes (such as quinacrine mustard) and identifies the so-called **Q bands.**

2. G banding. It uses Giemsa stain and identifies the G bands. With G banding three major types of chromatin can be recognized—euchromatin, centromeric and intercalary heterochromatin. The Q and G bands are generally similar and correspond to intercalary heterochromatin.

3. C banding. It stains specifically centromeric constitutive heterochromatin.

4. R banding. It gives a pattern that is the reverse of that of Q and G banding.

5. T banding. It stains telomeres of chromosomes.

Other banding techniques uses the Feulgen stain (**F bands**) and one selectively stains the nucleolar organizers (**N bands**) which are localized in the satellite of chromosomes 13, 14, 15, 21 and 22. G banding has become important tool in the analysis of mammalian, avian, reptilian and amphibian chromosomes; distinct G bands have not been found in plant chromosomes.

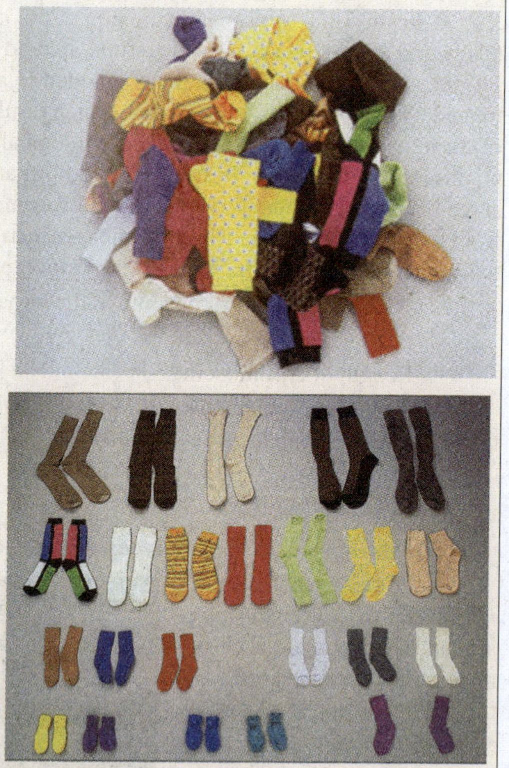

The 46 chromosomes in the human karyotype come in pairs, like socks.

Clinical importance of chromosome banding. Since banding patterns are unique and constant for each normal chromosomes, in case of a large number of chromosomal abnormalities, such as loss of a very small part, insertion of an additional segment and addition of whole chromosome can be easily recognized, *e.g.,* cat-cry syndrome due to loss of small part of chromosome 5 ; Down syndrome due to an extra chromosome 21.

SEX DETERMINATION

In human beings, the presence of Y chromosome determines maleness and its absence determines femaleness. So, males are XY and females are XX in human beings. However, in 1986, certain peculiar cases have been reported which were found to be males with XX chromosomes and females with XY chromosomes. These can be due to any one of the following two reasons : (i) A sex reversal gene SRY located on the Y chromosome leads to XX males and XY females. (ii) Translocation of a small segment of the Y chromosome to an X chromosome in XX males and its deletion from the Y chromosome results in the XY females, (For more details of sex determination, sex reversal and sex differentiation see Chapter 13).

Human Sex Anomalies

1. Turner's Syndrome (XO Females). A female with 44 autosomes and only with one X chromosome in her body cells exhibits symptoms of Turner's syndrome. Such females are sterile and have short stature, webbed neck, a low hairline on the nape of the neck, broad shield-shaped chest, low intelligence, under developed breasts, poorly developed ovaries, sparse pubic hairs and no axillary hair (Fig. 18.9).

2. Poly-X Females (XXX Females). Such females are called super females because they possess an extra X chromosome (44 autosomes+3X chromosomes). Some females may have 4 or 5X chromosomes besides the normal autosomes. All such poly-X females are mentally retarded and sterile showing abnormal sexual development.

3. Klinefelter's Syndrome (XXY Males). When an abnormal egg with XX chromosomes is fertilized by a sperm carrying Y chromosome, a zygote having three sex chromosomes (XXY chromosomes) is formed. The resulting young one is an abnormal sterile male showing the following features : small testicles, mental retardation, longer arms, feeble breasts, higher pitched voice and sparse body hairs (Fig. 18.10).

4. XYY males. Presence of an extra Y chromosome in males (XYY) results in their unusual height, mental retardation, severe facial acne during adolescence and criminal bent of mind. Their genitals are affected by developmental abnormalities.

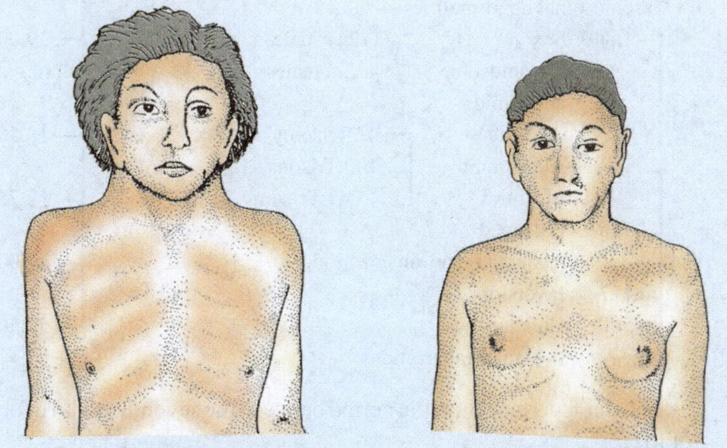

Fig. 18.9. XO female (Turner's syndrome)

Fig. 18.10. XXX male (Klinefelter's syndrome).

5. Hermaphroditism. True hermaphrodites are individuals that possess both ovarian and testicular tissue. The external genitalia are ambiguous, but often more or less masculinized; secondary sex characters vary from more or less male to more or less female. Ordinarily, true hermaphrodites are sterile because of rudimentary ovotestes. However, **J. Brazzel** and coworkers

(1978) reported a case of pregnancy in a hermaphrodite that terminated in delivery of a stillborn child. This individual was found to be engaged in male sexual activity in the early years but gradually shifted to a preference for the female role. Karyotype of true hermaphrodites may be 46, XX or 46, XY.

Pseudohermaphrodites have either testicular or ovarian tissue, generally rudimentary, but not both. On the basis of genetic sex, two major classes, **male** and **female**, are identified. The former are most often 46, XY and external genitalia are ambiguous. A penis-like organ of variable size is present. In adolescence pubic and axillary hair develops and the voice deepens. Often some breast development, as in females, occurs. **Female pseudohermaphrodites** are 46, XX (hence females) but present a more or less masculine phenotype. The external genitalia are ambiguous; ovaries are present but are immature or rudimentary.

SEX LINKAGE

In Chapter 12 we have discussed several cases of inheritance of X-linked recessive alleles, such as, of colour blindness and haemophilia and Y-linked alleles such as hairs on the ears.

CHROMOSOMAL ABERRATIONS

In human beings various types of chromosomal variations, both numerical as well as morphological, have been reported (Fig. 18.11). Most of autosomal aberrations have been discussed in Chapters 14 and 15.

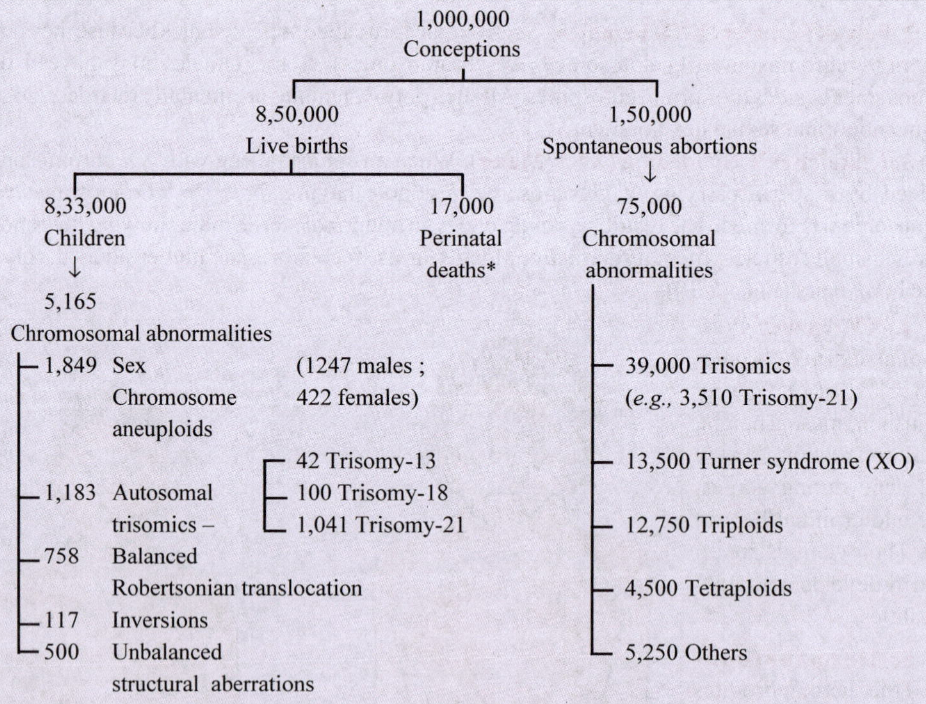

* perinatal = pertaining to the period between the seventh month of pregnancy and the first week of life.

Fig. 18.11. Occurrence of various numerical and structural chromosomal variations in human beings (based on data computated by **K. Sankaranarayana**, 1979) (after Suzuki *et al.,* 1986).

REVISION QUESTIONS

1. Describe some specific human traits controlled by heredity.
2. What is pedigree analysis? How are pedigree charts are used in human genetics? Explain.
3. Explain why, in human families, many traits, *e.g., albinism, blue eye,* and *phenylketonuria,* skip generations, while traits such as *polydactyly, free earlobes,* and *A* and *B blood groups* do not ?
4. (a) What are the major difficulties encountered in the study of the genetics of man ?

   (b) If you had several human pedigrees for a simply inherited trait, what criteria would you use to determine whether the trait was due to a dominant or a recessive allele ?
5. Describe the method of karyotyping human chromosomes.
6. What is the number of autosomes present in a human male and female ?
7. Suggest a method of detection of chromosome abnormalities at the foetal stage.
8. How the sex is determined in human beings ? Describe various types of syndromes related to variation in number of sex chromosomes.
9. Explain with illustrations the genotypes of the progeny expected in matings between a normal parent and one in which non-disjunction occurs at the time of gamete formation.
10. Suggest a mechanism by which individuals possessing Turner's syndrome and Klinefelter's syndrome may originate.
11. Write short notes on the following :

    (i) Amniocentesis ; (ii) Monozygotic twins ; (iii) Brachydactyly ; (iv) Albinism ; (v) Phenylketonuria; (vi) Sickle-cell anaemia ; (vii) Human chromosomes ; (viii) Turner's syndrome ; (ix) Klinefelter's syndrome.

CHAPTER

19

Eugenics, Euphenics and Genetic Engineering

Human life balanced against scientific progress.

Eugenics deals with the application of the laws of genetics to the improvement of humen race. The term eugenics (Gr. *eugenes* = well born) was coined by an English scientist **Francis Galton** in 1885. The science of eugenics can be defined as a science of well born, improving the inborn qualities of race and obtaining the better heritage by judicious breeding.

EUGENICS AND EUTHENICS

The betterment of human society can be achieved by following two inter-related methods :

1. By one of the method we can deal with the already existing human beings. The improvement of already existing human beings can be achieved by improving the environmental conditions, *e.g.,* by subjecting them to better nutrition, better unpolluted ecological conditions, better education and sufficient amount of medical facilities. This type of method of improving the human race is known as **euthenics**.

2. By another method we can improve the future generations by improving the germplasm of existing individuals. The

type of method is known as **eugenics**. Eugenics believes in artificial seleciton of physically and mentally sound individuals and discouragement of defective individuals for the inheritance of their defective germplasm to the future generations. In other words, eugenics seeks the measures to preserve the best type of germplasm and to eliminate defective germplasm from the human society by applying the laws of inheritance to man.

HISTORY

The primary aim of many of the ancient systems of eugenics was to produce a race of physically perfect human beings. The Greeks had definite ideas regarding eugenics. In Sparta a physically perfect manhood was the chief aim, where as the Atheniens carried more for the intellectual achievements.

Eugenics was the invention of the British scientist Sir Francis Galton.

Following the doctrine of Greeks and until the 19th century there was little interest in the eugenics. Of the particular importance was the eugenics movement in England in the last part of the nineteenth century. The movement, spearheaded by persons of outstanding intellect such as **Francis Galton** and **Karl Pearson**, had its objective in the application of biologically sound principles to human populations. Since the biological basis of heredity was unknown, the first objective was to establish the nature of heredity.

Galton and **Pearson** choose to work with human beings and with what they considered important human traits such as intelligence, stature and special abilities. We now know that these characters are very complex traits and are under the control of many genes interacting with environmental variations. So, quite naturally the early investigators made little progress. However, **Francis Galton** should be credited with being the real founder of the modern movements of eugenics. He defined the eugenics as the study of all the agencies under social control which may improve or impair the inborn qualities of future generatins of man either physically or mentally. **Darwin** also attached great importance to the eugenics and he compared it to a signpost with three directions. One of these indicates the influence of heredity on the fate of nations. Another point to the rules that an individual should strive to carry out in regard to parenthood based on the law of human heredity. The third arm indicates the regulations to be adopted by the society to encourage racial progress.

NEED FOR EUGENICS

The development of all organisms including the human individuals depends on both heredity and environmental factors. For the best development both good heredity and good-environment are essential. Even in the best environment there is little possibility for change of defective hereditary traits. The need of eugenics is apparent from the stand point of education, sociology and civilization. The aim of geneticists should be of increasing the normal and gifted population and at the same time decreasing the abnormal and deficient populations.

EUGENICS AND HUMAN BETTERMENT

Both hereditary and environmental factors play a significant role in the development of the organisms. For better type of development both good heredity and suitable environment are necessary. When we consider the future welfare of the human race then the following two factors alarm us greatly: (i) the declining birth rate among the normal and superior people (those having best germplasm) (ii) a relative rapid increase of the abnormal and defective individuals (those having defective germplasm). For the betterment of future generation, it is necessary to increase the populaiton of outstanding people and to decrease the population of abnormal and defective people by applying the principle of eugenics.

The eugenics can be applicable by adopting following two methods :

(A) By encouraging the marriages between desirable persons (constructive method or **positive eugenics**).

(B) By discouraging the marriages between undesirable persons (restrictive method or **negative eugenics**).

A. Positive Eugenics

The positive eugenics attempts to increase consistantly better or desirable germplasm and, thus, to preserve best germplasm of the society. The percentage of desirable traits can be increased by adopting following measures :

1. Early marriage of those having desirable traits. It is most commonly observed fact that the highly placed persons of the society often have great ambitions for the future life. In achieving their ambitious goals, they often devote the best part of their youth and they are able to marry in their mature age (*e.g.*, 30 to 35 years). The biological and psychological investigation have revealed that the aged persons often lack in necessary amount of emotional warmth for the sexual activities and moreover, their germplasm also lost its vigour. Therefore, some laws should be formulated to prevent the late marriages of highly endowed persons by applying high taxation on them and at the same time the young persons having best hereditary traits should be encouraged for early marriage.

2. Subsidizing the fit. Because the highly endowed persons lead a well-planned life and to avoid unnecessary difficulties in nursing the children they often prefer to have small number of children. Therefore, the selected young men and women of best eugenic value should be encouraged to increase their birth rate. Moreover, **H.J. Muller** has suggested that such persons not only should increase their family size but through artificial insemination the outstanding man can serve as fathers to many more children than would be otherwise possible. The artificial insemination is already widely practiced to permit those women whose husbands are sterile or have some serious hereditary afflictions to bear children. The sperms and eggs of outstanding persons can be stored for future use by quick freezing and storing them in deep freeze. These germ cells, thus, can be stored for 100 or more years.

Embryos being stored at –190°C in liquid nitrogen. At this low temperature, embryos can be preserved almost indefinitely.

Very recently the scientists felt the urgent need of establishment of **sperm** and **egg banks** to protect these precious germ cells from the radiation. The germ cells could be collected during early adulthood and stored in lead lined containers in the deep freeze. In this state the germ cells would not be subjected to radiaiton expossure which might affect the donors.

3. Education. For the eugenically oriented reforms in the society, the people should be educated about the basic principles of human biology, human genetics, eugenics and sex. The children should be instructed about basic laws of health and they should be encouraged to develop a physically and mentally healthy body. Moreover, the sex should be free from the wide spread confusion, narrow-minded concepts and religious and ethical bindings because that is a natural biological instinct. The children ignorant about the facts of sex may do more harm to society then otherwise.

4. By avoiding germinal waste. The wastage of best type of germplasm can be avoided by adopting following measures :

(i) The selection of marriage partners should be made with intelligence.

(ii) The social hinderance which do not allow the teachers, nuns and priests to get married, must be removed. By adopting such measures the wastage of best type of germplasm due to lack of opportunities can be prevented.

(iii) The wars must be avoided because in wars the best germplasm of the society is wasted.

The selection of marriage partners should be made with intelligence.

5. Genetic counselling. Genetic counselling can do great benefit to human society. The role of the genetic counsellor is to inform concerned individuals of the nature of the mutant condition that concern them. If it is inherited in a Mendelian fashion, then probability of producing affected offspring can be calculated. The final decision for taking a risk, is entirely the responsibility of the individual involved, it cannot be made by the counsellor.

6. Improvement of environmental conditions. Both heredity and environment have inter-related role in the development of eugenically better persons. Therefore, every person should get better food, living conditions, education and medical guidance, etc., so his or her hereditary traits can do their best development.

7. Promotion of genetic research. Our knowledge about the genetics is not sufficinet enough because we still have little informations about various human diseases and metabolic disorders which are generally related with the genes. Therefore, the research in the field of cytogenetics should be increased so that we can learn more and more about the man.

B. Negative Eugenics

The negative eugenics attempts to eliminate the defective germplasm of the society by adopting following measures :

1. Sexual separation of the defective. The defective persons may have various sex-linked diseases such as night blindness, haemophilia, colour blindness, etc., and various other defective traits which may be regulated by dominant or recessive genes. The increase of germplasm of the persons having such defective traits in the populaiton can be checked by keeping them away and separated from the society. Different states have wisely adopted the restricted measures in segregation the mental defectives from the society and to place them in mental hospitals.

2. Sterilization. The sterilization is the best means to deprive an individual from his power of reproduction without interfering with any of his normal funcitons. The sterilization method is based on surgical operation of sperm duct or vas deferens in males and oviducts or fallopian tubes in females. The former is known as **vasectomy** and the later is **splengectomy**. The family planning movement in India has adopted the sterilization as the tool for controlling the rate of vastly increasing population and in that case the sterilization is euthenical in its application than the eugenical.

3. Control of immigration. Through immigration there are enough chances that undesirable or defective genes of different races and nationalities may intermingle with the normal germplasm of the population. Therefore, the immigration rules must be strict and the persons with undesirable hereditary traits must not be allowed to migrate from one place to another.

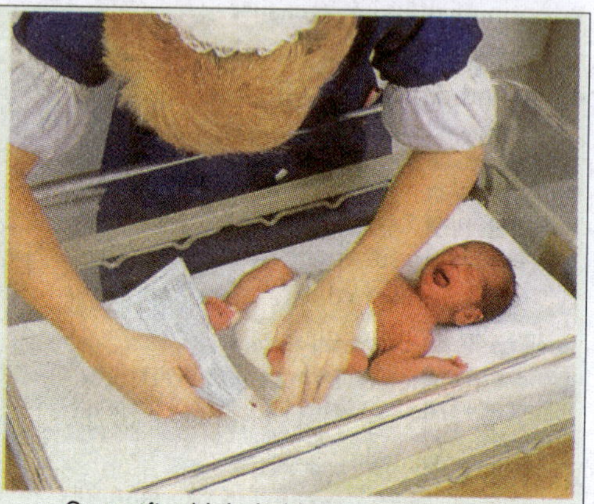

Soon after birth, babies have the Guthrie test which screens for phenylketonuria.

4. Regulation of marriage. Presently most human societies are money-minded and for the marriage relationship the wealthy or highly placed persons which, however, may contain several defective genetical traits, are more preferred than those which have eugenically sound hereditary traits but having no money and which fail to reach to highest status of the society due to lack of opportunities. Some rule must be enacted to encourage the marriages among desirable mates.

Recently human geneticists have devised following two eugenically sound methods (*i.e.*, euphenics and genetic engineering) for the improvement of human race.

EUPHENICS

The symptomatic treatment of genetic diseases of man is called **euphenics** (see **A.C. Pai**, 1974). The euphenics deals with the control of several inherited human diseases, especially inborn errors of metabolism in which the missing or defective enzyme has been identified. One example of this is the condition known as **phenylketonuria** or **pku**, determined by an autosomal recessive gene. Babies with this defect are unable to properly metabolize an amino acid, phenylalanine, the resulting chemical imbalance causes severe mental retardation. Now it is possible to distinguish pku homozygotes from normal individuals by testing the urine of all newborn babies with ferric chloride. In affected children, the metabolic imbalance caused by the mutation will turn the urine green. Once such a child is detected, a diet free of phenylalanine is imposed and the child can develop normally.

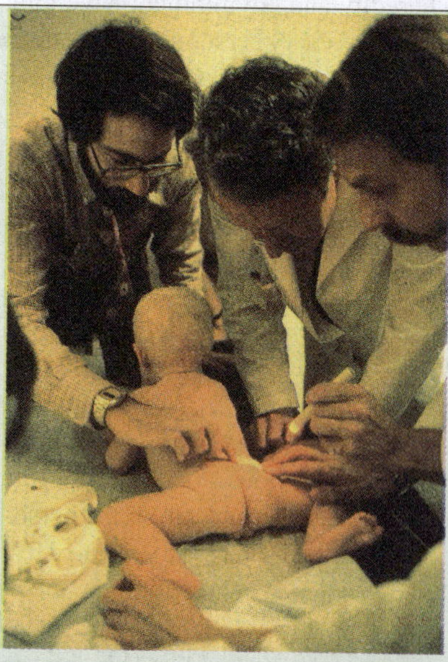

In modern healthcare the emphasis is on prevention rather than cure. A baby's DNA holds information about any genetic disease. DNA analysis should be a boon for medicine. But control over the use of genetic information will be needed to prevent its abuse.

Although a number of inherited diseases can be treated in a similar euphenic manner, but these constitute only a small fraction of known inherited disease. For the most part, biochemical geneticists could not identify the biochemical errors of many genetic diseases. In other cases, such as **albinism**, even though the metabolic block leading to an abnormality is known, but, it is not possible to correct it. However, in future following euphenic measures can get rid of man from certain fatal genetic diseases :

1. Intake of missing enzyme. One possible euphenic measure for the future would be to supply the known missing enzyme to individuals that would allow their cells to complete the required biochemical reaction. Some attempts to do this have been made without much success. Immunological difficulties are encountered, since the enzymes being supplied are antigenic and the body produces antibodies against them.

2. Cure for inherited anemia. Scientists studying the two anemias of man, **Cooley's** and **Lepore anemia**, resulting from abnormally low haemoglobin production in individuals homozygous for mutations at the beta-chain locus, hope that they may someday discover the factors regulating the beta gene activity and thereby increase the amount of beta chain synthesis. Furthermore, if the mechanism regulating haemoglobin synthesis in the foetus and the adult can be discerned, they might be able to cure lethal conditions, such as **thalassemia major** and **sickle cell anemia**, by suppressing synthesis of the abnormal beta chains and allowing foetal or gamma chains to be produced instead. In fact, individuals have been discovered whose β chain locus for some unknown reason never becomes active. Such individuals function normally with fetal haemoglobin even as adults.

3. Increasing role of genetics to medicine. The increasing number of human diseases that are being discovered to have a genetic basis lend great importance to the development of such euphenic measures. Three per cent of all humans have hereditary diseases which are transmitted in a Mendelian fashion. Two diseases which account for the deaths of hundreds of thousands a year, **cancer** and **heart disease**, are thought to have some heritable component. The future works of immunogenetics can suggest the ways by which an individual having genes for cancer may develop resistance for this diseases.

A study has shown that some forms of heart diseases may be inherited as an autosomal dominant trait (**L.K. Altman**, 1972). An understanding of the genetic basis of such heart diseases would alert persons from families with an incidence of the disease to the possibility of incurring heart conditions and perhaps cause them to alter their diets and life habits accordingly. For example, those who may inherit genes for heart trouble would then refrain from smoking and avoid high fat diets.

Harmful evolutionary effect of euphenics. To restore affected individuals to normalcy by euphenic measures is the only compassionate goal for scientists to maintain but to do so is to counter the forces of natural selection that are the basis for the evolutionary strength of a species. Pku homozygotes, for example, would normally not reproduce and transmit the harmful mutations to future generations. Selection aganist them would be total; however, when they do develop normally and produce progeny, every member of the next generation would be a carrier of the mutaiton, assuming that the wife of such person being normal. This has to add to our genetic load and to weaken the human species from the evolutionary point of view.

GENETIC ENGINEERING AND GENE THERAPY

During the late 1970s, the science of genetics entered a new era dominated by the use of **recombinant DNA technology** or **genetic engineering** (or **biotechnology**) to produce novel life forms not found in nature. Through this technology, it has been possible to transfer genes from mammals into bacteria, causing the microbes to become tiny factories for making (in relatively large quantities) proteins of great economic importance such as hormones (including growth hormone) and interferons (lymphocyte proteins that prevent replication of a wide variety of viruses). These proteins are produced in such small quantities in humans that the cost of their extraction and purification from tissues has been very expensive, thus, limiting their medical use in prophylaxis (prevention) and therapeutics (treatment) of disease. By genetic engi-

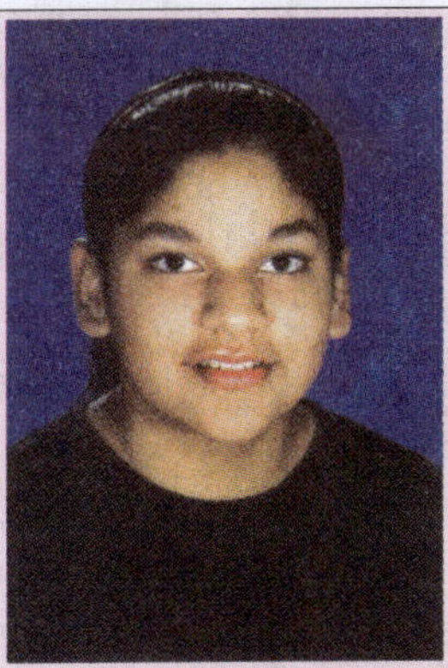

The most successful example of gene therapy. In 1990, researchers at the National Institutes of Health treated Ashanthi Desilva for the genetic disease SCID (severe combined immunodeficiency disease).

neering, it has become possible to produce various blood clotting factors, complement proteins (part of immune system) and other substances for the correction of genetic deficieny diseases (euphenics).

Recently, experiments have been conducted in which human cells deficient in the synthesis of purines have been obtained from the patients with **Lesch-Nyhan syndrome** and grown in culture; these cells have been converted to normal cells by transformation with recombinant DNA. The exciting potential of this technique lies in the possibility of correcting genetic defects—for example, restoring the ability of a diabetic individual to make insulin or correcting immunological dificiencies. This technique is called **gene therapy**.

Gene therapy basically involves the following two methods : (1) **Targeted gene modification.** This method is well demonstrated in mammalian systems. In this technique, genes have been introduced by any of the traditional gene transfer methods such as **calcium phosphate mediated gene transfer (or transfection)**, **electroporation** or **microinjection**. This is followed by **site specific mutagenesis** (*e.g.,* mutation of **HGPRT** or **hypoxanthine guanine phosphoribosyl transferase** locus and **int-2** loci in mouse embryonal stem cells). (2) **Gene augmentation method.** In this method, normal foreign gene sequences for the defective gene are introduced. A number of copies of the desired gene are introduced in the cell and are made to express at high level. Expression and transfer **vectors**, in the form of a number of viruses are now available to achieve this goal. Once the *in vitro* gene correction or gene augmentation has been achieved at the cellular level (in the cells obtained from affected organs depending on the disease), the modified cells can be implanted into a suitable region either in an organ of the patient or in the embryo. *In vivo,* direct delivery of the DNA is carrid by the vector into the living cells of the body.

Thus far, only a small number of human genes (Table 19-1) have been isolated and cloned for possible use in gene thereapy experiments. Gene therapy can be performed at two different levels :

(i) Patient therapy. In patient therapy, cells with healthy genes may be introduced in the affected tissue, so that the healthy gene overcomes the defect without affecting the inheritance of the patient. Patient therapy include the following steps :

(1) identification of a defective gene ; (2) isolation or synthesis of normal healthy gene; (3) isolation of cells of the tissue, where the normal healthy gene has to function; (4) introduction of healthy gene into the cell. During early 1990s, in the routine exercise of patient therapy any gene is isolated and this isolated gene is either directly injected into the cell or be carried by a virus (vector) to which it is linked by recombinant DNA technique. After entering the cell, the gene may become a part of nuclear DNA or remain free in cytoplasm like extra-chromosomal DNA. However, in each case RNA is synthesized only at the rate of few copies per cell in comparison to normal cells where thousands of copies are made.

Table 19-1. List of human genetic diseases for which genes have been isolated and cloned for gene therapy (Source : Burns and Bottino, 1989).

1. Alpha-Thalassemia	9. Haemophilia
2. Beta-Thalassemia	10. Glucose-6-phosphate dehydrogenase deficiency
3. Collagen disorder	11. Ornithine transcarbamoylase deficiency
4. Dwarfism	12. Antithrombin - 3 deficiency
5. Emphysema	13. Cholesterol metabolism/heart disease
6. Lesch - Nyhan syndrome	14. Cancer
7. Phenylketonuria	15. Leukemia, lymphomas
8. Christmas disease	16. Immune deficiencies

(ii) Embryo therapy. In embryo therapy the genetic constitution of embryo at the post-zygotic level is altered so that the inheritance is altered. This technique involves the following steps : (i) *in vitro* fertilization of the egg; (2) insertion of normal gene into embryo at post-zygotic level, either with viruses or directly by microinjection; and (iii) integration of inserted gene in host DNA, where it may or may not function.

REVISION QUESTIONS

1. Do you think that all effective eugenic programmes must based on detailed knowledge of the genetics of the trait being selected ? Discuss.
2. Distinguish between positive eugenics and negative eugenics.
3. Discuss the purpose and limitations of genetic counselling.
4. How can genes be transferred from one cell into another in mammalian systems ? How can this be utilized for therapeutic treatment of certain diseases ?
5. Discuss the role of human genetics in medical science with reference to (i) genetic counselling; (ii) antenatal diagnosis, and (iii) gene therapy.
6. Write short notes on the following :

   (a) Patient and embryo therapy ; (b) Euphenics ; and (c) Genetic engineering.

CHAPTER

20

Transposable Genetic Elements

(Jumping or Mobile Genes)

Sectors of purple and yellow tissue in the endosperm of maize kernels resulting from the presence of the transposable elements *Ds* and *Ac*.

Much of the investigations of the classical genetics have been devoted to the localization of genes on chromosomes. As discussed in Chapter 9 of Chromosome Mapping, genetic mapping depends on the assumption that genes do not move from one position to another. To a great extent, this hypothesis has been found correct. In fact, most genes occupy fixed sites on the chromosomes, and the overall structure of the genetic map is practically invariant. However, in early 1940's researchers have found that some DNA sequences (*viz*., genes) can actually change their position.

These mobile elements have been variously called '**jumping genes**', '**mobile elements**', '**cassettes**', '**insertion sequences**' and '**transposons**'. The formal name for this family of mobile genes is **transposable elements**, and their movement is called **transposition**. The term transposon was coined by **Hedges** and **Jacob** in 1974 for a DNA segment which could move from one DNA molecule (or chromosome) to other and carried resistance for antibiotic ampicilin. Transposable elements can be defined as small, mobile DNA sequences that move around chromosomes with no regard for homology and insertion of these elements may produce deletions, inversions, chromosomal fusions and even more complicated rearrangements.

MODE OF DISCOVERY OF TRANSPOSABLE ELEMENTS

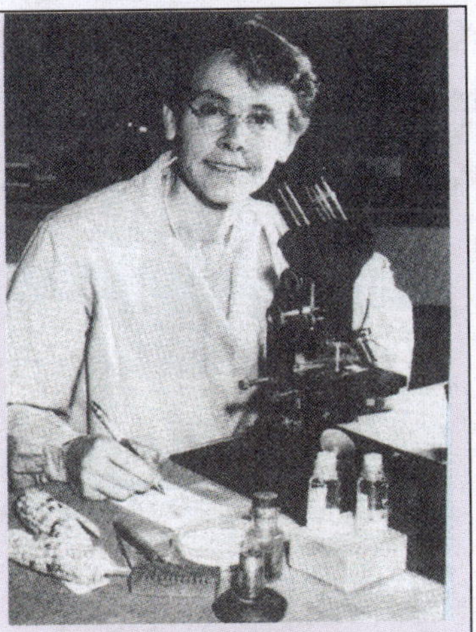

Barbara McClintock

Transposable elements were discovered by **Barbara McClintock** (1965) through an analysis of genetic instability in maize (corn). The instability involved chromosome breakage and was found to occur at sites where transposable elements were located. In McClintock's analysis, the events of breakage were detected by following the loss of certain genetic markers. In some experiments, McClintock used a marker (gene) that controlled the deposition of pigmentation in the aleurone which is the outer most layer of endosperm of maize kernels. Recall that the endosperm is **triploid**, being produced by the union of two maternal nuclei and one paternal nucleus. McClintock's marker was an allele of the *C* locus on the short arm of chromosome 9. Since this allele, called C^I, is a dominant inhibitor of aleurone colouration, any kernel possessing it should be colourless. McClintock fertilized *CC* ears with pollen from C^IC^I tassels, producing kernels in which the endosperm was C^ICC. Although many of these kernels were colourless, as expected, some showed patches of brownish-purple pigment. **McClintock** (1951) guessed that in such mosaics, the inhibitory C^I allele had been lost sometime during endosperm development, leading to a clone of tissue that was capable of producing pigment. The genotype in such a clone would be –*CC*, where the dash indicates loss of the C^I allele. Further study demonstrated that C^I allele had been lost through chromosome breakage.

As shown in Fig. 20.1, a break at the site labelled by the arrow would detach a segment of the chromosome from its centromere, creating an **accentric fragment**. Such a chromosomal fragment tends to be lost during cell division, so all the descendants of this cell would lack part of the paternal chromosome. Since the lost fragment carried the C^I allele, none of the cells in this clone would be inhibited from forming pigment, and if any of them produced a part of the aleurone, a patch of colour would appear.

Mutation in different kernels of Indian corn. It is caused due to jumping gene.

McClintock established that such kernel mosaics frequently resulted from breaks at a particular site on chromosome 9. She named the factor that produced these breaks, **Ds**, for "**Dissociation**". In her experiments, the chromosome that carried the C^I allele also carried the *Ds* factor. However, by itself, the *Ds* factor was not capable of inducing chromosome breakage. She found that *Ds* had to be activated by another factor, called **Ac**, for "**activator**". The *Ac* factor was present in some maize stocks, but absent in others. By crossing different stocks, *Ac* could be combined with *Ds*, creating the condition that led to chromosome breakage.

This two-factor system (*i.e.*, *Ac-Ds* family) provided an explanation for the genetic instability that McClintock had observed on chromosome 9 of maize. Her further studies have established that both *Ac* and *Ds* are members of a family of transposable elements. These elements are structurally related to each other and can insert at many different sites on the chromosomes. In fact, there often are multiple copies of the *Ac* and *Ds* elements present in the maize genome. Through genetic analysis, McClintock demonstrated that both *Ac* and *Ds* elements could move. When one of these elements was

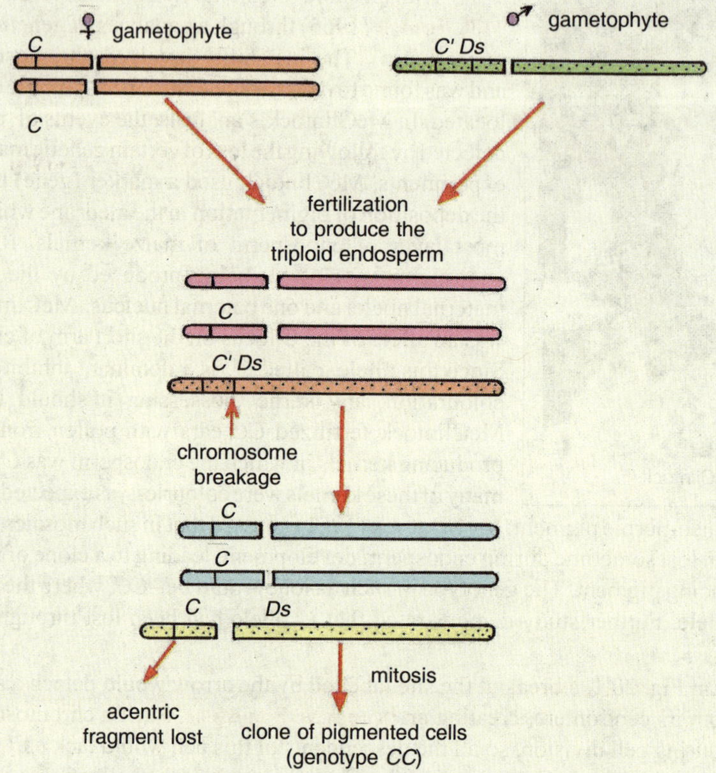

Fig. 20.1. Chromosome breakage caused by the transposable element *Ds* in maize. The allele *C′* on the short arm of chromosome 9 produces normal pigmentation in the aleurone (of maize kernel) ; the allele *C′* inhibits this pigmentation (after Gardner *et al.*, 1991).

inserted in or near a gene, McClintock sometimes found that the gene's function was altered. In extreme cases, the function was completely suppressed. Because of this effect on gene expression, **McClintock** (1956) referred to *Ac* and *Ds* as **controlling elements**.

CHARACTERISTICS OF TRANSPOSABLE ELEMENTS

Some salient features of the transposable elements are the following:

1. They were found to be DNA sequences that code for enzymes which bring about the insertion of an identical copy of themselves into a new DNA site.

2. Transposition events involve both recombination and replication processes which frequently generate two daughter copies of the original transposable elements. One copy remains at the **parent site** while the other appears at the **target site** (on the host chromosome).

3. The insertion of transposable elements invariably disrupts the integrity of their target genes. For example, if an IS (= insertion sequence) becomes inserted into an operon, it interrupts the coding sequence and inactivates the expression of the target gene into which it inserts as well as any gene downstream in that same operon. This is because an IS contains transcription and/or translation termination signals that block the expression of other genes downstream in an operon. This "one-way" mutational effect (or polarity) is referred to as a **polar mutation**. **Ghosal** (1979) reported that DNA sequence of IS1 and IS2 of *E. coli* contain nonsense codons in all reading frames. IS2 appears to have a chain termination codon.

4. Since transposable elements carry signals for the initiation of RNA synthesis, they sometimes activate previously dormant genes.

5. A transposable element is not a replicon (a sequence that contains a site for the origin of replication), thus, it cannot replicate apart from the host chromosome the way that plasmids and phage can.

6. No homology exists between the transposon and the target site for its insertion. Many transposons can insert at virtually any position in the host chromosome or into a plasmid. Some transposons seem to be more likely to insert at certain positions (hot spots), but rarely at base-specific target sites.

A single purple flower has appeared in the middle of this double pink African violet as the result of a 'jumping gene'.

TYPES OF TRANSPOSABLE ELEMENTS

Transposable elements are of the following types:

1. Insertion Sequences (IS) or Simple Transposons

The insertion sequences (IS) are shorter (800 to 1500 bp) and do not code for proteins. In fact, IS carry the genetic information necessary for their transposition (*i.e.*, the gene for the enzyme **transposase**) (Fig. 20.2). There are different insertion sequences such as IS1, IS2 , IS3, IS4 and so on in *E.coli* (Table 20-1). Recently, **IS21** has been reported in bacteria by **Willetts** *et al.,* (1981). Insertion sequences have been found in bacteriophages, in F factor plasmid and many bacteria.

The wrinkled trait in garden pea is caused by the insertion of transposable element into the structural gene for the starch-branching enzyme. (see W.S. Klug and M.R. Cummings 2003).

2. Transposons (Tn) or Complex Transposons

Transposons (Tn) are several thousand base pair long, and have genes coding for one or more proteins (including resistance factors in bacteria which act against antibiotics). The hallmark of a transposon is the presence of identical, **inverted terminal repeat (IR) sequences** of 8 to 38 base pairs (b.p.) (Fig. 20.2 B). Each type of transposon has its own unique inverted repeat. On either side of a transposon is a short (less than 10 b.p.) **direct repeat**. If a transposon exists in multiple copies, these direct repeats are of different base composition at each site where the transposon exists in the chromosome; the inverted terminal repeats, however, remain the same for a given transposon. The sequence into which a transposable element inserts is called the target sequence. During insertion of a transposon, the singular target sequence becomes duplicated and, thus, appears as direct repeats flanking the inserted transposable element. The direct repeats are not considered part of the transposon. These repeat sequences themselves act like IS or IS-like segments.

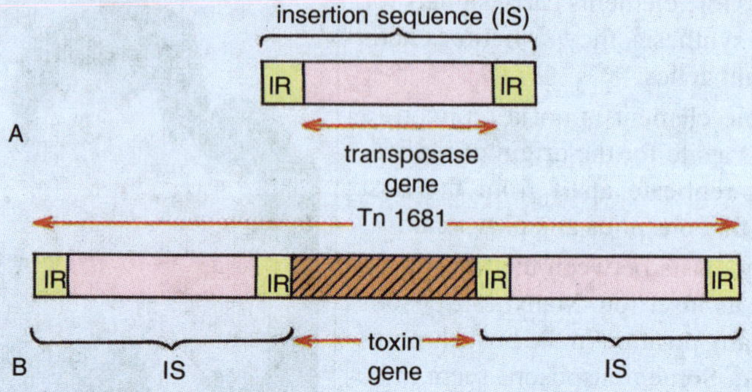

Fig. 20.2. A–A simple transposon (IS) carries only the genetic information necessary for transposition which is often enclosed by short regions of inverted repeat (IR) sequences ; B– The complex transposon (Tn) is composed of two insertion sequences (IS) and contains genetic information in addition to that needed for transposition. In Tn 1681 transposon carries the gene for a heat-labile toxin.

Table 20-1. Some examples of insertion sequences (IS) and complex transposons of *E.coli* (bacterium).

Type of transposable element	Size base pairs (bp)	Target DNA repeat (bp)	Known functions or proteins encoded (besides transposase)
A. Insertion sequences (IS)			
IS 1	768	9	—
IS 2	1327	5	—
IS10-R	1329	9	—
B. Complex Transposons (Tn)			
Tn 3	4957	5	Ampicillin resistance
Tn 5	5700	9 (IS 50 at end)	Kanamycin resistance
Tn 10	9300	9 (IS 10 at end)	Tetracycline resistance
Tn 1681	2100	9 (IS 1 at end)	Heat stable enterotoxin
Tn 2571	23000	9 (IS 1 at end)	Resistance to chloramphenicol, fusidic acid, streptomycin, sulphoamides and mercury

1 2 3 4 5 a b c d e.....................................e' d' c' b' a' 1 2 3 4 5
1' 2' 3' 4' 5' a' b' c' d' e'.....................................e d c b a 1' 2' 3' 4' 5'
Target sequence ← Transposon → Target sequence

Fig. 20.3. Diagram of a simple transposon. In the diagram, letters and numbers have been used instead of nucleotides to make repeats easier to read. Primed and unprimed symbols of the same kind (*e.g.*, 3 and 3', d and d') represent complementary base pairs. The ends of a transposon consist of inverted repeats (represented by letters). Direct repeats of the target gene(represented by numbers) flank the transposon. The dotted central region of simple transposons contains only genes necessary for translocation, but in complex transposons it may contain one or more additional genes (consult Fig. 20.2B) (after Stansfield, 1991).

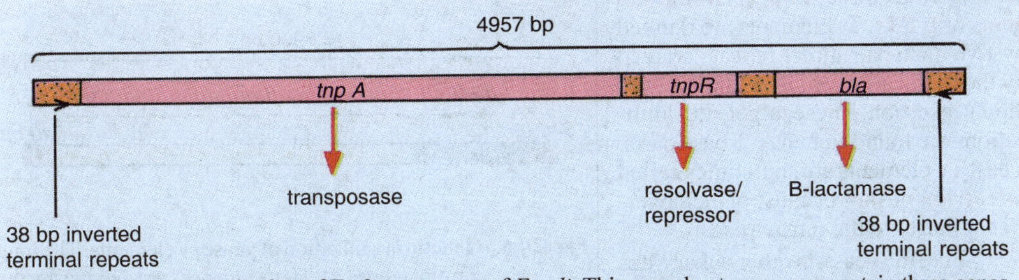

Fig. 20.4. Genetic organisation of Tn 3 transposon of *E.coli*. This complex transposon contain three genes that produce the protein indicated. Lengths of DNA sequences are given in base pairs (bp) (after Gardner *et al.*, 1991).

Examples of Transposons

1. Tn 3 transposon of E. coli. The molecular structure of transposon Tn 3 of *E.coli* has been worked out (Fig. 20.4). Tn 3 has 4957 bp and contains three genes such as *tnp A*, *tnp R* and *bla*, coding respectively for the following proteins : 1. **transposase** having 1015 amino acids and required for transposition; 2. a **repressor** (also called **resolvase**) containing 185 amino acids which regulates the transposase ; and 3. **β-lactamase** enzyme which confers resistance to the antibiotic ampicillin. On both sides of the Tn 3 is an inverted repeat of about 38 bp.

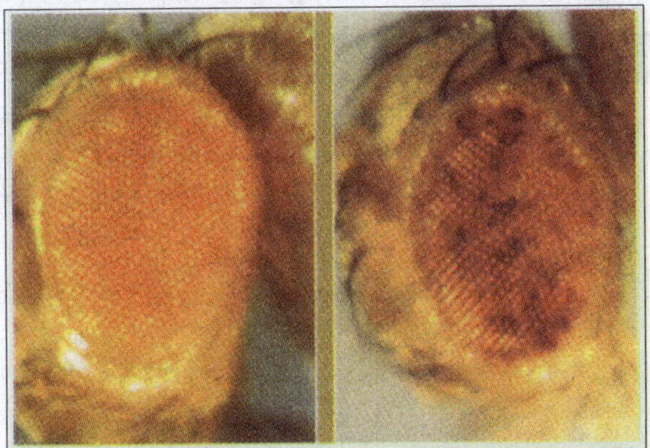

In these *Drosophila* eyes, the patches of red are caused by excision of a transposable element restoring normal eye colour to an otherwise peach-coloured eye.

2. Bacteriophage Mu. The bacteriophage *Mu* (*Mu* = mutator) is a temperate bacteriophage having typical phage properties and could be regarded as a giant transposon. Phage *Mu* was thought to behave in a unique manner since it does not multiply in a lytic way. It always starts its life-cycle by inserting itself into the *E.coli* chromosome at random locations. The resulting *prophage* often mutates the genes into which they become inserted (Fig. 20.5). Thus, phage *Mu* was originally known for its mutagenic properties.

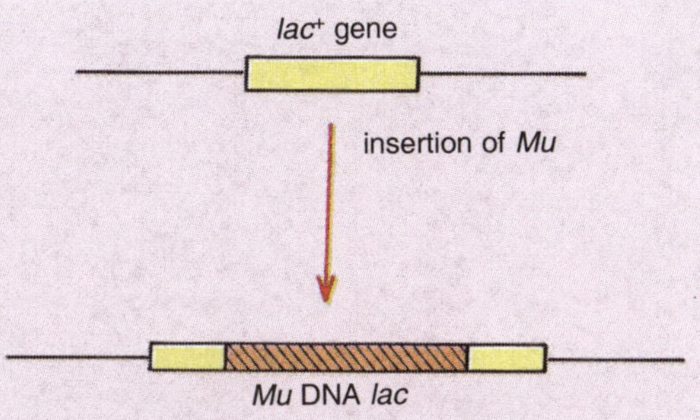

Fig. 20.5. Insertion of phage *Mu* into a *lac* gene renders the gene inactive (lac⁻).

During later part of phage life cycle, upon receipt of an appropriate signal, the *Mu* prophage somehow generates its transposition, but also activates the genes encoding for structural proteins which package its DNA. *Mu* phage is unique also in transposing much often than other transposons, providing nearly a hundred new copies of *Mu* during its hour-long life-cycle.

Like other transposons, *Mu* phage has inverted repeats (IR) at or near the ends of its DNA and it makes a short duplication with 5 base pairs of the bacterial DNA into which it is inserted.

3. Yeast *Ty* elements. The yeast *Saccharomyces cerevisiae* carries about 35 copies of a transposable element called ***Ty*** in its haploid genome. These transposons are about 5900 base pairs long and are bounded at each end by a DNA segment called the **delta (δ) sequence**, which is approximately 340 base-pairs long. Each δ sequence is oriented in the same direction, forming what are known as direct **long terminal repeats** or **LTRs**. *Ty* elements are flanked by five base-pair direct repeats created by the duplication of DNA at the site of the *Ty* insertion. These target site duplications are found rich in A-T base pairs. Yeast **Ty** elements are sometimes called **retrotransposons** because of their overall similarity to the retroviruses.

Like the yeast, in other eukaryotes such as *Zea mays*, *Drosophila*, mice and snapdragons (*Antirrhinum majus*) transposable elements of one or other type are found.

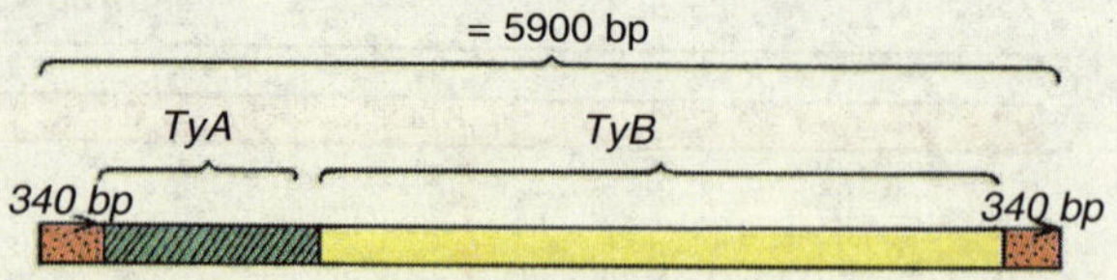

Fig. 20.6. Genetic organization of yeast *Ty* elements. The long terminal repeat (LTR) sequences are denoted by the Greek letter delta (δ). *Ty* elements apparently contain two genes *TyA* and *TyB*. Sequence length are shown in base pairs (after Gardner *et al.*, 1991).

REVISION QUESTIONS

1. Comment on **Ms. Barbara McClintock's** investigations regarding the controlling elements of maize.
2. Define the transposable elements. Describe different types of complex transposons of the eukaryotes.
3. Describe the transposable elements of *E.coli* and bacteriophage *Mu*.

MOLECULAR BIOLOGY

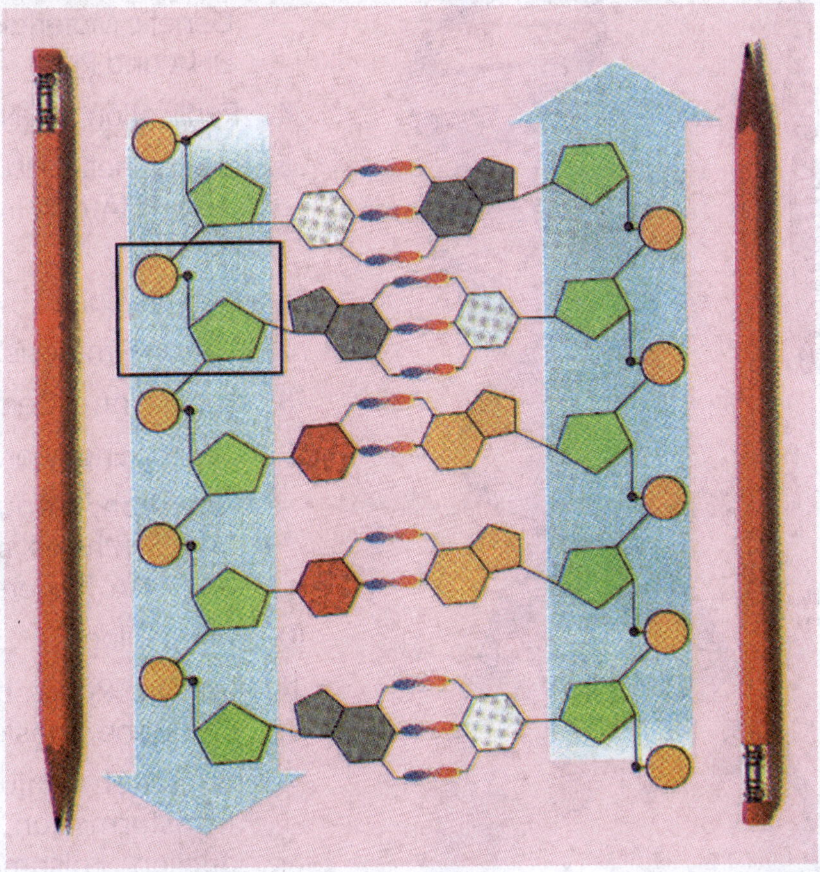

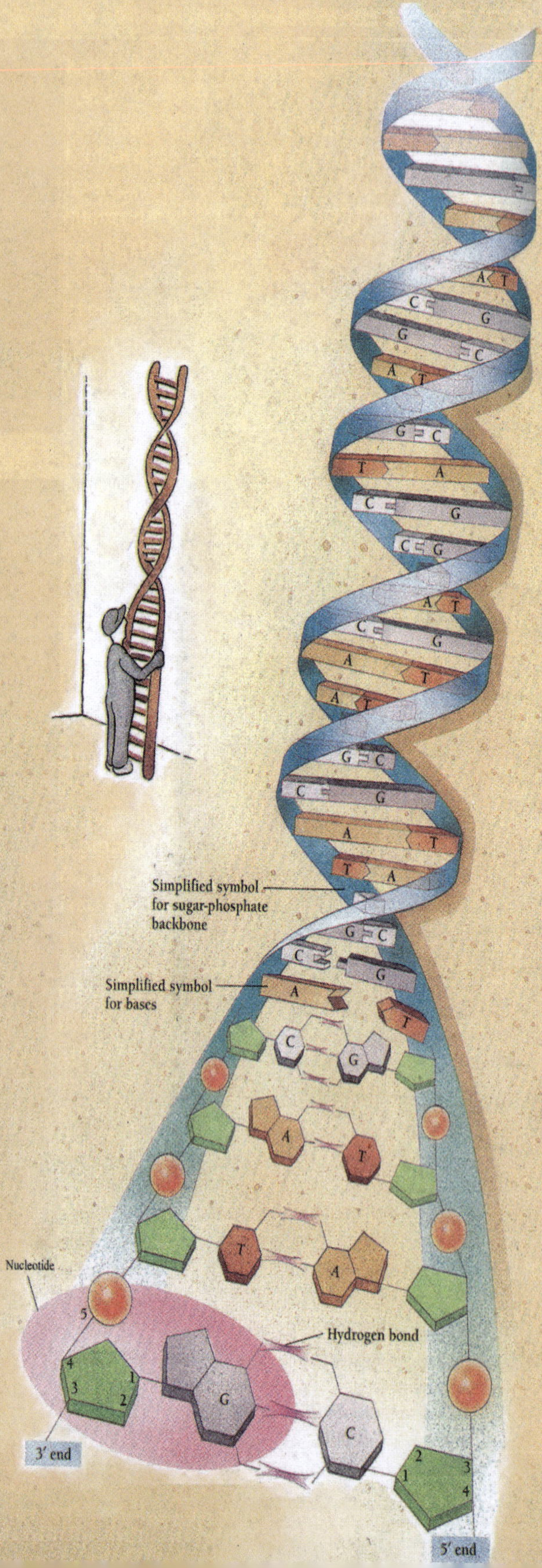
Simplified symbol for sugar-phosphate backbone
Simplified symbol for bases
Nucleotide
Hydrogen bond
3′ end
5′ end

CHAPTERS

CHAPTER

1

Introduction

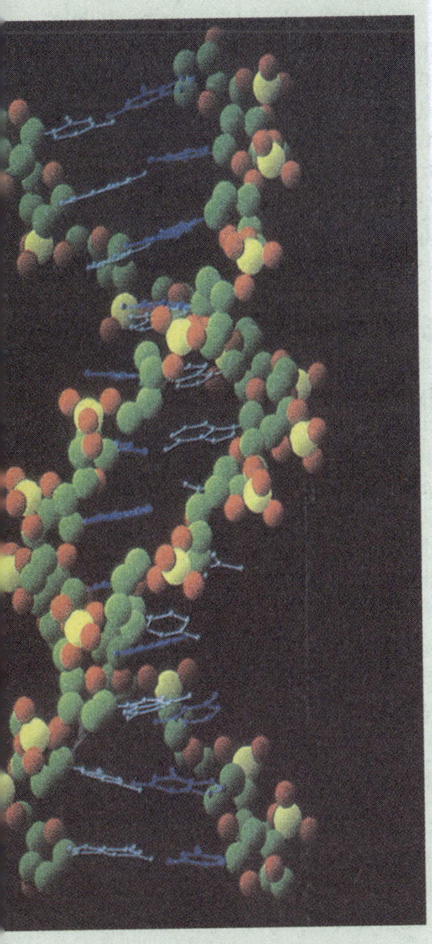

DNA is the genetic material of almost all organisms.

The term **molecular biology** was first used in 1945 by **William Astbury** who was referring to the study of the chemical and physical structure of biological macromolecules. By that time, biochemists had discovered many fundamental intracellular chemical reactions. The importance of specific reactions and of protein structure in defining the numerous properties of cells was also appreciated. However, the development of molecular biology had to await the understanding that the most advantageous approaches would be made by studying "simple" systems such as bacteria and bacteriophages which yield information about the basic biological processes more readily than animal cells. In fact, the faith in the basic uniformity of life processes was an

Francis Crick and James Watson.

important factor in rapid growth of molecular biology. That is, it was believed that fundamental biological principles that govern the activity of simple organisms, such as bacteria and viruses, must apply to more complex cells; only the details should vary. This faith has been amply justified by experimental results.

The roots of molecular biology were established in 1953 when an Englishman, **Francis Crick** and a young American, **James Watson** working at Medical Research Council Unit, Cavendish Laboratory, Cambridge, proposed a double helical model for the structure of DNA (deoxyribonucleic acid) molecule which was well known as the chemical bearer of genetic informations of certain microorganisms (bacteria, bacteriophages, etc.) due to pioneer discoveries made by **Grifith** (1928), **Avery**, **Macleod** and **McCarthy** (1944) and **Hershey** and **Chase** (1952). This discovery was followed by a thorough search of occurrence of DNA as the genetic material in other microorganisms, plants and animals and also by investigations of the molecular and atomic nature of different reactions of living cells. From all these studies has emerged the realization that the basic chemical organization and the metabolic processes of all living things are remarkably similar despite their morphological diversity and that the physical and chemical principles governing living systems are similar to those governing non-living systems.

The present understanding of molecular biology is that in most organisms the phenotype or the body structure and function ultimately depend for their determination on the structural and functional (*i.e.,* enzymatic) proteins or polypeptides. The synthesis of polypeptides is specified, directed and regulated by self-duplicating genes which are borne within molecules of DNA which is the universally accepted chemical bearer of genetic informations of most living organisms except certain viruses in which this function is carried by RNA, another nucleic acid. The genetic informations for polypeptide synthesis are initially dictated by the disposition of nitrogen bases in DNA molecule and are copied down by the process of **transcription**. During transcription stage copies (that is, transcripts) of an individual gene or genes are synthesized. These copies are molecules of RNA that include such familiar classes as ribosomal RNA, messenger RNA and transfer RNA. The biochemical interplay of these RNA copies which leads to the synthesis of a polypeptide chain, is called **translation**, meaning, literally, that the genetic message encoded in a messenger RNA molecule is translated into the linear sequence of amino acids in a polypeptide. The polypeptide in its turn determines the phenotype of the organism.

HISTORICAL BACKGROUND

The molecular biology is a very young biological discipline and has a very short history. Certain notable accomplishments of molecular biologists can be summarized as follows :

1928 **F. Grifth** discovered the phenomenon of transformation in bacteria.

1934 **M. Schlesinger** demonstrated that the bacteriophages are composed of DNA and protein.

1941 **G.W. Beadle** and **E.L.Tatum** published their classical study on the biochemical genetics of *Neurospora.*

1944 **O.T. Avery**, **C.M. MacLeod** and **M. McCarthy** recognized the DNA nature of transforming principle of pneumococcus bacteria. The fact suggested that it is DNA and not protein which is the hereditary chemical.

1948 **A. Boivin**, **R.Vendrely** and **C. Vendrely** showed that in the different cells of an organism the quantity of DNA for each haploid set of chromosome is constant.

1950 **E. Chargaff** demonstrated that in DNA the numbers of adenine and thymine groups are always equal and so are the numbers of guanine and cytosine groups.

1952 **A.D. Hershey** and **M. Chase** demonstrated that only the DNA of T_2 bacteriophage enters the host, the bacterium *Escherichia coli*, whereas the protein remains behind.

1953 **J. D. Watson** and **F.H.C. Crick** proposed a model for DNA comprising of two helically intertwined chains tied together by hydrogen bonds between the purines and pyrimidines.

1956 **A. Gierer** and **G. Schramm** demonstrated that RNA is the genetic material of tobacco mosaic virus (TMV).

1957 **H. Fraenkel-Contrat** and **B.Singer** separated RNA from the protein of TMV viruses, produced hybrid RNA viruses and confirmed the view that RNA is the genetic material of some viruses.

Mathew Meselson and **Franklin W. Stahl** performed a density-gradient experiment (using heavy isotope of nitrogen, ^{15}N) in bacteria to confirm the Watson and Crick's **semiconservative theory** of DNA replication.

George W. Beadle (1903-1989).

Edward L. Tatum (1909-1975).

1958 **G. Beadle** and **E. Tatum** received Nobel Prize for their contribution in biochemical genetics of fungus.

J. Laderberg got Nobel Prize for the discovery of bacterial recombination.

1959 **R.L. Sinsheimer** isolated single-stranded DNA from a small virus ϕ-X-174 which attacks *Escherichia coli*.

S. Ochoa ; **A. Kornberg** received Nobel Prize for artificial synthesis of nucleic acids.

1961 **M.W. Nirenberg** and **J.H. Matthaei** cracked the messenger RNA code.

F.H.C.Crick and his colleagues showed that the genetic language is made up of three-letter words (*i.e.*, triplet codons).

F.Jacob and **J. Monod** put forward the operon concept.

Maurice H.F. Wilkins (1916–).

1962 **J. Watson** and **F. Crick** ; **M. Wilkens** got Nobel Prize for the discovery of molecular nature of DNA.

1963 **J.P. Waller** reported that nearly one-half of all proteins in *E.coli* cells have the amino acid **methionine** in the N-terminal position.

1964 **K.A. Marcker** and **F. Sanger** discovered a peculiar aminoacyl-tRNA in *E.coli*, called **N-formyl- methionyl - tRNA** and suggested that this molecule may play a role in the special mechanism of chain elongation.

R.W. Holley and his colleagues gave detailed structure of alanyl tRNA ($tRNA^{ala}$) from yeast. Holley died in 1993.

H.G. Khorana.

1965 **F.H.C. Crick** proposed the **wobble hypothesis** for anticodons of tRNA and explained how several codons meant for same amino acid are recognized by same tRNA.

H. Wallace and **M.L. Birnstiel** isolated ribosomal RNA genes in *Xenopus*.

F. Jacob., **A. Lwoff**, and **J. Monod** received Nobel Prize for the discovery of protein synthesis mechanism in virus.

1968 **R.W. Holley** ; **H.G. Khorana** and **M.W. Nirenberg** got Nobel Prize for deciphering the genetic code.

1969 **A.D. Hershey**, **M. Delbruck** and **S.E. Luria** shared Nobel Prize in medicine for their contribution to replication and recombination in viruses (bacteriophages).

Britten and **Davidson** proposed the **gene-battery model** for regulation of protein synthesis in eukaryotes.

1970 **Howard Temin** and **David Baltimore** demonstrated the synthesis of DNA on RNA template tumour viruses. Both were awarded Nobel Prize in 1975 for the discovery of an enzyme called **RNA directed DNA polymerase** (or **reverse transcriptase**) which is present in the core of virus particle (rous sarcoma virus).

Howard Temin (1934–94).

David Baltimore (1938–).

Biotechnology emerged as a new discipline due to marriage of biological science with technology (see **Dubey**, 1995).

Knippers ; **Kornberg** and **Gefter** ; **Moses** and **Richardson** isolated DNA-polymerase-II enzyme.

1972 **Mertz** and **Davis** in 1972 demonstrated that cohesive termini of cleaved DNA molecule could be covalently sealed with *E.coli* DNA ligase and were able to produce recombinant DNA molecules.

Cohen *et al*., for the first time reported the cloning of DNA by using plasmid as vector.

R. Porter; **G.M. Edelman** received Nobel Prize (physiology and medicine) for the discovery of chemical structure of antibodies.

C.B. Anfinsen; **S. Moore** and **W.H. Stein** got Nobel Prize (chemistry) for the discovery of chemical structure and activity of the enzyme ribonuclease.

1973 **S.H. Kim** suggested three dimensional structure, *i.e*., L-shaped model, of tRNA.

1975 **E.M. Southern** developed a method, called **Southern blotting technique** for analysing the related genes in a DNA restriction fragment.

D. Pribnow discovered **Pribnow box** or **minus ten sequence** in *E. coli* genome.

1977 **P.A. Sharp** and **R.J. Roberts** discovered split genes of adenovirus.

D.S. Hogness, **I.B. David** and **N. Davidson** studied split genes for 28 S rRNA in *Drosophila*.

P. Chambon, **P. Leder** and **R.A. Flavell** studied split genes of B'globin, ovalbumin and tRNA.

Itakura *et al*., first of all produced human insulin (humulin) by means of recombinant technology.

1978-79 **W. Gilbert** first of all used the terms **exon** and **intron** (for split genes).

Walter Gilbert (1932–).

1978 **Hinnen** *et al*., first of all described the transformation of yeast (*Saccharomyces cervisae*) by the help of plasmid of *E.coli*.

1979 **Khorana** reported completion of the total synthesis of a biologically functional gene.

Alwine *et al*., developed **northern blotting technique** in which mRNA bands are blot transferred from the gel onto chemically reactive paper.

Towbin *et al*., developed the **western blotting technique** to find out the newly encoded protein by a transformed cell.

1980 **Fredrick Sanger** got the second Nobel Prize for discovering complete sequence of 5400 nucleotides of single stranded DNA of $\phi \times 174$ bacteriophage.

Frederick Sanger (1918–).

1982 **A. Klug** was awarded Nobel Prize in chemistry for providing three-dimensional structure of tRNAs.

Rubin and **Spradling** for the first time introduced *Drosophila* gene of xanthine dehydrogenase into a P-element (= parental element) which then was microinjected into embryo deficient for this gene.

R.D. Palmiter and **R.L. Brinter** produced **transgenic mice** by genetic engineering.

1983 **Marilyn Kozak** proposed the **scanning hypothesis** for initiation of translation by eukaryotic ribosomes.

1984 **Robert Tijan** identified a DNA-binding protein called **SP1** which is involved in eukaryotic gene regulation.

Thomas Cech (1947–).

1985 **Karry Mullis** discovered **polymerase chain reaction** (**PCR**) which is widely exploited in gene cloning for genetic engineering. He made use of a thermostable enzyme (acts best on 72^0 C temperature), called **Taq DNA polymerase**, isolated from *Thermus aquaticus*.

1984,86 **Alec Jeffreys** discovered the technique of **DNA fingerprinting**.

1987 **S. Tonegawa** was awarded Nobel Prize for discovering the mode of rearrangements of DNA sequences of mammalian immunoglobulin genes to produce a large variety of antibodies.

Stanford and coworkers developed the **particle bombardment gun** which shooted foreign DNA into plant cells or tissues at a very high speed.

1988 **J.W. Black**, **G.B. Elion** and **G.H. Hitchings** were awarded Nobel Prize for formulating drugs such as 6- mercaptopurine and thioguanine, which lead to inhibition of DNA synthesis and of cell division. This proved effective in cancer chemotherapy. They also designed drugs for treating gout, malaria and viral infections such as herpes.

1989 **T. Cech** and **S. Altman** awarded Nobel Prize for showing enzymatic role of some RNA molecules, called **ribozymes**.

1991 **Dr. Lalji Singh** at CCMB, Hyderabad has developed a new technique of DNA fingerprinting by using **BKM-DNA probe** (BKM = banded krait minor satellite). He discovered this probe while he was working on sex determination in snake, the banded krait (*Bungarus fasciatus*) for his Ph.D. work.

1992 **Edwin G. Krebs** and **Edmond H. Fisher** were awarded Nobel Prize for the pioneering work on "reversible protein phosphorylation as a biological regulator mechanism." Phosphorylation of proteins is shown to affect transcription, translation, cell division and many other cellular processes.

Sidney Altman (1939–).

Prof. Asis Dutta of JNU, New Delhi, was selected for the **Birla Award for Science and Technology** for cloning and characterization of two novel genes–gene for **oxalate decarboxylase** from *Lathyrus sativus* (in 1991) and gene for a **seed specific nutritionally balanced protein** from *Amaranthus* (in 1992).

1993 **M.J. Chamberlain** proposed the **inchworm model** for elongation of transcript of DNA template.

This year's Nobel Prize in chemistry was shared by **Kary Mullis** (for the discovery of PCR) with **Michael Smith** (for site directed mutagenesis).

MATERIAL AND METHODS IN MOLECULAR BIOLOGY

Different molecular biologists have made an intensive use of a variety of the microorganisms such as bacteriophages and other viruses; *Escherichia coli* and other bacteria; unicellular green algae, yeast, *Neurospora* and other fungi ; protozoans, etc., in their investigations of varied nature. However, many significant discoveries of molecular biology have been made by working on larval salivary gland chromosomes of Dipteran insect (*Drosophila*) and lampbrush chromosomes of oocytes of different amphibians and other higher animals. To understand different biochemical events of prokaryotic and eukaryotic cells at molecular level, a wide array of bio-physico-chemical techniques such as electron microscopy, ultracentrifugation, colorimetry, spectrophotometry, chromatography, isotopic tracers, X-ray crystallography, electrophoresis, etc., are used in molecular biology.

BASIC REQUIREMENTS TO BE MET BY GENETIC MATERIAL

According to the molecular biologists certain requirements must be met by any molecule if it is to be qualified as the substance that transmits genetic information from one generation to next. These requirements extend directly from what is known about the continuity of species and the process of evolutionary change.

1. The genetic material must contain biologically useful information that is maintained in a stable form.

2. The genetic information must be reproduced and transmitted faithfully from cell to cell or from generation to generation.

3. The genetic material must be able to express itself so that other biological molecules, and ultimately cells and organisms, will be produced and maintained. Implicit in this requirement is that some mechanism be available for decoding, or translating, the information contained in the genetic material into its "productive" form. A narrow, but important, distinction is, thus, made between a molecule that can generate only its own kind and a molecule that can also generate new kinds of molecules. For example, a salt crystal can "seed" a salt solution so that new salt crystals are formed, but this is the extent of its influence over its surroundings.

4. Genetic material must be capable of variation. Two sources of change have been recognized in present day genetic systems : mutation and recombination. A mutation changes the nature of information transmitted from parent to offspring, and, thus, it represents a relatively drastic way of bringing about variation. Recombination is a more moderate way of producing variation. It occurs during the course of some sort of sexual process, and it involves the precise shuffling of parental genetic information so that new combinations of genes are produced. These are then inherited by the offspring.

With these four requirements in mind we can study in the forthcoming chapters — How the genetic material is discovered ?, What is the physical and chemical properties of it ?, How it carries genetic information from one generation to next ?, How a genetic material determines the phenotype of an organism? and many problems of similar nature.

REVISION QUESTIONS

1. What is molecular biology? Discuss its importance for the humans.
2. Describe the historical growth of molecular biology.
3. Describe the contribution of the following instrumention and techniques in the growth of molecular biology :

   (i) Electrophoresis; (ii) Chromatography;

   (iii) Denaturation and annealing of DNA;

   (iv) Radioactive labelling; and (v) Autoradiography.

CHAPTER

2

Identification of the Genetic Materials

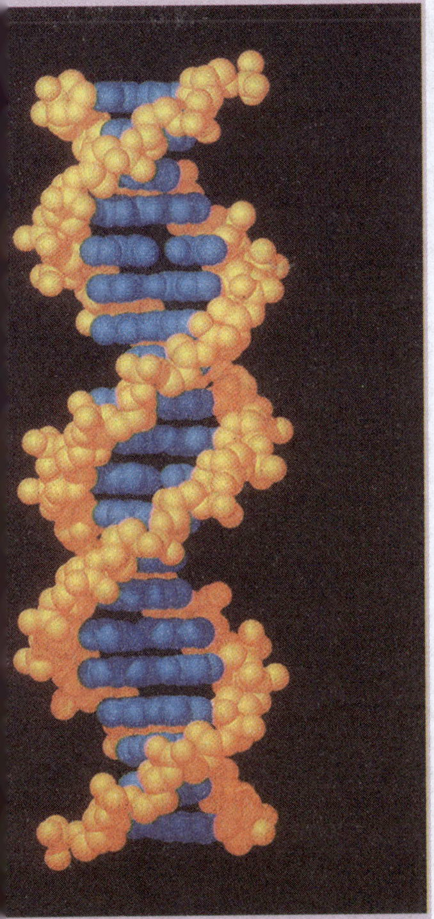

Computer-generated model of DNA molecule.

When it became evident that the chromosomes were the organs of heredity, because (i) they formed the only link between two generations, (ii) they carried linearly arranged genes, (iii) they occurred in every organism in specific number and had specific morphology for a particular species, and (iv) any variation in their number or morphology affected the phenotype of the species, then, various attempts were made by early molecular geneticists to identify the physical and chemical nature of genes. But the genes were found so minute structures that their physical identity remained almost impossible. However, the extensive chemical analysis of chromosomes of different organisms have revealed that chromosomes contain proteins and nucleic acids (DNA and RNA), and it was thought that genes might have either proteins or nucleic acids as their component molecules. Early molecular geneticists have assigned the informational roles of genes to the chromosomal proteins because, they found nucleic acids too simple to carry genetic informations. (This is called **tetranucleotide hypothesis**).

The controversy about the assignment of genetic role either to chromosomal proteins or to chromosomal DNA, existed up to 1949 when **A. Mirsky** and **H. Ris** had found that all cells of an organism appeared to contain the same amount of DNA, whereas different cell types contained quite different amounts and kinds of protein. Its constancy, therefore, favoured

DNA as the genetic material. Around 1953, it was universally accepted that DNA is the genetic substance (*i.e.*, chemical of which genes are composed) of most microorganisms and higher organisms. Later on, RNA was found to be the genetic material of some viruses. The concept that DNA or RNA is the genetic material of most organisms has been developed and supported by the following evidences :

A. DIRECT EVIDENCES FOR DNA AS THE GENETIC MATERIAL

The most conclusive evidences in support of DNA as the genetic material came from the following three avenues of approach on microorganisms : transformation of bacteria, mode of infection of bacteriophages and conjugation of bacteria.

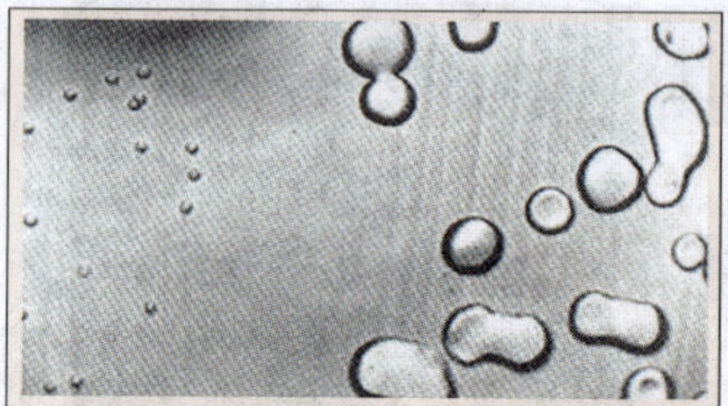

Petri plate with smooth and rough colonies of *Streptococcus pneumoniae*. R (rough) on the left and S(smooth) on the right.

1. The Transformation Experiments

In 1928, **Frederick Griffith** encountered a phenomenon now known as genetic transformation. Colonies of virulent strain (pathogenic) of pneumonia causing bacterium, *Diplococcus pneumoniae* grown on nutrient agar, have a **smooth** (**S**) glistering appearance owing to the presence of a type specific, polysaccharide (a polymer of glucose and glucuronic acid) capsule. The avirulent (non-pathogenic) strains, on the other hand, lack this capsule and they produce dull, **rough**(**R**) colonies. Smooth (S) and rough (R) characters are directly related to the presence or absence of the capsule and this trait is known to be genetically determined. Both S and R forms occur in several types and are designated as S-I, S-II, S-III, etc., and R-I, R-II, R-III, etc., respectively. All these subtypes of S and R bacteria differ with each other in the type of antigens, they produce. The kind of antigen produced is likewise genetically determined. Smooth (S) forms sometimes mutate to rough (R) forms, but this change has not been found reversible.

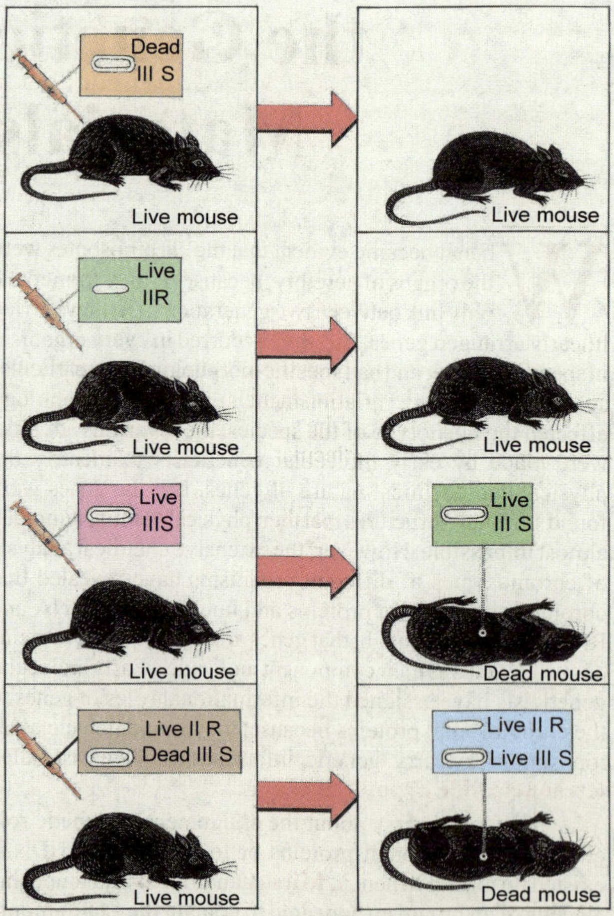

Fig. 2.1. Griffith experiment that demonstrated the principle of transformation.

In the course of his work, **Griffith** injected laboratory mice with live R-II pneumococci; the mice suffered no illness because R-II pneumococci was avirulent. But, when the mice were injected with virulent S-III pneumococci, the mice suffered from pneumonia and died. However, when he injected the heat killed S-III bacteria in mice, they did not suffer from pneumonia. But, when the mice were injected with the mixture of living avirulent R-II and heat killed S-III virulent, the unexpected symptoms of pneumonia appeared and high mortality resulted in them. By postmorteming the dead mice, it was found that their heart blood had both R-II and S-III pneumococci. From these results, Griffith concluded that the presence of the heat-killed S-III bacteria must have caused a **transformation** of the living R-II bacteria, so as to restore to them the capacity for capsule formation they had earlier lost by gene mutation. This was called "**Griffith effect**" or more popularly "**bacterial trans-formation**".

Identification of the "transforming principle or substance". Griffith could not understand the cause of bacterial transformation and that is first of all identified by **Oswald Avery**, **Colin MacLeod** and **Maclyn McCarty** (1944). They partially purified the **transforming principle** from the cell extract (*i.e.*, cell free extract of S-III bacteria) and demonstrated that it was DNA. These workers modified the known schemes for isolating DNA and prepared samples of DNA from S-III bacteria. They added this DNA to a live R bacterial culture; after a period of time they placed a sample of S-III containing R-II bacterial culture on an agar surface and allowed it to grow to form colonies. Some of the colonies (about 1 in 10^4) that grew were S-III type (Fig. 2.3). To show that this was a permanent genetic change, they dispersed many of the newly formed S-III colonies and placed them on a second agar surface. The resulting colonies were again S-III type. If an R-II colony arising from the original mixture was dispersed, only R-II bacteria grew in subsequent generations. Hence, the R-II colonies retained the R-II character, whereas the transformed S-III colonies bred true as S-III.

Further, because S-III and R-II colonies differed by a polysaccharide coat around each S-III bacterium, the ability of purified polysaccharide to transform was also tested, but no transformation was observed. The evidence that extracted trans-forming principle contain DNA, was provided by the following four methods:

1. Chemical analysis showed that the major component was deoxyri-bose containing nucleic acid.

2. Physical measurement showed that the sample con-tained a highly viscous substance having the properties of DNA.

3. Experiments demonstrated that trans-forming activity is not lost by reaction with either (a) purified proteolytic (protein-hydrolyzing) enzymes trypsin, chymo-trypsin, or a mixture of both or (b) ribonuclease (an enzyme that depoly-merizes RNA). The lipase too has no effect.

4. It was demon-strated that the treatment

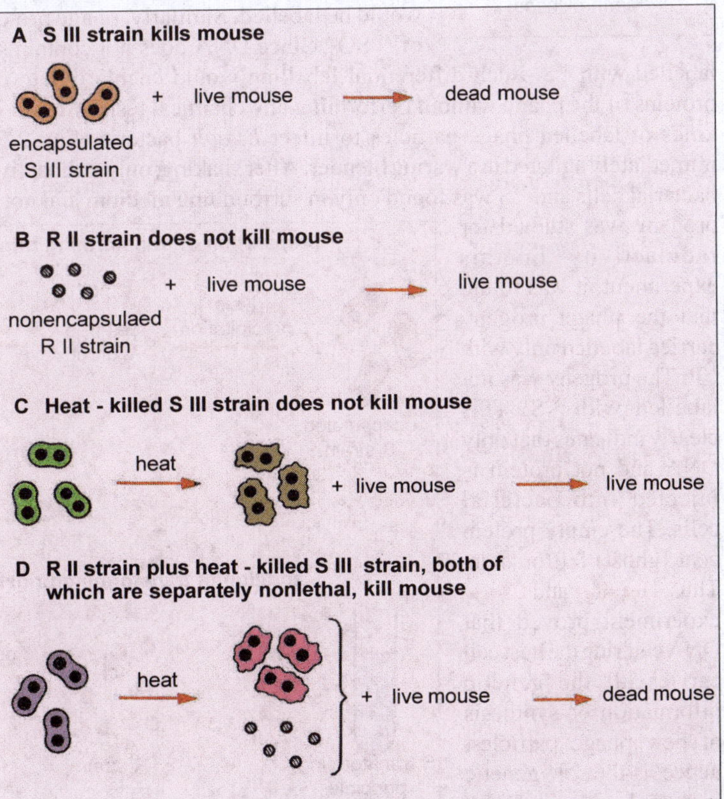

Fig. 2.2. Explanation of Griffith experiment showing conversion of a non-lethal bacterial strain to a lethal form by a cell extract (after Freifelder, 1985).

with materials, known to contain DNA-depolymerizing activity (DNAase enzyme), inactivated the transforming principle. Later on, such a transforming substance, the DNA, was found in a variety of bacteria (*e.g., Hemo-philus influenzae, Bacillus subtilus, Shigella para-dysenteriae,* etc.) and several other organisms.

2. The Blender Experiment

An elegant confirmation of the genetic nature of DNA came from an experiment with *E.coli* phage T_2. This experiment known as the **blender experiment** because a kitchen waring blender was used as a major piece of apparatus, was performed by **Alfred Hershey** and **Martha Chase** in 1952. They demonstrated that the DNA injected by a phage particle into a bacterium contains all the information required to synthesize progeny phage particles.

A.D. Hershey (1908–)

Martha Chase.

A single particle of phage T_2 consists of DNA (now known to be a single molecule) encased in a protein shell (Fig. 2.4). The DNA is the only phosphorus containing substance in the phage particle; the proteins of the shell, which contain the amino acids methionine and cysteine, have the only sulphur atoms. In these experi-ments, phage DNA was made radio-active by growing infected bacteria on a medium containing radioactive phosphate ($^{32}PO_4$). Since phage proteins do not contain phosphorus, only DNA would be labelled. Similarly, phage proteins were labelled with the help of $^{35}SO_4$. Since DNA does not contain sulphur, only protein would be labelled with ^{35}S. Such differential labelling would enable one to distinguish between DNA and proteins of the phage without performing any chemical tests. **Hershey** and **Chase** then allowed both kinds of labelled phage particles to infect *E. coli* bacteria (Fig. 2.5). The infected bacteria were immediately agitated in a waring blender. After shaking, only radioactive ^{32}P was found associated with bacterial cells and ^{35}S was found only in surrounding medium and not in bacterial cells. When phage progeny was studied for radioactivity in this experiment, it was found that the phage progeny carried labelled only with ^{32}P. The progeny was not labelled with ^{35}S. This clearly indicates that only DNA and not protein is injected into bacterial cells. The empty protein coat (ghost) left outside. Thus, **Hershey** and **Chase** experiment proved that DNA entering the host cell carries all the genetic information or synthesis of new phage particles, hence, is the sole genetic material in DNA bacteriophages (*e.g.*, T_2).

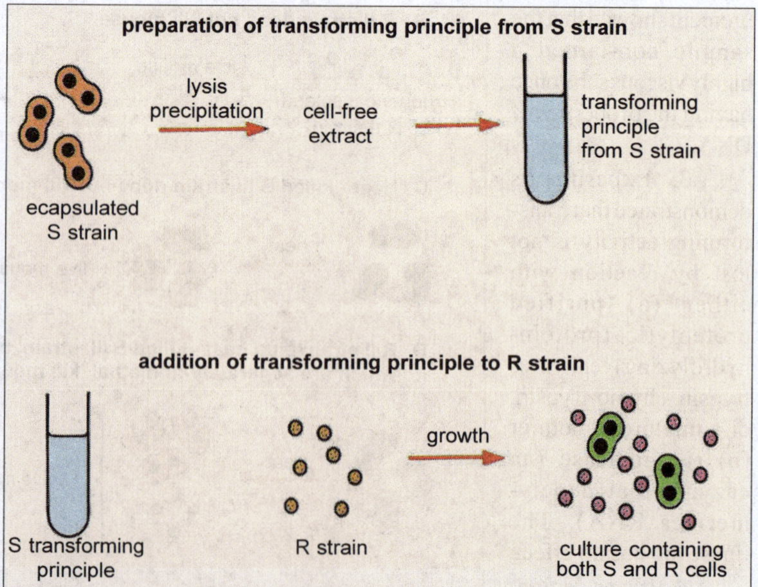

Fig. 2.3. Transformation experiment of **Avery, MacLeod** and **McCarthy** (after Freifelder, 1985).

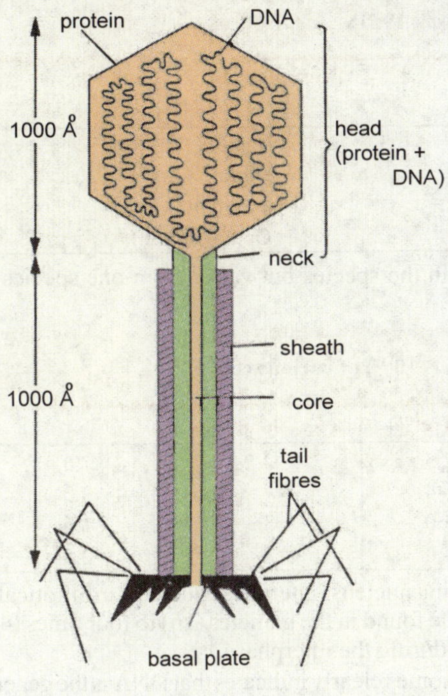

Fig. 2.4. Detailed structure of a T_2 bacteriophage.

3. Bacterial Conjugation

Another conclusive evidence for DNA as the genetic material came from the phenomenon of conjugation of bacteria. **Laderberg** and **Tatum** (1946) found that when an F^+ ('male') *E. coli* cell conjugated with an F^- ('female') *E. coli* cell, unidirectional transfer of F^+ factor of 'male' cell to F^- or 'female' cell took place, so that the latter was converted into a F^+ or 'male' strain. The F^+ factor was found to be a fragment of DNA molecule which occurred in the cytoplasm of bacterial cell.

From left : Dr. E. Tatum, Dr. Joshua Lederberg, Dr. Sol Spiegelman.

B. INDIRECT EVIDENCES FOR DNA AS THE GENETIC MATERIAL

The indirect evidences in favour of DNA as the genetic material came from higher organisms which are not as easy to manipulate as bacteria and viruses. The fact that DNA is the genetic material of higher organisms has been supported by the following facts :

1. The Feulgen techniques have shown that DNA entirely remains restricted to the chromosomes and it forms one of the major components of chromosomes.

2. Various quantitative measurements of amount of DNA in different cells have shown that there is a correlation between the amount of DNA and the number of chromosome sets (ploidy). Diploid cells contain twice as much DNA as do haploid cells of the same species as shown in the Table 2-1.

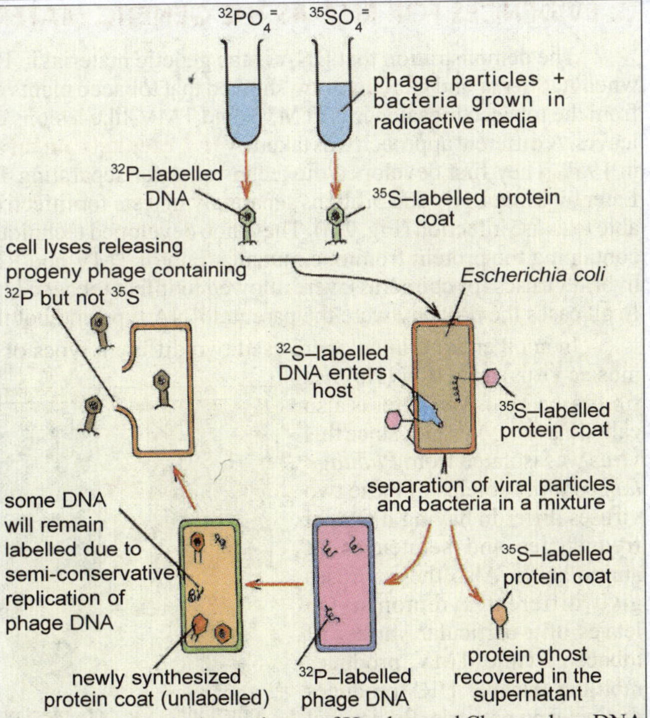

Fig. 2.5. A classical experiment of Hershey and Chase, where DNA of T_2 phage is labelled with ^{32}P and the protein is labelled with ^{35}S.

Table 2-1. DNA content and ploidy (Pollister *et al.*, 1951).

Cells	Mean DNA Feulgen content (Picograms)	Presumed chromosome set (Ploidy)
Spermatids	1.68	Haploid (2n)
Liver	3.16	Diploid (2n)
Liver	6.30	Tetraploid (4n)
Liver	12.80	Octoploid (8n)

3. The diploid amount of DNA is constant within the species but varies from one species to another, as shown in the Table 2-2.

Table 2-2. DNA contents in picograms (gram × 10^{-12}) of various cells.

S.No.	Organism	Kidney	Liver	Erythrocytes	Sperm
1.	Chicken	2.4	2.5	2.5	1.3
2.	Bovine	6.4	6.4	—	3.3
3.	Carp	—	3.0	3.3	1.6
4.	Human	5.6	5.6	—	2.5

4. In 1950 **Swift** found the amount of DNA per resting nucleus (interphase nucleus) in mitotically active tissues of animals ranged from twice (2n) the value found in the gametes (1n) to four times (4n) this value. He concluded that DNA synthesis occurred during the interphase.

This parallelism of behaviour in DNA and chromosomes clearly indicates that DNA is the genetic material of higher animals.

C. EVIDENCES FOR RNA AS THE GENETIC MATERIAL OF SOME VIRUSES

The demonstration that RNA is the genetic material in RNA-containing viruses came in 1956, when **A. Gierer** and **G. Schramm** showed that tobacco plants could be inoculated with purified RNA from the tobacco mosaic virus (TMV), and TMV-like lesions could later be identified on the tobacco leaves. A different approach was taken by **H. Fraenkel Conrat** and **B. Singer** in experiments published in 1957. They first developed the techn-iques for separating TMV particles into RNA and proteins. Later by using RNA and proteins separately in tests for infectivity, it was shown that RNA alone was able to cause infection (Fig. 2.6). They then developed techniques for forming "reconstituted" viruses containing the protein from one mutant strain of TMV and RNA from another, or vice-versa. Such hybrid viruses (or chimeras) were allowed to infect tobacco leaves, and the progeny were examined. In all cases the progeny were the parental RNA type and not the parental protein type.

In another experiment were used two different types of RNA viruses, namely TMV (tobacco mosaic virus) and HRV (Holmes rib-grass virus). The latter is also called **Plantago strain**, since this virus was isolated from *Plantago lanceolata*. Proteins of these two viruses differ in having different frequencies and sequences of amino acids. Also these viruses give different symptoms. On leaves of a particular variety of tobacco, while TMV produces mottling of leaves, HRV produces distinct ring patterns. Reciprocal chimeras using RNA of one strain and protein of other strain were

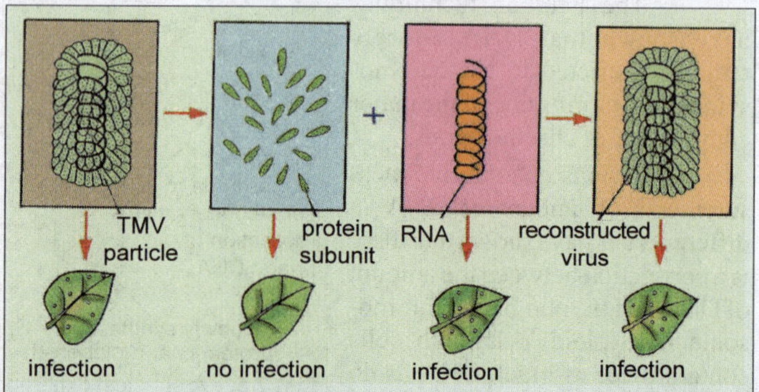

Fig. 2.6. Experiment of Fraenkel-Conrat on TMV showing that RNA and not the protein can cause infection.

obtained. It was found that when these chimeras were used for infection, the progeny had proteins which correspond to the virus from which RNA of the infecting virus particle was derived (Fig. 2.7). This proved that specificity of virus proteins was determined by RNA alone and that proteins carried no genetic information.

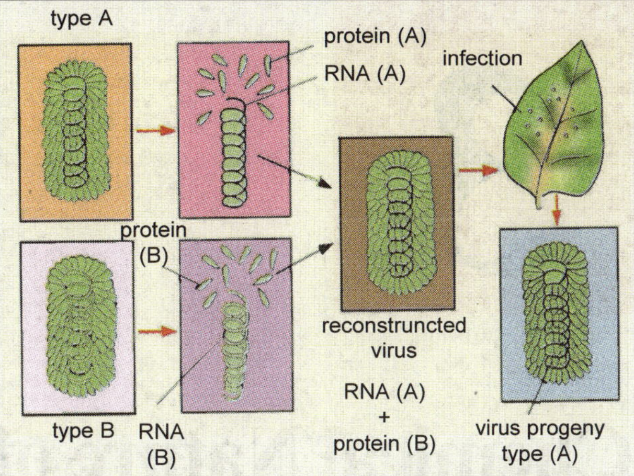

Fig. 2.7. Reconstruction of virus, showing that reconstructed virus gives progeny which resembles the virus from which RNA was derived.

In a much impro-ved technique **N. Pace** and **S. Spiegelman** in 1966 purified RNA from two different mutant strains of the RNA Phage QB which had quite distinct base compositions. The isolated RNAs were then incubated separ-ately in the presence of an *E. coli* cell extract containing an enzyme capable of RNA replication. The new RNA synthesized was in each case identical in base composition to the particular phage RNA presented to the *in vitro* system, thus, indicating that the phage RNA can serve as a template for its self-replication.

The genetic RNA is found to be either single-stranded or double-stranded (see Table 2-3).

Table 2-3. Different RNA viruses and the nature of genetic RNA associated with them.

Virus	Types of RNA
1. **Plant viruses**	
TMV	Single stranded
Wound tumour	Double stranded
2. **Animal viruses**	
Influenza virus	Single stranded
Rous sarcoma	Single stranded
Poliomyelitis	Single stranded
Reovirus	Double stranded
3. **Bacteriophages**	
MS2, F2, r17	Single stranded

REVISION QUESTIONS AND PROBLEMS

1. Describe two classical experiments which demonstrated that DNA is the genetic material.
2. Describe an experiment which established the genetic nature of RNA in certain viruses.
3. Write about the contribution of the following molecular biologists regarding the identification of the genetic material:
   (i) F. Griffth; (ii) O.T. Avery, C.M. Macleod and M. McCarthy;
   (iii) A.D. Hershey and M.J. Chase; (iv) Fraenkel and Conrat.
4. Differentiate between transformation and transduction.

ANSWERS TO PROBLESM

4. Transformation involves naked DNA from one cell becoming incorporated into another's DNA, whereas in transduction a virus serves as the vector transferring DNA derived from one cell into another.

CHAPTER

3

Chemical Nature of Genetic Materials (*i.e.*, DNA and RNA)

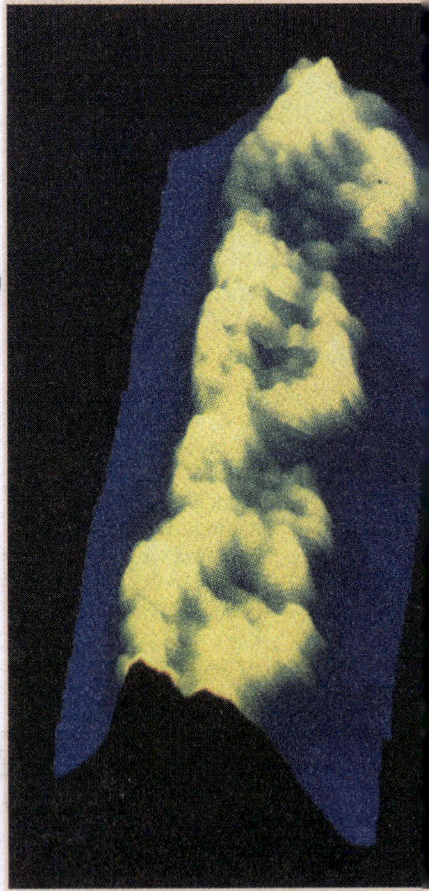

DNA magnified twenty-five million times by scanning tunneling microscopy.

DNA (deoxyribonucleic acid) and RNA (ribonucleic acid), the principal genetic materials of living organ isms, are chemically called **nucleic acids** and are complex molecules larger than most proteins and contain carbon, oxygen, hydrogen, nitrogen and phosphorus.

HISTORICAL

In 1869, **Friedrich Miescher**, a 22-year old Swiss physician (working in the laboratory of Felix Hoppe-Seyler in Tubingen) had isolated from pus cells obtained from discarded bandages in the Franko- Prussian War, and from salmon sperm, a previously identified macromolecular substance, to which he gave the name **nuclein**. Although he was unaware of the structure and function of nuclein, he submitted his findings for publication. The editor who received the paper was doubtful about some aspects of the Miescher's report and delayed publication for two years while he tried repeating some of the more questionable aspects of Miescher's work. Finally, in 1871, Miescher's report was published, but it made little immediate impact. He continued his careful work up to his death in 1895, recognizing (with the help of his student **Richard Altmann** in 1889) that nuclein was of higher molecular weight and was associated in some way with a basic protein, to which he gave the name **protamine**.

In 1880, **Emil Fischer** identified pyrimidines and purines. The biochemist, **Albrecht Kossel** identified the constituent nitrogenous bases of nuclein as well as its 5-carbon sugar and phosphoric acid. It was **Altmann** who first suggested, in 1899, the use of the term **nucleic acid** to describe phosphorus-containing nuclein. **Kossel** was awarded the 1910 Nobel Prize for demonstrating the presence of two pyrimidines (cytosine and thymine) and two purines (adenine and guanine) in nucleic acids. Kossel's work and later investigations of **Ascoli**, **Levine** and **Jones** during the first quarter of the 1900s disclosed the two kinds of nucleic acids, deoxyribonucleic acid (DNA) and ribonucleic acid (RNA). Development of DNA-specific staining techniques by **Feulgen** and **Rossenbeck** in 1924 enabled Feulgen to demonstrate in 1937 that most of the DNA content of the cell is located in the nucleus. It was not until the 1950s that the inter-nucleotide bond was established by **Todd** (**Judson**, 1979). Because DNA is the only genetic material of most living organisms, so deserves priorty in discussion.

DEOXYRIBONUCLEIC ACID OR DNA

Deoxyribose sugar molecule.

Highly purified DNA, extracted from a wide variety of plants, animals, bacteria and viruses, has been found to be complex macro-molecular or polymeric chemical compound which contains four kinds of smaller building blocks (monomers) called **deoxyribotids** or **deoxyribonucleotides**. Each deoxyribonucleotide is made up of three moieties: a **phosphoric acid molecule** (biologically called phosphate); a pentose sugar called **2 deoxyribose**; and **pyrimidine** and **purine** nitrogenous bases. Four major kinds of nitrogenous bases have been found in four kinds of deoxyribonucleotides of DNA : two are heterocyclic and two-ringed purines, **adenine** (**A**) and **guanine** (**G**), and two are one ringed pyrimidines, **cytosine** (**C**) and **thymine** (**T**). The pyrimidine ring can be numbered in two

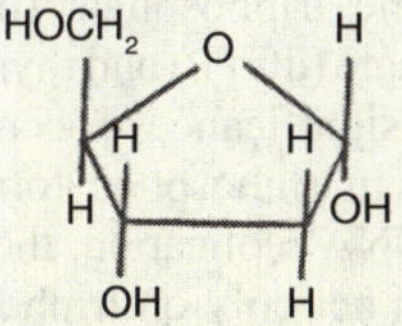

Fig 3.1. Chemical formula of deoxyribose.

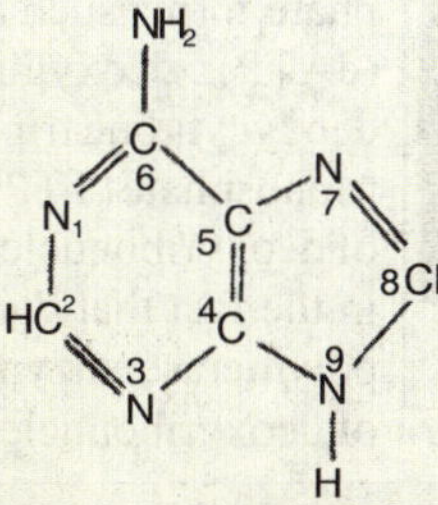

Fig 3.2. Chemical formula of adenine (6-amino-purine, mw=135.13 daltons).

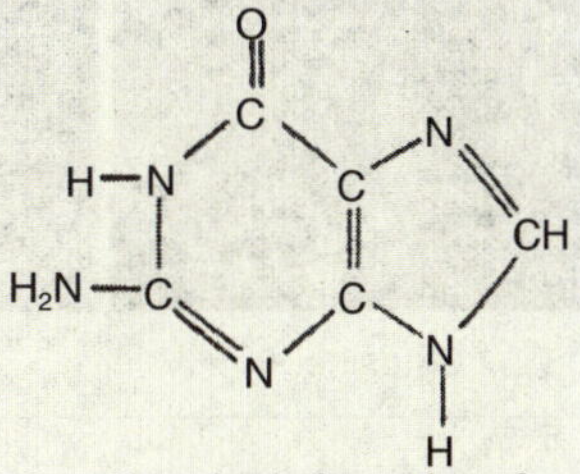

Fig 3.3. Chemical formula of guanine (2-amino– 6–hydroxypurine, mw= 151.13 daltons).

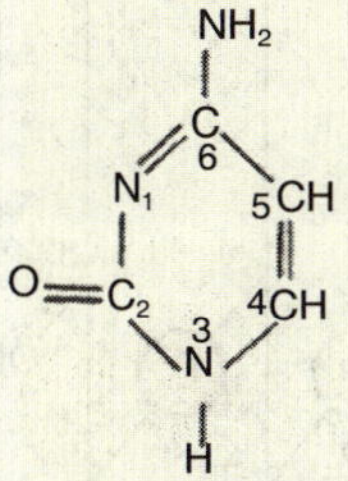

Fig. 3.4. Chemical formula of cytosine (2-hydroxy-6-aminopyrimidine, mw=111.10 daltons)

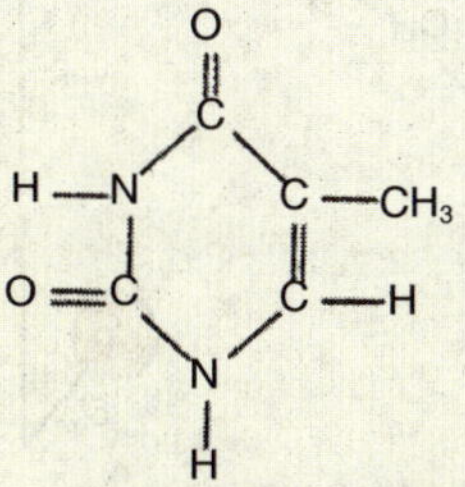

Fig. 3.5. Chemical formula of thymine (2, 6-dihydroxy–5–methylpyrimidine, mw =126.12 daltons.

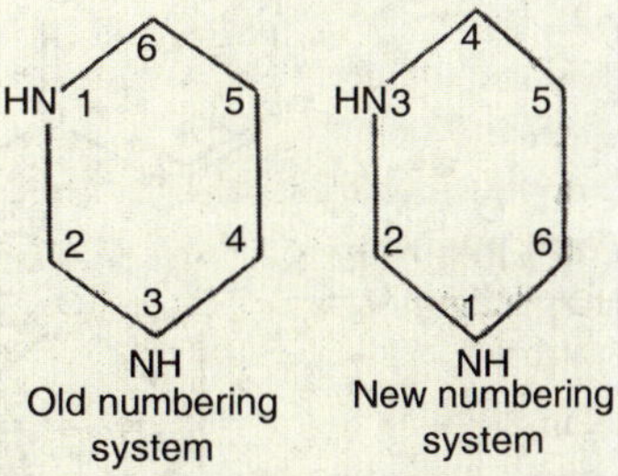

Fig. 3.6. Numbering of carbon atoms in a pyrimidine ring.

different ways (Fig 3.6). In this book, we will follow the old numbering system.

That part of each nucleotide which contains a nitrogenous base and deoxyribose is called **deoxyribonucleoside**. The four kinds of deoxyribonucleosides and deoxyribonucleotides can be tabulated as follows :

Table 3-1. Four nitrogen bases, nucleosides and nucleotides of DNA molecule.

Nitrogen base	Base + deoxyribose = deoxyribonucleoside	Deoxyribonucleoside + phosphoric acid = deoxyribonucleotide	Abbreviation for nucleotide
1. Adenine (A)	Deoxyadenosine	Deoxyadenylic acid (Deoxyadenosine monophosphate)	dAMP
2. Guanine (G)	Deoxyguanosine	Deoxyguanylic acid (Deoxyguanosine monophosphate)	dGMP
3. Cytosine (C)	Deoxycytidine	Deoxycytidylic acid (Deoxycytidine monophosphate)	dCMP
4. Thymine (T)	Thymidine	Thymidylic acid (Thymidine monophosphate)	TM

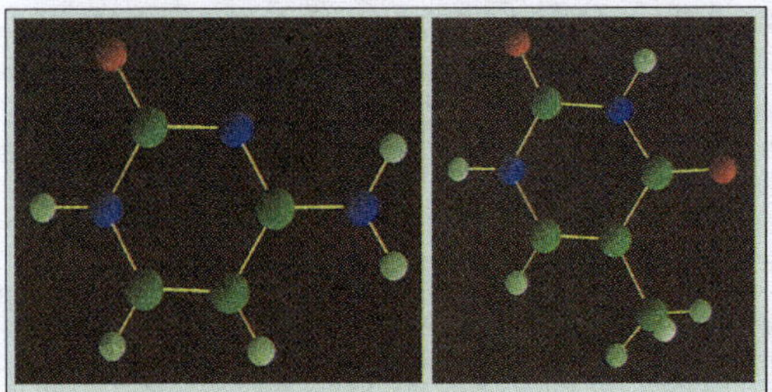
Cytosine. Thymine.

The four deoxyribonucleotides besides occurring in DNA molecule, occur also in nucleoplasm and cytoplasm, but in their triphosphate forms such as deoxyadenosine triphosphate (dATP), deoxyguanosine triphosphate (dGTP), deoxycytidine triphosphate (dCTP) and thymidine triphosphate (TTP). The significance of occurrence of deoxyribonucleotides in triphosphate forms lies in the fact that during DNA replication, the DNA polymerase enzyme can act only on triphosphate of deoxyribonucleotides.

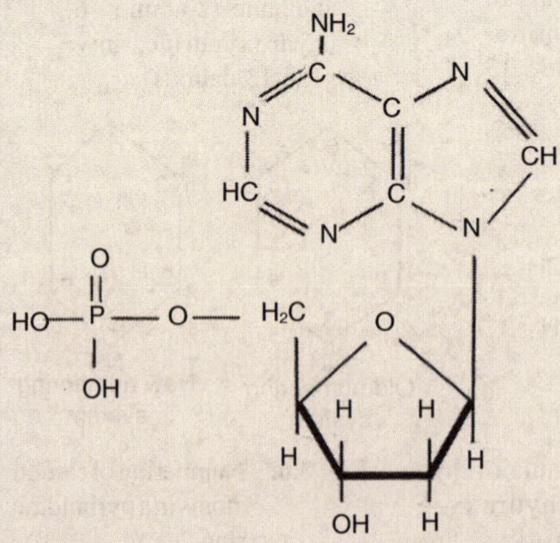

Fig. 3.7. Chemical formula of deoxycytidylic acid.

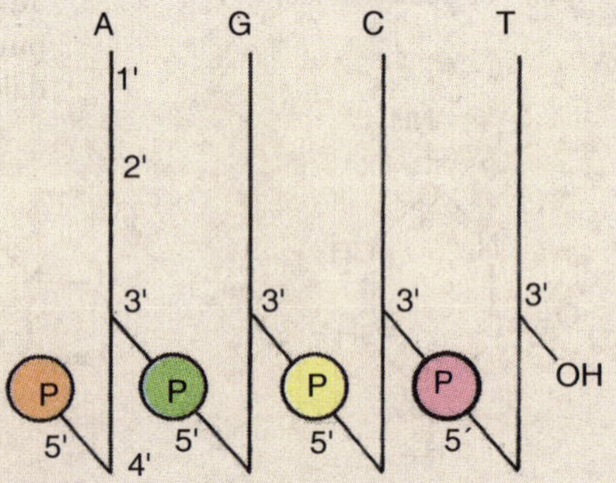

Fig 3.8. A short hand system of representation of a polynucleotide with four nucleotides.

Molar Ratios of Nitrogen Bases in DNA Molecule

Erwin Chargaff.

When many samples of DNA were isolated, purified and analysed by various techniques, such as paper chromatography, etc., it was found by **Hotchkiss** and **Chargaff** in 1948 that contrary to Levene's (1920's) tetranucleotide theory which considered DNA as a monotonous polymer having four DNA bases in approximately equal molar proportions, the four nucleotide bases are not necessarily present in DNA in exactly equal proportions. Thus, **Chargaff** reported that the DNA extracted from calf-thymus nuclei contains the four bases in the following molar proportions: 28% adenine, 24% guanine, 20% cytosine and 28% thymine. When he analyzed the DNA samples of varied animals, he found that the exact base composition of DNA differs according to its biological source. The relative amounts of purines and pyrimidines in samples of DNA of different living organisms are tabulated as follows :

Table 3-2. Relative amounts of nitrogen bases in different samples of DNA (Chargaff and Davidson, 1955).

S.N.	Source	Adenine	Guanine	Cytosine	Thymine	$\frac{A+T}{G+C}$
1.	Beef sperm	28.7	22.2	22.0	27.2	1.26
2.	Human thymus	30.9	19.9	19.8	29.4	1.52
3.	Human liver	30.3	19.5	19.9	30.3	1.53
4.	Human sperm	30.7	19.3	18.8	31.2	1.62
5.	Hen red cells	28.8	20.5	21.5	29.2	1.38
6.	Rat bone marrow	28.6	21.4	21.5	28.4	1.33
7.	Herring sperm	27.8	22.2	22.6	27.5	1.23
8.	*Paracentrotus lividus* (sea urchin) sperm	32.8	17.7	18.4	32.1	1.85
9.	Salmon	29.7	20.8	20.4	29.1	1.43
10.	Wheat germ	26.5	23.5	23.0	27.0	1.19
11.	Yeast	31.3	18.7	17.1	32.9	1.79
12.	*Diplococcus pneumoniae*	29.8	20.5	18.0	31.6	1.59
13.	*K-12 Escherichia coli*	26.0	24.9	25.2	23.9	1.00
14.	*Mycobacterium tuberculosis*	15.1	34.9	35.4	14.6	0.42
15.	Bacteriophage T_2	32.5	18.2	16.7	32.6	1.86

These different values of A, G, C, and T in different samples of DNA are suggesting that DNA rather than being a monotonous polymer, carries genetic information in the form of specific nucleotide base sequences.

The equivalence rule. Chargaff, in 1950, discovered the equivalence rule which suggested that despite wide compositional variations exhibited by different types of DNA, the total amount of purines equaled the total amount of pyrimidines (A+G=T+C); the amount of adenine equaled the amount of thymine (A=T) and the amount of guanine equaled the amount of cytosine (G=C). **Chargaff's** equivalence rule has been found to apply almost universally in different organisms (viruses, bacteria, plants and animals). However, DNA isolated from higher plants and animals was in general rich in adenine and thymine (A:T) and relatively poor in guanine and cytosine (G : C) (*e.g.*, AT/GC ratio of DNA of man was 1.40:1); whereas DNA isolated from microorganisms (viruses, bacteria and lower plants and animals) was in general rich in guanine and cytosine and relatively poor in adenine and thymine (*e.g.*, AT/GC ratio of DNA of *Mycobacterium tuberculosis* was 0.60 : 1). These differences in AT/GC ratios of microorganism and higher organisms undoubtedly reflect the difference in genetic information carried by these hereditary molecules and also phylogenetic, evolutionary and taxonomical significance of them.

Physical, Molecular or Geometrical Organization of DNA

Rosalind Franklin (1920–1958).

The first person to give any thought to the three dimensional structure of DNA was **W.T. Astbury** who by his X-ray crystallographic studies of DNA molecule concluded in 1938 that because DNA has high density, so, its polynucleotide was a stack of flat nucleotides, each of which was oriented perpendicularly to the long axis of the molecule and was situated every 3.4 A° along the stack. The X-ray crystallographic studies of Astbury were continued by **Wilkins** and his associates who managed to prepare highly oriented DNA fibres that allowed them to obtain an X-ray diffraction photograph. One of his female associates **Rosalind Franklin** obtained a superior X-ray diffraction photograph of DNA which confirmed Astbury's earlier inference of 3.4 internucleotide distance. Such studies demonstrated that DNA was a helical structure with a diameter of 20 A° and a pitch (one round) of about 34 A°. **Watson** and **Crick** who were already engaged in constructing some suitable model for DNA structure, when observed Franklin's picture of DNA molecule, they immediately utilized that information in constructing a molecular model for DNA. In April 1953, **Watson** and **Crick** published their conclusions about the structure of the DNA in the same issue of '**Nature**', in which **Wilkins** and his colleagues presented the X-ray evidence for that structure.

Considerations of Watson and Crick in the construction of double helical structure of DNA molecule. **Watson** and **Crick** concluded directly from the X-ray diffraction photograph of DNA taken by **Franklin** that (1) the DNA polynucleotide chain has the form of a regular helix, (2) the helix has a diameter of about 20A° and, (3) the helix makes one complete turn every 34A° along its length, and hence, since the internucleotide distance is 3.4A°, consists a stack of ten nucleotides per turn. Considering the known density of the DNA molecule, **Watson** and **Crick** next concluded that the helix must contain two polynucleotide chains, or two stacks of ten nucleotides each per turn, since the density of a cylinder 20A° in diameter and 34A° long would be too low if it contained but a single stack of ten, and too high if it contained three or more stacks of ten nucleotides each. Before trying to arrange these two polynucleotide chains into a regular helix of the required dimensions, however, **Watson** and **Crick** placed a further restriction on their model—a restriction that derived from their knowledge that DNA is, after all, the genetic material. If DNA is to contain hereditary information as they reasoned, and if that information is inscribed as a specific sequence of the four bases along the polynucleotide chain, then the molecular structure of DNA must be able to accommodate any arbitrary sequence of bases along its polynucleotide chains. Otherwise, the capacity of DNA as an information carrier would be too severely limited. Hence, they felt the need of construction of such a regular helix that though is composed of two polynucleotide chains containing an arbitrary sequence of nucleotide bases every 3.4A° along their length, it would, nevertheless, have a constant diameter of 20A°. Since the dimension of the purine ring is greater than that of the pyrimidine ring, **Watson** and **Crick** hit upon the idea that the two-chain helix could have a constant diameter if there existed a complementary relation between the two nucleotide stacks, so that at every level one stack harbours a purine base and the other a pyrimidine base. Finally, to endow the helix with thermodynamics stability, the structure would have ample opportunities for the formation of hydrogen bonds between amino or hydroxyl-hydrogens and keto-oxygens or immino-nitrogens of the purines and pyrimidine bases. These considerations led them to construct a double helix model for the molecular structure of DNA molecule.

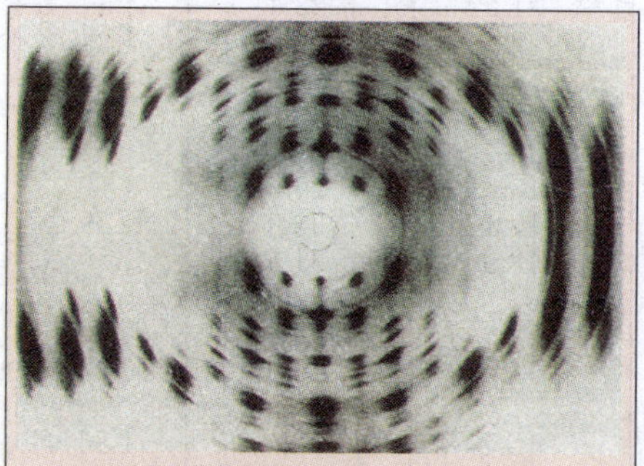

An X-ray diffraction image of DNA.

Watson and Crick's Structural Model of DNA

In DNA molecule the adjacent deoxyribonucleotides are joined in a chain by phosphodiester bridges or bonds which link the 5' carbon of the deoxyribose of one mononucleotide unit with the 3' carbon of the deoxyribose of the next mononucleotide unit. According to **Watson** and **Crick** DNA molecule consists of two such polynucleotide chains wrapped helically around each other, with the sugar-phosphate chain on the outside (forming ribbon-like backbone of double helix) and purines and pyrimidines on the inside of the helix (projecting between two sugar phosphate backbones as transverse bars). The two polynucleotide strands are held together by hydrogen bonds between specific pairs of purines and pyrimidines.

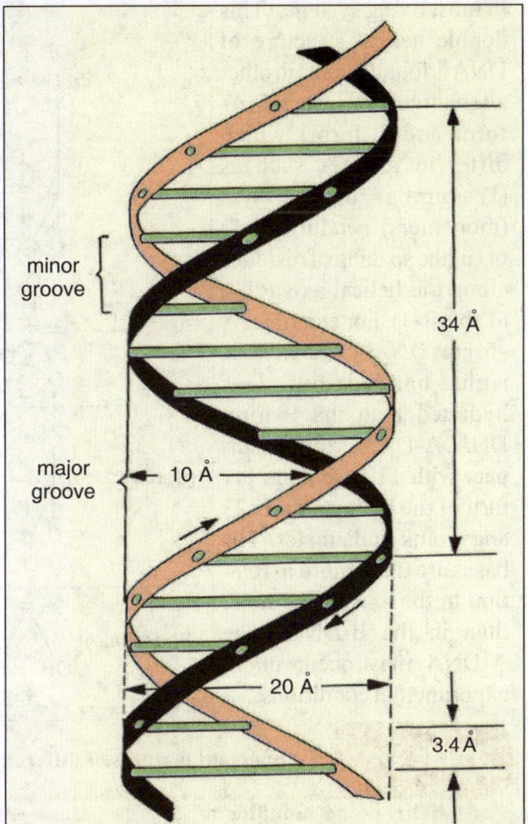

Fig. 3.9. Watson-Crick double helical DNA molecule.

The hydrogen bonds between purines and pyrimidines are such that adenine can bond only to thymine by two hydrogen bonds, and guanine can bond only to cytosine by three hydrogen bonds and no other alternative is possible between them. The specificity of the kind of hydrogen bonds that can be formed assures that for every adenine in one chain there will be thymine in the other. For every guanine in first chain there will be a cytosine in the other and so on. Thus, the two chains are complementary to each other; that is, the sequence of nucleotides in one chain dictates the sequence of nucleotides in the other. The two strands run anti-parallely—that is, have opposite directions. One strand has phosphodiester linkage in 3'→ 5' direction, while other strand has phosphodiester linkage in just reverse or 5'→3' direction. Further, both polynucleotides strands remain separated by 20 A° distance. The coiling of double helix is right handed and a complete turn occurs every 34 A°. Since each nucleotide occupies 3.4 A° distance along the length of a polynucleotide strand, ten mononucleotides occur per complete turn. (The base pairs are rotated 36° with respect to each adjacent pair). The helix has two external grooves, a deep wide one, called **major groove** and a shallow narrow one, called **minor groove** : both of these grooves are large enough to allow protein molecules to come in contact with the bases.

Polymorphism of DNA Helix

(Or Alternative Forms of DNA Double Helices)

For about 20 years after the discovery of the DNA double helix in 1953, DNA was thought to have the same monotonous structure, with exactly 36° of helical twist between its adjacent base pairs (10 nucleotide pairs per helical turn) and a uniform helix geometry. Subsequent experiments have shown that DNA is much more **polymorphic** than expected. Thus, DNA can be of the following types :

The above described DNA molecule contains the right handed helical coiling and has been called **B-form** or **B-DNA**. It is a biologically important form of DNA that is commonly and naturally found in most living systems. This double helical structure of DNA is found to exist in other alternative forms (such as A-form and C-form) which differ in features such as (i) number of residues (monomers) per turn ("n") or (ii) the spacing of residues along the helical axis ("h") (Table 3-3). For example, **A-form DNA** (A-DNA) is right handed but less hydrated than the B-form DNA. A-DNA is more compact with 11 base pairs per turn of the helix and it is 23 angstroms in diameter. The bases are tilted more in relation to the axis of the helix than in the B-DNA. The A-DNA may occur under experimental conditions.

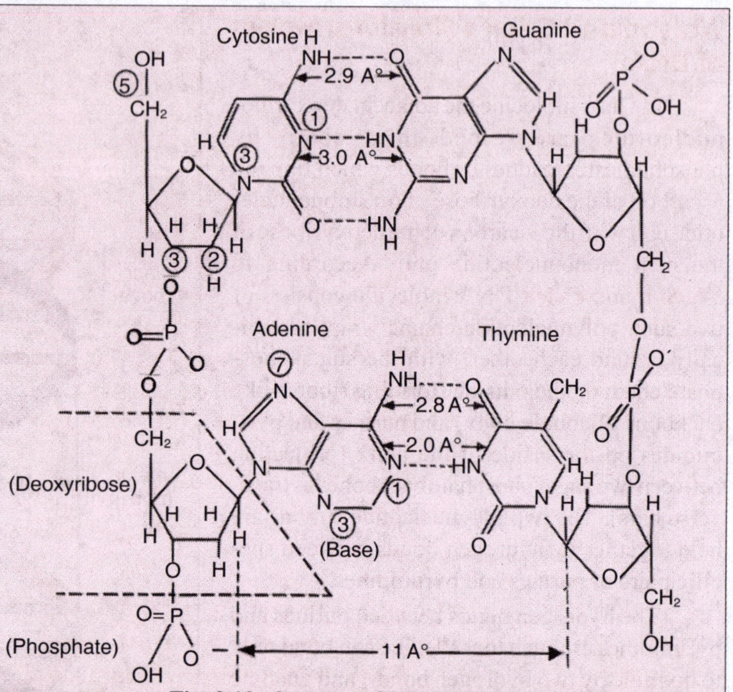

Fig. 3.10. Segment of a DNA molecule.

Table 3-3. **Important features of different forms of DNA double helical structures.**

Helix type	Conditions	Base pair per turn	Rotation per bp	Vertical rise per bp	Helical diameter
A	75 per cent relative humidity; Na^+, K^+, Cs^+, ions	11	+ 32.7° (right handed)	2.56 A°	23A°
B	92 per cent relative humidity, low ionic strength	10	+ 36.0° (right handed)	3.38A°	19A°
C	66 per cent relative humidity; Li^+ ions	9.33	+ 38.6° (right handed)	3.32A°	19A°
Z	Very light salt concentration	12	– 30.0° (left handed)	3.71 A°	18A°

The B-DNA is found in fibres of living cells at a very high 92 per cent relative humidity and low ionic strength. Likewise, A-form of DNA is found at 75 percent humidity in the presence of high ionic strength of NA^+, Ka^+ or Cs^+ ions. C-form of DNA is found at 66 per cent relative humidity in the presence of lithium (Li^+) ions. These three forms are assumed to be found in all DNAs.

There are certain other forms of DNA such as **D-form** and **E-form**, both of which are found as rare extreme variants and contain only 8 and 7½ base pairs per turn respectively. These rare DNA variants are found only in some DNA molecules which lack guanine.

Z-DNA (or Left-handed DNA). Crystallographic studies on synthetic nucleotides consisting of alternating purines and pyrimidines such as GCGCGCGC have shown that left-handed DNA can also exist. This DNA is called **Z-DNA** because of its zigzag structure. Z-DNA can also be found in solutions of high-ionic strength, such as 2M NaCl. In Z-form of DNA, the molecule still consists of two anti-parallel chains, but, otherwise, it is quite different from the A or B form (Fig. 3.11). The helix of Z-DNA is 18 angstrom in diameter, containing 12 base pairs per turn. The differences and similarities of Z-DNA and B-DNA have been summarized in Table 3-3; some other features are the following :

A. Similarities between Z-DNA and B-DNA

(1) Both are double helical.

(2) In both DNAs, two polynucleotide strands of double helix are antiparallel.

(3) Both forms exhibit G≡C pairing.

B. Differences between Z-DNA and B-DNA

(i) Z-DNA has left-handed helical sense, while B-DNA has right-handed helical sense (Fig. 3.12).

(ii) The phosphate backbone of Z-DNA follows a zigzag course, while in B-DNA this backbone is regular.

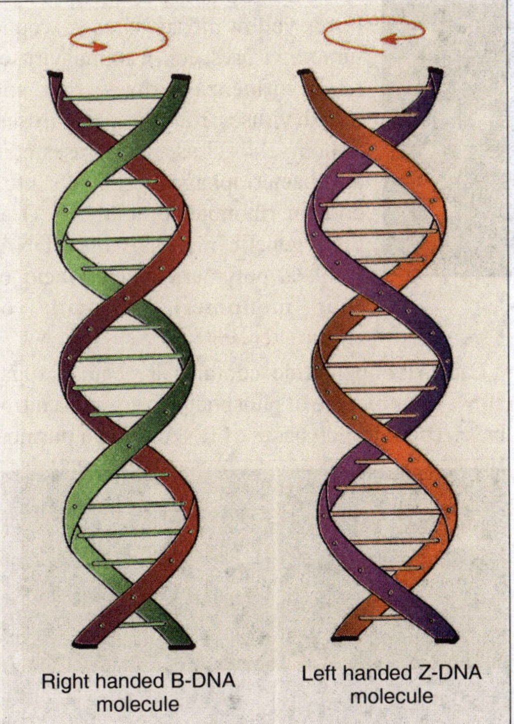

Fig. 3.12. The normal right handed helical sense of B-DNA (A) and rare left handed helical sense of Z-DNA(B).

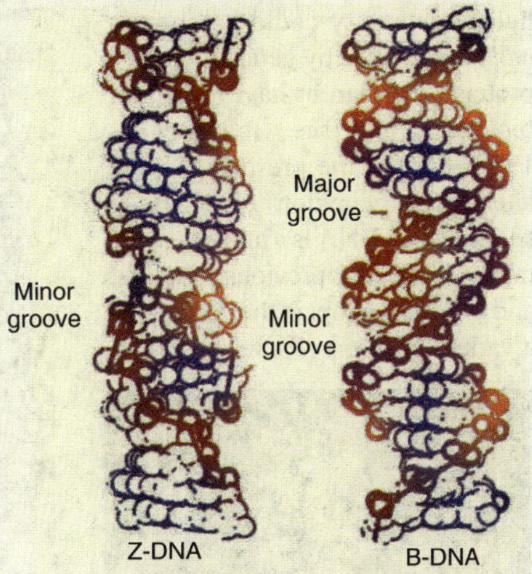

Fig. 3.11. Two molecular forms of DNA double helix, right-handed B-DNA and left-handed Z-DNA. The heavy black line in each molecule goes from phosphate group to phosphate group, indicating a smooth right-handed helix in B-DNA and irregular or zigzag left-handed helix in Z-DNA (after De Robertis and De Robertis Jr., 1987).

(iii) In Z-DNA, the adjacent sugar residues have opposite orientation, while in B-DNA they have same orientation. Due to this, repeating unit is a dinucleotide in Z-DNA as against a mononucleotide unit in B-DNA (Fig. 3.13).

(iv) In Z-DNA, one complete helix (*i.e.*, a twist through 360°) has twelve base pairs or six repeating dinucleotide units, while in B-DNA one complete helix has only ten base pairs or 10 repeating units.

(v) The angle of twist (rotation) per repeating unit (dinucleotide) in Z-DNA is 60° than the 36° of mononucleotide in B-DNA.

(vi) In Z-DNA, one complete turn of helix is 45A° long, while in B-DNA it is 34A° long.

(vii) Since bases get more length spread out in Z-DNA and since the angle of tilt is 60°, they are more closer to the axis. Due to this fact the diameter of Z-DNA molecule is 18 A° than the 20 A° diameter of B-DNA.

Regions of Z-DNA may be involved in gene regulation in the cells of "higher" organisms. Short regions of such drastically altered helical geometry could be specifically recognized by gene regulatory proteins, and thereby have important biological roles (see **Alberts** *et al.*, 1989). Lastly, the presence of any alternate configuration (*e.g.*, Z-DNA) suggests that DNA is a more flexible molecule than was previously thought and that it can adopt in the genome of a variety of forms (**Rich**, 1980).

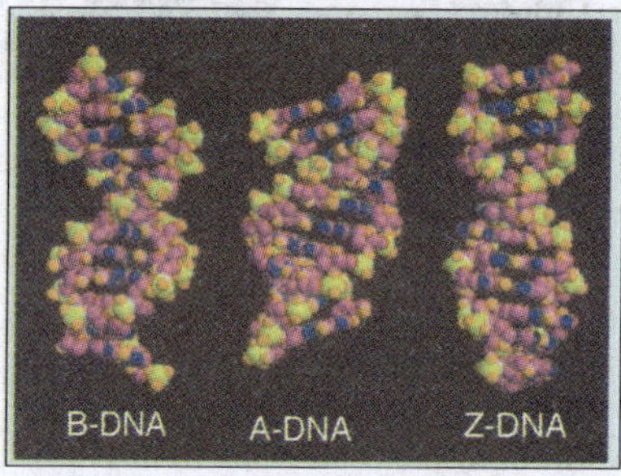

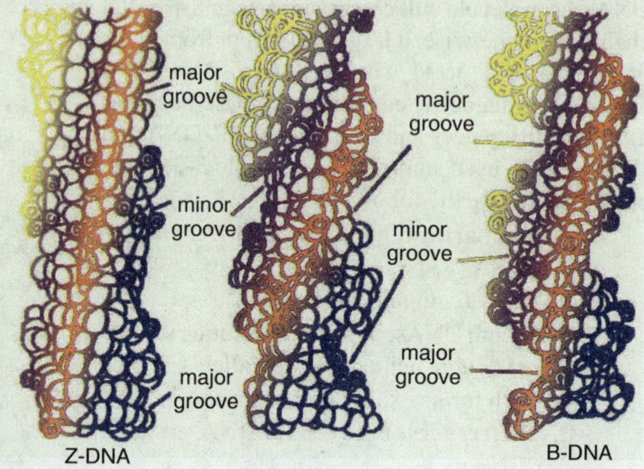

Fig. 3.13. Mode of orientation of adjacent sugar residues in Z-DNA and B-DNA. Z-DNA has opposite orientation resulting in dinucleotide units, while B-DNA has same orientation resulting in mononucleotide units.

RIBONUCLEIC ACID (RNA)

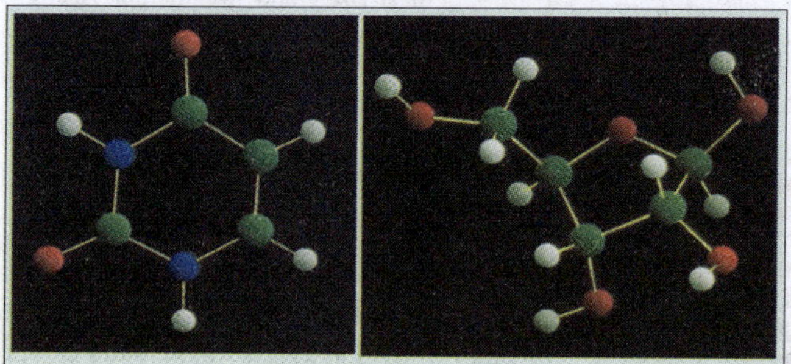

Uracil. Ribose.

Some plant viruses (*e.g.*, TMV, turnip yellow mosaic viruses, wound tumour viruses, etc.,), animal viruses (*e.g.*, influenza viruses, foot and mouth viruses; rous sarcoma viruses, poliomyelitis viruses, reoviruses, etc.) and bacteriophages (*e.g.*, MS_2 etc.) contain ribonucleic acid (RNA) as their genetic material. Like DNA, RNA is polymeric nucleic acid of four monomeric **ribotids** or **ribonucleotids**.

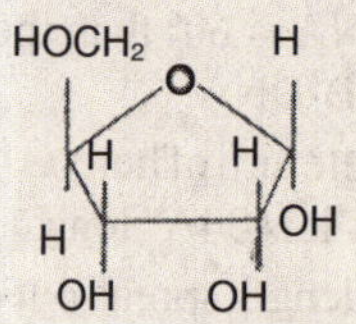

Fig. 3.14. Chemical formula of D-ribose.

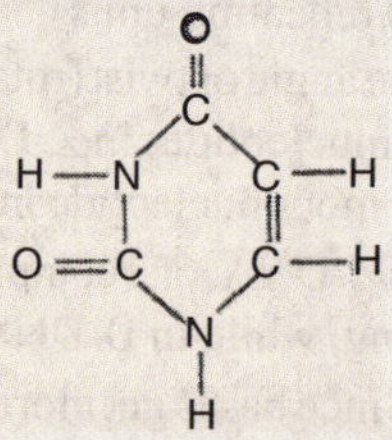

Fig. 3.15. Chemical formula of Uracil.

Each ribonucleotide contains a pentose sugar (**D-ribose**); a molecule of phosphate group and a nitrogen base. The nitrogen bases of RNA are two purines, **adenine** and **guanine** and two pyrimidines, **cytosine** and **uracil**. The four bases, ribonucleosides and ribonucleotides of RNA can be tabulated as follows :

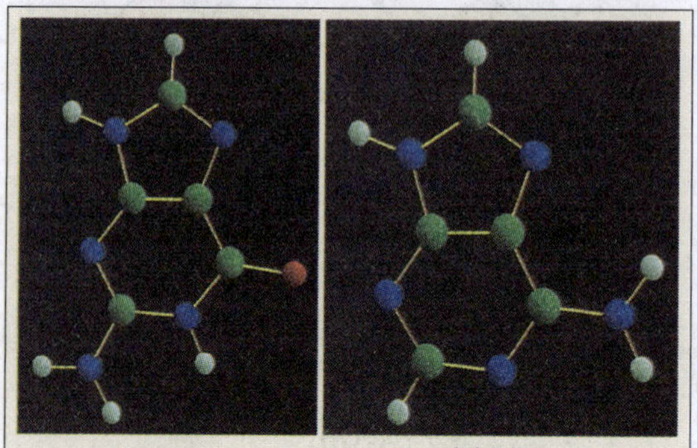

Guanine. Adenine.

Table 3-4. Four components of RNA.

Base	Ribonucleoside	Ribonucleotide	Abbreviation for ribonucleotide
1.Adenine(A)	Adenosine	Adenylic acid (Adenosine monophosphate)	AMP
2. Guanine(G)	Guanosine	Guanylic acid (Guanosine monophosphate)	GMP
3. Cytosine (C)	Cytidine	Cytidylic acid (Cytidine monophosphate)	CMP
4. Uracil (U)	Uridine	Uridylic acid (Uridine monophosphate)	UMP

The four ribonucleotides also occur freely in nucleoplasm but in the form of triphosphates of ribonucleosides such as adenosine triphosphate (ATP), and uridine triphosphate (UTP).

Molecular Structure of RNA

RNA molecule may be either single stranded or double stranded but not helical like DNA molecule. Single stranded RNA occurs as genetic material in plant viruses (*e.g.*, TMV, TYM), animal viruses (*e.g.*, influenza viruses, foot and mouth viruses, rous sarcoma viruses, poliomyelitis viruses) and bacteriophages (*e.g.*, MS_2). The non-genetic RNAs except tRNA of prokaryotes and eukaryotes, also have single stranded RNA molecules. The double stranded but non-helical RNA occurs as the genetic material in some plant viruses (*e.g.*, reoviruses). The transfer or soluble RNA (tRNA or sRNA) which is non-genetic RNA of prokaryotes and eukaryotes, is double stranded but non-helical structure.

Each strand of RNA is polynucleotidic, that is, made up of many ribonucleotides. In the polynucleotide strand of RNA, the ribose and phosphoric acids of nucleotides remain linked by phosphodiester bonds. The organisms which have only RNA, is called **genetic RNA**. While the organisms which have DNA along with RNA, they use the RNA in carrying the orders of DNA and in them because RNA has no genetic role, so called **non-genetic RNA**. The non-genetic RNA is heterogeneous and includes the following three genera : ribosomal RNA (rRNA), transfer RNA (tRNA) and messsenger RNA (mRNA). Each genera of non-genetic RNA has **DNA-dependent replication** of itself, that is, it is not self-replicating like DNA and is transcribed by DNA.

Replication of Genetic RNA

The genetic RNA of viruses is self-replicating, that is, it can produce its own replica by itself. So, its mode of replication is called **RNA-dependent RNA synthesis**. The genetical research on genetic RNA has revealed the following facts about it.

(1) The viral RNA functions directly as a messenger RNA which, in association with the ribosomal apparatus of the host, directs the synthesis of both the **RNA polymerase enzyme** (required for RNA replication) and the proteins of the viral coat.

(2) With the mediation of RNA polymerase and on the standard base-pairing principles, the viral RNA serves as a template in the synthesis of a complementary RNA chain, and thus a double stranded structure is produced.

REVISION QUESTIONS AND PROBLEMS

1. Discuss, in brief, the structure of deoxyribonucleic acid and compare it with that of ribonucleic acid.
2. Give an account of Watson and Crick's double-stranded model of DNA molecule.
3. Do the two strands of DNA helix carry the same genetic information? Explain.

4. What kind of evidence indicates (1) that DNA can reproduce itself and (2) that nucleotides occur in matched pairs in DNA molecule?
5. Write short notes on the following:
   (i) Z-DNA), (ii) B-DNA, (iii) Tetranucleotide hypothesis, (iv) Left- handed DNA versus right-handed DNA.
6. Assume the following base sequence was found in a 20 base DNA strand :
   3' ATT CGA CCT TAT TAC TGC AC 5'
   (a) What would be the first 5 bases in the 3′ end of the complementary strand?
   (b) What would be the 10 bases of the 5' end of the complementary strand?
   (c) Assuming the presence of complementary strands, what is the per cent composition of the polymer with respect to AT base pairs and with respect to GC base pairs.
7. Analysis of four double-stranded DNA samples yielded the following information:
   1.15% cytosine ; 2.12% guanine;
   3.35% thymine ; 4.28% adenine.
   (a) What would be the percentage of the other bases in each sample?
   (b) Could any of these samples have been obtained from the same organism? If so, which ones?
8. Four samples of nucleic acid were analyzed for the proportion of the different bases present, with the following results:
   (1) A = 30%, C = 30%, G = 20%, T = 20%.
   (2) A = 27.5%, C = 22.5%, G = 22.5%, T = 27.5%.
   (3) A = 18%, C = 32%, U = 32%, G = 18%.
   (4) A = 18%, C = 32%, U = 18%, G = 32%.
   Which of these samples were DNA and which were RNA? Which double-stranded?
9. Ratios of the bases present in different samples of nucleic acid yielded the following results:
   (1) (A+C) / (T+G) = 1, (2) (A+C) / (U+G) = 0.8, (3) (A+G) / (T+C) = 1.5.
   Which were RNA and which were DNA? Which were single and double-stranded?
10. If one DNA sample had a melting temperature of 85.5^{0}C and another showed a melting temperature of 88^{0}C. What might you conclude concerning the base composition of the two samples?
11. Assume an average-sized gene consisting of a linear sequence of 1000 bases and there were 1000 genes in a bacterial chromosome.
    (a) How many bases would such a chromosome contain in each strand of the double helix?
    (b) If 10 nucleotides = 34A°, how long would this chromosome would be in millimeters ?

ANSWERS TO PROBLEMS

6. (a) Bases at the 3′ end of the opposite strand will be complementary to those at the 5' end of the polymer given. Therefore, the first 5 bases will be 3' GTGCA.
   (b) 5' TAA GCT GGAA
   (c) A-T = 60 per cent, G-C = 40 per cent.
7. (a) 1.15% C, 15% G, 35% T, 35% A.
   2.12% G, 12% C, 38% T, 38% A.
   3.35% T, 35% A, 15% C, 15% G.
   4.28% A, 28% T, 22% G, 15% C.
   (b) Yes, samples 1 and 3.
9. (1) Double-stranded DNA , (2) Single-stranded RNA , (3) Single- stranded DNA.
10. The second sample has a higher G-C content.
11. (a) 1 million , (b) 3.4 mm.

CHAPTER

4

Replication of DNA

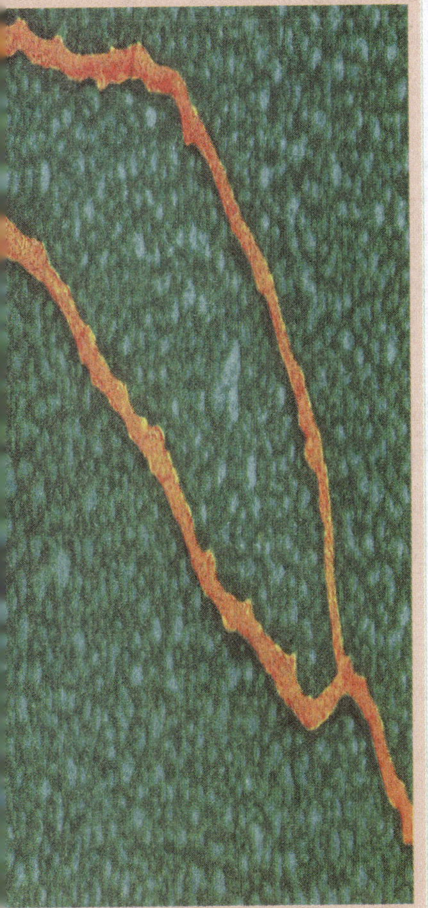

of human DNA from a HeLa cell, rating the replication bubble that acterizes DNA replication within a e replicon.

As a carrier of genetic information DNA has the following two important functions: **1. Heterocatalytic function** – When DNA directs the synthesis of chemical molecules other than itself (*e.g.,* synthesis of RNA, proteins, etc.), then such functions of DNA are called heterocatalytic functions **2. Autocatalytic functions** – The functions of DNA which directs the synthesis of DNA itself, are called autocatalytic functions. Here, we are concerned only with autocatalytic function of DNA.

WATSON AND CRICK'S MODEL FOR DNA REPLICATION

The double helix model of DNA molecule of Watson and Crick beautifully embodied a built-in template system for self-replication or autocatalytic function. Because of the specificity of base pairing, the sequence of base along one chain automatically determines the base sequence along the other. Thus, each chain of the double helix can serve as **template** for the synthesis of the other. For the replication of **DNA** molecule, **Watson** and **Crick** proposed that replication involved the disruption of hydrogen bonds followed by a rotation and separation of the two polynucleotide strands. Each purine and pyrimidine base of each polynucleotide strand is thought to attract a complementary free muleotide available for polymerization in cell and to hold it in place means of the specific hydrogen bonds. Once held in place on the parent template chain, the free mucleotides were sewn together by formation of the phosphate diester bonds that linked adjacent deoxyribose residues, forming a new polynucleotide molecule of predetermined base sequence. Thus, two double helical molecules, identical with each other, are formed.

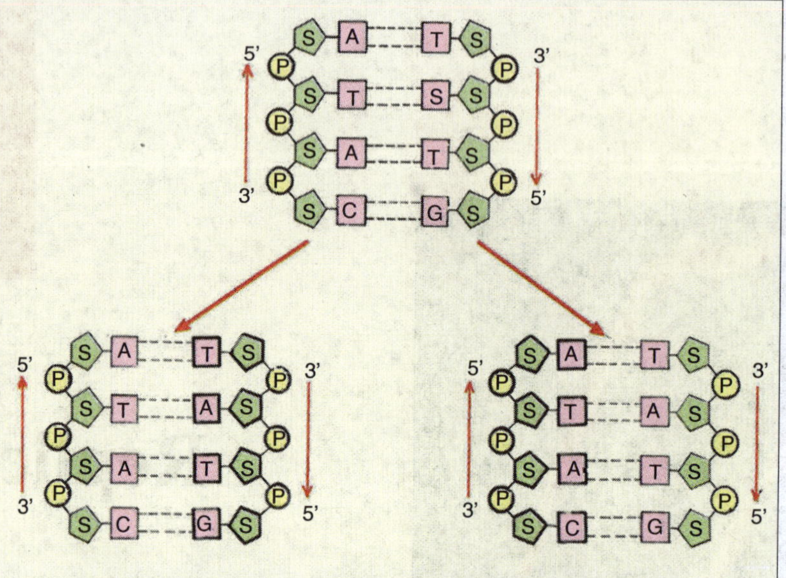

Fig. 4.1. Semi-conservative model of DNA replication.

1. Experimental Evidence for Semiconservative DNA Replication in *E. coli*

The Watson-Crick's model of DNA structure and replication suggested that once DNA replication is initiated, the two original polynucleotide strands of the duplex or helix will unwind, at least locally, so that each can serve as a template for a new strand. An immediate prediction follows from this proposal: both duplexes that result from replication should be **hybrid** in nature, each containing an old strand derived from the original molecule and a new strand which has been formed during the replication process. Since each of the two double helices or duplexes conserves only one of the parent polynucleotide strands, the process is said to be **semiconservative.** This prediction is diagrammed in Fig. 4.2 in which old DNA is shown in black and new DNA in white. Figure 4.2 also outline what would be

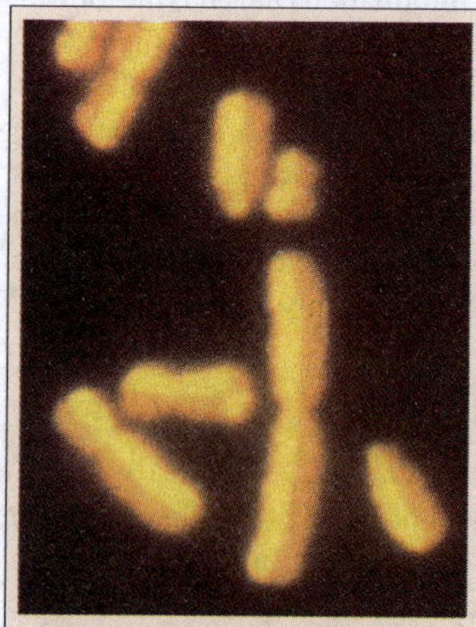

"Harlequin" chromosomes prove DNA replication is semiconservative.

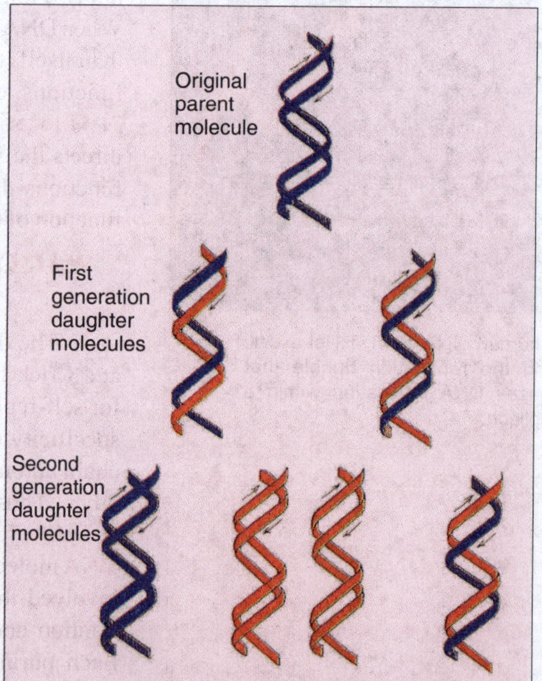

Fig. 4.2. Semi-conservative model of DNA replication.

predicted if these two hybrid duplexes went on to replicate themselves. Four duplexes would result, two of which would contain a single strand derived from the original chromosome and two of which would contain totally new DNA strands.

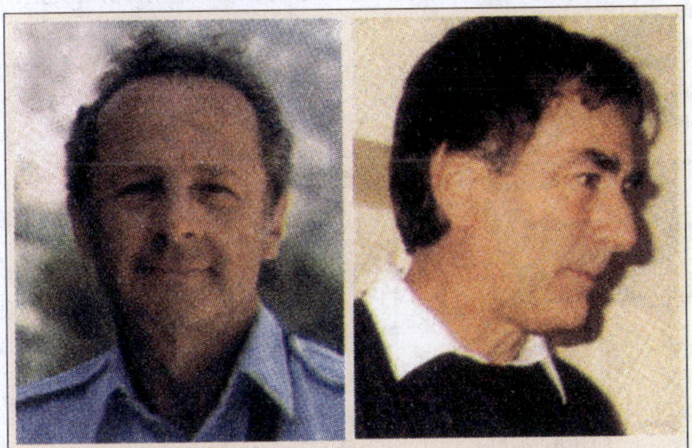

M. Meselson. F.W.Stahl.

Besides the semiconservative mode of DNA replication, the following two methods of DNA replication were deemed equally feasible: (i) **conservative replication,** in which both strands of parent double helix would be conserved and the new DNA molecule would consist of two newly synthesized strands; and (ii) **dispersive replication,** in which replication would involve fragmentation of the parent double helix and the intermixing of pieces of the parent strands with newly synthesized pieces, thereby forming the two new double helices.

(i) Meselson and Stahl's experiment. M. Meselson and **F.W. Stahl** (1958) verified the semiconservative nature of DNA replication in a series of elegant experiments using isotopically labelled DNA and a form of isopycnic density gradient centrifugation. They cultured *Escherichai coli* cells in a medium in which the nitrogen was ^{15}N (a 'heavy' isotope of nitrogen, but not a radioisotope) instead of commonly occurring and lighter ^{14}N. In time, the purines and pyrimidines of DNA in new cells contained ^{15}N (where ^{14}N normally occurs) and, thus, the DNA molecules were denser. DNA in which the nitrogen atoms are heavy (^{15}N) can be distinguished from DNA containing light nitrogen (^{14}N), because during isopycnic centrifugation, the two different DNAs band at different density positions in the centrifuge tube.

Depending on its content of ^{15}N and ^{14}N, the DNA bands at a specific position in the density gradient. Because the DNA synthezised by *E.coli.* cells grown in ^{15}N would be denser than ^{14}N-containing DNA, it would band further down the tube.

E. coli cells grown for sometime in the presence of ^{15}N-medium were washed free of the medium and transferred to ^{14}N-containing medium and allowed to continue to grow for specific lengths of time (*i.e.*, for various numbers of generation time). DNA isolated from cells grown for one generation of time in the ^{14}N medium had a density intermediate to that of the DNA from cells grown only in ^{15}N-containing medium (identified as *generation O;* Fig. 4.3) and that of DNA from cells grown only in ^{14}N-containing medium (*the controls*). Such a result immediately ruled out the possibility that DNA replication was conservative, because the conservative replication would have yielded two DNA bands in the density gradient for **generation 1** (*i.e.*, F_1 cells). The single band of intermediate density (identified as "**hybrid DNA**") consisted of DNA molecules in which one strand contained ^{15}N and the other contained 14N. When the incubation in the ^{14}N-medium was carried out for two generations of time (*i.e.* **generation 2**), two DNA bands were formed — one at the same density position as the DNA from cells grown exclusively in ^{14}N medium (*i.e.,* light controls) and the other of intermediate density. Subsequent generations produced greater numbers of DNA molecules

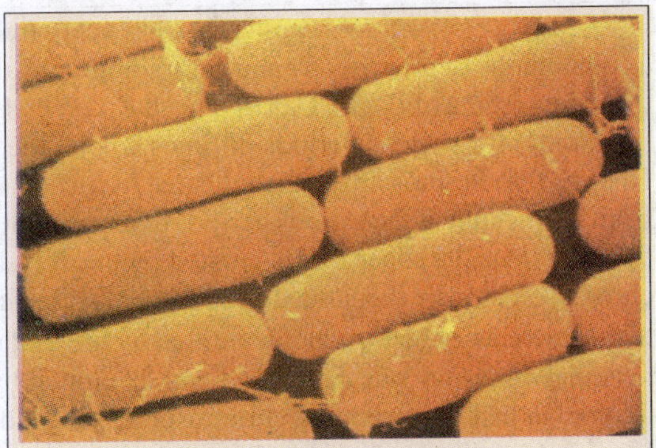

Cells from a clone of *E.coli.*

that banded at the ''light'' (^{14}N-containing DNA) position in the density gradient. These results are consistent only with the model of semiconservative replication. Studies using other prokaryotes as well as eukaryotes indicate the semiconservative replication of DNA is probably a universal mechanism.

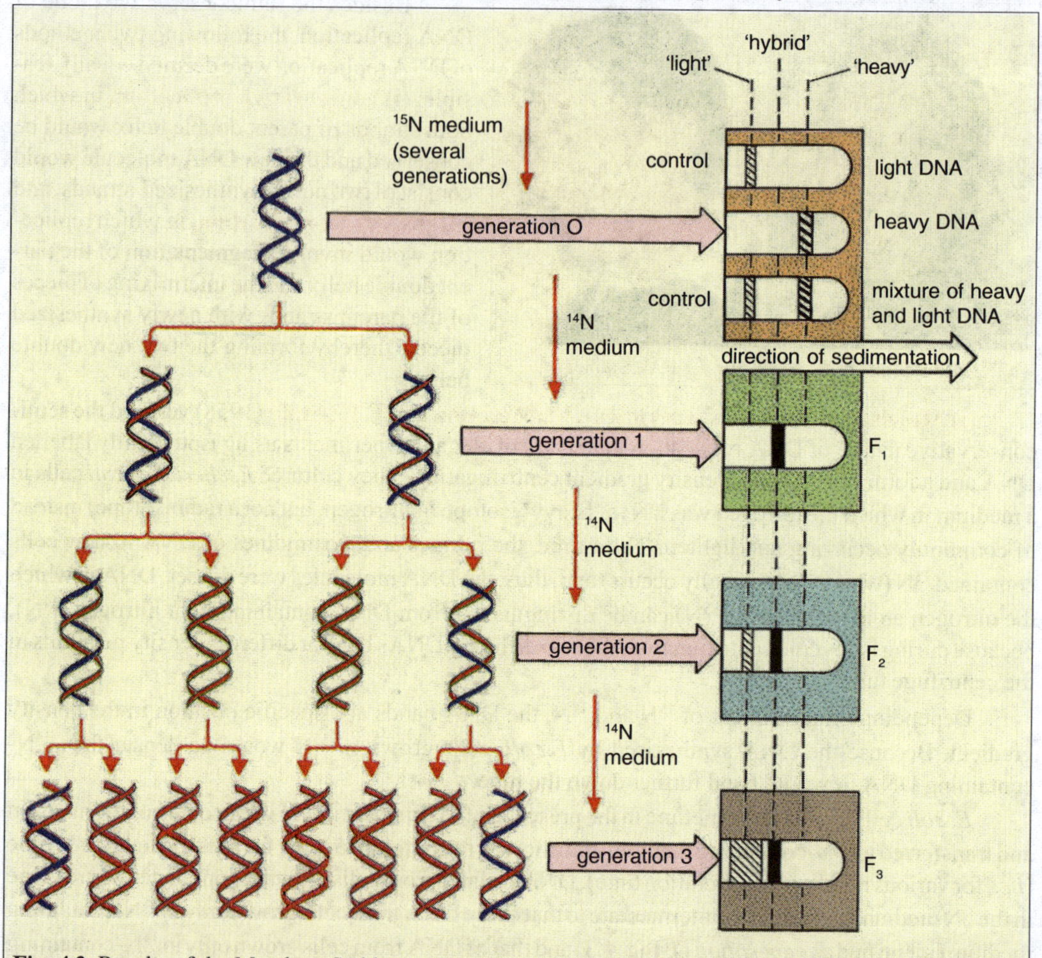

Fig. 4.3. Results of the Meselson-Stahl experiments (right) and their interpretation (left) (after Sheeler and Bianchi, 1987).

(ii) Visualization of replication in E. coli. In 1963, **J. Cairns** developed a technique employing a combination of microscopy and autoradiography that made it possible to visualize the replication of the chromosome of *E.coli.* Cairns placed *E. coli* cells in a medium containing ^3H-thymidine (tritiated-thymidine) for various periods of times so that the radioactive thymidine was incorporated into the DNA as the chromosome was replicated in successive generations of cells. *Ecoli* cells were removed from the medium after various periods of incubation and gently lysed to release the chromosome from the cell (since the shear forces created by harsh lysis break the chromosome into small pieces). The chromosomes were then transferred to glass slides and coated with a photographic emulsion sensitive to the low-energy beta-particles emitted by the ^3H-thymidine. After exposing the emulsion to the beta rays, the emulsion was developed and examined by light microscopy. Wherever decay of labelled thymidine had occurred in a chromosome, the emulsion was exposed and created visible grains. A chromosome not engaged in replication appeared as a circular structure formed from a close succession of exposure spots. Chromosomes ''caught in the act'' of replication gave rise to what are called **theta**

configurations because they have the appearance of the Greek letter theta (*i.e.*, q fig. 4.4). The theta structures reveal the positions of the replication forks in the circular chromosomes. Cairns' observations also clearly supported the semi-conservative nature of replication.

The rate at which the replication proceeds could also be worked out by measuring the length of DNA undergoing replication in a known interval of time. Cairns worked out generation time of *E. coli* as 30 minutes. The length of the chromosome was worked out to about 1 mm. The rate of replication, thus, would be approximately 30 µm - 40 µm per minute (1 mm = 1000 µm).

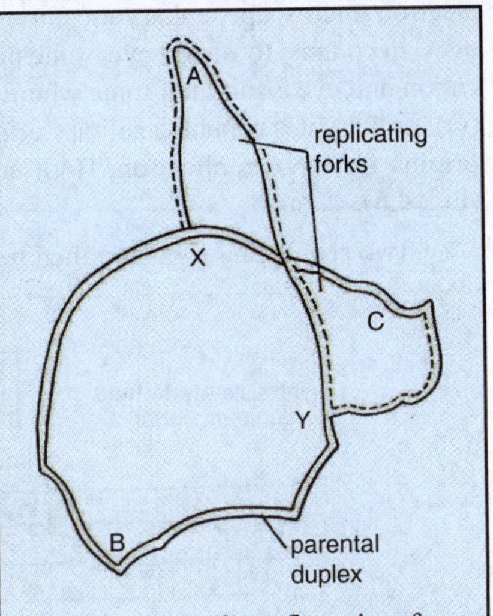

Fig. 4.4. Theta (θ) configuration of replicating *E.coil* chromosome obtained by Cairns. Loops A and B have completed replication, whereas loop C remains to be replicated. X and Y and the replication forks.

2. Evidence for Semi conservative Replication of Chromosomes (or DNA) in Eukaryotes

J.H. Taylor and P. Woods (1957) provided evidence in support of semi conservative mode of DNA replication in eukaryotes by using the technique of autoradiography and light microscopy in dividing root tip cells of the bean, *Vicia faba.* After incorporation of tritiated thymidine, when root tips were transferred to unlabelled culture medium (and colchicine was added to the medium to prevent anaphase separation of sister chromatids), in the first generation of duplication both chromatids were labelled (this is interpreted as one DNA double helix in each chromatid and only one of the two strands labelled). In the second cycle of duplication (in the unlabelled medium) in each chromosome, one of the two chromatids was found to be labelled (Fig. 4.5). This was interpreted as showing semiconservative mode of duplication.

3. Semidiscontinuous DNA Replication

Various experimental evidences have suggested that DNA synthesis is continuous on one strand (3' to 5' strand), called **leading strand** and discontinuous on the other strand (5' to 3' strand), called lagging strand. Since DNA synthesis always proceeds in 5' to 3' direction, so, on the lagging strand, synthesis takes place discontinuously in pieces, called **okazaki fragments** (after the name of discoverer **R. Okazaki**, 1968). Later on, these pieces are fused with the help of ligase enzyme to form an intact lagging strand. Such a DNA replication, where the leading strand is synthesized continuously and the lagging strand is synthesized discontinuously, is called semidiscontinuous replication.

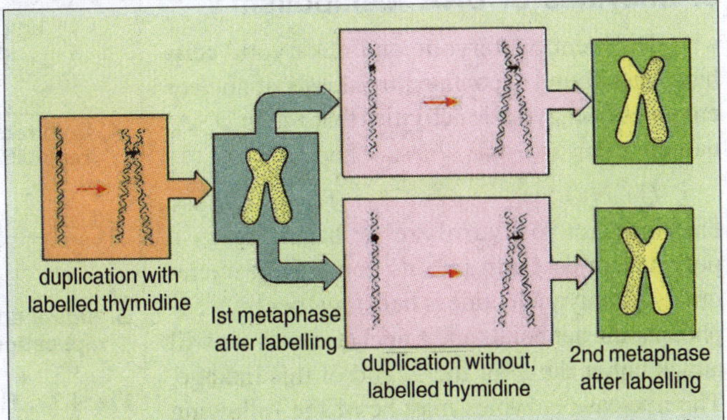

Fig. 4.5. Taylor and Wood's experiment on *Vicia faba* root tips using autoradiography technique (after Sheeler and Bianchi, 1987).

4. Unidirectional and Bidirectional DNA Replication

Regarding the direction of replication, DNA replication may be of the following two types:

(a) Bidirectional replication. All known DNA molecules, with only few exceptions, replicate as circles (or dubbles/eyes) and, hence, initiate within the helix. In the electron microscope, eukaryotic chromosomes are found to contain multiple expanding replication eyes, in contrast to single eye of the prokaryotic DNA. Further DNA synthesis within a given replication unit eye is initiated some where at a near the midpoint of the unit at a site termed as the origin (O); prokaryotes contain a solitary origin, while eukaryotes have multiple (up to several thousands) origins for DNA replication. Both ends of the eye are moving and serve as **replication forks** (Fig. 4.6).

Two replicating forks are then believed to travel in opposite directions until they reach either end of that unit, the two end points being called **termini (T).** A given replication unit may or may not undergo bidirectional synthesis in synchrony with contiguous units. In either case the newly replicated strands in adjacent units will eventually meet. These strands are then linked, perhaps, by a DNA ligase to form long, continuous daughter DNA strands (Fig. 4.8.).

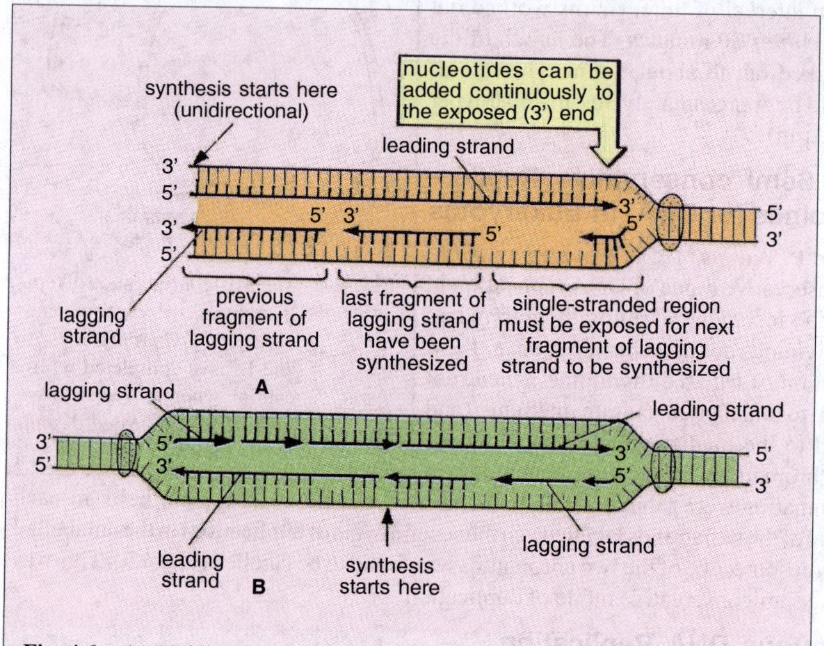

Fig. 4.6. A – Unidirectional DNA replication; B – Bidirectional DNA replication.

(b) Unidirectional replication. In case of this type of DNA replication, one of two ends of the replication eye remains stationary and the other ends serves as the replication fork and moves with replication. An example of unidirectional replication is the replication of mitochondrial DNA (mt DNA) by **D-loop** (or displacement loop) in vertebrates.

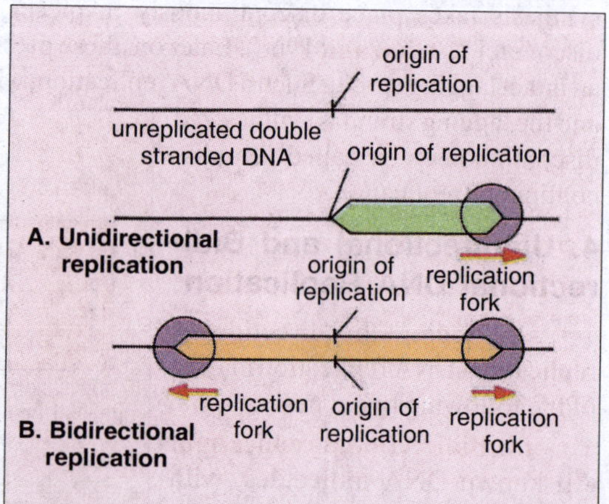

Fig. 4.7. The difference between unidirectional and bidirectional replications in the movement of replication forks.

5. Enzymes of DNA Metabolism

Different prokaryotic and eukaryotic cells have been found to contain three kinds of nuclear enzymes or enzymatic activities that act on DNA, namely **nuclease, polymerases** and **ligases.**

(I) Nuclease enzymes. The nuclease enzymes act to hydrolyze or break down a polynucleotide chain into its component nucleotides. A polynucleotide is held together by 3′, 5′ phosphodiester bonds and a nuclease enzyme will attack either the 3′ or the 5′ end of this linkage. The nuclease enzymes may be of the following two kinds :

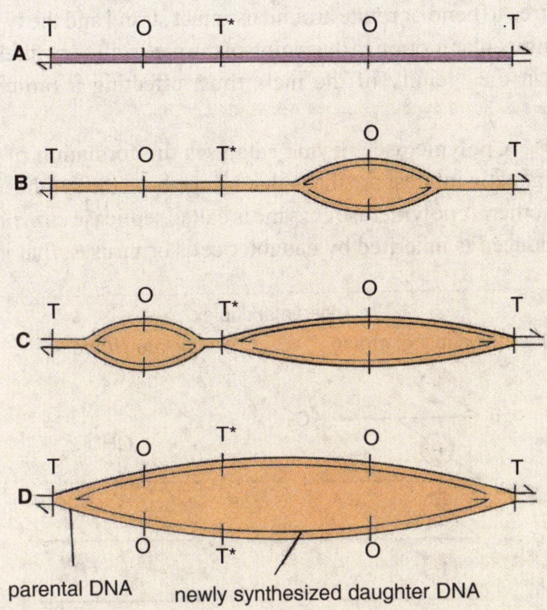

Fig. 4.8. Bidirectional model for mammalian chromosomal DNA replication. Origins are indicated by O and B – replication started in right-hand replication unit; C – replication started in left-hand replication unit and completed at termini of right-hand unit; D – replication completed, sister duplexes joined at common terminus T.

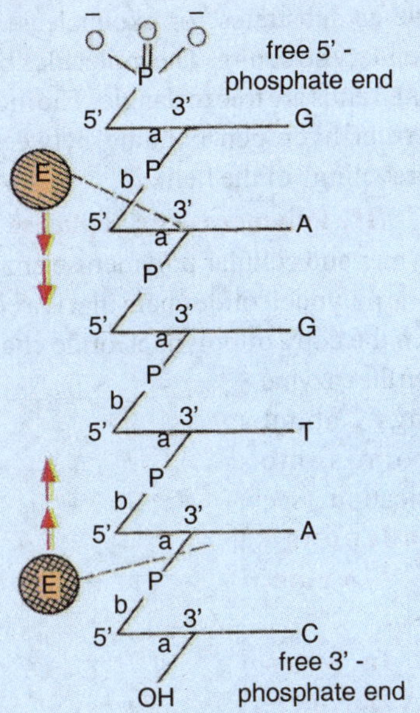

Fig. 4.9. Exonuclease action on a polynucleotide chain. The 3' side of each phosphodiester linkage is labelled *a*, the 5' side is labelled b, an exonuclease that cleaves *a* linkages, starting from a free 3' -OH end (*e.g.*, snake venom phosphodiesterase) is shown in check. An exonuclease that cleaves *b* linkages, starting from a free 5'-P end (*e.g.*, bovine spleen phosphodiestrase) is shown in line shade (after Lehninger, 1970).

(a) Exonuclease enzymes. A nuclease enzyme which begins its attack from a tree end of a polynucleotide is called exonuclease. Therefore, depending on the specificity of the enzyme, an exonuclease will either begin at a tree 3'-OH end of a polynucleotide and progressively cleave the bonds on the 3'-OH side of the phosphodiester backbone or it will begin at a free 5'-P end and digest the polynucleotide in a 5' $\rightarrow$ 3' direction. In' both cases the enzyme travels along the chain in a stepwise manner, liberating single nucleoside monophosphate molecules and eventually digesting the entire polymer.

(b) Endonuclease enzyme. Endonuclease enzyme also attacks one of the two sides of phosphodiester linkages, but they react only with those bonds that occur within the interior of a polynucleotide chain. If the polynucleotide chain is single-stranded *(e.g.,* viral DNA), such an attack will obviously cut the chain into two pieces. If, however, the polynucleotide strand is a member of a DNA helix (*e.g.,* prokaryotic and eukaryotic DNA), a single endonucleolytic cut will create a nick in the

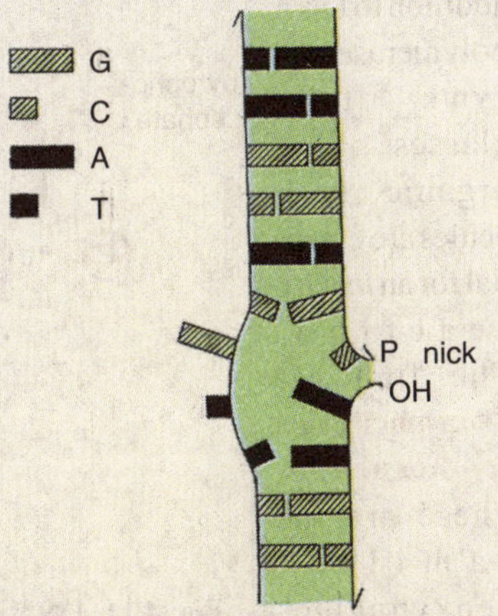

Fig. 4.10. A nicked duplex DNA molecule such as is formed by an endonuclease enzyme.

helix; the helix remains in one piece but it now possesses a gap that contains two tree ends, which can serve as substrates for exonucleases. A nicked double helix suffers a localized disruption of its secondary structure. The molecules becomes tree to bend or rotate around its intact strand and the two broken ends are tree to dangle. The increased molecular motion in the region of the nick will very likely interrupt hydrogen bonding between bases in the vicinity of the nick, thus, effecting a limited "unraveling" of the helix.

(II) Polymerase or replicase enzymes. A polymerase enzyme catalyses the formation of a polymer and cellular polymerase enzymes of genetic interest are those that bring about the synthesis of one polynucleotide chain that is a copy of another. A polymerase enzyme is called replicase enzyme when the copy of polynucleotide chain so produced is inherited by daughter cells or viruses, that is, when the enzyme brings about chromosome replication (see **Goodenough** and **Levine,** 1974).

In vitro DNA polymerization. To understand the mechanism of DNA-replication inside the living cell *(in vivo),* molecular biologists tried *in vitro* polymerization of DNA and found that in addition to DNA polymerase enzyme, three classes of organic molecules are essential for an *in vitro* reaction. The first are **deoxynucleoside triphosphates.** These are the familiar deoxynucleotide monophosphates *(i.e.,* d AMP, d CMP, d

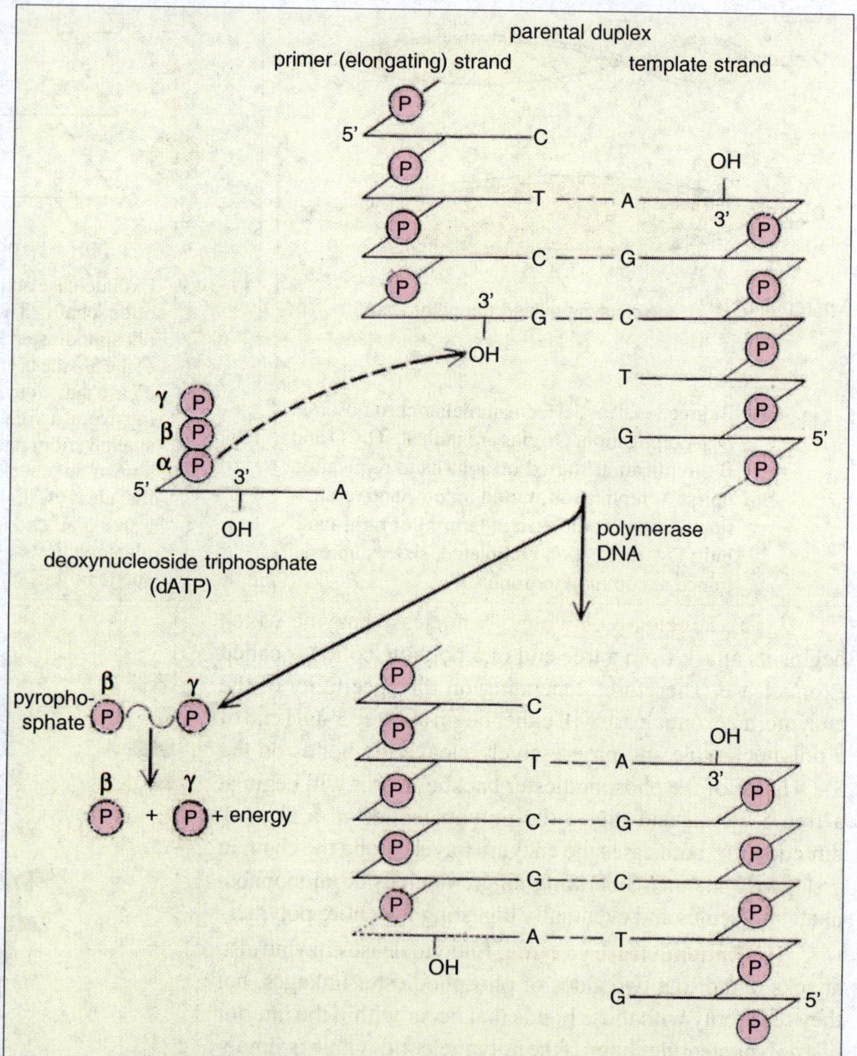

Fig. 4.11. DNA polymerization. A d ATP molecule is shown being added to a parental DNA duplex or helix at its 3'-OH end. As DNA polymerase catalyzes the formation of the phosphodiester bond, pyrophosphate (P~)) is hydrolyzed into two phosphate molecules.

GMP, and d TMP) with two additional phosphate groups attached to the initial or a. - phosphate group (*e.g.,* dATP, dCTP, dGTP and d TIP). The second class of essential molecules for many DNA polymerase enzymes are polynucleotide chains with tree 3'-08 ends (often called primer strands), meaning that they cannot initiate the *de novo* synthesis of a new strand. The third class of essential molecules for DNA polymerase enzymes are template strands. All biologically important DNA polymerases possess a critical property: they will add nucleotides to a primer strand only in response to the base sequence found on a second template strand. Just as **Watson** and **Crick** suggested in their original model of DNA replication, the polymerase enzymes observe the rule of complementary base pairing. Thus, if a tree **3'-OH** group on a primer strand lies opposite a thymine on a template strand (see Fig. 4.11), a polymerase enzyme will add only an adenine group to the primer, even when d CTP, dTTP, dGTP are also present in the reaction mixture.

Once the appropriate molecules are present and the appropriate ionic conditions are maintained, *in vitro* polymerization reaction summarized in Fig. 4.11 will occur. The a. - phosphate of a nucleoside triphosphate molecule forms a 3', 5' phosphodiester bond with a free 3'-OH in the growing polynucleotide chain, and a molecule or **phyrophosphate** (P~P) is simultaneoulsy release. Pyrophosphate contains a ''high-energy'' or ''~'' bond, meaning that when the released pyrophosphate is hydrolyzed into two phosphate molecules, energy is liberated which drives the polymerization process for ward. The resultant polymerization will always proceed in a net 5′ → 3′ direction, meaning that the nucleotide at the 3′ end is always the most recently added to the chain.

(1) Prokaryotic DNA polymerases. Three different DNA polymerases are known in *E. coli* and other prokaryotes (see Table 4-1), of which **DNA polymerase I** and **II** are meant for DNA repair and **DNA polymerase III** is meant for actual DNA replication.

A.Kornberg.

1. DNA polymerase I. This enzyme was isolated round 1960 by **Arthur Kornberg** and was the first enzyme suggested to be involved in DNA replication. It is also called **Kornberg enzyme.** DNA polymerase I enzyme is now considered to be a DNA repair enzyme rather than a replication enzyme. This enzyme is known to have five active sites, namely **template site, primer site, 5′ → 3′ cleavage or exonuclease site, nucleoside triphosphate site and 5′ → 3′ cleavage site (or 5′ → 3′ exonuclease site)** (Fig. 4.13). DNA polymerase I is mainly involved in removing RNA primers from okazaki or precursor fragments and filling the resultant gaps due to its 5′ → 3′ polymerizing capacity. DNA polymerase I enzyme can also remove thymine dimers produced due to UV-irradiation and fill the gap due to excision. Both polymerization (=chain elongation) and exonuclease activity of DNA polymerase I have been shown in Fig. 4.13. This is called **proof reading or editing function** of this enzyme.

2. DNA polymerase-II. This enzyme resembles DNA polymerase-I in its activity, but is a DNA repair enzyme. It brings about the growth in 5′ → 3′ direction, using free 3′-OH groups.

3. DNA polymerase-III. DNA polymerase-III or Poll II enzyme plays an essential role in DNA replication. It is a multimeric enzyme or holoenzyme having ten subunits such as alpha (α), beta (β), epsilon (ε), theta (θ), tau (τ) gamma (γ), delta (δ), delta dash (δ′), chi (χ), and psi (ψ). All these ten subunits are needed for DNA replication *in vitro*; however, all having different functions. For example, a subunit has 5′→ 3′ exonuclease proof-reading or editing activity. The **core enzyme** comprises three subunits- α, β and θ. Remaining seven subunits increase processivity (processivity means rapidity and efficiency with which a DNA polymerase extends growing chain).

Table 4-1. **Comparison of different characters of thre types of DNA polymerases of *E. coli*.**

Character	DNA polymerases		
	I	II	II
1. Polymerization 5′→ 3′	Yes	Yes	Yes
2. Exonuclease 5′→ 3′	Yes	Yes	Yes
3. Exonuclease 5′→ 3′	Yes	No	No
4. Use of primer single strands	Yes	No	No
5. Use of nicked duplex or helix	Yes	No	No
6. Molecular weight	109,000 (single chain)	120,000 (single chain)	>250,000 (heteromultimeric chain)
7. Molecules per cell	400	Not known	10-20
8. Gene	pol A	Pol B	pol C (dna E), dna N, dna ZX, dna Q, dna T
9. Nucleotides polymerized at 37°C, molecules/minute	up to 1,000	up to 50	up to 15,000
10. Affinity for nucleoside triphosphate (TPs) precursors	low	low	high

(2) Eukaryotic DNA polymerases. Eukaryotes (*e.g.,* yeast, rat liver, human tumour cells) are found to contain the following five types of DNA polymerases:

(i) DNA polymerase a (= alpha). This relatively high molecular weight enzyme is also called **cytoplasmic polymerase** or **large polymerase.** It is found both in nucleus and cytoplasm.

(ii) DNA polymerase β (=beta). This enzyme is also called **nuclear polymerase or small polymerase** and is found only in vertebrates.

(iii) DNA polymerase γ (= gamma). This enzyme is called **mitochondrial polymerase** and is encoded in the nucleus.

(iv) DNA polymerase δ (=delta). This enzyme is found in mammalian cells and is PCNA dependent for DNA-synthesis processivity (PCNA=proliferating cell nuclear antigen).

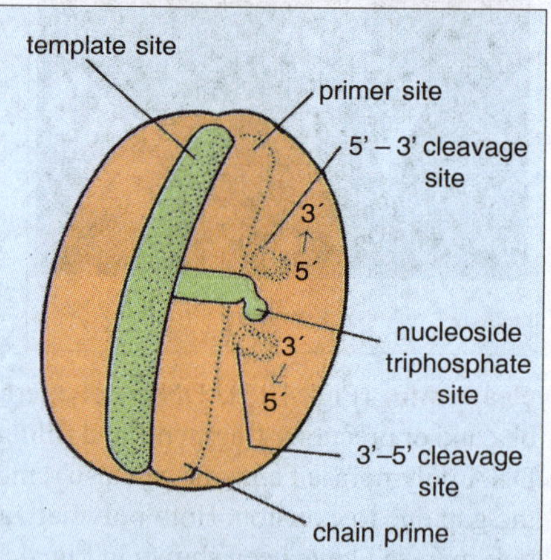

Fig. 4.12. A model of DNA polymerase-I enzyme, showing its five-different active sites.

(v) DNA polymerase ε (= epsilon). It was previously known as **DNA polymerase δ II.** This enzyme is PCNA independent and occurs in mammalian HeLa cells and budding yeast.

The large DNA polymerase a is the predominant DNA polymerase enzyme in eukaryotic cells and was belived for long time to be only enzyme involved in DNA replication. But now one more polymerase, namely, DNA polymerase δ is also found to be involved in eukaryotic DNA replication.

(III) DNA ligases. DNA ligase enzymes are capable of catalyzing phosphodiester bond formaion between free 3′-OH and free 5′-P groups of a nick of DNA which is created by endonuclease enzyme, thereby restoring an antact DNA duplex. Many DNA ligases have already been discovered.

The ligase enzyme from *E.coli* requires the presence of oxidized nicotinamide adenine dinucleotide (NAD+) a cofactor, whereas the ligase enzyme specified by T_4 bacteriophage requires ATP to bring about the joining reaction.

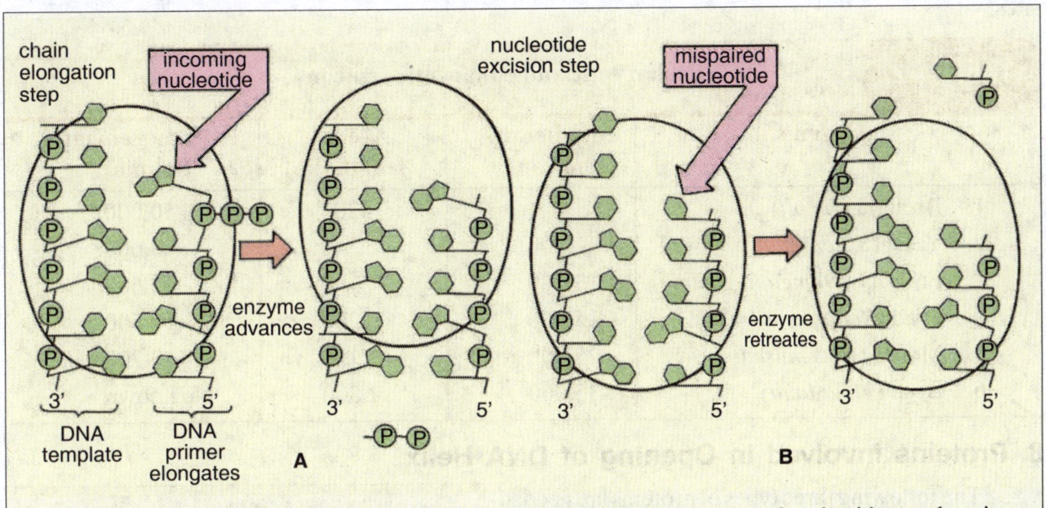

Fig. 4.13. Functions of DNA polymerase-I enzyme. A-Mode of polymerization of nucleotides, on the primer site of the enzyme, B-Removal of mispaired nucleotide by exonuclease activity of the enzyme.

6. Roles of RNA Primers in DNA Replication

No known DNA polymerase can initiate synthesis of DNA without the availability of a **primer RNA strand.** So, before actual DNA replication starts, short RNA oligonucleotide segments, called **RNA primers** or simply the **primers,** have to be synthesized by **DNA primase** enzyme utilizing ribonucleoside triphosphates. This RNA primer is synthesized by copying a particular base sequence from once DNA strand and differs from a typical RNA molecule in that after the synthesis the primer remains hydrogen-bonded to the DNA template **(Freifelder,** 1985). The primers are about 10 nucleotides long in eukaryotes and they are made at intervals on the lagging strand where they re elongated by the DNA polymerase enzyme to begin each okazaki fragment. These RNA primers are later excised and filled with DNA with the help of DNA repair system in eukayrotes (or DNA polymerase I in *E.coli).*

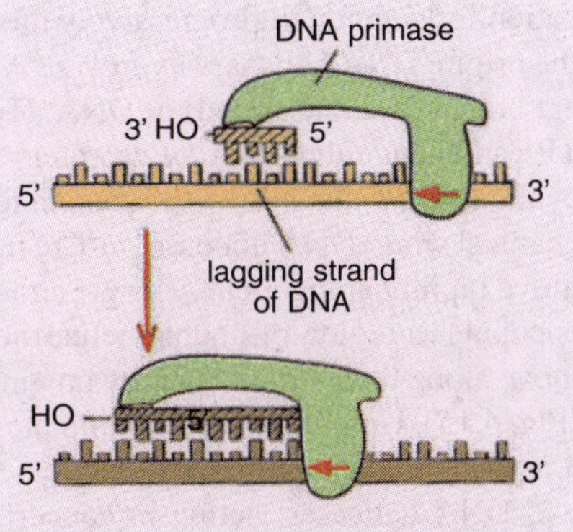

Fig. 4.14. Formating of RNA primer on lagging strand by DNA primase enzyme. Unlike DNA polymerase, this enzyme can start a new polynucleotide chain by joining two nucleotide triphosphates together (after Alberts *et al.,* 1989).

In bacteria, two different enzymes are known to synthesize primer RNA oligonucleotides – **RNA polymerase** (on the leading strand) and **DNA primase** (on the lagging strand).

7. Replicons

DNA replication in prokaryotes and eukaryotes is attained in discrete units, called **replicons.** The number of replicons may vary in a genome from one in bacteria (*E.coli*) and 500 in yeast to several thousands in plants and animals (Table 4-2). For example, in the *E. coli* there is single replicon with the **origin,** identified as a genetic locus *ori* C (245 bp). The origin is A: T rich, a feature that is related to unwinding of DNA to initiate replication. In *E. coli,* there are also termination sites (ter A-F), each consisting of ~ 23 bp. The process of termination of DNA replication requires the product of *tus* gene (**Tus protein** or *TBP, i.e.*, ter binding protein) which recognizes *ter* or termination sites.

Table 4-2. Prokaryotic and eukaryotic repicons.

Organism	Number of replicons	Average length (kb)	Fork movement (bp/min)
1. Bacteria (*E.coli*)	1	4200	50,000
2. Yeast (*S. cervisiae*)	500	40	3,600
3. Fruit fly (*D. melanogaster*)	3,500	40	2,600
4. Toad (*Xenopus laevis*)	15,000	200	500
5. Mouse (*Mus musculus*)	25,000	150	2,200
6. Bean (*Vicia faba*)	35,000	300	No known

8. Proteins involved in Opening of DNA Helix

The following three types of proteins are needed to help the DNA double helix to open and to provide exposed DNA template for the DNA polymerase to copy:

(i) DNA helicases. DNA helicases are A TP-dependent unwinding enzymes which promote separation of the two parental strands and establish replication forks that will progressively move away form the origin. DNA helicases hydrolyze ATP when they are bound to single strands of DNA. Hydrolysis of A TP can change the shape of a protein molecule in a cyclic manner that allows the protein to perform mechanical work. DNA helicases utilize this principle to move rapidly along a DNA single strand; when they encounter a region of double helix, they continue to move along their strand, thereby unwinding the helix (Fig. 4.15). Unwinding of the template DNA helix at a replication fork could in principle be catalyzed by two DNA helicases, acting in concert, one running along the leading strand and the other along the lagging strand.

(ii) Helix-destabilizing strand. (also called **single strand DNA-binding proteins** or **SSBPs**). Behind the replication fork, the single DNA strands are prevented from rewinding about one another (or forming doublestranded hair-pin loops in each single strands) by the action of SSB proteins. SSB proteins bind to exposed DNA strands without covering the bases, which, therefore, remain available for the templating process.

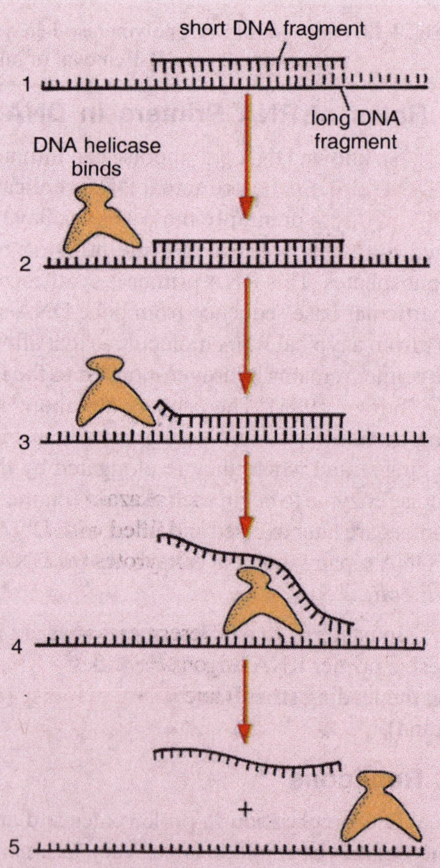

Fig. 4.15. Mode of action of DNA helicase (after Alberts *et. al.,* 1989).

(iii) Topoisomerases (DNA gyrases). The action of a helicase introduces a positive supercoil into the duplex DNA ahead of the replication fork. Enzymes, called **topoisomerases**, relax the supercoil by attaching to the transiently supercoil duplex, nicking one of the strands and rotating it through the unbroken strand. the nick is then resealed. thus, a DNA topoisomerase can be viewed as a "reversible nuclease" that adds itself covalently to a DNA phosphate, therey breking a phosphodiester bond in the DNA strand. Because the covalent linkage that joints a topoisomerase to DNA phosphate retains the energy of the broken phosphodiester bond, the brekage reaction is reversible; resealing is rapid and does not require additional energy input. The rejoining mechanism is different from DNA ligase enzyme.

One type of topoisomerase (*i.e.*, **topoisomerase I)** cause a single-strand break or nick which allows the two sections of DNA helix on either side of nick to rotate freely relative to each other, using the phosphodiester bond in the strand opposite the nick as a swivel point. A second type of topoisomerase (*i.e.*, **topoisomerase II)** forms a covalent bond to both strands of DNA helix at the same time, making transient **double-strand break** in the helix.

9. Replisome and Primosome

N.K. Sinha and **A.K.ornberg** have suggested that the DNA polymerases, RNA primases and helicases may be associated with one another to form a multienzyme complex — the **replisome** that carries out the synthesis of leading and lagging strands in **a coordinated fashion.** Such a complex would be highly processive and assure rapid replication of the DNA (Fig. 4.16).

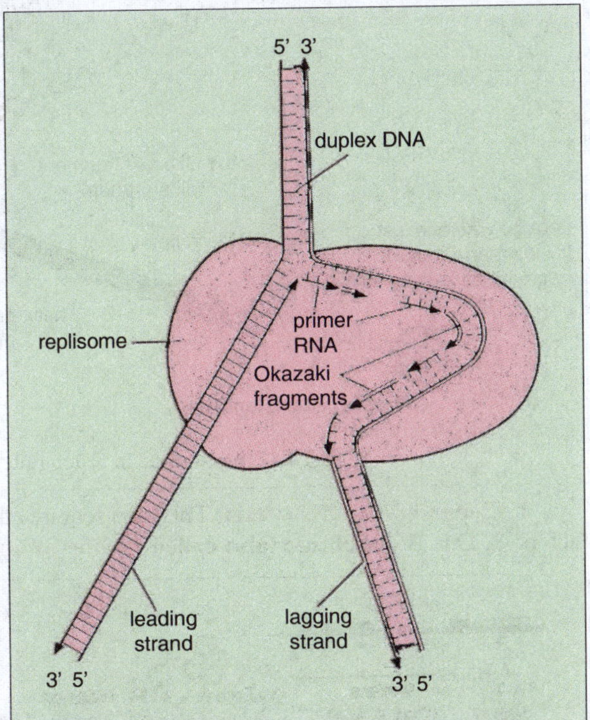

Fig. 4.16. Model of the replisome (after Sheeler and Bianchi, 1987).

Likewise, the proteins at a replication fork cooperate to form a replication machine, *e.g.*, the primase molecule is linked directly to the DNA helicase to form a unit on the lagging strand called a **primosome** which moves with the fork, synthesizing RNA primers as it moves (Fig. 4.17).

MECHANISM OF DNA REPLICATION IN PROKARYOTES

In vitro DNA replication has been extensively studied in *E.coli* and in the phages and plasmids of *E.coli.* In *E.coli,* the process of DNA replication involves the following three main steps:

1. Initiation of DNA replication. This process comprises three steps: (i) recognition of the origin (O), (ii) opening of DNA duplex to generate a region of single stranded DNA, and (iii) capture of **Dna B protein** (*i.e.*, 5′→3′ helicase; also acts as the activator of primase). thus, Dna-A (or initiator protein)-ATP complex binds at 9 bp inverted repeat regions (R_1, R_2, R_3, R_4) of ori C of *E. coli* and promotes opening of the DNA duplex in a region of three direct repeats of 13-bp sequence (called **13-mers).** The opening occurs from right 13-mer leftwards and requires negatively supercoiled DNA and HU or IHF initiator proteins (Fig. 4.18). Dna B (=helicase) is transferred to exposed single stranded DNA (Fig. 4.19) and causes unwinding of the DNA in the presence of ATP, SSB protein and **DNA gyrase** (a topoisomerase). This results in unwinding of DNA duplex and the replication from ori C proceeds in both directions (bidirectional); SSB binding occurs on single stranded regions and two Dna B complexes (= primosomes are loaded one on each strand.

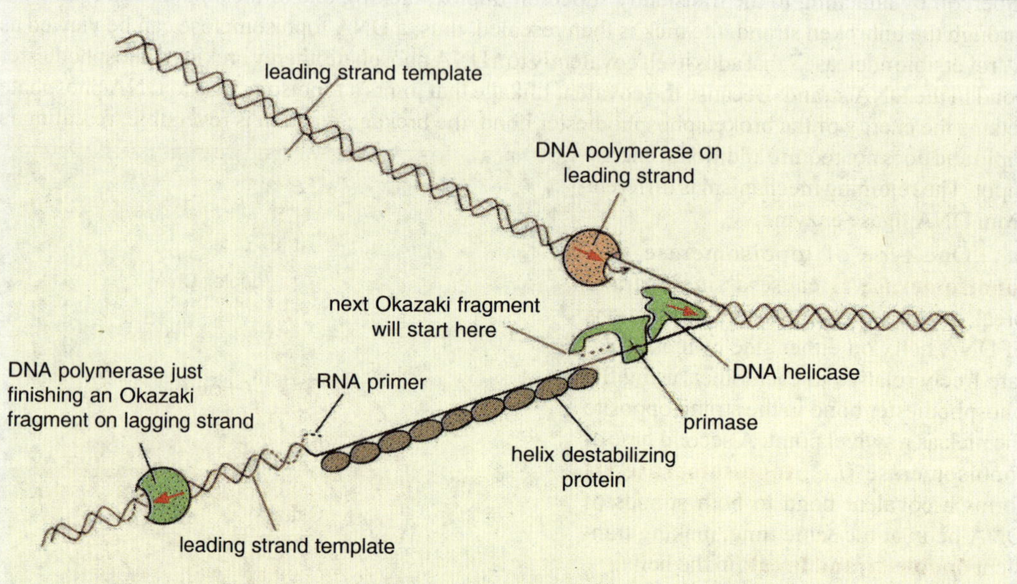

Fig. 4.17. primosome in action (after Alberts *et al,* 1989).

2. Elongation of DNA chain. This step requires the presence of the following enzymes and factors: 1. Dna B or helicase (also called **mobile promoter);** 2. primase (Dna G); 3. DNA polymerase holoenzyme (or DNA pol III HE); 4. SSB protein; 5. RNAse H which removes RNA rpimers; 6. DNA polymerase I which is used for filling the gap created due to RNA primers and 7. DNA ligase (which converts pimerless okazaki fragments into continuous strand). During initiation to elongation transition, the following envents occur:

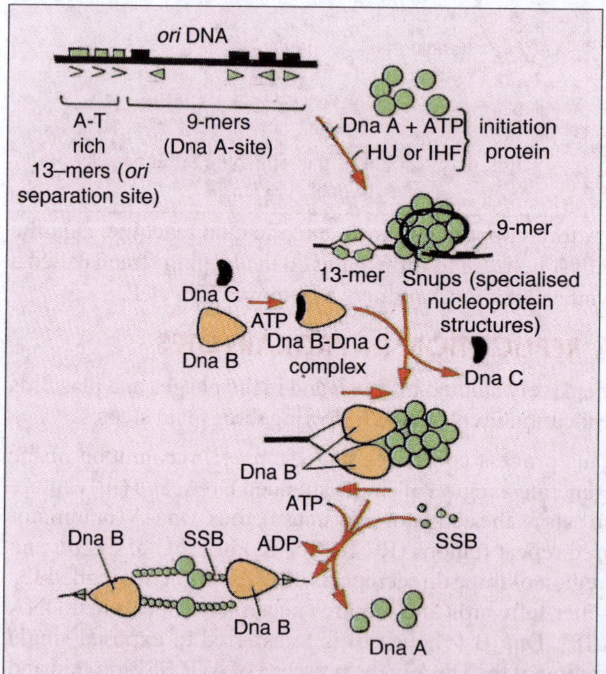

Fig. 4.18. Mode of the initiation of DNA replication at ori C of *e. coil.*

1. As helicase (or Dna B) travels in 5′ → 3′ direction, it generates a replication fork by opening the DNA duplex.

2. The DNA strand having helicase becomes the lagging strand. DNA primase associates with DNA B helicase, forming the primosome which synthesizes multiple primers for lagging strand and single RNA primer for the leading strand.

3. For the synthesis of lagging strand, the DNA poll III He has to work on the same strand to which DNA B helicase is bound, but it travels in opposite direction.

4. Dna B helicase, Dna G primase and DNA poll III HE work together in strand elongation. Helicase and DNA polymerase assembly remains *processive, i.e.,* they remain tighgly bount to the fork and stay bound throughout the reaction.

Synthesis (= elongation) of lagging and leading strands takes place by somewhat different methods; it is far more complex for lagging strand than for the leading strand.

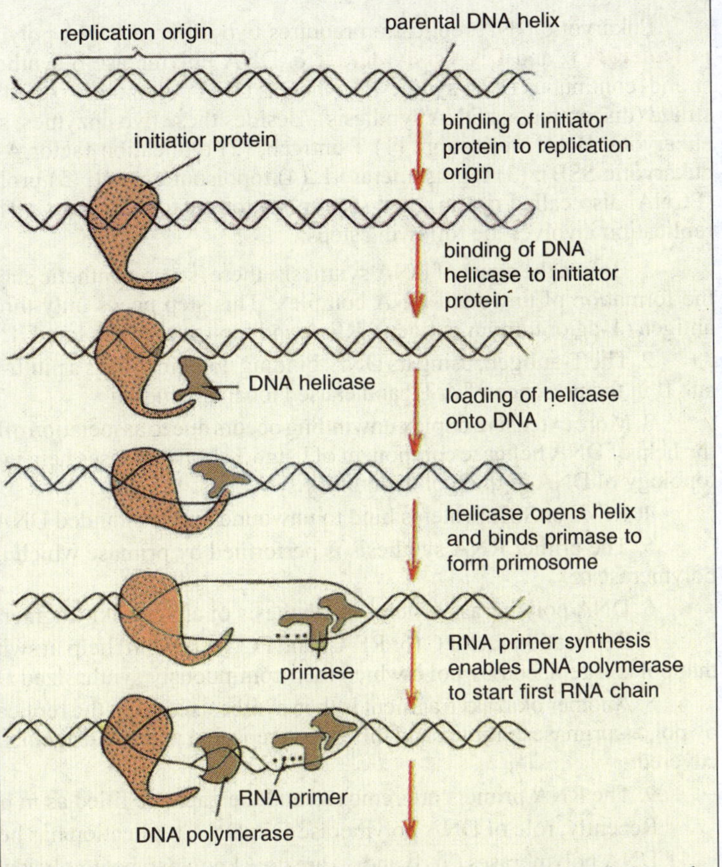

Fig. 4.19. Initial steps leading to formation of replication forks at the *E.coli* (after Alberts *et at.,* 1989).

(A) Discontinuous synthesis on lagging strand. 1. Primase is taken up from solution and is activated by helicase (DnaB) to synthesize a RNA primer (10 to 20nt or nucleotides long) on the lagging strand. 2. The RNA primers are recognized by DNA poll III HE on the lagging strand and are utilized for synthesis of precursor or okazaki fragments. In fact, each new RNA primer is recognized by the gamma (g) subunit of DNA pol III HE and loaded with b subunit of the same polymerase. This preloaded b subunit may then capture the *core* of DNA poly III HE when it becomes available after finishing its synthetic job on the preceding okazaki fragment. 3. On completion of the okazaki fragments, the RNA primers are excised by DNA polymerase I, which then fills the resulting gaps with DNA. 4. After DNA **polymerase I** adds the final deoxyribonucleotides in the gap left by the excised primer, the enzyme **DNA ligase** forms the phosphodiester bond that links the free 3′ end of the primer replacement of the 5′ end of the okazaki fragment.

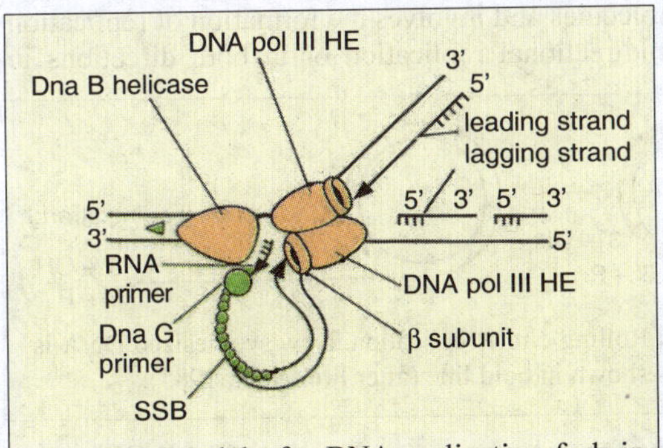

Fig. 4.20. Model of a DNA replication fork in prokaryotes during elongation.

B. Continuous synthesis on leading strand. (1) In bidirectional DNA replication, the leading strand is primed once on each of the parental strands. (2) The RNA primer of the leding strand is synthesized by RNA polymerase enzyme. (3) DNA poll III HE causes elongation of the leading strand and finally DNA pol I and ligase enzymes give final touch to the leading strand as in case of the lagging strand.

DNA Replication in Eukaryotes

Eukaryotic DNA replication requires two different DNA polymerase enzymes, namely **DNA polymerase** α and **DNA polymerase** δ. DNA polymerase δ synthesizes the DNA on the leading strand (continuous DNA synthesis), whereas DNA polymerase α synthesizes the DNA on the lagging strand (discontinuous DNA synthesis). Besides these two enzymes, six more factors are involved in eukaryotic DNA replication: (1) T antigen; (2) replication factor A or RF-A (also called RP-A or eukaryotic SSB); (3) topoisomerase I; (4) topoisomerase II; (5) proliferating - cell nuclear antigen (PCNA, also called cyclin), and (6) replication factor Cor RF-C. The process of eukaryotic DNA replication involves the following steps:

1. Before the onset of DNA synthesis, there is a presynthetic stage of 8-1 0 minutes duration for the formation of unwound DNA complex. This step needs only three purified proteins, namely T antigen (T-ag or tumour antigen), RF-A and topiosomerases I and II.
2. The T-antigen, using its DNA-binding domain, forms a multi-subunit complex with site I and site II in the presence of A TP and caused local unwinding.
3. More extensive duplex unwinding occurs due to association ofRF-A and a topoisomerase with the help of DNA helicase component ofT-ago Topoisomerases help in unwinding of DNA by altering topology of DNA at the replication fork.
4. RF-A or SSB proteins bind to unwound single stranded DNA.
5. The primer RNA synthesis is performed by primase which is tightly associated with DNA polymerase α.
6. DNA polymerase α helps in synthesis of an okazaki fragment in 5' to 3' direction.
7. Replication factor C (or RF-C) and PCNA (cyclin) help in switching of DNA polymerases so that pol α is replaced by pol δ which then continuously synthesized DNA on the leading strand.
8. Another okazaki fragment is then synthesized from the replication fork on the lagging strand by pol α - primase complex and this step is repeated again and again, till the .entire DNA molecule is covered.
9. The RNA primers are removed and the gaps are filled as in prokaryotic DNA replication.

Recently, role of DNA polymerase E in DNA replication has been stressed upon, so that three DNA polymerases (α, β and ε) are now known to be involved in eukaryotic DNA replication. **A. Sugino** and coworkers have proposed that DNA polymerase a might function at both the leading and lagging strands (since polymerase a has a primase activity), whereas polymerase E and polymerase δ are involved in elongation of the leading and lagging-strands respectively.

MODELS OF DNA REPLICATION

The following three models have been proposed for DNA replication in different organisms:

1. Replication Fork Model

It occurs both in linear and circular DNA molecules and involves the formation of replication forks which either move in one direction in unidirectional replication or in both directions in bidirectional replication.

2. Rolling Circle Model

It occurs during viral DNA replication and DNA replicating in *E. coli* during mating. According to this model, by some initiation event, a nick is made in the duplex circle and this nick has 3'-OH and 5' -ptermini (Fig. 4.21). Under the influence of a helicase and SSB protein a replication fork is generated. Synthesis of a primer is unnecessary because of the 3'-OH group, so leading-strand synthesis proceeds by elongation from this terminus. At the same time, the parental

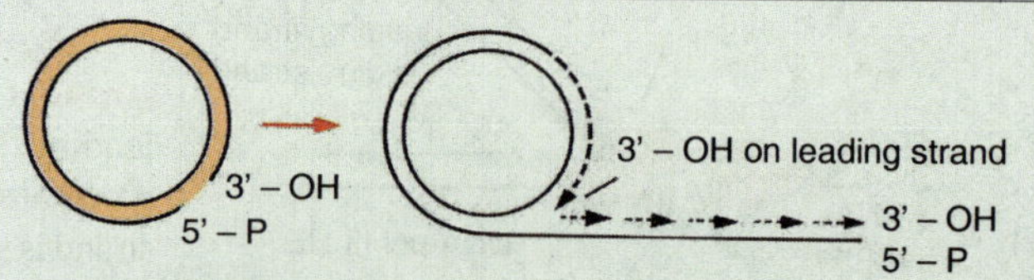

Fig. 4.21. Rolling circle replication. New synthesized DNA is shown in bold line (after Freifelder, 1985).

template for lagging-strand synthesis is displaced. The polymerase used for this synthesis is polymerase III. The displaced parental strand is replicated in the usual way by means of precursor fragments. The result of this mode of replication is a circle with a linear branch; it resembles the Greek letter sigma and is called σ **repliation** or **rolling circle replication.**

REVISION QUESTIONS AND PROBLESM

1. Describe the two classical experiments which demonstrated the semi conservative mode of DNA replication.
2. From what substrates is DNA polymerized? What properties do all known DNA polymerase share?
3. State whether each of the following is true or false:
   (a) In the synthesis of DNA the covalent bond forms in between a 3'-OH and 5'-P group.
   (b) In general, the DNA replicating enzyme in *E. coli* is DNA polymerase I.
   (c) A single strand of DNA can be copied ifthe four nucleoside triphosphates and polymerase I are provided.
   (d) If polymerase I is added to the four nucleoside triphosphates without a DNA template, DNA is synthesized but with a random base sequence.
   (e) An RNA primer must be complementary in base sequence to some region of the DNA to initiate DNA synthesis.
4. What is meant by the terms primer and template?
5. What is the role of RNA in DNA replication?
6. Distinguish the roles ofhelicases and SSB proteins in DNA replication.
7. Describe the different steps involved in the initiation of DNA synthesis in *E.coli.* Discuss the roles of different enzymes or proteins in this process.
8. Describe various steps of DNA replication in eukaryotes. Elaborate the specific role oftwo different DNA polymerases for the leading and the lagging strands.
9. Describe rolling circles and D-Ioops for DNA replication.
10. If one strand of DNA is found to have the sequence 5' A A C G T ACT G C 3\ what is the sequence of nucleotides on the 3',5' strand?
11. For a molecule of n-deoxyribonucleotide pairs, give a mathematical expression that can be used to calculate the number of possible sequences of those nucleotide pairs if only the "usual" bases are present.
12. Develop a formula for determining the length in micrometers of a DNA molecule whose number of deoxyribonucleotide pairs is known.
13. Phage T2 DNA is estimated to consist of about 200,000 deoxyribonucleotide pairs. What is the length in micrometers of its DNA complement?
14. Write short notes on the following:
    (i) Density centrifugation; (ii) Meselson and Stahl experiment; (iii) DNA polymerases; (iv) Unwinding proteins; (v) Replicons; (vi) Reverse transcription; (vii) Theta configuration; and (viii) Okazaki fragments.

ANSWER TO PROBLEMS

2. Deoxynucleoside triphosphates. Addition to 3'-OH group; requirement for a template.
3. (a) and (e) are true.
4. Primer means a nucleotide bound to DNA and having a3'-OH group; template means a polynucleotide strand whose base sequence can be copied.
5. RNA serves as a primer.
6. A helicase unwinds a helix. The SSB proteins prevent the helix from rewinding and prevent intramolecular base pairing from occurring.
10. 3' TIGCA TGACG 5' because of pairing qualities of deoxyribonucleotides.
11. 4^n.
12. $L\ \mu m = 3.4 \times 10^{-4}$ P, where $L\ \mu m$ represents the length in micrometers, and P representsthe number of nucleotide pairs.
13. 68 μm.

CHAPTER

5

Non-Genetic Ribonucleic Acid (RNA) and Transcription

In the organisms (*viz.*, prokaryotes and eukaryotes), where coded genetic informations are contained in the DNA molecule, different genetically controlled functions of their cells are performed by a different kind of nucleic acid, called **non-genetic ribonucleic acid (RNA).** In the cells of such organisms, the DNA molecule occurs in the chromosomes of nucleus or nucleoids of eukaryotes or prokaryotes respectively, and the process of protein synthesis occurs in the cytoplasm. It is investigated that a DNA molecule does not leave the nucleus to participate directly in the process of protein synthesis but employs different types of non-genetic ribobucleic acid molecules for carrying its genetic informations from the nucleus to the site of protein synthesis, *i.e.*, ribosomes. Thus, gene expression is accomplished by the transfer of genetic information from DNA to RNA molecules and then from RNA to protein molecules. RNA molecules are synthesized by using the base sequence (triplet codons) of one strand of DNA as a template in a polymerization reaction that is catalyzed by enzymes called **DNA-dependent RNA poly-**

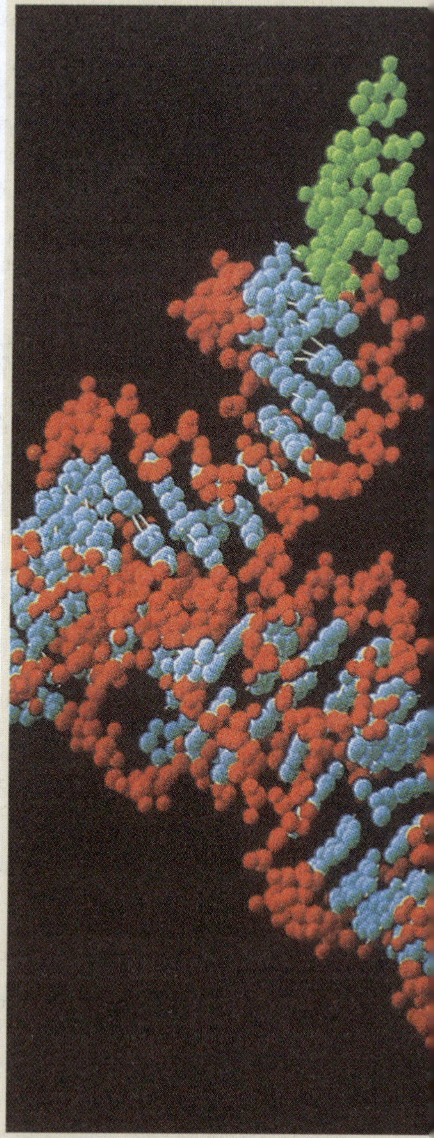

The molecular adaptor – t RNA.

merases or simply **RNA polymerases**. The process by which RNA molecules are initiated, elongated, and terminated is called **transcription**.

CHEMICAL COMPOSITION OF NON-GENETIC RIBONUCLEIC ACID (RNA)

Chemically, the non-genetic RNA is closely related with DNA. However, the non-genetic RNA is a single stranded polymer of ribonucleotide units and is composed of phosphoric acid, ribose (pentose), sugar and nitrogen bases, which are purines (adenine and guanine) and pyrimidines (cytosine and uracil). The molecular weights of the various RNA molecules vary from about 25,000 to over one million. Its base composition does not follow the base composition of DNA, although, in general, the proportion of A+C roughly equals G+U, as has been shown in Table 5-1.

Table 5-1. **Base ratios of RNA from various sources (as molar percentages) (From, Hall *et al.*, 1974).**

Source	Adenine	Guanine	Cytosine	Uracil
1. *Allium cepa* (onion seed)	24.9	29.8	24.7	20.6
2. *Phaseolus vulgaris* (bean seed)	24.9	31.4	24.1	19.6
3. *Cucurbita pepo* (pumpkin seed)	25.2	30.6	24.8	19.4
4. Ox liver	17.1	27.3	33.9	21.7
5. Yeast	25.4	24.6	22.6	27.4
6. *E. coli*	25.3	28.8	24.7	21.2

Further, some RNA molecules contain significant proportions of some methylated bases and an unusual nucleoside known as pseudouridine (ψU), in which the glycosidic bond is associated with position 5 of uracil rather than position 3.

COMPARISON BETWEEN DNA REPLICATION AND TRANSCRIPTION

The replication and transcription are two chief activities of DNA and they should be compared at the outset in the following manner :

1. In DNA replication an enzyme (or enzyme complex) matches up complementary deoxyribonucleoside triphosphates with a DNA template according to Watson-Crick pairing rules and the bases are polymerized to form a daughter DNA molecule. In DNA transcription a different enzyme, called **DNA-dependent RNA polymerase** mediates a similar process, except that complementary ribonucleoside-triphosphates are matched with the DNA template and RNA polymers are formed. The base uracil (U) replaces thymine in RNA but, otherwise, the DNA strand is faithfully copied. The RNA copy has the opposite polarity from the template, so that, for example, the sequence $\xrightarrow[\text{TACAAC}]{}$ in DNA is transcribed as $\xleftarrow[\text{AUGUUG}]{}$ in RNA.

2. The RNA transcripts do not ordinarily remain hydrogen-bonded with their DNA templates as DNA daughter strands do. Instead, they "peel off" the template as they are formed, thus, becoming available to participate in protein synthesis.

3. DNA of prokaryotes (*E.coli*) and eukaryotes has one or few initiation points for DNA replication, so that, at least, large portions of genome are copied into single, enormous daughter DNA molecules. In contrast, the synthesis of RNA molecules is initiated at close intervals along the DNA and, thus, relatively short RNA copies are produced.

4. Lastly, replication and transcription have to serve two different functions. The purpose of replication is to conserve the entire genome for next generation, whereas the purpose of transcription is to make RNA copies of individual genes that the cell can use in the biochemistry. The copies are themselves not endowed with the same kind of permanence as the genetic material; instead, they are typically degraded by cellular nucleases, once their functional usefulness has been spent.

MECHANISM OF PROKARYOTIC TRANSCRIPTION

Transcription involves the following three aspects : 1. The enzymatic synthesis of RNA; 2. The signals that determine at what points on a DNA molecule transcription starts and stops; and 3. The types of transcription products and how they are converted to the RNA molecules needed by the cell.

1. Enzymatic Synthesis of RNA

The essential chemical characteristics of the synthesis of RNA are the following :

(1) The precursors in the synthesis of RNA are the four ribonucleotide 5'-triphosphates (rNTP)—ATP, GTP, CTP and UTP. On the ribose portion of each NTP, there are two OH groups—one each on the 2'- and 3'- carbon atoms.

(2) In the polymerization reaction a 3'-OH group of one nucleotide reacts with the 5'-triphosphate of a second nucleotide; a pyrophosphate is removed and a phosphodiester bond results (Fig. 5.1)

(3) The sequence of bases in RNA molecule is determined by the base sequence of the DNA. Each base, added to the growing end of the RNA chain, is chosen by its ability to base-pair with the DNA strand used as template; thus, the bases C,T,G and A in a DNA strand cause G, A, C, and U respectively, to appear in the newly synthesized RNA molecule.

(4) The DNA molecule being transcribed is double-stranded, yet in many particular regions only one strand serves as a template (Fig. 5.2).

(5) The RNA chain grows in 5'→3' direction; that is, nucleotides are added only to the 3'-OH end of the growing chain. (This is the same as the direction of chain growth in DNA synthesis). The RNA molecule is terminated by a 5'-triphosphate at the non-growing end. The RNA molecule is antiparallel to the DNA strand being copied. Once initated, RNA chains grow at a rapid rate—40 to 50 nucleotides per second in *E. coli* at 37ºC.

(6) RNA polymerases, in contrast with DNA polymerases, are able to initiate chain growth; that is, no primer is needed.

(7) Only ribonucleoside 5'-triphosphates participate in RNA synthesis and the first base to be laid down

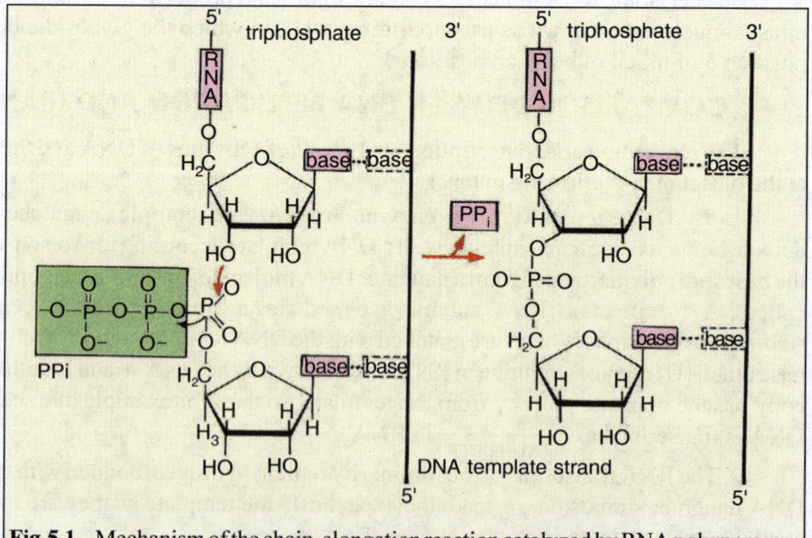

Fig 5.1. Mechanism of the chain-elongation reaction catalyzed by RNA polymerase. The broken arrows join the reacting groups. The pyrophosphate group (PP) and the bold hydrogen do not appear in the RNA strand. The DNA template and the RNA strands are antiparallel as in double-stranded DNA (after Freifelder, 1985).

in the initiation event is a triphosphate. Its 3'-OH group is the point of attachment of the subsequent nucleotide. Thus, the 5' end of a growing RNA molecule terminates with a triphosphate.

The overall polymerization reaction can be written as follows :

$$_{n}\text{NTP} + \text{XTP} \xrightarrow[\text{Mg}^{2+}]{\text{DNA, RNA-P}} (\text{NMP})_n\,\text{XTP} + \text{n PP}_1$$

in which XTP represents the first nucleotide at the 5' terminus of the RNA chain, NMP is a mononucleotide in the RNA chain. RNA-P is RNA polymerase, and PP_1 is the pyrophosphate released each time a nucleotide is added to the growing chain. The Mg^{2+} is required for all nucleic acid polymerization reactions.

The RNA Polymerase Enzyme

In *E. coli* and other prokaryotes a single RNA polymerase (RNA-P) enzyme is responsible for the synthesis of all kinds of RNAs (such as mRNA, tRNA and rRNA). RNA-P is, in fact, one of the largest enzymes known (MW 490,000). It consists of six subunits (*i.e.*, polypeptide chains)—two identical **alpha** (**α**) subunits and one chain of each of **beta (β)**, **beta dash (β′)**, **omega** (**ω**) and **sigma** (σ) subunits (Fig. 5.3). Some characteristics of each subunit of RNA-P has been tabulated in Table 5-2.

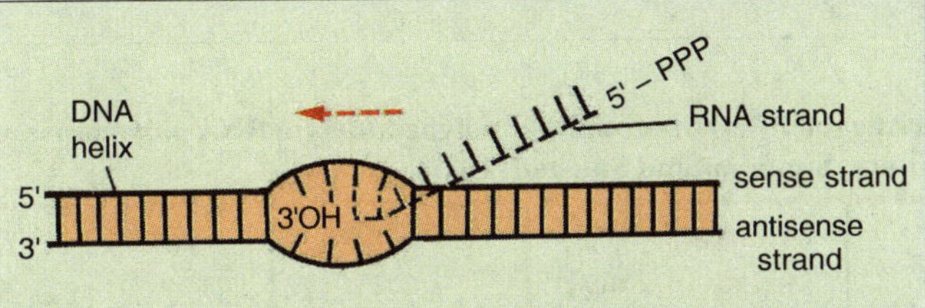

Fig. 5.2. An RNA strand is copied only from strand A (sense strand) of a segment of a DNA molecule. No RNA is copied from strand B (antisense strand) in that region of the DNA molecule. However, elsewhere, for example, in a different gene, strand B might be copied; in that case strand A would not be copied in that region of the DNA. The arrow shows the direction of RNA chain growth (after Freifelder, 1985).

The complete RNA polymerase enzyme is termed **holoenzyme** and can be represented as $\alpha\,_2\beta\beta'\omega\sigma$ in which attachment of sigma (σ) subunit (or factor) is not very firm, but resultant **core enzyme** ($\alpha_2\beta\beta'$) does not lose its catalytic activity of transcription. The active sites of core enzyme have been shown in Figure 5.4. Functions of different polypeptide subunits of RNA polymerase are now known, but not in any detail. Thus, β and β′ subunits form the catalytic centre of RNA-P and help RNA polymerase in unwinding of DNA molecule for the purpose of transcription. The sigma (σ) factor helps in the recognition of start signals on DNA molecule and directs RNA polymerase in selecting the initiation sites (promoter). Once RNA synthesis is initiated and RNA molecule becomes 8-9 bases long, the σ factor dissociates from the holoenzyme and then the core enzyme brings about elongation of mRNA (or any other RNA).

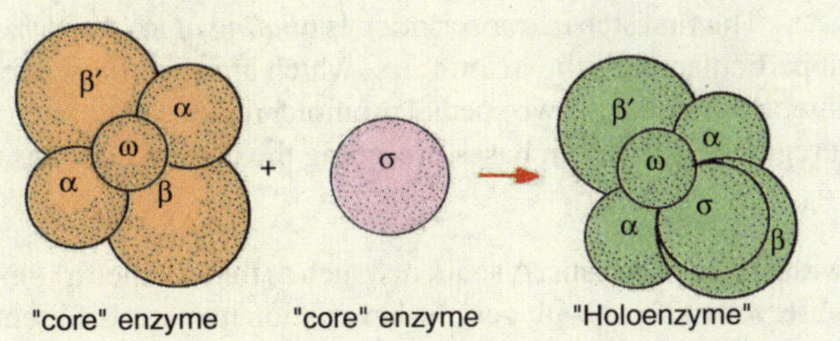

Fig. 5.3. A model of the structure of prokaryotic RNA polymerase showing association of 6 subunits or polypeptides (after Gardner *et al.*, 1991).

2. Binding of RNA Polymerase to Promoter, Initiation, Elongation and Termination

Unlike replication, transcription does not progress along the entire length of a chromosome. Instead, certain parts of the chromosome are transcribed. Only one of the two strands of a DNA duplex is transcribed; this strand is called the **sense strand**. The other strand of DNA which is not transcribed at the moment, is called **anti-sense strand**. Like DNA, RNA is synthesized in the 5→3' direction from the single-stranded region of the DNA template. This localized unwinding moves along the molecule followed by recoiling of the helix behind the newly synthesized RNA. The region of sense strand of DNA which is actually transcribed into RNA, is called the **coding region**.

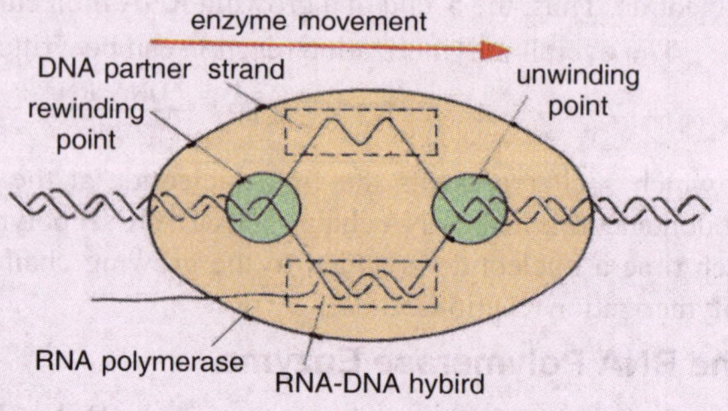

Fig. 5.4. Active centres in the core enzyme of bacterial RNA polymerase enzyme.

Table 5-2. Some basic characteristics of different subunits (= polypeptides) of RNA polymerase of *E.coli.* (Source : Gardner, Simmons and Snustad, 1991).

Subunit		Gene	Molecular weight (daltons)	Location	Function
1.	α_2	*rpo A*	41,000 each	Core enzyme	Promoter binding
2.	β	*rpo B*	155,000	Core enzyme	Nucleotide binding
3.	β'	*rpo C*	165,000	Core enzyme	DNA template binding
4.	ω	-	12,000	Core enzyme	—
5.	σ	*rpo D*	95,000	Sigma factor	Initiation

The first step in transcription is *binding of RNA polymerase to a DNA molecule*. Binding occurs at particular sites, the **promoters**, which are specific sequences of 20 to 200 bases at which several interactions occur. Two special promoter regions have been identified that appear in all organisms. In a region of five to ten bases preceding the coding region is a sequence of seven bases that reads :

TATAATG

with minor variations. A sequence such as this is called a **consensus sequence**, because it is a sequence observed to occur with very little variation in many different organisms. In bacteria (*e.g.*, *E.coli*) this region is called the **Pribnow box** and in the eukaryotes the same region has the sequence :

TATAAAT

and is called the **Hogness box**, each after the person who originally described the region. This region is generally referred to as the **TATA box** and is believed to orient the RNA polymerase enzyme, so that synthesis proceeds from left to right. It is also the region at which the double helix opens to form the **open promoter complex**. Further, the centre of pribnow box lies usually 10 bp upstream of coding region, it contains T and is called **–10 sequence** which is recognized by RNA polymerase during the binding reaction.

Another important region, further upstream from TATA box, is located approximately thirty-five bases upstream from the coding region (= mRNA start). This is called **–35 sequence** (also called

recognition sequence) and it consists of a nine-base consensus sequence, considered to be actual site of binding of the RNA polymerase. It seems likely that the sigma (σ) subunit first binds to the –35 sequence in a highly specific interaction and then, the appropriate region of this huge enzyme can come in contact with the –10 sequence of Pribnow box (Fig. 5.5).

CCAGGCTTTACACTTTATGCTTCCGGCTCGTATGATTGTGTGG**A**ATTG
CTTTTTGATGCAATTCGCTTTGCTTCTGACTATAATAGACAG**G**GTAA
GGCGGTGTTGACATAAATACCACTGGCGGTGATACTGAGCAC**A**TCAG
GTGCGTGTTGACTATTTTACCTCTGGCFGGTGATAATGGTTGC**A**TGTA
ATTGTTGTTGTTAACTTGTTTATTGCAGCTTATAATGGTTAC**A**AATA
GCTAACACTTTACAGCGCCGCGTCATTTGATATGATGCGCCCC**G**CTT

–35 Sequence **mRNA start**

Fig. 5.5. Base sequences in the noncoding strand of six different *E. coli* promoters, showing the three important regions: mRNA start, Pribnow box (including –10 sequence) and –35 sequence (after Freifelder, 1985).

The open-promoter complex is a highly stable complex and is the active intermediate in chain initiation. In this complex a local unwinding ("melting") of the DNA helix occurs starting about ten base pairs from the left end of the Pribnow box and extending to the end of the position of the first transcribed base. This melting is necessary for pairing of the incoming rebonucleotides. The base composition of the sequence of Pribnow box (which is A+T rich) makes the DNA strand open to denaturation. Apparently, RNA polymerase induces this conformational change.

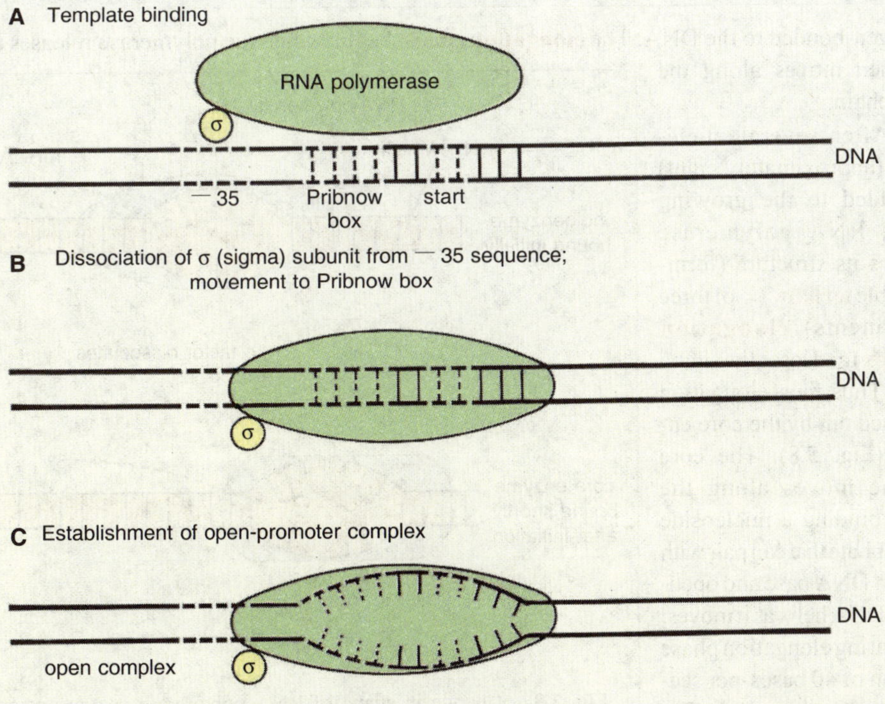

Fig. 5.6. A model for the binding of RNA polymerase to a promoter to form an open-promoter complex; PB=Pribnow box (after Freifelder, 1985).

Once an open-promoter complex has formed, RNA polymerase is ready to initiate RNA synthesis. RNA polymerase contains two nucleotides binding sites, called the **initiation site** and the **elongation site**. The initiation site binds only purine triphosphates, namely ATP and GTP, and one of these (usually ATP) is the first nucleotide in the growing RNA chain. Thus, the first DNA base that is transcribed is usually thymine (T). The initiating nucleoside triphosphate binds to the enzyme in the open-promoter complex and forms a hydrogen bond with the complementary DNA base (Fig. 5.7). The elongation site is then filled with a nucleoside triphosphate that is selected strictly by its ability to form a hydrogen bond with the next base in the DNA strand. The two nucleotides are then joined together, the first base is released from the initiation site, and initiation is completed. The dinucleotide remains hydrogen-bonded to the DNA. The **elongation phase** begins when the polymerase releases the base and then moves along the DNA chain.

RNA polymerase binds to promoter, slides into place, and forms an open complex. ATP in initiation site binds to T on coding strand.

II. A NTP is added to elongation site and is covalently linked to the A.

III. RNA polymerase moves to the next DNA base. The initiating dinucleotide is released. A NTP enters the elongation site and is covalenty linked to the dinucleotide. Then movement of RNA polymerase continues.

Fig. 5.7. Method of initiation of transcription. The enzyme is drawn without the sigma factor (after Freifelder, 1985).

After several nucleotides (approximately eight) are added to the growing chain, RNA polymerase changes its structure (forming stable **ternary** (= of three components) **elongation complex**) and loses the sigma factor. Thus, most elongation is carried out by the core enzyme (Fig. 5.8). The core enzyme moves along the DNA, binding a nucleoside triphosphate that can pair with the next DNA base and opening the DNA helix as it moves; thus, during elongation phase addition of 40 bases-per second at 37°C takes place. The open region extends only over a few base pairs; that is, the

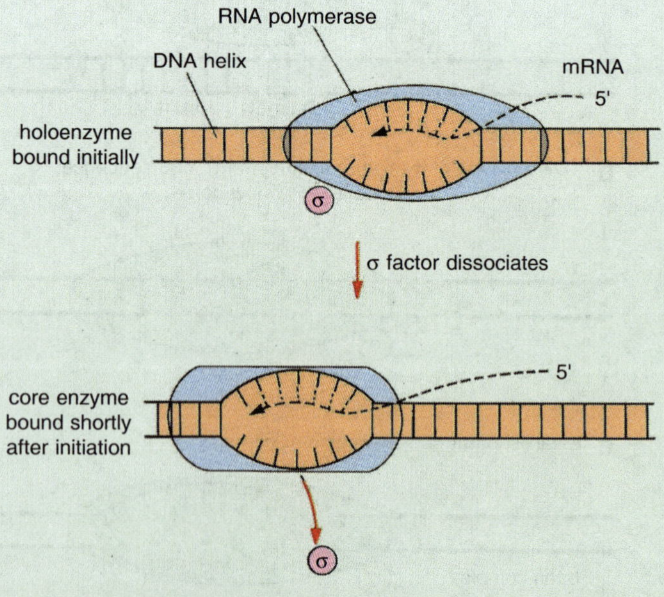

Fig. 5.8. Change in shape of RNA polymerase enzyme when the sigma factor gets dissociated from the core enzyme (after Freifelder, 1985).

DNA helix recloses just behind the enzyme. The newly synthesized RNA is released from its hydrogen bonds with the DNA as the helix reforms; however, a few RNA bases remain paired with the DNA template during RNA synthesis. It should be noted that the promoter itself is not transcribed.

Termination of RNA synthesis (or transcription) occurs at specific base sequences in the DNA molecule. Twenty termination sequences have so far been determined and each has the characteristics shown in Figure 5.9. Termination region consists of the following three important regions : 1. First, there is an inverted repeat base sequence containing a central non-repeating segment: that is, the sequence in one DNA strand would read like –

ABCDEF—XYZ—F′E′D′C′B′A′

in which A and A′, B and B′ and so on are complementary bases. Thus, this sequence is capable of intrastrand base pairing, forming a "stem-and-loop" configuration in the transcript (RNA) and possibly in the DNA strands. 2. The second region is near the loop end of the presumed stem (sometimes totally within the stem) and is a sequence having a high G+C content. 3. A third region (sometimes absent) is a sequence of A.T pairs that yields in the RNA a sequence of six to eight uracils (U) often followed by adenine.

In fact, there are two types of termination events : those that depend only on the DNA base sequence and those that require the presence of termination protein, called *rho* (ρ ; discovered by **J. Roberts**, 1969).

The final step in the termination process is dissociation of the core enzyme from the DNA. Following this event, the core enzyme interacts with a free σ (sigma) factor to reform the holoenzyme which becomes available for initiating RNA synthesis again.

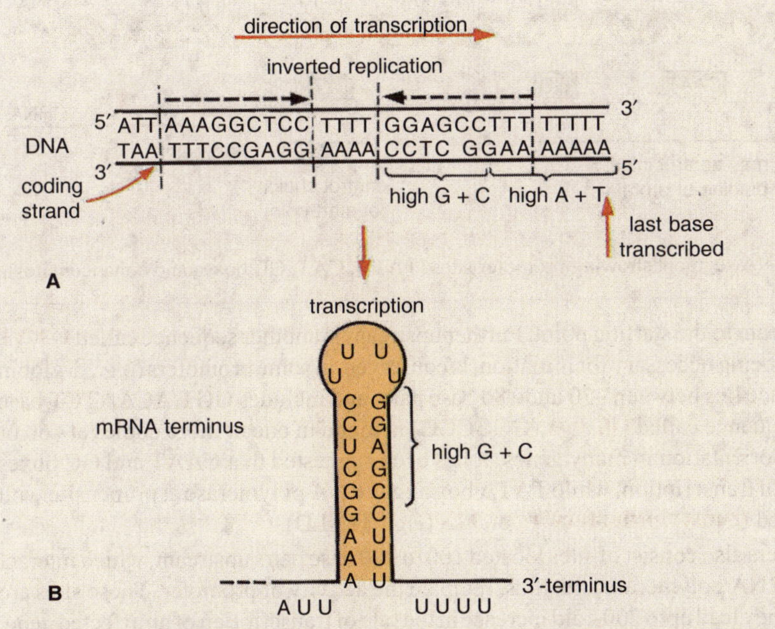

Fig. 5.9. Base sequence of (A) the DNA of the *E. coli trap* operon at which transcription termination occurs and of (B) the 3′ terminus of the mRNA molecule. The mRNA molecule is folded to form a stem-and-loop structure thought to exist (after Freifelder, 1985).

3. Classes of RNA Molecules and Processing

This step of prokaryotic transcription will be explained later on along with that of eukaryotes.

MECHANISM OF EUKARYOTIC TRANSCRIPTION

In eukaryotes, there are three major classes of RNA polymerases which are designated as I, II and III and are found in the nucleus. The three polymerases have different properties (see Table 5-3) and can be distinguished by the ions required for their activity, the optimal ion strength and their sensitivity to inhibition by various antibiotics (*e.g.*, α -amanitin). Each of these enzymes is a large protein (~ 500,00 daltons), with two large and several (8 to 10) smaller subunits.

Table 5-3. Proper.ties of three different eukaryotic nuclear RNA polymerases.

Enzymes	Location	Product and abundance	Sensitivity to α - amanitin
1. RNA polymerase I	Nucleolus	rRNA (50–70%) (except 5S rRNA)	Not sensitive
2. RNA polymerase II	Nucleoplasm	hnRNA (mRNA) (20–40%)	Sensitive
3. RNA polymerase III	Nucleoplasm	tRNA(~ 10%) (and 5S rRNA)	Inhibited in animals at high levels; not in yeast and insects.

Promoter, enhancer and silencers. The nature and function of RNA polymerase II (for hn RNA) is well studied than other two types of RNA polymerases. The promoters of genes of RNA polymerase II contain three distinct regions which are centred at sites lying between –25 bp and –100 p (Fig. 5.10). The least effective of these three regions is the TATA or Goldberg-Hogness box (7 bp

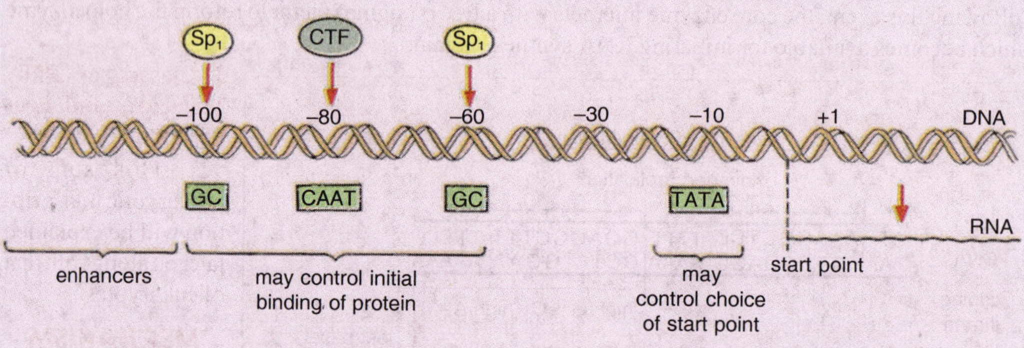

Fig. 5.10. An eukaryotic DNA segment showing promoter sites (TATA, CAT, GC boxes and enhancer sites).

long) located 20 bp upstream to the starting point. Further upstream is another sequence called **CAAT box** (or **CAT box**) which being necessary for initiation, is conserved in some promoters (*e.g.*, β-globin gene). CAAT box sequence lies between –70 and –80 base pairs and includes GGT/ACAATCT base composition. Another sequence called **GC box** (GGGCGG) is found in one or more copies at –60 or –100 bp upstream in any orientation in many genes. It has been suggested that CAAT and GC boxes determine the efficiency of transcription, while TATA box aligns RNA polymerase at proper site with the help of proteins, called **transcription factors** or **TFs** (*e.g.*, TF II D).

Eukaryotic promoters also consist of sites located 100 to 200 base pairs upstream, which interact with proteins other than RNA polymerase and, thus, regulate the activity of promoter. These sites are called **enhancers**, since they lead upto 200-fold increase in the rate of transcription of an affected gene. Examples of enhancers are known in the genome of viruses (SV40) and eukaryotes (*i.e.*, in the genes for immunoglobulin, insulin, alpha amylase, etc.; **Picard**, 1985). In the spacer regions between *Xenopus* large ribosomal genes are multiple regions 60–80 bp long that confer a 20-fold increase in transcription rate compared with genes lacking them (see **Maclean** and **Hall**, 1987).

There are other regulatory sites known as **silencers** which repress gene expression. Both enhancers and silencers can function at great distance (often many kilobases) from the genes they enhance and repress respectively. Silencers are known to occur in yeast and repress expression of *HML* and *HMR* loci involved in switching of mating types.

On the other hand, in each of the 5S RNA genes which is transcribed by RNA polymerase III, the promoter lies in the middle of transcription unit, 50 bp downstream from the start point (Fig. 5.11). Such internal promoters also occur in the genes for different types of RNA polymerase III enzymes. The polymerase III enzyme is big enough to occupy start point (+1bp) and the promoter region (+55 and +80) simultaneously and starts transcribing the start point without any apparent difficulty. The downstream promoters sequences of 5S rRNA and tRNA genes have subsequently been characterized more distinctly into **box A**, **box B** and **box C**. In 5S RNA genes box A and box C are found at +59 to +69 and +80 to +90 sequences, respectively. Likewise, in tRNA genes, box A and box B are located at +8 to +30 and +51 to +72 sequences, respectively. Box A of 5S RNA and tRNA genes contain similar conserved sequences and is recognized by the same transcription factor (TF III B).

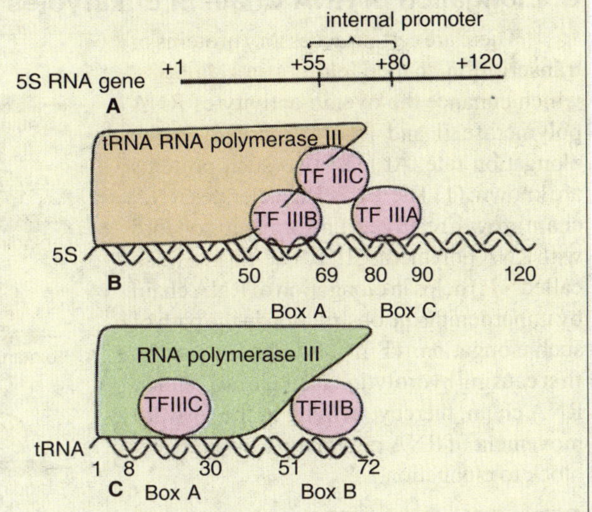

Fig. 5.11. A– Structure of internal promoter of a 5S RNA gene. B–Binding of TF III B to downstream promoters (box A), located at different positions in 5S and tRNA genes.

The eukaryotic transcription too involves the following three main steps: initiation, elongation and termination.

A. Initiation of Eukaryotic Transcription

For the eukaryotic transcription the regulatory DNA sequences (such as promoters, enhancers and silencers) for genes transcribed by each of the three RNA polymerases differ. Various transcription factors are also involved in the formation of a transcription complex which are needed for initiation of transcription. Generally, each of RNA polymerase is believed to have its own set of transcription factors, however, TF II D or a part of it (*e.g.*, TBP=TATA binding protein) is required for all the three RNA polymerases. The **transcription factors** (**TFs**) can be defined as proteins, which are needed for initiation of transcription, but are not part of the RNA polymerase. They help in DNA binding of a RNA polymerase to constitute the so-called **pre-initiation complex** or **transcription complex**. After the formation of this complex initiation of transcription occurs. All known transcription factors may recognize either DNA sequences, another factor or RNA polymerase.

Formation of transcriptosome with RNA pol II. A promoter sequence which is responsible for constitutive expression of common genes (also **called house keeping genes**) in all cells, is called **generic promoter.** The generic promoter cannot bring about regulated expression (i.e., tissue or stimulus specific expression of genes, called **luxary genes**). Initiation of transcription on the generic promoter by RNA polymerase II requires the action of diverse transcription factors (TFs) in the following order : (i) **TF II D** binds at TATA box; (ii) the step (i) permits the association of **TF IIA** and **TF IIB**; (iii) TF II B forms the so-called DB complex and RNA polymerase II associates to promoter site; (iv) RNA pol II is accompanied to the promoter by **TF II F** to form a transcription complex ; (v) orderly addition of **TF II E, TF II H** and **TF II J** helps the initiation process.

B. Elongation of RNA Chain in Eukaryotes

There are certain accessory proteins of transcription, called **elongation factors**, which enhance the overall activity of RNA polymerase II and lead to increase in the elongation rate. At least two such proteins are known: (1) The **TF II F** accelerates RNA chain growth relatively uniformly in concord with RNA polymerase II. 2. The **TF II S** (also called **S II**) helps in elongation of RNA chain by unburdening the obstruction in the path of such elongation. TF II S is known to act by first causing hydrolytic cleavage at 3′ end of RNA chain, thereby, helping in the forward movement of RNA polymerase through any block to elongation.

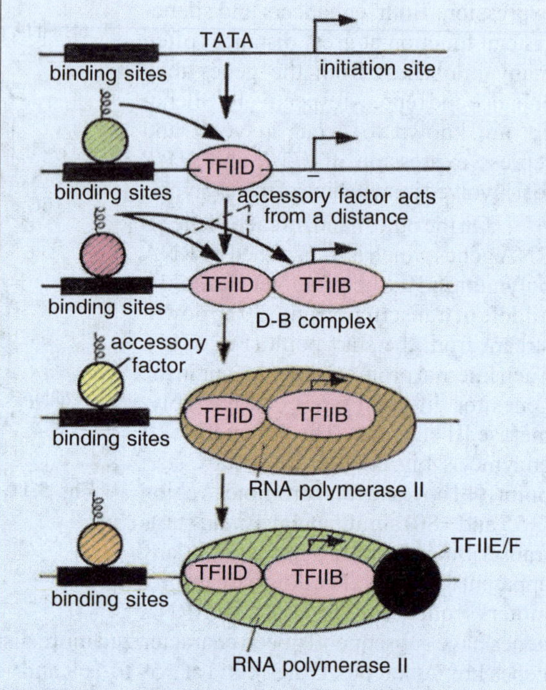

Fig. 5.12. The mode of formation of transcription complex at the promoter for RNA polymerase II.

C. Termination of Eukaryotic Transcription

In eukaryotes, the actual termination of RNA polymerase II activity during transcription may take place through termination sites similar to those found in prokaryotes. However, the nature of individual sites is not known. Such termination sites are believed to be present away (sometimes up to one kilobase away from the site of the 3′end of mRNA).

Chromatin Structure and Transcription

Decondensation of large chromosomal domains (coiled and supercoiled) is a prerequisite for transcripition. This process involves the following steps: 1. Some **activator proteins** bind and bring about the formation of nucleosome-free regions. These activator proteins gain access to specific DNA sequence in chromatin by the help of non-histone proteins in an unknown manner. The amino-terminal region of H3 and H4 histones may also undergo **acetylation**, which is often correlated with great amount of transcription. 2. In the so-generated nucleosome-free DNA, additional transcription factors and RNA polymerase bind. Transcription is, thus, initiated.

TYPES OF NON-GENETIC RNA AND PROCESSING

According to their specific functions during the process of protein synthesis, the following kinds of non-genetic RNA molecules have been recognized in prokaryotic and eukaryotic cells :

(1) Ribosomal RNA (rRNA)

Ribosomal RNA (rRNA), stable or **insoluble RNA** constitutes the largest part (up to 80%) of the total cellular RNA. It is found primarily in the ribosomes although, since it is synthesized in the nucleus it is also detected in that organelle. It contains four major RNA bases with a slight degree of methylation and shows differences in the relative proportions of the bases between species. Its molecules appear to be single polynucleotide strands which are unbranched and flexible. At low ionic strength, rRNA behaves as a random coil, but with increasing ionic strength the molecule shows helical regions produced by base pairing between adenine and uracil and guanine and cytosine.

Types and synthesis of rRNA. The eukaryotic cells have four kinds of rRNA molecules, namely **28S rRNA** (the sedimentation constant varies between 25S and 30S depending on the species), **18S rRNA, 5.8 S** and **5S rRNA**. The 28S rRNA, 5.8 S and 5S rRNA occur in 60S ribosomal subunit, while 18S rRNA occurs in 40S ribosomal subunit of 80S ribosomes of eukaryotes. The prokaryotic cells contain three kinds of rRNA molecules, namely **23S rRNA, 16S, rRNA** and **5S rRNA**. The 23S rRNA and 5S rRNA occur in 50S ribosomal subunit, while 16S rRNA occurs in 30S ribosomal subunit of 70S ribosomes of prokaryotes.

In bacteria, the genes for 5S, 23S and 16S rRNAs are clustered in one region to form a single operon working as a functional unit. The synthesis of rRNA molecules is initiated at promoter and completed at a terminator sequence.

In eukaryotes, rRNA genes (which belong to multigene families) occur in the region of nucleolar organizer (NO). The number of these genes may vary from 50 to 30,000 in a cell and this number may be unequally distributed on NOs, if more than such loci are present. The DNA comprising these genes is called **rDNA** (=ribosomal DNA) which is repetitive in nature. Each repeat unit has (i) a **coding region**, in which the genes for the 18S, 5.8S and 28S rRNA molecules exist next to each other in the order mentioned; (ii) a spacer region called **intergenic spacer** (**IGS**) and (iii) **internal transcribed spacers** (**ITS**), one each between 18S and 5.8S genes and another between 5.8S and 28S genes. For the synthesis of above mentioned rRNAs, a transcription initiation factor (TIFI) is needed which brings RNA polymerase I to the promoter region. A primary transcript is made by all three rRNA genes, including spacer regions between the genes. The transcript is then processed into the functional rRNA molecules.

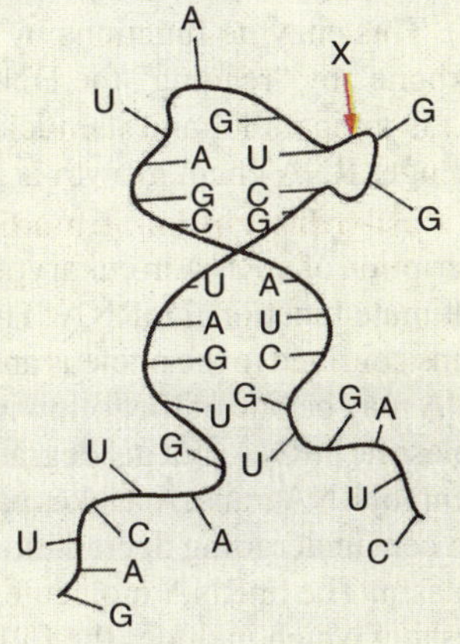

Fig. 5.13. The molecular (*i.e.*, secondary) structure of rRNA showing a helical region with complementary base pairing and a looped outer region of the helix at X (Davidson, 1972).

Further, the 5S RNA genes are located outside the nucleolar organizer. However, in prokaryotes and yeast, 5S RNA genes are present in close vicinity of rDNA. The 5S RNA genes are also organized in tandem repeats, each repeat consisting of a gene 120 bp long and a spacer region. The length of the complete repeat is 375 bp in *Drosophila*.

(II) Messenger RNA (mRNA)

The RNA molecules which are transcribed from large number of genes of the total genome (*i.e.*, 99 per cent genes of the total genome of *E. coli*) and have base sequence complementary to DNA, carry DNA's genetic informations for the assembly of amino acids into the polypeptide chains (protein molecules), to the cytoplasmic sites of protein synthesis, the ribosomes, to which they become associated to participate in codon-anticodon interaction with tRNA, are called **informational** or **messenger RNAs** (**mRNA**). The name messenger RNA has been proposed by **Jacob** and **Monod** (1961). The molecule of a mRNA is single-stranded like the rRNA molecule and it is DNA-like in its base composition so that GC contents of mRNA correspond to the GC contents of the genomes total DNA.

mRNA synthesis in bacteria. Messenger RNA is **complementary** to chromosomal DNA; it forms RNA-DNA hybrids after separation of the two DNA strands. Synthesis of mRNA is accomplished with only one of the two strands of DNA, which is used as template. The enzyme RNA polymerase joins the ribonucleotides, thus, catalyzing the formation of 3'-5'-phosphodiester bonds that form the RNA backbone. In this synthesis the AU/GU ratio of RNA is similar to the AT/GC ratio of DNA. The mRNA synthesis is initiated at 5' end and direction of growth is from the 5' end to 3' end. In bacteria, the RNA polymerase attaches to an initiator site of the structural gene, in the promoter and it catalyzes mRNA synthesis until termination site is reached.

In bacteria, the process of transcription of mRNA is simultaneous with translation, *i.e.*, as soon as the mRNA is being transcribed by RNA polymerases the ribosomes become attached to the mRNA to initiate protein synthesis.

mRNA synthesis in eukaryotes. Transcription of eukaryotic DNA to produce mRNA begins with the synthesis of long precursor molecules by RNA polymerase II from the template strand of DNA. In an average cell nucleus, there is only one molecule of RNA polymerase II per 750 nucleosomes - worth of DNA, *i.e.*, one enzyme molecule exists per 150,000 base pairs of DNA (**Maclean** and **Hall**, 1987). This enzyme functions by catalyzing formation of 5'→3' phosphodiester bonds of the RNA "backbone" by "reading" the DNA template in the 3'→5' direction. The developing mRNA (or hn RNA) is antiparallel and its nucleotides are complementary to those of the DNA template strand. Messenger RNA chain growth is rapid—from 15 to 100 nucleotides per second *in vitro*.

Post-transcriptional modification of processing of mRNA. The immediate product of transcription of mRNA in eukaryotes is a molecule of many more ribonucleotides than that comprising the ultimate functional mRNA. This primary transcript may range from 500 to 50,000 nucleotides; it remains confined to the nucleus and is called **heterogeneous nuclear RNA** (**hnRNA**). The fate of this hnRNA may be one of the followings : 1. RNA transcripts of some genes do not seem to give rise any cytoplasmic mRNA, but get degraded within the nucleus. 2. For each gene, only a small proportion (25 per cent) of RNA transcript takes part in RNA processing leading to formation of mRNA, the remaining 75 per cent undergoing degradation in the nucleus. Thus, only 5 per cent hn RNA (by mass) enters the cytoplasm. The hnRNA molecules which are destined to produce functional mRNA, undergo RNA processing which includes the following steps :

1. Addition of a cap of 7-MeG or m7G. During capping process, a cap of a methylated guanosine, called 7-methylguanosine (7-MeG or m^7G) , is added to 5' end of primary transcript (*i.e.*, hnRNA) in a rare 5'-5' linkage. Sometimes, this cap also includes methylation of additional sugars of both the 5' nucleotides: (7-MeG)-5′PPP-5' (G or A, with possibly methylated ribose) -3'-P— in which P and PPP refer to mono- and triphosphate groups, respectively.

Capping occurs shortly after initiation of synthesis of the mRNA, possibly before RNA polymerase II leaves the initiation site, and precedes all excision and splicing events. The biological significance of capping is that the cap may protect the mRNA from degradation by nucleases and may provide a feature for recognition by the protein-synthesizing machinery (*i.e.*, cap helps in recognition of ribosomes and thereby facilitates translation of mRNA). The 7-MeG caps are absent in mRNAs of histone proteins.

2. Addition of tail of poly-A. The 3′end of mRNA is generated in two steps (Fig. 5.14): 1. Endonuclease enzyme cuts the primary transcript at an appropriate location. 2. Poly (A) is added to the newly generated end by an enzyme, called **poly (A) polymerase**, utilizing ATP as a substrate. This step is called **polyadenylation**. Studies have shown that ordinarily ~ 30 per cent of hn RNA and ~ 70 per cent of mRNA are poly-adenylated. In a region 11 to 30 nucleotides upstream of the site of poly(A) addition, there is a sequence AAUAAAA (in all higher eukaryotes except yeast) which perhaps provides a signal for nuclease cleavage.

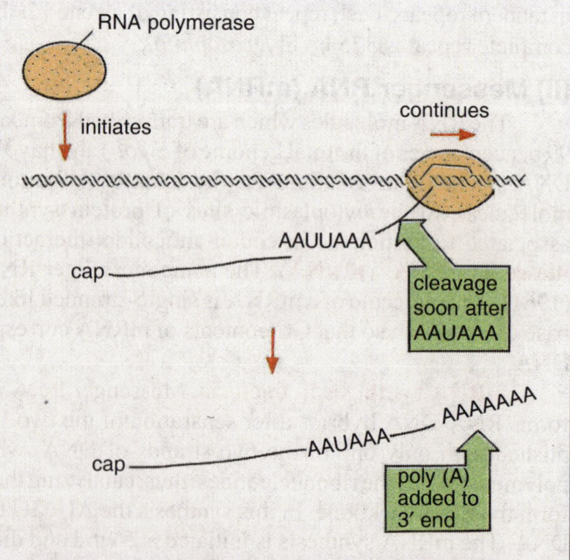

Fig. 5.14. Cleavage at AAUAAA sequence and poly adenylation at the cleaved 3' end.

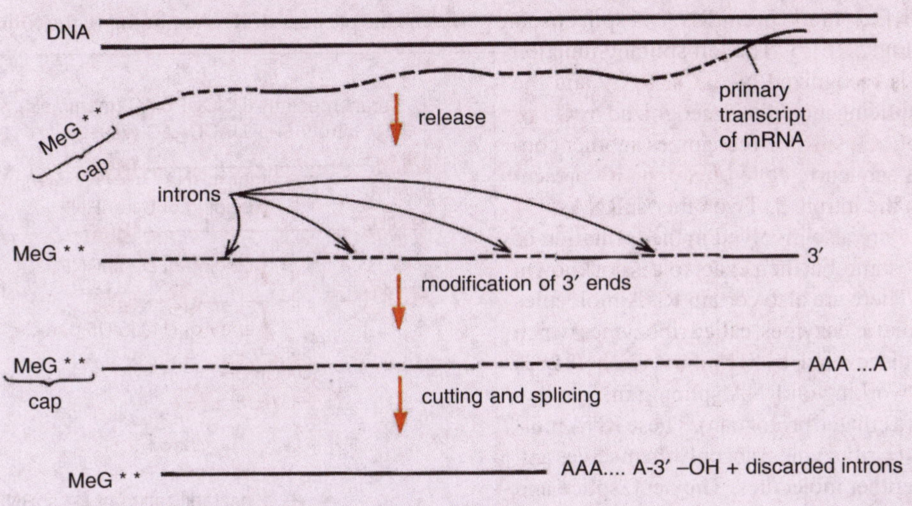

Fig. 5.15. Summary of various steps of production of mRNA. The primary transcript is capped before it is released. Then, its 3'-OH end is modified, and finally introns are excised and exons are rejoined to form functional mRNA. (MeG=7-methyl guanosine; * = nucleotide whose ribose is methylated) (after Freifelder, 1985).

In mRNAs of most histone proteins, no polyadenylation occurs at 3' ends, so that 3' ends are processed without addition of poly (A). The **U7 snRNA** (56 bases long) is involved in this processing through extensive complementary base pairing with histone hn RNAs.

3. RNA splicing. This is the controlled excision of large **intervening sequences** or **introns** from the transcript and rejoining of the remaining fragments, called **coding sequences** or **exons**, together to produce the finished mRNA. The number of introns per gene varies greatly (Table 5-5) and for a given protein is not the same in all organisms.

Table 5-5. Translated eukaryotic genes in which introns have been demonstrated (Source: Freifelder, 1985).

Gene	Number of introns
1. α- Globin	2
2. Immunoglobulin L chain	2
3. Immunoglobulin H chain	4
4. Yeast mitochondria cytochrome b	6
5. Ovomucoid	6
6. Ovalbumin	7
7. Ovotransferrin	16
8. Conalbumin	17
9. α-collagen (procollagen α)	52

The actual mechanism of cutting and splicing is not completely understood; however, it is known that the border regions of each intron usually contain similar sequences (called **consensus sequences**), usually started with a GU and ending with an AG (called **GU-AG rule**); and that a small nuclear RNA-protein particle, called the **snRNP particle**, is involved in RNA splicing. Small nuclear RNAs (sn RNAs) present in the sn RNP particles are designed as U1, U2, U4, U5 and U6. The sn RNPs form a macromolecular complex (called **spliceosome**; 40 to 60 nm in size) in association with other essential protein factors and pre-mRNA. A spliceosome performs the process of RNA splicing in the following way (Fig. 5.16): 1. The left splicing junction (GU) is recognized by **U1 snRNA** and the right splicing junction is recognized by **U5 sn RNA**. 2. **U2 snRNA** recognizes another consensus sequence, called **branch site** present within the intron. 3. Two other snRNAs, **U4** and **U6** are also involved in the formation of spliceosome, but their exact role is not known.

There are also certain RNA molecules which act as enzymes, called ribozymes which were discovered by **Thomas Cech** (1982) while working on RNA splicing in *Tetrahymena* (a ciliated protozoan). These RNA molecules (= ribozymes) act only themselves and not on other molecules. They cut, splice and assemble (= do processing of) precursor hn RNA into mRNA, or precursor rRNA into mature rRNA.

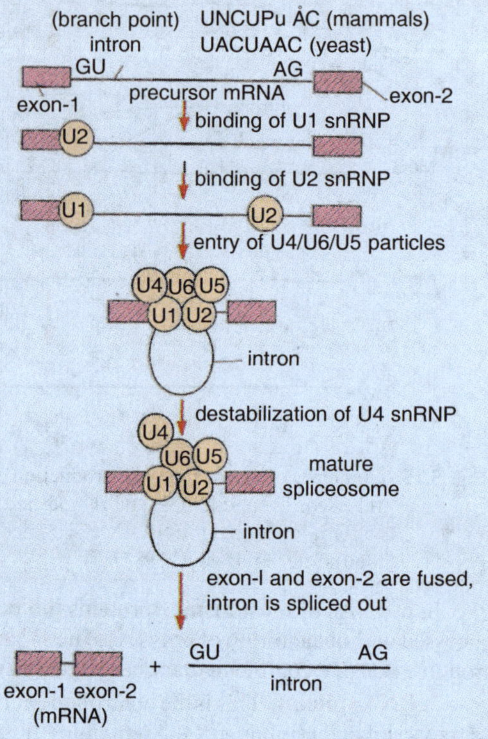

Fig. 5.16. Formation of a spliceosome and stages involved in RNA splicing.

Heterogeneity and types of mRNA. When the total mRNA population of an organism is considered, it is found to be heterogeneous in size, showing a wide range of S values of 6 to 30. This property of mRNA reflects the fact that the size or length of the mRNA molecule is directly related with the size of the codons for different protein molecules, the sizes of which may be quite variable. According to the size, the following two types of mRNA molecules can be recognized.

(a) Monocistronic mRNA. Mostly the mRNA carries the codons of single cistron (*i.e.*, codes for one complete protein molecule) of the DNA. Such mRNA molecule is called **monocistronic mRNA**. For example, for the synthesis of a polypeptide chain of 300 to 500 amino acid residues, a monocistronic mRNA of *E. coli* contains 900 to 1500 nucleotides in its molecule.

(b) Polygenic or polycistronic mRNA. Sometimes a mRNA molecule carries the codes from several adjacent DNA cistrons and become much longer in size. This type of mRNA is called **polygenic** or **polycistronic mRNA**. For example, for the metabolism of the histidine protein the cell synthesizes about 10 specific enzymatic proteins and a mRNA in this case may carry codons for all the 10 enzymes.

Life-Span of mRNA

In most prokaryotic and eukaryotic cells, mRNA has short life time. For example, the average life of mRNA of *E. coli* is about 2 minutes, because it is attacked by the cytoplasmic ribonuclease enzyme. So that, at most times, mRNA makes up only 5% of the total cellular RNA. Likewise, in most eukaryotes the average life span of mRNA is one to four hours. However, in both bacteria and eukaryotes mRNAs are known that are apparently resistant to nucleases and survive for long period of time. For example, mRNA with life time of six hours has been detected in the bacterium *Bacillus cerus* at a time when the cells are induced to become long lived spores. Likewise, in differentiating eukaryotic cells mRNAs with a life time of days have been detected. For example, in the immature red blood cells (reticulocytes) of the mammals the mRNA is synthesised originally by the nucleus in early stages and expelled to the cytoplasm. In later stages, the nuclei of maturing reticulocytes degenerate but the mRNA exists up to 2 days for prolong utilization in the synthesis of globin protein of haemoglobin. Further, in extreme cases, such as in the state of dormancy adopted by many animal eggs and plant seeds, mRNA is maintained in a stable form for months or even years.

Informosomes. In the eukaryotic cells, the stability to the mRNA is provided by certain proteins. Several investigators, *e.g.*, **Spirin**, **Beltisina** and **Lerman** (1965), **Perry** and **Kelley** (1968) and **Henshaw** (1968) have reported that in certain eukaryotic cells the mRNA does not enter in the cytoplasm as a naked RNA strand but often remains ensheathed by certain proteins. **Spirin** has coined a new term **informosome** to this mRNA and protein complex. The informosome is used by the cell when there is a delay in the translation. For instance, in the embryo the genetic expression is manifested late during organogenesis. In such cases, mRNA occurs in the form informosomes. The proteins of informosomes protect the mRNA from the degrading action of the enzyme, ribonuclease. These proteins may also control the synthesis at the level of the translation and, thus, may regulate or modulate the protein synthetic process.

(III) Transfer RNA (tRNA)

The RNA which possesses the capacity to combine specifically with only one amino acid in a reaction mediated by a set of amino acid-specific enzymes, called **aminoacyl-tRNA synthetases**; transfers that amino acid from the "amino acid pool" to the site of protein synthesis and recognises the codons of the mRNA is known as the **soluble RNA** (**sRNA**) or **transfer RNA** (**tRNA**). Thus, tRNA molecule has to perform several highly complex functions during protein synthesis—it interacts with a specific synthetase enzyme, possesses a site for binding an amino acid, possesses a second site for interacting with a ribosome, and contains an anticodon that must be exposed to the codons of mRNA.

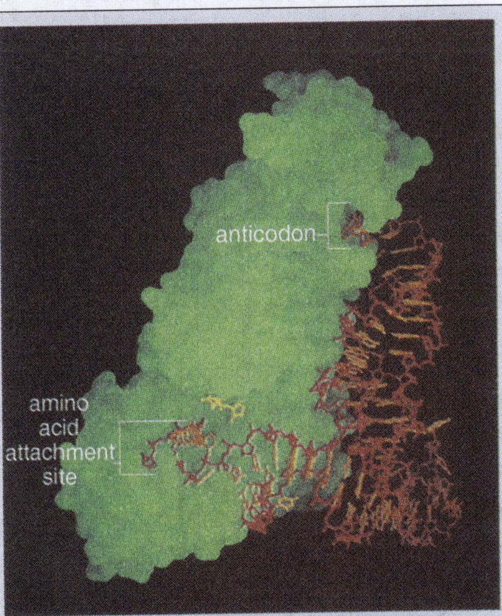

Computer-generated, three-dimensional model of one type of tRNA molecule. The tRNA (*reddish brown*) is shown attached to a bacterial enzyme (green) along with an ATP molecule (gold).

Structure of tRNA. Robert Holley (1965) and his colleagues reported the complete nucleotide sequence of alanine tRNA of yeast (**Holley** received the Nobel Prize in 1968 for his work along with **Khorana** and **Nirenberg**). Nucleotide sequences are now known for more than 100 different "species" of tRNA.

Transfer RNA has several unique characteristics: 1. It is a relatively small molecule of 75 to 90 ribonucleotides and is, thus, smaller than either mRNA or any of the rRNAs, and has a sedimentation coefficient of 4S.

2. The ratios of A:U and G:C are near unity which suggests the formation of DNA-like double helical segments (secondary structure). In these double helical segments, G:C base pairs are more common than A:U as suggested by the ratio AU:GC = 0.7

3. All tRNA molecules have a tertiary structure, the details for which are now known and Mg^{2+} ion concentration is important for its stabilization.

4. A number of "unusual" nucleotides are found in tRNA (*e.g.*, pseudouridine ψ or psi), inosine (I), dihydroxyuridine (DHU), etc.). Many of "unusual" nucleotides are methylated derivatives of common ones (*e.g.*, 1-methylguanylic acid, 1-methyladenylic acid, ribothymidylic acid and 5-methylcytosine).

The significance of these unusual bases of tRNA was understood well by molecular biologists during the construction of two-dimensional model from the primary-sequences of nucleotides of known tRNA. Thus, it was realized that most bases of tRNA pair according to Watson-Crick's pairing rule, but unusual bases fail to do so because they carry substitutions or alterations in those positions that usually participate in hydrogen bonding. Consequently, the presence of these bases forces the model builder to construct several non-base-paired loops in the tRNA molecule. By working on these lines, **R. Holley** (1965) first of all proposed a **clover leaf model** for yeast $tRNA^{ala}$. The clover leaf model of tRNA because accommodated several of the known functions of tRNA, so, it gained general acceptance. A typical clover-leaf model tRNA (Fig. 5.17) depicts the following structural peculiarities :

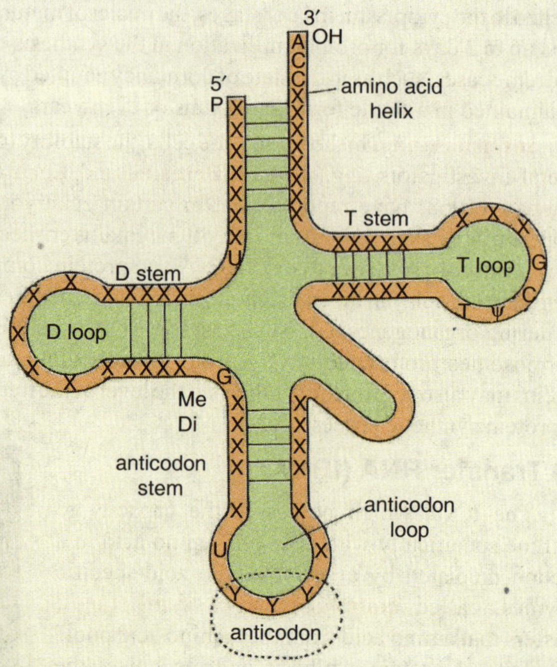

Fig 5.17. Generalized two-dimensional clover-leaf model of tRNA, based on analyses of several yeast tRNA molecules by various investigators. Note the common 3' terminal -CCA, the TCG in the T–loop, the anticodon–U (here YYYU) in the anticodon loop, the DiMeG between the anticodon loop and D loop and the U as the first unpaired base on the 5' strand. As diagrammed here, the anticodon is read from right to left 3'→5' (A=adenosine, C=cytidine, G=guanosine, T=ribothymidine, U=uridine, ψ = pseudouridine, DiMeG=dimethylguanosine, Y=any base of the anticodon (after Burns and Bottino, 1989).

(i) All tRNA molecules have guanine residue G at the 5' terminal end and unpaired (single stranded) C-C-A sequence at the 3′end. This is called **amino acid attachment site**, because the amino acid becomes covalently attached to adenylic acid or A of CCA sequence during polypeptide synthesis.

(ii) The **amino acid stem** or helix consists of seven paired bases.

(iii) The **T-stem** is composed of five paired bases—the last (*i.e.*, nearest the **T-loop** or **TψC loop** is C-G. T-loop contains seven unpaired bases and is involved in the binding of tRNA molecules to the ribosomes.

(iv) The **anticodon stem** includes five paired bases. The **anticodon loop** consists of seven unpaired bases, the third, the fourth and the fifth of which (from the 3' end of the molecule) constitute the **anticodon**. The anticodon permits temporary complementary pairing with three bases (triplet codon) on mRNA.

(v) The base on the 3' side of anticodon is a purine.

(vi) Immediately adjacent to the 5' side of the anticodon, uracil and another pyrimidine occurs.

(vii) A purine, often dimethylguanylic acid, is located in the "corner" between the anticodon stem and the D stem.

(ix) The **D- stem** is composed of three or four base pairs (depending on the "species" of tRNA). The **DHU-loop** or **D-loop** is also variable in size containing 8 to 12 unpaired bases. The D-loop helps in binding of amino-acyl synthetase.

(x) The extra arm is variable in nucleotide composition and is lacking entirely in some tRNA.

Three-dimensional structure of tRNA. In order to understand the structure-function relationship of tRNA, its three dimensional structure (TDS) was worked out by the help of X-ray crystallography study. **A.Klug**, the Nobel laureate of 1982, has contributed much to the TDS of tRNAs. **S.H.Kim** (1973) proposed a most acceptable TDS model of tRNA (*i.e.*, phenyl-alanine tRNA of yeast cells). According to **Kim**, TDS of tRNA takes the shape of letter L with a thickness of 20A^0. Each arm of the L doubled over by bonds holding complementary base together. Such an L-shape can also easily derived from two dimensional clover-leaf model.

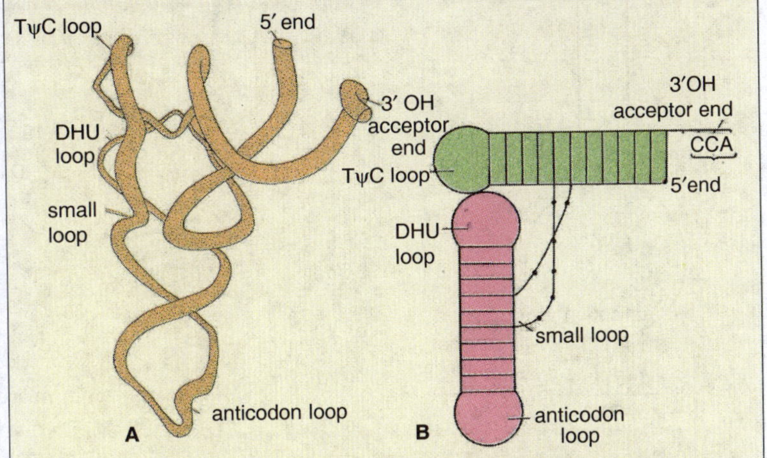

Fig. 5.18. Three dimensional structure of yeast phenylalanine tRNA. A—Actual appearance of the folding of the molecule. The polynucleotide chain is represented as a continuous coiled tube; B—The way that the clover leaf representation must be transformed in order to show the physical connections between various parts of the molecule. Two double-stranded helical regions are seen, each oriented at right angles to the other produces a L-shaped structure.

Extended anticodon hypothesis. Recently it is reported that the performance of anticodon, when isolated from tRNA, is weak and inaccurate. However, the performance of this anticodon triplet is enhanced if a matching sequence is present on the anticodon loop and on the stem on either side of anticodon triplet. As a result, an **extended anticodon hypothesis** has been proposed which suggests that the structures of anticodon loop and that of the proximal anticodon stem are related to the sequence of anticodon (**Michael Yarus**, 1982). Thus, the anticodon is extended into the nearby sequence and consists of in all 12 nucleotides arranged in the following way: (i) two nucleotides (*i.e.*, CU, ψU or UU) at the 5' side of anticodon loop; (ii) three nucleotides of anticodon (3' nucleotide of anticodon being very important and termed **cardinal nucleotide**); (iii) two nucleotides at the 3' side of anticodon loop; and (iv) five pairs of nucleotides in the anticodon stem which can be conveniently written by giving only the bases on 3' side of the stem (Fig. 5.19).

Processing of tRNA. Transfer RNA is transcribed from several particular sites on template DNA and comprised of about 15 per cent of the RNA present at any one time in an *E. coli* cell. Like the synthesis of rRNA and mRNA, synthesis of tRNA molecules is initiated at a promoter site and completed at a terminator sequence. Transfer RNAs are also processed from larger precursors. In prokaryotes, tRNA processing includes removal of nucleotides from the precursor or primary transcript and modification of some internal nucleotides. For example, in *E.coli*, the precursor for the amino acid tyrosine (pre tRNA^{tyr}) consists of 128 bases; in processing to tRNA ^{tyr} 41 nucleotides are removed from the 5' end and two are removed from the 3' end (Fig. 5.21).

In addition to processing, methylation and inclusion of other "unusual" intercalary bases takes place after transcription, as does the addition of the 3' terminal –C-C-A. The final functional tRNA molecule has 85 bases. Three-dimensional shape is achieved through hydrogen bonding. The

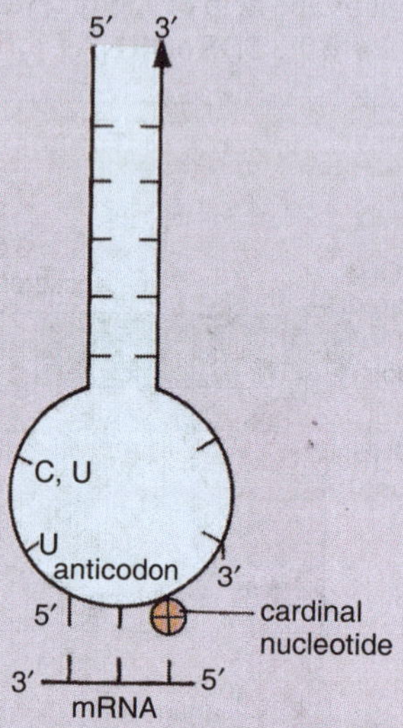

Fig. 5.19. General pattern of organization of nucleotide sequence around the anticodon, *i.e.,* in anticodon and loop and stem of tRNA.

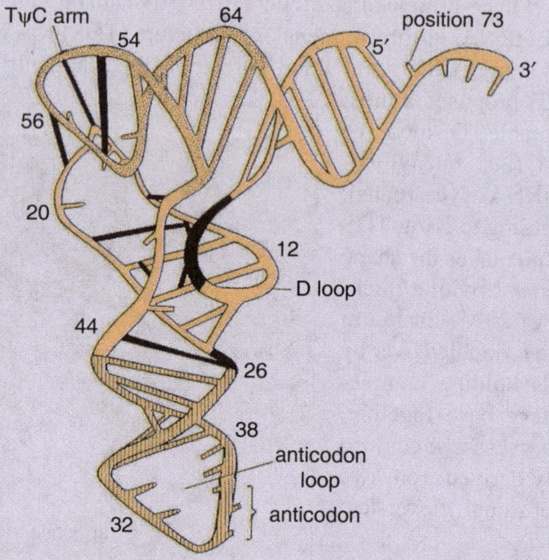

Fig. 5.20. Three dimensional model proposed by **Kim** *et al.*, (1974) for yeast phenylalanine tRNA molecule (after Burns and Bottino, 1989).

processing of eukaryotic tRNAs resembles with that of prokaryotes. However, in them it is far more complex.

In yeast, the process of tRNA splicing involves the following two steps : 1. Cutting reaction in which phosphodiester bond cleavage occurs by an **endonuclease** and this step does not require ATP. 2. Ligation reaction which requires ATP and involves bond formation with the help of **RNA ligase** (Fig. 5.22).

Genes for tRNA. There are, probably, at least 30 to 40 different tRNA genes and tRNA molecules in *E.coli.* Higher organisms are found to contain 60 tRNA molecules and 60 tRNA genes. Since the cell uses only 20 amino acids in protein synthesis (and probably only 20 synthetase enzymes), it follows that several tRNA will often have an affinity for the same amino acid. For example, *E. coli* cells contain five species of tRNA for leucine amino acid.

All the tRNA genes constitute far less than 1% of total genome in both *E.coli* and eukaryotic cells, yet some 10 to 15% of each cell's RNA may be in the form of tRNA. This discrepancy between the number of tRNA genes and gene transcripts occurs because of the following facts—(1) The tRNA molecules are

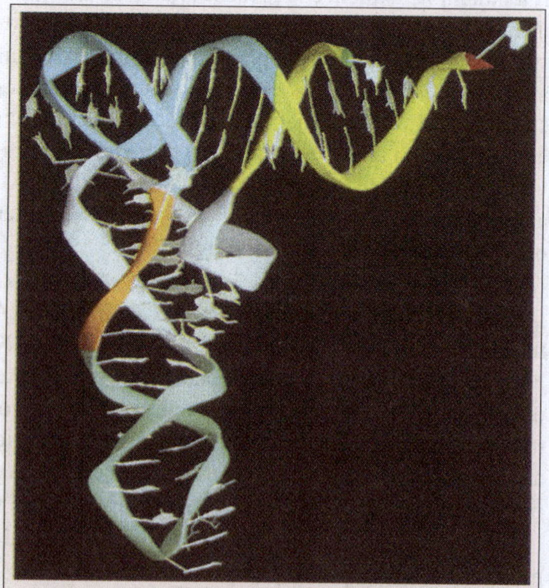

3-dimensional structure of tRNA.

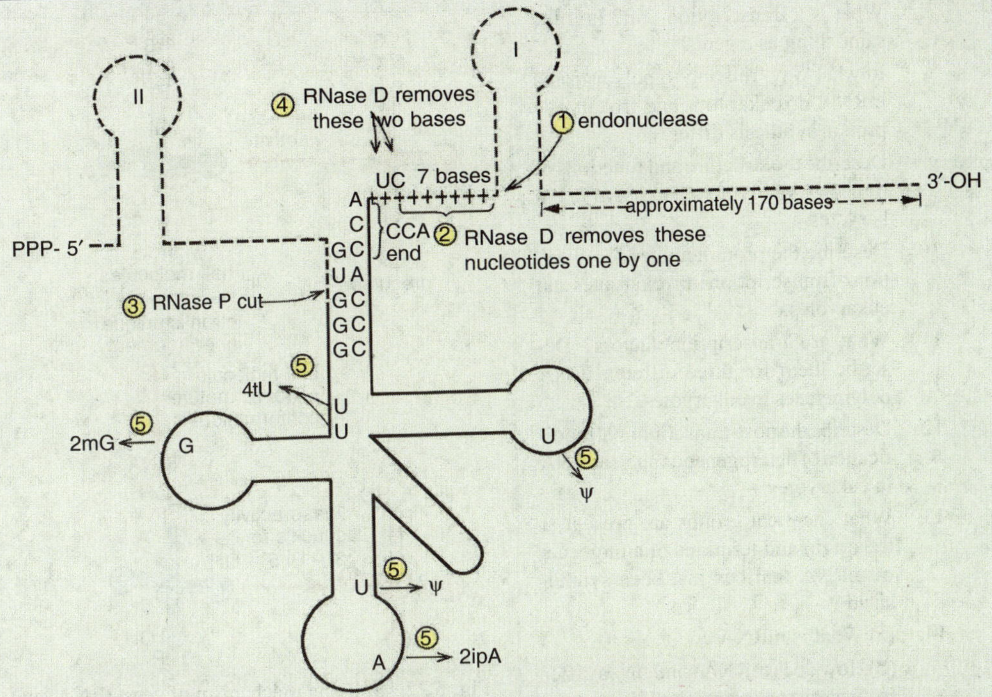

Fig. 5.21. Different stages in processing of the *E.coli* $tRNA^{Tyr}$ gene transcript. The five stages have been given Arabic numbers. Step 3 generates the 5'-P end. Step 4 generates the 3'-OH end (the CCA end). In step 5 six bases, all in or near the loops of tRNA molecule, are modified to form pseudouridine (ψ, psi), 2-isopentenyladenosine (2 ip A), 2-O-methylguanosine (2mg), and 4-thiouridine (4tU). The continuous sequence that forms the final tRNA molecule is given in black (after Freifelder, 1985).

relatively stable compared with many kinds of RNA. (2) The tRNA molecules are transcribed continuously and more quickly by tRNA genes than other RNAs because they are needed in plentiful amounts.

REVISION QUESTIONS AND PROBLEMS

1. (a) What genetic attributes does RNA share with DNA?
   (b) RNA is of three types in eukaryotes and bacteria. What are they? Where are they located in the cell? Where are they produced? What are their characteristics and functions?
2. Given a single strand of DNA.....3' TACCGAGTAC 5'...., construct (a) the complementry DNA chain, (b) the mRNA chain which would be made from strand.
3. (a) From what substrates is RNA made?
   (b) On what template?
   (c) With what enzyme?
   (d) Is a primer required?
4. Describe the differences, if any, between the chemical reactions catalyzed by DNA polymerase and RNA polymerase.
5. Describe the structure of *E.coli* RNA polymerase and discuss the roles of different components of this enzyme in RNA synthesis on DNA template.
6. What are the functions of the core enzyme and the holoenzyme *in vivo*.?

7. What is a transcription unit? Is it the same thing as a gene?
8. In what way, relating to polycistronic mRNA, do eukaryotic and prokaryotic protein synthesis differ?
9. Describe the structure and functions of three RNA polymerases known in eukaryotes.
10. Describe the promoter sites for initiation of transcription in prokaryotes and eukaryotes.
11. What are transcription factors? Describe them for three different RNA polymerases in eukaryotes.
12. Describe the post-transcriptional modification of heterogeneous nuclear RNA in eukaryotes.
13. What chemical groups are present at the origin and terminus of a molecule of mRNA that has just been synthesized ?
14. (a) What is mRNA ?
    (b) How does mRNA sometimes differ from a primary transcript?
    (c) Define coding strand and antisense strand.
    (d) Define cistron and polycistronic mRNA.
    (e) What parts of an mRNA molecule are not translated?
15. What is meant by the terms "upstream" and "downstream"?
16. What is Pribnow box? Describe its evolutionary and biochemical significance.
17. An RNA molecule is isolated having a 3'-OH terminus and a 5'-P terminus. What information does this fact provide?
18. These questions refer to eukaryotic RNA :
    (a) What is a cap?
    (b) At which end of mRNA is the poly (A)?
    (c) Are there eukaryotic mRNA molecules that do not contain either feature?
19. (a) What are intervening sequences or introns?
    (b) What is mRNA splicing?
20. Describe the structure and processing of tRNA.
21. Write short notes on the following :
    (1) Sigma factor, (2) Pribnow box, (3) TATA box, (4) Snurps and post-transcriptional cleavage, (5) Antiterminators, (6) Silencer sites, (7) Enhancer sites, (8) Ribozymes, (9) Polyadenylation, (10) Spliceosome, (11) Smart genes.

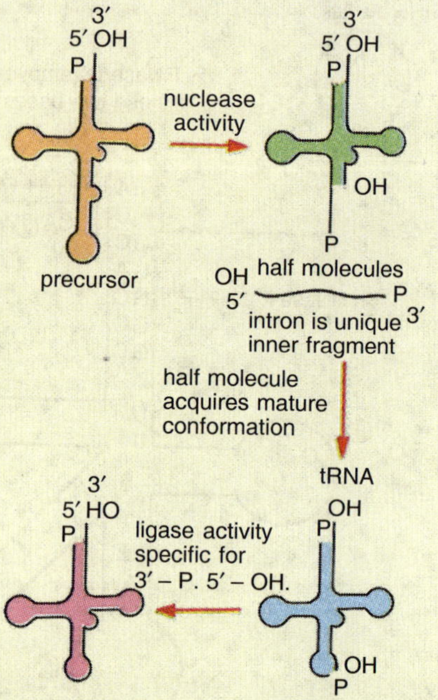

Fig. 5.22. Splicing mechanism of yeast tRNA, due to nuclease and ligase activities.

ANSWERS TO PROBLEMS

3. (a) Ribonucleoside 5'- triphosphates.
   (b) Double-stranded DNA.
   (c) RNA polymerase.
   (d) NO.

4. The reactions are identical, however, substrates are different, *i.e.*, DNA polymerase joins deoxynucleotides and RNA polymerase joins ribonucleotides.

7. A transcription unit is a section of DNA extending from a promoter to an RNA polymerase termination site. It is usually not a gene, but typically includes many genes.

8. In eukaryotes all mRNA is monocistronic.

13. A 5'-triphosphate and a 3'-OH.

14. (b) A primary transcript is a complementary copy of a DNA strand. It may contain mRNA, tRNA or rRNA and may be processed before translation can occur.

    (e) Leaders, spacers and the unnamed regions following the last stop codon of a mRNA are untranslated regions.

15. Upstream and downstream usually refer to regions in the 5' and 3' directions respectively, from a particular site that is being discussed.

17. Since the triphosphate of the primary transcript is absent, the molecule has been processed.

18. (a) A terminal structure in which a methylated guanosine is in 5'-5'- triphosphate linkage at the 5' terminus of mRNA.

    (b) The 3'-OH end.

    (c) All mRNA molecules (except those of several viruses) are capped. Some mRNA molecules lack the poly (A) tail.

19. (a) Untranslated sequences that interrupt the coding sequence of a transcript and that are removed before translation begins.

    (b) Removal of introns.

CHAPTER

6

Genetic Code

As DNA is a genetic material, it carries genetic informa tions from cell to cell and from generation to genera tion. .At this stage, an attempt will be made to determine that in what manner the genetic informations are existed in DNA molecule? Are they written in articulated or coded language on DNA molecule? If in the language of codes what is the nature of genetic code?

A DNA molecule is composed of three kinds of moieties: (i) phosphoric acid, (ii) deoxyribose sugar, and (iii) nitrogen bases. The genetic informations may be written in any one of the three moieties of DNA. But the poly-sugar-phosphate backbone is always the same, and it is, therefore, unlikely that these moiteies of DNA molecule carry the genetic informations. The nitrogen bases, however, vary from one segment of DNA to another, so the informations might well depend on their sequences. The sequences of nitrogen bases of a given segment of DNA molecule, actually, has been found to be identical to linear sequence of amino acids in a protein molecule. The proof of such a **colinearity** between DNA nitrogen base sequence and amino acid sequence in protein molecules has first obtained from an analysis of mutants of head protein of bacteriophage T_4 (**Sarabhai** *et al.*, 1964) and the A protein of tryptophan synthetase of *Escherichia coli* (**Yanofski** *et al.*, 1964). The colinearity of protein molecules and DNA polynucleotides has given the clue that the specific arrangement of four nitrogen bases (*e.g.*, A, T, C and G) in DNA polynucleotide chains, somehow, determines the sequence of amino acids in protein molecules. Therefore, these four DNA bases can be considered as four alphabets of DNA molecule. All the genetic informations, therefore, should be written by these four alphabets of DNA. Now the question

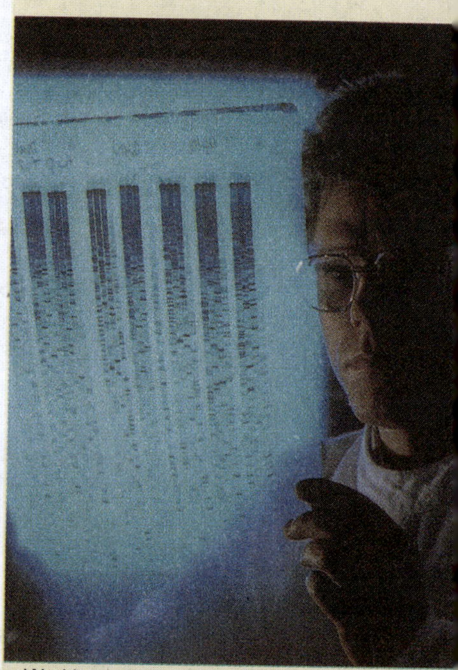

Working with X-ray pictures of gels, a scientist compares the relative positions of bands of DNA in sequencing "ladders". This enables him to establish the order of bases in a piece of DNA.

arises that whether the genetic informations are written in articulated language or coded language? If genetic informations might have occurred in an articulated language, the DNA molecule might require various alphabets, a complex system of grammar and ample amount of space on it. All of which might be practically impossible and troublesome too for the DNA. Therefore, it was safe to conclude for molecular biologists that genetic informations were existed in DNA molecule in the form of certain special language of code words which might utilize the four nitrogen bases of DNA for its symbols. Any coded message is commonly called **cryptogram**.

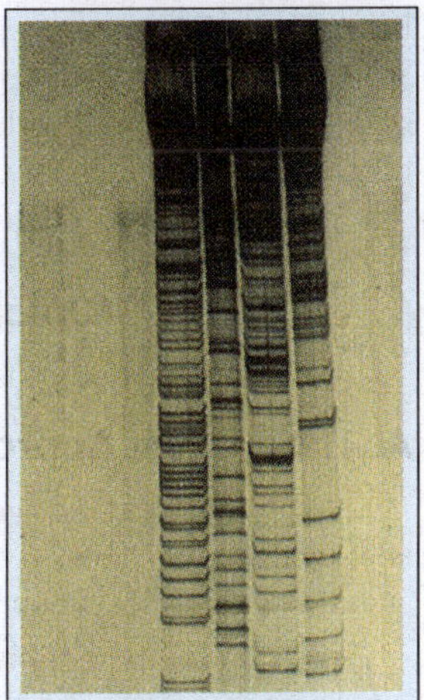

X-rays of DNA-containing gels, such as this one, allow scientists to determine the sequence of pieces of DNA containing 300 or so bases.

BASIS OF CRYPTOANALYSIS

The basic problem of such a genetic code is to indicate how information written in a four letter language (four nucleotides or nitrogen bases of DNA) can be translated into a twenty letter language (twenty amino acids of proteins). The group of nucleotides that specifies one amino acid is a **code word** or **codon**. By the **genetic code** one means the collection of base sequences (codons) that correspond to each amino acid and to translation signals. We can consider here the classical but logical reasoning done by **George Gamov** (1954) about the possible size of a codon. The simplest possible code is a **singlet code** (a code of single letter) in which one nucleotide amino acid could be specified. A **doublet code** (a code of two letters) is also inadequate, because it could specify only sixteen (4×4) amino acids, whereas a **triplet code** (a code of three letters) could specify sixty four (4×4×4) amino acids. Therefore, it is likely that there may be 64 triplet codes for 20 amino acids. The possible singlet, doublet and triplet codes, which are customarily represented in terms of "**mRNA language**" [mRNA is a complementary molecule which copies the genetic informations (cryptogram of DNA) during its transcription], have been illustrated in Table 6-1.

Table 6-1. Possible singlet, doublet and triplet codes of mRNA.

Singlet code (4 words)	Doublet code (16 words)				Triplet code (64 words)			
					AAA	AAG	AAC	AAU
					AGA	AGG	AGC	AGU
					ACA	ACG	ACC	ACU
					AUA	AUG	AUC	AUU
					GAA	GAG	GAC	GAU
					GGA	GGG	GGC	GGU
A	AA	AG	AC	AU	GCA	GCG	GCC	GCU
G	GA	GG	GC	GU	GUA	GUG	GUC	GUU
C	CA	CG	CC	CU	CAA	CAG	CAC	CAU
U	UA	UG	UC	UU	CGA	CGG	CGC	CGU
					CCA	CCG	CCC	CCU
					CUA	CUG	CUC	CUU
					UAA	UAG	UAC	UAU
					UGA	UGG	UGC	UGU
					UCA	UCG	UCC	UCU
					UUA	UUG	UUC	UUU

The first experimental evidence in support to the concept of triplet code is provided by **Crick** and coworkers in 1961. During their experiment, when they added or deleted single, or double base pairs in a particular region of DNA of T_4 bacteriophages of *E.coli*, they found that such bacteriophages ceased to perform their normal functions. However, bacteriophages with addition or deletion of three base pairs in DNA molecule, had performed normal functions. From this experiment, they concluded that a genetic code is in triplet form, because the addition of one or two nucleotides has put the reading of the code out of order, while the addition of third nucleotide resulted in a return to the proper reading of the message.

CODON ASSIGNMENT

(Cracking the Code or Deciphering the Code)

The genetic code has been cracked or deciphered by the following kinds of approaches :

A. Theoretical Approach

The physicist **George Gamow** proposed the **diamond code** (1954) and the **triangle code** (1955) and suggested an exhaustive theoretical framework to the different aspect of the genetic code. **Gamow** suggested the following properties of the genetic code :

(i) A **triplet codon** corresponding to one amino acid of the polypeptide chain.
(ii) **Direct template translation** by codon-amino acid pairing.
(iii) Translation of the code in an **overlapping** manner.
(iv) **Degeneracy** of the code, *i.e.*, an amino acid being coded by more than one codon.
(v) **Colinearity** of nucleic acid and the primary protein synthesized.
(vi) **Universality** of the code, *i.e.*, the code being essentially the same for different organisms.

Some of these Gamow's proposals have been contradicted by the molecular biologists. For example, **Brenner** (1957) showed that the overlapping triplet code is an impossibility, and subsequent work has shown that the code is a **non-overlapping** one. Similarly, Gamow's idea of direct template relationship between nucleic acid and polypeptide chain was challenged when **Crick** proposed his **adopter hypothesis**. According to this hypothesis, **adaptor molecules** intervene between nucleic acid and amino acids during translation. In fact, it is now known that tRNA molecules act as adaptors between codons of mRNA and amino acids of the resulting polypeptide chain.

B. The *in vitro* codon Assignment

1. Discovery and use of polynucleotide phosphorylase enzyme. Marianne Grunberg-Manago and **Severo Ochoa** isolated an enzyme from the bacteria (*e.g.*, *Azobacter vinelandii* or *Micrococcus lysodeikticus*) that catalyzes the breakdown of RNA in bacterial cells. This enzyme is called **polynucleotide phosphorylase**. **Manago** and **Ochoa** found that outside of the cell (*in vitro*), with high concentrations of ribonucleotides, the reaction could be driven in reverse and an RNA molecule could be made (see **Burns** and **Bottino**, 1989). Incorporation of bases into the molecule is random and does not require a DNA template. Thus, in 1955 **Manago** and **Ochoa** made possible the artificial synthesis of polynucleotides (=mRNA) containing only a single type of nucleotides (U,A, C, or G respectively) repeated many times.

Polynucleotide	Configuration
1. Polyuridylic acid or poly (U)	UUUUUU
2. Polyadenylic acid or poly (A)	AAAAAA
3. Polycytilic acid or poly (C)	CCCCCC
4. Polyguanidylic acid or poly (G)	GGGGGG

Thus, the action of polynucleotide phosphorylase can be represented in the following way :

$$(RNA)_n + \text{Ribonucleoside diphosphate} \xrightleftharpoons[]{\text{polynucleotide phosphorylase}} (RNA)_{n+1} + Pi$$

The polynucleotide phosphorylase enzyme differs from RNA polymerase used to transcribe mRNA from DNA polymerase used to transcribe mRNA from DNA in that : (i) it does not require a template or primer; (ii) the activated substrates are ribonucleoside diphosphates (*e.g.*, UDP, ADP, CDP and GDP) and not triphosphates; and (iii) orthophosphate (Pi) is produced instead of pyrophosphates (PPi).

The deciphering of the genetic code was made possible by the use of synthetic (or artificial) polynucleotides and trinucleotides. The different types of techniques used include the use of polymers containing a single type of nucleotide (called **homopolymers**), the use of mixed polymers (**copolymers**) containing more than one type of nucleotides (**heteropolymers**) in random or defined sequences and the use of trinucleotides (or "**minimessengers**") in ribosome-binding or filter-binding.

2. Codon assignment with unknown sequence. (i) Codon assignment by homopolymer. The first clue to codon assignment was provided by **Marshall Nirenberg and Heinrich Matthaei** (1961) when they used in vitro system for the synthesis of a polypeptide using an artificially synthesized mRNA molecule containing only one type of nucleotide (i.e., homopolymer). Prior to performing the actual experiments, they tested the ability of a cell-free protein synthesizing system to incorporate radioactive amino acids into newly synthesized proteins. Their cell-free extracts of E.coli contained ribosomes, tRNAs, aminoacyl-tRNA synthetase enzymes, DNA and mRNA. The DNA of this extract was eradicated by the help of **deoxyribonuclease** enzyme, thus, the template which might synthesize new mRNA was destroyed. When twenty amino acids were added to this mixture along with ATP, GTP, K^+ and MG^{2+}, they were incorporated into proteins. This incorporation continued so long as mRNA was present in such a cell-free suspension. It also continued in the presence of synthetic polynucleotides (mRNAs) which could be made with the help of polynucleotide phosphorylase enzyme.

M.W. Nirenberg.

The first successful use of this technique was made by **Nirenberg** and **Matthaei** who synthesized a chain of uracil molecules (poly U) as their synthetic mRNA (homopolymer). Poly (U) seemed a good choice, because there could be no ambiguity in a message consisting of only one base. Poly(U) was a good choice for other reasons : it binds well to ribosomes and, as it turned out, the product protein was insoluble and easy to isolate. When poly (U) was presented as the message to the cell-free system containing all the amino acids, one amino acid was exclusively selected from the mixture for incorporation into the polypeptide, called **polyphenylalanine**. This amino acid was phenylalanine and it could be concluded that some sequence of UUU coded for phenylalanine. Other homogeneous chains of nucleotides (Poly A, Poly C and Poly G) were inactive for phenylalanine incorporation. The mRNA code word for phenlalanine was, therefore, shown to be UUU. The corresponding DNA code word for phenylalanine can be deduced to be AAA. Thus, the first code word to be deciphered was UUU.

This discovery was extended in the laboratories of **Nirenberg** and **Ochoa**. The experiment was repeated using synthetic **poly (A)** and **poly (C)** chains, which gave **polylysine** and **polyproline** respectively. Thus, AAA was identified as the code for lysine and CCC as the code for proline. A poly (G) message was found non-functional *in vitro*, since it attains secondary structure and, thus, could not attach to the ribosomes. In this way three of 64 codons were easily accounted for.

(ii) Codon assignment by heteropolymers (Copolymers with random sequences). Further exposition of the genetic code took place by using synthetic messenger RNAs containing two kinds of bases. This technique was used in the laboratories of **Ochoa** and **Nirenberg** and led the deduction of the composition of codons for the 20 amino acids. The synthetic messengers contained the bases at random (called **random copolymers**). For example, in a random copolymer using U and A nucleotides eight triplets are possible, such as UUU, UUA, UAA, UAU, AAA, AAU, AUU and AUA. Theoretically, eight amino acids could be coded by these eight codons. Actual experiments, however, yielded only six phenylalanine, leucine, tyrosine, lysine, asparagine and isoleucine. By varying the relative compositions of U and A in the random copolymer and determining the percentage of the different amino acids in the proteins formed, it was possible to deduce the composition of the code for different amino acids.

3. Assignment of codons with known sequences. (i) Use of trinucleotides or minimessengers in filter binding (Ribosome-binding technique). Ribosome binding technique of **Nirenberg and Leder** (1964) made use of the finding that aminoacyl- tRNA molecules specifically bind to ribosome-mRNA complex. This binding does not require the presence of a long mRNA molecule; in fact, the association of a **trinucleotide** or **minimessenger** with the ribosome is sufficient to cause aminoacyl-tRNA binding. When a mixture of such small mRNA molecules-ribosomes and amino acid-tRNA complexes are incubated for a short time and then filtered through a nitrocellulose membrane, then the mRNA-ribosome-tRNA-amino acid complex is retained back and rest of the mixture passes through the filter. By using a series of 20 different amino acid mixtures, each containing one radioactive amino acid at a time, it is possible to find out the amino acid corresponding to each triplet by analysing the radioactivity absorbed by the membrane, e.g., the triplet GCC and GUU retain only alanyl-tRNA and valyl-tRNA respectively. All 64 possible triplets have been synthesized and tested in this way. Forty five of them have given clear-cut results. Later on, with the help of longer synthetic messages it has been possible to decipher 61 out of the possible 64 codons.

Table 6-2. The genetic dictionary. The trinucleotide codons are written in the 5'→3' direction.

First base	Second base: U	C	A	G	Third base
U	UUU	UCU	UAU	UGU	U
	UUC } Phe	UCC	UAC } Tyr	UGC } Cys	C
	UUA	UCA } Ser	UAA } Non-sense codon	UGA } Non-sense codon	A
	UUG } Leu	UCG	UAG	UGG } Trp	G
C	CUU	CCU	CAU	CGU	U
	CUC	CCC	CAC } His	CGC	C
	CUA } Leu	CCA } Pro	CAA	CGA } Arg	A
	CUG	CCG	CAG } Gln	CGG	G
A	AUU	ACU	AAU	AGU	U
	AUC } Ileu	ACC	AAC } Asn	AGC } Ser	C
	AUA	ACA } Thr	AAA	AGA	A
	AUG* Met	ACG	AAG } Lys	AGG } Arg	G
G	GUU	GUC	GAU	GGU	U
	GUC	GCC } Ala	GAC } Asp	GGC	C
	GUA } Val	GCA	GAA	GGA } Gly	A
	GUG	GCG	GAG } Glu	GGG	G

* AUG—Met or chain initiation codon.

Table 6-3. Amino acids and their messenger RNA codons.

Alanine (Ala)	GCA, GCG, GCC, GCU	Glycine (Gly)	GGA, GGG, GGC, GGU	Leucine (Leu)	CUA, CUG, CUC, CUU, UUA, UUG	Threonine (Thr)	ACA, ACG, ACC, ACU
Arginine (Arg)	AGA, AGG, CGA, CGG, CGC, CGU	Glutamine (Gln ; Glun)	CAA, CAG	Methionine (Met) (Starting codon)	AUG	Tryptophan (Trp)	UGG, UAU
Aspartic acid (Asp)	GAC, GAU	Glutamic acid (Glu)	GAA, GAG	Phenylalanine (Phe)	UUC, UUU	Tyrosine (Tyr)	UAC
Asparagine (Asn; Aspn)	AAC, AAU	Histidine (His)	CAC, CAU	Proline (Pro)	CCA, CCG, CCC, CCU	Valine (Val)	GUA, GUG*, GUC, GUU
Cysteine (Cys)	UGC, UGU	Isoleucine (Ile)	AUC, AUU, AUA	Serine (Ser)	AGC, AGU, UCA, UCG, UCC, UCU	Terminator (Nonsense codons)	UAA (ocher), UAG (amber), UGA (opal)
		Lysine (Lys)	AAA, AAG				

* GUG is also used as a start codon for some proteins.

P.Leder.

C. The *in vivo* Codon Assignment

The cell free protein synthetic systems, though have proved of great significance in decipherment of the genetic code, but they could not tell us whether the genetic code so deciphered is used in the living systems of all organisms also. Three kinds of techniques are used by different molecular biologists to determine whether the same code is also used *in vivo* (a) amino acid replacement studies (*e.g.*, tryptophan synthetase synthesis in *E.coli* (**Yanofsky** *et al.* 1963) and haemoglobin synthesis in man), (b) frameshift mutations (*e.g.*, investigations of **Terzaghi** *et al.* 1966, on lysozyme enzyme of T_4 bacteriophages, and (c) comparison of a DNA or mRNA polynucleotide cryptogram with its corresponding polypeptide clear text (*e.g.*, comparison of amino acid sequence of the R_{17} bacteriophage coat protein with the nucleotide sequence of the R_{17} mRNA in the region of the molecule that dictates coat-protein synthesis by **S. Cory** *et al.,* 1970).

Thus, *in vitro* and *in vivo* studies, so far described, gave the way to formulate a code table for twenty amino acids (see Table 6-2 and Table 6-3).

CHARACTERISTICS OF GENETIC CODE

The genetic code has the following general properties :

1. The code is a triplet codon. The nucleotides of mRNA are arranged as a linear sequence of codons, each codon consisting of three successive nitrogenous bases, *i.e.*, the code is a triplet codon. The concept of triplet codon has been supported by two types of point mutations: frameshift mutations and base substitution.

(i) Frameshift mutations. Evidently, the genetic message once initiated at a fixed point is read in a definite frame in a series of three letter words. The framework would be disturbed as soon as there is a deletion or addition of one or more bases. When such frameshift mutations were intercrossed, then in certain combinations they produce wild type normal gene. It was concluded that one of them was deletion and the other an addition, so that the disturbed order of the frame due to mutation will be restored by the other (Fig. 6.1).

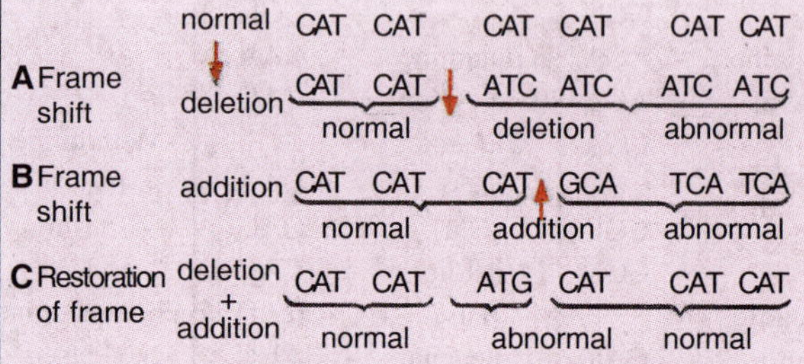

Fig. 6.1. Frame-shift mutations: A—Deletion; B—Addition; C—Restoration of frame.

(ii) Base substitution. If in a mRNA molecule at a particular point, one base pair is replaced by another without any deletion or addition, the meaning of one codon containing such an altered base will be changed. In consequence, in place of a particular amino acid at a particular position in a polypeptide, another amino acid will be incorporated. For example, due to substitution mutation, in the gene for tryptophan synthetase enzyme in *E.coli*, the GGA codon for glycine becomes a missence codon AGA which codes for arginine. **Missence codon** is a codon which undergoes an alteration to specify another amino acid.

A more direct evidence for a triplet code came from the finding that a piece of mRNA containing 90 nucleotides, corresponded to a polypeptide chain of 30 amino acids of a growing haemoglobin molecule. Similarly, 1200 nucleotides of "satellite" tobacco necrosis virus direct the synthesis of coat protein molecules which have 372 amino acids.

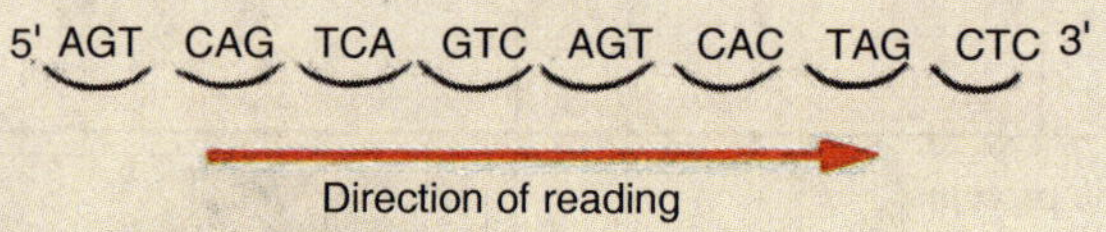

Fig. 6.2. Bases in mRNA are read sequentially in the 5' to 3' direction, in group of three bases.

2. The code is non-overlapping. In translating mRNA molecules the codons do not *overlap* but are "read' sequentially (Fig. 6.2). Thus, a **non-overlapping code** means that a base in a mRNA is not used for different codons. In Figure 6.3., it has been shown that an overlapping code can mean coding for four amino acids from six bases. However, in actual practice six bases code for not more than two amino acids. For example, in case of an overlapping code, a single change (of substitution type) in the base sequence will be reflected in substitutions of more than one amino acid in corresponding protein. Many examples have accumulated since 1956 in which a single base substitution results into a single amino acid change in insulin, tryptophan synthetase, TMV coat protein, alkaline phosphatase, haemoglobin, etc.

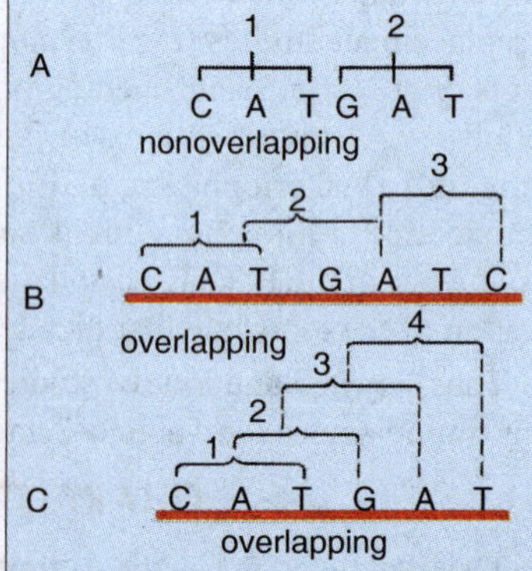

Fig. 6.3. A—Non-overlapping codons; B—Overlapping of codon due to one base; C—Overlapping of codon due to two bases.

Recently, however, it has been shown that in the bacteriophage $\phi \times 174$ there is a possibility of overlapping of genes and codons (**Barrel** and coworkers, 1976; **Sanger**, *et al.*, 1977).

3. The code is commaless. The genetic code is commaless, which means that no codon is reserved for punctuations. It means that after one amino acid is coded, the second amino acid will be automatically coded by the next three letters and that no letters are wasted as the punctuation marks (Fig. 6.4)

4. The code is non-ambiguous. Non-ambiguous code means that a particular codon will always code for the same amino acid. In case of ambiguous code, the same codon could have different meanings or in other words, the same codon could code two or more than two different amino acids. Generally, as a rule, the same codon shall never code for two different amino acids. However, there are some reported exceptions to this rule: the codons AUG and GUG both may code for *methionine* as initiating or starting codon, although GUG is meant for *valine*. Likewise, GGA codon codes for two amino acids *glycine* and *glutamic acid*.

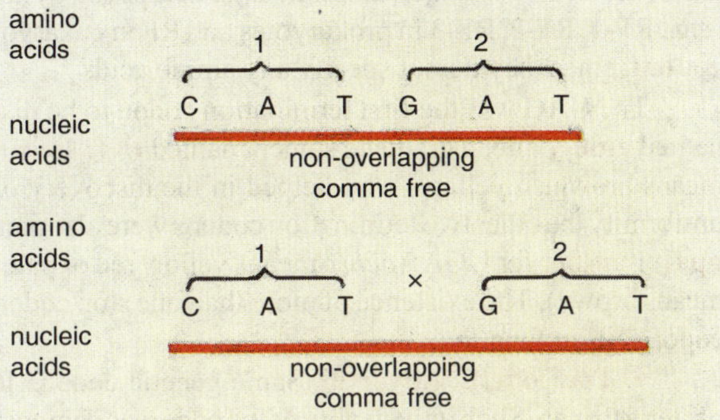

Fig. 6.4. Two forms of genetic code. A—Genetic code without comma; B—Genetic code with comma.

5. The code has polarity. The code is always read in a fixed direction, *i.e.*, in the 5'→3' direction. In other words, the codon has a **polarity**. It is apparent that if the code is read in opposite directions, it would specify two different proteins, since the codon would have reversed base sequence :

Codon :	UUG	AUC	GUC	UCG	CCA	ACA	AGG
Polypeptide: →	Leu	Ile	Val	Ser	Pro	Thr	Arg
	Val	Leu	Leu	Ala	Thr	Thr	Gly ←

6. The code is degenerate. More than one codon may specify the same amino acid; this is called **degeneracy** of the code. For example, except for *tryptophan* and *methionine*, which have a single codon each, all other 18 amino acids have more than one codon. Thus, nine amino acids, namely *phenylalanine*, *tyrosine*, *histidine*, *glutamine*, *asparagine*, *lysine*, *aspartic acid*, *glutamic acid* and *cysteine,* have two codons each. Isoleucine has three codons. Five amino acids, namely *valine*, *proline*, *threonine*, *alanine* and *glycine,* have four codons each. Three amino acids, namely *leucine*, *arginine* and *serine,* have six codons each (see Table 6-3).

The code degeneracy is basically of two types : partial and complete. **Partial degeneracy** occurs when first two nucleotides are identical but the third (*i.e.*, 3' base) nucleotide of the degenerate codons differs, *e.g.*, CUU and CUC code for leucine. **Complete degeneracy** occurs when any of the four bases can take third position and still code for the same amino acid (*e.g.*, UCU,UCC, UCA and UCG code for serine).

Degeneracy of genetic code has certain biological advantages. For example, it permits essentially the same complement of enzymes and other proteins to be specified by microorganisms varying widely in their DNA base composition. Degeneracy also provides a mechanism of minimizing mutational lethality.

7. Some codes act as start codons. In most organisms, AUG codon is the **start** or **initiation codon**, *i.e.*, the polypeptide chain starts either with **methionine** (eukaryotes) or **N-formylmethionine** (prokaryotes). Methionyl or N-formylmethionyl-tRNA specifically binds to the **initiation site** of mRNA containing the AUG initiation codon. In rare cases, GUG also serves as the initiation codon, *e.g.*, bacterial protein synthesis. Normally, GUG codes for valine, but when normal AUG codon is lost by deletion, only then GUG is used as initiation codon.

8. Some codes act as stop codons. Three codons UAG, UAA and UGA are the chain **stop** or **termination** codons. They do not code for any of the amino acids. These codons are not read by any tRNA molecules (via their anticodons), but are read by some specific proteins, called **release factors** (*e.g.*, RF-1, RF-2, RF-3 in prokaryotes and RF in eukaryotes). These codons are also called **nonsense codons**, since they do not specify any amino acid.

The UAG was the first termination codon to be discovered by **Sidney Brenner** (1965). It was named **amber** after a graduate student named **Bernstein** (= the German word for 'amber' and amber means brownish yellow) who helped in the discovery of a class of mutations. Apparently, to give uniformity the other two termination codons were also named after colours such as **ochre** for UAA and **opal** or **umber** for UGA. (*ochre* means yellow red or pale yellow; *opal* means milky white and *umber* means brown). The existence of more than one stop codon might be a safety measure, in case the first codon fails to function.

9. The code is universal. Same genetic code is found valid for all organisms ranging from bacteria to man. Such universality of the code was demonstrated by **Marshall**, **Caskey** and **Nirenberg** (1967) who found that *E.coli* (bacterium), *Xenopus laevis* (amphibian) and guinea pig (mammal) amino acyl-tRNA use almost the same code. **Nirenberg** has also stated that the genetic code may have developed 3 billion years ago with the first bacteria, and it has changed very little throughout the evolution of living organisms.

Recently, some differences have been discovered between the universal genetic code and mitochondrial genetic code (Table 6-4).

Table 6-4. **Differences between the 'universal genetic code' and two mitochondrial genetic codes (Source: Maclean and Hall, 1987).**

Codon	Mammalian mitochondrial code	Yeast mitochondrial code	"Universal code"
1. UGA	*Trp* *	*Trp*	Stop
2. AUA	*Met*	*Met*	Ile
3. CUA	Leu	*Thr*	Leu
4. AGA } 5. AGG }	*Stop*	Arg	Arg

* Italic type indicates that the code differs from the 'universal' code.

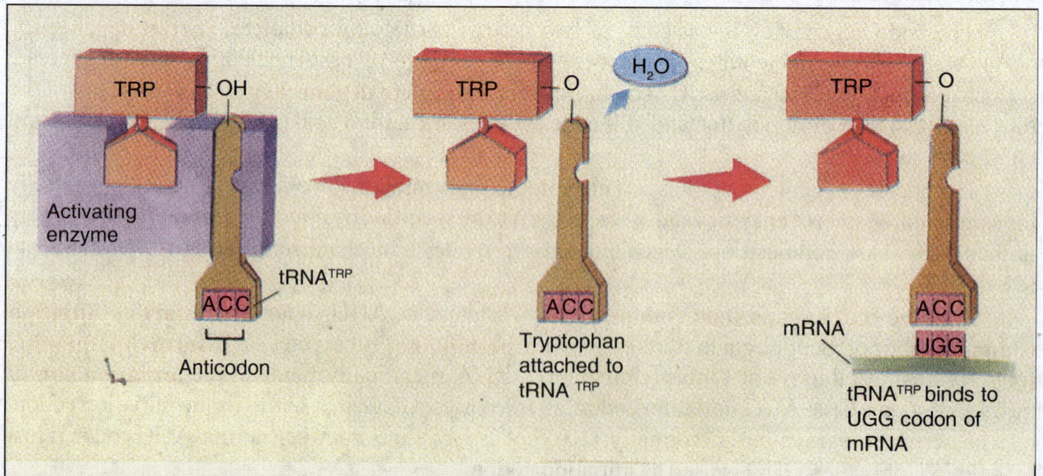

Each kind of activating enzyme recognizes and binds a specific amino acid such as tryptophan (TRP), and also recognizes and binds the tRNA molecules with anticodons specifying that amino acid, such as ACC for tryptophan.

Codon and Anticodon

The codon words of DNA would be complementary to the mRNA code words (*i.e.*, DNA codes run in 3'→5' direction and mRNA code words run in 5'→3' direction) and so thereby the three bases forming the **anticodon** of tRNA (*i.e.*, bases of anticodons run in 3'→5' direction). Three bases of anticodon pair with the mRNA on the ribosomes at the time of aligning the amino acids during protein synthesis (*i.e.*, translation of mRNA into proteins which proceeds in N_2→COOH direction). For example, one of two mRNA and DNA code words for the amino acid phenylalanine is UUC and AAG respectively, and the corresponding anticodon of tRNA is CAA. This indicates that codon and anticodon pairing is antiparallel. In this case, C pairs with G and U pairs with A.

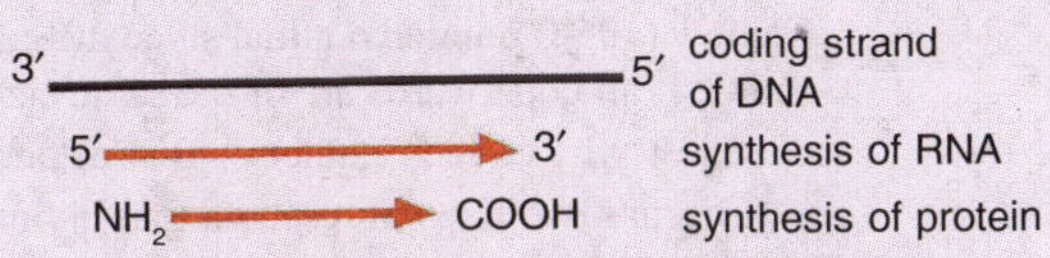

Fig. 6.5. Directions of synthesis of RNA and protein with respect to the coding strand of DNA (after Freifelder, 1985).

WOBBLE HYPOTHESIS

To explain the possible cause of degeneracy of codons, **Crick** (1966) proposed the **wobble hypothesis** (wobble means to sway or move unsteadily). Since there are 61 codons specifying amino acids, the cell should contain 61 tRNA molecules, each with a different anticodon. Actually, however, the number of tRNA molecule types discovered is much less than 61. This implies that the anticodons of tRNAs read more than one codon on mRNA. For example, yeast $tRNA^{ala}$ with anticodon bases 5' IGC 3' (where I stands for inosine, a derivative of adenine or A) could bind to three codons in mRNA such as 5' GCU 3', 5'GCC3' and 5' GCA3'. Inosine is frequently found as the 5' base of the anticodon; at the time of pairing with the base of the codons it wobbles and can pair either with U, C or A of three codons (see Table 6-5). Thus, Crick's wobble hypothesis states that the *base at 5' end of the anticodon is not spatially confined as the other two bases allowing it to form hydrogen bonds with any of several bases located at the 3' end of a codon* (Fig. 6.6).

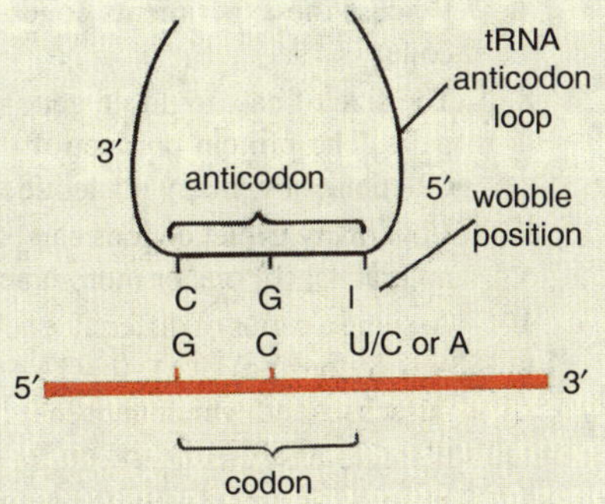

Fig. 6.6. Wobble hypothesis. In the third site (5' end) of the anticodon I can take either of three wobble posi

Table 6-5. Allowed base-pairing combinations according to the wobble hypothesis (Source: King, 1986).

5' base of codon	3′ base of anticodon
C	G
A	U
U	A or G
G	U, C or A
I	U, C or A

REVISION QUESTIONS AND PROBLEMS

1. What do you understand by 'genetic code'? How will you show that the minimum size of a code word should be triplet ?
2. A basic concept of molecular biology is the collinearity of gene, DNA, RNA and protein. Discuss the various lines of evidence supporting or providing this concept.
3. Gamow (1954) pointed out that since the genetic language contains only four letters, A, U (=T), C and G, if all code words are of the same size, codons must be at least three bases long.
   a. Show why codons cannot consist of one or of two bases.
   b. Discuss two lines of research that indicate that codons are three base long.
   c. Since codons are three base long, 64 different triplets can exit. Illustrate these using the branching method.
   d. How many of the 64 triplets will contain (1) no adenine, (2) at least one adenine?
4. (a) Explain what is meant by a degenerate code and illustrate your answer to show degeneracy in translation.
   (b) Which are the nonsense triplets and why are they so termed?
   (c) Do codons have the same meaning *in vitro* as *in vivo*?
5. Describe the general properties of the genetic code.
6. Give a brief account of the experiments which helped in deciphering the genetic code.
7. Discuss the experiments conducted by H.G. Khorana and his coworkers for cracking the genetic code.
8. The size of haemoglobin gene in humans is estimated to consist of approximately 450 nucleotide pairs. The protein product of the gene is estimated to consist of about 150 amino acid residues. Calculate how many nucleotides are present in a genetic codon for each amino acid.
9. How many triplet codons can be made from the four ribonucleotides A, U, G and C containing (a) no uracils; (b) one or more uracils?
10. Assume a series of different one-base changes in the codon GGA, which produces following several new codons: (a) UGA; (b) GAA; (c) GGC, (d) CGA. Which of these represent(s) degeneracy, which missense, and which nonsense?
11. If single-base changes occur in DNA (and, therefore, in mRNA), which amino acid, tryptophan or arginine, is most likely to be replaced by another in protein synthesis?
12. Write short notes on the following:
    (1) Homopolymers; (2) Copolymers of repetitive sequences; (3) Poly U; (4) Frameshift mutation; (5) Overlapping genes; (6) Wobble hypothesis; (7) Split genes.

ANSWERS TO PROBLEMS

8. Approximately 3 nucleotides code for each amino acid.
9. (a) Since uracil represents 1 among 4 nucleotides; the probability that uracil will be the first letter of the codon is 1/4 and the probability that U will not be the first letter is 3/4. The same reasoning holds true for the second and third letters of the codon. The probability that none of the three letters of the codon are uracils is $(3/4)^3 = 27/64$.
   (b) The number of codons containing at least one uracil is $1–27/64 = 37/64$.
10. (a) Nonsense; UGA is a stop signal and does not translate.
    (b) Missense; GGA codes for glycine, GAA for glutamic acid.
    (c) Degeneracy; both GGA and GGC code for glycine.
    (d) Missense; GGA codes for glycine, CGA for arginine.
11. Tryptophan, which has only one codon. Arginine has six codons.

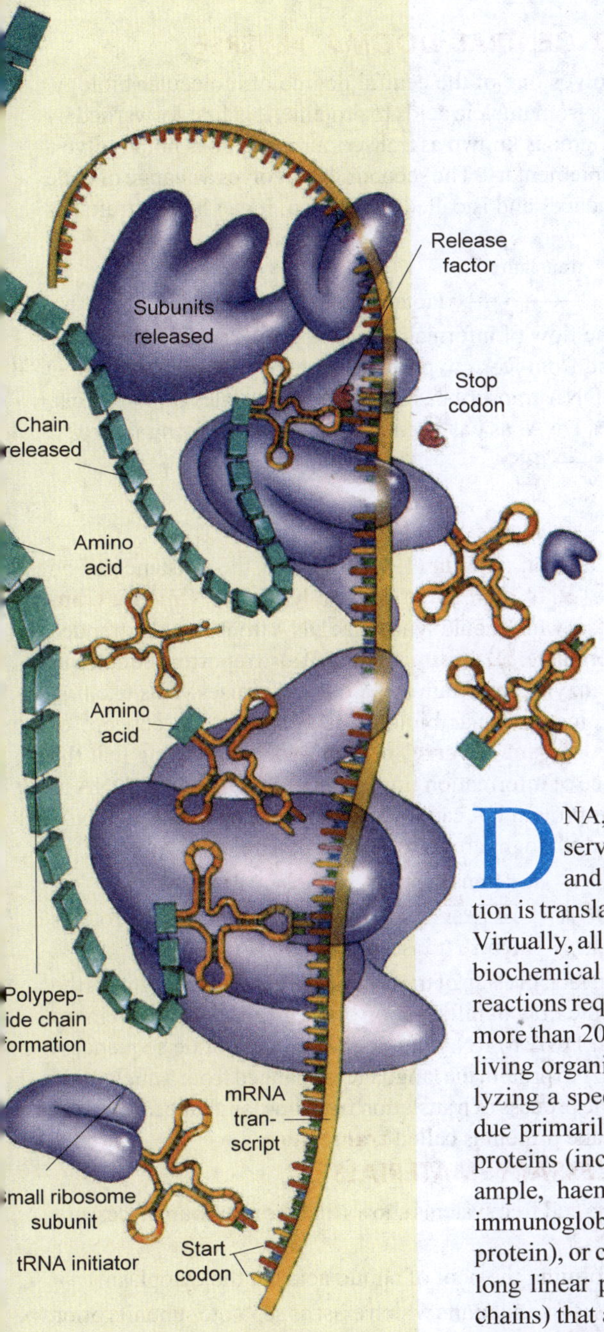

The process of protein synthesis.

CHAPTER

7

Protein Synthesis

DNA, with its correct mechanism of replication, serves to carry genetic information from cell to cell and from generation to generation. This information is translated into proteins that determine the phenotype. Virtually, all the phenotypes examined so far are the result of biochemical reactions that occur in the cell. All of these reactions require enzymes and enzymes are proteins. In fact, more than 2000 types of enzymes have been identified in the living organisms. Each enzyme is a unique molecule catalyzing a specific chemical reaction. Other phenotypes are due primarily to the kinds and amounts of non-enzymatic proteins (including the structural proteins) present, for example, haemoglobin, myoglobin, gamma globulin (*e.g.*, immunoglobins or IG), insulin, cytochrome C, fibroin (silk protein), or collagen. Proteins are composed of one or more, long linear polymers of amino acid residues (polypeptide chains) that are synthesized almost exclusively in the cytoplasm. The topic of this chapter is how the information present in the sequence of bases (=triplet codons) of the mRNA is translated into a sequence of amino acids in proteins. However, at the outset one point should be clear that the genetic information of mRNA is unidirectionally read beginning at the 5'- hydroxyl end by one or more ribosomes (polysomes) and that the 5'-end of the mRNA correspond to the N-terminal amino acid in the completed protein.

CENTRAL DOGMA AND CENTRAL DOGMA REVERSE

The process of synthesis of protein involves one of the central dogma of molecular biology, which postulates that genetic information flows from nucleic acids to protein. (It is first forwarded by **Crick** in 1958). The first step of this central dogma is known as transcription and does not involve a change of code since DNA and mRNA are complementary. The second step involves a change of code from nucleotide sequences to amino acid sequences and is called translation. It can be illustrated as follows :

$$\text{Duplication} \longrightarrow \text{DNA} \xrightarrow{\text{transcription}} \text{RNA} \xrightarrow{\text{translation}} \text{Proteins}$$

Thus, according to this central dogma, the flow of information is one way, *i.e.,* from DNA the information is transferred to RNA (mRNA) and from RNA to proteins. In 1968, **Barry Commoner** suggested a circular flow of information, *i.e.,* DNA transcribes RNA, RNA translates into proteins, proteins synthesize RNA and RNA synthesizes DNA, as has been illustrated in following figure :

$$\begin{array}{ccc} \text{DNA} & \longrightarrow & \text{RNA} \\ \uparrow & & \downarrow \\ \text{RNA} & \longleftarrow & \text{Proteins} \end{array}$$

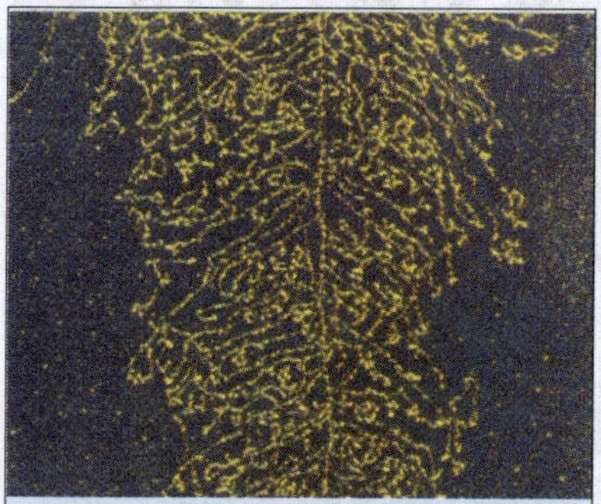
Clusters of mRNA molecules form a fern like structure around a backbone of DNA molecule which is undergoing transcription.

Later on, **Temin** (1970) reported the existence of an enzyme "**RNA dependent DNA polymerase**" (**inverse transcriptase**) which could synthesize DNA from a single stranded RNA template. **D. Baltimore** (1970) also reported the activity of this enzyme in certain RNA tumour viruses. This exciting finding in molecular biology gave rise to the concept of "**central dogma reverse**" or **teminism**, suggesting that the sequence of information flow is not necessarily from DNA to RNA to protein but can also take place from RNA to DNA. The central dogma reverse can be illustrated as follows :

$$\text{DNA} \underset{\text{inverse transcription}}{\overset{\text{transcription}}{\rightleftarrows}} \text{RNA} \xrightarrow{\text{translation}} \text{Proteins}$$

In fact, the step of translation of this central dogma is the time in the flow of information between genes and proteins to change the language being used. In going from DNA to RNA the language (nucleotide sequences) remained the same. However, in going from RNA to protein the language is changed from a nucleotide sequence to an amino acid sequence. Just as in the process of translation from one language to another, this process of using information in RNA to make protein is called **translation**.

MINIMUM NECESSARY MATERIALS

Success in polypeptide synthesis in *in vitro* cell free systems shows that the minimum necessary materials are the following :

1. Amino acids (*i.e.,* 20 amino acids forming the pool of amino acids in the cytoplasm)
2. Ribosomes (each of which comprises two subunits which exist as separate subunits prior to the translation of mRNA and contain two tRNA binding sites : the **P site** or **peptidyl site** and an **A** or **aminoacyl site**. One more site called **E** or **exit site** has been recognized in the ribosomes by some workers but this three site model (*i.e.,* A, P and E) still is not popular).

Mussels, busily demonstrating the importance of proteins for survival. Mussels produce the world's best underwater adhesive from a mix of proteins. The adhesive anchors them to substrates in their wave swept habitat.

3. mRNA
4. tRNA of several kinds
5. Enzymes
   (a) Amino-acid activating system (*e.g.*, aminoacyl-tRNA-synthetase).
   (b) Peptide polymerase system
6. Adenosine triphosphate (ATP) as an energy source
7. Guanosine triphosphate (GTP) for synthesis of peptide bonds
8. Soluble protein initiation and transfer factors
9. Various inorganic cations (*e.g.*, K^+, NH^+ Mg^{2+}).

MECHANISM OF PROTEIN SYNTHESIS

Protein synthesis is the most complex biochemical transformation which cells perform, and at least 200 different proteins are required for the protein synthesis itself. Both in prokaryotes and eukaryotes, the mechanism of protein synthesis can be divided into the following three main steps : 1. **initiation**; 2. **elongation** ; and 3. **termination**. The main features of the **initiation step** are the binding of mRNA to the ribosomes, selection of the initiation codon, amino acid activation, transfer of activated amino acid to tRNA and binding of acylated tRNA bearing the first amino acid. In the step of **elongation** there are two processes : joining together two amino acids by peptide bond formation, and moving the mRNA and ribosomes with respect to one another so that the codons can be translated successively. In the **termination** stage the completed polypeptide or protein is dissociated from the synthetic machinery and the ribosomes are released to begin another cycle of synthesis.

Various events of protein synthesis can be studied under the following headings :

A. Aminoacylation of tRNA (Formation of Aminoacyl-tRNA)

Amino acids alone do not come to the ribosome to be incorporated into protein. Instead, they are brought to the ribosome by their appropriate tRNA. The first step in incorporating an amino acid into a protein involves the amino acids attachment to its correct tRNA. This involves the following two steps:

(a) Activation of amino acids. Each of the 20 amino acids occur in the cytoplasm in an inactive state. Each amino acid before its attachment with its specific tRNA is activated by a specific activating enzyme known as the **aminoacyl synthetase** and ATP. The free amino acids react with ATP, resulting in the production of aminoacyl adenylate and pyrophosphate :

AA	+	ATP	+	Enzyme	⟶	AA ~ AMP—Enzyme	+	PP
Amino acid				Aminoacyl synthetase		Aminoacyl adenylate enzyme complex		Pyrophosphate

The reaction product aminoacyl adenylate (or aminoacyl adenosine monophosphate) is bound to the enzyme in the form of a monocovalent complex. This aminoacyl adennylate enzyme complex then esterifies to specific tRNA molecule.

The cell has at least 20 aminoacyl synthetase enzymes for the 20 amino acids. Each enzyme is specific and it attaches with the specific amino acid without any error.

(b) Attachment of activated amino acid to tRNA. The aminoacyl adenylate remains bounded with enzyme until it collides with the specific tRNA molecule and its synthetase is recognized by dihydrouridine (DHU) loop of specific tRNA. Then, amino acid residue of aminoacyl adenylate is transferred to amino acid attachment site of tRNA where its carboxyl group forms bond or linkage with the 3-OH group of the ribose of the terminal adenosine at CCA end of tRNA. As a result AMP and enzyme are released and a final product **aminoacyl tRNA** is formed by the following method :

AA—AMP—Enzyme + tRNA	⟶	AA—tRNA + AMP + Enzyme
Aminoacyl adenylate and enzyme		Amioacyl—tRNA

The aminoacyl-tRNA moves towards the site of protein synthesis, *i.e.,* ribosomes with mRNA.

In fact, the attachment of the amino acid to its tRNA is the only *step in protein synthesis in which the identity of the amino acid* (*i.e., R group*) *plays a part*. This was clearly shown in 1962 by **Chapeville**. Cysteine was attached to its tRNA (tRNA ^{cys}) by cysteinyl tRNA synthetase to form cysteinyl tRNA (cys-tRNA). The attached cysteine residue was then chemically reduced to alanine and the altered molecule (*i.e.,* alanyl tRNA^{cys}) added to a cell-free protein synthesizing system. The resultant polypeptide contained alanine residues in place of the normal cysteine residues (see **king**, 1986).

Further, recent evidence suggests that the aminoacyl synthetase enzyme checks for correct binding before the release of the charged tRNA (*i.e.,* tRNA-AA). If an error has been made, the wrong amino acid is removed and the correct one is attached.

B. Stages of Polypeptide Synthesis in Prokaryotes

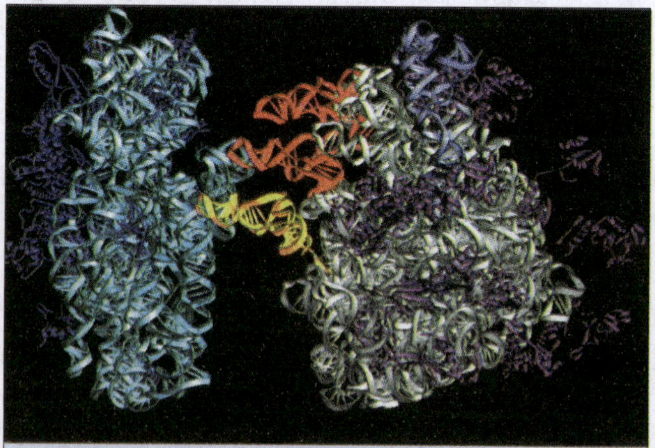

Crystal structure of 70s ribosome containing 3 bound tRNA.

Polypeptide synthesis in prokaryotes and eukaryotes follow the same overall mechanism though there are differences in detail (Table 7-1), the most important being the mechanism of initiation :

1. Initiation. An important feature of initiation of polypeptide synthesis is the use of specific initiating tRNA molecule. In prokaryotes this tRNA molecule is acylated with the modified amino acid **N-formyl methionine (f Met)**; the tRNA is often designated tRNA^{fMet}. Both tRNA^{fMet} and tRNA^{Met} recognize the codon AUG, but only tRNA$^{f\,Met}$ is used for initiation. The tRNA$^{f\,Met}$ molecule is first acylated (or charged) with methionine and an enzyme (found only in prokaryotes) adds a formyl group to the amino group of the methionine.

In eukaryotes also, the initiating tRNA molecule is charged with methionine, but formylation does not occur. The use of these initiator tRNA molecules means that while being synthesized, all prokaryotic proteins have N-formylmethionine at the amino terminus and all eukaryotic proteins have methionine at the amino terminus. However, these amino acids are frequently altered or removed later by the activity of a hydrolytic enzyme (this is called **processing**).

Since in protein synthesis the peptide chain always grows in a sequence from the free terminal amino ($-NH_2$) group towards the carboxyl (–COOH) end, so the function of formylmethionine–tRNA is to ensure that proteins are synthesized in that direction. In the formylmethionine–tRNA, the amino – (NH_2) group is blocked by the formyl group leaving only the —COOH group available to react with the $-NH_2$ group of the second amino acid (AA_2). In this way the synthesis of protein chain follows in the correct sequence.

Table 7.1. Differences between prokaryotes and eukaryotes in the mechanism of translation (*i.e.,* polypeptide synthesis).

Prokaryotes	Eukaryotes
1. Initiating amino acid, the methionine, needs to be formylated (Due to this reason there are present two tRNAs for methionine, *i.e.,* $tRNA^{fMet}$ and $tRNA^{Met}$).	1. Initiating amino acid, the methionine is not formylated (There occurs only one tRNA for methionine, *i.e.,* $tRNA^{Met}$).
2. Ribosomes enter the mRNA at AUG codon or at near by Shine-Delgarno site.	2. Ribosomes enter at the capped 5' end of mRNA and then advacne to AUG codon by linear scanning.
3. No initiation factors are needed for the initial contact between ribosomes and mRNA.	3. ATP and many protein factors are needed for ribosomes to engage the mRNA.
4. Small 30S ribosomal subunit can engage mRNA before binding of initiator met $tRNA^{fMet}$.	4. Small 40S ribosomal subunit binds stably to mRNA only after initiator met-$tRNA^{Met}$ has bound to it.

Protein synthesis in bacteria (*e.g., E. coli*) begins by the association of one **30S ribosomal subunit** (not the entire 70S ribosome), an mRNA molecule, f Met-tRNA, three proteins known as initiation factors (**IFs** such as **IFl**, **IF2**, and **IF3**, see Table 7-2) and guanosine 5'-triphosphates (GTP). These molecules constitute the 30S preinitiation complex (Fig. 7.1). Since protein synthesis begins at an AUG start codon and AUG codons are also found within coding sequences (that is, methionine occurs within a polypeptide chain), some signal must be present in the base sequence of the mRNA molecule to identify a particular AUG codon as an initiation or start signal. The means of selecting the correct AUG sequence differs in prokaryotes and eukaryotes. In prokaryotic mRNA molecules a particular base sequence—AGG AGGU, called **ribosome binding site**—is a part of untranslated segment of mRNA, called **leader sequence** containing 8 to 13 nucleotides. This site is also called **Shine-Dalgarno sequence** and occurs near the AUG codon used for initiation purpose and base pairs with a complementary sequence near the 3' terminus of the 16S rRNA molecule of the ribosome (Fig. 7.2). The $tRNA^{fMet}$ charged with the first amino acid (N-formylmethionine), binds to the 30S subunit at the P site at three ribonucleotides. During the next and succeeding step of translation the 50S subunit joins the 30S initiation complex to give the **70S initiation complex** (*i.e.,* a complete 70S ribosome). Energy for this union is supplied by the hydrolysis of a molecule of GTP which was originally brought to the initiation complex in conjunction with the initiation factor IF2. After union, the two parts of the ribosome hold between them the ribbon of mRNA with its attached N-formylmethionine tRNA.

Table 7.2. Properties of three initiation factors (IFs) of *E.coli* (a prokaryote).

Factor	Mass (dalton)	Functions
1. IF1	8,100	Stimulates activity of IF2 and IF3; increases the affinity of the 30S subunit for other factors.

2. IF2	97,300	Kinetic effector of 30S initiation complex formation; favours binding of aminoacyl tRNAs with blocked αNH_2 groups (*e.g.,* formylmethionine - $tRNA^{fMet}$); positions f-Met—$tRNA^{fMet}$ in P site of ribosome.
3. IF3	20,700	Kinetic effector of 30S initiation complex formation; shifts equilibrium $70S \rightleftharpoons 30S + 50S$ to the right side, thereby ensuring provision of free 30S ribosomal subunit for initiation.

There are several features of the initiation process which deserve comment. Firstly, the binding of f Met-tRNA is not **codon directed**, it is in position on the 30S subunit before mRNA bonds onto the ribosome. Secondly, f Met-tRNA enters directly into the 'P' site on the ribosome, whereas all other aminoacyl tRNAs enter the 'A' site and are subsequently translocated to the 'P' site (see **King**, 1986).

2. Elongation. The elongation of polypeptide chains is achieved by the following steps :

(i) The second charged tRNA binds (due to a codon-directed binding) to the first ribosome at the latter's 'A' site with the help of the proteins, called **elongation factors** (*e.g.,* **EF-Tu**). EF-Tu carries a molecule of GTP. Correct hydrogen bonding with the mRNA template dictates the selection of a new tRNA, and the activity of the EF-Tu ensures the proper positions of the tRNA in the A site. Such a placement activity needs energy that is provided by the hydrolysis of the molecule of GTP to GDP and phosphate. After performing its function, the EF-Tu protein dissociates from the ribosome, and in the cytoplasm is subsequently regenerated to its active form by another elongation factor, the **EF-Ts**. At this point, both sites of the ribosome are occupied by tRNA's, each of which carries an amino acid, and each of which is hydrogen bonded to the template mRNA (Fig. 7.3).

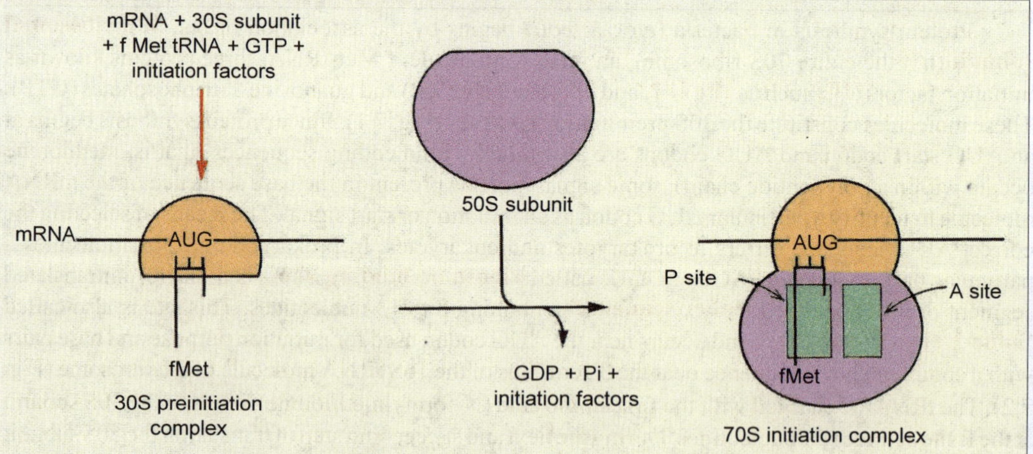

Fig. 7.1. Initiation of protein synthesis in prokaryotes: formation of the 30S preinitiation complex and of the 70S initiation complex (after Freifelder, 1985).

(ii) The next step is the formation of a peptide bond between the two amino acids. To accomplish this job, the first amino acid (N-formylmethionine) is removed from its attachment to its tRNA and transferred to the free–NH_2 terminus of the second amino acid. The first amino acid is, thus, placed "on top of" the second. The ensuing peptide bond, thus, joins the carboxyl group of the first amino acid with the amino group of the second amino acid (Fig. 7.3). The resulting compound is a **dipeptide** whose carboxyl end is still bonded to the second tRNA, but whose amino end is free. The reaction is catalyzed by an enzyme associated with 50S subunit and called **peptidyl transferase**. The energy for

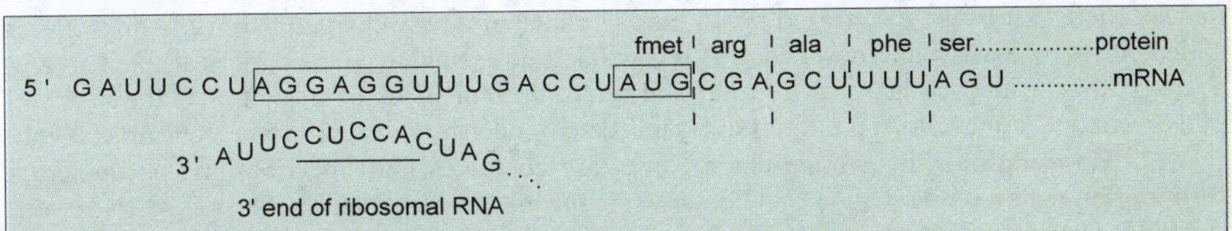

Fig. 7.2. Initiation of translation in prokaryotes. Base-pairing between the Shine-Dalgarno sequence (in box) in the mRNA and the complementary region (underlined) near the 3' terminus of 16S rRNA. The AUG initiation codon is hatched (after Freifelder, 1985).

peptide bond formation is supplied by the dissolution of the aminoacyl bond between the first amino acid and its carrier tRNA, this energy having originally been donated by ATP.

(iii) Although thus far a dipeptide has been generated, continued synthesis requires that the next codon be made available and that the next tRNA be admitted to the A site on the ribosome; this site being still occupied by dipeptide-carrying tRNA. The problem is solved by a movement of the entire ribosome relative to the mRNA strand. This is called **translocation** and consists of the following three steps :

(a) Ejection of discharged $tRNA^{fMet}$ from the 'P' site.

(b) Movement or physical shifting of tRNA dipeptide from the 'A' site to the 'P' site.

(c) Movement of the mRNA is such that the effect is the apparent movement of the ribosome in

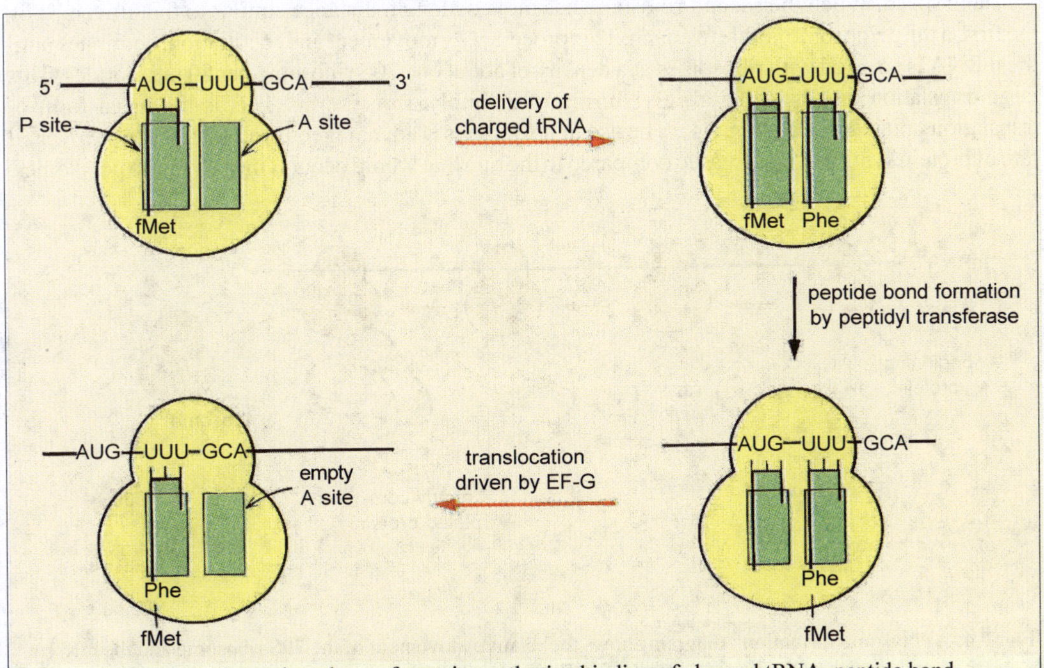

Fig. 7.3. Elongation phase of protein synthesis : binding of charged tRNA, peptide bond formation, and translocation (after Freifelder, 1985).

5'→3' direction by the length of one codon (three nucleotides). This step requires the presence of an elongation factor **EF-G** (called **translocase**) and GTP.

The sequential formation of a polypeptide continues in the manner described above. A tRNA in the 'P' site shifts its burden of growing polypeptide to the next succeeding tRNA, followed by translocation, exit of the discharged tRNA, and entrance of a new charged tRNA (having correct anticodon) to base pair with a new codon at 'A' site. Thus, the growing polypeptide is adopted in turn, by each tRNA, with each successive amino acid being added in effect, to the bottom of the stack. As process continues, the mRNA is progressively translated, codon by codon, from the 5' end to the 3' end.

3. Termination. Polypeptide chain elongation continues until the ribosome encounters a **termination codon** (either UAA, UAG or UGA) in the mRNA template. There are no tRNAs with anticodons complementary to the three termination codons. When the ribosome meets a termination codon, this codon is recognized by protein **release factors** (**RFs**) which bind to the 'A' (or aminoacyl) site of the ribosome. The factor (**RF1**) identifies termination codons UAA and UAG, while factor **RF2** recognizes codon UGA. Thus, the release factors help the ribosome to recognize these triplet codons. The function of **RF3** seems to stimulate the action of RF1 and RF2. The activity of these factors causes the termination of translation, the release of the polypeptide from the tRNA (this is specifically done by the peptidyl transferase enzyme which now serves as a hydrolase, see **King**, 1986), and dissociation of the 70S ribosome into its 30S and 50S subunits which leave the mRNA (This step is performed by IF3 factor).

Polysomes and Coupled Transcription – Translation

The unit of translation is never simply a ribosome traversing a mRNA molecule, but is a more complex structure. After about 25 amino acids have been joined in a polypeptide chain, the AUG initiation site of the encoding mRNA molecule is completely free of the ribosome. A second initiation complex then forms. The overall configuration is of two 70S ribosomes moving along the mRNA at the same speed. When the second ribosome has moved along a distance similar to that traversed by the first, a third ribosome is able to attach. This process, *i.e.,* movement and reinitiation, continues until the mRNA is covered with ribosomes at a density of about one 70S ribosome per 80 nucleotides. This large translation unit is called a **polyribosome** or simply a **polysome**. This is the usual form of translation unit in all cells (Fig.7.4). The use of polysomes is advantageous to a cell, since the over all rate of protein synthesis is increased compared to the rate that would occur if there were no polysomes.

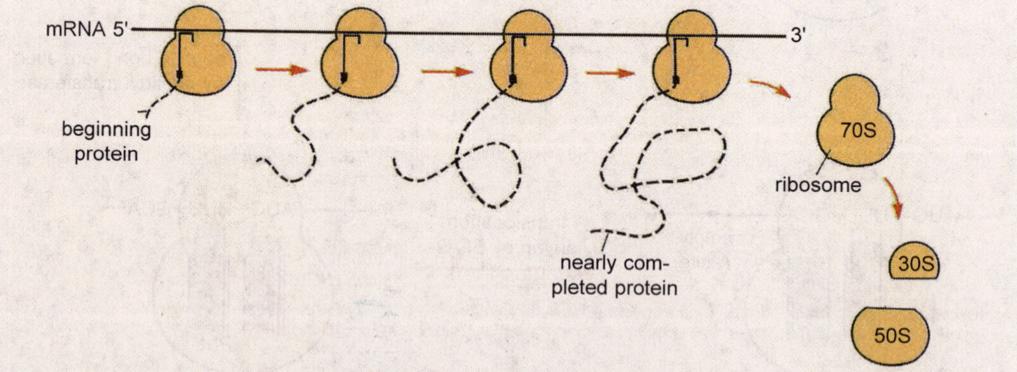

Fig. 7.4. Polysome formation : diagram shows the relative movement of the 70S ribosome and the mRNA, and growth of the protein chain (after Freifelder, 1985).

Further, a mRNA molecule being synthesized has a free 5' terminus and translation also occurs in the 5'→3' direction, so each cistron contained in the mRNA immediately starts its translation. As a result, the ribosome binding site is transcribed first, followed in order by the AUG codon, the region encoding the amino acid sequence, and finally the stop codon. Thus, in bacteria in which no nuclear membrane separates the DNA and the ribosomes, there is no obvious reason why the 70S initiation

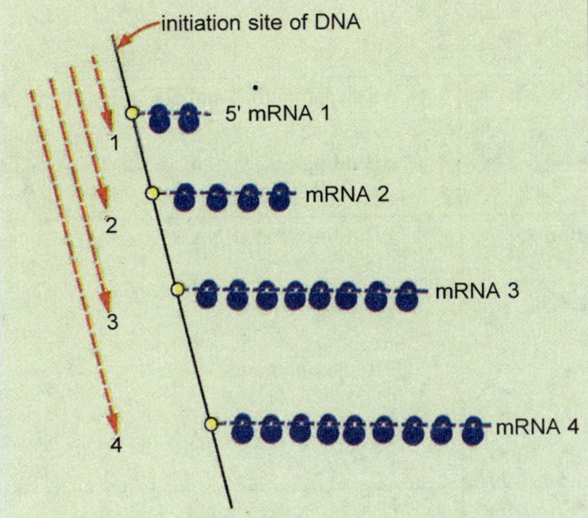

Fig. 7.5. Coupled transcription-translation. Transcription of a section of the DNA of *E. coli* and translation of the nascent mRNA. The dashed arrows show the distances of each RNA polymerase from the transcription initiation site. mRNA4 probably becomes shorter due to partial digestion of its 5' end by a RN ase (after Freifelder, 1985).

complex should not form before the mRNA is released from the DNA. With prokaryotes (*E. coli*) this does indeed occur; this process is called **coupled transcription-translation**. This coupled activity does not occur in eukaryotes, because the mRNA is synthesized and processed in the nucleus and later on transported through the nuclear membrane to the cytoplasm where the ribosomes are located. Coupled transcription-translation, too, speeds up protein synthesis in the sense that translation does not have to await release of mRNA from the DNA.

C. Stages of Polypeptide Synthesis in Eukaryotes

The process of polypeptide (protein) synthesis in eukaryotes follows the same general pattern as that described for the prokaryotes, with the following exceptions:

1. In eukaryotes there are more initiation factors and at least ten initiation factors have been identified in red blood cells (= reticulocytes). They are named by putting a prefix 'e' to signify their eukaryotic origin. These factors are **eIF1**, **eIF2**, **eIF3**, **eIF4A**, **eIF4B**, **eIF4C**, **eIF4D**, **eIF5** and **eIF6** (Table 7-3).

2. Although the initiation codon of mRNA is AUG (methionine) as in bacteria, the initiating methionine tRNA contains no formyl group.

3. In contrast to the often polycistronic mRNA of bacteria, the mRNA of eukaryotes are monocistronic, containing the coding sequence only for one polypeptide. The initiation codon for this sequence is located near the 5' end of the message. Thus, the association of mRNA takes place on 5' end and not at the initiation codon AUG as in prokaryotes.

Table 7.3. Structure and function of nine initiation factors (eIFs) isolated from the reticulocytes (an eukaryotic cell).

Factor	Structure and mass in daltons	Function
1. eIF3	Multimer; 7500,000	Binding mRNA
2. eIF4F	Multimer; 200,000	Binding mRNA 5' end; unwinding
3. eIF1	Monomer; 15,000	Assists mRNA binding
4. eIF4B	Monomer; 15,000	Assists mRNA binding and unbinding
5. eIF4A	Multimer; 15,000	Assists mRNA binding (also binds ATP)

6. eIF6	Monomer; 23,000	Prevents 40S-60S joining
7. eIF5	Monomer; 150,000	Releasing eIF2 and eIF3
8. eIF4C	Monomer; 15,000	Binding 60S subunit
9. eIF2	Trimer containing three chains :	Binding Met-tRNA
	(i) α-chain 35,000	Binds to GTP
	(ii) β-chain 38,000	Recycling factor
	(iii) γ-chain 55,000	Binds to Met-tRNA
10. eIF4D	Monomer	Unknown

4. In eukaryotes, smaller ribosomal subunit (40S) associates with initiator tRNA known as $tRNA^{Met}$ without the help of mRNA, while in prokaryotes, generally the 30S—mRNA complex is first formed, which then associates with f-Met-$tRNA^{fMet}$.

Now, let us discuss, in brief, the polypeptide synthesis in eukaryotes :

1. Initiation. This process involves the following steps : (i) GTP binds to eIF2 and to this complex becomes associated Met-$tRNA^{Met}$ to form a ternary complex, *i.e.,* Met-$tRNA^{Met}$ -eIF2-GTP. The protein eIF2 in mammals and wheat plant has three subunits, namely α, β and γ; eIF2 α binds to GTP; eIF2γ binds to Met-tRNA and eIF2β may be a recycling factor.

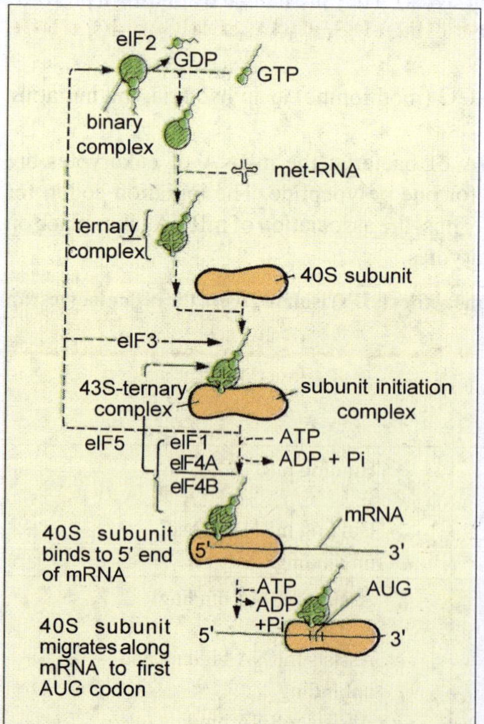

Fig. 7.6. Different steps of initiation of protein synthesis in eukaryotes.

(ii) The ternary (=of three components) complex associates with 40S subunit to form **43 initiation complex**.

(iii) The mRNA at its 5' end binds with 43 initiation complex. This reaction depends on eIF3 and the binding of mRNA is assisted by eIF4F, eIF4A, eIF4B and a high energy bond of ATP.

(iv) After association of the 5' end of mRNA, initiation complex moves towards 3' end in search of initiation codon AUG, and then also associates with 60S subunit. Association of 60S subunit with the initiation complex requires the factor eIF5, because it helps in releasing eIF2 and eIF3 (60S subunit cannot join, if these two initiation factors are not released). The eIF2 is released as a binary complex, eIF2-GDP. The 40S-60S joining reaction really depends on eIF4C. The GTP of the initiation complex is hydrolyzed, when 60S subunit joins (Fig. 7.6).

2. Elongation. For the process of elongation of the peptide chain the next (second) codon in mRNA is base paired by the appropriate AA-tRNA at the A site. This requires the presence of EF1 (elongation factor1) and energy (*i.e.*, one GTP is hydrolyzed to GDP). A peptide is then formed to join the amino group of aa-tRNA with the carboxyl group of p-tRNA. This is achieved by peptidyl transferase enzyme present in the 60S ribosomal subunit. The discharged tRNA in the P site is immediately ejected out. There upon EF2 (also called **translocase** or **G factor**) causes the translocation of newly formed P-tRNA and its codon from A to P site; again one GTP is hydrolyzed to GDP to provide the necessary energy. In this way the polypeptide chain continues to be elongated.

3. Termination. The termination of growing polypeptide chain occurs when a termination codon comes to occupy the 'A' site. There is no tRNA for these codons. The releasing factor 1 or 2 with the help of peptidyl transferase releases the polypeptide from the P-tRNA. This also causes the release of tRNA from the 'P' site and dissociation of 80S ribosome into 60S and 40S subunits.

In eukaryotic systems, only one release factor is known, *i.e.*, **eRFI**. GTP seems to be necessary for binding of this factor to ribosome. GTP is cleaved after termination step has occurred which may be essential for the release of eRFI from the ribosome.

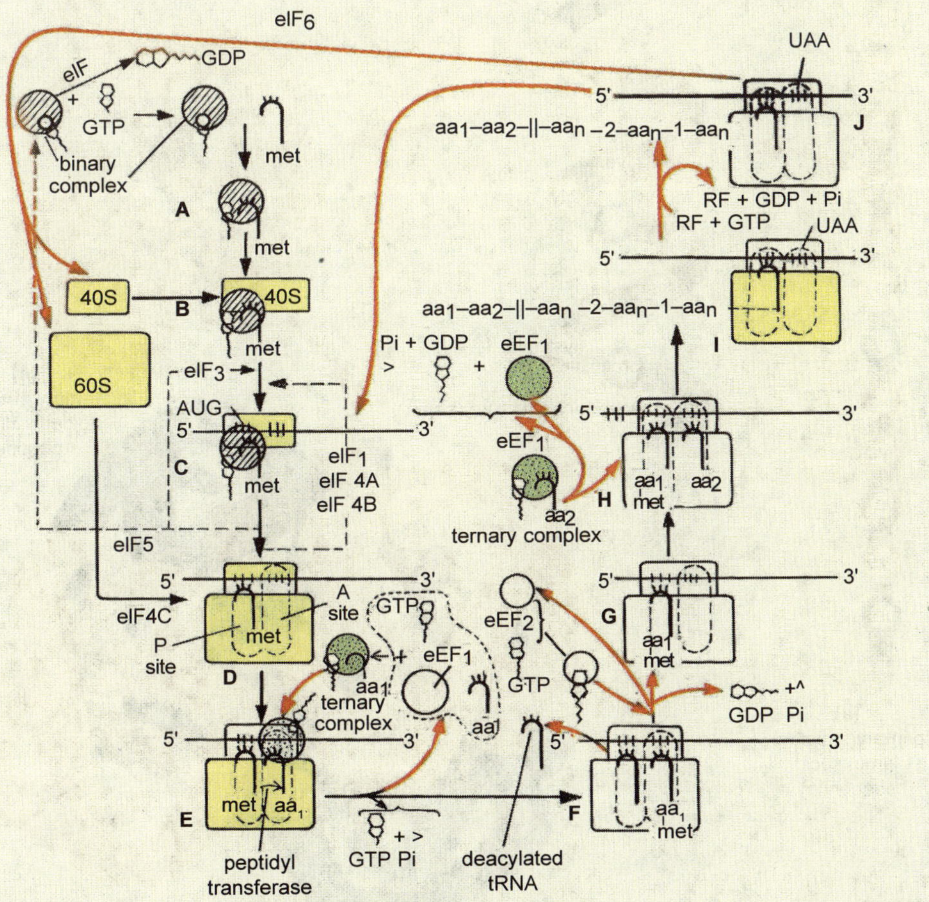

Fig. 7.7. Translation of mRNA in eukaryotes (reticulocytes).

Modification of Released Protein

The released polypeptide chain contains the formylated methionine at its one end. An enzyme **deformylase** removes the formyl group of methionine. The **exopeptidase** enzyme may remove some amino acids from N-terminal end or the C-terminal end of polypeptide chain. At this stage the polypeptide (protein) possesses its primary and probably its secondary structures. The linear sequence of amino acids forms the primary structure; at least some portion of many proteins have a secondary structure in the form of an alpha-helix. The protein chain may then fold back upon itself, forming internal bonds (including hydrogen bonds, ionic bonds, and strong disulphide bonds) which stabilize its tertiary structure into a precisely and often intricately folded pattern. Two or more tertiary structures may unite into a functional quarternary structure. For example, haemoglobin consists of four polypeptide chains, two identical α-chains and two identical β-chains. A protein does not become an active enzyme until it has assumed its tertiary or quarternary pattern. Some quaternary proteins consist of identical subunits (*e.g.,* the enzyme β-galactosidase consists of four identical polypeptide chains).

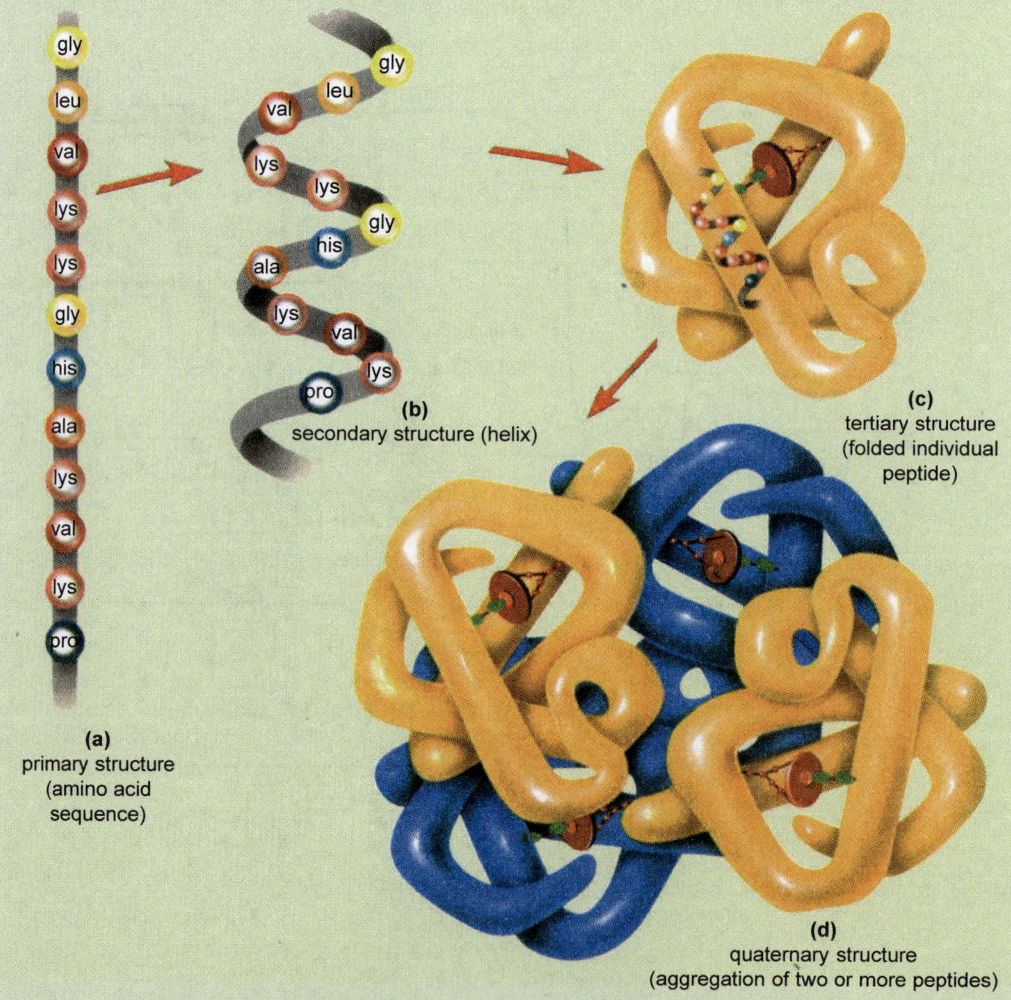

The four levels of protein structure.

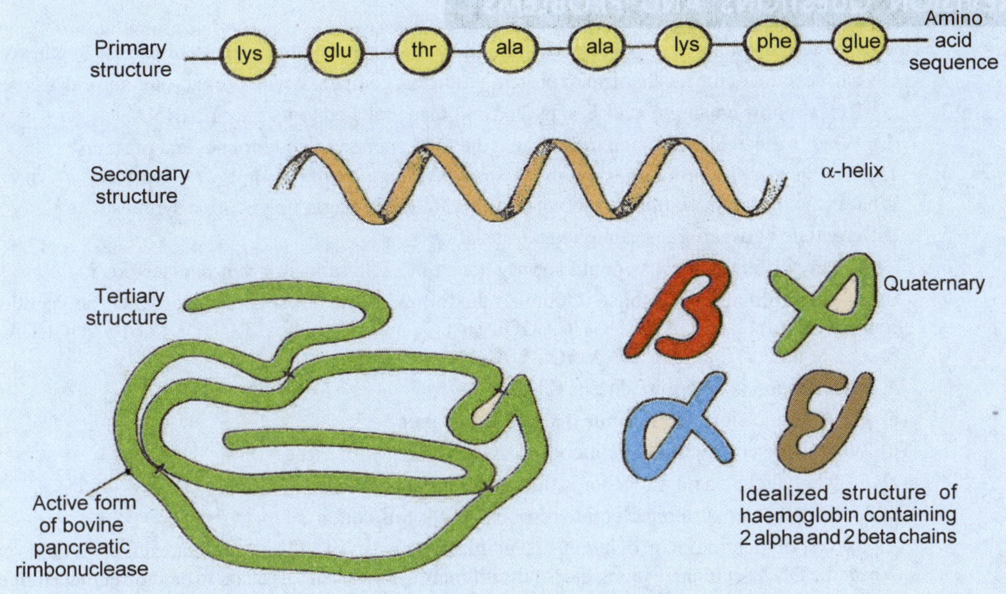

Fig. 7.8. Stages in the development of a functional protein.

Such proteins are called **homopolymers**. Other quaternary proteins, such as haemoglobin, consists of non-identical subunits and are called **heteropolymers**. Some polypeptide chains are subject to posttranslational modification before they assume their biologically active form. For example, chymotripsinogen is cut at one position by an enzyme and the active split product is chymotrypsin.

Role of chaperones in protein folding. Molecular chaperones represent a diverse class of protein molecules which play a significant role in folding of individual polypeptides and in assembly of multimeric proteins. They act by inhibiting incorrect molecular interactions which are not possible in their absence (In fact, chaperone is an English word which means a 'person', usually a married or elderly woman who, for the sake of propriety, accompanies a young unmarried lady in public as guide and protector). One example of chaperone is ATPase.

ANTIBIOTICS AND PROTEIN SYNTHESIS

Many antibacterial agents (called **antibiotics**) have been isolated from fungi. Most of these are inhibitors of protein synthesis. For example, **streptomycin** and **neomycin** bind to a particular protein in the 30S particle and thereby prevent binding of $tRNA^{fMet}$ to the 'P' site; the **tetracyclines** inhibit binding of charged tRNA; **lincomycin** and **chloramphenicol** inhibit the peptidyl transferase; and **puromycin** causes premature chain termination; **erythromycin** binds to a free 50S particle and prevents formation of the 70S ribosome. A particular antibiotic has clinical value only when it acts on bacteria and not on animal cells; the clinically useful antibiotics usually either fail to pass through the cell membrane of animal cells or do not bind to eukaryotic ribosomes, because of some unknown feature of their structure.

Some disease-causing bacteria exert their pathogenic effect, because they excrete inhibitors of mammalian protein synthesis. The agent causing **diphtheria** is an example; it binds to a factor necessary for movement of mammalian ribosomes along the mRNA.

REVISION QUESTIONS AND PROBLEMS

1. How do the functions of rRNA, mRNA and tRNA differ during the process of protein synthesis ?
2. Give an account of the mechanism of protein synthesis; compare it with prokaryotes and eukaryotes.
3. (a). Explain how an amino acid is activated and then attached to its specific tRNA.
   (b). What is the evidence that a mRNA may be long enough to code for several proteins?
4. The first step in protein synthesis is the formation of an initiation complex (**Nomura** *et.al.,* 1967). What is the initiation complex, and what events follow to begin polypeptide synthesis ?
5. Differentiate between transcription and translation.
6. How many different mRNAs could specify the amino acid sequence met-phe-ser-pro ?
7. Using the information in Table 6-2, convert the following mRNA segments into their polypeptide equivalents : (a)........5' GAA AUG GCA GUU UAC 3'..........; (b)............3' UUUUCGAGAUGUCAA 5'............; and (c) ....5' AAA ACC UAG AAC CCA3'........
8. A certain codon is determined to be AUG.
   (a) Of what nucleic acid molecule is this codon a part ?
   (b) What is the corresponding anticodon ?
   (c) Of what nucleic acid molecule is this anticodon a part ?
   (d) What is the deoxyribonucleotide responsible for this codon ?
9. Alpha (α) polypeptide chain of human haemoglobin consists of 141 amino acid residues. Would you expect the DNA segment responsible for the ultimate synthesis of this chain to be shorter than, longer than, or about the same length as the functional mRNA molecule for this chain ?
10. An antibiotic is a microbial product of low molecular weight that specifically interferes with the growth of microorganisms when it is present in exceedingly small amounts. Specify some of the physiological activities that might be interrupted by an appropriate antibiotic and the reason why human cells are not harmed.
11. Which processes in protein synthesis require hydrolysis of GTP ?
12. Write short notes on the following :
    (i) Central dogma : (ii) Teminism; (iii) Reverse transcription; (iv) Kozak's scanning hypothesis; (v) Shine-Dalgarno sequence; (vi) rho factor; (vii) Aminoacyl RNA synthetase; (viii) Elongation factors; (ix) Chaperones ; (x) Peptidyl transferase.

ANSWERS TO PROBLEMS

6. $1 \times 2 \times 6 \times 2 = 48$.
7. (a) - glu-met-ala-val-tyr-.
   (b) - phe - ala- arg- cys- asn - (since genetic code is always read in 5'→3' direction).
   (c) - lys - thr - (nonsense), chain terminates prematurely.
8. (a) By definition, codon occurs only in mRNA.
   (b) UAC
   (c) By definition, anticodons occur only in tRNA.
   (d) TAC.
9. Longer.
10. Cell wall formation is interfered with by penicillins and cephalosporins. DNA replication is prevented by the bleomycins and anthracyclines. Rifamycins interrupt the transcription of DNA into RNA. Translation is disrupted by erythromycin, tetracyclines, chloramphenicol, neomycin, lincomycin, puromycin, and streptomycin. Antibiotic are toxic for microorganisms but safe for humans, because all of these metabolic processes are subtly different in bacteria and humans.
11. Formation of the 70S initiation complex and translocation.

CHAPTER

8

Regulation of Gene Action

Many invertebrate animals – and some vertebrates – change form as they mature. The free swimming larva (top) grows into a slow – moving bottom – dwelling starfish (above). Changes in gene activities cause this metamorphosis. The pattern of change is itself controlled by other genes.

The synthesis of particular gene products is controlled by mechanisms collectively called **gene regulation**. Evidently, though cells contain the genetic capacity for the synthesis of an enormous number of different products (proteins), not all of these products are present at any given time, many being selectively activated only upon special occasion and in response to some environmental stimulus. For example, in prokaryotes some enzymes are synthesized **constitutively** (*i.e.,* continuously), indicating that transcription of mRNA is constantly occurring in them. However, other enzymes are synthesized only when a need for their action arises, and when this need has been fulfilled, enzyme synthesis stops. Transcription of mRNA in this case is evidently initiated only on demand and must, therefore, be subject to regulation. Exhaustive investigations have established that regulation of gene activity both in prokaryotes and eukaryotes may occur at three levels: transcription, translation and post-translation (*i.e.,* folding and processing of proteins).

The gene regulatory systems of prokaryotes and eukaryotes are slightly different from each other. Prokaryotes are generally free-living unicellular organisms that grow and divide continually as long as environmental conditions are suitable and the supply of nutrients is adequate. Thus, their gene regulatory systems are adapted to provide the maximum growth rate in a particular environment, except when such growth would be deterimental. This procedure seems to apply

DNA swquences involved in gene regulation have been identified.

to the free-living unicellular eukaryotes such as yeast, algae and protozoa.

The demands of tissue-forming eukaryotes are quite different from those of prokaryotes. For example, in a developing organism, in an embryo, a cell must not only grow and produce many progeny cells but also must undergo considerable change in morphology and biochemistry (*i.e.*, become differentiated) and then maintain the changed state. Also, during the growth and cell-division phases of the organism, these cells are challenged less by the environment (in contrast to the bacteria), because the composition and concentration of their growth media do not change drastically with time. Some examples of such media are blood, lymph, or other body fluids, or in the case of marine animals, the sea water. Lastly, in an adult organism, growth and cell division in most cell types have stopped, and each cell needs only to maintain itself and its properties. Many other examples could be quoted, the main point is that because a typical eukaryotic cell faces different emergencies than a bacterium does, the gene regulatory mechanisms of eukaryotes and prokaryotes are not the same.

REGULATION OF GENE ACTION IN PROKARYOTES

In prokaryotes (*e.g., E.coli*), the activities of genes are regulated according to the following mechanisms :

1. Transcriptional Control Mechanisms

In bacteria, there occur several mechanisms of gene regulation at the level of transcription. A notable method depends on whether the enzymes being regulated act in catabolic (degradative) or anabolic (synthetic) metabolic pathways. For example, in a multistep catabolic system the availability of the molecule to be degraded commonly determines whether the enzymes in the pathway will be synthesized. In contrast, in a biosynthetic pathway the final product is often the regulatory molecule. Even when a single protein molecule is translated from a monocistronic mRNA molecule, the protein may be **autoregulated**, *i.e.,* the protein itself may inhibit initiation of transcription and high concentrations of the protein may cause less transcription of the mRNA that encodes the protein. The molecular mechanisms for each of the regulatory patterns differ greatly and are of the following two types— negative regulation and positive regulation (Fig. 8.1). In a **negative regulated system**, an inhibitor is present in the cell and prevents transcription. An antagonist of the inhibitor, called an **inducer**, is needed to allow initiation of transcription. In a **positively regulated system**, an **effector** molecule (which may be protein, a small molecule, or a molecular complex) activates a promoter; no inhibitor must be countermanded. Negative and positive regulation are not mutually exclusive, and some systems are both positively and negatively regulated, utilizing two regulators to respond to different conditions in the cell. Thus, a catabolic system may be regulated positively or negatively. In a biosynthetic (anabolic) pathway, the final product usually regulates negatively its own synthesis; in the simplest type of negative regulation, absence of the product increases its synthesis and presence of the product decreases its synthesis.

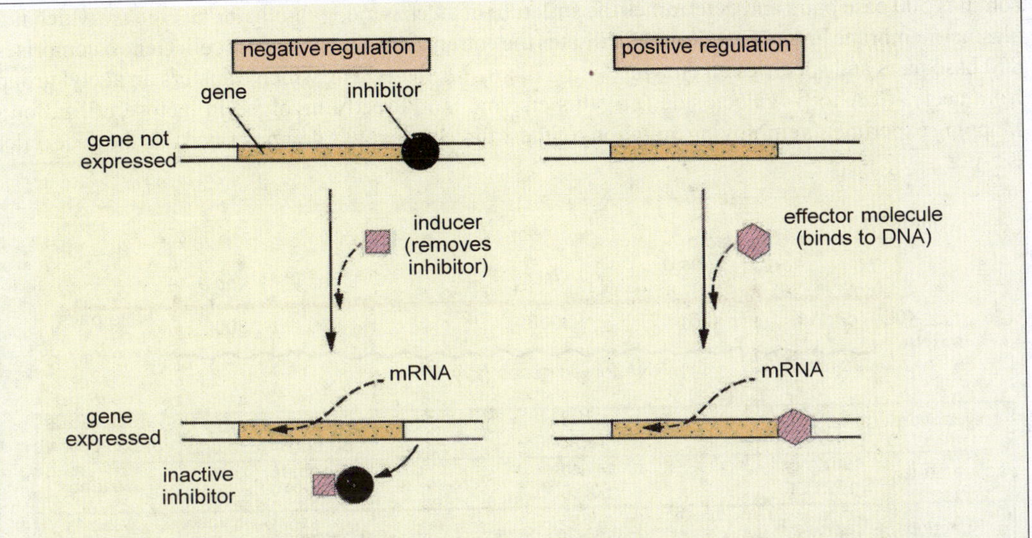

Fig. 8.1. The distinction between negative and positive regulation. In negative regulation an inhibitor, bound to the DNA, must be removed before transcription can occur. In positive regulation an effector molecule must bind to the DNA (after Freifelder, 1985).

A. Negative Control

(i) Inducible Operons (Inducible Systems)

An inducible enzyme is produced only when its substrate (inducer) is present in the environment (*i.e.,* active repressor + inducer = inactive repressor). Most enzymes in this category are catabolic in their activity. **F. Jacob** and **J. Monod** in 1961, on the basis of their study on the inducible system for the synthesis of β- galactosidase enzyme in *E.coli*, proposed a model in order to explain the induction or repression of enzyme synthesis. This model is popularly known as **operon model** and has been variously modified, ever since it was originally proposed by **Jacob** and **Monod**. According to this model, an **operon** was defined as a unit of coordinated control of protein synthesis which consisted of (i) an **operator gene** which controlled the activity of (ii) a number of **structural genes** which took part in the synthesis of protein(s). This means that the structural genes will synthesize mRNA under the operational control of an operator gene. The operator gene, in its turn, is under the control of a repressor molecule synthesized by a **regulator gene**, which is not a part of the operon. Thus, the members of an operon are transcribed coordinately a single, long, polycistronic mRNA molecule. One such operon in *E.coli*, called the **lactose** or **lac operon**, has provided a model system for the study of gene regulation.

Fancois Jacob and Jacques Monod.

Mechanism of lac operon. A *lac* operon contains three structural genes or cistrons, namely *z*, *y* and *a*, whose products (=enzymes) are involved in the breakdown (catabolism) of the sugar lactose (Fig. 8.2). Gene *z* contains 3063 base pairs and codes for an enzyme, **β- galactosidase**, which converts lactose into glucose and galactose; while gene *y* contains 500 base pairs and determines the structure of an enzyme, **galactoside permease** which is a plasma-membrane bound protein and facilitates the entrance of lactose into the cell. Gene *a* comprises 800 base pairs and specifies an enzyme, **thiogalactoside acetylase**, which transfers an acetyl group from acetyl-CoA to β- galactoside (*i.e.,* this enzyme is indirectly involved in lactose utilization). Mapping experiments employing mutations of these three genes have demonstrated that the gene order is *z-y-a*.

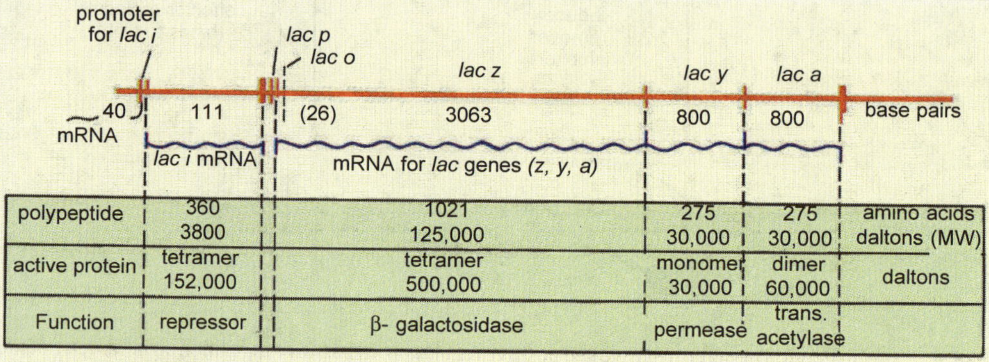

Fig. 8.2. The lactose operon of *E.coli* and its regulatory gene (after Lewin, 1990).

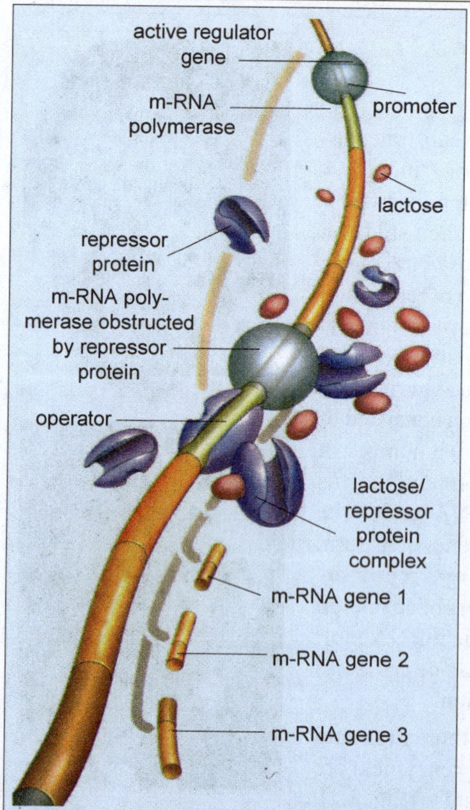

Breakdown of the sugar lactose in bacteria controlled by a segment of DNA called the lactose operon.

Normally, the synthesis of these three enzymes is not constitutive and in the absence of lactose only a few molecules of each enzyme are present. However, when lactose is provided, all three enzymes are synthesized rapidly and simultaneously as co-ordinated response to the presence of this substrate. Thus, lactose acts to **induce** the production of the enzymes needed for its catabolism. Since all three enzymes are synthesized through the translation of a single polycistronic mRNA, it follows that the entire operon is responding as a unit to the presence of inducer.

A clue to the cause of this controlled response on the part of the *lac* operon was provided by a mutation in whose presence all three enzymes were produced constitutively regardless of the presence or absence of lactose. Since the enzymes themselves were normal, it was concluded that this mutation had occurred in a so far unidentified controlling element which was given the name **operator (o)**. By mapping experiments the operator element was found to be situated immediately adjacent to gene *z*, so that the overall gene order was *o-z-y-a*. The operator is found to comprise 26 base pairs.

Another evidence to the mechanism controlling the *lac* operon was provided by the discovery of a gene called **inducer (i)** or **regulatory gene** whose locus was closely linked to the operon, but was separated from that of the operator, gene order being *i-o-z-y-a*.

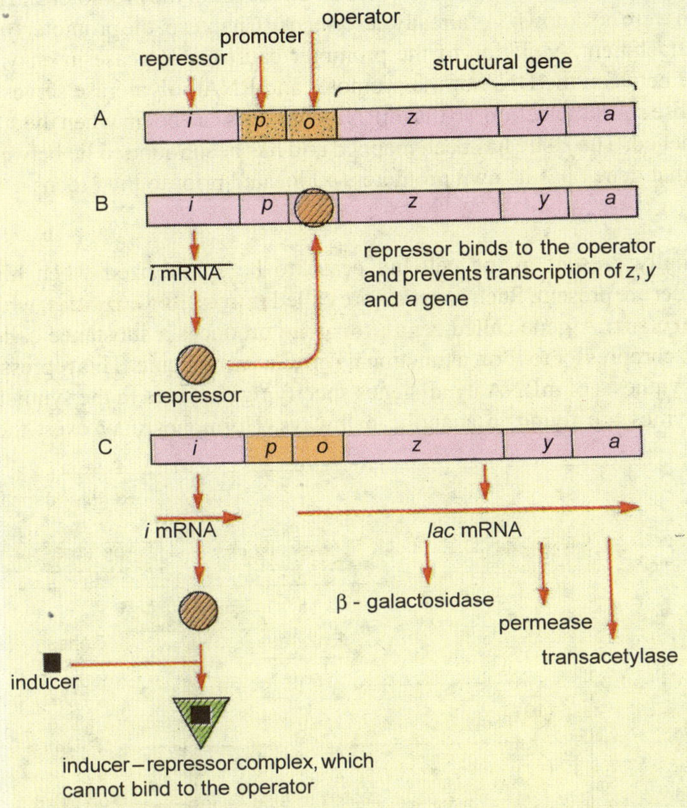

Fig. 8.3. A–Genetic map of the *lac* operon, not drawn to scale: the *p* and *o* sites are actually much smaller than the genes; B–*Lac* operon in repressed state; C–*Lac* operon in induced state. The inducer (lactose) changes the shape of repressor, so repressor can no longer bind to the operator (after Freifelder, 1985).

The *i* gene specified a product (called **repressor**) that could diffuse from the site of transcription and translation to the altogether different region of the genome and there influences the function of the wild type operator ($o+$) and $z+$ gene. Gene *i* is 1111 base pairs long and is transcribed separately from the genes of the *lac* operon as a monocistronic mRNA which is translated on the ribosomes into repressor protein. The repressor protein is a diffusible tetramer protein having 152,000 dalton M.W.

To clarify the respective roles of these elements, **F. Jacob** and **J. Monod** in 1961 proposed that a distinction be made between structural genes such as *z*, *y* and *a* that designate proteins required for cell metabolism, and regulatory genes such as *i*, whose products participate in control mechanisms imposed on structural genes. The operator element was designated a **controlling site** which governed the transcription of physically adjacent structural genes. **Jacob** and **Monod** also proposed a theory to explain the interaction of these loci in the regulation of coordinate enzyme synthesis. According to this theory, the regulatory gene *i* specified a repressor protein which in the absence of the inducer (lactose), was bound to the operator (*o*), thereby inactivating the operator and preventing transcription of the three *lac* cistrons (Fig. 8.3). Induction of transcription was explained as the result of the binding of the inducer (= lactose) to the repressor protein such that the repressor dissociated from the operator. Upon this release from repression, the operator would permit the transcription of the adjacent operon and coordinate enzyme synthesis would follow. The repressor protein was visualized as a molecule with two different, non-overlapping binding sites, one for the operator and the other for the inducer. Union of repressor with inducer would cause a change in the conformation of the repressor protein which rendered the binding site for the operator non-functional. However, when the supply of inducer was depleted through the activity of *lac* enzymes, dissociation of the inducer-repressor complex would occur, permitting a reverse change in conformation so that the repressor could once more bind to the operator to shut down transcription. Such a kind of reversible change in the conformation of a molecule is called **allostery**.

When **Jacob** and **Monod** conducted their studies in 1960s, the existence of the promoter and the mechanism of repressor function were not known. As already described in Chapter 5, the promoter (p) is the site of RNA polymerase attachment. Mutation of the promoter causes a decrease in enzyme synthesis. The complex formed between the DNA of the promoter and RNA polymerase serves to initiate strand separation and sense strand selection, so that mRNA synthesis can begin when the start codon within the operator is reached. The p site has been mapped and has been found to lie between i and o genes. We should note that gene i has its own promoter (p^i) located prior to the i locus.

(ii) Repressible System

Some enzymes are normally present in the cell but cease to be synthesized when high concentrations of their end product are present. Such enzymes are called **repressible enzymes**; while the end product is called **corepressor**. A gene called **regulator gene** produces a substance called **aporepressor** which unites with corepressor to form a functional **repressor** molecule. This repressor molecule or substance inhibits synthesis of mRNA by all genes specifying enzymes in the synthetic pathway. Most repressible enzymes are found in anabolic pathways. Hormones may exert their phenotypic effects by derepressing gen-es previously repressed.

Example. In *Salmonella ty-phimurium* when the amino acid, histidine occurs in high concentration in the medium, that starts to act as corepressor. As a corepressor, amino acid terminates the synthesis of 10 enzymes which are required in pathway to histidine. This kind of repression is called **coordinate repression** (Fig. 8.4).

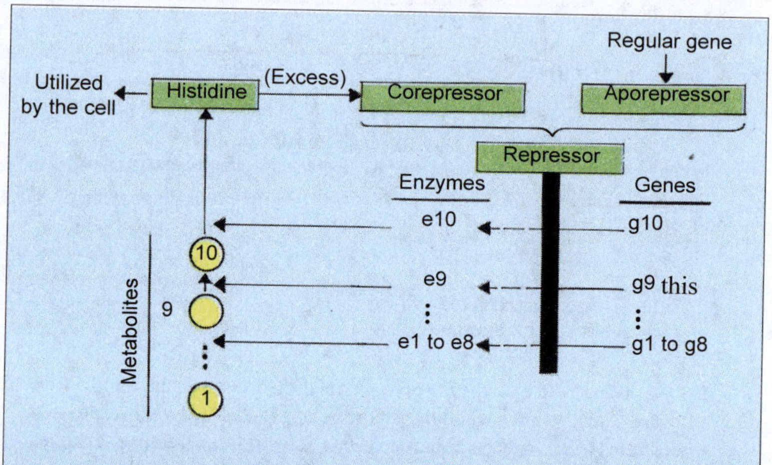

Fig. 8.4. Coordinate repression in *Salmonella typhimurium*.

B. Positive Control

Many bacterial genes are under positive control. This mode of gene regulation is attributed to the presence of factors that enhance the attachment of RNA polymerase to the promoters and initiation of mRNA synthesis.

(i) Effects of glucose on lac operon (Catabolic repression) or Glucose effect. The function of β-galactosidase enzyme in lactose metabolism is to form glucose by cleaving lactose. (The other cleavage product, galactose, is also ultimately converted to glucose by the enzymes of galactose operon). Thus, if both glucose and lactose are present in the growth medium, activity of *lac* operon is not needed, and indeed, no β- galactosidase is formed until virtually all of the glucose in the culture medium is consumed. The lack of synthesis of β-galactosidase is a result of lack of synthesis of *lac* mRNA. No *lac* mRNA is made in the presence of glucose, because in an addition of an inducer to inactivate the *lac i* repressor, another element (*i.e.*, cAMP-CAP) is needed for initiating *lac* mRNA synthesis; the activity of this element is regulated by the concentration of glucose. However, the inhibitory effect of glucose on expression of the *lac* operon is quite indirect.

Small molecules, the **cyclic AMP** (**cAMP**), are present in *E.coli* and many other bacteria. Cyclic AMP is synthesized enzymatically by **adenyl cyclase** and its concentration is regulated indirectly by glucose metabolism. When bacteria are growing in culture medium containing glucose, the cAMP concentration in the cells is quite low. In a medium containing glycerol or any carbon source that cannot enter the biochemical pathway used to metabolize glucose (the glycolytic pathway), the cAMP concentration becomes high. The mechanism by which glucose controls the cAMP concentration is poorly understood; the significant point is that cAMP regulates the activity of the *lac* operon (and other several operons as well). *E. coli* contains a protein called the **catabolic activator protein** (**CAP**), which is encoded in a gene called *crp*. Mutants of either *crp* or the adenyl cyclase gene are unable to synthesize *lac* mRNA, indicating that both CAP and cAMP are required for *lac* mRNA synthesis. CAP and cAMP bind to one another to form a unit, called **cAMP-CAP** which is an active regulatory element of the *lac* system. The cAMP-CAP complex must be bound to a base sequence in the DNA in the promoter region in order for transcription to occur (Fig. 8.5). Thus, cAMP-CAP is a positive regulator, in contrast with the repressor, and *lac* operon is independently regulated both positively and negatively.

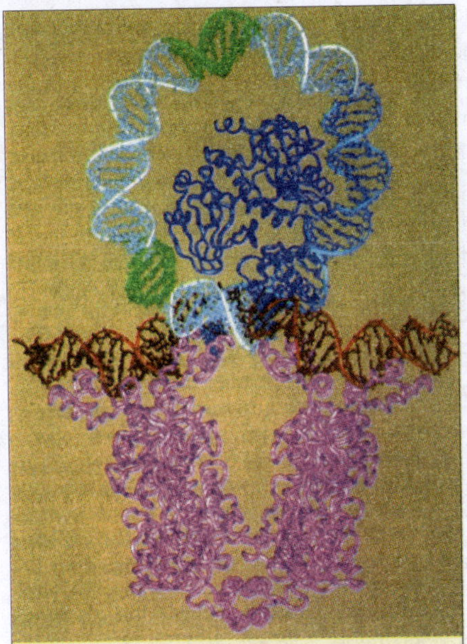

Structure of the *lac* operon repression loop. The *lac* repressor, shown in violet, binds to two DNA regions (red) and immediately upstream from the CRP binding site, within the loop is the CRP binding site (medium blue), shown bound with CAP protein (dark blue).

(ii) Tryptophan operon. The tryptophan (*trp*) operon of *E.coli* is responsible for the synthesis of the amino acid tryptophan. Regulation of this operon occurs in such a way that when tryptophan is present in the growth medium, *Trp* operon is not active. That is, when adequate tryptophan is present, transcription of the operon is inhibited; however, when its supply is insufficient, transcription occurs. The *Trp* operon is quite different from the *lac* operon in that tryptophan acts directly in the repression system rather than as an inducer. Moreover, since the *Trp* operon encodes a set of biosynthetic (or anabolic) rather than cata-

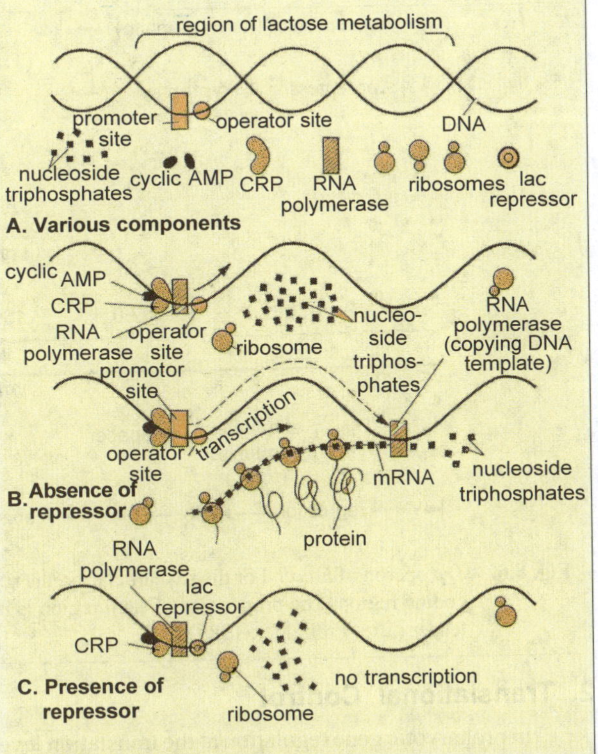

Fig. 8.5. Mechanism involved in the positive control system for the regulation of gene activity in *E.coli lac* operon. It should be noted that only in absence of the repressor, RNA polymerase enzyme can travel and transcribe *lac* operon as shown in B. The repressor when present on the operator site (C), it is an obstacle in the path of RNA polymerase.

bolic enzymes, neither glucose nor cAMP-CAP has a role in the operon activity.

A simple on-off system, as in the *lac* operon, is not ideal for a biosynthetic pathway; a situation may arise in nature in which some tryptophan is available, but not enough to allow normal growth if synthesis of tryptophan was totally shut down. Tryptophan starvation is prevented by a *modulating system in which the amount of transcription in the derepressed state is determined by the concentration of tryptophan*. Such a mechanism is found in many operons responsible for amino acid biosynthesis.

Tryptophan is synthesized in five steps, each requiring a particular enzyme. In the *E.coli* chromosome the genes encoding these enzymes are located adjacent to one another in the same order as they are used in the biosynthetic pathway; they are translated from a single polycistronic mRNA molecule. These genes are called *TrpE*, *TrpD*, *TrpC*, *TrpB* and *TrpA*. The *TrpE* gene is the first one translated. Adjacent to the *TrpE* gene are the promoter, the operator and two regions called the **leader** and the **attenuator**, which are designated *TrpL* and *Trpa* (not TrpA) respectively (Fig. 8.6). The repressor gene *TrpR* is located quite far from gene cluster.

The regulatory protein of the repression system of the *Trp* operon is the *Trp R*-gene product. Mutations either in this gene or in the operator cause constitutive initiation of transcription of trp-mRNA as the *lac* operon. This regulatory protein is called *Trp* **aporepressor** and it does not bind to the operator unless tryptophan is present. The aporepressor and the tryptophan molecule join together to form the active *Trp* repressor which binds to the operator. The reaction scheme is as follows :

Aporepressor (no tryptophan) → Active operator (Transcription occurs)

Aporepressor + Tryptophan → Active repressor
+
operator
↓
Inactive operator
(Transcription does not occur)

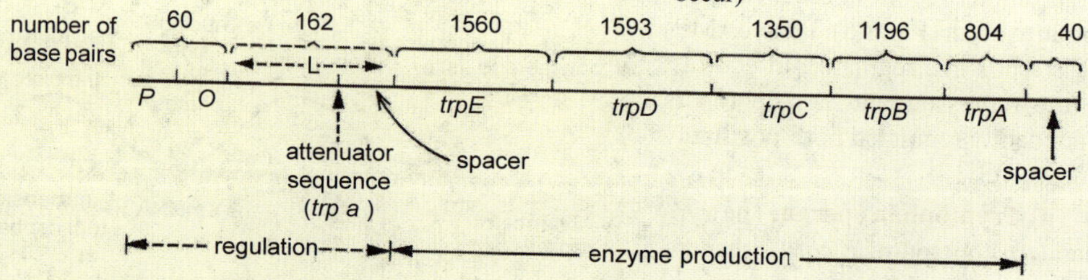

Fig. 8.6. *Trp* operon of *E.coli.* For the sake of clarity, the regulatory region is enlarged with respect to the coding region. The proper size of each region is indicated by the number of base pairs. L is the leader (after Freifelder, 1985).

2. Translational Control

In prokaryotic gene regulation at the translation level, the lifetime of a mRNA molecule may be genetically determined. Enzymatic degradation of mRNA is from the 5' to the 3' end, *i.e.*, the end of the RNA that is first synthesized is also the end that is first degraded. The average lifetime of many mRNA molecules of *E.coli* is only two minutes at 37°C. The specific nucleotide sequences at the 5' end may influence its susceptibility to enzymatic digestion. Further, catabolic enzymes are denied access to the mRNA when the ribosome coated at their 5' ends (*i.e.*, in case of polyribosomes). Hence, the lifetime of mRNAs may also be correlated with the number of free ribosomes available at any given moment to translate mRNA molecules. Bacteria vary their rates of protein synthesis by varying their ribosomal content rather than by varying the translational rate.

Example. In the lactose system of *E.coli*, there are three structural genes under control of a common operator locus determining production of (1) β- galactosidase, (2) galactoside permease and (3) galactoside acetylase. These three proteins are produced in the respective ratios 1:1/2:1/5, reflecting their respective locations relative to the 5' (operator) end of the polycistronic mRNA in which they are coded (these differences are the examples of translation regulations). Thus, there is a *polarity gradient* within the polycistronic mRNA that reduces the probability of cistron translation as a function of its distance fro the 5' end. It is hypothesized that ribosomes attach to different starting points (ribosome-binding sites) along the polycistronic mRNA at different rates as reflected by the relative amounts of the three proteins synthesized.

3. Post-translation Control (Feedback Inhibition or End Product Inhibition)

The expression of genes also can be regulated after proteins have been synthesized. This is called **post-translational control of gene action**. Feedback inhibition is a regulatory mechanism which does not affect enzyme synthesis, but rather inhibits enzyme activity (Fig. 8.7). The end product of a biosynthetic pathway may combine loosely (if in high concentration) with the first enzyme in the pathway. This union does not occur at the catalytic site, but it does modify the tertiary structure of an enzyme and, hence, inactivates the catalytic site. This **allosteric transition** of protein molecule blocks its enzymatic activity and prevents overproduction of end products and their intermediate metabolites.

Example. The studies on isoleucine synthesis in *E.coli* (**Umbarger** 1961) demonstrated that addition of isoleucine (the end product of a five step conversion of threonine) to a culture of the bacteria resulted in immediate blocking of the threonine→isoleucine pathway. In the presence of added isoleucine, the cells preferentially use this **exogenous** end product (*i.e.*, isoleucine) and their own isoleucine synthesis becomes ceased. Moreover, the production of each of the five enzymes is not interfered with, but action of an enzyme responsible for deamination of theronine to α-ketobutyrate is inhibited by the end product, isoleucine.

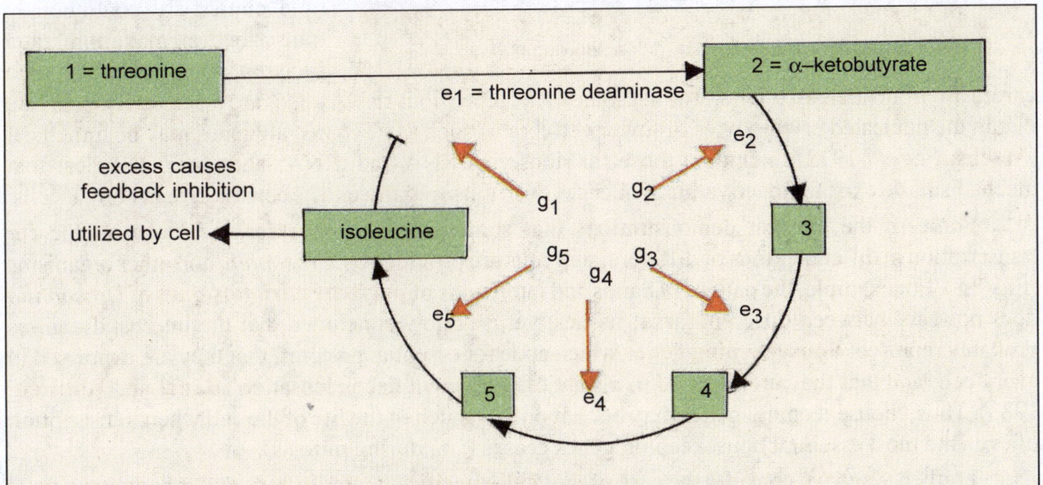

Fig. 8.7. Feedback inhibition in *E.coli.*

REGULATION OF GENE ACTION IN EUKARYOTES

In eukaryotes the following two kinds of controls or regulations of gene expression occur : 1. **Short-term** or **reversible regulation** corresponds to the kind of regulation we studied in bacteria and it represents a cell's response to fluctuations in the environment, specifically, it involves changes in activities or concentrations of enzymes as particular substrates and or hormone levels rise and fall. The changes a cell experiences during a cell cycle, particularly the fluctuations in rates of DNA, RNA, and protein synthesis that regularly occur with respect to the time of mitosis can also be placed in this category. 2. **Long-term** or **irreversible regulation** includes the phenomena associated with **determination**, **differentiation**, or more generally development: it is involved in the numerous steps by which a fertilized egg becomes an organism of, perhaps, trillions of cells with diverse and ultimately quite permanent roles to play in the maintenance of the whole. Short-term regulations also occur in developing and differentiating eukaryotic cells side by side of long-term regulation.

Both of these types of regulations of gene activities in eukaryotes, now, are considered to occur at the following levels involving diverse mechanisms: 1. Regulation at the level of DNA; 2. Regulation at the level of transcription; 3. Regulation at the level of translation; and 4. Regulation at the level of post-translation.

A. Regulation of Gene Action at the Level of Genome

In eukaryotic cells, it seems that certain classes of genes are transcribed more or less continuously, and only in extreme situations their activities are repressed. For example, genes coding for larger ribosomal RNA (28 S or 18 S rRNA) or transfer RNA (tRNA) are present as multiple copies forming **simple multigene family**. These genes are transcribed uniquely by RNA polymerase I for the larger ribosomal RNA or by RNA polymerase III for tRAN and 5S RNA. Although the products of some of these genes, the ribosomes, are used continuously in all cells, it does not conform that all of these multiple copies are continuously transcribed at maximum rate. Electron micrographs of spread chromatin from nucleoli often show that some of the repetitious rRNA genes are inactive. It is also true that in the nucleated erythrocytes of lower vertebrates such as *Xenopus*, all genes may be turned off (**Maclean** *et al*., 1972), including those for ribosomal RNA and tRNA. Therefore, it is clear that mechanisms do exist for inactivating sequences even those regarded to be constitutive in normal cells.

A giant chromosome of a midge showing puff.

Some of the clearest demonstrations that some specific genes are at least available for transcription in different kinds of differentiated cells are provided by *Drosophila* and other organisms (Fig. 8.8). For example, the pattern of bands and interbands of **polytene chromosomes** of *Drosophila* does not vary between different larval tissues, yet it is now concluded that the interband regions probably represent **housekeeping genes** which code for essential proteins, that they are expressed in every cell, and that they are retained in a state of permanent decondensation (**Bautz** and **Kabisch**, 1983). Thus, 'housekeeping' genes may be 'left on' for much of the life of the cell when transcription of even the most essential housekeeping genes ceases (*i.e.*, during mitosis).

Further, when we consider the case of the "**cell-specific**" **genes**, **luxury genes** or **smart genes**, which code for the products only found in specialized tissues, it becomes immediately clear that differential expression is the rule. Whether expression of gene is measured at the level of the messenger RNA or the protein, genes coding for products such as globin, crystallin, fibroin, ovalbumin, casein and immunoglobulin give every indication of complete repression in all but the specialized tissue characterized by their presence.

Thus, at the level of genome (*i.e.*, DNA), the following five modes of regulation are operative:

(i) Situations of total genetic shutdown. (a) During mitotic phase of the cell cycle, chromatin is highly condensed to form chromosomes, and transcriptional activity of all genes is suspended.

(b) During meiotic division of germ cells a somewhat similar situation to (a) is evident, although in some rare cases, such as **lampbrush chromosomes** of meiotic diplotene in vertebrates (**Vlad**, 1983), transcription proceeds very actively.

(c) The nucleus of mature nucleated erythrocytes of amphibians is transcriptionally inactive. Chromatin in these cells is highly condensed but not organized into discrete chromosomes (**Chegini** *et al.*, 1981). However, transcription can be partially reactivated in these nuclei by transferring them into new cytoplasm or exposing them *in vitro* to altered environmental conditions.

(d) In mammalian females, one of the two X chromosomes present in somatic cells undergoes condensation in early embryonic stages to become heterochromatic **sex chromatin** or **Barr body** (Dosage compensation). A variety of experiments indicate that most, though not all, genes of the condensed X chromosome are turned off. In developing oocytes, as opposed to somatic cells, Barr bodies are not present, the activities of both X chromosomes being required for normal oogenesis.

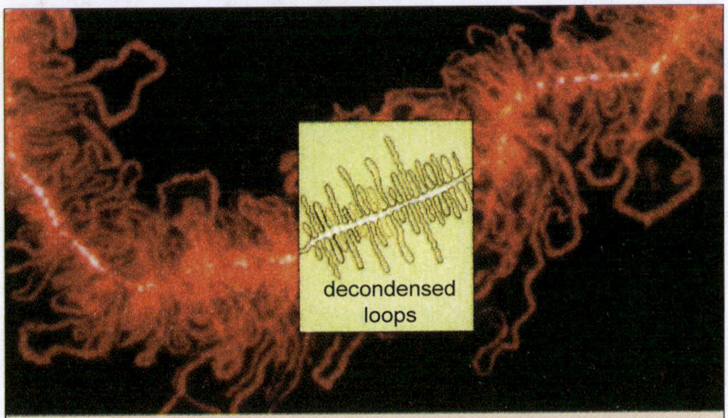

Lampbrush chromosome of a newt.

In somatic cells of normal XY males, genes of the single X chromosome remain active and Barr bodies are not found. However, in germ cells, the X chromosome is inactivated prior to spermatogenesis, otherwise, it may prevent sperm maturation and lead to sterility. In one extreme case, that of the creeping vole, *Microtus oregoni*, the X chromosome is eliminated from the germ cells of males by a special process of nondisjunction (see **Farnsworth**, 1988).

(e) Sperm cells clearly contain a complete genetic endowment but no transcription occurs until the sperm nucleus is activated within the egg cytoplasm.

(f) Complete suspension of transcriptional activity is also known in the following cases: cells of some plant seeds; cells within diapausing *Artemia* gastrulae; cells within inactive organisms such as desiccated *Tardigrada*; nuclei within bacterial and fungal spores; and nuclei within desiccated amoeba cells, as for example in the slime mould *Dictyostelium*.

(ii) Evidence for constitutive expression of some genes. (a) If the interbands of *Drosophila* polytene chromosomes are correctly interpreted as being loci for "housekeeping" genes, then the evidence is that such chromatin is permanently decondensed and is transcribed at a low but constant rate (**Semeshin** *et al.*, 1979).

(b) Electron microscopy of spread films of DNA extracted from nucleoli of *X. laevis* oocytes reveals tandemly arranged sequences coding for the 45S precursor of ribosomal RNA, each gene adorned with a Christmas-tree arrangement of RNA in the process of synthesis (*i.e.*, transcription).

(c) There is a constant and universal requirement for the products of certain genes in all cells and at all times. These include products such as the four kinds of rRNA—28S, 18S, 5.8S and 5S; tRNA of 20 basic types, and a few hundred proteins such as histones, ubiquitin and lactate dehydrogenase, RNA polymerase, and the like.

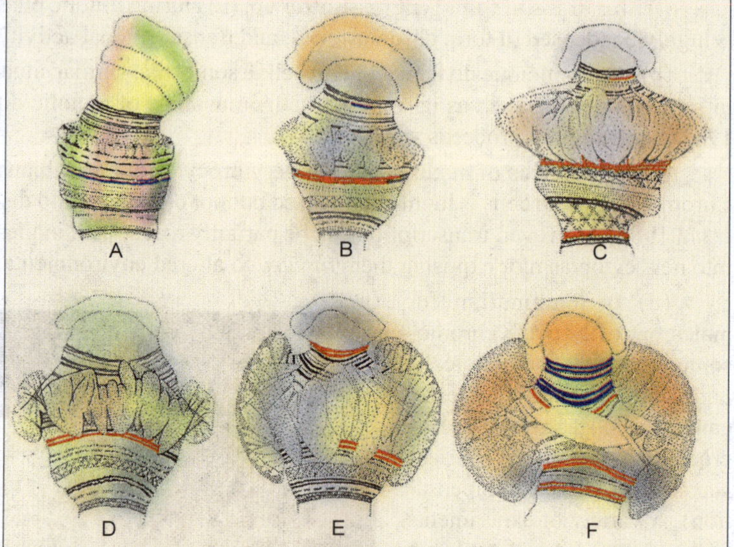

Fig. 8.8. The development of a chromosomal puff in a larval salivary gland cell nucleus of *Chironomus tentans*.

(iii) Many genes are expressed only in certain tissues. (a) *Xenopus* provides a good example of regulation of 5S genes. *Xenopus borealis* possesses 19,000 copies of the oocyte-specific 5S rRNA genes, and these genes are active only in the oocyte and in no other cell.

(b) The enzyme lactate dehydrogenase (LDH) is coded by a small family of genes, each gene determining the structure of a subunit. Subunits A and B are expressed in almost all mammalian cells, but one of the genes in the family, coding for subunit C, is active only in spermatocytes within the developing testes.

(c) The **puffing** (*i.e.*, chromatin decondensation) of restricted segments of the polytene chromosomes of *Drosophila, Chironomus* (Fig. 8.8), etc., provides visible evidence of the activity of genes coding for cell-specific products. Certain puffs, known as **heat-shock puffs**, can be induced to appear specifically when salivary glands are exposed to heat shock either *in vivo* or *in vitro*. The correlation between such chromatin decondensation and transcription activity is readily proved by autoradiography using tritiated precursors of RNA.

(iv) Some DNA is never transcribed in any cell. Analysis of various types of DNA sequences existing in eukaryotic cells reveals that some DNA is comprised of tandemly repeated short sequences that are concentrated in **heterochromatin** such as centromeres of chromosomes and the Y chromosome. Current evidence indicates that much of this DNA is never transcribed in any cell. Some spacer sequences occur between genes, for example, between multiple copies of genes for ribosomal RNA. Such spacer sequences are often taken to be untranscribed, but now it is found that some of the spacer DNA may transcribe nuclear RNA molecules.

The very large size of the genomes of the higher eukaryotes certainly indicates that much of the DNA is redundant (= repetitious) and probably not utilized as coding or regulatory sequence. The **pseudogenes** found in many gene families, often presumed to have arisen as cDNA (=complementary DNA) copies of reverse transcribed message. They lack introns and contain many stop signals so that RNA polymerase molecules fail to move very far along them.

(v) Some DNA is spliced to cause gene rearrangement. Such a mechanism occurs during expression of immunoglobulin (Ig) genes.

B. Regulation of Gene Action at the Level of Transcription

(a) Chromatin reconstitution experiments. Chromatin has three main components—DNA, histones and non-histones. While it is known since early 1960s that histones may be involved in repressing gene activity, the specific regulation by non-histones was shown only during 1970s. **Gilmour** and **Paul** (1973) performed a chromatin reconstitution experiment to demonstrate the positive role of non-histones in regulation of gene activity (Fig. 8.9). They isolated the chromatin from different tissues separately and then dissociated into DNA, histones and non-histones. This is followed by **chromatin reconstitution** using either the three components derived from the chromatin of the same tissue or by combining the non-histones of one tissue with the DNA and histones of another tissue (Fig. 8.9). From such experiments, it was demonstrated that the mRNA which is synthesized *in vitro* from reconstituted chromatin, mainly depended on the source of non-histone proteins (see **O'Malley** *et al.*, 1977).

(b) Change in chromatin conformation. Evidently nucleosomes continue to be present on most transcriptionally active DNA sequences, but they are probably reduced in number. Thus, although evidence from some laboratories suggests that the ribosomal genes of the amphibian nucleolus lack nucleosomes, active gene loci in *Drosophila* give a positive reaction to antibodies against H3 and H4, indicating that at least these subunits of the nucleosome persist on such DNA. In fact, in some active genes the nucleosomes are displaced or "phased" in these regions (**Samal** *et. al.*, 1981).

In this experiment plants treated with compounds that interfere with DNA methylation produced a greatly increased number of flower stalks. Moreover, flowers that developed showed altered morphology. Control plants are shown in parts A, A and E whereas antisense – treated plants are shown in B, D and F.

(c) Modification of DNA sequences : DNA methylation. The genomic DNA of higher eukaryotes is modified following replication so that a large proportion of the cytosine (C) residues are present as **5-methylcytosine (5mC)**. However, such methylation has not been detected in the DNA of lower eukaryotes such as yeast and *Dictyostelium*, nor in *Drosophila*. The percentage of methylated C residues in DNA relative to unmethylated C residues is highly variable, from less than 1 per cent in some insects to over 50 per cent in some higher plants and vertebrates. A much greater correlation exists between the methylation or under-methylation of sequence in the vicinity of gene promoters. For example, DNA of sperm is highly methylated, as in the DNA of the oocyte-specific 5S rRNA genes in adult tissues, whereas the sites around the coding regions of genes such as adult globin, ovalbumin, and immunoglobulin are under methylated in tissues in which they are expressed but are largely methylated in other cells in which they are not expressed.

There is evidence that the cytosine methylation in DNA alters the structure of the double helix in a fundamental way and favours the transition from B-form to Z-form DNA (**Bele** and **Felsenfeld**, 1981). It is possible that B–Z transition is itself involved in gene regulation and this may be the way in which DNA methylation has its effects on transcription.

(d) Modifications of histones. Histone component of chromatin is subject to three different post-synthetic modifications which have either direct or indirect effect on eukaryotic gene regulation : 1. **Histone methylation** affects only histones H3 and H4 and involves the irreversible methylation of a few lysine residues which alters the hydrophobic nature of the side chain of these histones. 2. **Histone phosphorylation** involves histone H1. Phosphorylation affects **serines** and **threonines**, changing them from a state of neutral charge to one of negative charge and is a reversible reaction. The state of phosphorylation of H1 protein varies through the eukaryotic cell cycle, and after H1 phosphorylation,

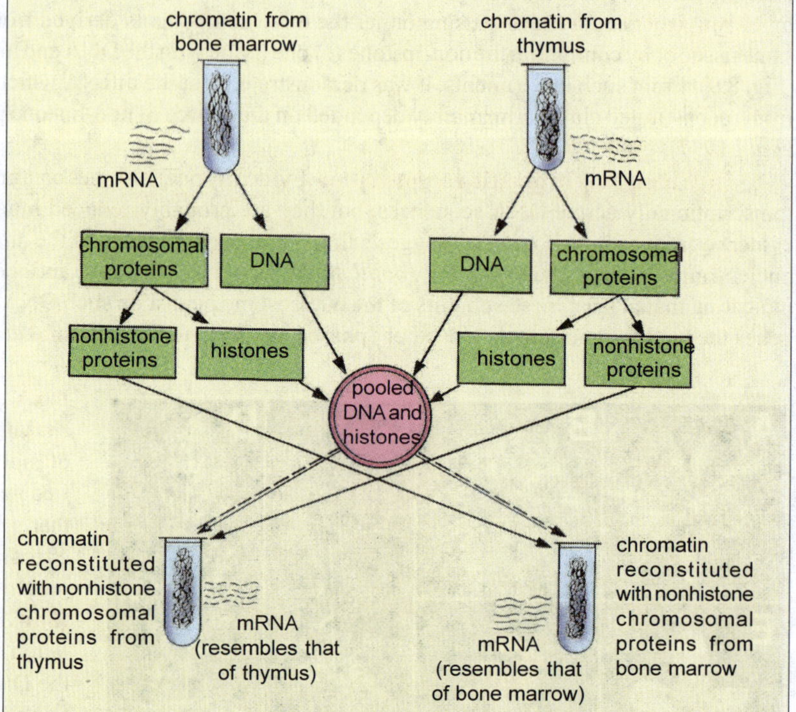

Fig. 8.9. Chromatin reconstitution experiment demonstrating the positive role of non-histone proteins in transcription.

chromatin becomes much more strongly condensed, as it does in mitotic chromosomes. Evidently, activation of the **histone kinase** enzyme that is responsible for H1 phosphorylation may be the first step in the chain of events that leads to eventual chromatin condensation prior to mitotic cell division. 3. **Histone acetylation** is of two types. The first is the irreversible acetylation of the amino-terminal **serines** of histones H1, H2A and H4. These modifications seem to be associated with histone synthesis. The second is reversible acetylation of **lysine** residues in the amino terminal regions of histones H2A, H2B, H3, and H4. The acetylation converts the normally basic lysine side chain to a neutral acetyl lysine, and, thus, reduces the net basic charge of the amino-terminal ends of the affected histones. Both H3 and H4 can have up to four lysines in the acetyl form, and there is a strong correlation between this type of histone acetylation, especially **tetracetylation** of all available lysines, and transcriptionally active chromatin. Evidently, the acetylation of the core histone lysines would tend to loosen the nucleosomal structure which brings about the transition from a silent condensed gene to a transcriptionally active and extended one.

(e) Transcriptional regulation by protein A24. The **A24** is an unusual hybrid protein, being a complex of histone **H2A** and the non-basic protein **ubiquitin**. The ubiquitin is covalently bound via the side chain amino group of lysine 19 of the histone. Some 10 per cent of H2A molecules are in the form of A24 and these specialized histones seem to be confined to interphase chromatin, disappearing as the chromosomes condense. A24 is found highly abundant in the chromatin of active genes.

(f) Gene regulatory molecules. Transcription of the eukaryotic genome is believed to be regulated by a variety of specific gene regulatory molecules which are produced by specific regulatory genes or by cytoplasm/cell surface. Examples of such gene regulatory molecules are the following :

(i) RNA polymerases. These enzymes are necessary for transcription and if they are short in supply, they tend to affect it. For example, there is a possible competition for type II polymerase (which is meant for hnRNA and mRNA) by the various promoter sequences that lie upstream of protein coding sequence.

(ii) Endonucleases. These enzymes are likely to affect the transcription, especially *in vitro* cell-free systems, by introducing into DNA nicks that may serve as initiation sites for some polymerases.

(iii) Topoisomerases, helicases and other DNA helix-destabilizing proteins. Various proteins are known that alter the three-dimensional structure of DNA and render it more available for processing which may affect transcription.

(iv) DNA methylase. This enzyme is likely to make DNA less available for transcription, and factors that antagonize methylation would enhance transcription.

(v) Histone acetylase and deacetylases. Such enzymes influence the rate of transcription by modulating acetylation.

Roy J. Britten.

(vi) Factors such as ATP. Such molecules may influence the transcription rate by changing the available energy.

(vii) Ions and small molecules. Many ions such as those of **calcium**, **magnesium** and **manganese** directly affect chromatin conformation, which modulates gene activity.

Britten-Davidson model or gene-battery model of transcription regulation. This model of eukaryotic gene regulation at the level of transcription was proposed by **Britten** and **Davidson** in 1969 and later on elaborated by them in 1973. The gene-battery model assumes the presence of four classes of sequences : 1. **producer gene** which is comparable to a structural gene of prokaryotic operon; 2. **receptor site** is located adjacent to each producer gene and is comparable to operator gene of prokaryotic operon; 3. **integrator gene** which is comparable to regulator gene and is responsible for synthesis of an **activator RNA** that may or may not give rise to proteins before it activates the receptor site; 4. **sensor site** regulates the activity of integrator gene, which can be transcribed only when the sensor site is activated. The sensor sites are recognized by agents which, like hormones and proteins, change the pattern of gene expression. For instance, hormone-protein complex or a transcription factor may bind to a sensor site and çause the transcription of integrator.

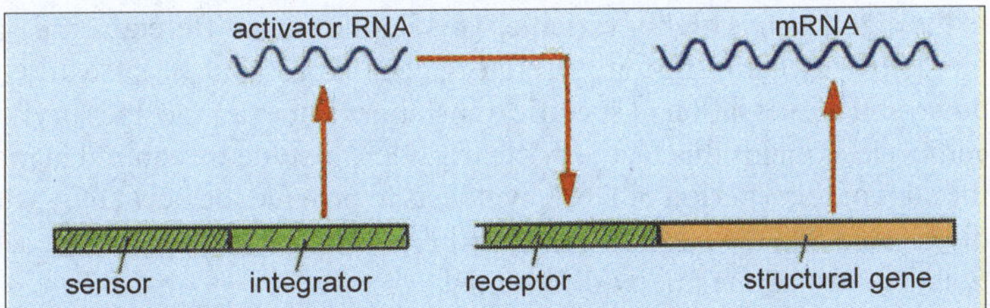

Fig. 8.10. Various components of Britten-Davidson's model for transcription regulation.

In this model, the genes (producer gene and integrator gene) are those sequences which are involved in RNA synthesis, while the receptor and sensor sites help only in recognition without taking part in RNA synthesis. Lastly, in Britten-Davidson's model, a set of structural genes controlled by one sensor site is called the **battery**.

C. Post-transcriptional Regulation

Steps that come between transcription and translation are described as **post-transcriptional** and are the following :

1. Some RNA is capped and tailed. The precise functions of capping and tailing of mRNA are not known, but they seem to serve to identify a message or potential message, and tailing may also help in the final export of this message from the nucleus.

2. RNA is processed to remove intron sequences. Introns removal and **splicing** together of the remaining exons during processing of hn RNA must be absolutely precise. This is in part engineered by a distinct group of nuclear particles (Sn RNPs containing U1, U2, U3, U4, U5 and U6 sn RNAs).

For example, **differential splicing** is used in different lymphocyte cells to produce different proteins from the same hn RNA molecule. As originally discovered by **Early** *et al*., (1980), the two types of mRNA molecules are produced by part of an intron being omitted from one of the mRNAs but included in the exon splice used to produce the other mRNA. This allows production of two distinct proteins both immunoglobulins (Ig), but *one* with a long strand of hydrophobic amino acids at its carboxyl terminus, and the *other* with only a short length of relatively hydrophilic amino acids. The Ig molecule with the long hydrophobic peptide is membrane bounded within a lymphocyte, whilst the molecule with the terminal hydrophilic peptide is secreted from the cell. This change in splicing takes place within the life of a single lymphocyte cell and clearly explains the following observation. Immature lymphocytes retain antibody and simply insert the Ig molecules into their plasma membranes, whereas following stimulation with antigen, the same lymphocyte becomes secretory, releasing antibody molecules into circulation.

3. Most RNA is never exported from the nucleus. About 5 per cent of total transcribed RNA never leaves the nucleus. This is explained partly by removal of intron RNA and also by many RNA molecules which break up within the nucleus. The significance of this process is not clear, but some clues about the identification of RNA for export are coming to light. Although not all genes contain introns, most do, and it seems that the presence of some of these introns is essential for RNA export. In other words, introns are used as a means of identifying or ticketing the molecules that are to be passed out of the nucleus (see **Maclean** and **Hall**, 1987).

4. Message degradation rates are significant. The rate at which eukaryotic mRNA is degraded in the cytoplasm is highly variable. This implies that differential message breakdown is an important method of regulating not only the rate of gene expression, but also the lag between transcriptional shut-down and the cessation of specific translation. For example, the survival of histone mRNA during the cell cycle explains this fact very clearly. New histone is required in massive amounts immediately at the start of the S period of DNA synthesis to provide the new DNA with nucleosomes. Recently, it is discovered that the restricted availability of histone message is not achieved as a result of transcriptional control alone but by differential breakdown rates for histone message.

D. Translational Control

In bacteria, most mRNA molecules are translated about the same number of times with only fairly small variation from gene to gene. In eukaryotes translational regulation occurs in which a mRNA molecule is not translated at all until a signal is received. Translational control may involve the following mechanisms :

1. Extension of lifetime of the mRNA. An important example of translational regulation is that of **informosomes** or **masked mRNA**. Unfertilized eggs are biologically static, but shortly after fertilization many new proteins must be synthesized, for example, the proteins of the mitotic apparatus, the cell membranes, histones for nucleosome formation as well as others.

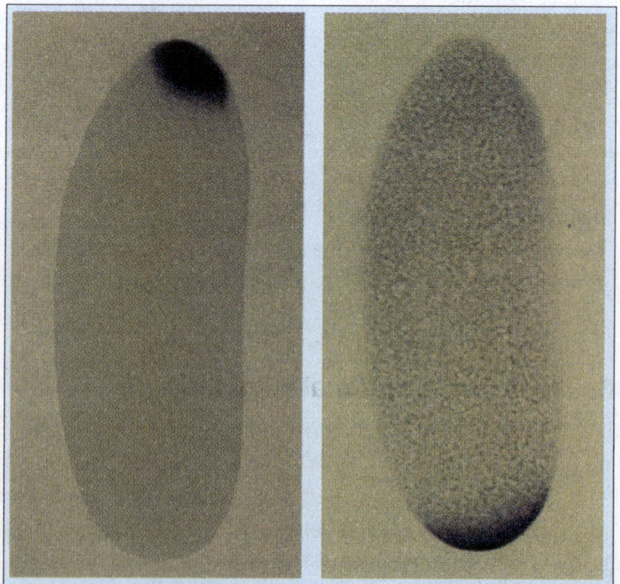

Localization of mRNA in the cytoplasm of a *Drosophila* egg.
Left – *bicoid* mRNA at anterior pole;
Right – *oskar* mRNA at posterior pole.

Unfertilized sea urchin eggs store large quantities of mRNA for many months in the form of mRNA-protein particles (= masked mRNA) made during formation of the egg. This mRNA is translationally inactive, but within minutes after fertilization, translation of these molecules begins. Here, the timing of translation is regulated; the mechanism for stabilizing the mRNA, for protecting it against RNases, and for activation are still unknown.

2. Regulation of rate of protein synthesis. This type of regulation also occurs in mature unfertilized eggs. These cells need to maintain themselves but do not have to grow or undergo a change of state. Thus, the rate of protein synthesis in eggs is generally low. This is not due to inadequate supply of mRNA but of a limitation of an as-yet-unidentified element, called the **recruitment factor** which apparently interferes with formation of the ribosome-mRNA complex. A good example of translational control is the extension of the lifetime of silk fibroin mRNA in the silkworm *Bombyx mori.* During cocoon formation the silk gland of the silkworm predominantly synthesizes a single type of protein, **silk fibroin**. Since the worm takes several days to construct its cocoon, it is the total amount and not the rate of fibroin synthesis that must be great; the silk worm achieves this by synthesizing a fibroin mRNA molecule that is very long lived.

Transcription of the fibroin gene is initiated at a strong promoter by an unknown signal and about 10^4 fibroin mRNA molecules are made in a period of several days (such a synthesis forms an example of transcriptional regulation). A typical eukaryotic mRNA molecule has a life-time of about three hours before it is degraded. However, the fibroin mRNA survives for several days during which each mRNA is translated repeatedly to yield 10^5 fibroin molecules. Thus, each gene is responsible for the synthesis of 10^9 protein molecules in four days. Altogether the silk gland makes 300 μg or 10^{15} molecules of fibroin during this period. If the lifetime of mRNA were not extended, either 25 times as many genes would be needed or synthesis of the required fibroin would take about 100 days.

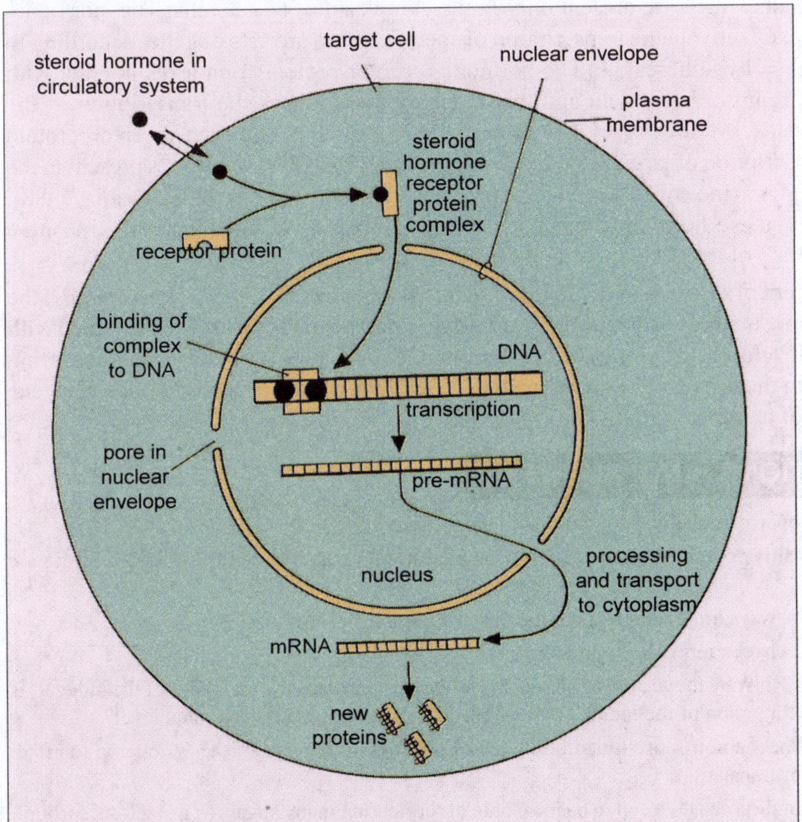

Fig.8.11. Mode of action of steroid hormones on the gene expression (*i.e.*, transcription) (after Gardner, *et al.*, 1991).

E. Post-translational Modification of Proteins to Make Them Active Ones

Some proteins are altered after synthesis, usually by partial degradation or trimming, as for example, by the enzymatic removal of the central section of the **proinsulin** molecule to yield the active protein, **insulin**. For their activity many proteins also depend on being complexed into compound proteins together with other subunits, either the same or different in nature. Such post-translational control mechanisms do play a significant role in determining the activities of differential cells. For example, **haemoglobin** production is highly dependent on the availability of haem to complex with globin protein subunits and may be deficient in cases of iron-dependent anaemia.

HORMONAL CONTROL OF GENE EXPRESSION

In higher plants and animals, intercellular communication is a very important phenomenon. Signals originating in various glands and/or secretory cells somehow stimulate **target tissue** or **target cells** to undergo dramatic changes in their metabolic patterns. These changes frequently include altered pattern of differentiation that are generally dependent on altered patterns of gene expression. **Peptide hormones** such as insulin, epinephrine, etc., and **steroid hormones** such as estrogen, progesterone, testosterone (in higher animals, *e.g.*, mammals) and ecdysone (in insects). In higher animals, hormones are synthesized in various specialized secretory cells (*i.e.*, endocrine cells) and are released into the blood stream. The peptide hormones do not normally enter cells because of their relative large size. Their effects appear to be mediated by receptor proteins located in target-cell membranes and by the intracellular levels of **cyclic AMP (cAMP)** (called **secondary messenger**). The cAMP activates a protein kinase (e.g., A-kinase) which phosphorylates (activates) many specific enzymes. The steroid hormones, on the other hand, are small molecules that readily enter cells through the plasma membrane. Once inside the appropriate target cells, the steroid hormones become tightly bound to **specific receptor proteins** which are present only in the cytoplasm of target cells. The hormone-receptor protein complexes activate the transcription of specific genes are sets of genes according to following two methods (= hypotheses): 1. The hormone receptor protein complexes interact with specific non-histone chromosomal protein and this interaction stimulates the transcription of the correct genes (**J. Stein**, **G. Stein** and **L. Kleinsmith**, 1975). 2. The hormone receptor protein complexes activate transcription of target genes by binding to specific DNA sequences present in the *cis*-acting regulatory regions (the enhancers and promoter regions) of the genes (**R.M. Evans**, 1988). In both of these cases, these hormone- receptor protein complexes would function as positive regulators (or "activators") of transcription much like the CAP-cAMP complexes in prokaryotes.

During development of dipteran flies (*Drosophila melanogaster* and *Chironomus tentans*) the steroid hormone **ecdysone** is released and triggers moulting. If larvae of these insects are treated with ecdysone at stages of development prior to or between moultings, patterns of chromosome puffing occur that are identical to those occurring during natural moultings. Ecdysone tends to affect the gene expression at the level of transcription.

REVISION QUESTIONS AND ANSWERS

1. Describe the operon model for regulation of gene activity.
2. Define the positive control of gene action. Compare it with negative control provided in the *lac* operon.
3. In what way or ways are operator and regulator sites similar? Dissimilar?
4. Differentiate between repressors and inducers.
5. Synthesis of mRNA by the *lac* operon of *Escherichia coli* increases with addition of the inducer. Is this evidence for action of the inducer at the transcriptional or at the translational level?
6. Would you expect a non-sense mutation in one of the structural genes of the *his* operon to affect transcription or translation?
7. A bacterial mutation causes a cell to be incapable of fermenting many sugars (*e.g.*, lactose, sorbitol, xylose) simultaneously. The operons of genes specifying the respective catabolic enzymes are wild type (*i.e.*, unmutated). Provide an explanation for this phenomenon.

8. The entry of lactose in a bacterial cell is mediated by a permease enzyme. In cells that have not previously been exposed to lactose, how can lactose enter an uninduced $i+$ $z+$ $y+$ cell to affect induction of synthesis of β- galactosidase enzyme ?

9. Describe the Britten-Davidson's model of regulation of gene activity in eukaryotes.

10. Describe regulation of gene action in eukaryotes at the level of DNA or genetic code.

11. Discuss the relative roles of histones and non-histone proteins in regulation of gene activity in the eukaryotes.

12. In an eukaryotic system, what barriers would an extra cellular repressor have to pass through before ultimately binding to DNA?

13. What kinds of mutations would completely eliminate translation of the entire sequence encoding a polypeptide?

14. In what way is the regulation of a gene that is active in a differentiating cell inherently different from regulation of a bacterial gene.

15. Write short notes on the following :

    (i) *lac* operon; (ii) repression; (iii) steroid hormones and gene expression; (iv) role of interferon in gene regulation; (v) heterochromatin.

ANSWERS TO PROBLEMS

3. Both sites are similar in that each is composed of a given segment of deoxyribonucleotides and each participates in regulatory control over cistrons. On the other hand, although the operator is transcribed in large part, it is not translated. The regulator is responsible for production of a regulatory protein (either a repressor or an activator).

5. Transcriptional level.

6. Translation; a nonsense mutation causes both chain termination and release. Termination and release function in translation, not in transcription.

7. The mutation could be in genes for adenyl cyclase or in the gene for catabolic activator protein (CAP).

8. Occasionally a repressor molecule will momentarily become dissociated from the operator and RNA polymerase will attach and begin transcription of the β- galactosidase and permease genes before the repressor reattaches. This so-called "**sneak synthesis**" endows the cell with enough enzymes to transport a few lactose molecules through the plasma membrane; these will be catabolized to the true inducer (*i.e.*, allo-lactose) so that derepression can occur.

13. A promotor mutation, which would prevent transcription; or a mutation in the start codon, which would prevent initiation.

CHAPTER

9

Genetic Engineering

(Isolation, Sequencing, Synthesis of Gene and DNA Fingerprinting)

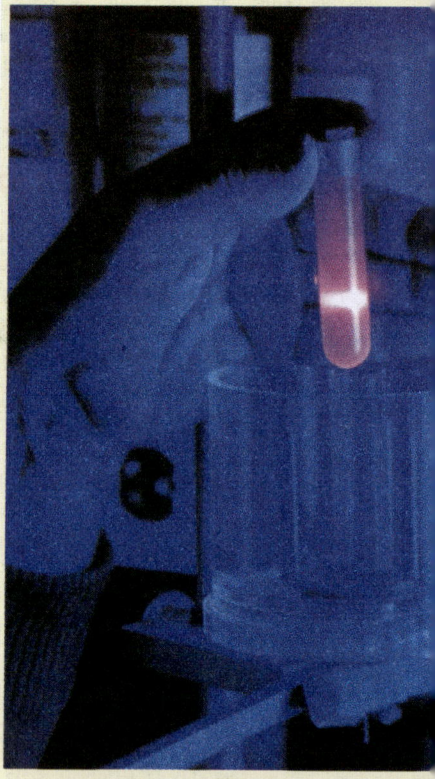

The future of the Earth's great diversity of life may eventually depend on genetic engineering, which has made it possible to extract and store the DNA of whole organisms indefinitely.

Allaby (1995) has defined the genetic engineering as the modification of the genetic information of living organisms by direct manipulation of their DNA (rather than by the more indirect method of breeding). Thus, a gene of known function (or economic importance) can be transferred from its normal location into a cell (which originally lack it) via a suitable mobile genetic element, called **vector** (such as plasmid, viruses (phages), etc.). The transferred gene replicates normally and is handed over to next progeny. On confirmation for its presence through biochemical procedures, the replica of the same cell (*i.e.*, clones) can be produced. With genetic engineering (also called **gene cloning**, **recombinant DNA technology** or **gene manipulation**), thus, genes can be isolated, cloned and characterized. More recently, **polymerase chain reaction** (**PCR**) which involves a thermostable DNA polymerase enzyme (*e.g.*, **Taq polymerase**) has also been used to obtain millions of copies of DNA segments (or genes) of choice.

The techniques of recombinant DNA and gene cloning are most powerful tools ever developed in the field of biology. The technique of designed genetic engineering of living cells has many potential applications. If used wisely, it promises to enhance the quality of human life. However, if used hapazardly and carelessly, genetic engineering could have negative impact on the quality of our life.

Genetic engineering is the 'hot cake' of todays' high-tech world; it has been applied for the production of valuable polypeptides, insulin, interferon, growth hormones and of course in the transfer of *Nif* (=nitrogen fixing) genes and control of genetic diseases (*e.g.*, cancer). Genetic engineers have promised a free agriculture from constraining requirements for fertilizers and pesticides. In this chapter some of the important techniques of genetic engineering will be described.

TOOLS OF GENETIC ENGINEERING

There are various biological tools which are used to carry out manipulation of genetic materials and cells as well. Some of them have been described as follows :

1. Enzymes such as exonucleases, endonucleases, restriction enzymes (=restriction endonucleases), SI enzymes (to change cohesive ends of single stranded DNA fragments into blunt ends), DNA ligases, alkaline phosphatase, reverse transcriptase, DNA polymerases.

2. Foreign DNA/Passenger DNA. It is a fragment of DNA molecule which is enzymatically isolated and cloned. The gene is identified on a genome and pulled out from it either before or after cloning. The cloned foreign DNA fragment expresses normally as in parent cell.

3. Cloning vectors. Vectors or vehicle DNA are those DNA that can carry a foreign DNA fragment when inserted into it. Based on the nature and sources, the vectors are grouped into bacterial plasmids, bacteriophages, cosmids and phasmids.

Plasmids have various curious properties : 1. The genes they carry may not be absolutely essential for life and so a plasmid can sometimes leave one bacterial cell and enter another, thereby transferring genetic traits between cells. 2. The plasmid can reproduce itself inside the bacterium independently of the main bacterial DNA. 3. A plasmid can sometimes fuse with the main DNA and later on can depart from the main genome, but in such a manner as to drag a piece of the main DNA with it. Nature seems to have evolved plasmid as an efficient way of exchanging gene between bacterial cells (see **Nossal** and **Coppel**, 1989).

4. cDNA bank or cDNA library. DNA copy of an mRNA molecule is known as **copy DNA** or **cDNA** (also called **complementary DNA**). The well characterized cDNA molecule is allowed to bind with a suitable vector which then **transforms** a bacterial cell in such a way that it does not disrupt its normal function. The transformed bacterial cell containing a plasmid with DNA copy of an mRNA molecule is known as **cDNA clone**. Since it is difficult to get cDNA from the double stranded DNA molecules, therefore, most of the cDNA clones are prepared by the use of **reverse transcriptase enzyme** from mRNA sequences of eukaryotic cells (Fig. 9.1).

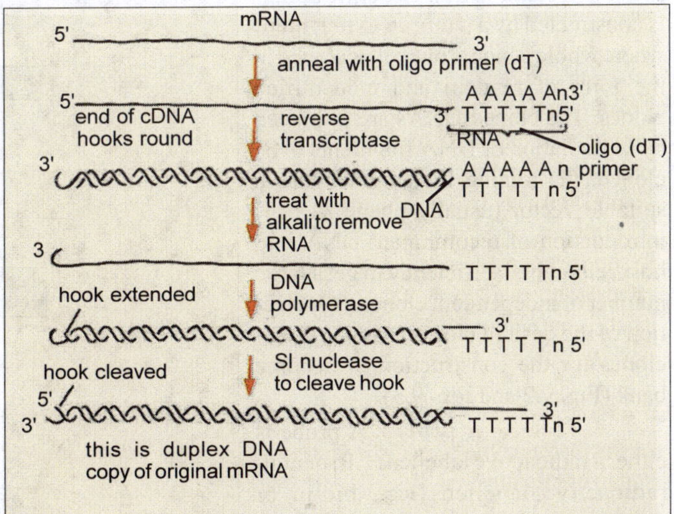

Fig 9.1. Synthesis of cDNA from mRNA, using reverse transcriptase enzyme.

A researcher analyzes the results of a DNA sequencing experiment and enters them directly into a computer. These are submitted to gene bank so that fellow researchers can have ready access to it.

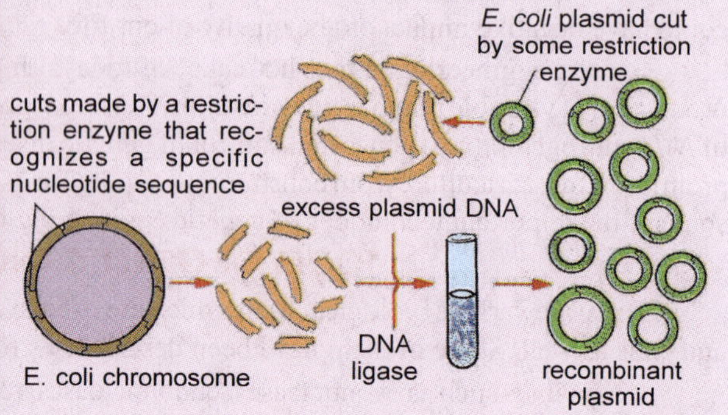

Fig. 9.2. Technique of formation of a genomic library using recombinant DNA technique.

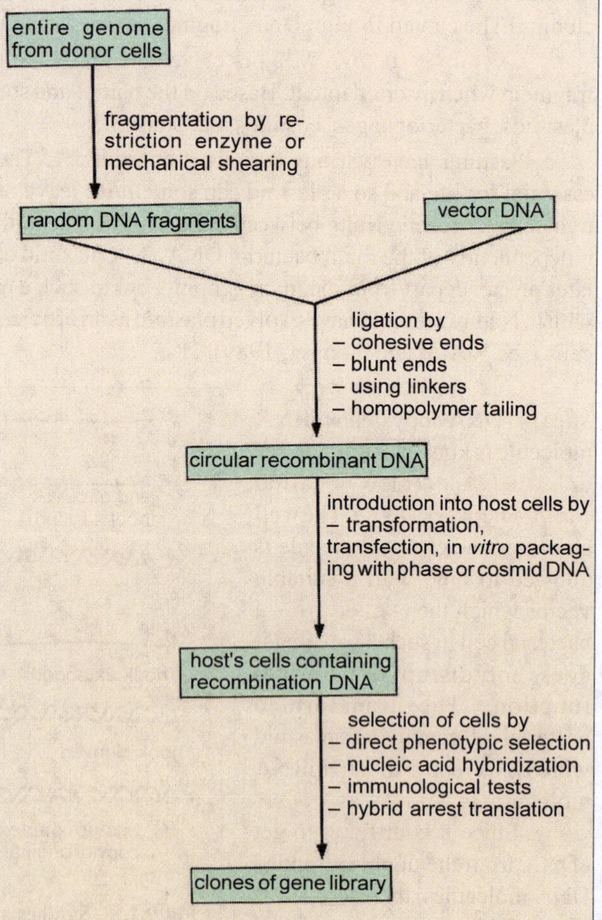

Fig. 9.3. Methods of construction of gene library.

5. Genomic library. Gene bank or **genomic library** is a complete collection of cloned DNA fragments which comprises the entire genome of an organism (**Dahl** *et al*., 1981). Genomic library is constructed by a **shotgun experiment** where whole genome of a cell is cloned in the form of random and unidentified clones. The cloned DNAs are produced by (1) isolation of DNA fragments to be cloned; (2) joining the fragments to a suitable vector (usually phage λ) ; (3) introduction of recombinant DNA into host cells at high efficiency to get a large number of independent clones; (4) selection of the desired clones; and (5) use of clones for the construction of genome bank (Fig. 9.2 and Fig. 9.3).

6. Molecular probes. A probe is either a radioactive labelled (^{32}P) or non-radioactive labelled (*viz*., biotin or digoxigenin), single stranded nucleic acid (20–40 nucleotide long) with a sequence complementary to at least one part of the desired DNA. The probe may be partially pure mRNA, a chemically synthesized oligonucleotide or a related gene which identifies the corresponding recombinant DNA. DNA/RNA probes have been commercially exploited in the diagnosis of infectious diseases, a variety of microbiological tests, identification of food contaminants and in forensic tests (*e.g.*, fingerprinting of murderers and rapists). Antibodies are also occasionally used as probes to recognize specific protein sequences (see **Dubey**, 1995).

CERTAIN GENERAL TECHNIQUES OF GENETIC ENGINEERING

The opening move of the genetic engineer is to break the DNA up into small, manageable bits, each containing one or just a few genes. Each little bit, one at a time, is stitched into a special, virus-like piece of DNA gifted with the ability for self-replication. These virus-like, recombined DNA molecules now invade rapidly-dividing host cells, again one at a time. Each host cell (*e.g.,* a bacterium or yeast) thereby becomes a factory for one pure gene. Clever tricks allow the genetic engineer to pick out the host cell carrying the gene wanted for that particular experiment. By isolating that one special cell and growing it up to any desired quantity, the one desired gene (or its protein product) can be obtained.

Now let us consider some of the general methods of genetic engineering as follows :

1. Isolation and Use of Restriction Enzymes

Recombinant DNA technology makes a frequent use of restriction endonucleases which cut the DNA double helix in very precise ways. They have the capacity to recognize specific base sequences on DNA and then to cut each strand at a given place. These enzymes are called **restriction enzymes** because they restrict infection of bacteria by certain viruses (*i.e.*, bacteriophages), by degrading the viral DNA without affecting the bacterial DNA. Thus, their function in the bacterial cell is to destroy foreign DNA that might enter the cell. The restriction enzyme recognizes the foreign DNA and cuts it at several sites along the molecule. Each bacterium has its own unique restriction enzymes and each enzyme recognizes only one type of sequence. As already described elsewhere, the DNA sequences recognized by restriction enzymes are called **palindromes**. Palindromes are the base sequences that read the same on the two strands but in opposite directions. For example, if the sequence on one strand is GAATTC read in 5'→3' direction, the sequence on the opposite strand is CTTAAG read in the 3'→5' direction, but when both strands are read in the 5'→ 3' direction the sequence is the same. The palindrome appears accordingly —

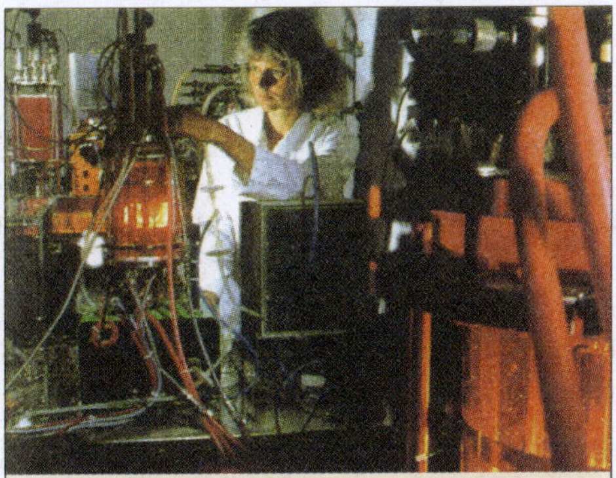

In recombinant protein research, a scientist takes a sample in which genetically engineered mammalian cells are being cultured.

5' GAATTC 3'
3' CTTAAG 5'

In addition, there is a point of symmetry within the palindrome. In our example, this point is in the centre between the AT/AT. The value of restriction enzymes is that they make cuts in the DNA molecule around this point of symmetry. Some enzymes cut straight across the molecule at the symmetrical axis producing **blunt ends**. Of more value, however, are the restriction enzymes that cut between the same two bases away form the point of symmetry on two strands, thus, producing a staggered break.

G AATTC
C TTAA G

G
C TTAA

AATTC
G

In this example, we have used the palindrome sequence recognized by one of the most popular restriction enzymes, called **Eco RI** from *E.coli* (bacterium). Hundreds of other restriction enzymes with different sequence specificities have been isolated from several bacteria and are commercially available.

The most useful aspect of restriction enzymes is that each enzyme recognizes the same unique base sequence regardless of the source of the DNA. It means that these enzymes establish fixed landmark along an otherwise very regular DNA molecule. This allows dividing a long DNA molecule into fragments that can be separated from each other by size with the technique of gel electrophoresis (*e.g.*, **agarose** or **polyacrylamide gel electrophoresis**, **pulsed field gel electrophoresis** or **PFGE**). Each fragment is, thus, also available for further analysis, including the sequencing.

One value of cutting DNA molecule up into discrete fragments is being able to locate a particular gene on the fragment where it resides. This is done by the general technique of **Southern blotting** (developed by **E.M. Southern**, 1975).

1. Southern blotting technique. In this technique, a DNA molecule is cut into discrete fragments by a restriction enzyme. It is electrophoresed through an agarose gel which separates the various fragments according to size. The DNA is then denatured into single strands by exposing the gel to NaOH. A few pieces of filter paper soaked in buffer are placed under the gel. A large piece of nitrocellulose paper is laid over the agarose gel, followed by several layers of absorbent material such as filter paper. This dry absorbent material pulls the buffer up through the gel from the lower layer (Fig. 9.4). This washes the DNA off the gel and on to the filter, where it covalently binds to the filter.

The positions of the DNA molecules on the filter paper are identical to their position in the gel. The nitrocellulose filter containing the DNA is first dried and then exposed to a solution of ^{32}P labelled mRNA called **molecular probe** from the gene to be isolated. The radioactive mRNA hybridizes (*i.e.*, establishes the hydrogen bonds) only with the single-stranded DNA in restriction fragments that contain complementary sequences. The nitrocellulose filter is then removed and placed in contact with photographic film that when developed will reveal fragments from the original gel containing complementary sequences to the mRNA used in the assay. The procedure allows specific identification of restriction fragments containing DNA sequences to specific RNA molecules.

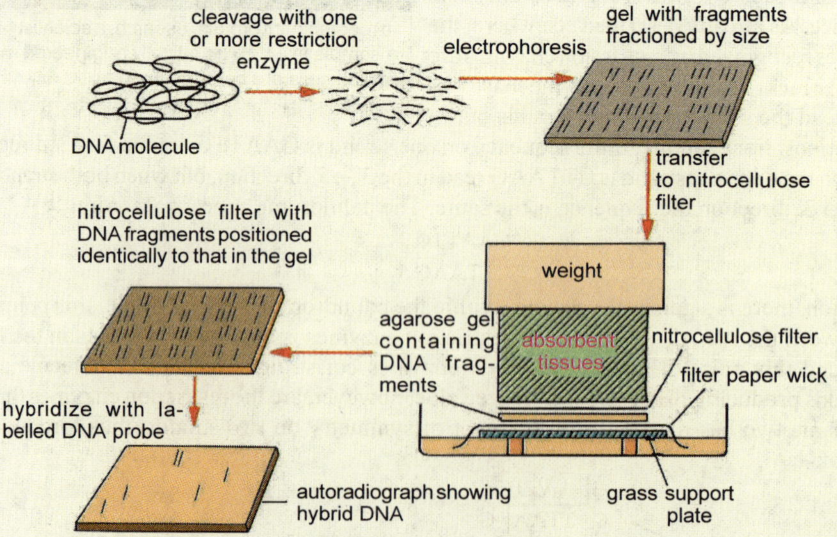

Fig. 9.4. Various steps of Southern blotting technique.

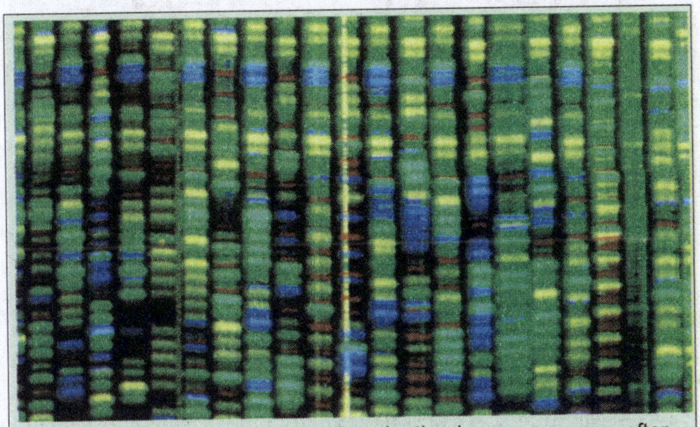

An autoradiogram is used to decode the base sequence after electrophoresis.

2. Northern blotting technique. **Alwine** *et al*., (1979) devised a technique, nicknamed **northern blotting** in which RNA bands are blot transferred from the gel onto chemically reactive paper. An **amino- benzyloxymethyl paper** which is prepared from Whatman filter paper No. 540 after a series of simple reactions, is **diazotized** (Gr., *di*= two; Fr., *azote* = nitrogen; to introduce the diazo group into a chemical compound, usually through the treatment of an amine with nitrous acid) and made into the reactive paper and, therefore, becomes available for hybridization with radiolabelled DNA probes. The hybridized bands are found out by autoradiography.

Later, it was shown (**Tomas**, 1980) that mRNA bands can be blotted directly onto nitrocellulose membrane, a technique which has been widely adopted. The mRNA bands blotted onto nitrocellulose membrane can be hybridized with a labelled DNA or RNA probe. The single stranded regions of probe are removed by nuclease (*e.g.* mung bean nuclease or S-1 nuclease), so that quantitative estimation of hybridized mRNA can also be made.

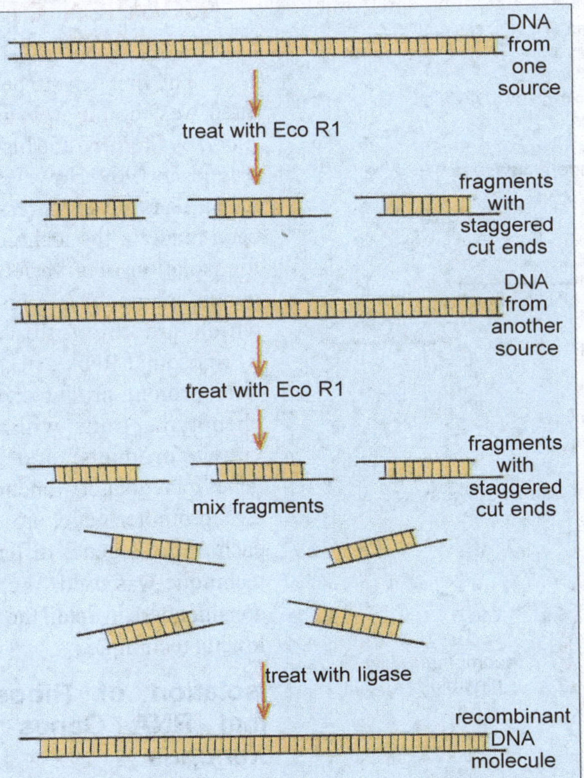

Fig. 9.5. Method of construction of a recombinant DNA molecule using a restriction enzyme.

3. Western blotting technique. **Towbin** *et al*., (1979) developed the **western blotting technique** to find out the newly encoded protein by a transformed cell. The extracted proteins are subjected to **polyacrylamide gel electrophoresis** (**PGE**) and are then transferred onto nitrocellulose to which they bind. Nitrocellulose membrane is then used for probing with a specific labelled antibody (Antibody tends to bind with a protein; it does not hybridize with protein). The antibody may be labelled with ^{125}I and the signal is detected again with autoradiography.

2. Vectors, Transformation and Molecular Cloning

Plasmids are extra-chromosomal DNA elements found mostly in bacteria. These plasmids contain DNA sequences coding for drug resistance, sex factor (F factor) etc., and probably has arisen from chromosomal DNA. When the bacterium multiplies, the plasmid DNA will also multiply along with chromosomal DNA. It is possible to isolate these plasmids in large quantity. In recombinant DNA technology, prokaryotic and eukaryotic DNAs as well as the plasmid DNAs are cut into specific fragments with restriction enzymes. The foreign DNA fragment (prokaryotic and eukaryotic) can be made to recombine with the plasmid DNA and the product is referred to as **recombinant DNA**. The plasmid DNA carrying the foreign DNA fragment can be put back into a suitable recipient bacterium. This bacterium can be grown in large quantities and the recombinant plasmids are isolated from such bacteria. (This is called **molecular cloning**). The foreign genes then can be released from recombinant plasmids once again by the use of restriction enzymes. Thus, large number of foreign genes can be isolated by this technology.

The foreign genes can also be introduced into viral DNAs and when such recombinant viruses (phages) infect and multiply, once again large quantities of desired genes can be isolated. Plasmids and viruses which are used as carriers of foreign DNAs are referred to as **vectors** or **vehicle DNAs**. By interchanging plasmid DNA and viral DNA fragments, several new vectors have been synthesized which carry new genes into bacteria, yeast, insect, plant and animal cells. When such new (foreign) DNA fragments are introduced into relevant host cells, such cells are said to be **transformed** (in case of bacteria) or **trangenic** (in case of plants and animals) and the process is called **transformation**. In animals the term **transfection** is used in place of transformation. For example, **D.W.OW** *et al*., (1986) produced a transgenic tobacco plant harbouring the luciferase gene of the firefly (see **Gardner** *et al*., 1991).

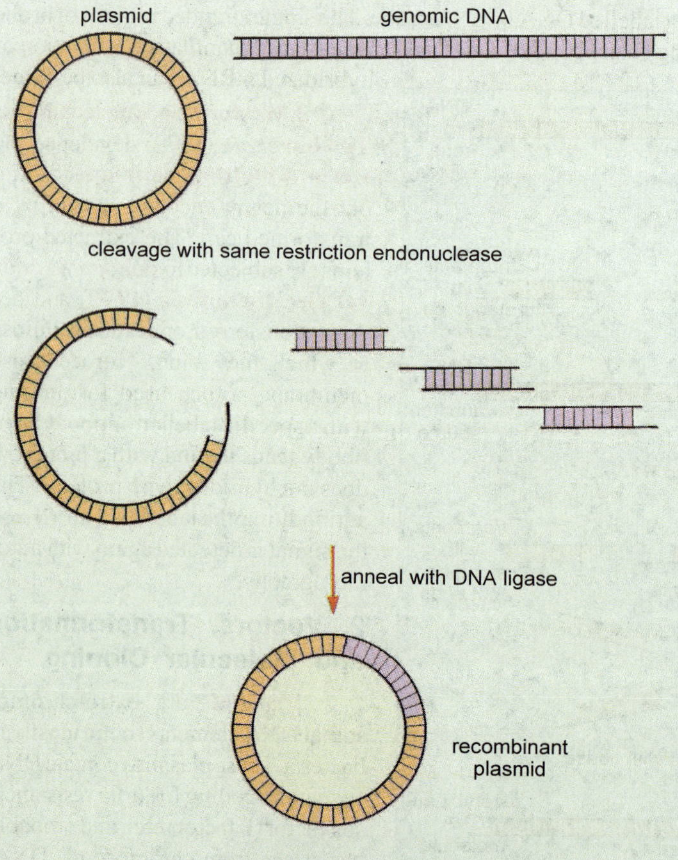

Fig. 9.6. Method of construction of a recombinant plasmid between a bacterial plasmid and genomic DNA from another organism.

ISOLATION OF GENES

The first gene to be isolated was the **lac operon** of *E.coli* by **Shapiro** and his colleagues in 1969. However, in recent years great progress has been made in the techniques for isolation of a variety of genes, some examples of which are the following: (i) ribosomal RNA ; (ii) specific protein products; (iii) phenotypic traits with unknown products; and (iv) genes for regulatory functions, *e.g.*, promoter gene, etc. For each of these genes different technique was used. We can consider here in detail the following technique :

Isolation of Ribosomal RNA Genes in *Xenopus*

As already described elsewhere, ribosomes consist of ribosomal proteins and ribosomal RNA (rRNA). Ribosomal RNA makes 80 per cent of cellular RNA and occurs in four sizes namely 28S, 18S, 5.8S and 5S. The rRNA is synthesized on ribosomal genes which have been isolated. Isolation of rRNA genes have been found easy due to the following three reasons: 1. availability of homogeneous rRNA; 2. differences between ribosomal RNA genes and other genes, *i.e.*, rRNA has high G+C content, *i.e.*, rRNA has 45 to 60 per cent G+C; while the remaining RNA has only 40 per cent G+C; 3. rRNA genes are present in multiple copies. Due to these facts, rRNA genes were isolated in 1965 in *Xenopus* by **Hugh Wallace** and **Max L. Birnstiel**. The technique of isolation of rRNA genes involved the following steps: (1) The rRNA was isolated from ribosomes of *Xenopus* and made radioactively labelled due to its replication in a medium having tritiated uridine. (2) Ribosomal DNA was isolated by density gradient centrifugation followed by its denaturation (since G+C content of rDNA differs from that of bulk DNA, it helps in its separation by centrifugation). (3) Single-stranded DNA was fixed on a filter paper. (4) Labelled rRNA was added on filter paper carrying single stranded DNA. (5) DNA-RNA hybridization was allowed to take place. (6) Excess labelled RNA was washed. (7) Radioactivity was measured and duplex hybrids isolated, which on denaturation, gave single stranded DNA, which could be made double stranded.

SEQUENCING OF GENE

Once a gene or DNA fragment is cloned, its further study involves DNA sequencing. The following three methods are used for the determination of DNA sequences :

1. Maxam and Gilbert's Chemical Degradation Method

As illustrated in Figure 9.7, this technique involves the following steps : 1. The 3' ends of DNA are labelled with ^{32}P. 2. The two strands of this radioactively labelled DNA are separated. 3. The mixture is divided into four samples, each treated with a different reagent having the property of destroying either only G, or only C, or A and G or T and C. The concentration of the reagent is adjusted in such a way that 50 per cent of target base is destroyed, so that fragments of different sizes having ^{32}P are produced. 4. Each of the four samples is electrophoresed in four different lanes of the gel. 5. The gel is autoradiographed to determine the sequence from positions of bands in four lanes.

2. Sanger's Dideoxynucleotide Synthetic Method

Fred Sanger had initially developed a method for DNA sequencing, which utilized DNA polymerase to extend DNA chain length. This was termed as **plus-minus method**. Later on, **Sanger** (1986) developed a more powerful method, utilizing single-stranded DNA as the template for DNA synthesis, in which **2', 3' dideoxynucleotides** were incorporated leading to termination of DNA synthesis. These dideoxynucleotides are used as triphosphates (ddNTP) and can be incorporated in a growing chain, but they terminate synthesis (Fig. 9.8), since they fail to form a phosphodiester bond with next incoming deoxynucleotide triphosphate (dNTP). Thus, Sanger's dideoxy methods includes the following steps : 1. Four reaction tubes are set up, each containing single stranded DNA sample (cloned in M13 phage) to be sequenced, all four dNTPs (radioactively labelled) and an enzyme for DNA synthesis (*i.e.*, DNA polymerase I). Each tube also contains a small amount of one of the four ddNTPs, so that each tube has a different ddNTP, bringing about termination at a specific base — adenine (A), cytosine (C), guanine (G) and thymine (T). 2. The DNA fragments which are generated by random incorporation of ddNTP leading to termination of reaction are then separated by electrophoresis on a high resolution polyacrylamide gel. This is done for all the four reaction mixtures on adjoining lanes in the gel. 3. The gel is used for autoradiography so that the position of different bands in each lane can be visualized. 4. The bands on the autoradiogram can be used for getting the DNA sequence.

3. Direct DNA Sequencing Using PCR

Polymerase chain reaction (PCR) has also been used for sequencing the amplified DNA product (Fig. 9.9). This method of DNA sequencing is faster and more reliable. It can utilize either the whole genomic DNA or the cloned fragments for sequencing a particular DNA segment. The DNA sequencing using PCR involves two main steps : 1. generation of sequencing templates (double stranded or single stranded) using PCR and; 2. sequencing of PCR products either with the thermostable Taq DNA polymerase. PCR is discovered by **Kary Mullis**, 1985, and nicknamed **people's choice reaction** in which instead of RNA primer, a deoxyolig onucleotide is used. (In PCR reaction, the normal DNA polymerase enzyme is replaced by **Taq DNA polymerase** (= an enzyme isolated from *Thermus aquaticus* growing in hot springs; this enzyme acts best at 72^{0}C and the denaturation temperature of 90^{0}C does not destroy its enzymatic activity, see **Kary Mullis**, 1990). An unlimited supply of amplified DNA is obtained by repeating the reaction, which is made possible by regular denaturation of freshly synthesized double stranded DNA molecules by heating it to 90° – 98°C. At this high temperature the two DNA strands separate. Once the double stranded DNA is made single stranded by heating up to 90 – 98°C, the mixture with two primers (= deoxyoligonucleotides) recognizing the two strands and bordering the sequence to be amplified, is cooled to 40 – 60°C. This permits the primers to bind to their complementary strands through renaturation. The presence of Taq DNA polymerase enzyme and all four essential nucleoside triphosphate in the 'eppendorf tube' allows synthe-

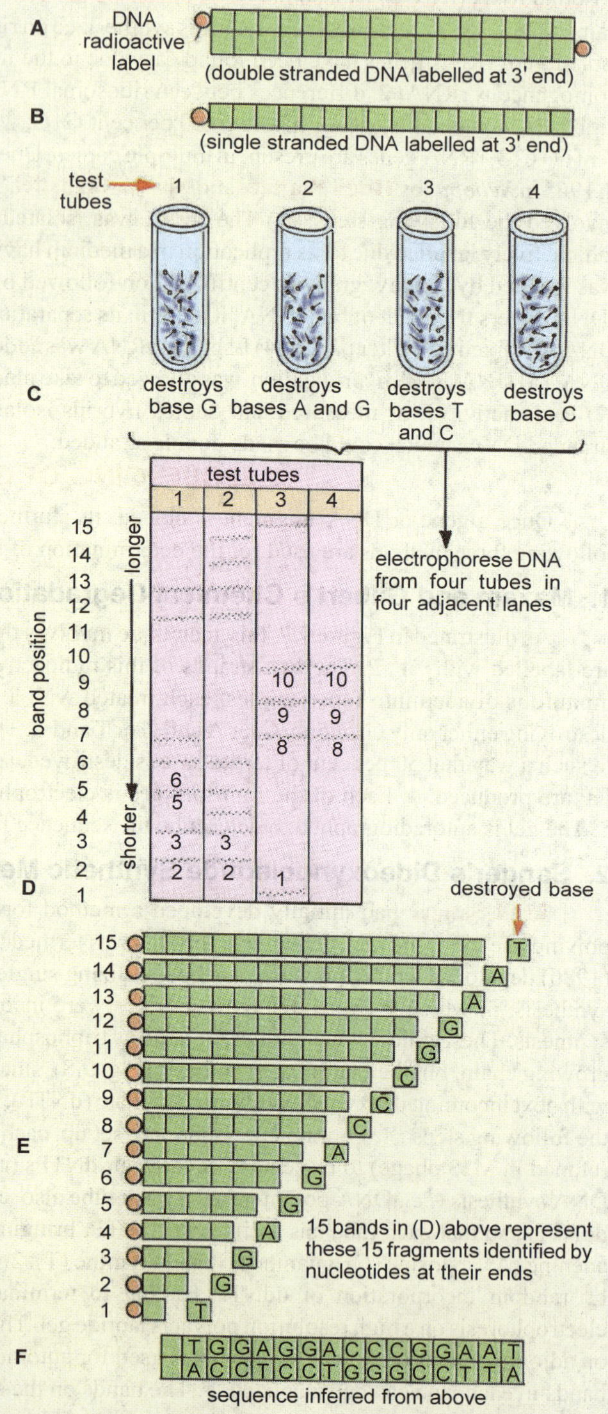

Fig. 9.7. Maxam and Gilbert's chemical degradation method for sequencing of DNA.

sis of complementary strands in the usual manner. In an automatic thermal cycler, this process is automatically repeated 20 – 30 times (as predetermined by a computer device), so that in a single afternoon a billion copies of the sequence flanked by the left and right primers, can be produced (Fig. 9.9). In order to continue the synthesis, the temperature of the mixture is alternately increased (for denaturation) and decreased (for renaturation) once every 1-3 minutes as fixed by the computer device). The use of Taq DNA polymerase and some other DNA polymerases have allowed automation of the entire PCR reaction.

In fact, the DNA sequencing method using PCR eliminates the need of cloning the DNA in single stranded DNA phage vector, *i.e.*, M13.

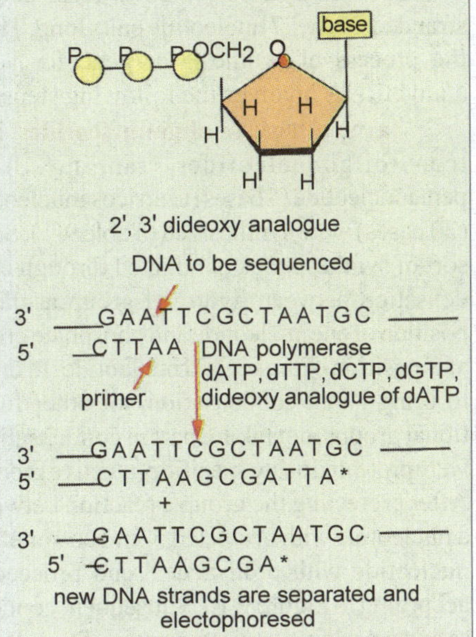

Fig. 9.8. A– Dideoxynucleotide; B– Technique involved in Sanger's chain termination method for sequencing of DNA.

SYNTHESIS OF GENE

The genes can be synthesized by the following two methods :

1. Organochemical Synthesis of Polynucleotides (or Chemical Synthesis of tRNA Genes)

When the detailed structure of a gene becomes procurable, then such a gene can be synthesized by a purely chemical method. The structure of gene could be inferred from its product. For instance, if a gene is responsible for giving rise to a polypeptide chain and the structure of the chain is known, then from the genetic code dictionary, structure of the gene could be easily inferred. Such genes were initially considered to be too long to be synthesized, because an average gene contains about 1500 base pairs. But since tRNA molecules are fairly small in size (about 80 nucleotides), a gene responsible for giving rise to a tRNA molecule was found to be within the reach of synthesis.

Gene machines are just automated chemistry sets. The machine pumps a precise amount of one of four solutions of the bases of DNA contained in the jar along fine pipes to a reaction chamber. A computer controls which base is added. The base is added chemically to a growing chain of DNA. The addition cycle repeats until the entire sequence is made.

(i) Synthesis of gene for yeast alanyl tRNA. R.W. Holley (who died in 1993) and his coworkers (1965), first of all gave the detailed structure of yeast **alanyl tRNA** (containing 77 nucleotides; see Chapter 5). **Khorana** and his coworkers had vast experience of synthesizing DNA of known base sequences. They found that such a long chain (*i.e.*, 77 base pairs of DNA of yeast alanyl tRNA) could not be synthesized by adding a single base each time, therefore, they decided that small oligo-deoxyribonucleotides ranging in length from 5 to 20 nucleotides should first be synthesized. These segments would be single stranded and would cover the whole length of both the strands of DNA.

These would then be joined to form double stranded DNA, 77 nucleotide pairs long. Thus, the process of synthesis of gene for yeast alanyl tRNA involves the following steps :

(a) Synthesis of oligonucleotides. Fifteen oligonucleotides ranging from pentanucleotide (5 bases) to an icosanucleotide (20 bases) were synthesized (Table 9-1). Such sort of synthesis was conducted through condensation between hydroxyl group at the 3' position of one nucleotide and phosphate group at 5' position of the second nucleotide. In order to bring about condensation, all other functional groups, not taking part in condensation, were protected using specific protective groups. After protecting the groups, reaction between a nucleotide with protected 5' end and another nucleotide with protected 3' end proceeded according to Figure 9.10. Subsequent condensation was done between groups of two, three or four nucleotides.

(ii) Synthesis of three duplex fragments. Fifteen single-stranded oligonucleotides were used to prepare three duplex fragments, each containing a single stranded end (Fig. 9.14). These three fragments are characterized as follows :

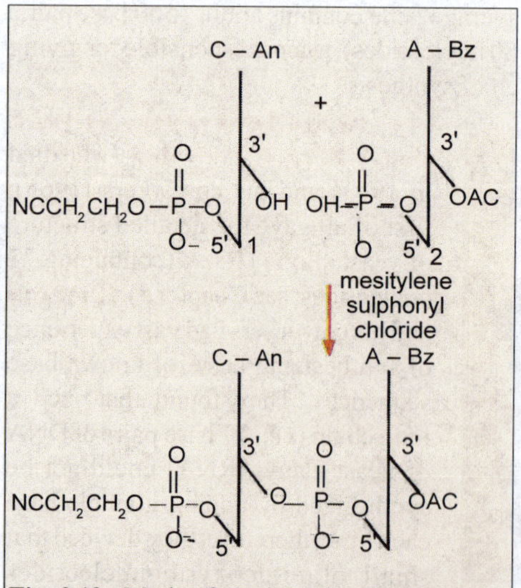

Fig. 9.10. Mode of condensation between two nucleotides with protected 5' OH and 3′ OH groups in the sugars and protected amino groups in the nitrogen bases (AC=acetyl, An=anisoyl and B_3= benzoyl protective groups).

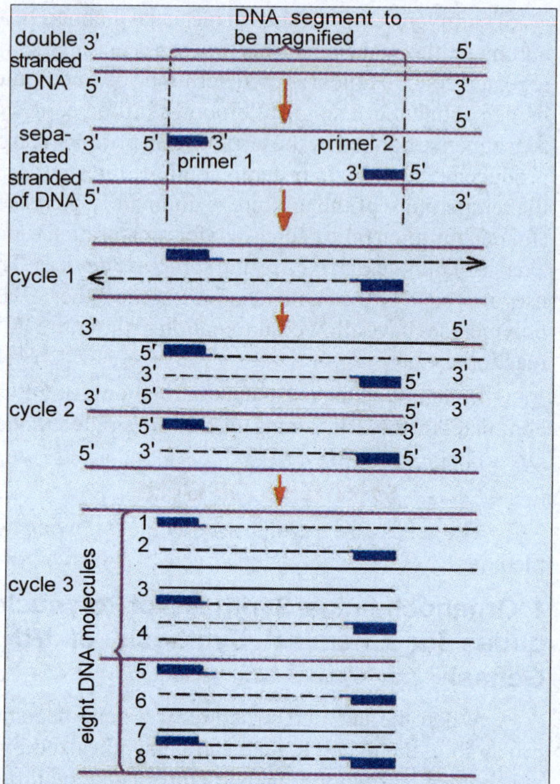

Fig. 9.9. Steps involved in basic polymerase chain reaction (PCR); only three cycles of PCR are shown; in each cycle primers are shown by solid boxes, template strands are shown by continuous lines and newly synthesized strands are shown by broken lines.

1. **Fragment A** consisted of the first 20 nucleotides in which nucleotides 17–20 being single stranded. 2. **Fragment B** contained nucleotide residues from 17 to 50, in which single stranded region being 17–20and 46–50. 3. **Fragment C** included nucleotide residues 46–77 with single stranded region 46–50.

(iii) Synthesis of tRNA gene from three duplex fragments. In the concluding step three duplex fragments A, B and C were joined (linked by ligase enzyme) to give complete gene in each of the following two ways :

1. In one scheme, A fragment was joined to B fragment taking advantage of the overlap in residues 17–20; fragment C was then added, with the overlap in the region of 46–50 residues. The complete double stranded DNA with 77 base pairs representing the gene was, thus, prepared. 2. In the alternative scheme

B fragment was first added to C fragment and to this A fragment was added in the end to get the complete gene. This gene was synthesized in 1970. Since such a synthetic gene could replicate and make its own copies, so, was used for subsequent work.

In 1977, **Riggs** and his colleagues have been able to synthesize a DNA piece (*i.e.*, a gene) which codes for the polypeptide (containing 14 amino acids) of **somatostatin**. This hormone regulates body growth as well as the production of insulin and glucagon hormones and also inhibits the release of other pituitary hormones in mammals.

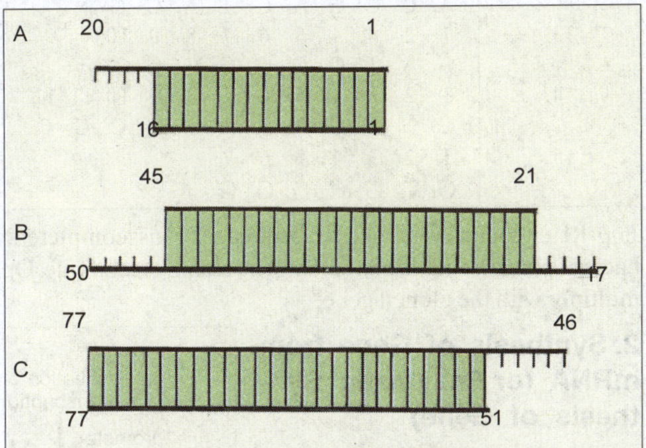

Fig. 9.11. Three duplex DNA fragments with single stranded sticky ends, produced for the synthesis of yeast alanyl tRNA gene by H.G. Khorana.

Synthesis of complete gene. During the synthesis of the gene or yeast alanyl tRNA, it became clear that natural tRNA was not the direct product of transcription. Instead a precursor molecule is synthesized which subsequently, after losing segments of RNA by cleavage, gives rise to tRNA. This has shown that the real natural gene for yeast alanyl tRNA was longer than the DNA duplex synthesized by Khorana. In view of this **Khorana** started the synthesis of a gene for *E.coli* **tyrosine suppressor tRNA precursor**. In 1979, he reported the completion of total synthesis of a biologically functional tyrosine suppressor transfer RNA gene carrying all regulatory sequences (Fig. 9.12). This gene was 207 base pairs long and included the following components : (i) a 51 base pairs long DNA promoter region; (ii) a 126 base pairs long DNA corresponding to the precursor tRNA and (iii) 25 base pairs long DNA corresponding to 16 base pairs adjoining CCA end of tRNA and the remainder, a modified sequence including

Table 9.1. **Base sequences and lengths of the single-stranded oligonucleotides synthesized for the construction of yeast alanine tRNA gene.**

Serial number	Base sequence	Length of nucleotides
1.	1 2 3 4 5 6 7 8 9 10 11 12 T G G T G G A C G A G T	12
2.	1 2 3 4 5 6 A C C A C C	6
3.	1 2 3 4 5 6 7 8 9 10 T G C T C A G G C C	10
4.	1 2 3 4 5 6 7 8 C C G G A A T C	8
5.	1 2 3 4 5 6 7 8 9 10 11 T T A G C T T G G C C	11
6.	1 2 3 4 5 6 7 8 9 10 11 12 13 14 15 16 17 18 19 20 T A A C C G G A G A G A C T C C C A T G	20
7.	1 2 3 4 5 6 7 T C T C T G A	7
8.	1 2 3 4 5 6 7 8 9 10 11 12 13 14 15 16 G G G T A C G A A A C C C T C G	16
9.	1 2 3 4 5	5

	C T T T G	
10.	1 2 3 4 5 6 7 8 9 10 G G A G C G C G C T	10
11.	1 2 3 4 5 6 7 8 9 10 C G C G A T G G C T	10
12.	1 2 3 4 5 6 7 8 9 10 A C C G A C T A C G	10
13.	1 2 3 4 5 6 7 8 9 10 G A T G C G C G G T	10
14.	1 2 3 4 5 6 7 8 9 10 11 12 C G C C A C A C G C C C	12
15.	1 2 3 4 5 6 7 G T G C G G G	7

Eco RI endonuclease specific sequence. This complete synthetic gene was cloned in the vector bacteriophage (lambda phage) by gene cloning technique. On transformation of *E. coli*, the phage could multiply with the cloned gene.

2. Synthesis of Gene from mRNA (or Enzymatic Synthesis of Gene)

RNA directed DNA polymerase enzyme, which was discovered by **Temin** and **Baltimore** (1970), can synthesize DNA from RNA template. This enzyme has enabled the molecular biologists to synthesize complementary (or copy) DNA (cDNA) using mRNA as a template. If mRNA transcribed from a specific gene is made available in purified form, it can be used in the production of cDNA which will represent the synthetic gene. Thus, by copying eukaryotic purified mRNA, several genes have been artificially synthesized. Most important of these genes are the genes for sea urchin histone proteins, ovalbumin gene in chicken and globin genes in mammals.

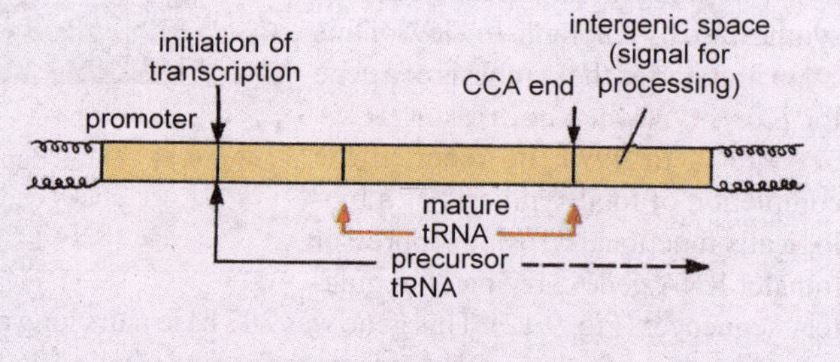

Fig. 9.12. Different parts of the complete *E. coli* tyrosine suppressor tRNA gene, synthesized by Khorana.

One of the leading discoveries, in which the technique of enzymatic synthesis of gene has been utilized, is the synthesis of rat **insulin cDNA** segments on insulin mRNA isolated from pancreatic islets of Langerhans (**Ullrich** *et al.*, 1977). This cDNA has been replicated to get multiple copies of the rat insulin gene. Further, after isolation and purification of mRNA from the cultured rat pituitary cells, **Seeberg** *et al.* (1977) has synthesized cDNA on mRNA that codes for a **rat growth hormone** (**RGH**) which comprises 190 amino acids. Similarly, a cDNA molecule has been synthesized on mRNA which codes for the **human chorionic somatomammotropin** (HCS), a placental lactogen (**Shine** *et al.*, 1977).

APPLICATION OF GENETIC ENGINEERING

The foremost application of genetic engineering has been in understanding the structure of eukaryotic genes such as exons, introns, promoters, enhancers, silencers, etc. Some other modern applications of genetic engineering include — (i) engineering of bacteria to carry out specific processes or to produce important molecules such as hormones or antibiotics; (ii) altering the genotypes of plants as an aid in plant breeding, *e.g.*, formation of insect resistant tomato plants; (iii) altering genotypes of animals to correct their genetic defects, *e.g.*, production of human growth hormones.

DNA Fingerprinting : The Ultimate Identification Test

Every year in court cases all over the world the ability to establish a person's identity is essential for a just decision. Genetics has come to the rescue of the courts and now the following new questions are routinely asked in the courts : (1) Is the drop of blood found at the crime scene from suspect on trial? Who is the child's father ? Until recently, there was no foolproof test. In a criminal case, if there was no identifiable fingerprint left behind at the crime scene, there was no case. Blood tests can determine who is not the parent, not who is. A test has now been developed that provides hundred per cent positive identification. The test is called **DNA fingerprinting**. The test of DNA fingerprinting can show conclusively whether the genetic material in a drop of blood matches that of the suspect, or it can be used to solve paternity case.

The technique of DNA fingerprinting relies on developments from recombinant DNA technology and allows an examination of each individual's unique genetic blueprint–DNA. The technique was discovered in England by **Alec Jeffreys**. It is based on the fact that the DNA of each individual is interrupted by a series of identical DNA sequences called **repetitive DNA** or **tandem repeats**. The pattern, length, and number of these repeats are

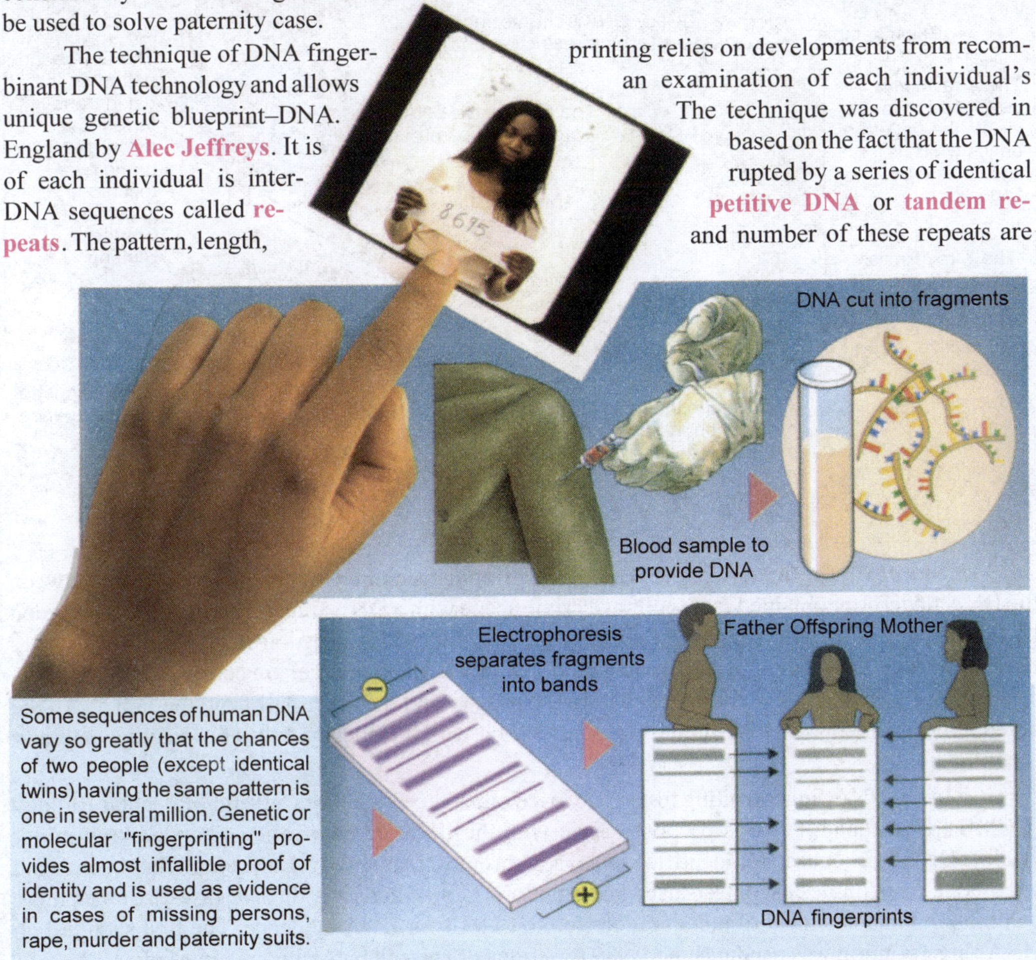

Some sequences of human DNA vary so greatly that the chances of two people (except identical twins) having the same pattern is one in several million. Genetic or molecular "fingerprinting" provides almost infallible proof of identity and is used as evidence in cases of missing persons, rape, murder and paternity suits.

unique for each individual. **Jeffreys** developed a series of DNA probes, which are short pieces of DNA that seek out any specific sequence they match, and base pair with that sequence. Such molecular probes are used to detect the unique repetitive DNA patterns characteristic of each individual. The procedure of DNA fingerprinting has the following steps : 1. DNA is purified from a small sample of blood, semen, or other DNA-bearing cells, and digested into smaller fragments with restriction endonucleases. 2. The fragments are separated by agarose gel electrophoresis. 3. The separated fragments are transferred to a nylon membrane by the technique of Southern blotting. 4. The DNA probes labelled with radioactive material are added to a solution containing the nylon membrane. 5. Wherever the probes fit a band containing repetitive DNA sequences, they attach. 6. The X-ray film is pressed against the nylon filter and exposed at bands carrying the radioactive probes attached to the fragments. 7. The patterns of bands obtained on the film is 100 per cent unique for each person, except for identical twins who would have the same pattern.

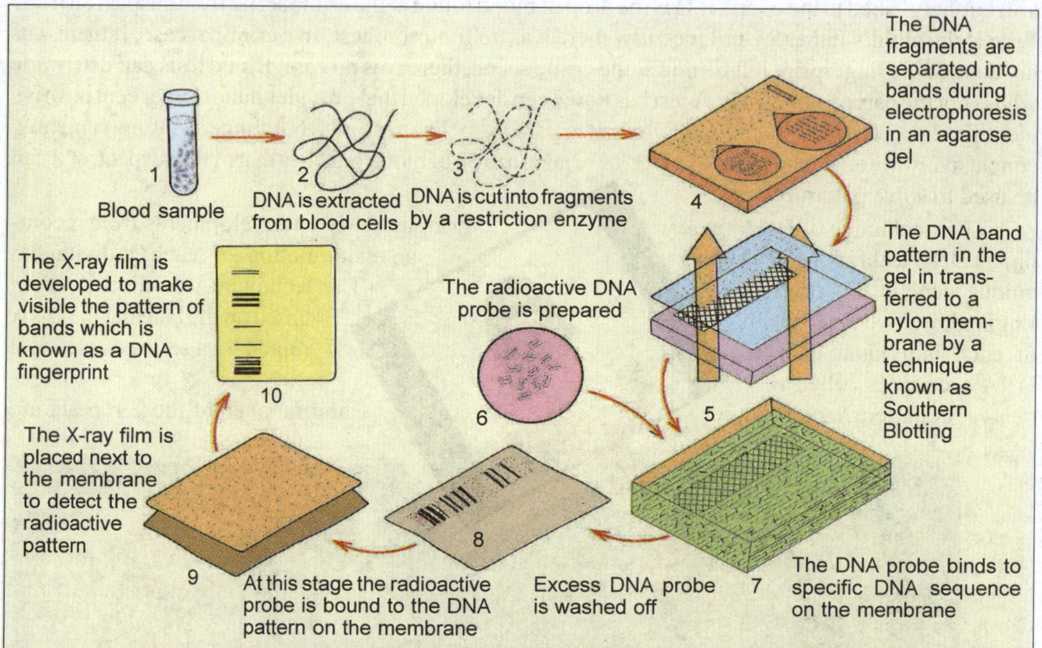

Fig. 9.13. Various steps of DNA fingerprinting process (after Burns and Bottino, 1989).

The **forensic application** of the DNA fingerprinting technique involves a comparison between the DNA fingerprint obtained from cells at a crime scene with a DNA fingerprint from cells provided by the suspect. If the DNA pattern matches exactly, certain identification is made. For **paternity determination**, DNA fingerprints of the mother, child and alleged father are compared. In this case, one-half of the bands in the child comes from the mother and the other half from the father. All the paternal bands in child's DNA fingerprint must match with the alleged father for positive paternity identification.

In India, DNA fingerprinting tests are carried out at the Centre for Cell and Molecular Biology (CCMB), Hyderabad. For this purpose, a test with the **BKM-DNA probe** (= banded krait minor satellite DNA) earlier used for identification of sex chromosomes (by **Dr. Lalji Singh**) has been found to cost one-tenth of the cost of tests used in Europe and U.S.A. Paternity dispute cases are much more common in India and most of them are referred to CCMB for DNA evidence. The first such test on DNA fingerprinting was used in June, 1989 to settle a drawn-out paternity case in Madras.

REVISION QUESTIONS AND PROBLEMS

1. What is genetic engineering ? Describe in brief various essential techniques of genetic engineering.
2. What is recombinant DNA and how is it made ?
3. Why did H.G. Khorana select the gene for yeast alanyl-tRNA for artificial synthesis ? Give a brief account of different steps involved in the artificial synthesis of the gene for yeast alanyl tRNA.
4. Give a brief account of the methods for the isolation of genes in eukaryotes.
5. Enumerate various methods of synthesis of a gene. How can a gene be synthesized from an mRNA molecule ?

6. What potential benefits and/or dangers for the human race do you see in recombinant DNA? Evaluate those benefits and/or dangers critically.
7. Given two DNA molecules, the overall composition of which is represented by the segments shown below, determine which molecule would have the higher melting temperature. Explain.

   (a) TTCAGAGAACTT
   AAGTCTCTTGAA

   (b) CCTGAGAGGTCC
   GGACTCTCCAGG
8. The genome of *E. coli* contains about 4000 kilobase pairs (kb; kilo = 1000); there are about 1.5 kb in 16S rRNA. If 0.14% of the genome forms hybrid double helices with RNA complementary to one strand of DNA, estimate the number of genetic loci encoding 16S rRNA.
9. Restriction nuclease Eco RI makes staggered cuts in a six-nucleotide DNA palindrome; Hae III nuclease cleaves at one point in the middle of a four-nucleotide DNA palindrome. If different aliquotes (*i.e.,* something which will divide a number without a remainder) of purified DNA preparation are treated with these enzymes, which one would be expected to contain more restriction fragments ?
10. Write short notes on the following : (i) cDNA library ; (ii) genomic library ; (iii) Southern blotting technique : (iv) complementary DNA (cDNA); (v) gene sequencing ; (vi) restriction enzyme; (vii) northern blotting technique; (viii) PCR; (ix) gene tagging; and (x) DNA fingerprinting.

ANSWERS TO PROBLEMS

7. (b) because it has a higher (G+C) (A+T) ratio.
8. (0.0014) (4000)/1.5 = 4.
9. More fragments are expected from Hae III, because the probability of a specific four base sequence is greater than the probability of a specific six-base sequence if the nucleotides are distributed along a chain in essentially a random order.

CHAPTER

10

Immunology

Immunology, the study of the immune system, grew out of the common experience that people who recover from certain infections become there-after "immune" to the disease again. *Immunity is highly specific* : an individual who recovers from measles is protected against the measles virus but not against other common viruses such as cold, chicken-pox or mumps. Normally, many of the responses of the immune system initiate the destruction and elimination of invading organisms and any toxic molecules produced by them. Because these immune reactions are destructive in nature, it becomes necessary that they be made in response only to molecules that are foreign to the host and not to those of the host itself. *This ability to distinguish foreign molecules from self molecules is another fundamental feature of the immune system.* However, occasionally it fails to make this distinction and reacts destructively against the host's own molecules; such **autoimmune diseases** can be fatal to the organism.

Almost any macromolecule (*e.g.,* protein, most polysaccharides and nucleic acids), as long as it is foreign to the recipient, can induce an immune response; any substance capable of eliciting an immune response is called an **antigen** (**anti**body **gen**erator). There are two broad classes of immune responses : (1) antibody responses and (2) cell-mediated immune responses. **Antibody-mediated (humoral) responses** involve the production of antibodies which are proteins, called **immunoglobulins** (**Ig**) (Fig. 10.1). The antibodies circulate in the blood stream and permeate the other body fluids, where they bind specifically to the foreign antigen that have induced them.

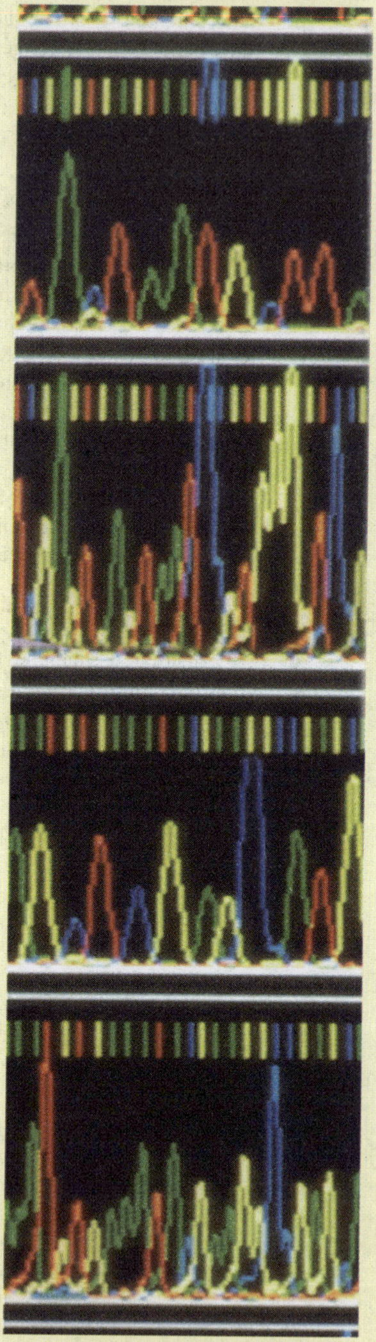

A computer displays an analysis of the sequence of nucleotide bases in a segment of DNA – in this case, the human gene cluster HL-A, which plays an important role in immunology.

Pollens provokes an allergic response in hayfever sufferers, whose immune systems make antibodies in response to normally harmless substances (antigens).

Generally, the antibodies either adhere to the surface of the microorganisms, making them clump together (**agglutination**), or they may cause them to disintegrate (**lysis**). Binding with antibody (called **antitoxin**) inactivates viruses and bacterial toxins (such as tetanus or botulism toxins) by blocking their ability to bind to receptors on the target cells. Antibody binding also marks or 'tags' invading microorganisms for destruction, either by making it easier for a phagocytic cell to ingest them (such antibodies are called **opsonins**) or by activating a system of blood proteins, collectively called **complement** that kills the invaders. **Precipitin** type antibodies cause aggregation of antigen molecules leading to the formation of a precipitate.

Cell mediated immune responses involve the production of specialized cells that react with foreign antigens on the surface of other host cells. The reacting cell can kill a virus-infected host cell that contains viral proteins on its surface, thereby eliminating the infected cell before the virus has replicated. In other cases the reacting cell secretes chemical signals that activate macrophages to destroy invading microorganisms.

Further, once specific antibodies have been produced against a particular microbe, defence against the disease is set up (at least for the time being) and *immunity is acquired*. This kind of immunity is called **active immunity**, because the body makes its own antibodies in response to the arrival of an antigen. However, during the development of a mammal a certain number of antibodies (*e.g.*, Ig G) pass from the mother to the foetus via the placenta or after birth via the milk (*e.g.*, Ig A). This confers **passive immunity** on the young animal, at any

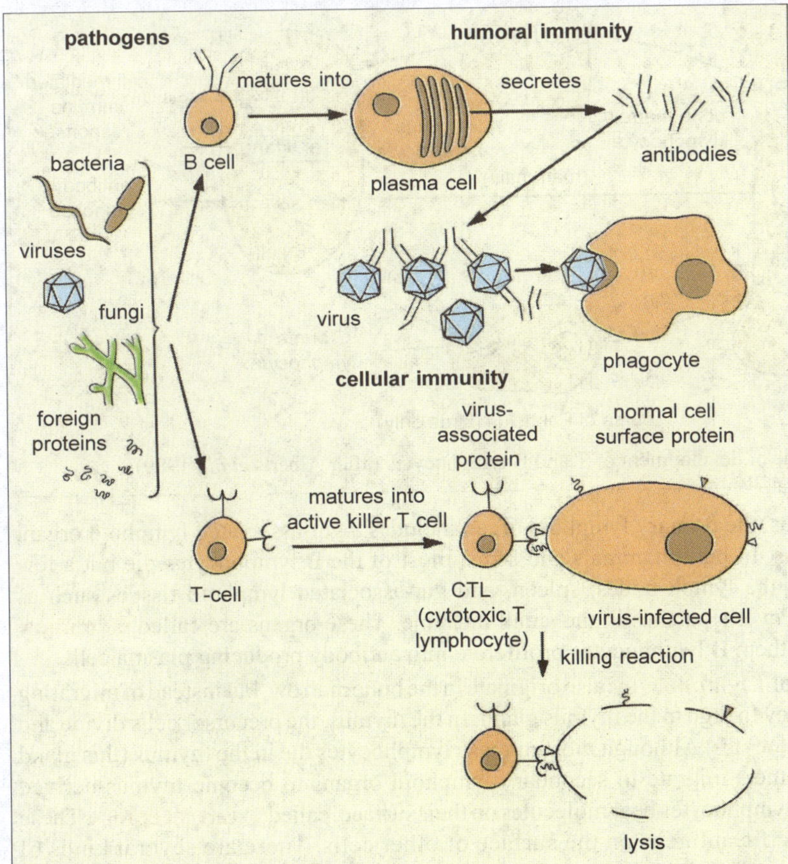

Fig. 10.1. Humoral immunity and cellular immunity (after Darnell *et al.*, 1986).

rate for a short time after birth. **Active artificial immunity** can also be obtained by injecting a small quantity of antigens, the **vaccine**, into the body (this process is called **immunization**). The immune response appears to be of rather recent evolutionary origin, because antibodies production is the characteristic only of vertebrates.

CELLULAR BASIS OF IMMUNITY

The cells responsible for immune specificity are a class of white blood cells known as **lymphocytes**. Lymphocytes occur in large numbers (in trillions in man) in the blood, lymph and lymphoid organs such as thymus, lymph nodes, spleen and appendix. During the 1960s, it was discovered that the two major classes of immune responses are mediated by different classes of lymphocytes : B lymphocytes and T lymphocytes. In mammals, **B lymphocytes** originate from the haemopoietic tissue of bone narrow (in adult) or foetal liver both of which are called **primary**

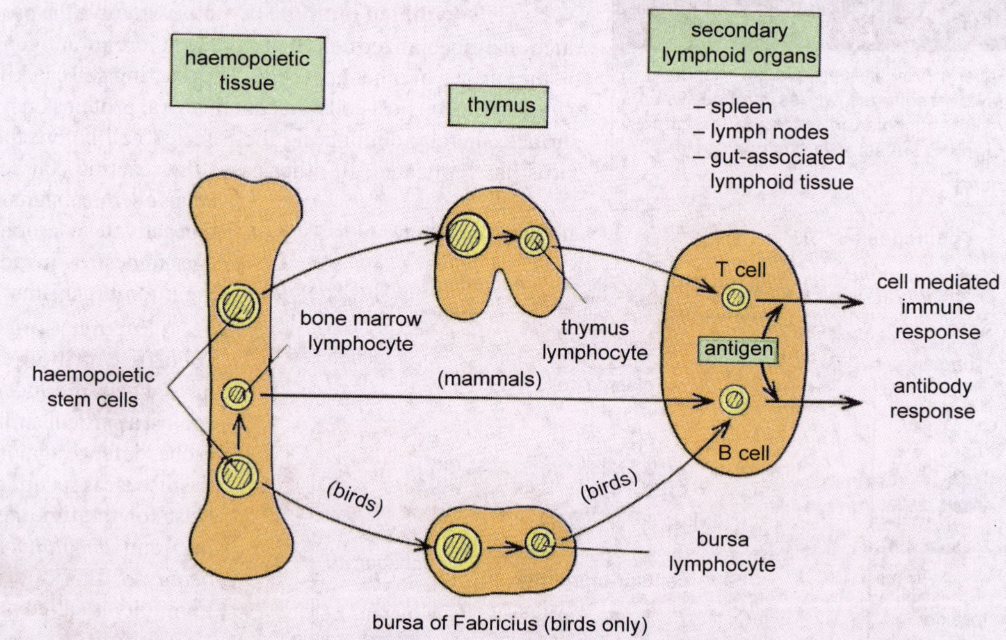

Fig. 10.2. Mode of development of T and B lymphocytes (after Alberts *et al.*, 1989).

lymphoid organs. In birds, the primary lymphoid organ includes a gut-associated lymphoid organ, called **bursa of Fabricius**. In both mammals and birds, most of the B lymphocytes die but a few migrate via the blood to the lymph nodes, spleen, and gut-associated lymphoid tissues such as appendix, adenoids and Peyer's patches in the small intestine. These organs are called **secondary lymphoid organs** and in them B lymphocytes proliferate into antibody producing plasma cells.

The precursor cells of **T lymphocytes** also originate in the bone marrow, but instead of migrating straight to lymph nodes they first go to the thymus gland. In the thymus, the precursor cells divide and differentiate into **T lymphocytes**. Although most of these lymphocytes die in the thymus (this gland too regresses in adult), others migrate to secondary lymphoid organs to become thymus-derived lymphocytes or **T cells**. T lymphocytes bear molecules on their surface, called **T-cell receptors**. These receptors react with specific antigens on the surface of other cells. There are several kinds of T-lymphocytes with different functions. Some of them destroy cancer cells, cells infected with viruses or intracellular bacteria, and cells which have been grafted into the body as, for example, in a heart transplant. They are called **cytotoxic T cells** or **killer T cells**. Since these cells are involved directly in defense against infection, they are also called **effector cells**. Other T lymphocytes, called **regulatory cells**, regulate the body's defence mechanism by controlling the phagocytes and B lymphocytes. Some of these regulatory lymphocytes help the B lymphocytes in their action (*i.e.*, they enhance B cell's responses) and are called **helper T cells**; others suppress them and are instrumental in preventing antibodies being produced against the individuals own cells. They are called **suppressor T cells.** The helper T lymphocytes achieve their effects by producing hormone-like chemical substances (local chemical mediators), called **lymphokines** or **interleukins**, that help B cells to make antibody responses, stimulate activated T cells to proliferate and activate macrophages (see **Alberts** *et al.*, 1989).

MOLECULAR STRUCTURE OF IMMUNOGLOBULINS OR ANTIBODIES

The human body is capable of synthesizing more than a million different kinds of immunoglobulin molecules, each capable of reacting with a different antigen, but all of them appear to share the same fundamental quaternary (globular) structure. Typically, an immunoglobulin molecule is a Y-shaped heteromer and composed of two identical **heavy (H) polypeptide chains** and two smaller identical **light (L) chains**. Each arm of the Y contains a complete L chain and a part of an H chain and the leg of the Y contains the remaining parts of the H chains. Near its C-terminus, each L chain is linked to an H chain by disulphide bridge, and two additional disulphide bridges link the H-chains together (Fig. 10.3). The H chains possess antigenic determinants in the "tail" segments by which they can be classified as **Ig G**, **Ig M**, **Ig A**, **Ig D** or **Ig E**, each with its own class of H chain, such as, γ **(gamma)**, μ **(mu)**, α **(alpha)**, δ **(delta)** and ε **(epsilon)** respectively (see Table 10-1.) Light chains can likewise be typed as **kappa (k)** or **lambda (λ)**. Within a H chain or L chain, C-termini segments exhibit very little variation in primary structure from one individual to another and are called **constant regions** (**C**). The amino ends or N-termini of both heavy and light chains, however, are extremely diverse in primary structure, even within a class and are called **vari-**

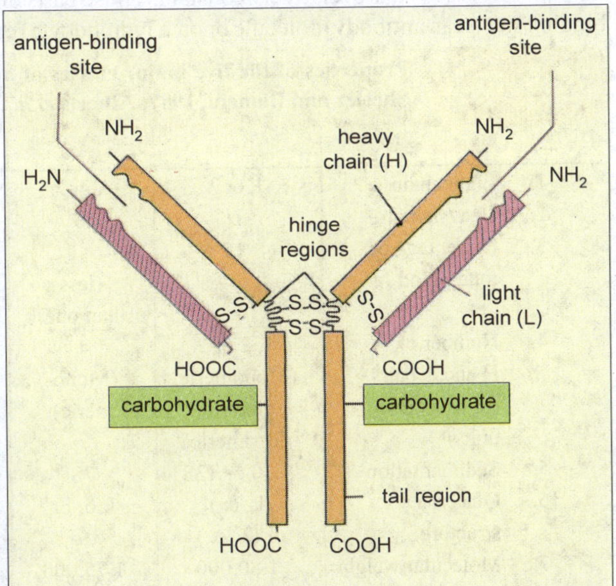

Fig. 10.3. Generalized chemical structure of an antibody molecule (after Alberts *et al.*, 1989).

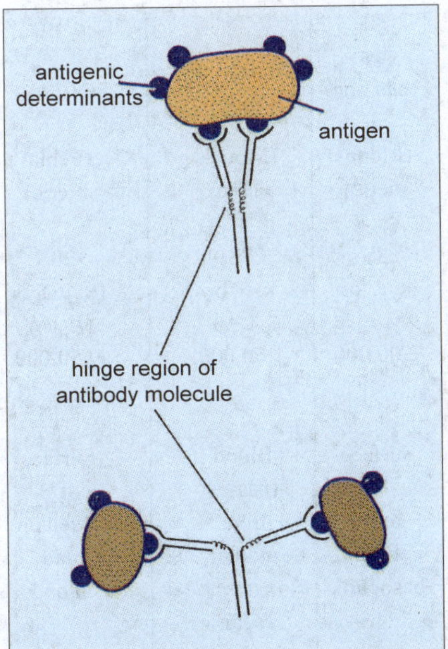

Fig. 10.4. The hinge region of an antibody molecule improves the efficiency of antigen binding (A) and cross-linking (B) (After Alberts *et al.*, 1989).

able (V) regions. The V_H and V_L regions together form antibody-combining site for specific interaction with a homologous antigen molecule. Thus, each Y-shaped antibody has two identical **antigen-binding sites**, one at the tip of each arm of the Y. Because of their two-antigen-binding sites, antibodies are said to be **bivalent**. The efficiency of antigen binding and cross-linking of antibodies is greatly increased by the flexible **hinge regions** in antibody molecules, which allow the distance between the two antigen binding sites to vary (Fig. 10.4).

Further, the proteolytic enzyme **papain** splits antibody molecule into different characteristic fragments : two separate and identical **Fab** (= **f**ragment **a**ntigen **b**inding) **fragments**, each with antigen-binding site and one **Fc fragment** (so called because it is readily crystallizes). Each of the four polypeptide chains of an immunoglobulin is also divided into repeating segments, called **domains**, each of which folds independently to form a compact functional unit. Thus, there are two domains in the L chains (*i.e.*, V_L and C_L) and four in the H chains (*i.e.*, V_H, C_H1, C_H2 and C_H3) (Fig. 10.5A).

Recently, it has been found that the antigen binding site of the antibodies is formed by only about 20 to 30 of the amino acid residues in the variable regions of both L and H chains. In fact, the variability in the variable regions of both L and H chains is for the most part restricted to three small **hyper variable regions** in each chain (Fig. 10.5B). The remaining parts of the variable regions, known as **framework regions**, are relatively constant. Those parts of an antigen that combine with the antigen-binding site on an antibody molecule or on a lymphocyte receptor, are called **antigenic determinants**

Table 10.1. **Properties of the five major classes of human antibodies (immunoglobulins) (after Sheeler and Bianchi, 1987; Alberts *et al.*, 1989).**

Characteristic	Ig A	Ig D	Ig E	Ig G	Ig M
1. Light chain	k or λ	k or λ	k or λ	k or λ	k or λ
2. Heavy chain	α	δ	ε	γ	μ
3. Percentage of total blood Ig	15	<1 (less than one)	<1 (less than one)	75	10
4. Number of H and L pairs (or of 4 chain units)	1, 2 or 3 (Monomeric, dimeric, or trimeric)	1 (Mono-meric)	1 (Mono-meric)	1 (Mono-meric)	5 (Penta-meric)
5. Sedimentation	7,10 or 13S	7S	8S	7S	20S
6. Quaternary structure	$(k_2\ \alpha_2)_n$ or $(\lambda_2\ \alpha_2)_n$	$k_2\delta_2$ or $\lambda_2\delta_2$	$k_2\ \varepsilon_2$ or $\lambda_2\varepsilon_2$	$k_2\ \gamma_2$ or $\lambda_2\gamma_2$	$(k_2\ \mu_2)_5$ or $(\lambda_2\ \mu_2)_5$
7. Molecular weight	1,80,000 to 5,00,000	1,75,000	2,00,000	1,50,000	9,50,000
8. Occurrence	Milk, saliva, tears and respiratory and intestinal secretions	Cell surface of B cells (Virgin B lymphocytes)	Surface of mast cells and basophils	Blood (passes from mother to foetus via placenta)	Surface of B cells and blood

or **epitopes**. Molecules that bind specifically to such an antigen-binding site but cannot induce immune responses, are called **haptens**. Haptens are usually small organic molecules, they become antigenic if they are coupled to a suitable macromolecule, called **carrier**. Haptens such as dinitrophenyl (DNP) group have been important tools in experimental immunology.

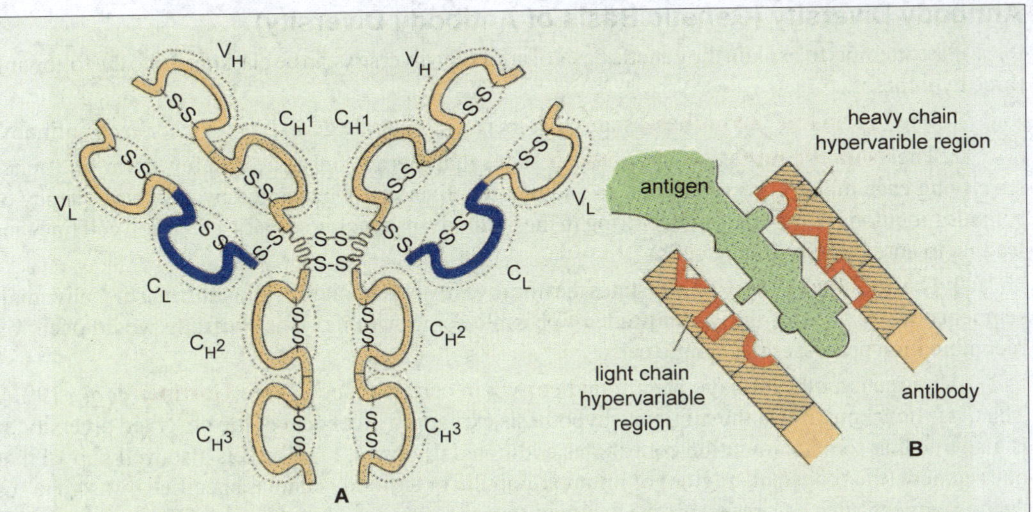

Fig. 10.5. A—The L-and H-chains in antibody molecule are each folded into repeating domains that are similar to one another. B—Hypervariable regions of each variable domain of an antibody and antigen (after Alberts *et al.*, 1989; Sheeler and Bianchi, 1987).

During the early stages of an infection, the response to the antigen involves the production of a specific class of immunoglobulins, called **IgM**. IgM is the first class of antibody which appears on the surface of a developing B lymphocyte and is secreted into the blood during the **primary antibody response**. In its secreted form, IgM is composed of five four-chain units and has a total of 10 antigen-binding sites. It also contains one copy of another polypeptide chain, called **J chain** (J=joining) which is produced by secretory B lymphocyte. Later on, the amount of Ig M gradually declines as another class or isotype of immunoglobulin, called **IgG**, appears. IgG represents the most abundant form of body's antibodies (75 per cent) and is one of the most extensively studied immunoglobulins.

IgG has a molecular weight of 1,50,000 and is composed of four polypeptide chains in its Y- shaped quaternary molecule—two identical H chains and two identical L chains. The L chains have the molecular weight of 20,000 to 25,000 and consist of about 450 amino acids. Each L chain is covalently linked to an H chain by a disulphide bridge, and two light chain-heavy chain pairs are covalently linked by two disulphide bridges (near the hinge). There are also 12 intra-chain disulphide bridges, four in each H chain and two in each L chain. Lastly, an aspargine residue in each H chain is bonded to carbohydrate, since an immunoglobulin is also a glycoprotein.

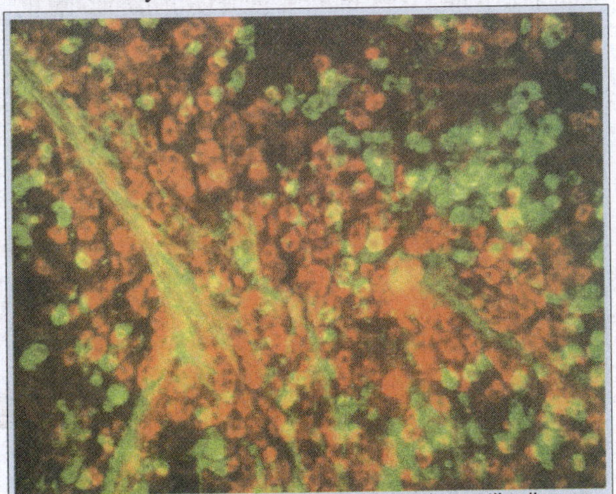

Micrograph of plasma cells producing antibodies IgM (coloured red) and IgG (green).

IgG is produced in large quantities during **secondary responses** and is involved in antibody dependent cell-mediated killing. IgG molecules are the only antibodies that can pass from mother to foetus via the placenta. The cells of the placenta that are in contact with F_C receptors bearing IgG molecules mediate their passage to the foetus. The antibodies are first ingested by receptor-mediated endocytosis and then transported across the cell in vesicles and released by exocytosis into the foetal blood. This process is called **transcytosis**.

Antibody Diversity (Genetic Basis of Antibody Diversity)

Past attempts to explain the genetic basis of antibody diversity can be classified into the following three hypotheses :

1. The **"germ line" hypothesis** states that there is a separate germ line gene for each antibody.

2. The **"somatic mutation" hypothesis** states that there is only one or a few germ line genes specifying each major class of antibodies and that the diversity is generated by a high frequency of somatic mutation (*i.e.*, mutations occurring in the antibody-producing somatic cells or in cell lineages leading to antibody-producing cells).

3. The **"minigene" hypothesis** states that the divesity is generated by the shuffling of many small segments of a few genes into a multitude of possible combinations. This shuffling would occur by recombination processes in somatic cells.

All three hypotheses have been found correct in certain respects (see **Gardner** *et al.*, 1991). Thus, it is now known that the minigene hypothesis explains a great deal of the observed diversity. It is also true that somatic mutation contributes additional diversity. Finally, it is also well known that one segment (the "constant" region) of immunoglobulin or antibody chain is specified by a "gene" or "gene segment" that is present in the genome in only a few copies. During differentiation of B lymphocytes, antibody diversity can be originated by the following three ways : 1. by genome rearrangement (or DNA splicing); 2. by alternate pathways of transcript or RNA splicing; and 3. by variable joining sites and somatic mutation. The following brief discussion regarding mode of origin of antibody diversity would illuminate our readers :

Here, we will consider how the enormous number of amino sequences of the V regions (variable chains) are formed from a comparatively small numbers of genes. To be clear, we will only describe how the kappa type light (L) chain is produced : Many genes can be used to form a kappa-type L chain; they are of three types — *V, J* and *C.* There are roughly 300 different *V* genes (which are responsible for the synthesis of the first 95 amino acids of the variable region), 4 different *J* genes (which encode the final 12 amino acids of the variable region and join the *V* and *C* regions; *J* = junction of joining sequences) and 1 copy of the *C* gene (which encodes the constant region). In an embryonic cell the *V* genes form a tight cluster, the *J* genes form a second tight cluster quite far from the *V* gene cluster, and the *C* gene follows not far after the *J* gene cluster. All of the genes are on the same chromosome as shown in Figure 10.6. Note that each *V* gene is preceded by leader regions where transcription can be initiated; the *J* and *C* genes are not preceded by leaders.

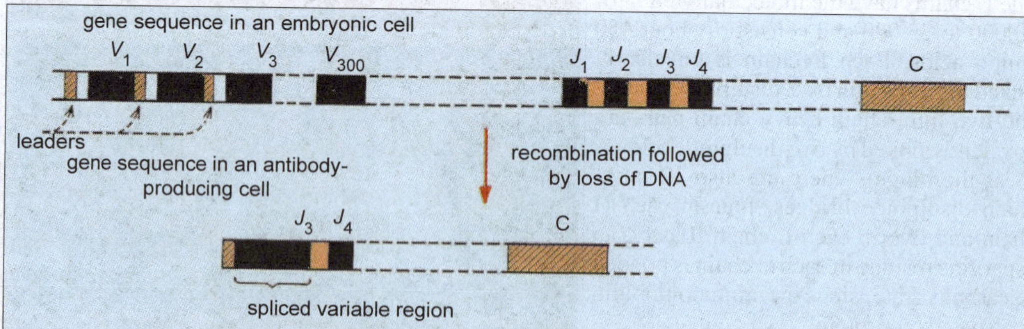

Fig. 10.6. DNA splicing in the formation of the variable region of an L chain. The V_{16}–J_3 joint has been removed, presumably by a site-specific recombination event. The *V* genes to the left of V_{16} have been removed either by homologous recombination or by site–specific recombination within the leader sequences (after Freifelder, 1985).

Regions encoding particular IgG molecules have been cloned (by recombinant DNA techniques) from various mouse cell lines, each producing a particular IgG molecule. The *V-J-C* region has also been cloned from mouse embryo cells, which have not yet been committed to antibody synthesis and, thus, presumably contain an unaltered master set of genes for antibody synthesis. For each clone obtained from an antibody-producing cell line it has been found that a large segment of the embryonic DNA sequence is not present and that the missing segment is always a sequence between the particular *V* gene that encodes the first 95 amino acids of the *V* region and the *J* gene that encodes the last 12 amino acids of the *V* region. This is explained by a gene rearrangement in which DNA between the particular *V* and *J* genes is deleted (*i.e.*, looping out of the intervening segment as in interstitial deletion). An example of one such rearrangement is shown in Figure 8.11. Many different gene sequences encoding particular IgG proteins have been cloned and it has been found that in each clone a different segment of DNA is not present. For example, in the figure the DNA between V_{16} and J_3 is absent; in another clone there might be a V_{210} - J_1 junction instead. In both cases a *V* gene and a *J* gene have become **spliced** together to form a complete gene for the variable region. Note that this is **DNA splicing** and not the RNA splicing that has been described in Chapter 7.

Studies of the base sequence of cloned IgG DNA sequences show that the junction between a particular *V* gene and a particular *J* gene is not always the same. That is, the two terminal triplets of juxtaposing *V* and *J* genes can exchange at any one of four sites that yields a triplet. The meaning of this statement is shown in Figure 10.7. In the example shown, there are three possible amino acids at the joint, which add diversity to the number of possible regions.

Since there are 300 different *V* genes, 4 different *J* genes, and (on the average) 2.5 different amino acids at the junction, there are, then, $300 \times 4 \times 2.5 = 3000$ different variable regions. The H chain genes are organized in a similar but not identical way (there are four different types of genes) and there are 5000 variable regions in the H chains. Thus, since each IgG molecule contains two identical H chains and two identical L chains, there are about $3000 \times 5000 = 1.5 \times 10^7$ different IgG molecules can be formed from the V_L, J_L, V_H and J_H genes. This is sufficient reason to explain the diversity of antibody molecules.

As shown in Figure 10.8, DNA splicing does not fully generate a L chain sequence, since (1) the spacer between the *J* and *C* genes remains and (2) the actual L chain has amino acids derived from only one *V* gene and one *J* gene and the spliced DNA usually contains many *V* and *J* genes. The correct amino acid sequence is obtained by a final RNA– splicing event (see Fig. 10.8). Note how the particular *V-J* joint determines the RNA splicing pattern. For example, the RNA removed is always a segment between the leader and the *V* gene and between *C* gene and the right end of the *J* gene in the *V-J* joint.

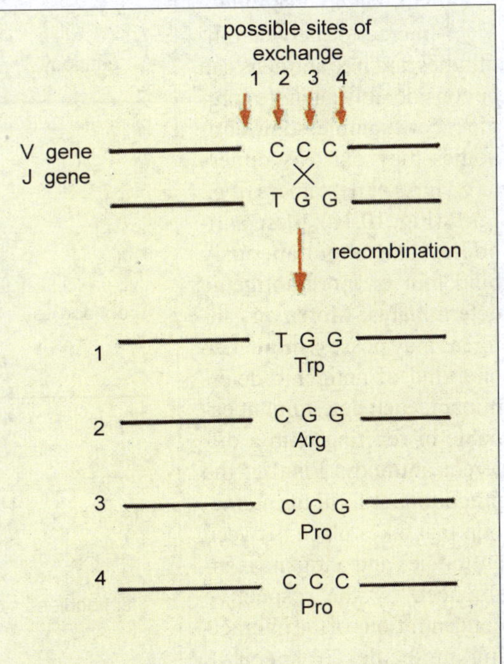

Fig. 10.7. Four possible junctions at a *V* - *J* joint giving rise to three different amino acids (after Freifelder, 1985).

B. LYMPHOCYTES AND THE IMMUNE RESPONSE

Antibodies secreted by plasma cells may have several different effects : (1) they may interact with free (*i.e.*, soluble) antigens causing **precipitation**; (2) they may interact with surface antigens of the pathogen (*i.e.*, particulate antigens) causing **agglutination**; or (3) they may promote **complement fixation**.

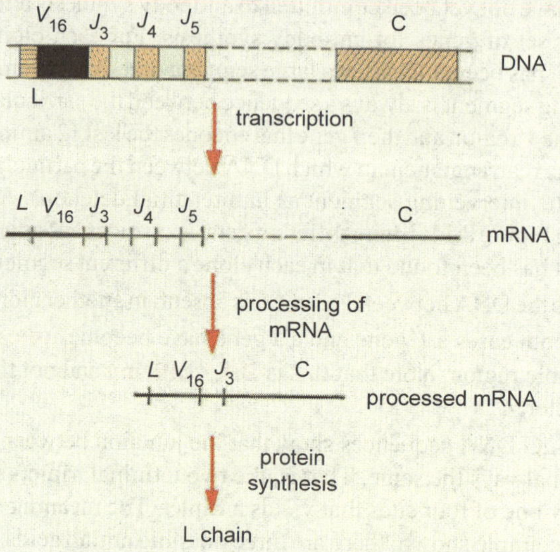

Fig. 10.8. Production of the L chain of a particular IgG molecule. Solid regions indicate coding sequences used to generate the final L chain (after Freifelder, 1985).

1. Precipitation of Soluble Antigens

Antigens may have one or more antigenic determinants (Fig. 10.9). If antigens contain one determinant, they are called **monodeterminants**; if they contain two determinants, they are called **bideterminants** and so on. Most antibodies are bivalents, meaning that they can simultaneously combine with two antigenic determinants. The products formed by interaction of antibody and antigen depend on the number of antigenic determinants that are present. For example, two monodeterminant antigens can be cross-linked by a single antibody, but the product is not usually insoluble unless the antigen itself is very large. However, if two antigenic determinants are present, cross-linking by the antibody can produce chains of antigens that are insoluble and form **precipitates**. Multideterminant antigens react with antibody to produce cross-linked network or lattices that are insoluble.

Interaction between antibodies and free antigens can be considerably far more complex. For example, some antibodies may exist as dimers (*e.g.*, IgA) or pentamers (*e.g.*, IgM) (Fig. 10.10); these antibodies can simultaneously bind four or more antigenic determinants. Moreover, antigens may possess more than one kind of antigenic determinant, each determinant capable of reacting with a different antibody. Finally, the predominant form of interaction that takes place between antibodies and antigens is influenced by the respective concentrations of the interacting molecules (= species). Thus, small soluble complexes are favoured when there is an excess of antibody;

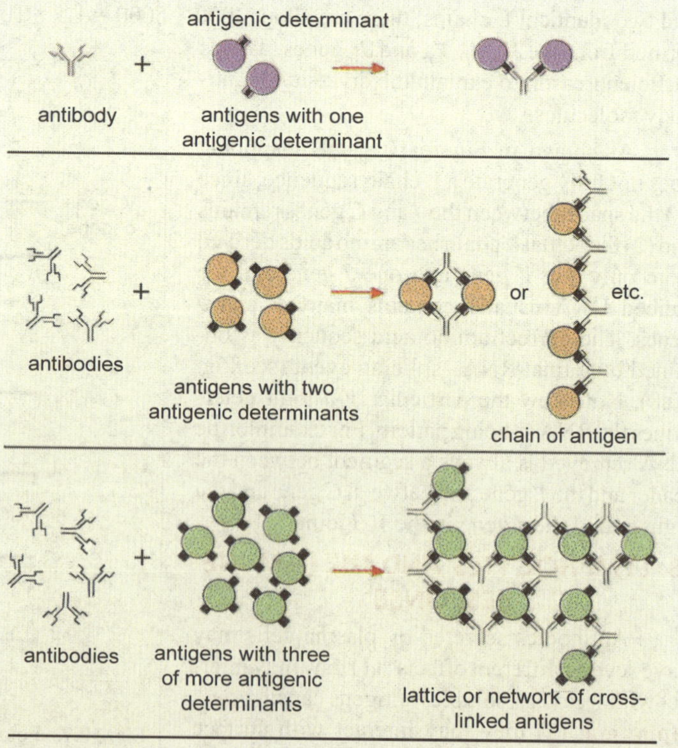

Fig. 10.9. Formation of antigen–antibody complexes (after Sheeler and Bianchi, 1987).

chains of cross-linked antigens are favoured when there is an antigen excess; and cross-linked lattices are favoured by nearly equal amounts of antibody and antigen. Regardless of the nature of the products formed, antigen-antibody complexes are eventually eliminated by the phagocytic action of macrophages.

2. Agglutination. Antibodies that interact with antigens present in the surfaces of invading microorganisms or other foreign particles cause agglutination (Fig. 10.11). During agglutination the particles become cross-linked to form small masses, and the masses are limited by the phagocytic action of macrophages.

The plasma membranes of macrophages possess receptors that recognise and bind the C-terminal or F_C regions of heavy chains of immunoglobulin. As a result, the macrophage receptors are called **FC receptors**. Because, the F_C regions of immunoglobulins include constant domains, macrophage F_C receptors can bind a variety of different antibodies. Interaction between a macrophage and a mass of agglutinated cells is followed by phagocytosis.

Foreign cells that have attached antibodies can be destroyed by **K** (or **Killer**) **cells**. Killer cells bind the agglutinated mass by interacting with the F_C regions of an antibody but do not internalize it. Instead, it is thought that there is the transfer of toxic substances from the K cells to the pathogen.

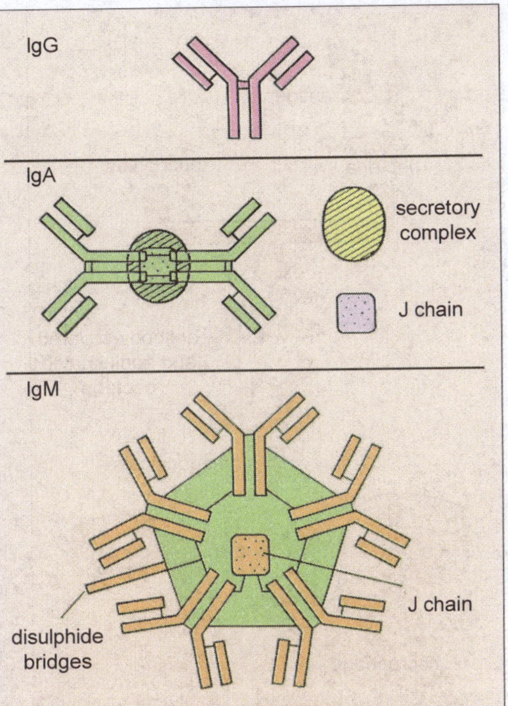

Fig. 10.10. Schematic illustration of IgG, IgA and IgM antibodies. IgG (and IgD and Ig E) molecules occur as monomers only, but IgA may form dimers and trimers, and IgM may form pentamers (after Sheeler and Bianchi, 1987).

3. Complement fixation (or The complement system). Complement, so called since it *complements* and amplifies the action of antibody, is the principal means by which antibodies defend vertebrates against most bacterial infections. It consists of a system of serum proteins that can be activated by antibody-antigen complexes or microorganisms to undergo a cascade of proteolytic reactions whose end result is the assembly of **membrane attack complexes** (or **lytic complex**). These complexes form holes in a microorganism and thereby destroy it. At the same time, proteolytic fragments released during the activation process promote the defense response by dilating blood vessels and attracting phagocytic cells to sites of infection. Complement also amplifies the ability of phagocytic cells to bind, ingest and destroy the microorganisms being attacked.

Individuals with a deficiency in one of the central components of complement system (*e.g.*, C_3) are subject to repeated bacterial infections, just as are individuals deficient in antibodies themselves. Complement-deficient individuals may also suffer from **immune-complex diseases**, in which antibody-antigen complexes precipitate in small blood vessels in skin, joints, kidney and brain, where they cause inflammation and destroy tissues; this suggests that complement normally helps to solubilize such complexes when they form during an immune response.

Mode of complement activation. Complement consists of about 20 interacting proteins, of which reacting components are designated **C1—C9**, **factor B** and **factor D**—the rest comprising a

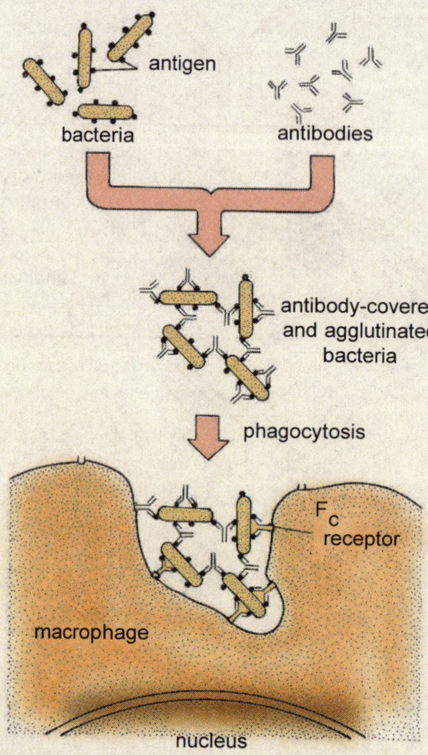

Fig. 10.11. Agglutination by antibodies and phagocytosis by macrophages (after Sheeler and Bianchi, 1987).

variety of regulatory proteins. The complement components are all soluble proteins. They are made mainly by the liver and circulate in the blood and extracellular fluid. Most are inactive unless they are triggered directly by an invading microorganisms or indirectly by an immune response. The ultimate result of complement activation is the assembly of the **late complement components** such as C5, C6, C7, C8 and C9, into a large protein complex, the membrane attack complex, that mediates microbial cell lysis.

Because its function is to attack the plasma membrane of microbial cells, the activation of complement is focused on the microbial plasma membrane, where it is triggered either by antibody bound to the microorganism or by microbial envelope polysaccharides. Both of these activate the early **complement components**. There are two sets of early complement components belonging to two distinct pathways of complement activation : C_1, C_2 and C_4 belong to the **classical pathway**, which is triggered by antibody binding; factor B and factor D belong to the **alternative pathway** (Fig. 10.12), which is triggered by microbial polysaccharides. The early complement components of both pathways ultimately act on C3, the most important complement component. The early complement components and C3 are **proenzymes** that are activated sequentially by limited proteolytic cleavage : as each proenzyme in the sequence is cleaved, it is activated to generate a serine protease, which cleaves the next proenzyme in the sequence and so on. Many of these cleavages liberate a small peptide fragment and expose a membrane-binding site on the large fragment. The larger fragment binds tightly to the target cell membrane by its newly exposed membrane-binding site and helps to carry out the next reaction in the sequence. In this way complement activation is confined largely to the cell surface where it began. The smaller fragment often acts independently as a diffusible signal that promotes an inflammatory response.

The activation of C3 by cleavage is the central reaction in the complement-activation sequence, and it is here that the classical and alternative pathways converge. In both pathway, C3 is cleaved by an enzyme complex called a **C3 convertase**. A different C3 convertase is produced by each pathway, formed by the spontaneous assembly of two of the complement components activated earlier in the cascade. Both types of C3 convertase cleave C3 into two fragments. The larger of these (C3b) binds covalently to the target-cell membrane and binds C5. Once bound, the C5 protein is cleaved by the C3 convertase (now acting as C5 convertase) to initiate the spontaneous assembly of the late components— (C5 through C9) – that creates the membrane attack complex.

Since each activated enzyme cleaves many molecules of the next protoenzyme in the chain, the activation of the early components consists of an amplifying **proteolytic cascade** : each molecule activated at the beginning of the sequence leads to the production of many membrane attack complexes.

The **classical pathway** is usually activated by IgG or IgM antibodies bound to antigens on the surface of a microorganism (or target cell). The binding of antigen by these antibodies enables their constant regions to bind in turn to the first component in the classical pathway, **C1**, which is a large complex composed of three sub components – **C1q**, **C1r** and **C1s** (Fig. 10.13). The molecule of C1q protein is large sized (~ 450,000 daltons) and made up of six identical subunits, each composed of three different polypeptide chains. The carboxyl-terminal halves of each of the three polypeptide chains in a subunit are folded into a globular structure; the amino-terminal halves have a typical collagen amino acid sequence and are wound together to form a collagen-like triple stranded helix. The six subunits are linked together by disulphide bonds between their triple-helical stems, forming a structure that resembles a bunch of tulips (Fig. 10.13).

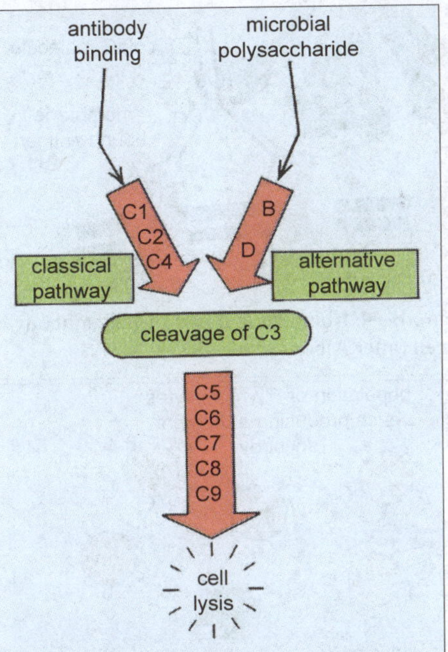

Fig. 10.12. The principal stages in complement activation by the classical and alternative pathways. Such activation occurs at the surface of an invading bacterium (after Alberts *et al.*, 1989).

The binding of a globular head of C1q to an IgG or IgM antibody bound to antigen activates C1q to start the early proteolytic cascade of the classical pathway. Activation of the C1q activates C1r to become proteolytic, and C1r in turn cleaves and activates C1s. Activated C1s then cleaves C4 into two fragments C4a and C4b and the smaller is designated as a. C4b immediately binds covalently to the membrane and then binds C2. Once bound, C2 is also cleaved by activated C1s.

There are some microbes which are ingested but not killed by macrophages. For example, *Mycobacterium leprae*, the causative agent of leprosy, and parasite of genus *Leishmania* (a flagellated protozoon causing leishmaniasis) actually grow only in the endocytic vesicles of macrophages, and for this reason they are extremely difficult to kill.

Clonal Selection Theory

The fact that specific antibodies are produced on demand by the arrival of particular antigen, led to the development of the **clonal selection theory** (first proposed in 1960 by **F.M.Burnet** and later developed by **G. L. Ada** in 1987). This theory is based on the idea that an enormously wide range of B lymphocytes, each potentially capable of producing a specific kind of antibody, is present in the body before birth. When an antigen gets into the body after birth, it 'selects' a lymphocyte of the appropriate type, adheres to its surface and causes it to proliferate into a clone of cells, all of which proceed to manufacture the correct antibodies (To recall, a clone is a population of identical cells all derived from one original cell). The lymphocyte is already programmed by its genes to make the right antibodies.

Sir Peter Medawar and **Sir Frank Macfarlane Burnet** shared the Nobel Prize in Physiology in 1960 for their research into how an organism's immune system learns how to recognize "self" and differentiate it from "non - self".

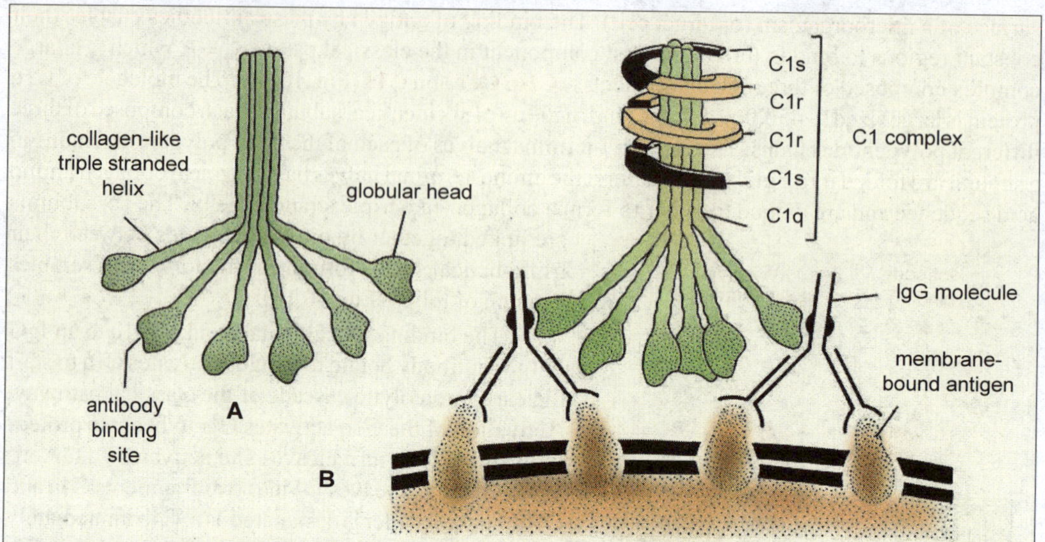

Fig. 10.13. A—Bunch of tulips-like molecule of C1q protein ; B—C1 binding to a pair of IgG antibody molecule bound to antigens of surface of target cell (after Alberts *et al.*, 1989).

The antigen simply triggers it into action, probably by switching on the appropriate part of its DNA. The theory is based on the lymphocyte recognizing its particular antigen. **Burnet** suggested that this is achieved by the antigen fitting into receptor site on the cell surface (antigen never gets inside the lymphocyte).

Allelic Exclusion

Now the question arises, why does each B lymphocyte make only one type of antibody ? Mammalian cells are diploid; they carry two sets of genetic information coding for each of the antibody chains. But only one productive genome rearrangement of light chain coding sequences and one productive genome rearrangement of heavy chain coding sequence occur in each B lymphocyte. This phenomenon is called **allelic exclusion**, because one of the "alleles" is excluded from being expressed. The exact mechanism involved in this phenomenon is not known. However, it appears that there must be some type of a feedback mechanism that arrests the recombination process(es) involved in these antibody gene rearrangement, once a productive rearrangement has occurred and the cell has started to synthesize a functional antibody. The simplest mechanism would involve inhibition of this process by the mature antibody itself (**Gardner** *et al.*, 1991).

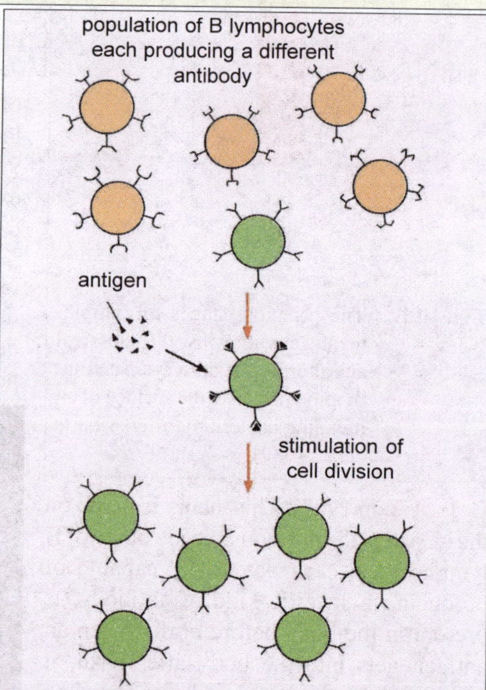

Fig. 10.14. Clonal selection theory. When a particular antibody binds to an antigen, the cell producing that particular antibody is stimulated to divide. After a series of cell divisions, a clonal population of cells is produced, with all cells producing the same antibody (after Gardner *et al.*, 1991).

Immunologic Memory

A relationship has been observed between time and the appearance of antibodies in response to a **first exposure** to a given antigen (Fig. 10.16). Following a short lag period, antibodies begin to appear in the blood, rising to maintain a plateau level for some time before falling again. This characteristic response curve is called a **primary immune response**. As long as the antibody content of the blood remains at its plateau level, a condition of **active immunity** exists. The response to a second exposure to the same antigen—the **secondary immune response**—is much more dramatic. In this case, the lag period is shorter, the response is more intense (*i.e.*, greater quantities of antibody are produced) and the elevated antibody level is maintained for a longer period of time. The difference between the two responses indicates that the body has "remembered" its earlier exposure to the antigen. This is called **immunological memory**.

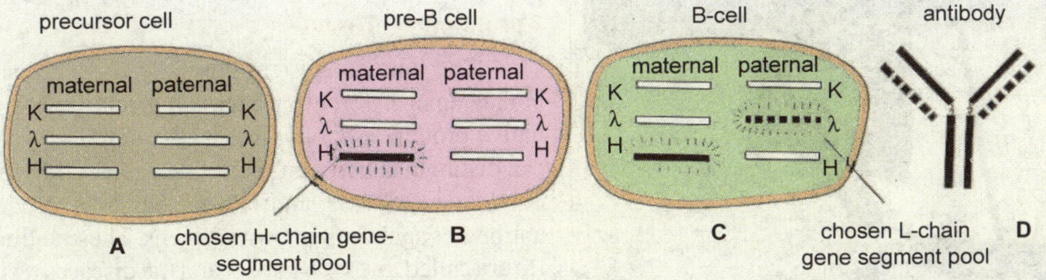

Fig. 10.15. The process of allelic exclusion. The sequential choices in Ig gene activation that a developing B cell must make in order to produce antibodies with only one type of antigen-binding site (*viz.*, B lymphocytes are called **monospecifics**). The cell must choose one of four L-chain gene-segment pools and one of two H-chain gene-segment pools. During development, a precursor cell first activates one H-chain gene pool to become a pre-B cell, making only μ heavy chains. After a period of extensive proliferation, a pre -B cell activates one kappa (k) or lambda (λ) light-chain pool to become a B cell that makes a unique IgM molecule (after Alberts *et al.*, 1989).

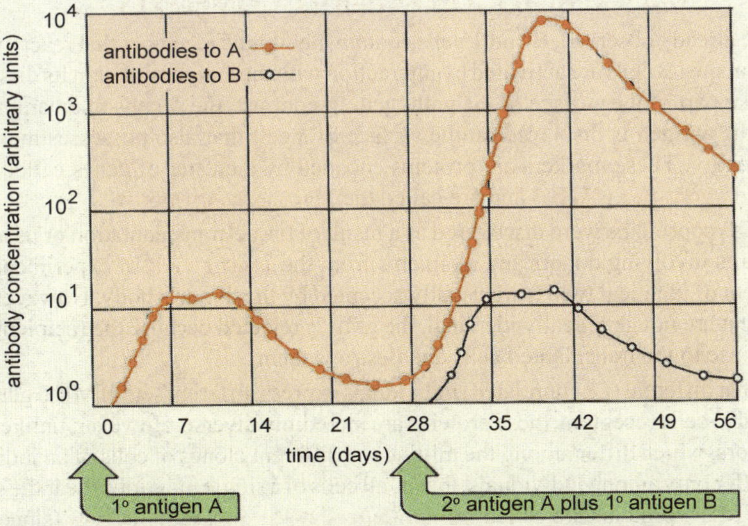

Fig. 10.16. Immunological memory. During a primary immune response the antibody level of the blood rises to a modest plateau and eventually declines. A second exposure to the same antigen results in a secondary immune response that is quicker and in which greater quantities of antibody are produced. Simultaneous exposure to different antigen elicits a primary immune response (after Sheeler and Bianchi, 1987).

People with rheumatoid arthritis have high levels of an antibody that binds to the body's own antibodies as to foreign antigens and deposits them on joint membranes.

Immunological memory may be explained in the following way. The initial exposure to antigen causes differentiation of B lymphocytes into **memory cells** as well as plasma cells. Whereas the plasma cells have a relatively short life span in which they are actively engaged in antibody secretion, memory cells do not secrete antibody and continue to circulate in the blood and lymph for months or years. These memory cells are able to respond more quickly to the reappearance of the same antigen than undifferentiated B lymphocytes. Memory cells are also produced by the multiplication and differentiation of T lymphocytes.

Autoimmune Disease

The immune system normally produces antibodies against foreign proteins but not against the native proteins of body, that is, the immune system can distinguish between "self" and "non-self." However, in rare cases, individuals begin to produce antibodies against their own antigens. These antibodies are called **autoantibodies** and the diseases resulting from their presence are the **autoimmune diseases**. Among these diseases are **paroxysmal cold haemoglobinuria** (antibodies against one's own red blood cells), **myasthenia gravis** (antibodies against one's own muscle cell acetylcholine receptors) and **systemic lupus erythematosus** (antibodies against one's own nuclear DNA).

MAJOR HISTOCOMPATIBILITY COMPLEXES

As we have already observed, B and T cells contain membrane receptors that react with specific antigenic determinants. B cells are activated by interaction with the antigen either in its dissolved form or while it is still a part of the surface of the pathogen. In contrast, the T cells of an animal are only activated when the antigen is displayed on the surface of a cell that also possesses markers of the animal's own identity. These markers are proteins encoded by a cluster of genes called the **major histocompatibility complex** (**MHC**) and are called the **MHC glycoproteins**.

The MHC glycoproteins were discovered as a result of tissue-transplantation or tissue-grafting experiments. Grafts involving donors and recipients from the **same strain** of experimental animals (*i.e.*, the equivalent of identical twins) are usually accepted by the donor's body. However, when the donor and recipient are not genetically identical, the graft is rejected because the recipient mobilizes an immune response to the transplanted cells and destroys them.

It appears that different (*i.e.*, unrelated) individuals express different sets of MHC genes and like the antibodies and T-cells receptors, these proteins are incredibly diverse. However, unlike antibodies and T-cell receptors, which differ among the millions of different clones of cells of an individual, the MHC proteins differ only among individuals, that is, all cells of a single individual bear the same MHC proteins. Human MHC proteins are called **human leucocyte-associated antigens** (since they were first identified in leucocytes or white blood cells) and can be divided into two major classes : **Class I** MHC antigens and **Class II** MHC antigens (Fig. 10.17).

1. Class I MHC Antigen

They are found in the surfaces of nearly all cells. These antigens consist of two polypeptide chains : a large **A chain** that is similar in size and organization to an immunoglobulin heavy chain and a smaller

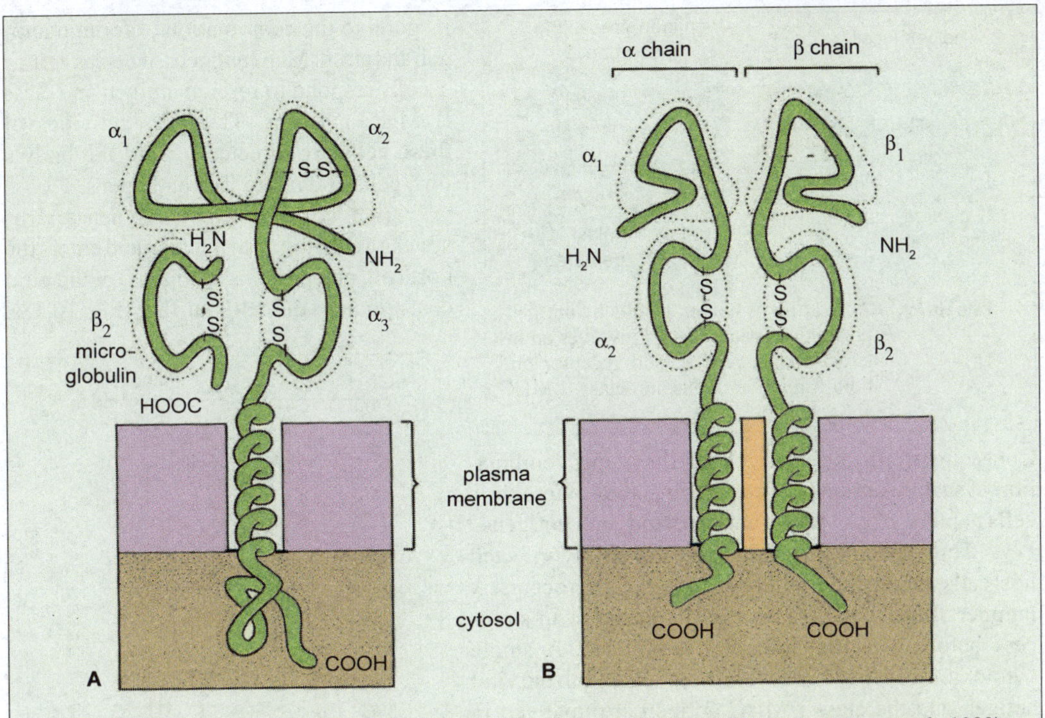

Fig. 10.17. A—Class I MHC glycoprotein; B—Class II MHC glycoprotein (after Albert *et al.*, 1989).

chain called **B2-microglobulin** (Fig. 10.17A).

2. Class II MHC Antigen

They are not found in the surfaces of all types of cells; rather, they are limited to a few types of cells that play a role in the immune response. For example, they are found in most B cells, some T cells and some antigen-presenting macrophages. Class II antigens are composed of two polypeptide chains : an **alpha chain** and a **beta chain**, each about the same size as an immunoglobulin light chain (Fig. 10.17B). MHC antigens bear some homology with the immunoglobulins and play a role in determining the action of T cell.

T LYMPHOCYTES AND THE IMMUNE RESPONSE

T lymphocytes do not interact with free antigens or with antigenic sites in the surfaces of foreign microorganisms. Instead, T cells respond only to cells bearing both a self MHC antigen and an antigenic determinant from a foreign source (*i.e.*, from bacteria, viruses, etc.). Thus, two stimuli are needed to trigger the proliferation and terminal differentiation of the required T-cell clones. **Cytotoxic T cells**

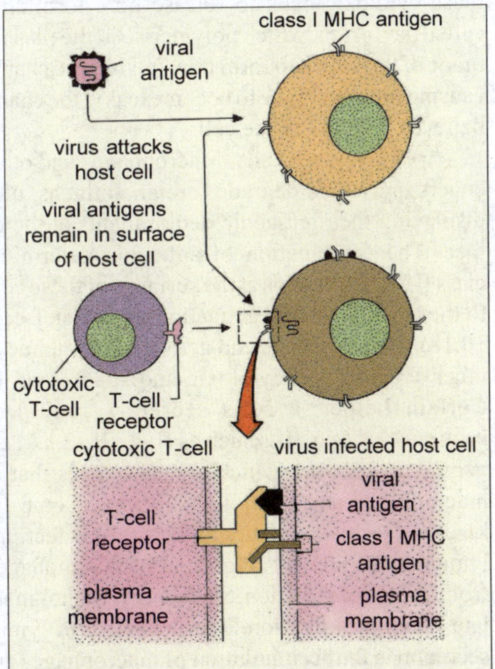

Fig. 10.18. Action of cytotoxic T cells during a response to viral infection (after Sheeler and Bianchi, 1987).

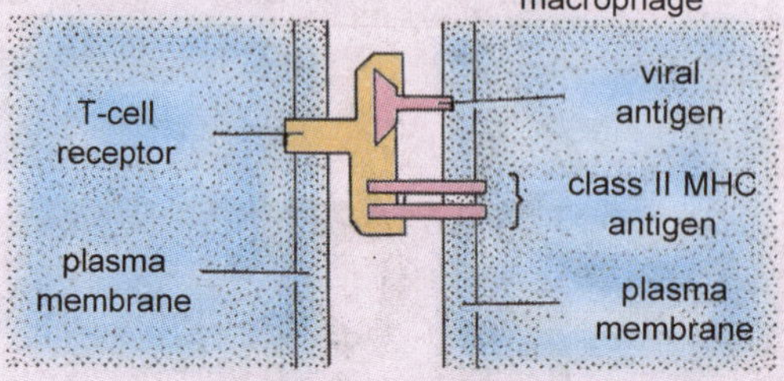

Fig.10.19. Activation of helper T cells by antigen-presenting macrophages involves an interaction between T cell receptor and both foreign antigens and class II MHC antigens.

respond to the combination of foreign antigen and class I MHC antigens, whereas **helper T cells** respond to foreign antigen and class II MHC antigens. Thus, the activities of these cells are directed toward the body's own cells and not to free pathogens.

(i) Cytotoxic T cells. When a virus attacks a cell, the viral nucleic acid enters the host cell, and proteins (antigens) of the viral coat remain at the cell's surface (Fig. 10.18). Consequently, the infected cell has the proper combination of surface antigens to be recognised by cytotoxic T cells namely, class I MHC antigens and viral antigens.

Thus, cytotoxic T cells attach to newly infected host cells, killing them before virus replication occurs. A number of host cells are necessarily killed by this process before the virus infection is reduced. A single cytotoxic T cell may kill several host cells. Both the viral antigen and the class I MHC antigen are involved in attachment of the T cell. Binding to the target stimulates these cytotoxic cells to release pore-forming protein called **perforins**, which polymerise in the plasma membrane of target cell to form transmembrane channels. By causing the membrane to become leaky, the channels are believed to help kill the cell.

(ii) Helper T cells. Macrophages and other scavengers ingest and degrade foreign antigens, ultimately displaying their antigenic determinants at the cell surface. The combination of antigenic determinant and class II MHC antigens at the surfaces of these cells leads to the attachment and activation of helper T cells (Fig. 10.18). Helper T cells can activate B lymphocytes and other T cells (*e.g.*, cytotoxic and suppressor T cells). Certain helper T cells secrete **lymphokines** or **interleukins** (*e.g.*, ILs such as IL-1 , IL-2........IL-6 and gamma interferon), which are substances that activate macrophages and other lymphocytes. Some lymphokines attract macrophages to the site of infection. Other lymphokines prevent migration of macrophages away from the site of infection. Still other lymphokines stimulate T cell proliferation. The net effect of lymphokine secretion is the accumulation of macrophages and lymphocytes in the region of an infection and is characterized by the inflammation that typically exists there.

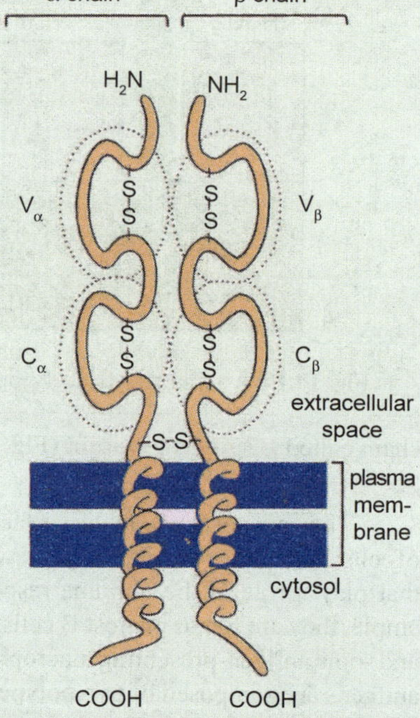

Fig. 10.20. A T cell receptor is a heterodimer comprising an α and a β polypeptide chain, both of which are glycosylated. Each chain is about 280 amino acid residues long, and its large extracellular part is folded into two immunoglobulin-like domains–one variable (V) and one constant (C). An antigen binding site is formed by V_α and V_β domain which is similar in its overall dimensions and geometry to the antigen-binding site of an antibody molecule. Unlike antibodies, however, T cell receptors have only one antigen binding site (after Alberts *et al.*, 1989).

Helper T-cells do not confine their help to lymphocytes. Those helper T cells that secrete interleukin-2 (IL-2) when stimulated by antigen also secrete other interleukins such as **gamma-interferon** that attract macrophages and activate them to become more efficient at phagocytosing and destroying invading microorganisms (*e.g.*, tuberculosis bacterium which can survive simple phagocytosis) (**Moller**, 1987).

(iii) Suppressor T cells. The discovery that T lymphocytes can help B cells make antibody responses was followed several years later by the discovery that they can also **suppress** the responses of B cells or other T cells to antigens. Such T cell suppression was first demonstrated in mice that had been made specifically unresponsive (tolerant) to sheep red blood cells (SRBC) by repeated injections of large numbers of SRBC. When T cells from tolerant mice were injected into normal mice, the latter also became specifically unresponsive to SRBC antigens. This implies that the tolerant state in this case is due to suppression of the response by T cells (**Harris** *et al.*, 1982). Subsequent experiments using surface antigenic markers suggested that cells responsible are a specialised class of T lymphocytes, called **suppressor T cells**. Thus, helper T cells act directly on the effector cells (*i.e.,* B cells and cytotoxic T cells) and suppressor T cells are thought to act indirectly by inhibiting the helper T cells on which the effector cells depend, although the mechanism of inhibition is unknown (**Asherson** *et al.*, 1986).

T cell receptors. Like the plasma membrane of B cells, T cell surfaces also contain antigen-binding receptors. T-cell receptors are a distinct class of proteins which have many properties common with antibodies (Fig. 10.19). The T cell receptor is about two-third the size of an antibody and consists of two subunit polypeptides connected by a disulphide bridge. One polypeptide is called an **alpha chain** and the other a **beta chain**. Each chain is composed of a constant domain and a variable domain. The disulphide bridge that connects α and β chains occurs between constant domains. At the end of each constant domain, there is a region rich in hydrophobic amino acids; this region anchors the receptor in the plasma membrane. A typical T cell has 20,000 to 40,000 α/β receptor proteins on its surface (**Allison** and **Lanier**, 1987).

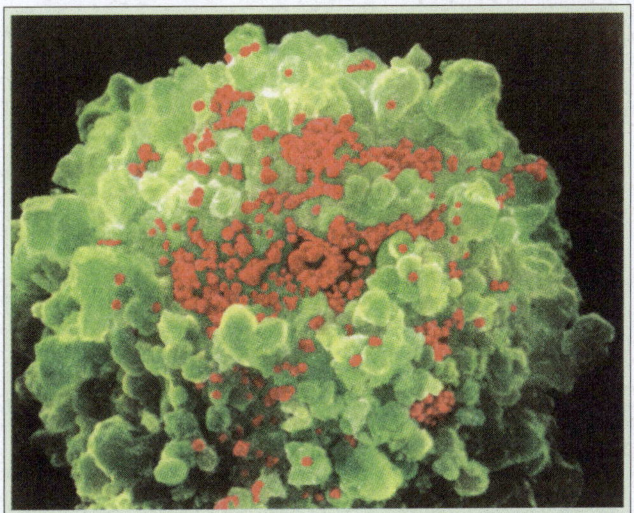

SEM of a T-lymphocyte (green) infected with the AIDS virus. Small structures (red) on the surface of the cell are new virus particles budding off.

AIDS (Acquired Immune Deficiency Syndrome)

The recent most dreaded disease of AIDS is caused by a virus called **human T-cell lymphotropic virus-III** (or **HTLV-III**). The HTLV-III is a retrovirus, that is, it is an RNA virus that induces its host cell to proliferate new viral genes by reverse transcription. This virus critically injures the immune system by infecting and eventually killing T cells. As a result of the progressive destruction of its T cells, the body is easily ravaged by a number of common infectious agents. Unable to battle infections in the normal manner, victims that develop a "full-blown" case of AIDS eventually succumb. In AIDS patients the HTLV-III virus has been shown to be present in semen as well as in the blood.

REVISION QUESTIONS

1. Describe the molecular structure of immunoglobulins.
2. How many polypeptide chains are present in each antibody molecule ? How many antigen-binding sites are present per antibody ? How many different antibodies are produced in each mature B lymphocyte ?
3. What are the three different sources of antibody variability ?
4. In what ways are the structures of antibodies and T- cell antigen receptors similar ?
5. Differentiate between the following :

   (a) Humoral immunity and cellular immunity ;

   (b) B lymphocytes and T lymphocytes.
6. Write short notes on the following :

   (a) Clonal selection theory ; (b) T-cell receptors ;

   (c) Complement systems ; (d) Immunological memory ;

   (e) Autoimmune disease ; (f) AIDS.

CHAPTER

11

Genetic Recombination and Gene Transfer

(Bacterial Conjugation, Transformation, Transduction, Episomes and Plasmids)

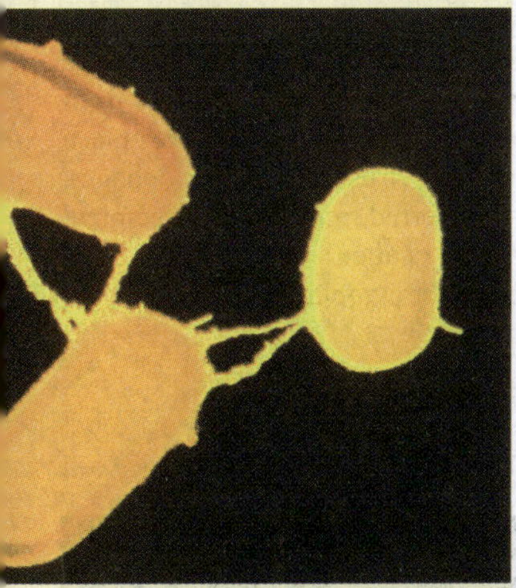

Bacteria undergoing conjugation.

Genetic recombination involves some kind of movement of genes (or genetic markers) between two distinct chromosomes derived from two different sources. The event of recombination may occur during the process of crossing over between meiotic chromosomes of eukaryotes, bacteriophage infection (*i.e.*, transduction), bacterial conjugation and bacterial transformation. Genetic recombination is characterised by high degree of precision ensuring neither gain nor loss of the genes (or genetic material). This process is generally accompanied by DNA synthesis in both prokaryotic and eukaryotic cells.

Joshua Lederberg and **Edward Tatum** (1946) showed that bacteria undergo conjugation, a parasexual process in which the genetic information from one bacterium is transferred to and recombined with that of another bacterium. Like meiotic cross-

Joshua Lederberg.

ing over in eukaryotes, genetic recombination in bacteria provided the basis for the development of methodology for chromosome mapping. It should be noted that the term genetic recombination, as applied to bacteria and bacteriophages, leads to *the replacement of one or more genes present in one strain with those from a genetically distinct strain.* While this is somewhat different from our use of genetic recombination in eukaryotes, where the term describes crossing over that results *in reciprocal exchange events*, the overall effect is the same : genetic information is transferred from one organism to another, resulting in an altered genotype. Two other phenomena, **transformation** and **transduction**, also result in the transfer of genetic information from one bacterium to another and have also served as a basis for determining the arrangement of genes on the bacterial chromosomes (**Klug** and **Cummings**, 1997).

Thus, in bacteria, the process of recombination can be performed by several mechanisms. Naked DNA molecules from one bacterial strain can be recombined into the DNA of another strain to produce a new phenotype. This phenomenon is called **transformation** (and it provided landmark support in identifying DNA as the genetic substance of pneumonia causing bacteria, see Chapter 2 Identification of Genetic Material). The unidirectional transfer of DNA from one cell to another through a cytoplasmic bridge is called **conjugation**. **Transduction** takes place when gene transfer from one bacterium to another is mediated by a virus (*i.e.*, bacteriophage). **Sexduction** is referred to as the phenomenon when the genes are transferred from donor to recipient cell on a molecule of DNA which acts as a **sex factor** or **fertility factor** called **F**. This sex gene can reside in bacterial chromosome or it may exist as an autonomous unit in the cytoplasm, called **episome** (**plasmid**).

CONJUGATION

Wild type of *E.coli* (= *Escherichia coli*, a colon bacillus of humans) can grow on **minimal medium** containing glucose, ammonium salts and a few additional substances needed in traces as well as on **complete medium**. A biochemically mutant strain can grow on complete medium but not on minimal medium because it cannot synthesize a particular amino acid or a vitamin essential for its growth from the minimal medium. Complete medium has that vitamin or amino acid, so the mutant strain has an allele of the wild gene which has arisen by mutation. The wild gene is able to direct the synthesis of the required amino acid or vitamin from the minimal medium but the mutant gene cannot.

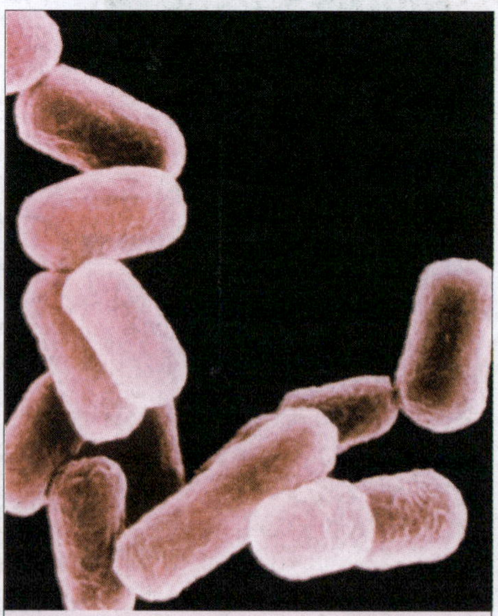

SEM of *E. coli*, magnified 18000 X.

Lederberg started his experiments with two double mutant strains. One strain was found unable to synthesize the amino acids threonine (thr^-) and leucine (leu^-) . The other strain was able to synthesize these two amino acids but could not make vitamins biotin (bio^-) and amino acid methionine ($meth^-$). The genotypes of the strains can be represented as :

$bio^+\ meth^+\ thr^-\ len^-$, and

$bio^-\ meth^-\ thr^+\ len^+$

The sign of plus (+) indicates the wild gene

which is able to synthesize these substances. Both the strains cannot grow on the minimal medium. These mutant strains are called the **auxotrophs** and the wild strain from which they are derived is known as **prototroph**.

In an experiment, when the above two auxotrophs were plated together on minimal medium, some growth was noticed. When the growing cells were analyzed, they proved to be the wild type cells with the genotype *bio*$^+$ *meth*$^+$ *thr*$^+$ *leu*$^+$. The frequency of such cells was 1 for every 10^7 cells plated. **Lederberg** and **Tatum** concluded that the growing cells of the genotype *bio*$^+$ *meth*$^+$ *thr*$^+$ *leu*$^+$ are the recombinants which were produced by the combination of the genes of the two auxotrophs. The two types of mutants got mated and produced zygotes. The zygotes after some sort of division (similar to meiosis) produce some recombinant cells.

Examples of Conjugation

A. F element and $F^+ \rightarrow F^-$ transfer. Conjugation is a one way transfer of DNA (*i.e.*, plasmid called **sex element** or **F factor**) from a donor cell (male) to a recipient cell (female). In *Escherichia coil* strain K-12, the cells that carry the sex factor F are known as **F^+** or male and such cells are found, albeit quite rarely, among natural populations of the bacterium. In F^+ cells perhaps 2 per cent of the cell's DNA is found in the F element which takes the form of a covalently closed circular molecule with a molecular weight of approximately 35×10^6. Females do not have the sex factor and are symbolized **F^-**.

An F^+ cell will usually ignore another F^+ cell, whereas it will readily establish contact, probably via a long appendage that extend from the surface of F^+ cells and is called **F pilus**, with a F^-cell. Once contact is made the pilus is believed to become modified and serve as a protoplasmic channel called **conjugation tube** between the two cells. Ordinarily, the only genetic element transferred through the tube is the F element itself. In a mixture of F^+ and F^-cells each F^+ donor will pass a copy of F factor on to an F^-recipient while retaining at least one copy for itself, so that eventually virtually, every cell in the population becomes an F^+ cell.

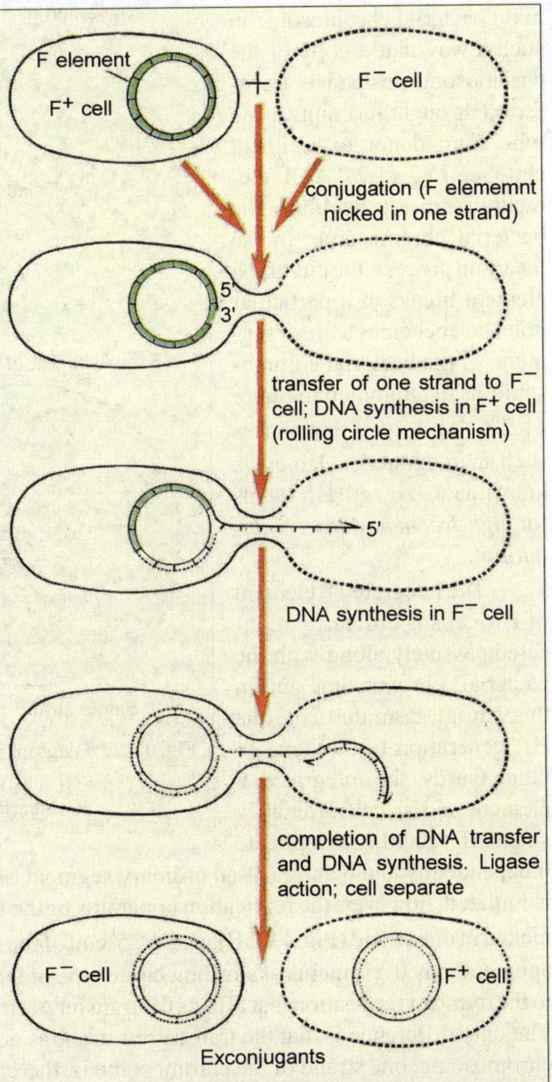

Fig. 11.1. Diagram illustrating transfer of the F element from an F^+ to an F^- cell during conjugation in *E. coli* (after Goodenough and Levine. 1974).

The transfer of F element is thought to proceed by asymmetric DNA replication of the rolling circle type. As shown in Fig. 11.1, the 5′ end of one strand of the F element is thought to be drawn into the recipient or female cell, where it is copied in the 5′→3′ direction. Meanwhile the second strand of F element remains in the donor or male and serves as a template for its own replication (**Goodenough** and **Levine**, 1974).

B. Formation of Hfr cells and Hfr → F– transfer. In certain F^+ bacterial cells, the F element infrequently (about once in every 10,000 F^+ cells) becomes associated with the main bacterial chromosome in such a way that a copy of the chromosome instead is transferred through the conjugation tube from donor to recipient cell (see Fig. 11.2). An F element first inserts itself into the bacterial chromosome. In the insertion process, the circular F element breaks at a particular point and becomes a linear segment of the bacterial chromosome, as diagrammed in figure 11.2 A. An F^+ cell that carries such an integrated F element is known as an **Hfr cell** (Hfr stands for *high frequency of recombination*).

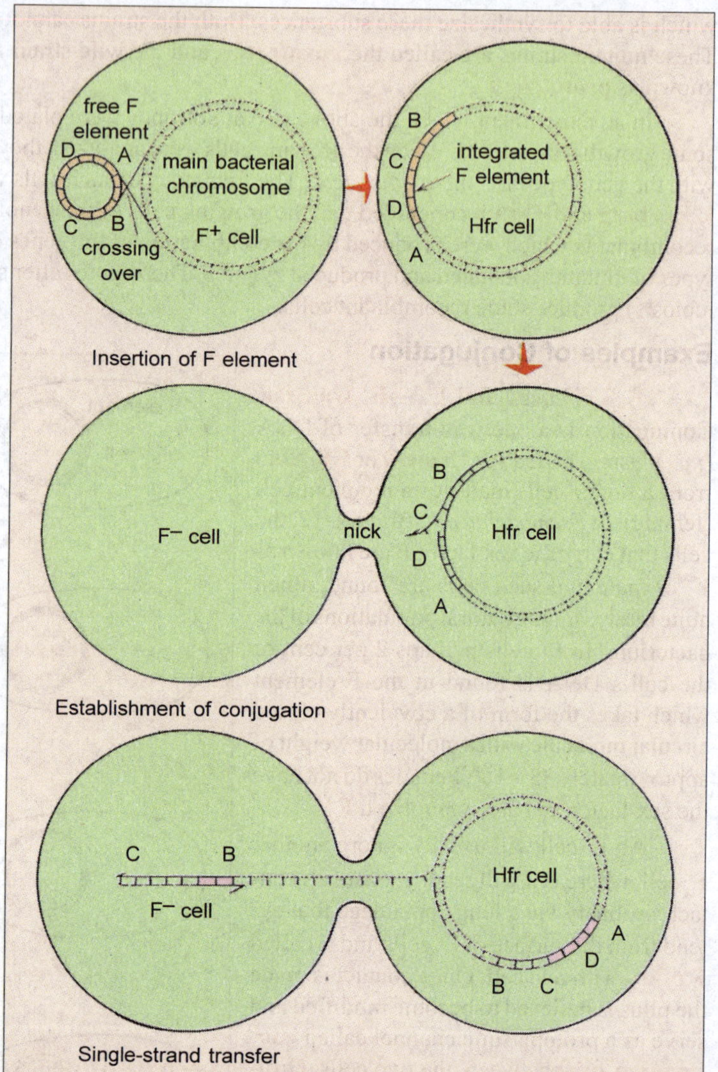

Fig, 11.2. Diagram illustrating integration of the F element to form Hfr cell and transfer of Hfr chromosome to an F^- cell during conjugation in *E.coli* (after Goodenough and Levine. 1974).

The integrated F element of Hfr cells is ordinarily replicated passively along with the bacterial chromosome and in this way is transmitted from one Hfr generation to the next. In other words, the integrated F element of Hfr cell ordinarily shunt off it stability to replicate independently and behaves as an ordinary segment of the bacterial chromosome. When conjugation is initiated, however, the replication apparatus of the F element is somehow activated. The F DNA is nicked in one strand (Fig. 11.2 B), and the 5′ end of the nicked strand is drawn into the conjugation tube apparently by the impetus of a rolling circle type of DNA replication. This mode of transfer is similar to the round of replication that affects the transfer of an F element in an $F^+ \rightarrow F^-$ conjugation (Fig. 11.1). The only difference is that the transferred strand is now covalently linked to the enormous bacterial chromosome; one strand of the chromosome is, therefore, drawn into the F^- cell as well (Fig. 11.2C) and is copied in the 5′→3′ direction as it enters the recipient.

Further, the F element carries some arbitrary gene marker called A, B, C and D (Fig. 11.2 A) and the nick that initiates rolling-circle replication occurs between gene markers C and D (Fig. 11.2 B). This means that the F element is split during transfer, and only some of its genes enter the F^- cell at the start of conjugation. The remaining F element genes (*i.e.*, D–A segment in Fig.11.2 C) will be transferred to the recipient cell only after approximately 1200μ of chromosomal DNA has passed through the conjugation tube. Usually the entering chromosome breaks at some intermediate position during transfer. Therefore, the F^- recipient cell usually inherits an incomplete copy of the F element during Hfr × F^+ conjugation and remains F^-. If transfer is complete, the recipient will inherit a complete F element and the Hfr property; its descendants can act as Hfr cells. Those F^- cell that have received a part of the donor chromosome are called **partial zygote** or **hemizygotes**, since they possess copies of one or more genes that derive from two different parents. A partial zygote is initially diploid for those genes that have been transferred, but the cells do not remain diploid. It soon takes part in non-reciprocal genetic exchanges so that some of the donor DNA is included in the recipient chromosome. Lastly, when donor recipient chromosomes carry different genetic markers, the emergent cells are frequently recombinant. It is for this reason that cells capable of donating chromosomal material to recipient cells have come to be called Hfr for **high frequency of recombination**.

Mapping the Bacterial Chromosomes

Mapping of bacterial chromosome involves the following methods : 1. interrupted conjugation mapping ; 2. uninterrupted conjugation mapping ; 3. recombination mapping ; 4. complementation mapping ; 5. mapping by deletion mutants. Here, we will consider only interrupted conjugation mapping method.

Elie Wollman.

Interrupted conjugation mapping. In conjugation experiments, it was reported that the number of genes transmitted from donor to recipient was directly proportional to the time interval for which conjugation was permitted. Based on this aspect, **Jacob** and **Wollman** (1958) developed a technique, called **interrupted conjugation mapping technique** for the bacterial chromosome. In this technique the donor Hfr and recipient F^- strains are mixed and allowed to conjugate for a short period of time. The conjugation is stopped at any desired time by subjecting the mixture to shearing forces of a waring blender (violent agitation) which artificially disrupts the conjugation tube. The sample is diluted immediately and plated at selective media, incubated, and then scored for recombinants. The length of donor chromosome transmitted could then be determined and mapped in terms of time units required for the transfer. It is known that 8 minutes are needed for conjugation to begin and then chromosome is transferred slowly in terms of time units, one time unit being equal to one minute. The complete chromosome of *E.coli* is transferred in about 89 minutes and, therefore, the bacterial chromosome is 89 time unit long (Fig. 11.3). The genes which are 2–3 times units apart can be precisely mapped by this method.

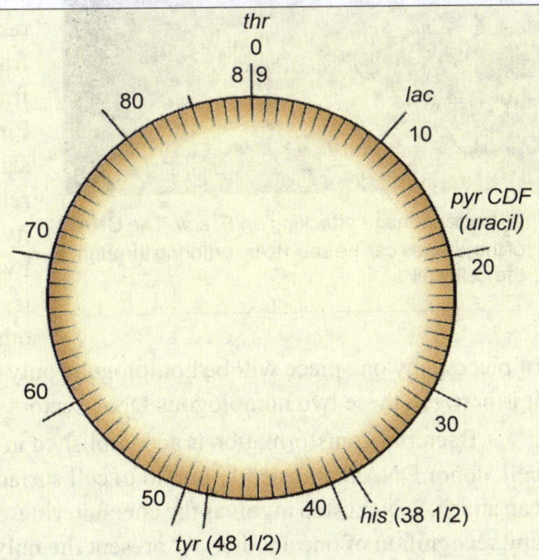

Fig, 11.3. Circular recombination map of *E.coli* chromosome.

Example. An Hfr bacterial strain carrying the protrophic markers such as a^+, b^+

and c^+ is mixed with a F^- strain carrying the auxotrophic alleles a, b, and c conjugation was interrupted at 5 minute intervals and plated on media which revealed the presence of recombinants:

Time (minutes)	Recombinants
5	a b^+ c
10	a b^+ c^+
15	a^+ b^+ c^+

The order of genes in the Hfr donor strain is $b^+ - c^+ - a^+$; b is less than 5 time units from the origin, c is less than 10 time units from b ; a is less than 10 time units from c.

TRANSFORMATION

It has been found that aging cultures of some bacteria spontaneously incur cell ruptures (autolysis) which releases DNA into the environment with other cells, and thereby offers the opportunity for transformation to occur. In *Bacillus subtilis*, however, living cells occasionally extrude large pieces of DNA into the medium and this DNA is taken up by recipients for transformation. Presumably under natural conditions, such as the infection of a host by two different bacterial strains, similar transformation events can and do occur.

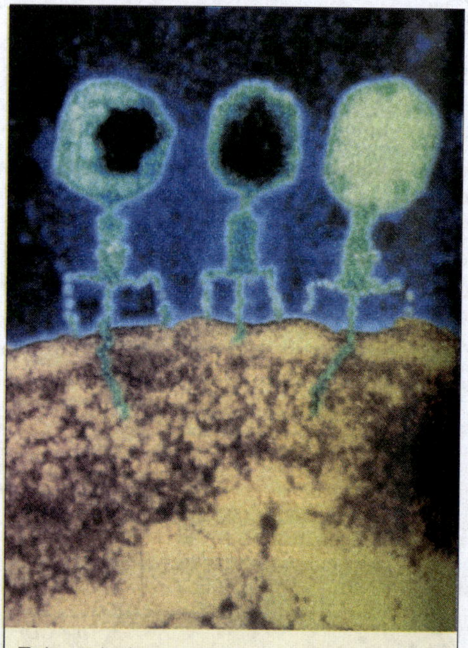
T_2 bacteriophage attacking an *E. coli*. The DNA of the phages can be seen to be entering through the cell wall.

Somewhat stringent conditions must be met before successful transformation comes about. Cells of a transformable strain must be **competent**, meaning that they must be in a particular physiological receptive state which occurs only during a fraction of its growth cycle when their walls and membranes are permeable to DNA. In addition, the **donor DNA** must be presented to the **host cell** (receptive cell) in the form of large, relatively intact, double-stranded pieces with molecular weights in the 0.3 to 8×10^6 dalton range; under most conditions single-stranded DNA fragments have no transforming activity, nor do DNA fragments with fewer than about 450 base pairs. Finally, the donor DNA must be obtained from the same species as the host; DNA from even a closely related species is generally without transforming ability. On a molecular level this probably means that the two DNAs must contain similar base sequences; that is, they should be generally **homologous** to one another. Obviously, since the donor DNA is in the form of pieces, any one piece will be homologous only to a certain segment of the host chromosome, and it is between these two homologous DNA sectors that recombination will occur.

Bacterial transformation is accomplished in the following steps–1. During competence of host cell, donor DNA is transiently bound to cell surface. Each cell has a number of sites at which DNA can attach. 2. Next step involves the encountering of homologous DNAs (donor DNA and host DNA) and recognition of one another. At present the only known basis for sufficiently reliable recognition between two DNA molecules rests on the formation of complementary base pairs. Base pairs can form only between single polynucleotide chains, yet the host chromosome and the transforming fragments are, at least initially, duplex structures. Consequently, there should be some means for the creation of local single-strandedness in both DNAs. Two possible means can be envisaged for creating these single-stranded regions. *A localized denaturation* could occur within an otherwise intact helix, an event that many investigators consider unlikely under physiological conditions. Alternatively, *one strand could be nicked by an endonuclease and then digested by an exonuclease to expose the complementary strand.* An additional wedging protein might participate here to maintain the single strand in an extended state, nucleotides exposed, so that the strand does not fold back on itself and form internal base-pairs.

Recombination after Transformation

Competence is the ability of a cell to incorporate naked DNA in the process of transformation. Most bacterial cells are only competent during a restricted part of their life cycle. During the competent state, the cell produces one or more proteins called **competence factors** that modify the cell wall so it can bind exogenous (foreign) DNA fragments. Thus, receptor sites are present only during the competent state. Absorbed DNA fragments are then reduced in size to molecular weights of about 4 or 5 million by enzymatic cleavage. As the double stranded DNA fragments penetrate the bacterial cell wall, one of the strands is degraded. Any fragment of DNA that has been translated (by transformation or some other method) from a donor cell to a recipient cell is referred to as an **exogenote** (incomplete genome) ; the **endogenote** (complete genome) is the native DNA of the recipient cell. A bacterial cell that has received an exogenote is initially **diploid** for part of its genome, and is called **merozygote** (partial diploid). However, single-stranded exogenotes are unstable and will usually be degraded unless they are integrated into the endogenote. Any process of genetic exchange that transfers only part of the genetic material from one cell to another is called **meromixis**. It is believed that the single stranded exogenote of transformation gets coated with a protein (such as the Rec A–protein of *E.coli*) that helps the exogenote to find a complementary region on the endogenote, to invade the double helix, to displace one of its strands, and to base-pair with the other strand. The displaced strand is enzymatically removed as the endogenote replaces by homologous base pairing (a phenomenon called **branch migration**). Triming enzymes remove the free ends (either donor or recipient) and ligase enzyme seals the nicks. Once the exogenote is integrated into the endogenote and the displaced strand is degraded, the cell is no longer regarded as a **merozygote** (Fig. 11.4)

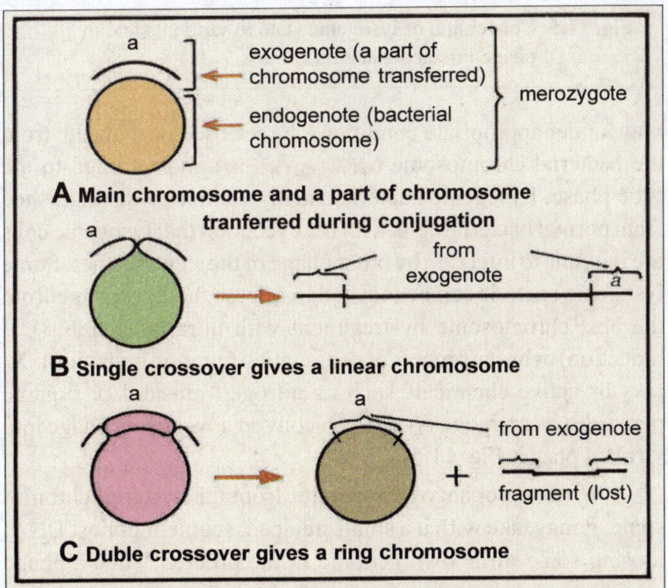

Fig. 11.4. Recombination in bacteria after gene transfer (single and double crossovers give linear and ring chromosomes).

Further, if the exogenote contains an allele of endogenote, the resulting recombinant double helix of DNA would contain one or more mismatched base pairs, and is called a **heteroduplex**. If progeny cells have to receive the new allele, **mismatch repair** must occur by excising a segment of the endogenote strand and using the exogenote strand as a template for its replacement. Since incorporation of the exogenote into the endogenote requires homologous recombination, the donor cell would normally belong to either the same species as the recipient cell or to a closely related one. Two or more closely linked genes may reside on the same transforming piece of DNA. If two or more genes are incorporated together into the endgenote, the recipient cell would be **cotransformed**. The frequency of cotransformation is a function of the linkage distance between the respective genes (**Stans- field**, 1991).

TRANSDUCTION AND RECOMBINATION OF VIRUSES

Transduction involves the carrying over of DNA (or gene transfer) from one organism to another by an intermediate agent, which is usually a bacteriophage. The phenomenon of transduction was discovered in 1952 by **N.D. Zinder** and **J.Lederberg**. When a bacteriophage infects a bacterial cell, it has two mutually exclusive choices : (1) It can act as virulent phage and enter the vegetative (**lytic**) cycle, or (2) it can act as a non-virulent (**temperate**) phage, becoming integrated into the host chromosome as a **prophage** and replicating in synchrony with the bacterial chromosome. A cell which harbours a prophage is said to be **lysogenic** because the prophage DNA may, under appropriate conditions, be released or induced from the bacterial chromosome (**deintegration**) and returned to the lytic phase. Lysogenic bacteria usually cannot be distinguished from normal bacterial cells with the exception that lysogenic cells are immune to infection by other phage of the same species. Some lysogenic bacteria can be induced to release their prophage from the host chromosome by treatment with ultraviolet light (UV induction) or by the process of conjugation (zygotic induction). X-rays or active chemicals such as nitrogen mustard or organic peroxide (= carcinogens) also can convert a lysogenic phage into virulent phage (Fig. 11.5).

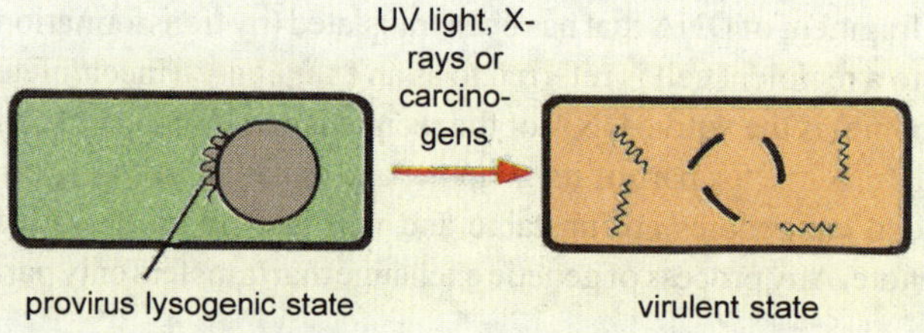

Fig. 11.5. Conversion of lysogenic state to virulent state in phage infected bacteria.

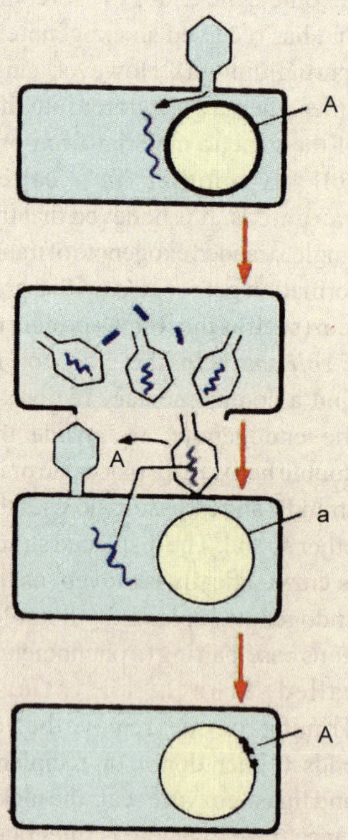

Fig. 11.6. Transduction involving transfer of A gene from one strain to another through phage particle.

When a prophage deintegrates from the bacterial chromosome; it may take with it a small adjacent segment of host DNA, loosing some of its own genome in the process. Such a phage particle can mediate the transfer of a bacterial gene or a portion of a bacterial gene from one cell to another in a process called **transduction**. All transducing phages are defective for a portion of their own genome.

Recombination in Viruses

The phenomenon of recombination in viruses was described

for the first time by **A.D. Hershey** and **R. Rotman** in 1949 and has been utilised for the preparation of linkage maps of viruses. Viruses are reported to undergo recombination in the host cells. For example, if the two strains of a virus having the genotypes $A^+ B^-$ and $A^- B^+$ are allowed to infect the host cells in such a way that both strains infect the same individual host cell, then the resulting progeny of virus particles after lysis may have the recombinants $A^+ B^+$ and $A^- B^-$. Like the *E. coli*, the linkage maps of T_2 and T_4 bacteriophage are found to be circular.

EPISOMES AND PLASMIDS

Episome is a closed, circular molecule of DNA that may be present in a given (bacterial) cell, either separately (autonomously) in the cytoplasm or integrated into the chromosome. The term episome was originally introduced by **Jacob** and his colleagues in 1960 to designate genetic material such as F or fertility factor of *E. coli*. The term **plasmid**, on the other hand, was reserved for those bits of genetic material that exist *only* extra chromosomally and cannot be integrated into the nucleioid (the DNA). Thus, episomes were said to be integrable, whereas plasmids were not. Currently, all types of extra-chromosomal genetic material, whether integrable or not are termed as plasmids (see **Burns** and **Bottino**, 1989).

1. Episomes

Episomes are characterised by the following features :

1. Episomes are DNA molecules which are not essential, therefore, they may or may not be present.

2. When absent, episomes cannot originate *de novo* and may be acquired from other strains either due to infection or conjugation.

3. When present, episomes may be in autonomous (free) or integrated state (attached to chromosomes). They may change their state from autonomous to integrated or *vice versa.*

4. When present, episomes may be lost finally.

On the basis of above criteria, elements like sex factor (F factor), bacteriophages and colcinogenic factors are included in the class of episomes.

2. Plasmids

The term plasmid was originally introduced by **Lederberg** in 1952 to describe any extra-chromosomal hereditary determinant. Currently, however, the term plasmid is restricted only to those accessory DNA circles which are found in bacteria in addition to the main chromosome. Plasmids being accessory DNA molecules are considered extranuclear genetic systems. They have the following properties :

Plasmid is a circle of DNA.

1. They are genetic elements which are made up of DNA.

2. They are smaller than and separate from the chromosome.

3. They are capable of replication.

Types of plasmids. Plasmids are of three main types : 1. F or sex factor responsible for transfer of genetic material from one strain to another ; 2. R plasmid responsible for drug resistance, and 3. Col factor responsible for colicin production. A plasmid may sometimes share more than one of the above properties.

1. Fertility (F) factor. F factor is also included in the category of episome and its nature has already been described, while discussing conjugation in *E. coli*.

2. R plasmid. R factor or R plasmids (R stands for resistance) when present imparts on bacteria (*e.g.*, *Shigella*, causing dysentery) resistance to various antibiotics. This drug resistance may be transferred from one strain to another and the R factor may also facilitate transfer of bacterial genes from one strain to another like the F factor. R factor is represented by a small DNA circle carrying genes for multiple drug resistance (for antibiotics such as penicillin, tetracycline, sulphonilamide, streptomycin, and chloramphenicol) or conjugation or both.

3. Col factor. The col factors are responsible for the production of **colicins** which are toxic antibiotic lipocarbohydrate-proteins killing bacteria other than those producing them. Like the F factor, the col factor can be transferred from one strain to another and may also facilitate transfer of bacterial genes.

Replication and recombination in plasmids. The plasmids being genetic elements undergo both replication and recombination. The replication of plasmids is carried on in a semi-conservative manner (Fig. 11.7). The DNA of plasmid may undergo recombination either with another plasmid DNA or with bacterial DNA. Certain sequences, called **insertion sequence (IS) elements** found in the bacterial chromosomes are homologous to plasmid DNA and help in recombination.

Uses of plasmids in genetic engineering and biotechnology. Due to recent advent of techniques of gene cloning (genetic engineering) and its use in biotechnology, plasmids have assumed great significance as *vector* or *carrier molecules*. Thus, genes can be isolated from DNA of living

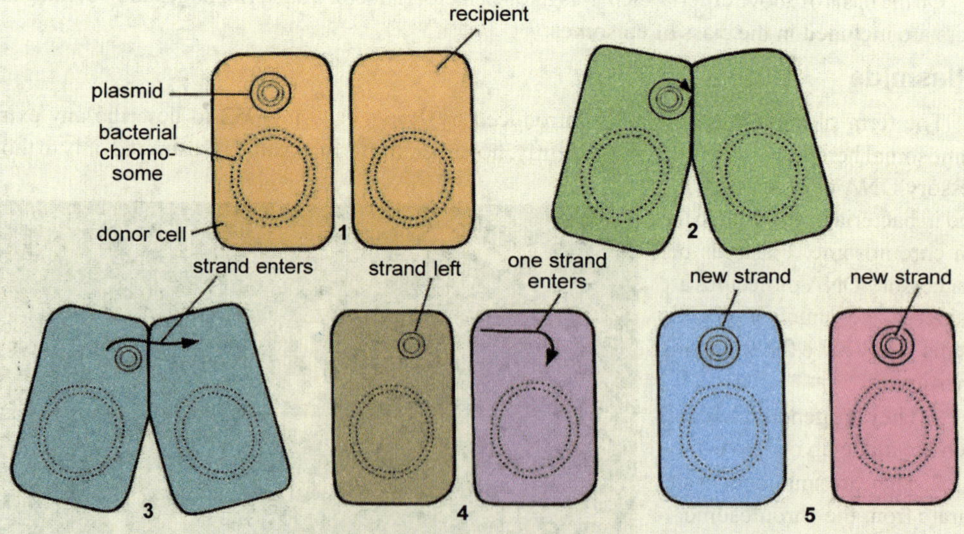

Fig. 11.7. Mode of transfer of a plasmid from one bacterial cell to another, involving replication. Only one strand of DNA is transferred (2, 3), the single strands, one in each donor and recipient cells, will then synthesize their complementary strands to form double stranded plasmid DNAs.

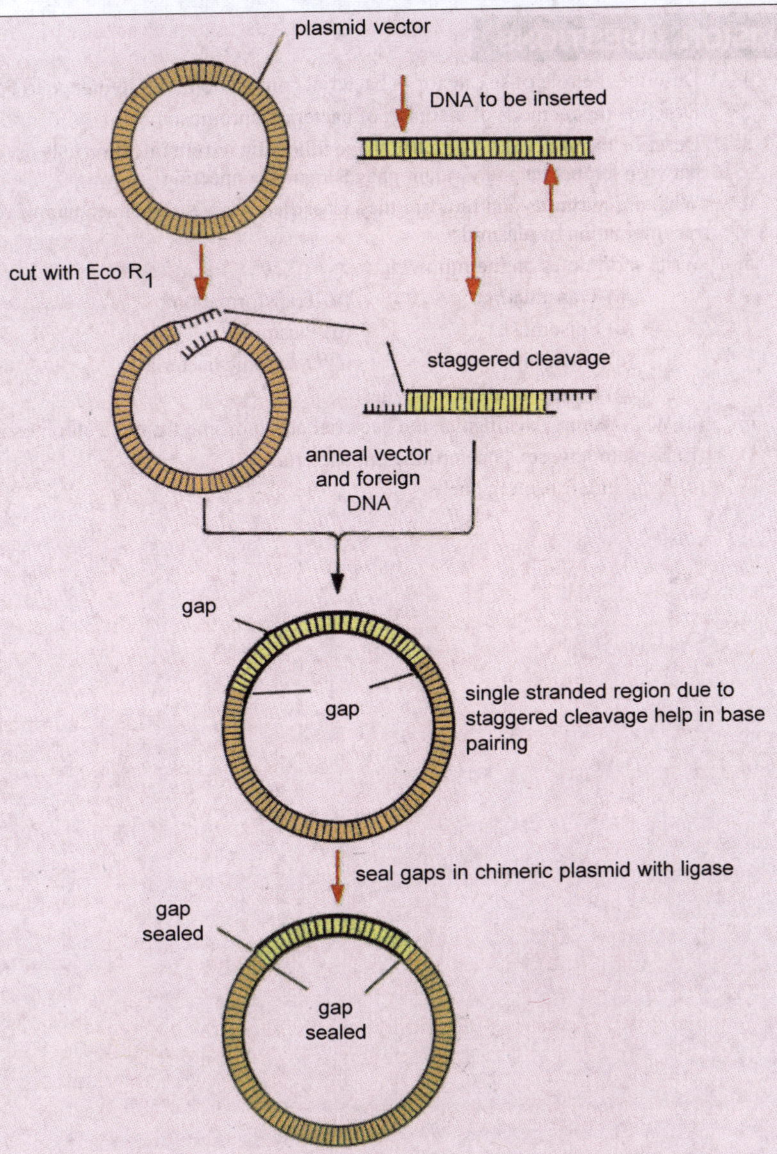

Fig.11.8. The technique of insertion of a DNA segment (which is to be cloned) in a plasmid vector.

organisms or may be artificially synthesized from mRNA (= messenger RNA) or from the nucleotide. These genes can multiply when they are inserted into bacteria with the help of vector which can live and multiply in bacterial cells. Plasmids are used as one class of such vector molecules (Fig.11.8) (For details, see Chapter 8 of Molecular Biology).

REVISION QUESTIONS

1. Describe the role of sex factor in bacterial conjugation with reference to F^+ and Hfr strains.
2. Describe the methods of mapping of bacterial chromosomes.
3. Describe the life cycle of bacteriophage following it from infection to lysis. Discuss the differences between temperate and virulent phages in this connection.
4. What are plasmids and how are they classified ? Give a brief account of replication, transfer and recombination of plasmids.
5. Write short notes on the following :
   (a) Transduction, (b) Transformation ;
   (c) Episomes ; (d) Plasmids ;
   (e) Hfr strains ; (f) Lysogenic bacteria ;
   (g) Recombination in phages.
6. (a) What features distinguish the bacterial chromosome from the chromosomes of eukaryotes ?
   (b) Explain how sex is determined in bacteria.
   (c) Explain Hfr bacterial cells.

EVOLUTION

(Evolutionary Biology)

CHAPTERS

CHAPTER

1

Introduction

Nature's variety is boundless.

We can make the following two clear-cut observations regarding the living world : (1) The living world comprises a rich **diversity** of animal, plant and microbial life. (2) All living organisms appear to be **well-fitted** (or **adapted**) for the problems posed by the environments in which they live. It is possible that either the characters of organisms are fixed and have remained so since the origin of life, or that they are mutable and their diversity and adaptability have unfolded progressively with time. The dynamic process is called **evolution** (see **Calow**, 1983).

The term *evolution* is derived from two Latin words – *e* = from; *volvere* = to roll and means the act of unrolling or unfolding, *i.e.*, the doctrine according to which higher forms of life have gradually arisen out of lower. This term was first used by **Charles Bonnet** (1720–1793), extrapolating from progressive embryogenesis (in individuals) to the development of species (see **Calow**, 1983). However, according to **Savage** (1969), the term evolution was first used by English philosopher **Herbert Spencer**.

Herbert Spencer (1820-1903)

Evolution has been defined as a gradual orderly change from one condition to another. The **principle of evolution** implies the development of an entity in the course of time through a gradual sequence of changes, from a simple to a more complex state. In other words, it maintains that the different kinds of organisms that we see today have evolved from common ancestors over millions of years. **Charles Darwin** (1859) has defined evolution as *"descent with modification"*, *i.e.*, closely related species resembling one another because of their inheritance and differing from one another because of the hereditary differences accumulated during the separation of their ancestors. In other words, evolution is the process by which related populations diverge from one another, giving rise to new species (or higher groups) (**Dodson** and **Dodson**, 1976).

Now, most of the evolutionary biologists envisage that the present-day varied forms of life of the planet Earth have evolved during the past history of earth by undergoing the following two steps of evolutionary process :

1. Inorganic, Chemical or Molecular Evolution and Origin of Life

In the first step of evolution took place the development of non-living matter into living matter or evolution of non-life into life. Thus, during chemical evolution, that occurred about 4600 millions of years ago, subatomic particles aggregated to form atoms of different elements; elements evolved into simple inorganic molecules; simple inorganic molecules gave origin to organic micromolecules such as amino acids and micromolecules developed into macromolecules such as proteins by **self-assembly process**. Certain macromolecules developed the ability of self-duplication and aggregated to form first primitive living systems, called **eobionts**. Eobionts evolved into prokaryotic cells which formed the kingdom Monera.

2. Organic and Biological Evolution and Origin of Species

The second step of evolution was started from the culmination point of chemical evolution (*i.e.*, origin of life). The concept of organic evolution holds that all the varied kinds of animals and plants which are now known, have developed out of earlier types by completely natural, gradual but continuous changes during the passage of time. According to **G.L. Stebbins** (1976), organic evolution is a series of partial or complete and irreversible transformations of the genetic composition of populations, based principally upon altered interactions with their environment. It consists chiefly of adaptive radiations into new environments, adjustments to environmental changes that take place in a particular habitat, and the origin of new ways for exploiting existing habitats. These adaptive changes occasionally give rise to greater complexity of developmental pattern, of physiological reactions and of interactions between populations and their environment.

The product of organic evolution is the **origin of species**, or evolution of old species into new species. For example, during organic evolution under certain environmental stresses certain autotrophic, heterotrophic, anaerobic and aerobic prokaryotic cells of Kingdom Monera started symbiotic relationship and evolved into eukaryotic plant and animal cells to form Kingdom **Protista** (**Margulis**, 1970). Later on, Kingdom Protista evolved and gave origin to three Kingdoms–**Plantae** (Metaphyta), **Fungi** and **Animalia** (Metazoa) by adaptive radiation or cladogenesis (Fig. 1.1).

The branch of biology which incorporates in it the studies concerning the problems of chemical evolution and origin of life and organic evolution and origin of man and other present-day organismal species of the planet Earth, is called **evolutionary biology** (see **Mayr**, 1970; **Salthe**, 1972).

FACT OF EVOLUTION

The concept of evolution is based on detailed comparisons of the structures of living and fossil forms, on the sequences of appearance and extinction of species in past ages, on the physiological and biochemical similarities and differences between species and an analysis of the genetic constitution of present-day plants and animals. In fact, various branches of biology such as palaentology, comparative anatomy, physiology, biochemistry, embryology, taxonomy, geographic distribution, ecology, animal behaviour, domestication techniques and population genetics have furnished strong support to the idea of evolution.

Now most biologists accept the fact that organisms have evolved and are still evolving, though no biologist has actually seen the origin of a major group of organism by evolution. In fact, certain races and species have been produced by duplicating in the laboratory some of the evolutionary processes such as mutation, selection, isolation, etc., which are known to take place in nature. The main reason that major steps in evolution have never been observed is that they require millions of years to be completed. The evolutionary processes which gave rise to major groups of organisms, such as

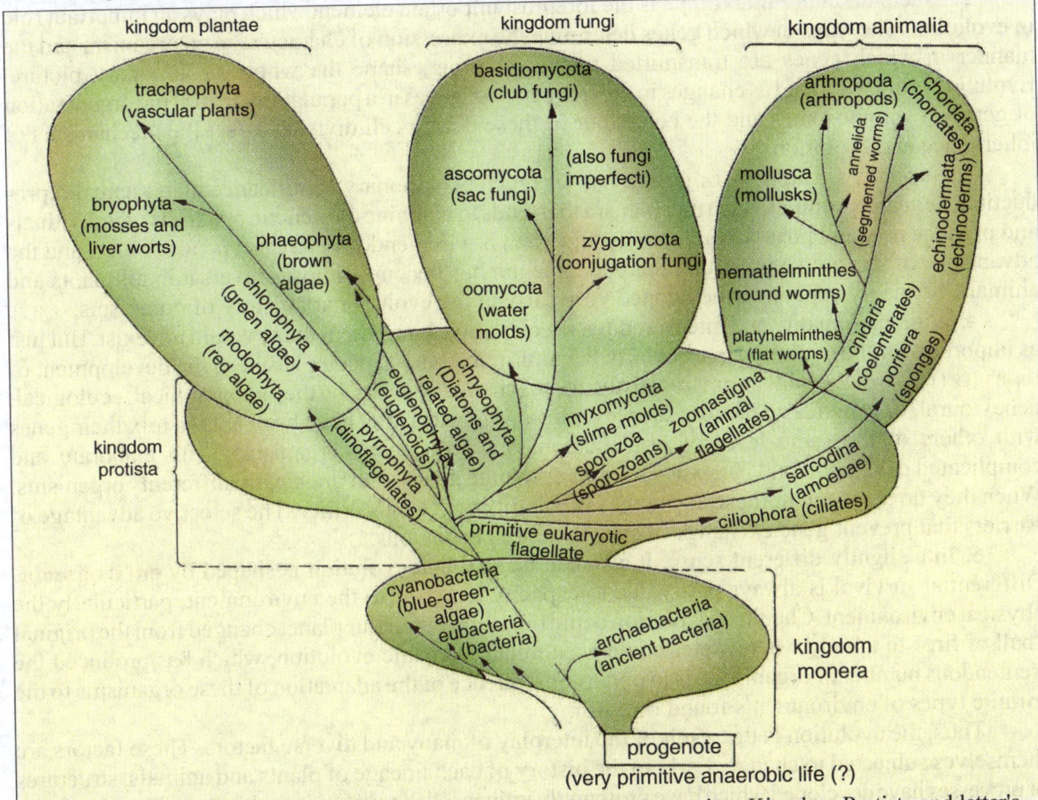

Fig. 1.1. Family tree showing organic evolution of Kingdom Monera into Kingdom Protista and latter's evolution into Kingdoms-Plantae, Fungi and Animalia.

genera and families, took place in the remote past, long before there were people to observe them. Nevertheless, the facts (*i.e.*, evidences of organic evolution) we know about these origins provide very strong circumstantial evidence to indicate that the processes which brought them about were very similar to those found in modern groups of animals and plants.

EVOLUTION COMPARED WITH ANCIENT HISTORY

Certain authors of evolutionary biology such as (1971), have compared the evolution with the ancient history. According to them, the state of our knowledge of the major steps of evolution which took in the past is much like that of our knowledge of ancient history. We do not have any reliable eye-witness account of the events which produced the rise and fall of the civilizations of ancient India, Egypt, Sumeria, Babylon and Crete, and the contemporary written records of these ancient times are fragmentary and unreliable. Nevertheless, the indirect evidences that these civilization existed are so strong that it is accepted by scholars and laymen alike. Moreover, we teach as historical fact many of the probable events connected with their rise and fall. The evidence which biologists now have about the rise and extinction of major groups of animals in past geological eras is of a very similar nature, and carries with it about the same degree of high probability.

A PREVIEW OF EVOLUTION

Evolution as a process of gradual development of organisms through time includes the following characteristics :

1. Basically evolution is the result of the **differential survival** in each generation of the progeny of individuals with certain special characteristics. In turn, these adaptive characteristics in part account for the differential survival.

2. **Mechanism of inheritance** is the foremost important element which plays an important role in evolution. The ways in which genes determine the expression of characters in an organism and the manner in which genes are transmitted to the offspring, shape the whole evolutionary picture. Evolution is often defined as changes in the frequency of genes in a population. In turn the organization of genes in chromosomes and the behaviour of these during cell division affect the mechanisms of inheritance and evolution.

3. Normally **sexual reproduction** has far much evolutionary significance than asexual reproduction. Sexual reproduction is a mechanism that tends to combine the genetic materials of individuals and produce new and novel combinations. Its effect is a tremendous increase in **variability**, and the advantages of this are so great that the phenomenon has become almost universal in all plants and animals. Sexuality apparently developed very early in the evolutionary history of organisms.

4. Without sexuality and interbreeding, species as we know them today would not exist. But just as important for the evolution, particularly the multiplication of species has been the development of **barriers** (*viz.*, **isolating mechanisms**) to the free exchange of genes, be they geographical, ecological, behavioural, or genetical. The very simple earliest organisms may have been able to mix their genes with others of the same level of organization, but present-day organisms, with elaborate and complicated developmental pathways, cannot exchange genes with drastically different organisms. When they do the result of these exchanges is lethality or at best sterility. The selective advantage of barriers that prevent gene exchange in such instances is obvious.

5. In a slightly different sense, it also can be said that evolution is shaped by **environment**. Differential survival is always partly due to capacity to **adapt** to the environment, particularly the physical environment. Chemical evolution could occur only after our planet changed from the original "ball of fire" to a body where water could accumulate. Organic evolution, which has produced the tremendous number of organisms, is in part a consequence of the adaptation of these organisms to the infinite types of environments found on earth.

Thus, the evolution is the result of the interplay of many and diverse factors. These factors are themselves subjected to change. Early in the history of each lineage of plants and animals, structures or processes have developed which have profoundly influenced the evolutionary history of that group. So, for example, a segmented body and an exoskeleton have been major factors in the success of the insects, but in turn these same factors have restricted the size and habits of the members of the class Insecta.

Recently, evolution is studied in the terms of genes and population genetics. The genes have been recognised as the ultimate causes of existent unity and diversity of various populations. The genes change by **point mutations** which are primarily alterations of base pairs of DNA molecules that find their initial phenotypic expressions as amino acids substitutions in proteins. Mutations are raw materials of evolution. Some genotypes survive and reproduce better than others and, hence, gain in relative gene frequency; this is, by definition, the process of evolution by **natural selection**. In simple terms, evolution is the shift in the frequency of genes within total populations and most such shifts of a sustained nature are due to natural selection.

Natural selection is never perfect in its operation. Many unfavourable genes are always being either phased out or renewed by mutation. Some genes are selected against in the homozygous state but favoured when in heterozygous state. These elements of imperfection constitute the **"genetic load"**. While reducing the immediate adaptiveness of organisms, they increase adaptability of future generations by providing genetic flexibility to cope with changes in the environment. Some evolutions also occur by random fluctuations in the gene frequency (called **genetic drift**) at least temporarily outside the influence of natural selection.

In spite of the irregularities of the evolutionary process, natural selection always tends to make the populations fit more precisely to the existing environment. But the environment is constantly shifting and populations are consequently forced to track it. They are always little behind; if they fall too far behind, they become **extinct. Species** are populations that diverge enough from each other during evolution to become **reproductively isolated**. At this critical juncture in their history, they are freed to evolve along their own paths and to disperse to new distributions that are different from those of all other genetically similar populations.

CERTAIN MISCONCEPTIONS OF EVOLUTIONARY BIOLOGY

The idea of evolution is very popular one, but unfortunately various misconceptions have crept in the minds of biologists and laymen about it during its extension and popularization. Some of these misconceptions and subsequent clarification of each of them are the following :

1. If the layman is asked the meaning of the word evolution, he is likely to reply, "Man came from monkeys". The shortcoming of the layman's definition lies in the fact that he pictures one modern form as descended from another modern form. Man and monkeys are contemporaries, both are products of long evolution. The relationship existing between monkey and man, rather than being a father-to-son relationship, is more comparable to cousin-to-cousin relationship. As one and his cousin have a pair of grandparents in common, similarly modern man and modern monkey are thought as having shared a common ancestor in the distant past. From this common ancestor both inherited some characteristics in which they still resemble each other. Was this common ancestor a man or a monkey? He was neither. He was a form that had the potentiality to give rise to a monkey on the one hand or to give rise to a man, on the other hand. There is no evidence that any of the modern monkeys have that potentiality (**Moody,** 1970).

2. The phrase "survival of the fittest" is Darwin's, but it is clear from his writings that he did not mean it literally. It was the British philosopher **Herbert Spencer** (1820–1903) who popularized the idea in his works and in his concept of social evolution. In the historical context of the times, the idea of struggle between individuals with the winner, the fittest, surviving to get the spoils was appealing to certain groups. In translating organic evolution to the social scene, Spencer misinterpreted the main idea–that is, differential survival of the progeny. He replaced it with one element of natural selection, differential mortality due to better adaptations to the environment, and interposed the idea of direct competition as responsible for differential mortality. The ideas as presented are entirely his, and have no scientific backing.

We must keep always in mind that by the "fittest" **Darwin** meant the one with the largest surviving progeny. This can be and often is a comparatively weak individual. In this sense rabbits are "fitter" than lions, in spite of man, than lions.

3. Another important point to understand is that no moral judgement of any sort can be read into evolution. Evolved organisms are not "better" in a moral sense, they are only better adapted to the environment they occupy compared to their extinct ancestors. Evolution is as blind as justice is supposed to be : those with the largest surviving progeny will multiply regardless of how good or bad humans may consider them (**O.T. Solbrig,** 1966).

SIGNIFICANCE OF EVOLUTIONARY BIOLOGY

The theory of evolution is quite rightly called the greatest unifying theory in biology. The diversity of organisms, similarities and differences between various kinds of organisms, patterns of distribution and behaviour, adaptation and interaction, all this was merely bewildering chaos of facts until given meaning by the evolutionary theory. There is no area of biology in which that theory does not serve as an ordering principle.

The distinguished evolutionist **Theodosius Dobzhansky** (1973) has aptly said : "*Nothing in biology makes sense except in the light of evolution*". For more than a century, the theory of evolution has exerted a very strong influence on our thinking about biology, also on developments in other disciplines such as sociology, politics, economics and religion.

REVISION QUESTIONS

1. What is evolution? Define the evolutionary biology.
2. Why is the theory of evolution considered as unifying theory of biology ?
3. Discuss some of the important misconceptions of evolution.
4. What is the significance and scope of evolutionary biology ?

CHAPTER

2

Development of the Idea of Organic Evolution

A view of the Galapagos Island.

The idea that the present forms of life have arisen from the earlier, simpler ones was far from new when **Charles Darwin** published his book "*The Origin of Species*" in 1859. The development of the idea of organic evolution through pre-Darwinian, Darwinian, and post-Darwinian eras can be studied under the following periods :

1. Period of Obscurity

The oldest speculations about evolution are found in the classical mythologies of different centres of ancient cultures, such as India, China, Babylon and Egypt. For instance, the classical Indian (Hindu) mythology has forwarded the idea that universe and the body of all living beings are made up of five elements or "**Panch Bhuta**", namely **kshiti** (earth), **ap** (water), **teja** (fire), **marut** (air) and **vyoum** (void or sky). Further, the **Padam Puran** conceived ten incarnations or "**Dashvatarahs**" for explaining the evolution of vertebrates. The sequence of '**Avatarahs**, includes **Matsya**, fish-like appearance; **Kurma**, reptile-like; '**Varaha**', mammal-like; '**Narasimha**', half-man and half-lion; "**Vamana**", the dwarf man-like appearance (anthropoid); "**Parshurama**", a *Homo sapiens* of Pleistocene; "**Shri Rama**", "**Balrama**", "**Shri Krishna**" and "**Kalpi**"–the latter four incarnations depict the formation of clans and communities.

Similar to ancient Indians, the ancient Chinese conceived the world at the beginning as made up of five primary elements—water, wood, fire, soil, and the precious metal, gold. According to their mythology, the positive and negative interactions of these primary five elements were responsible for the origin of life and different kinds of adaptations and diversities of organisms or "things" were originated from a single source through gradual unfolding and branching.

The Egyptian and the Babylonians gave the idea of cosmic evolution which included the following aspects : (i) a Divine Intelligence and creator, (ii) water as a primitive and boundless substance, (iii) the potentiality of all things in the Divine Intelligence, and (iv) creation as the process of becoming actual by stages.

During this period, because the speculation on evolution idea was much shrouded in mystery and mythology, so, this period has come to be known as the **period of obscurity**.

2. Period of Ancient Greeks and Romans

The classical writings of certain Greek philosophers contain the vague ideas and speculations about the organic evolution. Thus, **Thales** (624–548 B.C.) suggested aquatic origin of life, *i.e.*, all life originated from oceans. **Anaximander** (611–547 B.C.) explained the origin of universe. According to him, living things came from primordial fluid and men were first formed as fishes; eventually they cast off their fish skin and took up life on dry land. **Anaximenes** (382–524 B.C.) believed that life came from air and he proposed abiogenesis. **Xenophanes** (576–480 B.C.) is credited with being the first person to recognise that fossils, such as petrified shells embedded in rocks, represent the remains of animals that once lived. He also realised that the presence of fossils of marine animals on what is now dry land indicates that the sea once covered the area. He believed that in ancient times earth was covered with water where life began. **Empedocles** (495–435 B.C.) is considered as the "**Father of evolution idea**" by many evolutionists for his ideas which he forwarded in his "*Poem of Nature*". According to him, life arose spontaneously but gradually and change (evolution) would result when twin forces, the attraction and repulsion, interact with each other. Based on it, he further explained that first living things were cast out in the form of separate parts and by chance these parts were united by attraction force and as a result plants and animals were born. The sex originated later and through reproduction the developing embryo resulted from both paternal and maternal substances and in that fashion distinct species were formed. He believed that when reproduction was established, earth felt no need of special creation of organism. According to Empedocles plants came earlier than animals and perfect forms replaced imperfect types.

Aristotle discarded Empedocles's views and he observed that organisms could be arranged in graded series from lower to higher. The lowest stage is the inorganic. Organic beings arose from the inorganic by direct metamorphosis. He conceived the organic world to consist of three states: 1. plants, 2. plant-animals, a transitional group in which he included sponges and sea anemones; 3. animals, characterised by feeling or sensibility. Within the animal groups he constructed a genetic series leading from the lowest forms up to man, placed at the apex. **Aristotle** is considered as the **father of "family trees"**. From this taxonomic fact, he drew the correct inference that one higher group evolved from other lower one in the taxonomic series. However, he had the metaphysical belief that the gradual evolution of living things occurred because nature strives to change from the simple and imperfect to the more complex and perfect. An evolutionary explanation of the origin of plants and animals was given by the Roman poet **Lucretius** (99–55 B.C.) in his poem *De Rerum Natura.*

Aristotle.

3. Pre-Darwinian Period

Charles Darwin (1809-1882).

The medieval period is characterized as dark age. During this period, superstition and blind following of ancient thinkers prevailed and Aristotelian reasoning and the spirit of scientific inquiry were died out. Thus, no progress occurred in the idea of evolution during medieval period.

With the Renaissance, interest in the natural sciences quickened and the increasing knowledge of many kinds of animals led more and more scientists to consider the concept of evolution favourably. Thus, before **Charles Darwin**, several possible mechanisms have been proposed to account for the evolution; some of them have been considered here as follows :

(i) Fixity, design and creation. *Nullae speciae novae* (= No species new) was the catch-phrase for the early systematists, including **Linnaeus** (1707–1778). Species were the units of creation as prescribed in *Genesis* (The first book of the Bible) and, therefore, immutable. Moreover, the reason they were so well-fitted for the challenges presented by everyday life was that they had been designed by God for specific functions in nature. **John Ray** (1627–1705), clergyman, naturalist and early systematist, saw the fitness of species as evidence for the existence of a *Designer*, and this **Argument from Design** was made even more clear later by **William Paley** (1743–1805). According to early systematists, animals and plants are wonderful bits of machinery, more wonderful than any man-made machine, and so they must be the product of an **intelligence** more wonderful than that of Man. There was a later, more subtle, version of this argument which insisted on the **harmony of form** rather than utility. These are described as the **idealists** and **utilitarian** positions respectively.

This **creationist concept** held certain biological, theological and philosophical problems. For example, many fossilised organisms had been discovered that no longer existed on earth, and yet it was unbelievable that the perfect products of an omnipotent designer could ever have become obsolete. **Georges Cuvier** (1769–1832), a very influential French biologist and founder of palaeontology, offered **catastrophic hypothesis** as a way out. A series of catastrophes, fires and floods (*e.g.,* Noah's Flood) had removed some of the species initially created by God. Apparently useless characters were another worry because inferior design could hardly be attributed to a super-intelligent designer. To explain certain useless characters (*e.g.,* toes of pig), **Buffon** (1707–1788) suggested that the Supreme Being had created perfectly designed types embodied in the original species, but that new species arose from them by a process of hybridization and degeneration. Thus, the ass was supposed to be a degenerate horse and the ape a degenerate man. But this meant that the assumption of strict fixity of species had to be relaxed and the concept of *nullae speciae novae* disappeared from the last, revised edition of the book–*Systema naturae* of **Linnaeus**. **Linnaeus** though was not an evolutionist, nor was he greatly concerned with explanations of nature, but he devoted his life industriously in arranging the kinds of animals and plants known to him in a convenient system. This system he took to be actual and fixed; to discover it and use it was the major task of natural history. He broke the tradition in placing man, "the knowing one, *Homo sapiens*" in the animal kingdom, the class Mammalia and the order Primates. Of many categories employed today in classification, it is interesting that **Linnaeus** used only the kingdom, class, order, genus, and species, but other naturalists soon found it

Linnaeus (1707-1778).

advantageous to divide the animal and plant kingdoms into phyla and to divide orders into families. However, when Linnaeus established his idea of species, he created a dilemma that a species is real, objective and retains its identity in nature instead of shifting haphazardly or spontaneously changing its characteristics. Thus, his idea of special creation and of immutability of species caused a serious hindrance to the development of the evolution idea.

(ii) Programmed evolution. Some of the liberal medieval scholars considered the story of creation in *Genesis* to be a myth. For example, **Augustine** (353–430) likened the work of the Creator to the progressive growth of a tree, and **Aquinas** (1225–1274) similarly saw creation as a process whereby the powers given to the matter by God progressively unfolded. It was in this sense that **Charles Bonnet** (1720–1793) and **Treviranus** (1776–1853), were of the opinion that the *design* of all life-forms was contained as germs in the first being. However, **Bonnet** conceived of this unfolding as the working out of a rigidly determined programme of change, whereas **Treviranus** saw evolution as the calling into play of an endless variety of form assumptions (programmed into each organism by the Creator) according to the needs of survival in a changing world. Thus, organisms and environment interacted to generate all extant and extinct species.

These views of the world, as a giant organism endowed with a mysterious developmental impulse, can be traced back to **Plato** (427–347 B.C.) and were carried forward into the eighteenth and even nineteenth century by evolutionists as **Goethe** (1749–1831), **Kant** (1724–1804) and **Chambers** (1802–1871). These workers are called **natur philosophers**.

(iii) Less rigidly programmed evolution. Erasmus Darwin (1731–1807) – Charles Darwin's grandfather–wrote about the process of evolution in his book *Zoonomia* (1794) and poem *Temple of Nature.* He proposed a less rigid system of determined evolution based on the acquisition of new characters by the organisms themselves. Erasmus was the first to express clearly that millions of years have been required for the process of organic evolution and that all life arose from one primordial protoplasmic mass. **Jean Baptiste de Lamarck** (1744–1820) formulated a strikingly similar theory in his **Philosophie Zoologique** (1809). Both of these theories involved the following three main principles:

Jean Baptiste de Lamarck (1744-1820).

1. All organisms were supposed to have an **innate power** to progress towards a more complex and perfect form. For **Erasmus** this faculty to improve by its own inherent activity was God-given, but for Lamarck 'innate power' was supposed to emerge out of the organization of living beings. Subtle fluids – caloric and electricity – flowed through the organism not only maintaining order but also establishing more order.

2. All organisms had an inner disposition (what **Lamarck** called a *sentiment interieur*) which caused the performance of actions sufficient to meet the needs (survival) created by a changing environment. The environment does not cause change, it causes the need for change, which is recognised and acted upon by the organism. A mysterious principle did the job for **Erasmus**, but the subtle fluid did it for **Lamarck**.

3. Characters acquired through new use were transmissible from one generation to another by a process which was never clearly defined by **Lamarck**. Similarly, characters lost through disuse in one generation did not reappear in future generations.

Using these principles, **Lamarck** explained the evolution of tentacles in the head of snail in the following way. As the snail crawls along it finds the need to touch objects in front of it and makes efforts to touch them with the foremost part of its head. In so doing, it sends subtle fluids into these extensions. Ultimately nerves grow into them and the tentacles are made permanent. These acquired tentacles are transmitted from parent to offsprings, *i.e.,* they are heritable.

Lamarckism was supported by **E. Geoffroy St. Hilaire** (1772–1844), **Robbert Chambers** (1802–1871) and **Herbert Spencer** (1802–1903), but bitterly opposed by **G. Cuvier** (1769–1832) and **Weismann** (1887).

4. Darwinian Period

British naturalist **Alfred Wallace** (1823-1913) wrote to Darwin from Malaysia, asking advice on a short article he had written. To Darwin's dismay it contained the idea of natural selection. Wallace did not know that Darwin had been working on this idea for 20 years. A joint publication was hastily arranged.

Charles Darwin (1809–1882) and **Alfred Russel Wallace** (1823–1913), both spent many years in collecting and exploring South America and certain islands of its sea and both independently drew similar conclusions about the possible mechanism of organic evolution due to natural selection, after studying the essay on population by **Malthus** (1789) which suggested that human populations continually tend to outstrip the availability of food supply, but are kept in check by natural diseases, and famine, and man-made wars. **Darwin-Wallace theory of natural selection** was published in the Journal of the Linnaean Society in July, 1858.

Darwin seemed to agree with Lamarck's idea that characters acquired during the life of an organism might be transmissible, in a heredity sense, to its offspring (*e.g.,* Darwin's theory of pangenesis). However, it was on the mechanism of evolution that **Darwin** parted company, with both Lamarck and his own grandfather.

In his book, *Origin of Species* (first published in 1859), **Darwin** replaced the idea of **directed variation** (brought about by a sentiment interieur) with **random variation** that was heritable. Regarding the origin of this variation, he was a little vague – he thought that it might be due to environmental influences on the gonads and he was equally vague about the mechanism of inheritance. What he was perfectly clear about was that no matter how the variation originated, it did so without any reference to the needs of the organism. Moreover, for **Darwin** the variation upon which evolution was based involved very small continuous changes away from the original form. He accepted that large abrupt changes might occur, but claimed these were rare and most led to monstrosities. Finally, instead of Lamarck's mysterious struggling for perfection, Darwin explained evolution in terms of a completely ordinary and non-mysterious process. **Darwin** observed that animal and plant breeders had, by the selective breeding of variants, been able to affect considerable changes in the form of domestic plants and animals. This was **artificial selection**. Similarly, he thought that the struggle between organisms in an overcrowded world by finite resources would ensure that only the fittest survive and this would lead to a kind of **natural selection** of variants.

The gist of **Darwin-Wallace theory** of natural selection was as follows: 1. Individuals within species show considerable but continuous variation in the form and physiology. 2. This variation arises in a random fashion and is heritable. 3. The potential for increase within populations of animals and plants is considerable. 4. Since resources are limited, so individuals in a population struggle for their own existence and that of their offsprings. 5. Hence, only some (what **Darwin** called the **fittest**) survive and leave offsprings with the same traits. 6. Through this **natural selection** of the fittest, species become represented by individuals which are better and better adapted.

5. Post-Darwinian Period

The history of evolutionary thought subsequent to the publication of the book '*The Origin of Species*' may be divided into the following three periods :

(i) The romantic period. The romantic period extended from 1860 to about 1903 and was characterised by extreme enthusiasm for Darwinism, together with an uncritical acceptance of whatever data were claimed to support Darwinism. Negative evidence was given little weight, while illogical extremes of interpretation in order to make observed facts fit Darwinian theory were quite common. Leaders of this group in England included **T.H. Huxley** (1825–1895), **Herbert Spencer** (1820–1903) and **George Romanes**, while in the United States **David Starr Jordan** and **ASA Gray** (1810–1888) were the leaders. As a group, they went to interpretative extremes, reading adaptive significance into every organic structure, even on the most imaginary evidence. They often cited excellent anatomical and taxonomic evidence, but experiments to test adaptive values were unusual.

Thomas H. Huxley (1825-1895), nicknamed "Darwin's Bulldog", was a young energetic English scientist who took up the cause of evolution in many public debates. He fought on behalf of Darwin, who preferred to keep out of the public eye. However, Huxley was inspired by Haeckel, and he distorted Darwin's ideas with notions of "progress".

It should not be thought that these were second-rate biologists who were blinded by the brilliance of a great man like **Darwin**, on the contrary, they were excellent men in their respective fields. **Huxley** made splendid contributions to the development of invertebrate zoology, taxonomy and vertebrate anatomy. **Spencer** was one of the leading philosophers of his time. **Romanes** started his career as an invertebrate neurologist, but he soon became mainly engrossed in evolutionary problems. **Jordan** was one of the best ichthyologists of all times. **Gray** was a botanist of great repute. Nor should it be thought that they never ventured to differ from **Darwin,** for these men were independent thinkers. Yet the atmosphere of approval was extraordinary.

It has been said that evolution was born in England, but found its home in Germany. In fact, the German evolutionists of the Romantic period were more strictly Darwinian than their English and American colleagues in the sense that they were generally more thorough and careful collector of data. The leaders in Germany were **Von Baer**, **Carl Gegenbaur**, **Ernst Haeckel** and **August Weismann. Von Baer** (1792–1876) was a great embryologist and he tried to deduct some evolutionary meaning from embryology of different animals. His basic principles of embryology states that 1. general characters appear before special characters; 2. from the more general characters, less general and finally the special characters develop, 3. an animal during development departs progressively from the forms of other animals; 4. the young stages of animals are like the young (or embryonic) stages of other animals lower in the scale, but not like the adults of those animals. **Gegenbaur** (1826–1903) was one of the greatest and most influential comparative anatomists. He and his collaborators made exhaustive studies, in complete detail, upon all classes of vertebrates, and used the data so obtained in support of Darwinian theory. Much of the phylogeny (*i.e.,* the evolutionary or ancestral history of organisms) of the vertebrates in current textbooks of zoology is taken from the works of **Gegenbaur** and his collaborators.

Ernst Haeckel (1834-1919) developed the idea of "evolution a progress" to its fullest extent. He believed that nature had been deliberately moving towards find goal: human beings.

Ernst Haeckel (1834–1919) did significant work in anatomy, embryology and taxonomy. His studies in comparative embryology led him to broaden the principles of **Von Baer** into the **theory of recapitulation** or **biogenetic law,** in support of which he published extensively. In one of his writings '*Generelle Morphologie*', **Haeckel** (1866) defined the biogenetic law in the following way —an individual organism in its development (**ontogeny**) tends to recapitulate (or repeat) the stages passed through by its ancestor (**phylogeny**), or that *ontogeny repeats phylogeny.*

August Weismann (1834–1914) was a classical geneticist. For explaining the mode of inheritance in living organisms, he rejected outrightly the Lamarckian ideology of inheritance of acquired characters and Darwin's pangenesis hypothesis by demonstrating a well known experiment in which he cut the tails of mice for twenty-two generations, yet each new lot of offspring consisted only of animals with tails. He forwarded an entirely speculative and theoretical hypothesis of "the continuity of germplasm" or as most popularly known "**germplasm theory**". The main drawbacks of **Weismann** were that he was unaware of work of his contemporary geneticist **Mendel** and his speculations rested on some data on chromosome behaviour during mitosis which were often reported to him by his graduate students as he gradually became blind before he completed his major works. He correctly predicted the occurrence of reduction division (or meiosis) and the chemical nature of the hereditary units (see **Dodson** and **Dodson**, 1976). **Weismann** imagined that natural selection had operated at three levels : between individuals (**Darwin** and **Wallace**); between parts, organs, tissues, etc., of the same individual (**Roux**; according to **Roux**, 1881, the struggle is waged among the molecules, tissues and organs; what is at stake is the intake of food which, according to its quality, ensures greater or lesser growth, and consequently differentiation; see **Grasse**, 1977); and between germinal **determinants** (now called genes; **Weismann**).

Thus, according to **Weismann**, the most important struggle is between the "vital, invisible" particles (genes), whether they reside in the body cell (soma) or the sex cells. These **determinants** are never identical with one another and their ability to "assimilate" food varies from one another. The strongest take the most food, and their descendants prevail over the others. The result is that by the second generation the parts of the organism represented in the ovum by the strongest determinants, are the most highly developed. This explains the dominance of certain characteristics; the determinants of organs favoured by natural selection are better fed than the rest and are the winners in the competition, hence, the strengthening of change in that direction (see **Grasse**, 1977).

Also prominent during the romantic period was **Karl Pearson** and **Francis Galton** who laid the foundation of the new sciences of statistics and biometry which play a prominent role in modern evolutionary studies.

(ii) The agnostic period. The agnostic period extended from 1903 to 1935 and involved a lot of skepticism and disillusionment. Many factors converged to cause this change. One factor was the rediscovery of **Mendel's laws of heredity** in 1900. Today Mendelism is the foundation of most studies in evolution, but then the permanence of the gene seemed to raise fearful obstacles to the origin of new species. In consequence, genetics was regarded as a sort of blind path at the end of which stood the sign : **The gene**, **the dead end**. Another factor was the conclusion which was drawn by **Johannsen** from his hereditary experiment of the size in beans, that selection could be effective only in a stock (or organism) with hereditary variability. Variations produced by the environment (including nutrition, sunlight, temperature and moisture) were unimportant for evolution.

Finally, besides natural selection, many additional mechanisms and causes were sought by different evolutionists for explaining the evolution. For instance, **Moritz Wagner** (1868) emphasized the effects of different environments on isolated forms. **Gulik** pointed out the unavoidability of divergence even without difference in environment. **Karl Nageli** suggested that an inner directive guides the course of evolution mainly independent of the environment.

Hugo de Vries
(1848–1935).

Hugo de Vries (1848–1935) in his book *The Mutation Theory* (1909) distinguished between continuous individual variation and discrete, saltational variations. He applied the term **mutation** to the latter only. **De Vries** experimented with the plant evening primrose *Oenothera lamarckiana*. For **de Vries** it was these mutations that controlled evolutionary change and which were more important than natural selection itself.

The early Mendelians, such as **de Vries**, therefore, played down the ability of natural selection and emphasized the overriding importance of Mendelian ratios and mutations in moulding evolution. Some evolutionists recognized the importance and challenge of Mendelism and set out to reconcile it with the Darwinian theory. Because the results of their efforts went beyond the mechanism that Darwin had proposed, the synthesis that emerged is often referred to as **neo-Darwinism** (see **Calow**, 1983). Further, interest of some neo-Darwinist shifted from considering the consequences of breeding from pairs of individuals to the consequences of breeding between individuals in populations. The Mendelian ratio then became frequencies of occurrence of genotypes in populations and genetic and evolutionary changes were defined in the terms of changes in these frequencies. In 1908, **G.H. Hardy** and **W. Weinberg** independently concluded that Mendelian mechanisms themselves do not cause changes in gene frequencies and they cannot be an important element in directing evolutionary change. Moreover, two books—**J.B.S. Haldane's** *Genetical Theory of Evolution* (1930) and **R.A. Fisher's** *The Cause of Evolution* (1932)— were greatly important in shaping the subsequent development of neo-Darwinism. Both **Fisher** and **Haldane** expressed the intensity of selection in terms of changes in gene frequencies. Those alleles that increase in frequency relative to others are said to be fitter, so the change in these relative frequencies measures neo-Darwinian fitness. Fisher's *Fundamental Theorem of Natural Selection, i.e.,* the rate of increase of fitness of any organism at any time is equal to its genetic variance in fitness at that time.

Sewall Wright, another pioneer of neo-Darwinism, in 1931 developed a conceptual framework for considering the interaction between genes and their influence on fitness. This is referred to as an *adaptive landscape* or *fitness space.* Wright's emphasis was on **populations of genes**. However, **R.B. Goldschmidt** (1878–1958) who had wide experience in the study of geographic variations, taxonomy and physiological genetics, is the principal proponent of the main alternative to the dominant neo-Darwinian theory.

Punctuated equilibrium concept of evolution. This idea was proposed in 1972 by two American palaentologists, **Dr. S.J. Gould** and **Dr. N. Eldredge** and has been widely accepted as the best way to interpret evolutionary patterns over geological time (see **Henry Gee,** 1993). This concept contradicted Darwin's idea that *species evolve gradually one into other by insensible degree.* According to **Gould** and **Eldredge**, species evolve by *sudden bursts of rapid change that 'punctuate' long periods of static equilibrium.* The punctuated equilibrium concept showed that Darwinian gradualism should have been supported by fossil record, *i.e.,* transitional forms should make up the bulk of fossils. But fossil record shows that species seem to persist with little change over many millions of years. Thus, there is no use to try to find evidence of gradual change in fossil record, because it is not there to begin with (see **Gee**, 1993). According to **Gould** and **Eldredge**, species evolve very rapidly, within a few thousand years – too short a time for preservation of the fossil record to be likely. But once evolved species remain unchanged for many times, longer than it makes them to evolve. Thus, stasis (stoppage) of species becomes data of evolution.

Stephen Jay Gould (b . 1942) is among the scientists who are continually testing and questioning the details of evolution theory.

To support their idea, **Gould** and **Eldredge** took help of the idea of mutation. Thus, according to the process of speciation change, in the form of the success of mutations, it tends to happen most efficiently in small peripheral populations of a species rather than in the general mass in which such mutations might be swamped. Likewise, individuals from small populations in transition have even less chance of fossilisation than those from large one. Of course, natural selection can be a force for change in small as well as large populations, but a *chance* and *accident* assume more significance in smaller populations. Once formed the new species expand into new habitats and settles down to comfortable statis. The result is not one but two populations—exactly the pattern observed in the fossil record in many cases.

(iii) The modern synthesis period. The synthesis of a coherent theory of evolution which takes into account all the pertinent facts of modern biology, has been the work of several biologists during

Theodosius Dobzhansky (1900-1975) worked with T.H. Morgan on fruit flies and helped in the synthesis of genetics and evolutionary theory. He coined the term "**isolating mechanisms**" for the bioloical barriers that discourage crossing between different species.

the past fifty years. The first edition of **Theodosius Dobzhansky's** now classic book, '*Genetics and the Origin of Species*', which appeared in 1937, set the stage, and stimulated biologists in several fields to contribute to the synthetic evolutionary theory. The term **modern synthesis** was first used by **Julian Huxley** in 1942. As presented in the famous book – *Evolution, The Modern Synthesis* by **Huxley** (1942), the modern synthesis or synthetic theory was an amalgamation of the data and concepts of genetics, taxonomy, embryology, biogeography, and other disciplines (**Stanley**, 1979). Likewise books by zoologist **Ernst Mayr** (1942, 1963) showed how the modern theory could explain the origin of variation patterns in higher animals and **G.L. Stebbins** (1950) attempted to do the same for higher plants. **E.B. Babcock's** (1947) study of the genus *Crepis*, carried over a period of more than thirty years, is one of the most thorough and comprehensive studies ever made upon a single genus of plants, and it is one of the major supports of the neo-Darwinian theory. **Clausen** and coworkers in 1951 have published extensive studies on the behaviour of plants when grown in widely different habitats. A leading palaeontologist **G.G. Simpson** showed in his two books – *Tempo and Mode in Evolution* (1944) and *The Major Features of Evolution* (1953) that the fossil record of higher animals is best explained by assuming that throughout the evolutionary history of living things those same processes took place which were being experimentally demonstrated by many workers in populations of contemporary animals. In 1960, **Bernhard Rensch** has made a strong case for the belief that the evolution of genera, families and higher categories of animals has taken place through an extension into long periods of time of those same processes which at any one time level govern the origin of races and species.

According to **Mayr** (1970), most of the earlier evolutionary theories were characterized by heavy emphasis on a single factor, as has been tabulated below in Table 2-1.

Table 2-1. Theories of evolutionary changes (E. Mayr, 1970).

A. **Monistic** (single-factor explanations)
  1. **Ectogenetic** : Change directly induced by the environment.
     (a) Random response (for example, radiation effects)
     (b) Adaptive response (Geoffroyism)
  2. **Endogenetic** : Changes controlled by intrinsic forces
     (a) Finalistic (orthogenetic)
     (b) Volitional (genuine Lamarckism)
     (c) Mutational limitations
     (d) Epigenetic limitations.
  3. **Random events** ("accidents")
     (a) Spontaneous mutations
     (b) Recombination
  4. **Natural selection**

B. **Synthetic** (multiple-factor explanations)
  1b+2a+2b = most "Lamarckian-type" theories
  1b+2b+2c+4 = some recent "Lamarckian" theories

1b+3+4 = Late Darwinian, most non-mutationists during first three decades of 20th century
3+4 = early "modern synthesis"
1a+2c+2d+3+4 = recent "modern synthesis"

The modern synthetic theory selected the best aspects of the earlier hypotheses and combined them in a new and original manner. In essence a two-factor theory, it regards the diversity and harmonious adaptation of the organic world as the result of a steady production of variation and of the selective effects of the environment.

Attempting to explain evolution by a single-factor theory was the main weakness of the pre-Darwinian and most nineteenth-century evolutionary theories. For instance, Lamarckism with its internal self-improvement principle, Geoffroyism with its induction of genetic change by the environment, Cuvier's catastrophism, Wagner's isolation, De Vries' mutationism, all tried to explain evolution by a single principle, excluding all others. Even **Charles Darwin** occasionaly fell into this error, yet on the whole he was the first to make a serious effort to present evolutionary events as due to a balance of conflicting forces. The current theory of evolution the "**modern synthesis**", owes more to Darwin than to any other evolutionist and is built around Darwin's essential concepts. Yet it incorporates much that is distinctly post-Darwinian. The concepts of mutation, variation, population, inheritance, isolation, and species, all of which were quite nebulous in Darwin's days, but now are much better defined and understood, have been incorporated in recent "modern synthesis" (see **Mayr**, 1970).

PRESENT STATE OF EVOLUTION IDEA

The current synthetic theory of evolution recognizes five basic types of processes — **Gene mutation**, **Changes in chromosome number and structure**, **Genetic recombination, Natural selection**, and **Reproductive isolation**. The first three provide the genetic variability without which change cannot take place; natural selection and reproductive isolation guide populations of organisms into adaptive channels. In addition, three accessory processes affect the working of these five basic processes. **Migration** of individuals from one population to another, as well as **Hybridization** between races or closely related species both increase the amount of genetic variability available to a population. The effects of **chance**, acting on small populations, may alter the way in which natural selection guides the course of evolution.

REVISION QUESTIONS

1. Describe the evolutionary ideas of early Greek philosophers.
2. Discuss how did the idea of evolution mature during the 19th century?
3. Discuss the main points of pre-Darwinian period.
4. Discuss the main features of post-Darwinian romantic period and agnostic period.
5. What is the theory of modern synthesis? Describe its origin, growth and essence.
6. Discuss critically the following statement :
   '*Evolution was born in England, but found its home in Germany*'.
7. Write short notes on the following :
   (i) Father of family trees; (ii) Argument of design;
   (iii) Lamarck's evolutionary idea; (iv) Natural selection;
   (v) Weismann's determinants.

CHAPTER

3

Direct Evidences of Evolution : Fossils

(a)

(b)

(c)

Fossils of extinct organisms (a) trilobites, (b) seed ferns and (c) dinosaurs provide strong support that today's organisms were not created all at once but arose over time by the process of evolution.

Darwinian evolution permits two general predictions – (1) if there were an evolution from simple forms to more complex ones, there must be certain structural, developmental and chemical similarities between different forms of life, and (2) there must be a means by which variation in populations arise, and are transmitted from generation to generation. Chapters 3 and 4 will deal with the evidences that have been found to support the idea of a relationship between existing organisms and also of a relationship between extinct and existing organisms, while certain forthcoming chapters will deal with the way in which change can initiate and bring about an evolutionary process such as that proposed by Darwin.

The evidence that organic evolution has occurred is so overwhelming that no one who is acquainted with it has any doubt that new species are derived from previously existing one by descent with modification. The fossil record (palaeontology) provides **direct evidence** of organic evolution and gives the details of the evolutionary relationships of many lines of descent, while different biological disciplines such as comparative anatomy, taxonomy, embryology, physiology, biochemistry, genetics and biogeography provide **indirect evidence**, in support of biological evolution.

The direct evidences for biological evolution can be discussed as follows :

PALAEONTOLOGICAL EVIDENCES

Most convincing evidence that evolution has occurred comes from the fossil record of ancient creatures. The study of ancient life is called **palaeontology** (Gr., *paliaos* = ancient + *onta* = existing things + *logos* = discourse). The science of palaeontology links biology with geology and is concerned with the finding, cataloguing and interpreting of fossils.

Georges Cuvier (1769–1832).

Even before the Renaissance men had discovered shells, teeth, bones, and other parts of animals buried in the ground (*i.e.,* fossils). Some of these fossils corresponded to the parts of familiar living animals, but others were strangely unlike any known form. Many of the objects (fossils) found in rocks high in the mountains, far from the sea, resembled parts of marine animals. In the fifteenth century, the versatile artist and scientist, **Leonardo da Vinci** (1452–1519) gave the correct explanation of these curious finds, and gradually his conclusion that they were the remains of animals that had existed at one time but had become extinct, was accepted. This evidence of former life led to the formulation of the **theory of catastrophism** by the founder of palaeontology and a versatile French naturalist – **Georges Cuvier** (1769–1832). The theory of catastrophism holds that a succession of catastrophes, fires and floods, have periodically destroyed all living things, followed each time by the origin of new and higher types by the acts of special creation. **Cuvier** was also responsible for developing the **law of correlation** which states that certain structures are invariably found together. **Cuvier** is referred to as **Founder** or **Father of Modern Palaeontology**.

American scientist **Louis Agassiz** (1807-1873), a follower of Cuvier added a new type of catastrophe – the Ice Ages.

The concept of catastrophism was replaced by the theory of **uniformitarianism** due to the landmark investigations of three geologists, namely **James Hutton**, **John Playfair** and **Sir Charles Lyell** in the eighteenth and early nineteenth centuries. **James Hutton** (1785) developed the concept that the geologic forces at work in the past were the same as those operating now. He arrived at this conclusion after a careful study of the erosion of valleys by rivers and the formation of sedimentary deposits at the mouth of rivers. He demonstrated that the processes of erosion, sedimentation, disruption and uplift, carried on over long periods of time, could account for the formation of fossil-bearing rock strata. The publication of the book – *"Illustrations of the Huttonian theory of the Earth"* by **John Playfair** (1802) – gave further explanation and examples of the idea of uniformitarianism in geologic processes. Finally, the publication of **C. Lyell's** book – "*Principles of Geology*" (1932), had firmly established the concept of uniformitarianism.

In India, palaeontological work has been started as early as 1851 by the setting up of the *Geological Survey of India* in that year. The *India Museum* started in 1866 helped in collection of fossils and the spreading of palaeontological knowledge. Two other institutions which contributed to the progress of Indian palaeontology are the *Mining and Geological Institute* founded in 1906 at Asansol and the *Institute of Palaeobotany* established at Lucknow in 1946 by the initiative of **Prof. Birbal Sahani**. In 1950, *Palaeontological Society of India* was established and in 1965 *Palynological Society of India* was laid down. **P.N. Bose** was the first Indian palaeontologist whose paper was published in 1880 on a fossil carnivore from Siwalik hills.

In India, there are many good fossil collection spots both in the Himalayan regions and in the Deccan. In the foothills of Himalayas, the fossils have been collected from places such as Siwalik hills,

Spiti, Punjab and Kumaon. Fossils have also been collected from Sikkim. In the Deccan, there are fossil collection spots at Jabalpur and in many parts of Gujarat. In South India fossils have been collected at Rajamundri, Trichinapally and Quilon.

Branches of Palaeontology

Palaeontology is subdivided into the following branches :

1. Palaeobotany which deals with the study of plant fossils. It includes **palynology** which is the study of fossil spores and pollen grains.

2. Palaeozoology which deals with the study of animal fossils and includes **invertebrate palaeontology** and **vertebrate palaeontology**.

3. Micropalaeontology which deals with study of small microscopic fossils (microfossils) and their fragments, *e.g.,* foraminifera, fusulinids and ostracodes.

4. Palaeoecology which is the study of ancient organisms and their environment.

Fossil of arrow - headed amphibian *Diplocaulus* (Permian).

Fossils of Triassic nautiloids—ammonite and surviving nautiloid.

Fossil frog (Jurassic).

FOSSILS

Fossils (Latin *fossilis* or *fodere* = to dig) are the remains or impressions of organisms preserved from the geologic past. They are virtually anything that is formed by or derived from a prehistoric organism. This includes bones, wood, shells, teeth, skin, pollen, tracks, burrows, and even faeces or dung. The fossil record provides us with evidence that there were organisms that have become extinct.

As an archaeologist reconstructs, from imperfect relics occasionally preserved, something of the nature and customs of ancient people, so the palaeontologist obtains from the fossils and the sediments containing them, dim views of far earlier scenes in the life of animals and plants. These he fits together into a steadily growing history, but the record as a whole will always lie beyond his reach.

How Fossils are Formed ?

The animals or plants are only preserved and fossilized when they are suddenly buried in the silt of water, lava, ice or sand. As soon as the animal, plant or any other organism dies, its body decomposes sooner or later. Decomposition is caused by several agents. Scavengers of all sorts from vultures to bacteria and fungi take care of the soft parts very rapidly. Harder tissues such as shell, bones and wood are more resistant. Nevertheless, enzymes secreted by certain fungi and bacteria, acids occurring naturally in the soil, and the combined action of environmental agents such as water, wind, and temperature destroys even the hardest organic remains with time. However, under very special circumstances some parts may be preserved, with varying amounts of modifications, to form fossils (Fig. 3.1).

Thus, shells which are inorganic in nature are usually preserved completely unaltered; bones and wood, on the other hand, are often **mineralized** or **petrified** (*i.e.,* turned to stone). The organic matter in wood or bone gradually disintegrates, leaving the structure somewhat porous. Water seeps into the

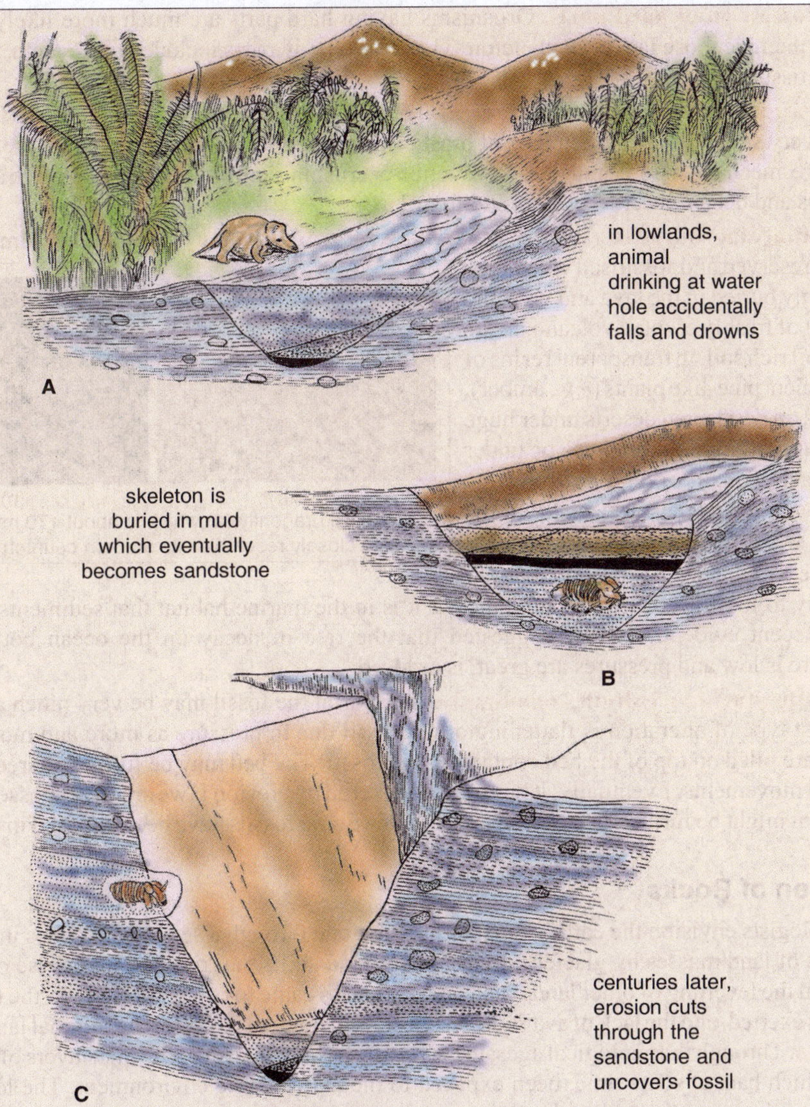

Fig. 3.1. Schematic representation of the process of fossilisation (after Solbrig, 1966).

interior of the bone and minerals dissolved in the water are slowly deposited there. Thus, the porosities gradually become filled with deposits of such materials as lime or silica. The portions of the original structure composed of inorganic materials may remain substantially as they were in life, or they also may be dissolved away and replaced by minerals. The replacing material may preserve the details of the original structure with great fidelity, or on the other hand, it may preserve only the general form of the original. Yet fossile may contain some of the original material found in the living organism; for example, amino acids have been found in fossils of millions of years old.

Conditions of Fossilisation

Fossilisation is the sum of the phenomena by which the remains of animals or plants are preserved. The process of fossilisation requires the occurrence of the following fundamental conditions :

1. Possession of hard parts. Organisms having hard parts are much more likely to become fossilised than are those lacking a skeleton. These hard body parts include bones, teeth, shell, chitin or woody tissues of plants.

2. Escape from immediate destruction. The possibility that any given animal will leave a fossil record is very small. Since the remains of most of these organisms are destroyed by the work of the atmosphere, mechanical forces such as wave action or crushing and biological agents such as predators, scavengers and decomposers (bacteria and fungi).

3. Immediate burial in a medium capable of retarding decomposition. The animals or plants are only preserved and fossilised when they are suddenly buried in the lava and volcanic ash or dust of freshly erupted volcano, in the ice, in an oil rich soil, in transparent resins of certain ancient pine-like plants (*e.g.,* amber), in swamps, in dessicated deserts under huge piles of sand at high temperature or under water in layers of silt piling one above the other (in seas and oceans, especially in the shallower regions and at the river deltas).

(a) (b)

The fossil dragonfly (a) which is about 170 million years old closely resembles its modern counterpart (b).

For example, organisms in the sea are more likely to fossilise than those on land, for it is in the marine habitat that sediments build most rapidly. Recent evidence further suggested that the rate of decay on the ocean bottom, where temperature is low and pressures are great, is reduced.

4. Subsequent vicissitudes (changes). After burial the fossil may be very much altered. The commonest type of alteration is flattening of the fossil due to pressure as more and more layers of sediment are piled on top of the bed containing the fossil. The bed may be further altered by folding and lateral movements. Eventually, it has to be uncovered by erosion if we are to get possession of the fossil, and it might be further damaged in this process. Some fossils have been exposed in the past and reburied.

Formation of Rocks

Geologists envision the earth's crust as having gone through a series of changes involving the alterations of land masses by glacier movements or land shifts that resulted in the rise of mountain ranges, and the levelling of other land when one land mass came to rest upon another, the tremendous pressure is exerted, and the lack of available water, caused the conversion of the original land mass into **rock strata**. Through the geological ages, then, land mass over land mass formed layers of rock strata, each of which had, at some time, been exposed to the atmospheric environment. The lowest strata solidified some two to three billion years ago.

The rock strata of earth includes **sedimentary** and **igneous rocks**, the formation of both of which remains significant in the process of fossilization. To understand the nature of the formation of sedimentary rocks, one must take into account certain natural processes that are occurring in the world today and that have been occurring since the crust of the earth was formed. By the process of **erosion**, through the action of wind and rain, and of freezing and thawing, rocks are gradually broken into the small particles that form **soil**. Through the action of rain, the particles of the soil are carried into streams and rivers and ultimately into lakes or oceans. This sedimentary material carries with it the bodies of many aquatic organisms and also the bodies of terrestrial forms that happen to be swept along by the streams and rivers. The hard parts of some of these organisms may be preserved, and in the course of time the sedimentary deposit, owing to the pressure of water above it and also to chemical reactions, is converted into sedimentary rock. The nature of the sedimentary material determines the kind of sedimentary rock formed – **limestone**, **sandstone**, and **shale** are familiar kinds. The method of formation of sedimentary rocks clearly distinguishes them from **igneous rocks**, which were formed by the solidification of molten material when the earth cooled, and are being formed by cooling of magma expelled from active volcanoes.

Other natural processes called **submergence** and **emergence** have occurred in the past and are still occurring between land and sea. Slow and gradual changes in level between land and sea are now in process. For instance, it is now recognized that the land on the Eastern coast of United States is gradually sinking into the sea, whereas the land on the Pacific coast is gradually rising. The sinking of the land below the sea is called **submergence**, and the rising of land above the sea is called **emergence**. In the geologic past, many regions of the earth have undergone a series of submergences and emergences. As a consequence, in many places (*e.g.,* Strata of Grand Cranyon of the Colorado), a whole series of different layers of sedimentary rocks is found. The chief fossil bearing sedimentary rocks are sandstone, shale, limestone, gypsum, phosphate rock, coal, iron formation, etc.

The walls of Grand Canyon have been formed of sedimentary rocks which cover more than a billion years of evolutionary history.

Further, the **law of superposition** of Geology states that the deepest layer of earth strata was deposited first and other layers in succession at later periods of time. Consequently, more superficially located fossils will be considered to be of recent origin. Moreover, if each formed layer of rock was left undisturbed until another layer was deposited on top of it, one would have perfect series for the study of fossil forms. The rock record of the earth, however, is not so complete. Layers deposited under water can emerge as land and be partially or completely eroded away. If a new submergence then occurs and a new layer of sedimentary rock is formed, there will be an unconformity between the two strata. This lack of sequence between layers makes the study and identification of strata difficult. Also, fossils that were formed may be destroyed in the formation of **metamorphic rocks**. As a result of great pressure and heat, deep layers may melt. When this material later solidifies again, the fossils originally present will usually be lost. Limestone is a form of sedimentary rock rich in fossils, when limestone melts and crystallizes into metamorphic rock, the result is marble. Despite the paucity of fossil formation and destruction of fossils by several reasons, the story of rocks is a very convincing one with reference to evolution.

Determination of Age of Rocks and Fossils

In the past, geologists and palaeontologists were able to make fairly accurate estimations of the age of different rock strata and their fossil record by using the known rate of the accumulation of salt in the oceans. Presently, rock deposits are dated largely by taking advantage of the fact that certain radioactive elements are transformed into other elements at rates which are slow and essentially unaffected by the pressures and temperature to which the rock has been subjected. For example, geologists standardly use **uranium-lead dating** (devised by **Boltwood**, 1907) to estimate the time of solidification of rock. Radioactive uranium decays spontaneously to lead. The half-life of uranium is about 4,500,000,000 years.

Since the uranium of the earth was formed four to five billion years ago when the great pressure of a contracting dust cloud created thermonuclear heat, why should not all uranium measurement give the same age, four to five billion years? The answer lies in the difference between the radioactive decay in a solid and in a liquid. In a liquid, the uranium is diluted, washed away, etc. But when liquid solidifies, the uranium it contains is not free to move, and undergoes its decay into lead in a highly localized region. The ratio of uranium to its stable product, lead, indicates the time of solidification of the liquid material and, therefore, the time at which it was added to the earth's crust. According to uranium-lead method, the precambrian rocks are the oldest and date back to about 3300 million years.

For the determination of the age of fossils, other radioactive materials are analysed. **Radioactive carbon** (^{14}C) is a natural radioactive form, produced in the atmosphere from the contact of naturally occurring ^{12}C with UV light. It passes down as $^{14}CO_2$ and enters plants and then animal material. The ratio of ^{14}C to ^{12}C remains constant during life, because of the constant interaction of biological organisms with the environment. Upon death and fossilization, this ratio decreases as the ^{14}C undergoes decay. It is possible, therefore, to determine when the fossilized individual lived by comparing its present ^{14}C to ^{12}C ratio with that usually maintained during life. Radioactive carbon has a half-life of about 5,568±30 years, and can only be measured up to 25,000 years or about 5 half-lives. This limitation results because the amount of ^{14}C in organisms is so small to begin with. **Radioactive-carbon dating** (it was devised by **Libby**, 1949) is excellent for the anthropological studies of early tribal civilizations, but not for the earlier strata.

Recently, the transformation of radioactive potassium ^{40}K to argon and rubidium to strontium has been used in a similar way for dating fossil-bearing rocks of any age and type. ^{40}K has a half-life of 1.3 billion years. Also because of its greater concentration in most rocks, it is more accurate method of dating fossils than uranium, a relatively rare element. A bed of prehuman fossils in South Africa dated by rock composition gave a result of 500,000 years, whereas **radioactive potassium dating method** indicated an age of 1,750,000 years – a most important difference, considering the nature of the fossils.

Relatively short periods of geologic time are estimated by measuring the rate at which waterfalls recede upstream as they wear away the rocks over which they tumble or by counting the annual deposits of clay on the bottom of ponds and lakes.

Nature of Fossils

A fossil is a record of an organism that lived in the past, whose remains have come into comparative equilibrium with the sediments in which it was buried. Fossils are formed in a variety of ways, depending upon the organic material involved and environmental conditions. They fall into the following two broad classes :

1. Unaltered fossils. Some specimens are preserved relatively unchanged from their original condition; teeth, bones and shells occasionally are found virtually intact buried in sediments. Even more dramatically, remains of organisms have been frozen in Arctic ice fields. Mammoths (an extinct form related to elephant) have been found in their hairy entirety throughout Siberia and have served as an extensive source of fossil ivory. Their flesh was so well preserved that it was eaten by dogs.

Preserved mammoth from Siberia.

Creatures (*e.g.,* largest mammals such as elephants, mastodons and paramylodons) have been trapped in asphalt or tar-pits at Rancho La Bera in Los Angeles (California). Likewise, in Poland two skeletons of the woolly rhinoceros, with some of the flesh and skin preserved, have been found buried in oil-soaked ground. The remains of great Iris deer are found in peat bogs of Ireland.

In the same way, the human bodies, their clothings and food, who lived in the plateau of Arizona (USA) and New Mexico, have been preserved in the dry air and remained as such for hundreds of years. Many tertiary insects have been preserved in the amber (fossil resin) exuded from the pine-like trees in Germany.

2. Altered fossils. The organisms become more or less completely changed by the infiltration of minerals from surrounding rocks, *e.g.,* bones and wood. For such fossils the term **petrifaction** is applied, *i.e.,* turned to stone. The petrified forests of Yellowstone National Park and the Arizona Painted Desert of the USA are well known examples of petrifaction. In Yellowstone, volcanic ash covered the forest about 40 to 60 million years ago. This was followed by partial mineralisation of the plant tissues by silica and quartz which came from the volcanic debris and was circulated through the plants by ground water. Millions of years later the forest was exposed when the surrounding volcanic matter was eroded away. In some regions as many as 27 layers of petrified forests have been discovered.

Fossil parks. India has rich deposits of fossil plants spanning a gap of 3500 million years. Birbal Sahani Institute of Palaeobotany, Lucknow, has discovered and studied 20 million years old fossil forests. These forests require to be systematically studied and conserved for scientific understanding. Palaeontologists have recognised the following localities which can be raised to the status of national fossil parks :

1. Fifty million years old fossil forests preserved in the intertrappean sediments between the streaming lava flows that poured out into the Deccan region and Mandla district, Madhya Pradesh.
2. One hundred million years old fossil forest localities in Rajmahal Hills, Bihar.
3. Two hundred and sixty million years old coal-forming fossils in Orissa.

Types of Fossils

Fossils are formed in a variety of ways, depending upon the organic materials involved and the environmental conditions.

1. Entire organism preserved. In a continuous dry or cold regions, all organic remains, even the softer body parts, may remain unchanged for a long time under exceptionally favourable conditions. Organisms may be preserved intact in a medium that protects them from decay by bacterial action. Among the most perfect fossils known are the insects preserved in **amber**, a fossil resin from pines, especially *Picea succinifera.* Millions of years ago insects such as bugs, aphids, caddice-fly, may-fly, etc., became entangled and entombed in soft, sticky resin exuding from pine trees. The resin hardened and eventually changed to amber, preserving the minutest details of structure of the contained insects. A few extinct animals are known from frozen specimens in which flesh and bones have been preserved in remarkable fresh condition for thousands of years. For example, woolly mammoth has occurred principally in Arctic Tundra (Siberia). Likewise, intact fossils of *Elaphas* and *Mastodon* have been obtained from oil-rich swampy soils from Los Angeles (California).

A ground sloth was found well preserved in an extinct volcanic crater in New Mexico in 1928. The skeleton of this Pleistocene mammal is complete.

2. Original hard parts of invertebrates preserved. In case of invertebrate animals and angiospermic plants the harder parts such as shells, chitinous exoskeleton or spicules of animals and woody tissue of plants get preserved and fossilised.

3. Skeletons preserved. The skeletons of vertebrates make perfect fossils because they could be well preserved in their original shapes and structure.

4. Altered hard parts of organism preserved. Sometimes the original hard parts of an organism get completely altered due to various factors :

Fossil perch.

(i) Carbonisation. Occasionally the disintegrating soft parts of a body leave behind a thin **film of carbon**. Due to carbonisation, have been obtained exact body outlines of the extinct swimming reptiles such as *Ichthyosaurus* and of other organisms such as jelly fishes, worms, leaves, etc.

(ii) Petrifaction. When the mineral rich water penetrates through the pores left behind by decomposition of organic matter of the harder parts, then minerals get deposited in these pores; the process is called **petrifaction** or **permineralisation**. The deposition of minerals makes these fossils of harder parts more resistant to weathering agents. Sometimes, the histological or finer structures such as fibres, wood or the pores and lamellae in shells are perfectly preserved; this is called **histometabasis**.

(iii) Mineralisation. Sometimes certain specific structures of the fossil remains are replaced by a specific mineral. This process is called **mineralisation** or **replacement**, *e.g.*, 1. mineralisation of valves of vessels by silica instead of collagen; 2. mineralisation of sieve tubes, vessels and cells of woody tissues of plants. Mineralisation involves more than 50 minerals; some common examples are the following :

1. Carbonate of lime or calcium carbonate ($CaCO_3$) replaces siliceous parts of sponge spicule.

2. Silica or silicon oxide (SiO_2) is deposited in rings around a central nucleus in hair or nails, etc.. The rings are known as **beekite rings**.

3. Iron oxide replaces the original material and converts it into haematite (Fe_2O_3) or limonite ($Fe_2O_3 . nH_2O$).

4. Iron sulphide replaces calcareous parts of molluscs, crustaceans and brachiopods and converts them into pyrites (iron disulphide, FeS_2) and marcasites.

(iv) **Iron silicate** replaces the tests of Foraminifera and shells of molluscans. These replaced fossils are called **glauconites** (silicates of iron).

(v) Carbonates of Ca and Mg alter the original skeleton of corals, brachiopods and echinoderms. Such fossils are called **dolomites**.

5. Traces of organisms preserved. Occasionally neither an entire organism nor a part of it is fossilized, instead the organism left certain impressions or marks in certain hard media. They too bear great palaeontological significance and are of the following types :

(i) Moulds and casts. A **mould** is an impression of some hard parts, *i.e.,* shell of an organism buried in the sediment or mud, which, in course of time, hardened to become a rock. Later on, the organism decays, the organic material is removed by percolating acidified water and a cavity (pit or depression) is formed that shows the external configurations and surface marking of the original material. It is called **external mould**. Sometimes, however, the internal cavity of part (*e.g.,* shell) is filled with mud or sediment or mineral deposits before the shell is dissolved by the action of percolating water. In such cases these moulds are called **internal moulds**. These represent internal structure of the shell and the exact shape of the animal. The moulds are available largely for molluscan shells, Foraminifera and Radiolaria.

If the mould showing external form, later on, is filled with mineral matter, **natural cast** of the original object is produced which reflects only its external form. Natural moulds and casts form a very large proportion of the fossils that are found in sandstones and limestones. For the preparation of **artificial casts**, latex (rubber) solution, melted wax or plaster of Paris is filled in the external mould. Natural moulds of several men and dogs, whose artificial casts have been produced by plaster of Paris, have been found in the ancient ash-covered town of Pompeii where they were mass buried by the eruption of the Mt. Vesuvius in 79 A.D.

(ii) Tracks and trails. Tracks (footprints) are formed over dryland or sea bottom or in muddy environments, where wet sand and mud receive impressions of the feet of passing animal and they are covered by the sediment before they are disturbed or eroded. Such fossils of footprints are very common for Dinosaur reptiles from Triassic rocks of Connecticut Valley. The tracks of Missippian amphibians form another example of track. With the help of these tracks palaeontologists gather information regarding the size and shape of foot, length of limb and posture and gait of the animal concerned.

Trails are the irregular markings of moving animals on the sedimentary rocks, such as those formed due to the crawling of a worm or snail, the dragging tentacles of a jelly-fish, the impressions of the fins of fish, and markings of the movements of crustaceans or urchins.

(iii) Burrows and borings. Some animals live in burrows, tubes and holes in the ground, wood or rocks either for taking shelter or in search of food. The presence of such burrows or tubes in the sedimentary rocks of ancient past are called **fossil burrows**. They belong to bivalve and other molluscs, worms and sponges. Fossil burrows provide us information regarding behaviour of their resident animals.

Borings are holes made by animals for the sake of food and shelter. Such fossil holes occur on fossil shells, wood and organic objects.

(iv) Gastroliths and coprolites. Gastroliths are hard stony pieces found from stomachs of ancient reptiles and fishes. **Coprolites** are fossilised faeces of ancient animals. Their study provides valuable information about the food and feeding habits of fossil forms to whom the faeces belong.

Microfossils. One should not get the impression from our discussion that fossils are all large, for many specimens are microfossils. Indeed, some fossils (*e.g.,* spores, pollen) are only observable by light or electron microscope.

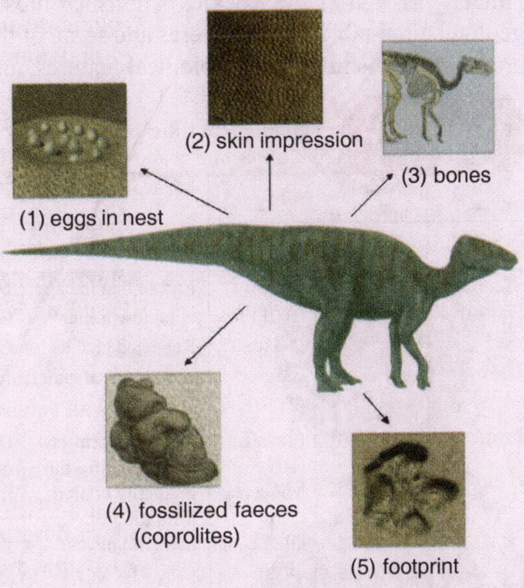

Various types of fossils of the dinosaurs.

Chemical fossils. In recent years biochemists have joined forces with geochemists and palaeontologists to track down a new variety of fossils, called **chemical fossils**. These are organic molecules that have survived relatively unmodified since they were formed by ancient organisms.

Pseudofossils. Many objects of inorganic origin closely resemble the forms of organic origin and are found in the sedimentary rocks. They are referred to as **pseudofossils**.

Significance of Fossils

The study of fossils has great significance. It helps in understanding the prehistoric forms, process of organic evolution and in reconstructing palaeographic (It deals with study of ancient geography during different geological eras and epochs) maps. Fossils are extensively used as indicators for prehistoric climate (salinity, sunlight and depth of water and availability of oxygen, etc.) and as stratigraphic indicators as well. Recently, fossils are used commercially to detect petroleum reserves, coal reserves, gas reserves and reserves of various metal ores. The fossils which indicate the age of the rock in which they are found, are called **index fossils.**

THE GEOLOGICAL TIME TABLE

Geologists, as a result of their studies of the strata of sedimentary rocks in the different regions of the world, have classified geologic history into six **eras.** The oldest era with fossils is the **Archeozoic** (era of primitive life) and this is followed in turn by the **Proterozoic** (era of early life), the **Paleozoic** (era of ancient life), the **Mesozoic** (era of medieval life), and the **Cenozoic** (era of modern life). The Paleozoic, Mesozoic, and Cenozoic eras are divided into **periods** and the periods of the Cenozoic into **epochs**. There is evidence that between the different eras there were widespread geologic disturbances, called **revolutions** which raised or lowered vast regions of the earth's surface and created or eliminated shallow island seas. These revolutions produced great changes in the distribution of sea and land organisms and wiped out many of the previous forms of life. For instance, the Paleozoic era ended with the revolution that raised the Appalachian mountain and it is believed, killed all but 3 per cent of the forms of life existing then. The **Rocky mountain revolution** (which raised the Andes, Alps and Himalay as well as the Rockies) abolished most reptiles of the Mesozoic. The following table of geologic time-table shows the eras and some of their subdivisions, the approximate duration of each era, some of the important geological features, and the characteristic animals and plants.

Table 3-1. **Geological time-table.**

Millions of years ago[+]			Geological feature	Plants	Invertebrates	Vertebrates
Eras	Periods	Epochs				
Cenozoic (70)	Quaternary (2)	Recent (0.011)	End of last ice age; climate warmer.	Decline of woody plants; rise of herbaceous ones.	–	Age of man
		Pleistocene (2)	Repeated glaciation; four ice ages.	Great extinction of species.	–	Extinction of great mammals; first human social life.
	Tertiary (68)	Pliocene (10) Miocene (25) Oligocene (35) Paleocene (70)	Climate warm in the beginning but gradually cooled, formation of Alps and Himalayas.	Development and spread of modern flowering plants. Rise of grasses and of herbs.	Arthropods and molluscs most abundant. Appearance of modern invertebrates type.	Archaic mammals declined after Eocence. Modern mammals evolved in the latter epochs. Rise of anthropoids.
Rocky Mountain Revolution (Little destruction of fossils)						
Mesozoic (230)	Cretaceous (135)		Great swamps in early part, Rocky Mountain and Andes formed.	Rapid development of angiosperms (first monocotyledons), gymnosperms declined.	Extinction of ammonites, spread of insects.	Extinction of dinosaurs and toothed birds; spread of birds: rise of primitive mammals.
	Jurassic (180)		Continents fairly high, shallow seas over some of Europe and Western U.S.	Increase of dicotyledons; conifers and cycads dominant.	Maximum of ammonites. Insects abundant, including social insects.	Dominance of dinosaurs, first toothed birds; early mammals.
	Triassic (230)		Climate warm; great desert areas.	Spread of cycads and conifers; seeds ferns disappear.	Limulus present; Marine invertebrates decline in numbers.	First dinosaurs. Mammal-like reptiles.
Appalachian Revolution (Some loss of fossils)						
Paleozoic (600)	Permian (280)		Appalachians and Urals formed. Glaciation and aridity.	First cycads and conifers; decline of lycopods and horse tails.	Last of trilobites; expansion of ammonites; modern insects arose.	Expansion of reptiles; mammal-like reptiles arose.
	Pennsyl-		Mountain build-	Great forests of seeds ferns	Insect common;	First reptiles;

Eras	Periods	Epochs	Geological feature	Plants	Invertebrates	Vertebrates
	vanian (320)		ing. Great coal swamps.	and gymnosperms.	first insects fossils.	spread of ancient amphibians.
	Mississippian (345)		Warm humid climate. Shallow island seas.	Lycopsids, horsetails, and seed ferns dominant. First coal deposits.	Culmination of crinoids (echinoderms)	Spread of sharks. Rise of amphibians.
	Devonian (405)		Emergence of land; some arid regions and glaciation.	First forests; lands plants well established; first gymnosperms.	Brachiopods flourishing; decline of trilobites.	First amphibians. Rise of fishes–lungfishes, shark abundant.
	Silurian (425)		Extensive continental seas; lowlands increasingly arid as land rose.	First definite evidence of land plants, algae dominant.	Corals, brachiopods, eurypterids. Marine arachnids dominant; first (wingless) insects.	Rise of ostracoderms (primitive fishes).
	Ordovician (500)		Great submergence of land; warm climates even in Arctic.	Marine algae abundant.	Trilobite abundant; diversified molluscs.	First vertebrates armored fishes.
	Cambrian (600)		Lands low, climate mild; earliest rocks with abundant fossils.	Algae, especially marine forms.	Trilobites, brachiopods dominant; all phyla represented.	
Second Great Revolution (Considerable loss of fossils)						
Proterozoic (1200)			Great sedimentation; volcanic activity later; extensive erosion, repeated glaciations.	Few fossils (sponges, radiolarian protozoans, worm burrows and algae). Thallophyta evolved. Fossils of blue green algae.	Most invertebrate phyla probably evolved.	–
First Great Revolution (Considerable loss of fossils)						
Archeozoic (3500)			Great volcanic activities; some sedimentary deposition; extensive erosion; rocks mostly igneous or metamorphosed.	Indirect evidence of life from graphite and limestone, but no recognizable fossils except bacteria, (microfossils).	–	–
Azoic (4500)			Origin of earth. Igneous rocks.	Organic material found in rock, origin of life.		

+ Numbers given in parentheses indicate approximate time since beginning of era, period or epoch.

Conclusions Drawn from Fossil Record

The most important features of the fossil record which is tabulated in Table 3-1 can be summarized as follows :

(i) There existed a multitude of diversified animal groups, similar to the animal groups that exist today.

(ii) All fossils did not appear at a time, but appeared during different great spans of times.

1. PALAEOZOIC

This piece of rock is from the Silurian period in the Palaeozoic era. This era ended 248 million years ago, when more than 90 percent of species became extinct. The latest theory is that the amount of oxygen in the air fell sharply, suffocating most large, active animals.

2. MESOZOIC

Triassic rock, containing ammonites, belongs to the Mesozoic era, the age of dinosaurs. Both dinosaurs and ammonites became extinct at the end of this era, 65 million years ago, when some 25 per cent of species died out.

3. CENOZOIC

This rock, bearing fossil fish, comes from the Tertiary period, part of the Cenozoic era. During the Cenozoic, which is still in progress, mammals and birds have taken over the many vacant slots left by the disappearance of the dinosaurs. After mass extinctions, some surviving species evolve into new forms that repopulate the Earth.

(iii) The most primitive forms of life are found in the oldest rocks.

(iv) Moving up through the various strata, from older to more recent formations, there is a succession of higher and more complex forms of life (geologic succession).

(v) In many instances, a group arose in one period or era and remained scarce, but in the next period or era it became dominant after undergoing adaptive radiation.

(vi) There have been many extinction of large groups, but after the establishment of all major phyla, some species of each phylum have persisted down to the present.

(vii) None of the past forms of life is exactly like any of those now living.

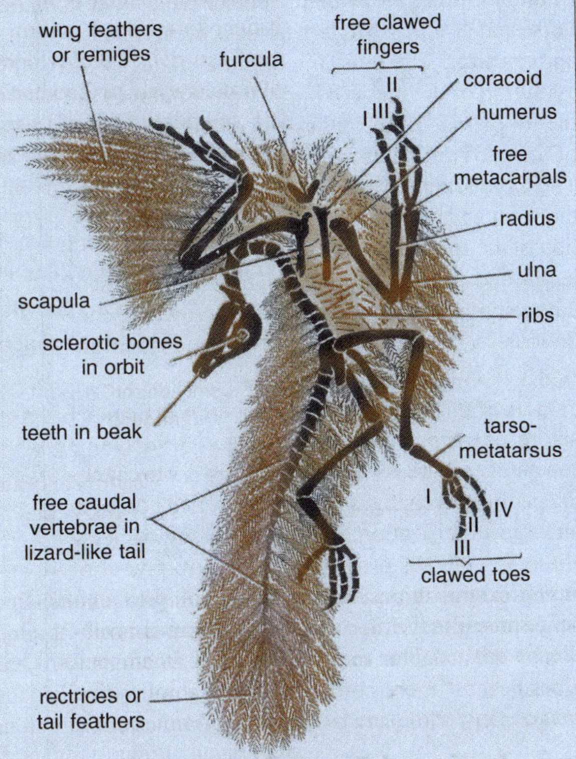

Fig. 3.2. Fossil of *Archaeopteryx*.

All these significant aspects of fossil record substantiate Darwin's theory of evolution by showing the increasing complexity of organismal structures with time. Besides many other interesting features, the fossil record helps in drawing correct conclusions about the possible origin of certain modern vertebrates (birds and horse) as follows :

1. Transitional forms. The critics of palaeontological evidence for the idea of evolution, sometimes maintain that the fossil record lacks transitional forms between one major group and another. In fact fossil record does contain some fossil transitional link forms. One such link is *Archaeopteryx,* transitional fossil between birds and reptiles. *Archaeopteryx* was found in the rocks of the Jurassic Period and it was about the size of a crow and in certain respects was quite like a reptile. Like birds it had wings with feathers, but they were small in size. Its tail, quite unlike that of modern birds, was long with a row of feathers on each side. Further, its jaws were equipped with conical teeth and three of the digits of each forelimb persisted, armed with claws. It is probable that the forelimbs were used for climbing as well as for flying. However, skeleton framework of *Archaeopteryx* resembled more with the reptiles than the birds. Thus, *Archaeopteryx* provides a connecting link between the reptiles and birds, suggesting that birds were evolved from animals like reptiles.

2. Forms showing serial evolution. Fossil records, sometimes, allow us to construct the phylogeny of an animal. Such phylogeny or serial evolution of an animal such as horse, elephant, camel, etc., provides a strong evidence in favour of idea of biological evolution.

Phylogeny of horse. The history of horse dates back to about 60 million years in the Eocene period, involving about 20 genera. The horse phylogeny starts with *Eohippus* (*Hyracotherium*) a member of Equidae. The *Eohippus* lived in the Eocene Epoch. It was about the size of a fox terrier dog, but with a longer head. Its legs were short, with four toes on each front foot and three on each hind foot. The third digit was somewhat longer than the others. All of the toes were placed on the ground and used in walking. *Eohippus* was a forest dweller and browser, feeding on soft vegetation. The lines of descent perhaps passed from *Eohippus* through *Mesohippus*, *Miohippus, Parahippus, Merychippus,* and

Pliohippus to *Equus,* which represents the modern horse and its allies. Modern horse is plains dweller and a grazer. The most important changes in the evolution of the modern horse from *Eohippus* are the following–1. The enlargement and elongation of the third digit, with a loss of the other digits; 2. The elongation of distal parts of the limbs so that the wrist and the ankle are high up; 3. The elongation of the fore part of the skull; 4. The increase of length and mobility of neck. 5. The development of the premolars and molars into high-crowned, continuously growing grinders, and 6. A general increase in body size.

The phylogeny of horses is not a freak of nature, as opponents of evolution idea think, but a constantly progressive adaptation history to changing geological changes which compelled horses to undergo modification. Thus, phylogeny of horse and other animals provides evidence in favour of biological evolution. Further, paleontological evidence provides the most direct indication of evolution by showing that through the years organismal types became more complex, it only implies that the simpler forms gave rise to more complex ones.

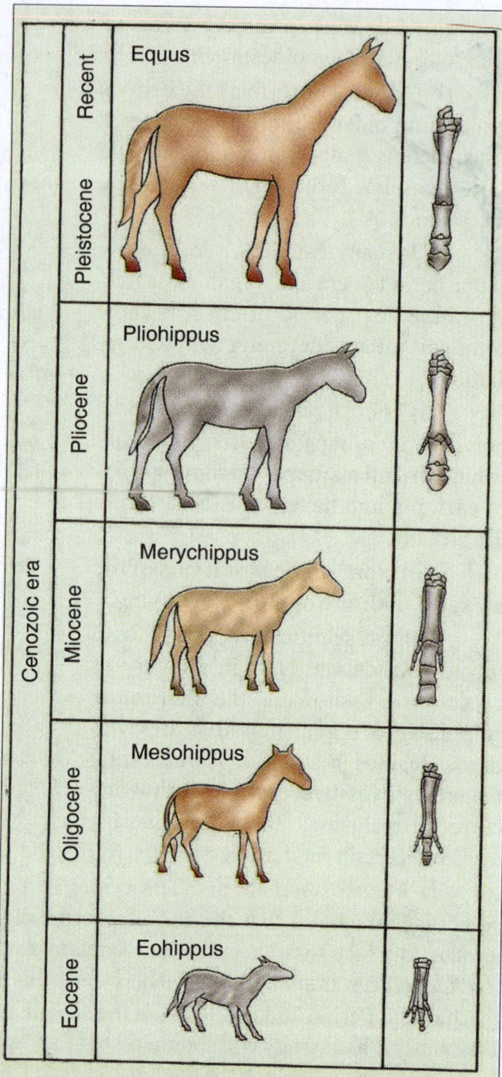

Fig. 3.3. Some stages in evolution of horse : 1. *Hyracotherium* (*Eohippus*), 2. *Mesohippus*. 3. *Merychippus*, 4. *Pliohippus*, 5. *Equus*.

Imperfection of Fossil Record

It is true that fossils provide direct and strong evidence of occurrence of organic evolution, yet, such a valuable evidence is generally marred by the incompleteness and discontinuity of the fossil record. Thus, about 1.5 million species of living organisms are known today, yet only 130,000 fossil species have been described. Many major groups are not represented as fossils at all. For this **Darwin** said that *fossil records are like a book whose many pages are missing or torn and whose all letters are not legible.* The following factors have been attributed for this discontinuity or imperfection of the fossil record :

1. Problems in excavation of fossils. Since fossils are present in deeply situated rocks, so it often becomes difficult to dig them out safely and in intact form. Further, since sedimentary rocks, in which fossils are formed, are mainly developed under sea and major portion of earth is occupied by sea, therefore, it is not possible to dig out these rocks under the sea. Moreover, it is also not possible to successfully excavate the high mountains, ice covered areas and very thick jungles, where there are great chances of finding the fossils. Because of these drawbacks a large number of fossils could not be excavated.

2. Problems of fossilisation. Another important factor responsible for discontinuity of fossil record is that of difficulties in the process of fossilisation, since majority of animals do not get proper environment for the fossilisation due to either of the following reasons :

(i) Animals are completely destroyed by predators, oxidizing agents or other factors.

(ii) The hard parts are either not present in most animals (*e.g.*, invertebrates) or get destroyed by factors such as acidity, etc. Due to this reason a large number of invertebrates such as protozoans, worms, helminths and arthropods could not be fossilised.

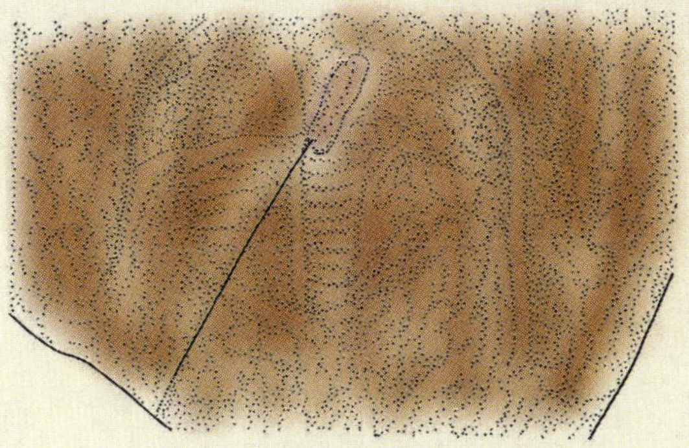

Fig. 3.4. Fossil of trilobite.

(iii) Animals living on land normally get minimum chances of being fossilised, because majority of them do not get any chance to find their way to the sea.

(iv) The occurrences of entirely unfavourable conditions for the fossilisation during various geological periods have contributed to the complete absence of fossils from certain geological periods.

(v) Population size also affects the fossil record. Thus, species with large populations are more likely to be preserved in the fossil record than population with low densities.

(vi) Even different amounts of calcium deposited in the skeleton modify the likelihood of preservation. Some experts suggest that the sudden appearance of **trilobites** in the fossil deposits is not because the creatures suddenly evolved but simply because they started using calcium to their shells.

A Trilobite (Silurian period).

3. Natural destruction of fossils. Lastly fossils of a variety of organisms are being destroyed due to the following natural causes :

(i) Enormous amount of heat and pressure results in alteration or complete destruction of the fossils present in deeper rocks.

(ii) Natural upheavals result in the destruction of fossils.

(iii) Changes in course of rivers, landslides, water erosion and floods result in the destruction of rocks and earth crusts containing fossils.

(iv) Flattening of fossils occurs due to pressure as more and more layers of sediment are piled on top of the bed containing the fossil.

REVISION QUESTIONS

1. Define fossils. How are they formed ? Comment on their significance.
2. Write about the significance of fossils. Describe their importance in organic evolution.
3. Write an essay on "fossils and fossilisation".
4. What do you understand by imperfection of geological records ?
5. How can you determine the age of fossils ?

CHAPTER

4

Indirect Evidences of Evolution

In the last chapter, we have studied direct evidences of organic evolution. In this chapter, we will discuss the following indirect evidences :

EVIDENCES FROM CLASSIFICATION (TAXONOMY)

Taxonomy is the science of classification of organisms. It began long before the doctrine of evolution was accepted. Indeed the founders of scientific taxonomy, **Ray** and **Linnaeus**, were firm believers in the fixity, the unchangingness, of species. **Linnaeus**' classification was based on the principle of **like with like**. Since the development of the theory of descent with modifications by **Lamarck** (1809) and **Darwin** (1859), the organisms were classified on the basis of their inter-relationships. Present-day taxonomists are concerned with the naming and describing of species primarily as a means of discovering or clarifying evolutionary relationships. This is based upon the assumption that the degree of resemblance in homologous structures is a measure of the degree of relationship. The characteristics of living things are such that they can be fitted into a hierarchical scheme of categories, each more inclusive

(a) (b) (c) (d)

Analogous structures.
Wings of (a) insects, (b) birds, (c) seals and (d) penguins.

Species is a population of closely similar individuals which form the basic unit of taxonomy.

than the previous one – species, genera, families, orders, classes and phyla. Thus, very closely related animals or plants are kept in the same **species** which shows that they descended from a common ancestor. The species which are much alike, are included in one **genus** which is distinct from other genera. In the same manner, all the like genera are included in one family which shows again a sign of affinity. Likewise, similar families form **orders**, and orders are grouped into **classes**, in which diverse members share only very fundamental characteristics. Lastly, all the related classes are included in a phylum. This system of classification of organisms can be best interpreted as proof of evolutionary relationship. If the kinds of animals and plants were not related by evolutionary descent, their characteristics would be present in a confused, random pattern and no such hierarchy of forms could be established.

Further, the scheme of classification is itself a proof of descent from a common ancestor as shown by the phylogenetic tree. The forms which are associated together in species, genera, classes or phyla are supposed to have descended from a single common ancestor. For example, the vertebrates like fish, frog, lizard, pigeon, rabbit and man look very different from each other, but they all resemble in having dorsal hollow central nervous system, a notochord and gill-slits. This similarity in the outwardly different vertebrate animals indicates their common ancestry.

The basic unit of taxonomy is the **species**, a population of closely similar individuals, which are alike in their morphologic, embryologic, and physiologic characters, which in nature breed only with each other and which have a common ancestry. It is difficult to give a definition of species that is universally applicable. The definition must be modified slightly to include species whose life cycles includes two or more quite different morphological forms (many coelenterates, parasitic worms, larval and adult insects and amphibians). A population that is spread over a wide territory may show local or regional differences which may be called **subspecies**. Many examples are known in which a species is broken up into a chain of subspecies, each of which differs slightly from its neighbours but interbreeds with them. The subspecies at the two ends of the chain, however, may be so different that they cannot interbreed. Such a series of geographically distributed subspecies is called a **Rassenkreis (German, racecircle)**.

Lastly, the classification of living organisms into well-defined groups is possible because most of the intermediate forms have become extinct. If representatives of every type of animal and plant that have ever lived were still living today, there would be many series of intergrading forms and the division of these into neat taxonomic categories would be difficult. Certain taxonomists have compared the present-day species to the terminal twigs of a tree whose main branches and trunk have disappeared. The main problems of modern taxonomists are to reconstruct the missing branches and to put each twig on the proper branch.

EVIDENCES FROM COMPARATIVE ANATOMY

A comparative anatomical study of any particular organ system in the diverse members of a given phylum reveals a basic similarity of form which is varied to some extent from one class to another. The skeletal, muscular, circulatory, and excretory systems of the vertebrates provide especially clear illustrations of this principle, but this is generally true of all systems in all phyla. Such studies in comparative anatomy provide many evidences of biological evolution by showing that (i) anatomical similarities become more and more complex progressively as one proceeds from a lower animal to a higher animal; and (ii) all the diverse animals inherited the anatomical similarity from a common ancestor.

Examples of anatomical similarities could be multiplied largely, but we restrict ourselves to the few following examples :

1. Connecting Link

Animals standing between two groups of animals are called **connecting links**, *i.e.*, *Archaeopteryx* and *Archaeornis* ; these extinct animals exhibit both reptilian and avian characters, suggesting evolutionary origin of birds from the reptiles. Among the existing mammals, Monotremata (duck-billed platypus and the spiny ant-eater) occupy an intermediate position between reptiles and typical mammals. For like the mammals they possess hair and suckle their youngs. Reptilian features of monotremes are: They lay large heavily yolked eggs, have a cloaca into which urinogenital ducts and alimentary canal discharge; the shoulder girdle is more like reptiles (presence of interclavicle which is not found in other mammals).

Duck-billed platypus is a connecting link between reptiles and mammals.

Thus, connecting links indicate the path along which the more highly organized groups have progressed during their evolution from more lowly organized ancestral groups.

2. Homology

Homology is the similarity between various organs of different organisms and it is based on common embryonic origin or common ancestry. **Homologous organs** are those which have the common origin and are built on the same fundamental pattern. But they perform different functions and have different appearance. These organs are variously modified in adaptation to different functions. The theory of evolution suggests that hereditary characters become gradually modified and these modifications make the organism better suited for the changed conditions of life. Homology is found in every organ system from pisces to mammals. For example, a whale's or seal's front flipper, a bat's wing, a cat's paw, a horse's front leg, a human hand and even a bird's wing, though superficially dissimilar and adapted for quite different functions (*viz.*, grasping, running, flying, swimming, etc.), nevertheless, are homologous structures. Each consists of almost the same number of bones, muscles, nerves, and blood vessels arranged in the same pattern, and their mode of development is very similar. The existence of such homologous organs implies a common evolutionary origin of birds and mammals from some ancient reptilian ancestor.

All insect legs are composed of five parts that always occur in the same order. Starting at their attachment on the thorax, these parts are : coxa, trochanter, femur, tibia, and tarsus. These parts are single segments except the tarsus, which varies from one to five segments. There have been many adaptive changes in insect legs during the evolution of the group. Many insects have a simple type of walking legs, such as the first and second legs of the grasshopper. The hindleg of the grasshopper, however, is modified for jumping. The forelimbs of a mole cricket are modified for digging; the legs of a hog louse are adapted for clinging to hairs; the legs of a diving beetle are constructed for swimming; and the claws, pulvilli, and hairs of the tarsi of the common housefly make it possible for the fly to walk upside down.

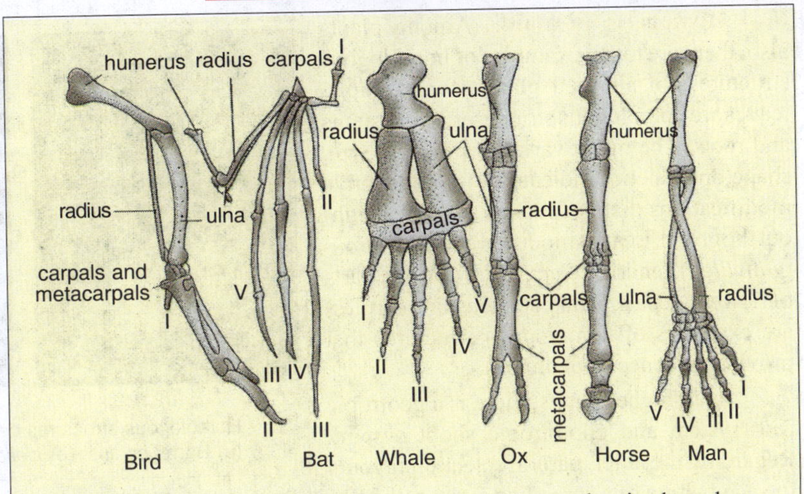

Fig 4.1. Comparison of forelimbs of some vertebrates showing homology.

The brain of vertebrates presents another good example of homology. The brains of vertebrates ranging from fishes to mammals, are constructed of similar parts : olfactory lobes, cerebral hemispheres, optic lobes, cerebellum, medulla and so on. As one moves higher through the series some lobes become more prominent than others. This happens due to adaptations and evolutionary change. Thus, the cerebral hemispheres are much smaller than the optic lobes in fishes; in mammals they become more prominent, hiding the traces of optic lobes beneath them. This is due to the difference of intelligence between the fish and mammals that the cerebral hemispheres become more developed in the latter.

Categories of homology. These are three categories of homology : (i) **Phylogenetic homology** is that existing between different species, *e.g.,* pentadactyl limbs of air-breathing vertebrates. (ii) **Sexual homology** exists in between sexes of the same species, *e.g.,* testes of man and ovaries of a woman ; penis of man and clitoris of woman. (iii) **Serial homology** is that existing between organs of the same individuals occupying different levels of the body, *e.g.,* arm and leg of man and appendages of crustacea. In a typical crustacean, there is one pair of appendages borne by each segment of the body and all have a single structural plan, consisting of a basal two segmented portion the **protopodite**, bearing two lateral outgrowths, the **exopodite** and **endopodite**. The appendages of various body segments perform different functions and accordingly they show modifications in the basic structural plan. This similarity is called **serial homology**.

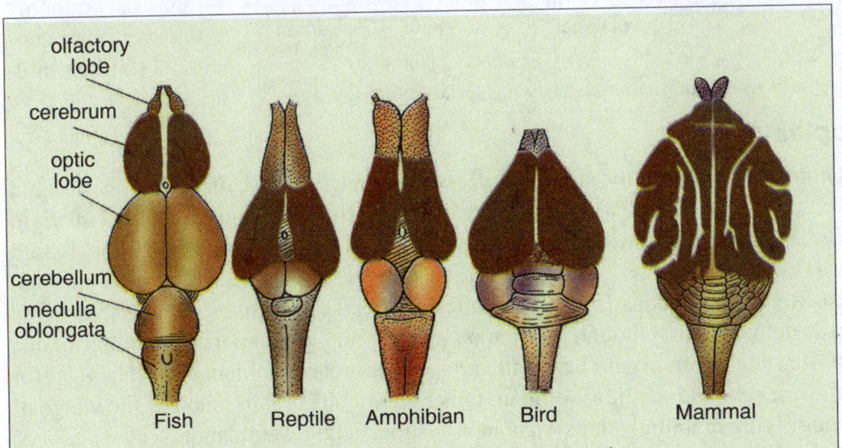

Fig 4.2. Brains of vertebrates showing homology.

Homology in plants. Among plants also, there are many examples of homologous structures. In all seed plants, for example, leaves are homologous structures, as are stems and roots. The many variations in their size, shape, and function indicate the many adaptive modifications that have taken place throughout the group. Leaves modified as thorns (*Bougainvillea*), tendrils (*Cucur-bita*) and the parts of flowers (sepals, petals, stamens, and carples) are examples of homologous structures that provide evidence of evolution.

(a) (b)
Homologous structures derived from leaves.
(a) Barrel cactus; (b) *Heliconia rostrata*.

In all higher plants, there is a geotropic root system and phototropic shoot system. Regarding the latter, natural selection favours the acquisition of a spreading form, rather than a compact form. This is due to positive advantage of trapping the maximum light energy for photosynthetic use. Likewise, the similarities that exist in many groups of families of plants in regard to floral parts and symmetry are related to common descent through organic evolution. Thus, in monocots the floral parts are arranged in odd number of 3 or its multiples (*e.g.,* sepals, petals, stamens and lobed pistil). In dicots, the parts of the flowers are commonly in the odd number of 5; very rarely in 3-s and less commonly in the even number of 2 and 4 (or their multiples).

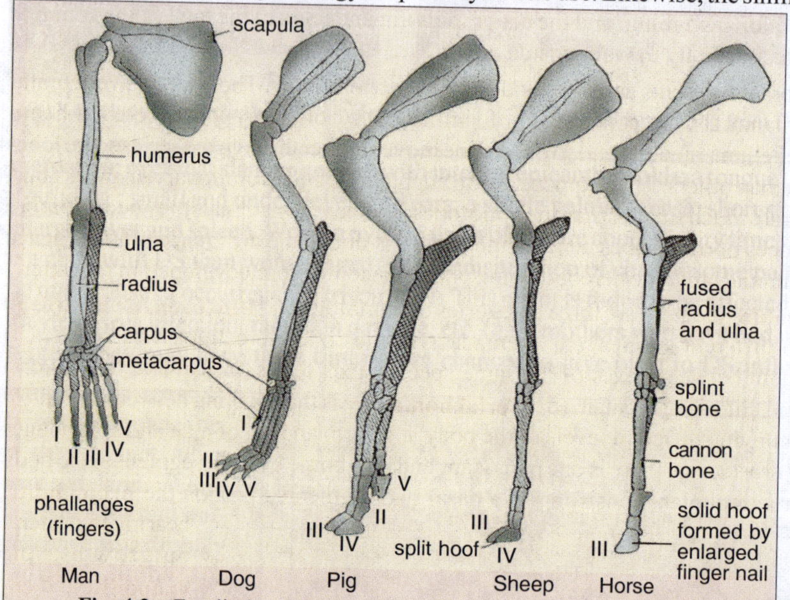

Fig. 4.3. Forelimbs of certain mammals showing serial homology.

Analogy (Homoplasy)

In the study of comparative anatomy, structures are often found that have the same function and are superficially alike, such as the wing of a bird and wing of a butterfly, yet are quite different in origin and in structural design. Such structures which are non-homologous but with similar functions are said to be **analogous organs** or **structures**. They have arisen in the evolutionary process through adaptations of quite different organisms to similar modes of life. The **analogy**, thus, refers to the relationship between structures which though differ anatomically but have superficial similarity due to similar functions. Often analogous organs have little gross structural resemblance to each other; for example, the gills of *Palaemon* and the lungs of man. Other analogous organs, such as the wings of insects, aves (birds) and flying mammals (bats) may have a superficial resemblance.

Their common function, flight, has only one structural solution the evolution of some sort of an airfoil. Analogous organs with a structural resemblance are said be **homoplastic**.

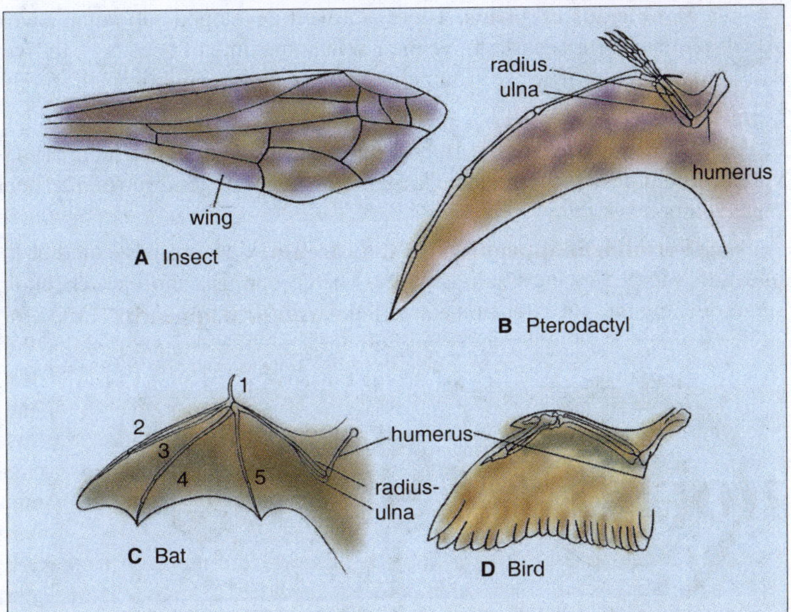

Fig. 4.4. Analogy and homology. The bones of the wings of the pterodactyle, bird and bat are homologous to each other. The insect's wing is similar to vertebrate wings in function, but it has only a superficial structural resemblance to them. It is, therefore, analogous and homoplastic.

There are many instances where different organisms that are not very closely related have developed similar adaptations for life in similar habitats. This phenomenon is often referred to as **convergent evolution**. The development of same fusiform type of body for life in the water by *Palaemon* (arthropod), shark (fish), *Ichthyosaurus* (reptile), and a dolphin (mammal) illustrates the convergent evolution well. The development of a limbless, cylindrical snake-like body by *Ophiosaurus* (lizard), *Uraeotyphlus* (caecilian amphibian) and *Typhlops* (blind snake) for life in the burrows, is another interesting example of convergent evolution in animals. Further, the development of a cactus-like body by members of several different families of plants is another good example of convergent evolution.

Among plants, a stem may be modified to appear like leaves and may perform the functions of leaves as in *Ruscus*. This is a case of analogy.

Vestigial Organs

Another source of evidence stemming from the studies of comparative anatomy is the presence of organs in various animals that have no apparent function and are useless and often small or lacking some essential part; however, in related organisms, these organs remain full-sized, complete and functional. Such organs are called **vestigial** or **rudimentary organs**. The vestigial organs are believed to be remnants of organs which were complete and functional in their ancestors. There are more than 100 vestigial organs in human body; some common examples of them are the following :

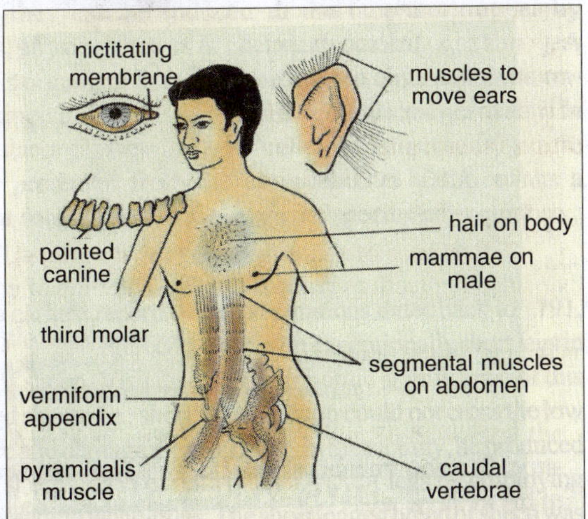

Fig. 4.5. Some vestigial organs of man.

1. Muscles of the external ears. They are well developed and functional in quadrupeds, but in man they are useless structures and cannot move the external ears (pinna).

2. Panniculus carnosus. There is a well developed sub-cutaneous skeletal muscle layer by which some quadrupeds can move their skin at the time of need (*i.e.,* to fly off the irritating insects). This muscle layer is of no use in humans. However, it is functional in the forehead by which it can move the eye-brows.

3. Coccyx. Coccyx is the much reduced string of caudal vertebrae and is present at the end of the vertebral column. It is a remnant of the tail present in early embryo of man but it is shed off much before the adulthood is attained.

4. Vermiform appendix and the caecum. Caecum is located at the point where the large intestine is joined by the small intestine. The caecum is a short pouch, ending blindly except for the small opening into its extension called the **vermiform appendix** (Fig. 4.6). The appendix of man is thought to be a remnant of the large caecum which is the storage organ for cellulose digestion in herbivorous mammals. In most primates except man, it is of large size. In human beings it has no function and is rudimentary. It indicates that ancestors of man (early primates) had a much coarse diet due to which caecum was functional.

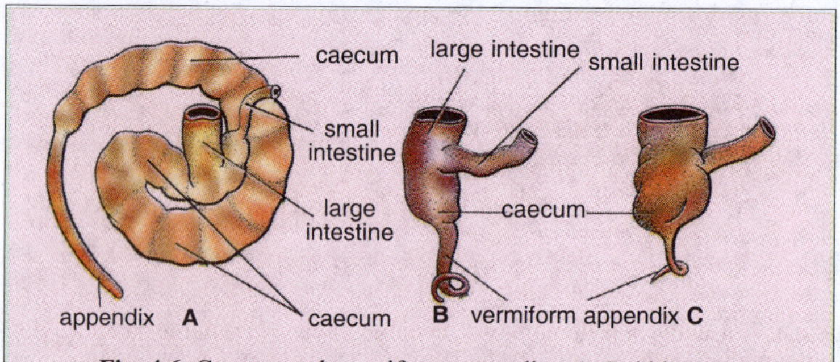

Fig. 4.6. Caecum and vermiform appendix : A–Rabbit; B–Man embryo; C–Man.

5. Presence of vestigial mammary glands in male human beings. This might suggest that one time males also suckled the young.

6. Presence of vestigial nictitating membranes. In the inner angle of each of our eyes is present a little fold of flesh called **semilunar fold** (plica semilunaris). This corresponds to a structure which in many lower animals is movable third eyelid, called **nictitating membrane**, lying under the outer eyelids and sweeping across the eye from the inner angle outwards.

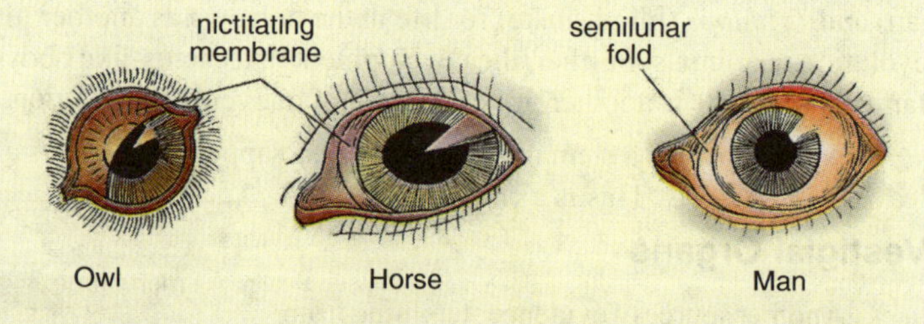

Fig. 4.7. Eyes of owl horse and man showing nictitating membrane or semilunar fold.

7. Presence of wisdom teeth (third molars). These are posteriormost teeth which are variable than the other teeth as regards the size and time of eruption. These are vestigial structures, since they are last to erupt or even fail to erupt. In other primates, third molars are as sound and as fully developed as the rest of the dentition.

All these vestigial organs have provided evidences that man has descended from simple primates.

Certain other animals also possess the vestigial organs which are as follows :

1. Tiny rudiments of the hindlimbs in *Python* and vestiges of pelvic girdle (ileum) and leg bones (femur) in the whale.
2. Wings of *Apteryx* (kiwi) of New Zealand (flightless birds).
3. Reduced third metacarpal (splint bone) in the feet of horse.

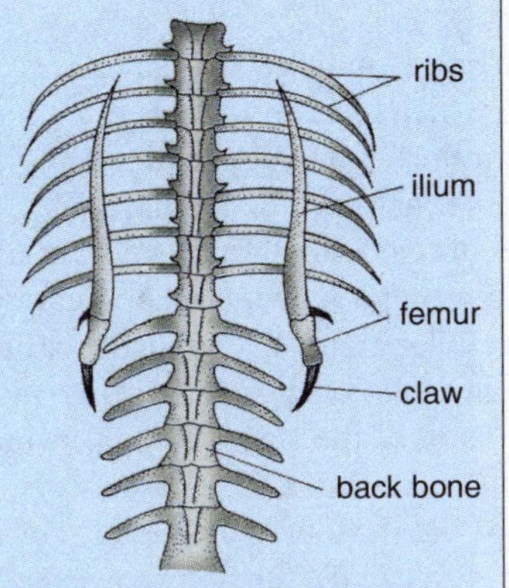

Fig. 4.8. Vestigial hindlimbs of Python (snake).

All the aforesaid examples provide conclusive evidence in support of the theory of organic evolution as follows:

1. Organisms having homologous and vestigial organs have developed from a common ancestor.

2. Organisms with analogous organs had different ancestors.

3. The present-day similarity and dissimilarity of organs is a sort of adaptive radiation to cope with new needs of the environment.

Darwin pointed out that both frigate birds and upland geese have webbed feet, yet neither goes into water. He explained their feet as a leftover from their past, both being descended from waterbirds.

EVIDENCES FROM COMPARATIVE EMBRYOLOGY

One of the most fascinating sources of evidence for the validity of Darwin's theory of evolution has come from studies of comparative embryology. The suspicions of a common ancestry for very diverse sets of organisms from comparative anatomical studies have been supported by embryological data.

As Darwinian evolution suggested that more complex organisms would have achieved their state by elaborating on the existing developmental patterns of more primitive forms, so one would expect to find that certain relatively simple organisms and more complex ones have many initial developmental steps in common. In fact, the more developmental steps two species have in common, the more closely related they are to a common ancestral form.

Comparative embryological studies have revealed that there was one developmental pattern that could be viewed as having undergone a series of branchings. All multicellular animals start their development as a single **zygote**, and through a series of mitotic divisions, increase in cell number until **a blastula** is formed. The developing embryo elaborates upon the blastula stage by forming two fundamental germ layers, **ectoderm** and **endoderm**, during the course of **gastrulation**. After the differentiation of the ectoderm and endoderm in the gastrula, the third germ layer, **mesoderm**, is formed. There are three distinct patterns by which developing embryo of different species produce a mesodermal layer. For instance, in annelids, molluscs and certain other invertebrates, the mesoderm develops from special cells which are differentiated early in cleavage. These cells migrate to the interior and come to lie between the ectoderm and endoderm. They then multiply to form two longitudinal cords of cells which develop into sheets of mesoderm between the ectoderm and endoderm. In primitive chordates (*i.e., Amphioxus,* etc.) the mesoderm arises as a series of bilateral pouches from the endoderm. These lose their connections with gut and fuse one with another to form a connected layer. The mesoderm in amphibia, reptilia, birds and mammals is formed from the cells which remain associated with either ectoderm or endoderm or both and which get disassociated from them to migrate and come to lie in between both germinal layers in the form of mesoderm.

Despite some differences in developmental patterns, comparative embryological studies overwhelmingly support the concept of Darwinian evolution. The species that follow the same pathway for several steps after the gastrula produce adult forms that are more similar than two species whose pathways diverge after that stage. The different developmental patterns are thought to be indications of different lines of evolution.

Another feature of embryological development that serves to link all multicellular animals is the embryological source of organ systems. Regardless of the way in which the developing animals are programmed for further differentiation of mesoderm they exhibit similar adult structures derived from the two primary germ layers. The outer covering of all multicellular animals, be it skin, scales, or gelatinous material, is derived from ectoderm. The universal features of ectodermally and endodermally derived tissues also indicate the presence of a common ancestral type early in evolutionary history.

Besides these basic embryological similarities in all multicellular animals, there exist certain embryological aspects which may compel one to believe in Darwinian evolution. For instance, it is most commonly observed that the embryos of higher animals repeat many of the stages through which embryos of higher animals have passed. This has been referred to as **recapitulation**. This concept, as originally used by **von Baer** (1792–1876) indicated that some of the developmental stages of an organism are similar to some of the developmental stages of its ancestors. Unfortunately, however, **E. Haeckel** in 1866, modified the concept into a **"biogenetic law"** which stated that *"ontogeny recapitulates phylogeny"*, *i.e.,* in its development, the individual passes through stages like the adult stages of its ancestors. The Haeckelian idea is wrong in the eyes of modern embryologists yet, it has stimulated research in embryology and focussed attention on the general resemblance between

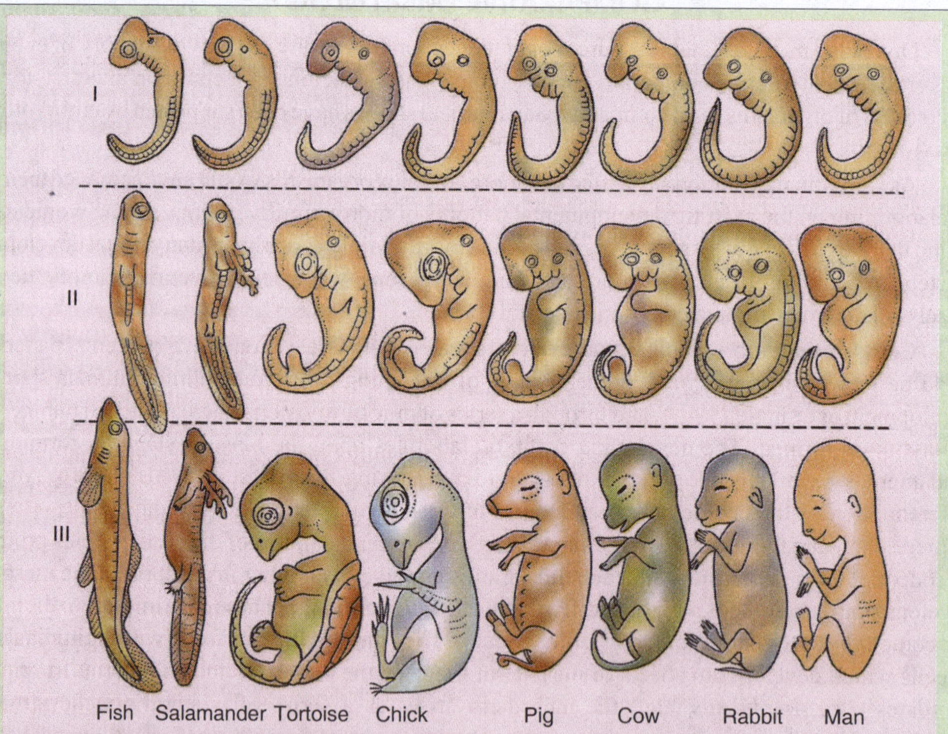

Fig. 4.9. Comparison of three embryonic stages from fish to man.

embryonic development and the evolutionary processes. According to biogenetic law of Haeckel, the fertilized egg can be compared to the hypothetical single-celled flagellate ancestor of all animals, and the blastula can be compared to a colonial protozoan or to some hypothetic blastula-like animal which has been postulated to be the ancestor of all Metazoa. Haeckel believed that the ancestor of coelenterates and all the higher animals were a gastrula-like organism with two layers of cells and a central cavity connected by a blastopore to the outside. Modern scientists prefer von Baer's recapitulation concept than biogenetic law of Haeckel.

There are numerous examples of recapitulation phenomenon during embryonic development, but one of the best is afforded by a comparison of different vertebrate embryos at comparable stages in development (see Fig. 4.9). During the early embryonic stage in the series, all the embryos look very much alike. All have similar **pharyngeal arches** and **pharyngeal clefts**. At somewhat later stage in the series, the **limb-bud** primordia of fore-and hindlimbs are forming in all embryos in a similar way and all of them have embryonic tails. The embryos of the lizard, chick, opossum, monkey, and man have strong resemblances, yet those of the fish and salamander are beginning to assume recognizable forms. At this stage, gills have formed from the tissues lining the gill-clefts of both the fish and salamander. Later, each embryo has developed features that indicate fairly clearly its definitive nature.

Further, in the development of any mammalian embryo, the heart is a four-chambered in series structure, as it is in fish embryo; then it develops partitions of the auricle (atria) similar to those of amphibian embryos, followed by ventricular division that is incomplete for a period, as it is in the embryos of reptiles. A similar example of recapitulation is found in mammalian embryos in the formation of pronephric, mesonephric and metanephric kidneys. The existence of a common trochophore larval stage in the development of many molluscs and many annelids is also typical of this process.

Plants provide a few examples of recapitulation. One good example is found in the *Acacia* tree. This tree has well-developed compound leaves, yet the seedling has simple leaves like those found in all stages of development of its ancestors. Further, it is generally believed that mosses and ferns are more highly evolved than algae. Protonema of mosses which is a green, filamentous structure that originates from an asexual spore, resembles certain green algae. This provides a clue to their evolutionary relationship. Both bryophytes and pteridophytes have ciliated sperms and require water for fertilization. Gymnosperms do not need water for fertilization, but *Cycads* and *Gingko,* the primitive gymnosperms, have ciliated sperms like the pteridophytes. This suggests that gymnosperms have descended from pteridophyte-like ancestors.

Genetic Basis of Recapitulation

Our increasing understanding of physiological genetics provides us with an explanation of the phenomenon of recapitulation. All chordates have in common a certain number of genes which regulate the processes of early development. As mammalian ancestors evolved from fish through amphibian and reptilian stages, they accumulated mutations for new characteristics but kept some of the original "fish" genes, which still control early development. Later in development the genes which the human or any other mammal shares with amphibians influences the course of development so that the embryo resembles a frog embryo. Subsequently, some of the genes which have in common with reptiles come into control. Only after this stage, most of the peculiarly mammalian genes exert their influence, and these are followed by the action of genes human has in common with other primates. The anthropoid apes, which have the most immediate ancestors in common with man, have the most genes in common with him and their development is identical with him (human) except for some fine details. In general, during development the general characteristics that distinguish phyla and classes appear before the special characteristics that distinguish genera and, finally, species. Within each phylum, the higher forms pass through a sequence of developmental stages which are similar to those of lower forms but achieve a different final form by adding changes at the end of the original sequence and by altering certain of the earlier embryonic stages they share with the lower forms (**Villee** *et al*, 1973).

EVIDENCES FROM COMPARATIVE PHYSIOLOGY AND BIOCHEMISTRY

Basically, evolution is biochemical phenomenon and it is, therefore, natural that physiology and biochemistry have given some of the following most important and dependable evidences to support the idea of evolution :

1. Protoplasm Chemistry

Biochemical analysis of the living matter in the protoplasm which is considered as "the physical basis of life", suggests that protoplasm from a variety of sources (*i.e.,* bacteria, blue green algae, plants and animals) has the same biochemical constitution. It mainly consists of substances like proteins, lipids, carbohydrates, water, etc. This would suggest that during evolution the most fundamental property of living things has remained intact, while variations in certain essential respects produced the variability according to the needs of differential forms.

2. Chromosome Chemistry

Like protoplasm, another remarkable similarity at the biochemical level is found in the chemistry of chromosomes. The chromosomes of all living organisms basically consist of nucleic acids (DNA and RNA) and proteins (histones and protamines). The molecules of these chemical substances remain arranged in all chromosomes in an almost identical fashion. Such a uniformity in the composition of chromosome again suggests a common origin of most living beings.

3. Enzyme Similarities

A large number of animals and plants contain identical enzymes. Several enzymes found in the digestive tract are common in a variety of animals. For example, trypsin and amylase are found from sponges to mammals. A number of enzymes used in photosynthesis are common in a variety of green plants. Such common enzymes and consequently a common mechanism of process of photosynthesis suggest a common ancestry of green plants.

4. Hormonal Similarities

Like enzymes, hormonal similarities are also found in all vertebrates. For example, thyroid hormone is commonly found in all vertebrates and this hormone from one class of organism can be substituted for that in another class of organisms. For example, in frogs deficiency of thyroid hormone can be corrected by feeding them on mammalian thyroid tissues. Likewise, another commonly occurring hormone of vertebrates is melanophore expanding hormone. It is concerned with the pigmentation of the skin to expand, thus, rendering the skin colour dark. This hormone is found in amphibians and mammals. In the latter it is a vestigial hormone, but if it is grafted into the amphibian skin, the skin pigmentation expands. The presence of these hormones in vertebrates is understandable only on the basis of descent from an ancestor to whom these hormones were useful.

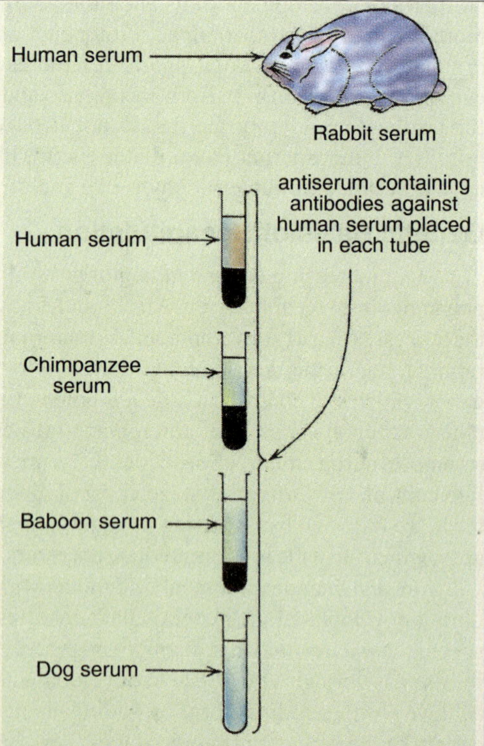

Fig. 4.10. Principle of precipitin test applied to investigation of animal relationship.

5. Comparative Serology

When a foreign protein is inoculated into the blood of an animal, the latter produces a complex protein compound against that foreign protein inoculated. These compounds are familiarly known as **antibodies** and the foreign inoculated protein is known as **antigen**. When a reaction occurs between antibody and antigen, a soft white precipitate will be formed. The strength of precipitate depends upon the concentration of antigen. The precipitate is the **precipitin** and the test is **precipitation test**. One of the remarkable features of this test is that the antibodies formed against one antigen, can also react with antigens of other source, provided the latter is chemically similar to the first antigen. Antibodies containing serum is known as antiserum. Antiserum of antigen of one animal can be tested with antigens of other animals in order to show their relationships. The test can be interpreted that if precipitate results with more diluted antigen of one animal against the test animal, then the former is more closely related to the latter ; if precipitate results with less diluted antigen of that animal, then it is distantly related to the test animal. Such precipitin tests have been conducted to resolve the disputed relationships of organisms in recent years. Of scores of examples, here we give two illustrations.

Till recently, it is believed that whales have relationship with fishes. It is because, almost all of their anatomy are so strongly modified to aquatic fish-like life. Only few anatomical clues to show their relationships to other mammals, remained. However, comparative serology of whales with other mammalian groups indicates that their serum proteins are most like those of the even-toed hoofed (Order Artiodactyla) mammals. This might suggest that whales sprang from primitive artiodactyl stock.

The same serological tests when performed in slightly modified form among the members of primates – man, an anthropoid ape, an old world-monkey, a new world-monkey and a lemur – the amount of precipitate of serum proteins would decrease in descending order, that is, the anthropoid ape is more closely related to man than other organisms. Even among anthropoid primates, tests done according to **Ouchterlony technique (Goodman**, 1962, 1967) reveal that the serum proteins of chimpanzee are more alike to man's serum proteins than the serum proteins of asiatic apes, gorilla and baboon.

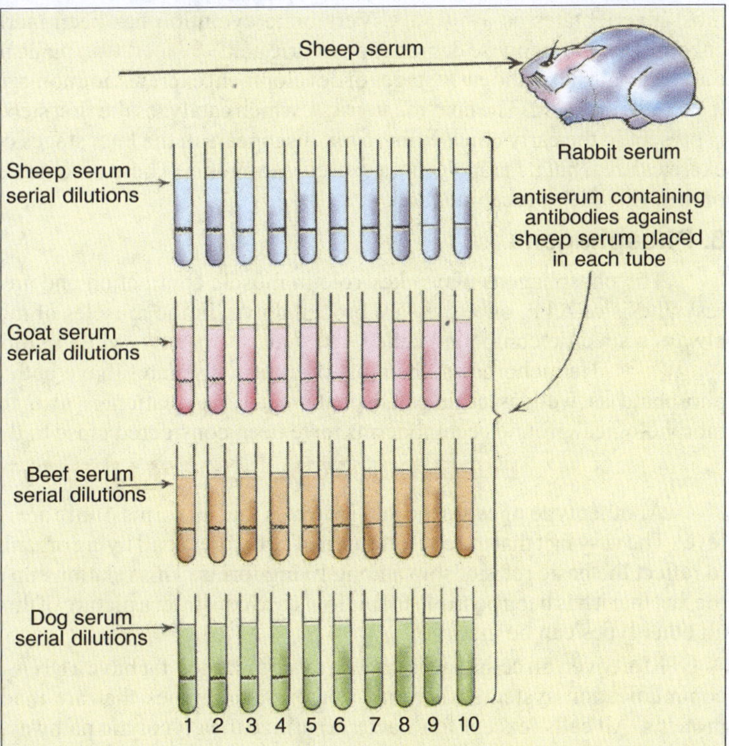

Fig. 4.11. Principle of precipitin test employing serial dilutions of antigen and the interfacial reaction.

Similar comparative serological tests reveal the fact that cats, dogs, and bears are closely related. Cows, sheep, goats, deer and antelopes constitute another closely related groups in terms of "blood relation." Serological tests also suggest that there is a closer relationship among the modern birds than among the mammals, for all of the several hundred species of birds tested give strong and immediate reactions with serum containing antibodies for chicken serum. From other tests it was concluded that birds are more closely related to the crocodile line of reptiles than to the snake-lizard line, which corroborates the pal-aeontological evidence. Similar tests of the sera of crustaceans, insects and molluscs have shown that forms regarded as being closely related from morphologic and palaeontologic evidence also show similarities in their serum proteins.

6. Amino-Acid Sequence Analyses

Molecular biological investigations of the sequence of amino acids in the α and β **chains of haemoglobins** from different species have revealed great similarities, of course, and specific differences, the pattern of which demonstrates the order in which the underlying mutations, the changes in nucleotide base pairs, must have occurred in evolution. The evolutionary relationships inferred from these studies agree completely with those based on morphologic studies. Analyses of the amino acid sequence in the protein portion of the **cytochrome enzyme** provide further concurring evolutionary relationships. Further the pattern and rates of reactions of lactate dehydrogenase and certain other enzymes with the normal pyridine nucleotide coenzyme (NAD) and with analogues of NAD can be used to demonstrate evolutionary relationships.

7. Excretory Product Analyses

An analysis of the urinary wastes of different species provides the evidence of evolutionary relationship. The kind of nitrogenous excretory waste depends upon the particular kinds of enzymes present, and the enzymes are determined by genes, which have been selected in the course of evolution. The waste products of the metabolism of purines, adenine and guanine are excreted by man and other primates as **uric acid,** by other mammals as **allantoin**, by amphibians and most fishes as **urea,** and by most invertebrates as **ammonia**. Vertebrate evolution has been marked by the successive loss of enzymes for the stepwise degradation of uric acid. **J. Needham** made the interesting observation that the chick embryo in the early stages of development excretes ammonia, later it excretes urea and finally it excretes uric acid. The enzyme **uricase**, which catalyzes the first step in the degradation of uric acid, is present in the early chick embryo but disappears in the later stages of development. The adult frog excretes urea, but its tadpole larva excretes ammonia. These biochemical examples are the repetition of the principle of recapitulation.

8. Phosphagens

The phosphagens play a key role in muscle contraction and are the sources of energy for the resynthesis of ATP, once they are broken down. In the muscles of most vertebrates, phosphagen is always a specific compound called **creatine phosphate**, while in most invertebrates it is **arginine phosphate**. Hemichordates, the most primitive chordates, have both the phosphagens, the creatine phosphate as well as argine phosphate. Such a situation is also found in echinoderms and on morphological grounds echinoderms have been considered close to the ancestor of chordates.

EVIDENCES FROM COMPARATIVE CYTOLOGY

Another type of evidence that indicates that all forms of life are related comes from the cellular level. The very fact that the cell is the unit of structure for all living organisms (except viruses) is thought to reflect the basic relationship among living forms. This relationship is even further emphasized by the fact that it has been possible for biologists to construct a picture of the "generalized" cell from which all other types can be inferred.

Moreover, all cells that have been examined thus far have a DNA-RNA- protein information and communication system. All forms contain membranes that are made up of double-layered lipo-proteins. All cells (except few bacteria) utilize the glycolytic pathway. Most bacterial forms, and all uni - and-multicellular organisms, have a Krebs cycle and an electron transport system. All are based on ATP as an energy donor. Certainly these factors provide an overwhelming demonstration of the interrelatedness of biological forms.

EVIDENCES FROM GENETICS

Genetics, the science of heredity, deals with the variability of plants and animals. Hereditary variations provide the raw material of evolution. There are mainly two sources of hereditary variations namely **recombination** and **mutation**. While recombinations after hybridization yield new combinations, mutations will create new genetic material which never existed earlier.

For the past several thousand years man has been selecting and breeding (*i.e.,* hybridizing) animals and plants for his own uses, and a great many varieties, adapted for different purposes, have been established. These results of artificial selection provide striking models of what may be accomplished by natural selection. All of our breeds of dogs have descended from one or perhaps a very few species of wild dog or wolf, yet they vary so much in colour, size and body proportions that if they occurred in the wild they would undoubtedly be considered separate species. They are all interfertile and are known to come from common ancestors, so they are regarded as varieties of a single species. A comparable range of varieties has been produced by artificial selection in cats, chickens, sheep, cattle and horses. Plant breeders have established by selective breeding a tremendous variety of plants. From the cliff cabbage, which still grows wild in Europe, have come cultivated cabbage, cauliflower, Kohlrabi, Brussels, sprouts, brocoli and kale.

Mutation seen in zinnia petals (a) and *Drosophila* (b).

Further, cytogeneticists have been able to trace the ancestry of certain modern plants by a combination of cytologic techniques in which the morphology of the chromosomes is compared and by breeding techniques which compare the kinds of genes and their order in particular chromosomes in a series of plants. In this way, the present cultivated tobacco plant, *Nicotiana tabacum*, was shown to have arisen from two species of wild tobacco, and corn was traced to teosinte, a grass-like plant which grows wild in the Andes and Mexico. Moreover, the cytologic details of the structure of the giant chromosomes of the salivary glands of fruit flies have been of prime importance in unrevealing the evolutionary history of many species of *Drosophila*.

Mutations have also played a very significant role in evolution. Different kinds of mutations, namely gross mutation, due to variation in chromosome number (polyploidy and aneuploidy), and point mutations (see chapter of mutation) introduce different kinds of variations in plants and animals and consequently result in speciation. Plant breeders have employed induced polyploidy methods in producing numerous economically important varieties of plants. Geneticists have produced many new strains of microorganisms, plants and animals (*Drosophila*) by artificially inducing point mutations in them.

Thus, the artificial selection methods due to recombination or polyploidy and induced methods of mutations have suggested the fundamental processes which may be involved in organic evolution.

EVIDENCES FROM BIOGEOGRAPHICAL RELATIONS

Biogeography deals with the manner in which plants and animals are distributed over our planet. On the basis of similarities in the existing fauna, found in different regions of the earth, the following six biogeographical regions have been distinguished :

(a) Nearctic. North America down to the Mexican Plateau.

(b) Palearctic. Asia North of the Himalayas, Europe and Africa, North of Sahara Desert.

(c) Neotropical. Central South of America.

(d) Oriental. Asia, south of the Himalayas.

(e) Ethiopian. Africa, south of the Himalayas.

(f) Australian. Australia and the associated islands.

Any one who is familiar with biogeography of South American and African regions, immediately be convinced by the fact that similar habitats are populated with similar animals. These two are extensive tropical regions and crossed by the equator. They both have similar habitats in that–have lowland jungles; extensive river systems and mountain regions. Both regions extend southward to temperate zones. Yet, surprisingly, they have dissimilarities in fauna more greatly than similarities. The characteristic fauna of Africa includes Lions, Elephants, Rhinoceroses, Hippopotami, many kinds of Antelopes, Giraffes, Zebras, Hyenas, Lemurs, Baboons, Monkeys with narrow noses and non-prehensile tailed Chimpanzees and Gorillas.

Alpacas.

Agouti.

In South America, Monkeys occur with broad noses, Tapirs, Odd-toed hoofed mammals, rodents like Capybara, Agouti, Chinchilla and Paca, Mountain lions like panthers – Ocelots Jaguars of cat family, well known Uamas, Gaunacos, Vicunas, and Alpacas, Armadillos, many opossums, Giant Anteaters, Raccoons, spectacled Bears and Sloths and many others.

Thus, if both Africa and S. America have similar varying habitats, then how could it have happened to them to possess different fauna instead of similar fauna ? We cannot simply satisfy with such an answer that suitability of habitat is important. Because we do find similarities in fauna of both regions. They both have such widely ranging animals, bats, rats, sq9 etc. The other alternative answer might be the accessibility of these two regions to the animals found in them. Let us examine this point here more briefly.

Vicunas.

Simpson has established a relationship "faunal stratification" between the separateness of animals and their period of length of appearance in S. America. Based on it, he concludes that armadillos and sloths are the oldest inhabitants of S. America since early coenozoic times. They have not appeared in other part of the world. Monkeys and field mice have formed part of S. American fauna during mid and late coenozoic times, while the rest of the fauna have independently evolved. These facts have been ascertained by fossil record.

The flightless Kiwi is found only in New Zealand, as were the giant flightless birds called Moas. Moas are now extinct. Moas and Kiwis evolved from flying birds that colonise the islands of New Zealand millions of years ago. Early in the Tertiary period, when mammals were still all fairly small, Moas did infect dominite. Moas died out quite recently because of overhunting.

The record reveals that during long periods S. American forms have no contact with those on other continents. Occasionally animals might have reached the continent by island hopping across the intervening sea. This type of dispersal is best seen in fauna on continental islands. For the most part the isthums of Panama that links S.

America to N. America, was submerged and thereby the two continents are isolated. When once geographical isolation takes place, local fauna along with invaders might evolve into distinct species in many different times and have become unique to S. America.

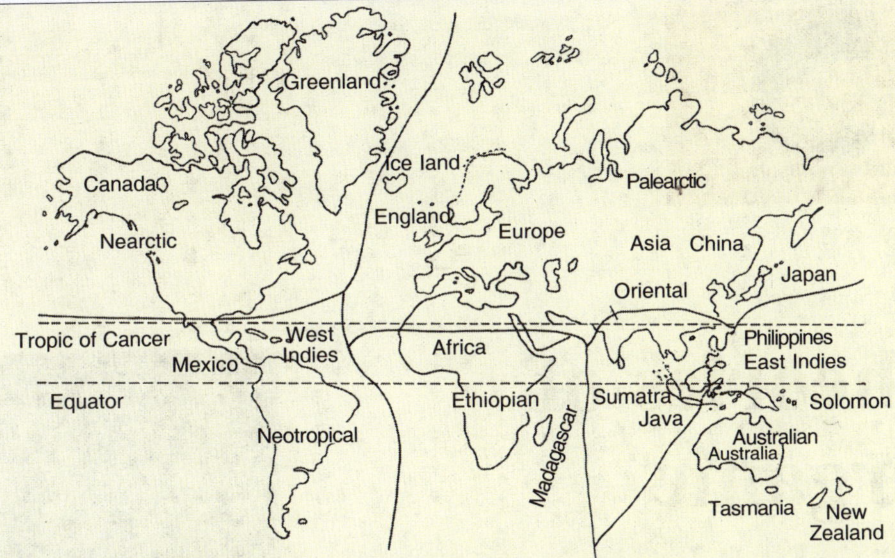

Fig. 4.12. Different zoogeographical regions of world.

Thus, this is the only possibility for the same habitats in different parts of earth have been populated by different animals and plants. This fact suggests that the animals today we see in different biogeographical regions, have descended from their predecessors with different structures, and have migrated from their place of origin to their new areas, but have failed to return to their place of origin, because it is separated geologically.

Continental Islands

These islands are located on continental shelves of continents. These islands sometimes are connected to mainland when ocean water recedes or the level of island rises.

The fauna and flora of continental islands characteristically resembled those of continents to which they were formerly joined. One interesting feature is the presence of amphibians and mammals on continental islands. Island hopping in these animals is less likely and, hence, they might have arrived on these islands through land connections.

Thus, these evidences have established that biological evolution is a fact but not a dogma. Consequently, the biologist's conviction that, through a series of changes resulting from natural selective processes, life came to the state known today, is of such magnitude that the entire science of biology has been oriented according to the evolutionary doctrine. Organisms have been reclassified according to proposed evolutionary relationships. Geneticists interpret their results as possible mechanisms of evolution or sources of variation. So powerful is this idea today that only a text organization based on evolutionary doctrine would truly represent the science of biology.

REVISION QUESTIONS

1. Discuss the indirect evidences of organic evolution.
2. Define the term organic evolution. Describe various evidences from comparative morphology (anatomy) or from physiology, biochemistry, serology and embryology or from vestigial organs and connecting links.
3. Explain the following terms : homology, analogy and connecting links.

CHAPTER

5

Theories of Organic Evolution

(Lamarckism, Darwinism, Modern Synthetic Theory, Germplasm Theory and Mutation Theory)

Competition between animals for food, space, light and moisture is an aspect of struggle for existence.

Two explanations have been offered for the origin of the different forms of life— special creation and evolution. Many evidences have been presented in previous chapters to support the concept of evolution, and almost without exception modern biologists are convinced of evolution. Uptill now, the following convincing theories have been forwarded to explain the mechanism of evolution. The first of these theories, which is now only of historical significance, was that of **Jean Baptiste de Lamarck** (1744–1829).

1. THEORY OF INHERITANCE OF ACQUIRED CHARACTERS (LAMARCKISM)

The theory of inheritance of acquired characters states that *modifications which the organism acquires in adaptation to the environments which it meets during its lifetime are automatically handed down to its descendants, and so become part of heredity.* This theory was propounded by a renowned French naturalist, **Lamarck,** shortly after the occurrence of French

Revolution. Lamarck spent the early part of his life as a botanist. Then at the age of 50 he turned his attention to zoology, particularly to the study of invertebrates (The terms "invertebrate" and "biology" have been coined by Lamarck). As a result of his systematic studies he became convinced that species were not constant but rather were derived from pre-existing species. This idea was in total conflict with the view of the period — that of fixity of species. As a result Lamarck's views were challenged by most of the biologists of that time, particularly by **Georges Cuvier**. In 1809, Lamarck published **Philosophie Zoologique,** which included his theory explaining the changes that occur in the formation of new types. Although, his views on evolutionary mechanism are outmoded now, he still occupies a very important place in the history of evolutionary thought. He was the first evolutionist to conclude that evolution is a general fact covering all forms of life. His evolutionary ideas can be discussed in brief as follows :

1. Internal forces of life tend to increase the size of the organism. New structures appear because of an "inner want" of the organism, *i.e.,* the internal forces of life tend to increase continuously the size of an organism and its component parts.

2. Direct environmental effect over living organisms. The organs of an animal became modified in appropriate fashion in direct response to a changing environment.

3. Use and disuse. The various organs became greatly improved through use or reduced to vestiges through disuse.

4. Inheritance of acquired characteristics. Such bodily modifications, in some manner, could be transferred and impressed on the germ cells to affect future generation. Thus, **inheritance** was viewed by **Lamarck** simply as the direct transmission of those superficial bodily changes that arose within the lifetime of the individual owing to use or disuse (**Volpe**, 1985).

Thus, Lamarck believed that organic changes seen in animals were resulted by the influence of environment on the gradual changes of species due to their tendency to become more and more perfect. According to him, when an animal's environment changes, its needs change, and this leads to special demands on certain organs. Organs used more extensively would enlarge and become more efficient. Conversely, an organ or organs, no longer used, would degenerate and atrophy. He postulated that such changed characteristics (acquired traits) would be transmitted to the offspring.

Examples of Lamarckism

1. The deer-like ancestor of giraffe lived in places (Africa) where the ground was almost invariably parched and without grass. Obliged to browse upon trees, it was continually forced to stretch upwards. This habit maintained over long periods of time by every individual of the race had resulted in the forelimbs becoming longer than the hind ones, and neck so elongated that a modern giraffe can raise his head to a height of eighteen feet without taking his forelimbs off the ground.

Fig. 5.1. Evolution of giraffe according to Lamarck.

2. Ducks and other aquatic birds invaded waters from land in search of enough food, because food was scarce on land

and these birds did not had power to fly. In water, the duck would stretch its toes apart to give more push during swimming. This new characteristic would be inherited, and the subsequent generation of duck would upon stretching their toes form a more defined web. Each generation would do the same until the webbed foot seen on ducks today was fully formed. This would then be passed on from generation to generation, essentially unchanged once the perfected state was attained.

3. Flat fishes (deep sea fishes) present at the bottom of sea where there is no sunlight, led an inactive life, lying on one side of the body. The eye of that side (lying towards bottom) migrated towards upper side and, thus, both eyes are on one side of the body.

4. The whales lost their hindlimbs as the consequence of the inherited effect of disuse.

5. The wading birds (*e.g.*, Jacana) developed its long legs through generations of sustained stretching to keep the body above the water level.

6. Snakes have elongated body accompanied by loss of limbs. The continuous creeping through holes and crevices made limbs continuously useless for locomotion with the result that limbs become completely lost in snakes.

7. Eyes are reduced in moles since they live underground. In cave animals also, eyes might become functionless and might even disappear.

Purple heron. According to Lamarck's theory, by trying to keep its belly out of the water a wading bird (purple heron) "acquires the habit of stretching and elongating its legs".

8. For plants Lamarck accepted the theory of his compatriot **St. Helaire**, who developed the notion that plant form is shaped by the combined effects of the environment.

Trofim Lysenko (1898-1976), Soviet geneticist, favoured Lamarckian ideas, since they agreed with communist ideology and he rose to power under Stalin. Lysenko banished Mendelian geneticists and dominated Soviet genetics for many years. Eventually, however, the evidence against Lamarckian inheritance was so strong that ideology had to give way to science. Lysenko was discredited and forced to resign.

There are other effects such as **vestigial organs** in living animals due to disuse. **Claws** in Carnivora, **sensitive skin** and **tactile points** on the ventral side of the body and **callosities** of palm in hard workers, exemplifying Lamarckian theory.

Significance of Lamarckism. Lamarckian theory was simple and it had some appeal, as it provided a **way** in which changes in organisms could come about. It was the first completely comprehensive mechanistic theory that was offered. Furthermore, it was the theory that lent itself to predictions and, therefore, to testing. Thus, Lamarckian theory enjoyed popular acceptance for near about 70 years, because it was exemplified by many common examples. Most persons know that exercise results in larger muscles.

Critical Analysis of Lamarck's Propositions

Lamarck defended his evolutionary theory vigorously until his death. For it, he suffered both social and scientific ostracism, but he had the courage of his convictions. However, he was criticized for the following reasons by the contemporary scientists during his lifetime and afterwards.

1. The first proposition of Lamarck suggests the tendency to increase in size. While the evolutionary trend in a certain groups of organisms may be associated with the increase in size, there are many cases, where evolution proceeded not only without any increase in size but rather through a reduction in size. Many plants contradict this Lamarckian principle by showing such a reduction in

size during their evolution. Many ferns and conifers which became extinct were gigantic trees and the more highly evolved flowering plants are really much smaller in size.

2. The second Lamarckian principle that new organs result from new needs, is quite manifestly false. In the case of plants, Lamarck believed that the environment acted directly upon the plant, causing the production of such new characters as might adapt the plant to its environment. In the case of animals, he believed that the environment acted through the nervous systems; in other words, the desire of the animal leads to the formation of new structures. In its crudest form this would mean that the man who mused "*Birds can fly, so why can't I*? should have sprouted wings and taken to the air.

3. The third Lamarckian principle that organs will develop due to use and degenerate due to disuse, may be correct as far as growth of an organ within the lifetime of an individual is concerned. For example, it is a commonly observed fact that if muscles are put to use these would develop. However, this principle is meaningful only when it is studied in relation to the following fourth principle.

4. The fourth and final proposition of Lamarck was that the inheritance of characters acquired during the lifetime of the individual. This principle has been tested by many biologists who have devised many types of experiments for it and have found it entirely incorrect. Certain experiments which have discredited it are the following :

August Weismann (1834-1914).

1. The noted German scientist **August Weismann** was the first person who for the first time made a definite distinction between heritable changes and those which cannot be inherited. In 1890, he performed some experiments to test if characters may disappear due to disuse. This he did by cutting the tails (mutilation) of white mice for more than 20 generations to see if this has any effect on tail length. The measuring of tail length of the offsprings of 20 successive generations, revealed that on average, the tails were not shorter. It means that, acquired character (cut tail) was not inherited.

2. Castle and **Phillips** performed transplantation experiments to show that environment has no effect on heredity. In one of the experiments they transplanted the ovary of a black female guinea pig into the body of white female guinea pig and the recipient female was mated with a white male guinea pig. They found that all the individuals from this pair were black. This shows that the environment does not affect the heredity as has been suggested by Lamarck.

3. Loeb artificially fertilized the eggs of sea urchin by certain chemical stimuli and these parthenogenetically fertilized eggs produced the generations, the members of which possessed no parthenogenesis.

4. Boaring of ears and nostrils in Indian women has been continued as the tradition from centuries but their offsprings do not show any trace of holes in ears and nostrils.

5. The Chinese bound the feet of their women for many generations, yet this has not resulted in any modification of the feet of present-day Chinese women.

6. Jewish boys (and also Muslim boys) have been circumcised for thousands of years, but this has not resulted in a tendency toward the reduction of the prepuce in them.

All the aforesaid cases and experiments clearly showed that Lamarckian theory is not tenable.

Neo-Lamarckism

The evolutionists who support the Lamarckian doctrine of inheritance of acquired characters come under the heading of **neo-Lamarckian**. Among the neo-Lamarckians notable supporters are

Cope (1840–1897), **Giard** (1846–1908), **Packard**, **Spencer** and **McBride** tried to modify Lamarckism in order to make it acceptable. These neo-Lamarckians considered that adaptation is universal. It arises as a result of casual relationship of structure, function and environment. Changed environmental conditions alter habits of organisms, hence, in response to new habits, organisms acquire new structures in place of old structures. Consequently, variations among animals have become distinct. These variations have finally become engrained in the heredity of the race. The kind of argument is, in fact, a modified version of Lamarckian principles, because it has omitted Lamarck's view that of a general perfecting tendency in evolution. It stressed direct action of environment on organic structure. For example, according to neo-Lamarckians, the development of fur on skin by animals as protection against cold weather, is the consequence of changed environment from warmer to colder state only. But if the environment reverts to normal state, the fur would also disappear. To account it, neo-Lamarckism included also the effects of use and disuse. Based on this, neo-Lamarckians have rejected natural selection as the sole mechanism of evolution. On the other hand, they believed in the interplay of structure, function and environment as the whole truth of evolution. However, no evolutionist of today is adherent to neo-Lamarckism.

The following are the evidences of transmission of acquired characters :

1. McDougall conducted learning experiments in rats and from the results obtained he tried to suggest that learning as an acquired character can be inherited. During his experiments, the rats were dropped into a tank of water from which there were two exits, one lighted and one dark, but not always the same one. A rat leaving by the lighted exit received an electric shock, while one leaving by the dark exit received no shock. Thus, the number of trials required for a particular rat to learn always to select the dark exit constituted a measure of the speed of learning. These rats were then bred and their descendants were similarly studied. It appeared that the speed of learning increased from generation to generation and, so, **McDougall** concluded that learning, an acquired trait *par excellence,* is inherited.

Some serious criticisms have been raised against McDougall's experiments. It is suggested that genetic constitution of rats was not controlled. Moreover, in the progeny of control rats not subjected to these learning tests, the learning habit was found to change the same way as in the treatments. So, it was said that some unanalysed changes in the technique may be responsible for the recorded increase in the speed of learning. Lastly, the most strange aspect about McDougall's experiment is, that when the same experiments, when repeated in other laboratories, never gave similar results.

2. F.B. Summer reared the white mice at 20^{0}C to 30^{0}C, because of which they developed longer bodies, tail and feet. When these mice were bred at lower temperatures their offsprings showed their normal proportions of size.

3. Lindsey subjected certain cold-blooded animals, warm-blooded animals and plants to unusual environmental conditions and found that environment affects certain changes in these organisms and these acquired modifications were inherited up to some extent in their offsprings.

4. Guver and **Smith** induced the hereditary changes in the eyes of foetuses of rabbit by simply destroying lens of living female with a needle *in situ.* The antilens serum has been produced in the blood of these animals.

5. Griffith and **Detlefson** reared rats in cages placed on rotating table for several months. Consequently, they became adapted to the rotating condition to such an extent that when rotation was stopped, they showed signs of **nystagmus** (dizziness) and other physiological conditions. This condition was inherited for several generations.

6. Various acquired diseases described by **Brown Sequard** are inherited from generation to generation. For example, **exophthalmia** was caused in parents by injuring in brain the restiform body. This disease was, later on, inherited to several generations. The other diseases such as **haematoma** and **dry gangrene** are produced by injury to the restiform body near the rib of calamus. Later, these were found to be inherited.

7. Kammerer worked on the tailed amphibian *Proteus anguinus* which lives in complete darkness in the water of underground cave. It was blind and colourless. He brought *Proteus* in daylight, due to which it became coloured (brown and black), which passed on to its progeny. Eyes of *Proteus* also developed normally in daylight. This showed that any change in environmental conditions, induced changes in the animals and the acquired characters were inherited.

Further, **Winterbert** (1962) has tried, with some success, to give Lamarckism a chemical basis by involving the capacity of every living being to react to environmental changes and aggressions. According to **Winterbert**, nothing is acquired by the living organism that is not the response of internal factors to an outside influence or ethological change. Living means reacting, never undergoing. Above all, it means not waiting around for a fortunate chance occurrence to save the situation (**Grasse**, 1977).

2. THEORY OF NATURAL SELECTION (DARWINISM)

More than a century ago, in 1859, **Charles Robert Darwin** (1809–1822) gave the biological world the master key that unlocked all previous perplexities regarding the mechanism of organic evolution. His natural selection theory can be compared only with such revolutionary ideas as Newton's law of gravitation and Einstein's theory of relativity (**Volpe**, 1985). The concept of natural selection was explained clearly and convincingly by Darwin in his masterpiece – *The Origin of Species* (The full title of the book was *On the Origin of Species by Means of Natural Selection, or The Preservation of Favoured Races in the Struggle for Life*). This epoch-making book was the fruition of more than twenty years of meticulous accumulation and analysis of facts. The first edition of the book, some 1,250 copies, was sold out on the very day it appeared, November 24, 1859. This book had opened a Pandora's box; it was immediately both acidly attacked and effusively praised.

Charles Robert Darwin was born on February 12, 1809 in Shrewsbury, England. His personality was not the type that one associates with a successful man, let alone a great one. While his family held a distinguished reputation (his grandfather **Erasmus Darwin** was a noted poet scientist, his father **Robert**, was an eminent physician, and his mother was a Wedgwood, the family of pottery fame), Charles was a singularly undistinguished individual whose contemporaries viewed him as just a normal, gentle, cautious, morally upright English lad. After completing his early education at Shrewsbury, Charles Darwin was sent at the age of fifteen to study medicine at the University of Edinburg. His medical interests soon dissipated, and he became something of an academic rover, studying a little science, but finding it too formal for his tastes.

His father then suggested him to become a clergyman, which appealed him. So, he got transferred, after two years, to Christ's college, Cambridge University, to study theology. There too he spent his energy in card-playing and drinking. He was obviously not ready for clerical commitments.

He did manage, however, to find some constructive interest in natural history and in collection of natural objects of all kinds. His friendship with Professor **John Stevens Henslow**, botanist at Cambridge, stimulated an interest in botany, while his exposure to the geologist **Sedgwick** stimulated an interest in geological explorations. Through the recommendation of **Dr. Henslow**, Darwin accepted a most inauspicious position as a naturalist without pay aboard the H.M.S. *Beagle.* The ship was to spend five years in exploration, mostly of South America. Darwin was to provide a collection of material representing the natural composition of the areas visited.

The ship *Beagle* left Plymouth on December 27, 1831 and visited many islands of the Atlantic ocean, some coasts of South America and some islands of south Pacific, of which Galapagos Islands are the most important. During his five years (from 1831 to 1836) on the *Beagle,* Darwin recorded almost everything he observed and sent an enormous amount of material back to England.

Upon his return home in 1836, he spent two years writing a book of his experiences, The *Voyage of the Beagle.* In 1838, he got married and later had two daughters and five sons. From 1838 to 1841,

Darwin acted as the secretary of Geological society, where he came in contact with an eminent geologist, **Charles Lyell**. It was not until 1844 that Darwin developed his idea of natural selection in an essay, but not for publication. He showed the manuscript to **Lyell** who encouraged him to prepare a book. Darwin still took no steps toward publishing his views. Perhaps he was well aware of how Lamarck's earlier theory had been received and wanted all possible supporting evidence before publishing his own new theory. It appears that Darwin might not have prepared his famous volume had not a fellow naturalist in the Dutch East Indies, **Alfred Russel Wallace** (1823 – 1913),

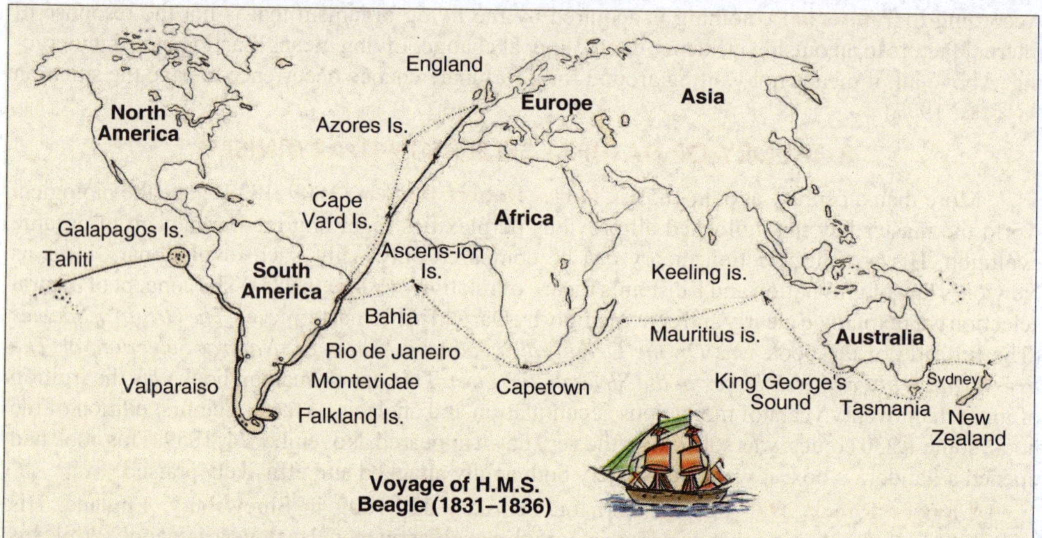

Fig.5.2. Five-years (from 1831 to 1836) world voyage of *H.M.S. Beagle*. Darwin's observations on this voyage convinced him of the reality of evolution (after Volpe, 1985).

independently conceived of the idea of natural selection. **Wallace** had travelled widely in tropical South America and Southeastern Asia, for studying the flora and fauna of these regions. **Wallace** also inspired by reading Malthus's essay, and the idea of natural selection came to him in a flash of insight during a sudden fit of malarial fever (This happened in February, 1858, when he was working in the island of Ternate in Indonesia). In June of 1858, Wallace sent Darwin a short essay *"On the tendency of varieties to depart indefinitely from the original type,"* and asked him if he thought of sufficient interest to present it to the Linnaean Society. We can easily imagine Darwin's amazement upon receiving Wallace's essay. Upon the insistence of the geologist **Charles Lyell** and the botanist **Joseph Hooker**, Darwin prepared an abstract of his conclusions for joint publication with Wallace's essay. Wallace's essay and a portion of Darwin's manuscript, each containing remarkably similar views, were read simultaneously before the Linnaean Society in London on July 1, 1858. The joint reading of the papers stirred little interest.

Darwin then laboured for eight months to compress his voluminous notes into a single book, which he modestly called " only an Abstract". Wallace shares with Darwin the honour of establishing the mechanism by which evolution is brought about, but it was the monumental *The Origin of Species* with its impressive weight of evidence and argument that left its mark on mankind. The fact that this book, written by Darwin alone, exposed the public at large to the theory of natural selection, and that Wallace was still in Indonesia at the time of the controversy and could not champion Darwin's defence by adding his own views, brought the public to associate only Darwin with this theory. Thus the theory is called the **Darwinian**, rather than the **Darwin-Wallace theory of evolution**.

Darwin died on April 19, 1882, when he was 73 years old. During his life, he wrote numerous books, journals, etc. Some of them can be listed in Table (5 - 1).

Table 5-1. **Some publications of Charles Darwin.**

	Title of book	Year of Publication
1.	Journal of researches	1839
2.	The structure and distribution of coral reefs	1844
3.	Geological observations on South America	1846
4.	A monograph on Cirripedia (Barancles, living and fossil, 4 volumes)	1851, 1854
5.	Origin of species	1859
6.	The fertilization of orchids	1862
7.	The variation of plants and animals under domestication	1868
8.	The descent of man	1871
9.	The expression of the emotions in men and animals	1872
10.	Insectivorous plants	1875
11.	The effects of cross and self-fertilization in the vegetable kingdom	1876
12.	Different forms of flowers on plants of the same species	1877
13.	The power of movement in plants	1880
14.	The formation of vegetable mould through the action of worms	1881

Facts that influenced Darwin's Thoughts

During the period in which Darwin massed his evidence and developed his natural selection theory, many things affected his thinking, of which three, in particular, deserve mention here :

1. During his voyage, Darwin took only a few books on board. One of them was the newly published first volume of **Charles Lyell's** *Principles of Geology*, a parting gift from his Cambridge mentor, **John Henslow**. In this book, **Lyell** challenged the prevailing belief that the earth has been created by a divine plan merely 6,000 years ago. On the contrary, the earth's age could be measured in hundreds of millions of years. Lyell asserted that the earth's mountains, valleys, rivers, and coastlines were shaped not by Noah's Flood but by the ordinary action of the rains, the winds, earthquakes, volcanoes, and other natural forces. Darwin was impressed by Lyell's emphasis on the great antiquity of the earth's rocks, and gradually came to conclusion that the characteristics of organisms as well as the face of the earth could change over a vast span of time.

The living and extinct organisms that Darwin observed in the flat plains of the Argentine pampas (Pampas are vast grassy treeless plains in South America) and the Galapagos Islands sowed the seeds of Darwin's views on evolution. From old river beds in the Argentine pampas, he dug up bony remains (fossils) of extinct mammals such as *Toxodon, Macrauchenia, Pyrotherium* and *Thoatherium* (Fig. 5.3). The presence of *Thoatherium* testified that a horse had been among the ancient inhabitants of the continent. It was the Spanish settlers who reintroduced the modern horse, *Equus*, to the continent of South America in the sixteenth century. Darwin wondered that a native horse should have lived and disappeared in South America. This was one of the first indications that *species gradually became modified with time, and that not all species survived through the ages.*

Further, when Darwin collected the remains of giant armadillos and sloths on an Argentine pampa, he revealed the fact that, although they clearly belonged to extinct forms, they were constructed on the same basic plan as the small living armadillos and sloths of the same region. This observation initiated him to think of the fossil sequence of a given animal species through the ages and causes of extinction. Moreover, what Darwin had appreciated was that *living species have ancestors.* On travelling from the north to the south of South America, Darwin observed that one species was replaced by similar, but slightly different, species. In the southern part of Argentina, Darwin caught a rare species of ostrich that was smaller and differently coloured from the more northern common American ostrich, *Rhea americanus.* This rare species of bird was later named *Rhea darwini.* It appeared to Darwin that *species change not only in time but also with geographical distance.*

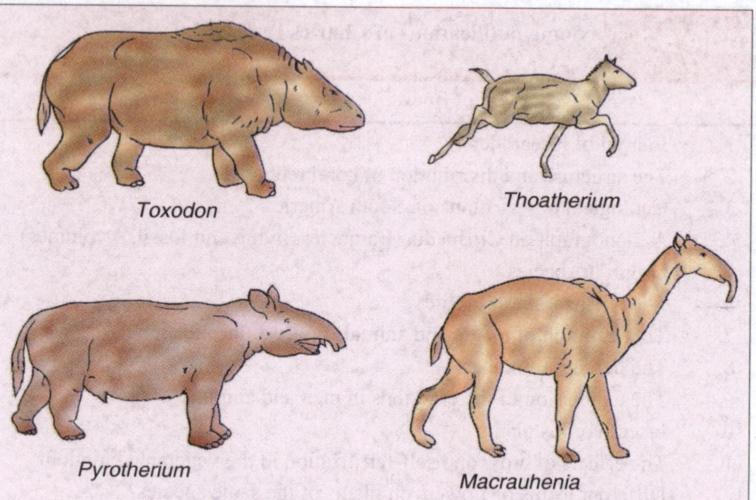

Fig. 5.3. Curious extinct hoofed mammals (ungulates) of South America. Bones of these great mammals were found by Darwin on the flat treeless plains of Argentina (after Volpe, 1985).

In the Galapagos Islands Darwin's scientific curiosity was greatly motivated by the many distinctive forms of life. The Galapagos consists of an isolated cluster of rugged volcanic islands in the eastern Pacific on the equator about 600 miles west of Ecuador. One of the most unusual animals is the giant land-dwelling tortoise, which may weigh as much as 275 kg, grow to 183 cm in length, and attain an age of 200 to 250 years. The Spanish word for tortoise, *galapago,* gives the islands their name. Darwin noticed that the tortoises were clearly different from island to island, although the islands were only a few miles apart. In isolation, Darwin reasoned, each population had evolved its own distinctive features. Yet, all the island tortoises showed basic resemblances not only to each other but also to relatively large tortoises on the adjacent mainland of South America. All this communicated to Darwin that island tortoises shared a common ancestor with the mainland forms. The same was true of a group of small black birds known today as **Darwin's finches**. Darwin observed that the finches were different on the various islands, yet they were obviously closely related to each other. Darwin concluded that finches were derived from an ancestral stock that had emigrated from the mainland to the volcanic islands and had undergone profound changes under the different conditions of the individual islands. Apparently, *a single ancestral group can give rise to several different varieties of species.*

(a) (b)

(a) Galapagos tortoises feed on prickly pear cactuses. (b) On islands with tortoises, a young cactus quickly grows a tall trunk, which lifts the succulent pads beyond the reach of the tortoises.

2. In the late 1830s, Darwin attended the meetings of animal breeders and intently read their publication. Animal breeders were conversant with the variability in their pet animals, and dwelled on the technique of **artificial selection**. Thus, the breeders selected and perpetuated those variant types that interested them or seemed useful to them. The breeders, however, had only vague notions as to the origin, or inheritance, of the variable traits.

Dog diversity illustrates artificial selection.

Darwin acknowledged the unlimited variability in organisms, but was never able to explain satisfactorily how a variant trait was inherited. Although Darwin criticized many aspects of Lamarck's theory, he did not deny that acquired characteristics can be transmitted. He realized that the nature of inheritance was unknown, but he devised a working hypothesis, called **hypothesis of pangenesis** (= origin from all) in 1868 to explain how acquired variations may be transmitted.

Pangenesis hypothesis. Darwin's pangenesis hypothesis assumed that all the organs, and perhaps all the cells, in the body of an animal produced miniatures of themselves. These miniatures, called **gemmules** or **pangenes**, were shed into the blood stream and carried to the sex glands (*viz.,* reproductive organs, the testes and ovaries), where they were assembled to form sex cells— the eggs or sperms. Later, when a fertilized egg undergoes development, according to the concept, the pangenes present are responsible for the particular features of the new individuals. In this manner, environmen-

Fig. 5.4. Darwin's finches.

tal changes would produce modified organs, which in turn would produce modified pangenes that would transmit the change to the next generation.

In year 1875, **Galton** made pangenesis hypothesis untenable by presenting several experimental proofs. He and others at later time made a series of experiments involving blood transfusions and later, transplants of ovaries between black and white varieties of rabbits and chickens. Gametes produced by the transplanted ovaries were consistent with the phenotype of the individual in which the ovary originated and not with the animal currently carrying the ovary. Blood transfusions had no effect on the gametes produced. These experiments readily demonstrated that the pangenesis hypothesis was incorrect.

3. Having explained the origin of variation (although incorrectly), Darwin wondered how artificial selection (a term familiar then only to animal breeders) could be carried in nature. There was no breeder in nature to pick and choose. In 1838, Darwin found the solution in **Thomas Robert Malthus** book *'An Essay on the Principle of Population'* in which Malthus asserted that the reproductive capacity of mankind far exceeds the food supply available to nourish an expanding human population. Humans compete among themselves for the necessities of life.

Thomas Malthus (1766-1834).

It, thus, occurred to Darwin that **competition** existed among all living things. Darwin then envisioned that the **"struggle for existence"** might be the means by which the well-adapted individuals survive and ill-adjusted are eliminated. Darwin was the first to realize that perpetual selection existed in nature in the form of **natural selection**. In natural selection, as contrasted to artificial selection, the animal breeder or horticulturist is replaced by the conditions of the environment that prevent the survival and reproduction of certain individuals. Thus, natural selection is a term serving to inform us that *some individuals leave more offspring than others.* It is not purposeful or guided by a specific aim; it does not seek to attain a specific end.

Darwin-Wallace Theory of Natural Selection

Darwin-Wallace explanation of the way in which evolution occurs may be generalized as follows– *"The change in species by the survival of an organismal type exhibiting a natural variation that gives it an adaptive advantage in an environment, thus, leading to a new environmental equilibrium, is evolution by natural selection."* Thus, natural selection is a continuous process of trial and error on a gigantic scale, for all of living matter is involved. It includes the following elements :

1. The universal occurrence of variation. Variation is the characteristic of every group of animals and plants and there are many ways in which organisms may differ. (Darwin and Wallace did not understand the cause of variation, and assumed it was one of the innate properties of living things. We now know that inherited variations are caused by mutations.)

2. An excessive natural rate of multiplication. Every species, in the absence of environmental checks, tends to increase in a geometrical manner. If a population of a given species doubles in one year and if there are no checks on its increase, it will quadruple the next year, and so on. Such a great reproductive potential of different species may be easily observed in nature. It has been estimated that a common Atlantic coast oyster may shed as many as 80 million eggs in one season. A salmon produces 28,000,000 eggs in a season. A single pair of English sparrows would be the ancestors of over 275 billion individuals in 10 years if they and their descendants could reproduce at their natural rate without any check. Darwin calculated that even a pair of elephants which are about the slowest breeding animals known, could, in the absence of any checks, have 29 million descendants at the end of 800 years.

Thus, more organisms of each kind are born than can possibly obtain food and survive. Since the number of each species remains fairly constant under natural conditions, it must be assumed that most of the offsprings in each generation perish. If all the offsprings of any species remained alive and reproduced they would soon crowd all other species from the earth.

Every species has an excessive rate of multiplication.

3. Struggle for existence. Since more individuals are born than can survive there is an *intraspecific or interspecific* or environmental struggle for survival, a competition for food, mates and space. This contest may be an active kill-or-be-killed struggle, or one less immediately apparent but no less real, such as the struggle of plants or animals to survive drought or cold.

4. The consequent elimination of the unfit and the survival of only those that are satisfactorily adapted. Some of the variations exhibited by living things make it easier for them to survive; others are handicaps which bring about the elimination of their possessors. This idea of "**The survival of the fittest**" is the core of the theory of natural selection.

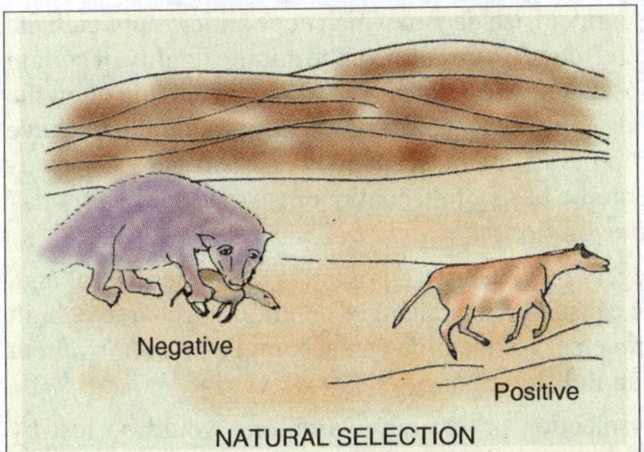

Fig. 5.5. Diagram showing negative and positive natural selection.

5. The inheritance of the mutations or recombinations that make for success in the struggle for existence. The surviving individuals will give rise to the next generation and, in this way, the "successful" variations are transmitted to the succeeding generations. The less fit will tend to be eliminated before they have reproduced.

Successive generations in this way tend to become better adapted to their environment; as the environment changes, further adaptations occur. The operation of natural selection over many generations may produce descendants which are quite different from their ancestors, different enough to be separate species. Furthermore, certain members of a population with one group of variations may become adapted to the environment in one way, while others, with a different set of variations, become adapted in a different way, or become adapted to a different environment. In this way two or more species may arise from a single ancestral stock.

Critical Analysis of Darwinism

With the appearance of Darwin's book in 1859 a veritable storm of controversy broke out. Many people became interested in his presentation; the arguments for and against that developed were often heated. The concept of natural selection was opposed primarily on ethical and religious grounds. But the sheer weight of evidence and the logic of Darwin's presentation finally convinced a majority of the educated people. The concept of evolution affected many aspects of man's thinking and it will continue to do so. However, the idea of "The survival of the fittest" of Darwinism has been erroneously interpreted by many to mean survival only by "Tooth and claw," an interpretation that has led in many cases to rationalizations for the attitude of "Every man for himself" in social and economic affairs. In fact, the process of evolution is not all "Tooth and claw". Plants have evolved too, and there is no bloody competition between them. In all forms, plants and animals alike, a part of the selective process involves the action of inanimate nature – factors such as drought, storm, moisture, temperature, etc. Active *competition* between organisms is a part of the process, but as **Walder Allee** and others have pointed out, *cooperation* has, in many cases, been involved in survival. Natural selection results in the survival of those forms that are best integrated with the various factors of the environment in which they live.

The poet **Alfred Lord Tennyson** wrote his poem *In Memoriam* in 1833, 25 years before Darwin published *The Origin of Species.* It includes the memorable line **"Nature red in tooth and claw".**

Certain other demerits of Darwinism. Besides its numerous merits, Darwinism has following demerits :

1. Darwinism or natural selection theory does not account for the beginning of organs, which may appear at first as the veriest rudiments having as yet no selection value. In other words, it remains concerned with the survival of the fittest, but not for the arrival of the fittest. Thus, to give rise to such specializations as elaborate mimicry, or the electric organ of the torpedo, etc., which are of apparent advantage only in the perfected state, natural selection, acting only upon minute gradations toward perfection, seems inadequate. The same is true of so complex a specialization as the eye and its function in the vertebrates or in the insects and crustaceans.

2. Over-specialization in certain cases like extinct Irish deer in which huge antlers outweigh the entire skeleton, or the immense spiral tusks of the Jefferson mammoth, or the minute fidelity of certain mimicking insects such as *Kallima,* or huge dinosaurs of Mesozoic – all cannot be explained on the basis of continuous variations and natural selection. These organs or body structures should not have reached such a harmful stage, if natural selection was operating. However, such cases of over-specializations have been explained by Darwin on the basis of discontinuous variations or "sports" which, according to him, do not play any role in evolution.

3. Natural selection cannot account for degeneracy. To say an organ is no longer useful and, hence, disappears, is to state the effect and not the cause. If under changed conditions a character built up by natural selection becomes a menace, the reversal of selection can accomplish its removal but this will not suffice where the characteristic is an indifferent one.

4. One of classical objections to natural selection is that new variations would be lost by "dilution" as the individuals possessing them bred with others without them. We now know that although the phenotypic expression of a gene may be altered when it exists in combination with certain other genes, yet the gene itself is not altered and is transmitted to succeeding generations.

5. Darwin indirectly accepted the lamarckian idea of inheritance of acquired characters in the form of **pangenesis hypothesis**, which cannot be accepted in the light of knowledge of genetics made available in the present century.

In year 1875, **Galton** made pangenesis hypothesis untenable by presenting several experimental proofs. He and others at later time made a series of experiments involving blood transfusions and later, transplants of ovaries between black and white varieties of rabbits and chickens. Gametes produced by the transplanted ovaries were consistent with the phenotype of the individual in which the ovary originated and not with the animal currently carrying the ovary. Blood transfusions had no effect on the gametes produced. These experiments readily demonstrated that the pangenesis hypothesis was incorrect.

6. Lastly, some persons have objected natural selection because it is essentially a materialistic doctrine, depending as it does purely on the laws of chance.

Neo-Darwinism

Neo-Darwinism is a modified form of Darwinism. The Neo-Darwinians like **T.H. Huxley** and **Herbert Spencer** of England, **D.S. Jordan** and **Asa Gray** of United States, and **E. Haeckel** and **A. Weismann** of Germany believed that natural selection has accounted everything that is involved in evolution. In their dogmatic belief they went further to secure enough proof towards natural selection.

Certain Neo-Darwinians, such as **Weismann** and his followers rejected Darwin's theory except its principal element of natural selection. These Neo-Darwinians, though distinguished between germplasm and somatoplasm of living organisms in their germplasm theory, yet they could not appreciate the role of mutations in evolution. While Darwin believed that the adaptations result mainly by a single source, *i.e.*, natural selection, Neo-Darwinians thought that adaptations result from multiple forces and natural selections is only one of these many forces. Neo-Darwinians also believed that characters are not inherited as such but there are character determiners, the **determinants** or **biophores**, which control only the development. The ultimate character would result out due to the interaction of the determiners, activity of the organism and the environment during development. Thus, Neo-Darwinism was incomplete and partly wrong because it lacked present understanding of genetics.

Maturation of Neo-Darwinism into Modern Synthesis

During later half of 19th century and at the beginning of 20th century, many important ideas have been forwarded for explaining the inheritance of variations from one generation to successive generations and also for explaining evolution. Thus, the rediscovery of **Mendel's laws** of heredity by **Correns**, **de Vries**, and **Tschermak** in 1900 made it clear that (1) the factors given to the offspring by the parents do not "mix" but are segregated, and (2) in more than one pair of contrasting characters are considered in the same cross, the factors responsible for these are inherited independently. Likewise, **mutation theory of evolution** of **Hugo de Vries** (1886, 1887) stated that new species arise by sudden changes or steps called **mutations** rather than by gradual processes. According to **de Vries**, it was mutations and not selection that should be considered as the primary factor in evolution. Further, the studies of **Wagner** (1868) suggested the role of **geographic** or **spatial isolation** in the formation of every species, race, or tribe of animals or plants on the earth. Even more, certain **population geneticists** realized that the actual physical struggle between animals for survival or the competition between plants for space, sun and water is much less important as an evolutionary force than Darwin believed. The evolution of any given kind of organism occurs over many generations during which individuals are born and die, but the population has a certain continuity. Thus, the unit in evolution is recognised not the individual but rather a **population** of individuals.

A population of similar individuals living within a circumscribed area and interbreeding is termed as a **deme** or a **genetic population**. The next larger unit of population in nature is the **species**, composed of a series of inter-grading demes. Further, in a population, the relative frequencies of the genes will remain constant from one generation to the next (1) if the population is large, (2) if there is no selection for or against any specific gene or allele, *i.e.,* if mating occurs at random, (3) if no mutations occur, and (4) if there is no migration of individuals into or out of the population. The operation of the Hardy-Weinberg Principle will result in maintaining a given gene frequency in a population. Thus, a population undergoing evolution is one in which the gene pool is changing from generation to generation. The gene pool is the sum total of all the allelic genes in a population, and the gene pool of a given population may be changed (1) by mutation, (2) by hybridization, that is, by introduction into the population of gene from some outside population, or (3) by natural selection. Recombination brought about by crossing over and by the assortment of chromosomes in meiosis may also lead to new combinations of genes and phenotypes with some specific advantage or disadvantage for survival that would be reflected in a change in the gene pool.

All these modern understandings in cytology, genetics, cytogenetics, population genetics, and evolution gave a way for the formulation of a coherent theory called **modern synthesis** around 1930s, by **S. Wright**, **H.J. Muller**, **Th. Dobzhansky**, **R.B. Goldschmidt**, **J.S. Huxley**, **R.A. Fisher**, **J.B.S. Haldane**, **Ernst Mayr** and **G.L. Stebbins**.

3. MODERN SYNTHETIC THEORY

The modern synthetic theory of evolution involves five basic processes—mutations, variations, heredity, natural selection and isolation. In addition, three accessory processes affect the working of these five basic processes. **Migration** of individuals from one population to another as well as **hybridization** between races or closely related species both increase the amount of genetic variability available to a population. The effects of **chance** acting on small populations, may alter the way in which natural selection guides the course of evolution (**Stebbins**, 1971).

1. Mutation. Alteration in the chemistry of gene (DNA) is able to change its phenotypic effect (*i.e.*, nature of polypeptide) is called **point mutation** or **gene mutation**. Mutation can produce drastic changes or can remain insignificant. There are equal chances of a gene to mutate back to normal. Most of the mutations are harmful or deleterious and lethal but not all. Most of the mutant genes are recessive to normal gene and these are able to express phenotypically only in homozygous condition. Thus, point mutations tend to produce variations in the offspring.

2. Variation (Recombination). The nature of genetic variations caused by reshuffling of genes during sexual reproduction (recombination) was very little known at the time of Darwin. **Recombination** — that is, new genotypes from already existing genes — is of several kinds : 1. the production of gene combinations containing in the same individual two different alleles of the same gene, or the production of heterozygous individuals (meiosis); (2) the random mixing of chromosomes from two parents to produce a new individual (sexual reproduction); (3) the mixing of a particular allele with a series of genes not previously associated with it, by an exchange between chromosomal pairs during meiosis, called **crossing over**, to produce new gene combinations. **Chromosomal mutations** such as polyploidy, deletion, duplication, inversion and translocation also result in variation.

3. Heredity. The transmission of characteristics or variations from parent to offspring, is an important mechanism of evolution. Organisms possessing hereditary characteristics that are helpful, either in the animal's native environment or in some other environment that is open to it, are favoured in the struggle for existence. As a result, the offsprings are able to benefit from the advantageous characteristics of their parents.

4. Natural selection. Natural selection brings about evolutionary change by favouring differential reproduction of genes. Differential reproduction of genes produces change in gene frequency from one generation to the next. Natural selection does not produce genetic change, but once genetic change has occurred it acts to encourage some genes over others. Further, natural selection creates new adaptive relations between population and environment, by favouring some gene combinations, rejecting others and constantly moulding and modifying the gene pool.

The workings of natural selection are exceedingly complex because of the range of organizational levels at which it functions. Selection discriminates among available reproducible biotic entities to produce more efficiently adapted units. Natural selection operates upon every stage in the life history of an organism. It produces non-random differential reproduction of biological units and may affect any biotic entity from the molecular to the community level. Examples of levels at which natural selection makes differential discrimination are the following : **intermolecule**, **intergene**, **interchromosome**, **intergamete**, **interindividual** (Darwinian selection), **interdemic**, **interracial**, **interspecific** and **intercommunity**. Darwinian selection may result from differential natality among others.

5. Isolation. Isolation of organisms of a species into several populations or groups under psychic, physiological or geographical factors is supposed to be one of the most significant factors responsible for evolution. **Geographical isolation** includes physical barriers such as high mountains, rivers, oceans and long distances preventing interbreeding between related organisms. **Physiological barriers** help in maintaining the individuality of the species, since these isolations do not allow the interbreeding amongst the organisms of different species. This is called **reproductive isolation**.

Speciation (Origin of new species). The populations of a species present in the different environments and are segregated by geographical and physiological barriers, accumulate different genetic differences (variations) due to mutations (both point and chromosomal), recombination, hybridization, genetic drifts and natural selection. These populations, thus, become different from each other morphologically and genetically, and they become reproductively isolated, forming new species.

4. WEISMANN'S GERM PLASM THEORY

August Weismann (1834–1914) was the German neo-Darwinian biologist who proposed the **germ plasm theory** which was published in the book *Das Keimplasma.* He asserted the continuity of germ plasm as the main criterion for inheritance of characters. All the heritable variations have their origin in germ cells and a new type of organisms arise only from changed type of germ cells. Thus, Weismann's germ plasm theory rejected outrightly the Lamarckian concept of inheritance of acquired characters and Darwin's pangenesis hypothesis. While Darwin's pangenesis hypothesis is **centripetal** in nature, germ plasm theory is **centrifugal**. The main points of germ plasm theory are the following:

1. Existence of somatoplasm and germ plasm. All substances of an organism can be divided into two parts, the **germ plasm** (which is the protoplasm of germ cells such as sperms and ova) and **somatoplasm** (which is the protoplasm of somatic or body cells). The germ plasm is thought to be the actual vehicle of heredity ; it is not affected by any influence either from the body or the external environment. The somatoplasm is thought not to play any role in heredity.

2. Continuity of germ plasm. In reproduction, a portion of germ plasm is derived from each parent, the paternal sperm and maternal ovum, each of which has an equivalent share in the inheritance. These combine to form the fertilized egg. The contents of egg divide and redivide to produce both body (soma) and the germ plasm of new individual. When the egg divides in the first or subsequent cleavages each daughter cell receives an equal share of germ plasm, and this holds true for all cells which go to form the adult body. None of the somatic cells, however, has any share in subsequent generations but only the germ cells which are derived directly from the original egg. It follows, therefore, that there is a continuous stream of germ plasm from generation to generation, to which the somatic cells contribute nothing. Hence, only those mutations which are germinal in their origin can possibly be handed down, and as the hereditary stream of germ plasm is already set apart before the adult body comes in use, one cannot see how any modification impressed upon the body through use or disuse can possibly become a part of the organism's heritage. Thus, germ plasm is immortal since it is perpetuated from generation to generation through the meiotic cell division. Each germ cell is the product of the division of previous germ cells. Soma (somatoplasm) is mortal, it perishes with the death of an organism. Germ plasm can produce the somatoplasm but somatoplasm cannot produce the germ plasm.

3. Architecture of germ plasm (Concept of determinants). In 1904, **Weismann** proposed that every distinct part of an organism is represented in the sex cell by a separate particle— the **idioplasm** or **determinant**. Each determinant is supposed to be made up of still smaller units called **biophores**. The sum total of determinants would represent the parts of the adult organism with all their peculiarities. The complete set of determinants would be handed down from generation to generation, which would account for hereditary transmission of characters. The determinants, according to **Weismann**, are localized in the chromosomes of the nucleus, just as are the genes of modern genetics.

Further, during the cleavage of the egg, the various determinants become segregated into different cleavage cells or blastomeres. Eventually each blastomere would contain only determinants of one kind and it would be forced to develop or differentiate in a specific way in accordance with the determinants present in it. Only the cells having the sex cells among their descendants tend to preserve the complete set of determinants, since these would be necessary for directing the development of next generation.

4. Parallel induction. In apposition of Lamarck's inheritance of acquired characters, **Weismann** had introduced the idea of **parallel induction**. According to this concept, the stimulus affects simultaneously the germ plasm and the body (soma). He proposed the occurrence of an internal stimulus which affects germ cells and results in heritable variations. The stimulus, according to Weismann, is the nourishment which is necessary for determinants and biophores (since they are living units). Those determinants or particles which obtain better nourishment are fast-growing and stronger than those getting less nourishment. Correspondingly, these particles tend to produce either strong or weaker part or organ in organisms. Thus, Weismann assumed a struggle for existence between better nourished and less nourished determinants, and there lies the causes of appearance and disappearance of variations.

Objections to Weismann's Germ plasm Theory

Germ plasm theory is criticised mainly for its speculativeness (*i.e.,* it lacks any experimental support), and also for its idea of determinants and their segregation during cleavage and for its failure for explaining causes of regeneration and asexual reproduction.

Significance of Weismann's Germ plasm Theory

1. It made ground for the understanding of the concept of particulate inheritance of **Mendel**.
2. It provides some clue about genes (determinants) which reside in chromosomes and represent some part of animal body.
3. According to the embryologists, the greatest contribution of this theory is that it proposes the division of germ plasm and somatoplasm during cleavage of the zygote during the embryogenesis (*e.g.,* cleavage of *Ascaris*).
4. The idea of continuity and immortality of germ plasm prepares the ground for the continuity of chromosome or DNA from one to next generation.

5. MUTATION THEORY

Darwin recognised two types of variations in nature, *viz.,* 1. **minor** and **continuous** and 2. **major** and **discontinuous** ones. He considered minor and continuous variations important in the origin of new species. Darwin considered major variations quite insignificant and for them he coined the terms **sports** or **saltatory** (L. *saltare* = to leap) variations. In England, **William Bateson** collected a large number of species showing discontinuous or saltatory variations. In Holland, **Hugo de Vries** (1901) gave much importance to these variations and proposed that new species arise not by the accumulation of minor and continuous variations through natural selection, but by the suddenly appearing saltatory variations for which he coined the term **mutation** (L. *mutare* = to change). **De Vries's** mutation theory states that species have not arisen through gradual selection accumulated for hundreds or thousands of years but have appeared by **sudden jumps** (saltations) and transformation.

Oenothera lamarckiana.

Dutch botanist **Hugo de Vries** (1848 – 1935) was born at Haarlem; he studied at

Leiden, Heidelberg and also Wurgberg. He was a university lecturer at Amsterdam and also held the professorship of Plant Physiology. His book entitled "*Die Mutation Theorie*" was published in 1901. In this book, **de Vries** proposed the mutation theory in order to explain the mechanism of evolution. This theory was based on his observations on evening primrose, *Oenothera lamarckiana* grown in a field near Amsterdam. He studied this plant in wild form for many years continuously and observed certain spontaneous changes in some of these wild plants. These plants differed considerably in stem height, flower colour and leave's shapes. He observed that these changes were heritable and ultimately led to several new varieties. He succeeded in cultivating all these new varieties and named them as **mutant varieties**. In fact, he selected for his breeding experiments two mutant varieties—*Oenothera laevifolia*, characterized by smooth leaves and *O. brevistylis* characterised by short styles. And he observed that these features were breeding true and so, he regarded these mutant strains as the distinct species. Thus, **de Vries** recognised the following seven distinct species of *Oeonothera:*

1. **O. gigas.** Giant and stout plants with large flowers and deep green leaves.
2. **O. rubrinervis.** Fruit red veined, leaves pale green and stem slender and brittle.
3. **O. oblonga.** Dwarf and weak plants with oblong leaves.
4. **O. albida.** Weak plants with whitish-pale leaves.
5. **O. laevifolia.** Leaves narrow and smooth and pale flowers.
6. **O. brevistylis.** Round leaves, flowers with very short style and flattened stigma.
7. **O. vannilla.** Very short or dwarf variety (20–30 cm); leaves sessile.

Of these varieties which arose suddenly from the parental variety *O. lamarckiana*, the *O. gigas* and *O. rubrinervis* were **progressive varieties**; the *O. laevifolia*, *O. brevistylis* and *O. vanniella* were **retrograde varieties** and *O. oblonga* and *O. albida* were weak and inconstant forms and could be grown only under strict protection.

Characteristics of Mutation Theory

De Vries's mutation theory has the following characteristics :

1. Mutations appear from time to time among the organisms of a naturally breeding species or populations. The organisms with mutations are called **mutants**. These mutants are clearly distinct from their parents.
2. Mutations are heritable and form new species. They do not disappear by crossing.
3. Mutations arise suddenly in one step, *i.e.*, new species arise suddenly in one step and not gradually.
4. Mutations occur in all possible directions and may be advantageous or disadvantageous.
5. Unsuitable mutants are destroyed by natural selection.
6. Mutations appear full-fledged and, hence, there is no question of incipient stages in the development of an organ.

Types of Mutation

Mutation in a broad sense include all those heritable changes, which alter the phenotype of an individual. In genetics, much investigations have been done in the phenomenon of mutation by **Morgan**, **Muller**, **Goldschmidt**, **Spencer**, etc. According to their work, following two types of mutations have been recognised :

1. Gene mutation (Point mutation). Any chemical change in a particular gene leading to the appearance of different alleles and affecting the viability or phenotype of the organism (For details see Chapter 16 in Part II – Genetics).

2. Chromosome mutation. Any change in the number and structure of the chromosome or chromosome sets, leading to **ploidy** (*e.g.*, polyploidy, trisomy) and chromosomal aberrations (*e.g.*, translocation) respectively (For details see Chapters 14 and 15 in Part II– Genetics).

Advantages of Mutation Theory

The advantages of de Vries's mutation theory are the following :

1. The mutation theory describes the importance of mutation in selective value of organisms.
2. The mutation theory explains the occurrence of evolutionary changes within short period in contrast to natural selection (which describes slow and continuous variations).
3. Mutation theory explains the absence of connecting links as no criteria against evolution but its possibility exist.
4. Occurrence of mutations in large and divergent direction removes the possibility of species disappearance by crossing.
5. Since mutations appear fully formed from the beginning, there is no difficulty in explaining the incipient stages in the development of organs.
6. Mutation is of great service to breeders in developing new useful varieties.

Objections to Mutation Theory

1. The mutation theory was unable to explain the presence of flightless birds on oceanic islands.
2. It could not explain the existence of discontinuity in distribution among individuals.
3. Many mutations, described by **de Vries** in *Oenothera lamarckiana,* are now known to be due to certain numerical and structural changes in the chromosomes. For instance, *"gigas"* mutant of *O. lamarckiana* was later found to be due to polyploidy.
4. Mutation theory alone could not explain evolution. It, however, provided raw material for other forces to act upon it and bring about evolutionary changes.

REVISION QUESTIONS

1. Define and describe the hypothesis of "Inheritance of acquired characters".
2. Give an account of Lamarckism. Give various evidences in support and criticism of it.
3. What is Lamarckism ? How does it differ from Darwinism ?
4. Write an essay on the theory of natural selection.
5. Describe Darwinism and explain the postulates of Neo-Darwinism.
6. Describe those circumstances which helped Darwin in the formulation of natural selection theory.
7. Explain the modern concept of evolution and discuss how does it support Darwinism.
8. What is modern synthetic theory ? Explain it.
9. Write short notes on the following :
   (i) De Vries's mutation theory ;
   (ii) Germ plasm theory ;
   (iii) Neo-Darwinism; and
   (iv) Modern synthetic theory.

CHAPTER

6

Selection in Action

(Examples and Types of Natural Selection)

Epigamic selection by peahen.

Darwin clearly saw the importance of selection as a prime evolutionary force in the natural world. He cites countless examples of adaptation through selection, and thousands of new cases are recognized each year. But one problem in discussing selection has usually been the correlation of experimental studies under controlled laboratory conditions with what actually happens in nature. During the last forty years a series of studies combining experiment and observation in the field have been carried out on the problem of industrial melanism in moths; they provide an exciting insight into the operation of selection under natural conditions. These brilliant investigations were undertaken originally by **R.A. Fisher** (1929) and **E.B. Ford** (1940, 1964, 1971) and more recently by **H.B.D. Kettlewell** (1956, 1961, 1973), all of Britain.

1. Melanism in Moths or Industrial Melanism

With increasing industrialization due to Industrial Revolution in Britain, entomologists began to record the replacement of light or grey colours in many different species of moths with dark or black ones (called **melanics**). A particularly good example is the peppered moth, *Biston betularia*, where the dark form is referred to as **carbonaria**. The following facts were discovered : (1) The earliest records of dark forms (*i.e.,* 1% in

1845 and 99% in 1895) came from sites near heavily industrialized areas (Manchester, Birmingham); (2) The highest frequency of dark forms was found in sites near industrial centres (hence, the phenomenon was described as **industrial melanism**); (3) Melanics usually occurred in night-flying species. The possible evolutionary explanation is that light forms are more conspicuous to predators (birds such as English robin, *Erithacus rubecula*) when they rest on sooty branches and tree trunks in industrial sites in the day time (Under unpolluted conditions light moths were almost mingled with lichened tree trunks and become well protected, Fig. 6.1). There are, of course, possible alternative explanations such as that the melanism might be phenotypic, *e.g.,* due to uptake of the industrial pollutants. However, breeding experiments established genotypic control and indicated that 'carbonaria' trait usually segregated as if it was a **dominant** at a single locus. There are other possible explanations, *e.g.,* a dark colour might protect against the direct effects of pollution. To analyse the situation further, therefore, Professor **H.B.D. Kettlewell** (1956, 1973) performed the following experiments :

Fig. 6.1. Typical (light-coloured) and carbonaria (dark-coloured) forms of *Biston betularia* (moth) on an unpolluted, lichen covered tree (after Hamilton, 1967).

1. To determine whether 'carbonaria' (*i.e.,* melanic moth) had a greater fitness than typical light peppered moth, **Kettlewell** released marked individuals into both industrialized and rural areas. More marked melanic forms were recovered in the industrial sites and more grey forms in the rural areas.

2. Direct observations were made on predation rates. Similar numbers of melanic and typical light moths were put on tree trunks in polluted and non-polluted areas. More melanics were eaten by birds in the rural areas and more non-melanics in the industrial areas. Obviously natural selection in terms of bird predators plays an enormously significant part in industrial melanism.

3. One bird, the creeper, did not distinguish between two moths (*i.e.,* melanics and grey). **Kettlewell** noticed that it feeds by creeping up and down the trunks of trees, and is, therefore, more apt to see the silhouette of the moth resting upon the bark of the tree than the colour of the wings.

Fig. 6.2. Typical (light-coloured) and carbonaria (dark-coloured) forms of *Biston betularia* (moth) on blackened bark in a polluted region of England (after Hamilton, 1967).

Points numbers 2 and 3 are particularly powerful supports for the idea that melanism has arisen in response to predation pressure. Thus, one of the many impressive features of Kettlewell's studies lies in the clear-cut identification of the selecting agent. Selection, we may recall, has been defined as differential reproduction. The act of selection in itself does not reveal the factors or agencies that enable one genotype to leave more offsprings than another. We may demonstrate the existence of selection, yet not know the *cause* of that selection. We might have reasonably suspected that predatory birds were directly responsible for the differential success of the melanic forms in survival and reproduction, but Kettlewell's laboriously accumulated data provided that all-important ingredient : the 'proof'.

Lastly, gradual replacement of soft, high-sulphur (sooty) coal to less smoky fuel or by oil and electricity, reduced the soot deposition on the trees. Conditions then became more suitable for the survival of grey moths, consequently their frequency once again increased (**Bishop** and **Cook**, 1975).

2. Australian Rabbits

In 1859, a small colony of 24 wild rabbits (*Oryctolagus cuniculus*) was brought from Europe to an estate in Victoria in the south eastern corner of Australia. From such modest beginnings, the rabbits multiplied enormously and by 1928 had spread over the greater part of the Australian continent. According to an estimate, the number of adult rabbit was over 500 million in an area of about 1 million square miles. The rabbits caused extensive damage to sheep-grazing pastures and to wheat cropfields.

For controlling the population explosion of these prolific rabbits, the Australian government spent huge sums of money for many years. Trapping, rabbit-proof fencing, poisoning of water holes, and fumigation all proved to be largely inadequate. Then, beginning in 1950, outstanding success in reducing the rabbit population was achieved by a biological control method, *i.e.,* inoculating rabbits with a virus that causes the fatal disease **myxomatosis**. The deadly myxoma virus was implanted into the tissues of rabbits in the southern area of Australia. In a remarkably short period of time, the virus had made its way, aided by insect carriers (mosquitoes), into most of the rabbit-infested areas of the continent. By 1953, more than 95 per cent of the rabbit population in Australia had been eradicated.

But, after their drastic decline in the early 1950s, the rabbit populations began to build up again. Evolutionary changes have occurred in both the **pathogen** (*i.e.,* myxoma virus) and the **host** (rabbit). Mutations conferring resistance to the myxoma virus have selectively accumulated in the rabbit populations. At the same time, the viruses themselves have undergone genetic changes; less virulent strains of the virus have evolved (**Frank J. Fenner**, 1959).

3. Resistance of Insects to Pesticides

The evolution of resistance of insect species to pesticides has in recent decades been a fantastic evidence of the ability of living species to undergo genetic changes in response to challenges of the environment (see **Brown**, 1967, **Georghiou**, 1972). The story is the same : when a new insecticide is introduced, a certain relatively small amount of concentration is sufficient to achieve a satisfactory control of the insect pest abundance; the necessary concentration gradually increases, until it becomes totally ineffective or economically impractical. In 1947 **Sacca** was the first to report that a population of the housefly, *Musca domestica*, had become resistant to **DDT** (dichloro-diphenyl-trichloroethane). Since then, resistance to one or more insecticides has been recorded in at least 225 species of insects (including the malaria-carrying mosquitoes and typhus-carrying lice) and other arthropods. Usually resistance is specific to one or a group of chemically related insecticides, but occasionally instances of multiple resistance are reported. An interesting parallel is found in a mammalian species, the grey rat. A substance known under the trade name of **warfarin** was destroying rats, chiefly through its action on a blood anticoagulant and interference with vitamin K. Warfarin-resistant rats (mutant strains) appeared first in a locality in England and have since been spreading and replacing the ordinary warfarin-sensitive rats. Our struggle against diseases and insect pests is unending as long as the mutation process continually produces variants that may have survival value in an altered environment.

4. Antibiotic Resistance in Bacteria

Penicillin, sulphonamides (sulpha drugs), streptomycin and other modern antibiotic agents made front-page headlines in the newspapers when first introduced. These wonder drugs were exceptionally effective against certain disease-producing bacteria and contributed immeasurably to the saving of human lives in World War II. However, the effectiveness of these drugs has been reduced by the emergence of resistant strains of bacteria. Medical authorities consider the rise of resistant bacteria as the most serious development in the field of infectious diseases over the past two decades. Bacteria now pass on their resistance to antibiotics faster than people spread the infectious bacteria.

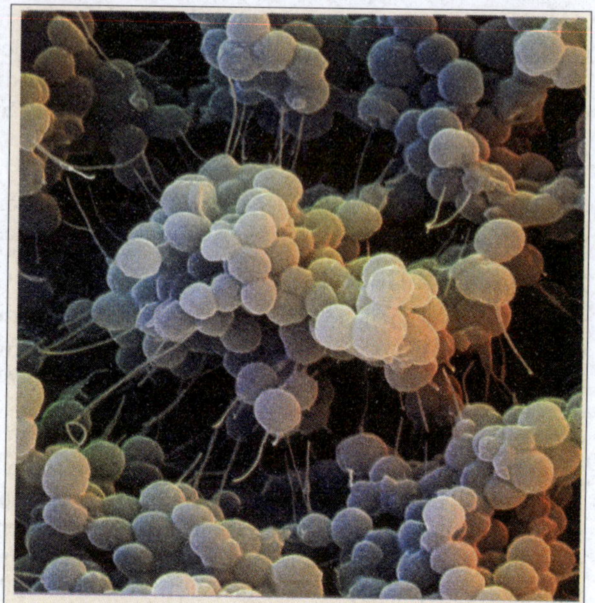

Staphylococcus aureus is among the many bacterial species that have evolved resistance to antibiotics. The evolution of such supergerms threatens to reverse our success in the battle against infectious disease.

Apparently, mutations have occurred in bacterial populations that enable the mutant bacterial cells to survive in the presence of the drug. In them, thus, mutations furnish the source of evolutionary changes and the fate of the mutant gene is governed by selection. In a normal environment, mutations that confer resistance to a drug are rare or undetected. In an environment changed by the addition of a drug, the drug-resistant mutants are favoured and replaced the previously normal bacteria.

It might be thought that the mutations conferring resistance are actually caused or induced by the drug. But this is not true. Drug-resistant mutations arise in bacterial cells irrespective of the presence or absence of the drug. An experiment devised by the Stanford University geneticist **Joshua Lederberg** (and his wife **Esther Lederberg**) in 1952 provided evidence that the drug (penicillin or streptomycin) permitted pre-existing mutations to express themselves. As shown in Figure 6.3, colonies of bacteria were grown on a streptomycin-free agar medium in a petri-plate. This is called **master plate**. When the agar surface of this master plate was pressed gently on a piece of sterile velvet, some bacterial cells from each bacterial colony clung to the fine fibres of the velvet. The imprinted velvet could now be used to transfer the bacterial colonies onto a second agar plate. In fact, more than one replica of the original bacterial growth can be made by pressing several agar plates on the same area of velvet. This ingenious technique has been aptly called **replica plating**.

In preparing the replicas **Lederberg** used agar plates containing streptomycin. Of course, on these agar plates, only bacterial colonies resistant to streptomycin grew. In the case depicted in Figure 6.3, one colony was resistant to the drug. Significantly, this one resistant colony was found in the same exact position in all replica plates. If mutations arose in response to exposure to a drug, it is hardly to be expected that mutant bacterial colonies would arise in precisely the same site on each occasion. In other words, a hapazard or random distribution of resistant bacterial colonies, without restraint or attention to location in the agar plate, would be expected if the mutations did not already exist in the original bacterial colonies.

In second set of experiment, **Lederberg** tested samples of original bacterial colonies in a test tube for sensitivity or resistance to streptomycin (Fig. 6.3). It is remarkable that the bacterial colonies on the original plate had not been previously in contact with the drug. When these original colonies

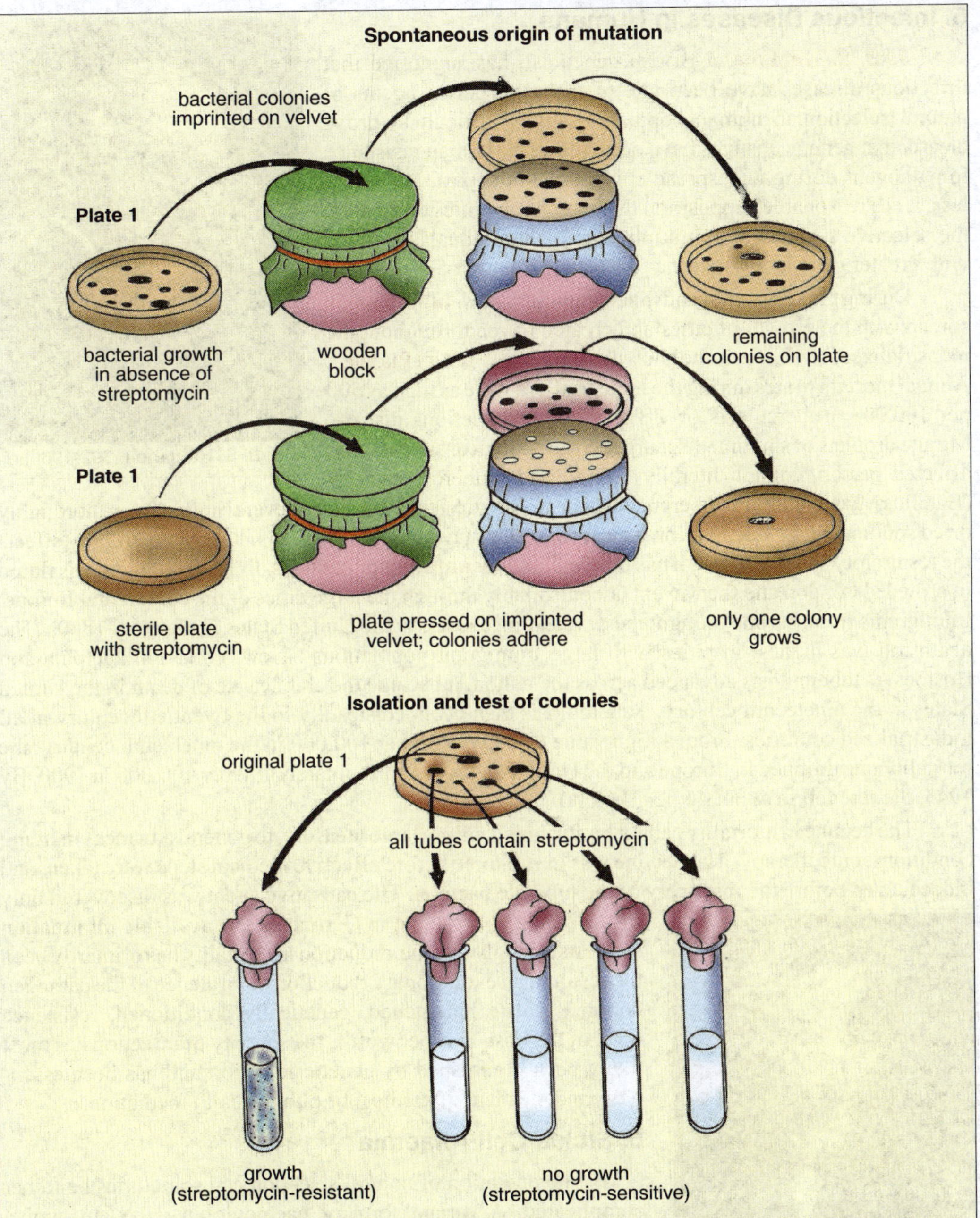

Fig. 6.3. Experiment of **Joshua Lederberg** revealing that drug-resistant mutations in bacterial cells had not been induced by the drug but had already been present prior to exposure of the bacteria to the drug (after Volpe, 1985).

were isolated and tested for resistance to streptomycin, only one colony proved to be resistant. This one colony occupied a position on the original plate identical with the site of the resistant colony on the replica plates. This experiment demonstrated conclusively that the mutation had not been induced by streptomycin but had already been present before exposure to drug.

5. Infectious Diseases in Humans

J. B. S. Haldane (1892-1964).

J. B. S. Haldane, a British geneticist, has suggested that infectious diseases have been one of the most potent agents of natural selection in human populations. It is difficult to prove clearly that genetic changes have occurred in the human resistance to pathogens during widespread epidemics in the past. Nevertheless, it is a reasonable supposition that major epidemics eliminated the selective survival and multiplication of individuals provided with greater genetic resistance.

During the eighteenth and nineteenth centuries, **tuberculosis** spread with the growth of cities and created havoc throughout the industrialized world. It became known as the "**Great White Plague**". Annual mortality rates during the industrial era were as high as 500 per 100,000. Tuberculosis or TB is a highly infectious disease. Minute droplets of sputum, discharged by the cough or sneeze of an affected person, contain literally thousands of tubercle bacteria (Bacillus Mycobacterium tuberculosis) which can survive in the air for several hours. Even thoroughly dried sputum, when inhaled, can be infective. TB may invade any organ but most commonly affects the respiratory system, where it has been called **consumption** or **phthisis**. In 1750, tuberculosis flared in crowded London and then swept uncontrollably through the large cities of the continental Europe. Immigrants from Europe brought the devastating disease to the United States in the early 1800s. The death toll was highest in cities with large immigrant populations—New York, Philadelphia and Boston. As tuberculosis advanced across the nation, it became the chief cause of death in the United States in the nineteenth century. Tuberculosis then declined steadily in the twentieth century in all industrialized countries. From a high value of 500 deaths per 100,000 in the nineteenth century, the mortality rate dropped in Europe and the United States to approximately 190 per 100,000 in 1900. By 1945, the rate fell to about 40 per 100,000.

The decline in mortality cannot be attributed, in any large measure, to general advances in living conditions, or to therapy. The decline was in full force before effective medical steps were taken, and indeed, even before the discovery of the tubercle bacteria. The causative agent was discovered only in 1882 by **Robert Koch** in Germany. The available information suggests strongly that the reduction in mortality has primarily been the result of an evolutionary reduction of virulence of the pathogen coupled with a heightened, genetically conditioned, resistance level of the host. In other words, the severity of infection has most likely been diminished by genetic selection and has been essentially independent of medical or public health mediation.

Robert Koch (1843-1910).

6. Sickle Cell Anaemia

The operation of natural selection can sometimes be rather complicated. A variant form of haemoglobin (HbS) in human beings is responsible for the genetic disorder or **molecular disease** (**Volpe**, 1985), called **sickle-cell anaemia** (discovered by American physician **James B. Herrick** in 1949). Individuals homozygous for the abnormal haemoglobin die at an early age. Even in the heterozygous condition, the red blood cells containing abnormal haemoglobin become sickle-shaped, and are unable to bind oxygen efficiently. Why has natural selection not eliminated the abnormal allele?

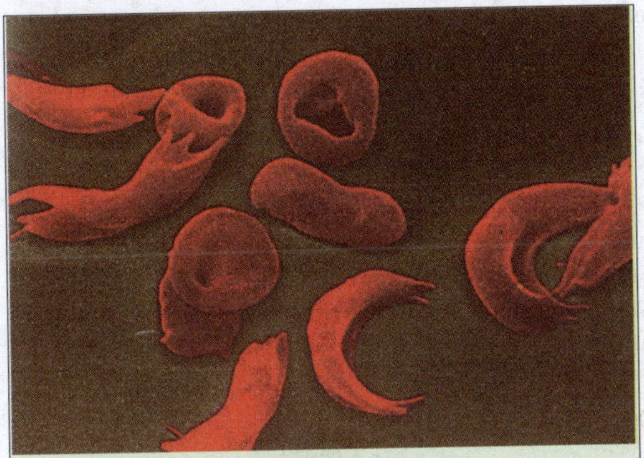
Sickle-shaped RBCs.

A clue to this puzzle is given by the geographical distribution of the abnormal haemoglobin. This variant is often found in areas where malaria is common. Detailed studies have shown that the malarial parasite spends a part of its life cycle in the human red blood cells. When a RBC in heterozygous individual becomes sickle-shaped, it effectively kills the parasite. As a result, individuals heterozygous for the abnormal haemoglobin are able to cope with malarianl infection much better than individuals homozygous for the normal haemoglobin (HbA). Thus, the natural selection tends to maintain the abnormal form of haemoglobin along with the normal form in a region where malaria is common.

7. Heavy Metal Resistance in Plants

Plants have also produced some clear-cut instances of action of selection. **Bradshaw** (1971) and **Gregory**, **Antonovics** and coworkers have studied varieties of several species of plants able to grow on soils contaminated with heavy metals (such as copper, zinc, lead). Such contaminated soils are often found near old mines where ores of these metals were exploited. The resistant varieties maintain themselves on soils with concentrations of the pollutants that are lethal or at least stressful for ordinary members of the same species. A conspicuous feature is that the resistant and non-resistant varieties may grow at distances only a few metres from each other, but meticulously confined to their respective contaminated and uncontaminated soils.

TYPES OF SELECTION

Some individuals in a population are more successful than others in passing along their genes into the next generation. This may occur for many reasons. Some organisms die before they reach reproductive age. Some organisms fail to find a mate. Among those individuals that do breeding, there are differences in the number of offsprings produced. The result of these differences means that for one reason or the other there is a differential rate of reproduction among members of a population, we call this **selection**.

Both in **natural selection** and **artificial selection** (discussed in Chapter 2), the principle is the same: some organisms breed more prolifically than others, thus, increasing the frequency of some genes and decreasing the frequency of others. Both types of selection operate on the phenotype, not directly on the genotype.

The following three types of natural selection have been categorized by population geneticists: directional, stabilizing and disruptive (diversifying).

1. Directional Selection

The directional selection describes the change that occurs when a population shows a particular trend through time. In the foregoing discussion (*i.e.,* Examples 1-7 of Natural Selection), we have seen that the selection process favours individuals that are best adapted to new situations or to new ecological opportunities. Such selection is said to be directional since the norm for the population is shifted with time in one direction (Fig. 6.4B). The selection for low virulence that occurred over time among the myxoma viruses in the host rabbits exemplifies directional selection. The raw materials for selection were a series of gene mutations, each of which acted to reduce the virus's virulence. The curve of distribution shifted to the right, as the less harmful strains of viruses displaced the virulent strains.

In fact, directional selection transforms the gene pool of a species or population toward the highest level of adaptedness that can be reached in the new environment. Apart from environmental change, origin of new and favourable mutant genes or super genes brings into action directional selection.

2. Stabilizing Selection

It describes the change when extreme individuals are eliminated from the population. So the intermediate values for a given trait are favoured over the extreme values. The result of this process is a reduced variability in the population. Most selection that occurs in populations is stabilizing and homeostatic, because it tends to maintain the status quo. As seen in Figure 6.4 A, the shape of the curve tends to narrow through continual elimination of the less adapted individuals. The birth weight of

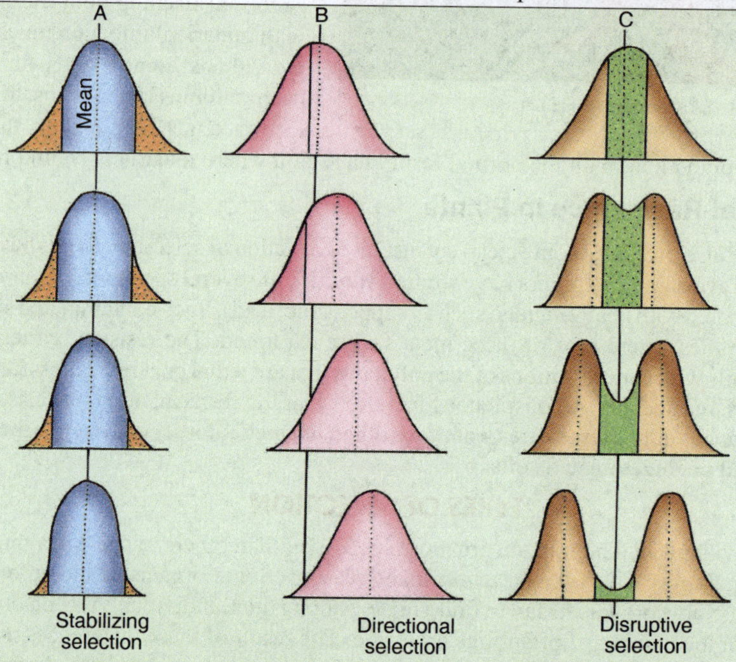

Fig. 6.4. Schematic representation of three types of selection and their effects. Each curve represents the normal distribution of the character in a population. From top to bottom, the lower curves show the expected distribution after the impact of selection. A–Stabilizing selection. The intermediate values for a given trait are favoured and preserved and the shape of the curve narrows through the elimination of the carriers of the extreme value; B–Directional selection. The adaptive norm changes as less adapted genotypes are replaced by better adapted genotypes; C–Disruptive or diversifying selection. Two adaptive norms are generated when the population exists in a heterogenous environment (after Volpe, 1985).

newborns provides a good example of a human character that has been subjected to stabilizing selection (Fig. 6.5). The optimum birth weight is 7.3 pounds; newborn infants less than 5.5 pounds and greater than 10 pounds have the highest probability of mortality. Given the strong stabilizing influence of weeding out the extremes, the optimum birth weight is associated with the lowest mortality. The curve for mortality is virtually the complement of the curve for survival.

3. Disruptive as Diversifying Selection

This form of selection occurs when the extreme values have the highest fitness and the intermediate values are relatively disadvantageous in terms of reproductive effectiveness. It is, essentially, selection for diversification with respect to a character. For example, shell patterns of limpets (marine molluscs) form a continuum ranging from pure white or dark tan. Limpets typically dwell in one of two distinct habitats, attaching either to white goose-neck barnacles, or to tan-coloured rocks. As might be expected, the light-coloured limpets seek the protection of the white barnacles, whereas tan limpets live by choice almost exclusively on the dark rocks. Limpets of intermediate shell patterns are conspicuous and are intensely selected against by predatory shore birds. If this disruptive type of selection (favouring the extremes) were to be accompanied by the sexual isolation of the two types of limpets then two new species could arise.

Evolutionary biologists also recognize the following two types of selection :

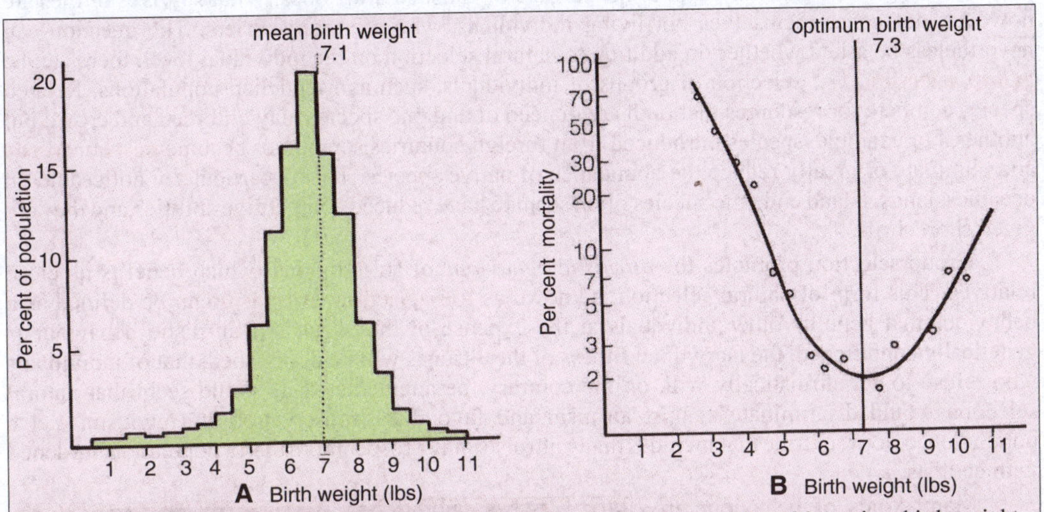

Fig. 6.5. The distribution of birth weights of human newborns and the mortality of the various birth-weight classes. A–The histogram shows the proportions of the population falling into the various birth-weight classes. The mean birth-weight is 7.1 pounds; B–The curve of mortality in relation to birth-weight reveals that the lowest mortality is associated with the optimum birth- weight (7.3 pounds) (after Volpe, 1985).

A. Sexual Selection

In many species of animals females and males are strikingly dimorphic. Males are often larger, stronger, more brightly coloured, have specialized song patterns and possessing various "adornments", some of which may seem to be handicaps rather than advantages to their possessors. Females and males always differ in behaviour to some extent, and sometimes strikingly so. Could sexual dimorphism originate in evolution by natural selection ? **Darwin** found this improbable as a general explanation, and proposed in 1871 that "sexual selection depends not on struggle for existence in relation to other organic beings or to external conditions, but on a struggle between the individuals of one sex". Thus, sexual selection is based on reproduction rather than survival.

During the century since it was proposed by Darwin, the theory of sexual selection has met with even more opposition than the theory of natural selection. At present, sexual selection is that form of natural selection directly related to differential reproduction, including the finding and acquisition of a mate, copulation, fertilization and parental care. **Huxley** (1938) had recognized the following two forms of sexual selection :

1. Epigamic selectin. It is based on choices made between males and females. The selection by a female of a particular male as a mating partner as a function of his large rack of antlers or bright colouration would represent epigamic selection.

2. Intrasexual selection. It is selection based on interactions between animals of the same sex, usually between males. An example of intrasexual selection would be competition among males for a territory in a breeding ground.

In practice, these two forms of selection overlap and distinctions are not clear, *e.g.,* a given territory may contain better resources such as food (survival selection) as well as being more attractive to females (sexual selection) (see **Dewsbury**, 1978).

B. Group and Kin Selection

In a population under selection the frequencies of some genes increase and those of others decrease as generation follows generation. The reproductive efficiency of individuals who carry the "successful" genes higher than that of the carriers of "unsucessful" ones. What survives or dies are however, not genes themselves but living individuals, who are their carriers. The question can nevertheless be asked whether, in addition to natural selection on the individual level, there can be **group selection**, *i.e.,* selection of groups of individuals, such as Mendelian populations. Related species compete for resources that both are in need of and one species may outbreed and crowd out another. For example, species introduced from foreign countries sometimes become **pests** or **weeds** and eliminate or greatly reduce the abundance of native species. This is particularly noticeable on oceanic islands. Island endemic species often seem to have reduced competitive abilities,and they are given short shrift.

Group selection promotes the *altruistic behaviour* of an individual which benefits its close relatives. This form of natural selection is known as **kin selection**. **Altruism** can be defined as a behaviour that benefits other indivduals at the expense of the altruist. If altruistic behaviour is genetically conditioned, the Darwinian fitness of the altruists will be diminished, that of individuals who refuse to act altruistically will, on the contrary, be augmentated. It would seem that natural selection would discriminate against altruism and favour self-preservation and egotism. Yet a population, colony, or tribe that includes many altruists may prosper more than a population burdened with egotists.

Many kinds of behaviour are equivalent to altruism, whereas they are really automatic and stereotyped. The behaviour of honeybees, ants, termites, and other social insects has provided good example of altruism (**Wilson**, 1971). Individual "workers" toil selflessly and are ready to sacrifice themselves for the benefit of the colony to which they belong. The evolutionary development of this apparent altruism is easily understood. The "workers" and "soldiers" do not reproduce. They are sterile females or males with underdeveloped sex organs. The perpetuation of an insect society is the task of a sexual caste, the so-called queens and fertile males, which are a minority, sometimes single individuals, in the colony. Sexual individuals do little or no work after the colony is founded and display no particular readiness for self-sacrifice. The workers share genes with the sexual members produced in the colony. The result of this arrangement is that natural selection operates not in individuals but in colonies consisting of reproductive and sterile casts as units. The apparent altruism of the workers enhances the probability of transmission to the next generation of the genes these workers share with the sexual caste of their colony.

The behaviour of honeybees provides good example of altruism.

REVISION QUESTIONS

1. Give some examples which support the phenomenon of natural selection.
2. Describe various types of selection.
3. Write short notes on the following : 1. Group selection; 2. Directional selection; 3. Industrial melanism; 4. Insecticide resistance in insects.

CHAPTER

7

Population Genetics and Evolution

These lady-bird beetles from the Chiricahua Mountains in Arizona show considerable phenotypic variation.

All the individuals of a species constitute a **population**. The genetical studies for the inheritance of phenotypic traits in a given population is called **population genetics**. The population genetics is a quantitative science. To calculate the results of the mode of inheritance of genes in a given population various statistical and mathematical models are employed in it. Certain fundamental aspects of population genetics are the following :

MENDELIAN POPULATION

A population of a particular species includes many inbreeding groups. The inbreeding groups may form a community within defined geographical boundaries and are called "**Mendelian population**". A Mendelian population, thus, is a group of sexually reproducing organisms with a relatively close degree of genetic relationship (such as species, subspecies, breed, variety, strain, etc.) residing within defined geographical boundaries where interbreeding occurs.

GENE POOL AND GENE FREQUENCY

To get a F_2 3 : 1 phenotypic ratio of a monohybrid cross, we began with two homozygous parental strains, such as AA and aa : that is, we introduced the alleles A and a in **equal frequency** (A **frequency** is the ratio of the actual number of a individuals falling in a single class to the total number of individuals; see **Gardner**, 1972). But in a Mendelian population, frequencies of alleles may vary considerably. For example, in Mendelian population of man, the gene for polydactyly is dominant, yet the polydactylous phenotype is fairly infrequent among infants. This indicates that frequency of the dominant allele in population is lower than that of its recessive allele and that both alleles do not exist in population in the 1:1 ratio like the individuals of monohybrid cross.

Further, if all the gametes produced by a Mendelian population are considered as a hypothetical mixture of genetic units from which the next generation will arise, we have the concept of a **gene** or **gamete pool**.

The percentages of gametes in the gene pool for a pair of alleles (A and a) depend upon the genotypic frequencies of the parental generation whose gametes form the pool. Thus, if a population is of dominant genotype AA, then the frequency of dominant alleles in the gene pool will be relatively high and the percentage of gametes bearing the recessive (a) allele will be correspondingly low.

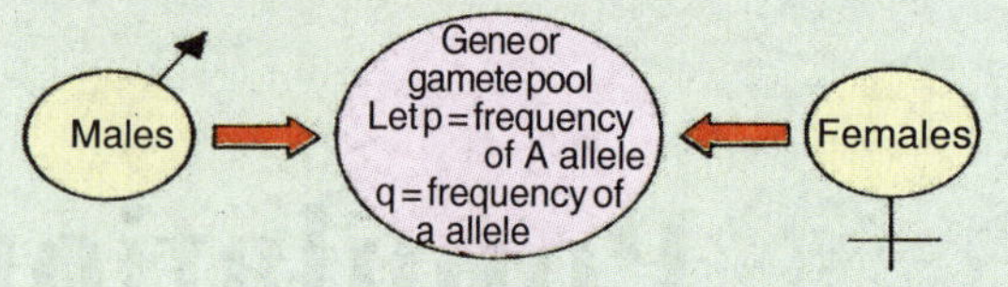

Fig. 7.1. The concept of gene pool (after Stansfield, 1969).

Two Models of Gene-pool Structure

Regarding the genetic structure of the population, the following two hypotheses have been proposed :

1. Classical hypothesis. It was developed by **T.H. Morgan** (1932) and supported by **H.J. Muller** and **Kaplan** (1966). The classical hypothesis proposes that the gene pool of a population consists at each gene locus of a wild-type allele with a frequency approaching one. Mutant alleles in very low frequencies may also exist at each locus. A typical individual would be homozygous for the wild-type allele at most gene loci; at a very small proportion of its loci the individual would be heterozygous for a wild and a mutant allele. Except in the progenies of consanguineous matings, individuals homozygous for a mutant allele would be extremely rare. The "normal" ideal genotype would be an individual homozygous for the wild-type allele.

According to classical hypothesis, mutant alleles are continuously introduced in the population by mutation pressure, but are generally deleterious and, thus, are more or less gradually removed from the population by natural selection. Periodically a beneficial mutant allele might arise, conferring higher fitness upon its carriers than the pre-existing wild-type allele. This beneficial allele would gradually increase in frequency by natural selection to become the new wild type allele, while the former wild type allele would be eliminated. Evolution, thus, consists of the replacement at an occasional locus of the pre-existing wild-type allele by a new wild-type allele (see **Ayala**, 1976).

2. Balance hypothesis. This hypothesis was proposed by **Dobzhansky** (1970) and **E.B. Ford** (1971). This hypothesis was derived by direct study of natural populations. According to the balance model, there is generally no single wild-type or "normal" allele. Rather, the gene pool of a population is envisioned as consisting at most loci of an array of alleles in moderate frequencies. A typical individual is heterozygous at a large proportion of its gene loci. There is no "normal" or ideal genotype, only an adaptive norm consisting of an array of genotypes that yield a satisfactory fitness in most environments encountered by the population.

The proponents of the balance hypothesis argue that the ubiquitous allelic polymorphisms are maintained in populations by various forms of balancing natural selection. The fitness conferred on its carriers by an allele depends on what other alleles exist in the genotype at that and other gene loci. It also depends, of course, on the environment. Gene pools are coadapted systems; the sets of alleles favoured at one locus depend on the sets of alleles that exist at other loci. Evolution occurs by gradual change in the frequencies and kinds of alleles at many gene loci. As the configuration of the set of alleles changes at one locus, it also changes at many other loci.

The balance model of genetic structure of populations has now become definitely established, although some controversy remains regarding the process maintaining the ubiquitous polymorphisms (Ayala, 1974; Lewontin, 1974).

CHANCE MATING OR PANMIXIS

Every member of Mendelian population depends on chance or random mating.

Just as the gene frequency is controlled in the genetic laboratory, so is the mating pattern, generation after generation. But outside the laboratory, mating is a chance or random affair. Every member of Mendelian population, thus, depends on chance or random matings. In other words, every male gamete in the gene pool has an equal opportunity of uniting with every female gamete. Zygotic frequencies expected in the next generation from such random gametic unions may be predicted from a knowledge of the gene (allelic) frequencies in the gene pool of the parental population. For example, the expected zygotic frequencies for allele A and a of a gene pool can be determined by chance mating or panmixis. If p stands for percentage of A alleles in the gene pool and q stands for the percentage of a alleles, the checkerboard of both alleles may predict possible chance combinations of A and a gametes as follows:

Male gametes → / Female gametes ↓	p (A)	q (a)
p(A)	(AA) p^2	(Aa) pq
q(a)	(Aa) pq	(aa) q^2

or
$$(p+q)^2 = p^2 + 2pq + q^2$$

Thus, p^2 is the fraction of the next generation expected to be homozygous (AA), 2pq is the fraction expected to be heterozygous (Aa) and q^2 is the fraction expected to be recessive (aa).

The zygotic combinations predicted in a randomly mating population may be represented by $p^2 : 2pq : q^2$, where p^2 represents the AA genotype, 2pq the Aa, and q^2 the aa genotype ; or, in the form of equation, $p^2+2pq+q^2=1$. When only two alleles are involved, and, therefore, p and q represent the frequencies of all of the alleles concerned $p+q=1$. Since $p+q=1$, $p=1-q$. Now if $1-q$ is substituted for p, all relations in the formula can be represented in terms of q as follows :

$$(1-q)^2 + 2q(1-q) + q^2 = 1.$$

HARDY-WEINBERG LAW

The formula $(p+q)^2=p^2+2pq+q^2$ is expressing the genotypic expectations of progeny in terms of gametic or allelic frequencies of the parental gene pool and is originally formulated by a British mathematician **Hardy** and a German physician **Weinberg** (1908) independently. Both forwarded the idea, called **Hardy-Weinberg law equilibrium** after their names, that *both gene frequencies and genotype frequencies will remain constant from generation to generation in an infinitely large interbreeding population in which mating is at random and no selection, migration or mutation occur.* Should a population initially be in disequilibrium, one generation of random mating is sufficient to bring it into genetic equilibrium and thereafter the population will remain in **equilibrium** (unchanged in gametic and zygotic frequencies) as long as Hardy-Weinberg condition persists. Hardy-Weinberg law depends on the following kinds of genetic equilibriums for its full attainment :

1. The population is infinitely large and mate at random.
2. No selection is operative.
3. The population is closed, *i.e.,* no immigration or emigration occurs.
4. No mutation is operative in alleles.
5. Meiosis is normal so that chance is the only factor operative in gametogenesis.

The significance of Hardy-Weinberg equilibrium was not immediately appreciated. A rebirth of biometrical genetics was later brought about with the classical papers of **R.A. Fisher**, beginning in 1918 and those of **Sewall Wright**, beginning in 1920. Under the leadership of these mathematicians, emphasis was placed on the population rather than on the individual or family group, which had previously occupied the attention of most Mendelian geneticists. In about 1935, **T. Dobzhansky** and others started to interpret and to popularize the mathematical approach for studies of genetics and evolution.

Genetic Equilibrium

As shown by **Hardy** and **Weinberg**, alleles segregating in a population tend to establish an equilibrium with reference to each other. Thus, if two alleles should occur in equal proportion in a large, isolated breeding population and neither had a selective or mutational advantage over the other, they would be expected to remain in equal proportion generation after generation. This would be a special case because alleles in natural populations seldom if ever occur in equal frequency. They may, however, be expected to maintain their relative frequency, whatever it is, subject only to such factors as chance, natural selection, differential mutation rates or mutation pressure, meiotic drive and migration pressure, all of which alter the level of the allele frequencies. A genetic equilibrium is maintained through random mating.

APPLICATION OF HARDY-WEINBERG LAW IN CALCULATING GENE FREQUENCIES IN A POPULATION

The gene frequencies for the autosomal and sex-chromosomal allele can be determined by the help of Hardy-Weinberg law by the following method :

Calculation of Gene Frequencies of Autosomal Genes

An autosomal gene locus may have codominant alleles, dominant and recessive alleles or multiple alleles. If one desires to determine the gene frequencies for each of these kinds of autosomal alleles in a given population, he has to adopt the different methods.

Calculation of gene frequencies for codominant alleles. When codominant alleles are present in a two-allele system, each genotype has a characteristic phenotype. The numbers of each allele in both homozygous and heterozygous conditions may be counted in a sample of individuals from the population and expressed as a percentage of total number of alleles in a sample. If the sample is representative of the entire population (containing proportionately same numbers as found in the entire population) then we can obtain an estimate of the allelic frequencies in the gene pool. If in a given sample of N individuals of which D are homozygous for one allele (A^1A^1), H are heterozygous (A^1A^2), and R are homozygous for the allele (A^2A^2), then N=D+H+R. Since each of the N individuals are diploid at this locus, there are 2N alleles represented in the sample. Each A^1A^1 genotype has two A^1 alleles. Heterozygotes have only one A^1 allele. Letting p represents the frequency of the A^1 allele and q the frequency of the A^2 allele, we have

$$P = \frac{2D+H}{2N} = \frac{D+1/2H}{N}, \quad q = \frac{H+2R}{2N} = \frac{1/2H+R}{N}$$

Example. The M-N blood type furnishes a useful example of a series of phenotypes due to a pair of codominant alleles. None of three possible phenotypes, M, MN and N, appears to have any selection value. The frequencies of the two alleles (*viz.*, L^M and L^N) for a sample from a group of white Americans living in New York City, Boston, and Columbus, Ohio, can be calculated by the following methods:

The sample of 6,129 Caucasian people includes the following three groups according to phenotypes and genotypes on M-N system :

Table 7-1.

Phenotype	Genotype	Number
1. M	L^ML^M	1,787
2. M N	L^ML^N	3,039
3. N	L^NL^N	1,303
		Total=6,129

To calculate frequencies of the two codominant alleles, L^M and L^N, it should be kept in mind that these 6,129 persons possess a total of 6,129 × 2 = 12,258 genes. The number of L^M alleles, for example, is 1,787+1,787+3,039. Thus, calculation of the frequency of L^M and L^N alleles is worked out in this way.

(i) $L^M = \frac{1{,}787, \ 1{,}787 \ 3{,}039}{12{,}258} \quad \frac{6{,}613}{12{,}258} \quad 0.5395$

(ii) $L^N = \frac{1{,}303 \ 1{,}303 \ 3{,}039}{12{,}258} \quad \frac{5{,}645}{12{,}258} \quad 0.4605$

Thus, the frequencies of the two codominant alleles in this sample are almost equal, and this is reflected in the close approximation to a 1 : 2 : 1 ratio, which is a simple monohybrid ratio for codominant alleles in Mendelian genetics.

Gene frequencies expressed as decimals may be used directly to state probabilities (A probability is a function that represents the likelihood of occurrence of any particular form of an event). If we can assume this sample to be representative of the population, then there is a probability of 0.5395 that of the chromosomes bearing this pair of alleles, any one selected randomly will bear gene L^M, and 0.4605 that it will bear L^N. Let p represents genotypic frequency of L^M allele and q represents frequency of L^N allele, then the frequencies of three phenotypes to be expected in the population are as follows:

Table 7-2. **Expansion of Hardy-Weinberg formula $(p+q)^2=p^2+2pq+q^2$, according to L^M and L^N gene frequencies in a population sample.**

Genotype	Phenotype	Genotype frequencies
L^ML^M	M	p^2 = 0.5395 × 0.5395 = 0.2911
L^ML^N, L^NL^M	MN	2pq = 2 (0.5395 × 0.4605) = 0.4968
L^NL^N	N	q^2 = 0.4605 × 0.4605 = 0.2121
		1.0000

FACTORS INFLUENCING ALLELE FREQUENCY OR DEVIATIONS FROM HARDY-WEINBERG EQUILIBRIUM

The Hardy-Weinberg explanation of equilibrium in the allele frequency pattern of a population required three assumptions–(1) Individuals with each genotype must be as reproductively fit as those of any other genotype in the population; (2) The population must consist of a large number of individuals; and (3) Random mating must occur throughout the population. The Hardy-Weinberg theorem with its assumptions does not account for any change in allele frequency within populations. That is just what Hardy and Weinberg intended because their formula described the statics of a Mendelian population. Something more was required to formulate a mathematical explanation of change or dynamics in terms of allele frequencies. This need was filled by **Fisher**, **Wright** and **J.B.S. Haldane**, who provided additional theoretical models and superimposed the mechanisms for change in allele frequencies upon the Hardy-Weinberg equilibrium. Population statics was, thus, extended to become population dynamics.

The shifts or changes in gene frequencies can be produced by a reduction in population size, selection, mutation, chance (genetic drift), meiotic drive and migration.

A. Selection

1. Evolution. Evolution has been described by **Lewontin** (1967) as a process that converts variation within a population into variation between populations, both in space (race formation and speciation) and in time (the evolution of phyla). The most prominent theory regarding the driving force behind evolution is of course the Darwin-Wallace theory of **natural selection** proposed in 1858 by **C. Darwin** and **A. Wallace.** The natural selection theory holds that as genetic variants arise within a population the fittest will be at a selective advantage and will be more likely to produce offspring than the rest. As the fit continue to enjoy greater survival, reproductively new species will eventually evolve.

2. Natural selection. Natural selection is generally believed to be the prominent agent for determining the relative frequency of alleles in a population. Natural selection differentiates between phenotypes in a population with respect to their ability to produce offspring. One phenotype may better survive endemic onslaughts of parasites or predators than another, one may penetrate new habitats more effectively than another; one may mate more efficiently than other; one may even prey on the other. The important point is that some natural situation or environmental feature selectively allows one organism to develop and to progpagate more efficiently, and one genotype is thereby afforded greater representation in the populations gene pool. If this selective process continues over many generations, allelic frequencies will change significantly and the potential will arise for evolutionary change. Further, the natural selection process, while acting on the total phenotype, will, in fact, influence only the heritable portion of the phenotype. If a trait has a high heritability, selection can rapidly affect its frequency within a population, whereas selection will take a great deal longer to have any effect on a trait with a low heritability.

Natural selection includes the following three parameters—**survival rate, relative fitness** and the **selection coefficient**. To calculate values for these parameters and, thus, best understand their meaning, we can consider the data from a particular population. For simplicity we shall consider only a single gene locus in the population defined by the A and a alleles and we shall assume incomplete dominance, so that A/a heterozygote can be distinguished phenotypically from A/A, and a/a homozygotes. We then count the number of A/A, A/a, and a/a individuals in a given generation immediately before and immediately after some selective event—the introduction of a parasite, a change in temperature—has occurred. The representative numbers have been given in Table 7-3. From these we can calculate a survival rate, relative fitness and selection coefficient, as shown in the table.

Table 7-3. Relative fitness and selection coefficients calculated from data taken in the same generation of population.

Data : Number of individuals in the population according to the genotype

	A/A	A/a	a/a
Before selection	4100	5000	2200
After Selection (same generation)	3900	4000	1200

Calculations :

(a) **Survival rate**

$$A/A = \frac{3900}{4100} = 0.95$$

$$A/a = \frac{4000}{5000} = 0.80$$

$$a/a = \frac{1200}{2200} = 0.55$$

(b) **Relative fitness W** (compared with A/A' s maximum survival rate)

$$W_{AA} = \frac{0.95}{0.95} = 1.00$$

$$W_{Aa} = \frac{0.80}{0.95} = 0.84$$

$$W_{aa} = \frac{0.55}{0.95} = 0.58$$

(c) **Selection coefficients s**

$$s_{AA} = 1-W_{AA} = 0$$

$$s_{Aa} = 1-W_{Aa} = 0.16$$

$$s_{aa} = 1-W_{aa} = 0.42$$

The genotype with the largest survival rate is defined as the fittest, and it is used as the standard for the **relative fitness** (W) of all other genotypes. Specifically, in our example, A/A has a survival rate of 0.95 and is, thus, fitter than A/a or a/a (Table 7-3). In determining values for W, therefore, all survival rates are divided by 0.95 so that the relative fitness of A/A becomes 0.95/0.95=1, the relative fitness of A/a becomes 0.80/0.95=0.84, and that of a/a becomes 0.55/0.95=0.58.

The **selection coefficient** (s) is simply 1–W. Just as W reflects the chances of an organism's reproductive success, so does s reflect the chances of its reproductive failure due to selection. They are simply two sides of the same coin.

3. Directional selection. A principal pattern followed by selection in natural populations is that of **directional selection**. Its effect is to eliminate or to reduce the reproductive potential of particular phenotypes in a population. Specifically, if the variance in a particular trait is normally distributed in a population before directional selection begins to operate, it might become markedly skewed in one direction afterwards (Fig.7.2).

Directional selection in its extreme form eliminates systematically the recessive homozygotes (a/a) from the population. In such cases the relative fitness of a/a is set at 0 and its selection coefficient, therefore, becomes 1–0=1. If we assume, for simplicity, that A/A and A/a have identical fitnesses and that no other alleles at the locus are involved, then by definition A/A and A/a each have fitness of 1 and selection coefficients of 0. Further, when directional selection remains in its incomplete form against homozygous recessives, then selection coefficients is less than 1 and the relative fitness W of the homozygous recessives is 1–s, a number greater than zero.

4. Artificial selection. The artificial selection is simply a selection programme devised by humans rather than managed by natural situations. The process of enriching a population for a given trait normally occurs far more rapidly under the artificial situation, since all but the desired phenotypes can be prevented from reproducing. Countless experiments have shown that it is possible to select for just about any trait present in a population, a major limitation being the heritability of the trait. Strains with particular shapes, sizes, behaviour patterns, temperature optima, sexual preferences, and so on have been selected. **S. Spiegelman** and colleagues have even subjected isolated RNA molecules from RNA phage Oβ to artificial selection and they reported that a rigorous selection for one trait can lead to the drastic exclusion of many other traits.

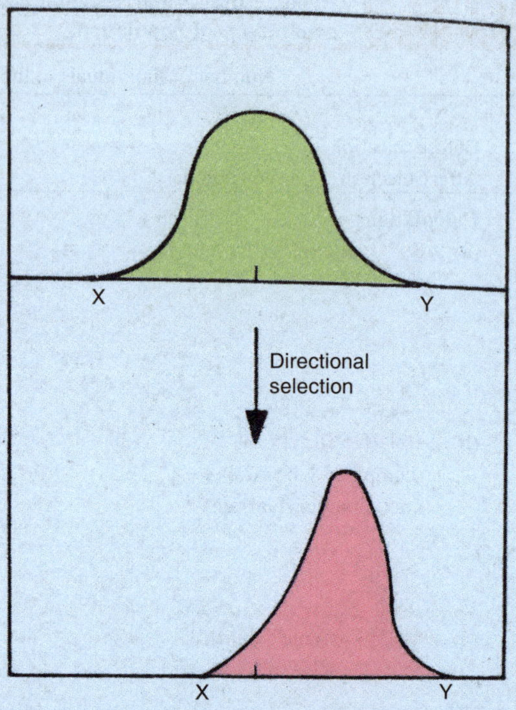

Fig. 7.2. The effects of directional selection on a population whose phenotypes are normally distributed between two extremes, X and Y. Directional selection acting against X and in favour of Y changes the distribution of the phenotypic variation (after Goodenough and Levine, 1974).

5. Significance of the heterozygote. When the frequency of an abnormal recessive gene becomes very low, most affected offspring (aa) will come from matings of two heterozygous carriers (Aa). For example, in the human population, the vast majority of newly arising albino individuals (*aa*) in a given generation (more than 99 per cent of them) will come from normally pigmented heterozygous parents.

Detrimental recessive genes in a population are obviously harboured mostly in the heterozygous state. The frequency of heterozygous carriers is many times greater than the frequency of homozygous individuals afflicted with the trait. Thus, an extremely rare disorder such as **alkaptonuria** (blackening of urine) occurs in 1 in one million persons. This detrimental gene, however, is carried in the hidden state by 1 out of 500 persons. There are two thousand as many genetic carriers of alkaptonuria. For another recessive trait, **cystic fibrosis**, 1 out of 1,000 individuals is affected with this homozygous trait. One out of 16 persons is a carrier of cystic fibrosis (see Table 7-4).

Genetic load (Concealed variability) and price of evolution. Genetic load can be defined as the relative deficiency in viability or fecundity or both viability and fecundity as compared to that which would occur if the population were all of the most fit phenotype. Thus, genetic load is the difference between the actual fitness and an optimum fitness, this difference being caused by the presence of suboptimal genetic variants, concealed or not, within the population.

In fact, mutants arise in any species and many or most of them are deleterious. Some of these mutants may persist in the populations, *i.e.,* one or several generations intervene between the origin and the populations of any species, including the best adapted and most successful ones, carry burdens or loads of genetic defects. One can distinguish between **expressed** (overt) or **concealed genetic loads**.

Table 7-4. Frequencies of recessive homozygotes and heterozygous carriers (Source Volpe, 1985).

Frequency of homozygotes (aa)	Frequency of heterozygous carriers (Aa)	Ratio of carriers to homozygotes
1. 1 in 500 (Sickle cell anaemia)	1 in 10	50 : 1
2. 1 in 1,000 (Cystic fibrosis)	1 in 16	60 : 1
3. 1 in 6,000 (Tay-Sachs disease)	1 in 40	150 :1
4. 1 in 20,000 (Albinism)	1 in 70	285 : 1
5. 1 in 25,000 (Phenylketonuria)	1 in 80	310 : 1
6. 1 in 50,000 (Acatalasia)	1 in 110	460 : 1
7. 1 in 1,000,000 (Akaptonuria)	1 in 500	2,000 : 1

The former comprises individuals with diseases, malformations, or constitutional weaknesses that are manifestly genetic. Concealed genetic burdens are due to recessive genes or gene complexes carried in heterozygous conditions in individuals who are themselves healthy or "normal". When homozygous, however, these genes too become lethal.

Types of genetic loads. According to **Fransworth** (1988), the genetic loads are of the following three types :

1. Mutational load. It refers to the deleterious effects on a population caused by recurrent mutation.

2. Segregational load. It arises when the **heterozygote** is the favoured genotype. The formation of heterozygotes is necessarily accompanied by the generation of the less well-adapted homozygous genotypes whose lowered viability and fecundity cause the segregational load. Segregational load is, thus, the loss in average fitness of a population caused by the segregation of the less fit homozygous genotypes. Because selection acts to preserve the heterozygous genotype, neither allele, even if one or both are lethal, is eliminated from the population. As an example, in malarial regions of Africa the loss or lowered fecundity of HbS and HbA homozygotes as compared to heterozygotes (HbA/HbS) constitute the segregational load for that locus.

3. Substitutional or transient load. It occurs when a change in the environment results in directional selection in favour of a previously neutral or unfavourable allele whose effects in the altered environment now confer greater fitness. The substitutional load, thus, consists of the relative number of deaths and lack of birth which the population must allow in order to progress from the original new state of fitness. Substitutional load is transient in that it occurs only during the replacement of one allele by another. For example, in the United States where falciparum malaria is absent, HbA homozygotes possess the best adapted phenotype, and directional selection in favour of the normal allele is occurring. As a result, the lowered fecundity and viability of HbS/HbA heterozygotes and HbS homozygotes, as compared to normal individuals contribute a transient substitutional load to the black population of the United States.

The substitutional genetic load has been called the **price of evolution** and **J.B.S. Haldane** (1957) has made some theoretical calculations of the extent of this price in terms of reduced fitness or loss of individuals. He has estimated that a selection intensity of 0.1, approximately 300 generations will be required to replace a gene at a cost in members lost by death or sterility of 30 times the average number of individuals present in a single generation. Thus, according to these calculations, approximately 10 per cent of a population would be lost per generation if gene replacement occurred over a time span of 300 generations.

Because of the enormous amount of protein, and, therefore, genetic polymorphism recently revealed at the molecular level by immunological and electrophoretic methods, these calculations pose a dilemma. Evolution requires the concurrent replacement of many genes, and to the substitutional load borne by a population must be added to the mutational load as well as segregational load contributed by stabilizing selection. It would appear that no population could withstand such a drain on its members, especially if the effects of individual subvital genes were independent and, therefore, multiplicative. The problem, then, is how are these numerous genic polymorphisms maintained without increasing a genetic load so heavy as to cause the extinction of a population ? The following two hypotheses have been proposed to resolve this problem :

(1) Selection hypothesis of genetic load. This hypothesis proposes that observed polymorphisms are each due to heterozygote advantage and are maintained in a population by various types of stabilizing or balancing selection. Examples include balanced polymorphism, such as sickle cell heterozygotes, as well as cases where one or another allele or chromosomal inversion favoured in different seasons, regions or stages of life cycle or in different sexes. Frequency-dependent selection also serves to maintain polymorphism, and at equilibrium the alleles involved are selectively neutral, contributing no genetic load.

Supporters of the selection theory suggest that the primary cause of the extensive heterozygosity observed in nature is heterosis at hundreds or even thousands of gene loci, the cumulative effects conferring fitness. In 1967, various workers such as **Sved**, **King**, **Milkman** and others have proposed that the genetic load should not be estimated on the assumption that each locus contributes to this load independently and that the individual effects are multiplicative. In view of the known interaction between genes they suggest that the values of the sums of these effects exhibit a normal distribution in a population. *Since selection scans the entire organism and not single gene loci, only those individuals that fall below a certain threshold of fitness, that is, those with fewer heterotic loci than a minimum number required for average fitness, are collected from the population.* As a result, members of a population are divided into two classes, those above and those below a given threshold and under these circumstances it is estimated that perhaps 1,000 selectively maintained polymorphisms, each conferring an advantage to heterozygotes of around 1 per cent, could persist in a population with only minimal effects as genetic load.

2. Neutralist theory of genetic load. This theory was proposed and developed by **J.F. Crow**, **K. Kimura**, **T. Ohta**, and **T.H. Jukes** during 1968 and 1983. These workers suggest that the majority of thousands of genic polymorphisms discovered at the biochemical level have no significant effect on fitness and, therefore, selectively neutral, making no contribution to the genetic load. They propose that adaptation does occur through natural selection, but that selection acts primarily to remove deleterious mutations from a population. Proponents of the neutralist theory indicate that while mutations which result in the substitution in a protein of a compatible amino acid are detectable by sophisticated methodology, they may not be detected by natural selection because the function of the protein will remain essentially unchanged. Under these circumstances the frequencies of polymorphic genes conferring no advantage or disadvantage will be determined primarily by random genetic drift rather than by selective processes.

Genetic load in human populations. Estimates of the genetic load in the human population have been based principally on the incidence of defective offspring from marriages of close relatives (consanguineous marriages). It can be safely stated that every human individual contains at least one newly mutated gene. It can also be accepted that any crop of gametes contains, in addition to one or more mutations of recent origin, at least 10 mutant genes that arose in the individuals of preceding generations and which have accumulated in the population. The average person is said to harbour four concealed lethal genes, each of which, if homozygous, is capable of causing death between birth and maturity. The most conservative estimates place the incidence of deformities to detrimental mutant genes in the vicinity of 2 per 1,000 births.

B. Mutation

Mutation is an evolutionary agent and mutability is a required property of the genetic material if evolution is to occur. One can visualize several ways that mutation might bring about evolutionary changes. Mutation may be highly directed at a particular locus such that allele a_1 is selectively driven to the a_2 form. Alternatively, the mutation process might be **random** but with time the a_2 form would come to predominate over a_1. Finally, mutation might simply provide a population with new alleles (new mutational "currency") on which any and all evolutionary agents (natural selection, for example) can act.

C. Meiotic Drive and Migration Pressure

Besides directional selection (natural and artificial) and mutation pressure, meiotic drive and migration pressure are two agents that can shift gene frequencies in a population out of a Hardy-Weinberg equilibrium.

1. Meiotic drive. The Hardy-Weinberg concept of an equilibrium population assumes that alleles will segregate in a 1 : 1 fashion at meiosis and that all gametes in the pool have an equal probability of fertilizing one another. Clearly, if gametes of genotype a_1 have greater success in fertilization than gametes of genotype a_2 the frequency of a_1 in the population should increase as a_2 decreases. Such a mode of bringing about potential evolutionary change is in many respects a form of natural selection but it has earned the specific designation of **meiotic drive**.

Normal segregation ratios can be biased by a number of factors. The most extensively studied example is the segregation distorter (SD) locus in *D. melanogaster* : a +/SD heterozygous male transmits to his offspring many SD alleles than + alleles. A high prevalence of SD might, therefore, be expected in natural populations, but the allele is, in fact, uncommon, its frequency being less than 0.1. Some additional selective pressures, thus, appear to counteract meiotic drive at this particular locus presumably by agents that act against SD-carrying flies. The existence of meiotic drive is not easy to demonstrate experimentally. However, its specific role in evolution is still unassessed.

2. Migration pressure. Perhaps the most obvious way that gene frequencies in a population can be changed is the introduction of new individuals with new genotypes, a process known as **migration pressure**. The effect of migration pressure can be expressed mathematically as follows. In a population we can call X a particular gene is present with a frequency q_x (Fig. 7.3). A certain number of individuals from this population now migrate and join population Y in which the gene in question is present at frequency q_Y. The immigrant, thus, comes to represent some fraction m of the total number of individuals present in the expanded version of population Y and a new gene frequency, q'_Y, becomes established. The value of q'_Y, will be equal to the contribution made by the immigrants ($q_x m$) plus the contribution made by the original population [q_Y (1–m)], where 1–m represents the proportion of non-migrants. Therefore,

$$q'_Y = q_X m + q_Y (1-m)$$

and the change in q after one generation in such a population becomes

$$\begin{aligned} \Delta q &= q'_Y - q_Y \\ &= q_Y - mq_Y + mq_X - q_Y \\ &= -m(q_Y - q_X) \end{aligned}$$

when numerical values are substituted into this equation, it becomes clear that significant changes in the value of q can result if these two populations differ only slightly in the frequency of a given gene and if only a moderate degree of migration occurs between them.

D. Random Genetic Drift

Random fluctuation in allele frequencies, called **genetic drift**, also occurs in breeding populations. The effect of genetic drift is negligible in large populations but in small breeding populations all the limited number of progeny might be of the same type with respect to certain gene pairs because of chance alone. Should this happen, fixation or homozygosity will have occurred at the locus concerned. Fixation is defined as gene frequency reaching $p = 1.00$ or $q = 1.00$. Chance fluctuations may or may not lead to fixation.

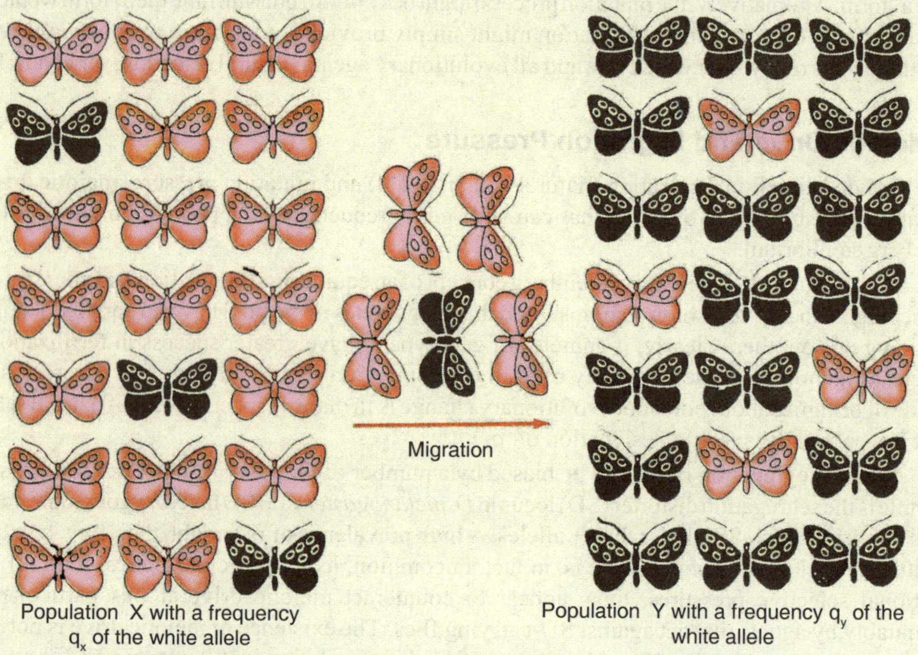

Fig. 7.3. Migration pressure as an evolutionary agent. The Y population of moths receives some immigrants from the X population, an event that changes its frequency of the white allele (after Goodenough and Levine, 1974).

E. Founder Principle

When a few individuals or a small group migrate from a main population, only a limited portion of the parental gene pool is carried away. In the small migrant group, some genes may be absent or occur in such low frequency that they may be easily lost. The unique frequency of genes that arise in population derived from small bands of colonizers or "founders", has been called the **founder effect** or **founder principle**. This principle was proposed in 1956 by Harvard evolutionist, **Ernst Mayr**. The founder principle essentially emphasizes the conditions or circumstances that support the operation of Sewall Wright's genetic drift. For example, North American Indian tribes, for the most part, lack the gene I^B that governs type B blood. However, in Asia, the ancestral home of the American Indians, the I^B gene is widespread. The ancestral population of Mongoloids that migrated across the Bering Strait to North America might have been very small. Accordingly, the possibility exists that none of the prehistoric immigrants happened to be blood group B. It is also likely that a few individuals of the migrant band did carry the I^B gene but they failed to leave descendants.

Evolutionary geneticists interpret this peculiar feature in terms of genetic drift. Most of the North American Indians possess only blood group O, or stated another way, contain only the blood allele *i*. With few exceptions, the North American Indian tribes have lost not only blood group allele I^B but also the allele that controls type A blood (I^A). The loss of both alleles, I^A and I^B, by sheer chance perhaps defies credibility. Indeed, many modern students of evolution are convinced that some strong selective force led to the rapid elimination of the I^A and I^B genes in the American Indian populations. If this is true, it would provide an impressive example of the action of natural selection in modifying the frequencies of the genes in a population.

GENETIC POLYMORPHISM

One of the most characteristic features of any natural population is its diversity. This diversity is obvious when we consider the human species, for we are attuned to sensing differences in human appearance, personality, sexuality, and so on. In the populations of flies or dandelions such well marked diversity does not occur but it exists nevertheless. In genetic terminology natural populations are said to be **polymorphic**.

Polymorphism is most apparent when it affects a visible or behavioral phenotype, but is not at all restricted to such traits. **R. Lewontin** and **J. Hubby**, in 1966, undertook the first extensive analysis of protein polymorphisms in natural population of *Drosophila pseudoobscura* by subjecting extracts of individual flies to get electrophoresis and observing the rates of migration of various proteins, which represented 18 gene loci. They found, quite unexpectedly, that many of the proteins existed in the population in the form of **isoelectric variants**, meaning that for a given type of protein some individuals possessed a fast-migrating species and others a slow-migrating species. Numerous subsequent studies of such diverse species as barley, wild oat, horse-shoe crab, mouse and man have all produced the same result : an abundance of **protein polymorphism** is found wherever it is sought. Protein polymorphism signals the existence of **allelism**, and it has been estimated that 20 to 50 per cent of all structural gene loci in a given species exist in two or more allelic forms in any given population.

The polymorphism may arise in a population by the following three basic avenues—**transient polymorphism**, **balanced polymorphism** and **random fixations of natural mutations**.

1. Transient polymorphism. Transient polymorphism is a by-product of directional natural selection. If we imagine that allele a_1 has a selective advantage over a_2, then with time a_1 should proceed toward fixation at $p = 1$, and a_2 should proceed toward elimination at $q = 0$. While this process is occurring, both a_1 and a_2 will be present in the gene pool and a_1/a_2 heterozygotes will be present in the population. As the name implies, transient polymorphism represents a temporary situation. For example, during the course of industrial melanization, both dark and light moths would be expected to cohabit the Manchester trees for the interim, but the proportion of light moths would be seen to diminish with time as dark moths gradually predominated.

2. Balanced polymorphism. Balanced polymorphism is also relatively permanent kind of equilibrium in which alleles a_1 and a_2 are present in the population at some steady-state frequencies. Balanced polymorphism is originated by **disruptive** or **diversifying selection** and **heterosis**.

3. Random fixation of natural fixation. The random fixation of natural fixation method of origin of polymorphism is also called **Neutral Mutation–Random Genetic Drift hypothesis** or "**Non-Darwinian Evolution**" and this idea has been developed by **S. Wright** and **Kimura**. This hypothesis is based on the following two assumptions : (1) The first assumption states that selectively neutral mutations can occur in genes that code for proteins. This will clearly be true in the case of "synonymous" mutations in which one codon is replaced by another codon dictating the same amino acids, but it is also proposed to be true in the case of mutations that lead to amino acid substitutions. The idea is that an acidic amino acid might occur in an "unimportant" region of the protein, with the result that the emergent mutant protein is identical to the original in all functional aspects. (ii) The second assumption states that neutral alleles, being neither selectively advantageous nor disadvantageous, simply drift in the gene pool. Thus, if a neutral mutation arises in woman's germ cell and this germ cell gives rise to a female child, the probability is about 0.5 that the mutant allele will be transmitted to a grandchild and 0.5 that it will not. If it is not, then q becomes equal to zero and the allele is lost. Polymorphism of cytochrome c of all eukaryotes and haemoglobin protein of all vertebrates strongly support the Wright-Kimura hypothesis.

POPULATION GENETICS AND EVOLUTION

Thus, population genetics has provided great support to the idea of organic evolution. Various aspects of evolution can be reviewed in terms of population genetics as follows :

Speciation

The process of formation of new species is called **speciation**. Two genetically divergent populations can form new species only when they become geographically isolated from each other. If a large population is fragmented into two or more units which are geographically isolated from one another, each independent unit follows different evolutionary paths for the following reasons :

1. Each isolated unit of a population may have its own type of mutation which provides raw materials for organic diversity.

2. The mutations and gene combinations which appear in different isolated population units will have different adaptive values in the new environments.

3. The organisms which originally colonize a certain geographical area and form an isolated population may not be representative of the group from which they came so that different gene frequencies exist from the beginning.

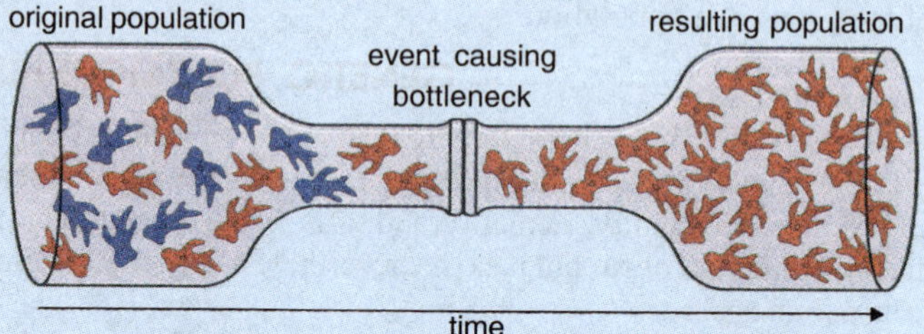

A In a population bottleneck, a population undergoes a drastic reduction in size — as a result of a natural catastrophe or overhunting. Then only a few individuals are left to contribute genes to entire future population.

B The cheetah passed through a population bottleneck in the recent past, resulting in an almost loss of genetic diversity.

4. The size of the new population may become quite small at various time so that a genetic "bottleneck" is formed, from which all subsequent organisms will arise.

5. During the period of small population size, the gene frequencies will fluctuate in unpredictable directions. The fluctuation in gene frequency is called **genetic drift**.

REVISION QUESTIONS

1. (a) What is Hardy-Weinberg equilibrium ?
   (b) What are the conditions necessary for the maintenance of this equilibrium in any population? Discuss.
   (c) What evidence is required before concluding that an allele pair is in *Hardy-Weinberg equilibrium* ?
   (d) What kinds of sexually reproducing species are likely to show this type of equilibrium ?
   (e) Clearly distinguish between allelic, genotypic and Hardy-Weinberg equilibrium, using a specific example to illustrate your answer.
2. Describe the significance of heterozygotes in the process of evolution.
3. What is genetic load ? Discuss it in relation of evolution.

CHAPTER

8

Evolution above Species Level

(Adaptation, Adaptive radiation, Microevolution, Macroevolution, Megaevolution, Punctuated equilibria and related phenomena)

Genetic changes in a species helps it to adjust to new or altered environmental conditions.

Adaptation can be defined as occurrence of genetic changes in a population or species as a result of natural selection so that it adjusts to new or altered environmental conditions. Thus, any characteristic that is advantageous to a particular organism or population is called an **adaptation (Savage**, 1969). For example, the lungs of land vertebrates make possible gaseous exchange between these organisms and the air and form an adaptation for terrestrial life. The fact of adaptation is confirmed by the diversity of life and the thousands of environmental situations inhabited by living organisms. (For various examples of adaptation see Part-V Ecology, Chapters 3 and 18).

ADAPTIVE RADIATION

The term **adaptive radiation** has been coined by **H.F. Osborn** (1902) for explaining evolution, from a single ancestor, of a number of descendants with a great variety of adaptations to different niches. More aptly stated the phenomenon of adaptive radiation is the diversification of a dominant evolutionary group into a large number of subsidiary types adapted to more restrictive modes of life (different adaptive zones) within the range of the larger group. According to **George Gaylord Simpson** (1940, 1953), adaptive radiation is the rapid proliferation of new taxa (species) from a single ancestral group. Certain authors of evolution biology such as **Savage** (1969), **Stanley** (1979) and **Volpe** (1985) have used an entirely new term **macroevolution** for the Osborn's law of adaptive radiation.

Examples of Adaptive Radiation

1. Adaptive radiation in Darwin's finches. A classical example of speciation involving the interplay of complex forces (such as isolation, competition, adaptation, etc.) leading to adaptive radiation at the species level is provided by Darwin's finches. These birds belong to largest family of birds, the Fringillidae and live in Islas Encantadas (Galapagos Islands) in the Pacific ocean. Significantly, the study of these small dark birds in their native habitat gave Darwin his first insight into evolutionary processes. **David Lack** (1947, 1953, 1969) has analysed the phenomenon of adaptive radiation of Darwin's finches very carefully.

The Galapagos Islands lie on the equator about 600 miles west of South America (*i.e.,* Ecuador). In the archipelago, there are 5 large islands in the group, with 19 smaller ones and 47 rocks. These islands are of volcanic origin (Fig. 8.1). They were never connected with the mainland of South America. The rugged shoreline cliffs are of grey lava and the coastal lowlands are parched, covered with cacti and thorn bushes. In the humid uplands trees flourish in rich black soil.

The present-day assemblages of Darwin's finches descended from small sparrow-like birds that once inhabited the mainland of South America. The ancestors of Darwin's finches were early migrants to the Galapagos Islands and probably the first land birds to reach the islands. These early colonists have given rise to 14 distinct species, each well adapted to a specific niche. Thirteen of these species occur in the Galapagos, one is found in the small isolated Cocos Island, northeast of Galapagos. Not all 13 species are found on each island. These 14 species belong to 4 genera: 1. *Geospiza;* 2. *Camarhynchus*; 3. *Certhidea;* and 4. *Pinaroloxias*.

The most conspicuous difference among the species are in the sizes and shapes of the beak, which are correlated with marked differences in feeding habits. Six of the species are ground finches, with heavy beaks specialized for crushing of seeds. Some of the ground finches live mainly on a diet of seeds found on the ground. For example, there are 3 seed-eater ground finches—small-beaked (*Geospiza fuliginosa*), medium-beaked (*Geospiza fortis*) and large-beaked (*Geospiza magnirostris*)— occur together in the coastal lowlands of several islands. Each species, however, is specialized in feeding on a seed of a certain size. The small-beaked finch feeds on small grass seeds, whereas large-beaked finch eats large, hard fruit. Other ground finches feed primarily on the flowers of prickly pear cacti. The cactus eaters possess decurved, flower-probing beaks. Their beaks are thicker than those of typical flower-eating birds.

All other species are tree finches, the majority of which feed on insects in the moist forests. One of the most remarkable of these tree dwellers is the woodpecker finch (Fig. 8.2). It possesses a stout, straight beak, but lacks the long tongue characteristic of the woodpecker. Like a woodpecker, it bores into wood in search of insect larvae, but then it uses a cactus spine or twig to probe out its insect prey from the excavated crevice. Equally extraordinary is the warbler finch which resembles in form and habit the true warbler. Its slender, warbler-like beak is adapted for picking small insects off bushes. Occasionally, like a warbler, it can capture an insect in flight.

Thus, the original ancestral stock of finches on the Galapagos diverged along several different paths. The pattern of divergence is reflected in the biologist's scheme of classification of these birds. All the finches are related to one another, but the various species of ground finches evidently are more related by descent to one another than to the members of the tree-finch assemblage. As a measure of evolutionary affinities, the ground finches are grouped together in one genus (*Geospiza*) and the tree finches are clustered in another genus (*Camarhynchus*). The different lineages of finches have been portrayed in Figure 8.3.

Darwin's finches like adaptive radiations within ecologic islands of any sort are called **ecoinsular radiations (Stanley**, 1969). Darwin's finches provide circumstantial evidence for the origin of a new species by means of geographical isolation.

2. Adaptive radiation in penguins. G.G. Simpson (1953) has cited evidence that when a group of organisms enters a new adaptive zone previously unoccupied by the group, there may be rapid bursts of speciation and adaptive divergence into a variety of ecological niches. In Fig. 8.4., we see Simpson's visualization of such adaptive radiation for penguins. Note that there are three zones for penguins : (1) aerial flight, (2) aerial and submarine flight and (3) submarine flight. These zones were sequentially invaded by penguins in their evolution and they are now extinct in the aerial flight zone or subzone.

3. Adaptive radiation in reptiles. The class Reptilia first appeared in the fossil record in Pennsylvanian times (250 million years ago). The ancestral reptilian stock initiated one of the most spectacular adaptive radiations in life's history. Adaptive radiation of the reptiles occurred between Permian and Cretaceous times, and living reptiles are derived from Cretaceous ancestors. The Mesozoic era, during which the reptiles thrived, is often referred to as the "**Age of Reptiles**".

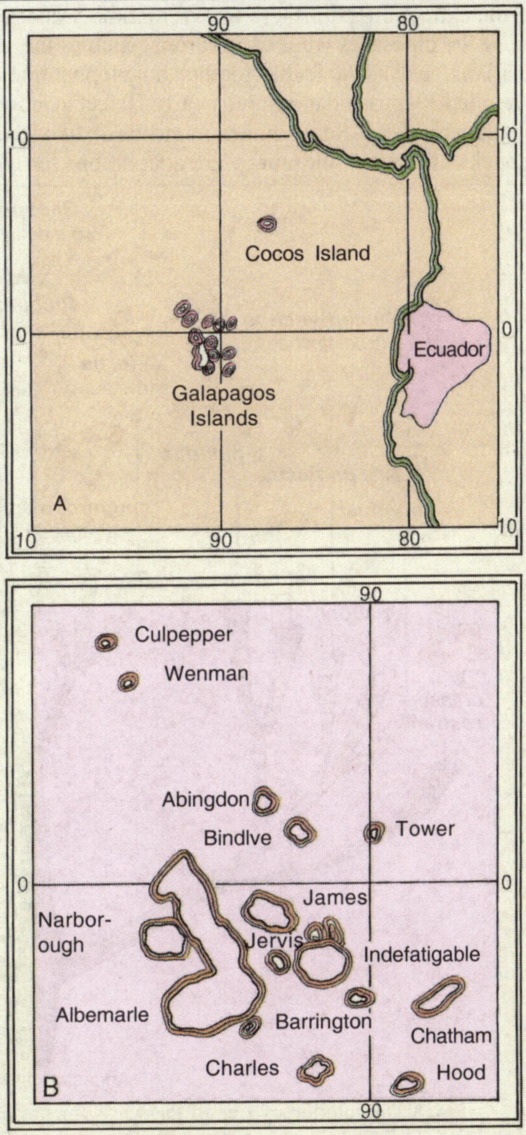

Fig. 8.1. A—Position of Galapagos Islands; B—Map of Galapagos Archipelago; Cocos Island is not in the group (after Lack, 1953).

Fig. 8.2. The woodpecker-finch. It is shown to carry a twig or cactus spine to dislodge insects from bark crevices (after Lack, 1953).

The initial success of the reptiles stem from a megaevolutionary shift from aquatic to completely terrestrial development, *i.e.*, the **cleidoic** or **amniotic** eggs of reptiles, like bird eggs, do not need to be immersed in water to survive. The basic stock of reptiles were the Cotylosauria (stem reptiles) from which a variety of reptiles were blossomed (Fig. 8.5). The **dinosaurs** were by far the most awe inspiring and famous. They ruled over the lands until the close of the Mesozoic era before suffering extinction. Dinosaurs were remarkably diverse; they varied in size, bodily form and habits. Some of the dinosaurs were carnivorous, such as the gigantic *Tyrannosaurus*, whereas others were vegetarians, such as the feeble-toothed but elephantine *Brontosaurus*. The bulky body of *Brontosaurus* weighed 30 tons and measured nearly 70 feet in length. Not all dinosaurs were huge; some were no bigger than chickens. Some dinosaurs strode on two feet; others had reverted to four. The exceedingly long necks of certain dinosaurs were adaptations for feeding on the foliage of tall coniferous trees.

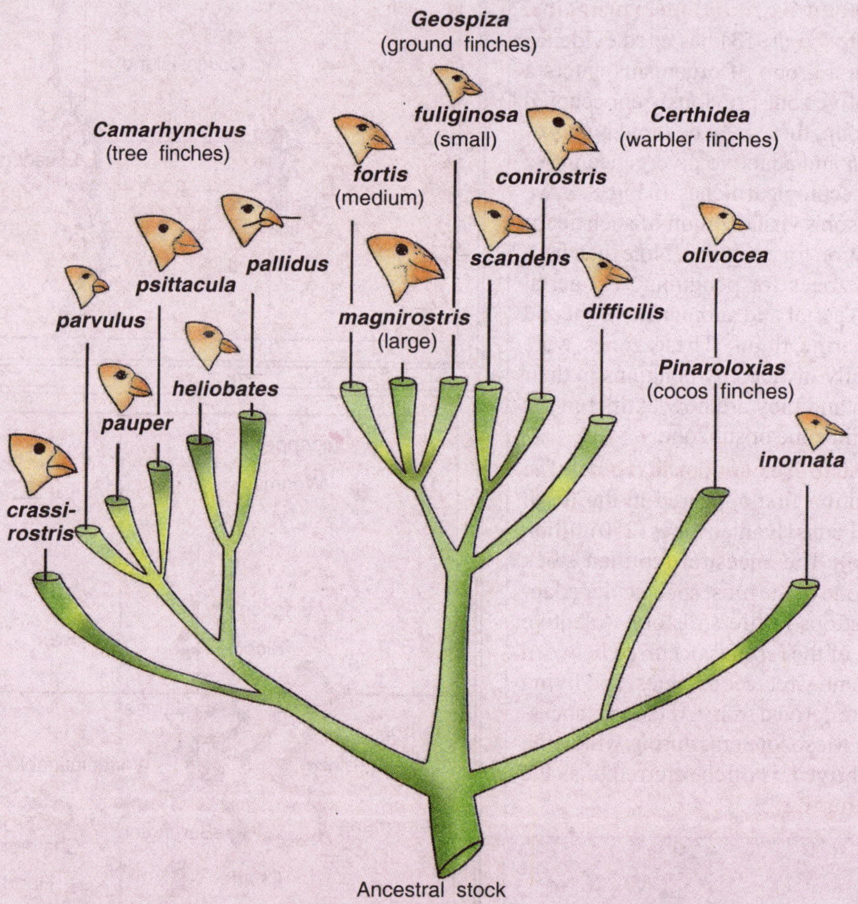

Fig. 8.3. Evolutionary tree of Darwin's finches based on Lack's findings. Darwin's finches evolved from a common stock and are represented today by 14 species. The species are assembled in four genera (after Volpe, 1985).

The dinosaurs were descended from the **thecodonts**—slender, fast-running lizard-like creatures. The thecodonts gave rise also to bizarre reptiles that took to the air, the **pterosaurs**. These "dragons of the air" possessed highly expansive wings and disproportionately short bodies. The winged pterosaurs extincted before the end of Mesozoic era. Another independent branch of the thecodonts led to exceptionally more successful flyers, the birds. The origin of birds from reptiles is revealed by the celebrated *Archaeopteryx*, a Jurassic form. The feathered creatures possessed a slender lizard-like tail and a scaly head equipped with reptilian teeth.

Certain mes-ozoic reptiles returned to water. The streamlined, dolphin-like **ichthyosaurs** and long-necked, short-bodied **plesiosaurs** were marine, fish-eating reptiles. The ichthyosaurs were expert swimmers; their limbs were fin-like and their tails were forked. The plesiosaurs were efficient predators, capable of swinging their heads 40 feet from side to side and seizing fish in their long, sharp teeth. These aquatic reptiles breathed by means of lungs; they did not redevelop the gills of their very distant fish ancestors. Indeed, it is axiomatic that *a structure once lost in the long course of evolution cannot be regained.* This is the doctrine of irreversibility of evolution, or **Dollo's Law**, after **Louis Dollo** (1895) the eminent Belgian palaeontologist to whom the principle is ascribed.

Further, among the early reptiles present before Mesozoic days (*i.e.,* Permian of Palaezoic era) were the **pelycosaurs** (*e.g., Dimetrodon*), notable for their peculiar sail-like extensions of the back. The function of the gaudy sail is unknown, but it should not be thought that this structural feature was merely ornamental or useless. It appears that the sail of pelycosaurs was a functional device to achieve some degree of heat regulation. Pelycosaurs gave rise to an important group of reptiles, the **therapsids**. These mammal-like forms bridged the structural gap between the reptiles and the mammals.

The history of the reptiles attests to the ultimate fate to many groups of organisms—extinction. The reptilian dynasty collapsed before the close of the Mesozoic era. Of the vast host of Mesozoic reptiles, relatively few have survived to modern times; the ones that have include the lizards, snakes, crocodiles and turtles.

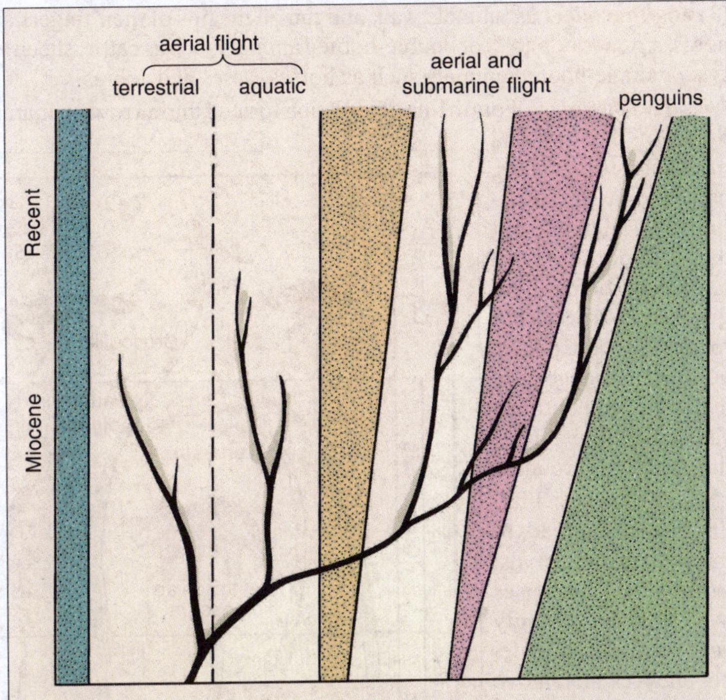

Fig. 8.4. Adaptive radiation (evolution) of penguin (after Hamilton, 1967).

Some authors maintain that the active thecodonts and dinosaurs were already warmblooded (**endothermic**) rather than, as often assumed, cold-blooded (**ectothermic**). An endothermic animal is not dependent on external sources of heat, but has the capacity to produce and control its own internal heat. The long reign of the dinosaurs could reflect, in part, the capacity for endothermy.

4. Adaptive radiation in eutherian mammals. The eutherian (placental) mammals provide another classical example of the process of adaptive radiation. In this case, from a primitive, insect-eating (insectivorous), short-legged, rat-like terrestrial creature that walked with the soles of its feet flat on the ground have evolved all the present day types of mammals. Thus, in respect to limb structure, among eutherian mammals, adaptive radiation occurred in the following five different lines:

Arboreal forms such as monkeys have adapted limbs for life in trees.

(i) Arboreal. One evolutionary line radiates to form **arboreal** forms which have adapted limbs for life in trees, *e.g.,* squirrels, sloth, monkeys, etc. The legs which are adapted for climbing are called **scansorial**.

(ii) Volant or aerial. Another line leads to

aerial or **volant**, representing mammals adapted for flight (*e.g.,* bats). Somewhere along this line we can place the **gliding mammals** such as "flying squirrel". The arboreal and aerial forms not arose independently from the terrestrial forms, but perhaps through semi-arboreal or climbing ancestor (Fig. 8.6).

(iii) Cursorial. Third line of radiation gives rise to cursorial forms such as horses and antelopes. They have developed limbs suitable to rapid movement. The cursorial mammals have the following three types of adaptation in their foot-postures, such as (1) **plantigrade**, *i.e.,* walking with whole sole of the foot touching the ground, *e.g.,* bears and primates including human beings; (2) **digitigrade**, *i.e.,* digits touch the ground and are provided with pads on their ventral side which absorb the shock and help in making stealthy approach towards the prey, *e.g.,* lion, tiger, leopard, cat, wolf and dog; and (3) **unguligrade**, *i.e.,* animals walk and run on the tips of their fingers and toes which are shielded by hoofs, *e.g.,* **artiodactyls** or double-hoofed animals such as cattle, sheep and buffaloes and **perissodactyls** or single-hoofed animals such as horses, asses and zebras.

(iv) Fossorial. Fourth line of radiation formed the burrowing mammals, the **fossorial** mammals. Some of the fossorial mammals, such as moles, have modified their forelimbs for digging but they are poorly adapted for locomotion on the ground. While other fossorials such as pocket gophers and badgers are expert diggers but they can also move readily on the surface of ground.

(v) Aquatic. Fifth line of radiation leads to the aquatic mammals. Whales and porpoises having limbs strongly adapted for aquatic life, but they cannot move about on land. While seals, sea lions and walruses have also strongly modified limbs for aquatic life but they are also able to move about on land. The third group includes accomplished swimmers such as otters and polar bears which are equally at home in water or on land.

Lastly, animal's legs modified for walking are called **ambulatory** and those that are adapted for jumping are called **saltatory**.

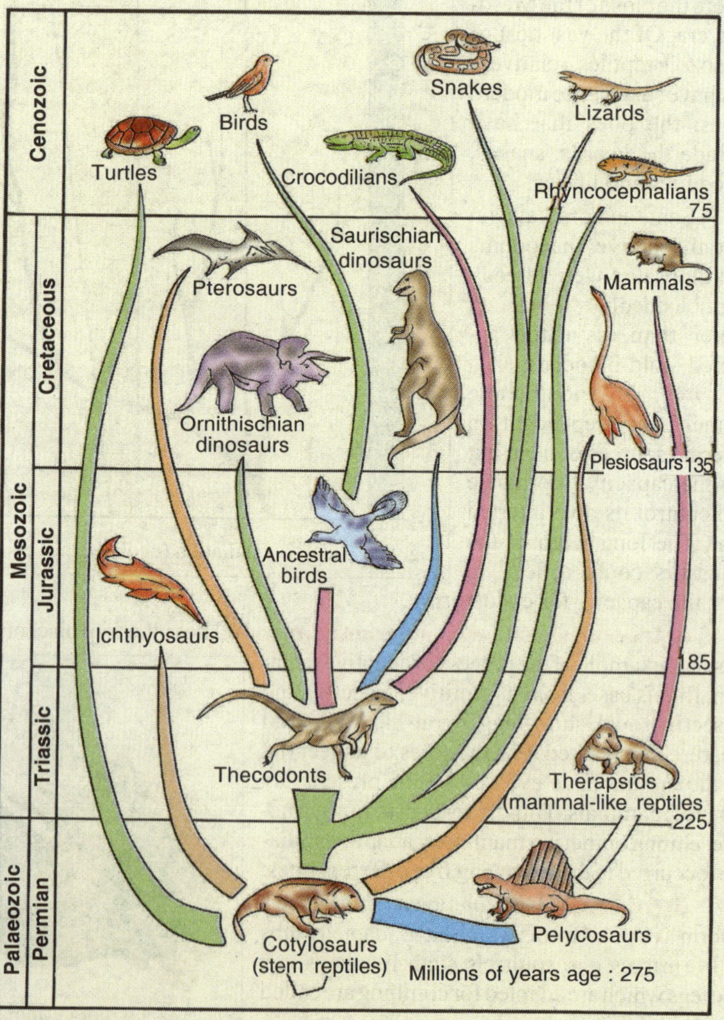

Fig. 8.5. Adaptive radiation of reptiles (after Volpe, 1985).

Among eutherial mammals, adaptive radiation is also applied to teeth (or mode of feeding) (Fig. 8.7).

5. Adaptive radiation in the marsupial (meta-therian) mammals. A remarkable example of adaptive radiation in animals is the tremendous diversity of marsupials in Australia. In many ways they parallel the adaptations of placental mammals in the rest of the world, affording striking examples of **parallel evolution**. Thus, in the absence of placental mammals, marsupials developed grazing forms (some kangaroos), burrowing forms (marsupial moles); forms resembling tree and flying squirrels (flying phalangers, pigmy gliders), forms resembling arboreal eutherians (teddy bear or koala and tree kangaroo); forms containing rodent-like dentition (wombats, marsupial mouse), rabbit-like forms (hare wallabies), wolf-like carnivores (Tasmania wolves), badger-like carnivores (Tasmania devil) and ant-eating carnivores (banded - anteater) (Fig. 8.8).

SIMPSON'S ADAPTIVE GRID AND MACROEVOLUTION

G. G. Simpson (1953) has developed a handy conceptual framework for describing major evolutionary patterns. We already know that at any moment in time the interaction of organisms and

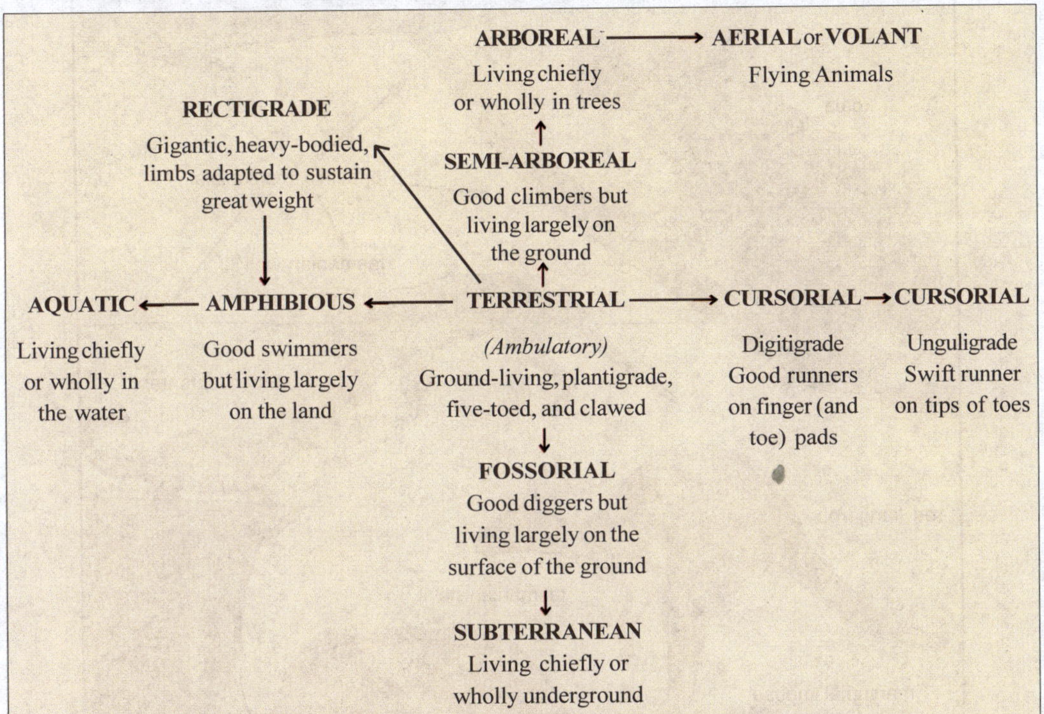

Fig. 8.6. Adaptive radiation in hindlimbs of eutherian mammals (after Lull, 1957).

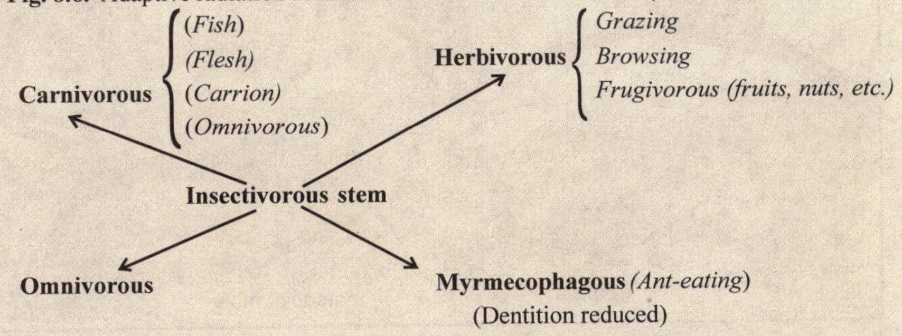

Fig. 8.7. Adaptive radiation in teeth of eutherian mammals (after Lull, 1957).

Fig. 8.8. Adaptive radiation of marsupials in Australia (after Hamilton, 1967).

their environment define a series of broad or narrow **adaptive fields** or **zones**. All members of the major group (for example the crustaceans) share one major adaptive zone because of their common possession of a complex general adaptations. Within the broad zone, each species of crustacean occupies a distinct but narrow field of its own because of the species' peculiar combination of special and general adaptations. Each kind of organism is an adaptive type discontinuous from other adaptive types; for example, a crab is distinct from a barnacle (both are crustaceans), a snake from a man or a sunflower. **Simpson** has suggested that for purposes of discussion the adaptive types or zones may be represented diagrammatically as bands or pathways on an **adaptive grid** (Fig. 8.9). The discontinuities between the major zones (A, B, C, D) are ecologic discontinuities or **unstable ecologic zones**. The adaptive zone itself is an ecologic role or characteristic relationship between organism and environment. Although the actual grid is of course very complex comprising many subzones (1, 2, 3) a simplified form is satisfactory for our discussion. The diagram is extremely useful in attempting to describe adaptive evolutionary change.

The evolution of adaptation may be summarized in terms of grid as follows : changes in adaptation involve movement of evolutionary lines within subdivisions of the subzones (microevolution), either from one major zone or subzone into others (macroevolution), or from one major set of zones into another (megaevolution). The basic feature of evolution above the species level is the movement of a group of organisms into a new adaptive zone. In order to move across the zone of ecologic instability (*i.e.*, an environment of strong negative selection) into the new adaptive zone, an organism must have evolutionary and ecologic access to it. In other words, the group must already have some characteristics adaptive to the new zone, and the zone must be unoccupied by a strong competitor. An example is provided by the evolution of terrestrial plants (Fig. 8.10).

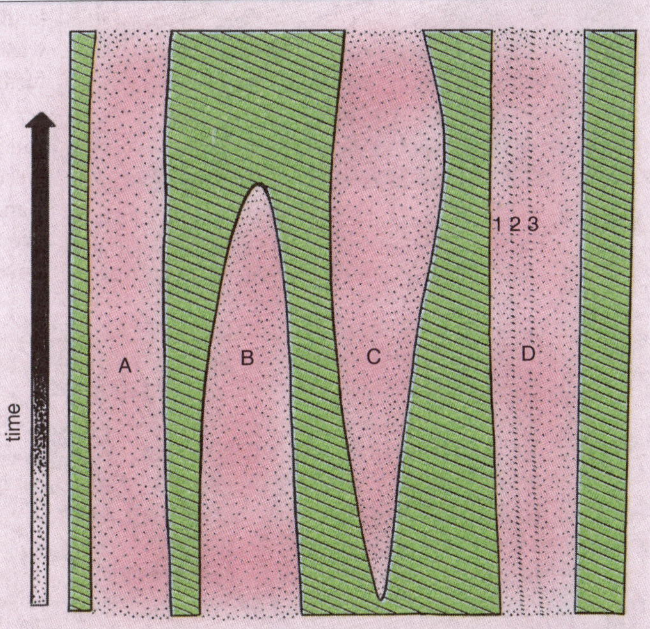

Fig. 8.9. Simpson's adaptive grid. Major adaptive zones are indicated by letters and subzones by arabic numbers; ecologically unstable zones are shaded (after Savage, 1969).

A good example of adaptive grid is provided by the evolution of lungless salamanders.

Evolution of lungless salamanders. Plethodontidae family of salamanders (amphibians) is the most successful having numerous groups of living salamanders found in Europe, North and tropical America. The adult of the ancestral stock probably occurred in and around the margins of small mountain streams. They apparently laid delicate eggs in the water which hatched into a free-living aquatic larva with functional gills. After undergoing further development, the larva changed into semi-aquatic salamander. A number of living genera of salam-anders such as *Desmognathus*, *Eurycea, Gyrinophilus, Pseudo- triton* and *Typhlotriton,* retain these habits and continue to occupy the ancestral semi-aquatic adaptive subzone. From this several descendant evolutionary lines have become completely aquatic in habits. In one genus, *Leurognathus*, which is closely allied to *Desmognathus* in the semi-aquatic group, the adult is strictly aquatic hides under rocks on the bottom of fast moving mountain streams. A second group of completely aquatic forms never transform, but become sexually mature and

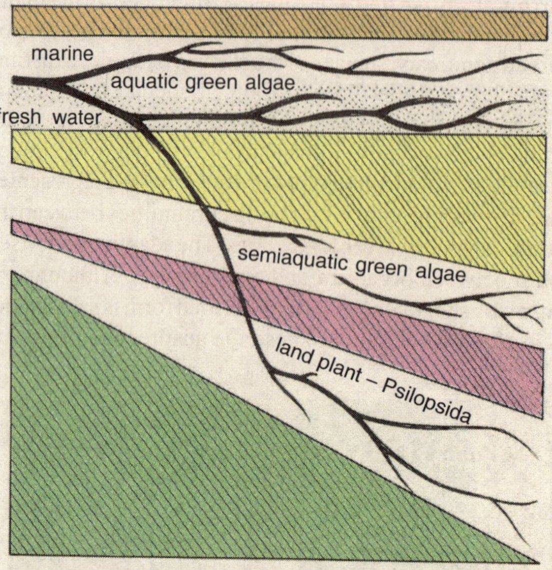

Fig. 8.10. An adaptive grid diagram of the evolution of terrestrial plants, indicating major breakthroughs and invasions of new adaptive zones (after Savage, 1969).

breed while still in a gilled larval state. Some species of *Eurycea* and *Gyrinophilus* have evolved by this means, while two distinctive genera allied to *Eurycea* also live in **paedogenetic subzone** (*i.e.,* larval have developed sex organs and behave like the adults: Fig. 8.11).

One species of *Desmognathus* and the genus *Hemidactylium* occupy the semiterrestrial subzone. The eggs of these salamanders are unspecialized aquatic types which are deposited in moist places on land. When the larvae hatch they are washed into streams where they undergo further development.

Some Plethodontidae have also invaded a second major adaptive zone, the **terrestrial zone**. Eggs of lungless salamanders in this zone have a special protective capsule and are deposited on land. Development proceeds directly and a small

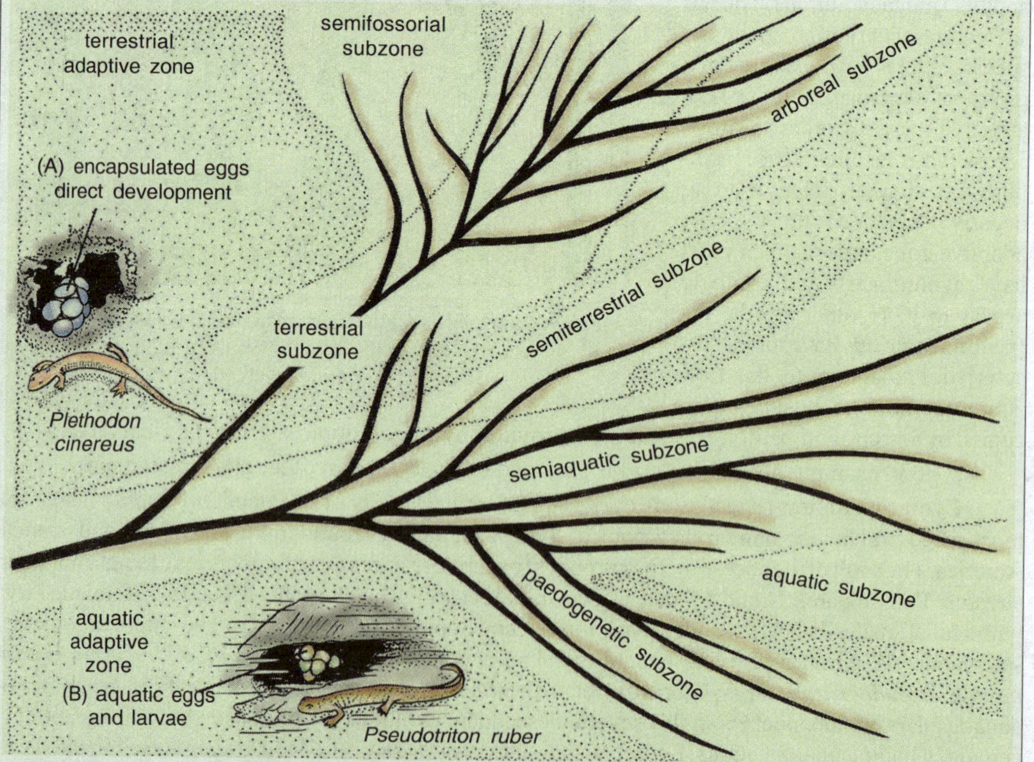

Fig. 8.11. Evolution in the lungless salamanders. Family Plethodontidae expressed on an adaptive grid diagram (after Savage, 1969).

salamander hatches out of the egg. No free-living larval stage occurs in the life history of these forms. These salamanders have invaded the terrestrial zone three times. The earliest of these invasions has resulted in adaptive radiation into 12 genera. The terrestrial forms or their descendants later occupied the semifossorial (burrowing) and arboreal (free-dwelling) subzones.

At much late date one genus (*Phaeognathus*) and a species of a related genus (*Desmognathus*) each seem to have independently penetrated the terrestrial adaptive zone. These forms also possess capsulated eggs with direct development.

Thus, when a new major zone or group of zones is occupied, the first zone entered is usually the widest and requires the least special adaptation. Later, more special zones will be occupied. The initial breakthrough to the new zone is due to general adaptation. Subsequent evolution leads to specialization. This pattern also fits the macroevolutionary speciation (adaptive radiation) of Darwin's finches.

MIVART'S DILEMMA AND PREADAPTATION

St. Geogre Jackson Mivart (1871) wrote in his book, '*On the Genesis of Species*', that he could not see how evolution would proceed by the gradual accumulation of small changes, since most intermediate character states would be maladaptive. For example, what use is a partially formed wing to a reptile. He, therefore, suggested that evolution would proceed in jumps, and that basis of these was the acquisition of useful characters.

Mivart's dilemma has been explained by invoking **preadaptation**. The intermediate structures act in a peculiar way for which they are well suited, but by chance they are appropriate for other roles which they are able to play after further elaboration. For example, a flap of skin between the fore-limb and the trunk might have evolved in reptiles as a thermoregulatory device and then have been found of use in gliding away from enemies, ultimately giving rise to a wing. Similarly, the air-bladder of fishes was not evolved so that they might one day invade land but as an accessory respiratory organ for use in the aquatic environment. However, once evolved, air-bladders were of use as lungs when new environmental problems, such as the drying of ponds, appeared (see **Calow**, 1983).

MICROEVOLUTION, MACROEVOLUTION, MEGAEVOLUTION AND HYPOTHESIS OF PUNCTUATED EQUILIBRIA

Goldschmidt (1940) divided evolution into **microevolution** (that of subspecies) and **macroevolution** (that of species and genera and perhaps also of higher categories). **Simpson** (1953) has proposed the additional term "**megaevolution**" for really large-scale evolution, such as that of families, orders, classes and phyla. Evolution at these levels receives much attention from students of the fossil record. **B. Rensch** (1959) has modified the terms microevolution into **infra-specific evolution** and macroevolution into **trans-specific evolution. Calow** (1983) has regarded macroevolution and cladogenesis as synonyms.

For several years (1976 to 1982) **Stephen Jay Gould** and **Niles Eldredge** of the American Museum of Natural History have questioned the conventional view that evolutionary changes in the distant past are principally the outcome of the gradual accumulation of slight inherited variations. They advocated that most evolutionary changes have consisted of rapid bursts of speciation alternating with long periods in which the individual species remain virtually unmodified. **Gould** and **Eldredge** maintain that most lineages display such limited morphological changes for long intervals of geologic time as to remain in stasis, or in "**equilibria**". Conspicuous or prominent evolutionary changes are concentrated in those brief periods ("**punctuations**") when the lineages actually split or branch. This is called **hypothesis of punctuated equilibria**.

Further, branching from a single lineage (or a single line of descent) during macroevolution, will ultimately produce a cluster of lineages known as **clade**. Thus, branching sometimes is called **cladogenesis**. When referring to a clade or to a group of related clads, it is common to use the adjective

phylogenetic (not to be confused with the similar word **phyletic**). Phylogenetic pertains to a portion of phylogeny consisting of more than one lineage, whereas phyletic refers to a single lineage. Moreover, a lineage reconstructed from fossil data may exhibit sufficient evolutionary change that a taxonomist deems it appropriate to divide it into two intergrading species. Such species are known as **chronospecies**, **successional species**, **palaeospecies** or **evolutionary species** (see **Stanley**, 1979).

1. Microevolution

Evolutionary changes in populations are ordinarily visualized as gradual, built upon many small genetic variations that arise and are passed on from generation to generation. The shifting gene frequencies in local populations may be thought of as microevolution. The progressive replacement of light-coloured moths by dark (melanic) moths in industrial regions in England (*i.e.,* industrial melanism) exemplifies the microevolution. Most population geneticists endorse the view that the same microevolutionary processes have been involved in the major transformations of organisms over longer span of geologic time (macroevolution). The traditional outlook is that small variations gradually accumulate in evolving lineages over periods of millions of years.

2. Macroevolution

The most significant feature of macroevolution is a progressive, sustained tendency for certain characters to develop along an evolutionary line. Trends (or directions of evolution) of this sort are numerous in the fossil record. Long-term progressing trends rarely appear in only one structure, but almost always involve a complex of different features. In fact, trends are produced by the driving force of natural selection operating within the limits of a particular adaptive zone or subzone. Evolution is not random, although certain elements in the process are random, and trends leading to greater efficiency are to be expected. Evolutionary trends are generally adaptive movements along one pathway, but they are never exclusively sequential and always involve divergent and repeated taking on of one or the other characters important in the trend. The populations undergoing change are constantly experimenting within the adaptive pathway, and parallel probings of a new subzone by related but different lines are the rule.

The classical example of evolutionary trends in macroevolution is provided by the horse family, Equidae.

Evolution of horses. In reconstructing the phylogeny of an animal group, it has been the standard practice to show the emergence from a central, generalized stock of a large number of divergent branches or lineages. Not all branches persist; indeed the general rule is that all but a few perish. The disappearance of many branches in the distant past might lead the observer today to the mistaken impression that the evolution of a particular group was not at all elaborately forked. Thus, the evolution of horses could be erroneously depicted as an undeviating straight-line progression from the small, terrier (small dog of various breeds) sized *Hyracotherium* (*Eohippus*) to the large modern horse, *Equus*. On the contrary, a wealth of evidence has revealed convincingly a pattern of many divergent lineages (Fig. 8.12).

Figure 8.12 demonstrates very nicely the gradual change from a dog-like, browsing creature with padded feet to a horse with grazing habits and hoofed springing feet. In fact, the appearance of the one-toed condition was a land mark in horse history. The fleet-footed, one-toed *Pliohippus* emerged during the Pliocene epoch from the slow-footed, three-toed *Merychippus*. This major event is usually interpreted as a gradual change within a single lineage. In other words, a direct ancestral-descendant relationship between *Merychippus* and *Pleiohippus* has been widely accepted. The paleontologist **Simpson** (1951) departed from the view of slow and steady change when he suggested that such a major transition might have involved the accumulation of small genetic changes in relatively rapid succession. This accelerated pace of phyletic gradualism was called "**quantum evolution**" by **Simpson**.

Proponents of hypothesis of punctuated equilibria explained the sudden appearance of *Pliohippus* in a quite contrast but convincing way. Thus, one might argue that *Merychippus* existed for a long period with little or no evolutionary change (**stasis**), and eventually suffered extinction. *Pliohippus* is not a linear descendant of *Merychippus*, but a dramatically new form that had earlier arisen as a small peripheral population of the parent species. It may be noticed (Fig. 8.12) that not all offshoots of *Merychippus* evolved the progressive single-toed condition. Several lines of Pliocene horses, such as *Hipparion* and *Nannippus*, retained the conservative three-toed pattern. These conservative Pliocene horses were evolutionary blind alleys, and may well have represented "punctuations" that failed.

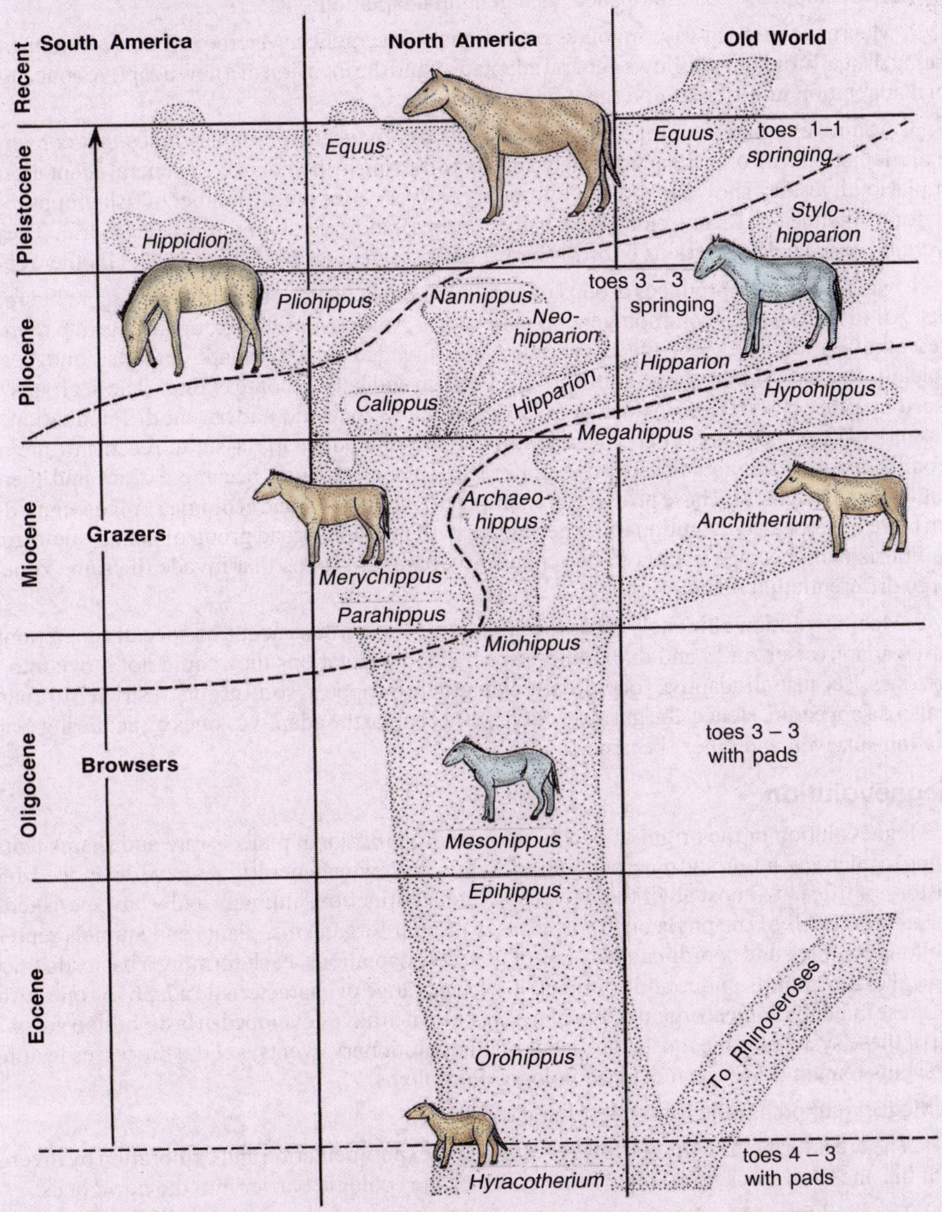

Fig. 8.12. Phylogenetic tree of horses through time, with its many divergent branches. All branches died out, save one which eventually culminating in the modern group of horses, *Equus*. The history of horses dates back to early Cenozoic times, some 60 million years ago. The Cenozoic era is divided into periods, which in turn are divided into epochs (after Savage, 1969).

General principles of macroevolution. The following common features have been derived from the phylogeny of horses regarding macroevolution :

1. Macroevolution follows the acquisition of new general adaptation or entrance into a new adaptive zone. For example, Darwin's finches radiated after an apparently generalized finch ancestor arrived to occupy the previously unoccupied Galapagos Islands. Reptiles radiated after completely terrestrial development was established as a general adaptation.

2. Macroevolution always involves evolutionary divergence. Macroevolution is not linear but radiating. Usually radiation follows general adaptation and the invasion of a new adaptive zone through special adaptation in different divergent descendant lines.

3. Adaptive radiation or macroevolution tends to produce evolutionary lines that converge in special adaptation with other distantly related groups differing in their matrix of general adaptation. For example, ichthyosaurs show a marked evolutionary **convergence** with a number of fish groups in their body form, manner of locomotion, food habits (fishes) and free-swimming pelagic life. They are convergent in special adaptations but still share a group of general adaptations with all other reptiles.

4. Macroevolution produces groups of parallel special adaptations among divergent but related stocks. All of them have a common general adaptation. Among reptiles, group after group representatives of every order except the cotylosaurs had invaded aquatic adaptive zones and become completely aquatic. Among these, some groups flourish and later become extinct. These groups were replaced by other parallel groups which invaded the same zone and underwent differentiation. The phytosaurs (Triassic) of the order Thecodontia and the crocodiles (Triassic to recent) of the order Crocodilia invaded the aquatic adaptive zone. But later phytosaurs became extinct and then the crocodiles replaced them. These provide an example of **parallelism** and **ecologic replacement**. Thus, where there are repeated evolutionary experiments with the same broad group of zones, some groups arise, flourish and become extinct to be replaced by parallel groups that invade the same zone and undergo differentiation in their turn.

5. Macroevolution ultimately leads to extinction. Some groups acquired special adaptations for a narrow adaptive subzones and due to these specialized adaptations they could not move into new major zones. Because all adaptive zones finally change and disappear, so all groups restricted to a narrow zone also disappeared. Hence, the groups which could change the adaptive zones by acquiring general adaptation survived and others became extinct.

3. Megaevolution

Megaevolution or the origin of new biological organizational plans is rare and shares features common to microevolution and macroevolution. Only a few major general types have developed during the history of life, but almost all of them persist without extinction, although a few have perished and others are relict. All of the phyla and most classes of microorganisms, plants and animals represent marvellous complex and coordinated groups of general adaptations, each forming a basic, distinctive biological organization, unique and dominant in a broad range of characteristic adaptive zones. About 200 of these major biological organization plans are known to have developed in last 3 billion years. The origin of these systems is the most significant of all evolutionary events, yet the processes leading to these events remain the least studied of biological problems.

Megaevolution has the following characteristics :

1. The breakthrough always follows evolutionary experimentation and exploration by divergent lines of the ancestral stock, until one of them crosses the ecologic barrier into the new zones.

2. The breakthrough and shift are always rapid; otherwise they fail because of the extreme negative selection in unstable ecologic zones.

3. The new zone is always ecologically accessible, is devoid of competition and requires a new general adaptive type for its invasion.

4. Adaptive radiation always follows the initial shift.

The Process involved in Macroevolution and Megaevolution

The major feature of organic evolution is divergence guided by the moulding force of natural selection. Evolution at the populational level is driven by the elemental forces of **mutation**, **selection** and **drift**. Basically population evolution may be sequential, but it is always divergent in the end. Speciation, macroevolution and megaevolution represent states or levels in a continuum of evolution; all are driven by the elemental forces but are subject to increasingly complicated effects from less understood forces as well. Selection becomes of greater significance above the microevolutionary level.

Of the greatest importance in speciation is the **isolation**. Selection may act to produce divergence in this process whenever fragments of an originally interbreeding gene pool become spatially or reproductively isolated. New species originated from old ones through the origin of reproductive isolation and independent microevolution. Thus, in addition to elementary forces, isolation, microevolution and macroevolution, megaevolution involves the following processes (Fig. 8.13) :

(1) The possession of new general adaptation or occupancy of a new adaptive zone.

(2) The breakthrough into new zones or subzones within the new adaptive zone by development of special adaptation.

(3) The loss of evolutionary flexibility, and channelization into greater and greater specialization within subzones.

(4) The ecological reinvasion of a zone or subzone when it becomes partially unoccupied because its original occupiers are now specially adapted (ecologic replacement).

(5) The irreversibility of evolution. Since each step is dependent upon the previous progressive changes, once a group is on an adaptive road, it is usually trapped in an adaptive zone and cannot reverse evolution against the direction of selection.

4. Doctrine of Punctuated Equilibria

A single line of descent or lineage may persist for long reaches of geologic time. As small changes accumulate over periods of millions of years within one lineage, the descendant populations may eventually be recognized as a **species** distinct from the antecedent population. The persistent

Interbreeding blurs the distinction between species.
(a) The myrtle warbler and ***(b)*** Audubon's warbler were formerly thought to be two separate species but are now considered to be merely local varieties of one widespread species.

accumulation of small changes within a lineage has been termed **phyletic gradualism**, and the transformation of a lineage over time has been termed as **anagenesis**. How organisms come to terms with and keep in tune with their environment, is called anagenesis (see **Calow**, 1983). As depicted in Figure 8.14 A, a new species (labelled "B") arises from the slow and steady transformation of a large antecedent population ("A"). If only transformation occurred, then life would cease as single lineages became extinct. Hence, a new species ("C") can also arise by the splitting or branching of a lineage. As already described, the splitting of one phyletic lineage into two or more lines is termed **cladogenesis** (*viz.*, the evolutionary process of genesis of variety). But here again the splitting is believed as proceeding slowly and gradually, with the two branching lineages progressively diverging, without significant reduction in population size. Thus, palaeontologists tend to view lineage splitting in terms of gradual morphological divergence.

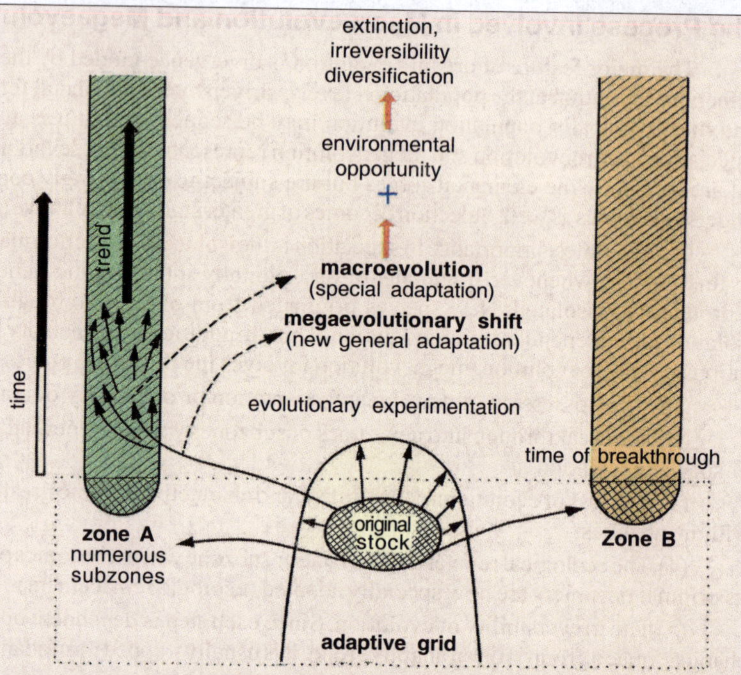

Fig. 8.13. The process of evolution above the species level — a diagrammatic representation of macroevolution and megaevolution (after Savage, 1969).

Gould and **Eldredge** maintain that phyletic gradualism is too slow to produce the major events of evolution. According to their theory of punctuated equilibria, the prominent episodes of evolution in the history of life are associated with the splitting of lineages, but the splitting is not seen as slow and steady cladogenesis. The new species arises through rapid evolution when a small local population becomes isolated at the margin of the geographic range of the parent species (Fig. 8.14 B). Indeed, the successful branching of a small isolated population from the periphery of the parental range virtually assures the rapid origin of a new species. In geologic time, the branching is sudden—thousands of years or less, compared to the millions of years of longevity of the species itself.

The punctuational change does not cause any unconventional evolutionary phenomenon. The sudden appearance of a species is fully consistent with **Ernst Mayr's rapid selection** in peripheral isolates. The morphological differentiation of geographical isolates occurs early and rapidly, especially when the population is small and adapting to the new abnormal environment at the periphery of the range. Once established as a new species, the assumption is that little evolutionary change will occur in that species over geologic time.

A well-studied example of punctuational mode of evolution is provided by **Peter G. Williams** in 1981. He studied the fossilized freshwater snails (molluscs) in the Lake Turkana's basin region of Africa (Northern Kenya). The thick, undisturbed fossil beds contain millions of preserved shells representing at least 19 species of snails. Several lineages of Cenozoic snails were exceptionally stable for 3–5 million years. When morphological changes in shell shape did occur, they were concentrated in brief periods of 5,000 — 50,000 years. These rapidly evolved new populations then persisted virtually unmodified until they became extinct. Some authors have convincingly commented that so-called sudden changes within 50,000 years might appear to be instantaneous to a palaeontologist but would be almost an infinity to an experimental geneticist. In fact, to many geneticists, an interval of 50,000 years (or 20 generations in snails) would be sufficient time for the morphological changes in the snail populations, to accumulate gradually rather than dramatically.

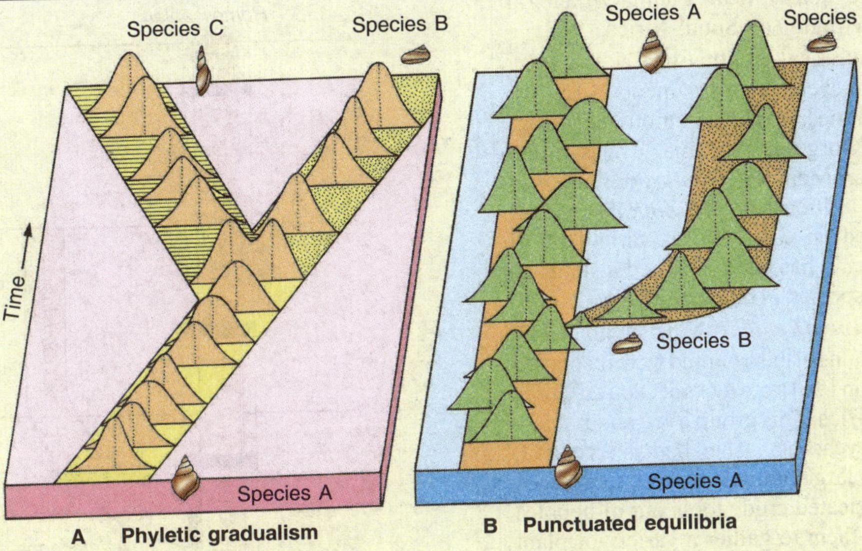

Fig. 8.14. A–According to phyletic gradualism, species A gradually transforms into species B, and may split into a slowly evolving species C; B– According to theory of punctuated equilibria a small isolate from the parental population (species A) evolves rapidly into a new entity (species B). Species A undergoes a long period of stasis during which little or no morphological change takes place. Species B does undergo minor structural modification with time (after Volpe, 1985).

Whether Human Evolution is Gradual or Punctuated ?

On the basis of phylectic gradualism mode of evolution, it has been standard practice to put *Homo sapiens* at the end of a continuous series tracing back through at least two other species (Fig. 8.15). According to certain modern authors such as **Stanley** (1979), **Calow** (1983) and **Volpe** (1985) various taxa involved in human phylogeny are the following :

1. *Ramapithecus*. This group also includes *Sivapithecus*. These ancestral forms (apes) lived approximately 8–14 My (=million years) ago and contained a flattened face with a short muzzle. First fossil (a fragile upper jaw) of *Ramapithecus* was recovered by **G. Edward Lewis** (1930) in Shivalik Hills of Northern India. Another fossil of this group was recovered from Lake Victoria in Africa by **Leaky** in 1955. The fossil evidence clearly indicated that such genera as *Ramapithecus* (from *Rama* = Hindu god + *pithekos* = Greek for ape) and *Sivapithecus* (from Siva=Hindu God) which lived in Africa and Eurasia were the forerunners of Hominids.

2. *Australopithecus afarensis*. (L. *austral* = southern). The fossil of this female ape-man (called **Lucy**) was recovered from Afar locality in Hadar, Ethiopia by **Donald Johanson** (1973). This fossil was probably the oldest among all the primitive man described. The *afarensis* remains dated between 3 and 4 million years (My) ago.

3. *Australopithecus*. This group lived 1.5—5 My ago. They had a flattened face. The orientation of the skull relative to the spinal cord, and the bones of the hind-limbs relative to the pelvis suggest an upright gait. There are following two groups of these ape-men :

(i) Robust australopithecines (*Australopithecus robustus*) which had a heavy skull and dentition suggesting a vegetation diet. The fossil remains of this group was recovered by **Robert Broom** and **Raymond A. Dart** (1938) from caves in Sterkfontein, Kromdraai and Swartkrans in South Africa.

(ii) Gracile australopithecines (*Australopithecus africanus*) which has a more delicate skeleton (jaw was light and slender) and supplemented their diet with animal food (*i.e.,* they had smaller molars than the *A. robustus*). First fossil remains of *A. africanus* were obtained by **R.A. Dart** in 1924 from a limestone quarry near Taung village in Transvaal region of South Africa.

Louis Leaky and his wife **Mary Leaky** (1959) uncovered bony fragments of a robust australopithecine with enormous teeth from Olduvai Gorge in Tanzania. It was called Nutcracker Man or *Zinjanthropus* and was about 1.8 million years old. *Zinjanthropus* is considered an east African variety of *A. robustus* and has been assigned a status of separate species—*Australopithecus boisei.*

4. *Homo habilis* ("handy man"). The fossil remains of this hominid were recovered by **Leaky** in 1964 from rocks of Olduvai Gorge in East Africa. This group lived about 1.5—2 million years ago. They had an erect gait, relatively large brains and were tool-users. They fabricated crude tools out of pebbles and used them to gather a variety of plant materials (nuts, seeds, fruits, tubers and roots) and small animals (lizards and rodents). Cooperative behaviour evolved in them and promoted the sharing of the food for the first time.

5. *Homo erectus* ("upright walking man"). This group lived 1 million years ago. They had a gait even more like ours and larger brains than the habilines. *H. erectus* used tools of bone and stone and used fire. Fossil remains of *H. erectus* were discovered at Trinil, Java, in 1894 by **Eugene Dubois**. They have been called Java men and had a cranial capacity of 775 to 1000 cubic centimetres (or cc). Another type of fossils of *H. erectus*, called **peking fossils** (*i.e.*, Peking men), were discovered by **Davidson Black** in 1920s. Peking men have the cranial capacity of 900 to 1200 cc.

6. *Homo sapiens* ("thinking man"). The grade of *sapiens* contains at least two contrastingly different anatomical types — the bulkily built and heavily muscled Neanderthal and slim-bodied Cro-Magnons.

The skeleton of Neanderthal man (*Homo sapiens neanderthalensis*) was first unearthed in 1856 in a lime stone cave in the Neander ravine near Dusseldorf, Germany. The *Neanderthals* were 1.5 lakh years old cave dwellers, short (about 5 feet) but powerfully built, with prominent facial brow ridges. They had large brains with an aver-

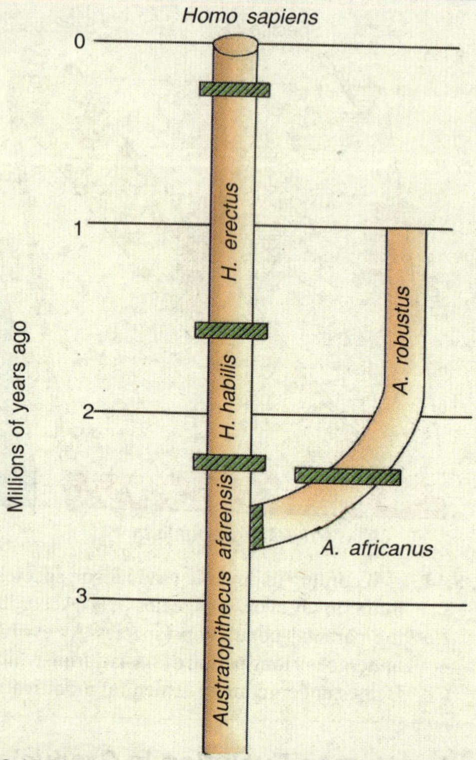

Fig. 8.15. Phylogeny of human. *Australopthecus afarensis* followed by *Homo habilis* and *Homo erectus* are believed to have existed successively in geological time and to form a lineage leading to *Homo sapiens*. In the distant past, the *Homo* assemblage co-existed with the *Australopithecus* groups. This phylogeny is based on the views by **Donald Johanson** (1979) (after Volpe, 1985).

(A) A fossil skull of ***Homo erectus*** show that the human brain gradually grew larger, beginning about 2.5 million years ago. At the same time the ridges above the eyes became smaller, and the face and jaw began to jut out less. Increasing intelligence and the ability to make better tools apparently went together. (B) The earliest stone tools, made in Africa about 2.5 million years ago, were simpler than the hand axe shown here.

age capacity of 1,450 cc as opposed to 1,350 cc in modern humans. Neanderthals made complex stone tools, were accomplished hunters of large game, used animal hides as clothing and withstood the rigors of the bitter cold climate of the last glaciation. There is evidence that the Neanderthals buried their dead with various ritual objects.

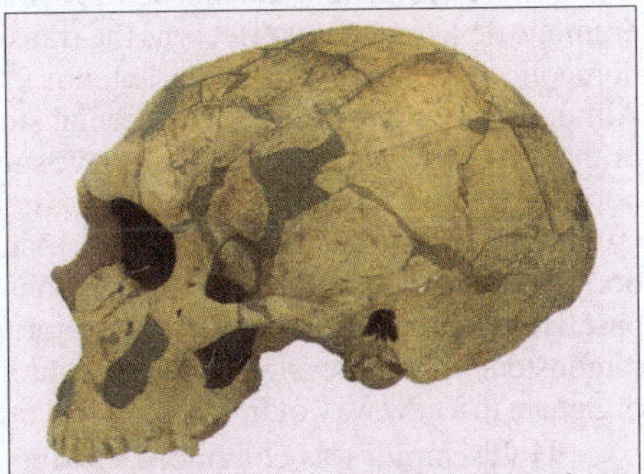

The skull of Neanderthals suggests a brain at least as large as that of modern humans. They were very strong and stocky and seem to have been well suited to the Ice Age climate.

The fossil remains of Cro-Magnons (*Homo sapiens sapiens*), a representative of our species, *Homo sapiens*, were unearthed from a French rock-shelter in the village of Les Ezgies. They were about 35,000 years old. The Cro-Magnon people occurred in Europe and left behind very elaborate cave paintings showing attainment of a form of culture not unlike our own. We do not know about the birthplace of modern *Homo sapiens*. Modern humans may have been cradled in Asia or Africa.

In the past three decades, Africa has become increasingly the focus for investigations into human origins and differentiation. Several sites in Africa have yielded fossils that reveal different phases of the human story from the earliest stages. The differences of opinion that exist concerning the phylogeny of the Hominidae are vividly revealed by the ever-changing reconstructions of the human and evolutionary tree, almost on a yearly basis. One of the most widely held views is that a gradual ancestral-descendant relationship existed for each successive hominid group. Thus, the australopithecines shaded faintly into *Homo habilis*, who in turn graded slowly into *Homo sapiens*. The theory of punctuated equilibria claims (see **Stanley**, 1979) that the evolutionary changes in hominids occurred in rapid bursts, concentrated in speciation events (Fig. 8.16). Such

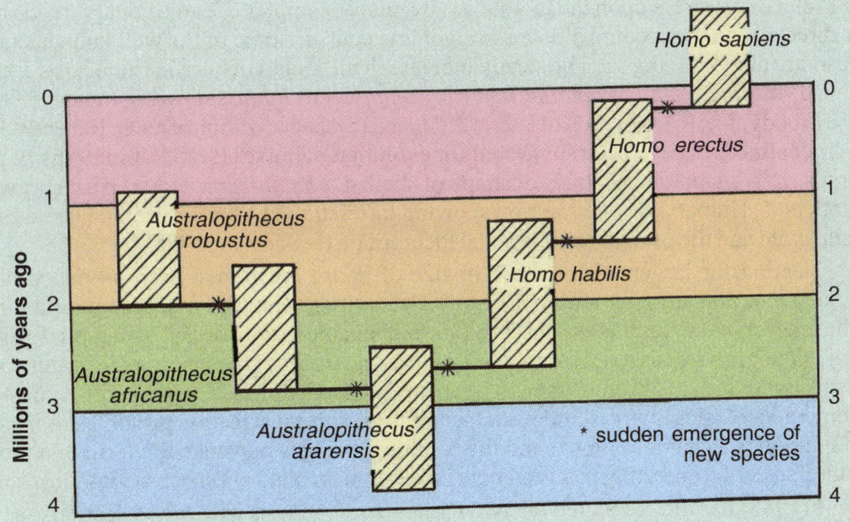

Fig. 8.16. The rectangular pattern of hominid phylogeny as expressed by the controversial theory of punctuated equilibria (after Volpe, 1985).

punctuated episodes were separated by long periods of stasis in hominid lineage during which little or no morphological change took place. Such long periods of stasis occurred within both the australopithecine and hominid groups and this is rather startling considering the high net rate of evolution which is characteristic of human evolution. Figure 8.16 shows that this pattern of evolution appears "rectangular" rather than tree-like.

SIMPSON'S HOPEFUL MONSTER

There are genetic changes, known as **homeotic mutations**, which are responsible for such dramatic developmental effects as the transformation of one major organ into another. In the fruit fly, homeotic mutations transform an antenna into a leg, or a balancer into a wing, or an eye into an antenna. Although fascinating for developmental studies, most evolutionists would not even consider that the origin of a new species could ever involve such drastic major morphological changes produced by single homeotic gene changes or **macromutation**. However, in 1940s, the prominent geneticist **Richard Goldschmidt** asserted that homeotic mutants do illustrate a possible mode of origin of novel types of body structure, and, hence, could be responsible for the rapid emergence of a new species in the distant past. He acknowledged that the vast majority of macromutation would have disastrous consequences ("monsters"). An occasional macromutation, however, might conceivable immediately adapt the organism to a new way of life, a "**hopeful monster**" in Goldschmidt's terminology.

Goldschmidt was convinced that the gradual accumulation of small mutations was too slow a process to account for broad macroevolutionary trends. He favoured the idea of a dramatic, abrupt transformation of a population by a favourable, instantaneous macromutation, which could take the form of a radical rearrangement of genetic material (*viz.*, a gross chromosomal change). The concept of punctuated equilibria does not require or imply chance favourable macromutations. Rather, in the process of allopatric speciation in peripheral isolates, the incipient species develop their distinctive features rapidly. However, the significance of one aspect of Goldschmidt's views cannot be denied : structural chromosomal rearrangements may play a primary role in the initiation or achievement of reproductive isolation between two diverging populations.

ORTHOGENESIS AND ORTHOSELECTION

The observed tendency of a part or organ to change progressively in size is called **orthogenesis** or **evolution in a straight line** (**Moody**, 1970). The theory of orthogenesis which was originally proposed by **Eimer** (1897), lays stress upon the fact shown by many examples of evolutionary trends which are gradually directed and proceeding along a straight line course. Some of the well known examples of orthogenesis are the following : 1. The steady increase in the length of horn in rhinoceros-like animal. 2. Fossil 'line' leading from *Hyractotherium* (or *Eohippus*) to *Equus*, showing trends in changes in (a) increase in body size; (b) reduction of lateral digits and (c) specialization of teeth. Recently, however, many divergent lineages have been suggested for evolution of horses (see **Volpe**, 1985). 3. *Paludina* (snail) fossils showing graded changes of shape of shell. 4. *Megaloceros* (fossil Irish elk) which had gigantic antlers. 5. Human line of evolution showing three straight trends, *viz.*, the increase in size of cranium and brain and the perfection of bipedal locomotion (see **Nayar**, 1985).

The selection for largeness (increase in size of body) must have been produced due to the advantage of becoming stronger and resisting predators (*e.g.*, elephants, ungulates and dinosaurs). Some of the progressive series that are observed are explainable as the result of differential growth rates (**allometry**). Other progressive series are explicable as the result of the operation of natural selection on organisms living in a stable environment or an environment that is changing with a constant trend (*e.g.*, becoming increasingly dry). Under such conditions natural selection promotes more and more perfect adaptation to that environment and the resulting changes may take the form of a progressive series. Natural selection operating in this manner is sometimes called **orthoselection** (**Simpson**, 1953).

Nayar (1985) has differentiated the two terms—orthogenesis and orthoselection—as follows: Thus, according to orthoselection the change is dependent upon the constancy of the favouring environmental opportunity. However, orthogenesis may involve straight line progress which may even go beyond even after the selection has passed its useful stage.

REVISION QUESTIONS

1. What is adaptive radiation ? Explain this phenomenon by citing example of mesozoic reptiles.
2. Describe the evolution of horse.
3. Give an account of phylogeny of man.
4. Write short notes on the following : (1) Macroevolution; (2) Punctuated equilibria ; (3) Orthogenesis.

CHAPTER

9

Isolation

Ecological isolation: A dramatic example is provided by many species of fig wasps which breeds in (and pollinates) the fruits of a particular species of fig and each fig species hosts one and only species by the pollinating wasp.

Isolation is an important factor contributing to the process of evolution. It is the segregation of the population of a particular species into smaller units which prevent interbreeding between them. Isolation aids in splitting of the species into separate groups. In consequence of being separated from one another, organisms have developed different characteristics, due to which they have become separate species. Hence, isolation is essential for the formation of new species, *i.e.*, speciation.

In fact, besides the variation and heredity, **segregation** or **isolation** forms an essential basic factor for the evolutionary process. Isolation is the physical or biotic barrier which prohibits accidental interbreeding. Forms with similar variational tendencies should interbreed to perpetuate them, and dissimilar forms should not, otherwise the new variations would be swamped and unless they were of dominant character would instantly disappear.

Thus, gene flow (the sharing of genes) between populations might be suppressed by a physical or geographical barrier. This process is called **allopatric speciation**. Alternatively, in principle, gene flow can be suppressed in populations which exist in the same area–if, for example, the bearers of favourable mutations mate preferentially with each other and if the selection of these mutations is sufficiently intense to overcome any gene flow from matings with more parental forms. This is **sympatric speciation**. Finally, even if there is some gene flow between partially isolated populations in adjacent areas there can, again in principle, be a divergence of characters under the influence of selection if gene flow is relatively low and the differences in selection pressure are relatively high. This is known as **parapatric speciation** and is likely to be common in organisms with low vagility (dispersal powers) such as plants and land snails (see **Calow**, 1985).

Whereas few biologists doubt the importance of the allopatric process in evolution, there is much debate on the importance of the sympatric and parapatric processes.

Isolation is accomplished by two means—**physical** (*e.g.* geographical isolation) or **biotic** (*e.g.*, reproductive isolation) **isolations**. Geographical isolation is one aspect of species multiplication which **Darwin** and **Wallace** failed to understand completely. Its theoretical development starts with the early studies of **Moritz Wagner** (1868,1889), **H. Seebohm**, **K. Jordan** (1905) and **E.B. Poulton** and passes through the subsequent work of **Rensch** (1929), **Dobzhansky** (1937, 1953), **Huxley** (1942, 1954), **Stebbins** (1950) and **Ford** to its culmination by **Mayr** (1942, 1949, 1963, 1971).

TYPES OF ISOLATION

Evolutionary biologists have recognised the following types of isolations which cause speciation.

1. Isolation by Time

The palaeontological history of animals and plants suggests that the population at one time is always the descendant of the one living earlier. In case of straight evolution one species simply gets transferred gradually into a new species through accumulating genetic differences. If the changes are great enough then the biologists called the population at two different times as two different species. This may well be illustrated with the sequential fossil records of the evolution of horse family. According to **G.G. Simpson** horse evolved some 16 million years ago in Eocene. It underwent successive changes produced by thousands of favourable mutations of every gene involved in the evolutionary trend of each characteristic.

2. Isolation by Distance (Spatial Isolation)

Sheer distance apart may also act as an isolating factor for a species which occupies a great range of area, which is unbroken by effective barriers. Its example is wrens (birds) of South America. Wrens are found all over the continent but the wrens of one region differ from those of the other in colour patterns, size, proportions and habits. It shows that without any barrier, sheer distance apart tends to produce local races (subspecies).

3. Geographical Isolation

It is the most common type of isolation and occurs when an original population is divided into two or more groups by geographical barriers such as a river, desert, glacier, mountain or ocean, all of which prevent interbreeding between them, then in the course of time different mutations may become incorporated in the gene pools of the different groups. Often these differences are of such a nature that the separated groups, when they come in contact again, do not interbreed, thus, species have been formed by geographical isolation. Certain additional kinds of geographical barriers are the volcanic formation of a mountain on land, mountain ranges with deep valleys between and land masses as islands in sea.

Geographical isolation

(a) the Kaibab squirrel lives only on the north rim of the Grand Canyon and ***(b)*** the Abert squirrel lives exclusively on the south rim. The two populations are geographically separated but still quite similar.

The classic example of continuing evolution by geographical isolation is that described by Darwin for finches. Darwin found that there were 26 groups of finches among the Galapagos Islands (which lie a few hundred miles west of South America). Only five of these groups were the same as the finches found on the mainland. The other twenty-one were types peculiar to the groups of islands. Some of the twenty-one groups interbred quite freely, while others did not. Apparently, each of the groups became isolated by migration. Each group evolved separately from the continental forms as well as from other isolated groups on the islands, forming a series of species and subspecies.

The saola, unknown to science until 1992, is one of a number of previously undiscovered species recently found in the mountains of Vietnam. The area's distinctive assemblage of species probably arose during a past period of geographic isolation.

Another interesting example of geographical isolation is provided by elephant seals. The Southern elephant seal, *Mirounga leonina*, occurs in the cool waters of the Southern Hemisphere around Antarctica, the Southern Coasts of South America, South Africa, Australia, New Zealand, and many of the antiboreal islands. A close ally, the northern elephant Seal, *Mirounga angustirostris*, is found in cool waters along the coasts of western North America. The two forms are very similar to each other and can be distinguished only with difficulty. However, the breeding populations of the two forms are separated by about 3000 miles of warm tropical seas, and, hence, are not capable of genetic exchange. Where forms occupy discrete geographic or ecologic ranges separated from one another by spatial barriers they are called **allopatric populations**. The two kinds of elephant seals are **allopatric species** genetically isolated by an eco-geographic barrier.

Other examples of geographical isolation can be observed anywhere. It is well established, for example, that there are more different species of the same genus in mountainous country than in plain regions. For instance, in the eastern part of the U.S.A., there are eight species of cotton-tail rabbits, whereas in the mountainous regions of the west of the U.S.A., there are 23 species of the rabbits. Often, in mountainous country many of the plants and animals found in deep valleys, which are separated by high peaks but which may be only a few miles apart, are peculiar to those valleys. Likewise, the isthmus of Panama provides another striking example of geographical isolation. On either side of the isthmus the phyla and classes of marine invertebrates are made up of different but closely related species. For some 16,000,000 years during the tertiary period, there was no connection between North and South America, and marine animals could migrate freely between what is now the Gulf of Mexico and the Pacific ocean. When the isthmus of Panama re-emerged, the closely related groups of animals were isolated and the differences between the fauna in the two regions represent the subsequent accumulation of hereditary differences.

Over the long haul, geographic isolation is seldom permanent. Changes in geography, migrations resulting from great population pressure, or chance dispersal during storms may bring separated groups in contact again. If they then interbreed and have fertile offspring, speciation has not occurred; but if they do not interbreed or if they interbreed and have sterile offsprings, new species have been produced, reproductive isolation has occurred.

4. Reproductive Isolation

In sexually reproducing organisms species can be defined as Mendelian populations between which the gene exchange is prevented by **reproductive isolation**. An **isolating mechanism** is any genetically conditioned barrier to gene exchange between populations (**Dobzhansky**, 1976). The term *isolating mechanism* was coined by **Dobzhansky** in 1937. In simple terms isolating mechanisms are those which prevent successful reproduction between members of two or more populations (*viz.*, closely related species) that have descended from the same original population.

TYPES OF ISOLATING MECHANISMS

Most modern evolutionists such as **Mecham** (1961), **Mayr** (1948, 1970), **Stebbins** (1966,1971), etc., have classified the reproductive isolating mechanisms into two classes, namely **premating** or **prezygotic isolating mechanisms** and **postmating** or **postzygotic isolating mechanisms**. Both types of isolating mechanisms differ fundamentally from each other in the following manner: premating or prezygotic isolating mechanisms prevent wastage of gametes (germ cells) and so are highly susceptible to improvement by natural selection; postmating or postzygotic isolating mechanisms do not prevent wastage of gametes and their improvement by natural selection is indirect (**Mayr**, 1970). Both types of isolating mechanisms included many subtypes, all of which have been listed in Table 9–1.

Table 9-1. Types of isolating mechanisms that prevent gene exchange between population of related species of organisms (Source: Stebbin's 1976, Mayr 1971).

A. Premating or prezygotic isolating mechanisms	— Mechanisms that prevent interspecific crosses (*i.e.*, fertilization and zygote formation).
1. Habitat isolation :	The populations live in the same regions but occupy different habitats, so that potential mates do not meet.
2. Seasonal or temporal isolation :	The populations exist in the same regions but are sexually mature at different times, so that potential mates remain unable to mate.
3. Ethological isolation :	The populations are isolated by different and incompatible behaviour (in animals only) before mating, so that potential mates meet but do not mate.
4. Mechanical isolation :	Cross fertilization or pollination is prevented or restricted by differences in structure of reproductive organs (genitalia in animals, flowers in plants), so that copulation is attempted but no transfer of sperm takes place.
B. Postmating or postzygotic isolating mechanisms	— Fertilization takes place, hybrid zygotes are formed, but these are inviable, or give rise to weak or sterile hybrids.
1. Gametic mortality :	Sperm transfer takes place but egg is not fertilized.
2. Zygotic mortality:	Egg is fertilized but zygote dies.
3. Hybrid inviability or weakness:	Zygote produces an F_1 hybrid of reduced viability.
4. Developmental hybrid sterility:	Hybrids are sterile because gonads develop abnormally, or meiosis breaks down before it is completed.
5. Segregational hybrid sterility:	Hybrids are sterile because of abnormal segregation to the gametes of whole chromosomes, chromosome segments, or combination of genes.
6. F_2 breakdown:	F_1 hybrids are normal, vigorous and fertile, but F_2 contains many weak or sterile individuals.

A. PREMATING OR PREZYGOTIC ISOLATING MECHANISMS

Premating or prezygotic isolating mechanisms are those that prevent contact between the species when they are reproductively active, or which prevent or restrict the union of gametes after mating or cross-pollination has occurred. It may be caused by the following reasons :

1. Habitat Isolation (Ecological Isolation)

Often two closely related species will thrive in different ecological conditions (habitats) within the same territory, but no hybrid between them will be found. However, when the members of the representative species are taken into the laboratory, hybrids between them can be obtained. Thus, their respective gene pools are isolated physically, but not physiologically. Again, because of the ecological separation that inhibits the sharing of traits, the incorporation of new adaptations may lead to functional separation of their gene pools, and speciation.

Habitat or ecological isolation is most common in plants because of their sedentary nature. In them it operates in two ways. The species may live in the same general area but have such different habitat preferences that their populations are rarely close enough together to cross fertilize each other frequently. In addition, if hybrids are occasionally formed, they may be either unable to grow maturity or leave few progeny under natural conditions, because no site to which they are adapted is available. This is well illustrated by several species of oaks. Many oaks are interfertile and hybrids are occasionally found. Such is the case with four Texas species: *Quercus mohriana*, *Q. havardi*, *Q. grisea* and *Q. stellate*, studied by **C.H. Muller** (1952). These four species are kept distinct because they grow in very distinct soil types—namely, limestone, sand, igneous outcrops and clay respectively—and the few hybrids between them are confined to the zones where the soil types come into contact. Furthermore, two plant species of scarlet oak (*Quercus coccinea*) and black oak (*Q. velutina*) are sympatric throughout most of the Eastern United States and both are distinguishable by shapes of their leaves and acorns. **Sympatry** is the phenomenon of occurrence of two or more populations in the same area; more precisely, the existence of a population in breeding condition within the cruising range of individuals of another population. Such populations which are related and share a portion of their ecologic ranges but remain isolated from the another not by space but through the physiologic expression of genetic difference, are called **sympatric populations** or **species**.

Habitat isolation is not a very effective isolating mechanism in mobile animals and has been observed most frequently in mouse, river fishes and certain water snakes. For example, two subspecies of American mouse (*Peromyscus maniculatis*) inhabit regions that share a common boundary. While laboratory hybrids between these subspecies exist, none has been found in nature. The explanation for this isolation is that one subspecies lives in the forests and the other on sandy beaches. Apparently, members of the two groups rarely encounter each other.

The same situation has been found for another subspecies of mouse in Oregon as well as for the water snake *Natrix sipedon*. One subspecies of snake is a freshwater race, and the other is a saltwater race.

2. Seasonal Isolation (Temporal Isolation)

Two groups may exist in exactly the same ecological area, but do not interbreed because they become sexually mature at different times of the year or under different conditions. Further, seasonal barriers are particularly frequent among aquatic animals, because water temperatures are most stable than air temperatures and embryonic development is more closely harmonised to definite temperatures. These two factors combined and permit a close regulation of the breeding seasons.

Seasonal isolation is common in plants and occurs frequently among insects and other invertebrates and some vertebrates. For example, over ten subspecies of Cypress trees (*Cupressus*) are found in California. They do not interbreed because each race produces pollen at a different time of the year.

In animals it may be the time or condition for mating which may vary. In the northeastern United States, three species of frog, *Rana pipiens*, *R. sylvatica* and *R. clamitans* all mate in the same ponds at different times. *R. sylvatica* begins breeding when the water is 44^0 F and completes its mating season before the temperature reaches 55^0 F, the breeding temperature necessary for *R. pipiens*. *R. clamitans* must have temperatures of at least 60^0 F. By this time the temperature is reached, both *R. sylvatica* and *R. pipiens* are past their breeding stages. Further, most closely related species of anurans (frogs and toads) have overlapping breeding seasons, but in them, the isolation is provided by acoustic stimuli.

3. Ethological or Behavioral Isolation (Sexual Selection)

Ethological (Gr, *ethos*=habit, custom) refers to behaviour pattern. Ethological isolating mechanisms are barriers to mating due to incompatibilities in behaviour. It is based on the production and reception of stimuli by the sex partners. The males of every species have specific courtship behaviour and females of the same species are receptive. Courtship involves an exchange of stimuli (visual, auditory, tactile, olfactory or chemical) between male and female continuing until both have reached a state of physiological readiness in which successful copulation can occur. The species specific stimuli which help two sexes for the simultaneous recognition of each other for courtship forms the **ethological isolation**.

For example, birds and insects embark on mating after highly standardized ritual dances. In some species of birds, plumage of the male stimulates the female sexual interest. Thus, the different colour patterns of insects; different colour patterns of plumages of birds; light signals sent out by male fire flies (beetles of family *Lampyridae*); melodious songs produced by the males of insects, frogs, birds, etc., sex attractants (ectohormones or pheromones) produced by most insects, mammals including human female; are some of the examples of stimuli which have a significant role in ethological isolation.

In most animals it is the male that actively searches for a mate. He is usually rather easily stimulated to display to objects. Sometimes quite inappropriately, when he does not receive adequate response from his display partner, or is actively repulsed, his display derive eventually becomes exhausted. Consequently, if such a displaying male encounters an individual of a different species, or a male of his own species, he will break off his courtship sooner or later. If the male is displaying to a non-receptive female, the same will happen, perhaps, after longer intervals.

The English naturalist **Gilber White** (1720-1793) was the first to notice that the chiff - chaff, the willow warbler, and the wood warbler were three different species, and not just one. The wood warbler is slightly larger and brighter in colour, but the chiff-chaff and willow warbler look almost exactly the same. The songs of these three, however, are all distinctly different. For the birds, the songs are used by the female to select a mate, so in this way they act as an isolating mechanism, separating the otherwise similar species.

In *Drosophila*, for example, it is shown that males act quite specifically to remove inhibitions, so that females will copulate with males of her own species. These actions of courtship include orientation of the body, display of wings and licking with tongue. Further, when the antennae, the most sensitive tactile organs of females, are etherised, the males can easily bring about copulation with females of different species.

4. Mechanical Isolation

Mechanical isolation is provided due to differences in the structure of flower of flowering plants or in the genital organs of different species in animals, so that cross-pollination or copulation may not be easily brought about. In case of higher plants mechanical isolation provides the most effective and a major isolating mechanism. It is more effective in those plants where flower structures remain well specialized for insect pollination. For example, different species of *Asclepias* of *Asclepiadaceae* differ in the shape of pollinia (sacs carrying pollen) and that of the stigmatic slits, so that with the help of particular insect bringing about pollination, pollinia of one species cannot bring about pollination on the stigma of another species. Such kind of mechanical isolation is also common in the other members of family *Asclepiadaceae* and *Orchidaceae*.

The orchid family has evolved flowers having an architecture that is particularly susceptible to large numbers of variations on a single adaptive theme. They are bilaterally symmetrical or **zygomorphic**, because one of the six perianth members or tepals is modified into an elaborate structure known as the **lip**; the stamens and stigma are united to form a single structure, the **column**, which in each species produces a very specific relationship between the pollen-releasing anthers and the pollen-receiving stigma; and their sticky pollen mass cannot be transported by any means other than a particular vector (Fig. 9.1). Unless the flower is visited by an insect or other visitor particularly adopted to one or a few related species, no seed is formed (see **G.L. Stebbins**, 1976). Flowers of some orchids even mimic in shape the females of certain species of wasps and bees (**Dodson**, 1967). Male wasps or bees are attracted to these flowers and engage in "**pseudo-copulation**" with them, thus, receiving and transferring the pollen sacs.

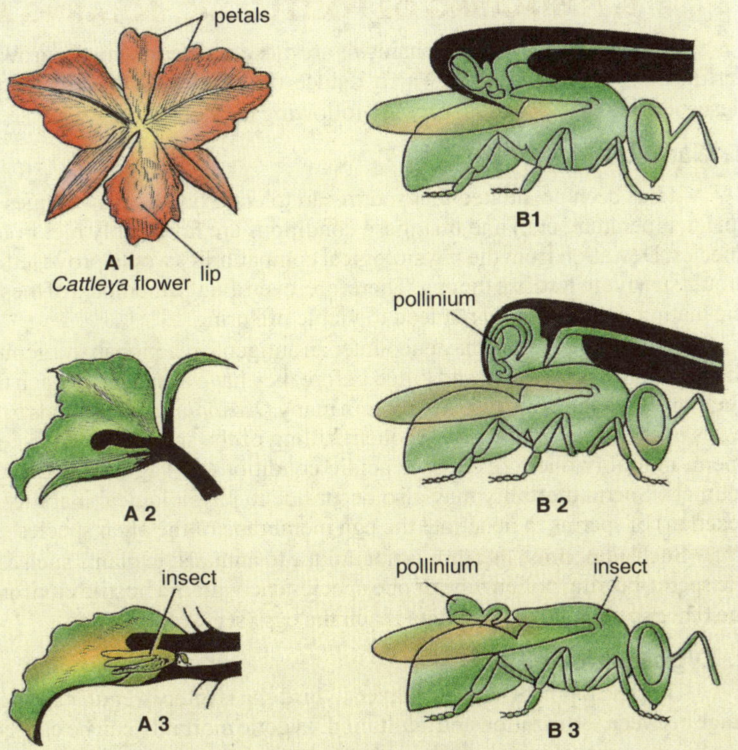

Fig. 9.1. Pollination of an orchid (*Cattleya*) flower, a highly specialized and vector specific flower. A— A–1. Face view of the flower showing the three narrower outer petals and three broader inner petals, the lowest of which is the lip. A–2. L.S. of the flower. Partly enclosed in the lip is the stamen that contains pollina, the stigma that receives pollen and the column that support them. A–3. L.S. flower showing position of an insect visitor in the act of delivering pollen and receiving a new pollinium; B—Enlarged sectional drawings showing how the pollinium becomes stuck to the bee's thorax (after Dobzhansky, *et al*.., 1976).

The occurrence of

mechanical isolation among animals was recognized quite early. As, soon after the discovery of the manifold structural differences in the genital armatures of different species of insects, it was asserted by **Dufour** (1844) that these genital armatures act like lock and key, preventing hybridization between individuals of different species. For example, interspecific crosses in *Drosophila* may cause injury or even death to the participants, and the same is true in *Glossina* (tsetse fly) and in pulmonate snails of subfamily *Polygyrinae*. However, **Karl Jordan** (1905) contradicted Dufour's hypothesis by showing that out of 698 species of *Sphingidae* family of *Insecta*, 48 were not different in their genitalia from other species of the family, while in about 50% of the species with geographic variation in colour, there was geographic variation in the structure of the genitalic armatures. Since that time much additional information has accumulated indicated the slight importance of the genitalic armatures as isolating mechanisms in animals.

According to **Mayr** (1970), the genitalic apparatus of animals is a highly complicated structure and is the pleiotropic by-product of very many genes of the species. Any change in the genetic constitution of the species may result in an incidental change in the structure of genitalia. As long as this does not interfere with the efficiency of fertilization, it will not be selected against.

B. POSTMATING OR POSTZYGOTIC ISOLATING MECHANISMS

Postzygotic isolating mechanisms are those which prevent the growth of hybrid individuals after fertilization has occurred or which reduce the fertility of the F_1 hybrids or the viability of their descendants. They are caused by the following methods :

1. Gametic Mortality

It has been postulated that sperm can tolerate only narrow ranges of physiological conditions (pH, temperature, etc.) and that these conditions are favourably met in female members of the same species. Deviation from the physiological compatibility would prove lethal to sperm, and, few, if any, would survive to fertilize the egg. Therefore, even though members of these two groups could copulate, the mating process would not lead to viable offspring.

Further, the sperms may encounter an antigenic reaction in the genital tract of the female, so that they become immobilized and killed before they have a chance to reach the eggs. **Patterson** reported such "insemination reaction" to occur in many *Drosophila* which leads to an enormous swelling of the walls of the vagina and the subsequent killing of the spermatozoa. In certain sterile human females sperm mortality due to excessive acidic conditions of vagina has been reported by many medical journals. Sperm mortality may also occur due to physiological inability (*i.e.*, fertilizin-antifertilizin reaction) of sperms to penetrate the egg membrane of the alien species.

Such **gametion isolation** is not unique to animals; in plants such as the Jimson weed (*Datura*), the sperm-bearing pollen tube of one species encounters a hostile environment in the flower tissue of the other species and is unable to reach the egg (see **Volpe**, 1985).

2. Zygotic Mortality

The development of the fertilized hybrid egg is often irregular and development may cease at any stage between fertilization and adulthood, zygotic mortality can be caused due to several cytological, genetical, embryological, molecular, biological, biochemical and physiological reasons in most animals. For example, certain hybrid zygotes of *Ambystoma mexicanum* lack nucleolus and in the absence of nucleolus, such zygotes die without undergoing cleavage.

Certain authors have included the isolations due to sperm mortality and zygotic mortality under **physiological isolation**. Such physiological isolation also can occur as an incompatibility between the embryo and the female parent. Interspecific hybrids between two species of Jimson weed (*Datura*) result in the death of the embryo at the eight-cell stage. By removing the embryos and culturing them artificially, viable seeds can be produced that germinate and grow into healthy offspring.

3. Hybrid Inviability

Many naturally occurring animal hybrids though have somatic hybrid vigour and fertility, but leave no offspring. The reason for the reproductive failure of fully fertile species hybrids is perhaps that they are less well adjusted to available ecological niches than individuals of the parental species. Also, species hybrids are usually less successful than individuals of pure species in courtship, when definite behaviour patterns and species-specific stimuli play an important role. Ecological and ethological inferiority, thus, reduces their chances of leaving offspring.

Rana catesbeiana

Rana clamitans

For example, two species of the chicory plant, *Crepis tectorum* and *Crepis capillaris*, can be crossed, but the hybrid seedlings die in early development. Crosses between the bullfrog, *Rana catesbeiana* and the green frog, *Rana clamitans*, results in inviable embryos.

4. Developmental Hybrid Sterility

Sometimes postzygotic isolation involves in the production of vigorous but sterile species hybrids. This type of hybrid sterility involves the abnormal development of gonads, or abnormal meiosis or abnormality in gamete formation. It is usually most common in male animals and plants. For instance, inviable F_1 hybrids between different species of *Drosophila* and flies, also between various species of mammals such as cattle × yalk, cattle × buffalo, horse × zebra, and between some species of birds such as mallard and muscory ducks. Thus, in certain hybrid crosses, such as between females of the toad species *Bufo fowleri* and males of *Bufo valliceps*, the hybrids may survive but are completely sterile.

Bufo valliceps

Bufo fowleri

The abnormality occurs due to poor growth and a low rate of mitosis in the cells of seminiferous tubules. Further, developmental abnormality also occurs at the time of meiosis of spermatocytes due to which either no sperm is produced or if any sperm is produced that remains undifferentiated, to be unable for fertilization. **A. Muntzing** (1930) has used the terms, **haplontic sterility**, where the sterility is set in haploid parts (gametes) and **diplontic sterility**, where the sterility is brought about due to events before meiosis or after fertilization in diploid tissues.

5. Segregational Hybrid Sterility

In both plants and animals hybrid sterility may result from abnormal segregation at meiosis of either whole chromosomes or blocks of genes contained in chromosomal segments. If the chromosomes of the parental species are not homologous (*i.e.*, chromosomes of both parents of a hybrid contain individual gene dissimilarities), they cannot pair at all. Further, segregational hybrid sterility or **genetic isolation** may also be caused due to genetically controlled spindle abnormalities, asynapsis or desynapsis and similar other abnormalities. For instance, in the hybrid (*Raphanobrassica*) between the radish and cabbage, the nine chromosomes derived from the radish may not pair at all with the nine chromosomes derived from the cabbage, so that at meiosis one observes 18 single chromosomes instead of nine pairs found in the parental species. Since these unpaired chromosomes are unable to line up on the meiotic metaphase spindle, they are distributed irregularly to the poles caused daughter cells with unbalanced complements of chromosomes, the genes of which are unable to direct the development of the pollen grains or embryo sacs.

Further, sometimes chromosomal sterility may result when the chromosomes from two parents do not lack homology but differ due to structural changes like translocation or inversions (chromosomal aberrations). In such a case the pairing of chromosomes will be imperfect and results in the segregation to the gametes of abnormal disharmonious combinations of genes (*i.e.*, sterile gametes). For instance, in hybrids in between *Primula verticillata* and *P. floribunda*, the sterility in hybrid is known due to small structural difference between parental chromosomes. Likewise, the functional offsprings, mules of horse and donkey, are sterile. The horse and donkey each contributes 33 chromosomes to the offspring or (species hybrid) mule. During meiosis of gametogenic cells of the mule, however, the chromosomes fail to pair. This failure is thought to be the result of the degree of genic dissimilarity between the horse and donkey chromosomes, so that they do not "recognize" one another at homologues. Since the orderly separation of chromosomes into gametes is necessary for fertilization to take place, the mule is sterile.

Primula verticillata.

Primula floribunda.

There is evidence that genetic isolating mechanisms are occasionally modified or reinforced by conditioning. Young birds can sometimes be imprinted on a foster species, when raised by foster parents. In parasitic birds, cuckoos and cowbirds, such conditioning is absent, and species recognition is rigid. Young cowbirds, for instance, leave the company of their foster parents and flock together as soon as they are independent. In the human species, however, conditioning is an important isolating mechanism. The free interbreeding of individuals coexisting in a geographical region is strongly influenced by religious, economic and cultural barriers.

6. F_2 Break Down

In both animals and plants, there are examples of hybrids which are highly or at least partly fertile, but which give rise to weak, abnormal, or sterile progeny in the second (F_2) generation. This phenomenon is called **hybrid breakdown**. Classical examples of hybrid breakdown are provided by the F_1 hybrids, *Gossypium arboreum* x *G.herbaceum* and *Gossipium barbadense* x *G. hersutum*. In both of these cases, F_1 progeny is either missing or is very weak.

From the developmental point of view, F_2 breakdown is similar to segregational sterility, except that the effects of the segregation of disharmonious gene combinations are delayed until after fertilization.

THE COACTION OF ISOLATING MECHANISMS

The interbreeding of closely related sympatric species of animals is prevented by a whole series of ecological, behavioral and cytogenetical factors, usually several for each species pair. One factor is often dominant, such as acoustic stimuli in certain frogs, grasshoppers, cicadas and mosquitoes; chemical contact stimuli in some insects; sex stuffs in Lepidoptera and certain marine organisms (*Bonellia*, etc.), and visual stimuli in certain birds, fishes and insects. The sterility barrier may be strong or weak but is rarely tested except when the other isolating mechanisms break down. Thus, the isolating mechanisms are arranged like a series of hurdles, if one breaks down, another must be overcome. For example, if the habitat barrier is broken, individuals of two species may still be separated from each other by behaviour patterns. If these also fail, the mates may be unable to produce viable hybrids, or if hybrids are produced they may be sterile (see **Mayr**, 1970).

THE GENETICS OF ISOLATING MECHANISMS

From the multiplicity of isolating mechanisms that protect every species, one can conclude that a considerable number of genes are involved. Almost any gene that changes the adaptation of a population may have an incidental effect on the interaction between male and female. Conversely, a mating advantage of a male will become established in a population provided it does not seriously lower the fitness of his offspring.

Detailed analysis of the genetics of isolating mechanisms other than sterility exist only for the genus *Drosophila*. Many mutations in *D. melanogaster*, such as yellow, white, and bar influence the mating success of the mutant individuals. Further studies have established now that behavioural isolation has a highly complex genetic basis. In addition to genes controlling the specificity and variation of the signals exchanged between males and females, there are others that control mating drive and levels and rhythms of activity.

ROLE OF ISOLATING MECHANISMS

One of the evident functions of isolating mechanisms is to increase the efficiency of mating. Where other closely related species do not occur, courtship signals can "afford" to be general, non-specific and variable. Where other related species coexist, however, non-specificity of signals may lead to wasteful courtship and delays, even where no heterospecific hybridization occurs. Under these circumstances there will be a selective premium on precision and distinctiveness of signals. This is exemplified by *Parus* bird of Tenerife island and *Regulus* song bird of Azores.

Moreover, each species is a delicately integrated genetic system that has been selected through many generations to fit into a definite niche in its environment. Hybridization usually leads to break-down of this system and results in the production of disharmonious types. It is the function of isolating mechanisms to prevent such a breakdown and to protect the integrity of the genetic system of species. Any attribute of a species that would favour the production of inferior hybrids is selected against, since it results in wastage of gametes. Such selection maintains the efficiency of the isolating mechanisms and indeed helps to perfect them. Isolating mechanisms are among the most important biological properties of species.

ORIGIN OF ISOLATION

The following two theories have been proposed about the origin of isolation :

Wallace has proposed the first theory. According to him, isolation mechanisms are selected by natural selection to root out lowly fitted hybrids between species and allow only those genotypes that have developed better isolation mechanism. There are instances where hybrids survive better under vigorous environmental stresses than the pure races. The serious objection to this theory principally comes from Darwin. If the back-crossing and introgression by a certain hybrid population would lead to the steady state deterioration of the pre-existing mechanism. Thus, Wallace's argument of improvement of isolation is failed.

Poulton and **Muller** put forward the second idea regarding origin of isolating mechanism. According to their idea, isolation mechanisms originate as a byproduct of genetic divergence of isolated populations. Genetical studies reveal that hybrids between two new species must be effectively sterile or inviable. Natural selection then picks out such instances, improves accessory isolating mechanisms, such as failure of sterile species survival. Such a selection ultimately renders hybrid production lesser and lesser in nature.

REVISION QUESTIONS

1. What is isolation? Describe various premating and postmating isolating mechanisms.
2. Write an essay on isolation.
3. Explain the role of isolation in organic evolution.

CHAPTER

10

Speciation

These species of cichlid fisher have descended from a single ancestral population within a million years.

The entire course of evolution depends upon the origin of new populations (species) that have greater adaptive efficiency than their ancestors. The study of population divergence or speciation is, therefore, crucial to an understanding of evolution. Before the discussion of speciation process, a brief discussion of the term species is desirable :

Species, Race and Deme

A **species** may be defined as an interbreeding population, which is reproductively isolated from other similar but morphologically distinguishable populations. In other words, a species may be envisioned as an isolated pool of genes flowing through space and time, constantly adapting to changes in its environment as well as to the new environments encountered by its extensions into other geographical regions. At various intervals of time and in different environments of this gene pool may become isolated. Such isolates may (1) become extinct, (2) reunite with the parental species, or (3) differentiate during isolation to form new species which in turn pass through space and time. Certain general characteristics of animal species are the following :

(i) Each species is an isolated pool of genes possessing regional (racial, populational) characteristics in gene complexes which are inter-connected by gene flow.

(ii) Each species fills an ecological niche not exactly utilized by another species.

(iii) Each species remains in process of continually adjusting to its environment.

(iv) Each species possesses a constellation isolating mechanisms that directly or indirectly prevent exchange of genes with related species.

(v) Each species has the capacity to give rise to new species provided some form of geographical or spatial isolation gives its isolates an opportunity to develop a unique gene pool without being swamped by gene flow from the parental species.

Each species remains in process of continually adjusting to its environment.

Lastly, a species is a population of individuals with similar structural and functional characteristics, which have a common ancestry and in nature breed only with each other. It is a collection of **demes**, the smallest population units which are the groups of genetically similar individuals bearing an intimate temporal and spatial relation to one another. Each deme is isolated to some extent from adjacent demes but there is always the possibility of genetic exchange between them. Even distantly located demes may contribute genetic material to one another over a period of time by the gradual passage of certain genotypes from one deme to another. Thus, demes are open genetic systems that are affected by gene flow from adjacent populations, that is, they are only partially isolated populations.

An intermediate population unit which exists in between species population and local population (deme) is **subspecies**, or **geographical race** or **race**. A subspecies is an aggregate of local populations of a species inhabiting a geographic subdivision of the range of the species and differing taxonomically from other populations of the species.

The nature of species, subspecies and deme now arms us to solve the following questions: How from a single stock of population arise different species? or, in other words, how does speciation occur?

NATURE OF SPECIATION

Speciation means the formation of species (**Mayr**, 1970). Ever since the speciation process became a separate topic in the 1930s (**Mayr**, 1931) alternative methods of speciation have been proposed. The principal ones are **quantum evolution** proposed by **Simpson** (1944), **quantum speciation** defined and discussed by **Grant** (1963, 1971) and **Carson** (1973, 1975), and called **saltational speciation** by **Ayala** (1975); **parapatric speciation** discussed by **Murray** (1972) and **Bush** (1975); and **sympatric speciation** described by **Stebbins** (1950), **Clausen** (1951), **Mayr** (1963, 1970) and **Grant** (1971).

The true meaning of the term "origin of species" used by **Darwin** was understood only rather recently. The pre-Darwinian evolutionists did not understand its exact significance, even Darwin himself seems to have considered "origin of species" the same as "evolution". He confused two essentially different problems, subsuming them under the single heading "origin of species". One of these is evolutionary change, in which Darwin was primarily interested and for which Darwin used the following terms— "phyletic evolution" "modification in time," or "decent with modification".

Phyletic evolution means evolution in a line or lineage. Species A living in a certain region in the course of time undergoes change so that the descendants are sufficiently unlike their distant ancestors to be considered a different species. *Mesohippus* evolving into *Miohippus* may be taken as an example of such phyletic evolution in which species A has become species B. Typically, species A will have disappeared in the process, leaving species B in place. Similarly, as time goes on species B may evolve into and be replaced by species C and so on. This phyletic pattern of evolution along a time axis is abundantly evident in the sequences of fossil forms.

Phyletic evolution.

According to **Mayr** (1970), such evolutionary transformation of a species does not necessarily lead to multiplication of species or speciation. Because an isolated population on an island might change in the course of time from species A through B and C into species D without ever splitting into several species. Consequently, in the end there will be only one species on the island, just as at the beginning. The essential aspects of phyletic evolution is the continuous genetic and evolutionary change within the populations composing the species, without the development of reproductive isolation between subspecies or demes of a species and, consequently, without its breaking into several species.

Speciation in its restricted modern sense, however, means the splitting of a single species into several, that is **multiplication of species**.

The opposite of splitting of a single species into several daughter species is the **complete fusion of two species**. Since the product of such a process would be an entirely new unit, truly a new biological species, it is legitimate to consider the fusion of two species into one as a form of speciation, which is the reverse of a multiplication of species. Since species are defined as reproductively isolated populations, the fusion of two species is, on the whole, a logical contradiction. Yet occasionally in nature a previously existing reproductive isolation breaks down and two previously distinct sympatric species merge. This happens most easily in species in which the isolating mechanisms are primarily ecological. Such fusion method of speciation has been reported in two species of birds, towhees *Pipilo erythrophthalmus* and *P.ocai*, both of which fuse because of breakdown of isolation and form a new species.

In this chapter, a detailed discussion will be made only for speciation by multiplication of species as other methods of speciation namely, transformation of species (phyletic speciation) and reduction in number of species (fusion of two species) are of little significance.

POTENTIAL MODES OF SPECIATION

According to **Mayr** (1970), true speciation or multiplication of species may occur by the following agencies :

A. Instantaneous speciation (through individuals)
  1. Genetically
    (a) By single mutation in asexual "species"
    (b) By macrogenesis
  2. Cytologically, in partially or wholly sexual species
    (a) By chromosomal mutations or aberrations (translocation, etc.)
    (b) By polyploidy
B. Gradual speciation (through populations)
  1. Geographical speciation
  2. Sympatric speciation

A. INSTANTANEOUS SPECIATION

The process of instantaneous speciation may be defined as the production of a single individual (or the offspring of a single mating) that is reproductively isolated from the species to which the parental stock belongs and that is reproductively and ecologically capable of establishing a new species population. It can be achieved by the following methods :

1. Instantaneous Speciation through Ordinary Mutation

Mutation is a genetic phenomenon of such relatively high frequency that in a higher animal with more than 10,000 gene loci almost every individual will be the carrier of a new mutation. Such mutations merely increase the heterozygosity of a population; they do not lead to the production of new species. Any mutation drastically affecting reproductive behaviour or ecology will be selected against, if it lowers the viability, or will displace the original allele if it is of higher viability. In neither case will there be any origin of discontinuities. It is evident that ordinary mutation cannot produce new species in sexually reproducing species.

Certain phenomena among rotifers, cladocerans and nematodes suggest the occasional occurrence of **asexual speciation by mutation**. For example, in the rotifer order *Bdelloidea*, there occurs no male individual in entire order and so it is assumed that the ancestral species of this order was parthenogenetic. This ancestral species evolved into over 200 species, about 20 genera and 4 families only due to asexual speciation by mutation.

Coelenterates (hydroids).

Vegetative reproduction by growth and splitting, thus, might offer a favourable condition for instantaneous speciation. This form of reproduction, however, occurs only in some of the lowest groups of animals, such as sponges, coelenterates (hydroids), turbellarians and bryozoans, groups in which the taxonomy is as yet too uncertain for a study of speciation.

2. Instantaneous Speciation through Macrogenesis

The sudden origin of new species, new higher taxa, or quite generally of new types by some sort of **saltation** (a change by leap across a discontinuity) has been termed **macrogenesis**. The macrogenesis theory of speciation has been supported by some geneticists such as **Goldschimdt** (1940, 1948) and **Schindewolf** (1936,1950). According to this theory, the production of a new type by a complete genetic reconstruction (macrogenesis) or by a major "systemic mutation" or **macromutation** is the crucial event in speciation. Such an event will produce a "**hopeful monster**" (as **Goldschmidt**, 1940 called it) which will become the ancestor of a new evolutionary lineage. The believers of macrogenesis theory also proposed that all new types appear in the fossil record suddenly and abruptly. These types are not connected with the ancestral types by intermediates, they claim, and cannot be derived from them by gradual evolution.

All these views of macrogenesis theory have been bitterly criticized by many modern evolutionists. The basic objection to macrogenesis theory is that it can never be proved by any kind of evidences, because it is obviously impossible to witness the occurrence of a major jump, particularly one that achieves simultaneously reproductive isolation and ecological compatibility. Further, **Simpson** (1953) and **Rensch** (1960) have criticized this theory by showing palaeontological evidences which clearly suggest that one 'type' can be derived from a previously existing one. Moreover, modern genetic studies have shown that the production of a monster or freak individual due to mutation does not provide the raw material for speciation as such freak individuals have low viability, do not get any mate for sexual reproduction and subsequently fail to establish reproductive isolation from normal members of the parental population. Therefore, macrogenesis is a poor source of speciation (**Mayr**, 1963, 1970).

3. Instantaneous Speciation through Chromosomal Aberrations

Closely related species often differ more conspicuously in their karyotype than in their morphology. Among aspects of the karyotype that differ are chromosome number; the number of metacentric or acrocentric chromosomes; the presence and kind of paracentric or pericentric inversions; or of supernumerary chromosomes, and just about every aspect of chromosomal aberrations. Visualizing this fact most cytogeneticists believed that chromosomal mutations have a significant role in the causing of instantaneous speciation. This belief is based on two assumptions : (1) that the degree of difference displayed by two species requires a speciation process of such drastic dimensions that only chromosomal mutation can qualify, and (2) that reproductive isolation between two species cannot be achieved without chromosomal reorganization. Now, it is known that chromosomal rearrangements in most cases does not produce new species, however, only in few cases it produces speciation.

(i) Chromosomal rearrangement without speciation. Except deleterious chromosomal mutations, most kinds of chromosomal aberrations (mutations) lead to **chromosomal polymorphism** rather than to the development of isolating mechanisms. The paracentric inversions of *Drosophila* are a well-known example of such chromosomal polymorphism. Yet each species has its own species-specific polymorphism, and only very rarely do even the most closely related species share the same chromosomal polymorphism. This fact underlines the drastic nature of the chromosomal aberration during much of speciation. This phenomenon suggests that there is no necessary correlation between chromosomal mutations and speciation, since either can occur without the other.

(ii) Speciation coinciding with a chromosomal mutation. If the chromosomal mutations during speciation would not have some selective (evolutionary) advantage, they would not have occurred so frequently in nature. Consequently, they have the following two advantages in chromosomal speciation—(a) chromosomal mutations have the potential to serve as (or contribute to) isolating mechanisms, and (b) the locking up and protection of a particularly favourable gene complement through a chromosomal mutation may create a new **supergene** as **Wallace** (1959) first of all recognized it. Both of these components of chromosomal speciation can subsequently be improved by natural selection, either during a period of segregation in a geographic isolate or during subsequent **parapatric speciation** or by both processes. The term **parapatric** stands for the population or species which remain geographically in contact but not overlapping and rarely or never interbreeding (**Mayr**, 1970).

(a) Chromosomal mutations as potential new isolating mechanisms. Any change in the structure of the chromosomes is called chromosomal mutations or chromosomal aberrations, whether it is an **inversion**, **translocation**, **duplication**, or any other change in the linear sequence of the genes or in the mechanics of the chromosome (for instance, spindle attachment). Chromosomal mutations (mostly inversions) are estimated to occur at a rate of 1 in 1000. Most of these are sufficiently deleterious to be eliminated at once, that is, before the mutation's carrier can reach reproductive age. Other kinds of chromosomal mutations are capable of giving rise to a system of balanced polymorphism and they will be retained in the population. In addition, there is a third class of chromosomal mutations which appears to reduce the **fecundity** (a term used for reproductive potential as measured by the quantity of gametes, particularly eggs, produced) of the heterozygotes to some extent. The heterozygotes containing some kind of chromosomal mutation that seems important in speciation, usually encounters the following difficulties during meiosis (gametogenesis) : (i) meiotica synapsis (partial failure of chromosome pairing), or (ii) malorientation of multivalents at the first meiotic metaphase, or both. Both of these difficulties lead to the production of gametes carrying chromosomal deletion or duplication or broken or acentric chromosomes and, thus, lead to a significant reduction in the fecundity of male hybrids.

Unlike **gene** or **point mutations**, chromosomal mutations usually have no effect on the visible phenotype of the individual, so they are difficult to detect. Further, unlike the heterozygotes containing point mutations, the heterozygotes of chromosomal mutations are not shielded by dominance and such heterozygotes with their meiotic difficulties are the prime target of selection (**John** and **Lewis**, 1969). Finally, chromosomal mutation is not a change in the DNA programme, but a change in the linear sequence of genes (duplication or a deletion). The most frequent chromosomal change is the **fusion** of two acrocentric chromosomes into a single metacentric chromosome. The result is the reduction in chromosome number. A trend toward such a reduction is widespread among animal taxa. For instance, the most primitive isopods have 28 haploid chromosomes and this number is independently reduced through fusion to the lower number of 8 in several unconnected lines. It is now clear that most differences in chromosome numbers between closely related species of animals are the result of such fusions.

If only fusions have occurred during evolution, soon all species would have only a single pair of chromosomes. Obviously, there should be some opposing mechanism which may increase the chromosome number. The seemingly simplest such process would be the exact reverse of fusion, that is **fission** (or disassociation) of a metacentric chromosome into two acrocentrics. Most cytologists doubted such fission to occur. However, increase in the chromosome number (not due to polyploidy) whatever the mechanism, is frequent in animal evolution and may well play an important role in speciation (see **Mayr**, 1970).

(b) Chromosomal mutation and the production of new supergenes. It becomes now clear that gene contents rather than mechanical qualities, as discussed in preceding heading, determine in many cases whether or not a new gene arrangement can establish itself in a population. Most structural rearrangements of chromosomes inhibit or prevent crossing over in heterozygotes. A new gene arrangement may "lock up" a coadapted gene sequence and, by protecting it from crossing over, create a new **supergene** (**Mayr**, 1963). **Wallace** (1959) pointed out that bearers of such protected chromosomes or chromosome segments being members of peripheral populations, are better adapted for the marginal environment of the species border than are genotypes from centre of the species range. To him, the most important aspect of the chromosomal reorganization is the protection from disruptive recombination that it affords certain new supergenes. Chromosomal mutations, thus, is an instrument of ecotypic adaptation.

It is evident that these two aspects of chromosomal mutation—the production of mechanical incompatibilities and development of new supergenes—reinforce each other. There will be a steady selection for an improvement of the adaptation supergene and this will tend to produce an increase in genic heterozygote inferiority. This, in turn, will strengthen the effectiveness of the cyto-mechanic isolating mechanisms.

4. Instantaneous Specification through Polyploidy

Polyploidy is a multiplication of the normal chromosome number. It is very widespread among plants and is an important mechanism of speciation in the plant kingdom (**Stebbins**, 1950, **Grant**, 1963). Except conifers, one-third of all species of plants have arisen by polyploidy. Among animals the polyploidy is much rarer. Among animals it occurs only in those groups which reproduce parthenogenetically. In few parthenogenetically reproducing animal groups such as lumbricid earthworms, turbellarians and in certain groups of weevils polyploidy is the principal method of speciation.

B. GRADUAL SPECIATION

Gradual speciation is the gradual divergence of populations until they have reached the levels of specific distinctness. Two modes of gradual speciation have been postulated—(i) one involving geographical separation of the diverging populations (**geographic speciation**), and (2) the other involving divergence without geographic separation (**sympatric speciation**).

1. Geographic or Allopatric Speciation

Geographic or allopatric speciaton states that in sexually reproducing animals a new species develops when a population that is geographically isolated from the other populations of its parental species acquires during this periods of isolation characters that promote or guarantee reproductive isolation after the external barriers break down (**Mayr**, 1942). The geographic speciation is the almost exclusive mode of speciation among animals, and most likely the prevailing mode even in plants, is now quite generally accepted.

For example, suppose there is a large assemblage of land snails subdivided in three geographical aggregations or races *A*, *B* and *C*, each adapted to local environmental conditions (Fig. 10.1). Initially there exists no barrier separating the populations from each other and where race *A* meets race *B* and race *B* meets race *C*, interbreeding occurs. Zones of interbreeding individuals are, thus, established between the races, and the width of these zones depends on the extent to which the respective populations intermingle.

Now suppose, some striking physical feature (barrier) such as great river, forging its way through the territory and effectively isolating the land snails of race *C* from those of race *B*. These two population may be spatially separated from each other for an indefinitely long period of time, affording an opportunity for race *C* to pursue its own evolutionary course. Two populations that are geographically separated, such as *B* and *C* in our pictorial model, are said to be allopatric. After eons of time, the river may dry up and the hollow bed may eventually become filled in with land. Now, if the members of population *B* and *C* were to extend their ranges and meet again, one of the two things might happen : 1. The snails of the two populations may freely interbreed and establish once again a zone of intermediate individuals. 2. The two populations may no longer be able to interchange genes. If two assemblages can exist side by side without interbreeding, then the two groups have reached the evolutionary status of separate species. A **species** is a breeding community that preserves its genetic identity by its inability to exchange genes with other such breeding communities. In our pictorial model (Fig.10.1) race *C* has become transformed into a new species *C′*.

2. Sympatric Speciation

The majority of authors until fairly recently considered sympatric speciation, that is, speciation without geographic isolation to be the prevailing mode of speciation. Such a speciation is based on two postulates : (a) the establishment of new populations of a species in different ecological niches within the normal cruising range of individuals of the parental population, and (b) the reproductive isolation of the founders of the new population from individuals of the parental population. Gene flow between daughter and parental population is postulated to be inhibited by intrinsic rather than extrinsic factors. The concept of sympatric speciation is far older than that of geographic speciation and goes back to pre-Darwinian days.

Definition of sympatric speciation. The method of origin of reproductive isolating mechanisms within the dispersal area of the offspring of a single deme is called sympatric speciation. The size of the dispersal area is determined, for instance, in marine organisms by the dispersal of the larval stages. In most insects it is determined in the adult state by the more mobile sex. Since there are normally a great many ecological niches (= the constellation of environmental factors into which a species (or other taxon) fits within the dispersal area of a deme, niche specialization is impossible without continued new pollution by immigrants in every generation.

Reasons for postulating sympatric speciation. Numerous biological phenomena suggest, at first sight, the occurrence of sympatric speciation. One is the abundance of finely adapted sympatric species in local fauna. For instance, there are several hundred (perhaps 500) species of bees of the genus *Perdita* in North America. All are oligolectic, confining their visits to the flowers of a single species or of a group of closely allied species. Many species of *Perdita* may be found at a single locality but never together because they occur on different plants or in some cases at different times of the year. A similar ecological specificity characterizes the species of many other genera of animals. All such species are perfectly adapted to the particular environment in which they occur. Could such a perfect fitting of species into their ecological niches have evolved as a purely accidental byproduct of genetics changes accumulated during geo-

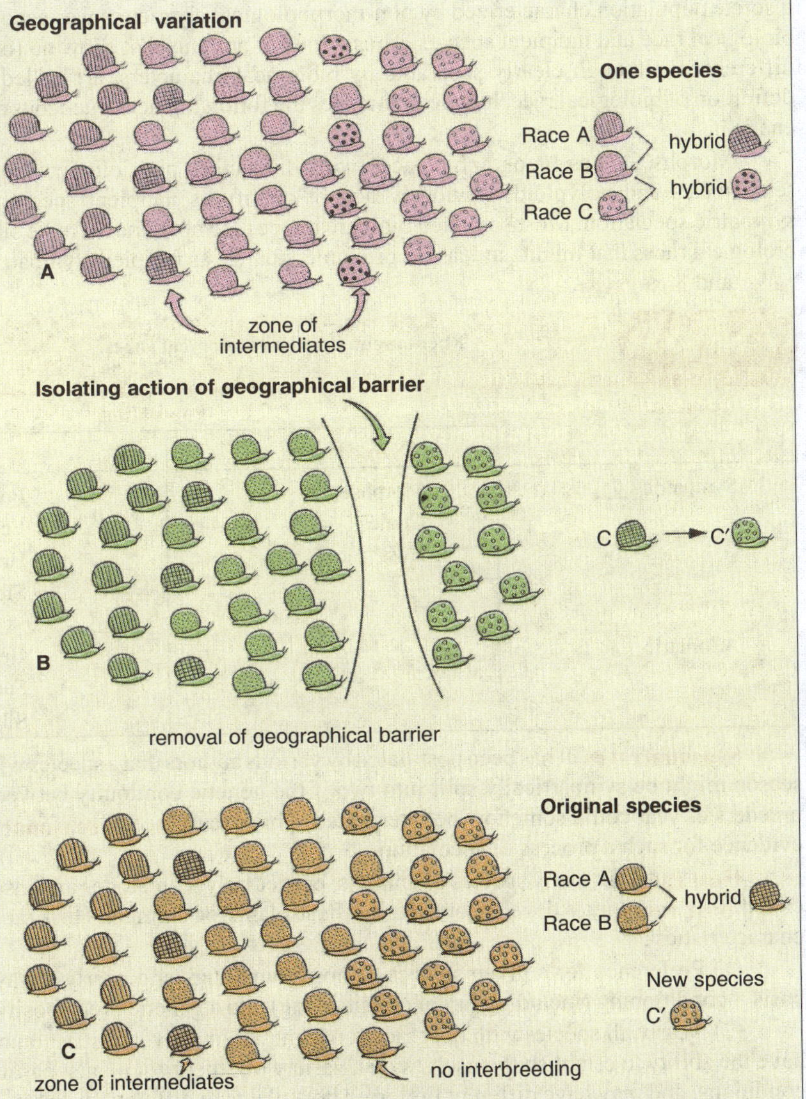

Fig.10.1. Model for the process of geographic speciation. Members of population or race *C* had diverged genetically during geographical isolation in ways that have made them reproductively antagonistic with race *B* when they met again. Race *C* has, thus, transformed into new species *C′* (after Volpe, 1985).

graphic isolation? Recent understanding of genetics of geographic speciation has shown that ecological differences can be acquired in geographic isolation and can be increased (by "character displacement") after the secondary establishment of sympatry.

Biological and host races. A necessary corollary of any theory of gradual speciation is that there should exist in nature some 'forms' or 'varieties' or 'populations' that are **incipient species**. The supporters of sympatric speciation have always cited the occurrence of biological races as examples of such incipient species. Most of the early supporters of sympatric speciation adhered to a morphological species definition and assumed that the acquisition of ecological or other non-morphological species characters preceded the acquisition of morphological species characters. Any discrete population characterized by non-morphological characters was accordingly interpreted as a biological race and incipient species. Thus, kinds of animals that show no (or only slight) structural differences, although clearly separable by biological characters, are called biological races. This definition of biological race, however, includes the following heterogeneous assemblage of phenomena in it.

Morphs, clones, geographic races with distinctive physiological or ecological properties, semispecies and polyploids obviously do not qualify as incipient species under the concept of sympatric speciation. Most so-called biological races are now known to be sibling species. The only biological races that might , at least in principle, qualify as incipient sympatric species are **seasonal races** and **host races**.

Table 10-1. Phenomena listed as biological races.

Occurrence	Population	
	Same	Different
Sympatric	Morphs Clones Host "races"	Polyploids Ecological races Host races Sibling species
Allopatric		Geographical races Semispecies Sibling species

Seasonal races. It has been postulated by various authors that a species with a very long breeding season might be sympatrically split into two if the genetic continuity between the earliest and last breeders of year could somehow be interrupted. This occurrence of **seasonal races** is often cited as evidence for such a process of speciation.

Host races. In many species of animals, particularly, nematodes and insects, temporary strains called **host races** may develop on specific host plants or animals. Host races have the following characteristics :

(1) Preference for a given species of host plants may and nearly always does have a double basis—conditioning (including larval conditioning) and a genetic predisposition.

(2) Nearly all species with host races concentrate in a given district on one host : yet they also have the ability to establish themselves on a variety of other host plants, particularly under crowded conditions, and may have different preferred host plants in different districts.

(3) Local populations (demes) of insects often have considerable genetic variability with respect to host specificity. Any monophagous (*e.g.*, Microlepidoptera, solitary bees, beetles (*Acmaeodera*, *Agrilus*), chrysomelid (*Calligrapha*, *Arthrochlamys*, leaf-Diptera and other groups) or oligophagous species of insect will come in contact, during its dispersal phase, with numerous plant species other than its normal host. If the species has the appropriate genetic constitution it will establish itself on the new

host and this, according to the Ludwig theorem, leads to an expansion of the food niche of the species.

(4) It is also found that if such a mixing of demes or local populations is prevented artificially and an inbred strain is selected experimentally on a single one of several original hosts, such a strain may become progressively less tolerant ecologically until a stage is reached when it may be difficult to re-establish it on any of the other original host species.

(5) Mortality occurs whenever a strain is established on a new host. The more usual the host, the heavier the initial mortality.

Among the many kinds of so-called biological races, it is the host races that have the best claim to the designation of "biological race" (**Mayr**, 1970).

Means of sympatric speciation. Thus, we can conclude that the sympatric speciation is caused by the following natural situations—sibling species; monophagous species; parasites; species swarms; and the instantaneous splitting of fossil lineages.

Hypothesis of Sympatric Speciation

Certain evolutionists have forwarded the following hypothesis for sympatric speciation, but all of these are not supported by evidences :

1. Homogamy. The concept of homogamy was postulated under the pre-Mendelian theory of blending inheritance because without its random mating would soon have led to complete elimination of all genetic variability. According to this concept, the most similar individuals of a population tend to mate with each other. Further, it postulated that monogamy would lead to homogamy. Both of these assumptions are supported by facts (**Mayr**, 1947) except that in a single case of birds where monogamy is a rule but never leads to homogamy. In case of two kinds of geese, namely snow goose and blue goose, monogamy leads to homogamy and both are found to belong to a single species, (*Anser coerulescens*) and are found to have differences in colour pattern due to a single gene, semidominant for blue (**Cooke** and **Cooch,** 1968). However, a mild form of homogamy occurs in many animals with highly variable adult size but that is quite insufficient for sympatric speciation.

2. Conditioning. Many workers such as **Thorpe** (1945) and others have suggested that the establishment of a new sympatric species population (*e.g.*, insects) in a new niche might be achieved through conditioning. But **Mayr** (1947) contradicted it by showing that complete isolation of the two populations was never achieved by any of the conditioning experiment of **Thorpe** with insects.

3. Preadaptation and niche selection. Most hypotheses of sympatric ecological speciation postulate that dispersing individuals search actively for that particular niche to which they are best adapted on the basis of their particular genotype. There is some evidence for the validity of this assumption, since a given species usually has well-defined species-specific habitat preferences and these are not necessarily identical for all the various genotypes of a species. Most habitat selection is, however, species-specific only in a very generalized way and is also often affected by non-genetic influences (conditioning and so forth). Only in an exceptional case will an individual search out that particular subniche for which it is specifically preadapted by its genetic constitution.

The concept of sympatric speciation by preadaptation is quite typological in making the assumption that a single gene preadapts an individual for a new niche. Indeed it would require a veritable systemic mutation to achieve the simultaneous appearance of a genetic preference for a new niche, a special adaptedness for this niche, and a preference for mates with a similar niche preference. The known facts do not support these assumptions.

Sympatric Speciation by Disruptive Selection

The method of natural selection for phenotypic extremes in a population until a discontinuity is achieved is called **disruptive selection** (**Mather**, 1955) or **diversifying selection** (**Dobzhansky**). **Muller** (1940) was first to suggest that the accumulation of different sets of specific modifiers might lead to sympatric speciation. **Thoday** and **Gibson** (1962) recently have studied different aspects of disruptive selection and they have found that when simultaneous selection for high and low bristle number was carried on in a population of *Drosophila melanogaster*, it responded with a strongly bimodel distribution within 12 generations of selection. However, when these selection experiments were repeated by other workers, similar results were not obtained. When the selection was phenotypically successful, one of the resulting morphs was of lowered viability and could be maintained only with difficulty in competition with the more successful morphs. Further, a selection pressure that would permit the survival of only two opposite extremes is unlikely under the variable conditions of natural environment.

A polymorphism established by diversified selection is unlikely to lead speciation. A species would lose all the advantages of improved utilization of the environment acquired through adaptive polymorphism if it were to split into a series of narrowly specialized species. **K. Jordan** suggested that selection of different physiological varieties within a population can have only two outcomes, either polymorphism or extinction of inferior types. Jordan's conclusion is well illustrated by the North American butterfly *Limenitis arthemis*.

According to **Mayr** (1970), geographic (allopatric) speciation is the only important mean for gradual multiplication of species because sympatric speciation is not supported by strong evidences.

Differences between allopatric (geographic) and sympatric speciation. The theories of sympatric and allopatric speciation agree in their emphasis on the importance of ecological factors in speciation. They differ in the sequence in which the steps of the speciation process follow each other.

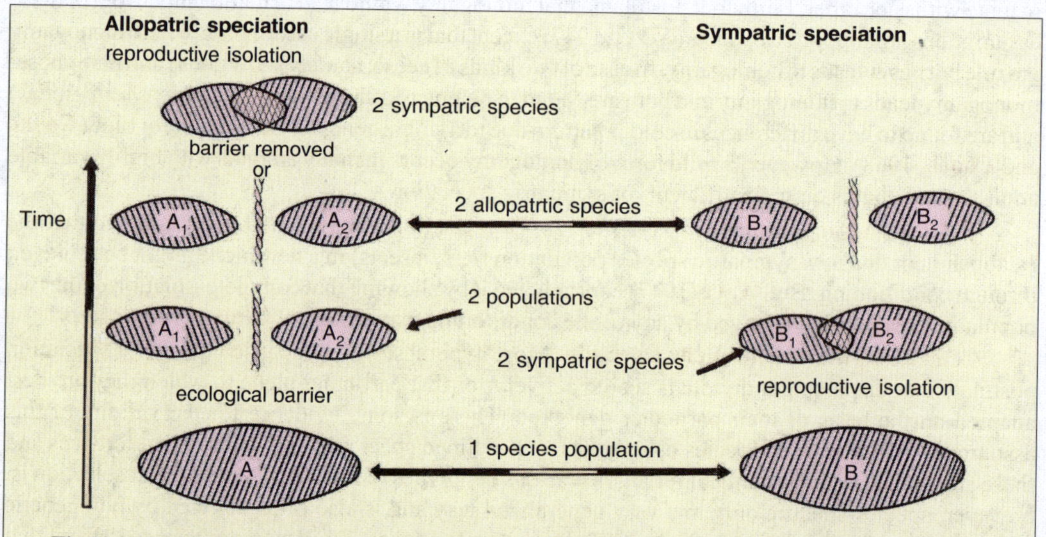

Fig 10.2 Comparison of the process of allopatric and sympatric speciation (after Savage, 1969).

The theory of allopatric speciation lets an extrinsic event which separates the single gene pool into several gene pools, with the ecological factors playing their major role after the populations have become geographically separated. According to the theory of sympatric speciation, the splitting of the gene pool is itself caused by ecological factors, and any spatial isolation of the populations formed thereby is a secondary, later phenomenon.

QUANTUM SPECIATION

Quantum speciation can be defined as the budding off of a new and very different daughter species from a semi-isolated peripheral population of the ancestral species in a cross fertilizing organism (**Grant**, 1963, 1971). The process of quantum speciation by which rapid evolutionary transition is achieved in small populations, is not fully understood, but chromosomal rearrangement (translocations) and change in gene regulation often play important roles (**Stanley**, 1979). Geographically, quantum speciation often occurs during adaptive radiation, by the release of genetic variability within ecologic islands. For example, most characteristic example in plant is *Clarkia lingulata* a narrowly endemic species of the Central Sierra Nevada California, clearly derived from the more widespread *C. biloba*, and differing from it in both floral morphology and karyotype (see **Stebbins**, 1976). **Hampton Carson** (1968) reported how eco-insular radiations (leading to quantum speciation) has produced about 238 species of *Drosophila* on the Hawaiian islands during about 5.6 million years.

Clarkia biloba. *Clarkia lingulata.*

The quantum speciation differs from the conventional speciation in the following ways :

1. Quantum speciation is rapid, requiring only a few generations.

2. The ancestors of new species do not include a large proportion of the populations belonging to the pre-existing one and may consist of only one or a few individuals. Conventional speciation is a process of splitting, quantum speciation is a budding process.

3. Conventional speciation may be promoted by drastic reduction in population size, but this is not necessary as it is for quantum speciation.

4. Conventional selection in its entirety is either guided by or is the by-product of natural selection. Quantum selection usually includes one or more chance events.

DIFFERENCES BETWEEN SPECIATION IN ANIMALS AND IN PLANTS

The pattern of speciation in animals and in plants has the following five significant differences :

1. A plant is a much less complex organism than an animal (**Grant**, 1971). One way of quantifying this difference is to compare the number of different kinds of differentiated cells found in the adult body. Estimates made by **Stebbins** (1976) indicate that this number varies form 47 to 52 kinds of cells in flowering plants, about 66 different kinds in an earthworm, 100 to 150 in an insect, and 200 to 250 in a mammal such as *Homo sapiens*. Moreover, the integration and delicate balance between organs required for motility and sense perception in animals are far greater than anything existing in plants. Finally, plants have an **open system of growth**, based upon embryonic tissue or meristems that occur at the ends of their shoots or roots and are potentially immortal, while animal integration requires a finite (fixed) body size and compactness, and youth, maturity, old age and death are necessary consequences. This difference affects speciation because when developmental patterns are relatively simple, the possibility that elements from two different patterns can be combined to make a functional intermediate is much greater than when they are highly complex.

2. Because of their highly developed sense organs, animals can easily develop barriers of reproductive isolation based upon divergent behaviour patterns, *i.e.*, ethological isolation. In many groups of plants, on the other hand, particularly those having accidental cross pollination by the wind, the possibility for building up prezygotic barriers of reproductive isolation is much less (**Baker**, 1959).

3. Because of their great capacity for vegetative growth, sterile hybrid plants in many groups may be virtually immortal. For instance, the Canadian pond weed (*Elodea canadensis*) was introduced into

Europe as female plant about 1840; but no males of this dioecious species arrived. Nevertheless, between 1840 and 1880 it spread by vegetative means throughout the inland waters of Europe (**Gustafsson**, 1946–47). Similar is true for the land plant, the grass red fescue (*Festuca rubra*; **Harberd**, 1961).

Thus, sexual sterility does not cause any hindrance to occupation of habitat in plants as it in animals. In perennial plant species, particularly those equipped with rhizomes or bulbs, a sterile hybrid can occupy indefinitely whatever habitat may be open to it. Moreover, even if it has a very low level of fertility, such that less than one per cent of the flowers can produce seed by sexual means, it can eventually produce many offspring that share its adaptive vegetative characteristics (*e.g.*, two grass species—*Elymus condensatus* and *E. triticoides;* **Stebbins**, 1959).

Elodea canadensis.

4. Sterile hybrids of plants may acquire fertility by chromosome doubling or polyploidy more easily than animals. For this reason alone, **reticulate evolution** is likely to be much more common in plants than in animals. Reticulate evolution can be defined as evolutionary change resulting from genetic recombination between strains is an inter-breeding population.

5. Many plant species are both hermaphroditic and capable of self-pollination, so that uniparental reproduction by autogamy is for them a normal method of reproduction.

REVISION QUESTIONS

1. Write an essay on the process of speciation.
2. Write short notes on the following:
   (i) Parapatric speciation ; (ii) Allopatric speciation ; (iii) Sympatric speciation ;
   (iv) Quantum speciation.
3. Describe the differences between the following :
   1. Conventional speciation and quantum speciation;
   2. Processes of speciation in plants and animals.

CHAPTER

11

Barriers

Mountain ranges are topographic barriers as they tend to limit the distribution of terrestrial animals.

Any physical, biological or climatic factor that restricts the migration or free movement of individuals or populations is called **barrier**. For example, the physical barriers include high mountain ranges, large water bodies between two land masses and large land masses between two water reservoirs, the difference in salinity of water, the depth of oceans, presence of very thick forests and absence of vegetation. According to **R.S. Lull** (1957) and **P.A. Moody** (1970), there are the following types of barriers :

1. Topographic Barriers

High and extensive **mountain ranges** act as barriers. They tend to limit the distribution of terrestrial animals. They are more effective if they are parallel with the equator as in Asia and Europe. In consequence, there exists a marked difference between the species occupying the northern and those occupying the southern slopes. For example, the Himalayan ranges of India are parallel with the equator and are covered by snow. On the south of it, there are hot moist plains of India having tropical fauna, which resembles that of Africa. It contains elephants, tigers and rhinoceros. In the north of Himalayas, the climate and other conditions are changed and as a result the animals of that region are comparable to those of Europe. It contains oxen and gibbons. In the New World (*i.e.,* North and South America), the mountain ranges run North to South and their influence as barriers upon the distribution of animals is very less.

The mountain ranges of western United States of America exercise an inescapable influence on the control of the humidity. The moisture-laden winds from Pacific Ocean never cross the mountains and are precipitated on the western side. Due to this action dry climate prevails on the eastern side. This has a great effect upon the nature of the vegetation and also upon plant-eating (or herbivorous) animals.

One important topographic barrier in North America is that which limits the Mexican plateau. The plateau itself has a temperate climate, and the fauna is similar to that of the region to the north and northwest. From the edge of the plateau the land drops away abruptly and conditions rapidly change to that of a hot, moist, tropical region, with a corresponding change in the plant and animal life, which is now that of tropical central and South America.

2. Climatic or Ecological Barriers

They include the abiotic ecological factors such as temperature, moisture and light which control the survival and dispersal of animals.

(i) Temperature. The temperature acts as an important barrier for both cold-blooded as well as warm-blooded animals, because none of the two can survive the extremes of temperature changes. The effect of temperature is much more marked in limiting distribution of cold-blooded animals. For example, the amphibians and reptiles are tropical and temperate in their distribution; they rapidly diminish towards North and South Poles. Among reptiles, crocodiles are tropical or subtropical. The northern limit of turtles is 50^0 latitude. The lizards are very few beyond 40^0 latitude. The serpents (snakes) have the widest range, but only three species in Europe are found beyond 55^0 latitude and one common viper extends up to the Arctic circle. Birds and mammals (*e.g.,* tigers, elephants, etc.) have no such limitations.

(ii) Moisture. The moisture is another very powerful factor which acts as barrier, because its lack and excess both produce different types of habitats. **Lack of moisture** produces desert conditions. Some forms can bear the drought conditions and some not and, hence, act as an efficient barrier. The most notable **desert barrier** is that of the **Sahara** in Africa which limits the distribution of deer. Deers are found in Eurasia but are absent from Africa except in its northern part (Straits of Gibralter to Tripoli). The **Arabian desert** also acts as a barrier for such forms. Animals such as *Amphibia*, which require moisture for their larval life and have no great migratory powers, find even a small arid area an impassable obstacle.

Lack of moisture produces desert conditions.

Increase of moisture in a region makes it unsuitable due to swampy conditions or excessive growth of trees . For certain creatures (*e.g.,* Permian reptiles) it had become a formidable place.

3. Vegetative Barriers

The forestation and the food supply, both, tend to limit the distribution of animals. The **luxuriance of vegetation** depends very largely upon the climatic conditions, *i.e.,* the temperature and degree of moisture. Its prevalence is favourable to the distribution of certain types of animals and unfavourable to others. The influence of vegetation is both direct and indirect. Its direct effect is the **arboreal** animals (tree dwelling), which cannot cross the regions where forests do not occur (*e.g.,* prairies). In the same way, the larger terrestrial animals cannot penetrate through the dense forest of tropics. Thus, in Pleistocene times mastodons of the genus *Dibelodon* migrated from North to South America across the newly established isthmian **land bridge**, but during the Pleistocene, while there was several species of magnificent elephants (*Elephas* and *Mastodon*) in North America, none of them apparently succeeded in penetrating south of the southern limits of Mexican plateau. Primates were abundant in west North America in Eocene but by the end of Eocene they became extinct due to the change of tropical forest by deciduous forests.

The **lack of vegetation** also affects the distribution of certain animals; for example, the primates, which are mostly dwellers of the tropical forests, depend directly upon the huge vegetation for movement and also for food, because they eat fruits, nuts, buds, etc.

Further, there is a necessity of trees and shrubs for animals, which have short crowned browsing teeth and pasturage of harsh grasses for those who have long-crowned grazing teeth. Each type of vegetation being unadapted to the other type of animals. Distribution of certain types of insects and caterpillars is also dependent upon the vegetation of an area. They feed upon certain plants and, hence, if these plants are absent from that area, those insects and caterpillars will not exist. Insectivorous birds are also dependent upon the availability of specific insects.

4. Large Bodies of Water as Barriers

Large bodies of water such as large rivers and oceans act as barriers, especially for the terrestrial vertebrates such as amphibians, reptiles and mammals but not for the aerial creatures such as birds and bats. Mozambique channels (400 km wide), separating the Madagascar island from the main land of Africa and sea, separating New Zealand from Australia. The fauna found in Madagascar is quite different from Africa. The river Amazon, Brahmaputra and Ganga with their tributaries limit the distribution of forest animals.

Large water bodies act as barriers for terrestrial vertebrates but not for aerial creatures.

Freshwater fishes, *e.g.,* carps, garpikes, catfishes, bowfin, etc., are unable to migrate through the large bodies of salt water. For amphibians, salt water even of 1% is poisonous. They are completely absent from southern hemisphere, *e.g.,* Australia, South America and Africa. Burrowing caecilians are confined to South America and Africa. For crocodiles and marine turtles, the seas do not act as barriers. Giant tortoises are confined to certain oceanic islands, *e.g.,* Galapagos Islands and the islands of the western Indian oceans and are totally absent on the mainlands of South America, Africa and Eurasia.

The serpents (snakes), except the sea snakes, cannot swim through the large bodies of sea water. Lizards too cannot cross these oceans but their eggs may be transported to oceanic islands by the birds. Birds, except flightless birds, *e.g., Rhea*, ostrich, cassowary can do trans-oceanic migrations. Among mammals, except whales and seals, all cannot pass through the large oceans. Bats, however, are capable to cross the large oceans and are found in oceanic islands.

On the other hand, large land masses act as barriers for the aquatic animals.

5. Lack of Salinity of Sea-water as Barrier

Salinity of sea water remains constant and it acts as an effective barrier for certain kinds of sea animals, such as the brachiopods, echinoids, crinoids, starfishes, squids, foraminiferans, corals and sponges. These animals tend to live in pure saline water and are totally absent from those regions where rivers enter the sea.

The corals are mostly marine, and few are found in brackish and fresh water. With few exceptions, all corals require water of maximum purity and salinity. They are never found near the mouths of large rivers such as the Amazon (South America), the Orinoco (Venezuela) and the Mississippi (USA), where the salinity is greatly changed. Great Barrier Reef on the east coast of Australia acts as barrier. It hinders the dispersal of animals.

Great barrier reef.

Further, like the salinity the **depth of ocean** acts as an effective barrier. Thus, animals living on the surface and mid-water cannot move to deeper waters because their bodies would not be able to withstand the increase in atmospheric pressure and decrease in light. Similarly, animals from depth cannot survive in surface waters because their bodies will not be able to withstand the decrease in pressure and increase in light intensity.

6. Biological Barriers

The biological barriers include preferential type of food. Certain animals depend on particular type of plants for their food, shelter and breeding and if those plants are not present or removed, then that particular species would gradually perish. However, if such plants are available in some nearby area they would migrate. Similarly, certain animals act as parasites and others as predators. Of these animals availability of the host animals and the prey respectively, is most essential for the survival. If the host species or prey are destroyed or removed from an area, the parasites and predators would either disperse or become extinct.

REVISION QUESTIONS

1. Describe the various barriers, which restrict the zoogeographical distribution of animals.
2. What are the barriers ? How do they affect the distribution of animals in space ?
3. Discuss the role of barriers in the dispersal of animals.

CHAPTER

12

Origin of Life

The outcome of chemical evolution was the origin of life.

The kind of evolution which includes evolution from atoms and molecules to simple and then complex sub stances and from these to still more complex ones capable of self-duplication, has been termed as **inorganic**, **molecular** or **chemical evolution**. The most significant outcomes of chemical evolution are the origin of biologically important macromolecules such as proteins and nucleic acids (RNA and DNA), **origin of life** and an environment to sustain life on the planet earth.

HISTORICAL AND THEORIES

The question of origin of life on our planet, the earth, has remained a most complicated and puzzling problem for thinkers, philosophers and naturalists of modern as well as ancient times. Different views have been forwarded concerning the origin of life on earth by thinkers of different ages. Most of them are quite bizarre and merely having a historical significance. Certain most important hypotheses and theories concerning the phenomenon of origin of life are the following:

1. Special Creation Theory

According to the special creation theory, the life is mysterious force or *vital spirit* that set living things completely apart from the non-livings and it is originated on our planet due to some supernatural event, which cannot be studied scientifically but must be accepted on faith. Most religions of different human cultures accepted the concept of special creation in one way or other. The followers of this theory argue that all living beings on the earth were originally created by a super natural power–God. According to Bible, for instance, world was created within six natural days. Plants were created on third day, fish and fowl on the fifth day, and animals on the sixth day. Lastly human beings were created, first man then woman. **Adam**, the first man, was moulded by God from inanimate matter—clay, which he furnished with a soul, thus, breathing life into him.

Objections to special creation theory. The special creation theory lacks sound logic and scientifically sound evidences, therefore, it could not convince the scientists. **Charles Darwin** condemned the special creation theory by saying that earth has not always been inhabited by the plants and animals as we know now.

2. Hindu Concept of Origin of Life

Different schools of Indian philosophy and Hindu religion have elaborated the problem of origin of life in sufficient details, but that is beyond the scope of this book. However, it will not be out of context to discuss here some ideas of Upanishad-philosophy. For instance, the cosmology elaborated in the "**Taittiriya Upanishad**" describes the origin of life in the following way—from brahman sprang ether, from ether air, form air fire, from fire water, from water earth, from earth herbs, from herbs food, from food seed and from seed man." Likewise, according to the "**Mundaka Upanishad**", life arose from food and from food arose the mind. Most of the early Upanishadic philosophers considered Brahman as not being something transcendental, superhuman, supernatural or otherworldly. They did not recognise any God as the creator, controller and designer of the world. To them the Brahman was not a god, but the infinite totality of existence (for details see, **K. Damodaran**, 1970— "*Man and Society in Indian Society*", People's Publishing House, New Delhi, India).

3. Theories of Spontaneous Generation or Abiogenesis

For centuries, the concept that life arose spontaneously from inanimiate material was a principal doctrine of how life originated. The spontaneous generation or abiogenesis of life was visualized as beginning with either inorganic materials or with putrefying organic matter. In China, for example, even from the earliest times, there was belief in the spontaneous generation of aphids or other insects under the influence of heat and moisture. In sacred books of India there are indication of the sudden emergence of various parasites, flies, and beetles from sweat and manure. It has been deciphered from Babylonian cuneiform texts that worms and other creatures were formed from the mud of canals. In ancient Egypt the conviction prevailed that the layer of humus deposited by the Nile in its flood stage could give birth to living creatures. When it was warmed by the sun : frogs, toads, snakes and mice were formed thus.

Apparently, before the 1700s in Europe, it was common among laymen and scientists alike to believe that living matter could spontaneously develop from non-living materials. Infestations of worms, flies, fungi and other organisms appeared regularly in food and organic matter without any apparent explanation. People were unaware of the life history of most of these organisms and could not know how easily eggs and spores were transferred from place to place.

Disproof of the spontaneous generation theory. During the late seventeenth century, an Italian doctor **Francesco Redi** (1626–1698) performed what are now considered classical experiments to test the validity of the theory of spontaneous generation and had the honour of being the first to come forward with experimental refutation of spontaneous generation so firmly established for many centuries. In his paper "*Esperienze intorno alla generazione degli netti* (1668), he described a series of experiments in which he placed meat or fish (eel) under clean muslin coverings and demonstrated that while flies laid eggs on muslin, maggots or larvae appeared only when those eggs were transferred to the meat and allowed to hatch. He concluded that maggots came only from pre-existing flies and were not spontaneously generat-ed by any other form of material.

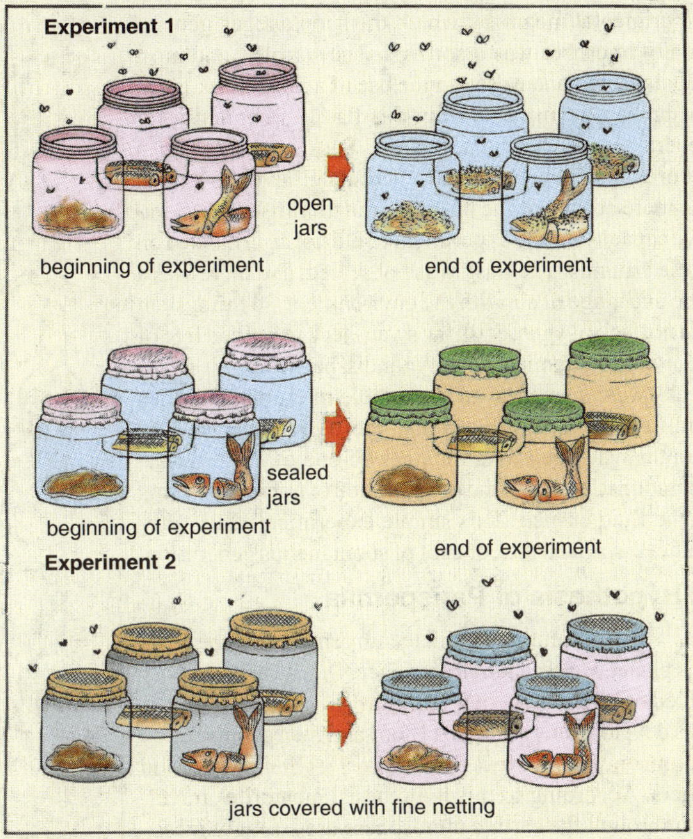

Fig 12.1. Redi's experiments.

While Redi's experiment proved that even the simple forms of life known in the seventeenth century did not arise by spontaneo us generation, the discovery of microbes by **Van Leeuwenhoek** (1632–1723) brought the question into the forefront of biological thought once again. After all, maggots are far more complex than those little spherical, rod, or corkscrew-shaped bacterial forms observed under the microscope. So it was considered that microorganisms arose by spontaneous generation. This idea was very authoritatively maintained by the noted German philosopher **Leibniz** (1646–1716). French naturalist **Buffon** (1708–1788) and Englishman **John Nedham** (1713–1781).

The Decline and Fall of the Theory of Spontaneous Generation

An Italian scholar **Abbe Spallanzani** in 1765 tested the theory of spontaneous generation of microbes. He prepared flasks of meat broth which were boiled for several hours and then sealed. The broth remained clear for months, and when the seals were broken and the broth tested, it was shown to be free of microbes.

Spallanzani's experiments were neither conclusive nor satisfying to many of his contemporary scientists. Some claimed that by boiling he had driven out a "vital force" necessary for spontaneous generation. When **Priestly** proclaimed oxygen as a necessary factor for life and experiments of **Gay-Lussac** demonstrated that such boiled and sealed flasks contained no oxygen, than these were sized upon as evidence that the vital principle of life required oxygen. Many of Spallanzani's successors modified the experimental procedure in attempt to quell the criticisms, but each new technique gave rise to new criticisms. The death blow to the theory of spontaneous generation of microbes was finally dealt with **Louis Pasteur** (1822–1895) in the nineteenth century, some two hundreds years after the initial experiments of Spallanzani.

The period of Pasteur. Pasteur devised several experimental means by which the spontaneous generation of microbes was disproved. The simplest and most sophisticated one was with the use of a swan-neck flask. He prepared a meat broth in this flask and boiled it for several hours. He then left the flask **unsealed** on a laboratory bench. Previous experiments have shown that the mere boiling of the broth did not diminish its generative capacity, so the experiment could not be criticized on these grounds. The flask was not sealed, and there was a free exchange of air with the environment, so the system did not lack oxygen. Still, the swan-neck remained free of microbial contamination for months, because their swan-necks were shaped so to trap viable microbial particles and to allow only air to enter the flask. After several months when he broke the neck of one of these flasks, contamination by air and proliferation of microorganisms in the fluid ensued. This simple experiment, thus, altogether disproved the concept of spontaneous generation.

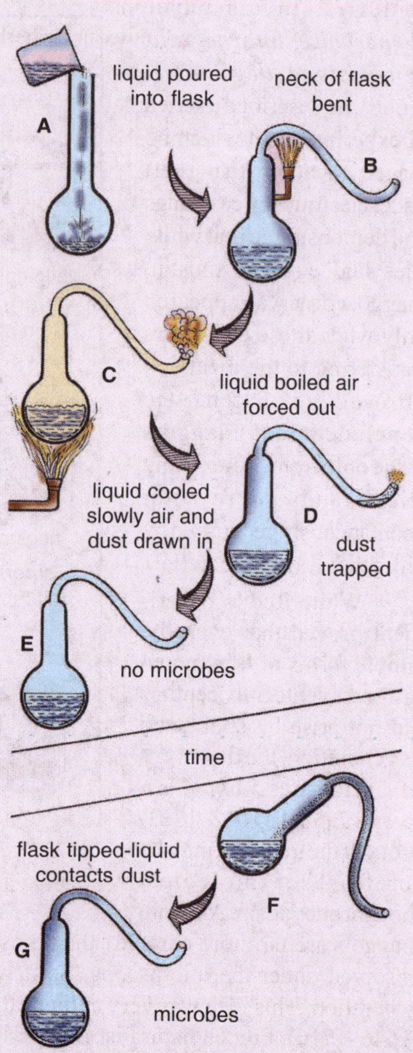

Fig. 12.2. Pasteur's experiment.

4. Hypothesis of Panspermia

Some scientists of nineteenth century assumed that this planet was 'seeded' from space. One of such a theory called **cosmozoic theory** or **hypothesis of panspermia** was developed by **Richter** (1865) and then supported by **Thomson**, **Helmholtz** (1884), **Van Tieghem** (1891) and others. According to this hypothesis, meteorites travelling through the earth's atmosphere are strongly incandescent only on the surface, while the interior remains cold. Thus, the embryos of organisms inhabiting meteorites or the planets from which they were formed were preserved alive in the interior of meteorites, finding fertile soil on earth, and grew and evolved to produce all the species now present.

Arrhenius (1911)

Another hypothesis of panspermia which was developed by **Arrhenius** (1911) holds that spores of life together with particles of cosmic dust could be transformed from one heavenly body to another under the pressure of stellar rays. But it is doubtful that whether such spore remained viable because as especially serious danger for living embryos are short wavelength ultraviolet radiations which penetrate inter-planetary and interstellar space and are destructive to all life.

From 1870 to 1940, most areas of science progressed toward a level of maturity where their firmly established principles collectively created a unified picture of the world. Biology had brought life down to the chemical level through physiology and biochemistry. Chemistry had raised the elementary structure of matter to large synthetic molecules. Physics had provided the energy considerations on which reactions were based. Geology had measured the age of earth as a few billion years and that of the universe as close to ten billion years. The stage was set for the idea of spontaneous generation to be given a more rigorous inspection. The first form of life was not to be a maggot, or bacterium, or even a virus but a collection of biologically important molecules, capable of producing organized forms of life through processes of natural selection. The laboratory was the entire surface of earth, the time was in terms of several billion years, and the experimental procedure was natural selection.

5. Theory of Chemical Evolution and Spontaneous Origin of Life at Molecular Level

Since Pasteur's time, countless generations of students have been taught not to believe in spontaneous generation. However, Pasteur's experiments revealed only that life cannot arise spontaneously under conditions that exist on earth today. In fact, a Russian biochemist **Alexander I. Oparin** in 1924 has proposed that conditions on the primeval earth billions of years ago were definitely different from present conditions, and the first form of life, or self-duplicating particles, did arise spontaneously from chemical inanimate or abiotic substances. Thus, Oparin was the first to suggest that a long evolution of chemical substances occurred before life actually originated. In his book "*The Origin of life on Earth*", in 1924 he provided a biochemical explanation of origin of life. According to him, origin of life occurred along the origin and evolution of earth and its atmosphere. The primordial atmosphere of earth had water, methane and ammonia from which as a result of a series of changes, a colloidal body called **coacervate** and containing a mixture of biologically important macromolecules like proteins, lipids, nucleic acids, etc., were formed. These coacervates though were not living, but they behaved in a manner similar to biological systems by being subjected to natural selection, by being chemically directive and by having the capacity of reproduction by fragmentation.

Synthesis of coacervate droplets. The coacervate droplets have been made typically by combining solutions of oppositely charged colloids, *e.g.*, gelatin + gum arabic, or, serum albumin + gum; or clupein+gum, or histone + gelatin, histone + RNA, or gelatin + gum + DNA. Each solution is uniform, but when the two are brought together they interact to form '**swarms**' or '**clusters**'. When these clusters reach a certain size they separate from the solution in the form of coacervate droplets.

Characters of coacervate droplets. Oparin's coacervate droplets can take up materials from the external medium and some of them show enzymatic activities. For example, Oparin and his associates have made coacervate droplets from gum arabic and histone at pH 6.2 and have included phosphorylase enzyme during the formation of droplets. When glucose-1-phosphate is dissolved in the equilibrium liquid, the droplets are found to store starch. In another experiment, coacervate droplets of some sort have been prepared with phosphorylase and β-amylase. In this case, the starch that formed from the entering glucose-1-phosphate then decomposed into maltose which diffused into the external medium where it was detected.

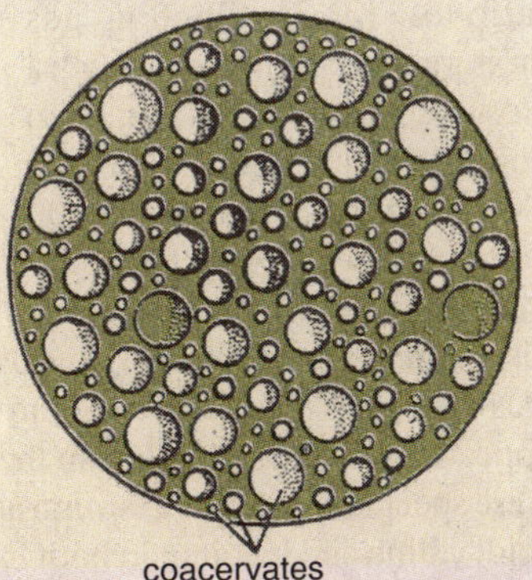

Fig. 12.3. Coacervate droplets.

Similarly, polynucleotides were synthesized enzymatically within coacervate droplets. These experiments with coacervate droplets, thus, show some sense of the special benefit of encapsulation of enzymes in living entities. Oparin proposed that although the coacervates which might be originated during chemical evolution, were not living, they behaved in a manner similar to biological systems (1) by being subjected to natural selection (in terms of a positive selection for stability as expressed through persistent size and constant chemical properties, (2) by being chemically directive (selective accumulation of material) and (3) by reproduction (by fragmentation).

But since coacervate droplets lack stability, they fall apart on standing and they were synthesized from polymers obtained from living systems, they no longer served as the models of protocells.

Laderberg has recognised the following three steps in the Oparin's thesis: (1) **Chemogeny** (Chemical evolution); (2) **Biogeny** (Formation of self-reproducing biological units) and (3) **Cognogeny** (Evolution of mechanisms of expressions, communication, perception and feeding). Oparin like observations were made by an English biologist **J.B.S. Haldane** (1929) about the origin of life. Both of these scientists got the ideas from the studies of biochemistry made by Sir **F.G. Hopkins**, and both were of the opinion that the evolution is purely chemical, the gradual transmutation of inorganic compounds into organic ones.

Experimental Support of Oparin's Hypothesis

Oparin's ideas created a rash of excitement, but it was not until 1953 that some experimental evidence was gathered to support Oparin's hypothesis of spontaneous generation of life at molecular level.

Miller's Experiment

In 1953, **Stanley Miller** then at the University of Chicago and a student of Nobel Laureate **Harold Urey**, synthesized organic compounds under conditions resembling the primitive atmosphere of the earth. At that time it was believed that the primitive atmosphere had probably been composed chiefly of **water vapour**, **methane**, **hydrogen** and **ammonia** (latter three in the ratio of 1:2:2). Such a gas mixture was circulated through a closed apparatus by steam from boiling water and subjected to an electric spark discharge (75,000 volts) between tungsten electrodes (Fig. 12.4). The electric discharge duplicated the effects of violent electrical storms in the primitive universe. The gas mixture was then condensed and added to the boiling water for recirculation; any non-volatile substance that appeared would accumulate in the water. This apparatus was permitted to run for a week. The results were spectacular. Several amino acids were synthesized (*e.g.*, alanine, glycine, glutamic acid, aspartic acid, valine and leucine; in both D- and L- forms) and a number of other organic compounds, such as hydrogen cyanide (HCN), aldehyde and cyano compound (– C ≡ N), appeared; about 15 per cent of the carbon in the gas mixture became incorporated into these compounds.

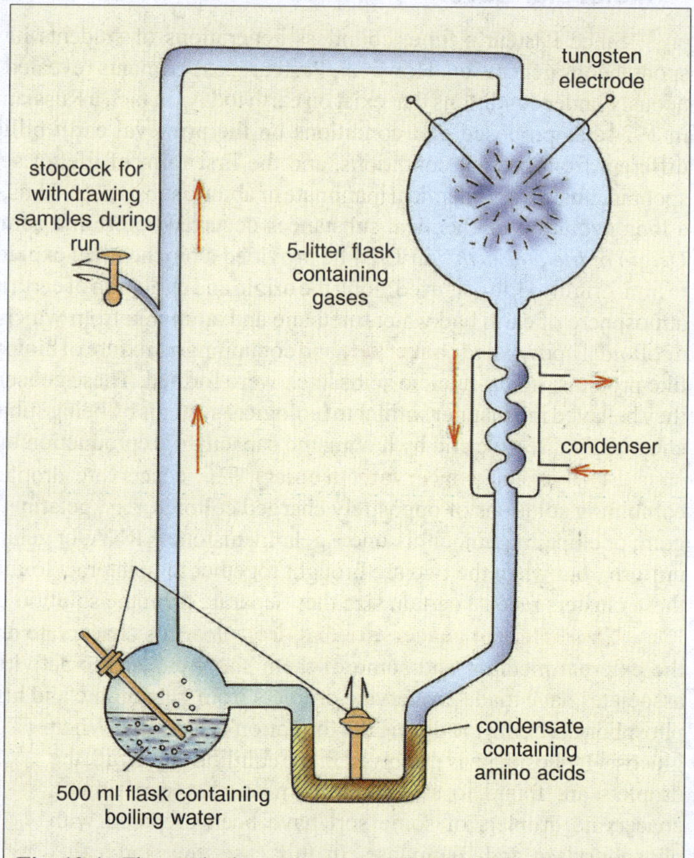

Fig. 12.4. The spark-charge apparatus of Miller for producing organic compounds from a reducing atmosphere.

Miller's instructive experiment has been successfully repeated by a number of investigators; amino acids can also be generated by irradiating a similar gas mixture with ultraviolet light. Both purines and pyrimidines can be formed when HCN and cyanoacetylene (HC ≡ C – C ≡ N) are present in the gas phase of the Miller-type reaction. HCN is reactive in the presence of UV light alone and purines (mainly **adenine**) can be regarded as condensation products of five HCN molecules, although the route of synthesis is still not clear (**Miller** and **Orgel**, 1973). Polymerization of HCN also gives rise to arotic acid, which can be photochemically decarboxylated by sunlight to pyrimidine (**uracil**). Even ATP can be formed under prebiotic conditions, especially in the presence of the common mineral apatite (calcium phosphate).

Protenoid Microspheres

The synthesis of monomers (*e.g.*, amino acids) is only a small step toward the synthesis of a living cell. In 1964, **Sidney W. Fox**, then at Florida State University and later at the University of Miami, reasoned that **proteins** were synthesized from amino acids in the primitive earth by thermal energy or heat. **Fox** accordingly heated a mixture of 18 amino acids to temperatures of 160–200° C for varying periods of time. He obtained stable, protein-like macromolecules, which he termed **protenoids**. These thermally produced proteinoids are similar to natural proteins in many respects. For instance, bacteria can actually utilise the protenoids in a culture medium, degrading them enzymatically into individual amino acids. Equally important, when the protenoid material was cooled and examined under a microscope, **Fox** observed small, spherical, cell-like units that has arisen from aggregations of the protenoids. These molecular aggregates were called **protenoid microspheres**.

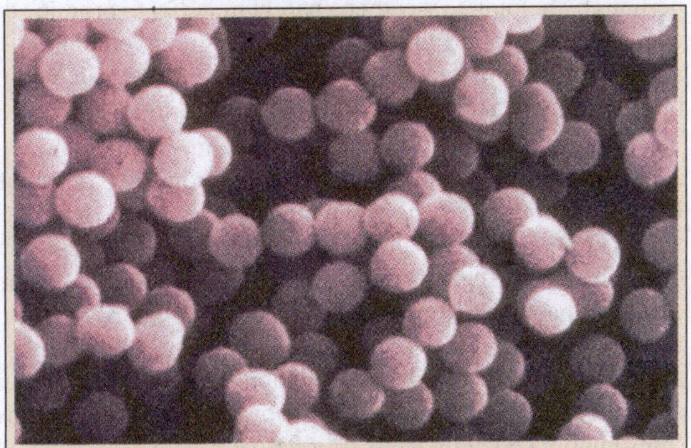

Do microspheres resemble the earliest cells ?
Cell-like microspheres can be formed by agitating proteins and lipids in a liquid medium. Such microspheres can take in material from the surrounding solution, grow, and even "reproduce".

Physical properties of protenoid microspheres. The protenoid microspheres were spherical, mircroscopic and of uniform diameter. Their size remained in between the range of 0.5–0.7mμ diameter. They had a number of properties in common with coccoid bacteria and microfossils found in ancient geological strata (**Fox**, 1969), in containing uniformity in size, stability in centrifugation, electron microscopy, extreme pH variations, etc.

Structural properties of protenoid microspheres. Electron microscopic studies of protenoid microspheres have revealed concentric double layered boundaries around them through which diffusion of material from the interior to the exterior occurs. Further, protenoid microspheres has the ability of motility, growth, binary fission into two particles, and a capacity of reproduction by budding and fragmentation. Further, some of them are stainable by gram-positive stains others by gram-negative stain.

Enzyme-like activities of protenoid microspheres. They are found to have catalytic activity, such as degradation of glucose and this enzymatic activity of them is partially lost during heating.

The model of protenoid microsphere for protocells is widely accepted because of its following significances— (1) Such protenoid microspheres arise from monomers, rather than from polymers obtained from organisms already in biota, as is true for the usual experiments with coacervate droplets. (2) This model suggests that protenoids are informational and it shows the origin of communication which may be intercellular or intergenerational communication. In fact, it suggests the simultaneous origin of both types of communication.

Since both proteins and nucleic acids (plus other simpler substances) are required to develop and reproduce organisms living today, an obvious question is which of these substances arose first. No clear answer is available. To date protein-like materials have proven easiest to synthesize prebiotically; perhaps they arose first. This implies that there was some sort of replicating genetic material associated with early proteins upon which selection could act. This replicating material could have been a protein itself. It is difficult to say that this is impossible, but no self-replicating proteins are known (**J.W. Valentine**, 1976). Nucleic acids seem difficult to form, but they do replicate; perhaps the earliest gene was a simple nucleic acid (*i.e.*, RNA). The trouble with nucleic acids is that they do not do anything themselves. This paradox has been recently resolved with the discovery of a catalytic RNA by **Cech** (1982–1986). At this stage, let us analyse the following models regarding the first genetic substance:

1. Cairns-Smith's model. The model of primitive genetic system proposed by **Cairns-Smith** (1971) postulates that *the first genes were minerals*, minute crystals associated with proteins as concentration sites and to some extent as templates. Certain minerals, *e.g.*, clay or other layered silicates, were in very large supply as products of early rock weathered. They might have favoured the formation of classes of polymers for which they provided good templates. Natural selection would favour a polymer whose activities preserved or even increased the supply of its mineral template, perhaps by helping to weather a parent material to produce the **clay gene**. The association of two (and eventually many) different polymers might have aided in mineral formation, while others provide adhesion for the polymer colony, and still others cleaved prebiotic organic compounds to provide building block materials for the colony, and so forth.

The polymer in any such early colonies would have **proteins** but eventually a **nucleic acid** was added to a colony, presumably for some property other than protein-coding or even self-replication. Perhaps its original function was mechanical, acting as a fibre or in some other structural capacity. This would be a very early example of **preadaptation**. The ability of the nucleic acid to code for proteins that were themselves useful to the colony eventually became manifest. Once the nucleic acid was contributing to colony fitness, any process that would replicate it *in situ* would be considerably favoured. Thus, a nucleic acid (probably a simple RNA at first) would replace minerals as genetic material. Presumably, the nucleic acid gradually took over the task of coding for all the proteins employed in the colony, including those associated with its own replication. Eventually DNA evolved (presumably coded by RNA originally) and an essentially modern nucleic acids—protein system of gene—enzyme was present. At some stage **membranes** were developed around the colonies and those membrane-bound units became **cells**. Thus, according to this model genes evolved in the following way:

Clay genes having proteins → **RNA genes** → **DNA genes**

Cairn-Smith scenario is highly speculative. Evidence are lacking which may prove that life followed such a protoadaptive pathway from prebiotic materials to living cells (see **Valentine,** 1976).

2. RNA first model. In recent years, evidences are in favour of RNA to be material of first formed gene (**Woese**, 1967; **Crick**, 1968; **Orgel**, 1973, 1986; **Watson** *et al.*, 1986; **Darnell** *et.al.*, 1986). Thus, a number of protein-free (*i.e.*, non-enzymatic) reactions that could have been decisive in the development of the first biologically important polymers have been discovered by molecular biologists. These reactions include (i) formation of RNA oligonucleotide from activated monomeric precursors; (ii) template-directed RNA synthesis; (iii) site-specific RNA cleavage (with and without the aid of second RNA molecule), and (iv) RNA-RNA splicing. Thus, *RNA may have been the first polymer and some form of reverse transcription may have given rise later to DNA*. How the coordination between the RNA oligonucleotides and amino acids or peptides occurred remain unknown, but the result was a single **universal three-letter** code that is still used by all cells today.

Further, comparison between the rRNAs of many different cell types as well as comparison between bacterial rRNAs and rRNAs of organelles (mitochondria and chloroplasts) have suggested two important conclusions about evolution: (i) the overall two-subunit structure of ribosomes is universal; (2) secondary (stem-loop) structure is extremely similar in all rRNAs. Thus, ribosomes can be considered as highly preserved but fairly complicated early cellular structure.

Catalytic RNA. The precursor of rRNA of the ciliate protozoon *Tetrahymena* was found to contain a **self-splicing intron** (containing 413 nucleotides) which acts as an enzyme, *i.e.*, it excises itself from the rRNA precursor without the help of a protein enzyme (**Cech** and coworkers, 1982, 1983, 1986). The ability of rRNA intron to function as enzyme implied that the very first living molecule might have been an **RNA replicase** that catalysed its own replication without the help from a protein (see **Watson** *et al*., 1986).

The discovery in 1983 of a second catalytic RNA species (377 nucleotides long RNA component in RNP or ribonucleoprotein) of *E.coli* lent strong support to this notion. In streak virusoid (*Lucerne transient*) too RNA enzymes work like protein enzymes (**Pace** and **Marsh**, 1985). Catalytic RNA are now considered as **molecular fossils** whose history dates back to the earliest life forms.

Thus, the discovery of intervening sequences, the **introns** or **junk DNA** (mainly in the eukaryotic genomes) has given rise to much speculation about the nature of the earliest genes. Because protein-free RNA reactions include **cutting** and **splicing**, and because spliced gene structures are relatively well preserved over enormous evolutionary intervals (*i.e*., more than a billion years), it is possible that the first gene had non-contiguous coding regions and that splicing was a prominent feature of the most primitive genomes. According to this view, cells lacking introns may have lost them due to selective pressures associated with **rapid growth**, whereas eukaryotic precursors were slower growers that did not lose their introns. The antiquity of the eukaryotic lineage is supported by the rRNA lineage studies (see **Darnell** *et al*.,1986).

Why RNA and not DNA was the first living molecule?

New examples of RNA molecules with enzymatic activity are constantly being discovered, but no enzymatic activity has ever been attributed to DNA. Further, ribose is much more readily synthesized than deoxyribose under simulated prebiotic conditions.

PROCESS OF ORIGIN OF LIFE

With such a historical background about the speculated modes of protobiogenesis, now we are in better position to discuss the present (current) understanding of molecular evolution and its culmination into the origin of life. The pre-requisite for the treatment of the problem of protobiogenesis is the origin of the raw materials (matter, elements, stars, planets, etc.) for the molecular evolution, *viz*., a brief introduction of **cosmology**.

1. Structure of Cosmos

The **universe or cosmos** is made up of countless aggregation of stars organized into systems of various shapes and types, called **galaxies**. These galaxies, each of which includes hundreds billions of stars with their accompanying **planets** and **satellites**, are distributed into the vast reaches of **space**. **Milky way** is a galaxy which is composed of 100 billion stars besides **sun** and its **planets** (*viz*., **solar system**). The sun is a star being five billion years old and remains in a rapid state of motion. The planets of solar system are **Mercury**, **Venus**, **Earth**, **Mars**, **Jupiter**, **Saturn**, **Uranus**, **Neptune** and **Plato**. All the planets of solar system rotate around their axes and revolve around the sun in elliptical paths determined by the gravitational force.

2. Primitive Earth

It is a tenable hypothesis that the cosmos as we know it originated in gigantic explosion (**Big bang hypothesis** of **Abbe Lemaitre**, 1950) which blew matter from a primordial fireball, scattering outwards the materials of the universe in an ever-expanding volume. Clouds of dust and gases, chiefly **helium** and **hydrogen**, spread through the space, thinned in some regions and dense in others. **Stars** can be formed from the condensation of such clouds; our own **sun** probably formed from a rotating mass of dust and gas that was flattened by the spin and which collapsed upon a central region due to the force of gravity. As the gaseous disc condensed inwards, the spin of the central region must have increased in order to conserve **momentum**. At some point the spin could have become so great that material would be ejected owing to **centrifugal force**. However, as such a point was approached, and the centrifugal force increased, matter in the outer portions of the condensing disc would tend to be more attracted to local dense patches than do the central region. Thus, at some distances from the **proto-sun**, matter would be condensed into separate bodies. These would be the **planets**, which lie in the plane of the sun's rotation. The solar **satellites**, including the earth and presumably the other **planets** and **asteroids** and the earth's **moon** as well, appear to date from about 4.6 billion years ago (see **Valentine**, 1976).

3. Prebiotic Synthesis

The **core** of the primitive earth melted from **heat of condensation**, but the outer **crust** cooled quickly except where convection or volcanic eruptions brought heat and mass from the molten core to the surface. The **primitive atmosphere** was formed as various gases escaped from the interior of the earth (**outgassing**). The early gas cloud was especially rich in **hydrogen** (H_2). The hydrogen of the primitive earth chemically united with carbon to form **methane** (CH_4), with nitrogen to form **ammonia** (NH_3), and with oxygen to form **water** vapour (H_2O). Thus, the early atmosphere had a strongly reducing (non-oxygenic) character, containing primarily hydrogen, methane, ammonia and water. **Darnell** *et al.*, (1986) has, however, argued that these gases (or compounds) may not have existed free in the atmosphere in large quantities for extended times. **Carbon dioxide** (CO_2) and **hydrogen sulphide** (H_2S) were probably more permanent constituents of the early atmosphere.

The crust and atmosphere of the primitive earth cooled slowly, and the first shallow seas were formed as the less volatile gases including water, rained back down to earth. The oceans formed as a result of millions of years of torrential rains. The rains eroded the rocks and washed minerals (such as chlorides and phosphates) into the seas. The stage was set for the combination of the varied chemical elements. Chemicals from the atmosphere mixed and reacted with those in the waters to form a wealth of hydrocarbons (that is, compounds of hydrogen and carbon). Water, hydrocarbons and ammonia are the **raw materials** of **amino acids** which, in turn, are the building blocks for the larger protein molecules. Thus, in the **primitive seas**, amino acids accumulated in considerable quantities and became linked together to form proteins or oligopeptides (**prebiotic synthesis**).

Energy sources for prebiotic synthesis. Complex carbon compounds, such as amino acids and proteins, are termed as **organic** because they are made by living organisms. Our present-day green plants use the energy of sunlight to synthesize the organic compounds from simple molecules. What, then was the energy sources in the primitive earth, and how was synthesis of organic compounds effected in the absence of living things? It is generally held that **ultraviolet (UV) rays** from the intense solar radiations penetrating through the thin ozone-free atmosphere, electric discharges such as **lighting** from rainstorms and intense dry heat from **volcanic activity** furnished the energy for prebiotic synthesis such as joining the simple carbon compounds and nitrogenous substances into amino acids (**Orgel**, 1973). Superheated hot springs were also formed along rifts in the sea floor. Alternating cycles of cold nights and hot days too might have affected the prebiotic synthesis (see **Watson**, *et al.*, 1986).

Evidence that oxygen was absent in primitive atmosphere. The sedimentary rocks that remain from 3.8 to about 1.9 billion years ago, which appear to have been deposited chiefly in shallow epicontinental seas, generally suggest an **anoxic atmosphere** of the primitive earth; the minerals are mostly in low states of oxidation relative to today (although more highly oxidized layers do appear locally). The most famous examples are the banded iron formations which contain vast amounts of **ferrous iron**. This indicates that free oxygen was not available in the atmosphere; otherwise the iron would be in a **ferric state** (**Cloud**, 1968).

There are certain other considerations which show why the primitive atmosphere lacked free oxygen? Free oxygen is available in quantity from only two sources. One is **photodissociation** of water vapour, which occurs high in atmosphere. Hydrogen, freed from water molecules, leaks out into space, while the heavier oxygen is retained. The extent of this process in the past, and the rate at which photodissociated oxygen could accumulate, are much debated. Oxygen is also produced from water via **photosynthesis**. It is commonly believed that the bulk of free oxygen was produced by plants. Before the advent of life, and indeed until plant biomass became large, this source of oxygen was absent or unimportant.

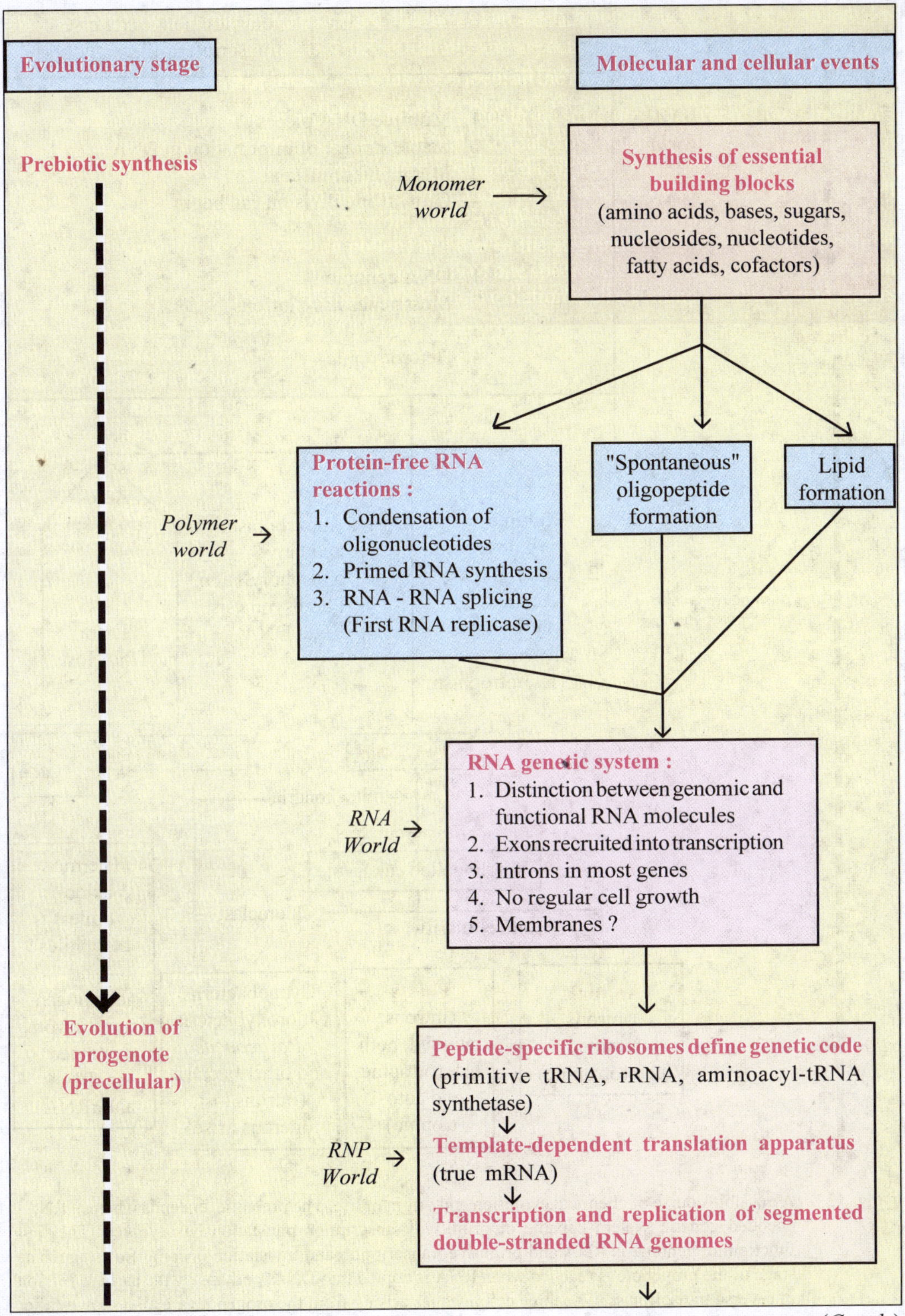
Evolutionary stage
Molecular and cellular events
Prebiotic synthesis
Monomer world
Synthesis of essential building blocks
(amino acids, bases, sugars, nucleosides, nucleotides, fatty acids, cofactors)
Polymer world
Protein-free RNA reactions :
1. Condensation of oligonucleotides
2. Primed RNA synthesis
3. RNA - RNA splicing (First RNA replicase)
"Spontaneous" oligopeptide formation
Lipid formation
RNA World
RNA genetic system :
1. Distinction between genomic and functional RNA molecules
2. Exons recruited into transcription
3. Introns in most genes
4. No regular cell growth
5. Membranes ?
Evolution of progenote (precellular)
RNP World
Peptide-specific ribosomes define genetic code
(primitive tRNA, rRNA, aminoacyl-tRNA synthetase)
Template-dependent translation apparatus
(true mRNA)
Transcription and replication of segmented double-stranded RNA genomes

(*Contd.*)

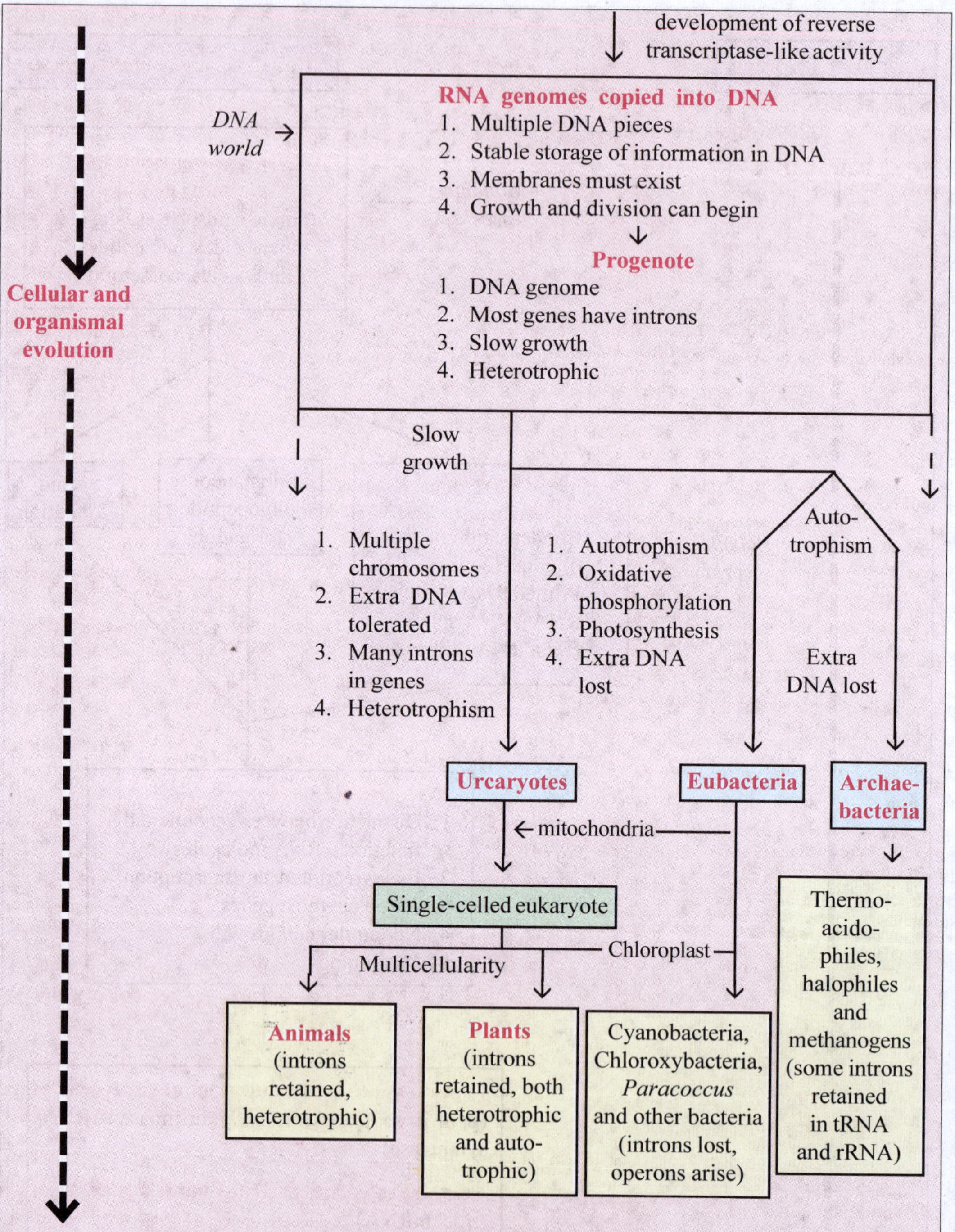

Fig. 12.5. A possible course of chemical evolution and origin of life. The prebiotic era ends when an RNA-encoded genetic system capable of primitive transcription-translation has evolved. The first functioning genome is RNA and primitive transcription and translation occur at this stage. The stage of the progenote is reached when RNA is copied into DNA, perhaps by the earliest version of reverse transcriptase. The three cell lineages arising from the progenote are also shown (after Darnell *et al.*, 1986; Watson *et al.*, 1986).

One final evidence that oxygen was not ordinarily present in the early atmosphere and, therefore, was not dissolved in early oceans, is that organic compounds are stable in the presence of hydrogen (that is, under reducing conditions) but are quickly destroyed in the presence of oxygen (**Haldane**, 1933). This indicates that compounds required as precursors or building blocks for living systems must have originated and accumulated in a reducing environment.

Primitive sea served as broth for the prebiotic synthesis. Classical experiment of **Miller** (1953) has established that organic compounds can be formed without the intervention of living organisms. Thus, it appears likely that the sea of the primitive earth spontaneously accumulated a rich mixture of organic molecules. In the absence of living organisms and oxygen, the organic compounds would have been stable and would persist for countless years. The sea became a sort of dilute organic soup or broth (or **aquatic Garden of Eden** of **Volpe**, 1985 or **chemical cauldron** of **Watson** *et al.*, 1986). In the organic soup of cauldron the molecules collided and reacted to form new molecules of increasing levels of complexity. Proteins capable of catalysis or enzymatic activity, had to evolve, and nucleic acid molecules capable of self-replication must also have developed.

4. Evolution of Progenote

The precellular evolution involved the following three main steps :

(i) Origin and evolution of RNA world. The living cell is an orderly system of chemical reactions (which are directed by enzymes) that has the ability of reproduction. Current consensus is that the machinery for self-replication (that is self-duplicating nucleic acids) evolved before the development of the metabolic (enzymatic) machinery of the cell. It is likely that self-replicating polynucleotides (nucleic acids) slowly became established in the primordial earth some 3.5 billion years ago. In the absence of proteinaceous enzymes in the prebiotic environment, less efficient catalysts in the form of minerals or metal ions would be sufficient to promote the reactivity of the nucleotide precursors. It is postulated that errors continually occurred in the copying process and that sequence of nucleotides in the original polynucleotide molecule became altered on many occasions. Large numbers of polynucleotide variations undoubtedly maladapted. Just as organisms today compete for available resources, molecules with different nucleotide sequences competed for the available nucleotide precursors in promoting copies of themselves. Any new mutant sequence with a replication rate higher than the antecedent sequence could be "fittest" and prevail. *Natural selection (or differential reproduction), thus, operated on population of molecules as it did later on population of organism.*

The highlights of RNA would include (i) distinction between genomic and functional RNA molecules, (ii) recruitment of **exons** into transcription; (iii) inclusion of introns in most genes; (iv) no regular cell growth; and (v) formation of membranes (Fig. 12.5).

(ii) Origin and evolution of ribnucleoprotein (RNP) world. Certain polynucleotides not only specified their own sequences, but apparently directed the synthesis of macromolecules of an entirely different type—namely, polypeptides (proteins). In present-day organisms, RNA guides protein synthesis, which suggests that the nucleotides of RNA (not DNA) were the first carriers of genetic information. Thus, RNA established a primitive genetic code for ordering amino acids into proteins. Evidently, the genetic code— the translation of nucleotide sequences into amino acid sequences— became established at a very early stage of organic evolution. The universality of the genetic code in present-day organisms attests to the early origin of genetic code.

The highlights of RNP world include (i) origin of primitive tRNA, rRNA and aminoacyl-tRNA synthetase enzyme; (ii) evolution of template-dependent translation apparatus (true mRNA) and transcription and replication of segmented double-stranded RNA genomes.

(iii) Origin of plasma membrane. A selective advantageous RNA molecule would be one that directs the synthesis of a protein that accelerates the replication of that particular RNA (*i.e.*, RNA polymerase). However, a protein specified by a particular variant of RNA could not foster the reproduction of that kind of RNA unless it is retained in the immediate vicinity of the RNA. The free diffusion of proteins could be prevented if some form of a compartment evolved to enclose or circumscribe the specific protein made by a particular RNA. All present-day cells have a limiting plasma membrane composed principally of phospholipids and proteins. Most probably the **first cell** was formed when polarised films of phospholipids formed soap-like bubbles enclosing aggregations of complex macromolecules. Specifically, a membranous structure arose that encircled a self-replicating aggregation of RNA and protein molecules. Once bounded by a limiting membrane, a given RNA molecule could be assured of propagating its own special protein molecules.

(iv) DNA world. At some later stage in the evolutionary process, DNA took the place of RNA as the repository of the genetic information. This step may have occurred due to evolution of an enzyme called inverse transcriptase. In other words, the genetic code was stored in DNA rather than RNA. Today, RNA molecules serve essentially their primal function : directing protein synthesis. Thus, in modern cells, genetic information is stored in DNA, transcribed into RNA, and translated into protein. An endless permutation of nucleic acids and proteins has fostered an enormous richness of life.

Thus, evolutionary highlights of DNA world include (i) evolution of multiple pieces of DNA, (ii) stable storage of information in DNA; (iii) sure existence of plasma membrane; and (iv) growth and division.

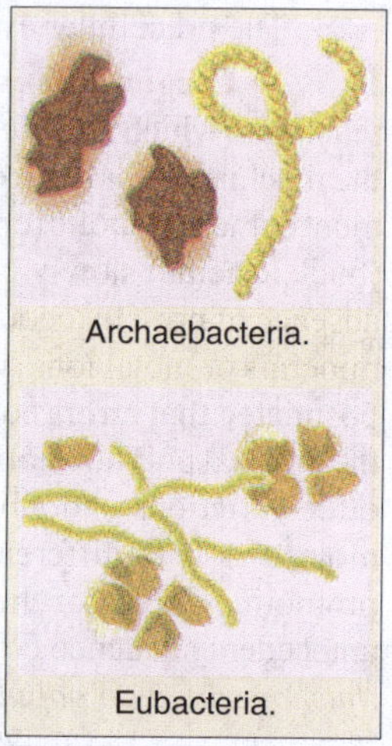
Archaebacteria.
Eubacteria.

(v) Origin of progenote. The precellular stage of organization has been termed **progenote**. The nature of progenote is uncertain. It might have been a functional, slow replicating, independently evolving organism, or it might have been an amorphous group of primitive transcription units plus a primitive transcription and translation apparatus from which many different cell lineages evolved; three of which have survived to the present day (*i.e.*, archaebacteria, urkaryote and eubacteria). The progenote contained DNA, most genes with introns, a slow growth and heterophic nutrition.

The first living system (the progenote) drew upon the wealth of organic materials in the sea broth. Organisms that are nutritionally dependent on their environment for ready-made organic substances are called **heterotrophs**. The primitive one-celled heterotroph probably had little more than a few genes, a few proteins and a limiting plasma membrane. The heterotrophs multiplied rapidly in an environment with a copious supply of dissolved organic substances. However, the ancient heterotrophs could survive only as long as the existing store of organic molecules lasted. Eventually, living systems evolved the ability to synthesize their own organic requirements from simple inorganic substances. Thus, in the course of time, **autotrophs** arose, which were able to manufacture organic nutrients from simpler molecules.

However, primitive autotrophs would require a whole array of enzymes to direct a multistep chain of metabolic reactions involved in the synthesis of a protein. Indeed, it would be too much to expect that all the necessary enzymes evolved at the same time. **Norman H. Horowitz** (1959) has proposed that the chain of steps in a complicated chemical pathway evolved backward. Let us analyse his concept in the following way :

Retrograde evolution. According to **Horowitz**, an organism might acquire, by successive mutations, the enzymes required in a biosynthetic pathway in the reverse order from that in which they normally appear in a sequence. In other words, evolution began with the end product of the pathway and worked backward, one step at a time, toward the beginning of the reaction chain. The total pathway became constructed when the organism had tapped from its surroundings all of the intermediate precursors of the reaction.

For instance, let us suppose that an organic compound O is synthesized through the following steps, where A through D represent precursors :

$$A \rightarrow B \rightarrow C \rightarrow D \rightarrow O$$

The first primitive heterotroph, lacking synthetic ability, would require the presence of *O* in its environment. This essential organic compound, continually being used by the heterotroph, would ultimately become rare, if not exhausted. If a mutation occurred that granted the heterotroph an ability to catalyze the reaction from *D* to *O*, then the organism would no longer remain dependent on the availability of *O* in the environment. Indeed, in an environment where *O* had become depleted, the new mutant would have a survival or selective advantage over the ancestral type and replace it. As the new mutant reproduced, *D* in turn would become scarce. Another mutation might occur in the mutant, converting it to a still newer form capable of catalyzing the reaction *C* to *D*. At this juncture, both the first mutant and the newly arisen mutant could live together in a close mutual (symbiotic) relationship establishing the first two-gene systems.

The intimate two-gene combination would be able to survive and reproduce in the presence of *C* and *D*. As other compounds in synthetic pathway become progressively rarer, additional mutations for their synthesis would be favourably selected. Ultimately, a multigenic system would evolve capable of directing the synthesis of *O* from inorganic substance *A* by way of intermediate products *B*, *C*, *D*. At first glance, it might appear that an unreasonable number of mutational events has been involved. But it should be recalled that mutations occur normally and continually in living organism. In fact, the capacity of mutation should be regarded as an essential property of life.

(vi) Adaptive radiation of progenote. The first simple **autotroph** arose in an anaerobic world, one in which little free oxygen was available. The primitive autotrophs obtained their energy from the relatively inefficient process of **fermentation** (the breakdown of organic compounds in the absence of oxygen). In addition to provide available energy in the form of ATP, such fermentative reactions could also be developed to provide reducing power in the form of NADH that could be used to derive biosynthetic reaction (Fig. 12.6) :

(i) Glucose + ADP	→	Lactate + ATP (*e.g., Lactobacillus*)
(ii) Glucose + ADP	→	Ethanol + Lactate + CO_2 + ATP (*e.g., Leuconostoc*)
(iii) Glucose + ADP	→	Ethanol + CO_2 + ATP (*e.g., Zymomonas*)
(iv) Lactate + ADP	→	Propionate + Acetate + ATP + CO_2 (*e.g., Clostridium*)

Fig. 12.6. Some fermentation reactions producing ATP and examples of prokaryotes which perform them (after King, 1986).

Thus, the early fermentative autotrophs were much like our present-day anaerobic bacteria (*Lactobacillus*, etc.) and yeast. The metabolic processes of the anaerobic autotrophs resulted in the liberation of large amount of carbon dioxide (CO_2) into the atmosphere. Once this occurred, the way was paved for the evolution of organisms that could use CO_2 as the sole source of carbon in synthesizing organic compounds and could use sunlight as the only source of energy. Such organisms would be the photosynthetic cells.

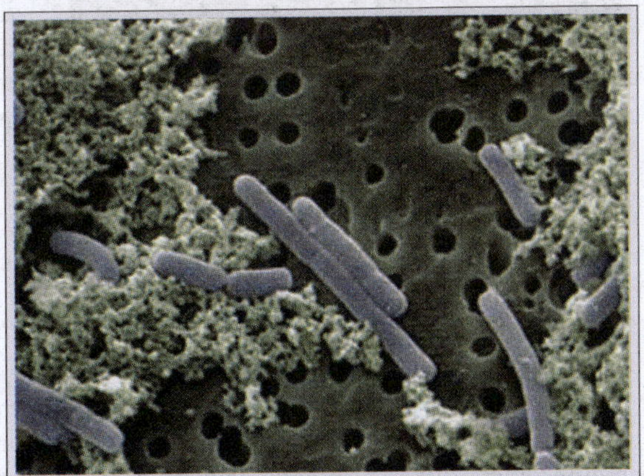

Electron micrograph of *Lactobacillus*.

Early photosynthetic cells probably split hydrogen containing compounds, such as **hydrogen sulphide** (**H_2S**). In other words, as is still performed today by anaerobic green/purple sulphur bacteria, hydrogen sulphide is cleaved into hydrogen and sulphur. The hydrogen is used by the cell to synthesize organic compounds, and the sulphur is released as a waste product (as evidenced by the earth's great sulphur deposits). In time, the process of photosynthesis was refined so that water (H_2O) served as the source of hydrogen. The result was the release of oxygen (O_2) as a waste product. At this stage, free oxygen became established for the first time in the atmosphere.

The first organisms to use water as the source of hydrogen in photosynthesis were the blue-green algae (cyanobacteria). Since blue-green algae were active photosynthesizers, atmospheric oxygen accumulated in increasing amounts converting reducing primitive atmosphere into oxidizing atmosphere. Oxygen would be **toxic** to anaerobic bacteria. Many primitive anaerobic bacteria, incapable of adapting to free oxygen, remained in portions of environment that were anaerobic, such as sulphur springs and oxygen-free muds. Today, there are bacterial types that are anaerobic as well as aerobic.

Geologic evidence indicates that **free oxygen** began to accumulate in the atmosphere about two billion years ago. The rising levels of atmospheric oxygen set the stage for the appearance of **one-celled eukaryotic organisms**, which arose at least one billion years ago. Then, within the comparatively short span of last 600 million years, the one-celled eukaryotes evolved in various directions to give rise wealth of **multicellular life forms** inhabiting the earth.

Evolution of Eukaryotes

The simplest cells in nature—the bacteria and blue green algae (cyanobacteria)—have been classified as **prokaryotic cells**. These cells have no nuclear membrane by which the hereditary materials (DNA) are set apart from the cytoplasm, and lack specialized cytoplasmic bodies (organelles) such as mitochondria and chloroplasts. In a bacterial cell, the DNA molecule, which does not contain introns but does have operons, forms a simple closed loop that is attached to the inside of the plasma membrane. The replication of a bacterial cell occurs rapidly by the simple division of the cell into two.

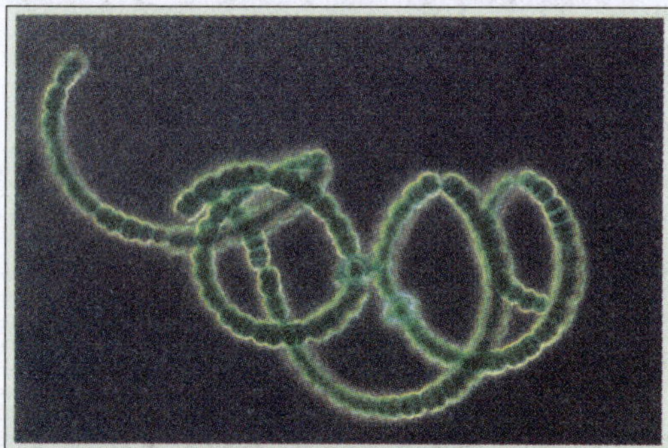

A cyanobacterium, *Anabaena*.

To recall, phylogenetically, the progenote evolved into three types of cells: 1. Archaebacteria, 2. Eubacteria and 3. Urkaryote. **Archaebacteria** are the prokaryotes which appear to have diverged from the true bacteria (or eubacteria) very early in evolution (Fig. 12.7). The strange class of archaebacteria consists of bacteria that are tolerant of acid (**thermophiles** or **thermoacidophiles**), bacteria that thrive in salt (**halophiles**) and bacteria that generate methane (**methanogens**). Recent developments in molecular biology reveal that these bacteria share some striking biochemical features (such as 16 S rRNA sequence) that are absent from other bacteria. **Eubacteria** includes **cyanobacteria** (aerobic, H_2O photosynthetic); **chloroxybacteria** (aerobic, H_2O photosynthetic); **paracoccus** (aerobic, heterotrophic), **non-sulphur bacteria** (aerobic, H_2S photosynthetic), sulphur bacteria (anaerobic, H_2S photosynthetic), **green-filamentous bacteria** (aerobic, H_2S photosynthetic), **green sulphur bacteria** (anaerobic, H_2S photosynthetic), **spirochaetes** (anaerobic, heterotrophic), *Clostridia* (anaerobic, heterotrophic) and *Desulphovibrio* (sulphate respiration) (see **Sleigh**, 1986). The urkaryote is the ancestor of modern eukaryote.

In contrast to the prokaryotes, the cells of the morphologically more complex plants and animals, or **eukaryotic cells**, have a distinct nuclear membrane that encloses strands of DNA, or discrete chromosomes. Eukaryotic cells also have an elaborate system of membrane-bound cytoplasmic organelles. The enzymes that are responsible for releasing energy from complex organic molecules are packed inside sausage-shaped organelles called **mitochondria**. In the eukaryotic plant cell, a prominent organelle is the **chloroplast**, the photosynthesizing particle that converts light energy into

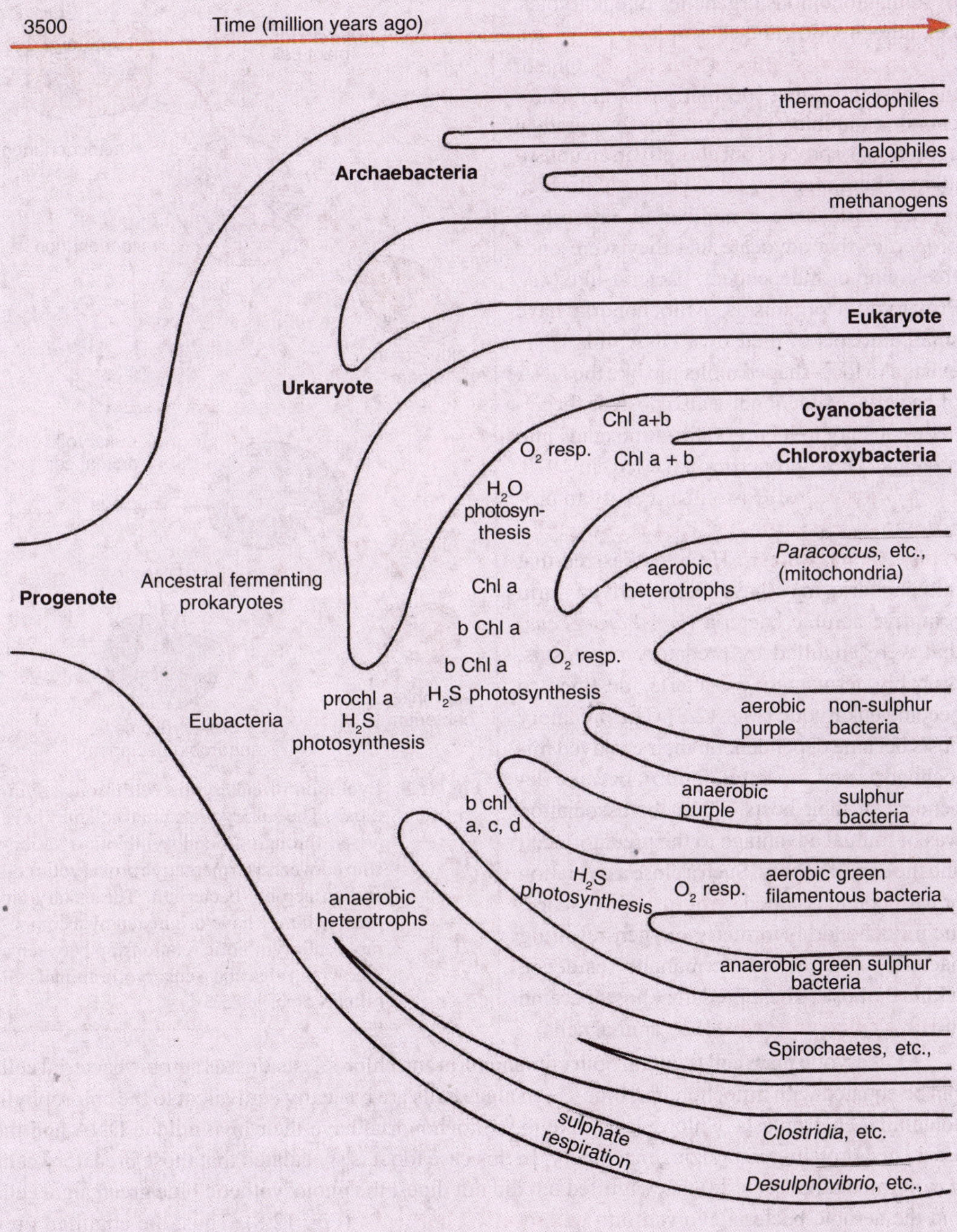

Fig. 12.7. A phylogenetic tree of the main groups of prokaryotes mentioned in the text. ch1 is chlorophyll; bch1, bacterio-chlorophyll; pb, phycobilin; prochl a, photo chlorophyll a and resp., is respiration (after King, 1986).

chemical energy. The following two hypotheses have been proposed regarding the origin of semiautonomous organelles of eukaryotes, *i.e.*, mitochondria and chloroplasts :

A. Endosymbiotic hypothesis. One of the most intriguing speculations is that mitochondria and chloroplasts arose not by a gradual evolutionary process but abruptly in an unusually striking manner, *i.e.*, **symbiosis**. In fact, mitochondria have a number of interesting properties that advocate that they were once free-living or independent, bacteria-like (*i.e.*, Prokaryotic) organisms. Mitochondria have small amounts of their own DNA; this DNA exists as a loop-shaped molecule like the DNA of bacteria. Mitochondria also possess the genetic capacity to incorporate amino acids into proteins. These properties have led, in 1970, **Lynn Margulis** of Boston University to propose the **serial symbiotic hypothesis** or **endosymbiotic hypothesis**. He hypothesized that mitochondria may have been derived form primitive aerobic bacteria (*i.e.*, *Paracoccus*) that were engulfed by predatory organisms, probably fermentative bacteria, destined to become eukaryotic (Fig. 12.8). The predatory hosts became dependent on their enslaved mitochondria and the latter, in turn, became dependent of their hosts. Thus, the association was of mutual advantage to the predatory cell and the engulfed prey. Such a close association or partnership is called **symbiosis**. In essence, the mitochondria (formerly oxygen-respiring bacteria) established permanent residence within the hosts. These predatory hosts became the **first eukaryotic cells** (*i.e.*, animal cells).

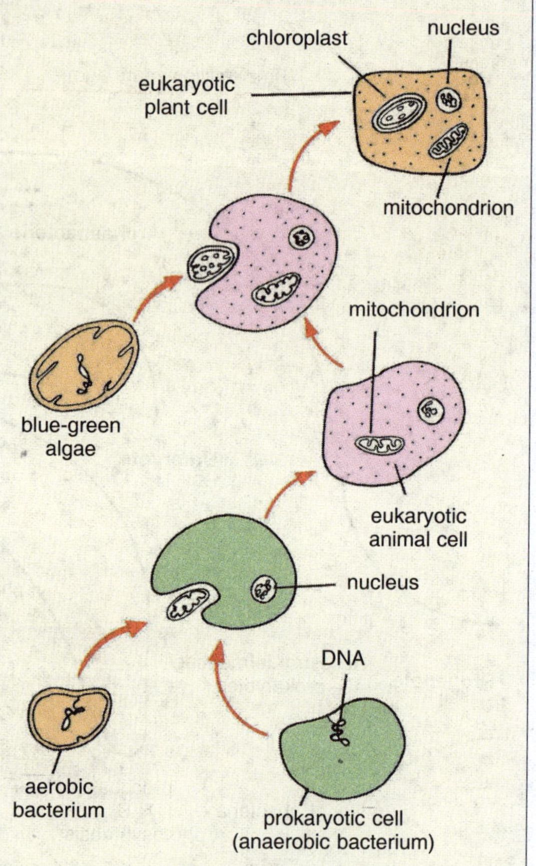

Fig. 12.8. Evolution of eukaryotic cell through symbiosis. The eukaryotic animal cell may have arisen through an endosymbiotic relationship between a fermentative prokaryotic cell and an aerobic bacterium. The eukaryotic plant cell may have originated by a comparable endosymbiotic relationship between a blue-green alga and a eukaryotic animal cell (after Volpe, 1985).

Eukaryotic plant cells contain both mitochondria and chloroplasts. Just as aerobic bacterial cells can be equated with mitochondria, blue-green algae cells are basically equivalent to the chlorophyll-containing chloroplasts. Chloroplasts, like the mitochondria, have their own unique DNA and the associated protein-synthesizing machinery. In this case too it is speculated that those predatory cells (*i.e.*, fermentative bacteria) that engulfed but did not digest the photosynthetic blue green algal cells and the aerobic bacteria evolved into **eukaryotic plant cells** (Fig. 12.8). Thus, the engulfed prey became permanent symbiotic residents—either as mitochondria or chloroplasts—within the predatory cell.

B. Invagination of surface membrane hypothesis. Many authors are not convinced with the hypothesis suggesting the relatively abrupt origin of the organelles of eukaryotes by a process of symbiosis. According to them, the evolution of living organisms normally proceeds through a series of gradual changes. Thus, continual modification of the surface membrane (plasma membrane) may have been the basic evolutionary mechanism in the differentiation of eukaryotic cell from prokaryotic cell. The organelles of eukaryotic cells may have evolved by the invagination (or drawing inward) of the surface membrane of a primitive cell. Figure 12.9 explains the postulated mechanism for the origin of a mitochondrion from the cell surface membrane. In support of this scheme, it is interesting that the enzymes for the breakdown of carbohydrates for energy production in bacteria are incorporated in the structure of the cell surface membrane. In similar manner other specialized internal structures of eukaryotes, including the membrane enclosed **nucleus**, may have differentiated from invaginated cell surface (Fig. 12.9).

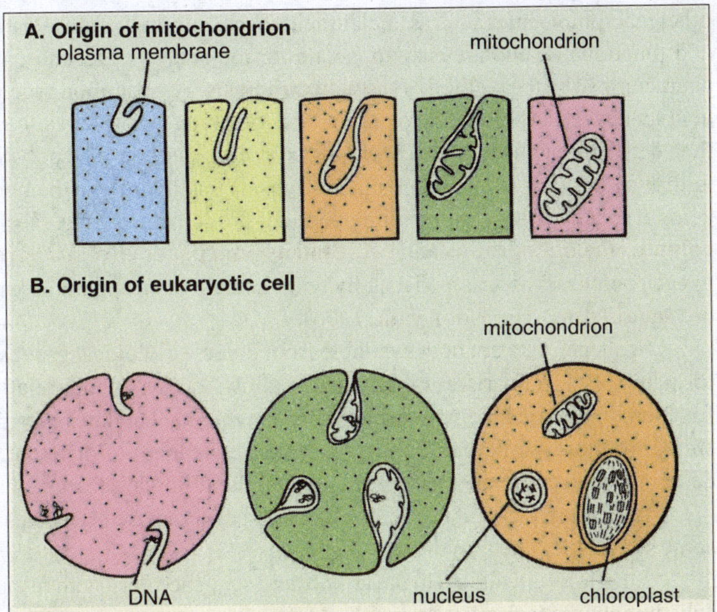

Fig. 12.9. Hypothetical scheme of the origin of eukaryotic organelles from the surface membrane. A—Development of the mitochondrion from the surface membrane; B—Origin of a eukaryotic cell by the invagination of the surface membrane in several places (after Volpe, 1985).

Finally, one of the new twists contributed by modern molecular biology is the possibility that today's prokaryotes are very highly evolved cells whose evolution has made them the most flexible and efficient cells on the planet.

MOLECULAR EVOLUTION

Evolution is a process of change. At the molecular level, this process involves the insertion, deletion, or substitution of nucleotides in the DNA. If the DNA encodes a polypeptide, these events may cause a change in the amino acid sequence. Over time, such changes can accumulate, leading to a molecule that bears little resemblance to its ancestor. Comparisons of the amino acid compositions of proteins in present-day organisms enable us to assume the molecular events that occurred in the past. The more distant in the past that ancestral stock diverged into two present-day species, the more changes will be evident in the amino acid sequences of the proteins of the two contemporary species. Viewed another way, the number of amino acid modifications in the lines of descent can be used as a measure of time since the divergence of the two species from a common ancestor. In principle, protein molecules incorporate a record of their evolutionary history that can be just as informative as the fossil record. Thus, proteins can be considered as '**chemical fingerprints**' of evolutionary history, because they bear amino acid sequences that have changed as a result of genetic changes.

The Evolution of Proteins

A few thousand kinds of proteins are required for the functioning of an *E.coli* cell and about a million kinds are required by human beings. They offer a preferable field for evolutionary study than other macromolecules such as nucleic acids because proteins are more heterogeneous both structurally and functionally, and are easy to isolate for the analysis. Recently a number of proteins have been characterized by the method of **sequence analysis**. The common procedure for establishing the amino acid sequence of a polypetide requires, first, breaking the protein into small fragments of peptides, and then determining the amino acid sequence in each peptide. The polypeptide is broken into fragments with enzymes (*e.g.*, trypsin and chymotrypsin) that hydrolyze the bonds between contiguous amino acids at specific sites. The resulting peptides are separated from each other using such procedures as **column chromatography** and **two-dimensional paper chromatography**. The amino acid sequence in each peptide is ascertained usually with the help of an apparatus known as a "**protein sequencer**" or "**sequinator**" (see **F.J. Ayala**, 1976).

Sequence data are now available from diverse biological groups ranging from microorganisms to mammals. Comparison of sequences allows study of interrelationship between structure and function, and to deduce how proteins have evolved. The laws governing evolution of proteins are similar to those for heritable traits, and interrelatedness of different organisms points toward their descent from a common ancestor. In general the following features have been noted :

1. Identical proteins are not found among different living species. But homologous proteins with some similarities occur in diverse organisms.

2. Different positions in the amino acid sequence vary with respect to the number of amino acid substitutions that can take place without impairing the function of the polypeptide chain. Thus, some positions allow more substitutions than others.

3. Due to the tightly packed three dimensional structure of the protein, a change undergone by an amino acid situated in the interior of the chain is likely to affect residues in neighbouring positions. Such evolutionary changes, therefore, occur in pairs.

4. Enzymes performing similar functions have similar structures.

Examples of Protein Evolution

1. Insulin. The first protein to be sequenced was **insulin** (Insulin is a protein hormone produced by the beta cells of the islets of Langerhans of pancreas which participates in carbohydrate and fat metabolism) which composed of only 51 amino acids (**Sanger** and **Thompson**, 1953). Comparison of the amino acid sequence of insulin in a variety of mammals, including cattle, pigs, horses, sperm whales and sheep, shows that with the exception of a stretch of three amino acids, the protein is identical in each of these species.

2. Haemoglobin. In comparison between haemoglobin of vertebrates, **E. Zuckerkandl** (1965), calculated an average of 22 differences between human haemoglobin chains (α and β) and the similar haemoglobin of horse, pig, cattle and rabbit. It was, therefore, concluded that human haemoglobin diverged from those of animals by about one amino acid change per 7 million years. The rate of change of amino acids is, however, different for different proteins.

3. Cytochrome c. Cytochrome c is an ancient, evolutionarily conservative molecule that serves as an essential enzyme in respiration. It is relatively small protein of a chain length slightly over 100 amino acid residues, which apparently exists in all eukaryotic organisms. This ubiquitous protein has been extracted, purified and analyzed in more than 80 eukaryotic species (**Fitch** and **Margoliash**, 1967; **Dickerson**, 1971). Cytochrome c has 104 amino acids in vertebrates and a few more in certain species lower in the phylogenetic tree. Since cytochrome c performs similar vital functions in all organisms, fairly rigid evolutionary limitations have been placed on the acceptance of chance alterations (or mutations) of the primary amino acid sequence (*i.e.*, an evolutionary conservatism is maintained).

Some amino acid residues of cytochrome c have not varied at all throughout time. In particular,

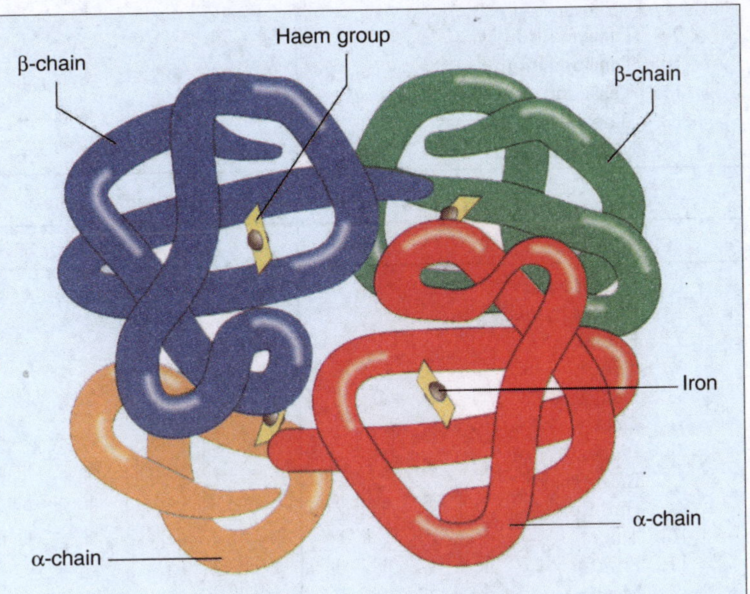

Structure of haemoglobin molecule.

the same amino acids have been found in 20 positions in all organisms tested, from moulds to humans. Obviously, the 20 amino acids at specific positions are essential, and any substitutions at these sites are likely to interfere with the function of the enzyme. In fact, there is one region of cytochrome c formed by a sequence of 11 amino acids (positions 70 through 80) that has remained intact in all organisms.

In Table 12-1., the number of amino acid replacements in cytochrome c of several species have been compared (**Dickerson**, 1971). Firstly, it is evident that the chromosomes of closely related vertebrates differ in only a few residues, or not at all. For example, there is no difference at all in the composition of the amino acid residues between humans and the chimpanzee. Secondly, the greater the phylogenetic differences, the greater the likelyhood that the cytochrome composition differ. This becomes clear when one compares a vertebrate with an insect (*i.e.*, tuna and moth) or an insect with yeast.

Neutral Theory of Protein Evolution

Evolution of proteins have also been studied in certain other proteins such as myoglobin, α-globin, histone, carbonic anhydrases, fibrinopeptides, etc. Analyses of the amino acid sequences of polypeptides suggest that some evolutionary changes may be random. This fact led to the formulation of a theory by **Kimura** (1968) called **neutrality theory of protein evolution**. This controversial theory assumes that for any gene, a large proportion of all possible mutants are harmful to their carriers; these mutants are eliminated or kept at very low frequencies by natural selection. A large fraction of mutations, however, are assumed to be adaptively equivalent (*i.e., the rate of evolution is equal to the rate of mutation*).

Table 12-1. Evolution of cytochrome c (Source : Volpe, 1985)

A. Comparison of cytochrome c between different organisms.

Organisms	Number of variant amino acid residues
1. Cow and sheep	0
2. Cow and whale	2
3. Horse and cow	3
4. Rabbit and pig	4
5. Horse and rabbit	5
6. Whale and kangaroo	6
7. Rabbit and pigeon	7
8. Shark and tuna	19

9. Tuna and fruitfly	21
10. Tuna and moth	28
11. Yeast and mould	38
12. Wheat germ and yeast	40
13. Moth and yeast	44

B. Comparison with human cytochrome c

Organisms	Number of differences of amino acid residues
1. Chimpanzee	0
2. Rhesus monkey	1
3. Rabbit	9
4. Cow	10
5. Kangaroo	10
6. Duck	11
7. Pigeon	12
8. Rattle snake	14
9. Bull frog	18
10. Tuna	20
11. Fruitfly	24
12. Moth	26
13. Wheat germ	37
14. Mould (*Neurospora*)	40
15. Baker's yeast	42

Although the neutral theory does admit a role for natural selection as a purifying agent, that is, as a force that eliminates harmful mutations, it has little or no place for selection in the positive Darwinian sense. In the neutral theory species do not get better by fixing beneficial mutations; they simply do not get worse by fixing only neutral or nearly neutral mutations. For this reason, this theory is also called **non-Darwinian theory of evolution** (see **Gardner** *et al.*, 1991).

REVISION QUESTIONS

1. Describe how life originated on earth.
2. Describe different theories for the origin of life on earth. Discuss in detail the Oparin's theory regarding the origin of life.
3. What is chemical evolution and origin of life? Discuss.
4. Describe **Urey** and **Miller** experiment. How does this experiment support Oparin's hypothesis?
5. Write short notes on the following:

   (i) Coacervates; (ii) Microspheres ; (iii) Origin of eukaryotic cells; (iv) Endosymbiotic hypothesis; and (v) Current postulated role of RNA in origin of life.

ECOLOGY

(Environmental Biology)

CHAPTERS

CHAPTER 1

Introduction

Ecology is the study of inter-relationship of organisms with their physical and biotic environment.

No microorganism, plant or animal species including man is an isolated organism living in a void. Each of them is surrounded by a host of physical conditions that can be measured in terms of chemical composition, texture, pressure, temperature, and humidity, as well as being surrounded by a host of other living organisms which can be described in such terms as microorganism, plants, animals, food, parasites, and enemies. Studies of the inter-relationships of organisms with their physical and biotic environments are termed **environmental biology** or **ecology**. These words are very much in the public consciousness today as we become aware of some of the past and current ecological malpractices. It is important for everyone to know and appreciate the principles of this aspect of biology so that one can form an intelligent opinion regarding topics such as insecticides, detergents, mercury pollution, sewage disposal, power dams, urbanization and their effects on mankind, on human civilization, and on the world we live in.

The term **ecology** (*oekologie*) is derived from two Greek words–*oikos*=means 'house' or 'place to live' and–*logos* means 'a discussion or study'. Literally, ecology is the study of organisms 'at home', in their native environment. The term was first of all introduced by **Reiter** in 1868, but because the German biologist **Ernst Haeckel** (1869) first of all fully defined this term and made an extensive use of this term in his writings, he is therefore, usually falsely credited for the coinage of the term ecology (see **C.B. Knight**, 1965).

DEFINITION OF ECOLOGY

The traditional definition of ecology is '*the study of an organism and its environment*', however, different ecologists have defined it variously. **Ernst Haeckel** (1869) defined ecology as, '*the total relation of the animal to both its organic and its inorganic environment.*' In 1936, **Taylor** defined ecology as 'the *science of all the relations of all organisms to all their environments.*' **Charles Elton** (1947) in his pioneer book *Animal Ecology* defined ecology as '*scientific natural history.*' Although this definition does point out the origin of many of our ecological problems, yet it is much broad and vague like Haeckelian definition of ecology. **Allee** *et al.*, (1949), in their definition of ecology, clearly emphasize the all-encompassing character of this field of study. According to them ecology may be defined broadly as '*the science of the interrelation between living organisms and their environments, including both the physical and biotic environments, and emphasizing interspecies as well as intra-species relations.*' Though, **F.J. Vernberg** and **W.B. Vernberg** (1970) completely agree with **Allee** *et al's* definition, yet there are certain ecologists which are not satisfied with this definition and have provided their own definitions of ecology. For instance, **Andrewartha** (1961) defined ecology as '*the scientific study of the distribution and abundance of organisms.*' **G.A. Petrides** (1968) has defined ecology as '*the study of environmental interactions which control the welfare of living things, regulating their distribution, abundance, production and evolution.*' **Eugene Odum** (1963, 1969 and 1971) has defined ecology as '*the study of the structure and function of ecosystems*' or '*structure and function of nature.*' The Indian ecologist **R. Misra** (1967) defined ecology as the '*interactions of form, functions and factors.*' These three interacting aspects, *i.e.*, form, functions and factors integrate together to construct the **triangle of nature**.

Eugene Odum.

C.J. Krebs (1972) has shown his satisfaction over the definition of **Odum** for its emphasizing the form and function idea that permeates biology, but, he considered it vague. He, however, proposed the modified version of Andrewartha's definition of ecology, according to which '*ecology is the scientific study of the interactions that determine the distribution and abundance of organisms.*' According to **M.E. Clark** (1973) '*Ecology, the science of the environment, is a study of ecosystems, or the totality of the reciprocal interactions between living organisms and their physical surroundings.*' Two Indian authors **Ananthakrishnan** and **Viswanathan** (1976) have defined ecology as '*the study of the ways by which individual organisms, populations of species and communities of populations respond to diverse environmental pressures, i.e., physical and biotic pressures.*' This definition is merely the modified version of the definition of ecology which has been originally proposed by **T. Lewis** and **L.R. Taylor** (1967) in their book, *Introduction to Experimental Ecology.* According to **Lewis** and **Taylor** '*Ecology is the study of the way in which individual organisms, populations of same species and communities of populations respond to these changes.*' Lastly to avoid this ecological jargon we can rely upon the simple definition of ecology which has been provided by **Charles H. Southwick** (1976). According to him '*ecology is the scientific study of the relationships* *of living organisms with each other and with their environments'*. Thus, it is the science of biological interactions among individuals, populations, and communities. Ecology is also the science of ecosystems–the interrelations of biotic components with their non-living environments.

Ecology versus Environmental Studies

The continuing increase of human population and the associated destruction of natural environments with deforestation, pesticides and pollutants have awakened the awareness of the public to the word of ecology. However, much of this recent interest centres on the human environment and human ecology. Unfortunately the word ecology became identified in the public mind with the much broader problems of the human environment, and "ecology" came to mean everything and anything about the environment. The science of ecology is concerned with the environments of all plants and animals and is not solely concerned with humans. As such, ecology has much to contribute to some of the broad questions about humans and their environments. Apparently, ecology should be to environmental science as physics is to engineering. Just as humans are constrained by the law of physics when we build aeroplanes and bridges, so also we should be constrained by the principles of ecology when altering the environment (see **C. J. Krebs**, 1985).

HISTORICAL BACKGROUND OF ECOLOGY

The roots of ecology lie in natural history, which is as old as man himself. The vague beginning of ecology may be traced back to prehistoric man, who utilized environmental information to hunt food; fishing, trap animals, find edible vegetation, and locate shelter to survive the hardships imposed by nature. An increased knowledge of the importance of environmental conditions led quite naturally to religious rituals, worship of weather gods, and rain dances, all of which are quite prevalent in different social groups and tribes of Indian civilization. The establishment of agriculture increased the need to learn about the practical ecology of plants and domestic animals.

Further, the science of ecology has had a gradual but spasmodic development during the recorded history of human civilization. Consequently, the historical background of ecology can be traced as follows :

1. Ecology in Indian Classics

Early Indian philosophers and thinkers had some knowledge of ecology as has been revealed by the Indian classical writings, in the Vedic, Upanishadic, Pauranic, and Epic literature. **Charaka** considered the important factors of *Vayu* (air and gases), *Jala* (water), *Desha* (topography) and *time* in regulating the life of plants (see **R. Misra**, 1962). The great classical Indian Poet **Kalidasa** has displayed his ecological faculty in his classics– *'Ritu Sanhar'*, *'Meghdoot'* and *'Abhigyan Sakuntlam.'*

2. Early Greek's Ideas of Ecology

The early Greek philosophers too were well aware of the importance of environmental studies. One of the papers entitled *Airs, waters and places* of **Hippocrates** stressed on the need of ecological background for medical students. According to him–"whoever wished to investigate medicine properly, should proceed thus : in the first place to consider the seasons of the year, and what effects each of them produces........" In the fourth century B.C. **Aristotle** made references about the habits of animals and environmental conditions prevailing in certain areas, in his writings on natural history. One of the Aristotle's friend and associate **Theophrastus** (370–250 B.C.) who is regarded as the first true ecologist, wrote about the association of plant communities and the relation of plants to each other and the non-living environment.

3. Ecology in 18th Century

After a considerable long gap of many centuries, the first naturalist who made a serious attempt to systematise knowledge concerning relations of animals to environment was **Buffon** (1707–1788). Though such works as *Historia animalium* by **Konard Gesner** (1519–1565); 14 volumes of collected works of natural history by **Ulisse Aldrovandi** (1522–1605); the natural histories of fishes and birds by **Pierre Belon** (1517–1564); *The plants* by **Andrea Caesalpino** (1519–1603), the botanical work of **Gaspard Bauhin** (1550–1624); demographical studies about human ecology by **Graunt** (1664); studies of the sexuality of plants by **Rudolph Camerarius** (1665–1721); the works of English naturalist **John Ray** (1627–1705); studies on natural history of insects by French naturalist **Rene Reaumur** (1683–1757) and the works of many other naturalists doubtlessly laid the foundation upon which **Buffon** built his remarkable work. **Buffon** considered that animals and plants developed adaptations which enabled them to favourably respond to their environment and termed this **environmental induction**. He touched on many of our modern ecological problems and recognised that population of man, other animals and plants are subjected to the same processes. **Buffon** also explained how the great fertility of every species was counter-balanced by innumerable agents of destruction. He, thus, dealt with problems of population regulation.

Animals and plants develop adaptations which enables them to favourably respond to their environment.

Anton Van Leeuwenhoek, best known as a pioneer microscopist of the early 1700s, also pioneered the study of "food chains" and "population regulation", which are two important areas of modern ecology (see **Odum**, 1971).

Malthus published one of the earliest controversial book on demography. In his *Essay on Population* (1798) he calculated that although the number of organisms can increase geometrically (1,2,4,8,16,32.......); their food supply may never increase faster than arithmetically (1, 2, 3, 4, 5....). Such a great disproportion between these two powers of increase led **Malthus** to infer that reproduction must eventually be checked by food production. Many workers questioned the ideas of **Malthus**. For example, in 1841 **Doubleday** brought out his true law of population. To contradict the **Malthus** idea, he believed that whenever a species was threatened, nature made a corresponding effort to preserve it by increasing the fertility of its members. Human populations that were undernourished had the highest fertility; those that were well fed had the lowest fertility. **Doubleday** explained these effects by the oversupply of mineral nutrients in well-fed populations.

Further, many of the early developments in ecology came from the applied fields of agriculture, fisheries, and medicine. For example, work on the insect pests of crops has been one of the important source of ideas. The regulation of population size of obnoxious insects is a basic problem that has been under study. In 1762 the mynah bird was introduced from India to the island of Mauritius to control the red locust. Due to such **biological control**, by 1770 the locust threat was a negligible problem. **Forska** wrote in 1775 about the introduction of predatory ants from nearby mountains into date palm orchards to control other species of ants feeding on the palms in South Western Arabia. In subsequent years an increasing knowledge of insect parasitism and predation led to many such introductions all over the world in the hope of controlling introduced and native agricultural pests.

Similarly, medical work on infectious diseases such as malaria around the 1890's gave rise to the study of epidemiology and interest in the spread of disease through a population. Before malaria could be controlled adequately it was necessary to know in detail the ecology of mosquitoes. Likewise, the production ecology had its beginnings in agriculture, and **Egerton** (1969) has traced this back to the eighteenth-century botanist **Richard Bradley**. Bradley recognized the fundamental similarities of animal husbandary and production, and he proposed methods of maximizing agricultural yields (and hence profit) for vineyards, trees, poultry, rabbits and fish.

Production ecology enables maximum agriculture yields.

4. Ecology in 19th Century

In 1807, **Humboldt** after extensive travelling of tropical and temperate South America, discussed the geographical distribution of plants and animals in relation to climate. **Edward Forbes** in 1844 described the distribution of animals in British coastal waters and part of the Mediterranean sea, and he wrote of zones of differing depths which were distinguished by the associations of species they contained. He noted that some species are found only in one zone and that other species have a maximum of development in one zone but occur sparsely in other adjacent zones. Mingled in are stragglers that do not, fit in zonation pattern. **Forbes** recognized the dynamic aspect of the interrelations between these organisms and their environment. As the environment change, one species might die out, another might increase its abundance. English naturalist, **St. George Jackson Mivart** coined the term **hexicology** around 1859 and in 1894 he defined it as "devoted to the study of the relations which exist between the organisms and their environment as regards to the nature of the locality they frequent, the temperatures and the amounts of light which suit them, and their relations to other organisms as enemies, rivals or accidental and involuntary benefactors." **Saint-Hilaire** (1859) used the term "**ethology**" for the study of relations of organisms to environment and outlined a volume, which he never wrote, on instincts, habits, and other ecological matters. The term ethology, though was not accepted by early ecologists, but this term is now used as an important ecological discipline dealing with animal behaviour. In 1869, **Haeckel** proposed "**oecologie**" for the relations of organisms to organic and inorganic environments. The modern term *ecology* is derived from the word *oecologie*. The observations of **Priestley** and **Scheele** that plants produce oxygen led to an understanding of the inter-relationships of plants and animals. The discovery that green plants and animals use carbon dioxide and water to form organic matter for use by animals and result in the release of oxygen and water as wastes, provided new understanding of ecological food chains. **Justus von Liebig** (1803–1873) initiated the idea of carbon and nitrogen cycles in nature and his work (1840) led to **ecoclimatology** and **physiological ecology**. Further, out of such studies in aquatic environment developed the field of **ecological energetics**. **Louis Agassiz** (1807–1873) initiated the research in marine ecology by establishing many laboratories for marine ecological research. **Karl Mobius** (1877) introduced the word "**biocoenosis**" to designate a group of organisms as an ecological unit. **Charles Darwin** (1809–

Poultry has a economic base.

1882) and **Alfred R. Wallace** (1823–1915) have increased our knowledge of island of insular life with their characteristically different types of environmental conditions. **Hensen** (1887) in a classical paper on "*The Lake as a Microcosm*" suggested that the species assemblage in a lake was an organic complex and that by affecting one species we exerted some influence on the whole assemblage. Thus each species maintains a 'community of interest' with the other species, and we cannot limit our studies to a single species. **Forbes** believed that there was a steady balance of nature, which held each species within limits year after year, even though each species was always trying to increase its number. The Russian pioneer ecologist **V.V. Dokuchaev** (1846–1903) and his chief disciple **G.F. Morozov** (who specialized in forest ecology) placed great emphasis on the concept of the "**biocoenosis**", a term later expanded by Russian ecologist to "**geobiocoenosis**", a term equivalent to ecosystem.

Ethology deals with particular behaviour patterns of animals.

Schroter (1896) introduced the terms "**autecology**" and "**synecology**". **Herdman** (1896) indicated the economic possibilities of scientific agriculture. Danish botanist **Warming** (1895, 1909) raised questions about the structure of plant communities and the associations of species in these communities. In 1899, **H.C. Cowles** described *plant succession* on the sand dunes at the Southern end of Lake Michigan.

Thus, as a recognized distinct field of biology, the science of ecology dates from 19th century, and only in the mid 20th century has the word become the part of the general vocabulary.

5. Ecology in the 20th Century

At the beginning of the twentieth century ecology was a young, but an established science and such eminent ecologists as **Wasmann** (1901), **Dahl** (1901) and **Wheeler** (1902) were discussing whether **Saint Hilaire's** ethology or **Haeckel's** ecology should be used to designate the science of relations of organisms to environments. The latter term has gradually come into general usage. **Case** (1905) discussed environmental conditions of past geological periods and has continued to make contributions in that field. Two botanists, **Clements**, in his "*Research Methods in Ecology*" (1905) and subsequent writings, and **Warming** (1909, 1925) in his "*Oecology of Plants*" have done much to clarify ideas and contribute terms. **Adams** (1913) gave an excellent review of significant ecological literature, urged the usefulness of ecological surveys, and discussed methods to be used in making them, **Shelford** (1912) demonstrated excellent examples of ecological succession and has made notable contributions concerning animal communities (1913) and environmental factors (1929). In 1920 the German limnologist **Thienemann** introduced the concept of trophic levels in terms of producers and consumers. The theoretical approaches in population dynamics of **Lotka** (1925) and **Volterra** (1926) stimulated the experimental approaches by biologists. **Gleason** (1926) studied the development and dynamics of plant communities. **Shelford** (1913, 1937), **Adams** (1909) and **Dice** (1943) in America and **Elton** (1927) in England investigated the interrelations of plants and animals. In 1935 **A.G. Tansley** first proposed the term **ecosystem**.

Gause (1935) investigated the interactions of predators and prey and the competitive relationships between species. At the same time **Nicholson** studied interspecific competition. In 1954 **Andrewartha** and **Birch** and **Lack** provided a broader foundation for the study of regulation of populations. The discovery of the role of territory in bird life by **H.E. Howard** in 1920 led to further studies by **Nice** in the 1930s and 1940s. Out of such studies came the field of **behavioural ecology**.

In the 1940s and 1950s **Lorenz** and **Tinbergen** developed concepts of instinctive and aggressive behaviour. In 1962, **Wynne-Edwards** of England explored the role of social behaviour in the regulation of populations.

Starling chicks forage for food without leaving the nest showing begging behaviour.

In the field of ecological energetics, in 1920 the German limnologist **Thienemann** introduced the concept of trophic levels in terms of producers and consumers. Two American limnologists, **Birge** and **Juday** in the 1940s, through their measurements of the energy budgets of lakes, developed the idea of primary production. Out of their studies came the trophic-dynamic concept of ecology. Introduced by **Lindemann** in 1942, this concept marked the beginning of modern ecology. Out of **Lindemann's** study came further pioneering work on energy flow and energy budgets by **Hutchinson** and **H.T.** and **E.P. Odum** in the 1950s in America. Early work on the cycling of nutrients was done by **Ovington** (1957) in England and Australia and by **Rodin** and **Bazilevig** (1967) in Russia.

One of the interesting aspects of development of ecology during the first third of this century remains that a tendency towards increased specialization in ecology has occurred as is evidenced by the establishment of such disciplines as **palaeoecology**, a study of environmental conditions and life as it existed in past ages. Pollen analysis, radioactive dating, and paleontology have aided the palaeoecologists. Other specialized fields of ecology which took their birth during this time, are **zoogeography**, the scientific study of the geographic distribution of animals; **oceanography**, the study of the biotic and physical conditions existing on oceans, bays, and estuaries, and **limnology**, the study of the living and non-living components of inland waters. With the establishment of specialized disciplines of ecology, the value of application of ecology to such fields as agriculture, forestry, wildlife conservation, fisheries management, and pest control became apparent, and, thus, a new discipline called **applied ecology** came into its existence.

Further, in the second decade of 20th century great emphasis was laid on statistical studies on populations, sampling techniques, and community studies. During this period several important texts published–**Lotka's** *Elements of Physical Biology; **Elton's** Animal Ecology; **Volterra's** Animal Ecology; **Gause's** The Struggle for Existence,* and **Bodenheimer's** *Problems of Animal Ecology.* Certain distinguished ecologists of earlier half and present decade of this century are **Clements**, **Cowles**, **Shelford**, **Pearse**, **Oosting**, **A.M. Woodbury**, **Hutchinson**, **Deevey**, **Lindeman**, **F.E. Smith**, **Slobodkin**, **Hairston**, **T. Park**, **O.Park**, **Emerson**, **MacArthur**, **G.L. Clark**, **C.B. Knight**, **H.T.** and **E.P. Odum**, **R. L. Smith**, **F.J. Vernberg**; **C.J. Krebs**, **R. Misra**, **M.S. Mani**, etc.

Limnology is the study of living and non-living components of inland waters e.g., pond, river.

Development of ecology during this century have been greatly stimulated by organization of ecologists. The British Ecological Society was founded in 1913 and this society sponsored two Journals namely *Journal of Ecology* (started in 1913) and *Journal of Animal Ecology* (started in 1932). Following the British, the Ecological Society of America was established in 1915 which also sponsored two publications, *Ecology* in 1920 and *Ecological Monographs* in 1931.

BRANCHES OF ECOLOGY

The science of ecology is often divided into autecology and synecology. **Autecology** deals with the study of the individual organism or an individual species. In this, life histories and behaviour as a means of adaptation to the environment are usually emphasized. **Synecology** deals with the study of groups of organisms which are associated together as a unit (*i.e.,* community). Thus, if a study is made of the relation of a white oak tree (or of white oak trees in general) to the environment, the work would be autecological in nature. If the study concerned the forest in which the white oak lives, the approach would be synecological (**Odum**, 1971). While autecology is experimental and inductive, the synecology is philosophical and deductive (**Smith**, 1974).

Synecology is often further subdivided into **aquatic** and **terrestrial ecology**. The aquatic ecology includes **freshwater ecology**, **eustuarine ecology** and **marine ecology**. Terrestrial ecology, subdivided further into such areas as **forest ecology**, **grassland ecology**, **cropland ecology** and **desert ecology**, is concerned with terrestrial ecosystems—their microclimate, soil chemistry, nutrient and hydrological cycle and productivity.

Further, **demecology** is that branch of ecology which deals with the ecology of populations. Early ecologists have recognized two major subdivisions of ecology in relation to plants and animals—**plant ecology** and **animal ecology**. But when it was found that in the ecosystems plants and animals are very closely associated and interrelated, then, both of these major subdivisions of ecology into plant ecology and animal ecology became vague. Besides these major subdivisions the ecology has been classified in the following branches according to the level of organization, kind of environments or habitats and taxonomic position :

1. Habitat ecology. It deals with the study of different habitats of the biosphere. According to the kind of habitat, ecology is subdivided into **marine ecology**, **freshwater ecology**, and **terrestrial ecology**. The terrestrial ecology too is further subdivided into **forest ecology**, **cropland ecology**, **grassland ecology**, etc., according to the kind of study of its different biomes.

2. Ecosystem ecology. It deals with the analysis of ecosystem from structural and functional point of view including the interrelationship of physical (abiotic) and biological (biotic) components of environment.

3. Conservation ecology. It deals with methods of proper management of natural resources such as land, water, forests, sea, mines, etc., for the benefit of human beings.

Conservation of natural resources can be done through proper management.

4. Production ecology. It is the modern subdivision of ecology which deals with the gross and net production of different ecosystems such as freshwater, sea water, cropfields, orchards, etc., and tries to do proper management of these ecosystems so that maximum yield can be obtained from them.

5. Radiation ecology. It deals with the study of gross effects of radiations and radioactive substances over the environment and living organisms.

6. Taxonomic ecology. It is concerned with the ecology of different taxonomic groups and eventually includes following subdivisions of ecology–**plant ecology**, **insect ecology**, **invertebrate ecology**, **vertebrate ecology**, **microbial ecology** and so on.

7. Human ecology. It deals with the study of relationship of man with his environment.

8. Space ecology. It is a modern subdivision of ecology which remains concerned with the development of partially or completely regenerating ecosystems for supporting life of man during long space flights or during extended exploration of extra-terrestrial environments.

9. Systems ecology. The systems ecology is the most modern branch of ecology which is concerned with the analysis and understanding of the function and structure of ecosystem by the use of applied mathematics such as advanced statistical techniques, mathematical models and computer science.

RELATIONSHIP OF ECOLOGY WITH OTHER DISCIPLINES

Modern ecology is a multidisciplinary science which depends on a variety of disciplines such as **physics**, **chemistry**, **mathematics**, **statistics**, **meterology**, **climatology**, **geology**, **geography**, **economics**, **sociology**, **agriculture science**, **forestry**, **horticulture**, **genetics**, **physiology**, etc. All these disciplines have helped in the better understanding of many ecological principles.

For instance, meteorological and climatological data for certain geographic localities allow for a more implied interpretation of results. A basic knowledge of forestry can be invaluable for a forest-ecologist to understand forest type distribution, floristic composition and prevalent environmental factors. Likewise, statistical data helps in interpreting the reasons for activity, population increases, migrations, probability of ecological events occurring in a particular area, sampling techniques and reliability of results. Palaeontology (geology) provides information about the ancestral organisms and environmental situations prevalent in past. Evolution and genetics are utilized to interpret the reason for organic changes when linked with environmental conditions, establishment of new populations and species, environmental effects on genetic populations and species and, so on. Such interdisciplinary approaches to ecology, consequently have given rise to following subdivisions of ecology :

1. Ecological genetics. An ecologist recognised a kind of genetic plasticity in the case of every organism. In any environment only those organisms that are favoured by the environment survive. The branch of ecology dealing with genetics in relation to ecology is called ecological genetics.

Pedology is the study of soil and their influence on the organisms.

2. Palaeoecology. It deals with the movements of biotic elements based on palaentological evidence, which provides information about ancestral organisms and environmental conditions existing in the past.

3. Ecophysiology. The factors of environment have a direct bearing on the functional aspects of organisms. The ecophysiology deals with the survival of populations as a result of functional adjustments of organisms with different ecological conditions of the ecosystem.

4. Chemical ecology. It deals with the adaptations of animals or preferences of particular organisms like insects to particular chemical substances.

5. Pedology. It is a branch of terrestrial ecology and it deals with the study of soil, in particular their acidity, alkalinity, humus content, mineral contents, soil type, etc., and their influence on the organisms.

6. Ecogeography. It deals with the study of the role of the environment in animal distribution. It is related with **biogeography** which is concerned with the structural and functional relations of living organisms in space, which form the immediate environment of the individuals as well as populations. **Ecofloras** and **ecofaunas** are the lowest units of which a biogeographic flora or fauna is made up of.

7. Ecological energetics. It deals with energy conservation and its flow in the organisms within the ecosystem. In it thermodynamics has its significant contribution.

Because of its far-flung involvements with so many fields, ecology is often regarded as a generality rather than a speciality. Visualizing this fact, an ecologist, **A. Macfadyen** (1957) wrote in his book *Animal Ecology : Aims and Methods*—*"The ecologist is something of a chartered libertine. He roams at will over the legitimate preserves of the plant and animal biologist, the taxonomist, the physiologist, the behaviourist, the meterologist, the geologist, the physicist, the chemist, and even the sociologist; he poaches from all these and from other established and respected disciplines. It is indeed a major problem for the ecologist in his own interest, to set bounds to his divagations"*.

ECOLOGICAL TOOLS AND TECHNIQUES

Ecology has been studied on three broad lines—the **field**, **laboratory** and **mathematics**. Most of the ecologists have studied ecology from either a **descriptive** or **analytical** point of view. The descriptive point of view remains mainly natural history. A descriptive ecologist as a natural historian has asked the question *"What is present ?"* He has described the life histories and characteristics of organisms and also has done a correlated study of many of the responses of an organism, whether it was responding as an individual or as a member of a population, to the presence of physical forces in its habitat, such as weather, food, space and others. According to **Moen** (1973), a descriptive ecologist has to proceed a bit further than the natural historian in relating observed characteristics of the organism to observed characteristics of the habitat in a quantitative way. This has resulted in the formation of many ecological rules. **Bergman's** rule is an example : animals living further north tend to be larger. These types of rules are generally applicable, although exceptions can be found in looking at detailed relationships. The analytical ecologist ask the question *"why"?* He is interested in the mechanism operating in the natural world. The recognition of simple relationship such as the condition of the range and conditions of the animal are persued further by analyzing the requirements of the animal through time and the ability of the range to satisfy these requirements.

An analytical ecologist has to face several practical problems because he has to work with living systems and more often he comes across the variables which are numerous and normally highly complex. Consequently the tool and techniques used by the physical scientists could not be easily applied in ecological investigations, nor were the results of ecological experiments so precise as those obtained in physics and chemistry. Despite these problems certain modern ecologists have made an active use of tools and techniques of chemistry, physics and mathematics in their ecological investigations. Physical and chemical measurements were taken to measure the various parameters of the environment. These may range from simple chemical determination of the various elements to the use of such sophisticated apparatuses such as **paper chromatogram**, **infrared gas analyzer**, **recording spectrophotometer**, and **microbomb calorimeter**. The use of statistical procedures such as **correlation**, **multiple regression**, and **matrix algebra** and the application of modern algebra, calculus and computer science to mathematical models simulating field conditions

Gas analyzer for H_2O and CO_2.

have provided new insights into population interactions and ecosystem functioning. With the use of **electronic equipment** and **biotelemetric techniques**, ecologists can sample and measure plant and animal populations without destroying them. **Radioisotopes** have enabled the investigators to follow the pathways of nutrients through ecosystem and to determine the time and extent of transfer. **Laboratory microcosms**—samples of both aquatic and soil microsystems—are useful in determining the rates of nutrient cycling and other parameters of ecosystems functioning.

SIGNIFICANCE OF ECOLOGY FOR MAN

Man is himself an organism within an environment. Like other animals man is influenced by the physical features of his environment, he is absolutely dependent upon other species for his food, clothing, medicine, and other similar aspects and he has to adjust to other individuals of his own species. Therefore, the basic laws of ecology apply well to him and its fundamental knowledge is must for man for his own existence on this planet (Earth).

Man almost always has a modifying influence, and without proper regulation he often has a destructive effect. For instance, by applying certain ecological principles to such fields as agriculture, biological surveys, game management, pest control, forestry, horticulture, and fishery biology, he has received tremendous economic gains. Its knowledge is found critically important for intelligent conservation whether in relation to soil, forest, wildlife, water supply or fishery resources. Further, though man has been known to control his environment successfully to meet his needs, but indiscriminate control of different pathogens or pests such as bacteria, fungi, weeds, insects, rodents, etc., through use of different chemical-poisons or pesticides (bacteriocides, fungicides, insecticides, etc.), the release of massive quantities of radioactive debris, **xenobiotics** (*i.e.*, chemical compounds synthesized by humans which are not naturally found in living organisms and cannot normally be metabolised (broken down) by them; see **Green** *et al.*, 1990), discarded chemicals and industrial wastes into rivers result in atmospheric and aquatic pollution, which may have short and long term ecological effects. Further, rapid growth of urbanization and fast rate of multiplication of human population have resulted in fatal threat of scarcity of wild life, food, open space, and of survival.

There are certain other ecological problems. Agriculture and now forestry are concentrating upon monoculture–**single species ecosystems**–in spite of the difficulties and dangers associated with unnaturally simplified ecosystems that lack a diversity of species. Over much of the world, especially in the grasslands, we continue to disrupt the energy balance through overgrazing and end up with eroded mountain sides, silt-clogged streams and lakes, and a scarcity of water. Therefore, future of human life on earth demands more knowledge about the ecosystems and other ecological problems.

REVISION QUESTIONS

1. Define the term ecology. Why it has attracted the attention of modern man ? Explain.
2. Give the detailed account of historical background of ecology.
3. Enumerate different branches of ecology and write short account about each branch of ecology.
4. Ask your nonbiological friends how they would define *ecology*, and discuss the distinction between *ecology* and *environmental studies*.

CHAPTER

2

Ecology in India

Growth of ecology in India is based on the earliest contributions (1875 to 1929) of some forest officers who provided purely descriptive accounts of forests. In fact, the first exhaustive ecological contribution was made by **Winfield Dudgeon** (1921) who published an ecological account of the Upper Gangetic Plains. He employed in his studies the concept of seasonal succession. **Saxton** (1922) elaborated this concept, however, **Misra** (1946, 1958, 1959) contradicted this view of succession. Instead of it, he concluded that the processes mentioned in Dudgeon's work might be better referred to as seasonality of communities rather than ecological succession.

Physical and biotic conditions influences the responses of the organisms.

GROWTH OF PLANT ECOLOGY

The plant communities were studied with phytosociological approach first time by **Agharkar** (1924) who studied the grasslands. **Bharucha** and co-workers (1930, 1937, 1957) at Bombay worked out phytosociology of grasslands. The study of forest vegetation was conducted by **Champion** (1929, 1935, 1937), **Troup** (1925), **Bor** (1947, 1948), etc. Ecological studies of autecological nature were performed for a variety of forest trees by **Champion** and **Pant** (1931), **Jagat Singh** (1925), **Phadnis** (1925) and **Champion** and **Griffith** (1947). Indeed, comprehensive autecological studies in India was initiated by **Prof. R. Misra** who worked out the autecology of herbaceous plants of different habitats. **Misra** and **Rao** (1948) studied the autecology of *Lindenbergia polyantha,* revealing various significant aspects regarding the distribution of this species. Further growth of Indian Ecology can be contributed to extensive investigation into phytosociology of grasslands and mangroves by **Bharucha** (1941), forests by **G.S. Puri** (1950, 1951, 1960) and deserts by **Sarup** and co-workers.

Two pioneer centres of ecological research were established in 1930's in India, first by **R. Misra** at the Department of Botany, Benaras Hindu University, and second by **F.R. Bharucha** at the Institute of Science, Bombay. From 1937 onwards, **R. Misra** did extensive work at Benaras (Varanasi) on the ecology of grasslands, wastelands, playgrounds, lakes and ponds, low-lying lands, riparian eroded slopes and ecotypic differentiation. In 1950, Prof. Misra established another Ecological Research Institute at Sagar University (M.P.). In 1956, **Misra** moved to Varanasi and developed an internationally recognized centre of Ecological Investigation. At this centre, in between 1955–1962, investigations were carried on in the diverse areas such as ecophysiology (**V.Kaul**), root-soil relationships of grasslands (**S.S. Raman**) and riparian lands (**R.S. Ambasht**), ecotypic differentiation (**P.S. Ramakrishnan**) and reproductive capacity of herbage species (**H.R. Sant** and **V.G. Nelvigi**). During 1963–1971, Benaras centre of ecology carried mainly autecological work on the medicinal plants and the weeds (**R.S. Tripathi**), grassland productivity (**J.S. Singh**) and forest litter decomposition and productivity (**K.P. Singh**).

Ecologists have conducted studies on several herbaceous plants of grasslands and waste lands.

Autecological and synecological studies of forest communities were conducted by **Sharma** (1955), **Bhatia** (1954, 1955, 1956), **Misra** and **Joshi** (1952), **Arora** (1961–1964), **Puri** (1949, 1950), **Rao** (1967), etc. **Meher-Homji** (1989) did vegetation mapping in India. **Dhani** and **Behera** (1990) did quantitative analysis of forest community of Boudh-Khondmals (Phulbani, Orissa). **Joshi** and **Behera** (1990) did qualitative analysis of vegetation from a mixed tropical forest of Orissa. Productivity studies of forests have been performed by **Misra** (1969), **Raman** (1970), **Singh** (1971), **Bandhu** (1971), **Sharma** (1972), **Farugui** (1972), and **Pandeya** *et al.*, (1967, 1969, 1971). **Rajvanshi** and **Gupta** (1980) studied decomposition of litter in a tropical and deciduous forest. **Rajvanshi** and **Gupta**, in 1985, worked on mineral cycling in a tropical deciduous *Dalbergia sissoo* Roxb., forest. **Chaturvedi** and **Singh** (1978, 1987) described dry matter dynamics and nutrient dynamics of pine forests of Central Himalaya. **Ambasht** (1988) studied biomass productivity of some tropical grazing lands and plantation forests.

Autecological and phytosociological studies on several herbaceous plants of grasslands and wastelands have been made by **Pandeya** (1953, 1960, 1964, 1968), **Tiwari** (1955), **Ramakrishnan** (1960), **Ambasht** (1963, 1964), **Tripathi** (1965, 1977), **Kaul** (1959), **Sant** (1962, 1965), etc. Autecology of some medicinal plants have been made by **Pathak** (1967), **Kaul** (1965), **Shetty** (1967), etc. **Vimal** and **Talshikar** (1983) suggested various methods of recycling of agricultural wastes. Ecological investigations of specialized habitats were also done, such as of walls (**Varshney**, 1963), of eroded river banks (**Ambasht**, 1968), of deserts (**Joshi**, 1956, **Mulay** and **Joshi**, 1964), freshwaters (**Misra**, 1946, **Das**, 1968, **Gupta**, 1968, **Jha**, 1968, **Gopal**, 1968 and **Sinha**, 1969).

Dalbergia sissoo.

Ecological studies of fungi were initiated at School of Ecology at Benaras by 1960's under the leadership of **R.Y. Roy**. His team made synecological studies on the rhizosphere mycoflora of different kinds of plants. His team mate, **R.S. Dwivedi** has also contributed to ecology of fungi active in the decay of herbaceous plants, forest litter, etc. **P.K. Khanna**, **P.D. Sharma** and **B. Rai** investigated the phenomenon of succession of air-borne fungi on decaying grasses. Some other active centres of fungal ecology were established at Madras (by **T.S. Sadasivan**), at Sagar (by **S.B. Saksena**) and at Lucknow (by **J.N. Rai**). In recent years (*i.e.,* 1972–1985), interaction of soil microfungi and earthworms has been investigated by **M.C. Dash** and his team of researchers comprising **J.B. Cragg**, **Senapati**, **Nanda**, **Behera**, **P.C. Mishra**, **H.K. Dash**, etc., at Sambalpur, Orissa.

GROWTH OF ANIMAL ECOLOGY

In comparison to plant ecology, relatively little work has so far been done for the enhancement of Indian animal ecology. Most animal ecologists of India have dealt with the ecology of regional fauna. Thus, **Baker** (1921, 1935) carried out some ecological studies on birds of India, Burma and Ceylon. **Salim Ali** had watched the Indian birds very closely and his work greatly contributed to the ethology and ecology of Indian birds. **Dharmkumar-Sinhji** (1937–1975) made extensive ecological studies on water fowls and other birds of Gujarat. **Mani** (1953–1974) worked on the ecology of some insects of Himalaya. In 1974, he also has given historical account of animal ecology of India. **Dr. Ishwar Prakash** at the Central Arid Zone Research Institute (CAZRI) has exhaustively dealt with ecology and conservation of wild life (vertebrate fauna such as reptiles, birds, and mammals including bats) of Indian hot deserts during span of more than 30 years (1958–1988). Ecology of mammals of Indian deserts were also made by **Purohit** (1968). **Choudhuri** (1989) worked on the conservation of Indian tigers.

Salim Ali —a well known ornithologist of india.

Ramalingum (1961) made ecological observations on the larvae of trematode. **Southwick** *et al.,* (1961) carried a survey of populations of Rhesus monkeys in villages, towns and temples of North India. **Ananthasubramanian** and **Ananthkrishnan** (1963) studied ecology of microarthropods of pasture soil. **Nayar** and **Prabhu** (1965) studied the soil fauna of tea plantation of Kerala. Studies of ecological nature were conducted on village dwelling langurs (primates) of West Bengal by **Oppenheimer** (1973). **Singh** *et al.,* (1961) and **Bains** and **Shukla** (1975) made autecological studies on *Chilo partellus*. Ecological studies on the population of Thar in Nilgiri, Tamil Nadu were made by **Davidar** (1976).

In recent years, ecology of earthworms, soil nematodes and certain amphibians and reptiles has been extensively worked out. **M.C. Dash** (1978–1993) working at the School of Life Sciences, Sambalpur University, Orissa, has worked on diverse ecological aspects of earthworms and soil nematodes of the tropical pasture. He also worked on the energetics of anuran larvae and turtle conservation. **Dash** and **Senapati** (1986) published an extremely useful paper entitled *"Vermitechnology : an option for organic waste management in India"*. **Dash** and **Kar** (1987, 1990) suggested methods of conservation and management of sea turtle resources at the Orissa coast. **Mohapatra** and **Dash** (1987) studied density effect on growth and metamorphosis of *Bufo stomaticus* larvae. **Rao** (1984) studied diapause induced physiological and biochemical changes in a tropical earthworm, *Octochaetoma surensis*. **B.K. Senapati** *et al.,* (1979–1985) studied various other ecological aspects of earthworms in the tropical pastures. **Sahu** and **Senapati** (1988) studied secondary production and energy utilization strategy of *r* and *k* selected earthworms from tropical pastures of Orissa.

GROWTH OF DESERT ECOLOGY

The hot arid zone occupies 2.86 lac km^2 or 8.7 per cent of the geographic area of India. It is characterized by distinct ecological features such as low and fluctuating rainfall, high solar radiation, low atmospheric humidity, strong wind regime and dominance of dunes and sandy plains amongst the landforms. Despite being the zones of scarcity and hardships, the arid zones of India exhibit a spectacular and vivid biotic diversity. Indian hot deserts have attracted the attention of several ecologists and significant contributions have been made in the field of desert ecology. **Krishnan** (1952) described geological history of Rajasthan to trace its present day conditions. **Wadia** (1960) studied the post glacial dessication of central Asia. **Lahiri** and coworkers (1961–1988) worked on the adaptations of arid zone plants to soil water deficit. **Abi Chandani** (1964) studied genesis, morphology and management of arid zone soils of India. **Sankhala** (1964) described wildlife sanctuaries of Rajasthan. **Roy** and **Sen** (1968) prepared a soil map of Rajasthan. **Mann** (1971) presented a report on operational research project on arid land management. **Seth** *et al.* (1971) studied micronutrient status of Rajasthan soil. **Gupta** and **Saxena** (1972) described potential grassland types and their geological succession in Rajasthan. **Mann** *et al.,* (1974) did a qualitative analysis of arid zones of India and analyzed the problem of spread of Indian desert. **Roonwal** (1975) described termites of Thar Desert ; in 1982, he studied field biology and morphology of some central India grasshoppers. **Shankar** *et al.,* (1976) studied effects of Khejri tree (*Prosopis cineraria)* on the productivity of range grasses growing in its vicinity. **Shankar** (1977) suggested the means of measuring desertification through plant indicators. In 1983, he studied the depleted vegetation of the desert habitats and studied the process of natural regeneration. **Saxena** (1977) described vegetation and its succession in the Indian desert. **Shankarnaryan** (1977) analyzed impact of overgrazing on grasslands. In 1985, he and **Sen** suggested the means of combating desertification. **Dhir** (1977) analyzed the problem of soil degradation due to over exploitative human efforts. **Mann** and **Saxena** (1980) released monograph entitled "*Khejri* (*Prosopis cineraria*) *in the Indian Desert—its role in agro-forestry*". **Dhabriya** (1982–84) performed drought monitoring/green biomass mapping in the arid and semi-arid parts of Rajasthan and Gujarat. He also suggested ways to check the desertification process. **Mathana** *et al.,* (1984) described root system of desert tree species. **Bhardwaj** *et al.,* (1984) reviewed ecological study of Rajasthan desert during years 1964–1984. **Bhandari** (1988) described floral wealth and plant adaptation of the Indian desert. **Mathur** and **Govil** (1988) suggested methods of greening of the Indian Desert. **Professor Ishwar Prakash** (1988) has published a research paper entitled *Wildlife, human-animal interactions and conservation in the Rajasthan desert.*

Indian deserts exhibit vivid biotic diversity.

GROWTH OF OCEANOGRAPHY AND LIMNOLOGY

The ecology of Indian seas and oceans (Indian oceanography) has been studied extensively, since it has an importance in defence. A variety of ships and vessels have been used from time to time to record data on physical, chemical and biological characteristics of Indian seas and oceans. Thus, various forms of life, meteorology, geology, and geophysics of the Indian seas and oceans have been studied. During 1960–1964, an International Indian Ocean Expedition (IIOE) was organized to study the Indian ocean in which India collaborated with other 20 countries. In 1966, the National Institute of Oceanography (NIO) was set up in 1966 with its Head Quarter at New Delhi as national laboratory under the Council of Scientific and Industrial Research (CSIR). In 1969, NIO's Head quarter was shifted to Dona Paula (Goa). NIO and its regional centres at Bombay and Cochin consist of seven divisions such as Indian Ocean Biological Centre (Cochin), Physical Oceanography, Chemical Oceanography, Geological Oceanography, Biological Oceanography, Data and Documentation and Oceanographic Instrumentation. Some noted oceanographers of India are **Jayaraman**, **Ramasastry**, **Mahadevan**, **Nair**, **Qasim**, **Gopinathan**, **Prasad**, etc.

Indian sea.

Pruthi (1933) reported seasonal changes in the physical and chemical conditions of the water in the tank in Indian Museum compound. **Philipose** (1940) studied ecology and seasonal succession in a permanent pool at Madras city. In the year 1959, **Philipose** made observations on the ecology of fresh water phytoplanktons of Inland fisheries. **Ganapati** (1941, 1943, 1947) made divergent limnological investigations on the freshwater ponds of Madras City. **Banerjea** and coworkers (1954–1970) studied limnology of various fish ponds of India. **Chacko** and **Krishanamoorthy** (1954) studied ecology of planktons of three freshwater ponds of Madras City. **Biswas** and **Calder** (1955) made floral study on certain common water and marsh plants of India and Burma. **Mitra** (1955, 1966) studied autecology of certain aquatic plants such as *Limnanthemum cristatum, L. indicum* and *Hydrilla verticillata*. **Das** and **Shrivastava** (1955–1959) studied the limnology of various freshwater ponds of Lucknow. **Das** and **Moitra** (1955) worked on the feeding ecology of some common fishes of U.P. and classified them into surface feeders, mid feeders, and bottom feeders. **Mitra** and coworkers (1956, 1965) worked on feeding ecology of the Indian major carp fry. **Alikunhi** (1956) made certain ecological observation such as fecundity on *Labeo bata.* **Singh** (1956) studied limnological relations of Indian inland waters with special reference to algal blooms. **R.S. Ambasht**, Professor of Botany, Benaras Hindu University, Varanasi, (1958–1990) and his team of researchers have published more than 120 research papers dealing with primary production; forest ecology; aquatic ecology—ecology of pond ecosystem; Ganga river ecosystem (*e.g.,* riparian ecology), grassland ecology, aquatic pollution, weed's ecology and conservation of riparian wasteland. **Das** (1959–1967) studied ecological effects of micronutrients, vitamin B_{12}, density, etc., on the Indian freshwater carps.

Hydrilla verticillata.

George (1961–1966) made ecological observations on the physico-chemical nature of water and zooplanktons and rotifers of certain shallow ponds of Delhi. **Subhramanyam** (1962) published a monograph on certain common aquatic angiosperms of India. **Saigal** *et al.,* (1962) studied ecology of *Mystus seenghala* of Ganga river system. **Bhatt** *et al.,* (1963) reported mineral contents of freshwater and freshwater organisms of Kalyan, Kolhapur and Jaduguda areas. **Michael** (1964) studied limnology of planktons and macrofauna of a freshwater pond of West Bengal. **Sreenivasan** (1964) studied hydrology of tropical impoundment, Bhavanisagar Reservoir,

Madras state for the years 1956–1961. In 1968, **Sreenivasan** undertook the studies on the limnology of two upland impoundments of Nilgiris of Madras state. **Banerjee** and **Raychoudhury** (1966) worked out the physico-chemical nature of Chilka lake of Orissa. **Krishnamurthy** (1966) made preliminary studies on the macrofauna of the Tungabhadra Reservoir. **Verma** (1967) studied diurnal variation in a fish pond in Seoni. **Vasisht** (1968) worked out ecology of a pond ecosystem of Benaras. **Moitra** and **Bhowmik** (1961) studied seasonal cycles of rotifers in a freshwater fish pond in Kalyani, West Bengal. **David** *et al.,* (1969) studied limnology and fishery of the Tungabhadra reservoir. **Nayar** (1970) studied ecology of rotifer populations of two ponds at Pilani, Rajasthan. **Khan** and **Siddiqui** (1970–1974) investigated the diurnal and seasonal variations in the limnological features of a perennial fish moat of Aligarh, Uttar Pradesh. **Kaul** (1971) studied the production and ecology of some macrophytes of Kashmir lakes. **Patnaik** (1971) worked on the seasonal abundance and distribution of benthos of Chilka lake, Orissa. **Ganapati** (1972) compared the organic production of various types of aquatic ecosystems of India. **Kant** and **Kachroo** (1975) recorded the diurnal changes in the temperature and pH of water and diurnal movement of planktons in Dal lake of Shrinagar. **Pillai** and **Sreenivasan** (1975) estimated the carbon and nitrogen status of some lakes, reservoirs and ponds of Tamil Nadu. **Gupta** (1976) studied the limnology of macrobenthic fauna of Loni reservoir of Rewa of Madhya Pradesh. **Sreenivasan** (1977) studied the limnology and fisheries of Tirmoorthy reservoir, Tamil Nadu. **Rao** (1977) investigated the ecology of certain phytoplanktons such as diatoms, Euglenineae and Myxophyceae of three fresh water ponds of Hyderabad.

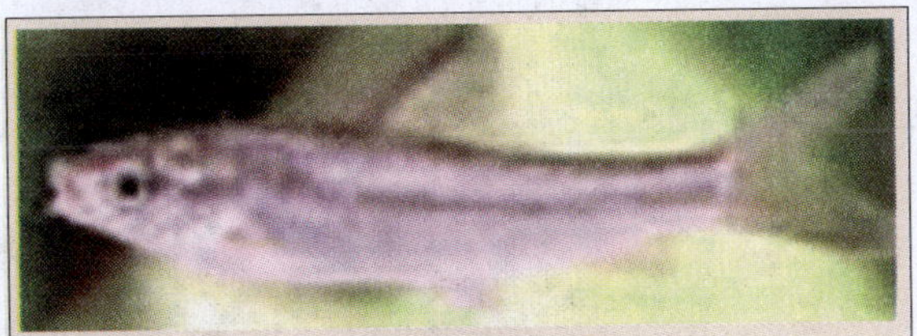

Labeo bata.

GROWTH OF POLLUTION BIOLOGY

C. K. Varshney at Delhi worked on pollution ecology with special reference to SO_2 pollution. **D.N. Rao** (1979) used plant leaf as a pollution monitoring device. **L. K. Dadhick** (1982) described certain indicator plants which may be useful in the recognition and monitoring of air pollutants. **Ambasht** *et al.,* (1979) made an assessment of effluents of a chemical and fertilizer factory for irrigation of agricultural lands. **Ambasht** (1982) wrote an essay entitled "*Responses of aquatic plants to pollution.*" **V.Venkateswarlu** (1982) studied the role of algae as indicators of river water quality and pollution. **Mahajan** (1982) described zooplankton as indicators for assessment of water pollution. **Kar** *et al.,* (1987) studied the problem of pollution in river Ib and described plankton population and primary productivity. **Palharya** and **Malvia** (1988) studied pollution of the Narmada river at Hoshangabad in Madhya Pradesh and suggested measures for its control. **Tiwari** and **Ali** (1988) prepared a water quality index for Indian rivers. **Shardendu** and **Ambasht** (1988) studied impact of urban waste waters on primary production and nutrient status of *Hydrilla verticillata*. **Ambasht** *et al.,* (1984) performed ecological investigation of conservation nature on the plant biomass and energy used for the burning of human dead bodies at Varanasi Ghat.

REVISION QUESTIONS

1. Describe the historical growth of plant ecology in India.
2. Write an essay on 'growth of animal ecology in India'.
3. Trace the historical growth of oceanography in India.
4. Describe the history of limnology in India.
5. Describe the historical growth of desert ecology of India.

CHAPTER

3

Environment

Each and every living organism has its specific surrounding, medium of environment to which it continuously interacts and remains fully adapted. The **environment** is the sum total of physical and biotic conditions influencing the responses of the organisms (**S.C. Kendeigh**, 1974). The life supporting environment of planet earth—the **biosphere** is composed of following three chief media—**air**, **water** and **soil**, which are the components of three major sub-divisions of the biosphere—**atmosphere**,

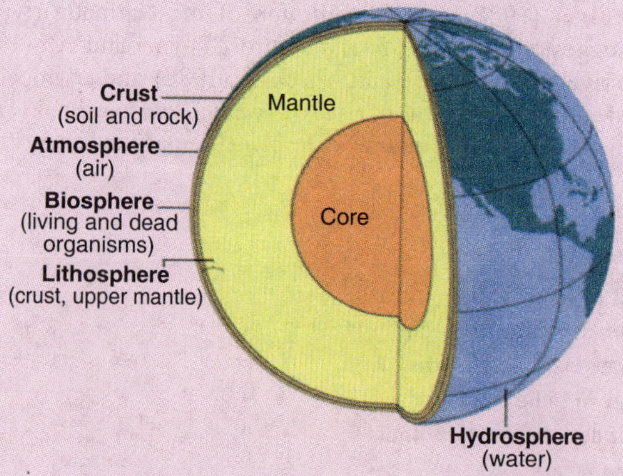

Our life-support system: the general structure of the earth.

Physical and biotic conditions influences the responses of the organisms.

hydrosphere and **lithosphere**, respectively. These media are not completely isolated from each other, however, some of the atmospheric gases are dissolved in all natural waters, and some moisture is present almost everywhere in the atmosphere. Each of these media can be discussed separately in the following manner :

A. ATMOSPHERE (AIR)

The multilayered gaseous envelope surrounding the planet earth is called **atmosphere**. The atmosphere remains in contact with all the major types of environment of earth, interacting with them and greatly affecting their ability to support life. It filters sunlight reaching the earth, affect climate, and is a reservoir of several elements essential for life.

Various Zones of Atmosphere

The atmosphere is divided into five distinct layers or zones : troposphere, stratosphere, mesosphere, ionosphere and exosphere. The **tropopause** separates the stratosphere from the troposphere and **stratopause** separates mesosphere from stratosphere.

1. Troposphere. It is the lowest region of atmosphere which subjects to differential heating, temperature inversions and convection currents and which extends from the surface of the earth up to a height of 8 to 10 km at polar latitudes (poles), 10 to 12 km at moderate latitudes and 16 to 20 km at the equator. For the organisms, troposphere forms a most important zone of atmosphere. Many important climatic events such as cloud formation, lightening, thundering, thunder storm formation, etc., all take place in troposphere. In this zone the percentage concentration of different gases in air does not vary with an increase in height. But the water vapour content in air depends upon the weather (*e.g.,* part of the troposphere over an ocean carries more moisture than that over a land surface) and it decreases sharply with an increase in height as does the air temperature. Air temperature in this zone gradually decreases with height at the rate of about 6.5°C per km (more specifically 5°C per km (more specifically 5°C per km in the lower troposphere and 7°C per km in the upper troposphere). In fact, towards the upper layers of troposphere, the temperature may decrease up to – 60°C. Upper region of the troposphere has a narrow boundary called the **tropopause** which has a constant temperature.

Troposphere.

Further, the non-uniform heating of the ground surface by sun's radiations produces ascending and descending air currents, which cause turbulence and mixing of air masses vertically. Moreover, the average air pressure at the earth's surface is 1,014 millibars (*viz.,* 1 millibar= 1/1000 bar; 1 bar = 1.019 kg per cm^2 close to 1 atmosphere, 1.332 millibars=1mm of mercury (Hg). At an altitude of 5 km, the air pressure is half that at the surface; at 11 km it is 225 millibars and at 17 km it is only 90 millibars.

2. Stratosphere. Next to troposphere is the second zone, called **stratosphere**, which is about 30 km in height. This zone is free from clouds and aeroplanes usually fly in its lower zone. The temperature of stratosphere increases up to 90°C and such an increase in temperature is due to ozone formation under the influence of ultraviolet component of sunlight. Such a layer of ozone is called **ozonosphere**. In ozonosphere, the sunlight ionize oxygen to ozone by photochemical dissociation. The ozonosphere completely absorbs solar radiation, ultraviolet radiation from the sun and also a lot of the solar infra-red, thus, becoming warmer than adjacent layers above or below. There is a serious threat to this ozone layer now due to the harmful effects of gaseous pollutants. A big hole (*i.e.,* thinning of the O_3 layer) has occurred in it, above the antarctic region. Upper layers of stratosphere form **stratopause**.

3. Mesosphere. Stratosphere is followed by next zone called **mesosphere** which is 40 km in height. In mesosphere, temperature shows again a decrease up to –80°C. Upper layers of this zone form the **mesopause.**

4. Ionosphere. The remaining part of atmosphere above the mesosphere, up to the height of about 300 km above earth's surface, is called **ionosphere**. Ionosphere contains several layers of ionized air. Thus, most of the gaseous components which become ionized under the influence of radiant energy, remain as ions. Ionosphere reflects short radio waves, making telecommunication possible over long distances.

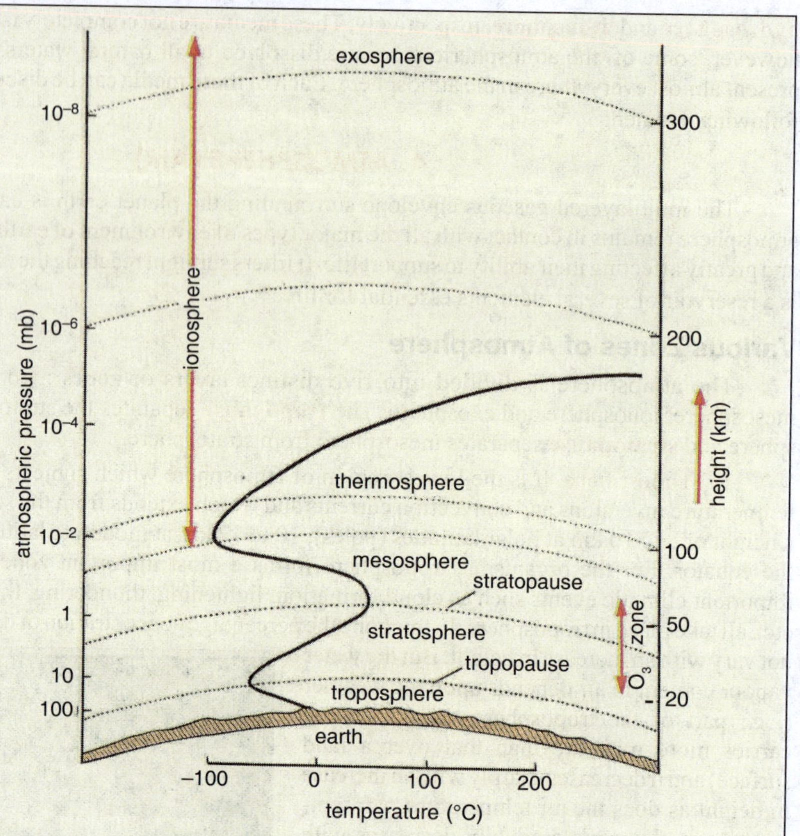

Fig. 3.1. Diagrammatic sketch showing the principal zones of atmosphere and variations in temperature and pressure in them.

5. Exosphere. The exosphere is the outer fringe of earth's atmosphere and outer space begins after it. The air density is very low in this zone; hydrogen being dominant element of it.

Air

The gaseous mixture of troposphere, is utilized by most organisms in respiration to liberate energy from food during oxidation and is called air. In atmosphere, about 95 per cent of the total air is present up to the height of about 20 km above earth's surface and the remaining 5 per cent in the rest, of about 280 km height. In the gaseous mantle, there is found a mixture of gases in different proportions (Table 3-1). Of these various gases nitrogen and oxygen are the major components of air.

Table 3-1. Composition of atmospheric air.

Component	% by volume	% by weight
Nitrogen (N_2)	78.09	75.54
Oxygen (O_2)	20.93	23.14
Argon (A)	0.93	1.27
Carbon dioxide (CO_2)	0.032	0.46
Miscellaneous	0.02	0.02

Miscellaneous component includes traces of hydrogen, ozone, radon, helium, neon, krypton, xenon, sulphur dioxide, hydrogen sulphide, ammonia, methane, etc.. Besides these gases, air may carry suspensions of liquids such as water in clouds and solids such as dust from the ground (soil), smoke from fires or salt from ocean spray. Air also contains microorganisms (viruses, bacteria, etc.), pollen grains and fungal spores, all forming biological constituents of the atmosphere.

In general, the composition of these gases in air and in other media on earth, such as water or soil, is in equilibrium with the atmosphere. In special ecological habitats such as the anaerobic regions of deep lakes or sand-mud flats, however, the composition of gases of air is altered dramatically. Slight differences occur in the atmosphere at different latitudes and, at places where gases are entering or leaving the atmosphere, such as volcanoes, fires, smelters, cities, metropolitan areas, and vegetation.

Physiologic-ecologic inter-relationships of gases and animals. The atmospheric composition is not always the most important factor to organism; rather, it is the partial pressure of a gas, especially oxygen and CO_2, which influences the existence of life, as illustrated by altitudinal studies on organisms. The **partial pressure** of a gas is the product of the total barometric pressure times the concentration of gas in dry air.

The solubility characteristics of oxygen and carbon dioxide in water are different, for carbon dioxide is about 200 times more soluble than oxygen. Both temperature and salinity have evident effects on the solubility of these two gases; with either increased temperature or increased salinity, there is a decrease in their solubility in water as has been shown in table 3-2.

Table 3-2. Coefficients of saturation of CO_2 in water at different temperatures and salinity (ml/litre in equilibrium with 760 mm Hg).

Temperature, °C	Salinity, %		
	0	28.91	36.11
0	1715	1489	1438
12	1118	980	947
24	782	695	677

In the process of photosynthesis, green plants use CO_2 and release O_2 during the day time; oxygen is used in respiration by all the organisms all the time. Photosynthesis, thus, regulates the oxygen and carbon dioxide balance in nature. Nitrogen which is abundantly present in the atmosphere is not used directly by plants or animals except by some bacteria and blue green algae. If gases are to be functionally important to an organism, there must be a mechanism whereby gases can enter and leave the body. Apparently the passage of oxygen across a membrane is accomplished by **diffusion**. The rate of diffusion can be calculated by **Fick's Law**, which is based on concentration coefficient, amount of surface thickness of the membrane, and time, and is expressed as the diffusion coefficient (**Davson**, 1959). The diffusion coefficient value is not the same for different animal tissues. For example, some representative values of diffusion coefficient of O_2 are : muscle, 0.000014; connective tissue, 0.000011; water, 0.000034; air, 11.0;

Green plants regulate the oxygen and carbon dioxide balance in the atmosphere by the process of photosynthesis.

chitin, 0.000013. Not only is the diffusion rate different in various tissues, but the behaviour of each gas also differs. Carbon dioxide, for example, diffuses through water and animal tissues 25 times faster than oxygen.

Air as a medium for living organism. Air is not an easy and suitable medium to support life (biota) and actually no organism ever originated in air, though, certain aquatic and terrestrial organisms have become secondarily adapted for aerial existence. Air has so much less **buoyancy** than water that organisms emerging from water or land are immediately subjected to a stronger pull of **gravity** which holds them to the earth. They are also exposed to **evaporation** of water from their bodies, which threatens loss of water from the protoplasm and death by desiccation. Changes in **temperature** are much more drastic in air and there is great danger of chilling or overheating. Further, periods of light exposure are much longer and much more intense in air than in water or upland. The problem of **mineral supply** becomes acute in the air.

Draco volans.

Biota of the air. Only a few microorganisms, plants and animals have invaded the air. The most common examples of aerially adapted animals are insects, flying fish (*Exocoetus volitans*), flying frog, flying lizard (*Draco volans*), flying phalangers (*Petaurus*), bats (*Eptesicus*), and birds (For volant adaptation see Chapter 18).

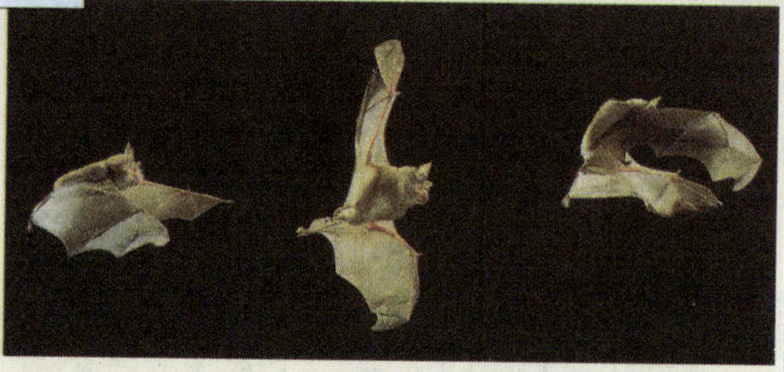

Bats.

B. HYDROSPHERE (WATER)

The oceans, rivers, streams, lakes, ponds, pools, polar ice caps, water vapour, etc., form the **hydrosphere**. About three-fourth of the earth's surface (75%) is covered with hydrosphere, the main component of which is **water**. Water is one of the most unusual natural compounds found on earth, and it is also one of the most important. The water remains in solid (snow), liquid (water) and gaseous (water vapour) forms. Life on earth began in the seas, and water in some form or the other is absolutely essential for the maintenance of all life. Water is one of the main agent in pedogenesis (soil formation) and is the medium for several different ecosystems. It permeates the atmosphere and the outer layers, of lithosphere and has uneven distribution on earth, so that, some of the great ocean depths are approximately six or seven miles (9750 meters). Further water in its two forms, salt water and freshwater, forms two chief aquatic environments—namely **marine environment** and **fresh water environment** of earth. The oceans holding marine environment are two and one half times more extensive than land and provide over 300 times the living space, since they are habitable throughout their entire depth by certain groups of organisms. Water is obviously heavier than air which imparts a greater buoyancy to the aquatic medium enabling organisms to float at variable levels. The most unique features of water concern its physical properties.

Physical Properties of Water

Henderson (1913, 1924) listed in his stimulating book, *The Fitness of the Environment,* the following characteristics of water which are favourable to biological systems :

1. A tremendous quantity of water exists on earth in three forms : gaseous, liquid, and solid. Not only is 75 per cent of the earth's surface covered by water but the atmosphere also contains an abundance of aqueous vapour, and the polar region is ice-covered. Moreover, at least 60 per cent of active protoplasm is water.

2. Water is an extremely inert body in relation to most other chemical substances.

3. It has unique thermal properties such as heat capacity, latent heat and higher freezing point.

(a) Heat capacity. Water has high heat capacity and it can withhold large amounts of heat. Because of the high heat capacity of water, oceans and lakes tend to maintain a relatively constant temperature, and therefore, the temperature of the biosphere is relatively stable. This property of water is functionally important to animal life. For example, a 165 lb (about 75 kg) man at rest produces sufficient heat to raise his body temperature more than 32°C, but if the heat capacity of his body were to correspond to that of many other substances, this amount of heat production would raise his temperature 100–150°C.

(b) Latent heat of melting and evaporation. The latent heat of melting is the number of calories required to convert 1 gram of solid at the freezing point into 1 gram of liquid at the same temperature. This value is about 90 for water, and means that the amount of heat necessary to melt ice is the same as that required to raise the temperature of the resulting ice water to 80°C. The latent heat of evaporation is defined as the number of calories required to change 1 gram of liquid into vapour. For water this value is 536; therefore, as much heat is required to boil away 1 gram of water as to raise the temperature of 536 grams through 1°C. These properties of water are important not only because they moderate the temperature of the biosphere, but also because they play a basic role in the evaporation of water and its precipitation as rain and as dew in the **hydrological (water) cycle**.

(c) Thermal conductivity. Although water is a poor thermal conductor compared to metals, among the common liquids it is excellent : for example, the conductivity value for silver is 1.10, for water, 0.0125, for alcohol, 0.00048, and for benzene, 0.00033.

(d) Expansion before freezing. The relationship between temperature and density, or mass per unit volume, is very unusual. When water is cooled from room temperature, it contracts, becoming denser until it reaches a maximum density at 3.94°C. If cooled further, it begins to expand again. At the freezing point (0°C) it expands markedly, unlike almost all other substances. Thus, ice will always float on the top of a lake or stream, and it is very unusual for an aquatic ecosystem ever to freeze solid, unless it is very small.

4. No other compound compares to water as a **solvent**. So many different substances can be dissolved in it that it is known as the **universal solvent**. More things, in fact, can be dissolved in water than in any other liquid. This is especially true for inorganic chemicals which split, or dissociate to form electrically charged entities termed **ions**. Ionization influences most electrical phenomena and many chemical phenomena of solutions. It is probable that all natural elements are soluble in water, at least in trace amounts, and that they are all found in natural water at some place or other on the earth's surface. In addition, many organic chemicals are water-soluble. Thus, water is the main medium by which chemical constituents are transported from one part of an ecosystem to the other. It is the only medium by which these constituents can pass from the abiotic portion of the ecosystem into the living portion. Even in the driest of terrestrial environments, nutrient materials pass into the roots of plants in aqueous solution ; when air is breathed by animals, oxygen is dissolved in water at the surface of the lung before it can cross the mucous membrane and be absorbed by the blood.

5. Water has the greatest **surface tension** of all common liquids, except mercury. The role of surface tension is most obvious in the way it allows certain things, such as pollen, dust, and water striders, to remain at the surface of a water body even though they are denser than the water. More important, however, the high surface tension of water allows soils to contain a significant amount of water through capillary attraction and to make it available to terrestrial plants.

Visualizing these unique physical properties of water, **Henderson** (1913) concluded that no other known substance could be substituted for water as a basic abiotic environmental factor. Now, ecologists have known certain other important physical properties of water which affect life in someway. Some of the additional important physico-chemical properties of water are the following :

(i) Viscosity. Water is a fairly viscous liquid. Animals that live and move in water need to be much more streamlined than those that move through air, because the resistance to motion in a viscous medium is high. But at the same time, the viscosity of water allows organisms to swim using relatively simple movements. Further, high viscosity of water protects the aquatic animals and plants from the mechanical disturbances.

(ii) Buoyancy. Water is buoyant medium. Organisms can exist in it without specialized supportive structures such as those that are needed by organisms that inhabit terrestrial environments.

(iii) Transparency. Water is transparent medium. Its transparency enables the penetration of light to the depths where it is ultimately absorbed. Different wavelengths are absorbed at different depths. The long heat waves are stopped near the surface. Shorter waves with more energy penetrate successively farther. The ultraviolet rays penetrate beyond 100 meters. The zone up to which light rays penetrate is called as **photic zone** and below this zone there is complete darkness and organisms that require light cannot live. The transparency of water is greatly affected by the presence of suspended particles, phytoplankton, etc., which absorb light and so penetration of light in turbid water is less.

(iv) Pressure. Organisms living at sea level experience a pressure of about 15 psi, which is defined as 1 atm (760 mm of Hg). Pressure increases with increased depth of water at the rate of one atmosphere (1 atm) for every 10 meters of descent. Organisms inhabiting the floor of deep-sea areas at depth of 10,500 meters are exposed to hydrostatic pressure of about one ton per square centimeter. Pressure influences solubility, ionic dissociation, and surface tension, and water is slightly compressible with increased pressure.

(v) Salinity. Salinity has been defined as "the total amount of solid material in grams contained in one kilogram of the water, when all the carbonate has been converted into oxide, bromine and iodine replaced by chlorine and all organic matter completely oxidized." All types of natural waters contain various amounts of different salts (ions) such as Na, K, Mg, Cl, SO_4, PO_4, CO_3, HCO_3, NO_3, etc., and all these salts are responsible for the **saltiness**, **salinity** or **salt content** of water (Table 3-3). The salinity of marine water is rather constant being about 3.5%. The salinity of fresh water varies greatly. Some salt lakes may have a salinity of 25% to 30% which greatly restricts life in them.

Table 3-3. Comparison of some of the principal ions found in different kinds of water.

Water	Na	K	Ca	Mg	Cl	SO_4	CO_3	Total per litre
1. Soft water	0.016	–	0.01	0.0005	0.019	0.007	0.012	0.065
2. Hard water	0.021	0.016	0.62	0.014	0.041	0.025	0.119	0.301
3. Sea	10.56	0.30	0.40	1.27	18.98	2.65	0.71	34.85
4. Great salt lake	65.54	3.76	0.065	4.47	110.08	13.04	–	197.51

Salinity of water acts as an important limiting factor for the distribution of a number of species of plants and animals. Certain animals such as, spider crab, *Maia*, etc., can tolerate only narrow fluctuations in salinity of water and are known as **stenohaline** animals. While some animals such as *Mytilus, Aplysia*, etc., can withstand wider ranges of salinity and are called **euryhaline** animals. However, there are certain animals, such as, *Anguilla, Salmon,* etc., which are both stenohaline and euryhaline.

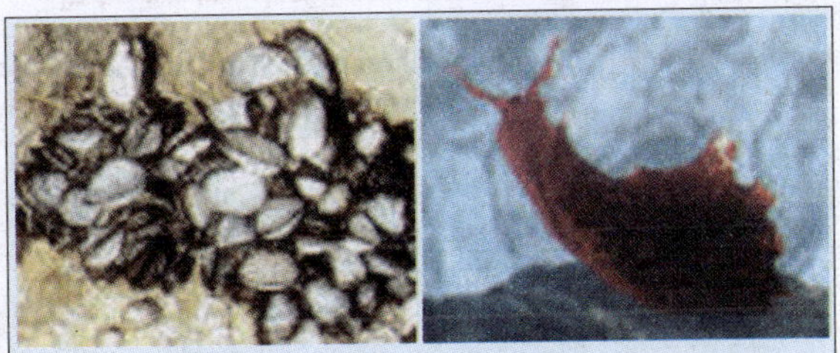

Mytilus. *Aplysia.*

Chemical Properties of Water

Water consists primarily of a single compound, H_2O. It is a universal solvent and most chemical compounds ionize readily in water and provide many radicals and considerable versatility in the rearrangement of chemical substances. It has following chemical properties :

Fishes use oxygen dissolved in water for respiration.

(a) Solubility of gases in water. Most gases dissolve readily in water, most notably those that are essential for life. The concentration of any gas in water generally varies between zero and a theoretical maximum or **saturation** (Table 3-4). The latter is the amount of gas that can be dissolved in water when the atmosphere and the water are in equilibrium with one another. Except for water falls and very turbulent streams, the water in natural ecosystem is seldom in equilibrium with the atmosphere. A gas may show a **deficit**, if it is being utilized in the ecosystem faster than it is going into solution across the air water interface, or it may be **supersaturated**, if it is being produced in the ecosystem faster than it is being released from solution across the interface. The concentration of important gases may vary widely in any ecosystem, both horizontally and vertically.

The saturation level of any gas in water depends on several variables, most notably temperature, salinity, (see Table 3-2) the concentration of the gas in the atmosphere, and its relative solubility in water. The greater the concentration of a gas in the atmosphere, the greater its concentration in the water will tend to be, depending on its relative solubility in water.

1. Oxygen. One of the most critical factor in an aquatic environment is the amount of oxygen in the water because most living organisms (excepting anaerobic forms) require this gas for respiration. In contrast to atmosphere, the oxygen becomes limiting factor for aquatic animals as the saturation concentration of oxygen in water is governed by temperature and salinity. As is evident in table 3-5, the lower the temperature, the greater the oxygen retaining capacity of water, whether it is freshwater or sea water.

Table 3-4. Comparison between equilibrium concentrations of important gases in the atmosphere and in water (Clapham, Jr. 1973).

Gas	Atmospheric concentration	Saturation in water
Oxygen	210 cc./1. (21%)	7 cc./1. (32.9%)
Nitrogen	780 cc./1. (78%)	14 cc./1 (65.7%)
Carbon dioxide	0.3 cc./1. (0.03%)	0.3 cc./1. (1.4%)

Table 3-5. **Comparison of the saturation concentration of oxygen in freshwater and salt water environment with varying temperature.**

Water	Temperature, °C	Saturation point in millilitre per litre
Freshwater	0	10.27
Saltwater	0	8.08
Freshwater	30	5.57
Saltwater	30	4.52

2. Nitrogen. Nitrogen is significantly less soluble in water than oxygen. But because it constitutes 78 per cent of the atmosphere, it still accounts for about 65 per cent of the dissolved gases at equilibrium. It is fairly inert chemically and does not react with water, although some bacteria, fungi, blue-green algae, and so on, can use it to satisfy their nitrogen requirements; and other bacteria can produce it through reduction of nitrate under conditions of very low oxygen concentration.

3. Carbon dioxide. The decomposition of organic matter and the respiratory activity of aquatic plants and animals produce carbon dioxide. This gas is one of the essential raw materials necessary for photosynthetic activity by green plants. Carbon dioxide combines chemically with water to produce carbonic acid (H_2CO_3), which influences the hydrogen ion concentration (pH) of water. Carbonic acid dissociates to produce hydrogen (H^+) and bicarbonate (HCO_3^-) ions. The bicarbonate radical may undergo further dissociation forming more hydrogen (H^+) and carbonate ($CO_3^=$), as represented in the following reactions :

$$\underset{\text{atmospheric}}{CO_2 + H_2O} \underset{\text{interface}}{\overset{\text{air-water}}{\rightleftharpoons}} \underset{\text{dissolved}}{CO_2 + H_2O} \rightleftharpoons \underset{\text{carbonic acid}}{H_2CO_3} \rightleftharpoons \underset{\text{bicarbonate}}{H^+ + HCO_3^-}$$

$$\underset{\text{bicarbonate}}{HCO_3^-} \rightleftharpoons 2H^+ + \underset{\text{carbonate}}{CO_3^=}$$

The amount of free or uncombined carbon dioxide in water is of ecological importance : it governs the precipitation of calcium in the form of calcium carbonate ($CaCO_3$). Calcium precipitates when temperature and salinity are high and the amount of uncombined carbon dioxide is low. This means more carbonate ($CO_3^=$) is present to combine with the calcium cation (Ca^{++}). These conditions exist in shallow tropical waters, where evaporation is high. This raises the salinity and photosynthetic activity of plants and reduces the quantity of free carbon dioxide in water. The precipitation of calcium carbonate in tropical areas as the Bahamas explains the preponderance of thick calcareous shells of shallow water tropical molluscs, plankton and algae. In deep oceanic water, temperature is low and there are no photosynthetic plants, consequently, the carbon dioxide content of the water is high. Deep water fauna (molluscs, crustaceans) possess very fragile skeleton because the precipitation of calcium carbonate is minimum.

4. Hydrogen sulphide. The deeper layers of many bodies of water, including ponds, lakes, and some estuaries, may contain significant amounts of the toxic gas, hydrogen sulphide, which is released by decaying organic matter. If concentration of the gas builds up, all life but anaerobic bacteria excluded from the area (*e.g.,* deeper strata of Black sea).

Effect of Factor of Aquatic Environment on Aquatic Organisms

The aquatic environment is subject to water movements ranging from small vertical circulations to strong currents. The streams have a unidirectional movement and in seas the movement is reversible. Many aquatic animals have accordingly taken to sedentary or sessile lives depending on water movements. Radial symmetry is a characteristic of such animals. Transformation from a sessile to a locomotive existence favours a bilateral symmetry. The water currents of water often abrade (= rub off) the inhabiting flora and fauna and varied modifications are encountered to withstand this abrasive action. Thick scales, strong shells and many attachment devices such as the holdfasts and suckers all are the results of this environmental stress. The ability to breathe air dissolved in water, at times even resorting to anaerobic existence, the modification of various senses to respond to stimuli characteristic of aquatic environments, the phenomenon of osmoregulation, and above all the phenomenon of external fertilization are other remarkable physiological adaptations to live in an exclusively aquatic medium.

Water and Ecological Adaptations

Water makes up a large proportion of the bodies of plants and animals, whether they live on land or in water. Active cytoplasm holds about 70–90 per cent of water. It has several important physiological properties. There exists a strong relationship between the water status of soil, plant and atmosphere. The rooting zone of the soil (zone of soil in which the water absorbing organs, roots, root hairs are present), the plant body and the lower layer of atmosphere behave as a continuum, called **spac** or **soil plant atmosphere continuum** in relation to water transfer (**Phillip**, 1966). Solar radiation is the primary energy source for the water transport process in the SPAC. On the other hand, animals obtain water (i) by drinking (ii) by absorbing it through their skin from contact with some damp ground, (iii) directly from their food or (iv) from water produced by metabolism. The method of obtaining water and the relation to the supply of liquid water as well as resistance to the drying effects of the surrounding atmosphere are important in determining the distribution of animals.

The scarcity or abundance of water brings about adaptations in living organisms. Plants which grow in areas where water is available in plenty, are classified as **mesophytes** and terrestrial animals living under such conditions are called **mesocoles**. Plants growing in water are called **hydrophytes**, while animals that live in the aquatic environment are called **aquatic animals** or **hydrocoles**. Some plants can grow in ecosystems where water is scarce and where the day temperature is very high. These plants are called **xerophytes** and the animals living in such xeric conditions are called **desert animals** or **xerocoles**. Xerophytes living in physiologically dry soils, *i.e.*, saline soils with high concentrations of salts such as NaCl, $MgCl_2$ and $MgSO_4$, are called **halophytes**. Based on their specific habitat, halophytes can be further classified into **lithophilous**, **psammophilous**, **pelophilous** and **helophilous** plants growing on rock and stones, sand, mud and swamp, respectively. Helophilous helophytes include mangroves of sea shores of Bombay such as *Rhizophora mucronata* and *Sonneratia*.

1. Hydrophytes and hydrocoles and their adaptations. Hydrophytes include : (a) **free-floating hydrophytes** (*e.g., Wolffia, Lemna, Spirodella, Azolla, Eichhornia crassipes, Salvinia* and *Pistia*), (b) **rooted hydrophytes with floating leaves** (*e.g., Trapa, Nelumbo, Nymphaea, Marsilea,* etc.), (c) **Submerged floating hydrophytes** (*e.g., Ceratophyllm, Utricularia, Najas,* etc.), (d) **rooted submerged hydrophytes** (*e.g., Hydrilla, Chara, Vallisneria,* etc.), (e) **rooted emergent hydrophytes** or amphibious plants (*e.g., Sagittaria, Ranunculus,* etc.). **Tenagophytes** are amphibious plants—they grow in water bodies as well as in water logged soil.

Vallisneria.

These hydrophytes grow in hydric conditions and show the following general adaptive features: They possess poor mechanical, absorbing, conductive and protective tissues. They also contain an extensive development of air spaces (aerenchyma) in the tissues. Roots are either absent (*e.g., Wolffia*) or poorly developed (*e.g.,*

Hydrilla). Roots may not have root hairs, root cap (instead of root cap *Eichhornia* has root pockets) and vascular tissue. Roots of hydrophytes are generally fibrous and adventitious, when present. The stem of hydrophytes is weak, slender and spongy. In some it is like a horizontal rhizome covered with mucilage, while it may be hard, as in *Nelumbo*. The aerial leaves may be broad but the submerged leaves are thin, long or ribbon-shaped. Stomata are completely absent in submerged leaves (*e.g.*, *Anacharis*), but in floating forms, stomata are confined only to the upper surfaces of leaves as in *Nymphaea*.

Nelumbo.

Nymphaea.

Aquatic animals or hydrocoles in general exhibit an elongated stream-lined body having a compressed head, body and tail. Hydrocoles include fishes, sea turtles, mammals such as whales, and many others. There are also amphibious forms such as frog, toad, crocodile, etc., and many birds which visit water bodies either for reproduction or for collection of food. (For further details of aquatic adaptations in animals see Chapter 18).

2. Xerophytes and xerocoles and their adaptations. Xerophytes grow in conditions of water scarcity, high temperature, strong winds, high transpiration rate and evaporation higher than precipitation. The soil is very dry and porous. The essential adaptations of xerophytes involve increased water absorption by roots, storing of water and retardation of transpiration. Thus, in search of water, xerophytic trees may go very deep in the soil and have extensive root hairs to absorb it. The roots of plants such as *Calotropis procera, Ficus,* and *Acacia nilotica* may go as deep as 10 to 16 metres and may reach the water table. As a consequence, these plants survive in deserts or arid conditions even if their rate of transpiration is higher. The storage of water is facilitated either by modifications of leaves, as in *Mesembryanthemum* and in the malacophyllous xerophytes (in which leaves contain turgescent parenchymatous cells) as *Aloe, Begonia, Bryophyllum, Agave, Yucca,* etc., or modification of stems, as in cacti such as *Opuntia* (phylloclade) and *Euphorbia.* In some xerophytes the water is stored in their roots as in *Asparagus* and *Ceiba parvifolia*. All these xerophytes are called **succulents** because they possess thick, fleshy, water storage organs such as stems, leaves and roots. Non-succulent xerophytes such as *Calotropis*, *Prosopis, Acacia, Zizyphus, Casuarina, Nerium, Saccharum* and *Pinus* possess other sort of xerophytic adaptations, *viz.*, extensive root system, high osmotic pressure and other modifications in the leaves. Reduced transpiration is achieved by decreasing the leaf surface, as in *Casurina, Acacia* and *Asparagus* or by modifying the leaves into spines and barbed bristles, as in cacti, or by having thick, leathery, thick cuticle or wax-coating bearing leaves with well-developed hypoderma and sunken stomata to reduce transpiration, as in *Calotropis* and *Nerium*. Halophytes (*e.g.*,

Nerium.

mangroves) resemble xerophytes and have high osmotic pressure; succulent organs; thin, evergreen, small leathery leaves with water storing tissues and thick cuticles and special air-breathing roots called **pneumatophores**.

Different animals have evolved the following adaptive features to live in arid environment :

1. Nocturnal life style. Most desert animals are nocturnal and seek shade or burrow deep in the soil in the day time to avoid excessive heat and dryness. Some xerocole rodents passively lose heat through conduction by pressing their bodies against the burrow walls.

2. Deceptive colouration. Desert animals are usually grey, brown or red matching with the colour of the sand or rock.

3. Suspended animation. Certain animals, usually with simpler organization, such as rotifers, nematodes, tardigrades, desert snails, etc., retain their vitality in long dry environment. Other forms (frogs, toads, etc.) **aestivate** during droughts and are active during moist season of the year.

4. Fast movement. Desert animals move much faster than other land animals, since they have to travel long distances in search of food and water.

5. Migration. Many birds and mammals of arid zones migrate when water becomes scarce or as a result of drought or for other reasons, the food supply is less.

6. Heat loss by radiation. Animals such as jack-rabbits (*Lepus*) and fox (*Vulpes velox*) have large ears that reduce the need of water evaporation to regulate the body temperature. Their ears function as efficient radiators to the cooler desert sky, which on clear days may have a radiation temperature 25°C below than that of the animal body. By seeking shade and sitting in depressions, *Lepus* could radiate 5 kcal/day through its two large ears (400 cm^2).

7. Impervious skin. The drier habitats (deserts, etc., are invaded by only those animals which contain a thick impermeable body covering. Such integuments occur in many insects, birds and mammals. Some mammals such as men, apes and horses lose much water (and salt) through sweat glands in heat regulation. Most rodents and some ruminants such as antelope nearly or completely lack sweat glands. Moist skinned forms (most amphibians and earthworms), certain mites and soft-bodied insects are restricted to swamps, stream margins, moist soils and other similarly damp places.

This nocturnal kangaroo rat in a California desert is a master of water conservation. Instead of drinking water it gets the water it needs from its food and cellular respiration. It also conserves water by excreting dry faeces and thick nearly solid urine.

8. Upturned nostrils. Desert animals have nostrils directed upwards; this may provide a protection from clogging by wind blown sand.

9. Water from food and from metabolism. Most herbivores and carnivores live on the moisture obtained with food. Many insects utilize the high water contents of plants to meet their water requirements. In fact, most animals make use of water released during metabolism when fats and carbohydrates are broken.

Kangaroo rat, which seals its burrow by day to keep its chamber moist, can live throughout year without drinking water. It obtains its water from its own metabolic processes and from hygroscopic water in its food.

10. Internal lungs or tracheal system. Mode of respiration has some correlation with water. Crustaceans, with their gills covered by a water-retaining carapace, carry with them a liquid environment for their gills. The scaly body covering of a fish may be practically impermeable to water and exchange of gases may be limited to gills and gut. Internal lungs, whether in pulmonate snails, land isopods, spiders or higher vertebrates (amphibians, reptiles, birds and mammals) together with the internal tracheal system of insects are water saving. Much water is lost in breathing even in animals having internal lungs.

11. Dry excretion. A further water-saving device is the excretion of concentrated, relatively dry nitrogenous and faecal waste materials. Water-saving insects, reptiles and birds excrete nitrogenous waste as solid uric acid.

Oryx–a drought-resistant animal.

Camel and African antelope, *Oryx* provide good examples of xerocoles or drought-resistant animals. Camel can go without food and water as long as 10 days at a stretch. For water conservation, camel has following adaptations : (1) When camel drinks water, it can take in up to 50 litres in one gulp and this water is evenly distributed all over its body in tissues and not in specific pockets or organs (*e.g.,* in stomach) as the common misconception holds. (2) Camel excretes highly concentrated urine. Its dung also contains very little water, compared to the dung of the donkey or the cow (both of which are xerocoles). (3) It perspires very little and its breathing rhythm is very slow. (4) Camel can withstand dehydration up to 25 per cent of body weight and it loses water from body tissues rather than from blood. (5) Body temperature of camel is labile, dropping to 33.8°C overnight and raising to 40.6°C by day, at which point it begins to sweat. Due to such thermoregulation, the amount of water loss by perspiration and in other ways is greatly reduced. (6) Camel accumulates its fat in the hump rather than all over body. This speeds heat flow away from the body and its thick coat prevents the flow of heat inward towards the body.

Snow as Habitat

Snow which is the solid state of liquid water, is an integral factor in the physical environment. It is important particularly in the higher latitudes and on the higher parts of mountains where snow cover may last throughout the year. Its significance is increased by the large amount of land surface in northern Eurasia and North America. Snow covers much of the available food for many animals and is a factor causing migration, both of birds and mammals, towards areas that are somewhat free from its direct influence. Certain mammals (rodents, etc.) tunnel through the snow near the underlying mosses and get access of food and often protection from winter cold. In the Tundra, the non-hibernating lemmings (*Dicrostonyx*) breed during winter beneath the snow. Snow prevents the locomotion of mammals and of running birds, and animals with long feet only are well adapted for moving on snow cover.

Lemming.

C. LITHOSPHERE (SOIL) (Edaphic Factors)

The earth is a cooled, spherical, solid planet of solar system, which rotates on its axis and revolves around the sun at a certain constant distance. The solid component of earth is called **lithosphere**. The lithosphere is multilayered including following main layers—**crust**, **mantle**, and **outer** and **inner core**. The **core** is the central fluid or vapourised sphere having diameter of about 2500 kms from the centre and is possibly composed of Nickel-Iron. The **mantle** extends about 2900 kms above the core. This is in molten state. The outer most solid zone of the earth is called **crust** which is about 8 to 40 kms above the mantle. The crust is very complex and its surface is covered with **soil** supporting rich and varied biological communities, for living organisms find in the soil an environment providing food, shelter, and concealment from predators (**Wallwork**, 1970).

Soil

The word soil is derived from a Latin word *'solum'* meaning earthy material in which plants grow. The science which deals with the study of soil is called **Soil Science, Pedology** (*pedos*=earth) or **Edaphology** (*edaphos*=soil). The study of soil is significant for us due to various reasons. Soil is a natural habitat for microorganisms, plants and animals. Its knowledge is helpful in practices of agriculture, horticulture, forestry, etc., such as cultivation, irrigation, artificial drainage, and use of fertilizers. Pedology is important also for geology, petrology, mineralogy, palaeobotany and palaeozoology.

The soil can be defined as the weathered surface of the earth's crust which is mixed with organic material and in which microorganisms live and plants grow. Soil consists of the inorganic materials (the mineral matter) derived from parent rocks; the organic materials (the humus) derived from dead organisms; the air and water occupying the pores between the soil particles which are loosely-packed (the soil water and the soil atmosphere); small organisms (the biological systems) such as bacteria, fungi, algae, protozoa, rotifers, mematodes, oligochaetes, molluscs and arthropods, and higher plants which live in it. Soil is the ultimate source of all food production since plants form the base of all ecological pyramids and plants grow in soil and derive nutrition from them. It provides mechanical anchorage to plants, besides serving as a reservoir of food materials and water. It is the site where nutrient elements are brought into biological circulation by mineral weathering. The soil harbours the bacteria which incorporate atmospheric nitrogen into the soil. Plant's roots occupy a considerable portion of the soil; tie the vegetation to the soil; and pump water and its dissolved minerals to other parts of the plant for photosynthesis and other biochemical processes (see **Smith**, 1974).

Further, a **mature soil** is that state of soil that has assumed the profile features (*i.e.,* succession of natural layers) characteristic of predominant soils on the smooth uplands within the general climatic and botanic regions in which it is found (**Marbut**, 1926).

Soil formation or Pedogenesis

Soil is a stratified mixture of inorganic and organic materials, both of which are decomposition products. The mineral constituents of soil are derived from some parent material, the **soil forming rocks** by fragmentation or weathering, while, organic components of soil are formed either by decomposition (or transformation) of dead remains of plants or animals or through metabolic activities of living organisms present in the soil. Before discussing the soil-forming processes, let us describe in brief the nature of soil forming rocks as follows :

Soil forming rocks. Basically, there are following three kinds of soil forming rocks :

1. Igneous rocks which are formed due to cooling of molten megma or lava, *e.g.,* granite, diorite and basalt.

2. Sedimentary rocks which are formed by deposition of weathered minerals which are derived from igneous rocks, *e.g.,* shales, sandstone and limestone.

3. Metamorphic rocks which are formed by change of pre-existing rocks (*e.g.,* igneous or sedimentary rocks) through heat and pressure, *e.g.,* gneiss, schist, slate, quartzite and marble.

Chemistry of minerals of soil-forming rocks. Rocks are the chemical mixture of numerous kinds of minerals. The chemical nature of certain most common and abundant minerals of soil-forming rocks has been listed in table 3-6.

Table 3-6. Chemical composition of some common soil minerals.

Minerals	Chemical constituents
A. Sand and silt minerals	
1. Quarts or silica	SiO_2
2. Feldspars	
a. Orthoclase	$K_2Al_2Si_6O_{16}$
b. Plagioclase	$NaAlSi_3O_8$
c. Calcium feldspar	$CaAl_2Si_2O_8$
3. Micas	
a. Muscovite	$K(OH)_2Al_2(AlSi_3)O_{10}$
b. Biotite	K, Mg, Fe, Al silicate
4. Pyroxene	(Mg, Fe) SiO_3
5. Amphibole	$(Mg, Fe)_7 (Si_4O_{11})_2 (OH)_2$
6. Olivine and serpentine	$(Mg, Fe)_2 SiO_4$
7. Calcite; magnesite; and dolomite	$CaCO_3$, $MgCO_3$; and ($CaCO_3$, $MgCO_3$)
8. Iron oxides	
a. Haematite	Fe_2O_3
b. Magnetite	Fe_3O_4
c. Limonite	FeO (OH), xH_2O
B. Clay minerals	
1. Kaolin	Al_2O_3, $2SiO_2$, $2H_2O$
2. Montmorillonite	$(Ca,MgO)Al_2O_3,5SiO_3,5H_2O$

Process of Soil Formation

The processes which are involved in the formation of mature soil can be studied under the following heads :

A. Weathering of Soil Forming Rocks

Soil formation is started by **disintegration** or **weathering** of parent rocks by some physical, chemical and/or biological agents, because of which the soil-forming rocks are broken down in small particles called **regoliths**. Regoliths are the basic materials which under the influence of various other pedogenic processes finally develop into mature soil.

(a) Physical weathering. The physical weathering agents are primarily climatic in character, exerting a mechanical effect on the substratum with the result that fragments are pulverised into progressively decreasing particle sizes (*i.e.,* regoliths). Such climatic weathering of rocks does not cause any chemical transformation of rock-minerals and commonly occurs in deserts, in high altitudes, in high latitudes, and in localities with marked topographic relief and sparse vegetation cover. The agents which are involved in climatic weathering of rocks, are **temperature**, **water**, **ice**, **gravity** and **wind** (see **Wallwork**, 1970).

The **temperature** causes break down of those rocks which have heterogeneous structure, due to the fact of differential expansion and contraction coefficient of materials composing the rocks. Minerals composing the rocks have got different degrees of expansion. These minerals expand in the high temperature of day and contract when the temperature falls. The differential expansion and contraction of different minerals of rocks set up internal tensions and produce cracks in the rocks and consequently, the rocks weather into finer particles.

In its liquid state **water** causes mechanical weathering of rocks by following methods : (i) **Rain water**. Natural water falling either in the form of rain drops or hail storm on the surface of rock with beating effect bring about abrasion of massive rocks into smaller particles. (ii) **Torrent water**. Rapidly flowing water rolls the heavy rock masses such as rock boulders along the bottom of stream and grinds them into finer particles. (iii) **Wave action**. The wave actions are most effective in sea shores. The rapidly striking water waves dislocate solid particles of varying diameters from sea shore rocks and the debris is then settled at the sea bottom to form **marine soil**. Water also acts as a mechanical carrier. Nearly 5 billion tons of mineral matter is annually carried away in solution from land to sea (see **Pearse**, 1939).

Torrent waterfall rolls the heavy rock masses and grinds them into finer particles.

In its freezing and ice-melting states water causes rock-weathering by **frost action** and **glacier formation**. Water in the form of frost or ice, is an extremely effective physical weathering agent of rocks. It seeps into rock crevices, freezes due to sudden fall of temperature of rocks, expands about nine per cent of its original volume, exerts a pressure (expansion force) of approximately 150 ton/ft^2 and eventually cracks the rock into smaller pieces. Likewise, in summer when ice at mountain tops starts melting and glaciers (huge sliding masses of ice) move downwardly on the slope, then during the glacier movement, the rock over which they move is gradually worn down to produce fine particles which are deposited as **drift** or **till**, when glacier finally retreats.

Gravitation weathering action is most effectively demonstrated by land slides and rock slippages caused by earth quakes and faulting during which the rock is fragmented by abrasion and the forces of impact. Lastly, the rapid stormy **wind** carrying suspended sand particles causes the abrasion of exposed rock. It acts like a mechanical carrier in moving the particles over the surface of earth as **dunes** or **drifts** and in transporting large quantities of fine suspended particles long distances.

Biological weathering.

(b) Chemical weathering. The physical weathering produces a greater surface area of rock exposed to the **chemical weathering**, which occurs simultaneously with physical weathering and continues much beyond that. During chemical weathering, a chemical transformation or decomposition of parental mineral materials into new mineral complexes or secondary products occurs. For example, primary minerals that contain aluminium and silicon, such as feldspar, are converted to secondary mineral such as clay. Because chemical weathering requires the presence of moisture and air as essential factors, therefore, chemical weathering is not effective in deserts. It includes reactions such as solution, hydrolysis, oxidation, reduction, carbonation and hydration.

(c) Biological weathering. Though the surface of bare rock is unsuitable for many forms of life, even then a number of microorganisms (bacteria, protozoans, fungi, nematodes, etc.), lichens and mosses can gain a foothold. These early colonizers transform the rock into a dynamic system, storing energy and synthesizing organic material (**Jacks**, 1965). Their activities alter the mineral composition as well as the physical structure of the rock. For example, lichens are present in the initial stages of biological succession and their growth may cause cracking or flaking, exposing greater area of rock to further weathering. The lichens and mosses extract mineral nutrients such as P, S, Ca, Mg, K, Na, Fe, Si, and Al, from the rock. These elements are combined with organic complexes, and eventually return to the developing soil when the vegetation decomposes.

Products of Weathering and Soil Types

The soil which is formed by weathering of soil-forming rocks is called **embryonic** or **primary soil**. The **secondary** or **transported** soils are those which are carried to other places by carriers such as gravity, water, glacier and wind. Soil material transported from one area to another is known as **loess** or **eolian**; that transported by water as **alluvial**, **lacustrine** (lake soil) and **marine deposits**; that transported by gravity as **colluvial soil** and that transported by sand storms as **sand dunes**. In a few places soil material comes from accumulated organic matter as **peat**. The soil which develops *in situ* above parent bedrock is called **residual** or **sedentary soil**.

B. Mineralization and Humification

The weathering of inorganic soil matter is followed by decomposition of organic matter, which starts at an early stage of pedogenesis, and continues up to much later stages. During early stages of pedogenesis, the organic contents of the embryonic soil are not very high, as the vegetation and its associated fauna are not richly developed. Also, the products of organic composition frequently occur in combination with those of inorganic decomposition as complexes, not easily separable into component parts in these initial stages. However, as both the size of soil particles and the spaces between them become smaller with the result that the water holding capacity increases. This together with the increasing amount of plant nutrients and organic material, allows the soil to support higher plant life, such as grasses; the protective covering effect and binding action of root systems so provided, promotes a greater stability of soil structure. Thus, soil development begins under some influence of plants. They root, draw nutrients from mineral matter, reproduce and die. Their roots penetrate and further breakdown the regolith. The plants pump up nutrients from its depth and add them to the surface, and in doing so recapture minerals carried deep into the material by weathering processes. By photosynthesis plants capture the sun's energy and add a portion of it as organic carbon to the soil each year. This energy source, the plant debris, enables bacteria, fungi (such as *Rhizoctonia solani, Armiliaria mellea,* etc.), earthworms and other soil organisms to colonize the area.

The breakdown of organic debris into humus is accomplished by decomposition and finally mineralization. Higher organisms in the soil—millipedes, centipedes, earthworms, mites, springtails, grasshoppers, and others—consume fresh material and leave partially decomposed products in their excreta. This is further decomposed by microorganisms, the bacteria and fungi, into various compounds of carbohydrates, proteins, lignins, fats, waxes, resins and ash. These compounds are broken down into simpler products such as carbon dioxide, water, minerals and salts. This latter process is called **mineralization**. The residual amorphous, incompletely decomposed black coloured organic matter which undergoes mineralization is called **humus**. The process of humus formation is called **humification**. **Muller** (1879, 1884) has recognized two kinds of humus–mor and mull. **Mor humus** is acidic and supports an abundant fungal growth and low number of soil bacteria. Fungal mycelia may help to bind together particles of humus and decomposition litter into matted layers and in a well developed mor, three such layers can be distinguished, namely a surface **litter** or **L layer**, comprising undecomposed leaves and twigs, below which is the **fermentation** or **F layer**, in which decomposition has proceeded same way toward the development of humus, and beneath this, the **humus** or **H layer**, in which degraded humus fractions accumulate. Mor has low calcium content and developed on sandy soils under conifers. **Mull humus** is neutral or slightly alkaline and contain rich microflora of bacteria. It lacks distinct layering of the mor, largely because the presence of calcium compound favours the development of a rich earthworm fauna which promotes a greater mixing of organic and mineral materials. The mull humus develops in brown forest soils under tree species having a relatively high calcium content, such as alder, elm, bass wood, aspen and few conifers (red cedar). Intermediate between the two types of humus is the '**moder**' which has richer and varied fauna.

In order to emphasize the part played by living plants and animals in soil formation, **Taylor** (1930) proposed the following formula :

$$S = M\,(C+V+VA+A)\ t+D$$

where, A signifies animals ; C = climate; D = erosion or deposition ; M = parent material, S = soil; t = time and V = vegetation.

Role of earthworms in soil formation and soil fertility. Earthworms play a significant role in soil formation. They may burrow two metres into the soil, make numerous transverse furrows and produce **worm casts** by eating the soil and soil organic matter (litter). These casts are very rich in nitrogen, water soluble aggregates and mineral substances and increase soil fertility. The burrows facilitate aeration and increase water holding capacity of the soil. **Dash** and **Patra** (1979) have calculated 77 tonnes of dry weight of earthworm cast production per hectare per year in an Indian grassland site. It is estimated that soil turn over due to earthworm action provides a stone-free layer about 15 cm deep on the surface.

C. Formation of Organo-Mineral Complexes

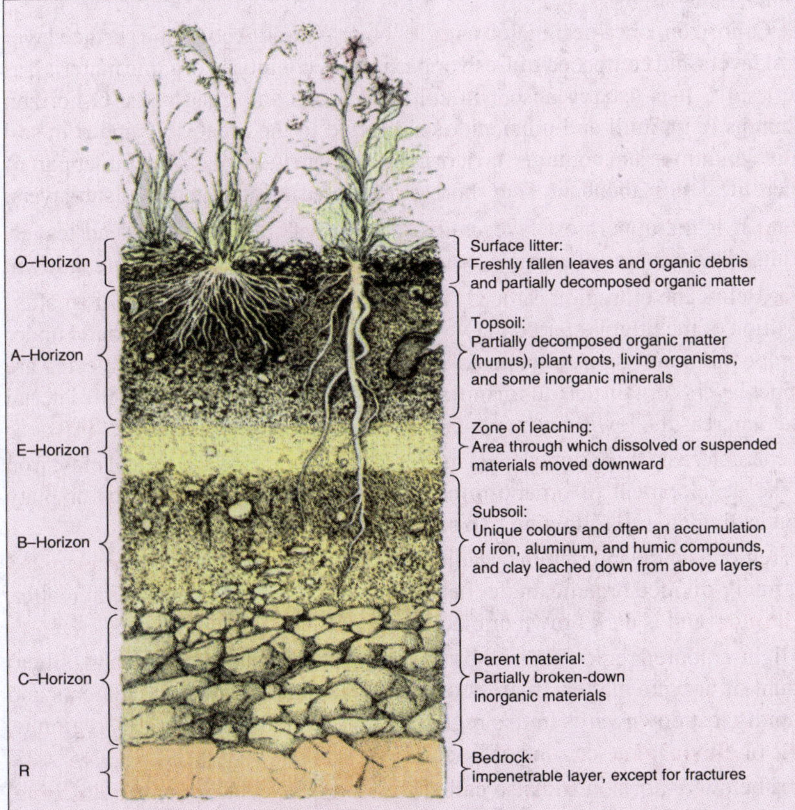

Fig. 3.2. A Generalized profile of soil. Zones vary in number, composition, and thickness, depending on the type of soil.

During final stage of pedogenesis, colloidal particles which are formed due to weathering, humification and mineralization accumulate and may aggregate into **crumbs** or **concretions**. Some colloidal humus particles may become associated with mineral particles to form organo-mineral complexes. According to **Wallwork** (1970) the crumbs or organo-mineral complexes are formed by following two mechanisms—**electro chemical bonding** and **cementing**. In electrochemical bonding method of crumb formation, aggregation of negatively charged colloidal clay and/or humus particles of water molecules and metallic ions, particularly calcium, takes place. The cementing mechanism of crumb formation involves the action of substances absorbed on the surface of soil particles which effectively glues them together.

The crumbs increase the total pore space in the soil, allowing good aeration and drainage. Eventually, a characteristic profile develops under the influence of climate, vegetation, parent material and the activities of the soil communities. The mature soil, thus, becomes a complex system of living and non-living materials not inert but active.

Soil Profile

Soil profile is the term used for the vertical section of earth crust generally up to the depth of 1.83 meter or up to the parent material to show different layers or horizons of soil for the study of soil in its undisturbed state. It is made up of a succession of horizontal layers or **horizons**, each of which varies in thickness, colour, texture, structure, consistency, porosity, acidity and com-position. A **pedon** is the smallest three-dimensional volume of soil needed to give full representation of horizontal variability of soil.

In general, soils have following four horizons an **organic** or **O-horizon** and three **mineral (A,B,C) horizons**. Some workers recognized a **D-horizon**, in which rocks are in active weathering state, in between C and R-horizons. **R-horizon** is the consolidated bed rock on which a soil profile rests. A and B-horizons form the **true soil** or **solum**. Each horizon of soil profile is further subdivided. Horizon subdivisions are indicated by arabic numbers, *e.g,* O_1, O_2, A_1, A_2, etc. Different horizons of soil-profile have following characteristics :

1. O-horizon. The O-horizon, once designated as **L**, **F**, **H**, or Ao and Aoo, is the surface layer forming above the mineral layers and composed of fresh or partially decomposed organic material, as found in temperate forest soils. It is usually absent in cultivated soils and grasslands. O-horizon contains both kinds of humus (*e.g.,* mull and mor) and is subjected to the greatest changes in soil temperatures and moisture conditions and contains most organic carbon. O-horizon and upper part of A horizon is the region where life is most abundant. The O horizon is divided into following two sub-layers:

(i) O_1 (Aoo) region. It is the uppermost layer which consists of freshly-fallen dead leaves, branches, flowers and fruits, dead remains of animals, etc. All these do not show evident breakdown.

(ii) O_2 (Ao) region. Below the O_1 region is the O_2 layer of partly decomposed organic matter. The process of decomposition of the litter is started in O_2 region. Thus, organic matter is found under different stages of decomposition and microorganisms such as bacteria, fungi, actinomyeetes are frequently found in it. Upper layers contain detritus in initial stage of decomposition, in which material can be faintly recognized, whereas the lower layers contain fairly decomposed matter, the **duff**.

2. A-horizon. It is characterized by major organic matter accumulation, by the loss of clay, iron and aluminium and by the development of organo-mineral complexes, granular crumbs or platy structures. The A-horizon is divided into following two sub-layers :

(i) A_1 region. This region is dark and rich in organic matter and is called **humic** or **melanized region**. The amorphous, finely-divided organic matter here becomes mixed with the mineral matter, which is now known as **humus** and is dark brown or black-coloured.

(ii) A_2 region. It is light-coloured region where the mineral particles of large size as sand (silica) are more with little amount of organic matter. In areas of heavy rainfall, the mineral elements and organic chemicals are rapidly lost downwards in this region, making it light-coloured. A_2 region is, thus, also called **podsolic** or **eluvial zone** or zone of leaching.

3. B-horizon. It lies below A-horizon and also called **sub-soil** or **illuviation** or **illuvial zone**, since, the nutrients received from A-horizon due to leaching are accumulated in this region. B-horizon is dark-coloured and coarse textured due to the presence of silica rich clay, organic compounds, hydrated oxides of aluminium, iron, etc. It is poorly developed in dry areas. B-horizon can be divided into B_1 (A_3), B_2 and B_3 regions, depending upon the stages of soil development in the area.

4. C-horizon. Below B-horizon and above the surface of weathered parent rock, is the zone of regolith or C-horizon. It is a light coloured horizon containing weathered parent material.

5. R-horizon. Below all these horizons may lie the **R-horizon**, which is the parent, unweathered bedrock. The percolated soil water tends to collect at the surface of the bedrock.

Climate and Soil Types

Variation in climatic factors and rock properties, therefore, lead to different types of soil development in different climatic conditions, as follows :

1. Laterization. Laterite (L. *latus*=brick) soils are reddish brown in colour and are found in the warm and humid (=heavy rainfall) tropics. In the process of laterization, iron and aluminium oxides which are resistant to decomposition, do not leach down but remain in the surface soil. Silicic acid (*i.e.*, colloidal form of silica) on the other hand leaches to lower horizons. Laterites are comparatively less fertile soils and are found in Australia, India, some warm parts of Europe and North America.

Lotsol is a wider term than laterite and currently used to refer to soils of warm humid belt of the earth where silica usually leaches to lower horizons while oxides of iron and aluminium remain on the surface. Lotosol soil is also called **oxide** and **ferrasol** and is poor in bases and contains low organic content. It is rich in free quartz grains and kaolinite clays.

Vertisols are the other characteristic soils where there is a long dry season. These soils are formed from limestones, marl or ferromagnesium rocks and rich in montmorillonite clays. During dry summer, deep cracks are formed in vertisols and these cracks get filled with clay due to dust blow which in the rainy season on wetting swells so much as to form small bumps. Thus, vertisols are characterized by cracks in summer and bumps in rains.

2. Podsolization. It is the most actively studied soil forming process. It takes place in cold humid climate under forest vegetation (chiefly of genera of Ericaceae and conifers) where leaf litter and other organic matter decompose slowly, chiefly through fungal activity. Litter contains high lignin and low nutrient (particularly calcium) contents; it is also rich in phenolic compounds which inhibit microbial activities. In consequence, such a litter is acidic in nature. The water percolating through such litter, being acidic dissolves out with it minerals and humus contents (such as carbonates, sulphates and iron and aluminium compounds) from A-horizon of soil. These leached materials reach to the lower horizons, being collected in the form of a hard, distinct layer in the B-horizon. Leaching results into a grey ash- like surface called **podsol** (Russian *pod* = under, *zole* = ash). In the Indian conifer forests of the Himalayas true podosols are not met with, mainly due to alternate phases of wet and dry seasons.

3. Gleization. In humid climates, due to high ground level or retention of surface water in the soil, water-logging and consequent reducing conditions are fairly common. Under such reduced conditions firstly due to ferrous compounds, the soil colour becomes blue-grey or grey, and secondly the rates of decomposition of the organic matter are slow. These together result into the accumulation of a sticky compact layer of blue-grey colour at the bottom of B horizon. This process in called **gleization** and the soil as **gleys**. Gleization also takes place in ice covered regions of arctics (also tundra) where soil is not saline.

4. Melanization. The process is very common in regions of low humidity where humus is formed from the organic matter. As a result of melanization, the black humus along with water is mixed in the A-horizon of the soil which turn dark-coloured. Such a blackish brown top soil is called **chernozem** and it has a minimum leaching, so it is rich in organic detritus.

5. Calcification. It is a common process found in North India, where calcium in the form of calcium bicarbonate dissolves in water and leaches down escaping adsorption around dry particles. The leached $CaCO_3$ gets precipitated at the depths of about half to three metres, depending on the quantity of rainfall and depth of infiltration. The precipitated calcium remains in the form of nodules or *kankar* and form a hard pan. This gradually reduces soil fertility. In the Northern India the kankar nodules are unearthed and are used in the building of roads and in the preparation of lime. Calcification is typical of grassland and a layer of calcium carbonate generally occurs at the base of B horizon. Sometimes, a layer of more soluble calcium sulphate (gypsum) is also present.

There are some other types of soils such as a hydromorphic, halomorphic and azonal. **Hydromorphic soils** are formed on poorly drained water filled regions with poor aeration, *i.e.,* they are highly reducing soils in which ferric ions are reduced to ferrous. But in dry seasons the soil becomes aerated and ferrous is again converted into ferric. The hydromorphic soils of tropical paddy fields are called **grey soil**. **Halomorphic soils** are characterized by the presence of high salt contents on the surface. Salinity and alkalinity are commonly found in big patches in the North Indian states resulting into the formation of *usar* or *reh* soils. In fact, such soils result due to upward capillary movement of dissolved salts from the rising water table in too frequently irrigated and fertilized regions, or from nutrient rich water received through stream flow from the surroundings during the periods of high rainfall. On evaporation, salts in different forms are left on the surface. Grazing animals often lick such salt deposits to fulfil their metabolic needs. **Azonal soils** lack horizon B and the horizon A is also thin. Among azonal soils, the common are **lithosols** found on mountain tops where parent rock is quite hard and resistant to weathering. **Regosols** are formed from volcanic deposits.

Morphology of Soil

In the field, differences among soils and among horizons within a soil are primarily reflected by variations in texture, arrangement, structure and colour.

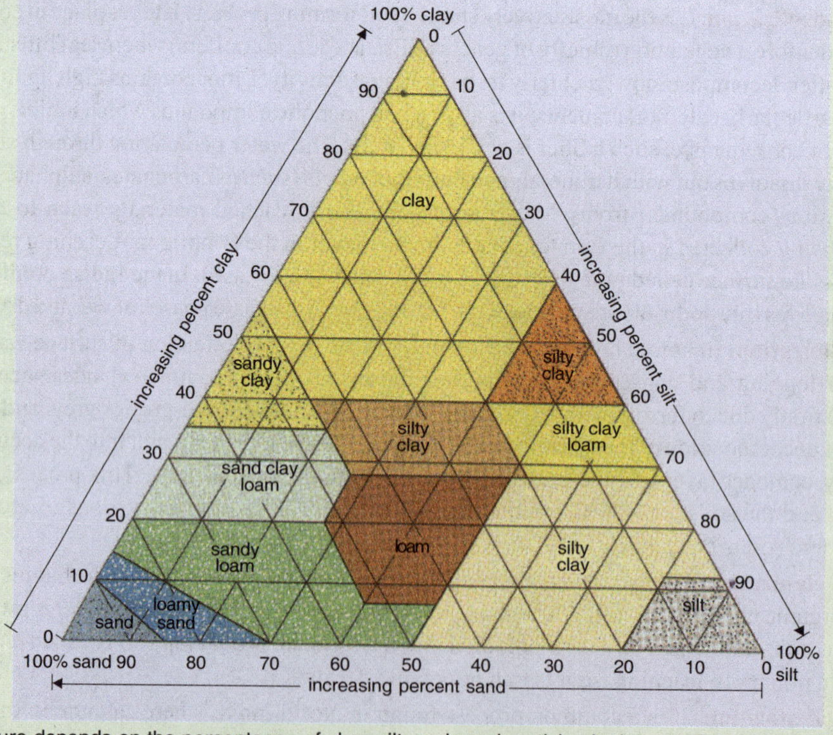

Soil texture depends on the percentages of clay, silt, and sand particles in the soil. Soil texture affects soil porosity—the average number and spacing of pores in a volume of soil. Loams are the best soils for growing most crops.

1. Texture of soil. The texture of a soil is determined by the production of different sized soil particles. Mineral fractions or particles of soil are called "**soil fractions**" or "**soil separates**". The soil particles have been classified into **gravel**, **sand**, **silt** and **clay** on the basis of their size differences. Gravel consists of coarse particles larger than 2.0 mm. Sand ranges from 0.02 to 2.0 mm in diameter and is easily seen and feels gritty. Silt consists of particles from 0.002 to 0.02 mm in diameter, which scarcely can be seen by the naked eye, feels and looks like flour. Clay particles range below 0.002 mm in diameter, are too fine to be seen even under the ordinary microscope, and are colloidal in nature. Clay controls the most important properties of soils, including plasticity and exchange of ions between soil particles and soil solution.

Most soils are mixture of these various particles. Based on the proportions of the various particles contained in them, soils can be grouped into 5 textural groups : **(1) Coarse textured soils** are loose, consist mainly of sand and gravel. They retain very little moisture and supply some plant nutrients. **(2) Moderately coarse soils** include sandy loam to very fine sandy loam. **(3) Medium textured soils** are mixture of sand, silt and clay, high enough to hold water and plant nutrients. **(4) Moderately fine textured soils** are high in clay. They are moderately sticky and plastic when wet, they may form a crust on the surface if organic matter is low. They have a high moisture-holding capacity. **(5) Fine textured soils** contain more than 40 per cent clay, may be sticky and plastic when wet and hold considerable water and plant nutrients, but may have restricted internal drainage. All these 5 textural soil-groups include 12 textural classes (see Table 3-7) which have been recognized on the basis of relative percentage of soil particles (*i.e.,* sand, silt and clay).

Table 3-7. Textural classes of soils.

Soil classes or Textural names	Range in relative percentage of soil separates		
	Sand	Silt	Clay
1. Sandy soil	85-100	0-15	0-10
2. Loamy sand	70-90	0-30	0-15
3. Sandy loam	43-80	0-50	0-20
4. Loam	23-52	28-50	7-27
5. Silt loam	0-50	50-88	0-27
6. Silt	0-20	8-10	0-12
7. Sandy clay loam	45-80	0-28	20-35
8. Clay loam	20-45	15-53	27-40
9. Silty clay loam	0-20	40-73	27-40
10. Sandy clay	45-65	0-20	35-45
11. Silt clay	0-20	40-60	40-60
12. Clay	0-45	0-40	40-100

2. Structure of soil. Soil separates (sand, silt and clay) are held together in clusters or shapes of various sizes, called **aggregates** or **peds**. The arrangement of these peds in earth crust is called **soil structure**. Like texture, there are many types of soil structure. Soil peds may be classified as **granular**, **crumb-like**, **blocky plate-like** or **platy**, and **prismatic** or **columnar**. The properties of different soil peds have been tabulated in Table 3-8.

Table 3-8. **Characteristics of peds.**

Types of ped	Properties
1. Granular	Small, spherical and non-porous.
2. Crumb-like	Small, porous and spheroidal.
3. Agranular blocky	Block-like with sharp ends, one end may be pointed.
4. Sub-granular blocky	Block-like but bounded by other aggregates.
5. Platy	Plate-like, sometimes plates are overlapped.
6. Prismatic or columnar	Prism-like but without rounded surface.

Structureless soil can be either single-grained or massive soil aggregates (peds) tend to become larger with increasing depth. Structure is influenced by texture, air, moisture, organic nutrients, micro-organisms, root-growth and soil chemical status.

3. Soil colour. Soils exhibit a variety of colour. Soil colour may be inherited from the parental material (*i.e.*, **lithochromic**) or sometimes it may be due to soil forming processes (**acquired** or **genetic colour**). Though colour has little direct influence on the functions of a soil, but is important in the identification of soil type. In temperate regions dark-coloured soils generally are higher in organic matter than light coloured ones. Well-drained soil may range anywhere from very pale brown to dark brown and black, depending upon the organic matter-content. Red and yellow soils are the result of iron oxides, the bright colours indicating good drainage and good aeration. Other red soils obtain their colour from parent material and not from soil-forming processes. Well-drained yellowish sands are white sands containing a small amount of organic matter and such colouring matter as iron oxide. Red and yellow colours increase from the cool regions to the equator. Quartz, kaolin, carbonates of lime and magnesium, gypsum and various compounds of ferrous iron give whitish and grayish colours to the soil. The grayest are permanently saturated soils in which the iron is in the ferrous form.

Soil colour and temperature. Soil colour influences the soil temperature. The dark coloured soils absorb heat more readily than light coloured soils. **Ramdas** *et al.*, (1936) showed that black cotton soils of Poona absorbed 86% of the total solar radiations falling on the soil surface as against 40% by the grey alluvial soil.

Physical Properties of Soil

Soil possesses many characteristic physical properties such as density, porosity, permeability, temperature, water and atmosphere, each of which can be studied under following separate headings:

1. Soil density. Average density of soil is 2.65 gms. per ml. Density of soil varies greatly depending upon the degree of weathering.

2. Porosity. The spaces present between soil particles in a given volume of soil are called **pore spaces**. The percentage of soil volume occupied by pore space or by the interestitial space is called **porosity** of the soil. Porosity of soil depends upon the texture and structure compactness and organic contents of soil. Porosity of the soil increases with the increase in the percentage of organic matter in the soil.

The pore spaces are of two types–**(1) Micro-pore spaces** (capillary pore spaces) and **(2) Macro-pore spaces** (non-capillary pore spaces). Capillary pore spaces can hold more water and restrict the free movement of water and air to a considerable extent, whereas macro-pore spaces have little water holding capacity and allow free movement of moisture and air in the soil under normal conditions.

3. Permeability of soil. The characteristic of soil that determines the movement of water through pore spaces is known as **soil permeability**. Soil permeability is directly dependent on the pore size, therefore, it is higher for the loose soil with large number of macro-pore spaces than it is for compact soil with numerous micro-pore spaces.

4. Soil temperature. Soil gets heat energy from different sources such as solar radiation, decomposing organic matter, and heat formed in the interior of earth. The temperature of soil is affected by its colour, texture, water content, slope, altitude of the land and also by climate and vegetational cover of the soil. Evaporation of water from soil makes it cooler. Black soils absorb more heat than white soils. Sandy soils absorb more heat and radiate it out quickly at night than clay or loam soils.

The soil temperature greatly affects the physico-chemical and biological processes in the soil. For example, the germination of seeds, normal growth of roots and biological activity of soil-inhabiting micro-and macro-organisms which require proper and specific temperature.

5. Soil water. In soil, water is not only important as a solvent and transporting agent but in various ways it maintains soil texture, arrangement and compactness of soil particles and makes soil habitat livable for plants and animals. It comes in soil mainly through infiltration of precipitated water (dew, rain, sleet, snow, and hail) and irrigation. According to improved classification method of **Bouyoucos** (1920), soil water is classified into following types— gravitational and ground water, capillary water, hygroscopic water, combined water and water vapour.

(i) Gravitational water. In a well-saturated soil, the accessory (extra) amount of water displaces air from the pore spaces between soil particles and percolates downwardly under gravitational influence and finally it is accumulated in the pore spaces. This accumulated excess water of large soil spaces is called **gravitational water**. When this gravitational water further percolates down and reaches to the level of parent rock, it is called **ground water**. Both kinds of these soil water are ecologically important in the leaching of nutrients.

(ii) Capillary water. The water which is held by capillary forces (*i.e.,* surface tension and attraction forces of water molecules) in smaller soil channels, when the gravitational water and ground water have been drained, is called **capillary water**. Capillary water occurs as a thin-film around soil particles in the capillary spaces and represents the normal available water to the plants. It remains in soil for long periods and carries with it nutrients in solution. Humus has more capillary water than soil minerals.

(iii) Hygroscopic water. Soil particles retain some water so firmly that the plants cannot absorb this. Such soil water is called hygroscopic water.

(iv) Combined water. Combined water is the water of chemical compounds held by chemical forces of molecules such as hydroxides of silicon, iron and aluminium. It is of no ecological significance.

(v) Water vapour. Some soil water occurs as moisture or water vapours in the soil atmosphere.

Further, the total amount of water present in the soil is called **holard**. The quantity of water that plant-roots can absorb out of holard is called **chresard** and that amount of soil water which cannot be absorbed by plant-roots is called **echard**. Moreover, there are some terms that reflect the water status of soil and are generally used for comparative studies of different soils.

(a) Soil water potential. This is an expression of the total reduction of water potential in the soil, due to mineral matrix, solubility, external pressure and gravitation effects.

(b) Field capacity. When a soil holds all the water it can, but no gravitational water, it is said to be at its field capacity. Field capacity is generally defined as the water content of an undisturbed soil (% oven-dry weight) after it is saturated by rainfall and drainage of gravitational water has completely stopped. Field capacity of soil may be taken as the total amount of capillary, hygroscopic and combined water plus water vapour.

(c) Moisture equivalent. It is defined as the water content (% oven dry weight) retained by the undisturbed soil.

(d) Water holding capacity (or storage capacity). This is the extent to which a soil can hold capillary water against gravity. It is equal to field capacity less hygroscopic water.

(e) Hygroscopic-coefficient. It is defined as the water vapour (%) absorbed by unit weight of dry soil, when placed in an atmosphere completely saturated with water vapours.

(f) Permanent wilting percentage (or wilting coefficient). It is the amount of water (% oven dry weight) that remains in a soil when permanent wilting is present in the plants growing in soil.

The status of soil water is also represented in terms of **capillary potential** (= metric potential). The relationship that exists between soil water content and water potential is usually determined by a pressure membrane apparatus, called **tensiometer**.

6. Soil atmosphere. Gases found in pore spaces of soil profiles form the **soil atmosphere**. The soil atmosphere contains three main gases namely O_2, CO_2, and N_2. Soil air differs from atmospheric air in having more of moisture and CO_2 and less of O_2. The soil atmosphere is affected by temperature, atmospheric pressure, wind, rainfall, etc. Loam soils with humus contain a normal proportion of air and water (about 34% air and 66% water) and therefore are good for majority of crops.

Course-textured or well-structured soils contain higher gaseous diffusive transfer rates than fine-textured or poorly-structured soils under **wet conditions**. This is because they contain many large pores which remain gas-filled. The water-filled pores of the fine-textured soils form potent barriers of gaseous diffusion; O_2 diffuses ten thousand times more slowly in water than in gas. In **dry soil**, the situation is reversed, as the fine textured soils have a greater total pore space and provide a larger gas-filled cross sectional area for diffusion. Soil aeration is very important in growth of roots, seed germination and microbial activity. Poor soil aeration suppresses root hair development, and may reduce rate of absorption of water and nutrients.

7. Soil solution. There exists a weak solution of various salts, alongwith other liquids and gases in the soil mass. This soil solution contains almost all the essential minerals. Complex mixture of minerals such as carbonates, sulphates, nitrates, chlorides, and organic salts of Ca, Mg, Na, K, etc., are found as dissolved in water. The chemical nature of soil solution depends on the nature of parent matter, chemical nature of organic matter and climatic factors and other factors involved in pedogenesis. The soil solution is the primary source of inorganic nutrients for plant roots. Soils with optimal concentration of various nutrient solutes are called **eutrophic**, whereas those with suboptimal concentration of these nutrient salts are called **oligotrophic**.

Chemical Properties of Soil

Soil is a mixture of various inorganic and organic chemical compounds and exhibits certain significant chemical properties, all of which can be discussed as follows :

(a) Inorganic elements and compounds of soil. The chief inorganic constituents of soils are the compounds of following elements—A1, Si, Ca, Mg, Fe, K and Na. Soil also contains smaller amounts of compounds of following inorganic elements—B, Mn, Cu, Zn, Mo, Co, I, F, etc.

(b) Organic matter of soil. The chief organic component of soil is **humus** which chemically contains amino acids, proteins, purines, pyrimidines, aromatic compounds, hexose sugars, sugar alcohols, methyl sugars, fats, oils, waxes, resins, tannins, lignin and some pigments. Further humus is black coloured, odourless, homogeneous complex substance.

(c) Colloidal properties. As soil is composed of crystalloids and colloids, therefore, it exhibits all the physico-chemical properties which are related with these two soil particles. Colloids for example exhibit absorption, electrical properties, coagulation, Tyndal phenomenon, Brownian movement, dialysis, etc.

(d) Soil pH. Many chemical properties of soils centre around soil reaction. As regards their nature, some soils are **neutral**, some are **acidic** and some **basic**. The acidity, alkalinity and neutrality of soils are described in terms of **hydrogen-ion concentrations** or **pH** values. A pH value of 7.0 indicates neutrality, a value above this figure (7.1–14.00) indicates alkaline condition and a value below (0–6.9) indicates acid conditions. Normally, the pH value of soils lies between 2.2 and 9.6. In

India, acidic soils (pH below 5.5 to 5.6) occur in the high rainfall areas of Western ghats, Kerala, Eastern Orissa, West Bengal, Tripura, Manipur and Assam. The saline, alkaline or basic soils (called *'Usar'*, contain pH upto 8.5) of India, occur in U.P., West Bengal, Punjab, Bihar, Orissa, Maharashtra, Madras, M.P., A.P., Gujarat, Delhi and Rajasthan.

Some pH values are ecologically significant for the plants (**Pearsall**, 1952); plants regarded as **calcicoles** usually occur in soil with pH 6.5, whereas **calcifuges** occur in soil with pH value below 3.8–4.0. Soils above pH 6.5 are generally cation-saturated (those containing free $CaCO_3$ called **calcareous soils**) while soil below pH 3.8-4.0 contain a considerable content of exchangeable hydrogen. These limits of pH value are also reflected in the nature of the soil's organic matter: raw humus or mor which is associated with soils below of pH 3.8, while mull is characteristic of the more cation saturated soils of pH 4.8–5.0 and above.

Highly acidic and highly saline or alkaline soils often remain injurious for plant growth, micro-organisms, etc. Soil pH strongly affects the microbial activities, as at below pH 5.0 bacterial as well as fungal activities are reduced. Neutral or slightly acidic soil, however, remains best for the growth of majority of plants.

Soil as Habitats for Animals

Soil is used by various animals as a hideout from enemies, to escape desiccation and extremes of temperature; as material for abodes, and as a highway. Clay is cold, dense, poorly aerated and though it holds much water, does not give up water readily to plants and animals. Sand is dry, loose (cave in readily) and has a variable temperature. Rock heats and cools quickly, has little available water and practically no air, and is usually too resistant to allow penetration by burrowing animals. Alkaline soils are "**physiologically dry**" because the salts present hold the water so that it is not readily available to organisms. Humus has a great water capacity and gives up water readily, is usually well-aerated and furnishes nitrogen as well as mineral salts and water. It remains rich in biota. Peaty soils are raw humus and are uninhabitable for most animals because they are poorly aerated and have an acid reaction. In general, the character of the vegetation cover serves as a good index of the habitability of soils for animals; conditions that make plants flourish are usually favourable for animals.

Soil Fauna and Soil Flora

Soil supports a wide array of organisms of different body-sizes and taxonomic groups. Generally, soil organisms are classified into three major groups namely **microfauna** and **microflora**, **mesofauna**, and **macrofauna**. Mesoflora and macroflora because occur above the surface of soil (land-surface), therefore, are excluded from this discussion :

1. Microfauna. It includes animals with body size within the range of 20μm to 200μm. It includes all Protozoa and small-sized mites, nematodes, rotifers, tardigrades and copepode Crustacea. Soil inhabiting **protozoans** such as amoeba, ciliates, zoomastigine flagellates occur near the surface soils, while the testate forms such as *Thecamoeba, Euglypha* and *Difflugia,* have a wider vertical distribution. The common terrestrial **polyclad** is *Bipalium.* The **nematodes** such as *Rhabditis, Diplogaster, Tylenchus, Heterodera, Aphelenchoides, Mononchus, Pratylenchus, Xiphinema* and *Criconemoides,* abound by as much as 1-3 million in raw humus soils to 20 million/m^2 in grassland soils.

2. Microflora. The microflora of soil includes bacteria, soil fungi, soil actinomycetes, blue green algae and algae. In soil, microflora bacteria form about 90 per cent of the total population. Fungi and algae together represent one per cent and actinomycetes cover only 9 per cent.

Soil bacteria grow fairly well in the neutral soils richly supplied with organic nutrients. Soil inhabitant bacteria fall into categories namely—**autotrophic bacteria** and **heterotrophic bacteria**. The autotrophic bacteria derive their energy from the oxidation of simple carbon compounds or from inorganic substances and their carbon from the atmospheric CO_2. The common autotrophic bacteria of soil are nitrifying bacteria, hydrogen bacteria, sulphur bacteria, iron bacteria, manganese bacteria, carbon monoxide bacteria and methane bacteria. Most of soil bacteria are heterotrophic bacteria depending upon the organic matter of soil for their energy source and are primarily concerned with the decomposition of cellulose, and other carbohydrates, proteins, fats and waxes. They bring about mineralization of organic matter of soil and release considerable amount of nitrogen, phosphorus and other nutrients for plants. The common nitrogen-fixing bacteria of soil are *Rhizobium* (occurs in root nodules of leguminous plants); *Azotobacter* and *Clostridium pasteurianum* (the latter two are free occurring in soil).

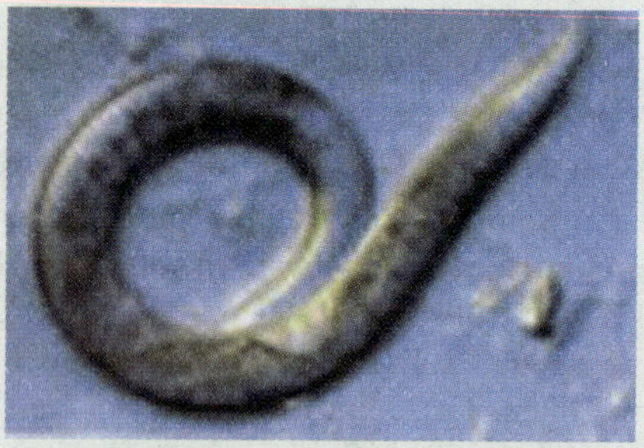

Nematode such as *Rhabditis* are found in raw humus soils.

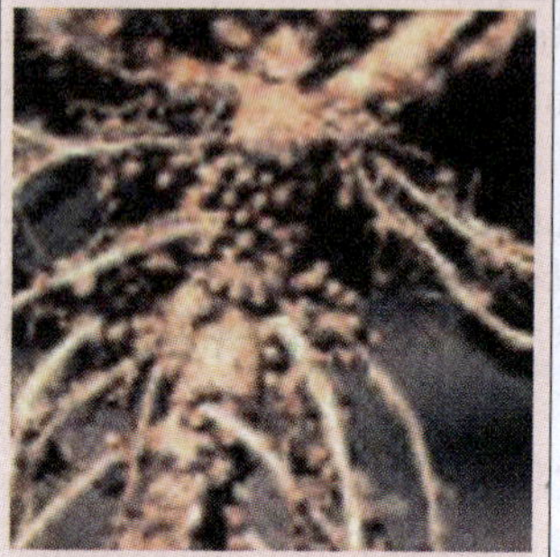

Root nodules of leguminous plants have *Rhizobium* which are nitrogen - fixing bacteria.

Majority of **soil fungi** are found in acidic soils. They may be parasitic, saprophytic and symbiotic. **Parasitic fungi** of soil infect roots of plants and cause plant diseases such as cotton root rot and many kinds of wilts, rusts, blights and smuts. Certain wilt-forming fungi produce toxins which are harmful, for example, *Fusarium lini,* which causes wilt of flax (*Alsi*) and secretes HCN and *Fusarium udum*, a fungus causing wilt of pigeon pea (*Arhar*) secretes fusaric acid in the roots of host plants. However, certain parasitic fungi produce growth stimulating substances for host plant. *Fusarium* sp., for example, have been found to secrete gibberellin and gibberellic acid ($C_{19}H_{22}O_6$). **Symbiotic fungi** of soil live on the roots of certain plants and both fungus and plants are benefited. **Saprophytic fungi** depend on dead organic matter of soil and derive the energy from decomposition of the latter. They break down cellulose, lignin, and gum, sugars, starch, protein, etc., into simple gradients to be utilized by higher plants as nutrients.

Actinomycete fungi prefer saline soils and bring about the decomposition of organic matter such as cellulose. They produce a variety of antibiotics of great economic significance for man. The most important blue green algae of soil are those which fix nitrogen in soil. *Anabaena, Nostoc, Microcystis* are important nitrogen fixing blue green algae of soil. They also make soil aggregates because of having mucilage.

3. Mesofauna. Mesofauna include animals with body size within the range 200μm to 1 cm. The micro-arthropods Acari (mites) and Collembola (spring tails) are important members of this group which also include the larger nematodes, rotifers, and tardigrades, together with most of the isopods, Arachnida (spiders), Chelognathi (pseudoscorpions), Opiliones (harvestmen), Enchytraeidae (pot-worms), insect larvae and small millipedes (Diplopoda), isopods and molluscs.

Among **annelids** the microscopic enchytraeids are represented by *Enchytraeus fridericia* and *Achaeta lumbricellus*, which are more abundant in organic soils and forests than in grassland. *Oniscus, Porcellio,* and *Armadillidum* are the most common isopods (**crustacean**) of the tropics in the humid zone. Among the soil **arachnids**, mites are the most predominant. Mites flourish in moist organic soils and certain mites such as *Galumna, Cepheus,*

Oniscus.

Hemorobates occur in lichens and mosses. Certain mites, such as *Schelorbates* and *Brachychthornus* live in humus. The mites are saprophagous, predatory and phytophagous and are involved in the process of organic decomposition and its resultant processes. Certain arachnids such as scorpions, *Thelyphonus, Galeodes;* and some spiders are crepuscular, hiding under rocks or in crevices in soil and in loose litter, and has no ecological significance in decomposition like other arachnids. Many **opiliones** or harvestmen occur in forest litter, frequently preying upon soil organisms.

Besides mites, only the pseudoscorpions or **chelognathi** occur in surface soils and most decaying vegetations. Of the xerophil litter inhabitants are *Stenatemnus indicus, Dhanus indicus, Fealla indica* and the hygrophil inhabitants living in the litter and under stones are–*Comsaditha indica, Tyrannochthonius madrasensis, Tyrannochithonius chelatus* and *Hygrochelifer indicus* (**Murthy**, 1964), feeding on Collembola, enchytraeids, etc.

The common millipedes or diplopodes of forest soil which are chief decomposers of soils are *Spirostreptus, Thyropygus, Glomeris, Arthrosphaera, Polydesmus, Iulis,* etc. **Tardigrades** or bear animalcules occur in surface layer of most soils in grassland being represented by the *Macrobiotus* and *Hypsibius.*

Among **insects**, apterygote Collembola form numerically the most important groups of soil insects. Other insects such as Dermaptera, Psocoptera, Dictyoptera, Isoptera, Coleoptera, a few Hymenoptera and some Diptera also occur in soil, sometimes as juveniles. The termites such as *Reticulotermes* and *Odontotermes* are important soil-dwellers of tropics and play an important role in the break up of organic materials and their mixing up with mineral soils. Among the Hymenoptera, ants are the most important soil dwelling forms. Among Collembola, Onchuridae, Isotomidae, Poduridae and Entomobrydae are richly represented in the soil both in number and species composition. Large-sized Collumbola such as *Tomo-cerus, Entomobrya* and *Orchesella* occur in surface layer, while the smaller *Onychurus, Tullbergia,* etc., occur in deeper layers of soils.

A dung beetle rolling away a ball of dung. This will provide the beetle and its young with food.

Among Diplura, *Anajapyx, Japyx* and *Campodea* are often found in small numbers in moist soils under stones and in humus. Proturans are more abundant than Diplura and very common in moist forest and grassland soils which abound in species of *Eosentomon, Acerentomon* and *Acerentulus.*

4. Macrofauna of soil. Macrofauna of soil includes those animals whose body size is greater than 1 cm. Here belong the majority of Lumbricidae, the Mollusca, the largest insects and arachnids and the soil-dwelling vertebrates.

Earthworms usually occur in abundance in alkaline and moist soils and sparse in acid soils. They have been proverbial for their influence on the process of decomposition of organic materials, breaking up litter fragments and mixing them thoroughly with mineral soils resulting in the formation of organic soils. Some of the

Earthworm (A) and wood ants (B) help make the soil fertile. They do this by constantly breaking up debris on the forest floor.

common Indian annelidan species of soil are *Megascolex, Pheretima, Octochaetus, Drawida* and *Moniligaster*. Among chilopods the carnivorous *Scolopendra* and *Lithobius* are common in moist soils feeding on leaf litter inhabitants.

Among soil vertebrates, following animals are well adapted for burrowing life in soils– *Ichthyophis, Cacopus systema, Breviceps* (Amphibia), *Sphenodon, Uromastix,* limbless lizards and snakes (Reptilia), *Talpa, Dasyurus, Notoryctes* and various insectivores and rodents (Mammalia).

Adaptations of soil animals. Animals which are adapted for digging the burrows and for subterranean mode of life are called **fossorial animals**. These animals may dig either for their food or simply for retreat. Zoologically they are primitive, defenseless and unambitious animals. They have following adaptations : 1. The body contour is either cylindrical (*e.g., Ichthyophis,* limbless lizards, snakes, earthworms, *Scolopendra,* etc.), or spindle-shaped or fusiform (e.g., *Talpa, Echidna,* etc.), so as to offer least resistance to subterranean passage. 2. The head tapers anteriorly to form a sort of snout for burrowing. 3. The tail is short or vestigial. 4. The eyes tend to become vestigial as they are of no use in dark habitat. 5. The external ears also tend to disappear since they would be obstructing in burrowing. 6. For digging, many structures may be found in different fossorial animals, *e.g.*, hands are well adapted for digging. In the insect *Gryllotalpa*, the fore-legs are modified for digging purpose.

Mole is almost blind. Sensitive whiskers and a good sense of smell are more useful than eyesight underground. Powerful feet and claws are used by mole to dig through soil.

The ground-dwelling animals which may be **cur-sorial** (running), such as ostrich, rhea, ungulates, wolves, cats, bears, hyaenas, etc., **saltatory** (jumping) such as rodents, rabbits, wallabies, kangaroos, etc., or **graviportal** (heavy) such as turtles, armadillos, elephants, hippopotami, etc., exhibit different kinds of adaptations for different kinds of soils. For example, if the soil is firm and hard, the large animals inhabiting the ecosystem tend to have small hooves or paws; if the soil is wet and spongy, they tend to have broad hooves or paws.

Ostrich is a ground dwelling cursorial (running) bird.

REVISION QUESTIONS

1. Describe physico-chemical nature of the atmosphere.
2. Define the following terms : environment ; atmosphere; ozonosphere; ionosphere; air; wind.
3. Describe physical and chemical properties of water and show, how they affect the organisms.
4. Describe different ecological adaptations of plants and animals in relation to water.
5. What is soil ? Describe its formation, morphology and profile (*i.e.,* soil profile).
6. Discuss the role of edaphic factor in ecological studies.
7. Describe the soil-flora and soil-fauna of Indian soils.
8. Describe the influence of climate, vegetation and parent rock on soil development.
9. Describe the role of living organisms in soil formation and in increasing soil fertility.

CHAPTER

4

Abiotic Environmental Factors

Non-living elements in an environment comprise the abiotic components of our ecosystem.

Environment of an organism has two components — **abiotic** and **biotic**. The first includes the **atmosphere** (air), **hydrosphere** (water) and **lithosphere** (land including soil). The abiotic components are characterized by physical and chemical factors such as light, temperature, rainfall, pressure, pH, the content of oxygen and other gases, and so on. These factors exhibit diurnal, nocturnal, seasonal, and annual changes. The biotic components include all living organisms which interact with each other and with the abiotic components. Earth's living organisms interacting with their physical or abiotic environment (including air, land and water) form a giant and vast ecosystem, called **ecosphere** or **biosphere** which is largest and most nearly self-sufficient biological system.

In physical terms, the biosphere is a relatively thin and incomplete envelope covering most of the world. It represents a mosaic of different biotic communities from simple to complex, aquatic to terrestrial, and tropical to polar. It does not exist in the extremities of the polar regions, the highest mountains, the deepest ocean troughs, the most extreme

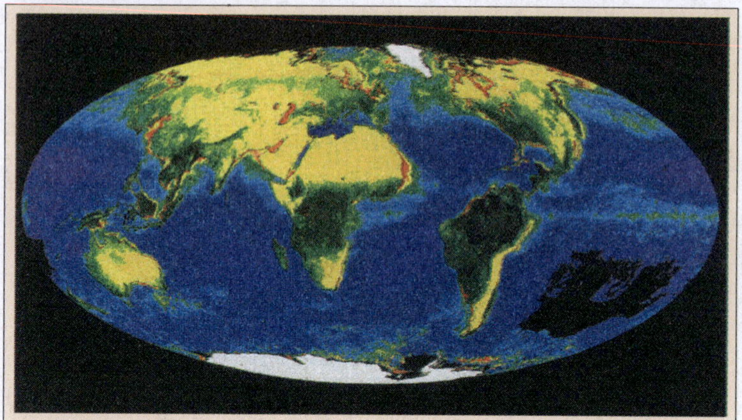

The biosphere. Three years of satelite data were combined to produce this picture of the earth's biological productivity. Rain forests and other highly productive areas appear as dark green, deserts as yellow. The concentration of phytoplankton, a primary indicator of ocean productivity, is represented by a scale that runs from red (highest) to orange, yellow, green, and blue (lowest).

deserts, or the most highly polluted areas of land and water. Its total thickness, including all portions of the earth where living organisms can exist, is less than 26 kilometres. Its zone of active biological production, in terms of photo-synthesis, is much narrower, and varies from a few centimetres to over 100 metres. This zone would, for instance, be only a few centimetres in muddy or turbid water, whereas in very clear ocean water, it could be more than 100 metres in thickness. On land, the zone of biological production might be only a few millimeters in a desert or rock environment, whereas it might again be more than 100 metres in a sequoia or tropical rain forest.

TYPES OF ABIOTIC ENVIRONMENTAL FACTORS

The distribution, abundance, growth and reproduction of the organisms comprising the individual members of populations are controlled by certain **environmental** or **ecological factors**. An environmental factor is any external force, substance or condition which surrounds and affects the life of an organism in any way. Abiotic environmental factors are customarily classified as follows :

1. Climatic factors

(i) Light ;

(ii) Temperature ;

(iii) Water (including atmospheric water, rainfall or precipitation, soil moisture, etc.) ;

(iv) Atmosphere (gases and wind) ;

(v) Fire.

2. Topographic or **physiographic factors**

(i) Altitude ;

(ii) Direction of mountain chains and valleys ;

(iii) Steepness and exposure of slopes.

3. Edaphic factors (soil formation, physical and chemical properties of soil, nutrients).

ESSENTIAL ELEMENTS AND LIMITING FACTORS

The individual organisms of species population, in order to grow and multiply, must be supplied with certain essential materials. Of the hundred four (*i.e.,* 104) naturally occurring chemical elements on the earth, all living organisms are believed to utilize only 16 different chemical elements for their survival. These are called **essential elements**. Several other elements are needed in small quantities by some species. They are summarized in Table 4-1.

Table 4-1. Essential elements (Source : Clapham, 1973).

Macronutrients : Elements used in relatively large quantities	Micronutrients : Elements generally needed in relatively small quantities	Micronutrients : Elements needed by certain species in relatively small quantities
Carbon	Iron	Sodium
Hydrogen	Manganese	Vanadium
Oxygen	Boron	Cobalt
Nitrogen	Molybdenum	Iodine
Phosphorus	Copper	Selenium
Calcium	Zinc	Silicon
Magnesium	Chlorine	Fluorine
Sulphur		Barium

Some of these tabulated elements are used in relatively large quantities as the fundamental building blocks of organic tissues and are called **macronutrients**. Others, the **micronutrients**, are used in much smaller quantities. The micronutrients are also known as **trace elements**. For instance, a crop such as corn will remove over 45 kg. of nitrogen per acre from the soil but only 10–15 g. of boron. In addition to the nutrients required by all species, some organisms have special nutrient requirements.

Liebig-Blackman Law of Limiting Factors

An organism is seldom, if ever, exposed solely to the effect of a single factor in its environment. On the contrary, an organism is subjected to the simultaneous action of all factors in its immediate surroundings. However, some factors exert more influence than do others, and the attempt to evaluate their relative roles had led to the development of the law of the minimum by the German biochemist **Justus Liebig** in 1840.

Corn.

Liebig while investigating the relationship between the available amounts of essential elements and plant growth, discovered that crop yield was frequently limited by elements other than those utilized in the largest quantity. Freely translated, a part of his statement on his experimental results is that "*growth is dependent on the amount of food stuff that is present in minimum quantity.*" This statement has come to be known as **Liebig's law of minimum**. It is now usually incorporated with a **law of limiting factors** developed by a British physiologist **F. F. Blackman** (1905), who at the beginning of this century investigated the factors affecting the rate of photosynthesis. He listed five factors involved in controlling the rate of photosynthesis : amount of CO_2 available, amount of H_2O available, intensity of solar radiation, amount of chlorophyll present, and temperature of the chloroplast. **Blackman** discovered that the rate of photosynthesis is governed by the level of the factor that is operating at a limiting intensity. Further work on limiting factors and substances has shown that a high level of one factor will modify the limiting effect of a second, a process described as **factor interaction** (**Shelford**, 1932). The same principle of limiting factors applies well to animal functions.

The law of minimum has been restated by **Taylor** (1934) in broad ecological terms, as follows : *"The functioning of an organism is controlled or limited by that essential environmental factor or combination of factors present in the least favourable amount. The factors may not be continuously effective but only at some critical period during the year or perhaps only during some critical year in a climatic cycle."*

THRESHOLD AND RATE

Every environmental factor varies through a wider range of intensity than any single organism can tolerate characteristically. For each individual organism there is present a lower and an upper limit in the range of an environmental factor between which it functions efficiently. For any one factor, different organisms find optimal conditions for existence at different points along the range, hence their segregation into different habitats.

The **threshold** is the minimum quantity of any factor that produces a precipitable effect on the organism. It may be the lowest temperature at which an animal remains active, the least amount of moisture in the soil that permits growth of a plant, the minimum intensity of light at which photoreceptor is stimulated, and so forth. Above the threshold, the rate of a function increases more or less rapidly as the quantity of heat, moisture, light or other environmental factor is increased, until a maximum rate is attained. Above the maximum, there is usually a decline in the rate of a process either because of deleterious effect produced, the interference of some other factor or exhaustion.

SHELFORD'S LAW OF TOLERANCE

Organisms may be limited in their growth and their occurrence not only by too **little** of an element or too **low** an intensity of a factor but also by too **much** of the element or too **high** intensity of the factor. For example, carbon dioxide is necessary for the growth of all green plants, small increase in

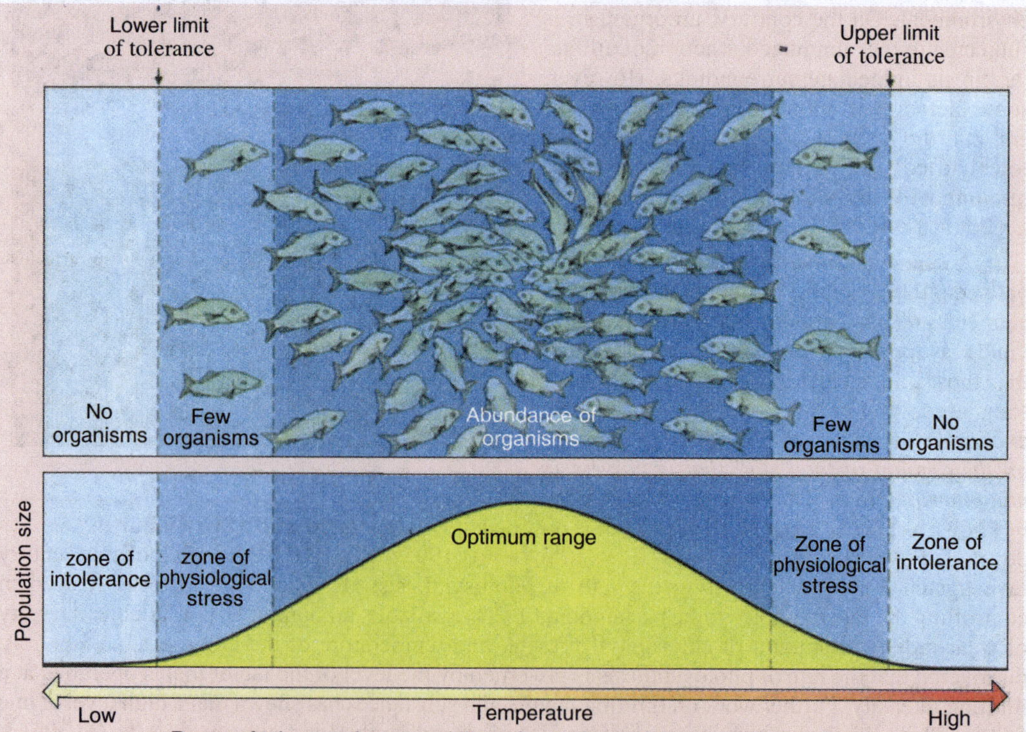

Range of tolerance for a population of organisms of the same species to an abiotic environmental factor –in this case, temperature.

concentration of carbon dioxide in the atmosphere will, under certain circumstances, increase the rate of plant growth, but very considerable increases become toxic. Likewise, small additions of arsenic to the human diet actually have a tonic effect, further increase in the dosage, however, soon proves fatal.

The idea that factors could be limiting at their maximum as well as minimum quantities was incorporated in **law of tolerance** formulated by **V.E. Shelford** in 1913. This law postulates that each ecological factor to which an organism responds has maximum and minimum limiting effect between which lies a range or gradient that is now known as the **limits of tolerance**.

Between the **lower** and **upper** limits of tolerance lies a broad middle sector of a gradient which is called the **zone of compatibility**, the **zone of tolerance**, the **biokinetic zone** or the **zone of capacity adaptation**. The region at either end of the zone of compatibility is called the **lethal zone** or the **zone of resistance** or **zone of intolerance**. The zone of compatibility too includes a broad **range of optimum** and narrow **zones of physiological stresses** in between the range of optimum and lethal zones.

Upper and **lower limits of tolerance** are intensity levels of a factor at which only half of the organisms can survive. These limits are sometimes difficult to determine, as for example with low temperature, organisms may pass into an inactive, dormant, or hibernating state from which they may again become functional when the temperature rises above a threshold. At high temperatures, there may be similar inactivation or aestivation before the lethal level is attained. Even without dormancy occurring, there are normally **zones of physiological stresses** before the limits of tolerance are reached.

The species as a whole is limited in its activities more by conditions that produce physiological discomforts or stresses than it is by the limits of tolerance themselves. Death comes close to the limits of tolerance, and the existence of the species would be seriously risked if it was frequently exposed to these extreme conditions. Therefore, in retreat before conditions of physiological stress there is a margin of safety, and the species adjusts its activities so that limits of tolerance are avoided. There is a variation in hardiness of individuals within a species, so that some hardy individuals find existence possible under conditions that disturb other individuals. The population level of a species becomes reduced before the limits of its range are actually reached.

Further, species vary in their limits of tolerance to the same factor. For example, the Atlantic salmon spends most of its adult life in the sea, but goes annually into freshwater streams to breed. Most other marine fishes are killed quickly when placed in fresh water, as are freshwater fishes when placed in salt water. The following terms (Table 4-2) are used to indicate the relative extent to which organisms can tolerate variation in environmental factors. The prefix *steno*—means that the species, population, or individual has a narrow range of tolerance and the prefix *eury*—indicates that it has a wide range.

Table 4-2. **Terminology for ranges of certain abiotic environmental factors.**

Stenothermal–Eurythermal	Pertaining to temperature
Stenohaline–Euryhaline	Pertaining to salinity
Stenohydric–Euryhydric	Pertaining to water
Stenophagic–Euryphagic	Pertaining to food
Stenobathic–Eurybathic	Pertaining to depth
Stenoecious–Euryecious	Pertaining to niche or habitat selection

Moreover, a plant or animal may have a wide range of tolerance for one factor in the environment, but a relatively narrow range of tolerance for another condition. Thus, we find that some species of freshwater fishes are eurythermal but they are stenohaline.

Lastly, though climatic, edaphic and biotic factors affect plants/animals, their populations and community growth and dynamics take place in a holistic manner; but, it is difficult to understand the mechanism of environmental influences unless we study the different components of environment separately as follows.

1. LIGHT AND RADIATIONS

The radiant energy from the sun is the basic requirement for the existence of life on the earth. This source of energy is of fundamental importance to the photosynthetic production of food by plants and as mentioned previously, the heat budget of the world is dependent on solar radiation. Although we generally, think only in terms of visible light, the sun emits other radiations of different wavelengths—cosmic rays, gamma rays, X-rays, ultraviolet rays, infra-red rays, heat waves, spark discharges, radar waves, radio waves, slow electro-magnetic waves. Biologists have been primarily interested in only three regions near the centre of the electromagnetic spectrum—(1) the infra-red, (2) the visible light and (3) ultraviolet regions. The **infra-red wavelengths**, which are the longest of these three, are not visible to the human eye ; they contribute to the warmth of the earth at the high altitudes in the terrestrial atmosphere. **Visible light**, is only a small fraction of the radiation spectrum and contains the frequency of wavelengths ranges from 390 to 700 millimicrons (mμ). It is made up of a series of colours ranging from violet through indigo, blue, green, yellow, orange and red, all constituting the **visible spectrum**. Light energy, thus, reaches the earth as electro-magnetic waves of solar radiation with tremendous velocity and supplies most of the warmth the earth receives from the sun and also supplies the main source of energy which is utilized in photosynthesis of plants, oriented and rhythmic behaviour of ani-

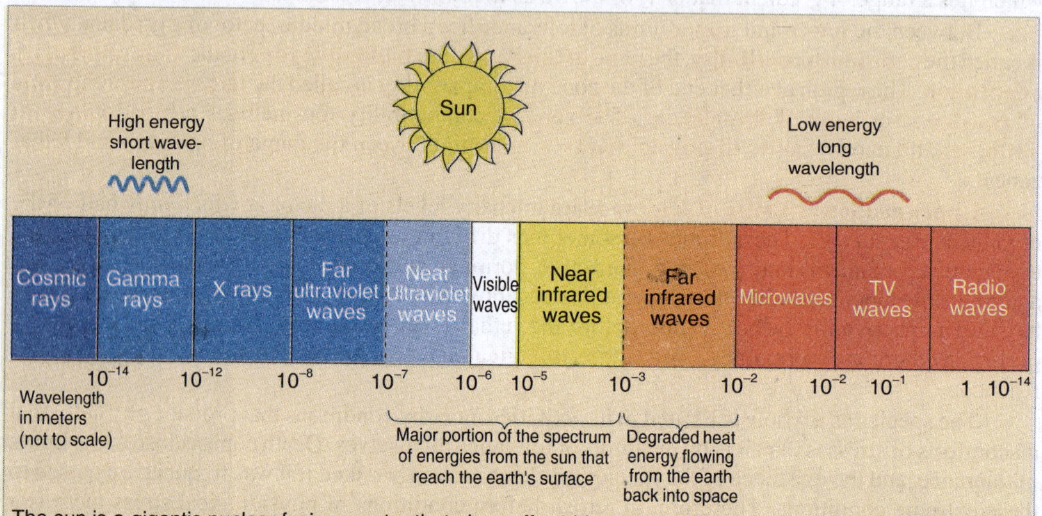

The sun is a gigantic nuclear fusion reactor that gives off a wide spectrum of radiant energy into space. When this energy reaches the earth, much of it is either reflected or absorbed by chemicals in the atmosphere, which prevents most of the harmful, high-energy cosmic rays, gamma rays, X-rays, and far-ultraviolet ionizing radiation from reaching the earth's surface.

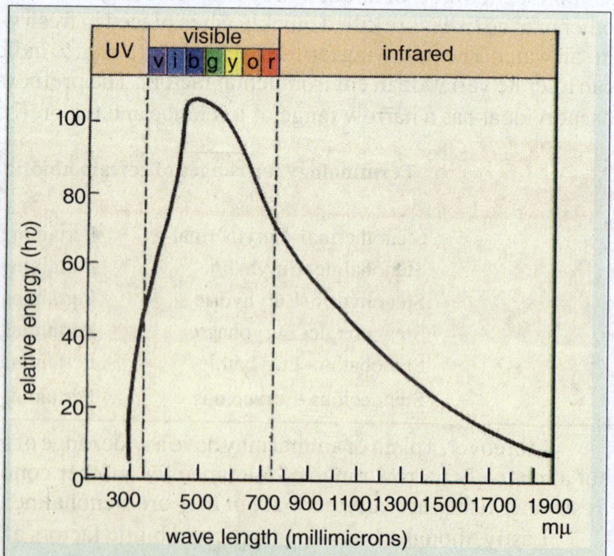

Fig. 4.1. Spectral distribution of ultraviolet light, visible light and infra-red radiation at the earth's surface.

mals, bioluminescence, periodicities of occurrence and periods of inactivity. Unlike temperature, light is a non-lethal ecological factor and it has a specific direction in its flow. The wavelength of **ultraviolet light** from the sun is shorter than that of visible light, and it produces the upper levels of the earth known as the ionosphere.

Light Receptors of Animals

Since animals depend on light for orientation, diurnal migrations and synchrony of rhythmic activities, light reception probably is the most important sensory modality in the exploration of the environment. It is not surprising, therefore, that receptors for light are common to almost all animals : in those with better developed photoreceptors, light greatly influences behaviour. Light receptors may be well-defined organs such as the vertebrate eyes, the compound eyes of Crustacea and insects, the simple eyes or ocelli of other arthropods and invertebrates, and the dermal light receptors.

Primates can distinguish colours.

Many homeotherm animals with normal vision do not commonly respond to ultraviolet or infra-red rays, since their retinas are not stimulated. They also do not properly distinguish between the different colours of the visible spectrum. Dogs, cats, hamsters, and oppossums are colour blind, while horned cattle cannot identify red light. Horses, deer, sheeps, pigs and squirrels cannot distinguish close to red and green. Primates can distinguish colours. Some insects such as honey bee can see ultraviolet radiation.

Light Variations in Different Environments

Light energy varies with different media. The transparency of air and water is important in regulating the amount and quantity of light that may be available in particular habitats. For example, the intensity of light reaching the earth's surface varies with the angle of incidence, degree of latitude and altitude, season, time of day, amount absorbed and dispersed by atmosphere and a number of climatic and topographical factors such as fog, clouds, suspended water drops, dust particles, etc., when the angle of incidence is smaller, light rays have to travel by a longer distance through the atmosphere, which results into relative reduction in intensity. Likewise, sun's altitude changes due to differences in latitude, changes in the season and in the time of day. When sun remains overhead, the intensity of sunlight over the earth's surface will be greatest. At higher latitudes the intensity of light becomes correspondingly reduced. The **illumination** or intensity of daylight is greatly diminished by moisture, clouds, and dust in the atmosphere and also by forest vegetation. The direction and slope of the mountain also affect light intensity. There will be no light on the one side of the slope. Illumination is measured in **lux** or **foot candle**. One foot candle is 10.7 lux. One lux is equal to one **lumen** per square meter. A lumen is equal to 1/620 watt. Most organisms respond to moonlight particularly on a full moon, which has an illumination of 0.25 lux in cloudless weather.

Phyllophora.

The light which enters in the aquatic media, comes from sun by passing through the atmosphere existing above the water surface and, hence, that is subjected to all kinds of atmospheric factors like that of terrestrial media. About 10% of the sunlight which falls over the water surface, is reflected back and rest 90% of that pass downward in the water and is modified in respect to intensity, spectral composition, angular distribution (refraction) and time distribution. The phytoplankton, zooplankton, suspended organic and inorganic particles either reflect or absorb the light rays. Further, in water there is a selective absorption of light at various depths. The longer light rays are absorbed near the surface and in general the shortest light rays penetrate deepest. Thus, long infra-red rays are absorbed in the upper layers of water (about 4 metres); red and orange rays are completely absorbed up to the depth of 20 metres; yellow rays penetrate up to 50 metres and green and blue rays penetrate up to 80 to 100 metres deep. Violet and ultraviolet rays penetrate beyond 100 metres and no light ray penetrates beyond 200 metres depth. Depending upon the penetration of light, oceans are divided into **euphotic zone** (up to 50 metres depth), **disphotic zone** (up to 80 to 200 meters depth) and **aphotic zone** (below 200 meters of depth). In the ocean, algae are distributed according to length of light rays that their colours are best suited to absorb and to utilize. Photosynthesis in deeper waters occurs with blue and green rays, which are absorbed by the brown and red pigments of red algae (*Phyllophora*). The red and blue-green algae use pigments of photocyanin and photoerythrin for photosynthesis. Each plant performs optimum photosynthesis in complementary colours, *e.g.,* green pigment for red rays, brown pigment for green rays, and red pigment for blue rays. Further, since the photosynthesis is a function of the illumination, a variation in the amplitude of which depends on the earth's rotation, altitude, season, presence or absence of clouds and dust and so on.

Effect of Light on the Plants

Light energy influences almost all the aspects of plant life directly or indirectly. Thus, it controls plant's structure, form, shape, physiology, growth, reproduction, development, local distribution, etc. On the basis of light factor certain ecologists have classified plants into **sciophytes**, **shade loving**, or **photophobic plants** which have best growth under lower intensities of light, and **heliophytes** or **photophilous plants** which have best growth in full sunlight. The light factor affects following aspects of plant life :

A. Direct effects of light on plants. Light affects directly the following physiological processes of the plants— 1. It is an essential factor in the formation of **chlorophyll** pigment in chlorophyllous plants. 2. It has a very strong influence on the number and position of **chloroplasts**. The upper part of the leaf which receives full sunshine has larger number of chloroplast which are arranged in line with the direction of light. In leaves of plants which grow under shade, chloroplasts are very few in number and are arranged at right angle to the light rays, thus, increasing the surface of light absorption. 3. Light has its most significant role in **photosynthesis**. During photosynthesis, the green plants which are the "primary producers" of an ecosystem, synthesize their carbohydrate food from water and CO_2 in the presence of sunlight. Thus, during photosynthesis, the solar radiant energy is transformed into the chemical or molecular energy which remains stored in chemical bonds of carbohydrates and this chemical energy is utilized by other chlorophyllous and non-chlorophyllous parts of plants, all animals, bacteria and viruses in their different life activities. The rate of photosynthesis is greater in intermittant light than in the continuous light. At high intensity of light a photo-oxidation of chlorophylls and other enzymes takes place, which consequently reduces the rate of synthesis of carbohydrates and proteins. However, high intensity of light results in the formation of **anthocyanin** pigments. It is for this reason alpine plants have beautifully coloured flowers. 4. Light inhibits the production of **auxins** or **growth hormones** as a result of which it influences the shape and sizes of plants. Plants grown in insufficient light or in the total darkness, produce maximum amount of growth hormones, as a result of which they are elongated with weak pale yellow stems with very few branches. 5. Light also influences certain chemical compounds of plants which affect the differentiation of specialized tissues and organs. 6. Leaf structure too is influenced by the intensity of light. 7. The development of flowers, fruits and seeds is greatly affected by light intensity. Diffused light or reduced light promotes the development of vegetative structures and causes delicacy. For example, vegetative crops such as turnips, carrots, potato and beets give highest yield in regions with high percentage of cloudy days. Intense light favours the development of flowers, fruits and seeds. 8. Duration of light is also very important. Actual duration or length of the day (**photoperiod**) is a significant factor in the growth and flowering of a wide variety of plants. The controlling effect of **photoperiod** is called **photoperiodicity**. According to the response to length of photoperiods, the plants have been classified into following three groups :

(i) Short day plants. Which bloom when the light duration is less than 12 hours per day *e.g., Nicotiana tabacum* (tobacco), *Dahlia variabilis, Chrysanthemum indicum, Cosmos bipinnatus, Cannabis sativa* (hemp), etc.

(ii) Long day plants. Which bloom when the light duration is more than 12 hours per day, *e.g., Allium cepa* (onion), *Beta vulgaris* (beet root), *Daucus carota* (carrot), *Papaver somniferum* (opium poppy), *Vicia faba* (broad bean), *Brassica rapa* (turnip), *Avena sativa* (oat), *Secale cereale* (rye), *Sorghum vulgare* (sorghum), etc.

(iii) Day neutral plants. Which show little response to length of day light, *e.g., Cucumis sativus* (cucumber), *Gossypium hirsutum* (cotton), *Solanum tuberosum* (potato), etc.

9. Light also affects the movement in some plants. The effect of sunlight on the plant movement is called **heliotropism** or **phototropism**. The stems elongate towards light (positive phototropism) and the roots are negatively phototropic. The leaves grow transversely to the path of light. 10. The seeds when moist are very sensitive to light. In some cases the **germination of seeds** is delayed in light. 11. Light is an important factor in the distribution of plants. Some plants grow in full sunlight, while others prefer to grow in the shades.

Roots are negatively phototropic.

Stems are positively phototropic.

B. Indirect effects of light on plants. Light affects opening and closing of stomata, influences the permeability of plasma membrane and has heating effect. All these in turn affect **transpiration** which in turn affects **absorption** of water. Light affects **respiration** of plants indirectly, as in the presence of light the respiratory substrates are synthesized. In many plants the respiratory rate increases with the increase in the light intensity (*e.g., Canna, Nerium, Bougainvillea*). However, in certain plants respiration rate is decreased slightly in intense light.

Effects of Light on Animals

Light affects divergent aspects of animal's life. It influences cellular metabolism, growth, pigmentation, locomotion, reproduction, ontogenetic development, and also controls the periodicity and biological clocks of animals. Some of its significant effects can be discussed as follows :

1. Effect of light on protoplasm. Though the bodies of most animals remain protected by some sort of body covering which save animal tissues from the lethal effects of solar radiations. But, sometimes sun rays penetrate such covers and cause excitation, activation, ionization and heating of protoplasm of different body cells. Ultraviolet rays are known to cause mutational changes in the DNA of various organisms.

2. Effect of light on metabolism. The metabolic rate of different animals is greatly influenced by light. The increased intensity of light results in an increase in enzyme activity, general metabolic rate and solubility of salts and minerals in the protoplasm. Solubility of gases, however, decreases at high light intensity. Cave dwelling animals are found to be sluggish in their habits and to contain slow rate of metabolism.

3. Effect of light on pigmentation. Light influences pigmentation in animals. Cave animals lack skin pigments. If they are kept out of darkness for a long time, they regain skin pigmentation. The darkly pigmented skin of human inhabitants of the tropics also indicate the effect of sunlight on skin pigmentation. The skin pigment's synthesis is dependent on the sunlight. Light also determines the characteristic patterns of pigments of different animals which serve the animals in sexual dimorphism and protective colouration. Animals that dwell in the depths of the ocean where the environment is monotonous, though pigmented do not show patterns in their colouration.

4. Effect of light on animal movements. The influence of light on the movement of animals is evident in lower animals. Oriented locomotory movements towards and away from a source of light is called **phototaxis**. Positively phototactic animals such as *Euglena, Ranatra*, etc., move towards the source of light, while, negatively phototactic animals such as planarians, earthworms, slugs, copepodes, siphonophores, etc., move away from the source of light.

The light directed growth mechanisms are called **phototropisms** which occur in sessile animals. Phototropisms also include responsive movement of some body part of some active animals to the light stimulus, such as the movement of flagellum of *Euglena* towards light and movements of polyps of many coelenterates.

The velocity or speed of the movement of certain animals is also regulated by light. It has been observed that animals when responding to light reduce their velocity of movement and these movements which are non-directional are called **photokinesis**. Photokinesis may be a change in linear velocity (**rheokinesis**) or in the direction of turning (**klinokinesis**). During photokinesis when only a part of the body of an animal deviates away from the source of light, the reaction is termed **photoklinokinesis**. Larvae of *Musca domestica* show such movements. When animals are confronted with two lights of equal brightness they move towards or away to a position that is distanced between the two lights. This is termed **phototropotaxis**. Attraction of males towards the flash of the female is called **telotaxis**. Movement of animals at a constant angle towards the source of light is called **light compass reaction** or **celestial orientation**.

Celestial orientation. Some organisms, particularly arthropods, birds and fish, utilize their time sense as an aid to find their way from one area to another. To orient themselves, the animals use the sun, moon, or stars as a compass. To do this, they utilize both their biological clock and observations on the azimuthal position of the sun in relation to an established direction. The **azimuth** is the angle between a fixed line on the earth's surface and a projection of the sun's direction on the surface. Using the sun as a reference point involves some problems for animals because the sun moves. The target angle changes throughout the day. But animals which use the sun as a reference, correct their orientation somehow. Such celestial orientation has been observed in fishes, turtles, lizards, most birds, and such invertebrates as ants, bees, wolf spiders and sand hoppers.

5. Photoperiodism and biological clocks (Biorhythms). During evolution, organisms have acquired a variety of endogenous rhythms, their periods are matched with the rhythmic events in the environment. A **rhythm** is a recurring process which is wave like in character, because maximum and minimum states appear at identical intervals of time. The time taken between two maxima (peaks) or two minima (troughs) is called a **period** or **cycle** and consists of two phases, a rise and fall in the biological process. The **amplitude** is the range of fluctuations from an average value. The response of different organisms to environmental rhythms of light and darkness is termed **photoperiodism**. Each daily cycle inclusive of a period of illumination followed by a period of darkness is called the **photoperiod**. The term **photophase** and **scatophase** are sometimes used to denote the period of light and the period of darkness respectively.

Some rhythms of organisms are matched to **24 hours** cycle of light and dark (**circadian**), other to the **12.4** or **24.8 hours** tide cycle (**circatidal**), the **29 days** lunar cycle (**circalunar** or **circasynodic**), the **yearly seasons** (**circanular**) or the time between successive spring low waters (**14.7 days**; **semilunar** or **circasyzygic**). Prefixing word *circa* (which means about) with most types of biological rhythms is very necessary because all these internal clocks are only approximately matched, *e.g.,* 24 hours cycle or rhythm does not mean 24 hours by wrist watch, it means about 24 hour or circadian (L. *circa* = about + *di*(*em*) = day). This term was coined by an American ethologist **Franz Halberg** in 1959.

Now it is believed that all animals (and also plants) possess an internal (endogenous) and automatic **clock** that controls the rhythm of behaviour and keeps it going with the help of internal stimuli. This clock is entrained (*i.e.,* set and reset) by external environmental stimuli, called **zeitgebers** or **entrainers** such as day length, lunar phases, tides, temperature, humidity, etc. The biological clocks show following general characteristics: 1. They keep almost the same timings at high and low temperatures, *i.e.,* they are unaffected by metabolic inhibitors which are known to block biochemical reactions. 3. They are controlled by the physiological processes of body. 4. They keep normal cycle even in the absence of environmental cues and are self sustained in nature. 5. They function through nerves (brain) and hormones.

Evening primrose.

Chicory.

Types of biorhythms. (1) Circadian rhythms. Operating on an approximately **24-hour day-night cycle** of the earth's rotation, they are the most studied and well understood biological clocks or biorhythms. They are found in almost all the major taxonomic groups of animal kingdom. In their simplest form, circadian rhythms are reflected in the alternating periods of **activity** and **sleep** which correlate with the **light/dark cycle**. In this cycle some animals remain most active at sunrise and sunset times, such animals are known as **crepuscular**, some animals are active during the night (**nocturnal**) but most animals are active during the day time, they are called **diurnal**. In the diurnal rhythm, the animal activity stage includes small cycles of flitting, flying, running and rest. Among plants showing diurnal rhythm, the flowers of some plants open up before the break of dawn (*e.g.,* at 4 to 5 a.m. in Dogrose, Chicory and Poppy) and in some others, with the onset of dusk (*e.g.,* evening primrose, *Matronalis,* etc.). In fact, when a flower closes up, it does so to protect its inner organs from the cold of the night and against extra moisture. In the day, flowers open up for pollination under most favourable conditions. Such periodic change in the position of the organs (*viz,* petals and leaves) is called **sleep**. Such sort of movements of plant organs are called **nyctinastic movements**. Circadian rhythms also have been observed in metabolic rates, cell division, growth, heart beat, rate of photosynthesis, cellular enzyme activity and a host of other activities.

Poppy.

2. Circatidal rhythms. The biological rhythms synchronized with the **low** and **high tides** (*i.e.,* the alternate rise and fall of the sea due to gravitational pull exerted by moon) in the sea are called **circatidal rhythms**. Thus, the organisms living in the intertidal zone of the sea shore are alternately submerged in water and exposed to air and in doing this, these animals become exposed to various ecofactors such as pressure, salinity, food supply, temperature and risk of predation. Animals inhabiting tidal areas show behavioural periodicity associated with the tides, for example, bivalve molluscs (*e.g.,* pearl oysters) increase filteration rates when high tide brings in more food and sea anemones increase the rate of body expansion and contraction during high tide, waving its tentacles actively to trap more food.

3. Circalunar rhythms. Biological rhythms which are synchronized with the phases of moon are called **circalunar rhythms.** For example, a marine polychaete *Platynereis dumerii* is a long worm-like non sexual creature which lives in a burrow. During its breeding season, this seaworm becomes brilliantly colourful, gets bigger in size, its parapodia, eyes and tentacles also get enlarged and it becomes a sexual form called **heteronereis**. Many *Heteronereis* swarm actively at sea shores during the **full moon**, perform nuptial dance and spawn.

4. Semilunar rhythms. The biological rhythms which are synchronized with the fortnightly cycle of **spring tide** (*i.e.,* high tide occurring a day or two after the new or full moon) and **neap tide** (*i.e.,* low tide which occurs in the middle of the second and fourth quarters of the moon) are called **semilunar rhythms**. For example, the gastropod mollusc pariwinkle, *Littorina rudis* shows a marked 14.5 day periodicity in its locomotory activity; the species lives high up on the shore and is only covered by the high water of spring tides, when it comes out of its burrow to move around and to feed.

5. Circannual rhythms. The activity of some animals and plants is influenced by the **seasons** occurring once in a year. They show circannual rhythms. For example, for most birds the height of the breeding season is the spring; for the deer the mating season is the fall. Likewise, brook trout spawns in the fall; bass and blue gills in late spring and summer. Many animals have a part of their life in the resting stage when they exhibit reduced metabolic activity. They either hibernate, aestivate or migrate from the place to avoid rigours of temperature. Migratory birds undertake long distance migrations every year and show cyclic changes in their body weight, gonad size, plumage and colouration. The larvae of many insects (*e.g.,* parasitic wasp, flesh fly, etc.) show two phases every year (i) **active** (when the insect does work actively) and (ii) **inactive** or **diapause** (a period of arrested growth and development in insects which is under the control of endocrine system. This is an adaptation to avoid adverse conditions).

Mode of function of biological clocks. It is believed that in the course of evolution, DNA evolved the biological clock regulating growth, cell division, diurnal rhythmicity of metabolic processes and finally the activity pattern of organisms. Biological clocks show ecophysiological integrity of the DNA-cell/organism; the evolutionary mechanism has brought stability to these clocks through the heredity.

Circadian rhythm in *Arabadopsis.*

It is hypothesized that the intracellular (=endogenous) clocks are located in the nucleus of the cell and can be compared with a **spring** which is tightened in one phase and gradually released in another phase without ever reaching a zero level. Oscillatory motion occurs between the winding (tension) and unwinding (relaxed) phases. The winding of the spring requires energy, which is received in plants from reaction initiated by light and in mammals from oxidation and reduction reactions. Many biochemical studies suggest that in the **spring model**, the winding phase corresponds to the synthesis of RNA and enzymes and unwinding phase to the breakdown of proteins in the cytoplasm. Such a hypothesis may work well in case of unicellular organisms, but in the case of multicellular and complex organisms various processes occur simultaneously in different tissues and organ systems. It appears that in them a central mechanism may operate through the genes–hormones–organ systems.

Recent findings suggest that in invertebrates both **neural** and **neuroendocrine products** (hormones) are responsible for biological rhythms. In higher vertebrates, the central regulation of rhythms is largely performed by the **hypothalamus-pituitary** complex which is genetically controlled and influenced by environmental factors such as photoperiod. For example, in birds and mammals, the **pineal gland** receives light stimuli and entrains to show biological rhythms. In amphibians, reptiles and birds the pineal gland is located under the skull but in some terrestrial forms of vertebrates, it is placed over the brain in the form of the third eye. The information of photoperiod is received first by the eye and then via neural pathways reaches the pineal body which secretes hormones such as **melatonin** (**Reiter**, 1980) which has antigonadotropic effects in various organisms such as rats, mice and many other mammals. Melatonin is believed to pass to the anterior **pituitary gland** and decrease secretion of gonadotropic hormones (see **Guyton** and **Hall**, 1996). Pineal body partly controls circannual rhythms of reproduction. In case of turtles, **serotonin** is synthesized during the day whereas melatonin during the night but this mechanism is completely lost during hibernation period.

6. Effect of light on reproduction. In many animals (*e.g.,* birds) light is necessary for the activation of gonads and in initiating annual breeding activities. The gonads of birds are found to become active with increased illumination during summer and to regress during shorter periods of illumination in winter.

7. Effect of light on development. Light in some cases (*e.g., Salmon* larvae) accelerates development, whereas, in other (*e.g, Mytilus* larvae) it retards it.

Further, occasionally the output of sunlight is increased by the development of **sunspots**. As a result of this excess energy is radiated to space and this naturally increases the output of solar energy near the earth. A direct consequence of this is the greater evaporation of water which results in cloud formation to prevent more exposure to sunshine and thus to equalize temperature and modify climate.

2. TEMPERATURE

Temperature is one of the essential and changeable environmental factor. It penetrates into every region of the biosphere and profoundly influences all forms of life by exerting its action through increasing or decreasing some of the vital activities of organism, such as behaviour, metabolism, reproduction, ontogenetic development (*viz.,* embryogenesis and blastogenesis) and death. Temperature is a universal influence and is frequently a limiting factor for the growth or distribution of animals and plants. Normal life activities go on smoothly at a specific temperature or at a specific range of temperatures. This is called the **optimum temperature** or the optimum range of temperature. Organisms react to any rise or fall of the optimum temperature range and biotic communities more often encounter alterations only due to extremes of temperature (*viz.,* **minimum** or **maximum** temperatures). The interaction of temperature with certain other abiotic environmental factors such as humidity, etc., cause into many other climatic changes which influence the living organisms in one way or another.

Nature of Temperature

Temperature is a measure of the intensity of **heat** in terms of a standardized unit, and is commonly expressed as degrees on either the **Fahrenheit** or **Celsius scale**. Heat is a form of energy, and as such is necessary for the very existence of life. Heat may be received or lost by animals as molecular vibrations transmitted from one part of an environment to another by radiation, convection, or conduction. It may be received as radiant energy, ofcourse largely from the sun. Heat may be produced by mechanical work as in Hawk moth, the heat is generated by vibrating the wings before take-off; or by chemical reactions.

Heat Budget

The temperature at the earth's surface is governed by the brightness of the sun; the constancy of brightness of sun has remained virtually unchanged for about 3 billion years. In fact, average terrestrial temperatures now probably do not differ radically from those at the earth's beginning (**Schwarzschild**, 1967). The total amount of heat entering the biosphere from the sun must balance the amount lost per unit time if temperatures are to remain unchanged, since the flow of geothermal heat from the interior of the earth is small by comparison and probably has been negligible for at least the past 500 million years. The estimate of this energy flow is referred to as the **heat budget** (**Vernberg** and **Vernberg**, 1970). Heat budget has been projected for the total surface of the earth as well as for special environments.

Temperature Fluctuations in Different Environments

Environmental temperatures fluctuate both daily and seasonally. Different environments such as freshwater, marine and terrestrial environments, are subjected to varied responses to fluctuating temperature. Temperature fluctuations are comparatively less in the aquatic environment than in the terrestrial environment. The increase in depth of aquatic medium often increases the temperature fluctuations. There exists a distinct difference in the response of living organisms of the freshwater and sea to temperature fluctuations because of the presence of dissolved salts in sea water. The minimum temperature in the sea is–3°C, while in freshwater pond it never goes below 0°C. The maximum temperature of ocean generally goes up to 36°C, but in shallow pools of freshwater and tide of pools littoral zone, temperature may go higher. In deeper bodies of water, heating and cooling are restricted to the surface strata. But the deeper layers also get a lot of heat as a result of what is usually termed **vertical circulation**, where in, due to circulatory movement of water, surface waters are brought to the deeper regions, and vice versa. Studies on the vertical changes of temperature have led to the hypothetical classification of the freshwater media into three strata. The superficial layer of freshwater is constantly stirred by wind and is called **epilimnion**. It is the layer of warmer water and its temperature may rise up to 22°C during summers. The stagnant water of the bottom constitutes the **hypolimnion**. The hypolimnion has temperature of 5°C to 9°C. In between epilimnion and hypolimnion occurs an intermediate zone called **thermocline** or **metalimnion**, which has rapid vertical temperature changes (Fig. 4.2 B). During winter, the temperature of epilimnion of a freshwater lake becomes 0°C and the lake becomes ice covered. The process of differentiation of freshwater habitat into these three strata is called **thermal stratification**.

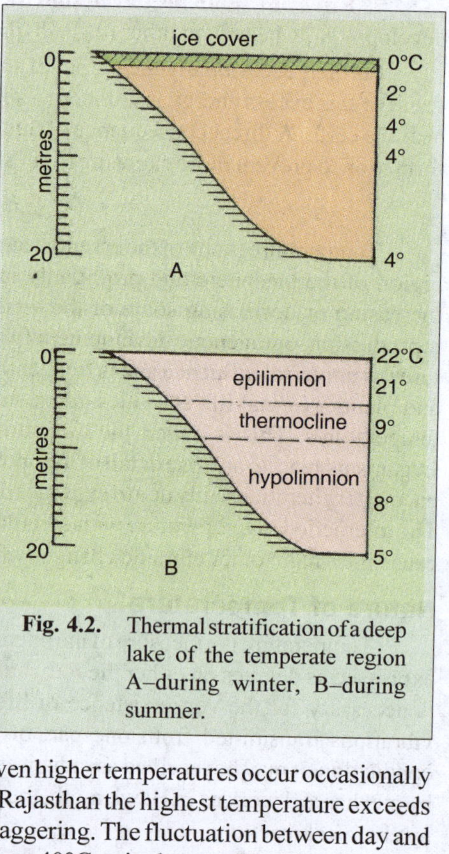

Fig. 4.2. Thermal stratification of a deep lake of the temperate region A–during winter, B–during summer.

In the terrestrial environment the seasonal and daily fluctuations in temperature are varied and marked. The lowest temperature recorded for any land mass is –70°C (Siberia in 1947). Higher temperatures may likewise go often 85°C as in certain deserts at noon. However, the water in hot springs and geysers may approach 100°C, and even higher temperatures occur occasionally in the very special situation presented by volcanic areas. In Rajasthan the highest temperature exceeds 50°C. On land, diurnal variations of temperature are quite staggering. The fluctuation between day and night temperatures may be 17°C as in ordinary land masses or 40°C as in deserts. The Thar Desert of South Rajasthan (India) shows a diurnal range of 20–30°C for all seasons. Difference in latitudes often cause variation in the annual temperature cycles with visible effects on the organisms. Further, with the increased altitude for every 150 meters, the decrease in temperature of 1°C takes place.

Range of Temperature Tolerance

Life in this universe exists within a range of—200 to 100°C. Though normal life persists within narrow temperature limits of about—10 to + 50°C. Individual species survive in a smaller range and are active within even narrower limits. Many nematodes, rotifers and tardigrades have shown to withstand cooling to –272°C without ill effects. Larvae of chironomids and certain other Diptera have been found to thrive at temperatures near 55°C. Praying mantis is reported to live in bare grounds at a temperature of 62°C in deserts. Some algae and bacteria are reported to live in hot springs at 88°C. Non-photosynthetic bacteria inhabiting hot springs can actively grow at temperatures greater than 90°C (**Bott** and **Brock**, 1969). On the other land, some arctic algae can complete their life cycles in places where the temperatures barely rise above 0°C. Further, the eggs of the acanthocephalan *Macracanthorhynchus hirudinaceus* have been known to withstand temperatures from – 10 to – 45°C for about 140 days, and desiccation at temperatures up to 39°C for a period of 265 days.

The organisms (microbes, plants and animals) which can tolerate very large fluctuations in temperature are called **eurythermal organisms**. For example, cyclops, toad, wall lizard, grass snake, man, etc., are the eurythermal animals. The organisms which can tolerate only a small variation in temperature are termed **stenothermal organisms**. The common stenothermal animals are fishes, snails, coral reefs, etc. In organisms all metabolic processes necessary for life start at a certain **minimum temperature** and increase with rise in temperature until they reach the maximum level at a temperature called **optimum temperature**. Further rise in temperature beyond optimum brings about decrease in metabolic rate, until it ceases at a temperature called **maximum temperature**.

Thus, the favourable temperature range for any particular species is determined by the prevailing temperature at which normal physiological activities of the animals take place, as in the distribution of the rotifer *Keratella procura* (**Nayar**, 1970). This rotifer species is known to appear in the ponds of Pilani, Rajashthan (India), when the temperature is below 24°C and to disappear when it rises above 24°C, the frequency of distribution reaching a peak during months of October to March, with the fall in temperature.

Poikilotherms and Homeotherms

An animal's response to wide ranges in temperature is influenced by its physiology. All invertebrates, lower chordates (hemichordates, urochordates and cephalochordates), cyclostomes, fishes, amphibians and reptiles have no internal mechanism for temperature regulation, and their body temperatures vary with the surrounding environmental temperatures. Such animals are called **cold blooded**, **ectothermic** or **poikilothermic** organisms. The temperature range that a poikilothermic animal tolerates can be correlated closely with the environmental temperatures normally encountered in nature : tropical animals cannot withstand low temperatures; polar animals cannot withstand high temperature; temperate-zone animals survive a wide temperature range. Many poikilothermic animals show a rather precise discrimination of temperature. If the temperature decreases or increases appreciably, these animals may tolerate and adjust to change, may avoid the change by seeking a less exposed environment or may become inactive. Some animals respond by becoming dormant during periods of extreme temperatures. Such dormancy is called a **estivation** at high temperatures and **hibernation** at low temperatures. For example, most poikilotherms become inactive when the temperature of their surroundings goes below 8°C or rises to 42°C.

The dormouse hibernates in winter when food is scarce.

A few exceptional poikilotherms, especially insects, certain amphibians and reptiles, exercise a degree of thermoregulation by either physiological or behavioural mechanisms. For example, Hawkmoths can rise the temperature of their flight muscle to 32°–36°C by vibrating the wings before take-off and gregarious butterfly larvae may raise their temperature 1½–2°C when clustered together. Locust, and grasshoppers may increase their temperature 10°C by basking, sideways in the sun. Ants move their larvae to warm or cool places within the nest and bees maintain temperatures within their hives between 13° and 25°C by fanning with their wings to evaporate water droplets when it is too hot, or releasing body heat through increased metabolic activity, when too cold.

When temperature drops, lizards bask in the sun to achieve the desired body temperature; once this thermal level is attained, they will divide their time between sun and shade to maintain it. Poikilotherms such as desert dwellers that live in environments where the temperature is apt to be very high are often nocturnal and, thus, avoid the highest temperatures of the day. Some poikilotherms, both vertebrates and invertebrates, lower their body temperatures slightly by evaporative cooling. In frogs and reptiles, evaporative cooling can occur through the skin or via the respiratory tract by panting (**Warburg**, 1967). Among the invertebrates, evaporative cooling has been reported in tropical intertidal-zone animals (**Lewis**, 1963).

In contrast to poikilotherms, birds and mammals can, within limits, maintain constant body temperatures, regardless of temperature variations of air and water. Such animals are termed **warm blooded**, **homeothermic** or **endothermic animals**. The life processes are adjusted to function at the animals's normal temperature, averging a little less than 38°C in mammals and 3 to 4° higher in birds. If its temperature control fails, the animal dies. The homeothermic animals are able to maintain the constancy of body temperature by a combination of several factors : (1) a thermoregulating centre in the brain (hypothalamus) (2) insulation (3) a peripheral vascular response to ambient temperature; and (4) metabolic compensation.

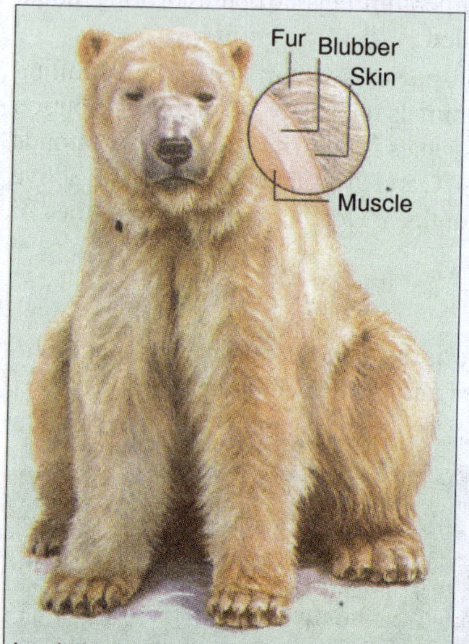

Insulation by fur and blubber in polar bear.

Among these factors, **insulation** is highly important particularly in enabling large arctic animals to withstand very low temperature. If the temperature is lowered, the oxygen consumption rate of tropical mammals increases, whereas that of arctic mammals remains basal. This response pattern is due to differences in insulation; some arctic animals are fur-insulated, whereas others are insulated with a layer of blubber. Many small animals, however, are not well insulated and must seek heat retention by other means, often by huddling together.

Homeothermic animals living in very hot environments cannot tolerate greatly elevated body temperatures and they utilize methods that facilitate heat transfer to the environment, including an increased peripheral blood flow and surface cooling by sweating or panting. Birds may accomplish this by rearranging their plumage so that more skin is exposed. There is however, no reduction in metabolic rate. Animals subjected to high temperature may exhibit diurnal patterns of behaviour–that is, they may reduce locomotor activity during the heat of the day or may move into the shade to avoid direct sunlight. Some animals such as monotremes and marsupials have a limited power of temperature regulation, they are called **heterothermic animals**; *e.g.*, the pigmy mouse and the little pocket mouse, which respond to temperature extremes by a estivating or hibernating, and others such as the humming bird, which experience a nocturnal drop in temperature.

Plants too can be divided into the following three categories on the basis of their heat tolerating capacity : **megatherms**, **microtherms** and **mesotherms**. Megatherm plants occur in warm habitat (*e.g.* desert vegetation). Microtherm plants occur in cold habitat, (*e.g.*, plants of high altitudes). Mesotherm plants are the plants of the habitat which is neither very hot nor very cold (*e.g.*, aquatic plants). **Hekistotherms** include alpine vegetation which tolerate very low temperatures.

Factors or Variables Affecting Organismal Response to Temperature

The response of an organism to temperature is affected by a number of factors such as **thermal history**, **genetic differences**, **diet**, **size**, **stage in life cycle**, **sex**, **moulting**, **parasitism**, **hormones**, etc. 1. Generally, animals found in warmer environments can withstand higher temperatures than animals from colder situations. It is also generally true that animals from colder environments tolerate lower temperatures better than those from the warmer climates. These differences between warm-climate and cold-climate populations may reflect basic genetic differences or may be phenotypic expressions resulting from different thermal histories. 2. The thermal resistance of a species may be closely identified with the genetic composition of the parent stock, as illustrated by hybrid development in two anuran species, *Bufo valliceps* and *B. luetkeni* (**Ballinger** and **Mckinney**, 1966). The lower lethal temperature for development of *B. valliceps* is 18°C ; for *B. luetkeni,* 22°C. In their hybrid, the lower lethal limit is found to be in between that of the two parents at 19.5°C. 3. In some animals a relationship between the total intake of food and resistance to thermal stress has been demonstrated. Some species are more sensitive to elevated temperatures when they are starved for even short periods of time. Goldfish showed increased resistance to high temperature when placed on a high fat diet (**Hoar** and **Cottle**, 1952). 4. The size of animal body is found to have some correlation with thermal lethal limits of animals. In some species, the smaller animals are more resistant to higher temperatures than the larger ones; in some species the reverse is noted, and in other species size is not an apparent variable. Further, some smaller animals die faster at low temperatures than do larger animals of the same species, but at higher temperatures body size is not a factor. 5. Many species of animals have exceedingly complex life cycles during which the larval stages are not only morphologically dissimilar from the adult but also occupy different ecological niches. In the wharf crab, *Sesarma cinereum*, thermal requirements of the planktonic zoeal stages are different from those of adults because these larvae are limited to a smaller temperature range than are the adults. 6. The female sex is found to be more tolerant than males to the temperature fluctuations. 7. Among certain invertebrates, such as, in some crabs, moulting adversely affected heat resistance, but had no effect on other species. 8. Mud-flat snails, *Nassarius obsoleta*, which are heavily infected with trematode larvae (*i.e.,* parasites) cannot withstand high temperatures like the non-parasitized snails. 9. Hormones have been shown to influence cold and warm resistance as well as acclimatization (*viz.,* temperature adaptation) to temperature, especially in mammals and a few other invertebrates.

Effect of Temperature on Plants and Animals

Temperature has been found to affect the living organisms in various ways. Some of well-studied effects of temperature on living organisms are the following :

1. Temperature and cell. The minimum and maximum temperatures have lethal effects on the cells and their components. If too cold, cell proteins may be destroyed as ice forms, or as water is lost and electrolytes become concentrated in the cells; heat coagulates proteins (**Lewis** and **Taylor**, 1967).

2. Temperature and metabolism. Most of metabolic activities of microbes, plants and animals are regulated by varied kinds of enzymes and enzymes in turn are influenced by temperature, consequently increase in temperature, up to a certain limit, brings about increased enzymatic activity, resulting in an increased rate of metabolism. For instance, the activity of liver arginase enzyme upon arginine amino acid, is found to increase gradually and gradually, with the simultaneous increase in the temperature from 17°C to 48°C. But an increase in temperature beyond 48°C is found to have an adverse effect on the metabolic rate of this enzymatic activity which retards rapidly.

In plants, the absorption rate is retarded at low temperature. Photosynthesis operates over a wide range of temperature. Most algae require lower temperature range for photosynthesis than the higher plants. The rate of respiration in plants, however, increases with the rise of temperature, but beyond the optimum limit high temperature decreases the respiration rate. The rate of respiration become doubled (like in animals) at the increase of 10°C above the optimum temperature, provided other factors are favourable (**Vant Hoff's law**). However, optimum temperature for photosynthesis is lower than that for respiration (**Smith**, 1974).

3. Temperature and reproduction. The maturation of gonads, gametogenesis and liberation of gametes takes place at a specific temperature which varies from species to species. For example, some species breed uniformly throughout the year, some only in summer or in winter, while some species have two breeding periods, one in spring and other in fall. Thus, temperature determines the breeding seasons of most organisms.

Temperature also affects fecundity of animals. **Fecundity** of an animal is defined as its reproductive capacity, *i.e.,* the total number of young ones given birth during the life time of the animal. For example, females of the insect, acridid *Chrotogonus trachypterus* became sexually mature at 30°C and 35°C than at 25°C, and the highest number of eggs per female was laid at temperature of 30°C. The number of eggs decreased from 243 to 190 when the temperature was raised to 30–35°C (**Grewal** and **Atwal**, 1968). Likewise, in grasshopper species—*Melanoplus sanguinipes* and *Camnula pellucida* when reared at 32°C produce 20–30 times as many eggs than those reared at 22°C (see **Ananthakrishnan** and **Viswanathan**, 1976). On the other hand, the fecundity of certain insects such as cotton stem weevil (*Pempherulus affinis*) was found to decline with an increase in temperature beyond 32.8°C (**Ayyar** and **Margabandhu**, 1941).

4. Temperature and sex ratio. In certain animals the environmental temperature determines the sex ratio of the species. For example, the sex ratio of the copepod *Macrocyclops albidu* is found to be temperature dependent. As the temperature rises there is a significant increase in number of males. Similarly in plague flea, *Xenopsylla cheopis*, males outnumbered females on rats, on days when the mean temperature remains in between 21–25°C. But the position becomes reverse on more cooler days.

Xenopsylla cheopis.

5. Temperature and ontogenetic development. Temperature influences the speed and success of development of poikilothermic animals. In general, complete development of eggs and larvae is more rapid in warm temperatures. Trout eggs, for example, develop four times faster at 15°C than at 5°C. The insect, chironomid fly *Metriocnemus hirticollis*, requires 26 days at 20°C for the development of a full generation, 94 days at 10°C, 153 days at 6.5°C, and 243 days at 2°C (**Andrewartha** and **Birch**, 1954). However, the seeds of many plants will not germinate and the eggs and pupae of some insects will not hatch or develop normally until chilled. Brook trout grows best at 13°C to 16°C, but the eggs develop best at 8°C.

The ermine, also called the stoat, grows a white coat in winter. In its summer coat it would show against the snow.

6. Temperature and growth. The growth rate of different animals and plants is also influenced by temperature. For example, the adult trouts do not feed much and do not grow until the water is warmer than 10°C. Likewise, in the oyster *Ostraea virginica*, the length of the body increase from 1.4 mm to 10.3 mm when temperature is increased from 10°C to 20°C. Sea urchin *Echinus esculentus* shows maximum size in warmer waters. Corals flourish well in those waters which contain water below 21°C.

7. Temperature and colouration. The size and colouration of animals are subject to influence by temperature. In warm humid climates many animals such as insects, birds and mammals bear darker pigmentation than the races of some species found in cool and dry climates. This phenomenon is known as **Gloger rule**. In the frog *Hyla* and the horned toad

Phrynosoma, low temperatures have been known to induce darkening. Some prawn (crustacean invertebrates) turn light coloured with increasing temperature. The walking stick *Carausius* has been known to became black at 15°C and brown at 25°C.

8. Temperature and morphology. Tempera-ture also affects the absolute size of an animal and the relative properties of various body parts (**Bergman's Rule**). Birds and mammals, for example attain a greater body size when they are in cold regions than in warm regions, and colder regions harbour larger species. But poikilotherms tend to be smaller in colder regions. Body size has played a significant role in adaptation to low temperature because it has influenced the rate of heat loss. According to **Brown** and **Lee** (1969), larger wood rats have a selective advantage in cold climates, apparently because their surface to air ratio and greater insulation permit them to conserve metabolic heat. For opposite reasons small-sized animals are favoured in deserts.

Further, the extremities of organism such as tail, ears and legs of mammals often appear to be shorter in colder climate (**Allen's Rule**). Mice reared at 31° to 33.5°C have longer tails than those of the same strain reared at 15.5° to 20°C. Moreover, the races of birds with relatively narrow and more acuminate (*i.e.,* tapered to a slender point) wings tend to occur in colder regions, while those in warmer climates tend to be broader (**Rensch's Rule**). Temperature also influences the morphology of certain fishes and is found to have some relation with the number of vertebrae (**Jordon's Rule**). Cod which hatches off New Foundland at a temperature between 4° and 8°C has 58 vertebrae, while that hatches East of Nantucket at a temperature between 10° and 11°C has 54 vertebrae.

9. Temperature and cyclomorphosis. The relation between seasonal changes of temperature and body form is manifested in a remarkable phenomenon termed **cyclomorphosis** exhibited by certain cladocerans such as *Daphnia* during the warm months of summer. These crustaceans show a striking variation in the size of their helmet or head projection between the winter and summer month (**Coker**, 1931). The helmet develops on the *Daphnia* head in spring, it attains its maximum size in summer and disappears altogether in winter to provide usual round shape to the head. Such a kind of cyclomorphosis in the terms of size of the helmet is clearly showing a correlation to the degree of warmth of different seasons. These prolongations of the helmet have been interpreted as an adaptation aiding floatation since the buoyancy of water becomes reduced as the temperature increases (the **buoyancy hypothesis**). According to other interpretation (*viz.,* **stability hypothesis**), the helmet acts like the rudder and gives greater stability to the animal. Besides temperature such structural polymorphism can be caused by other environmental factors including the food.

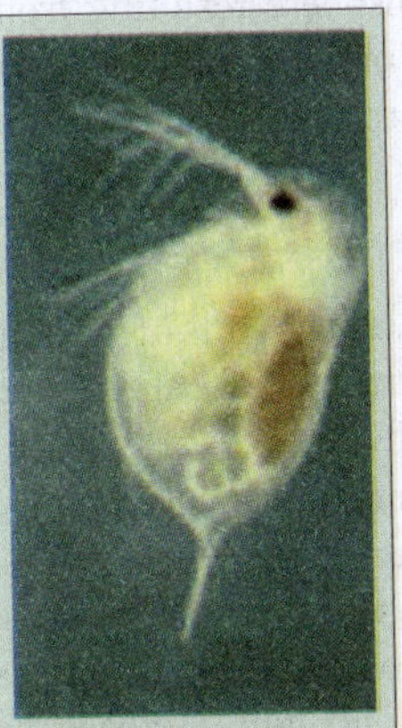

Daphnia.

10. Temperature and animal behaviour. Temperature generally influences the behavioural pattern of animals. In temperate waters the influence of temperature on the behaviour of wood borers is profound. For example, in the winter months in general, both *Martesia* and *Teredo* occur in smaller numbers in comparison with *Bankia campanulata* whose intensity of attack is maximum during the winter months. Further, the advantage gained by certain cold blooded animals through thermotaxis or orientation towards a source of heat are quite interesting. Ticks locate their warm blooded hosts by a turning reaction to the heat of their bodies. Certain snakes such as rattle snake, copper heads, pit vipers are able to detect mammals and birds by their body heat which remains slightly warmer than the surroundings. Even in the dark these snakes strike on their prey with an unnerving accuracy, due to heat radiation coming from the prey. The arrival of cold weather in temperate zones causes the snakes to coil up and huddle together.

11. Temperature and animal distribution. Because the optimum temperature for the completion of the several stages of the life cycle of many organisms varies, temperature imposes a restriction on the distribution of species. Generally the range of many species is limited by the lowest critical temperature in the most vulnerable stage of its life cycle, usually, the reproductive stage. Although the Atlantic lobster will live in water with a temperature range of 0° to 17°C, it will breed only in water warmer than 11°C. The lobster may live and grow in colder water but a breeding population never becomes established there. Not only temperature affects on breeding in the geographical distribution but also temperature affects on survivality (*i.e.,* lethal effect of temperature), feeding, and other biological activities which are responsible in geographic distribution of animals. As noted earlier in this chapter, the animals from colder geographic regions are generally less heat tolerant and more cold tolerant than those animals from warmer regions ; for example, members of *Aurelia,* a jelly fish from Nova Scotia die at a water temperature of 29°–30°C, while *Aurelia* from Florida can tolerate temperatures up to 38.5°C. Thus, lethal limit of temperature may regulate the range of distribution of *Aurelia.*

Terrestrial invertebrates, particularly arthropods generally are distributed in all thermal environments where life is found. Many arthropods that have invaded the colder areas have one stage in their life cycle which is very resistant to cold, enabling them to over-winter until warmer weather returns (**Salt**, 1964). Birds and mammals are also adapted to live in nearly all thermal environments. The distribution of amphibians and reptiles, however, is limited to the relatively warmer thermal climates. **Hock** (1964) has listed three factors that limit the invasion of reptiles into cold environments : the daily ambient (=atmospheric) temperature must be high enough and long enough to allow breeding and to allow adults and youngs to acquire food for "overwintering" and there must be adequate sites for hibernation.

12. Temperature and moisture. The differential heating of the atmosphere resulting from temperature variation over the earth's surface produces a number of ecological effects, including local and trade winds and hurricanes and other storms, but more importantly it determines the distribution of precipitation.

Thermal Adaptations of Plants and Animals

Most animals and plants of different ecological habitats have developed various sorts of thermal adaptations during the course of evolution to overcome the harmful effects of extremes of temperature. Some of the significant thermal adaptations of plants and animals are the following :

1. Formation of heat resistant spores, cysts, seeds, etc. Some of the animals and plants produce heat resistant cysts, eggs, pupae, spores and seeds which can tolerate extremes of temperatures. *Amoeba* in encysted conditions, can tolerate temperature below 0°C. Similarly, *rye* seeds remain active even at 0°C and can germinate at that temperature. As an adaptation against frost the starch of plants changes to fats or oils in the autumn. The fatty oils diminish the freezing points and, thus, increase the power of resistance in plants against frost. Many leaves, that grow in the coldest lands, store fats. Pentosans mucilage and pectic substances which have high moisture retaining power are abundant in many plants. They decrease the danger of plants from desiccation during extremes of heat and save them from death.

North African jerboa, or desert rat, lives underground by day to protect themselves from the burning rays of sun. They come out to hunt in the cool of the night.

2. Removal of water from tissue. Dried seeds, spores and cysts avoid freezing because there remains no liquid in them that can freeze. Due to removal of water from seeds, the cold resistance of seeds of certain plants increases up to the extent that their exposure for 3 weeks to 190°C, does not diminish their germinating capacity.

3. Dormancy. Dormancy includes two already discussed phenomena namely **hibernation** and **aestivation**. During both kinds of dormancies metabolic rate becomes reduced, body temperature becomes low and heart beat rate is also reduced.

4. Thermal migration. Thermal migrations occur only in animals. The journeys taken by animals that enable them to escape from extremely hot or cold situations are referred to as **thermal migrations**. For example, desert animals move to shaded places to avoid burning heat of noon and some animals such as desert reptiles and snakes become nocturnal to avoid heat of the day. The frogs, toads, other amphibians, turtles, etc., make short trips into or out of water (or moist places) and this provides desired cooling and warming to the animal.

3. PRECIPITATION (RAIN FALL)

The moisture falling on an area in liquid, vapours or frozen form is termed as **precipitation**. Thus, precipitation includes all moisture that comes to earth in the form of rain, snow, hail and dew. Precipitation is the chief source of soil water. The water available to plants and animals from soil comes as a result of rainfall. Due to **water cycle** or **hydrological cycle**, there occurs an interchange of water between the earth's surface and the atmosphere. In this cycle following, two important events are involved : **precipitation** and **evapotranspiration**. The ecofactor of precipitation depends upon season, wind, air pressure and temperature.

Evapotranspiration.

Precipitation occurs as a result of the cooling and condensation of water vapour at high altitudes. The low temperature at high altitudes cools the air, which gets saturated and loses its water-holding capacity. As the temperature starts falling, the water vapour condenses and falls as rain due to gravity. Depending on the environmental conditions, precipitation falls as hail, snow or rain. In winter the ground temperature falls and as a result, atmospheric vapour gets condensed as **dew** or **frost** on the surfaces of objects, plants, animals, soils, etc. Dew becomes an important source of moisture to plants in the winter season. **Drizzle** involves minute drops appearing as to float in air. **Rain** is the drops of liquid water, which are larger than drizzle and also heavier. **Snow** is the moisture as solid state. **Sleet** is the form of small grains or pellets of ice, whereas **hail** consists of balls or lumps of ice.

Dew.

Light drizzle is of little importance as very little moisture penetrates the soil because much of it evaporates rapidly. Snow is injurious to plants, breaking tender branches, flowers and fruits. Hail and sleet also cause similar damage. Some sedges grow in snow patches. Of all the above forms of precipitation, the rain is most important. It is the source of soil water and also affects humidity of atmosphere. The total precipitation of the world is about 4.46×10^{20} g per year and of this amount 0.99×10^{20} g fell on land and the rest on oceans (**Hutchinson**, 1957). In India, rains are caused by monsoons; **monsoon** is a special pattern of moist air movement prevalent in India. During monsoon, the air masses moving from the Arabian sea to the West coasts of India and from the Bay of Bengal to the eastern part of country, become extremely moist. About 45 per cent of the water available during annual precipitation flows into river, 20 per cent percolates in the ground and the remaining 35 per cent is lost by evaporation. Gentle steady rains are most effective because much of it penetrates the soil. Torrential rains on the land are most disastrous because they lead to flooding, soil erosion, destruction of vegetation and of animals.

The quantity, duration and intensity of rainfall profoundly regulate the vegetation of any place. For example, in tropical areas with heavy rainfall throughout the year, vegetation mainly include **evergreen forests**. In countries, with heavy rainfall during winter and low during summer, there are present **sclerophyllous forests**. The plants are shrubs, stunted in height, with leathery, thick, evergreen leaves. The areas with heavy rainfall during summer and low during winter are characterized by the presence of **grasslands**. The regions where rainfall is scanty, are seen with deserts and xerophytic vegetation.

An evergreen forest.

The average annual rainfall of India is about 117 cm, the highest in the world. However, there exists great variations between the different regions in India with regard to rainfall. Thus, the average annual rainfall in Assam and the north-east is about 250 cm. In Cherrapunji in Meghalaya, the annual average of rainfall is about 1,100 cm, whereas in Jaisalmer in Rajasthan it is only 20 cm. The distribution of rainfall over India depends largely on the position of hills and mountains and the forest cover. Thus, monsoon winds strike the Western Ghats from a southwest direction and shed most of their water on the windward side. Bombay being on the windward side, receives 200 cm of rainfall annually during the monsoon, while, Pune being on the leeward side (*i.e.*, side of anything away from wind) receives only about 75 cm. If the rainfall to evaporation ratio is zero or less, deserts develop. Grasslands develop when this ratio is more than 0.2 and less than 1 and forests develop when the ratio is more than 1 (around 1.6 to 2). Often in forests, the floor receives rain from trees in the form of drips from tree leaves due to accumulation of condensed water on winter mornings.

4. HUMIDITY OF AIR

Atmospheric moisture in the form of invisible vapour is known as **humidity**. The humidity of air is expressed in terms of values of **relative humidity** which is the amount of moisture in air as percentage of the amount which the air can hold at saturation at the existing temperature. Relative humidity is measured by the instrument called **psychrometer** or by paper strip **hygrometer** or a **thermo-hydrograph**. Humidity is greatly affected by intensity of solar radiation, temperature, altitude, wind exposure, cover and water status of soil. High temperature increases the capacity of the air to retain moisture and results in lower relative humidity. Low temperatures result in higher relative humidity by decreasing the capacity of air for moisture and reaches 100% or saturation point. Further cooling results in condensation of vapour into water and this temperature/moisture point is called the **dew point**. Daily variations in relative humidity values depend upon the type of habitat conditions. In plains and deserts, it may show variations during day, whereas in oceanic islands there is little variation, being same throughout the year.

Humidity plays an important role in the life of plants and animals. It affects the life processes such as transpiration, absorption of water, etc. The rate of evapo-transpiration is modified by the factors as saturation deficit (*i.e.*, degree of wetness or dryness), temperature and wind velocity. Parasitic fungi becomes abundant in moist weather and reduces primary production (see **Ambasht**, 1990). Some plants such as orchids, lichens, mosses, etc., make direct use of atmospheric moisture. In fungi, and other microbes, humidity plays an important role in germination of spores and subsequent stages in life cycle.

5. FIRE

Fire is an interesting ecofactor. Fire is of a common occurrence in natural vegetation all over the world; it is more common in drier habitats than the wet. Lightening is the commonest natural cause of fire initiation. Our earth's surface is hit by lightening every second in one or another part of the globe and many of these are of great magnitude. Other causes of fire are abrassive effects of falling rocks or of dried plant material such as bamboos, or spontaneous combination of very dry and hot material or by volcanic activities. Most forest fires are now man-caused, *i.e.,* by incendiarists (such as poachers), debris burners, smokers, campers, short-circuiting of high-tension electric lines, and nearby railway lines. In the Meghalay region, tribals and even modern farmers practice the *Jhum* cultivation by periodically slashing the forest, burning them and raising crops on the ash enriched soil for a few years and moving to another segment of forest as the earlier one becomes eroded and nutrient deficient.

slash-and-burning subsistence farming in a small patch of cleared tropical rain forest in Costa Rica.

Surface fire in Ocala National Forest in Florida. Occasional fires like this burn deadwood, undergrowth, and litter and help prevent more destructive crown fires in some types of forests.

Types of Fire

Fires are generally classified as (i) **Ground fires** which develop in such conditions where organic matter (litter) accumulates richly as heaps and they catch fire which generally smoulder for longer periods. Thus, in dry litter, fire is rapid and extinguishes quickly, while in somewhat moist litter, the fire is slow and with its heat the inner parts of litter-heap also get dried and fire continues for a longer period. Ground fires are flameless and subterranean and kill almost all herbaceous plants rooted in the burning material except some woody species (shrubs and trees). (ii) **Surface fires** which sweep over the ground surface rapidly and their flames consume the litter, living herbaceous vegetation and shrubs and also scorching the tree bases if comes in contact. (iii) **Crown fires** which are most destructive, burning the forest canopy and are common in dense woody vegetations. They spread in the top layers from the canopy of one tree to the canopy of another and so on. Canopy fires produce a temperature up to 704.5°C, killing the trees, shrubs and herbs. However, in moist soil surface, the underground plant parts and buried seeds escape death.

Destructive crown fire in Yellowstone National Park during the summer of 1988. Wildlife that can't escape are killed and wildlife habitats are destroyed. Severe erosion can occur.

Effects of Fire

Fire has direct (*e.g.,* lethal) as well as indirect effects on plants and wild-life. Some well confirmed indirect effects of fire on plants are as follows :

1. Fire causes injury to some plants, resulting large scars on their stems. Such scars may serve as suitable avenues of entry of parasitic fungi and insects.

2. Fire arrests the course of succession and modify the edaphic environment very much.

3. Fire brings about distinct changes of such ecofactors as light, rainfall, nutrient cycles, fertility of soil, litter and humus contents of soil, pH, water holding capacity and soil fauna (earthworms, nematodes, arthropods, etc.) Soil fungi are reduced while bacteria increase due to post-fire changes in the soil.

The microclimate too is greatly changed due to addition of ash, loss of shade, loss of raindrop interception, accelerated erosion, etc.

Epilobium angustifolium.

4. Fire plays an important role in the removal of competition for surviving species. **Fire tolerant plant** species generally increase in abundance at the expense of those killed by fire (**fire-sensitive plants**) due to considerable reduction in competition and possibly due to alteration in other conditions.

5. Some plants such as *Populus tremuloides* and *Epilobium angustifolium* are stimulated to growth by fires. A number of such grasses as *Aristida stricta, Cynodon dactylon, Paspalum notatum*, are stimulated by fire to produce large quantities of seeds. In some grasses and legumes, the seeds would germinate only after these get fire treatment (*e.g.,* seeds of *Themeda, Heteropogon, Andropogon, Rheus, Tephrosia* and *Stipa*).

6. Some fungi, mainly some ascomycetes, grow in soils of burnt areas. Such fungi are known as **pyrophilous** (*e.g., Pyrophilous confluens*).

Adaptations to Fire

Quercus.

In frequent fire prone areas certain plants develop the following adaptations :

1. Certain trees, particularly conifers such as *Pinus* and *Larix* and dicots such as *Quercus* develop fire resistant bark with insulating effect against heat. These trees also have tall trunk with the crown restricted to upper zone only. This helps in escaping the destructions against surface and ground fires.

2. In some plants leaves are fire-resistant due to poor contents of such compounds such as resin or oil, and may check surface fires.

3. Some plants as *Pinus rigida* and species of *Eucalyptus* have adventitious or latent axillary buds which may develop into new branches. Similarly *Betula papyrifera* and *Vaccinium* sp., may develop new shoots after fire kills the older ones.

4. *Epilobium anguistifolium* acts as a fire indicator species. It grows in patches and in dormant condition. In case of fire, these plants rapidly grow while other plants die due to fire.

6. WIND FACTOR

The strong moving current of air is called **wind**. It is an important ecological factor of the atmosphere affecting variously the plant life on flat plains, along sea coasts and at high altitudes in mountains. Wind is directly involved in transpiration, in causing several types of mechanical damage and in dissemination of pollen, seeds and fruits. Wind also modifies the water relations and light conditions of a particular area. The movement or velocity of wind is affected by such factors such as temperature, atmospheric pressure, geographical features (including topography), vegetation masses and position with respect to sea shores. Air moves from a region of high pressure to low pressure. The pressure differences are mainly due to differential heating of atmosphere. The equatorial regions receive more heat than north or south regions, thus, low pressure occurs at lower latitude. The air generally moves from poles towards equator. Winds result in various physical, anatomical and physiological effects on plants: 1. **Physical effects** such as breakage and uprooting; deformation; lodging or flattening of her-baceous plants such as wheat, maize, sugarcane, etc.; abrasion of buds of plants (by wind-carried soil particles), erosion and deposition of soil around plant roots; plant injury due to salt spray along sea coasts 2. **Anatomical** and **physiological effects** such as formation of dense, reddish xylem called **compression wood** on the compressed side of wind deformation, desiccation due to increased rates of evaporation and transpiration which are caused by strong winds and **dwarfing** of trees on sea coasts, arctic or alpine timberline due to prolonged dehydration and consequent loss of turgidity under the influence of drying winds.

Strong prevailing winds can leave trees permanently bent.

7. PHYSIOGRAPHIC FACTORS

Climate at any point on the globe depends upon the combined effects of three factors, the amount of solar heat (solar radiation), atmospheric circulation and physiographic factors. **Physiographic factors** are those which are introduced by the structure, conformity and behaviour of the earth's surface, by topographic or orographic features, such as elevation and slopes, by geodynamic processes such as silting and erosion and consequently by local geology. Mountains, hills, hillocks, valleys, etc., result from irregularities in earths's surface. These distinct topographic reliefs tend to produce marked local climate or microclimates. For example, in respect to the microclimate, the summits are different from the sides of mountains and narrow valleys are different from the open plains. Different physiographic factors can be grouped under the following headings :

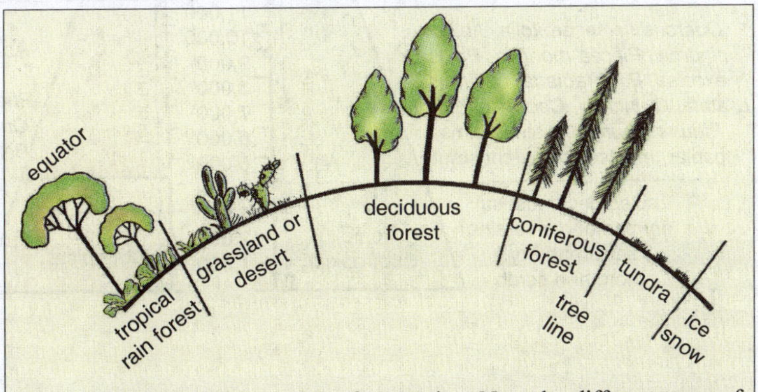

Fig. 4.3. Latitudinal zonation of vegetation. Note the different types of vegetation from equator towards poles (increasing latitude).

1. Latitudes and Altitudes

Latitude represents distance from the equator. Temperature values are maximum at the equator, decreasing gradually towards poles. Marked variations in temperature at different latitudes result in the division of earths's vegetation into different zones such as equatorial, tropical rain forests, desert or grasslands, deciduous forests, coniferous forests, alpine forests, tundra, ice and snow of poles (Fig. 4.3).

Height above the sea level forms the **altitude**. At high altitudes, the velocity of wind remains high, temperature and air pressure decrease, humidity as well as intensity of light increases. Due to these factors, vegetation at different altitudes is different, showing distinct zonation (Fig. 4.4).

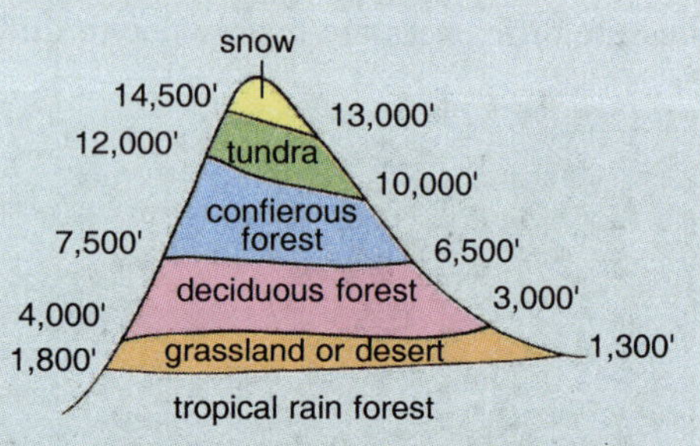

Fig. 4.4. Altitudinal zonation of vegetation on mountains. Note the different types of vegetation with increasing altitude.

Generally, mean temperature of air decreases about 10°F (–12.2°C) for every degree of latitude north of equator and for every 300' (or 91.44 metres) of altitude. Altitudinally, this temperature fall is more rapid on the leeward side of a mountain than on the windward side. For example, the fall in temperature in the Nilgiri hills is 1°C for every 100 metre rise on the leeward side and 122 metre rise on the windward side. Further, valleys and lowlands sometimes become much cooler than the mountain tops due to sinking of the heavier cold air.

Climatic changes in both, the latitude and altitude, exhibit identical effects upon the type of major vegetations of the world. Thus, communities at altitudes (alpine) roughly resemble those of polar regions (arctic tundra). Below the arctic line and the trees or timber line exists the coniferous forests or taiga which are followed by temperate zones which have quite cold conditions and support deciduous forests.

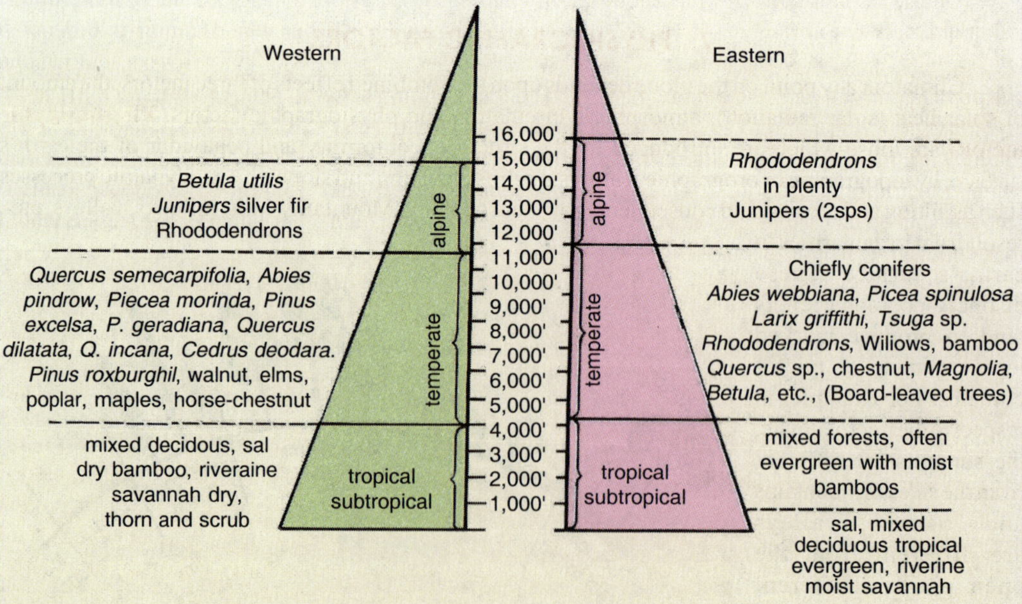

Fig. 4.5. Altitudinal zonation of Himalayas. Note the distribution of various species of plants on the Western and Eastern Himalayas.

Lower to the temperate zone, the sunlight falls nearly at an angle of 90°C, making the climate warm and dry. These places are known as **tropical regions** and lie on either side of the equator.

Temperature and rainfall are commonly said to determine the type of vegetation at different altitudes of mountain. In addition to these two factors edaphic conditions (soil characteristics) are also found important. At higher altitudes, generally temperature decreases and rainfall increases, and these certainly affect the development of soil and vegetation. Due to low temperature and high rainfall the organic matter content of the soil increases at high altitudes, resulting an increase in soil nitrogen and a decrease in soil's pH value. As a result, xerophytes are more common at lower altitudes and chamaephytes occur at higher altitudes.

Moreover, at sea level, the atmospheric pressure is 760mm Hg. The partial pressure of oxygen remains approximately one-fifth (*i.e.,* about 160 mm Hg) of the atmospheric pressure. With increasing altitudes, the atmospheric pressure and partial pressure of O_2 goes on decreasing. Since oxygen is thinner at higher altitudes, it becomes a stress factor for many animals, including mammals.

Lastly, in Himalayan mountains, temperature variations are quite evident, and there is a general zonation of vegetation from lower to higher altitudes. The successive zones of vegetation from base upwards are tropical and subtropical, temperate and alpine. But, there are practically no sharp boundaries between these vegetational zones due to differences in topography, soil and geology. Thus, after studying the vegetational zones at different altitudes on Himalayas, western as well as eastern, the effect of temperature together with altitude and other factors becomes quite apparent (Fig. 4.5).

2. Height of Mountain Chains

Effect of different heights (altitude) can be better seen on mountains. With an increase in altitude above sea level, there are changes in values of temperature, pressure, wind velocity, humidity, intensity of solar radiation, etc. Due to these changes, vegetation at different altitudes differ much.

3. Direction of Mountains and Valleys

The directions of mountain chains or ranges and high mountains act as wind barriers and affect the climate. Mountains also steer or deflect winds into different directions and capture moisture from wind on certain sides to cause precipitation. Thus, moisture-bearing winds may not be able to cross a mountain range and may discharge rain near it. The far side of the mountain may not get much rain and develop into an arid zone (having xeric conditions). For example, the Himalayas act as great barriers for moisture bearing sea winds, therefore, the ocean side of the land in the north eastern region and the Gangetic delta get good rains and have luxuriant vegetation, while, the middle and inner Himalayas are dry with poor vegetation. The absence of a high land masses may be one of the reasons for the development of desert conditions in Rajasthan. Similarly, southern side of Kulu Valley (in Himachal) is moist with rich vegetation, whereas at lower level into Lahul Valley, there is very poor vegetation.

High mountains not only act as climatic barriers to air masses between neighbouring areas or zones but also cause local air circulation, giving rise to mountain winds. The most common winds in such areas are **slope winds** which blow down-hill during night and up hill during the day. Such winds are frequent in Antarctica. Further, the relative humidity (*i.e.,* dryness, humidity or severity) of a climate depends on whether the air comes from the north or the south, from the sea, or deserts, mountains or forests and so on. The heating of earth's surface is directly proportional to the angle at which solar rays arrive on the ground. The closer their incidence to perpendicular, the more solar heat is delivered per square centimetre. Hence, mountain ranges affect the climate through rainfall and other factors which have a significant affect on the growth of vegetation and the distribution of animals.

4. Steepness of Slope

Slope is the characteristic feature of mountains. The steepness of a slope has a distinct effect on the climate of the area, *i.e.,* incidence of solar radiation, rainfall, wind velocity and the temperature of the region.

Generally, southern and western slopes of mountain display higher temperature than do the northern or eastern slopes. Most probably these differences arise due to different solar radiation received, differences in the amount of rainfall, snowfall, relative humidity and wind movement on the two slopes. In India, all south-facing slopes are in general exposed to more solar radiation than north facing slopes. North facing slopes are generally cooler than south facing ones. This is due to the fact that the steep southern slopes receive the rays of the mid-day sun almost at right angles, whereas the northern slopes receive only oblique rays during morning and evening hours and sometimes none at all except for a short period during summer (*i.e.,* the duration of sunlight on the south facing slope is longest). East and west-facing slopes receive equal amounts of sunlight, but in the former, the angle of incidence of the sun's rays is greatest in the morning hours when the air is cool and the dew has not fully disappeared. In west-facing slopes, on the other hand, the sun's rays become scorching as the air is already heated up, causing a desiccation effect on the environment. The differences in solar radiation and consequent temperature values bring about changes in the vegetation on the slopes. For example, northern slope (which is protected from sun) support virgin forests with hygrophilous ground vegetation, whereas the southern slopes (which is heated by sun) have only a xerophytic vegetation. Various climatic factors of the slopes also affect the distribution of consumers (animals).

Steepness of slope decides the rapidity, with which water flows away from the surface and determines the characteristic of the soil. Thus, loss of water as run off is more with increase in steepness of the slope. This is why having same amount of rainfall bear different types of vegetations. On a steep slope, plants are unable to establish, even if there is sufficient rainfall. Any little vegetation occurring there will be xerophytic in nature. Soil erosion is more prevalent on steep slopes. The water moving over the slopes causes erosion of the top soil and consequent disappearance of vegetation from the areas. Eroded hills, steep slopes and rock may produce a special habitat where only particular species of plants can adapt themselves. Gravel, sand or clay eroded by water, wind, etc., from rock surfaces, are generally brought down by rivers and streams and these are ultimately deposited as silt at their mouths, providing new habitat (*e.g.,* salt marshes) which may support different kinds of plants. The areas subjected to quick and continuous erosion are completely devoid of vegetation.

Exposure of slope. A slope often remains exposed to sun and wind and this affects greatly the kind of plants growing there. Generally, the slope exposed to the sun and wind supports vegetation which may be entirely different from that which is less or not at all, exposed to sun and/or wind. Visualizing this fact, green houses and hot beds in temperate climates are always built in a way so as to face the sun or on southern slopes, which receive much heat from the sun.

REVISION QUESTIONS

1. Write short notes on the following :
   (i) Essential elements,
   (ii) Limiting factors,
   (iii) Liebig's law of minimum,
   (iv) Threshold,
   (v) Law of tolerance,
   (vi) Cyclomorphosis,
   (vii) Ciradian rhythms,
2. Discuss the role of temperature as an ecological factor. Give an account of different morphological and physiological adaptations of plants and animals in relation to temperature.
3. Describe the role of light as an ecological factor. Also add a note about photoperiodism.
4. Describe the role of fire as an ecofactor.
5. With suitable examples, discuss the effect of topographic factors on distribution of plants.
6. Explain the following :
   (a) Outer Himalayas have luxuriant vegetation than inner Himalayas.
   (b) Vegetation differs at various altitudes of Himalayas.
7. Describe the role of the following ecofactors :
   (a) Precipitation ; (b) Wind ;
   (c) Humidity ; (d) Steepness of slope.
8. Write an essay on photoperiodism and biological clocks.
9. Describe how the chemical factor such as O_2, CO_2 and pH regulate distribution of the organisms.

CHAPTER

5

Biotic Environmental Factors

All organisms are inerdependent on each other.

Organisms do not exist alone in nature but in a matrix of other organisms of many species. Many species in an area will be unaffected by the presence or absence of one another (This is often termed **neutralism**), but in some cases two or more species will interact. The evidence for such interaction is quite direct : populations of one species are different in the absence and in the presence of a second species.

INTERSPECIFIC INTERACTIONS

The interactions between species may have positive or negative results. Following six general types of interactions have been described :

A. Positive interactions

1. Mutualism ;
2. Commensalism ;
3. Protocooperation.

B. Negative interactions

4. Exploitation ;
   (i) Social parasitism ;
   (ii) Parasitism ;
   (iii) Predation.
5. Amensalism and antibiosis ;
6. Competition.

A. Positive Interactions

In case of positive interactions, populations help one another and either one or both the species are benefited. This benefit may be regarding the food, shelter, substratum or transport. Further, such an association may be continuous or transitory, obligate or facultative and the two partners may be in close contact (*i.e.*, their tissues remain intermixed with each other) or one of them may live within a specific area of the other or attached to its surface. Moreover, for the positive interactions, how does one determine that what is the benefit to the species ? Individuals that participate in the interaction should have a higher relative fitness than those that do not participate. At the population level, the growth rate of a population containing mutalistic individuals should be higher than that of a population lacking such individuals (**Boucher** *et al.*, 1982). Positive interactions may be of the following types :

I. Mutualism. Mutualism is an obligatory positive interspecific interaction that is strongly beneficial to both species. In past, it was termed **symbiosis** (see **Dash**, 1993). Such mutually beneficial interactions are more common in the tropics than elsewhere. In this case, both of the species derive benefit and there exists a close and often permanent and obligatory contact which is more or less essential for survival of each. In mutualism, two populations enter into some sort of physiological exchange and resulted in coevolution of both species (see **Krebs**, 1985). Some common examples of mutualistic association are the following :

1. Pollination by animals. Certain insects such as bees, moths, butterflies, etc., and birds derive food from the nectar, pollen or other plant products, and in return bring about cross pollination (Fig. 5.1A). To ensure the success of this function, various structural adaptations have occurred in both plants and animals, leading to coevolution.

2. Dispersal of fruits and seeds. Birds and mammals are of great importance as agents of plant distribution. Seeds, fruits, even entire plants become attached to feathers or fur or ingested seeds are eaten and eliminated unharmed with the faeces.

3. Mutual defence in ants and acacias. A mutualistic system of defence has been achieved by the swollen thorn acacias and their ant inhabitants in the New World Tropics. The ants (*Pseudomyrmex* sp.) depend on acacia (*Acacia* sp.) tree for food and a place to live and the acacia depends on the ants for protection from herbivores and neighbouring plants (**Janzen**, 1966). Swollen-thorn acacias (*e.g., Acacia cornigera*) have large, hollow thorns in which ants live (Fig. 5.2). The ants feed on modified leaflet tips, called **Beltian bodies**, which are the primary source of protein and oil for the ants, and also on enlarged nectaries, which supply sugars. Swollen thorn acacias maintain year-around leaf production, even in the dry season, to provide food for the ants. The acacia ants continually patrol the leaves and branches of the tree and immediately attack any herbivore that attempts to eat acacia leaves or bark. The ants also bite and sting any foreign vegetation that touches an acacia, and they clear all the vegetation from the ground beneath the acacia tree. Some of the species of ants that inhibit acacia thorns are obligate acacia ants and live nowhere else.

Beltian bodies.

4. Lichens. They form the examples of mutualism where contact is close and permanent as well as obligatory. The body of lichens is made up of a matrix formed by a fungus, within the cells of which an alga is embedded (Fig. 5.1B). The fungus makes available the moisture and minerals to the algae, which prepare food by photosynthesis. In nature, neither of the two can grow alone independently. Lichens tend to grow abundantly on bare rock surfaces.

5. Symbiotic nitrogen fixation. Here a bacterium *Rhizobium* forms nodules in the roots of leguminous plants and lives symbiotically with the host (Fig. 5.1 C). Bacteria get a protective space to live in and derive prepared food from the roots of higher plants and in return fix gaseous nitrogen, making it available to the plants. The leguminous plants use this nitrogen in the protein synthesis and so legumes are rich in proteins. Nitrogen fixation-like association also occurs in root nodules of *Alnus, Alopecurus, Casuarina, Cycadacease, Myrica, Podocarpus, etc.*, and leaves of about 400 species of non-leguminous plants.

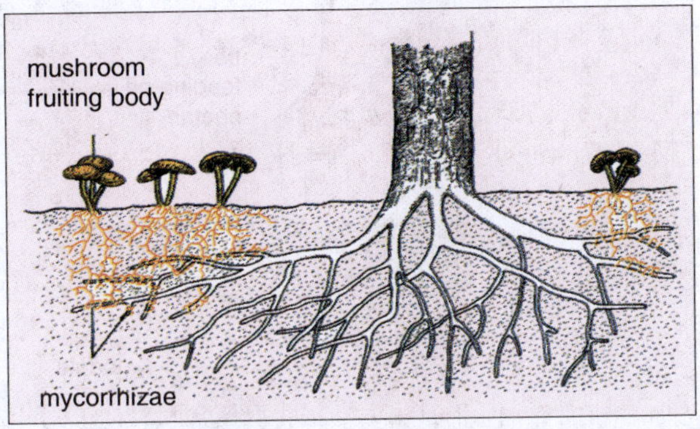

6. Mycorrhizae. In mycorrhizal associations, tree roots become infested with fungal hyphae (Fig. 5.1D). The fungi derive their food from the tree roots and in return their hyphae supply water and minerals that they absorb from the soil much like the root hairs of trees. It is believed that the fungus also regulates the pH and sugar level for a good growth of roots in acidic soils (*e.g.*, conifers). Mycorrhizae may be on the surface of roots (**ectotrophic**) or inside between the cells of the roots (**endotrophic**). Ectotrophic mycorrhizae are common in nature on pines, oaks, hickories and beech and endotrophic ones occur in red maple and are common in roots and other tissues of many orchids and members of Ericaceae.

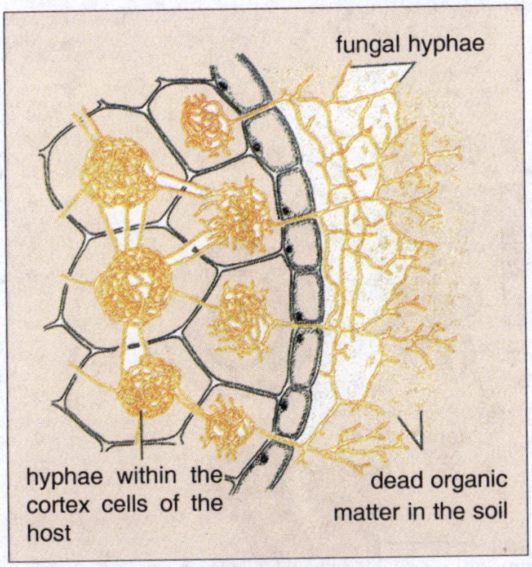

The other similar root associations are met with actinomycetes (actinorhizal) in form of nodules and with some blue-green algae (cyanobacteria) such as *Anabaena* and *Nostoc* forming coralloid roots of *Cycas*. Both of these kinds of associations are connected with nitrogen fixation by microscopic organisms present in the roots.

7. Zoochlorellae and Zooxanthellae. Some unicellular photosynthetic plants, especially algae, known as *zoochlorellae*, live symbiotically in the outer tissues of certain sponges, coelenterates, molluscs and worms (Fig. 5.1E). Some brown or yellow cells, probably flagellates (zooxanthellae) are also present. Algae are photosynthetic and produce oxygen and nitrogenous compounds beneficial to hosts and in exchange, they obtain materials such as carbon dioxide and nitrogenous wastes released by metabolism of host animals. The unicellular green alga, *Chlorella vulgaris* lives within the gastrodermal cells of *Hydra*. Likewise, alga *Zoochlorella* lives in a planarian, *Convoluta roscoffensis*.

8. Microorganisms and cellulose digestion. Interspecific mutualism is nicely demonstrated by the flagellate protozoan, *Trichonympha* an obligate anaerobe in the gut of several species of wood-eating termites where it digests cellulose. *Trichonympha* also occurs in the alimentary canal of wood-eating roach *Cryptocerus*. The termite and roach reduce the wood to small fragments, passing them through the alimentary canal to hind gut, where the protozoans digest the cellulose, changing it into sugar. The host benefits the protozoa by removing harmful metabolic waste products and maintaining anaerobic conditions in the intestine.

9. Fungus gardens. There exist certain mutualistic associations between insects and fungi. For example, the insect-fungus-galls of gall insects (*e.g.*, midge, *Lasioptera*) are lined by a fungus which is parasitic on the host plant. It is believed that the female midge deposits spores of fungus at the time she lays her eggs in the plant or when insect sucks plant sap. Gall fungus seems to assist the insect indirectly by partly breaking down the tissues so that the insect can digest it (Fig. 5.3). Besides these

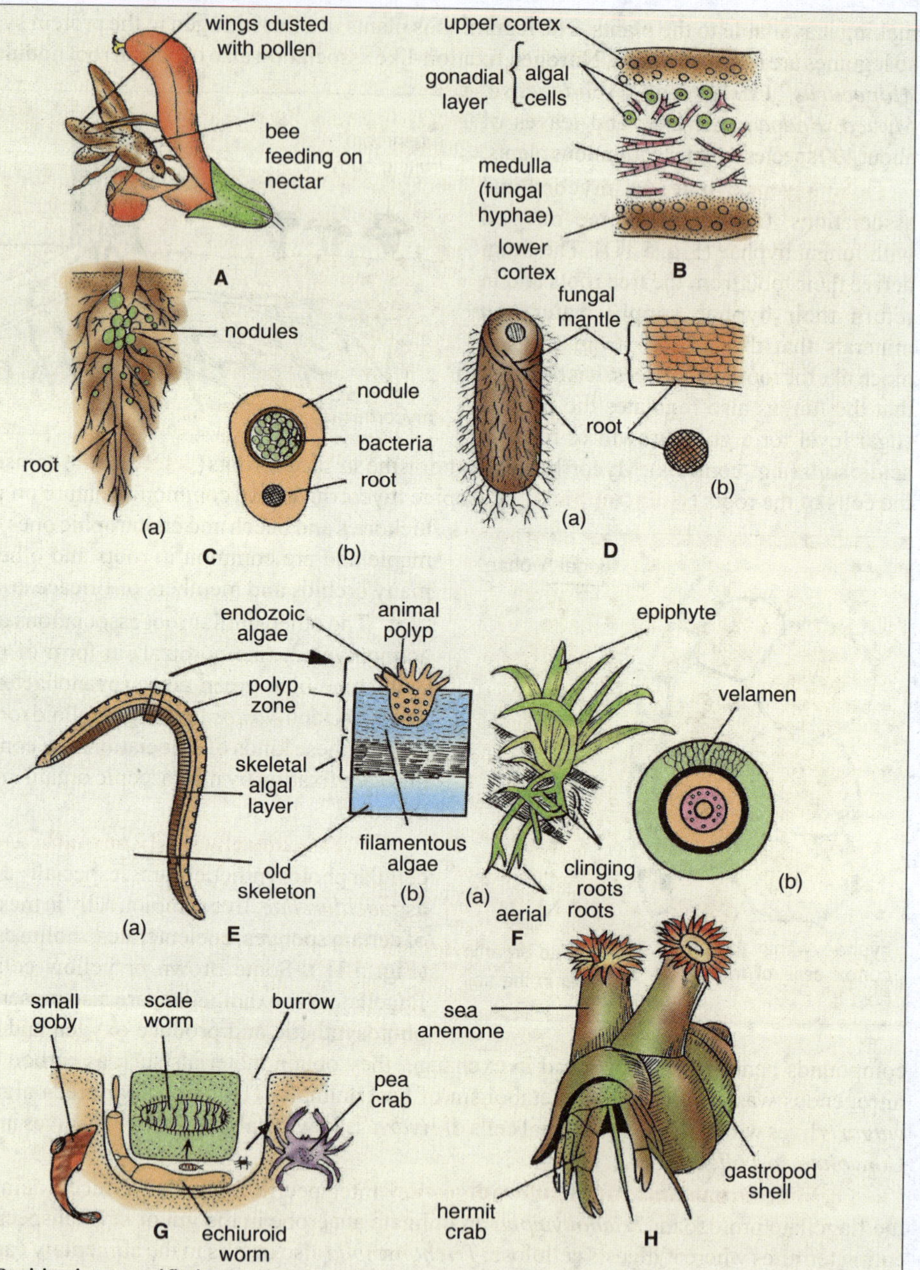

Fig. 5.1. Positive interspecific interactions. A—Mutalism, a bee feeding on nectar and also is dusted with pollen; B—Mutalism, a part of T.S. of lichen thallus showing algal and fungal partners; C—Mutalism, (a) root system of a legume showing nodule, (b) T.S. through the root and nodule; D—Mutalism, mycorrhizae, *i.e.*, association between root of higher plant and fungal hyphae, (a) A part of mycorrhizal root coverd by fungal hyphae, (b) A part of mycorrhizal root showing fungal mantle, E—Mutualism, a relationship between the polyp (animal) and several algae, (a) Cross section of "head" of coral (polyp) showing algal layer, (b) A part of C.S. of coral head magnified; F—Commensalism, epiphyte orchid growing perched on the branch of another plant, (b) T.S. of aerial root of orchid showing velamen; G—Commensalism, a diagrammatic sketch showing three commensals, the small goby, scale worm and pea crab, living in the burrow of the echiuriod worm; H—Protocooperation, association between sea anemone and hermit crab.

gall insects, there are various cases of insect gardeners such as ambrosia beetles, tropical termites and American ants which do gardening of fungi in their nests to get a regular supply of their food directly or indirectly.

II. Commensalism. Commensalism defines the coaction in which two or more species are associated and one species at least, derives benefit from the association, while the other associates are neither benefited nor harmed. The concept of commensalism has been broadened in recent years, to apply to coactions other than those centering on food; cover, support, protection, and locomotion are now frequently included (**Bear**, 1951; **Kendeigh**, 1974). Some common examples of commensalism are the following :

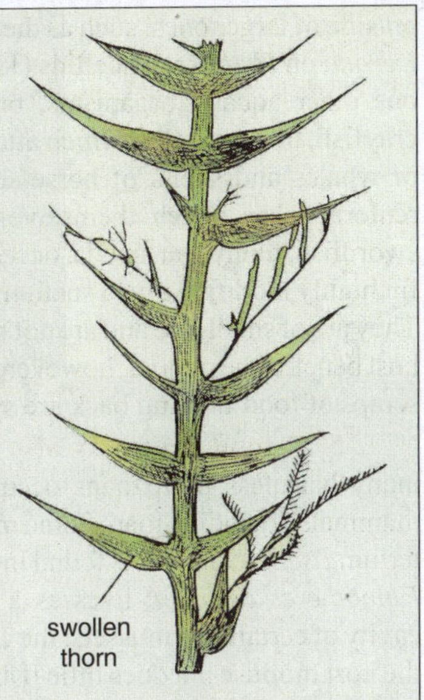

Fig. 5.2. Swollen thorns of *Acacia cornigera* on a lateral branch. Each thorn is occupied by 20-40 immature ants and 10 to 15 workers ants. All the thorns on the tree are occupied by one ant colony (after Krebs, 1985).

1. Lianas. These plants are common in tropical rain forests where light at ground level is scarce because of the dense and multistoreyed growth of vegetation. Lianas are vascular (woody) plants rooted in the ground but which climb up with the support of other trees and reach almost to the top of the forest canopy. The woody stem of lianas is closely attached to the supporting tree but it is not involved with it in any nutritional relationship. Thus, with much economy of mechanical tissues, lianas are able to get better light. On the basis of the type of device used for climbing their supports, lianas are classified as leaners, thorn lianas, twiners or tendril lianas. In North Indian tropical deciduous forests *Bauhinia vahili* is a common liana. Other common lianas are species of *Ficus* and *Tinospora*.

Lianas.

2. Epiphytes. Epiphytes are the plants growing on other plants (Fig. 5.1F). They use other plants only as support and not for water or food supply. Epiphytes differ from the lianas in that they are not rooted into the soil. They may grow on trees, shrubs or larger submerged plants. They grow either on the trunks or leaves. Certain common examples of epiphytes are orchids, bromeliads, hanging "mosses", *Usnea* and *Alectoria*. Epiphytes are most common in tropical rain forests. They derive their supply of moisture and nutrient from the frequent rains and debris accumulated in bark crevices. Dust is also a source of the nutrients. Roots of epiphytes often store water in a special tissue, called **velamen**. For meeting water requirements, in the epiphyte *Dischidia*, the leaves fold to form a jug-like structure having a narrow mouth where water accumulates; roots of the epiphyte enter and grow in these specialized cavities of leaf. Many epiphytes develop a thick network of roots upon which wind blown dust accumulates and provides the necessary edaphic environment.

3. Epizoans. Some plants grow on the surfaces of animals. For example, some green algae grow on the long, grooved hair of sloth. *Basicladia* (Cladophoraceae) grows on the backs of fresh water turtle.

4. Epizoite. Likewise, ectocommensals or **epizoite** animals are associated with another animal for the purpose of anchorage and protection. Many small animals, for example, become attached to the outside of larger ones, such as the protozoans *Trichodina* and *Kerona* on *Hydra,* vorticellids (*Vorticella, Epistylis*) or various other aquatic organisms, branchiobdellid annelids on crayfish, the barnacles, which attach themselves to the backs of whales and shells of horse-shore crabs and so on. The remora fishes attach themselves to the bellies of sharks, swordfish, tunny, barracuda, or sea turtle by means of a dorsal fin highly modified into a suction disk on the top of the head. They are of small size and are not burdensome to the host. The host benefits the remora, however, for when the host feeds, the scraps of food floating back are swept up by the remora.

5. Commensals may also be internal. For example, many harmless protozoans occur in the intestinal tract of mammals, including man. Some microorganisms such as bacterium *Escherchia coli* is found in human colon. The pea crab *Pinnotheres ostreum*, lives as a commensal in the mantle cavity of certain sea mussels; the crab steals food collected by the host mollusc but does little if any other known injury. The hermit crab normally lives inside gastropod shell. The pitcher of the pitcher plant found in bogs furnishes a breeding site or home for certain species of midge flies, mosquitoes, and tree toads. Many kinds of microorganisms, plants and animals (such as small crabs, shrimps, polychaets, ophiuroids, ascidians, etc.) live in the canal system of sponges.

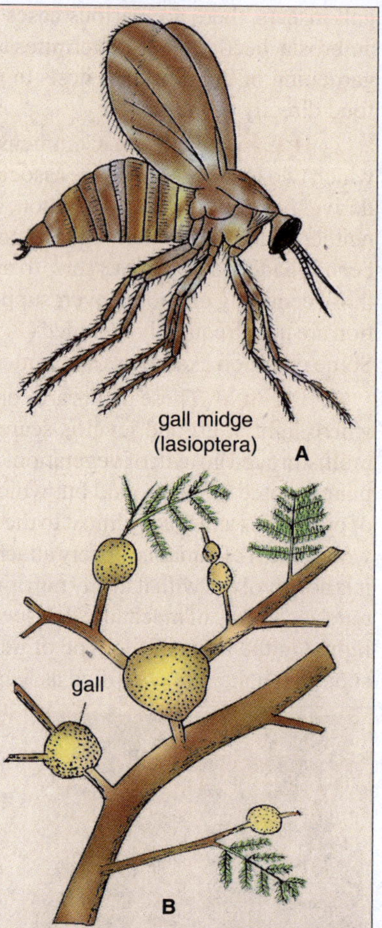

Fig. 5.3. A—Gall midge; B—Plant galls.

The nest of one species often furnishes shelter and protection for other species as well. Ant nests may contain guest species of various other insects; some birds place their nests close to wasps, bees or ants for the protection offered by these insects. Woodchuck burrows are used also by rabbits, skunks and racoons, especially in winter.

6. Commensalism-like associations also exist between a variety of microorganisms and higher plants. For example, the zone of soil around the roots of higher plants characterized by intense microbial activity, called the **rhizosphere**; the surface proper of the roots growing in soil—the **rhizoplane**; the boundary layer of air over the green leaves with active microbes—the **phyllosphere**; and the leaf surface—**phylloplane**, constitute important ecological niches, where occur rich populations of microorganisms including fungi, actinomycetes and bacteria. The living roots and leaves of higher plants are known to continuously diffuse various metabolic products, mainly sugars and amino acids. These are sources of nutrition for microorganisms present therein.

III. Protocooperation. Protocooperation is less extreme sort of interaction than mutualism in which the interaction is clearly beneficial to both species, allowing the equilibrium population levels of both to be higher than they otherwise would be. However, it is not obligatory for either species. Examples of protocooperation are many, including the relationship between water moccasins and large birds such as herons and ibises on several islands off the west coast of Florida. The birds nest in the lower branches of relatively unprotected trees, while the snakes congregate around the bases. This protects the birds from tree climbing predators such as racoons. In turn, the snake feed, in part, on fish dropped by the birds and the occasional baby birds that fall out of the nest (**Ehrenfeld**, 1970).

Another good example of protocooperation is of coelenterate, sea anemone (*Adamsia palliata*) attached to the empty molluscan shell harbouring hermit crab (*Eupagurus prideauxi*; Fig. 5.1H). The sea anemone is carried by the crab to fresh feeding grounds and crab in turn is said to be protected from its enemies by sea anemone. Some ecologists prefer to include this association under commensalism.

B. Negative Interactions

In case of negative interactions, one or both species are harmed in any way during their life period. Some authors such as **Clarke**, 1954, prefer to call these types of interactions as **antagonism**. The negative interactions include the following three broad categories :

1. Exploitation. In exploitation, one species harm the other by making its direct or indirect use for support, shelter or food. It is of following types :

(i) Social parasitism. Social parasitism describes the exploitation of one species by another, for various advantages. It is a kind of parasitism in which the parasite foists the rearing of its young into the host. Social parasitism in various stages of development is found among some higher vertebrates and insects. For example, there occurs an **egg parasitism** in two species of birds— old world cuckoos and the brown headed cowbirds of North America, both of which do not build nests of their own, rather they deposit their eggs in nests of other species, abandoning eggs and young to the care of foster parents. Similarly, there occurs a **brood parasitism** in between two Indian birds—Indian koel and crow. The Indian koel (*Eudynamys scolopacea*), like many of its cuckoo cousins (Family Cuculidae), neither incubates its eggs nor feed its young. But these functions are secretly forced upon the house crow (*Corvus splendens*) and jungle crow (*Corvus macrorhynchus*) (**Davis**, **De**, and **Pal**, 1977). One species of ant attacks foraging workers of another species and snatches away the food they are transporting or the robber species may deliberately rob another nest for food. Some species of ants make slaves of the workers of other species. Such kind of dependency of one species on another also occurs in other social insects such as termites, wasps and bees. Social insects are apparently the only animals other than man to have succeeded in domesticating other species, and of cultivating plants particularly fungi for food.

(ii) Parasitism. Parasitism is a kind of harmful coaction (disoperation) between two species. It is the relation between two individuals wherein one individual called **parasite** receives benefit at the expense of other individual called **host**. Parasitism is mainly a food coaction, but the parasite derives shelter and protection from the host, as well. A parasite usually parasitizes a host which is larger in body size than it. Further, a parasite does not ordinarily kill its host, at least not until the parasite has completed its reproductive cycle. However, a host may die due to some secondary infection or suffer from stunted growth, emaciation, or sterility. The balance between parasite and host is upset if the host produces antibodies or other substances which hamper normal development of the parasite. In general the parasite derives benefit from the relation while the host suffers harm. But this does not mean that all parasites are harmful. The former category constitutes the **pathogenic parasites**.

Classification of parasites. Parasites exhibit a tremendous diversity in ways and adaptations to exploit their hosts. The parasites may be **viral parasites** (*e.g.,* bacterial, plant and animal viruses), **microbial parasites** (*e.g.,* bacteria, protozoa, fungi, etc.), **phytoparasites** (*e.g.,* plant parasites) and **zooparasites** (*e.g.,* animal parasites such as platyhelminthes, nematodes, arthropodes, etc.). They may parasitize microorganisms, plants and animals. They may occur on the outside of the host (**ectoparasites**) or live within the body of the host (**endoparasites**). The endoparasites usually live in the alimentary tract, body cavities, various organs or blood or other tissues of host. Ectoparasites may be parasitic only in the immature stages—the hairworm larvae, parasitic in aquatic insects; only the adult parasitic— fleas, on birds and mammals; or both larvae and adults may be parasitic—the blood sucking lice and flies, biting lice, mites, and ticks that occur on birds, mammals, and sometimes reptiles, and the monogenetic trematodes on fish. Similar relation occurs in endoparasites but in them usually all the stages of life are parasitic—entozoic amoebae, trichomonad flagellates, opalinid ciliates, sporozoans, pentastomids, nematodes, digenetic trematodes, acanthocephalons, cestodes, and some copepodes.

Further, endoparasites rely on various means of transport from one host to another so that their survival, range of distribution, and life cycle are unaffected. For the endoparasite, the interior of its host is its **microhabitat** and **microenvironment**, while the outside environment is the **macroenvironment**.

The host acts as a buffer between the parasite and the outside environment. Changes in the microenvironment directly influence the parasite while alterations in the macroenvironment influence it directly. Thus, mutual adaptations and tolerance between the host and parasite are important factors facilitating successful transmission of endoparasites, and climatic fluctuations tend to alter the physiology of the parasite, host or vector. Mutual adaptations are the parasite's abilities to establish and maintain itself in a favourable location within the host and exist in a proper form for sufficient length of time enabling subsequent transport to another host and continuation of species.

Animals may also be parasitic to plants. Nematodes infest the roots of plants. Wasps or gnats form galls on plants such as oaks, hickories, willows, roses, goldenrods, and asters. Mites stimulate formation of witches brooms in hackberry. A variety of insects the larvae of which are leaf miners, wood borers, cambium feeders, and fruit eaters should be included here.

Lastly, parasites may be **full-time** (or **permanent**) **parasites** or **part-time** (or **temporary**) **parasites**. Mosquitoes and bugs that suck the blood of their hosts are temporary parasites. Some temporary parasites spend only a part of their life cycle as parasites. For example, glochidium larva of *Anodonta* (freshwater mussel) attaches itself to the body of the fish by means of its hooks and penetrates inside the fish integument to remain buried there for several weeks and finally emerges out as the young mussel to lead an independent existence. Permanent parasites, however, spend their life completely on other organisms. The common examples of permanent parasites are—*Plasmodium, Entamoeba histolytica,* and other protozoan pathogens, different platyhelminthes, nematodes, arthropods, etc.

Parasitic adaptations. While studying parasitism as a biotic factor influencing the life and activities of parasites and hosts we often encounter manifold adaptations, both offensive and defensive, being developed by the host, as well as by the parasite. An endoparasite entering a host often meets with the antibodies or phagocytic cells produced by the host. The ectoparasites, and endoparasites have following parasitic adaptations :

1. In parasitic animals, a reduction of organs of special sense of nervous system, and of locomotory organs occur.

2. Most ectoparasites develop some clinging organs such as hooks, suckers, etc., to get attached with the body of their hosts. They also develop special piercing and sucking organs to suck the blood of animals or sap of plants. Certain blood-sucking ectoparasites such as leeches, blood-sucking mosquitoes, etc., contain certain anticoagulant enzymes in their salivary secretions.

3. Most endoparasites exhibit anaerobic respiration, high rate of reproduction, parthenogenesis, hermaphroditism, polyembryony, intermediate host and a complicated life-cycle (**Lapage**, 1951). Some endoparasites, such as tapeworm, have become so adapted to the host that they no longer require a digestive system. They simply absorb their food directly through their body wall. Further, parasites that live within the bodies of plants and animals possess cuticles or develop cysts resistant to the digestive enzymatic action of the host.

4. Many parasites pass their entire existence in a single host; others require one, two, even three intermediate hosts. It is of ecological significance that both primary and intermediate hosts of a parasite occur in the same habitat or community. Even then the hazards to successful passage from one host to another are so great and mortality so high that enormous quantities of offspring are produced to ensure that at least a few individuals will complete the cycle.

5. Parasites are transferred from one host to another by active locomotion of the parasite itself; by ingestion, as one animal sucks the blood of or eats another; by ingestion as an animal takes in eggs, spores, or encysted stages of the parasite along with its food or drinking water; as a result of bodily contact between hosts; or by transportation from host by way of vectors. For example, the bacteria that cause tularemia in man are carried from rabbit to rabbit by ticks. Man contacts the disease when he handles infected rabbits but the incidence of infection is greatly reduced in the autumn when cold weather forces the ticks (*i.e.,* the vectors of bacteria) to leave the rabbits and go into hibernation.

Host specificity of parasites. Certain parasites, such as copepodes, infect a wide range of hosts and they are ubiquitously present in various invertebrates and fishes. Most parasitic genera, however, are adapted to hosts of one phylum only. For example, each order of birds possesses its own particular species of tapeworms. This is true even when several order of birds live in the same habitats as do, for, instance, grebes, loons, herons, ducks, waders, flamingoes and cormorants (**Bear**, 1951). Species of flagellate protozoans that occur in termite alimentary canals are largely host-specific. Some species of gall wasps attack only one species of oak. Where a single species parasitizes two or more host species, the shape and structure of the gall formed around the egg and larva on both hosts is essentially similar. When several insects are found on the same oak, each kind of parasite produces its own characteristic gall form. The parasite specific gall formation is found to be related with parasite-specific enzyme (**Kinsey**, 1930).

Effects of parasite on the host : disease. Parasites may not cause immediate mortality but they cause damage to body structures, should it become excessive, may cause death. Because of these parasitically caused anatomical damages, the finely adjusted balance of different vital processes of host's body become disturbed and host is said to be diseased. There are a number of causes of diseases. Parasites are one; physiological stress, nutritional deficiency and poisoning are others. Some of the common parasitic agents of disease and consequent mortality of animals are following :

1. Viruses are the potent agents of several disastrous diseases of plants and animals including man. For instance, some viruses are the agents of hoof and mouth disease in deer, spotted fever in rabbit, encephalitis and distemper in foxes.

2. Bacteria may produce localized inflamatory changes in tissue, enter the blood stream, or produce powerful poisons known as **toxins**. They cause a variety of diseases, notably tularemia, paratyphoid, and tuberculosis among birds and mammals, as well as other diseases in lower types of organisms.

3. Fungus spores of *Aspergillus* may be drawn into the lungs of ground-feeding birds, where they germinate and grow, causing **aspergillosis disease**. Fungus may also develop on the external surface of animals.

4. Protozoan parasites are especially important in the alimentary tract and in the blood. A sporozoan species of *Eimeria* damages the walls of the intestine in upland game birds and causes **coccidiosis disease**; *Toxoplasma* becomes encysted in the brain of rodents; *Leucocytozoon* is a common blood parasite of waterfowl and game birds.

5. Worm parasites, such as tapeworms, nematodes, and acanthocephalans may wander through the host's body doing mechanical injury as well as destroying and consuming tissues. The host may respond by forming a fibrous capsule or cyst around an embedded parasite.

6. External parasites such as ticks, fleas, lice, mites, and flies do not commonly produce serious mortality by themselves, but they are often vectors, transmitting protozoa, bacteria and viruses from one animal to another. Heavy infestations of external parasites however lower the vitality of an animal and cause diseases of fur or feathers.

7. Nutritional deficiency in vitamins or minerals, or improper balance among carbohydrates, proteins, and fats may produce malformations, lack of vigour, or even death.

8. Food poisoning, **botulism**, occurs when certain food become contaminated with the toxins released by the bacterium *Clostridium botulinum.*

9. **Physiological stress** (**Selye**, 1955) is a term that has come to be applied to changes produced in the body non-specifically by many different agencies which may accompany any disease. Effects of stress include loss of appetite and vigour, ache and pains, and loss of weight. Internally, physiological stress causes following abnormalities, such as, acute degeneration of the lymphatic organs, decrease of the blood eosinophils, enlargement and increased secretory activity of the adrenal cortex, and a variety of changes in the chemical constitution of the blood and tissues.

Parasitism in the plants. There are some parasitic vascular plants. Species of *Cuscuta* (doddar) which are **total stem parasites**, grow on other plants on which they depend for nourishment. Young stem twines around the host stem (Fig. 5.4) and it develops adventitious roots which finally penetrate the stem of host, establishing relationship with its conducting elements. The specialized roots are called **haustoria**. Other examples of such association are **total root parasites** as *Orabanche, Conopholis, Epifagus, Balanophora* and *Striga* which are found on roots of higher plants. *Orabanche* affects the yield of mustard and brinjal and is parasitic to the roots of Solanaceae, Cruciferae and Gramineae. *Rafflesia* is root parasite of *Vitis*. Partial root parasites belong to family Loranthaceae (*Viscum album, Loranthus* sp.) They grow rooted in branches of host tree. Examples of partial root parasites include *Santalum album* and *Thesium*. Fungi, bacteria, mycoplasmas, rickettsias and viruses are other common examples of plant parasites. Thus, fungi such as rusts and smuts are parasitic to many crop plants and reduce their production.

Striga.

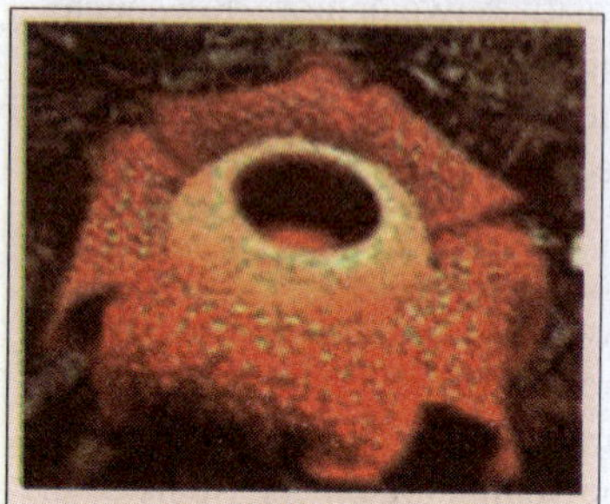
Rafflesia.

Parasitism may occur within the species. For example, in higher plants the growing pollen tube may be considered as a male plant parasitic on the tissues of the stigma and style of the flower (female plant).

Parasitodism. Some Diptera and Hymenoptera deposit their eggs in the immature stages of other insects; the larvae on hatching feed on the host until they are fully grown. In such type of parasitism in which host generally dies of the larval destruction before the larva emerges and the parasitoid larva generally lives in spite of the host's death, is called **parasitodism**. It stands in between parasitism and predation. The parasitoid may in turn be infested with **hyperparasitoids**. In the chicago, *Samia cecropia*, a saturniid moth, suffers the destruction of nearly 23% of its cocoon by an ichneumonid parasitoid, *Spilocryptus extrematis*, which deposits an average of 33 eggs on the inside of each cocoon or on the surface of larva. The host larva dies in a few hours after the parasitoid hatches and the ichneumonid larva moves about freely, feeding on the cuticle and burrowing into the tissues to drink the body fluids. *S. extrematis* is infested by another hyperparasitoid, *Aenoplex smithii* which causes about

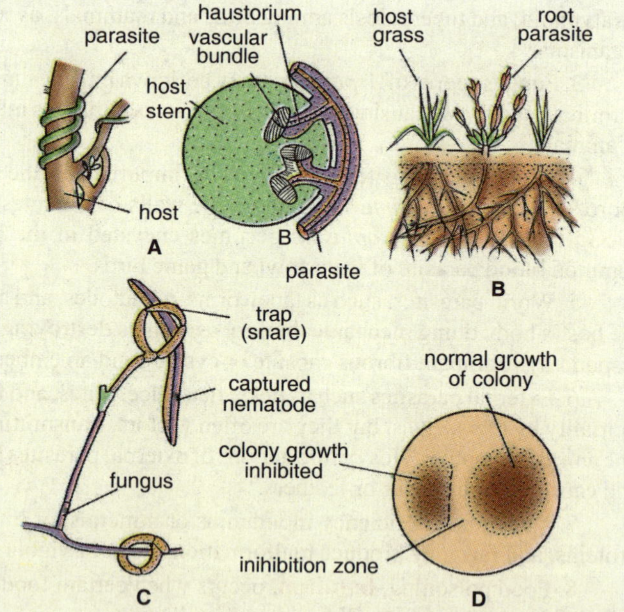

Fig. 5.4 Negative interactions. A—Parasitism, total stem parasites : a–*Cuscuta* twinning the host stem, b–T.S. of stems of host and parasite to show haustoria. B—A partial root parasite with its haustoria attached to the roots of host. C—Predation, a part of the mycelium of a nematophagous fungus (*Arthrobotrys dactyloides*) capturing the nematode as its prey. D—Antibiosis, inhibition of growth of one microbe colony by other colony.

13% destruction of *S.extrematis* cocoons. A chalcidid, *Dibrachys boucheanus*, feed both upon *S. extermatis* and as a tertiary parasitoid, upon *A. smithii.* A pycnidial fungus, *Cicinnobolus cesatii* is found as hyperparasite on a number of powdery mildew fungi.

(iii) Predation. Predation occurs when members of one species eat those of another species. Often, but not always, this involves the killing of the prey. Most of the predatory organisms are animals, but there are some plants (carnivores) also. Four types of predation have been recognized (see **Krebs**, 1985) : **Herbivores** are animals that prey on green plants or their seeds or fruits; often the plants eaten are not killed but may be damaged. Typical predation occur when **carnivores** prey on herbivores or on other carnivores. **Insect parasitism** is another form of predation in which the insect parasite lays eggs on or near the host insect, which is subsequently killed and eaten. Finally, **cannibalism** is a special form of predation in which the predator and prey are from the same species.

Predation is an important ecological process from three points of view. One effect of predation on a population may be to restrict distribution or reduce abundance. If the affected animal is a pest, predation appears useful. If it is a valuable resource such as caribou, the predation seems undesirable. Second, along with competition, predation is the second major process which affects community structure. Third, predation is a major selective force and many adaptations we see in organisms, such as warning colouration, have their explanation in predator-prey coevolution (Evolutionary changes in two or more interacting species are called **coevolution**.)

(A) Predation in animals. A predator animal tends to be larger than its prey. The main components of predation are predator and prey. A successful **predator** has following characteristics:

1. The hunting ability of a predator remains well developed.

2. By their hunting activities predators can be regarded as specialized or generalized. **Specialized predators** are those adapted to hunt only a few species. They are forced to move where the vulnerability of a staple prey item drops to a point where the predator population cannot support itself. For example, Peale's falcon shows a marked preference for ducks and pheasants. Deer exhibits a pronounced preference for certain species of browse (**Klein**, 1970). **Generalized predators**, not so restricted in diet, adjust to other food sources. The horned owl and buted hawk have a large range of collective prey available. Foxes can shift to vegetable and carrion diet, should conditions require it.

3. Hunting ability and success of predator involve the development of a searching image on the part of the predator. Once it has secured a palatable item of prey, the predator finds it progressively easier to find others of the same kind.

4. Though a predator may have strong preference for a particular prey, it can turn in the time of relative scarcity to an alternate, more abundant species that provides more profitable hunting. For example, if rodents are more abundant than rabbits and quail, foxes and hawks will concentrate on them instead of game animals.

5. Habitat preferences or overlapping territories can bring predator and prey into close contact, increasing prey risks. For example, predatory rainbow trout in Paul Lake, British Columbia, moves into the shoals when their prey, the red side shiner, are most heavily concentrated there.

6. Age, size, and strength of prey influence the direction that predation takes. Predators select food on the basis of size. Mountain lions, for example, avoid attacking large healthy elk, which they cannot successfully handle, and concentrate instead on deer and young or feeble elk (**Hornocker**, 1970).

7. Predators hunt only when it is necessary for them to procure food. The searching rate of predator is influenced by the speed of the predator relative to the speed and escape reactions of the prey, to the distance at which predators first notice and attack the prey, and to the proportion of attacks that result in successful capture.

Like the predators, the prey has its certain defensive specializations. The prey risk is determined by density of prey population, availability of food and protective cover (concealment place), movement, activity, habits, size, age, strength and escape reactions of prey. Besides these striking morphological features, prey often develop the following two antipredator defence strategies : aposematic colouration and group living.

(i) Aposematic colouration. Many animals have evolved distastefulness as a means of predator defence. Toxins may be obtained from food or synthesized by the prey species. Many of these toxic prey species feature bright colours, termed **aposematic colouration** and these colours are usually regarded as a signal to predators. For example, Monarch butterflies advertise their noxiousness to blue jays by their large size and distinct colouration.

Monarch butterflies.

(ii) Group living. Group living is normally not advantageous for the individuals. Then why do birds form blocks and ungulates go about in herds ? One possible reason for living in groups is to reduce predation losses (**Bertram**, 1978). There are three main advantages a prey organism can obtain by living in a group. First, early detection of predators may be facilitated if many eyes are looking. Prey in groups may, thus, be able to spend more time feeding and less time looking for potential predators. Second, if the prey are not too much smaller than the predator, several of them acting together may be able to deter the predator from attacking. One possible example of this is mobbing in birds (*e.g.*, crow). Primate troops will also mob predators in some circumstances and musk-oxen threatened by wolves gather into a circular defence formation. Third, if the predator is still able to attack a group, it must select one individual in the group to capture. By fleeing in confusion, the prey may confuse the predator, who may not be able to concentrate on any one individual. Alternatively, by being in the centre of a group, an individual may reduce its chances of being eaten.

Evolution of prudent predators. Coevolution of predator-prey system has produced **prudent predators** (a term coined by **Slobodkin**, 1961). If one predator is better than other at catching prey, the first individual will probably leave more descendants to subsequent predator generations. Thus, predators should be continually selected (= evolved) to become more efficient at catching prey. The problem, of course, is that by becoming too efficient, the predator will exterminate the prey and then suffer starvation. So a prudent predator should harvest prey with some constraint against overharvesting.

For example, a prudent predator would not eat prey individuals in their peak reproductive ages because that type of mortality would reduce the productivity of the prey population. Prudence dictates eating of those individuals that would die in the normal course and which contribute little to productivity. Often these are the oldest and the youngest individuals in a prey population which are hunted by the predator (**Slobodkin**, 1974). Old individuals may be post-reproductive, and young individuals often have a high death rate due to other cause.

Regulating effect of predation. It is commonly concluded that if the predator–prey interaction causes a reduction in the prey population, it is deteremental in same fashion to the prey. This belief has led to extensive "predator control" efforts conducted in the name of wildlife conservation. But the coevolution of species within natural ecosystems has in fact led to a dynamic balance between the populations in a community, so that the population sizes of predator and prey species are inter-regulated by feed back mechanisms that effectively control the population of both.

(B) Predation in plants. Plants provide the following examples of their predation :

1. Carnivorous fungi. A number of fungi such as species of *Dactylella, Dactylaria, Arthrobotrys, Zoophagus,* etc., capture insects, nematodes and other worm-like animals. Such fungi use specialized structures such as **traps** or **snares** (Fig. 5.5C) which are formed on their mycelia to capture the nematodes.

Nepenthes.

2. Carnivorous plants. A number of highly specialized plants such as *Nepenthes, Utricularia, Drosera, Dionaea, Darlingtonia, Sarracenia,* etc., are dependent for part of their nitrogen requirements upon small animals such as insects. These **insectivorous plants** have some specialized structures and mechanisms to attract, trap and ingest insects and then secrete some proteolytic enzymes to "digest" the protein contents of the insect body.

(C). Plants as prey (food) of animals: Browsing, grazing and scraping. Herbivores such as cattle and wild herbivores kill the plants and use unharvested herbs, shrubs or even trees as their food and sometimes pose great problem of management of natural and artificial vegetations (forests and grasslands). Different plants receive varying degree of damage as a result of browsing and grazing. Many insects and ruminants browse lightly over the vegetation. Cattle, camels, goats, etc., frequently browse the tender shoots of shrubs and trees and sheep graze the grasses. Generally, annuals suffer more due to grazing than the perennials, shrubs are damaged less than herbs.

Grazing has a profound effect on the composition, structure and physiognomy of vegetation. Heavy grazing reduces the photosynthetic parts so much that many plants succumb or their population decreases. This is true for palatable species. The unpalatable species avoided by grazing animals multiply and increase in number in absence of competition due to decrease in the population of palatable species. In a pasture with *Cynodon dactylon* and *Eragrostis amabilis* as dominant grass species, **Senapati** and **Dash** (1981) reported that a grazing pressure of 1.3 livestock/acre caused the accumulation of higher proportion of plant biomass (68%) underground compared to 46% in an ungrazed (*i.e.,* protected) pasture. Plant diversity was more in the grazed plot and about 18% of NP (= nitrogen and phosphorus) was removed by grazing. Grazing also stimulated higher secondary production by soil invertebrate animals, particularly oligochaetes, but it also increased mortality due to hoof action and trampling.

Cynodon dactylon.

In *Dichanthium annulatum*, there is a considerable loss of seeds and herbage due to grazing. The reduced reproductive capacity due to loss of seeds is compensated by a remarkable adaptation of increased reproductive buds on rhizomes (**Ambasht** and **Maurya**, 1970). The indirect effect of grazing is on the rate of soil erosion. This is certainly the greatest menace of grazing aspect of biotic factor. The decreased shoot cover ultimately affects root growth and exposes the soil surface to erosion due to rain and run off water and also to wind erosion.

Ambasht (1974) has indicated that grazing involves the large scale consumption of seeds, due to their high nutritive value, which results in poor successive crops, unless the plants adapt to vegetative propagation. Rats selectively remove seeds from grasses and store them in their burrow in large quantities. Fruit-eating birds and animals have some adavantageous effects also. Some seeds fail to germinate unless they have passed through the gut of birds or other animals. The seed coat in these seeds are hard and impermeable to water or oxygen. The seed coat is softened and often digested by the animals and the rest of the seeds come out and germinate quickly.

Further, in the outskirts of Indian forests, villagers tend to scrape grasses to feed their cattle in summer. **Scraping** is usually done right at the ground level. Completely exposed and loosened top soil is blown by strong summer wind. This is followed by rainy season, during which initial heavy showers result in great soil erosion.

2. Amensalism. Amensalism is a comprehensive term in biotic interactions to denote the adverse or depressing effect of one of the competing population or species while the other, remains stable. Thus, amensalism is a situation in which one population definitely inhibits the other while remaining unaffected itself. By so modifying the environment, the organism improves its own chance of survival. Ammensalism which commonly involves some type of chemical interaction, by which the organisms of one species affects the well-being and growth of individuals or the population biology of other, is called **allelochemic** (see **Whittaker**, 1970, **Smith**, 1974; **Kendeigh**, 1974). The inhibitor substance may be inorganic chemicals such as acids or bases which are produced by pioneering organisms (inhibitor species) and which reduce the competition for nutrients, light, and space, between the amensal species and inhibitor species. The production of relatively simple organic toxins is another source of chemical inhibitors. These toxins inhibit seedling growth in the vicinity. This may affect succession of plant species, especially important in the early stages (**Muller**, 1966). A third type of inhibitor chemical is the **antibiotic**–a potent antimicrobial agent. It is a substance produced by an organism, which, in low concentrations, can inhibit or kill the growth of another organism.

Salvia leucophylla.

The majority of inhibiting chemicals are produced as secondary substances by plants and released into the soil through the roots or leaf wash. The suppression of growth through the release of chemicals by a higher plant is known as **allelopathy**. Thus, allelopathy (*allelon* = each other + *pathy* = suffering) means chemical control of distribution among plants. For example, walnut tree produces a non-toxic substance, juglone (**Bode**, 1958) which is found in its leaves, fruits and other tissues. When the leaves or fruits fall on the ground, juglone is released to soil, where it is oxidized to a substance that inhibits the growth of certain under story species and garden plants such as heaths and broad-leaf herbs, and favours others such as bluegrass and blackberries. In Southern California, *Salvia leucophylla* leaves emit some volatile oils which reach the soil surface and inhibit the germination of seeds of other species and the inhibitory effects persist for several months (**Muller**, 1966). Similar allelopathic effects are also reported to be exerted by *Adenostoma fasciculatum* (**Christensen** and **Muller**, 1975). **Datta** and **Chakrabarti** (1982) prepared weed plant extracts such as of *Croton bonplandianum* and *Clerodendrum viscosum* in water and tested their effects under different dilutions on seed germination and growth of crops as pea, mustard, rice and lettuce. *C. bonplandianum* inhibits germination of seeds of above crops except of pea at high concentration. They found that the active ingredient in *C. bonplandianum* leaves showing allelopathic potential are **abscisic** and **phaseic acids**, while in *C. viscosum* inhibiting mustard seed germination may be a **terpene**. *Clerodendrum* is also reported to exert allelopathic potential on several weeds such as *Abutilon indicum, Amaranthus spinosus, Cassia tora,* etc. (see **Ambasht, 1990**).

Agriculturists have recognized the action of **smother crops** as weed suppressors. The smother crops include barley, rye, sorghum, millet, sweet clover, alfalfa, soyabeans and sunflowers. Their inhibition of weed growth was assumed to be due to competition for water, light or nutrients. However,

Overland showed that barley (*Hordeum vulgare*) inhibited the germination and growth of several weeds, even in the absence of competition for nutrients or water. He found that the adverse effect of barley is partly due to the secretion by its roots of chemicals (alkaloid) that reduce growth and germination of nearby weeds.

Amaranthus spinosus.

The term **antibiosis** generally refers to the complete or partial inhibition or death of one organism by another through the production of some substance or environmental conditions as a result of metabolic pathway. These substances and/or conditions are harmful (antagonistic) to other organisms. Antibiosis is more commonly referred to secretions by microorganisms that check the growth of others. Production of chemicals that are antagonistic to microbes—the **antibiotics** is well known. Antibiotics produced by bacteria, fungi, actinomycetes, and lichens are widespread in nature and may be one of the reasons why bacteria pathogenic to man cannot multiply well in soils. *Penicillium*, a fungus found in soil produces antibiotic substances that check the growth of a large variety of bacteria. A number of antibiotics such as penicillin, have been used extensively in human medicine.

Antagonistic substances are also reported in some algae, as for example in culture of *Chlorella vulgaris*, some substance accumulates which inhibits the growth of the diatom, *Nitzschia frustrulum*. Substances produced by senescent cultures of *Chlorella* and of the diatoms, *Navicula* and *Scendesmus* inhibit the filter-feeding of *Daphnia* in laboratory. Pond "blooms" of blue-green algae especially *Microcystis* are known to produce toxins such as hydroxylamine which causes death of fish and cattle. In marine waters, populations of some microbes, popularly known as **red tide**, cause catastrophic destruction of fish and other aquatic animals.

Allelochemic effects are of great variety in both plants and animals : repellants, escape substances, suppressants, venoms, inductants, counteractants, attractants, signals, stimulants, autotoxins, autoinhibitors, and so on. **Pheromones** (ectohormones) are chemical messages between members of a species especially important in reproductive behaviour, social regulation and recognition, alarm and defense, territory or trail marking, food location, and so on. Many of these effects are beneficial to the individual, while other serve for competitive purposes (**Whittaker** and **Feeny**, 1971).

3. Competition. The presence of other organisms may limit the distribution of some species through **competition**. Such competition can occur between any two species that use the same sorts of places. Note that two species do not need to be closely related to be involved in competition. For example, birds, rodents and ants may compete for seeds in desert environments. Competition among animals is often over food, water and mates. Animals may also do competition for space, *i.e.,* nesting sites, wintering sites, safer sites from predators. Plants can compete for light, water, nutrients, or even for pollinators and/or attachment sites. Competition is an important process affecting the distribution and abundance of plants and animals.

There are two different types of competition, defined as follows (**Birch**, 1957) :

(a) Resource competition (also called **scramble competition**) occurs when a number of organisms (of the same or of different species) utilise common resources that are in short supply.

(b) Interference competition (also called **contest competition**) occurs when the organisms seeking a resource harm one another in the process, even if the resource is not in short supply.

Note that the competition may be **interspecific** (between two or more different species) or **intraspecific** (between members of the same species).

(A) Intraspecific competition. It is an important density-dependent factor regulating populations.

The wildebeest (gnu) population is thought to be regulated by intraspecific scramble competition for a limited supply of grass of adequate quality in dry season.

Intraspecific competition is responsible for levelling off or fluctuations around a certain density of a population. Exact mechanism do vary in different species. Intraspecific competition is worked out in the beetle, the azuki bean weevil (*Callosobruchus chinensis*) which feed on stored legume seeds. Figure 5.5 shows results from a series of experiments begun with different numbers of adult weevils. Thus, at high density, female fecundity is reduced because overcrowding leads to fewer successful matings due to mechanical interference between individuals. These conflicts are the main factors which result in the decrease in population density. Minor effects include increased egg mortality and increased level competition for food at high density. The egg mortality is due to adult knocking eggs from the sites at which they were laid.

Dash and **Hota** (1980), **Misra** and **Dash** (1984) and **Mohapatro** and **Dash** (1984) studied the density effect on the growth and metamorphosis of three species of tropical anuran tadpoles *Rana tigrina (=tigerina)*, *Polypedates* (*=Rhacophorus*) *maculatus* and *Bufo stomaticus.* For each of these tadpoles, they found the competition for food due to crowding which affects growth rate and the threshold size for metamorphosis. In very high density populations (with severe competition) some larvae were not able to attain the threshold size and could not metamorphose (see **Dash**, 1990).

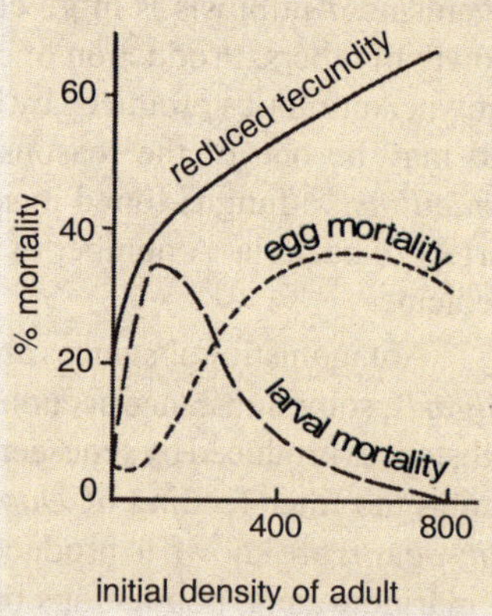

Fig. 5.5. Intraspecific competition between azuki bean weevils change as population density is increased in laboratory cultures. Fecundity is reduced and mortality among younger shows more complex and contrasting patterns.

(B) Interspecific competition. A.J. Lotka and **V. Volterra** in 1920 developed theoretical or mathematical models to explain prey-predator interactions which produce oscillatory patterns. They also discussed the problem of competition. If two different species populations require a common resource for living and if this resource is potentially or actually limiting, then the two species are said to be in competition for it. These models predict the following three types of outcomes :

1. One species only survives, it being the one with the greatest negative effect on its competitor. Growth of the surviving population to its carrying capacity is slower than if the second population had been absent.

Thus, only one species will survive and there will be an exclusion or displacement of one of the populations by competition. This is called **Gause's principle of competitive exclusion** which states that complete competitors cannot exist.

2. Both species coexist indefinitely. This occurs when interspecific competition is less intense than intraspecific one in both species. Neither population reaches the carrying capacity it would have in the absence of other species.

3. The species beginning at higher density persists, and the other is eliminated. This is a special case when the populations have equally negative effects on the growth of each other, but interspecific competition is stronger than intraspecific one.

Example. (i) **Gause** studied the growth patterns of two species of *Paramecium* (*P. aurelia* and *P. caudatum*) when cultured together and separately (Fig. 5.6). Each species population has a sigmoidal growth pattern when grown independently. However, when both species grown together, their growth patterns were sigmoid in the first week, but later there was a gradual increase of *P. aurelia* and gradual decrease of *P. caudatum. P. aurelia* too did not grow to the level it had done when grown separately. Thus, this experiment confirmed the competitive exclusion principle.

(ii) The principle of competitive exclusion applied well to comparative studies of **Thomas Park** (1948-1962) using the grain beetles, *Tribolium confusum* and *Tribolium castaneum* under different environmental conditions.

(iii) Introduction of wasp parasite, *Opius oophilus* to control a fruit fly pest in Hawaii, apparently displaced the other two wasp species from the field.

(iv) The tramp ant, *Wasmannia auropunctata* has displaced all the native ants in some areas of Santa Cruz Island in the Galapagos group.

(v) **Misra** and **Dash** (1979) studied interspecific interactions in four species (*Lampito mauritii, Drawida calebi, Octochaetona surensis* and *Peryonix millardi*) of tropical earthworms. The rate of oxygen consumption of the different species of earthworms taken in mixed species pairs, showed significant reduction when compared to the sum of their individual rates of respiration.

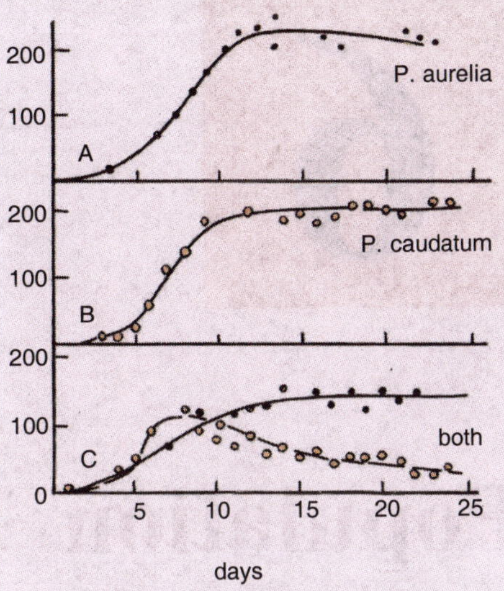

Fig. 5.6. Population growth and interspecific competition between two species of *Paramecium* in laboratory cultures. When cultured separately (A) and (B), the two species approximate to logistic growth. However, in mixed cultures (C) *P. aurelia* survives and *P.caudatum* dies out.

(vi) **Ramakrishnan** and **Gupta** (1980) worked on the competition in related species of *Argemone* (*A. mexicana* and *A. ochroleuca*). They found that *A.mexicana* dominates over *A. ochroleuca* in mixed populations if seedling establishment occurs in the later part of the growing season and *A. ochroleuca* dominates if the seedling establishment occur in the early part. Thus, environmental conditions influence the competitive ability of the species.

REVISION QUESTIONS

1. Describe the phenomenon of mutualism by providing examples both from plants and animals.
2. What is exploitation ? Describe, with suitable examples, this negative ecological interaction.
3. Discuss the role of parasitism as a biotic ecological factor and compare it with predation.
4. Write an essay on 'prey-predation interaction.'
5. Describe ecological effects of browsing, grazing and scrapping.
6. Describe with appropriate examples the ecological aspects of competition.
7. Discuss the phenomenon of ammensalism.
8. Discuss with suitable examples the positive and negative interactions between two different populations.
9. Write short notes on the following :
   (i) Mutualism ; (ii) Commensalism ; (iii) Protocooperation ; (iv) Mycorrhizae ; (v) Epiphytes ; (vi) Social parasitism ; (vii) Fungus gardens ; (viii) Prudent predators ; (ix) Aposematic colouration ; (x) Antibiosis.

CHAPTER

6

Population

(Population Ecology)

The term population has its origin in the Latin word *populus*, meaning people. In ecology, a **population** may be defined as a group of organisms of the same species occupying a particular space. Thus, we may speak of the deer population of Corbett National Park, the deer population of Kaziranga Wild Life Sanctuary, the human population of Delhi, or the human population of India. The ultimate constituents of the population are **individual organisms** that can potentially interbreed. The populations may be subdivided into **demes** or local populations, which are groups of interbreeding organisms, the smallest collective unit of a plant or animal population. Individuals in demes, thus, share a common gene pool. The boundaries of a population both in space and in time are vague and in practice are usually fixed arbitrarily by the investigator.

Some ecologists recognized following two types of populations : 1. **Monospecific population** is the population of individuals of only one species; and 2. **Mixed** or **polyspecific population** is the population of individuals of more than one species. Often ecologists use the term **community** for polyspecific population.

The population has various **group characteristics**, which are statistical measures that cannot be applied to individuals. These group characteristics are of three general types. The basic characteristic of a population is its **size** or **density** which is

Similar species of organisms constitute a population.

affected by four primary population parameters such as **natality** (births), **mortality** (deaths), **immigration** and **emigration** (Fig. 6.1). In addition to these attributes, one can derive secondary characteristics of a population, such as its **age distribution**, **genetic composition**, and **pattern of distribution** (distribution of individuals in space). These population parameters result from a summation of individual characteristics. Thus, group attributes of population are different from those of an individual, for instance an individual cannot have a density, it cannot have a rate of birth or death, although it occupies a place, is born and dies. The group attributes of a population with respect to rates of birth, death, reproduction, etc., form the science of **demography**. However, when they are studied from a ecological point of view, this forms a **population ecology** or **democology**. In other words, population ecology is the study of individuals of the same species where the processes such as aggregation, interdependence between individuals, etc., and the various factors governing such processes are emphasized.

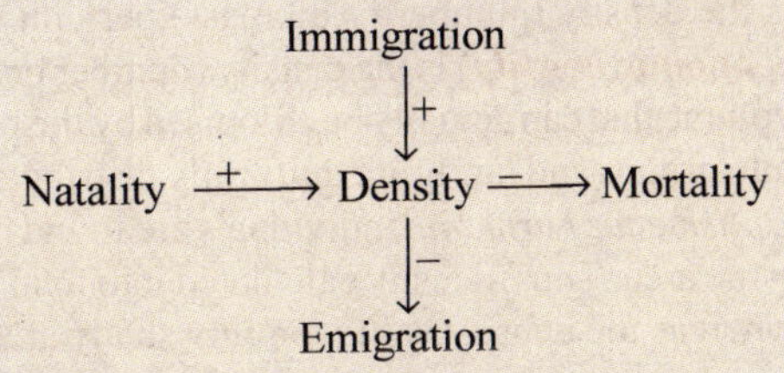

Fig. 6.1. The primary population parameters (after Krebs, 1985).

The ecological study of populations include the following three main aspects :

1. Population characteristics;
2. Population dynamics;
3. Regulation of population.

POPULATION CHARACTERISTICS

The population has the following characteristics :

A. Population Size and Density

Total **size** of population is generally expressed as number of individuals in a population. The population size (N) at any given place is determined by the processes of **birth** (B), **death** (D), new arrivals from outside or **immigration** (I) and going out or **emigration** (E). Therefore, change in population size between an interval of time Nt + 1 is Nt (initial stage) + B – D + I – E.

Population density is defined as numbers of individuals per unit area or per unit volume of environment. Larger organisms as trees may be expressed as 600 trees per hectare, whereas smaller ones such as phytoplanktons (as algae) as 2 million cells per cubic metre of water. In terms of weight it may be 100 kilograms of fish per hectare of water surface. Density may be **numerical density** (number of individuals per unit area or volume) or **biomass density** (biomass per unit area or volume). When the size of individuals in the population is relatively uniform, as mammals, birds or insects, then density is expressed in terms of number of individuals (numerical density). But, when the size of individuals is variable, such as true of fishes, trees or mixed populations (biomass density). To measure density of a population in terms of measurement of biomass will be satisfactory as a measure of density biomass one can take wet weight, dry weight, volume and carbon and nitrogen weight. Population density is also measured as **abundance** or absolute number of population. Generally, in an area smaller animals are more abundant than larger animals. But birds are less abundant than mammals of equivalent size (**Krebs**, 1985).

Cassia tora.

Further, since the pattern of distribution of organisms in nature is different it becomes important to distinguish between crude density and ecological density. **Crude density** is the density (number or biomass) per unit total space. **Ecological density** (also called **specific** or **economic density**) is the density (number or biomass) per unit of habital space, *i.e.,* available area or volume that can actually be colonised by the population. In nature, organisms grow generally clumped into groups and rarely are uniformly distributed. For example, in plant species such as *Cassia tora* and *Oplismenus burmanni* individuals are found more crowded in shady patches, and few in other parts of same area. Thus, density calculated into total area (shady as well as exposed) would be crude density, whereas the density value for only shady area (where the plants actually grow) would be ecological density.

Fine distinction between the crude density and ecological density becomes more clear by the Figure 6.2, which is based upon **Kahl's** (1964) study on fish density in a water body. As shown in the figure the crude density of small fish in the area as a whole goes down as the water level drops during the winter dry season, but the ecological density in the contracting pool of water increases as fish are crowded into smaller and smaller water area. Thus, with changing time, water level and total number of fish, the two densities become different.

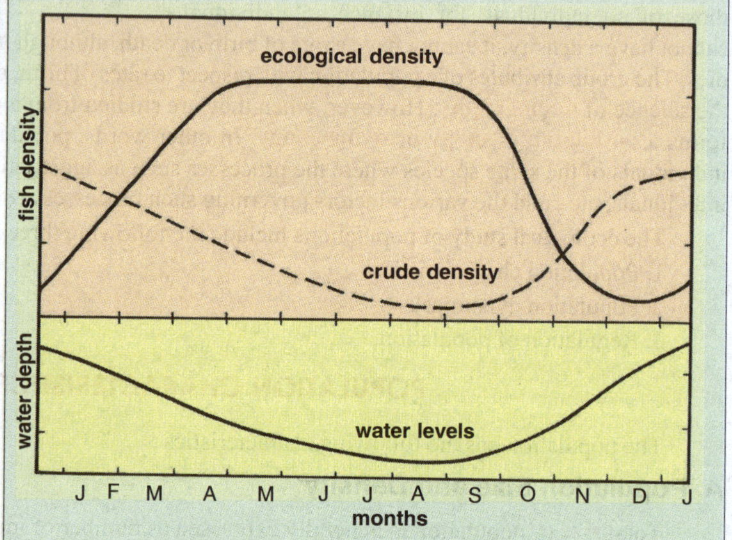

Fig. 6.2. Crude density and ecological density of fish prey in relation to the breeding of stork predator (after Kahl, 1964).

B. Patterns of Population Dispersion (or Spatial Distribution)

Individuals of a population arrange themselves in a manner that is specific for each population and these arrangements in space appear to be of great significance in the studies of the dynamics of the ecosystems. In nature, due to various biotic interactions and influence of abiotic factors, the following three basic population distributions can be observed : random distribution, uniform or regular dispersion, and clumped dispersion (Fig. 6.3).

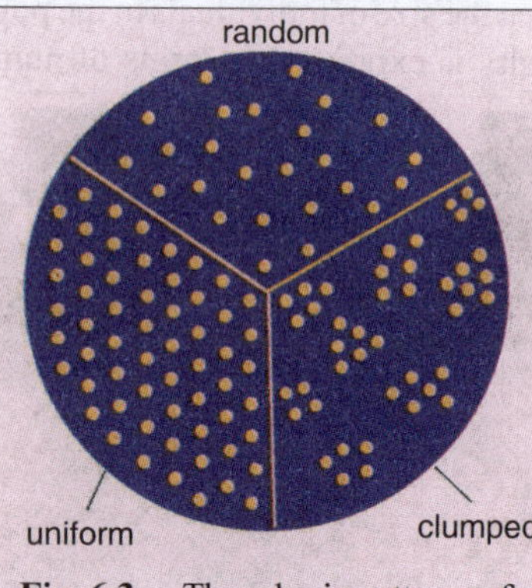

Fig. 6.3. Three basic patterns of population dispersion.

C. Age Structure

In most types of populations, individuals are of different age. The proportion of individuals in each age group is called **age structure** or **age distribution** of the population. Age distribution of the population influences natality (birth rate) and mortality (death rate). Reproductive ability is determined by age of the females and, hence, natality denotes the numerical representation of young ones produced in unit time by the same age group. Similarly chances of death are more towards the earlier and later periods of life span and so mortality is under the control of age. Therefore, the ratio of the various age groups in a population determines the current reproductive status of the population and helps in anticipating its future. From an ecological point of view there are three major **functional** or **ecological ages** (age groups) in any population. These are (i) **pre-reproductive** (or juvenile or dependent phase), (ii) **reproductive** (or adult phase) and (iii) **post-reproductive** (or old age). The relative duration of these age groups in proportion to the life span varies greatly with different organisms. In human beings, the "three" ages are relatively equal in length, about a third of his life falling in each class. Many plants and animals have very long pre-reproductive period. For example, insects have extremely long pre-reproductive periods, a very short reproductive period and no post-reproductive period.

Age pyramids. A model representing geometrically the proportion of different age groups in the population of any organism is called **age pyramid** (or **age-sex pyramid**); (see **Ananthkrishnan** and **Viswanathan**, 1983). An age pyramid is a vertical bar graph in which the number or proportion of individuals in various age ranges at any given time is shown from youngest at the bottom of the graph to oldest at the top. It is very useful in monitoring the commercial exploitation of food species by man among other things. According to **Bodenheimer** (1938), there are following three basic types of age-sex geometric figures (Fig. 6.4) :

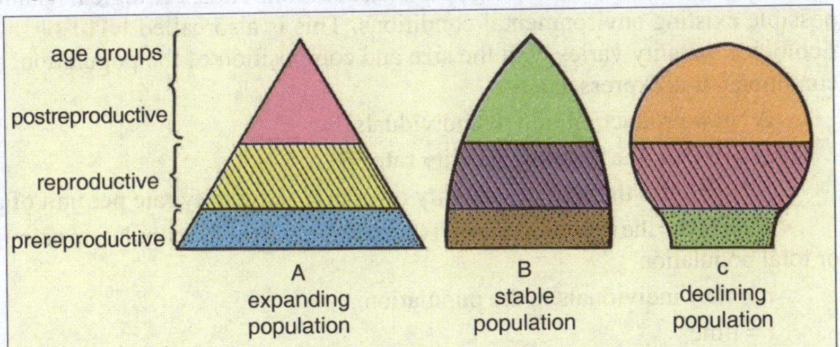

Fig. 6.4. Age structure in different types of population (after Kormondy, 1969).

1. Pyramid with broad base (or Triangular structure). It indicates a rapidly expanding population with a high percentage of young individuals and only few old individuals. Thus, in rapidly growing young population, birth rate is high and population growth may be exponential as in yeast, housefly, *Paramecium*, etc.

2. Bell-shaped polygon. It indicates a stationary population having an equal number of young and middle-aged individuals. As the rate of growth becomes slow and stable, *i.e.*, the pre-reproductive and reproductive age groups become more or less equal in size, post-reproductive group remaining as the smallest.

3. Urn-shaped structure. It indicates a low percentage of young individuals. It shows a declining population. Such an urn-shaped figure is obtained when the birth rate is drastically reduced, the pre-reproductive group dwindles in proportion to the other two age groups of the population.

D. Natality

Population increases because of natality. **Natality** is a broader term covering the production of new individuals by birth, hatching, germination, or fission. The **natality rate** may be expressed as the number of organisms born per female per unit time. In human population, the natality rate is eqivalent to the 'birth-rate'.

At this stage, one should be clear regarding two essential aspects of the reproduction which forms main component of natality : 1. **Fertility** is the physiological notion which indicates that *an organism is capable of breeding*. 2. **Fecundity** is an ecological concept that is based *on the numbers of offspring produced during a period of time.* Fecundity is of two types—**potential fecundity** and **realized fecundity**. For example, the realized fecundity rate for an actual human population may be only one birth per 8 years per female in the childbearing ages. Whereas the potential fecundity rate for human is one birth per 9 to 11 months per female in the childbearing ages (see **Krebs**, 1985).

Natality is of following two types :

1. Maximum natality. (Also called **absolute**, **potential** or **physiological natality**). It is theoretical maximum production of new individuals under ideal conditions which simply means that there are no ecological limiting factors and that reproduction is limited only by physiological factors. Hence, absolute natality is constant for a species population. This is also called **fecundity rate** (see **Sharma**, 1994).

2. Ecological natality. (Also called **realized natality** or simply **natality**). It refers to population increase under an actual, existing specific condition. Thus, ecological natality takes into account all possible existing environmental conditions. This is also called **fertility rate** (see **Sharma**, 1994). Ecological natality varies with the size and composition of the population, and with environmental conditions. It is expressed as—

Δ Nn = production of new individuals

Δ Nn/ Dt = the absolute natality rate (B)

Δ Nn/NDt = the specific natality rate (b) (*i.e.,* natality rate per unit of population)

where, N = the reproductive part (mature female) of the population or initial number of organisms or total population

n = new individuals in the population

t = time

Δ = delta : a change in value.

Further, the rate at which females produce offsprings is determined by the following three population characteristics :

1. **Clutch size** or the number of young produced on each occasion;
2. the time between one reproductive event and the next, and
3. the age of first reproduction.

Thus, natality usually increases with the period of maturity and then falls again as the organism gets older (Fig. 6.5). But there are some trees which continue to increase fruit production as they get older. Natality patterns differ in tropical and temperate populations.

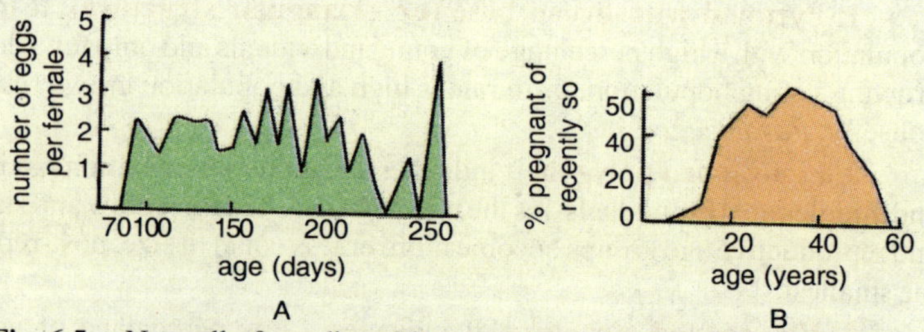

Fig. 6.5. Normally fecundity of female animals is low at the onset of the sexual maturity, then reaches a lengthy plateau, followed by a decline in old age. A—Milkweed bug (*Oncopeltus unifasciatellus*) cultured in laboratory; B—African elephants (*Loxodonta africans*) in a National Park, Uganda.

Fecundity. Species vary greatly in the characteristic number of generations, broods or litters produced per year, and in the size of them. Protozoans often divide so rapidly that they produce a new generation every few hours. Plankton organisms, less fecund, may produce a new generation every few days. Many vertebrates breed but once a year, some large animals only once every two or three years. Several species of small birds and mammals have two or more broods per year. Rodents may continue to breed throughout the winter under favourable environmental conditions, so that their reproductive potential is enormous.

The maximum size of a litter is determined by the physiological and morphological characteristics of the species. In mammals, which produce viviparous young, the size of uterus and body cavity as well as the number of mammary glands for suckling the young after birth acts as limiting factors for the size of the litter. In birds there is a limit on the number of eggs that one individual can cover and successfully incubate. However, in species that do not take care of their eggs after laying, the number produced may be limited only by the energy resources of the parent.

The number of eggs or young produced per litter is correlated inversely with the amount of attention (*viz.*, parental care) that they require. When parental care is altogether lacking, invertebrates may lay 1,000 to 500,000,000 eggs at one maturation; where there is some protection afforded by brood pouches, 100 to 1,000 eggs may be laid; with a high degree of brood protection, 1 to 10 or more eggs may be laid. Mammals seldom have more than a dozen young in a single litter and in larger species usually only one. Characteristic clutch size among birds varies from 1 to 15; rarely, 20.

Further, there is a limit on the size of the brood or litter that adult warm blooded animals can successfully feed and raise to maturity. There is no advantage, for example, for starlings to have broods larger than five. In larger broods, each individual receives less food, and, hence, has less vigour and weight on leaving the nest. In such weak individuals, mortality increases either before fledging or in subsequent months (**Perrins**, 1965). In those species that feed their young in the nest, the clutch size has evolved through natural selection to the greatest number that can be hatched and raised successfully through efforts of the adults (**Lack**, 1967). In those species whose young leave the nest and feed themselves at hatching, the clutch size depends in large part on the capability of the female to mobilize energy in her body to produce eggs of a particular size (**Ryder**, 1970). The variability in clutch and litter size for most species allows them to take advantage of temporarily improved conditions.

White tail deer.

The size of brood, litter or clutch is also regulated by weather. For example, the clutches laid by birds during periods of hot weather are usually smaller than those laid when temperature is moderate (**Kendeigh**, 1941). Clutches laid by related species in temperate latitudes tend to be larger than those laid in the tropics. Likewise, the fecundity of white-tailed deer is higher with good forage than with poor forage. The reproduction is generally more successful after periods of high mortality than during years of abundance.

E. Mortality

Mortality means the rate of death of individuals in the population. It is a negative factor for population growth. Like natality, mortality may be of following types :

1. Minimum mortality. It is also called **specific** or **potential** mortality. Minimum mortality represents the theoretical minimum loss under ideal or non-limiting conditions. It may be constant for a population.

Senescence in plant leaves.

2. Ecological mortality. It is also called **realised mortality**. It is the actual loss of individuals under a given environmental condition. Ecological mortality is not constant for a population and varies with population and environmental conditions, such as predation, disease and other ecological hazards.

Aging.

All organisms have a **physiological** or **potential longevity** representing the age (called **old age**) up to which the organisms can live under ideal conditions without stresses or diseases and death occurs purely due to aging or **senescence**.

Like natality, mortality may be expressed as the number of individuals dying in a given period (death per time), or as specific rate in terms of units of the total population or any part there of . The death rates vary among species and are correlated with rates of reproduction (natality). The death rate of a species is influenced by a number of factors such as destruction of nests, eggs or young by storms, wind, floods, predators, accidents, and desertation of parents, but of fundamental importance is the number of young that are born in relation to the carrying capacity of the habitat. When more young are born than the habitat can support, the surplus must either die or leave the area. Mortality rate is usually expressed as a percentage of individuals dying within a given time. The mortality rate in many species varies from one age level to another; thus, a mean death rate has only general significance. In birds, however, the death rate is nearly constant once they become adult, and it is then apparent that it varies inversely with adult longevity. In adult penguins, pelicans, shore birds, gulls, and swifts, the annual mortality rate is commonly between 12 and 30 per cent; in herons, hawks, and owls it is about 30 per cent; in ducks, doves, and song birds it is between 40 and 68 per cent, while in gallinaceous birds it is the highest, 60 to 80 per cent. As mortality varies positively with age in the majority of organisms, the specific mortalities at particular ages can be illustrated in the form of life table.

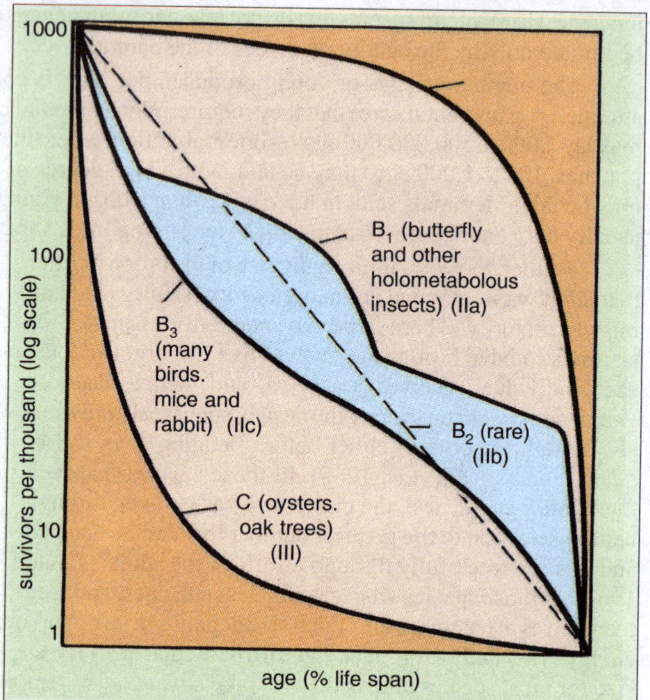

Fig. 6.6. Different types of survivorship curves plotted on the basis of survivors and age. Examples of each has been given within the brackets.

Vital index and survivorship curves. A birth-rate ratio (100 × births/death) is called **vital index**. For a population the surviving individuals are more significant for a population than the dead ones, the death rate can be represented by **survivorship curves** or **life tables**, both of which provide an estimate of organisms surviving at various ages.

(i) Survivorship curves. The pattern of mortality with age is best illustrated by survivorship curves which plot the numbers surviving to a particular age. There are following three types of survivorship curves (Fig. 6.6) which represent the different nature of survivors in different types of populations :

(a) Highly convex curve. This type of curve (pattern I) in the figure is the characteristic of the species in which the mortality rate of the population is low until near the end of the life span. Thus, such species tend to live throughout their life span, with low mortality rate in young and adult. Many species of large animals as deer, mountain sheep, man and small rotifers show such curves. Among various large-sized perennial plants, generally the plant die after reaching reproductive phase within a period of old age.

(b) Highly concave curve. This type of curve (pattern III) is the characteristic of such species where mortality rate is high during the young stages. Some birds, blacktail deer, oysters, shell fish, oak trees, short-lived weedy annuals, etc., exhibit pattern III type curve.

(c) Diagonal straight-line curve. This type of curve (pattern II b) indicates an age-specific constant survivorship, *i.e.,* a constant rate of mortality occurs at every age. Some animals such as hydra, gull, American robin, etc., exhibit this type of curve.

In fact, no population in the real world has a constant age-specific survival rate throughout the whole life span. Thus, a slightly concave or sigmoid curve (pattern IIc) is characteristic of many birds, mice and rabbits. In them, the mortality rate is high in the young but lower and almost constant in the adult (1 year or older). In some holometabolous insects (*i.e.,* insects with complete metamorphosis), such as butterflies, the survival rate differs in successive life-history stages and the curve becomes the **stair-step** type survivorship curve (curve IIa), the initial, middle and final steep segments represent the egg population and short lived adult stages and the two middle flatter segments represent the larval and pupal stages which exhibit less mortality. Some earthworms also exhibit stair-step type survivorship curve (**Dash**, 1987). Lastly, crowding (high density) in certain populations such as black tail deer population may influence the shape of survivorship curve (**Taber** and **Dasmann**, 1957).

The survivorship curve of human population is highly convex. This has become possible because of increased medical care, better hygiene, improved nutrition, and so on. The average life span of an Indian male has gone up from 30 years (before 1947) to 58 years now. But maximum longevity has not gone up very much in any society.

(ii) Life tables. Species differ widely in the number of young produced each year, in the average age to which they live and in their average rate of mortality. When sufficient facts about a species are known, a **life table** that tabulates the vital statistics of mortality and life expectancy for each age

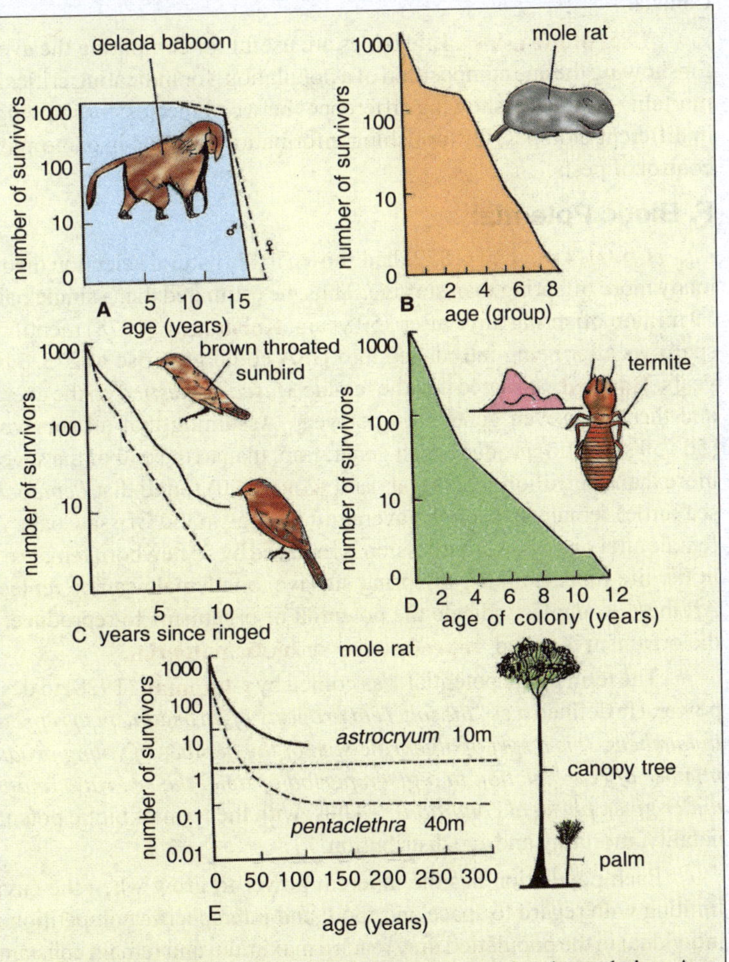

Fig. 6.7. Some examples of survivorship curves of real population. A—Convex curve in gelada baboons, (montane Ethiopia); B—Diagonal curve of mole rat (Nairobi, Kenya); C—Diagonal curve of brown-throated sunbird (forests of south-east Asia); D—Concave curve of whole termite colonies (Nigerian savanna); E—Concave curves of two rainforest trees (solid line is for understory palm in Mexico; broken line for a canopy tree in Costa Rica).

group in the population may be formulated. In such tables, age is usually represented by the subscript index x and is some convenient fraction of a species mean life span, such as year or stage of development. The life table is set up on the basis of an initial cohort of 100, 1000, or 100,000 individuals; and the number living to the beginning of each successive age interval is symbolized as lx. Plotting these data gives a **survivorship curve** for the species and a number of survivorship curves have been worked out for different species. The number dying with each age interval is designated as dx and gives a mortality rate. The rate of mortality during each age interval is commonly expressed as the percentage of the number at the beginning of the interval 100 (dx/lx) and is indicated as qx. Survival rate is the difference between the mortality rate and 100 per cent (100–qx) and is expressed as sx. Life expectancy (ex) is the mean time that elapses between any specified age and the time of death of all animals in the age group. For example we can consider in Table 6-1 the life table for budworms of Canada.

Uses of life tables. Life tables are useful for computing the average longevity of a population, for showing the age composition of a population, for indicating critical stages in the life cycle at which mortality is high, for showing differences between species, for showing the success of the same species in different biotopes, for furnishing information of value in game and fish exploitation (yield) and in control of pests.

F. Biotic Potential

Charles Darwin (1859) had shown in his natural selection theory that all organisms reproduce many more offspring than survive. Thus, he estimated that a single pair of elephants would have over 19 million offspring alive after 750 years. **Kormondy** (1978) mentioned that a single pair of English sparrows after being introduced into USA could give rise to 275, 716, 983, 698 descendants in ten years. **Howard** estimated that the female *Musca domestica* on the average produces 120 eggs at a time and there are seven generation per year. Assuming that all survive one year and all the females (50 : 50 sex ratio) produce each generation, than at the end of the seventh generation, there would be more than six trillion flies. **Dash** and **Kar** (1990) found that *Lepidochelys olivacea* (the olive ridley sea turtle) female lays, on the average 110 eggs in the Orissa Coast. Assuming a 50 : 50 sex ratio, a female after one egg-laying session is replaced by 55 newborn females. A female lays eggs many times in her life time. If all the offspring survive, one female can be replaced by a few hundred females. All these examples indicate the potential of organisms to reproduce. This reproductive capacity of the organism is called **reproductive** or **biotic potential**.

The term biotic potential was coined by **Chapman** (1928) to designate maximum reproductive power. He defined it as "*the inherent property of an organism to reproduce to survive, i.e., to increase in numbers. It is a sort of algebraic sum of the number of young produced at each reproduction, the number of reproduction in a given period of time, the sex ratio and their general ability to survive under given physical conditions.*" Thus, with the term of biotic potential, one is able to put together natality, mortality and age distribution.

Each population has the inherent power to grow when the environmental conditions are not limiting with regard to space and food; and interspecific competition is absent. The growth rate per individual in the population may reach a maximum and remain constant under specific environmental conditions. In a population with a stable age distribution, the reproductive capacity or growth rate is called the **intrinsic rate of natural increase**, represented by the symbol r. The r is the exponent in the differential equation for population growth in an unlimited environment under specific physical conditions. In fact, the index r is the difference between the instantaneous specific natality rate (*i.e.*, rate per time per individual) and the instantaneous specific death rate and may be expressed as :

$$r = \mathrm{b} - \mathrm{d}$$

Table 6-1. Life table for the 1952-53 generations of the spruce budworm in New Brunswick, Canada (after Morris and Miller, 1954, source Kendeigh, 1974).

x	lx	Factors responsible for dx	dx	qx
Eggs	1000	Parasites	17	2(—)
		Predators	86	9(—)
		Others	6	1(—)
			Total = 109	12
Instar I	891	Dispersal, etc.	428	48
Hibernacula	463	Winter	79	17
Instar II	384	Dispersal	242	63
Instars III-VI	142	Parasites	51	36
		Disease	3	2
		Birds	20	14
		Others	61	43
			Total = 135	95
Pupae	7	Parasites	0.6	8
		Predators	0.7	10
		Others	1.3	18
			Total = 2.6	36
Moths	4.3	Total for generation	= 995.7	99.5

The *r* depends on the age composition and the specific growth rates due to reproduction of component age groups. Thus, there may be several values of *r* for a species depending upon population structure (Table 6-2) :

Table 6-2. Intrinsic rate of natural increase.

Organism	Approximate biotic potential (r) per year
1. Large mammals	0.02 – 0.5
2. Birds	0.05 – 1.5
3. Small mammals	0.3 – 8
4. Larger invertebrates	10 – 30
5. Insects	4 – 50
6. Small invertebrates (including large protozoans)	30 – 800
7. Protozoa and unicellular algae	600 – 2,000
8. Bacteria	3,000 – 20,000

Under natural conditions biotic potential is a rare phenomenon, since environmental conditions do not permit unlimited growth of any population. The difference between the maximum *r* (biotic potential) and the rate of increase which occurs in an actual laboratory or field conditions is often taken as a measure of the **environmental resistance**, which is the sum total of environmental limiting factors which prevent the biotic potential from being realized. Environmental resistance is often low when a species is first introduced into a new territory so that the species increases in number at a fantastic rate as when the rabbit was introduced into Australia and English sparrow and Japanese beetle were brought into the United States. But as a species increases in number the environmental resistance to it also increases, in the form of organisms that prey upon it parasitize it, and the competition between the members of the species for food and living space, and consequently, population decreases. An equilibrium will be reached either by decreasing the birth rate (natality) or by increasing the mortality rate.

POPULATION DYNAMICS

Populations have characteristic pattern of increase which are called **population growth forms**. Such growth forms represent the interaction of biotic potential and environmental resistance. The study of population dynamics is done by three approaches (1) mathematical models, (2) laboratory studies and (3) field studies.

The growth is the most fundamental dynamic feature that a species population displays. Populations characteristically increase in size in a sigmoid, S-shaped or logistic fashion. When a few organisms are introduced into an unoccupied area, the growth of the population is at first slow (**positive acceleration phase**), then becomes very rapid (**logarithmic phase**) and finally slows down as the environmental resistance increases (**the negative acceleration phase**) until an equilibrium level is reached around which the population size fluctuates more or less irregularly according to the constancy or variability of given environment. The level beyond which no major increase can occur represents the saturation level or **carrying capacity**. The carrying capacity or equilibrium density is represented by the letter **K**. It is often useful to define the maximum rate of growth of the population. This parameter, generally termed the **intrinsic rate of natural increase**, is symbolized **r0** and represents the growth rate of a population that is infinitely small. Accordingly such type of population growth can be described by following **logistic equation** :

$$dN/dt = r_0 N (K - N) / K$$

where r_0 = innate capacity of population to increase (birth rate without resource limitation), N = population size and K = highest population density that can be maintained in real environment, *i.e.*, at carrying capacity.

There are two main types of population growth forms. (1) J-shaped and (2) S- shaped or sigmoid forms (Fig. 6.8). The growth forms are due to the nature of species and prevailing environmental conditions. In **J-shaped curve** there is a rapid increase in density with the passage of time (called **exponential growth**). The density values when plotted against time give a J-shaped growth curve and at the peak the population growth ceases abruptly due to environmental resistance. For example, the population growth curve in human populations and growth of yeast, *Drosophila* and rabbit under laboratory conditions show an initial slow rate and then it accelerates and finally slows giving the growth curve which is **sigmoid** or **S-shape**. The peak constant level represented by *K* or upper level (called **asymptote**) of the sigmoid curve is called the **maximum carrying capacity**. It marks the limit to which the environment can support the population.

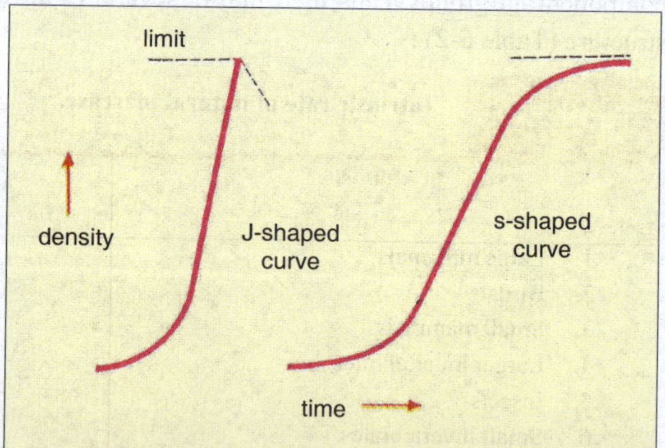

Fig. 6.8. J-shaped and S-shaped population growth curves.

Plant Population Dynamics

In many respects, plant populations behave like the animal population, but, they have some unique features such as follows : Most higher plants are **modular** organisms, developing from a single zygote but producing an indeterminate number of repeatitive structures, called **modules** vegetatively. A clump of herbs, grasses or trees may be product of one zygote. Plants cannot move to mate or disperse.

Thus, they have evolved means as gravity, wind, water flow or animals for dispersal of pollen, seed, vegetative parts, etc. The **seed population** present in the soil for different species are referred to as **seed bank** or **seed pool** (**Silvertown**, 1987). All these seeds do not germinate or all the seedlings do not establish. Some die due to environmental stresses and this is called **environmental sieve** which allows only the stronger individuals to survive. In most cases, the seeds germinate in batches and seedlings of one lot is known as **cohort**. In this way, from a huge seed bank through ecological selection, cohorts are formed and these in turn result into adult population. This process is referred to as **recruitment**. Further, a plant may originate from a vegetative part, called **ramete** (or **tiller**) or from seed called **genet**. Thus, ramete and genet form two levels of population structure. The term **clone** is normally used to designate the population derived from ramete of the same parent plant.

The 3/2 thinning law. Most aspects of growth of population are density related. One important generalisation applied is **3/2 thinning law**. If we plot the relationship between the dry weight and density of shoots (known number of individuals of) in plant population, the line relating weight of each individual to density has a slope of –1.5 (or – 3/2). The slope would be –1, if increasing density has been exactly compensated by reduction in weight of individuals. Thinning is normally inversely density dependent, but does not always occur if the growth of the plants is extremely plastic. The 3/2 is universal and applied well in a wide variety of plants from mosses to trees. Exact reason of its occurrence is still not clear.

GROWTH RATE OF POPULATION

The rate of growth of a population is expressed as the number of individuals by which the population increases divided by the amount of time that elapses while this population increase is taking place :

$$\text{Growth rate } (r) = \frac{\text{number of births } (b) - \text{number of deaths } (d)}{\text{average population in time interval}}$$

The actual change in population number (ΔN) over any span of time (Δt) is equal to rN (Δ is the entity that is changing). This can be written $\Delta N/\Delta t = rN$ or, using the symbology of the calculus, the rate of change of the population at any instant time (dN/dt) can be expressed $dN/dt = rN$. This is equivalent to saying that the number of individuals at any arbitrary time t, or Nt, is related to the number of individuals at the beginning, N_0, by the equation $Nt = N_0e^{rt}$, where e= 2.71828........,the base of the natural logarithms.

If r is constant, the growth of population will be exponential. If r is positive ($b>d$), the population shows an exponential increase to indefinite density and if r is negative ($b<d$), it shows an exponential decay to extinction. It is impossible for a population to change at an exponential rate indefinitely. However, there are many cases in which conditions are such that b is substantially large than d for a period of time, following which conditions change so that d becomes much larger than b. The responses of populations to variations of this sort is an exponential "**population explosion**" during favourable conditions, followed by a "**crash**" when conditions change. Diatom populations in Lake Michigan, USA, for example, undergo such exponential increases at different times of years, triggered by variations in abiotic factors within the lake, followed by equally rapid declines.

POPULATION DISPERSION

Populations have a tendency to disperse, or spread out in all directions, until some barrier is reached. Accordingly, population dispersion is the movement of individuals into or out of the population area. It takes three forms : **emigration**—one-way outward movement; **immigration**—one-way inward movement; and **migration**—periodic departure and return. Dispersion supplements natality and mortality in shaping population, growth form and density, and also it plays a significant role in the distribution of plants and animals even to the areas previously unoccupied by the members of the population. Most types of population dispersion occur due to a number of reasons such as for obtaining food, avoiding predators, preventing overcrowding, result of action of wind and water, environmental factors as light and temperature, breeding behaviour, physiological reasons as secretion of some hormone or for interchange of genetic material between populations.

Emigration

Emigration under natural conditions occurs when there is overcrowding in the migratory locust, lemming, grouse, snowy owl, snowshoe rabbit, Arctic fox, gray squirrel and occasionally in other species. This is generally regarded as an adaptive behaviour that regulates the population on a particular site and prevents over-exploitation of the habitat. Further, it leads to occupation of new areas elsewhere. By dispersing into new localities, there is opportunity gained for interbreeding with other populations leading to more genetic heterozygosity and adaptability.

It is, of course, population pressure that is responsible in large part for the dispersal of the young and extension of ranges into new areas. Under normal conditions adult animals, especially among the higher vertebrates, are well established on their territories and the youngs are forced to seek homes elsewhere. Among insects, there is a relation between emigration and inherited behaviour tendencies. Individual tent caterpillars, both larvae and adults, differ innately in the extent to which they show activity even within same colony. In the development of populations of excessive size, spread of infestations of the insect into new regions is largely by the more active individuals. The outbreak finally terminates when the proportion of sluggish individuals comes to predominate in the population (**Wellington**, 1966). Continuous emigrations are rare and when they occur, result in **depopulation**. Equilibrium of populations is maintained in such circumstances by enhancing the reproductive ability as well as by decreased mortality among the populations.

Immigration

Immigration leads to a rise in population level, causing an **overpopulation** which may lead to an increase beyond the carrying capacity. These immigrations result in increased mortality among the immigrants or decreased reproductive capacity of the individuals. Both emigration and immigration are initiated by weather and other abiotic and biotic environmental factors.

Migration

Migration is a peculiar kind of population dispersion which involves the mass movement of entire population. This can occur only in mobile organisms and best developed in insects such as desert locust, *Schistocerca gregaria,* migratory locust, *Locusta migratoria*, butterfly, *Danaus plexippus* and in the migratory dragonflies, *Libellula quadrimaculata* and *Pantala flavescens*; in fishes like eels, in birds and in certain mammals. Most two-way migratory movements are rhythmic processes of population and regular periodicity is a common feature. Very often, environmental periodicities control these migratory movements, as for example day and night rhythm, lunar periods, tides and changing seasons. The monarch butterflies, *Danaus plexippus*, travel very long distances and their migrations are found to be pathed every year through their same conventional routes, and their migratory movements are initiated by the oncoming winter and the return trip being influenced by spring.

Migration of population may occur for food, shelter, or reproduction.

In most cases, migration of population may occur for food, shelter, or reproduction. Better utilization of uninhabited or hitherto untouched habitats and their resources are the greatest benefits derived from the migratory movements. However, during migration of population, mortality of numerous individuals may occur due to different ecological hazards such as temperature fluctuations, scarcity of food, predation, etc. Anyhow, migration has certain benefits for populations—as it enables wider dispersion of populations; it avoids intraspecific competition for food, shelter or any other means.

REGULATION OF POPULATION SIZE

The inherent tendency of all animal populations is to increase in number. But this increase in number is, however, not infinite since the carrying capacity of the environment always imposes a restriction upon it. Thus, after reaching the carrying capacity level, the population density tends to fluctuate above and below this level and such fluctuation in population between upper and lower limits tend to give some stability to the population. The nature of the processes that regulate the numbers of plants and animals is a major problem of population ecology.

Many groups of animals and plants are provided with unique and intrinsic self-regulatory mechanism such as failure of reproduction and self-inflicted mortality for controlling the size of the population. Populations grow when natality exceeds mortality, and they decline when mortality exceeds natality. However, limitation of animal number in a population is brought about by the action and interaction of two basic regulatory processes namely **density independent** and **density dependent factors**. Density independent factors are the extrinsic factors which tend to regulate the density of a population under different conditions, appearing to act on the population and inflict loss of individuals irrespective of the population density. Variations in space or cover, favourable weather and food occur independently of population densities and may cause drastic changes in the abundance of animals. Such ecological or environmental factors influence negatively or positively all the individuals of a population irrespective of density.

The density dependent factors are intrinsic or biotic factors and they depend on coaction between individuals within the same population or between populations of different species. Density-dependent factors may stabilize populations at an asymptote, the level of which is determined by the carrying capacity of the environment. Some of the important density dependent factors are competition, reproductivity, predation, emigration and disease. The combination of factors or any specific factor involved in the density dependent action may vary from species to species. Further these species have been known to display the property of intercompensation. According to intercompensation, if there is a change in the environment to relieve the population from the pressure of an existing effect, then the population increases till it reaches a level when the second effect takes over. Thus, if the predators that normally keep herbivorous animals down are removed, the population of herbivores may increase to become overcrowded and result is starvation. A supply of abundant food at this junction would make the individuals susceptible to diseases due to the intensity of crowding (**Wilson** and **Bossert**, 1971). Generally, the population fluctuations controlled by extrinsic factors tend to be irregular and correlated with the variation in one or more major physical limiting factors such as temperature, food, water, etc., while fluctuations of populations controlled by intrinsic factors exhibit regularity and population cycles.

Population Cycles

Populations are said to be cyclic when they alternatively errupt and subside in a more or less uniform manner between high and low levels of density. Different animals exhibit population cycles different times. The best established cycles of population density of fluctuations are those of periodicities of 3–4 years and 9–10 years. The 3–4 year cycles are most commonly observed in many birds such as snowy owl, willow ptarmigan, capercaillie, Blackgame, Hazel grouse etc.; mammals such as lemmings, voles, arctic foxes, etc., and fishes. 9–10 years cycles of population density fluctuations have been observed in birds such as ruffed-grouse, sharp-tailed grouse, willow ptarmigan, etc., and mammals such as snowshoe rabbit, muskrats, Canada lynx, etc. Among invertebrates, insect pests of coniferous forests in Germany fluctuate in periods variously from 6 through 18 years. *Asterias forkesi* is found to contain periodicity of 14 years.

Attempts to explain these vast fluctuations in numbers on the basis of climatic changes have been unsuccessful. At one time it was believed that these were caused by sunspots, and the sunspots and lynx cycles do appear to correspond during the early part of the nineteenth century. However, the cycles are of slightly different lengths and by 1920 were completely out of phase, sunspot maxima corresponding to lynx minima. Attempts to correlate these cycles with other periodic weather changes and with cycles of disease organisms have been unsuccessful.

The snowshoe hares die off cyclically even in the absence of predators and in the absence of known disease organisms. The animals apparently die of "shock" characterized by low blood sugar (hypoglycemia), exhaustion, convulsions and death symptoms which resemble the "alarm responses" induced in laboratory animals subjected to physiological stress. Visualizing this similarity, **J.J. Christian** (1950) proposed that their death, like the alarm response is the result of some upset in the adrenal-pituitary system. As population density increases; there is increasing physiological stress on individual hares owing to crowding and competition for food. Some individuals are forced into poorer habitats where food is less abundant and predators more abundant. The physiological stresses stimulate the adrenal medulla to secrete **epinephrine** or **adrenaline** which stimulates pituitary via the hypothalamus to secrete more **ACTH** (adrenocorticotropic hormone). This in turn, stimulates the adrenal cortex to produce corticosteroids, an excess or imbalance of which produces the alarm response or physiological shock.

In the latter part of the winter of a year of peak abundance, with the stress of cold weather, lack of food and the onset of the new reproductive season putting additional demands on the pituitary to secrete gonadotropins, the adrenal-pituitary system fails, becomes unable to maintain its normal control on carbohydrate metabolism, and low blood sugar (hypoglycemia), convulsion and death follows.

According to **Kendeigh** (1974) several instrinsic factors such as disease, predation, food factor and natural selection affect animal cycles in addition to several extrinsic factors such as weather, solar radiation, and so on.

POPULATION ECOLOGY AND EVOLUTION

Population parameters have been used by population ecologists to draw conclusions about evolution of species :

r-selection and k-selection. In this regard an important synthesis was done by **MacArthur** and **Wilson** (1967) who declared that populations are outcome or *r*- or *k*- selection. The **r-selected** populations have a high intrinsic rate of growth (*r*) and tend to "boom" when environmental conditions are favourable and "best" when these conditions deteriorate. In consequence, they show great fluctuations in density, and incidentally have the potential for large genetic change through the founder effect. The **k-selected populations** have relatively constant density at or near the carrying capacity (*K*) of the environment.

REVISION QUESTIONS

1. What is population ? Describe its different characteristics with suitable examples.
2. Describe the *r* and *k*- selection.
3. Describe in brief various techniques of measurements of population density of animals.
4. Discuss that in what way a population is a self-regulating system.
5. Write short notes on the following :
   (i) Fecundity, (ii) Natality, (iii) Mortality of population, (iv) Survivorship curve, (v) Age pyramids, (vi) Population dispersion.

CHAPTER

7

Biotic Communities

(Community Ecology : Communities, Niche, and Bioindicators)

Living creatures of an ecosystem form a biotic community.

All the living entities of an ecosystem form a single biotic component, the **community** or **biotic community**. All the organisms of a community live together, share same habitat, influence each other's life directly or indirectly and have reached a survival level within a given radiant energy. Thus, a community is any assemblage of populations of living organisms in a prescribed area of habitat. A community is claimed to have one or more of the following attributes : (1) co-occurrence of species, (2) recurrence of groups of the same species, and (3) homeostasis or self-regulation (**Krebs**, 1985).

Community is a larger unit than the population and it achieves many characteristics that are not found in its constituents, *i.e.,* the organisms and the populations. Communities may have a wide range of sizes, ranging from a small patch of land or water-body to extensive forests. **Minor communities** are greatly influenced by inputs from adjacent communities, while **major communities** are relatively independent and self-sufficient of their habitat. Communities differ from place to place and at the same place at different times.

The approach of botanists and zoologists to community studies is quite different. While the zoologist is mainly concerned with the functional relationships within a community, involving both the plants and animals, the botanists are generally concerned with community structure and the changes that undergo in time and space.

Characteristics of a Community

Like a population, a community has a series of characteristics such as follows :

1. Species diversity. Various species of plants and animals live in a community and exhibit species richness or species diversity. The study of species diversity is an essential component of community study. For animal communities, a study of age structure and growth pattern is important, while for plant communities, floristics, study of taxonomy, life forms such as herbs, shrubs, climbers, trees are important. Since seasonal changes occur in the appearance of plant structure and growth, periodicities and phenology (= seasonal succession in natural communities) are significant parameters.

2. Growth form and structure. The type of the community is described by major categories of growth forms (*e.g.,* trees, shrubs, herbs, mosses, etc.) These different growth forms determine the stratification, or vertical layering of the community.

3. Dominance. Among several species present in a community, a few exert a major controlling influence by virtue of their size, numbers, or activities. These are called as **ecological dominants** or **dominant species**.

4. Relative abundance. Different populations in a community exist in relative proportions and this idea is called as **relative abundance**.

5. Trophic structure. Who eats whom ? The feeding relations of the species in the community will determine the flow of energy and materials from plants to herbivorous animals to carnivorous animals.

CLASSIFICATION OF THE COMMUNITIES

Communities have been classified by different ecologists from different view points. In terms of the general growth, composition, shape, etc., of vegetation, and organisms associated with them, communities may be classified as **forests**, **deserts**, **grasslands**, **tundra** and so on. Likewise, according to the amount of water in the habitat, communities may be divided as **hydrophytic** in predominantly aquatic habitats, **mesophytic** in moderately moist soils and **xerophytic** in arid or dry conditions. Communities growing in condition of abundant light are called **heliophytic** and those growing in shade are called **sciophytic**. **Clements** (1916) recognized the fact that plant communities are not always the same at any place and he classified the communities on two parallel lines : One in the process of change which are called **seral communities** and the others are called **stable** or **climax communities**.

Further, the **global community** is an enormous mass of life, comprising all the plants and animals in the world. The global community is further divided into : **continental communities** and **oceanic communities**. Since due to great variability in climatic factors, an exhaustive study in such vast areas is practically impossible, therefore, communities are often studied as **biotic province**. A biotic province can be defined as a considerable geographic area, over which the climate is relatively uniform though often modified by physiographic features (**Dice**, 1952). Since biotic province is an abstract community, so based on ecological criteria, the associations are studied as **concrete communities**. A concrete community can be defined as a specific area which can be observed directly and which is an assemblage of plants and animals that actually exists and from which some ecological data can be collected.

Lastly, a **stand** is the largest concrete community, for example, a particular forest, river, swamp, meadow or lake that can be seen, observed, measured and worked over by the ecologist. A **microstand** is a small localized area within stand. Thus, each individual plant or animal with its associated parasites, epiphytes and commensals are good examples of the microstands.

COMPOSITION OF COMMUNITY

Composition of community comprises the following parameters :

1. Size. Communities may be large or small. Larger one extends over areas of several thousands of square kilometers, as forests, others such as deserts, etc., are comparatively smaller with dimensions in hundreds of kilometers, and still others such as meadows, rivers, ponds, rocky plateaus, etc., covering a more restricted area. Very small-sized communities are the groups of microorganisms in such microhabitats as leaf surface, fallen log, litter, soil, etc.

2. Number of species. The number of species and population abundance in communities vary greatly. **Charles Elton** (1927) calculated that the communities of certain British rivers included 131 species of invertebrates, in addition to fish, amphibians, populations of algae, protozoa, bacteria, and rooted aquatic plants. In contrast to this, he found a community on a sandy beach consisting of only five species of invertebrates. A meadow (=a rich pasture ground especially beside a stream) on clay near Oxford, England included 93 species of invertebrates in the soil and on above ground vegetation. In India too, various ecologists recently have estimated the number of species in the composition of various Indian communities. **M.C. Dash** and his coworkers (1980, 1981 and 1984) studied the composition of a biotic community in some tropical pastures of Sambalpur, India and found that it consisted of 22 grass and other herbaceous species (the primary producers), 5 species of earthworms, 25 of test-bearing protozoa, 15 of nematodes and 7 of microarthropods. The team of **Dash** (1980, 1984) reported 9 species of grasses and other herbaceous species, 1 of earthworms, 11 of testate protozoa and 16 of nematodes in a tropical hill ecosystem of Sambalpur. At one site of river Ib, Orissa, the community consisted of 1 species of rooted plant, 32 of phytoplanktons, 20 of zooplanktons and 7 of fish (**Kar** *et al.*, 1987; **Dash** *et. al.*, 1988).

3. Dominants. In each community there occurs diverse species. All these species are not equally important but there are only a few over topping species which by their bulk and growth modify the habitat and control the growth of other species of the community, thus, forming a kind of characteristic nucleus in the community. These species are called the **dominants**. Generally in most of the communities, only a single species is particularly conspicuous and dominant and in such case the community is called by the name of dominant species, *e.g.*, spruce forest community. In other cases, there may be more than one dominants in a community, *e.g.*, oak-hickory forest community.

According to **Clements** and **Shelford** (1939), dominance is most commonly expressed in the reactions of an organism on its habitat. Community dominants sustain the full impact of the climate or the environment but modify this effect for other organisms within the community by tempering light, moisture, space, and other conditions. Only those other organisms that find these modified physical conditions tolerable can exist within the community. Furthermore, dominants are ordinarily the most prominent species in the community, make up its greatest mass of living material (**biomass**), and serve as the major source of food, substrate, and shelter for the animals that are present. In a forest community, trees are dominant. They decrease light intensity, increase the relative humidity, intercept precipitation, monopolize most of the moisture and nutrients in the soil, decrease wind velocity and furnish shelter and food for animals. Grasses play a similar, though less conspicuous, role in prairie communities; sedges, rushed and cattails in marsh communities : sagebrush in the arid habitat of the Great Basin; mussels and barnacles on a rocky seashore; and so forth.

5. Ecological amplitude. The range of environmental conditions which a taxon can tolerate is called **ecological amplitude**. The composition of a biotic community in any habitat is dependent upon the frequency of environmental conditions in that habitat and the ecological amplitude of species populations. Thus, the climate and other abiotic and biotic conditions of a habitat determine the type of community which survives and develops.

STRUCTURE OR STRATIFICATION OF COMMUNITY

The communities exhibit a **structure** or recognizable pattern in the spatial arrangement of members of the communities. **Stratification** forms one of the outstanding general principles of community structure. Organisms are distributed unevenly throughout the biotic community, resulting in stratification. The stratification in a community may be resulted by the following factors : 1 Specific tolerations and adaptations. 2. Chemical reactions taking place between the by-products and physical stratification. 3. Other organisms take up temporary or permanent residence as a direct response to the presence of initial residents. The stratification in community may be horizontal or vertical and is different in aquatic and terrestrial habitats.

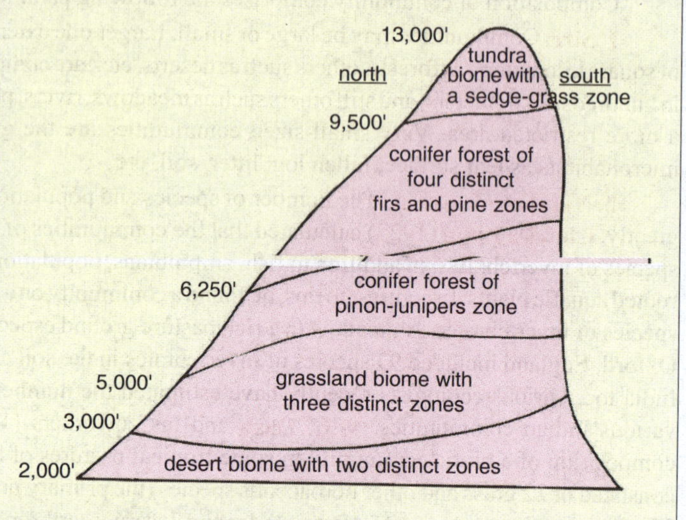

Fig. 7.1. Horizontal vegetational zones (*i.e.,* zonation) on one of the mountains of western North America.

1. Horizontal Stratification

A community may be divided horizontally into **subcommunities** which are units of homogeneous life forms and ecological relations. This horizontal division constitutes the **zonation** in the community. Figure 7.1 shows the zonation of different distinct vegetational types on a mountain. Latitudinal as well as altitudinal zonations of vegetation have also been shown in relation with climatic factors described in Chapter 4.

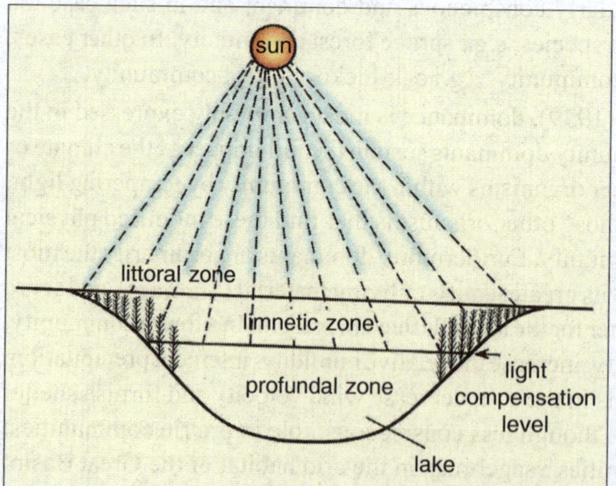

Fig. 7.2 Three major zones of a freshwater lake.

In shallow ponds, the zonation is very little. However, in deep ponds and lakes, there may be recognized three zones, *viz.,* **littoral zone**, **limnetic zone** and **profundal zone** (Fig. 7.2). In each zone, organisms differ from each other. Further, in case of animal communities of rocky sea shore, there are following four distinct zones : 1. **Balanus zone**. This zone includes the animals such as *Balanus* and *Patella* which live on exposed rocks. 2. **Littoral zone** or **intertidal zone** (*i.e.,* zone between high and low tides). This habitat develops in the sheltered niche beneath loose stones and contain a variety of crustaceans and collembola (springtail). Lower on the shore, where the tides always flood are living beneath the stone some annelids and nemertine worms (*e.g., Lineus*). Adjacent to low water mark are certain species of *Neries*, shore crabs, etc. 3. Holes and cracks in the rocks are occupied by various annelid worms, crustaceans, small sea-cucumbers and rock-boring bivalves. 4. There are rock-pools that enclose those animals which escape direct current of water.

Sandy-shore dwelling marine animals are in great part burrower such as spinoid (*Prionospio, Aricia*), annelids (*Glycera, Nerina*), archiannelids (*Protodrilus*) and crustacean (*Emerita, Albunea*). Populations which inhabit this substrata adjust to both sand and mud particles. This becomes evident in their respiratory adaptations and locomotory and feeding behaviour.

2. Vertical Stratification

Vertical gradients in environmental factors such as the availability of sunlight, the temperature and so on bring about a recognizable stratification in water bodies, particularly in marine ecosystem (Fig. 7.3) such as deep water lakes. The **oceanic region** (*i.e.,* the region of open sea beyond continental shelf) comprises two zones : 1. **bathyal zone** (*i.e.*, region of continental slope and rise) and 2. **abyssal region** (*i.e.,* area of ocean "deeps"). In terms of light penetration, the ocean region is vertically also divided into an upper thin **euphotic zone** (= the light compensation zone) and a vastly thicker permanently darker zone, the **aphotic zone** (including bathyal and abyssal zones). The oceanic region also includes two more vertical zones : the **benthic** (bottom) and the **pelagic** (whole body of water). The community of pelagic zone includes (1) **phytoplanktons** (the producers such as diatoms, dinoflagellates, flagellates, green algae, brown algae such as *Sargassum*); (2) **zooplanktons** (the consumers such as Foraminifera, Radiolaria, annelid worms, swimming snails, jelly fishes, ctenophores, etc., (3) **nektons** (these are free swimming marine animals which are essentially independent of water movements). Nektons are consumers such as cephalopods (*Sepia, Loligo, Nautilus*) bony fishes, sharks, sea turtles, whales, seals and sea birds. Minute organisms that float or swim on surface water are called **neustons**. **Benthos** are bottom dwellers of benthic environment. They are sessile forms such as limpets, chitons, mussels, oysters, sponges, corals, hydroids, anemones, bryozoans, crabs, lobsters, some echinoderms and flat fishes.

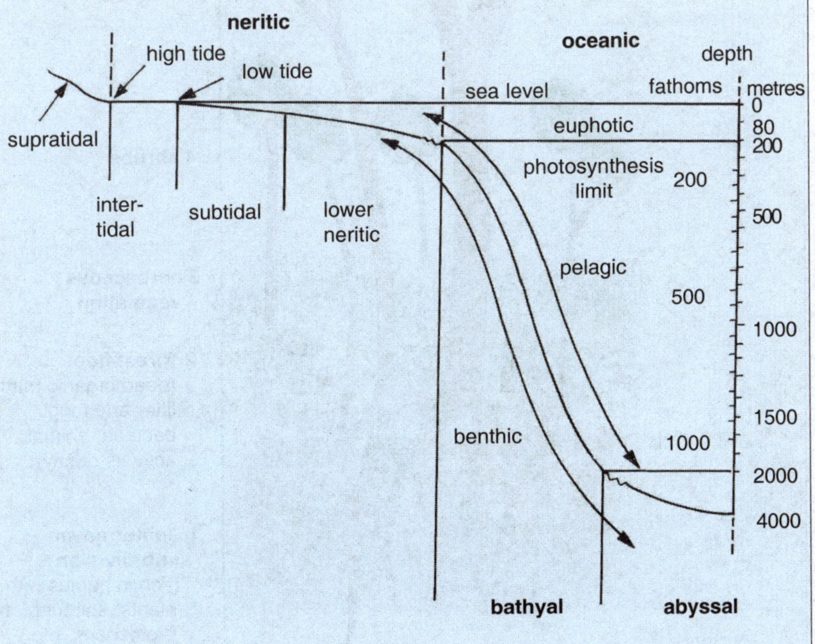

Fig. 7.3. Zonation in the sea.

Vertical stratification in communities is also observed in terrestrial ecosystems. For example, in a **grassland community**, the euphotic zone consists of the **herbaceous stratum** consisting of grasses and other herbs as producers and rodents, cattle and insects as consumers. The **subterranean stratum** consists of different layers of soil containing litter, humus, mineral fraction of soil, plant roots, rhizomes and organisms such as bacteria, fungi, protozoa, microarthropods, insect larvae, pupae, oligochaetes and molluscs. In **forest community**, however, the vertical stratification is most complicated and includes following five subdivisions: 1. **subterranean subdivision**; 2. **forest floor**; 3. **herbaceous vegetation**, 4. **shrubs** and 5. **trees** (Fig. 7.4). In some **tropical rain-forests** three more subdivisions exist in the forest floor such as leaf litter layer (**L layer**), partially decomposed leaf litter layer (**F layer**) and a humus layer. Thus, based upon the light and relative humidity requirements, we find stratification in above ground parts.

Ecological potential of a zone or strata determines the distribution of community. **Fenton** (1947) has classified these communities as (1) **Micro-biota** which include bacteria, algae, fungi, protozoans, rotifers, nematodes and mites. (2) **Meso-biota** which include many nematodes, centipedes, millipedes, spiders, insects, etc. (3) **Macro-biota** which include plant roots, earthworms, toads, lizards, snakes, rodents and large mammals including man. He also classified these communities according to their feeding habits such as **chemophagous**, **ectophagous**, **entophagous**, **predators** and **shelterers**.

CHARACTERS USED IN COMMUNITY STRUCTURE

Structure of plant communities can be determined by quantitative and qualitative features as follows :

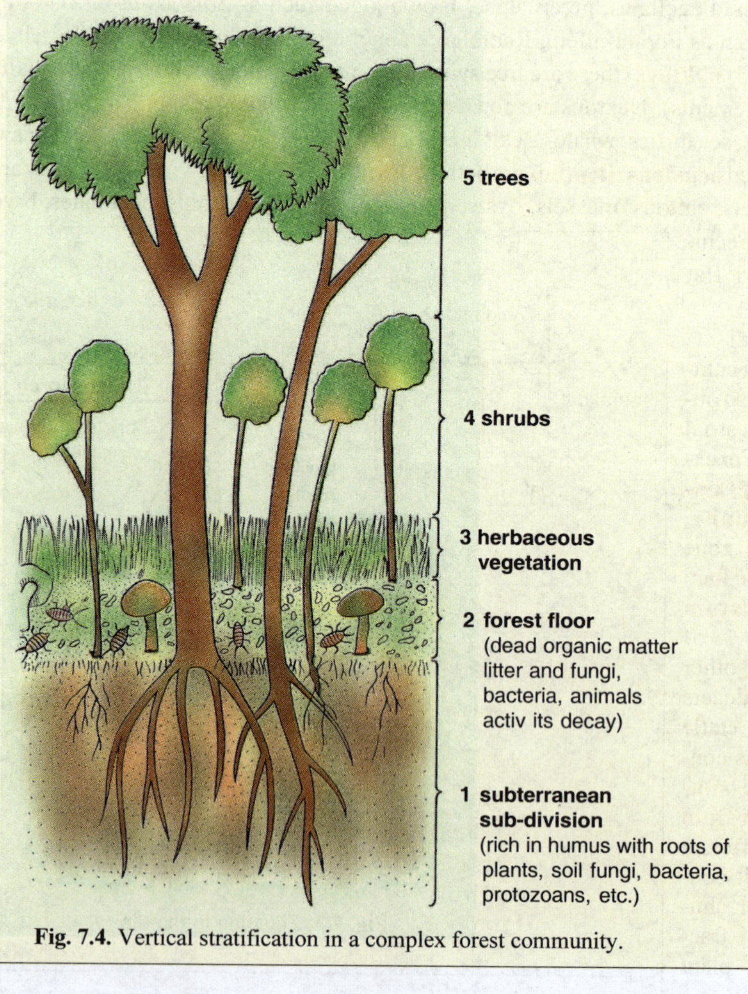

Fig. 7.4. Vertical stratification in a complex forest community.

A. Quantitative Structure of Plant Communities

Quantitative characters of the community include such characters as frequency, density, abundance, cover and basal area, etc.

1. Frequency. Various species of the community are recorded by different phytosociological methods, by taking any sampling unit such as **transect**, **bisect**, **quadrat**, **point centre** or **point quarters**, etc. Frequency is the number of sampling units (as %) in which a particular species occurs. Thus, the frequency of each species is calculated in the following way :

$$\text{Frequency (\%)} = \frac{\text{Number of sampling units in which the species occurred}}{\text{Total number of sampling units studied}} \times 100$$

After determining the percentage frequency of each species, various species are distributed among **Raunkaier's** (1934) five frequency classes depending upon their frequency values (Table 7-1)

Table 7-1. **Raunkaier's frequency classification of plant communities.**

Frequency class	Species with frequency range (%)
A	0–20
B	21–40
C	41–60
D	61–80
E	81–100

The **relative frequency** (**RFR**) is determined by the use of the following formula (data collected using the quadrant method).

$$\text{Relative frequency (RFR) of the species} = \frac{\text{Number of occurrence of a species}}{\text{Number of occurrence of all species}} \times 100$$

Since environmental conditions vary, the frequency values differ in different communities. **Raunkaier** observed the following relationships between different frequency classes : A = 53, B = 14, C = 9, D = 8 and E = 16. He generalized his findings in the **law of frequency**, which showed that

$$A > B > C \gtreqless D < E$$

Such a relationship occurs only in undisturbed communities or communities where disturbance is minimum.

2. Density. Density represents the numerical strength of a species in the community. The number of individuals of the species in any unit area is its density. It gives an idea of degree of competition. Samples are taken by the help of different types of samples and sampling techniques (**Southwood**, 1966) and estimation are made of the numerical strength of a species population and of the total community. The data can be multiplied with an appropriate factor to express the total number of organisms per unit of area. The density of a species is calculated as follows :

$$\text{Density} = \frac{\text{Total number of individuals of the species in all the sampling units}}{\text{Total number of sampling units studied}}$$

The **relative density (RDE)** of a species is calculated in the following way :

$$\text{RDE} = \frac{\text{Number of individuals of the species in all quadrates}}{\text{Number of individuals of all species in all quadrates}} \times 100$$

3. Abundance. This is the number of individuals of any species per sampling unit of occurrence. Abundance is calculated as follows :

$$\text{Abundance} = \frac{\text{Total number of individuals of the species in all the sampling units}}{\text{Number of sampling units in which the species occurred.}}$$

4. Cover and basal area. The above ground parts of plants (such as leaves, stems and inflorescence) cover a certain area— if this area is demarcated by vertical projections, the area of the ground covered by the plant canopy is called **canopy cover**, **foliage cover** or **herbage cover**. Canopy cover is a good measure of the herbage availability and is estimated by chart quadrat, line intercept or point frame.

Basal area refers to the ground actually penetrated by all stems and is readily seen when the leaves and stems are clipped at the ground surface. It is either measured 2.5 cm above ground or actually on the ground level. Basal area of a plant is measured by callipers, line-interception or point-centered quadrat method. It is one of the chief characteristics to determine dominance.

Since plants differ in their growth form, so the relationship between herbage cover and basal area differs in different types of plants as shown in Figure 7.5.

For the measurement of crown cover, the diameter is taken at ground level, across the canopy perimeter passing by the main stem. The cover is expressed as a percentage. Based on their cover values, species are grouped into the following six groups : Group 1 (less than 5%), Group 2 (5–25%), Group 3 (25–50%), Group 4 (50–75%), Group 5 (75–95%) and Group 6 (95–100%).

B. Qualitative Characteristics of Plant Communities

These involves parameters such as physiognomy, phenology, stratification, abundance, sociability or gregariousness, vitality and vigour, life form (growth form), etc. All these parameters can be described and may be grouped in point scales.

1. Physiognomy. Physiognomy is the general appearance of vegetation as determined by the growth form of dominant species. Such a characteristic appearance is often expressed by single term. For example, observation of a plant community having trees and some shrubs as the dominants, clearly forced us to conclude that it is a forest. Similarly on the basis of appearance of a community, it may be a grassland, desert, etc.

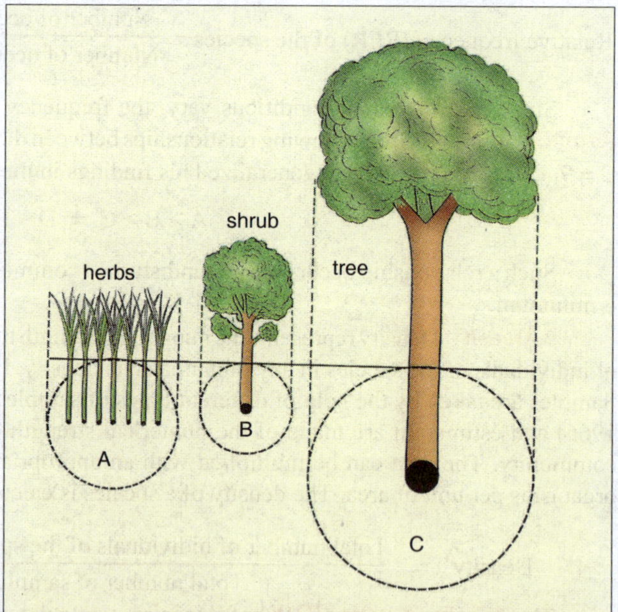

Fig. 7.5. Diagram showing relationship between herbage cover and basal area in different types of growth forms (herbs, shrubs and trees) of plants.

2. Phenology. The life history of a plant species involves seed germination, vegetative growth, flowering, fruit formation, seed maturation, leaf fall, seed dispersal and death. These events are different for different species and are recorded for the individual species. A study of the date and time of occurrence of these events is called **phenology**. In other words, phenology is the calender of events in the life history of the plant. Environmental factors tend to influence the phenological behaviour of a species population. These phenological events are customarily recorded diagrammatically (Fig. 7.6), monthwise and seasonwise and provide valuable information. These diagrams are called **phenograms**. It will be significant to apply phenology to those ecosystems where seasonal changes are not sharply defined, *e.g.,* deserts. Phenology also includes the periodicity and aspection of the community.

Periodicity of communities. As already described, all the major phenomena on earth are cyclic. The communities being dynamic entities respond constantly to seasonal changes and exhibit a variety of periodicities such as seasonal rhythms, lunar and tidal rhythms and many periodicities associated with the cycle of day and night (e.g., circadian rhythms). Ecologists tend to recognize six phases (seasons) in the annual cycle—prevernal, vernal, extival, aestivation, autumnal and hiemal periods. Such cyclic fluctuations can be observed in both abiotic and biotic factors of a particular environment such as follows : 1. Abiotic variability , i.e., cycle changes in temperature, moisture, wind speed and a number of other factors. 2. Biotic variability, i.e., flora and fauna influenced by the immediate abiotic environment.

3. Stratification. As described elsewhere, stratification of communities is the way in which plants of different species are arranged in different vertical layers in order to make full use of the available physical and physiological requirements.

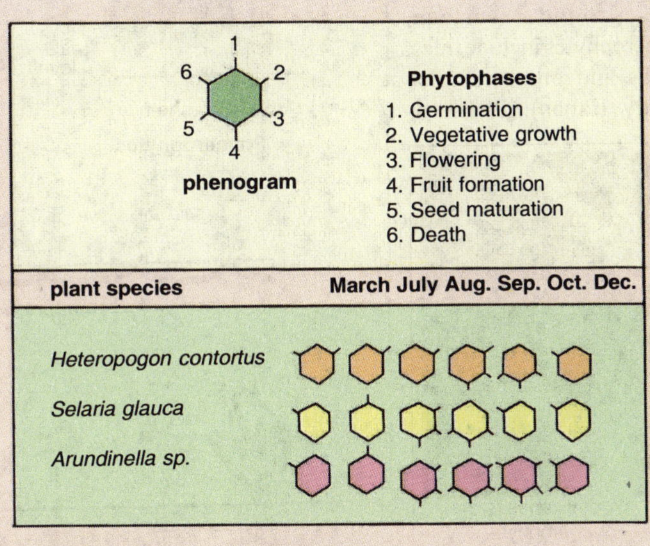

Fig. 7.6. A phenogram showing phenological events of three Indian herbaceous species.

4. Abundance. Plants are not found uniformly distributed in an area. They are found in smaller patches or groups, differing in number at each place. Abundance is divided into following five arbitrary groups, depending upon the number of plants : 1. **very rare**, 2. **rare**, 3. **common**, 4. **frequent**, and 5. **very much frequent**.

5. Sociability. Sociability or gregariousness expresses the degree of association between species. It denotes the proximity of plants to one another. Plants generally grow as isolated individual, in patches, colonies or groups. **Braun -Blanquet** (1951) classified plants into the following five sociability groups :

(i) S_1 plants. Plants (stems/shoots) found quite separately from each other, growing singly.

(ii) S_2 plants. Plants growing in small groups (4 to 5 plants).

(iii) S_3 plants. Plants growing in small scattered patches.

(iv) S_4 plants. Several bigger groups of many plants at one place.

(v) S_5 plants. Pure populations of plants covering large areas.

Interdependence is another community feature which includes inter-and intraspecific competition, parasitism, symbiosis and commensalism, all of which have been discussed else-where in this book.

6. Vitality. It is the capacity of normal growth and reproduction, which are important for successful survival of a species. In plants, the vitality is determined by the following parameters : stem height, root length, leaf area, leaf number, number and weight of flowers, fruits, seeds, etc.

7. Life-form (Growth form). Ecologists generally use **Christen Raunkiaer's** classification (1934) of plant life forms. A **life form** is the sum of the adaptation of the plant to the community (it is determined by the stature of the plant species, their spread and characters of life-form). **Raunkiaer** classified plants into following five broad life-form categories (Fig. 7.7) ; his classification is based on the position of perennating buds on plants and the degree of their protection during adverse conditions.

(i) Phanerophytes. The growing buds of these plants are naked or covered with scale (*i.e.,* they are not well protected), and situated in upright shoots much above the ground surface. Phanerophytes include trees, shrubs and climbers (also woody lianas). Phanerophytes are found mostly in tropical regions and decrease progressively from the tropics to the temperate to the polar regions.

Woody lianas is a phanerophyte.

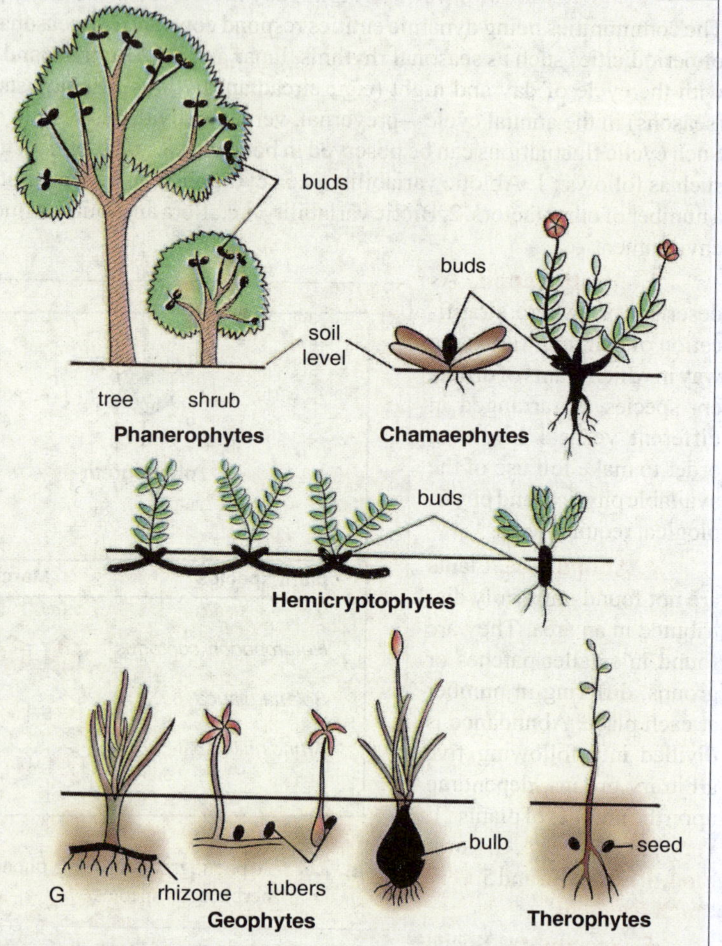

Fig. 7.7. Diagrammatic representation of Raunkiaer's life-forms. The barred regions survive during adverse periods of the growing season. Note the successive increasing degree of protection of renewal or perennial buds, organs or seeds from phanerophytes to therophytes.

(ii) Chamaephytes. In these plants the buds are situated close to the ground surface and these buds get protection from fallen leaves and snow cover. Chamaephytes commonly occur in high altitudes and latitudes, *e.g., Trifolium repens* which is found in North America.

(iii) Hemicryptophytes. These are mostly found in cold temperate zone. Their buds are hidden under soil surface, protected in soil itself. In the warm season the growth of aerial parts is marked. Their shoots generally die each year, *e.g.,* most of the biennial and perennial herbs including grasses.

(iv) Cryptophytes or geophytes. In these plants, the buds are usually buried in the soil or in bulbs and rhizomes where food is stored to withstand long periods of adverse climatic conditions (*i.e.,* freezing and drying). Cryptophytes include the hydrophytes (buds remaining under water), helophytes (marsh plants with rhizomes under the soil) and geophytes (terrestrial plants with underground rhizomes or tubers).

(v) Therophytes. These are seasonal plants, completing their life cycle in a single favourable season, and remain dormant throughout the rest unfavourable period of year in the form of seeds. Therophytes are commonly found in dry, hot or cold environments (*i.e.,* deserts).

Life-forms of animals. There have been several attempts to classify the life-forms of animals, but no definite system has resulted (**Remane**, 1952 ; **Krivolutskii**, 1972). The major life-forms more often agree with their taxonomy than do plants, but some life-forms include representatives from several different taxonomic groups. These can be recognized **encrusting forms**, such as the freshwater bryozoan *Plumatella* and some sponges ; **coral forms**, including grass, leaf, or shrub forms; **radiate forms**, such as coelenterates and echinoderms; **bivalve forms**; **snail forms**; **slug forms**; **worm forms**; **crustaecean forms**; **insect forms** ; **fish snake**, **bird** and **four-footed forms**. Each of these major types may be divided into narrower structural or behavioural types, for example, **Osburn** *et al.,* (1903) subdivided four-footed mammalian forms into following types :

1. **Aquatic (swimming).** Seal, whale, walrus.
2. **Fossorial (burrowing).** Mole, shrew, pocket gopher.
3. **Cursorial (running).** Deer, antelope, zebra.
4. **Saltatorial (leaping).** Rabbit, kangaroo, jumping mouse.
5. **Scansorial (climbing).** Squirrel, opossum, monkey.
6. **Aerial (flying).** Bat.

C. Synthetic Characters

Synthetic characters are determined after computing the data on analytical characters (*i.e.,* quantitative and qualitative characters) of the community. For comparing the vegetation of different areas, community comparison needs the calculation of their synthetic characters. Synthetic characters are determined in terms of the following parameters :

1. Presence and constance. The presence expresses the extent of occurrence of the individuals of a particular species in the community. The species on the basis of its percentage frequency may belong to any of the following five presence classes that were first proposed by **Braun Blanquet** :

I. **Rare**—present in 1 to 20% of the sampling units.

II. **Seldom present**—present in 21–40% of the sampling units.

III. **Often present**—present in 41–60% of the sampling units.

IV. **Mostly present**—present in 61–80% of the sampling units.

V. **Constantly present**—present in 81–100% of the sampling units.

A species with high frequency is considered constant. The term constancy is used when equal sample areas are taken in each stand, and presence is used when the area of the sample varies from stand to stand.

2. Fidelity. Fidelity or "faithfulness" is the degree with which a species is restricted in distribution to one kind of community. Such species are sometimes known as **indicators**. The species have been grouped into the following fidelity classes which were first formulated by **Braun Blanquet**:

(i) Fidelity 1 or Strangers. Rarely occurring species which either have arrived from other communities or are relicts from earlier stages of succession.

(ii) Fidelity 2 or Indifferents (or **ubiquitous**). Species which occur in any community without exhibiting any preference to any particular kind of community.

(iii) Fidelity 3 or Preferentials. Species which occur in several kinds of communities but are predominant in one.

(iv) Fidelity 4 or Selectives. Species occurring frequently in one kind of community and sometimes also in others. Sometimes preferential and selective species are considered together as **characteristic**.

(v) Fidelity 5 or Exclusive or True. Species which are completely restricted to one kind of community.

3. Dominance. It is used as a synthetic as well as analytical character (**Daubenmire**, 1959). The number of organisms sometimes may not give correct idea of the species. If one bases his conclusion on number, a single or few trees in a grassland, or few grasses in a forest should be of little value. But if he considers the species on the basis of area occupied or weight (biomass), the situation may be different. Thus, in dominance, cover is included as an important character. The single tree in grassland may occupy fairly a large area and may have much mass. Relative dominance (cover; RDO) is calculated in the following way :

$$\text{Relative dominance (cover, RDO)} = \frac{\text{Dominance (cover) of the species}}{\text{Total dominance (cover) of all the species}} \times 100$$

4. Importance value index (IVI) and polygraph construction. This index is used to determine the overall importance of each species in the community structure. In calculating this index, the percentage values of the relative frequency, relative density and relative dominance are summed up together and this value is designated as IVI or importance value index of the species. The IVI provides an idea of the sociological structure of a species in its totality in the community but does not indicate its position separately with regard to other aspects, such as frequency, density, and so on. In consequence, plant ecologists often use **polygraphs** to show the individual and combined aspects of the position of each species in the structure of the community.

The method of polygraph preparation involves drawing a circle and dividing it into four equal segments, by drawing two lines at right angles to each other passing through its centre. Three of the radii are divided into 100 equal parts from the centre to the circumference and the fourth is divided into 300 equal parts to mark the IVI value. On the 0–100 scale in the three radii, the values of relative frequency, relative density and relative dominance are marked. Figure 7.8 shows some of the polygraphs (or phytographs) which are generally used by plant ecologists (*e.g.*, **Dani** and **Behera**, 1990).

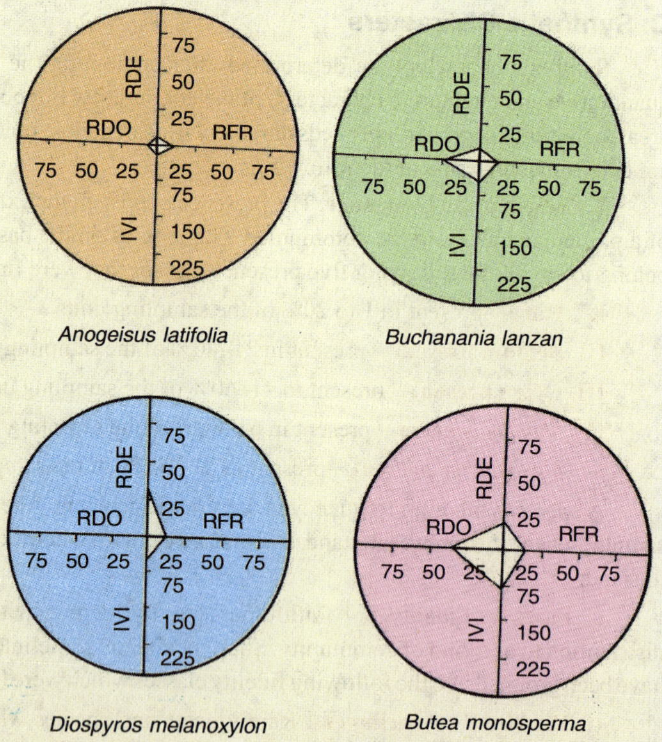

Fig. 7.8. Polygraphs prepared by **Dani** and **Behera** (1990) to show the sociological characters of individual species in a montane deciduous forest ecosystem (foot hill) of Sambalpur, India. IVI = importance value index, RDO = relative dominance, RDE = relative density, RFR = relative frequency.

Animal ecologists use an **importance value** (IV) as follows (**Sajise** *et al.*, 1976) :

$$IV\ (\%) = \frac{Wa}{Wt} \times 100$$

where Wa is the average dry weight (dry weight in grams per square meter per month) of the species and Wt is the total dry weight of all species of the community (of the area).

HABITAT AND NICHE

The term **niche** was for the first time used by **Joseph Grinnel** (1971) to explain microhabitats of California thrashers. According to him, "niche is the ultimate distributional unit, within which each species is held by its structural and instinctive limitations......no two species in the same general territory can occupy for long identically the same ecological niche." The **area** of a species or any other taxon refers to the total extent of its geographical range of dispersal; this can be plotted on a map. The **habitat** of a species described in a single word or phrase is the totality of abiotic factors to which the species is exposed in this area. Thus, one can talk of marine habitats, coastal habitats, marsh habitats, forest habitats, disturbed habitats or even dry habitats.

In many cases, however, the habitat of the species is highly specialized. Certain species of leaf miners, for example, live only in the upper photosynthetic layer of the leaves of certain species of plants, while other species live in the lower cell layer. These patterns of location are so consistent that it is clear that the habitats of the species are different. Subdivisions of the environment on this scale are commonly called **microhabitats**. Thus, the leaf constitutes a microhabitat for leaf miners within the total forest, and the different cell layers of leaf constitute different microhabitats within the leaf for different species of leaf miner. Conditions within the leaves are quite different from the general conditions in the forest. The specific environmental variables in the microhabitat of a population is called **microenvironment** or the **microclimate** (**Clapham**, **Jr**., 1973).

Though habitats and microhabitats indicate the places where organisms of a community live, but, if one wishes to understand how a species fit into the fabric of the ecosystem, it is not enough to merely describe the creature's habitat or microhabitat, he must also understand its functional role within the community—"address" of the organism but its "profession" as well. This includes what it eats, what it does, where it lives—everything about it that influences the community. Thus, the ecological niche is the property of the community and it represents the place of the species in the formal community structure.

In fact, the ecological niche is an inclusive term that involves not only the physical space occupied by an organism, but also its functional role in the community (*i.e.,* trophic position occupied) and its position in environmental gradients including other conditions of existence. These three aspects of ecological niche are generally designated as (i) the **spatial** or **habitat niche** (the physical space occupied); (ii) the **trophic niche** (functional role, *i.e.,* trophic position), and (iii) the multidimensional or **hypervolume niche** (position in the environmental gradients).

1. Spatial or Habitat Niche

This type of niche pertains to the physical space occupied by an organism. **O΄ Neill** (1967) has cited a good example of spatial niche in the distribution of seven species of millipedes in forest floor of a maple-oak forest. All the seven species occur in the same general habitat, *i.e.,* forest floor and all belong to the same basic trophic level (*i.e.,* all are decomposers in their role as detritivores). Each species of millipedes predominates in its own specific different microhabitat. In fact, there exist several gradients in decomposition stage from the centre of log to the position underneath the leaf litter, and these gradients constitute distinct microhabitats in the same general habitat, the forest floor. Thus, *Euryurus erythropygus* predominates in the heartwood at centre of logs, *Pseudopolydesmus serratus* in superficial wood of logs, *Narceus americanus* in the outer surface of the logs beneath the bark, *Scytonotus granulatus* under log but on the log surface, *Fontaria virginiensis* under the log but on ground surface, *Cleidogonia caesioannularis* within leaves of litter, and *Abacion lacterium* beneath litter on ground surface.

2. Trophic Niche

In this case two species live in the same habitat but they occupy different trophic niches, because of differences in food habits. For example, two aquatic bugs, *Notonecta* and *Corixa* live in the same pond but occupy different trophic niches. The former is an active predator that swims about grasping and eating other animals, but the latter feeds mainly on decaying vegetation.

Notonecta.

Finches of Galapagos Islands in America provide another good example of trophic niche. These birds belong to three genera, *viz., Geospiza* (ground finche), *Camarhynchus* (tree finche) and *Certhidia* (warbler finch). These all birds live in the same general habitat, but differ in terms of their trophic position. For example, in tree finches, one of the species, *Camarhynchus crassirostris* has a parrot-like beak and is basically vegetarian, feeding on buds and fruits. Rest of the tree finches are insect eaters, three of which feed on insects of different sizes. Another species, *C. heliobates* feeds on insects in mangrove swamps. The most interesting is wood-pecker finch, *C. pallidus* which climbs tree in search of insects in the cracks of bark. Although, it has wood-pecking beak, it lacks long extensive tongue of wood-pecker.

Corixa.

3. Multifactor or Hypervolume Niche

This concept, where niche is considered as an abstract n-dimensional inhabitated hypervolume, was developed by **G.E.Hutchinson** (1965) of Yale University. Suppose that we measure the range of some environmental factor (*e.g.,* temperature) over which a particular species can live and reproduce (in effect, its range of tolerance) and we put this on a graph. Suppose that we then do the same for another environmental factor (*e.g.,* humidity) and put this on the second axis of the graph. The space that is enclosed will represent the niche of the species. If the effects of two variables (or factors) (*i.e.,* temperature and humidity) are independent, the space would be two-dimensional box (Fig. 7.9 A). However, in fact, temperature and relative humidity are not independent in their biological effects. Thus, tolerance to higher temperature may be related with increase in relative humidity. Under such condition of interacting variables or factors the niche instead of being the box-like would be an oval in shape (Fig. 7.9 B). Now suppose that the tolerance to levels of a third factor (*i.e.,* available phosphorus) was affected by interactions with both temperature and relative humidity. We would now have a niche representing three variables or three-dimensional, volumetric figure (Fig. 7.9 C). If there is present a fourth variable, the space enclosed would be a hyper-volume with four dimensions. Since there are a large number (n) of other environmental factors, both abiotic and biotic, that affect the population, the niche is n-dimensional hypervolume, an abstraction since one is able to draw only with respect to three dimensions. This is the **fundamental niche** of the species. When the fundamental niches of two species overlap them, the two species are competing. Competitive exclusion would be called into play only in the complete overlapping of n-dimensional niches (Fig. 7.9 D). Thus, this concept refers to the totality of abiotic and biotic factors to which a given species is uniquely adapted.

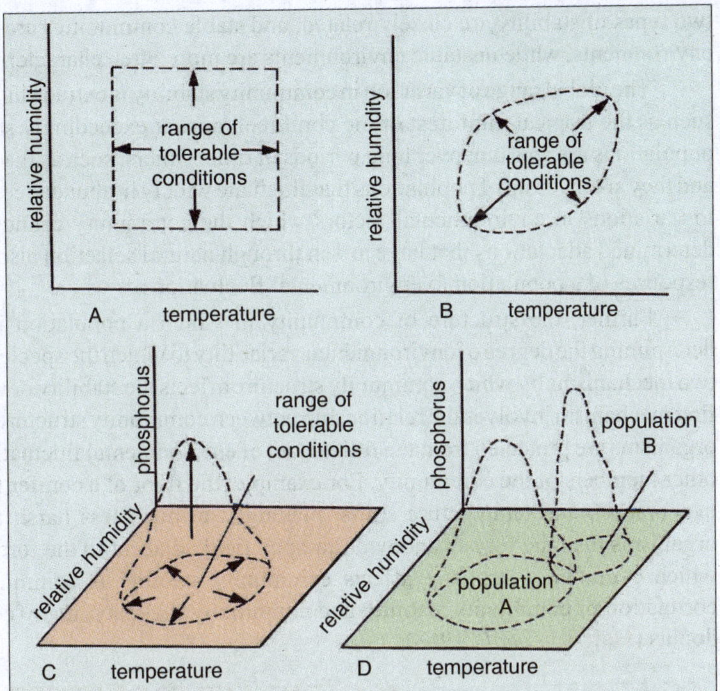

Fig. 7.9. Multifactor or hypervolume niche. A—A two-dimensional niche with independent variables or factors ; B—A two-dimensional niche with interdependent factors; C—A three-dimensional niche ; D—Two populations with overlapping niches.

Advantages of niche segregation. The major advantage which organisms gain by occupying different niches is escape from continuous intense competition. In fact, the niche occupied is favourable to the species physically in furnishing suitable substratum and microclimate, although many species have the ability to live elsewhere where competition is not involved. Automatic segregation of a species into its niche through inherited behaviour patterns or imprinting avoids the great expenditure of energy and loss of time that would be required if this segregation had to be worked out anew each year or each generation. Segregation into niches also avoids confusion of activities between organisms in the community and permits a more orderly and efficient life cycle on the part of each species. Furthermore, the segregation of each species into different niches permits the occupancy of the area by a large number of species, since they will better divide the available resources between them. Similarly, the more distinct the niche of a species is, the more it can avoid conflict with its neighbours and lead a life that is orderly, productive and quite efficient (**Kendeigh**, 1974).

COMMUNITY METABOLISM

Once a basic idea of community structure is gained it becomes necessary to ask the question how does a community function? The study of functional organisation or community metabolism is fundamentally a study of food relationship within the community. Such studies involve investigations on the flow of energy and cycling of nutrients within a community. Through the **food web** one can make the graphic depiction of the actual flow of food materials from population to population through the community.

COMMUNITY STABILITY

One of the most important considerations about communities is their stability, or lack of variation in time. There are two discrete senses in which this term is used. The first, **community stability**, refers to the degree of fluctuation in population size of the populations comprising the community. The second, **environmental stability**, refer to the fluctuations in the abiotic factors of the ecosystem. The two types of stability are closely related, and stable communities are generally to be found in stable environments, while unstable environments are more often characterized by unstable communities.

The global range of variation in community stability is extraordinarily broad. Some communities such as the tropical rainforest or the coral reef, appear exceedingly stable, and the densities of their populations are constant over long periods of time. Others, such as the tundra, are extremely variable, and they are made up of populations that fluctuate widely in abundance. Much of this fluctuation is due to variations in environmental factors which the community cannot avoid. However, genetically determined adaptations that have arisen through natural selection also play a part in determining the responses of a population to environmental fluctuations.

Further, the structure of community in which a population is found, has a major role in determining the degree of environmental variability to which the species is subjected. There are at least two mechanisms by which community structure affects the stability or variability of a community. The first mechanism involves the relationship between community structure and microenvironment. Many organisms are protected from the full impact of environmental fluctuations because of the presence of other members of the community. For example, the floor of a conifer forest is protected from winds, evaporation, and temperature shifts, making it a much less harsh and variable environment for organisms than the forest canopy or an open field adjacent to the forest. The second mechanism by which community structure affects community stability is community diversity. But about the correlation of community stability and community diversity, many ecologists have expressed their doubts (**Hairston** *et al.*, 1968).

ECOTONE AND EDGE EFFECT

In the ecocline (*i.e.*, gradient of ecosystems), the line of demarcation (boundary-line) between two communities is often very difficult in view of the chances of overlapping of one community over another. Such demarcation will be conspicuous only when the dominants of the adjacent communities show clear and characteristic differences. The transition zone between two or more diverse communities is called the **ecotone**. The common examples of ecotone are following—the border between forest and grassland, the bank of a stream running through a meadow or between a soft bottom and hard bottom marine community. An ecotone, thus, presents conditions that are intermediate to the communities which are on either side of it. The ecotones may be narrow or very wide, extending to large areas. The community on either side of the ecotone may have a typical structure, but the ecotone is strikingly different. The ecotone has a higher diversity than either of the main communities, a diversity that is not directly controlled by the climate or further fundamental environmental factors, but because of the migrations of individuals of different species-populations from both communities. Further, a number of special populations can become adapted specifically to the ecotone, even when both of the major communities are too simple or are otherwise unsuited for successful colonization by the species. This potential for the ecotone to act as habitat for species found in neither major community is termed as the **edge effect**. A common example of the edge effect in action can be seen in those species of owl that live in or near ecotones between forests and grassland. They depend on forest trees for nesting, and they do their hunting in the grassland, where they depend on field rodents for food. In man-made communities such as agricultural fields, the

Ecotone.

transition zone (ecotone) between the field and the forest may act as a refuge for animal species formerly found in the plowed area, as well as for other plants such as weeds. Ecotones of this type are also the prime habitat of many species of insect, game bird, and mammal.

FACTOR COMPENSATION AND ECOTYPES

Organisms are not just 'slaves' to the physical environments, they adapt themselves and modify the physical environment so as to reduce the limiting effects of temperature, light, water, and other physical conditions of existence. Such **factor compensation** is particularly effective at the community level of organization, but also occurs within the species. Species with wide geographical ranges almost always develop locally adapted populations called **ecotypes** that have different genetically based tolerance ranges and optima to correspond to the environments within which they normally live (**Odum**, 1971; **Clapham Jr.**, 1973). The term ecotype was coined by a Swedish botanist, **Gote Turesson** in 1912 and in 1922 he defined it as the product arising as a result of genotypical response of an ecospecies to a particular habitat. Ecotypes are especially common in plants, but are also shown by animals and microorganisms.

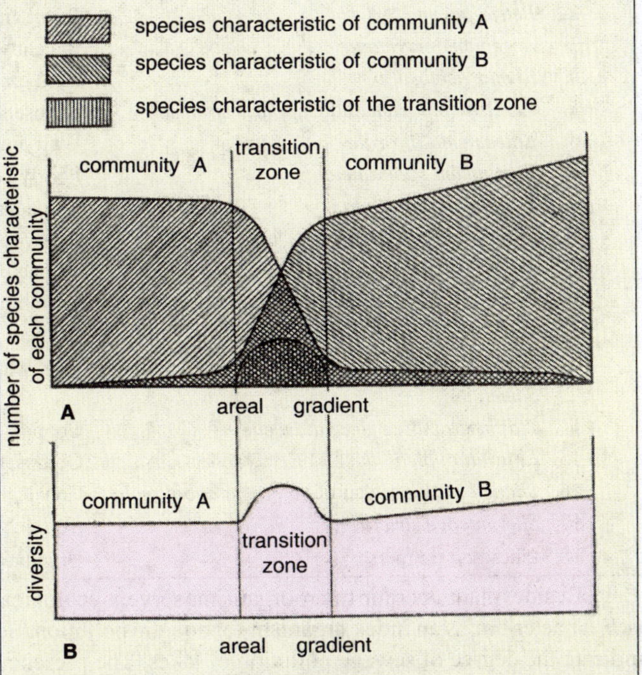

Fig. 7.10. Diagram of the species distribution across a transition zone or ecotone between two communities, labelled A and B (after Clapham, Jr., 1973).

The ecotypes of the plant species though remain genetically distinct, but are interfertile and are put into one taxonomic species. The different ecotypes of a particular species may differ in their edaphic, biotic, or microclimatic requirements. In India **Misra** and **Rao** (1948) first time reported ecotypic differentiation in *Lindenbergia polyantha.* Ecotypes have also been reported in following Indian plant species as *Euphorbia hirta, Euphorbia thymifolia, Cassia tora, Xanthium strumarium,* etc.

Among animals, ecotypes have been reported in jelly fish, *Aurelia aurita* (**Bullock**, 1955), European oyster, *Ostrea edulis* (**Korringa**, 1957), and frog. Some frogs, for example, undergo rapid development at lower temperatures at one extreme of their range of distribution, while others show a more retarded development rate at higher temperatures at another extreme of their range.

ECOLOGICAL INDICATORS

Since, specific factors often determine rather precisely what kinds of organisms will be present in a particular habitat, we can turn the situation around and judge the kind of physical environment from the organism present. It is found that certain species of microorganisms, plants and animals have one or more specific requirements and they become very much limited in their distribution. Thus, the occurrence of such species in a particular area indicates special habitat conditions, and such species are called **biological** or **ecological indicators**, since they indicate some very specific condition of the environment. Some of the Indian plant species which serve as ecological indicators have been tabulated in Table 7-2.

Table 7-2. Certain ecological indicators of plant species.

	Name of species	Indicator (s) of —
1.	*Utricularia, Chara, Wolffia, Ottelia alismoides*	Water pollution
2.	*Petridium* sp.	Burnt and highly disturbed coniferous forest
3.	*Argemone maxicana*	Recently disturbed or flooded soil
4.	*Rumex acetosella*	Acid grassland soil
5.	*Carissa spinarum* and *Capparis spinosa*	Intense soil erosion
6.	*Enicostema littorale*	Well drained soil
7.	*Saccharum spontaneum*	Poorly drained soil
8.	*Zizyphus rotundifola*	Soil deposition
9.	*Andropogon scoparium*	Sandy loam type soil
10.	*Lippia nodiflora* and *Rumex* sp.	Nitrate rich soil
11.	*Woodfordio foribunda* and *Chloris virgata*	High lime content of soil
12.	*Atriplen, Salsola,* and *Saueda*	Saline water condition
13.	*Hydrilla verticillata* and *Ceratophyllum demersum*	Hard water
14.	*Waltheria indica sterculiasceae*	Copper, lead and zinc rich soil
15.	*Equisetum plebejum* and *E. arvense*	Gold rich soil
16.	*Silene cobalticola* and *Crotalaria cobalticola*	Cobalt rich soil
17.	*Spermacoce stricta*	Iron rich soil
18.	Pines and junipers	Uranium rich soil

Besides plants, certain microorganisms serve as ecological indicators. For example, *Escherichia coli,* a bacterium, is an index organism for aquatic pollution. Some species of diatoms (Araphidinae) indicate the degree of sewage pollution of lakes. The presence of fusilinids (protozoans) in the soil indicate the petroleum deposits.

Similarly, certain animal species serve as ecological indicators. Chironomid larvae and maggots increase in number as the pollution increases. Burrowing may fly *Hexagenia* serves as indicator for well oxygenated aquatic environments. Emigration and consequent disappearance of certain Indian fish species such as *Catla catla, Labeo gonius, Labeo bata, Labeo rohita, Notopterus,* etc., from Kalinadi river near Mansurpur Sugar Factory, U.P. indicate industrial pollution of water, as these fishes are found to be very sensitive to aquatic pollution (**Verma**, 1971).

REVISION QUESTIONS

1. Define biotic community and describe in detail its different characteristics.
2. What is community stratification ? Explain the phenomenon with suitable examples of plants and animals.
3. Write short notes on the following :
   (i) Ecotone,
   (ii) Edge effect,
   (iii) Ecotype,
   (iv) Ecological indicators,
   (v) Ecological niche,
   (vi) Life or growth forms.

CHAPTER

8

Ecological Succession

The process of directional change in vegetation is termed as succession.

When stripped of its original vegetation by fire, flood, or glaciation, an area of bare ground does not remain devoid of plants and animals. Beginning with plants, area is rapidly colonized by a variety of both plant and animal species that subsequently modify one or more environmental factors. This modification of the environment may in turn allow additional species to become established. The transitional series of communities which develop in a given area are called **sere** or **seral stages**, while the final stable and mature community is called the **climax**. The development of the community by the action of vegetation on the environment leading to the establishment of new species is termed **succession**. Succession is the universal process of directional change in vegetation during ecological time. It can be recognized by the progressive change in the species composition of the community. Retrogression in community development does not occur unless succession is disturbed or halted by fire, grazing, scraping or erosion.

CAUSES OF SUCCESSION

Since succession involves a series of complex processes, so there exist many causes of its occurrence. Ecologists have recognized the following three primary causes of succession :

1. Initial or Initiating causes. These are climatic as well as biotic in nature. The climatic causes include factors such as erosion and deposits, wind, fire, etc., which are caused by lightening or volcanic activity. The biotic causes include various activities of organisms. All these causes produce the bare areas or destroy the existing populations in an area.

2. Ecesis or Continuing causes. These are processes as migration, ecesis, aggregation, competition, reaction, etc., which cause successive waves of populations as a result of changes, chiefly in the edaphic (soil) features of the area.

3. Stabilising causes. These include factors such as climate of the area which result in the stabilisation of the community.

TRENDS OF SUCCESSION (Functional Changes)

Trends of changes during ecosystem development from a young to mature stages include the following features :

1. A continuous change occurs in the kinds of plants and animals.
2. An increase in the diversity of species takes place. The general appearance of the community or the physiognomy keeps on becoming more and more complex as the succession proceeds.
3. There is a progressive increase in the amount of living biomass and dead organic matter. Such an increase occurs in gross as well as net primary production in the initial and seral stages. Thus, there is more biomass accumulation, gradually reaching a huge biomass structure in the climax.
4. Green pigment (Chlorophyll) go on increasing during the early phase of primary succession. The ratio of yellow/green pigments remains around 2 in the early stages and increases to 3 to 5 in the climax stage. Pigment diversity also increases.
5. The community respiration increases but the P/R (*i.e.,* Production/Respiration) ratio remains more than 1 in the seral stages. The huge living biomass respires a lot in the climax stage and the P/R ratio equals 1 (*i.e.,* $P/R = 1$). Thus, in the early stages $P > R$ and in the climax stage $P = R$.
6. The food chain relationships become more complex as succession proceeds.
7. Nutrients in the young stage are allocated mostly in the soil, but as the seral stages advance, nutrients get allocated more in the vegetation and less in soil. Further the nutrient cycling becomes more closed or intrabiotic with an efficient cycling mechanism whereas in the young stage the nutrients easily leak out from the system, *i.e.,* the cycling is more of an open type.
8. The role of detritus becomes progressively more and more important.
9. The quality of the habitat gets progressively modified to a more mesic condition from either too dry or too wet condition, in the early seral stage.
10. The niche specialization increases, *i.e.,* different functions are more effectively performed by specialist species in mature seral stage, whereas in early stage many functions are performed but less efficiently by a few species.
11. The life cycle of mature community species are longer and more complex.
12. The importance of macroenvironment becomes less in later stages.
13. Relationship among component species becomes more and more mutalistic or helpful even though there exist enough competition and allelochemic activities to prevent invasion of outside elements.
14. Dispersal of seeds and propagules is by wind in young stage, while by animals in mature stage.

BASIC TYPES OF SUCCESSION

Based on different criteria, there are following kinds of succession :

1. Primary succession. If an area in any of the basic environments (such as terrestrial, freshwater or marine) is colonized by organisms for the first time, the succession is called **primary succession**. Thus, primary succession begins on a sterile area (an area not occupied previously by a community), such as newly exposed rock or sand dune where the conditions of existence may not be favourable initially.

2. Secondary succession. If the area under colonization has been cleared by whatsoever agency (such as burning, grazing, clearing, felling of trees, sudden change in climatic factors, etc.) of the previous plants, it is called **secondary succession**. Usually the rate of secondary succession is faster than that of primary succession because of better nutrient and other conditions in area previously under plant cover.

Primary succession.

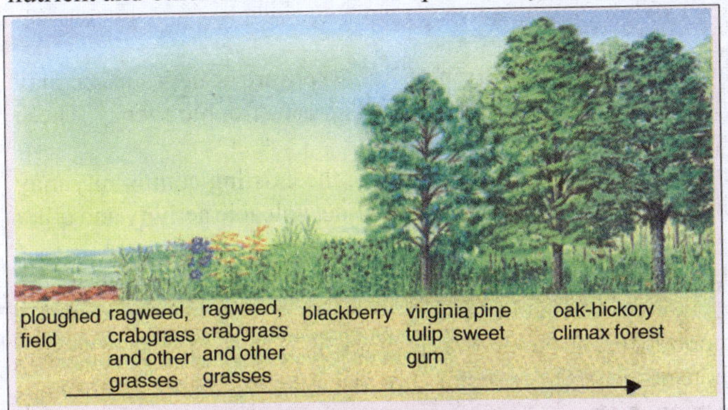

Secondary succession.

3. Autogenic succession. After the succession has begun, in most of the cases, it is the community itself which, as a result of its reactions with the environment, modifies its own environment and, thus, causing its own replacement by new communities. This course of succession is known as **autogenic succession**.

4. Allogenic succession. In some cases replacement of one community by another is largely due to forces other than the effects of communities on the environment. This is called **allogenic succession** and it may occur in a highly disturbed or eroded area or in ponds where nutrients and pollutants enter from outside and modify the environment and in turn the communities.

5. Autotrophic succession. It is characterized by early and continued dominance of autotrophic organisms such as green plants. It begins in a predominantly inorganic environments and the energy flow is maintained indefinitely. There is gradual increase in the organic matter content supported by energy flow.

6. Heterotrophic succession. It is characterized by early dominance of heterotrophic organisms such as bacteria, actinomycetes, fungi and animals. It begins in a medium which is rich in organic matter such as small areas of rivers, streams; these are polluted heavily with sewage or in small pools receiving leaf litter in large quantities.

7. Induced succession. Activities such as overgrazing, frequent scraping, shifting cultivation or industrial pollution may cause deterioration of an ecosystem. Agricultural practices are retrogression of a stable state to a young state by man's deliberate action.

8. Retrogressive succession. It means a return to simpler and less dense or even impoverished form of community from an advanced or climax community. In most cases, the causes are allogenic, *i.e.,* forces from outside the ecosystem become severe and demanding. For example, most of our natural forest stands are degrading into shrubs, savanna or impoverished desert-like stands by the severity of grazing animals brought from surrounding villages. Excessive removal of wood, leaf and twig litter also leads to retrogressive succession.

9. Cyclic succession. It is of local occurrence within a large community. Here *cyclic* refers to repeated occurrence of certain stages of succession whenever there is an open condition created within a large community.

Further, in ecological literature, there are mentioned still so many other kinds of succession, depending mainly upon the nature of the environment where the process of succession has begun. Thus, it may be a **hydrosere** or **hydrarch** when succession occurs in regions where water is in plenty, as ponds, lakes, streams, swamps, bogs, etc., a **mesarch** when there are present adequate moisture conditions; and a **xerosere** or **xerarch** when there is present minimum amount of moisture, *e.g.*, deserts, rocks, etc. Sometimes, few more categories are recognized such as **lithosere** – succession initiating on rocks, **psammosere** –succession initiating on sand and **halosere** –succession occurring in saline water or saline soil.

GENERAL PROCESS OF SUCCESSION

The entire process of primary autotrophic succession is completed through the following sequential steps :

1. Nudation

This is the development of a bare area without any form of life. Exposure of new surface may occur due to several causes such as landslide, erosion, deposition, or other catastrophic agency. These causes of nudation are of three main types :

(i) Topographic. Due to soil erosion by gravity, water or wind, the existing community may disappear. Other topographic causes include deposition of sand, landslide, volcanic activity and other factors.

(ii) Climatic. Glaciers, dry period, hails and storm, frost, fire, etc., may also destroy the community.

(iii) Biotic. Man forms a most important biotic factor; he is responsible for destruction of forests, grasslands for industry, agriculture, housing, etc. Other factors are disease epidemics due to fungi, viruses, etc., which destroy the whole population.

2. Invasion

Invasion is the successful establishment of a species in a bare area. The species actually reaches this new site from any other area. Invasion includes the following three steps :

(i) Migration (dispersal). The seeds, spores, or other propagules of the species reach the bare area. This process is called **migration** and is generally brought about by air, water, etc.,

(ii) Ecesis (establishment). After reaching to new area, the process of successful establishment of the species, as a result of adjustment with the conditions prevailing there, is known as **ecesis**. Ecesis is followed by full-scale **colonization**. In plants, after migration, seeds or propagules germinate, seedlings grow, and adults start to reproduce.

(iii) Aggregation. Colonization by successive offspring and new migrants help increase the population, a process called **aggregation**. Thus, after ecesis, as a result of reproduction, the individuals of the species increase in number and they come close to each other. Plants or autotrophic organisms which are the first to colonize and aggregate are called **pioneers**. The pioneer communities are likely to be more dynamic and have low-nutrient requirements and to take minerals in comparatively more complex forms. They are small-sized and make less demand from environment.

3. Competition and Coaction

Due to aggregation of a large number of individuals of the species at the limited place, there develops **competition** (*i.e.*, interspecific and intraspecific competition) for space and nutrition. Individuals of a species affect each other's life in various ways and this is called **coaction**. The species which fail to compete with other species are ultimately discarded. The reproductive capacity, wide ecological amplitude, etc., help the species to withstand the competition.

4. Reaction

Reaction includes mechanism of the modification of the environment through the influence of living organisms on it. Due to this very significant stage, changes take place in soil, water, light conditions, temperature, etc., of the environment. As a result of reaction, the environment is modified and become unsuitable for the existing community which sooner or later is replaced by another community (seral community). The whole sequence of communities that replaces one another in the given area is called a **sere**, and different communities constituting the sere are called **seral communities**, **seral stages** or **developmental stages**.

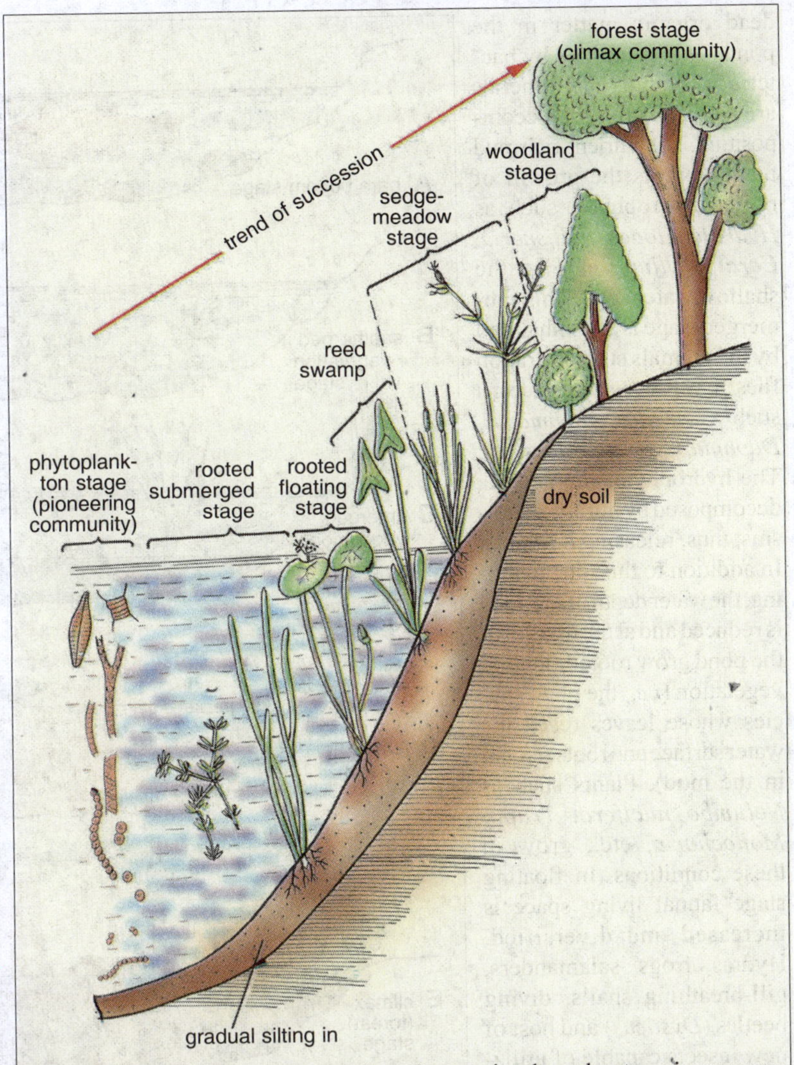

Fig. 8.1. Diagrammatic representation of succession through autogenic processes. Seral communities modify the environmental complex which in turn changes the vegetation and the process continues until the dynamic equilibrium is reached. Note the pioneer, seral and climax communities.

5. Stabilization (Climax)

Finally, there occurs a stage in the process, when the final terminal community becomes more or less established for a longer period of time and it can maintain itself in equilibrium with the climate of the area. The final community is not replaced and is known as **climax community** and the stage as **climax stage**.

SOME EXAMPLES OF SUCCESSION

1. Hydrosere. A good example of succession is the **hydrarch succession** or **hydrosere** (Fig. 8.2), in which a pond and its community are converted into a land community. In the initial stage, phytoplankton (*e.g.*, some blue green algae (cyanobacteria), green algae (*e.g., Spirogyra, Oedogonium*), diatoms and bacteria) are the pioneer colonizers. They are consumed by zooplankton (*e.g.*, protozoans such as *Amoeba*, *Paramecium*, *Euglena*, etc.), fish such as blue gill fish, sun fish, large mouth, etc. Gradually these organisms die and increase the content of dead organic matter in the pond. This is utilized by bacteria and fungi, and minerals are released after the decomposition. The nutrient-rich mud then supports the growth of rooted hydrophytes such as *Hydrilla, Elodea, Vallisneria, Ceratophyllum,* etc., in the shallow water zone. This submerged stage is also inhabited by the animals such as dragon flies, may flies and crustaceans such as *Asellas, Gammarus, Daphnia, Cypris, Cyclops,* etc. The hydrophytes die and are decomposed by mircroorganisms, thus, releasing nutrients. In addition to this, due to silting, the water depth of the pond is reduced and at the margin of the pond grow rooted floating vegetation (*i.e.,* the plant species whose leaves reach the water surface and roots remain in the mud). Plants such as *Nelumbo nucifera, Trapa, Monochoria,* etc., grow in these conditions. In floating stage faunal living space is increased and diversified. Hydras, frogs, salamanders, gill-breathing snails, diving beetles (*Dysticus*) and host of new insects capable of utilizing the under surfaces of floating leaves appear. Some turtles and snakes also invade the pond.

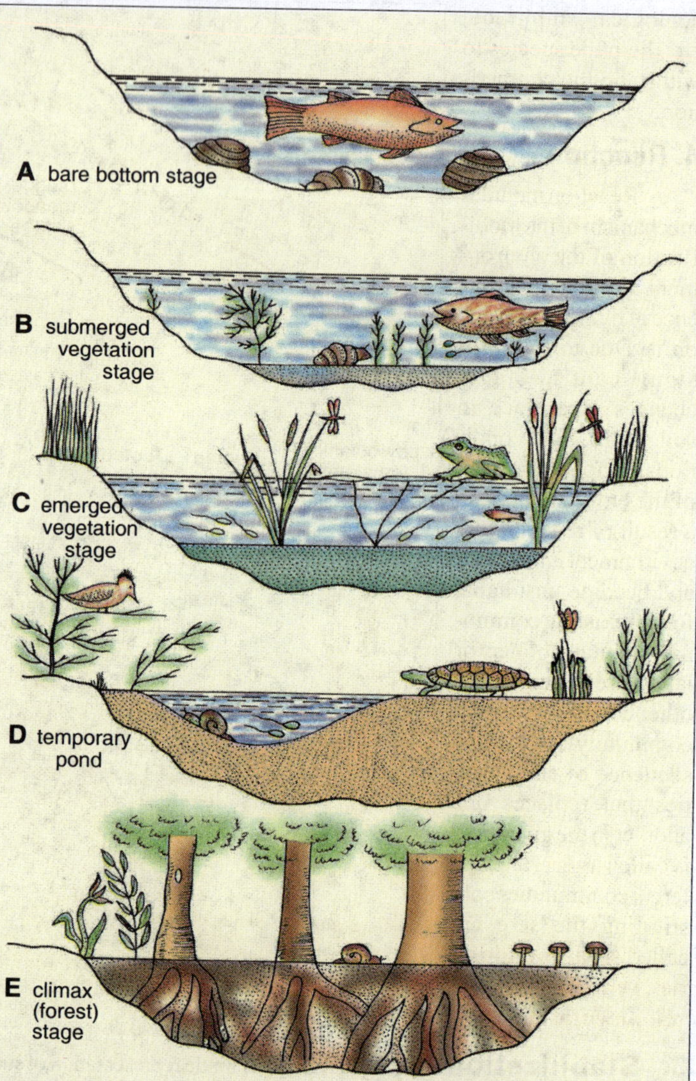

Fig. 8.2. Diagram showing ecological succession from a pond to a climax forest (after Cockrum and McCauley, 1965).

Azolla.

Gradually the water depth in the pond decreases due to evaporation and the deposition of organic matter and the concentration of the nutrients increases. Free-floating plants such as *Azolla, Lemna, Pistia, Wolffia, Spirodella,* etc., increase in number because of the high nutrient availability. Gradually their dead parts fill up the pond ecosystem, resulting in the further build up of the substratum.

At this stage, the pond becomes a swampy ecosystem. The reed swamp species (such as *Scirpus* or bulrushes, *Typha* or cattail, *Phragmites* (reed grass), *Rumex,* etc.) and sedges (*e.g., Carex, Juncus, Cyperus,* etc.) invade the pond and the latter are gradually replaced by mesic communities as the water depth is reduced greatly. Gradually land plants, such as, shrubs (*Salix, Cornus*) and trees (*Populus, Almus*) invade ending in the climax community such as deciduous forest (Fig. 8.2). In association with the changes in water depth and vegetation, the aquatic fauna also change and ultimately gets replaced by land animals.

Thus, possible trend of succession in the aquatic environment is as follows :

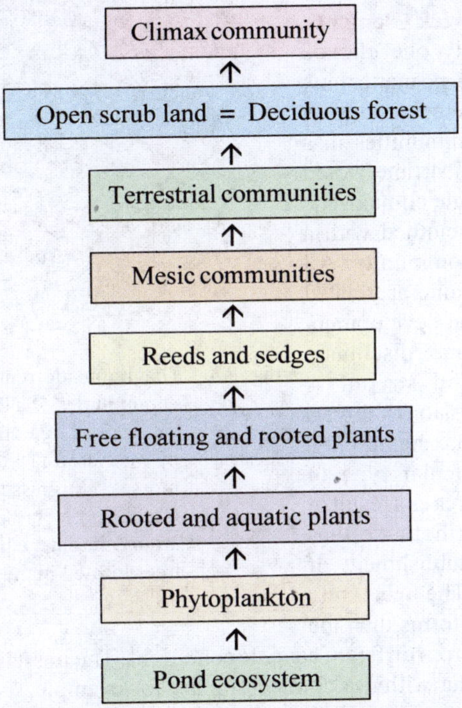

Climax communities vary from place to place. For example, in low lying lands in some parts of Kashmir, the climax community has trees such as *Salix* (**Ambasht,** 1988). In Indian upland plateaus, the climax woody species consist of *Diospyros, Butea* and *Zizyphus* and ground vegetation of *Eragrostis, Sporobolus, Bothriochloa,* etc. Climax vegetation of lowlands and valleys which provide a mesic environment includes *Terminalia, Ficus, Sterculia, Salix,* etc.

2. Succession in xeric habitat. Xerosere or xerarch succession begins on exposed parent rocks (**lithosere**) or dry sand (**psammosere**). A lithosere (Fig. 8.3) involves the following stages : crustose lichens stage (pioneers) → foliose lichens stage → moss stage → herbs stage → shrub stage → forest stage (climax stage). Thus, pioneer plants are lichens, mosses and *Selaginella*, which help in soil formation by accelerating erosion. In course of time grasses, annuals and herbaceous vegetation grow on the soils deposited on rocks.

Like the hydrosere, the lithosere involves successive changes in animal life. The pioneer animals of the lichen stages are few species of mites, ants and spiders. These animals are exposed to harsh environment such as extreme fluctuations in temperature. During the moss stage, many new species of mites, small spiders, tardigrades and springtails invade the community. The herb stage is characterized by nematodes, mites, collembola, ants and various insect larvae. During the shrub and forest stages great qualitative and quantitative modifications occur in the fauna. Thus, there occur numerous kinds of animals such as slugs, snails, wire worms, millipedes, centipedes, mites, ants, sow bugs, springtails, amphibians such as salamanders, frogs, etc., reptiles such as turtles, skinks and other lizards, snakes, birds such as flycatcher and grouse, and mammals such as mole, mouse, shrews, squirrels, chipmunk and fox.

CONCEPT OF CLIMAX

The concept of climax has since long been a subject of much controversy and discussion. There are following three theoretical approaches to the climax :

1. Monoclimax theory. This theory is developed largely by **Frederick Clements**. This theory recognizes only one climax, determined solely by climate, no matter how great the variety of environmental conditions is at the start. All seral communities in a given region, if allowed sufficient time, would ultimately converge to a single climax. The whole landscape would be clothed with a uniform plant and animal community. All other communities than the climax are related to the climax by successional development and are recognized as subclimax, disclimax, preclimax, postclimax and so on. A **subclimax** is a stage in succession of forests just preceding the climate climax community. The **disclimax** is the particular type of vegetation maintained in an area as a result of recurrent disturbance, chiefly the biotic, thus preventing a successful establishment of climate climax community. The **preclimax** is a vegetation of lower-life forms than the one adjacent to it and results from different edaphic conditions. The **postclimax** is a strip of vegetation of higher life-forms occurring within a climate climax, for example, a forest along a stream in a grassland community constitutes postclimax community. Later on numerous other terms such as coclimax, superclimax, quasiclimax, anticlimax, pseudoclimax, etc., were coined by some post-Clements ecologists to describe specific situations.

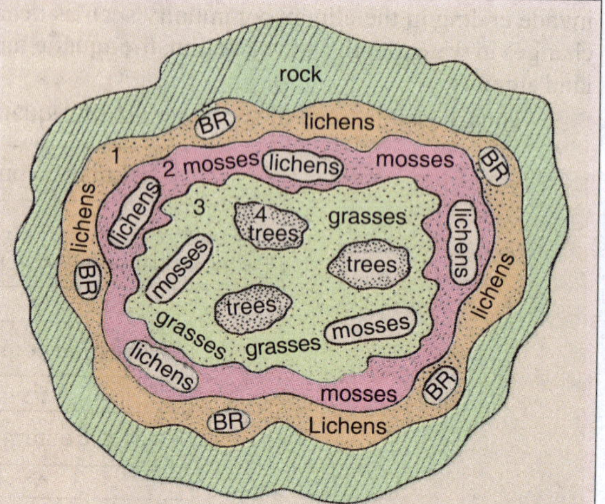

Fig. 8.3. Diagrammatic representation of different plant communities of a lithosere appearing on a rock. The various zones from outside towards the centre of rock are (1) lichens (pioneers), (2) ring of mosses, (3) grasses (broad zone) and (4) trees (seedlings). Note the pioneer community of lichens around the outer edge and more advanced stage of trees located in the centre. BR=bare rock.

The monoclimax theory is supported by **Cowles**, **Ranganathan**, and **Puri**, but strongly objected by **Daubenmire** (1968).

2. Polyclimax theory. This theory was developed by **Tansley**. This theory considers that the climax vegetation of region consists of not just one type but a mosaic of vegetational climaxes controlled by soil moisture, soil nutrients, topography, slope exposure, fire and animal activity. The advocates of polyclimax theory preferred to call each stable community as a climax and described them with a prefix as edaphic climax, topographic climax, biotic (zootic, grazing or anthropogenic) climax and fire climax. For example, grassland communities of central India, Sri Lanka and parts of California have developed under the influence of fire, grazing and other biotic factors and so have been considered as biotic climaxes by **Misra**, **Pandeya** and **Holmes**.

3. Climax pattern hypothesis. This theory was developed by **Whittaker**, **MacIntosh** and **Sellack**. According to this theory, the composition, species structure and balance of a climax community are determined by the total environment of the ecosystem and not by one aspect, such as climate alone. Involved are the characteristics of each species population, their biotic interrelationships, availability of flora and fauna to colonize the area, the chance dispersal of seeds and animals, and the soils and climate. The pattern of climax vegetation will change as the environment changes. Thus, the climax community presents a pattern of populations that corresponds with and changes with the pattern of environmental gradients, intergrading to form ecoclines. The central and most widespread community in the pattern is the prevailing or climatic climax. It is the community that most clearly expresses the climate of the area.

4. Information theory. This theory was proposed by **Leith**, **Odum** and **Golley**. It considered succession and climax in terms of ecosystem development. In autotrophic succession (ecosystem development), diversity of species tends to increase with an increase in organic matter content and biomass supported by the available energy. Thus, in a climax community, the available energy and biomass, which is called **information content**, increase. In contrast to it, in a heterotrophic succession occurs a gradual depletion of energy, because the rates of respiration always exceed production rates. However, in an ecosystem both the autotrophic and heterotrophic successions operates in a co-ordinate manner. The autotrophic individuals derive mineral elements from the soil and atmosphere, while the heterotrophic individuals carry on the return of the nutrients to soil and atmosphere, through decomposition of complex dead organic matter. Thus, succession reaches a stage, the climax stage, when the amount of energy and nutrients received from the environment by the plants is again returned in more or less similar amount to the environment by decomposition through heterotrophs.

Certain Recent Models of Succession

Connell and **Slatyer** (1977) have proposed the following three models to accommodate different possible pathways of succession :

1. Facilitation model. It is based on the Clements ideas of relay communities in which the seral community is believed to modify. According to this model each new community in course of time prepares suitable ground to facilitate its own replacement by another better suited community. So each community like a 'relay process' delivers the habitat to next or higher status community. This model has not been supported by any proper evidence.

2. Tolerance model. This model is based on the concept of **IFC** or **initial floritic composition** which suggests that arrival of new invaders of higher life form types necessarily does not eliminate pioneers. According to this model, only such higher succession or climax species are able to join which can be tolerated by the early settlers. So, the tolerance model differs from Clement's relay succession model in the sense that as the succession proceeds, more and higher life form plants are tolerated to join and co-exist, than necessarily replacing the earlier component species. With the passage of time, species which mutually tolerate each other gain control over the habitat to form the climax vegetation.

Clements gaue the idea of relay community.

3. Inhibition model. According to this model, the early arrived species (populations) on a new habitat may develop counter-mechanism to normal replacement process. For example, the allelopathy may be the common counter-acting adaptation to thwart or inhibit the entry of late arriving species. This kind of highly adaptated early stage communities may not be common on a wide range of habitats. In such a case, relay succession gains control only after the death of the allelopathic plants. Since inhibition model lacks universality, so it was also criticised.

Resource-Ratio Hypothesis of Succession

This model was proposed by **Tilman** (1985) who was agreed to Clement's idea of replacement of communities. According to him, the resources of the habitat are regarded as the key factor. Limited resource leads to competition. Due to competition a resource ratio is created for each kind of resource. Depending on the newly created ratio level, new species adapted to it succeeds.

In conclusion, succession is directional and progresses towards the climax. In some cases, it may be non-directional, *i.e.*, it returns to a particular stage and it is cyclic. It take hundreds and thousands of years to complete primary succession, while the secondary succession is quicker (takes 10 to 200 years for its completion). In some cases, the initial stage community may inhibit altogether the process of succession by allelochemic inhibition of new arrivals.

COMMUNITY EVOLUTION

Like the responses of communities to changing abiotic conditions, community evolution involves progressive changes in climax communities. Because the evolution is exceedingly slow, it cannot be observed in operation, and few instances from the fossil record are sufficiently complete to show the process in action.

The example that best demonstrates evolution of the basic structure of the community is that of the development of a terrestrial community of a modern type by early reptiles some 250 million years ago. Between the time when vertebrates (amphibians) first became able to lead a predominant terrestrial existence some 350 million years ago and the establishment of an essentially modern type food web some 100 million years later, the structure of terrestrial community was decidedly different from what it is now. Development of the modern type of community structure required not only a complete rearrangement of the niche structure of the community but also the evolution of new species that could fill the new niches (**Olson**, 1961, 1966).

Attainment of the adaptations needed for terrestrial life by the first amphibians did not in itself establish a land-based vertebrate community. These early amphibians were carnivores, and the only animals inhabiting the land environment were insects. It is unbelievable that the clumsy locomotor system of early amphibians would have allowed them to prey effectively on animals such as insects. Thus, the first communities inhabited by terrestrial vertebrates are best regarded as extensions of aquatic communities, with the land habit as an adaptation to improve the capabilities of organisms whose prime food supply was aquatic invertebrates and fish.

By some 300 million years ago, reptiles had evolved that could feed effectively on terrestrial invertebrates. An entirely land-based community was theoretically possible in which all herbivore niches were assumed by invertebrates and some of the carnivore niches by vertebrates. However, the palaeoecological evidences suggest that most contemporary carnivorous vertebrates were unable as yet to realize an entirely terrestrial carnivore niche, so that the great majority of the energy flow through the community continued to pass through the aquatic route. The typical food chain to the highest terrestrial vertebrate carnivore was plant → aquatic invertebrate → aquatic-invertebrate-feeding vertebrate → semi-aquatic predator → terrestrial predator.

By 250 million years ago terrestrial herbivorous vertebrates had evolved and a fully terrestrial vertebrates community could come into being. From this time onward the basic structure of the terrestrial community was of an essentially modern sort, with all consumer trophic levels occupied by a wide range of animals, both vertebrates and invertebrates.

Such evolutionary changes in the structure of communities are caused by a large number of factors. One factor is changes in the regional climate. It became progressively drier during the period under consideration, and the development of a land-based community reasonably responded to this sort of change. Indeed many evolutionary changes in community structure can be explained on the basis of responses of major changes in the regional abiotic factors of the environment (**Axelrod**, 1950, 1958). But other chief factors of evolutionary change in community include reorganization of the community's structure in response to the realization of niches that had not previously existed in the community.

REVISION QUESTIONS

1. What is ecological succession ? Describe the process of ecological succession in pond or on a bare rock. Add a note about climax.
2. Describe the causes, trends and basic types of succession.
3. Write short notes on :
   (i) Ecesis ; (ii) Climax ; (iii) Allogenic succession ; (iv) Retrogressive succession.
4. Describe various models explaining the nature of climax communities.
5. Give an illustrated account of the sequential stages of a typical hydrosere.

CHAPTER

9

Ecosystem : Structure and Function

Our earth is a giant ecosystem where abiotic and biotic components react with each other bringing functional changes.

Any ecological unit that includes all the organisms (*i.e.,* the communities in a given area) which interact among themselves and with the physical environment, so that a flow of energy leads to clearly defined trophic structure, biotic diversity and material cycle (*i.e.,* exchange of materials) within the system, is known as **ecological system** or **ecosystem**. There exist nutritional relationships (or food links) amongst the living organisms of such a system. Keeping this in view, the earth can be considered as a giant ecosystem where abiotic and biotic components are constantly acting and reacting upon each other bringing forth structural and functional changes in it. This vast ecosystem — the **biosphere** is however, difficult to handle and, thus, for the sake of convenience, we generally study nature by making its artificial subdivisions into units of smaller ecosystems (such as **terrestrial** — forest, desert, grassland; **aquatic** — fresh water, marine; and **man-made** — cropland, etc.). An ecosystem may, thus, be as small as a single log, a pond, a cropland, or as large as an ocean, desert or forest. Though these unit ecosystems are separated from each other with time and space, but functionally they all are linked with each other, forming an integrated whole.

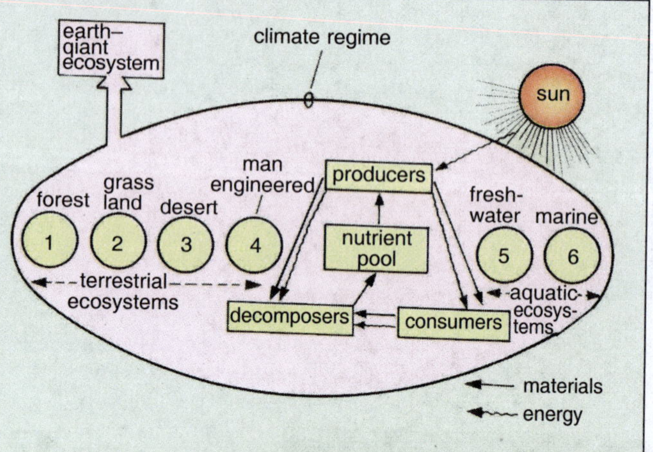

Fig. 9.1. Diagrammatic representation of the basic types of ecosystems all of which together constitute the giant ecosystem— the biosphere.

The term **ecosystem** was proposed by **A.G. Tansley** in 1935. There are many other parallel terms or synonyms for the ecosystem which have been proposed by various ecologists, *e.g.*, **biocoenonsis** (**Karl Mobius**, 1877), **microcosm** (**S.A. Forbes**, 1887), **holocoen** (**Friederichs**, 1930), **biosystem** (**Thienemann**, 1939), **geobiocoenosis** (**Sukhachev**, 1944), **bioenert body** (**Vernadsky**, 1944) and **ecosom**, etc.

In recent years, ecological studies of ecosystems undertake besides structure, the similarities and differences in food and energy relationships among living components of ecosystem. This is called **bioenergetic approach** of modern ecology.

KINDS OF ECOSYSTEM

An ecosystem can be natural or artificial, temporary or permanent and large or tiny. Thus, various constituent ecosystems of the biosphere fall into the following categories :

1. Natural ecosystems. These types of ecosystems operate by themselves without any major interference by man. Based upon the particular kind of habitat, these are further classified as :

(i) Terrestrial ecosystems such as forests, grasslands, deserts, a single log, etc.

(ii) Aquatic ecosystems which may be further distinguished as follows :

(a) Fresh water ecosystems. These may be **lotic** (running water as spring, brook, stream or river) or **lentic** (standing water as lake, pond, pool, puddle, ditch, swamp, etc.).

(b) Marine ecosystems. These include salt water bodies which may be deep bodies as an ocean or shallow ones as a sea or estuary.

2. Artificial ecosystems. These are also called **man-made** or **man-engineered ecosystems**. These are maintained artificially by man where, by addition of energy and planned manipulations, natural balance is disturbed regularly, *e.g.*, croplands such as sugarcane, maize, wheat, rice-fields ; orchards, gardens, villages, cities, dams, aquarium and manned spaceship.

Microecosystems

Natural ecosystems are usually large in size. In them numerous variables (factors) operate at one time leading to great complexity. As a result, it becomes usually difficult to study them with the normal scientific methods. Thus, to reduce the number of variables and to work in a system with a discrete boundary, ecologists are trying to simulate microecosystems in the laboratory, which can be replicated and manipulated at will. The microsystems can be laboratory systems build by taking one or very few species, at a time, from axenic cultures (An **axenic culture** is a bacterial culture that consists of only one species) and studying them in desired combinations. Recently, the concept of polyaxenic cultures has been developed. **Odum** (1971) and others have elucidated the microecosystem concept.

STRUCTURE OF ECOSYSTEM

The structure of an ecosystem is basically a description of the species of organisms that are present, including information on their life histories, populations and distribution in space. It is a guide to who's who in the ecosystem. It also includes descriptive information on the non-living (physical) features of environment, including the amount and distribution of nutrients. An ecosystem typically has two major components :

A. Abiotic or Non-living Components

Abiotic component of the ecosystem comprises three sort of components : (1) **Climatic condition** and **physical factors** of the given region such as air, water, soil, temperature, light (*i.e.,* its duration and intensity), moisture (relative humidity), pH, etc. (2). **Inorganic substances** such as water, carbon (C), nitrogen (N), sulphur (S), phosphorus (P) and so on, all of which are involved in cycling of materials in the ecosystem (*i.e.,* biogeochemical cycles). The amount of these inorganic substances, present at any given time in an ecosystem, is designated as the **standing state** or **standing quality**. (3). **Organic substances** such as proteins, carbohydrates, lipids, humic substances, etc., present either in the biomass or in the environment, *i.e.,* **biochemical structure** that link the biotic and abiotic components of the ecosystem.

B. Biotic or Living Components

In the trophic structure of any ecosystem, living organisms are distinguished on the basis of their nutritional relationships, which are discussed as follows :

1. Autotrophic component. Autotrophic (*auto* = self ; *trough* = nourishing) component of ecosystem includes the **producers** or **energy transducers** which convert solar energy into chemical energy (that becomes locked in complex organic substances such as carbohydrate, lipid, protein, etc.) with the help of simple inorganic substances such as water and carbon dioxide and organic substances such as enzymes. Autotrophs fall into following two groups : (i) **photoautotrophs** which contain green photosynthetic pigment **chlorophyll** to transduct the solar or light energy of sun, *e.g.,* trees, grasses, algae, other tiny phytoplanktons and photosynthetic bacteria and cyanobacteria (=blue green algae). (ii) **Chemoautotrophs** which use energy generated in oxidation - reduction process, but their significance in the ecosystem as producers is minimal, *e.g.,* microorganisms such as *Beggiatoa,* sulphur bacteria, etc.

2. Heterotrophic component. In the heterotrophic (*hetero* = other; *trophic* = nourishing) organisms predominate the activities of utilization, rearrangement and decomposition of complex organic materials. Heterotrophic organisms are also called **consumers**, as they consume the matter built up by the producers (autotrophs). The consumers are of following two main types :

(a) Macroconsumers. These are also called **phagotrophs** (phago = to eat) and include mainly animals which ingest other organisms or chunks of organic matter. Depending on their food habits, consumers may either be **herbivores** (plant eaters) or **carnivores** (flesh eaters). Herbivores live on living plants and are also known as **primary consumers**, *e.g.,* insects, zooplanktons and animals such as deer, cattle, elephant, etc. **Secondary** and **tertiary consumers**, if present in the food chain of the ecosystem, are carnivores or omnivores, *e.g.,* insects such as preying mantis, dragon flies; spiders and large animals such as tiger, lion, leopard, wolf, etc. Secondary consumers are the carnivores which feed on primary consumers or herbivores. Carnivores are, often, recognized as carnivore order - 1 (C_1), carnivore order - 2 (C_2) and so on, depending on their food habits.

Ticks and mites, leeches and blood-sucking insects (mosquito, bed-bug) are dependent on herbivores, carnivores and omnivores.

(b) Microconsumers. These are also called **decomposers**, **reducers**, **saprotrophs** (*sapro* = decompose), **osmotrophs** (osmo = to pass through a membrane) and **scavengers**. **Wiegert** and **Owen** (1971) have coined the term, **biophages** for heterotrophic decomposers which feed on the dead organic matter. Microconsumers include microorganisms such as bacteria, actinomycetes and fungi. Microconsumers breakdown complex organic compounds of dead or living protoplasm, absorb some of the decomposition or breakdown products and release inorganic nutrients in the environment, making them available again to autotrophs or producers. Some invertebrate animals such as protozoa, oligochaeta such as earthworms, etc., use the dead organic matter for their food, as they have the essential enzymes and, hence, can be classified as decomposer organisms. Some ecologists believe that micro-organisms are **primary decomposers**, while invertebrates are **secondary decomposers**.

Detritivores act on organic detritus and breaks it down to release inorganic nutrient.

The disintegrating dead organic matter is also known as **organic detritus** (Latin word *deterere* means to wear away). By the action of **detritivores** (=decomposers), the disintegrating detritus result into particulate organic matter (POM) and dissolved organic matter (DOM) which play important role in the maintenance of the edaphic environment.

EXAMPLE OF ECOSYSTEM

A pond as a whole serves as a good example of an aquatic and freshwater ecosystem (Fig. 9.2). In fact, it represents a self-sufficient and self-regulating system. It has following components:

1. Abiotic Component

The chief non-living or abiotic substances are heat, light, pH value of water, and the basic inorganic and organic compounds, such as water itself, carbon dioxide gas, oxygen gas, calcium, nitrogen, phosphates, amino acids, humic acid, etc. Inorganic salts occur in the form of phosphates, nitrates and chlorides of sodium, potassium and calcium. Some proportion of nutrients exist in solution state but most of them are present as stored in particulate matter as well as in living organisms.

2. Biotic Component

It includes various organisms which are classified into the following types :

(a) Producers. These are photoautotrophic green plants and photosynthetic bacteria. The producers fix radiant energy of sun and with the help of minerals derived from water and mud, they manufacture complex organic substances as carbohydrates, proteins and lipids. Producers of pond are of following types :

(i) Macrophytes. These include mainly the rooted large-sized plants which comprise three types of hydrophytes : partly or completely submerged, floating and emergent aquatic plants. The common plants are species of *Trapa, Typha, Eleocharis, Sagittaria, Nymphaea, Potamogeton, Chara, Hydrilla, Vallisneria, Utricularia, Marsilea, Nelumbo,* etc. Besides these plants, some free floating forms also occur in the pond ecosystem, *e.g., Azolla, Salvinia, Wolffia, Eichhornia, Spirodella, Lemna,* etc.

(ii) Phytoplanktons. These are microscopic (minute), floating or suspended lower plants (algae) that are distributed throughout the water, but mainly in the photic zone. Most of them are filamentous algae such as *Spirogyra, Ulothrix, Zygnema, Cladophora* and *Oedogonium.* There also occur some chlorococcales (*e.g., Chlorella*), *Closterium, Cosmarium, Eudorina, Pandorina, Pediastrum, Scendesmus, Volvox,* Diatoms, *Anabaena, Gloeotrichia, Microcystis, Oscillatoria, Chlamydomonas, Spriulina,* etc., and some flagellates.

(b) Macroconsumers. They are phagotrophic heterotrophs which depend for their nutrition on the organic food manufactured by producers, the green plants. Macroconsumers are of following three types :

(i) Herbivores (Primary consumers). These animals feed directly on living plants (producers) or plant remains. They may be large or minute in size and are of following two types : 1. **Benthos** which are the bottom dwelling forms such as fish, insect larvae, beetles, mites, molluscs (*e.g., Pila,*

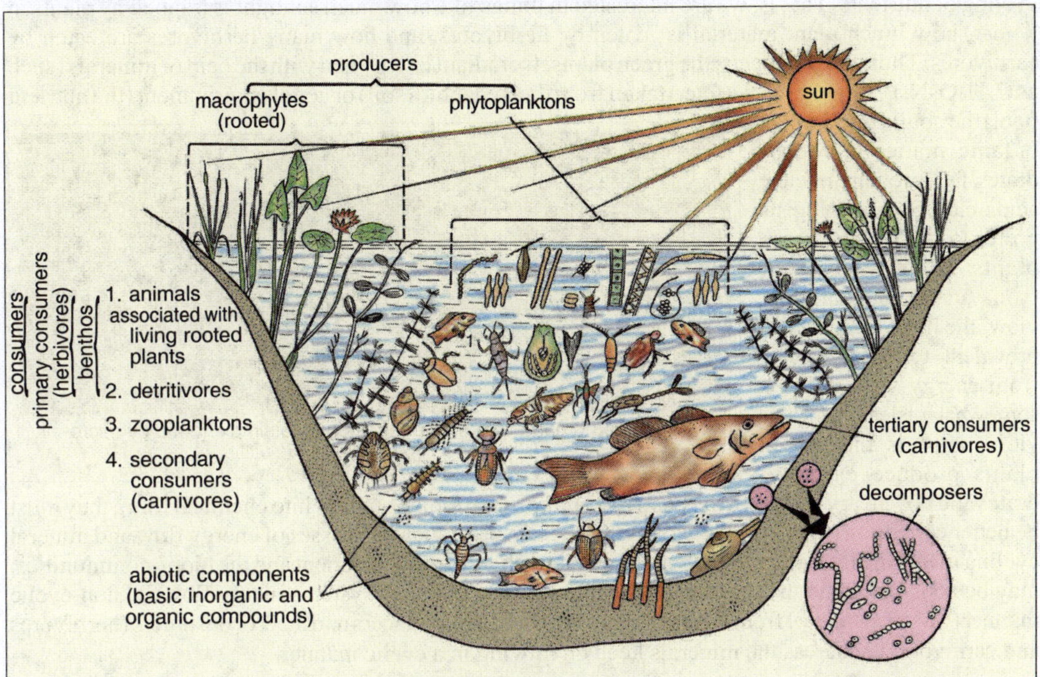

Fig. 9.2. A pond ecosystem showing its basic structural units – the abiotic (inorganic and organic compounds) and biotic (producers, and consumers – herbivores, carnivores and decomposers) components.

Planorbis, Unio, Lamellidens, etc.), crustaceans, etc. 2. **Zooplanktons** which feed chiefly on phytoplanktons and are chiefly the rotifers as *Brachionus, Asplanchna, Lecane,* etc., although some protozoans as *Euglena, Coleps, Dileptus,* etc., and crustaceans such as *Cyclops, Stenocypris,* etc., are also present in the pond.

Besides these small-sized herbivores, some mammals such as cow, buffaloes, etc., also visit the pond casually and feed on marginal rooted macrophytes. Some birds also regularly visit the pond to feed on some hydrophytes.

(ii) Carnivore order-1 (Secondary consumers). These carnivores feed on the herbivores and include chiefly insects, fish and amphibians (frog). Most insects are water beetles which feed on zooplanktons; some insects are the nymphs of dragonflies which feed upon aquatic insects.

(iii) Carnivore order-2 (Tertiary consumers). These are some large fish as game fish that feed on the smaller fish and, thus, become the tertiary (top) consumers.

(c) Decomposers. They are also called microconsumers, since they absorb only a fraction of the decomposed organic matter. They bring about the decomposition of dead organic matter of both producers (plants) as well as macroconsumers (animals) to simple forms. Decomposers help in returning of mineral elements again to the medium of the pond and in running biogeochemical cycles. Decomposers of pond ecosystem include chiefly bacteria, actinomycetes and fungi. Among fungi, species of *Aspergillus, Cephalosporium, Cladosporium, Pythium, Rhizopus, Penicillium, Thielavia, Alternaria, Trichoderms, Circinella, Fusarium, Curvularis, Paecilomyces, Saprolegnia,* etc., are most common decomposers in water and mud of the pond.

FUNCTION OF AN ECOSYSTEM

When we consider the function of an ecosystem, we must describe the flow of energy and the cycling of nutrients. That is, we are interested in things like how much sunlight is trapped by plants in a year, how much plant material is eaten by herbivores, and how many herbivores are eaten by carnivores. Thus, the producers, the green plants, fix radiant energy and with the help of minerals (such as C, H, O, N, P, Ca, Mg, Zn, Fe, etc.) taken from their edaphic (soil) or aerial environment (the nutrient pool) they build up complex organic matter (carbohydrates, fats, proteins, nucleic acids, etc.). Some ecologists prefer to refer to the green plants as **converters** or **transducers**, since in their view, the most popular and prevalent term 'producer' from energy view point is somewhat misleading. Their view point is that green plants produce carbohydrates and not energy and since they convert or transduce radiant energy into chemical form, they must be better called **converters** or **transducers**. The two ecological processes of energy flow and mineral cycling involving interaction between the physico-chemical environment and the biotic communities, may be considered the 'heart' of ecosystem dynamics. In an ecosystem, energy flows in non-cyclic manner (**unidirectional**) from sun to the decomposers via producers and macroconsumers (herbivores and carnivores), whereas the minerals keep on moving in a cyclic manner.

The dead organic matter is decomposed to simple forms by micro-organisms.

Productivity of Ecosystem

The productivity of an ecosystem refers to the rate of production, *i.e.,* the amount of organic matter accumulated in any unit time. It is of following types :

1. Primary productivity. It is defined as the rate at which radiant energy is stored by photosynthetic and chemosynthetic activity of producers. Primary productivity is of following types:

(i) Gross primary productivity. It refers to the total rate of photosynthesis including the organic matter used up in respiration during the measurement period. GPP depends on the chlorophyll content. The rate of primary productivity are estimated in terms of either chlorophyll content as chl/g dry weight/ unit area or photosynthetic number, *i.e.,* amount of CO_2 fixed/g chl/hour.

(ii) Net primary productivity. It is the rate of storage of organic matter in plant tissues in excess of the respiratory utilization by plants during the measurement period.

Primary production is measured by following methods—harvest method, oxygen measurement method (or light or dark method), oxygen diurnal curve method, carbon dioxide measurement method (enclosure method), the aerodynamic method, the pH method, radioisotope method, chlorophyll estimation method (see **Dash**, 1993).

2. Secondary productivity. It is the rate of energy storage at consumer's levels—herbivores, carnivores and decomposers. Consumers tend to utilise already produced food materials in their respiration and also convert the food matter to different tissues by an overall process. So, secondary productivity is not divided into 'gross' and 'net' amounts. Due to this fact some ecologists such as **Odum** (1971), prefer to use the term **assimilation** rather than production at this level – the consumers level. Secondary productivity, in fact, remains mobile (*i.e.,* keeps on moving from one organism to another) and does not live *in situ* like the primary productivity.

3. Net productivity. It is the rate of storage of organic matter not used by the heterotrophs or consumers, *i.e.,* equivalent to net primary production minus consumption by the heterotrophs during the unit period as a season or year, etc.

Food Chains in Ecosystems

In an ecosystem one can observe the transfer or flow of energy from one trophic level to other in succession. A **trophic level** can be defined as the number of links by which it is separated from the producer, or as the *n*th position of the organism in the food chain. The patterns of eating and being eaten forms a linear chain called **food chain** which can always be traced back to the producers. Thus, primary producers trap radiant energy of sun and transfer that to chemical or potential energy of organic compounds such as carbohydrates, proteins and fats. When a herbivore animal eats a plant (or when bacteria decompose it) and these organic compounds are oxidized, the energy liberated is just equal to the amount of energy used in synthesizing the substances (first law of thermodynamics), but some of the energy is heat and not useful energy (second law of thermodynamics). If this animal, in turn, is eaten by another one, along with transfer of energy from a herbivore to carnivore a further decrease in useful energy occurs as the second animal (carnivore) oxidizes the organic substances of the first (herbivore or omnivore) to liberate energy to synthesize its own cellular constituents. Such transfer of energy from organism to organism sustains the ecosystem and when energy is transferred from individual to individual in a particular community, as in a pond or a lake or a river, we come across the food chains. The number of steps in a food chain are always restricted to four or five, since the energy available decreases with each step. For example, in a typical food chain of an Indian river, a diatom may be eaten by a copepod which is eaten by a small fish, which forms the food source of large fish and so on (**Dash**, 1993) :

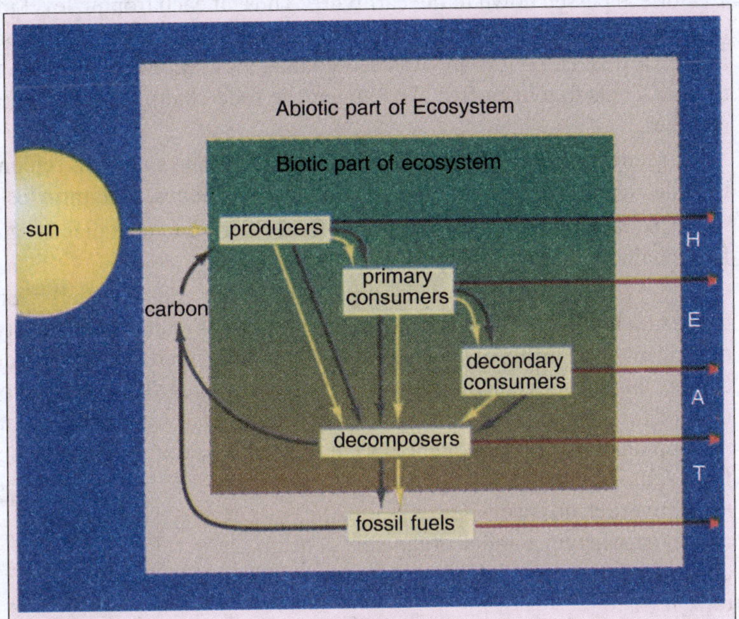

Food Chain.

Scenedesmus boligues (phytoplankton) → *Brachionus falcatus* (zooplankton)
↓
Homo sapiens (man) ← *Wallago attu* (a large fish) ← *Amblypharyngodon* sp. (a small fish)

In an Indian pasture, the following food chain operates :

Cynodon dactylon (a grass species) → *Melanoplus differentialis* (a grasshopper)
↓
Hawk ← *Zamensis mucosus* (a rat snake) ← *Bufo melanostictus* (a toad).

One may ask—why is the number of trophic levels in a food chain limited ? In a simple food chain (Fig. 9.3), out of 1000 calories of energy reaching a plant only 10 calories (1%) are stored by the plant. The remaining calories of energy (99%) are lost to the environment or for plant's own maintenance. Of the 10 calories which are available to the herbivore, 9 calories (99%) are lost at its level and only 1 calorie is passed down to the carnivore. Thus, at each trophic level in a food chain, a large portion of energy is used for its own maintenance and ultimately lost as heat. Consequently, organisms in each trophic level pass on less and less energy than they receive. This tends to limit the number of steps or trophic levels to four or five. The longer the food chain, the less is the energy available to the final member.

Many direct or indirect methods are employed to study food chain relationships in nature. They include gut content analysis, use of radioactive isotopes, precipitin test, etc.

In nature, basically two types of food chains are recognized—grazing food chain and detritus food chain.

1. Grazing food chain. This type of food chain (Fig. 9.4) starts from the living green plants, goes to grazing herbivores and on to the carnivores. Ecosystems with such type of food chain are directly dependent on an influx of solar radiation. Thus, this type of food chain depends on autotrophic energy capture and the movement of this energy to herbivores. Most of the ecosystems in nature follow this type of food chain. These chains are very significant from energy standpoint. The phytoplanktons → zooplanktons → fish sequence or the grasses → rabbit → fox sequence are the examples of grazing food chain. Further the producer → herbivore → carnivore chain is a **predator chain**. **Parasitic chains** also exist wherein smaller organisms consume larger ones without outright killing as the case of the predators.

2. Detritus food chain. The organic wastes, exudates and dead matter derived from the grazing food chain are generally termed **detritus**. The energy contained in this detritus in not lost to the ecosystem as a whole; rather it serves as the source of energy for a group of organisms (**detritivores** that are separate from the grazing food chain, and generally termed as the **detritus food chain** (Fig. 9.5). The detritus food chain represents an exceedingly important component in the energy flow of an ecosystem. Indeed in some ecosystems, considerably more energy flows through the detritus food chain than through the grazing food chain. In the detritus food chain the energy flow remains as a continuous passage rather than as a stepwise flow between discrete entities. The organisms of the detritus food chain are many and include algae, bacteria, slime molds, actinomycetes, fungi, Protozoa, insects, mites, Crustacea, centipedes, molluscs, rotifers, annelid worms, nematodes and some vertebrates. Some species are highly specific in their food requirements and some can eat almost anything. Many Protozoa, for instance, need certain specific organic acids, vitamins, and other nutrients before they can thrive; on the other hand, the guts of small Collembola (a group of tiny soil insects) have been reported to contain decaying plant material,

Fig. 9.3. Mode of energy flow through a simple food chain.

fungal fragments, spores, fly pupae, other Collembola, parts of decaying earthworms, and cuticle from their own faecal casting (**Hale**, 1967). In contrast to the grazing food chain, in which energy storage is entirely within the tissues of living organisms, energy storage for the detritus food chain may be largely external to the organisms, and in the detritus itself.

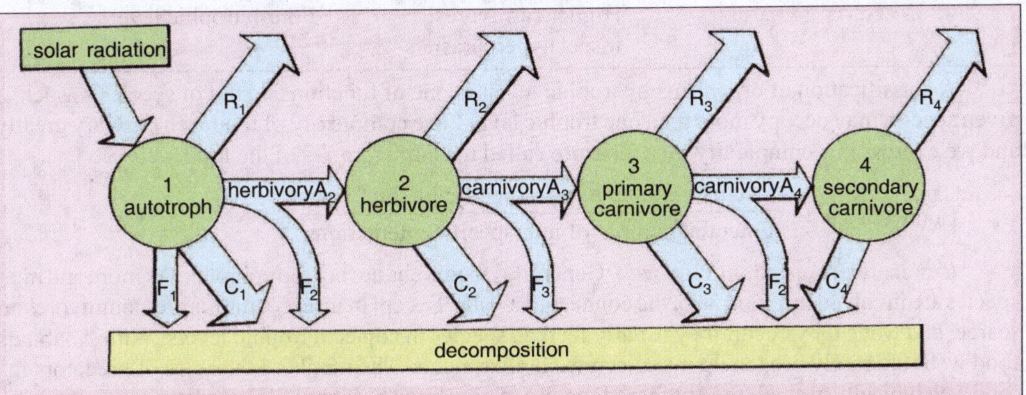

Fig. 9.4. Diagrammatic representation of a grazing food chain. Showing inputs and losses of energy at each trophic level. Trophic levels are numbered and used as subscripts to letters indicating energy transfer. A–assimilation of food by the organisims at the trophic level; F–energy lost in the form of faeces and other excretory products; C–energy lost through decay and R–energy lost to respiration (after Clapham, Jr., 1973).

Significance of food chain. The food chain studies help understand the feeding relationships and the interaction between organisms in any ecosystem. They also help us to appreciate the energy flow mechanism and matter circulation in ecosystem, and understand the movement of toxic substances in the ecosystem and the problem of biological magnification (*e.g.,* DDT; for details see Chapter 14).

Food web. In nature simple food chains occur rarely. The same organism may operate in the ecosystem at more than one trophic level, *i.e.,* it may derive its food from more than one source. Even the same organism may be eaten by several organisms of a higher trophic level or an organism may feed upon several different organisms of a lower trophic level. Usually the kind of food changes with the age of the organism and the food availability. Thus, in a given ecosystem various food chains are linked together and intersect each other to form a complex network called **food web**.

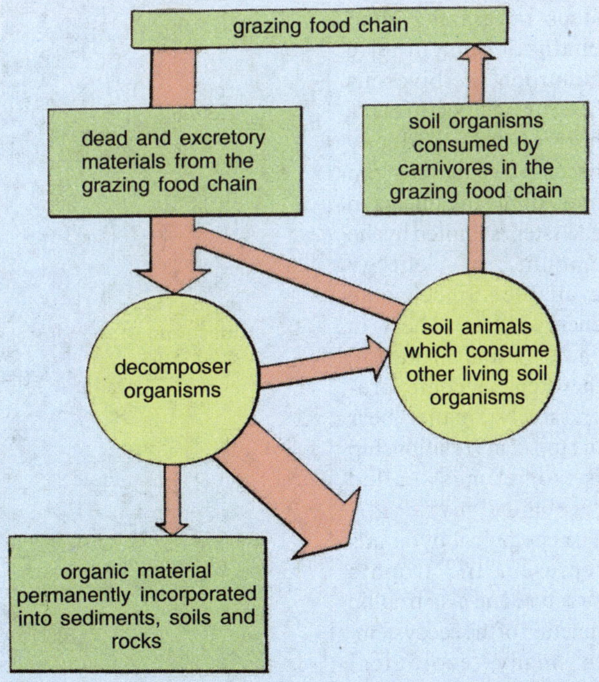

Fig. 9.5. Diagrammatic representation of the detritus food chain, showing energy transfers between it and the grazing food chain, as well as energy losses to the detritus food chain, (after Clapham, Jr., 1973).

In any complex food web, one can recognize several different trophic levels :

1. **Producers**	Green plants	First trophic level
2. **Primary consumers**	Herbivores	Second trophic level
3. **Secondary consumers**	Carnivores, insectivores	Third trophic level
4. **Tertiary consumers**	Higher carnivores, insect hyperparasites	Fourth trophic level

A classification of organisms by trophic levels is one of function and not of species as such. A given species may occupy more than one trophic level. The complexity of food web can vary greatly, and we express this complexity by a measure called the **connectance** of the food web :

$$\text{Connectance} = \frac{\text{Actual number of interspecific interactions}}{\text{Potential number of interspecific interactions}}$$

Generalizations about food web. Generally, food webs are not too complex. As more and more species are involved in a food web, the connectance falls. Except in insect communities, omnivores are scarce, and when they occur, they usually feed on species in adjacent trophic levels. Within habitats, food webs are rarely broken up into discrete compartments. The number of species of predators in a food web typically exceeds the number of species of prey by an average of 1.3 predator species per prey species.

ECOLOGICAL PYRAMIDS

In the successive steps of grazing food chain–photosynthetic autotroph, herbivorous heterotroph, carnivores heterotroph, decay bacteria–the number and mass of the organisms in each step is limited by the amount of energy available. Since some energy is lost as heat, in each transformation the steps become progressively smaller near the top. This relationship is sometimes called "**ecological pyramid**". The ecological pyramids represent the trophic structure and also trophic function of the ecosystem. In many ecological pyramids, the producer form the base and the successive trophic levels make up the apex.

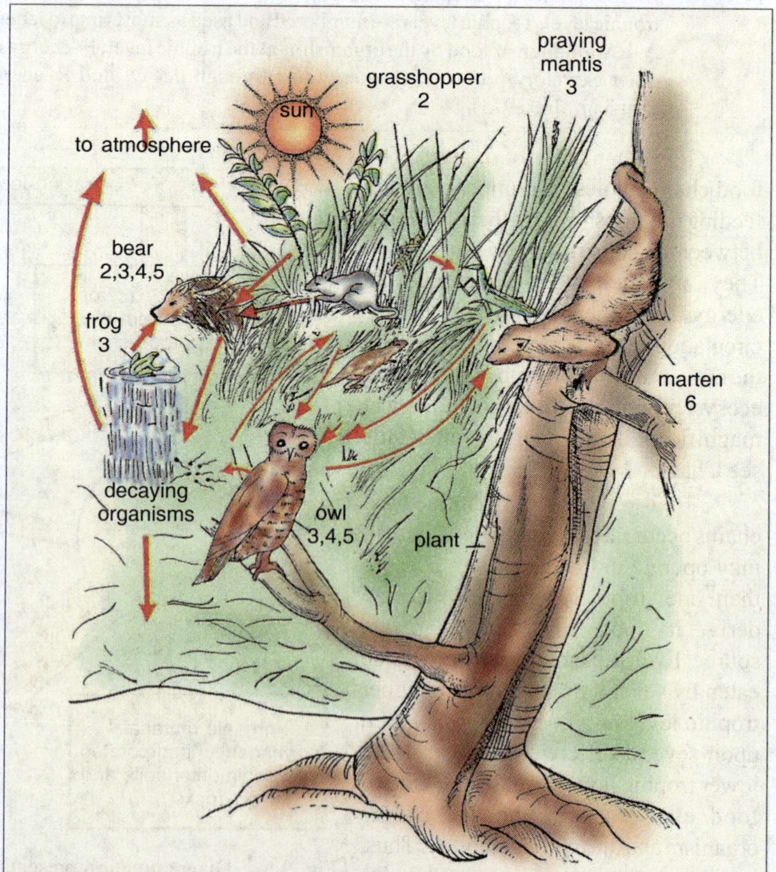

Fig. 9.6. Food web showing 1–6 trophic levels : 1–Plants ; 2–Grasshopper, mouse and bear ; 3–Praying mantis, frog, owl and bear ; 4–Owl, shrew and bear ; 5–Owl and bear ; 6–Marten.

Thus, communities of terrestrial ecosystems and shallow water ecosystems contain gradually sloping ecological pyramids because these producers remain large and characterized by an accumulation of organic matter. This trend, however, does not hold for all ecosystems. In such aquatic ecosystems as lakes and open sea, primary production is concentrated in the microscopic algae. These algae have a short-cycle, multiply rapidly, accumulate little organic matter and are heavily exploited by herbivorous zooplankton. At any one point in time the standing crop is low. As a result, the pyramid of biomass for these aquatic ecosystems is inverted: the base is much smaller than the structure it supports.

Types of Ecological Pyramids

The ecological pyramids may be of following three kinds :

1. Pyramid of number. It depicts the number of individual organisms at different trophic levels of food chain. This pyramid was advanced by **Charles Elton** (1927), who pointed out the great difference in the number of the organisms involved in each step of the food chain. The animals at the lower end (base of pyramid) of the chain are the most abundant. Successive links of carnivores decrease rapidly in number until there are very few carnivores at the top. The pyramid of number ignores the biomass of organisms and it also does not indicate the energy transferred or the use of energy by the groups involved. The lake ecosystem provides a typical example for pyramid of number.

2. Pyramid of biomass. The biomass of the members of the food chain present at any one time forms the pyramid of the biomass. Pyramid of biomass indicates decrease of biomass in each trophical level from base to apex. For example, the total biomass of the producers ingested by herbivores is more than the total biomass of the herbivores in an ecosystem. Likewise, the total biomass of the primary carnivores (or secondary consumer) will be less than the herbivores and so on.

3. Pyramid of energy. When production is considered in terms of energy, the pyramid indicates not only the amount of energy flow at each level, but more important, the actual role the various organisms play in the transfer of energy. The base upon which the pyramid of energy is constructed is the quantity of organisms produced per unit

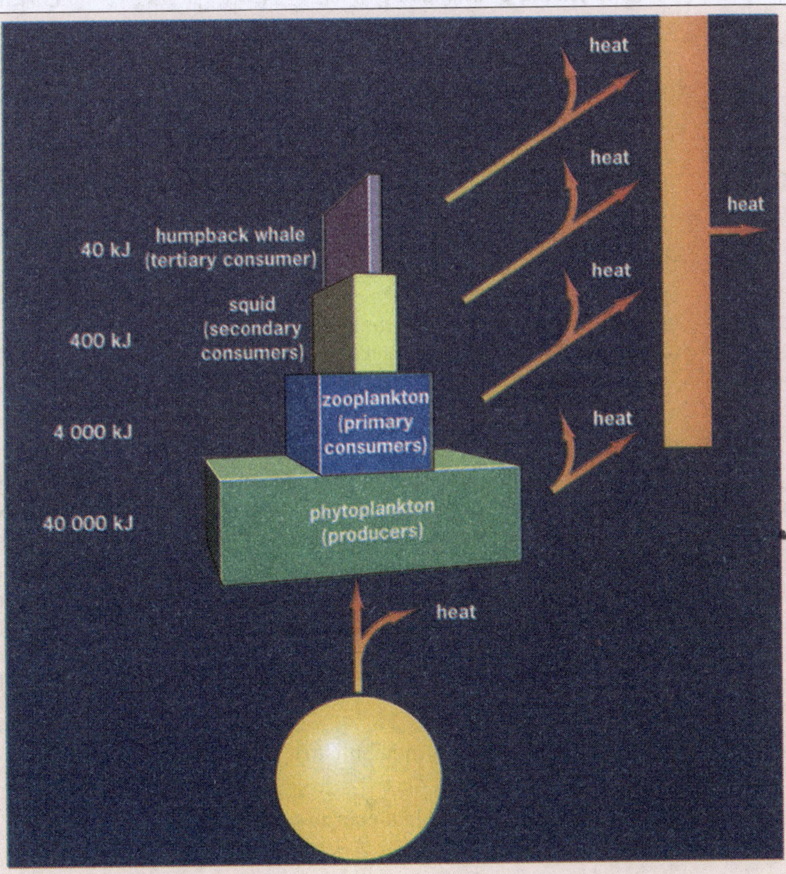

Pyramid of energy.

time, or in other words, the rate at which food material passes through the food chain. Some organisms may have a small biomass, but the total energy they assimilate and pass on, may be considerably greater than that of organisms with a much larger biomass. Energy pyramids are always slopping because less energy is transferred from each level than was paid into it. In cases such as in open water communities the producers have less bulk than consumers but the energy they store and pass on must be greater than that of the next level. Otherwise the biomass that producers support could not be greater than that of the producers themselves. This high energy flow is maintained by a rapid turn over of individual plankton, rather than an increase of total mass.

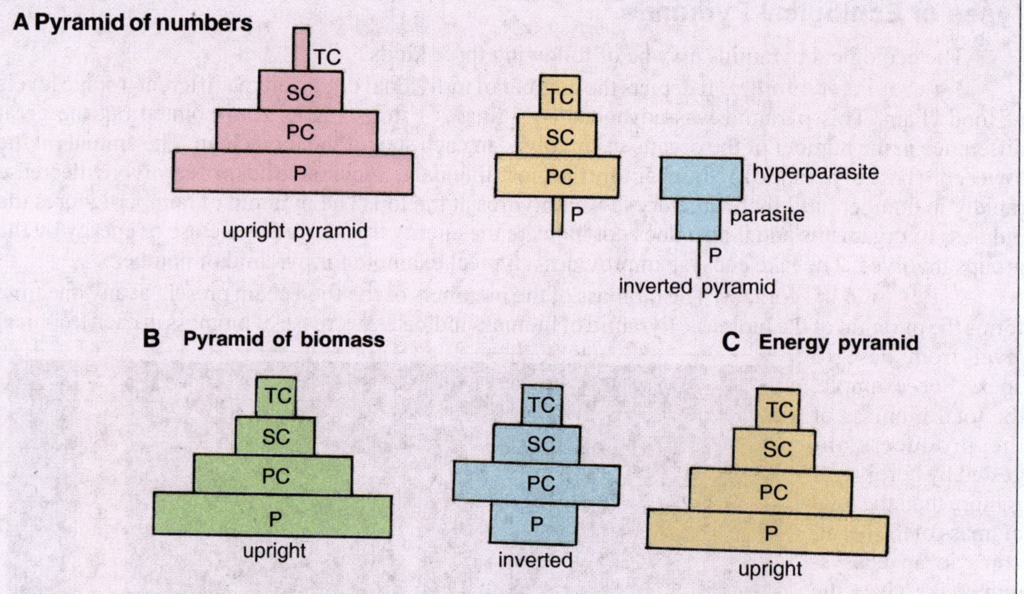

Fig. 9.7. Ecological and Eltonian pyramids : pyramid of numbers can be both upright and inverted, pyramid of biomass can also be both upright and inverted but the pyramid of energy is always upright (P=producers; PC=primary consumers (herbivores) ; SC=secondary consumers (carnivores) ; TC=tertiary consumers (carnivores).

ENERGY FLOW IN ECOSYSTEMS

1. Concept of energy. Energy is the capacity to do work. Biological activities require consumption of energy which ultimately comes from the sun. Radiant energy of sun (or **solar energy**) is transformed into **chemical energy** by the process of photosynthesis—this is stored in plant tissues and then transformed into **mechanical** and **heat** form of energy during metabolic activities. In the biological world, the energy flows from the sun to plants and then to all heterotrophic organisms, such as microorganisms, animals and man in the following manner :

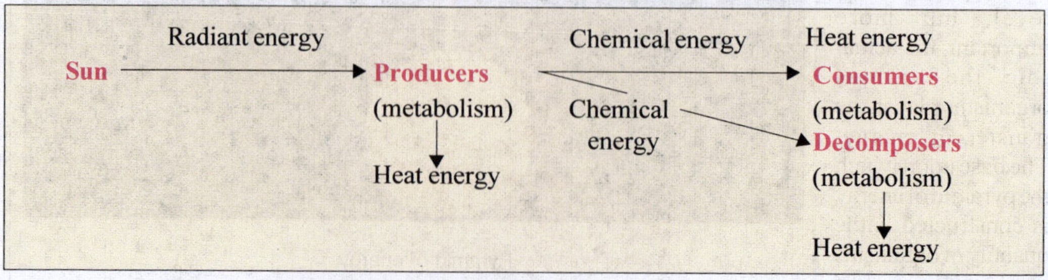

Mechanical energy has two forms, namely kinetic or free energy and potential energy. The energy a body possess by virtue of its motion is called **kinetic energy**, and is measured by the amount of work done in bringing the body at rest. **Potential energy** is stored energy (the energy at rest) and becomes useful after conversion into kinetic energy. All organisms require a source of potential energy, which is found in the chemical energy of food. The oxidation of food releases energy which is used to do work. Thus, chemical energy is converted into mechanical energy. Food means material containing energy that organism can use. Food is the means to transfer of both matter and energy in the living world. Plants synthesize food with the help of solar energy and inorganic substances such as nutrients, CO_2 and H_2O in a biochemical process called **photosynthesis** :

$$6CO_2 + 6H_2O \xrightarrow[\text{Enzymes, etc.}]{\text{Solar energy}} C_6H_{12}O_6 + 6O_2$$

(Chlorophyll) Glucose; food

2. Unit of energy. The unit of measurement of energy is **erg**; the work done in lifting 1 gram of weight to a height of 1 cm against the force of gravity is equal to 981 ergs. One crore ergs (10^7 ergs) is equal to one **Joule**. All forms of energy can be completely converted into heat energy. For a better and uniform expression in ecology, therefore, energy is measured not in terms of ergs but joules or units of heat measurement. Heat is measured in calories. One calorie is equal to the heat energy required to raise the temperature of 1 gram of water from 14.5°C to 15.5°C, and one calorie is equal to 4.2 joules or 4.2×10^7 ergs. One thousand calories (10^3) makes one kilo calories or a kilogram calories (Kcal or Cal). Now, there is a trend of expressing energy in ecological literature in terms of kilojoules.

3. Ecological energetics. Ecological energetics includes energy transformation which occur within ecosystems. In ecological energetics, we consider (i) quantity of energy reaching an ecosystem per unit of area (say a square metre) per unit of time (say one hour, day or year); (ii) quantity of energy trapped by green plants and converted to a chemical form (photosynthesis) and (iii) the quantity and path of energy flow from green plants to organisms of different trophic levels over a period of time in a known area (*i.e.*, energy flow from producers to consumers).

Only about 0.02% of the sunlight reaching the atmosphere is used in the process of photosynthesis.

The energy used for all plant life processes is derived from solar radiations. A fraction, *i.e.*, about 1/50 millionth of the total solar radiation reaches the earth's atmosphere. Solar radiation travels through the space in the form of waves, their wavelength ranging from 0.03 A° to several kilometres. While most radiations are lost in space, those ranging from 300 mμ to 10 mμ and above 1 cm (radiowaves) enter the earth's outer atmosphere (which is about 28 km altitude). The energy reaching the earth's surface consists mainly of visible light (390–760 mμ) and infrared component. On a clear day radiant energy reaching the earth's surface is about 10% UV, 45% visible and 45% infra-red. Green plants absorb strongly the blue and red light (400 to 500 mμ and 600 to 700 mμ respectively).

About 34% of the sunlight reaching the earth's atmosphere is reflected back (by clouds and dust), 10% is held by ozone layer, water vapour and other atmospheric gases. The rest, 56% reaches the earth's surface. Only a fraction of the energy reaching earth's surface (1 to 5%) is used by green plants for photosynthesis and rest is absorbed as heat by ground vegetation or water. In fact, only about 0.02% of the sunlight reaching the atmosphere is used in the process of photosynthesis.

The amount of radiant energy of all wavelengths that cross unit area per unit time is called **solar flux**. The solar flux is about 8.368 J (2 cal)/cm^2 min. At a given place, it varies diurnally because of the earth's rotation on its axis.

4. Laws governing energy transformation. Energy transformation in ecosystems can also be explained in relation to the laws of thermodynamics, which are usefully applied to closed systems. The **first law of thermodynamics** is the law of conservation of energy, which says that *energy may be transformed from one form into another but is neither created nor destroyed.* If an increase or decrease occurs in the internal energy (E) of the system itself, work (W) is done and heat (Q) is either evolved or absorbed. Thus,

ΔE	=	W	+	Q
Decrease in the internal energy of the system		Work done by the system		Heat given off by the system

The sign Δ refers to a change in quantity.

The total amount of heat produced or absorbed in a chemical reaction, either occurring directly or in stages, always remains the same. This is called the **specific law of constant heat sums** and included in the first law.

For example, combustion (a direct chemical reaction) of food release the following amount of energy :

$$C_6H_{12}O_6 + 6\,O_2 \rightarrow 6\,H_2O + 6\,CO_2 + 673 \text{ kcal } (2816 \text{ kJ})$$

The fermentation is a two-step reaction, but releases the same amount of energy :

1. $C_6H_{12}O_6 \rightarrow 2\,C_2H_5OH + 2\,CO_2 + 18 \text{ kcal } (75 \text{ kJ})$
2. $2\,C_2H_5OH + 6\,O_2 \rightarrow 6\,H_2O + 4\,CO_2 + 655 \text{ kcal } (2740 \text{ kJ})$

(1+2) $C_6H_{12}O_6 + 6O_2 \rightarrow 6H_2O + 6CO_2 + 673 \text{ kcal } (2815 \text{ kJ})$

This law explains the interconvertibility of all forms of energy but does not refer to the efficiency of transformation or conversion. In ecological systems solar energy is converted into chemical energy stored in food materials which is ultimately converted into mechanical and heat energy. Thus, in ecological systems, the energy is neither created nor destroyed but is converted from one form into another. Thus, when wood is burned the potential energy present in the molecules of wood equals the kinetic energy released, and heat is evolved to the surroundings. This is an **exothermic reaction**. In an **endothermic reaction**, energy from the surrounding may be paid into a reaction. For example, in photosynthesis, the molecules of the products store more energy than the reactants. The extra energy is acquired from the sunlight, but even then there is no gain or loss in total energy.

The second law of thermodynamics states that *processes involving energy transformation will not occur spontaneously unless there is degradation of energy from a non-random to a random form.* In man-made machines (closed systems), heat is the simplest and most familiar medium of energy transfer. But in biological systems, it is not a useful medium of energy transfer, as living systems are essentially isothermal and there are no significant differences in temperature between different parts of a cell or between different cells in a tissue.

5. Concept of free energy, enthalpy and entropy. Free energy is that component of the total energy of a system which can do work under isothermal conditions. All physical and chemical processes proceed with a decline in free energy until they reach an equilibrium where the free energy of the system is at a minimum.

$$\Delta G = \Delta H - T\Delta S$$

Where ΔG = Change in the free energy of the system

ΔH = Change in **enthalpy** (*i.e.,* a change in the amount of energy in the form of heat liberated or absorbed by the system during physical or chemical changes)

ΔS = Entropy change of the system (**Entropy** is the name of a quantity in thermodynamics representing the degree of disorder in a physical system or the extent to which the energy in a system is available for doing work)

T = Absolute temperature

Thus, a decline in G is accompanied by an increase in $T\Delta S$. These are equal if there is no heat transfer between the system and the surrounding. If a reaction proceeds with a decline in free energy, we call it spontaneous.

6. Lindeman's trophic-dynamic concept. As **Lindeman** (1942) pointed out, the amount of energy reaching each trophic level is determined by the net primary production (NPP) and the efficiency with which food energy is converted to biomass energy within each trophic step. Of the light energy assimilated by plants, 15 to 70 per cent is used for maintenance and, therefore, is unavailable to consumers. Most herbivores and carnivores are more active than plants and spend correspondingly more of their assimilated energy on maintenance. As a result productivity of each trophic level is usually no more than 5 to 20 per cent that of the level below it. The percentage transfer of energy from one trophic level to the next is called both the ecological efficiency and food chain efficiency.

7. Maintenance cost of secondary producers. In general, 55 to 75% of the assimilated energy is spent on maintenance of secondary producers. Temperature, moisture conditions of the habitat and the type of species determine the maintenance cost. The drier and hotter the habitat, the higher is the maintenance cost, irrespective of species. For example, the average maintenance cost of some Indian earthworm species was found to be 6.48, 9.96 and 20.54 kJ/g dry tissue/month in the winter, rainy and summer seasons respectively in tropical pastures (**Dash**, 1987). The maintenance cost varies seasonally and is three times more in summer than in winter.

8. Assimilated energy and respiration energy. Once food is eaten, its energy follow a variety of paths through the organisms (Fig. 9.8). Regardless of an organism's source of food, what it digests and absorbs is referred to as **assimilated energy**, which supports maintenance, builds tissues, or is excreted in unusable metabolic byproducts. The energy used to fulfill metabolic needs, most of which is lost as heat, is known as **respired energy**. A smaller fraction of assimilated energy is excreted in the form of

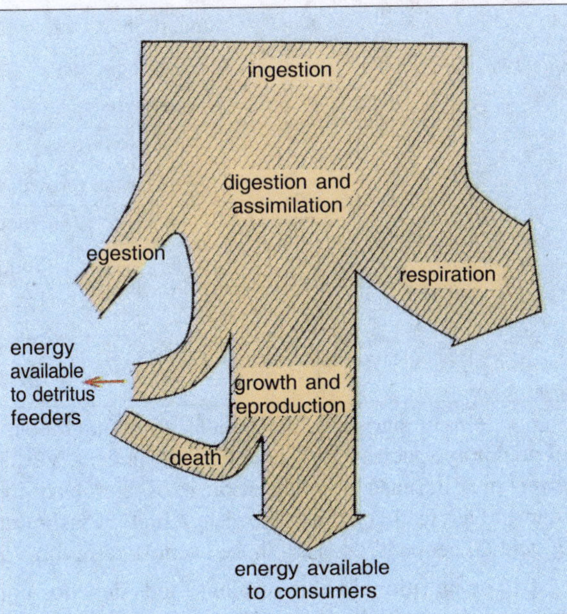

Fig. 9.8. Partitioning of energy within a link of the food chain.

organic, nitrogen –containing wastes (primarily ammonia, urea, or uric acid) produced when the diet contains an excess of nitrogen. Assimilated energy retained by the individual organism is available for the synthesis of new biomass (**production**) through growth and reproduction, which may then be consumed by herbivores, carnivores, and detritivores.

Further, not all food can be fully assimilated; hair, feathers, insect exoskeletons, cartilage and bone in animal foods, and cellulose and lignin in plant foods resist digestion by most animals. These materials are egested either by defecation or by regurgitation of pellets of undigested remains. Some egested wastes are substances that have been relatively unaltered chemically during their passage through an organism, but nearly all have been mechanically broken up into fragments by chewing and by contractions of the stomach and intestines, which makes them more readily usable by detritus feeders (**Edwards** and **Heath**, 1963).

9. Ecological efficiency. Ecological efficiency is the product of efficiencies with which organisms exploit their food resources and transform them into biomass which becomes available to the next higher trophic level. Because most biological production is consumed, **exploitation efficiency** is 100 per cent overall, and ecological efficiency depends on two factors : the proportion of assimilated energy incorporated in growth, storage and reproduction. The first proportion is called the **assimilation efficiency** and the second, the **net production efficiency** (Table 9-1). The product of the assimilation and net production efficiencies is the **gross production efficiency** : the proportion of food energy that is transformed into consumer biomass energy.

Net production efficiency for the plants is defined as the ratio of net to gross production. This index has been found to vary between 30 and 85 per cent, depending on habitat and growth form. Rapidly growing plants in temperate zones, whether trees, old-field herbs, crop species, or aquatic plants, have constantly high net production efficiencies (75 to 85 per cent). Similar vegetation types in the tropics exhibit lower net production efficiencies, perhaps 40 to 60 per cent respiration increases relative to photosynthesis at low latitudes.

Table 9-1. **Definition of several energetic efficiencies.**

1. Exploitation efficiency = $\dfrac{\text{Ingestion of food}}{\text{Prey production}}$

2. Assimilation efficiency = $\dfrac{\text{Assimilation}}{\text{Ingestion}}$

3. Net production efficiency = $\dfrac{\text{Production (growth and reproduction)}}{\text{Assimilation}}$

4. Gross production efficiency = (2) X (3) = $\dfrac{\text{Production}}{\text{Ingestion}}$

5. Ecological efficiency = (1) X (2) X (3) = $\dfrac{\text{Consumer production}}{\text{Prey production}}$

In comparison to the plant food, animal food is more easily digested. Assimilation efficiency of predatory species vary from 60 to 90 per cent. Vertebrates prey are digested more efficiently than insect prey because the indigestible exoskeletons of insects constitute the larger proportion of body than the hair, feathers and scales of vertebrates. Assimilation efficiencies of insectivorous vary between 70 and 80 per cent, whereas those of most carnivores are about 90 per cent.

The nutritional value of plant foods depends upon the amount of cellulose, lignin, and other indigestible materials present. Herbivores assimilate as much as 80 per cent of the energy in seeds, and 60 to 70 per cent of that in young vegetation (**Chew** and **Chew**, 1970). Most grazers and browsers (*e.g.*, cattle, elephants and grasshoppers) assimilate 30 to 40 per cent of the energy in their food. Millipedes, which eat decaying wood composed mostly of cellulose and lignin (and the microorganisms that occur in decaying wood), assimilate only 15 per cent (**O'Neil**, 1968).

Maintenance, movement and in warm-blooded animals, heat production require energy that otherwise could be utilized for growth and reproduction. Active warm-blooded animals (homeotherms) exhibit low net production efficiency—birds less than 1 per cent, small mammals with high reproductive rates up to 6 per cent. More sedentary, cold-blooded animals (poikilotherms), particularly aquatic species, channel as much as 75 per cent of their assimilated energy into growth and reproduction (**Welch**, 1968). The extreme high value approaches the biochemical efficiency of egg production and tissue growth, between 70 to 80 per cent in domesticated animals (**Ricklefs**, 1974).

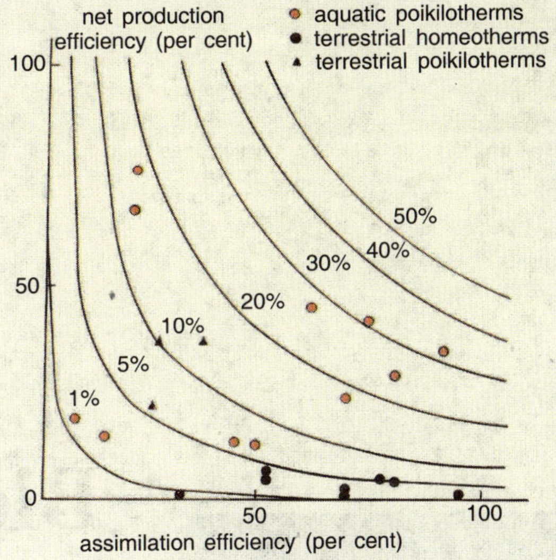

Fig. 9.9. Relationship between assimilation efficiency and net production efficiency for a variety of animals. Gross production efficiencies are indicated by the curved lines on the graph (after Ricklefs, 1990).

The **gross production efficiency** (*i.e.,* biomass production efficiency within a trophic level) is the product of assimilation efficiency and net production efficiency. Gross production efficiencies of warm-blooded terrestrial animals rarely exceed 5 per cent, and those of some birds and large mammals fall below 1 per cent (**Turner**, 1970). Gross production efficiencies of insects lie within the range of 5 to 15 per cent, and those of some aquatic animals exceed 30 per cent (Fig. 9.9).

REVISION QUESTIONS

1. Describe the structure of an ecosystem by taking an example of pond ecosystem.
2. Describe the function of ecosystem.
3. Into how many types the ecosystems can be classified ?
4. How does energy flow in an ecosystem ?
5. Describe different kinds of ecological pyramids.
6. What is food chain ? Describe different food chains with examples.
7. What do you mean by ecological efficiency ? Discuss it briefly.

CHAPTER

10

Biogeochemical Cycles

There is a constant movement of chemical elements of the biosphere between the organism and the environment.

The total mass of all the organisms that have lived on the earth in the past 1.5 billion years is much greater than the mass of carbon and nitrogen atoms present. According to the **law of conservation of matter**, matter is neither created nor destroyed; obviously the carbon and nitrogen must have been used over and over again in the course of time. The earth neither receives any great amount of matter from other parts of the universe nor

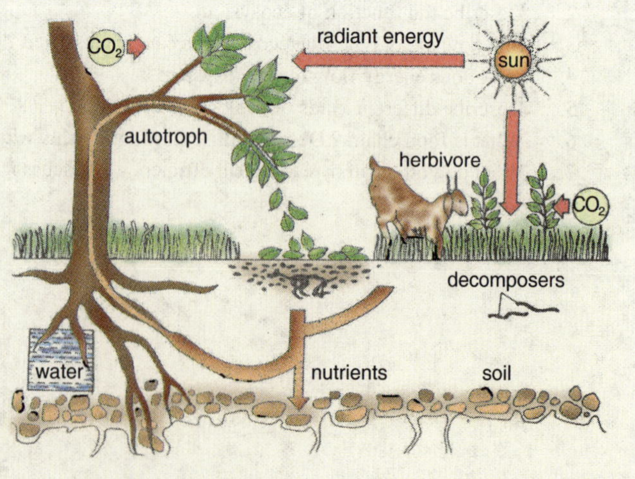

Fig. 10.1. Biogeochemical or nutrient cycle in a forest ecosystem.

does it lose significant amount of matter to outer space. The atoms of each element as carbon, hydrogen, oxygen, nitrogen, phosphorus, calcium, and the rest are taken from the environment, made a part of some cellular component of an organism and finally, perhaps by a quite circuitous route involving several other organisms, are returned to the environment to be used over again. The cyclic movements of chemical elements of the biosphere between the organism and the environment are referred to as **biogeochemical cycles**, after **Vernadsky** (1934). "*Bio*" refers to living organisms and "*geo*" to the rocks, soil, air and water of the earth.

Organic and Abiotic Phases of Geochemical Cycles

Since an element is necessary for the maintenance of life, its movement through biotic communities (organisms) can be viewed in the terms of food chain. The flow of a chemical element through the food chain can be viewed as the **organic phase** of the biogeochemical cycle. Further, the biogeochemical cycles also include **abiotic phases** which are the functions of the chemistry of the elements in question. These abiotic phases are of critical importance to the ecosystem, as the major reservoirs for all nutrient elements are external to the food chains, and flow in the abiotic phases tends to be much slower than in the organic phase. The rapidity and direction of nutrient cycling through the abiotic phases determine not only the distribution of the element in the total environment, but also its availability of living systems. There are two classes of abiotic phases in biogeochemical cycles, a **sedimentary phase**, which is part of all cycles, and an **atmospheric phase**, which is possessed by some. In some cycles, such as nitrogen, the atmospheric phase is more important than the sedimentary. In others, such as phosphorus, the atmospheric phase is essentially non-existent. In still others, such as sulphur, both phases are present and their relative importance depends on other environmental factors. Biogeochemical cycles that have dominant atmospheric phases are often called **atmosphere-reservoir cycles**; those whose sedimentary phase is dominant are termed **sediment-reservoir cycles**.

TYPES OF BIOGEOCHEMICAL CYCLES

There are two types of biogeochemical cycles, the **gaseous** and the **sedimentary**. In gaseous cycles the main reservoir of nutrients is the atmosphere and the ocean. In sedimentary cycles the main reservoir is the soil and the sedimentary and other rocks of the earth's crust. Both involve biotic and abiotic agents, both are driven by the flow of energy and both are tied to the **water cycle**.

A. Water Cycle

Living organisms, atmosphere and earth maintain between them a circulation of water and moisture, which is referred to as **water cycle** or **hydrologic cycle**. As we have already discussed in Chapter 3, water forms a very significant factor of environment and without the cycling of water, biogeochemical cycles could not exist, ecosystems could not function, and life could not be maintained. Water is important for an ecosystem for several reasons— it is the medium by which nutrients are introduced into autotrophic plants; it is an important part of living tissue, either as liquid water or as part of essential organic molecules; it serves as a means of thermal regulation for both plants and animals; it is the medium by which sediments — a prime source of mineral nutrients— are removed from or added to local ecosystem; it covers the great majority of the earth's surface, and is the dominant feature of all aquatic ecosystems.

The hydrologic cycle is driven by solar energy and gravity. More than 80 per cent of the total insolation that is not lost immediately as electromagnetic radiation goes to **evaporate** water. The atmospheric water vapour produced by this means can then condense around particles of dust in the atmosphere, often called **nucleation particles**. The atmosphere possesses a limited capacity for holding water vapour, thus, the droplets formed by this means are heavy enough to fall as precipitation, under the influence of gravity. Eventually, the hydrologic cycle can be defined as an alteration of evaporation and precipitation, with the energy used to evaporate the water being dissipated as heat in the atmosphere as the water condenses.

Distribution of water in earth's surface. Water is not evenly distributed throughout the earth. Almost 95 per cent of the total water on earth is chemically bound into rocks and does not cycle. Of the remainder, about 97.3 percent is in the ocean, about 2.1 per cent exists as ice in the polar caps and permanent glaciers, and the rest is fresh water, present in the form of atmospheric water vapour, ground water, soil water, or inland surface water (Nace, 1967).

Table 10-1. Distribution of water in the earth's crust and surface (Clapham, Jr., 1973).

A. Chemically bound water of rocks : Does not cycle	
1. Crystalline rocks	$250{,}000 \times 10^{17}$ Kg.
2. Sedimentary rocks	$21{,}00 \times 10^{17}$ Kg.
B. Free water : Moves via hydrologic cycle	
1. Oceans	$13{,}200 \times 10^{17}$ Kg.
2. Ice caps and glaciers	292×10^{17} Kg.
3. Ground water to a depth of 4000 m	83.5×10^{17} Kg.
4. Freshwater lakes	1.25×10^{17} Kg.
5. Saline lakes and inland seas	1.04×10^{17} Kg.
6. Soil moisture	0.67×10^{17} Kg.
7. Atmospheric water vapour	0.13×10^{17} Kg.
8. Rivers	0.013×10^{17} Kg.

The rate of cycling of water. The rate of cycling between surface and atmosphere is very rapid. The amount of water vapour in the atmosphere is sufficient, on the average, to cover the entire earth to a depth of 2.25 cm. But the average annual rainfall for the earth is about 81.1 cm (Furon, 1967) and in some places it ranges up to 1200 cm. This means that the average turnover time for atmospheric water is about 11.4 days, or that the equivalent of all the water vapour in the entire atmosphere falls as precipitation and is re-evaporated more than 32 times per year.

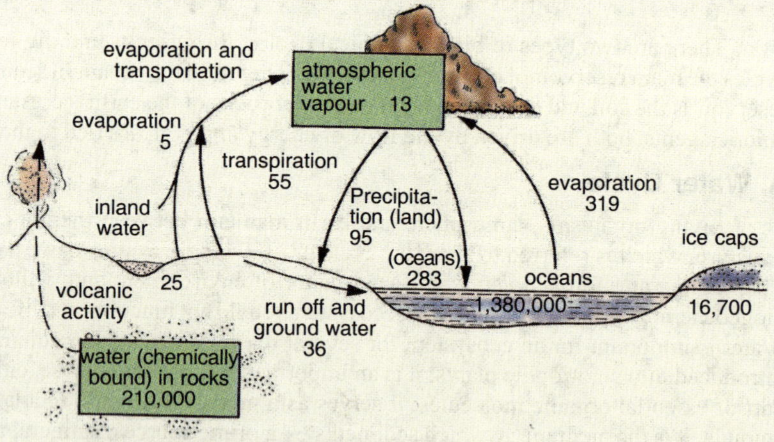

Fig. 10.2. Global water or hydrological cycle. Most of the storage of water is in the oceans but most of the flux is to and from the atmosphere units: stores, 10^{18}g : fluxes, 10^{18} g.

Further, the distribution of evaporation and rainfall is quite uneven. If we compare the annual evaporation and precipitation over land and sea, we find that relatively more water precipitates on land than evaporates from land. This is fortunate from the point of view of terrestrial organisms. Even so, the amount of rainfall on the open ocean is proportionately greater than that on land, taking into account the relative percentages of the earth's surface covered by land and sea.

Nature of hydrologic cycles. The hydrologic cycle over the oceans is extremely simple—the water is evaporated from the surface of the ocean and water vapours form the clouds which when cool down precipitate the water as rain fall. But several routes are open to precipitation that falls on land—direct evaporation, transpiration, entry of water into ground water system and runoff. Consequently, the routes of hydrologic cycles on land can be divided into following three main categories–the rapidly cycling portion, or **evapotranspiration**, which includes the evaporation and transpiration, the less rapidly cycling water, or surface runoff, and very slowly cycling ground water that seeps into the soil can end up in any one of these three categories.

1. Evapotranspiration. Evapotranspiration includes evaporation and transpiration. **Evaporation** refers to water that is evaporated directly from any surface other than a plant, such as a lake, soil surface, or animal skin. In most cases, the main effects of direct evaporation are to moderate the temperature of local area and to allow the hydrologic cycle to continue. In some ecosystems, evaporation also leads to a concentration of salts in the water of soil which may be a critical environmental factor. **Transpiration** is water that evaporates from the surface of leaves of plants. Transpiration acts to move the biogeochemical cycles for all mineral nutrients that enter the food chain via the roots of plants.

2. Surface runoff. If transpiration is related to the mechanism of nutrient uptake, the gross movement of soluble and solid particles in the ecosystem is accomplished largely by runoff. Nutrients that have accumulated in sediments or soils can be eroded by streams and removed altogether from a local ecosystem, or soluble nutrients may be carried by soil seepage into surface waters, where they are removed from the area. Streams may carry sediment particles which can be chemically altered through additional weathering so that the nutrient elements they contain may be utilized by organisms. Finally moving water acts as an agent of **erosion** which removes soil and allows weathering of the underlying rock to make their nutrients available to plants.

3. Ground water. Ground water is water that saturates either sediment or rock below the water table. In general, it is not trapped by plants for transpiration and it is too deep to be directly evaporated from the soil surface. It is an exceedingly important reservoir for water which moves from one place to another under the influence of gravity. The area where the net water movement is from the surface into the ground water systems is termed a **catchment area**; areas where ground water reaches the surface and runs off are termed **springs**. A rock body through which ground water flows is called an **aquifer**. A well drilled into an aquifer that has sufficient hydrostatic pressure to force water up into it is called an **artesian well**.

The hydrologic cycle on land, thus, includes evapotranspiration of water from earth's surface and leaf surface → formation of clouds → precipitation → surface runoff + accumulation of water as ground water → return of water to sea via streams or direct evaporation and cloud formation, and so on.

B. Gaseous Cycles

The gaseous geochemical cycles are of following types :

1. The oxygen cycle. Oxygen (O_2), the by-product of photosynthesis, is involved in the oxidation of carbohydrates with release of energy, carbon dioxide and water. Its primary role in biological oxidation is that of a hydrogen acceptor. The break-down and decomposition of organic molecules proceeds primarily by dehydrogenation. Hydrogen is removed by enzymatic activity from organic molecules in a series of reactions and is finally accepted by the oxygen, forming water. Though oxygen is necessary for life, but being very active chemically, molecular O_2 may be toxic to living

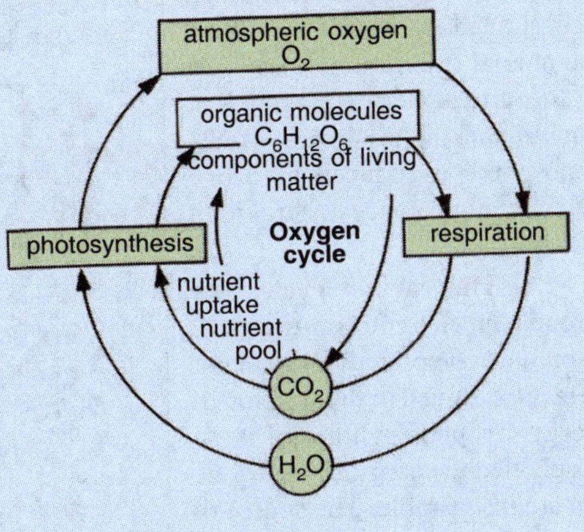

Fig. 10.3. The oxygen cycle.

body cells. Therefore, for the protection from toxic effects of molecular O_2, cells possess the cellular organelles called **peroxisomes** which mediate oxidative reactions resulting in the production of hydrogen peroxide which in turn is used through the mediation of other enzymes as an acceptor in oxidizing other compounds.

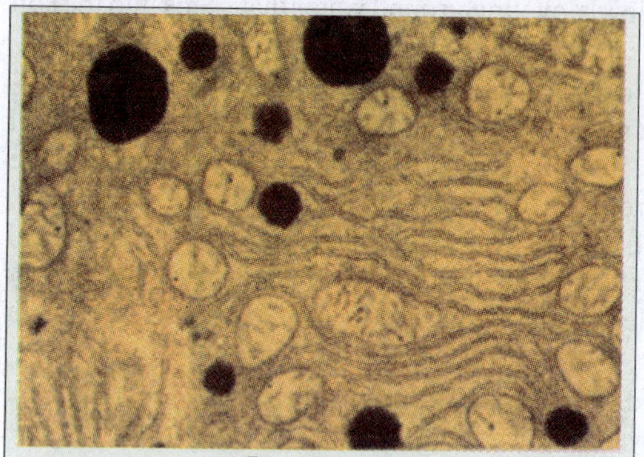

Peroxisomes.

The major supply of free oxygen which supports life occurs in the atmosphere. There are two significant sources of atmospheric oxygen. One is the **photodissociation** of water vapour in which most of the hydrogen released escapes into outer space. The other source is **photosynthesis**, active only since life began on earth. Because photosynthesis and respiration are cyclic, involving both the release and utilization of oxygen, one would seem to balance the other, and no significant quantity of oxygen would accumulate in the atmosphere. However, at some time in the earth's history the amount of oxygen introduced into the atmosphere had to exceed the amount used in the decay of organic matter and that tied up in the oxidation of sedimentary rocks. Part of the atmospheric oxygen represents that portion remaining from the unoxidized reserves of photosynthesis–coal, oil, gas, and organic carbon in sedimentary rocks. The amount of stored carbon in the earth suggests that 150×10^{20}g of oxygen has been available to the atmosphere, over 10 times as much as now present, 10×10^{20}g (**Johnson**, 1970). The main non-living (abiotic) oxygen pool consists of molecular oxygen, water, and carbon dioxide, all intimately linked to each other in photosynthesis and other oxidation-reduction reactions, and all exchangeable in such compounds as nitrates and sulphates utilized by organisms that reduce them to ammonia and hydrogen sulphide.

The cycling of oxygen is very complex. As a constituent of CO_2, it circulates freely throughout the biosphere. Some carbon dioxide combines with calcium to form carbonates. Oxygen combines with nitrogen compounds to form nitrates, with iron to ferric oxides, and with many other minerals to form various other oxides. In these states oxygen is temporarily withdrawn from circulation. In photosynthesis the oxygen freed is split from the water molecule. This oxygen is then reconstituted into water during plant and animal respiration. Some part of the atmospheric oxygen that reaches the higher levels of the troposphere is reduced to ozone (O_3) by high energy ultraviolet radiation.

2. The carbon cycle. The carbon being a basic constituent of all organic compounds and a major element involved in the fixation of energy by photosynthesis, is so closely tied to energy flow that the two are inseparable. The source of all the fixed carbon both in living organisms and fossil deposits is

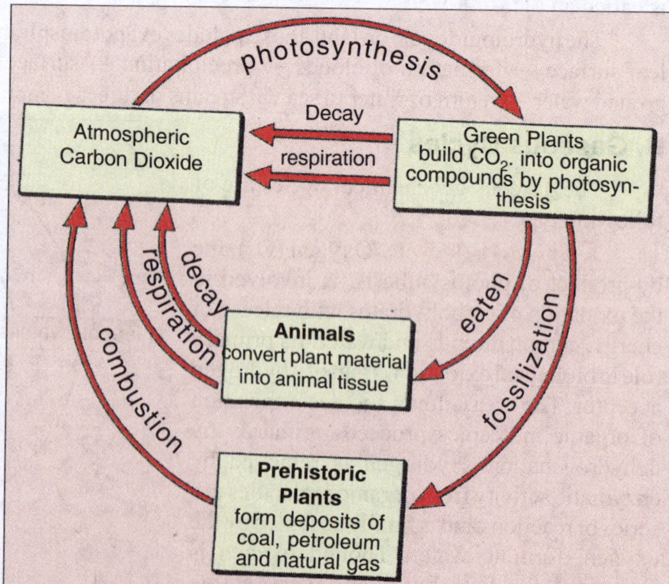

Fig. 10.4. The carbon cycle.

carbon dioxide CO_2, found in the atmosphere and dissolved in the waters of the earth. During photosynthesis, carbon from atmospheric CO_2 is incorporated into the production of the carbohydrate, glucose, $C_6H_{12}O_6$, that subsequently may be converted to other organic compounds such as polysaccharides (sucrose, starch, cellulose, etc.), proteins and lipids. All the polymeric organic compounds containing carbon are stored in different plant-tissues as food and from them the carbon is passed on to the trophic levels of herbivores or phytoparasites, or retained by the plant until it serves as food for decay organisms (*viz.,* decomposers). Some of the carbon is returned to the atmosphere (or the enveloping aqueous medium) in the form of CO_2 a by-product of plant respiration, in which, a considerable portion of glucose is oxidized to yield CO_2, H_2O and energy as follows :

$$1/6C_6H_{12}O_6 + O_2 \rightarrow CO_2 + H_2O + \text{Energy}$$

The CO_2 which is released as the by-product of plant respiration is again used by plants in photosynthesis. This is the basic carbon cycle which is simple and complete. Decomposing micro-organisms are important in breaking down dead material with the release of carbon back into the carbon cycle.

Similarly, carbon taken up by herbivores or phytoparasites may travel a number of routes. It may be incorporated into protoplasm (assimilation) and stored until the organism dies, where upon it is utilized by decomposers; it may be released through animal respiration; it may serve as live food for other organisms; or finally it may be stored in the environment as CO_2. Similar fates await carbon at the carnivore trophic levels. In fact, all the carbon of plants, herbivores, carnivores and decomposers is not respired but some is fermented and some is stored. The carbon compounds that are lost to the food chain after fermentation, such as methane, are readily oxidized to carbon dioxide by inorganic reactions in the atmosphere. As for the storage of carbon in sediments, just as deposition works to store materials, erosion may uncover them, and inorganic chemical weathering of rock can oxidize the carbon contained there. Some carbon is permanently stored in sediments and not uncovered by weathering; it may be replaced by carbon dioxide released from volcanoes and other similar examples of intense geological activity. In modern age, man has greatly increased the rate at which carbon is passing from sedimentary form to carbon dioxide. The combustion of fossil fuels is a significant means of recycling sedimentary carbon much faster than natural weathering.

Small portion of carbon, especially in the sea, is found not as organically fixed carbon, but as carbonate ($CO_3^{=}$), especially calcium carbonate ($CaCO_3$). $CaCO_3$ is very commonly used for shell construction by such animals as clams, oysters, some protozoa, and some algae. Carbon dioxide reacts with water to form carbonate in the following three step reaction :

$$\underset{}{CO_2 + H_2O} \rightleftarrows \underset{\text{Carbonic acid}}{H_2CO_3} \rightleftarrows \underset{\text{Bicarbonate}}{H^+ + HCO_3^-} \rightleftarrows \underset{\text{Carbonate}}{2H^+ + CO_3^{=}}$$

The precise amount of each of these constituents in the water depends on the pH of the water. Organisms such as clams can combine bicarbonate or carbonate with calcium dissolved in the water to produce calcium carbonate. After the death of the animal, this calcium carbonate may either dissolve or remain in sedimentary form.

Certain control mechanisms are inherent in the carbon cycle. The rate of carbon utilization is dependent on its availability. If excessive amounts of carbon are taken up in any one phase of the cycle, other phases of activity may be inhibited or slowed down. For example, if the pH of water is alkaline, more carbon is tied up in a carbonate and less is in solution. This removal of carbon in solution would upset the equilibrium established between the atmospheric and the dissolved CO_2 and the net effect would be a movement of CO_2 into solution until equilibrium was reached.

Peculiarities of carbon cycle. Though carbon-cycle exhibits basic similarity with other biogeochemical cycles, yet it is unusual in that the organic phase is not essentially a complete cycle within itself. The organic (biotic) and atmospheric (abiotic) phases, however, are so closely intertwined that the rapid cycling typical of the organic phase is present. The multiplicity of paths along which carbon can flow is typical of biogeochemical cycles in general, and provides a well-buffered system with adequate feedback mechanisms to insure an adequate supply of the carbon. It is significant that all phases of the cycle yield carbon dioxide at some time, and carbon dioxide is the raw material for them. Thus, despite its relative low concentration in the atmosphere (0.03 per cent), carbon in a form in which it can be used by living organisms is virtually always present.

3. The nitrogen cycle. Nitrogen is an essential constituent of different biologically significant organic molecules such as amino acids and proteins, pigments, nucleic acids and vitamins. It is also the major constituent of the atmosphere, comprising about 79 per cent of it. The paradox is that in its gaseous state, N_2 is abundant but is unavailable to most life. Before it can be utilized it must be converted to some chemically usable form.

To be used biologically, the free molecular nitrogen has to be fixed and fixation requires an input of energy. In the first step molecular nitrogen, N_2 has to be split into two atoms : $N_2 \rightarrow 2N$. The free nitrogen atoms then must be combined with hydrogen to form ammonia, with the release of some energy:

$$2N + 3H_2 \rightarrow 2NH_3$$

This fixation comes about in two ways. One is by high-energy fixation such as cosmic radiation, meteorite trails, and lightning that provide the high energy needed to combine nitrogen with oxygen and hydrogen of water. The resulting ammonia and nitrates are carried to the earth in rain water. The second method of nitrogen-fixation which contributes about 90 per cent of fixed nitrogen of earth, is biological. Some bacteria, fungi, and blue-green algae can extract molecular nitrogen from the atmosphere and combine it with hydrogen to form ammonia. Some of this ammonia is excreted by the nitrogen-fixing organism, and, thus, becomes directly available to other autotrophs. Some of these nitrogen-fixing organisms may be free-living, either in the soil (*e.g.,* bacteria — *Azotobacter* and *Clostridium*) or in water (*e.g.,* blue-green algae—*Nostoc, Calothrix* and *Anabaena*) and produce vast quantities of fixed nitrogen. In other cases, certain symbiotic bacteria of genus *Rhizobium*, although unable to fix atmospheric nitrogen themselves, can do this when in combination with cells either from the roots of legumes (*e.g.,* peas, beans, clover and alfalfa) and of other angiosperms such as *Alnus, Ceanothus, Shepherdia, Elaeagnus* and *Myrica*, or from the leaves of African genera of *Rubiaceae* and *Pavetta.* The bacteria invade the roots or leaves and stimulate the formation of root-nodules or leaf-nodules, a sort of harmless tumor. The combination of symbiotic bacteria and host cells remains able to fix atmospheric nitrogen and for this reason legumes are often planted to restore soil fertility by increasing the content of fixed nitrogen. Nodule bacteria may fix as much as 50 to 100 kilograms of nitrogen per acre per year, and free soil bacteria as much as 12 Kilograms per acre per year. Further both free soil bacteria (*Azotobacter* and *Clostridium*) produce ammonia as the first stable product and like

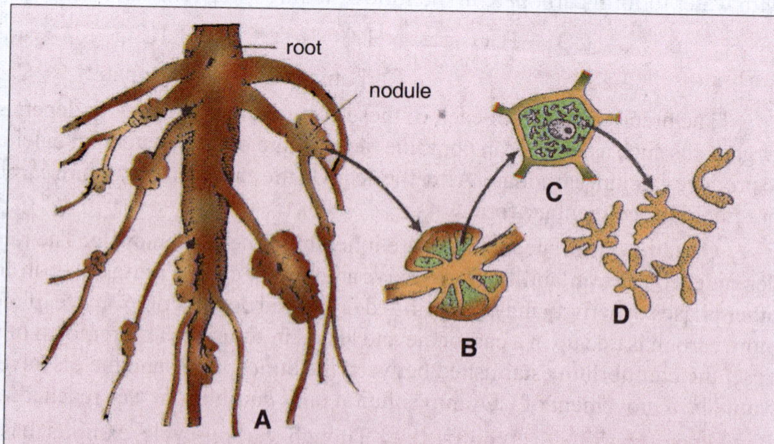

Fig. 10.5. Root-nodules of a legume plant. A—Legume root with root nodules. B—T.S. of root nodule. C—Single cell of nodule. D— Nitrogen-fixing bacterium, *Rhizobium.*

the symbiotic bacteria, they require molybdenum as an activator and are inhibited by an accumulation of nitrates and ammonia in soil.

Recently, certain lichens (*Collema tunaeforme* and *Peltigera rufescens*) were also implicated in nitrogen fixation (**Henriksson**, 1971). Lichens with nitrogen-fixing ability possess nitrogen-fixing blue green species as their algal component.

Nitrogen fixed by symbiotic and non-symbiotic microorganisms in soil and water is one source of nitrogen. Another source is organic matter. The nitrogenous wastes and carrion of animals are degraded by the detritus organisms, nitrogen is converted to the amino form (*e.g.,* L-Alanine). The amino group ($-NH_2$) is liberated from organic molecules to form ammonia; this process is called **deamination**. Certain specific bacteria, most notably of the genus *Nitrosomonas*, can oxidize ammonia to nitrite (NO_2) by the reaction.

$$2NH_3 + 3O_2 \rightarrow 2NO_2^- + 2H_2O + 2H^+$$

This reaction takes place in the soil, in lake or sea water or sediments, and whenever ammonia is being released and oxygen is present. As fast as nitrite is produced, other bacteria, such as *Nitrobacter*, can combine nitrite with oxygen to form nitrate (NO_3) by the reaction :

$$2NO_2^- + O_2 \rightarrow 2NO_3^-$$

Both of these reactions which are performed by two nitrifying bacteria — *Nitrosomonas* and *Nitrobacter* are the parts of a single biological process called **nitrification**. In nitrification process, thus, ammonia is oxidized to nitrate and nitrite yielding energy. This energy is used by the bacteria to make their organic materials directly from carbon dioxide and water. Nitrate can be taken up by autotrophs at the beginning of food chain.

Under certain circumstances, nitrate is either not produced in the nitrogen cycle or it is degraded before it can be utilized by autotrophs. Degradation of nitrate is called **denitrification**, and may be important when oxygen concentration is low. Denitrifying bacteria such as *Pseudomonas* can use the energy of the nitrate ion to drive their metabolism, and in so doing, they break the nitrate down to nitrite, ammonia, or molecular nitrogen :

$$C_6H_{12}O_6 + 12NO_3^- \rightarrow 12NO_2 + 6CO_2 + 6H_2O$$
$$C_6H_{12}O_6 + 8NO_2 \rightarrow 4N_2 + 2CO_2 + 4CO_3^= + 6H_2O$$
$$C_6H_{12}O_6 + 3NO_3^- \rightarrow 3NH_3 + 6CO_2 + 3OH^-$$

If denitrification is significant in an ecosystem, nitrite is transitory and is also degraded into either ammonia or molecular nitrogen.

Cycling of nitrogen in the ecosystem. The sources of inputs of nitrogen under natural conditions are the bacterial fixation of atmospheric nitrogen, addition of inorganic nitrogen in rain from such sources as lightning fixation and fixed "juvenile" nitrogen from volcanic activities, ammonia absorption from the atmosphere by plants and soil, and nitrogen accretion from windblown aerosols, which contain both organic and inorganic forms of nitrogen. In terrestrial ecosystems, nitrogen, largely in the form of ammonia and nitrates is taken up by plants, which convert it into amino acids and proteins. Animals (primary macro-consumers) may eat the plants and utilize the amino acids from the plant proteins in the synthesis of their own proteins and other cellular constituents. When animals and plants die, the decay bacteria convert the nitrogen of their proteins and other compounds into ammonia. Animals excrete several kinds of nitrogen-containing wastes—urea, uric acid, creatinine, and ammonia and the decay bacteria converts these wastes to ammonia. Ammonia may be lost as gas to the atmosphere, may be acted upon by nitrifying bacteria, or may be taken up directly by plants. The nitrates may be utilized by plants, immobilized by microbes, stored in decomposing humus, or leached away. This material is carried to streams, lakes, and eventually the sea, where it is available for use in aquatic ecosystems. There nitrogen is cycled in a similar manner, except that the large reserves contained in the soil humus are largely lacking. Life in the water contributes organic matter and dead organisms that undergo decomposition and subsequent release of ammonia and ultimately nitrates. In aquatic ecosystem atmospheric nitrogen is fixed by numerous blue-green algae.

Under natural conditions nitrogen lost from ecosystems by denitrification, volatilization, leaching, erosion, wind blown aerosols, and transportation out of the system is balanced by biological fixation and other sources. Both chemically and biologically, terrestrial and aquatic ecosystems constitute a dynamic equilibrium system in which a change in one phase affects the other.

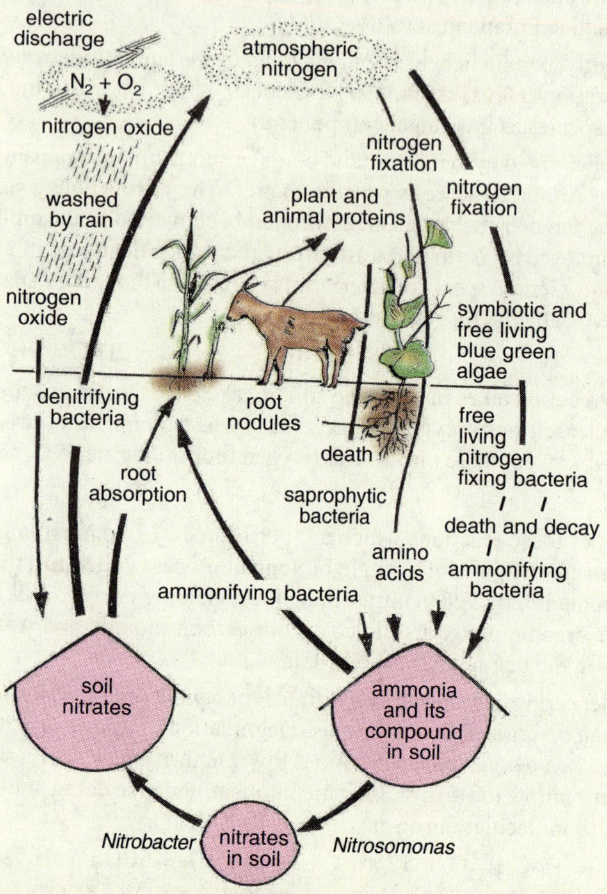

Fig. 10.6. The nitrogen cycle.

C. Sedimentary Cycles

Mineral elements required by living organisms are obtained initially from inorganic sources. Available forms occur as salts dissolved in soil water or lakes, streams, and seas. The **mineral cycle** varies from one element to another, but essentially it consists of two phase; the **salt-solution phase** and the **rock phase**. Mineral salts come directly from the earth's crust by weathering. The soluble salts then enter the water-cycle. With water they move through the soil to streams and lakes and eventually reach the sea, where they remain indefinitely. Other salts are returned to the earth's crust through sedimentation. They become incorporated into salt beds, silts and limestones; after weathering they again enter the cycle.

Plants and animals, the living components of the ecosystems, fulfil their mineral requirements from mineral solutions in their environments. Other animals acquire the bulk of their minerals from plants and animals they consume. After the death of living organisms the minerals are returned to the soil and water through the action of the organisms and process of decay.

There are different kinds of sedimentary or mineral cycles, depending on the kinds of elements, but following two cycles are very significant for a ecosystem :

(i) Sulphur cycle. Sulphur, like nitrogen, is an essential part of protein and amino acids and is characteristic of organic compounds. It exists in a number of states–elemental sulphur, S, sulphides, sulphur monoxide, sulphite and sulphates. Of these three are important in nature : elemental sulphur, sulphides and sulphates.

The sulphur cycle is both sedimentary and gaseous (*i.e.,* it includes gaseous phase and sedimentary phase). The sedimentary phase of sulphur cycle is long-termed and in it sulphur is tied up in organic and inorganic deposits. From these deposits, it is released by weathering and decomposition, and is carried to terrestrial and aquatic ecosystems in a salt solution. Atmospheric (gaseous) phase of sulphur-cycle is less pronounced and it permits circulation on a global scale.

Sulphur enters the atmosphere from several sources—the combustion of fossil fuels, volcanic eruption, the surface of the oceans and gases released by decomposition. Initially sulphur enters the atmosphere as hydrogen sulphide, H_2S, which quickly oxidizes into another volatile form, sulphur dioxide, SO_2. Atmospheric sulphur dioxide, soluble in water, is carried back to earth in rainwater as weak sulphuric acid, H_2SO_4. Whatever its source, sulphur in a soluble form, mostly as sulphate ($SO_4^{=}$) is absorbed through plant roots, where it is incorporated into certain organic molecules, such as some amino acids (*e.g.,* cystine) and proteins. From the producers the sulphur in amino acids is transferred to the consumer animals, with excess being excreted in the faeces.

Excretion and death carry sulphur in living material back to the soil and to the bottoms of ponds, lakes, and seas where the organic material is acted upon by bacteria of detritus food chain. Within the detritus food chain, the sulphydryl group (–SH) of amino acids (*e.g.,* L-cysteine) is separated from the rest of the molecule as hydrogen sulphide (H_2S) by most decomposing bacteria as a normal part of the degradation of proteins. In an aerobic environment, the hydrogen sulphide is oxidized to sulphate by bacteria specially adapted to perform this conversion :

$$H_2S + 2O_2 \rightarrow SO_4^{=} + 2H^+$$

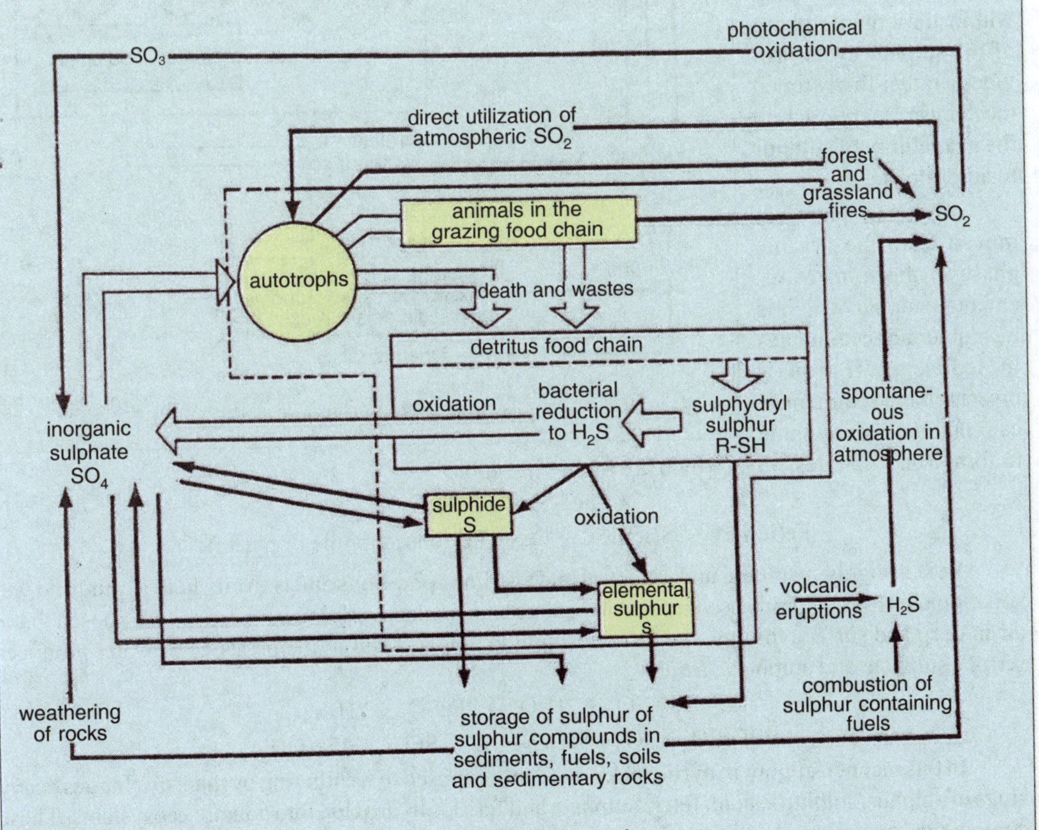

Fig. 10.7. Sulphur cycle.

The sulphate produced then can be reused by the autotrophs. In anaerobic environments, such as bottom of certain lakes, it is impossible to oxidize sulpide by this means, because the process of oxidation requires oxygen. But if infrared radiation is present in these environments, there are photosynthetic bacteria that can use it to manufacture carbohydrates and oxidize sulphide either to elemental sulphur or to sulphate :

$$6CO_2 + 12H_2S + hv \rightarrow C_6H_{12}O_6 + 6H_2O + 12S$$

$$6CO_2 + 12H_2O + 3H_2S + hv \rightarrow C_6H_{12}O_6 + 6H_2O + 3SO_4^{=} + 6H^+$$

Elemental sulphur can also be utilized by other bacteria to form sulphate. If oxygen is present, the reaction is quite rapid.

$$2S + 3O_2 + 2H_2O \rightarrow 2SO_4^{=} + 4H^+$$

Under anaerobic conditions, elemental sulphur can still be oxidized to sulphate by certain bacteria if nitrate is present :

$$6NO_3^- + 5S + 2CaCO_3 \rightarrow 3SO_4^{=} + 2\,CaSO_4 + 2CO_2 + 3N_2$$

None of these bacterial reactions is unidirectional; under certain conditions, sulphate can also be reduced either to sulphide or to elemental sulphur by bacteria. This series of reactions operating within the organic phase of the sulphur cycle provides a rather finely tuned mechanism for regulating the availability of sulphur to autotrophs.

The sulphur is removed from the organic phase in the form of elemental sulphur which is insoluble and accumulates in sediments. If iron is present in the sediment, it can combine with sulphide to form iron sulphides, all of which are highly insoluble :

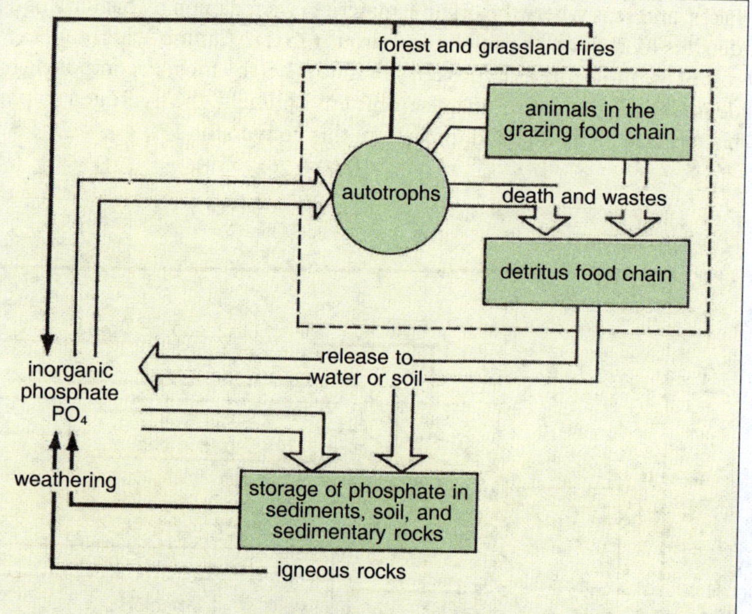

Fig. 10.8. Phosphorus cycle.

$$Fe^{++} + S^{=} \rightarrow FeS$$

$$Fe(ionic) + 2S(ionic) \rightarrow FeS_2 \text{ (Ferrous sulphide or pyrite)}$$

FeS_2 is highly insoluble under neutral and alkaline conditions and is firmly held in mud and wet soil. Some ferrous sulphide is contained in sedimentary rocks overlying coal deposits. Exposed to the air in deep and surface mining, the ferrous sulphide oxidizes and in the presence of water produces ferrous sulphate and sulphuric acid :

$$2FeS_2 + 7O_2 + 2H_2O \rightarrow 2FeSO_4 + 2H_2SO_4$$

$$12FeSO_4 + 3O_2 + 6H_2O \rightarrow 4Fe_2(SO_4)_3 + 4Fe(OH_3)$$

In this manner sulphur in pyrite rocks, suddenly exposed to weathering by man, discharges heavy slugs of sulphur, sulphuric acid, ferric sulphate and ferrous hydroxide into aquatic ecosystems. These compounds destroy aquatic life and cause acidic water.

(ii) Phosphorus cycle. Phosphorus cycle has no atmospheric phase. It occurs naturally in environment as phosphate (PO_4^-, or one of its analogues, HPO_4^- or $H_2PO_4^-$), either as soluble inorganic phosphate ions, as soluble organic phosphate (*i.e.,* as a part of a soluble organic molecule), as particulate phosphate (*i.e.,* as part of an insoluble organic or inorganic molecules) or as mineral phosphate (*i.e.,* as part of a mineral grain as found in a rock or sediment). The ultimate source of phosphate in the ecosystem is crystalline rocks. As these are eroded and weathered, phosphate is made available to living organisms, generally as ionic phosphate. This is introduced into autotrophic plants through their roots, where it is incorporated into living tissues. From autotrophs, it is passed along the grazing food chain in the same fashion as nitrogen and sulphur, with excess phosphate being excreted in the faeces. An extreme example of faecal phosphate is the tremendous guano deposits built up by birds on the desert west coast of South America. Phosphates can also be released as particulate matter from forest and grassland fires.

In the detritus food chain, as large organic molecules containing phosphate are degraded, the phosphate is liberated as inorganic ionic phosphate. In this from it can be immediately taken up by autotrophs, or it can be incorporated into a sediment particle, either in the soil of a terrestrial ecosystem or in a sediment of an aquatic ecosystem. The sedimentary phase of phosphorous cycle remains comparatively slow than the organic phase.

Besides phosphorus, there are biogeochemical cycles for all the other nutrients (minerals) used by living organisms, as well as some that are not. Most of them has complete cycles in sedimentary phase. The availability depends on their solubility in water and availability of water as solvent.

Thus, the geochemical cycles of different chemical substances are closed : the atoms are used over and over again. To keep the cycles going does not require new matter but it does require **energy**, for the energy cycle is not a closed one. Further, the patterns of flow, both of energy and of chemical substances, are of great significance. The simpler pat-

Fig. 10.9. Forest ecosystem dynamics. Arrows indicate the flows of matter and energy.

terns involve energy, as the sources of energy are external to the ecosystem, and flow is undirectional through it. Chemical substances, on the other hand, are finite and have their origin inside the ecosystem, thus, they must continuously cycle within the system.

Biogeochemical Cycle of Micronutrients

Many other micronutrients (elements) such as calcium, potassium, sodium, etc., have a similar pathway of cycling like the other nutrients. They also involve the abiotic (soil, water, etc.) and biotic phases (living organisms and their decomposition). These elements usually cycle within the same ecosystem and in small quantities globally. Usually, the rocks are their reserve pools and the weathering process enriches the soil and water with these elements. Though, the dust of these nutrients may be blown away by the wind and carried across ecosystems and some of their amount is lost in the runoff to lakes and oceans and then to deep sediments. For example, in a tropical rain forest on average, more than 80% of total store of these micronutrients is found in the soil, with progressively less in vegetation, litter and the fauna (Fig. 10.9). Though, the concentration of trace elements tends to increase along the food chains in the sequence such as soil – plants – herbivores – carnivores and litter – detritivores. Thus, concentration of iron in a rain forest were 6 ppm in soil, 218 ppm in vegetation, 3054 ppm in herbivores and 3387 ppm in the carnivores. Many of these micronutrients become toxic to plants in high concentration (see **Sharma**, 1994).

The nutrient concentration of the leaf litter of some plants have been tabulated in Table 10-2.

Table 10-2. Concentration of nutrients (in %) in the leaf litter of some plant species (Source : Vyas and Galley, 1975 ; Rajvanshi and Gupta, 1985).

Species	Calcium (Ca)	Magnesium (Mg)	Potassium (K)	Sodium (Na)	Nitrogen (N)	Phosphorus (P)
1. *Terminalia arjuna*	4.42	0.37	0.54	0.06	0.85	0.24
2. *Terminalia tomentosa*	3.73	0.54	0.40	0.06	0.93	0.52
3. *Tectona grandis*	2.54	0.25	0.24	0.20	0.78	0.18
4. *Diospyros melanoxylon*	1.90–2.10	0.51–0.69	0.57–0.90	0.05–0.10	0.56–0.61	0.008–0.02
5. *Butea monosperma*	1.85	0.56	0.75	0.06	2.26	0.25
6. *Shorea robusta*	1.03–1.58	0.36–0.48	0.26–0.59	0.05–0.08	0.71–0.81	0.41–0.50
7. *Dalbergia sissoo*	1.02	0.96	0.25	0.07	2.23	0.025

REVISION QUESTIONS

1. What are biogeochemical cycles ? What role they play in the ecosystem ? Discuss.
2. Describe nitrogen and phosphorus biogeochemical cycles with suitable examples.
3. Write short notes on the following :
   (a) Carbon cycle ;
   (b) Water cycle ;
   (c) Sedimentary biogeochemical cycles ;
   (d) Gaseous biochemical cycles.

CHAPTER

11

Aquatic Ecosystems : Freshwater Communities

Aquatic ecosystems are the source of life on earth.

The **habitat approach**, describes the distinctive fea tures of the major habitats and their subdivisions, how they are organized, the organisms present in each and the ecologic role of these organisms in that region (*i.e.,* the identity of the major producers, consumers and decomposers). According to **Elton** (1949) habitat is definable as an area possessing uniformity of physiography, vegetation, climate or any other quality the investigator assumes important. Four major habitats can be distinguished : **marine**, **estuarine**, **fresh-water** and **terrestrial**. No plant or animal is found in all four major habitats and, indeed, no animal or plant is found every-where within any one of these. According to recent ecological trend only two major habitats or more specifically ecosystems namely **aquatic ecosystems** and **terrestrial ecosystems**, are recognized.

AQUATIC ECOSYSTEMS

Liquid water covers about three quarters of the earth's surface either as oceans or as freshwater. Virtually all these waters contain life in one form or other; hence, aquatic ecosystems would be important for their sheer volume, if for nothing else. But aquatic ecosystems are historically the source of life on earth. Even now, tiny single-celled marine plants are the main source of the earths' oxygen supply, because most photosynthetic oxygen is derived from these plants (**Clapham, Jr.**, 1973). In addition, aquatic ecosystems are simpler in many ways than terrestrial ecosystems. The reason for this is that the omnipresent environmental factor which sets the tone for all aquatic systems, regardless of their biotic complexity, is water. It is the medium within which all aspects of the ecosystem coexist, both living and non-living; it is the source of all nutrients for aquatic life, including the gaseous nutrients such as oxygen and carbon dioxide; it is the medium by which organic and inorganic wastes and sediments are distributed throughout the ecosystem. The amount of light energy reaching the community is determined by the way in which light is absorbed by water; the heat properties of water determine much of the circulation patterns within the ecosystem and have a major control over the structure of the aquatic community that exists in any one place.

SUBDIVISIONS OF AQUATIC ECOSYSTEMS

Global aquatic system fall into two broad classes definable by salinity, or amount of material dissolved in water—the **freshwater ecosystems** and **salt-water ecosystems**. The latter may include inland brackish water, as well as marine and estuarine habitats. In fact, estuary represents a transitional zone between a river and the sea and it contains dissolved solid content intermediate between those of fresh and marine waters. Freshwater ecosystems, the study of which is known as **limnology**, are conveniently divided into two groups—**lentic**, **standing** or **still water habitats** and **lotic** or **running water habitats**. Both can be considered on an environmental gradient. The lotic follows a gradient from springs to mountain brooks to streams to rivers. The lentic involves a gradient from lakes to ponds to bogs, swamps and marshes.

FRESHWATER ECOSYSTEMS : PHYSICO-CHEMICAL NATURE OF FRESHWATER

The freshwater of both kinds—lentic and lotic, has low percentage of dissolved salts and is subjected to the influence of a wide array of physical and chemical factors. The rise and fall of these factors very frequently affect the fauna, altering their number and diversity. Some of the important factors of freshwater environment are following :

1. Pressure, Density and Buoyancy

The pressure imposed on a lake-dwelling organism is the weight of the column of water above it plus the weight of the atmosphere. In all freshwater environments maximum pressure is much less than in the ocean, and organisms appear to adjust to them readily. The absence of animal life from deep water is ordinarily a consequence of low oxygen supply, or low temperature, rather than pressure.

The density of water varies inversely with temperature and directly with the concentration of dissolved substances. Water is most dense at approximately 4°C and becomes progressively less dense as it cooled below +4°C. Ice also expands markedly the colder it gets. It is because the coldest water is at the surface in winter that ice forms there, rather than at the bottom of a lake. In summer, the coldest waters of deep lakes are at the bottom. Dissolved salts increase the density of water; the density of most inland water-bodies is much less than that of the ocean. However, when great evaporation occurs in a lake having no outlet, the lake may come to contain a higher percentage of salts (*i.e.,* hypersalinity) than the ocean. The few species capable of living in these very salty lakes (*e.g.,* Great Basin of USA) include some algae and Protozoa, the brine shrimp *Artemia gracilis* and the immature stages of two brine flies, *Ephydra gracilis* and *E. hians*.

According to the **law of Archimedes** the buoyancy of an object is equal to the weight of the water it displaces. Buoyancy varies with the density of water, and is influenced by the factors that affect density. Viscosity, the measure of the internal friction of water, varies inversely with temperature and also influence buoyancy. Most aquatic organisms keep stations by swimming movements or have special adaptations to decrease the specific gravity of the body and take advantage of any turbulence in the water. For this purpose freshwater aquatic organisms have some swimming adaptations, clinging organs (in case of animal of lotic habitat) or following adaptations—absorption of large amounts of water to form jelly-like tissues; storage of gas or air bubbles within the body; formation of light-weight fat deposits within the body or oil droplets within the cell; increase of surface area in proportion to body mass, which increase frictional resistance (**David**, 1955). When an organism so equipped dies, the special mechanism quickly cease to function, and it sinks to the bottom.

2. Temperature

The unique thermal properties of water are best demonstrated by freshwater environment. Diurnal and seasonal variations of temperatures are very much evident in these environments than in marine environments. A diurnal variation range of 4.8–5.0°C has been recorded by **Sreenivasan** (1964) in a tropical pond, with an average depth of 3.0 meters. In shallow water habitats, difference between day and night temperatures remain more conspicuous. For example, in a polluted moat with an average depth of 1.5 meters, the lowest night time temperature was 26.6°C, the highest day time temperature was 32°C with a variation of 5.4°C (**Sreenivasan**, 1964). However, the Kodaikanal lake in South India showed a diurnal variation of only 2.8°C (**Sreenivasan**, 1964). Flowing lotic waters of streams and rivers lack such wide fluctuations in temperature.

Further, the lentic water of lakes and ponds undergo thermal stratification phenomenon according to the seasons. Thermal stratification has been reported most frequently in the lakes of tropical countries such as Java, Sumatra, etc. In fact, according to their temperature relations, lakes have been classified into three types : (1) **Tropical lakes** in which surface temperatures are always maintained above 4°C ; (2) **Temperate lakes** in which surface temperature vary above and below 4°C and (3) **Polar lakes** in which surface temperatures never go above 4°C. The seasonally regulated thermal stratification of lentic habitats has a significant influence on their inhabitant biotic communities. Decreasing temperatures often cause a fall in metabolism, resulting in a lower rate of food consumption. The extremes of lower and higher temperature have lethal effects on the aquatic organisms. So fluctuations in temperature of aquatic media regulate the breeding periods, initiate hibernation, gonadial activitation and a number of other biological phenomena such as thermally oriented migration, etc., of freshwater biota. On the basis of their ability to tolerate thermal variations, most freshwater organisms are **stenothermic** with a narrow range temperature tolerance, but some are **euthermic** with a wide range of temperature tolerance. For example, the stenothermic oligochaets includes **steno-minimothermal forms** (narrow range of temperature, *e.g., Aeolosoma, Megascolex mauritii*), **steno-maximothermal** (*e.g., Dero limosa*) or **stene-optimothermal forms** (*e.g., Branchiodrilus semperi and B. menoni*) (**Sitaramiah**, 1966).

3. Light

Light influences freshwater ecosystems greatly. The freshwaters often have a lot of suspended material. While affording protection to the light sensitive species, these substances more often obstruct the light that normally reaches the water. The degree of such obstruction of light influence the productivity of the freshwater ecosystems. A shallow lake receive light to its very bottom resulting in an abundant growth of vegetation both phytoplankton and rooted vascular plants. These plants in living or dead states form nice food for consumers of grazing food chain or organisms of detritus food chain, respectively. The running water contains little plant or animal plankton not due to the lack of sunlight but because of the action of the currents in washing it away. Further, light controls the orientation and changes in position of attached species and their nature of growth and it also causes the diurnal migration of planktonic species of freshwater.

4. Oxygen

Chemically pure water is biologically uninhabitable and all freshwaters contains an array of chemical substances. The oxygen, which is a most essential chemical component of life processes, remain dissolved in freshwaters. The aquatic environments which remain in close proximity with atmosphere contain an abundance of oxygen, that reaches the water either by direct diffusion or by movements of water such as wave action or water circulation. Lotic (moving) water of streams and rivers often have a high percentage of oxygen. Aquatic plants supply water with oxygen that is formed as a product of photosynthesis. Rooted vegetation of shallow water zones and floating phytoplankton of open waters also produce oxygen. The amount of photosynthetically produced oxygen remain high at warm temperatures and at greater light intensities.

Oxygen the most essential chemical component is dissolved in fresh waters.

The oxygen level in a tropical pond exhibits diurnal variation—it remains at peak between 14.00 and 17.00 hours of day (Sreenivasan, 1964). Oxygen contents of a freshwater body are depleted in numerous ways. Primarily oxygen is utilised in the respiration of organisms and decomposition of dead organisms in the aquatic environment. While photosynthesis remains restricted to the surface layer of water containing phytoplankton and exposed regions of rooted vascular plants, respiration and decomposition occur at all levels. In stagnant pools with a lot of decaying vegetation oxygen content often reaches a stage of complete depletion. The reduction in dissolved oxygen is magnified by the release of many gases as end products of decomposition or by the mixing up of waters of low oxygen content reaching the habitat as an inflow. Aquatic animals with very few exception, *i.e.*, those that breath air, utilize the oxygen dissolved in water. Certain freshwater inhabitants such as many anaerobic bacteria and insect larvae of chironomids perform anaerobiosis and requires no oxygen.

5. Carbon dioxide

Aquatic vegetation and phytoplankton require carbon dioxide for photosynthetic activity. The carbon dioxide of freshwater environments is produced as the end product of respiration and of decomposition. Carbon dioxide also diffuses directly from the atmosphere and is readily dissolved in water to result in carbonic acid (H_2CO_3) which affects the pH of water. It is also present in the freshwater as carbonates and bicarbonate of calcium, magnesium and other minerals. The growing plants and lime deposition bacteria and other animals may cause a depletion in carbon dioxide resources. Photosynthesis is the major cause for its drain. The high saturation levels of O_2 and CO_2 have been found to have toxic effects on aquatic biota.

6. Other Gases

Streams and lakes contaminated by sewage and stagnant pools with decaying vegetation show an abundance of the gas, hydrogen sulphide which is a decomposition product. This gas is highly toxic to living organisms and results into complete denudation of bottom fauna. Methane and carbon monoxide are other toxic gases which are the products of decomposition. Nitrogen, hydrogen, sulphur dioxide and ammonia are some of the other gases which are found dissolved in freshwaters.

7. Dissolved Salts and Salinity

Freshwater being efficient solvent contains many solutes in solution, but even then its salt contents remain under 1/5% than marine water which contains about thirty-five parts per thousand (%) dissolved salts (see **Clapham, Jr.**, 1973). Different dissolved salts reach the water by erosion, inflow and decay of aquatic forms. Dissolved substances have peculiar significance for floating aquatic vegetation and phytoplankton, since these organisms do not depend on the substratum for the supply of nutrients. Compounds of nitrogen, phosphorus and silicon are most important substances found dissolved in freshwater. Nitrate, nitrites and ammonium salts are essential for the food of aquatic vegetation such as algae and water weeds. Nitrate always remain available due to nitrogen cycle occurring between nitrogen fixing bacteria and nitrogen consuming plants. Ammonium salts in excess have a lethal effect on the fauna. Dissolved silicates of freshwater are readily utilized by diatoms and sponges in constructing their body structures such as shell in case of diatoms and spicules in case of sponges.

All freshwater environments also contain small amounts of phosphorus which more often acts as a limiting factor. Utilization of phosphorus by plankton during the periods of abundance may result in a total elimination of other plants that require the element (phosphorus). Many other elements such as calcium, magnesium, manganese, iron, sodium, potassium, sulphur, and zinc are found dissolved in water and influence the fauna variously. Iron being a growth promoting element for plants exists as the compound of oxygen (ferrous oxide) or sulphur (ferrous sulphide) in different freshwater bodies. Its influence is often modified by the pH of water. Calcium is an essential element for plants. The abundance and scarcity of carbonate of calcium determine the faunal composition. Deposition of calcium carbonate in water called **marl** is produced by the activity of plants. External coverings of arthropods and the shell of molluscs and tubes of some worms need calcium carbonate. Snails are found to develop a heavy shell if the waters in which they lived contained excess of calcium. Bryozoans, sponges, and cladocerans prefer an increased calcium content.

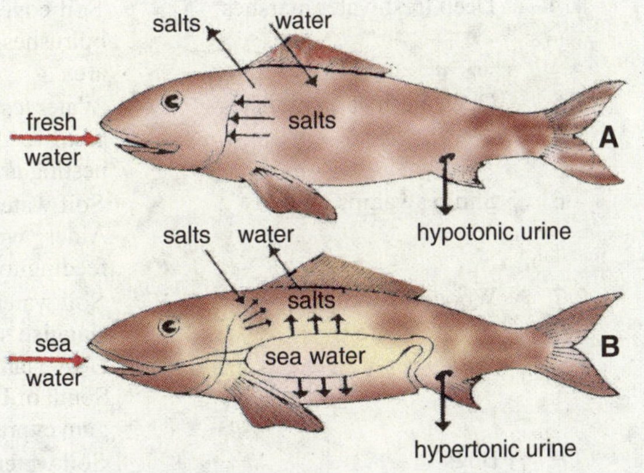

Fig. 11.1. Osmoregulation in (A) fresh water fish and (B) marine fish.

Due to low salinity of freshwater, animals face the problem of **osmoregulation**. Because the salt concentration of body fluids of animals remain higher than the freshwater, hence, the water continuously tend to enter the body which should be readily removed. Most aquatic animals (*e.g.,* Protozoa and fishes) have the means to excrete extra amount of water of body by osmoregulation. For this purpose Protozoa employ contractile vacuoles and other multicellular invertebrates and chordates use excretory organs, such as nephridia, kidney, etc.

8. pH or Hydrogen Ion Concentration

In freshwater environments pH is a determining factor for the biota by becoming a limiting factor. The pH value of different freshwater bodies may fluctuate seasonally and annually. The pH of surface waters and deeper waters exhibit marked differences. **Sreenivasan** (1968) has reported a marked pH variation of 2.2 units between surface water and deeper water in the Sandynulla reservoir in the Nilgiris (India). Though pH range is species specific, yet lower aquatic forms in general showed little reaction to alterations in pH, while higher aquatic organisms (*e.g.,* fishes) responded quickly to little pH variations.

LENTIC ECOSYSTEMS

Lentic ecosystems include all standing water (freshwater) habitats such as lakes, ponds, marshes, swamps, bogs, meadows, etc. Some of the important lentic ecosystems or wetlands have been tabulated in Table 11-1.

Table 11-1. Classification of wetlands or lentic ecosystems (excluding lakes and ponds) (Smith 1974).

A.	**Inland freshwater areas**	
1.	Seasonally flooded	Soil covered with water or waterlogged during variable periods but well-drained during much of the growing season. In upland depressions and bottomlands. Bottomland hardwoods to herbaceous growth.
2.	Freshwater meadows	Without standing water during growing season ; water logged to within few inches of surface. Grasses, rushes, sedges, broadleaf plants.
3.	Shallow freshwater marshes	Soil waterlogged during growing season ; often covered with about 15 cm or more of water. Grasses, bulrushes, spike rushes, cattails, arrowhead, smartweed, pickerelweed. A major water-fowl-production area.
4.	Deep freshwater marshes	Soil covered with 15 cm to 91 cm of water. Cattails, reeds, bulrushes, spike rushes, wild rice. Principal duck-breeding area.
5.	Open freshwater marshes	Water less than 3 m deep. Bordered by emergent vegetation : pondweed, naiads, wild celery, water lily. Brooding, feeding, nesting areas of ducks.
6.	Shrub swamps	Soil waterlogged; often covered with 15 cm or more water. Alder, willow, buttonbush, dogwoods. Ducks nesting and feeding to limited extent.
7.	Wooded swamps	Soil waterlogged; often covered with 30 cm of water. Along sluggish streams, flat uplands shallowlakes, basins. North of USA : tamarack, arborvitae, spruce, red maple, silver maple. South of USA : water oak, overcup oak, tupelo, swamp black gum cypress.
8.	Bogs	Soil waterlogged; spongy covering of mosses. Heath shrubs, sphagnum, sedges.
B.	**Coastal freshwater areas**	
9.	Shallow freshwater marsh	Soil waterlogged during growing season ; at high tides as much as 15 cm of water on landward side, deep marshes along tidal rivers, sounds, deltas. Grasses and sedges. Important water-fowl areas.
10.	Deep freshwater	At high tide covered with 15 cm to 1.8 m of water. Along tidal rivers and bays. Cattails, wild rice, giant cutgrass.
11.	Open freshwater	Shallow portions of open water along fresh tidal rivers and sounds. Vegetation scarce or absent. Important water-fowl areas.

1. Lakes and Ponds

Lakes are inland depressions containing standing water. They may vary in size from small ponds of less than a hectare to large seas covering thousands of square kilometers. They may range in depth from a few centimeters to over 1666 meters. **Ponds**, however, are considered as small bodies of standing water so shallow that rooted plants can grow over most of the bottom. Most ponds and lakes have outlet streams and both are more or less temporary features on the landscape because their filling is inevitable.

The aquatic habitats of lake and pond remain vertically stratified in relation to light intensity, wave length absorption, hydrostatic pressure, temperature, etc. In a lake, for example there are three well recognized horizontal strata namely ; **(i)** Shallow water near the shore forms the **littoral zone**. It contains upper warm and oxygen rich circulating water layer which is called **epilimnion**. The littoral zone includes rooted vegetation. **(ii) Sublittoral zone** extends from rooted vegetation to the non-circulating cold water with poor oxygen zone, *i.e.,* **hypolimnion**. **(iii) Limnetic zone** is the open water zone away from the shore. It is the zone upto the depth of effective light penetration where rate of photosynthesis is equal to the rate of respiration. **(iv) Profundal zone** is the deep-water area beneath limnetic zone and beyond the depth of effective light penetration. **(v) Abyssal zone** is found only in deep lakes, since it begins at about 2,000 meters from the surface. Ponds have little vertical stratification. In them littoral zone is larger than the limnetic zone and profundal zone. In a small pond the limnetic and profundal zone are not found. Lakes, thus, differ from ponds in having relatively larger limnetic zone and profundal zone than littoral zone.

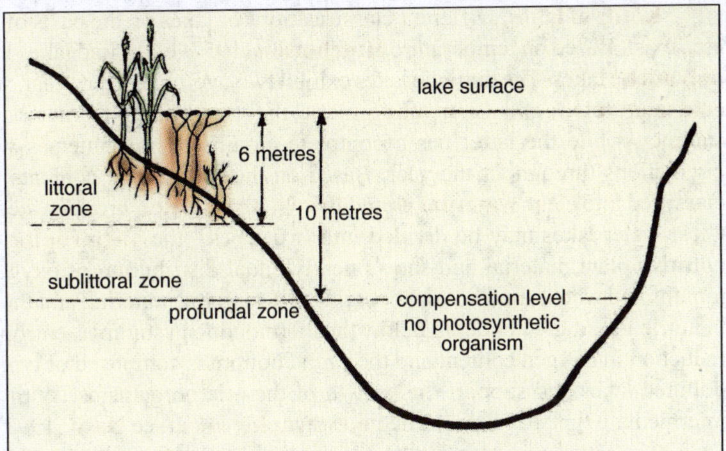

Fig. 11.2. Different zones of a deep freshwater lake.

Further, lentic water of a lake or pond is also classified on the basis of the depth of light penetration enabling photosynthesis into **trophogenic zone** (includes littoral plus sublittoral zones), and a **tropholytic zone** (upper part of profundal zone). The former is often distinguished by abundant plant growth and dependent fauna, while the latter denotes a general absence of vegetation and harbours mostly saprobes. In between the two zones is the **compensation level** which forms a boundary between two zones. It exhibits perfect equilibrium between respiration and photo-synthesis.

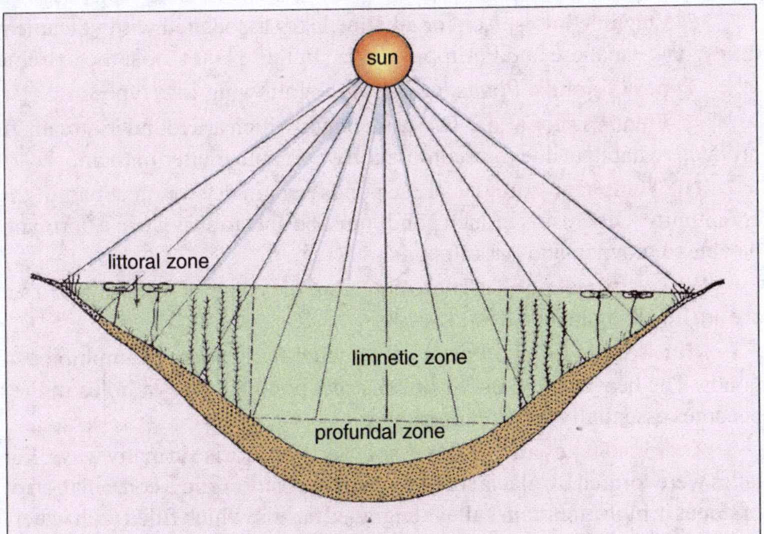

Fig. 11.3. Different zones of a freshwater pond.

Physico-chemical properties of lakes and ponds. Lakes have the tendency to become thermally stratified during summer and winter to undergo definite seasonal periodicity in depth distribution of heat and oxygen. Light too penetrates only to a certain depth, depending upon turbidity. These gradations of oxygen, light and temperature profoundly influence the life in the lake, its distribution and adaptation.

Kinds of lakes. Different classifications of lakes on the basis of physical factors, productivity, etc., exist. Based on temperature **Hutchinson** (1957) classified lakes into **dimictic**, **monomictic** and **polymictic** lakes. The dimictic lakes exhibit two overturns every year, while monomictic lakes may be **cold monomictic** and **warm monomictic**, the former being characterized by a circulation only during summer, while the latter has a complete circulation in winter as well. Polymictic lakes present circulations throughout the year. Based on the humic acid contents the lakes of world have been classified into **clear water lakes** and **brown water lakes**, brown water contains high humus content. Clear water lakes may be divided into two types—the **oligotrophic type**, whose water is poor in nutritive plant material and shows nearly equal distribution of oxygen during summer and winter months and whose mud bottom contains little organic material; and the **eutrophic type**, which is rich in nutrients. At greater depth below the thermocline in summer, eutrophic lakes show a considerable reduction in oxygen content and their mud bottom is composed of typical muck. Eutrophic lakes are polluted lakes and support the growth of the microorganisms. With regards to above features, an intermediate type of lake is called **mesotrophic lake** (see **Dash**, 1993).

Special types of lakes. There also occur the following five types of lakes :

1. Dystrophic lakes. These lakes contain high concentration of humic acid in water. The margins of these lakes have a low pH and develop into peat bogs.

2. Deep ancient lakes. Lake Baikal in Russia is the most famous of ancient lakes. It is the deepest lake in the world. This lake is often nicknamed as "**Australia of Freshwater**", since it contains various unique animals. It contains about 384 species of arthropods, 98 per cent of which are **endemic** (*i.e.*, found no where else). Likewise, 81 per cent of its 36 species of fish are also endemic.

3. Desert salt lakes. These lakes exist in sedimentary drainages in arid climates where evaporation exceeds precipitation, *e.g.*, Great Salt Lake of Utah, USA. Biotic community of these lakes comprises only those species which can tolerate high salinity, *e.g.*, *Artemia* (Brine shrimp).

4. Desert alkali lakes. These lakes occur in igneous drainages in arid climates, *e.g.*, Pyramid lake, Navada, USA. High pH and concentration of carbonates are the characteristic features of these lakes.

5. Volcanic lakes. Acid or alkaline lakes associated with volcanic regions are called **volcanic lakes**, *e.g.*, Japanese and Philippine lakes. In these lakes exist the extreme chemical conditions.

Types of ponds. Ponds too are of the following four types :

(i) Flood plain ponds. These are ponds which are found in stream flood plains. These are quite productive habitats due to accumulation of organic matter in them.

(ii) Temporary ponds. These ponds remain dry for most part of the year but support a unique community. Organisms in such ponds must be able to survive in a dormant stage during dry periods or be able to move in and out of ponds.

(iii) Artificial ponds. These ponds are the result of damming of a stream or basin by man. They are artificially managed habitats.

(iv) Beaver ponds. Ponds constructed by beavers (=amphibious rodents) are called beaver ponds. The beaver often does not construct pond but live in holes in the bank of streams and, thus, becomes essentially the stream animals.

Origin of lakes and ponds. Lakes and ponds arise in many ways. For example, North American lakes were formed by glacial erosions and deposition and a combination of the two. Glacial abrasions of slopes in high mountain valleys engraved basins, which filled with water from rain and melting snow to produce tarns *i.e.*, small mountain lakes. Retreating valley glaciers left behind crescent-shaped ridges of rock debris, which dammed up water behind them. Lakes are also formed by the deposition of silt, driftwood, and other debris in the beds of slow-flowing streams. Craters of extinct volcanos may fill with water and landslides can block off streams and valleys to form new lakes and ponds. Further man intentionally creates artificial lakes by damming rivers and streams for power, irrigation, and water storage, or by constructing small ponds and marshes for water, fishing and wild life. Man made lakes are often called **impoundments**. They may be a 'close' type or 'open' type depending on the discharge of water into it. Sandynulla is a closed impoundment, while Pykara is an open impoundment (**Sreenivasan**, 1957). Impoundments differ from natural lakes in thermal and oxygen variations and are often characterized by a low percentage of bottom dwellers or benthos.

Ponds occur in most regions of adequate rainfall. Temporary ponds, which are often shallow and quick drying contain a characteristic lentic environment. Permanent ponds contain water throughout the year, while flood-plain ponds are formed when a stream shifts its position, leaving the formed bed isolated as a body of standing water.

Biotic Communities of Lakes and Ponds

Different organisms of the lentic environment can be ecologically classified based on whether they are dependent on the substratum or free from it. Organisms depending on the substratum are called **pedonic forms** and those that are free from it are the **limnetic forms**. Further, the aquatic organisms may also be classified into following groups depending upon their sizes and habits :

1. Neuston. These are unattached organisms that live at the air-water interface. They may include floating plants such as duckweed, as well as many types of animals. Animals that spend their lives on top of the air-water interface, such as water striders, are termed **epineuston**, while others, including

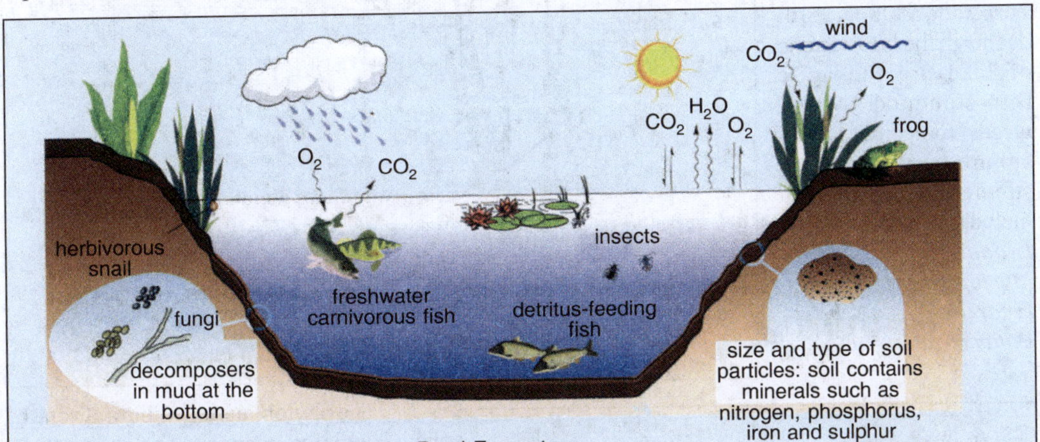

Pond Ecosystem

insects such as diving beetles and back swimmers, which spend most of their time on the underside of the air-water interface and obtain much of their food from within water, are termed **hyponeuston**.

2. Plankton. These are forms which are found in all aquatic ecosystems except for fast-moving rivers. They are small plants and animals whose powers of self-locomotion are so limited that they cannot overcome currents. Thus, their distribution is controlled largely by the currents in their ecosystems. Most planktons (phytoplanktons and zooplanktons) can move a bit, however, either, to control their vertical distribution or to seize prey. Certain animal planktons or zooplanktons are exceedingly active and move relatively great distances considering their small size, but they are so small that their range is still controlled largely by currents and such planktons are also known as **nektoplankton**.

3. Nekton. Nektonic animals are swimmers and are found in all aquatic systems except for fastest-moving rivers. In order to overcome currents, these animals are relatively large and powerful; they range in size from the swimming insects of quiet water, which may be only about 2 mm long, to the largest animal that has ever lived on earth, the blue whale.

4. Benthos. The benthos include the organisms living at the bottom of the water mass. They occur virtually in all aquatic ecosystems. The benthos organisms living above the sediment-water interface are termed **benthic epifauna** and those living in the sediment itself are termed **infauna**.

Neustons of lentic aquatic environment. The water surface of lake or pond contains certain free-floating hydrophytes such as *Wolffia, Lemna, Spirodella, Azolla, Salvinia, Pistia* and *Eichhornia*. These plants remain in contact with water and air, but not soil.

Biota of littoral zone. Lentic aquatic life is most prolific in the littoral zone. The littoral zone of a lake remains rich in pedonic flora especially up to the depth to which effective light penetration is possible facilitating the growth of rooted vegetation. At the shore proper are the **emergent vegetation** which remain firmly rooted in the shore substratum but their tops with thin chlorophyll bearing regions are exposed. Certain rooted emergent plant species of littoral zone are *Ranunculus, Monochoria, Cyperus* and *Rumex*. Interspersed with these plants are the cattails (*Typha*), bulrushes (*Scirpus*), arrowheads (*Sagittaria*) and pickerel-weeds. Slightly deeper are the rooted plants with floating leaves such as the water lilies–*Nymphaea, Nelumbo, Aponogeton, Trapa,* and *Marsilea.* Still deeper in the littoral zone of lake are the agile thin-stemmed water weeds, rooted but totally submerged. Such a submerged vegetation includes plants such as *Elodea, Vallisneria, Myriophyllum, Isoetes, Hydrilla, Chara, Potamo-geton,* etc.

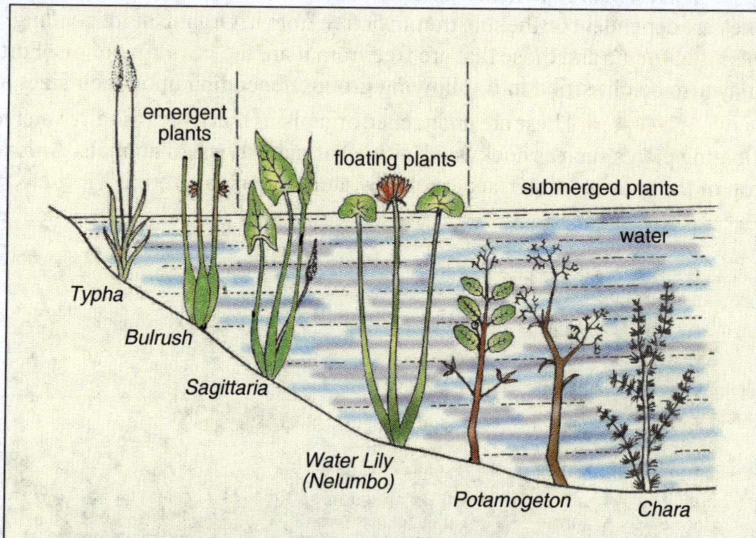

Fig.11.4. Flora of the lentic habitat.

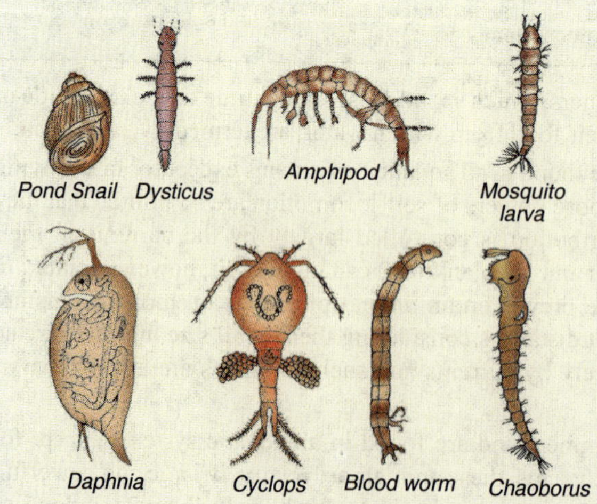

Fig. 11.5. Fauna of littoral zone of lentic habitat.

The phytoplankton of littoral zone of lake is chiefly composed of diatoms (*e.g., Navicula, Cyclotella*), blue green algae (*e.g., Microcystis, Oscillatoria*), green algae (*e.g., Cosmarium, Staurastrum*) and holophytic flagellates. **Sreenivasan** (1970) has observed the abundance of marine dinophycean flagellate *Ceratium hirudinella* in some impoundments of India.

The littoral zone also contains great concentration of animals, which remain distributed in recognizable communities. In or on the bottom are various dragonfly nymphs, crayfish, isopods, worms, snails and clams. Other animals live in or on plants and other objects projecting up from the bottom. These include protozoans such as *Vorticella, Stentor* ; larvae of *Dytiscus, Laccotrophes,* leach-like *Glossophonia,* climbing dragonfly, damsel fly nymphs, rotifers (*Brachionus*), flatworms, Bryozoa, Hydra, snails (*Lymnaea*) and other. With the exception of a few rotifers such as *Keratella, Brachionus, Diurella,* and *Trichocerca* and crustaceans such as *Cyclops, Daphnia* and *Moina,* most freshwater animals of lentic habitat depend in one way or other on the aquatic vegetation. Larvae of *Chironomous* are found underneath the leaves of floating plants .The zooplankton of littoral zone consists of water fleas such as *Daphnia,* rotifers and ostracods. The free swimming fauna (Nekton) includes *Paramecium, Euglena, Ranatra, Corixa, Dytiscus,* larvae of *Culex* and of *Chaoborus, Gerris, Gyrinus,* etc. Among the vertebrates frogs, salamanders, snakes and turtles are the nektons of littoral zone. Floating members of the community (neuston) are whirliging beetles, water spiders and numerous Protozoa. Many pond fishes such as sun-fish, top minnows, bass, pike and gar spend much of their time in the littoral zone.

Biota of limnetic zone. Limnetic zone is the region of rapid variation, with the water level, temperature and oxygen content varying from time to time. In certain Himalayan lakes which have a glacial origin, change in water level is a conspicuous phenomenon, the onset of summer bringing forth a rapid and sudden rise in water level. Various protozoans which are capable of encystment during adverse ecological conditions, tardigrades such as *Macrobiotus,* rotifers such as *Rotaria, Philodina,* copepods, snails and frogs occupy the limnetic zone. Sedentary and slow moving forms are excluded from this zone because of predators and lack of permanent substratum for attachment. Many microscopic plants (*Volvox, Euglena*) and fishes also occupy this zone. The limnetic zone has autotrophs in abundance.

Biota of profundal zone. The deep (profundal) life consists of bacteria, fungi, clams, blood worms (larvae of midges), annelids and other small animals capable of surviving in a region of little light and low oxygen. In the profundal zone, autotrophs cannot produce food, and the main source of energy is detritus that rains out of the limnetic zone. All the organisms of this zone are heterotrophs, either as detritus feeders or as carnivores. Most of our largest lake fishes inhabit the dark waters of hypolimnion for most of their lives. In addition to detritus from the limnetic zone, food washed into lakes by rivers can settle out and serve as food for profoundal animals.

Benthos of lake bottom. The lake bottom in young lakes may be of the original rock; in older lakes, it will have been covered with sediment to form a uniform substrate of mud or sand. The benthos community includes several species of insect larvae, including those of small mosquito such as midges, burrowing mayflies, clams, snails, and tubeworms.

Distribution of Oxygen and Dissolved Nutrients in Lakes

The productivity of a lake depends on the amounts of oxygen and available nutrients dissolved in the water. As a general rule, the littoral zone behaves as a normal ecosystem, with oxygen being produced by photosynthesis within the community, as well as being constantly mixed into the water at the surface. Productivity within the littoral zone is directly controlled by the nutrients cycling within the littoral zone ecosystem and biogeochemical cycles apparently operate as do in terrestrial ecosystems, with sediments serving as a storage place for inorganic nutrients such as phosphate, and some rooted plants being able to draw nutrients out of the sediments through their roots (**McRoy** and **Barsdate**, 1970). But in the pelagic zone (limnetic and profundal zones), the distribution of oxygen and nutrients is exceedingly uneven, and this has a profound effect on the behaviour of the lake as a whole, especially in those whose pelagic zone is much larger than the littoral.

The epilimnion of a lake is typically well oxygenated, both because of the production of photosynthetic oxygen in the epilimnion and because of the oxygen uptake at the surface. The amount of nutrients dissolved in the water is highly variable, mainly because most of the nutrients in the epilimnion have been incorporated into living tissue. Thus, production of epilimnion is controlled by amount of available nutrients, which, in turn is controlled primarily by the rate of cycling within the epilimnion, the rate of loss to the hypolimnion of nutrient-rich detritus, and the rate of return of soluble nutrients from the hypolimnion during overturn. The fact that only heterotrophic metabolism proceeds in the hypolimnion means that the level of production is determined by the amount of detritus raining into it. As the decomposition of detritus releases soluble nutrients, therefore, there is an inverse relationship between oxygen concentration and concentration of vital nutrients. It is paradoxical that the greater the productivity of a lake the lower the oxygen concentration in its hypolimnion. For example, oligotrophic lake which is virtually sterile biologically has low production but its water remain well saturated with oxygen from top to bottom (**orthograde oxygen curve**). Likewise, a eutrophic lake has high productivity, but there is substantial oxygen depletion from top of the hypolimnion to the bottom, with depletion most extreme in the bottom layers (**clinograde oxygen curve**). However, in most lakes, oxygen is thoroughly mixed and is essentially at saturation during overturn.

LOTIC ECOSYSTEMS

Running-or moving-water or lotic ecosystems include rivers, streams, and related environments. They are remarkably variable, ranging in size from Ganga, Yamuna, Hindon, Kali Nadi, Sutlaj, Gomti, Brahmaputra, Narmada, Mahanadi, Kaveri, Krishna, Godavari, etc., to the trickle of a small spring. They vary from raging torrents and waterfalls to rivers whose flow is so smooth as to be almost unnoticeable. A given river varies considerably over its length, as it changes from a mountain brook to a large river. This is most noticeable in the abiotic factors of the environment, but all features of the ecosystem vary in response.

Characteristics of Lotic Environment

Moving waters differ in the three major aspects from lakes and ponds : current is a controlling and limiting factor; land-water interchange is great because of the small size and depth of moving water systems as compared with lakes and oxygen is almost always in abundant supply except when there is pollution. Temperature extremes tend to be greater than in standing water. Besides these, the most distinctive features of moving water ecosystems are those related to their motion, *i.e.,* the **rate of flow** and the **stream velocity**. The rate of flow or discharge refers to the volume water passing a given observation point during a specific unit of time; it is measured in units such as m^3/sec., ft^3/sec., or acre-feet/sec., one acre-foot is equivalent to a volume of water 1 acre in area by 1 foot deep). It tends to increase steadily growing down stream, as tributaries join with the main river. The **velocity of flow** is the speed at which the water moves, and is measured in m./sec., ft./sec., or mi./hr. Velocity is variable, but it also tends to increase downstream with increasing discharge. A factor that may be even more important to the biotic community is the **turbulence**, or the irregularity of the motion of the water particles. Perfectly even flow, in which water particles move parallel to one another, is called **laminar flow**. It is contrasted with **turbulent flow**, in which the movement of water particles is highly irregular. In highly turbulent water the erosive power is great, the sheer forces at the water-sediment interface at the bottom of the river are powerful, and the amount of oxygen incorporated into the water is very high. In streams with laminar flow the erosive power is lower, the sheer forces at the bottom of the stream are lower, and relatively less oxygen is incorporated into the water than turbulent waters.

The moving-water ecosystems can be divided into following two ecosystems depending upon the velocity and rate of flow of water body :

1. Rapidly flowing water and 2. Slowly flowing water.

A. Rapidly Flowing Water

Rapidly flowing water (of fast moving torrential streams, etc.) can be defined as the portion of the stream in which the flow is both rapid and turbulent. Everything that is not attached or weighted is swept away by the current ; this includes organisms and sediment particles alike. The substrate tends to be rock or gravel, and the fragments are rounded and smoothed by the water. The habitat itself is an extremely diverse one. Physical parameters such as sheer force and rate of water movement tend to be quite different on top of a rock fragment, between rock fragments, or beneath rock fragments, and different species can exploit these differences in microhabitats. Aquatic plants, in addition, provide microhabitats for some torrential animals. The rarely occurring phytoplankton of **rapid** or **riffle zone** of stream includes diatoms, blue-green and green algae (*e.g., Cladophora, Ulothrix*) and water moss (*e.g., Fontinalis*).

1. Plant inhabiting torrential forms. Some animals of rapidly flowing water streams live among the mosses and flowering plants such as *Eriocaulon miserum, Hydrolyum lichenoides,* and *Duroea wallichii.* Animals inhabiting these plants have a torpedo-like body which enables them to offer minimum resistance to the current. Some animals such as funnel mouthed tadpoles of *Megalophrys* and tipulid larvae live entangled in the roots of these plants. They possess devices for fixation and have hydrostatic organs.

Some animals such as *Cephalopteryx, Helodes, Phalacrocera, Gammarus,* etc., live among leaves and stems of aquatic plants. All these animals possess hook-like structures on their body helping them in anchoring on leaves and stems. The larvae of simulids and chironomids, however, live on the exposed surfaces of plants.

2. Rock inhabitant forms. The animals that live on the tops of exposed rocks have an efficient mechanism for staying in one place, otherwise, they will be swept away. The organisms in this microhabitat are universally flattened. Some, like the fresh-water limpet, *Ferrissia,* are virtually flat, offering little resistance to the current they hold themselves in place with a very large powerful foot that extends over almost the entire area of the shell. Others, such as water penny, the larvae of the riffle beetle, and *Baetis* larvae, are not only almost flat, but each of their legs possesses hooked claws that enable them to hold onto the substrate more firmly. Larvae of *Simulium* and chironomids cling to rock top by grapling hooks at the posterior end of the body. Caddishfly worms contain both of these structures (claws on legs and hooks at posterior end of body). The mayfly nymphs *Iron* and *Psephenus* attach themselves to the rocks by means of functional pads. Some animals, including freshwater sponges, actually cement themselves to the surface of the rocks. Others, such as caddish flies, build "houses" out of sediment or wood fragments, which are then cemented firmly to the rocks. Exposed surfaces of rocks also contain few sessile algae and few plants. A significant amount of organic detritus is washed into rapid-water ecosystems from up stream and adjacent terrestrial areas, however. This is the source of the most of energy in fast water ecosystems, and is much more important as an energy source than primary production within the stream itself. In consequence, the majority of primary consumers of such a microhabitat are detritus-eaters.

3. Inhabitant forms of spaces between rocks. Numerous different kinds of animals live in the spaces between the rocks. Many of them such as mayfly and stonefly naiads, are flattened and have certain behavioural adaptations to hold them in place. These include **thigmotaxis**, by which they cling instinctively to any hard surface such as rock or another insect larva. **Rheotaxis**, by which the animals orient themselves to face into current and move upstream is another adaptation of this sort. In many of these naiads, the combination of the organism's shape and behaviour is such that the current presses the insect tightly against the rock, increasing the friction between the animal and its substrate. Other insect larvae, such as the hellgramite, are large and covered with spines. Their size make it somewhat harder for the current to sweep them away, while the spines help in holding the larvae in places between stones.

4. Inhabitants occurring beneath rocks. Many species such as flatworms, annelid worms, other insect larvae, clams and some species of snails, live beneath the rocks. The current is weaker here, and animals are less likely to be carried away. They lack any special adaptation.

Finally, if the current is sufficiently slow, certain swimming organisms such as fish, will be present. The fish of fast-water ecosystems tend to be stenothermal cold-water fish, such as trout. The fishes of fast-water ecosystems have small body size and streamlined body. Trout and masheer are capable of moving against the current or waterfall by muscular effort. Majority of torrential fishes, however, live on or among rocks and boulders, over which water flows very swiftly. Small loaches (*Noemacheilus*) or loach-like fishes (*Ambyceps olyra*) are met with at the bottom and limpet-like fishes (*e.g., Glyptosternum, Balitora*) or fishes with special modifications (*Garra, Glyptosternum, Balitora*) occur on the rocks. The loaches and carps adhere to rocks like the limpets, with the ventral surface, inclusive of paired fins forming one large sucker, *e.g., Gastromyzon* of Borneo, the Indian loach *Baletora* of the eastern Himalayas and Assam hills (**Hora**, 1947)

Physico-chemically, the fast-water ecosystems resemble with cold, deep lakes. Water temperatures tend to be quite low, productivity is also quite low, and diversity is high. In the fast-water ecosystem, the main control of productivity is the current, which seriously limits the amount and type of autotrophs production that can take place.

B. Slowly Flowing Water

A slowly flowing water ecosystem is a very different type of system from the fast streams. Because the flow is both slower and more likely to be laminar, the erosive power of the stream is greatly reduced, and smaller sediment particles (silt) and decaying organic debris, instead of being carried away by the stream, are deposited on the bottom. In addition to these the slow streams have a higher temperature, consequently, planktonic organisms, especially protozoans, occur in large numbers in this ecosystem. The detritus-feeding benthos of slow water ecosystems include those which either live on the bottom, such as isopods (sowbugs), molluscs (*Sphaenius, Pisidium,* and *Anodonta dominata*), and mayfly and damselfy naiads, or which burrow into the sediment, such as tubeworms, naiads of the burrowing mayflies, *Sialis* (alderfly), midge *Chironomus*, and several other insect larvae, as well as, clams, nematodes, snails and rotifers. Swimming organisms are also abundant, including not only fishes, such as carps, cat fishes, suckers, stingers, spoonbills, etc., which tend to be different species from those of fast-water areas, but also larger Crustacea such as amphipods (fast-water shrimps). Finally, several insects spend most of their time at the surface of the stream. These include forms such as watrer striders, water boatmen, backswimmers, and predaceous diving beetles. Zooplankton is abundant, including a rich assemblage of Protozoa and smaller Crustacea such as Cladocera (water-fleas) and copepods. The richness of food afforded by slow streams attract a large number of reptiles (water-snakes, crocodiles, turtles) and amphibians.

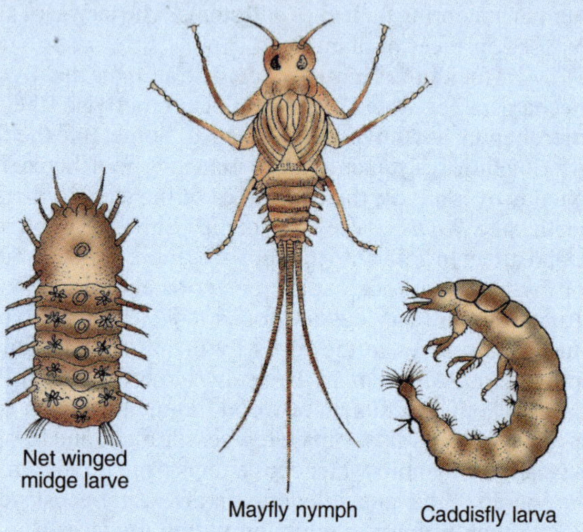

Fig. 11.6. Some inhabitants of lotic environment.

Plant life is also abundant in a slow-water ecosystem. It includes rooted vascular plants such as pond weeds and grasses, firmly attached aquatic mosses and large multicellular filamentous algae. Motile algae such as diatoms and flagellates may abound in the open water. In all, the primary productivity of slow-water ecosystem is higher than that of the rapid-water ecosystem, and the community is relatively less dependent on food materials from outside. Further, the detritus food chain consisting of organisms such as bacteria and fungi is much better developed in this community as the partially decomposed organic debris that comprises the main food source for these organisms collects in the mud bottom. In some slow-moving streams, in fact, the bottom muds contain more organic material than mineral fragments.

In slow-water streams oxygen concentration remains main limiting factor. The high level of animal activity coupled with an active detritus food chain, can withdraw a large amount of oxygen from water of a slowly moving stream. In addition, the low level of turbulence means that less oxygen is incorporated into the water at the surface. Thus, the dissolved oxygen content of a slowly moving stream is likely to be much lower than that of a fast-moving stream. This is commonly reflected in the fauna of the stream. Fishes tolerant of low oxygen levels such as carps and catfishes are the most common fishes in slow water, while species with high oxygen demands, such as trout are found in fast water.

REVISION QUESTIONS

1. Classify the freshwater environment. Describe lentic ecosystem with suitable example.
2. What is lotic habitat ? Describe its physico-chemical characteristics and flora and fauna.
3. Write short notes on the following :
   (a) Limnology ; (b) Lakes ; (c) Biotic zonation of a deep lake ; (d) Thermal stratification of lake.

CHAPTER

12

Aquatic Ecosystems : Estuaries and Marine Communities

Habitat of marine fish.

ESTUARINE ECOLOGY

Waters of all streams and rivers ultimately drain into the sea ; the place where this freshwater joins the salt water is called an **estuary**. An estuary (L. *aestus*—tide) is the wide lower tidal part of a river which is strongly affected by tidal action. Estuaries are the transitional zones (ecotones) between the sea and rivers and are the sites of unique ecological properties possessing a characteristic biological make up. They are semienclosed coastal bodies of water that have a free connection with the open sea and within which sea water is measurably diluted with freshwater from rivers. Examples of estuaries are river mouths, coastal bays, tidal marshes and water bodies behind barrier beaches. Not all rivers open into estuaries : some simply discharge their runoff into the ocean. Estuaries differ in size, shape and volume of water flow, all influenced by the geology of the region in which they occur. As the river reaches the encroaching sea, the stream carried sediments are dropped in the quite water. These accumulate to form **deltas** in the upper reaches of the mouth and shorten the estuary. When silt and mud accumulations become high enough to be exposed at low tide, **tidal flats** develop, which divide and braid the original channel of the estuary. At the same time, ocean currents and tides erode the coast line and deposit material on the seaward side of the estuary, also shortening the mouth. If more material is deposited than is carried away, then **barrier beaches**, **island** and **brackish lagoons** appear.

Type of Estuaries

Estuaries are classified in various ways depending on diverse criteria such as geomorphology, water circulation and stratification and systems energetics. Thus, on the basis of geomorphology, estuaries include (1) drowned river valleys, (2) ford-type estuaries, (3) bar-built estuaries, (4) estuaries produced by tectonic processes, (5) river delta estuaries, etc. Based on water circulation and stratification, estuaries are classified as (i) highly stratified or salt-wedge estuaries, (ii) partially mixed stratified estuaries, (iii) completely mixed or vertically homogeneous estuaries and (iv) hypersaline estuaries. On the basis of the ecosystem energetics, the estuaries have been classified by **Odum** *et al.*, (1969) as (1) physically stressed systems of wide latitudinal range, (2) natural arctic ecosystems with ice stress, (3) natural temperate ecosystem with seasonal programming, (4) natural tropical coastal ecosystems of high diversity, and (5) emerging new systems associated with man.

Physico-Chemical Aspects of Estuaries

Current and **salinity**, both very complex and variable, shape life in the estuary, where the environment is neither freshwater nor salt. Estuarine currents result from the interaction of a one-direction stream flow, which varies with the season and rain fall, with oscillating ocean tides, and the wind. The salinity of estuaries varies vertically and horizontally and fluctuates amazingly between 0.5–0.35 per cent. Due to low salinity freshwater has a lower density than sea water (± 1.00, as opposed to 1.03). Were there no tides in an estuary to mix fresh and salt-water the lighter freshwater would simply flow over the heavier sea water and dissipate in the ocean. However, the tidal action acts as a plunger to thoroughly flush the estuary and mix the fresh and salt water. This has several results. The water level in the estuary fluctuates regularly, unlike that of river. For this reason the habitat that is covered at high tide and uncovered at low tide is a prominent one in the estuary, and has no analogue in any purely freshwater ecosystem. Further, the salinity is exceedingly variable and may change by a factor of ten over the course of a day at any location. At low tide, most of water passing through the estuary is fresh-river water, and the salinity is correspondingly low. At high tide, most of the water may be of marine origin, and the salinity correspondingly high. Flow may be somewhat stratified, or mixing may be incomplete, and the salinity at any time may vary considerably from one location to another. The intensity of current, and, hence, of the degree of mixing, is a function of the intensity of the tides and of the rivers flow rate. In other words, an estuary is an exceedingly variable environment. Further, salinity of estuaries remains highest during the summer and during periods of drought, when less freshwater flows into the estuary. It is lowest during the winter and spring, when rivers and streams are discharging their peak loads. No stenohaline organism (such as Echinodermata, Cephalopoda, and other molluscs) could hope to survive in the estuaries. Likewise, no organism that could not tolerate strong currents and the turbid water that results from strong currents could live there.

The temperatures in estuaries fluctuate considerably diurnally and seasonally. Waters are heated by solar radiation and inflowing and tidal currents. High tide on the mud flats may heat or cool the water, depending on the season. The upper layer of estuarine water may be cooler in winter and warmer in summer than the bottom, a condition that, as in a lake, will result in a spring and autumn overturn.

Anyhow, all estuaries have high productivity. As estuary's high productivity is exceeded among aquatic communities only by coral reefs. The reason is that although an estuary may be a harsh environment in some ways it is not in others. Because the typical estuary is shallow and turbulent, the amount of dissolved oxygen tends to be fairly high although it may be low in bottom layers of those estuaries where water does not mix from top to bottom. More important, the tidal action acts to accumulate (concentrate) the nutrient and energy materials that wash in from upstream or, in some cases, that enter from the nutrient rich bottom waters of the sea.

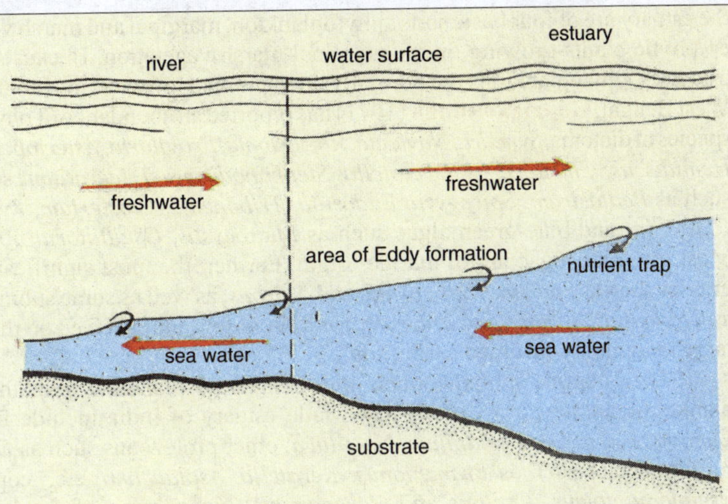

Fig. 12.1. Diagram showing formation of nutrient traps and horizontal turbulance or eddy.

The mechanism of nutrient concentration is very simple—particulate nutrient material enters the estuary at its upper end is carried seaward by the falling tide and is brought back through the estuary by the rising tide, and so on for several cycles. The length of time it takes from a nutrient particle to traverse the estuary is substantially greater than it would take for it to traverse a similar length of even the most slowly flowing river. Thus, the estuary acts as a nutrient trap, with an average nutrient level significantly higher than either the river or the sea that it connects. Likewise, the concentra-tion of energy-rich organic materials remain high in estuaries.

The primary result of the concentration of nutrients and fixed carbon is a very high level of production within the detritus food chain. This has two aspects. First the nutrient material is broken down bacterially at a very high rate, and recycled back into soluble form. This allows a very high rate of gross primary production for estuarine plants. More important, perhaps, the amount of nutrient-rich organic detritus allows a level of productivity for detritus-eating animals much higher than could be maintained on the basis of the primary production by estuarine plants.

Biotic Communities of Estuaries

Carikker (1967) has classified the regions of estuaries into the **head**, where freshwater enters the estuary, **upper**, **middle** and **lower reaches** with increasing range of salinities and the **mouth** with salinity nearly equal to the sea. He has also classified the animals inhabiting the estuarine region into **oligohaline** (0.5 to 5 per cent), **mesohaline** (5 to 18 per cent) in the upper reaches, **polyhaline** (18 to 25 per cent) in the lower reaches, 25 to 30 per cent in the middle and **euhaline** (30 to 50 per cent) in the mouth of the estuary. Communities of estuaries are a mixture of endemic species and those which come in from sea. **Krishnamoorthy** (1963) in fact reported that the extent of penetration of the polychaets, *Marphysa gravelyi, Diopatra uariabilis, Clymene insecta, Loimia medusa, Glycera embranchiata, Onuphis eremita* into the Adyar estuary in Madras from the Bay of Bengal vary with varying salinities at constant temperatures.

Furthermore a estuary being a transitional zone between the freshwaters and seas is an ecotone and, therefore, typical estuarine forms are unique in their habitat. Thus, the estuarine community is a mixture of three components: the marine, the freshwater and the brackish water. However, the diversity of both freshwater and marine components reaches a maximum. Total diversity of the estuarine community is lower than that of either of the more normal environments. The number of interactions between species is not high and some estuarine populations may fluctuate greatly in size. The plant of the estuary are of four basic sorts : phytoplankton, marginal and marsh vegetation, mud-flat algae, and epiphytic plants-growing on the marginal marsh vegetation. Because of the turbid water found in estuaries, phytoplanktons are normally uncommon. However, in Hooghly-Matla estuarine system in West Bengal, **Gopalakrishnan** (1971) has reported an abundance of phytoplanktonic forms—several species of diatoms, *Synedra, Navicula, Rhizostoma, Fragilaria, Asteriobella, Biddulphia, Planktoniella, Hemidiscus, Chaetoceros, Cyclotella, Stephanodiscus, Triceratium*; several species of green algae such as *Pediastrum, Spirogyra, Eudornia, Tribonema, Closterium, Zygonema, Pandorina, Volvox, Chlorella;* and blue-green algae such as *Microcystis, Oscillatoria, Anabaena,* and *Trichodesmus*. Most estuarine algae are of marine origin. Further, the most significant estuarine plants are marsh grasses such as *Spartina, Salicornia,* and *Scirpus*, as well as some submerged filamentous algae such as *Cladophora, Chara* and *Enteromorpha.* Very few animals feed on these plants directly, but a very large amount is consumed as detritus.

The estuarine animal communities include zooplanktons and other animals. For example, the planktonic animal forms of Hoogly-Matla estuary of India include flagellate protozoans such as *Euglena, Ceratium, Peridinium, Noctiluca*; other protozoans such as *Difflugia, Arcella, Vorticella;* rotifer species such as *Branchionus, Keratella, Asplanchna,* etc.; copepodes such as *Diaptomus, Pseudodiaptomus, Cyclops* and *Paracalanus;* cladocerans of the genus *Bosmina, Bosminopsis Ceriodaphnia, Moina,* besides a number of isopods (**Gopalakrishnan**, 1971). All these forms remain confined to water and lead a pelagic existence. The best known estuarine animals are detritus feeders such as oysters (*Ostrea* sp., of Lake Chilka), clams, lobsters, and crabs. Several insect larvae, annelid worms, molluscs, enter the estuary from the freshwater, while most marine phytoplankton, Crustacea, annelid worms, anemones and Bryozoa enter an estuary from marine ecosystems. For example, out of 130 species of fishes and 30 species of prawns of Hoogly-Matla estuary, several fishes such as *Hilsa, Harpodon, Mugil, Trachyurus,* etc., migrate from sea in to the mouth of the estuary, to form important fish catches, while some such as palaemonid prawns are freshwater inhabitants and come to live in estuaries.

Adyar estuary of Madras coast is found rich in invertebrate and vertebrate fauna (**Aiyar** and **Panikkar**, 1937). The common invertebrates of this estuary are sea anemones (*e.g., Phytocoetes gangeticus, Phytocoeteopsis ramunnii, Stephensonaetis ornata, Pelocoetes exul, Bolocractis gopalai,* etc.); hermaphrodite nereids (*e.g., Lycastis indica*); the tubicolous polychaet *Diopatra variabilis*; the burrowing polychaet *Marphysa gravelyi*, the oligochaet *Pontodrilus bermudensis*; the mysids *Rhopalophthalmus egregius* and *Mesopodopsis orientalis*; the crabs *Uca annulipes, Uca triangularis, Neptunus pelagieus, Scylla serrata, Veruna litterata, Sesarma tetragonum*; the hermit crabs — *Clibonarius olivaceus* and *Clibanarius padavensis*, and several molluscs (*e.g., Ostraea arkanensis, Modiolus undulatus, Meretrix casta, Cuspidaria cochinensis, Stenothyria blanfordiana, Amnicola stenothyroides, Potamides cingulatus, Styliger gopalai* and *Bursatella leachii*). The vertebrate fauna of Adyar estuary includes the fishes such as *Etroplus maculatus, Etroplus suratensis, Acentrogobius virdipunctatus, Acentrogobius neilli, Panchax parvus, Aplocheilus melastigma,* and mud skipper *Periophthalmus koelreuteri* and snakes such as *Natrix piscator* and *Cerberus rhyncops.*

Subsystems of Estuaries

An estuary consists of several basic subsystems linked together by the ebb and flow of water that is driven by the hydrological cycle and the tidal cycle. Some of its main subsystems are the following:

1. Shallow water production zones. In these zones, the rate of primary production exceeds the rate of community respirations. The producers are reefs, seaweed or sea grass beds, algal mats and salt marshes. This system exports energy and nutrients to deeper waters of the estuary and adjacent coastal shelf. Being rich in diverse types of producers these remain "programmed" for virtually year-round photosynthesis and derive benefit from tidal action in creating a subsidised fluctuating water level ecosystem. They have all the three types of producers such as **macrophytes** (*e.g.,* see weeds, sea grasses and marsh grasses), **benthic microphytes** and **phytoplanktons**. Certain significant macrophytes are *Spartina, Zostera* and *Thalassia*. Benthic algae grow on macrophytes, sessile animals, rocks, sand and mud. Some estuaries have distinct blooms such as **red tides** of large blooms of red pigmented dinoflagellates (*i.e., Gonyaulax* and *Gymnodinium*).

2. Sedimentary subsystems. These subsystems occur in the deeper channels, sounds (=straits) and lagoons in which respiration exceeds production and in which particulate and dissolved organic matter from the production zone is used. In these habitats, nutrients are regenerated, recycled, stored and vitamins and growth regulators are manufactured. The consumers of these zones are versatile in their feeding habits.

3. Plankton and nekton. These organisms move freely between the two abovesaid fixed subsystems. They continue producing, converting and transporting nutrients and energy while responding to diurnal, tidal and seasonal periodicities. Holoplanktons tend to comprise relatively few species, while the meroplanktons being more diverse reflecting a variety of benthic habitats.

MARINE ECOSYSTEMS

Marine ecosystem includes the salt water mass of seas and oceans that covers 70.8% of the earth's surface and contain about 97% of the planet's water; a volume of 1370×10^6 km^3. The average depth of oceans is 3730 m. Oceans strongly influence climate by acting as a reservoir of solar heat, thus, improving temperature extremes, and as the source of water which falls as precipitation (*i.e.,* hydrologic cycle). The **Pacific ocean** is the largest (179.7×10^6 km^2), coldest (average 3.36°C), least saline (34.6 parts per thousand) and deepest (average depth 4028m). The **Atlantic ocean** is warmer (average 3.73°C), more saline (average 34.9 parts per thousand), and relatively shallow (average depth 3310 m); it also has the longest coastline providing many natural harbours. The smallest and shallowest ocean is the **Arctic.** The **antarctic (Southern) ocean** shows wide variations in temperature (–1.8—10°C). The **Indian ocean** covers 77×10^6 km^2 to an average depth of 3872 m.

The volume of surface area of marine environment lighted by the sun is small in comparison to the total volume of water involved. This and the dilute solutions of nutrients limits production. All the seas are interconnected by currents, dominated by waves, influenced by tides and characterized by saline waters. Not only the seashore and banks which are the homes of many organisms but the open ocean, many hundreds of miles away from land, supports plant and animal communities of great diversity and complexity.

Physico-Chemical Aspects of Marine Environment

In the marine environment, the most important physical factors which influence marine life are light, temperature, pressure, salinity, tides and currents.

1. Light. Light is a very significant factor in regulating the pattern of distribution of marine animals and it contributes significantly to organic production. The autotrophic primary producers exploit the light energy in the photosynthetic production of food for primary macroconsumers of marine ecosystem. The amount of light exploited in photosynthesis depends on intensity of light and turbidity of water. For example, a ten meters deep turbid coastal zone receives an equal amount of light energy as a hundred meters deep clear oceanic zone, but the maximum intensity at a ten meters coastal water is greater. Light determines diurnal migrations of marine organisms, and it also regulates colour pattern of marine animals. The deep sea fauna which lives in total darkness exhibit either colourlessness or uniform colouration. It is also somehow related with the development of visual sense organs as shown by absence of functional eyes in deep-sea animals.

2. Temperature. The range of temperature in sea is far less than that on land. Arctic waters at 27°F (–2.8°C) are much colder than tropical waters at 81°F (27.2°C), and currents are warmer or colder than the waters through which they flow. Seasonal and daily temperature changes are larger in coastal waters than on the open sea. The surface of coastal waters is the coolest at dawn and the warmest at dusk. In general, sea water is never more than 2° to 3° below the freezing point of freshwater or higher than 81°F (27.2°C). At any given place the temperature of deep water is almost constant cold, below the freezing point of freshwater. Unlike freshwater, sea water does not have a density maximum at 4°C;

rather it becomes continuously denser as it gets colder.

3. Pressure. Pressure in the ocean varies from 1 atm at the surface to 1,000 atm at the greatest depth. Pressure changes are many times greater in the sea than in terrestrial environments and have a pronounced effect on the distribution of life. Certain organisms are restricted to surface waters when the pressure is not so great, whereas other organisms are adapted to life at great depths. Some marine organisms, such as the sperm-whale and certain seals, can dive to great depths and return to the surface without difficulty.

4. Zonation of marine environment. Just as lakes exhibit stratification and zonation, so do the seas. The ocean itself is divided into two main division, the **pelagic**, or whole body of water, and the **benthic**, or bottom region (Fig. 7.3). The pelagic region is further divided into two provinces : the **neritic** (near shore), water that overlies the continental shelf, and the **oceanic provinces**. The **continental shelf** is the underwater extension of the continent and it generally extends to a depth of roughly 125—200m. The edge of the continental shelf may be within a few kilometers of the shore, or it may be several hundred kilometers from shore. From the edge of the continental shelf there is a more rapid descent, the **continental slope**; to the broad flat **abyssal plain** that underlies most of the ocean at a depth of 4,000—5,000m. On the abyssal plain, there are extensive mountain ranges, or **midoceanic ridges**, some of which have tips projecting above sea level. In addition, there are very deep troughs, which drop down below 11,000 m., below sea level. Life exists from the very top of sea to the bottom in virtually all areas, although its abundance is exceedingly variable.

The benthic life zones are defined in terms of following physical subdivisions : The **littoral**, or **intertidal** zone, is the zone between high tide and low tide levels. The **sublittoral** extends from the low tide mark to the edge of the continental shelf, with the **bathyal zone** comprising the continental slope. The **abyssal zone** includes the abyssal plains, and the **hadal zone** includes any life in the deep trenches below 5,000 m.

The pelagic zone is divided into three vertical layers or zones : from the surface to about 200 m is the **photic zone**, in which there is sharp gradients of illumination, temperature, and salinity. The region below this zone is called **aphotic zone**. From 200 to 1,000m where very little light penetrates and where the temperature gradient is more even and gradual and without much seasonal variation, is the **mesopelagic zone** of aphotic zone. It contains an oxygen-minimum layer and often maximum concentrations of nitrate and phosphate. Below mesopelagic is the **bathypelagic zone**, where darkness is virtually complete except for bioluminescence, and where temperature is low and pressure is great.

5. Stratification of marine environment. The upper layers of ocean water exhibit a stratification of temperature and salinity. Depths below 300m are usually thermally stable. In high and low latitudes, temperatures remain fairly constant throughout the year. In middle latitudes temperatures vary with the season, associated with climatic changes. In summer the surface waters become warmer and lighter, forming a temporary seasonal thermocline. In subtropical regions the surface waters are constantly heated, developing a marked permanent thermocline. Between 500 and 1,500m a permanent but relatively slight thermocline exists.

Associated with a temperature gradient is a salinity gradient or **halocline**, especially at the higher latitudes. There the abundant precipitation reduces surface salinity and causes a marked change in salinity with depth. Thus, in the middle latitudes in particular, the two produce a marked gradient in density. Water masses form density layers with increasing depth. Because density of seawater does increase in depth and does not reach its greatest density at 4°C as with freshwater, there is no seasonal overturn. This results in a normally stable stratification of density known as the **pycnocline**. Because there are marked changes in temperatures with depth in the open ocean and, thus, with density, the pycnocline is the deep zone which is oxygen rich, cold and comprises 80 per cent of the ocean.

6. Salinity. The sea is salty, with an average salinity of 35 parts of salt (weight basis) per 1000 parts of water or 3.5 percent, that is usually written as 35%, *i.e,* parts per 1000 (in contrast the salinity of freshwater is 0.5%). The main salts are chloride, sulphates, bicarbonates, carbonates and bromides of sodium, magnesium, calcium and potassium. Of these salts, sodium chloride is present in maximum amount. The salinity of marine water fluctuates from place to place. The marine animal life has specific osmoregulatory adaptation for high saline sea water. The absence of many animal species in marine environment has been related to their inability to tolerate the high salt contents of sea water. Except few insects such as *Halobates* and shore collembolans such as *Isotoma* and *Sminthurus,* most insects do not occur in marine ecosystems.

7. Currents and tides. Sea water is never static and waves, currents, and tides are the regular features of sea water. All these phenomena are controlled by winds, cosmic forces and varying water densities. Most world-wide water currents occur at the surface and at great depths of sea water. These currents determine the interchange between the surficial and deep water masses, as well as horizontal movements. Both the horizontal and vertical movements of ocean water are significant, but for different reason. Within a given current, such as warm Gulf stream, for instance, the water mass retains its identity for great distances, and the warm-sea community ranges far north of what have been anticipated as its northernmost limits. In addition, the climate of terrestrial ecosystems is strongly affected by the nature of adjacent water masses. The prevailing winds blow across the ocean and are heated or cooled by the waters they traverse. In marine ecosystems, however, the oceanic current system controls the distribution of productivity of marine communities through vertical mixing of water masses. Because most of the ocean is permanently stratified, any essential nutrient that settles out of the upper layers, either as an inorganic mineral or as organic detritus, is not returned to the upper layers at the same location. This had led to such an improvisation of the surficial waters of the oceans that productivity in most parts of the open oceans is about the same as that of desert, because of the very low concentrations of nutrients, especially phosphorus. At the same time, the lowest waters of the ocean are nutrient-rich. There are certain regions in the world, notably at the west coasts of continents and in subpolar latitudes in both hemispheres, where deep currents rise to the surface. These are called **zones of upwelling**, and even though their waters are very cold, they are sites of intense production because of their relatively high concentration of nutrients. Nutrients are then cycled back to the rest of the ocean via surface currents.

Tides.

Tides are water movements which are caused by some astronomical factors (*i.e.,* due to a pull by moon and sun). They represent a rhythmic rise and fall of water and often waves of long wave lengths characterize the process. The chief astronomic force behind the tidal rhythm is the attraction of the moon and sun which depends on the movement of earth in relation to the moon and the sun. Tides influence the sea-shore fauna variously, as low tides expose the shore and high tides flood the substratum.

MARINE COMMUNITIES

The sea biota is not abundant but contains well marked diversity. Every major group of algae and almost every major group of animals can be found some where in the oceans. The only striking omissions are the vascular plants and insects, which are abundant in estuaries but which have few or no marine representatives. Communities of marine environment includes : (1) **producers** such as phytoplanktonic diatoms and dinoflagellates, green algae, brown algae, red algae and kelps; (2) **consumers** such as zooplanktons, benthos, nekton and neuston and bacteria. The marine biotic communities can be studied separately for different life zones of the ocean as follows :

A. Biotic Communities of Oceanic Region

The oceanic region or pelagic zone is less rich in species and numbers than the coastal areas, but it has its characteristic species. Many of these are transparent or bluish and since the sediment-free water of the open sea is marvelously transparent, these animals are nearly invisible. Animals that are too thick to be transparent frequently have smooth shiny and silvery bodies which make them invisible by mirroring the water in which they swim. Further, the animals of the pelagic zone of ocean encounter stable and uniform environmental conditions due to the continuity of the waters of the sea. This continuity of sea waters causes the individuals of the open ocean to stay away temporarily or permanently from any solid objects, and they are provided with different locomotor organs. However, the planktonic forms lack a locomotor organ and float freely in the vast expanse of the sea.

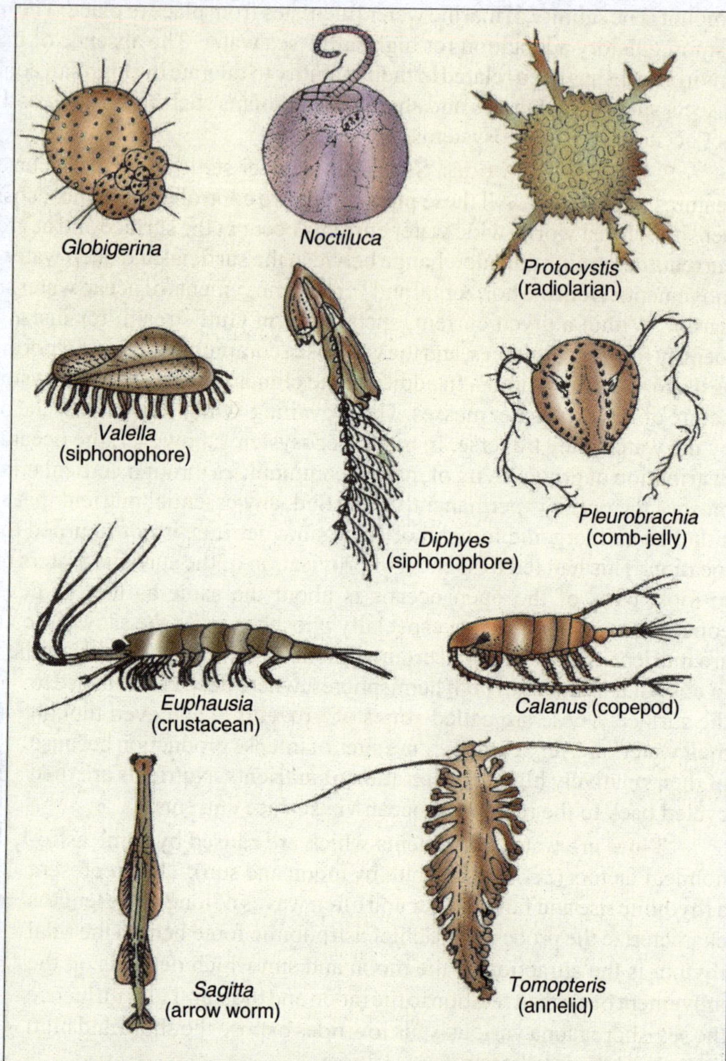

Fig. 12.2. Certain holoplanktons of pelagic zooplanktons.

(i) Biotic communities of pelagic zone. The pelagic plankton also called **epiplankton** are exceedingly diverse. **Pytoplankton** includes diatoms and dinoflagellates, which together produce most of the organic carbon in the sea (and most of the oxygen in the atmosphere), as well as other forms of golden-brown algae and flagellated green algae. Some seaweeds, such as the large brown algae *Sargassum*, may have a floating stage. The **zooplankton** of pelagic zone includes representatives of every major phylum and most minor phyla, either as permanent members of the plankton community (**holoplanktononic forms**) or as transients during their larval stages (**meroplanktonic forms**). Common among permanent planktonic forms are the Foraminifera and Radiolaria, arrow worm (*Sagitta*), certain annelid worms, swimming snails, jelly fishes and most abundant of all are the crustaceans such as shrimps, copepods and cladocerans. Among the temporary zooplanktons are the larva of large animals from all marine environments and even some freshwater environments. For example, in the zooplanktons of inshore water of Mandapam (South India) **Prasad** (1956) reported following animal species—flagellate protozoan *Noctiluca miliaris*; larval forms of coelenterates such as *Planula, Ephyra, Semper's* larva and *Cerianthus* larva; medusae of certain coelenterates such as *Bougainvillia, Obelia, Charybdea, Rhizo-stoma* and *Rhopalina;* cteno-phores such as *Pleurobrachia* and *Beroe;* larvae of many polychaet annelids and annelids such as *Tomopteris, Alciopa* and *Autolytus,* several species of chaetognaths such as *Sagitta* and *Spadella*; heteropod and pteropod molluscs such as *Carinaria, Pterotrachea, Cliona, Creseis;* larvae of

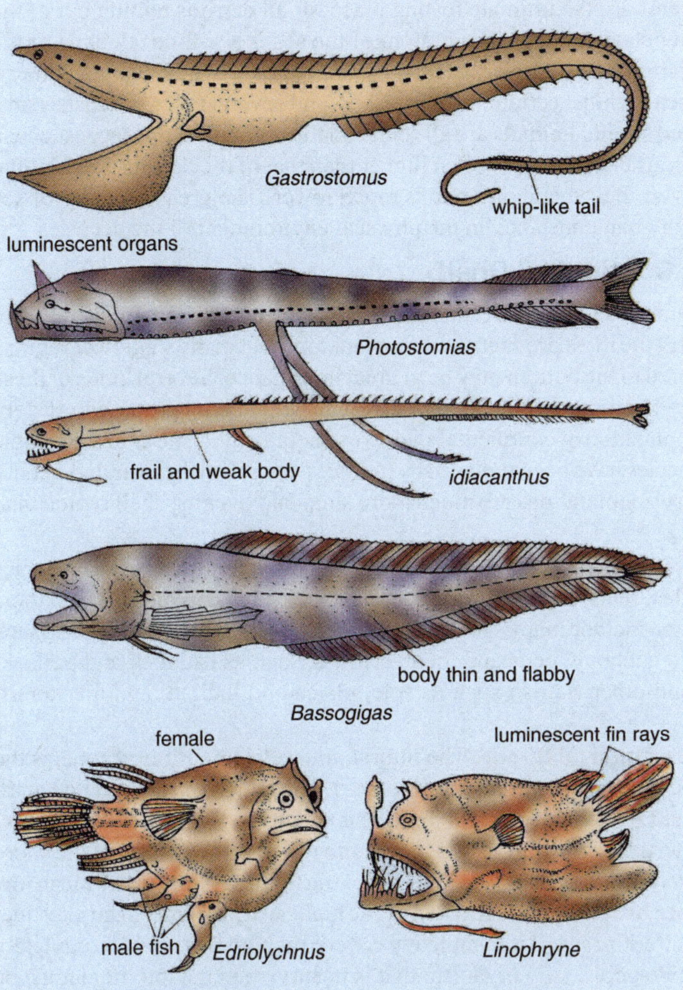

Fig. 12.3. Fishes of the deep water of the ocean.

echinoderms such as bipinnaria, ophio-pluteus, echinopluteus, auricularia; tornaria of *Balanoglossus*; species of copepods such as *Acartia, Calanopia, Labidocera, Centropages, Paracalanus, Canthocalanus, Acrocalanus, Eucalanus* and *Microsetella;* larvae of decapod *Lucifer*; protochordate species of Larvacea, Thaliaceae and Ascidiaceae such as *Thalia democratica, Jasis zonaria, Salpa cylindrica;* appendicularian *Oilopleura diocia, O. parava* and *O.cephalocera* and also cephalochordate *Amphioxus.*

The largest animals in the pelagic region are the nekton. These include cephalopods such as squid and nautili among the invertebrates, as well as many marine vertebrates, such as bony fishes, sharks, sea turtles and whales. In addition, sea birds also feed on many of the same food types as these large carnivores. Air-breathing nekton, such as the turtles, and whales, are found mainly in the photic zone, but fishes extend from the sea surface to the bottom. Certain fishes such as tuna, shark, sardine, mackerel, herring, bonito, and anchovy live near the surface. Those fishes which occur at greater depths are often bizarre and unlike any fish found at the surface. They tend to be small—15 to 20 m is large for the ocean deeps —and they are exceedingly dispersed. Many have luminescent appendages with which to lure prey, or mouth that looks too large for the rest of the body. Food is not plentiful in the deep waters of the sea, so these fishes must go for long periods of time without food, and then consume as much as they can when they have a chance.

(ii) Biotic communities of abyssal benthic zone. The abyssal benthic zone or deep sea of oceanic region is pitch dark and universal absence of light in this environment excludes the possibility of any growth of vegetation or other photosynthetic organisms. Consequently, most of the deep sea dwellers depend upon the detritus (dead body and excreta of surface forms) which sink to the sea bottom while a large number of species are active predators (carnivores). However the deep ocean benthos are surprisingly diverse. Food is a little more plentiful on the bottom than it is in the deep waters of the ocean, because the bottom sediments are the ultimate resting place for all detritus raining out of the upper layers (oceanic surface and pelagic zones). The bottom of the sea is a soft ooze, made of the organic remains and shells of Foraminifera, Radiolaria, and other animals and plants. Sea cucumbers, brittle stars, crinoids (sea lilies), sea urchins, certain benthic fishes, and several types of crustaceans, as well as sea anemones, clams, and similar animals are all found on the bottoms of the deepest ocean trenches. The biomass of carnivorous brittle stars is often higher than that of the detritus feeders that serve as their food source; however, their metabolic rate is much lower. The great diversity of sea bottom fauna has been related with great constancy in the physical environment of the deep sea.

B. Biotic Communities of Continental Shelf

The communities of continental shelf are both richer and more diverse than those of open ocean. Diatoms and dinoflagellates are still the most productive phytoplanktons, but in the shallower regions green, brown, and red algae anchored to the bottom may be of great importance. Several kinds of these seaweeds are harvested from rocky shores as human food or for some commercial purpose in eastern countries, USA and India. The zooplankton of continental shelf are generally the same as in the pelagic region, but some purely pelagic species have been replaced by neritic species, and the overall diversity is somewhat higher. The temporary zooplanktons are much more abundant over the shelf region than in the open ocean.

The necktons of the **neritic sea** are both diverse and well known. The significant neckton species of the region are large squid, whales, seals, sea otters, and sea snakes. The most numerous necktonic forms, of course, are the fishes, which include many species of shark as well as herring-like species such as menhaden, herring, sardine and anchovy; cods and their relatives, such as haddock and pollack; salmon and sea trout; flounder and other flatfish such as sole, plaice and halibut; and mackerels, including tuna and bonito.

Communities of sea shore or intertidal zone. The littoral, intertidal or eulittoral zone, is the region of sea shore which exists between the high and low tide lines. The region of the high tide mark is called **supratidal** or **supralittoral zone**. The region of low tide is called **subtidal zone. Davenport** (1903) has divided sea shore or intertidal zone into **submerged zone** which is the portion of the shore below the low tide exposed by very low tides, a **lower beach** occupying the area between the mean low tide and mean high tide and an **upper beach** which is reached by the highest of high tides. The intertidal zone is the most variable zone in the entire sea. It is completely covered at high tide, and is completely uncovered at low tide except for tide pools. Any organism that is to survive in the intertidal must be either resistant to periodic desiccation or able to burrow to water level. For example intertidal polychaets escape desiccation by resorting to a tubicolous mode of life whereby they can resist desiccation by retaining sufficient moisture till the occurrence of the next wetting. Barnacles can survive desiccation by remaining in a state of suspended activity during dry periods. Many plants of sea shore contain jellylike substances such as agar, which absorb large quantities of water and retain it while the tide is out. Further, the majority of shore animals utilize oxygen dissolved in water and when the low tide exposes them during the periods of drought, they utilize the oxygen stored in the blood plasma, *e.g.*, nereid *Arenicola.* Branacles achieve this by enclosing an air bubble between the shells.

One of the notable characteristics of this region is the ever present action of the waves, therefore, the organisms living on a sandy or rocky beach have had to evolve ways of resisting wave action. Many seaweeds have tough pliable bodies, able to blend with the waves without breaking, while the animals either are encased in hard calcareous shells, such as those of mollusks, Bryozoa, starfish, barnacles and crabs, or are covered by a strong leathery skin that can bend without breaking, like that of the sea anemone and octopus. Many sedentary shore animals (*e.g.*, sponges, tunicates, etc.) have special modes of adhesion which keep them firmly attached to the substratum. However, animals of sandy beaches escape wave action by taking burrowing mode of life, because, these habitats lack a hard substratum for the attachment of the animals.

The biota of intertidal zone also have to over come the wide fluctuations in salinity and temperature. Thus, like the estuary, this variable environment is a zone in which an organism must be adapted to a broad range of environmental conditions. Like the estuary also, it is an area of very high productivity with a simple community, many of whose members may be exceedingly abundant. The diversity of biota of intertidal zone is determined by nature of substratum whether loose sand or mud or rocky coast. Ecologists have recognized three major types of intertidal zones or seashores, namely, **rocky**, **sandy** and **muddy**. **Eltringham** (1971) has included a fourth type of sea shore, the **pabble shore** is this classification. Each type of sea shore possesses specific biota each of which remains specifically adapted for its peculiar habitats. Here we will discuss only rocky shore and sandy beach habitats.

(a) (b)

Features of (a) sandy shore and (b) rocky beach.

1. Biotic communities of rocky shore. The rocky shore presents solid substratum for the attachment of many sessile animals which often remain abundant here. The animals which are sedentary and inhabit rocky shore are limpets such as *Patella, Haliotis, Fissurella,* and oysters, barnacles, tunicates and bryozoans. Sessile organisms such as the sponges, the colonial hydrozoans and anthozoans such as *Gammaria* and *Zoanthus* attached to rocks. Certain animals such as some sponges, annelids and molluscs either bore the soft rocks like limestone or seek protection in the crevices of hard rocks. All these sedentary sessile animals are adopted for filter feeding.

Further, all these animals of rocky shore occurs in successive zones. At the uppermost end is **a zone of bare rock** marking the transition between land and sea. Next is a **spray zone** with dark patches of algae on which the periwinkles (*Littoria*) graze. Below this is the zone called **barnacle zone** regularly covered by the high tide; rocks in this zone are encrusted with acorn barnacles, limpets and mussels (*e.g., Mytilus*). This zone is a vulnerable zone being exposed to the stress of wave action and active predation by starfishes and gastropods. The mussels act like hard substratum for tunicates, sponges and small mussles. The next one is the **zone of oysters** which is less vulnerable and includes green algae (*Enteromorpha* and *Ulva*) and barnacles. The next zone includes mussels, *i.e.,* boring bivalves, which remain within crevices with exposed siphons only to be protected from waves and predation. The chief species of borer bivalves are *Pholas, Martesia, Hiatella, Tridacna,* etc. Besides bivalves a variety of filter feeder bryozoans, and brachiopods, and tubicolous polychaets such as *serpulids, terebellids, sabellids* and *cirratulids* which live in tubes

Green alga: *Ulva*.

are the chief inhabitants of this zone. The subtidal zone includes red algae, brown algae (Laminarians and fucoids), sea anemone, sea urchin, corals, etc.

For example, the Okha coast of Gujarat in India is a typical rocky sea coast which has well defined fauna and flora (**Gopalakrishnan**, 1970). Dominant weeds of subtidal zone of Okha coast are *Sargassum, Dictyopteris, Graciliaria, Padina, Ulva, Polysiphonia,* etc. This subtidal zone also contains sea urchins, serpulid tubes of species *Pomatoleius crosslandi,* and mollusc species such as *Trochus radiatus, Cypraea arabica, Thais alveolata.* The upper limit of the dense weed growth of subtidal zone marks the lower end of eulittoral or intertidal zone, where occurs a dense cluster of *Crassostrea cucullata* and *Littorina* sp. The intertidal zone includes three subzones, *i.e.,* 1. the subzone of patchy weed growth; 2. the subzone of limpets such as *Sellana radiata* and barnacles and 3. the subzones of oysters and barnacles. The supratidal zone is inhabited by green algae and molluscs. Chlorophyceae such as *Enteromorpha, Ulva, Bryopsis, Codium, Halimeda, Caulerpa, Dictyosphaeria* and *Cladophora;* Phaeophyceae such as *Padina, Sargassum, Colpomenia, Iyengaria, Spathoglossum, Dictyopteris* and *Turbinaria*; Rhodophyceae such as *Gracilaria, Hypnea, Acanthophora, Polysiphonia, Cryptonemia* are the chief algae of the area. The common animals of this area are coelenterate species *Stoichactis, Ixalactis,* and *Gyrostama;* Nereid (*e.g., Eunice, Tubifex*); molluscs such as *Chiton, Nerita rumphii, Cypraea integerrimus, Pilumnus vespertilio,* seaweed grazer (nudibranch) molluscs, *Aplysia benedicti, Onchidium, Elysia grandifolia, Eolis, Doris* and floating mollusc *Janthina*, etc. The midlittoral reef community of Okha coast consists of fishes such as *Epinephelus, Petrachus*; the crab *Pilumnus vespertilio*, the annelid *Eurythoe complanata* and molluscs such as *Murex, Bursa*, etc. Rock crevices of intertidal zone harbour within them crabs such as *Atergatis*, *Grapsus, Pilumnus, Matuta, Charybdea, Gelasimus,* squillids (*e.g., Sesarma* and *Gonodactylus*); coelenterates (*e.g., Stoichactis, Ixalactis, Zoanthus, Lobophytum*; the hydroid species such as *Lytocarpus, Sertularia* and *Plumularia*); and annelids such as *Sepula, Polynoe, Eurythoe, Sabella,* etc.

(ii) Biotic communities of sandy shore. The sandy shore may be even more harsh than the rocky shore. It is subjected to all the extremes of the latter (temperature, salinity, turbidity, wave action, etc.) plus inconvenience of a constantly shifting substratum. The last factor makes life on the surface almost impossible; life has retreated below the surface. Generally sandy beaches are characterized by gentler wave action. Because of the prolonged time taken for drying up, these beaches are suitable for animal life. Decaying seaweeds and dead remains of animals result in the addition of organic matter to the sand. The coastal subsoil water is the environment of a special interstitial brackish water fauna. Most of the interstitial organisms are small, elongate, vermiform with transparent bodies. They have no eye but possess well developed adhesive organs and sense organs. They are negatively phototactic and gregarious in habit. They may be herbivore, carnivore, omnivore or detritus feeder. Large animal forms which are few in number prefer coarse sand grains and smaller one which are very abundant, prefer finer sand. The principal animal groups of interstitial water are the Nematoda, Turbellaria, Annelida, Gastrotricha and Acraina. Certain animal groups such as Protozoa, occur sporadically. The top 20 cm of sand between low and midtide levels of the beach is occupied by the inhabitants such as Hydrozoa, Turbellaria, Nemertina, Rotifera, Archiannelida, Polychaeta, Ostracoda, Halacaridae and Nudibranchiata. Species of Gastrotricha, Kinorhyncha, and Isopoda occur at mid-tide levels preferably in deep layers of sand. Oligochaets are sporadic and are found at different levels. Harpactcoid copepods remain restricted to mid-tide level. Tardigrades occupy the region between mid-tide and high-tide. Thus, sandy beaches also possess characteristic successive zones or zonations of animals.

For example, according to studies of **Ganapathy** and **Rao** (1962) sandy beach fauna of Visakhapatnam coast though lacks or contains small number of tubicolous polychaets and Crustacea due to very unstable substratum, but include following animal species—spinoid *Prionospio krusadiensis, Aricia* and *Lumbriconeries*, all of which possess the capacity to secrete mucus and formation of cover of sand grain over the body for protection, *Glycera lancadivae, G. alba, Nerina bonnieri, Pisionidens* *indica, Pisione complexa* and archiannelids *Saccocirrus minor* and *Protodrilus* also occur. The crustacean fauna of this zone includes *Emerita asiatica, Albunea symnista* (burrowing), amphipod species *Harpinia* sp., isopod *Sphaeroma walkeri,* mysid shrimp *Gastrosaccus spinifer,* hermit crabs *Clibanarius aretheustus* living within the shells of *Thais* and *Cerithium*, the eight-oared swimming

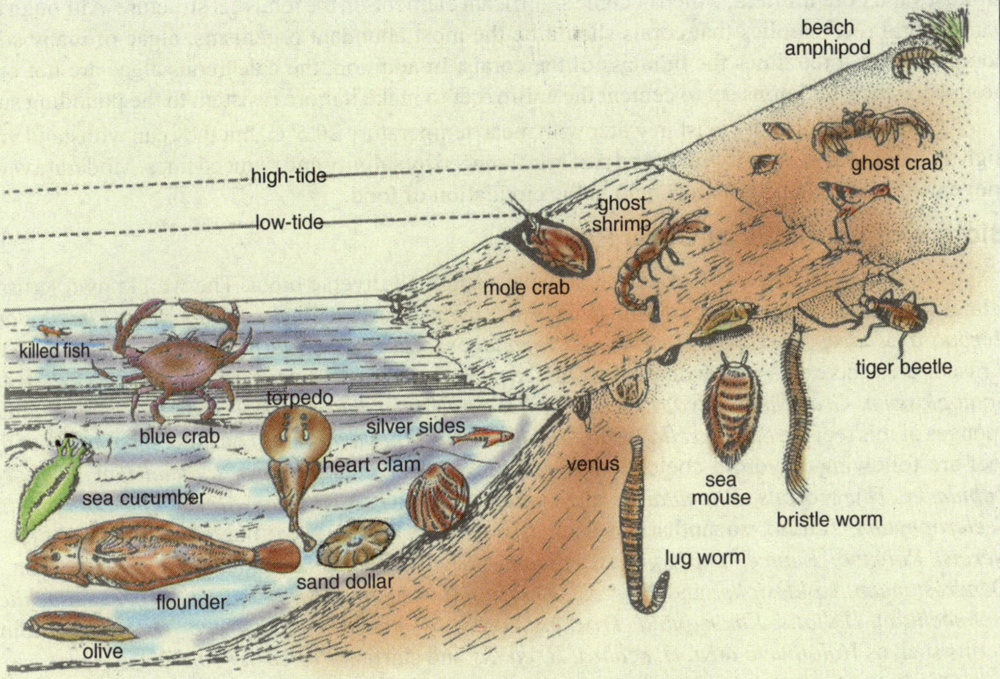

Fig. 12.4. Life on a sandy beach along the Atlantic coast.

crab *Matuta victor*, the spider crab, *Philyra scabruescula, Ocypoda platytarsis* and *O.macrocera.* Lamellibranchs such as *Donax cuneatus* and *Pamphia textile,* burrowing snails (*e.g., Sinum neritoideum*), common gastropods (*e.g., Oliva gibbosa* and *Terebra* sp.), *Bullia vittata* are also found along the coast of Visakhapatnam. This beach also contains eel *Ophichtys orientalis* and siphonophore coelenterates such as *Physalia, Velella* and *Porpita*; and medusae such as *Aequoria* and *Chiropsalmus.*

Coral Reef as a Specialized Oceanic Ecosystem

Coral reefs are the specialized ecosystems of ocean which are among the most productive of all ecosystems anywhere, with a diversity equalled only by tropical rainforests. The reef environment is very rugged one because it is constantly subjected to the pounding of the waves, and parts of the reef are often above water at low tide. But reef is also a very rich environment for those species that can withstand the physical pounding of water. The amount of oxygen is very high. During the day, it may reach 250 per cent of saturation because of the production of oxygen by algae in the reef structure. Likewise, the tremendous amount of CO_2 produced by the metabolism can be carried off before it becomes too high.

Reefs can be found wherever a suitable substrate exists in the lighted waters of the tropics, away from areas of continental sediments which would tend to bury the reef in mud or cold upwellings which would tend to lower the temperature too much. They may form **barrier reefs** (*e.g.,* barrier reef in Australia which contains canal or lagoon between them and coast) adjacent to continents; **atolls** or ring shaped or horse-shoe-shaped islands on the top of submarine heights, or **fringing reefs**, as in

Mandapam, Appa Islands and Hare Islands. In all modern reefs, the main reef structure is formed by madreporarian corals of following families—Poritidae (*Porites minicoiensis, Porites mannarensis*), Acroporitidae (*Acropora intermedia*), Astraeidae and Milliporidae. Secretions of calcareous algae go into the building up of reefs. Equally important are calcareous Bryozoa, lamellibranchs and gastropod molluscs and Foraminifera, which become significant elements in the total reef structure. Although the name 'coral reef' implies that corals should be the most abundant organisms, algae of many sorts comprise over three times the biomass of the corals. In addition, the calcareous algae are not only producers; their skeletons act to cement the entire reef to make it more resistant to the pounding surf.

Coral reefs normally exist in water with mean temperature 20.5°C, but they can withstand very high temperatures. The growth of coral reef is influenced by salinity and reduced tides. Moderate water movement around the coral reefs help in the circulation of food.

Biotic Communities of Coral Reef

Coral reef ecosystems harbour an abundant and well diverse biota. The well known Krusadi Island is the coral island of gulf of Manaar. It's dominant corals are *Porites, Montipora, Poecilopora, Acropora, Madrepora, Favia, Symphylia, Psammocora, Platygyra, Merulina* and *Leptastrea.* In the Krusadi area there is a luxuriant growth of sea weeds such as *Sargassum, Caulerpa, Turbinaria, Spatoglossum, Gracilaria, Ulva, Enteromorpha, Chaetomorpha* and *Polysiphonia.* The abundant sponges of this reef are *Spirastrella incorstans* and *Callyspongia fibrosa.* The other inhabitants of this reef are following—hydroid coelenterates such as *Syncoryne, Sertularia, Eudendrium, Pennaria, Tubularia, Thyroscypus, Plumularia;* alcyonarians such as *Clavularia, Alcyonium, Lobophytum, Scelerophytum, Telesto,* zooantharians such as *Zoanthus, Germmaria;* polychaet annelids such as *Nereis, Polynoe, Eunice, Lepidonotus, Iphion, Eurythoe, Hesione*; sipunculid species such as *Dendrostomum, Aspidosiphon* and *Phascolosoma;* molluscs such as *Doris, Dendradoris, Acanthochiton, Ischnochiton, Haliotis, Emarginula, Trochus, Turbo, Nerita, Natica, Cypraea, Thais*; and echinoderms such as *Holothuria atra, H. scabra, H. edulis* and starfish *Proteaster lincki.*

Beautifully pigmented fishes such as *Holocentrus, Lutianus* sp., *Pomacanthoides, Chaetodontopsis, Chaetodon, Linophora, Thallasoma* and *Zanclus* and the eels *Gymnothorax undulatus* and *G. punctatus* are commonly frequent in the coral reefs. These inhabitants of coral reef community, when exposed to low tides are subjected to predation by large number of birds such as the crested tern *Sterna bergii*, golden plover *Pluvialis dominica fulva*, etc.

According to **Clapham, Jr.,** (1973) the great productivity of reefs is obviously caused by something more than the abundance of light and oxygen. He considers nutrient cycles which runs very quickly and efficiently between the components of reef ecosystem to be main cause of great productivity of reef ecosystems. In fact, no organic matter is incorporated permanently into the sedimentary system portion of reef. The conditions of rapid nutrient cycling allow a mechanism by which the natural paucity of nutrients in the ocean can be effectively circumvented. It may also allow a gradual increase in nutrient availability by incorporating the nutrients of the non-reef organisms that wash into the reef, while still continuing to prevent the reef's own nutrients from being lost to the deeper seas.

REVISION QUESTIONS

1. What are the physico-chemical characteristics of marine environment ? Classify the marine environment and describe in detail the flora and fauna of the pelagic region.
2. Describe physico-chemical characteristics and biota of estuaries.
3. Describe the "coral reef as a specialized oceanic ecosystem."
4. Write short notes on the following ;
   (a) Sublittoral zone ;
   (b) Deep sea fauna ;
   (c) Intertidal sandy shore.

CHAPTER

13

Terrestrial Ecosystems

Terrestrial A Ecosystem.

Only about a quarter of the earth's surface is dry land, yet the complexity and variegation of terres trial ecosystems are much greater than those of aquatic ecosystems. There is no unifying theme in terrestrial ecosystems analogous to the physical properties of water. Rather, the variety of climates, the diversity of all lithosphere, and heterogeneity of terrestrial biotic communities all conspire to give rise to such a variety of themes that they are woven together in nature to give an enormous diversity of biotic communities and terrestrial ecosystems.

Physico-Chemical Nature of Terrestrial Ecosystems and their Comparison with Aquatic Ecosystems

An aquatic system is essentially a **single-phase system**, where water sets the tone for entire habitat. A terrestrial ecosystem, on the contrary, is a **three-phase system**, where the characteristics of the habitat are a function of the atmosphere and climate, the soil, and the biotic community itself. The atmosphere is the source of oxygen for animals and carbon dioxide for plants. Both are sufficiently common and well-mixed in the atmosphere that they are essentially never limiting. Air as a medium of support is much less buoyant than water. Thus, an organism need relatively little support to survive in water, but it needs a fairly rigid skeleton to live and move on land. Climate in a terrestrial ecosystem is much more variable than it is in water. The most important aspects of climate are the temperature and water relations of ecosystems. These relations include mean annual temperature, amount of temperature fluctuation, annual rainfall, degree and time of fluctuation in rainfall and potential evapotranspiration. Rainfall tends to be somewhat irregular even in the most predictable terrestrial ecosystems, and in arid regions or during periods of drought, lack of water may be a source of extreme stress. The temperature is much more variable on land than in water. This is caused by the fact that aquatic ecosystems are heated by absorption of radiation by water, and water has a high specific heat; terrestrial ecosystems, on the other hand, are heated by the absorption of heat by soil, rock, and vegetation, whose specific heat are much lower (Table 13-1). A given quantity of heat energy absorbed by solid material can change its temperature two to five times as much as the same energy absorbed by water. Thus, heat is gained and lost rapidly by terrestrial ecosystems, with a resulting wide fluctuations of temperature, both diurnally and seasonally.

Table 13-1. Specific heat of several common substances (Clapham, Jr. 1973).

1.	Freshwater	1.00
2.	Sea water	0.93
3.	Wet mud	0.60
4.	Moist sandy clay	0.33
5.	Solid rock	0.20
6.	Copper	0.09

In terrestrial ecosystems, the soil serves two major functions—it provides support for living organisms, and it is the source for all essential nutrients except for carbon, oxygen and hydrogen (for detail see Chapter 3). Even its supportive role is different from the way in which a lake sediment supports rooted vegetation. Because the entire weight of all living organisms is borne by the soil, there is no partition of support between the soil and the air. The role of soil as a source of nutrients is unique to terrestrial environments. It is the site of the entire detritus food chain (see Chapter 3) and, thus, is central to the biogeochemical cycling of nutrient materials. Different types of soils have different properties that affect the availability of nutrients to plants, so the productivity of the terrestrial ecosystems is very closely tied to the chemistry of the soil. Further, in terrestrial biotic communities interactions between species are real but they are simple, involving relationships such as predation, mutualism, parasitism and the like. In contrast to aquatic ecosystems, living organisms of terrestrial ecosystems leave a permanent mark on the system. For example, the role of plants in breaking down and weathering of rock and building soil is much more evident than it is in almost all aquatic ecosystems. In addition, the succession that takes place in a terrestrial community is almost entirely a function of the organisms within the ecosystem, unlike aquatic successions such as lake eutrophication, which are dependent in large part on materials being washed in from out side.

Lastly, in each of these two ecosystems, namely aquatic and terrestrial ecosystems, several types of major subdivisions can be recognized : thus, one can distinguish freshwater, estuarine and marine aquatic ecosystems, and several major types of terrestrial ecosystems such as prairies, forests, tundra, etc. The former are distinguished on the basis of a major chemical difference (*i.e.,* salt content), the latter generally on the basis of the predominant type of vegetation (grass, tree, etc. ;) see **Kormondy**, 1976).

CLASSIFICATION OF TERRESTRIAL ECOSYSTEMS

The earth surface— the continental land masses have been classified by biogeographers and ecologists into following regions :

(i) Biogeographic realms ; (ii) Biomes.

(I) Biogeographic Realms or Regions

Careful studies of the distribution of plants and animals over the earth have revealed the existence of six major biogeographic realms, namely **Palaearctic realm**, **Nearctic realm**, **Neotropical realm**, **Ethiopian realm**, **Oriental realm** and **Australian realm**, each characterized by the presence of certain unique organisms. Each of these realms embraces a major continental land mass and each remains separated by oceans, mountain ranges, or desert.

(II) Biomes

Within these biogeographic realms and established by a complex interaction of climate, other physical factors and biotic factors, are large community units called **biomes**. A biome is a large community unit characterized by the kinds of plants and animals present. In each biome the **kind** of climax vegetation is uniform—grasses, conifers, deciduous trees—but the particular **species** of plant may vary in different parts of the biome. The kind of climax vegetation depends upon the physical environment and the two together determine the kind of animals present. The definition of biome includes not only the actual climax community of a region, but also the several intermediate communities that precede the climax community. Further, there is usually no sharp line of demarcation between adjacent biomes; instead each blends with the next through a fairly broad transition region termed as **ecotone**. There is, for example, an extensive region in Northern Canada where the tundra and coniferous forests blend in the tundra-coniferous ecotone. Ecologists have identified the following biomes :

1. Tundra Biome

Tundra presents the most common example of "fragile ecosystem" (**Clapham**, **Jr.**, 1973). Tundra which means "marshy plain", lies largely north of latitude 60^0N (*i.e.,* between the Arctic ocean and polar ice-caps and the forests to the south) and is characterized by the absence of trees, the presence of dwarfed plants, and an upper ground surface that is wet, spongy and uneven, or hummocky, as a result of freezing and thawing of this poorly drained land (**Kormondy**, 1976). Some five million kilometers areas of Tundra stretch across Northern America, Northern Europe and Siberia. Although there is variation from place to place within the biome, temperature, precipitation, and evaporation are characteristically low, the warmest months averaging below 10°C and the wettest with about 25 mm of precipitation. Despite the small amount of precipitation, water is usually not a limiting factor because the rate of evaporation is also very low.

Tundra Biome

The ground usually remains frozen except for the uppermost 10 to 20 cm, which thaw during the brief summer seasons. The permanently frozen deeper soil layer is called **permafrost**. The permafrost line which may exist at a depth of a few centimeters to several meters, is the ultimate limit of plant root growth, but the immediate control is the depth to which soil is thawed in summer. The rather thin carpet of vegetation of tundra biome includes few species : grasses and sedges are characteristic of numerous marshes and poorly drained areas, but large areas consist of grasses, sedges, mosses and lichens, with occasional occurrence of dwarf birches (*Betula*) and willows (*Salix*). Perhaps the most characteristic arctic tundra plant is the lichen known as " reindeer moss" (*Cladonia*). Tundra is also characterized by heath-plants such as *Erica* and other members of Ericaceae family. The net primary productivity is extremely low.

The animals that have adapted to survive in the tundra are caribou or reindeer, musk ox, the arctic hare, arctic fox, polar bear, wolves, lemmings, snowy owls, ptarmigans and during the summer, swarms of flies (Dipteran black flies), mosquitoes and a host of migratory birds.

Tundra biomes cover large areas of the arctic zone. There are two tundra biomes one in the palaearctic, and another in the nearctic region. Tundra-like areas, called the **alpine tundra** are also quite similar to some of Arctic tundra, but in the absence of permafrost and in the growing season, mosses and lichens are less prominent, flowering plants more so.

2. High Altitude or the Alpine Biome

The region of mountain above the timber line contains a distinct flora and fauna and is referred to as the **alpine zone**. Alpine zone remains conspicuous in those mountains whose peaks reach up to the niveal or snow zone, as in the Himalayas. The alpine zone (zone which lies between timber line and snow zone) includes in the descending order, a **sub-snow zone** immediately below the snow zone, a **meadow zone** in the centre and a **shrub zone** which gradually merges into the timber zone. According to **Mani** (1957) snow zone of Himalayas lies over 5100 m above mean sea level and alpine zone exists at a height of 3600m. From an ecological view point, the zone above the limits of tree growth (timber line) exhibits extreme environmental conditions which greatly influence the biota of this region.

The characteristic features of high altitude environment are following—a low air density, low oxygen and carbon dioxide contents and water vapour, high ozone content, high atmospheric transparency affording greater penetration of light, cold, snow cover, increased rate of desiccation, high wind velocity, insolation of high intensity during the day and rapid radiation during the night, the high glare from the sky and snow, high intensity of ionizing radiation and the exclusion of trees.

In the western Himalayas of Jammu and Kashmir, Himachal Pradesh and U.P. and the eastern Himalayas of Assam and Arunachal Pradesh, **subalpine forests** dominated by trees of such species as *Abies, Pinus*, and so on, occur at altitudes of 2900 to 3500 meters, while **alpine scrub** forests are found at altitudes of 36,000 meters and more. These scrub forests contain the broad-leaved *Betula utilis.* Alpine zone of Himalayas is also characterized by a sparseness of animal groups, the scarcity being relatively important as far as tropical Indian, South Chinese, Indo-Chinese and Malayan derivatives are concerned. Important constituents of the fauna of this zone are cold-adapted palaearctic forms (**Mani**, 1968).

Many invertebrates of alpine zone are predatory and occur in lakes, streams and ponds. Among vertebrates fishes and amphibians are totally lacking and reptilian fauna is greatly impoverished. However, pelobatid frog, *Aleurophryne mammata* is found to occur at 4500 meters of height in Himalayas and viviparous *Mabuya varia* has been reported to live at 4000 meters of height in Mount Kilimanjaro. Some of the representative vertebrates of Himalayan alpine zone are—crow, *Corvus corax*; snow partridge, *Lerwa nivicola*; snow leopard, *Felis unica* and *Felis lynx*; European leech marten, *Mustela foina*; the vole, *Lagomys*; Tibetan yak, *Bos grunnius*; Tibetan sheep, *Ovis hodgsoni*; pamir sheep, *Ovis poli*; ibex, *Capra sibirica*; markhor, *Capra falconeri*; and Persian wild goat, *Capra aegagrus.*

A large number of insects and arachnids remain best adapted to Himalayan alpine zone and at 6900 meters height. The torrential streams which are the products of melting snow contain many mayflies, stoneflies and caddisflies. Some of common insects of this zone are—stonefly *Rhabdiopteryx lunata*; grasshoppers such as *Bryodema* and *Gomphomastax*; apterous species *Conophyma*; *Dicranophyma*; *Spingonotus*; tettigonid *Hypsinomus fasciata*; earwigs such as *Anechura*; bugs such as *Dolmacoris, Tibetocoris,* and *Nysius ericae;* aquatic beetle *Amphizoa;* Carabids such as *Bembidion, Carabus, Calosoma, Harpalus, Trechus, Calathus, Bradytus, Broscus,* and *Dyschirius;* staphylinids such as *Atheta, Aleochara, Geodromicus, Oxypoda, Philonthus* and *Tachinus;* tenebrionids such as *Bioramix* and *Chianalus;* ants such as *Formica picea*; bumble bees such as *Lapidariobombus separandus;* butterflies such as *Calias, Argynnis* and *Parnassius*; the fly *Deuterophlebia*; collembolans such as *Entomobrya, Isotoma, Proistoma, Hypogastrura, Isotomurus, Tomocerus* and *Onychiurus*; and numerous simulids blepharocerids, syrphids, anthomyids tachinids and sarcophagids (**Ananthankrishnan** and **Viswanathan**, 1976).

3. Forest Biomes

The word **forest** is derived from Latin '*Foris*' meaning outside, the reference being to village boundary fence, and must have included all uncultivated and uninhabitated land. Today a forest is any land managed for the diverse purposes of forestry whether covered with trees, shrubs, climbers, etc., or not (see **S.P. Sagreiya**, 1967). The forest biomes include a complex assemblage of different kinds of biotic communities. Optimum conditions of temperature and ground moisture responsible for the growth of trees contribute greatly to the establishment of forest communities. In addition, 50 mm rainfall is a pre-requisite for the trees. The nature of soil, wind and air currents determines the distribution (abundance or sparseness) of forest vegetation. Normally ecologists recognise among forest communities such features as their evergreen nature, whether deciduous or indeciduous, as well as their shape, whether broad-leaved as in temperate forests or more needle-like as in the conifers. On the basis of these features the forest biomes of the world have been classified into following biomes—**coniferous forest**, **tropical forest**, and **temperate forest**. All these forest biomes are generally arranged on a gradient from north to south or from high altitude to lower altitude.

(i) Boreal coniferous forest. Cold regions with high rainfall and strongly seasonal climates with long winters and fairly short summers are characterized by boreal coniferous forest which is transcontinental. For example, adjacent to the tundra region either at Latitude or high altitude is the **northern coniferous forest**, which stretches across both North America and Eurasia just south of the tundra (*i.e.,* Canada, Sweden, Finland, Siberia and Mussoorie). The term **taiga** is applied to the northern range of coniferous forests. This is characterized by evergreen plant species such as spruce (*Picea glauca*), fir (*Abies balsamea*) and pine trees (*Pinus resinosa, Pinus strobus*) and by animals such as snow shoe hare, the lynx, the wolf, bears, red fox, porcupines, squirrels, amphibians like *Hyla* and *Rana,* etc. Large size is common, both in the trees, which range upto some 40 m in height, and in vertebrates, which include the giants of several groups of animals, such as moose, caribou, elk, grizzly bear, wolverine, beaver, and several large species of birds. Species diversity is low, and pure stands of trees and shrubs are common, as are outbreaks of defoliating insects of several sorts (*e.g.,* bark beetles, saw flies, geometrid moths, etc.). Understory trees are uncommon as a result of the continual low light penetration. Among common understory associates are orchids and ericaceous shrubs like the blue berry. The thalloid mosses and lichens being very rich understory vegetation.

A boreal coniferous forest.

Boreal forest soils are thin podozols, and are rather poor both because the weathering of rock proceeds slowly in the cold environments, and because the litter derived from conifer needle is broken down very slowly and is not particularly rich in nutrients. These soils are acidic and mineral deficient, the result of the movement of a large amount of water through the soil; in the absence of a significant counter upward movement of evaporation, soluble essential nutrients such as calcium, nitrogen, and potassium are leached sometimes beyond the roots thereby leaving no alkaline-oriented cations to encounter the organic acids of the accumulating litter (**Kormondy**, 1976). The productivity and community stability of a boreal coniferous forest are lower than those of any other biome.

(ii) Temperate deciduous forest. The temperate forest biomes are characterized by a moderate climate and broad-leaved deciduous trees, which shed their leaves in fall, are bare over winter and grow new foliage in the spring. These forests are the characteristics of North America, Europe, Eastern Asia, Chile, part of Australia and Japan, with a cold winter and an annual rainfall of 75—150 cm and a temperature of 10—20°C. In these biomes the precipitation may be fairly uniform throughout the year. In India, at elevations of 9000' (=2743.2m)—12,000' (=3655.6m) in Himalayas occur temperate vegetation including pines, fir, yew and juniper trees with an undergrowth of scrubby rhododendrons.

Temperate deciduous forest in Rhode Island during winter, spring, summer and fall.

Soils of temperate forests are podozolic and fairly deep. Trees are quite tall—about 40—50m in height and their leaves are thin and broad. The predominant genera of this biome are maple (*Acer*), beech (*Fagus*), oak (*Quercus*), hickory (*Carya*), basswood (*Tilia*), chestnut (*Castnea*), cottonwood (*Populas*), sycamore (*Platanus*), elm (*Ulmus*) and willow (*Salix*). In some locations, coniferous vegetation may be quite predominant and that incudes white pine (*Pinus strubos*), hemlock (*Tsuga canadensis*) and red cedar (*Juniperus virginianus*). There are not many epiphytes or lianas save for some species of mosses, algae and lichens growing on tree trunks, and a few vines, notably *Vitis,* the grape. The understory of shrubs and herbs in the deciduous forest is typically well-developed and richly diversified, with a considerable portion of the photosynthesis and flowering attuned to the short day of the spring season, prior to the leafing out of and consequent shading by the tree canopy. Accordingly, there are often two separate herb assemblages. One consists of spring flowers, which bloom before the trees have expanded their leaves and are gone by summer, and the other is adapted to the low light levels of the forest floor and lasts into the fall.

The animals originally present in temperate forests are deer, bears, squirrels, gray foxes, bobcats, wild turkey and wood peckers. Other common animals of this region are invertebrates such as earthworms, snails, millipedes, Coleoptera and Orthoptera and vertebrates like amphibians such as newts, salamanders, toads, and cricket frogs; reptiles such as turtles, lizards and snakes; mammals such as racoon opossum, pigs, mountain lion, etc., and birds like horned owl, hawks, etc. All these animals and plants show a profound seasonality; some may even hibernate throughout the winter. The range of animal size and adaptations is wide; the largest animals include such forms as the deer and black bear. The dominant carnivores are large, including the wolf and mountain lion, although smaller carnivores such as fox and skunk are also common. Diversity of fauna is lower than in any of the rain forests and a few species seems clearly to be dominant.

(iii) Temperate evergreen woodland (Chaparral). Many parts of the world have a Mediterranean-type climate with warm, dry summers and cool, moist winters. These are commonly inhabited by low evergreen trees with small hard needles or slightly broader leaves. The most important area of tropical evergreen woodland in North America is the **'chaparral'** of the Pacific coast, the Mediterranean '**maquis**', Spanish '**encinar**' and '**melle scrab**' on Australia's South coast are the same type of community. In such a woodland, trees are essentially lacking, although shrubs may range upto 3—4 m in height. Species diversity is roughly intermediate between that of a temperate deciduous forest and a drier grassland. Fire is an important factor in this ecosystem, and the adaptations of the plants enable them to regenerate quickly after being burned. The characteristic animals of temperate evergreen woodland or chaparral are mule deer, brush rabbits, wood rats, chipmunks, lizards, wem-tits and brown towhees. Small-hooved cursorial ungulates are the dominant herbivores. Saltatorial (jumping) animals and many fast moving ungulates are also common in this fauna.

(iv) Temperate rainforests. The temperate rainforest is a colder ecosystem than any other rainforest. Such a forest has a definite seasonality, with both temperatures and rainfall varying throughout the year. Rainfall is high, but fog maybe very heavy and actually more important as a source of water than rainfall. The diversity is much lower, both in plants and animals, in comparison to warmer rainforests, yet it remains still higher than other temperate forest types. The diversity is much lower, both in plants and animals, in comparison to warmer rainforests, yet it remains still higher than other temperate forest type. The dominant trees (canopies) are coast redwood (*Sequoia sempervirens*) of the Pacific coast of North America and the alpine ash (*Eucalyptus regnans*) of Australia and Tasmania, both of which reach more than 100m in height. Epiphytes and lianas are common but are not abundant like those of other rainforest. The animals of temperate rainforests are similar to those of deciduous forests, but show a somewhat higher diversity.

(v) Tropical rainforests. Tropical rainforests occur near the equator in Central and South America, Central and Western Africa (Congo, Zambesi river), Southeast Asia (parts of India and Malaysia), Malaya, Borneo, New Guinea and Northwest Australia. Tropical rainforests are among the most diverse communities on earth. Both temperature and humidity are high and constant. The annual rainfall which exceeds 200 to 225 cm is generally evenly distributed throughout the year.

A Tropical rainforest.

The flora is highly diversified : a square mile may contain 300 different species of trees, a diversity unparalleled in any other biome. The extremely dense vegetation of the tropical rainforests remains vertically stratified with tall trees often covered with vines, creepers, lianas, epiphytic orchids and bromeliads. Under the tall trees is a continuous evergreen carpet, the canopy layer, some 25 to 35 meters tall. The lowest layer is an understory of trees, shrubs, herbs like ferns and palms, all of which become dense where there is a break in the canopy. Nearly all plants are evergreen, and those that do lose their leaves entirely do so at irregular intervals with no apparent regard to the climatic regime. The leaves of most plant are of moderate size, leathery and dark green in colour. Their roots are often shallow and have swollen bases or flying buttresses.

Soils of tropical rainforests are red latosols, and they may be exceedingly thick. The high rate of leaching makes these soils virtually useless for agriculture purposes, but if they are left undisturbed the extremely rapid cycling of nutrients within the litter layer (due to decomposition) can compensate for the natural poverty of the soil. It is the nature of the soil, both its potential for high leaching as well as its chemical composition that promotes a rock-like quality when exposed to air, that largely has prevented western style agriculture from being applied to the tropical forests (**Kormondy**, 1976).

Invertebrate density and abundance are very high in tropical rainforests, but while vertebrates are diverse, they are not as abundant as in many other communities. The common invertebrates of these forests are worms, snails, millipedes, centipedes, scorpions, isopods, spiders, insects, planarians and leeches. Amount insects, heteropterans, orthopterans, blattids, mantids, phasmids, bees, termites and ants are most common. The common vertebrates of tropical rainforests are the arboreal amphibian *Rhacophorus malabaricus*; aquatic reptiles, chameleons, agamids, geckoes, and many species of snakes; many species of birds, social birds being predominant; and a variety of mammals. Nocturnal and arboreal habits are most common in many mammals such as insectivores, leopards, jungle cats, ant-eaters, giant flying squirrels, monkeys and sloths. But in New Guinea and Northern Queensland, where monkeys are absent, there are arboreal kangaroos (*Dendrolagus* sp.), despite the fact that the basic anatomy of the kangaroo is not particularly well-suited to arboreal life. Further, in the foot hills of the forest zone of peninsular India covered with dense tropical vegetation, we have the tiger (*Panthera tigris*), the elephant (*Elephas maximus*), samber deer (*Rusa unicolor*), muntjac (*Muntiacus muntjak*), the gaur (*Bibos gaurus*), the chital or spotted deer (*Axis axis*), and the swamp deer (*Rucervus duraucelli*) as the major ground dwelling mammals (see **Ananthakrishnan** and **Viswanathan**, 1976).

(vi) Tropical seasonal forests. Tropical seasonal forests occur in region whose total annual rainfall is very high, but segregated into pronounced wet and dry periods. Tropical seasonal forests are found in Southeast Asia, Central and South America, Northern Australia, Western Africa and the tropical islands of the specific as well as India and Southeast Asia. In exceedingly wet tropical seasonal forests, commonly known as **monsoon forests**, the annual precipitation may be several times higher than that of the tropical rainforests. Trees may reach heights over 40 m, but are more commonly 20-30m high. Stratification is of a relatively simple type with a single understory tree layer, canopy is deciduous and understory is evergreen. Teak is often a major large tree in the best-known tropical seasonal forests, those of India (Central India) and Southeast Asia. Bamboo is also an important climax shrub in these areas although in other areas it is important only in earlier stages of the succession.

(vii) Subtropical rainforests. In regions of fairly high rainfall but where temperature differences between winter and summer are less marked, as in Florida (USA), the broad-leaved evergreen subtropical biome is found. The vegetation includes mahogany, gumbo limbo, bays, palms, oaks, magnolias, tamarinds, all laden with epiphytes (of pineapple and orchid families), ferns, vines and strangler fig (*Ficus aureus*). This stratification is simpler, with only one understory tree horizon. All these plants tend to be evergreen, but may lose their leaves during the dry season. Animal life of subtropical forest is very similar to that of tropical rainforests.

Rainforests of India. In India, patches of rainforests are found in Kerala, Assam and the Gandhamardan hills of Orissa. They are of following three types :

(a) Moist tropical forests. They include the southern tropical wet evergreen forests found in Assam and West Bengal, the northern semievergreen forests of Assam and Orissa, and the southern tropical semievergreen forests of Andamans.

(b) Montane subtropical forests. They include the northern subtropical broad-leaved wet hill forests of Assam and West Bengal, the southern subtropical broad-leaved hill forests of Orissa and Kerala and the subtropical pine forests of UP, Himachal Pradesh, Assam, Manipur, etc. The *Chir* forests of UP and HP are also example of this type of forests.

(c) Montane wet temperate forests. They include the forests of Kodaikanal and Udagamandalam in Tamil Nadu and Kerala, the northern wet temperate forest of the north-eastern region and West Bengal and the Himalayan moist temperate forests.

4. Tropical Savanna Biomes

Savannas are tropical grassland with scattered, drought resistant trees which generally do not exceed above ten meters in height and do not form a canopy. Thus, a savanna is an intermediate between a forest and a grassland. Savannas constitute extensive areas in Eastern Africa, which support the richest diversity of grazing mammals in the world, and also occur in Australia, South America and Asia. The climate is generally characterized by a rainy (May through October) and dry (November through April) season; in the llanos of Venezuela, for example, nearly 90 per cent of the annual rainfall of 130 cm falls during the rainy season. The latosol soils of savannas are nutrient poor owing a heavy leaching. Quite widely distributed in savannas are soils called laterite which when dried harden to rock-like consistency thereby precluding their use in western style agriculture (**Kormandy**, 1976). Although climate and soil are significant regulating factors in this biome, the controlling factor appears to be fire, which gives grass and certain species of trees a powerful advantage over other tree species. As a result of this latter factor, species diversity is quite low in comparison to adjacent tropical forests ; in some situations a single species of both grass and tree may be dominant over large areas.

A tropical savanna biome.

The trees of savannas are resistant to desiccation, and may be either deciduous or evergreen. Their leaves are often hard and drought-resistant. Grasses are the most conspicuous plants, and may reach heights of 1½–2m. Gigantism of certain animal groups in these tropical savannas is as pronounced as it is in the boreal forest including such giants as many antelopes, giraffes, elephants, buffalo and lions (**Clapham**, Jr., 1973). In addition, a rich insectan fauna especially those with strong mandibles capable of mastication are conspicuous in this biome. Grasshoppers and termites are encountered in large numbers.

Savannas of India. The savannas of India are dominated by grasses and sal trees and the consumers are cattle, rodents, insects, jackals, hyaenas, etc. They are classified into the following types :

(1) High savanna. It occurs in the Brahmaputra valley and constitutes open stands of low branching trees which are usually 2 to 3 metres tall, such as *Syzigium cerasoideum* and *Emblica officinalis.* Its common grasses are *Imperata, Saccharum spontaneum, Ophiurus* and *Vetiveria.*

(2) Moist sal savanna. It occurs in the Gangetic plain and consists of open forests of sal (*Shorea robusta*) having tall grasses. Its common grasses are *Imperata cylindrica, Themeda arundinacea, Cymbopogon nardus, Erianthus* and *Apluda.*

(3) Low alluvial savanna woodlands. It occurs in the Gangetic plain and on riverine flats. The soil of this savanna is sandy and alluvial and contains patches of clay in depressions. The common trees of this region are *Dalbergia sissoo, Butea monosperma, Albizia, Adina cordifolia* and *Zizyphus procerum, Arundinella, Themeda gigantea* and *Erianthus* sp.

(4) Dry savanna. This region is found in Punjab, Haryana, Bihar, Orissa and eastern Tamil Nadu. It is characterized by trees which stand far apart, singly or in small groups such as *Acacia lenticularis, Emblica officinalis, Gardenia turgida, Crotalaria hirta* and *Pterocarpus marsupium.* The common and abundant grasses of this region are *Themeda triandra, T.quadrivalves, Apluda mutica* and *Arundinella setosa.*

(5) Saline alkalina scrub savanna. It occurs throughout the Indo-Gangetic Plain. The common tree of this region are the *Phoenix sylvestris, Acacia* sp., *Tamarix* sp., *Calotropis procera, C. giganta* and *Kochia indica.*

5. Grassland Biomes

Sheep grazing on a temperate grassland (short grass prairie).

The grassland biome is found where rainfall is about 25 to 75 cm per year, not enough to support a forest, yet more than of a true desert. The seasonality of grasslands is pronounced, both with respect to rainfall, which is concentrated in the summer and to temperature. Grasslands typically occur in the interiors of continents and include the **tall grass prairies**, **short grass prairie** or great plains and **arid grassland** of North America as well as the **steppes** of Eurasia (Southern Russia, Siberia and Asia), **puszta** in Hungary, the **veldt** of Africa and the **pampas** of South America (Argentina). The grassland communities are open land communities with limited moisture conditions, irregular rainfall, sharp seasonal and diurnal variations and very high radiation. Vegetation of grassland is dominated by grasses, legumes and composites. Since tall trees or other thick vegetations are excluded from these communities there is a free movement of air. These winds carry the particles of sand or dust. These open habitats provide natural pasture for grazing animals (herbivore) which are excluded from predation by predators which hide in thick vegetation and prey upon them.

In grasslands, the stratification is reduced essentially to a single story, but within that level, species diversity may be as high as in a deciduous forest-especially for the tall grass prairies. Only along the streams are trees to be found, but the "gallery forests" within a few meters of stream bank are characteristics of grasslands. The grasses comprising of the grassland can be divided into two basic groups, the tall grasses more than 1m high, which are found in moist portions of the grassland, and the short grasses less than 1 m high which are found in the drier regions. For example, prior to its conversion to agriculture and urban development, the tall grass prairie toward east of North America was dominated by species of bluestern (*Andropogon*) forming dense covers 122 cm to 183 cm tall. Westward, Buffalo grass (*Buchloe dactyloides*) and other grasses but a few centimeters high dominated the landscape. Flowering herbs including many kinds of composites are common, but much less important than grasses.

Since the precipitation—evaporation ratio is below one in the grassland, leaching is considerably less. The soils of the grasslands are rich, fertile prairie soils (black-earths) and chernozems. Organic matter (humus) accumulates in the upper portion of the soil, rendering it dark; this upper portion remains neutral to slightly alkaline because of the continued replenishment of cations like calcium and potassium through the upward movement associated with evaporation.

Typical animals of grasslands tend to be quite small, with the exception of a few very large cursorial herbivore mammals such as the bison and pronghorn in North America, the wild horse, ass, and saiga antelope of Eurasia and zebra and antelops of Southern Africa. The large herbivores are no where near as diverse as they are in savanna areas. Likewise, carnivores are relatively small such as are coyotes, weasels, badgers, foxes, ferrets, owls and rattle snakes. Rodents such as prairie dogs, rabbits, and ground squirrels are common. Most herbivores characteristically aggregate into herds or colonies; this aggregation provides some protection against predators. The characteristic birds of grasslands are prairie chickens, meadow larks, and rodent hawks. Further, saltatorial motion is widespread, not only in mammals such as rabbits and kangaroo rats, but also in insects, such as grasshoppers and crickets. This probably has to do with increasing rate at which animal can move through or over the tall grasses, as well as making it harder for a predator to catch the animal. Visibility in the grasslands in exceedingly good and the premium placed on efficiency including predators is very high.

The common insects of steppes are termites, locusts (*Locusta migratoria, Schistocerca gregaria* and *Melanoplus* sp.), bees, burrowing wasps, mutilid wasps, bumble bees and blister bees. The steppes also harbour an undisturbed reptilian fauna. Lizards and snakes are met with in large numbers and possess, remarkable diversity. Most of them are fossorial, insectivorous and carnivorous.

Types of grasslands. Grasslands tend to occupy about 20% of the earth's land surface and are of three types :

(i) Tropical grasslands. These are situated 20 degrees away from the equator and the rainfall varies from 40 to 100 cm. Tall grasses rise to a height of about 1.5 to 3.5 metres. The tropical grasslands of Africa include ungulates, deer, antelopes, giraffes, lions, etc.

(ii) Temperate grasslands. They are found in Europe, Asia and North America. Their rainfall varies from 25 to 75 cm per year.

(iii) Alpine grasslands. They occur at higher latitudes. They are of the meadow type and many flowering herbs also grow here.

6. Desert Biomes

Deserts are the biomes formed in the driest of environments. Temperature may range from very hot as in **hot deserts** to very cold as in **cold deserts**. Major hot deserts of the world are situated near the tropics of cancer and capricon, with a rainfall of less than 10 mm. The most important hot deserts

(a) A cold desert. (b) A hot desert.

of world is the Sahara-Arabia-Gobi desert complex extending from Africa to Central Asia and contains highly irregular and very insignificant rainfall, and low humidity due to excessive evaporation. Fairly extensive hot deserts also occur in India (Sind-Rajasthan deserts), South America (Chile), North America ad Australia. The cold deserts occur at high elevations where the temperatures are low and rainfall scanty as the air losses all its moisture content as it ascends higher and higher. Cold deserts occur in Ladakh regions of Himalayas, Tibet, and Bolivia Arctic. The hot and cold deserts may also be distinguished by differences in plant population which are mostly succulent type (*e.g.,* cactus, palo verde trees, creosote bush, etc.). Most cold deserts have sage brush.

Low abnormal precipitation coupled with soil and air temperatures that are extremely high by day and drop abruptly by night, low humidity, and high insolation are the major desiccating environmental factors to which desert vegetation and animals have adapted. Desert plants which tend to be shrubs are adapted to drought conditions through reduced leaf size and the dropping of leaves in dry conditions, both reducing water loss via evapotranspiration. The roots of most desert plants remain well developed and occur in the top meter of the soil to take maximum possible advantage of any rainfall. Further, the root hairs on many desert plants are ephemeral, drying back under drought conditions and thereby reducing potential water loss by osmosis. Yet other species are short lived annuals that complete their life cycles during the short-moist period. In most hot deserts, there occur plants such as cacti, water-storing succulents such as acacias, euphorbias, cacti, prickly pears, etc., which are adapted by their protoplasmic colloids, which enable the accumulation of substantial water reserves, as well as by a reduced leaf surface, which obviates water loss via evapotranspiration.

The animals present in the desert are reptiles, insects and burrowing rodents. All these animals possess special morphological, physiological and ethological adaptations for deserts. In general, large animals are very uncommon except mule deer and some species of gazelle and all animals have cursorial, fossorial and/or saltatorial adaptations. Some desert animals are nicely adapted for high extremes of temperature. For example, the lethal temperature of different species of insects found to be following—for the canal spider *Galeodes granti* it is 50°C, for the *Gryllus domesticus* it is 40°C and for the forficulid *Labidura riparia* it is 38°C (Cloudsely-Thompson, 1962). Diurnal rhythms are perhaps the best method of avoiding the heat. While some desert plants close their petals at night, many blossom only at night. Some insects such as tenebrionid bettle *Akis spinosa* remain active during the day, and the centipede *Scolopendra clavipes* restricts its activities to night time.

Further, certain reptiles and certain insects are well adapted for survival in deserts because of their thick, impervious integuments and the fact they excrete dry waste matter. A few species of mammals have become secondarily adapted to the desert by excreting very concentrated urine. They avoid the sun by remaining in their burrows during the day. Kangaroo rat and pocket mouse, both are able to live without drinking water by extracting the moisture from the seeds and succulent cactus they eat. The camel and the desert birds (ostrich, etc.) must have an occasional drink of water but can go for long periods of time using the water stored in the body. Most insects of deserts are herbivores and as a correlation the number of small insectivorous lizards found in the desert is usually high.

Fauna of Indian deserts. Prakash (1974) have recognized four habitat types in the Indian desert based on land forms namely **aquatic**, **sandy**, **rocky** and, **ruderal habitats**. The aquatic habitats of Indian deserts which remain confined to the perennial lakes, are inhabited by various species of fishes such as *Labeo nigripinnis, Oxygaster clupeoides, Tor khurdee, Puntius amphibia, Noemacheilus denisonii.* Among amphibians of the Indian desert are a species of toad (*Bufo andersonii*) and five species of frogs (Indian bull frog *Rana tigrina,* cricket frog *Rana limnocharis* and burrowing frog *Rana breviceps* being dominant species). Among reptiles, there occur two species of testudines (Loricata), 18 species of lizards and 18 species of snakes. Of the lizards, some species such as *Acanthodactylus cantoris, Calotes versicolor, Uromastix hardwicki, Ophiomorus tridactylus* are predatory on the desert locust inhabiting localized areas in the Thar desert. The avian fauna of the Indian desert includes species such as painted partridge *Francolinus pictus pallidus,* grey partidge *Francolinus pondicerianus,* common quail *Coturnix coturnix,* black breasted quail *Coturnix coromandelicus,* rock bush quail, the little bustard quail *Lurnix sylvatica,* button quail *Turnix tanki,* great Indian bustard *Choriotes nigriceps,* florican *Sypheotides indica,* sandgrouses such as *Petrocles exustus erlangeri, Petrocles alchata, P. senegallus,* and *P. indicus indicus.* During winter numerous aquatic birds become abundant in Indian deserts. Such birds are white fronted goose *Anser albifrons,* wigeon *Anas penelope*, garganey *anas querguedrila,*red crested pochard *Netta fufina* and tufted duck *Aythya futigola.* Among the predominant predatory birds are two species of the vultures namely *Gyps bengalensis* and the white vulture *Neophron percnopterus.*

The mammalian fauna of Indian deserts includes many species, some of which have been tabulated in Table 13-2.

All these mammals have well specialized adaptations for survival in thermal extremes and low humidity. Many of them possess diurnal rhythmic activities.

7. Wetland Biomes

Wetlands may be defined as submerged or water saturated lands, natural and artificial, permanent or temporary with water that is static or flowing and fresh, brackish or saltish. They are characteristic transitional conditions between terrestrial land and deep water bodies and are represented by **swamps**, **marshes** (or palustrine lands), **fens** (*i.e.,* tract of marshy land), **Peatlands**, **lagoons** (*i.e.,* a shallow lake, especially one near and communicating with the sea or river), **lakes**, etc. Wetlands are also represented by paddy fields, fishery ponds, *Trapa* cultivation ponds, riverine flood plains, lacustrine (L.*lacus*= a lake) marshy lands, marine backwaters and coastal wetlands. Wetlands are often seasonal, *i.e.,* with free overlying water in the rainy season, marshy and often almost dry in

Table 13-1. **Certain common mammals of Indian deserts (Prakash, 1974).**

	Order	Species	Popular Name
1.	**Chiroptera**	*Rasetius arabicus*	Fruit bat
		Pteropus giganteus giganteus	Fruit bat
		Rhinopoma kinneari	Rat-tailed bat
		Rhinopoma hardwicki	The lesser rat-tailed bat
		Taphozour perforatus	Tomb bat
		Megaderma lyra	Indian false vampire
		Rhinolophus lepidus lepidus	Horse-shoe-bat
2.	**Insectivora**	*Hemiechinus auritus lepidus*	Long ear hedgehog
		Paraechinus micropus	Long ear hedgehog
		Suncus muinus sindensis	House Shrew
3.	**Pholidota**	*Manus crassicaudata*	Scaly ant eater
4.	**Lagomorpha**	*Lepus nigricollis caudatus*	Hare
		Lepus nigricollis dayanus	Hare
5.	**Rodentia**	*Funambulus pennanti*	Northern palm squirrel
		Hystrix indica indica	Indian crested porcupine
		Rattus rattus	House rat
		Mus musculus rufescens	House mouse
		M. musculus bactrianis	House mouse
		Gerbillus gleadovi	Indian hairy footed gerbil
		Tatera indica	Gerbil
		Merriones hurrianae	Gerbil
6.	**Artiodactyla**	*Sus scrofa cristata*	Wild boar
		Boselaphus tragocamelus	Blue bull
		Antilope cervicapra rajputani	Black buck
		Gazella gazella bennetti	Gazelle
7.	**Carnivora**	*Canis Iupus*	Wolf
		Canis aureus	Asiatic jackal
		Vulpes vulpes	Red fox
		Vulpes bengalensis	Bengal fox
		Lutra perspicillata	Indian otter
		Viverricula indica	Small Indian civet
		Herpestes aeropunctatus	Small Indian mongoose
		Felis chaus prateri	Jungle cats
		Felis libyes ornata	Jungle cats
		Panthera pardus sindica	Panther

summer. Wetlands support specialized vegetation and fauna; they are suitable niche for fish and other aquatic fauna, as breeding and nursery ground for water fowls and as filters for sediments and pollutants. Flood plains and marshes are sometimes not easily accessible and, hence, serve as a secure place for many kinds of wild life. Sunderbans is one such habitat in the coastal belt of the Bay of Bengal. Since wetlands contain supraoptimal water, they are quite productive. In tropics (*e.g.,* India), because of associated bright sunshine and warm condition their primary productivity becomes very high. Rice which is the major crop of India, is best grown in wetlands. Biomass of wetlands is well distributed between the above ground foliage and underground rhizomatous and cormous parts. *Typha* and *Phragmites* have very high production efficiency. *Euryale ferox*, a prized dry fruit crop is most extensively grown in wetlands of north Bihar. Wetlands are fast disappearing in many parts of world because of dumping of wastes and reclamation for agriculture, housing, forestation, etc. The flowing or lotic wetlands are most easily polluted by city sewage, industrial effluents and runoff pesticides and other agrochemicals applied in the drained upland crop fields.

8. Mangroves in India

Halophytes growing in the muddy swamps of the estuaries of tropical and subtropical regions and sea coasts flooded by rivulets and tides form the **mangroves**. Mangroves are, thus, the salt-tolerant forest ecosystems found mainly in tropical and subtropical inter-tidal regions. They comprise the swamps, forest land within and its water spread areas. In India, the total area of mangroves is about 6740 sq. km., which is nearly 7% of the world's mangroves. The mangrove areas of Sunderbans (W.Bengal) and Andman and Nicobar Islands constitute about 80% of the Indian mangroves, the rest are scattered in the states of Andhra Pradesh, Tamil Nadu, Orissa, Maharashtra, Gujarat, Goa and Karnataka.

Halophytes seen in a mangrove.

Mangroves are dense forests of low trees, especially adapted to (a) fixation, (b) respiration, (c) vivipary, (d) migration, and (e) xeric characters. In mangroves, the plant and animal diversity is very rich. Typical plant associations of Indian mangroves include *Avicennia officinalis, Avicennia alba, Aegiceras majus, Aegialitis rotundifolia, Acanthus ilicifolius, Kandelia rheedii, Bruguiera caryophyloides, Sonneratia griffithi* and *Carapa obovata.*

Mangroves are highly productive ecosystems. The net primary production is usually high. Secondary productivity and decomposition activity of the mangroves are also very high. The terrestrial consumers are snakes, lizards, turtles, jackals, wild boars, hyaena, tigers and so on. In the Bhittar Kanika sea coast of Orissa adjacent to the mangroves exists the largest sea turtle (*Lepidochelys olivacea*) rookery. Every year 300,000 to 500,000 female turtles come to the sea beach (arribada) in late December-January and again in March-April to lay about 50 million eggs (**Dash** and **Kar**, 1990). Birds are also the most colourful and conspicuous inhabitats of the mangroves. Mangroves of the Florida key support a fauna of mostly insects and spiders, along with scorpions, isopods and other arthropods (see **Krebs**, 1985). In recent years, the Indian mangroves have been subjected to reckless exploitation and other biotic interference.

REVISION QUESTIONS

1. Define biome and describe the physio-chemical characteristics and biota of tundra biomes.
2. Describe the biota of tundra biomes.
3. Write short notes on the following :
   (a) Alpine biome ; (b) Forest biome ;
   (c) Grassland biome ; (d) Wetlands ;
   (e) Mangroves.

CHAPTER

14

Pollution

(Environmental Pollutants and Toxicology)

Harmful materials introduced by man into the environment is the cause of pollution.

Recently, first time in his entire cultural history, man has faced one of the most horrible ecological crisis–the problem of pollution of his environment which sometime in past, was pure, virgin, undisturbed, uncontaminated and basically quite hospitable for him. **Pollution** is an undesirable change in the physical, chemical or biological characteristics of our air, land, and water that may or will harmfully affect human life or that of desirable species, our industrial processes, living conditions, and cultural assets (**Odum**, 1971). In other words, pollution is the unfavourable alteration of our environment, largely as a result of human activities (**Southwick**, 1976).

By and large, the term pollution is used to refer to harmful materials introduced by man into the environment. Thus, in a way, pollution is the release in excess of permitable limits of foreign substances within the environment (see **Michael**, 1984).

ORIGIN OF POLLUTION

Some authors (such as **Southwick**, 1976), associate the human population explosion with the pollution problem. They point out that with more people there is more sewage, more solid wastes, more fuel being burned, more fertilizers and insecticides being used to produce more food for hungry mouths. But, there are certain writers who have pointed out that in underdeveloped countries, pollution is not the severe problem as it is in technologically developed countries and yet the populations may be very dense. They feel that it is the wasteful aspects of our technology which strive always to produce more convenient products ("disposable" items) which pollute our environment.

There are certain modern ecologists such as **Odum** (1971), **Southwick** (1976), **Smith** (1977), etc., who sought many factors such as human population explosion, unplanned urbanization and deforestation, profit oriented capitalism and technological advancement, which may have originated pollution crisis on earth. In fact, in countries where there have been the greatest technological advances, the worst pollution occurs. In these countries, whether capitalist, socialist, or communist, 'there has been an emphasis on growth. Thus, in well-developed nations on a per capita basis, citizens consume more food , use more pesticides, fertilizers, fuel, minerals, cars, and other manufactured products of all kinds. Most of these products are manufactured in one or other kinds of industries, all of which in their turn add some pollutants in our environment and cause pollution.

POLLUTANTS: THE CREATORS OF POLLUTION

Every human society, be it rural, urban, industrial and most technologically advanced society, dispose of certain kinds of byproducts and waste products which when are injected into the biosphere in quantities so great that they affect the normal functioning of ecosystems and have an adverse effect on plants, animals, and man are collectively called **pollutants** (**Smith**, 1977). A pollutant is a constituent in the wrong amount, at the wrong place or at the wrong time. For example, nitrogen and phosphorus are essential nutrients for living organisms and are extensively used in agriculture to increase crop yields but they can also cause pollution in lakes and rivers when found in excess by promoting undue algal growth (eutrophication).

Types of Pollutants

Pollutants primarily are grouped into the following two types:

1. Natural pollutants. Certain pollutants such as carbon dioxide, carbon monoxide, sulphur dioxide, lead, mercury and other trace elements are the consequence of life processes being produced through respiration, faeces, urine and body decomposition. With an increase in human population, the pollutants are increasing with alarming rate.

2. Synthetic, man-made, anthropogenic or xenobiotic pollutants. A vast array of synthetic pollutants are increasing continuously with urbanization and industrial growth. They include pesticides, detergents, pharmaceuticals, cosmetic products, organic acids, aerosols, and metals, etc. Several of these compounds are extremely stable and persist in the environment for a considerable period posing serious environmental hazards.

From the ecosystem viewpoint, these pollutants can be classified into two basic types: **nondegradable pollutants and biodegradable pollutants** (**Odum**, 1971). The materials and poisons, such as aluminium cans, mercurial salts, long-chain phenolic chemicals and DDT that either do not degrade or degrade only very slowly in the natural environment, are called **nondegradable pollutants**. Such nondegradable pollutants not only accumulate but are often "biologically magnified" as they move in biogeochemical cycles and along food chains. Also they frequently combine with other compounds in the environment to produce additional toxins. **The biodegradable pollutants** include domestic sewage, heat, etc. The domestic sewage can be rapidly decomposed by natural processes or in engineered systems (such as a municipal sewage treatment plant) that enhance nature's great capacity to decompose and recycle. Problems arise with the biodegradable pollutants when their input into the environment exceeds the decomposition or dispersal capacity.

1. AIR POLLUTION

Air pollution is the presence in the atmosphere, or injection into it, of substances that are not present naturally, or present naturally but is in much smaller concentrations, and that may harm living organisms directly or indirectly (**Allaby** 1995). Thus, when due to some natural processes or human activities the amount of solid waste or concentration of gases other than O_2 increase in the air which normally has constant percentage of different gases in it, the air is said to be polluted and this phenomenon is referred to as **air** or **atmospheric pollution**. Air pollution is one of the most dangerous and common kind of environmental pollution that is reported in most industrial towns and metropolitans of India and abroad such as Delhi, Bombay, Calcutta, Kanpur, Madras, Hyderabad, Jaipur, Ahmedabad, Nagpur, Firozabad and also in London, New York, Tokyo, Pittsburg, etc.

Air Quality

In our country, data on air quality have been collected by NEERI (1978-79, 1980-81). Although there exist several parameters to determine air quality, but only three, *i.e.,* SO_2 NO_x and SPM (= suspended particulate matter) are used; these give fair idea of load of pollution carried by the air. NEERI has collected data from ten Indian cities/Metropolitans such as Delhi, Kanpur, Calcutta, Nagpur, Hyderabad, Madras, Cochin, Bombay, Ahmedabad and Jaipur to assess the extent and nature of degradation of air quality due to industrialization and urbanization. Each of these ten Indian cities was surveyed for residential, commercial and industrial situations.

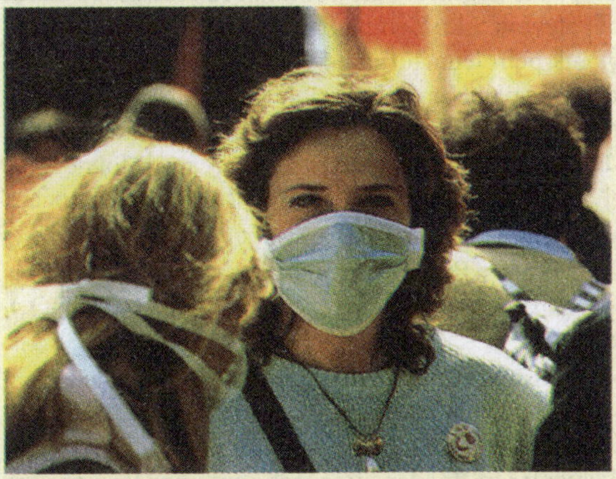

Vehicle exhausts release a collection of chemicals including carbon monoxide, sulfur dioxide, nitrogen oxides and hydrocarbons. Some gasoline also contains lead, which appears in the exhaust fumes and can have adverse effects on brain development in children. Throughout the world's cities, many people – such as cyclists, who have to breathe deeply while in close proximity to vehicle exhausts – have begun to wear masks to filter the air they breathe. Here, a Green Party protestor against the poor quality of city air in rome emphasizes the point.

Thus, for sulphur dioxide (SO_2), Calcutta is found to be most polluted city, followed in a descending order by Bombay, Delhi, Ahmedabad, Kanpur, Hyderabad, Madras, Nagpur and Jaipur. However, yearly averages did not exceed 80 $\mu g/m^3$. Levels of oxides of nitrogen (NO_x) ranged from 4 $\mu g/m^3$ in residential and industrial areas of Jaipur to the highest 40 $\mu g/m^3$ in industrial areas of Ahmedabad (1980) and commercial areas in Kanpur (1980). Level of SPM was highest in Delhi and Calcutta and the lowest in Madras and Bombay (coastal cities). In general, SPM levels in all cities were much above the international levels. Residents of Delhi are exposed to some of the highest levels of air pollution in the country and perhaps the world.

Methods of Detection and Measurement of Air Pollution

Air pollution is usually measured by sampling of air by *thermal* and by *electrostatic precipitation, Sonkin impactor* and *electrostatic dust collectors.* The particulate pollution is measured by the instrument called *deposit gauge* or by *Owen's dust counter.* The thickness of the smoke is measured by *Liegean sphere* and by *Ringelmann chart.* The rough estimation of SO_2 in air can be made by chemical analysis of the dust collected in a deposit gauge or by a *bubbler method.* Fluorides are estimated by colour reactions.

Sources of Air Pollution

1. Air Pollution by Natural Means

Nature adds few natural pollutants such as pollen, hydrocarbons released by vegetation, dusts from deserts, storms, and volcanic activity. Thus, **volcanic eruptions** may eject large amounts of gases and particulate matter. Settling volcanic ash can kill vegetation by coating leaves and preventing photosynthesis and transpiration (*e.g.,* following the 1980 eruption of Mt. St. Helens, USA). Fine particles, mainly of sulphates, may penetrate the stratosphere, spread widely and reflect significant amounts of solar radiation, leading to climatic cooling (*e.g.,* the 1991 eruption of Mt. Pinatubo, Phillippines). Likewise, **dust storms** sometimes carry fine sand for thousands of kilometers and favourable weather conditions stimulate the release of pollen, affecting people sensitive to it.

2. Air Pollution by Human Activities

(a) Industrial chimney wastes. There are a number of industries which are potent sources of air pollution. **Petroleum refineries** are the major sources of gaseous pollutants (*e.g.,* SO_2, NO_x, etc.) Mathura-based petroleum refinery has been accused to aggravate the pollution-related decay of Taj Mahal in Agra and other historical monuments of Fatehpur Sikri Complex. **Industrial processors** such as metallurgical plants and smelters, chemical plants, petroleum refineries, pulp and paper mills, sugar mills, cotton mills, and synthetic rubber manufacturing plants are responsible for about one fifth of the air pollution. **Cement factories** emit plenty of dust, which is potential health hazard. Stone crushers and hot mix plants also create a menace. The SPM levels in such stone crushing areas are found to be five time the industrial safety limits. **Chemical manufacturing industries** emit acid vapours in air.

(b) Thermal power stations. The coal consumption of thermal power stations of India (*e.g.,* Delhi has three thermal power stations, one at Indraprastha Estate, others at Rajghat and Badarpur) is several million tonnes. The chief pollutants of coal burning are fly ash, SO_2' and other gases (CO, NO_2), aldehydes and hydrocarbons (**Chaudhuri**, 1982).

(c) Automobiles. The transportation industry exclusive of automobiles and including railroads, ships, aircrafts, trucks, buses, tractors, etc., contribute the same type of pollutants as cars (see **Smith**, 1977). The vehicular exhausts are toxic being a source of considerable air pollution, next only to thermal power plants. The ever increasing vehicular traffic density poses continued threat to the surrounding air quality. At the global level, there are over 300 million cars, trucks and buses and their number is increasing rapidly. India too has millions of vehicles, of which more than 65% are two wheelers operating on petrol.

The sources of emission in the automobiles are (i) exhaust system; (ii) fuel tank and carburettor and (iii) crankcase (Fig. 14.1). The exhaust produces many air pollutants including unburnt hydrocarbons, CO, NO_x and lead oxides. There are also traces of aldehydes, esters, ethers, peroxides and ketones; these are chemically active and combine to form smog in presence of light. Due to volatile nature of petrol, evaporation from fuel tank goes on constantly and results in emission of hydrocarbons. The evaporation through carburettor occurs when engine is stopped and heat builds up and as much as 12 to 40 ml of fuel (petrol/ diesel) is lost during each long stop causing emission of hydrocarbons. Some gas vapour escapes between walls and the piston, which enters the crankcase and then

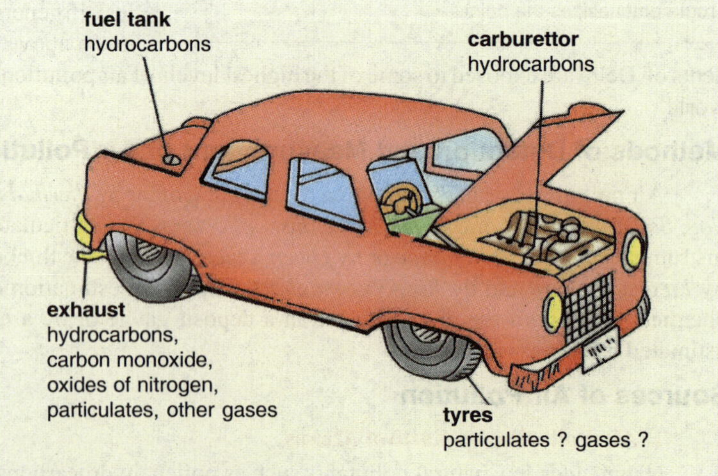

Fig. 14.1. Hydrocarbons and other particulate and gaseous emissions from automobile.

discharges into the atmosphere. Thus, the total hydrocarbon emission of an engine reaches upto 25%.

Other sources of air pollution are minor in quantities but bear significance due to the harmful substances they release, these are **agriculture,** which is responsible for pesticides, dust from agriculture practices and field burning, and the construction industry.

Types of air pollutants. All the just described sources of air pollution release the following air pollutants: 1. **Carbon compounds** (*e.g.,* CO_2, CO); 2. **Sulphur compounds** (*e.g.,* SO_2, H_2S and H_2SO_4) 3. **Nitrogen oxides** (*e.g.,* NO, NO_2 and HNO_3); 4. **Ozone** (O_3); 5. **Flurocarbons;** 6. **Hydrocarbons** (*e.g.,* benzene, benzypyrene, etc.); 7. **Metals** (*e.g.,* lead, nickel, arsenic, beryllium, tin, vanadium, titanium, cadmium, etc.); 8. **Photochemical products** (*e.g.,* olefins, aldehydes, photochemical smog, PAN, PB_2N, etc.); 9. **Particulate matter** (*e.g.,* fly ash, dust, grit and SPM); and 10. **Toxicants.**

Some of the most common air pollutants, their sources and their effects on human health have been tabulated in Table 14-1.

Table 14-1. Common air pollutants, their sources and pathological effects on man (Source: Southwick, 1976)

Pollutants	Where they come from (source)	Pathological effect on man
1. Aldehydes	Thermal decomposition of fats, oil, or glycerol.	Irritate nasal and respiratory tracts.
2. Ammonias	Chemical processes—dye-making; explosives, lacquer; fertilizer.	Inflame upper respiratory passages.
3. Arsines	Processes involving metals or acids containing arsenic soldering.	Break down red cells in blood, damage kidneys; cause jaundice.
4. Carbon monoxides	Gasoline motor exhausts; burning of coal.	Reduce oxygen-carrying capacity of blood.
5. Chlorines	Bleaching cotton and flour; many other chemical processes.	Attack entire respiratory tract and mucous membranes of eyes; cause pulmonary edema.
6. Hydrogen cyanides	Fumigation; blast furnaces, chemical manufacturing; metal plating.	Interfere with nerve cells; produce dry throat, indistinct vision, headache.
7. Hydrogen fluorides	Petroleum refining; glass etching; aluminium and fertilizer production.	Irritate and corrode all body passages.
8. Hydrogen sulphides	Refineries and chemical industries; bituminous fuels.	Smell like rotten eggs; cause nausea; irritate eyes and throat.
9. Nitrogen oxides	Motor vehicle exhausts; soft coal.	Inhibit cilia action so that soot and dust penetrate far into the lungs.
10. Phosgenes (carbonyl chloride, $COCl_2$)	Chemical and dye manufacturing.	Induce coughing, irritation, and sometimes fatal pulmonary edema.
11. Sulphur	Coal and oil combustion.	Cause chest constriction, headache, vomiting, and death from respiratory ailments.
12. Suspended particles (ash, soot, smoke)	Incinerators; almost any type of manufacturing.	Cause emphysema, eye irritations and possibly cancer.

Ecology of Air Pollution

Once injected into the atmosphere, pollutants enter the biogeochemical cycles by different routes. The air above many cities can assimilate and disperse great quantities of fine particulate and

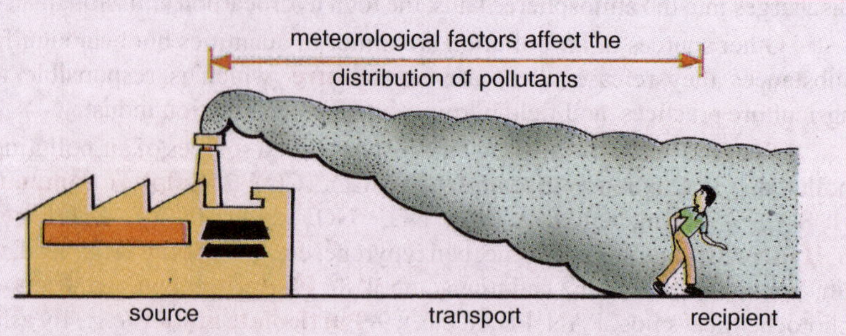

Fig. 14.2. Mode of movement of air pollutant from source to recipient.

gaseous pollutants as long as air can move and disperse (Fig. 14.2). But if air masses over cities become stagnant, pollutants accumulate quickly and deteriorate air quality which cause many respiratory diseases in man and other animals. Air pollutants also accumulate during **temperature inversions**, when cooler surface layers of air become trapped under warmer upper layers (Fig. 14.3). In these situations, the upper layers of warm air prevent the vertical rise and dispersal of pollutants which are held near the ground. Temperature inversions commonly occur in cities surrounded by mountains or bordered by mountains on the leeward side.

Further a portion of air pollutants reaches land as dry fallouts; it may then enter various nutrient cycles and food chains through water and soil. Other contaminants of air react chemically or photochemically with each other and produce such secondary pollutants as sulphuric acid, ozone, and, peroxyacetyl nitrate or PAN. Aerosols and other forms of fine particulate matter act as **condensation nuclei,** to which water vapours present in the air quickly surround to form droplets of fog or rain.

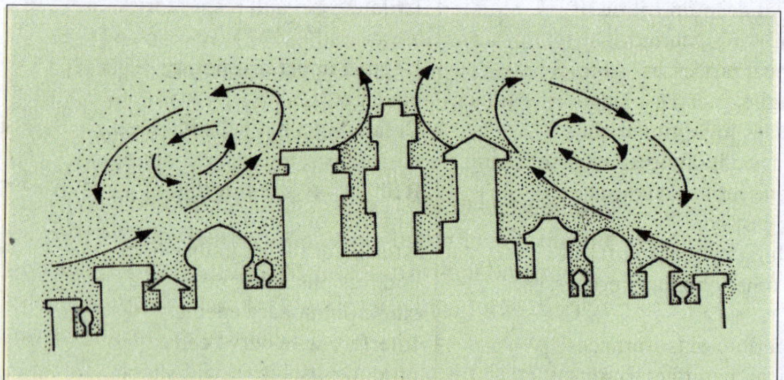

Fig. 14.3. Pattern of circulation of warm and cool air over a city and formation of the heat island.

Moreover, different air pollutants adversely affect flora, fauna and climate of a given area variously and some of the common air pollutants and their specific effects on man, vegetation, climate, etc., have been discussed as follows:

A. Gaseous Pollutants

1. Sulphur oxides and hydrogen sulphide. These gaseous pollutants are naturally released by the biological decomposition and from volcanic eruptions. They are also released artificially due to human activities such as smelting of sulphide-containing ores, combustion of sulphur-containing fuels such as coal and oil, petroleum refining and obtaining of geothermal energy.

According to a report the sulphur dioxide (SO_2) content of the atmosphere in Delhi city has reached the level of 0.233 ppm, whereas in USA and West Germany the permissible limit is only 0.1 and 0.05 ppm. Concentrations as low as 0.3 ppm may damage plants. Lichens are particularly sensitive to SO_2 and in polluted regions one does not find lichens growing on the tree trunks. Thus, lower quantities of sulphur dioxide suppresses the overall vegetative as well as reproductive growth and yield. Its high atmospheric concentrations produce various injuries to leaves such as interveinal and blade damage, necrosis of leaves and cellular collapse (Fig. 14.4). However, moderate SO_2 pollution results in chlorosis of leaves without cellular collapse. Pine trees are more susceptible than broad leaved trees and react by partial defoliation and reduced growth. Plant's exposure to hydrogen sulphide results in leaf lesions, mottling, defoliation and reduced growth.

Sulphur dioxide pollution causes in human beings various types of injuries such as eye irritation, chest constriction, headache, vomiting and death from respiratory ailment (Fig. 14.5). It paralyzes or destroys bronchial cilia in air passages of man, constricts bronchii, damages lungs, lowers resistance to pneumonia and influenza and causes bronchitis, emphysema and irritation of the mucous membranes *(i.e.,* an increase in cough and sputum). In fact, SO_2 and other pollutants bring about coalescence of alveoli (Fig. 14.6) and reduce the amount of surface area available for the transport of oxygen and also reduce the rate at which air is exchanged. When there occurs severe pollution of SO_2, the death rate and bronchial asthma are found to increase and in past it caused such disasters as Meuse Valley in Belgium in 1930; Donora, in 1938; London, in 1952; and New York and Tokyo in 1960s (see **Southwick**,1976).

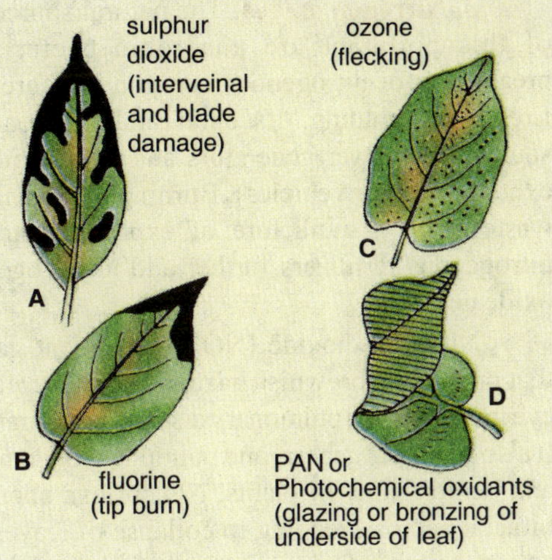

Fig. 14.4. Injuries caused to plant leaves by SO_2 (A), HF (B), O_3 (C) and PAN (D).

Further, in the atmosphere, SO_2 does not remain in the gaseous state for long time, but very soon it reacts with moisture to form sulphuric acid or H_2SO_4. Sulphuric acid causes many respiratory diseases in man and also produces **acid rainfalls** over parts of the earth. In Scandinavia, downwind from the industrial centers of Britain and the Ruhr Valley, the acidity of the rainfall has increased 200 fold since 1966, with pH values as low as 2.8 being recorded (**Oden** and **Ahl**, 1970). This acid rainwater has increased the acidity of Scandinavian streams, interfering with salmon reproduction and destroying salmons runs. It has reduced forest growth and increased the amount of calcium and other nutrients leached from agricultural soil. The acid rainfall is also involved in the erosion of building materials as limestone marble, the slate used in roofing, mortar and deterioration of statues.

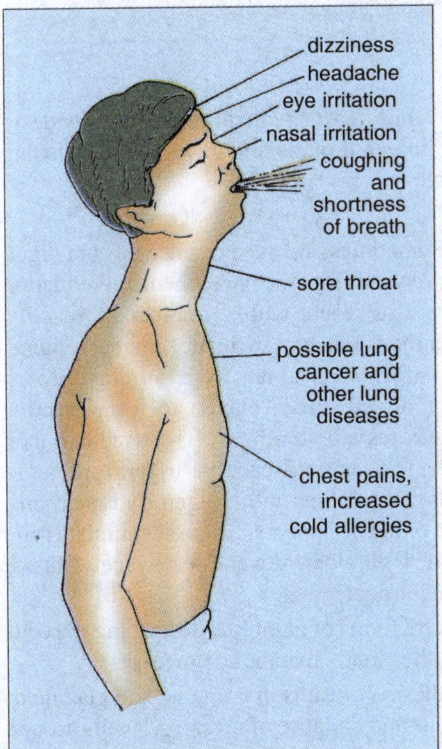

Fig. 14.5. Ailments caused in man by air pollution.

2. Carbon monoxide. It is released chiefly from gasoline engines and burning of coal in the defective furnaces. In man, CO produces headache, dizziness, inability to distinguish time intervals, nausea, ringing in the ears, heart palpitation, pressure in chest and difficulty in breathing. This gas combines with haemoglobin of blood to form carboxy-heamog-lobin in RBC which reduces its oxygen carrying capacity to all parts of body and, thus, causing asphyxia. The higher concentrations of carbon monoxide (CO) may be even fatal. CO and nicotine from the cigarette smoke increase the tendency for the blood to clot and so block the coronary arteries. Carbon monoxide also increases the rate at which the fatty materials are deposited in the arteries.

3. Nitrogen oxides. The natural sources of this pollutant are anaerobic bacterial breakdown of nitrogenous compounds, forest fires and lightning. Its chief anthropogenic sources are power generators and automobile exhausts (motor vehicles). Burning of organic wastes, and manufacture of explosives and nitrogenous fertilizers further add to nitrogen oxide pollution.

Nitrogen dioxide (NO_2), a pungent gas that produces a brownish haze, causes nose and eye irritations and pulmonary discomfort in man. It also produces general and pulmonary oedema and hemorrhage. In plants, NO_2 brings about bifacial necrosis leading to collapse of leaves, enhancement of green colour followed by chlorosis and extensive leaf drop. Ultimately there occur an increase in fruit drop and decrease in fruit crop.

Acid deposition of corrosive
This limestone statue at Rheims Cathedral in France is being dissolved away by acid deposition.

Photochemical smog. In the atmosphere nitrogen dioxide is reduced by ultraviolet light to nitrogen monoxide and atomic oxygen :

$$NO_2 \rightarrow NO + O$$

Atomic oxygen reacts with oxygen to form ozone :

$$O_2 + O \rightarrow O_3$$

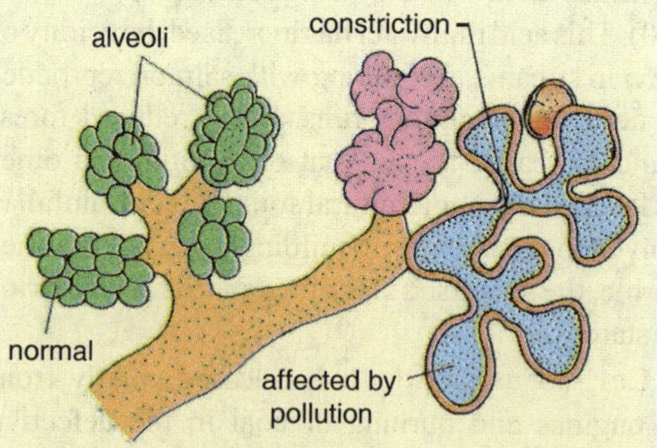

Fig. 14.6. Effect of sulphur dioxide and other air pollutants on the bronchial tubes of human lungs.

Ozone reacts with nitrogen monoxide to form nitrogen dioxide and oxygen, thus, closing the cycle :

$$NO + O_3 \rightarrow NO_2 + O_2$$

Sometimes, in the presence of sun light, atomic oxygen from the photochemical reduction of NO_2 also reacts with a number of reactive hydrocarbons (such as methane, ethane, toluene, etc., all of which originate from burning of fossil fuels or directly from plants) to form reactive intermediates called **radicals.** These radicals then take part in a series of reactions to form still more radicals that combine with oxygen, hydrocarbons, and NO_2. As a result nitrogen dioxide is regenerated, nitric oxide disappears, ozone accumulates and a number of secondary pollutants are formed such as formaldehyde, aldehydes and peroxyacetyl nitrate or PAN ($C_2H_3O_5N$). All of these collectively form **photochemical smog**.

Ozone, PAN and nitrogen dioxide severely injure many forms of plant life, destroying the cells of leaves, damaging the chloroplasts, and interfering with the plant's metabolic processes.

PAN is known to block "Hill reaction" of photosynthesis. It results in bronzing and glazing of abaxial leaf surface (Fig. 14.4D) which is due to plasmolysis and collapse of mesophyll cells around substomatal chambers (Fig. 14.8 B). Epidermal and guard cells are not injured by PAN.

4. **Ozone**. Levels of ozone (O_3) may rise in atmosphere due to human activities. It is also

Fig. 14.7. Formation of the smog an its climatic effects.

formed by NO_2 under UV-radiations effect. Minor amounts of ozone are also added to the atmosphere by electric discharges such as lightning flashes, by vertical flux of stratopheric ozone and by tropospheric storms.

Ozone near the earth's surface in the troposphere creates pollution problem. Increase in O_3 concentration near earth's surface is toxic to plants reducing crop yields significantly. It also has adverse effects on human health (see Table 14-2). Thus, while higher levels of O_3 in the atmosphere protects us, it is harmful when it comes in direct contact with us and plants at earth's surface.

Table 14-2. **Effects of ozone (O3) on human health.**

Concentration (ppm)	Effects observed
0.2	No ill effects
0.3	Nose and throat irritation
1.0–3.0	Extreme fatigue after 2 hrs.
9.0	Severe pulmonary oedema

In plants, O_3 enter through stomata and produces visible damage to leaves and results in decrease in yield and quality of plant products. Thus, O_3 results in necrotic flecking of upper surface of leaf (Fig. 14.4C), general chlorosis and bronzing, premature senescence of plants, precocious dropping of older leaves, reduced growth of shoots and roots, suppression of nodulation, reduction in seed set and yield. Ozone causes shrinkage of nuclei and cytoplasm or mesophyll cells which become granular and results in increased intercellular space (Fig. 14.8A). At 0.02

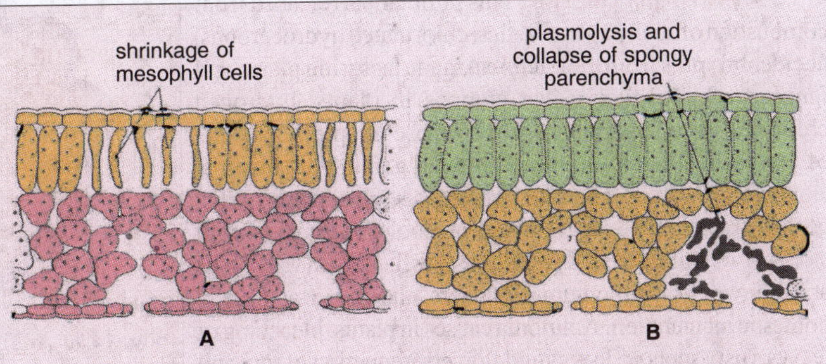

Fig. 14.8. A–Affect of ozone on the palisade layer of leaf ; B – collapse of spongy parenchyma in the region of substomatal cavity due to PAN.

ppm, O_3 damages tobacco, tomato, bean, pine and other plants. It is now known to be the cause of several widespread diseases of plants such as weather fleck of tobacco, leaf tip burn of carnations, tip burn of onions, bronzing of bean, speckle-leaf of potato and brown leaf of grapes. In human beings, O_3 at its 9.0 ppm concentration result in severe damage of pulmonary organs. Its lower concentrations result in nose and throat irritation and severe tired feeling (see Table 14-2).

5. Fluorocarbons (Hydrogen fluoride). Natural sources of fluorides in the atmosphere are active volcanoes. Their man-made or artificial sources are petroleum refining, aluminium, steel and electrochemical reduction plants, blast furnaces, brick-kilns, and tile, glass etching and superphosphate fertilizer industries and combustion of coal.

Fluoride burns the tip of plant leaves (Fig. 14.4B). It's low amounts impair plant growth, result in excessive dropping of bloom and young fruits, development of small, partially or completely seedless fruits and premature formation of soft red flesh and splitting of peach. In human being, it irritates and corrode all body passages.

6. Hydrocarbons. Biological decomposition of organic matter, spill and seepage from natural gas and oil fields and volatile emissions from plants are some major natural causes for the release of hydrocarbons such as methane, terpenes, ethylene and aniline. Incomplete combustion of fuels, automobile exhaust (Fig. 14.1), petroleum-refineries, agricultural burning, motor fuel marketing, manufacture of explosives and cracking of natural gas in petrochemical plants (as a blow-off emissions) constitute the man-made sources that emit hydrocarbons.

The hydrocarbon **ethylene** causes yellowing and occasional necrosis of leaves, chlorosis of floral buds, inhibition of terminal growth, epinasty of leaves, shortening of internodes, thickening of stems, lack of apical dominance, stunted growth, dry sepal disease of orchids and decrease in the amount of chlorophyll and carotenoids. Another hydrocarbon **aniline** results in the appearance of bands on leaves as if they are water-soaked, necrotic spots and abscission of leaves. In man, hydrocarbons bring about irritation of mucous membrane, bronchial constriction and eye irritation. One of the hydrocarbons released due to incomplete combustion, is **3-4 benzpyrene** which is said to cause lung cancer. **Methyl isocyanate** gas when accidentally leaked out from the storage tanks of the pesticide factory in Bhopal on December 2, 1984, had killed over 3,000 persons and seriously affecting lakhs of residents.

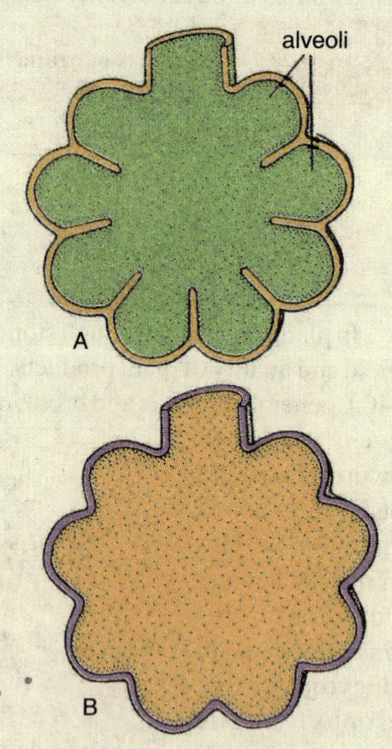

Fig. 14.9. A–The appearance of alveoli of a normal lung; B–The appearance of alveoli in the lung of the person suffering with emphysema.

7. Hydrogen chloride. This pollutant is released from combustion of coal, paper, plastics, chlorinated hydrocarbons, accidental spills from the chemical manufacturing plants and ignition of solid-fuel rocket engines in plants, hydrogen chloride causes plasmolysis and collapse of epidermal cells of leaves and thereby results in abaxial glazing of leaves.

8. Ammonia. The main anthropogenic sources of this gaseous pollutant are refrigerator, precooler systems of cold storages, manufacture of dyes, explosives, lacquers (varnishes) and anhydrous ammonium fertilizers and nitric acids and domestic incineration. Ammonia causes in plants, bleaching of leaves, rusty spots on leaves and flowers, reduction of root and shoot growth, browning and softening of fruits, development of dark, corky lenticels in apples, and reduction in the rate of seed germination. In man, it inflames upper respiratory passages.

9. Tobacco smoke. Tobacco smoke contains about 300 chemical compounds including nicotine and carcinogens such as tar ("aromatic hydrocarbon"). It is mainly produced by smoking cigarettes and bidis. It is gradually becoming a pollutant especially in closed atmospheres such as buses, trains, auditoria, discotheques and so on. **Nicotine** stimulates some types of synapse of nervous system, increase blood pressure and heart rate by the production of adrenaline, causes vasodilation in the muscles and vaso-constriction in the skin. When a person smokes, tiny particles in the smoke get caught on the lining of the windpipe and bronchial tubes. Extra mucus is produced and the cilia stop beating. The mucus collects in the bronchial tube and this gives rise to a "smoker's cough".

Repeated coughing may cause the delicate walls of the alveoli to break down into larger spaces. This cuts down the surface area over which gaseous exchange can take place, so the person gets very short of breath. Doctors call this condition **emphysema** (Fig. 14.9). Tobacco smoke also brings about thickening of bronchial epithelial layer, loss of ciliated cells and appearance of cells with bizarre nuclei, which are probably the precursors of cancerous cells (Fig. 14.10). Although smoking mainly affects the lungs, it can also cause cancer of other organs such as mouth, throat, oesophagus and bladder. It is also associated with heart disease and stomach ulcers and a woman who smokes while she is pregnant is more likely to have spontaneous abortion of still birth or to give birth to an under sized baby.

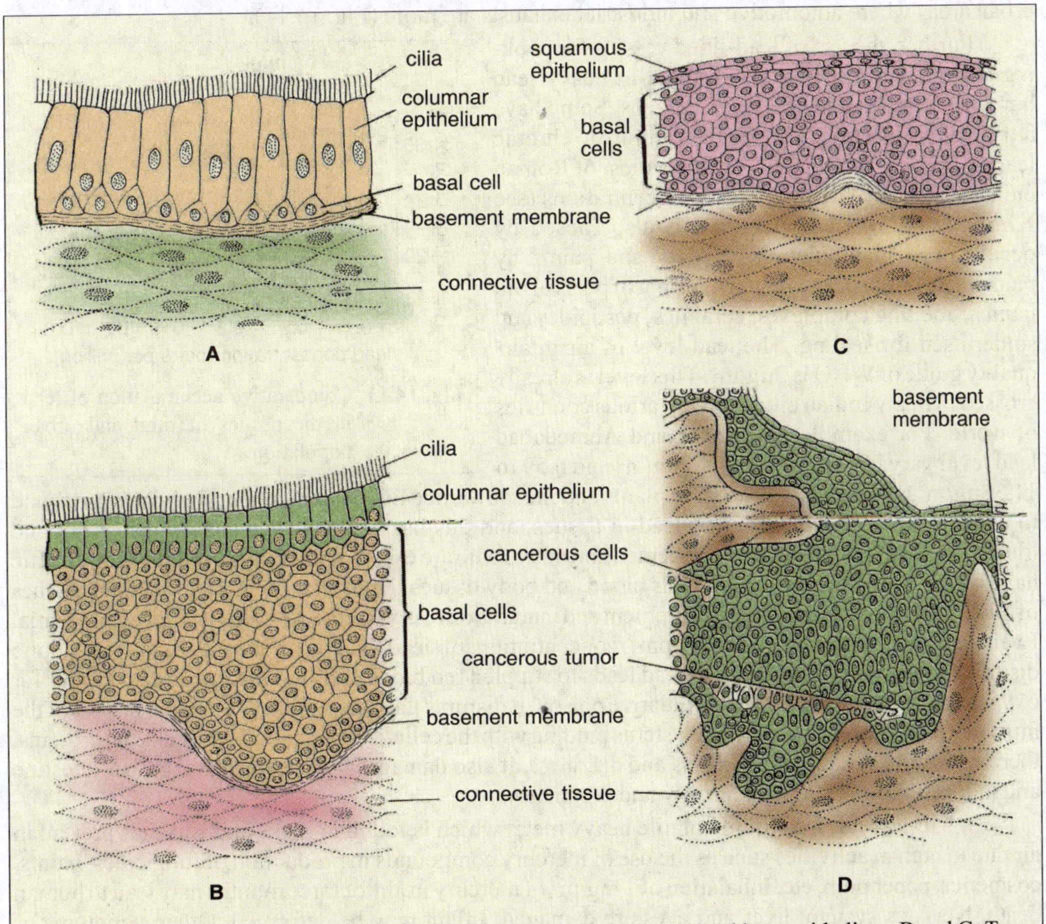

Fig. 14.10. Effect of smoking on the lung epithelium of smokers. A–Normal lung epithelium; B and C–Two stages of abnormal changes in the lung epithelium found in smokers ; D–An epithelial tumor growing in the lung tissue.

B. Particulate Pollutants

1. Fluorides. The particulate fluorides originate in the same way as the gaseous fluorides. They settle and accumulate on the grass and other vegetation. They are less toxic to these plants causing occasional leaf-tip burns. However, ingestion by cattle of various fluorine compounds falling on forage, causes **fluorosis,** a disease characterized by abnormal calcification of bones and teeth, eventually resulting in loss of teeth, body weight and in lameness.

Fluoride pollution in man and animals is mainly through water. In our country, fluorosis is a public health problem in states of Gujarat, Rajasthan, Punjab, Haryana, U.P., Andhra Pradesh, Tamil Nadu, Karnataka, and some areas of Delhi. Globally it is a problem of various other countries such as USA, Italy, Holland, France, Germany, Spain, Switzerland, China, Japan and some African and Latin American countries.

2. Lead. Lead, a heavy metal, is injected in the atmosphere mainly from automobile exhaust. Automobile gasoline contains tetraethyl lead that is used as an antiknock additive. Lead is emitted into the air with the exhaust as volatile **lead halides** (bromides and chlorides). About 75% of lead burnt in gasoline comes out as lead halides through tail pipe in exhaust gases. Of this about 40% settles immediately on the ground and the rest (60%) goes into air. That is why its concentration is higher in urban areas where automotive and industrial exhausts are more (Fig. 14.11).

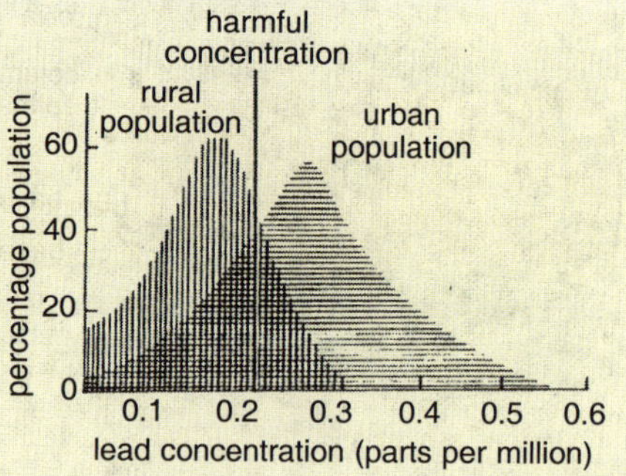

Fig. 14.11. Comparative accumulation of lead in the tissues of rural and urban population.

Further, the use of lead-lined vessels for cooking and the storage of wine resulted in heavy lead burdens in the bodies of Roman citizens. Some have attributed the decline of the Roman empire to chronic lead poisoning. In fact, analysis of bones, of Roman citizens have revealed high lead concentrations (see **Kimball**, 1975). Lead pollution is also caused by occupational manipulation of putty and paints by painters; habitual or accidental nibbling of peels of old paints; smelting complexes, ceramics, pesticides and solder used for sealing. The lead level of air in air-quality guide of WHO is 2μg/m^3. This level is already crossed in many Indian cities and in various countries of world. For example, in Kanpur and Ahmedabad lead level vary between 1.05 to 8.3μg/m^3 and 0.59 to 11.38μg/m^3, respectively. Roadside plants and meadow mice living along major highways are found to contain high concentrations of lead in tissues, and this has a sublethal effect on the health and longevity of the animal. Traffic policemen and others who are exposed for long periods to heavy traffic have higher than average lead in their blood and body tissues. High lead accumulation in the tissues of human body interferes with development and maturation of erythrocytes (RBC) and causes anaemia. Even 0.2 parts per million (0.2 ppm) concentration of lead in human body creates metabolic disturbances. Chronic exposure to lead leads to stippled red blood cells having impaired capacity for oxygen transportation. Being a cumulative poison, it disrupts the functioning of cells and organs of the muscular, circulatory and nervous systems binding with the cellular enzymes and coagulating proteins, thereby, results in nausea, weakness and dizziness. It also damages liver, kidney and gastro-intestine and induces abnormalities in fertility and pregnancy.

3. Mercury. It is a liquid volatile heavy metal which is found in rocks and soil. It is present in air due to human activities such as the use of mercury compounds in production of fungicides, paints, cosmetics, paper pulp, etc. Inhalation of 1 mg/m^3 of mercury in air for three months may lead to human death. Nervous system, liver and eyes are damaged. Infant may be deformed. Other symptoms of mercury toxicity are headache, fatigue, anxiety, lethargy, loss of appetite, etc.

4. Zinc. Zinc in air occurs mostly as white zinc oxide fumes and is toxic to man. It exists in air around zinc smelters and scrap zinc refineries.

5. Cadmium. This metal is emitted to air by human activities and industries (*e.g.*, electroplating and welding of cadmium containing materials; industries producing pesticides and phosphate fertilizers. Cadmium occurs in the air in the form of oxide, sulphate or chloride compounds. It is poisonous at very low levels and is known to accumulate in human liver and kidney, Cadmium causes hypertension, emphysema and kidney damage. It may also act as carcinogen in mammals.

6. Potassium salts. These particulate pollutants are derived mainly from potash mines and cause in plants abnormalities such as branch tip death, chlorosis and necrosis of leaves.

7. Sodium chloride. Certain de-icing salts such as sodium chloride, used to remove ice and snow from roads in winters, are found to cause multiple damage to the roadside trees such as leaf necrosis, defoliation, suppression of flowering, and dieback of terminal shoots in apple.

8. Agricultural chemicals. Several types of chemicals such as insecticides, herbicides, fungicides and pesticides, used widely in agriculture are found to result in foliar lesions, chlorosis and abscission of leaves and reduction in fruit set.

9. Particulate matter. The word particulate has been derived from particle and includes all solid or liquid substances primarily in the air. Particulate matter is usually divided into two categories : suspended particulates and dust fall.

(a) Suspended particulates. These are smaller than 10 microns. They are generated by various industrial processes, combustion processes and black soot. In polluted air, their concentration may reach up to 100,000 per cubic centimeter.

(b) Dust fall. Dust fall includes the particles larger than about ten microns. They are emitted into the air by physical processes such as grinding and abrasion, a soot and fly ash (or PFA = Pulverised fly ash) from fuel combustion. Dust particles settle at the ground quickly and cause nuisance to certain industries requiring aseptic conditions such as drug industries and food processing plants. The average level of dust and particulate matter in Dehi is about 600 mg/m^3 in Bombay about 200 mg/m^3, in Calcutta about 300mg/m^3, whereas it is just 150 mg/m^3 in other polluted western cities.

Dust may be inorganic and organic in nature: 1. **Inorganic dust** containing silica and trapped heavy metals is the main pollutant in mining, quarrying and stone cutting operations. 2. **Organic dust** raised in cotton textile mills, ginning plants, coir retting and processing, jute and hemp processing, saw mills and plywood industries are also potentially toxic with properties of sensitization of persons exposed to them.

Particulates emitted from cement manufacturing units cause in plants premature fall of needles, higher puberty of leaves, formation of more stomata and trichomes, reduction in number and size of cobs and weight of seeds and increase in number of infertile seeds. Dust from stone crushers, lime kilns and slate making units is also hazardous to plants and human beings. Particles of asbestos that are released from factories can wear off brakes and gears of cars. Coal dust and asbestos have been found to cause in plants, necrotic lesions, reduction in fruit sets and silicosis, asbestosis and lung cancer in human beings. The common dust particles sometime become health hazards as they may lead to diseases such as allergic asthma, bronchitis, emphysema and even fibrosis of lungs. Pulverised fly ash (PFA) is getting accumulated in such large quantities particularly in areas where thermoelectric generators are installed that it has proved to be one of the major sources of solid wastes. Since it contains boron, its deposition by wind in agricultural tracts often results in deleterious effects.

In fact, living organisms are rarely exposed to a single pollutant in nature. The influence of mixtures of pollutants is usually synergistic, *i.e*, greater than the additive effects of the different pollutants alone.

C. Effects of Air Pollution on Weather, Climate, and Atmospheric Processes

At gross level, air pollution causes two worldwide problems – contamination of the upper atmosphere and the alteration of weather and climate.

Air pollution also affects weather on a continental or global basis. Many gaseous pollutants and fine aerosols reach the upper atmosphere, where they have basic effects on the penetration and absorption of sunlight. **Brodine** (1973) and certain other modern environmental biologists feel that increasing particulate pollution may be reducing the amount of sunlight energy reaching the earth's surface, thereby, lowering solar radiation at the earth's surface and producing a cooling effect on world climates which could ultimately trigger another ice age. In fact, **Thompson** (1975) has reported a decrease in mean annual temperatures in the northern hemispheres and an increase in the north polar ice cap.

Green house effect. Carbon dioxide is a natural constituent of the atmosphere, but, its concentration is increasing in the air with an alarming rate. A byproduct of the burning of fossil fuel, it is not necessarily a pollutant. It produces adverse physiological effects only at very high levels. It is estimated that approximately one-half of the CO_2 input stays in the atmosphere and other half of it is removed by the oceans and by plants. The increased amount of CO_2 in atmosphere is found to increase the temperature of earth.

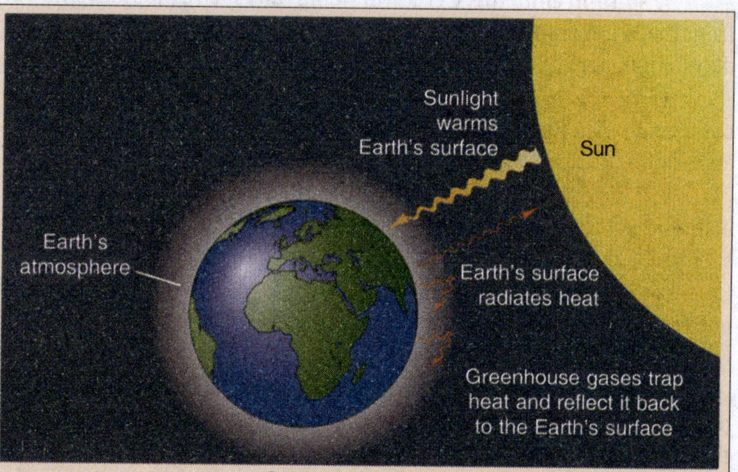

Green House Effect.

Energy from the Sun is either transmitted, reflected or trapped by the atmosphere. Energy radiated back into space is at longer wavelengths (in the infrared region) than the energy arriving from the Sun, and the consequence is that more heat enters than can escape.

Greenhouse gases (CO_2, CH_4, nitrous oxides, CFCs), together with water vapour in clouds, contribute to the 'blanketing' effect of the atmosphere.

The spectral properties of CO_2 in the atmosphere are such that it tends to prevent the long wave radiations (i.e., infra-red heat radiation) from earth from escaping into outer space and deflect it back to earth. The latter has an increased temperature at surface (**Turk** *et. al., 1974)*. This phenomenon is called **atmospheric effect (Lee,** 1974) or **greenhouse effect** (see **Southwick**, 1976, **Smith,** 1977).

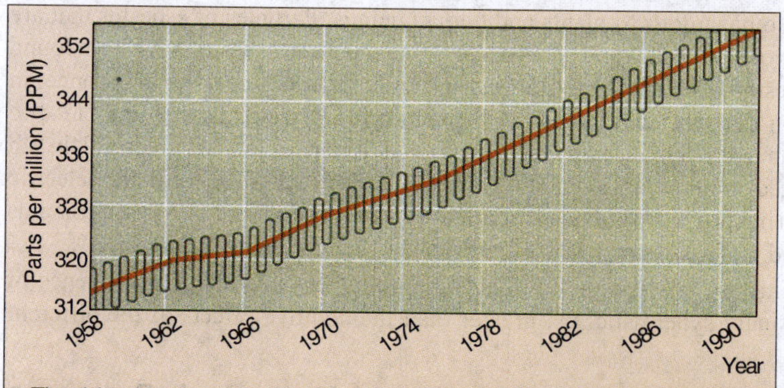

The rising level of carbon dioxide in the atmosphere. Data was collected at the Mauna Loa observatory in Hawaii. There is good evidence to suggest that changes in the average temperature at the surface of the Earth are closely linked to changes in the levels of greenhouse gases.

The simultaneous cooling and heating effects of air pollution on earth have increased variability in the world-wide weather patterns which may be a serious threat to global food production **(Thompson,** 1975). Recently, certain ecologists have tried to correlate air pollution with serious and prolonged droughts, heavier rains and floods, and more serious hurricanes and tornadoes (see **Southwick,** 1976).

Scientists fear that increased level of CO_2 will increase the greenhouse effect and thereby increase the temperature of the earth. Only a slight temperature rise would cause the polar ice caps to melt and to cause an enough rise in sea level submerging a number of major cities of the world, because many are along coasts.

Simultaneously, along with carbon dioxide man has been adding solid particles and droplets into the air which increase the **albedo** or shineness of the earth. This should act contrary to the greenhouse effect, reduce the sunlight reaching the earth and tend to lower the temperature of earth.

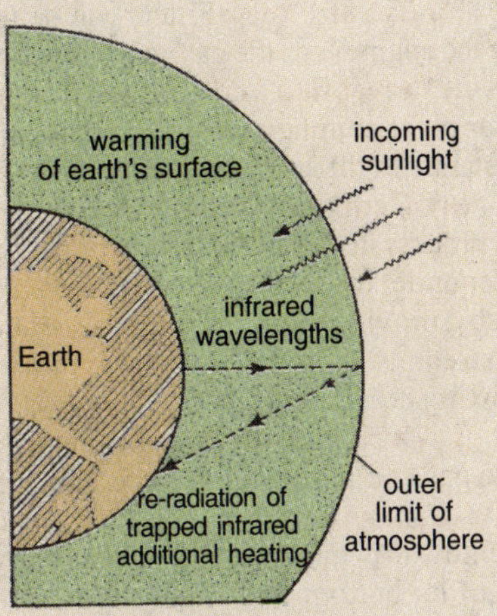

Fig. 14.12. The greenhouse effect of carbon dioxide

Peeling of ozone umbrella by CFMs. Certain fluorocarbon compounds which are called chlorofluoromethanes or CFMs or "freon" are used as propellants in pressurized aerosol cans. They are inert in normal chemical and physical reactions, but they get accumulated in greater amounts at high altitudes and there in the stratosphere these inert gaseous compounds (*i.e.*, CFMs) release chlorine atoms under the influence of intense short-wave ultraviolet radiation. Each atom of chlorine chain then reacts with more than 1,00,000 molecules of ozone, converting ozone to oxygen. The reduction in stratospheric ozone permits greater penetration of ultraviolet light, which intensifies UV radiation at the earth's surface. Some scientists such as **Ahmed** (1975), **Brodeur** (1975), and **Russell** (1975), feel that this intensified radiation will cause a significant increase in skin cancer and eventually have a lethal effects on many organisms, including man. In plants such enhanced level of UV radiations are feared to cause stunted growth, short, thick stems, smaller leaves, plasmolysis of cells, destruction of anthocyanin, chlorophyll and nuclei. They are known to cause bronzing of leaves, injury to various fruits, somatic aberrations, discolouring of staminal hairs, inhibition of pollen germination and pollen tube growth.

The protective ozone layer of the stratosphere is also considered by many ecologists to be endangered by supersonic jets, the SSTs. The jet engines of supersonic aircraft flying at high altitudes release nitrogen oxides (NO_x) which catalytically destroy ozone molecules (see **Southwick**, 1976).

Control of Air Pollution

Most kinds of air pollutions can be controlled by modern technology, but the costs ultimately be borne by the public in the form of higher prices for manufactured goods, higher taxes, reduced profit margins in industry, and more restrictions on individual activities such as burning leaves and trash and use of automobiles. Following steps have to be taken to control pollution at source (prevention) as well as after the release of pollutants in the atmosphere.

1. Prevention and control of vehicular pollution. For controlling the air pollution because of vehicles, the following measures are adopted:

(a) Curbing the pollutant emission from vehicular exhaust. This type of control can be attained by (i) using new proportion of gasoline and air; (2) more exact timing of fuel feeding; (3) using gas additives to improve combustion; (4) by injecting air into the exhaust to convert exhaust compounds into less toxic substances, and by (5) correcting the engine design and/or fixing cessation device to improve combustion with the existing design. Complete elimination of three main pollutants, namely CO, NOx and hydrocarbons can be attained by either updating the present design of engines or by making appropriate changes in devices for improving combustion. In recent years, **I.K.Bharati** of Mumbai has claimed to have devised a simple attachment (by the patent name **Thermoreactor**) to curb air pollution by motor vehicle. The reactor is fitted to the exhaust tail pipe and it converts carbon monoxide into oxygen. Various devices such as positive crankcase ventilation valve and catalytic convertor have been developed in USA to reduce exhaust emissions by automobiles.

(b) Control of evaporation from fuel tank and carburetter. This can be performed by (1) collection of vapours with activated charcoal when the engine is turned off; (2) subjecting the gasoline in the tank to slight pressure to prevent the gas from evaporation; and (3) developing low volatile gasoline that does not evaporate easily.

(c) Use of filters. Filters can be used to capture and recycle the escaped gases (hydrocarbons) from the engine (*i.e.*, the gas vapours which escape between walls and the piston and reach the crankcase).

2. Prevention and control of industrial pollution. To check the air pollution by industrial and power plant chimney wastes , measures are taken for the removal of particulate matter and gaseous pollutants. In different industries, air pollution can be checked at various steps (Fig.14.13).

(a) Removal of particulate matter. This process involves collection of particulate matter under the influence of different forces, thereby, moving them continuously out of the gas stream. This step involves the use of following equipments :

(i) Cyclone collector. In this case the waste gas containing particles is subjected to **centrifugation.** The suspended particles move towards the wall of cyclone body and then to its bottom and finally are discarded out. The cyclone collectors can remove upto 70 per cent of the particles.

(ii) Electrostatic precipitators (ESPs). To remove the suspended particles from gas stream, the **electrical forces** are applied within the chamber in the precipitator. The particles become charged or ionized, and they are attracted to charged electrodes and removed. ESPs can remove 99 per cent of the particulate pollutants from the chimney exhaust. ESPs work very efficiently in power plants, papermills, cement mills, carbon black plants, etc. Sometimes, high resistivity dust may make separation in an ESP difficult. To overcome this difficulty, **fibric filters** or **beg filters** are employed. Further, tall chimneys are used for vertical dispersion of air pollutants.

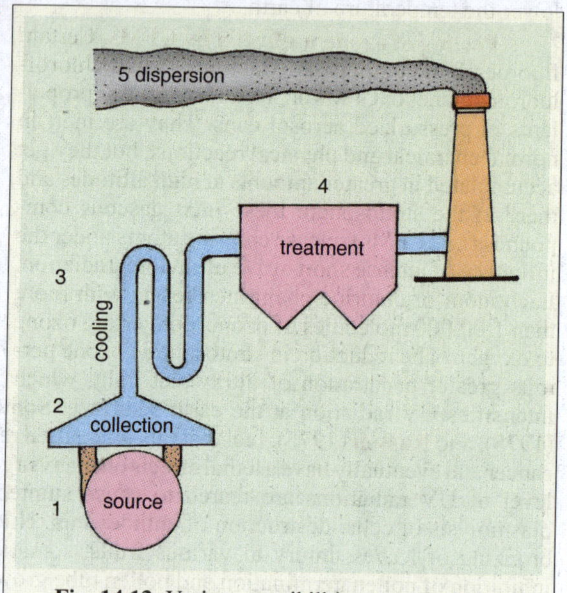

Fig. 14.13. Various possibilities to control the air pollution.

(b) Removal of gaseous pollutants. These are removed from the chimney exhaust by the following three methods:

(i) Wet systems. These are used as **washing towers** in which alkali fluid circulate continuously. This liquid react with SO_2 to produce a precipitate. In **gas scrubber**, a fine spray of water can effectively separate many gases such as ammonia and sulphur dioxide.

(ii) Dry systems. In this case, the gas pollutants are allowed to react with an **absorbent** under a dry phase. Absorbents such as dolomite, lime (CaO) and limestone (CaOH) are placed in the way of the flowing gas (SO_2). The process is cheap and does not involve any spray of water.

(iii) Wet dry systems. In these cases water in the absorbent reacts with the acid components. The absorbent calcium hydroxide slurry is spread into the hot gas stream in the form of small droplets. Calcium reacts with SO_2 and the hot gases cause the water to evaporate simultaneously. The end product is a dry powder containing mostly fly ash and salts.

Control of air pollution through law. In our country there have been several legislative measures both at state and central Govt., levels to prevent and control different types of air pollutions:

1. Bengal smoke Nuisance Act, 1905 ;
2. The Motor Vehicle Act, 1938;
3. The Gujarat Smoke Nuisance Act, 1953 ;
4. The Air (prevention and Control of Pollution Act, 1981) (It was amended in 1987);

5. The Environment (protection) Act, 1986 ;
6. The Motor Vehicles Act, 1988 (This Act came into force from 1.7.1989).

2. WATER POLLUTION

The great solvent power of water makes the creation of absolutely pure water a theoretical rather than a practical goal. Even the highest-quality distilled water contains dissolved gases and to a slight degree, solids. The problem, therefore, is one of determining what quality of water is needed to meet a given purpose and then finding practical means of achieving that quality. The problem is further compounded because every use to which water is put-washing, irrigation, flushing away wastes, cooling, making paper, etc., adds something to the water. In fact for centuries rivers and lakes have been used as dumping grounds for human sewage and industrial wastes of every conceivable kind, many of them are highly toxic. Added to this are the materials leached and transported from land by water percolating through the soil and running off its surface to aquatic ecosystems.

The term **water pollution** is referred to any type of aquatic contamination between following two extremes: (1) a highly enriched, over productive biotic community, such as a river or lake with nutrients from sewage or fertilizer (**cultural eutrophication**), or (2) a body of water poisoned by toxic chemicals which eliminate living organisms or even exclude all forms of life (see **Southwick**, 1976).

Normally water contains two types of impurities- dissolved and suspended. Dissolved impurities are gases (H_2S, CO_2, NH_3, etc.) and minerals (Ca, Mg, Na, salts). Suspended matter includes clay, silt and sand and even microbes. Polluted waters are turbid, unpleasant, foul smelling, unfit for drinking, bathing and washing or other purposes. They are harmful and means of many diseases as cholera, dysentery, typhoid, hepatitis, etc.

The Rio Tinto in Spain is so called because of the color of its waters (*tinto* means "dark red" or "dyed"). The color comes from what were once Europe's largest copper mines, close to the river. The mining produced a lot of spoil, which piled around the mine. Water running off the site picked up the copper, which colored it. Although the mine itself is now closed, much of the soil at the huge copper deposits in the river persist. Like other metals, copper is Present in many organisms in tiny amounts, but a large dose is extremely toxic.

Types of Water Pollution

Types of water pollution may be classified by the *medium* in which they occur, such as **surface water pollution, ground water pollution, soil water pollution,** etc.; the *habitat* in which they occur, such as **river pollution, lake pollution, estuarine pollution, coastal water pollution, open ocean pollution,** etc.; and *source* or *type* of contamination, such as **nutrient pollution, bacterial pollution, viral pollution, metallic pollution, petrochemical pollution, pesticide pollution, thermal pollution, radioactive pollution,** etc.

Kinds and Sources of Water Pollutants

Pollutants entering water sources are classified broadly into following categories: domestic sewage and oxygen-demanding wastes; infectious agents; plant nutrients; chemicals such as insecticides, herbicides, and detergents; other minerals and chemicals; sediment from land erosions; radioactive substances; and heat from power and industrial plants.

These aquatic pollutants come from many sources. Excessive nutrients, such as nitrates and phosphates, commonly originate in domestic sewage, run-off from agricultural fertilizer, waste materials from animal feed lots, packing plants, etc. Toxic chemicals as agents of water pollution originate in industrial operations, acid mine drainage, surface erosion from strip mines, washing of herbicides and insecticides, radioactive fall out from atomic explosion, and commercial accidents such as oil spills or the rupture of chemical tanks. Besides the pollutants which come from point sources such as sewage, factory or industry, there are many pollutants which come from watershed run-off. Urban and sub-urban run-off, for example, contains many pollutants such as oil, pesticides, radioactive dust, salt, fertilizers, miscellaneous chemicals, and nematodes, pathogenic protozoans, ,bacteria and viruses. However, **Kimbal** (1975) recognized only three major sources of pollution: *domestic, industrial* and *agricultural.*

Ecology of Water Pollution

Each type of water pollution affects the abiotic and biotic factors of different aquatic systems in different degrees and its ultimate effect on man remains quite drastic in medical, aesthetic, and economical sense. Some of the well known ecological effects of water pollution are following:

1. Sewage Pollution

Contamination of freshwaters and shallow offshore seas by sewage is a common occurrence. Domestic sewage and waste-water is about 99.9 per cent water and 0.02-0.04 per cent solids of which proteins and carbohydrates each comprise 40–50 per cent and fats 5–10 per cent (**Simmons**, 1974). In other words, sewage includes mostly biodegradable pollutants such as human faecal matter, animal wastes, and certain dissolved organic compounds *(e.g.,* carbohydrates, urea, etc.) and inorganic salts such as nitrates and phosphates of detergents and sodium, potassium, calcium and chloride ions. Under natural processes most of the biodegradable pollutants of sewage are rapidly decomposed, but, when they accumulate in large quantities, they create problem, *i. e.,* when their input into environment exceeds the decomposition or dispersal capacity of the latter. Most cities of well developed countries such as USA, Britain, etc., and some cities of developing countries such as India have evolved certain engineering systems, such as, **septic tanks, oxidation ponds, filter beds, waste water treatment plants and municipal sewage treatment plants** for the removal of many harmful bacteria and other microbes, organic wastes and other pollutants from the sewage, before it is tipped into river or sea.

Sewage treatment is usually performed in following three stages: (1) **Primary treatment**, which removes large objects and suspended undissolved solids of raw sewage and converts them into a biologically inactive and aesthetically inoffensive state, the **sludge,** a valuable fertilizer. (2) **Secondary treatment,** which supplies aeration and bacteriological action to decompose organic compounds into harmless substances such as CO_2, sulphate and water. During later stages of secondary treatment, whole waste water is chlorinated (*i.e.,* treated with chlorine) to reduce its content of bacteria. (3) **Tertiary treatment,** which removes nitrates and phosphates and releases pure water. These three stages of sewage treatment have become increasingly expensive and only in most advanced countries all the three treatments of sewage are done. However, most Indian cities either lack any sewage or waste-water treatment plant or have inadequate sewage treatment facilities. Consequently, normally and especially during heavy downpour and floods, raw sewage or incompletely treated sewage is dumped into rivers which cause severe water pollution problems in following ways:

(i) Bacterial and viral contamination. Sewage wastes may contain pathogenic bacteria and viruses which are a threat to human health. Waterborne diseases such as typhoid, bacillary dysentery, amoebic dysentery, botulism, poliomyelitis, and hepatitis all represent potential health hazards in sewage-contaminated waters. Due to such kinds of sewage pollution waters of many ponds, lakes, rivers, sea beaches in India and abroad have been prohibited for human use, whether for drinking, bathing, swimming or other sort of water recreation.

BOD test. BOD or biological oxygen demand is the amount of oxygen required for biological oxidation by microbes in any unit volume of water. The test is done at 20°C for at least five days. BOD value generally approximates the amount of oxidisable organic matter (such as sewage and other organic wastes, animals and human excreta, all of which are called **oxygen demanding wastes)** and is, therefore, used as a measure of degree of water pollution and waste level. Thus, due to addition of sewage and waste, oxygen levels are depleted which are reflected in terms of BOD values of water. The number of microbes as *Escherichia coli* (bacterium) also increase tremendously and these also consume most of the oxygen. The number of bacteria as *E.coli* in unit volume of water is also taken (called ***E.coli* index)** as a parameter of water pollution. The quantity of oxygen in water (called **dissolved oxygen** or **DO)** along with BOD is indicated by the kind of organism present in water. For example, fish become rare at DO value of 4 to 5 ppm of water. Further decrease in DO value may lead to increase in anaerobic bacteria. Typical BOD value for raw sewage run from 200 to 400 mg of oxygen per liter of water (therefore, 200—400 ppm). Water for drinking should have a BOD less than 1.

(ii) Eutrophication. According to **Hutchinson** (1969), the eutrophication is a natural process which literally means "well nourished or enriched." It is a natural state in many lakes and ponds which have a rich supply of nutrients, and it also occurs as part of the aging process in lakes, as nutrients accumulate through natural succession. Eutrophication becomes excessive, however, when abnormally high amounts of nutrients from sewage, fertilizer, animal wastes and detergents, enter streams and lakes, causing excessive growth or "bloom' of microorganisms and aquatic vegetation.

Nitrates stimulate algae growing in water as well as plants growing in soil. If runoff from fertilizer gets into a body of water, algae grow so profusely that they form a blanket over the surface. This usually happens in summer, when the light levels and warm temperatures favor growth.

Most secondary sewage treatment plants, though, precipitate solids and inactivate most bacteria in domestic sewage, yet they do not remove the basic nutrients such as ammonia, nitrogen, nitrates, nitrites and phosphates. These nutrients stimulate algal growth and lead to **plankton blooms.** Some plankton blooms, particularly those of blue-green algae produce obnoxious odours and tastes in waters. Others, such as the dinoflagellate blooms or **"the red tide"** of southern coastal regions, produce toxic metabolic products which can result in major fish kills.

Plankton blooms of green algae do not always produce undesirable odours or toxic products, but still create problems of oxygen supply in the water. While these blooms exist under abundant sunlight, they contribute oxygen to the water through photosynthesis, but under conditions of prolonged cloudiness, they begin to decay and consume more oxygen and with heavy load the oxygen contents of the water may diminish below the point where most fish can not survive. As the conditions in the water become anaerobic due to increased oxygen depletion by bacterial decomposition or planktonic blooms, the breakdown products become reduced rather than oxidized molecules, many of which (*e.g.,* hydrogen sulphide) produce -offensive odours and tastes.

Excessive nutrient levels in aquatic systems can also cause two other kinds of ecological problems. Primarily, they may lead to extensive growth of aquatic weeds such as Eurasion milfoil, water hyacinth, water chestnut, etc. Excessive growth of these weeds can impair fishing, bathing, fish spawning, shell fish production, and even navigation (**Sculthorpe,** 1967). Secondarily, nitrates can be converted in the human digestive tract by certain bacteria to nitrites, and the same transition may occur in opened cans of food even if they are subsequently refrigerated. Nitrites react with haemoglobin, forming methemoglobin which will not take up oxygen. Laboured breathing and occasional suffocation result most severely in human infants. Nitrites may also react with creatinine (present in the vertebrate muscles to form nitrosarcosine which can be carcinogenic (see **Simmons,** 1974; **Southwick**, 1976).

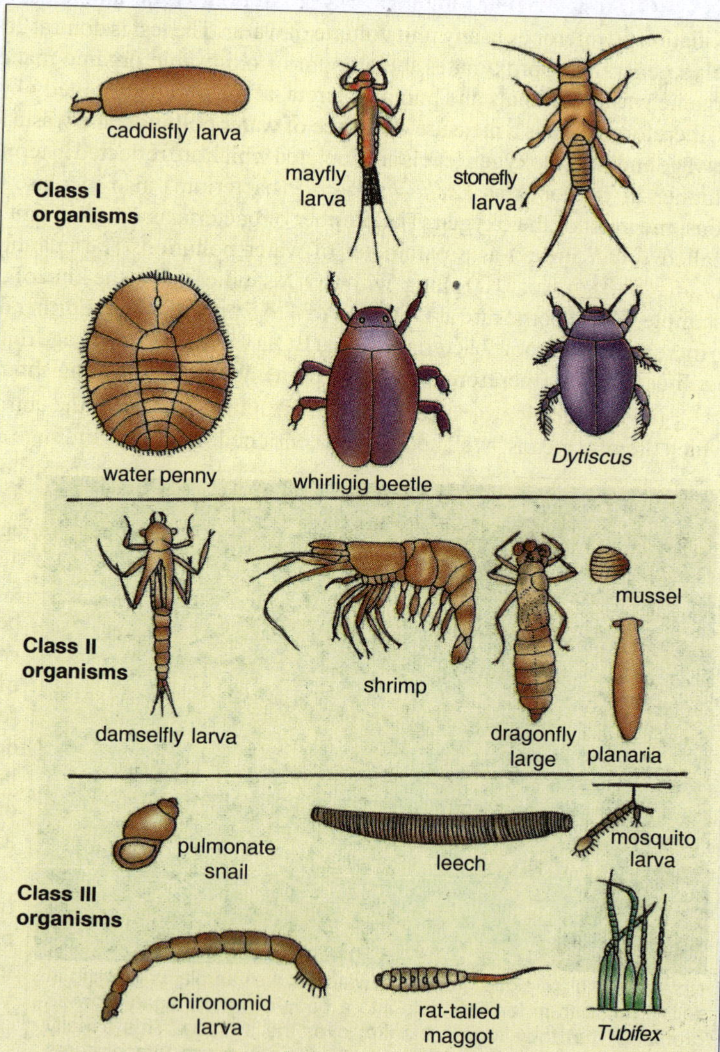

Fig. 14.14. Classes of stream fauna used as pollution indicators.

Effects of organic pollution on aquatic animal life. Organic pollution tends to bring about changes in faunal composition in a freshwater ecosystem. Nymphs of stone flies and may flies are the first to disappear from water which has high organic pollution. As pollution increases, caddishfly larvae and many fish which require high levels of environmental oxygen move into less polluted area of the stream. Shrimps, water fleas, leeches; snails and most of the fish vanish as the pollution becomes severe. At such levels of pollution there is very little of dissolved oxygen and the animals present are chironomid larvae (blood worms) and the oligochaete worms *Tubifex* (Fig. 14.14). Some decomposing plants are known to produce toxins as **strychnine** which kills animals including cattle.

2. Industrial Pollution

Most of the Indian rivers and freshwater streams are seriously polluted by industrial waste of effluents (see Table 14-3). **Effluents** are waste products in a liquid form resulting from industrial processes and domestic activities. They are released by different industries such as petro-chemical complexes; fertilizer factories; oil refineries; pulp, paper, textile, sugar and steel mills, tanneries, distilleries, coal washeries, synthetic material plants for drugs, fibres, rubber, plastics, etc. The industrial wastes of these industries and mills include metals (copper, zinc, lead, mercury, etc.), detergents, petroleum, acids, alkalis, phenols, carbamates, alcohols, cyanide, arsenic, chlorine and many other inorganic and organic toxicants. All of these chemicals of industrial waste are toxic to animals and many cause death or sublethal pathology of the liver, kidneys, reproductive systems, respiratory systems, or nervous systems in both invertebrate and vertebrate aquatic animals (**Wilbur**, 1969). Chlorine which is added to water to control growth of algae and bacteria in the cooling system of power station, may persist in streams to cause mortality of plankton and fish. Heavy fish mortality in river **Sone** near Dehri-on-sone in Bihar is reported to cause by free chlorine content of the chemical wastes discharged by factories near Mirzapur in U.P.

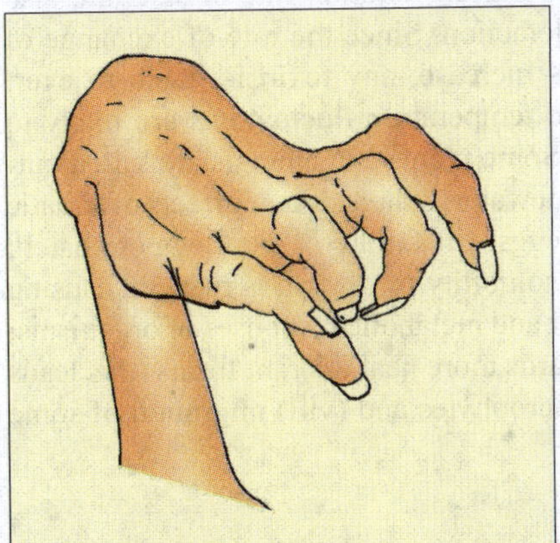

Fig. 14.15. Minamata disease caused by mercury pollution.

Mercury like other heavy metals such as lead and cadmium has cropped up as a toxic agent of serious nature. Mercury, a byproduct of the production of vinyl-chloride, is used in many chemical industries and it is also a byproduct of some incinerators, power plants, laboratories and even hospitals, (**Aaronson**, 1971). In Japan, illness and even death occurred in the 1950s among fishermen who ingested fish, crabs, and shell-fish contaminated with methyl mercury from Japanese coastal industries. This mercury poisoning produced a crippling and often fatal disease called **Minamata disease** (Fig.14.15). Initial symptoms of minimata disease included numbness of the limbs, lips, and tongue, impairment of motor control, deafness, and blurring of vision. Cellular degeneration occurred in the cerebellum, midbrain, and cerebral cortex and this led to spasticity, rigidity, stupor and coma. In Japan in 1953, due to Minamata disease 17 persons died and 23 became permanently disabled (see **Southwick**, 1976).

Quit recently, **Dash** and coworkers (1988) studied mercury pollution due to effluents of a paper mill in the river Ib of the Mahanadi water system in Orissa. The paper mill uses mercury as a cathode in chloroalkali electrostatic cells and mercury leakage can be as high as 200 mg/tonne of chlorine produced (**Fergusson**, 1982). An analysis of the mercury in some fish showed 500 times biomagnification (or bioamplification) over the downstream surface water. Generally benthic fish and those higher in the food chain accumulated more mercury. People of the villages of that region, who used river water for drinking, suffered from diseases of eye, skin rashes and anemia.

3. Thermal Pollution

Thermal pollution.

Various industrial processes may utilize water for cooling, and resultant warmed water has often been discharged into streams or lakes. Coal-or oil-fired generators and atomic energy plants cause into large amount of waste heat which is carried away as hot water and cause **thermal pollution or calefaction** (warming). Thermal pollution produces distinct charges in aquatic biota. A body of water at 30-35°C is essentially a biological desert and many game fish require temperatures of 10°C for successful reproduction, although they will survive above that temperature. A temperature rise of 10°C will double the rate of many chemical reactions and so the decay of the organic matter, the rusting of iron and the solution rate of salts are also accelerated by calefaction. Since the rate of exchange of salts in organisms increase, any toxin is liable to exert greater effects and temperature fluctuations are likely to affect organisms. Some plants and animals are killed outright by the very hot water. Other adverse effects of aquatic pollution on aquatic life include (i) early hatching of fish eggs, (ii) failure of trout eggs to hatch; (iii) failure of salmon to spawn, (iv) increase in BOD, *i.e.,* solubility of oxygen is reduced causing deoxygenation; (v) change in diurnal and seasonal behaviour and metabolic responses of organisms; (vi) significant shift in algal forms and other organisms towards more heat tolerant forms (this leads to decrease in species diversity); (vii) affect changes in macrophytes and (viii) migration of some aquatic forms.

4. Silt Pollution

A result of intensive agriculture, earth moving for construction projects, poor conservation practices and downpour with resultant floods, is the increased production of silt in streams and lakes. This load of particulate matter cuts down primary productivity by decreasing the depth of light penetration. Silt may also interrupt or prevent the reproduction of fish, by smothering eggs laid on the bottom.

5. Water Pollution by Agrochemicals

Water that flows on the surface of crop fields, where agrochemicals such as fertilizers, pesticides and herbicides are used, contributes to heavy water and soil pollution. Pesticides and weedicides are used by human beings to control crop diseases by the pests or to kill the weeds and, hence, to increase the productivity. The use of these toxic chemicals has created health hazards not only for livestock and wild life but also for fish, other, aquatic organisms, birds and mammals including man.

Apart from killing the living organisms present on the surface of the soil, they reach even the deeper layers through tilling and irrigation of the land, killing still more living forms which might be involved in soil formation or humus formation, *(e.g.,* earthworms, centipede, millipede, etc.). With their continuous use the soil microorganisms lose their ability of nitrogen fixation. Moreover, when these chemicals find their way into water supplies, they contaminate and disrupt the aquatic ecosystem as well.

Any substance or mixture of substances which prevents, repels, destroys or mitigates any pest *(e.g.,* bacteria, viruses, fungus, insects, nematodes, rodents, etc.) is called a **pesticide.** Pesticides have proved tremendously beneficial to human populations, in reducing or eliminating the target organisms such as insects, snails, rats, and other animals which transmit disease, destroy agricultural crops, damage homes and stored products, and directly or indirectly affect human health and welfare. Hence, the significance of chemical pesticides controlling mosquitoes, termites, houseflies, cockroaches, house crickets, weevils, locusts and grasshoppers, borers, snails, rats, rabbits, and a multitudes of other animals has been great, and it is difficult to imagine modern disease control and agricultural programmes without some forms of chemical control. Like pesticides, **herbicides** are specifically designed chemicals for the control of weed pests and unwanted plant growth.

Table 14-3. Some Indian rivers and their major sources of pollution.

Name of the river	Sources of pollution
1. Kali at Meerut (U.P.)	Sugar mills; distilleries; paint, soap, rayon, silk, yarn, tin and glycerine industries.
2. Jamuna near Delhi	D.D.T. factory, sewage, Indraprastha Power Station, Delhi.
3. Ganga at Kanpur	Jute chemical, metal, and surgical industries; tanneries, textile mills and great bulk of domestic sewage of highly organic nature.
4. Gomati near Lucknow (U.P.)	Paper and pulp mills; sewage.
5. Dajora in Bareilly (U.P.)	Synthetic rubber factories.
6. Damodar between Bokaro and Panchet	Fertilizers, fly ash from steel mills, suspended coal particles from washeries, and thermal power station.
7. Hooghly near Calcutta	Power Stations; paper pulp, jute, textile, chemical mills, paint, varnishes, metal, steel, hydrogenated vegetable oils, rayon, and soap, match, shellack, and polythene industries and sewage.
8. Sone at Dalmianagar (Bihar)	Cement, pulp and paper mills..
9. Bhadra (Karnataka)	Pulp, paper and steel industries.
10. Cooum, Adyar and Buckinghum canal (Madras)	Domestic sewage, automobile workshops.
11. Cauvery (Tamil Nadu)	Sewage, tanneries, distilleries, paper and rayon mills.
12. Godavari	Paper mills.
13. Siwan (Bihar)	Paper, sulphur, cement, sugar mills.
14. Kulu (between Bombay and Kalyan)	Chemical factories, rayon mills and tanneries.
15. Suwao (in Balrampur)	Sugar industries.

Ecologically pesticides and herbicides have created two major serious problems which were not previously anticipated. In the first place many of them have persisted and accumulated in the environment and have harmed or contaminated numerous animals or plants not intended to be targets. Secondly, many of them have directly or indirectly affected human health.

Biomagnification or bioamplification. Many of pesticides, such as DDT, aldrin and dieldrin, have a long life time in the environment. They are fat-soluble and generally not biodegradable. They get incorporated into the food chain and ultimately deposited in the fatty tissues of animals and man. In the food chain, because of their build up, they get magnified in the higher trophic levels (called **biological magnification** or **biological amplification).** The pesticides have been in use during last 50 years. Their targets are insect pests, fungi, nematodes and rodents which damage crops. But these pesticides have created great problems for non-target organisms consisting largely beneficial species such as earthworms, honeybees, fish, amphibia, some reptiles, birds, mammals and man.

The phenomenon of biological magnification is also reported for certain other pollutants such as heavy metals such as lead, mercury and copper and radioactive substances (or radionuclids) as strontium-90.

Toxicity of pesticides. The toxicity of organo-chlorine pesticides (*i.e.,* DDT or Dichlorodiphenyl trichloroethane, hexachlorocyclohexane, chlordane, aldrin, dieldrin, etc.) lies in their inhibiting Na+,

K+ and Mg+ adenosine triphosphatase activity in the nerve endings of animals particularly insects. It affects the sensory, motor nerve fibers and the motor cortex (**Matsuma** and **Patil,** 1961). In the giant axons of cockroach, DDT is known to influence the efflux of potassium ions from the axon.

DDT and other organochlorine pesticides are absorbed from the intestinal tract, from the alveoli of lungs and also through the skin, if the pesticides are in solution A high concentration of DDT causes brain damage, centrolobular necrosis of the liver, and liver enlargement in small mammals. Concentration as low as 5—10 ppm in diet cause liver damage. In some birds, DDT concentration of 1—3 ppm destroys the female sex hormone and the egg shell becomes so thin that the eggs break when the parents sit on them for hatching.

Organophosphorus insecticides are absorbed by the gastrointestinal and respiratory tracts and the skin. These insecticides inhibit the **acetylcholinesterase** (**Ach E**) enzyme. Thus, an abnormal accumulation of endogenous acetylcholine occurs. This cause excessive activity of the para-sympathetic system, in the form of sweating, abdominal cramp, chest discomfort, vomiting, overactivity of smooth muscles, and so on. It also causes headache, nervousness, and over activity of voluntary muscles. The nervous system is also adversely affected.

The carbamate insecticides are absorbed by all portals, including the skin. These pesticides are reversible inhibitors of Ach E.

6. Marine Pollution

The marine ecosystem (*i.e.,* seas, oceans and estuaries) is so grand and vast but interacting in various ways with human life. How mighty and majestic ocean may appear to us, it was once challenged and tamed by Lord Rama during his march towards Ravana's Lanka. The oceans are used for navigation, fisheries, aquaculture, mining, acquisition of water, naval and military exercises as well as for the discharge and dumping of a variety of wastes. The very vastness of the ocean has led to the assumption that all wastes dumped into them can be harmlessly absorbed. Such an assumption now belongs to the past. The ever increasing range and volume of polluting activities and tend to seriously affect marine production. Marine pollution is most evident in coastal waters and estuarine areas. The oceans have in fact, become the final settling basin for millions of tons of waste products from human activities. For example, in the late 1960s West Germany was dumping 375 tons of sulphuric acid, 750 tons of iron sulphate, 20 tons of chlorinated hydrocarbons and 16,000 tons of gypsum wastes into the North sea and North Altantic Ocean every day. Table 14-4 shows a few of the industrial and agricultural pollutants reaching the ocean annually in the mid- 1970s.

Thus, petroleum and its distillates, halogenated hydrocarbons including pesticides, organic compounds including detergents, domestic raw sewage, agricultural run-offs, inorganic chemicals including metals such as mercury, lead and copper, dumped containers, radioactive wastes, cooling waters from industries and hydel projects and coastal constructions are the major pollutants entering the sea. Often the noxious pollutants are not dispersed and diluted but tend to remain concentrated and localised, or when diluted may be reconcentrated by living organisms.

This tremendous burden of pollutants in evidently affecting the health and integrity of the world's oceans. Due to oceanic pollution the marine biota has been seriously affected. **Cousteau** (1974) reported a 30 to 40 per cent decline in the over all productivity of pelagic organisms from shrimps to whales in the past 25 years. He also observed a serious shrinkage of coral reefs in many tropical areas of world, and a displacement of these rich and varied communities with turbid and relatively barren waters. There has occurred a decline in populations of many fishes due to oceanic pollution. Oil spills have killed water birds, mammals, fish and vegetation (**Hunt,** 1965).

Oil is the most apparent pollutant of the ocean. It discharged into the sea either deliberately when oil tankers are washed out prior to reloading or in accidental spills when tankers are wrecked (during war or otherwise). In coastal areas when oil spreads on the water surface, it clogs the feathers of diving birds making their flight impossible. White preening themselves in an attempt to clear the plumage, they swallow enough oil to poison themselves. Apart from these, oil interferes with the insulation provided by the feathers and the birds die of cold or become susceptible to pneumonia. When oil covers the rocks and sea weeds, molluscs and crustaceans growing on them die. Oil-spills on the coasts of sea-side resorts, drive away holiday makers, affecting the economy of the place.

Table 14-4. **Examples of industrial and agricultural pollutants discharged annually into the world's oceans (Source: southwick, 1976).**

Pollutant	Estimated annual discharge, 1970-75 (in metric tons)	Source
1. Petroleum and industrial hydrocarbons	3,405,000	Offshore wells, oil tankers, industrial wastes.
2. Hydrocarbons (airborne)	15,000,000	Vehicles, industries, power plants.
3. Airborne lead	350,000	Vehicles.
4. Mercury	100,000	Industrial operations.
5. Aldrin-Toxaphene (converted to dieldrin)	25,000	Agricultural and public health operations.
6. Benzene hexachloride (BHC)	50,000	Agricultural and public health operations.
7. DDT	25,000	Agricultural and public health operations.
8. Polychlorinated biphenols (PCBs)	25,000	Plastic industries.

Control of Water Pollution

Most cities of world have evolved certain engineering systems such as **septic tanks, oxidation ponds, filter beds, waste water treatment plants** and **municipal sewage treatment plants** for the removal of various pollutants from the sewage before it is tripped into river or sea. It is essential to have modern sewage treatment plants for every town and city of India so that the biodegradable as well as non-biodegradable pollutants can be removed from it and pure water obtained for recirculation.

Water hyacinth plays an important role in controlling water pollution.

The urban sewerage condition of India is quite distressing. At present only 15 class-I cities out of 142 are fully sewered, whereas, only 7 of the 190 class-II towns claim to have this facility. Still the Indian cities continue to pour their waste water (286 thousand million cubic metre) into natural water sources.

Currently, water hyacinth (*Eichhornia crassipes*), an otherwise pernicious aquatic weed, has come into prominence for purifying domestic and industrial waste water. The plant regenerates rapidly and has a tremendous capacity to accumulate heavy and even radioactive metals. It is efficient in absorbing nitrogen, phosphorus and similar chemical pollutants. The polluted water fed into reservoirs or lagoons with water hyacinth, becomes markedly clean and free from 75-90 per cent of its pollutants. This plant has also been used as a new source of food, fertilizer and biogas (energy).

In India, the enactment of 'Prevention and Control of Water Pollution Act' in 1974, has helped to a certain extent to prevent water pollution. However, adequate legislative measures have to be adopted by every state to ensure: (a) proper disposal of sewage and industrial wastes; (b) prevention of abuse of water resources; (c) recycling of waste waters through proper methods of purifications; and (d) punishment of erring industries which do not install effluent devices in their factories. A large number of international and inter-governmental agencies are now busy in controlling marine pollution. A number of highly hazardous pollutants are **black listed** and their discharge in the sea is totally prohibited. These are halogenated organic compounds, tin, mercury and cadmium compounds, radioactive isotopes and carcinogenic material. Somewhat less toxic substances are **gray listed** such as compounds of arsenic, zinc, and antimony, cyanides, organosilicon compounds, crude oils, foam forming detergents and surfactants (see **Ambasht,** 1990).

Realizing the devastation caused by the population of cities, towns and villages of 8 states and the union territory located along the coasts of river Ganga, through its run of 2525 km from Gangotri (Himalayas) to Ganga Sagar (Bay of Bengal), the Government of India has launched Ganga Action Plan to restore its water quality. The first phase of this gigantic cleaning operation had been initiated in year 1985 and may take several years in its completion.

Further, since xenobiotics, *i.e.,* synthetic herbicides and pesticides and fertilizers pose a challenge to the natural degradative process. Therefore, we should encourage a judicious, efficient and optimum use of organic manures. Recently, certain microbes have been used for the breakdown of these synthetic compounds. For instance, specific microorganisms are being evolved by molecular breeding process to convert the chief ingredients of herbicide agents, 2, 4, 5-T, into carbon dioxide and chloride.

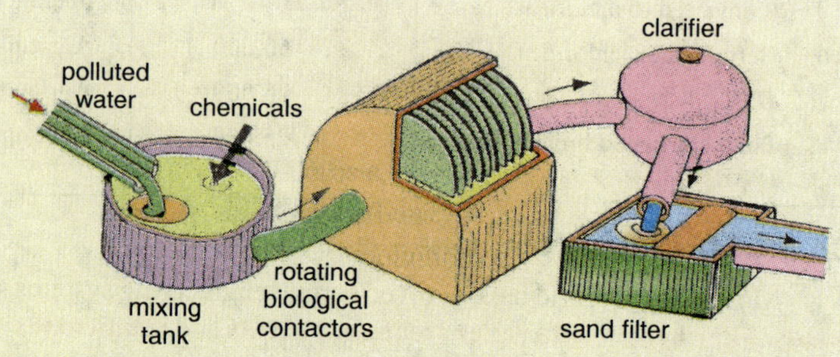

Fig. 14.16. The method of purification of water polluted with cyanide and heavy metals through bacteria.

The waste water polluted with cyanides and heavy metals can be purified by certain bacteria which cannot only withstand but are also able to grow at high level of cyanides. Millions of these bacteria are introduced in each of the several rotating discs of the plant's main processing units (Fig. 14.16). The stickly body surfaces of these bacteria pick up zinc, iron and other metals from the industrial waste water as it passes over the plates. These bacteria also absorb cyanide so toxic to the fish and other aquatic life. Subsequently this water is passed through clarifier, and filtered and then released into the streams.

In the past, attempts have been made to get rid of oil slicks by spraying them with **detergents** which break up the oil into drops. The trouble is that detergents are even more deadly than the oil and they kill many marine organisms that might have escaped the oil. Nowadays less destructive methods are used. Recently, a product **"bregoil"**, resembling sawdust from paper industry waste has been employed. It rapidly absorbs oil. The resulting residue can be conveniently handled and even burnt as a fuel. Bacteria which thrive on oil eventually break it down and this brings about a regeneration of fauna and flora even within a year. Recently a strain of the bacterium *Pseudomonas putida* has been created by genetic engineering which breaks down octane, xylene and camphor (see **Green** *et al.,* 1990).

3. LAND POLLUTANTS AND LAND POLLUTION

The land pollution is caused by solid wastes and chemicals. There are many examples of land that has been stripped of vegetation by industrial development and disposal of waste. Some common soil contaminants, their sources and wide range impacts on the biota have been summarized as follows:

The slag heaps from mines bear witness to the destructive effects which this can have on our environment. Areas around smelting and mining complexes are usually soiled by metals such as cadmium, zinc, lead, copper, arsenic and nickel. These are not only phytotoxic even in small amounts but also render plants unsafe for human and animal consumption. Zinc, often with cadmium, is released into the environment during the use or breakdown of lubricating oils, vehicle tyres, galvanized metals and fertilizers.

The major sources of land pollution are the industries such as pulp and paper mills, sugar mills, oil refineries, power and heating plants, chemicals and fertilizer manufacturing units, iron and steel plants, plastic and rubber producing complexes and so on. Huge amounts of solid wastes are either dumped, burnt, or emptied into rivers and seas. Most industrial furnaces produce a grey, powdery residue of unburnt material known as fly ash. The fly ash, cinders, solid wastes and litter all are thrown away by industries and form huge mounds which spoil the landscape.

Our households too contribute a large amount of solid rubbish in the form of domestic wastes. Some common examples are groceries, food scraps, vegetable remains, packing materials, cans, cardboard cartons, rags, paper, cinders, ash, broken gadgets, wood, worn-out furniture, metals, bones of dead animals, plastics, polythene bags, ceramics, glass, aluminium, rubber, leather, construction rubbish, brick, sand and other junk. Some man-made materials can be used again such as paper, scrap metal, glass, polythene, plastic, etc. But the majority of these cannot be reused and must be got rid of somehow. All these go to constitute heaps of municipal refuse. If not properly disposed, this rubbish can prove perilous, filthy and unhygienic. Such places often become a home for rats, flies, mosquitoes, bacteria and many other vectors, which may spread numerous human diseases.

Minimizing Land Pollution

It should be necessary for the industries to install collectors to remove the particulate wastes (fly ash) from the chimneys. Appropriate methods should be developed to dispose off or utilize the other types of pollutants. The garbage, instead of burning in the open, can be used not only to produce energy but also as filler for cement, bricks, asphalt and pavings. Some of these wastes, if properly sifted and separated, can even be recycled as raw material for other industries.

Another simplest method is crude **tipping** or **open dumping**, a common method used in most Indian cities. More satisfactory is **controlled tipping** or the **sanitary landfill,** which is recently adopted in Delhi for solid waste disposal. The **sanitary landfill** is a better remedy for larger objects because it brings about inexpensive biodegradation of such trash without causing much pollution, disease and ugliness. In sanitary landfill, a layer of about 2 meters of refuse is covered by at least 23 cm of earth, ash or other inert material, up to the level of the hole chosen. The chosen holes may be low lying watery areas and ditches. The land, thus, reclaimed can be used for making gardens, parks, playgrounds or apartment complexes. Before such filling, the wastes can be pulverized by machines to a uniform size— by this method the volume of refuse is reduced and some of the refuse is more quickly biodegraded. There are certain persons like **Nek Chand** who has found an aesthetic use of household refuse in the construction of world famous monuments such as "Rock Garden" of Chandigarh.

4. RADIOACTIVE POLLUTION

Radioactive isotopes, or radionuclides, are forms of elements with unstable atomic nuclei; that is, they decompose with ionizing radiation in the form of alpha or beta particles, or gamma rays. Many radioisotopes, such as radium-226, uranium-235 or238, thorium-232, potassium-40, or carbon-14, occur naturally in rocks and soil. Other radioisotopes such as those of cesium, cobalt, iodine, krypton, plutonium, and strontium, result primarily as fission products from atomic bomb fallout, nuclear reactors or other radiation sources. Of more than 450 radioactive isotopes which can occur as fission products, only a few are of major environmental con-cern. These are primarily argon-41, cobalt-60, cesium-137, iodine-131, krypton-85, strontium-90, tritium and plutonium-239 (**Bebbington**, 1973).

Within biotic communities and ecosystems, these radioactive elements may become dispersed or accumulated, depending upon the biological activity of the element and period of radioactivity of the isotope. **Strontium-90,** for example, normally occurs in radioactive fallout, has a half-life of 28 years and behaves like calcium in biogeochemical cycles. Thus it is absorbed by plants, ingested by animals and deposited in bone tissues close to blood forming tissue. Strontium-90 can also concentrate in natural biological systems in following method: water → bottom sediments → aquatic plants → freshwater clams → minnows and small fish → musk rats. It is demonstrated that due to this food chain musk rats concentrate strontium-90, 3500 times above the levels of the water in which they live. Grazing animals concentrate strontium-90 by ingesting it through grass and forage, and it can then be passed on to humans through milk.

Radioactive phosphate, cesium and iodine-132 also can readily accumulate in plants and animals through natural food chains. However, in food chains involving arthropods radioactive isotopes of potassium, sodium, and phosphorus accumulate, but isotopes of strontium and cobalt do not (**Reichle** *et.al.*, 1970).

Although isotopes may accumulate in human tissues as well as those of plants and animals, it is not established at the present time whether current levels of isotopes in human tissues represent serious health hazards to man (**Southwick**, 1976; **Smith**, 1977). Some medical scientists such as **Gofman** and **Tamplin** (1970) and **Sternglass** (1972), however, feel that man's radiation exposure from artificial sources is already sufficient to produce serious disease problems (leukemia and bone tumors), genetic damage, and infant mortality.

5. NOISE POLLUTION

Noise is primarily a feature of cities and is defined as 'sound without value' or 'any noise that is undesired by the recipient". Noise levels in many urban-industrialized situations are known to be deleterious to human health and efficiency, with effects on the sense organs, cardiovascular, glandular and nervous systems.

High intensity sopund or noise pollution is caused by many machines man has invented during his technological advancement. Thus, there exists a long list of sources of noise pollution including different machines of numerous factories, industries and mills, different kind of auto and motor vehicles such as scooters, motorbikes, cars, tempos, buses, trucks, tractors, aircrafts, motorboats, ships, loudspeakers, social gatherings, loud pop-music, supersonic alrcrafts, etc. Noise can be measured by a sound metre and is expressed in a unit called the **decibel (dB)**. The quietest sound that the human ear can detect (zero decibels) is called the **threshold of hearing**. Calcutta, Bombay and Delhi are regarded to be nosiest cities in the world, where the average noise level was 90 decibels in 1975.

Noise pollution has certain well evident ecological and pathological effects on biota and human beings. For instance, the sonic boom path associated with SST projects such as Concorde porduces noise of a very different order, in the form of sudden but repeated shock waves, which is suspected to cause disturbance to wild birds as well as domestic stock and buildings (see **Simmons**, 1974). Noise is not only annoying but if continued for a long time can also result in emotional and behavioural alterations in man.

Health Hazards of Noise Pollution

Noise causes disturbances in the atmosphere which in turn interferes with the systems of communication. It affects our peace of mind, health and behaviour. Sudden loud note can cause acute damage to the ear drum and the tiny hair cells in the internal ear, whereas prolonged noise results in temporary loss of hearing or even permanent impairment. It causes headache irritability and impairs focussing. Noise is known to flush the skin, constrict stomach muscles and produce ulcers, heart disease, high blood pressure, nervousness and other defects in sensory and nervous systems.

Reducing Noise Pollution

Though it is almost difficult to completely get rid of the malady of noise pollution of current electronic age. However, there are certain methods by which deleterious influences of noise pollution can be minimized and intensity of noise level can be curbed. The means of noise control are : (a) to manipulate the source so as to reduce the noise at its origin; (b) to interrupt the path of transmission and (c) to protect the recipient.

Legislation and public awareness are essential. Nobody should be permitted to create noise in silent zones or during night. Noise producing traffic vehicles should be prevented from plying on the roads and their use of pressure horns should be entirely checked. Standards for noise control measures should be set up for industry and community, and a comprehensive safety programme enforced.

The path of the sound can be interrupted by using various materials which absorb the sound energy. Horticulturists should suggest adequate varieties of vegetation which can be planted around factories, hospitals, educational institutions, public libraries and houses which may reduce sound pollution and also may minimize dust pollution. Acoustic materials and mufflers can also be used to protect oneself. Under noisy situations, one may hold his hands over his ears or may run away from the source, or may simply stuff a bit of cotton or ear plugs in the ears to reduce much hazards of noise pollution.

REVISION QUESTIONS

1. Define the term 'pollution,' Write an essay on the 'problem' of ecological pollution and its control methods.
2. What is 'air pollution' ? How it affects the biota including man? Describe certain recent methods to control the air pollution.
3. What is 'water pollution' ? Describe different kinds of water pollutions and their ecological effects on the aquatic life and also on man.
4. Write an essay on 'the eutrophication of aquatic habitats'.
5. Write short notes on the following :
   (i) Minamata disease; (ii) Greenhouse effect; (iii) Photochemical smog; (iv) Thermal pollution; (v) Pesticide and herbicide contamination; (vi) Noise pollution; (vii) Radioactive contamination.

CHAPTER

15

Ecology and Human Welfare

(Natural Resource Ecology : Natural Resources, Conservation and Management)

These wind pumps use the power of the wind as their source of energy.

Mother nature is quite generous. For thousands of years, it has provided us with all of our basic needs—air, water, sunlight, food, clothing and shelter. Up to the beginning of 20th century our thinking was that our natural treasures were eternal and virtually inexhaustible. However, by observing the present trend of ever increasing human population and mindless exploitation of these repositories, we can speculate for future that mankind seems to be heading for a grim time which might be full of shortages, scarcities and chaos. The fortune which the mankind had inherited from the nature, has been and is being exhausted so carelessly that it is not likely to last for very long, if not harnessed properly.

CLASSIFICATION OF NATURAL RESOURCES

We may classify resources into two major types according to how they are affected by their consumers :

1. Non-renewable resources. These resources once gone, have very little chance of recovery or resynthesis. Non-renewable resources are not altered by use. For example, space once occupied becomes unavailable; it is "replenished" only when the consumer (user of a commodity) leaves. Mineral deposits are formed slowly over million of years and once used cannot be regenerated, *e.g.,* fossil fuels, such as petrol and coal. Since the formation of soil takes thousands of years and is not renewable in the life span of many generations, it is thought of as a non-renewable resource.

2. Renewable resources. These resources are constantly regenerated or renewed. Births in a population of prey continually supply food items for predators. Plants (crops, forests, etc.) and animals (milk and meat production) form another good example of renewable resources; since their continued harvest is possible upon proper planning and management. Other examples of renewable resources are wild life, aquatic life, pastures and forests.

CONSERVATION OF NATURAL RESOURCES

Conservation has been defined as "the management for the benefit of life including human kind of the biosphere so that it may yield sustainable benefit to the present generation while maintaining its potential to meet the needs and aspirations of the future generations." Thus, conservation is one of the most significant application of ecology. It avoids unplanned development which breaks ecological as well as human laws. A conservationist has two folds basic aims (i) to insure the preservation of a quality environment that considers esthetic and recreational as well as product needs and (ii) to insure a continuous yield of useful plants, animals, and materials by establishing a balanced cycle of harvest and renewal (**Odum**, 1971). Thus, conservation process remains chiefly concerned with the use, preservation and proper management of the **natural resources** of the earth and their protection from the destructive influences, misuse, decay, fire, or waste.

In fact, the preservation of rare species is certainly not all that conservation means. Conservation is never a "hoarding" or even "rationed use" of a limited source but a constant effort to increase the resources and rapid resynthesis of the materials. Thus, conservation of living resources has following three objectives : (1) to maintain essential ecological processes and life support system; (2) to preserve biological diversity; and (3) to ensure that any utilization of species and ecosystems is sustainable. Conservation, therefore, makes significant contributions to social and economic development.

Different kinds of natural resources and their conservation and management practices can be studied under following heads :

1. MINERALS AND THEIR CONSERVATION

Minerals are often called "**stock**" or "**non-renewable**" resources, because, their 'new' materials can only be extracted from the earth's crust once. But even in the transformed state in which they are used, they are not lost to the planet and so are ideally available for reuse.

(a) Terrestrial Mineral Resources

For his industrial, technological and cultural growth man has required and still needs a great variety of inorganic materials, all of which come from the earth's crust. Chief among these are the ores which are used on a large scale to yield metals such as **iron**, **aluminium**, **copper**, **silver**, **gold**, **platinum**, etc. To them must be added elements which may not be needed in large quantities but which are indispensable in many modern industrial processes, as for example catalysts and hardeners such as **vanadium**, **tungsten**, and **molybdenum**. Finally there are non-metallic materials which are vital to industrialized nations such as sand and gravel, cement, fluxes, clay, salt, sulphur, phosphorus, diamonds, and the chemical byproducts of petroleum refining.

(b) Marine Mineral Resources

The mineral resources of seas can be divided into three categories : those which are dissolved in water itself; sediments present on the sea-bed at various depths; and those present at some depth below the sea-floor, beyond the sediments of relatively recent origin.

At present the utility of the dissolved elements is in the direct proportion to their abundance and to the relative cost from terrestrial sources. **Cloud** (1969) has enlisted commonest elements (*e.g.,* strontium, boron, silicon, fluorine, argon, nitrogen, lithium, rubidium, phosphorus, etc.) in sea water which could be extracted. Common salt is one of the resources that has been utilised since prehistoric times for its value in flavouring and meat preservation. At present only **salt**, **magnesium** and **bromine** are being extracted in commercial quantities and sea seems to remain inexhaustible for these elements. Except a few other elements, economically feasible recovery processes of most other elements from sea is very low.

Sediments and sedimentary rocks on the continental shelves are sources of certain materials. **Placer deposits** (= materials that has been concentrated in a particular place by mechanical action) contain workable quantities of **gold**, **tin**, and **diamonds**, and the other sediments which may be liable to exploitation include **sand**, **gravel** and **shells**. However, land -use problems by their extraction from the land would largely be prevented by the use of the sea as a source, provided that the ecosystems of the oceans were not too greatly damaged by the recovery processes, which create great quantities of silt and also eventuate imbalance in the sedimentary systems of the sea-floor. **Phosphates** are found as nodules and in crust. A further resource of the continental shelves is **freshwater**; large quantities of artesian water may be found in certain aquifers. Finally there are petroleum and natural gas.

The deep sea basins is found to contain enormous quantities of minerals such as **manganese**, **nickel**, **cobalt** and **copper**, but their extraction is not economically feasible (**Cloud**, 1969).

(c) Conservation of Terrestrial Mineral Resources

Until recently little attention was paid to conservation of terrestrial mineral resources because it was assumed that there were plenty for centuries to come and that nothing could be done to save them any way. But recently **Cloud** (1969) made it evident that both assumption are dead wrong. During his assessment of the situation, he has formulated following two situations about the exploitation of terrestrial mineral resources—(1) The first is the **demographic quotient** or **Q** :

$$Q = \frac{\text{total resources available}}{\text{population density} \times \text{per capita consumption}}$$

As this quotient goes down, so does the quality of modern life; it is going down at a frightening rate because available supplies can only go down as consumption goes up.

(2) The other concept introduced by **Cloud** is the graphic model of **depletion curve**, as shown in Figure 15.1. With the present procedure of "*mine, use and throw away*", a huge boom and bust is projected, as shown in curve A. The time scale is uncertain because lack of data, but the "bust" could begin within this century since certain key metals and fuels such as zinc, tin, lead, copper, uranium–235, natural gas could be mined or extracted out within 20 years in so far as the readily exploitable reserves are concerned. If a programme of mineral conservation, substituations (using less scarce minerals wherever possible), and partial recycle were to be started now, the depletion curve could be flattened as shown in curve B. Efficient recycling combined with stringent conservation and a reduction in per capita use could prolong depletion for a long time, as shown in curve C. It should be noted that even with prefect recycle, depeletion would still occur, because small amounts of most metals lose by friction, rust, etc.

Consequently, terrestrial mineral conservationists have suggested following measures for terrestrial mineral conservation : 1. Technological advancement for extraction of minerals from sea which is unlimited supply depot. 2. Extraction of metals from lean ores with the use of vast quantities of cheap atomic energy. **Granite** is suggested as a source for the extraction of many metals. 3. Recovery of mineral elements from scrap and waste (recycling).

Ecological Aspects of Mining

Most kinds of mining processes of man have side-effects of an ecological nature. Underground mining may include whole new towns among its surface installations (*e.g.*, the Khateri Copper Project, JhunJhunu, Rajasthan, India). Timber is cut in forested areas, often leading to soil erosion and the tailings and mine-waste have to be discarded. Large-size solid wastes can be used as backfill or sold for aggregates, but tailings usually yield silt particles to wind and water and are often chemically unstable; the only suitable treatment appears to be to "fix" them with vegetation. Mine waters are often heavily contaminated and have to be treated chemically and physically, or injected into "safe" rock strata.

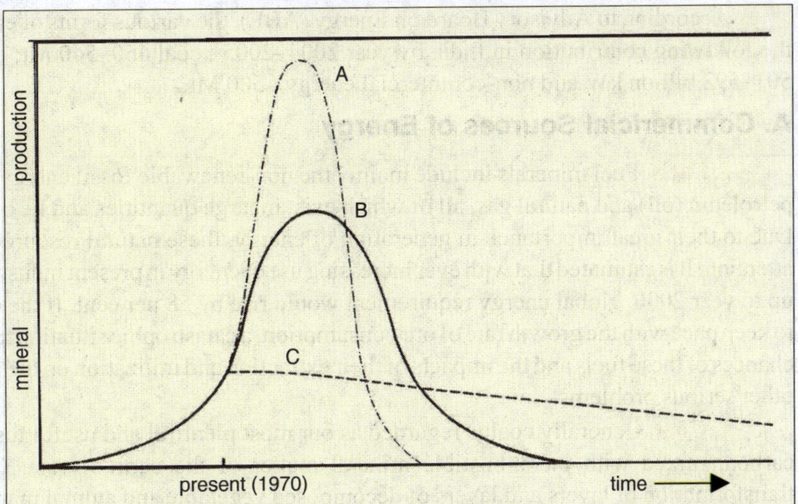

Fig. 15.1. Alternate depletion patterns for mineral resources. A—Pattern of rapid extraction and depletion of minerals that will occur under the present custom of unrestricted mining, use, and throw away ; B—Depletion time can be extended by partial recycle and less wasteful use ; C—Efficient recycle, combined with stringent conservation and substitutions can extend mineral depletion curves indefinitely (Cloud, 1969).

Further, in most countries open-pit mining is more widespread than extraction by shaft. In such mining waste disposal becomes a major problem to which back fill is the obvious solution ; if the topsoil is saved then restoration of agriculture or recreational use is often possible. The processes of concentration, beneficiation and refining may all create biological changes if various products are released into nearby ecosystems, *i.e.,* they may create environmental pollution which affect adversely the ecosystems.

2. ENERGY AND ITS CONSERVATION

Energy utilization has become the modern key to human progress. At the present time, modern nations differ markedly in their energy consumption, with industrialized nations such as European nations like Britain, Germany, Poland, etc., United States, Japan, Hong Kong and Singapore, typically using 10 to 50 times as much energy per capita as agricultural and developing nations such as India, Bangladesh, Indonesia and Ethiopia (**Southwick**, 1976). Higher energy consumption is often associated with higher gross national product and personal income per capita, but it is also associated with higher levels of industrial pollution.

Fossil fuels such as oil, gas and coal form the non-renewable sources of energy and account for 90% of the world's production of commercial energy. Hydroelectric and nuclear power accounting for only 10%. The figures are :

Oil	39.5% energy
Natural gas	19.6% energy
Coal	30.3% energy
Hydro-electric	6.7% energy
Nuclear	3.9% energy

According to Advisory Board on Energy (ABE), the various forms of energy are likely to make the following contribution in India by year 2004–2005 : coal 450–540 Mt; oil 90.11 Mt; electricity 501–592 billion kW and non-commercial energy : 500 Mt.

A. Commericial Sources of Energy

I. Fuels. Fuel minerals include mainly the non-renewable fossil energy resources such as coal, petroleum (oil) and natural gas, all of which exist in large quantities and lie buried under the ground. Due to their great importance in generation of energy, these natural resources are receiving utmost attention. It is estimated that with ever increasing use of energy in present industrial—urban complexes, up to year 2000, global energy requirement would rise by 58 per cent. If the energy production fails to keep pace with the growth rate of our consumption, a catastrophic situation may arise. Fast depletion chances of these fuels and the impacts of their extraction and utilization on the environment are certain other serious problems.

1. Coal. Generally coal is regarded as our most plentiful and useful fossil fuel. This is largely carbon, mixed with incombustible mineral matter of the earth's crust. Coal is formed by the transformation of layers and layers of decomposed vegetable and animal matter under the combined effect of heat and pressure of earth. Coal deposits are probably being formed even today as in the Ganga river delta of our country. Coal deposits occur in abundance and are fairly evenly distributed over the various continents of world. Soviet Union, China and United States contain about 80 per cent of the world's total reserves. The rate of coal consumption in the world as calculated in 1976, was 3.7 billion short tons per year. On this basis, our total reserves would last for 212 years and recoverable resources for approximately 1700 years. Coal is the only long run fossil fuel which is the long-run source of gas and many coal-tar and petroleum products that can be derived from it.

2. Petroleum, oil and natural gas. The organic deposits of these fossil fuels were probably formed from tiny microorganisms rather than from the debris of large plants as was the case with coal. When these microscopic creatures settled on ocean floor and were later covered with mineral sediments, tiny droplets of body oil were squeezed out of them. This oil was trapped into large deposits during formation and was altered chemically by heat and pressure.

According to an estimate, the recoverable amount of petroleum of the world is approximately 2000 billion barrels. Of these, 339 billion barrels have already been produced, 646 billion barrels are reserves, and slightly over 1000 billion barrels are to be discovered as yet. About 60 per cent of the total world's reserves are concentrated in Persian Gulf and South West Asian countries. The industrialized nations, excluding the USSR, with the highest rate of petroleum consumption have the smallest reserves and resources. In comparison to other countries of the world, India's reserves of oil and production of crude petroleum are quite meagre. India produced only 21.7 billion barrels of oil in 1976. It is estimated that if the present oil production rate remained constant, global supply would last about 30 years and the entire resources no longer than 77 years.

However, before we reach that stage, we will undoubtedly have ravaged the great oil-laden shale deposits. A wide variety of laminated solidified mixtures of inorganic sediments and organic matter, called **oil-shales**, have the property of yielding oil on destructive distillation. Certain countries such as Brazil, China, USSR, Zaire and Scotland exploit this resource to get oil. It is speculated that 15,200 billion barrels of high grade oil (25 to 100 gallons per ton shale) and 3,18,000 billion barrels of low grade oil (10 to 25 gallons per ton material) constitute the total resources. Likewise, **tar sands** (also known as oil sands, bituminous sand and bituminous rocks) contain heavy asphaltic, viscous substance called **bitumen** (petroleum). These occur in Canada and have 731 billion barrels of inplace oil.

Natural gas is another energy resource. USA is its largest producer as well as consumer. About 50 trillion cubic feet natural gas was obtained in 1976 and if this pace continues constantly in future, the world's reserve of natural gas would last about 45 years.

3. Nuclear fuels. The nuclear energy of the atoms has been harnessed by three types of reactors: fission reactor, fast breeder reactor and nuclear fusion reactor. While fission and fast breeder reactors are based on the splitting of uranium and plutonium, the nuclear fusion reactor is latest and most powerful and involves fusion of hydrogen nuclei. Nuclear reactors produce enormous energy for driving ships and generating electric power.

Our planet is estimated to have 4900 thousand short tons of uranium, out of which 3510 are known from USA. These can fuel 800 gigawatts (800 billion watts) of light water reactor plants. In comparison to oil and gas, uranium represents a resource of limited potential under present state of technologies. However, new enrichment technologies may prolong these supplies. In fact, utmost care need to be exercised to produce energy from the nuclear fuel, radioactive contamination is a significant biological hazard. Any accidental fall-out or escape from nuclear wastes can produce long lasting disasters often lethal effects on man and other biotic components of the biosphere.

II. Electric energy production. The various sources of electric energy are (i) atomic (nuclear) power reactors; (ii) lignite; (iii) lean gas; (iv) hydropower plants ; (v) coal plants, and (vi) small (micro-hydel plants).

1. Atomic power reactors. Atomic power has become a principal source of energy in recent years, since the fossil fuel reserves are depleting very fast. A small quantity of radioactive material can produce an enormous amount of energy. For example, one tonne of uranium –235 (U^{235}) would provide as much energy as by three million tonnes of coal or 12 million barrels of oil. Besides getting the electricity, atomic power is also used as fuel for marine vessels, heat generation for chemical and food processing plants and for spacecrafts.

An atomic power reactor.

For the generation of atomic energy one requires a **nuclear reactor**. The decay of fissionable matter produces great amount of heat. This heat is used to make steam which is channeled through a turbine connected to **electric generator**.

2. Hydroelectric energy. Moving water has been harnessed as a source of energy for many years. Waterfalls turned paddle wheels to grind corn or wheat in early Europe, America and India. Later, moving water was backed up behind great dams and then released to turn turbines of huge generators for **hydroelectric energy**. Hydroelectric power is the most important renewable energy source. It is a cheapest source of power generation, since land is free and environmental costs are seldom calculated. It is regenerative, thus, helps in the conservation of fossil fuels. Potential of such hydroelectric energy is being developed in India through large-scale projects in the Himalayan region in North and North-East regions. Recently, however, these megaprojects have been criticized and vehemently opposed by the environmentalists such as **Sunder Lal Bahuguna** and **Medha Patekar**, for these have resulted (during development of their infrastructure) into deforestation, wildlife destruction, increased soil erosion and other socio-political and economic problems.

Hydroelectric power is an important renewable energy source.

3. Micro-hydels. In India, there is great scope for the development of micro-and minihydel power schemes. Micro-hydels are proper means of generating electricity in the remote areas. Such istallations are usually developed on the run-off-the river schemes and do not require much infrastructure. They are more safe environmentally and are very useful for the local people. In fact, they do not involve approach roads, no pondage, no loss of vegetative cover, and are free from pollution hazards.

B. Non-commercial Sources of Energy

All materials which originate from photosynthesis form the **biomass**. It includes all new plant growth, residues and wastes (wood, short rotational trees) ; herbaceous plants, freshwater and marine algae, aquatic plants; agricultural and forest residues (straw husks, bagasse, corn cobs, bark, sawdust, wood shavings, roots, animal droppings); wastes (garbage, dung, night soil or faeces), industrial refuse, etc. Biodegradable organic effluents from industries such as cannaries, sugar mills, slaughter house, meat packing plants, breweries, distillaries, etc., are also included in this category. Biomass-based energy can be obtained by the following sources :

1. Fire wood. According to Advisory Board on Energy or ABE the demand for firewood is going to go up in order of 300 – 330 Mt in the year 2004/ 2005 against the present level of 120 – 130 Mt.

Recently, efforts are going on to improve the firewood utilisation to get more and more energy. Some new techniques are the following :

(i) Briquetting. The Indian Institute of Technology, New Delhi has developed the technology of briquetting saw dust into smokeless fuel.

(ii) Gasification. Gasification of biomass is an important means of harvest energy through thermo-chemical conversion. It yields **biogas**, **producer gas** and **pyrogas.**

(iii) Improved chullahs. Present day chullahs (which are about 112 million in number) have a very low, 2–10% efficiency. This results into wastage of wood, deforestation, air pollution and health problems. Department of Non-conventional Energy Sources (DNES) have designed certain improved stoves which contain thermal efficiency up to 15 to 20%.

2. Petroplants. There are attempts to identify potential plant species as sources of liquid hydrocarbons, a substitute for liquid fuels. In this respect, 15 species hold promise and they belong to families Euphorbiaceae, Asclepiadaceae, Apocynanceae, Urticaceae, Convolvulaceae and Sapotaceae. There is a need to increase the biomass of these plants, and conversion of their hydrocarbons into petroleum fractions. The Indian Institute of Petroleum, Dehra Dun done a remarkable job in this area, particularly on hydrocracking of the crude products. The products, thus, obtained are gases, naphtha, kerosene, gas oil, coke, etc.

3. Biogas. Biogas provides a significant solution to present energy crisis, especially in rural areas. This is an environmentally clean technology. The generation of biogas involves decomposition of animal dung (even human faecal matter), leafy and agricultural residues, under anaerobic conditions by bacterial action, in a suitable capacity digester. This results in the production of methane gas which is used for cooking and even for lighting purpose. Biogas is composed of methane, CO_2, H_2 and N_2. At 40% methane content, calorific value of biogas is 3,214 kcal/m^3, at 50% is 4,429 k cal/m^3 and at 55% is 4,713 k cal/m^3.

Dung of domestic herbivore is used in biogas production. According to **Dr. Archana Sharma** (1987), India annually produces about 1000 million tonnes of dung, from which about 22,425 million m^3 gas can be produced. Besides the cooking gas, slurry (*i.e.,* residual organic matter obtained from the digester) can produce 206 million tonnes of organic manure (or bio-fertilizer) every year that can replace 1.4 million tones of N_2, 1.3 million tonnes of P_2O_5 (phosphate) and 0.9 million tonnes of K_2O (potash). However, this source of bioenergy is not being fully tapped in India, though China has fully exploited this biotechnology.

Biogas can also be generated from sludge obtained from primary treatment of raw sewage and one such plant is in operation at Okhla. Water hyacinth, *Hydrilla*, duck weeds and algae can act as supplements for production of biogas.

At present, in India, there are about 6.1 lakh biogas plants. It is estimated that 1.5 lakh biogas plants will save 6 lakh tonnes of wood equivalent every year.

C. Non-conventional Renewable Sources of Energy

These sources can boost production in specific areas, in a decentralised manner and are of the following types :

1. Dendrothermal energy. In Philippines, energy is experimentally generated from energy plantation. There is a claim of 3.75 kw of electric energy production from wood growing on 7 hectares of land.

2. Solar Energy. Solar energy of sunlight is a highly potential renewable resource like the wind energy. It does not cause any pollution and can be manned to produce enough power by developing economical technologies. India receives abundant sun-shine about 1648–2108 kWh/m^2/ year with nearly 250-300 days of useful sun-shine in a year. The daily solar energy incidence is between 5 to 7 k Wh/m^2 at different parts of the country. This immense resource of solar energy can be converted into other form of energy through thermal or photovoltaic conversion routes. 1. The **solar thermal** uses solar radiation in the form of heat that in turn may be converted to mechanical, electric or chemical energy. Solar thermal devices such as solar cookers, solar water/air heaters, solar dryers, solar wood seasoning ovens or kilns and silicon systems have been developed. 2. The **photovoltaic conversion systems** convert solar radiation directly into electricity through **silicon solar cells**. These cells may be of various types such as single crystal silicon cell, polycrystalline cells, amorphic solar cells, etc. The systems have been exploited for community lighting, radio and TV sets, light houses, offshore platforms and installation in remote areas. A significant application of photovoltaic conversion systems is pumping of water for micro-irrigation and drinking purposes. These systems being decentralised on-the-spot electricity generation systems help to replace old, noisy and polluting diesel-utilising systems. They may be installed in remote areas such as forests, deserts and hills.

3. Wind energy. Wind has provided cheap and pollution free energy throughout much of western Europe, but windmills are presently often treated as quaint relics. Part of the problem is that wind is a rather local phenomenon and is extremely unpredictable in most places. In India, there are areas which are quite windy. Average annual wind density of 3 kW/m^2 /day are common at various places in peninsular India and also along coast line of Gujarat, Western Ghats and parts of central India. Wind densities exceeding 4 kW/m^2/day are available for 5–7 months in a year. During winter, wind densities become even more than 10 kW/m^2/day.

Wind energy can be converted into mechanical and electrical energy. National Aeronautical Limited, Bangalore is engaged in research and development of power generation through wind mills. Wind energy is useful in remote areas, helps saving fossil fuels, delivers on-the-spot small quantity of energy which is free from pollution and environmental degradation.

4. Ocean (tidal) energy. Tidal power is a form of hydroelectricity produced by harnessing the ebb and flow of the tides. Barriers containing reversible **turbines** are built across an estuary or gulf where the tidal range (*i.e.,* rise and fall of sea level) is great.

5. Geothermal energy. Geothermal energy provides the most promosing sources of power. Geothermal energy utilisation is possible in volcanic regions or where hot springs and geysers occur. A cold storage unit and 5 mW power plant using geothermal energy have been set up at Manikaran (H.P.). About 350 geothermal springs have been located in the country (in 46 of them the water temperature exceeds 150^{0}C; **Dash**, 1993). Data of the investigations on geothermal energy potential at Tapovan, Badrinath (U.P.) by Wadia Institute of Himalayan Geology and other agencies only are available (see **Dubey**, 1995).

3. FOOD, AGRICULTURE AND AQUACULTURE

Humans need food as a source of energy and for tissue replacement, like any other animal. The source of much of the food consumed by man is **terrestrial agriculture**. This represents the most manipulated of all the non-urban ecosystems, in which the energy and matter pathways are directed almost entirely to man and where he maintains a high level of input of matter and energy to keep the system stable in order to yield his preferred crop. Not only is the ecosystem man-made, but the plant and animal components of it have usually been genetically altered by man in the course of their domestication. There are two main types of agriculture; **crop agriculture**, in which plant production is harvested for use by man either directly or after processing and **animal agriculture**, where a crop from highly manipulated ecosystem is fed to domesticated animals. Ecologically, terrestrial agriculture presents man either as a herbivore or as a third trophic level carnivore.

Terrestrial agriculture systems today exhibit a major division into shifting and sedentary types:

(i) Shifting Cultivation

In shifting cultivation, total manipulation of the natural system is practised over a limited area but for only a short (1–5 years) period of time. Thus, the agricultural path is spatially and temporarily enclosed by wild vegetation. Today, shifting cultivation is largely confined to tropical forests, savannas, and grasslands. In this type of cultivation crops are planted in a mosaic of different heights and times of fruition so that the plant cover of the soil remains as complete as possible throughout the year in order to reduce the leaching effect of heavy rain fall. 'Slash and burn' techniques provide mineral nutrients for their uptake by the crops. The natural diversity of the forests is imitated by the variety of crops which are grown by some shifting cultivators. For example, the tropic crops of Hanunoo of the Philippines include rice, beans, root crops, shrub, legumes, tree crops, yam, taro, sweet potato, vines, bananas and sugarcane, together with European-contact crops such as maize, ground nuts, tomatoes, melons and pumpkins.

Ecologists have found shifting cultivation as a well-adapted system in forested lands where the trees regenerate easily when the plots are deserted, and where an equilibrium population has established.

(ii) Sedentary Cultivation

Sedentary cultivation represents the permanent manipulation of an ecosystem ; the natural biota are removed and replaced with domesticated plants and animals. Competition by the remnants of the original biota or by man-introduced organisms may still remain and considerable efforts may be needed to keep these weeds and pests at an acceptable level. In **dry land agriculture** the soil assumes an importance for it now becomes the long-term reservoir of all nutrients and is constantly depleted as crops are harvested and removed. The nutrients must be replenished either by the addition of organic excreta or chemical fertilizers. There is also the **paddy-culture** of rice, where the soil is very largely a mechanical rooting medium for the plants and the water supplies the essential mineral nutrients ; it often contains blue-green algae which fix nitrogen, for example. The variety of crops grown under the various forms of sedentary cultivation is very high and changing patterns of agriculture together with shifting trade flows and altered rates of consumption make a world kaleidoscope of infinite variety.

Ecological training has been widely used in agriculture. To select particular climate and edaphic conditions for a particular crop is primarily an ecologist's job. Ecology plays a vital role in weed control. The nitrogen and phosphorus requirements of soil, crop rotation practice, etc., have been learned from ecology. Due to application of ecologically sound principles agriculture has resulted into **Green Revolution**, *i.e.*, increased agriculture production. In fact, the Green Revolution of the late 1960s, was based on new genetic strain of rice and wheat, improved irrigation, and better application of fertilizers and it has achieved a doubling and tripling of crop yields in many tropical countries. But recently, Green Revolution has exhibited certain ecological problems—the breakdown of soil structure in European countries under a regime of continuous cereal cropping with the use of physically heavy machinery and its increase dependence upon petroleum and fertilizer. Further, new strains of agricultural crops which are high yielding, protein-rich, and weather and pest resistant, require more careful management, more water, more fertilizer, more insecticides, and more intensive care.

New Sources of Food

Most modern ecologists hold the view that even if the rapid development of conventional agriculture is sustained, protein deficiencies will continue to exist (**Simmons**, 1974). A search for supplementary sources of plant and animal proteins is, therefore, in progress. Animal proteins are more preferred ones because their amino-acid make-up is closest to man's requirements. Animal flesh has the further advantage that it is usually the more easily assimilable, since the plant proteins are locked away behind a cell wall of cellulose not easily broken down by the action of the human stomach. Animals, thus, have been a means of harvesting the plant protein in a digestible form, and so have considerable dietary advantages. If animals are to be avoided, a food source which will yield plant protein in a digestible form or from which the majority of cellulose has been removed is clearly attractive. Fungi appear to be easily assimilated and contain a good deal of protein by comparison with some other sources such as beef, fish, etc. Other advantages of mushrooms are that they do not absorb much human or fossil energy in production, can be grown independently of environmental factors in places such as caves and abandoned railway tunnels, can readily be stored in dried form and require little sophisticated knowledge or technology (**Pyke**, 1970)

Animals retain many desirable characteristics as cellulose-converters and as saliva-inducers. As very small number of wild animals have been domesticated by man for milk, wool, skin, food, transportation, etc., recently, certain other wild animals are increasingly domesticated. Birds such as young colonial seabirds, reptiles, amphibians, insects, crustaceans, molluscs, rodents, and many other animals can be utilized as food. Freshwater and brackish-water fish are other sources of animal proteins. Marine and freshwater fisheries provide a best sources of animal proteins, many nations such as Peru, Japan, USSR, China, Norway, U.S.A., UK, Burma, Thailand and India depend on them for food requirements. Due to overfishing of certain favourite fish species, recently following trends are adopted for continuous abundant supply of fishes :

(i) Extension of fisheries. The fisheries have been extended from three sources : the utilization of untapped species, the cropping of hitherto unattractive areas, and the development of more novel methods of culture and harvesting. During recent years many new fisheries have started to flourish, such as Peruvian anchoveta, Alaska pollock, Bering sea flatfishes and herring, and several more. In future, fishing method may be extended in the cool temperate parts of southern hemisphere.

(ii) Krill. Uneaten food of whales is also found suitable for human consumption as food. Approximately 80 per cent of the prey of blue and fin whales is **krill**, the shrimp *Euphausia superba,* each of which contains 7% fat and 16% protein.

(iii) Aquaculture. To increase the yield of certain edible species of freshwater and marine fishes, molluscs and crustaceans and other aquatic organisms, **fish farming** or **aquaculture** that includes their culturing and herding, has been employed and at a commercial scale in Hongkong, Philippines and Japan. Framework are lowered into shallow offshore waters and allowed to colonize with sedentary molluscs like oysters and mussels with sense species, the individuals grow on ropes that hang clear of the action so that they are out of the reach of predators such as starfish. Although productive, such systems are very vulnerable to contamination, and since the organisms filter large quantities of water their ability to concentrate substances toxic either to themselves or to consumers is very high.

Livestock as renewable resource. The branch of agriculture which is mainly concerned with the breeding, feeding and caring of domestic animals is called **animal husbandry**. When it incorporates the study of proper utilization of economically important domestic animals, it is called **livestock management**. Livestock (*viz.,* cattle, buffaloes, sheep, goats, pigs, camels, horses, donkeys, mules (ponies), etc.,) form an important renewable resource, as they provide milk, eggs, meat, skin (leather), horn, dung and other products. The total livestock in India amount to a huge population consisting of 185 million cattle, 97 million goats, 61 million buffaloes, 45 million sheep, 1 million horses and ponies, 1 million camels and about 1 million other livestock. There are also 156 million poultry (fowls) and other domestic birds (such as ducks, turkeys, etc.). All these provide 1 million tonnes of meat, 40 million tonnes of milk, 39 million kg of wool and 13 million eggs. The amount of dung produced is enormous and is used for biogas production and as manure. Livestock is also used for rural transport and agriculture. Therefore to get a sustained yield from them, their proper breeding, health care and management and diet are essential.

4. WASTE MANAGEMENT

(Recycling of Resources and Vermitechnology)

Some types of solid wastes can now be recycled by the help of certain new technologies. For example, the fermentation of organic waste such as dung (of cattle, buffalo, etc.), animal excreta (including faeces or nightsoil and urine), garbage, aquatic weeds, etc., can be utilised to produce **biogas** which can be used for cooking, streetlighting, industry and so on. The **slurry** produced by this method is used as manure. Likewise, waste paper can be converted into toilet paper. Some solid wastes can be utilised in vermitechnology for the production of earthworm cast, earthworm tissue protein and so on. The worm casts can be utilised for the production of mushrooms (**Dash** and **Das**, 1989) and vegetable crops (**Senapati**, 1989). The earthworm tissue is used as feed material for the poultry (**Das** and **Dash**, 1990), fish, amphibia and pigs.

A domestic 'biogas' digester in an Indian village, using animal manure and producing methane for cooking and heating water.

Further, after extraction of metal from the ore, a waste product, called **slag** remains. The slag can be crushed and added to cement for construction work. **Flyash** is also used as a cementing material. **Scrap** metal produced in mills and factories, used metals of discarded vehicles (trucks, car, scooters, cycles, etc.), machines, ships and aeroplanes can be melted and recycled for various purposes. Thus, metals such as copper, zinc, lead, aluminium and platinum can be recovered from the used materials by recycling.

Ecologists have also suggested the following means of resource-recycling from the industrial effluents : 1. Waste heat (*i.e.,* heated water from power plants may maintain constant water temperatures in coastal temperate zones and there certain tropical fishes of high productivity can be raised besides native fishes. 2. Sewage and eutrophicated waste products might be used as the basis for algal production which forms the food of oysters which then filter the water as well. 3. Carbon dioxide from a power station chimney is used to enhance production of algae which are fed to clams.

Vermitechnology (Organic Waste Management through Earthworms or Verms)

India tends to produce about 2,500 million tonnes of organic waste annually, which can be utilised for the recovery of essential resources such as fertilizer, fodder, fuel and food. Earthworms (*e.g., Dichogaster bolaui, Drawida willsi, Lampito mauritii* and *Perionyx excavatus*) are now known to be a good biological elements for the recovery of vermifertiliser and vermiprotein (called **vermitin**) for use in agroecosystems, aquaculture and poultry. The earthworm is physically an aerator, crusher and mixer, chemically a degrader and biologically a stimulator in the decomposer subsystem (**Mitchell**, 1978). The degradation of organic waste by earthworm consumption is known as **vermicomposting** (**Dash** and **Senapati**, 1986).

5. FOREST RESOURCES

India has a rich flora and fauna (*i.e.,* rich biological diversity), much of which is present in forest areas. Forests occupy a special place in the life and thought of the people. They form an important renewable natural resource. Forest ecosystem is dominated by trees, their species-content varying in different parts of the world. Forests contribute essentially to the economic development of our country by providing goods and services to the people and industry. They are intimately linked with our culture and civilization. The chief products that forests supply is **wood** which is used as fuel; raw materials for various industries as pulp, paper, news-print, board, plywood; timber for furniture items, toys, sports goods, musical instruments, wooden crates, boxes (for packing articles for fruit, tea, etc.), boats, truck bodies, carts, ploughs, railway sleepers, bridges, buildings; **fodder** for cattles, sheep, goats and camels and **bamboos**. Bamboos, called poor man's timber, are used in rafters, scaffolding, roofing, walling, flooring, basketry, cartwood and cordage. Industrially bamboos are used as a raw material in paper and rayon industry. Indian forests also supply **minor forest products** such as canes (rattans), gums, resins, rubber, dyes, tannins, fibres, flocs, medicines, katha, insecticides, camphor, essential oils (*e.g.,* rosha grasses, khas and sandal woods), cooking oils and spices. Tendu (kendu) leaves are used as wrappers for *bidis*, soap-substitutes such as *ritha* and *shikakai*, sola pith and ornamental seeds *rudraksha* are important commercial products of the forests. Lac, honey, wax, tusser or moga silk are obtained from forest insects. Feathers, horns, hides and ivory are also other significant forest product (obtained from forest wild life). For tribal people forests also provide food (tuber, roots, leaves, fruits, meat from birds and other animals).

Besides above discussed uses, forests are also a major factor of environmental concern, providing protection to wild life, help in gaseous (*i.e.,* CO_2, O_2) cycles of atmosphere, tend to enhance local rainfall and water holding capacity of soil, maintain the soil fertility, regulate the earth's temperature regimes and water cycle, check soil erosion, landslides, shifting of sand and silting and reduce the flood havoc. Forests play an important role in reducing atmospheric pollution by collecting the suspended particulate matter and by absorbing carbon dioxide. Lastly, forests have aesthetic and touristic values and serve as gene reserve of important species.

Forest Cover

According to **Brewbaker** (1984) the total forest area of the world in 1900 was nearly 7000 M ha(=million hectares). By 1975 it was reduced to 2890 M ha and if this trend continues by 2000 A.D., it would be merely 2370 M ha. It is also estimated that major reduction will be in tropics and subtropics (40.2%), being only 0.6% in temperate areas. In the tropics and subtropics, too, maximum reduction would be 50% in Asia and the Pacific. This occurs mainly due to population explosion in these areas.

According to **CFC** (=Central Forestry Commission), the forest cover of India in 1980 was around 74.8 Mha. (22.7% of the total land mass). This data have been contradicted by remote sensing surveys which have shown that it is hardly 14 per cent. Among 16 different forest types of India, the most common is the **tropical dry deciduous** (38.7%), followed by **tropical moist deciduous** (30.9%) both forming the major bulk (69.6%) of the forest area in India. Other forest types contain following percentage of forest cover : Tropical thorn (6.9%), Tropical dry evergreen (0.1%), Pure coniferous (or high mountain areas 6.3%), Sal forests (16%), Teak forests (13%), Broad leaved forests (excluding sal and teak, 55.8%), Bamboos (including in plantations, 8.8%). Further, according to the study made by the National Remote Sensing Agency (NRSA) in 1984, there has been a decrease in the forest area in India within a seven-year period, *i.e.,* between 1972–75 and 1980–82. In general, there has been a decrease in forest area in all states except in Sikkim and Arunachal Pradesh. Maximum decrease (11.3%) has been in Tripura, and least (less than 1%) is in Haryana, Delhi, Arunachal Pradesh and Nagaland. The overall forest area has decreased from 55.20 Mha to 45.70 Mha, *i.e.,* 2.89% during this period.

Deforestation (Destruction of Forests)

Deforestation is an alarming threat to the economy, quality of life and future of the environment. The main causes of the deforestation in India are : explosion of human and livestock population, increased require-ment of timber and fuel wood, expansion of croplands and enhanced grazing. Other prominent causes of forest destruction are construction of roads along the mountain (hill-roads), dam building (valley projects) ever migrating graziers, shifting (*jhum*) cultivation, agricultralisa-tion, urbanisation and industrialisation. Stress on revenue earning and resources regeneration for pulp and match industries seems to have accelerated soil erosion. Resin tapping from chir-pine trees has choked the trees to death.

Several trees are cut every year for firewood which is a major cause of deforestation.

India consumes nearly 170 million tonnes of firewood annually and 10–15 million hectares of forest cover is being stripped every year to meet fuel requirements. During a period of 20 years (1951 to 1971) forests have been cut for : agriculture (24.32 lakh hectares), river valley projects (4.01 lakh hectares), industrial uses (1.24 lakh hectares), road-construction (0.55 lakh hectares) and miscellaneous uses (3.88 lakh hectares). Thus, a total of 3.4 million hectares of forests were lost during this period. Due to deforestation, nearly 1 per cent of the land surface of India is turning barren and unproductive every year.

The devastating effects of deforestation in India include soil, water and wind erosions. Deforestation has a major impact on the productivity of our croplands. This happens in two ways : (i) soil erosion increases manifold and the soil actually gets washed, leading to an intensified cycle of floods and drought. (ii) the shortage of firewood has an obvious impact on the productivity of our croplands. Thus, when firewood becomes scarce, people begin to use cattle dung and crop wastes as fuel, mainly for cooking. In a way, every part of the plant gets used up gradually and goes back to the soil. Over a period of time, such a nutrient drain affects crop productivity due to loss in soil fertility. Further, the local cattles, goats, sheep, etc., not only destroy the vegetation but also pull out the roots of the plants.

Destruction of biotic potential of land also leads to **desertification**. Desertification is the process leading to desert formation. Removal of vegetal cover brings about marked changes in the local climates of the area. Thus, deforestation, overgrazing, etc., bring about changes in rainfall, temperature, wind velocity, etc., and also leads to soil erosion. Destruction often starts as patchy destruction of productive land. Increased dust particles in atmosphere lead to desertification and drought in margins of the zones that are not humid. For example, after denudation of Himalayas, the process of deforestation started in the Shivalik range (extending parallel to the Himalayas). Shivalik sal forests were exploited for industrial use (railway sleepers etc.). Foot hills of Shivaliks once covered with dense forests, are facing an acute water scarcity and semidesert conditions. When forests die, ecological balance maintained by nature breaks away; and floods and droughts are the dangerous consequences. In the Himalayan range, the rainfall has declined 3 to 4 per cent due to deforestation.

Afforestation

For the purpose of afforestation, following two types of methods of forestry are practiced :

A. Conservation or Protective Forestry

This includes the following three strategies :

(a) Conservation or reserve forests. These forests are the areas where our water regimes are located (*e.g.,* Himalayas and Western and Eastern Ghats together with catchment areas). They also include National Parks, Sanctuaries, Sacred Groves, Biosphere Reserves and all ecologically fragile areas. In these areas, no commercial exploitation can be allowed. These areas, in fact, need protection from fuel-starved villagers and fodder-starved cattle. The real goal of eco-development (conservation of reserve forests) cannot be achieved without getting support from public such as Chipko Movement and Honey Bee network of Prof. A.K. Gupta, AIIM, Ahmedabad.

Chipko movement. Chipko is a Hindi word which means to embrace or to hug. Hugging the trees is an age-old practice in the Himalayan region. In 1970, Gopeswar and some 20 villages of the Tehri-Garhwal district of Uttar Pradesh were devastated by flash flood in the Alakananda river. This flood occurred due to deforestation and was an eye-opener for the villagers. The people of these villages under the leadership of **Chandi Prasad Bhatt** pledged that they would not permit any more felling of trees. The people (mostly women) started hugging trees whenever forest contractors tried to cut them down. This movement became very successful and was popularised all over the world by **Sunderlal Bahuguna**. **Bahuguna** presented the plan of this chipko movement for the protection of soil and water through ban on tree felling in the Himalayas, at the meeting of UNEP (= United Nations Environment Programme) held in London in June 1982. According to him, every standing green tree in the forest is a sentry to protect us from avalanches and landslides to save our soils and conserve our water. Chipko movement advocates the slogan of planting five Fs—*food, fodder, fuel, fibre* and *fertiliser* trees—to make communities self-sufficient in all their basic needs.

Chipko movement.

(b) Limited production forests. These are less fertile areas at more than 1000 metres altitude with hilly topography. A part of the annual growth may be harvested in a very careful and controlled manner so as to avoid soil and tree damage.

(c) Production forests. Such types of forests lie on the flat land and are managed for high degree of production. Their working on scientific lines with proper logging methods does not pose environmental problems.

B. Commercial or Exploitative Forestry

The basic aim of commercial forestry is to supply goods and services and meet the needs of local people for firewood, fodder, food, fertilizer, fibre, timber, medicines, etc. These forests also supply material for industrial purposes as timber of all types, plywood, matchwood, fibre board, paper and pulp, rayon, silvichemicals, etc. This can be achieved by (i) intensive plantation and (ii) captive plantations.

(i) Intensive plantations. This type of forestry includes planting all the available land from villagers fields to commercial land, to road/rail sides and every available space. For removing pressure on the natural forests, indigenous and/or exotic species can be used for plantation. Intensive plantations include two types of programmes : social forestry and agro-forestry.

(a) Social forestry. This type of forestry started with NCA (National Commission on Agriculture) in 1976. **Social forestry** envisages use of community lands, individual holdings and other public lands, denuded/degraded lands for producing what the dependent rural and tribal communities need and for environmental purposes. There are two main objectives in social forestry : 1. use of public and common land to produce in a decentralised way fire wood, fodder and timber for the local poor men and also to manage soil and water conservation, and 2. to relieve pressure on conservation forests. This programme has been mainly aimed for the poor villagers and children who become involved in the intensification of nursery operation at village level for multipurpose species for firewood, fodder, fibre, pole, fruit, etc. In our country, social forestry programme become quite successful particularly in Gujarat. There are followed two systems internationally : Chinese system and South Korean system. The area under social forestry increased from 15 million hectare in First plan to 1524 million hectare in Sixth plan.

(b) Captive or Production plantations (Agro-forestry). This is absolutely commercial forestry which is developed to fulfil the needs of the various forest-based industries requiring large quantities of raw materials. The captive plantation is done on the fallow land which is not being used for agriculture, mostly on free grazing lands. A part of this plantation is used to produce fodder for the cattle. Further, in this type of plantation, short rotation of indigenous or exotic species are preferred over long duration sal or teak. In USA productivity of captive forests has been increased by adopting the following measures : 1. Proper manipulation of **silviculture** (The theory and practice of controlling the establishment, composition, and growth of stands of trees for any of the goods and benefits that they may be called upon to produce) and nutritional requirements such as use of fertilizers, irrigation, bacterial and mycorrhizal inoculations. 2. Management of diseases and pests. 3. Weed control 4. Adoption of advanced techniques in forest tree breeding for superior genetic strains. 5. Skilful use of tissue culture and cloning methods.

Few more measures for the conservation forestry. Following strategy can be adopted to prevent further depletion of tree cover in the conservation forests :

1. Extraction of timber should not interfere with watershed protection. Tree-felling should be matched with tree-planting programmes.

2. The use of firewood should be discouraged to reduce pressure on more valuable natural forests. Other sources of energy such as biogas, solar energy, etc., have to be provided to supplement practices.

3. A better understanding should be developed between the persons who manage and those utilising trees. Thus, there is a growing need for (i) creating a new cadre of forest managers and bare-foot foresters committed to conservation. (ii) involvement of rural/tribal people in forestry programmes and (iii) encouraging industries using forest products to obtain raw material at market price and help them to raise plantations much in advance to meet their demands.

4. Forest dwellers should have access to subsidised sources of fuel, fodder, building material, etc., so that they do not cut forest trees.

5. A ban or freeze (*i.e.,* moratorium) of 15 to 20 years should be imposed on commercial tree fellings in fragile areas of Himalayas and other hilly areas.

6. During moratorium, an extensive afforestation programme with people's participation should be followed.

7. Environmentally sound action plans based on scientific research should be adopted.

8. By the help of people's cooperatives, community forests should be developed around villages.

9. Protection of standing forests should be done.

10. Creation of new stock should be made.

11. Masses and voluntary agencies should be involved in the task of tree planting.

12. Building of information base have to be made.

Forest conservation through law. The National Forest policy 1952 stated that one third of the geographical area of the country should be under forest. Forest (Conservation) Act, 1980 was enacted with a view to check indiscriminate dereservation and diversion of forest land to non-forest purposes. This act was amended in 1988 to incorporate stricter panel provisions against violators.

6. RANGE MANAGEMENT (GRASSLAND MANAGEMENT)

Grassland biomes are important to maintain the crops of many domesticated and wild herbivorous mammals such as horses, mules, asses, cattle, pigs, sheeps, goats, buffaloes, camels, deers, zebras, etc., all of which provide food, milk, wool, hide, or transportation, etc., to man. The range management involves an important application of ecological principles in maintaining the grassland biomes. The objective here is to preserve grasslands for maximum forage, *i.e.,* food for cattle. In India, most of the grasslands represent the seral stages in succession and if they are not maintained by grazing and fire they would develop into forest communities. A team of scientists at Indian Grasslands and Fodder Research Institute, Jhansi and Central Arid Zone Institute, Jodhpur, is mainly entrusted with such responsibility.

During range management a suitable degree of grazing is to be maintained. Some species are more palatable to the stock and are very sensitive for grazing and, thus, overgrazing usually results into their disappearance from the area where now some unpalatable, annuals, weeds, shrubs start to grow and turn the area into man-made desert. The palatable plant species, which are sensitive to grazing, have been called the **decreasers** whose disappearance from the area is an indicator of grazing stress and warning signal for range managers.

Overgrazing has certain other ecological effects—reduction of the mulch cover of the soil occur, microclimate becomes more dry and severe and is readily invaded by xerophytic plants. Due to absence of humus cover, mineral soil surface is heavily trampled when wet and produces puddling of the surface layers, which in turn reduces the infiltration of water into the soil and accelerates its runoff, producing drought. These changes all contribute to the reduction of the rate of energy flow, and the disruption of the stratification and periodicity of the primary producers results in a breakdown of the biogeochemical cycles of water, carbon and nitrogen. Water and wind erosion completely breakdown a very dry grassland-microclimate. Further, intensive grazing which results in increased areas of bare soil, creates a new habitat for burrowing animals such as mice, jackrabbits, gophers, prairie dogs, locusts, etc., which render sterile much areas of forage lands.

In range management, fire, plays an important role. Under moist conditions fire favours grass over trees and under dry conditions fire is often necessary to maintain grasslands against the invasion of desert shrubs. Burning of *Cynodon dactylon* increases forage yields.

7. WILD-LIFE MANAGEMENT

Wild-life management involves the following three main steps : 1. the restoration of the habitat; 2. enforcement of the law to stop poaching completely ; and 3. prevention of competition from domestic livestock regarding the grazing activity; the latter step may result in avoidance of transmission of diseases. Management programmes also involve the establishment of more national parks, wild-life sanctuaries and biosphere reserves (For details see Chapter 16).

8. WATER RESOURCE AND ITS MANAGEMENT

Water occurs naturally in gaseous, solid and liquid phases; man's use of it is nearly all concerned with the last state and is also dominated by his demand for water relatively low in dissolved salts, *i.e.,* freshwater. Freshwater is a precious resource and is obtained from the following three natural sources:

A. Rain water. India receives about 3 trillion m^3 of water from rainfall, which amounts to about 105 to 117 cm annually. This is a huge resource and largest in the world. But almost 90% of this precipitation falls between mid-June and October. Of the total annual precipitation, India utilises only 10% which may increase about 26% by 2025 A.D. There is a need of good management of rainwater resource.

B. Surface flow. In India surface flow takes place through 14 major river systems namely 1. Brahmani, 2. Brahmaputra, 3. Cauvery, 4. Ganga, 5. Godavari, 6. Indus, 7. Krishna, 8. Mahanadi, 9. Mahi, 10. Narmada, 11. Perriar, 12. Sabarmati, 13. Subarnarekha and 14. Tapi. Between them they share 83% of the drainage basin, account for 85% of surface flow. Apart from them, there are 44 medium and 55 minor river systems—these are fast flowing, monsoon fed and originate in the coastal mountains. Of the major rivers, Brahmputra, Ganga and Indus basins along with Godavari cover more than 50% of the country. Only 4 rivers namely Brahmputra, Ganga, Mahanadi and Brahmani are **perennial rivers** with a minimum discharge of 0.47 M m^3/km^2/year.

C. Ground water. Underground reservoirs of freshwater are called **aquifers**; they are continuously recharged through infiltration, seepage and evapo-transpiration. The total volume of ground water found in the aquifers is estimated to be 42.3×10^{10} m^3. At present about 25% of the ground water is being used by man. Agriculture uses the maximum amount of water in the world. This amount is about 73% and leads to a lot of pressure on ground water. Excessive use of ground water depletes aquifers, lowers the water table and may lead to salinisation, waterlogging and alkalinisation of the soils. As the human population goes on increasing, so do agriculture and industrial activities. As a result, there is a considerable demand for water resources.

Water Resource Management

Water resource management should ensure that (i) there is no wastage or misuse of water ; (ii) pure water is made available to man for various purposes and (iii) water storage and distribution is done in a scientific way. Usually water is wasted by leaking taps and excessive irrigation. Public awareness should be made against using more water than is necessary.

There are central and corresponding state organisations which are concerned with specific aspects of water resource management. Some of them have been tabulated in Table 15.2.

Water resource management.

A. Quantitative management of water resource. Water is an integral part of land/soil productivity base and its misuse can cause soil degradation and soil erosion. So water management becomes necessary for crop yields and other activities. Primary channel flow originates in upper catchments and these watersheds are very important for future..

1. Watershed managements. Availability of water in a given soil environment is a critical factor and is related to erosion, siltation, loss of plant cover and productivity. In India, **floods** bring considerable damage resulting in loss of life and property each year. Due to floods, the plains have become silted with mud and sand and affect the cultivable lands. The National Commission on Floods has calculated that the land areas prone to floods has doubled from 20 million hectares in 1971 to 40 million hectares in 1980. The worst flood-suffering states are Assam, Bihar, Orissa, U.P., and West Bengal. The management of rainfall and resultant runoff is very important and found to depend on **water sheds**. To recall, a watershed is an area bounded by the divide line of water. Thus, it may be a drainage basin or stream. The Himalayas are one of the most vital watersheds in the world. These watersheds are threatened by deforestation and other ecological malpractices and have resulted in the depletion of water resources. Thus, for watershed managements the following measures are adopted: 1. soil and land use survey; 2. soil conservation in catchments of River Valley Projects and flood prone areas; 3. afforestation, social forestry programmes and Drought prone area Development Programmes, 4. desert development and 5. control of shifting cultivation.

Table 15-2. Some Indian organisations involved in water resource management.

Name of agency	Nature of management
1. Central Water Commission	Surface water
2. Central Ground Water Board	Ground water
3. Indian Meteorological Department	Precipitation
4. Central Pollution Control Board	Water quality
5. Ministry of Agriculture and ICAR	Water use for agriculture
6. Department of Environment, Forests and Wildlife (Ministry of Environment and Forests).	Environmental impact assessment
7. Central Public Health and Environmental Engineering (Ministry of Urban development)	Water supplies, sanitation and sewage disposal
8. Department of Power	Hydro-electric power
9. Department of Forests	Watershed management

2. River valley projects. The environmental side effects of river valley and hydel power projects can be classified in following three categories : 1. Impacts within and around the area covered by dam and reservoir, *e.g.,* changes in microclimate, loss of vegetal cover, soil erosion, variation in water table and enhanced seismic activities due to pressure of water ; 2. Down stream effects caused by alteration in hydraulic regime ; and 3. Regional effects in terms of overall aspects including resource use and socio-economic aspects.

Other impacts caused by construction of dams and reservoirs are the following : 1. Blasting operations in hilly tracts for road construction can cause considerable damage to the environment through loosening of hill sides and resultant landslides, sedimentation of reservoirs, drying up of springs and flash floods. The creation of new settlements for the workman and rehabilitation of project oustees (= expelled population) in the watershed areas may increase the ecological hazards. In recent years, environmentalists like **Sunder Lal Bahuguna**, **Medha Patekar** and others have opposed the construction of the following river projects : (1) Tehri Dam Project, U.P. ; (2) Narmada Sagar Project, Madhya pradesh; (3) Sardar Sarovar Project, Gujarat; (4) Bodhghat Project, M.P.

3. Irrigation (waterlogging and salinity). Rain water is conserved through wells, reservoirs, budding of streams and canal systems. Canal irrigation increased by 6.7 million ha., from 1950-51 to 1978-79. It is found harmful to soils which degrade due to water logging and salinity. In India, more than 20% of irrigated land is damaged due to water logging and salinity.

B. Qualitative management of water resource. Availability of potable or drinking water is becoming increasingly difficult. There is a lot of industrial liquid waste, there are very poor urban sewerage and sewerage treatment facilities and there is heavy pollution load on water. Measures of qualitative management of water fall in two categories : (1) **Short term measures.** These include the following measures : (i) pollution control of urban settlement sources, (ii) sewerage regulation (discharge of effluents only in municipal sewers, etc.), (iii) proper pattern of sewage collection systems, (iv) pollution control of industrial sources, (v) environmental planning guides for industrial estate, (vi) protection of drinking water resources, and (vii) coastal management. (2) **Long term measures.** These are planned for each river basin and are of following nature : (i) preparation of water use maps (these are to classify and zone river waters based on best uses); (ii) evaluation of pollution potential in the river basin, and (iii) preparation of water quality map on the basis of continuous water quality monitoring.

9. LAND USE PLANNING AND MANAGEMENT

Land is a precious resource, since it is put to diverse use by man. India with a land area of 32,88000 km^2 which is about 2.4% of the world supports 15% of the world's population. There were about 238 million people in India in the year 1901 and are now about 850 million. The per capita land resource available now in India is less than 0.4 hectares, in comparison to more than 0.9 hectare in China and about 8.4 hectares in the USSR. About 44% of our land is used for agriculture, 23% is covered with forests, 4% is used for pastures and grazing fields, 8% for housing, agroforestry, industrial areas, roads and so on. The 14% land is barren and about 8% is used for miscellaneous purposes. The rapid increase of urbanisation and migration of population from rural areas to towns and cities has created many problems. All this has led to the utilization of agricultural land for housing, construction of office buildings, industries, and so forth.

The rational use of land resource is possible by adopting an integrated land-use policy which involves prevention of land misuse and reclamation of degraded and under-utilised land, wastelands, fallows, etc. Reclamation of abandoned mines and brick kilns may yield some much required land. Fertile agricultural land should not be sacrificed for non-agricultural purposes, such as roadbuilding, development of industries or construction of water reservoirs. Urban areas should not be developed on agricultural lands.

10. SOIL EROSION AND SOIL CONSERVATION

It is a common experience that the soil is liable of removal from one place to another whenever there is physical force such as storm or flood (or running water). The dust blows along with wind or flows in runoff water, usually after rains. This is called **erosion**. The term 'erosion' is derived from the Latin word *Erodere* which means to 'gnaw away' or 'tear away'. **Rama Rao** (1962) called soil erosion as **creeping death of the soil**. Soil erosion refers to physically detaching soil particles from their original place and transporting them to some other place. Thus, though it takes a very long time to build the soil, its erosion by the forces of rain and runoff water, wind action, etc., is a rapid process. Volume and intensity of precipitation, slope conditions, vegetation cover and wind speed are some important factors of soil erosion.

Soil erosion.

Type of Soil Erosion

Based on various **agents** that bring about soil erosion and the **form** in which the soil is lost during erosion, there are following types of soil erosion :

1. Water erosion. Water erodes or cuts and removes soil chiefly in four ways : (i) **Sheet erosion.** The soil is removed in small but uniform amounts from all over and therefore, does not leave a mark behind. (ii) **Rill erosion.** The run off water moves rapidly and cuts small stream-like structures. (iii)

Gully erosion. Several rills converge towards the steep slopes and join to form broad channels of water called **gullies**. (iv) **Riparian erosion** or **stream bank erosion.** The rivers during floods splash their water against the banks and thus, cuts through them.

2. Wind erosion. Erosion due to wind is very common and includes saltation, suspension and surface creep. During dust storms, a huge quantity of dust is raised high and transported to great distances. The wind lifts finer particles high up (called **suspension**) whereas the coarser and heavier particles roll along the surface (called **surface creep**). The rolling particles rub the ground and due to abrasive action help in loosening the soil. The process continues and more dust particles gather as the storm advances.

In the arid regions, where rainflow is low, drainage is poor and high temperatures prevail, water evaporates quickly leaving behind the salts. Salt accumulation occurs mainly in lowlands around the oceans. The salts are normally chloride, sulphates, carbonates and nitrates of potassium, magnesium and sodium. The salts also include chlorides and nitrates of calcium. The major portion of such salty soil is carried by wind in the form of small leaps.

3. Landslide erosion. The hydraulic pressure caused by heavy rains increases the weight of the rocks at cliffs which come under the gravitation force and finally slip or fall off.

4. Erosion due to overhunting, overgrazing and deforestation. Human beings are involved directly or indirectly in accelerating the rate of erosion. In the normal course of development of communities the edaphic conditions and other environmental conditions attain some sort of dynamic equilibrium with the vegetation and other biota. As a result of this the soil is held in position by vegetation cover. Killing of lions, tigers and other carnivores by man results in increased population of herbivore. The natural balance gets upset; herbivores in greater numbers damage the natural vegetation (by grazing) in greater quantities. Natural vegetation is also damaged by scraping and deforestation. All these lead to the removal of protective plant cover from the soil surface. The beating of rain drops on the bare soil raises soil particles selectively; finer particles of clay and humus being lighter are raised to greater heights than sand particles. Thus, the finer components which are accidentally more fertile brought upon the surface and these seal and clog the pores. The accumulated water fails to percolate down and therefore, runs down the slope. The fertile top soil also gets washed away.

Soil Conservation

Soil conservation means protection, improvement and sustained renewal of the soil at any place.

1. Principles of soil conservation. The soil conservation depends on the following principles: (i) protection of soil from impact of rain drops, (ii) to prevent water from concentrating and moving down the slope, (iii) to slow down the water movement when it flows along the slope, (iv) to encourage more water to enter the soil, (v) to increase the size of soil particle, (vi) reduction in the wind velocity near the ground by growing vegetation cover, ridging the land, and (vii) to grow the strips of stubble or other vegetation cover which help to catch and hold the moving particles of soil.

2. Methods of soil conservation. For the conservation of soil (*i.e.,* to check the loss of soil during erosion) ecologists have devised the following methods :

A. Biological methods. In these methods, conservation is achieved by the use of plant vegetation cover : (i) **Agronomic practices.** Natural protection is provided to the soil by growing vegetation in a manner that reduces soil loss. These are the following : (a) **Contour farming.** This oldest method is found useful in those areas where the rainfall is low. It involves preparation of the field with alternate furrows and ridges (or contours). The water is caught and held in furrows and stored; this tends to reduce run off and erosion. Contour farming is also useful for lands with gentle slopes. In this method ploughing is done at right angles to the direction of the slope. (b) **Terrace cropping.** On hilly terrains where there is some good soil profile, terrace cropping is in practice since time immemorial to reduce soil erosion and obtain a reasonable productivity. In preparing **terrace**, one block of flat land abruptly ends and a new terrace begins at a lower level. The soil is cut and laid to make steps. On the margins

Terrace cropping.

between two steps the slope is vertical. This zone is planted with some perennial soil binding grass whereas the rest of the space is used for cultivation. The runoff of rainwater is greatly reduced and the soil becomes stable. (c) **Mulching.** In this method certain plants such as maize stalks, cotton stalks, potato tops, etc., are used as a mulch. Mulch is a protective layer formed by stubble (*i.e.,* the basal parts of herbaceous plants), especially cereals attached to the soil after harvest. Mulches (13 cm to 20 cm) bind the soil particles, reduce evaporation of soil moisture and increase the amount of soil moisture by adding organic matter to soil. (d) **Crop rotation.** This method decreases soil loss and preserves the productivity of land. The typical crop rotation involves one year of fallow, followed by winter heat. (e) **Strip cropping.** This involves the planting the crop in rows or strips to check flow of water. On more slopy lands for strip cropping, strips of land are ploughed at right angles of the direction of the slope and different crops are raised in adjacent strips. In different years the one crop of the strip is rotated with another crop.

Strip cropping

(ii) Dry farming. This method is found useful for the croplands existing in the areas of low or moderate rainfall ; in these areas ordinary farming is at risk. Thus, for checking the soil erosion, there are only three alternatives : crop production, animal husbandary and growing grazing fields. Dry farming methods differ in different areas and involves the fallowing the land, strip cropping, crop rotation, contour farming, etc.

(iii) Agrostological methods. In these methods, grasses such as *Cynodon dactylon* are used as erosion-resisting plants. They are grown in strips between the crops. They act as stabilisers when grown in gully and sodding (*i.e.,* to cover with sod). These methods involve lay farming, retiring lands to grass, etc.

Excess run off water is charelled through drains or diversions.

B. Mechanical methods. They are used as supplements to biological methods and are of following types : **(a) Basin listing.** It involves the construction of small basin along the contours to retain water which also reduces its velocity. **(b) Contour terracing.** This involves digging a channel along the slope to intercept and divert the run off water.

C. Other methods. These are of following types : (1) **Gully control.** To check the widening of gullies by constructing bunds, dams, drains or diversions through which excess run off water is channelled. (2) **Stream bank protection.** For preventing the cutting and caving of river banks, vegetation is grown alongside the river bank, drains are constructed and concrete or pitching is done. **(3) Afforestation.** Trees as windbreaks are planted in deserts which check the velocity of wind. Plantation of trees in short blocks are called **wind breaks**. Extensive plantation of trees are called **shelter belts**. Windbreaks are planted across the area at 90^0 to the prevailing wind. They check the spread of sand dunes or desert conditions or blowing away of the fertile top soil. Windbreaks are planted in several rows. In Indian deserts, following species have been planted under the afforestation—*Lowsonia alba, Agava americana, Thevetia neriifolia, Calotropis gigantea, Ricinus communis, Zizyphus jujuba, Acacia catechu, Cassia, Acacia nelotica, Dalbergia* *sissoo*, *Mangifera indica* and *Tamarindus indica.* Both types of plantations (wind breaks and shelter belts) have been grown in the region of Uttar Pradesh where desert is encroaching. The plantation is usually being done in two or three belts. Small sized plants planted on the wind-ward side in the region are *Saccharum munja*, *Calligonum polygonoides*, *Laptadenia spartium*, *Cenchrus ciliaris*, *Balanites roxburghii*, *Kochia indica*, *Panicum antidotale*, etc. On the lee ward are being planted the following – *Acacia leucophloea*, *Acacia senegal*, *Ricis communis*, *Tecoma undulata*, *Prosopis juliflora*, *Parkinshonia.* The roots of the plants (which are used in windbreaks or shelter belts) bind the soil and protect it from erosion. The shoot conopy acts as a physical barrier for wind and dust and greatly reduces the wind speed.

REVISION QUESTIONS

1. Write an essay on ecology and human welfare.
2. What are the natural resources ? Give an account of different natural resources and the means of their conservation.
3. Write short notes on the following :
   (i) Aquaculture ;
   (ii) Wildlife management ;
   (iii) Soil conservation ;
   (iv) Silviculture ;
   (v) Waste-management parks;
   (vi) Vermitechnology ;
   (vii) Social forestry ;
   (viii) Land use management.

CHAPTER

16

Wild Life Management

Management of wild life assures the maximum possible populations of game animals consistent.

Wild life management which is an important branch of conservation, is concerned with assuring the maximum possible populations of game animals consistent with other land uses in the same area and with the number that the given habitat will support. This is attained by manipulating the balance of nature in such a way that the desired game species are favoured. Many sportsmen have the idea that the way to increase game animals is to kill off all of their predators, Wild life managers, of course, are well aware that predator elimination may well be the very thing that should not be done. Successful game management depends upon a number of factors including an understanding of ecological principles, an appreciation of the conflicts between game species and agricultural uses of land, and educating both the landowner and the hunter concerning the objectives, techniques and limitations of game management.

WILD LIFE OF INDIA

Indian subcontinent is unique in having great natural beauty in its different biomes and also in possessing a rich and diverse wild life fauna. It includes about 123 families of terrestrial vertebrates. According to an estimate, there are 400 species of mammals, 1200 species of birds, 350 species of reptiles and more than 29,70,000 species of insects (**Khajuria**, 1957). The animals such as black buck, Nilgiri tahr, pigmy hog, golden langur, lion-tailed macaque, etc., are unique to India. The other typical wild

Blue bull (nilgai).

animals of India are elephant; rhinoceros; deer such as musk deer (kastura), barking deer (kotra), spotted deer (cheetal), hog deer (hoghiran), mouse deer (Indian chevrotain), swamp deer (Bara singha), dancing deer (sambhar), Thamin, and kashmir stag; antelopes such as black buck, four-horned antelope (chausingha), Indian gazelle (chinkara), and blue bull (nilgai); bison ; wild buffalo; Himalayan ibex or wild goat (*Capra siberica*); wild boar; wild ass; Nilgiri tahr; carnivores such as big cats such as lion, tiger and leopard, striped hyaena, jackal (*Canis aureus*) wolf (*canis lupus*), desert cat (*Felis lybica*), desert fox (*Vulpes v.pusilla*); bears such as black Himalayan bear and sloth bear, monkeys and apes such as Nilgiri langur (*Ceropithecus johni*), lion tailed monkey (*Macaca silensus*), rhesus monkey (*Macaca mulatta*), and Hanuman monkey (*Semnopithecus entellus*); Indian giant squirrel (*Ratufa indica*); desert hare (*Lepus nigricollis*) ; porcupine ; pangolin; birds such as peafowl (*Pavo cristatus*), jungle fowl, partridge, quail, great Indian bustard (*Choriotes nigriceps*), duck, pigeon, sandgrouse, storks and egrets, pelican, eagle, crane, owl, hornbill, etc., and reptiles such as crocodiles, gharials, lizards (*Uromastix*) and enormous varieties (about 216 species) of snakes. Wild life management requires autecological informations about game animals and some basic facts of game species are the following:

1. Deer

The deer belong to the family 'Cervidae' of the mammalian order 'Artiodactyla' and they are herbivores and form significant component of grazing food-chains of the Indian forests. They control the overgrowth of forest vegetation and themselves become food of carnivores. Due to limited shelter and food supply, deforestation, urbanisation, mass hunting, etc., there has been a great reduction in the number of deer population in India. Sangai of Manipur or brow-antlered deer (*Cervus eldi eldi*) is at the verge of extinction. Mouse deer, *Tragulus meminna* stands just 30 cm off the forest floor. It has no antlers but male has downward pointed canines that look like tusks. Its small size helps them to evade predators by hiding beneath low bushes. Timid and gentle, these dwarfs lend enchantment to Indian forests and are subject of Indian folk tales.

The musk deer (*Moschus moschiferus*) is found at 2500 to 4000 metres altitudes in Himalayas, Himachal Pradesh, Northern Uttar Pradesh, Nepal and Sikkim. It is about 50 cm high, has long ears, short tail and gall-bladder, but no antlers and no pit-glands (suborbital glands) below the eyes. Male or buck has downwardly projecting canines of the upper jaw. It is a shy animal and lives a solitary life. This deer has been almost hunted out mainly for musk which is a valuable product used as perfume fixative and as important ingredient of several Ayurvedic and Homeopathic drugs and is obtained from the musk pouch existing below the skin of abdomen of buck. For the conservation of musk deer a National Park of 1,000 sq. km has been established at Manali in Kulu Valley by Himachal Pradesh Government.

The barking deer or rib-faced deer (*Cervulus muntijac*) is found at the altitudes of 2000 to 3000 metres and has four subspecies in India. It is 61 cm in height and has suborbital glands below eyes. Buck has antlers and non-protruding canines. It lives in pairs or small families in thickets and being diurnal, grazes in the morning and evening hours. The dancing deer or sambhar (*Cervus unicolour*) is large-sized (150–160 cm high), browsing and gregarious deer, occurring in the herds of 4 to 12 animals and is found in forests all over the country. It has very long (up to 100 cm) antlers. The spotted deer or cheetal (*Axis axis*) is 76 to 91 cm high, grazing and gregarious deer, living in the herds of 20 to 30. It is spotted with white on rich brown coat. The buck has long slender antlers, each of which has three branches or tines and is shed annually. It is mostly found in the forests of Indian plains and in hill upto 1333 metres but are not found in Assam, Punjab and Rajasthan. It prefers open type forests with good grazing sites and running streams. The hog deer (*Axis porcinus*) is smaller and stouter in built and has hog-like appearance. It lives either solitary or in pair and prefers to live in grass patches bordering on forests. It is diurnal and comes out to feed only during morning and evening.

2. Antelopes and Other Herbivores

(1) Antelopes. In India there are four species of antelopes : the Indian gazelle, nilgai, black buck and chowsingha.

The Indian gazelle or chinkara (*Gazella dorcas bennetii*) is mostly found throughout desert region of southern part of Uttar Pradesh and Rajasthan. It lives in herds of 3 to 5 animals and has adaptations to live in extremely arid conditions, kilometres away from drinking water sources (**Prakash**, 1977). It prefers to live in open grasslands with thick bush cover. They feed on a variety of grasses and browse on shrubs.

The nilgai or blue bull (*Boselaphus tragocamelus*) occurs in plains and Rajasthan deserts but not in Bengal and Assam, living in small herds of 4 to 10. It prefers to stay near cultivated lands but is also found in desert grassland. It is 131–137 cm high, females being smaller. Young bulls and cows are tawny (= light yellowish-brown), but old males have iron gray colour. Both sexes have a short dark mane on the neck. Male has small cone-shaped curved horns. In Thar desert it is regarded sacred like 'cow', so is spared from hunting. However, it is killed under crop-protection act, because herds of nilgai inflict severe damage to crops.

The black buck has its four subspecies in India : *Antelope cervicapra rajputanae, Antelope cervicapra cervicapra, Antelope cervicapra rupicapra* and *Antelope cervicapra centralis.* Black buck is a graceful and fast-running animal which occurs in grassy plains of India and avoids forests. It inhabits regions having scrub vegetation in the vicinity of tanks and lakes where they pass the day hiding under bushes and become active during night. The male antelope is brown or black in colour while doe and fawns are yellowish brown in colour. Male has spirally ringed, unbranched, long (46 cm) and permanent horns. There occur large preorbital glands below the eyes of both sexes. Black buck is a polygamous and gregarious animal and its herd often include single male, several (50 or more) females and a few fawns. Almost all subspecies of black bucks have been hunted out from most part of India and now they occur in appreciable numbers only in certain pockets of Rajasthan, especially around the villages of 'Visnoi caste' of Hindus, who protect them from hunting.

The four-horned antelope or chausingha (*Tetracerus quadricornis*) has two pairs of horns, the front pair being shorter. It occurs in the foot hills of the Himalayas and in Bundel Khand region. It mostly leads a solitary life.

(2) Wild-buffalo. The wild buffalo (*Bubalus bubalis*) is a large-sized, robust animal having streamlined body. It prefers tall grassy forests close to marshy areas in the vicinity of rivers and lakes. It was once distributed over the grass jungles and river in forests of the Gangetic plain, the terai, Assam and eastern peninsular India down to the river Godavari. Great hunting of wild buffalo has restricted them now only to Assam (Manas and Kaziranga sanctuaries), the Nepal Terai and Bastar district in Madhya Pradesh. According to **Sheshadari** (1969), there are only 1425 wild buffaloes in India.

(3) Bison. Mithun, bull gaur or bison (*Bos gaurus*) is a relative of ox and is found in Andhra pradesh, Orissa, Madhya Pradesh and Bihar. Due to certain epidemic diseases such as rinderpest and enteric viral disease numerous individuals were wiped out some years ago and now a few bisons are mainly confined to Bandipur in Karnataka.

(4) Wild goat and sheep. In India, wild goat and sheep are restricted to north. Ibex is found in the Himalayas. The Nilgiri tahr (*Hemitragus hylocrius*) is the only one to be in south, in Nilgiris, Tamil Nadu. It moves about in herds and grazes on grasses. This species is greatly endangered due to habitat disturbance, predation by black panther and wild dogs (dhole) and hunting by poachers and only 450 tahr are present today (**Davidar**, 1975).

(5) Elephant. The Indian elephant, *Elephas maximus* is a largest terrestrial mammal and is confined to the Terai and the foot-hills because of its dependence on succulent grass, bamboo and plenty of water. It is gregarious and its herds move constantly in the search of new feeding grounds. Its population is hunted out mainly for tusk and now Periyar sanctuary (Madras) has some elephants.

(6) Rhinoceros. The Indian rhinoceros (*Rhinoceros unicornis*) is found in the grasslands and jungle area of the foot hills of the Himalayas (central Nepal) and certain pockets of plains of West Bengal and Assam. It prefers swamps and open savannah, covered with the tall elephant grass. A full grown rhinoceros is 1.8 metres in height and 3–9 metres in length. It has a single horn on the nose. The horn is made of keratinized skin and matted hair, it is said to have aphrodisiac and other medicinal values, so that large sums are paid for it, this results in indiscriminate shooting and its near extinction in India. It is a slow, solitary animal and is strictly territorial.

(7) Wild ass. The wild ass (*Equs hemionus khur*) had a fairly wide distribution in the dry regions of North-West India (Jaisalmer and Bikaner), but their distribution now become restricted to the Little Rann in the southern part of the Thar desert, where **Gee** (1963) estimated their number to be 870. In past, the population of wild ass severely reduced due to hunting, catching for breeding mules, surra disease and south African Horse sickness disease. However, people around the Little Rann are orthodox and vegetarian and do not molest the wild ass inspite of the fact they inflict serious damages to crops. Because it is the only true 'wild ass' of the world, so it needs protection and conservation (**Prakash,** 1977).

Wild boar.

(8) Wild boar. The wild boar, *Sus scrofa cristatus,* has a long mobile snout with terminal nostrils. The canines of both jaws grow continuously and form long triangular and up curved tusks for defence and digging roots, they are better developed in males. The wild boar or pig is omnivorous, it feeds on carrion, snakes, insects, roots, tubers and cultivated crops. It is gregarious living in families in grass or bush of marshy places and is now rare and is mostly found in small numbers along the Aravalli ranges in the eastern most part of Thar desert where they inhabit rocky slopes.

3. Big Cats and Other Carnivores

India has a variety of world-known large-sized cats and other carnivores which are as follows:

The Asiatic lion, *Panthera leo persica* was found in the Thar desert, in the arid plains of Sind, and Rajasthan and Punjab. It is on record that the last lions were shot at Anadra and at Jaswantpura of Rajasthan desert in about 1876. In India lions are now preserved only in the Gir sanctuary, in Gujarat, on the South-West of the Thar desert. Its extermination is well known.

The cheetah, *Acinonyx jubatus venaticus* was found in Central India, Deccan region, and in Thar desert upto Jaipur in Rajasthan. **Sheshadri** (1969) stated that it disappeared from Rajasthan about the turn of this century, from Central India in 1920's and in Deccan it was last reported in 1952. Now the cheetah, one of the most beautiful, agile and fastest animal, is regarded to be extinct from India.

The tiger *Panthera tigris tigris* is a gracious carnivore which is distributed in Uttar Pradesh from the Himalayas to the Vindhya forests in the South. In Himalayas it is found up to an elevation of about 2,000 metres. It lives in a variety of habitats, from thorn forests to the dense terai forests. According to an estimate, in 1948, the tiger population was 20,000 to 25,000, but in 1958, about 4000 and in 1970 less than 3000 (merely 1540 ; **Daniel**, 1970). To save the Tiger from extinction in India, "**Project tiger**" has been launched in 1972 ; this project planned to create Tiger reserves in selected areas of India. Due to this effort, considerable improvement was observed in tiger population in 1973 (Table 16-1).

Table 16-1. Increase in Tiger population due to 'Project Tiger' (News Letter, 1977, Vol. I, No.1).

Tiger Reserve	Population in 1972	Population in 1973
1. Manas	31	41
2. Palamaus	22	30
3. Simplipal	17	50
4. Corbett National Park	44	55
5. Ranthambore	14	20
6. Kanha National Park	43	48
7. Melghat	27	32
8. Bandipur	10	19

The leopard, *Panthera pardus* is the smaller spotted cousin of tiger which can live in all types of forests. Once it was found all over India but now its number has been drastically reduced. Leopard is given full protection in the Mount Abu sanctuary, however, can be hunted under licence from the Indian Wild Life Act. The snow leopard (*Panthera unica*) is found in the high Himalayas from Kashmir to Sikkim near snow line. It has a creamy grey coat with large black rings. It is greatly hunted for its fur and now is endangered species. Other important cats are clouded leopard (*Neofelis nebulosa*), leopard cat (*Felis bengalensis*), pallas's cat (*Felis manul*), caracal (*Felis caracal*), etc. Caracal once was common in Rajasthan deserts, now it has become extremely rare and is suspected to occur only in the Sirohi-Jalore and Bikaner regions (**Prakash**, 1977).

The sloth bear, *Melursuu ursinus*, is a stout carnivore which is found in the hilly forests like Mount Abu and Erinpura ranges. It lives mostly on fruits, honey and insects, it sucks up white ants from their hills which are broken by its claws with great speed. It is rapidly vanishing due to ruthless hunting and it needs conservation. The black Himalayan bear, has V-shaped white mark on the chest and is carnivorous.

Dolphin. In the Indus river, in the North-West and in the rivers Chenab and Sutlaj in the north of the Thar desert, the freshwater gangetic dolphin, *Platanista gangetica* (Cetacean mammal) was once found in good numbers (**Murray**, 1884), but is now totally extinct (**Prakash**, 1977).

4. Birds

The important wild bird-fauna of India include the following species : pea fowl (*Pavo cristatus*), grey jungle fowl (*Gallus sonnerati*), red spur fowl (*Galloperdix spadicea*), ducks (*Anser indicus Sarkidiornis melanotos, Anaspoecilorhyncha*), pigeons (*Columba livia*), sand grouses (*Pterocles exustus, P. orientalis, Streptopelia decaocta*), storks (*Leptotilos dubuis*), egrets (*Egretta garetta*), grey partridge (*Francolinus pondieirianus*), black partridge (*Francolinus francolinus*), golden eagle (*Aquilla chrysaetos*), vultures (*Gyps bengalensis, Neophron percnopterus*), spotted billed pelican (*Pelacanus philippensis*), sarus crane (*Grus antigone*), quails (*Coturnix coturnix, C. coromandelica*), and great Indian bustard (*Choriotes nigriceps*). Certain birds such as osprey, great Indian bustard, cranes, mountain quail, horned owl, red start avocet, pinkheaded duck, etc., are on the verge of extinction, if not saved. Great Indian Hornbill with its peculiar nesting behaviour is the pride of India and needs to be saved.

The great Indian bustard is a large Indian game bird having a heavy body, long neck and long bare running legs. It is essentially an inhabitant of wide, open dry scrubby plains and waste, broken undulating lands of the Rajasthan, West Punjab and Gujarat. It is ruthlessly hunted out for its delicious flesh and is almost at the verge of extinction. In 1980, an International Conference was held at Jaipur to assess its present position and to suggest conservation steps to save it from extinction.

Indian bustard is almost at the verge of extinction.

5. Crocodiles and Other Reptiles

The following three Indian aquatic reptiles have been found endangered : the estuarine crocodile (*Crocodilus porosus*) ; subnosed crocodile, freshwater or marsh crocodile or muggar (*Crocodilus palustris*) and gharial (*Gavialis gangeticus*). The estuarine crocodile is found in Bhitarkanika Island and adjacent mangrove areas of Orissa and in the Sunderbans of West Bengal. The marsh crocodile is widely distributed in India and has been greatly reduced in number in the past due to indiscriminate shooting for its valuable hide. *Gavialis* was once common in most large Indian rivers such as Indus, Ganga, Brahmputra and Mahanadi, but now has very sparse distribution in Ganga and larger feeders in the Nepal-Terai, Bihar and Bengal. According to **Bustard** (1974), Indian gharials are on the verge of extinction and there are only 143 living gharials (**Subba Rao**, 1977). Mugger is also fast depleting and may be endangered, if not saved. In 1974, under the guidance of **Dr. H.R. Bustard** the project called 'Save the Crocodiles' was started for the conservation and management of crocodiles, through the incubation of eggs, rearing and release in preserves. Projects for preservation of muggar, estuarine crocodile and gharial are now being undertaken in Andhra Pradesh, Bihar, Gujarat, Kerala, Orissa, Rajasthan, Tamil Nadu and Uttar Pradesh.

Gharial.

Sea turtles migrate to "**Gahirmatha**" in Orissa to carry on nesting popularly known as"**Arribada**". The deep sea fishing is a continuous threat to this reptilian species. When turtles float on surface during copulation, certain tribals which feed on them trap them and sell in the market. Their nesting grounds are also disturbed by man who collect turtle's eggs for consumption.

A rare limbless lizard is the *Barkudia insularis* which is mistook as the snake and killed indiscriminately. It is subterranean in habit and found at Barkudia inland of Chilka lake and needs protection.

In India, snakes are killed by some and worshipped by others. Generally they are feared. They are believed to be farmer's best friends as a single rat snake can eat hundreds of rats and mice per year. Indiscriminate killing of snakes for skin results in a loss of ten million rat snakes, cobras and pythons every year.

6. Frog

The "State of India's Environment" 1984–85 reports that every year Indians kill an estimated 100 million frogs to satisfy the palates of well fed westerners. In 1981, 4368 tonnes of frogs legs were exported from India; these have been finding their way to Canada, Saudi-Arabia, Japan and UAE. Two common species of frogs *viz., Rana tigrina* and *Rana hexadactyla* have been affected by this trade and deserve protection.

CONCEPT OF THREATENED SPECIES

The International Union of Conservation of Nature and Natural Resources (IUCN) has categorized the rare species of plants and animals for conservation purpose. This classification is based on (i) the present and past distribution of the species or taxa; (ii) abundance and quality of natural habitat; (iii) the decline in the density of population in course of time; and (iv) the biology and ecological value of species. According to this system the following categories have been identified:

1. Endangered (E) species. These are species or taxa, which are in danger of extinction and which may not survive if the adverse factors continue to operate. The species whose numbers have been reduced to a critical level or their habitats have been drastically reduced in such a way that they are in the instant danger of extinction.

2. Vulnerable (V) species. The species or taxa likely to move into the endangered category in the near future if the causal adverse factors continue to operate. These are species whose populations have been seriously decreased and whose ultimate security is not assured and also those whose populations are still abundant but are under threat throughout their range.

3. Rare (R) species. These are species with small populations in the world. At present, these species are not endangered and vulnerable but are at risk. These species are usually localized within restricted geographical areas or habitats or are thinly scattered over a more extensive range.

4. Threatened (T) species. The term 'threatened' is used in the context of conservation of the species which are in any one of the above three categories, *viz.,* endangered, vulnerable or rare.

The most threatened reptiles, birds and mammals of India have been listed in Table 16-2.

Table 16-2. **Table 16-2. Threatened reptiles, birds and mammals of India.**

Amphibians and Reptiles	Birds	Mammals
1. *Tylotriton varrncosus* (Himalayan Newt or Salamander) 2. *Varanus salvator* 3. *Crocodilus palustris* (Marsh crocodile) 4. *Crocodilus porosus* (Estuarine crocodile) 5. *Gavialis gangeticus* (Gharial)	1. *Choriotis nigriceps* (The great Indian bustard) 2. *Cairina cutalata* (The white winged wood duck)	1. *Presbytis pileatus* (Capped langur) 2. *Macaca silensus* (Lion tailed macaque) 3. *Hylobates hoolock* (White browed gibbon) 4. *Panthera tigris tigris* (Tiger) 5. *Panthera leo persica* (Lion) 6. *Panthera pardus* (Leopard) 7. *Panthera uncia* (Snow leopard) 8. Clouded leopard 9. *Felis marmorata* 10. *Felis temmincki* (Golden cat) 11. *Felis bengalensis* (Leopard cat) 12. *F. silvestris ornata* (Desert cat) 13. *F. viverrina* (Fishing cat) 14. *F. caracal* (Caracal) 15. *F. Lynx* (Lynx) 16. *Arctictis binturong* (Binturong) 17. *Prionodon pardicolar* (Spotted linsang) 18. *Canis lupus pallipes* (Indian wolf) 19. *Melursus ursinus* (Sloth bear)

REASONS FOR DEPLETION OF WILD LIFE

Many wild animals became extinct due to various human and natural activities : 1. Absence of cover or shelter to wild animals. 2. Due to deforestation for cultivation, road-building, railway routes, dam construction or for urbanisation, occurs reduction in the area for free movement of wild animals which retard reproductive capacity of certain wild animals such as deer, bison, rhino, tiger, etc. 3. Destruction of wild plants of forests for timber, charcoal and fire wood often deprive wild animals their most palatable food and affects their survival. 4. Noise pollution by different transporting media (trucks, buses, rails, aeroplanes, etc.) and polluting river water have adversely affected wild animals. 5. Various natural calamities such as floods, droughts, fires, epidemics, etc., have also caused great destruction of wild life. 6. Hunting methods of all kinds and for any purpose (*i.e.,* for food, recreation, hide, fur, plumage, musk, tusk, horn, etc.) have caused destruction of wild life.

NECESSITY FOR WILD LIFE CONSERVATION

The conservation of wild life is required for the following benefits; 1. The wild life helps us in maintaining the 'balance of nature'. Once this equilibrium is disturbed it leads to many problems. The destruction of carnivores or insectivores often leads to the increase in herbivores which in turn affects the forest vegetation or crops. 2. The wild life can be used commercially to earn more and more money. It can increase our earning of foreign exchange, if linked with tourism. 3. The preservation of wild life helps many naturalists and behaviour biologists to study morphology, anatomy, physiology, ecology, behaviour biology of the wild animals under their natural surroundings. 4. The wild life provides best means of sports and recreation. 5. The wild life of India is our cultural asset and has deep-rooted effect on Indian art, sculpture, literature and religion.

MODES OF WILD LIFE CONSERVATION

In India, the following measures have been under taken for the wild life management :

1. Protection by Law

India was probably the first country to enact a Wild Life Protection Act. The Wild Birds and Animals Protection Act was passed in 1887 and repealed in 1912. For game protection in the states, in 1927, the Forest Act XVI was enacted. Indian Board for Wild Life was established in 1952 and this was followed by setting up of Wild Life Boards in different states. In 1972, new Wild Life Protection Act was passed. Under this Act, possession, trapping, shooting of wild animals alive or dead; serving their meat in eating houses; their transport and export are all controlled and watched by special staff (Chief Wild Life Warden and authorized officers). This Act prohibits hunting of females and young ones. Under this Act threatened species are absolutely protected and the rest afforded graded protection according to their state of population size.

Pangolin.

2. Protected Species of Indian Wild Life

The following wild animals have been enlisted as threatened and protected species ; White eyed buck, Indian antelope or black buck, Four-horned antelope, Bharal, Swamp deer, Eastern pangolin, Elephant, Indian gazelle or Chinkara, Musk deer, Serow, Tahr, Golden cat, Snow leopard, Great Indian bustard, Pink headed duck, Peafowl, Monal pheasant, Koklas pheasant, Chir pheasant, Horned pheasant, Monitor lizard, Water lizard, Gharial, Marsh crocodile, Python, and Leathery turtle. Albino and melanic individuals of all species of wild animals are also protected.

Water lizard.

3. Establishment of Sanctuaries and National Parks

To protect, preserve and propagate rich natural bounty, the Government of India passed the Wild life (protection) Act, in 1972 under which national parks and sanctuaries could be created. Creation of biosphere reserves has also been put into practice since 1986. In India, nearly 434 national parks and sanctuaries have been established, out of which 17 have been selected as "Project Tiger" areas (on March 1, 1989, there were 445 national parks and sanctuaries, see Rana, 1994).

National parks are set up for preserving flora, fauna, landscapes and historic objects of an area. A national park is an area which is strictly reserved for the betterment of the wild life and where activities such as forestry, grazing or cultivation are not permitted. No private ownership right is allowed. There are 66 national parks in the country which are spread over an area of 33,988.14 square kilometres or nearly 1 per cent of the country's geographical area. **Sanctuaries** are places where the killing or capturing of any animal is prohibited except under orders of the authorities concerned. They provide protection and optimum living conditions to wild animals. Thus, in a sanctuary, protection is given only to the fauna and operations such as harvesting of timber, collection of minor forest products and private ownership rights are permitted so long as they do not interfere with the well-being of animals. There are 368 sanctuaries in India covering over 1,07,310,13 square kilometres, amounting to 3.2 per cent of India's total geographical area.

During the past three decades the concept of **Biosphere Reserves** has been evolved by the **Man and Biosphere (MAB) programme** of the UNESCO. India has identified 13 areas to be declared as biosphere reserves. Of these the Nilgiri Biosphere Reserve, including parts of Karnataka, Kerala and Tamil Nadu was declared in 1986 and the Nanda Devi Biosphere Reserve in 1988. Two more biosphere reserves, one at Uttarakhand (including the Valley of Flowers in North-Western Himalaya) and another at Nokrek (North-Himalaya) have also been recently included in the list of biosphere reserves. In a biosphere reserve, multiple land use is permitted by designating various zones. These zones are : the **core zone** (where no human activity is permitted), the **buffer zone** (where limited human activity is allowed) and the **manipulated zone** (where a large number of human activities would go on). In a biosphere reserve, wild populations as well as traditional life styles of tribals and varied domesticated plant and animals genetic resources are protected. Some of the well known wild life sanctuaries and national parks of India are as follows :

1. Kaziranga wild life sanctuary. It was established in 1926 in district Sibsagar, subdivision Jorhat (Assam) on the south bank of the Brahmputra river. It consists of 430 sq km of forest grasslands and swamps and supports a fauna of 700 rhinoceros, besides elephant, wild buffalo, bison, tiger, leopard, sloth bear, sambhar, swamp deer, hog deer, barking deer, wild boar, gibbon and birds such as pelican, stork and ring-tailed fishing eagles.

2. Manas wild life sanctuary and Tiger reserve. It is located in district Kamrup in Assam, has an area of 540 sq km and is situated at an altitude of 80 metres. River Manas passes through it. It contains the following wild animals: tiger, panther, wild dog, wild boar, rhinoceros, gaur, wild buffalo, sambhar, swamp deer and golden langur.

3. Jaldapara wild life sanctuary. It is situated in Jalpaiguri district of West Bengal and is a 65 sq km stretch of grassland. Its wild life fauna includes animals such as rhinoceros, gaur, elephant, tiger, leopard, deer, and a variety of birds and reptiles.

4. Palamau national park. It is situated in Daltongunj district in Bihar and has an area of 345 sq km. Its flora is thick tropical forests. The fauna of this park includes tiger, panther, sloth bear, elephant, chital, gaur, nilgai, chinkara, chowsingha and mouse-deer.

5. Hazaribagh national park. This national park was established in 1954 in Bihar. It has an area of 184 sq km of thick tropical forests. The typical fauna of this park includes wild boar, sambhar, nilgai, tiger, leopard, sloth bear, hyena and gaur.

6. Simlipal national park. It is situated in district Mayuri Bhanj in Orissa and has an area of 2750 sq. km. It is covered over by dense sal forests and is chosen for project tiger. Its typical fauna includes tiger, elephant, deer, pea fowl, talking mynas, chital, sambhar, panther, gaur, hyena and sloth bear.

7. Chilka lake. This largest inland lake has an area of 1000 sq km and is about 100 km from Bhubaneshwar (Orissa). Its typical fauna includes water fowls, ducks, cranes, ospreys, golden plovers, sandpipers, stone curlews, flamingoes, etc.

8. Kolameru bird sanctuary. This is a small bird sanctuary near Tadepallegudam in Andhra State. It is a breeding place for pelicans and many marine birds visit this place.

9. Vedanthangal bird sanctuary. This small-sized but very old sanctuary extend over a lake of about 0.30 sq km, 85 km south of Madras. Many migratory birds regularly visit this temporary lake. The regular visitors of this lake are birds such as spoon bill, open billed stork, egrets, ibis, cormorant, darter, grey heron, pelican, snipes and dab chick.

10. Guindy deer park. It is situated near Madras city and has mainly chitals and black bucks. A few albinos of black buck are also found here.

11. Point calimera wild life sanctuary. This is situated at the southern tip of the Thanjavur district of Tamil Nadu abutting the Palk strait. Its back water and lagoon is visited by flamingos and pelicans. Its nearby vedaranyam forests has a fauna of numerous black bucks, chitals and wild boars.

12. Mundanthurai sanctuary. It was established in 1962 in Tirunelveli district of Tamil Nadu and has an area of 520 sq km. It has evergreen forests and Tamaraparani river flows through this sanctuary. Its typical fauna includes panther, tiger, sambhar and chital.

13. Periyar wild life sanctuary. This is situated in Kerala State and has an area of 777 sq km. It was established in 1940 around the artificial lake which arose behind the dam built across the Periyar river is 1900. This sanctuary of great scenic beauty supports a fauna of wild elephants, gaurs, leopards, sloth bear, sambhars, barking deer, wild dogs, wild boars, black Nilgiri langur and water birds such as grey hornbills, egrets.

Periyar wild life sanctuary.

14. Mudumalai wild life sanctuary. This wild life sanctuary was established in 1940 in North Western part of Nilgiris in Tamil Nadu. It is known for its rich forests and diversity of fauna that includes wild elephant, gaur, sambhar, chital, barking deer, mouse deer, four-horned antelope, tiger, panther, bonnet monkey, langur, giant squirrel, flying squirrel, wild dog, jackal, wild cat, sloth bear, porcupine, pangolin, flying-lizard, monitor-lizard, rat snake, python and various birds.

15. Ranganthittoo bird sanctuary. It covers an area of 166 sq km and includes a series of islands in the Cauvery river 15 km off the Banglore-Mysore Road, near Shrirangapattnam. Its avial fauna includes open bill stork, white ibis, egret, spoon bill, wild duck, peafowl, night heron birds.

16. Bandipur wild life sanctuary. This sanctuary was established in 1941 by the then ruler of Mysore (Karnataka State). It is situated 80 km south of Mysore city enroute to Octacamund. It has an area of 874 sq km and is at an altitude of 1454.4 metres. Its forest is very thick and has plenty of rain fall. Its wild life fauna includes plenty of gaurs and animals such as elephant, leopard, sloth bear, wild dog, chital, panther, barking deer, porcupine and langur.

17. Cotigao wild life sanctuary. It is located in South Goa and has an area of 105 sq km. It has wet evergreen forest of Elve and Berla trees and supports rich fauna of the following animals : gaur, sambhar, chital, hog-deer, barking deer, wild boar, hyaena, panther, leopard, jackal, otter, porcupine and birds such as parakeet, lorikeet, woodpecker, kingfisher, bulbul, jungle fowl, egret, etc.

18. Bhagwan Mahadev wild life sanctuary. It is located in North Goa and has an area of 240 sq km and supports fauna almost similar to Cotigao wild life sanctuary.

19. Sesar Gir. This famous wild life sanctuary for the Asiatic lion is situated in Gujarat state, 468 km from Ahmedabad and 43 km from Veraval. It has an area of 1295 sq km of semiarid country with patches of thorn scrub and deciduous trees. Its fauna includes Asiatic lion, spotted deer, blue bull (nilgai), four-horned antelope, chinkara, striped hyena, wild boar, porcupine, langur, python, crocodiles and birds such as green pigeon, partridge, rock-grouse, etc.

20. Kanha national park. This national park was established in 1955 in former Banjar Valley Reserve (Madhya Pradesh). This park has an area of 939.94 sq km and includes hilly terrains and streams. It is 175 km away from Jabalpur and has forests of sal trees. Its typical fauna includes animals such as tiger, chital, panther, sambhar, black buck and bara singha.

21. Tandoba national park. It is located in Chandrapur district (Maharastra) and has an area of 116 sq km. Its typical fauna includes tiger, sambhar, sloth bear, bison, chital, chinkara, barking deer, blue bull, four horned deer, langur, peafowl and few crocodiles.

22. Sariska. This is one of the most beautiful wild life sanctuary of Rajasthan near Alwar. It has an area of 800 sq km and has dense Dohokra and Solar forests. Its typical fauna includes tiger, leopard, spotted deer, jungle cat, four-horned antelope, langur, porcupine, hedgehog, peafowl, etc.

23. Bharatpur bird sanctuary (Koeldeo Ghana). It is located at Bharatpur (Rajasthan), has an area of 29 sq km and harbour all kinds of indigenous nesting water birds, water side birds and migratory birds. More than 328 varieties of birds including cormorants, spoonbills, white ibis, Indian darters, egrets, painted storks, open billed storks, great black necked storks, etc., Many migratory birds such as ducks, geese, siberian cranes, etc., regularly visit this sanctuary. Drier parts of this marshy sanctuary have animals like spotted deer, black buck, sambhar, blue bull, wild boar and python.

Bharatpur bird sanctury.

24. Sultanpur lake bird sanctuary. This small sized (2 sq km area) bird sanctuary is located in Gurgaon district (Haryana) about 30 km from Delhi. Its typical avian fauna includes crane, sarus, spot-bill, rudyshel and drake.

25. Shikari Devi wild life sanctuary. It is located in Mandi district in Himachal Pradesh. It has an area of 213 sq km and has the following animals : snow leopard, flying fox, black bear, barking deer, musk deer, chakor, partridge, etc.

26. Bir Motibagh sanctuary. It is located near Patiala in Punjab. The typical fauna of this wild life sanctuary includes black buck, blue bull, hog deer, hare, jackal and birds such as peafowl, partridge, sparrow, babbler, myna, parakeet, pigeon, dove, etc.

27. Dachigam wild life sanctuary. It was established in 1951 in Kashmir 26 km away from Srinagar. It has an area of 89 sq km and has two levels : upper Dachigam at 3692.3 metres altitudes and Lower Dachigam at 1846.2 metres altitudes. It mainly preserves hangul or Kashmir stag, but also has animals such as musk deer, leopard, black bear, brown bear and baboon.

28. Corbett national park. It is one of India's most famous wild life sanctuary and was constituted in 1935 as the first national park of India. It is situated between Nainital and Garhwal districts in Uttar pradesh. It has an area of 525 sq km and is located within west-to-south bend of river Ramganga. It supports a rich and diverse fauna of the following animals : tiger, panther, sloth bear, hyaena, elephant, blue bull, swamp deer, barking deer, Indian antelope, porcupine, birds such as bulbul, woodpecker, barbet, babbler and bee eater, and reptiles such as crocodile, python, etc.

29. Shivpuri sanctuary. It is located in Madhya Pradesh and is an asylum for tigers.

30. Anamalai sanctuary. This sanctuary was established in 1972 in the southern part of Coimbatore District in Tamil Nadu. It has a vast area of 958 sq km and supports rich fauna of animals such as elephant, gaur, sambhar, spotted deer, barking deer, Nilgiri tahr, lion-tailed macaque, tiger, panther, sloth bear, langur, porcupine and pangolin.

Other Conservation Measures

Other important steps need to preserve wild life include the following measures : (i) For the preservation of species wild life management staff should have a correct idea about the exact habitat which the species under consideration needs. (ii) Natural habitat of wild animals should be carefully protected. (iii) Habitats of wild life should be improved by constructing water holes, saltlicks and by raising plantation of better and nourishing fodder grasses and trees. (iv) Effective means of census operation should be adopted to measure population sizes of various wild animals. (v) The enforcement of Wild Life Protection Acts should be observed strictly. (vi) Shooting and hunting of endangered species should be totally banned. (vii) To take care of wild life in case of epidemics, veterinary efforts should be made. (viii) To understand the biology and behaviour of wild animals research on wild life should be encouraged. (ix) Public should be educated about the advantages and disadvantages of wild life.

REVISION QUESTIONS

1. Write an essay on the wild life of India.
2. Describe, why should wild life of India be protected ?
3. Describe various measures of wild life management.
4. Define the terms wild life sanctuary and national park and describe some important wild life sanctuaries and national parks of Indian Subcontinent.

CHAPTER

17

Biogeography

(Distribution of Animals and Plants)

There is geographical distribution of species of all organisms on this earth.

No organism occurs in uniform numbers throughout the world. Rather, specific organisms are restricted to spe cific communities or groups of communities. Three aspects of the distribution of an organism are generally recognized : **geographic range**, or the specific extent of land or water area where the organism normally occurs ; the **geologic range**, or the distribution in time, past and present; and the **ecological distribution**, or the major biotic communities (*i.e.,* marine biome, freshwater biome and terrestrial biome) of which the organism is a member. Certain biologists have also made distinction between **geographic distribution** (horizontal or superficial distribution) and **bathymetric distribution** (vertical or altitudinal distribution). Bathymetric distribution includes the following three realms : (i) **Halobiotic**, or vertical distribution of organisms in marine (sea) habitat, (ii) **Limnobiotic**, or vertical distribution of organisms in freshwater habitat; and (iii) **Geobiotic** or altitudinal distribution of organisms on land. All the living organisms in a given region are termed the **biota** of that region. The animals of a given region are collectively termed the **fauna**, and the plants of a given region, the **flora** (*i.e.,* fauna+flora=biota). Here, a distinction can be made between the following two terms– flora and vegetation. **Flora** mainly refers to the botanical composition of a place, *i.e.,* the names of different plant species, while **vegetation** means the totality of forms in which the emphasis is not on names of different plants but their life forms, number and coverage. The studies of the distribution of biota are collectively called **biogeography** ; of animals only, **zoogeography** ; and of plants only, **plant geography** or **phytogeography**. There are two major approaches to the study of biogeography or **geographical ecology** (**Kendeigh**, 1974) : (i) **descriptive** or **static biogeography** which deals with the description of biota of different botanical and zoological areas of earth ; and (ii) **interpretative** or **dynamic biogeography** which describes the forces which have brought about plant and animal distribution.

DESCRIPTIVE PHYTOGEOGRAPHY

With a descriptive approach of phytogeography, will be discussed the major plant communities (biomes) of the world, and different vegetational belts of the earth constituted by these biomes together with characteristic climatic conditions of the area.

1. Major Plant Communities (Biomes) of the World

As already discussed in sufficient details major plant communities of the world are classified chiefly on the basis of the kinds of habitat and environmental conditions into the following chief types:

Plant communities
- Aquatic
  - Fresh
  - Marine
  - Estuarine
- Terrestrial (Biomes)
  - Tundra
  - Forest
  - Grassland
  - Desert

2. Phytogeographical Regions of World (or Chief Vegetational Belts of the Earth)

On the basis of climatic and geographical conditions, the earth is usually divided into the following four broad vegetational belts :

A. Arctic zone. This zone is divided into the following two types :

(1) Arctic proper. This zone occurs around the north pole and remains covered with ice throughout the year. This zone in fact is Tundra biome and its chief components of vegetation include some algae, annual flowering plants, mosses and lichens.

(2) Subarctic. This is a less defined zone which extends from southern arctic to the northern limits of temperate zone. This region is very cold, contains abundant bogs and have vegetation including small height trees, shrubs and herbs in the months of June and July. Ground vegetation often includes some pteridophytes, orchids, insectivorous plants, mosses and lichens.

B. North temperate zone. This zone extends between 30^0 N latitude and 55^0 N latitude and includes following two major zones :

1. North temperate of the eastern zone. This zone is further subdivided into the following four zones : (i) **Western and central Europe.** This zone is demarcated in north by the subarctics and in south by Alps and British Islands. Its forests are dominated by several gymnospermous tall trees such as *Pinus, Picea* and *Abies*, and angiospermous trees such as oaks, maple and chestnuts. Ground vegetation comprises orchids, wild roses, buttercups, *Viola, Salvia, Dianthus,* etc. At high altitudes of this zone trees are replaced by grassy vegetation with some herbaceous flowering plants. (ii) **Mediterranean**. This zone extends between 30^0N and 40^0N latitudes, south of mountain ranges in Europe and in Asia around Mediterranean sea and is characterized by warm temperate type climate. Vegetation is chiefly composed of fruit trees, olives, nut trees, oranges, and also some foreign palms, cacti, acacias, etc. In the Asian region of Mediterranean as in Arab countries, rainfall is low, deserts are common and sparse vegetation include following species *Atriplex, Alhagi, Polygonum* and *Phoenix dactylifera*. (iii) **Northern Africa**. This zone includes the northern parts of Morocco, Algeria, Libya and Egypt. It is characterized by scanty rainfall and sparse vegetation. In cooler areas (*i.e.*, mountains) some conifers and broad-leaved oaks are common. In deserts (including some portion of Sahara desert) some herbs, shrubs, woody acacias and succulent xerophytes occur. (iv) **Himalayas, eastern Asia** and **Japan**. Tibet, China and Japan have different type of vegetation. In China and Japan conifers such as *Cryptomria, Sciadopitys, Cephalotaxus, Ginkgo biloba* and *Cycas* and angiosperms such as *Rhododendrons, Cinnamomum camphora* and *Begonia* are common. The vegetation of Himalayas will be described later in this chapter.

(2) North temperate of the western hemisphere. It includes the parts of United States and Canada lying mostly between north latitudes 30^0 to 55^0. The eastern coastal regions of these countries in the temperate belt have some very characteristic plant species such as tropical fern (*Schizaea pusilla*). The forest communities are composed of conifers and deciduous trees. On lower altitudes some wild cherries, plums, roses and orchids are abundant. In the New England region trees of *Ulmus americana* and *Castania dentata* are abundant. Forests of conifers are common in southern parts and on western parts of Rocky mountains of USA. In north California grows *Sequoia sempervirens,* the tallest tree of world. Ground vegetation is composed of *Salicornia herbacea, Rumex maritima, Monotropa uniflora, Saxifragea, Primula,* etc. In the Colorado desert of Arizona and south-eastern California, there are several types of xerophytic plants.

C. Tropical zone. This zone is divided into two subzones :

(A) Palaetropic. It comprises old world or eastern tropics and has the following two botanical regions : **(1) Tropical Africa**. This is a large-sized landmass of varied topography. It includes high altitudes and Sahara deserts which have little or no rainfall. In Africa, most remarkable plant is *Welwitschia mirabilis,* Eastern part of Central Africa has India-like vegetation of *Borassus flabelliformis, Tamarindus indica, Ficus, Aspargus, Clematis, Phaseolus, Cassia fistula, Erythrina, Acacia, Albizzia, Zizyphus, Bauhinia,* etc. **(2) Tropical Asia**. It includes Arabia, Pakistan, India, Burma, Ceylon, Thailand, Indonesia, Philippines and Islands of Indian sea. In Arabia, the rainfall is low and most of the desert species are found, *Coffea arabica* is a native plant of Arabia. Ceylon is rich in species diversity and ferns are the chief components of its sparse natural vegetation. Malaya, Java and Sumatra are characterized by heavy rainfall and their vegetation comprises varied types of palms, some types of ferns, tall trees, lianas and insectivorous plants. Some important plants of Java are *Albizzia, Pterocarpus, Tamarindus, Bombax, Cassia, Dendrocalamus,* etc. The common trees of Burma and Thailand are jack fruit, orange, mango, banana, betelnut, etc.

Cassia fistula

(B) Neotropics. It comprises Mexico and major part of South America. The low-rainfall areas of Mexico are rich in xerophytes. At higher cooler altitudes there is a forest of conifers such as *Pinus, Spruce, Quercus* and *Populus.* On mountain peaks grasses are most common. In wet areas of Mexico, there are mosses, palms, bamboos, orchids, etc. In South America, there are extensive forests of flood-resistant trees such as *Bertholletia excelsa, Maximiliana regia* and also mangrove vegetation. There are also found many epiphytes.

D. South temperate zone. This zone includes extreme southern region of Africa, Australia and New Zealand. In African area, the vegetation is chiefly made up of ferns and gymnosperms. On the hills conifers are present. Its lower wet regions contain *Salix* and *Phragmites* and dry regions have grasses such as *Andropogon* and trees such as *Acacia.* The vegetation of northern part of Australia is similar to that of south east Asia and includes trees of palms, nuts, *Eucalyptus, Acacia* and *Casuarina.* Some petridophytes are also found in the ground vegetation. In the South Australia *Araucarias* are common.

New Zealand forests are mostly made up of conifers together with ferns, palm such as *Rhopalostylis*, many species for *Metrosideros*. New Zealand has richest bryophytic flora.

casuarina.

3. Phytogeography of India

Indian subcontinent lying between 8^0 and 37^0 N, and 68^0 and 97^0 E has its own peculiar physiographic, climatic and biotic features. It is surrounded on its south, east and west by oceans and in the north by mountainous chains highest in the world. The subcontinent stretches out between tropical and subtropical belts. Its climate is chiefly modified by oceans and mountains. In the south and in the far east the climate is typically tropical, while it is temperate in the north, and highly arid in the northwest. The general climate of India is of monsoon type.

Vegetation of India. The most important factors used in the classification of Indian vegetation are rainfall, temperature, edaphic factors, biotic influences and life forms. There are two most common types of plant formation in India—forest and grassland.

(a) Forest vegetation. Champion and **Seth** (1968) have recognised the following 16 major types of forests in India.

I. MOIST TROPICAL FORESTS

(1) Tropical wet evergreen forests. These forests occur on the Western Ghats, Assam, Cachar, parts of Bengal, parts of Mysore and Andamans. These forests are climatic climax forests with very dense growth of tall trees (more than 45 metres high). The shrubs, lianas, climbers and epiphytes are abundant, however, grasses and herbs are rare on the forest floor. The rainfall exceeds 250 cm in these regions. The characteristic flora of these forests include *Dipterocarpus indica, Hopea, Artocarpus, Mangifera, Emblica, Michelia, Ervatamia, Lagerstroemia, Ixora,* etc.

(2) Tropical moist semi-evergreen forests. These forests occur along Western Ghats, in parts of Upper Assam and Orissa. The rainfall of these regions is usually high, above 200 cm per year. These forests consist of dense and tall (25 to 35 metres) trees along with shrubs. In these forests, because deciduous species such as *Terminalia, Tetrameles* and *Shorea* grow intermixed with the evergreen species such as *Artocarpus, Michelia* and *Eugenia,* so they are called semi-evergreen forests.

(3) Tropical moist deciduous forests. These forests are distributed in a narrow belt along the foot of the Himalayas, on the eastern side of Western Ghats, Chhota Nagpur and Khasi Hills. These forests have high rainfall (150–200 cm) throughout the year except few very short dry periods. The trees are deciduous remaining leafless for one or two months. The most common species is teak (*Tectona grandis*) in the south and sal (*Shorea robusta*) in the north. Other common species of these forests are *Dalbergia, Cedrela, Salmalia, Albizzia, Terminalia, Cordia, Bombusa, Melia, Dillenia, Eugenia, Dendrocalamus,* etc.

(4) Littoral and swamp forests. These forests comprise mostly evergreen species and occur along sea coasts in wet and marshy areas and also in deltas of larger rivers on eastern coast. In saline swamps, mangroves are chiefly composed of *Rhizophora, Bruguiera, Ceriops, Nipa,* etc. In less saline areas, *Phoenix, Ipomoea, Phragmites, Casuarina,* etc., are found.

II. DRY TROPICAL FORESTS

(5) Tropical dry deciduous forest. These forests are distributed almost throughout the country except Kashmir, beyond Bengal, Rajasthan and Western Ghats. These occur in areas with annual rainfall ranging from 75 cm to 125cm and with a dry season of about 6 months. The trees are small (10–15 metres) but canopy is open. Shrubs are abundant. These forests include following common plant species *Shorea robusta* (in north), *Tectona grandis* (in south), *Anogeissus, Terminalia, Semecarpus, Carissa, Emblica, Acacia, Zizyphus, Dalbergia, Dendrocalamus,* etc.

Euphorbia is a common plant of tropical thorn forests.

(6) Tropical dry deciduous forests of South-East Daccan Region. In some parts of Tamil Nadu and the Karnatic coasts evergreen forests of short (10–15 metres) but dense trees are found though the annual rainfall is relatively small. There are no bamboos but grasses are abundant. The common flora of this region includes species of *Memecylon, Maba, Pavetta, Feronia, Ixora, Sterculia,* etc.

(7) Tropical thorn forests. These forests are distributed in areas with very low rainfall as in Western Rajasthan, parts of Southern Punjab, Bundelkhand, Gujarat, Maharashtra, Madhya Pradesh and Tamil Nadu. The trees are short (8–10 meters), sparsely distributed, mostly thorny ; shrubs are more common than trees. Plants remain leafless throughout the year. During rains, grasses and herbs become abundant. The common plants of these forests are *Acacia nilotica, Prosopis spicigera, Capparis, Albizzia, Zizyphus, Euphorbia, Calotropis, Madhuca, Tephrosia, Crotalaria,* etc.

III. MONTANE (MOUNTAINOUS) SUBTROPICAL FORESTS

(8) Subtropical broad-leaved hill forests. These forests are dense, have predominantly evergreen species, and occur in relatively moist areas as lower slopes of eastern Himalayas, Bengal, Assam and on hill ranges as Khasi, Nilgiri, Mahabaleshwar, etc. In the north occur *Quercus. Castanopsis, Schima* and some temperate species. In eastern Himalayas due to higher humidity, bamboos, many epiphytes including orchids and ferns become abundant. In the south common trees are *Eugenia, Actinodaphne, Randia, Ficus, Populus, Canthium, Mangifera* and climbers are *Piper, Gnetum, Smilax,* etc.

(9) Subtropical pine forests. In Western Himalayas and Khasi Jayantia hills of Assam, up to an altitude of 1875 metres, open forests composed mainly of pine species such as *Pinus roxburghi* and *P. khasya*.

(10) Subtropical dry evergreen forests. Lower elevations of Himalayas have xerophytic forests containing mainly thorny and small-leaved evergreen species. These areas are characterized by low rainfall and low temperature. The common species are *Acacia modesta, Dodonea viscosa, Olea cuspidata,* etc.

IV. MONTANE TEMPERATE FORESTS

These forests are found at 1800 to 3800 metres altitude in the Himalayas.

(11) Montane wet temperate forests. These forests are found in eastern Himalayas with high rainfall and also in Nilgiris in the south. The forests in the south are evergreen and are known as **Sholas**. The trees are 15 to 20 metres high with a dense growth forming closed canopy. Epiphytes are abundant. The common plant species of these forests are *Rhododendron nilagircum, Hopea, Artocarpus hirsuta, Salmalia,* etc.

In the north, the west temperate forests extend from eastern Nepal to Assam at an altitude of 1800 to 2900 metres. These forests are evergreen, dense and high (up to 25 metres). The common plant species include *Quercus, Acer, Prunus, Ulmus, Begonia, Loranthus,* etc. In the western Himalayas *Quercus* species dominate all over. Conifers as *Abies, Cedrus, Picea,* etc., and broad-leaved species as *Betula, Alnus, Ulmus, Acer, Populus, etc.,* are also widely distributed.

(12) Himalayan moist temperate forests. Central and Western Himalayas, have the forests of conifers or/and oaks at an altitude of 1500 to 3000 metres. The trees are tall (up to 45 metres), but undergrowth is thin and deciduous. These forests have plants such as *Cotoneaster, Berberis, Spiraea, Quercus, Cedrus, Pinus, Picea, Abies, Tsuga, Cupressus,* etc.

Rhododendron is widely found in alpine forests which occur throughout Himalyas.

(13) Himalayan dry temperate forests. These forests occur in a narrow belt in the western Himalayas extending from parts of U.P. through Himachal Pradesh and Punjab to Kashmir. They are open, evergreen forests with scrub undergrowth. Oaks and conifers are dominant. In western drier areas *Pinus gerardiana* and *Quercus ilex* are common, while in the eastern moist zone *Abies, Picea, Larix, Juniperus* are found. Other associated plant species are *Cedrus. Picea, Daphne, Artenisia, Fraxinus, Alnus,* etc.

V. ALPLINE FORESTS

(14) Sub-alpine forests. These evergreen forests are open and occur throughout the Himalayas at high altitudes (above 3000 metres up to tree line). They comprise mostly conifers such as *Abies spectabilis,* broad-leaved species such as *Betula* and *Rhododendron* and shrubs as *Cotoneaster, Rosa, Lonicera, Smilax,* etc.

(15) Moist-alpine scrub. Throughout the Himalayas, above the timber line up to the height of 5500 metres, dwarf evergreen shrubby growth of conifers (*Juniperus*) and broad-leaved species (*Rhododendron*) are found.

(16) Dry-alpine scrub. In areas with low rainfall (below 35 cm) open xerophytic scrubs such as *Juniperus, Caragana, Eurotia, Salix,* etc., are found upto 5500 metres.

(b) Grassland vegetation. In India, natural grasslands (as climax formation) are not present but occur only during succession. The following three types of grasslands occur in India : (1) **Xerophilous** that occur in dry regions of north-west India under semidesert conditions : (2) **Mesophilous** which are extensive grass flats or savannas and occur in moist deciduous forests of Uttar Pradesh ; and (3) **Hygrophilous** which are called wet **savannas**. These three basic types of Indian grasslands have been further subdivided into the following eight subtypes, each of which is named on its dominant species: *Sehima–Dichanthium, Dichanthium–Cenchrus, Phragmites–Saccharum, Bothriochloa, Cymbopogon, Arundinella, Deyeuxia–Arundinella,* and *Deschampsia–Deyeuxia.*

4. Floristic (Botanical) Regions (Provinces) of India

The Indian sub-continent is characterised with a variety of climate types and the flora of the country is also correspondingly of different types in its different parts. The country has been divided into the following nine **floristic regions** for the study of flora :

A. Western Himalayas. It extends from Kumaon to Kashmir and has annual rainfall up to 200 cm. Altitu-dinally there are following three zones of vegetation corresponding to three climatic belts:

(i) Submo-ntane zone. This extends up to 1500 metres altitude and comprises mostly of Siwalik ranges. The forests are tropical and subtropical having trees such as *Shorea robusta, Dalbergia sissoo,*

Cedrela toona, Ficus glomerata, Eugenia jambolano, Acacia catechu, Butea monosperma (dhak), *Zizyphus* and thorny succulent euphorbias on slopes.

(ii) Temperate zone. Above submontane zone extend montane temperate forests up to 3500 metres altitude. They are dominated by plant species such as *Quercus, Acer, Ulmus, Rhododendron, Betula* (birch), *Salix* (cane), *Populus, Cornus, Prunus, Fraxinus, Pinus, Cedrus, Picea* and *Taxus.*

(iii) Alpine zone. This zone extends from 3500 to 4500 metres altitudes (snow line) and is characterised with alpine forest vegetation with scrub and meadows. Most common tree species are *Abies, Betula, Juniperus* and *Rhododendrons.* The herbs which occur near the snow line include species such as *Primula, Potentilla, Polygonum, Geranium, Saxifraga, Aster,* etc.

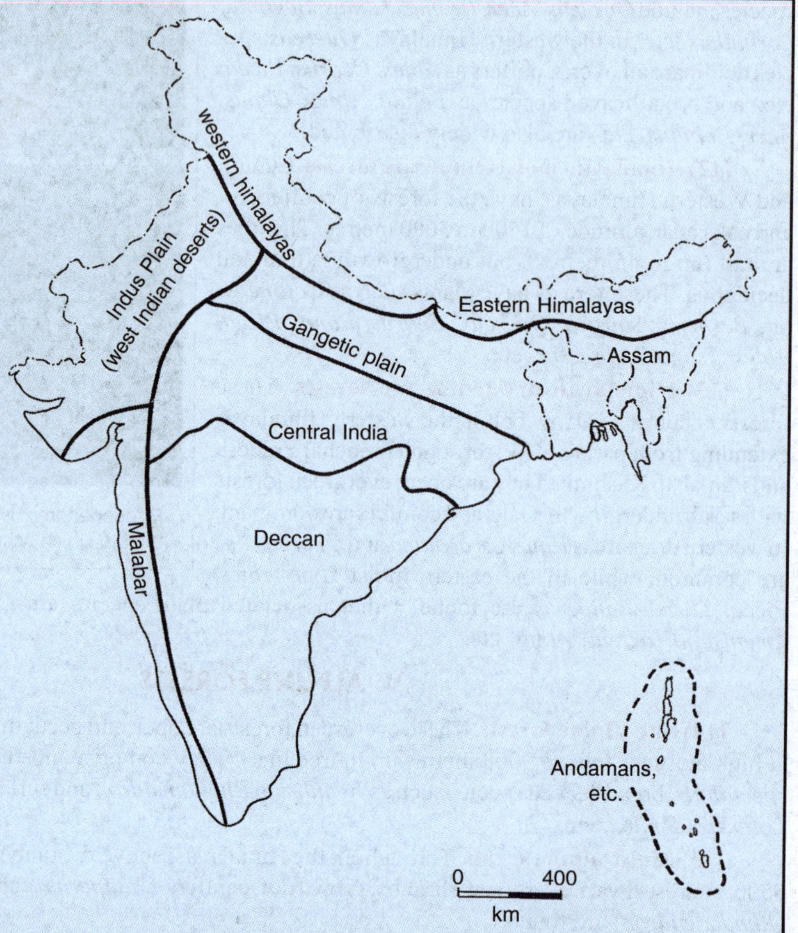

Fig. 17.1. Map showing different floristic regions of India.

B. Eastern Himalayas. It includes regions of Sikkim and NEFA and is characterised by more rainfall, less snow and higher temperature. This is also divided into the following three zones altitudinally :

(a) Tropical zone. Up to 1800 metres altitudes, this zone has tropical semi-evergreen or moist deciduous forests. These forests comprise the plants such as *Shorea robusta, Acacia catechu, Delbergia sissoo, Terminalia, Albizzia, Cedrela, Dendrocalamus* (bamboo), etc.

(b) Temperate zone. This zone extends between 1800 metres to 3800 metres altitudes and has typical montane temperate forests which are dominated by oaks such as *Michelia, Quercus, Pyrus, Symplocos, Eugenia,* etc., at lower levels and by conifers as *Juniperus, Cryptomeria, Abies, Pinus, Larix* and *Tsuga* and also *Salix, Rhododendron* and *Arundinaria* (bamboo) at higher cooler levels.

(c) Alpine zone. Beyond the temperate zone extends alpine zone up to 5000 metres altitudes. It has alpine vegetation including *Juniperus* and *Rhododendron* with its other typical flora.

C. Indus plains. This zone includes the arid and semiarid regions of Punjab, Rajasthan, Kutch, part of Gujarat and Delhi. The rainfall is less than 70 cm. The vegetation is tropical thorn forest in semiarid region and is typical desert in the arid region. The plants of this zone are primarily xerophytic. The common plant species of this zone are *Acacia nelotica, Prosopis* sp., *Salvadora, Tecomella, Capparis, Tamarix, Zizyphus, Calotropis, Panicum, Saccharum, Cenchrus, Euphorbia,* etc.

D. Gangetic plains. This region extends over Uttar Pradesh, Bihar, Bengal and part of Orissa and is characterised by moderate amount of rainfall and most fertile (*i.e.,* alluvial) soils. Vegetation of this zone is chiefly of tropical moist and deciduous and dry deciduous forest type. The common plants of this zone are *Dalbergia sissoo* (shesham), *Acacia nelotica* (babul), *Saccharum munja, Butea monosperma, Madhuca indica* (mahua), *Terminalia arjuna* (arjuna), *Buchanania lanzan* (chiraunji), *Diospyros melanoxylon* (tendu) , *Cordi myxa* (lisora), *Acacia catechu* (khair), *Azadirachta indica* (neem), *Ficus bengalensis* (bergad), *Ficus religiosa* (pipal), *Mangifera indica* (mango), and weeds and grasses such as *Xanthium, Cassia, Argemone, Amaranthus,* etc. In Gangetic delta (South Bengal) mangrove vegetation is common.

E. Central India. It comprises Madhya Pradesh, parts of Orissa and Gujarat. The rainfall is 150–200 cm and its vegetation is thorny, mixed deciduous and teak type. The chief plants of this region are *Tectona grandis, Madhuca, Diospyros, Butea, Dalbergia, Terminalia, Carissa, Zizyphus, Acacia, Mangifera,* etc.

F. Malabar (west coast). This region includes western coast of India from Gujarat to Cape Comorin and has heavy rainfall. The forests are tropical evergreen in west, semi-evergreen towards interior, subtropical or montane temperate evergreen forests in Nilgiris and mangroves near Bombay and Kerala coast.

G. Deccan Plateau. This region extends all over peninsular India (*i.e.,* Andhra Pradesh, Tamil Nadu and Karnataka) and has rainfall up to 100 cm. Its central hilly plateau has tropical dry deciduous forests of *Bowsellia serrata, Tectona grandis* and *Hardwickia pinnata,* while, the low eastern dry Coromandal Coast has tropical dry evergreen forests of *Santalum album* (chandan), *Cedrela toona* and plants such as *Acacia, Prosopis, Euphorbia, Capparis, Phyllanthus,* etc.

H. Assam. This region is characterised by heavy rainfall (200 to 1000 cm). The vegetation is either dense evergreen forest or sub-tropical. The evergreen forests include trees such as *Dipterocarpus macrocarpu, Mesua ferrea, Shorea robusta, Ficus elastica,* etc. bamboos as *Bambusa pallida, Dendrocalomus hamiltonii,* etc., grasses such as *Imperata cylindrica, Saccharum* sp., *Themeda* sp., insectivorous plants as *Nepenthes* sp., and also epiphytes (ferns and orchids). In the northern cooler regions, wet hill forests include plants such as *Alnus, Betula, Rhododendron, Magnolia,* etc., The hilly tracts also have pine forests of *Pinus khasiya* and *P. insularis.*

I. Andmans. This region possesses a varied type of vegetation : mangroves and beech forest at its coasts and evergreen forests of tall trees in the interior. Important plant species of this island are *Rhizophora, Mimusops, Calophyllum, Lagerstroemia,* etc.

PATTERNS OF DISTRIBUTION OF BIOTA

1. Distribution

The plants and animals exhibit specific patterns of distribution over the globe :

(i) Continuous distribution. A taxon distributed throughout the world in all climatic zones or in all the continents in at least one climatic zone is said to have continuous distribution. It is of the following types :

A. Cosmopolitan distribution. Such a taxon or a species occurs in all climatic zones. The common plants species of nearly cosmopolitan distribution are *Chenopodium album, Taraxcum officinale, Phragmites communis, Urtica dioica, Poa annua,* etc. Among the animals the following species have cosmopolitan distribution : *Tyto alba, Ardea cineria, Falco pereginus, Vanessa cardui, Nonnophila noctuella,* and *Macrobiotus hufelandi.*

B. Circumpolar distribution. Certain species such as *Saxiphraga oppositaefolia, Carex lapponica, Ranunculus nivalis,* etc., are distributed in a belt around the north-pole.

C. Circumboreal and circumaustral distribution. The species which are distributed in a near continuous belt in the temperate region of northern or southern hemisphere are said to have circumboreal and circumaustral distribution respectively. The genera *Alnus, Draba, Acer,* etc., are circumboreal in distribution and *Danthonia* is circumaustral in distribution.

D. Pantropical distribution. Certain plant species such as *Bauhinia, Dalbergia, Dioscorea, Corchorus, Ocimum, Cassia, Eugenia, Phyllanthus,* etc., are distributed throughout the tropical belt.

Bauhinia is a common plant of tropical belt.

(ii) Discontinuous distribution. A taxon distributed in two or more widely separated geographical areas, is said to have discontinuous distribution. Such taxa may occur in several small areas in the same continent or in two different continents of the same or different hemispheres. For example, two plant species—*Saxifraga* and *Silene* are distributed in Arctic region and high altitudes. Likewise, three genera of Dipnoi fish, *Epiceratodus, Protopterus* and *Lepidosiren* are distributed in Australia, Africa and South America. There are various other examples in plants and animals which show discontinuous distribution.

2. Endemism

Species which are spread over a wide area or in different ecological conditions are said to be **cosmopolitan**. Some of them, called **endemics** are restricted in their distribution to a small region. This phenomenon of restricted distribution associated with some geographical or ecological factors is called **endemism**. Numerous plant families, genera and species are found to be highly endemic. In the Himalayas, some 28% (3,169 dicot species) of all the dicots are endemic (**Chatterjee**, 1939). For example, a great proportion of the taxa of the flora of oceanic islands is endemic to those islands only. Some important endemic plant genera of Indian deserts are *Omania, Xerotia, Leptadenia, Daemia,* etc., and of India are *Amphicome, Dittoceras, Dodecania, Ulteria, Cruddasia, Heylandia, Lagenandra, Zeylandium, Hitchenia, Blepharistemma,* etc. Certain important examples of endemic Indian plant species include *Ficus religiosa, Ficus bengalensis, Aegle marmelos, Artocarpus nobilis, Crotalaria juncea, Datura metal, Indigofera tinctoria, Elettaria, Eleusine coracana, Piper nigrum, Piper longum, Sesamum indicus, Hibiscus abelmoschus, Butea monosperma, Beaumontia grandiflora, Memecylon umbellatum, Holmskioldia sanguinea, Feronia elephantum, Saraca indica, Shorea robusta* and *Caryota urens.*

Among endemics, some species exhibit very localised distribution and are called **local endemics**. Sometimes mutants appear and vanish without being able to compete with parental species, and are called **pseudo-endemics**. Some species may show a restricted distribution but cover large areas in course of time. This is called **expanding** or **progressive endemics**. Some old species may be restricted to a small region because of a severe decline in their population, a phenomenon called **contracting** or **retrogressive endemics**. Lastly, there are following two more types of endemics : **1. Palaeoendemic or epibiotic or relics**. These endemics are supposed to have been the remnants of a once widely distributed taxon in the past, e.g., *Gingko biloba, Sequoia semipervirens, Trapa natans.* **2. Neoendemics** or **microendemics**. These endemic taxa are supposed to have evolved only during the recent times and did not have sufficient time to extend their ranges of distribution.

Ficus.

3. Centre of Origin

It is maintained that for every species, there is one region from where it came into existence. Thus, some areas might have given rise to many species. These regions are, therefore, called the centres of origin. Plants evolved much earlier than man. Paleobotanical evidence helps us to arrive at conclusions on the centre of origin. For example, evidently cincona, tomato (*Lycopersicon esculentum*) and tobacco (*Nicotiana tabacum, N. rustica*) originated in South America. Cotton (*Gossipium* sp.) and chilly (*Capsicum* sp.) originated in Central America. Lichi (*Litchi chinensis*), tea (*Camellia* or *Thea sinensis*) and brinjal (*Solanum melongena*) originated in China. India is the centre of origin of many crop plants such as rice (*Oryza sativa*), sugarcane (*Saccharum officinarum*) and gram (*Cicer arietinum*).

DESCRIPTIVE ZOOGEOGRAPHY

A few groups of animals are cosmopolitan in having world-wide distribution and most animals are restricted in distribution because of some kind of barrier or because of past history of origin and dispersal. An understanding of their present-day distribution takes us into zoogeography. Major units of distribution are the **zoological region** or **zoogeographical regions**, areas defined largely by the past and present relations of the continent to each other. Each region is further subdivided into **faunal** or **ecological units**, depending on the criteria used.

Zoogeographical Regions

There are six zoogeographical regions, each more or less embracing a major continental land mass and all have long been separated from one another by water (oceans and fresh-water bodies), mountain ranges or desert, so that, each region has evolved its distinctive and characteristic orders and families of animals. These major distributional units were first recognised by **Clater** (1858), modified by **Huxley** (1868), extended by **Wallace** (1876) and best described in a modern way by **Darlington** (1957) and others. The six zoogeographical regions are the **Palaearctic**, the **Nearctic**, the **Neotropical**, the **Ethiopian**, the **Oriental** and the **Australian**. Because some zoogeographers consider the Neotropical and the Australian regions to be so different from the rest of the world, these two are often considered as regions or realms equal to the other four combined. They are classified as **Neogea** (the Neotropical), **Notogea** (the Australian) and **Metagea** (the Palaearctic, Nearctic, Ethiopian and Oriental). Moreover, two zoogeographic regions, namely the Palaearctic and the Nearctic are quite closely related, so the two often considered as one, the **Holarctic**. In fact, these two regions are similar in climate, vegetation, and in their faunal composition having animals such as wolf, hare, moose (called elk in North America), caribou, wolverine and bison.

A. Palaearctic region. This largest region comprises the whole of Europe, Soviet Russia, Northern China, Japan, Northern Arabia (Persia) and narrow strip of coastal North Africa. It is subdivided into **European**, **Mediterranean**, **Siberian** and **Manchurian** subregions. Its eastern parts have characteristic fauna of the following animal species : *Rhacophorus* (flying frog), *Bombinator* (fire-bellied frog), *Agkistrodon halys* (pit viper), *Strix uralensis, Grus japonensis* (crane), *Aix galericulata, Erinaceus* (hedgehog), *Podces panderi, Capricornis sumatraensis, Camelus ferus* (camel), *Bos mutus, Ailuropus* (great panda), *Panthera tigris altaica* (tiger), *Uncia uncia, Phoca sibirica* (fresh-water seal), *Pteromys volans, Equus hemionus* (donkey) and *Macaca fuscata* (rhesus monkey). The common fauna of Western Palaearctic include animal species such as *Clupea harengus* (herring), *Sardina pilchardus* (fish), *Gadus morrhua* (codfish), *Vipera ammodytes* (viper snake), *Chamaeleon chamaeleon* (arboreal lizard), *Pelecanus onocrotalus* (pelecan), *Fratercula arctica, Phoenicopterus ruber, Rangifer tarandus* (caribou), *Ovis ammon, Dama dama, Capra aegagrus, Odobenus rosmarus* (walrus), *Ursus arctos, Lynx lynx* (wild cat), *Castor fiber* (beaver), *Hystrix cristata* (porcupine), and *Panthera pardus* (leopard).

B. Nearctic region. This region comprises the North American continent south to the Tropic of Cancer (*i.e.,* Greenland and North America). It is subdivided into **Californian**, **Rocky mountain**,

Allegany and **Canadian** subregions. The Nearctic is the home of many reptiles and has more endemic families of vertebrates. Characteristic fauna of this region include animal species as *Ambystoma tigrinum, Crotalus adamanteus* (rattle snake), *Branta canadensis* (Canadian goose), *Larus schistisagus* (gull), *Bucephala albeola, Meleagris gallopavo* (turkey), *Didelphis marsupialis* (opossum), *Ovibos moschatus* (musk-ox), *Rangifer tarandus arcticus* (caribou) *Bison bison* (North American bison), *Odobenus rosmarus divergens, Cystophora cristata, Martes pennanti* (fisher), *Lutra canadensis* (otter), *Dasypus novemcinctus* (armadillo), *Canis latrans, Procyon lotor* (racoon), *Lagenorhynchus acutus, Delphinapterus leucas* and *Enhydra lutric*.

C. Neotropical region. It includes Central America, South America, part of Mexico and the West Indies. This region lacks well-developed ungulate fauna of the plains, but have rich, distinctive, varied endemic fauna of vertebrates. Characteristic animals of this region are *Lepidosiren* (lung-fish), *Lepidosteus* (Garpike), *Histrio histrio*, tortoises (*Dermatemys, Stourotypus, Peltocephalus*), *Heloderma horridum, Iguana iguana, Amazilia yucatanensis, Phoenicopterus ruber, Pelicanus occidentalis* (brown pelecan), *Phalacrocorax harrisi, Sarcoramphus papa, Amazona amazonica, Spheniscus magellanicus, Chironectes minimus, Desmodus rotundus, Myrmecophaga tridactyla* (giant anteater), *Tapirus terrestris* (tapir), *Lama vicugna, Lama guanicoe, Panthera onca, Arctocephalus australis, Cavia aperea, Bradypus tridactylus* (three-toed sloth), *Vultur gryphus, Pterocnemia pennata* and *Ara macao*.

The jaguar (*Panthera onca*) maintains the populations of many prey species at low levels, establishing a "balance of nature" in the Neotropical forest.

D. Ethiopian region. The old world counterpart of the Neotropical is the Ethiopian, which includes the continent of Africa, south of the Atlas mountain and Sahara Desert, Southern Arabia, Madagascar and Mauritius. It embraces tropical forests in central Africa and in the mountains of East Africa, Savanna, grasslands and deserts. It includes a varied vertebrate fauna and several endemic families. The characteristic animals of this region are *Protopterus* (lung fish), *Crocodilus niloticus, Sagittarius serpentarius, Struthio camelus* (ostrich), *Oryx gazella* (African antelope), *Giraffa camelopardalis* (giraffe), *Loxodonta africana* (Africana elephant), *Equus quagga, Hippopotamus amphibius, Daubentonia madagascariensis* (ayeaye), *Lemur catta, Macaca sylvana, Pan troglodytes* (chimpanzee), *Gorilla gorilla, Acinonyx jubatus, Panthera leo* (lion), *Hyaena hyaena*, and *Delphinus delphis*. It lacks deer and bear among mammals and salamanders and tree-frogs among amphibians and contain several endemic families of birds.

E. Oriental region. It includes India, Indochina, South China, Malaya, and the western islands of the Malay Archipelago. It is bounded on the north by the Himalayas and on the other sides by the Indian and Pacific oceans. On the southeast corner, where the islands of the Malay Archipelago stretch out toward Australia, there is no definite boundary, although **Wallace's line** is often used to separate the Oriental from the Australian regions. This line runs between the Philippines and the Moluccas in the north, then bends southwest between Borneo and the Celebes, then south between the islands of Bali and Lombok. A second line, **Weber's line**, has been drawn to the east of Wallace's line ; it separates the islands with a majority of Oriental animals from those with a majority of Australian ones. Since the islands between these two lines form a transition between the Oriental and the Australian regions, some zoogeographers call the area **Wallacea (Smith**, 1977).

The Oriental region is divided into the following four subregions : **Indian sub-region, Ceylonese sub-region**, **Indo-Chinese subregion** and **Indo-Malayan sub-region**. Its characteristic fauna includes the following animal species : *Rhacophorus pardalis, Gavialis niloticus* (gharial), *Calotes versicolor* (garden lizard), *Draco volans* (flying lizard), *Draco dussumieri* (flying lizard). *Python reticulatus, Naja naja* (Indian cobra), *Ophiophagus hannah, Gallus gallus* (jungle fowl), *Pavo cristatus* (peacock), *Milvus migrans* (pariah kite), *Eudynamys* (Koel), *Psittacula krameri* (rose-ringed parakeet), *Bubo bubo* (great horned owl), *Coracias benghalensis* (blue jay), *Dinopium benghalense* (golden-backed woodpecker), *Corvus splendens* (crow), *Argusianus argus. Elephas maximus* (elephant), *Rhinoceros unicornis, Antelope cervicapra* (black buk), *Axis axis* (spotted deer), *Boselaphus tragocamelus* (blue bull), *Selenarctos tibetans* (black Himalayan bear), *Melursus ursinus* (sloth bear) *Sus cristatus* (wild boar), *Hystrix leucura* (Indian porcupine), *Herpestes* (mongoose), *Manis* (pangolin), *Soriculus* (Indian shrew), *Hyaena striata, Panthera tigris tigris, Panthera pardus, Babyrousa babyrussa, Macaca mulatta, Presbytis entellus* (langoor), *Pongo pygmaeus* (organgutan), *Ailurus fulgens,* and *Hylobates lar* (gibbon).

F. Australian region. This region includes Australia, Tasmania, New Guinea, and a few smaller islands of the Malay Archipelago. New Zealand and the Pacific Islands are excluded, for these are regarded as oceanic islands separate from the major faunal regions. Partly tropical and partly south temperate, the Australian region is noted for its lack of a land connection with other regions; the poverty of freshwater fish, amphibians and reptiles; the absence of placental mammals and dominance of marsupials. Included are the egg-laying mammals (monotremes) and the spiny ant-eaters. The characteristic fauna of this region includes following animal species : *Neoceratodus* (fish), *Chelmon rostratus, Phyllopteryx eques, Ornithorhynchus anatinus* (platypus or duck mole), *Zagoglossus bruijni* (proechidna), *Techyglossus aculeatus* (echidna, spiny ant-eater), *Macropus giganteus* (kangaroo), *Notoryctes typhlops, Macrotis lagotis, Dromaius novaehollandiae, Petaurus austalis, Paradisaea rubra, Pteridophora alberti* and *Phaseolarctos cinereus* (koala).

New Zealand has *Varanus komodensis, Sphenodon punctatus, Abteryx owenii haasti* (kiwi), etc. Sometime, south polar region called the **Antarctic** or **Archinotic** is also included in these major distributional units. Antarctic region has the following characteristic animal species : *Balaenoptera musculus* (blue whale), *Orcinus orcam, Megadyptes antipodes, Aptenodytes patagonica* (penguin). *A. foresteri* (emperor penguin), *Pygoscelis adeliae, Diomedea exulans* (albatross), *Chionis alba,* and *Lobodon carcinophagus.*

REVISION QUESTIONS

1. Describe different patterns of distribution of biota.
2. What is phytogeography ? Describe chief vegetational belts of Earth.
3. Write an essay on the phytogeography of India.
4. What is zoogeography ? Describe different zoogeographical regions of Earth.
5. Describe the process of dispersal of organisms.
6. Write short notes on the following : Endemic species ; Oriental region ; range expansion ; broadcasting; barriers; continental drift theory; land bridges; migration; and ecesis.

CHAPTER

18

Adaptations

(Aquatic Adaptations, Volant Adaptations And Desert Adaptations)

Pit Viper has heat sensitive pits between each eye and nostril. This enables them to pick up changes in temperature around them.

The idea of adaptation maintains that organisms (animals and plants) are structurally and functionally designed for meeting the needs of life in the habitats in which they live. Thus, adaptations include adjustments by which an organism accommodates itself to its environment. These may occur by natural selection. In this chapter only following three types of adaptations of animals have been described : 1. Aquatic adaptations ; 2. Volant adaptations and 3. Desert adaptations. (For other sort of adaptations of plants and animals see Chapter 3).

AQUATIC ADAPTATIONS

Aquatic adaptations occur in those animals which live in water habitat, *viz.,* fresh, brackish or sea water. They are called **aquatic animals** or **hydrocoles**. Based upon the phylogenetic history of the aquatic animals, following two types of hydrocoles have been recognised :

1. Primary aquatic animals. The primarily aquatic animals are those in which the phylogenetic history is restricted to water as habitat. Therefore, all their adaptations are originally designed to meet the necessities of aquatic life. Generally, by primarily aquatic forms is meant the fishes, which have never had a terrestrial ancestry, but have evolved from more primitive aquatic progenitors. As a consequence their adaptation to a dense watery medium is perfect and they do not suffer as those secondarily adapted do through their inability to breathe water. They are, therefore, the primitive gill breathing vertebrates. Primary aquatic animals include protozoans, sponges, coelenterates, some annelids, molluscs and arthropods ; echinoderms ; and chordates such as cephalochordates (*Amphioxus*), urochordates (*Herdmania*, *Doliolum*, *Salpa*, etc.), cyclostomates (*Petromyzon, Myxine*), fishes, etc.

2. Secondary aquatic animals. These are those hydrocoles which have a record of terrestrial life in their phylogeny (i.e., they are descended from ancestors which led a life on land). Secondary aquatic animals are lung-breathers, mainly amphibious vertebrates, which through stress of circumstances such as inhospitable lands, where food was scarce or severe competitions were forced to return once more to the water. Consequently these animals show in their body structure, the evidences which speak of their ancestry from land-living animals, *e.g., Pila,* frog and other amphibians, swamp river turtles (*Notosaurs, Phytosaurs*), crocodiles, birds such as *Ichthyornis,* penguins, albatrosses, petrels, ducks, geese, etc., and mammals such as *Hippopotamus, Otter,* whales, porpoises, etc.

A. Primary Aquatic Adaptations

1. Body contour. The form of body depends upon the habits of life. The majority of fixed and partly sedentary forms have radially symmetrical body forms, *e.g.*, sponges (*Sycon, Euplectella, Hyalonema,* etc.,), *Hydra, Obelia, Aurelia, Metridium* (sea anemone), echinoderms such as *Holothuria, Echinus, Astropecten* or star fish. The active locomotor type have fusiform, spindle-shaped elongated and worm-like bodies. The spindle form is the characteristic of fishes and wavy, worm-like form is found in the annelids (*Nereis*).

The piscine body is designed for fast locomotion in water. There occurs a side to side compression of head, body and tail into a beautifully curved streamlined fish form. Head is sub-conical. There is no protuberance over the body, which would retard the swift passage of the animal through water. Further, since the relative weight is more on the upper half of the fish body (on account of the myotomes) a dead fish will float ventral side turned upwards.

Regarding the body form of planktonic organisms (which float passively at or about the surface), some are globular such as *Noctiluca,* some have umbrella shape (medusa of *Obelia, Aurelia*) and some have barrel-like shape (tunicates such as *Doliolum* and *Salpa*).

2. Swimming organs. Some aquatic animals float passively and do not possess much powers of movement, *e.g.,* dinoflagellates (Protozoa). The medusae and most siphonophores move by alternative contraction and expansion of their sub-umbrellar side.

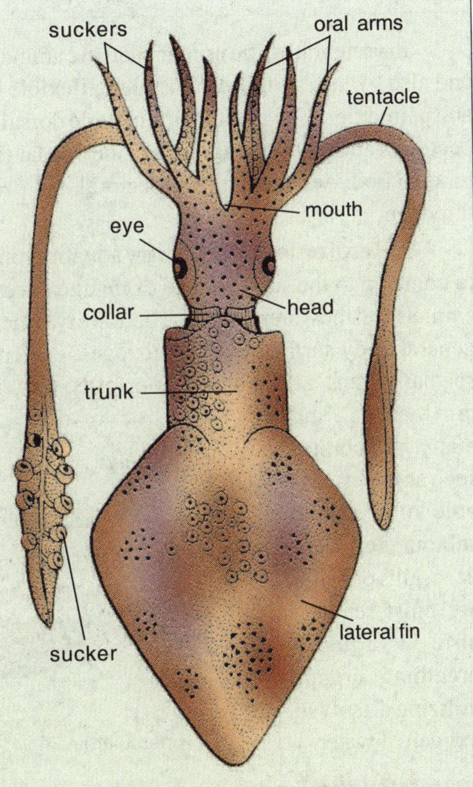

Fig. 18.1. *Loligo* (squid).

Organs for active swimming exist in arthropods (*e.g.,* prawn or *Palaemon*), annelids

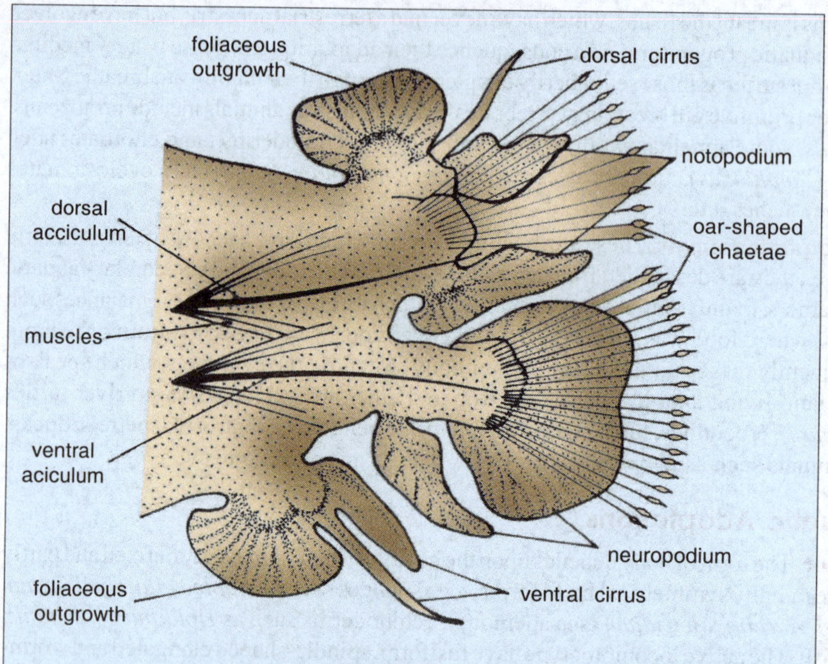

Fig. 18.2. Parapodium of *Heteronereis.*

(*e.g.*, *Heteronereis*), cephalopods (*e.g.*, *Sepia, Octopus* and *Loligo*) and vertebrates. For example, the **parapodia** of *Hetero-nereis* (Fig. 18.2) are well adapted for active swimming. Each parapodium has a flattened structure divided into the noto-podium and neuropodium bearing cirri and setae. The prawn has specially adapted abdominal appendages, called **pleopods** and **uropods** for swimming.

In vertebrates, the primary aquatic animals are the fishes. The fish move (swim) by the help of fins and also by lateral undulations of the flexible body. The fins of fish are of two types, the median fins and paired fins. The median fins include dorsal fins, caudal fins and ventral fins. The paired fins are the pectorals and pelvics (Fig. 18.3). The caudal fin is the chief propeller, the dorsal and ventral fins help to keep body vertical and the pectoral and pelvic fins help in propulsion and in making changes in direction.

3. Respiration. The primary aquatic animals are able to respire inside the water, without the need to come up to the surface. The exchange of respiratory gases takes place between the blood of these animals and the water outside. There are two methods of aquatic breathing : 1. through diffusion through general body surface, *e.g.*, protozoans, coelenterates and planktonic larvae, and 2. with the help of special organs called branchia or **gills**, *e.g.*, prawns, and other crustaceans; *Unio, Pila* and other molluscs ; and many vertebrates such as fishes, tadpole of frog and salamanders. Indeed, gills of fishes are most remarkable aquatic breathing organs utilizing dissolved oxygen of water.

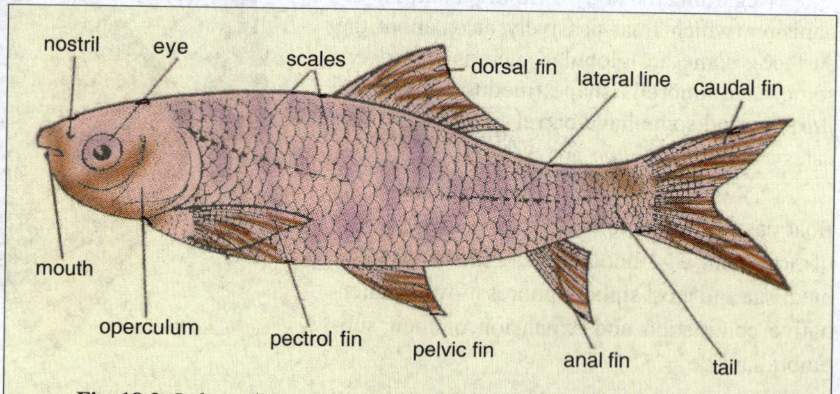

Fig. 18.3. *Labeo rohita.* A freshwater fish (called **carp** or **rohu**) of India.

4. Air bladder. Advanced

bony fishes contain air bladder (or swim bladder) which serve as an accessory respiratory organ and hydrostatic organ. Air bladder is a hollow outgrowth of the alimentary canal and is filled with gas or air.

5. Lateral line sense organs. Fishes have lateral line systems extending all over the body. It contains neuromast organs which act as rheoreceptors (*i.e.,* detect pressure changes in surrounding water).

6. Skin. Skin of fishes is rich in mucous glands and/or is protected with scales.

B. Secondary Aquatic Adaptations

1. Stream-lined body. The body shape is stream-lined like primarily adapted forms : neck constriction disappears and tail enlarges, *e.g.,* Ichthyosauria (extinct fish-lizards), Cetacea (whales, dolphins, and porpoises), Sirenia (manatees and dugongs), Pinnipedia (walrushes and seals). Frog also contains stream-lined body.

2. Enlargement of size. Aquatic vertebrates tend to be larger in size because in these creatures energy, which in terrestrial form is exhausted in gravitational forces, is turned into growth. For example, largest sulphur-bottom whale (*Balaenoptera musculus*) is several times bigger than the largest elephant. Other example include giant sharks and squids.

3. Submergence. All secondary aquatic animals need to develop capacity of submergence since swimming below water surface demands such an adaptation. For example, in whales the ribs are strongly arched, the lungs are massive, the external nostrils communicate with the median "blow hole" which is closable.

Certain adult aquatic insects too are able to increase their period of submergence by storing air inside the subelytral space, *e.g., Nepa.*

4. Shortening of neck. There occurs reduction of length and mobility of neck. In whales cervical vertebrae (which are seven in number like other mammals) are fused to form a solid and compressed mass of bone.

5. Disappearance of excrescences. The external ears (pinnae) which hinder water locomotion tend to disappear, since they collect sound waves in air medium and are useless in aquatic forms. Thus, ears are reduced in amphibious mammals and are lost in whales, true seals and walruses. The nostrils (nares) move towards the apex of head as in whales, ichthyosaurs, phytosaurs, etc. Nares are often capable of being closed (*e.g.,* otter). Likewise, eyes become water-adapted by shifting higher on the face as in hippopotamus.

Fig. 18.4. Marine turtle hawks bill (*Chelone imbricata*).

6. Occurrence of locomotory paddles (fins). There occur fleshy and fin-like expansions of the body wall in whales and ichthyosaurs which help in propulsion. These fins may be dorsal or caudal. Dorsal fin is present in killer whale, while absent in *Delphinopterus* and *Balaena.* Caudal fin (also called **caudal** or **tail fluke**) of marine mammals in horizontal (vertical in reptiles) and the bone divides the tail into two equal parts rather than running into one lobe. In turtles **oar propulsion** occurs by fin-like limbs (Fig. 18.4); but in whales, sirenians, etc., **tail propulsion** takes place as their hindlimbs become disappeared. Pectoral paddles of whales and sirenians exhibit the following adaptations : (1) the restriction of movements corresponding to the elbow and wrist joints ; (2) the fusion between digits ; (3) increase in the number of phalanges, called hyperphalangy, and (4) increase in the number of digits for increase of expanse of paddling surface, called hyperdactyly.

In *Nepa* (Insecta) legs are flattened and oar-like.

7. Disappearance of hairs, skin glands, etc. In whales and sirenians, the skin becomes naked due to loss of hairs. The hair loss is compensated by the formation of a fatty layer below the skin (**blubber**) for the retention of the bodily heat. The blubber also has a hydrostatic advantage (*e.g.,* it helps in floatation or to keep positions in the water and act in combination with the buoyancy of the aquatic medium).

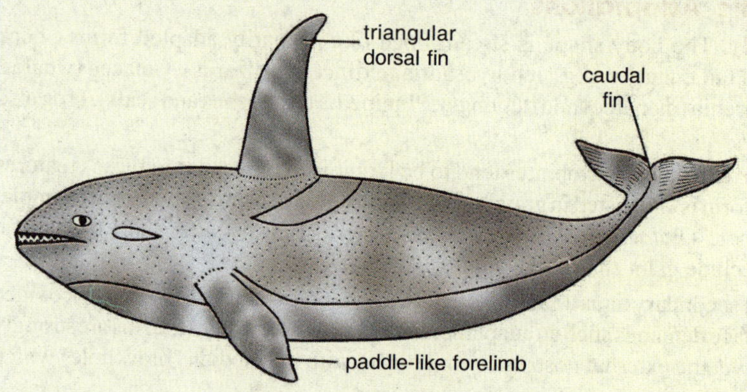

Fig. 18.5. Killer whale (*Orcinus*).

Sweat or oil glands disappear as they have nothing to do with the aquatic mode of life. Muscles and nerves also atropy from the integument due to its thickening and immobility.

8. Mouth armament. Since jaws are not used for mastication in whales they lost the power of movement. Teeth become simplified (homodont in dolphins) and greater in number. In sperm whale, teeth are present only on one jaw or entirely absent from both the jaws (*e.g.,* baleen whale).

9. Skull modification. In certain aquatic mammals (*e.g.,* Dolphins, porpoises) the cranium is shortened and front part of the skull becomes elongated to acquire the shape of a **rostrum**. In the skull of the cetacea, the zygomatic arch is reduced to a vestige.

10. Simplification of vertebrae. In secondarily aquatic forms (vertebrates) the vertebrae tend to be simple. In Ichthyosaurs, vertebrae are simple with biconcave centra like the fishes. Various secondary articulations or zygapophysis become reduced, as body weight is supported by water. The chest too become cylindrical. The rib articulations are modified and are central, *i.e.,* they are articulated to the centrum and are not articulated to the transverse processes.

Sacrum in cetaceans and sirenians is more or less reduced, since it does not withstand and transmit the supporting impact of the hindlimbs, as does in terrestrial forms.

11. Lightness of bones. The bones in aquatic forms are light and spongy. In whales, their interstices are filled with oil.

VOLANT ADAPTATIONS

The volant adaptations are concerned with the flight. The flight is a form of locomotion in the air under which the body has to be firstly prevented from falling down and secondly moved forwards, the speedier the better. Thus, volant adaptations must include modifications in the animals body for reducing the weight of the body and also for the formation of organs capable of executing the flight.

The flight may be of following two types :

1. Passive or gliding type flight. This type of movement involves no propulsion other than the initial force of jumping. Gliding is characterised by leaping or jumping from a high point and held up by some sustaining organs, then to glide to lower level. Thus, there is no locomotive force other than gravity. Here the 'wings' are made of **petagia** which do not flapped (*i.e.,* do not move up and down) by the muscular action.

The gliding flights are performed by various lizards (*e.g., Draco volans* or flying dragon, Fig. 18.6), fishes (*e.g., Exocoetus,* Fig. 18.7), birds (*e.g.,* ostriches, etc.,) mammals such as flying phalangers (*Petaurus sciurens,* a marsupial, Fig. 18.8), flying squirrel (Fig. 18.8) etc., and amphibians (*e.g., Rhacophorus,* Fig. 18.8). All these forms are found to possess the following adaptations for the gliding :

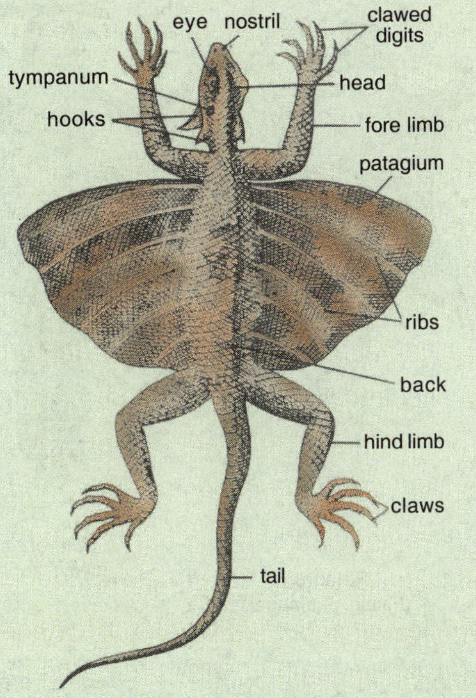

Fig. 18.6. *Draco volans.*

(i) Development of patagia. The sustaining surface ("wing") for the gliding is a fold or series of folds of skin, called **patagium**. The patagium lies between forelimbs and hindlimbs and can be folded like a fan against the body when is not in use. In *Draco,* the patagium is supported by ribs (Fig. 18.6). *Ptychozoon* (Fig. 18.8) is another gliding lizard which is commonly known as "the flying or fringed gecko" and in which lateral expansion of skin (patagium) extends along the side of neck, body, tail and limbs and between toes. Flying snakes (*e.g., Chrysopelea*) leap by the concave ventral side of body. The extinct reptiles pterodactyls (Fig. 18.8) of Mesozoic era were volant creatures akin to birds, containing true flight patagia. Their patagia were extensions between limbs supported by ribs.

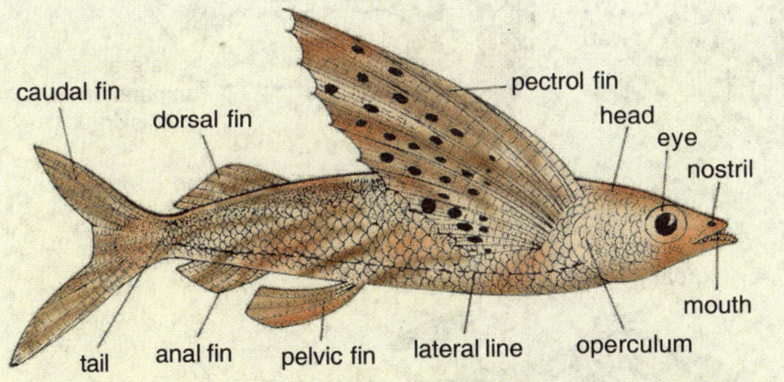

Fig. 18.7. *Exocoetus.*

Among mammals, flying squirrel, *Sciuropterus* (Fig. 18.9) has highly developed patagium. Its wide hairy tail is further supplemented by the hairy fringe on the patagium and along the rear of the thighs. In flying lemur (*Galeopithecus volans*; Fig. 18.9) the patagium extends from side of the neck to the tip of the tail even including digits, which are also webbed. In the bats, the patagium is supported mainly by the elongated forelimbs and the second, third and fifth digits. The first digit remains free.

Traces of patagia are also found in front and behind the arms (forelimbs) in birds which have adequate supporting function. Flying frog too contains rudiments of patagia in front and behind the limbs.

(ii) Enlargement and high insertion of pectoral fins. In flying fishes (*Exocoetus*) are trim-built creatures with large parachute-like pectoral fins which are highly inserted on the body. The pelvic fins are much smaller in size. The lower lobe of tail is also invariably longer, helping in leaping. The pectoral

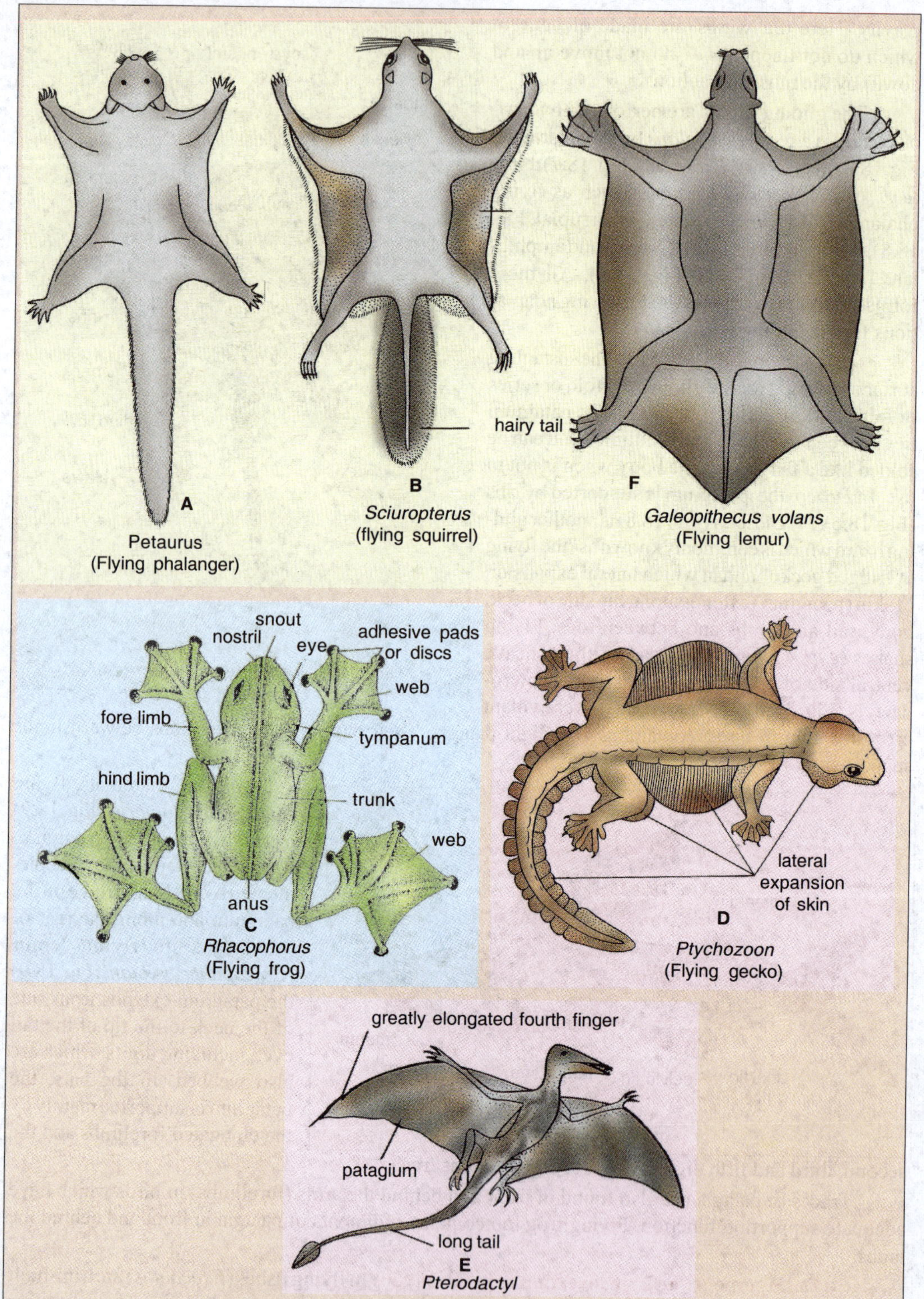

Fig. 18.8. Patagia of various animals.

fins do vibrate. *Exocoetus* can fly up to 200–300 metres to escape from the large fishes such as tunny and albicore). Other genera of flying fishes are *Dactylopterus, Pantodon* and *Pegasus. Pegasus* is a little marine fish found along the coasts of Japan, China, India and Australia. They skim along the surface of the water for 40 feet or more.

(iii) Webbing of feet. In flying frog (*Rhacophorus paradailis*, Fig. 18.10), the feet are webbed which sustain the prolonged leaps. Flying frog's digits terminate in adhesive pads which help in adhesion to trees.

2. Active or true flight. It is the aerial flight caused by the action of wings. True flight is found in insects, pterodactyls, birds and bats. In all of them the nature of development and structure of wings are quite different and their analogy suggest that the flight has evolved independently in different groups. In true flights the power is implied and the movement in air is sustained.

Varieties of Wings

1. Insect wings. The flying insects develop wings at their last moult. The mode of origin of the wing of the insects is different in the Exopterygota and Endopterygota. The wings are made up of cuticle and are strengthened by thickenings, called **veins**. The insect flight is affected by the flapping movements of the wings. Typically, there are two pairs of wings developed on the dorsolateral sides of the meso-and meta-thoracic segments (*e.g.,* Orthoptera, Isoptera, Coleoptera, Ephemeroptera, etc). In Diptera only the mesothoracic wings are developed. The wing muscles are highly striated and are rich in mitochondria.

In fact, insect's body is ideally suited for the evolution of aerostatic adaptations because of the system of tracheae which contain air and penetrate to the different parts of body. The highest degree of adaptation is met with in the dragon flies, may flies and damsel flies. In may flies, the adults are short lived and has its alimentary canal filled with air. In the honey-bees, there are air sacs associated with the tracheal system, these function in reducing the weight of the body.

2. Bat wing. In a bat wing, humerus bone is well developed, radius is long and curved and ulna is vestigial. The pollex (thumb) is free and clawed for crawling and climbing. In smaller insectivorous bats (Microptera; Fig. 18.9), the second finger is not

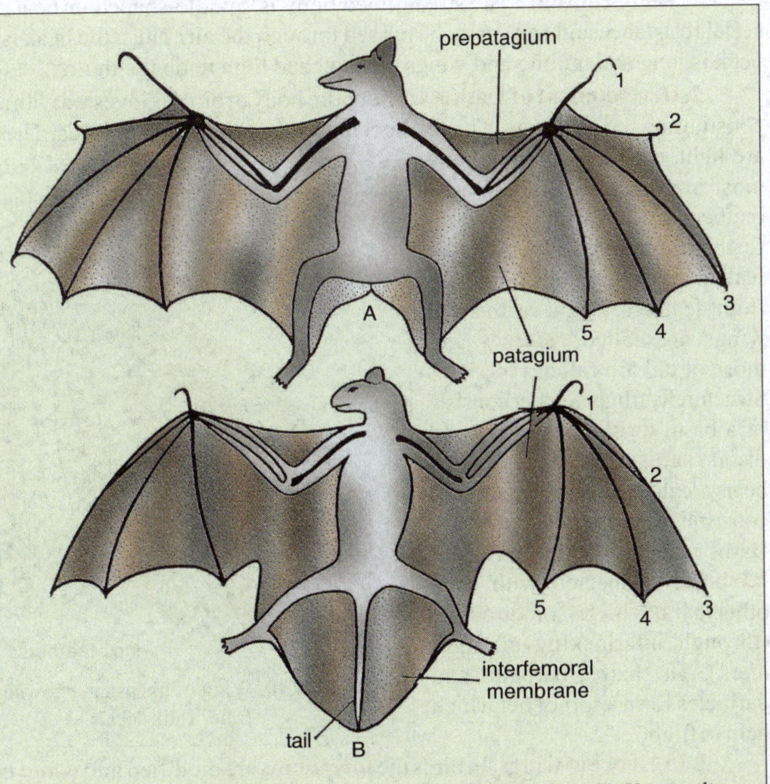

Fig. 18.9. A–*Pteropus* (frugivorous, Megachiroptera); B–*Vespertilia noctula* (insectivorous, Microchiroptera).

free from third, but is attached to it distally. These two support the anterior margin of the wing. The fourth and fifth digits are well developed. In frugivorous bats (Megachiroptera, Fig. 18.9), the second digit is clawed and free from the third.

3. Pterodactyl wing. In pterodactyl's wings, the radius and ulna bones are nearly equal. The next segment consists of a heavy fourth metacarpal bearing great wing finger and three small metacarpals supporting the first, second and third clawed digits. There is also present a **pteroid** bone which is directed towards the shoulder and supposed to support the anterior margin of a prepatagium which lies in front of arm from the wrist to the neck. The single wing finger (*i.e.,* fourth) is huge and forms the entire anterior support of the patagium beyond the wrist.

4. Bird wing. In birds forelimbs are modified into flight structures, the wings. The wings of birds are more specialized of all modern wings. In a wing of bird, digits are reduced to three and these are fused together to help in flight. The metacarpals are three which are unequally developed and co-ossified. The digits are represented by one or two phalanges which support the so-called **bastard quills** (*i.e.,* three, four or five feathers on the first digit (homologous of the thumb) of a bird's wings, (Fig. 18.10).

Volant adaptations of birds. Bird flight is characterized by the flapping of the wings. In flying, the bird lifts its body and drives forward by beating its wings in a characteristic way. An object moved swiftly in the air is affected by two forces : 1. upward pushing (**lifting**) and 2. downward pull (**drag**) due to the action of gravity. Hence, for successful flight, enough force must be applied to neutralize the drag and also to move forward. Birds wings are slightly concave, so are able to produce the air-current for producing the lifting force.

Birds have the following adaptations for true flight :

1. Body contour. The streamlined body is spindle-shaped or boat shaped, encountering least aerial resistance and can easily be passed through the air. Thus, the beak is pointed; head is compact; neck is long and mobile and wings are attached high upon the thorax.

2. Development of feathers. The entire body of birds is invested with a close covering of feathers, constituting the **plumage**. The feathers form the exoskeleton of birds. These nature's "master pieces" are light, elastic, waterproof and most important in flight. Bird's feathers are classified into quill feathers, contour feathers, down feathers and filoplumes. Quills are flight feathers. Flight feathers of wings are called **remiges** and those of tail are called **rectrices**. Structurally, flight feather consists of a basal **quill** or **calamus** and distal **rachis** or **shaft**. The rachis bears a leaf-like **vane** (or vexillum) consisting of series of lateral **barbs**. Barbs consist of double row of **barbules** connected with each other by **barbicles** or **hooklets** (through interlocking arrangement). The barbs, barbules and barbicles form a sort of net which help in flight.

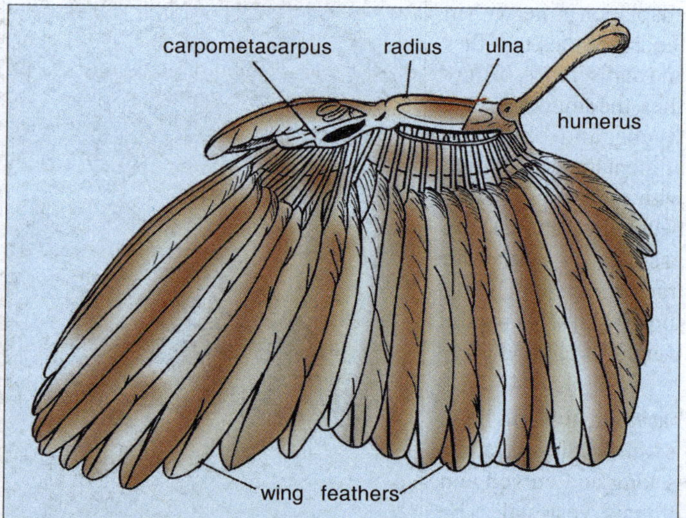

Fig. 18.10. Wing of pheasant, showing the two "bastard quills" (after Lull, 1957).

3. Presence of wings. In birds the **fore-limbs** are modified into **wings** helping in flying. The hind-limbs or legs are large and variously adapted for walking, running, scratching, perching, food-capturing, swimming.

4. Pneumatization of bones. The bones of birds are hollow and air filled. They also contain many air cavities. These add buoyancy during flight.

5. Occurrence of flight muscles and keeled sternum. In birds, specific flight muscles are developed which connect the wings with limb bones. Each wing is depressed or lowered by an enormous muscle called **pectoralis major**. It is elavated or raised by **pectoralis minor** (the tendon of which is inserted on the head of humerus). The **sternum** or breast bone is well developed and bears a median keel or carina for the attachment of pectoralis muscles.

6. Development of air sacs. They act as air reservoir during respiration and serve as balloons to provide lightness and buoyancy to the body. Air sacs also help in internal perspiration, thus, helping in the regulation of body temperature.

7. Brain and sense organ's specificity. Cerebrum is well developed (for controlling manoeuvrability) and optic lobes become enlarged (for controlling the great development of sight) and olfactory lobes are reduced (*i.e.,* power of smell is reduced). Bird's eyes are large and bear characteristic sclerotic plates to resist variable air pressure. Eyes also contain pectens) which are comb-like, vascular and pigmented structures) to regulate fluid pressure within the eye (*i.e.,* accomodation).

8. Beak. The conversion of forelimbs into wings is compensated by the presence of a **bill** or **beak**. The beak is horny and lacks teeth.

9. Mobile neck. The neck of birds is very long and flexible.

10. Single ovary. Presence of a single functional ovary of the left side in the female bird also leads to reduction of weight which is very essential for flight.

11. Absence of urinary bladder. Birds do not have a urinary bladder which is present to store the urine temporarily in other animals. Further, birds excrete a semisolid excreta which chiefly contains the insoluble uric acid and urates (**urecotelic excretion**). These features help in reducing the weight of body.

DESERT ADAPTATIONS

Desert animals or xerocoles have adaptations for following three sorts : 1. Moisture getting; 2. Moisture conservation and 3. Self-defence against physical and organic environment.

1. Moisture getting. As already described in Chapter 3, the securing of adequate quantitiy of moisture is the prime need of desert forms and its scarcity is the characteristic of the desert. At certain places of desert is found deep ground water, even then the surface sand is dry and burning. Hence, for getting the deep water, rain water from the superficial layers of sand, absorption of dew water, the xerophytic plants (*e.g.,* date palm) and xerocoles (*e.g., Moloch*) have varied adaptations. The date palm has long deep exploring and horizontally spreading roots. The sand lizard or *Moloch* has hygroscopic skin to absorb water like the

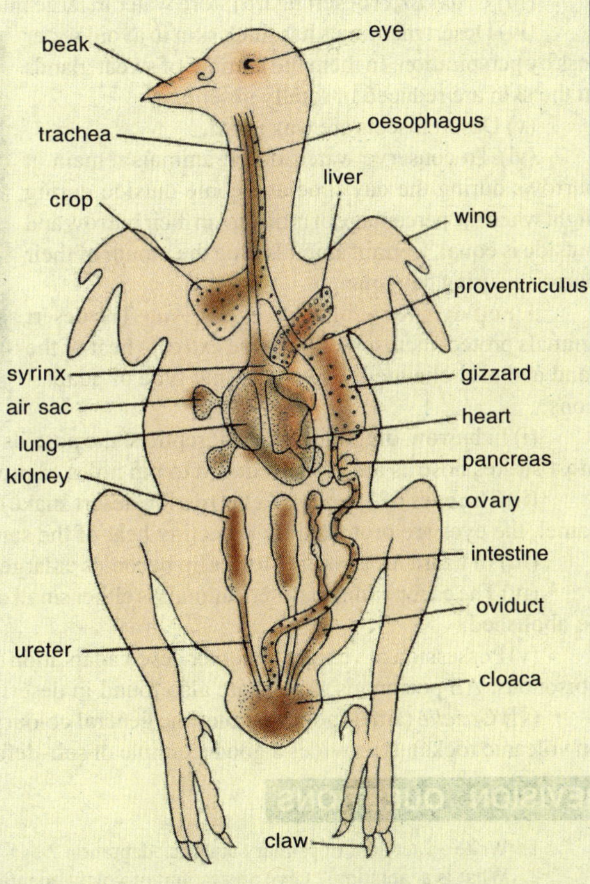

Fig. 18-11. Various volant adaptations of a female bird.

blotting paper. Its surface is covered with thorn-like scales (Fig. 18.12). A few animals absorb dew drops along with vegetation food. Barrel cactus affords drinkable water to thirsty desert people.

Desert rabbit, tortoises and wood rat (*Neotoma*) eat succulent plant for their water need. Jerboas, kangaroo rat (*Dipodomys*), pocket mouse (*Perognathus*) and certain other rodents eat dry seeds and vegetation and quench their thirst. Carnivorous animals feed on their prey for food and water. Grasshopper mice of western North America and long eared hedgehog of Sahara desert eat insects and small rodents for getting water.

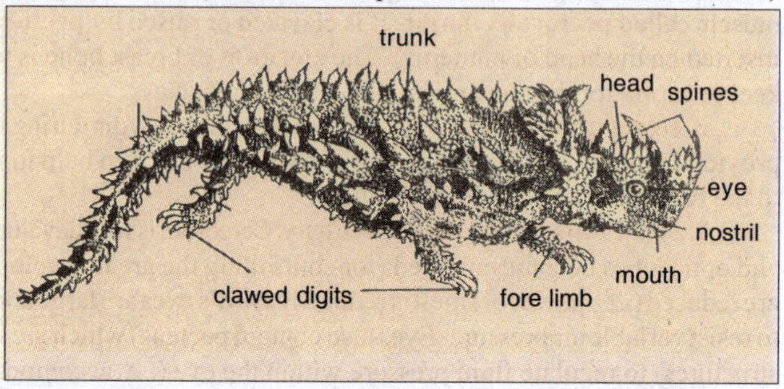

2. Moisture conservation. After getting moisture it becomes essential for the xerocoles to conserve it. As a result, desert animals contain the following types of adaptations for the water conservation.

(i) Camel exhibits many adaptations for water conservation (For details see Chatper 3).

(ii) To avoid water loss through skin, the horned toad (*Phrynosoma*,) has hard and rough skin and *Moloch* has scales and spines all over the body surface (*i.e.*, skin).

(iii) *Uromastix* (desert lizard) store water in large intestine.

(iv) Desert mammals has thick skin to avoid water loss by perspiration. In them the number of sweat glands in the skin are reduced or totally absent.

(v) Desert insects are wax proof.

(vi) To conserve water, desert animals remain in burrows during the day time and come outside during night when the percentage of moisture in their burrow and outside is equal. Certain animals plug the mouth of their burrows during day time.

Horned toad (*Phrynosoma*) has hard and rough skin to avoid water loss.

3. Self defence against scorching sun. The desert animals protect themselves from the extreme heat of the sand and arid climate by the following type of adaptations:

(i) In burrow-digging species of reptiles, the nostrils are directed upward instead of forward. In most snakes nostrils are either reduced to pin holes or protected by complicated valves.

(ii) The eyes of *Typhlops* (a burrowing desert snake) are overhung by the sheath. In ostrich and camel, the eyes are protected by reflecting heat of the sand by possessing the long neck.

(iii) In lizard *Mabuya*, the lower lip becomes enlarged with a transparent window in it.

(iv) The ear opening of desert animals is either small or protected by fringes or scales or they may be abolished.

(v) Possession of venom is another desert adaptation (self-defence). Almost all desert snakes are poisonous. All poisonous spiders are also found in desert. Scorpions too are common in deserts.

(vi) *Gazelle* (antelope) resembles the general colouring of landscape (white on sand, dark grey on volcanic rocks). It provides a good example of self-defence.

REVISION QUESTIONS

1. Write an account of primary aquatic adaptation ?
2. What is adaptation ? Give an account of volant adaptation.
3. Give an account of desert adaptations.

INDICES

- Cell Biology
- Genetics
- Molecular Biology
- Evolution
- Ecology

INDICES

CELL BIOLOGY

A

B

C

M

N

O

P

Q

R

S

T

U

V

X

GENETICS

A

B

C

D

E

X

Y

Z

MOLECULAR BIOLOGY

A

B

C

D

E

F

G

H

I

L

M

N

O

P

EVOLUTION

B

C

D

E

ECOLOGY

A

B

H

I

L

M

N

O

P

Q

R

S

T

U

V